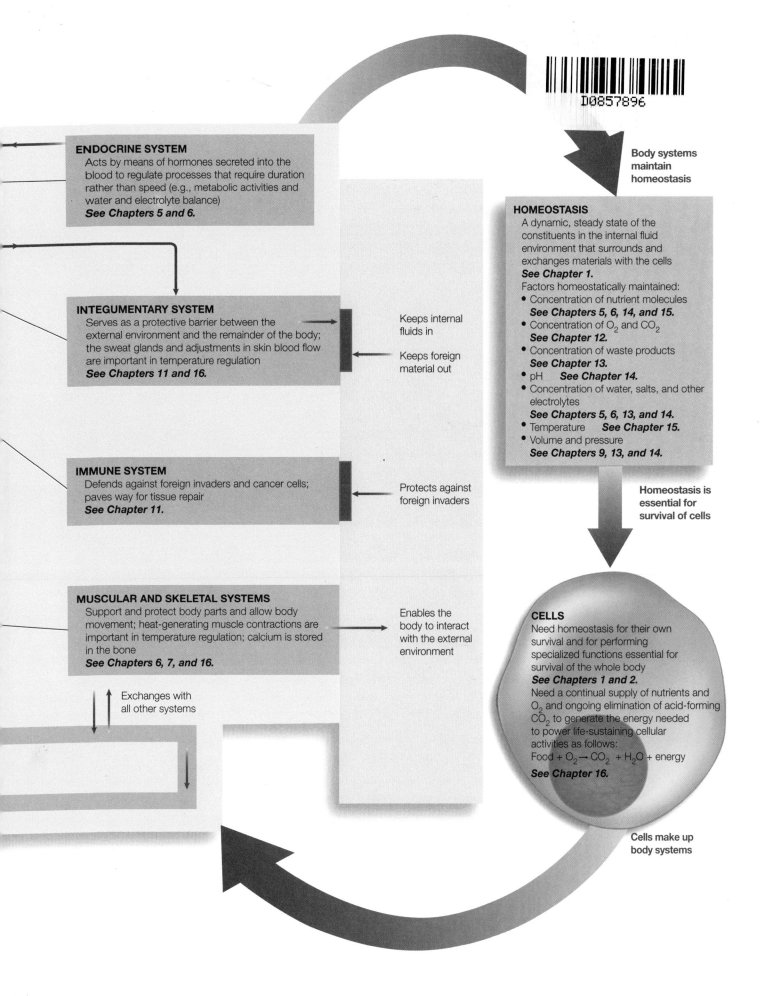

**ENDOCRINE SYSTEM**
Acts by means of hormones secreted into the blood to regulate processes that require duration rather than speed (e.g., metabolic activities and water and electrolyte balance)
*See Chapters 5 and 6.*

**INTEGUMENTARY SYSTEM**
Serves as a protective barrier between the external environment and the remainder of the body; the sweat glands and adjustments in skin blood flow are important in temperature regulation
*See Chapters 11 and 16.*

Keeps internal fluids in

Keeps foreign material out

**IMMUNE SYSTEM**
Defends against foreign invaders and cancer cells; paves way for tissue repair
*See Chapter 11.*

Protects against foreign invaders

**MUSCULAR AND SKELETAL SYSTEMS**
Support and protect body parts and allow body movement; heat-generating muscle contractions are important in temperature regulation; calcium is stored in the bone
*See Chapters 6, 7, and 16.*

Enables the body to interact with the external environment

Exchanges with all other systems

Body systems maintain homeostasis

**HOMEOSTASIS**
A dynamic, steady state of the constituents in the internal fluid environment that surrounds and exchanges materials with the cells
*See Chapter 1.*
Factors homeostatically maintained:
• Concentration of nutrient molecules
  *See Chapters 5, 6, 14, and 15.*
• Concentration of $O_2$ and $CO_2$
  *See Chapter 12.*
• Concentration of waste products
  *See Chapter 13.*
• pH    *See Chapter 14.*
• Concentration of water, salts, and other electrolytes
  *See Chapters 5, 6, 13, and 14.*
• Temperature    *See Chapter 15.*
• Volume and pressure
  *See Chapters 9, 13, and 14.*

Homeostasis is essential for survival of cells

**CELLS**
Need homeostasis for their own survival and for performing specialized functions essential for survival of the whole body
*See Chapters 1 and 2.*
Need a continual supply of nutrients and $O_2$ and ongoing elimination of acid-forming $CO_2$ to generate the energy needed to power life-sustaining cellular activities as follows:
Food + $O_2 \rightarrow CO_2$ + $H_2O$ + energy
*See Chapter 16.*

Cells make up body systems

FOURTH CANADIAN EDITION

# Human Physiology
## From Cells to Systems

**Lauralee Sherwood**
Department of Physiology and Pharmacology
School of Medicine
West Virginia University

**Christopher Ward**
Department of Biomedical and Molecular Sciences
Cardiovascular Physiology
Queen's University

**NELSON**

# NELSON

**Human Physiology, Fourth Canadian Edition**

by Lauralee Sherwood, Christopher Ward

**VP, Product Solutions, K–20:**
Claudine O'Donnell

**Senior Publisher, Digital and Print Content:**
Paul Fam

**Marketing Manager:**
Kim Carruthers

**Content Manager:**
Theresa Fitzgerald

**Photo and Permissions Researcher:**
Julie Pratt

**Senior Production Project Manager:**
Imoinda Romain

**Production Service:**
SPi Global

**Copy Editor:**
Carolyn Jongeward

**Proofreader:**
SPi Global

**Indexer:**
SPi Global

**Design Director:**
Ken Phipps

**Higher Education Design Project Manager:**
Pamela Johnston

**Interior Design Modifications:**
Brian Malloy

**Cover Design:**
Courtney Hellam

**Cover Image:**
ChrisChrisW/iStock

**Art Coordinator:**
Suzanne Peden

**Illustrator(s):**
Articulate Graphics, Crowle Art Group, SPi Global

**Compositor:**
SPi Global

**Library and Archives Canada Cataloguing in Publication**

Sherwood, Lauralee, author
    Human physiology : from cells to systems / Lauralee Sherwood, Department of Physiology and Pharmacology, School of Medicine, West Virginia University; Christopher Ward, Department of Biomedical and Molecular Sciences, Cardiovascular Physiology, Queen's University. — Fourth Canadian edition.

Includes index.
Issued in print and electronic formats.
ISBN 978-0-17-674484-7 (hardcover).
—ISBN 978-0-17-685380-8 (PDF)

    1. Human physiology—Textbooks. 2. Textbooks. I. Ward, Christopher, 1966–, author II. Title.

QP34.5.S54 2018        612
C2017-907805-4
C2017-907806-2

ISBN-13: 978-0-17-674484-7
ISBN-10: 0-17-674484-3

To my family for all they have done for me in the past, all they mean to me in the present, and all I hope will yet be in the future:

My parents, Larry and Lee Sherwood (both in memoriam)

My husband, Pater Marshall

My daughters and son-in-law, Melinda and Mark Marple, Allison Tadros and Bill Krantz

My grandchildren, Lindsay Marple, Emily Marple, Alexander Tadros, Lauren Krantz

**Lauralee Sherwood**

To my family for their love and support and to the students whose quest for knowledge has inspired me

**Christopher Ward**

# Brief Contents

# Contents

---

## Chapter 3 | The Central Nervous System 91

---

**Chapter 4 | The Peripheral Nervous
System: Sensory,
Autonomic, Somatic   141**

Chapter 7 | **Muscle Physiology    297**

Chapter 8 | **Cardiac Physiology    345**

---

Chapter 11 | Body Defences   459

---

Chapter 12 | The Respiratory System   499

## Chapter 13 | **The Urinary System   559**

# Preface

## Goals, Philosophy, and Theme

Our goal in writing physiology textbooks is to help students not only learn about how the body works but also to share our enthusiasm for the subject matter. We have been teaching physiology for over 40 years and remain awestruck at the intricacies and efficiency of body function. No machine can perform even a portion of natural body function as effectively. Most of us, even infants, have a natural curiosity about how our bodies work. When a baby first discovers it can control its own hands, it is fascinated and spends many hours manipulating them in front of its face. By capitalizing on students' natural curiosity about themselves, we try to make physiology a subject they can enjoy learning.

Even the most tantalizing subject can be difficult to comprehend if not effectively presented, however. Therefore, this book has a logical, understandable format with an emphasis on how each concept is an integral part of the entire subject. Too often, students view the components of a physiology course as isolated entities; by understanding how each component depends on the others, a student can appreciate the integrated functioning of the human body. The text focuses on the mechanisms of body function from cells to systems and is organized around the central theme of *homeostasis*—how the body meets changing demands while maintaining the internal constancy necessary for all cells and organs to function. The text is written in simple, straightforward language, and every effort has been made to ensure smooth reading through good transitions, common-sense reasoning, and integration of ideas throughout.

This text is designed for undergraduate students preparing for health-related careers, but its approach and depth are also appropriate for other undergraduates. Because this book is intended as an introduction and, for most students, may be their only exposure to a formal physiology text, all aspects of physiology receive broad coverage; yet depth, where needed, is not sacrificed. The scope of this text has been limited by judicious selection of pertinent content that a student can reasonably be expected to assimilate in a one-semester physiology course. Materials were selected for inclusion on a "need to know" basis, so the book is not cluttered with unnecessary detail. Instead, content is restricted to relevant information needed to understand basic physiological concepts and to serve as a foundation for later more in-depth studies of selected aspects of physiology and related health sciences, as well as in the health professions. Some controversial ideas and hypotheses are presented to illustrate that physiology is a dynamic, changing discipline.

To keep pace with today's rapid advances in the health sciences, students in the health professions must be able to draw on their conceptual understanding of physiology instead of merely recalling isolated facts that soon may be out of date. Therefore, this text is designed to promote understanding of the basic principles and concepts of physiology rather than memorization of details.

In consideration of the clinical orientation of most students, research methodologies and data are not emphasized, although the material is based on up-to-date evidence. New information based on recent discoveries has been included throughout. Students can be assured of the timeliness and accuracy of the material presented. To make room for new, pertinent information, we have carefully trimmed content while clarifying, modifying, and simplifying as needed to make this edition fresh, reader-friendly, and current.

Because the function of an organ depends on the organ's construction, enough relevant anatomy is integrated within the text to make the inseparable relation between form and function meaningful.

## Organization

There is no ideal organization of physiologic processes into a logical sequence. With the sequence we chose for this book, most chapters build on material presented in immediately preceding chapters, yet each chapter is designed to stand alone, allowing the instructor flexibility in curriculum design. This flexibility is facilitated by cross-references to related material in other chapters. The cross-references let students quickly refresh their memory of material already learned or to proceed, if desired, to a more in-depth coverage of a particular topic.

The general flow is from introductory information on cells and metabolism, to excitable tissue, to organ systems. We have tried to provide logical transitions from one chapter to the next. For example, Chapter 7, Muscle Physiology, ends with a discussion of cardiac muscle, which is carried forward in Chapter 8, Cardiac Physiology. Even topics that seem unrelated in sequence, such as Chapter 11, Body Defences, and Chapter 12, The Respiratory System, are linked together, in this case by discussion of the respiratory system's contribution to protection against inhaled viruses and particles.

Several organizational changes have been made, which we believe will enhance the students' understanding of human physiology and the readability of the text. For example, we have included three inserts called the Integration of Human Physiology. Each insert focuses on the integration of a number of body systems, to aid the instructor in teaching integrative physiology. Again, these inserts were designed to stand alone.

The first insert, for example, discusses the concept of locomotion, and contrasts locomotion in a healthy person with the problems associated with locomotion when disease disrupts normal neural communication. Each of the three integrative inserts is strategically placed, following the chapters upon which they build.

We have also placed the chapters discussing the nervous and endocrine systems near the beginning of the textbook. The decision to place these chapters in close proximity and near the start of the text was made based on the belief that it is important for the student to understand how communication takes place within the human body, as communication has a considerable influence on homeostasis.

More specific hormones are subsequently introduced in appropriate chapters, such as vasopressin and aldosterone in the chapters on the kidneys (Chapter 13) and fluid balance (Chapter 14). Intermediary metabolism (Chapter 2) of absorbed nutrient molecules is largely under endocrine control, providing a link from digestion (Chapter 15) and energy balance (Chapter 16) to the endocrine chapters. Chapter 6, The Endocrine Glands, pulls together the source, functions, and control of specific endocrine secretions discussed in Chapter 5, Principles of Endocrinology, and serves as a unifying capstone for homeostatic body function. Finally, building on the gonadotropic hormones introduced in Chapter 6, Chapter 17 diverges from the theme of homeostasis to focus on reproductive physiology.

Besides the novel handling of hormones and the endocrine system, other organizational features are unique to this book. Unlike other physiology texts, the skin is covered in the chapter on body defences (Chapter 11), in consideration of the skin's newly recognized immune functions. Bone is also covered more extensively in the endocrine chapter than in most undergraduate physiology texts, especially with regard to hormonal control of bone growth and bone's dynamic role in calcium metabolism.

A concerted effort was made to bolster specific topics in each chapter. The following are some examples of this. The cell physiology chapter (Chapter 2) has been expanded and includes an introduction to excitable cells and neurons. The central nervous system (Chapter 3) has an expanded discussion on somatosensory mapping and plasticity of the motor cortex, and it introduces the Babinski reflex. In the following chapter, The Peripheral Nervous System, the topic of somatosensory mapping is carried forward from Chapter 3, and a more detailed discussion of mechanoreceptors has been included, as well as an introduction to the fight-or-flight response.

Also new to this edition is the inclusion of a feature called **Clinical Connections**. The beginning of most chapters has an introduction to a clinical case that is related to the content of the chapter. As you move through the chapter, this case is revisited to put the underlying physiology into context.

Departure from traditional groupings of material in several important instances has permitted more independent and more extensive coverage of topics that are frequently omitted or buried within chapters concerned with other subject matter. For example, a separate chapter (Chapter 14) is devoted to fluid balance and acid–base regulation, topics often included in the kidney chapter. Further, we have grouped energy balance and temperature regulation into an independent chapter (Chapter 16).

Although there is a rationale for covering the various aspects of physiology in the order given here, it is by no means the only logical way to present each topic. Each chapter is able to stand alone, especially with the cross-references provided, so that the sequence of presentation can be varied at the instructor's discretion. Some chapters may even be omitted, depending on the students' needs and interests and the time constraints of the course. For example, a cursory explanation of the defensive role of the leukocytes appears in the chapter on blood, so an instructor can choose to omit the more detailed explanations of immunity in Chapter 11, Body Defences.

*Lauralee Sherwood and Christopher Ward*

## For the Canadian Edition

The fourth Canadian edition of *Human Physiology: From Cells to Systems* has been substantially revised to meet the needs of Canadian students and instructors in core physiology programs, as well as related programs such as kinesiology, nursing, physical therapy, and zoology, where physiology represents an important component of the curriculum. The reorganization of the text has allowed for the inclusion of integrative physiology sections to assist the instructor in addressing this component of their course. A continued effort has been made by the authors to highlight topics of research important to Canadians, and to include statistics, organizations, and researchers from Canadian schools, institutes, and hospitals. The Système international d'unités (SI) and Canadian spelling have been used throughout. In addition to these changes, this fourth Canadian edition also provides students and instructors with updated physiological concepts, new figures and tables, new trends in physiological research, and current and relevant examples of the body's function in disease, exercise, and health. The end-of-chapter review questions have been retooled to the individual chapter's objectives. As well, the Nelson Education Testing Advantage (NETA) program was used to ensure high-quality test banks.

## New to the Fourth Canadian Edition

### Global changes to this edition

- Updated references to Canadian research
- Timely material incorporated throughout
- Over 280 figures, which include examples of extensively revised, and newly conceptualized ideas
- Every opportunity used to make the writing as clear, concise, well-organized, and relevant for readers as possible
- The inclusion of Clinical Connections feature to most chapters

### Chapter 1—The Foundation of Physiology

- Updated references to Canadian research
- Updated information on stem cells

## Chapter 2—Cell Physiology

- Updated discussion of neurons
- Expanded discussion on membrane potential and cellular communication

## Chapter 3—The Central Nervous System

- Condensed and focused content on the CNS
- Reorganized the order of discussion of the CNS

## Chapter 4—The Peripheral Nervous System: Sensory, Autonomic, Somatic

- Enhanced clarity for some content

## Chapter 5—Principles of Endocrinology: The Central Endocrine Glands

- Streamlined content

## Chapter 6—The Endocrine Glands

- This chapter repositioned to follow the chapter on the central endocrine glands
- Updated content

## Chapter 7—Muscle Physiology

- Enhanced clarity of key concepts

## Chapter 8—Cardiac Physiology

- Included discussion on how afterload affects stroke volume

## Chapter 9—Vascular Physiology

- Updated discussion
- Clarified some key concepts

## Chapter 10—The Blood

- Updated and clarified content

## Chapter 11—Body Defences

- Updated content
- Streamlined content
- Enhanced clarity of discussion

## Chapter 12—The Respiratory System

- Streamlined content
- Enhanced clarity of discussion

## Chapter 13—The Urinary System

- Streamlined topic discussion

## Chapter 14—Fluid and Acid–Base Balance

- Clinical case on diabetic ketoacidosis (DKA) to highlight the clinical relevance of fluid balance and acid–base regulation

- New brief discussion on the two types of diabetes insipidus (nephrogenic and central) as they relate to vasopressin
- Addition of new figures, including $Na^+$–$H^+$ exchanger of the proximal nephron and a nomogram to help diagnose acid–base disorders

## Chapter 15—The Digestive System

- Updated content

## Chapter 16—Energy Balance and Temperature Regulation

- Updated to include current statistics

## Chapter 17—The Reproductive System

- Updated content

In addition, pedagogical features unique to the fourth Canadian edition of *Human Physiology: From Cells to Systems* include a new addition: **Clinical Connections** have been added to most chapters. This feature serves to relate diseases and their symptoms to what is happening in the underlying physiology. The **Further Reading** list of relevant research articles from international journals within the **Concepts, Challenges, and Controversies** boxes is intended to facilitate further study, and the **Chapter Terminology** list of relevant key terms at the end of each chapter is available for students to review. **Why It Matters** boxes have been added to some chapters to draw attention to the importance of material that at first glance may appear trivial. The **Table of Contents** has been simplified to allow for easier navigation of the text, as have the **section headings** within each chapter. **Appendixes**, which include Reference Values for Commonly Measured Variables in Blood and Commonly Measured Cardiorespiratory Variables, and the **Glossary** have been expanded to accommodate new physiology concepts.

# Text Features and Learning Aids

## Implementing the homeostasis theme

A unique, easy-to-follow schematic of the relations between cells and systems to illustrate homeostasis is developed in the introductory chapter and presented on the inside front cover as a quick reference. Each chapter begins with its own quick reference, accompanied by a brief, written introduction emphasizing how the body system considered in the chapter functionally fits in with the body as a whole. This opening feature is designed to orient students to the homeostatic aspects of the material that follows. Then, at the close of each chapter, the **Chapter in Perspective: Focus on Homeostasis** is intended to put into perspective how the part of the body just discussed contributes to homeostasis. This capstone feature, the opening homeostatic model, and the introductory comments are designed to work together to facilitate comprehension of the interactions and interdependency of body systems, even though each system is discussed separately.

## Analogies

Many analogies and frequent references to everyday experiences are included to help students relate to the physiology concepts presented. These useful tools have been drawn in large part from Lauralee Sherwood's four decades of teaching experience. Knowing which areas are likely to give students the most difficulty, she has tried to develop links that help them relate the new material to something with which they are already familiar.

## Pathophysiology and clinical coverage

 Another effective way to keep students' interest is to help them realize they are learning worthwhile and applicable material. Because most students using this text will have health-related careers, frequent references to pathophysiology and clinical physiology demonstrate the content's relevance to their professional goals. *Clinical Note* icons flag clinically relevant material featured throughout the text.

## Check Your Understanding

**Check Your Understanding** questions serve as study breaks for students to test their knowledge before continuing with more material. These questions are different than the **Written Questions** that cover the same content at the end of the chapter. Many of the new section questions involve doing something other than copying an answer from a text description, such as drawing and labelling, preparing a chart, predicting based on information provided, and so on.

## Boxed features

Two types of boxed features are integrated within the chapters, and in most cases they conclude with a list of journal articles and references pertinent to the topic. As much as possible, the journal articles and references are Canadian.

**Concepts, Challenges, and Controversies** boxes expose students to high-interest, tangentially relevant information on such diverse topics as stem cell research, acupuncture, new discoveries regarding common diseases such as strokes, historical perspectives, and body responses to new environments such as those encountered in space flight and deep-sea diving.

The **Why It Matters** boxes can be used by students and instructors to help glean the importance of, for example, a seeming simple reaction, shedding light on how something small and apparently insignificant can in actuality have an impact on large-scale system-wide function.

The **Integration of Human Physiology** pages are intended to enrich the understanding of the issues covered within the chapters. These pages are distinct and easy to find due to their purple background and dark purple edges. They build on the concepts introduced within the chapters and complement the physiological processes presented. These pages are included to highlight both the normal physiological responses and the abnormal physiological responses to stimuli, such as movement, altitude, and exercise.

## Pedagogical illustrations

The anatomic illustrations, schematic representations, photographs, tables, and graphs are designed to complement and reinforce the written material. Unique to this book are the numerous process-oriented figures with step-by-step descriptions, which provide a concise visual aid for understanding the processes.

Flow diagrams are used extensively to help students integrate the written information. In the flow diagrams, light and dark shades of the same colour denote a decrease or an increase in a controlled variable, such as blood pressure or the concentration of blood glucose.

Integrated, colour-coded figure/table combinations help students better visualize what part of the body is responsible for what activities. For example, the anatomic depiction of the brain is integrated with a table of the functions of the major brain components, with each component shown in the same colour in the figure and the table.

A unique feature of this book is that people depicted in the various illustrations are realistic representatives of a cross-section of humanity (they were drawn based on photographs of real people). Sensitivity to various races, sexes, and ages should enable all students to identify with the material being presented.

## Key terms and word derivations

In most cases, key terms are defined as they appear in the text. These terms are key to understanding the concepts outlined within each chapter. A chapter terminology section also appears at the end of each chapter. Many key terms are also found in the glossary.

## End-of-chapter learning and review

A **Chapter Terminology** list of relevant key terms appears at the end of each chapter for students to review. The **Review Exercises** at the end of each chapter include a variety of question formats for students to self-test their knowledge and application of the facts and concepts presented. Also available are **Quantitative Exercises** that provide the students with an opportunity to practise calculations that will enhance their understanding of complex relationships. A **Points to Ponder** section features thought-provoking problems that encourage students to analyze what they have learned, and the **Clinical Consideration**, a mini case history, challenges them to apply their knowledge to a patient's specific symptoms. Answers and explanations for all of these questions are found in Appendix E.

## Appendixes, Glossary, and Study Cards

The appendixes are designed for the most part to help students brush up on some foundation material they are assumed to have studied in prerequisite courses.

- *Appendix A*, **Système international d'unités (SI)**, is a conversion table between metric measures and their Imperial equivalents.
- Most undergraduate physiology texts have a chapter on chemistry, yet physiology instructors rarely teach basic chemistry concepts. Knowledge of chemistry beyond that introduced in secondary schools is not required for understanding this text. Therefore, to reserve valuable text space for discussion of physiology concepts, we have provided **Appendix B, A Review of Chemical Principles** as a handy reference for the review of basic chemistry concepts that apply to physiology.

- Likewise, **Appendix C, Storage, Replication, and Expression of Genetic Information**, can serve as a reference for students and, when appropriate, be used by the instructor as assigned material. It includes a discussion of DNA and chromosomes, protein synthesis, cell division, and mutations.
- **Appendix D, Principles of Quantitative Reasoning**, is designed to help students become more comfortable working with equations and translating back and forth between words, concepts, and equations. This appendix supports the Quantitative Exercises in the end-of-chapter material.
- **Appendix E, Answers to End-of-Chapter Objective Questions, Quantitative Exercises, Points to Ponder, and Clinical Considerations**, provides answers to all objective learning activities, solutions to the Quantitative Exercises, and explanations for the Points to Ponder and Clinical Considerations.
- **Appendix F, Reference Values for Commonly Measured Variables in Blood and Commonly Measured Cardiorespiratory Variables**, provides normal male and female values.
- **Appendix G, A Deeper Look into Chapter 11: The Respiratory System**, provides greater detail on the terminology and gas laws covered in Chapter 11.
- The **Glossary**, which offers a way to review the meaning of key terminology, includes phonetic pronunciations of the entries.
- The **Study Cards** present the major points of each chapter in concise, section-by-section bulleted lists, including cross-references for page numbers, figures, and tables. With this summary design, students can review more efficiently by using both written and visual information to focus on the main concepts.

## About the Nelson Education Teaching Advantage (NETA)

The **Nelson Education Teaching Advantage (NETA)** program delivers research-based instructor resources that promote student engagement and higher-order thinking to enable the success of Canadian students and educators. Be sure to visit Nelson Education's **Inspired Instruction** website at http://www.nelson.com/inspired/ to find out more about NETA. Don't miss the testimonials of instructors who have used NETA supplements and seen student engagement increase!

## Instructor Resources

### Downloadable instructor supplements

All NETA and other key instructor ancillaries can be accessed through www.nelson.com/instructor, giving instructors the ultimate tools for customizing lectures and presentations.

**NETA Test Bank**: This resource was written by Dr. Alice Khin, University of Alberta. It includes over 5000 questions; 1750 questions are multiple-choice questions written according to NETA guidelines for effective construction and development of higher-order questions. The Test Bank was copy-edited by a NETA-trained editor. Also included are fill in the blank, matching, and true-or-false test items as well as essay questions. The NETA Test Bank is available in a new, cloud-based platform. **Nelson Testing Powered by Cognero®** is a secure online testing system that allows you to author, edit, and manage test bank content from any place you have Internet access. No special installations or downloads are needed, and the desktop-inspired interface, with its drop-down menus and familiar, intuitive tools, allows you to create and manage tests with ease. You can create multiple test versions in an instant, and import or export content into other systems. Tests can be delivered from your learning management system, your classroom, or wherever you want. Nelson Testing Powered by Cognero® for *Human Physiology: From Cells to Systems,* Fourth Canadian Edition, also be accessed through www.nelson.com/instructor. Printable versions of the Test Bank in Microsoft® Word® and PDF formats are available by contacting your sales and editorial representative.

**NETA PowerPoint**: Microsoft® PowerPoint® lecture slides for every chapter have been created by Dr. Stephen Donald Turnbull, University of New Brunswick. There is an average of 100 slides per chapter, many featuring key figures, tables, and photographs from *Human Physiology: From Cells to Systems*, Fourth Canadian Edition. Incorporating animations and "build slides," these illustrations with labels from the book have been formatted to allow optimal display in PowerPoint®. NETA principles of clear design and engaging content have been incorporated throughout, making it simple for instructors to customize the deck for their courses.

**Image Library**: This resource consists of over 600 digital copies of figures, short tables, and photographs used in the book. Instructors may use these JPEG to customize the NETA PowerPoint or create their own PowerPoint presentations.

**NETA Instructor's Manual**: This resource was written by Dr. Coral Murrant, University of Guelph. It is organized according to the textbook chapters and addresses key educational concerns, such as typical stumbling blocks student face and how to address them. It also includes elements of a traditional Instructor's Manual, including lecture outlines, discussion topics, student activities, Internet connections, media resources, and testing suggestions that will help time-pressed instructors more effectively communicate with their students and also strengthen the coverage of course material.

**DayOne**: DayOne—Prof InClass is a PowerPoint presentation that instructors can customize to orient students to the class and their text at the beginning of the course.

## MindTap

⸭ MINDTAP

**MindTap for** *Human Physiology: From Cells to Systems*, Fourth Canadian Edition, is a personalized teaching experience with relevant assignments that guide students to analyze, apply, and elevate thinking, allowing instructors to measure skills and promote better outcomes with ease. A fully online learning solution, MindTap combines all student learning tools—readings, multimedia, activities, and assessments—into a single Learning Path that guides the student through the curriculum. Instructors personalize the experience by customizing the presentation of these learning tools to their students, even seamlessly introducing their own content into the Learning Path.

## Aplia

**Aplia**™ is a Cengage Learning online homework system dedicated to improving learning by increasing student effort and engagement. Aplia makes it easy for instructors to assign frequent online homework assignments. Aplia provides students with prompt and detailed feedback to help them learn as they work through the questions, and features interactive tutorials to fully engage them in learning course concepts. Automatic grading and powerful assessment tools give instructors real-time reports of student progress, participation, and performance, and while Aplia's easy-to-use course management features let instructors flexibly administer course announcements and materials online. With Aplia, students will show up to class fully engaged and prepared, and instructors will have more time to do what they do best… teach.

Jacqueline Carnegie of the University of Ottawa revised the MindTap and Aplia resources to ensure close alignment to the fourth Canadian edition.

## Student Ancillaries

### MindTap

⸭ MINDTAP

Stay organized and efficient with **MindTap**—a single destination with all the course material and study aids you need to succeed. Built-in apps leverage social media and the latest learning technology. For example,

ReadSpeaker will read the text to you.

Flashcards are pre-populated to provide you with a jump start for review—or you can create your own. You can highlight text and make notes in your MindTap Reader. Your notes will flow into Evernote, the electronic notebook app that you can access anywhere when it's time to study for the exam.

Self-quizzing allows you to assess your understanding.

Visit www.nelson.com/student to start using **MindTap**. Enter the Online Access Code from the card included with your text. If a code card is *not* provided, you can purchase instant access at NELSONbrain.com.

## Aplia

Founded in 2000 by economist and Stanford professor Paul Romer, **Aplia**™ is an educational technology company dedicated to improving learning by increasing student effort and engagement. Currently, Aplia products have been used by more than a million students at over 1300 institutions. Aplia offers a way for you to stay on top of your coursework with regularly scheduled homework assignments that increase your time on task and give you prompt feedback. Interactive tools and additional content are provided to further increase your engagement and understanding. See http://www.aplia.com for more information. If Aplia isn't bundled with your copy of *Human Physiology: From Cells to Systems*, Fourth Canadian Edition, you can purchase access separately at NELSONbrain.com. Be better prepared for class with Aplia!

## U.S. Acknowledgments

I gratefully acknowledge the many people who helped with the first eight editions or this edition of the textbook. Also, I remain indebted to four people who contributed substantially to the original content of the book: Rachel Yeater (West Virginia University), who contributed the original material for the exercise physiology boxes; Spencer Seager (Weber State University), who prepared Appendix A, "A Review of Chemical Principles"; and Kim Cooper (Midwestern University) and John Nagy (Scottsdale Community College), who provided the Solving Quantitative Exercises at the ends of chapters.

In addition to the 184 reviewers who carefully evaluated the forerunner books for accuracy, clarity, and relevance, I express sincere appreciation to the following individuals who served as reviewers for this edition:

Ahmed Al-Assal, West Coast University

Amy Banes-Berceli, Oakland University

Patricia Clark, Indiana University-Purdue University

Elizabeth Co, Boston University

Steve Henderson, Chico State University

James Herman, Texas A&M University

Qingsheng Li, University of Nebraska-Lincoln

Douglas McHugh, Quinnipiac University

Linda Ogren, University of California, Santa Cruz

Roy Silcox, Brigham Young University

Also, I am grateful to the users of the textbook who have taken time to send helpful comments.

I have been fortunate to work with a highly competent, dedicated team from Cengage Learning, along with other capable external suppliers selected by the publishing company. I would like to acknowledge all of their contributions, which collectively made this book possible. It has been a source of comfort and inspiration to know that so many people have been working diligently in so many ways to bring this book to fruition.

From Cengage Learning, Yolanda Cossio, Senior Product Team Manager, deserves warm thanks for her vision, creative ideas, leadership, and ongoing helpfulness. Yolanda was a strong advocate for making this the best edition yet. Yolanda's decisions were guided by what is best for the instructors and students who will use the textbook and ancillary package. Thanks also to Product Assistant Victor Luu, who coordinated many tasks for Yolanda during the development process.

I appreciate the efforts of Content Developer Alexis Glubka for helping us launch this project and for Managing Developer Trudy Brown for taking over and not missing a beat when Alexis moved to another position. Trudy facilitated most of the development process, which proceeded efficiently and on schedule. Having served as a Production Editor in the past, Trudy was especially helpful in paving the way for a smooth transition from development to production. I am grateful for the biweekly conference calls with Yolanda and Trudy. Their input, expertise, and support were invaluable as collectively we made decisions to make this the best edition yet.

I appreciate the creative insight of Cengage Learning Senior Art Director John Walker, who oversaw the overall artistic design of the text and found the dynamic cover image that displays simultaneous power, agility, and grace in a "body in motion," the theme of our covers. The upward and "over-the-hump" movement of the high jumper symbolizes that this book will move the readers upward and over the hump in their understanding of physiology. I thank John for his patience and perseverance as we sifted through and rejected many photos until we found just what we were looking for.

The technology-enhanced learning tools in the media package were updated under the guidance of Media Developer Lauren Oliveira. These include the online interactive tutorials, media exercises, and other e-Physiology learning opportunities at the CengageBrain website. Associate Content Developers Casey Lozier and Kellie Petruzzelli oversaw development of the multiple hard-copy components of the ancillary package, making sure it was a cohesive whole. A hearty note of gratitude is extended to all of them for the high-quality multimedia package that accompanies this edition.

On the production side, I would like to thank Senior Content Project Manager Tanya Nigh, who closely monitored every step of the production process while simultaneously overseeing the complex production process of multiple books. I felt confident knowing that she was making sure that everything was going according to plan. I also thank Photo Researcher Priya Subbrayal and Text Researcher Kavitha Balasundaram for tracking down photos and permissions for the art and other copyrighted materials incorporated in the text, an absolutely essential task. With everything finally coming together, Manufacturing Planner Karen Hunt oversaw the manufacturing process, coordinating the actual printing of the book.

No matter how well a book is conceived, produced, and printed, it would not reach its full potential as an educational tool without being efficiently and effectively marketed. Senior Market Development Manager Julie Schuster played the lead role in marketing this text, for which I am most appreciative.

Cengage Learning also did an outstanding job in selecting highly skilled vendors to carry out particular production tasks. First and foremost, it has been my personal and professional pleasure to work with Cassie Carey, Production Editor at Graphic World, who coordinated the day-to-day management of production. In her competent hands lay the responsibility of seeing that all copyediting, art, typesetting, page layout, and other related details got done right and in a timely fashion. Thanks to her, the production process went smoothly, the best ever. I also want to extend a hearty note of gratitude to compositor Graphic World for their accurate typesetting; execution of many of the art revisions; and attractive, logical layout. Lisa Buckley deserves thanks for the fresh and attractive, yet space-conscious, appearance of the book's interior and for envisioning the book's visually appealing exterior.

Finally, my love and gratitude go to my family for the sacrifices in family life as this ninth edition was being developed and produced. The schedule for this book was especially hectic because it came at a time when a lot else was going on in our lives. My husband, Peter Marshall, deserves special appreciation and recognition for assuming extra responsibilities while I was working on the book. I could not have done this, or any of the preceding books, without his help, support, and encouragement.

Thanks to all!

*Lauralee Sherwood*

# Canadian Acknowledgments

I am extremely thankful to all the reviewers who took part in the process of developing the fourth Canadian edition of *Human Physiology: From Cells to Systems*. Without your help the text would not be as thorough.

| | |
|---|---|
| Gary V. Allen | Dalhousie University |
| Darren S. DeLorey | University of Alberta |
| Iain McKinnell | Carleton University |
| Coral Murrant | University of Guelph |
| Kirsten Poling | University of Windsor |
| Ramona Stewart | Lethbridge College |
| Anita Woods | Western University |

## Reviewers

We would like to thank Nelson, and in particular Paul Fam and Theresa Fitzgerald, for making this project possible and enjoyable. We would also like to thank Imoinda Romain, Carolyn Jongeward, Kim Carruthers, and Udhaya Harisudan for their help in the success of this text. And to all of my colleagues and students, who contributed in many different ways, thanks so much.

*Christopher Ward*

# About the Authors

**Lauralee Sherwood, Ph.D.** Following graduation from Michigan State University in 1966, Dr. Lauralee Sherwood joined the faculty at West Virginia University, where she is currently a professor in the Department of Physiology and Pharmacology, School of Medicine. For the past 40 years, Professor Sherwood has taught an average of over 400 students each year in physiology courses for pharmacy, medical technology, physical therapy, occupational therapy, nursing, medical, dental, dental hygiene, nutrition, exercise physiology, and athletic training majors. She has authored three physiology textbooks: *Human Physiology: From Cells to Systems, Fundamentals of Human Physiology,* and *Animal Physiology: From Genes to Organisms,* all published by Cengage Brooks/Cole. Dr. Sherwood has received numerous teaching awards, including an Amoco Foundation Outstanding Teacher Award, a Golden Key National Honor Society Outstanding Faculty Award, two listings in *Who's Who among America's Teachers,* and the Dean's Award of Excellence in Education.

**Christopher Ward, Ph.D.** Dr. Ward is an associate professor in the Department of Biomedical and Molecular Sciences at Queen's University. Dr. Ward has taken a leadership role in the development of innovative teaching methodologies to enhance undergraduate, graduate, and medical student learning. He is credited with the launch of online courses and blended learning initiatives, and is part of a curricular redesign team responsible for pioneering facilitated small-group learning at Queen's Faculty of Medicine. A cardiac physiologist, Dr. Ward also maintains an active laboratory and graduate research team that focuses on studying the electrical activity of the heart and understanding how ion channel function can be modulated by disease states.

## About the Contributor
## (Chapter 14, Fluid and Acid–Base Balance)

**Joseph Smuczek, M.D.** Dr. Smuczek completed his medical training and postgraduate family medicine training at McMaster University. He is currently a practising physician with the division of Primary Care at Trillium Health Partners in Mississauga, Ontario. He has been actively involved in undergraduate education and medical education and has previously taught undergraduate-level human anatomy and human physiology at Brock University and University of Guelph-Humber.

# Homeostasis Systems and Cells

Body systems maintain homeostasis

## Homeostasis

Homeostasis is the maintenance by the highly coordinated, regulated actions of the body systems of relatively stable chemical and physical conditions in the internal fluid environment that bathes the body's cells.

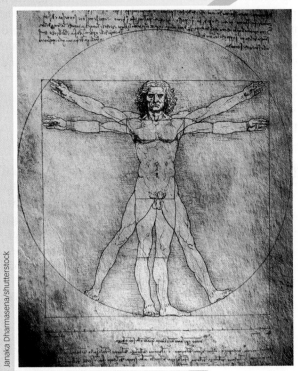

Da Vinci's *Vitruvian Man*

Homeostasis is essential for survival of cells

Cells make up body systems

Animal cell cut-away

# 1

# The Foundation of Physiology

## 1.1 | Introduction to Physiology

The human body is incredibly complex in that it is made up of a multitude of single cells that come together to form tissues and organ systems. These cells and systems depend on many daily processes that occur in our body to keep us alive. We usually take these life-sustaining activities for granted and don't really think about what makes us tick, but that's what physiology is about. Specifically, in this textbook we will study **human physiology**, the scientific study of the functions of our body.

We will also discuss other disciplines that rely on understanding human physiology, such as **nursing**, **medicine**, and **physical therapy**, to name but a few. In special topic boxes called Concepts, Challenges, and Controversies we address information related to human physiology that you may find interesting. As you read through this chapter and others, you will find another special topic box called Why It Matters, which conveys the importance and relevance of material covered in the chapter. New to this edition is a Clinical Connection feature that presents a clinical case at the beginning of the chapter and then revisits it throughout the chapter to highlight the relevant physiology. To bring it all together, you will also find Check Your Understanding sections that let you pause and see if you have understood the topic so far.

## Approaches to understanding

Two approaches are used to explain the events that occur in the body: one emphasizes the purpose, and the other examines the underlying mechanism. In response to the question, "Why do I shiver when I am cold?" one answer would be "to help warm up, because shivering generates heat." This approach—a **teleological approach**—explains body functions in terms of meeting a bodily *need*, without considering how this outcome is accomplished; it emphasizes the "why" or purpose of body processes. **Physiologists** (persons who study physiology; see Table 1-1 for a list of Canadian scientists who have received the Nobel Prize for Physiology or Medicine) primarily use a **mechanistic approach** to body function. They view the body as a machine whose mechanisms of action can be explained in terms of cause-and-effect sequences of physical and chemical processes—the same types of processes that occur in other components of the universe. Physiologists therefore explain the "how" of events that occur in the body. A physiologist's mechanistic explanation of shivering is that when temperature-sensitive nerve cells detect a fall in body temperature, they signal the temperature-sensitive area of the brain. In response, the brain activates nerve pathways that ultimately bring about involuntary, oscillating contraction-relaxation of skeletal muscle (shivering).

Because most bodily mechanisms do serve a useful purpose (naturally selected throughout evolution), it is helpful when studying physiology to predict what mechanistic process would be useful to the body under a particular circumstance. Thus, you can apply a certain amount of logical reasoning to each new situation you encounter in your study of physiology. *If you try to find the thread of logic in what you are studying, you can avoid a good deal of pure memorization, and, more importantly, you will better understand and assimilate the concepts being presented.*

## Structure and function

Physiology is closely related to **anatomy**, the study of the structure of the body. Physiological mechanisms are made possible by the structural design and relationships of the various body parts that carry out each of these functions. Just as the functioning of an automobile depends on the shapes, organization, and interactions of its various parts, the structure and function of the human body are inseparable. Therefore, as we explain how the body works, we will provide sufficient anatomic background for you to understand the function of the body part under discussion.

Some of the structure–function relationships are obvious. For example, the heart is well designed to receive and pump blood, the teeth to tear and grind food, and the hinge-like elbow joint to permit bending of the elbow. Other situations in which form and function are interdependent are more subtle but equally important. Consider the interface between the air and blood in the lungs: the respiratory airways that carry air from the outside into the lungs branch extensively when they reach the lungs. Tiny air sacs cluster at the ends of the huge number of airway branches. The branching is so extensive that the lungs contain about 300 million air sacs. Similarly, the blood vessels carrying blood into the lungs branch extensively and form dense networks of small vessels that encircle each of the air sacs (see Figure 12-2). Because of this structural relationship, the total surface area exposed to the air in the air sacs and the blood in the small vessels is much larger than a tennis court, depending on the size of the person. This large interface is crucial for the lung's ability to efficiently carry out the transfer of needed oxygen from the air into the blood and the unloading of the by-product of cellular respiration (carbon dioxide) from the blood into the air. The greater the available surface area for exchange, the faster the rate of movement (i.e., diffusion) of oxygen and carbon dioxide between the air and the blood. This large functional interface packaged within the confines of your lungs is possible only because both the air-containing and blood-containing components of the lungs branch extensively.

## 1.2 | Levels of Organization in the Body

We now turn our attention to how the body is structurally organized into a total functional unit, from the chemical level to the systems level to the whole body (⟩ Figure 1-1). These levels of organization make life as we know it possible.

### Chemical level

Like all matter on this planet, the human body is made up of a combination of specific chemicals. *Atoms* are the smallest building blocks of all nonliving and living matter. The most common atoms (or constituents) in the body—oxygen, carbon, hydrogen, and nitrogen—make up approximately 99 percent of total body chemistry. Calcium, phosphorus, and potassium contribute roughly 1 percent. These atoms combine to form the *molecules* of life, such as proteins, carbohydrates, fats, and nucleic acids (genetic material, such as deoxyribonucleic acid, or DNA). These important atoms and molecules are the raw ingredients

| ▍ **TABLE 1-1** Canadians Who Have Received Nobel Prizes in Physiology or Medicine | | |
|---|---|---|
| Year | Name | Discovery Awarded |
| 1923 | Frederick Banting | Insulin |
| 1966 | Charles B. Huggins | How hormones control the spread of some cancers |
| 1981 | David H. Hubel | Information processing in the visual system |
| 2009 | Jack W. Szostak | How chromosomes are protected by telomeres |
| 2011 | Ralph Steinman | The dendritic cell and its role in adaptive immunity |

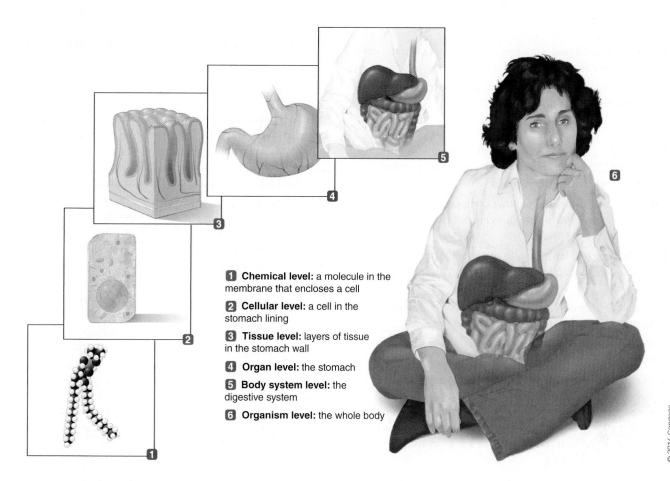

1. **Chemical level:** a molecule in the membrane that encloses a cell

2. **Cellular level:** a cell in the stomach lining

3. **Tissue level:** layers of tissue in the stomach wall

4. **Organ level:** the stomach

5. **Body system level:** the digestive system

6. **Organism level:** the whole body

> FIGURE 1-1 **Levels of organization in the body**

© 2016 Cengage

from which all living things arise. (See Appendix B for a review of this chemical level.)

## Cellular level

The mere presence of a particular collection of atoms and molecules does not confer the unique characteristics of life. Instead, these nonliving chemical components must be arranged and packaged in very precise ways to form a living entity. The **cell**, the basic or fundamental unit of both structure and function in a living being, is the smallest unit capable of carrying out the processes associated with life. Cell physiology is the focus of Chapter 2.

An extremely thin, oily barrier, the **plasma membrane**, encloses the contents of each cell, separating these chemicals from those outside the cell. Because the plasma membrane can control the movement of materials into and out of the cell, the cell's interior contains a combination of atoms and molecules that differs from the mixture of chemicals in the exterior environment surrounding the cell. Given the importance of the plasma membrane and its associated functions for carrying out life processes, Chapter 2 describes this structure in detail.

**Organisms** are independent living entities. The simplest forms of independent life are single-celled organisms, such as bacteria and amoebas. Complex multicellular organisms, such as trees and humans, are structural and functional aggregates of trillions of cells (*multi* means "many"). In multicellular organisms, cells are the living building blocks. In the simpler multicellular forms of life—for example, a sponge—the cells of the organism are all similar. However, more complex organisms, such as humans, have many different kinds of cells, such as muscle cells, nerve cells, and glandular cells.

Each human organism begins when an egg and sperm unite to form a single cell that starts to multiply and form a growing mass through myriad cell divisions. If cell multiplication were the only process involved in development, all the body cells would be essentially identical, as in the simplest multicellular life forms. During development of complex multicellular organisms, such as humans, however, each cell *differentiates*—that is, becomes specialized to carry out a particular function. Because of **cell differentiation**, your body is made up of many different specialized types of cells.

### BASIC CELL FUNCTIONS

All cells, whether they exist as solitary cells or as part of a multicellular organism, perform certain basic functions essential for survival of the cell. These basic cell functions can be described as metabolism, growth, and reproduction, and include the following:

1. Obtaining food (nutrients) and oxygen ($O_2$) from the environment surrounding the cell

2. Performing chemical reactions that use nutrients and $O_2$ to provide energy for the cells, as follows:

$$Food + O_2 \rightarrow CO_2 + H_2O + energy$$

3. Eliminating from the cell's internal environment carbon dioxide ($CO_2$) and other by-products, or wastes, produced during these chemical reactions

4. Synthesizing proteins and other components needed for cell structure, growth, and carrying out particular cell functions

5. Controlling to a large extent the exchange of materials between the cell and its surrounding environment

6. Moving materials from one part of the cell to another when carrying out cell activities. Some cells are even able to move through their surrounding environment

7. Being sensitive and responsive to changes in the surrounding environment

8. For most cells, reproducing. Some body cells, notably nerve cells and cardiac muscle cells, lose the ability to reproduce after they are formed during early development. This is why strokes, which result in losing nerve cells in the brain, and heart attacks, which kill heart muscle cells, can be so devastating.

Cells are remarkably similar in the ways they carry out these basic functions. Thus, they share many common characteristics.

### SPECIALIZED CELL FUNCTIONS

In multicellular organisms, each cell also performs a specialized function, which is usually a modification or elaboration of a basic cell function. Here are a few examples:

- The glandular cells of the digestive system take advantage of their protein-synthesizing ability and produce and secrete digestive enzymes (proteins) that break down ingested food.

- The specialized ability of the kidney cells enables them to control the exchange of materials between the cell and its environment to selectively retain needed substances while eliminating unwanted substances in the urine.

- Muscle contraction selectively moves internal structures to bring about the shortening of muscle cells, which shows the inherent capability of these cells to produce intracellular (within the cell) movement; this allows the organism to move through the external environment.

- Nerve cells generate, transmit, and store information in the form of changes in electrical impulse. For example, nerve cells in the ear relay information to the brain about sound in the external environment.

Each cell performs these specialized activities in addition to carrying on the unceasing, fundamental activities required of

all cells. The basic cellular functions are essential to the survival of each individual cell, whereas the specialized contributions and interactions among the cells of a multicellular organism are essential to the survival of the whole body.

### Tissue level

Just as a machine does not function unless all its parts are properly assembled, the body's cells must be specifically organized to carry out their life-sustaining processes as a whole, such as digestion, respiration, and circulation. Cells are progressively organized into tissues, organs, body systems, and, finally, the whole body.

Cells of similar structure and specialized function combine to form **tissues**, of which there are four primary types: muscle, nervous, epithelial, and connective. Each tissue consists of cells of a single specialized type, along with varying amounts of extracellular (outside the cell) material.

- **Muscle tissue** consists of cells specialized for contracting and generating force. There are three types of muscle tissue: *skeletal muscle*, which moves the skeleton; *cardiac muscle*, which pumps blood out of the heart; and *smooth muscle*, which encloses and controls movement of contents through hollow tubes and organs (e.g., food through the digestive tract).

- **Nervous tissue** consists of cells specialized for initiating and transmitting electrical impulses, which carry information over various distances: for example, one body part to another. Nervous tissue is found in the brain, spinal cord, and peripheral nerves.

- **Epithelial tissue** consists of cells specializing in exchange of materials between the cell and its environment. Any substance entering or leaving the body must cross an epithelial barrier. Epithelial tissue is organized into two general types: epithelial sheets and secretory glands.

  - Epithelial cells join tightly to form sheets of tissue that cover and line various parts of the body; for example, the outer layer of the skin is epithelial tissue, as is the lining of the digestive tract. Generally, these sheets serve as boundaries that separate the body from the external environment and from the contents of cavities that open to the external environment, such as the digestive tract lumen. (A **lumen** is the cavity within a hollow organ or tube.) Only selective transfer of materials is possible between regions separated by an epithelial sheet. The type and extent of controlled exchange vary, depending on the location and function of the epithelial tissue. For example, the skin can exchange very little between the body and external environment, whereas the epithelial cells lining the digestive tract are specialized for absorbing nutrients.

  - **Glands** are epithelial tissue derivatives specialized for secreting. **Secretion** is the release from a cell, in response to appropriate stimulation, of specific products that have been produced by the cell. Glands are formed during embryonic development by pockets of epithelial tissue that dip inward from the surface and develop secretory capabilities. There are two categories of glands: *exocrine* and *endocrine* (> Figure 1-2). If, during development, the connecting cells between the epithelial surface cells and the secretory gland cells within the depths of the invagination remain intact, as in a duct between the gland and the surface, an exocrine gland is formed. **Exocrine glands** secrete through ducts to the outside of the body (or into a cavity that communicates with the outside) (*exo* means "external"; *crine* means "secretion"). Sweat glands and glands that secrete digestive juices are examples of exocrine glands. In contrast, if the connecting cells disappear during development and the secretory gland cells are isolated from the surface, an endocrine gland is formed. **Endocrine glands** lack ducts and release their secretory products, known as *hormones*, internally into the blood (*endo* means "internal"); for example, the pancreas secretes insulin into the blood, which transports the hormone to its sites of action (e.g., muscle).

- **Connective tissue** is distinguished by having relatively few cells dispersed within an abundance of extracellular material. Connective tissue connects, supports, and anchors body parts. It includes diverse structures such as loose connective tissue that attaches epithelial tissue to underlying structures; tendons, which attach skeletal muscles to bones; bone, which gives the body shape, support, and protection; and blood, which transports materials (e.g., hormones, $O_2$) from one part of the body to another. Except for blood, the cells within connective tissue produce specific structural molecules that they release into the extracellular spaces between the cells. One such molecule is the rubber-band-like protein fibre *elastin*, whose presence facilitates the stretching and recoiling of structures such as the lungs, which inflate and deflate during breathing.

Muscle, nervous, epithelial, and connective tissue are the primary tissues in a classical sense, as each is an integrated collection of cells of the same specialized structure and function. The term *tissue* is also often used, as in clinical medicine, to mean the aggregate of various cellular and extracellular components that make up a particular organ (e.g., lung tissue or liver tissue).

## Organ level

**Organs** consist of two or more types of primary tissue organized together to perform a particular function or functions. The stomach is an example of an organ made up of all four primary tissue types. The tissues that make up the stomach function collectively to store ingested food, move it forward into the rest of the digestive tract, and begin the digestion of protein. The stomach is lined with epithelial tissue that restricts the transfer of harsh digestive chemicals and undigested food from the stomach lumen into the blood. Epithelial gland cells in the stomach include exocrine cells, which secrete protein-digesting juices into the lumen, and endocrine cells, which secrete a hormone that regulates the stomach's exocrine secretion and muscle

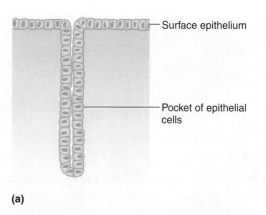

**(a)**

Surface epithelium

Pocket of epithelial cells

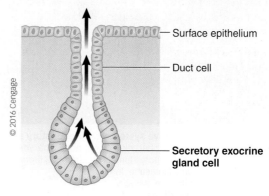

**(b)**

Surface epithelium

Duct cell

Secretory exocrine gland cell

© 2016 Cengage

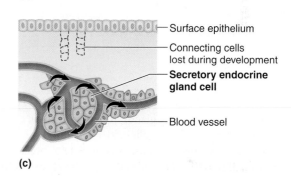

**(c)**

Surface epithelium

Connecting cells lost during development

**Secretory endocrine gland cell**

Blood vessel

> **FIGURE 1-2 Exocrine and endocrine gland formation.** (a) Glands arise during development from the formation of pocket-like invaginations of surface epithelial cells. (b) If the cells at the deepest part of the invagination become secretory and release their product through the connecting duct to the surface, an exocrine gland is formed. (c) If the connecting cells are lost and the deepest secretory cells release their product into the blood, an endocrine gland is formed.

contraction. The wall of the stomach contains smooth muscle tissue, whose contractions mix ingested food with the digestive juices and push the mixture out of the stomach and into the intestine. The stomach wall also contains nervous tissue, which, along with hormones, controls muscle contraction and gland secretion. Connective tissue binds together all of these various tissues.

## Body system level

Groups of organs are further organized into **body systems**. Each system is a collection of organs that perform related functions and interact to accomplish a common activity essential for

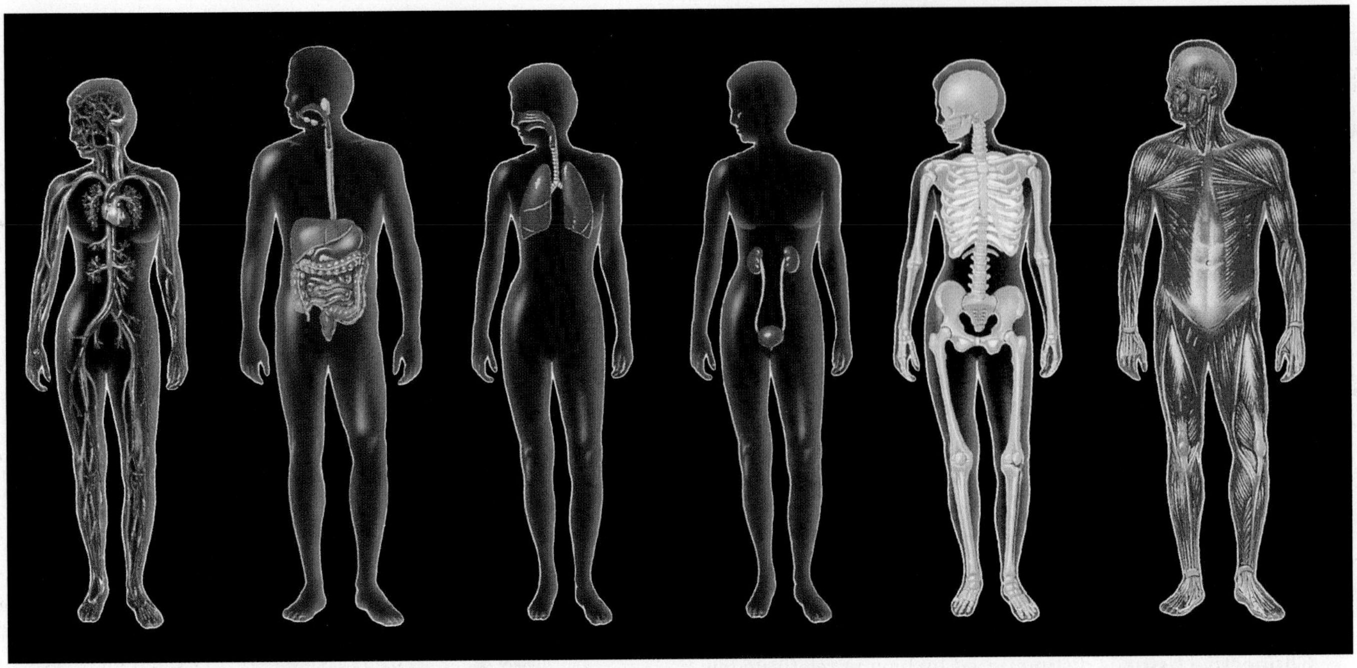

**Circulatory system**
heart, blood vessels, blood

**Digestive system**
mouth, pharynx, esophagus, stomach, small intestine, large intestine, salivary glands, exocrine pancreas, liver, gallbladder

**Respiratory system**
nose, pharynx, larynx, trachea, bronchi, lungs

**Urinary system**
kidneys, ureters, urinary bladder, urethra

**Skeletal system**
bones, cartilage, joints

**Muscular system**
skeletal muscles

© 2016 Cengage

> FIGURE 1-3 **Components of the body systems**

survival of the whole body. For example, the digestive system consists of the mouth, salivary glands, pharynx (throat), esophagus, stomach, pancreas, liver, gallbladder, small intestine, and large intestine. These digestive organs cooperate to break food down into small nutrient molecules that can be absorbed into the blood for distribution to all cells.

The human body has 11 systems: circulatory, digestive, respiratory, urinary, skeletal, muscular, integumentary, immune, nervous, endocrine, and reproductive (> Figure 1-3). Chapters 3 through 17 cover the details of these systems.

## Organism level

Each body system depends on the proper functioning of other systems to carry out its specific responsibilities. The whole body of a multicellular organism—a single, independently living individual—consists of the various body systems structurally and functionally linked as an entity that is separate from the external (outside the body) environment. Thus, the body is made up of living cells organized into life-sustaining systems.

The different body systems do not act in isolation from one another. Many complex body processes depend on the interplay among multiple systems. For example, regulation of blood pressure depends on coordinated responses among the cardiac, circulatory, urinary, nervous, and endocrine systems, as you will

learn later. Even though physiologists may examine body functions at any level, from cells to systems (as the title of this book makes clear), their ultimate goal is to integrate these mechanisms into the big picture: how the entire organism works as a cohesive whole.

Currently, researchers are pursuing several approaches for repairing or replacing tissues or organs that can no longer perform vital functions due to disease, trauma, or age-related changes (see Concepts, Challenges, and Controversies). A similar feature in each chapter explores interesting, tangential information on such diverse topics as environmental impact on the body, aging, ethical issues, new discoveries regarding common diseases, historical perspectives, and so on.

### Check Your Understanding 1.1

1. Define *physiology*.
2. List and describe the levels of organization in the body.
3. Compare basic cell functions and specialized cell functions.
4. Distinguish among external environment, internal environment, intracellular fluid, extracellular fluid, plasma, and interstitial fluid.

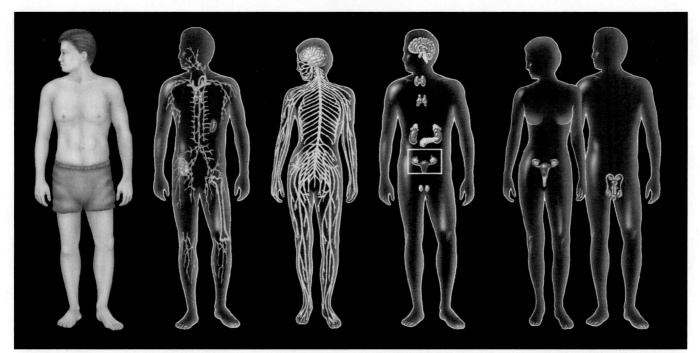

**Integumentary system**
skin, hair, nails

**Immune system**
lymph nodes, thymus, bone marrow, tonsils, adenoids, spleen, appendix, white blood cells (not shown), and mucosa-associated lymphoid tissue

**Nervous system**
brain, spinal cord, peripheral nerves, special sense organs (not shown)

**Endocrine system**
all hormone-secreting tissues, including hypothalamus, pituitary, thyroid, adrenals, endocrine pancreas, gonads, kidneys, pineal, thymus, and, not shown, parathyroids, intestine, heart, skin, and adipose tissue

**Reproductive system**
Male: testes, penis, prostate gland, seminal vesicles, bulbourethral glands, and associated ducts

Female: ovaries, oviducts, uterus, vagina, breasts

© 2016 Cengage

# 1.3 | Concepts of Homeostasis

## Homeostasis

**Homeostasis** (homeostatic) is the ability of a cell or organism to regulate its internal conditions (e.g., pH), typically using feedback systems (e.g., negative feedback) to minimize variation and maintain health regardless of changes in the external environment (e.g., cold temperatures). Accordingly, if the internal environment changes in some fashion, then purposeful adjustments will be initiated to move the internal environment back toward the point prior to the change. That is, if internal body temperature were to decrease, the body would enact processes (e.g., shivering) to increase internal body temperature to 37°C. *It should be noted that in many cases our body does not have a specific* **set point** *(e.g., 37°C) but more a* **physiological range** *(e.g., 36−37°C) that the body operates most efficiently within (steady state).* Examples addressing homeostasis are often more easily portrayed using a specified set point. However, it should be understood that physiological ranges are more accurate targets for homeostatic mechanisms than one specific value.

Each cell has basic survival skills, but why can't the body's cells live without performing specialized tasks and being organized according to specialization into systems that accomplish the functions essential for the whole body's survival? The cells in a multicellular organism cannot live without contributions from the other body cells because the vast majority of cells are not in direct contact with the external environment. A single-celled amoeba can directly obtain nutrients and oxygen from, and eliminate waste into, its immediate external surroundings. In contrast, a muscle cell in a multicellular organism needs nutrients and oxygen, yet the muscle cell cannot directly make these exchanges with the environment surrounding the body because the cell is isolated from this external environment. How does the muscle cell make these vital exchanges? The key is the presence of a watery **internal environment** with which the body cells are in direct contact and make life-sustaining exchanges.

## Body cells

The fluid collectively contained within all body cells is called **intracellular fluid (ICF)**; the fluid outside the cells is called **extracellular fluid (ECF)** (*intra* means "within"; *extra* means "outside of"). The extracellular fluid is the internal environment of the body, the fluid environment in which the cells live. Note that the internal environment is outside the cells but inside the body. By contrast, the intracellular fluid is inside of each cell, and the external environment is outside the body. You live in the external environment; your cells live within the body's internal environment.

# The Science and Direction of Stem Cell Research in Canada

## What Are Stem Cells?

Stem cells are undifferentiated cells that are found in all multicellular organisms (e.g., humans). These primal cells maintain the ability to renew themselves through cell division (mitosis) and can differentiate into a wide variety of specialized cell types. Human stem cell research grew from the findings of Canadian scientists Ernest McCulloch and James Till in the 1960s. There are two broad categories of mammalian stem cells: *embryonic stem cells*, which are derived from blastocysts; and *adult stem cells*, which are found in adult tissues. In the developing embryo, stem cells are able to differentiate into all of the specialized embryonic tissues. In adults, however, the stem cells and progenitor cells act as a repair system for the body. Because stem cells can be readily grown and transformed through cell culture into specialized cells with characteristics consistent with cells of various tissues, such as muscles or nerves, their use in medical therapies has been proposed. In particular, embryonic cell lines and highly plastic adult stem cells from umbilical cord blood or bone marrow are believed to be promising candidates for use in medical therapy.

## Medical Treatments

Researchers believe that stem cell treatments have the potential to change the way we treat human disease. A number of stem cell treatments already exist, although most are still experimental or very controversial. An example of one treatment that is not experimental is bone marrow transplants. Medical researchers predict the use of stem cells to treat spinal cord injuries, cancer, muscle damage, and even degenerative diseases such as Parkinson's and Alzheimer's. Technical difficulties remain, however, and slow the progress of adult stem cell therapeutics. For example, challenges exist in extracting sufficient adult stem cell populations from patients and efficiently growing enough stem cells to add corrective factors. These and other problems still need to be solved. There also still exists a great deal of social uncertainty surrounding embryonic stem cell research.

## Controversy over the Process

In general, the controversy over stem cell research centres on the techniques used in the creation and use of embryonic stem cells. The state of technology today requires the destruction of a human embryo and/or therapeutic cloning in order to establish a human stem cell line. Opponents of stem cell research argue that this practice is very similar to reproductive cloning, which devalues the worth of a potential human being. Proponents of stem cell research argue that the research is necessary because the results are expected to have significant medical potential and the human cost is minimal: the embryos used for stem cell research had been allocated for destruction anyway. The debate has prompted medical and governmental authorities worldwide to develop a regulatory framework that will speak to both social and ethical challenges.

In 2002, the Canadian Institutes for Health Research (CIHR) first developed guidelines governing the use of stem cells for research in Canada (*Human Pluripotent Stem Cell Research: Guidelines for CIHR-Funded Research*). Since then, these guidelines have continued to be updated as this field of scientific research continues to develop. These guidelines are based on six guiding principles:

- Research undertaken should have potential health benefit for all Canadians rather than just specific groups.
- Free and informed consent must be voluntarily provided; this includes consent from donors of both the sperm and the ova used to create embryos, which is highly relevant for those opting for in vitro fertilization if the surplus embryos may be used for research.
- Respect for privacy and confidentiality must be upheld.

---

The extracellular fluid (internal environment) is made up of two components: **plasma**, the fluid portion of the blood; and **interstitial fluid**, which surrounds and bathes the cells (*inter* means "between"; *stitial* means "that which stands") (⟩ Figure 1-4).

No matter how remote a cell is from the external environment, it can make life-sustaining exchanges with its own surrounding internal environment. In turn, particular body systems transfer materials between the external environment and the internal environment so that the composition of the internal environment is maintained (e.g., homeostasis; p. 9) to support the life and functioning of the cells. For example, the digestive system transfers the nutrients required by all body cells from the external environment into the plasma. Likewise, the respiratory system transfers oxygen from the external environment into the plasma. The circulatory system distributes these nutrients and oxygen throughout the body. Materials are thoroughly mixed and exchanged between the plasma and the interstitial fluid across the capillaries, the smallest and thinnest of the blood vessels. As a result, the nutrients and oxygen originally obtained from the external environment are delivered to the interstitial fluid surrounding the cells. The body cells, in turn, pick up these needed supplies from the interstitial fluid. Similarly, wastes produced by the cells are extruded into the interstitial fluid, picked up by the plasma, and transported to the organs that specialize in eliminating these wastes from the internal environment to the external environment. The lungs remove carbon dioxide from the plasma, and the kidneys remove other wastes for elimination in the urine.

- There can be no direct or indirect payment for tissues collected for stem cell research and no financial incentives.
- Embryos cannot be created for research purposes.
- Respect individual and community notions of human dignity and physical, spiritual, and cultural integrity.

In 2014, these guidelines were integrated into the 2nd Edition of the Tri-Council Policy Statement: Ethical Conduct for Research Involving Humans.

### The Search for Noncontroversial Stem Cells

In 2007, the Japanese research team led by Shinya Yamanaka announced that they had created pluripotent stem cells using human skin cells. This finding was a major step forward as it meant that researchers were able to obtain pluripotent stem cells for research without the need of embryos. This type of stem cell is what we now call induced pluripotent stem cells (iPSC or iPS). The technique involves using viral vectors to introduce stem cell–associated genes into nonpluripotent cells and then selectively choosing cells that have the characteristics of stem cells. Although this discovery increased the availability of stem cells for research, its therapeutic potential was limited as it was also prone to the formation of cancers.

### Canadian Efforts

Canadians have played many prominent roles in stem cell research. As early as 1961, Canadian scientists James Till and Ernest McCulloch were the first to demonstrate the existence of stem cells, for which they were awarded the 2005 Lasker Prize. In 1992, Dr. Sam Weiss, the 2008 Gairdner award–winning scientist at the University of Calgary, identified stem cells in the adult brain and discovered that the brain uses these stem cells to heal itself. At the University of Toronto in 1994, Derek van der Kooy identified retinal stem cells in mice, while John Dick isolated the first cancer stem cells from a leukemia patient. More recent advances include Andras Nagy's contributions, in collaboration with Scottish scientists, which allows the creation of iPSCs without the use of potentially cancer-causing viral vectors, and the work by Freda Miller's discovery of skin-derived progenitor cells, which are currently being investigated for use in spinal cord injury. These are just a few examples of important research by Canadians.

The **Stem Cell Network** (SCN) is a Canadian venture that fosters ethical stem cell research by bringing together more than 100 individuals to examine the therapeutic potential of stem cells. The network was started by a $21.1-million grant from the Networks of Centres of Excellence in 2001. It includes scientists, clinicians, engineers, and ethicists. Funding is made possible by the Natural Sciences and Engineering Research Council of Canada (NSERC), Canadian Institutes of Health Research (CIHR), and the Social Sciences and Humanities Research Council (SSHRC), in partnership with Industry Canada. The goal of SCN is "to be a catalyst for realizing the full potential of stem cell research for Canadians." More information about SCN can be found at http://www.stemcellnetwork.ca. Diseases that may benefit from stem cell research include cancer, HIV/AIDS, diabetes, and obesity.

### Further Reading

Hawke, T.J. (2005). Muscle stem cells and exercise training. *Exerc Sport Sci Rev*, 33(2): 63–68.

Price, F.D., Kuroda, K., & Rudnicki, M.A. (2007). Stem cell based therapies to treat muscular dystrophy. *Biochim Biophys Acta*, 1772(2): 272–83.

## Body systems

The body cells can live and function only when the extracellular fluid is compatible with their survival; thus, the chemical composition and physical state of this internal environment must be maintained within narrow limits, that is, a physiological range (e.g., pH). As cells take up nutrients and oxygen from the internal environment, these essential materials are continuously replenished. Similarly, wastes are constantly removed from the internal environment so that they do not reach toxic levels. Other aspects of the internal environment important for maintaining life, such as temperature and pH, must be kept relatively constant.

The functions performed by each body system contribute to homeostasis, thereby maintaining our internal environment and making survival of all the cells possible. Cells, in turn, make up body systems. This theme is central to physiology and this book: *Homeostasis is essential for the survival of each cell, and each cell, through its specialized activities, contributes as part of a body system to the maintenance of the internal environment shared by all cells* (> Figure 1-5).

Homeostasis is not a rigid, fixed state, or absolute setting, but rather a dynamic, steady state in which changes occur but are minimized by multiple dynamic equilibrium adjustment mechanisms (basic component parts are receptor, control centre, effector). Both external and internal factors continuously activate these adjustment mechanisms to maintain a relatively stable internal environment. The term *dynamic* refers to the fact that each homeostatically regulated factor is marked

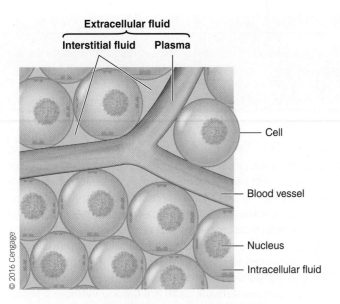

**Extracellular fluid**

**Interstitial fluid    Plasma**

Cell

Blood vessel

Nucleus

Intracellular fluid

© 2016 Cengage

❯ **FIGURE 1-4 Components of the extracellular fluid (internal environment).** A body cell takes in essential nutrients from its watery surroundings and eliminates wastes into these same surroundings, just as an amoeba does. The main difference is that each body cell must help maintain the composition of the internal environment so that this fluid continuously remains suitable to support the existence of all the body cells. In contrast, an amoeba does nothing to regulate its surroundings.

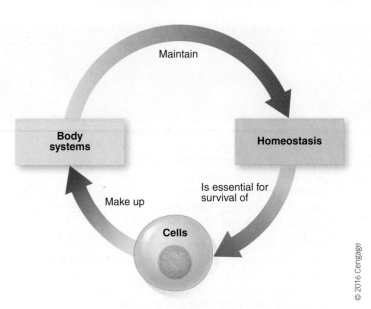

Maintain

**Body systems**

**Homeostasis**

Is essential for survival of

Make up

**Cells**

© 2016 Cengage

❯ **FIGURE 1-5 Interdependent relationship of cells, body systems, and homeostasis.** The depicted interdependent relationship serves as the foundation for modern-day physiology: homeostasis is essential for the survival of cells; body systems maintain homeostasis; and cells make up body systems.

by continuous change, whereas *steady state* implies that these changes do not deviate too far from a constant level. This situation is comparable to the minor steering adjustments made while driving a car along a straight course down the highway. Small fluctuations around the optimal level for each factor in the internal environment are normally kept, by carefully regulated mechanisms, within the narrow limits of a physiological range compatible with life.

Some of the dynamic equilibrium adjustments are immediate, transient responses to a situation that moves a regulated factor in the internal environment outside the desired range. Others are more long-term (i.e., chronic) adaptations that take place in response to prolonged or repeated exposure to a situation that disrupts homeostasis (e.g., exercise). Chronic adaptations make the body more efficient in responding to an ongoing or repetitive challenge. The body's reaction to exercise includes examples of both short-term (i.e., acute) compensatory responses and chronic adaptations among the different body systems: for example, the acute response of the body to environmental heat.

**FACTORS HOMEOSTATICALLY REGULATED**

Many factors of the internal environment must be homeostatically maintained, including the following:

1. *Concentration of nutrient molecules.* Cells need a constant supply of nutrient molecules for energy production. Energy, in turn, is needed to support life-sustaining and specialized cell activities.

2. *Concentration of oxygen and carbon dioxide.* Cells need oxygen to carry out energy-yielding chemical reactions. The carbon dioxide produced during these reactions must be removed so that acid-forming carbon dioxide does not increase the acidity of the internal environment.

3. *Concentration of waste products.* Some chemical reactions produce end products that exert a toxic effect on the body's cells if these wastes are allowed to accumulate.

4. *pH.* Changes in the pH (relative amount of acid) adversely affect nerve cell function and wreak havoc with the enzymatic activity of all cells.

5. *Concentration of water, salt, and other electrolytes.* Because the relative concentrations of salt and water in the extracellular fluid influence how much water enters or leaves the cells, these concentrations are carefully regulated to maintain the proper volume of the cells. Cells do not function normally when they are swollen or shrunken. Other electrolytes perform a variety of vital functions. For example, the rhythmic beating of the heart depends on a relatively constant concentration of potassium in the extracellular fluid.

6. *Volume and pressure.* The circulating component of the internal environment—the plasma—must be maintained at adequate volume and blood pressure to ensure body-wide distribution of this important link between the external environment and the cells.

7. *Temperature.* Body cells function best within a narrow temperature range. If cells are too cold, their functions slow down

too much; worse yet, if they get too hot, their structural and enzymatic proteins are impaired or destroyed.

## CONTRIBUTIONS OF THE BODY SYSTEMS TO HOMEOSTASIS

The 11 body systems contribute to homeostasis in the following important ways. (Note that the figure included on the inside front cover can be used as a handy reference to help you keep track of how all the pieces fit together.)

1. The *circulatory system* transports materials, such as nutrients, oxygen, carbon dioxide, wastes, electrolytes, and hormones, from one part of the body to another. It also assists with thermoregulation by moving heat to the periphery from the core.

2. The *digestive system* breaks down dietary food into small nutrient molecules that can be absorbed into the plasma for distribution to the body cells, and transfers water and electrolytes from the external environment into the internal environment. It eliminates undigested food residues to the external environment in the feces.

3. The *respiratory system*, consisting of the lungs and major airways, receives oxygen from the external environment and eliminates carbon dioxide from the internal environment. By adjusting the rate of removal of acid-forming carbon dioxide, the respiratory system is also important in maintaining the proper pH of the internal environment.

4. The *urinary system* removes excess water, salt, acid, and other electrolytes from the plasma and eliminates them in the urine, along with waste products other than carbon dioxide. This system includes the kidneys and associated "plumbing."

5. The *skeletal system* (bones and joints) provides support and protection for the soft tissues and organs. It also serves as a storage reservoir for calcium, an electrolyte whose plasma concentration must be maintained within very narrow limits. Furthermore, the bone marrow—the soft interior portion of some types of bone—is the ultimate source of all blood cells.

6. The *muscular system* (skeletal muscles) and skeletal system form the basis of movement. The skeletal muscles attach via tendons to bones. When the muscles contract, this enables the bones to move and lets us walk, grab, and jump. From a purely homeostatic view, this system enables an individual to move toward food or away from harm. Furthermore, the heat generated by muscle contraction is important in temperature regulation. In addition, because skeletal muscles are under voluntary control, a person can use them to accomplish many other movements. These movements, which range from the fine motor skills required for delicate needlework to the powerful movements involved in weight-lifting, are not necessarily directed toward maintaining homeostasis.

7. The *integumentary system* (skin and related structures) serves as an outer protective barrier that prevents internal fluid from being lost from the body and foreign microorganisms from entering (part of our external defence; p. 460). This system is also important in regulating body temperature. The amount of heat lost from the body surface to the external environment can be adjusted by controlling sweat production and by regulating the flow of warm blood through the skin.

8. The *immune system* (white blood cells and lymphoid organs) defends against foreign invaders and body cells that have become cancerous. It also paves the way for repairing or replacing injured or worn-out cells.

9. The *nervous system* (brain, spinal cord, and nerves) is one of the two major regulatory systems of the body. In general, it controls and coordinates bodily activities that require swift responses. The nervous system is especially important in detecting and initiating reactions to changes in the external environment. It is also responsible for higher functions that are not entirely directed toward maintaining homeostasis, such as consciousness, memory, and creativity.

10. The *endocrine system* is the other major regulatory system (and a communication system). Unlike the nervous system (our body's fast communication system), in general the hormone-secreting glands of the endocrine system regulate activities that require duration rather than speed, such as growth. This system is especially important in controlling the concentration of nutrients and, by adjusting kidney function, controlling the internal environment's volume and electrolyte composition.

11. The *reproductive system* is essential for perpetuating the species. Homeostatic mechanisms ensure that both the male and female reproductive systems are optimized to favour reproductive success.

As we examine each of these systems in greater detail, always keep in mind that the body is a well-coordinated and integrated whole, even though each system provides its own special contributions. It is easy to forget that all the body parts actually fit together into a functioning, interdependent whole body. Accordingly, each chapter begins with a figure and a discussion that focus on how the body system described in that chapter fits into the body as a whole. In addition, each chapter ends with a brief overview of the homeostatic contributions of the body system. You should also be aware that the whole functioning is greater than the sum of its separate parts. Through specialized, coordinated, and interdependent functions, cells combine to form an integrated, unique, single living organism with more diverse and complex capabilities than those possessed by any of the cells that make it up. For humans, these capabilities go far beyond the processes needed to maintain life. A cell, or even a random combination of cells, obviously cannot create an artistic masterpiece or design a spacecraft, but body cells working together permit an individual to create these things.

You have now learned what homeostasis is and how the functions of different body systems maintain it. Now let's look at the regulatory mechanisms by which the body reacts to changes and controls the internal environment.

# 1.4 | Homeostatic Control Systems

To maintain homeostasis you need a homeostatic control system. This is a functionally interconnected network of body components that operate to maintain a given factor (variable; e.g., temperature) in the internal environment within a narrow range to maintain a steady state. To maintain homeostasis, the control system must be able to (1) detect deviations from normal in the internal environment (receptor), (2) integrate this information with any other relevant information (control centre), and (3) trigger the needed adjustments responsible for restoring this factor within the normal range (effector).

## Local or body-wide control

Homeostatic control systems can be grouped into two classes—intrinsic and extrinsic controls. **Intrinsic (local) controls** are built into or are inherent in an organ (*intrinsic* means "within"). For example, as an exercising skeletal muscle rapidly uses up oxygen to generate energy to support its contractile activity, the oxygen concentration within the muscle falls. This local chemical change acts directly on the smooth muscle in the walls of the blood vessels that supply the exercising muscle, causing the smooth muscle to relax so that the vessels dilate (open wider). As a result, an increased amount of blood flows through the dilated vessels into the exercising muscle, bringing in more oxygen. This local mechanism helps maintain an optimal level of oxygen in the interstitial fluid around the exercising muscle's cells.

However, most factors in the internal environment are maintained by **extrinsic controls**, which are regulatory mechanisms initiated outside an organ to alter the activity of the organ (*extrinsic* means "outside of"). Extrinsic control of the organs and body systems is achieved by the two major regulatory systems, the nervous and endocrine systems. Extrinsic control permits synchronized regulation of several organs to a common goal; in contrast, intrinsic controls are self-serving for the organ in which they occur. Coordinated regulatory mechanisms are crucial to maintain the dynamic equilibrium associated with the steady state of the internal environment. To maintain blood pressure within the necessary range, the nervous system simultaneously acts on the heart and the blood vessels throughout the body to increase or decrease blood pressure.

To stabilize the necessary physiological factors, homeostatic control systems must be able to detect and make the necessary adjustments to various changes. In making these dynamic equilibrium adjustments, feedback and feedforward loops come into play. The term **feedback** refers to responses made after a change has been detected; the term **feedforward** describes responses made in anticipation of a change. Let's take a look at these mechanisms in more detail.

## Negative feedback

Homeostatic control mechanisms operate primarily on the principle of negative feedback, which attempts to provide stability within the internal environment. In **negative feedback**, a change in a homeostatically controlled factor triggers a response that seeks to maintain homeostasis by moving the factor in the opposite direction of its initial change. That is, a corrective adjustment opposes the original deviation from the homeostatic steady state.

For example, room temperature is a **controlled variable** (factor) held within a narrow range by a control system. The control system includes a thermostatic device, a furnace, and all their electrical connections. Room temperature is determined by the activity of the furnace. To switch on or switch off the furnace, the control system as a whole must be able to determine what the *actual* room temperature is, compare this with the *desired* room temperature, and adjust the output of the furnace to maintain the temperature in the desired level. A thermometer provides information about the actual room temperature, acting as a **sensor** (physiologically, a sensor is termed a *receptor*) to monitor the magnitude and direction of the controlled variable. The sensor typically converts the information regarding a change into a language the control system can understand. For example, the thermometer part of the thermostat converts the magnitude of the air temperature into electrical impulses. This message serves as the input into the control system. The thermostat setting provides the desired temperature steady state. The thermostat acts as a **control centre** (integrator), which compares the sensor's input with the optimal range and adjusts the furnace output to bring about the appropriate response, opposing the deviation outside the range, which provides stability. The furnace is the **effector**, the component of the control system that brings about the desired effect. These general components of a negative-feedback control system are summarized in › Figure 1-6a.

Let's look at a typical negative-feedback loop. When room temperature falls below the predetermined range, the thermostat activates the furnace, which produces heat to raise the room temperature to the selected steady state (› Figure 1-6b). Once the room temperature reaches the predetermined range, the thermometer no longer detects a deviation in temperature, and the thermostat no longer activates the furnace. In this way, the heat from the furnace counteracts, or is *negative* to, the original fall in temperature. If the heat-generating pathway were not shut off once the target temperature was reached, heat production would continue indefinitely increasing the temperature. Overshooting the optimal range does not occur, because the heat "feeds back" to shut off the thermostat and terminate the heat output. Thus, a negative-feedback control system detects any change in a controlled factor away from the steady-state value, initiates mechanisms to correct the situation, and then shuts itself off.

What happens when the original deviation involves a rise in room temperature above the selected steady state? A heat-producing furnace is of no use in returning the room temperature to the desired level. In this case, the thermostat activates a different effector, an air conditioner, to cool the room air. The effect is opposite that of the furnace. Using negative feedback, the optimal range is reached, and the air conditioner is turned off to prevent overcooling. However, if the house is equipped with only a furnace that produces heat to oppose a fall in room temperature, no mechanism is available to prevent the house from getting too hot. To keep room temperature at a steady state, two opposing mechanisms must be used: one that heats and one that cools.

Homeostatic negative-feedback systems in the human body operate in the same way. When temperature-monitoring nerve cells in the hypothalamus detect a decrease in body temperature below the physiological range, they signal other nerve cells in a control centre, which initiate shivering (and other responses) to generate metabolic heat and increase body temperature to within the physiological range (⟩ Figure 1-6c). When body temperature rises to within the physiological range, the temperature-monitoring nerve cells turn off the stimulatory signal to the skeletal muscles, stopping the shivering. Conversely, when the temperature-monitoring nerve cells in the hypothalamus detect a rise in body temperature above normal, cooling mechanisms, such as sweating, come into play to reduce the temperature to normal. When the temperature reaches the physiological range, the cooling mechanisms are shut off. As with body temperature, opposing mechanisms can move most homeostatically controlled variables in either direction as needed.

## Positive feedback

Negative feedback is used to resist change and add stability to the internal environment. With **positive feedback** the output enhances or amplifies a particular change so that the controlled factor continues to move in the direction of the initial change. This action is comparable to the heat generated by a furnace triggering the thermostat to call for even *more* heat output from the furnace, increasing room temperature even more.

Because the body's goal is to remain within a relatively stable homeostatic range, positive feedback occurs much less frequently than negative feedback. Positive feedback is used in certain circumstances, such as child birth. The hormone oxytocin causes powerful uterine contractions. These contractions push the baby against the cervix, stretching the cervix. This stretch activates sensors that lead to the release of more oxytocin, which causes stronger uterine contractions, triggering the release of more oxytocin, and so on. This positive-feedback cycle does not stop until the baby exits the cervix. Likewise, all other normal instances of positive feedback include some mechanism for stopping the cycle. However, some abnormal circumstances are characterized by runaway positive-feedback loops that continue to move the body farther and farther from homeostatic balance, potentially until death or a medical intervention stops the cycle.

Such an example of abnormal positive feedback occurs during heatstroke. When temperature-regulating mechanisms are unable to cool the body sufficiently in the face of pronounced environmental heat exposure, the body temperature may rise so high that the temperature control centre becomes impaired. Because

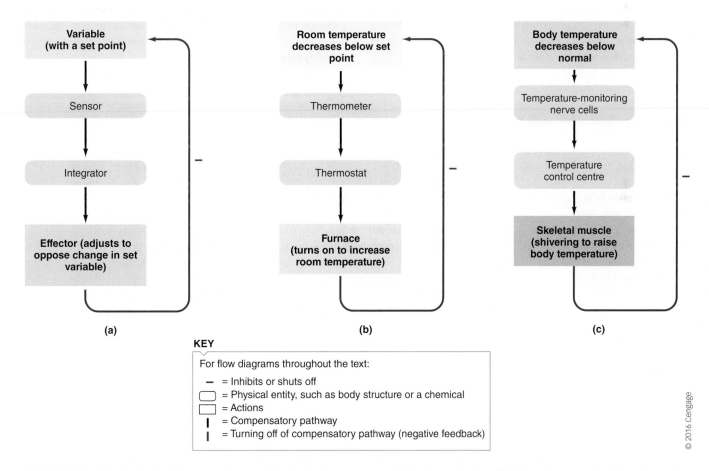

(a)                    (b)                    (c)

**KEY**

For flow diagrams throughout the text:

— = Inhibits or shuts off
⬭ = Physical entity, such as body structure or a chemical
▭ = Actions
❙ = Compensatory pathway
❙ = Turning off of compensatory pathway (negative feedback)

© 2016 Cengage

⟩ **FIGURE 1-6 Negative feedback.** (a) Components of a negative-feedback control system. (b) Negative-feedback control of room temperature. (c) Negative-feedback control of body temperature

the temperature control centre is dysfunctional, its ability to start the cooling mechanism is diminished and body temperature continues to increase, causing further damage to the control centre. As a result, body temperature spirals out of control.

## Feedforward mechanisms and anticipation

Feedback mechanisms bring about a reaction to a change in a regulated factor, whereas a feedforward mechanism anticipates change in a regulated factor. For example, when a meal is still in the digestive tract, a feedforward mechanism increases secretion of a hormone (insulin) that will promote the cellular uptake and storage of ingested nutrients after they have been absorbed from the digestive tract. This anticipatory response helps limit the rapid rise in nutrients (e.g., glucose) after absorption into the blood, assisting in maintenance of a more steady-state blood nutrient level.

## Disruptions in homeostasis

Despite control mechanisms, when one or more of the body's systems malfunction, homeostasis is disrupted, and the regulated factor moves outside the dynamic range in which it most effectively operates. In this situation, all cells suffer because they no longer have an optimal environment in which to live and function. Various pathophysiological states ensue, depending on the type and extent of homeostatic disruption. The term **pathophysiology** refers to the abnormal functioning of the body (altered physiology) associated with disease. The result of disease may be a homeostatic disruption that is so severe that it is no longer compatible with survival, and death results.

### Check Your Understanding 1.2

1. Distinguish between intrinsic controls and extrinsic controls.
2. Compare and contrast negative feedback and positive feedback.
3. Draw a diagram of negative feedback showing how the different components are connected.

# Chapter in Perspective: Focus on Homeostasis

In this chapter, you have learned what homeostasis is: a dynamic, steady state of the constituents in the internal fluid environment (the extracellular fluid) that surrounds and exchanges materials with the cells. Maintenance of homeostasis is essential for the survival and normal functioning of cells. Each cell, through its specialized activities, contributes as part of a body system to the maintenance of homeostasis.

This relationship is the foundation of physiology and the central theme of this book. We have described how cells are organized according to specialization into body systems. How homeostasis is essential for cell survival and how body systems maintain this internal constancy are the topics covered in the rest of this book. Each chapter concludes with this capstone feature to facilitate your understanding of how the system under discussion contributes to homeostasis, as well as of the interactions and interdependency of the body systems.

## CHAPTER TERMINOLOGY

anatomy (p. 4)
body systems (p. 7)
cell (p. 5)
cell differentiation (p. 5)
connective tissue (p. 7)
control centre (p. 14)
controlled variable (p. 14)
effector (p. 14)
endocrine glands (p. 7)
epithelial tissue (p. 6)
exocrine glands (p. 7)
extracellular fluid (ECF) (p. 9)
extrinsic controls (p. 14)
feedback (p. 14)
feedforward (p. 14)

glands (p. 6)
homeostasis (p. 9)
human physiology (p. 3)
internal environment (p. 9)
interstitial fluid (p. 10)
intracellular fluid (ICF) (p. 9)
intrinsic (local) controls (p. 14)
lumen (p. 6)
mechanistic approach (p. 4)
medicine (p. 3)
muscle tissue (p. 6)
negative feedback (p. 14)
nervous tissue (p. 6)
nursing (p. 3)
organisms (p. 5)

organs (p. 7)
pathophysiology (p. 16)
physical therapy (p. 3)
physiological range (p. 9)
physiologists (p. 4)
plasma (p. 10)
plasma membrane (p. 5)
positive feedback (p. 15)
secretion (p. 6)
sensor (p. 14)
set point (p. 9)
stem cell network (p. 11)
teleological approach (p. 4)
tissues (p. 6)

### Objective Questions (Answers in Appendix E, p. A-35)

1. Which of the following activities is NOT carried out by every cell in the body?
   a. Obtaining oxygen and nutrients
   b. Performing chemical reactions to acquire energy for the cell's use
   c. Eliminating wastes
   d. Controlling to a large extent exchange of materials between the cell and its external environment
   e. Reproducing

2. What is the proper progression of the levels of organization in the body?
   a. Chemicals, cells, organs, tissues, body systems, whole body
   b. Chemicals, cells, tissues, organs, body systems, whole body
   c. Cells, chemicals, tissues, organs, whole body, body systems
   d. Cells, chemicals, organs, tissues, whole body, body systems
   e. Chemicals, cells, tissues, body systems, organs, whole body

3. Which of the following is NOT a type of connective tissue?
   a. Bone
   b. Blood
   c. The spinal cord
   d. Tendons
   e. The tissue that attaches epithelial tissue to underlying structures

4. The term *tissue* can apply either to one of the four primary tissue types or to a particular organ's aggregate of cellular and extracellular components. *(True or false?)*

5. Cells in a multicellular organism have specialized to such an extent that they have little in common with single-celled organisms. *(True or false?)*

6. Cell specializations are usually a modification or elaboration of one of the basic cell functions. *(True or false?)*

7. The four primary types of tissue are _____, _____, _____, and _____.

8. The term _____ refers to the release from a cell, in response to appropriate stimulation, of specific products that have in large part been synthesized by the cell.

9. _____ glands secrete through ducts to the outside of the body, whereas _____ glands release their secretory products, known as _____, internally into the blood.

10. _____ controls are inherent to an organ, whereas _____ controls are regulatory mechanisms initiated outside an organ that alter the activity of the organ.

11. Match the following:

| | |
|---|---|
| ___ 1. circulatory system | (a) obtains oxygen and eliminates carbon dioxide |
| ___ 2. digestive system | (b) support, protect, and move body parts |
| ___ 3. respiratory system | (c) controls, via hormones it secretes, processes that require duration |
| ___ 4. urinary system | (d) acts as transport system |
| ___ 5. muscular and skeletal systems | (e) removes wastes and excess water, salt, and other electrolytes |
| ___ 6. integumentary system | (f) perpetuates the species |
| ___ 7. immune system | (g) obtains nutrients, water, and electrolytes |
| ___ 8. nervous system | (h) defends against foreign invaders and cancer |
| ___ 9. endocrine system | (i) acts through electrical signals to control body's rapid responses |
| ___ 10. reproductive system | (j) serves as outer protective barrier |

### Written Questions

1. Define *physiology*.

2. What are the basic cell functions?

3. Distinguish between the external environment and the internal environment. What constitutes the internal environment?

4. What fluid compartments make up the internal environment?

5. Define *homeostasis*.

6. Describe the interrelationships among cells, body systems, and homeostasis.

7. What factors must be homeostatically maintained?

8. Define and describe the components of a homeostatic control system.

9. Compare and contrast negative and positive feedback.

(Explanations in Appendix E, p. A-35)

1. Considering the nature of negative-feedback control and the function of the respiratory system, what effect do you predict that a decrease in carbon dioxide in the internal environment would have on how rapidly and deeply a person breathes?

2. Would the oxygen levels in the blood be (a) normal, (b) below normal, or (c) elevated in a patient with severe pneumonia resulting in impaired exchange of oxygen and carbon dioxide between the air and blood in the lungs? Would the carbon dioxide levels in the same patient's blood be (a) normal, (b) below normal, or (c) elevated? Because $CO_2$ reacts with $H_2O$ to form carbonic acid ($H_2CO_3$), would the patient's blood (a) have a normal pH, (b) be too acidic, or (c) not be acidic enough (i.e., be too alkaline), assuming that other compensatory measures have not yet had time to act?

3. The hormone insulin enhances the transport of glucose (sugar) from the blood into most body cells. Its secretion is controlled by a negative-feedback system between the concentration of glucose in the blood and the insulin-secreting cells. Therefore, which of the following statements is correct?
   a. A decrease in blood glucose concentration stimulates insulin secretion, which in turn further lowers blood glucose concentration.
   b. An increase in blood glucose concentration stimulates insulin secretion, which in turn lowers blood glucose concentration.
   c. A decrease in blood glucose concentration stimulates insulin secretion, which in turn increases blood glucose concentration.
   d. An increase in blood glucose concentration stimulates insulin secretion, which in turn further increases blood glucose concentration.

4. Given that most AIDS victims die from overwhelming infections or rare types of cancer, what body system do you think HIV (the AIDS virus) impairs?

5. Body temperature is homeostatically regulated around a physiological range. Given your knowledge of negative feedback and homeostatic control systems, predict whether narrowing or widening of the blood vessels of the skin will occur when a person exercises strenuously. (Hint: Muscle contraction generates heat. Narrowing of the vessels supplying an organ decreases blood flow through the organ, whereas vessel widening increases blood flow through the organ. The more warm blood flowing through the skin, the greater the loss of heat from the skin to the surrounding environment.)

**CLINICAL CONSIDERATION**

(Explanation in Appendix E, p. A-35)

Jennifer R. has a stomach flu that is going around campus. She has been vomiting profusely for the past 24 hours. Not only has she been unable to keep down fluids or food, but she has also lost the acidic digestive juices secreted by the stomach that are normally reabsorbed back into the blood farther down the digestive tract. In what ways might this condition threaten to disrupt homeostasis in Jennifer's internal environment? That is, what homeostatically maintained factors are moved away from normal by her profuse vomiting? What body systems respond to resist these changes?

# Body Systems

Body systems maintain homeostasis

## Homeostasis

The specialized activities of the cells that make up the body systems are aimed at maintaining homeostasis, a dynamic steady state of the constituents in the internal fluid environment.

Homeostasis is essential for survival of cells

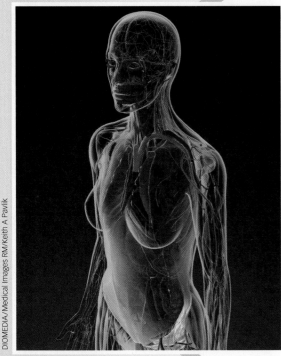

DIOMEDIA/Medical Images RM/Keith A Pavlik

Anatomy and orientation of body systems

Cells make up body systems

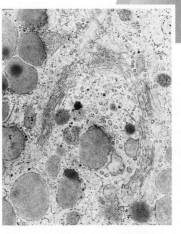

Dr. Kari Lounatmaa/Science Source

Electron micrograph of the Golgi apparatus

Cells are the body's living building blocks. Just as the body as a whole is highly organized, so too is a cell's interior. A cell has three major parts: a *plasma membrane* that encloses the cell; the *nucleus*, which houses the cell's genetic material; and the *cytoplasm*, which is organized into discrete, highly specialized *organelles* dispersed throughout a gel-like liquid, the *cytosol*. The cytosol is pervaded by a protein scaffolding, the cytoskeleton, that serves as the "bone and muscle" of the cell.

Through the coordinated actions of each of these cell components, every cell can perform certain basic functions essential to its own survival as well as a specialized task that helps maintain homeostasis. Cells are organized according to their specialization into body systems that maintain the stable internal environment essential for the whole body's survival. All body functions ultimately depend on the activities of the individual cells that compose the body.

# 2

# Cell Physiology

The same chemicals that make up living cells are also found in nonliving objects. Even though researchers have analyzed the chemicals of which cells are made, they have not been able to organize these chemicals into a living cell in a laboratory. Life stems from the complex organization and interaction of these chemicals within the cell. Groups of inanimate chemicals are structurally organized and function together in unique ways to form a cell, the smallest living entity. Cells, in turn, serve as the living building blocks for the immensely complicated whole body. Thus, cells are the bridge between chemicals and humans (and all living organisms). Furthermore, each new cell and all new life arise from the division of pre-existing cells, not from nonliving sources. Because of this continuity of life, the cells of all organisms are fundamentally similar in structure and function. ▌ Table 2-1 summarizes these principles, which are known collectively as the **cell theory**. By probing deeper into the molecular structure and organization of the cells that make up the body, modern physiologists are unravelling many of the broader mysteries of how the body works.

## TABLE 2-1 Principles of the Cell Theory

- The cell is the smallest structural and functional unit capable of carrying out life processes.
- The functional activities of each cell depend on the specific structural properties of the cell.
- Cells are the living building blocks of all plant and animal organisms.
- An organism's structure and function ultimately depend on the individual and collective structural characteristics and functional capabilities of its cells.
- All new cells and new life arise only from pre-existing cells.
- Because of this continuity of life, the cells of all organisms are fundamentally similar in structure and function.

© 2016 Cengage

# 2.1 Observation of Cells

Until the invention of the microscope in the middle of the 17th century, scientists did not know that the cell existed. It is not entirely clear who invented the microscope, but three people from the Netherlands are given credit: Hans Lippershey, and Hans and Zacharias Janssen (who were father and son). In the early part of the 19th century, with the development of better light microscopes (the lens quality improved greatly), researchers learned that all plant and animal tissues consist of individual cells. The cells of a hummingbird, a human, and a whale are all about the same size. Larger species have a greater number of cells, not larger cells. These early investigators also discovered that cells are filled with a fluid, which, given the microscopic capabilities of the time, appeared to be a rather uniform, soupy mixture believed to be the elusive "stuff of life." In the 1940s, when scientists first employed the technique of electron microscopy to observe living matter, they began to understand the great diversity and complexity of the internal structure of cells. Electron microscopes have approximately 1000 times more magnification as light microscopes. Modern microscopic techniques like confocal imaging even allow us to look into live cells. Now that scientists have even more sophisticated microscopes, biochemical techniques, cell culture technology, and genetic engineering, the concept of the cell as a microscopic bag of formless fluid has given way to the current understanding of the cell as a complex, highly organized, compartmentalized structure.

### Check Your Understanding 2.1

1. State the principles of the cell theory.
2. Compare the average size of cells in your body with those in a mouse and in an elephant.

# 2.2 An Overview of Cell Structure

The trillions of cells in a human body are classified into about 200 different cell types based on specific variations in structure and function. Even though there is no such thing as a typical cell, because of these diverse structural and functional specializations, different cells share many common features. For example, a single skeletal muscle cell is cylindrical and elongated, but it can vary in length, which depends on the recruitment of **myoblasts** into existing fibres. A muscle cell in the sartorius muscle of the thigh contains single cells that can be 300 mm (30 centimetres) in length. Every muscle cell contains multiple nuclei (i.e., they are multinucleated), as well as cytosol, ribosomes, and mitochondria, which are common to many cells. Depending on the type of muscle cell (type I or type II; p. 318), there may be differences in the number or volume of mitochondria. A series of common components are also directly associated with force generation and are more specific to the muscle cell. Examples of these components are regulatory proteins (troponin and tropomyosin), contractile proteins (actin and myosin), and structural proteins (titin, nebulin, and desmin). Thus, not all cells are the same, as we will see. However, most cells have three major subdivisions: the *plasma membrane*, which encloses the cells; the **nucleus**, which contains the cell's genetic material; and the **cytoplasm**, the portion of the cell's interior not occupied by the nucleus but containing numerous organelles, structural proteins, transport and secretory vesicles, and enzymes. A summary of cell structures and functions can be found in Table 2-2.

# 2.3 Cellular Metabolism

The term **intermediary metabolism** refers collectively to the large set of chemical reactions inside the cell that involve the degradation, synthesis, and transformation of small organic molecules, such as simple sugars, amino acids, and fatty acids. These reactions are critical for ultimately capturing energy to be used for cell activities and for providing the raw materials needed for maintaining the cell's structure and function and for the cell's growth. All intermediary metabolism occurs in the cytoplasm, with most of it accomplished in the cytosol. The cytosol contains thousands of enzymes involved in glycolysis and other intermediary biochemical reactions.

Cells are incredibly dynamic and are constantly doing things related to their primary functions as well as their housekeeping chores to keep the cells alive and well. In order for this to occur, cells must be both making things that are needed and breaking down things that are not. **Anabolic** processes are those that favour the synthesis of molecules for building up organs and tissues. In contrast, **catabolic** processes are those that favour the breakdown of complex molecules into more simple ones. It is important to understand that cells are simultaneously undergoing both anabolism and catabolism and that these processes may be balanced or that one of the states is favoured, depending on the demands of the cell at any given time. Anabolic processes,

**TABLE 2-2** Summary of Cell Structures and Functions

| Cell Part | Structure | Function |
|---|---|---|
| **Plasma membrane** | Lipid bilayer studded with proteins and small amounts of carbohydrate | Acts as selective barrier between cellular contents and extracellular fluid; controls traffic in and out of the cell |
| **Nucleus** | DNA and specialized proteins enclosed by a double-layered membrane | Acts as control centre of the cell, providing storage of genetic information; nuclear DNA provides codes for the synthesis of structural and enzymatic proteins and serves as blueprint for cell replication |
| **Cytoplasm** | | |
| *Organelles* | | |
| Endoplasmic reticulum | Extensive, continuous membranous network of fluid-filled tubules and flattened sacs, partially studded with ribosomes | Forms new cell membrane and other cell components and manufactures products for secretion |
| Golgi complex | Sets of stacked, flattened membranous sacs | Modifies, packages, and distributes newly synthesized proteins |
| Lysosomes | Membranous sacs containing hydrolytic enzymes | Serve as digestive system of the cell, destroying foreign substances and cellular debris |
| Centriole | Usually paired, small barrel-shaped organelles that consist of nine short triplet microtubules | Site of growth of new microtubules: both cytoplasmic transport microtubules and the microtubules that form the mitotic spindle |
| Peroxisomes | Membranous sacs containing oxidative enzymes | Perform detoxification activities |
| Mitochondria | Rod- or oval-shaped bodies enclosed by two membranes, with the inner membrane folded into cristae that project into an interior matrix | Act as energy-producing organelles; major sites of ATP production; contain enzymes for citric acid cycle and electron transport chain |
| Vaults | Shaped like hollow octagonal barrels | Serve as cellular trucks for transport from nucleus to cytoplasm |
| *Cytosol: gel-like portion* | | |
| Intermediary metabolism enzymes | Dispersed within the cytosol | Facilitate intracellular reactions involving the degradation, synthesis, and transformation of small organic molecules |
| Ribosomes | Granules of RNA and proteins—some attached to rough endoplasmic reticulum, some free in the cytoplasm | Serve as workbenches for protein synthesis |
| Transport, secretory, and endocytotic vesicles | Transiently formed, membrane-enclosed products synthesized within or engulfed by the cell | Transport and/or store products being moved within, out of, or into the cell, respectively |
| Inclusions | Glycogen granules, fat droplets | Store excess nutrients |
| *Cytosol: cytoskeleton portion* | | As an integrated whole, serves as the cell's "bone and muscle" |
| Microtubules | Long, slender, hollow tubes composed of secretory vesicles | Maintain asymmetric cell shapes and tubulin molecules; coordinate complex cell movements, specifically facilitating transport of secretory vesicles within cells, serving as main structural and functional component of cilia and flagella, and forming mitotic spindle during cell division |
| Microfilaments | Intertwined helical chains of actin molecules; microfilaments composed of myosin molecules also present in muscle cells | Play a vital role in various cellular contractile systems, including muscle contraction and amoeboid movement; serve as a mechanical stiffener for microvilli |
| Intermediate filaments | Irregular, threadlike proteins | Help resist mechanical stress |

in particular, put a demand on the cell. For a cell to make more complex molecules, the building blocks must be present. If a cell is not capable of creating these building blocks itself, it must transport such molecules from the extracellular environment into the cell. Once inside, the cell is then capable of combining these precursor molecules into more complex ones. However, in addition to these precursor molecules, anabolic metabolism has an additional requirement: it takes energy to combine simpler molecules into more complex molecules.

The source of energy for the body is the chemical energy stored in the carbon bonds of ingested food. Body cells are not equipped to use this energy directly; the energy must be extracted from the food nutrients and converted into a usable form of energy—namely, the high-energy phosphate bonds of **adenosine triphosphate (ATP)**. Adenosine triphosphate consists of adenosine with three phosphate groups attached (*tri* means "three") (see Appendix B, p. A-17). When a high-energy bond, such as that binding the terminal phosphate to adenosine, is split, a substantial amount of energy is released. Adenosine triphosphate is the universal energy carrier—the common energy "currency" of the body. Cells use ATP to pay for the cost of operating the body's cells (in essence, the cost of cellular work). To obtain immediate usable energy, cells split the terminal phosphate bond of ATP, which yields **adenosine diphosphate (ADP)**—adenosine with two phosphate groups attached (*di* means "two")—plus inorganic phosphate ($P_i$), plus energy:

$$ATP \overset{spiliting}{\rightarrow} ADP + P_i + energy \text{ for use by the cell}$$

In this energy scheme, food can be thought of as the crude fuel, and ATP the refined fuel for operating the body's machinery. In this fuel-conversion process, dietary food is digested, or broken down, by the digestive system into smaller absorbable units that can be transferred from the digestive tract lumen into the blood (see Chapter 15). For example, dietary carbohydrates are broken down primarily into glucose, which is absorbed into the blood. No usable energy is released during the digestion of food. When delivered to the cells by the blood, the nutrient molecules are transported across the plasma membrane into the cytosol (described in Sections 2.6–2.8) where they are available for ATP production.

The chemical pathways for the production of ATP involve three separate processes: *substrate-level phosphorylation*, *anaerobic glycolysis*, and *aerobic metabolism*. In most cells, the majority of ATP is generated from the sequential dismantling of absorbed nutrient molecules in four steps: (1) *glycolysis* (both aerobic and anaerobic), (2) the decarboxylation of pyruvate, (3) the *tricarboxylic acid cycle* (TCA cycle; aerobic), and (4) the *electron transport chain* (ETC; aerobic). We will use glucose as an example to describe these four steps in ATP production (> Figure 2-1). However, before focusing on glycolysis and the mitochondrial ATP-generating processes, we first discuss substrate-level phosphorylation, which is completed in the cytosol of skeletal muscle and brain cells for immediate ATP generation, such as at the onset of intense exercise.

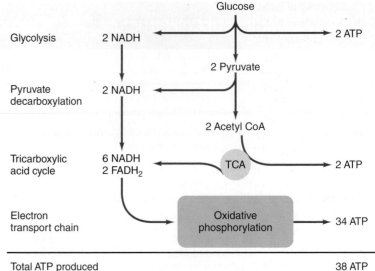

> **FIGURE 2-1 An overview of ATP production from glucose.** ATP production starts in the cytoplasm with the process of glycolysis to form pyruvate, which can be decarboxylated to form acetyl CoA. Acetyl CoA enters the mitochondrial tricarboxylic acid cycle. The NADH and FADH$_2$ molecules produced can enter the mitochondrial electron transport chain to produce even more ATP. Altogether, the maximum theoretical yield of ATP from 1 molecule of glucose is 38 ATP.

## Pathways for production of ATP

### SUBSTRATE-LEVEL PHOSPHORYLATION

In some tissues, such as skeletal muscle, ATP appears in relatively low concentrations. This means that even small decreases in ATP content can be problematic for cellular operation. Therefore, ATP production is closely regulated. The first step in ATP generation is the substrate-level phosphorylation of ADP using creatine phosphate (CP). **Creatine phosphate** (a.k.a. phosphocreatine) is the first energy store tapped at the onset of contractile activity. It is not stored in the mitochondria, but rather in the cytosol of the cell. Like ATP, CP contains a high-energy phosphate group. Just as energy is released when the terminal phosphate bond in ATP is spliced, similar energy is released when the bond between phosphate and creatine is broken. The energy released from the hydrolysis (a reaction using $H_2O$) of CP, along with the phosphate, can be donated directly to ADP to form ATP. This reaction, which is catalyzed by the skeletal muscle cell enzyme **creatine kinase**, is reversible; energy and phosphate from ATP can be transferred to creatine to form CP:

$$CP + ADP \xrightarrow{\text{creatine kinase}} creatine + ATP$$

In accordance with the law of mass action, as energy reserves are built up in a resting skeletal muscle, the increased concentration of ATP favours transfer of the high-energy phosphate group from ATP to form CP. By contrast, at the onset of muscle contraction, when myosin ATPase (the enzyme on the myosin head; see Chapter 7) splits the meagre reserves of ATP, the resultant fall in ATP concentration favours transfer of the high-energy phosphate group from stored CP to form more ATP. A rested muscle contains about five times as much CP

(18 mM per kg of muscle) as ATP. Thus, most energy stored in skeletal muscle is in the form of CP. When the skeletal muscle starts to vigorously contract using ATP to accomplish the cellular work, the CP replenishes the ATP, preserving the level of ATP in the working skeletal muscle. Therefore, CP levels tend to decrease more greatly than ATP levels. There tends to be a large change in CP levels with a minimal change in the level of ATP. Because only one enzymatic reaction is involved in this energy transfer, ATP can be formed rapidly (within a fraction of a second) by using CP.

### GLYCOLYSIS

Among the thousands of enzymes within the cytosol are those responsible for **glycolysis**, a chemical process involving ten separate sequential reactions that break down the simple sugar molecule, glucose, into two pyruvic acid molecules (*glyc-* means "sweet"; *lysis* means "break down"). During glycolysis, some of the energy from the broken chemical bonds of glucose is used to directly convert ADP into ATP (> Figure 2-2) while also producing NADH. However, glycolysis is not very efficient in terms of energy extraction: the net yield is only two molecules of ATP and two molecules of NADH per glucose molecule processed. Much of the energy originally contained in the glucose molecule is still locked in the chemical bonds of the pyruvic acid molecules. The low-energy yield of glycolysis is not enough to support the body's demand for ATP. This is where the mitochondria come into play.

*Clinical Note* There are numerous metabolic diseases that can affect glycolysis. Persons with McArdle disease, for example, are unable to break down glycogen to glucose and thereby produce energy. (This disease was named after Brian McArdle, a British physician who in 1951 discovered a muscle disorder that caused cramp-like pains yet was not associated with the normal production of lactic acid from exercise. The defect was later identified as an absence of phosphorylase, the enzyme involved in the first step in the splitting-off of the glucose-1-phosphate units from glycogen.) The symptoms of McArdle disease include muscle fatigue, pain, and cramps. These are due to a deficiency of the skeletal muscle form of glycogen phosphorylase (enzyme) and result in the unavailability of muscle glycogen as a source of energy for muscle contraction. Once diagnosed, McArdle disease may be treated by correcting the enzyme deficiency. One such treatment is gene-replacement therapy, using the adenovirus vector.

### PYRUVATE DECARBOXYLATION

Pyruvic acid, which is produced in the cytosol of the cell via glycolysis, enters the mitochondrial matrix through the carrier protein monocarboxylate transporter, which is located on the inner mitochondrial membrane. Once inside, pyruvate is metabolized by the pyruvate dehydrogenase complex and is decarboxylated. Decarboxylation is the removal of a carbon and the formation of carbon dioxide ($CO_2$), as well as the transfer of a hydrogen to nicotinamide adenine dinucleotide ($NAD^+$), forming NADH (discussed in the section Electron Transport Chain). Carbon dioxide is eliminated from the body via the cardiorespiratory system. Once the pyruvate is converted to acetyl coenzyme A (acetyl CoA), it is ready to enter the TCA cycle.

### TRICARBOXYLIC ACID CYCLE

Acetyl CoA enters the **tricarboxylic acid cycle** (also known as the **Krebs cycle** or the **citric acid cycle**), which consists of a series of eight separate biochemical reactions that are directed by the enzymes of the mitochondrial matrix. > Figure 2-3 is a schematic of a TCA cycle, which is one way to characterize a series of biochemical reactions. (The molecules themselves are not physically moved around in a cycle.) Acetyl CoA is a two-carbon molecule that enters the TCA cycle and combines with oxaloacetic acid (a four-carbon molecule) to form a six-carbon molecule, citric acid (hence the name citric acid cycle). Coenzyme A is removed from the acetyl CoA, allowing it to be reused in the conversion of more pyruvic acid into acetyl CoA. Next, the atoms of the citric acid are rearranged to become isocitric acid. Isocitric acid then becomes alpha-ketoglutaric acid, in a two-reaction process: first, a hydrogen is removed, and then a carbon, via $CO_2$. Then, hydrogen is again removed, and $CO_2$ forms. The new structure then attaches itself to coenzyme A, forming succinyl CoA. Succinyl CoA consists of four carbons.

The next molecular arrangement is the only reaction in the TCA cycle that forms ATP directly. Coenzyme A is removed and a phosphate group is added. This group is transferred later via guanosine triphosphate (GTP) to ADP, forming ATP. In the following steps, more hydrogen is removed (forming fumaric acid), water is added (forming malic acid), and then more hydrogen is again removed (forming oxaloacetic acid). This last removal of hydrogen and acceptance of hydrogen by $NAD^+$ forms the molecule that we initially began with: oxaloacetic acid. As the cycle progresses, different molecules are formed. These molecular alterations have the following important consequences:

1. Two carbons are sequentially removed from the six-carbon citric acid molecule, converting it back into the four-carbon oxaloacetic acid, which is now available at the top of the cycle to pick up another acetyl CoA for another revolution through the cycle.

2. The released carbon atoms, which were originally present in the acetyl CoA that entered the cycle, are converted into two molecules of $CO_2$. This $CO_2$, as well as the

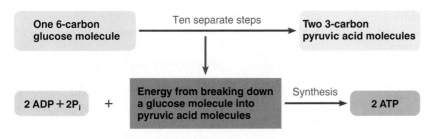

> **FIGURE 2-2 A simplified summary of glycolysis.** Glycolysis involves the breakdown of glucose into two pyruvic acid molecules, with a net yield of two molecules of ATP for every glucose molecule processed.

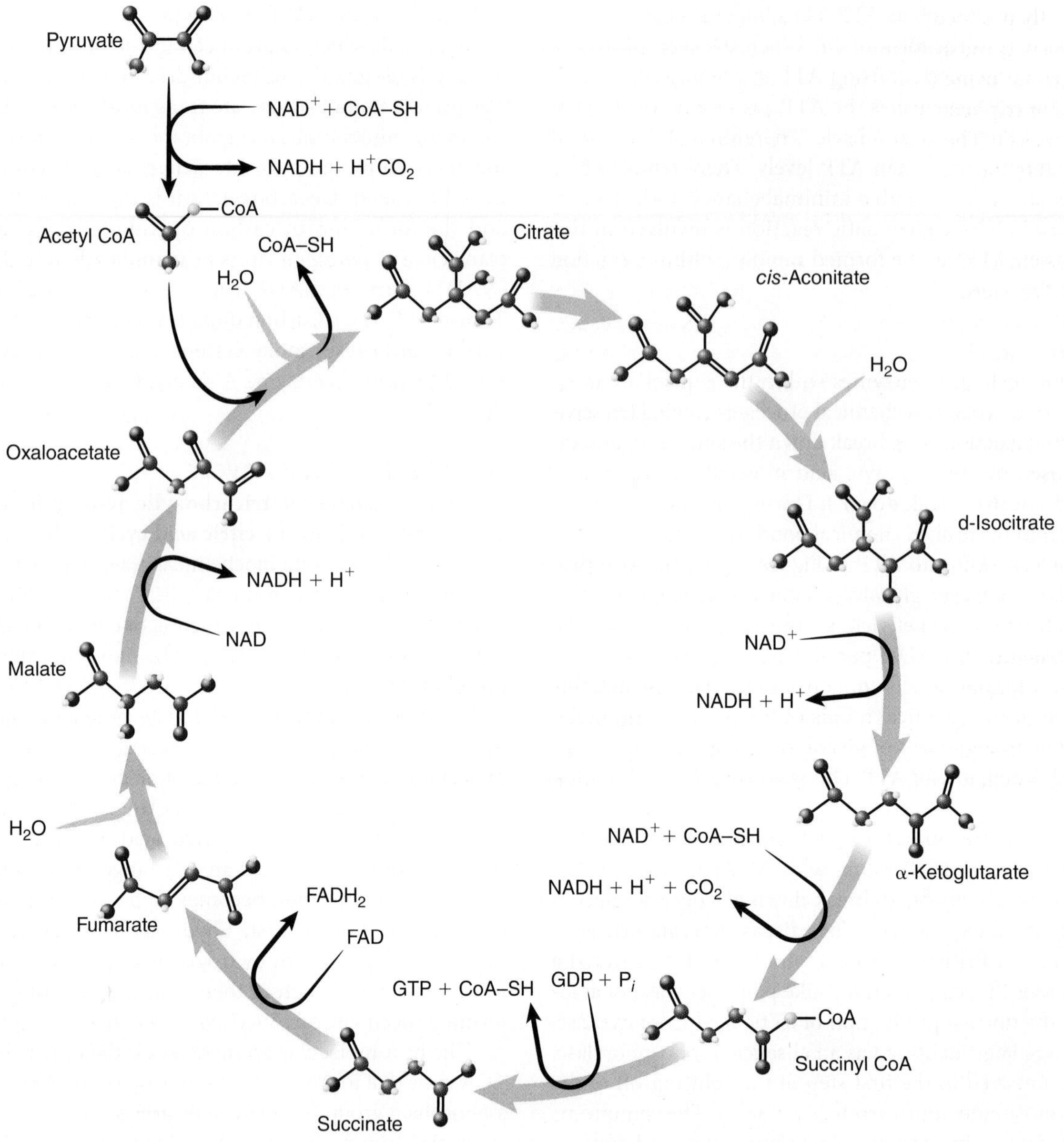

**Pyruvate**

$NAD^+ + CoA-SH$

$NADH + H^+ CO_2$

**Acetyl CoA** — CoA

$CoA-SH$

$H_2O$

**Citrate**

*cis*-Aconitate

$H_2O$

**Oxaloacetate**

**d-Isocitrate**

$NADH + H^+$

$NAD$

$NAD^+$

$NADH + H^+$

**Malate**

$H_2O$

$NAD^+ + CoA-SH$

$NADH + H^+ + CO_2$

**Fumarate**

$FADH_2$

$FAD$

α-Ketoglutarate

$GTP + CoA-SH$   $GDP + P_i$

— CoA

**Succinyl CoA**

**Succinate**

> **FIGURE 2-3 Tricarboxylic acid cycle.** This simplified version of the TCA cycle shows how the two carbons entering the cycle by means of acetyl CoA are eventually converted into carbon dioxide, and how oxaloacetic acid, which accepts acetyl CoA, is regenerated at the end of the cyclical pathway. Also denoted is the release of hydrogen atoms at specific points along the pathway. These hydrogens bind to the hydrogen carrier molecules $NAD^+$ and FAD for further processing by the electron transport chain. One molecule of ATP is generated for each molecule of acetyl CoA that enters the TCA cycle, yielding a total of two molecules of ATP for each molecule of processed glucose.

$CO_2$ produced during the formation of acetic acid from pyruvic acid, passes out of the mitochondrial matrix and subsequently out of the cell to enter the blood. In turn, the blood carries the $CO_2$ to the lungs, where it is finally eliminated into the atmosphere through the process of breathing (see Chapter 12). The oxygen used to make $CO_2$ from these released carbon atoms is derived from the molecules that were involved in the reactions, not from free molecular oxygen supplied by breathing.

3. Hydrogen atoms are also removed during the cycle at four of the chemical conversion steps. The key purpose of the

citric acid cycle is to produce these hydrogens for entry into the electron transport chain. These hydrogens are "caught" by two other compounds that act as hydrogen carrier molecules: **nicotinamide adenine dinucleotide ($NAD^+$)**, a derivative of the B vitamin niacin; and **flavine adenine dinucleotide (FAD)**, a derivative of the B vitamin riboflavin. The transfer (or addition) of hydrogen converts these compounds to NADH and $FADH_2$, respectively.

4. One more molecule of ATP is produced for each molecule of acetyl CoA processed. Actually, ATP is

not directly produced by the citric acid cycle. The released energy is used to directly link inorganic phosphate to **guanosine diphosphate (GDP)** to form **guanosine triphosphate (GTP)**, a high-energy molecule similar to ATP. The energy from GTP is then transferred to ATP as follows:

$$ADP + GTP = ATP + GDP$$

Because each glucose molecule is converted into two acetyl CoA molecules, permitting two turns of the TCA cycle, two more ATP molecules are produced from each glucose molecule.

So far the cell still does not have much of an energy profit. However, the citric acid cycle is important in preparing the hydrogen carrier molecules for their entry into the next step, the electron transport chain, which produces far more energy than the sparse amount of ATP produced by the cycle itself.

### ELECTRON TRANSPORT CHAIN

Most of the potential energy is still stored in the released hydrogen atoms, which contain electrons at high energy levels. The "big payoff" comes when NADH and FADH$_2$ enter the **electron transport chain**, which consists of electron carrier molecules located in the inner mitochondrial membrane lining the cristae (⟩ Figure 2-4). The high-energy electrons are extracted from the hydrogens held in NADH and FADH$_2$ and are transferred through a series of steps from one electron-carrier

molecule—complexes I–IV, the quinone pool (Q), and cytochrome c (C)—to another, within the cristae membrane, in a kind of assembly line (⟩ Figure 2-4). As a result of giving up hydrogen and electrons within the electron transport chain, NADH and FADH$_2$ are converted back to NAD$^+$ and FAD. These molecules are now free to pick up more hydrogen atoms released during glycolysis and the citric acid cycle. Thus, NAD$^+$ and FAD serve as the link between TCA and the electron transport chain.

During the process of moving the high-energy electrons down the electron transport chain, some of the energy is used to pump hydrogen ions across the inner mitochondrial membrane. This occurs at three different places and leads to the accumulation of hydrogen ions within the intermembrane space. It is this accumulation of hydrogen ions that leads to the generation of ATP by what is known as the **chemiosmotic mechanism**. The enzyme **ATP synthase** is also located within the inner mitochondrial membrane, and it is activated by the flow of hydrogen ions back into the matrix from the inner mitochondrial membrane. Once activated, ATP synthase creates ATP from ADP + P$_i$ in a process called **oxidative phosphorylation**. The amount of ATP that can be produced from high-energy electron carriers depends on where they enter the electron transport chain. For example, NADH enters the electron transport chain early, and each molecule of NADH can lead to the production of three ATP molecules, whereas FADH$_2$ enters the electron transport chain at a later stage and can only produce two ATP molecules.

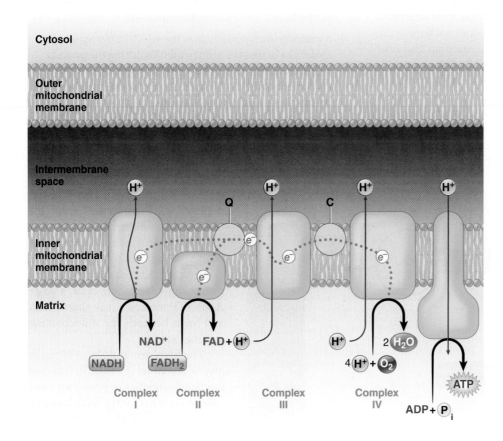

⟩ **FIGURE 2-4 Oxidative phosphorylation at the mitochondrial inner membrane.** ATP synthesis that results from the passage of high-energy electrons through the mitochondrial electron transport chain and the activation of ATP-synthase by movement of H$^+$.

### HOW MUCH ATP IS REALLY PRODUCED FOR EACH GLUCOSE?

Although the processes that lead to the formation of ATP seem relatively straightforward, defining the exact quantity of ATP that can be produced from each glucose is not. So far we have examined the theoretical maximum of the amount of ATP that can be produced, which is 38 ATP molecules (see ⟩ Figure 2-1). This theoretical maximum makes two primary assumptions (1) that the NADH produced within the cytosol during glycolysis can be transported without cost of energy into the mitochondria to enter the electron transport chain, and (2) that the electron transport chain always runs at 100 percent efficiency.

Although several carrier systems exist to move cytosolic NADH into the mitochondria, not all of them are able to directly shuttle NADH without expending energy. Some of these systems require active transport and actually consume ATP to move the NADH. Others result in the conversion of cytosolic NADH to mitochondrial FADH$_2$, both of which reduce the total number of ATP molecules that can be produced from each glucose.

The electron transport system itself does not operate at maximum efficiency. When NADH enters the electron transport chain, it ultimately leads to the production of two or three ATP molecules, so the value of 2.5 is generally used. Similarly, $FADH_2$ is said to produce 1.5 ATP molecules since it actually produces either one or two ATP molecules.

So taking altogether the amount of ATP, NADH, and $FADH_2$ produced from each glucose molecule (see ❯ Figure 2-1), we can determine that the actual amount of ATP produced for each glucose molecule is somewhere between 30 and 32, depending on how the energy stored in the NADH from glycolysis ultimately ends up entering the electron transport chain.

## Aerobic versus anaerobic conditions

All of our discussion so far on the production of ATP from glucose has focused on **aerobic** metabolism, which requires the presence of oxygen for these reactions to occur. What happens when $O_2$ is not present? When $O_2$ levels are insufficient to maintain aerobic metabolism, **anaerobic** (lack of air, specifically lack of $O_2$) metabolism becomes the predominant source of energy (❯ Figure 2-5). Recall that the first step of ATP production, glycolysis, takes place in the cytosol and involves the breakdown of glucose into pyruvic acid, producing a low yield of two molecules of ATP and one molecule of NADH per molecule of glucose. Glycolysis is always considered anaerobic since the conversion of glucose to pyruvate does not require the presence of $O_2$. However, the intermediate reactions that lead to the decarboxylation of pyruvate to acetyl CoA do require $O_2$ to proceed. The result is that in the absence of $O_2$, pyruvate is instead converted to lactic acid (❯ Figure 2-5), which does not enter the mitochondria or the TCA cycle for further ATP production. The conversion of pyruvate to lactic acid consumes NADH, negating the NADH produced during glycolysis, which further decreases the amount of ATP produced during anaerobic metabolism.

## ATP production from other dietary sources

Glucose, the principal nutrient derived from dietary carbohydrates, is the fuel preference of most cells. However, nutrient molecules derived from fats (fatty acids) and, if necessary, from protein (amino acids) can also participate at specific points in this overall chemical reaction to eventually produce energy. Amino acids are usually used for protein synthesis instead of energy production, but they can be used as fuel if insufficient glucose and fat are available. Details of these processes are discussed in Chapter 16.

Note that the oxidative reactions within the mitochondria generate energy, unlike the oxidative reactions controlled by the peroxisome enzymes. Both organelles use oxygen, but for different purposes.

## ATP for synthesis, transport, and mechanical work

Once formed, ATP is transported out of the mitochondria and is then available as an energy source as needed within the cell. Cell activities that require energy expenditure fall into three main categories:

1. *Synthesis of new chemical compounds*, such as protein synthesis by the endoplasmic reticulum. Some cells, especially cells with a high rate of secretion and cells in the growth phase, use up to 75 percent of the ATP they generate just to synthesize new chemical compounds.

2. *Membrane transport*, such as the selective transport of molecules across the kidney tubules during the process of urine formation. Kidney cells can expend as much as 80 percent of their ATP currency to operate their selective membrane-transport mechanisms.

3. *Mechanical work*, such as contraction of the heart muscle to pump blood or the contraction of skeletal muscles to lift an object. These activities require tremendous quantities of ATP.

As a result of cell energy expenditure to support these various activities, large quantities of ADP are produced. These energy-depleted ADP molecules enter the mitochondria for recharging by participating in oxidative phosphorylation before cycling back into the cytosol as energy-rich ATP molecules. In this recharging–expenditure cycle, a single ADP/ATP molecule may shuttle back and forth between the mitochondria and the cytosol.

The high demands for ATP make glycolysis alone an insufficient and inefficient supplier of power for most cells. Were it not for the mitochondria, which house the metabolic machinery for oxidative phosphorylation, the body's energy capability would be very limited. However, glycolysis does provide cells with a sustenance

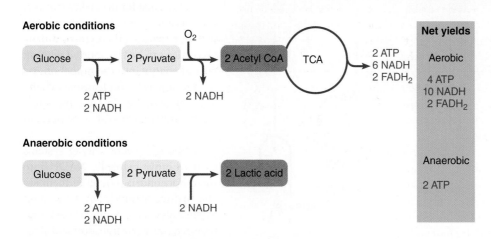

❯ **FIGURE 2-5 Comparison of high-energy products under anaerobic and aerobic conditions.** In the absence of $O_2$, pyruvate cannot be converted into acetyl CoA, and lactic acid is produced instead, dramatically reducing the number of high-energy products available for ATP production.

mechanism that can produce at least some ATP under anaerobic conditions. Skeletal muscle cells in particular take advantage of this ability during short bursts of strenuous exercise, when energy demands for contractile activity outstrip the body's ability to produce ATP at a fast enough rate through oxidative phosphorylation. Even though a single pass through glycolysis produces less ATP than oxidative phosphorylation, the high flux rate of glucose to either pyruvate or lactate can produce ATP much faster. As well, red blood cells, which are the only cells that do not contain any mitochondria, rely solely on glycolysis for their limited energy production. The energy needs of red blood cells are low, however, because they also lack a nucleus and therefore are not capable of synthesizing new substances, which is the biggest energy expenditure for most noncontractile cells.

## Check Your Understanding 2.2

1. State how much ATP can be generated for each glucose molecule.
2. Compare aerobic and anaerobic ATP production.

# 2.4 | The Plasma Membrane

The survival of every cell depends on the maintenance of intracellular contents unique for that cell type, despite the remarkably different composition of the extracellular fluid surrounding it. This difference in fluid composition inside and outside a cell is maintained by the plasma membrane, an extremely thin layer of lipids and proteins that forms the outer boundary of every cell and encloses the intracellular contents. In addition to serving as a mechanical barrier that traps needed molecules within the cell, the plasma membrane plays an active role in determining the composition of the cell by selectively permitting specific substances to pass between the cell and its environment. Besides controlling the entry of nutrient molecules and the exit of secretory and waste products, the plasma membrane maintains differences in ion concentrations between a cell's interior and exterior. These ionic differences, as you will learn, are important in the electrical activity of the plasma membrane. As well, the plasma membrane participates in the joining of cells to form tissues and organs. Furthermore, it plays a key role in the ability of a cell to respond to changes, or signals, in the cell's environment. This ability is important in communication between cells. For every cell type, these membrane functions are important to (1) the cell's survival, (2) its ability to maintain homeostasis, and (3) its ability to function cooperatively and in coordination with surrounding cells. Many of the functional differences between cell types are due to subtle variations in the composition of their plasma membranes, which in turn enable different cells to interact in different ways with essentially the same extracellular fluid environment. If problems occur in the membrane of the cell, the ramifications can be very serious. (To understand what can happen when there are problems with the cell membrane, see Concepts, Challenges, and Controversies; p. 32).

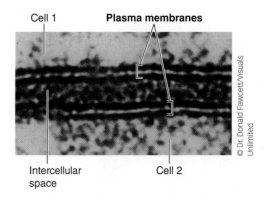

> FIGURE 2-6 **Trilaminar appearance of a plasma membrane in an electron micrograph.** Depicted are the plasma membranes of two adjacent cells. Note that each membrane appears as two dark layers separated by a light middle layer.

The plasma membrane is too thin to be seen under an ordinary light microscope, but with an electron microscope it appears as a **trilaminar structure** consisting of two dark layers separated by a light middle layer (*tri* means "three"; *lamina* means "layer") (> Figure 2-6). The specific arrangement of the molecules that make up the plasma membrane is responsible for this three-layered "sandwich" appearance.

## Membrane structure and composition

The plasma membrane consists of mostly lipids (fats), proteins, and some carbohydrates. The most abundant membrane lipids are phospholipids, with lesser amounts of cholesterol. Approximately a billion phospholipid molecules are present in the plasma membrane of a typical human cell. **Phospholipids** have a polar (electrically charged; see Appendix B, p. A-13) head containing a negatively charged phosphate group and two nonpolar (electrically neutral) fatty acid tails (> Figure 2-7a). The polar end is **hydrophilic** (water loving) because it can interact with water molecules, which are also polar; the nonpolar end is **hydrophobic** (water fearing) and will not mix with water. Such two-sided molecules self-assemble into a **lipid bilayer**, a double layer of lipid molecules, when in contact with water (> Figure 2-7b) (*bi* means "two"). The hydrophobic tails bury themselves in the centre away from the water, whereas the hydrophilic heads line up on both sides in contact with the water. The outer surface of the layer is exposed to extracellular fluid (ECF), whereas the inner surface is in contact with the intracellular fluid (ICF) (> Figure 2-7c).

The lipid bilayer is not a rigid structure but is fluid in nature. The phospholipids, which are not held together by strong chemical bonds, can twirl around rapidly as well as move about within their own half of the layer. This phospholipid movement accounts in large part for membrane fluidity.

**Cholesterol** also contributes to the fluidity as well as the stability of the membrane. The cholesterol molecules are tucked in between the phospholipid molecules, where they prevent the fatty acid chains from packing together and crystallizing, a process that would drastically reduce membrane fluidity.

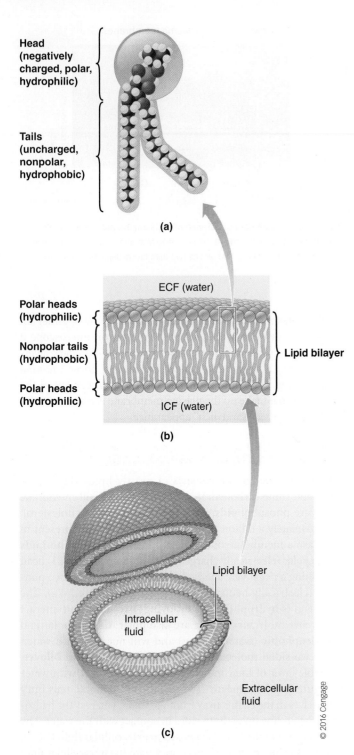

Head (negatively charged, polar, hydrophilic)

Tails (uncharged, nonpolar, hydrophobic)

**(a)**

ECF (water)

Polar heads (hydrophilic)

Nonpolar tails (hydrophobic)

Polar heads (hydrophilic)

**Lipid bilayer**

ICF (water)

**(b)**

Lipid bilayer

Intracellular fluid

Extracellular fluid

© 2016 Cengage

**(c)**

⟩ **FIGURE 2-7 Structure and organization of phospholipid molecules in a lipid bilayer.** (a) Phospholipid molecule. (b) In water, phospholipid molecules organize themselves into a lipid bilayer with the polar heads interacting with the polar water molecules at each surface and the nonpolar tails all facing the interior of the bilayer. (c) An exaggerated view of the plasma membrane enclosing a cell, separating the ICF from the ECF

Cholesterol's spatial relationship with phospholipid molecules helps stabilize the phospholipids' positions.

Because of its fluid nature, the plasma membrane has structural integrity but at the same time is flexible, enabling the cell to change shape. For example, muscle cells change shape as they contract, and red blood cells must change shape considerably as they travel through the capillaries, the tiniest of blood vessels.

The **membrane proteins** are attached to or inserted within the lipid bilayer (⟩ Figure 2-8). Some of these proteins extend through the entire thickness of the membrane. Other proteins stud only the outer or inner surface; they are anchored by interactions with a protein that spans the membrane or by attachment to the lipid bilayer. The plasma membrane contains approximately 50 times as many lipid molecules as protein molecules; however, protein accounts for approximately half of the membrane's mass, since protein molecules are so large. The fluidity of the lipid bilayer enables many membrane proteins to move within the membrane, although the cytoskeleton restricts the mobility of proteins that perform a specialized function in a specific area of the cell. This view of membrane structure is known as the **fluid mosaic model**, in reference to the membrane fluidity and the ever-changing mosaic pattern of the proteins embedded within the lipid bilayer. (A mosaic is a surface decoration made by inlaying small pieces of variously coloured tiles to form patterns or pictures.)

The small amount of membrane carbohydrate is located on the outer surface, making the plasma membrane "sugar coated." Short-chain carbohydrates protrude like tiny antennas from the outer surface, bound primarily to membrane proteins and, to a lesser extent, to lipids. These sugary combinations are known as *glycoproteins* and *glycolipids*, respectively (⟩ Figure 2-8).

The different components of the plasma membrane carry out a variety of functions, including the following: (1) the lipid bilayer acts as a barrier to diffusion, (2) proteins perform specialized functions, and (3) carbohydrates act as self-recognition molecules for cell-to-cell interaction. Let's look at these functions in more detail.

**LIPID BILAYER**

The lipid bilayer serves three important functions:

1. It forms the basic structure of the membrane. The phospholipids can be visualized as the "pickets" that form the "fence" around the cell.

2. Its hydrophobic interior serves as a barrier to passage of water-soluble substances between the ICF and ECF. Water-soluble substances cannot dissolve in and pass through the lipid bilayer. By means of this barrier, the cell can maintain different mixtures and concentrations of solutes inside and outside the cell.

3. It is responsible for the fluidity of the membrane.

**MEMBRANE PROTEINS**

Different types of membrane proteins serve the following specialized functions:

1. Some proteins span the membrane to form water-filled pathways, or channels, across the lipid bilayer (⟩ Figure 2-8). Water-soluble substances small enough to enter a channel can pass through the membrane this way without coming into direct contact with the hydrophobic lipid interior. Only small ions can fit through channels as

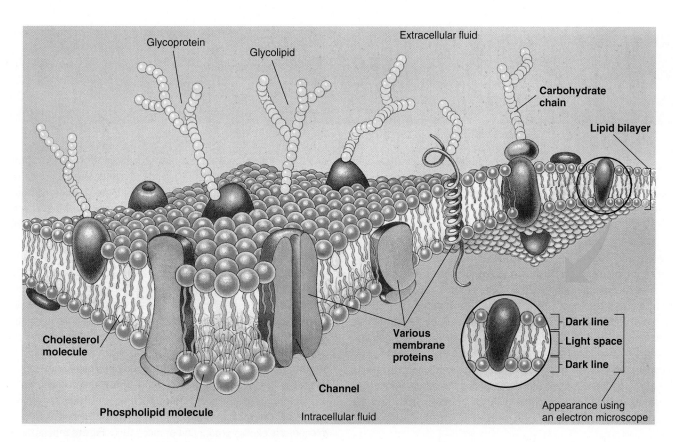

> **FIGURE 2-8 Fluid mosaic model of plasma membrane structure.** The plasma membrane comprises a lipid bilayer embedded with proteins. Some of these proteins extend through the thickness of the membrane, some are partially submerged in the membrane, and others are loosely attached to the surface of the membrane. Short carbohydrate chains are attached to proteins or lipids on the outer surface only.

their small diameter precludes passage of particles greater than 0.8 nm in diameter. Furthermore, a given channel can selectively attract or repel particular ions. For example, only sodium $(Na^+)$ can pass through $Na^+$ channels, whereas only potassium $(K^+)$ can pass through $K^+$ channels. This selectiveness is due to specific arrangements of chemical groups on the interior surfaces of the proteins that form the channel walls. A given channel may even be open or closed to its specific ion as a result of changes in channel shape in response to a controlling mechanism, for example, glucose uptake by muscle cells (p. 273). This is a good example of function depending on structural details. Cells vary in number, kind, and activity of channels they possess.

2. Another group of proteins that spans the membrane serves as **carrier molecules,** which transfer specific substances across the membrane that are unable to cross on their own. How carriers accomplish this transport will be described in Section 2.8. (Thus, channels and carrier molecules are both important in the transport of substances between the ECF and ICF.) Each carrier can transport only a particular molecule or closely related molecules. Cells of different types have different kinds of carriers. As a result, they vary as to which substances they can selectively transport across their membranes. For example, thyroid gland cells are the only cells in the body to use iodine. Appropriately, only the plasma membranes of thyroid gland cells have carriers for iodine, so these cells alone can transport this element from the blood into the cell interior.

3. Other proteins located on the inner membrane surface serve as **docking-marker acceptors** that bind in a lock-and-key fashion with the docking markers of secretory vesicles. Secretion is initiated when stimulatory signals trigger the fusion of the secretory vesicle membrane with the inner surface of the plasma membrane through interactions between these matching labels. The secretory vesicle subsequently opens up and empties its contents to the outside by exocytosis.

4. Another group of surface-located proteins function as **membrane-bound enzymes** that control specific chemical reactions at either the inner or the outer cell surface. Cells are specialized in the types of enzymes embedded within their plasma membranes. For example, the outer layer of the plasma membrane of skeletal muscle cells contains an enzyme that destroys the chemical messenger that triggers muscle contraction, thus enabling the muscle to relax.

5. Many proteins on the outer surface serve as **receptor sites** (or, simply, **receptors**) that "recognize" and bind with specific molecules in the cell's environment. This binding initiates a series of membrane and intracellular events that alter the activity of the particular cell. In this way,

# Cystic Fibrosis: A Fatal Defect in Membrane Transport

APPROXIMATELY ONE IN EVERY 3500 CHILDREN IS BORN with cystic fibrosis (CF). It is characterized by the production of abnormally thick, sticky mucus. Most dramatically affected are the respiratory airways and the pancreas.

## Respiratory Problems

The presence of the thick, sticky mucus in the respiratory airways makes it difficult to get adequate air in and out of the lungs. And because bacteria thrive in the accumulated mucus, CF patients suffer from repeated respiratory infections. They are especially susceptible to *Pseudomonas aeruginosa*, an "opportunistic" bacterium that is frequently present in the environment but usually causes infection only when some underlying problem handicaps the body's defences. Gradually, the involved lung tissue becomes scarred (fibrotic), making the lungs harder to inflate. This complication increases the work of breathing beyond the extra work required to move air through the clogged airways.

## Underlying Cause

During the last decade, researchers found that cystic fibrosis is caused by any one of several different genetic defects that lead to production of a flawed version of a protein known as *cystic fibrosis transmembrane conductance regulator (CFTR)*. CFTR normally helps form and regulate the chloride ($Cl^-$) channels in the plasma membrane. With CF, the defective CFTR "gets stuck" in the endoplasmic reticulum/Golgi system, which normally manufactures and processes this product and ships it to the plasma membrane (p. 48). That is, in CF patients, the mutated version of CFTR is only partially processed and never makes it to the cell surface. The resultant absence of CFTR protein in the plasma membrane's $Cl^-$ channels leads to membrane impermeability to $Cl^-$. Because $Cl^-$ transport across the membrane is closely linked to sodium ion ($Na^+$) transport, cells lining the respiratory airways cannot absorb salt (NaCl) properly. As a result, salt accumulates in the fluid lining the airways.

What has puzzled researchers is how this $Cl^-$ channel defect and resultant salt accumulation lead to the excess mucus problem. Two recent discoveries have perhaps provided an answer, although these proposals remain to be proven and research into other possible mechanisms continues to be pursued. One group of investigators found that the airway cells produce a natural antibiotic, *defensin*, which normally kills most of the inhaled airborne bacteria. It turns out that defensin cannot function properly in a salty environment. Bathed in the excess salt associated with CF, the disabled antibiotic cannot rid the lungs of inhaled bacteria. This leads to repeated infections. One of the outcomes of the body's response to these infections is excess production of mucus. In turn, this mucus serves as a breeding ground for more bacterial growth. The cycle continues as the lung-clogging accumulation of mucus and the frequency of lung infections grow ever worse. The excess mucus is also especially thick and sticky, making it difficult for the normal ciliary defence mechanisms of the lung to sweep up the bacteria-laden mucus from the lungs. The mucus is thick and sticky because it is underhydrated (has too little water), a problem believed to be linked to the defective salt transport.

The second new study found an additional complicating factor in the CF story. These researchers demonstrated that CFTR appears to

---

chemical messengers in the blood, such as water-soluble hormones, can influence only the specific cells that have receptors for a given messenger. Even though every cell is exposed to the same messengers via their widespread distribution by the blood, a given messenger has no effect on cells that do not have receptors for this specific messenger. For example, the anterior pituitary gland secretes into the blood thyroid-stimulating hormone (TSH), which can attach only to the surface of thyroid gland cells to stimulate secretion of thyroid hormone. No other cells have receptors for TSH, so only thyroid cells are influenced by TSH, despite its widespread distribution.

6. Still other proteins serve as **cell adhesion molecules (CAMs)**. Many CAMs protrude from the outer membrane surface and form loops or hooks that the cells use to grip each other and to grasp the connective tissue fibres that interlace between cells. For example, **cadherins**, on the surface of adjacent cells, interlock in a zipper-like fashion to help hold the cells within tissues and organs together. Some CAMs, such as the **integrins**, span the plasma membrane. Integrins not only serve as a structural link between the outer membrane surface and its extracellular surroundings but also connect the inner membrane surface to the intracellular cytoskeletal scaffolding. These CAMs mechanically link the cell's external environment and intracellular components. Furthermore, integrins can also relay regulatory signals through the plasma membrane in either direction. Although CAMs originally were believed to serve only as adhesive molecules, investigators have now learned that some of them also act as **signalling molecules**. These CAMs participate in signalling cells to grow and in signalling immune system cells to interact

serve a dual role as a Cl⁻ channel and as a membrane receptor that binds with *P. aeruginosa* (and perhaps other bacteria). CFTR subsequently destroys the captured bacteria. In the absence of CFTR in the airway cell membranes of CF patients, *P. aeruginosa* is not cleared from the airways as usual. In a double onslaught, these bacteria were shown to trigger the airway cells to produce unusually large amounts of an abnormally thick, sticky mucus. This mucus promotes more bacterial growth, as the vicious cycle continues.

Canadian researcher Dr. Han-Peter Hauber and associates from McGill University have suggested that the increased production of mucus may be associated with our immune system. Mucus overproduction by the epithelium in patients with cystic fibrosis could be regulated by increased expression of interleukin-9 (IL-9) and the IL-9 receptor (IL-9R), as well as upregulation of calcium-activated chloride channel hCLCA1, which may be a new way to control mucus overproduction in cystic fibrosis patients.

### Pancreatic Problems

Furthermore, in CF patients, the pancreatic duct, which carries secretions from the pancreas to the small intestine, becomes plugged with thick mucus. Because the pancreas produces enzymes important in the digestion of food, malnourishment eventually results. In addition, as the pancreatic digestive secretions accumulate behind the blocked pancreatic duct, fluid-filled cysts form in the pancreas, with the affected pancreatic tissue gradually degenerating and becoming fibrotic. The name "cystic fibrosis" aptly describes long-term changes that occur in the pancreas and lungs as the result of a single genetic flaw in CFTR.

### Treatment

Treatment consists of physical therapy to help clear the airways of the excess mucus and antibiotic therapy to combat respiratory infections, plus special diets and administration of supplemental pancreatic enzymes to maintain adequate nutrition. Despite this supportive treatment, most CF victims do not survive beyond their early 30s, with most dying from lung complications.

With the recent discovery of the genetic defect responsible for the majority of CF cases, investigators are hopeful of developing a means to correct or compensate for the defective gene. Another potential cure being studied is the development of drugs that induce the mutated CFTR to be "finished off" and inserted in the plasma membrane. Furthermore, several promising new drug-therapy approaches, such as a mucus-thinning aerosol drug that can be inhaled, offer hope of reducing the number of lung infections and extending the lifespan of CF victims until a cure can be found.

### Further Reading

Alothman, G.A., Ho, B., Alsaadi, M.M., Ho, S.L., O'Drowsky, L., Louca, E., et al. (2005). Bronchial constriction and inhaled colistin in cystic fibrosis. *Chest*, *127*(2): 522–29.

Durie, P.R. (2000). Pancreatic aspects of cystic fibrosis and other inherited causes of pancreatic dysfunction. *Med Clin N Am*, *84*(3): 609–20, ix.

Hauber, H.P., Manoukian, J.J., Nguyen, L.H., Sobol, S.E., Levitt, R.C., Holroyd, K.J., et al. (2003). Increased expression of interleukin-9, interleukin-9 receptors, and the calcium-activated chloride channel hCLCA1 in the upper airways of patients with cystic fibrosis. *Laryngoscope*, *113*(6): 1037–42.

with the right kind of other cells during inflammatory responses and wound healing, among other things.

7. Finally, still other proteins on the outer membrane surface, especially in conjunction with carbohydrates (as glycoproteins), are important in the cells' ability to recognize "self" (i.e., cells of the same type) and in cell-to-cell interactions.

### SELF-RECOGNITION

The short sugar chains on the outer membrane surface serve as self-identity markers that enable cells to identify and interact with one another in the following ways:

1. Different cell types have different markers. The unique combination of sugar chains projecting from the surface membrane proteins serves as the trademark of a particular cell type,

enabling a cell to recognize others of its own kind. Thus, these carbohydrate chains play an important role in self-recognition and in cell-to-cell interactions, which helps cells of the same type recognize one another and form tissues. This is especially important during embryonic development. If cultures of embryonic cells of two different types, such as nerve cells and muscle cells are mixed, the cells sort themselves into separate aggregates of nerve cells and muscle cells.

2. Carbohydrate-containing surface markers are also involved in tissue growth, which is normally held within certain limits of cell density. Typically, cells do not "trespass" across the boundaries of neighbouring cells. The exception to this rule is seen in the disease of cancer (p. 489), where cancer cells, with their abnormal cell surface carbohydrate markers, spread uncontrollably, invading the space of neighbouring cells.

## 2.5 | Cell-to-Cell Adhesions

In multicellular organisms, such as humans, plasma membranes not only serve as the outer boundaries of all cells but also participate in cell-to-cell adhesions. These adhesions bind groups of cells together into tissues and package them further into organs. The life-sustaining activities of the body systems depend not only on the functions of the individual cells of which they are made but also on how these cells live and work together in tissue and organ communities.

Organization of cells into appropriate groupings is at least partially attributable to the carbohydrate chains on the membrane surface. Once arranged, cells are held together by three different means: (1) the extracellular matrix, (2) cell adhesion molecules in the cells' plasma membranes, and (3) specialized cell junctions.

### Biological glue

Tissues are not made up solely of cells, and many cells within a tissue are not in direct physical contact with neighbouring cells. Instead, they are held together by the **extracellular matrix (ECM)**, an intricate meshwork of fibrous proteins embedded in a watery, gel-like substance composed of complex carbohydrates. The ECM serves as the biological glue. The watery gel provides a pathway for diffusion of nutrients, wastes, and other water-soluble traffic between the blood and tissue cells. It is usually called the *interstitial fluid*. Interwoven within this gel are three major types of protein fibres: collagen, elastin, and fibronectin.

1. *Clinical Note* **Collagen** forms cable-like fibres or sheets that provide tensile strength (resistance to longitudinal stress). In *scurvy*, a condition caused by vitamin C deficiency, these fibres are not properly formed. As a result, the tissues, especially those of the skin and blood vessels, become very fragile. This leads to bleeding in the skin and mucous membranes, which is especially noticeable in the gums.

2. **Elastin** is a rubber-like protein fibre most abundant in tissues that must be capable of easily stretching and then recoiling after the stretching force is removed. It is found, for example, in the lungs, which stretch and recoil as air moves in and out.

3. **Fibronectin** promotes cell adhesion and holds cells in position. Reduced amounts of this protein have been found within certain types of cancerous tissue, possibly accounting for the fact that cancer cells do not adhere well to each other but tend to break loose and metastasize (spread elsewhere in the body).

The ECM is secreted by local cells present in the matrix. The relative amount of ECM compared with cells varies greatly among tissues. For example, the ECM is scant in epithelial tissue but is the predominant component of connective tissue. Most of this abundant matrix in connective tissue is secreted by **fibroblasts** (fibre formers). The exact composition of ECM components varies for different tissues, thus providing distinct local environments for the various cell types in the body. In some tissues, the matrix becomes highly specialized to form such structures as cartilage or tendons or, on appropriate calcification, the hardened structures of bones and teeth.

Contrary to long-held belief, the ECM is not just passive scaffolding for cellular attachment but also helps regulate the behaviour and functions of the cells with which it interacts. Investigators have shown that cells are able to function normally and indeed even to survive only when in association with their normal matrix components. The matrix is especially influential in cell growth and differentiation. In the body, only circulating blood cells are designed to survive and function without attaching to the ECM.

### Cell junctions

In tissues where the cells lie in close proximity to one another, some tissue cohesion is provided by the cell adhesion molecules (CAMs). These special loop-and-hook-shaped surface membrane proteins function like Velcro to connect adjacent cells to each other. In addition, some cells within given types of tissues are directly linked by one of the three types of specialized cell junctions: (1) desmosomes (adhering junctions), (2) tight junctions (impermeable junctions), or (3) gap junctions (communicating junctions).

#### DESMOSOMES

**Desmosomes** act like spot rivets, anchoring together two closely adjacent but nontouching cells. A desmosome consists of two components: (1) a pair of dense, button-like cytoplasmic thickenings known as *plaque*, located on the inner surface of each of the two adjacent cells; and (2) strong glycoprotein filaments containing cadherins (a type of CAM) that extend across the space between the two cells and attach to the plaque on both sides (› Figure 2-9). These intercellular filaments bind adjacent plasma membranes together so that they resist being pulled apart. Thus, desmosomes are adhering junctions.

Desmosomes are most abundant in tissues that are subject to considerable stretching, such as those found in the skin, the heart, and the uterus. In these tissues, functional groups of cells are riveted together by desmosomes that extend from one cell to the next, then from that cell to the next, and so on. Furthermore, intermediate cytoskeletal filaments, such as tough keratin filaments in the skin, stretch across the interior of these cells and attach to the desmosome plaques located on opposite sides of the cell's inner surface. This arrangement forms a continuous network of strong fibres extending throughout the tissue, both through the cells and between the cells, much like a continuous line of people firmly holding hands. This interlinking fibrous network provides tensile strength, reducing the chances of the tissue being torn when stretched.

#### TIGHT JUNCTIONS

At **tight junctions**, adjacent cells firmly bind with each other at points of direct contact to seal off the passageway between the two cells. They are found primarily in sheets of epithelial tissue. Epithelial tissue covers the surface of the body and lines all its

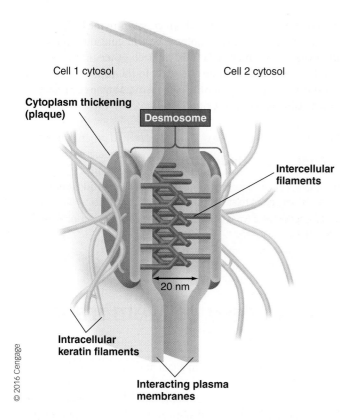

> FIGURE 2-9 **Desmosome.** Desmosomes are adhering junctions that spot-rivet cells, anchoring them together in tissues subject to considerable stretching.

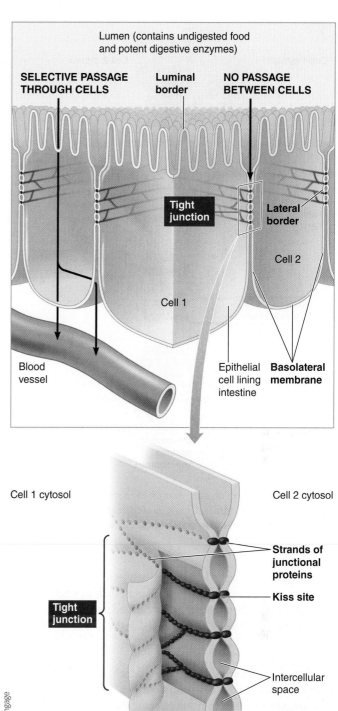

> FIGURE 2-10 **Tight junction.** Tight junctions are impermeable junctions that join the lateral edges of epithelial cells near their luminal borders, thus preventing materials from passing between the cells. Only regulated passage of materials can occur through these cells, which form highly selective barriers that separate two compartments of highly different chemical composition.

internal cavities. All epithelial sheets serve as highly selective barriers between two compartments that have considerably different chemical compositions. For example, the epithelial sheet that lines the digestive tract separates the food and potent digestive juices within the inner cavity (lumen) from the blood vessels that lie on the other side. It is important that only completely digested food particles, and not undigested food particles or digestive juices, move across the epithelial sheet from the lumen to the blood. Accordingly, the lateral (side) edges of the adjacent cells in the epithelial sheet are joined in a tight seal near their luminal border by "kiss" sites, sites of direct fusion of *junctional proteins* on the outer surfaces of the two interacting plasma membranes ( ) Figure 2-10). These tight junctions are impermeable and thus prevent materials from passing between the cells. Passage across the epithelial barrier, therefore, must take place *through* the cells, not *between* them. This traffic across the cell is regulated by means of the channels and carriers present. If the cells were not joined by tight junctions, uncontrolled exchange of molecules could take place between the compartments by unpoliced traffic through the spaces between adjacent cells. Tight junctions thus prevent undesirable leaks within epithelial sheets.

### GAP JUNCTIONS

At a **gap junction**, as the name implies, a gap exists between adjacent cells, which are linked by small, connecting tunnels formed by connexons. A **connexon** is made up of six protein subunits arranged in a hollow tube-like structure. Two connexons extend outward, one from each of the plasma membranes of two adjacent cells, and join end to end to form the connecting tunnel between the two cells ( ) Figure 2-11). Gap junctions are communicating junctions. The small diameter of the tunnels permits

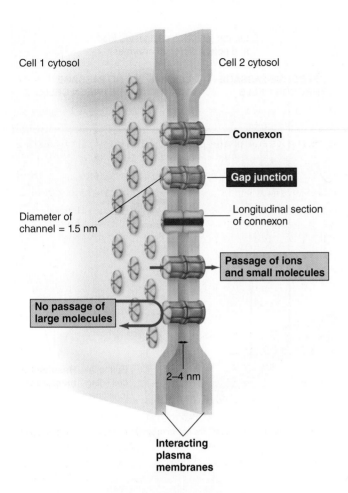

Cell 1 cytosol

Cell 2 cytosol

Connexon

Gap junction

Diameter of
channel = 1.5 nm

Longitudinal section
of connexon

Passage of ions
and small molecules

No passage of
large molecules

2–4 nm

Interacting
plasma
membranes

> FIGURE 2-11 Gap junction. Gap junctions are communicating junctions consisting of small connecting tunnels made up of connexons that permit movement of charge-carrying ions and small molecules between two adjacent cells.

small, water-soluble particles to pass between the connected cells but precludes passage of large molecules, such as vital intracellular proteins. Through these specialized anatomic arrangements, ions (electrically charged particles) and small molecules can be directly exchanged between interacting cells without ever entering the ECF.

Gap junctions are especially abundant in cardiac muscle and smooth muscle. In these tissues, movement of ions between cells through gap junctions transmits electrical activity throughout an entire muscle mass. Because this electrical activity brings about contraction, the presence of gap junctions enables synchronized contraction of a whole muscle mass, such as the pumping chamber of the heart.

Gap junctions are also found in some nonmuscle tissues, where they permit unrestricted passage of small nutrient molecules between cells. For example, glucose, amino acids, and other nutrients pass through gap junctions to a developing egg cell from cells surrounding the egg within the ovary, thus helping the egg stockpile these essential nutrients.

Gap junctions also serve as avenues for the direct transfer of small signalling molecules from one cell to the next. Such transfer permits the cells connected by gap junctions to directly communicate with each other. This communication provides

one possible mechanism by which cooperative cell activity may be coordinated. In later chapters, we will continue to examine other means by which cells "talk" to each other.

Here, we now turn our attention to the topic of membrane transport, focusing on how the plasma membrane can selectively control what enters and exits the cell.

## 2.6 | Overview of Membrane Transport

Anything that passes between a cell and the surrounding extracellular fluid must be able to penetrate the plasma membrane. If a substance can cross the membrane, the membrane is said to be **permeable** to that substance; if a substance cannot pass, the membrane is **impermeable** to it. The plasma membrane is **selectively permeable** in that it permits some particles to pass through while excluding others.

Two properties of particles influence whether they can permeate the plasma membrane without any assistance: (1) the relative solubility of the particle in lipid, and (2) the size of the particle. Highly lipid-soluble particles can dissolve in the lipid bilayer and pass through the membrane. Uncharged or nonpolar molecules (such as $O_2$, $CO_2$, and fatty acids) are highly lipid soluble and readily permeate the membrane. Charged particles (ions, such as $Na^+$ and $K^+$) and polar molecules (such as glucose and proteins) have low lipid solubility but are very soluble in water. The lipid bilayer serves as an impermeable barrier to particles poorly soluble in lipid. For water-soluble (and thus lipid-insoluble) ions less than 0.8 nm in diameter, the protein channels serve as an alternative route for passage across the membrane. Only ions for which specific channels are available and open can permeate the membrane.

Particles that have low lipid solubility and are too large for channels cannot permeate the membrane on their own. Yet some of these particles must cross the membrane for the cell to survive and function. Glucose is an example of a large, poorly lipid-soluble particle that must gain entry to the cell but cannot permeate by dissolving in the lipid bilayer or passing through a channel. Cells have several means of assisted transport to move particles across the membrane that must enter or leave the cell but cannot do so unaided, as you will learn shortly.

Even if a particle can permeate the membrane by virtue of its lipid solubility or its ability to fit through a channel, some force is needed to produce its movement across the membrane. Two general types of forces are involved in accomplishing transport across the membrane: (1) forces that do not require the cell to expend energy to produce movement (**passive forces**), and (2) forces that do require the cell to expend energy (ATP) to transport a substance across the membrane (**active forces**).

The next section describes the various methods of membrane transport, including whether each is an unassisted or assisted means of transport and whether each is a passive- or active-transport mechanism.

# 2.7 | Unassisted Membrane Transport

Molecules (or ions) that can penetrate the plasma membrane on their own are passively driven across the membrane by two forces: diffusion down a concentration gradient and/or movement along an electrical gradient. We first examine diffusion down a concentration gradient.

## Passive diffusion of particles

At temperatures above absolute zero, all molecules (or ions) are in continuous random motion due to heat energy. This motion is most evident in liquids and gases, where the individual molecules have more room to move before colliding with another molecule. Each molecule moves separately and randomly in any direction. As a consequence of this haphazard movement, the molecules frequently collide, bouncing off each other in different directions like billiard balls striking each other. The greater the molecular concentration of a substance in a solution, the greater is the likelihood of collisions. Consequently, molecules within a particular space tend to become evenly distributed over time. Such uniform spreading out of molecules because of their random intermingling is known as **diffusion** (*diffusere* means "to spread out"). To illustrate, in ⟩ Figure 2-12a, the concentration differs between area A and area B in a solution. Such a difference in concentration between two adjacent areas is called a **concentration gradient** (or **chemical gradient**). Random molecular collisions will occur more frequently in area A because of its greater concentration of molecules. For this reason, more molecules will bounce from area A into area B than in the opposite direction. In both areas, the individual molecules will move randomly and in all directions, but the net movement of molecules by diffusion will be from the area of higher concentration to the area of lower concentration.

The term **net diffusion** refers to the difference between two opposing movements. If ten molecules move from area A to area B while two molecules simultaneously move from B to A, the net diffusion is eight molecules moving from A to B. Molecules will spread in this way until the substance is uniformly distributed between the two areas and a concentration gradient no longer exists (⟩ Figure 2-12b). At this point, even though movement is still taking place, no net diffusion is occurring because the opposing movements exactly counterbalance each other—that is, are in equilibrium. Movement of molecules from area A to area B will be exactly matched by movement of molecules from B to A. This situation is an example of a **steady state**.

Many different factors influence how long it takes for molecules in solution to reach steady state, the two most prominent being temperature and molecular size. Temperature is directly proportional to the speed by which molecules move in solution, meaning that at higher temperatures, the molecules move faster and reach steady state faster. In contrast, molecular movement is inversely proportional to molecular size. The larger the molecule, the slower it moves in solution and, therefore, the longer it takes to reach steady state.

What happens if a plasma membrane separates different concentrations of a substance? If the substance can permeate the membrane, net diffusion of the substance will occur through the membrane down its concentration gradient from the area of high concentration to the area of low concentration (⟩ Figure 2-13a). No energy is required for this movement, so it is a passive mechanism of membrane transport. The process of diffusion is crucial to the survival of every cell and plays an important role in many specialized homeostatic activities. As an example, oxygen is transferred across the lung membrane by this means. The blood carried to the lungs is low in oxygen, having given up oxygen to the body tissues for cell metabolism. The air in the lungs, in contrast,

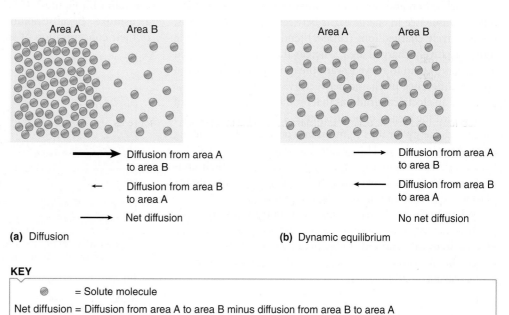

Diffusion from area A to area B

⟵ Diffusion from area B to area A

⟶ Net diffusion

**(a)** Diffusion

Diffusion from area A to area B

⟵ Diffusion from area B to area A

No net diffusion

**(b)** Dynamic equilibrium

**KEY**

⦿ = Solute molecule

Net diffusion = Diffusion from area A to area B minus diffusion from area B to area A

⟶ / ⟵ : Differences in arrow length, thickness, and direction represent the relative magnitude of molecular movement in a given direction.

⟩ **FIGURE 2-12 Diffusion.** (a) Diffusion down a concentration gradient. (b) Steady state, with no net diffusion occurring

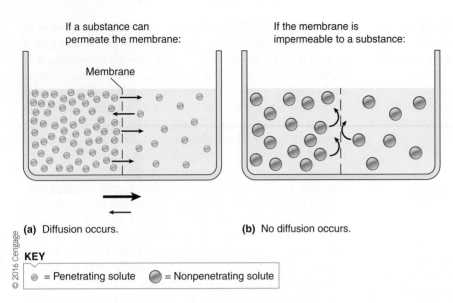

If a substance can permeate the membrane:

Membrane

If the membrane is impermeable to a substance:

**(a)** Diffusion occurs.

**(b)** No diffusion occurs.

KEY

○ = Penetrating solute  ● = Nonpenetrating solute

› **FIGURE 2-13 Diffusion through a membrane.** (a) Net diffusion across the membrane down a concentration gradient. (b) No diffusion through the membrane despite the presence of a concentration gradient

is high in oxygen because it is continuously exchanged with fresh air through the process of breathing. Because of this concentration gradient, net diffusion of oxygen occurs from the lungs into the blood as blood flows through the lungs. Thus, as blood leaves the lungs for delivery to the tissues, it is high in oxygen.

## Passive diffusion of ions

The movement of ions (electrically charged particles that have either lost or gained an electron) is also affected by their electrical charge. Like charges (those with the same kind of charge) repel each other, and opposite charges attract each other. If a relative difference in charge exists between two adjacent areas, the positively charged ions (*cations*) tend to move toward the more negatively charged area, whereas the negatively charged ions (*anions*) tend to move toward the more positively charged area. A difference in charge between two adjacent areas thus produces an **electrical gradient** that promotes the movement of ions toward the area of opposite charge. The cell does not have to expend energy for ions to move into or out of the cell along an electrical gradient. This method of membrane transport is passive, similar to particles moving passively down their concentration gradient. However, only ions that can permeate the plasma membrane can move along this gradient.

When both an electrical and a concentration (chemical) gradient act simultaneously on a specific ion, the result is an **electrochemical gradient**. In Section 2.10 you will learn how electrochemical gradients contribute to the electrical properties of the plasma membrane.

## Osmosis

Water can readily permeate the plasma membrane. It is small enough to slip between the phospholipid molecules within the lipid bilayer. As well, in some cell types, membrane proteins form

**aquaporins**, which are channels used for the passage of water (*aqua* means "water"). About a billion water molecules can pass single file through an aquaporin channel in one second. The driving force for movement of water across the membrane is the same as for any other diffusing molecule—namely, its concentration gradient. Usually the term *concentration* refers to the density of the solute (dissolved substance) in a given volume of water. It is important to recognize, however, that adding a solute to pure water in essence decreases the water concentration. In general, one molecule of a solute displaces one molecule of water.

Compare the water and solute concentrations in the two containers in › Figure 2-14. The container in part (a) of the figure is full of pure water, so the water concentration is 100 percent and the solute concentration is 0 percent. In part (b), solute has replaced 10 percent of the water molecules. The water concentration is now 90 percent, and the solute concentration is 10 percent; this is a lower water concentration and a higher solute concentration than in part (a). Note that as the solute concentration increases, the water concentration decreases correspondingly.

If solutions of unequal solute concentration (and hence unequal water concentration) are separated by a membrane that permits passage of water, such as the plasma membrane, water will move passively down its own concentration gradient from the area of higher water concentration (lower solute concentration) to the area of lower water concentration (higher solute concentration) (› Figure 2-15). This net diffusion of water is known as **osmosis**. Because solutions are always referred to in terms of their concentration of solute, *water moves by osmosis to the*

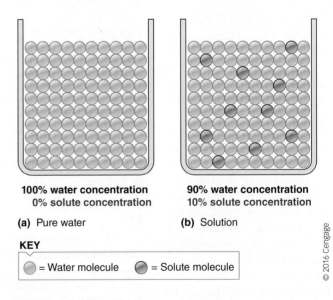

**100% water concentration**
**0% solute concentration**

**90% water concentration**
**10% solute concentration**

**(a)** Pure water

**(b)** Solution

KEY

○ = Water molecule  ● = Solute molecule

› **FIGURE 2-14 Relationship between solute and water concentration in a solution**

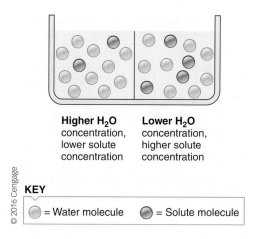

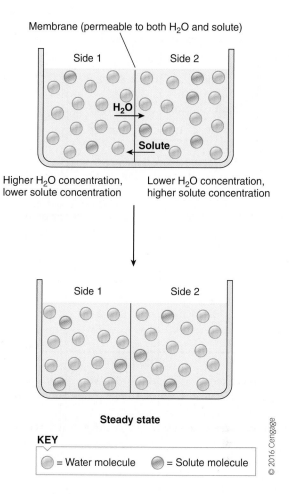

**Higher H₂O** concentration, lower solute concentration

**Lower H₂O** concentration, higher solute concentration

**KEY**

⬤ = Water molecule  ⬤ = Solute molecule

© 2016 Cengage

› **FIGURE 2-15 Osmosis.** Osmosis is the net diffusion of water down its own concentration gradient (to the area of higher solute concentration).

Membrane (permeable to both H₂O and solute)

Side 1                    Side 2

H₂O →

← Solute

Higher H₂O concentration, lower solute concentration

Lower H₂O concentration, higher solute concentration

Side 1                    Side 2

**Steady state**

**KEY**

⬤ = Water molecule  ⬤ = Solute molecule

© 2016 Cengage

› **FIGURE 2-16 Movement of water and a penetrating solute unequally distributed across a membrane**

*area of higher solute concentration.* Despite the impression that the solutes are "drawing"—that is, attracting—water, osmosis is nothing more than diffusion of water down its own concentration gradient across the membrane.

The discussion of osmosis so far has ignored any movement of solute. In the following sections we consider the movement of solute and compare the results of osmosis when the solute can and cannot permeate the membrane.

### OSMOSIS WHEN A MEMBRANE SEPARATES UNEQUAL SOLUTIONS OF A PENETRATING SOLUTE

Assume that solutions of unequal solute concentration are separated by a membrane that permits passage of both water and the solute. Because the membrane is permeable to the solute as well as to water, the solute can move down its own concentration gradient in the opposite direction of the net water movement (› Figure 2-16). This movement continues until both the solute and the water are evenly distributed across the membrane. With all concentration gradients gone, osmosis ceases. The final volume of the compartments when the steady state is achieved is the same as at the onset. Water and solute molecules have merely exchanged places between the two compartments until their distributions have equalized; that is, an equal number of water molecules have moved from side 1 to side 2 as solute molecules have moved from side 2 to side 1.

### OSMOSIS WHEN A MEMBRANE SEPARATES UNEQUAL SOLUTIONS OF A NONPENETRATING SOLUTE

For solutions of unequal solute concentrations separated by a membrane that is permeable to water but impermeable to the solute, the solute cannot cross the membrane to move down its concentration gradient (› Figure 2-17). At first, the concentration gradients are identical to those in the previous example. However, even though net diffusion of water takes place from side 1 to side 2, the solute cannot move. As a result of water movement alone, the volume of side 2 increases while the volume of side 1 correspondingly decreases. Loss of water from side 1 increases the solute concentration on side 1, whereas addition of water to side 2 reduces the solute concentration on that

side. Eventually, the concentrations of water and solute on the two sides of the membrane become equal, and net diffusion of water ceases. Unlike the situation in which the solute can also permeate, diffusion of water alone has resulted in a change in the final volumes of the two compartments. The side originally containing the greater solute concentration has a larger volume, having gained water.

### OSMOSIS WHEN A MEMBRANE SEPARATES PURE WATER FROM A SOLUTION OF A NONPENETRATING SOLUTE

What happens when a nonpenetrating solute is present on side 2 and pure water is present on side 1 (› Figure 2-18)? Osmosis occurs from side 1 to side 2, but the concentrations between the two compartments can never become equal. No matter how dilute side 2 becomes because of water diffusing into it, it can never become pure water, nor can side 1 ever acquire any solute. Because equilibrium is impossible to achieve, does net diffusion of water (osmosis) continue until all the water has left side 1? No. As the volume expands in compartment 2, a difference in hydrostatic pressure between the two compartments is created, and it opposes osmosis. **Hydrostatic (fluid) pressure** is the pressure exerted by a standing, or stationary, fluid on an object—in this case, the plasma membrane (*hydro* means "fluid";

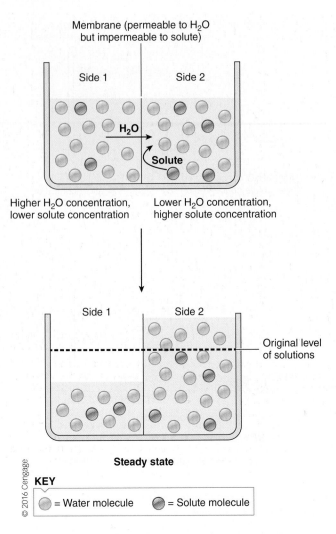

> FIGURE 2-17 Osmosis in the presence of an unequally distributed nonpenetrating solute

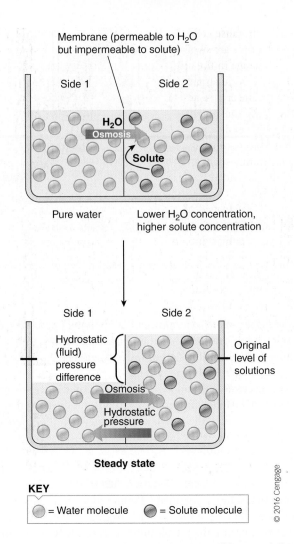

> FIGURE 2-18 Osmosis when pure water is separated from a solution containing a nonpenetrating solute

*static* means "standing"). The hydrostatic pressure exerted by the larger volume of fluid on side 2 is greater than the hydrostatic pressure exerted on side 1. This hydrostatic pressure difference tends to push fluid from side 2 to side 1.

The **osmotic pressure** of a solution is a measure of the tendency for water to move into that solution because of its relative concentration of nonpenetrating solutes and water. Net movement of water continues until the opposing hydrostatic pressure exactly counterbalances the osmotic pressure. The magnitude of the osmotic pressure is equal to the magnitude of the opposing hydrostatic pressure necessary to completely stop osmosis. The greater the concentration of nonpenetrating solute and the lower the concentration of water, the greater the drive for water to move by osmosis from pure water into the solution and, therefore, the greater the opposing pressure required to stop the osmotic flow and the greater the osmotic pressure of the solution. Therefore, a solution with a high concentration of nonpenetrating solute exerts greater osmotic pressure than a solution with a lower concentration of nonpenetrating solute does.

Osmosis is the major force responsible for the net movement of water into and out of cells. Approximately 100 times the volume

of water in a cell crosses the plasma membrane every second. Yet body cells normally do not experience any net gain (swelling) or loss (shrinking) of volume because the concentration of nonpenetrating solutes in the ECF is normally carefully regulated (e.g., by the kidneys; p. 560) so that the osmotic pressure in the ECF is the same as the osmotic pressure within the cells.

## TONICITY

When a solution surrounds the cell, the **tonicity** of the solution is the effect the solution has on cell volume—that is, whether the cell remains the same size, swells, or shrinks. The tonicity of a solution is determined by its concentration of nonpenetrating solutes. Solutes that can penetrate the plasma membrane quickly become equally distributed between the ECF and ICF, so they do not contribute to osmotic differences. An **isotonic solution** (*iso* means "equal") has the same concentration of nonpenetrating solutes as normal body cells do. When a cell is bathed in an isotonic solution, there is no net movement of water across the membrane, so cell volume remains constant. For this reason, the ECF is normally kept isotonic so that no net diffusion of water occurs across the plasma membranes of body cells. This is important because cells, especially brain cells, do not function properly if they are swollen or shrunken.

Any change in the concentration of nonpenetrating solutes in the ECF produces a corresponding change in the water concentration difference across the plasma membrane. The resultant osmotic movement of water brings about changes in cell volume. The easiest way to demonstrate this phenomenon is to place red blood cells in solutions with varying concentrations of nonpenetrating solutes. Normally the plasma in which red blood cells are suspended has the same osmotic activity as the fluid inside these cells, so the cells maintain a constant volume. If red blood cells are placed in a dilute or **hypotonic solution** (*hypo* means "below")—a solution with a below-normal concentration of nonpenetrating solutes (and therefore a higher concentration of water)—water enters the cells by osmosis. Net gain of water by the cells causes them to swell, perhaps to the point of rupturing. If, in contrast, red blood cells are placed in a concentrated or **hypertonic solution** (*hyper* means "above")—a solution with an above-normal concentration of nonpenetrating solutes (and therefore a lower concentration of water)—the cells shrink as they lose water by osmosis. Thus, it is crucial that the concentration of nonpenetrating solutes in the ECF quickly be restored to normal should the ECF become hypotonic (as with ingesting too much water) or hypertonic (as with losing too much water through severe diarrhoea). (See pp. 611–619 for further details about the important homeostatic mechanisms that maintain the normal concentration of nonpenetrating solutes in the ECF.)

## 2.8 | Assisted Membrane Transport

All the kinds of transport discussed so far—diffusion down concentration gradients, movement along electrical gradients, and osmosis—produce net movement of molecules capable of permeating the plasma membrane by virtue of their lipid solubility or small size. Large, poorly lipid-soluble molecules, such as proteins, glucose, and amino acids, cannot cross the plasma membrane on their own no matter what forces are acting on them. These molecules are too big for channels, and they cannot dissolve in the lipid bilayer. This impermeability ensures that the large polar intracellular proteins cannot escape from the cell. These proteins therefore stay in the cell where they belong and can carry out their life-sustaining functions. For example, they serve as metabolic enzymes (e.g., phosphofructokinase, used in glycolysis).

However, because large, poorly lipid-soluble molecules cannot cross the plasma membrane on their own, the cell must provide mechanisms for deliberately transporting these types of molecules into or out of the cell as needed. For example, the cell must usher in essential nutrients, such as glucose for energy and amino acids for the synthesis of proteins, and transport out metabolic wastes and secretory products, such as water-soluble protein hormones. Furthermore, passive diffusion alone cannot always account for the movement of small ions. Cells use two different mechanisms of membrane transport to accomplish these selective transport processes: (1) *carrier-mediated transport* for transfer of small, water-soluble substances across the membrane, and (2) *vesicular transport* for movement of large molecules and multimolecular particles between the ECF and ICF. We now examine each of these methods of assisted membrane transport.

### Carrier-mediated transport

Carrier proteins span the thickness of the plasma membrane and can reverse shape so that specific binding sites can alternately be exposed at either side of the membrane. That is, the carrier "flip-flops" so that binding sites located in the interior of the carrier are alternately exposed to the ECF and the ICF. › Figure 2-19 is a schematic representation of this **carrier-mediated transport**. As the molecule to be transported attaches to a binding site within the interior of the carrier on one side of the membrane (step 1), this binding causes the carrier to flip its shape so that the same site is now exposed to the other side of the membrane (step 2). Then, having been moved in this way from one side of the membrane to the other, the bound molecule detaches from the carrier (step 3). After the passenger detaches, the carrier reverts to its original shape (step 4).

Carrier-mediated transport systems display three important characteristics that determine the kind and amount of material that can be transferred across the membrane: *specificity*, *saturation*, and *competition*.

1. **Specificity**. Each carrier protein is specialized to transport a specific substance or, at most, a few closely related chemical compounds. For example, amino acids cannot bind to glucose carriers, although several similar amino acids may be able to use the same carrier. Cells vary in the types of carriers they have, thereby permitting transport selectivity among cells.

A number of inherited diseases involve defects in transport systems for a particular substance. *Cysteinuria* (cysteine in the urine) is such a disease, involving defective cysteine carriers in the kidney membranes. This transport system normally removes cysteine from the fluid destined to become urine and returns this essential amino acid to the blood. When this carrier malfunctions, large quantities of cysteine remain in the urine, where it is relatively insoluble and tends to precipitate. This is one cause of urinary stones.

2. **Saturation**. A limited number of carrier binding sites are available within a particular plasma membrane for a specific substance. Therefore, there is a limit to the amount of a substance a carrier can transport across the membrane in a given time. This limit is known as the **transport maximum ($T_m$)**. Until the $T_m$ is reached, the number of carrier binding sites occupied by a substance and, accordingly, the substance's rate of transport across the membrane are directly related to its concentration. The more substance available for transport, the more will be transported. When the $T_m$ is reached, the carrier is saturated (all binding sites are occupied), and the rate of the substance's transport across the membrane is maximal. Further increases in the substance's concentration are not accompanied by corresponding increases in the rate of transport (› Figure 2-20).

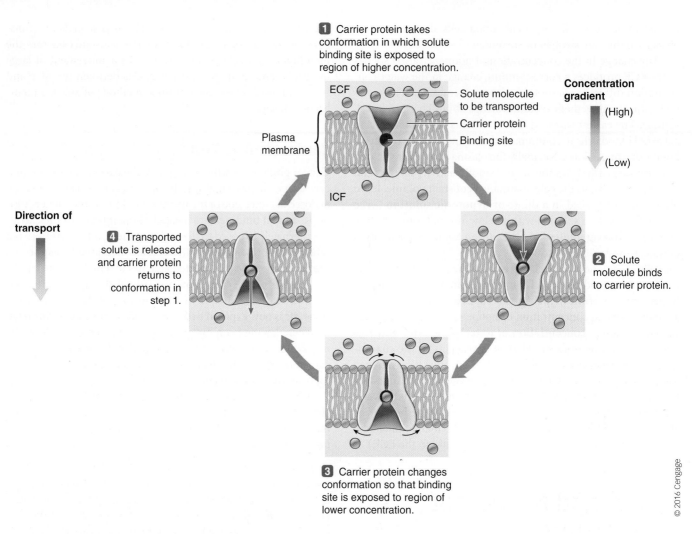

> FIGURE 2-19  **Model for facilitated diffusion, a passive form of carrier-mediated transport**

1. Carrier protein takes conformation in which solute binding site is exposed to region of higher concentration.

ECF

Solute molecule to be transported
Carrier protein
Binding site

Concentration gradient

(High)

(Low)

Plasma membrane

ICF

Direction of transport

4. Transported solute is released and carrier protein returns to conformation in step 1.

2. Solute molecule binds to carrier protein.

3. Carrier protein changes conformation so that binding site is exposed to region of lower concentration.

© 2016 Cengage

As an analogy, consider a ferry boat that can maximally carry 100 people across a river during one trip in an hour. If 25 people are on hand to board the ferry, 25 will be transported that hour. Doubling the number of people on hand to 50 will double the rate of transport to 50 people that hour. Such a direct relationship will exist between the number of people waiting to board (the concentration) and the rate of transport until the ferry is fully occupied (its $T_m$ is reached). The ferry can maximally transport 100 people per hour. Even if 150 people are waiting to board, still only 100 will be transported per hour.

Saturation of carriers is a critical rate-limiting factor in the transport of selected substances across the kidney membranes during urine formation and across the intestinal membranes during absorption of digested foods. Furthermore, it is sometimes possible to regulate (e.g., by hormones) the rate of carrier-mediated transport by varying the affinity (attraction) of the binding site for its passenger or by varying the number of binding sites. For example, the hormone insulin greatly increases the carrier-mediated transport of glucose into most cells of the body by promoting an increase in the number of glucose carriers in the cell's plasma membrane.

Deficient insulin action (*diabetes mellitus*) drastically impairs the body's ability to take up and use glucose as the primary energy source.

3. **Competition**. Several closely related compounds may compete for a ride across the membrane on the same carrier. If a given binding site can be occupied by more than one type of molecule, the rate of transport of each substance is less when both molecules are present than when either is present by itself. To illustrate, assume the ferry has 100 seats (binding sites) that can be occupied by either men or women. If only men are waiting to board, up to 100 men can be transported during each trip; the same holds true if only women are waiting to board. If, however, both men and women are waiting to board, they will compete for the available seats so that fewer men and fewer women will be transported than when either group is present alone. Fifty of each might make the trip, although the total number of people transported will still be the same, 100 people. In other words, when a carrier can transport two closely related substances, such as the amino acids glycine and alanine, the presence of both diminishes the rate of transfer for either.

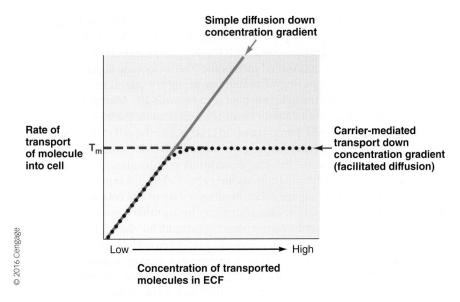

Simple diffusion down
concentration gradient

Rate of
transport
of molecule
into cell

$T_m$

Carrier-mediated
transport down
concentration gradient
(facilitated diffusion)

Low ——————————> High

Concentration of transported
molecules in ECF

© 2016 Cengage

> FIGURE 2-20 **Comparison of carrier-mediated transport and simple diffusion down a concentration gradient.** With simple diffusion of a molecule down its concentration gradient, the rate of transport of the molecule into the cell is directly proportional to the extracellular concentration of the molecule. With carrier-mediated transport of a molecule down its concentration gradient, the rate of transport of the molecule into the cell is directly proportional to the extracellular concentration of the molecule until the carrier is saturated, at which time the rate of transport reaches a maximum value (transport maximum, or $T_m$). The rate of transport does not increase with further increases in the ECF concentration of the molecule.

## Active or passive transport

Carrier-mediated transport takes two forms, depending on whether energy must be supplied to complete the process: *facilitated diffusion* (not requiring energy) and *active transport* (requiring energy). **Facilitated diffusion** uses a carrier to facilitate (assist) the transfer of a particular substance across the membrane "downhill"—that is, along a concentration gradient, from high to low concentration. This process is passive and does not require energy because movement occurs naturally down a concentration gradient. The unassisted diffusion described earlier is sometimes called *simple diffusion*, to distinguish it from facilitated diffusion. **Active transport**, in contrast, requires the carrier to expend energy to transfer its passenger "uphill"—that is, against a concentration gradient, from an area of lower concentration to an area of higher concentration. An analogous situation is a car on a hill. To move the car downhill requires no energy; it will coast from the top down. Driving the car uphill, however, requires the use of energy (gasoline or electricity).

### FACILITATED DIFFUSION

The most notable example of facilitated diffusion is the transport of glucose into cells. Glucose is in higher concentration in the blood than in the tissues. Fresh supplies of this nutrient are regularly added to the blood through eating and by using reserve energy stores in the body. Simultaneously, the cells metabolize glucose almost as rapidly as it enters the cells from the blood. As a result, a continuous gradient exists for net diffusion of glucose into the cells. However, glucose cannot cross cell membranes on its own. Because glucose is polar, it is not lipid soluble and it is also too large to fit through a channel. Without the glucose-carrier molecules to facilitate membrane transport of glucose, the cells would be deprived of glucose, their preferred source of fuel.

The carrier binding sites involved in facilitated diffusion can bind with their passenger molecules when exposed to either side of the membrane (see > Figure 2-19). Passenger binding triggers the carrier to flip its conformation and drop off the passenger on the opposite side of the membrane. Because passengers are more likely to bind with the carrier on the high-concentration side than on the low-concentration side, the net movement always proceeds down the concentration gradient from higher to lower concentration. As is characteristic of mediated transport, the rate of facilitated diffusion is limited by saturation of the carrier binding sites, unlike the rate of simple diffusion, which is always directly proportional to the concentration gradient (see > Figure 2-20).

### ACTIVE TRANSPORT

Active transport also involves the use of a protein carrier to transfer a specific substance across the membrane, but in this case the carrier transports the substance against its concentration gradient. For example, the uptake of iodine by thyroid gland cells necessitates active transport because 99 percent of the iodine in the body is concentrated in the thyroid. To move iodine from the blood, where its concentration is low, into the thyroid, where its concentration is high, requires expenditure of energy to drive the carrier. Specifically, in active transport, energy in the form of ATP is required to alter the relative affinity of the binding site when exposed on one side of the plasma membrane or to the other. By contrast, in facilitated diffusion, the affinity of the binding site is the same when exposed to either the outside or the inside of the cell.

With active transport, the binding site has a greater affinity for its passenger on the low-concentration side due to *phosphorylation* of the carrier. The carrier exhibits ATPase activity in that it splits the terminal phosphate from an ATP molecule to yield ADP plus a free inorganic phosphate. The phosphate group is then attached to the carrier. This phosphorylation and the binding of the passenger on the low-concentration side cause the carrier protein to flip its conformation so that the passenger is now exposed to the high-concentration side of the membrane. The change in carrier shape is accompanied by *dephosphorylation*; that is, the phosphate group detaches from the carrier. Removal of phosphate reduces the affinity of the binding site for the passenger, so the passenger is released on the high-concentration side. The carrier then returns to its original conformation. In this way, ATP energy is used in the phosphorylation–dephosphorylation cycle of the carrier. It alters the affinity of the carrier's binding sites on opposite sides of the membrane so that transported particles are moved uphill from an area of low concentration to an area of higher concentration. These active-transport mechanisms are frequently called **pumps**, analogous to water pumps that require energy to lift water against the downward pull of gravity.

Let us now apply these basic principles of active transport and look at the **$Na^+-K^+$ ATPase pump** (**$Na^+-K^+$ pump** for short), a very important pump found in the plasma membrane of all cells.

This carrier transports sodium ions ($Na^+$) out of the cell, concentrating them in the ECF, and picks up potassium ions ($K^+$) from the outside, concentrating them in the ICF (⟩ Figure 2-21). Splitting of ATP through ATPase activity and the subsequent phosphorylation of the carrier on the intracellular side increases the carrier's affinity for $Na^+$ and induces a change in carrier shape, leading to the drop-off of $Na^+$ on the exterior. The subsequent dephosphorylation of the carrier increases its affinity for $K^+$ on the extracellular side and restores the original carrier conformation, thereby transferring $K^+$ into the cytoplasm. There is not a direct exchange of $Na^+$ for $K^+$, however. The $Na^+-K^+$ pump moves three $Na^+$ out of the cell for every two $K^+$ it pumps in. (To appreciate the magnitude of active $Na^+-K^+$ pumping that takes place, consider that a single nerve cell membrane contains perhaps 1 million $Na^+-K^+$ pumps capable of transporting about 200 million ions per second.)

The $Na^+-K^+$ pump plays three important roles:

1. It establishes $Na^+$ and $K^+$ concentration gradients across the plasma membrane of all cells; these gradients are critically important in the ability of nerve and muscle cells to generate electrical signals essential to their functioning (this topic is discussed thoroughly in Sections 2.10–2.12).

2. It helps regulate cell volume by controlling the concentrations of solutes inside the cell, thereby minimizing osmotic effects that would induce swelling or shrinking of the cell.

3. The energy used to run the $Na^+-K^+$ pump also indirectly serves as the energy source for the cotransport of glucose and amino acids across intestinal and kidney cells. This process is known as *secondary active transport*.

### SECONDARY ACTIVE TRANSPORT

Unlike most cells of the body, the intestinal and kidney cells actively transport glucose and amino acids by moving them uphill from low to high concentration. The intestinal cells transport these nutrients from inside the intestinal lumen into the blood, concentrating them in the blood until none of these molecules are left in the lumen to be lost in the feces. Similarly, the kidney cells save these nutrient molecules for the body by transporting them out of the fluid that is to become urine, moving them against a concentration gradient into the blood. However, energy is not directly supplied to the carrier in these instances. The carriers that transport glucose against its concentration gradient from the lumens of the intestine and kidneys are distinct from the glucose facilitated–diffusion carriers. The luminal

carriers in intestinal and kidney cells are **cotransport carriers** in that they have two binding sites, one for $Na^+$ and one for the nutrient molecule. The $Na^+-K^+$ pumps in these cells are located in the basolateral membrane (the membrane at the base of the cell opposite the lumen and along the lateral edge of the cell below the tight junction; see ⟩ Figure 2-10). More $Na^+$ is present in the lumen than inside the cells because the energy-requiring $Na^+-K^+$ pump transports $Na^+$ out of the cell at the basolateral membrane, keeping the intracellular $Na^+$ concentration low. Because of this $Na^+$ concentration difference, more $Na^+$ binds to the luminal cotransport carrier when it is exposed to the outside (⟩ Figure 2-22). Binding of $Na^+$ to the cotransport carrier increases the carrier's affinity for its other passenger (e.g., glucose), so the carrier has a high affinity for glucose when exposed to the outside of the membrane. When both $Na^+$ and glucose are bound to the carrier, it undergoes a change in shape and opens to the inside of the cell. Both $Na^+$ and glucose are released to the interior: $Na^+$ because of the lower intracellular $Na^+$ concentration, and glucose because of the reduced affinity of the binding site on release of $Na^+$.

The movement of $Na^+$ into the cell by this cotransport carrier is downhill because the intracellular $Na^+$ concentration is low, but the movement of glucose is uphill because glucose becomes concentrated in the cell. The released $Na^+$ is quickly pumped out by the active $Na^+-K^+$ transport mechanism, keeping the level of intracellular $Na^+$ low. The energy expended in this process is not directly used to run the cotransport carrier, because there is no need for phosphorylation to alter the affinity of the binding site to glucose. Instead, the establishment of an $Na^+$ concentration gradient by a primary active-transport mechanism (the $Na^+-K^+$ pump) drives this secondary active-transport mechanism ($Na^+-$glucose cotransport carrier) to move glucose against its concentration gradient. With **primary active transport**, energy is *directly* required to move a substance uphill. The term *active transport* without a qualifier typically means primary active transport. With **secondary active transport**, energy is required in the entire process, but it is *not directly* required to run the pump. Rather, it uses second-hand energy stored in the form of an **ion concentration gradient** (e.g., a $Na^+$ gradient) to move the cotransported molecule uphill. This is very efficient, because $Na^+$ must be pumped out anyway to maintain the electrical and osmotic integrity of the cell.

The glucose carried into the cell across the luminal border by secondary active transport then passively moves out of the cell across the basolateral border by facilitated diffusion down its concentration gradient and enters the blood. This facilitated diffusion is mediated by a passive carrier in the basolateral membrane identical to the one that transports glucose into other cells, but in intestinal and kidney cells it transports glucose out of the cell. The difference depends on the direction of the glucose concentration gradient. In the case of intestinal and kidney cells, glucose is in higher concentration inside the cells.

Before leaving the topic of carrier-mediated transport, think about all the activities that rely on carrier assistance. All cells depend on carriers for the uptake of glucose and amino acids,

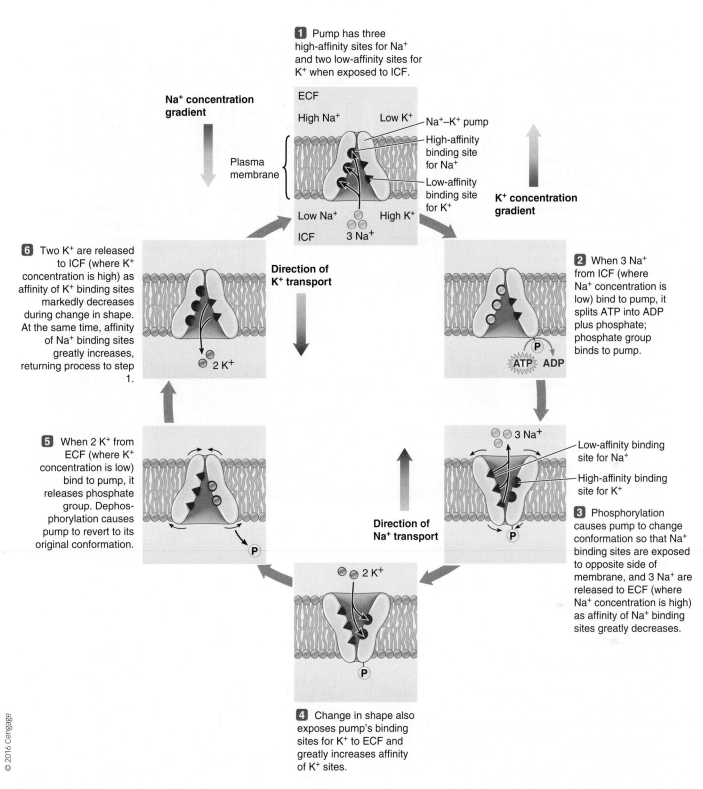

**1** Pump has three high-affinity sites for Na$^+$ and two low-affinity sites for K$^+$ when exposed to ICF.

**Na$^+$ concentration gradient**

ECF

High Na$^+$  Low K$^+$

Na$^+$–K$^+$ pump

Plasma membrane

High-affinity binding site for Na$^+$

Low-affinity binding site for K$^+$

Low Na$^+$  High K$^+$

ICF

3 Na$^+$

**K$^+$ concentration gradient**

**Direction of K$^+$ transport**

**6** Two K$^+$ are released to ICF (where K$^+$ concentration is high) as affinity of K$^+$ binding sites markedly decreases during change in shape. At the same time, affinity of Na$^+$ binding sites greatly increases, returning process to step 1.

2 K$^+$

**2** When 3 Na$^+$ from ICF (where Na$^+$ concentration is low) bind to pump, it splits ATP into ADP plus phosphate; phosphate group binds to pump.

P

ATP  ADP

**5** When 2 K$^+$ from ECF (where K$^+$ concentration is low) bind to pump, it releases phosphate group. Dephosphorylation causes pump to revert to its original conformation.

P

**Direction of Na$^+$ transport**

3 Na$^+$

Low-affinity binding site for Na$^+$

High-affinity binding site for K$^+$

P

**3** Phosphorylation causes pump to change conformation so that Na$^+$ binding sites are exposed to opposite side of membrane, and 3 Na$^+$ are released to ECF (where Na$^+$ concentration is high) as affinity of Na$^+$ binding sites greatly decreases.

2 K$^+$

P

**4** Change in shape also exposes pump's binding sites for K$^+$ to ECF and greatly increases affinity of K$^+$ sites.

⟩ **FIGURE 2-21 Na$^+$−K$^+$ ATPase pump.** The plasma membrane of all cells contains an active transport carrier, the Na$^+$−K$^+$ pump, which uses energy in the carrier's phosphorylation-dephosphorylation cycle to sequentially transport Na$^+$ out of the cell and K$^+$ into the cell against these ions' concentration gradients. This pump moves three Na$^+$ out and two K$^+$ in for each ATP split.

which serve as the major energy source and structural building blocks, respectively. Na$^+$−K$^+$ pumps are essential for generating cellular electrical activity and for ensuring that cells have an appropriate intracellular concentration of osmotically active

solutes. Active transport, either primary or secondary or both, is used extensively to accomplish the specialized functions of the nervous and digestive systems as well as the kidneys and all types of muscle.

© 2016 Cengage

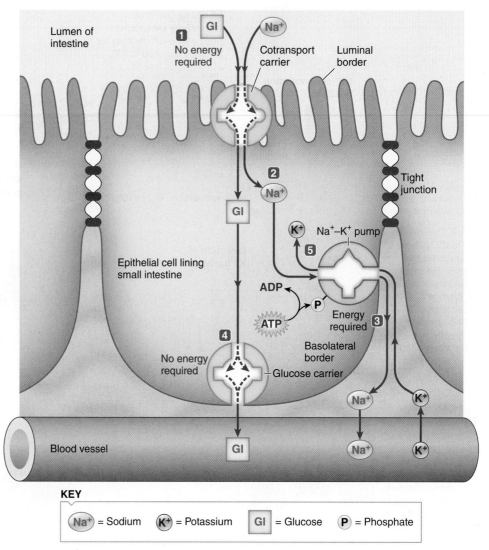

**1** A cotransport carrier at the luminal border simultaneously transfers glucose against a concentration gradient and Na⁺ down a concentration gradient from the lumen into the cell.

**2** No energy is directly used by the cotransport carrier to move glucose uphill. Instead operation of the cotransport carrier is driven by the Na⁺ concentration gradient (low Na⁺ in ICF compared with lumen) established by the energy-using Na⁺–K⁺ pump.

**3** The Na⁺–K⁺ pump actively transports Na⁺ out of the cell at the basolateral border keeping the ICF Na⁺ concentration lower than the luminal concentration.

**4** After entering the cell by secondary active transport, glucose is transported down its concentration gradient from the cell into the blood by facilitated diffusion, mediated by a passive glucose carrier at the basal border.

**5** The Na⁺–K⁺ pump also actively transports K⁺ into the cell, maintaining a high intracellular K⁺ concentration.

**KEY**

| Na⁺ = Sodium | K⁺ = Potassium | Gl = Glucose | P = Phosphate |
|---|---|---|---|

> **FIGURE 2-22 Secondary active transport.** Glucose (as well as amino acids) is transported across intestinal and kidney cells against its concentration gradient by means of secondary active transport.

---

### Check Your Understanding 2.4

1. Draw a graph comparing simple diffusion down a concentration gradient and carrier-mediated transport.
2. Describe what causes the carrier to change shape and thereby expose binding sites to passengers on opposite sides of the membrane in facilitated diffusion, primary active transport, and secondary active transport.

## Vesicular transport

Special carrier-mediated transport systems embedded in the plasma membrane can selectively transport ions and small polar molecules. But what happens in the case of large polar molecules or even multimolecular materials that must leave or enter the cell—such as during secretion of protein hormones (large polar molecules) by endocrine cells, or during ingestion of invading bacteria (multimolecular particles) by white blood cells? These materials are unable to cross the plasma membrane, even with assistance. They are much too big for channels, and no carriers exist for them (they would not fit into a carrier molecule). These large particles are transferred between the ICF and ECF not by crossing the membrane but by being wrapped in a membrane-enclosed vesicle, a process known as **vesicular transport**. Vesicular transport requires energy expenditure by the cell, so it is an active method of membrane transport. Energy is needed to accomplish vesicle formation and vesicle movement within the cell. Transport into the cell in this manner is termed **endocytosis**, whereas transport out of the cell is called **exocytosis**.

#### ENDOCYTOSIS
In endocytosis, the plasma membrane surrounds the substance to be ingested, then fuses over the surface, pinching off a membrane-enclosed vesicle so that the engulfed material is trapped within the cell (> Figure 2-23). There are three

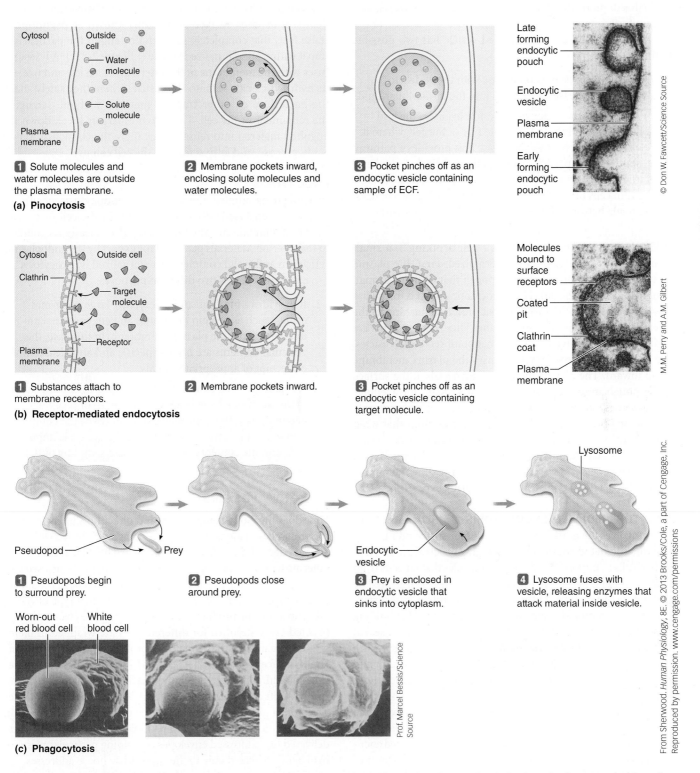

**(a) Pinocytosis**

1 Solute molecules and water molecules are outside the plasma membrane.

2 Membrane pockets inward, enclosing solute molecules and water molecules.

3 Pocket pinches off as an endocytic vesicle containing sample of ECF.

Cytosol
Outside cell
Water molecule
Solute molecule
Plasma membrane

Late forming endocytic pouch
Endocytic vesicle
Plasma membrane
Early forming endocytic pouch

© Don W. Fawcett/Science Source

**(b) Receptor-mediated endocytosis**

1 Substances attach to membrane receptors.

2 Membrane pockets inward.

3 Pocket pinches off as an endocytic vesicle containing target molecule.

Cytosol
Outside cell
Clathrin
Target molecule
Receptor
Plasma membrane

Molecules bound to surface receptors
Coated pit
Clathrin coat
Plasma membrane

M.M. Perry and A.M. Gilbert

**(c) Phagocytosis**

1 Pseudopods begin to surround prey.

2 Pseudopods close around prey.

3 Prey is enclosed in endocytic vesicle that sinks into cytoplasm.

4 Lysosome fuses with vesicle, releasing enzymes that attack material inside vesicle.

Pseudopod
Prey
Endocytic vesicle
Lysosome

Worn-out red blood cell
White blood cell

Prof. Marcel Bessis/Science Source

From Sherwood. *Human Physiology*, 8E. © 2013 Brooks/Cole, a part of Cengage, Inc. Reproduced by permission. www.cengage.com/permissions

> **FIGURE 2-23 Three forms of endocytosis.** (a) Diagram and electron micrograph of pinocytosis. The surface membrane dips inward to form a pouch, then seals the surface, forming an intracellular endocytic vesicle that nonselectively internalizes a bit of ECF. (b) Diagram and electron micrograph of receptor-mediated endocytosis. When a large molecule such as a protein attaches to a specific surface receptor, the membrane pockets inward with the aid of a coat protein, forming a coated pit, then pinches off to selectively internalize the molecule in an endocytic vesicle. (c) Diagram and scanning electron micrograph series of phagocytosis. White blood cells internalize multimolecular particles such as bacteria or old red blood cells by extending pseudopods that wrap around and seal in the targeted material. A lysosome fuses with and degrades the vesicle contents.

forms of endocytosis, depending on the nature of the material internalized: pinocytosis, receptor-mediated endocytosis, and phagocytosis.

Once inside the cell, an engulfed vesicle has two possible destinies:

1. In most instances, lysosomes fuse with the vesicle to degrade and release its contents into the intracellular fluid.

2. In some cells, the endocytotic vesicle bypasses the lysosomes and travels to the opposite side of the cell, where it releases its contents by exocytosis. This provides a pathway to shuttle intact particles through the cell. Such vesicular traffic is one way that materials can be transferred through the thin cells lining the capillaries, across which exchanges are made between the blood and surrounding tissues.

**PINOCYTOSIS**   With **pinocytosis** ("cell drinking"), a small droplet of extracellular fluid is internalized. First, the plasma membrane dips inward, forming a pouch that contains a small bit of ECF (❯ Figure 2-23a). The endocytotic pouch is formed as a result of membrane-deforming coat proteins attaching to the inner surface of the plasma membrane. These coat proteins are similar to those involved in the formation of secretory vesicles. The linking of the coat proteins causes the plasma membrane to dip inward. The plasma membrane then seals at the surface of the pouch, trapping the contents in a small, intracellular **endocytotic vesicle**. Dynamin, a protein molecule responsible for pinching off an endocytotic vesicle, forms rings that wrap around and "wring the neck" of the pouch, severing the vesicle from the surface membrane. Besides bringing ECF into a cell, pinocytosis provides a way to retrieve extra plasma membrane that has been added to the cell surface during exocytosis.

**RECEPTOR-MEDIATED ENDOCYTOSIS**   Unlike pinocytosis, which involves the nonselective uptake of the surrounding fluid, **receptor-mediated endocytosis** is a highly selective process that enables cells to import specific large molecules that it needs from its environment. Receptor-mediated endocytosis is triggered by the binding of a specific molecule, such as a protein, to a surface membrane receptor site specific for that protein. This binding causes the plasma membrane at that site to sink in, then seal at the surface, trapping the protein inside the cell (❯ Figure 2-23b). Cholesterol complexes, vitamin B12, the hormone insulin, and iron are examples of substances selectively taken into cells by receptor-mediated endocytosis.

Unfortunately, some viruses can sneak into cells by exploiting this mechanism. For instance, flu viruses and HIV, the virus that causes AIDS (p. 491) gain entry to cells via receptor-mediated endocytosis. They do so by binding with membrane receptor sites normally designed to trigger the internalization of a needed molecule.

**PHAGOCYTOSIS**   During **phagocytosis** ("cell eating"), large multimolecular particles are internalized. Most body cells perform pinocytosis, many carry out receptor-mediated endocytosis, but only a few specialized cells are capable of phagocytosis. The latter are the "professional" phagocytes, the most notable being certain types of white blood cells that play an important role in the body's defence mechanisms (see Chapter 11 to find

out more about immune cells). When a white blood cell encounters a large multimolecular particle, such as a bacterium or tissue debris, it extends surface projections known as **pseudopods** (false feet) that completely surround or engulf the particle and trap it within an internalized vesicle (❯ Figure 2-23c). A lysosome fuses with the membrane of the internalized vesicle and releases its hydrolytic enzymes into the vesicle, where they safely attack the bacterium or other trapped material without damaging the remainder of the cell. The enzymes largely break down the engulfed material into reusable raw ingredients, such as amino acids, glucose, and fatty acids that the cell can use.

### EXOCYTOSIS

In exocytosis, almost the reverse of endocytosis occurs. A membrane-enclosed vesicle formed within the cell fuses with the plasma membrane, then opens up, and releases its contents to the exterior. Materials packaged for export by the endoplasmic reticulum and Golgi complex are externalized by exocytosis.

Exocytosis serves two different purposes:

1. It provides a mechanism for secreting large polar molecules, such as protein hormones and enzymes that are unable to cross the plasma membrane. In this case, the vesicular contents are highly specific and are released only on receipt of appropriate signals.

2. It enables the cell to add specific components to the membrane, such as selected carriers, channels, or receptors, depending on the cell's needs. In such cases, the composition of the membrane surrounding the vesicle is important, and the contents may be merely a sampling of ICF.

## Secretory vesicles

How does the Golgi complex sort and direct finished proteins to the proper destinations? Finished products are collected within the dilated edges of the Golgi complex's sacs. The dilated edge of the outermost sac then pinches off to form a membrane-enclosed vesicle that contains the finished product. For each type of product to reach its appropriate site of function, each distinct type of vesicle takes up a specific product before budding off. Vesicles with their selected cargo destined for different sites are wrapped in membranes containing different surface protein molecules. Each different surface protein marker serves as a specific **docking marker** (like an address on an envelope). Each vesicle can "dock" and "unload" its cargo only at the appropriate "docking-marker acceptor," a protein located only at the proper destination within the cell (like a house address). Thus, Golgi products are sorted and delivered like addressed envelopes containing particular pieces of mail being delivered only to the appropriate house addresses.

Specialized secretory cells include endocrine cells, which secrete protein hormones, and digestive gland cells, which secrete digestive enzymes. In secretory cells, numerous large **secretory vesicles**, which contain proteins to be secreted, bud off from the Golgi stacks. The secretory proteins remain stored within the secretory vesicles until the cell is stimulated by a specific signal that indicates a need for release of that particular secretory product. On appropriate stimulation, the vesicles move to the cell's periphery. Vesicular contents are quickly released to

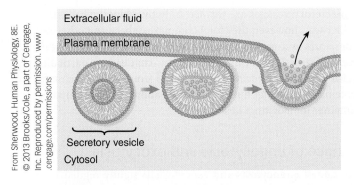

From Sherwood. Human Physiology, 8E. © 2013 Brooks/Cole, a part of Cengage, Inc. Reproduced by permission. www .cengage.com/permissions

> **FIGURE 2-24 Exocytosis of a secretory product.** A secretory vesicle fuses with the plasma membrane, releasing the vesicle contents to the cell exterior. The vesicle membrane becomes part of the plasma membrane.

the cell's exterior as the vesicle fuses with the plasma membrane, opens, and empties its contents to the outside (> Figure 2-24). Release of the contents of a secretory vesicle by means of exocytosis constitutes the process of secretion. Secretory vesicles fuse only with the plasma membrane and not with any of the internal

membranes associated with organelles, thereby preventing fruitless or even dangerous discharge of secretory products into the organelles.

We now examine how secretory vesicles take up specific products to be released into the ECF and then dock only at the plasma membrane. Before budding off from the outermost Golgi sac, the portion of the Golgi membrane that will be used to enclose the secretory vesicle becomes coated with a layer of specific proteins from the cytosol (> Figure 2-25). These and associated membrane proteins serve three important functions:

1. Specific proteins on the interior surface of the membrane facing the Golgi lumen act as *recognition markers* for the recognition and attraction of specific molecules that have been processed in the Golgi lumen. The newly finished proteins destined for secretion contain a unique sequence of amino acids known as a *sorting signal*. Recognition of the right protein's sorting signal by the complementary membrane marker ensures that the proper cargo is captured and packaged within a secretory vesicle that forms and buds off the outermost Golgi sac.

2. *Coat proteins* from the cytosol bind with another specific protein facing the outer surface of the membrane. The linking

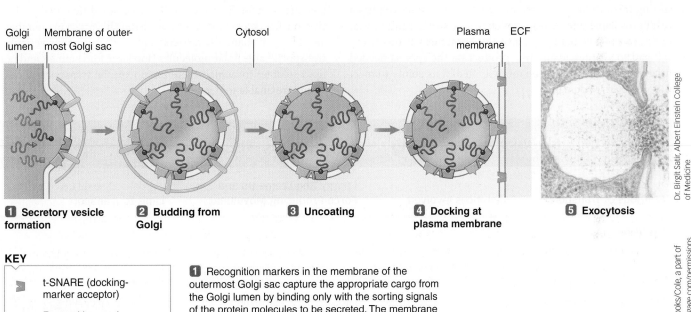

**1 Secretory vesicle formation**  **2 Budding from Golgi**  **3 Uncoating**  **4 Docking at plasma membrane**  **5 Exocytosis**

Dr. Birgit Satir, Albert Einstein College of Medicine

**KEY**

- ▸ t-SNARE (docking-marker acceptor)
- ⎰ Recognition marker
- ⎱ Coat-protein acceptor
- ⎰ v-SNARE (docking marker)
- Sorting signal
- ∿ Cargo proteins
- Coatomer (coat protein that causes membrane to curve)

**1** Recognition markers in the membrane of the outermost Golgi sac capture the appropriate cargo from the Golgi lumen by binding only with the sorting signals of the protein molecules to be secreted. The membrane that will wrap the vesicle is coated with coatomer, which causes the membrane to curve, forming a bud.

**2** The membrane closes beneath the bud, pinching off the secretory vesicle.

**3** The vesicle loses its coating, exposing v-SNARE docking markers on the vesicle surface.

**4** The v-SNAREs bind only with the t-SNARE docking-marker acceptors of the targeted plasma membrane, ensuring that secretory vesicles empty their contents to the cell's exterior.

From Sherwood. Human Physiology, 8E. © 2013 Brooks/Cole, a part of Cengage, Inc. Reproduced by permission. www.cengage.com/permissions

> **FIGURE 2-25 Packaging, docking, and release of secretory vesicles.** The diagram series illustrates secretory vesicle formation and budding, with the aid of a coat protein, and docking with the plasma membrane by means of v-SNAREs and t-SNAREs. The transmission electron micrograph shows secretion by exocytosis.

of these coat proteins causes the surface membrane of the Golgi sac to curve and form a dome-shaped bud around the captured cargo. Eventually, the surface membrane closes and pinches off the vesicle.

3. After budding off, the vesicle sheds its coat proteins. This uncoating exposes *docking markers*, which are other specific proteins facing the outer surface of the vesicle membrane. These docking markers, known as *v-SNAREs*, can link in a lock-and-key fashion with another protein marker, a *t-SNARE*, found only on the targeted membrane. In the case of secretory vesicles, the targeted membrane is the plasma membrane, the designated site for secretion to take place. In this way, the v-SNAREs of secretory vesicles are able to fuse only with the t-SNAREs of the plasma membrane. Once a vesicle has docked at the appropriate membrane by means of matching SNAREs, the two membranes completely fuse; then the vesicle opens up and empties its contents at the targeted site.

Note that the contents of secretory vesicles never come into contact with the cytosol. From the time these products are first synthesized in the endoplasmic reticulum until they are released from the cell by exocytosis, they are always wrapped in a membrane and thus isolated from the remainder of the cell. By manufacturing its particular secretory protein ahead of time and storing this product in secretory vesicles, a secretory cell has a readily available reserve from which to secrete large amounts of this product on demand (e.g., digestive hormones). If a secretory cell had to synthesize all of its product on the spot as needed for export, the cell would be more limited in its ability to meet varying levels of demand.

Secretory vesicles are formed only by secretory cells. In a similar way, however, the Golgi complex of these and other cell types sorts and packages newly synthesized products for different destinations within the cell. In each case, a particular vesicle captures a specific kind of cargo from among the many proteins in the Golgi lumen, and then addresses each shipping container for a distinct destination.

## Balance of endocytosis and exocytosis

The rate of endocytosis and exocytosis is tightly regulated in order to maintain a constant membrane surface area and cell volume. More than 100 percent of the plasma membrane may be used in an hour to wrap internalized vesicles in a cell actively involved in endocytosis, and this necessitates the rapid replacement of surface membrane by exocytosis. In contrast, when a secretory cell is stimulated to secrete, it may temporarily insert up to 30 times its surface membrane through exocytosis. This added membrane must be specifically retrieved by an equivalent volume of endocytotic activity. Thus, through exocytosis and endocytosis, portions of the membrane are constantly being restored, retrieved, and generally recycled.

Our discussion of membrane transport is now complete; Table 2-3 summarizes the pathways by which materials can pass between the ECF and the ICF. Cells are differentially selective in what enters or leaves because they have varying numbers and kinds of channels, carriers, and mechanisms for vesicular transport. Large polar molecules (too large for channels and not lipid soluble) for which there are no special transport mechanisms are unable to permeate.

---

**▌ TABLE 2-3 Characteristics of the Methods of Membrane Transport**

| Methods of Transport | Substances Involved | Energy Requirements and Force-Producing Movement | Limit to Transport |
|---|---|---|---|
| **Diffusion** | | | |
| Through lipid bilayer | Nonpolar molecules of any size (e.g., $O_2$, $CO_2$, and fatty acids) | Passive; molecules move down concentration gradient (from high to low concentration) | Continues until the gradient is abolished (steady state with no net diffusion) |
| Through protein channel | Specific, small ions (e.g., $Na^+$, $K^+$, $Ca^{2+}$, and $Cl^-$) | Passive; ions move down electrochemical gradient through open channels (from high to low concentration and by attraction of ion to area of opposite charge) | Continues until there is no net movement and a steady state is established |
| Special case of osmosis | Water only | Passive; water moves down its own concentration gradient (water moves to area of lower water concentration, i.e., higher solute concentration) | Continues until concentration difference is abolished or until stopped by an opposing hydrostatic pressure or until cell is destroyed |
| **Carrier-Mediated Transport** | | | |
| Facilitated diffusion | Specific polar molecules for which a carrier is available (e.g., glucose) | Passive; molecules move down concentration gradient (from high to low concentration) | Displays a transport maximum ($T_m$); carrier can become saturated |

| | | | |
|---|---|---|---|
| Primary active transport | Specific ions or polar molecules for which carriers are available (e.g., $Na^+$, $K^+$, and amino acids) | Active; ions move against concentration gradient (from low to high concentration); requires ATP | Displays a transport maximum; carrier can become saturated |
| Secondary active transport | Specific polar molecules and ions for which cotransport carriers are available (e.g., glucose, amino acids, and some ions) | Active; molecules move against concentration gradient (from low to high concentration); driven directly by ion gradient (usually $Na^+$) established by ATP-requiring primary pump | Displays a transport maximum; cotransport carrier can become saturated |
| **Vesicular Transport** | | | |
| *Endocytosis* | | | |
| Pinocytosis | Small volume of ECF fluid; also important in membrane recycling | Active; plasma membrane dips inward and pinches off at surface, forming an internalized vesicle | Control poorly understood |
| Receptor-mediated endocytosis | Specific, large polar molecule (e.g., protein) | Active; plasma membrane dips inward and pinches off at surface, forming an internalized vesicle | Necessitates binding to specific receptor site on membrane surface |
| Phagocytosis | Multimolecular particles (e.g., bacteria and cellular debris) | Active; cell extends pseudopods that surround particle, forming an internalized vesicle | Necessitates binding to specific receptor site on membrane surface |
| *Exocytosis* | Secretory products (e.g., hormones and enzymes) as well as large molecules passing through cell intact; also important in membrane recycling | Active; increase in cytosolic $Ca^{2+}$ induces fusion of secretory vesicle with plasma membrane; vesicle opens up and releases contents to outside | Secretion triggered by specific neural or hormonal stimuli; other controls involved in transcellular traffic and membrane recycling not known |

© 2016 Cengage

**2**

# 2.9 | Intercellular Communication and Signal Transduction

Coordination of the diverse activities of cells throughout the body to accomplish life-sustaining and other desired activities depends on the ability of cells to communicate with one another.

## Communication between cells

Intercellular communication takes place in the following ways (› Figure 2-26):

1. The most intimate means of intercellular communication is through gap junctions. Through these specialized anatomic arrangements, small ions and molecules are directly exchanged between interacting cells without ever entering the extracellular fluid.

2. The presence of identifying markers on the surface membrane of some cells permits them to directly link up transiently and interact with certain other cells in a specialized way. This is the means by which the phagocytes of the body's defence system specifically recognize and selectively destroy only undesirable cells, such as cancer cells, while leaving the body's own healthy cells alone.

3. The most common means by which cells communicate with one another is indirectly through extracellular chemical messengers, of which there are four types: *paracrines, neurotransmitters, hormones*, and *neurohormones*. In each case, a specific chemical messenger is synthesized by specialized cells to serve a designated purpose. On being released into the ECF by appropriate stimulation, these signalling agents act on other particular cells, the messenger's **target cells**, in a prescribed manner. To exert its effect, an extracellular chemical messenger must bind with target cell receptors specific for it.

The four types of chemical messengers differ in their source and the distance and means by which they get to their site of action as follows:

• **Paracrines** are local chemical messengers whose effect is exerted only on neighbouring cells in the immediate environment of their site of secretion. Because paracrines are distributed by simple diffusion, their action is restricted to short distances. They do not gain entry to the blood in

## DIRECT INTERCELLULAR COMMUNICATION

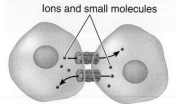

Ions and small molecules

**Gap junctions**

**Transient direct linkup of cells**

## INDIRECT INTERCELLULAR COMMUNICATION VIA EXTRACELLULAR CHEMICAL MESSENGERS

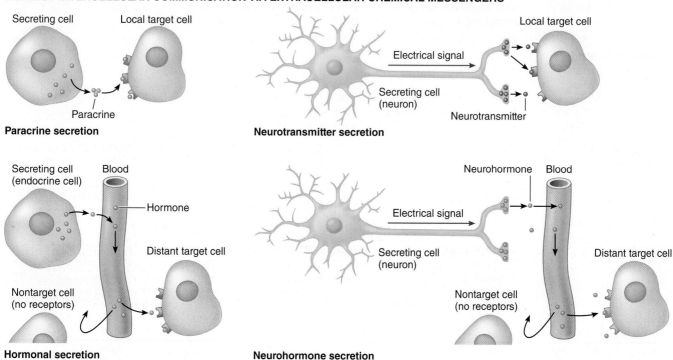

Secreting cell

Local target cell

Paracrine

**Paracrine secretion**

Electrical signal

Secreting cell (neuron)

Local target cell

Neurotransmitter

**Neurotransmitter secretion**

Secreting cell (endocrine cell)

Blood

Hormone

Distant target cell

Nontarget cell (no receptors)

**Hormonal secretion**

Neurohormone    Blood

Electrical signal

Secreting cell (neuron)

Distant target cell

Nontarget cell (no receptors)

**Neurohormone secretion**

> FIGURE 2-26 **Types of intercellular communication.** Gap junctions and transient direct linkup of cells are two means of direct communication between cells. Paracrines, neurotransmitters, hormones, and neurohormones are all extracellular chemical messengers that accomplish indirect communication between cells. These chemical messengers differ in their source and the distance they travel to reach their target cells.

any significant quantity because they are rapidly inactivated by locally existing enzymes. One example of a paracrine is *histamine*, which is released from a specific type of connective tissue cell during an inflammatory response within an invaded or injured tissue (p. 469). Among other things, histamine dilates (opens more widely) the blood vessels in the vicinity, which increases blood flow to the tissue. This action brings additional blood-borne combat supplies into the affected area. It is important to note that paracrines must be distinguished from chemicals that influence neighbouring cells after being nonspecifically released during the course of cellular activity. For example, an increased local concentration of carbon dioxide in an exercising muscle is among the factors that promote local dilation of the blood vessels supplying the muscle, but it is not specifically released to accomplish this particular response; therefore, carbon dioxide and

similar nonspecifically released chemicals are not considered paracrines.

- Neurons communicate directly with the cells they innervate (their target cells) by releasing *neurotransmitters*, which are very short-range chemical messengers, in response to electrical signals (action potentials). Like paracrines, neurotransmitters diffuse from their site of release across a narrow extracellular space to act locally on only an adjoining target cell, which may be another neuron, a muscle, or a gland.

- **Hormones** are long-range chemical messengers that are specifically secreted into the blood by endocrine glands in response to an appropriate signal. The blood carries the messengers to other sites in the body, where they exert their effects on their target cells some distance away from their site of release. Only the target cells of a particular hormone have membrane receptors for binding with this hormone. Nontarget cells are not influenced by any blood-borne hormones that reach them.

- **Neurohormones** are hormones released into the blood by *neurosecretory neurons*. Like ordinary neurons, neurosecretory neurons can respond to and conduct electrical signals. Instead of directly innervating target cells, however, a neurosecretory neuron releases its chemical messenger, a neurohormone, into the blood upon appropriate stimulation. The neurohormone is then distributed through the blood to distant target cells. Thus, like endocrine cells, neurosecretory neurons release blood-borne chemical messengers, whereas ordinary neurons secrete short-range neurotransmitters into a confined space. From now on in this textbook the general term *hormone* will tacitly include both blood-borne hormonal and neurohormonal messengers.

In every case, extracellular chemical messengers are released from a particular cell type and interact with specific target cells to bring about a desired effect in the target cells. We now turn our attention to how these chemical messengers bring about the desired cell response.

## Signal transduction

The term **signal transduction** refers to the process by which incoming signals (instructions from extracellular chemical messengers) are conveyed to the target cell's interior for execution. (A *transducer* is a device that receives energy from one system and transmits it in a different form to another system. For example, a radio receives radio waves sent out from the broadcast station and transmits these signals in the form of sound waves that can be detected by your ears.) Lipid-soluble extracellular chemical messengers, such as cholesterol-derived steroid hormones, gain entry into the cell by dissolving in and passing through the lipid bilayer of the target cell's plasma membrane. These extracellular chemical messengers bind to receptors inside the target cell and initiate the desired intracellular response themselves. By contrast, extracellular chemical messengers that are water soluble cannot gain entry to the target cell because they are poorly soluble in lipid and cannot dissolve in the plasma membrane. Protein hormones delivered by the blood and neurotransmitters released from nerve endings are the major water-soluble extracellular messengers. These messengers signal the cell to perform a given response by first binding with surface membrane receptors specific for that given messenger. These receptors are specialized proteins within the plasma membrane (p. 31). The combination of an extracellular messenger and a surface membrane receptor triggers a sequence of intracellular events that ultimately controls a particular cellular activity, such as membrane transport, secretion, metabolism, or contraction.

Despite the wide range of possible responses, the binding of an extracellular messenger (the **first messenger**) to its matching receptor brings about the desired intracellular response by only two general means: (1) by opening or closing channels, or (2) by activating second-messenger systems. We now examine these important cellular events more closely.

## Chemically gated channels

Some extracellular messengers accomplish the desired intracellular response by opening or closing specific chemically gated channels in the membrane, thereby regulating the movement of particular ions into or out of the cell. An example is the opening of chemically gated channels in the subsynaptic membrane of a neuron in response to the binding of a neurotransmitter. The resultant small, short-lived movement of charge-carrying ions across the membrane through these open channels generates electrical signals.

Another example is the stimulation of muscle cells to bring about contraction. This occurs when chemically gated channels in the muscle cells open in response to the binding of neurotransmitter released from the neurons supplying the muscle. (Chapter 7 on muscle physiology describes this mechanism in detail.) The control of chemically gated channels by extracellular messengers is an important regulatory mechanism in both nerve and muscle physiology.

On completion of the response, the extracellular messenger is removed from the receptor site, and the chemically gated channels close once again. The ions that moved across the membrane through opened channels to trigger the response are returned to their original location by special membrane carriers.

## Second-messenger pathways

Many extracellular chemical messengers that cannot actually enter their target cells bring about the desired intracellular response by means other than the opening of chemically gated channels. These first messengers issue their orders by triggering a "Psst, pass it on" process. The binding of the first messenger to a membrane receptor serves as a signal for activating an intracellular **second messenger**. The second messenger ultimately relays the orders through a series of biochemical intermediaries to particular intracellular proteins that carry out the dictated response, such as changes in cellular metabolism or secretory activity. The intracellular pathways activated by a second messenger in response to binding of the first messenger to a surface receptor are remarkably similar among different cell types despite the diversity of ultimate responses to that signal. The variability in response depends on the specialization of the cell, not on the mechanism used.

Some neurotransmitters function through intracellular second-messenger systems. Most, but not all, neurotransmitters function by changing the conformation of chemically gated channels, thereby altering membrane permeability and ion fluxes across the postsynaptic membrane. Synapses involving these rapid responses are considered fast synapses. However, another mode of synaptic transmission used by some neurotransmitters, such as *serotonin*, involves the activation of intracellular second messengers. Synapses that lead to responses mediated by second messengers are known as slow synapses, because these responses take longer and often last longer than those accomplished by fast synapses. For example, neurotransmitter-activated second messengers may trigger long-term postsynaptic cellular changes believed to be linked to neuronal growth and development, as well as possibly playing a role in learning and memory.

Second-messenger systems are widely used throughout the body, including being one of the key means by which most water-soluble hormones ultimately bring about their effects. Hormonal communication is discussed in Chapter 5, where we examine second-messenger systems in more detail.

## 2.10 | Membrane Potential

The plasma membranes of all living cells have a membrane potential; that is, they are polarized electrically. This potential allows for cellular communication in excitable cells such as nervous tissue and muscle.

### Separation of opposite charges

The term **membrane potential** quite simply refers to the difference in the electrical potential between inside and outside a cell. In Sections 2.4 and 2.6, we discussed the structure of the lipid bilayer and several methods of transport that are available for moving substances across the plasma membrane. One of the main reasons these transport mechanisms take place is that the intracellular and extracellular environments are not the same. In Section 2.8, we discussed how a particular substance such as glucose moves with or against its concentration gradient, but what happens in the case of charged ions?

When there is a difference in the relative number of cations (+) and anions (−) in the ICF (inside the cell) and ECF (outside the cell), a separation of charges is created. Opposite charges tend to attract each other, whereas like charges tend to repel. Work is performed (energy expended) to separate opposite charges after they have come together. Conversely, when oppositely charged particles have been separated, the electrical force of attraction between them can be harnessed to perform work when the charges are permitted to come together again. Because separated charges have the potential to do work, a separation of charges across the membrane is called a *membrane potential*. The potential is measured in units of volts (the same unit used for electrical devices), but because the membrane potential is relatively low, the unit used is the **millivolt (mV)** (1 mV = 1/1000 volt).

This membrane voltage is very important: if you change the voltage of the membrane, ions will move across the membrane to generate an electrical current. Likewise, if a cellular response is to move ions across the membrane, this will change the voltage or membrane potential. The relationship between voltage and current is described by Ohm's law:

$$V = IR$$

Where $V$ is voltage (potential across the plasma membrane), $I$ is *current* (ions moving across the membrane), and $R$ is *resistance* (intrinsic properties of the plasma membrane to resist the movement of ions). This relationship forms the basis for understanding the importance of electrical potentials in cells.

### Concentration and permeability of ions

The cells of *excitable tissues*—namely, nerve and muscle cells—have the ability to produce rapid, transient changes in their membrane potential when excited. These brief fluctuations in potential serve as electrical signals. The constant membrane potential present in the cells of nonexcitable tissues and those of excitable tissues when they are at rest—that is, when they are not producing electrical signals—is known as the **resting membrane potential**. We focus first on the generation and maintenance of the resting membrane potential and then on the changes that take place in excitable tissues during electrical signalling.

The unequal distribution of key ions between the ICF and the ECF and their selective movement through the plasma membrane are responsible for the electrical properties of the membrane. The ions primarily responsible for the generation of the resting membrane potential are sodium ($Na^+$), potassium ($K^+$), and anions ($A^-$; large, negatively charged intercellular proteins). Other ions (e.g., calcium, magnesium, chloride, bicarbonate, and phosphate) do not make a significant contribution to the resting membrane potential in most cells, but they do play other important roles in the body.

The concentrations and relative permeabilities of the ions critical to membrane electrical activity are compared in ▌ Table 2-4. Note that *$Na^+$ is in greater concentration in the extracellular fluid and $K^+$ is in much higher concentration in the intracellular fluid*. These concentration differences are maintained by the $Na^+-K^+$ pump with the expense of energy (ATP). Because the plasma membrane is virtually impermeable to $A^-$, these large, negatively charged proteins are found only inside the cell. They are synthesized from amino acids transported into the cell, and remain trapped within the cell once synthesized.

In addition to the active carrier mechanism, $Na^+$ and $K^+$ can passively cross the membrane through specific protein channels. It is usually much easier for $K^+$ than for $Na^+$ to get through the membrane, because typically the membrane has many more channels open for passive $K^+$ movement than for passive $Na^+$ movement. At resting membrane potential in a nerve cell, the membrane is about 50 to 75 times as permeable to $K^+$ as to $Na^+$.

Knowledge of the relative concentrations and permeabilities of these ions allows us to analyze the forces acting across the plasma membrane. This analysis includes the following aspects: (1) the effect that the movement of $K^+$ alone would have on membrane potential; (2) the effect of $Na^+$ alone; and (3) the situation that exists in the cells when both $K^+$ and N+ effects are taking place concurrently. Remember, throughout this discussion *the concentration gradient for $K^+$ is always outward* and *the concentration gradient for $Na^+$ is always inward*, because the $Na^+-K^+$ pump maintains a higher concentration of $K^+$ inside the cell and a higher concentration of $Na^+$ outside the cell. Note as well that because $K^+$ and $Na^+$ are both cations (+ charges), *the electrical gradient for both of these ions is always toward the negatively charged side of the membrane.*

**TABLE 2-4** Concentration and Permeability of Ions Responsible for Membrane Potential in a Resting Nerve Cell

### CONCENTRATION (millimoles/litre)

| Ion | Extracellular | Intracellular | Relative Permeability |
|-----|---------------|---------------|-----------------------|
| Na⁺ | 150 | 15 | 1 |
| K⁺ | 5 | 150 | 50–75 |
| A⁻ | 0 | 65 | 0 |

### EFFECT OF THE MOVEMENT OF POTASSIUM ALONE ON MEMBRANE POTENTIAL: K⁺ EQUILIBRIUM POTENTIAL

In our first hypothetical situation, we will examine the driving forces affecting the movement of $K^+$ across the plasma membrane. As ▍ Table 2-4 shows, the concentration of $K^+$ is higher inside the cell than outside; therefore, there is a concentration gradient that favours the movement of $K^+$ out of the cell. › Figure 2-27 illustrates what happens when $K^+$ leaves the cell, making the inside of the cell more negative relative to the outside and thereby establishing a membrane potential. The membrane potential, however, is negative, meaning the negatively charged interior favours the movement of cations such as $K^+$ to the inside of the cell. As a result, *two opposing forces would now be acting on $K^+$: the concentration gradient tending to move $K^+$ out of the cell, and the electrical gradient tending to move these same ions into the cell.*

Eventually, the net outward concentration gradient and the net inward electrical gradient become balanced, so there is no net movement of $K^+$ across the membrane. The potential that would exist at this equilibrium is known as the $K^+$ **equilibrium potential** $(E_{K^+})$. At this point, because of the exactly equal opposing electrical gradient, a large concentration gradient for $K^+$ would still exist, but no more net movement of $K^+$ would occur out of the cell. (› Figure 2-27).

The membrane potential at $E_{K^+}$ is −90 mV. By convention, *the sign always designates the polarity of the excess charge on the inside of the membrane.* A membrane potential of −90 mV means that the potential is of a magnitude of 90 mV, with the inside being negative relative to the outside; a +90 mV would have the same strength, but the inside would be more positive.

The equilibrium potential for a given ion of differing concentrations across a membrane can be calculated by means of the **Nernst equation**, as follows:

$$E = 61 \log \frac{C_o}{C_i}$$

where

$$E = \text{equilibrium potential for ion in mV}$$

61 = a constant that incorporates the universal gas constant (R), absolute temperature (T), the ion's valence ($z$) (when the valence is 1+, as for $K^+$ and $Na^+$), an electrical constant known as Faraday (F), along with the conversion of the natural logarithm (ln) to the logarithm to base 10 (log)

61 = RT / $z$F. For any ion with a valence other than 1+, 61 must be divided by $z$ to calculate a Nernst potential.

$C_o$ = concentration of the ion outside the cell in millimoles/litre (millimolars; mM)

$C_i$ = concentration of the ion inside the cell in mM

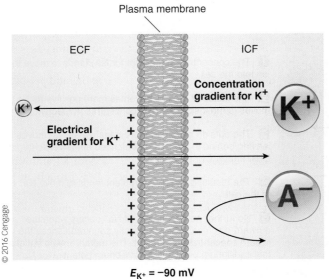

Plasma membrane

ECF · ICF

K⁺

Electrical gradient for K⁺

Concentration gradient for K⁺

K⁺

A⁻

$E_{K^+}$ = −90 mV

**1** The concentration gradient for K⁺ tends to move this ion out of the cell.

**2** The outside of the cell becomes more positive as K⁺ moves to the outside down its concentration gradient.

**3** The membrane is impermeable to the large intracellular protein anion (A⁻). The inside of the cell becomes more negative as K⁺ moves out, leaving behind A⁻.

**4** The resulting electrical gradient tends to move K⁺ into the cell.

**5** No further net movement of K⁺ occurs when the inward electrical gradient exactly counterbalances the outward concentration gradient. The membrane potential at this equilibrium point is the equilibrium potential for K⁺ ($E_{K^+}$) at −90 mV.

› **FIGURE 2-27** Equilibrium potential for K⁺

Given that the ECF concentration of $K^+$ is 5 mM and the ICF concentration is 150 mM,

$$E_{K^+} = 61 \log \frac{5 \text{ mM}}{150 \text{ mM}}$$

$$= 61 \log \frac{1}{30}$$

Because the log of $\frac{1}{30} = -1.447$,

$$E_{K^+} 61 \times (-1.477) = -90 \text{mV}$$

Because 61 is a constant, the equilibrium potential is essentially a measure of the membrane potential (i.e., the magnitude of the electrical gradient) that exactly counterbalances the concentration gradient (i.e., the ratio between the ion's concentration outside and inside the cell) that exists for the ion. Note that the larger the concentration gradient is for an ion, the greater the ion's equilibrium potential. A comparably greater opposing electrical gradient would be required to counterbalance the larger concentration gradient.

### EFFECT OF MOVEMENT OF SODIUM ALONE ON MEMBRANE POTENTIAL: Na+ EQUILIBRIUM POTENTIAL

❯ Figure 2-28 illustrates another hypothetical situation, this time for the movement of $Na^+$ alone across the membrane. In this case, the concentration gradient for $Na^+$ would move this ion into the cell, producing a buildup of positive charges inside of the membrane and leaving negative charges unbalanced outside—primarily in the form of chloride $(Cl^-)$. ($Na^+$ and $Cl^-$ [i.e., salt] are the predominant ECF ions.) Net inward movement would continue until equilibrium was established by the development of an opposing electrical gradient that exactly counterbalances the concentration gradient. This is the same principle as occurs with $K^+$, but in the opposite direction. Given the concentrations for $Na^+$, the $Na^+$ equilibrium potential $(E_{Na^+})$ would be +60 mV. In this case the inside of the cell would be positive. The

magnitude of $E_{Na^+}$ is somewhat less than that of $E_{K^+}$ (60 mV compared with 90 mV). Because the concentration gradient for $Na^+$ is not as large as that for $K^+$ (see ▮ Table 2-4), the opposing electrical gradient (membrane potential) for $Na^+$ is not as great as that for $K^+$ at equilibrium.

### CONCURRENT POTASSIUM AND SODIUM EFFECTS ON MEMBRANE POTENTIAL

Neither $K^+$ nor $Na^+$ exists alone in the body fluids, so equilibrium potentials are not present in body cells; they exist only in hypothetical or experimental conditions. In a living cell, the concurrent effects of both $K^+$ and $Na^+$ must be taken into account. *The greater the permeability of the plasma membrane for a given ion, the greater is the tendency for that ion to drive the membrane potential toward the ion's own equilibrium potential.* Because the membrane at rest is 50 to 75 times as permeable to $K^+$ as to $Na^+$, $K^+$ passes through more readily than $Na^+$; thus, $K^+$ influences the resting membrane potential to a much greater extent than $Na^+$ does. Recall that $K^+$ acting alone would establish an equilibrium potential of −90 mV. However, because membrane is somewhat permeable to $Na^+$, some $Na^+$ enters the cell in a limited attempt to reach its equilibrium potential. This neutralizes some of the membrane potential produced by $K^+$ alone.

To better understand this concept, assume that each separated pair of charges in ❯ Figure 2-29 represents 10 mV of potential. (Technically this is incorrect, because in reality many separated charges must be present to account for a potential of 10 mV.) In this simplified example, nine separated pluses and minuses, with the minuses on the inside, would represent the $E_{K^+}$ of −90 mV. Superimposing the slight influence of $Na^+$ on this $K^+$-dominated membrane, assume that two sodium ions enter the cell down the $Na^+$ concentration and electrical gradients. (Note that the electrical gradient for $Na^+$ is now inward, in contrast to the outward electrical gradient for $Na^+$ at $E_{Na^+}$. At $E_{Na^+}$, the inside of the cell is positive as a result of the inward movement of $Na^+$ down its concentration gradient. In a resting nerve cell, however,

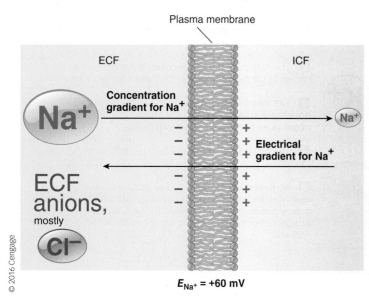

Plasma membrane

1 The concentration gradient for $Na^+$ tends to move this ion into the cell.

2 The inside of the cell becomes more positive as $Na^+$ moves to the inside down its concentration gradient.

3 The outside becomes more negative as $Na^+$ moves in, leaving behind in the ECF unbalanced negatively charged ions, mostly $Cl^-$.

4 The resulting electrical gradient tends to move $Na^+$ out of the cell.

5 No further net movement of $Na^+$ occurs when the outward electrical gradient exactly counterbalances the inward concentration gradient. The membrane potential at this equilibrium point is the equilibrium potential for $Na^+$ $(E_{Na^+})$ at +60 mV.

$E_{Na^+} = +60$ mV

❯ FIGURE 2-28 Equilibrium potential for Na+

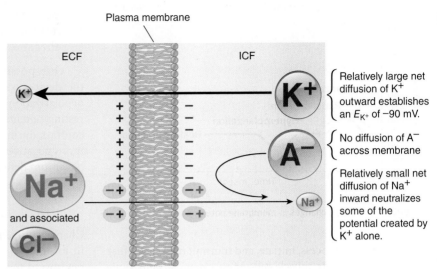

1. The Na$^+$–K$^+$ pump actively transports Na$^+$ out of and K$^+$ into the cell, keeping the concentration of Na$^+$ high in the ECF and the concentration of K$^+$ high in the ICF.

2. Given the concentration gradients that exist across the plasma membrane, K$^+$ tends to drive membrane potential to the equilibrium potential for K$^+$ (−90 mV), whereas Na$^+$ tends to drive membrane potential to the equilibrium potential for Na$^+$ (+60 mV).

3. However, K$^+$ exerts the dominant effect on resting membrane potential because the membrane is more permeable to K$^+$. As a result, resting potential (−70 mV) is much closer to $E_{K^+}$ than to $E_{Na^+}$.

4. During the establishment of resting potential, the relatively large net diffusion of K$^+$ outward does not produce a potential of −90 mV because the resting membrane is slightly permeable to Na$^+$ and the relatively small net diffusion of Na$^+$ inward neutralizes (in *grey* shading) some of the potential that would be created by K$^+$ alone, bringing resting potential to −70 mV, slightly less than $E_{K^+}$.

5. The negatively charged intracellular proteins (A$^-$) that cannot cross the membrane remain unbalanced inside the cell during the net outward movement of the positively charged ions, so the inside of the cell is more negative than the outside.

Plasma membrane

ECF          ICF

Relatively large net diffusion of K$^+$ outward establishes an $E_{K^+}$ of −90 mV.

No diffusion of A$^-$ across membrane

Relatively small net diffusion of Na$^+$ inward neutralizes some of the potential created by K$^+$ alone.

Resting membrane potential = −70 mV

© 2016 Cengage

⟩ **FIGURE 2-29** **Effect of concurrent K$^+$ and Na$^+$ movement on establishing the resting membrane potential**

the inside is negative because of the dominant influence of K$^+$ on membrane potential. Thus, both the concentration and the electrical gradients now favour the inward movement of Na$^+$.) The inward movement of these two positively charged sodium ions neutralizes some of the potential established by K$^+$, so now only seven pairs of charges are separated, and the potential is −70 mV. This is the *resting membrane potential* of a typical nerve cell. The resting potential is much closer to $E_{K^+}$ than to $E_{Na^+}$ because of the greater permeability of the membrane to K$^+$, but it is slightly less than $E_{K^+}$ (−70 mV is a lower potential than −90 mV) because of the weak influence of Na$^+$.

### CHLORIDE MOVEMENT AT RESTING MEMBRANE POTENTIAL

The Cl$^-$ ion, present in high concentrations in the ECF, is the principal ECF anion. Its equilibrium potential is −70 mV, exactly the same as the resting membrane potential. Movement alone of negatively charged Cl$^-$ into the cell down its concentration gradient would produce an opposing electrical gradient, with the inside negative compared with the outside. When physiologists were first examining the ionic effects that could account for the membrane potential, they were tempted to think that Cl$^-$ movements and establishment of the Cl$^-$ equilibrium potential could be solely responsible for producing the identical resting membrane potential. Actually, the reverse is the case. The membrane potential is responsible for driving the distribution of Cl$^-$ across the membrane.

Most cells are highly permeable to Cl$^-$ but have no active-transport mechanisms (no pumps) for this ion. With no active forces acting on it, Cl$^-$ passively distributes itself to achieve an individual state of equilibrium. In this case, Cl$^-$ is driven out

of the cell, establishing an inward concentration gradient that exactly counterbalances the outward electrical gradient (i.e., the resting membrane potential) produced by K$^+$ and Na$^+$ movement. This means the concentration difference for Cl$^-$ between the ECF and the ICF is brought about passively by the presence of the membrane potential rather than maintained by an active pump, as is the case for K$^+$ and Na$^+$. Therefore, in most cells Cl$^-$ does not influence resting membrane potential; instead, membrane potential passively influences the Cl$^-$ distribution. (Some specialized cells have an active Cl$^-$ pump, with subsequent movement of Cl$^-$ accounting for part of the potential.)

### Specialized use of membrane potential in nerve and muscle cells

Two types of cells—nerve cells and muscle cells—have developed a specialized use for membrane potential. These cells can rapidly and transiently alter their membrane permeabilities to specific ions in response to appropriate stimulation, bringing about fluctuations in membrane potential. These fluctuations in potential serve as electrical signals. The constant membrane potential that exists when a nerve or muscle cell is not displaying rapid changes in potential is referred to as the *resting potential*. All cells display a membrane potential, but the significance of this in other cells is uncertain. For example, the membrane potential of some secretory cells appears to be linked to the level of their secretory activity.

Nerve and muscle are considered **excitable tissues** because, when excited, they change their resting potential to produce electrical signals. Nerve cells, or *neurons*, use these electrical

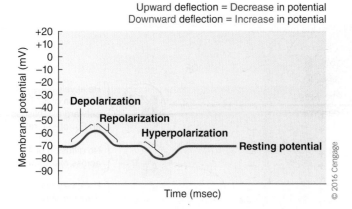

Upward deflection = Decrease in potential
Downward deflection = Increase in potential

> FIGURE 2-30 **Types of changes in membrane potential**

signals to receive, process, initiate, and transmit messages—that is, to communicate. In muscle cells, these electrical signals initiate muscular contraction, which is responsible for the movement involved in work and exercise. Consequently, electrical signals are critical to the function of the nervous system and all muscles. In the following section, we consider how neurons undergo changes in potential to accomplish their function. The functioning of muscle cells is discussed in Chapter 7.

## Depolarization and hyperpolarization

To help you understand the nature of electrical signals and how they are created, become familiar with the terms used to describe changes in potential: polarization, depolarization, repolarization, and hyperpolarization. (> Figure 2-30 represents these terms graphically.)

1. **Polarization**. Charges are separated across the plasma membrane, so the membrane has potential. Any time the value of the membrane potential is other than 0 mV, in either a positive or a negative direction, the membrane is in a state of polarization. Recall that the magnitude of the potential is directly proportional to the number of positive and negative charges separated by the membrane and that the sign of the potential (+ or −) always designates whether excess positive or excess negative charges are present, respectively, on the inside of the membrane. At resting potential, the membrane is polarized at −70 mV in a typical neuron.

2. **Depolarization**. The resting potential becomes less negative (e.g., a change from −70 to −60 mV), and fewer charges are separated than at resting potential. So depolarization is a movement in the positive (+) direction, or upward on a recording device.

3. **Repolarization**. The membrane returns to resting potential after having been depolarized. So repolarization is a movement in the negative (−) direction, or downward on a recording device.

4. **Hyperpolarization**. The membrane becomes more polarized than at resting potential. During hyperpolarization the membrane potential moves even farther from 0 mV, becoming more negative (for instance, a change from −70 to −80 mV); this means more charges are separated than at resting potential. Again, hyperpolarization is a movement in the negative (−) direction, or downward on the recording device.

One possibly confusing point should be clarified. On the device used for recording rapid changes in potential, a depolarization—when the inside becomes less negative than at resting potential—is represented as an *upward* deflection. By contrast, during a hyperpolarization—when the inside becomes more negative than at resting potential—is represented by a *downward* deflection.

## Electrical signals and ion movement

Changes in membrane potential are brought about by changes in ion movement across the membrane. For example, if the net inward flow of positively charged ions increases compared with the resting state, the membrane becomes depolarized (less negative inside). By contrast, if the net outward flow of positively charged ions increases compared with the resting state, the membrane becomes hyperpolarized (more negative inside).

Changes in ion movement, in turn, are brought about by changes in membrane permeability in response to *triggering events*. Depending on the type of electrical signal, a triggering event might be (1) a change in the electrical field in the vicinity of an excitable membrane; (2) an interaction of a chemical messenger with a surface receptor on a nerve or muscle cell membrane; (3) a stimulus, such as sound waves stimulating specialized nerve cells in the ear; or (4) a spontaneous change of potential caused by inherent imbalances in the leak–pump cycle. (You will learn more about the nature of these triggering events as our discussion of electrical signals continues.)

Because the water-soluble ions responsible for carrying charge cannot penetrate the plasma membrane's lipid bilayer, these charges can cross the membrane only through channels specific to them. Membrane channels may be either *leak channels* or *gated channels*. **Leak channels** are open all the time, thereby permitting unregulated leakage of their chosen ion across the membrane through the channels. **Gated channels**, in contrast, have gates that can alternately be open, permitting ion passage through the channels, or closed, preventing ion passage through the channels. The opening-and-closing of gates results from a change in the three-dimensional conformation (shape) of the protein that forms the gated channel. There are four kinds of gated channels, depending on the factor that induces the change in channel conformation: (1) **voltage-gated channels**, which open or close in response to changes in membrane potential; (2) **chemically gated channels**, which change conformation in response to the binding of a specific chemical messenger to a membrane receptor in close association with the channel; (3) **mechanically gated channels**, which respond to stretching or other mechanical deformation; and (4) **thermally gated channels**, which respond to local changes in temperature (heat or cold).

Thus, triggering events alter membrane permeability and consequently alter ion flow across the membrane by opening or closing the gates guarding particular ion channels. These ion movements redistribute charge across the membrane, causing membrane potential to fluctuate.

There are two basic forms of electrical signals: (1) *graded potentials*, which serve as short-distance signals; and (2) *action potentials*, which signal over long distances. Before we explore how nerve cells use these signals to convey messages, we will examine these types of signals in more detail, beginning with graded potentials.

### Check Your Understanding 2.5

1. Explain why the resting membrane potential of a cell is closer to the $K^+$ equilibrium potential.

2. Draw a graph depicting the changes in potential during depolarization, repolarization, and hyperpolarization as compared to resting membrane potential.

3. State the factor responsible for triggering gate opening.

## 2.11 | Graded Potentials

**Graded potentials** are local changes in membrane potential that occur in varying grades or degrees of magnitude or strength. For example, membrane potential could change from $-70$ to $-60$ mV (a 10 mV graded potential) or from $-70$ to $-50$ mV (a 20 mV graded potential).

### Triggering events

Graded potentials are usually produced by a specific triggering event that causes gated ion channels to open in a specialized region of the excitable cell membrane. In most cases, these are chemically gated or mechanically gated channels. Most commonly, gated $Na^+$ channels open. When $Na^+$ channels open, $Na^+$ permeability increases as $Na^+$ moves down its concentration and electrical gradients. As a result, the membrane potential moves toward the $Na^+$ equilibrium potential. The resultant depolarization—the graded potential—is confined to this small, specialized region of the total plasma membrane.

The magnitude of this initial graded potential (i.e., the difference between the new potential and the resting potential) is related to the magnitude of the triggering event: *The stronger the triggering event, the more gated channels that open, the greater the positive charge entering the cell, and the larger the depolarizing graded potential at the point of origin. Also, the longer the duration of the triggering event, the longer the duration of the graded potential* (> Figure 2-31).

### Graded potentials and passive currents

When a graded potential occurs locally in a nerve or muscle cell membrane, the remainder of the membrane is still at resting potential. The temporarily depolarized region is called an *active area*. Note from

> Figure 2-32 that inside the cell, the active area is relatively more positive than the neighbouring *inactive areas* that are still at resting potential. Outside the cell, the active area is relatively less positive than these adjacent areas. Because of this difference in electrical charges, passive flow between the active and adjacent resting regions on both the inside and the outside of the membrane begins. Any flow of electrical charges is called a **current**. By convention, the direction of current flow is always designated by the direction in which the positive charges are moving (> Figure 2-32b). On the inside, positive charges flow through the ICF away from the relatively more positive depolarized active region toward the more negative adjacent resting regions. Outside the cell, positive charges flow through the ECF from the more positive adjacent inactive regions toward the relatively more negative active region. Ion movement (i.e., current) is occurring *along* the membrane between regions next to each other on the same side of the membrane. This flow is in contrast to ion movement *across* the membrane through ion channels.

As a result of local current flow between an active depolarized area and an adjacent inactive area, the potential changes in the previously inactive area. Positive charges have flowed into this adjacent area on the inside, while simultaneously the positive charges have flowed out of this area on the outside. Thus, at this adjacent site the inside is more positive (or less negative), and the outside is less positive (or more negative) than before (> Figure 2-32c). Stated differently, the previously inactive adjacent region has been depolarized, so the graded potential has spread. This area's potential now differs from that of the inactive region immediately next to it on the other side, inducing further current flow at this new site, and so on. In this manner, current spreads in both directions away from the initial site of the potential change.

The amount of current that flows between two areas depends on the difference in potential between the areas and on the resistance of the material through which the charges are moving.

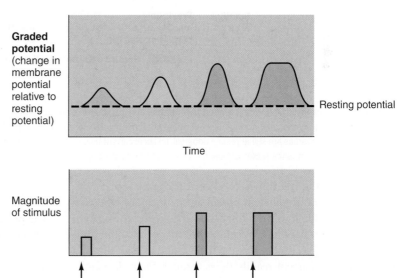

> FIGURE 2-31 **The magnitude and duration of a graded potential.** The magnitude and duration of a graded potential depend directly on the strength and duration of the triggering event, such as a stimulus.

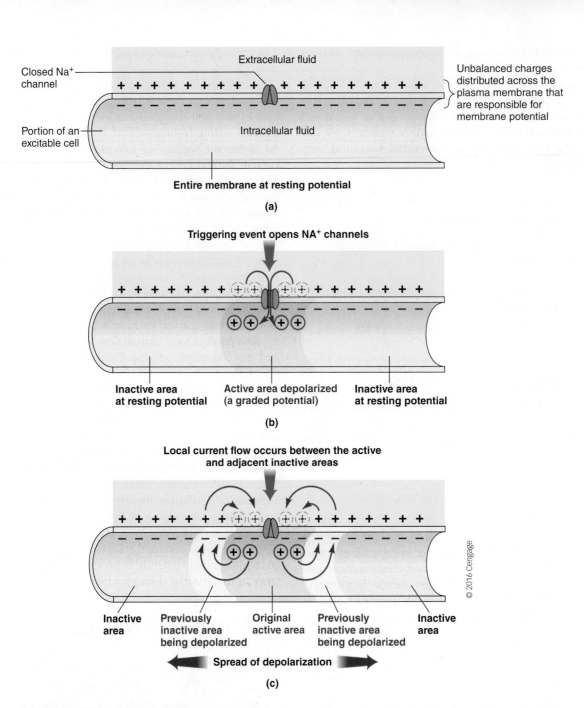

Extracellular fluid

Closed Na⁺ channel

Portion of an excitable cell

Intracellular fluid

Unbalanced charges distributed across the plasma membrane that are responsible for membrane potential

**Entire membrane at resting potential**

**(a)**

**Triggering event opens NA⁺ channels**

**Inactive area at resting potential** — **Active area depolarized (a graded potential)** — **Inactive area at resting potential**

**(b)**

**Local current flow occurs between the active and adjacent inactive areas**

**Inactive area** — **Previously inactive area being depolarized** — **Original active area** — **Previously inactive area being depolarized** — **Inactive area**

◄ **Spread of depolarization** ►

**(c)**

© 2016 Cengage

> **FIGURE 2-32 Current flow during a graded potential.** (a) The membrane of an excitable cell at resting potential. (b) A triggering event opens Na⁺ channels, leading to the Na⁺ entry that brings about depolarization. Note that the depolarization takes place at the Na⁺ channel and not just anywhere along the membrane. The adjacent inactive areas are still at resting potential. (c) Local current flow occurs between the active and the adjacent inactive areas. This local current flow results in depolarization of the previously inactive areas. In this way, the depolarization spreads away from its point of origin.

**Resistance** is the hindrance to electrical charge movement. The greater the difference in potential, the greater is the current flow; and the lower the resistance, the greater the current flow. *Conductors* have low resistance, providing little hindrance to current flow. Electrical wires and the ICF and ECF are all good conductors, so current readily flows through them. *Insulators* have high resistance and greatly hinder movement of charge. The plastic surrounding electrical wires has high resistance, as do body lipids. As a result, current does not freely flow across the plasma membrane's lipid bilayer. Current, carried by ions, can move across the membrane only through ion channels.

## Graded potentials and current loss

The passive current flow between active and adjacent inactive areas is similar to the means by which current is carried through electrical wires. We know from experience that people can get an electric shock if they touch a bare wire. This danger is due to the fact that current leaks from an electrical wire, unless it is covered with an insulating material, such as plastic. Likewise, current is lost across the plasma membrane as charge-carrying ions leak through "uninsulated" parts of the membrane—that is, through open channels. Because of this current loss, the magnitude of

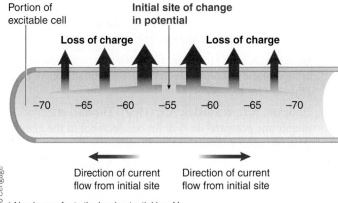

Portion of excitable cell

Initial site of change in potential

**Loss of charge**        **Loss of charge**

−70  −65  −60  −55  −60  −65  −70

Direction of current flow from initial site

Direction of current flow from initial site

© 2016 Cengage

\* Numbers refer to the local potential in mV at various points along the membrane.

> **FIGURE 2-33** **Current loss across the plasma membrane.** Leakage of charge-carrying ions across the plasma membrane results in progressive loss of current with increasing distance from the initial site of potential change.

the local current progressively diminishes with increasing distance from the initial site of origin (> Figure 2-33). Thus, the magnitude of the graded potential continues to decrease the farther it moves away from the initial active area. Another way of saying this is that the spread of a graded potential is *decremental* (gradually decreases). Note that in > Figure 2-33 the magnitude of the initial change in potential is 15 mV (a change from −70 to −55 mV), which decreases as it moves along the membrane to a change in potential of 10 mV (from −70 to −60 mV) and continues to diminish the farther it moves away from the initial active area, until there is no longer a change in potential. In this way, these local currents die out within a few millimetres from the initial site of change in potential and consequently can function as signals for only very short distances.

Graded potentials have limited signalling distance, but they are critically important to the body's function, as explained in this and later chapters. The following are all graded potentials: *postsynaptic potentials, receptor potentials, end-plate potentials, pacemaker potentials,* and *slow-wave potentials.* Although these terms may be unfamiliar now, you will become acquainted with them as we continue discussing nerve and muscle physiology. For now, this list simply introduces the many different types of graded potentials. Excitable cells usually produce one of these types of graded potentials in response to a triggering event. In turn, graded potentials can initiate *action potentials,* the long-distance signals, in an excitable cell.

## 2.12 | Action Potentials

Action potentials are brief, rapid, large changes in membrane potential during which the potential actually reverses, so that the inside of the excitable cell transiently becomes more positive than the outside. Like graded potentials, a single action potential involves only a small portion of the total excitable cell membrane. However, unlike graded potentials, action potentials are conducted, or propagated, throughout the entire membrane in *nondecremental* fashion—that is, they do not diminish in

strength as they travel from their site of initiation throughout the remainder of the cell membrane. Thus, action potentials can serve as faithful long-distance signals. Think about the nerve cell that brings about the contraction of muscle cells in your big toe. If you want to wiggle your big toe, commands are sent from your brain down your spinal cord to initiate an action potential at the beginning of this nerve cell, which is located in the spinal cord. This action potential travels, undiminished, all the way down the nerve cell's long axon, which runs through your leg and terminates in your big toe muscle cells. The signal does not weaken, but instead maintains its full strength from initiation in the spinal cord to muscle contraction of the toe.

Before we look at how action potentials spread throughout the cell in an undiminishing fashion, let's first consider the changes in potential during an action potential and the permeability and ion movements responsible for generating this change in potential.

### Reversal of membrane potential

If they have sufficient magnitude, graded potential changes can initiate an action potential before the graded change dies off. Typically, the portion of the excitable membrane where graded potentials are produced in response to a triggering event does not undergo action potentials. Instead, the graded potential, by electrical or chemical means, brings about depolarization of adjacent portions of the membrane where action potentials can take place.

To initiate an action potential, a triggering event causes the membrane to depolarize from the resting potential of −70 mV (> Figure 2-34). Depolarization proceeds slowly at first, until it reaches a critical level known as **threshold potential**, typically between −50 and −55 mV. At threshold potential, an explosive depolarization takes place. A recording of the membrane

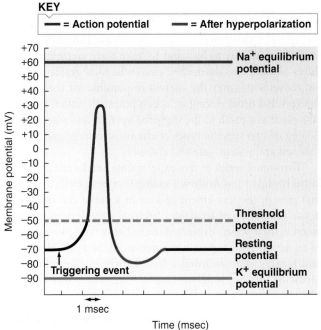

**KEY**

━━ = Action potential        ━━ = After hyperpolarization

Na⁺ equilibrium potential

Threshold potential

Resting potential

K⁺ equilibrium potential

Triggering event

1 msec

Time (msec)

Membrane potential (mV)

© 2016 Cengage

> **FIGURE 2-34** **Changes in membrane potential during an action potential**

potential at this time shows a sharp upward deflection to +30 mV; this coincides with the rapid reversal of the membrane potential, causing the inside of the cell to become positive compared to the outside. Just as rapidly, the membrane repolarizes, dropping back to resting potential. Often the forces that repolarize the membrane push the potential too far, causing a brief phase known as **after hyperpolarization**, during which the inside of the membrane briefly becomes even more negative than normal (e.g., −80 mV) before the resting potential is restored.

The entire rapid change in potential from threshold to peak and then back to resting is called the *action potential*. Unlike the variable duration of a graded potential, the duration of an action potential is always the same in a given excitable cell. In a nerve cell, an action potential lasts for only 1 millisecond (0.001 sec), but it lasts longer in muscle, with the duration depending on the muscle type. The portion of the action potential during which the potential is reversed (between 0 and +30 mV) is called the **overshoot**. Often an action potential is referred to as a **spike**, because of its spike-like recorded appearance. Alternatively, when an excitable membrane is triggered to undergo an action potential, it is said to **fire**. Thus, the terms *action potential*, *spike*, and *firing* all refer to the same phenomenon of rapid reversal of membrane potential. If threshold potential is not reached by the initial triggered depolarization, no action potential takes place. Therefore, threshold is a critical all-or-none point. Either the membrane is depolarized to threshold and an action potential takes place, or threshold is not reached in response to the depolarizing event, and no action potential occurs (all-or-none).

## Changes in membrane permeability

How is the membrane potential, which is usually maintained at a constant resting level, thrown out of balance to such an extent as to produce an action potential? Recall that K⁺ makes the greatest contribution to the establishment of the resting potential, because the membrane at rest is considerably more permeable to K⁺ than to Na⁺ (p. 54). During an action potential, marked changes in membrane permeability to Na⁺ and K⁺ take place, permitting rapid fluxes of these ions down their electrochemical gradients. These ion movements carry the current responsible for the potential changes that occur during an action potential. Action potentials take place as a result of the triggered opening and subsequent closing of two specific types of channels: voltage-gated Na⁺ channels and voltage-gated K⁺ channels.

Through a series of five experiments conducted by Alan Lloyd Hodgkin and Andrew Fielding Huxley in 1952, the rules that govern the movement of ions in a nerve cell resulting in the generation of an action potential were determined. In essence, their papers explained the initiation and propagation of an action potential within a nerve cell. In 1963, Hodgkin and Huxley were awarded a Nobel Prize in Physiology or Medicine for the Hodgkin–Huxley model, as it was known. The references for their five papers are as follows:

1. Hodgkin, A.L., Huxley, A.F., & Katz. B. (1952). Measurement of current-voltage relations in the membrane of the giant axon of *Loligo. J Physiol, 116*: 424–48.

2. Hodgkin, A.L., & Huxley, A.F. (1952). Currents carried by sodium and potassium ions through the membrane of the giant axon of *Loligo. J Physiol, 116*: 449–72.

3. Hodgkin, A.L., & Huxley, A.F. (1952). The components of membrane conductance in the giant axon of *Loligo. J Physiol, 116*: 473–96.

4. Hodgkin, A.L., & Huxley, A.F. (1952). The dual effect of membrane potential on sodium conductance in the giant axon of *Loligo. J Physiol, 116*: 497–506.

5. Hodgkin, A.L., & Huxley, A.F. (1952). A quantitative description of membrane current and its application to conduction and excitation in nerve. *J Physiol, 117*: 500–44.

### VOLTAGE-GATED Na⁺ AND K⁺ CHANNELS

Voltage-gated membrane channels consist of proteins that have a number of charged groups. The electrical field (potential) surrounding the channels can exert a distorting force on the channel structure because charged portions of the channel proteins are electrically attracted or repelled by charges in the fluids surrounding the membrane. Unlike the majority of membrane proteins, which remain stable despite fluctuations in membrane potential, the voltage-gated channel proteins are especially sensitive to voltage changes. Small distortions in channel shape induced by changes in potential can cause them to flip to another conformation—an example of how subtle changes in structure can profoundly influence function.

The voltage-gated Na⁺ channel has two gates: an *activation gate* and an *inactivation gate* (› Figure 2-35). The activation gate guards the channel by opening and closing like a hinged door. The inactivation gate consists of a ball-and-chain-like sequence of amino acids. This gate is open when the ball is dangling free on its chain and closed when the ball binds to its receptor located at the channel opening, thus blocking the opening. Both gates must be open to permit passage of Na⁺ through the channel, and closure of either gate prevents passage. This voltage-gated Na⁺ channel can exist in three different conformations: (1) *closed but capable of opening* (activation gate closed, inactivation gate open; › Figure 2-35a); (2) *open, or activated* (both gates open; › Figure 2-35b); and (3) *closed and not capable of opening, or inactivated* (activation gate open, inactivation gate closed; › Figure 2-35c). It is the membrane

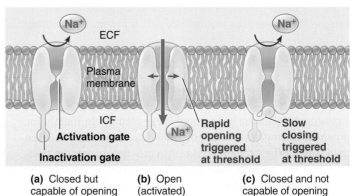

(a) Closed but capable of opening    (b) Open (activated)    (c) Closed and not capable of opening (inactivated)

› **FIGURE 2-35** Conformations of voltage-gated sodium channels

© 2016 Cengage

potential that determines which conformation the channel is in, which is why these are called voltage-gated channels.

Voltage-gated $K^+$ channels are similar to voltage-gated $Na^+$ channels in that they can exist in three conformations (open, closed, and inactivated); however, because most voltage-gated $K^+$ channels consist of four individual subunits, there are no distinct activation and inactivation gates. Rather, the influence of the electrical field changes the conformation of the subunits, which determines whether $K^+$ may flow through the channel.

## CHANGES IN PERMEABILITY AND ION MOVEMENT DURING AN ACTION POTENTIAL

At resting potential ($-70$ mV), all the voltage-gated $Na^+$ and $K^+$ channels are closed. Specifically, the $Na^+$ channels' activation gates are closed and their inactivation gates are open—that is, the voltage-gated $Na^+$ channels are in their "closed-but-capable-of-opening" conformation. Therefore, passage of $Na^+$ and $K^+$ does not occur through these voltage-gated channels at resting potential. However, due to the presence of many $K^+$ leak channels and very few $Na^+$ leak channels, the resting membrane is 50 to 75 times as permeable to $K^+$ as to $Na^+$.

When a membrane starts to depolarize toward threshold as a result of a triggering event, the activation gates of some of its voltage-gated $Na^+$ channels open. Because both the concentration and electrical gradients for $Na^+$ favour its movement into the cell, $Na^+$ starts to move in. The inward movement of positively charged $Na^+$ depolarizes the membrane further, thereby opening even more voltage-gated $Na^+$ channels and allowing more $Na^+$ to enter, and so on, in a positive-feedback cycle (> Figure 2-36).

This opening of channels leads to an explosive increase in **$Na^+$ permeability ($P_{Na^+}$)**; the membrane swiftly becomes 600 times as permeable to $Na^+$ as to $K^+$. $Na^+$ rushes into the cell, rapidly eliminating the internal negativity and even making the inside of the cell more positive than the outside in an attempt to drive the membrane potential to the $Na^+$ equilibrium potential (which is +60 mV; see p. 56) (> Figure 2-37). The potential reaches +30 mV, close to the $Na^+$ equilibrium potential. The potential does not become any more positive, because, at the peak of the action potential, the $Na^+$ channels start to close and inactivate, and $P_{Na^+}$ starts to fall to its low resting value.

What causes the $Na^+$ channels to close? When the membrane potential reaches threshold, two closely related events take place in the gates of each $Na^+$ channel. First, the activation gates are triggered to *open rapidly* in response to the depolarization, converting the channel to its open (activated) conformation (see > Figure 2-35b). Surprisingly, this channel opening initiates the process of channel closing. The conformational change that opens the channel also allows the inactivation gate's ball to bind to its receptor at the channel opening, thereby physically blocking the mouth of the channel. However, this closure process takes time, so the inactivation gate *closes slowly* compared with the rapidity of channel opening. Meanwhile, during the 0.5 millisecond delay after the activation gate opens and before the inactivation gate closes, both gates are open and $Na^+$ rushes into the cell through these open channels, bringing the action potential to its peak. Then the inactivation gate closes, membrane permeability to $Na^+$ plummets to its low resting value, and further $Na^+$ entry is prevented. The channel remains in this inactivated conformation until the membrane potential has been restored to its resting value.

Simultaneous with the inactivation of $Na^+$ channels, the voltage-gated $K^+$ channels start to slowly open at the peak of the action potential. Opening of the $K^+$ channel gate is a delayed voltage-gated response triggered by the initial depolarization to threshold. Thus, three action potential–related events occur at

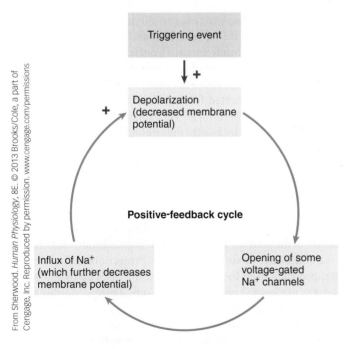

> **FIGURE 2-36** Positive-feedback cycle responsible for opening $Na^+$ channels at threshold

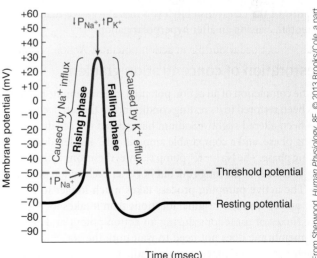

> **FIGURE 2-37** Permeability changes and ion fluxes during a neuronal action potential

threshold: (1) the rapid opening of the Na$^+$ activation gates, which permits Na$^+$ to enter, moving the potential from threshold to its positive peak; (2) the slow closing of the Na$^+$ inactivation gates, which halts further Na$^+$ entry after a brief time delay, keeping the potential from rising any further; and (3) the slow opening of the K$^+$ gates, which is in large part responsible for the potential plummeting from its peak back to resting potential. At this stage in the action potential, the membrane potential returns to resting potential after closure of the Na$^+$ channels, and K$^+$ continues to leak out, but no more Na$^+$ can enter. Furthermore, the return to resting potential is hastened by the opening of K$^+$ gates at the peak of the action potential. Opening of the voltage-gated K$^+$ channels greatly increases K$^+$ permeability $\left(P_{K^+}\right)$ to about 300 times the resting $P_{Na^+}$.

This marked increase in $P_{K^+}$ causes K$^+$ to rush out of the cell down its concentration and electrical gradients, carrying positive charges back to the outside. Note that at the peak of the action potential, the positive potential inside the cell tends to repel the positive K$^+$ ions, so the electrical gradient for K$^+$ *is* outward, unlike at resting potential. The outward movement of K$^+$ rapidly restores the negative resting potential.

To review (> Figure 2-37), *the rising phase of the action potential* (from threshold to +30 mV) *is due to* Na$^+$ *influx* (Na$^+$ entering the cell) induced by an explosive increase in $P_{Na^+}$ at threshold. The *falling phase* (from +30 mV to resting potential) is *brought about largely by K$^+$ efflux* (K$^+$ leaving the cell), which is caused by the marked increase in $P_{K^+}$ occurring simultaneously with the inactivation of the Na$^+$ channels at the peak of the neuronal action potential.

As the potential returns to resting, the changing voltage shifts the Na$^+$ channels to their closed-but-capable-of-opening conformation: that is, activation gate closed and the inactivation gate open. Now the channel is reset, ready to respond to another triggering event. The newly opened voltage-gated K$^+$ channels also close, so the membrane returns to the resting number of open K$^+$ leak channels. Typically, the voltage-gated K$^+$ channels are slow to close. As a result of this persistent increased permeability to K$^+$, more K$^+$ may leave than is necessary to bring the potential to resting. This slight excessive K$^+$ efflux makes the interior of the cell transiently even more negative than resting potential, causing an after hyperpolarization.

## Restoration of concentration gradient

At the completion of an action potential, the membrane potential has been restored to its resting condition, but the ion distribution has been altered slightly. Sodium has entered the cell during the rising phase, and a comparable amount of K$^+$ has left during the falling phase. The Na$^+$−K$^+$ pump restores these ions to their original locations in the long run, but not after each action potential.

The active pumping process takes much longer to restore Na$^+$ and K$^+$ to their original locations than it takes for the passive fluxes of these ions during an action potential. However, the membrane does not need to wait until the Na$^+$−K$^+$ pump slowly restores the concentration gradients before it can undergo another action potential. Actually, the movement of relatively few of the total number of Na$^+$ and K$^+$ ions present causes the large swings in potential that occur during an action potential.

Only about 1 out of 100 000 K$^+$ ions present in the cell leaves during an action potential, while a comparable number of Na$^+$ ions enters from the ECF. The movement of this extremely small proportion of the total Na$^+$ and K$^+$ during a single action potential produces dramatic 100 mV changes in potential (between −70 and +30 mV) but only infinitesimal changes in the ICF and ECF concentrations of these ions. Much more K$^+$ is still inside the cell than outside, and Na$^+$ is still predominantly an extracellular cation. Consequently, the Na$^+$ and K$^+$ concentration gradients continue to exist, and repeated action potentials can occur without the pump having to keep pace to restore the gradients.

Were it not for the pump, of course, even tiny fluxes accompanying repeated action potentials would eventually cause what is known as rundown of the concentration gradients, making further action potentials impossible. If the concentrations of Na$^+$ and K$^+$ were equal between the ECF and the ICF, changes in permeability to these ions would not bring about ion fluxes, so no change in potential would occur. Thus, the Na$^+$−K$^+$ pump is critical to maintaining the concentration gradients in the long run. However, it does not have to perform its role between action potentials, nor is it directly involved in the ion fluxes or potential changes that occur during an action potential.

### Check Your Understanding 2.6

1. Describe the relative contributions of K$^+$ and Na$^+$ to the resting membrane potential.

2. Outline the changes in ion fluxes that underlie the neuronal action potential.

## Propagation of action potential

A single action potential involves only a small patch of the total surface membrane of an excitable cell. But if action potentials are to serve as long-distance signals, they cannot be merely isolated events occurring in a limited area of a nerve or muscle cell membrane. Mechanisms must exist to conduct or spread the action potential throughout the entire cell membrane. Furthermore, the signal must be transmitted from one cell to the next cell (e.g., along specific nerve pathways). We explain these mechanisms in three stages: first, by briefly describing neuronal structure; second, by examining how an action potential (nerve impulse) is conducted throughout a nerve cell; and third, by describing how the signal is passed to another cell.

A single nerve cell, or **neuron**, typically consists of three basic parts: the *cell body*, the *dendrites*, and the *axon*—although there are variations in structure, depending on the location and function of the neuron. The nucleus and organelles are contained in the **cell body**, or soma. Numerous projections called dendrites protrude from the cell body; in some cases up to 400 000 dendrites are located on a single cell body. These dendrites are sometimes called the dendritic tree, because of their shape. The dendrites receive signals from other nerve cells (> Figure 2-38). In most cases, the cell body and dendrites have many protein receptors for binding the chemical messengers from other nerve cells. Together, the cell body and dendrites constitute the *input*

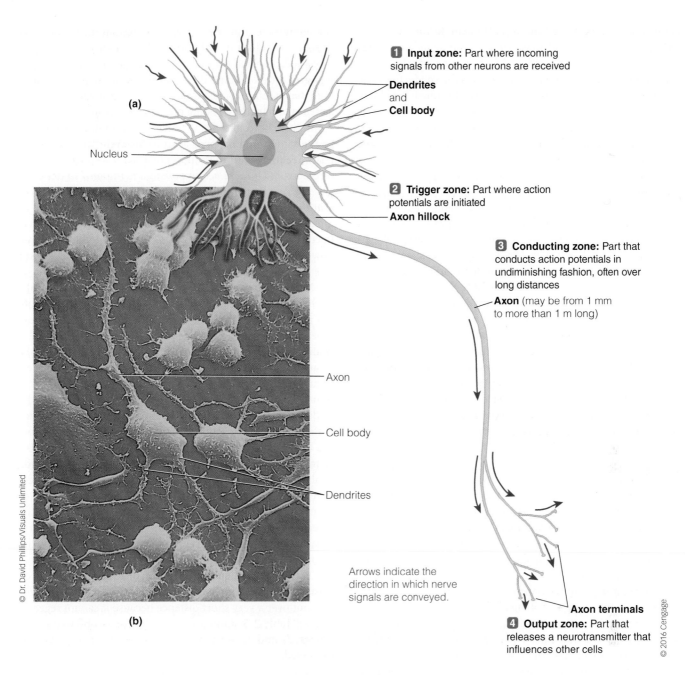

**1** **Input zone:** Part where incoming signals from other neurons are received

**Dendrites**
and
**Cell body**

(a)

Nucleus

**2** **Trigger zone:** Part where action potentials are initiated
**Axon hillock**

**3** **Conducting zone:** Part that conducts action potentials in undiminishing fashion, often over long distances

**Axon** (may be from 1 mm to more than 1 m long)

Axon

Cell body

Dendrites

© Dr. David Phillips/Visuals Unlimited

(b)

Arrows indicate the direction in which nerve signals are conveyed.

**Axon terminals**
**4** **Output zone:** Part that releases a neurotransmitter that influences other cells

© 2016 Cengage

› **FIGURE 2-38** **Anatomy of the most abundant structural type of neuron (nerve cell).** (a) Most, but not all, neurons consist of the basic parts represented in this figure. (b) An electron micrograph highlighting the cell body, dendrites, and part of the axon of this type of neuron within the central nervous system

*zone* for the nerve cell. This is the region where graded potentials are produced in response to triggering events: in this case, incoming chemical messengers.

The **axon**, or **nerve fibre**, is a single, elongated, tubular extension that conducts action potentials *away from* the cell body and eventually terminates at other cells. The axon can extend tens, hundreds, or even tens of thousands of times the diameter of the cell body in length (μm to m in length). For example, the axon of the nerve cell innervating your big toe must traverse the distance from the origin of its cell body within the spinal cord in the lower region of your back all the way down your leg to your

toe. The axon frequently gives off side branches, or **collaterals**, along its course. The first portion of the axon, in conjunction with the region of the cell body where the axon emerges from the soma, is called the **axon hillock**. The axon hillock is the part of the neuron that has the greatest density of voltage-dependent $Na^+$ channels, making it the most easily excitable portion of the neuron and the *trigger zone* of the axon (it also has the most hyperpolarized action potential threshold). The axon hillock is the zone where action potentials can be triggered by a graded potential of sufficient magnitude. The action potentials are then conducted along the axon from the axon hillock to the typically highly branched nerve cell ending at the **axon terminals**. These terminals release chemical messengers that simultaneously influence numerous other cells with which they come into close association. Functionally, therefore, the axon is the *conducting zone* (electrical wire) of the neuron, and the axon terminals constitute its *output zone* (electrical station). (The major exception to this typical neuronal structure and functional organization is found in neurons specialized to carry sensory information, a topic described in Chapter 4.)

Action potentials can be initiated only in portions of the membrane that have an abundance of voltage-gated $Na^+$ channels that can be triggered to open by a depolarizing event. Typically, regions of excitable cells where graded potentials take place do not undergo action potentials, because voltage-gated $Na^+$ channels are sparse there. Therefore, sites specialized for graded potentials do not undergo action potentials, even though they might be considerably depolarized. However, graded potentials can, before dying out, trigger action potentials in adjacent portions of the membrane by bringing these more sensitive regions to threshold through a local current flow spreading from the site of the graded potential. In a typical neuron, for example, graded potentials are generated in the dendrites and cell body in response to incoming chemical signals. If these graded potentials have sufficient magnitude by the time they have spread to the axon hillock, they initiate an action potential at this triggering zone.

## Conduction via a nerve fibre

Once an action potential is initiated at the axon hillock, no further triggering event is necessary to activate the remainder of the nerve fibre. The impulse is automatically conducted throughout the neuron without further stimulation; this is accomplished by one of two methods of propagation: *contiguous conduction* or *saltatory conduction.*

**Contiguous conduction** involves the spread of the action potential along every patch of membrane down the length of the axon (*contiguous* means "touching" or "next to in sequence"). This process is illustrated in › Figure 2-39, which is a schematic representation of a longitudinal section of the axon hillock and the portion of the axon immediately beyond it. The membrane at the axon hillock is at the peak of an action potential. The inside of the cell is positive in this active area, because $Na^+$ has already rushed into the nerve cell at this point. The remainder of the axon, still at resting potential and negative inside, is considered inactive. For the action potential to spread from the active to the inactive areas, the inactive areas must somehow be depolarized to threshold before they can undergo an action potential. This depolarization is accomplished by local current flow between the area already undergoing an action potential and the adjacent inactive area, similar to the current flow responsible for the spread of graded potentials. Because opposite charges attract, current can flow locally between the active area and the neighbouring inactive area on both the inside and the outside of the membrane. This local current flow in effect neutralizes or eliminates some of the unbalanced charges in the inactive area; that is, it reduces the number of opposite charges separated across

the membrane, thus reducing the potential in this area. This depolarizing effect quickly brings the involved inactive area to threshold, at which time the voltage-gated $Na^+$ channels in this region of the membrane are all thrown open, leading to an action potential in this previously inactive area. Meanwhile, the original active area returns to resting potential as a result of $K^+$ efflux.

In turn, beyond the new active area is another inactive area, so the same thing happens again. This cycle repeats itself in a chain reaction until the action potential has spread to the end of the axon. *Once an action potential is initiated in one part of a nerve cell membrane, a self-perpetuating cycle is initiated so that the action potential is propagated along the rest of the fibre automatically.* In this way, the axon is like a firecracker fuse that needs to be lit at only one end. Once ignited, the fire spreads down the fuse; it is not necessary to hold a match to every separate section of the fuse.

Note that the original action potential does not travel along the membrane. Instead, it triggers an identical new action potential in the adjacent area of the membrane, and this process is repeated along the axon's length. An analogy is the "wave" at a stadium. Each section of spectators stands up (the rising phase of an action potential), then sits down (the falling phase) in sequence, one after another, as the wave moves around the stadium. The wave, not individual spectators, travels around the stadium. Similarly, new action potentials arise sequentially down the axon. Each new action potential in the conduction process is a fresh local event that depends on the induced permeability changes and electrochemical gradients, which are virtually identical down the length of the axon. Consequently, the last action potential at the end of the axon is identical to the original one, no matter how long the axon. This is how an action potential is spread along the axon in an undiminished fashion, and, in this way, action potentials can serve as long-distance signals without attenuation or distortion.

This nondecremental propagation of an action potential contrasts with the decremental spread of a graded potential, which dies out over a very short distance because it cannot regenerate itself. ▌ Table 2-5 summarizes the differences between graded potentials and action potentials, some of which are yet to be discussed.

## One-way propagation

What ensures the one-way (unidirectional) propagation of an action potential away from the initial site of activation? Note in › Figure 2-40 that once the action potential has been regenerated at a new neighbouring site (now positive inside) and the original active area has returned to resting (once again negative inside), the close proximity of opposite charges between these two areas is conducive to local current flow in the backward direction, and also in the forward direction into as yet unexcited portions of the membrane. If such backward current flow were able to bring the just-inactivated area to threshold, another action potential would be initiated right there and spread both forward and backward. However, if action potentials could move in both directions, the situation would be chaotic, with numerous action potentials bouncing back and forth along the axon until the nerve cell

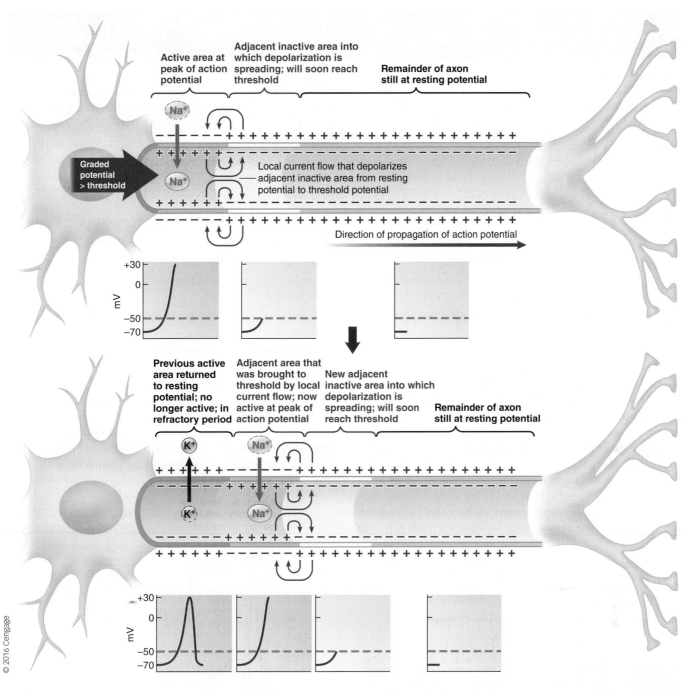

**Active area at peak of action potential**

**Adjacent inactive area into which depolarization is spreading; will soon reach threshold**

**Remainder of axon still at resting potential**

Na⁺

Graded potential > threshold

Na⁺

Local current flow that depolarizes adjacent inactive area from resting potential to threshold potential

Direction of propagation of action potential

+30
0
−50
−70
mV

**Previous active area returned to resting potential; no longer active; in refractory period**

**Adjacent area that was brought to threshold by local current flow; now active at peak of action potential**

**New adjacent inactive area into which depolarization is spreading; will soon reach threshold**

**Remainder of axon still at resting potential**

K⁺

Na⁺

K⁺

Na⁺

+30
0
−50
−70
mV

© 2016 Cengage

⟩ **FIGURE 2-39 Contiguous conduction.** Local current flow between the active area at the peak of an action potential and the adjacent inactive area still at resting potential reduces the potential in this contiguous inactive area to threshold, which triggers an action potential in the previously inactive area. The original active area returns to resting potential, and the new active area induces an action potential in the next adjacent inactive area by local current flow as the cycle repeats itself down the length of the axon.

eventually fatigued. Fortunately, neurons are saved from this fate of oscillating action potentials by the **refractory period**, during which a new action potential cannot be initiated by normal events in a region that has just undergone an action potential.

The refractory period has two components: the *absolute refractory period* and the *relative refractory period* (⟩ Figure 2-41). These two periods occur as a result of the changing status of the voltage-gated Na⁺ and K⁺ channels during and after an action potential. During the time that a particular patch of axonal membrane is undergoing an action potential, it cannot initiate another action potential, no matter how strongly a triggering event stimulates it. This time period when a recently activated patch of membrane is completely refractory (meaning "stubborn," or unresponsive) to further stimulation is known as the **absolute refractory period**. Once the voltage-gated Na⁺ channels have flipped to their open, or activated, state, they cannot be triggered to open again in response to another depolarizing triggering event, no matter how strong, until resting potential is

## ▌ TABLE 2-5 Comparison of Graded Potentials and Action Potentials

| Graded Potentials | Action Potentials |
|---|---|
| Graded potential change; magnitude varies with magnitude of triggering event | All-or-none membrane response; magnitude of triggering event coded in frequency rather than amplitude of action potentials |
| Duration varies with duration of triggering event | Constant duration |
| Decremental conduction; magnitude diminishes with distance from initial site | Propagated throughout membrane in undiminishing fashion |
| Passive spread to neighbouring inactive areas of membrane | Self-regeneration in neighbouring inactive areas of membrane |
| No refractory period | Refractory period |
| Can be summed | Summation impossible |
| Can be depolarization or hyperpolarization | Always depolarization and reversal of charges |
| Triggered by stimulus, by combination of neurotransmitter with receptor, or by spontaneous shifts in leak-pump cycle | Triggered by depolarization to threshold, usually through spread of graded potential |
| Occurs in specialized regions of membrane designed to respond to triggering event | Occurs in regions of membrane with abundance of voltage-gated Na⁺ channels |

restored and the channels are reset to their original positions. The channels are in their inactive state. Accordingly, the absolute refractory period lasts the entire time from the opening of the voltage-gated $Na^+$ channels' activation gates at threshold, through closure of their inactivation gates at the peak of the action potential, until the return to resting potential when the channels' activation gates close and inactivation gates open once again—that is, until the channels are in their "closed-but-capable-of-opening" conformation. Only then can they respond to another depolarization with an explosive increase in $P_{Na^+}$ to

initiate another action potential. Because of this absolute refractory period, one action potential must be over before another can be initiated at the same site.

Following the absolute refractory period is a **relative refractory period**, during which a second action potential can be produced only by a triggering event considerably stronger

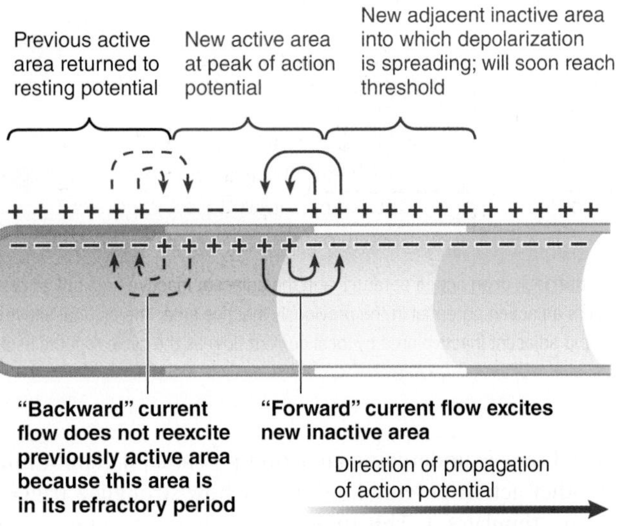

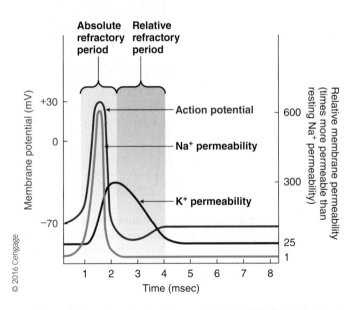

Previous active area returned to resting potential

New active area at peak of action potential

New adjacent inactive area into which depolarization is spreading; will soon reach threshold

"Backward" current flow does not reexcite previously active area because this area is in its refractory period

"Forward" current flow excites new inactive area

Direction of propagation of action potential

> FIGURE 2-40 **Value of the refractory period.** Backward current flow is prevented by the refractory period. During an action potential and slightly beyond, an area cannot be restimulated by normal events to undergo another action potential. Thus, the refractory period ensures that an action potential can be propagated only in the forward direction along the axon.

> FIGURE 2-41 **Absolute and relative refractory periods.** During the absolute refractory period, the portion of the membrane that has just undergone an action potential cannot be restimulated. This period corresponds to the time during which the Na⁺ gates are not in their resting conformation. During the relative refractory period, the membrane can be restimulated only by a stronger stimulus than is usually necessary. This period corresponds to the time during which the K⁺ gates that opened during the action potential have not yet closed.

than is usually necessary. The relative refractory period occurs after the action potential is completed because of a twofold effect: lingering inactivation of the voltage-gated $Na^+$ channels and slow closing of the voltage-gated $K^+$ channels that opened at the peak of the action potential. During this time, fewer-than-normal voltage-gated $Na^+$ channels are in a position to be jolted open by a depolarizing triggering event. Simultaneously, $K^+$ continues to leave through its slow-to-close channels during the after hyperpolarization. The less-than-normal $Na^+$ entry in response to another triggering event is opposed by a persistent hyperpolarizing outward leak of $K^+$ through its not-yet-closed channels. Therefore, a greater-than-normal depolarizing triggering event is needed to bring the membrane to threshold during the relative refractory period. If threshold is reached, action potentials stimulated during the relative refractory period may be of smaller amplitude, depending upon the number of voltage-gated $Na^+$ channels that have recovered from inactivation and are in the closed, ready-to-open conformation.

By the time the original site has recovered from its refractory period and is capable of being restimulated by normal current flow, the action potential has been rapidly propagated in the forward direction only and is so far away that it can no longer influence the original site. Thus, *the refractory period ensures the one-way propagation of the action potential down the axon away from the initial site of activation.*

## The refractory period and the frequency of action potentials

The refractory period is also responsible for setting an upper limit on the frequency of action potentials—that is, it determines the maximum number of new action potentials that can be initiated and propagated along a fibre in a given period of time. The original site must recover from its refractory period before a new action potential can be triggered to follow the preceding action potential. The length of the refractory period varies for different types of neurons. The longer the refractory period, the greater the delay before a new action potential can be initiated and the lower the frequency with which a nerve cell can respond to repeated or ongoing stimulation.

## All-or-none fashion

If any portion of the neuronal membrane is depolarized to threshold, an action potential is initiated and relayed along the membrane in undiminished fashion. Furthermore, once threshold has been reached, the resultant action potential generally goes to maximal height. The reason for this effect is that the changes in voltage during an action potential result from ion movements down concentration and electrical gradients, and these gradients are not affected by the strength of the depolarizing triggering event. A triggering event stronger than one necessary to bring the membrane to threshold does not produce a larger action potential. However, a triggering event that fails to depolarize the membrane to threshold does not trigger an action potential at all. Thus, *an excitable membrane either responds to a triggering event with an action potential that spreads nondecrementally throughout the membrane, or it does not respond with an action potential at all.* This property is called the **all-or-none law**.

This all-or-none concept is analogous to firing a gun. Either the trigger is not pulled sufficiently to fire the bullet (threshold is not reached), or it is pulled hard enough to elicit the full firing response of the gun (threshold is reached). Squeezing the trigger harder does not produce a greater explosion. Just as it is not possible to fire a gun halfway, it is not possible to cause a halfway action potential.

The threshold phenomenon allows some discrimination between important and unimportant stimuli or other triggering events. Stimuli too weak to bring the membrane to threshold do not initiate action potentials and therefore do not clutter up the nervous system by transmitting insignificant signals.

## The strength of a stimulus and the frequency of action potentials

How is it possible to differentiate between two stimuli of varying strengths when both stimuli bring the membrane to threshold and generate action potentials of the same magnitude? For example, how can one distinguish between touching a warm object or touching a very hot object if both trigger identical action potentials in a nerve fibre relaying information about skin temperature to the central nervous system? The answer lies in part on the *frequency* with which the action potentials are generated. A stronger stimulus does not produce a larger action potential, but it does trigger a greater *number* of action potentials per second. For an illustration, see > Figure 9-36 (p. 421), in which changes in blood pressure are coded by corresponding changes in the frequency of action potentials generated in the nerve cells monitoring blood pressure.

In addition, a stronger stimulus in a region causes more neurons to reach threshold, increasing the total information sent to the central nervous system. For example, lightly touch this page with your finger and note the area of skin in contact with the page. Now, press down more firmly and note that a larger surface area of skin is in contact with the page. Therefore, more neurons are brought to threshold with this stronger touch stimulus.

Once initiated, the velocity, or speed, with which an action potential travels down the axon depends on two factors: (1) whether the fibre is myelinated, and (2) the diameter of the fibre. Contiguous conduction occurs in unmyelinated fibres. In this case, as just described, each individual action potential initiates an identical new action potential in the next contiguous (bordering) segment of the axon membrane so that every portion of the membrane undergoes an action potential as this electrical signal is conducted from the beginning to the end of the axon. A faster method of propagation, *saltatory conduction*, takes place in myelinated fibres. We now examine how a myelinated fibre compares with an unmyelinated fibre, and then how saltatory conduction compares with contiguous conduction.

## Myelination and the speed of conduction

**Myelinated fibres**, as the name implies, are covered with myelin at regular intervals along the length of the axon (› Figure 2-42a). **Myelin** is composed primarily of lipids (phospholipids are 80 percent lipid and 20 percent protein). Because the water-soluble ions responsible for carrying current across the membrane cannot permeate this thick lipid barrier, the myelin coating acts as an insulator, just like plastic around an electrical wire, to prevent current leakage across the myelinated portion of the membrane. Myelin is not actually a part of the nerve cell but consists of separate myelin-forming cells that wrap themselves around the axon in a jelly-roll fashion (› Figures 2-42b and 2-42c). These myelin-forming cells are **oligodendrocytes** in the brain and spinal cord of the central nervous system and **Schwann cells** in nerves of the peripheral nervous system that run between the central nervous system and body regions. Between the myelinated regions are areas called the **nodes of Ranvier**, where the axonal membrane is bare and exposed to the ECF. The unmyelinated nodes of Ranvier are located at about 1 mm intervals along the length of the axon, with each nodal bare space about 2 μm long. Only at these bare spaces can current flow across the membrane to produce action potentials. Voltage-gated $Na^+$ channels are concentrated at the nodes, whereas the myelin-covered regions are almost devoid of these special passageways. By contrast, an unmyelinated fibre has a high density of voltage-gated $Na^+$ channels throughout its entire length. Recall that action potentials can be generated only at portions of the membrane furnished with an abundance of these channels.

The distance between the nodes is short enough that local current can travel from an active node to an adjacent inactive node before dying off. When an action potential occurs at one node, local current flow between this node and the oppositely charged adjacent node reduces the adjacent node's potential to threshold so that it undergoes an action potential, and so on. Consequently, in a myelinated nerve, the impulse "jumps" from node to node, like an athlete performing plyometric exercises jumping from the floor (node) over a box (myelin) to the floor (node), over and over again. This process is called **saltatory conduction** (*saltere* means "to jump"). Saltatory conduction propagates action potentials more rapidly than contiguous conduction does, because the action potential is regenerated only at the unmyelinated axonal nodes and not between. In myelinated fibres, local current generated at an active node travels a longer distance, depolarizing the next node instead of the next section. Myelinated fibres conduct impulses about 50 times faster than unmyelinated fibres of comparable size. Thus, the most urgent types of information are transmitted via myelinated fibres, and nerve pathways that carry less urgent information are unmyelinated.

In addition to permitting action potentials to travel faster, a second advantage of myelination is that it conserves energy. Because the ion fluxes associated with action potentials are confined to the nodal regions, the energy-consuming $Na^+ - K^+$ pump does not need to restore as many ions to their respective sides of the membrane following propagation of an action potential.

## Fibre diameter and the velocity of action potentials

Besides the effect of myelination, fibre diameter influences the speed with which an axon can conduct action potentials. The magnitude of current flow (i.e., the amount of charge that moves) depends not only on the difference in potential between two adjacent electrically charged regions but also on the resistance or hindrance to electrical-charge movement between the two regions. When fibre diameter increases, the resistance to local current decreases. Thus, the larger the fibre diameter, the faster action potentials can be propagated. For example, the giant squid has one of the largest axons (>1 mm diameter) known, and one of the fastest velocities of nerve impulses in the animal kingdom.

Large myelinated fibres, such as those supplying skeletal muscles, can conduct action potentials at a speed of up to 120 m/sec (268 mi/h), compared with a conduction velocity of 0.7 m/sec (2 mi/h) in small unmyelinated fibres, such as those supplying the digestive tract. This difference in speed of propagation is related to the urgency of the information being conveyed. A signal to skeletal muscles to execute a particular movement (e.g., to prevent you from falling as you trip on something) must be transmitted more rapidly than a signal to modify a slow-acting digestive process. Without myelination, axon diameters within urgent nerve pathways would have to be very large and cumbersome to achieve the necessary conduction velocities. Indeed, this is the case in many invertebrates. In the course of vertebrate evolution, the need for very large nerve fibres has been overcome by the development of the myelin sheath, allowing economic, rapid, long-distance signalling.

The presence of myelinating cells can be either of tremendous benefit or of tremendous detriment when an axon is cut, depending on whether the damage occurs in a peripheral nerve or in the central nervous system (CNS). Next we discuss the regeneration of damaged nerve fibres, a matter of crucial importance in spinal cord injuries or other trauma affecting nerves.

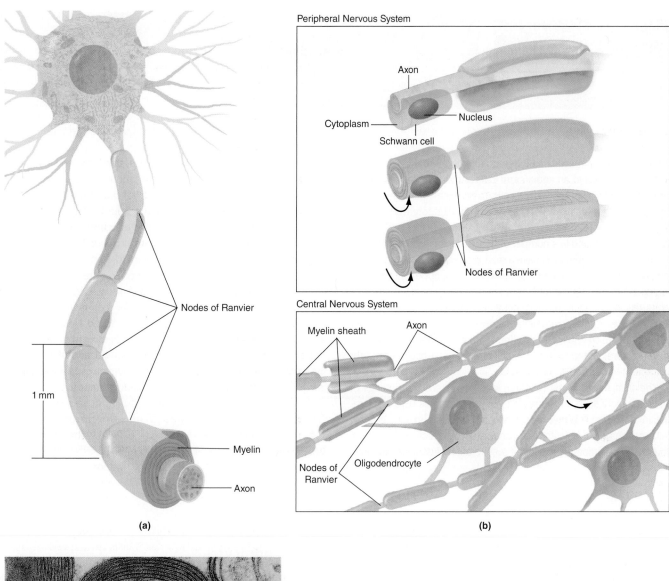

Peripheral Nervous System

Axon

Cytoplasm — Nucleus

Schwann cell

Nodes of Ranvier

Nodes of Ranvier

1 mm

Myelin

Axon

(a)

Central Nervous System

Myelin sheath

Axon

Nodes of Ranvier

Oligodendrocyte

(b)

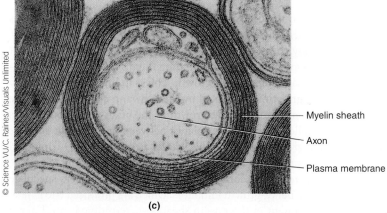

Myelin sheath

Axon

Plasma membrane

(c)

❯ **FIGURE 2-42 Myelinated fibres.** (a) A myelinated fibre is surrounded by myelin at regular intervals. The intervening unmyelinated regions are known as nodes of Ranvier. (b) In the peripheral nervous system, each patch of myelin is formed by a separate Schwann cell that wraps itself jelly-roll fashion around the nerve fibre. In the central nervous system, each of the several processes ("arms") of a myelin-forming oligodendrocyte forms a patch of myelin around a separate nerve fibre. (c) An electron micrograph of a myelinated fibre in cross-section

# Neural Stem Cells and Spinal Cord Repair

## Stem Cells

In the embryo, stem cells can differentiate into all of the specialized embryonic tissues. In an adult, stem cells act as a repair system for the body, replenishing specialized cells. They also maintain the normal turnover of regenerative organs, such as skin. Pluripotent adult stem cells (cells capable of having more than one potential outcome) are rare but can be found in a number of tissues, including umbilical cord blood. Most adult stem cells have a restricted lineage and as such are multipotent (they can generate several cell types, but those types are limited in number). The multipotent stem cell is generally referred to by its tissue of origin: the mesenchymal stem cell, adipose-derived stem cell, etc. Human stem cell research uses primarily multipotent cells that have the ability to proliferate in an undifferentiated state, to self-restore, and to form all the cell types of a particular tissue, for example, nervous tissue.

The field of stem cell research was born from the findings of Dr. Ernest McCulloch, a cellular biologist, and Dr. James Till, a biophysicist, in the 1960s. Together they conducted research on bone marrow cell injections into irradiated mice. Today, stem cell research holds great potential for therapeutic use in the regeneration of damaged organs.

## Neural Stem Cells

Neural stem cells come from the nervous system itself, and mature into central nervous system cells. The most productive source of nervous system cells (e.g., neurons, glia, astrocytes, and oligodendrocytes) is embryonic neural tissue. In the embryo, neuroepithelial cells of the neural tube generate a variety of lineage-restricted (multipotent) precursor cells that migrate and differentiate into neurons, astrocytes, and oligodendrocytes. Central nervous system (CNS) stem cells have now been discovered in humans, and these could potentially be used to promote neurogenesis following injury and disease. Research suggests that transplanted neural stem cells can alter their fate in response to the environment into which they are reintroduced. Neural stem cells also can be isolated from different areas and grown in culture for prolonged periods without losing their multipotentiality. When the cells are transplanted back into the CNS, they have the capacity to migrate, integrate, and respond to local cues for differentiation.

## Spinal Cord Repair

The findings from neural stem cell research point to a promising future for this cell in helping those with spinal cord injuries. One of the best

## 2.13 | Regeneration of Nerve Fibres

Whether or not a severed axon regenerates depends on its location. Cut axons in the peripheral nervous system can regenerate, whereas those in the central nervous vous system cannot.

### Schwann cells

In the case of a cut axon in a peripheral nerve, the portion of the axon farthest from the cell body degenerates, and the surrounding Schwann cells phagocytize the debris. The Schwann cells themselves remain and form a **regeneration tube** to guide the regenerating nerve fibre to its proper destination. The remaining portion of the axon connected to the cell body starts to grow and move forward within the Schwann cell column by amoeboid movement. The growing axon tip "sniffs" its way forward in the proper direction, guided by a chemical secreted into the regeneration tube by the Schwann cells. Successful fibre regeneration is responsible for the return of sensation and movement after a period of time following traumatic peripheral nerve injuries, although regeneration is not always successful.

### Oligodendrocytes

Fibres in the CNS, which are myelinated by oligodendrocytes rather than Schwann cells, do not have this regenerative ability. Actually, the axons themselves have the ability to regenerate, but the oligodendrocytes surrounding them synthesize certain proteins that inhibit axonal growth. This is in sharp contrast to how nerve growth is promoted by the Schwann cells that myelinate peripheral axons. Nerve growth in the brain and spinal cord is controlled by a delicate balance between *proteins that enhance nerve growth and those that inhibit nerve growth*. During fetal development, nerve growth in the CNS is possible when the brain and spinal cord are being formed. Researchers speculate that the nerve-growth inhibitors, which are produced late in fetal development in the myelin sheaths surrounding central nerve fibres, may normally serve as "guardrails" to keep new nerve endings from straying outside their proper paths. In this way, the growth-inhibiting action of oligodendrocytes may serve to stabilize the enormously complex structure of the CNS.

Growth inhibition is a disadvantage, however, when central axons need to be mended, as when the spinal cord has been severed accidentally. Damaged central fibres show immediate signs of repairing themselves after an injury, but within several weeks they start to degenerate, and scar tissue forms at the site of injury,

sources of cells for this treatment is neuronal-restricted precursors, from the developing spinal cord. These cells can be expanded in vitro and have the potential to differentiate into numerous neuronal types, such as motor neurons. Injury to the spinal cord is associated with damage to a particular region; the severity of the injury will dictate the size of the area affected. Function may be lost in the affected region because nerve fibres carrying information up and down the cord beyond that point are blocked.

For example, spinal cord injury to the cervical or thoracic region may result in the loss of control of the legs and bladder, because of the damage to nerve fibres that previously reached parts of the body below the injury. Researchers hope to use neural stem cells and precursors to replace the damaged or lost nerve cells.

Bridging the damaged area to allow axonal regeneration across and beyond the damaged area requires replacing the myelin as well. Thus, stem cells that mature into neurons and oligodendrocytes (which make myelin) are needed for most cellular repair in the injured spinal cord. Myelin is the biological insulation for nerve fibres. It is critical to the maintenance of electrical conduction in the CNS. When myelin is stripped away through disease or injury, sensory and motor deficiencies result, and paralysis is common. Animal research has demonstrated that when adult human neural stem cells are injected into mice with spinal cord injuries, the transplanted stem cells differentiate into new oligodendrocyte cells that restore myelin around the damaged axons. Moreover, transplanted cells can differentiate into new neurons that form synaptic connections with mouse neurons. Mice that received human neural stem cells within a few days of getting spinal cord injuries have shown improvements in mobility as compared with control mice. For an update on some of the most current research, visit www.icord.org, a Canadian-led initiative focused on spinal cord injury.

**Further Reading**

Khazaei, M., Ahuja, C.S., & Fehlings, M. G. (2017). Induced pluripotent stem cells for traumatic spinal cord injury. *Front Cell Dev Biol*, *19*(4): Article 152.

Cote, D.J., Bredenoord, A.L., Smith, T.R., Ammirati, M., Brennum, J., Mendez, I, Ammar, A.S., Bolles, G., Esene, I.N., Mathiesen, T., & Broekman, M.L. (2017). Ethical clinical translation of stem cell interventions for neurologic disease. *Neurology*, 88 (3): 322–28.

halting any recovery. Therefore, damaged neuronal fibres in the brain and spinal cord never regenerate. (For another example of nerve degeneration and the human body's inability to regenerate, see Concepts, Challenges, and Controversies; p. 72)

## Regeneration of cut central axons

In the future, with the help of exciting new findings, it may be possible to induce significant regeneration of damaged fibres in the CNS. Here are some current lines of research:

- Scientists have been able to induce significant nerve regeneration in rats with severed spinal cords by *chemically blocking the nerve-growth inhibitors*, thereby allowing the nerve-growth enhancers to promote abundant sprouting of new nerve fibres at the site of injury. One of the nerve-growth inhibitors, dubbed *Nogo*, was recently identified. Now investigators are trying to encourage axon regrowth in experimental animals with spinal cord injuries by using an antibody to Nogo.

- Other researchers are experimentally *using peripheral nerve grafts* to bridge the defect at an injured spinal cord site. These grafts contain the nurturing Schwann cells, which release proteins that enhance nerve growth.

- Another promising avenue under study involves *transplanting olfactory ensheathing glia* into the damaged site. Olfactory neurons, the cells that carry information about odours to the brain, are replaced on a regular basis, unlike most neurons. The growing axons of these newly generated neurons enter the brain to form functional connections with appropriate neurons in the brain. This capability is promoted by special olfactory ensheathing glia, which wrap around and myelinate the olfactory axons. Early experimental evidence suggests that transplants of these special myelin-forming cells might help induce axonal regeneration in the CNS.

- Another avenue of hope is the recent discovery of *neural stem cells* (pp. 10–11). These cells might someday be implanted into a damaged spinal cord and coaxed into multiplying and differentiating into mature, functional neurons that will replace those lost.

- Yet another new strategy under investigation is *enzymatically breaking down inhibitory components in the scar tissue* that naturally forms at the injured site and prevents sprouting nerve fibres from crossing this barrier.

- Still other investigators are exploring other promising avenues of spurring repair of central axonal pathways, with the goal of enabling victims of spinal cord injuries to walk again.

You have now seen how an action potential is propagated along the axon and learned about the factors that influence the speed of this propagation. But what happens when an action potential reaches the end of the axon? We now turn our attention to this topic.

# 2.14 | Synapses and Neuronal Integration

When the action potential reaches the axon terminals, they release a chemical messenger that alters the activity of the cells on which the neuron terminates. A neuron may terminate on one of the three structures: (1) a muscle, (2) a gland, or (3) another neuron. Therefore, depending on where a neuron terminates, it can cause a muscle cell to contract, a gland cell to secrete, another neuron to convey an electrical message along a nerve pathway, or some other function. When a neuron terminates on a muscle or a gland, the neuron is said to **innervate** the structure. The following section focuses on the junction between two neurons: the synapse. (Sometimes the term *synapse* is used to describe a junction between any two excitable cells, but we will reserve this term for the junction between two neurons.)

## Synapses

Typically, a synapse involves a junction between an axon terminal of one neuron, known as the *presynaptic neuron*, and the dendrites or cell body of a second neuron, known as the *postsynaptic neuron*. (*Pre* means "before," and *post* means "after"; the presynaptic neuron lies before the synapse, and the postsynaptic neuron lies after the synapse.) The dendrites and, to a lesser extent, the cell body of most neurons receive thousands of synaptic inputs, which are axon terminals from many other neurons. Some neurons within the central nervous system receive as many as 100 000 synaptic inputs (> Figure 2-43).

The anatomy of one of these thousands of synapses is shown in > Figure 2-44a. The axon terminal of the **presynaptic neuron**, which conducts its action potentials *toward* the synapse, ends in a slight swelling, the **synaptic knob**. The synaptic knob contains **synaptic vesicles**, which store a specific chemical messenger, a **neurotransmitter** that has been synthesized and packaged by the presynaptic neuron. The synaptic knob comes into close proximity to, but does not actually directly touch, the **postsynaptic neuron**, the neuron whose action potentials are propagated *away* from the synapse. The space between the presynaptic and postsynaptic neurons is called the **synaptic cleft**.

Current does not directly spread from the presynaptic to the postsynaptic neuron, because no channels are present in the presynaptic membrane for passage of charge-carrying $Na^+$ and $K^+$. Therefore, action potentials cannot electrically pass between the neurons. Instead, an action potential in the presynaptic neuron alters the postsynaptic neuron's potential by chemical means. Excitation in synapses moves in one direction only, from the presynaptic neuron to the postsynaptic neuron; the reason for this will become apparent as you read further.

## Neurotransmitters and signals

Here are the events that occur at a synapse (illustrated in > Figure 2-44):

1. When an action potential in a presynaptic neuron has been propagated along the axon to the axon terminal (step **1**), this local change in potential opens another kind of voltage-gated channel: the voltage-gated calcium ($Ca^{2+}$) channels in the synaptic knob. Voltage-gated $Ca^{2+}$ channels are a class of transmembrane ion channels that are activated by changes in electrical potential. $Ca^{2+}$ channels are critical in neurons, and are common in other cell types.

2. Because $Ca^{2+}$ is much more highly concentrated in the ECF and its electrical gradient is inward, this ion flows into the synaptic knob through the opened channels (step **2**).

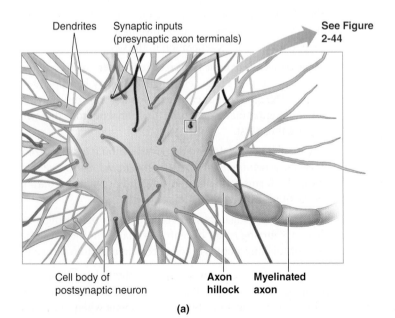

Dendrites  Synaptic inputs (presynaptic axon terminals)

See Figure 2-44

Cell body of postsynaptic neuron    **Axon hillock**  **Myelinated axon**

(a)

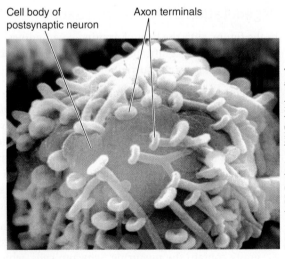

Cell body of postsynaptic neuron    Axon terminals

© E.R. Lewis; T.E. Everhart, and Y.Y. Zevi, University of California/Visuals Unlimited

(b)

> **FIGURE 2-43 Synaptic inputs to a postsynaptic neuron.** (a) Schematic representation of synaptic inputs (presynaptic axon terminals) to the dendrites and cell body of a single postsynaptic neuron. (b) Electron micrograph showing multiple presynaptic axon terminals to a single postsynaptic cell body

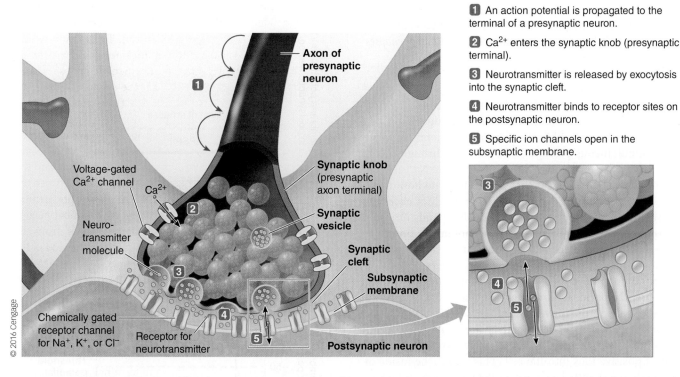

1. An action potential is propagated to the terminal of a presynaptic neuron.

2. Ca²⁺ enters the synaptic knob (presynaptic terminal).

3. Neurotransmitter is released by exocytosis into the synaptic cleft.

4. Neurotransmitter binds to receptor sites on the postsynaptic neuron.

5. Specific ion channels open in the subsynaptic membrane.

> FIGURE 2-44 **Synaptic structure and function.** Schematic representation of the structure of a single synapse. The circled numbers designate the sequence of events that take place at a synapse.

3. Ca²⁺ induces the release of a neurotransmitter from some of the synaptic vesicles into the synaptic cleft (step 3). A delay (i.e., synaptic delay) of 1–2 milliseconds occurs between the arrival of the action potential and the neurotransmitter release. This results from the conversion of the electrical impulse to the chemical signal. The more complex the pathway is, the greater the synaptic delay and the longer the *total reaction time (response to the stimulus)*. The neurotransmitter release is accomplished by exocytosis (p. 48), though the molecular mechanism is not completely understood. Also, it is believed that the old vesicles used during exocytosis become part of the presynaptic membrane, and new vesicles are made from adjacent membrane.

4. The released neurotransmitter diffuses across the cleft and binds with specific protein receptors on the **subsynaptic membrane**, which is the postsynaptic membrane directly below the synaptic knob (step 4). The protein receptor has recognition sites specific for that neurotransmitter.

5. This binding triggers the opening of specific ion channels in the subsynaptic membrane, changing the ion permeability of the postsynaptic neuron (step 5). The binding of the neurotransmitter to the receptor triggers a postsynaptic response specific for that receptor. These are chemically gated channels, in contrast to the voltage-gated channels responsible for the action potential and for Ca²⁺ influx into the synaptic knob.

Because the presynaptic terminal releases the neurotransmitter and the subsynaptic membrane of the postsynaptic neuron has specific receptor sites for the neurotransmitter, the synapse can operate only in the direction from presynaptic to postsynaptic neuron. Additionally, these responses can be either *excitatory* or *inhibitory*, depending on the properties of the receptor.

## Synapse behaviour

Neurons vary in the type of neurotransmitters they release, but does each neuron release only one type of neurotransmitter? No, not exactly. Neuroactive peptides and other neuroactive molecules can be contained in the same neuron. Mature neurons commonly contain one small-molecule transmitter and one or more peptides, and when released together by a presynaptic neuron they work synergistically on the same target cells. It also seems that the same neuron can release more than one neurotransmitter, depending on location. For example, it is known that motor neurons release acetylcholine (Ach; on muscle and on neurons in the spine), but it is also believed that they release the neurotransmitter glutamate on neurons in the spine. This means the concept of a neuron releasing only one neurotransmitter, or communication molecule, is too simplistic. However, here we focus on neurotransmitter release and its effects, not on all the potential neuroactive substances released by neurons, which is beyond the scope of this text.

Once the neurotransmitter binds with its subsynaptic receptor sites, a change in ion permeability ensues. There are two types of synapses, depending on the permeability changes induced in the postsynaptic neuron by the combination of a specific neurotransmitter with its receptor sites: *excitatory synapses* and *inhibitory synapses.*

### EXCITATORY SYNAPSES

At an **excitatory synapse**, the response to the binding of a neurotransmitter to the receptor is the opening of nonspecific cation channels within the subsynaptic membrane that permit simultaneous passage of Na$^+$ and K$^+$ through them. What makes these channels different from the Na$^+$ and K$^+$ channels described earlier is that permeability to both these ions is increased at the same time. How much of each ion diffuses through an open cation channel depends on their electrochemical gradients. At resting potential, both the concentration and electrical gradients for Na$^+$ favour its movement into the postsynaptic neuron, whereas only the concentration gradient for K$^+$ favours its movement outward. Therefore, the permeability change induced at an excitatory synapse results in the movement of a few K$^+$ ions out of the postsynaptic neuron, while a relatively larger number of Na$^+$ ions simultaneously enter this neuron. The result is net movement of positive ions into the cell. This makes the inside of the membrane slightly less negative than at resting potential, thereby producing a *small depolarization* of the postsynaptic neuron. Excitation and inhibition depend on the properties of the receptor, not the neurotransmitter. Note too that receptors coupled to Na$^+$ or Ca$^{2+}$ channels are excitatory and, when activated, produce a depolarization of the postsynaptic membrane.

Activation of one excitatory synapse can rarely depolarize the postsynaptic neuron sufficiently to bring it to threshold. Too few channels are involved at a single subsynaptic membrane to permit enough ion flow to reduce the potential to threshold. This small depolarization, however, does bring the membrane of the postsynaptic neuron closer to threshold, increasing the likelihood that threshold will be reached (but further excitatory input is needed) and an action potential will occur. That is, the membrane is now more excitable (easier to bring to threshold) than when at rest. Accordingly, this change in postsynaptic potential occurring at an excitatory synapse is called an **excitatory postsynaptic potential (EPSP)** (⟩ Figure 2-45a).

### INHIBITORY SYNAPSES

At an **inhibitory synapse**, the binding of a different released neurotransmitter with its receptor sites increases the permeability of the subsynaptic membrane to either K$^+$ or Cl$^-$. Receptors coupled to Cl$^-$ or K$^+$ channels are therefore inhibitory and produce a *small hyperpolarization* of the postsynaptic neuron— having greater internal negativity. In the case of increased P$_{K^+}$, more positive charges leave the cell via K$^+$ efflux, which leaves more negative charges behind on the inside. To hyperpolarize the membrane in the case of increased P$_{Cl^-}$, more negative charges enter the cell in the form of Cl$^-$ ions (because Cl$^-$ is in higher concentration outside the cell) than are driven out by the opposing electrical gradient established by the resting

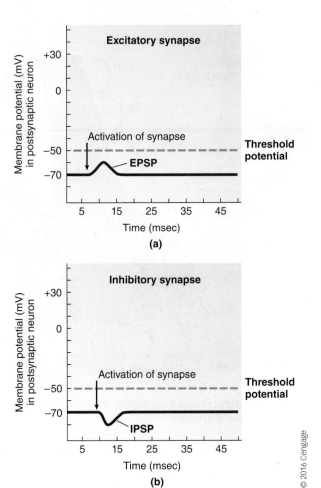

> FIGURE 2-45 **Postsynaptic potentials.** (a) Excitatory synapse. An excitatory postsynaptic potential (EPSP) brought about by activation of an excitatory presynaptic input brings the postsynaptic neuron closer to threshold potential. (b) Inhibitory synapse. An inhibitory postsynaptic potential (IPSP) brought about by activation of an inhibitory presynaptic input moves the postsynaptic neuron farther from threshold potential.

membrane potential (p. 57). In either case, this small hyperpolarization moves the membrane potential even farther away from threshold (⟩ Figure 2-45b), reducing the likelihood that the postsynaptic neuron will reach threshold and undergo an action potential. That is, the membrane is now less excitable (harder to bring to threshold by excitatory input) than when it is at resting potential. The membrane is said to be inhibited under these circumstances, and the small hyperpolarization of the postsynaptic cell is called an **inhibitory postsynaptic potential (IPSP)** (see Why It Matters).

In cells where the equilibrium potential for Cl$^-$ exactly equals the resting potential, an increased P$_{Cl^-}$ does not result in a hyperpolarization because there is no driving force to produce Cl$^-$ movement. Opening of Cl$^-$ channels in these cells tends to hold the membrane at resting potential, and this reduces the likelihood that threshold will be reached.

Note that EPSPs and IPSPs are produced by the opening of chemically gated channels, unlike action potentials, which are produced by the opening of voltage-gated channels.

IT IS NOT UNCOMMON FOR excitatory and inhibitory neurons to synapse on the same neuron. The arrival of an excitatory signal (EPSP) triggers a wave of depolarization along the membrane of a postsynaptic neuron. Inhibitory signals (IPSP) have an opposite effect. IPSP signals cause a wave of hyperpolarization along the membrane of a postsynaptic cell. When they arrive at the same neuron, these opposing signals travel along the postsynaptic membrane until they reach the axon terminal, and are integrated. If enough EPSPs arrive at the axon terminal simultaneously, the membrane potential rises above threshold, and an action potential forms. Arrival of the IPSPs lowers the membrane potential. If enough IPSPs have formed, these inhibitory signals stop the neuron from firing. During stressful times, excitatory neurons in the amygdala fire rapidly, sending excitatory signals to other areas of the brain. This type of firing leads to a feeling of panic or fear. The IPSP interneurons in the amygdala modulate these emotions by releasing GABA (g-aminobutyric acid). Its recognition by postsynaptic receptors (e.g., GABAA) inhibits excitatory signals that result in feelings of panic and fear; thus, the inhibitory neurotransmitter GABA has a calming effect on our emotions and prevents us from becoming overwhelmed in stressful situations. GABA's inhibitory role is key in maintaining normal firing rates in many physiological systems.

## Receptor combinations

Many different chemicals serve as neurotransmitters (❚ Table 2-6). Even though neurotransmitters vary from synapse to synapse, the same neurotransmitter is always released at a particular synapse. Furthermore, at a given synapse, the binding of a neurotransmitter with its appropriate subsynaptic receptors always leads to the same change in permeability and the resultant change in potential of the postsynaptic membrane. That is, the response to a given neurotransmitter–receptor combination is always constant and produces the same response. Some neurotransmitters (e.g., *glutamate*, the most common excitatory neurotransmitter in the brain) typically bring about EPSPs, whereas others (e.g., *gamma-aminobutyric acid*, or GABA, the brain's main inhibitory neurotransmitter) always cause IPSPs. Still other neurotransmitters (e.g., *norepinephrine*) are quite variable, producing EPSPs at one synapse and IPSPs at a different synapse. This means that changes in permeability in the postsynaptic neuron differ in response to the binding of the same neurotransmitter to the subsynaptic receptor sites of different postsynaptic neurons.

Most of the time, each axon terminal releases only one neurotransmitter. Recent evidence suggests, however, that in some cases two different neurotransmitters can be released simultaneously from a single axon terminal. For example, *glycine* and GABA, both of which produce inhibitory responses, can be packaged and released from the same synaptic vesicles. Scientists speculate that the fast-acting glycine and more slowly acting GABA may complement each other in the control of activities that depend on precise timing: for example, coordination of complex movements.

## Neurotransmitter removal

The alteration in membrane permeability responsible for the EPSP or IPSP continues as long as the neurotransmitter remains bound to the receptor sites. However, after it has produced the appropriate response in the postsynaptic neuron, the neurotransmitter must be inactivated or removed. This, so to say, wipes cleans the postsynaptic neuron, leaving it ready to receive additional messages from the same or other presynaptic inputs. Several mechanisms can remove the neurotransmitter: it may (1) diffuse away from the synaptic cleft, (2) be inactivated by specific enzymes within the subsynaptic membrane, or (3) be actively taken back up into the axon terminal by transport mechanisms in the presynaptic membrane. Once the neurotransmitter is taken back up, it can be stored and later released again (recycled) in response to a subsequent action potential, or it may be destroyed by enzymes within the synaptic knob. What actually occurs depends on the particular synapse.

## Grand postsynaptic potential

EPSPs and IPSPs are graded potentials. Unlike action potentials, which behave according to the all-or-none law, graded potentials can be of varying magnitude, have no refractory period, and can be summed (added on top of one another). What are the mechanisms and significance of summation?

The events that occur at a single synapse result in either an EPSP or an IPSP at the postsynaptic neuron. But if a single EPSP is inadequate to bring the postsynaptic neuron to threshold and an IPSP moves it even farther from threshold, how can an action potential be initiated in the postsynaptic neuron? The answer

### ❚ TABLE 2-6 Some Common Neurotransmitters

| Acetylcholine | Histamine |
|---|---|
| Dopamine | Glycine |
| Norepinephrine | Glutamate |
| Epinephrine | Aspartate |
| Serotonin | Gamma-aminobutyric acid (GABA) |

lies in the thousands of presynaptic inputs (sum total) that a typical neuronal cell body receives from many other neurons. Some of these presynaptic inputs may be carrying sensory information from the environment; some may be signalling internal changes in homeostatic balance; others may be transmitting signals from control centres in the brain; and still others may arrive carrying other bits of information. At a given time, any number of these presynaptic neurons (probably hundreds) may be firing and thus influencing the postsynaptic neuron's level of activity. The total potential in the postsynaptic neuron, the **grand postsynaptic potential (GPSP)**, is a sum total of all EPSPs and IPSPs occurring at approximately the same time.

The postsynaptic neuron can be brought to threshold in two ways: (1) *temporal summation* and (2) *spatial summation*. To illustrate these methods of summation, we examine the possible interactions of three presynaptic inputs—two excitatory inputs (Ex1 and Ex2) and one inhibitory input (In1)—on a hypothetical postsynaptic neuron (› Figure 2-46). The recording shown in the figure represents the potential in a single postsynaptic cell; typically there are thousands of synapses.

### TEMPORAL SUMMATION

Suppose that Ex1 has an action potential that causes an EPSP in the postsynaptic neuron. If later, after this EPSP has died off, another action potential occurs in Ex1, an EPSP of the same magnitude takes place (panel (a) in › Figure 2-46). Next, suppose that

Ex1 has two action potentials in close succession (panel (b)). The first action potential in Ex1 produces an EPSP in the postsynaptic membrane. While the postsynaptic membrane is still partially depolarized from this first EPSP, the second action potential in Ex1 produces a second EPSP. The second EPSP adds on to the first EPSP, bringing the membrane to threshold, so an action potential occurs in the postsynaptic neuron. Graded potentials do not have a refractory period, so this additive effect is possible.

The summing of several EPSPs occurring very close together in time because of successive firing of a single presynaptic neuron is known as **temporal summation** (*tempus* means "time"). In reality, the situation is much more complex than just described. The sum of up to 50 EPSPs might be needed to bring the postsynaptic membrane to threshold. Each action potential in a single presynaptic neuron triggers the emptying of a certain number of synaptic vesicles. The amount of neurotransmitter released and the resultant magnitude of the change in postsynaptic potential are thus directly related to the frequency of presynaptic action potentials. One way, then, in which the postsynaptic membrane can be brought to threshold is through rapid, repetitive excitation from a single persistent input.

### SPATIAL SUMMATION

Now consider the situation when both of the excitatory inputs are stimulated simultaneously in the postsynaptic neuron (panel (c) in › Figure 2-46). An action potential in either Ex1 or Ex2

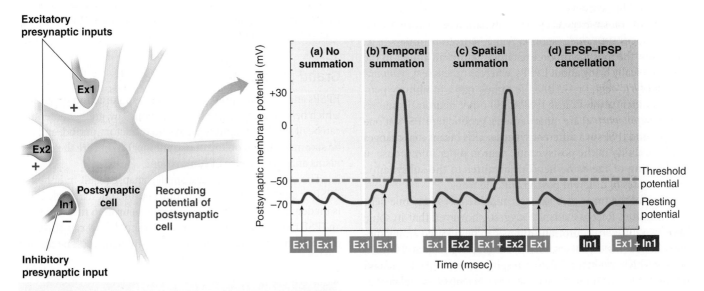

**(a)** If an excitatory presynaptic input (Ex1) is stimulated a second time after the first EPSP in the postsynaptic cell has died off, a second EPSP of the same magnitude will occur.

**(b)** If, however, Ex1 is stimulated a second time before the first EPSP has died off, the second EPSP will add onto, or sum with, the first EPSP, resulting in *temporal summation*, which may bring the postsynaptic cell to threshold.

**(c)** The postsynaptic cell may also be brought to threshold by *spatial summation* of EPSPs that are initiated by simultaneous activation of two (Ex1 and Ex2) or more excitatory presynaptic inputs.

**(d)** Simultaneous activation of an excitatory (Ex1) and inhibitory (In1) presynaptic input does not change the postsynaptic potential, because the resultant EPSP and IPSP cancel each other out.

› **FIGURE 2-46 Determination of the grand postsynaptic potential by the sum of activity in the presynaptic inputs.** Two excitatory (Ex1 and Ex2) and one inhibitory (In1) presynaptic inputs terminate on this hypothetical postsynaptic neuron. The potential of the postsynaptic neuron is being recorded.

will produce an EPSP in the postsynaptic neuron; however, neither of these alone brings the membrane to threshold to elicit a postsynaptic action potential. But simultaneous action potentials in Ex1 and Ex2 produce EPSPs that add to each other, bringing the postsynaptic membrane to threshold, and making it possible for an action potential to occur. The summation of EPSPs originating simultaneously from several different presynaptic inputs (i.e., from different places, or points in space) is known as **spatial summation**. A second way, therefore, to elicit an action potential in a postsynaptic cell is through concurrent activation of several excitatory inputs. Again, in reality up to 50 simultaneous EPSPs are required to bring the postsynaptic membrane to threshold.

Similarly, IPSPs can undergo temporal and spatial summation. As IPSPs add together, however, they progressively move the potential farther from threshold.

### CANCELLATION OF CONCURRENT EPSPS AND IPSPS

If an excitatory and an inhibitory input are simultaneously activated, the concurrent EPSP and IPSP more or less cancel each other out. The extent of cancellation depends on their respective magnitudes. In most cases, the postsynaptic membrane potential remains close to resting potential (panel (d) in › Figure 2-46).

### IMPORTANCE OF POSTSYNAPTIC NEURONAL INTEGRATION

The magnitude of the GPSP depends on the sum of activity in all the presynaptic inputs and, in turn, determines whether or not the neuron will undergo an action potential to pass information on to the cells on which the neuron terminates. The following oversimplified real-life example demonstrates the benefits of this neuronal integration. The explanation is not completely accurate technically, although the principles of summation are.

Assume for simplicity's sake that urination is controlled by a postsynaptic neuron that innervates the urinary bladder. When this neuron fires, the bladder contracts. (Actually, voluntary control of urination is accomplished by postsynaptic integration at the neuron controlling the external urethral sphincter, rather than the bladder itself.) As the bladder starts to fill with urine and becomes stretched, a reflex is initiated that ultimately produces EPSPs in the postsynaptic neuron responsible for causing bladder contraction. Partial filling of the bladder does not cause enough excitation to bring the neuron to threshold, so urination does not take place (panel (a) of › Figure 2-46). As the bladder becomes progressively filled, the frequency of action potentials progressively increases in the presynaptic neuron that signals the postsynaptic neuron of the extent of bladder filling (Ex1 in panel (b)). When the frequency becomes great enough that the EPSPs are temporally summed to threshold, the postsynaptic neuron undergoes an action potential that stimulates bladder contraction.

What if the time is inopportune for urination to take place? IPSPs can be produced at the bladder postsynaptic neuron by presynaptic inputs originating in higher levels of the brain responsible for voluntary control (In1 in panel (d)). These voluntary IPSPs in effect cancel out the reflex EPSPs triggered by stretching of the bladder. Because the postsynaptic neuron

remains at resting potential and does not have an action potential, the bladder is prevented from contracting and emptying even though it is full.

What happens when a person's bladder is only partially filled—so that the presynaptic input originating from this source is insufficient to bring the postsynaptic neuron to threshold to cause bladder contraction—but a urine specimen has to be provided for laboratory analysis? The person can voluntarily activate an excitatory presynaptic neuron (Ex2 in panel (c)). The EPSPs originating from this neuron and the EPSPs of the reflex-activated presynaptic neuron (Ex1) are spatially summed to bring the postsynaptic neuron to threshold. This achieves the action potential necessary to stimulate bladder contraction, even though the bladder is not full.

This example illustrates the importance of postsynaptic neuronal integration. Each postsynaptic neuron in a sense "computes" all the input it receives and makes a "decision" about whether to pass the information on (i.e., whether threshold is reached and an action potential is transmitted down the axon). In this way, neurons serve as complex computational devices, or integrators. The dendrites function as the primary processors of incoming information. They receive and tally the signals coming in from all the presynaptic neurons, providing a sum total. Each neuron's output in the form of frequency of action potentials to other cells (muscle cells, gland cells, or other neurons) reflects the balance of activity in the inputs it receives via EPSPs or IPSPs from the thousands of other neurons that terminate on it. Each postsynaptic neuron filters out and does not pass on information it receives that is not significant enough to bring it to threshold. If every action potential in every presynaptic neuron that impinges on a particular postsynaptic neuron were to cause an action potential in the postsynaptic neuron, the neuronal pathways would be overwhelmed with trivia. Only if an excitatory presynaptic signal is reinforced by other supporting signals through summation will the information be passed on. Furthermore, interaction of postsynaptic potentials provides a way for one set of signals to offset another set (IPSPs negating EPSPs). This allows a fine degree of discrimination and control in determining what information will be passed on.

Next we examine why action potentials are initiated at the axon hillock.

---

### Check Your Understanding 2.8

1. Compare and contrast excitatory synapses and inhibitory synapses.
2. Explain what is meant by temporal and spatial summation and why they are important.

---

### Action potentials at the axon hillock

Threshold potential is not uniform throughout the postsynaptic neuron. The lowest threshold is present at the axon hillock, because this region has a much greater density of voltage-gated $Na^+$ channels than anywhere else in the neuron. The greater density of these voltage-sensitive channels makes the axon hillock considerably more responsive than the dendrites or the remainder of the cell body to changes in potential. The latter regions have a significantly higher threshold than the axon hillock. Because of local current flow, changes in membrane potential (EPSPs or IPSPs) occurring anywhere on the dendrites or cell body spread throughout the dendrites, cell body, and axon hillock. When summation of EPSPs takes place, the lower threshold of the axon hillock is reached first, whereas the dendrites and cell body at the same potential are still considerably below their own, much higher thresholds. Therefore, an action potential originates in the axon hillock and is propagated from there to the end of the axon.

## Neuropeptides as neuromodulators

In addition to the classical neurotransmitters just described, some neurons also release *neuropeptides*. Neuropeptides differ from classical neurotransmitters in several important ways (▮ Table 2-7). **Classical neurotransmitters** are small, rapid-acting molecules that typically trigger the opening of specific ion channels to bring about a change in potential in the postsynaptic neuron (an EPSP or IPSP) within a few milliseconds or less. Most classical neurotransmitters are synthesized and packaged locally in synaptic vesicles in the cytosol of the axon terminal. These chemical messengers are primarily amino acids or closely related compounds.

**Neuropeptides** are larger molecules made up of anywhere from 2 to about 40 amino acids. They are synthesized in the neuronal cell body in the endoplasmic reticulum and Golgi complex and are subsequently moved by axonal transport along the microtubular highways to the axon terminal. Neuropeptides are not stored within the small synaptic vesicles with the classical neurotransmitter, but instead are packaged in large **dense-core vesicles**, which are also present in the axon terminal. The dense-core vesicles undergo $Ca^{2+}$-induced exocytosis and release neuropeptides at the same time as the neurotransmitter is released from the synaptic vesicles. An axon terminal typically releases only a single classical neurotransmitter, but the same terminal may also contain one or more neuropeptides that are cosecreted along with the neurotransmitter.

Even though neuropeptides are currently the subject of intense investigation, our knowledge about their functions and control is still limited. They are known to diffuse locally and act on other adjacent neurons at much lower concentrations than do classical neurotransmitters, and they bring about slower, more prolonged responses. Some neuropeptides released at synapses may function as true neurotransmitters, but most are believed to function as neuromodulators. **Neuromodulators** are chemical messengers that do not cause the formation of EPSPs or IPSPs, but rather bring about long-term changes that subtly *modulate*—depress or enhance—the action of the synapse. They bind to neuronal receptors at nonsynaptic sites (i.e., not at the subsynaptic membrane), and they do not directly alter membrane permeability and potential. Neuromodulators may act at either presynaptic or postsynaptic sites. For example, a neuromodulator may influence the level of an enzyme involved in the synthesis of a neurotransmitter by a presynaptic neuron, or it may alter the sensitivity of the postsynaptic neuron to a particular neurotransmitter by causing long-term changes in the number of subsynaptic receptor sites for the neurotransmitter. In this way, neuromodulators delicately fine-tune the synaptic response. The effect may last for days or even months or years. Whereas neurotransmitters are involved in rapid communication between

### ▮ TABLE 2-7 Comparison of Classical Neurotransmitters and Neuropeptides

| Characteristic | Classical Neurotransmitters | Neuropeptides |
| --- | --- | --- |
| **Size** | Small; one amino acid or similar chemical | Large; 2 to 40 amino acids in length |
| **Site of Synthesis** | Cytosol of synaptic knob | Endoplasmic reticulum and Golgi complex in cell body; travel to synaptic knob by axonal transport |
| **Site of Storage** | In small synaptic vesicles in axon terminal | In large dense-core vesicles in axon terminal |
| **Site of Release** | Axon terminal | Axon terminal; may be cosecreted with neurotransmitter |
| **Speed and Duration of Action** | Rapid, brief response | Slow, prolonged response |
| **Site of Action** | Subsynaptic membrane of postsynaptic cell | Nonsynaptic sites on either presynaptic or postsynaptic cell at much lower concentrations than classical neurotransmitters |
| **Effect** | Usually alter potential of postsynaptic cell by opening specific ion channels | Usually enhance or suppress synaptic effectiveness by long-term changes in neurotransmitter synthesis or postsynaptic receptor sites (act as neuromodulators) |

neurons, neuromodulators are involved with more long-lasting events, such as learning and motivation.

Interestingly, the synaptically released neuromodulators include many substances that also have distinctive roles as hormones, which are released into the blood from endocrine tissues. For example, *cholecystokinin (CCK)*, a well-known hormone released from the small intestine, is involved in digestive activities such as causing the gallbladder to contract and release bile into the intestine (see Chapter 15). However, CCK has also been found in axon terminal vesicles in the brain, where it is believed to cause the feeling of satiation. In many instances, neuropeptides are named for their first-discovered role as hormones, as is the case with cholecystokinin (*chole* means "bile"; *cysto* means "bladder"; *kinin* means "contraction"). It appears that a number of chemical messengers are quite versatile and can assume different roles, depending on their source, their distribution, and their interaction with different cell types.

## Presynaptic inhibition or facilitation

Besides neuromodulation, presynaptic inhibition or facilitation is another means of depressing or enhancing synaptic effectiveness. Sometimes a third neuron can influence activity between a presynaptic ending and a postsynaptic neuron. A presynaptic axon terminal (labelled A in ⟩ Figure 2-47) may itself be innervated by another axon terminal (labelled B). The neurotransmitter released from modulatory terminal B binds with receptor sites on terminal A. This binding alters the amount of neurotransmitter released from terminal A in response to action potentials. If the amount of neurotransmitter released from A is reduced, the phenomenon is known as **presynaptic inhibition**. If the release of neurotransmitter is enhanced, the effect is called **presynaptic facilitation**.

Here, we look more closely at how this process works. Recall that $Ca^{2+}$ entry into an axon terminal causes the release of neurotransmitter by exocytosis of synaptic vesicles. The amount of neurotransmitter released from terminal A depends on how much $Ca^{2+}$ enters this terminal in response to an action potential. $Ca^{2+}$ entry into terminal A, in turn, can be influenced by activity in modulatory terminal B. Consider the example of presynaptic inhibition represented in ⟩ Figure 2-47. The amount of neurotransmitter released from presynaptic terminal A, an excitatory input in this example, influences the potential in the postsynaptic neuron (C) at which it terminates. Firing of A, alone, brings about an EPSP in postsynaptic neuron C. Now consider that B is stimulated simultaneously with A. When neurotransmitter from terminal B binds to terminal A, $Ca^{2+}$ entry into terminal A is reduced. Less $Ca^{2+}$ entry means less neurotransmitter release from A. Note that modulatory neuron B can suppress neurotransmitter release from A only when A is firing. If this presynaptic inhibition by B prevents A from releasing its neurotransmitter, the formation of EPSPs on postsynaptic membrane C from input A is specifically prevented. As a result, no change in the potential of the postsynaptic neuron occurs despite action potentials in A.

Could the same inhibitory action result from the simultaneous production of an IPSP through the activation of an inhibitory input that negates an EPSP produced by the activation of A? Not quite. Activation of an inhibitory input to cell C would produce an IPSP in cell C, but this IPSP could cancel out not only an EPSP from excitatory input A but also any EPSPs produced by other excitatory terminals, such as terminal D. The entire postsynaptic membrane is hyperpolarized by IPSPs, thereby negating (cancelling) excitatory information fed into any part of the cell from any presynaptic input. By contrast, presynaptic inhibition (or presynaptic facilitation) works in a much more specific way than does the action of inhibitory inputs to the postsynaptic cell. Presynaptic inhibition provides a means by which certain inputs to the postsynaptic neuron can be *selectively* inhibited without affecting the contributions of any other inputs. For example,

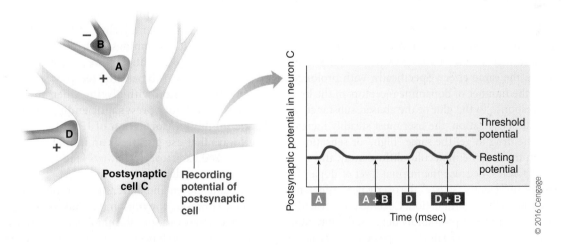

⟩ **FIGURE 2-47 Presynaptic inhibition.** A, an excitatory terminal ending on postsynaptic cell C, is itself innervated by inhibitory terminal B. Stimulation of terminal A alone produces an EPSP in cell C, but simultaneous stimulation of terminal B prevents the release of excitatory neurotransmitter from terminal A. Consequently, no EPSP is produced in cell C despite the fact that terminal A has been stimulated. Such presynaptic inhibition selectively depresses activity from terminal A without suppressing any other excitatory input to cell C. Stimulation of excitatory terminal D produces an EPSP in cell C, even though inhibitory terminal B is simultaneously stimulated, because terminal B only inhibits terminal A.

the firing of B specifically prevents the formation of an EPSP in the postsynaptic neuron from excitatory presynaptic neuron A, but does not have any influence on other excitatory presynaptic inputs. Excitatory input D can still produce an EPSP in the postsynaptic neuron even when B is firing. This type of neuronal integration is another means by which electrical signalling between nerve cells can be carefully fine-tuned.

## Drugs, diseases, and transmission

*Clinical Note* A drug is any biological substance that when consumed will in some way alter the biological function of the organism. The vast majority of drugs that influence the nervous system perform their function by altering synaptic mechanisms. Synaptic drugs may block an undesirable effect or enhance a desirable effect. Possible drug actions include (1) altering the synthesis, axonal transport, storage, or release of a neurotransmitter; (2) modifying neurotransmitter interaction with the postsynaptic receptor; (3) influencing neurotransmitter reuptake or destruction; and (4) replacing a deficient neurotransmitter with a substitute transmitter.

For example, cocaine blocks the reuptake of the neurotransmitter *dopamine* at presynaptic terminals. It does so by binding competitively with the dopamine reuptake transporter, which is a protein molecule that picks up released dopamine from the synaptic cleft and shuttles it back to the axon terminal. With cocaine occupying the dopamine transporter, dopamine remains in the synaptic cleft longer than usual and continues to interact with its postsynaptic receptor sites. The result is prolonged activation of neural pathways that use this chemical as a neurotransmitter. Among these pathways are those that play a role in emotional responses, especially feelings of pleasure. Thus, cocaine locks the switch in the neural pleasure pathway in the "on" position.

Cocaine is addictive because it causes the neurons involved to generate long-term molecular adaptations, such that they cannot transmit normally across synapses without increasingly higher doses of the drug. The postsynaptic cells become accustomed to high levels of stimulation; they become "hooked" on the drug. The term **tolerance** refers to this *desensitization* to an addictive drug so that the user needs greater quantities of the drug to achieve the same effect. Specifically, with prolonged use of cocaine, the number of dopamine receptors in the brain is reduced in response to the glut of the abused substance. As a result of this desensitization, the user must steadily increase the dosage of the drug to get the same "high," or sensation of pleasure. When the cocaine molecules diffuse away, the sense of pleasure evaporates, because the normal level of dopamine activity does not sufficiently "satisfy" the overly needy demands of the postsynaptic cells for stimulation. Cocaine users reaching this low become frantic and profoundly depressed. Only more cocaine makes them feel good again. But repeated use of cocaine modifies responsiveness to the drug. Over the course of abuse, the user often no longer can derive pleasure from the drug but suffers unpleasant *withdrawal symptoms* once the effect of the drug has worn off. Furthermore, the amount of cocaine needed to overcome the devastating crashes progressively increases.

The user typically becomes **addicted** to the drug, compulsively seeking out and taking the drug at all costs, first to experience the pleasurable sensations and later to avoid the negative withdrawal symptoms, even when the drug no longer provides pleasure. Cocaine is abused by millions who have become addicted to its mind-altering properties, with devastating social and economic effects.

Whereas cocaine abuse leads to excessive dopamine activity, **Parkinson's disease (PD)** is attributable to a deficiency of dopamine in the *basal nuclei*, a region of the brain involved in controlling complex movements. This movement disorder is characterized by muscular rigidity and involuntary tremors at rest, such as involuntary rhythmic shaking of the hands or head. The standard treatment for PD is the administration of *levodopa* (*L-dopa*), a precursor of dopamine. Dopamine itself cannot be administered because it is unable to cross the blood-brain barrier (discussed in Chapter 3), but L-dopa can enter the brain from the blood. Once inside the brain, L-dopa is converted into dopamine, thus substituting for the deficient neurotransmitter. This therapy greatly alleviates the symptoms associated with the deficit in most patients. This condition is described further in the discussion of the basal nuclei (Chapter 3, Section 3.9).

Synaptic transmission is also vulnerable to neural toxins, which may cause nervous system disorders by acting at either presynaptic or postsynaptic sites. For example, two different neural poisons, tetanus toxin and strychnine, act at different synaptic sites to block inhibitory impulses while leaving excitatory inputs unchecked. Tetanus toxin prevents the release of an inhibitory neurotransmitter, GABA, and therefore blocks the inhibitory presynaptic inputs to neurons that supply the skeletal muscles. Unchecked excitatory inputs to these neurons result in uncontrolled muscle spasms. These spasms occur especially in the jaw muscles early in the disease, giving rise to the common name of *lockjaw* for this condition. Later they progress to the muscles responsible for breathing, at which point death occurs.

Strychnine, in contrast, competes with another inhibitory neurotransmitter, glycine, at the postsynaptic receptor site. This poison combines with the receptor but does not directly alter the potential of the postsynaptic cell in any way. Instead, strychnine blocks the receptor so that it is not available for interaction with glycine when the latter is released from the inhibitory presynaptic ending. As a result, postsynaptic inhibition (formation of IPSPs) is abolished in nerve pathways that use glycine as an inhibitory neurotransmitter. Unchecked excitatory pathways lead to convulsions, muscle spasticity, and death.

Tetanus toxin and strychnine poisoning have similar outcomes. However, one poison (tetanus toxin) prevents the presynaptic release of a specific inhibitory neurotransmitter, and the other (strychnine) blocks specific postsynaptic inhibitory receptors. Other drugs and diseases that influence synaptic transmission are too numerous to mention, but as these examples illustrate, any site along the synaptic pathway is vulnerable to interference, either pharmacologic (drug induced) or pathologic (disease induced).

## Neurons: complex converging and diverging pathways

Two important relationships exist between neurons: convergence and divergence. A given neuron may have many other neurons synapsing on it. Such a relationship is known as **convergence** (› Figure 2-48). Through this converging input, a single cell is influenced by thousands of other cells. This single cell, in turn, influences the level of activity in many other cells by divergence of output. The term **divergence** refers to the branching of axon terminals so that a single cell synapses with and influences many other cells.

Note that a particular neuron is postsynaptic to the neurons converging on it but presynaptic to the other cells at which it terminates. So the terms *presynaptic* and *postsynaptic* refer only to a single synapse. Most neurons are presynaptic to one group of neurons and postsynaptic to another group.

There are an estimated 100 billion neurons and $10^{14}$ (100 quadrillion) synapses in the brain alone! When you consider the vast and intricate interconnections possible between these neurons through converging and diverging pathways, you can begin to imagine how complex the wiring mechanism of our nervous system really is. Even the most sophisticated computers are far less complex than the human brain. The "language" of the nervous system—that is, all communication between neurons—is in the form of graded potentials, action potentials, neurotransmitter signalling across synapses, and other nonsynaptic forms of chemical chatter. All activities for which the nervous system is responsible—every sensation you feel, every command to move a muscle, every thought, every emotion, every memory, every spark of creativity—depend on the patterns of electrical and chemical signalling between neurons along these complexly wired neural pathways.

# Chapter in Perspective: Focus on Homeostasis

The ability of cells to perform functions essential for their own survival as well as specialized tasks that help maintain homeostasis within the body ultimately depends on the successful, cooperative operation of the plasma membrane and the intracellular components. For example, to support life-sustaining activities, all cells must generate energy, in a usable form, from nutrient molecules. Energy is generated intracellularly by chemical reactions that take place within the cytosol and mitochondria. Additionally, many cellular functions require the movement of molecules either into or out of the cell, which can only be accomplished by the many different plasma membrane transport systems.

In addition to being essential for basic cell survival, many of the specialized tasks performed by cells that contribute to homeostasis require the production of cellular energy and the regulation of the intracellular environment. Here are some examples:

- Nerve and endocrine cells both release protein chemical messengers that are important in regulatory activities aimed at maintaining homeostasis. These protein chemical messengers (neurotransmitters in nerve cells and hormones in endocrine

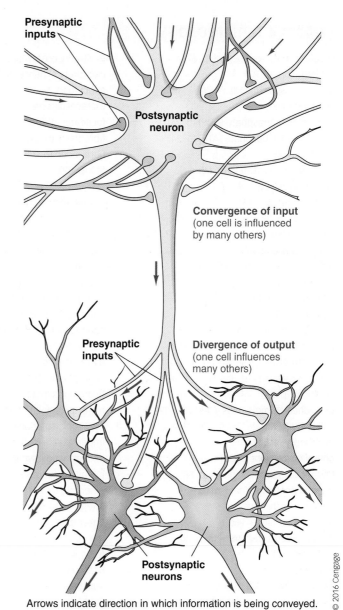

Arrows indicate direction in which information is being conveyed.

© 2016 Cengage

› **FIGURE 2-48** Convergence and divergence

cells) are all produced by the endoplasmic reticulum and Golgi complex and released by exocytosis from the cell when needed.

- The ability of muscle cells to contract depends not only on their highly developed cytoskeletal microfilaments sliding past one another but also on the vast amounts of ATP produced within their mitochondria. Muscle contraction is responsible for many homeostatic activities, including (1) contracting the heart muscle, which pumps life-supporting blood throughout the body; (2) contracting the muscles attached to bones, which enables the body to procure food; and (3) contracting the muscle in the walls of the stomach and intestine, which moves the food along the digestive tract so that ingested nutrients can be progressively broken down into a form that can be absorbed into the blood for delivery to the cells.

As we begin to examine the various organs and systems, keep in mind that proper cell functioning is the foundation of all organ activities.

absolute refractory period (p. 67)

active forces (p. 37)

active transport (p. 43)

addicted (p. 82)

adenosine diphosphate (ADP) (p. 24)

adenosine triphosphate (ATP) (p. 24)

aerobic (p. 28)

after hyperpolarization (p. 62)

all-or-none law (p. 69)

anabolic (p. 22)

anaerobic (p. 28)

aquaporins (p. 38)

ATP synthase (p. 27)

$Na^+-K^+$ ATPase pump (p. 44)

axon (p. 65)

axon hillock (p. 65)

axon terminals (66)

basal nuclei (p. 82)

cadherins (p. 32)

carrier-mediated transport (p. 41)

carrier molecules (p. 31)

catabolic (p. 22)

cell adhesion molecules (CAMs) (p. 32)

cell body (p. 64)

cell theory (p. 21)

chemical gradient (p. 37)

chemically gated channels (p. 58)

chemiosmotic mechanism (p. 27)

cholesterol (p. 29)

citric acid cycle (p. 25)

classical neurotransmitters (p. 80)

collagen (p. 34)

collaterals (p. 65)

competition (p. 42)

concentration gradient (p. 37)

connexon (p. 35)

contiguous conduction (p. 66)

convergence (p. 83)

cotransport carriers (p. 44)

creatine kinase (p. 24)

creatine phosphate (p. 24)

current (p. 59)

cytoplasm (p. 22)

dense-core vesicles (p. 80)

desmosomes (p. 34)

diffusion (p. 37)

divergence (p. 83)

docking marker (p. 31, 48)

elastin (p. 34)

electrical gradient (p. 38)

electrochemical gradient (p. 38)

electron transport chain (p. 27)

endocytosis (p. 46)

endocytotic vesicle (p. 48)

equilibrium potential ($E_{K^+}$) (p. 55)

excitable tissues (p. 57)

excitatory postsynaptic potential (EPSP) (p. 76)

excitatory synapse (p. 76)

exocytosis (p. 46)

extracellular matrix (ECM) (p. 34)

facilitated diffusion (p. 43)

fibroblasts (p. 34)

fibronectin (p. 34)

fire (p. 62)

first messenger (p. 53)

flavine adenine dinucleotide (FAD) (p. 26)

fluid mosaic model (p. 30)

gap junction (p. 35)

gated channels (p. 58)

glycolysis (p. 25)

graded potentials (p. 59)

grand postsynaptic potential (GPSP) (p. 78)

guanosine diphosphate (GDP) (p. 27)

guanosine triphosphate (GTP) (p. 27)

hormones (52)

hydrophilic (p. 29)

hydrophobic (p. 29)

hydrostatic (fluid) pressure (p. 39)

hypertonic solution (p. 41)

hypotonic solution (p. 41)

impermeable (p. 36)

inhibitory postsynaptic potential (IPSP) (p. 76)

inhibitory synapse (p. 76)

innervate (p. 74)

integrins (p. 32)

intermediary metabolism (p. 22)

ion concentration gradient (p. 44)

isotonic solution (p. 40)

Krebs cycle (p. 25)

leak channels (p. 58)

lipid bilayer (p. 29)

mechanically gated channels (p. 58)

membrane-bound enzymes (31)

membrane potential (p. 54)

membrane proteins (p. 30)

millivolt (mV) (p. 54)

myelin (p. 70)

myelinated fibres (p. 70)

myoblasts (p. 22)

$Na^+$ permeability ($P_{Na^+}$) (p. 63)

Nernst equation (p. 55)

nerve fibre (p. 65)

net diffusion (p. 37)

neurohormones (p. 53)

neuromodulators (p. 80)

neuron (p. 64)

neuropeptides (p. 80)

neurotransmitter (p. 74)

nicotinamide adenine dinucleotide ($NAD^+$) (p. 26)

nodes of Ranvier (p. 70)

nucleus (p. 22)

oligodendrocytes (p. 70)

osmosis (p. 38)

osmotic pressure (p. 40)

overshoot (p. 62)

oxidative phosphorylation (p. 27)

paracrines (p. 51)

Parkinson's disease (PD) (p. 82)

passive forces (p. 37)

permeable (p. 36)

phagocytosis (p. 48)

phospholipids (p. 29)

pinocytosis (p. 48)

postsynaptic neuron (p. 74)

presynaptic facilitation (p. 81)

presynaptic inhibition (p. 81)

presynaptic neuron (p. 74)

primary active transport (p. 44)

pseudopods (p. 48)

pumps (p. 44)

receptor sites (p. 31)

receptor-mediated endocytosis (p. 48)

receptors (p. 31)

refractory period (p. 67)

regeneration tube (p. 72)

relative refractory period (p. 68)

resistance (p. 60)

resting membrane potential (p. 54)

saltatory conduction (70)

saturation (p. 41)

Schwann cells (p. 70)

second messenger (p. 53)

secondary active transport (p. 44)

secretory vesicles (p. 48)

selectively permeable (p. 36)

signal transduction (p. 53)

signalling molecules (p. 32)

spatial summation (p. 79)

specificity (p. 41)

spike (p. 62)

steady state (p. 37)

subsynaptic membrane (p. 75)

synaptic cleft (p. 74)

synaptic knob (p. 74)

synaptic vesicles (p. 74)

target cells (p. 51)

temporal summation (p. 78)

thermally gated channels (p. 58)

threshold potential (p. 61)

tight junctions (p. 34)

tolerance (p. 82)

tonicity (p. 40)

transport maximum ($T_m$) (p. 41)

tricarboxylic acid cycle (p. 25)

trilaminar structure (p. 29)

vesicular transport (p. 46)

voltage-gated channels (p. 58)

## Objective Questions (Answers on p. A-37)

1. The barrier that separates and controls movement between the cell contents and the extracellular fluid is the _____.

2. The chemical that directs protein synthesis and serves as a genetic blueprint is _____, which is found in the _____ of the cell.

3. The cytoplasm consists of _____, which are specialized, membrane-enclosed intracellular compartments, and a gel-like mass known as _____, which contains an elaborate protein network called the _____.

4. The universal energy carrier of the body is _____.

5. The largest cells in the human body can be seen by the unaided eye. (*True or false?*)

6. The nonpolar tails of the phospholipid molecules bury themselves in the interior of the plasma membrane. (*True or false?*)

7. The hydrophobic regions of the molecules composing the plasma membrane correspond to the two dark layers of this structure visible under an electron microscope. (*True or false?*)

8. Using the following answer code, indicate which form of energy production is being described:

    ___ 1. takes place in the mitochondrial matrix

    ___ 2. produces $H_2O$ as a by-product

    ___ 3. results in a rich yield of ATP     (a) glycolysis

    ___ 4. takes place in the cytosol     (b) citric acid cycle

    ___ 5. processes acetyl CoA     (c) electron transport chain

    ___ 6. located in the mitochondrial inner-membrane cristae

    ___ 7. converts glucose into two pyruvic acid molecules

    ___ 8 uses molecular oxygen

9. Using the answer code on the right, indicate which membrane component is responsible for the function in question:

    ___ 1. channel formation

    ___ 2. barrier to passage of water-soluble substances

    ___ 3. receptor sites     (a) lipid bilayer

    ___ 4. membrane fluidity     (b) proteins

    ___ 5. recognition of "self"     (c) carbohydrates

    ___ 6. membrane-bound enzymes

    ___ 7. structural boundary

    ___ 8. carriers

10. Using the answer code on the right, indicate the direction of net movement in each case:

    ___ 1. simple passive diffusion     (a) movement from high to low concentration

    ___ 2. facilitated diffusion     (b) movement from low to high concentration

    ___ 3. primary active transport

    ___ 4. $Na^+$ during secondary active transport

    ___ 5. cotransported molecule during secondary active transport

    ___ 6. water with regard to the water concentration gradient during osmosis

    ___ 7. water with regard to the solute concentration gradient during osmosis

11. Using the answer code on the right, indicate the type of cell junction described:

    ___ 1. adhering junction     (a) gap junction

    ___ 2. impermeable junction     (b) tight junction

    ___ 3. communicating junction     (c) desmosome

    ___ 4. made up of connexons, which permit passage of ions and small molecules between cells

    ___ 5. consists of interconnecting fibres, which spot-rivet adjacent cells

    ___ 6. formed by an actual fusion of proteins on the outer surfaces of two interacting cells

    ___ 7. important in tissues subject to mechanical stretching

    ___ 8. important in synchronizing contractions within heart and smooth muscle by allowing spread of electrical activity between the cells composing the muscle mass

    ___ 9. important in preventing passage between cells in epithelial sheets that separate compartments of two different chemical compositions

12. Through its unequal pumping, the $Na^+ - K^+$ pump is directly responsible for separating sufficient charges to establish a resting membrane potential of $-70$ mV. (*True or false?*)

13. At resting membrane potential, there is a slight excess of negative charges on the inside of the membrane, with a corresponding slight excess of positive charges on the outside. (*True or False?*)

14. Following an action potential, there is more $K^+$ outside the cell than inside because of the efflux of $K^+$ during the falling phase. (*True or false?*)

15. Postsynaptic neurons can either excite or inhibit presynaptic neurons. (*True or false?*)

16. The one-way propagation of action potentials away from the original site of activation is ensured by the _____.

17. The _____ is the site of action potential initiation in most neurons because it has the lowest threshold.

18. A junction in which electrical activity in one neuron influences the electrical activity in another neuron by means of a neurotransmitter is called a _____.

19. Summing of EPSPs occurring very close together in time as a result of repetitive firing of a single presynaptic input is known as _____.

20. Using the answer code on the right, indicate which potential is being described:

_____ 1. behaves in all-or-none fashion      (a) graded potential

_____ 2. magnitude of the potential change varies with the magnitude of the triggering response

         (b) action potential

_____ 3. decremental spread away from the original site

_____ 4. spreads throughout the membrane in undiminishing fashion

_____ 5. serves as a long-distance signal

_____ 6. serves as a short-distance signal

## Written Questions

1. Compare exocytosis and endocytosis.

2. Define *secretion, pinocytosis, receptor-mediated endocytosis,* and *phagocytosis.*

3. Describe the structure of mitochondria, and explain their role in oxidative phosphorylation.

4. Describe the fluid mosaic model of membrane structure.

5. What are the functions of the three major types of protein fibres in the extracellular matrix?

6. What two properties of a particle influence whether it can permeate the plasma membrane?

7. List and describe the methods of membrane transport. Indicate what types of substances are transported by each method, and state whether each is a passive or active means of transport.

8. List and describe the types of intercellular communication.

9. Define the term *signal transduction.*

10. Distinguish between a first messenger and a second messenger.

11. What are the two types of excitable tissue?

12. Define the following terms: *polarization, depolarization, hyperpolarization, repolarization, resting membrane potential, threshold potential, action potential, refractory period,* and *all-or-none law.*

13. Describe the permeability changes and ion fluxes that occur during an action potential.

14. Compare the events that occur at excitatory and inhibitory synapses.

15. Discuss the possible outcomes of the grand postsynaptic potential brought about by interactions between EPSPs and IPSPs.

16. Discuss the possible outcomes of the grand postsynaptic potential brought about by interactions between EPSPs and IPSPs.

## Quantitative Exercises (Solutions on p. A-35 and A-36)

(See Appendix D, Principles of Quantitative Reasoning)

1. Each "turn" of the Krebs cycle
   a. generates 3 $NAD^+$, 1 FADH, and 2 $CO_2$
   b. generates 1 GTP, 2$CO_2$, and 1 $FADH_2$
   c. consumes 1 pyruvate and 1 oxaloacetate
   d. consumes an amino acid

2. Consider how much ATP you synthesize in a day. Assume that you consume 1 mol of $O_2$ per hour or 24 mol/day (a mole is the number of grams of a chemical equal to its molecular weight). About 6 mol of ATP are produced per mole of $O_2$ consumed. The molecular weight of ATP is 507. How many grams of ATP do you produce per day at this rate? Given that 1000 g equal 2.2 lbs., how many pounds of ATP do you produce per day at this rate? (This is under relatively inactive conditions!)

3. Under resting circumstances a person produces about 144 mol of ATP per day (73 000 g ATP/day). The amount of *free energy* represented by this amount of ATP can be calculated as follows: cleavage of the terminal phosphate bond from ATP results in a decrease of free energy by approximately 7300 cal/mol. This is a crude measure of the energy available to do work that is contained in the terminal phosphate bond of the ATP molecule. How many calories, in the form of ATP, are produced per day by a resting individual, crudely speaking?

4. Calculate the number of cells in the body of an average 68 kg (150 lb.) adult. (This will only be accurate to about 1 part in 10 but should give you an idea how this commonly quoted number is arrived at.) Assume all cells are spheres 20 μm in diameter. The volume of a sphere can be determined by the equation. (*Hint:* We know that about two-thirds of the water in the body is intracellular, and the density of cells is nearly 1 g/mL. The proportion of the mass made up of water is about 60 percent.)

5. If sucrose is injected into the bloodstream, it tends to stay out of the cells (cells do not use sucrose directly). If it doesn't go into cells, where does it go? In other words, how much "space" is in the body that is not inside some cell? Sucrose can be used to determine this space. Suppose 150 mg of sucrose is injected into a 55 kg woman. If the concentration of sucrose in her blood is 0.015 mg/mL, what is the volume of her extracellular space, assuming that no metabolism is occurring and that the blood sucrose concentration is equal to the sucrose concentration throughout the extracellular space?

6. When using the Nernst equation for an ion that has a valence other than 1, you must divide the potential by the valence. Thus, for $Ca^{2+}$ the Nernst equation becomes

$$E = \frac{61mV}{z} \log \frac{C_o}{C_i}$$

Use this equation to calculate the Nernst (equilibrium) potentials from the following sets of data:
a. Given $[Ca^{2+}]_o = 1$ mM, $[Ca^{2+}]_i = 100$ nM, find $E_{Ca^+}$,
b. Given $[Cl-]_o = 110$ mM, $[Cl-]_i = 10$ mM, find $E_{Cl^-}$,

7. One of the important uses of the Nernst equation is in describing the flow of ions across cell membranes. Ions move under the influence of two forces: the concentration gradient (given in electrical units by the Nernst equation), and the electrical gradient (given by the membrane voltage). This is summarized by *Ohm's law*, as follows:

$$l_x = G_x (V_m - E_x)$$

This equation describes the movement of ion x across the membrane. $I$ is the current in amperes (A); $G$ is the conductance, a measure of the permeability of x, in Siemens (S), which is $\Delta/I\Delta V$; $V_m$ is the membrane voltage; and $E_x$ is the equilibrium potential of ion x. Not only does this equation tell us how large the current is but it also tells us in what direction the current is flowing. By convention, a negative value of the current represents either a positive ion entering the cell or a negative ion leaving the cell. The opposite is true of a positive value of the current.
a. Using the following information, calculate the magnitude of $I_{Na^+}$

$$[Na^+]_o = 145 \text{ mM}, [Na^+]_i = 15 \text{ mM}$$
$$G_{Na^+} = 1 \text{ns}, \quad V_m = -70 \text{ mV}$$

b. Is $Na^+$ entering or leaving the cell?
c. Is $Na^+$ moving with or against the concentration gradient?
d. Is it moving with or against the electrical gradient?

8. Another important use of the Nernst equation is in determining the resting membrane potential of a cell. The cell resting membrane potential is a weighted average of the equilibrium potentials of all permeant ions. The weighting factor is the relative permeability (conductance) to that ion. For a cell permeable only to $Na^+$ and $K^+$, this equation is

$$V_m = \left\{ G_{Na^+} / G_T \right\} E_{Na^+} + \left\{ G_{K^+} / G_T \right\} E_{K^+}$$

In this equation, $G_T$ is the total conductance (in this case,
a. Given the following information, calculate $V_m$:

$$G_{Na^+} = 1 \text{ nS}, G_{K^+} = 5.3 \text{ nS}, E_{Na^+} = 59.1 \text{ mV}$$
$$E_{K^+} = -9.44 \text{mV}$$

b. What would happen to $V_m$ [$K^+$]$_o$ were increased to 1 50 mM?

---

## POINTS TO PONDER

(Explanations on pp. A-37)

1. The poison *cyanide* acts by binding irreversibly to one component of the electron transport chain, blocking its action. As a result, the entire electron-transport process comes to a screeching halt, and the cells lose more than 94 percent of their ATP-producing capacity. Considering the types of cell activities that depend on energy expenditure, what would be the consequences of cyanide poisoning?

2. Why do you think a person is able to only briefly perform anaerobic exercise (such as lifting and holding a heavy weight) but can sustain aerobic exercise (such as walking or swimming) for long periods? (*Hint*: Muscles have limited energy stores.)

3. Which of the following methods of transport is being used to transfer the substance into the cell in ❭ Figure 2-49?
a. diffusion down a concentration gradient
b. osmosis
c. facilitated diffusion
d. active transport
e. vesicular transport
f. It is impossible to tell from the information provided.

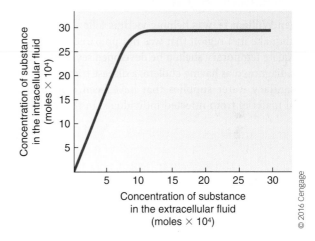

❭ FIGURE 2-49 **Concentration of substance**

4. Colostrum, the first milk that a mother produces, contains an abundance of antibodies, large protein molecules. These maternal antibodies help protect breastfed infants from infections until they are capable of producing their own antibodies. By what means do

you suspect these maternal antibodies are transported across the cells lining a newborn's digestive tract into the bloodstream?

5. Which of the following would occur if a neuron were experimentally stimulated simultaneously at both ends?
   a. The action potentials would pass in the middle and travel to the opposite ends.
   b. The action potentials would meet in the middle and then be propagated back to their starting positions.
   c. The action potentials would stop as they meet in the middle.
   d. The stronger action potential would override the weaker action potential.
   e. Summation would occur when the action potentials meet in the middle, resulting in a larger action potential.

6. Compare the expected changes in membrane potential of a neuron stimulated with a *subthreshold stimulus* (a stimulus not sufficient to bring a membrane to threshold), a *threshold stimulus* (a stimulus just sufficient to bring the membrane to threshold), and a *suprathreshold stimulus* (a stimulus larger than that necessary to bring the membrane to threshold).

Assume you touched a hot stove with your finger. Contraction of the biceps muscle causes flexion (bending) of the elbow, whereas contraction of the triceps muscle causes extension (straightening) of the elbow. What pattern of postsynaptic potentials (EPSPs and IPSPs) do you expect would be initiated as a reflex in the cell bodies of the neurons controlling these muscles to pull your hand away from the painful stimulus?

7. Now assume your finger is being pricked to obtain a blood sample. The same *withdrawal reflex* would be initiated. What pattern of postsynaptic potentials would you voluntarily produce in the neurons controlling the biceps and triceps to keep your arm extended in spite of the painful stimulus?

8. Assume presynaptic excitatory neuron A terminates on a postsynaptic cell near the axon hillock and presynaptic excitatory neuron B terminates on the same postsynaptic cell on a dendrite located on the side of the cell body opposite the axon hillock. Explain why rapid firing of presynaptic neuron A could bring the postsynaptic neuron to threshold through temporal summation, thus initiating an action potential, whereas firing of presynaptic neuron B at the same frequency and the same magnitude of EPSPs may not bring the postsynaptic neuron to threshold.

## CLINICAL CONSIDERATION

### (Explanation on p. A-37)

When William H. was helping victims following a devastating earthquake in a region that was not prepared to swiftly set up adequate temporary shelter, he developed severe diarrhoea. He was diagnosed as having cholera, a disease transmitted through unsanitary water supplies that have been contaminated by fecal material from infected individuals. In this condition, the toxin produced by the cholera bacteria leads to opening of the $Cl^-$ channels in the luminal membranes of the intestinal cells, thereby increasing the secretion of $Cl^-$ from the cells into the intestinal tract lumen. By what mechanisms would $Na^+$ and water be secreted into the lumen in accompaniment with $Cl^-$ secretion? How does this secretory response account for the severe diarrhoea that is characteristic of cholera?

## Body Systems

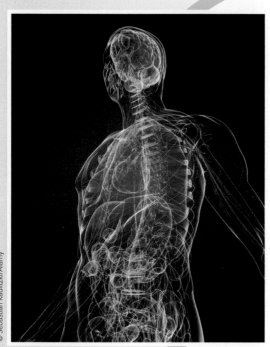

Central nervous system

Body systems maintain homeostasis

### Homeostasis

To maintain homeostasis, cells must work together in a coordinated fashion toward common goals. The two major regulatory systems of the body that help ensure life-sustaining coordinated responses are the nervous and endocrine systems.

Homeostasis is essential for survival of cells

Cells make up body systems

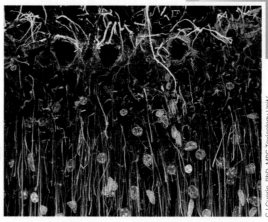

Cerebellum tissue

To maintain homeostasis, cells must work together in a coordinated fashion toward a common purpose.

The nervous system is one of the two major regulatory systems of the body; the other is the endocrine system. A complex interactive network of three basic functional types of nerve cells—afferent neurons, efferent neurons, and interneurons—constitutes the excitable cells of the nervous system.

The central nervous system (CNS) consists of the brain and spinal cord, which receive input about the external and internal environments from the afferent neurons. The CNS sorts and processes this input, then transmits appropriate directions to the efferent neurons, which carry the instructions to glands or muscles to bring about the desired response—some type of secretion or movement. Many of these neurally controlled activities are directed toward maintaining homeostasis. In general, the nervous system acts by means of its electrical signals (action potentials) to control the rapid responses of the body.

# 3

# The Central Nervous System

The **nervous system** is a complex system with two primary divisions, the central nervous system and the peripheral nervous system. The central nervous system is the focus of this chapter and builds on the principal function of the nervous system, communication, which was introduced in Chapter 2.

**▌Clinical Connections**

Jackie is a 40-year-old Caucasian woman. She is married, has two children, full-time employment, and enjoys a very active lifestyle. Over the past several months she has progressively been feeling more tired and occasionally noticed that she doesn't seem as strong as she once did. At work, she also noticed that she just didn't seem to be as mentally sharp as she used to be and started having some vision problems. She attributed these symptoms to just getting older. Her symptoms gradually worsened and started to include occasional loss of bladder control. Not until her family noticed her speech was slurred did she seek medical help. Fearing that she may have had a stroke, she went to her local emergency room.

She underwent a medical resonance imaging (MRI) test, and the results showed that although she did not have a stroke, there were several areas of scar tissue in her brain. After additional MRIs were done, more of these lesions were found in her spinal cord. Then a sample of cerebrospinal fluid was taken, and she was diagnosed with multiple sclerosis, an immune disease that affects the central nervous system.

# 3.1 Organization of the Nervous System

## The central nervous system and the peripheral nervous system

The nervous system is organized into the **central nervous system (CNS)**, consisting of the brain and spinal cord, and the **peripheral nervous system (PNS)**, consisting of nerve fibres that carry information between the CNS and other parts of the body (the periphery) (⟩ Figure 3-1).

The PNS is further subdivided into afferent and efferent divisions. The **afferent division** carries information to the CNS, informing it of changes in the external environment (e.g., ambient temperature) and internal environment (e.g., hunger). The **efferent division** communicates instructions from the CNS to **effector organs**—the muscles or glands that carry out the instructions to bring about the desired effect. The efferent nervous system has two subdivisions: the somatic and autonomic nervous systems. The **somatic nervous system** consists of the nerve fibres of the motor neurons that supply the skeletal muscles.

The **autonomic nervous system (ANS)** consists of nerve fibres that innervate smooth muscle, cardiac muscle, and glands. The ANS is further subdivided into the **sympathetic nervous system** and the **parasympathetic nervous system**, both of which innervate most of the organs supplied by the autonomic system.

Recognize that all these types of nervous systems refer to subdivisions of a single integrated system. Each subdivision is based on differences in the structure, location, and functions of the various diverse parts of the whole nervous system.

## Three functional classes of neurons

Three functional classes of neurons make up the nervous system: *afferent neurons, efferent neurons,* and *interneurons.* The afferent division of the peripheral nervous system consists of afferent neurons, which are shaped quite differently from efferent neurons and interneurons (⟩ Figure 3-2). At its peripheral ending, a typical afferent neuron has a **sensory receptor** that generates action potentials in response to a particular type of stimulus (e.g., touch). (Do not confuse this type of stimulus-sensitive afferent receptor with the protein receptors on all cell surfaces that bind chemical

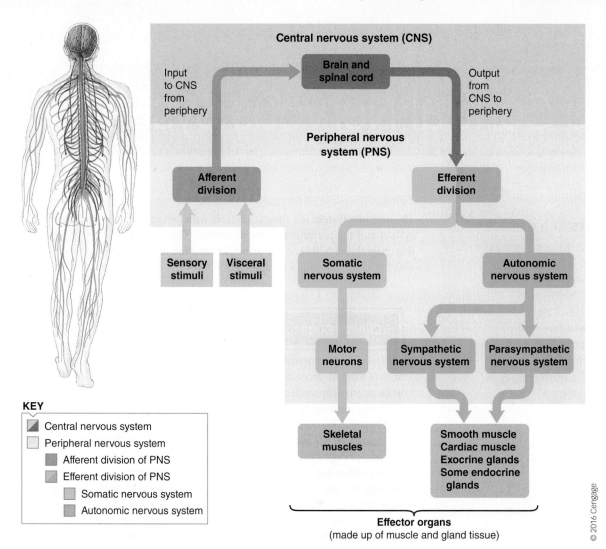

**KEY**
- ▨ Central nervous system
- ☐ Peripheral nervous system
- ▮ Afferent division of PNS
- ▨ Efferent division of PNS
- ☐ Somatic nervous system
- ▮ Autonomic nervous system

⟩ **FIGURE 3-1 Organization of the nervous system**

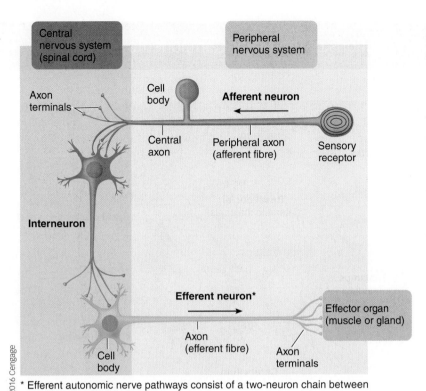

**Central nervous system (spinal cord)**

**Peripheral nervous system**

Axon terminals

Cell body

**Afferent neuron**

Central axon

Peripheral axon (afferent fibre)

Sensory receptor

**Interneuron**

**Efferent neuron***

Effector organ (muscle or gland)

Axon (efferent fibre)

Axon terminals

Cell body

© 2016 Cengage

* Efferent autonomic nerve pathways consist of a two-neuron chain between the CNS and the effector organ.

> **FIGURE 3-2** Structure and location of the three functional classes of neurons

messengers.) The afferent neuron cell body, which is devoid of dendrites and presynaptic inputs, is located adjacent to the spinal cord. A long *peripheral axon*, commonly called the *afferent fibre*, extends from the receptor to the cell body, and a short *central axon* passes from the cell body into the spinal cord. Action potentials are initiated at the receptor (distal end) of the peripheral axon in response to a stimulus and are propagated along the peripheral axon and central axon toward the spinal cord (proximal end). The terminals of the central axon diverge and synapse with other neurons within the spinal cord, thus disseminating information about the stimulus. Afferent neurons lie primarily within the PNS, and only a small portion of their central axon endings project into the spinal cord and relay peripheral signals.

**Efferent neurons** also lie primarily in the peripheral nervous system (Figure 3-2). Their cell bodies originate in the CNS, where many centrally located presynaptic inputs converge on them to influence their outputs to the effector organs. Efferent axons (efferent fibres) travel out of the CNS to meet the muscles or glands they innervate. They convey their integrated output for the effector organs to put into effect. (An autonomic nerve pathway consists of a two-neuron chain between the CNS and the effector organ.)

**Interneurons** lie entirely within the CNS. About 99 percent of all neurons belong to this category. The human CNS is estimated to have more than 100 billion interneurons! These neurons serve two main roles. First, as their name implies, they lie between the afferent and efferent neurons (connectors) and are important in integrating peripheral responses to peripheral information (*inter* means "between"). For example, on receiving information through afferent neurons that you are touching a hot object,

appropriate interneurons signal efferent neurons that transmit to your hand and arm muscles the message "Pull the hand away from the hot object!" The more complex the required action, the greater the number of interneurons interposed between the afferent message and the efferent response. Second, interconnections between interneurons themselves are responsible for abstract phenomena associated with the mind, such as thoughts, emotions, memory, creativity, intellect, and motivation. These activities are the least understood functions of the nervous system.

With this brief introduction to the types of neurons and their location in the various divisions of the nervous system, we now turn our attention to the CNS.

## 3.2 | Structure and Function of the Central Nervous System

The central nervous system consists of the brain and spinal cord. The estimated 100 billion neurons in your brain are assembled into complex networks that enable you to (1) subconsciously regulate your internal environment by neural means, (2) experience emotions, (3) voluntarily control your movements, (4) perceive (be consciously aware of) your own body and your surroundings, and (5) engage in other higher cognitive processes, such as thought and memory. The term **cognition** refers to the act or process of knowing, including both awareness and judgment.

No part of the brain acts in isolation from other brain regions, because networks of neurons are anatomically linked by synapses, and neurons throughout the brain communicate extensively with each other by electrical and chemical means. However, neurons that work together to ultimately accomplish a given function tend to be organized within a discrete location. Thus, although the brain functions as a whole, it is organized into different regions. The parts of the brain can be arbitrarily grouped in various ways based on anatomical distinctions, functional specialization, and evolutionary development. In this text we use the following grouping (see ▌ Table 3-1):

1. Brain stem
2. Cerebellum
3. Forebrain
a. Diencephalon
   (1) Hypothalamus
   (2) Thalamus
b. Cerebrum
   (1) Basal ganglia
   (2) Cerebral cortex

The Central Nervous System **93**

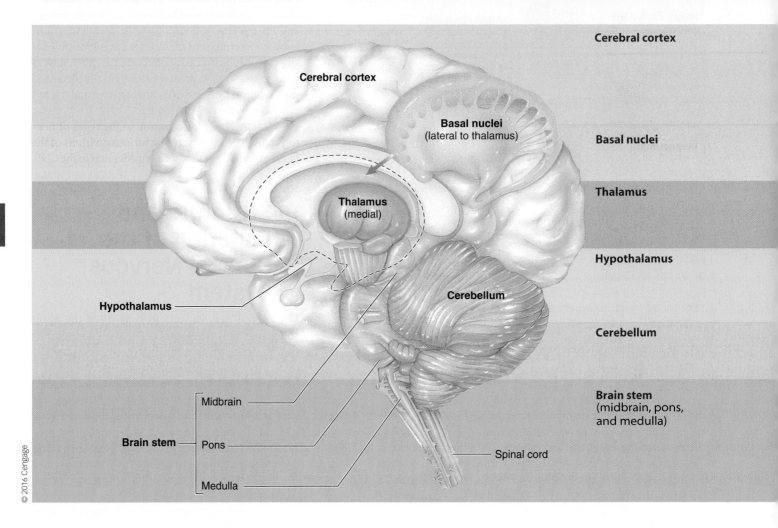

© 2016 Cengage

The order in which these components are listed generally represents both their anatomic location (from bottom to top) and their complexity and sophistication of function (from the least specialized, oldest level to the newest, most specialized level).

A primitive nervous system consists of comparatively few interneurons interspersed between afferent and efferent neurons. During evolutionary development, the interneuronal component progressively expanded, formed more complex interconnections, and became localized at the head end of the nervous system, forming the brain. Newer, more sophisticated layers of the brain were added on to the older, more primitive layers. The human brain represents the present peak of development.

The *brain stem*, the oldest region of the brain, is continuous with the spinal cord (Table 3-1; and see Figure 3-21b). It consists of the midbrain, pons, and medulla. The brain stem controls many of the life-sustaining processes, such as respiration, circulation, and digestion, which are common to many of the lower vertebrate forms. These processes are often referred to as *vegetative functions*, meaning functions performed unconsciously, as

the person has no awareness or control. When a person loses higher brain functions, these lower brain levels—in accompaniment with appropriate supportive therapy, such as adequate nourishment—can still sustain the functions essential for survival.

Attached to the top rear portion of the brain stem is the *cerebellum*, which is concerned with maintaining proper position of the body in space (proprioception; p. 117) and subconscious coordination of motor activity (movement). The cerebellum also plays a key role in learning skilled motor tasks, such as a dance routine.

On top of the brain stem, tucked within the interior of the cerebrum, is the *diencephalon*. It houses two brain components: the *hypothalamus*, which controls many homeostatic functions important in maintaining stability of the internal environment; and the *thalamus*, which performs some primitive sensory processing.

On top of the lower brain regions is the **cerebrum**. The cerebrum is progressively larger and more highly convoluted (i.e., it has tortuous ridges delineated by deep grooves or folds) in the

## Major Functions

1. Sensory perception
2. Voluntary control of movement
3. Language
4. Personality traits
5. Sophisticated mental events, such as thinking, memory, decision making, creativity, and self-consciousness

1. Inhibition of muscle tone
2. Coordination of slow, sustained movements
3. Suppression of useless patterns of movement

1. Relay station for all synaptic input
2. Crude awareness of sensation
3. Some degree of consciousness
4. Role in motor control

1. Regulation of many homeostatic functions, such as temperature control, thirst, urine output, and food intake
2. Important link between nervous and endocrine systems
3. Extensive involvement with emotion and basic behavioural patterns
4. Role in sleep–wake cycle

1. Maintenance of balance
2. Enhancement of muscle tone
3. Coordination and planning of skilled voluntary muscle activity

1. Origin of majority of peripheral cranial nerves (midbrain, pons, and medulla)
2. Cardiovascular, respiratory, and digestive control centres
3. Regulation of muscle reflexes involved with equilibrium and posture
4. Reception and integration of all synaptic input from spinal cord; arousal and activation of cerebral cortex
5. Role in sleep–wake cycle

more advanced vertebrate species. The cerebrum is most highly developed in humans, where it constitutes about 80 percent of the total brain weight. The outer layer of the cerebrum is the highly convoluted *cerebral cortex*, which caps an inner core that houses the *basal ganglia*. The myriad convolutions of the human cerebral cortex give it the appearance of a much-folded walnut (see Figure 3-21a). In many lower mammals, the cortex is perfectly smooth. Without these surface wrinkles, the human cortex would take up to three times the area it does, and, accordingly, would not fit like a cover over the underlying structures. The increased neural circuitry housed in the extra cerebral cortical area not found in lower species is responsible for many of our unique human abilities. The cerebral cortex plays a key role in the most sophisticated neural functions, such as voluntary initiation of movement, fine sensory perception, conscious thought, language, personality traits, and other factors we associate with the mind or intellect. It is the highest, most complex, integrating area of the brain.

Before we describe each region of the CNS in detail, we are going to look at how the brain is protected and nourished.

## 3.3 | Protection of the Central Nervous System

### Glial cells

About 90 percent of the cells within the CNS are not neurons but **glial cells**, or **neuroglia**. Despite their large numbers, glial cells occupy only about half the volume of the brain, because they do not branch as extensively as neurons do.

Unlike neurons, glial cells do not initiate or conduct nerve impulses. However, they do communicate with neurons and among themselves by means of chemical signals, and they maintain homeostasis within the nervous system. *Glia* means "glue," and since the discovery of glial cells in the 19th century, scientists thought of these cells as passive "mortar" that physically supported the functionally important neurons. In the last decade the roles of these dynamic cells have become apparent. Glial cells serve as the connective tissue of the CNS and, as such, help support the neurons both physically and metabolically. They homeostatically maintain the composition of the specialized extracellular environment surrounding the neurons within the narrow limits optimal for normal neuronal function. Furthermore, they actively modulate synaptic function and are now considered nearly as important as neurons to learning and memory. We now look at the specific roles of the four major types of glial cells in the CNS: *astrocytes, oligodendrocytes, microglia, and ependymal cells* (❯ Figure 3-3 and ▮ Table 3-2).

### Astrocytes

Named for their starlike shape (*astro* means "star"; *cyte* means "cell") (❯ Figure 3-4), **astrocytes** are the most abundant glial cells. They fulfill seven critical functions:

1. As the main connective tissue of the CNS, astrocytes hold the neurons together in proper spatial relationships.

2. Astrocytes serve as a scaffold to guide neurons to their proper final destination during fetal brain development.

3. These glial cells induce the small blood vessels of the brain to undergo the anatomic and functional changes that are responsible for establishing the blood–brain barrier, a highly selective barricade between the blood and brain that we will soon describe in greater detail.

4. Astrocytes are important in the repair of brain injuries and in neural scar formation.

5. They play a role in neurotransmitter activity. Astrocytes take up and degrade glutamate and gamma-aminobutyric acid (GABA), excitatory and inhibitory neurotransmitters, respectively, thus bringing the actions of these chemical messengers to a halt.

6. Astrocytes take up excess K$^+$ from the brain ECF when high action potential activity outpaces the ability of the K$^+$ pump to return the effluxed K$^+$ to the neurons. (Recall that K$^+$ leaves a neuron during the falling phase of an action potential; p. 63) By taking up excess K$^+$, astrocytes help maintain the proper brain–ECF ion concentration to sustain

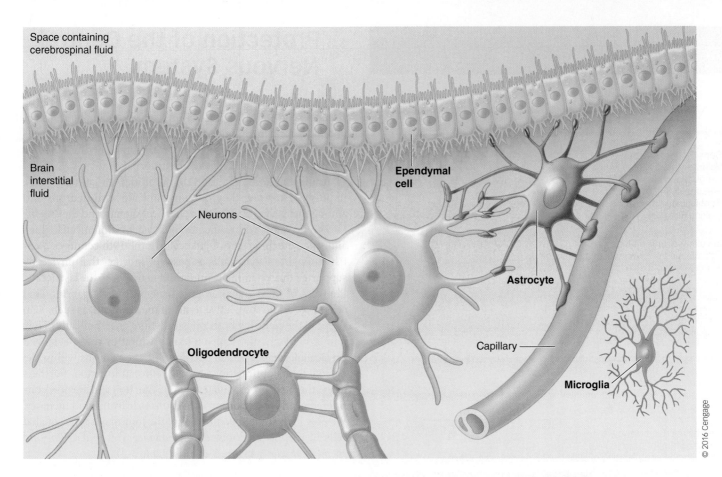

Space containing cerebrospinal fluid

Brain interstitial fluid

Neurons

Ependymal cell

Astrocyte

Oligodendrocyte

Capillary

Microglia

© 2016 Cengage

> **FIGURE 3-3** **Glial cells of the central nervous system.** The glial cells include the astrocytes, oligodendrocytes, microglia, and ependymal cells.

## ▌ TABLE 3-2 Functions of Glial Cells

| Type of Glial Cell | Functions |
|---|---|
| **Astrocytes** | Physically support neurons in proper spatial relationships |
| | Serve as a scaffold during fetal brain development |
| | Induce formation of blood–brain barrier |
| | Form neural scar tissue |
| | Take up and degrade released neurotransmitters into raw materials for synthesis of more neurotransmitters by neurons |
| | Take up excess $K^+$ to help maintain proper brain–ECF ion concentration and normal neural excitability |
| | Enhance synapse formation and strengthen synaptic transmission via chemical signalling with neurons |
| **Oligodendrocytes** | Form myelin sheaths in CNS |
| **Microglia** | Play a role in defence of brain as phagocytic scavengers |
| **Ependymal cells** | Line internal cavities of brain and spinal cord |
| | Contribute to formation of cerebrospinal fluid |
| | Serve as neural stem cells with the potential to form new neurons and glial cells |

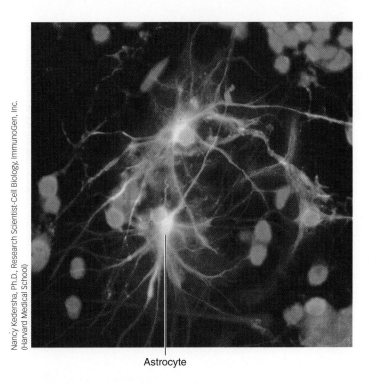

Astrocyte

> **FIGURE 3-4  Astrocytes.** Note the starlike shape of these astrocytes, which have been grown in tissue culture.

normal neural excitability. If $K^+$ levels in the brain ECF were allowed to rise, the resultant lower $K^+$ concentration gradient between the neuronal ICF and surrounding ECF would reduce the neuronal membrane closer to threshold, even at rest. This would increase the excitability of the brain. In fact, an elevation in brain–ECF $K^+$ concentration may be one of the factors responsible for the brain cells' explosive convulsive discharge that occurs during epileptic seizures.

1.  In recent discoveries, astrocytes along with other glial cells are now known to enhance synapse formation and to modify synaptic transmission. Astrocytes communicate with neurons and with one another by means of chemical signals in two ways. First, gap junctions (pp. 34–35) have been identified between the astrocytes themselves and between astrocytes and neurons. Chemical signals can pass directly between cells through these small connecting tunnels without entering the surrounding ECF. Second, astrocytes have receptors for the neurotransmitter glutamate released by neurons. Furthermore, the firing of neurons in the brain in some instances triggers the release of ATP along with glutamate from the axon terminal. Binding of glutamate to an astrocyte's receptors and/or detection of extracellular ATP by the astrocyte leads to $K^+$ influx into this glial cell. The resultant rise in intracellular $K^+$ prompts the astrocyte itself to release ATP, thereby activating adjacent glial cells. In this way, astrocytes share information about action potential activity in a nearby neuron. Thus, astrocytes can communicate among themselves by means of coupling of the astrocytes at gap junctions and by $K^+$ wave propagation. Astrocytes and other glial cells can also release

the same neurotransmitters as neurons do, as well as other chemical signals. These extracellular chemical signals released by glial cells can affect neuronal excitability and strengthen synaptic activity, such as by increasing neuronal release of a neurotransmitter or promoting the formation of new synapses. Glial modulation of synaptic activity is likely important in memory and learning. Scientists are trying to sort out the two-directional chatter that takes place between and among the glial cells and neurons, because this dialogue plays an important role in the processing of information in the brain.

Some neuroscientists believe that synapses should be considered "three-party" junctures involving the glial cells as well as the traditional synapse members, the presynaptic and postsynaptic neurons. This point of view is indicative of the increasingly important role being placed on astrocytes in synapse function.

## Oligodendrocytes

As mentioned in Chapter 2 (p. 70), oligodendrocytes form the insulating myelin sheaths around axons in the CNS. An oligodendrocyte has several elongated projections, each of which is wrapped jelly-roll fashion around a section of an interneuronal axon to form a patch of myelin (see > Figures 2-43b and 3-3).

| **Clinical Connections** | Multiple sclerosis is a chronic disease that leads to nerve damage in the brain, spinal |

cord, and optic nerves. It is an immune disease and causes inflammation that destroys the myelin of neurons and leaves the scar tissue that can be seen on an MRI. Recall from Chapter 2 that signal conduction in nerves is affected by myelin. The pattern of demyelination is not the same in all patients. Demyelination can occur in different nerves, and this leads to the variable symptoms of multiple sclerosis.

## Microglia

**Microglia** are the immune cells of the CNS. These scavengers are "cousins" of monocytes, a type of white blood cell that leaves the blood and sets up residence as a front-line defence agent (i.e., has innate immunity) in various tissues throughout the body. Microglia are derived from bone marrow tissue that gives rise to monocytes. During embryonic development, microglia migrate to the CNS, where they remain stationary until activated by an infection or injury.

In the resting state, microglia are wispy cells with many long branches that radiate outward. Resting microglia don't just watch and wait until activated, they also release low levels of growth factors, such as *nerve growth factor*, that help neurons and other glial cells survive and thrive. When trouble occurs in the CNS, microglia retract their branches, round up, and become highly mobile, moving toward the affected area to remove any foreign invaders or tissue debris. Activated microglia release destructive chemicals for assault against their target.

 Researchers increasingly suspect that excessive release of these chemicals from overzealous microglia may damage the neurons they are meant to protect, thereby contributing to the insidious neuronal damage seen in stroke, Alzheimer's disease, multiple sclerosis, dementia (mental failing) in AIDS patients, and other *neurodegenerative diseases.*

## Ependymal cells

**Ependymal cells** line the internal, fluid-filled cavities of the CNS. As the nervous system develops embryonically from a hollow neural tube, the original central cavity of this tube is maintained and modified to form the ventricles and central canal. The **ventricles**, which consist of four interconnected chambers within the interior of the brain, are continuous with the narrow, hollow **central canal** that tunnels through the middle of the spinal cord (❯ Figure 3-5). The ependymal cells lining the ventricles help form the cerebrospinal fluid that surrounds and cushions the brain and spinal cord (a topic to be discussed shortly). Ependymal cells are one of the few cell types to bear cilia: short, brushlike projections from the cell surface, which are continuous with the cell's surface membrane and beat rhythmically back and forth. Microtubules inside the cilia provide the structure as well as the rhythmic movement driven by the protein dynein. Beating of ependymal cilia contributes to the flow of cerebrospinal fluid throughout the ventricles.

New research has identified a different role for the ependymal cells. According to Dr. Jonas Frisén and colleagues, these cells serve as neural stem cells with the potential of forming not only other glial cells but new neurons as well. The traditional view has long held that new neurons are not produced in the mature brain. Then, in the late 1990s, scientists discovered that new neurons are produced in one restricted site: namely, a specific part of the hippocampus, a structure important for learning and memory (p. 126). Neurons in the rest of the brain are considered irreplaceable. But the discovery that ependymal cells are precursors for new neurons suggests that the adult brain has more potential for repairing damaged regions than previously assumed. Currently, there is no evidence that the brain spontaneously repairs itself following neuron-losing insults, such as head trauma, strokes, and neurodegenerative disorders. Apparently, most brain regions cannot activate this mechanism for replenishing neurons, probably because the appropriate "cocktail" of supportive chemicals is not present.

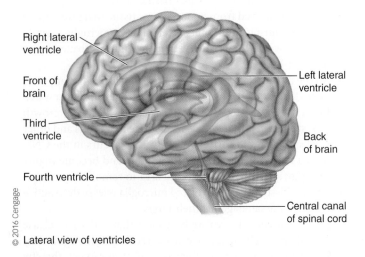

Right lateral ventricle
Front of brain
Third ventricle
Fourth ventricle

Left lateral ventricle
Back of brain
Central canal of spinal cord

Lateral view of ventricles

© 2016 Cengage

❯ **FIGURE 3-5 The ventricles of the brain**

Researchers hope that probing into why these ependymal cells are dormant and how they might be activated will lead to the possibility of unlocking the brain's latent capacity for self-repair.

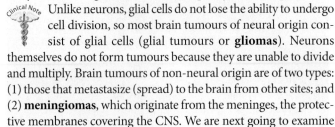

*Clinical Note* Unlike neurons, glial cells do not lose the ability to undergo cell division, so most brain tumours of neural origin consist of glial cells (glial tumours or **gliomas**). Neurons themselves do not form tumours because they are unable to divide and multiply. Brain tumours of non-neural origin are of two types: (1) those that metastasize (spread) to the brain from other sites; and (2) **meningiomas**, which originate from the meninges, the protective membranes covering the CNS. We are next going to examine the meninges and other means by which the CNS is protected.

---

### Check Your Understanding 3.1

1. Draw a flow diagram showing the organization of the subdivisions of the human nervous system.
2. Compare the structure, location, and function of the functional classes of neurons. Do the same for glial cells.

---

## Protection of the delicate central nervous tissue

Central nervous tissue is very delicate. This characteristic, coupled with the fact that damaged nerve cells cannot be replaced, makes it imperative that this fragile, irreplaceable tissue be well protected. Four major features help protect the CNS from injury:

1. Hard, bony structures enclose the CNS; the *cranium* (skull) encases the brain, and the *vertebral column* surrounds the spinal cord.
2. Three protective and nourishing membranes, the *meninges*, lie between the bony covering and the nervous tissue.
3. The brain "floats" in a special cushioning fluid, the *cerebrospinal fluid (CSF).*
4. A highly selective *blood–brain barrier* limits the access of blood-borne materials into the vulnerable brain tissue.

The role of the first of these protective devices, the bony covering, is self-evident. The latter three protective mechanisms warrant further discussion.

## Meningeal membranes: Wrapping, protecting, and nourishing

The **meninges**—three membranes that wrap the CNS—are, from the outermost to the innermost layer, the dura mater, the arachnoid mater, and the pia mater (❯ Figure 3-6). (*Mater* means "mother," indicative of the protective and supportive role of these membranes.)

The **dura mater** is a tough, inelastic covering consisting of two layers (*dura* means "tough"). Usually, these layers adhere closely, but in some regions they are separated to form blood-filled cavities, **dural sinuses**, or, in the case of the larger cavities, **venous sinuses**. Venous blood draining from the brain empties into these sinuses to be returned to the heart. Cerebrospinal fluid also re-enters the blood at one of these sinus sites.

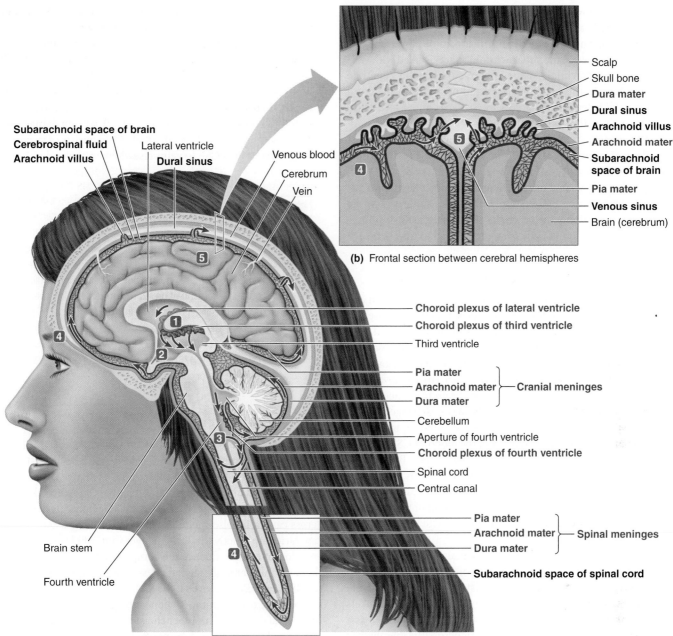

**Subarachnoid space of brain**
**Cerebrospinal fluid**
**Arachnoid villus**

Lateral ventricle

**Dural sinus**

Venous blood

Cerebrum

Vein

Scalp
Skull bone
**Dura mater**
**Dural sinus**
**Arachnoid villus**
Arachnoid mater
**Subarachnoid space of brain**
Pia mater
**Venous sinus**
Brain (cerebrum)

**(b)** Frontal section between cerebral hemispheres

**Choroid plexus of lateral ventricle**
**Choroid plexus of third ventricle**
Third ventricle

Pia mater
Arachnoid mater  } **Cranial meninges**
Dura mater

Cerebellum
Aperture of fourth ventricle
**Choroid plexus of fourth ventricle**
Spinal cord
Central canal

Pia mater
Arachnoid mater  } **Spinal meninges**
Dura mater

**Subarachnoid space of spinal cord**

Brain stem

Fourth ventricle

**(a)** Sagittal section of brain and spinal cord

Cerebrospinal fluid

**1** is produced by the choroid plexuses,

**2** circulates throughout the ventricles,

**3** exits the fourth ventricle at the base of the brain,

**4** flows in the subarachnoid space between the meningeal layers, and

**5** is finally reabsorbed from the subarachnoid space into the venous blood across the arachnoid villi.

⟩ **FIGURE 3-6  Relationship of the meninges and cerebrospinal fluid to the brain and spinal cord.** (a) Brain, spinal cord, and meninges in sagittal section. The arrows and squared numbers with accompanying explanations indicate the direction of flow of cerebrospinal fluid (in yellow). (b) Frontal section in the region between the two cerebral hemispheres of the brain, depicting the meninges in greater detail

The **arachnoid mater** gets its name from the cobweb appearance of this delicate, richly vascularized layer (*arachnoid* means "spider-like"). The space between the arachnoid layer and the underlying pia mater, the **subarachnoid space**, is filled with CSF. Protrusions of arachnoid tissue, the **arachnoid villi**, penetrate through gaps in the overlying dura and project into the dural sinuses. CSF is reabsorbed across the surfaces of these villi into the blood circulating within the sinuses.

The innermost meningeal layer, the **pia mater**, is the most fragile (*pia* means "gentle"). It is highly vascular and closely adheres to the surfaces of the brain and spinal cord, following every ridge and valley. In certain areas it dips deeply into the brain to bring a rich blood supply into close contact with the ependymal cells lining the ventricles. This relationship is important in the formation of CSF, the topic we turn to now.

## Cerebrospinal fluid: A cushion for the brain

**Cerebrospinal fluid (CSF)** surrounds and cushions the brain and spinal cord. The CSF has about the same density as the brain itself, so the brain essentially floats, or is suspended, in this special fluid environment. The major function of CSF is to serve as a shock-absorbing fluid to prevent the brain from bumping against the interior of the hard skull when the head is subjected to sudden, jarring movements.

In addition to protecting the delicate brain from mechanical trauma, the CSF plays an important role in the exchange of materials between the neural cells and the surrounding interstitial fluid. The brain interstitial fluid—not the blood or CSF—comes into direct contact with the neurons and glial cells. Because the brain interstitial fluid directly bathes the neural cells, its composition is critical. The composition of the brain interstitial fluid is influenced more by changes in the composition of the CSF than by alterations in the blood. Accordingly, the CSF's composition must be carefully regulated.

Cerebrospinal fluid is formed primarily by the **choroid plexuses** found in particular regions of the ventricle cavities of the brain. Choroid plexuses consist of richly vascularized, cauliflower-like masses of pia mater tissue that dip into pockets formed by ependymal cells. Cerebrospinal fluid is formed as a result of selective transport mechanisms across the membranes of the choroid plexuses. The composition of CSF differs from that of plasma. For example, CSF is lower in $K^+$ and higher in $K^+$, thus making the brain interstitial fluid an ideal environment for the movement of these ions down concentration gradients, a process essential for conduction of nerve impulses (p. 34).

Once CSF is formed, it flows through the four interconnected ventricles within the interior of the brain and through the spinal cord's narrow central canal, which is continuous with the last ventricle. Cerebrospinal fluid escapes through small openings from the fourth ventricle at the base of the brain, enters the subarachnoid space, and subsequently flows between the meningeal layers over the entire surface of the brain and spinal cord (see ⟩ Figure 3-6). When the CSF reaches the upper regions of the brain, it is reabsorbed from the subarachnoid space into the venous blood through the arachnoid villi.

Flow of CSF through this system is facilitated by ciliary beating along with circulatory and postural factors that result in a CSF pressure of about 10 mmHg. Reduction of this pressure by removal of even a few millilitres of CSF during a spinal tap for laboratory analysis may produce severe headaches.

Through the ongoing processes of formation, circulation, and reabsorption, the entire CSF volume of about 125–150 mL is replaced more than three times a day. If any one of these processes is defective so that excess CSF accumulates, **hydrocephalus** ("water on the brain") occurs. The resulting increase in CSF pressure can lead to brain damage and mental retardation if untreated. Treatment consists of surgically shunting the excess CSF to veins elsewhere in the body.

> **■ Clinical Connections**
>
> In many diseases that affect the CNS, the CSF can sometimes be used for diagnostic testing, which explains why Jackie's CSF was sampled after doctors identified multiple lesions in her brain and spinal cord. In her case, the CSF was clear and colourless and had normal cell counts, indicating there was no active inflammation, as might be seen in other CNS diseases, such as meningitis. However, in the case of multiple sclerosis—although the CSF visually appears normal—the CSF frequently has elevated levels of the antibody IgG, compared to IgG levels in the blood (see Chapter 11, p. 475). This is a sign that a recurring immunological response has occurred.

## The blood–brain barrier

The brain is carefully shielded from harmful changes in the blood by a highly selective **blood–brain barrier (BBB)** consisting of endothelial cells. Throughout the body, exchange of materials between blood and surrounding interstitial fluid can take place only across the walls of capillaries. Unlike the rather free exchange across capillaries elsewhere, permissible exchanges across brain capillaries are strictly limited. Changes in most plasma constituents do not easily influence the composition of brain interstitial fluid, as only carefully regulated exchanges can be made across the BBB. For example, even if the $K^+$ level in the blood is doubled, little change occurs in the $K^+$ concentration of the fluid bathing the central neurons. This is beneficial because alterations in interstitial fluid $K^+$ would be detrimental to neuronal function.

The BBB consists of both anatomic and physiological factors. Capillary walls throughout the body are formed by a single layer of cells. Usually, all plasma components (except the large plasma proteins) can be freely exchanged between the blood and the surrounding interstitial fluid through holes or pores between the cells making up the capillary wall. In brain capillaries, however, the cells are joined by *tight junctions* (p. 34), which completely seal the capillary wall so that nothing can be exchanged across the wall by passing between the cells. The only possible exchanges are through the capillary cells themselves. Lipid-soluble substances, such as oxygen, carbon dioxide, alcohol, and steroid hormones, penetrate these cells easily by dissolving in their lipid plasma membrane. Small water molecules also diffuse through readily, apparently by passing between the phospholipid molecules that compose the plasma membrane. All other substances exchanged between the blood and brain interstitial fluid—including such essential materials as glucose, amino acids, and ions—are transported by highly selective membrane-bound carriers. Therefore, transport across brain capillary walls *between* the wall-forming cells is *anatomically prevented*, and transport *through* the cells is *physiologically restricted*. Together, these mechanisms constitute the BBB.

By strictly limiting exchange between the blood and the brain, the BBB protects the delicate brain from chemical fluctuations in the blood and minimizes the possibility that potentially harmful blood-borne substances might reach the central neural tissue. It further prevents certain circulating hormones that could also act as neurotransmitters from reaching the brain, where they could produce uncontrolled nervous activity. On the negative side, the BBB limits the use of drugs for the treatment of brain and spinal cord disorders, because many drugs are unable to penetrate this barrier.

Brain capillaries are surrounded by astrocyte processes, which at one time were erroneously thought to be physically responsible for the BBB. Scientists now know that astrocytes have two roles regarding the BBB: (1) They signal the cells forming the brain capillaries to "get tight." Capillary cells do not have an inherent ability to form tight junctions; they do so only at the command of a signal within their neural environment. (2) Astrocytes participate in the cross-cellular transport of some substances such as K⁺.

For functional reasons, certain areas of the brain called the **circumventricular organs** are not subject to the BBB. The lack of a BBB in these areas allows these organs to "sample" the blood and adjust their output accordingly to maintain homeostasis. Part of this output is in the form of hormones that must enter the capillaries to be transported to their sites of action. Appropriately, these capillaries are not sealed by tight junctions.

## The role of oxygen and glucose

Even though many substances in the blood never actually come in contact with the brain tissue, the brain, more than any other tissue, is highly dependent on a constant blood supply. Unlike most tissues, which can resort to anaerobic metabolism to produce ATP in the absence of oxygen for at least short periods (p. 28), the brain cannot produce ATP in the absence of oxygen. Scientists recently discovered an oxygen-binding protein, **neuroglobin**, in the brain. This molecule is similar to haemoglobin, the oxygen-carrying protein in red blood cells (p. 436), and is thought to play a key role in oxygen handling in the brain, although its exact function remains to be determined. As well, in contrast to most tissues, which can use other sources of fuel for energy production in lieu of glucose, the brain normally uses only glucose, but does not store any of this nutrient. Therefore, the brain absolutely depends on a continuous, adequate blood supply of oxygen and glucose.

*Clinical Note* Brain damage results if this organ is deprived of its critical oxygen supply for approximately 5 minutes or if its glucose supply is cut off for more than 15 minutes. The most common cause of inadequate blood supply to the brain is a stroke. (See Concepts, Challenges, and Controversies for details.)

## 3.4 | Spinal Cord

The **spinal cord** is a long, slender cylinder of nerve tissue that extends from the brain stem. It is about 45 cm long and 2 cm in diameter (about the size of your thumb), and is protected by the vertebral column.

### The vertebral canal

Exiting through a large hole in the base of the skull, the spinal cord descends through the vertebral canal and is surrounded by the vertebral column (› Figure 3-7). Paired **spinal nerves** emerge from the spinal cord through spaces formed between the bony, winglike arches (pedicle and lamina) of adjacent vertebrae. The spinal nerves are named according to the region of the vertebral column from which they originate (› Figure 3-8): there are eight pairs of *cervical (neck) nerves* (namely C1–C8), twelve *thoracic (chest) nerves*, five *lumbar (abdominal) nerves*, five *sacral (pelvic) nerves*, and one *coccygeal (tailbone) nerve*.

During development, the vertebral column grows about 25 cm longer than the spinal cord. Because of this differential

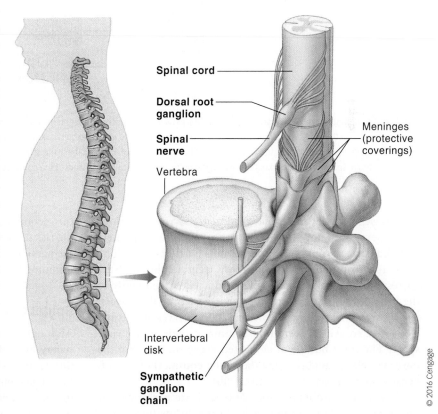

© 2016 Cengage

› **FIGURE 3-7** Location of the spinal cord relative to the vertebral column

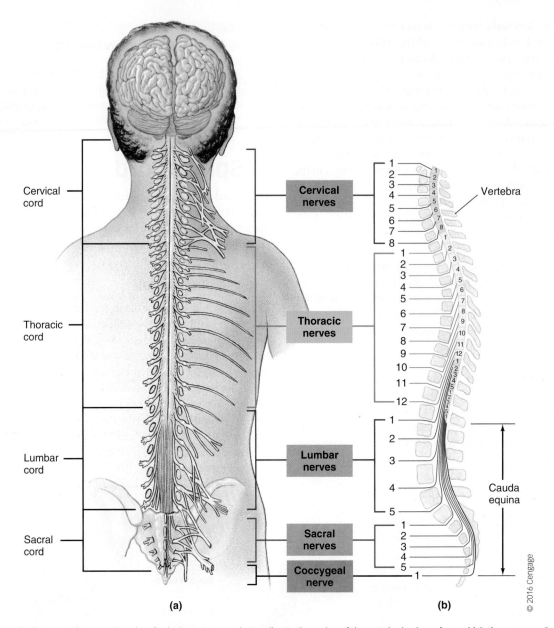

Cervical cord

Thoracic cord

Lumbar cord

Sacral cord

Cervical nerves

Thoracic nerves

Lumbar nerves

Sacral nerves

Coccygeal nerve

Vertebra

Cauda equina

© 2016 Cengage

(a)

(b)

> **FIGURE 3-8 Spinal nerves.** There are 31 pairs of spinal nerves named according to the region of the vertebral column from which they emerge. Because the spinal cord is shorter than the vertebral column, spinal nerve roots must descend along the cord before emerging from the vertebral column at the corresponding interverte-bral space, especially those beyond the level of the first lumbar vertebra (L1). Collectively these rootlets are called the cauda equina, literally, horse's tail. (a) Posterior view of the brain, spinal cord, and spinal nerves (on the right side only). (b) Lateral view of the spinal cord and spinal nerves emerging from the vertebral column

growth, segments of the spinal cord that give rise to various spinal nerves are not aligned with the corresponding inter-vertebral spaces. Most spinal nerve roots must descend along the cord before emerging from the vertebral column at the corresponding space. The spinal cord itself extends only to the level of the first or second lumbar vertebra (about waist level), so the nerve roots of the remaining nerves are greatly elongated, to exit the vertebral column at their appropriate space. The thick bundle of elongated nerve roots within the lower vertebral canal is called the **cauda equina** (horse's tail) because of its appearance (Figure 3-8b).

*Spinal taps* to obtain a sample of CSF are performed by inserting a needle into the vertebral canal below the level of the second lumbar vertebra. Insertion at this site does not run the risk of penetrating the spinal cord. The needle pushes aside the nerve roots of the cauda equina so that a sample of the sur-rounding fluid can be withdrawn safely.

## Spinal cord white matter

Although there are some slight regional variations, the cross-sectional anatomy of the spinal cord is generally the same throughout its length (> Figure 3-9). In contrast to the grey matter forming an outer shell that caps an inner white core in the brain, the grey matter in the spinal cord forms an inner butterfly-shaped region surrounded by the outer white matter. The cord **grey matter** consists primarily of neuronal cell bodies and their dendrites, short interneurons, and glial cells. The **white matter** is organized into tracts, which are bundles of nerve fibres (axons of long interneurons) with a similar function. The bundles are grouped into columns that extend the length of the cord. Each of these tracts begins or ends within a particular area of the brain, and each transmits a specific type of information. Some are **ascending tracts** (cord to brain) that transmit to the brain signals derived from afferent input. Others are **descending tracts** (brain to cord) that relay messages from the brain to efferent neurons (> Figure 3-10).

The tracts are generally named for their origin and termination. For example, the **ventral spinocerebellar tract** is an ascending pathway that originates in the spinal cord, runs up the ventral (toward the front) margin of the cord, with several synapses along the way, and eventually terminates in the cerebellum (> Figure 3-11a). This tract carries information—derived from muscle stretch receptors and delivered to the spinal cord by afferent fibres—for use by the spinocerebellum. In contrast, the **ventral corticospinal tract** is a descending pathway that originates in the motor region of the cerebral cortex, then travels down the ventral portion of the spinal cord, and terminates in the spinal cord on the cell bodies of efferent motor neurons supplying skeletal muscles (> Figure 3-11b). Because various types of signals are carried in different tracts within the spinal cord, damage to particular areas of the cord can interfere with some functions, whereas other functions remain intact.

## Spinal cord grey matter

The centrally located grey matter is also functionally organized (> Figure 3-12). The central canal, which is filled with CSF, lies in the centre of the grey matter. Each half of the grey matter is divided into a **dorsal (posterior) horn**, a **ventral (anterior) horn**, and a **lateral horn**. The dorsal horn contains cell bodies of interneurons on which afferent neurons terminate. The ventral horn contains cell bodies of the efferent motor neurons supplying skeletal muscles. Autonomic nerve fibres supplying cardiac and smooth muscle and exocrine glands originate at cell bodies found in the lateral horn.

## Spinal nerves

Spinal nerves connect with each side of the spinal cord by a **dorsal root** and a **ventral root** (see Figure 3-9). Afferent fibres carrying incoming signals from peripheral receptors enter the spinal cord through the dorsal root. The cell bodies for the afferent neurons at each level are clustered in a **dorsal root ganglion**. (Note that a collection of neuronal cell bodies located outside the CNS is called a *ganglion*, whereas a functional collection of cell bodies within the CNS is referred to as a *centre* or a *nucleus*.) The cell bodies for the efferent neurons originate in the grey matter and send axons out through the ventral root. Therefore, efferent fibres carrying outgoing signals to muscles and glands exit through the ventral root.

The dorsal and ventral roots at each level join to form a *spinal nerve* that emerges from the vertebral column (see Figure 3-9). A spinal nerve contains both afferent and efferent fibres that traverse between a particular region of the body and the spinal cord. Note the relationship between a *nerve* and a *neuron*. A **nerve** is a bundle of peripheral neuronal axons, some afferent and some efferent, enclosed by a covering of connective tissue

and following the same pathway (> Figure 3-13). A nerve does not contain a complete nerve cell, only the axonal portions of many neurons. (By this definition, there are no nerves in the CNS! Bundles of axons in the CNS are called *tracts*.) The individual fibres within a nerve generally do not have any direct influence on one another. They travel together for convenience, just as many individual telephone lines are carried within a telephone cable, yet any particular phone connection can be private without interference or influence from other lines in the cable.

The 31 pairs of spinal nerves, along with the 12 pairs of cranial nerves that arise from the brain, constitute the *peripheral nervous system*. After they emerge, the spinal nerves progressively branch to form a vast network of peripheral nerves that supply the tissues. Each segment of the spinal cord gives rise to a pair of spinal nerves that ultimately supply a particular region of the body with both afferent and efferent fibres. Consequently, the location and extent of sensory and motor deficits associated with spinal cord injuries can be clinically important in determining the level and extent of the cord injury.

> **3**

> **▌Clinical Connections**
>
> In the case of Jackie, a patient with multiple sclerosis, several regions of scarring were seen in the spinal cord, and this provides an example of how sensory and motor defects can be identified. Jackie was experiencing muscle weakness and bladder-control problems that were due to an interruption of the nerve signals. In patients with multiple sclerosis, spinal cord lesions are frequently observed in the cervical, or neck, area, and these affect the corticospinal tract. What is different in multiple sclerosis compared to a trauma that severs the spinal cord is that afferent and efferent nerve signals still occur, but they are impaired.

 With reference to sensory input, each specific region of the body surface supplied by a particular spinal nerve is called a **dermatome**. These same spinal nerves also carry fibres that branch off to supply internal organs, and sometimes pain originating from one of these organs is "referred" to the corresponding dermatome supplied by the same spinal nerve. **Referred pain** originating in the heart, for example, may appear to come from the left shoulder and arm. The mechanism responsible for referred pain is not completely understood. Inputs arising from the heart presumably share a pathway to the brain in common with inputs from the left, upper extremity. The higher perception levels, being more accustomed to receiving sensory input from the left arm than from the heart, may interpret the input from the heart as having arisen from the left arm.

## Basic reflexes

The spinal cord is strategically located between the brain and the afferent and efferent fibres of the peripheral nervous system. This location enables the spinal cord to fulfill its two primary functions: (1) serving as a link for transmission of information between the brain and the remainder of the body, and (2) integrating reflex activity between afferent input and efferent output without involving the brain. This type of reflex activity is called a *spinal reflex*.

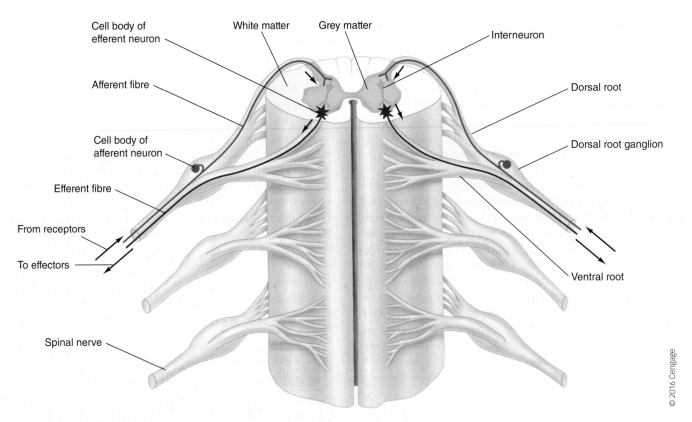

> **FIGURE 3-9 Spinal cord in cross section**. Schematic representation of the spinal cord in cross section showing the relationship between the spinal cord and spinal nerves. The afferent fibres enter through the dorsal root, and the efferent fibres exit through the ventral root. Afferent and efferent fibres are enclosed together within a spinal nerve.

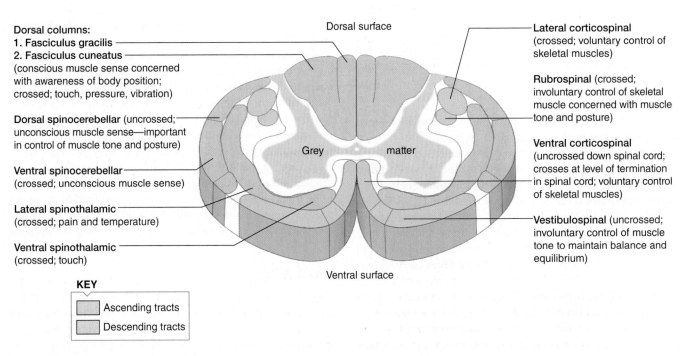

> **FIGURE 3-10 Ascending and descending tracts in the white matter of the spinal cord in cross section**

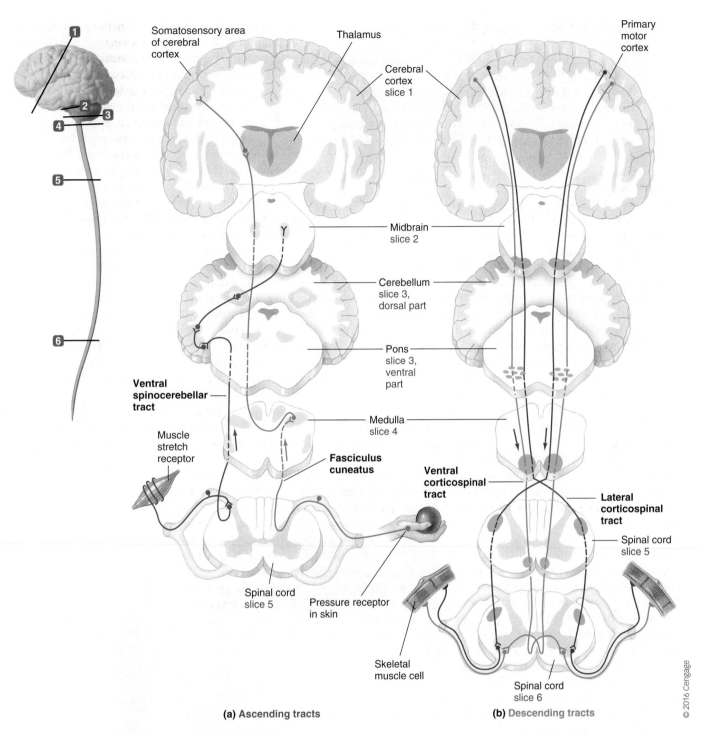

**(a) Ascending tracts**

**(b) Descending tracts**

Somatosensory area of cerebral cortex

Thalamus

Primary motor cortex

Cerebral cortex slice 1

Midbrain slice 2

Cerebellum slice 3, dorsal part

Pons slice 3, ventral part

**Ventral spinocerebellar tract**

Medulla slice 4

Muscle stretch receptor

**Fasciculus cuneatus**

**Ventral corticospinal tract**

**Lateral corticospinal tract**

Spinal cord slice 5

Spinal cord slice 5

Pressure receptor in skin

Skeletal muscle cell

Spinal cord slice 6

© 2016 Cengage

> **FIGURE 3-11 Examples of ascending and descending pathways in the white matter of the spinal cord.** (a) Cord-to-brain pathways of several ascending tracts (fasciculus cuneatus and ventral spinocerebellar tract). (b) Brain-to-cord pathways of several descending tracts (lateral corticospinal and ventral corticospinal tracts)

A **reflex** is any response that occurs automatically without conscious effort and is part of a biological control system that links stimulus and response. There are two types of reflexes: (1) **simple (basic) reflexes**, which are built-in, unlearned responses, such as pulling the hand away from a burning hot object; and (2) **acquired (conditioned) reflexes**, which are a result of practice and learning, such as a pianist striking a particular key upon seeing a certain note on a music staff. The musician reads music and plays it automatically, but only after considerable conscious training effort.

**REFLEX ARC**

The neural pathway involved in accomplishing reflex activity is known as a **reflex arc**, which typically includes five basic components:

1. Receptor
2. Afferent pathway
3. Integrating centre
4. Efferent pathway
5. Effector

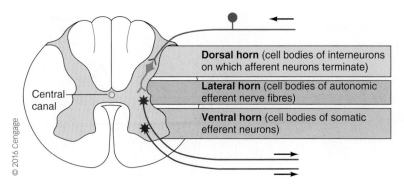

**Dorsal horn** (cell bodies of interneurons on which afferent neurons terminate)

**Lateral horn** (cell bodies of autonomic efferent nerve fibres)

**Ventral horn** (cell bodies of somatic efferent neurons)

Central canal

© 2016 Cengage

> **FIGURE 3-12** Regions of the grey matter

The receptor responds to a **stimulus**, which is a detectable physical or chemical change in the environment of the receptor. In response to the stimulus, the receptor produces an action potential that is relayed by the **afferent pathway** to the **integrating centre** for processing. Usually the integrating centre is the CNS. The spinal cord and brain stem integrate basic reflexes, whereas higher brain levels usually process acquired reflexes. The integrating centre processes all information available to it from this receptor as well as from all other inputs, then makes a decision about the appropriate response. The instructions from the integrating centre are transmitted via the **efferent pathway** to the effector—a muscle or gland—which carries out the desired response. Unlike conscious behaviour, in which any one of a number of responses is possible, a reflex response is predictable, because the pathway between the receptor and the effector is always the same.

### STRETCH REFLEX

The **stretch reflex** involves an afferent neuron originating at a stretch-detecting receptor in a skeletal muscle. This afferent neuron terminates directly on the efferent neuron supplying the same skeletal muscle and, when activated, causes it to counteract the stretch by stimulating the muscle to contract. The stretch reflex is a **monosynaptic reflex** (one-synapse), because the only synapse in the reflex arc is the one between the afferent neuron and the efferent neuron. The withdrawal reflex and all other reflexes are **polysynaptic** (many synapses), because interneurons are interposed in the reflex pathway and, therefore, a number of synapses are involved.

Whenever a whole muscle is passively stretched, its muscle spindle intrafusal fibres are likewise stretched, which increases the firing rate in the afferent nerve fibres whose sensory endings terminate on the stretched spindle fibres. The afferent neuron directly synapses on the alpha motor neuron that innervates the extrafusal fibres of the same muscle, resulting in contraction of that muscle.

This stretch reflex serves as a local negative-feedback mechanism to resist any passive changes in muscle length so that optimal resting length can be maintained.

The classic example of the stretch reflex is the **patellar tendon (knee-jerk) reflex** (Figure 3-14). The extensor muscle of the knee is the *quadriceps femoris*, which forms the anterior (front) portion of the thigh and is attached just below the knee to the tibia (shinbone) by the *patellar tendon*. Tapping this tendon with a rubber mallet passively stretches the quadriceps muscle, activating its spindle receptors. The resulting stretch reflex brings about contraction of this extensor muscle, causing the knee to extend and raise the foreleg in the well-known knee-jerk fashion.

This test is routinely done as a preliminary assessment of nervous system function. A normal knee jerk indicates that a number of neural and muscular components—muscle spindle, afferent input, motor neurons, efferent output, neuromuscular junctions, and the muscles themselves—are functioning normally. It also indicates an appropriate balance of excitatory and inhibitory input to the motor neurons from higher brain levels. Muscle jerks may be absent or depressed with loss of higher-level excitatory inputs, or may be greatly exaggerated with loss of inhibitory input to the motor neurons from higher brain levels.

The primary purpose of the stretch reflex is to resist the tendency for the passive stretch of extensor muscles by gravitational forces when a person is standing upright. Whenever the knee joint

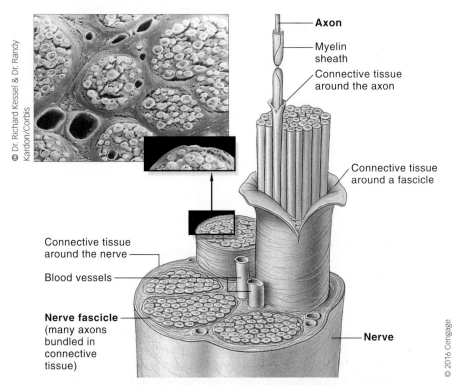

© Dr. Richard Kessel & Dr. Randy Kardon/Corbis

**Axon**

Myelin sheath

Connective tissue around the axon

Connective tissue around a fascicle

Connective tissue around the nerve

Blood vessels

**Nerve fascicle** (many axons bundled in connective tissue)

**Nerve**

© 2016 Cengage

> **FIGURE 3-13 Structure of a nerve.** Diagrammatic view of a nerve, showing neuronal axons (both afferent and efferent fibres) bundled together into connective tissue–wrapped fascicles. A nerve consists of a group of fascicles enclosed by a covering of connective tissue and following the same pathway. The photograph is a scanning electron micrograph of several nerve fascicles in cross section.

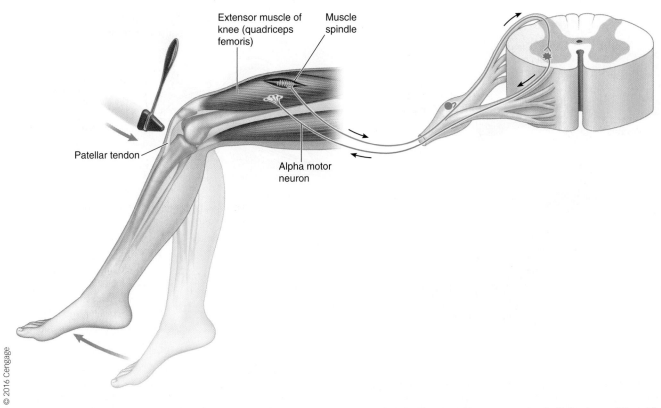

Extensor muscle of
knee (quadriceps
femoris)

Muscle
spindle

Patellar tendon

Alpha motor
neuron

© 2016 Cengage

> **FIGURE 3-14** **Patellar tendon reflex (a stretch reflex).** Tapping the patellar tendon with a rubber mallet stretches the muscle spindles in the quadriceps femoris muscle. The resultant monosynaptic stretch reflex results in contraction of this extensor muscle, causing the characteristic knee-jerk response.

tends to buckle because of gravity, the quadriceps muscle is stretched. The resulting enhanced contraction of this extensor muscle brought about by the stretch reflex quickly straightens out the knee, holding the limb extended so that the person remains standing.

### WITHDRAWAL REFLEX

A basic **spinal reflex** (escape reflex) is one integrated by the spinal cord; that is, all components necessary for linking afferent input to efferent response are present within the spinal cord. The **withdrawal reflex** can serve to illustrate a basic spinal reflex (Figure 3-15). When a person touches a hot stove (or receives a painful stimulus), a withdrawal reflex is initiated to pull the hand away from the stove (to withdraw from the pain). The skin has different receptors for warmth, cold, light touch, pressure, and pain. Even though all information is sent to the CNS by way of action potentials, the CNS can distinguish between various stimuli because different receptors, and, consequently, different afferent pathways are activated by different stimuli. When a receptor is stimulated enough to reach threshold, an action potential is generated in the afferent sensory neuron. The stronger the stimulus, the greater the frequency of action potentials generated and propagated to the CNS. Once the afferent neuron enters the spinal cord, it diverges to synapse with the following different interneurons (the numbers correspond to those in Figure 3-15, step 3).

(3a) An excited afferent neuron stimulates excitatory interneurons that in turn stimulate the efferent motor neurons supplying the biceps, the muscle in the arm that flexes (bends) the elbow joint. By concentrically contracting (shortening muscle fibres, p. 303), the biceps pulls the hand away from the hot stove.

(3b) The afferent neuron also stimulates inhibitory interneurons that in turn inhibit the efferent neurons supplying the triceps to prevent it from contracting. The triceps is the muscle in the arm that extends (straightens out) the elbow joint. When the biceps is contracting to flex the elbow, it would be counterproductive for the triceps to be contracting concentrically. Thus, the triceps muscle relaxes (lengthens), allowing for a smooth, controlled motion. Therefore, built into the withdrawal reflex is inhibition of the muscle that antagonizes (opposes) the desired response. This type of neuronal connection involving stimulation of the nerve supply to one muscle and simultaneous inhibition of the nerves to its antagonistic muscle is known as **reciprocal innervation**.

(3c) The afferent neuron stimulates still other interneurons that carry the signal up the spinal cord to the brain via an ascending pathway. Only when the impulse reaches the sensory area of the cortex is the person aware of the pain, its location, and the type of stimulus. As well, when the impulse reaches the brain, the information can be stored as memory, and the person can start thinking about the situation—how it happened, what to do about it, and so on. All this activity at the conscious level is above and beyond the basic reflex.

As is characteristic of all spinal reflexes, the brain can modify the withdrawal reflex. Impulses may be sent down descending pathways to the efferent motor neurons supplying the involved muscles to override the input from the receptors, actually preventing the biceps from contracting in spite of the painful stimulus. When your finger is pricked to obtain a blood sample, pain receptors are

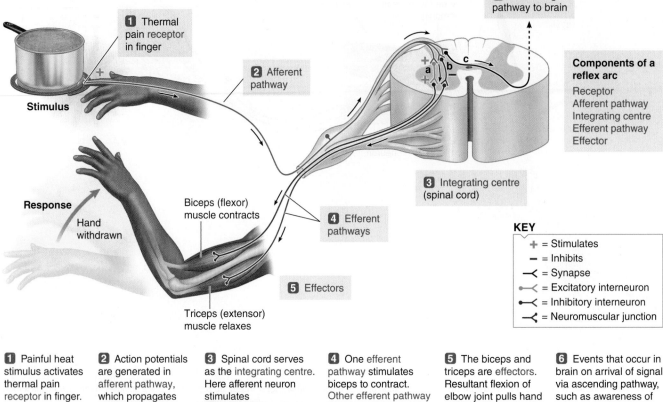

**1** Thermal pain receptor in finger

**2** Afferent pathway

**Stimulus**

**6** Ascending pathway to brain

**Components of a reflex arc**
Receptor
Afferent pathway
Integrating centre
Efferent pathway
Effector

**3** Integrating centre (spinal cord)

**Response**

Hand withdrawn

Biceps (flexor) muscle contracts

**4** Efferent pathways

**5** Effectors

Triceps (extensor) muscle relaxes

**KEY**

| + | = Stimulates |
|---|---|
| − | = Inhibits |
| | = Synapse |
| | = Excitatory interneuron |
| | = Inhibitory interneuron |
| | = Neuromuscular junction |

**1** Painful heat stimulus activates thermal pain receptor in finger.

**2** Action potentials are generated in afferent pathway, which propagates impulses to the spinal cord.

**3** Spinal cord serves as the integrating centre. Here afferent neuron stimulates
**3a** excitatory interneurons, which stimulate motor neurons to biceps;
**3b** inhibitory interneurons, which inhibit motor neurons to triceps;
**3c** interneurons that are part of ascending pathway to brain.

**4** One efferent pathway stimulates biceps to contract. Other efferent pathway leads to relaxation of triceps by preventing counterproductive excitation and contraction of this antagonistic muscle.

**5** The biceps and triceps are effectors. Resultant flexion of elbow joint pulls hand away from painful stimulus. This response completes the withdrawal reflex.

**6** Events that occur in brain on arrival of signal via ascending pathway, such as awareness of pain, memory storage, and so on, are above and beyond reflex arc.

› **FIGURE 3-15 The withdrawal reflex**

stimulated, initiating the withdrawal reflex. Knowing that you must be brave and not pull your hand away, you can consciously override the reflex by sending IPSPs via descending pathways to the motor neurons supplying the biceps and EPSPs to those supplying the triceps. The activity in these efferent neurons depends on the sum of activity of all their synaptic inputs. Because the neurons supplying the biceps are now receiving more IPSPs from the brain (voluntary) than EPSPs from the afferent pain pathway (reflex), these neurons are inhibited and do not reach threshold. Therefore, the biceps is not stimulated to contract and withdraw the hand. Simultaneously, the neurons to the triceps are receiving more EPSPs from the brain than IPSPs via the reflex arc, so they reach threshold, fire, and consequently stimulate the triceps to contract. Thus, the arm is kept extended despite the painful stimulus. In this way, the withdrawal reflex has been voluntarily overridden.

### THE BABINSKI ANKLE JERK REFLEX
Babinski reflex occurs when the big toe dorsiflexes (straightens) and the other toes fan outward after the lateral portion of the sole of the foot has been stimulated by a sweeping pressure (Figure 3-16). Babinski reflex is an infantile reflex, which is

normal up to two years of age. It disappears as the child ages and the nervous system develops, being replaced with a plantarflexion (bending) response. The presence of the Babinski reflex after age two is a sign of neurological issues associated with the CNS, and more specifically the corticospinal tract. As the corticospinal tract runs down both sides of the spinal cord, Babinski reflex can occur on one or both sides of the body. An abnormal Babinski reflex can be temporary or permanent. Some commonly associated diseases are amyotrophic lateral sclerosis, head injury, meningitis, multiple sclerosis, spinal cord injury, stroke, tuberculosis, and rabies.

### OTHER REFLEX ACTIVITY
Spinal reflex action is not necessarily limited to motor responses on the side of the body to which the stimulus is applied. Assume that a person steps on a tack instead of burning a finger. A reflex arc is initiated to withdraw the injured foot from the painful stimulus, while the opposite leg simultaneously prepares to suddenly bear all the weight so that the person does not lose balance or fall (› Figure 3-17). Unimpeded bending of the injured extremity's knee is accomplished by concurrent

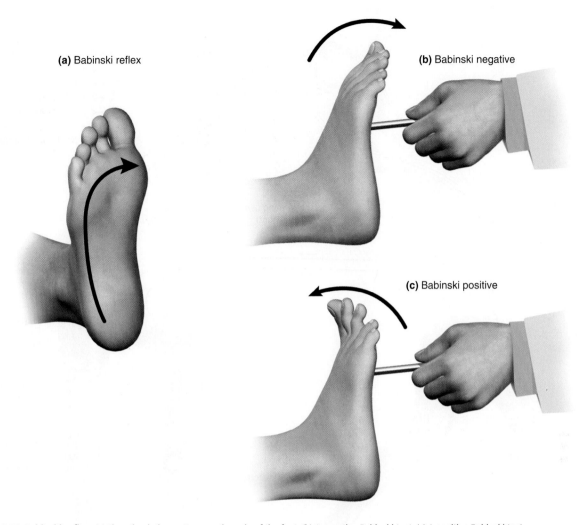

(a) Babinski reflex

(b) Babinski negative

(c) Babinski positive

> **FIGURE 3-16** **Babinski reflex.** (a) The stimulation pattern on the sole of the foot. (b) A negative Babinski test. (c) A positive Babinski test

reflex stimulation of the muscles that flex the knee and inhibition of the muscles that extend the knee. This response is a typical withdrawal reflex. At the same time, unimpeded extension of the opposite limb's knee is accomplished by activation of pathways that cross over to the opposite side of the spinal cord to stimulate this knee's extensors and inhibit its flexors. This **crossed extensor reflex** ensures that the opposite limb will be in a position to bear the weight of the body as the injured limb is withdrawn from the stimulus.

Besides protective reflexes (such as the withdrawal reflex) and simple postural reflexes (such as the crossed extensor reflex), basic spinal reflexes mediate the emptying of pelvic organs (e.g., urination, defecation, and expulsion of semen). All spinal reflexes can be voluntarily overridden at least temporarily by higher brain centres.

Not all reflex activity involves a clear-cut reflex arc, although the basic principles of a reflex (i.e., an automatic response to a detectable change) are still present. Pathways for unconscious responsiveness digress from the typical reflex arc in two general ways:

1. *Responses mediated at least in part by hormones.* A particular reflex may be mediated solely by either neurons or hormones or may involve a pathway using both.

2. *Local responses that do not involve either nerves or hormones.* For example, the blood vessels in an exercising muscle dilate because of local metabolic changes, thereby increasing blood flow to match the active muscle's metabolic needs.

## Check Your Understanding 3.3

1. Draw a cross section of a spinal cord and a pair of spinal nerves, showing the location of an afferent neuron, efferent neuron, and interneuron. Label the grey matter, white matter, dorsal root, ventral root, and spinal nerve.

2. Distinguish among a *tract, ganglion, nucleus, centre,* and *nerve.*

## 3.5 | Brain Stem

The **brain stem** is inferior to (beneath) the occipital lobe and anterior (in front) of the cerebellum. It is an adjoining structure continuous with the spinal cord. The brain stem, consists of the *pons, medulla oblongata,* and *midbrain.* It has five functions, which will be discussed shortly (see Figure 3-21b).

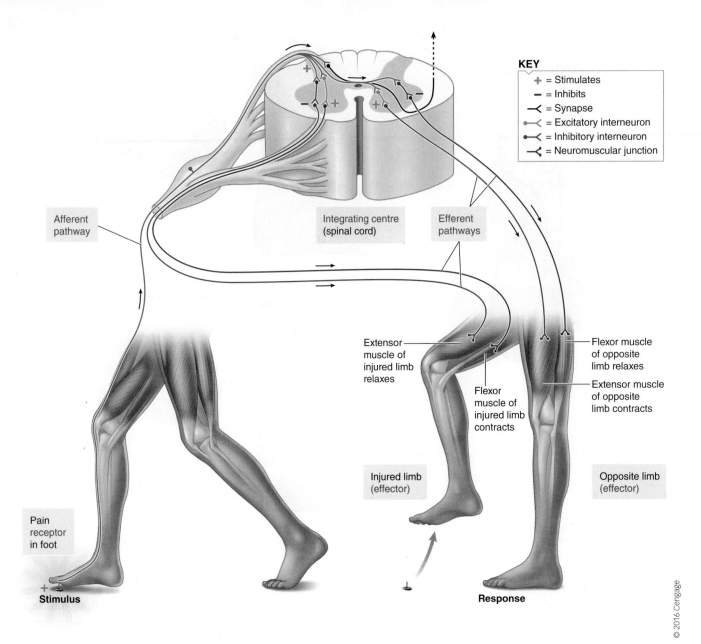

**KEY**
- **+** = Stimulates
- **−** = Inhibits
- ⌁ = Synapse
- ●⌁ = Excitatory interneuron
- ●⌁ = Inhibitory interneuron
- ⌁ = Neuromuscular junction

Afferent pathway

Integrating centre (spinal cord)

Efferent pathways

Extensor muscle of injured limb relaxes

Flexor muscle of injured limb contracts

Flexor muscle of opposite limb relaxes

Extensor muscle of opposite limb contracts

Injured limb (effector)

Opposite limb (effector)

Pain receptor in foot

Stimulus

Response

© 2016 Cengage

> **FIGURE 3-17 The crossed extensor reflex coupled with the withdrawal reflex.** The withdrawal reflex, which causes flexion of the injured extremity to withdraw from a painful stimulus. The crossed extensor reflex, which extends the opposite limb to support the full weight of the body

## A vital link

All incoming and outgoing fibres traversing between the periphery and higher brain centres must pass through the brain stem. Incoming fibres relay sensory information to the brain, and outgoing fibres carry command signals from the brain for efferent output. A few fibres merely pass through, but most synapse within the brain stem for important processing. Thus, the brain stem is a critical connecting link between the rest of the brain and the spinal cord.

The functions of the brain stem include the following:

1. The majority of the 12 pairs of **cranial nerves** arise from the brain stem (> Figure 3-18). With one major exception, these nerves supply structures in the head and neck with both sensory and motor fibres. They are important in sight, hearing, taste, smell, sensation of the face and scalp, eye movement, chewing, swallowing, facial expressions,

and salivation. The major exception is cranial nerve X, the **vagus nerve**. Instead of innervating regions in the head, most branches of the vagus nerve supply organs in the thoracic and abdominal cavities. The vagus is the major nerve of the parasympathetic nervous system.

2. Within the brain stem are neuronal clusters, called centres, that control heart and blood vessel function, respiration, and many digestive activities.

3. The brain stem plays a role in regulating muscle reflexes involved in equilibrium and posture.

4. A widespread network of interconnected neurons, called **the reticular formation**, runs throughout the entire brain stem and superiorly into the thalamus. This network receives and integrates all incoming sensory synaptic input. Ascending fibres originating in the reticular

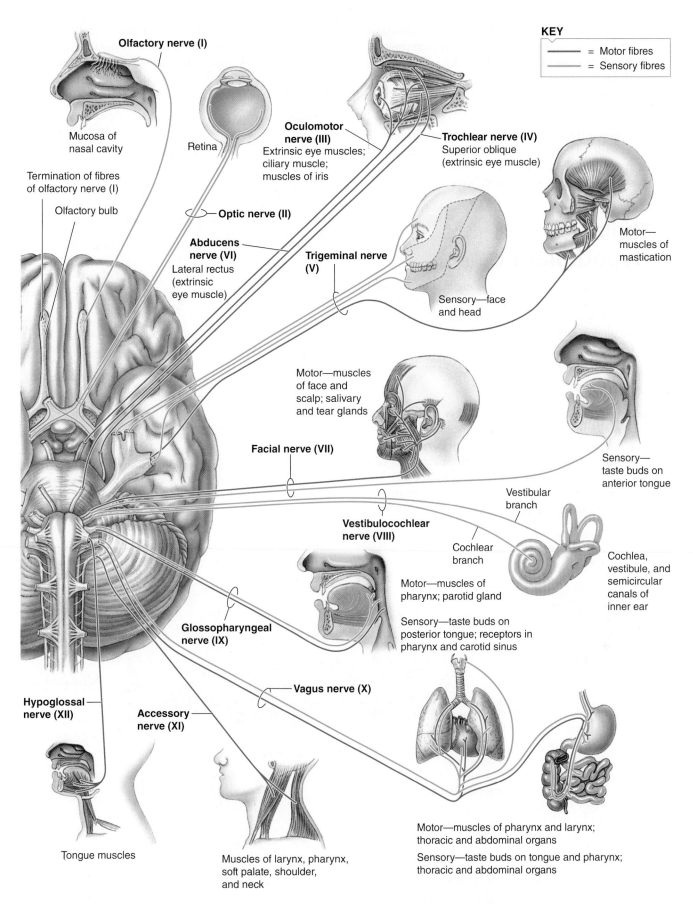

**KEY**

= Motor fibres

= Sensory fibres

**Olfactory nerve (I)**

Mucosa of nasal cavity

Termination of fibres of olfactory nerve (I)

Olfactory bulb

Retina

**Oculomotor nerve (III)**
Extrinsic eye muscles; ciliary muscle; muscles of iris

**Trochlear nerve (IV)**
Superior oblique (extrinsic eye muscle)

**Optic nerve (II)**

**Abducens nerve (VI)**
Lateral rectus (extrinsic eye muscle)

**Trigeminal nerve (V)**

Motor— muscles of mastication

Sensory—face and head

Motor—muscles of face and scalp; salivary and tear glands

**Facial nerve (VII)**

Sensory— taste buds on anterior tongue

Vestibular branch

**Vestibulocochlear nerve (VIII)**

Cochlear branch

Cochlea, vestibule, and semicircular canals of inner ear

Motor—muscles of pharynx; parotid gland

Sensory—taste buds on posterior tongue; receptors in pharynx and carotid sinus

**Glossopharyngeal nerve (IX)**

**Vagus nerve (X)**

**Hypoglossal nerve (XII)**

**Accessory nerve (XI)**

Tongue muscles

Muscles of larynx, pharynx, soft palate, shoulder, and neck

Motor—muscles of pharynx and larynx; thoracic and abdominal organs

Sensory—taste buds on tongue and pharynx; thoracic and abdominal organs

› **FIGURE 3-18 Cranial nerves.** Inferior (underside) view of the brain, showing the attachments of the 12 pairs of cranial nerves to the brain and many of the structures innervated by those nerves

3

formation carry signals upward to arouse and activate the cerebral cortex (> Figure 3-19). These fibres compose the **reticular activating system (RAS)**, which controls the overall degree of cortical alertness and is important in the ability to direct attention. In turn, fibres descending from the cortex, especially its motor areas, can activate the RAS.

5. The centres of the brain that govern sleep traditionally have been considered to be housed within the brain stem, although recent evidence suggests that the centre that promotes slow-wave sleep lies in the hypothalamus (see Section 3.12).

# 3.6 | Thalamus and Hypothalamus

Deep within the brain near the basal ganglia is the **diencephalon**, a midline structure that forms the walls of the third ventricular cavity, one of the spaces through which CSF flows. The diencephalon consists of two main parts, the *thalamus* and the *hypothalamus* (> Figure 3-20; see also Figure 3-21b).

## The thalamus

The **thalamus** serves as a "relay station" and synaptic integrating centre for preliminary processing of all sensory input on its way to the cortex. It screens out insignificant signals and routes the important sensory impulses to appropriate areas of the somatosensory cortex, as well as to other regions of the brain. Along with the brain stem and cortical association areas, the thalamus is important in the ability to direct attention to stimuli

of interest. For example, parents can sleep soundly through the noise of outdoor traffic but become instantly aware of their baby's slightest whimper. The thalamus is also capable of crude awareness of various types of sensation but cannot distinguish their location or intensity. Some degree of consciousness resides

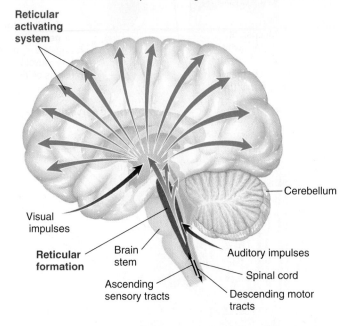

> **FIGURE 3-19 The reticular activating system.** The reticular formation, a widespread network of neurons within the brain stem (in red), receives and integrates all synaptic input. The reticular activating system, which promotes cortical alertness and helps direct attention toward specific events, consists of ascending fibres (in blue) that originate in the reticular formation and carry signals upward to arouse and activate the cerebral cortex.

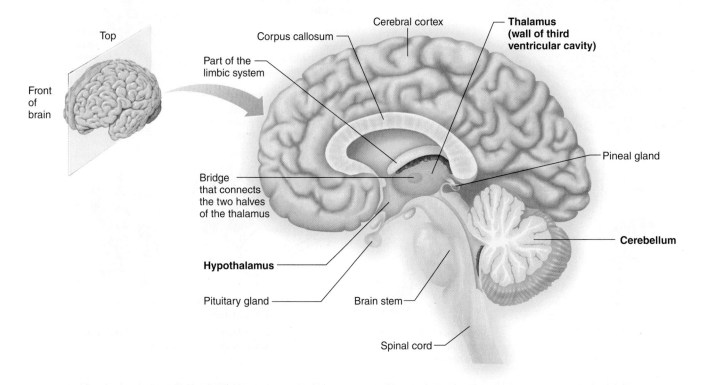

> **FIGURE 3-20 Location of the thalamus, hypothalamus, and cerebellum in sagittal section**

here as well. As described in Section 3.5, the thalamus also plays an important role in motor control by positively reinforcing voluntary motor behaviour initiated by the cortex.

## The hypothalamus

The **hypothalamus** is a collection of specific nuclei and associated fibres that lie inferior to (beneath) the thalamus. It is an integrating centre for homeostatic functions and is an important link between the autonomic nervous system and the endocrine system. Specifically, the hypothalamus (1) controls body temperature; (2) controls thirst and urine output; (3) controls food intake; (4) controls anterior pituitary hormone secretion; (5) produces posterior pituitary hormones; (6) controls uterine contractions and milk ejection; (7) serves as a major autonomic nervous system coordinating centre, which in turn affects all smooth muscle, cardiac muscle, and exocrine glands; (8) plays a role in emotional and behavioural patterns; and (9) participates in the sleep–wake cycle.

The hypothalamus is the brain area most involved in directly regulating the internal environment. For example, when the body is cold, the hypothalamus initiates internal responses to increase heat production (such as shivering) and to decrease heat loss (such as constricting blood vessels in the skin to reduce the flow of warm blood to the body surface, where heat could be lost to the external environment). Other areas of the brain, such as the cerebral cortex, act more indirectly to regulate the internal environment. For example, a person who feels cold is motivated to voluntarily put on warmer clothing, close the window, turn up the thermostat, and so on. Even these voluntary behavioural activities are strongly influenced by the hypothalamus, which, as a part of the limbic system, functions together with the cortex in controlling emotions and motivated behaviour. We now turn our attention to the cerebral cortex.

### Check Your Understanding 3.4

1. Describe how the thalamus serves as a sensory relay station.
2. Name the brain area most involved directly in regulating homeostatic functions.

## 3.7 | Cerebral Cortex

The cerebrum, by far the largest portion of the human brain, is divided into two halves, the right and left **cerebral hemispheres** (Figure 3-21a). They are connected to each other by the **corpus callosum**, a thick band consisting of an estimated 300 million neuronal axons travelling between the two hemispheres

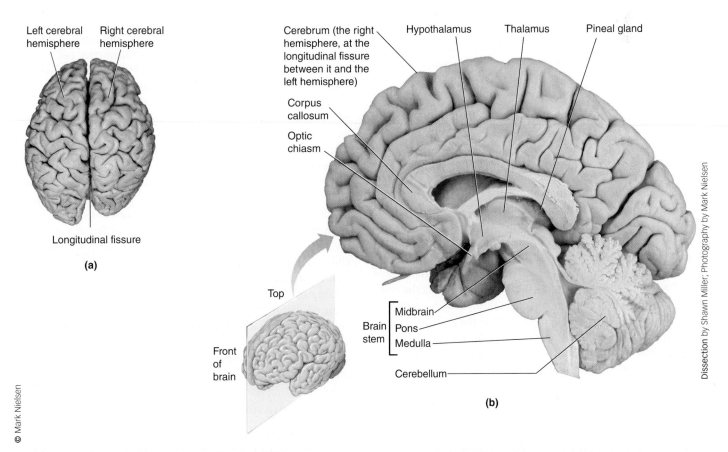

Left cerebral hemisphere
Right cerebral hemisphere
Longitudinal fissure
**(a)**

Cerebrum (the right hemisphere, at the longitudinal fissure between it and the left hemisphere)
Corpus callosum
Optic chiasm
Hypothalamus
Thalamus
Pineal gland
Top
Front of brain
Brain stem { Midbrain Pons Medulla
Cerebellum
**(b)**

© Mark Nielsen

Dissection by Shawn Miller; Photography by Mark Nielsen

> **FIGURE 3-21 Brain of human cadaver.** (a) Dorsal view looking down on the top of the brain. Note that the deep longitudinal fissure divides the cerebrum into the right and left cerebral hemispheres. (b) Sagittal view of the right half of the brain. All major brain regions are visible from this midline interior view. The corpus callosum serves as a neural bridge between the two cerebral hemispheres.

# Strokes: A Deadly Domino Effect

**T**HE MOST COMMON CAUSE OF BRAIN DAMAGE is a **cerebrovascular accident (CVA)**, or stroke. When a brain (cerebral) blood vessel is blocked by a clot or ruptures, the brain tissue supplied by that vessel loses its vital oxygen and glucose supply. The result is damage and usually death of the deprived tissue. New findings show that neural damage (and the subsequent loss of neural function) extends well beyond the blood-deprived area as a result of a neurotoxic effect that leads to the death of additional nearby cells. Whereas the initial blood-deprived cells die by *necrosis* (unintentional cell death), the doomed neighbours undergo *apoptosis* (deliberate cell suicide). The initial oxygen-starved cells release excessive amounts of glutamate, a common excitatory neurotransmitter. Glutamate or other neurotransmitters are normally released in small amounts from neurons as a means of chemical communication between brain cells. The excitatory overdose of glutamate from the damaged brain cells binds with and overexcites surrounding neurons. Specifically, glutamate binds with excitatory receptors known as NMDA receptors, which function as calcium $K^+$ channels. As a result of toxic activation of these receptor channels, they remain open for too long, permitting too much $K^+$ to rush into the affected neighbouring neurons. This elevated intracellular $K^+$ triggers these cells to self-destruct. During this process, **free radicals** are produced. These highly reactive, electron-deficient particles cause further cell damage by snatching electrons from other molecules. Adding to the injury, researchers speculate that the $K^+$ apoptotic signal may spread from the dying cells to abutting healthy cells through gap junctions—the cell-to-cell conduits that allow $K^+$ and other small ions to diffuse freely between cells. This action kills even more neuronal victims.

(Figure 3-21b; see also > Figure 3-15). The corpus callosum is the body's "information superhighway." The two hemispheres communicate and cooperate with each other by means of constant information exchange through this neural connection.

## The cerebral cortex: Grey matter and white matter

Each hemisphere is composed of a thin outer shell of *grey matter*, the **cerebral cortex**, covering a thick central core of *white matter* (see > Figure 3-28). Another region of grey matter, the basal ganglia, is located deep within the white matter. Throughout the entire CNS, grey matter consists predominantly of densely packaged neuronal cell bodies and their dendrites as well as most glial cells. Bundles or tracts of myelinated nerve fibres (axons) constitute the white matter; its white appearance is due to the lipid composition of the myelin. The grey matter can be viewed as the computers of the CNS and the white matter as the wires that connect the computers to one another. Integration of neural input and initiation of neural output take place at synapses within the grey matter. The fibre tracts in the white matter transmit signals from one part of the cerebral cortex to another, or between the cortex and other regions of the CNS. Such communication between different areas of the cortex and elsewhere facilitates integration of their activity, which is essential for even a simple task, such as picking a flower. Vision of the flower is received by one area of the cortex, reception of its fragrance takes place in another area, and movement is initiated by still another area. More subtle neuronal responses, such as appreciation of the flower's beauty and the urge to pick it, are poorly understood but undoubtedly extensively involve interconnecting fibres between different cortical regions.

## Layers and columns in the cerebral cortex

The cerebral cortex is organized into six well-defined layers based on varying distributions of several distinctive cell types. These layers are organized into functional vertical columns that extend perpendicularly about 2 mm from the cortical surface down through the thickness of the cortex to the underlying white matter. The neurons within a given column function as a team, with each cell being involved in different aspects of the same specific activity: for example, perceptual processing of the same stimulus from the same location.

The functional differences between various areas of the cortex result from different layering patterns within the columns and from different input–output connections, not from the presence of unique cell types or different neuronal mechanisms. For example, those regions of the cortex responsible for perception of senses have an expanded layer, a layer rich in **stellate cells**, which are responsible for initial processing of sensory input to the cortex. In contrast, cortical areas that control output to skeletal muscles have a thickened layer, which contains an abundance of large **pyramidal cells**. These cells send fibres down the spinal cord from the cortex to terminate on efferent motor neurons that innervate skeletal muscles.

## Lobes in the cerebral cortex

We now consider the locations of the major functional areas of the cerebral cortex. Throughout this discussion, keep in mind that even though a discrete activity is ultimately attributed to a particular region of the brain, no part of the brain functions in isolation. Each part depends on a complex interplay among

Thus, the majority of neurons that die following a stroke are originally unharmed cells that commit suicide in response to the chain of reactions unleashed by the toxic release of glutamate from the initial site of oxygen deprivation.

Until the last decade, physicians could do nothing to halt the inevitable neuronal loss following a stroke, leaving patients with an unpredictable mix of neural deficits. Treatment was limited to rehabilitative therapy after the damage was already complete. In recent years, armed with new knowledge about the underlying factors in stroke-related neuronal death, the medical community has been seeking ways to halt the cell-killing domino effect. The goal, of course, is to limit the extent of neuronal damage and thus minimize, or even prevent, clinical symptoms such as paralysis. In the early 1990s doctors started administering clot-dissolving drugs within the first three hours after the onset of a stroke to restore blood flow through blocked cerebral vessels. Clot busters were the first drugs used to treat strokes, but they are only the beginning of new stroke therapies. Other methods are currently under investigation to prevent adjacent nerve cells from succumbing to the neurotoxic release of glutamate. These include (1) blocking the NMDA receptors that initiate the neuronal death-wielding chain of events in response to glutamate, (2) halting the apoptosis pathway that results in self-execution, and (3) blocking the gap junctions that permit the $K^+$ death messenger to spread to adjacent cells. These tactics hold much promise for treating strokes, which are the most prevalent cause of adult disability and the third leading cause of death in Canada. However, to date no new neuroprotective drugs have been found that do not cause serious side effects.

numerous other regions for both incoming and outgoing messages.

The anatomic landmarks used in cortical mapping are certain deep folds that divide each half of the cortex into four major lobes: the *occipital, temporal, parietal,* and *frontal lobes* (Figure 3-22). To accompany the following discussion of the major activities attributed to various regions of these lobes, study the basic functional map of the cortex in Figure 3-23a.

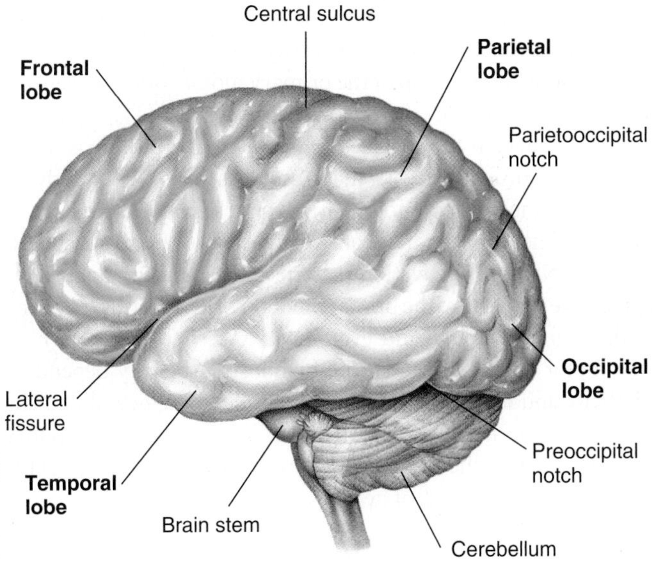

The **occipital lobes**, located posteriorly (at the back of the head), carry out the initial processing of visual input. Sound sensation is initially received by the **temporal lobes**, located laterally (on the sides of the head) (Figures 3-23a and 3-23b). You will learn more about the functions of these regions in Chapter 4 when we discuss vision and hearing.

The frontal lobes and parietal lobes, located on the top of the head, are separated by a deep infolding, the **central sulcus**, which runs roughly down the middle of the lateral surface of each hemisphere. The **parietal lobes** lie to the rear of the central sulcus on each side, and the **frontal lobes** lie in front of it. The frontal lobes are responsible for three main functions: (1) voluntary motor activity, (2) speaking ability, and (3) elaboration of thought. The parietal lobes are primarily responsible for receiving and processing sensory input.

### The frontal lobes

The area in the rear (dorsal) portion of the frontal lobe, immediately anterior of the central sulcus and adjacent to the somatosensory cortex, is the **primary motor cortex** (Figure 3-24a; also see Figure 3-23a). It works in coordination with the premotor cortex (a few millimetres anterior of the primary motor cortex) to plan and execute movements. The primary motor cortex contains large pyramidal cell neurons (Betz cells, located in the grey matter) with long axons that descend the spinal cord and synapse (neurotransmitter glutamate) with the alpha-motor neurons of skeletal muscles. The premotor cortex also works in concert with the basal ganglia to plan movements or actions, and then receives input from the cerebellum to refine these movements (Section 3.8).

> FIGURE 3-22 **Cortical lobes.** Each half of the cerebral cortex is divided into the occipital, temporal, parietal, and frontal lobes, as depicted in this schematic lateral view of the brain.

Labels: Central sulcus, Frontal lobe, Parietal lobe, Parietooccipital notch, Lateral fissure, Temporal lobe, Brain stem, Cerebellum, Preoccipital notch, Occipital lobe

© 2016 Cengage

As in sensory processing, the motor cortex on each side of the brain primarily controls muscles on the opposite side of the body. Neuronal tracts originating in the motor cortex of the left hemisphere cross over before passing down the spinal cord to terminate on efferent motor neurons that stimulate skeletal muscle contraction on the right side of the body (see Figure 3-11b). Some fibres cross over to the opposite (contralateral) side in the medulla oblongata, and then the axons travel down the spinal cord as the lateral corticospinal tract. Other fibres travel down separately in ventral corticospinal tract with most crossing over to the contralateral side in the spinal cord just before reaching their associated motor neurons. Accordingly, damage to the motor cortex on the left side of the brain produces paralysis on the right side of the body, and the converse is also true.

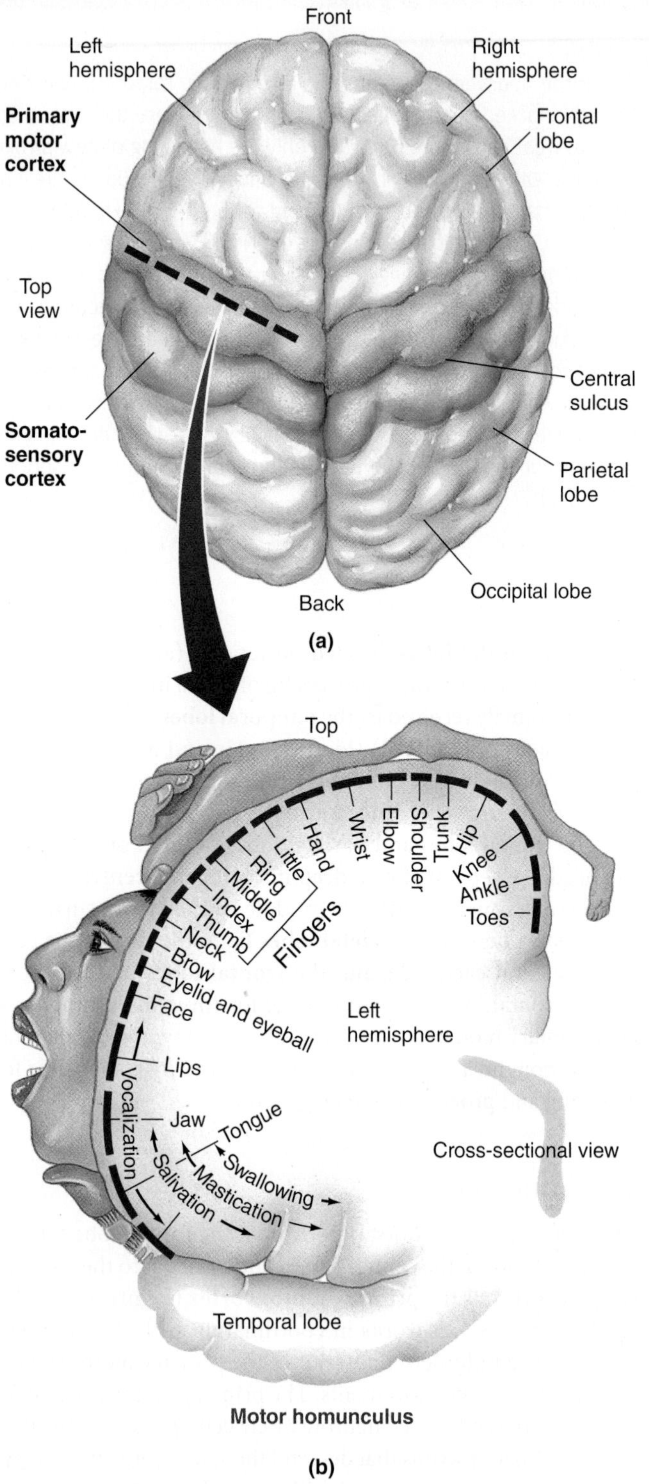

> ■ **Clinical Connections** Given that the cerebral cortex is involved in many functions, including cognition, speech, vision, and motor control, it is no surprise that any lesions in this area due to multiple sclerosis could present in numerous ways. The majority of Jackie's symptoms could be the result of scarring in this region of the brain, but it is important to note that multiple sclerosis generally leads to multiple lesion formations throughout the entire CNS, which suggests that any body function could be affected at multiple levels.

Stimulation of different areas of the primary motor cortex brings about movement in different regions of the body. Like the sensory homunculus for the somatosensory cortex, the **motor homunculus**, which depicts the location and relative amount of motor cortex devoted to output to the muscles of each body part, is upside down and distorted (Figure 3-24b). The fingers, thumbs, and muscles important in speech, especially those of the lips and tongue, are grossly exaggerated, indicating the fine degree of motor control these body parts have. Compare this with how little brain tissue is devoted to the trunk, arms, and lower extremities, which are not capable of such complex movements. This illustrates that the extent of representation in the motor cortex is proportional to the precision and complexity of motor skills required of the respective part.

The somatotopic map of the primary motor cortex is relatively stable as it pertains to the major body divisions (e.g., arm, leg). However, beneath the surface each of these map subdivisions is a distinct neuronal network that is not as stable. The underlying network of the primary motor cortex is not simply a static setup of horizontal neuronal connections; change and/or reorganization is possible. Neuronal activation patterns viewed through neuroimaging have shown that the primary motor cortex is plastic following both traumatic injury and everyday experiences (e.g., learning new motor skills). These neurons demonstrate activity-dependent plasticity and modify in association with learning new skills. The reorganization is likely dependent on excitatory and inhibitory influences within the horizontal neural connections, changes at the synapse, and the release of neurotransmitters (e.g., GABA).

## The parietal lobes

Sensations from the surface of the body, such as touch, pressure, heat, cold, and pain, are collectively known as **somaesthetic sensations** (*somaesthetic* means "body feelings"). The

> ❯ **FIGURE 3-23 Somatotopic map of the primary motor cortex.** (a) Top view of cerebral hemispheres. (b) Motor homunculus showing the distribution of motor output from the primary motor cortex to different parts of the body. The distorted graphic representation of the body parts indicates the relative proportion of the primary motor cortex devoted to controlling skeletal muscles in each area.

© 2016 Cengage

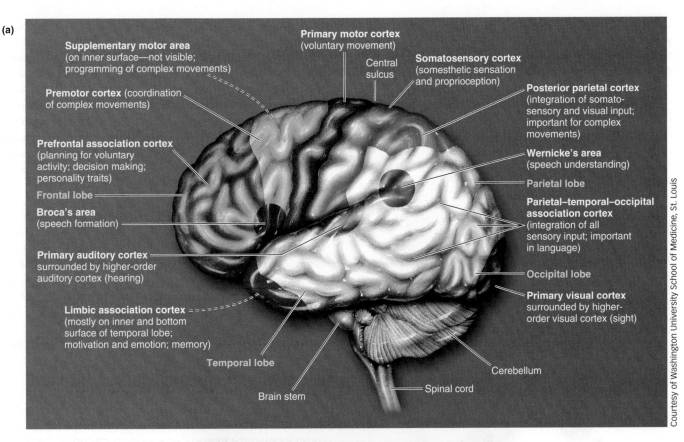

**(a)**

Supplementary motor area (on inner surface—not visible; programming of complex movements)

Primary motor cortex (voluntary movement)

Central sulcus

Somatosensory cortex (somesthetic sensation and proprioception)

Premotor cortex (coordination of complex movements)

Posterior parietal cortex (integration of somato-sensory and visual input; important for complex movements)

Prefrontal association cortex (planning for voluntary activity; decision making; personality traits)

Wernicke's area (speech understanding)

Frontal lobe

Parietal lobe

Broca's area (speech formation)

Parietal–temporal–occipital association cortex (integration of all sensory input; important in language)

Primary auditory cortex surrounded by higher-order auditory cortex (hearing)

Occipital lobe

Primary visual cortex surrounded by higher-order visual cortex (sight)

Limbic association cortex (mostly on inner and bottom surface of temporal lobe; motivation and emotion; memory)

Temporal lobe

Cerebellum

Brain stem

Spinal cord

Courtesy of Washington University School of Medicine, St. Louis

**3**

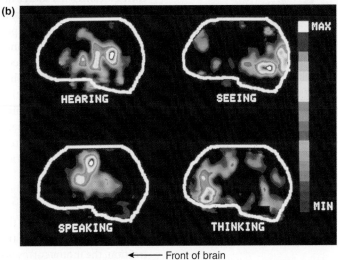

**(b)**

MAX

HEARING

SEEING

SPEAKING

THINKING

MIN

← Front of brain

› **FIGURE 3-24 Functional areas of the cerebral cortex.** (a) Various regions of the cerebral cortex are primarily responsible for various aspects of neural processing, as indicated in this schematic lateral view of the brain. (b) Different areas of the brain "light up" on positron-emission tomography (PET) scans as a person performs different tasks. PET scans detect the magnitude of blood flow in various regions of the brain. Because more blood flows into a particular region of the brain when it is more active, neuroscientists can use PET scans to "take pictures" of the brain at work on various tasks.

means by which afferent neurons detect and relay information to the CNS about these sensations will be covered in Chapter 4, where we explore the afferent division of the PNS in detail and compare the perceptual maps of stimuli that activate A-beta (touch), A-delta (fast, pain), and C fibres (slow, heat) on the skin's surface. Within the CNS, this information is *projected* (i.e., transmitted along specific neural pathways to higher brain levels) to the **somatosensory cortex**. The somatosensory cortex is located in the front (anterior) portion of each

parietal lobe immediately behind (posterior) the central sulcus (Figures 3-23a and 3-25a). It is the site for initial cortical processing and perception of somaesthetic input as well as proprioceptive input. **Proprioception** is the awareness of body position.

Each region within the somatosensory cortex receives somaesthetic and proprioceptive input from a specific area of the body. This distribution of cortical sensory processing is depicted in Figure 3-25b. The mapping of the somatosensory cortex using electrical stimulation by Penfield and colleagues (in the 1950s and earlier) gave us two important findings. First, representations of the skin surface have a generally somatotopic organization; for example, the hand representation is next to the arm

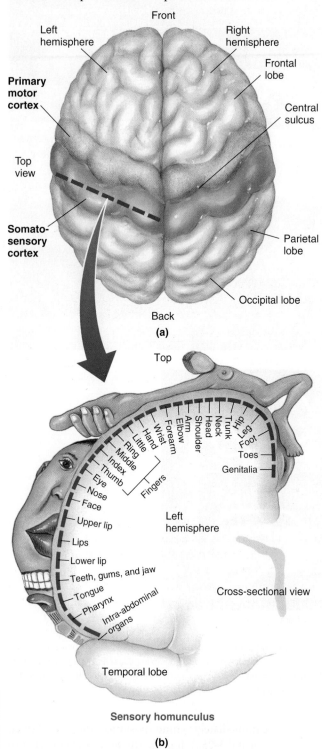

Front

Left hemisphere

Right hemisphere

**Primary motor cortex**

Frontal lobe

Central sulcus

Top view

**Somato-sensory cortex**

Parietal lobe

Occipital lobe

Back

**(a)**

Top

Little
Ring
Middle
Index
Thumb
Hand
Wrist
Forearm
Elbow
Arm
Shoulder
Head
Neck
Trunk
Hip
Leg
Foot
Toes

Eye
Nose
Face
Fingers
Genitalia

Upper lip

Left hemisphere

Lips

Lower lip

Teeth, gums, and jaw

Cross-sectional view

Tongue

Pharynx

Intra-abdominal organs

Temporal lobe

Sensory homunculus

**(b)**

> **FIGURE 3-25 Somatotopic map of the somatosensory cortex.** (a) Top view of cerebral hemispheres. (b) Sensory homunculus showing the distribution of sensory input to the somatosensory cortex from different parts of the body. The distorted graphic representation of the body parts indicates the relative proportion of the somatosensory cortex devoted to reception of sensory input from each area.

© 2016 Cengage

representation. Second, the cortical representations of body parts are larger for those that have higher sensitivity and use. This orderly somatotopic representation is named the **sensory homunculus** (*homunculus* means "little man"). The body is represented upside down on the somatosensory cortex, and as mentioned, different parts of the body are not equally represented. The size of each body part in this homunculus indicates the relative proportion of the somatosensory cortex devoted to that area. The exaggerated size of the face, tongue, hands, and genitalia indicates the high degree of sensory perception and/or use (e.g., touch) associated with these body parts.

For the most part, the somatosensory cortex on each side of the brain receives sensory input from the opposite side of the body, because most of the ascending pathways carrying sensory information up the spinal cord cross over to the opposite side before eventually terminating in the cortex (see Figure 3-11a). Therefore, damage to the somatosensory cortex in the left hemisphere produces sensory deficits on the right side of the body, whereas sensory losses on the left side are associated with damage to the right half of the cortex.

Simple awareness of touch, pressure, temperature, or pain is detected by the thalamus, a lower level of the brain, but the somatosensory cortex goes beyond pure recognition of sensations to fuller sensory perception. The thalamus makes you aware that something hot versus something cold is touching your body, but it does not tell you where or of what intensity. The somatosensory cortex localizes the source of sensory input and perceives the level of intensity of the stimulus. It also is capable of spatial discrimination, as it can discern shapes of objects being held, subtle differences in similar objects, and changes in limb position.

The somatosensory cortex, in turn, projects this sensory input via white matter fibres to adjacent higher sensory areas for even further elaboration, analysis, and integration of sensory information. These higher areas are important in perceiving complex patterns of somatosensory stimulation—for example, simultaneous appreciation of the texture, firmness, temperature, shape, position, and location of an object you are holding.

## Other brain regions and motor control

Even though signals from the primary motor cortex terminate on the efferent neurons that trigger voluntary skeletal muscle contraction, the motor cortex is not the only brain region involved in motor control. First, lower brain regions and the spinal cord control involuntary skeletal muscle activity, such as in maintaining posture. Some of these same regions also play an important role in monitoring and coordinating voluntary motor activity that the primary motor cortex has set in motion. Second, although fibres originating from the motor cortex can activate motor neurons to bring about muscle contraction, the motor cortex itself does not *initiate* voluntary movement. The motor cortex is activated by a widespread pattern of neuronal discharge, the **readiness potential**, which occurs about 750 milliseconds before specific electrical activity is detectable in the motor cortex. Three higher motor areas of the cortex are involved in this voluntary decision-making period. These higher areas, which all command the primary motor cortex, include the *supplementary motor* *area*, the *premotor cortex*, and the *posterior parietal cortex* (see Figure 3-23a). Furthermore, a **subcortical region** of the brain, the *cerebellum*, plays an important role, specifically, by sending input to the motor areas of the cortex regarding the planning, initiating, and timing certain kinds of movement.

The three higher motor areas of the cortex and the cerebellum carry out different, related functions that are all important in programming and coordinating complex movements that involve simultaneous contraction of many muscles. Even though electrical stimulation of the primary motor cortex brings about contraction of particular muscles, no purposeful coordinated movement can be elicited—just as pulling on isolated strings of a puppet does not produce any meaningful movement. A puppet displays purposeful movements only when a skilled puppeteer manipulates the strings in a coordinated manner. In the same way, these four regions (and perhaps other areas as yet undetermined) develop a **motor program** for the specific voluntary task and then "pull the appropriate pattern of strings" in the primary motor cortex to bring about the sequenced contraction of appropriate muscles to accomplish the desired complex movement.

The **supplementary motor area** lies on the medial (inner) surface of each hemisphere anterior (in front) of the primary motor cortex. It plays a preparatory role in programming complex sequences of movement—for example, those requiring simultaneous use of both hands and feet. Stimulation of various regions of this motor area brings about complex patterns of movement. Lesions here do not result in paralysis, but they do interfere with performance of more complex, useful, integrated movements.

The **premotor cortex**, located on the lateral surface of each hemisphere in front of the primary motor cortex, is important in orienting the body and arms (trunk muscles) toward a specific target. To command the primary motor cortex to bring about the appropriate skeletal muscle contraction to accomplish the desired movement, the premotor cortex must be informed of the body's momentary position in relation to the target. The premotor cortex is guided by sensory input processed by the **posterior parietal cortex**, a region that lies posterior (behind) to the primary somatosensory cortex. These two higher motor areas have many anatomic interconnections and are closely interrelated functionally. When either of these areas is damaged, the person cannot process complex sensory information to accomplish purposeful movement in a spatial context; for example, the person cannot successfully manipulate eating with utensils. However, these higher motor areas command the primary motor cortex and are important in preparing for execution of deliberate, meaningful movement. Researchers cannot say, however, that voluntary movement is actually initiated by these areas.

Think about the neural systems called into play, for example, during the simple act of picking up an apple to eat. Your memory tells you that the fruit is in a bowl on the kitchen counter. Sensory systems, coupled with your knowledge based on past experience, enable you to distinguish the apple from the other kinds of fruit in the bowl. On receiving this integrated sensory information, motor systems issue commands to the exact muscles of the body in the proper sequence to enable you to move to the fruit bowl and pick up the targeted apple. During execution of this act, minor adjustments in the motor command are made as needed, based on continual updating provided by sensory input about the position of your body relative to the goal. Then there is the issue of motivation and behaviour. Why are you reaching for the apple in the first place? Is it because you are hungry (detected by a neural system in the hypothalamus) or because of a more complex behavioural scenario unrelated to a basic hunger drive, such as the fact that you started to think about food because you just saw someone on television eating? Why did you choose an apple rather than a banana when both are in the fruit bowl and you like the taste of both, and so on? Consequently, initiating and executing purposeful voluntary movement actually include a complex neuronal interplay involving output from the motor regions that is guided by integrated sensory information and ultimately depends on motivational systems and elaboration of thought. All this plays against a background of memory stores from which you can make meaningful decisions about desirable movements.

## Somatotopic maps

Although the general organizational pattern of sensory and motor somatotopic (body representation) maps of the cortex is similar in all people, the precise distribution is unique for each individual. Just as each of us has two eyes, a nose, and a mouth, and yet no two faces have these features arranged in exactly the same way, so it is with brains. Furthermore, an individual's somatotopic mapping is not "carved in stone" but is subject to constant subtle modifications based on use.

When Penfield and colleagues did the original research on somatotopic mapping, they did not explore the manner in which the map could be altered by time and experience. Since then, research has shown that the primary somatosensory representations are plastic—that is, capable of changing in response to central or peripheral adjustments and experiences. For example, after an injury, tactile reduction occurs, altering the somatosensory maps. In contrast, following an amputation there is geographical expansion of the representations of neighbouring areas into the somatosensory cortex previously represented by the amputated body part.

It is known that altering body position results in changes in the perceived sensation and localization of touch. Localization is believed to occur in two stages: (1) the somatotopic stage in which a stimulus (e.g., pain) location is identified relative to its position on the skin surface, and (2) it takes into account limb position in external space. Thus, we must also know the location of our body to effectively localize the tactile stimuli relative to our environment.

Consequently, mapping is governed by both genetic and developmental processes, but the individual cortical setup can be influenced by **use-dependent competition** for cortical space. Here is another example of the adaptation of somatosensory mapping: When monkeys were encouraged to use a specific finger (middle finger), instead of other fingers to press a bar to release food, only after thousands of bar presses was the "middle finger area" of the motor cortex greatly expanded, to the point of intruding upon space previously devoted to the other fingers.

Other regions of the brain besides the somatosensory cortex and motor cortex can also be modified by experience. We are now going to turn our attention to this plasticity of the brain.

## Plasticity and neurogenesis of brain tissue

From the previous discussion, you can see that the brain displays a degree of **plasticity**—that is, an ability to change or be functionally remodelled in response to the demands placed on it. The ability of the brain to modify as needed is more pronounced in the early developmental years, but even adults retain some plasticity (e.g., sensory and motor mapping). When an area of the brain associated with a particular activity is destroyed, other areas of the brain may gradually assume some or all of the functions of the damaged region. Researchers are only beginning to unravel the underlying molecular mechanisms responsible for the brain's plasticity. Current evidence suggests that the formation of new neural pathways (not new neurons, but new connections between existing neurons) in response to changes in experience are mediated in part by alterations in dendritic shape resulting from modifications in certain cytoskeletal elements (p. 130). As its dendrites become more branched and elongated, a neuron becomes able to receive and integrate more signals from other neurons. In this way, the precise synaptic connections between neurons are not fixed but can be modified by experience. The gradual modification of each person's brain by a unique set of experiences provides a biological basis for individuality.

In contrast to the growth of new neuronal connections, recently, adult neurogenesis (the creation of new brain cells) has been proposed. New neurons integrate themselves into the brain, providing the potential for self-healing. Neurogenesis could lead to improved treatments for disorders or damaged tissue (e.g., traumatic brain injury), and to ways to help keep our mind sharp. For example, hippocampal neurogenesis was demonstrated in humans, which may be important for learning and memory. Support for adult neurogenesis and improved memory include new neurons to increase memory capacity, reduction of interference between memories, and addition of temporal information to the memories.

## Electroencephalograms

Extracellular current flow arising from electrical activity within the cerebral cortex can be detected by placing recording electrodes on the scalp to produce a graphic record known as an **electroencephalogram (EEG)**. These "brain waves" for the most part are not due to action potentials but instead represent the momentary collective postsynaptic potential activity (i.e., EPSPs and IPSPs; p. 76) in the cell bodies and dendrites located in the cortical layers under the recording electrode.

Electrical activity can always be recorded from the living brain, even during sleep and unconscious states, but the waveforms vary, depending on the degree of activity of the cerebral cortex. Often the waveforms appear irregular, but sometimes distinct patterns in the wave's amplitude and frequency can be observed. A dramatic example of this is illustrated in Figure 3-26, in which the EEG waveform recorded over the occipital (visual) cortex changes markedly in response to simply opening and closing the eyes.

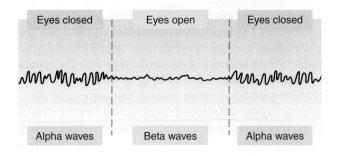

> FIGURE 3-26 **Replacement of an alpha rhythm on an EEG with a beta rhythm when the eyes are opened**

The EEG has three major uses:

1. The EEG is often used as a *clinical tool in the diagnosis of cerebral dysfunction*. Diseased or damaged cortical tissue often gives rise to altered EEG patterns. One of the most common neurologic diseases accompanied by a distinctively abnormal EEG is **epilepsy**. Epileptic seizures occur when a large collection of neurons abnormally undergo synchronous action potentials that produce stereotypical, involuntary spasms and alterations in behaviour. Different underlying problems, including genetic defects and traumatic brain injuries, can lead to the neuronal hyperexcitability that characterizes epilepsy. Typically, there is too little inhibitory compared with excitatory activity, as with compromised functioning of the inhibitory neurotransmitter GABA or prolonged action of the excitatory neurotransmitter glutamate. The seizures may be partial or generalized, depending on the location and extent of the abnormal neuronal discharge. Each type of seizure displays different EEG features.

2. The EEG finds further use in the *legal determination of brain death*. Even though a person may have stopped breathing and the heart may have stopped pumping blood, it is often possible to restore and maintain respiratory and circulatory activity if resuscitative measures are instituted soon enough. Yet because the brain is susceptible to oxygen deprivation, irreversible brain damage may have already occurred before lung and heart function have been re-established, resulting in the paradoxical situation of a dead brain in a living body. The determination of whether a comatose patient being maintained by artificial respiration and other supportive measures is alive or dead has important medical, legal, and social implications. The need for viable organs for modern transplant surgery has made the timeliness of such life/death determinations of utmost importance. Physicians, lawyers, and the Canadian public in general have accepted the notion of brain death—that is, a brain that is not functioning and has no possibility of recovery—as the determinant of death under such circumstances. Brain-dead people make good organ donors because the organs are still being supplied by circulating blood and thus are in better condition than those obtained from a person whose heart has stopped beating. The most widely accepted indication of brain death is

*electrocerebral silence*—an essentially flat EEG. This must be coupled with other stringent criteria, such as absence of eye reflexes, to guard against a false terminal diagnosis in individuals with a flat EEG that is due to causes that can be reversed, as in certain kinds of drug intoxication.

3. The EEG is also used to *distinguish various stages of sleep*, as described in Section 3.12.

## Neurons

Most information about the electrical activity of the brain has been gleaned not from EEG studies but from direct recordings of individual neurons in experimental animals engaged in various activities. Following surgical implantation of an extremely thin recording microelectrode into a single neuron within a specific region of the cerebral cortex, scientists have been able to observe changes in the electrical activity of the neuron as the animal engages in particular motor tasks or encounters various sensations. Investigators have concluded that neural information is coded by changes in action potential frequency in specific neurons—the greater the triggering event, the greater the firing rate of the neuron.

Even though this finding is important, single-neuron recordings have not been able to identify concurrent changes in electrical activity in a group of neurons working together to accomplish a particular activity. As an analogy, consider if you tried to record a concert by using a single microphone that could pick up only the sounds produced by one musician. You would get a very limited impression of the performance by hearing only the changes in notes and tempo as played by this one individual. You would miss the richness of the melody and rhythm being performed in synchrony by the entire orchestra. Similarly, by recording from single neurons and detecting their changes in firing rates, scientists were unable to observe a parallel information mechanism that involves changes in the relative timing of action potential discharges among a functional group, or *assembly*, of neurons. By taking into account simultaneous recordings from multiple neurons, some researchers now suggest that interacting neurons may transiently fire together for fractions of a second. Many neuroscientists believe that the brain encodes information not just by changing the firing rates of individual neurons but also by changing the patterns of these brief neural synchronizations. That is, groups of neurons may communicate, or send messages about what is happening, by changing their pattern of synchronous firing.

Neurons within an assembly that fire together may be widely scattered, but integrated. For example, when you view a bouncing ball, different visual units initially process different aspects of this object—its shape, its colour, its movement, and so on. Somehow all these separate processing pathways must be integrated, or bound together, for you to see the bouncing ball as a whole unit without stopping to contemplate its many separate features. The solution to the mystery of how the brain accomplishes this task might lie in the synchronous firing of neurons in separate regions of the brain that are functionally linked by virtue of being responsive to different aspects of the same objects—such as the bouncing ball.

### Check Your Understanding 3.5

1. Draw a lateral view of the left cerebral cortex, and then label the location of each of the following: frontal lobe, parietal lobe, occipital lobe, temporal lobe, primary motor cortex, somatosensory cortex, primary visual cortex, auditory cortex, and prefrontal association cortex.

2. Define *plasticity*.

## 3.8 | Cerebellum

The **cerebellum** means "little brain" in Latin. It is a baseball-sized region located inferiorly to the occipital lobe and posteriorly to the brain stem. The cerebellum is central to the integration of motor output and sensory perception. It is linked by neural pathways to the motor cortex, which sends information to the skeletal muscles, causing movement, and to the spinocerebellar tract, which provides feedback on body position (proprioception) (see Figures 3-21b and 3-20).

### Balance and planning

More individual neurons are found in the cerebellum than in the rest of the brain, indicative of the importance of this structure. The cerebellum consists of three functionally distinct parts, each with different roles concerned primarily with subconscious control of motor activity (› Figure 3-27). Specifically, the different parts of the cerebellum perform the following functions:

1. The **vestibulocerebellum** is important for maintaining balance and controls eye movements.

2. The **spinocerebellum** enhances muscle tone and coordinates skilled, voluntary movements. This brain region is especially important in ensuring the accurate timing of various muscle contractions to coordinate movements involving multiple joints. For example, the movements of your shoulder, elbow, and wrist joints must be synchronized even during the simple act of reaching for a pencil. When cortical motor areas send messages to muscles for executing a particular movement, the spinocerebellum is informed of the intended motor command. This region also receives input from peripheral receptors that inform it about the body movements and positions that are actually taking place. The spinocerebellum essentially acts as middle management, comparing the intentions or orders of the executive management (motor cortex) with the performance of the lower levels of the organization (muscles) and then correcting any errors or deviations from the intended outcomes (movement). The spinocerebellum even seems able to predict the position of a body part in the next fraction of a second during a complex movement and to make adjustments accordingly. If you are reaching for a pencil, for example, this region "puts on the brakes" soon enough to stop the forward movement of your hand

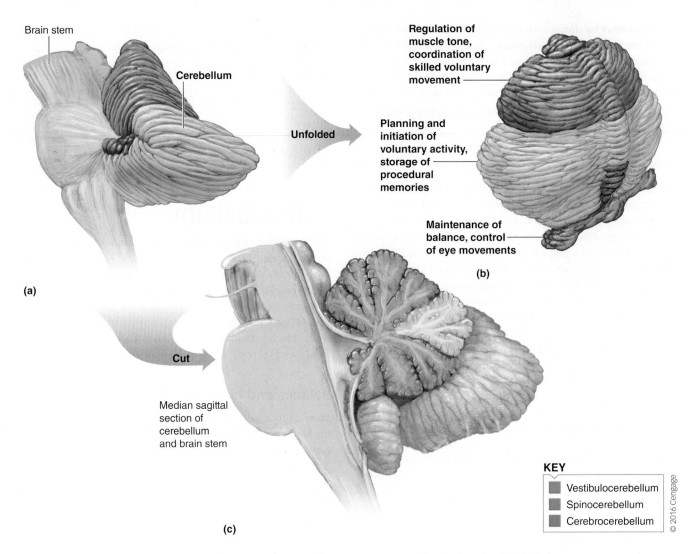

Brain stem

Cerebellum

Unfolded

**Regulation of muscle tone, coordination of skilled voluntary movement**

**Planning and initiation of voluntary activity, storage of procedural memories**

**Maintenance of balance, control of eye movements**

(b)

(a)

Cut

Median sagittal section of cerebellum and brain stem

(c)

**KEY**

▨ Vestibulocerebellum
▨ Spinocerebellum
▨ Cerebrocerebellum

© 2016 Cengage

❯ **FIGURE 3-27 Cerebellum.** (a) Gross structure of the cerebellum. (b) Unfolded cerebellum, revealing its three functionally distinct parts. (c) Internal structure of the cerebellum

at the intended location rather than letting you overshoot your target. These ongoing adjustments, which ensure smooth, precise, directed movement, are especially important for rapidly changing (phasic) activities, such as typing, playing the piano, or running.

3. The **cerebrocerebellum** plays a role in planning and initiating voluntary activity by providing input to the cortical motor areas. This is also the cerebellar region that stores procedural memories.

Recent discoveries suggest that in addition to these well-established functions, the cerebellum has even broader responsibilities, such as perhaps coordinating the brain's acquisition of sensory input. Researchers are currently trying to make sense of new and surprising findings that do not fit with the cerebellum's traditional roles in motor control.

The following symptoms of cerebellar disease can be referred to a loss of the cerebellum's established motor functions: poor balance, nystagmus (rhythmic, oscillating eye movements),

reduced muscle tone but no paralysis, inability to perform rapid movements smoothly, and inability to stop and start skeletal muscle action quickly. The latter gives rise to an *intention tremor* characterized by oscillating to-and-fro movements of a limb as it approaches its intended destination. In trying to pick up a pencil, a person with cerebellar damage may overshoot the pencil and then rebound excessively, repeating this to-and-fro process until successful. No tremor is observed except in performing intentional activity, in contrast to the resting tremor associated with disease of the basal ganglia, most notably Parkinson's disease.

The cerebellum and basal ganglia both monitor and adjust motor activity commanded from the motor cortex, and, like the basal ganglia, the cerebellum does not directly influence the efferent motor neurons. Although they perform different roles (e.g., the cerebellum enhances muscle tone, whereas the basal ganglia inhibit it), both function indirectly by modifying the output of major motor systems in the brain. The motor command for a particular voluntary activity arises from the motor cortex, but the actual execution of that activity is coordinated subconsciously by these subcortical regions. To illustrate, you can voluntarily decide to get up and walk, but you do not have to consciously think about the specific sequence of movements needed to accomplish this intentional act. Accordingly, much voluntary activity is actually involuntarily regulated. You will learn more about motor control when we discuss muscle physiology in Chapter 7.

## 3.9 | Basal Ganglia

The **basal ganglia** consist of several masses of grey matter located deep within the cerebral white matter (⟩ Figure 3-28) and are associated with a variety of functions, including motor control, cognition, emotions, and learning.

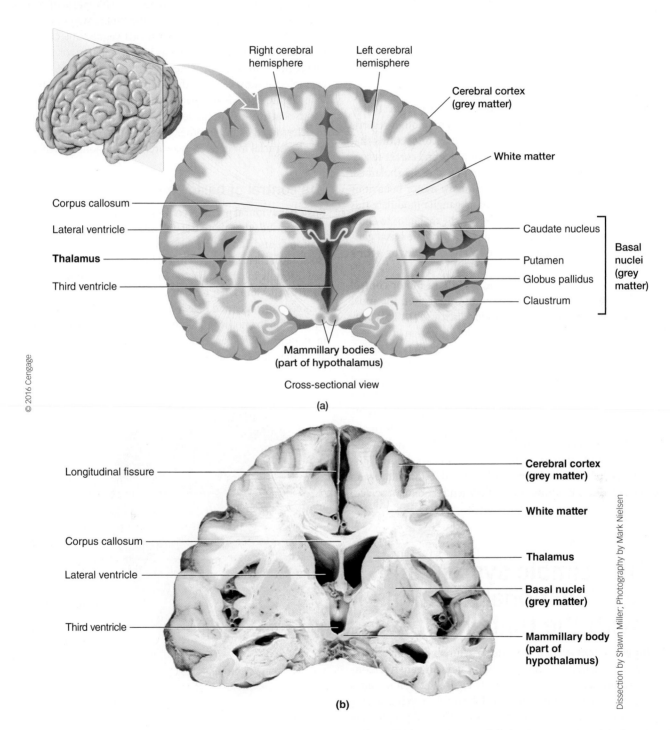

⟩ **FIGURE 3-28 Frontal section of the brain.** (a) Schematic frontal section of the brain. The cerebral cortex, an outer shell of grey matter, surrounds an inner core of white matter. Deep within the cerebral white matter are several masses of grey matter, the basal ganglia. The ventricles are cavities in the brain through which the cerebrospinal fluid flows. The thalamus forms the walls of the third ventricle. (b) Photograph of a frontal section of the brain of a cadaver

© 2016 Cengage

Dissection by Shawn Miller; Photography by Mark Nielsen

The basal ganglia play a complex role in controlling movement, in addition to having nonmotor functions that are less understood. More specifically, the basal ganglia are important in (1) inhibiting muscle tone throughout the body (proper muscle tone is normally maintained by a balance of excitatory and inhibitory inputs to the neurons that innervate skeletal muscles); (2) selecting and maintaining purposeful motor activity while suppressing useless or unwanted patterns of movement; and (3) helping monitor and coordinate slow, sustained contractions, especially those related to posture and support. The basal ganglia do not directly influence the efferent motor neurons that bring about muscle contraction but do modify ongoing activity in motor pathways.

To accomplish these complex integrative roles, the basal ganglia receive and send out much information, as is indicated by the tremendous number of fibres linking them to other regions of the brain—for example, the strategic interconnections that form a complex feedback loop linking the cerebral cortex (especially its motor regions), the basal ganglia, and the thalamus. The thalamus positively reinforces voluntary motor behaviour initiated by the cortex, whereas the basal ganglia modulate this activity by exerting an inhibitory effect on the thalamus to eliminate antagonistic or unnecessary movements. The basal ganglia also exert an inhibitory effect on motor activity by acting through neurons in the brain stem.

*Clinical Note* The importance of the basal ganglia in motor control is evident in diseases involving this region, the most common of which is *Parkinson's disease (PD)*. This condition is associated with a deficiency of dopamine, an important neurotransmitter in the basal ganglia (see Integration of Human Physiology: Movement and Parkinson's Disease, p. 339). Because the basal ganglia lack enough dopamine to exert their normal roles, three types of motor disturbances characterize PD: (1) increased muscle tone, or rigidity; (2) involuntary, useless, or unwanted movements, such as *resting tremors* (e.g., hands rhythmically shaking, making it difficult or impossible to hold a cup of coffee); and (3) slowness in initiating and carrying out different motor behaviours. People with PD find it difficult to stop ongoing activities. If sitting down, they tend to remain seated, and if they get up, they do so very slowly.

# 3.10 | The Limbic System and Its Functional Relations with the Higher Cortex

The **limbic system** is not a separate structure but a ring of forebrain structures that surround the brain stem and are interconnected by intricate neuron pathways (> Figure 3-29). It includes portions of each of the following: the lobes of the cerebral cortex (especially the limbic association cortex), the basal ganglia, the thalamus, and the hypothalamus. This complex interacting network is associated with emotions, basic survival and sociosexual behavioural patterns, motivation, and learning. Let's examine each of these brain functions further.

## The limbic system and emotion

The concept of **emotion** encompasses subjective emotional feelings and moods (such as anger, fear, and happiness) plus the overt physical responses associated with these feelings. These responses include specific behavioural patterns (e.g., preparing for attack or defence when angered by an adversary) and observable emotional expressions (e.g., laughing, crying, or blushing). Evidence points to a central role for the limbic system in all aspects of emotion. Stimulating specific regions within the limbic system of humans during brain surgery produces various vague subjective sensations that the patient describes as joy, satisfaction, or pleasure in one region and discouragement, fear, or anxiety in another. For example, the **amygdala**, on the interior underside of the temporal lobe (Figure 3-29), is an especially important region for processing inputs that give rise to the sensation of fear. In humans and to an undetermined extent in other species, higher levels of the cortex are also crucial for conscious awareness of emotional feelings.

## The control of basic behavioural patterns

**Basic behavioural patterns** controlled at least in part by the limbic system include those aimed at individual survival (attack, searching for food) and those directed toward perpetuating the species (sociosexual behaviours conducive to mating). In experimental animals, stimulating the limbic system brings about complex and even bizarre behaviours. For example, stimulation in one area can elicit responses of anger and rage in a normally docile animal, whereas stimulation in another area results in placidity and tameness, even in an otherwise vicious animal. Stimulation in yet another limbic area can induce sexual behaviours, such as copulatory movements.

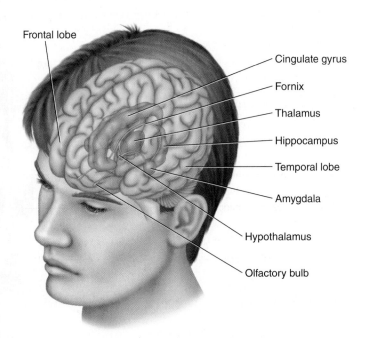

> FIGURE 3-29 **Limbic system.** This partially transparent view of the brain reveals the structures composing the limbic system.

**Source:** From Sherwood. *Human Physiology*, 8E. © 2013 Brooks/Cole, a part of Cengage, Inc. Reproduced by permission. www.cengage.com/permissions

## ROLE OF THE HYPOTHALAMUS IN BASIC BEHAVIOURAL PATTERNS

The relationships among the hypothalamus, limbic system, and higher cortical regions regarding emotions and behaviour are still not well understood. Apparently the extensive involvement of the hypothalamus in the limbic system governs the involuntary internal responses of various body systems in preparation for appropriate action to accompany a particular emotional state. For example, the hypothalamus controls the increase of heart rate and respiratory rate, elevation of blood pressure, and diversion of blood to skeletal muscles that occur in anticipation of attack or when angered. These preparatory changes in the internal state require no conscious control.

## ROLE OF THE HIGHER CORTEX IN BASIC BEHAVIOURAL PATTERNS

In executing complex behavioural activities, such as attack, flight, or mating, the individual (animal or human) must interact with the external environment. Higher cortical mechanisms are called into play to connect the limbic system and hypothalamus with the outer world so that appropriate overt behaviours are manifested. At the simplest level, the cortex provides the neural mechanisms necessary for implementing the appropriate skeletal muscle activity required to approach or avoid an adversary, participate in sexual activity, or display emotional expression. For example, the stereotypical sequence of movement for the universal human emotional expression of smiling is apparently preprogrammed in the cortex and can be called forth by the limbic system. One can also voluntarily call forth the smile program, as when posing for a picture. Even individuals who are blind from birth have normal facial expressions; that is, they do not learn to smile by observation. Smiling means the same thing in every culture, despite widely differing environmental experiences. Such behavioural patterns shared by all members of a species are believed to be more abundant in lower animals.

Higher cortical levels also can reinforce, modify, or suppress basic behavioural responses so that actions can be guided by planning, strategy, and judgment based on an understanding of the situation. Even if you were angry at someone and your body were internally preparing for attack, you probably would judge that an attack would be inappropriate and could consciously suppress the external manifestation of this basic emotional behaviour. Thus, the higher levels of the cortex, particularly the prefrontal and limbic association areas, are important in conscious, learned control of innate behavioural patterns. Using fear as an example, exposure to an aversive experience calls two parallel tracks into play for processing this emotional stimulus: a fast track in which the lower-level amygdala plays a key role, and a slower track mediated primarily by the higher-level prefrontal cortex. The fast track permits a rapid, rather crude, instinctive response—a gut reaction—and is essential for the feeling of being afraid. The slower track involving the prefrontal cortex permits a more refined response to the aversive stimulus, based on a rational analysis of the current situation compared with stored past experiences. The prefrontal cortex formulates plans and guides behaviour, thereby suppressing amygdala-induced responses that may be inappropriate for the situation at hand.

## REWARD AND PUNISHMENT CENTRES

An individual tends to reinforce behaviours that have proved gratifying and to suppress behaviours that have been associated with unpleasant experiences. Certain regions of the limbic system have been designated as **reward centres** and **punishment centres**, because stimulation in these respective areas gives rise to pleasant or unpleasant sensations. When a self-stimulating device is implanted in a reward centre, an experimental animal will self-deliver up to 5000 stimulations per hour and will even shun food when starving, in preference for the pleasure derived from self-stimulation. In contrast, when the device is implanted in a punishment centre, animals will avoid stimulation at all costs. Reward centres are found most abundantly in regions involved in mediating the highly motivated behavioural activities of eating, drinking, and sexual activity.

## Motivated behaviours

**Motivation** is the ability to direct behaviour toward specific goals. Some goal-directed behaviours are aimed at satisfying specific identifiable physical needs related to homeostasis. **Homeostatic drives** represent the subjective urges associated with specific bodily needs that motivate appropriate behaviour to satisfy those needs. For example, the sensation of thirst accompanying a water deficit in the body drives an individual to drink to satisfy the homeostatic need for water. However, whether water, a soft drink, or another beverage is chosen as the thirst quencher is unrelated to homeostasis. Much human behaviour does not depend on purely homeostatic drives related to simple tissue deficits, such as thirst. Human behaviour is influenced by experience, learning, and habit, shaped in a complex framework of unique personal gratifications blended with cultural expectations. To what extent, if any, motivational drives unrelated to homeostasis, such as the drive to pursue a particular career or win a certain race, are involved with the reinforcing effects of the reward and punishment centres is unknown. Indeed, some individuals motivated toward a particular goal may even deliberately "punish" themselves in the short term to achieve their long-range gratification (e.g., the temporary pain of training in preparation for winning a competitive athletic event).

## Neurotransmitters in pathways for emotions and behaviour

The underlying neurophysiological mechanisms responsible for the psychological observations of emotions and motivated behaviour largely remain a mystery, although the neurotransmitters *norepinephrine, dopamine,* and *serotonin* all have been implicated. Norepinephrine and dopamine, both chemically classified as *catecholamines*, are known transmitters in the regions that elicit the highest rates of self-stimulation in animals equipped with do-it-yourself devices. A number of **psychoactive drugs** affect moods in humans, and some of these drugs have also been shown to influence self-stimulation in experimental animals. For example, increased self-stimulation is observed after the administration of drugs that increase catecholamine synaptic activity, such as *amphetamine*, an "upper" drug.

Although most psychoactive drugs are used therapeutically to treat various mental disorders, others, unfortunately, are abused. Many abused drugs act by enhancing the effectiveness of dopamine in the pleasure pathways, initially giving rise to an intense sensation of pleasure. As you learned in Chapter 2, one example is cocaine, which blocks the reuptake of dopamine at synapses (p. 82).

**Depression** is among the mental disorders associated with defects in limbic system neurotransmitters. A functional deficiency of serotonin or norepinephrine or both is implicated in depression, a disorder characterized by a pervasive negative mood accompanied by a generalized loss of interests, an inability to experience pleasure, and suicidal tendencies. All effective antidepressant drugs increase the available concentration of these neurotransmitters in the CNS. *Prozac*, one of the most commonly prescribed drugs in Canadian psychiatry, is illustrative. It blocks the reuptake of released serotonin, thus prolonging serotonin activity at synapses. Serotonin and norepinephrine are synaptic messengers in the limbic regions of the brain involved in pleasure and motivation, suggesting that the pervasive sadness and lack of interest (no motivation) in depressed patients are related at least in part to disruption of these regions by deficiencies or decreased effectiveness of these neurotransmitters.

Researchers are optimistic that as understanding of the molecular mechanisms of mental disorders is expanded in the future, many psychiatric problems can be corrected or better managed through drug intervention, a hope of great medical significance.

---

### Check Your Understanding 3.6

1. State the brain functions associated with the limbic system.
2. Name the brain area most important in processing the sensation of fear.

---

## 3.11 | Learning, Memory, and Language

### Learning

**Learning** is the acquisition of knowledge or skills as a consequence of experience, instruction, or both. It is widely believed that rewards and punishments are integral parts of many types of learning. If an animal is rewarded on responding in a particular way to a stimulus, the likelihood increases that the animal will respond in the same way again to the same stimulus as a consequence of this experience. Conversely, if a particular response is accompanied by punishment, the animal is less likely to repeat the same response to the same stimulus. When behavioural responses that give rise to pleasure are reinforced or those accompanied by punishment are avoided, learning has taken place. Housebreaking a puppy is an example. If the puppy is praised when it urinates outdoors but scolded when it wets the carpet, it soon learns the acceptable place to empty its bladder. In

this regard, learning involves a change in behaviour that occurs as a result of experiences. It is highly dependent on the organism's interaction with its environment. The only limits to the effects that environmental influences can have on learning are the biological constraints imposed by species-specific and individual genetic endowments.

### Memory

**Memory** is the storage of acquired knowledge for later recall. Learning and memory form the basis by which individuals adapt their behaviour to their particular external circumstances. Without these mechanisms, it would be impossible for individuals to plan for successful interactions and to intentionally avoid predictably disagreeable circumstances.

The neural change responsible for retention or storage of knowledge is known as the **memory trace**. Generally, concepts—not verbatim information—are stored. As you read this page, you are storing the concept discussed, not the specific words. Later, when you retrieve the concept from memory, you will convert it into your own words. It is possible, however, to memorize bits of information word by word.

Storage of acquired information is accomplished in at least two stages: short-term memory and long-term memory (Table 3-3). **Short-term memory** lasts for seconds to hours, whereas **long-term memory** is retained for days to years. The process of transferring and fixing short-term memory traces into long-term memory stores is known as **consolidation**.

A recently developed concept is that of **working memory**, or what has been called "the erasable blackboard of the mind." Working memory temporarily holds and interrelates various pieces of information relevant to a current mental task. Through your working memory, you briefly hold and process data for immediate use—both newly acquired information and related, previously stored knowledge that is transiently brought forth into working memory—so that you can evaluate the incoming data in context. This integrative function is crucial to your ability to reason, plan, and make judgments. By comparing and manipulating new and old information within your working memory, you can comprehend what you are reading, carry on a conversation, calculate a restaurant tip in your head, find your way home, and know that you should put on warm clothing if you see snow outside. In short, working memory enables people to string thoughts together in a logical sequence and plan for future action.

Although strong evidence is still lacking, new findings hint that once an established memory is actively recalled, it becomes labile (unstable or subject to change) and must be **reconsolidated** back into a restabilized, inactive state. According to this controversial proposal, new information may be incorporated into the old memory trace during reconsolidation.

#### COMPARISON OF SHORT-TERM AND LONG-TERM MEMORY

 Newly acquired information is initially deposited in short-term memory, which has a limited capacity for storage. Information in short-term memory has one of two eventual fates. Either it is soon forgotten (e.g.,

## ▮ TABLE 3-3 Comparison of Short-Term and Long-Term Memory

| Characteristic | Short-Term Memory | Long-Term Memory |
| --- | --- | --- |
| **Time of storage after acquisition of new information** | Immediate | Later; must be transferred from short-term to long-term memory through consolidation; enhanced by practice or recycling of information through short-term mode |
| **Duration** | Lasts for seconds to hours | Retained for days to years |
| **Capacity of storage** | Limited | Very large |
| **Retrieval time (remembering)** | Rapid retrieval | Slower retrieval, except for thoroughly ingrained memories, which are rapidly retrieved |
| **Inability to retrieve (forgetting)** | Permanently forgotten; memory fades quickly unless consolidated into long-term memory | Usually only transiently unable to access; relatively stable memory trace |
| **Mechanism of storage** | Involves transient modifications in functions of pre-existing synapses, such as altering amount of neurotransmitter released | Involves relatively permanent functional or structural changes between existing neurons, such as formation of new synapses; synthesis of new proteins plays a key role |

forgetting a telephone number after you have looked it up and finished dialling), or it is transferred into the more permanent long-term memory mode through *active practice* or *rehearsal*. The recycling of newly acquired information through short-term memory increases the likelihood that the information will be consolidated into long-term memory. (Therefore, when you cram for an exam, your long-term retention of the information is poor!) This relationship can be likened to developing photographic film. The originally developed image (short-term memory) rapidly fades unless it is chemically fixed (consolidated) to provide a more enduring image (long-term memory). Sometimes only parts of memories are fixed, whereas others fade away. Information of interest or importance to the individual is more likely to be recycled and fixed in long-term stores, whereas less important information is quickly erased.

The storage capacity of the long-term memory bank is much larger than the capacity of short-term memory. Different informational aspects of long-term memory traces seem to be processed and codified, then stored with other memories of the same type; for example, visual memories are stored separately from auditory memories. This organization facilitates future searching of memory stores to retrieve desired information. For example, in remembering a woman you once met, you may use various recall cues from different storage pools, such as her name, her appearance, the fragrance she wore, an incisive point she made, or the song playing in the background.

Stored knowledge is of no use unless it can be retrieved and used to influence current or future behaviour. Because long-term memory stores are larger, it often takes longer to retrieve information from long-term memory than from short-term memory. *Remembering* is the process of retrieving specific information from memory stores; *forgetting* is the inability to retrieve stored information. Information lost from short-term memory is permanently forgotten, but information in long-term storage is frequently forgotten only transiently. Often you are only temporarily unable to access the information—for example, not being able to remember an acquaintance's name, then having it suddenly occur to you later.

Some forms of long-term memory involving information or skills used on a daily basis are essentially never forgotten and are rapidly accessible, such as knowing your own name or being able to write. Even though long-term memories are relatively stable, stored information may be gradually lost or modified over time, unless it is thoroughly ingrained as a result of years of practice.

### AMNESIA

*Clinical Note* Occasionally, individuals suffer from a lack of memory that involves whole portions of time rather than isolated bits of information. This condition, known as **amnesia**, occurs in two forms. The most common form, *retrograde* (going backward) *amnesia*, is the inability to recall recent past events. It usually follows a traumatic event that interferes with electrical activity of the brain, such as a concussion or stroke. If a person is knocked unconscious, the content of short-term memory is essentially erased, resulting in loss of memory about activities that occurred within about the last half hour before the event. Severe trauma may interfere with access to recently acquired information in long-term stores as well.

*Anterograde* (going forward) *amnesia*, conversely, is the inability to store memory long term for later retrieval. It is usually associated with lesions of the medial portions of the temporal lobes, which are generally considered critical regions for memory consolidation. People suffering from this condition may be able to recall things they learned before the onset of their problem, but they cannot establish new permanent memories. New information is lost as quickly as it fades from short-term memory. In one case study, the person could not remember where the bathroom was in his new home but still had total recall of his old home.

## Memory traces

What parts of the brain are responsible for memory? There is no single memory centre in the brain. Instead, the neurons involved in memory traces are widely distributed throughout the subcortical and cortical regions of the brain. The regions of the brain most extensively implicated in memory include the hippocampus and associated structures of the medial (inner) temporal lobes, the limbic system, the cerebellum, the prefrontal cortex, and other regions of the cerebral cortex.

### THE HIPPOCAMPUS AND DECLARATIVE MEMORIES

The **hippocampus** is the elongated, medial portion of the temporal lobe that is part of the limbic system (see Figure 3-29). It plays a vital role in short-term memory involving the integration of various related stimuli and is also crucial for consolidation into long-term memory. The hippocampus is believed to store new long-term memories only temporarily and then transfer them to other cortical sites for more permanent storage. The sites for long-term storage of various types of memories are only beginning to be identified by neuroscientists.

The hippocampus and surrounding regions play an especially important role in **declarative memories**—the "what" memories of specific people, places, objects, facts, and events. This type of memory often results after only one experience and can be declared in a statement such as, "I saw the CN Tower last summer" or conjured up in a mental image. Declarative memories require conscious recall. The hippocampus and associated temporal/limbic structures are especially important in maintaining a durable record of the everyday episodic events in our lives.

*Clinical Note* People with hippocampal damage are profoundly forgetful of facts critical to daily functioning. Declarative memories typically are the first to be lost. Interestingly, extensive damage in the hippocampus region is evident during autopsy in patients with Alzheimer's disease.

### THE CEREBELLUM AND PROCEDURAL MEMORIES

In contrast to the role of the hippocampus and surrounding temporal/limbic regions in declarative memories, the cerebellum and relevant cortical regions play an essential role in **procedural memories**—the "how to" memories involving motor skills gained through repetitive training, such as memorizing a particular dance routine. The cortical areas important for a given procedural memory are the specific motor or sensory systems engaged in performing the routine. In contrast to declarative memories, which are consciously recollected from previous experiences, procedural memories can be brought forth without conscious effort. For example, during a competition, an ice skater typically performs best by "letting the body take over" the routine instead of thinking about exactly what needs to be done next.

*Clinical Note* The distinct localization in different parts of the brain of these two types of memory is apparent in people who have temporal/limbic lesions. They can perform a skill, such as playing a piano, but the next day they have no recollection that they did so.

### THE PREFRONTAL CORTEX AND WORKING MEMORY

The major orchestrator of the complex reasoning skills associated with *working memory* is the prefrontal association cortex. The prefrontal cortex not only serves as a temporary on-line storage site for holding relevant data but also is largely responsible for the so-called executive functions involving manipulation and integration of information for planning, juggling competing priorities, problem solving, and organizing activities. The prefrontal cortex carries out these complex reasoning functions in cooperation with all the brain's sensory regions, which are linked to the prefrontal cortex through neural connections. Researchers have identified different storage bins in the prefrontal cortex, depending on the nature of the current relevant data. For example, working memory involving spatial cues is in a prefrontal location distinct from working memory involving verbal cues or cues about an object's appearance. A recent fascinating proposal suggests that level of intelligence may be determined by the capacity of a person's working memory to temporarily hold and relate a variety of relevant data.

## Short-term and long-term memory: Different molecular mechanisms

As well as the question of where memory is stored, there is the question of how it is stored. Despite a vast amount of psychological data, only a few tantalizing scraps of physiological evidence concerning the cellular basis of memory traces are available. Obviously, some change must take place within the neural circuitry of the brain to account for the altered behaviour that follows learning. A single memory does not reside in a single neuron but rather in changes in the pattern of signals transmitted across synapses within a vast neuronal network.

Different mechanisms are responsible for short-term and long-term memory. Short-term memory involves transient modifications in the function of pre-existing synapses, such as a temporary change in the amount of neurotransmitter released in response to stimulation or a temporary increased responsiveness of the postsynaptic cell to the neurotransmitter within affected nerve pathways. Long-term memory, in contrast, involves relatively permanent functional or structural changes between existing neurons in the brain. Let's look at each of these types of memory in more detail.

## Short-term memory

Experiments in the sea snail *Aplysia* have shown that two forms of short-term memory—habituation and sensitization—are due to modification of different channel proteins in presynaptic terminals of specific afferent neurons involved in the pathway that mediates the behaviour being modified. This modification, in turn, brings about changes in transmitter release. **Habituation** is a decreased responsiveness to repetitive presentations of an indifferent stimulus—that is, one that is neither rewarding nor punishing. **Sensitization** is increased responsiveness to mild stimuli following a strong or noxious stimulus. *Aplysia* reflexly withdraws its gill when its siphon, a breathing organ at the top of its gill, is touched. Afferent neurons responding to touch of the siphon (presynaptic neurons) directly synapse on efferent motor neurons (postsynaptic neurons) controlling gill withdrawal. The snail becomes habituated when its siphon is repeatedly touched; that is, it learns to ignore the stimulus and no longer withdraws its gill in response. Sensitization, a more complex form of learning, takes place in *Aplysia* when it is given a hard bang on the siphon. Subsequently, the snail withdraws its gill more vigorously in response to even mild touch. Interestingly, these different forms of learning affect the same site—the synapse between a siphon afferent and a gill efferent—in opposite ways. Habituation depresses this synaptic activity, whereas sensitization enhances it (> Figure 3-30). These transient modifications persist for as long as the time course of the memory.

### MECHANISM OF HABITUATION

In habituation, the closing of $K^+$ channels reduces $K^+$ entry into the presynaptic terminal, which leads to a decrease in neurotransmitter release. As a consequence, the postsynaptic potential is reduced compared with normal, resulting in a decrease or absence of the behavioural response controlled by the postsynaptic efferent neuron (gill withdrawal in *Aplysia*). Thus, the memory for habituation in *Aplysia* is stored in the form of modification of specific $K^+$ channels. With no further training, this reduced responsiveness lasts for several hours. A similar process is responsible for short-term habituation in other species studied. This suggests that $K^+$ channel modification is a general mechanism of habituation; however, in higher species the involvement of intervening interneurons makes the process somewhat more complicated. Habituation is probably the most common form of learning and is believed to be the first learning process to take place in human infants. By learning to ignore indifferent stimuli, the animal or person is free to attend to other more important stimuli.

### MECHANISM OF SENSITIZATION

Sensitization in *Aplysia* also involves channel modification, but a different channel and mechanism are involved. In contrast to what happens in habituation, $K^+$ entry into the presynaptic terminal is enhanced in sensitization. The subsequent increase in neurotransmitter release produces a larger postsynaptic potential, resulting in a more vigorous gill-withdrawal response. Sensitization does not have a direct effect on presynaptic $K^+$ channels. Instead, it indirectly enhances $K^+$ entry via presynaptic facilitation (p. 81). The neurotransmitter serotonin is released from a facilitating interneuron that synapses on the presynaptic terminal to bring about increased release of presynaptic neurotransmitter in response to an action potential. It does so by triggering activation of a cyclic AMP (adenosine monophosphate) second messenger within the presynaptic terminal, which ultimately brings about an inhibition of $K^+$ channel function in the terminals. Cyclic AMP is a second messenger used in many

biological processes, and is derived from adenosine triphosphate (ATP) and used for intracellular signal transduction (see more information on cyclic AMP in Chapter 5, p. 217). This blockage prolongs the action potential in the presynaptic terminal. Remember that $K^+$ efflux through opened $K^+$ channels hastens the return to resting potential (repolarization) during the falling phase of the action potential. Because the presence of a local action potential is responsible for the opening of $K^+$ channels in the terminal, a prolonged action potential permits the greater $K^+$ influx associated with sensitization.

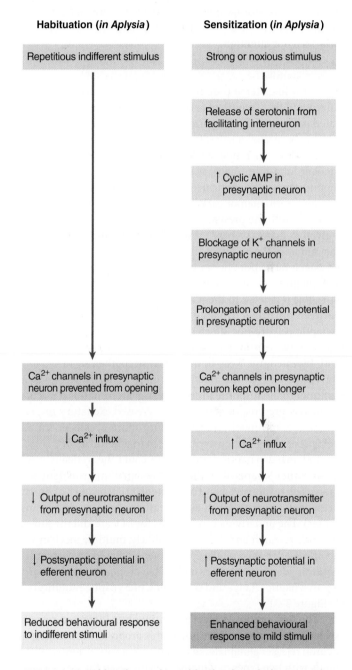

> **FIGURE 3-30 Habituation and sensitization in Aplysia.** Researchers have shown that in the sea snail Aplysia, two forms of short-term memory—habituation and sensitization—result from opposite changes in neurotransmitter release from the same presynaptic neuron, caused by different transient channel modifications.

Thus, existing synaptic pathways may be functionally interrupted (habituated) or enhanced (sensitized) during simple learning. Scientists speculate that much of short-term memory is similarly a temporary modification of already existing processes. Several lines of research suggest that the cyclic AMP cascade, especially activation of protein kinase, plays an important role, at least in elementary forms of learning and memory.

Further studies have revealed that declarative memories, which involve conscious awareness and are more complex than habituation and sensitization, are initially stored by means of more persistent changes in activity of existing synapses. Specifically, initial storage of declarative information appears to be accomplished by means of long-term potentiation, to which we now turn our attention.

### MECHANISM OF LONG-TERM POTENTIATION

The term **long-term potentiation (LTP)** refers to a prolonged increase in the strength of existing synaptic connections in activated pathways following brief periods of repetitive stimulation. LTP has been shown to last for days or even weeks—long enough for this short-term memory to be consolidated into more permanent long-term memory. LTP is especially prevalent in the hippocampus, a site critical for converting short-term memories into long-term memories. When LTP occurs, simultaneous activation of both the presynaptic and postsynaptic neurons across a given excitatory synapse results in long-lasting modifications that enhance the ability of the presynaptic neuron to excite the postsynaptic neuron. Because both the presynaptic and postsynaptic neurons must be active at the same time, the development of LTP is restricted to the pathway that is stimulated. Pathways between other inactive presynaptic inputs and the same postsynaptic cell are not affected. With LTP, signals from a given presynaptic neuron to a postsynaptic neuron become stronger with repeated use. Keep in mind that strengthening of synaptic activity results in the formation of more EPSPs in the postsynaptic neuron in response to chemical signals from this particular excitatory presynaptic input. This increased excitatory responsiveness is ultimately translated into more action potentials being sent along this postsynaptic cell to other neurons.

Enhanced synaptic transmission could theoretically result from either changes in the postsynaptic neuron (such as increased responsiveness to the neurotransmitter) or in the presynaptic neuron (such as increased release of neurotransmitter). The underlying mechanisms for LTP are still the subject of much research and debate. Most likely, multiple mechanisms are involved in this complex phenomenon. Based on current scientific evidence, here are two plausible proposals, one involving a postsynaptic change and the other a presynaptic modification (> Figure 3-31).

The first proposed mechanism involves increased responsiveness of the postsynaptic cell. In this proposal, the presynaptic neuron releases glutamate, an excitatory neurotransmitter, which binds with two types of glutamate receptors in the postsynaptic neuron's plasma membrane—**NMDA receptors** and *AMPA receptors*. Because *NMDA receptors* act as nonselective cation channels, their activation and opening on binding with glutamate leads to $K^+$ as well as $K^+$ entry into the postsynaptic cell.

The entering calcium activates a $K^+$ second-messenger pathway in the postsynaptic neuron. This second-messenger pathway leads to the physical insertion of additional AMPA receptors in the postsynaptic membrane. **AMPA receptors** are primarily responsible for generating EPSPs in response to glutamate activation. Because of this increased availability of AMPA receptors, the postsynaptic cell exhibits a greater EPSP response to subsequent release of glutamate from the presynaptic cell. This heightened sensitivity of the postsynaptic neuron to glutamate from the presynaptic cell helps maintain LTP.

A second proposed mechanism involves increased neurotransmitter release from the presynaptic cell as a major contributor to LTP. In this proposed mechanism, activation of the $K^+$ second-messenger pathway in the postsynaptic neuron causes this postsynaptic cell to release a retrograde factor that diffuses to the presynaptic neuron. Here, the retrograde factor activates a second-messenger system in the presynaptic neuron, ultimately enhancing the release of glutamate from the presynaptic neuron. This positive feedback strengthens the signalling process at this synapse, helping sustain LTP. Note that in this mechanism for the development of LTP, a chemical factor from the postsynaptic neuron influences the presynaptic neuron— exactly the opposite direction of neurotransmitter activity at a synapse. Synapses traditionally operate unidirectionally, with the presynaptic neuron releasing a neurotransmitter that influences the postsynaptic neuron. The cell body and dendrites of the postsynaptic neuron do not contain any transmitter vesicles. Therefore, the retrograde factor is distinct from classical neurotransmitters or **neuropeptides**. Most investigators believe that the retrograde messenger is **nitric oxide**, a chemical that recently has been found to perform a bewildering array of other functions in the body. These other functions range from dilation of blood vessels in the penis during erection to destruction of foreign invaders by the immune system (p. 463).

Studies suggest a regulatory role for the cAMP second-messenger pathway in the development and maintenance of LTP, in addition to the $K^+$ second-messenger pathway. Participation of cAMP may hold a key to linking short-term memory to long-term memory consolidation.

## Long-term memory

Whereas short-term memory involves transient strengthening of pre-existing synapses, long-term memory storage requires the activation of specific genes that control the synthesis of proteins needed for lasting structural or functional changes at specific synapses. Examples of such changes include the formation of new synaptic connections or permanent changes in presynaptic or postsynaptic membranes. Thus, long-term memory involves permanent physical changes in the brain.

The brains of experimental animals reared in a sensory-deprived environment are observably different at a microscopic level from animals raised in a sensory-rich environment. The animals afforded more environmental interactions—and supposedly, therefore, more opportunity to learn—displayed greater branching and elongation of dendrites in nerve cells in regions of the brain suspected to be involved with memory storage. Greater

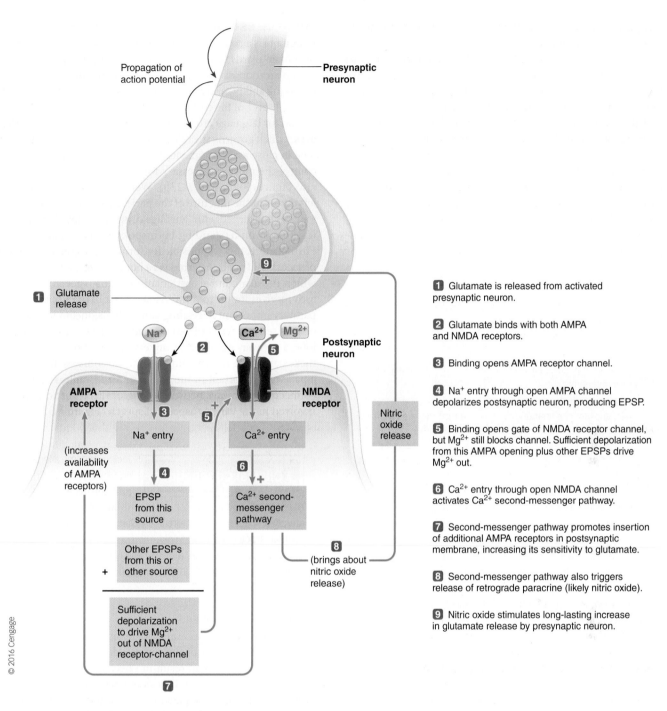

Propagation of action potential

**Presynaptic neuron**

1 Glutamate release

Na⁺  Ca²⁺  Mg²⁺

**Postsynaptic neuron**

2  5

**AMPA receptor**  **NMDA receptor**

3  5

Nitric oxide release

(increases availability of AMPA receptors)

Na⁺ entry  Ca²⁺ entry

4  6

EPSP from this source

Ca²⁺ second-messenger pathway

Other EPSPs from this or other source

(brings about nitric oxide release)

+

Sufficient depolarization to drive Mg²⁺ out of NMDA receptor-channel

7

9  +

© 2016 Cengage

1 Glutamate is released from activated presynaptic neuron.

2 Glutamate binds with both AMPA and NMDA receptors.

3 Binding opens AMPA receptor channel.

4 Na⁺ entry through open AMPA channel depolarizes postsynaptic neuron, producing EPSP.

5 Binding opens gate of NMDA receptor channel, but Mg²⁺ still blocks channel. Sufficient depolarization from this AMPA opening plus other EPSPs drive Mg²⁺ out.

6 Ca²⁺ entry through open NMDA channel activates Ca²⁺ second-messenger pathway.

7 Second-messenger pathway promotes insertion of additional AMPA receptors in postsynaptic membrane, increasing its sensitivity to glutamate.

8 Second-messenger pathway also triggers release of retrograde paracrine (likely nitric oxide).

9 Nitric oxide stimulates long-lasting increase in glutamate release by presynaptic neuron.

› FIGURE 3-31 **Possible pathways for long-term potentiation**

dendritic surface area presumably provides more sites for synapses. This means long-term memory may be stored at least in part by a particular pattern of dendritic branching and synaptic contacts.

A positive regulatory protein, **CREB**, is the molecular switch that activates (turns on) the genes important in long-term memory storage. Another related molecule, **CREB2**, is a repressor of CREB-facilitated protein synthesis. Formation of enduring memories involves not only the activation of positive regulatory factors (CREB) that favour memory storage but also the turning off of inhibitory constraining factors (CREB2) that prevent memory storage. The shifting balance between positive

and repressive factors is believed to ensure that only information relevant to the individual, not everything encountered, is put into long-term storage.

The mechanism that activates CREB and inhibits CREB2 is still unknown, but may be associated with cAMP activating CREB. This second messenger plays a regulatory role in LTP as well as in simpler forms of short-term memory, such as sensitization. Furthermore, cAMP can activate CREB, which ultimately leads to new protein synthesis and the consolidation of long-term memory. Consolidation of both declarative and procedural memories depends on CREB (and presumably CREB2). CREB serves as a common molecular switch for converting both these categories of memory from the transient short-term to the permanent long-term mode.

The CREB regulatory proteins regulate a group of genes, the **immediate early genes (IEGs)**, which play a critical role in memory consolidation. These genes govern the synthesis of the proteins that encode long-term memory. The exact role that these critical new long-term memory proteins might play remains speculative. They may be needed for structural changes in dendrites or used for synthesis of more neurotransmitters or additional receptor sites. Alternatively, they may accomplish long-term modification of neurotransmitter release by sustaining biochemical events initially activated by short-term memory processes.

To complicate the issue even further, numerous hormones and neuropeptides are known to affect learning and memory processes.

---

### Check Your Understanding 3.7

1. Define *consolidation*.
2. Compare the molecular mechanisms for short-term and long-term memories.
3. Distinguish among declarative memories, procedural memories, and working memory, and indicate the brain area primarily associated with each.

---

## Language

Language ability is an excellent example of early cortical plasticity coupled with later permanence. Unlike the sensory and motor regions of the cortex, which are present in both hemispheres, in the vast majority of people the areas of the brain responsible for language ability are found in only one hemisphere—the left.

*Clinical Note* However, if a child under the age of two accidentally suffers damage to the left hemisphere, language functions are transferred to the right hemisphere with no delay in language development, but at the expense of less obvious nonverbal abilities for which the right hemisphere is normally responsible. Up to about the age of 10, after damage to the left hemisphere, language ability can usually be re-established in the right hemisphere following a temporary period of loss. If damage occurs beyond the early teens however, language ability is permanently impaired, even though some limited restoration may be possible. The regions of the brain involved in comprehending and expressing language apparently are permanently assigned before adolescence.

Even in normal, healthy individuals, there is evidence for early plasticity and later permanence in language development. Infants can distinguish between and articulate the entire range of speech sounds, but each language uses only a portion of these sounds. As children mature, they often lose the ability to distinguish between or express speech sounds that are not important in their native language. For example, Japanese children can distinguish between the sounds of "r" and "l," but many Japanese adults cannot perceive the difference between them.

### ROLES OF BROCA'S AREA AND WERNICKE'S AREA

**Language** is a complex form of communication in which written or spoken words symbolize objects and convey ideas. It involves the integration of two distinct capabilities—namely, *expression* (speaking ability) and *comprehension*—each of which is related to a specific area of the cortex. The primary areas of cortical specialization for language are Broca's area and Wernicke's area. **Broca's area**, which governs speaking ability, is located in the left frontal lobe in close association with the motor areas of the cortex that control the muscles necessary for articulation (Figure 3-32; see also Figures 3-23a and 3-23b). **Wernicke's area**, located in the left cortex at the juncture of the parietal, temporal, and occipital lobes, is concerned with language comprehension. It plays a critical role in understanding both spoken and written messages. Furthermore, it is responsible for formulating coherent patterns of speech that are transferred via a bundle of fibres to Broca's area, which in turn controls articulation of this speech. Wernicke's area receives input from the visual cortex in the occipital lobe, a pathway important in reading comprehension and in describing objects seen, as well as from the auditory cortex in the temporal lobe, a pathway essential for understanding spoken words. Wernicke's area also receives input from the somatosensory cortex, a pathway important in the ability to read Braille. Precise interconnecting pathways between these localized cortical areas are involved in the various aspects of speech (Figure 3-32).

---

**Clinical Connections**

Jackie became alerted that she had a serious medical problem when her speech became slurred. However, since she had no problem understanding speech, it was more likely that one of her lesions was in or near Broca's area. That said, the control of speech integrates many areas of the brain, and you always need to think of other causes that could result in the same effect.

---

### LANGUAGE DISORDERS

*Clinical Note* Because various aspects of language are localized in different regions of the cortex, damage to specific regions of the brain can result in selective disturbances of language. Damage to Broca's area results in a failure of word formation, although the patient can still understand the spoken and written word. Such people know what they want to say but cannot express themselves. Even though they can move their lips and tongue, they cannot establish the proper motor command to articulate the desired words. In contrast, patients with a lesion in Wernicke's area cannot understand words they see or hear. They can speak fluently, even though their perfectly articulated words make no sense. They cannot attach meaning to words or choose appropriate words to convey their thoughts. Such language disorders caused by damage to specific cortical areas are known as **aphasias**, most of which result from strokes. Aphasias should not be confused with **speech impediments**, which are caused by a defect in the mechanical aspect of speech, such as weakness or incoordination of the muscles controlling the vocal apparatus.

**Dyslexia**, another language disorder, is a difficulty in learning to read because of inappropriate interpretation of

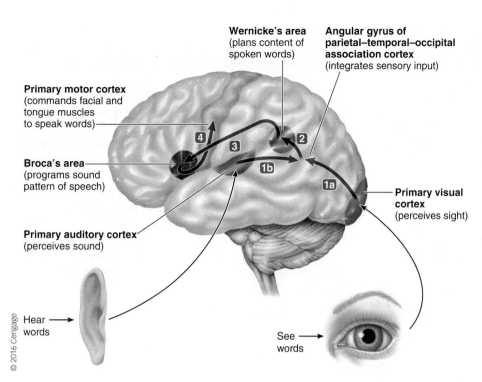

Wernicke's area
(plans content of
spoken words)

Angular gyrus of
parietal–temporal–occipital
association cortex
(integrates sensory input)

Primary motor cortex
(commands facial and
tongue muscles
to speak words)

Broca's area
(programs sound
pattern of speech)

Primary auditory cortex
(perceives sound)

Primary visual
cortex
(perceives sight)

Hear
words

See
words

© 2016 Cengage

**1a** To speak about something seen,
the brain transfers the visual information
from the primary visual cortex to the angular
gyrus of the parietal–temporal–occipital
association cortex, which integrates
inputs such as sight, sound, and touch.

**1b** To speak about something heard,
the brain transfers the auditory
information from the primary auditory
cortex to the angular gyrus.

**2** The information is transferred to
Wernicke's area, where the choice
and sequence of words to be spoken
are formulated.

**3** This language command is then
transmitted to Broca's area, which
translates the message into a
programmed sound pattern.

**4** This sound program is conveyed
to the precise areas of the primary
motor cortex that activate the appropriate
facial and tongue muscles for causing
the desired words to be spoken.

**3**

> **FIGURE 3-32 Cortical pathway for speaking a written word or naming a visual object.** The grey arrows and squared numbers with accompanying explanation indicate the pathway used to speak about something seen. Similarly, appropriate muscles of the hand can be commanded to write the desired words.

words. The problem arises from developmental abnormalities in connections between the visual and language areas of the cortex or within the language areas themselves; that is, the person with dyslexia is born with altered neuronal pathways within the language-processing system. Emerging evidence suggests that dyslexia stems from a deficit in phonological processing, meaning an impaired ability to break down written words into their underlying phonetic components. Dyslexics have difficulty decoding and thus identifying and assigning meaning to words. The condition is in no way related to intellectual ability.

## The association areas

The motor, sensory, and language areas account for only about half of the total cerebral cortex. The remaining areas, called **association areas**, are involved in higher functions. There are three association areas: (1) the *prefrontal association cortex*, (2) the *parietal–temporal–occipital association cortex*, and (3) the *limbic association cortex* (see Figure 3-23a). At one time the association areas were called "silent areas" because stimulation does not produce any observable motor response or sensory perception. (During brain surgery, typically the patient remains awake and only local anaesthetic is used along the cut scalp. This is possible because the brain itself is insensitive to pain. Before cutting into this precious, nonregenerative tissue, the neurosurgeon explores the exposed region with a tiny stimulating electrode. The patient is asked to describe what happens with each stimulation—the flick of a finger, a prickly feeling on the bottom of the foot, nothing? In this way, the surgeon can ascertain the appropriate landmarks on the neural map before making an incision.)

The **prefrontal association cortex** is the front portion of the frontal lobe, just anterior of the premotor cortex. This is the part

of the brain that "brainstorms." Specifically, the roles attributed to this region are (1) planning for voluntary activity, (2) decision making (i.e., weighing consequences of future actions and choosing between different options for various social or physical situations) (see Figure 3-23b), (3) creativity, and (4) personality traits. To carry out these highest of neural functions, the prefrontal cortex is the site of operation of working memory, where the brain temporarily stores and actively manipulates information used in reasoning and planning.

*Clinical Note* Stimulating the prefrontal cortex does not produce any observable effects, but deficits in this area change personality and social behaviour. Because damage to the prefrontal lobe was known to produce these changes, about 70 years ago prefrontal lobotomy (surgical removal) was used for treating violent individuals or others with so-called bad personality traits or social behaviour, in the hopes that their personality change would be for the better. Of course, the other functions of the prefrontal cortex were lost in the process (fortunately, the technique was used for only a short time).

The **parietal–temporal–occipital association cortex** lies at the interface of the three lobes for which it is named. In this strategic location, it pools and integrates somatic, auditory, and visual sensations projected from these three lobes for complex perceptual processing. It enables you to "get the complete picture" of the relationship of various parts of your body with the external world. For example, it integrates visual information with proprioceptive input to let you place what you are seeing in proper perspective, such as realizing that a bottle is in an upright position despite the angle from which you view it (i.e., whether you are standing up, lying down, or hanging upside down from a tree branch). This region is also involved in the language pathway connecting Wernicke's area to the visual and auditory cortices.

The **limbic association cortex** is located mostly on the bottom and adjoining the inner portion of each temporal lobe. This area is concerned primarily with motivation and emotion and is extensively involved in memory.

The cortical association areas are all interconnected by bundles of fibres within the cerebral white matter. Collectively, the association areas integrate diverse information for purposeful action. An oversimplified basic sequence of linkage between the various functional areas of the cortex is schematically represented in › Figure 3-33.

### The cerebral hemispheres

The cortical areas described so far appear to be equally distributed in both the right and left hemispheres, except for the language areas, which are found only on one side, usually the left. The left side is also most commonly the dominant hemisphere for fine motor control. Most people are right-handed, because the left side of the brain controls the right side of the body. Furthermore, each hemisphere is somewhat specialized in the types of mental activities it carries out best. The **left cerebral hemisphere** excels in logical, analytic, sequential, and verbal tasks, such as math, language forms, and philosophy. In contrast, the **right cerebral hemisphere** excels in non-language skills, especially spatial perception and artistic and musical talents. Whereas the left hemisphere tends to process information in a fine-detail, fragmentary way, the right hemisphere views the world in a big-picture, holistic way. Normally, much sharing of information occurs between the two hemispheres so that they complement each other, but in many individuals the skills associated

with one hemisphere are more strongly developed. Left cerebral hemisphere dominance tends to be associated with "thinkers," whereas the right hemispheric skills dominate in "creators."

## 3.12 | Sleep: An Active Process

The term **consciousness** refers to subjective awareness of the external world and self, including awareness of the private inner world of one's own mind—that is, awareness of thoughts, perceptions, dreams, and so on. Even though the final level of awareness resides in the cerebral cortex and a crude sense of awareness is detected by the thalamus, conscious experience depends on the integrated functioning of many parts of the nervous system.

The following states of consciousness, are listed in decreasing order of arousal level, based on the extent of interaction between peripheral stimuli and the brain:

- Maximum alertness
- Wakefulness
- Sleep (stages 1–4 and paradoxical)
- Coma

Maximum alertness depends on attention-getting sensory input that stimulates the reticular activating system (RAS) and subsequently the activity level of the CNS as a whole. At the other extreme, coma is the total unresponsiveness of a living person to external stimuli, caused either by brain stem damage that interferes with the RAS or by widespread depression of the cerebral cortex, for example, as a result of oxygen deprivation.

The **sleep–wake cycle** is a normal cyclic variation in awareness of surroundings. In contrast to being awake, sleeping people are not consciously aware of the external world, but they do have inward conscious experiences, such as dreams. Furthermore, they can be aroused by external stimuli, such as an alarm going off.

**Sleep** is an active process, not just the absence of wakefulness. Sleep is characterized by a decreased reaction to external stimuli, reduced voluntary movement, increased rate of anabolism, and a decreased rate of catabolism. The brain's overall level of activity is not reduced during sleep. During certain stages of sleep, oxygen uptake by the brain is even increased above normal waking levels.

There are two types of sleep, characterized by different EEG patterns and different behaviours: *slow-wave sleep* (delta-wave; non-rapid eye movement, NREM) and *paradoxical sleep* (rapid eye movement, REM) (▮ Table 3-4).

### EEG patterns during sleep

**Slow-wave sleep** occurs in four stages, each displaying a progressively lower frequency of EEG wave but higher in amplitude (› Figure 3-34). At the onset of sleep, you move through the stages of slow-wave sleep from light sleep (stage 1) to deep sleep (stage 4) during about a 30- to 45-minute period; you then reverse through the same stages in the same amount of time. A 10- to 15-minute episode of **paradoxical sleep** punctuates the end of each slow-wave sleep cycle. Paradoxically, your EEG

| Sensory input | Relayed from afferent neuronal receptors |
| Primary sensory areas (somatosensory, 1° visual, 1° auditory cortices) | Initial cortical processing of specific sensory input |
| Higher sensory areas | Further elaboration and processing of specific sensory input |
| Association areas | Integration, storage, and use of diverse sensory input for planning of purposeful action |
| Higher motor areas | Programming of sequences of movement in context of diverse information provided |
| Primary motor cortex | Commanding of efferent motor neurons to initiate voluntary movement |
| Motor output | Relayed through efferent motor neurons to appropriate skeletal muscles, which carry out desired action |

For simplicity, a number of interconnections have been omitted.

› **FIGURE 3-33** Schematic linking of various regions of the cortex

## ▌TABLE 3-4 Comparison of Slow-Wave and Paradoxical Sleep

### TYPE OF SLEEP

| Characteristic | Slow-Wave Sleep | Paradoxical Sleep |
|---|---|---|
| **EEG** | Displays slow waves | Similar to EEG of alert, awake person |
| **Motor activity** | Considerable muscle tone; frequent shifting | Abrupt inhibition of muscle tone; no movement |
| **Heart rate, respiratory rate, blood pressure** | Minor reductions | Irregular |
| **Dreaming** | Rare (mental activity is extension of waking-time thoughts) | Common |
| **Arousal** | Sleeper easily awakened | Sleeper hard to arouse but apt to wake up spontaneously |
| **Percentage of sleeping time** | 80 percent | 20 percent |
| **Other important characteristics** | Has four stages; sleeper passes through this type of sleep first | Rapid eye movements |

pattern during this time abruptly becomes similar to that of a wide-awake, alert individual, even though you are still asleep (hence, *paradoxical* sleep) (Figure 3-34). After the paradoxical episode, the stages of slow-wave sleep repeat.

A person cyclically alternates between the two types of sleep throughout the night. In a normal sleep cycle, you always pass through slow-wave sleep before entering **paradoxical** sleep. On average, paradoxical sleep occupies 20 percent of total sleeping time throughout adolescence and most of adulthood. Infants spend considerably more time in paradoxical sleep. In contrast, paradoxical as well as deep, stage 4, slow-wave sleep declines in the elderly. People who require less total sleeping time than normal spend proportionately more time in the lighter stages of slow-wave sleep.

## Behavioural patterns during sleep

In addition to distinctive EEG patterns, the two types of sleep are distinguished by behavioural differences. It is difficult to pinpoint exactly when an individual drifts from drowsiness into slow-wave sleep. In this type of sleep, the person still has considerable muscle tone and frequently shifts body position. Only minor reductions in respiratory rate, heart rate, and blood pressure occur. During this time the sleeper can be easily awakened and rarely dreams. The mental activity associated with slow-wave sleep is less visual than dreaming. It is more conceptual and plausible—like an extension of waking-time thoughts concerned with everyday events—and it is less likely to be recalled. The major exception is nightmares, which occur during stages 3 and 4. People who walk and talk in their sleep do so during slow-wave sleep.

The behavioural pattern accompanying paradoxical sleep is marked by abrupt inhibition of muscle tone throughout the body. The muscles are completely relaxed, with no movement taking place. Paradoxical sleep is further characterized by *rapid eye movements*, hence the alternative name, REM sleep. Heart rate and respiratory rate become irregular, and blood pressure may fluctuate. Another characteristic of REM sleep is *dreaming*. The rapid

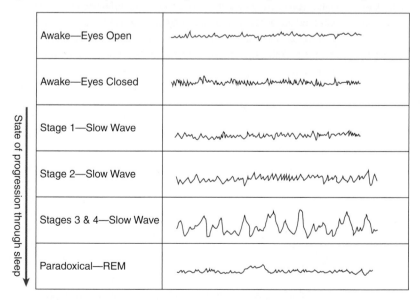

State of progression through sleep

| Awake—Eyes Open |
| Awake—Eyes Closed |
| Stage 1—Slow Wave |
| Stage 2—Slow Wave |
| Stages 3 & 4—Slow Wave |
| Paradoxical—REM |

❯ **FIGURE 3-34 EEG patterns during wake and different types of slow-wave and paradoxical sleep.** Note that the EEG pattern during paradoxical (REM) sleep is similar to that of an alert, awake person, whereas the pattern during slow-wave sleep displays distinctly different waves.

eye movements are not related to "watching" the dream imagery. The eye movements are driven in a locked, oscillating pattern, uninfluenced by dream content.

Brain imaging of volunteers during REM sleep shows heightened activity in the higher-level visual-processing areas and limbic system (the seat of emotions), coupled with reduced activity in the prefrontal cortex (the seat of reasoning). This pattern of activity lays the groundwork for the characteristics of dreaming: internally generated visual imagery reflecting activation of the person's "emotional memory bank" with little guidance or interpretation from the complex thinking areas. As a result, dreams are often charged with intense emotions, a distorted sense of time, and bizarre content, which, at the time of dreaming, is uncritically accepted as real and involves little reflection about all the strange happenings.

**3**

### Check Your Understanding 3.8

1. Define *consciousness*.
2. Discuss the location and functions of the three neural systems that play a role in the sleep–wake cycle.

# Chapter in Perspective: Focus on Homeostasis

To interact in appropriate ways with the external environment to sustain the body's viability, such as in acquiring food, and to make the internal adjustments necessary to maintain homeostasis, the body must be informed about any changes taking place in the external and internal environments, be able to process this information, and send messages to various muscles and glands to accomplish the desired results. The nervous system, one of the body's two major regulatory systems, plays a central role in this life-sustaining communication. The central nervous system (CNS), which consists of the brain and spinal cord, receives information about the external and internal environments by means of the afferent peripheral nerves. After sorting, processing, and integrating this input, the CNS sends directions, by means of efferent peripheral nerves, to bring about appropriate muscular contractions and glandular secretions.

With its swift electrical signalling system, the nervous system is especially important in controlling the rapid responses of the body. Many neurally controlled muscular and glandular activities are aimed toward maintaining homeostasis. The CNS is the main site of integration between afferent input and efferent output. It links the appropriate response to a particular input so that conditions compatible with life are maintained in the body. For example, when informed by the afferent nervous system that blood pressure has fallen, the CNS sends appropriate commands to the heart and blood vessels to increase the blood pressure to normal. Likewise, when informed that the body is overheated, the CNS promotes secretion of sweat by the sweat glands. Evaporation of sweat helps cool the body to normal temperature. Were it not for this processing and integrating ability of the CNS, maintaining homeostasis in an organism as complex as a human would be impossible.

At the simplest level, the spinal cord integrates many basic protective and evacuative reflexes that do not require conscious participation, such as withdrawing from a painful stimulus and emptying of the urinary bladder. In addition to serving as a more complex integrating link between afferent input and efferent output, the brain is responsible for the initiation of all voluntary movement, complex perceptual awareness of the external environment and self, language, and abstract neural phenomena, such as thinking, learning, remembering, consciousness, emotions, and personality traits. All neural activity—from the most private thoughts to commands for motor activity, from enjoying a concert to retrieving memories from the distant past—is ultimately attributable to propagation of action potentials along individual nerve cells and chemical transmission between cells.

During evolutionary development, the nervous system has become progressively more complex. Newer, more complicated, and more sophisticated layers of the brain have been piled on top of older, more primitive regions. Mechanisms for governing many basic activities necessary for survival are built into the older parts of the brain. The newer, higher levels progressively modify, enhance, or nullify actions coordinated by lower levels in a hierarchy of command, and they also add new capabilities. Many of these higher neural activities are not aimed toward maintaining life, but they add immeasurably to the quality of being alive.

acquired (conditioned) reflexes (p. 105)
afferent division (p. 92)
afferent pathway (p. (p. 106))
AMPA receptors (p. 130)
amygdala (p. 124)
amnesia (p. 127)
aphasias (p. 132)
arachnoid mater (p. 100)
arachnoid villi (p. 100)
ascending tracts (p. 103)
association areas (p. 133)
astrocytes (p. 95)
autonomic nervous system (ANS) (p. 92)
basal ganglia (p. 123)
basic behavioural patterns (p. 124)
blood–brain barrier (BBB) (p. 100)
brain stem (p. 109)
Broca's area (p. 132)
cauda equina (p. 102)
central canal (p. 98)
central nervous system (CNS) (p. 92)
central sulcus (p. 115)
cerebellum (p. 121)
cerebral cortex (p. 114)
cerebral hemispheres (p. 113)
cerebrocerebellum (p. 122)
cerebrospinal fluid (CSF) (p. 100)
cerebrovascular accident (CVA) (p. 114)
cerebrum (p. 94)
choroid plexuses (p. 100)
circumventricular organs (p. 101)
cognition (p. 93)
consciousness (p. 134)
consolidation (p. 126)
corpus callosum (p. 113)
cranial nerves (p. 110)
CREB (p. 131)
CREB2 (p. 131)
crossed extensor reflex (p. 109)
declarative memories (p. 128)
depression (p. 126)
dermatome (p. 103)
descending tracts (p. 103)
diencephalon (p. 112)
dorsal (posterior) horn (p. 103)
dorsal root (p. 103)
dorsal root ganglion (p. 103)
dura mater (p. 98)
dural sinuses (p. 98)
dyslexia (p. 132)
effector organs (p. 92)
efferent division (p. 92)
efferent neurons (p. 93)
efferent pathway (p. 106)
electroencephalogram (EEG) (p. 120)

emotion (p. 124)
ependymal cells (p. 98)
epilepsy (p. 120)
free radicals (p. 114)
frontal lobes (p. 115)
glial cells (p. 95)
gliomas (p. 98)
grey matter (p. 103)
habituation (p. 128)
hippocampus (p. 128)
homeostatic drives (p. 125)
hydrocephalus (p. 100)
hypothalamus (p. 113)
immediate early genes (IEGs) (p. 132)
integrating centre (p. 106)
Interneurons (p. 93)
language (p. 132)
lateral horn (p. 103)
learning (p. 126)
left cerebral hemisphere (p. 134)
limbic association cortex (p. 134)
limbic system (p. 124)
long-term memory (p. 126)
long-term potentiation (LTP) (p. 130)
memory (p. 126)
memory trace (p. 126)
meninges (p. 98)
meningiomas (p. 98)
microglia (p. 97)
monosynaptic reflex (p. 106)
motivation (p. 125)
motor homunculus (p. 116)
motor program (p. 119)
nerve (p. 103)
nervous system (p. 91)
neuroglia (p. 95)
neuroglobin (p. 101)
neuropeptides (p. 130)
nitric oxide (p. 130)
NMDA receptors (p. 130)
occipital lobes (p. 115)
paradoxical (p. 135)
paradoxical sleep (p. 134)
parasympathetic nervous
   system (p. 92)
parietal lobes (p. 115)
parietal–temporal–occipital association
   cortex (p. 133)
patellar tendon (knee-jerk) reflex (p. 106)
peripheral nervous system (PNS) (p. 92)
pia mater (p. 100)
plasticity (p. 120)
polysynaptic (p. 106)
posterior parietal cortex (p. 119)
prefrontal association cortex (p. 133)

premotor cortex (p. 119)
primary motor cortex (p. 115)
procedural memories (p. 128)
proprioception (p. 117)
psychoactive drugs (p. 125)
punishment centres (p. 125)
pyramidal cells (p. 114)
readiness potential (p. 118)
reciprocal innervation (p. 107)
reconsolidated (p. 126)
referred pain (p. 103)
reflex (p. 105)
reflex arc (p. 105)
reticular activating system (RAS) (p. 112)
reticular formation (p. 110)
reward centres (p. 125)
right cerebral hemisphere (p. 134)
sensitization (p. 129)
sensory homunculus (p. 118)
sensory receptor (p. 92)
short-term memory (p. 126)
simple (basic) reflexes (p. 105)
sleep (p. 134)
sleep–wake cycle (p. 134)
slow-wave sleep (p. 134)
somaesthetic sensations (p. 116)
somatic nervous system (p. 92)
somatosensory cortex (p. 117)
speech impediments (p. 132)
spinal cord (p. 101)
spinal nerves (p. 101)
spinal reflex (p. 107)
spinocerebellum (p. 121)
stellate cells (p. 114)
stimulus (p. 106)
stretch reflex (p. 106)
subarachnoid space (p. 100)
subcortical region (p. 119)
supplementary motor area (p. 119)
sympathetic nervous system (p. 92)
temporal lobes (p. 115)
thalamus (p. 112)
use-dependent competition (p. 119)
vagus nerve (p. 110)
venous sinuses (p. 98)
ventral (anterior) horn (p. 103)
ventral corticospinal tract (p. 103)
ventral root (p. 103)
ventral spinocerebellar tract (p. 103)
ventricles (p. 98)
vestibulocerebellum (p. 121)
Wernicke's area (p. 132)
white matter (p. 103)
withdrawal reflex (p. 107)
working memory (p. 126)

**3**

## Objective Questions (Answers in Appendix E, p. A-38)

1. The major function of the CSF is to nourish the brain. (*True or false?*)

2. In emergencies when oxygen supplies are low, the brain can perform anaerobic metabolism. (*True or false?*)

3. The hands and structures associated with the mouth have a disproportionately large share of representation in both the sensory cortex and the motor cortex. (*True or false?*)

4. The left cerebral hemisphere specializes in artistic and musical ability, whereas the right side excels in verbal and analytical skills. (*True or false?*)

5. The specific function a particular cortical region will carry out is permanently determined during embryonic development. (*True or false?*)

6. _____ is a decreased responsiveness to an indifferent stimulus that is repeatedly presented.

7. The process of transferring and fixing short-term memory traces into long-term memory stores is known as _____.

8. Afferent fibres enter through the _____ root of the spinal cord, and efferent fibres leave through the _____ root.

9. Using the answer code on the right, indicate which neurons are being described (a characteristic may apply to more than one class of neurons):

    ___ 1. have receptor at peripheral endings    (a) afferent neurons

    ___ 2. lie entirely within the CNS    (b) efferent neurons

    ___ 3. lie primarily within the peripheral nervous system    (c) interneurons

    ___ 4. innervate muscles and glands

    ___ 5. cell body is devoid of presynaptic inputs

    ___ 6. predominant type of neuron

    ___ 7. responsible for thoughts, emotions, memory, etc.

10. Using the answer code on the right, match the following:

    ___ 1. consists of nerves carrying information between the periphery and the CNS    (a) somatic nervous system

    ___ 2. consists of the brain and spinal cord    (b) autonomic nervous system

    ___ 3. division of the peripheral nervous system that transmits signals to the CNS    (c) central nervous system

    ___ 4. division of the peripheral nervous system that transmits signals from the CNS    (d) peripheral nervous system

    ___ 5. supplies skeletal muscles    (e) efferent division

    ___ 6. supplies smooth muscle, cardiac muscle, and glands    (f) afferent division

## Written Questions

1. Discuss the function of each of the following: astrocytes, oligodendrocytes, ependymal cells, microglia, cranium, vertebral column, meninges, cerebrospinal fluid, and blood–brain barrier.

2. Compare the composition of white and grey matter.

3. Draw and label the major functional areas of the cerebral cortex, indicating the functions attributable to each area.

4. Discuss the function of each of the following parts of the brain: thalamus, hypothalamus, basal ganglia, limbic system, cerebellum, and brain stem.

5. Discuss the roles of Broca's area and Wernicke's area in language.

6. Compare short-term and long-term memory.

7. What is an electroencephalogram?

8. Distinguish between a monosynaptic and a polysynaptic reflex.

1. Special studies designed to assess the specialized capacities of each cerebral hemisphere have been performed on *split-brain* patients. In these people, the corpus callosum—the bundle of fibres that links the two halves of the brain—has been surgically cut to prevent the spread of epileptic seizures from one hemisphere to the other. Even though no overt changes in behaviour, intellect, or personality occur in these patients, because both hemispheres individually receive the same information, deficits are observable with tests designed to restrict information to one brain hemisphere at a time. One such test involves limiting a visual stimulus to only half of the brain. Because of a crossover in the nerve pathways from the eyes to the occipital cortex, the visual information to the right of a midline point is transmitted to only the left half of the brain, whereas visual information to the left of this point is received by only the right half of the brain. A split-brain patient presented with a visual stimulus that reaches only the left hemisphere accurately describes the object seen, but when a visual stimulus is presented to only the right hemisphere, the patient denies having seen anything. The right hemisphere does receive the visual input, however, as demonstrated by nonverbal tests. Even though split-brain patients deny having seen anything after an object is presented to the right hemisphere, they can correctly match the object by picking it out from among a number of objects, usually to the their own surprise. What is your explanation of this finding?

2. What symptom is most likely to occur as the result of a severe blow to the back of the head?
   a. Paralysis
   b. Hearing impairment
   c. Visual disturbances
   d. Burning sensations
   e. Personality disorders

3. Give examples of conditioned reflexes you have acquired.

4. The hormone insulin enhances the carrier-mediated transport of glucose into most of the body's cells but not into brain cells. The uptake of glucose from the blood by neurons does not depend on insulin. Knowing the brain's need for a continuous supply of blood-borne glucose, predict the effect that insulin excess would have on the brain.

3

Julio D., who had recently retired, was enjoying an afternoon of playing golf when suddenly he experienced a severe headache and dizziness. These symptoms were quickly followed by numbness and partial paralysis on the upper right side of his body, accompanied by an inability to speak. After being rushed to the emergency room, Julio was diagnosed as having suffered a stroke. Given the observed neurologic impairment, what areas of his brain were affected?

# The Peripheral Nervous System

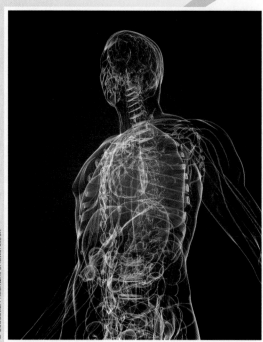

Peripheral nervous system

Body systems maintain homeostasis

## Homeostasis

The nervous and endocrine systems, as the body's two major regulatory systems, regulate many body activities aimed at maintaining a stable internal fluid environment.

Homeostasis is essential for survival of cells

Cells make up body systems

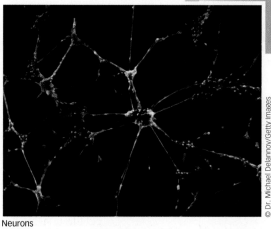

Neurons

The nervous system is one of two major regulatory systems of the body. It consists of the central nervous system (CNS), composed of the brain and spinal cord, and the peripheral nervous system (PNS), composed of the afferent and efferent fibres that relay signals between the CNS and the periphery (other parts of the body).

The afferent division of the PNS detects, encodes, and transmits peripheral signals to the CNS, thus informing the CNS about the internal and external environment. This afferent input (i.e., feedback) to the controlling centres of the CNS is essential to making adjustments in effector organs to maintain homeostasis. The efferent division of the PNS enables the CNS to make adjustments by controlling the activities of effector organs (e.g., muscles and glands) through transmitting signals *from* the CNS to these organs. Unrelated to homeostasis, the PNS allows for planned voluntary actions, such as skeletal muscle contraction.

# The Peripheral Nervous System: Sensory, Autonomic, Somatic

## 4.1 | Introduction

The peripheral nervous system (PNS) is the link between the periphery and the central nervous system (CNS). It consists of nerve fibres that transmit information via myelinated and unmyelinated neurons. The afferent division of the PNS senses (via various receptors) and sends (via neurons) information about both the internal and external environment to the CNS. The CNS receives this input (i.e., information), interprets it, and initiates a response that is carried out through the efferent division. The efferent division, under the direction of the CNS, serves to consciously control the activity of some effector organs (e.g., skeletal muscles), and also to subconsciously control various tissues and organs (e.g., smooth muscle in blood vessels).

The conscious control is exerted by the efferent division through the somatic nervous system, whereas the subconscious control is exerted through the autonomic nervous system. Neither is more important than the other.

> ▮ **Clinical Connections**
>
> Deborah is a 60-year-old female. She is obese and was diagnosed with type 2 diabetes mellitus at the age of 40. Despite recommendations to alter her diet and to exercise, she claimed that standing has become painful, and it is easier for her to order in food than to walk to the grocery store a few blocks away. Deborah also struggles to remember to take her mediations. Due to all these factors, her blood sugar levels are chronically elevated. Deborah used to be a very active knitter, but recently she had to stop because she struggles

*Continued*

with the fine coordination. Interestingly, she increasingly noticed that when she accidentally poked herself with knitting needles, she didn't feel it. Deborah also realized she had progressive development of numbness in her fingers and toes, mild muscle weakness, more frequent unsteadiness on her feet, and problems with her vision. She attributed all these symptoms to aging and ignored them until she started to have occasional sharp pain in her toes while sleeping. Then she went to see her physician.

Given her symptoms, her physician suspected and confirmed diabetic peripheral neuropathy. Her elevated blood sugar levels had led to microvascular injury. When this occurs, the blood supply to nerves can become compromised and result in nerve injury or abnormal nerve function.

## 4.2 | The Peripheral Nervous System: Afferent Division

The afferent neurons, also called sensory or receptor neurons, carry nerve impulses (action potentials) from receptors or sense organs toward the CNS.

Sensory information (e.g., pain) is transmitted via afferent pathways from a peripheral receptor to the CNS with no noticeable delay. The stimulus is sensed in the brain and perceived (as pain) based on the location of the brain that is stimulated. The structure of the afferent neurons is such that they have a single long dendrite, a short axon, and a smooth rounded cell body. The dendrite is structurally and functionally similar to an axon. Immediately external to the spine are thousands of afferent cell bodies that are aggregated in a swell, the *dorsal root ganglion*.

This chapter opens with a description of the fight-or-flight response, as it pertains to sensation, to demonstrate the robust nature of our body, and how the divisions of the nervous system work in symphony to prepare for stressful circumstances.

### Fight-or-flight response

When you perceive a threatening situation, your body readies itself to either fight or flee the situation. An example: While walking in the woods you hear the rustling of leaves and branches behind you. You stop and look—it's a bear. Do you run, back up slowly, or confront the bear, making yourself seem as threatening as possible? During this acute (short-term) stress response, unnecessary functions are temporarily slowed, and energy is diverted to other functions vital to survival. In general, stress can be physical (an altercation), psychological (unpleasant event), or emotional (anger). Afferent input is necessary to respond to stress, for arousal, perception, and determination of efferent output. Using the example of seeing a bear, the physiological adjustments to the stress associated with the afferent division of the PNS are as follows:

- The afferent division provides information (sound, sight, smell) to the CNS about the external environment that influences conscious activity (i.e., decision making). Arousal occurs, which adjusts efferent output (e.g., muscles, heart rate).

- Hearing the rustling of leaves and branches, you quickly turn your head in the direction of the sounds. This requires the ears to sense and transmit the sound waves more quickly on the side of the body closest to the sound.

- To focus on the bear, your eyes use accommodation to provide a clear, crisp image.

- Pupils dilate to allow in more light.

- The decision to fight or flee in this situation is most likely based on your ability to judge distance. If the bear is close, you may not run, but if the bear is far off, you may decide to run. This requires cognitive processing and also good eye sight.

We return to this example in our discussion of the autonomic and efferent divisions' contributions to the fight-or-flight response later in this chapter (p. 187). Staying with the afferent division, we now examine somatosensation in detail.

### Somatosensation

The somatosensory system consists of receptors (PNS) and processing centres (CNS) that integrate and create the sensory modalities (e.g., pain, temperature). The modality of the stimulus is determined by which sensory neuron is activated and where the stimulus terminates in the brain. The primary somatosensory area is located in the parietal lobes of the cerebral cortex. The somatosensory system reacts to diverse stimuli by means of four primary types of receptors: thermoreceptors, mechanoreceptors, photoreceptors, and chemoreceptors. These receptors are located throughout the body. Sensory information is transmitted in the form of an action potential from the receptors via sensory nerves (afferents) through tracts in the spinal cord to the brain. The system works when a receptor is stimulated (e.g., heat) and information is passed along the sensory neuron to an area in the brain uniquely attributed to that area on the body.

### Sensation: Internal and external

Afferent information about the internal environment, such as blood pressure and concentration of carbon dioxide in the body fluids, never reaches the level of conscious awareness, but this input is essential for determining the appropriate efferent output to maintain homeostasis. The incoming pathway for information derived from the internal viscera (*viscera* are organs in the body cavities, such as the abdominal cavity) is called a **visceral afferent**. Even though mostly subconscious information is transmitted via visceral afferents, people do become aware of pain signals arising from the viscera.

In contrast, afferent input derived from receptors located at the body surface or in the muscles or joints typically reaches the level of conscious awareness. Final processing of sensory input by the CNS not only is essential for interaction with the environment for basic survival (e.g., fight-or-flight) but also adds immeasurably to the richness of life. This input is known as *sensory information*, and the incoming pathway is considered a **sensory afferent**. Sensory information is categorized as either (1) **somatic sensation** (body sense) arising from the body surface, including *somaesthetic sensation* from the skin and *proprioception* from the muscles, joints, skin, and inner ear or (2) **special senses**, including *vision, hearing, taste, smell,* and *equilibrium*.

## Perception: Conscious awareness of surroundings

**Perception** is the conscious interpretation of the external world as created by the brain from a pattern of nerve impulses delivered to it from sensory receptors. Is the world, as we perceive it, reality? The answer is a resounding no. Our perception is different from what is really "out there," for several reasons. First, humans have receptors that detect only a limited number of existing energy forms. We perceive sounds, colours, shapes, textures, smells, tastes, and temperature but are not informed of magnetic forces, polarized light waves, radio waves, or X-rays because we do not have receptors to respond to these other energy forms. What is not detected by receptors, the brain will never know. There are several reasons for this. First, our response range is limited even for the energy forms for which we do have receptors. For example, dogs can hear a whistle whose pitch is above our level of detection. Second, the information channels to our brains are not high-fidelity recorders. During precortical processing of sensory input, some features of stimuli are accentuated, and others are suppressed or ignored. Third, the cerebral cortex further manipulates the data—comparing the sensory input with other incoming information as well as with memories of past experiences—to extract the significant features: for example, sifting out a friend's words from the hubbub of sound in a school cafeteria. In the process, the cortex often fills in or distorts the information to abstract a logical perception; that is, it "completes the picture." As a simple example, in ⟩ Figure 4-1a you "see" a white square even though there is no white square, but merely right-angle wedges taken out of four red circles. Optical illusions illustrate how the brain interprets reality according to its own rules. Do you see two faces in profile or a wine glass in ⟩ Figure 4-1b? You can alternately see one or the other out of identical visual input. Thus, our perceptions do not replicate reality. Other species, equipped with different types of receptors and sensitivities and with different neural processing, perceive a markedly different world from what we perceive.

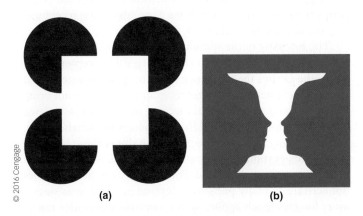

⟩ **FIGURE 4-1 Optical illusions.** (a) Do you "see" a white square that is not really there? (b) Variable perceptions from the same visual input. Do you see two faces in profile, or a wine glass?

© 2016 Cengage

An afferent neuron has a receptor at its peripheral ending that responds to a stimulus in both the external and internal environments. A stimulus is a change detectable by the body that meets a minimum threshold. Stimuli exist in a variety of energy forms, such as heat, light, sound, pressure, and chemical changes. Receptors vary widely in complexity, and their axons can be myelinated or unmyelinated. **Somatosensory receptors** such as free nerve endings consist of a neuron with an exposed receptor, whereas the **special senses receptor** in the ear turns a mechanical stimulation (non-neural) into a neural signal by synapsing onto a sensory neuron. Because the only way afferent neurons can transmit information about these stimuli to the CNS is via action potential propagation, receptors must convert these other forms of energy into electrical signals. This energy-conversion process, known as **transduction**, converts the mechanical or chemical stimulation into an electrical signal through changes in ion permeability at controlled ion channels.

As all stimuli are converted to action potentials in sensory neurons, this would seem to present a problem for the CNS in distinguishing the various forms of stimuli. However, by using four basic properties of the stimuli, the CNS is able to accurately differentiate the incoming stimuli from the PNS. These properties include modality, intensity, location, and duration.

### Adequate stimuli and threshold

Each type of receptor is specialized to respond best to one type of energy, known as an **adequate stimulus**. For example, receptors in the eye are most sensitive to light, receptors in the ear to sound waves, and warmth receptors in the skin to heat energy. Because of this differential sensitivity of receptors, we cannot see with our ears or hear with our eyes. Some receptors can respond weakly to stimuli other than their adequate stimulus, but even when activated by a different stimulus, a receptor still gives rise to the sensation usually detected by that receptor type. As an example, the adequate stimulus for eye receptors (photoreceptors) is light, to which they are exquisitely sensitive, but these receptors can also be activated to a lesser degree by mechanical stimulation. When hit in the eye, a person often "sees stars," because the mechanical pressure stimulates the photoreceptors. Thus, the sensation perceived depends on the **modality** of receptor stimulated rather than on the type of stimulus. The minimum stimulus required to activate the receptor is known as the **threshold**.

#### TYPES OF RECEPTORS ACCORDING TO THEIR ADEQUATE STIMULUS

Depending on the type of energy to which they ordinarily respond, sensory receptors are categorized as follows:

- **Photoreceptors** are responsive to visible wavelengths of light.
- **Mechanoreceptors** are sensitive to mechanical energy. Examples include skeletal muscle receptors sensitive to stretch; the receptors in the ear that contain fine hair cells, which bend as a result of sound waves; blood pressure—monitoring baroreceptors; and osmoreceptors, which respond to stretch. Other stimuli sensed are vibration and acceleration.

- **Thermoreceptors** are sensitive to varying amounts of heat.
- **Chemoreceptors** are sensitive to specific chemicals. Chemoreceptors include the receptors for smell and taste, the chemical content of the digestive tract, and oxygen and carbon dioxide in the blood.

Some sensations are compound sensations, in that their perception arises from the central integration of several simultaneously activated primary sensory inputs. For example, the perception of wetness comes from touch, pressure, and thermal receptor input; there is no such thing as a "wetness receptor."

### USES OF INFORMATION DETECTED BY RECEPTORS

Information detected by receptors and conveyed via afferent neurons to the CNS is used for a variety of purposes:

- It is essential for the control of efferent output, both for regulating motor behaviour based on external and internal stimuli and also to direct and maintain homeostasis. At the most basic level, afferent input provides information (for use—consciously and subconsciously) for the CNS to use in directing activities necessary for survival (e.g., fight-or-flight response). We could not be aware of and interact successfully with our environment or with one another without sensory input.
- Processing of sensory input by the reticular activating system in the brain stem is critical for cortical arousal and consciousness (p. 112).
- Central processing of sensory information gives rise to our perceptions of the world around us.
- Selected information delivered to the CNS may be stored for future reference.
- Sensory stimuli can have a profound impact on our emotions. The smell of just-baked apple pie, the sight of a loved one, hearing bad news—sensory input can gladden, sadden, arouse, calm, anger, frighten, or evoke a wide range of emotions.

We now examine how adequate stimuli initiate action potentials that ultimately are used for these purposes.

## Stimuli and receptor permeability

A receptor may be either (1) a specialized ending of the afferent neuron or (2) a separate receptor cell closely associated with the peripheral ending of the neuron. Stimulation of a receptor alters its membrane permeability, usually by causing a nonselective opening of all small ion channels. The means by which this permeability change takes place is individualized for each receptor type. Because the electrochemical driving force is greater for sodium ($Na^+$) than for other small ions at resting potential, the predominant effect is an inward flux of $Na^+$, which depolarizes the receptor membrane (p. 59). (There are exceptions; for example, photoreceptors are hyperpolarized upon stimulation.) This local depolarizing change in potential is known as a **generator potential**, if the receptor is a specialized ending of an afferent neuron, or as a **receptor potential**, in the case of a separate receptor cell. The receptor (or generator) potential is a graded potential whose amplitude and **duration** can vary,

depending on the strength and the rate of application or removal of the stimulus (p. 59). The stronger the stimulus, the greater is the change in ion permeability, and the larger the receptor potential. As is true of all graded potentials, receptor potentials have no refractory period, so summation in response to rapidly successive stimuli is possible. Because the receptor region has a very high threshold, action potentials do not take place at the receptor itself. For long-distance transmission, the receptor potential must be converted into action potentials that can be propagated along the afferent fibre.

## Receptor potentials and action potentials

If a receptor (or generator) potential has sufficient magnitude, it may initiate an action potential in the afferent neuron membrane next to the receptor by triggering the opening of $Na^+$ channels in this region. The means by which the $Na^+$ channels are opened differ depending on whether the receptor is a separate receptor cell or a specialized afferent ending.

- In the case of a separate receptor cell, a receptor potential triggers the release of a chemical messenger that diffuses across the small space separating the receptor cell from the ending of the afferent neuron, similar to a synapse (⟩ Figure 4-2a). Binding of the chemical messenger with specific protein receptor sites on the afferent neuron membrane opens chemically gated $Na^+$ channels (p. 58).
- In the case of a specialized afferent ending, local current flow between the activated receptor ending undergoing a generator potential and the cell membrane adjacent to the receptor brings about opening of voltage-gated $Na^+$ channels in this adjacent region (⟩ Figure 4-2b).

In either case, if the magnitude of the resulting ionic flux is big enough to bring the adjacent membrane to threshold, an action potential is initiated and self-propagates along the afferent fibre to the CNS. (For convenience, from here on we will refer to both receptor potentials and generator potentials as receptor potentials.)

Note that the initiation site of action potentials in an afferent neuron differs from the site in an efferent neuron or interneuron. In the latter two types of neurons, action potentials are initiated at the axon hillock located at the start of the axon next to the cell body (p. 64). By contrast, action potentials are initiated at the peripheral end of an afferent nerve fibre next to the receptor, a long distance from the cell body (⟩ Figure 4-3).

The **intensity** of the stimulus is reflected by the magnitude of the receptor potential. In turn, the larger the receptor potential is, the greater the frequency of action potentials generated in the afferent neuron. A larger receptor potential cannot bring about a larger action potential (because of the all-or-none law), but it can induce more rapid firing of action potentials (p. 59). Stimulus strength is also reflected by the size of the area stimulated. Stronger stimuli usually affect larger areas, so correspondingly more receptors respond. For example, a light touch does not activate as many pressure receptors in the skin as does a more forceful touch applied to the same area. Stimulus intensity is therefore distinguished both by the frequency of action potentials generated in the afferent neuron and by the number of receptors activated within the area (see Why It Matters).

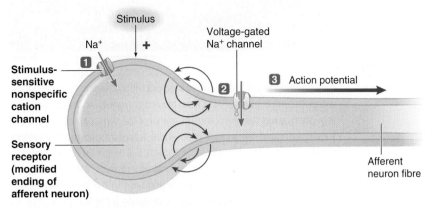

**Stimulus**

Na⁺

**1** Voltage-gated Na⁺ channel

**Stimulus-sensitive nonspecific cation channel**

**2** **3** Action potential

**Sensory receptor (modified ending of afferent neuron)**

Afferent neuron fibre

**(a)** Receptor potential in specialized afferent ending

1 In sensory receptors that are specialized afferent neuron endings, stimulus opens stimulus-sensitive channels, permitting net Na⁺ entry that produces receptor potential.

2 Local current flow between depolarized receptor ending and adjacent region opens voltage-gated Na⁺ channels.

3 Na⁺ entry initiates action potential in afferent fibre that self-propagates to CNS.

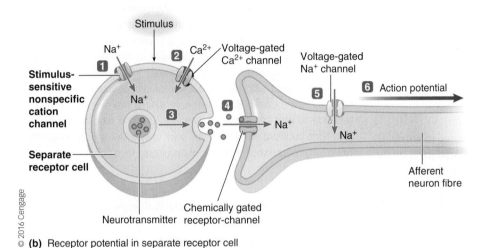

**Stimulus**

Na⁺ **1** **2** Ca²⁺ Voltage-gated Ca²⁺ channel

Voltage-gated Na⁺ channel

**Stimulus-sensitive nonspecific cation channel**

Na⁺ **3** **4** **5** **6** Action potential

Na⁺

Na⁺

**Separate receptor cell**

Afferent neuron fibre

Neurotransmitter Chemically gated receptor-channel

© 2016 Cengage

**(b)** Receptor potential in separate receptor cell

1 In sensory receptors that are separate cells, stimulus opens stimulus-sensitive channels, permitting net Na⁺ entry that produces receptor potential.

2 This local depolarization opens voltage-gated Ca²⁺ channels.

3 Ca²⁺ entry triggers exocytosis of neurotransmitter.

4 Neurotransmitter binding opens chemically gated receptor channels at afferent ending, permitting net Na⁺ entry.

5 Resultant depolarization opens voltage-gated Na⁺ channels in adjacent region.

6 Na⁺ entry initiates action potential in afferent fibre that self-propagates to CNS.

**4**

> **FIGURE 4-2 Conversion of receptor potential into action potentials.** (a) Specialized afferent ending as sensory receptor. Local current flow between a depolarized receptor ending undergoing a receptor potential and the adjacent region initiates an action potential in the afferent fibre by opening voltage-gated Na⁺ channels. (b) Separate receptor cell as sensory receptor. The depolarized receptor cell undergoing a receptor potential releases a neurotransmitter that binds with chemically gated channels in the afferent fibre ending. This binding leads to a depolarization that opens voltage-gated Na⁺ channels, initiating an action potential in the afferent fibre.

**▌ Clinical Connections** Recall that Deborah was experiencing numbness in her fingers and toes. In diabetic peripheral neuropathy, the longer afferent nerves are affected first, which is why she was experiencing numbness in the extremities. In her case, some afferent nerves in her fingers and toes were injured and were unable to initiate or propagate action potentials in response to touch. While this is an example of mechanoreceptors being affected, those with diabetic neuropathy can also experience injury to other receptor types. For example, when thermoreceptor afferent neurons are damaged, the sense of heat or cold can be lost. When this happens, someone who touches a hot surface may not withdraw their hand in time to prevent serious burning.

## Adaptation to sustained stimulation

Stimuli of the same intensity do not always bring about receptor potentials of the same magnitude from the same receptor. Some receptors can diminish the extent of their depolarization despite sustained stimulus strength—a phenomenon called **adaptation**. Subsequently, the frequency of action potentials generated in the afferent neuron decreases. That is, the receptor "adapts" to the stimulus by no longer responding to it to the same degree.

### TYPES OF RECEPTORS ACCORDING TO THEIR SPEED OF ADAPTATION

There are two types of receptors—*tonic receptors* and *phasic receptors*—based on their speed of adaptation. **Tonic receptors** do not adapt at all, or they adapt slowly (> Figure 4-4a). These receptors are important in situations where it is valuable to maintain information about a stimulus. Examples of tonic receptors are muscle stretch receptors, which monitor muscle length, and joint proprioceptors, which measure the degree of joint flexion. To maintain posture and balance, the CNS must continually get information about the degree of muscle length and joint position. It is important, therefore, that these receptors do not adapt to a stimulus but continue to generate action potentials to relay this information to the CNS.

**Phasic receptors**, in contrast, are rapidly adapting receptors. The receptor rapidly adapts by no longer responding to a maintained stimulus, but when the stimulus is removed, the receptor typically responds with a slight depolarization called the **off response** (> Figure 4-4b). Phasic receptors are useful in

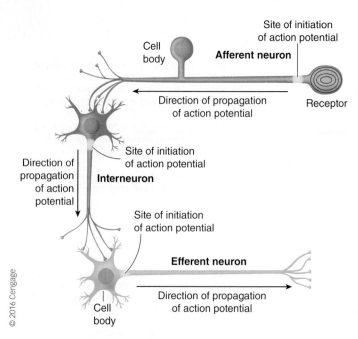

> FIGURE 4-3 **Comparison of the initiation site of an action potential in the three types of neurons.**

A receptor potential is a type of graded potential (Chapter 2). It is generally a depolarizing event resulting from inward current flow. The influx of current often brings the membrane potential of the sensory receptor toward threshold, triggering an action potential. Receptor potentials therefore vary in size and, if frequent or large enough, will trigger an action potential. An example of a receptor is a taste bud, where taste, a chemical signal, is converted into an electrical signal and sent to the brain (specifically, the primary gustatory region). When stimulated, the taste bud triggers the release of a neurotransmitter through exocytosis of synaptic vesicles from the presynaptic membrane. The neurotransmitter molecules diffuse across the synaptic cleft to the postsynaptic membrane, where the excitatory stimulus continues to transmit the signal to the primary gustatory region of the brain. The sense of taste is equivalent to the excitation of the taste receptors, and receptors for a large number of specific chemicals have been identified. Researchers have grouped taste into five basic sensations: sweet, salty, sour, bitter, and umami (savoury or meaty). All basic tastes are classified as either appealing or aversive, which can be beneficial to your health. If, for example, a serving of food has gone bad, your taste buds will often warn you prior to swallowing it. This warning will greatly reduce your likelihood of developing food poisoning.

situations where it is important to signal a change in stimulus intensity rather than to relay status quo information. Rapidly adapting receptors include *tactile (touch) receptors* in the skin that signal changes in pressure on the skin surface. Because these receptors adapt rapidly, you are not continually conscious of wearing your watch, rings, and clothing. For example, when you put something on, you soon become accustomed to it, due to these receptors' rapid adaptation. When you take the item off, you are aware of its removal, because of the off response.

## Labelling somatosensory pathways

On reaching the spinal cord, afferent information has two possible destinies: (1) it may become part of a reflex arc, bringing about an appropriate effector response, or (2) it may be relayed

upward to the brain via ascending pathways for further processing and possible conscious awareness. Pathways conveying conscious somatic sensation, the **somatosensory pathways**, consist of discrete chains of neurons, or **labelled lines**, synaptically interconnected in a particular sequence to accomplish progressively more sophisticated processing of the sensory information.

### LABELLED LINES AND LOCATION OF STIMULUS

The afferent neuron with its peripheral receptor that first detects the stimulus is known as a **first-order sensory neuron**. It synapses on a **second-order sensory neuron**, either in the spinal cord or in the medulla, depending on which sensory pathway is involved. This neuron then synapses on a **third-order sensory neuron** in the thalamus, and so on. With each step, the input is processed further. A particular sensory modality (e.g., touch, sight, sound, pain) detected by a specialized receptor type is sent over a specific afferent and ascending pathway—a pathway committed to that modality—to excite a defined area in the somatosensory cortex. That is, a particular sensory input is projected to a specific location of the cortex (see ⟩ Figure 4-24a, for an example). For example, if a free nerve ending associated with pain is stimulated, it does not matter if the stimulus is thermal, pressure, or chemical; the person perceives it as pain. In this way, different types of incoming information are kept separate within specific labelled lines between the periphery and the cortex. In this way, even though all information is propagated to the CNS via the same type of signal (action potentials), the brain can decode the type and location of the stimulus. ▌ Table 4-1

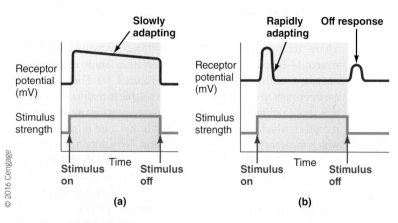

> FIGURE 4-4 **Tonic and phasic receptors.** (a) Tonic receptor. This receptor type does not adapt at all or adapts slowly to a sustained stimulus, and thereby provides continuous information about the stimulus. (b) Phasic receptor. This receptor type adapts rapidly to a sustained stimulus and frequently exhibits an off response when the stimulus is removed. Thus, the receptor signals changes in stimulus intensity rather than relaying status quo information.

© 2016 Cengage

| ■ TABLE 4-1 Coding of Sensory Information | |
|---|---|
| **Stimulus Property** | **Mechanism of Coding** |
| **Type of Stimulus** (stimulus modality) | Distinguished by the type of receptor activated and the specific pathway over which this information is transmitted to a particular area of the cerebral cortex |
| **Location of Stimulus** | Distinguished by the location of the activated receptive field and the pathway that is subsequently activated to transmit this information to the area of the somatosensory cortex representing that particular location |
| **Intensity of Stimulus** (stimulus strength) | Distinguished by the frequency of action potentials initiated in an activated afferent neuron and the number of receptors (and afferent neurons) activated |

© 2016 Cengage

summarizes how the CNS is informed of the type (what), location (where), and intensity (how much) of a stimulus.

### PHANTOM PAIN

*Clinical Note* Activation of a sensory pathway at any point gives rise to the same sensation that would be produced by stimulation of the receptors in the body part itself. This phenomenon has served as the traditional explanation for **phantom pain**—for example, pain perceived as originating in the foot by a person whose leg has been amputated at the knee. Irritation of the severed endings of the afferent pathways in the stump can trigger action potentials that, on reaching the foot region of the somatosensory cortex, are interpreted as pain in the missing foot. New evidence suggests that, in addition, the sensation of phantom pain may arise from extensive remodelling of the brain region that originally handled sensations from the severed limb. This remapping of the vacated area of the brain is speculated to somehow lead to signals from elsewhere being misinterpreted as pain arising from the missing extremity.

## Acuity, receptive field size, and lateral inhibition

Each somatosensory neuron responds to stimulus information only within a circumscribed region of the skin surface surrounding it; this region is called its **receptive field**. The size of a receptive field varies inversely with the density of receptors in the region; the more closely receptors of a particular type are spaced, the smaller the area of skin each monitors. The smaller the receptive field is in a region, the greater its **acuity**—that is, its **discriminative ability**. Compare the tactile discrimination in your fingertips with that in your elbow by "feeling" the same object with both. You can sense more precise information about the object with your richly innervated fingertips because the receptive fields

there are small; as a result, each neuron signals information about small, discrete portions of the object's surface. An estimated 17 000 tactile mechanoreceptors are present in the fingertips and palm of each hand. In contrast, the skin over the elbow is served by relatively few sensory endings with larger receptive fields. Subtle differences within each large receptive field cannot be detected. The distorted cortical representation of various body parts in the sensory homunculus (p. 118) corresponds precisely with the innervation density; more cortical space is allotted for sensory reception from areas with smaller receptive fields and, accordingly, greater tactile discriminative ability.

Besides receptor density, a second factor influencing acuity is **lateral inhibition**. You can appreciate the importance of this phenomenon by slightly indenting the surface of your skin with the point of a pencil (> Figure 4-5a). The receptive field is excited immediately under the centre of the pencil point where the stimulus is most intense, but the surrounding receptive fields are also stimulated, only to a lesser extent because they are less distorted. If information from these marginally excited afferent fibres in the fringe of the stimulus area were to reach the cortex, localization of the pencil point would be blurred. To facilitate localization and sharpen contrast, lateral inhibition occurs within the CNS (> Figure 4-5b). The most strongly activated signal pathway originating from the centre of the stimulus area inhibits the less excited pathways from the fringe areas. This occurs via inhibitory interneurons that pass laterally between ascending fibres serving neighbouring receptive fields. Blockage of further transmission in the weaker inputs increases the contrast between wanted and unwanted information so that the pencil point can be precisely localized. The extent of lateral inhibitory connections within sensory pathways varies for different modalities. Those with the most lateral inhibition—touch and vision—bring about the most accurate localization.

## Mechanoreceptors

Mechanoreceptors come in different forms, but in essence are sensitive to pressure, stretch, vibration, acceleration, and sound. Some mechanoreceptors are discussed in other sections and chapters. (Section 4.6 of this chapter describes mechanoreceptors for sound, and Chapter 9 examines baroreceptors for pressure.) But here we focus on five types of mechanoreceptors that are important to the somatic sensation of touch: (1) Pacinian corpuscles, (2) Meissner's corpuscles, (3) Merkel's discs, (4) Ruffini corpuscles, and (5) free nerve endings. Touch (tactile) receptors are the most common in the body; they respond to various forms of physical contact such as pressure, vibration, stretch, texture, and caressing movements. They are found in the skin, deep tissues, muscle, joints, and organs; they are also difficult to study due to their small size.

**Pacinian corpuscles** are located in the skin and respond to touch and deep pressure, are useful in detecting moderately rough to rougher surfaces, and are very sensitive to vibration. Their distribution ranges from approximately 350 per finger to 800 in the palm. The Pacinian corpuscle is oval shaped, ~1 mm in length, and wrapped by a layer of connective tissue. They are classified as a phasic tactile mechanoreceptor. Their optimal

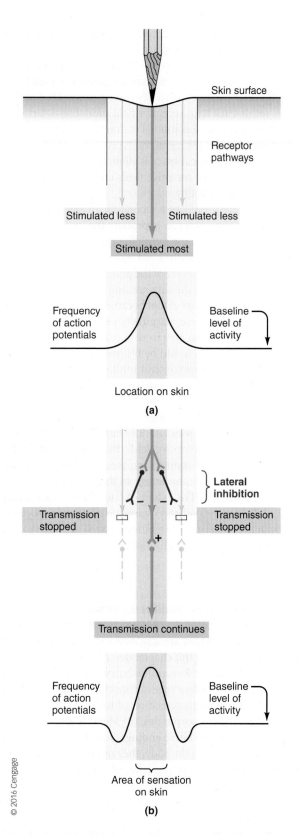

**Location on skin**

**(a)**

**Area of sensation on skin**

**(b)**

© 2016 Cengage

> **FIGURE 4-5 Lateral inhibition.** (a) The receptor at the site of most intense stimulation is activated to the greatest extent. Surrounding receptors are also stimulated but to a lesser degree. (b) The most intensely activated receptor pathway halts transmission of impulses in the less intensely stimulated pathways through lateral inhibition. This process facilitates localization of the site of stimulation.

sensitivity is about 250 Hz, and this is the frequency range generated upon fingertips by textures made of very small features (~200 μm). They respond to transient touches rather than sustained pressure, because they are rapidly adapting receptors. (The mechanism by which adaptation is accomplished varies for different receptors and has not been fully elucidated for all receptor types.)

Adaptation in a Pacinian corpuscle involves both mechanical and electrochemical components. The mechanical component depends on the physical properties of this receptor. The receptor ending consists of concentric layers of connective tissue (20–60 concentric lamellae), resembling layers of an onion wrapped around the peripheral terminal of a myelinated afferent neuron. When pressure is first applied to the Pacinian corpuscle, the underlying terminal responds with a receptor potential that reflects the intensity of the stimulus. As the stimulus continues, the pressure energy is dissipated because it causes the receptor layers to slip (just as the layers of an onion slip under pressure). Because this physical effect filters out the steady component of the applied pressure, the underlying neuronal ending no longer responds with a receptor potential and adaptation occurs.

**Meissner's corpuscles** are a phasic, tactile, myelinated mechanoreceptor, consisting of nerve endings located in the skin that are responsible for sensing light touch. They have a low threshold when sensing vibration (<50 Hz), and they rapidly adapt. Their receptive field is about 3–5 mm. Structurally, they are encapsulated nerve endings surrounded by a connective tissue capsule. The corpuscle is 30–140 μm in length, 40–60 μm in diameter, and has a density of approximately 150 per cm² at the fingertip. They are distributed on various areas of the skin, but concentrated in areas sensitive to light touch such as the fingers, lips, and nipples.

Of all the mechanoreceptors, **Merkel's discs** are the most sensitive to low frequency vibrations (5–15 Hz), having a receptive field of about 2–3 mm. They are classified as a slow-adapting, type I myelinated mechanoreceptor that provides touch information. Merkel's discs are found in the superficial layers of the skin and mucosa. They are found clustered beneath the ridges of the fingertips that make up fingerprints. Merkel nerve endings are extremely sensitive to tissue displacement, and may respond to less than 1 μm of displacement. Merkel nerve endings are clustered into specialized epithelial structures called *touch domes* or *hair disks*. They are also located in the mammary glands.

**Ruffini corpuscles** are slow-adapting, myelinated nerve endings found in the deep layers of the skin (subcutaneous tissue) and joints. They are encapsulated, tied into the local collagen matrix, and respond to stretch (deformation) and torque. They register mechanical deformation within joints (sensitivity of 2 degrees), as well as continuous pressure.

Free nerve endings are found in the skin, around hair roots, in the eyes, and many other tissues. They may be myelinated or unmyelinated, and are the most abundant of the skin receptors. Free nerve endings specialize in detecting touch and pressure, but are also used for temperature and nociception (sensing pain). When you touch your eye while putting in your contact lens, or rub your eye to remove debris, the free nerve endings associated with the cornea make you aware of the pressure and touch. Also, the sensations of tickling and itching are detected by free nerve endings, considered noxious stimuli—most commonly found in the superficial skin—and transmitted via unmyelinated slow fibres.

Having completed our discussion of receptor physiology associated with the somatic sense touch, we are now going to examine one important somatic sensation in greater detail—pain.

### Check Your Understanding 4.1

1. Define *stimulus, receptor potential, labelled line,* and *perception.*
2. Draw the response of a tonic receptor and of a phasic receptor to a stimulus of sustained strength.
3. Compare the receptive field size for a sensory neuron on your tongue and a sensory neuron on your back.

## 4.4 | Pain

Pain is an unpleasant sensation, associated with **nociceptor** stimulation (nociception is the perception of physiological pain) and is a measurable physiological event. "Pain" comes from the Latin *poena,* meaning punishment or penalty. Pain can also include nociception from both external, perceived events (e.g., seeing something) and internal, cognitive events (e.g., phantom limb pain; p. 147). Nociception is a critical component of the body's defence system because it is part of a rapid-warning system that relays instructions to the CNS to initiate an efferent motor response, thereby triggering cognitive processes to reduce or remove the physical harm.

### Stimulation of nociceptors

Unlike other somatosensory modalities, the sensation of pain is accompanied by motivated behavioural responses such as the withdrawal reflex, as well as emotional reactions such as crying. As well, the subjective perception of pain can be influenced by other past or present experiences—for example, heightened pain perception accompanying fear of the dentist.

#### CATEGORIES OF PAIN RECEPTORS

Nociceptors form late during neurogenesis, developing from neural crest stem cells, which are responsible for development of the peripheral nervous system. There are three categories of nociceptors classified by modality. **Mechanical nociceptors** respond to mechanical damage, such as cutting, crushing, or pinching. **Thermal nociceptors** respond to temperature extremes, especially heat. **Chemical nociceptors** respond equally to many kinds of irritating chemicals released from damaged tissue. The nociceptor free nerve endings are found in skin and deep tissue, joints, muscle, and bone. Because of their value to survival, nociceptors do not adapt to sustained or repetitive stimulation.

A good example of nociception is the noxious response to heat (temperature). Since the discovery of the **capsaicin** (active ingredient in hot chili peppers) receptor (TRPV1), the mechanism for thermal pain transduction has been well documented. Thermal pain is detected via a channel-gated nociceptor. Six types of thermo-TRP receptors are known: Transient Receptor Potential Vanilloid (TRPV) 1, TRPV2, TRPV3, TRPV4, TRPM8, and TRPA1. TRPV1 and TRPV2 are sensitive to noxious heat and warmth, whereas TRPV3 and TRPV4 are sensitive to cool and cold, respectively. Still other transient receptor potential (TRP) receptors are undetermined.

*Clinical Note* Nociceptors can be modulated by the presence of chemicals such as *prostaglandins,* which greatly enhance the receptor response to noxious stimuli. Prostaglandins are a special group of fatty acid derivatives that are cleaved from the lipid bilayer of the plasma membrane and act locally once released (p. 741). Tissue injury, among other things, can lead to the local release of prostaglandins. These chemicals act on the nociceptors' peripheral endings to lower their activation threshold. Aspirin-like drugs inhibit the synthesis of prostaglandins, which accounts at least in part for the analgesic (pain-relieving) properties of these types of drugs. Other chemical substances that influence nociceptors are histamine, potassium, serotonin, and substance P (peptides). These chemicals also affect the inflammatory response associated with damaged tissue.

#### FAST AND SLOW AFFERENT PAIN FIBRES

Nerve impulses vary in conduction speed, and the conduction velocity is based on myelination and nerve fibre size. We judge and categorize pain as either **fast pain** or **slow pain**. Pain impulses from nociceptors are transmitted to the CNS via one of two types of afferent fibres, which range in size and conduction speed ( Table 4-2): **A-delta (δ) fibres,** which are largest and fastest, and **C fibres,** which are smallest and slowest. Signals resulting from cold, warmth, and mechanical stimuli are

## ▌ TABLE 4-2 Characteristics of Pain Nerve Fibres

| Fibre Types | General Characteristics | Speed | Diameter | Stimuli | Description | Receptor Classification |
|---|---|---|---|---|---|---|
| **Aδ** | Small and myelinated | 6–30 m/sec | 1–5 μm | Fast pain, cold, warmth, mechanical | Sharp, stabbing, or acute | Free nerve ending |
| **C Fibres** | Small and unmyelinated | <1–2 m/sec | 0.5–2 μm | Slow pain, heat, cold, mechanical | Burning, aching, throbbing | Free nerve ending |

transmitted via fast myelinated A δ fibres at rates of 6–30 m/sec. Impulses carried via unmyelinated slow C fibres carry slow pain, heat, cold, and mechanical stimuli at a much slower rate of <1–2 m/sec.

Think about the last time you cut or burned your finger. You undoubtedly felt a sharp twinge of pain at first, with a more diffuse, disagreeable pain commencing shortly thereafter. Pain typically is perceived initially as a brief, sharp, prickling sensation that is easily localized; this is the fast pain pathway originating from specific mechanical or heat nociceptors. This feeling is followed by a dull, aching, poorly localized sensation that persists for a longer time and is more unpleasant; this is the **slow pain pathway**, which is activated by chemicals, especially **bradykinin**, a normally inactive substance that is activated by enzymes released into the ECF from damaged tissue. Bradykinin and related compounds not only provoke pain—presumably by stimulating the **polymodal nociceptors**—but also contribute to the inflammatory response to tissue injury (Chapter 11). Polymodal receptors are small-diameter receptors that respond to more than one type of noxious (painful) stimulus (e.g., heat or mechanical stimuli). The persistence of these chemicals might explain the long-lasting, aching pain that continues after removal of the mechanical or thermal stimulus that caused the tissue damage.

### HIGHER-LEVEL PROCESSING OF PAIN INPUT

Multiple structures are involved in pain processing. The primary afferent pain fibres synapse with specific second-order interneurons in the dorsal horn of the spinal cord. In response to stimulus-induced action potentials, afferent pain fibres release neurotransmitters that influence these next neurons in line. The two best known of these pain neurotransmitters are *substance P* and *glutamate*. **Substance P** activates ascending pathways that transmit nociceptive signals to higher levels for further processing (> Figure 4-6a). Ascending pain pathways have different destinations in the *cortex*, the *thalamus*, and the *reticular formation*. Cortical somatosensory processing areas localize the pain, whereas other cortical areas participate in other conscious components of the pain experience, such as deliberation about the incident. In the absence of the cortex, pain can still be perceived, presumably at the level of the thalamus. The reticular formation increases the level of alertness associated with the noxious encounter. Interconnections from the thalamus and reticular formation to the hypothalamus and limbic system elicit the behavioural and emotional responses accompanying the painful experience. The limbic system appears to be especially important in perceiving the unpleasant aspects of pain.

Glutamate, the other neurotransmitter released from primary afferent pain terminals, is a major excitatory neurotransmitter (p. 77). Glutamate acts on two different types of plasma membrane receptors on the dorsal horn neurons, with two different outcomes (p. 130). First, binding of glutamate with its *AMPA receptors* leads to permeability changes that ultimately result in the generation of action potentials in the dorsal horn cell. These action potentials transmit the pain message to higher centres. Second, binding of glutamate with its *NMDA receptors* leads to calcium $(Ca^{2+})$ entry into the dorsal horn cell. This pathway is not involved in the transmission of pain messages. Instead, $Ca^{2+}$ initiates second-messenger systems that make the dorsal horn

neuron more excitable than usual. This hyperexcitability contributes in part to the exaggerated sensitivity of an injured area to subsequent exposure to painful, or even normally nonpainful, stimuli, such as a light touch. Think about how exquisitely sensitive your sunburned skin is, even to clothing. Other mechanisms in addition to glutamate-induced hyperexcitability of the dorsal horn neurons also contribute to supersensitivity of an injured area. For example, responsiveness of the pain-sensing peripheral receptors can be boosted so that they react more vigorously to subsequent stimuli. This exaggerated sensitivity presumably serves a useful purpose by discouraging activities that could cause further damage or interfere with healing of the injured area. Usually this hypersensitivity resolves as the injury heals.

Clinical Note Chronic pain, sometimes excruciating, occasionally occurs in the absence of tissue injury. In contrast to the pain accompanying peripheral injury, which serves as a normal protective mechanism to warn of impending or actual damage to the body, abnormal chronic pain states result from damage within the pain pathways in the peripheral nerves or the CNS. That is, pain is perceived because of abnormal signalling within the pain pathways in the absence of peripheral injury or typical painful stimuli. For example, strokes that damage ascending pathways can lead to an abnormal, persistent sensation of pain.

> **Clinical Connections**
>
> This abnormal chronic pain is sometimes categorized as neuropathic pain, which is what someone with diabetic peripheral neuropathy can experience. At night, Deborah was experiencing sharp pains in her toes. Her diabetes had led to injury of Aδ fibres that resulted in the inappropriate signalling of pain to the CNS. As well, this could be related to why she was experiencing pain when standing. Although pain receptors may not have been directly activated by standing, because they were injured, they may have misinterpreted the pressures on the lower extremities during standing and initiated pain signalling.

### The brain's built-in analgesic system

In addition to the chain of neurons connecting peripheral nociceptors with higher CNS structures for pain perception, the CNS contains a built-in pain-suppressing or **analgesic system** that suppresses transmission in the pain pathways as they enter the spinal cord. Two regions are known to be a part of this descending analgesic pathway. Electrical stimulation of the *periaqueductal grey matter* (grey matter surrounding the cerebral aqueduct, a narrow canal that connects the third and fourth ventricular cavities) results in profound analgesia, as does stimulation of the *reticular formation* within the brain stem. This analgesic system suppresses pain by blocking the release of substance P from afferent pain-fibre terminals (> Figure 4-6b).

The analgesic system depends on the presence of **opiate receptors**. We have known for many years of the effect of morphine. Since the discovery of morphine as an analgesic, scientists began the search for endogenous opiates (that act like morphine) that bind these receptors and dull pain. They found that endorphins, enkephalins, and dynorphins were a part of the body's natural

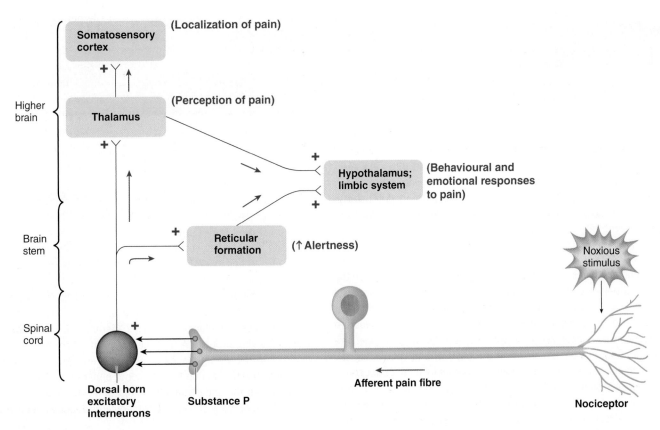

**(a)** Substance P pain pathway

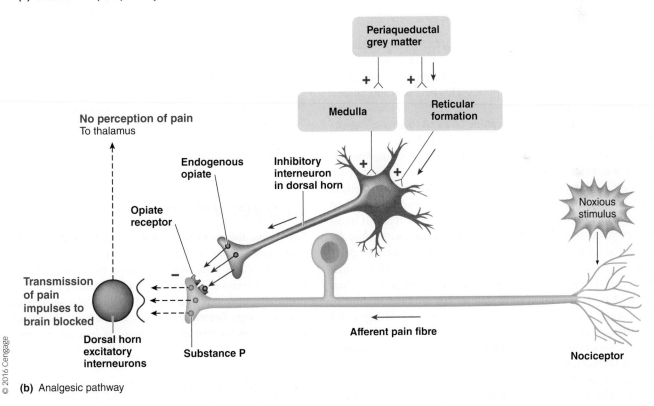

**(b)** Analgesic pathway

› **FIGURE 4-6 Substance P pain pathway and analgesic pathway.** (a) When activated by a noxious stimulus, some afferent pain pathways release substance P, which activates ascending pain pathways that provide various brain regions with input for processing different aspects of the painful experience. (b) Endogenous opiates released from descending analgesic (pain-relieving) pathways bind with opiate receptors at the synaptic knob of the afferent pain fibre. This binding inhibits the release of substance P, thereby blocking transmission of pain impulses along the ascending pain pathways.

The Peripheral Nervous System: Sensory, Autonomic, Somatic

## Acupuncture: Is It for Real?

**I**T SOUNDS LIKE SCIENCE FICTION. How can a needle inserted in the hand relieve a toothache? Acupuncture analgesia (AA), the technique of relieving pain by inserting and manipulating threadlike needles at key points, has been practised in China for more than 2000 years but is relatively new to Western medicine.

### Brief History

Traditional Chinese teaching holds that disease can occur when the normal patterns of flow of healthful energy (called *Qi*; pronounced "chee") just under the skin become disrupted, and acupuncture is able to correct this imbalance and restore health. Many Western scientists have been skeptical because, until recently, the phenomenon could not be explained on the basis of any known, logical, physiological principles, although a tremendous body of anecdotal evidence in support of the effectiveness of AA existed in China. From the point of view of Western medicine, the success of acupuncture was considered a placebo effect. The term *placebo effect* refers to a

chemical or technique that brings about a desired response through the power of suggestion or distraction rather than through any direct action.

Because the Chinese accepted anecdotal evidence for the success of AA, this phenomenon did not come under close scientific scrutiny until European and American scientists started studying it in the last several decades. As a result, an impressive body of rigorous scientific investigation supports the contention that AA works based on a physiological rather than a placebo/psychological effect. In controlled clinical studies, 55–85 percent of patients were helped by AA. Pain relief was reported by only 30–35 percent of placebo controls (people who thought they were receiving proper AA treatment, but in whom needles were inserted in the wrong places or not deep enough). Furthermore, its mechanisms of action have become apparent. Indeed, more is known about the underlying physiological mechanisms of AA than about those of many conventional medical techniques, such as gas anesthesia.

---

analgesic system. These endogenous opiates serve as analgesic neurotransmitters; they are released from the descending analgesic pathway and bind with opiate receptors on the afferent pain-fibre terminal. This binding suppresses the release of substance P via presynaptic inhibition, thereby blocking further transmission of the pain signal (p. 81). Morphine binds to these same opiate receptors, which accounts for its analgesic properties.

It is not clear how this natural pain-suppressing mechanism is normally activated. Factors known to modulate pain include exercise, stress, and acupuncture. Researchers believe that endorphins are released during prolonged exercise and presumably produce the "runner's high." Some types of stress also induce analgesia. It is sometimes disadvantageous for a stressed organism to display the normal reaction to pain. For example, when two male lions are fighting for dominance of the group, withdrawing, escaping, or resting when injured would mean certain defeat. For an examination of how acupuncture relieves pain, see Concepts, Challenges, and Controversies.

We have now completed our coverage of somatic sensations, which included touch, temperature, and pain. As you are now aware, somatic sensation is detected by various types of widely distributed receptors that provide information about the body's interactions with the environment. In contrast, each of the special senses has highly localized, extensively specialized receptors that respond to unique environmental stimuli. The special senses include *vision*, *hearing*, *taste*, *smell*, and *equilibrium* to which we now turn our attention, starting with vision.

> **Check Your Understanding 4.2**
>
> 1. Compare and contrast the classes of pain receptors
> 2. Compare the type of pain signals transmitted via A-delta fibres and C fibres.
> 3. Describe the role of endogenous opiates in the body's natural analgesic system.

## 4.5 | Eye: Vision

For **vision**, the eyes capture the patterns of illumination in the environment as an "optical picture" on a layer of light-sensitive cells, the *retina*. This coded image on the retina is transmitted through a series of progressively more complex steps of visual processing until it is finally perceived as a visual likeness of the original image. Before considering the steps involved in the process of vision, we first examine how the eyes are protected from injury.

### Protective mechanisms and eye injuries

Several mechanisms help protect the eyes from injury. The bony socket in which the eye is positioned shelters the eyeball, except for its most anterior portion. The **eyelids** act like shutters to protect the front of the eye from environmental insults. They close reflexly to cover the eye under threatening circumstances,

### Mechanism of Action

An overwhelming body of evidence supports the *acupuncture endorphin hypothesis* as the primary mechanism of AA's action. According to this hypothesis, acupuncture needles activate specific afferent nerve fibres, which send impulses to the central nervous system. Here, the incoming impulses activate three centres (a spinal cord centre, a midbrain centre, and a hormonal centre—the anterior pituitary unit of the hypothalamus) to cause analgesia. Researchers have shown that all three centres block pain transmission by means of endorphins and closely related compounds. Several other neurotransmitters, such as serotonin and norepinephrine, as well as cortisol, the major hormone released during stress, are implicated as well. (Pain relief in placebo controls is believed to occur as a result of placebo responders subconsciously activating their own built-in analgesic system.)

### Acupuncture in Canada

In Canada, AA has not been used in mainstream medicine, even by physicians who have been convinced by scientific evidence that the technique is valid. AA methodology has traditionally not been taught in Canadian medical colleges, and the techniques take time to learn. Also, AA is much more time consuming than the use of drugs. Western physicians who have been trained to use drugs to solve most pain problems are generally reluctant to give up their known methods for an unfamiliar, time-consuming technique. However, acupuncture is gaining favour as an alternative treatment for relief of chronic pain, especially because analgesic drugs can have troublesome side effects. After decades of being spurned by most of the Canadian medical community, acupuncture started gaining respectability after the publication of scientific research. The research has suggested that acupuncture is effective as an alternative or adjunct to conventional treatment for many kinds of pain and nausea. Now that acupuncture has gained acceptance, some medical insurers now pay for this scientifically legitimate treatment, and some of the nation's medical schools have incorporated the technique into their curricula. There are now many accredited acupuncture schools for nonphysicians, and thousands of accredited acupuncturists.

---

such as rapidly approaching objects, dazzling light, and instances when the exposed surface of the eye or eyelashes are touched. Frequent, spontaneous blinking of the eyelids helps disperse the lubricating, cleansing, bactericidal (germ-killing) **tears**. Tears are produced continuously by the **lacrimal gland** in the upper lateral corner under the eyelid. This eye-washing fluid flows across the anterior surface of the eye, drains into tiny canals in the corner of each eye (⟩ Figure 4-7a), and empties into the back of the nasal passageway. This drainage system cannot handle the profuse tear production during crying, so the tears overflow from the eyes. The eyes are also equipped with protective **eyelashes**, which trap fine, airborne debris, such as dust, before it can fall into the eye.

### The eye: A fluid-filled sphere

Each **eye** is a spherical, fluid-filled structure enclosed by three layers. From outermost to innermost, these are (1) the *sclera/cornea*; (2) the *choroid/ciliary body/iris*; and (3) the *retina* (Figure 4-7b). Most of the eyeball is covered by a tough outer layer of connective tissue, the **sclera**, which forms the visible white part of the eye (Figure 4-7a). Anteriorly, the outer layer consists of the transparent **cornea**, through which light rays pass into the interior of the eye. The middle layer underneath the sclera is the **choroid**, which contains many blood vessels that nourish the retina. The choroid layer becomes specialized anteriorly to form the *ciliary body* and *iris*, which will be described shortly. The innermost coat under the choroid is the **retina**, which consists of an outer layer and an inner nervous-tissue layer. The latter contains the **rods** and **cones**, the photoreceptors that convert light energy into nerve impulses. The choroid and the outer layer of the retina are highly pigmented to prevent reflection, or scattering, of light within the eye.

The interior of the eye consists of two fluid-filled cavities, separated by an elliptical **lens**, all of which are transparent to permit light to pass through the eye from the cornea to the retina. The larger cavity between the lens and retina contains a semifluid, jelly-like substance, the **vitreous humour**. The vitreous humour is important in maintaining the spherical shape of the eyeball. The anterior cavity between the cornea and lens contains a clear, watery fluid, the **aqueous humour**. The aqueous humour carries nutrients for the cornea and lens, both of which lack a blood supply. Blood vessels in these structures would impede the passage of light to the photoreceptors.

Aqueous humour is produced at a rate of about 5 mL/day by a capillary network within the **ciliary body**, a specialized anterior derivative of the choroid layer. This fluid drains into a canal at the edge of the cornea and eventually enters the blood (⟩ Figure 4-8). If aqueous humour is not drained as rapidly as it forms (e.g., because of a blockage in the drainage canal), the excess accumulates in the anterior cavity, causing the pressure to rise within the eye. This condition is known as **glaucoma**. The excess aqueous humour pushes the lens backward into the vitreous humour, which in turn is pushed against the inner neural layer of the retina. This compression causes retinal and optic nerve damage that can lead to blindness if the condition is not treated.

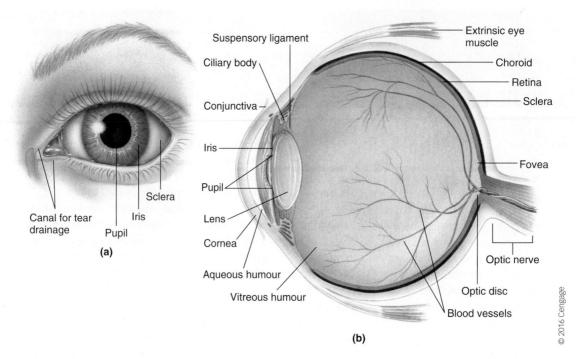

> **FIGURE 4-7 Structure of the eye.** (a) External front view. (b) Internal sagittal view

## The iris

Not all the light passing through the cornea reaches the light-sensitive photoreceptors. This is due to the iris, a thin, pigmented smooth muscle that forms a visible ringlike structure within the aqueous humour (see › Figure 4-7a and › 4-7b). The pigment in the iris is responsible for eye colour. The varied flecks, lines, and other nuances of the iris are unique for each individual, making the iris the basis of the latest identification technology. Recognition of iris patterns by a video camera that captures iris images and translates the landmarks into a computerized code is more foolproof than fingerprinting or even DNA testing.

The round opening in the centre of the iris through which light enters the interior portions of the eye is the **pupil**. The size of this opening can be adjusted by variable contraction of the iris muscles to admit more or less light as needed, much as the diaphragm controls the amount of light entering a camera. The iris contains two sets of smooth muscle networks, one *circular* (the muscle fibres run in a ringlike fashion within the iris) and the other *radial* (the fibres project outward from the pupillary margin like bicycle spokes) (› Figure 4-9). Because muscle fibres

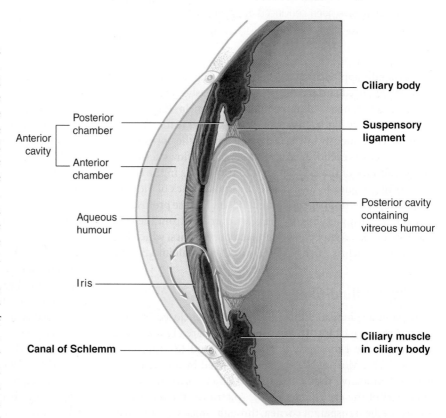

> **FIGURE 4-8 Formation and drainage of aqueous humour.** Aqueous humour is formed by a capillary network in the ciliary body, then drains into the canal of Schlemm, and eventually enters the blood.

**Source:** From Sherwood. *Human Physiology*, 8E. © 2013 Brooks/Cole, a part of Cengage, Inc. Reproduced by permission. www.cengage.com/permissions

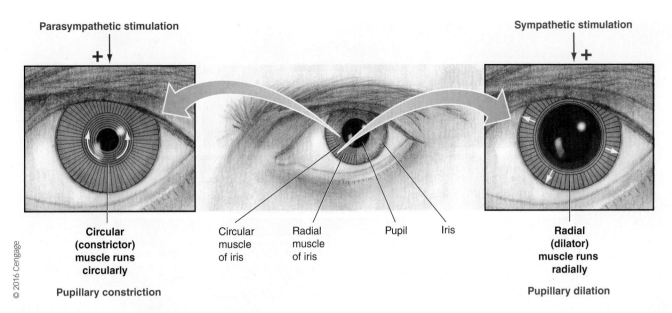

**Parasympathetic stimulation**

**Sympathetic stimulation**

Circular
(constrictor)
muscle runs
circularly

**Pupillary constriction**

Circular
muscle
of iris

Radial
muscle
of iris

Pupil

Iris

Radial
(dilator)
muscle runs
radially

**Pupillary dilation**

© 2016 Cengage

› FIGURE 4-9 **Control of pupillary size**

shorten when they contract, the pupil gets smaller when the **circular (constrictor) muscle** contracts and forms a smaller ring. This reflex pupillary constriction occurs in bright light to decrease the amount of light entering the eye. When the **radial (dilator) muscle** shortens, the size of the pupil increases. Such pupillary dilation occurs in dim light to allow the entrance of more light. Iris muscles are controlled by the autonomic nervous system (ANS). Parasympathetic nerve fibres innervate the circular muscle (causing pupillary constriction), whereas sympathetic fibres supply the radial muscle (causing pupillary dilation).

## Properties of light waves

**Light** is a form of electromagnetic radiation composed of particle-like individual packets of energy called **photons** that travel in wavelike fashion. The distance between two wave peaks is known as the **wavelength** (› Figure 4-10). The wavelengths in the electromagnetic spectrum range from $10^{-14}$ m (quadrillionths of a metre, as in the extremely short cosmic rays) to $10^4$ m (10 km, as in long radio waves) (› Figure 4-11). The visible light range is between 400 and 700 nm (a billionth of a metre), a small portion of the total spectrum. This range, however, is sufficient to allow us to perceive different colour sensations, since different wavelengths correspond to different colours. The shorter visible wavelengths are sensed as violet and blue; the longer wavelengths are interpreted as orange and red.

In addition to having variable wavelengths, light energy also varies in intensity; that is, the amplitude, or height, of the wave (› Figure 4-10). Dimming a bright red light does not change its colour; it just becomes less intense or less bright.

Light waves *diverge* (radiate outward) in all directions from every point of a light source. The forward movement of a light wave in a particular direction is known as a **light ray**.

Divergent light rays reaching the eye must be bent inward to be focused back into a point (the **focal point**) on the light-sensitive retina to provide an accurate image of the light source.

Light travels faster through air than through other transparent media, such as water and glass. When a light ray enters a medium of greater density, it is slowed down (the converse is also true). The course of direction of the ray changes if it strikes the surface of the new medium at any angle other than perpendicular. The bending of a light ray is known as **refraction**. With a curved surface, such as a lens, the greater the curvature is, the greater the degree of bending and the stronger the lens. When a light ray strikes the curved surface of any object of greater density, the direction of refraction depends on the angle of the curvature (› Figure 4-12). A **convex** surface curves outward (like the outer surface of a ball), whereas a **concave** surface curves inward (like a cave). Convex surfaces

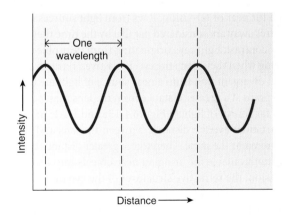

› FIGURE 4-10 **Properties of an electromagnetic wave.** A *wavelength* is the distance between two wave peaks. The *intensity* is the amplitude of the wave.

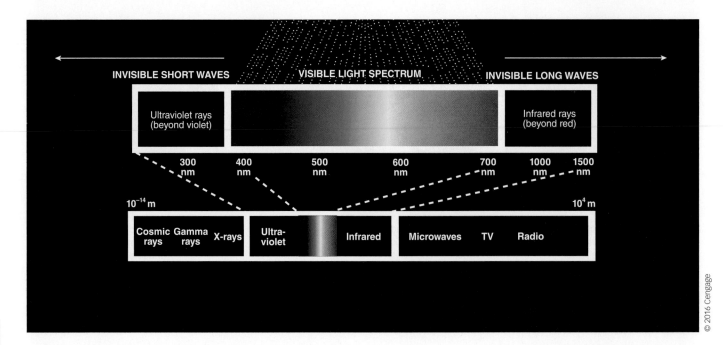

> **FIGURE 4-11 Electromagnetic spectrum.** The wavelengths in the electromagnetic spectrum range from less than $10^{-14}$ m to $10^4$ m. The visible spectrum includes wavelengths ranging from 400 to 700 nm.

converge light rays, bringing them closer together. Concave surfaces diverge light rays (spread them farther apart).

## The eye's refractive structures

The cornea and lens are the structures most important to the eye's refractive ability. The first structure, the curved corneal surface, contributes most extensively to the eye's total refractive ability because the difference in density at the air–cornea interface is much greater than the differences in density between the lens and the fluids surrounding it. In **astigmatism** the curvature of the cornea is uneven, so light rays are unequally refracted. The refractive ability of a person's cornea remains constant, because the curvature of the cornea never changes. In contrast, the refractive ability of the lens can be adjusted by changing its curvature as needed for near or far vision. Rays from light sources more than six metres away are considered parallel by the time they reach the eye. By contrast, light rays originating from near objects are still diverging when they reach the eye. For a given refractive ability of the eye, divergent rays from a near source of light are brought to a focal point at a greater distance from the lens than parallel rays from a far source of light (› Figure 4-13a and › 4-13b). However, in a particular eye the distance between the lens and the retina always remains the same. Therefore, a greater distance beyond the lens is not available for bringing near objects into focus. Yet for clear vision, the refractive structures of the eye must bring both near and far light sources into focus on the retina. If an image is focused before it reaches the retina or is not yet focused when it reaches the retina, it will be blurred. To bring both near and far light sources into focus on the retina (i.e., in the same distance), a stronger lens must be used for the near source (› Figure 4-13c). Let's see how the strength of the lens can be adjusted as needed.

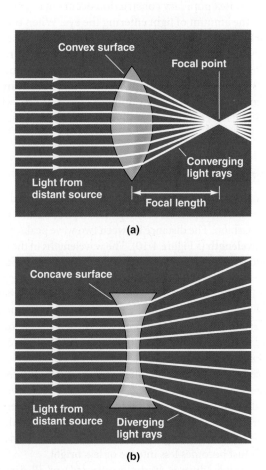

> **FIGURE 4-12 Refraction by convex and concave lenses.** (a) A lens with a convex surface, which converges the rays (brings them closer together). (b) A lens with a concave surface, which diverges the rays (spreads them farther apart)

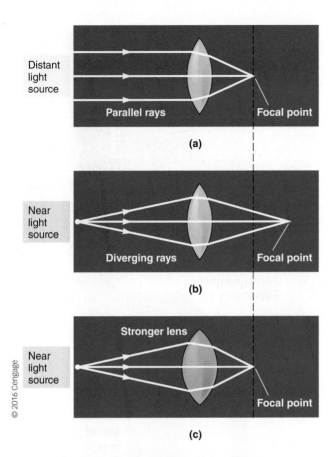

© 2016 Cengage

**› FIGURE 4-13 Focusing of distant and near sources of light.** (a) The rays from a distant (far) light source (more than six metres from the eye) are parallel by the time the rays reach the eye. (b) The rays from a near light source (less than six metres from the eye) are still diverging when they reach the eye. A longer distance is required for a lens of a given strength to bend the diverging rays from a near light source into focus compared with the parallel rays from a distant light source. (c) To focus on both distant and near light sources in the same distance (i.e., the distance between the lens and retina), a stronger lens must be used for the near source.

## Accommodation

The ability to adjust the strength of the lens is known as **accommodation**. The strength of the lens depends on its shape, which in turn is regulated by the ciliary muscle.

The **ciliary muscle** is part of the ciliary body, an anterior specialization of the choroid layer. The ciliary body has two major components: the ciliary muscle and the capillary network that produces aqueous humour (see Figure 4-8). The ciliary muscle is a circular ring of smooth muscle attached to the lens by **suspensory ligaments** (› Figure 4-14a and › 4-14b).

When the ciliary muscle is relaxed, the suspensory ligaments are taut, and they pull the lens into a flattened, weakly refractive shape (› Figure 4-14c). As the muscle contracts, its circumference decreases, slackening the tension in the suspensory ligaments (› Figure 4-14d). When the suspensory ligaments subject the lens to less tension, it becomes more spherical, because of its inherent elasticity. The greater curvature of the more rounded lens increases its refractive abilities, further bending light rays. In the normal eye, the ciliary muscle is relaxed, and the lens is flat for far vision, but the muscle contracts to let the lens become more convex and

stronger for near vision. The ciliary muscle is controlled by the ANS, with sympathetic stimulation causing its relaxation and parasympathetic stimulation causing its contraction.

*Clinical Note* The lens is made up of about 1000 layers of cells that destroy their nucleus and organelles during development, so the cells are perfectly transparent. Lacking DNA and protein-synthesizing machinery, mature lens cells cannot regenerate or repair themselves. Cells in the centre of the lens are in double jeopardy. Not only are they oldest, but also they are farthest away from the aqueous humour, the lens's nutrient source. With advancing age, these nonrenewable central cells die and become stiff. With loss of elasticity, the lens can no longer assume the spherical shape required to accommodate for near vision. This age-related reduction in accommodative ability, **presbyopia**, affects most people by middle age (45–50), requiring them to resort to corrective lenses for near vision (reading).

The elastic fibres in the lens are normally transparent. These fibres occasionally become opaque so that light rays cannot pass through—a condition known as a **cataract**. The defective lens can usually be surgically removed and vision restored by an implanted artificial lens or by compensating eyeglasses.

Other common vision disorders are *nearsightedness (myopia)* and *farsightedness (hyperopia)*. In a normal eye (**emmetropia**) (› Figure 4-15a), a far light source is focused on the retina without accommodation, whereas the strength of the lens is increased by accommodation to bring a near source into focus. In **myopia** (› Figure 4-15b1), because the eyeball is too long or the lens is too strong, a near light source is brought into focus on the retina without accommodation (even though accommodation is normally used for near vision), whereas a far light source is focused in front of the retina and is blurry. Thus, a myopic individual has better near vision than far vision, a condition that can be corrected by a concave lens (› Figure 4-15b2). With **hyperopia** (› Figure 4-15c1), either the eyeball is too short or the lens is too weak. Far objects are focused on the retina only with accommodation, whereas near objects are focused behind the retina even with accommodation and, accordingly, are blurry. Thus, a hyperopic individual has better far vision than near vision, a condition that can be corrected by a convex lens (› Figure 4-15c2). Instead of using corrective eyeglasses or contact lenses, many people are now opting to compensate for refractive errors with laser eye surgery (such as LASIK) to permanently change the shape of the cornea.

## The retinal layers

The primary responsibility of the eye is to focus light rays on to the rods and cones, which transform the light energy into electrical signals that are sent to the CNS.

The receptor-containing portion of the retina is actually an extension of the CNS, not a separate peripheral organ. During embryonic development, the retinal cells "back out" of the nervous system, so the retinal layers, surprisingly, are facing backward! The neural portion of the retina consists of three layers of excitable cells (› Figure 4-16): (1) the outermost layer (closest to the choroid) containing the rods and cones, whose light-sensitive ends face the choroid (away from the incoming light); (2) a middle layer of **bipolar cells**; and (3) an inner layer of **ganglion cells**. Axons of

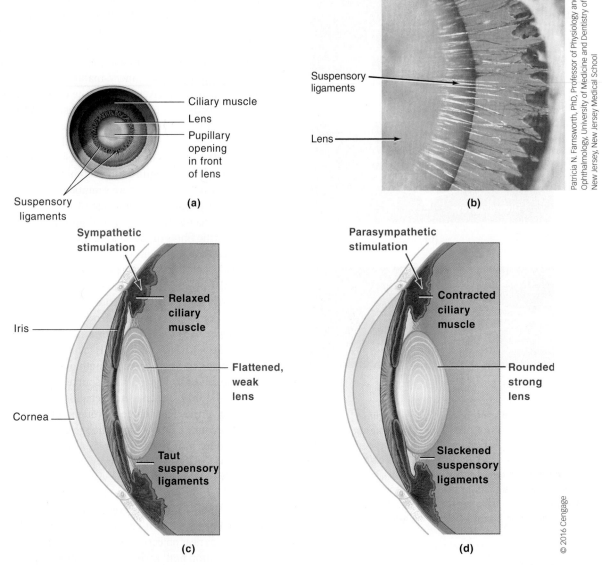

Ciliary muscle
Lens
Pupillary opening in front of lens
Suspensory ligaments

**(a)**

Suspensory ligaments
Lens

**(b)**

Patricia N. Farnsworth, PhD, Professor of Physiology and Ophthalmology, University of Medicine and Dentistry of New Jersey, New Jersey Medical School

**Sympathetic stimulation**
Iris
**Relaxed ciliary muscle**
**Flattened, weak lens**
Cornea
**Taut suspensory ligaments**

**(c)**

**Parasympathetic stimulation**
**Contracted ciliary muscle**
**Rounded strong lens**
**Slackened suspensory ligaments**

**(d)**

© 2016 Cengage

› **FIGURE 4-14 Mechanism of accommodation.** (a) Schematic representation of suspensory ligaments extending from the ciliary muscle to the outer edge of the lens. (b) Scanning electron micrograph showing the suspensory ligaments attached to the lens. (c) When the ciliary muscle is relaxed, the suspensory ligaments are taut, putting tension on the lens so that it is flat and weak. (d) When the ciliary muscle is contracted, the suspensory ligaments become slack, reducing the tension on the lens. The lens can then assume a stronger, rounder shape because of its elasticity.

the ganglion cells join to form the **optic nerve**, which leaves the retina slightly off-centre. The point on the retina at which the optic nerve leaves and through which blood vessels pass is the **optic disc** (see Figure 4-7b). This region is often called the **blind spot**; no image can be detected in this area because it has no rods and cones (› Figure 4-17). We are normally not aware of the blind spot, because central processing somehow "fills in" the missing spot. You can discover the existence of your own blind spot by a simple demonstration (› Figure 4-18).

Light must pass through the ganglion and bipolar layers before reaching the photoreceptors in all areas of the retina except the fovea. In the **fovea**, which is a pinhead-sized depression located in the exact centre of the retina, the bipolar and ganglion cell layers are pulled aside so that light strikes the photoreceptors directly (see Figure 4-7b). This feature, coupled with the fact that

only cones (which have greater acuity or discriminative ability than the rods) are found here, makes the fovea the point of most distinct vision. In fact, the fovea has the greatest concentration of cones in the retina. Thus, we turn our eyes so that the image of the object at which we are looking is focused on the fovea. The area immediately surrounding the fovea, the **macula lutea**, also has a high concentration of cones and fairly high acuity (› Figure 4-17). Macular acuity, however, is less than that of the fovea, because of the overlying ganglion and bipolar cells in the macula.

*Clinical Note* **Macular degeneration** is the leading cause of blindness in the western hemisphere. This condition is characterized by loss of photoreceptors in the macula lutea in association with advancing age. Its victims have "doughnut" vision. They suffer a loss in the middle of their visual field, which normally has the highest acuity, and are left with only the less distinct peripheral vision.

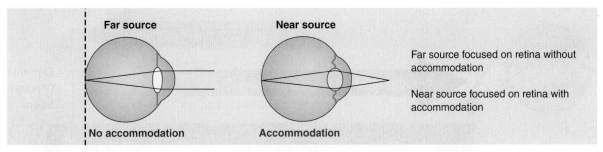

**(a) Normal eye (Emmetropia)**

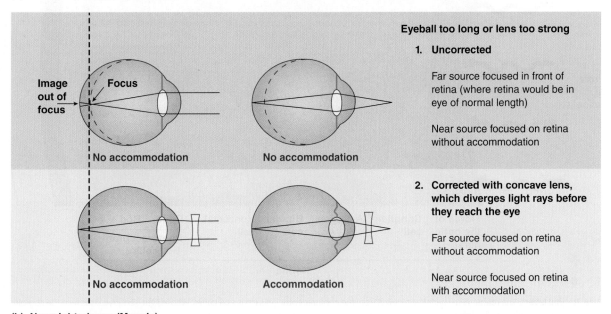

**(b) Nearsightedness (Myopia)**

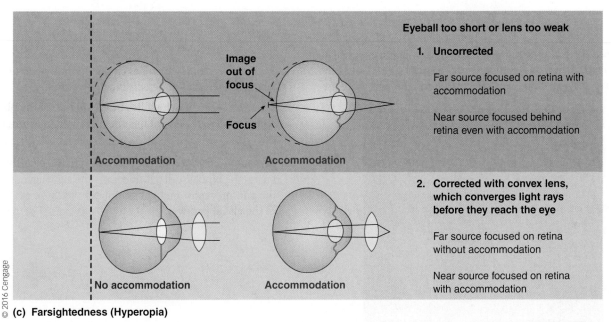

© 2016 Cengage

**(c) Farsightedness (Hyperopia)**

› **FIGURE 4-15 Emmetropia, myopia, and hyperopia.** This figure compares far vision and near vision (a) in the normal eye with (b) nearsightedness and (c) farsightedness in both their (1) uncorrected and (2) corrected states. The vertical dashed line represents the normal distance of the retina from the cornea, that is, the site at which an image is brought into focus by the refractive structures in a normal eye.

The Peripheral Nervous System: Sensory, Autonomic, Somatic

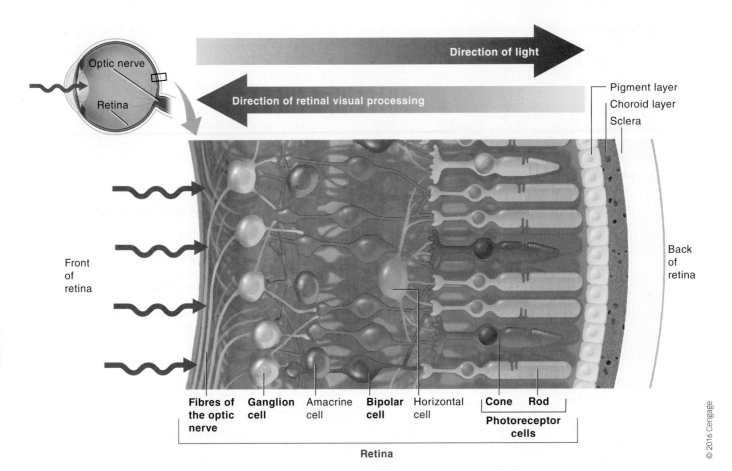

> **FIGURE 4-16  Retinal layers.** The retinal visual pathway extends from the photoreceptor cells (rods and cones, whose light-sensitive ends face the choroid away from the incoming light) to the bipolar cells, and then to the ganglion cells. The horizontal and amacrine cells act locally for retinal processing of visual input.

## Photoreceptors

A photoreceptor is a specialized type of neuron in the eye's retina. The two types of photoreceptor cells are rods and cones. Cones are adapted to detect colours and work well in bright light. Rods are more sensitive but do not detect colour well; they are adapted for low light. The photoreceptor consists of three parts (> Figure 4-19a):

1. An *outer segment*, which lies closest to the eye's exterior, facing the choroid. It detects the light stimulus.

2. An *inner segment*, which lies in the middle of the photoreceptor's length. It contains the metabolic machinery of the cell.

3. A *synaptic terminal*, which lies closest to the eye's interior, facing the bipolar cells. It transmits the signal generated in the photoreceptor on light stimulation to these next cells in the visual pathway.

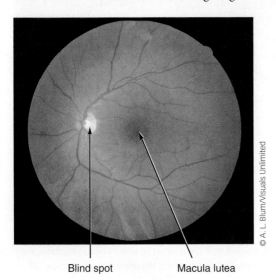

Blind spot          Macula lutea

> **FIGURE 4-17  View of the retina seen through an ophthalmoscope.** With an ophthalmoscope, a lighted viewing instrument, it is possible to view the optic disc (blind spot) and macula lutea within the retina at the rear of the eye.

> **FIGURE 4-18  Demonstration of the blind spot.** Find the blind spot in your left eye by closing your right eye and holding the book about 10 cm from your face. While focusing on the cross, gradually move the book away from you until the circle vanishes from view. At this time, the image of the circle is striking the blind spot of your left eye. You can similarly locate the blind spot in your right eye by closing your left eye and focusing on the circle. The cross will disappear when its image strikes the blind spot of your right eye.

The outer segment, which is rod shaped in rods and cone shaped in cones (› Figure 4-19a), consists of stacked, flattened, membranous discs containing an abundance of light-sensitive *photopigment* molecules. Each retina has about 150 million photoreceptors, and more than a billion photopigment molecules may be packed into the outer segment of each photoreceptor.

**Photopigments** undergo chemical alterations when activated by light. Through a series of steps, this light-induced change and subsequent activation of the photopigment bring about a receptor potential that ultimately leads to the generation of action potentials, which transmit this information to the brain for visual processing. A photopigment consists of two components: **opsin**, a protein that is an integral part of the disc membrane; and **retinene**, a derivative of vitamin A that is bound within the interior of the opsin molecule (› Figure 4-19b). Retinene is the light-absorbing part of the photopigment.

There are four different photopigments: one in the rods and one in each of three types of cones. These four photopigments differentially absorb various wavelengths of light. **Rhodopsin**, the rod photopigment, absorbs all visible wavelengths. Using visual input from the rods, the brain cannot discriminate between various wavelengths in the visible spectrum. Therefore, rods provide vision only in shades of grey by detecting different intensities, not different colours. The photopigments in the three types of cones—**red, green, and blue cones**—respond selectively to various wavelengths of light, making colour vision possible.

**Phototransduction**, the process of converting light stimuli into electrical signals, is basically the same for all photoreceptors, but the mechanism is contrary to the usual means by which receptors respond to their adequate stimulus. Specifically, receptors typically *depolarize* when stimulated, whereas photoreceptors *hyperpolarize* on light absorption. Let's first examine the status of the photoreceptors in the dark, and then consider what happens when they are exposed to light.

### PHOTORECEPTOR ACTIVITY IN THE DARK

The plasma membrane of a photoreceptor's outer segment contains chemically gated $Na^+$ channels. Unlike other chemically gated channels that respond to extracellular chemical messengers, these channels respond to an internal second messenger: **cyclic GMP (cGMP)** (cyclic guanosine monophosphate). The binding of cGMP to these $Na^+$ channels keeps them open. In the absence of light, the concentration of cGMP is high (Figure 4-20a). Therefore, the $Na^+$ channels of a photoreceptor, unlike most receptors, are open in the absence of stimulation—that is, in the dark. The resultant passive, inward $Na^+$ leak depolarizes the photoreceptor. The passive spread of this depolarization from the outer segment (where the $Na^+$ channels are located) to the synaptic terminal (where the photoreceptor's neurotransmitter is stored) keeps the synaptic terminal's voltage-gated $Ca^{2+}$ channels open. $Ca^{2+}$ entry triggers the release of neurotransmitter from the synaptic terminal while in the dark.

### PHOTORECEPTOR ACTIVITY IN THE LIGHT

On exposure to light, the concentration of cGMP is decreased through a series of biochemical steps triggered by photopigment activation (› Figure 4-20b). Retinene changes shape when

it absorbs light (see › Figure 4-19b). This change in conformation activates the photopigment. Rod and cone cells contain a G protein (p. 217) called **transducin**. The activated photopigment activates transducin, which in turn activates the intracellular enzyme phosphodiesterase. This enzyme degrades cGMP, thus decreasing the concentration of this second messenger in the photoreceptor. During the light excitation process, the reduction in cGMP permits the chemically gated $Na^+$ channels to close. This channel closure stops the depolarizing $Na^+$ leak and causes membrane hyperpolarization. This hyperpolarization, which is the receptor potential, passively spreads from the outer segment to the synaptic terminal of the photoreceptor. Here the potential changes lead to closure of the voltage-gated $Ca^{2+}$ channels and a subsequent reduction in neurotransmitter release from the synaptic terminal. Thus, photoreceptors are *inhibited by their adequate stimulus* (hyperpolarized by light) and *excited in the absence of stimulation* (depolarized by darkness). The hyperpolarizing potential and subsequent decrease in neurotransmitter release are graded according to the intensity of light. The brighter the light is, the greater the hyperpolarizing response and the greater the reduction in neurotransmitter release.

### FURTHER RETINAL PROCESSING OF LIGHT INPUT

How does the retina signal the brain about light stimulation through such an inhibitory response? The photoreceptors synapse with bipolar cells. These cells, in turn, terminate on the ganglion cells, whose axons form the optic nerve for transmission of signals to the brain. The neurotransmitter released from the photoreceptors' synaptic terminal has an *inhibitory* action on the bipolar cells. The reduction in neurotransmitter release that accompanies light-induced receptor hyperpolarization decreases this inhibitory action on the bipolar cells. Removal of inhibition has the same effect as direct excitation of the bipolar cells. The greater the illumination is on the receptor cells, the greater the removal of inhibition from the bipolar cells and the greater the excitation of the visual pathway to the brain.

Bipolar cells, similar to the photoreceptors, display graded potentials and do not conduct action potentials. This is in contrast to the ganglion cells, the first neurons in the chain, which propagate the visual message over long distances to the brain through the generation of action potentials.

The altered photopigments are restored to their original conformation in the dark by enzyme-mediated mechanisms (see › Figure 4-19b). Subsequently, the membrane potential and rate of neurotransmitter release of the photoreceptor are returned to their basal state, and no further action potentials are transmitted to the visual cortex (› Figure 4-20a).

 *Clinical Note* Researchers are currently working on an ambitious and still highly speculative microelectronic chip that would serve as a partial substitute retina. Their hope is that the device will be able to restore at least some sight in people who are blinded by loss of photoreceptor cells, but whose ganglion cells and optic pathways remain healthy. For example, if the research is successful, the chip could benefit people with macular degeneration. The envisioned "vision chip" would bypass the photoreceptor step altogether: images received by means of a camera mounted on eyeglasses would be translated by the chip

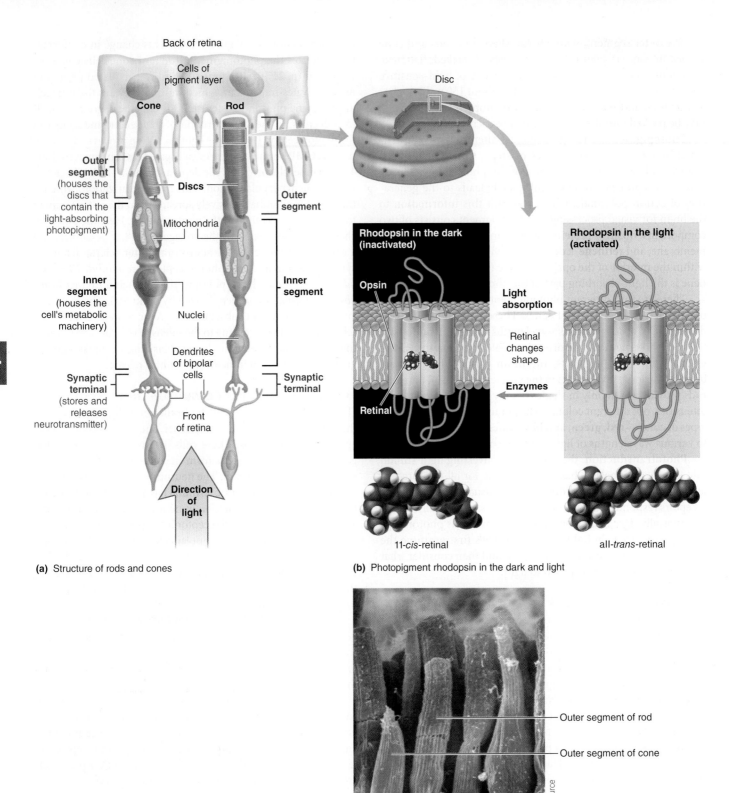

**(a)** Structure of rods and cones

**(b)** Photopigment rhodopsin in the dark and light

**(c)** Outer segments of rods and cones

> **FIGURE 4-19 Photoreceptors. (a)** The three parts of the rods and cones, the eye's photoreceptors. Note in the outer segment of the rod and cone the stacked, flattened, membranous discs, which contain an abundance of photopigment molecules. **(b)** A photopigment, such as rhodopsin, depicted here and found in rods, consists of opsin, a plasma-membrane protein, and retinal, a vitamin A derivative. In the dark, 11-*cis*-retinal is bound within the interior of opsin, and the photopigment is inactive. In the light, retinal changes to all-*trans*-retinal, activating the photopigment. **(c)** Scanning electron micrograph of the outer segments of rods and cones. Note the rod shape in rods and the cone shape in cones.

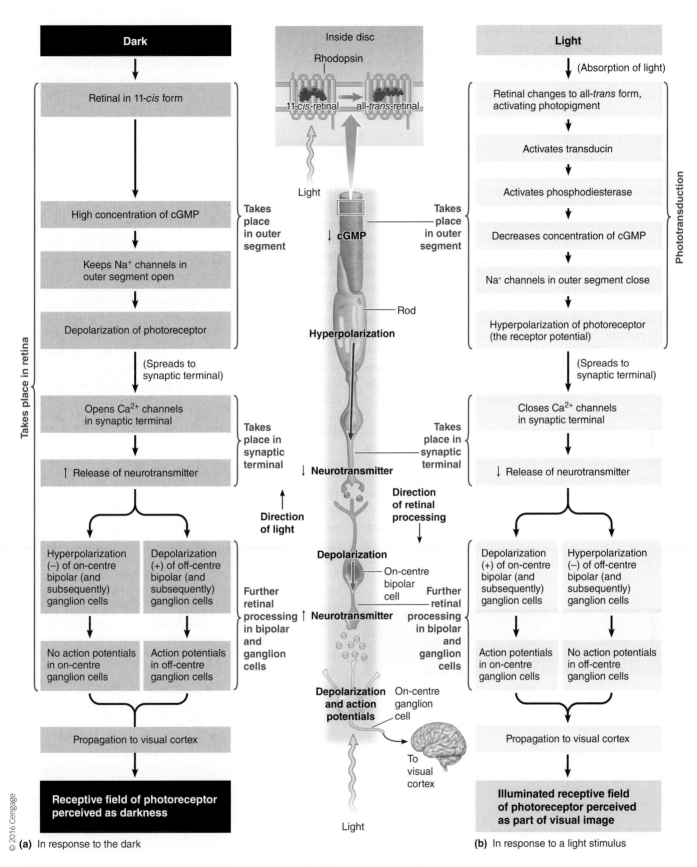

> FIGURE 4-20 **Phototransduction, further retinal processing, and initiation of action potentials in the visual pathway.** (a) Events occurring in the retina and visual pathway in response to the dark. (b) Events occurring in the retina and visual pathway in response to a light stimulus

into electrical signals detectable by the ganglion cells and transmitted on for further optical processing. Another promising avenue under investigation for halting or even reversing the loss of sight in degenerative eye diseases involves regenerating the retina through use of fetal retinal cell transplants.

## Properties of rods and cones

The retina contains more than 30 times as many rods as cones (100 million rods compared with 3 million cones per eye). Cones are most abundant in the macula lutea in the centre of the retina. From this point outward, the concentration of cones decreases and the concentration of rods increases. Rods are most abundant in the periphery. We have examined the similar way in which phototransduction takes place in rods and cones. Now we focus on the differences between these photoreceptors. You already know that rods provide vision only in shades of grey, whereas cones provide colour vision. The capabilities of the rods and cones also differ in other respects because of a difference in the "wiring patterns" between these photoreceptor types and other retinal neuronal layers ( Table 4-3). Cones have low sensitivity to light, being "turned on" only by bright daylight, but they have high acuity (sharpness; ability to distinguish between two nearby points). Thus, cones provide sharp vision with high resolution for fine detail. Humans use cones for day vision, which is in colour and distinct. Rods, in contrast, have low acuity but high sensitivity, so they respond to the dim light of night. You can see at night with your rods but at the expense of colour and distinctness. Let us see how wiring patterns influence sensitivity and acuity.

There is little convergence of neurons in the retinal pathways for cone output (p. 83). Each cone generally has a private line connecting it to a particular ganglion cell. In contrast, there is much convergence in rod pathways. Output from more than 100 rods may converge via bipolar cells onto a single ganglion cell.

Before a ganglion cell can generate an action potential, the cell must be brought to threshold through the influence of the graded potentials in the receptors to which it is wired. Because a single-cone ganglion cell is influenced by only one cone, only bright daylight is intense enough to induce a sufficient receptor potential in the cone to ultimately bring the ganglion cell to threshold. The abundant convergence in the rod visual pathways, in contrast, offers good opportunities for summation of subthreshold events in a rod ganglion cell (p. 78–79). Whereas a small receptor potential induced by dim light in a single cone would not be sufficient to bring its ganglion cell to threshold, similar small receptor potentials induced by the same dim light in multiple rods converging on a single ganglion cell would have an additive effect to bring the rod ganglion cell to threshold. Because rods can bring about action potentials in response to small amounts of light, they are much more sensitive than cones. However, because cones have private lines into the optic nerve, each cone transmits information about an extremely small receptive field on the retinal surface. Cones are thus able to provide highly detailed vision at the expense of sensitivity. With rod vision, acuity is sacrificed for sensitivity. Because many rods share a single ganglion cell, once an action potential is initiated, it is impossible to discern which of the multiple rod inputs were activated to bring the ganglion cell to threshold. Objects appear fuzzy when rod vision is used, because of this poor ability to distinguish between two nearby points.

## Sensitivity of the eyes to dark and light

The eyes' sensitivity to light depends on the amount of light-responsive photopigment present in the rods and cones. When you go from bright sunlight into darkened surroundings, you cannot see anything at first, but gradually you begin to distinguish objects as a result of the process of **dark adaptation**. Breakdown of photopigments during exposure to sunlight tremendously decreases photoreceptor sensitivity. For example, a reduction in the content of inactivated rhodopsin by only 0.6 percent from its maximum value decreases rod sensitivity approximately 3000 times. In the dark, the photopigments broken down during light exposure are gradually regenerated. As a result, the sensitivity of your eyes gradually increases so you can begin to see in the darkened surroundings. However, only the highly sensitive, rejuvenated rods are "turned on" by the dim light.

Conversely, when you move from the dark to the light (e.g., leaving a movie theatre and entering the bright sunlight), at first your eyes are very sensitive to the dazzling light. With little contrast between lighter and darker parts, the entire image appears bleached. As some of the photopigments are rapidly broken down by the intense light, the sensitivity of the eyes decreases and normal contrasts can once again be detected, a process known as **light adaptation**. The rods are so sensitive to light that enough rhodopsin is broken down in bright light to essentially "burn out" the rods; that is, after the rod photopigments have already been broken down by the bright light, they no longer can respond to the light. Furthermore, a central neural adaptive mechanism switches the eyes from the rod system to the cone system on exposure to bright light. Therefore, only the less sensitive cones are used for day vision.

**TABLE 4-3 Properties of Rod Vision and Cone Vision**

| Rods | Cones |
| --- | --- |
| 100 million per retina | 3 million per retina |
| Vision in shades of grey | Colour vision |
| High sensitivity | Low sensitivity |
| Low acuity | High acuity |
| Night vision | Day vision |
| Much convergence in retinal pathways | Little convergence in retinal pathways |
| More numerous in periphery | Concentrated in fovea and macula lutea |

© 2016 Cengage

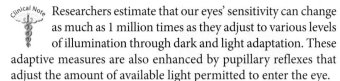

Researchers estimate that our eyes' sensitivity can change as much as 1 million times as they adjust to various levels of illumination through dark and light adaptation. These adaptive measures are also enhanced by pupillary reflexes that adjust the amount of available light permitted to enter the eye.

Because retinene, one of the photopigment components, is a derivative of vitamin A, adequate amounts of this nutrient must be available for the ongoing resynthesis of photopigments. **Night blindness** occurs as a result of dietary deficiencies of vitamin A. Although photopigment concentrations in both rods and cones are reduced in this condition, there is still enough cone photopigment to respond to the intense stimulation of bright light, except in the most severe cases. However, even modest reductions in rhodopsin content can decrease the sensitivity of rods so much that they cannot respond to dim light. In this condition, a person can see in the day by using cones but cannot see at night because the rods are no longer functional. So carrots are "good for your eyes" because they are rich in vitamin A.

## Colour vision

Vision depends on stimulation of retinal photoreceptors by light. Certain objects in the environment, such as the sun, fire, and light bulbs, emit light. But how do you see such objects as chairs, trees, and people, which do not emit light? The pigments in various objects selectively absorb particular wavelengths of light transmitted to them from light-emitting sources, and the unabsorbed wavelengths are reflected from the objects' surfaces. These reflected light rays enable you to see the objects. An object perceived as blue absorbs the longer red and green wavelengths of light and reflects the shorter blue wavelengths, which can be absorbed by the photopigment in the eyes' blue cones, thereby activating them.

Each cone type is most effectively activated by a particular wavelength of light in the range of colour indicated by its name—blue, green, or red. However, cones also respond in varying degrees to other wavelengths (› Figure 4-21). **Colour vision**, the perception of the many colours of the world, depends on the three cone types' various *ratios of stimulation* in response to different wavelengths. A wavelength perceived as blue does not stimulate red or green cones at all but excites blue cones maximally (the percentage of maximal stimulation for red, green, and blue cones, respectively, is 0:0:100). The sensation of yellow, in comparison, arises from a stimulation ratio of 83:83:0—that is, red and green cones are stimulated 83 percent of maximum, and blue cones are not excited at all. The ratio for green is 31:67:36, and so on, with various combinations giving rise to the sensation of all the different colours. White is a mixture of all wavelengths of light, whereas black is the absence of light.

The extent to which each of the cone types is excited is coded and transmitted in separate, parallel pathways to the brain. A distinct colour vision centre in the primary visual cortex combines and processes these inputs to generate the perception of colour, taking into account the object in comparison with its background. The concept of colour is therefore in the mind of the beholder. Most of us agree on what colour we see because we have the same types of cones and use similar neural pathways

for comparing their output. Occasionally, however, individuals lack a particular cone type, so their colour vision is a product of the differential sensitivity of only two types of cones—a condition known as **colour blindness**. Not only do those with colour blindness perceive certain colours differently, they also cannot distinguish as many varieties of colours (› Figure 4-22). For example, people with certain defects in colour perception cannot distinguish between red and green, so at a traffic light they can tell which light is "on" by its intensity, but they must rely on the position of the bright light to know whether to stop or go. Colour blindness is often genetic in nature, but also may result from eye, nerve, or brain damage.

Although the three-cone system has been accepted as the standard model of colour vision for more than two centuries, new evidence suggests that perception of colour may be more complex. DNA studies have shown that men with normal colour vision have a variable number of genes coding for cone pigments. For example, many had multiple genes (from two to four) for red-light detection, and these men could distinguish more subtle differences in colours in this long-wavelength range than could those with single copies of red-cone genes. This finding will undoubtedly lead to a re-evaluation of how the various photopigments contribute to colour vision.

## Visual information

The field of view that can be seen without moving the head is known as the **visual field**. The information that reaches the visual cortex in the occipital lobe is not a replica of the visual field for several reasons:

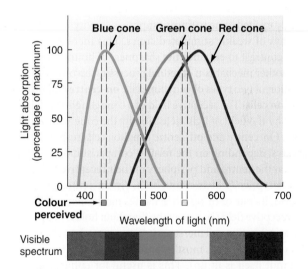

| Colour perceived | Percent of maximum stimulation | | |
|---|---|---|---|
| | Red cones | Green cones | Blue cones |
| | 0 | 0 | 100 |
| | 31 | 67 | 36 |
| | 83 | 83 | 0 |

© 2016 Cengage

› **FIGURE 4-21 Sensitivity of the three types of cones to different wavelengths.** The ratios of stimulation of the three cone types are shown for three sample colours.

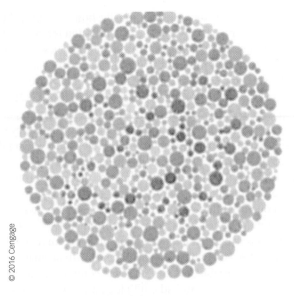

> FIGURE 4-22 **Colour blindness chart**. People with red–green colour blindness cannot detect the number 29 in this chart.

> FIGURE 4-23 **Example of the outcome of retinal processing by on-centre and off-centre ganglion cells**. Note that the grey circle surrounded by black appears brighter than the one surrounded by white, even though the two circles are identical (same shade and size). Retinal processing by on-centre and off-centre ganglion cells is largely responsible for enhancing differences in relative (rather than absolute) brightness, which helps define contours.

The image detected on the retina at the onset of visual processing is upside down and backward because of bending of the light rays through the lens of the eye. Once projected to the brain, the inverted image is interpreted as being in its correct orientation.

The information transmitted from the retina to the brain is not merely a point-to-point record of photoreceptor activation. Before the information reaches the brain, the retinal neuronal layers beyond the rods and cones reinforce selected information and suppress other information to enhance contrast. One mechanism of retinal processing is lateral inhibition, by which strongly excited cone pathways suppress activity in surrounding pathways of weakly stimulated cones. This increases the dark–bright contrast to enhance the sharpness of boundaries.

Another mechanism of retinal processing involves differential activation of two types of ganglion cells: **on-centre and off-centre ganglion cells**. The receptive field of a cone ganglion cell is determined by the field of light detection by the cone with which it is linked. On-centre and off-centre ganglion cells respond in opposite ways, depending on the relative comparison of illumination between the centre and periphery of their receptive fields. Think of the receptive field as a doughnut. An on-centre ganglion cell increases its rate of firing when light is most intense at the centre of its receptive field (i.e., when the doughnut hole is lit up). In contrast, an off-centre cell increases its firing rate when the periphery of its receptive field is most intensely illuminated (i.e., when the doughnut itself is lit up). This is useful for enhancing the difference in light level between one small area at the centre of a receptive field and the illumination immediately around it. By emphasizing differences in relative brightness, this mechanism helps define contours of images, but in so doing, information about absolute brightness is sacrificed (> Figure 4-23).

Various aspects of visual information, such as form, colour, depth, and movement, are separated and projected in parallel pathways to different regions of the cortex. Only when these separate bits of processed information are integrated by higher visual regions do we perceive a reassembled picture of the visual scene. This is similar to the blobs of paint on an artist's palette versus the finished portrait; the separate pigments do not represent a portrait of a face until they are appropriately integrated on a canvas.

*Clinical Note* Patients with lesions in specific visual-processing regions of the brain may be unable to completely combine components of a visual impression. For example, a person may be unable to discern movement of an object but have reasonably good vision for shape, pattern, and colour. Sometimes the defect can be remarkably specific, like being unable to recognize familiar faces, while retaining the ability to recognize inanimate objects.

Because of the pattern of neural wiring between the eyes and the visual cortex, the left half of the cortex receives information only from the right half of the visual field as detected by both eyes, and the right half receives input only from the left half of the visual field of both eyes.

As light enters the eyes, light rays from the left half of the visual field fall on the right half of the retina of both eyes (the medial or inner half of the left retina and the lateral or outer half of the right retina) (> Figure 4-24a). Similarly, rays from the right half of the visual field reach the left half of each retina (the lateral half of the left retina and the medial half of the right retina). Each optic nerve exiting the retina carries information from both halves of the retina it serves. This information is separated as the optic nerves meet at the **optic chiasm** located underneath the hypothalamus (*chiasm* means "cross") (see Figure 3-7b). Within the optic chiasm, the fibres from the medial half of each retina cross to the opposite side, but those from the lateral half remain on the original side. The reorganized bundles of fibres leaving the optic chiasm are known as **optic tracts**. Each optic tract carries information from the lateral half of one retina and the medial half of the other retina. Therefore, this partial crossover brings together from the two eyes fibres that carry information from the same half of the visual field. Each optic tract, in turn, delivers to the half of the brain on its same side information about the opposite half of the visual field.

*Clinical Note* Knowledge of these pathways can facilitate diagnosis of visual defects arising from an interruption of the visual pathway at various points (> Figure 4-24b).

Before we move on to how the brain processes visual information, take a look at ▌ Table 4-4, which summarizes the functions of the various components of the eyes.

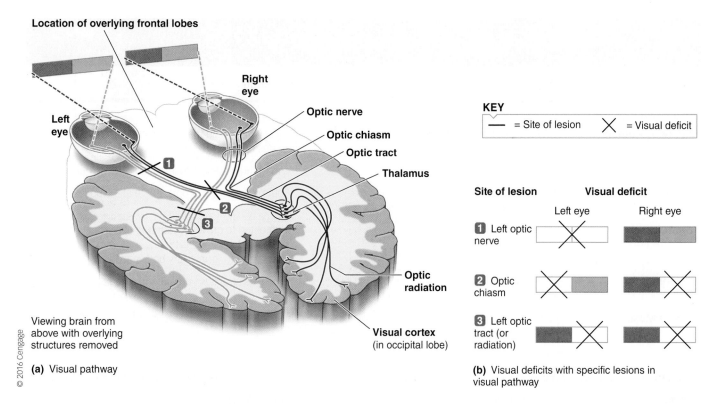

Location of overlying frontal lobes

Left eye

Right eye

Optic nerve

Optic chiasm

Optic tract

Thalamus

**KEY**

— = Site of lesion    ✕ = Visual deficit

Site of lesion    Visual deficit

Left eye    Right eye

**1** Left optic nerve

**2** Optic chiasm

**3** Left optic tract (or radiation)

Optic radiation

Visual cortex (in occipital lobe)

Viewing brain from above with overlying structures removed

© 2016 Cengage

**(a)** Visual pathway

**(b)** Visual deficits with specific lesions in visual pathway

> **FIGURE 4-24 The visual pathway and visual deficits associated with lesions in the pathway.** (a) Note that the left half of the visual cortex in the occipital lobe receives information from the right half of the visual field of both eyes (in blue), and the right half of the cortex receives information from the left half of the visual field of both eyes (in red). (b) Each visual deficit illustrated is associated with a lesion at the corresponding numbered point in the visual pathway in part (a).

**▌Clinical Connections**

Because Deborah is diabetic, her reported visual disturbances are likely due to diabetic retinopathy. In retinopathy, the microvasculature supplying the retina can weaken and leak blood, causing swelling in the retina. This can cause the blurred vision and minor vision loss that is experienced in about 25 percent of those with diabetes mellitus. As the disease progresses, the blood vessel injury gets worse, and decreases blood supply to the retina, in a manner similar to that underlying the neuropathy already discussed. Eventually this leads to severe vision loss and even blindness.

## The thalamus and visual cortices

The first stop in the brain for information in the visual pathway is the lateral geniculate nucleus in the thalamus (› Figure 4-24a). It separates information received from the eyes and relays it via fibre bundles—known as **optic radiations**—to different zones in the cortex, each of which processes different aspects of the visual stimulus (e.g., colour, form, depth, movement). This sorting process is no small task, because each optic nerve contains more than a million fibres carrying information from the photoreceptors in one retina. This is more than all the afferent fibres carrying somatosensory input from all other regions of the body! Researchers estimate that hundreds of millions of neurons, occupying about 30 percent of the cortex, participate in visual processing, compared with 8 percent devoted to touch perception and 3 percent to hearing. Yet the connections in the visual pathways are precise. The lateral geniculate nucleus and each of

the zones in the cortex that processes visual information have a topographical map representing the retina point for point. As with the somatosensory cortex, the neural maps of the retina are distorted. The fovea, the retinal region capable of greatest acuity, has much greater representation in the neural map than do the more peripheral regions of the retina.

### DEPTH PERCEPTION

Although each half of the visual cortex receives information simultaneously from the same part of the visual field as received by both eyes, the messages from the two eyes are not identical. Each eye views an object from a slightly different vantage point, even though the overlap is tremendous. The overlapping area seen by both eyes at the same time is known as the **binocular field of vision** (*binocular* means "two-eyed"), which is important for **depth perception**. Like other areas of the cortex, the primary visual cortex is organized into functional columns, each processing information from a small region of the retina. Independent, alternating columns are devoted to information about the same point in the visual field from the right and left eyes. The brain uses the slight disparity in the information received from the two eyes to estimate distance, allowing you to perceive three-dimensional objects in spatial depth. Nevertheless, some depth perception is possible using only one eye, based on experience and comparison with other cues. For example, if your one-eyed view includes a car and a building and the car is much larger, you correctly interpret that the car must be closer to you than the building is.

| Structure | Location | Function |
|---|---|---|
| *(in alphabetical order)* | | |
| **Aqueous Humour** | Anterior cavity between cornea and lens | Clear watery fluid that is continually formed and carries nutrients to the cornea and lens |
| **Bipolar Cells** | Middle layer of nerve cells in retina | Important in retinal processing of light stimulus |
| **Blind Spot** | Point slightly off centre on retina where optic nerve exits; is devoid of photoreceptors (also known as *optic disc*) | Route for passage of optic nerve and blood vessels |
| **Choroid** | Middle layer of eye | Pigmented to prevent scattering of light rays in eye; contains blood vessels that nourish retina; anteriorly specialized to form ciliary body and iris |
| **Ciliary Body** | Specialized anterior derivative of the choroid layer; forms a ring around the outer edge of the lens | Produces aqueous humour and contains ciliary muscle |
| **Ciliary Muscle** | Circular muscular component of ciliary body; attached to lens by means of suspensory ligaments | Important in accommodation |
| **Cones** | Photoreceptors in outermost layer of retina | Responsible for high acuity, colour, and day vision |
| **Cornea** | Anterior clear outermost layer of eye | Contributes most extensively to eye's refractive ability |
| **Fovea** | Exact centre of retina | Region with greatest acuity |
| **Ganglion Cells** | Inner layer of nerve cells in retina | Important in retinal processing of light stimulus; form optic nerve |
| **Iris** | Visible pigmented ring of muscle within aqueous humour | Varies size of pupil by variable contraction; responsible for eye colour |
| **Lens** | Between aqueous humour and vitreous humour; attaches to ciliary muscle by suspensory ligaments | Provides variable refractive ability during accommodation |
| **Macula Lutea** | Area immediately surrounding the fovea | Has high acuity because of abundance of cones |
| **Optic Disc** | *(see entry for blind spot)* | |
| **Optic Nerve** | Leaves each eye at optic disc (blind spot) | First part of visual pathway to the brain |
| **Pupil** | Anterior round opening in middle of iris | Permits variable amounts of light to enter eye |
| **Retina** | Innermost layer of eye | Contains the photoreceptors (rods and cones) |
| **Rods** | Photoreceptors in outermost layer of retina | Responsible for high-sensitivity, black-and-white, and night vision |
| **Sclera** | Tough outer layer of eye anteriorly specialized to form cornea | Protective connective tissue coat; forms visible white part of eye |
| **Suspensory Ligaments** | Suspended between ciliary muscle and lens | Important in accommodation |
| **Vitreous Humour** | Between lens and retina | Semifluid, jelly-like substance that helps maintain spherical shape of eye |

*Clinical Note* Sometimes the two views are not successfully merged. This condition may occur for two reasons: (1) the eyes are not both focused on the same object simultaneously, because of defects of the external eye muscles that make fusion of the two eyes' visual fields impossible; or (2) the binocular information is improperly integrated during visual processing. The result is double vision, or **diplopia**, a condition in which the disparate views from both eyes are seen simultaneously.

## HIERARCHY OF VISUAL CORTICAL PROCESSING

Within the cortex, visual information is first processed in the primary visual cortex and then sent to higher-level visual areas for even more complex processing and abstraction. The cortex contains a hierarchy of visual cells that respond to increasingly complex stimuli. Three types of visual cortical neurons have been identified based on the complexity of stimulus required for the cell to respond; these are called **simple cells**, **complex cells**, and **hypercomplex cells**. Simple and complex cells are stacked on top of one another within the cortical columns of the primary visual cortex, whereas hypercomplex cells are found in the higher visual-processing areas. Unlike a retinal cell, which responds to the amount of light, a cortical cell fires only when it receives information regarding a particular pattern of illumination for which it is programmed. These patterns are built up by converging connections that originate from closely aligned photoreceptor cells in the retina. For example, some simple cells fire only when a bar is viewed vertically in a specific location, others when a bar is horizontal, and others at various oblique orientations. Movement of a critical axis of orientation becomes important for responses generated by some of the complex cells. Hypercomplex cells add a new dimension to visual processing by responding only to particular edges, corners, and curves.

Each level of cortical visual neurons has increasingly greater capacity for abstraction based on information built up from the increasing convergence of input from lower-level neurons. In this way, the cortex transforms the dot-like pattern of photoreceptors stimulated to varying degrees by varying light intensities in the retinal image into information about depth, position, orientation, movement, contour, and length. Other aspects of this information, such as colour perception, are processed simultaneously. How and where the entire image is finally put together is still unresolved. Only when the higher visual regions integrate these separate bits of processed information do we perceive a reassembled picture of the visual scene.

## Visual input to other areas of the brain

Not all fibres in the visual pathway terminate in the visual cortices. Some are projected to other regions of the brain for purposes other than direct vision perception. Examples of nonsight activities dependent on input from the rods and cones include (1) contribution to cortical alertness and attention, (2) control of pupil size, and (3) control of eye movements. Each eye is equipped with a set of six **external eye muscles** that position and move the eye so that it can better locate, see, and track objects. Eye movements are among the fastest, most discretely controlled movements of the body.

### MULTIPLE SENSORY-PROCESSING AREAS IN THE BRAIN

Before concluding this section on the eye, consider a controversial new theory regarding the senses—one that challenges the prevailing view that the separate senses feed into distinct brain regions that handle only one sense. A growing body of evidence suggests that the brain regions devoted almost exclusively to a certain sense, such as the visual cortex for visual input and the somatosensory cortex for touch input, actually receive a variety of sensory signals. Therefore, tactile and auditory signals also arrive in the visual cortex. For example, one study using new brain-imaging techniques showed that people who are blind from birth use the visual cortex when they read Braille, even though they are not "seeing" anything. The tactile input from their fingers reaches the visual area of the brain as well as the somatosensory cortex. This input helps them "visualize" the patterns of the Braille bumps.

Also reinforcing the notion that central processing of different types of sensory input overlaps to some extent, scientists recently discovered *multisensory neurons*—brain cells that react to multiple sensory inputs instead of just to one. No one knows whether these cells are rare or commonplace in the brain.

Next, we concentrate on the mainstream functions of the other special senses. Let's shift attention from the eyes to the ears.

---

### Check Your Understanding 4.3

1. Draw two sagittal sections of an emmetropic eye, one for far vision and one accommodated for near vision.
2. Explain how light absorption by a photopigment leads to a hyperpolarizing receptor potential.
3. Compare rod and cone vision.

---

## 4.6 | Ear: Hearing and Equilibrium

Each **ear** consists of three parts: the *external*, the *middle*, and the *inner ear* (› Figure 4-25). The external and middle portions of the ear transmit airborne sound waves to the fluid-filled inner ear, amplifying the sound energy in the process. The inner ear houses two different sensory systems: the *cochlea*, which contains the receptors for conversion of sound waves into nerve impulses, making hearing possible; and the *vestibular apparatus*, which is necessary for the sense of equilibrium.

### Sound waves

**Hearing** is the neural perception of sound energy. Hearing involves two aspects: identification of the sounds (what) and their localization (where). We first describe the characteristics of sound waves, and then we examine how the ears and brain process sound input to accomplish hearing.

**Sound waves** are travelling vibrations of air that consist of regions of high pressure, caused by compression of air molecules, alternating with regions of low pressure, caused by rarefaction of the molecules (› Figure 4-26a). Any device capable of producing such a disturbance pattern in air molecules is a source of sound. A simple example is a tuning fork. When a tuning fork is struck, its prongs vibrate. As a prong of the fork moves in one direction (› Figure 4-26b), air molecules ahead of it are pushed closer together, or compressed, increasing the pressure in this area. Simultaneously, as the prong moves forward, the air molecules behind the prong spread out, or are rarefied, lowering the pressure in that region. As the prong moves in the opposite direction, an opposite wave of compression and rarefaction is created. Even though individual molecules are moved only short distances as the tuning fork vibrates, alternating waves of compression and rarefaction spread out considerable distances in a rippling fashion. Disturbed air molecules disturb other molecules in adjacent regions, setting up new regions of compression and rarefaction, and so on (› Figure 4-26c). Sound energy

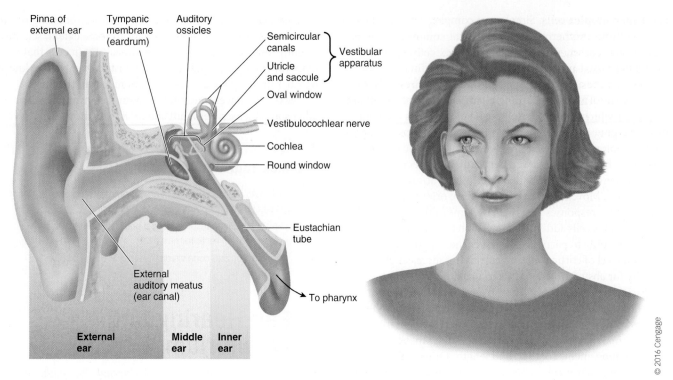

> FIGURE 4-25 **Anatomy of the ear**

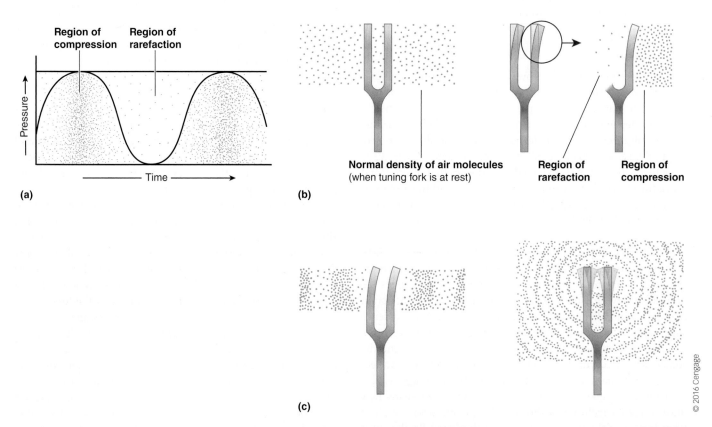

> FIGURE 4-26 **Formation of sound waves.** (a) Sound waves are alternating regions of compression and rarefaction of air molecules. (b) A vibrating tuning fork sets up sound waves as the air molecules ahead of the advancing arm of the tuning fork are compressed and the molecules behind the arm are rarefied. (c) Disturbed air molecules bump into molecules beyond them, setting up new regions of air disturbance more distant from the original source of sound. In this way, sound waves travel progressively farther from the source, even though each individual air molecule travels only a short distance when it is disturbed. The sound wave dies out when the last region of air disturbance is too weak to disturb the region beyond it.

is gradually dissipated as sound waves travel farther from the original sound source. The intensity of the sound decreases, until it finally dies out when the last sound wave is too weak to disturb the air molecules around it.

Sound waves can also travel through media other than air, such as water. They do so less efficiently, however; greater pressures are required to cause movements of fluid than movements of air because of the fluid's greater inertia (resistance to change).

Sound is characterized by its pitch (tone), intensity (loudness), and timbre (quality) (❭ Figure 4-27):

- The **pitch**, or **tone**, of a sound (e.g., whether it is a C or a G note) is determined by the frequency of vibrations. The greater the frequency of vibration is, the higher the pitch. Human ears can detect sound waves with frequencies from 20 to 20 000 cycles per second but are most sensitive to frequencies between 1000 and 4000 cycles per second.

- The intensity, or **loudness**, of a sound depends on the amplitude of the sound waves, which is the pressure difference between a high-pressure region of compression and a low-pressure region of rarefaction. Within the hearing range, the greater the amplitude is, the louder the sound. Human ears can detect a wide range of sound intensities, from the slightest whisper to the painfully loud takeoff of a jet. Loudness is measured in decibels (dB), which are a logarithmic measure of intensity compared with the faintest sound that can be heard: the **hearing threshold**. Because of the logarithmic relationship, every 10 dB indicates a tenfold increase in loudness. A few examples of common sounds illustrate the magnitude of these increases (❚ Table 4-5). Note that the rustle of leaves at 10 dB is 10 times as loud as hearing threshold, but the sound of a jet taking off at 150 dB is a quadrillion (a million billion) times, not 150 times, as loud as the faintest audible sound. Sounds greater than 100 dB can permanently damage the sensitive sensory apparatus in the cochlea.

- The **timbre**, or **quality**, of a sound depends on its overtones, which are additional frequencies superimposed on the fundamental pitch or tone. A tuning fork has a pure tone, but most sounds lack purity. For example, complex mixtures of overtones impart different sounds to different instruments

| ❚ TABLE 4-5 Relative Magnitude of Common Sounds | | |
|---|---|---|
| **Sound** | **Loudness (in dB)** | **Comparison with Faintest Audible Sound (Hearing Threshold)** |
| **Rustle of leaves** | 10 | 10 times as loud |
| **Ticking of watch** | 20 | 100 times as loud |
| **Hush of library** | 30 | 1 thousand times as loud |
| **Normal conversation** | 60 | 1 million times as loud |
| **Food blender** | 90 | 1 billion times as loud |
| **Loud rock concert** | 120 | 1 trillion times as loud |
| **Takeoff of jet plane** | 150 | 1 quadrillion times as loud |

**4**

playing the same note (a C note on a trumpet sounds different from C on a piano). Overtones are likewise responsible for characteristic differences in voices. Timbre enables the listener to distinguish the source of sound waves, because each source produces a different pattern of overtones.

## The external ear

The specialized receptors for sound are located in the fluid-filled inner ear. Airborne sound waves must therefore be channelled toward and transferred into the inner ear, compensating in the process for the loss in sound energy that naturally occurs as sound waves pass from air into water. This function is performed by the external ear and the middle ear.

The **external ear** (see ❭ Figure 4-25) consists of the *pinna* (ear), *external auditory meatus* (ear canal), and *tympanic membrane* (eardrum). The **pinna**, a prominent skin-covered flap of cartilage, collects sound waves and channels them down the external ear canal. Many species (dogs, for example) can cock their ears in the direction of sound to collect more sound waves, but human ears are relatively immobile. Because of its shape, the pinna partially shields sound waves that approach the ear from the rear, changing the timbre of the sound and thus helping a person distinguish whether a sound is coming from directly in front or behind.

Sound localization for sounds approaching from the right or left is determined by two cues. First, the sound wave reaches the ear closer to the sound source slightly before it arrives at the farther ear. Second, the sound is less intense as it reaches the farther ear, because the head acts as a sound barrier that partially disrupts the propagation of sound waves. The

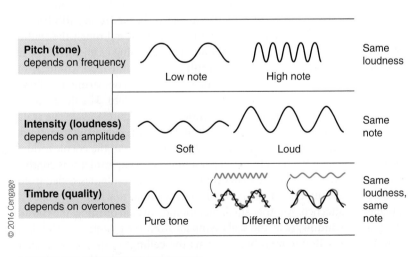

| | | |
|---|---|---|
| **Pitch (tone)** depends on frequency | Low note · High note | Same loudness |
| **Intensity (loudness)** depends on amplitude | Soft · Loud | Same note |
| **Timbre (quality)** depends on overtones | Pure tone · Different overtones | Same loudness, same note |

❭ **FIGURE 4-27 Properties of sound waves**

**The Peripheral Nervous System: Sensory, Autonomic, Somatic**

auditory cortex integrates all these cues to determine the location of the sound source. It is difficult to localize sound with only one ear. Recent evidence suggests that the auditory cortex pinpoints the location of a sound by differences in the timing of neuronal firing patterns, not by any spatially organized map, such as the one projected on the visual cortex point for point from the retina that enables the location of a visual object to be identified.

The entrance to the **ear canal** is guarded by fine hairs. The skin lining the canal contains modified sweat glands that produce **cerumen** (earwax), a sticky secretion that traps fine foreign particles. Together, the hairs and earwax help prevent airborne particles from reaching the inner portions of the ear canal, where they could accumulate or injure the tympanic membrane and interfere with hearing.

## The tympanic membrane

The **tympanic membrane**, which is stretched across the entrance to the middle ear, vibrates when struck by sound waves. The alternating higher- and lower-pressure regions of a sound wave cause the exquisitely sensitive eardrum to bow inward and outward in unison with the wave's frequency.

For the membrane to be free to move as sound waves strike it, the resting air pressure on both sides of the tympanic membrane must be equal. The outside of the eardrum is exposed to atmospheric pressure that reaches it through the ear canal. The inside of the eardrum facing the middle ear cavity is also exposed to atmospheric pressure via the **eustachian (auditory) tube**, which connects the middle ear to the **pharynx** (back of the throat) (see › Figure 4-25). The eustachian tube is normally closed, but it can be pulled open by yawning, chewing, and swallowing. Such opening permits air pressure within the middle ear to equilibrate with atmospheric pressure so that pressures on both sides of the tympanic membrane are equal. During rapid external pressure changes (e.g., during air flight), the eardrum bulges painfully as the pressure outside the ear changes while the pressure in the middle ear remains unchanged. Opening the eustachian tube by yawning allows the pressure on both sides of the tympanic membrane to equalize, relieving the pressure distortion as the eardrum "pops" back into place.

*Clinical Note* Infections originating in the throat sometimes spread through the eustachian tube to the middle ear. The resulting fluid accumulation in the middle ear not only is painful but also interferes with sound conduction across the middle ear.

## The middle ear bones

The **middle ear** transfers the vibratory movements of the tympanic membrane to the fluid of the inner ear. This transfer is facilitated by a movable chain of three small bones, or **ossicles** (the **malleus**, **incus**, and **stapes**), that extend across the middle ear (› Figure 4-28a). The first bone, the malleus, is attached to the tympanic membrane, and the last bone, the stapes, is attached to the **oval window**, the entrance into the fluid-filled cochlea. As the tympanic membrane vibrates in response to sound waves, the chain of bones is set into motion at the same frequency, transmitting this frequency of movement from the tympanic membrane to the oval window. The resulting pressure on the oval window with each vibration produces wavelike movements in the inner ear fluid at the same frequency as the original sound waves. However, as noted earlier, greater pressure is required to set fluid in motion.

The ossicular system amplifies the pressure of the airborne sound waves by two mechanisms to set up fluid vibrations in the cochlea. First, because the surface area of the tympanic membrane is much larger than that of the oval window, pressure is increased as the force exerted on the tympanic membrane is conveyed by the ossicles to the oval window (pressure 5 force/ unit area). Second, the lever action of the ossicles provides an additional mechanical advantage. Together, these mechanisms increase the force exerted on the oval window by 20 times what it would be if the sound wave struck the oval window directly. This additional pressure is sufficient to set the cochlear fluid in motion.

Several tiny muscles in the middle ear contract reflexively in response to loud sounds (more than 70 dB), causing the tympanic membrane to tighten and limiting movement of the ossicular chain. This reduced movement of middle ear structures diminishes the transmission of loud sound waves to the inner ear to protect the delicate sensory apparatus from damage. This reflex response is relatively slow, however, happening at least 40 milliseconds after exposure to a loud sound. It thus provides protection only from prolonged loud sounds, not from sudden sounds like an explosion. Taking advantage of this reflex, World War II antiaircraft guns were designed to make a loud pre-firing sound to protect the gunner's ears from the much louder boom of the actual firing.

## The cochlea

The pea-sized, snail-shaped **cochlea**, the "hearing" portion of the inner ear, is a coiled tubular system lying deep within the temporal bone (see › Figure 4-25). It is easier to understand the functional components of the cochlea by "unrolling" it, as shown in › Figure 4-28a. The cochlea is divided throughout most of its length into three fluid-filled longitudinal compartments. A blind-ended **cochlear duct**, also known as the **scala media**, constitutes the middle compartment. It tunnels lengthwise through the centre of the cochlea, almost but not quite reaching its end. The upper compartment, the **scala vestibuli**, follows the inner contours of the spiral, and the **scala tympani**, the lower compartment, follows the outer contours (› Figure 4-28a and 4-28b). The fluid within the cochlear duct is called **endolymph** (Figure 4-29a). The scala vestibuli and scala tympani both contain a slightly different fluid, the **perilymph**. The region beyond the tip of the cochlear duct where the fluid in the upper and lower compartments is continuous is called the **helicotrema**. The scala vestibuli is sealed from the middle ear cavity by the oval window, to which the stapes is attached. Another small membrane-covered opening, the **round window**, seals the scala tympani from the middle ear. The thin **vestibular membrane** forms the ceiling of the cochlear duct and separates it from the scala vestibuli. The **basilar membrane**

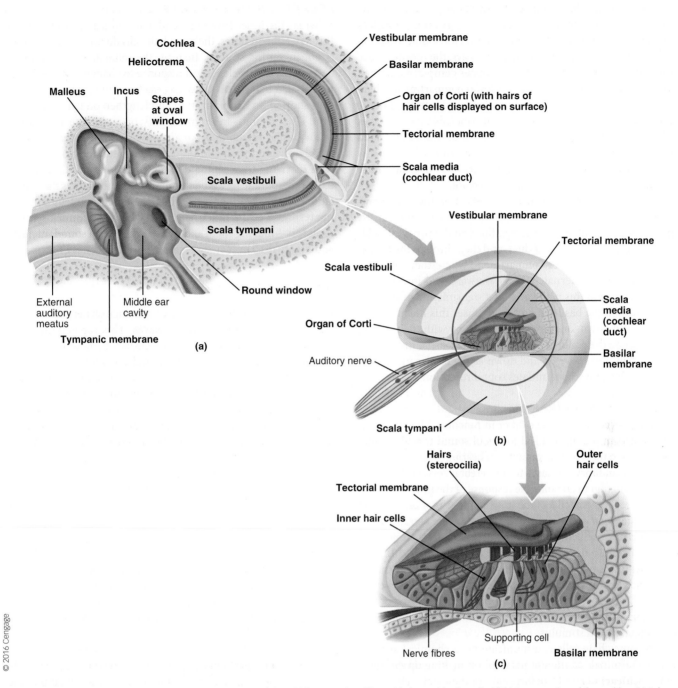

> **FIGURE 4-28 Middle ear and cochlea.** (a) Gross anatomy of the middle ear and cochlea, with the cochlea "unrolled." (b) Cross-section of the cochlea. (c) Enlargement of the organ of Corti

forms the floor of the cochlear duct, separating it from the scala tympani. The basilar membrane is especially important because it bears the **organ of Corti**, the sense organ for hearing.

## Hair cells

The organ of Corti, which rests on top of the basilar membrane throughout its full length, contains **hair cells** that are the receptors for sound. The 16 000 hair cells within each cochlea are arranged in four parallel rows along the length of the basilar membrane: one row of **inner hair cells**, and three rows of **outer hair cells** (> Figure 4-28c). Protruding from the surface of each hair cell are about 100 hairs known as **stereocilia**, which are actin-stiffened microvilli. Hair cells generate neural signals when their surface hairs are mechanically deformed in association with fluid movements in the inner ear. These stereocilia contact the **tectorial membrane**, an awning-like projection overhanging the organ of Corti throughout its length (> Figure 4-28b and > 4-28c).

The piston-like action of the stapes against the oval window sets up pressure waves in the upper compartment. Because fluid is incompressible, pressure is dissipated in two ways as the stapes causes the oval window to bulge inward: (1) displacement of the round window, which dissipates sound energy; and (2) deflection of the basilar membrane, which allows for sound reception ( > Figure 4-29a). In the first of these pathways, the pressure wave pushes the perilymph forward in the upper compartment, then around the helicotrema, and into the lower compartment, where it causes the round window to bulge outward into the middle ear cavity to compensate for the pressure increase. As the stapes rocks backward and pulls the oval window outward toward the middle ear, the perilymph shifts in the opposite direction, displacing the round window inward. This pathway does not result in sound reception; it just dissipates pressure.

The second pathway involves the deflection the basil membrane and results in sound reception. Pressure waves in the upper compartment are transferred through the thin vestibular membrane, into the cochlear duct, and then through the basilar membrane into the lower compartment, where they cause the round window to alternately bulge outward and inward. The main difference in this pathway is that transmission of pressure waves through the basilar membrane causes this membrane to move up and down, or vibrate, in synchrony with the pressure wave. Because the organ of Corti rides on the basilar membrane, the hair cells also move up and down as the basilar membrane oscillates; this movement initiates sound reception.

### ROLE OF THE INNER HAIR CELLS

The inner and outer hair cells differ in function. The inner hair cells transform the mechanical forces of sound (cochlear fluid vibration) into the electrical impulses of hearing (action potentials propagating auditory messages to the cerebral cortex). The outer hair cells function to enhance the response of the inner hair cells.

Because the stereocilia of the inner hair receptor cells contact the stiff, stationary tectorial membrane, they are bent back and forth when the oscillating basilar membrane shifts their position in relationship to the tectorial membrane ( > Figure 4-30). This back-and-forth mechanical deformation of the hairs alternately opens and closes mechanically gated ion channels in the hair cell, resulting in alternating depolarizing and hyperpolarizing potential changes—the receptor potential—at the same frequency as the original sound stimulus.

The inner hair cells communicate via a chemical synapse with the terminals of afferent nerve fibres making up the **auditory (cochlear) nerve**. Depolarization of these hair cells (when the basilar membrane is deflected upward) increases their rate of neurotransmitter release, which steps up the rate of firing in the afferent fibres. Conversely, the firing rate decreases as these hair cells release less neurotransmitter when they are hyperpolarized upon displacement in the opposite direction.

Thus, the ear converts sound waves in the air into oscillating movements of the basilar membrane that bend the hairs of the receptor cells back and forth. This shifting mechanical deformation of the hairs alternately opens and closes the receptor cells' channels, bringing about graded potential changes in the receptor that lead to changes in the rate of action potentials propagated to the brain. In this way, sound waves are translated into neural signals that can be perceived by the brain as sound sensations ( > Figure 4-31).

### ROLE OF THE OUTER HAIR CELLS

Whereas the inner hair cells send auditory signals to the brain through afferent fibres, the outer hair cells do not signal the brain about incoming sounds. Instead, the outer hair cells actively and rapidly change length in response to changes in membrane potential, a behaviour known as *electromotility*. The outer hair cells shorten on depolarization and lengthen on hyperpolarization. These changes in length amplify, or accentuate, the motion of the basilar membrane. An analogy would be a person deliberately pushing the pendulum of a grandfather clock in time with its swing to accentuate its motion. Such modification of basilar membrane movement improves and fine tunes the stimulation of the inner hair cells. Thus, the outer hair cells enhance the response of the inner hair cells—the real auditory sensory receptors—making them exquisitely sensitive to sound intensity and highly discriminatory between various pitches of sound.

## Pitch discrimination

**Pitch discrimination** is the ability to distinguish between various frequencies of incoming sound waves. This capacity depends on the shape and properties of the basilar membrane, which is narrow and stiff at its oval window end and wide and flexible at its helicotrema end (see > Figure 4-29b). Different regions of the basilar membrane naturally vibrate maximally at different frequencies; that is, each frequency displays peak vibration at a different position along the membrane. The narrow end nearest the oval window vibrates best with high-frequency pitches, whereas the wide end nearest the helicotrema vibrates maximally with low-frequency tones (see > Figure 4-29c). The pitches in between are sorted out precisely along the length of the membrane from higher to lower frequency. When a sound wave of a particular frequency is set up in the cochlea by oscillation of the stapes, the wave travels to the region of the basilar membrane that naturally responds maximally to that frequency. The vigorous membrane oscillation dissipates the energy of the pressure wave, so the wave dies out at the region of maximal displacement.

The hair cells in the region of peak vibration of the basilar membrane undergo the most mechanical deformation and accordingly are the most excited. This information is propagated to the CNS, which interprets the pattern of hair cell stimulation as a sound of a particular frequency. Modern techniques have determined that the basilar membrane is so fine-tuned that the peak membrane response to a single pitch probably extends no more than the width of a few hair cells.

## Loudness discrimination

**Intensity (loudness) discrimination** depends on the amplitude of vibration. As sound waves originating from a loud sound source strikes the eardrum, they cause it to vibrate more vigorously (i.e., bulge in and out to a greater extent) but at the same frequency as a softer sound of the same pitch. Consequently, the greater tympanic membrane deflection is converted into a greater amplitude of basilar membrane movement in the region of peak responsiveness. The CNS interprets greater basilar membrane oscillation as a louder sound.

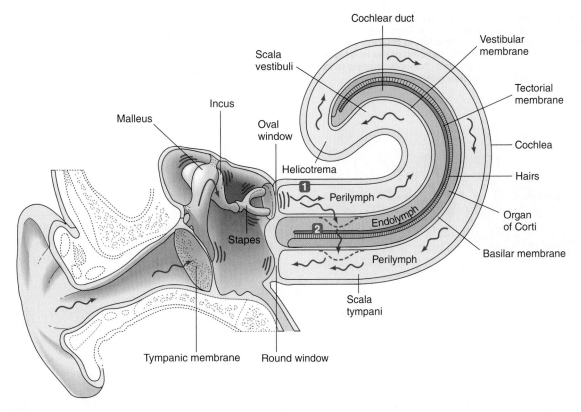

Fluid movement within the perilymph set up by vibration of the oval window follows two pathways:

**Pathway** **1** Through the scala vestibuli, around the helicotrema, and through the scala tympani, causing the round window to vibrate. This pathway just dissipates sound energy.

**Pathway** **2** A "shortcut" from the scala vestibuli through the basilar membrane to the scala tympani. This pathway triggers activation of the receptors for sound by bending the hairs of hair cells as the organ of Corti on top of the vibrating basilar membrane is displaced in relation to the overlying tectorial membrane.

(a)

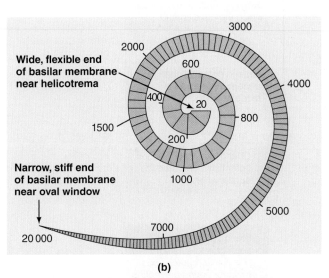

(b)

The numbers indicate the frequencies in cycles per second with which different regions of the basilar membrane maximally vibrate.

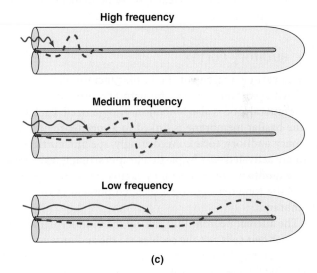

(c)

© 2016 Cengage

> **FIGURE 4-29 Transmission of sound waves.** (a) Fluid movement within the cochlea set up by vibration of the oval window follows two pathways, one dissipating sound energy and the other initiating the receptor potential. (b) Different regions of the basilar membrane vibrate maximally at different frequencies. (c) The narrow, stiff end of the basilar membrane nearest the oval window vibrates best with high-frequency pitches. The wide, flexible end of the basilar membrane near the helicotrema vibrates best with low-frequency pitches.

The stereocilia (hairs) from the hair cells of the basilar membrane contact the overlying tectorial membrane. These hairs are bent when the basilar membrane is deflected in relation to the stationary tectorial membrane. This bending of the inner hair cells' hairs opens mechanically gated channels, leading to ion movements that result in a receptor potential.

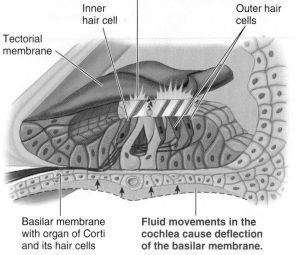

© 2016 Cengage

> FIGURE 4-30 **Bending of hairs on deflection of the basilar membrane**

The auditory system is so sensitive and can detect sounds so faint that the distance of basilar membrane deflection is comparable to only a fraction of the diameter of a hydrogen atom, the smallest of atoms. No wonder that very loud sounds—which cannot be sufficiently attenuated by protective middle ear reflexes (e.g., the sounds of a typical rock concert)—can set up such violent vibrations of the basilar membrane that irreplaceable hair cells are actually sheared off or permanently distorted, leading to partial hearing loss (› Figure 4-32).

### The auditory cortex

Just as various regions of the basilar membrane are associated with particular tones, the **primary auditory cortex** in the temporal lobe is also *tonotopically* organized. Each region of the basilar membrane is linked to a specific region of the primary auditory cortex. Accordingly, specific cortical neurons are activated only by particular tones; that is, each region of the auditory cortex becomes excited only in response to a specific tone detected by a selected portion of the basilar membrane.

The afferent neurons that pick up the auditory signals from the inner hair cells exit the cochlea via the auditory nerve. The neural pathway between the organ of Corti and the auditory cortex involves several synapses en route, the most notable of which are in the brain stem and *medial geniculate nucleus* of the thalamus. The brain stem uses the auditory input for alertness and arousal. The thalamus sorts and relays the signals upward. Unlike signals in the visual pathways, auditory signals from each ear are transmitted to both temporal lobes because the fibres partially cross over in the brain stem. For this reason, a disruption of the auditory pathways on one side above the brain stem does not affect hearing in either ear to any extent.

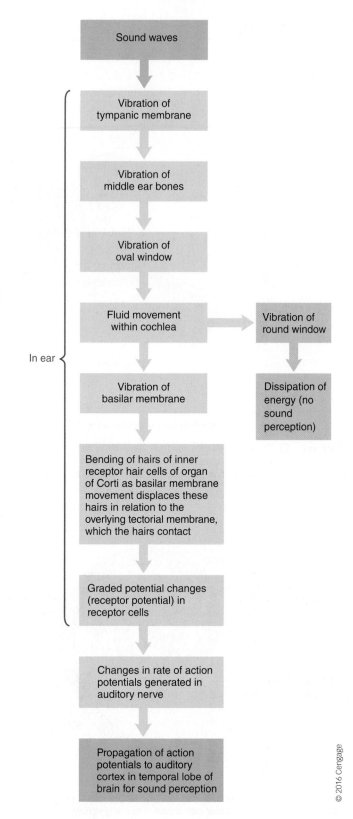

> FIGURE 4-31 **Sound transduction**

The primary auditory cortex appears to perceive discrete sounds, whereas the surrounding higher-order auditory cortex integrates the separate sounds into a coherent, meaningful pattern. Think about the complexity of the task accomplished by your auditory system. When you are at a concert, your organ of Corti responds to the simultaneous mixture of the instruments, the applause and hushed talking of the audience, and the background noises in the theatre. You can distinguish these separate parts of the many sound waves reaching your ears and can pay attention to those of importance to you.

## Deafness

Loss of hearing, or **deafness**, may be temporary or permanent, partial or complete. Hearing loss, which affects about 8 percent of all Canadians, is the second most common physical disability in the United States. Deafness is classified into two types—*conductive deafness* and *sensorineural deafness*—depending on the part of the hearing mechanism that fails to function adequately. **Conductive deafness** occurs when sound waves are not adequately conducted through the external and middle portions of the ear to set the fluids in the inner ear in motion. Possible causes include physical blockage of the ear canal with earwax, rupture of the eardrum, middle ear infections with accompanying fluid accumulation, or restriction of ossicular movement because of bony adhesions between the stapes and oval window. In **sensorineural deafness**, sound waves are transmitted to the inner ear, but they are not translated into nerve signals that are interpreted by the brain as sound sensations. The defect can lie in the organ of Corti or the auditory nerves or, rarely, in the ascending auditory pathways or auditory cortex.

One of the most common causes of partial hearing loss, **neural presbycusis**, is a degenerative, age-related process that occurs as hair cells "wear out" with use. Over time, exposure to even ordinary modern-day sounds eventually damages hair cells so that, on average, adults have lost more than 40 percent of their cochlear hair cells by age 65. Unfortunately, partial hearing loss caused by excessive exposure to loud noises is affecting people at younger ages than in the past, not only because we live in an increasingly noisy environment but also because of the increased use of earbuds and earphones associated with mobile phones and music devices like MP3 players. Hair cells that process high-frequency sounds are the most vulnerable to destruction.

**Hearing aids** are helpful in conductive deafness but are less beneficial for sensorineural deafness. These devices increase the intensity of airborne sounds and may modify the sound spectrum and tailor it to the person's particular pattern of hearing loss at higher or lower frequencies. For the sound to be perceived, however, the receptor cell–neural pathway system must still be intact.

In recent years, **cochlear implants** have become available. These electronic devices, which are surgically implanted, transduce sound signals into electrical signals that can directly stimulate the auditory nerve, thus bypassing a defective cochlear system. Cochlear implants cannot restore normal hearing, but they do permit recipients to recognize sounds. Success ranges from an ability to "hear" a phone ringing to being able to carry on a conversation over the telephone.

Exciting new findings suggest that in the future it may be possible to restore hearing by stimulating an injured inner ear to repair itself. Scientists have long considered the hair cells of the inner ear irreplaceable. Thus, hearing loss resulting from hair cell damage caused by the aging process or exposure to loud noises is considered permanent. Encouraging new studies suggest, on the contrary, that hair cells in the inner ear have the latent ability to regenerate in response to an appropriate chemical signal. Researchers are currently trying to develop a drug that will spur regrowth of hair cells, thereby repairing inner ear damage and hopefully restoring hearing. Other investigators are using neural growth factors to coax auditory nerve cell endings to resprout in the hopes of re-establishing lost neural pathways.

**Spoken languages** are developed in the early years of life. A hearing impairment can prevent not only the ability to talk but also the ability to understand the spoken word. By the time severely hearing-impaired children are diagnosed, they may have already developed problems with communication that hinder family socialization and social skills. Early diagnosis means that sign language can be taught at a young age, thus facilitating communication. In industrialized countries, hearing is evaluated during the newborn period in an attempt to prevent the isolation that is associated with deafness. Although sign language is a full communication system, literacy (the ability to read and write)

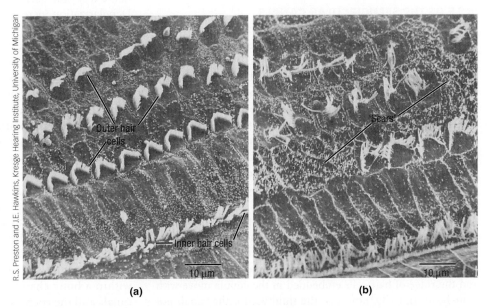

R.S. Preston and J.E. Hawkins, Kresge Hearing Institute, University of Michigan

(a)　(b)

❯ **FIGURE 4-32 Loss of hair cells caused by loud noises.** Injury and loss of hair cells caused by intense noise. Portions of the organ of Corti, with its three rows of outer hair cells and one row of inner hair cells, from the inner ear of (a) a normal guinea pig and (b) a guinea pig after a 24-hour exposure to noise at 120 dB SPL (sound pressure level), a level approached by loud rock music

depends on understanding spoken language. In most written languages, the sound of the word is coded in symbols. However, a person who can hear early in life and learns to speak and read will retain the ability to read, even if significant hearing loss occurs later. In contrast, a person who had significant hearing impairment from a very early age will rarely be able to read well. Listening also plays an important role in learning a second language. Thus, early identification of a hearing impairment is key to learning spoken language.

## The vestibular apparatus

In addition to its cochlear-dependent role in hearing, the inner ear has another specialized component, the **vestibular apparatus**. This part of the inner ear provides essential information for the sense of equilibrium and also for the coordination of head movements with eye and postural movements. The vestibular apparatus consists of two sets of structures lying within a tunnelled-out region of the temporal bone near the cochlea: the *semicircular canals* and the *otolith organs*, namely the *utricle* and *saccule* (› Figure 4-33a).

The vestibular apparatus detects changes in position and motion of the head. As in the cochlea, all components of the vestibular apparatus contain endolymph and are surrounded by perilymph. Also, similar to the organ of Corti, the vestibular components each contain hair cells that respond to mechanical deformation triggered by specific movements of the endolymph. And like the auditory hair cells, the vestibular receptors may be either depolarized or hyperpolarized, depending on the direction of the fluid movement. Unlike information from the auditory system, much of the information provided by the vestibular apparatus does not reach the level of conscious awareness.

### ROLE OF THE SEMICIRCULAR CANALS

The **semicircular canals** detect rotational or angular acceleration or deceleration of the head, such as when starting or stopping spinning, somersaulting, or turning the head. Each ear contains three semicircular canals arranged three-dimensionally in planes that lie at right angles to each other. The receptive hair cells of each semicircular canal are situated on top of a ridge located in the **ampulla**, a swelling at the base of the canal (Figure 4-33a and 4-33b). The hairs are embedded in an overlying, caplike, gelatinous layer, the **cupula**, which protrudes into the endolymph within the ampulla. The cupula sways in the direction of fluid movement, much like seaweed leaning in the direction of the prevailing tide.

Acceleration or deceleration during rotation of the head in any direction causes endolymph movement in at least one of the semicircular canals, because of their three-dimensional arrangement. When you start to move your head, the bony canal and the ridge of hair cells embedded in the cupula move with your head. Initially, however, the fluid within the canal, not being attached to your skull, does not move in the direction of the rotation but lags behind because of its inertia. (Due to inertia, a resting object remains at rest, and a moving object continues to move in the same direction, unless the object is acted on by some external force that induces change.) When the endolymph is left behind as you start to rotate your head, the fluid in the same plane as the head movement is in effect shifted in the opposite direction of the head movement (similar to your body tilting to the right as the car in which you are riding suddenly turns to the left) (› Figure 4-34). This fluid movement causes the cupula to lean in the opposite direction from the head movement, bending the sensory hairs embedded in it. If your head movement continues at the same rate in the same direction, the endolymph catches up and moves in unison with your head so that the hairs return to their unbent position. When your head slows down and stops, the reverse situation occurs. The endolymph briefly continues to move in the direction of the rotation while your head decelerates to a stop. As a result, the cupula and its hairs are transiently bent in the direction of the preceding spin, which is opposite to the way they were bent during acceleration.

The hairs of a vestibular hair cell consist of one cilium, the **kinocilium**, along with a tuft of 20 to 50 microvilli—the *stereocilia*—arranged in rows of increasing height (see › Figure 4-33c and › 4-33d). The stereocilia are linked at their tips by **tip links**, which are tiny molecular bridges between adjacent stereocilia. When the stereocilia are deflected by endolymph movement, the resultant tension on the tip links pulls on mechanically gated ion channels in the hair cell. Depending on whether the ion channels are mechanically opened or closed by hair bundle displacement, the hair cell either depolarizes or hyperpolarizes. Each hair cell is oriented so that it depolarizes when its stereocilia are bent toward the kinocilium; bending in the opposite direction hyperpolarizes the cell. The hair cells form chemically mediated synapses with terminal endings of afferent neurons whose axons join with those of the other vestibular structures to form the **vestibular nerve**. This nerve unites with the auditory nerve from the cochlea to form the **vestibulocochlear nerve**. Depolarization increases the release of neurotransmitter from the hair cells, thereby bringing about an increased rate of firing in the afferent fibres. Conversely, hyperpolarization reduces neurotransmitter release from the hair cells, in turn decreasing the frequency of action potentials in the afferent fibres. When the fluid gradually comes to a halt, the hairs straighten again. Thus, the semicircular canals detect changes in the rate of rotational movement (rotational acceleration or deceleration) of your head. They do not respond when your head is motionless or when it is moving in a circle at a constant speed.

### ROLE OF THE OTOLITH ORGANS

The **otolith organs** provide information about the position of the head relative to gravity (i.e., static head tilt) and also detect changes in the rate of linear motion (moving in a straight line regardless of direction). The otolith organs, the **utricle** and the **saccule**, are saclike structures housed within a bony chamber situated between the semicircular canals and the cochlea (see › Figure 4-33a). The hairs (kinocilium and stereocilia) of the receptive hair cells in these sense organs also protrude into an overlying gelatinous sheet, whose movement displaces the hairs and results in changes in hair cell potential. Many tiny crystals of calcium

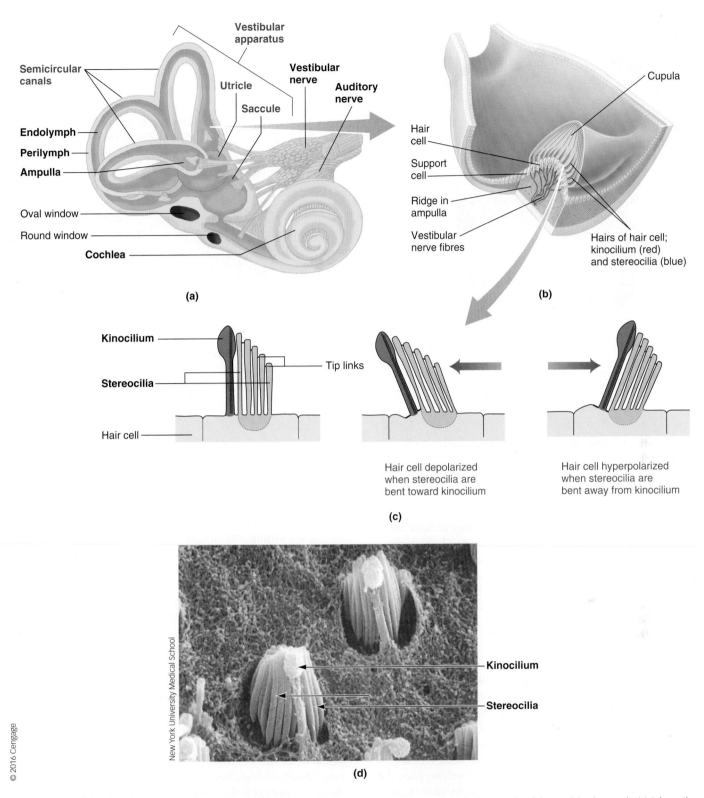

Semicircular canals

Endolymph

Perilymph

Ampulla

Oval window

Round window

Cochlea

Vestibular apparatus

Utricle

Saccule

Vestibular nerve

Auditory nerve

**(a)**

Cupula

Hair cell

Support cell

Ridge in ampulla

Vestibular nerve fibres

Hairs of hair cell; kinocilium (red) and stereocilia (blue)

**(b)**

Kinocilium

Stereocilia

Tip links

Hair cell

Hair cell depolarized when stereocilia are bent toward kinocilium

Hair cell hyperpolarized when stereocilia are bent away from kinocilium

**(c)**

New York University Medical School

Kinocilium

Stereocilia

**(d)**

> FIGURE 4-33 **Vestibular apparatus.** (a) Gross anatomy of the vestibular apparatus. (b) Receptor cell unit in the ampulla of the semicircular canals. (c) Schematic representation of the "hairs" on the sensory hair cells of the semicircular canals. (d) Scanning electron micrograph of the kinocilium and stereocilia on the hair cells within the vestibular apparatus

carbonate—the otoliths (ear stones)—are suspended within the gelatinous layer, making it heavier and giving it more inertia than the surrounding fluid (>Figure 4-35a). When a person is in an upright position, the hairs within the utricles are oriented vertically, and the saccule hairs are lined up horizontally.

Let's look at the *utricle* as an example. It's otolith-embedded, gelatinous mass shifts positions and bends the hairs in two ways:

The Peripheral Nervous System: Sensory, Autonomic, Somatic    **179**

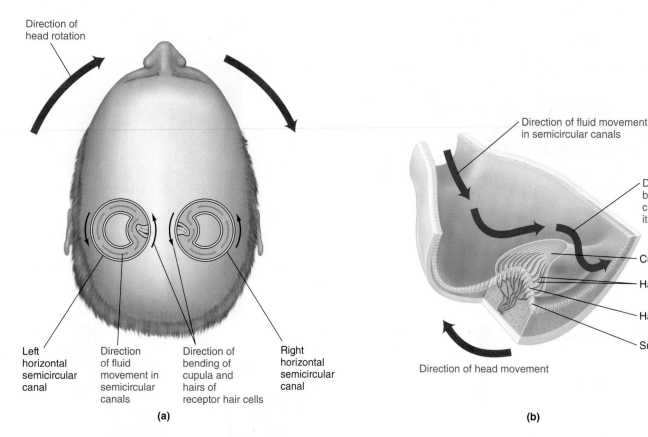

Direction of head rotation

Direction of fluid movement in semicircular canals

Direction of bending of cupula and its hairs

Cupula

Hairs

Hair cell

Support cell

Left horizontal semicircular canal

Direction of fluid movement in semicircular canals

Direction of bending of cupula and hairs of receptor hair cells

Right horizontal semicircular canal

Direction of head movement

© 2016 Cengage

(a)

(b)

› **FIGURE 4-34 Activation of the hair cells in the semicircular canals**

in the direction of the tilt of the head and in the opposite direction to the tilt.

1. The utricle hairs are displaced by any change in head movement away from the vertical. When you tilt your head in any direction other than vertical (i.e., other than straight up and down), the hairs are bent in the direction of the tilt because of the gravitational force exerted on the top-heavy gelatinous layer (› Figure 4-35b). This bending produces depolarizing or hyperpolarizing receptor potentials depending on the tilt of your head. The CNS thus receives different patterns of neural activity depending on head position with respect to gravity.

2. The utricle hairs are also displaced by any change in horizontal linear motion (such as moving straight forward, backward, or to the side). When you start to walk forward (› Figure 4-35c), the top-heavy otolith membrane at first lags behind the endolymph and hair cells because of its greater inertia. The hairs are thus bent to the rear, in the opposite direction of the forward movement of your head. If you maintain your walking pace, the gelatinous layer soon catches up and moves at the same rate as your head so that the hairs are no longer bent. When you stop walking, the otolith sheet continues to move forward briefly as your head slows and stops, bending the hairs toward the front. In this way, the hair cells of the utricle detect horizontally directed linear acceleration and deceleration, but they do not provide information about movement in a straight line at constant speed.

The *saccule* functions similarly to the utricle, except that it responds selectively to tilting of the head away from a horizontal

position (such as getting up from bed) and to vertically directed linear acceleration and deceleration (such as jumping up and down or riding in an elevator).

Signals arising from the various components of the vestibular apparatus are carried through the vestibulocochlear nerve to the **vestibular nuclei**—a cluster of neuronal cell bodies in the brain stem—and to the cerebellum. Here, the vestibular information is integrated with input from the skin surface, eyes, joints, and muscles to accomplish three important functions: (1) maintaining balance and desired posture; (2) controlling the external eye muscles so that the eyes remain fixed on the same point, despite movement of the head; and (3) perceiving motion and orientation (› Figure 4-36).

*Clinical Note* Some people, for poorly understood reasons, are especially sensitive to particular motions that activate the vestibular apparatus and cause symptoms of dizziness and nausea; this sensitivity is called **motion sickness**. Occasionally, fluid imbalances within the inner ear lead to **Ménière's disease**. Not surprisingly, because both the vestibular apparatus and the cochlea contain the same inner ear fluids, both vestibular and auditory symptoms occur with this condition. An afflicted individual suffers transient attacks of severe vertigo (dizziness), accompanied by pronounced ringing in the ears and some loss of hearing. During these episodes, the person cannot stand upright and reports feeling as though self or surrounding objects in the room are spinning around.

▌ Table 4-6 summarizes the functions of the major components of the ear.

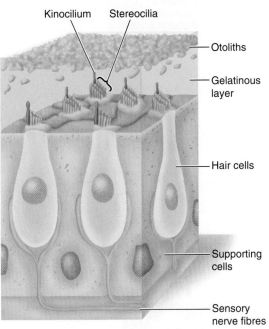

Kinocilium    Stereocilia

Otoliths

Gelatinous layer

Hair cells

Supporting cells

Sensory nerve fibres

**(a)** Receptor cell unit in utricle

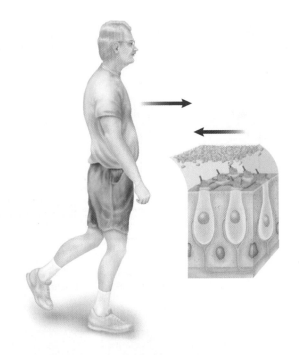

**(c)** Activation of utricle by horizontal linear acceleration

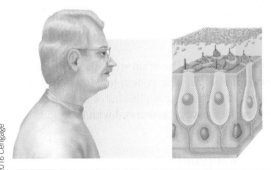

**(b)** Activation of utricle by change in head position

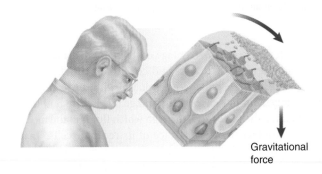

Gravitational force

© 2016 Cengage

> **FIGURE 4-35** **Utricle**

### Check Your Understanding 4.4

1. Describe the function of the middle ear.
2. Compare the mechanisms for discrimination of pitch, loudness, and timbre.
3. Schematically draw one semicircular canal on each side of the head (viewed from above), showing the direction of fluid movement in the canals and the direction of bending of the cupula and hairs of the receptor hair cells when the head is rotating clockwise.

## 4.7 | Chemical Senses: Taste and Smell

Unlike the eyes' photoreceptors and the ears' mechanoreceptors, the receptors for taste and smell are chemoreceptors, which generate neural signals on binding with particular chemicals in their environment. The sensations of **taste** and **smell**

in association with food intake influence the flow of digestive juices and affect appetite. Furthermore, stimulation of taste or smell receptors induces pleasurable or objectionable sensations and signals the presence of something to seek (a nutritionally useful, good-tasting food) or to avoid (a potentially toxic, bad-tasting substance). In this way, the chemical senses provide a "quality-control" checkpoint for substances available for ingestion. In lower animals, smell also plays a major role in finding direction, in seeking prey or avoiding predators, and in sexual attraction to a mate. The sense of smell is less sensitive in humans and much less important in influencing our behaviour (although millions of dollars are spent annually on perfumes and deodorants to make us smell better and thereby more socially attractive).

We first examine the mechanism of taste (**gustation**) and then turn our attention to smell (**olfaction**).

### Taste receptor cells

The chemoreceptors for taste sensation are packaged in taste buds, about 10 000 of which are present in the oral cavity and

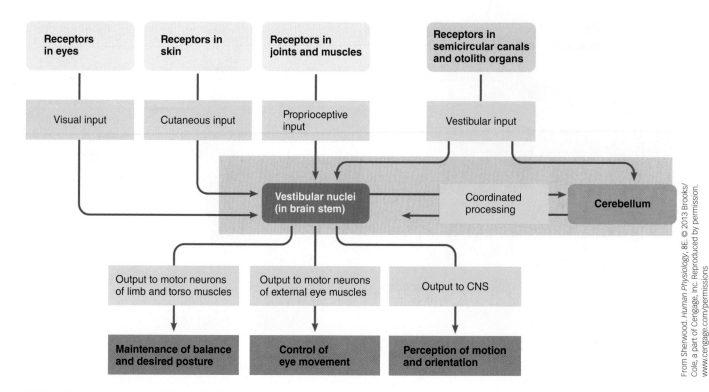

From Sherwood. *Human Physiology*, 8E. © 2013 Brooks/ Cole, a part of Cengage, Inc. Reproduced by permission. www.cengage.com/permissions

> FIGURE 4-36  Input and output of the vestibular nuclei

throat, with the greatest percentage on the upper surface of the tongue (› Figure 4-37). A taste bud consists of about 50 long, spindle-shaped *taste receptor cells* packaged with *supporting cells* in an arrangement like slices of an orange. Each taste bud has a small opening, the **taste pore**, through which fluids in the mouth come into contact with the surface of its receptor cells. **Taste receptor cells** are modified epithelial cells, with many surface folds (microvilli) that protrude slightly through the taste pore, greatly increasing the surface area exposed to the oral contents. The plasma membrane of the microvilli contains receptor sites that bind selectively with chemical molecules in the environment. Only chemicals in solution—either ingested liquids or solids that have been dissolved in saliva—can attach to receptor cells and evoke the sensation of taste. Binding of a taste-provoking chemical, a **tastant**, with a receptor cell alters the cell's ionic channels to produce a depolarizing receptor potential. This receptor potential, in turn, initiates action potentials within terminal endings of afferent nerve fibres with which the receptor cell synapses.

Most receptors are carefully sheltered from direct exposure to the environment, but the taste receptor cells, by virtue of their task, frequently come into contact with potent chemicals. Unlike the eye or ear receptors, which are irreplaceable, taste receptors have a life span of about 10 days. Epithelial cells surrounding the taste bud differentiate first into supporting cells and then into receptor cells to constantly renew the taste bud components.

Terminal afferent endings of several cranial nerves synapse with taste buds in various regions of the mouth. Signals in these sensory inputs are conveyed via synaptic stops in the brain stem and thalamus to the **cortical gustatory area**, a region in the parietal lobe adjacent to the "tongue" area of the somatosensory cortex. Unlike most sensory input, the gustatory pathways are

primarily uncrossed. The brain stem also projects fibres to the hypothalamus and limbic system to add affective dimensions, such as whether the taste is pleasant or unpleasant, and to process behavioural aspects associated with taste and smell.

## Taste discrimination

We can discriminate among thousands of different taste sensations, yet all tastes are varying combinations of five **primary tastes**: *salty, sour, sweet, bitter*, and *umami*. Umami, a meaty or savoury taste, has recently been added to the list of primary tastes.

Each receptor cell responds in varying degrees to all five primary tastes but is generally preferentially responsive to one of the taste modalities. The richness of fine taste discrimination beyond the primary tastes depends on subtle differences in the stimulation patterns of all the taste buds in response to various substances, similar to the variable stimulation of the three cone types that gives rise to the range of colour sensations.

Receptor cells use different pathways to bring about a depolarizing receptor potential in response to each of the five tastant categories:

- **Salty taste** is stimulated by chemical salts, especially NaCl (table salt). Direct entry of positively charged $Na^+$ ions through specialized $Na^+$ channels in the receptor cell membrane, a movement that reduces the cell's internal negativity, is responsible for receptor depolarization in response to salt.

- **Sour taste** is caused by acids containing a free hydrogen ion ($H^+$). The citric acid content of lemons, for example, accounts for their distinctly sour taste. Depolarization of the receptor cell by sour tastants occurs because $H^+$ blocks

| Structure | Location | Function |
|---|---|---|
| **External Ear** | | Collects and transfers sound waves to the middle ear |
| *Pinna (ear)* | Skin-covered flap of cartilage located on each side of the head | Collects sound waves and channels them down the ear canal; contributes to sound localization |
| *External auditory meatus (ear canal)* | Tunnels from the exterior through the temporal bone to the tympanic membrane | Directs sound waves to tympanic membrane; contains filtering hairs and secretes earwax to trap foreign particles |
| *Tympanic membrane (eardrum)* | Thin membrane that separates the external ear and the middle ear | Vibrates in synchrony with sound waves that strike it, setting middle ear bones in motion |
| **Middle Ear** | | **Transfers vibrations of the tympanic membrane to the fluid in the cochlea** |
| *Malleus, incus, stapes* | Movable chain of bones that extends across the middle ear cavity; malleus attaches to the tympanic membrane, and stapes attaches to the oval window | Oscillate in synchrony with tympanic membrane vibrations and set up wavelike movements in the cochlear perilymph at the same frequency |
| **Inner Ear: Cochlea** | | **Houses sensory system for hearing** |
| *Oval window* | Thin membrane at the entrance to the cochlea; separates the middle ear from the scala vestibuli | Vibrates in unison with movement of stapes, to which it is attached; oval window movement sets cochlear perilymph in motion |
| *Scala vestibuli* | Upper compartment of the cochlea, a snail-shaped tubular system that lies deep within the temporal bone | Contains perilymph that is set in motion by oval window movement driven by oscillation of middle ear bones |
| *Scala tympani* | Lower compartment of the cochlea | Contains perilymph that is continuous with the scala vestibuli |
| *Cochlear duct (scala media)* | Middle compartment of the cochlea; a blind-ended tubular compartment that tunnels through the centre of the cochlea | Contains endolymph; houses the basilar membrane |
| *Basilar membrane* | Forms the floor of the cochlear duct | Vibrates in unison with perilymph movements; bears the organ of Corti, the sense organ for hearing |
| *Organ of Corti* | Rests on top of the basilar membrane throughout its length | Contains hair cells, the receptors for sound; inner hair cells undergo receptor potentials when their hairs are bent as a result of fluid movement in cochlea |
| *Tectorial membrane* | Stationary membrane that overhangs the organ of Corti and contacts the surface hairs of the receptor hair cells | Serves as the stationary site against which the hairs of the receptor cells are bent to generate receptor potentials as the vibrating basilar membrane moves in relation to this overhanging membrane |
| *Round window* | Thin membrane that separates the scala tympani from the middle ear | Vibrates in unison with fluid movements in perilymph to dissipate pressure in cochlea; does not contribute to sound reception |
| **Inner Ear: Vestibular Apparatus** | | **Houses sensory systems for equilibrium and provides input essential for maintenance of posture and balance** |
| *Semicircular canals* | Three semicircular canals arranged three-dimensionally in planes at right angles to each other near the cochlea | Detect rotational or angular acceleration or deceleration |
| *Utricle* | Saclike structure in a bony chamber between the cochlea and semicircular canals | Detects (1) changes in head position away from vertical and (2) horizontally directed linear acceleration and deceleration |
| *Saccule* | Lies next to the utricle | Detects (1) changes in head position away from horizontal and (2) vertically directed linear acceleration and deceleration |

4

© 2016 Cengage

The Peripheral Nervous System: Sensory, Autonomic, Somatic

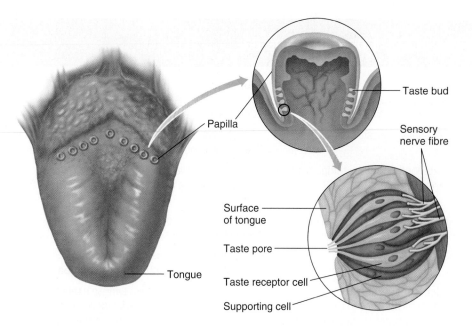

Papilla

Taste bud

Sensory
nerve fibre

Surface
of tongue

Taste pore

Taste receptor cell

Supporting cell

Tongue

> **FIGURE 4-37 Location and structure of the taste buds.** Taste buds are located primarily along the edges of moundlike papillae on the upper surface of the tongue. The receptor cells and supporting cells of a taste bud are arranged like slices of an orange.

- **Umami taste**, which was first identified and named by a Japanese researcher, is triggered by amino acids, especially glutamate. The presence of amino acids, as found in meat for example, serves as a marker for a desirable, nutritionally protein-rich food. Glutamate binds to a G protein–coupled receptor and activates a second-messenger system, but the details of this pathway are unknown. In addition to giving us our sense of meaty flavours, this pathway is responsible for the distinctive taste of the flavour additive monosodium glutamate (MSG), which is especially popular in Asian dishes.

Taste perception is also influenced by information derived from other receptors, especially smell (odour). When you temporarily lose your sense of smell because of swollen nasal passageways during a cold, your sense of taste is also markedly reduced, even though your taste receptors are unaffected by the cold. Other factors affecting taste include temperature and texture of the food as well as psychological factors associated with past experiences with the food. How the cortex accomplishes the complex perceptual processing of taste sensation is currently not known.

potassium $(K^+)$ channels in the receptor cell membrane. The resultant decrease in the passive movement of positively charged $K^+$ out of the cell reduces the internal negativity, producing a depolarizing receptor potential.

- **Sweet taste** is evoked by the particular configuration of glucose. From an evolutionary perspective, we crave sweet foods because they supply necessary calories in a readily usable form. However, other organic molecules with similar structures but no calories, such as saccharin, aspartame, sucralose, and other artificial sweeteners, can also interact with "sweet" receptor binding sites. Binding of glucose or another chemical with the taste cell receptor activates a G protein, which turns on the cAMP second-messenger pathway in the taste cell (p. 217). The second-messenger pathway ultimately results in phosphorylation and blockage of $K^+$ channels in the receptor cell membrane, leading to a depolarizing receptor potential.

- **Bitter taste** is elicited by a more chemically diverse group of tastants than the other taste sensations. For example, alkaloids (such as caffeine, nicotine, strychnine, morphine, and other toxic plant derivatives), as well as poisonous substances, all taste bitter, presumably as a protective mechanism to discourage ingestion of these potentially dangerous compounds. Taste cells that detect bitter flavours possess 50 to 100 bitter receptors, each of which responds to a different bitter flavour. Because each receptor cell has this diverse family of bitter receptors, a wide variety of unrelated chemicals all taste bitter despite their diverse structures. This mechanism expands the ability of the taste receptor to detect a wide range of potentially harmful chemicals. The first G protein in taste—**gustducin**—was identified in one of the bitter-signalling pathways. This G protein, which sets off a second-messenger pathway in the taste cell, is very similar to the visual G protein, transducin.

## Olfactory receptors in the nose

The **olfactory mucosa** (smell), a $3 \, cm^2$ patch of mucosa in the ceiling of the nasal cavity, contains three cell types: *olfactory receptor cells*, *supporting cells*, and *basal cells* (> Figure 4-38). The supporting cells secrete mucus, which coats the nasal passages. The basal cells are precursors for new olfactory receptor cells, which are replaced about every two months. An **olfactory receptor cell** is an afferent neuron whose receptor portion lies in the olfactory mucosa in the nose and whose afferent axon traverses into the brain. The axons of the olfactory receptor cells collectively form the **olfactory nerve**.

The receptor portion of an olfactory receptor cell consists of an enlarged knob bearing several long cilia that extend like a tassel to the surface of the mucosa. These cilia contain the binding sites for attachment of **odourants**, which are molecules that can be smelled. During quiet breathing, odourants typically reach the sensitive receptors only by diffusion, because the olfactory mucosa is above the normal path of airflow. The act of sniffing enhances this process by drawing the air currents upward within the nasal cavity so that a greater percentage of the odoriferous molecules in the air come into contact with the olfactory mucosa. Odourants also reach the olfactory mucosa during eating by wafting up to the nose from the mouth through the pharynx (back of the throat).

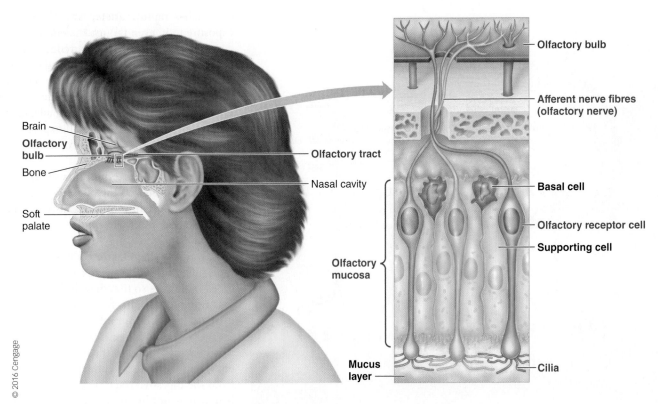

Brain
Olfactory bulb
Bone
Soft palate
Olfactory tract
Nasal cavity
Olfactory mucosa
Mucus layer

Olfactory bulb
Afferent nerve fibres (olfactory nerve)
Basal cell
Olfactory receptor cell
Supporting cell
Cilia

© 2016 Cengage

> FIGURE 4-38  Location and structure of the olfactory receptors

## Detecting and sorting odour components

To be smelled, a substance must be (1) sufficiently volatile (easily vaporized) that some of its molecules can enter the nose in the inspired air, and (2) sufficiently water soluble that it can dissolve in the mucus that coats the olfactory mucosa. As with taste receptors, to be detected by olfactory receptors, molecules must be dissolved.

### Detecting and sorting odour components

The human nose contains 5 million olfactory receptors, of which there are 1000 different types. During smell detection, an odour is "dissected" into various components. Each receptor responds to only one discrete component of an odour rather than to the whole odourant molecule. Accordingly, each of the various parts of an odour is detected by one of the thousand different receptors, and a given receptor can respond to a particular odour component shared in common by different scents. Compare this to the three cone types for coding colour vision and to the taste buds that respond differentially to only five primary tastes to accomplish coding for taste discrimination.

Binding of an appropriate scent signal to an olfactory receptor activates a G protein, triggering a cascade of cAMP-dependent intracellular reactions that leads to opening of Na$^+$ channels. The resultant ion movement brings about a depolarizing receptor potential that generates action potentials in the afferent fibre. The frequency of the action potentials depends on the concentration of the stimulating chemical molecules.

The afferent fibres arising from the receptor endings in the nose pass through tiny holes in the flat bone plate separating the olfactory mucosa from the overlying brain tissue (Figure 4-38). They immediately synapse in the **olfactory bulb**, a complex neural structure containing several different layers of cells that are functionally similar to the retinal layers of the eye. The twin olfactory bulbs, one on each side, are about the size of small grapes. Each olfactory bulb is lined by small ball-like neural junctions known as **glomeruli** (little balls) (Figure 4-39). Within each glomerulus, the terminals of receptor cells carrying information about a particular scent component synapse with the next cells in the olfactory pathway, the **mitral cells**. Because each glomerulus receives signals only from receptors that detect a particular odour component, the glomeruli serve as "smell files." The separate components of an odour are sorted into different glomeruli, one component per file. Thus, the glomeruli, which serve as the first relay station in the brain for processing olfactory information, play a key role in organizing scent perception.

The mitral cells on which the olfactory receptors terminate in the glomeruli refine the smell signals and relay them to the brain for further processing. Fibres leaving the olfactory bulb travel in two different routes:

- A subcortical route going primarily to regions of the limbic system, especially the lower medial sides of the temporal lobes (considered the **primary olfactory cortex**). This route, which includes hypothalamic involvement, permits close coordination between smell and behavioural reactions associated with feeding, mating, and direction orienting.

- A route through the thalamus to the cortex. As with other senses, the cortical route is important for the perception and fine discrimination of smell.

**The Peripheral Nervous System: Sensory, Autonomic, Somatic**

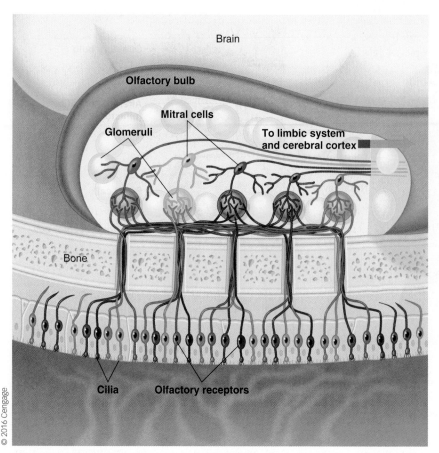

Brain

Olfactory bulb

Mitral cells

Glomeruli

To limbic system
and cerebral cortex

Bone

Cilia    Olfactory receptors

© 2016 Cengage

> **FIGURE 4-39 Processing of scents in the olfactory bulb.** Each of the glomeruli lining the olfactory bulb receives synaptic input from only one type of olfactory receptor, which, in turn, responds to only one discrete component of an odourant. Thus, the glomeruli sort and file the various components of an odoriferous molecule before relaying the smell signal to the mitral cells and higher brain levels for further processing.

diminishes rapidly after a short period of exposure to it, even though the odour source continues to be present. This reduced sensitivity does not involve receptor adaptation, as researchers thought for years; actually, the olfactory receptors themselves adapt slowly. It apparently involves some sort of adaptation process in the CNS. Adaptation is specific for a particular odour, and responsiveness to other odours remains unchanged.

What clears the odourants away from their binding sites on the olfactory receptors so that the sensation of smell doesn't linger after the source of the odour is removed? Several "odour-eating" enzymes recently discovered in the olfactory mucosa serve as molecular janitors, clearing away the odoriferous molecules so that they do not continue to stimulate the olfactory receptors. Interestingly, these odourant-clearing enzymes are very similar chemically to detoxification enzymes found in the liver. (These liver enzymes inactivate potential toxins absorbed from the digestive tract; p. 671) This resemblance may not be coincidental. Researchers speculate that the nasal enzymes may serve the dual purpose of clearing the olfactory mucosa of old odourants and transforming potentially harmful chemicals into harmless molecules. Such detoxification would serve a very useful purpose, considering the open passageway between the olfactory mucosa and the brain.

## Odour discrimination

Because each odourant activates multiple receptors and glomeruli in response to its various odour components, odour discrimination is based on different patterns of glomeruli activation achieved by various scents. In this way, the cortex can distinguish more than 10 000 different scents. This mechanism for sorting out and distinguishing different odours is very effective. A noteworthy example is our ability to detect methyl mercaptan (garlic odour) at a concentration of 1 molecule per 50 billion molecules in the air! This substance is added to odourless natural gas to enable us to detect potentially lethal gas leaks. Despite this impressive sensitivity, humans have a poor sense of smell compared with other species. By comparison, a dog's sense of smell is hundreds of times more sensitive than that of a human. Bloodhounds, for example, have about 4 billion olfactory receptor cells, compared with our 5 million such cells, accounting for their superior scent-sniffing ability.

## Adaptability of the olfactory system

Although the olfactory system is sensitive and highly discriminating, it is also quickly adaptive. Sensitivity to a new odour

## The vomeronasal organ

In addition to the olfactory mucosa, the nose contains another sense organ, the **vomeronasal organ (VNO)**, which is common in other mammals but until recently was thought nonexistent in humans. The VNO is located about 15 mm inside the human nose next to the vomer bone—hence, its name. It detects **pheromones**, nonvolatile chemical signals passed subconsciously from one individual to another. In animals, binding of a pheromone to its receptor on the surface of a neuron in the VNO triggers an action potential that travels through nonolfactory pathways to the limbic system, the brain region that governs emotional responses and sociosexual behaviours. These signals never reach the higher levels of conscious awareness. In animals, the VNO is known as the "sexual nose" for its role in governing reproductive and social behaviours, such as identifying and attracting a mate and communicating social status.

Some scientists now claim the existence of pheromones in humans, although many skeptics doubt these findings. Although the role of the VNO in human behaviour has not been validated, some researchers suspect that it is responsible for spontaneous "feelings" between people, either "good chemistry," such as "love at first sight," or "bad chemistry," such as "getting bad vibes" from someone you just met. They speculate that pheromones in humans subtly influence sexual activity, compatibility with others, or group behaviour, similar to the role they play in other mammals, although this messenger system is nowhere as powerful or important in humans as in animals. Because messages conveyed by the VNO seem to bypass cortical consciousness, the response to the largely odourless pheromones is not a distinct, discrete perception, such as smelling a favourite fragrance, but more like an inexplicable impression.

This completes our discussion of the afferent nervous system, which is responsible for sensing (via receptors) and transmitting information from the periphery to the CNS (spinal cord and higher brain centres). We now address the transmission of information from the CNS to the periphery (e.g., skeletal muscle, heart, lungs). The information transmitted is used to effect change, for example, increasing or decreasing heart rate.

### Check Your Understanding 4.5

1. List the five established primary tastes and the stimuli that evoke each of these taste sensations.
2. Describe how odour discrimination is accomplished.

## 4.8 | The Peripheral Nervous System: Efferent Division

Once the afferent division of the PNS communicates with the CNS, the efferent division then communicates with the effector organs (muscles and glands) to carry out the intended effects. Therefore, the action potential flow is in the opposite direction to that of the afferent neuron.

The efferent nerves, known as motor or effector neurons, carry nerve action potentials away from the CNS to muscles and glands and other structures (e.g., cells of the inner ear). Structurally, the cell body, oval in shape, of the efferent neuron is connected to a single, long axon and several shorter dendritic projections from the cell body. This axon then forms a neuronal junction with the effector. The efferent neuron is present in the grey matter of the spinal cord and medulla oblongata.

Only two neurotransmitters are used by effector neurons: acetylcholine and norepinephrine. These neurotransmitters bring about diverse effects such as salivary secretion, bladder contraction, and voluntary motor movements. These effects are a prime example of how the same chemical messenger may elicit a multiplicity of responses from various tissues, depending on the specialization of the effector organs.

Cardiac muscle, smooth muscle, most exocrine glands, some endocrine glands, and adipose tissue (fat) are innervated by the autonomic nervous system (ANS), the involuntary branch of the peripheral efferent division. Skeletal muscle is innervated by the somatic nervous system, the branch of the efferent division subject to voluntary control. Efferent output typically influences either movement (e.g., walking) or secretion (e.g., catecholamines). We examine these two branches of the efferent division of the PNS in Sections 4.9 and 4.10.

Much of the efferent output is directed toward maintaining homeostasis, but it is also important in responding to stresses, such as the fight-or-fight response. Let's revisit the example of fight-or-flight response introduced at the beginning of this chapter and, in doing so, outline the contributions of the efferent division of the PNS.

### Fight-or-flight response

The fight-or-flight example asked you to envision walking in the woods and hearing a rustling of leaves and branches behind you; you stop to look and see a bear. Do you run, back up slowly, or confront the bear by making yourself as threatening as possible? During this acutely stressful situation, unnecessary functions are reduced, and energy is diverted to other functions vital to survival. Some of the physiological adjustments to stress are as follows (for further discussion, see Times of Sympathetic Dominance , p. 191):

- The cardiorespiratory system responds by increasing heart rate, blood pressure, and rate and depth of respiration.
- Energy stores are mobilized for use by working tissues, for example, skeletal muscles.
- Redistribution of resources (via vasodilation and vasoconstriction) takes place, which is based on priority need. Blood flow and oxygen delivery to the digestive tract is constricted, and saliva production in the mouth is reduced, as these are nonessential at this time. In contrast, blood flow and oxygen delivery is increased to skeletal muscles, heart, lungs, and skin.
- Secretion of epinephrine, norepinephrine, and other hormones is increased, which assist in accelerating heart rate, mobilizing energy, and if necessary, reducing pain.

Now let's look at a general overview of the autonomic nervous system (ANS).

## 4.9 | Autonomic Nervous System

The ANS influences smooth muscles, glands, and the heart through its two subdivisions: the sympathetic and the parasympathetic systems. These subdivisions maintain a dynamic equilibrium aimed at maintenance of homeostasis. The ANS also includes the enteric nervous system that functions within the gastrointestinal tract and influences the pancreas, liver, and gallbladder, thereby controlling gastrointestinal motility, secretion, and blood flow. The hypothalamus, brain stem, and spinal cord are the hub of the neural output of the ANS. The motor and premotor cortex and the hypothalamus adjust much of the output from the ANS. These central regulators of the ANS also adjust the secretion of hormones that influence blood volume and total peripheral resistance.

The cerebral cortex incorporates all inputs of sight, smell, hearing, and equilibrium, as well as somatosensory information (e.g., touch and pain). The integration of these inputs into an appropriate response takes place largely in the hypothalamus and medulla. The efferent signals are sent to the periphery via the sympathetic and parasympathetic pathways. The central regulators of the ANS also coordinate the stress response (e.g., fight-or-flight response), reproduction, and thermoregulation.

## Autonomic nerve pathways

Each autonomic nerve pathway extending from the CNS to an innervated organ is a two-neuron chain (⟩ Figure 4-40). The cell body of the first neuron in the series is located in the CNS. Its axon, the **preganglionic fibre**, synapses with the cell body of the second neuron, which lies within a ganglion. (Recall that a ganglion is a cluster of neuronal cell bodies outside the CNS.) The axon of the second neuron, the **postganglionic fibre**, innervates the effector organ (smooth muscle, heart).

The ANS has two subdivisions—the sympathetic nervous system and the parasympathetic nervous systems (⟩ Figure 4-41). Sympathetic nerve fibres originate in the thoracic and lumbar regions of the spinal cord (p. 102). Most sympathetic preganglionic fibres are very short; they synapse with cell bodies of postganglionic neurons within ganglia that lie in a **sympathetic ganglion chain** (also called the **sympathetic trunk**) located along either side of the spinal cord (see Figure 3-24). Long postganglionic fibres originating in the ganglion chain end on the effector organs. Some preganglionic fibres pass through the ganglion chain without synapsing. Instead, they end later in sympathetic **collateral ganglia** about halfway between the CNS and the innervated organs, and postganglionic fibres travel the rest of the distance.

Parasympathetic preganglionic fibres arise from the cranial (brain) and sacral (lower spinal cord) areas of the CNS. These fibres are longer than sympathetic preganglionic fibres because they do not end until they reach **terminal ganglia** that lie in or near the effector organs. Very short postganglionic fibres end on the cells of an organ itself.

## Parasympathetic and sympathetic postganglionic fibres

Sympathetic and parasympathetic preganglionic fibres release the same neurotransmitter, **acetylcholine (ACh)**, but the postganglionic endings of these two systems release different neurotransmitters (the neurotransmitters that influence the effector organs). Parasympathetic postganglionic fibres release acetylcholine. Accordingly, they, along with all autonomic preganglionic fibres, are called **cholinergic fibres**. Most sympathetic postganglionic fibres, in contrast, are called **adrenergic fibres** because they release **noradrenaline**, commonly known as **norepinephrine**.[1] Both acetylcholine and norepinephrine also serve as chemical messengers elsewhere in the body (▌Table 4-7).

Postganglionic autonomic fibres do not end in a single terminal swelling like a synaptic knob. Instead, the terminal branches of autonomic fibres have numerous swellings, or **varicosities**, that simultaneously release neurotransmitter over a large area of the innervated organ rather than on single cells (see ⟩ Figure 4-40, and Figure 7-29). This diffuse release of neurotransmitter, coupled with the fact that any resulting change in electrical activity is spread throughout a smooth or cardiac muscle mass via gap junctions (p. 35), means that autonomic activity typically influences whole organs instead of discrete cells.

## Innervations of visceral organs

Afferent information coming from the viscera (internal organs) usually does not reach the conscious level. Likewise, visceral activities, such as circulation, digestion, sweating, and pupillary size, are regulated outside the realm of consciousness and voluntary control via autonomic efferent output. Most visceral organs are innervated by both sympathetic and parasympathetic nerve fibres (⟩ Figure 4-42). ▌Table 4-8 summarizes the major effects of these autonomic branches. Although the details of this wide array of autonomic responses are described more fully in the later chapters that discuss the individual organs involved, you can consider several general concepts now. As you can see from Table 4.8, the sympathetic and parasympathetic nervous systems generally exert opposite effects in a particular organ.

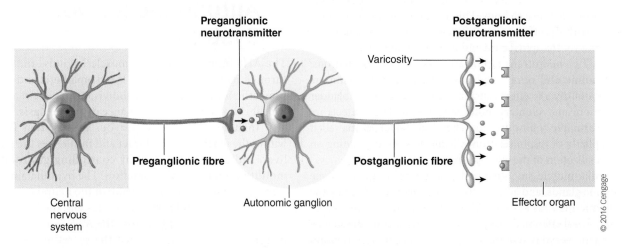

⟩ **FIGURE 4-40 Autonomic nerve pathway**

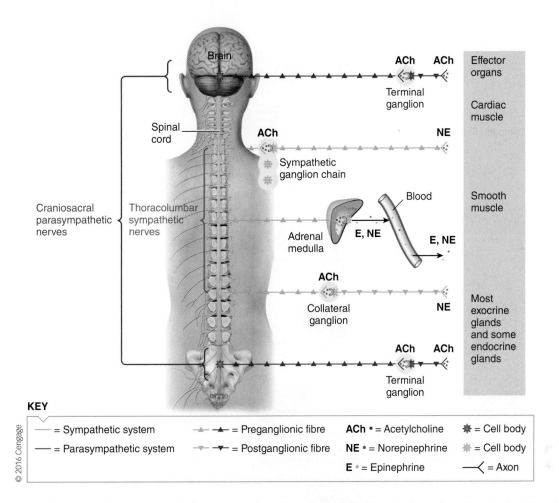

KEY

| | |
|---|---|
| ⎯⎯⎯ = Sympathetic system | ▲▲ = Preganglionic fibre | **ACh** • = Acetylcholine | ✳ = Cell body |
| ⎯⎯ = Parasympathetic system | ▼▼ = Postganglionic fibre | **NE** • = Norepinephrine | ✳ = Cell body |
| | | **E** • = Epinephrine | ⟨ = Axon |

© 2016 Cengage

> **FIGURE 4-41 Autonomic nervous system.** The sympathetic nervous system, which originates in the thoracolumbar regions of the spinal cord, has short cholinergic (acetylcholine-releasing) preganglionic fibres and long adrenergic (norepinephrine-releasing) postganglionic fibres. The parasympathetic nervous system, which originates in the brain and sacral region of the spinal cord, has long cholinergic preganglionic fibres and short cholinergic postganglionic fibres. In most instances, sympathetic and parasympathetic postganglionic fibres both innervate the same effector organs. The adrenal medulla is a modified sympathetic ganglion, which releases epinephrine and norepinephrine into the blood.

For example, sympathetic stimulation increases the heart rate, whereas parasympathetic stimulation decreases it; sympathetic stimulation slows down movement within the digestive tract, whereas parasympathetic stimulation enhances digestive motility. Note that both systems increase the activity of some organs and reduce the activity of others.

Rather than memorize a list, such as that in ▌Table 4-8, it is better to logically deduce the actions of the two systems by first understanding the circumstances under which each system dominates. Usually both systems are partially active; that is, normally some level of action potential activity exists in both the sympathetic and the parasympathetic fibres supplying a particular organ. This ongoing activity is called **sympathetic or parasympathetic tone**, or tonic activity. Under specific circumstances, the activity of one ANS subdivision can dominate the other. *Sympathetic dominance* to a particular organ exists when the sympathetic fibres' rate of firing to that organ increases above tonic level, coupled with a simultaneous decrease below tonic level in the parasympathetic fibres' frequency of action potentials to the same organ. The reverse situation is true for *parasympathetic dominance*. The balance between sympathetic and parasympathetic activity can be shifted

▌ **TABLE 4-7 Sites of Release for Acetylcholine and Norepinephrine**

| Acetylcholine | Norepinephrine |
|---|---|
| All preganglionic terminals of the autonomic nervous system | Most sympathetic postganglionic terminals |
| All parasympathetic postganglionic terminals | Adrenal medulla |
| Sympathetic postganglionic terminals at sweat glands and some blood vessels in skeletal muscle | Central nervous system |
| Terminals of efferent neurons supplying skeletal muscle (motor neurons) | Central nervous system |

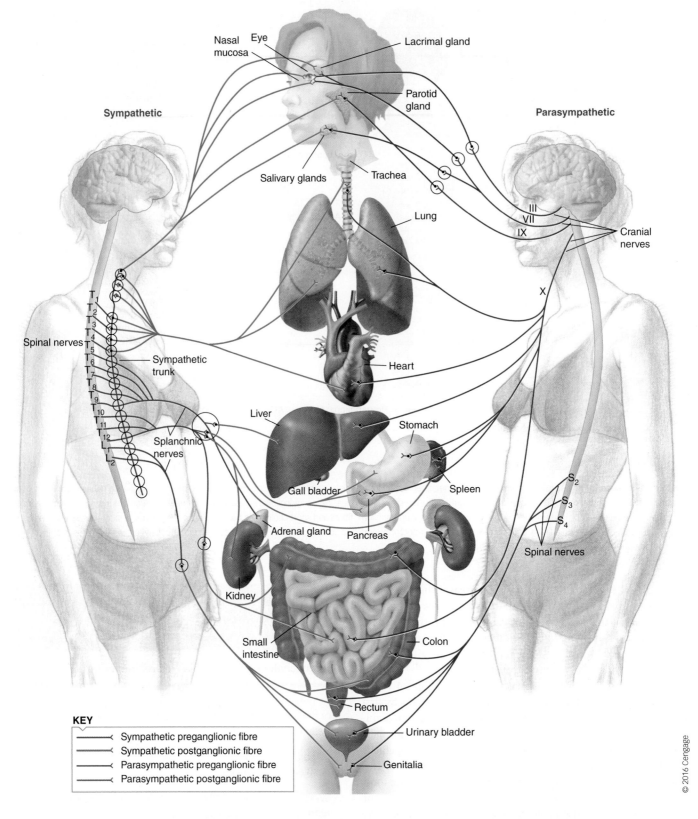

**Sympathetic**

- Nasal mucosa
- Eye
- Lacrimal gland
- Parotid gland
- Salivary glands
- Trachea
- Lung
- Heart
- Liver
- Stomach
- Gall bladder
- Spleen
- Adrenal gland
- Pancreas
- Kidney
- Small intestine
- Colon
- Rectum
- Urinary bladder
- Genitalia

Spinal nerves
- $T_1$
- $T_2$
- $T_3$
- $T_4$
- $T_5$
- $T_6$
- $T_7$
- $T_8$
- $T_9$
- $T_{10}$
- $T_{11}$
- $T_{12}$
- $L_1$
- $L_2$

Sympathetic trunk

Splanchnic nerves

**Parasympathetic**

- III
- VII
- IX
- X
- Cranial nerves
- $S_2$
- $S_3$
- $S_4$
- Spinal nerves

**KEY**

| | |
|---|---|
| ⊰ | Sympathetic preganglionic fibre |
| ⊰ | Sympathetic postganglionic fibre |
| ⊰ | Parasympathetic preganglionic fibre |
| ⊰ | Parasympathetic postganglionic fibre |

› **FIGURE 4-42** Schematic representation of the structures innervated by the sympathetic and parasympathetic nervous systems

© 2016 Cengage

in two ways: (1) separately for individual organs to meet specific demands (e.g., sympathetically induced dilation of the pupil in dim light; p. 154–155) and (2) by a more generalized, widespread discharge of one autonomic system in favour of the other to control body-wide functions. Such massive, widespread discharges take place more frequently in the sympathetic system. The value of massive sympathetic discharge is clear, considering the circumstances during which this system usually dominates.

| Organ | Effect of Sympathetic Stimulation | Effect of Parasympathetic Stimulation |
|---|---|---|
| **Heart** | Increased rate, increased force of contraction (of whole heart) | Decreased rate, decreased force of contraction (of atria only) |
| **Blood Vessels** | Constriction | Dilation of vessels supplying the penis and clitoris only |
| **Lungs** | Dilation of bronchioles (airways) | Constriction of bronchioles |
| | Inhibition of mucous secretion | Stimulation of mucous secretion |
| **Digestive Tract** | Decreased motility (movement) | Increased motility |
| | Contraction of sphincters (to prevent forward movement of contents) | Relaxation of sphincters (to permit forward movement of contents) |
| | Inhibition of digestive secretions | Stimulation of digestive secretions |
| **Urinary Bladder** | Relaxation | Contraction (emptying) |
| **Eye** | Dilation of pupil | Constriction of pupil |
| | Adjustment of eye for far vision | Adjustment of eye for near vision |
| **Liver** (glycogen stores) | Glycogenolysis (glucose released) | None |
| **Adipose Cells** (fat stores) | Lipolysis (fatty acids released) | None |
| **Exocrine Glands** | | |
| *Exocrine pancreas* | Inhibition of pancreatic exocrine secretion | Stimulation of pancreatic exocrine secretion (important for digestion) |
| *Sweat glands* | Stimulation of secretion by most sweat glands | Stimulation of secretion by some sweat glands |
| *Salivary glands* | Stimulation of small volume of thick saliva rich in mucus | Stimulation of large volume of watery saliva rich in enzymes |
| **Endocrine Glands** | | |
| *Adrenal medulla* | Stimulation of epinephrine and norepinephrine secretion | None |
| *Endocrine pancreas* | Inhibition of insulin secretion; stimulation of glucagon secretion | Stimulation of insulin and glucagon secretion |
| **Genitals** | Ejaculation and orgasmic contractions (males); orgasmic contractions (females) | Erection (caused by dilation of blood vessels in penis [male] and clitoris [female]) |
| **Brain Activity** | Increased alertness | None |

**TIMES OF SYMPATHETIC DOMINANCE**

The sympathetic nervous system promotes responses that prepare the body for strenuous physical activity in emergency or stressful situations, such as a physical threat from the environment. This response is typically referred to as a **fight-or-flight response**, because the sympathetic nervous system readies the body to fight against or flee from the threat (as in the chapter opening example of encountering a bear). Think about the body resources needed in such circumstances. The sympathetic outflow (from the hypothalamus, brain stem, spinal cord) to the heart increases heart rate and force of myocardial contraction (contractility), and global vasoconstriction (narrowing of blood vessels) occurs, both of which increase mean arterial and systolic blood pressure. Respiratory airways dilate (open wide) to permit increased ventilation (airflow); energy stores (i.e., carbohydrate and fat) are mobilized and broken down to release additional fuel into the blood for use by active tissues (e.g., skeletal muscle); and local vasodilation occurs in skeletal muscles to supply the additional oxygen and carbohydrates, and to remove by-products (e.g., hydrogen ions). All these responses are aimed at providing an increase in oxygen and nutrient availability via the blood to the active tissues (e.g., heart, lungs, skeletal muscles, and brain) in anticipation of strenuous physical activity. And because digestion and urination are not essential functions at this time, the sympathetic nervous system inhibits them, saving the additional blood flow for the active tissue. Sweating is promoted in anticipation of excess heat production by the physical exertion. Furthermore, vision and hearing are heightened to allow the person to accurately assess the entire threatening scene.

### TIMES OF PARASYMPATHETIC DOMINANCE

In quiet and restful circumstances, the parasympathetic system dominates, allowing the body to focus on its own "general housekeeping" activities (e.g., digestion). The parasympathetic system promotes these **"rest-and-digest"** types of bodily functions while slowing down those activities that are enhanced by the sympathetic system. There is no need, for example, to have the heart beating rapidly and forcefully when the person is in a tranquil setting.

### ADVANTAGE OF DUAL AUTONOMIC INNERVATION

The dual control exerted by the ANS provides more precise control, similar to the accelerator-and-brake system in an automobile. This type of duel control mechanism is termed **antagonistic**, while the unopposed sympathetic on-off control of vasoconstriction is termed **tonic**. If an animal suddenly darts across the road as you are driving, you could eventually stop if you just take your foot off the accelerator (tonic control), but you might stop too slowly to avoid hitting the animal. If you simultaneously apply the brake as you lift up on the accelerator (antagonistic control), however, you can come to a more rapid, controlled stop. In a similar manner, a sympathetically accelerated heart rate could gradually be reduced to normal following a stressful situation by decreasing the firing rate in the cardiac sympathetic nerve (letting up on the accelerator). However, the heart rate can be reduced more rapidly by simultaneously increasing activity in the parasympathetic supply to the heart (applying the brake). Indeed, the two divisions of the ANS are usually reciprocally controlled; increased activity in one division is accompanied by a corresponding decrease in the other.

There are several exceptions to the general rule of dual reciprocal innervations by the two branches of the ANS; the most notable are the following:

- *Innervated blood vessels* (most arterioles and veins are innervated; arteries and capillaries are not) receive only sympathetic nerve fibres. Regulation is accomplished by increasing or decreasing the firing rate above or below the tonic level in these sympathetic fibres. The only blood vessels to receive both sympathetic and parasympathetic fibres are those supplying the penis and clitoris. The precise vascular control this dual innervation affords these organs is important in accomplishing erection.

- Most *sweat glands* are innervated only by sympathetic nerves. The postganglionic fibres of these nerves are unusual because they secrete acetylcholine rather than norepinephrine.

- *Salivary glands* are innervated by both autonomic divisions, but, unlike elsewhere, sympathetic and parasympathetic activity is not antagonistic. Both stimulate salivary secretion, but the saliva's volume and composition differ, depending on which autonomic branch is dominant.

You will learn more about these exceptions in later chapters. We are now going to turn our attention to the adrenal medulla, a unique endocrine component of the sympathetic nervous system.

## The adrenal medulla

There are two *adrenal glands*, one lying above the kidney on each side. The adrenal glands are endocrine glands, each with an outer portion, the *adrenal cortex*, and an inner portion, the *adrenal medulla*. The **adrenal medulla** is a modified sympathetic ganglion that does not give rise to postganglionic fibres. Instead, on stimulation by the preganglionic fibre that originates in the CNS, it secretes hormones into the blood (› Figure 4-43; also see › Figure 4-41). Not surprisingly, the hormones are identical or similar to postganglionic sympathetic neurotransmitters. About 20 percent of the adrenal medullary hormone output is norepinephrine, and the remaining 80 percent is the closely related substance **epinephrine (adrenaline)**. These hormones, in general, reinforce activity of the sympathetic nervous system.

## Different receptor types

Because each autonomic neurotransmitter and medullary hormone stimulates activity in some tissues but inhibits activity in others, the particular responses must depend on specialization of the tissue cells rather than on properties of the chemicals themselves. Responsive tissue cells have one or more of several different types of plasma membrane receptor proteins for these chemical messengers. Binding of a neurotransmitter to a receptor induces the tissue-specific response.

### CHOLINERGIC RECEPTORS

Researchers have identified two types of acetylcholine (cholinergic) receptors—*nicotinic* and *muscarinic receptors*—on the basis of their response to particular drugs. **Nicotinic receptors** are activated by the tobacco plant derivative nicotine, whereas **muscarinic receptors** are activated by the mushroom poison muscarine.

Nicotinic receptors are found on the postganglionic cell bodies in all autonomic ganglia. These receptors respond to acetylcholine released from both sympathetic and parasympathetic preganglionic fibres. Binding of acetylcholine to these receptors brings about the opening of cation channels in the postganglionic cell, which permits passage of both $Na^+$ and $K^+$. Because of the greater electrochemical gradient for $Na^+$ than for $K^+$, more $Na^+$ enters the cell than $K^+$ leaves, bringing about a

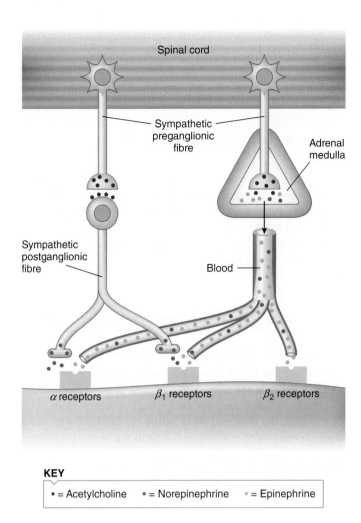

Spinal cord

Sympathetic preganglionic fibre

Adrenal medulla

Sympathetic postganglionic fibre

Blood

$\alpha$ receptors    $\beta_1$ receptors    $\beta_2$ receptors

**KEY**

• = Acetylcholine    • = Norepinephrine    • = Epinephrine

> **FIGURE 4-43 Comparison of the release and binding to receptors of epinephrine and norepinephrine.** Norepinephrine is released both as a neurotransmitter from sympathetic postganglionic fibres and as a hormone from the adrenal medulla. Beta-1 $(\beta_1)$ receptors bind equally with both norepinephrine and epinephrine, whereas beta-2 $(\beta_2)$ receptors bind primarily with epinephrine. Alpha $(\alpha)$ receptors of both subtypes have a greater affinity for norepinephrine than for epinephrine.

depolarization that leads to initiation of an action potential in the postganglionic cell.

Muscarinic receptors are found on effector cell membranes (smooth muscle, cardiac muscle, and glands). They bind with acetylcholine released from parasympathetic postganglionic fibres. There are five subtypes of muscarinic receptors, all of which are linked to G proteins that activate second-messenger systems that lead to the target cell response.

### ADRENERGIC RECEPTORS

There are two primary classifications of adrenergic receptors for epinephrine and norepinephrine (catecholamine receptors): alpha $(\alpha)$ and beta $(\beta)$ receptors. The $\alpha$ and $\beta$ receptors are further subclassified as $\alpha_1$ and $\alpha_2$, and $\beta_1$, $\beta_2$, and $\beta_3$. Catecholamine receptors act through secondary messengers to transfer the signal from the cell surface into the cytoplasm, in order to influence metabolic processes and thus cellular function. These various receptor types are distinctly distributed among the

effector organs. Receptors of the $\beta_2$ type bind primarily with epinephrine, whereas $\beta_1$ receptors have about equal affinities for norepinephrine and epinephrine, and $\alpha$ receptors of both subtypes have a greater sensitivity to norepinephrine than to epinephrine (Figure 4-43).

All adrenergic receptors are coupled to G proteins, but the ensuing pathway differs for the various receptor types, which can have either stimulatory or inhibitory influences. Activation of both $\beta_1$ and $\beta_2$ receptors brings about the target cell response by means of the cyclic AMP second-messenger system (p. 217). Stimulation of $\alpha_1$ receptors elicits the desired response via the $Ca^{2+}$ second-messenger system (p. 218). By contrast, binding of a neurotransmitter to an $\alpha_2$ receptor blocks cyclic AMP production in the target cell.

Activation of $\alpha_1$ receptors usually brings about an excitatory response in the effector organ—for example, arteriolar constriction caused by increased contraction of smooth muscle in the walls of these blood vessels. The $\alpha_1$ receptors are present in most sympathetic target tissues. Activation of $\alpha_2$ receptors, in contrast, brings about an inhibitory response in the effector organ, such as decreased smooth muscle contraction in the digestive tract.

Stimulation of $\beta_1$ receptors, which are found primarily in the heart, causes an excitatory response, namely, increased rate and force of cardiac contraction. The response to $\beta_2$ receptor activation is generally inhibitory, such as arteriolar or bronchiolar (respiratory airway) dilation caused by relaxation of the smooth muscle in the walls of these tubular structures. Salbutamol is a short-lived bronchodilator that acts through the $\beta_2$-adrenergic receptor to provide relief from bronchospasm, but it also has implications for aerobic sports as an ergogenic aid. Salbutamol's function as an ergogenic aid is currently under investigation by researchers B. Sporer and D. MacKenzie at the University of British Columbia. While $\beta_1$ and $\beta_2$ receptors are more commonly found within the body, $\beta_3$ is less common. An example of a tissue associated with the $\beta_3$ receptor is adipose tissue (i.e., fat cells). The sympathetic nervous system stimulates adipose tissue to break down fat molecules via both the $\beta_3$ and $\alpha_2$ receptors.

### AUTONOMIC AGONISTS AND ANTAGONISTS

*Clinical Note* Drugs are available that selectively alter autonomic responses at each of the receptor types. **Agonists** bind to the same receptor as the neurotransmitter and elicit an effect that mimics that of the transmitter. **Antagonists**, by contrast, bind with the receptor and block the neurotransmitter's response. Some of these drugs are only of experimental interest, but others are very important therapeutically. For example, *atropine* blocks the effect of acetylcholine at muscarinic receptors, but does not affect nicotinic receptors. Because acetylcholine released at both parasympathetic and sympathetic preganglionic fibres combines with nicotinic receptors, blockage at nicotinic synapses would knock out both these autonomic branches. By acting selectively to interfere with acetylcholine action only at muscarinic junctions—the sites of parasympathetic postganglionic action—atropine effectively blocks parasympathetic effects but does not influence sympathetic activity at all. Doctors use this principle to suppress salivary and bronchial secretions before surgery, to reduce the risk of a patient inhaling these secretions into the lungs.

Likewise, there are widely used drugs that act selectively at α- and β-adrenergic receptor sites to either activate or block specific sympathetic effects. *Salbutamol* is an excellent example. It selectively activates $\beta_2$-adrenergic receptors at low doses, making it possible to dilate the bronchioles in the treatment of asthma without undesirably stimulating the heart (the heart has mostly $\beta_1$ receptors).

## Control of autonomic activities

Messages from the CNS are delivered to cardiac muscle, smooth muscle, and glands via the autonomic nerves, but which regions of the CNS regulate autonomic output? The following summarizes these regions and their regulatory functions:

- The spinal cord integrates some autonomic reflexes, such as urination, defecation, and erection, but all these spinal reflexes are subject to control by higher levels of consciousness.
- The medulla within the brain stem is most directly responsible for autonomic output, and includes centres for controlling cardiovascular, respiratory, and digestive activity via the autonomic system.
- The hypothalamus plays an important role in integrating the autonomic, somatic, and endocrine responses that automatically accompany various emotional and behavioural states. For example, the increased heart rate, blood pressure, and respiratory activity associated with anger or fear are brought about by the hypothalamus acting through the medulla.
- The prefrontal association cortex can also influence autonomic output through its involvement with emotional expression characteristic of the individual's personality. An example is blushing when embarrassed, which is caused by dilation of

blood vessels supplying the skin of the cheeks. Such responses are mediated through hypothalamic-medullary pathways.

Table 4-9 summarizes the main distinguishing features of the sympathetic and parasympathetic nervous systems.

### Check Your Understanding 4.6

1. Illustrate the origin, termination, fibre length, and neurotransmitter released for parasympathetic and sympathetic preganglionic fibres and postganglionic fibres.
2. Compare the times of sympathetic and of parasympathetic dominance.
3. Discuss the relationship of the adrenal medulla to the autonomic nervous system.

## 4.10 | Somatic Nervous System

### Motor neurons and skeletal muscle

Skeletal muscle is innervated by **motor neurons**, the axons of which constitute the somatic nervous system. The cell bodies of almost all motor neurons are within the ventral horn of the spinal cord; the only exception is that the cell bodies of motor neurons supplying muscles in the head are in the brain stem. Unlike the two-neuron chain of autonomic nerve fibres, the axon of a motor neuron is continuous from its origin in the CNS to its ending on skeletal muscle. Motor neuron axon terminals release acetylcholine, which brings about excitation and contraction of

**TABLE 4-9** Distinguishing Features of Sympathetic Nervous System and Parasympathetic Nervous System

| Feature | Sympathetic System | Parasympathetic System |
|---|---|---|
| **Origin of Preganglionic Fibre** | Thoracic and lumbar regions of spinal cord | Brain and sacral region of spinal cord |
| **Origin of Postganglionic Fibre (location of ganglion)** | Sympathetic ganglion chain (near spinal cord) or collateral ganglia (about halfway between spinal cord and effector organs) | Terminal ganglia (in or near effector organs) |
| **Length and Type of Fibre** | Short cholinergic preganglionic fibres | Long cholinergic preganglionic fibres |
| | Long adrenergic postganglionic fibres | Short cholinergic postganglionic fibres |
| **Effector Organs Innervated** | Cardiac muscle, almost all smooth muscle, most exocrine glands, and some endocrine glands | Cardiac muscle, most smooth muscle, most exocrine glands, and some endocrine glands |
| **Types of Receptors for Neurotransmitters** | For preganglionic neurotransmitter: nicotinic | For preganglionic neurotransmitter: nicotinic |
| | For postganglionic neurotransmitter: $\alpha_1, \alpha_2, \beta_1, \beta_2$ | For postganglionic neurotransmitter: muscarinic |
| **Dominance** | Dominates in emergency fight-or-flight situations; prepares body for strenuous physical activity | Dominates in quiet, relaxed situations; promotes "general housekeeping" activities, such as digestion |

the innervated muscle cells. Motor neurons can only stimulate skeletal muscle, and not inhibit it, unlike the autonomic fibres and their effector organs. Inhibition of skeletal muscle activity can be accomplished only within the CNS through inhibitory synaptic input to the dendrites and cell bodies of the motor neurons supplying that particular muscle.

## Motor neurons: The final common pathway

Motor neuron dendrites and cell bodies are influenced by many converging presynaptic inputs, both excitatory and inhibitory. Some of these inputs are part of spinal reflex pathways originating with peripheral sensory receptors. Others are part of descending pathways originating within the brain. Areas of the brain that exert control over skeletal muscle movements include the motor regions of the cortex, the basal nuclei, the cerebellum, and the brain stem; see Figure 7-22 for a summary of motor control, and Figure 3-28b, for specific examples of these descending motor pathways).

Motor neurons are considered the **final common pathway**, because the only way any other parts of the nervous system can influence skeletal muscle activity is by acting on these motor neurons. The level of activity in a motor neuron and its subsequent output to the skeletal muscle fibres it innervates depend on the relative balance of EPSPs and IPSPs (pp. 76) brought about by its presynaptic inputs originating from these diverse sites in the brain.

The somatic system is under voluntary control, but much of the skeletal muscle activity involving posture, balance, and stereotypical movements is subconsciously controlled. You may decide you want to start walking, but you do not have to consciously bring about the alternate contraction and relaxation of the involved muscles because these movements are involuntarily coordinated by lower brain centres.

*Clinical Note* The cell bodies of the crucial motor neurons may be selectively destroyed by **poliovirus**. The result is paralysis of the muscles innervated by the affected neurons. **Amyotrophic lateral sclerosis (ALS)**, also known as **Lou Gehrig's disease**, is the most common motor-neuron disease. ALS is associated with pathological changes in (1) neurofilaments that block axonal transport of crucial materials, (2) extracellular accumulation of toxic levels of the excitatory neurotransmitter glutamate, (3) aggregation of misfolded intracellular proteins, or (4) mitochondrial dysfunction leading to reduced energy production.

Before turning our attention to the junction between a motor neuron and the muscle cells it innervates, we are going to pull together in tabular form two groups of information we have been examining. Table 4-10 summarizes the features of the two branches of the efferent division of the peripheral nervous system—the ANS and the somatic nervous system. Table 4-11 compares the three functional types of neurons—afferent neurons, efferent neurons, and interneurons.

## ■ TABLE 4-10 Comparison of the Autonomic Nervous System and the Somatic Nervous System

| Feature | Autonomic Nervous System | Somatic Nervous System |
|---|---|---|
| Site of Origin | Brain or lateral horn of spinal cord | Ventral horn of spinal cord for most; those supplying muscles in head originate in brain |
| Number of Neurons from Origin in CNS to Effector Organ | Two-neuron chain (preganglionic and postganglionic) | Single neuron (motor neuron) |
| Organs Innervated | Cardiac muscle, smooth muscle, exocrine and some endocrine glands | Skeletal muscle |
| Type of Innervation | Most effector organs dually innervated by the two antagonistic branches of this system (sympathetic and parasympathetic) | Effector organs innervated only by motor neurons |
| Neurotransmitter at Effector Organs | May be acetylcholine (parasympathetic terminals) or norepinephrine (sympathetic terminals) | Only acetylcholine |
| Effects on Effector Organs | Either stimulation or inhibition (antagonistic actions of two branches) | Stimulation only (inhibition possible only centrally through IPSPs on dendrites and cell body of motor neuron) |
| Types of Control | Under involuntary control | Subject to voluntary control; much activity subconsciously coordinated |
| Higher Centres Involved in Control | Spinal cord, medulla, hypothalamus, prefrontal association cortex | Spinal cord, motor cortex, basal nuclei, cerebellum, brain stem |

| Feature | Afferent Neuron | EFFERENT NEURON Autonomic Nervous System | EFFERENT NEURON Somatic Nervous System | Interneuron |
|---|---|---|---|---|
| **Origin, Structure, Location** | Receptor at peripheral ending; elongated peripheral axon, which travels in peripheral nerve; cell body located in dorsal root ganglion; short central axon entering spinal cord | Two-neuron chain; first neuron (preganglionic fibre) originating in CNS and terminating within a ganglion; second neuron (postganglionic fibre) originating in the ganglion and terminating on the effector organ | Cell body of motor neuron lying in spinal cord; long axon travelling in peripheral nerve and terminating on the effector organ | Various shapes; lying entirely within CNS; some cell bodies originating in brain, with long axons travelling down the spinal cord in descending pathways; some originating in spinal cord, with long axons travelling up the cord to the brain in ascending pathways; others forming short local connections |
| **Termination** | Interneurons* | Effector organs (cardiac muscle, smooth muscle, exocrine and some endocrine glands) | Effector organs (skeletal muscle) | Other interneurons and efferent neurons |
| **Function** | Carries information about the external and internal environment to CNS | Carries instructions from CNS to effector organs | Carries instructions from CNS to effector organs | Processes and integrates afferent input; initiates and coordinates efferent output; responsible for thought and other higher mental functions |
| **Convergence of Input on Cell Body** | No (only input is through receptor) | Yes | Yes | Yes |
| **Effect of Input to Neuron** | Can only be excited (through receptor potential induced by stimulus; must reach threshold for action potential) | Can be excited or inhibited (through EPSPs and IPSPs at first neuron; must reach threshold for action potential) | Can be excited or inhibited (through EPSPs and IPSPs; must reach threshold for action potential) | Can be excited or inhibited (through EPSPs and IPSPs; must reach threshold for action potential) |
| **Site of Action Potential Initiation** | First excitable portion of membrane adjacent to receptor | Axon hillock | Axon hillock | Axon hillock |
| **Divergence of Output** | Yes | Yes | Yes | Yes |
| **Effect of Output on Structure on Which It Terminates** | Only excites | Postganglionic fibre either excites or inhibits | Only excites | Either excites or inhibits |

© 2016 Cengage

* Except in stretch reflex where afferent neuron terminate directly on efferent neurons; see p. 106.

## Check Your Understanding 4.7

1. Compare the effector organs innervated by the autonomic nervous system and by the somatic nervous system.

2. Explain why motor neurons are considered the final common pathway.

# 4.11 Neuromuscular Junction

## Linkage of motor neurons and skeletal muscle fibres

An action potential in a motor neuron is rapidly propagated from the cell body within the CNS to the skeletal muscle along the large myelinated axon (efferent fibre) of the neuron. As the axon approaches a muscle, it divides into many terminal branches and loses its myelin sheath. Each of these axon terminals forms a special junction, a **neuromuscular junction**,[2] with one of the many muscle cells (**muscle fibres**) that compose the whole muscle (e.g., biceps muscle) (› Figure 4-44). A single muscle cell is long and cylindrical with bands of light and dark areas. The axon terminal is enlarged into a knoblike structure, the **terminal button**, which fits into a shallow depression, or cleft, in the underlying muscle fibre (› Figure 4-45). Some scientists alternatively call the neuromuscular junction a motor end plate. However, we will reserve the term **motor end plate** for the specialized portion of the muscle cell membrane immediately under the terminal button.

## Acetylcholine

Nerve and muscle cells do not actually come into direct contact at a neuromuscular junction. The space (i.e., synaptic cleft) between these two structures is too large to permit electrical transmission of an impulse between them; thus, the action potential cannot jump that far. Furthermore, there are no channels for exit of charge-carrying current from the terminal button. Therefore, just as occurs at a neuronal synapse (p. 75), a chemical messenger carries the signal between the neuron terminal and the muscle fibre. This neurotransmitter is acetylcholine (ACh).

### RELEASE OF ACH AT THE NEUROMUSCULAR JUNCTION

Each terminal button contains thousands of vesicles that store ACh. Propagation of an action potential to the axon terminal (› Figure 4-45, step 1) triggers the opening of voltage-gated $Ca^{2+}$ channels in the terminal button (p. 76). Opening of

$Ca^{2+}$ channels permits $Ca^{2+}$ to diffuse into the terminal button down its concentration gradient (step 2), which in turn causes the release of ACh by exocytosis from several hundred vesicles into the cleft (step 3).

### FORMATION OF AN END-PLATE POTENTIAL

The released ACh diffuses across the cleft and binds with specific receptor sites, which are specialized membrane proteins unique to the motor end-plate portion of the muscle fibre membrane (step 4). (These cholinergic receptors are of the nicotinic type.) Binding of ACh with these receptor sites induces the opening of chemically gated channels in the motor end plate. These channels permit a small amount of cation traffic through them (both $Na^+$ and $K^+$), but no anions (step 5). Because the permeability of the end-plate membrane to $Na^+$ and $K^+$ on opening of these channels is essentially equal, the relative movement of these ions through the channels depends on their electrochemical driving forces. Recall that at resting potential, the net driving force for $Na^+$ is much greater than that for $K^+$, because the resting potential is much closer to the $K^+$ equilibrium potential. Both the concentration and electrical gradients for $Na^+$ are inward, whereas the outward concentration gradient for $K^+$ is almost, but not quite, balanced by the opposing inward electrical gradient. As a result, when ACh triggers the opening of these channels, considerably more $Na^+$ moves inward than $K^+$ outward, depolarizing the motor end plate. This potential change is called the **end-plate potential (EPP)**. It is a graded potential similar to an EPSP (p. 76), except that an EPP is much larger for the

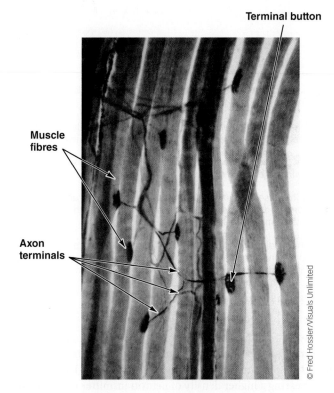

**Terminal button**

**Muscle fibres**

**Axon terminals**

© Fred Hossler/Visuals Unlimited

› **FIGURE 4-44 Motor neuron innervating skeletal muscle cells.** When a motor neuron reaches a skeletal muscle, it divides into many terminal branches, each of which forms a neuromuscular junction with a single muscle cell (muscle fibre).

---

[2]Many scientists refer to a **synapse** as any junction between two cells that handle information electrically. According to this broad point of view, *chemical synapses* include junctions between two neurons as well as those between a neuron and an effector cell (such as muscle cells of any type or gland cells), and *electrical synapses* include gap junctions between smooth muscle cells or between cardiac muscle cells. We narrowly reserve the term *synapse* specifically for neuron-to-neuron junctions and use different terms for other types of junctions, such as the term *neuromuscular junction* for a junction between a motor neuron and a skeletal muscle cell.

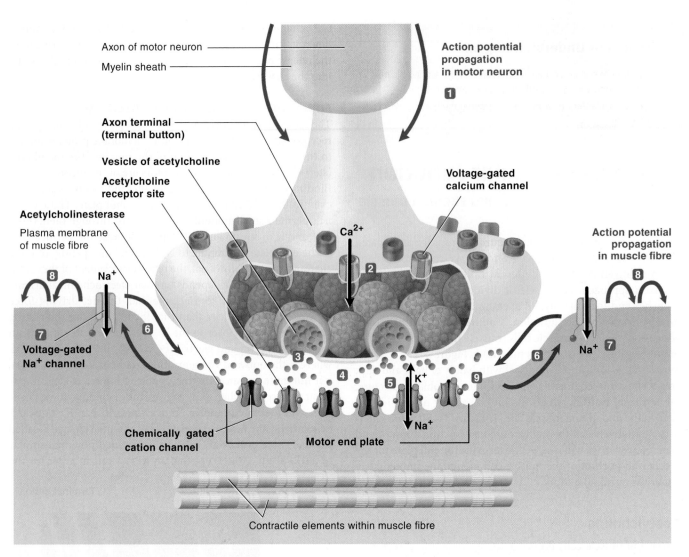

Axon of motor neuron

Myelin sheath

**Action potential propagation in motor neuron** ❶

**Axon terminal (terminal button)**

**Vesicle of acetylcholine**

**Acetylcholine receptor site**

**Acetylcholinesterase**

Plasma membrane of muscle fibre

$Ca^{2+}$

**Voltage-gated calcium channel**

**Action potential propagation in muscle fibre**

❽

$Na^+$ ❼

$Na^+$

❷

❸

❻

**Voltage-gated $Na^+$ channel**

❼

❻

❹

❺

$K^+$

$Na^+$

❾

**Chemically gated cation channel**

**Motor end plate**

Contractile elements within muscle fibre

❶ An action potential in a motor neuron is propagated to the axon terminal (terminal button).

❷ This local action potential triggers the opening of voltage-gated $Ca^{2+}$ channels and the subsequent entry of $Ca^{2+}$ into the terminal button.

❸ $Ca^{2+}$ triggers the release of acetylcholine (ACh) by exocytosis from a portion of the vesicles.

❹ ACh diffuses across the space separating the nerve and muscle cells and binds with receptor channels specific for it on the motor end plate of the muscle cell membrane.

❺ This binding brings about the opening of these nonspecific cation channels, leading to a relatively large movement of $Na^+$ into the muscle cell compared to a smaller movement of $K^+$ outward.

❻ The result is an end-plate potential. Local current flow occurs between the depolarized end plate and the adjacent membrane.

❼ This local current flow opens voltage-gated $Na^+$ channels in the adjacent membrane.

❽ The resultant $Na^+$ entry reduces the potential to threshold, initiating an action potential, which is propagated throughout the muscle fibre.

❾ ACh is subsequently destroyed by acetylcholinesterase, an enzyme located on the motor end-plate membrane, terminating the muscle cell's response.

> **FIGURE 4-45 Events at a neuromuscular junction**

© 2016 Cengage

following reasons: (1) more neurotransmitter is released from a terminal button than from a presynaptic knob in response to an action potential; (2) the motor end plate has a larger surface area bearing a higher density of neurotransmitter receptor sites, and accordingly, has more sites for binding with neurotransmitter than a subsynaptic membrane has; and (3) many more ion channels are opened in response to the neurotransmitter-receptor complex in the motor end plate. This permits a greater net influx of positive ions and a larger depolarization. As with an

EPSP, an EPP is a graded potential, whose magnitude depends on the amount and duration of ACh at the end plate.

### INITIATION OF AN ACTION POTENTIAL

The motor end-plate region itself does not have a threshold potential, so an action potential cannot be initiated at this site. However, an EPP brings about an action potential in the rest of the muscle fibre, as follows. The neuromuscular junction is usually in the middle of the long, cylindrical muscle fibre. When an

EPP takes place, local current flow occurs between the depolarized end plate and the adjacent, resting cell membrane in both directions (step 6), opening voltage-gated $Na^+$ channels and thus reducing the potential to threshold in the adjacent areas (step 7). The subsequent action potential initiated at these sites propagates throughout the muscle fibre membrane by contiguous conduction (step 8). The spread runs in both directions, away from the motor end plate toward both ends of the muscle fibre, triggering the contraction of the muscle fibre. Thus, by means of ACh, an action potential in a motor neuron brings about an action potential and subsequent contraction in the muscle fibre.

Unlike synaptic transmission, the magnitude of an EPP is normally enough to cause an action potential in the muscle cell. Therefore, one-to-one transmission of an action potential typically occurs at a neuromuscular junction; one action potential in a nerve cell triggers one action potential in a muscle cell that it innervates. Other comparisons of neuromuscular junctions with synapses can be found in ▌ Table 4-12.

## Acetylcholinesterase and acetylcholine activity

To ensure purposeful movement, a muscle cell's electrical and resultant contractile response to stimulation by its motor neuron must be switched off promptly when there is no longer a signal from the motor neuron. The muscle cell's electrical response is turned off by an enzyme in the motor end-plate membrane; this enzyme is **acetylcholinesterase (AChE)**, which inactivates ACh.

As a result of diffusion, many of the released ACh molecules come into contact with and bind to receptor sites on the surface of the motor end-plate membrane. However, some of the ACh molecules bind with AChE, which is also at the end-plate surface. Being quickly inactivated, this ACh never contributes to the end-plate potential. The acetylcholine that does bind with receptor sites does so very briefly (for about 1 millionth of a second), then detaches. Some of the detached ACh molecules quickly rebind with receptor sites, keeping the end-plate channels open, but some randomly contact AChE instead this time around and are inactivated (Figure 4-45, step 9 ). As this process repeats, more and more ACh is inactivated until it has been virtually removed from the cleft within a few milliseconds after its release. ACh removal ends the EPP, so the remainder of the muscle cell membrane returns to resting potential. Now the muscle cell can relax. Or, if sustained contraction is essential for the desired movement, another motor neuron action potential leads to the release of more ACh, which keeps the contractile process going. By removing contraction-inducing ACh from the motor end plate, AChE permits the choice of allowing relaxation to take place (no more ACh released) or keeping the contraction going (more ACh released), depending on the body's momentary needs.

▌ **Clinical Connections**  So far we have discussed how diabetes has affected Deborah's afferent nerves. However, she was also experiencing mild muscle weakness, becoming unsteady on her feet, and having problems with fine motor control, which prevented her from knitting. Surprisingly, such symptoms are very common in diabetes, yet there is little medical literature to indicate what the exact causes are. Such changes are potentially caused by a combination of damage to motor neurons, damage to motor end plates, and damage to the muscles themselves. Neuropathy due to mechanisms already discussed could account for damaged neurons and motor end plates, but more research is needed. Within the muscles themselves, the weakness can be caused by muscle atrophy—a loss of the muscle mass.

## Vulnerability of the neuromuscular junction

Several chemical agents and diseases are known to affect the neuromuscular junction by acting at different sites in the transmission process. Two well-known toxins—*black widow spider venom* and *botulinum toxin*—alter the release of ACh, but do so in opposite directions.

### BLACK WIDOW SPIDER VENOM CAUSES EXPLOSIVE RELEASE OF ACH

The venom of black widow spiders exerts its deadly effect by triggering explosive release of ACh from the storage vesicles, not only at neuromuscular junctions but at all cholinergic sites. All cholinergic sites undergo prolonged depolarization, the most harmful result of which is respiratory failure. Breathing is accomplished by alternate contraction and relaxation of skeletal muscles, particularly the diaphragm. Respiratory paralysis occurs as a result of prolonged depolarization of the diaphragm. During this so-called **depolarization block**, the voltage-gated $Na^+$ channels are trapped in their inactivated state (p. 62), thereby prohibiting the initiation of new action potentials and resultant contraction of the diaphragm. Thus, the victim cannot breathe.

### BOTULINUM TOXIN BLOCKS RELEASE OF ACH

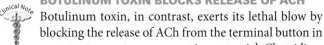

Botulinum toxin, in contrast, exerts its lethal blow by blocking the release of ACh from the terminal button in response to a motor neuron action potential. *Clostridium botulinum* toxin causes **botulism**, a form of food poisoning. It most frequently results from improperly canned foods contaminated with clostridial bacteria, which survive and multiply, producing their toxin in the process. When this toxin is consumed, it prevents muscles from responding to nerve impulses. Death is due to respiratory failure caused by inability to contract the diaphragm. Botulinum toxin is one of the most lethal poisons known; ingesting less than 0.0001 mg can kill an adult human. (See Concepts, Challenges, and Controversies, p. 200), to learn about a surprising new wrinkle in the botulinum toxin story.)

### CURARE BLOCKS ACTION OF ACH AT RECEPTOR SITES

Other chemicals interfere with neuromuscular junction activity by blocking the effect of released ACh. The best-known example is the antagonist **curare**, which reversibly binds to the ACh receptor sites on the motor end plate. Unlike ACh, however, curare does not alter membrane permeability, nor is it inactivated by AChE. When curare occupies ACh receptor sites, ACh cannot combine with these sites to open the channels that would permit the ionic movement responsible for an EPP.

# Botulinum Toxin's Reputation Gets a Facelift

THE POWERFUL TOXIN PRODUCED by *Clostridium botulinum* causes the deadly food poisoning botulism. Yet this dreaded, highly lethal poison has been put to use as a treatment for alleviating specific movement disorders and, more recently, has been added to the list of tools that cosmetic surgeons use to fight wrinkles.

During the last decade, botulinum toxin, marketed in therapeutic doses as *Botox*, has offered welcome relief to people with a number of painful, disruptive neuromuscular diseases known categorically as **dystonias**. These conditions are characterized by spasms (excessive, sustained, involuntarily produced muscle contractions) that result in involuntary twisting or abnormal postures, depending on the body part affected. For example, painful neck spasms that twist the head to one side result from *spasmodic torticollis* (*tortus* means "twisted"; *collum* means "neck"), the most common dystonia. The problem is believed to arise from too little inhibitory compared with excitatory input to the motor neurons that supply the affected muscle. The reasons for this imbalance in motor-neuron input are unknown. The end result of excessive motor-neuron activation is sustained, disabling contraction of the muscle supplied by the overactive motor neurons. Fortunately, injecting minuscule amounts of botulinum toxin into the affected muscle causes a reversible, partial paralysis of the muscle. Botulinum toxin interferes with the release of muscle-contraction-causing acetylcholine from the overactive motor neurons at the neuromuscular junctions in the treated muscle. The goal is to inject just enough botulinum toxin to alleviate the troublesome spasmodic contractions but not enough to eliminate the normal contractions needed for ordinary movements. The therapeutic dose is considerably less than the amount of toxin needed to induce even mild symptoms of botulinum

Consequently, because muscle action potentials cannot occur in response to nerve impulses to these muscles, paralysis ensues. When enough curare is present to block a significant number of ACh receptor sites, the person dies from respiratory paralysis caused by inability to contract the diaphragm. In the past, some people used curare as a deadly arrowhead poison.

### ORGANOPHOSPHATES PREVENT INACTIVATION OF ACH

**Organophosphates** are a group of chemicals that modify neuromuscular junction activity in yet another way—namely, by irreversibly inhibiting AChE. Inhibition of AChE prevents the inactivation of released ACh. Death from organophosphates is also due to respiratory failure, because the diaphragm cannot

---

### ■ TABLE 4-12 Comparison of a Synapse and a Neuromuscular Junction

| Similarities | Differences |
|---|---|
| Both consist of two excitable cells separated by a narrow cleft that prevents direct transmission of electrical activity between them. | A synapse is a junction between two neurons. A neuromuscular junction exists between a motor neuron and a skeletal muscle fibre. |
| Axon terminals of both store chemical messengers (neurotransmitters) that are released by the $Ca^{2+}$-induced exocytosis of storage vesicles when an action potential reaches the terminal. | One-to-one transmission of action potentials occurs between motor neurons and muscle fibres, whereas one action potential in a presynaptic neuron usually cannot by itself bring about an action potential in a postsynaptic neuron. An action potential in a postsynaptic neuron typically occurs only when summation of EPSPs brings the membrane to threshold. |
| In both, binding of the neurotransmitter with receptor sites in the membrane of the cell underlying the axon terminal opens specific channels in the membrane, permitting ionic movements that alter the membrane potential of the cell. | A neuromuscular junction is always excitatory (an EPP); a synapse may be either excitatory (an EPSP) or inhibitory (an IPSP). |
| The resultant change in membrane potential in both cases is a graded potential. | Inhibition of skeletal muscles cannot be accomplished at the neuromuscular junction; it can occur only in the CNS through IPSPs at the dendrites and cell body of the motor neuron. |

**Source:** From Sherwood. *Human Physiology*, 8E. © 2013 Brooks/Cole, a part of Cengage, Inc. Reproduced by permission. www.cengage.com/permissions

poisoning. Botulinum toxin is eventually cleared away, so its muscle-relaxing effects wear off after three to six months, at which time the treatment must be repeated.

The first dystonia for which Botox was approved as a treatment by the U.S. Food and Drug administration (FDA) was *blepharospasm* (*blepharo* means "eyelid"). In this condition, sustained and involuntary contractions of the muscles around the eye nearly permanently close the eyelids.

Botulinum toxin's potential as a treatment option for cosmetic surgeons was accidentally discovered when physicians noted that injections used to counter abnormal eye muscle contractions also smoothed the appearance of wrinkles in the treated areas. It turns out that frown lines, crow's feet, and furrowed brows are caused by facial muscles that have become overactivated, or permanently contracted, as a result of years of performing certain repetitive facial expressions. By relaxing these muscles, botulinum toxin temporarily smooths out these age-related wrinkles. Botox has recently received FDA approval as an antiwrinkle treatment. The agent is considered an excellent alternative to facelift surgery for combating lines and creases. This treatment is among the most rapidly growing cosmetic procedures in the United States, especially in the entertainment industry and in high-fashion circles. However, as with its therapeutic use to treat dystonias, the costly injections of botulinum toxin must be repeated every three to six months to maintain the desired effect in appearance. Furthermore, Botox does not work against the fine, crinkly wrinkles associated with years of excessive sun exposure, because these wrinkles are caused by skin damage, not by contracted muscles.

repolarize and return to resting conditions, then contract again to bring in a fresh breath of air. As well, many important biochemicals are organophosphates, including DNA and RNA, and many cofactors vital to life. Organophosphates are the source of many insecticides and military nerve gases.

### MYASTHENIA GRAVIS INACTIVATES ACH RECEPTOR SITES

**Myasthenia gravis**, a disease involving the neuromuscular junction, is characterized by extreme muscular weakness (*myasthenia* means "muscular weakness"; *gravis* means "severe"). It is an autoimmune condition (*autoimmune* means "immunity against self") in which the body erroneously produces antibodies against its own motor end-plate ACh receptors. Thus, not all the released ACh molecules can find a functioning receptor with which to bind. As a result, AChE destroys much of the ACh before it ever has a chance to interact with a receptor and contribute to the EPP. Treatment consists of administering a drug, such as **neostigmine**, that inhibits AChE temporarily (in contrast to the toxic organophosphates, which irreversibly block this enzyme). This drug prolongs the action of ACh at the neuromuscular junction by permitting it to build up for the short term. The resultant EPP has sufficient magnitude to initiate an action potential and subsequent contraction in the muscle fibre, as it normally would.

---

### Check Your Understanding 4.8

1. Discuss the role of ACh and of AChE at a neuromuscular junction.
2. Compare the magnitude of an EPP and an EPSP, and explain the functional significance of this difference.

---

# Chapter in Perspective: Focus on Homeostasis

### AFFERENT DIVISION

To maintain a life-sustaining stable internal environment, the body must constantly make adjustments to compensate for myriad external and internal factors that continuously threaten to disrupt homeostasis, such as external exposure to cold or internal acid production. Many of these adjustments are directed by the nervous system, one of the body's two major regulatory systems. The CNS, the integrating and decision-making component of the nervous system, must continuously be informed of "what's happening" in both the internal and the external environment so that it can command appropriate responses in the organ systems to maintain the body's viability. In other words, the CNS must know what changes are taking place before it can respond to these changes.

The afferent division of the peripheral nervous system is the communication link by which the CNS is informed about the internal and external environment. It detects, encodes, and transmits peripheral signals to the CNS for processing. Afferent input is necessary for arousal, perception, and determination of efferent output.

Afferent information about the internal environment, such as the carbon dioxide level in the blood, never reaches consciousness, but this input to the controlling centres of the CNS is essential for maintaining homeostasis. Afferent input that reaches the level of conscious awareness, called sensory information, includes somatosensory sensation (body sense) and special senses (vision, hearing and equilibrium, taste, and smell).

The somatosensory receptors are distributed over the entire body surface as well as throughout the joints and muscles. Afferent signals from these receptors provide information about what's happening directly to each specific body part in relation to the external environment (i.e., the "what," "where," and "how much" of stimulatory inputs to the body's surface and the momentary position of the body in space). In contrast, each special sense organ is restricted to a single site in the body. Rather than provide information about a specific body part, a special sense organ provides a specific type of information about the external environment that is useful to the body as a whole. For example, through their ability to detect, extensively analyze, and integrate patterns of illumination in the external environment, the eyes and visual processing system enable you to see your surroundings. The same integrative effect could not be achieved if photoreceptors were scattered over your entire body surface, as are touch receptors.

Sensory input (somatosensory and special senses) enables a complex multicellular organism, such as a human, to interact in meaningful ways with the external environment in procuring food, defending against danger, and engaging in other behavioural actions geared toward maintaining homeostasis. In addition to providing information essential for interactions with the external environment for basic survival, the perceptual processing of sensory input adds immeasurably to the richness of life, such as enjoyment of a good book, concert, or meal.

## EFFERENT DIVISION

While the afferent division of the PNS detects and carries information to the CNS for processing and decision making, the efferent division of the PNS carries directives from the CNS to the effector organs (muscles and glands), which carry out the intended response. Much of this efferent output is directed toward maintaining homeostasis.

The ANS is the efferent branch that innervates smooth and cardiac muscle, and glands. Its role in homeostatic activities is as follows:

- regulating blood pressure
- controlling digestive tract secretions, contractions, and mixing of ingested food
- controlling sweating to help maintain body temperature

The somatic nervous system is the efferent branch that innervates skeletal muscle and contributes to homeostasis by stimulating the following:

- Skeletal muscle contractions that enable body movement and postural control, which contribute to homeostasis by moving the body toward food or away from harm
- Skeletal muscle contractions that enable breathing to maintain required levels of oxygen and carbon dioxide
- Shivering thermogenesis by skeletal muscles, which helps maintain body temperature

In addition, efferent output to skeletal muscles accomplishes many movements that enrich our lives and enable us to engage in activities that contribute to society, such as dancing, building bridges, or performing surgery.

## CHAPTER TERMINOLOGY

accommodation (p. 157)
acetylcholine (ACh) (p. 188)
acetylcholinesterase (AChE) (p. 199)
acuity (p. 147)
adaptation (p. 145)
A-delta (δ) fibres (p. 149)
adequate stimulus (p. 143)
adrenal medulla (p. 192)
adrenergic fibres (p. 188)
agonists (p. 193)
ampulla (p. 178)
amyotrophic lateral sclerosis (ALS) (p. 195)
analgesic system (p. 150)
antagonists (p. 193)
antagonistic (p. 192)
aqueous humour (p. 153)
astigmatism (p. 156)
auditory (cochlear) nerve (p. 174)
basilar membrane (p. 172)
binocular field of vision (p. 167)
bipolar cells (p. 157)
bitter taste (p. 184)
blind spot (p. 158)

botulism (p. 199)
bradykinin (p. 150)
capsaicin (p. 149)
cataract (p. 157)
cerumen (p. 172)
C fibres (p. 149)
chemical nociceptors (p. 149)
chemoreceptors (p. 144)
cholinergic fibres (p. 188)
choroid (p. 153)
ciliary body (p. 153)
ciliary muscle (p. 157)
circular (constrictor) muscle (p. 155)
cochlea (p. 172)
cochlear duct (p. 172)
cochlear implants (p. 177)
collateral ganglia (p. 188)
colour blindness (p. 165)
colour vision (p. 165)
complex cells (p. 168)
concave (p. 155)
conductive deafness (p. 177)
cones (p. 153)

convex (p. 155)
cornea (p. 153)
cortical gustatory area (p. 182)
cupula (p. 178)
curare (p. 199)
cyclic GMP (cGMP) (p. 161)
dark adaptation (p. 164)
deafness (p. 177)
depolarization block (p. 199)
depth perception (p. 167)
diplopia (p. 168)
discriminative ability (p. 147)
duration (p. 144)
dystonias (p. 200)
ear (p. 169)
ear canal (p. 172)
emmetropia (p. 157)
endolymph (p. 172)
end-plate potential (EPP) (p. 197)
epinephrine (adrenaline) (p. 192)
eustachian (auditory) tube (p. 172)
external ear (p. 171)
external eye muscles (p. 169)

eye (p. 153)
eyelashes (p. 153)
eyelids (p. 152)
fast pain (p. 149)
fight-or-flight response (p. 191)
final common pathway (p. 195)
first-order sensory neuron (p. 146)
focal point (p. 155)
fovea (p. 158)
ganglion cells (p. 157)
generator potential (p. 144)
glaucoma (p. 153)
glomeruli (p. 185)
gustation (p. 181)
gustducin (p. 184)
hair cells (p. 173)
hearing (p. 169)
hearing aids (p. 177)
hearing threshold (p. 171)
helicotrema (p. 172)
hypercomplex cells (p. 169)
hyperopia (p. 157)
incus (p. 172)
inner hair cells (p. 173)
intensity (loudness) discrimination (p. 174)
kinocilium (p. 178)
labelled lines (p. 146)
lacrimal gland (p. 153)
lateral inhibition (p. 147)
lens (p. 153)
light (p. 155)
light adaptation (p. 164)
light ray (p. 155)
loudness (p. 171)
Lou Gehrig's disease (p. 195)
macula lutea (p. 158)
macular degeneration (p. 158)
malleus (p. 172)
mechanical nociceptors (p. 149)
mechanoreceptors (p. 143)
Meissner's corpuscles (p. 148)
Ménière's disease (p. 180)
Merkel's discs (p. 148)
middle ear (p. 172)
mitral cells (p. 185)
modality (p. 143)
motion sickness (p. 180)
motor end plate (p. 197)
motor neurons (p. 194)
muscarinic receptors (p. 192)
muscle fibres (p. 197)
myasthenia gravis (p. 201)
myopia (p. 157)
neostigmine (p. 201)
neural presbycusis (p. 177)
neuromuscular junction (p. 197)
nicotinic receptors (p. 192)

night blindness (p. 165)
nociceptor (p. 149)
noradrenaline (p. 188)
norepinephrine (p. 188)
odourants (p. 184)
off response (p. 145)
olfaction (p. 181)
olfactory bulb (p. 185)
olfactory mucosa (p. 184)
olfactory nerve (p. 184)
olfactory receptor cell (p. 184)
on-centre and off-centre ganglion cells
    (p. 166)
opiate receptors (p. 150)
opsin (p. 161)
optic chiasm (p. 166)
optic disc (p. 158)
optic nerve (p. 158)
optic radiations (p. 167)
optic tracts (p. 166)
organ of Corti (p. 173)
organophosphates (p. 200)
ossicles (p. 172)
otolith organs (p. 178)
outer hair cells (p. 173)
oval window (p. 172)
Pacinian corpuscles (p. 147)
perception (p. 167)
perilymph (p. 172)
phantom pain (p. 147)
pharynx (p. 172)
phasic receptors (p. 145)
pheromones (p. 186)
photons (p. 155)
photopigments (p. 161)
photoreceptors (p. 143)
phototransduction (p. 161)
pinna (p. 171)
pitch (p. 171)
pitch discrimination (p. 174)
poliovirus (p. 195)
polymodal nociceptors (p. 150)
postganglionic fibre (p. 188)
preganglionic fibre (p. 188)
presbyopia (p. 157)
primary auditory cortex (p. 176)
primary olfactory cortex (p. 185)
primary tastes (p. 182)
pupil (p. 154)
quality (p. 171)
radial (dilator) muscle (p. 155)
receptive field (p. 147)
receptor potential (p. 144)
red, green, and blue cones (p. 161)
refraction (p. 155)
"rest-and-digest" (p. 192)
retina (p. 153)

retinene (p. 161)
rhodopsin (p. 161)
rods (p. 153)
round window (p. 172)
Ruffini corpuscles (p. 148)
saccule (p. 178)
salty taste (p. 182)
scala media (p. 172)
scala tympani (p. 172)
scala vestibuli (p. 172)
sclera (p. 153)
second-order sensory neuron (p. 146)
semicircular canals (p. 178)
sensorineural deafness (p. 177)
sensory afferent (p. 142)
simple cells (p. 168)
slow pain (p. 149)
slow pain pathway (p. 150)
smell (p. 181)
somatic sensation (p. 142)
somatosensory pathways (p. 146)
somatosensory receptors (p. 143)
sound waves (p. 169)
sour taste (p. 182)
special senses (p. 143)
special senses receptor (p. 143)
spoken languages (p. 177)
stapes (p. 172)
stereocilia (p. 173)
substance P (p. 150)
suspensory ligaments (p. 157)
sweet taste (p. 184)
sympathetic ganglion chain (p. 188)
sympathetic or parasympathetic
    tone (p. 189)
sympathetic trunk (p. 188)
synapse (p. 197)
tastant (p. 182)
taste (p. 181)
taste pore (p. 182)
taste receptor cells (p. 182)
tears (p. 153)
tectorial membrane (p. 173)
terminal button (p. 197)
terminal ganglia (p. 188)
thermal nociceptors (p. 149)
thermoreceptors (p. 144)
third-order sensory neuron (p. 146)
threshold (p. 143)
timbre (p. 171)
tip links (p. 178)
tone (p. 171)
tonic (p. 192)
tonic receptors (p. 145)
transducin (p. 161)
transduction (p. 143)
tympanic membrane (p. 172)

**4**

## REVIEW EXERCISES

### Objective Questions (Answers in Appendix E, p. A-38)

1. Conversion of the energy forms of stimuli into electrical energy by the receptors is known as_____.

2. The type of stimulus to which a particular receptor is most responsive is called its_____.

3. All afferent information is sensory information. (*True or false?*)

4. Off-centre ganglion cells increase their rate of firing when a beam of light strikes the periphery of their receptive field. (*True or false?*)

5. During dark adaptation, rhodopsin is gradually regenerated to increase the sensitivity of the eyes. (*True or false?*)

6. An optic nerve carries information from the lateral and medial halves of the same eye, whereas an optic tract carries information from the lateral half of one eye and the medial half of the other. (*True or false?*)

7. Displacement of the round window generates neural impulses perceived as sound sensations. (*True or false?*)

8. Hair cells in different regions of the organ of Corti and neurons in different regions of the auditory cortex are activated by different tones. (*True or false?*)

9. Each taste receptor responds to just one of the five primary tastes. (*True or false?*)

10. Rapid adaptation to odours results from adaptation of the olfactory receptors. (*True or false?*)

(l) sclera

___ 1. layer that contains photoreceptors
___ 2. point from which optic nerve leaves retina
___ 3. forms white part of eye
___ 4. thalamic structure that processes visual input
___ 5. coloured diaphragm of muscle that controls amount of light entering eye
___ 6. contributes most to eye's refractive ability
___ 7. supplies nutrients to lens and cornea
___ 8. produces aqueous humour

11. Match the following:

(a) choroid
(b) aqueous humour

(c) fovea
(d) lateral geniculate nucleus
(e) cornea

(f) retina

(g) lens

___ 9. contains vascular supply for retina and a pigment that minimizes scattering of light within eye
___ 10. has adjustable refractive ability
___ 11. portion of retina with greatest acuity
___ 12. point at which fibres from medial half of each retina cross to opposite side

(h) optic disc; blind spot
(i) iris

(j) ciliary body

(k) optic chiasm

12. Using the answer code on the right, indicate which properties apply to taste and/or smell:
   a. applies to taste
   b. applies to smell
   c. applies to both taste and smell

___ 1. Receptors are separate cells that synapse with terminal endings of afferent neurons.
___ 2. Receptors are specialized endings of afferent neurons.
___ 3. Receptors are regularly replaced.
___ 4. Specific chemicals in the environment attach to special binding sites on the receptor surface, leading to a depolarizing receptor potential.
___ 5. There are two processing pathways: a limbic system route and a thalamic-cortical route.
___ 6. The discriminative ability is based on patterns of receptor stimulation by five different modalities.
___ 7. A thousand different receptor types are used.
___ 8. Information from receptor cells is filed and sorted by neural junctions called glomeruli.

13. Sympathetic preganglionic fibres begin in the thoracic and lumbar segments of the spinal cord. (*True or false?*)

14. Action potentials are transmitted on a one-to-one basis at both a neuromuscular junction and a synapse. (*True or false?*)

15. Which statement correctly describes the sympathetic nervous system?
   a. It is always excitatory.
   b. It innervates only tissues concerned with protecting the body against challenges from the outside environment.
   c. It has short preganglionic and long postganglionic fibres.
   d. It is part of the afferent division of the peripheral nervous system.
   e. It is part of the somatic nervous system.

16. Which statement correctly describes acetylcholinesterase?
    a. It is stored in vesicles in the terminal button.
    b. It combines with receptor sites on the motor end plate to bring about an end-plate potential.
    c. It is inhibited by organophosphates.
    d. It is the chemical transmitter at the neuromuscular junction.
    e. It paralyzes skeletal muscle by strongly binding with acetylcholine receptor sites.

17. The two divisions of the ANS are the _____ nervous system, which dominates in fight-or-flight situations, and the _____ nervous system, which dominates in rest-and-digest situations.

18. The _____ is a modified sympathetic ganglion that does not give rise to postganglionic fibres, but instead secretes hormones similar or identical to sympathetic post-ganglionic neurotransmitters into the blood.

19. Using the answer code on the right, identify the autonomic transmitter being described:

    ___ 1. secreted by all preganglionic fibres      (a) acetylcholine
    ___ 2. secreted by sympathetic postganglionic fibres      (b) norepinephrine
    ___ 3. secreted by parasympathetic postganglionic fibres
    ___ 4. secreted by the adrenal medulla
    ___ 5. secreted by motor neurons
    ___ 6. binds to muscarinic or nicotinic receptors
    ___ 7. binds to $\alpha$ or $\beta$ receptors

20. Using the answer code on the right, indicate which type of efferent output is being described:

    ___ 1. composed of two-neuron chains      (a) characteristic of the somatic nervous system
    ___ 2. innervates cardiac muscle, smooth muscle, and glands      (b) characteristic of the autonomic nervous system
    ___ 3. innervates skeletal muscle
    ___ 4. consists of the axons of motor neurons
    ___ 5. exerts either an excitatory or an inhibitory effect on its effector organs
    ___ 6. dually innervates its effector organs
    ___ 7. exerts only an excitatory effect on its effector organs

## Written Questions

1. List and describe the receptor types according to their adequate stimulus.

2. Compare tonic and phasic receptors.

3. Explain how acuity is influenced by receptive field size and by lateral inhibition.

4. Compare the fast and slow pain pathways.

5. Describe the built-in analgesic system of the brain.

6. Describe the process of phototransduction.

7. Compare the functional characteristics of rods and cones.

8. What are sound waves? What is responsible for the pitch, intensity, and timbre of a sound?

9. Describe the function of each of the following parts of the ear: pinna, ear canal, tympanic membrane, ossicles, oval window, and the various parts of the cochlea. Include a discussion of how sound waves are transduced into action potentials.

10. Discuss the functions of the semicircular canals, the utricle, and the saccule.

11. Describe the location, structure, and activation of the receptors for taste and smell.

12. Compare the processes of colour vision, hearing, taste, and smell discrimination.

13. Distinguish between preganglionic and postganglionic fibres.

14. Compare the origin, preganglionic and postganglionic fibre length, and neurotransmitters of the sympathetic and para-sympathetic nervous systems.

15. What is the advantage of dual innervation of many organs by both branches of the ANS?

16. Distinguish among the following types of receptors: nicotinic receptors, muscarinic receptors, $\alpha_1$ receptors, $\alpha_2$ receptors, $\beta_1$ receptors, and $\beta_2$ receptors.

17. What regions of the CNS regulate autonomic output?

18. Why are motor neurons called the "final common pathway"?

19. Describe the sequence of events that occurs at a neuromus-cular junction.

20. Discuss the effect each of the following has at the neuro-muscular junction: black widow spider venom, botulinum toxin, curare, myasthenia gravis, and organophosphates.

## Quantitative Exercises (Solutions in Appendix E, p. A-39)

1. Calculate the difference in the time it takes for an action potential to travel 1.3 m between the slow (12 m/sec) and fast (30 m/sec) pain pathways.

2. Have you ever noticed that humans have circular pupils, whereas cats' pupils are more elongated from top to bottom? For simplicity in calculation, assume the cat pupil is rectangular. The following calculations will help you understand the implication of this difference. For sim-plicity, assume a constant intensity of light.
    a. If the diameter of a human's circular pupil were decreased by half on contraction of the constrictor muscle of the iris, by what percentage would the amount of light allowed into the eye be decreased?
    b. If a cat's rectangular pupil were decreased by half along one axis only, by what percentage would the amount of light allowed into the eye be decreased?
    c. Comparing these calculations, do humans or cats have more precise control over the amount of light falling on the retina?

3. A decibel is the unit of sound level, β, defined as follows:

$$\beta = (10\ dB)\log_{10}\left(\frac{l}{l_0}\right),$$

where $l$ is *sound intensity*, or the rate at which sound waves transmit energy per unit area. The units of $l$ are watts per square metre $(W/m^2)$. $l_0$ is a constant intensity close to the human hearing threshold, namely, $10^{-12}\ W/m^2$.

a. For the following sound levels, calculate the corresponding sound intensities:

   (1)  20 dB (a whisper)

   (2)  70 dB (a car horn)

   (3)  120 dB (a low-flying jet)

   (4)  170 dB (a space shuttle launch)

b. Explain why the sound levels of these sounds increase by the same increment (i.e., each sound is 50 dB higher than the one preceding it), yet the incremental increases in sound intensities you calculated are so different. What implications does this have for performance of the human ear?

4. When a muscle fibre is activated at the neuromuscular junction, tension does not begin to rise until about 1 millisecond after initiation of the action potential in the muscle fibre. Many things occur during this delay, one time-consuming event being diffusion of acetylcholine across the neuromuscular junction. The following equation can be used to calculate how long this diffusion takes:

$$t = \frac{x^2}{2D}$$

In this equation, $x$ is the distance covered, $D$ is the diffusion coefficient, and $t$ is the time it takes for diffusion of the substance across the distance $x$. In this example, $x$ is the width of the cleft between the neuronal axon terminal and the muscle fibre at the neuromuscular junction (assume 200 nm), and $D$ is the diffusion coefficient of acetylcholine (assume $1 \times 10^{-5}\ cm^2/sec$). How long does it take the acetylcholine to diffuse across the neuromuscular junction?

## POINTS TO PONDER

### (Explanations in Appendix E, p. A-39)

1. Patients with certain nerve disorders are unable to feel pain. Why is this disadvantageous?

2. Ophthalmologists frequently instill eye drops in their patients' eyes to bring about pupillary dilation, which makes it easier for the physician to view the eye's interior. In what way would the drug in the eye drops affect ANS activity in the eye to cause the pupils to dilate?

3. A patient complains of not being able to see the right half of the visual field with either eye. At what point in the patient's visual pathway does the defect lie?

4. Explain how middle ear infections interfere with hearing. Of what value are the tubes that are sometimes surgically placed in the eardrums of patients with a history of repeated middle ear infections accompanied by chronic fluid accumulation?

5. Explain why your sense of smell is reduced when you have a cold, even though the cold virus does not directly adversely affect the olfactory receptor cells.

6. Explain why epinephrine, which causes arteriolar constriction (narrowing) in most tissues, is frequently administered in conjunction with local anesthetics.

7. Would skeletal muscle activity be affected by atropine? Why or why not?

8. Considering that you can voluntarily control the emptying of your urinary bladder by contracting (preventing emptying) or relaxing (permitting emptying) your external urethral sphincter, a ring of muscle that guards the exit from the bladder, of what type of muscle is this sphincter composed and what branch of the nervous system supplies it?

9. The venom of certain poisonous snakes contains α bungarotoxin, which binds tenaciously to acetylcholine receptor sites on the motor end-plate membrane. What would the resultant symptoms be?

10. Explain how destruction of motor neurons by poliovirus or amyotrophic lateral sclerosis can be fatal.

## CLINICAL CONSIDERATION

### (Explanation in Appendix E, p. A-40)

Suzanne J. complained to her physician of bouts of dizziness. The physician asked her whether by "dizziness" she meant a feeling of lightheadedness, as if she were going to faint (a condition known as *syncope*), or a feeling that she or surrounding objects in the room were spinning around (a condition known as *vertigo*). Why is this distinction important in the differential diagnosis of her condition? What are some possible causes of each of these symptoms?

# The Endocrine System

Body systems maintain homeostasis

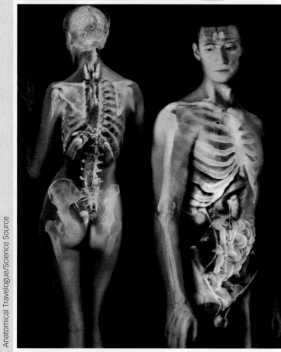

Anatomical Travelogue/Science Source

Endocrine system

## Homeostasis

The endocrine system, one of the body's two major regulatory systems, secretes hormones that act on their target cells to regulate the blood concentrations of nutrient molecules, water, salt, and other electrolytes, among other homeostatic activities. Hormones also play a key role in controlling growth and reproduction and in stress adaptation.

Homeostasis is essential for survival of cells

Cells make up body systems

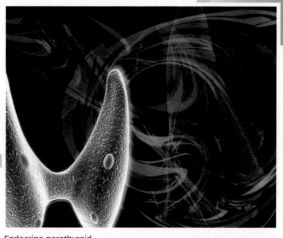

© O2creationz/Shutterstock

Endocrine parathyroid

The endocrine system regulates activities that require duration rather than speed. Endocrine glands release hormones: the blood-borne chemical messengers that act on target cells located a long distance from the endocrine gland. Most target-cell activities under hormonal control are directed toward maintaining homeostasis. The central endocrine glands, which are in or closely associated with the brain, include the hypothalamus, the pituitary gland, and the pineal gland. The hypothalamus (a part of the brain) and posterior pituitary gland act as a unit to release hormones essential for maintaining water balance, giving birth, and breastfeeding. The hypothalamus also secretes regulatory hormones that control the hormonal output of the anterior pituitary gland, which secretes six hormones that, in turn, largely control the hormonal output of several peripheral endocrine glands. One anterior pituitary hormone, growth hormone, promotes growth and influences nutrient homeostasis. The pineal gland is a part of the brain that secretes a hormone important in establishing the body's biological rhythms.

# 5

# Principles of Endocrinology: The Central Endocrine Glands

**❚ Clinical Connections**

Michael is a 13-year-old male of African Canadian decent. As a younger child Michael always seemed happy and healthy and engaged in school and social life typical for his age. However, as he started getting older, his childish face and chubby body stayed looking young. In contrast, his friends were growing taller and showing signs of puberty. Michael started to withdraw from his friends, stopped playing sports, and when not at school, spent all his time at home. Concerned for his well-being, his mother took him to their family physician.

At the physician's office, Michael said that he is teased for being the shortest boy in his gym class. When his height was measured and compared to the growth charts, his height showed up as the lowest for his age. An X-ray of his hand revealed the bone age of a 10-year-old boy. Blood tests taken to measure the various hormones involved in growth showed he had normal levels of the growth hormone–releasing hormone, very low levels of growth hormone, and low levels of insulin-like growth factor. Michael was diagnosed with pituitary dwarfism.

## 5.1 | General Principles of Endocrinology

The **endocrine system** consists of the ductless endocrine glands (p. 7) scattered throughout the body (❭ Figure 5-1). Even though the endocrine glands for the most part are not connected anatomically, they constitute a system in a functional sense. They all accomplish their functions by secreting hormones into the blood, and many functional

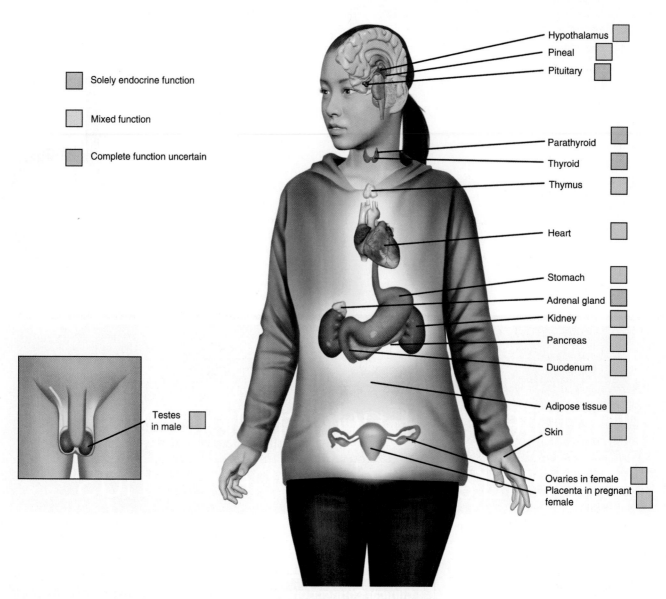

Solely endocrine function

Mixed function

Complete function uncertain

Hypothalamus
Pineal
Pituitary

Parathyroid
Thyroid
Thymus

Heart

Stomach
Adrenal gland
Kidney
Pancreas
Duodenum

Adipose tissue

Skin

Ovaries in female
Placenta in pregnant female

Testes in male

> FIGURE 5-1 **The endocrine system**

interactions take place among the various endocrine glands. Once secreted, a hormone travels in the blood to its distant target cells, where it regulates or directs a particular function. **Endocrinology** is the study of the homeostatic chemical adjustments and other activities that hormones accomplish.

A hormone is a chemical substance that is secreted in low quantities into the blood by a cell or grouping of cells and exerts a physiological effect on specific target tissue (or cells). Although the blood distributes hormones throughout the body, only specific target cells can respond to each hormone, because only the target cells have receptors for binding with the particular hormone (p. 32). The binding of a hormone with its specific target-cell receptors initiates a chain of events within the target cells to bring about the hormone's final effect. Pages 217–221 explain in detail how most hormones initiate their effect by first combining with a receptor that is located either on the cell membrane or inside the cell. A cell may have several different receptors that recognize the same hormone and activate different signalling pathways, or, in contrast, different hormones and their receptors may summon the same biochemical pathway.

## Hormones

The endocrine system is one of the body's two major regulatory systems; the other is the nervous system (Chapters 3 and 4). Recall that the nervous system is specialized for controlling different types of activities. In general, the nervous system coordinates rapid, precise responses and is especially important in mediating the body's interactions with the external environment. The endocrine system, by contrast, primarily controls activities that require duration rather than speed. It regulates, coordinates, and integrates cellular and organ function at a distance.

### OVERALL FUNCTIONS OF THE ENDOCRINE SYSTEM

1. Regulating organic metabolism and water and electrolyte balance, which are important collectively in maintaining a constant internal environment

2. Inducing adaptive changes to help the body cope with stressful situations

3. Promoting smooth, sequential growth and development

4. Controlling reproduction

5. Regulating red blood cell production

6. Along with the autonomic nervous system, controlling and integrating both circulation and the digestion and absorption of food

## TROPIC HORMONES

Some hormones regulate the production and secretion of another hormone. A hormone that has as its primary function the regulation of hormone secretion by another endocrine gland is classified functionally as a **tropic hormone** (*tropic* means "nourishing"). Tropic hormones stimulate and maintain their endocrine target tissues. For example, the tropic hormone thyroid-stimulating hormone (TSH), from the anterior pituitary, stimulates thyroid hormone secretion by the thyroid gland and also maintains the structural integrity of this gland. In the absence of TSH, the thyroid gland atrophies (shrinks) and produces very low levels of its hormone.

## COMPLEXITY OF ENDOCRINE FUNCTION

The following factors add to the complexity of the system:

- A single endocrine gland may produce multiple hormones. For example, the thyroid gland produces thyroxine, triiodothyronine, and calcitonin.

- A single hormone may be secreted by more than one endocrine gland. For example, the hypothalamus, pancreas, and stomach all secrete the hormone somatostatin, and the stomach also releases somatostatin that acts as a paracrine (a chemical released to adjoining cells or surrounding tissues).

- Frequently, a single hormone has more than one type of target cell and therefore can induce more than one type of effect. An example of this is oxytocin, which contracts the uterus during birth and also acts on the myoepithelial cells in the breasts to assist in expelling milk when the baby suckles.

- The rate of secretion of some hormones varies considerably over time in a cyclic pattern. Therefore, endocrine systems also provide temporal (time) coordination of function. This is particularly apparent in endocrine control of reproductive cycles, such as the menstrual cycle, in which normal functioning requires highly specific patterns of change in the secretion of various hormones.

- A single target cell may be influenced by more than one hormone. Some cells contain an array of receptors for responding in different ways to different hormones. To illustrate, insulin promotes the conversion of glucose into glycogen within liver cells by stimulating one particular hepatic enzyme; whereas another hormone, glucagon, by activating yet another hepatic enzyme, enhances the degradation of glycogen into glucose within liver cells.

- The same chemical messenger may be either a hormone or a neurotransmitter, depending on its source and mode of delivery to the target cell. Norepinephrine, which is secreted as a hormone by the adrenal medulla and released as a neurotransmitter from sympathetic postganglionic nerve fibres, is a prime example.

- Some organs are exclusively endocrine in function (they specialize in hormone secretion alone, the anterior pituitary being an example), whereas other organs of the endocrine system perform nonendocrine functions in addition to secreting hormones. For example, the testes produce sperm and also secrete the male sex hormone testosterone.

## Plasma concentration

The primary function of most hormones is the regulation of various homeostatic activities. Because hormones' effects are proportional to their concentrations in the plasma, these concentrations are subject to control according to homeostatic need. The effective plasma concentration of a free, biologically active hormone—and thus the hormone's availability to its receptors—depends on several factors: (1) the hormone's rate of secretion into the blood by the endocrine gland; (2) for a few hormones, its rate of metabolic activation; (3) for lipophilic hormones, its extent of binding to plasma proteins; and (4) its rate of removal from the blood by metabolic inactivation and excretion in the urine. Furthermore, the magnitude of the hormonal response depends on the availability and sensitivity of the target cells' receptors for the hormone. However, the quantity of hormones (concentration) in the blood is small: ranging from 1 picogram (1 millionth of a millionth of a gram) per millilitre of blood to a few micrograms (1 millionth of a gram) per millilitre of blood. The rate of secretion, also small, is typically measured in micrograms or milligrams per day.

We first examine the factors that influence the plasma concentration of the hormone and then turn our attention to the target cells' responsiveness to the hormone.

Normally, the effective plasma concentration of a hormone is regulated by appropriate adjustments in the rate of its secretion. Endocrine glands do not secrete their hormones at a constant rate; the secretion rates of all hormones vary, subject to control often by a combination of several complex mechanisms. The regulatory system for each hormone is considered in detail in later sections. For now, we will address the general mechanisms of controlling secretion that are common to many different hormones: negative-feedback control, neuroendocrine reflexes, and diurnal (circadian) rhythms.

## NEGATIVE-FEEDBACK CONTROL

Negative feedback is a prominent feature of hormonal control systems. Stated simply, *negative feedback exists when the output of a system counteracts a change in input*, thereby maintaining a controlled variable within a narrow range around a set level (p. 14). Negative feedback maintains the plasma concentration of a hormone at a given level, similar to the way a home heating system maintains the room temperature at a given set point. Control of hormonal secretion provides some classic physiological examples of negative feedback. For example, when the plasma concentration of free circulating thyroid hormone falls below a given set point, the anterior pituitary secretes thyroid-stimulating hormone (TSH), which stimulates the thyroid to increase its secretion of thyroid hormone (› Figure 5-2). Thyroid hormone, in turn, inhibits further secretion of TSH by the anterior pituitary. Negative feedback ensures that once thyroid gland secretion has been "turned on" by TSH, it will not continue unabated, but will be "turned off" when the appropriate level of free circulating thyroid hormone has been achieved. Thus, the effect of a particular hormone's actions can inhibit its own secretion. The feedback loops

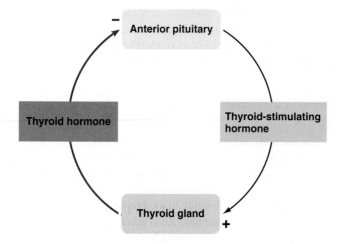

> FIGURE 5-2 **Negative-feedback control**

often become quite complex to the extent that several endocrine tissues and hormones, or other messengers, may be involved.

Although the majority of endocrine feedback systems are negative, it is important to mention that some feedback loops are **positive-feedback** loops. Simplistically, positive feedback means that the actions of a hormone cause the further release of the hormone. Such mechanisms tend to be self-perpetuating and out of control, which is contrary to the concept of homeostasis; this is why such systems are not common. An excellent example of a positive-feedback loop is the role of oxytocin in child birth (parturition) (see Figure 17-30).

### NEUROENDOCRINE REFLEXES

Many endocrine control systems involve **neuroendocrine reflexes**, which include neural as well as hormonal components. The purpose of such reflexes is to produce a sudden increase in hormone secretion (i.e., "turn up the thermostat setting") in response to a specific stimulus, frequently a stimulus external to the body. In some instances, neural input to the endocrine gland is the only factor regulating secretion of the hormone. For example, secretion of epinephrine by the adrenal medulla is solely controlled by the sympathetic nervous system. Some endocrine control systems, in contrast, include both feedback control (which maintains a constant basal level of the hormone) and neuroendocrine reflexes (which cause sudden bursts in secretion in response to a sudden increased need for the hormone). An example is the increased secretion of cortisol, the "stress hormone," by the adrenal cortex during a stress response (see Figure 6-8).

### DIURNAL (CIRCADIAN) RHYTHMS

The secretion rates of many hormones rhythmically fluctuate up and down as a function of time. The most common endocrine rhythm is the **diurnal (circadian) rhythm** (*diurnal* means "day–night"; *circadian* means "around a day"), which is characterized by repetitive oscillations in hormone levels that are very regular and cycle once every 24 hours. This rhythmicity is caused by endogenous oscillators similar to the self-paced respiratory neurons in the brain stem that control the rhythmic motions of breathing, except that the time-keeping oscillators cycle on a much longer time scale. Furthermore, unlike the rhythmicity of breathing, endocrine rhythms are locked

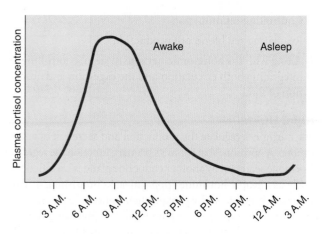

> FIGURE 5-3 **Diurnal rhythm of cortisol secretion**

on, or **entrained**, to external cues, such as the light–dark cycle. That is, the inherent 24-hour cycles of peak and ebb of hormone secretion are set to "march in step" with cycles of light (period of activity) and dark (period of inactivity). For example, cortisol secretion rises during the night, reaches its peak secretion in the morning before a person gets up, then falls throughout the day to its lowest level at bedtime (> Figure 5-3). Inherent hormonal rhythmicity and entrainment are not accomplished by the endocrine glands themselves but result from the central nervous system changing the set point of these glands. We discuss the master biological clock further in Section 5.7. Negative-feedback control mechanisms operate to maintain whatever set point is established for that time of day. Some endocrine cycles operate on time scales other than a circadian rhythm; a well-known example is the monthly menstrual cycle.

## Transport, metabolism, and excretion

Even though the effective plasma concentration of a hormone is normally regulated by adjusting its rate of secretion, alterations in its transport, metabolism, or excretion can also influence the hormone's plasma concentration, sometimes inappropriately (Appendix F lists reference values). For example, because the liver synthesizes plasma proteins, liver disease may result in abnormal endocrine activity by altering the balance between free and bound pools of lipophilic hormones.

Eventually, all hormones are metabolized by enzyme-mediated reactions that modify the hormonal structure in some way. In most cases, this process *inactivates* the hormone. However, in some cases, metabolism *activates* a hormone; that is, the hormone's product has greater activity than the original hormone. For example, after the thyroid hormone thyroxine is secreted, it is converted to a more powerful hormone by enzymatic removal of one of its iodine atoms. Usually the rate of such hormone activation is itself under hormonal control. The liver is the most common site for metabolic hormone inactivation, but some hormones are also inactivated in the blood, kidneys, or target cells.

Hormones and their metabolites are typically eliminated from the blood by urinary excretion. In contrast to the tight controls on hormone secretion, hormone inactivation and excretion are not regulated.

The amount of time after a hormone is secreted before it is inactivated, and the means by which this takes place, differ for different classes of hormones. In general, the hydrophilic peptides and catecholamines are easy targets for blood and tissue enzymes, so they remain in the blood only briefly (a few minutes to a few hours) before being enzymatically inactivated. In the case of some peptide hormones, such as insulin, the target cell actually engulfs the bound hormone by endocytosis and degrades it intracellularly. In contrast, binding of lipophilic hormones to plasma proteins makes them less vulnerable to metabolic inactivation and keeps them from escaping into urine. Therefore, lipophilic hormones are removed from plasma much more slowly. They may persist in the blood for hours (steroids) or up to a week (thyroid hormone). Lipophilic hormones typically undergo a series of reactions that reduce their biological activity and make them more water soluble so they can be freed from their plasma protein carriers and be eliminated in the urine.

*Clinical Note* When liver and kidney function are normal, the measuring of urinary concentrations of hormones and their metabolites provides a useful, noninvasive way to assess endocrine function, because the rate of excretion of these products in the urine directly reflects their rate of secretion by the endocrine glands. Because the liver and kidneys are important in removing hormones from the blood, patients with liver or kidney disease may suffer from excess activity of certain hormones solely because hormone elimination is reduced.

## Endocrine disorders

*Clinical Note* Abnormalities in a hormone's effective plasma concentration can arise from a variety of factors (▌Table 5-1). Endocrine disorders most commonly result from abnormal plasma concentrations of a hormone caused by inappropriate rates of secretion—that is, too little hormone secreted (**hyposecretion**) or too much hormone secreted (**hypersecretion**). Occasionally, endocrine dysfunction arises because target-cell responsiveness to the hormone is abnormally low, even though plasma concentration of the hormone is normal.

▌ **TABLE 5-1** Means by Which Endocrine Disorders Can Arise

| Too Little Hormone Activity | Too Much Hormone Activity |
|---|---|
| Too little hormone secreted by the endocrine gland (hyposecretion)* | Too much hormone secreted by the endocrine gland (hypersecretion)* |
| Increased removal of the hormone from the blood | Reduced plasma protein binding of the hormone (too much free, biologically active hormone) |
| Abnormal tissue responsiveness to the hormone | Decreased removal of the hormone from the blood |
| Lack of target-cell receptors | Decreased inactivation |
| Lack of an enzyme essential to the target-cell response | Decreased excretion |

© 2016 Cengage

* Most common causes of endocrine dysfunction

### HYPOSECRETION

*Primary hyposecretion* occurs when an endocrine gland is secreting too little of its hormone because of an abnormality within that gland. *Secondary hyposecretion* takes place when an endocrine gland is functioning normally but is secreting too little hormone because of a deficiency of its tropic hormone.

The following are among the many different factors (each listed with an example) that may cause primary hyposecretion: (1) genetic (inborn absence of an enzyme that catalyzes synthesis of the hormone); (2) dietary (lack of iodine, which is needed for synthesis of thyroid hormone); (3) chemical or toxic (certain insecticide residues may destroy the adrenal cortex); (4) immunological (autoimmune antibodies may destroy the body's own thyroid tissue); (5) other disease processes (cancer or tuberculosis may coincidentally destroy endocrine glands); (6) *iatrogenic* (physician induced, such as surgical removal of a cancerous thyroid gland); and (7) *idiopathic* (meaning the cause is not known).

The most common method of treating hormone hyposecretion is to administer a hormone that is the same as (or similar to, such as from another species) the deficient or missing one. Such replacement therapy seems straightforward, but the hormone's source and means of administration involve some practical problems. The sources of hormone preparation for clinical use include (1) endocrine tissues from domestic livestock, (2) placental tissue and urine of pregnant women, (3) laboratory synthesis of hormones, and (4) "hormone factories," or bacteria into which genes coding for the production of human hormones have been introduced. The method of choice for a given hormone is determined largely by its structural complexity and degree of species specificity.

### HYPERSECRETION

Like *hypo*secretion, *hyper*secretion by a particular endocrine gland is designated as primary or secondary depending on whether the defect lies in that gland or is due to excessive stimulation from the outside, respectively. Hypersecretion may be caused by (1) tumours that ignore the normal regulatory input and continuously secrete excess hormone and (2) immunological factors, such as excessive stimulation of the thyroid gland by an abnormal antibody that mimics the action of TSH, the thyroid tropic hormone. Excessive levels of a particular hormone may also arise from substance abuse, such as the outlawed practice among athletes of using certain steroids that increase muscle mass by promoting protein synthesis in muscle cells.

There are several ways of treating hormonal hypersecretion. If a tumour is the culprit, it may be surgically removed or destroyed with radiation treatment. In some instances, hypersecretion can be limited by drugs that block hormone synthesis or inhibit hormone secretion. Sometimes the condition may be treated by giving drugs that inhibit the action of the hormone without actually reducing the excess hormone secretion.

### ABNORMAL TARGET-CELL RESPONSIVENESS

Endocrine dysfunction can also occur because target cells do not respond adequately to the hormone, even though the effective plasma concentration of a hormone is normal. This unresponsiveness may be caused, for example, by an inborn lack of receptors for the hormone, as in *testicular feminization syndrome*. In this condition, receptors for testosterone, a masculinizing hormone produced by the male testes, are not produced because of a specific genetic defect. Although adequate testosterone is available, masculinization does not take place, just as if no testosterone were present. Abnormal responsiveness may also occur if the target cells for a particular hormone lack an enzyme essential to carrying out the response.

## Target cells

In contrast to endocrine dysfunction caused by *unintentional* receptor abnormalities, the target-cell receptors for a particular hormone can be *deliberately altered* as a result of physiological control mechanisms. A target cell's response to a hormone is correlated with the number of the cell's receptors occupied by molecules of that hormone, which in turn depends not only on the plasma concentration of the hormone but also on the number of receptors in the target cell for that hormone. Thus, the response of a target cell to a given plasma concentration of hormone can be fine-tuned up or down by varying the number of receptors available for hormone binding. The number of receptors in a target cell does not remain constant from minute to minute because receptors are destroyed and new ones are manufactured by the cell.

### DOWN REGULATION

As an illustration of this fine-tuning, when the plasma concentration of insulin is chronically elevated, the total number of target-cell receptors for insulin is reduced as a direct result of the effect an elevated level of insulin has on the insulin receptors. This phenomenon, known as **down regulation**, constitutes an important locally acting negative-feedback mechanism that prevents the target cells from overreacting to the high concentration of insulin; that is, the target cells are *desensitized* to insulin, helping blunt the effect of insulin hypersecretion.

Down regulation of the insulin receptor is accomplished by the following mechanism. The binding of insulin to its surface receptors induces endocytosis of the hormone-receptor complex, which is subsequently attacked by intracellular lysosomal enzymes. This internalization serves a twofold purpose: it provides a pathway for degrading the hormone and helps regulate the number of receptors available for binding on the target cell's surface. At high plasma insulin concentrations, the number of surface receptors for insulin is gradually reduced by the accelerated rate of receptor internalization and degradation, brought about by increased hormonal binding. The rate of synthesis of new receptors within the endoplasmic reticulum and their insertion in the plasma membrane do not keep pace with their rate of destruction. Over time, this self-induced loss of target-cell receptors for insulin reduces the target cell's sensitivity to the elevated hormone concentration.

### PERMISSIVENESS, SYNERGISM, AND ANTAGONISM

A given hormone's effects are influenced not only by the concentration of the hormone itself but also by the concentrations of other hormones that interact with it. Because hormones are widely distributed through the blood, target cells may be simultaneously exposed to many different hormones, giving rise to numerous complex hormonal interactions on target cells. Hormones frequently alter the receptors for other kinds of hormones as part of their normal physiological activity. A hormone can influence the activity of another hormone at a given target cell in one of three ways: permissiveness, synergism, and antagonism.

- With **permissiveness**, one hormone must be present in adequate amounts for the full exertion of another hormone's effect. In essence, the first hormone, by enhancing a target cell's responsiveness to another hormone, "permits" this other hormone to exert its full effect. For example, thyroid hormone increases the number of receptors for epinephrine in epinephrine's target cells, increasing the effectiveness of epinephrine. In the absence of thyroid hormone, epinephrine is only marginally effective.

- **Synergism** occurs when the actions of several hormones are complementary and their combined effect is greater than the sum of their separate effects. An example is the synergistic action of follicle-stimulating hormone and testosterone, both of which are required for maintaining the normal rate of sperm production. Synergism results from each hormone's influence on the number or affinity of receptors for the other hormone.

- **Antagonism** occurs when hormones have opposing effects. For example, in regulating plasma calcium ($Ca^{2+}$) concentrations, the hormone calcitonin promotes the incorporation of $Ca^{2+}$ into bone, whereas the hormone parathyroid hormone stimulates the release of $Ca^{2+}$ from bone. Another example of antagonism is the actions of insulin and glucagon on regulating both cellular metabolism and plasma glucose concentrations.

## 5.2 | Principles of Hormonal Communication

### Hydrophilic or lipophilic

Hormones are not all similar chemically but instead fall into two distinct groups based on their solubility properties: hydrophilic or lipophilic hormones. Hormones within each group are further classified according to their biochemical structure and/or source as follows (Table 5-2):

1. **Hydrophilic hormones** (water-loving) are highly water soluble and have low lipid solubility. Most hydrophilic hormones are peptide or protein hormones consisting of specific amino acids arranged in a chain of varying length. The shorter chains are peptides, and the longer ones are proteins. For convenience, we will refer to this entire category as *peptides*. An example is insulin from the pancreas. Another group of hydrophilic hormones are the *catecholamines*, which are derived from the amino acid tyrosine and are specifically secreted by the adrenal medulla. The adrenal gland consists of an inner adrenal medulla surrounded by an outer adrenal cortex. (You will learn more about the location and structure of

## ❚ TABLE 5-2 Chemical Classification of Hormones

| | | AMINES | | |
|---|---|---|---|---|
| **Properties** | **Peptides** | **Catecholamines** | **Thyroid Hormone** | **Steroids** |
| **Solubility** | Hydrophilic | Hydrophilic | Lipophilic | Lipophilic |
| **Structure** | Chains of specific amino acids | Tyrosine derivative | Iodinated tyrosine derivative | Cholesterol derivative |
| **Synthesis** | In rough endoplasmic reticulum; packaged in Golgi complex | In cytosol | In colloid, an inland extracellular site | Stepwise modification of cholesterol molecule in various intracellular compartments |
| **Storage** | Large amounts in secretory granules | In chromaffin granules | In colloid | Not stored; cholesterol precursor stored in lipid droplets |
| **Secretion** | Exocytosis of granules | Exocytosis of granules | Endocytosis of colloid | Simple diffusion |
| **Transport in Blood** | As free hormone | Half bound to plasma proteins | Mostly bound to plasma proteins | Mostly bound to plasma proteins |
| **Receptor Site** | Surface of target cell | Surface of target cell | Inside target cell | Inside target cell |
| **Mechanism of Action** | Channel changes or activation of second-messenger system to alter activity of target proteins that produce the effect | Activation of second-messenger system to alter activity of target proteins that produce the effect | Activation of specific genes to make new proteins that produce the effect | Activation of specific genes to make new proteins that produce the effect |
| **Hormones of this Type** | Majority of hormones | Only hormones from the adrenal medulla | Only hormones from the thyroid follicular cells | Hormones from the adrenal cortex and gonads |



the endocrine glands in Chapter 6.) Epinephrine is the primary catecholamine secreted by the adrenal medulla.

2. **Lipophilic hormones** (lipid-loving) have high lipid solubility and are poorly soluble in water. Lipophilic hormones include *thyroid hormone* and the *steroid hormones.* Thyroid hormone, as its name implies, is secreted exclusively by the thyroid gland. Thyroid hormone is an iodinated tyrosine derivative. Even though catecholamines and thyroid hormone behave very differently, they are sometimes grouped together as *amine hormones* because of their common tyrosine derivation. Steroids are neutral lipids derived from cholesterol. The hormones secreted by the adrenal cortex, such as cortisol, and the sex hormones (testosterone in males and estrogen in females) secreted by the reproductive organs are all steroids.

Minor differences in chemical structure between hormones within each category often result in profound differences in biological response. Comparing two steroid hormones in › Figure 5-4, for example, note the subtle difference between testosterone, the male sex hormone responsible for inducing the development of masculine characteristics, and estradiol, a form of estrogen, which is the feminizing female sex hormone.

The solubility properties of a hormone determine the means by which (1) the hormone is processed by the endocrine cell,

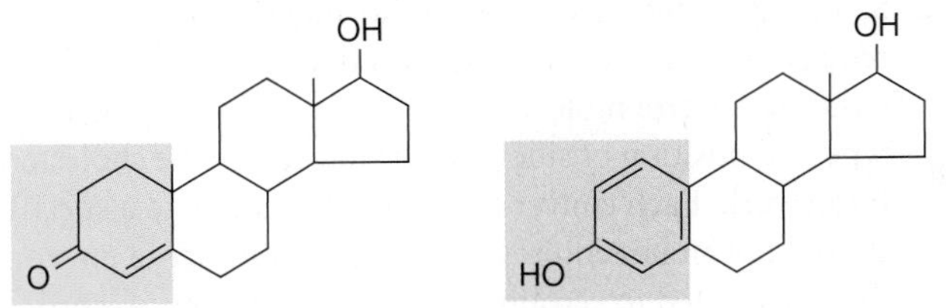


Testosterone, a masculinizing hormone

Estradiol, a feminizing hormone

› **FIGURE 5-4 Comparison of two steroid hormones, testosterone and estradiol.**

(2) the way the hormone is transported in the blood, and (3) the mechanism by which the hormone exerts its effects at the target cell. First we consider the different ways these hormone types are processed at their site of origin—the endocrine cell—and then we compare their means of transport and their mechanisms of action.

## The mechanisms of synthesis, storage, and secretion

Because of their chemical differences, the means by which the various types of hormones are synthesized, stored, and secreted differ.

### PROCESSING OF HYDROPHILIC PEPTIDE HORMONES

Peptide hormones are synthesized and secreted by the same steps used for manufacturing any protein that is exported from the cell. From the time peptide hormones are synthesized until they are secreted, they are always segregated from intracellular proteins through containment in membrane-enclosed compartments. Here is a brief overview of the steps:

1. Preprohormones, or precursor proteins, are synthesized by ribosomes on the rough endoplasmic reticulum (ER). They then migrate to the Golgi complex in membrane-enclosed vesicles that pinch off from the smooth ER.

2. During their journey through the ER and Golgi complex, the large preprohormone precursor molecules are processed to become active hormones.

3. The Golgi complex then packages the finished hormones into secretory vesicles that are pinched off and stored in the cytoplasm until an appropriate signal triggers their secretion.

4. On appropriate stimulation, the secretory vesicles fuse with the plasma membrane and release their contents to the outside by exocytosis (p. 49). Such secretion usually does not go on continuously; it is triggered only by specific stimuli. The blood then picks up the secreted hormone for distribution.

### PROCESSING OF LIPOPHILIC STEROID HORMONES

All steroidogenic (steroid-producing) cells perform the following steps to produce and release their hormonal product:

1. Cholesterol is the common precursor for all steroid hormones.

2. Synthesis of the various steroid hormones from cholesterol requires a series of enzymatic reactions that modify the basic cholesterol molecule—for example, by varying the type and position of side groups attached to the cholesterol framework. Each conversion from cholesterol to a specific steroid hormone requires the help of a number of enzymes that are limited to certain steroidogenic organs. Accordingly, each steroidogenic organ can produce only the steroid hormone or hormones for which it has a complete set of appropriate enzymes. For example, a key enzyme necessary for producing cortisol is found only in the adrenal cortex, so no other steroidogenic organ can produce this hormone.

3. Unlike peptide hormones, steroid hormones are not stored. Once formed, the lipid-soluble steroid hormones immediately diffuse through the steroidogenic cell's lipid plasma membrane to enter the blood. Only the hormone precursor cholesterol is stored in significant quantities within steroidogenic cells. Accordingly, the rate of steroid hormone secretion is controlled entirely by the rate of hormone synthesis. In contrast, peptide hormone secretion is controlled primarily by regulating the release of presynthesized stored hormone.

The adrenomedullary catecholamines and thyroid hormone have unique synthetic and secretory pathways; these will be described in Chapter 6.

## Dissolving of hydrophilic hormones; transporting of lipophilic hormones

All hormones are carried by the blood, but they are not all transported in the same manner:

- The hydrophilic peptide hormones are transported simply dissolved in the plasma.

- Lipophilic steroids and thyroid hormones cannot dissolve to any extent in the watery plasma; instead, the majority circulate in the blood, bound to plasma proteins. Some are bound to specific plasma proteins designed to carry only one type of hormone, whereas other plasma proteins, such as albumin, indiscriminately pick up any "hitchhiking" hormone.

- Only the small, unbound, freely dissolved fraction of a lipophilic hormone is biologically active (i.e., free to cross capillary walls and bind with target-cell receptors to exert an effect). Once a hormone has interacted with a target cell, it is rapidly inactivated or removed, so it is no longer available to interact with another target cell. Because the carrier-bound hormone is in dynamic equilibrium with the free hormone pool, the bound form of steroid and thyroid hormones provides a large reserve of these lipophilic hormones that can be called on to replenish the active free pool. To maintain normal endocrine function, the magnitude of the small, free, effective pool, rather than the total plasma concentration of a particular lipophilic hormone, is monitored and adjusted.

- Catecholamines are unusual in that only about 50 percent of these hydrophilic hormones circulate as free hormone; the other 50 percent are loosely bound to the plasma protein albumin. Because catecholamines are water soluble, the importance of this protein binding is unclear.

*Clinical Note* The chemical properties of a hormone dictate not only the means by which blood transports it but also how it can be artificially introduced into the blood for therapeutic purposes. Because the digestive system does not secrete enzymes that can digest steroid and thyroid hormones, when taken orally, these hormones, such as the sex steroids contained in birth control pills, can be absorbed intact from the digestive tract into the blood. No other type of hormones can be taken orally, because protein-digesting enzymes would attack and convert them into inactive fragments. Therefore, these hormones must be administered by non-oral routes; for example, insulin deficiency is treated with daily injections of insulin.

We now examine how the hydrophilic and lipophilic hormones vary in their mechanisms of action at their target cells.

## Hormones and intracellular proteins

To induce their effect, hormones must bind with target-cell receptors specific for them. Each interaction between a particular hormone and a target-cell receptor produces a highly characteristic response that differs among hormones and among different target cells influenced by the same hormone. Both the location of the receptors within the target cell and the mechanism by which binding of the hormone with the receptors induces a response vary, depending on the hormone's solubility characteristics.

### LOCATION OF RECEPTORS FOR HYDROPHILIC AND LIPOPHILIC HORMONES

Hormones can be grouped into two categories based on the primary location of their receptors:

1. The hydrophilic peptides and catecholamines, which are poorly soluble in lipid, cannot pass through the lipid membrane barriers of their target cells. Instead, they bind with specific receptors located on the *outer plasma membrane surface* of the target cell.

2. The lipophilic steroids and thyroid hormone easily pass through the surface membrane to bind with specific receptors located *inside* the target cell.

### GENERAL MEANS OF HYDROPHILIC AND LIPOPHILIC HORMONE ACTION

Even though hormones elicit a wide variety of biological responses, all hormones ultimately influence their target cells by altering the cell's proteins through three general means:

1. A few hydrophilic hormones, on binding with a target cell's surface receptors, *change the cell's permeability* (either opening or closing channels to one or more ions) by *altering the conformation (shape) of adjacent channel-forming proteins already in the membrane.*

2. Most surface-binding hydrophilic hormones function by *activating second-messenger systems* within the target cell. This activation directly *alters the activity of intracellular target proteins, usually enzymes,* to produce the desired effect.

3. All lipophilic hormones function mainly by *activating specific genes* in the target cell *to cause formation of new intracellular proteins,* which in turn produce the desired effect. The new proteins may be enzymatic or structural.

Let's examine the two major mechanisms of hormonal action (activation of second-messenger systems and activation of genes) in more detail.

## Hydrophilic hormones and target proteins

Most hydrophilic hormones (peptides and catecholamines) bind to surface membrane receptors and produce their effects in their target cells by acting through a second-messenger system to alter the activity of preexisting target proteins. There are two major second-messenger pathways: One uses **cyclic adenosine monophosphate (cyclic AMP or cAMP)** as a second messenger, and the other employs calcium in this role.

### CYCLIC AMP SECOND-MESSENGER PATHWAY

In the following description of the cAMP pathway, the numbered steps correlate to the numbered steps in › Figure 5-5.

1. Binding of an appropriate extracellular messenger (a first messenger) to its surface membrane receptor eventually activates the enzyme **adenylyl cyclase** (step1), which is located on the cytoplasmic side of the plasma membrane. A membrane-bound "middleman," a **G protein**, acts as an intermediary between the receptor and adenylyl cyclase. G proteins are found on the inner surface of the plasma membrane. An unactivated G protein consists of a complex of alpha ($\alpha$), beta ($\beta$), and gamma ($\gamma$) subunits. A number of different G proteins employing varying subtypes of these subunits have been identified. The different G proteins are activated in response to the binding of various first messengers to surface receptors. When a first messenger binds with its receptor, the receptor attaches to the appropriate G protein, resulting in activation of the $\alpha$ subunit. Once activated, the $\alpha$ subunit breaks away from the G protein complex and moves along the inner surface of the plasma membrane until it reaches an effector protein. An effector protein is either an ion channel or an enzyme within the membrane. The $\alpha$ subunit links up with the effector protein and alters its activity. In the cAMP pathway, adenylyl cyclase is the effector protein that is activated. Researchers have identified more than 300 different receptors that convey instructions of extracellular messengers through the membrane to effector proteins by means of G proteins.

2. Adenylyl cyclase induces the conversion of intracellular ATP to cAMP by cleaving off two of the phosphates (step 2). (This is the same ATP used as the common energy currency in the body.)

3. Acting as the intracellular second messenger, cAMP triggers a preprogrammed series of biochemical steps within the cell to bring about the response dictated by the first messenger. To begin, cyclic AMP activates a specific intracellular enzyme, **protein kinase A** (step 3).

4. Protein kinase A, in turn, phosphorylates (attaches a phosphate group from ATP) a specific intracellular target

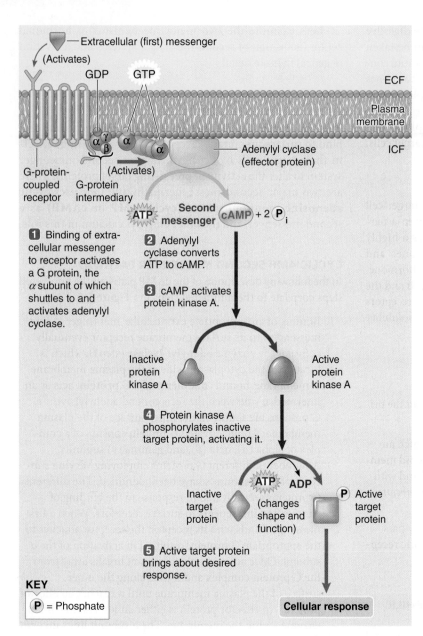

**KEY**

$\boxed{\text{P}}$ = Phosphate

1. Binding of extra-cellular messenger to receptor activates a G protein, the α subunit of which shuttles to and activates adenylyl cyclase.

2. Adenylyl cyclase converts ATP to cAMP.

3. cAMP activates protein kinase A.

4. Protein kinase A phosphorylates inactive target protein, activating it.

5. Active target protein brings about desired response.

> FIGURE 5-5 **Mechanism of action of hydrophilic hormones via activation of the cyclic AMP second-messenger system**

Note that in this signal-transduction pathway, the steps involving the extracellular first messenger, the receptor, the G protein complex, and the effector protein occur *in the plasma membrane* and lead to activation of the second messenger. The extracellular messenger cannot gain entry into the cell to "personally" deliver its message to the proteins that carry out the desired response. Instead, it initiates membrane events that activate an intracellular messenger, cAMP. This second messenger then triggers a chain reaction of biochemical events *inside the cell* that leads to the cellular response.

Different types of cells have different pre-existing proteins available for phosphorylation and modification by protein kinase A. Therefore, a *common second messenger, cAMP, can induce widely differing responses in different cells,* depending on what proteins are modified. Cyclic AMP may be viewed as an on-off switch used by many different cells, triggering different cell events depending on the type of protein activity in the target cell. The type of proteins altered by a second messenger depends on the unique specialization of a particular cell type. This can be likened to being able to either illuminate or cool down a room depending on whether the wall switch you flip on is wired to a device specialized to light up (a chandelier) or one specialized to create air movement (a ceiling fan). In the body, the variable responsiveness once the switch is turned on is due to the genetically programmed differences in the sets of proteins within different cells. For example, activating the cAMP system brings about modification of heart rate in the heart, stimulation of the formation of female sex hormones in the ovaries, breakdown of stored glucose in the liver, control of water conservation during urine formation in the kidneys, creation of some simple memory traces in the brain, and perception of a sweet taste by a taste bud.

protein (step 4), such as an enzyme important in a particular metabolic pathway.

5. Phosphorylation causes the protein to change its shape and function (either activating or inhibiting it) (step 5).

The resultant change is the target cell's ultimate physiological response to the first messenger. For example, the activity of a particular enzymatic protein that regulates a specific metabolic event may be increased or decreased.

After the response is accomplished and the first messenger is removed, the α subunit rejoins the β and γ subunits to restore the inactive G protein complex. Cyclic AMP and the other participating chemicals are inactivated so that the intracellular message is "erased" and the response can be terminated. Otherwise, once triggered, the response would go on indefinitely until the cell ran out of necessary supplies.

**Check Your Understanding 5.1**

1. Compare and contrast the synthesis, storage, secretion, and transport in the blood of peptide hormones and steroid hormones.

2. Explain how a common second messenger such as cAMP can induce widely differing responses in different cells.

**CALCIUM SECOND-MESSENGER SYSTEM**

Some cells use calcium instead of cAMP as a second messenger. In such cases, binding of the first messenger to the surface receptor eventually leads by means of G proteins to activation of the enzyme **phospholipase C**, a protein effector that is bound to the inner side of the membrane (> Figure 5-6, step 1). This enzyme breaks down **phosphatidylinositol bisphosphate (PIP₂)**, a component of the tails of the phospholipid molecules within the membrane itself. The products of (PIP₂) breakdown are **diacylglycerol (DAG)** and **inositol trisphosphate (IP₃)** (step 2). (IP₃) is the fragment responsible for mobilizing intracellular $Ca^{2+}$ stores to increase cytosolic $Ca^{2+}$ (step 3a). Calcium then takes over the role of second messenger, ultimately bringing

about the response dictated by the first messenger. Many of the $Ca^{2+}$-dependent cellular events are triggered by activation of **calmodulin**, an intracellular $Ca^{2+}$-binding protein (step 4a). The activated calmodulin complex activates $Ca^{2+}$-calmodulin dependent protein kinase (CaM kinase) (step 5a). Activation of CaM kinase is similar to activation of protein kinase A by cAMP. From here, the patterns of the two pathways are similar, meaning that

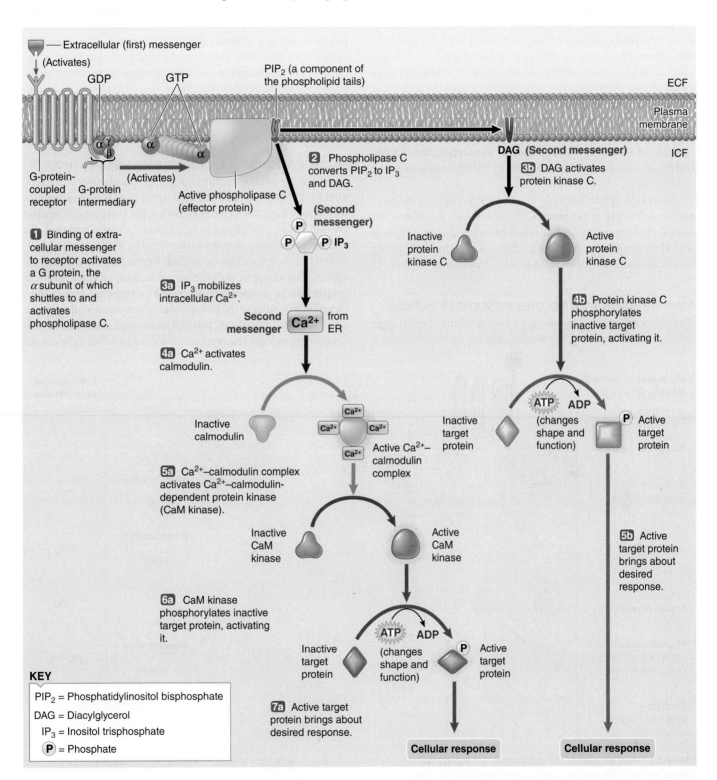

> FIGURE 5-6 Mechanism of action of hydrophilic hormones via concurrent activation of the (IP₃)–Ca²⁺ second-messenger pathway and the DAG pathway

CaM kinase phosphorylates the inactive target protein (step 6a), which alters its shape and function to either activate or inhibit it. This altered protein brings about the ultimate desired cellular response (step 7a). Simultaneously, the other ($PIP_2$) breakdown product, DAG, sets off another second-messenger pathway. DAG activates *protein kinase C* (PKC), which in turn brings about a given cellular response by phosphorylating particular cellular proteins (steps 3b–5b).

The cAMP and $Ca^{2+}$ pathways frequently overlap in bringing about a particular cellular activity. For example, cAMP and $Ca^{2+}$ can influence each other. Calcium-activated calmodulin can regulate adenylyl cyclase and thus influence cAMP, whereas protein kinase A may phosphorylate and thereby change the activity of $Ca^{2+}$ channels or carriers.

Although the cAMP and $Ca^{2+}$ pathways are the most prevalent second-messenger systems, they are not the only ones. For example, in a few cells **cyclic guanosine monophosphate (cyclic GMP)** serves as a second messenger in a system analogous to the cAMP system.

Many hydrophilic hormones use cAMP as their second messenger. A few use intracellular $Ca^{2+}$ in this role; for others, the second messenger is still unknown. Remember that activation of second messengers is a universal mechanism employed by a variety of extracellular messengers in addition to hydrophilic hormones.

### AMPLIFICATION BY A SECOND-MESSENGER PATHWAY
Amplifiers allow small signals to be converted into much larger signals; second messengers work in a similar fashion through a cascade of events that amplify the initial stimulus (⟩ Figure 5-7). The binding of one extracellular chemical-messenger molecule to a receptor activates a number of adenylyl cyclase molecules (let us arbitrarily say 10), each of which activates many (in our hypothetical example, let's say 100) cAMP molecules. Each cAMP molecule then acts on a single protein kinase A, which phosphorylates and thereby influences many (again, let us say 100) specific proteins, such as enzymes. Each enzyme, in turn, is responsible for producing many (perhaps 100) molecules of a particular product, such as a secretory product. The result of this cascade of events, with one event triggering the next event in sequence, is a tremendous amplification of the initial signal. In our hypothetical example, one chemical-messenger molecule has been responsible for inducing a yield of 10 million molecules of a secretory product. In this way, very low concentrations of hormones and other chemical messengers can trigger pronounced cell responses.

### MODIFICATIONS OF SECOND-MESSENGER PATHWAYS
Although membrane receptors serve as links between extracellular first messengers and intracellular second messengers in the regulation of specific cellular activities, the receptors themselves are also frequently subject to regulation. In many instances, the number and affinity (attraction of a receptor for its chemical messenger) of receptors can be altered, depending on the circumstances. For example, the number of receptors for the hormone insulin can be deliberately decreased in response to a chronic elevation of insulin in the blood. Later, when we cover the specific endocrine glands in detail, you will learn more about this mechanism to regulate the responsiveness of a target cell to its hormone.

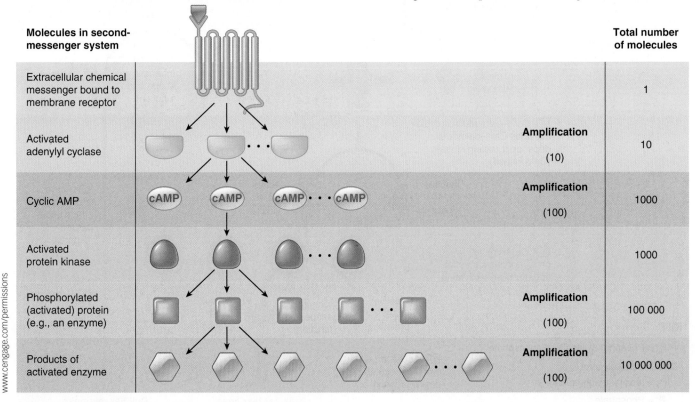

⟩ **FIGURE 5-7 Amplification of the initial signal by a second-messenger pathway.** Through amplification, very low concentrations of extracellular chemical messengers, such as hormones, can trigger pronounced cellular responses.

Many disease processes can be linked to malfunctioning receptors or defects in one of the components of the ensuing signal-transduction pathways. For example, defective receptors are responsible for *Laron dwarfism*. In this condition, the person is abnormally short despite having normal levels of growth hormone, because the tissues cannot respond normally to growth hormone. This is in contrast to the more usual type of dwarfism in which the person is abnormally short because of growth hormone deficiency.

## Lipophilic hormones and protein synthesis

All lipophilic hormones (steroids and thyroid hormone) bind with intracellular receptors and primarily produce their effects in their target cells by activating specific genes to cause the synthesis of new enzymatic or structural proteins (see Why It Matters). The following steps outline the process, as shown in ⟩ Figure 5-8:

1. Free lipophilic hormone (hormone not bound with its carrier) diffuses through the plasma membrane of the target cell and binds with its specific receptor. Depending on the lipophilic hormone, this receptor is either in the cytoplasm or within the nucleus.

2. Each receptor has a specific region for binding with its hormone and another region for binding with DNA. Once the hormone is bound to the receptor, the hormone receptor complex binds with DNA at a specific attachment site on DNA known as the **hormone response element (HRE)**. Different steroid hormones and thyroid hormone, once bound with their respective receptors, attach at different HREs on DNA. For example, the estrogen receptor complex binds at DNA's estrogen response element.

3. Binding of the hormone receptor complex with DNA ultimately turns on a specific gene within the target cell. This gene contains a code for synthesizing a given protein. The code of the activated gene is transcribed into complementary messenger RNA.

4. The new messenger RNA leaves the nucleus and enters the cytoplasm.

5. In the cytoplasm, messenger RNA binds to a ribosome, the "workbench" that mediates the assembly of new proteins (Appendix C, p. A-21). Here, messenger RNA directs the synthesis of the designated new proteins according to the DNA code in the activated genes.

6. The newly synthesized protein, either enzymatic or structural,

**❚ Why It Matters**

Lipophilic compounds are significant due to their ability to cross biological membranes. Some of the most highly studied groups of lipophilic compounds are estrogen and estrogen-like (phytoestrogens) compounds. The presence of estrogens in our waste waters has been linked to the disruption of reproduction in several fish species through the feminization of the males. Although this is a specific example using estrogen, similar principles apply to any lipophilic hormone that can persist in the environment. By crossing biological membranes they can activate genes to cause the production of potentially inappropriate proteins and enzymes. The long-term effects on the environment and the human population are largely unknown but are of significant concern.

produces the target cell's ultimate physiological response to the hormone.

By means of this mechanism, different genes are activated by different lipophilic hormones, resulting in different biological effects.

Even though most steroid actions are accomplished by hormonal binding with intracellular receptors that leads to gene activation, recent studies have unveiled another mechanism by which steroid hormones induce effects that occur too rapidly to be mediated by gene transcription. Researchers have learned that some steroid hormones, most notably some of the sex hormones, bind with unique steroid receptors in the plasma membrane, in addition to

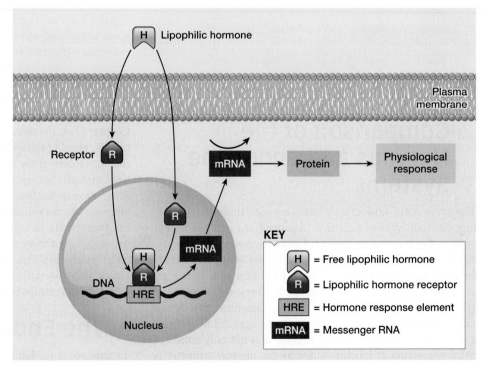

⟩ **FIGURE 5-8 Activation of genes by lipophilic hormones**

binding with the traditional steroid receptors in the nucleus. This membrane binding leads to *nongenomic steroid receptor actions*—that is, actions accomplished by something other than altering gene activity, such as by inducing changes in ionic flux across the membrane or by altering activity of cellular enzymes.

### Hormonal responses versus neural responses

Compared with neural responses, which are brought about within milliseconds, hormone action is relatively slow and prolonged; it takes minutes to hours after the hormone binds to its receptor for the response to take place. The variability in time of onset for hormonal responses depends on the mechanism employed. Hormones that act through a second-messenger system to alter a pre-existing enzyme's activity elicit full action within a few minutes. In contrast, hormonal responses that require the synthesis of new protein may take up to several hours before any action is initiated.

As well, in contrast to neural responses, which are quickly terminated once the triggering signal ceases, hormonal responses persist for a period of time after the hormone is no longer bound to its receptor. Once an enzyme is activated in response to hydrophilic hormonal input, it no longer depends on the presence of the hormone. Thus, the response lasts until the enzyme is inactivated. Likewise, once a new protein is synthesized in response to lipophilic hormonal input, it continues to function until it is degraded. As a result, a hormone's effect usually lasts for some time after its withdrawal. Predictably, the responses that depend on protein synthesis last longer than do those stemming from enzyme activation.

---

### Check Your Understanding 5.2

1. Describe the three general mechanisms of controlling hormone secretion that are common to many different hormones.
2. Explain down regulation, permissiveness, synergism, and antagonism.
3. Describe the processes by which lipophilic hormones initiate cellular responses.

---

## 5.3 | Comparison of the Nervous and Endocrine Systems

The nervous and endocrine systems are specialized for controlling different types of activities. In general, the nervous system governs the coordination of rapid, precise responses. It is especially important in the body's interactions with the external environment. Neural signals in the form of action potentials are rapidly propagated along nerve cell fibres, resulting in the release at the nerve terminal of a neurotransmitter that has to diffuse only a microscopic distance to its target cell to generate a response. A neurally mediated response is not only rapid but brief; the action is quickly halted as the neurotransmitter is swiftly removed from the target site. This permits either ending the response, almost immediately repeating the response, or

rapidly initiating an alternative response, as circumstances demand (e.g., the swift changes in commands to muscle groups needed to coordinate walking). This mode of action makes neural communication extremely rapid and precise. The target tissues of the nervous system are the muscles and glands, especially exocrine glands, of the body.

The endocrine system, in contrast, is specialized to control activities that require duration rather than speed, such as regulating organic metabolism and water and electrolyte balance; promoting smooth, sequential growth and development; and controlling reproduction. The endocrine system responds more slowly to its triggering stimuli than does the nervous system, for several reasons. First, the endocrine system must depend on blood flow to convey its hormonal messengers over long distances. Second, hormones' mechanisms of action at their target cells typically are more complex than that of neurotransmitters and thus require more time before a response occurs. The ultimate effect of some hormones cannot be detected until a few hours after they bind with target-cell receptors. As well, because of the receptors' high affinities for their respective hormones, hormones often remain bound to receptors for some time, thus prolonging their biological effectiveness. Furthermore, unlike the brief, neurally induced responses that stop almost immediately after the neurotransmitter is removed, endocrine effects usually last for some time after the hormone's withdrawal. Neural responses to a single burst of neurotransmitter release usually last only milliseconds to seconds, whereas the alterations that hormones induce in target cells range from minutes to days or, in the case of growth-promoting effects, even a lifetime. Thus, hormonal action is relatively slow and prolonged, making endocrine control particularly suitable for regulating metabolic activities that require long-term stability.

Although the endocrine and nervous systems have their own areas of specialization, they are intimately interconnected functionally. Some nerve cells do not release neurotransmitters at synapses but instead end at blood vessels and release their chemical messengers (neurohormones) into the blood, where these chemicals act as hormones. A given messenger may even be a neurotransmitter when released from a nerve ending and a hormone when secreted by an endocrine cell. The nervous system directly or indirectly controls the secretion of many hormones (see Chapter 6). At the same time, many hormones act as neuromodulators, altering synaptic effectiveness and thereby influencing the excitability of the nervous system. The presence of certain key hormones is even essential for the proper development and maturation of the brain during fetal life. Furthermore, in many instances the nervous and endocrine systems both influence the same target cells in a supplementary fashion. For example, these two major regulatory systems both help regulate the circulatory and digestive systems. Thus, many important regulatory interfaces exist between the nervous and endocrine systems. The study of these relationships is known as **neuroendocrinology**.

## 5.4 | The Endocrine Tissues

To this point we have provided an overview of the general principles and functions of the endocrine system. ▌ Table 5-3 summarizes the most important specific functions of the

| Endocrine Gland | Hormones | Target Cells | Major Functions of Hormones |
|---|---|---|---|
| **Hypothalamus** | Releasing and inhibiting hormones (TRH, CRH, GnRH, GHRH, GHIH, PRH, PIH) | Anterior pituitary | Controls release of anterior pituitary hormones |
| **Posterior Pituitary (hormones stored in)** | Vasopressin | Kidney tubules | Increases $H_2O$ reabsorption |
| | | Arterioles | Produces vasoconstriction |
| | Oxytocin | Uterus | Increases contractility |
| | | Mammary glands (breasts) | Causes milk ejection |
| **Anterior Pituitary** | Thyroid-stimulating hormone (TSH) | Thyroid follicular cells | Stimulates $T_3$ and $T_4$ secretion |
| | Adrenocorticotropic hormone (ACTH) | Zona fasciculata and zona reticularis of adrenal cortex | Stimulates cortisol secretion |
| | Growth hormone | Bone; soft tissues | Essential but not solely responsible for growth; stimulates growth of bones and soft tissues; metabolic effects include protein anabolism, fat mobilization, and glucose conservation |
| | | Liver | Stimulates somatomedin secretion |
| | Follicle-stimulating hormone (FSH) | *Females*: ovarian follicles | Promotes follicular growth and development; stimulates estrogen secretion |
| | | *Males*: seminiferous tubules in testes | Stimulates sperm production |
| | Luteinizing hormone (LH) (interstitial cell-stimulating hormone—ICSH) | *Females*: ovarian follicle and corpus luteum | Stimulates ovulation, corpus luteum development, and estrogen and progesterone secretion |
| | | *Males*: interstitial cells of Leydig in testes | Stimulates testosterone secretion |
| | Prolactin | *Females*: mammary glands | Promotes breast development; stimulates milk secretion |
| | | *Males* | Uncertain |
| **Pineal Gland** | Melatonin | Brain; anterior pituitary; reproductive organs; immune system; possibly other systems | Entrains body's biological rhythm with external cues; inhibits gonadotropins; its reduction likely initiates puberty; acts as an antioxidant; enhances immunity |
| **Thyroid Gland** | Tetraiodothyronine ($T_4$ or thyroxine); triiodothyronine ($T_3$) | Most cells | Increases the metabolic rate; essential for normal growth and nerve development |
| **Thyroid Gland C Cells** | Calcitonin | Bone | Decreases plasma $Ca^{2+}$ concentration |
| **Adrenal Cortex** | | | |
| *Zona glomerulosa* | Aldosterone (mineralocorticoid) | Kidney tubules | Increases $Na^+$ reabsorption and $K^+$ secretion |
| *Zona fasciculata and zona reticularis* | Cortisol (glucocorticoid) | Most cells | Increases blood glucose at the expense of protein and fat stores; contributes to stress adaption |
| | Androgens (dehydroepiandrosterone) | *Females*: bone and brain | Responsible for the pubertal growth spurt and sex drive in females |

*(continued)*

| Endocrine Gland | Hormones | Target Cells | Major Functions of Hormones |
|---|---|---|---|
| **Adrenal Medulla** | Epinephrine and norepinephrine | Sympathetic receptor sites throughout the body | Reinforces the sympathetic nervous system; contributes to stress adaptation and blood pressure regulation |
| **Endocrine Pancreas (Islets of Langerhans)** | Insulin ($\beta$ cells) | Most cells | Promotes cellular uptake, use, and storage of absorbed nutrients |
| | Glucagon ($\alpha$ cells) | Most cells | Important for maintaining nutrient levels in blood during postabsorptive state |
| | Somatostatin (D cells) | Digestive system | Inhibits digestion and absorption of nutrients |
| | | Pancreatic islet cells | Inhibits secretion of all pancreatic hormones |
| **Parathyroid Gland** | Parathyroid hormone (PTH) | Bone, kidneys, intestine | Increases plasma $Ca^{2+}$ concentration; decreases plasma $PO_4^{3-}$ concentration; stimulates vitamin D activation |
| **Gonads** | | | |
| *Female: ovaries* | Estro gen (estradiol) | Female sex organs; body as a whole | Promotes follicular development; governs development of secondary sexual characteristics; stimulates uterine and breast growth |
| | | Bone | Promotes closure of the epiphyseal plate |
| | Progesterone | Uterus | Prepares for pregnancy |
| *Male: testes* | Testosterone | Male sex organs; body as a whole | Stimulates sperm production; governs development of secondary sexual characteristics; promotes sex drive |
| | | Bone | Enhances pubertal growth spurt; promotes closure of the epiphyseal plate |
| *Testes and ovaries* | Inhibin | Anterior pituitary | Inhibits secretion of follicle-stimulating hormone |
| **Placenta** | Estrogen (estriol); progesterone | Female sex organs | Help maintain pregnancy; prepare breasts for lactation |
| | Chorionic gonadotropin | Ovarian corpus luteum | Maintains corpus luteum of pregnancy |
| **Kidneys** | Renin ($\rightarrow$ angiotensin) | Zona glomerulosa of adrenal cortex (acted on by angiotensin, which is activated by renin) | Stimulates aldosterone secretion |
| | Erythropoietin | Bone marrow | Stimulates erythrocyte production |
| **Stomach** | Gastrin | Digestive tract exocrine glands and smooth muscles; pancreas; liver; gallbladder | Controls motility and secretion to facilitate digestive and absorptive processes |
| **Duodenum** | Secretin; cholecystokinin | | |
| | Glucose-dependent insulinotropic peptide | Endocrine pancreas | Stimulates insulin secretion |
| **Liver** | Somatomedins (insulin-like growth factors [IGF]) | Bone; soft tissues | Promotes growth |
| **Skin** | Vitamin D | Intestine | Increases absorption of ingested $Ca^{2+}$ and $PO_4^{3-}$ |
| **Thymus** | Thymosin | T lymphocytes | Enhances T lymphocyte proliferation and function |
| **Heart** | Atrial natriuretic peptide | Kidney tubules | Inhibits $Na^+$ reabsorption |
| **Adipose Tissue** | Leptin | Hypothalamus | Suppresses appetite; important in long-term control of body weight |
| | Other adipokines | Multiple sites | Play role in metabolism and inflammation |

major hormones for both the central and the peripheral endocrine glands. As extensive as ▌Table 5.3 appears, it leaves out a variety of potential hormones that have not yet been fully recognized as hormones, either because they do not quite fit the classic definition of a hormone or because they have been discovered so recently that their hormonal status has not yet been conclusively documented. The table also excludes the cytokines secreted by the effector cells of the defence system (white blood cells and macrophages; p. 485) and a variety of recently revealed and poorly understood growth factors that promote growth of specific tissues, such as *epidermal growth factor* and *nerve growth factor*. Furthermore, new hormones are likely to be discovered, and additional functions may be found for known hormones. As an example, vasopressin's role in conserving water during urine formation was determined first, followed later by the discovery of its constrictor effect on arterioles. More recently, vasopressin has also been found to play roles in fever, learning, memory, and behaviour.

Some of the hormones listed in the table are not discussed further in this chapter but are described in other chapters, as follows: the renal hormones (erythropoietin in Chapter 10 and renin in Chapter 13), thrombopoietin from the liver (Chapter 10), the gastrointestinal hormones (Chapter 15), thymosin from the thymus (Chapter 11), atrial natriuretic peptide from the heart (Chapter 13), and leptin and other adipokines from adipose tissue (Chapter 16). The remainder of the hormones are described in detail in the following sections of this chapter and in Chapter 6, where the peripheral endocrine glands are discussed.

We now focus on the central endocrine glands—those in the brain itself or in close association with the brain—namely, the hypothalamus, the pituitary gland, and the pineal gland.

## 5.5 | Hypothalamus and Pituitary

The **pituitary gland (hypophysis)** (1 cm in diameter, 1 g in weight) is a small endocrine gland located in a bony cavity at the base of the brain just below the hypothalamus (❯ Figure 5-9). The pituitary is connected to the hypothalamus by a thin connecting stalk. If you point one finger between your eyes and another finger toward one of your ears, the imaginary point where these lines would intersect is about where your pituitary is located.

### The anterior and posterior glands

The pituitary has two anatomically and functionally distinct lobes, the **posterior pituitary gland** and the **anterior pituitary gland**. The posterior pituitary is composed of nervous tissue and thus is also termed the **neurohypophysis**. The anterior pituitary consists of glandular epithelial tissue and accordingly is called

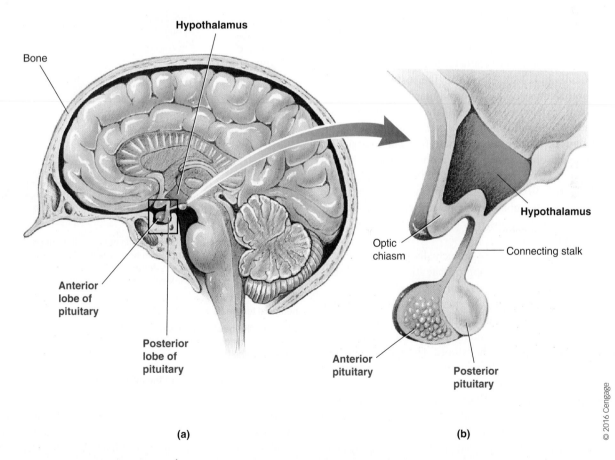

**Hypothalamus**

Bone

Anterior lobe of pituitary

Posterior lobe of pituitary

(a)

Hypothalamus

Optic chiasm

Connecting stalk

Anterior pituitary

Posterior pituitary

(b)

© 2016 Cengage

❯ **FIGURE 5-9 Anatomy of the pituitary gland.** (a) Relation of the pituitary gland to the hypothalamus and to the rest of the brain. (b) Schematic enlargement of the pituitary gland and its connection to the hypothalamus

the **adenohypophysis** (*adeno* means "glandular"). The anterior and posterior pituitary have only their location in common.

In some species, the adenohypophysis also includes a third, well-defined *intermediate lobe*, but humans lack this lobe. In lower vertebrates, the intermediate lobe secretes several **melanocyte-stimulating hormones (MSH)**, which regulate skin colouration by controlling the dispersion of granules that contain the pigment **melanin** (p. 462). By causing variable skin darkening in certain amphibians, reptiles, and fishes, MSH plays a vital role in the camouflage of these species.

In humans, a part of the anterior pituitary that briefly exists as a separate intermediate lobe during fetal development secretes a small amount of MSH. However, although excessive MSH activity does darken skin, MSH is not involved in differences in skin colouration associated with the amount of melanin deposited in the skin nor in the process of skin tanning. Instead, MSH in humans plays a totally different role in helping control food intake. It also appears to influence excitability of the nervous system, perhaps improving memory and learning. Also, MSH has been shown to suppress the immune system, perhaps helping to serve in some way as a check-and-balance that prevents excessive immune responses.

## The hypothalamus and posterior pituitary

The release of hormones from both the posterior and the anterior pituitary is directly controlled by the hypothalamus. However, the nature of the relationship between the hypothalamus and each of these parts of the pituitary is entirely different. The posterior pituitary connects to the hypothalamus by a neural pathway, whereas the anterior pituitary connects to the hypothalamus by a unique vascular link. First, let's look at the posterior pituitary.

The hypothalamus and posterior pituitary form a neuroendocrine system that consists of a population of neurosecretory neurons whose cell bodies lie in two well-defined clusters in the hypothalamus (the **supraoptic nuclei** and **paraventricular nuclei**). The axons of these neurons pass down through the thin connecting stalk and terminate on capillaries in the posterior pituitary (> Figure 5-10). The posterior pituitary consists of these neuronal terminals plus glial-like supporting cells. Functionally as well as anatomically, the posterior pituitary is simply an extension of the hypothalamus.

The posterior pituitary does not actually produce any hormones. It simply stores and, on appropriate stimulation, releases into the blood two small peptide hormones, *vasopressin* and *oxytocin*, which are synthesized by the neuronal cell bodies in the hypothalamus. These hydrophilic peptides are made in both the supraoptic and the paraventricular nuclei, but a single neuron can produce only one of these hormones. The synthesized hormones are packaged in secretory granules that are transported down the cytoplasm of the axon and stored in the neuronal terminals within the posterior pituitary. Each terminal stores either vasopressin or oxytocin, but not both. Thus, these hormones can be released independently as needed. On stimulatory input to the hypothalamus, either vasopressin or oxytocin is released into the systemic blood from the posterior pituitary by exocytosis of the appropriate secretory granules. This hormonal release is

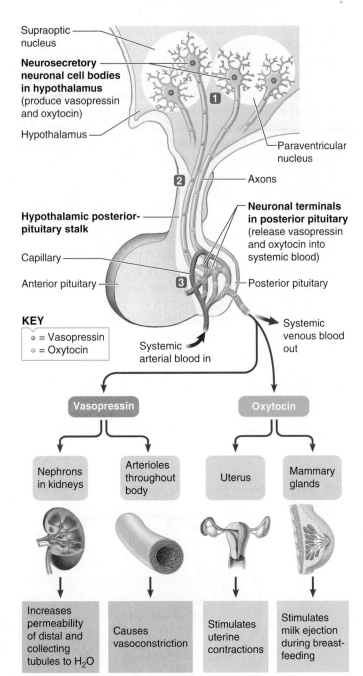

**KEY**
- = Vasopressin
- = Oxytocin

**1** The paraventricular and supraoptic nuclei both contain neurons that produce vasopressin and oxytocin. The hormone, either vasopressin or oxytocin depending on the neuron, is synthesized in the neuronal cell body in the hypothalamus.

**2** The hormone travels down the axon to be stored in the neuronal terminals within the posterior pituitary.

**3** When the neuron is excited, the stored hormone is released from the terminals into the systemic blood for distribution throughout the body.

> FIGURE 5-10 **Relationship of the hypothalamus and posterior pituitary**

triggered in response to action potentials that originate in the hypothalamic cell body and sweep down the axon to the neuronal terminal in the posterior pituitary. As in any other neuron, action potentials are generated in these neurosecretory neurons in response to synaptic input to their cell bodies.

## Relationship of the hypothalamus and posterior pituitary

The actions of vasopressin and oxytocin are summarized here and more thoroughly described in later chapters—vasopressin in Chapters 13 and 14 and oxytocin in Chapter 17.

### VASOPRESSIN

Vasopressin (antidiuretic hormone, ADH) has two major effects that correspond to its two names: (1) it enhances the retention of water by the kidneys (an antidiuretic effect), and (2) it causes contraction of arteriolar smooth muscle (a vessel pressor effect). The first effect has more physiological importance. Under normal conditions, vasopressin is the primary endocrine factor that regulates urinary water loss and overall water balance. When vasopressin is absent from the collecting tubes and ducts, very little water is reabsorbed and the loss of water in the urine is substantially greater. In contrast, typical levels of vasopressin play only a minor role in regulating blood pressure by means of the hormone's pressor effect.

The major control for hypothalamic-induced release of vasopressin from the posterior pituitary is input from hypothalamic osmoreceptors, which increase vasopressin secretion in response to a rise in plasma osmolarity. However, the exact mechanism through which the osmotic concentration of ECF controls the release of vasopressin is unclear. A less powerful input from the left atrial volume receptors increases vasopressin secretion in response to a fall in ECF volume and arterial blood pressure (p. 617).

### OXYTOCIN

Oxytocin stimulates contraction of the uterine smooth muscle to help expel the infant during childbirth, and it promotes ejection of the milk from the mammary glands (breasts) during breastfeeding. Appropriately, oxytocin secretion is increased by reflexes that originate within the birth canal during childbirth and by reflexes that are triggered when the infant suckles the breast. Oxytocin is important to the birthing process. Animal studies have demonstrated that if oxytocin is removed during the latter part of gestation, labour becomes prolonged.

In addition to these two major physiological effects, oxytocin has recently been shown to influence a variety of behaviours, especially maternal behaviours. For example, this hormone fittingly facilitates bonding, or attachment, between a mother and her infant.

## Anterior pituitary hormones: Mostly tropic

Unlike the posterior pituitary, which releases hormones synthesized by the hypothalamus, the anterior pituitary itself synthesizes the hormones it releases into the blood. Different cell populations within the anterior pituitary secrete six major peptide hormones. The actions of each of these hormones are described in detail in later sections. For now, here is a brief statement of their primary effects to provide a rationale for their names › (Figure 5-11):

1. **Growth hormone (GH, somatotropin)**, the primary hormone responsible for regulating overall body growth, is also important in intermediary metabolism.

2. **Thyroid-stimulating hormone (TSH, thyrotropin)** stimulates secretion of thyroid hormone and growth of the thyroid gland.

3. **Adrenocorticotropic hormone (ACTH, adrenocorticotropin)** stimulates cortisol secretion by the adrenal cortex and promotes growth of the adrenal cortex.

4. **Follicle-stimulating hormone (FSH)** in females stimulates growth and development of ovarian follicles, within which the ova, or eggs, develop. It also promotes secretion of the hormone estrogen by the ovaries. In males, FSH is required for sperm production.

5. **Luteinizing hormone (LH)** in females is responsible for ovulation and luteinization (i.e., the formation of a hormone-secreting corpus luteum in the ovary following ovulation). LH also regulates ovarian secretion of the female sex hormones, estrogen and progesterone. In males, the same hormone stimulates the interstitial cells of Leydig in the testes to secrete the male sex hormone, testosterone, giving rise to its alternative name of **interstitial cell-stimulating hormone (ICSH)**.

6. **Prolactin (PRL)** enhances breast development and milk production in females. Its function in males is uncertain, although evidence indicates that it may induce the production of testicular LH receptors. Furthermore, recent studies suggest that prolactin may enhance the immune system and support the development of new blood vessels at the tissue level in both sexes—both actions totally unrelated to its known roles in reproductive physiology.

TSH, ACTH, FSH, and LH are all tropic hormones, because they each regulate the secretion of another specific endocrine gland. FSH and LH are collectively referred to as **gonadotropins** because they control secretion of the sex hormones by the gonads (ovaries and testes). Because growth hormone exerts its growth-promoting effects indirectly by stimulating the release of liver hormones, the *somatomedins*, it too is sometimes categorized as a tropic hormone. Among the anterior pituitary hormones, prolactin is the only one that does not stimulate secretion of another hormone. Of the tropic hormones, FSH, LH, and growth hormone exert effects on nonendocrine target cells in addition to stimulating secretion of other hormones.

## Hypothalamic releasing and inhibiting hormones

None of the anterior pituitary hormones are secreted at a constant rate. Even though each of these hormones has a unique control system, there are some common regulatory patterns. The two most important factors that regulate anterior pituitary hormone secretion are (1) hypothalamic hormones and (2) feedback by target-gland hormones.

Because the anterior pituitary secretes hormones that control the secretion of various other hormones, it long held the undeserved title of "master gland." Scientists now know that the release of each anterior pituitary hormone is largely controlled by still other hormones produced by the hypothalamus.

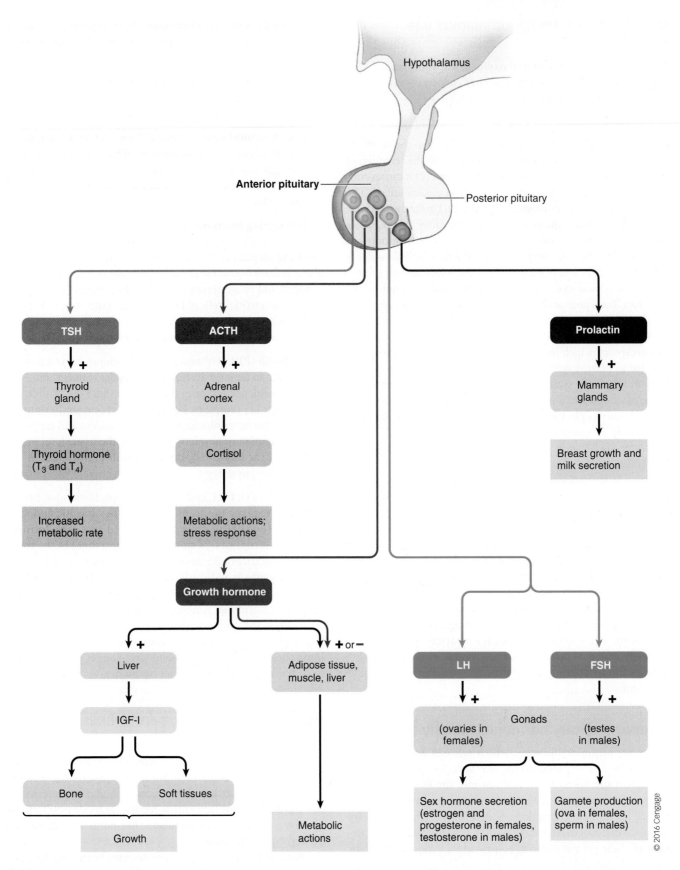

> **FIGURE 5-11 Functions of the anterior pituitary hormones.** Five different endocrine cell types produce the six anterior pituitary hormones—TSH, ACTH, growth hormone, LH and FSH (produced by the same cell type), and prolactin—which exert a wide range of effects throughout the body.

The secretion of these regulatory neurohormones, in turn, is controlled by a variety of neural and hormonal inputs to the hypothalamic neurosecretory cells.

### ROLE OF THE HYPOTHALAMIC RELEASING AND INHIBITING HORMONES

The secretion of each anterior pituitary hormone is stimulated or inhibited by one or more of the seven hypothalamic **hypophysiotropic hormones** (*hypophysis* means "pituitary"; *tropic* means "nourishing"). These small peptide hormones are listed in ▌ Table 5-4. Depending on their actions, these hormones are called **releasing hormones** or **inhibiting hormones**. In each case, the primary action of the hormone is apparent from its name. For example, **thyrotropin-releasing hormone (TRH)** stimulates the release of TSH (alias thyrotropin) from the anterior pituitary, whereas **prolactin-inhibiting hormone (PIH)** inhibits the release of prolactin from the anterior pituitary. Note that hypophysiotropic hormones in most cases are involved in a three-hormone hierarchic chain of command (⟩ Figure 5-12): The hypothalamic hypophysiotropic hormone (*hormone 1*) controls the output of an anterior pituitary tropic hormone (*hormone 2*). This tropic hormone, in turn, regulates secretion of the target endocrine gland's hormone (*hormone 3*), which exerts the final physiological effect.

Although endocrinologists originally speculated that there was one hypophysiotropic hormone for each anterior pituitary hormone, many hypothalamic hormones have more than one effect, so their names indicate only the function first identified. Moreover, a single anterior pituitary hormone may be regulated by two or more hypophysiotropic hormones, which may even exert opposing effects. For example, **growth hormone–releasing hormone (GHRH)** stimulates growth hormone secretion, whereas **growth hormone–inhibiting hormone (GHIH)**, also known as **somatostatin**, inhibits it. The output of the anterior pituitary growth hormone–secreting cells (i.e., the rate of growth hormone secretion) in response to two such opposing inputs depends on the relative concentrations of these hypothalamic hormones as well as on the intensity of other regulatory inputs.

Chemical messengers that are identical in structure to the hypothalamic releasing and inhibiting hormones and to vasopressin are produced in many areas of the brain outside the hypothalamus. Instead of being released into the blood, these messengers act locally as neurotransmitters and as neuromodulators in these other sites. For example, PIH is identical to *dopamine*, a major neurotransmitter in the basal nuclei and elsewhere (p. 124). Others are thought to modulate a variety of functions that range from motor activity (TRH) to libido (GnRH) to learning (vasopressin). These examples further illustrate the multiplicity of ways chemical messengers function.

### ROLE OF THE HYPOTHALAMIC-HYPOPHYSEAL PORTAL SYSTEM

The hypothalamic regulatory hormones reach the anterior pituitary by means of a unique vascular link. In contrast to the direct neural connection between the hypothalamus and posterior pituitary, the anatomic and functional link between the hypothalamus and anterior pituitary is an unusual capillary-to-capillary

| Hormone | Effect on the Anterior Pituitary |
|---|---|
| Thyrotropin-Releasing Hormone (TRH) | Stimulates release of TSH (thyrotropin) and prolactin |
| Corticotropin-Releasing Hormone (CRH) | Stimulates release of ACTH (corticotropin) |
| Gonadotropin-Releasing Hormone (GNRH) | Stimulates release of FSH and LH (gonadotropins) |
| Growth Hormone–Releasing Hormone (GHRH) | Stimulates release of growth hormone |
| Growth Hormone–Inhibiting Hormone (GHIH) | Inhibits release of growth hormone and TSH |
| Prolactin-Releasing Hormone (PRH) | Stimulates release of prolactin |
| Prolactin-Inhibiting Hormone (PIH) | Inhibits release of prolactin |

**▌ TABLE 5-4** Major Hypophysiotropic Hormones

© 2016 Cengage

**5**

connection: the **hypothalamic-hypophyseal portal system**. A portal system is a vascular arrangement in which blood flows directly from one capillary bed through a connecting vessel to another capillary bed. The largest and best-known portal system is the hepatic portal system, which drains intestinal blood directly into the liver for immediate processing of absorbed nutrients (p. 671). Although much smaller, the hypothalamic-hypophyseal portal system is no less important, because it provides a critical link between the brain and much of the endocrine system. It begins in the base of the hypothalamus with a group of capillaries that recombine into small portal vessels, which pass down through the connecting stalk into the anterior pituitary. Here, the portal vessels branch to form most of the anterior pituitary capillaries, which in turn drain into the systemic system (⟩ Figure 5-13).

As a result, almost all the blood supplied to the anterior pituitary must first pass through the hypothalamus. Because materials can be exchanged between the blood and surrounding tissue only at the capillary level, the hypothalamic-hypophyseal portal system provides a route where the releasing and inhibiting hormones can be picked up at the hypothalamus and delivered immediately and directly to the anterior pituitary at relatively high concentrations, completely bypassing the general circulation. If the portal system did not exist, once the hypophysiotropic hormones were picked up in the hypothalamus, they would be returned to the heart by the systemic venous system. From here, they would travel to the lungs and back to the heart through the pulmonary circulation, and finally enter the systemic arterial system for delivery throughout the body, including the anterior pituitary. This process would not only take longer, but the

Principles of Endocrinology: The Central Endocrine Glands

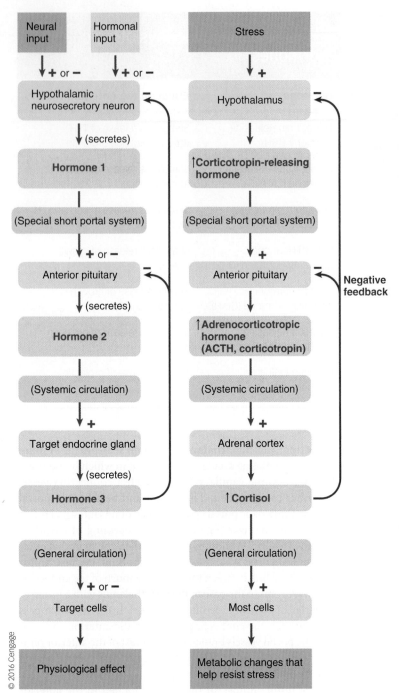

> FIGURE 5-12 **Hierarchic chain of command and negative feedback in endocrine control.** The general pathway involved in the hierarchic chain of command among the hypothalamus, anterior pituitary, and peripheral target endocrine gland is depicted on the left. The pathway on the right leading to cortisol secretion provides a specific example of this endocrine chain of command. The hormone ultimately secreted by the target endocrine gland, such as cortisol, acts in negative-feedback fashion to reduce secretion of the regulatory hormones higher in the chain of command.

secrete their hormones in the same way as the hypothalamic neurons that produce vasopressin and oxytocin. The hormone is synthesized in the cell body and then transported to the axon terminal. It is stored there until its release into an adjacent capillary on appropriate stimulation. The major difference is that the hypophysiotropic hormones are released into the portal vessels, which deliver them to the anterior pituitary, where they control the release of anterior pituitary hormones into the general circulation. In contrast, the hypothalamic hormones stored in the posterior pituitary are themselves released into the general circulation.

## CONTROL OF HYPOTHALAMIC RELEASING AND INHIBITING HORMONES

What regulates secretion of these hypophysiotropic hormones? As in other neurons, the neurons secreting these regulatory hormones receive abundant input of information (both neural and hormonal and both excitatory and inhibitory) that they must integrate. Studies are still in progress to unravel the complex neural input from many diverse areas of the brain to the hypophysiotropic secretory neurons. Some of these inputs carry information about a variety of environmental conditions. One example is the marked increase in secretion of corticotropin-releasing hormone (CRH) in response to stress (see › Figure 5-12). Numerous neural connections also exist between the hypothalamus and the portions of the brain concerned with emotions (the limbic system; p. 124). Thus, emotions greatly influence secretion of hypophysiotropic hormones. Menstrual irregularities sometimes experienced by women who are emotionally upset are a common manifestation of this relationship.

In addition to being regulated by different regions of the brain, the hypophysiotropic neurons are controlled by various chemical inputs that reach the hypothalamus through the blood. Unlike other regions of the brain, portions of the hypothalamus are not guarded by the blood–brain barrier, so the hypothalamus can easily monitor chemical changes in the blood. The most common blood-borne factors that influence hypothalamic neurosecretion are the target-gland hormones that produce negative-feedback effects, to which we now turn our attention.

## Target-gland hormones

In most cases, hypophysiotropic hormones initiate a three-hormone sequence: (1) hypophysiotropic hormone, (2) anterior pituitary tropic hormone, and (3) hormone from the peripheral target endocrine gland. Typically, in addition to producing its physiological effects, the target-gland hormone suppresses secretion of the tropic hormone that is driving its secretion. This negative feedback is accomplished by the target-gland hormone acting either directly on the pituitary itself or on the release of hypothalamic hormones, which in turn regulate anterior pituitary function (see › Figure 5-12). As an example,

hypophysiotropic hormones would be considerably diluted by the much larger volume of blood flowing through this usual circulatory route.

The axons of the neurosecretory neurons that produce the hypothalamic regulatory hormones terminate on the capillaries at the origin of the portal system. These hypothalamic neurons

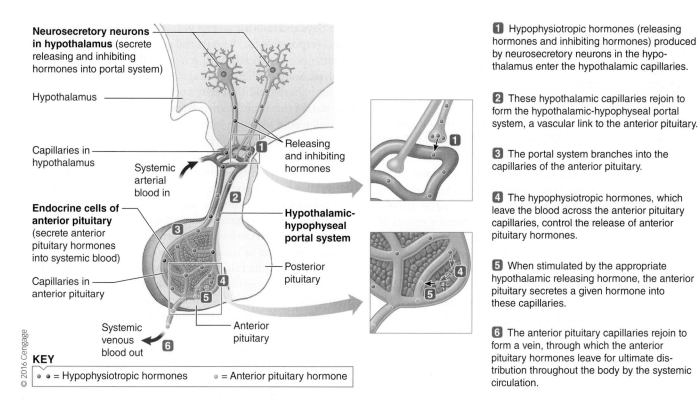

Neurosecretory neurons in hypothalamus (secrete releasing and inhibiting hormones into portal system)

Hypothalamus

Capillaries in hypothalamus

Systemic arterial blood in

Endocrine cells of anterior pituitary (secrete anterior pituitary hormones into systemic blood)

Capillaries in anterior pituitary

Releasing and inhibiting hormones

Hypothalamic-hypophyseal portal system

Posterior pituitary

Anterior pituitary

Systemic venous blood out

© 2016 Cengage

**KEY**
● ● = Hypophysiotropic hormones    ● = Anterior pituitary hormone

**1** Hypophysiotropic hormones (releasing hormones and inhibiting hormones) produced by neurosecretory neurons in the hypothalamus enter the hypothalamic capillaries.

**2** These hypothalamic capillaries rejoin to form the hypothalamic-hypophyseal portal system, a vascular link to the anterior pituitary.

**3** The portal system branches into the capillaries of the anterior pituitary.

**4** The hypophysiotropic hormones, which leave the blood across the anterior pituitary capillaries, control the release of anterior pituitary hormones.

**5** When stimulated by the appropriate hypothalamic releasing hormone, the anterior pituitary secretes a given hormone into these capillaries.

**6** The anterior pituitary capillaries rejoin to form a vein, through which the anterior pituitary hormones leave for ultimate distribution throughout the body by the systemic circulation.

> FIGURE 5-13 **Vascular link between the hypothalamus and anterior pituitary**

consider the CRH–ACTH–cortisol system. Hypothalamic CRH (corticotropin-releasing hormone) stimulates the anterior pituitary to secrete ACTH (adrenocorticotropic hormone, alias corticotropin), which in turn stimulates the adrenal cortex to secrete cortisol. The final hormone in the system, cortisol, inhibits CRH secretion from the hypothalamus and also reduces the sensitivity of the ACTH-secreting cells to CRH by acting directly on the anterior pituitary. Through this double-barrelled approach, cortisol exerts negative-feedback control to stabilize its own plasma concentration. If plasma cortisol levels start to rise above a prescribed set level, cortisol suppresses its own further secretion by its inhibitory actions at the hypothalamus and anterior pituitary. This mechanism ensures that once a hormonal system is activated, its secretion does not continue unabated. If plasma cortisol levels fall below the desired set point, cortisol's inhibitory actions at the hypothalamus and anterior pituitary are reduced, so the driving forces for cortisol secretion (CRH–ACTH) increase accordingly. The other target-gland hormones act by similar negative-feedback loops to keep their plasma levels relatively constant at the set point.

Diurnal rhythms are superimposed on this type of stabilizing negative-feedback regulation; that is, the set point changes as a function of the time of day. Furthermore, other controlling inputs may break through the negative-feedback control to alter hormone secretion (i.e., change the set level) at times of special need. For example, stress raises the set point for cortisol secretion.

The detailed functions and control of all the anterior pituitary hormones except growth hormone are discussed in other chapters in conjunction with the target tissues that they

influence. For example, thyroid-stimulating hormone is covered in Chapter 6 with the discussion of the thyroid gland. Accordingly, growth hormone is the only anterior pituitary hormone we elaborate on at this time.

### Check Your Understanding 5.3

1. Draw a flow diagram showing the hierarchic chain of command and negative feedback in the hypothalamic–anterior pituitary–peripheral target endocrine-gland axis.

2. List the posterior pituitary hormones, the anterior pituitary hormones, and the hypophysiotropic hormones.

3. Compare the means by which the hypothalamus controls hormonal output from the posterior pituitary and from the anterior pituitary.

## 5.6 | Endocrine Control of Growth

In growing children, continuous net protein synthesis occurs under the influence of growth hormone as the body steadily gets larger. Weight gain alone is not synonymous with growth, because weight gain may occur from retaining excess water or storing fat without true structural growth of tissues. Growth requires net synthesis of proteins and includes lengthening of the long bones (the bones of the extremities) as well as increases in the size and number of cells in the soft tissues.

## Growth hormone

Although, as the name implies, growth hormone (GH) is absolutely essential for growth, it is not wholly responsible for determining the rate and final magnitude of growth in a given individual. The following factors affect growth:

- *Genetic determination* of an individual's maximum growth capacity. Attaining this full growth potential further depends on the other factors listed here.

- *An adequate diet*, including enough total protein and ample essential amino acids to accomplish the protein synthesis necessary for growth. Malnourished children never achieve their full growth potential. The growth-stunting effects of inadequate nutrition are most profound when they occur in infancy. In severe cases, the child may be locked into irreversibly stunted body growth and brain development. About 70 percent of total brain growth occurs in the first two years of life. By contrast, a person cannot exceed his or her genetically determined maximum by eating a more than adequate diet. The excess food intake produces obesity instead of growth.

- *Freedom from chronic disease and stressful environmental conditions.* Stunted growth under adverse circumstances is due in large part to the prolonged stress-induced secretion of cortisol from the adrenal cortex. Cortisol exerts several potent antigrowth effects, such as promoting protein breakdown, inhibiting growth in the long bones, and blocking the secretion of GH. Even though sickly or stressed children do not grow well, if the underlying condition is corrected before adult size is achieved, they can rapidly catch up to their normal growth curve through a remarkable spurt in growth.

- *Normal levels of growth-influencing hormones.* In addition to the absolutely essential GH, other hormones—including thyroid hormone, insulin, and the sex hormones—play secondary roles in promoting growth.

The rate of growth is not continuous, neither are the factors responsible for promoting growth the same throughout the growth period. *Fetal growth* is promoted largely by certain hormones from the placenta (the hormone-secreting organ of exchange between the fetal and maternal circulatory systems; p. 760), with the size at birth being determined principally by genetic and environmental factors. GH plays no role in fetal development. After birth, GH and other nonplacental hormonal factors begin to play an important role in regulating growth. Genetic and nutritional factors also strongly affect growth during this period.

Children display two periods of rapid growth—*a postnatal growth spurt* during their first two years of life and a *pubertal growth spurt* during adolescence (› Figure 5-14). From age 2 until puberty, the *rate* of linear growth progressively declines, even though the child is still growing. Before puberty there is little sexual difference in height or weight. During puberty, a marked acceleration in linear growth takes place because the long bones lengthen. Puberty begins at about age 11 in girls and

13 in boys and lasts for several years in both sexes. The mechanisms responsible for the pubertal growth spurt are not clearly understood. Apparently both genetic and hormonal factors are involved. Some evidence indicates that GH secretion is elevated during puberty and thus may contribute to growth acceleration during this time. Furthermore, **androgens** (male sex hormones), whose secretion increases dramatically at puberty, also contribute to the pubertal growth spurt by promoting protein synthesis and bone growth. The potent androgen from the male testes, testosterone, is of greatest importance in promoting a sharp increase in height in adolescent boys. The less potent adrenal androgens from the adrenal gland, which also show a sizable increase in secretion during adolescence, are most likely important in the female pubertal growth spurt. Although estrogen secretion by the ovaries also begins during puberty, it is unclear what role this female sex hormone may play in the pubertal growth spurt in girls. Testosterone and estrogen both ultimately act on bone to halt its further growth so that full adult height is attained by the end of adolescence.

As previously discussed, pituitary hormones primarily regulate the function of the target cells (glands). However, the exception to this rule is growth hormone, which exerts its effects on most tissues of the body. GH is the most abundant hormone produced by the anterior pituitary, even in adults in whom growth has already ceased, although GH secretion typically starts to decline after middle age. The continued high secretion of GH beyond the growing period implies that this hormone has important influences other than on growth. In addition to promoting growth, GH has important metabolic effects and enhances the immune system. We will briefly describe GH's metabolic actions before turning our attention to its growth-promoting actions.

### METABOLIC ACTIONS OF GH UNRELATED TO GROWTH

The specific metabolic effects of growth hormone are (1) increased rate of protein synthesis in all body cells, (2) increased fatty acid mobilization from adipose tissue and increased fatty acid use by body tissues, and (3) decreased

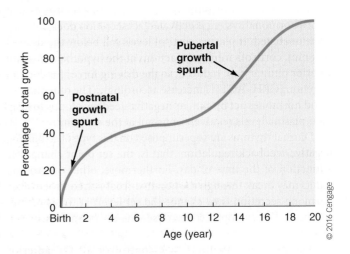

› FIGURE 5-14 **Normal growth curve**

rate of glucose/glycogen use by body tissues. Thus, the overall metabolic effect of GH is to mobilize fat stores as a major energy source while conserving glucose for glucose-dependent tissues, such as the brain. This metabolic pattern is suitable for maintaining the body during prolonged fasting or other situations when the body's energy needs exceed available glucose stores.

### GROWTH-PROMOTING ACTIONS OF GH ON SOFT TISSUES

When tissues are responsive to its growth-promoting effects, GH stimulates growth of both soft tissues and the skeleton. GH promotes growth of soft tissues by (1) increasing the number of cells (**hyperplasia**) and (2) increasing the size of cells (**hypertrophy**). GH increases the number of cells by stimulating cell division and by preventing apoptosis (programmed cell death; p. 114). GH increases the size of cells by favouring synthesis of proteins, the main structural component of cells. GH stimulates almost all aspects of protein synthesis while it simultaneously inhibits protein degradation. It promotes the uptake of amino acids (the raw materials for protein synthesis) by cells, decreasing blood amino acid levels in the process. Furthermore, it stimulates the cellular machinery responsible for accomplishing protein synthesis according to the cell's genetic code.

Growth of the long bones resulting in increased height is the most dramatic effect of GH. Before you can understand the means by which GH stimulates bone growth, you must first become familiar with the structure of bone and how growth of bone is accomplished.

## Bone thickness and length

**Bone** is a living tissue. As a form of connective tissue, it consists of cells and an extracellular organic matrix produced by the cells. Bone also has a high degree of vascularization, which allows delivery of nutrients hormones. The bone cells that produce the organic matrix are known as **osteoblasts** (bone formers). Osseous tissue is the primary tissue of bone, a lightweight composite material made primarily of *calcium phosphate*. It is the osseous tissue that gives bone its rigidity and compressive strength. Although bones are relatively brittle, they do have a high degree of elasticity that is provided by collagen (p. 34). If bones consisted entirely of inorganic crystals, they would be brittle, like pieces of chalk. Bones have structural strength approaching that of reinforced concrete, yet they are not brittle and are much lighter in weight, because they have the structural blending of an organic scaffolding hardened by inorganic crystals. **Cartilage** is similar to bone, except that living cartilage is not calcified.

A long bone basically consists of a fairly uniform cylindrical shaft, the **diaphysis**, with a flared articulating knob at either end, an **epiphysis**. In a growing bone, the diaphysis is separated at each end from the epiphysis by a layer of cartilage known as the **epiphyseal plate** (> Figure 5-15a). The central cavity of the bone is filled with bone marrow, the site of blood cell production.

### BONE GROWTH

Growth in *thickness* of bone is achieved by adding new bone on top of the outer surface of already existing bone. This growth is produced by osteoblasts within the **periosteum**, a connective tissue sheath that covers the outer bone. As osteoblasts deposit new bone on the external surface, other cells within the bone, the **osteoclasts** (bone breakers), dissolve the bony tissue on the inner surface next to the marrow cavity. In this way, the marrow cavity enlarges to keep pace with the increased circumference of the bone shaft.

Growth in *length* of long bones is accomplished by a different mechanism. Bones grow in length as a result of activity of the cartilage cells, or **chondrocytes**, in the epiphyseal plates (> Figure 5-15b). During growth, cartilage cells on the outer edge of the plate next to the epiphysis divide and multiply, temporarily widening the epiphyseal plate. As new chondrocytes are formed on the epiphyseal border, the older cartilage cells toward the diaphyseal border are enlarging. This combination of proliferation of new cartilage cells and hypertrophy of maturing chondrocytes temporarily widens the epiphyseal plate. This thickening of the intervening cartilaginous plate pushes the bony epiphysis farther away from the diaphysis. Soon the matrix surrounding the oldest hypertrophied cartilage becomes calcified. Because cartilage lacks its own capillary network, the survival of cartilage cells depends on diffusion of nutrients and oxygen through the matrix, a process prevented by the deposition of calcium salts. As a result, the old nutrient-deprived cartilage cells on the diaphyseal border die. As osteoclasts clear away dead chondrocytes and the calcified matrix that imprisoned them, the area is invaded by osteoblasts, which swarm upward from the diaphysis, trailing their capillary supply with them. These new "tenants" lay down bone around the persisting remnants of disintegrated cartilage until bone entirely replaces the inner region of cartilage on the diaphyseal side of the plate. When this **ossification** (bone formation) is complete, the bone on the diaphyseal side has lengthened and the epiphyseal plate has returned to its original thickness. The cartilage that bone has replaced on the diaphyseal end of the plate is as thick as the new cartilage on the epiphyseal end of the plate. Thus, bone growth is made possible by the growth and death of cartilage, which acts to push the epiphysis farther out while it provides a framework for future bone formation on the end of the diaphysis.

### MATURE, NONGROWING BONE

As the extracellular matrix produced by an osteoblast calcifies, the osteoblast—like its chondrocyte predecessor—becomes entombed by the matrix it has deposited around itself. Unlike chondrocytes, however, osteoblasts trapped within a calcified matrix do not die, because they are supplied by nutrients transported to them through small canals that the osteoblasts themselves form by sending out cytoplasmic extensions around which the bony matrix is deposited. Thus, within the final bony product a network of permeating tunnels radiates from each entrapped osteoblast, serving as a lifeline system for nutrient delivery and waste removal. The entrapped osteoblasts, now called **osteocytes**, retire from active bone-forming duty, because their imprisonment prevents them from laying down

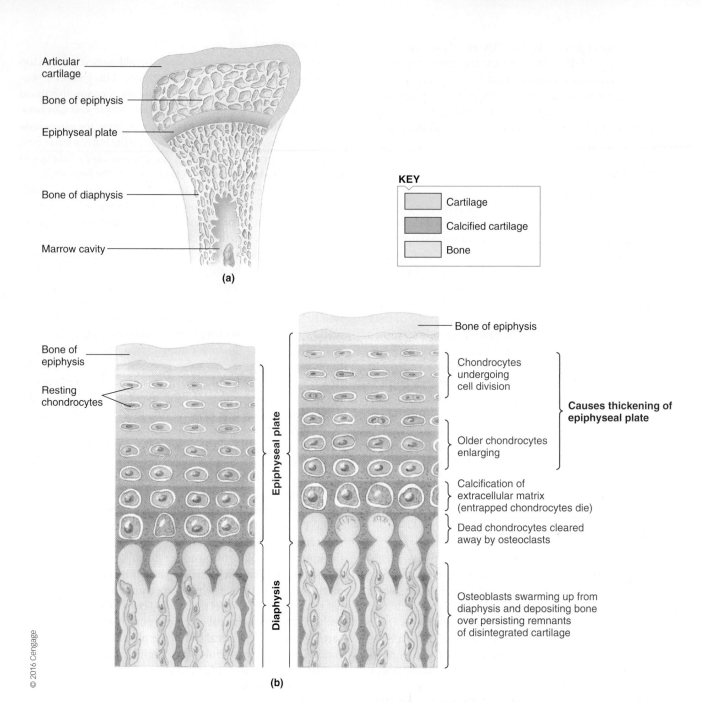

Articular cartilage

Bone of epiphysis

Epiphyseal plate

Bone of diaphysis

Marrow cavity

**(a)**

**KEY**

☐ Cartilage

☐ Calcified cartilage

☐ Bone

Bone of epiphysis

Bone of epiphysis

Resting chondrocytes

Epiphyseal plate

Diaphysis

Chondrocytes undergoing cell division

**Causes thickening of epiphyseal plate**

Older chondrocytes enlarging

Calcification of extracellular matrix (entrapped chondrocytes die)

Dead chondrocytes cleared away by osteoclasts

Osteoblasts swarming up from diaphysis and depositing bone over persisting remnants of disintegrated cartilage

**(b)**

© 2016 Cengage

> **FIGURE 5-15 Anatomy and growth of long bones.** (a) Anatomy of long bones. (b) Two sections of the same epiphyseal plate at different times, depicting the lengthening of long bones

new bone. However, they are involved in the hormonally regulated exchange of calcium between bone and the blood. This exchange is under the control of parathyroid hormone (discussed in Chapter 6), not GH.

### GROWTH-PROMOTING ACTIONS OF GH ON BONES

GH promotes growth of bone in both thickness and length. It stimulates osteoblast activity and the proliferation of epiphyseal cartilage, thereby making space for more bone formation. GH can promote lengthening of long bones as long as the epiphyseal plate remains cartilaginous, or is "open." At the end of adolescence, under the influence of the sex hormones these plates completely ossify, or "close," so that the bones cannot lengthen any further despite the presence of GH. Thus, after the plates are closed, the individual does not grow any taller.

**▌Clinical Connections** Michael was much shorter than other children his age because his anterior pituitary was releasing very little growth hormone, which severely decreased the rate of long-bone elongation. Treatment with hGH will allow his bones to continue to elongate but he will remain shorter than his peers.

## Somatomedins

According to the traditional view of endocrinologists, GH does not act directly on its target cells to bring about its growth-producing actions (increased cell division, enhanced protein synthesis, and bone growth). When GH is supplied directly to chondrocytes outside the body (i.e., cultured), there appears to be no effect (e.g., enlargement). But when GH is provided in vivo, the same chondrocytes begin to proliferate and enlarge. This shows that the effects of GH are indirectly brought about by peptide mediators known as **somatomedins**. These peptides are also referred to as **insulin-like growth factors (IGF)** because they are structurally and functionally similar to insulin. Two somatomedins—IGF-I (IGF-1) and IGF-II (IGF-2)—have been identified. However, by far the most important somatomedin is IGF-I. Recent studies suggest that in addition to working together, GH and IGF-I each act independently to promote growth.

### IGF-I

**IGF-I** synthesis is stimulated by GH and mediates much of this hormone's growth-promoting actions. The concentration of IGF-I in the blood plasma normally mimics the rate of secretion of GH. However, GH's duration of action is short, because GH attaches weakly to the plasma proteins in the blood and it is quickly released into the tissues. Its half-life in the blood is about 20 minutes. In contrast, IGF-I's half-life is approximately 20 hours, because it tightly binds the blood plasma proteins. This association with IGF-I allows the effects of GH to be more prolonged. The major source of circulating IGF-I is the liver, which releases this peptide product into the blood in response to GH stimulation. IGF is also produced by most other tissues, although they do not release it into the blood to any extent. Researchers propose that IGF-I produced locally in target tissues may act through paracrine means (p. 51). For example, the injection of GH into epiphyseal cartilage of bone causes growth within that specific cartilage, which indicates that the localized release of IGF-I by the tissue has a significant effect. Such a mechanism could explain why blood levels of GH are no higher, and indeed circulating IGF-I levels are lower, during the first several years of life compared with adult values, even though growth is quite rapid during the postnatal period. Local production of IGF-I in target tissues may possibly be more important than delivery of blood-borne IGF-I or GH during this time.

Production of IGF is controlled by a number of factors other than GH, including nutritional status, age, and tissue-specific factors, as follows:

- IGF-I production depends on adequate nutrition. Inadequate food intake reduces IGF-I production. As a result, changes in circulating IGF-I levels do not always coincide with changes in GH secretion. For example, fasting decreases IGF-I levels even though it increases GH secretion.

- Age-related factors influence IGF-I production. A dramatic increase in circulating IGF-I levels accompanies the moderate increase in GH at puberty, which may, of course, be an impetus to the pubertal growth spurt.

- Finally, various tissue-specific stimulatory factors can increase IGF-I production in particular tissues. To illustrate, the gonadotropins and sex hormones stimulate IGF-I production within reproductive organs, such as the testes in males and the ovaries and uterus in females.

Thus, control of IGF-I production is complex and subject to a variety of systemic and local factors. (For further discussion of the interaction and influence of GH and IGF-I, see Concepts, Challenges, and Controversies, p. 236.)

### IGF-II

In contrast to IGF-I, the production of IGF-II is not influenced by GH. **IGF-II** is primarily important during fetal development. Although IGF-II continues to be produced during adulthood, its role in adults remains unclear.

The following are the basic physiological actions of GH—accomplished either directly or through somatomedins (IGFs):

- Decreased glycogen synthesis
- Reduced glucose use
- Increased lipolysis (breakdown of stored fat)
- Increased use of fatty acids
- Metabolic sparing of glucose and amino acids
- Increased amino acid transport across cell membrane
- Increased protein synthesis
- Increased collagen synthesis
- Increased cartilage growth
- Promotion of hypertrophy and hyperplasia of tissues

## Secretion

The control of GH secretion is complex, with two hypothalamic hypophysiotropic hormones playing a key role.

### GROWTH HORMONE–RELEASING HORMONE AND GROWTH HORMONE–INHIBITING HORMONE

Two antagonistic regulatory hormones from the hypothalamus are involved in controlling growth hormone secretion: growth hormone–releasing hormone (GHRH), which is stimulatory, and growth hormone–inhibiting hormone (GHIH or somatostatin), which is inhibitory (› Figure 5-16). (Note the distinctions among *somatotropin*, alias growth hormone; *somatomedin*, a liver hormone that directly mediates the effects of GH; and *somatostatin*, which inhibits GH secretion.) Any factor that increases GH secretion could theoretically do so either by stimulating GHRH release or by inhibiting GHIH release. Endocrinologists do not know which of these pathways is used in each specific case.

As with the other hypothalamus–anterior pituitary axes, negative-feedback loops participate in the regulation of GH secretion. Both GH and the somatomedins inhibit pituitary secretion of GH, presumably by stimulating GHIH release from the hypothalamus. The somatomedins may also directly influence the anterior pituitary to inhibit the effects of GHRH on GH release.

# Human Growth Hormone

**G**ROWTH HORMONE (GH) IS A HORMONE that stimulates growth and cell reproduction in humans and other animals. It is a single-chain polypeptide (191-amino acid) hormone that is synthesized, stored, and secreted by the anterior pituitary gland (somatotroph cells). Growth hormone is now synthetically produced via recombinant DNA (rhGH) technology.

GH is used for the treatment of several diseases (e.g., GH-replacement therapy in adults). The genes for human growth hormone are located on chromosome 17 and are closely related to human chorionic somatomammotropin genes. GH, human chorionic somatomammotropin, and prolactin (PRL) are a group of homologous hormones with growth-promoting and lactogenic activity. The two major regulators of GH are peptides: growth hormone–releasing hormone, which is stimulating; and somatostatin, which is inhibiting. Released into the portal vein by the hypothalamus, these peptides bathe the anterior pituitary gland. However, many physiological factors influence these stimulating and inhibiting peptides, including the following:

- Growth hormone–releasing hormone (arcuate nucleus)
- Ghrelin (peptide hormone secreted by the stomach)
- Sleep
- Exercise
- Low levels of blood sugar (hypoglycaemia)
- Dietary protein
- Estradiol
- Arginine

Inhibitors of GH secretion include the following:

- Somatostatin (peptide produced by the periventricular nucleus and other tissues)
- GH and IGF-I concentrations
- Hyperglycaemia
- Glucocorticoids

The secretion of GH occurs each day in several large pulsatile peaks, with a typical duration lasting from 10 to 30 minutes, and changes in the plasma concentration of GH ranging from approximately 5 to >35 ng/mL (between peaks basal GH level is usually >3 ng/mL over a 24-hour period). There is wide disparity in GH peaks between days and among individuals, but the most predictable of these peaks occurs about an hour after the onset of sleep. Additionally, the amplitude and pattern of GH secretion change throughout the lifetime. Basal levels are greatest during childhood, the amplitude and frequency of peaks are largest during puberty, and basal levels and the amplitude and frequency of peaks decline throughout adulthood.

### The Effects of Growth Hormone

The effects of growth hormone on human tissues are anabolic in nature. Growth hormone acts by interacting with a specific receptor on the surface of cells. For example, GH acts directly on adipocytes, stimulating the breakdown of triglycerides, while suppressing the ability to uptake and accumulate circulating lipids. One of the primary roles of GH is to stimulate the liver and other tissues (e.g., skeletal muscle) to secrete insulin-like growth factor-I (IGF-I). IGF-I stimulates proliferation of chondrocytes (cartilage cells), resulting in bone growth as well as skeletal muscle growth. The influence of GH on skeletal muscle development is via its inhibitory influence on myostatin.

Myostatin is a cytokine that inhibits myogenesis (formation of muscle tissue); thus, the cytokine myostatin is catabolic in nature. GH has an inhibitory effect on myostatin mRNA expression. It is suspected (by in vitro studies) that skeletal muscle cells may demonstrate a significant reduction in myostatin expression by myotubes in response to GH. The result (by in vivo studies) is an increase in lean body mass (e.g., skeletal muscle tissue), which translates into improved athletic performance (e.g., aerobic performance, strength, and power). As long

## FACTORS THAT INFLUENCE GH SECRETION

A number of factors influence GH secretion by acting on the hypothalamus. GH secretion displays a well-characterized diurnal rhythm. Through most of the day, GH levels tend to be low and fairly constant. About an hour after the onset of deep sleep, however, GH secretion markedly increases, up to five times the daytime value, and then rapidly drops over the next several hours.

Superimposed on this diurnal fluctuation in GH secretion are further bursts in secretion that occur in response to exercise, stress, and low blood glucose, the major stimuli for increased secretion. The benefits of increased GH secretion during these situations when energy demands outstrip the body's glucose reserves are presumably that glucose is conserved for the brain and fatty acids are provided as an alternative energy source for muscle.

Because GH uses up fat stores and promotes synthesis of body proteins, it encourages a change in body composition away from adipose deposition toward an increase in muscle protein. Accordingly, the increase in GH secretion that accompanies exercise may at least in part mediate the effects of exercise in reducing the percentage of body fat while increasing the lean body mass.

A rise in blood amino acids after a high-protein meal also enhances GH secretion. In turn, GH promotes the use of these amino acids for protein synthesis. GH is also stimulated by a decline in blood fatty acids. Because GH mobilizes fat, such regulation helps maintain fairly constant blood fatty acid levels.

Note that the known regulatory inputs for GH secretion are aimed at adjusting the levels of glucose, amino acids, and fatty acids in the blood. No known growth-related signals influence

as GH levels are elevated, the inhibitory influence on myostatin can be sustained, maintaining lean body mass. GH assists in the following:

- Strengthens and increases the mineralization of bone
- Increases skeletal muscle mass
- Promotes lipolysis (fat breakdown)
- Increases protein synthesis
- Stimulates the growth of internal organs
- Reduces liver uptake of glucose
- Promotes gluconeogenesis in the liver
- Contributes to the maintenance and function of pancreatic islets
- Stimulates the immune system

### Use of Growth Hormone

Because of GH's influence on bone and skeletal muscle, it has been studied as a potential anti-aging treatment since the 1990s. Some research has demonstrated that GH supplementation in elderly men and women can significantly increase lean body mass, bone mineral density, and upper and lower body strength, as well as decrease fat mass. However, other research has found no improvement in bone density, cholesterol levels, maximal oxygen consumption, or muscle strength.

As well, habitual use of GH has demonstrated several negative side effects, such as joint swelling, joint pain, and an increased risk of diabetes. It is because of the potential positive side effects that athletes have become interested in the potential benefits of using rhGH as an ergogenic aid. The scientific evidence to validate the effectiveness of rhGH in athletic populations is lacking as a result of prudent scientific ethical guidelines. The doses of rhGH (anecdotally) used by athletes would generally exceed what is considered acceptable and safe by Research Ethics Boards in Canada and the United States, making verification of the ergogenic benefits in athletic populations difficult.

The scientific community must base its educated opinion on research involving more therapeutic doses, which is more speculative in nature. A critical review of the current literature pertinent to rhGH's effectiveness in benefiting athletic performance by increasing muscular size and strength was undertaken by Dr. Dean at the University of Manitoba. His literature review (2002) suggested that administration of rhGH in athletes did not significantly improve muscular strength and thus athletic performance. However, more research is needed in this area to better determine the influence of rhGH on athletic performance, but as previously mentioned this will be a challenging task.

### Regulation of rhGH in Athletic Competition

The use of rhGH by athletes for ergogenic purposes is banned by the World Anti-Doping Agency (WADA). WADA's position on rhGH is clear and can be found on its website (http://www.wada-ama .org). WADA outlines the properties of HGH (same as rhGH), the side effects, the existence of a test, the reliability of testing, and other factors pertinent to informing the athlete of their responsibility and WADA's position. Despite the unproven benefits of rhGH on athletic performance, and its ban by WADA, some athletes continue to use it.

### Further Reading

Dean, H. (2002). Does exogenous growth hormone improve athletic performance? *Clin J Sport Med*, 12(4): 250-53.

Liu, W., Thomas, S.G., Asa, S.L., Gonzalez-Cadavid, N., Bhasin, S., & Ezzat, S. (2003). Myostatin is a skeletal muscle target of growth hormone anabolic action. *J Clin Endocr Metab*, 88(11): 5490-96.

Woodhouse, L.J., Mukherjee, A., Shalet, S.M., & Ezzat, S. (2006). The influence of growth hormone status on physical impairments, functional limitations, and health-related quality of life in adults. *Endocr Rev*, 27(3): 287-317.

GH secretion. The whole issue of what really controls growth is complicated by the fact that GH levels during early childhood— a period of quite rapid linear growth—are similar to those in normal adults. As mentioned earlier, the poorly understood control of IGF-I activity may be important in this regard. Another related question is, Why aren't adult tissues still responsive to GH's growth-promoting effects? We know we do not grow any taller after adolescence because the epiphyseal plates have closed, but why don't soft tissues continue to grow through hypertrophy and hyperplasia under the influence of GH? One speculation is that levels of GH may be high enough to produce its growth-promoting effects only during the secretion bursts that occur in deep sleep. It is interesting to note that time spent in deep sleep is greatest in infancy and gradually declines with age. Still, even as we age, we spend some time in deep sleep, yet we do not

gradually get larger. Further research is needed to unravel these mysteries.

**❚ Clinical Connections**

Recall that Michael was found to have normal levels of growth hormone–releasing hormone yet low levels of both growth hormone and insulin-like growth factor (IGF-1). Further tests did not show any kind of tumour or lesion within the anterior pituitary. When treated with human growth hormone (hGH), his IGH-1 levels also increased. Taken together, it can be concluded that the cause of Michael's dwarfism is that the anterior pituitary gland does not release growth hormone. Review ❯ Figure 5-16 to understand the role of IGF-1 in the diagnosis.

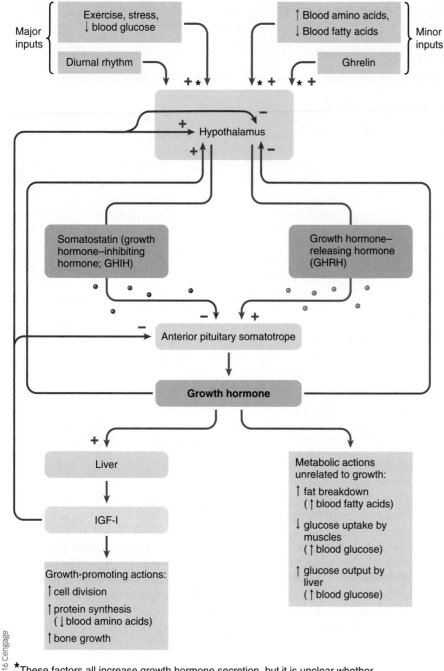

*These factors all increase growth hormone secretion, but it is unclear whether they do so by stimulating GHRH or inhibiting somatostatin (GHIH), or both.

> FIGURE 5-16 **Control of growth hormone secretion**

## Aberrant growth patterns

 Diseases related to both deficiencies and excesses of growth hormone can occur. The effects on the pattern of growth are much more pronounced than the metabolic consequences.

### GROWTH HORMONE DEFICIENCY

GH deficiency may be caused by a pituitary defect (lack of GH) or may occur secondary to hypothalamic dysfunctions (lack of GHRH). Hyposecretion of GH in a child is one cause of **dwarfism**. The predominant feature is short stature caused by retarded skeletal growth (> Figure 5-17). Less obvious characteristics include poorly developed muscles (reduced muscle protein synthesis) and excess subcutaneous fat (less fat mobilization).

In addition, growth may be thwarted because the tissues fail to respond normally to GH. An example is **Laron dwarfism**, which is characterized by abnormal GH receptors that are unresponsive to the hormone. The symptoms of this condition resemble those of severe GH deficiency even though blood levels of GH are actually high.

In some instances, GH levels are adequate and target-cell responsiveness is normal, but somatomedins are lacking. African Pygmies are an interesting example; their short stature is attributable to a congenital incapacity to produce a normal amount of IGF-I.

The onset of GH deficiency in adulthood after growth is already complete produces relatively few symptoms. GH-deficient adults tend to have reduced skeletal muscle mass and strength (less muscle protein) as well as decreased bone density (less osteoblast activity during ongoing bone remodelling). Furthermore, because GH is essential for maintaining cardiac muscle mass and performance in adulthood, GH-deficient adults may be at increased risk of developing heart failure.

### GROWTH HORMONE EXCESS

Hypersecretion of GH is most often caused by a tumour of the GH-producing cells of the anterior pituitary. The symptoms depend on the age of the individual when the abnormal secretion begins. If overproduction of GH begins in childhood before the epiphyseal plates close, the principal manifestation is a rapid growth in height without distortion of body proportions. Appropriately, this condition is known as **gigantism** (see > Figure 5-17). If not treated by removal of the tumour or by drugs that block the effect of GH, the person may reach a height of 2.4 metres or more. All the soft tissues grow correspondingly, so the body is still well proportioned.

If GH hypersecretion occurs after adolescence when the epiphyseal plates have already closed, further growth in height is prevented. Under the influence of excess GH, however, the bones become thicker, and the soft tissues, especially connective tissue and skin, proliferate. This disproportionate growth pattern produces a disfiguring condition known as **acromegaly** (*acro* means "extremity"; *megaly* means "large"). Bone thickening is

© 2016 Cengage

> **FIGURE 5-17 Examples of the effect of abnormalities in growth hormone secretion on growth.** The man on the left displays pituitary dwarfism resulting from underproduction of growth hormone in childhood. The man at the centre of the photograph has gigantism caused by excessive growth hormone secretion in childhood. The woman at the right is of average height.

most obvious in the extremities and face. The gradual thickening of the facial bones and the skin result in the jaws and cheekbones becoming more prominent (> Figure 5-18). The hands and

feet enlarge, and the fingers and toes become greatly thickened. Peripheral nerve disorders often occur as nerves are entrapped by overgrown connective tissue or bone.

## Other hormones

Several other hormones in addition to GH contribute in special ways to overall growth.

- *Thyroid hormone* is essential for growth but is not itself directly responsible for promoting growth. It plays a permissive role in skeletal growth; specifically, the actions of GH fully manifest only when enough thyroid hormone is present. As a result, growth is severely stunted in hypothyroid children, and hypersecretion of thyroid hormone may result in excessive skeletal growth, which increases the child's height.

- *Insulin* is an important growth promoter. Insulin deficiency often blocks growth, and hyperinsulinism frequently spurs excessive growth. Because insulin promotes protein synthesis, its growth-promoting effects should not be surprising. However, these effects may also arise from a mechanism other than insulin's direct effect on protein synthesis. Insulin structurally resembles the somatomedins and may interact with the somatomedin (IGF-I) receptor, which is very similar to the insulin receptor.

- *Androgens*, which are believed to play an important role in the pubertal growth spurt, powerfully stimulate protein synthesis in many organs. Androgens stimulate linear growth, promote weight gain, and increase muscle mass. The most potent androgen, testicular testosterone, is responsible for the general development of heavier musculature in men than in women. These androgenic growth-promoting effects depend on the presence of GH. Androgens have virtually no effect on body growth in the absence of GH, but in its presence they

> **FIGURE 5-18 Progressive development of acromegaly.** In this series of photos from childhood to early adulthood, note that the brow bones, cheekbones, and jawbones become more prominent. This results from progressive thickening of the bones and skin due to excessive GH secretion.

synergistically enhance linear growth. Although androgens stimulate growth, they ultimately stop further growth by promoting closure of the epiphyseal plates.

- *Estrogens*, like androgens, ultimately terminate linear growth by stimulating complete conversion of the epiphyseal plates to bone. However, the effects of estrogen on growth prior to bone maturation are not well understood. Some studies suggest that large doses of estrogen may even hinder further body growth by inhibiting chondrocyte proliferation while the epiphyseal plates are still open.

Several factors contribute to the average height differences between men and women. First, because puberty occurs about two years earlier in girls than in boys, on average boys have two more years of prepubertal growth than girls do. As a result, boys are usually several inches taller than girls at the start of their respective growth spurts. Second, as already mentioned, boys experience a greater androgen-induced growth spurt than girls before their respective gonadal steroids seal their long bones from further growth; this results in greater heights in men than in women, on average. Third, the pubertal rise in estrogen may reduce the pubertal growth spurt in females. Fourth, recent evidence suggests that androgens "imprint" the brains of males during development, giving rise to a secretory pattern of GH typical in males, characterized by higher cyclic peaks, which are speculated to contribute to the greater height of males.

In addition to these hormones that exert overall effects on body growth, a number of poorly understood peptide *growth factors* have been identified that stimulate mitotic activity of specific tissues (e.g., epidermal growth factor).

We now shift our attention to the other central endocrine gland—the pineal gland.

---

### Check Your Understanding 5.4

1. Describe GH's metabolic effects.
2. Explain the relationship between GH and IGF-I in promoting growth.
3. What factors can influence the release of growth hormone?

---

## 5.7 | Pineal Gland and Circadian Rhythms

The **pineal gland**, a tiny, pinecone-shaped structure located in the centre of the brain (see > Figure 5-1), secretes the hormone **melatonin**. (Do not confuse melatonin with the skin-darkening pigment, *melanin*.) Although melatonin was discovered in 1959, investigators have only recently begun to unravel its many functions. One of melatonin's most widely accepted roles is helping to keep the body's inherent circadian rhythms in synchrony with the light–dark cycle. We first examine circadian rhythms in general and then look at the role of melatonin in this regard and in other functions.

## The suprachiasmatic nucleus

Hormone secretion rates are not the only factor in the body that fluctuates cyclically over a 24-hour period. Humans have similar biological clocks for many other bodily functions, ranging from gene expression to physiological processes, such as temperature regulation (p. 710), to behaviour. The master biological clock that serves as the pacemaker for the body's circadian rhythms is the **suprachiasmatic nucleus (SCN)**. It consists of a cluster of nerve cell bodies in the hypothalamus above the optic chiasm, the point at which part of the nerve fibres from each eye cross to the opposite half of the brain (*supra* means "above"; *chiasm* means "cross") (p. 166; see Figure 3-7 and > Figure 5-9). The self-induced rhythmic firing of the SCN neurons plays a major role in establishing many of the body's inherent daily rhythms.

### ROLE OF CLOCK PROTEINS

Scientists have now unravelled the underlying molecular mechanisms responsible for the SCN's circadian oscillations. Specific self-starting genes within the nuclei of SCN neurons set in motion a series of events that brings about the synthesis of **clock proteins** in the cytosol surrounding the nucleus. As the day wears on, these clock proteins continue to accumulate, finally reaching a critical mass, at which time they are transported into the nucleus. Here, they block the genetic process responsible for their own production. The level of clock proteins gradually dwindles as they degrade within the nucleus, thus removing their inhibitory influence from the clock-protein genetic machinery. No longer blocked, these genes once again rev up the production of more clock proteins, as the cycle repeats itself. Each cycle takes about a day. The fluctuating levels of clock proteins bring about cyclic changes in neural output from the SCN that, in turn, lead to cyclic changes in effector organs throughout the day. An example is the diurnal variation in cortisol secretion (see > Figure 5-3). Circadian rhythms are therefore linked to fluctuations in clock proteins, which use a feedback loop to control their own production. In this way, internal timekeeping is a self-sustaining mechanism built into the genetic makeup of the SCN neurons.

### SYNCHRONIZATION OF THE BIOLOGICAL CLOCK WITH ENVIRONMENTAL CUES

On its own, this biological clock generally cycles a bit slower than the 24-hour environmental cycle. Without any external cues, the SCN sets up cycles that average about 25 hours. The cycles are consistent for a given individual but vary somewhat among different people. If this master clock were not continually adjusted to keep pace with the world outside, the body's circadian rhythms would become progressively out of sync with the cycles of light (periods of activity) and dark (periods of inactivity). As a result, the SCN must be reset daily by external cues so that the body's biological rhythms are synchronized with the activity levels driven by the surrounding environment. The effect of not maintaining the internal clock's relevance to the environment is well known by people who experience **jet lag** when their inherent rhythm is out of step with external cues. The SCN works in conjunction with the pineal gland and its hormonal

# Tinkering with Our Biological Clocks

RESEARCH SHOWS THAT THE HECTIC PACE OF MODERN LIFE— the stress, noise, pollution, and the irregular schedules many workers follow—can upset internal rhythms, illustrating how a healthy external environment affects our own internal environment and our health.

Dr. Richard Restak, a neurologist and author, notes that the "usual rhythms of wakefulness and sleep … seem to exert a stabilizing effect on our physical and psychological health." The greatest disrupter of our natural circadian rhythms is the variable work schedule, surprisingly common in industrialized countries. Today, one out of every four working men and one out of every six working women has a variable work schedule—that is, shifting frequently between day and night work. In many industries, to make optimal use of equipment and buildings, workers are on the job day and night. As a spin-off, more restaurants and stores stay open 24 hours a day and more health-care workers must be on duty at night to care for accident victims.

To spread the burden, many companies that maintain shifts around the clock alter their workers' schedules. One week, employees work the day shift. The next week, they move to the "graveyard shift" from midnight to 8 a.m. The next week, they work the night shift from 4 p.m. to midnight. Many shift workers feel tired most of the time and have trouble staying awake at the job. Work performance suffers because of the workers' fatigue. When workers arrive home for bed, they're exhausted but can't sleep, because they're trying to doze off at a time when the body is trying to wake them up. Unfortunately, the weekly changes in schedule never permit workers' internal alarm clocks to fully adjust. Most people require 4–14 days to adjust to a new schedule.

Workers on alternating shifts suffer more ulcers, insomnia, irritability, depression, and tension than workers on unchanging shifts. Their lives are never the same. To make matters worse, tired, irritable workers whose judgment is impaired by fatigue pose a threat to society as a whole. Consider an example.

At 4 a.m. in the control room of the Three Mile Island nuclear reactor in Pennsylvania, three operators made the first mistake in a series of errors that led to the worst nuclear accident in U.S. history. The operators did not notice warning lights and failed to observe that a crucial valve had remained open. When the morning-shift operators entered the control room the next day, they quickly discovered the problems, but it was too late. Pipes in the system had burst, sending radioactive steam and water into the air and into two buildings. John Gofman and Arthur Tamplin, two radiation health experts, estimate that the radiation released from the accident will cause at least 300 and possibly as many as 900 additional fatal cases of cancer in the residents living near the troubled reactor, although other experts (especially in the nuclear industry) contest these projections, saying the accident will have no noticeable effect. Whatever the outcome, the 1979 accident at Three Mile Island cost several billion dollars to clean up.

Late in April 1986, another nuclear power plant ran amok. This accident, in Chernobyl in the former Soviet Union, was far more severe. In the early hours of the morning, two engineers were testing the reactor. Violating standard operational protocol, they deactivated key safety systems. This single error in judgment (possibly caused by fatigue) led to the largest and most costly nuclear accident in world history. Steam built up inside the reactor and blew the roof off the containment building. A thick cloud of radiation rose skyward and then spread throughout Europe and the world. While workers battled to cover the molten radioactive core that spewed radiation into the sky, the whole world watched in horror.

The Chernobyl disaster, like the accident at Three Mile Island, may have been the result of workers operating at a time unsuitable for clear thinking. One has to wonder how many plane crashes, auto accidents, and acts of medical malpractice can be traced to judgment errors resulting from our insistence on working against inherent body rhythms.

Thanks to studies of biological rhythms, researchers are finding ways to reset biological clocks, which could help lessen the misery and suffering of shift workers and could improve the performance of the graveyard-shift workers. For instance, one simple measure is to put shift workers on three-week cycles to give their clocks time to adjust. And instead of shifting workers from daytime to graveyard shifts, transfer them forward, rather than backward (e.g., from a daytime to a nighttime shift). It's a much easier adjustment. Bright lights can also be used to reset the biological clock. It's a small price to pay for a healthy workforce and a safer society. Furthermore, use of supplemental melatonin, the hormone that sets the internal clock to march in step with environmental cycles, may prove useful in resetting the body's clock when that clock is out of synchrony with external cues.

### Further Reading

Boivin, D.B., & James, F.O. (2005). Light treatment and circadian adaptation to shift work. *Ind Health*, 43(1): 34-48.

Boivin, D.B., Tremblay, G.M., & James, F.O. (2007). Working on atypical schedules. *Sleep Med*, 8(6): 578-99.

Mrosovsky, N. (2003). Beyond the suprachiasmatic nucleus. *Chronobiol Int*, 20(1): 1-8.

product melatonin to synchronize the various circadian rhythms with the 24-hour day–night cycle. For a discussion of problems associated with being out of sync with environmental cues, see Concepts, Challenges, and Controversies, p. 241.

## Melatonin

Daily changes in light intensity are the major environmental cue for the body to adjust the SCN master clock. Special photoreceptors in the retina pick up light signals and transmit them directly to the SCN. These photoreceptors are distinct from the rods and cones used to perceive, or see, light (p. 161). Scientists recently discovered that **melanopsin**, a protein found in a special retinal ganglion cell (p. 157), is the receptor for light that keeps the body in tune with external time. The vast majority of retinal ganglion cells receive input from the rod and cone photoreceptors. The axons of these ganglion cells form the *optic nerve* that carries information to the visual cortex in the occipital lobe (p. 166). Intermingled among the visually oriented retinal ganglion cells, about 1 to 2 percent of the retinal ganglion cells instead form an entirely independent light-detection system that responds to levels of illumination, like a light meter on a camera, rather than the contrasts, colours, and contours detected by the image-forming visual system. The melanopsin-containing, illumination-detecting retinal ganglion cells cue the pineal gland about the absence or presence of light by sending their signals along the **retino-hypothalamic tract** to the SCN. This pathway is distinct from the neural systems that result in vision perception. The SCN relays the message regarding light status to the pineal gland. This is the major way the internal clock is coordinated to a 24-hour day. Melatonin is the hormone of darkness. Melatonin secretion increases up to tenfold during the darkness of night and then falls to low levels during the light of day. Fluctuations in melatonin secretion, in turn, help entrain the body's biological rhythms with the external light–dark cues.

Proposed roles of melatonin besides regulating the body's biological clock include the following:

- Melatonin induces a natural sleep without the side effects that accompany hypnotic sedatives.
- Melatonin is believed to inhibit the hormones that stimulate reproductive activity. Puberty may be initiated by a reduction in melatonin secretion.

- In a related role, in some species, seasonal fluctuations in melatonin secretion associated with changes in the number of daylight hours are important triggers for seasonal breeding, migration, and hibernation.
- In another related role, melatonin is being used in clinical trials as a method of birth control, because at high levels it shuts down ovulation (egg release). A male contraceptive using melatonin to stop sperm production is also under development.
- Melatonin appears to be a very effective **antioxidant**, a defence tool against biologically damaging free radicals. *Free radicals* are very unstable electron-deficient particles that are highly reactive and destructive. Free radicals have been implicated in several chronic diseases, such as coronary artery disease (p. 374) and cancer, and are believed to contribute to the aging process.
- Evidence suggests that melatonin may slow the aging process, perhaps by removing free radicals or by other means.
- Melatonin appears to enhance immunity and has been shown to reverse some of the age-related shrinkage of the thymus, the source of T lymphocytes (p. 474), in old experimental animals. Melatonin taken in coordination with calcium is a potent stimulator of the T cell response.

*Clinical Note* Because of melatonin's many proposed roles, use of supplemental melatonin for a variety of conditions is very promising. However, most researchers are cautious about recommending supplemental melatonin until its effectiveness as a drug is further substantiated. Meanwhile, many people are turning to melatonin as a health food supplement; as such, it is not regulated by Health Canada for safety and effectiveness. The two biggest self-prescribed uses of melatonin are as a prevention for jet lag and as a sleep aid.

# Chapter in Perspective: Focus on Homeostasis

The endocrine system is one of the body's two major regulatory systems; the other is the nervous system. Through its relatively slowly acting hormone messengers, the endocrine system generally regulates activities that require duration rather than speed. Most of these activities are directed toward maintaining homeostasis. The specific contributions of the central endocrine organs to homeostasis are as follows:

- The hypothalamus–posterior pituitary unit secretes vasopressin, which acts on the kidneys during urine formation to help maintain water balance. Control of water balance, in turn, is essential for maintaining ECF osmolarity and proper cell volume.
- For the most part, the hormones secreted by the anterior pituitary do not directly contribute to homeostasis. Instead, most are tropic; that is, they stimulate the secretion of other hormones.
- However, growth hormone from the anterior pituitary, in addition to its growth-promoting actions, also exerts metabolic effects

that help maintain the plasma concentration of amino acids, glucose, and fatty acids.

- The pineal gland secretes melatonin, which helps entrain the body's circadian rhythm to the environmental cycle of light (period of activity) and dark (period of inactivity).

The peripheral endocrine glands further help maintain homeostasis in the following ways:

- Hormones help maintain the proper concentration of nutrients in the internal environment by directing chemical reactions involved in the cellular uptake, storage, and release of these molecules. Furthermore, the rate at which these nutrients are metabolized is controlled in large part by the endocrine system.
- Salt balance, which is important in maintaining the proper ECF volume and arterial blood pressure, is achieved by hormonally controlled adjustments in salt reabsorption by the kidneys during urine formation.
- Likewise, hormones act on various target cells to maintain the plasma concentration of calcium and other electrolytes. These electrolytes, in turn, play key roles in homeostatic activities.

For example, maintenance of calcium levels within narrow limits is critical for neuromuscular excitability and blood clotting, among other life-supporting actions.

- The endocrine system orchestrates a wide range of adjustments that help the body maintain homeostasis in response to stressful situations.

- The endocrine and nervous systems work in concert to control the circulatory and digestive systems, which in turn carry out important homeostatic activities.

Unrelated to homeostasis, hormones direct the growing process and control most aspects of the reproductive system.

## CHAPTER TERMINOLOGY

acromegaly (p. 238)
adenohypophysis (p. 226)
adenylyl cyclase (p. 217)
adrenocorticotropic hormone (ACTH, adrenocorticotropin) (p. 227)
androgens (p. 232)
antagonism (p. 214)
anterior pituitary gland (p. 225)
antioxidant (p. 242)
bone (p. 233)
calmodulin (p. 219)
cartilage (p. 233)
cascade (p. 220)
chondrocytes (p. 233)
clock proteins (p. 240)
cyclic adenosine monophosphate (cyclic AMP or cAMP) (p. 217)
cyclic guanosine monophosphate (cyclic GMP) (p. 220)
diacylglycerol (DAG) (p. 219)
diaphysis (p. 233)
diurnal (circadian) rhythm (p. 212)
down regulation (p. 214)
dwarfism (p. 238)
endocrine system (p. 209)
endocrinology (p. 209)
entrained (p. 212)
epiphyseal plate (p. 233)
epiphysis (p. 233)
follicle-stimulating hormone (FSH) (p. 227)
G protein (p. 217)
gigantism (p. 238)

gonadotropins (p. 227)
growth hormone (GH, somatotropin) (p. 227)
growth hormone–inhibiting hormone (GHIH) (p. 229)
growth hormone–releasing hormone (GHRH) (p. 229)
hormone response element (HRE) (p. 221)
hydrophilic hormones (p. 214)
hyperplasia (p. 233)
hypersecretion (p. 213)
hypertrophy (p. 233)
hypophysiotropic hormones (p. 229)
hyposecretion (p. 213)
hypothalamic-hypophyseal portal system (p. 229)
IGF-I (p. 235)
IGF-II (p. 235)
inhibiting hormones (p. 229)
inositol trisphosphate ($IP_3$) (p. 219)
insulin-like growth factors (IGF) (p. 235)
interstitial cell-stimulating hormone (ICSH) (p. 227)
jet lag (p. 240)
Laron dwarfism (p. 238)
lipophilic hormones (p. 215)
luteinizing hormone (LH) (p. 227)
melanin (p. 226)
melanocyte-stimulating hormones (MSH) (p. 226)
melanopsin (p. 242)
melatonin (p. 240)

neuroendocrine reflexes (p. 212)
neuroendocrinology (p. 222)
neurohypophysis (p. 225)
ossification (p. 233)
osteoblasts (p. 233)
osteoclasts (p. 233)
osteocytes (p. 233)
paraventricular nuclei (p. 226)
periosteum (p. 233)
permissiveness (p. 214)
phosphatidylinositol bisphosphate ($PIP_2$) (p. 219)
phospholipase C (p. 218)
pineal gland (p. 240)
pituitary gland (hypophysis) (p. 225)
positive-feedback (p. 212)
posterior pituitary gland (p. 225)
prolactin (PRL) (p. 227)
prolactin-inhibiting hormone (PIH) (p. 229)
protein kinase A (p. 217)
releasing hormones (p. 229)
retino-hypothalamic tract (p. 242)
somatomedins (p. 235)
somatostatin (p. 229)
suprachiasmatic nucleus (SCN) (p. 240)
supraoptic nuclei (p. 226)
synergism (p. 214)
thyroid-stimulating hormone (TSH, thyrotropin) (p. 227)
thyrotropin-releasing hormone (TRH) (p. 229)
tropic hormone (p. 210)

## Objective Questions (Answers in Appendix E, p. A-40)

1. One endocrine gland may secrete more than one hormone. *(True or false?)*

2. One hormone may influence more than one type of target cell. *(True or false?)*

3. All endocrine glands are exclusively endocrine in function. *(True or false?)*

4. A single target cell may be influenced by more than one hormone. *(True or false?)*

5. Hyposecretion or hypersecretion of a specific hormone can occur even though its endocrine gland is perfectly normal. *(True or false?)*

6. Growth hormone levels in the blood are no higher during the early childhood growing years than during adulthood. *(True or false?)*

7. A hormone that has as its primary function the regulation of another endocrine gland is classified functionally as a _____ hormone.

8. Self-induced reduction in the number of receptors for a specific hormone is known as _____.

9. Activity within the cartilaginous layer of bone known as the _____ is responsible for lengthening of long bones.

10. The _____ in the hypothalamus is the body's master biological clock.

11. Indicate the relationships among the hormones in the hypothalamic/anterior pituitary/adrenal cortex system by using the following answer code to identify which hormone belongs in each blank:

(a) cortisol
(b) ACTH
(c) CRH

(1) _____ from the hypothalamus stimulates the secretion of (2) _____ from the anterior pituitary. (3) _____, in turn, stimulates the secretion of (4) _____ from the adrenal cortex. In negative-feedback fashion, (5) _____ inhibits secretion of (6) _____ and furthermore reduces the sensitivity of the anterior pituitary to (7) _____.

## Written Questions

1. List the overall functions of the endocrine system.

2. How is the plasma concentration of a hormone normally regulated?

3. List and briefly state the functions of the posterior pituitary hormones.

4. List and briefly state the functions of the anterior pituitary hormones.

5. Compare the relationship between the hypothalamus and posterior pituitary with the relationship between the hypothalamus and anterior pituitary. Describe the role of the hypothalamic-hypophyseal portal system and the hypothalamic releasing and inhibiting hormones.

6. Describe the actions of growth hormone that are unrelated to growth. What are growth hormone's growth-promoting actions? What is the role of somatomedins (IGF)?

7. Discuss the control of growth hormone secretion.

8. What are the source, functions, and stimulus for secretion of melatonin?

## POINTS TO PONDER

### (Explanations in Appendix E, p. A-40)

1. Would you expect the concentration of hypothalamic releasing and inhibiting hormones in a systemic venous blood sample to be higher, lower, or the same as the concentration of these hormones in a sample of hypothalamic-hypophyseal portal blood?

2. Thinking about the feedback control loop among TRH, TSH, and thyroid hormone, would you expect the concentration of TSH to be normal, above normal, or below normal in a person whose diet is deficient in iodine (an element essential for synthesizing thyroid hormone)?

3. A patient displays symptoms of excess cortisol secretion. What factors could be measured in a blood sample to determine whether the condition is caused by a defect at the hypothalamic/anterior pituitary level or the adrenal cortex level?

4. Why would males with testicular feminization syndrome be unusually tall?

5. A black market for growth hormone abuse already exists among weightlifters and other athletes. What actions of growth hormone would induce a full-grown athlete to take supplemental doses of this hormone? What are the potential detrimental side effects?

(Explanation in Appendix E, p. A-41)

At 18 years of age and 2.4 metres tall, Anthony O. was diagnosed with gigantism caused by a pituitary tumour. The condition was treated by surgically removing his pituitary gland. What hormonal replacement therapy would Anthony need?

5

## The Endocrine Glands

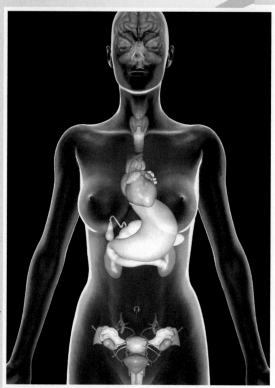

The female endocrine system

Body systems maintain
homeostasis

### Homeostasis

The endocrine system, one of the body's two major regulatory systems, secretes hormones that act on their target cells to regulate the blood concentrations of nutrient molecules, water, salt, and other electrolytes, among other homeostatic activities. Hormones also play a key role in controlling growth and reproduction and in stress adaptation.

Homeostasis is essential
for survival of cells

Cells make up body systems

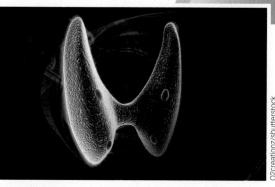

Endocrine gland

The endocrine system, by means of the blood-borne hormones it secretes, generally regulates activities that require duration rather than speed. The peripheral endocrine glands include the thyroid gland, which controls the body's basal metabolic rate; the adrenal glands, which secrete hormones important in metabolizing nutrient molecules, in adapting to stress, and in maintaining salt balance; the endocrine pancreas, which secretes hormones important in metabolizing nutrient molecules; and the parathyroid glands, which secrete a hormone important in calcium metabolism.

# The Endocrine Glands

**❚ Clinical Connections**

Megan, a 19-year-old female, away at university, was studying for mid-term exams when she suddenly experienced a pounding headache, blurred vision, sweating, and a rapid heartbeat. She assumed she was experiencing an anxiety attack due to the upcoming exam so she ignored it and went to bed. In class a few weeks later, she suddenly experienced similar symptoms, but much worse this time. Being away from her family, she decided to go to the hospital emergency room close to campus. Upon presentation, her blood pressure was 170/120 mmHg and her heart rate was 90 bpm, but within the hour her symptoms had vanished, and her blood pressure returned to the higher end of normal. Standard blood work came back normal. She was released and told to follow up with her family physician.

Later that evening she called her parents to tell them what had happened. Her father told her that his brother frequently had symptoms like that for years before he passed away at an age of 35 from what everyone assumed was a heart attack. The next day the symptoms returned, and she went again to the emergency room. Upon hearing her updated family history, the doctor ordered additional tests, including blood work, urine analysis, and an MRI. The urine analysis showed high concentrations of catecholamine metabolites, and an MRI found a tumour in the celiac ganglia. Ultimately she was diagnosed with pheochromocytoma due to an extra-adrenal paraganglioma.

# 6.1 | Thyroid Gland

The **thyroid gland** consists of two lobes of endocrine tissue joined in the middle by a narrow portion of the gland, giving it a bow-tie shape (> Figure 6-1a). The gland is even located in the appropriate place for a bow tie, lying over the trachea just below the larynx.

## Secretion of the thyroid hormone

The major thyroid secretory cells, known as **follicular cells**, are arranged into hollow spheres, each of which forms a functional unit called a **follicle**. On a microscopic section (> Figure 6-1b), the follicles appear as rings of follicular cells enclosing an inner lumen filled with **colloid**—a substance that serves as an extracellular storage site for thyroid hormone. Note that the colloid within the follicular lumen is extracellular (i.e., outside the thyroid cells), even though it is located within the interior of the follicle. Colloid is not in direct contact with the extracellular fluid that surrounds the follicle, similar to an inland lake that is not in direct contact with the oceans that surround a continent.

The chief constituent of the colloid is a large protein molecule known as **thyroglobulin (Tg)**, within which are incorporated the thyroid hormones in their various stages of synthesis. The follicular cells produce two iodine-containing hormones derived from the amino acid tyrosine: **tetraiodothyronine** ($T_4$ or thyroxine) and **triiodothyronine** ($T_3$). The prefixes *tetra* and *tri* and the subscripts *4* and *3* denote the number of iodine atoms incorporated into each of these hormones. Approximately 90 percent of the hormone secreted by the thyroid gland is in the form of $T_4$, whereas about 10 percent is in the form of $T_3$. These two hormones, collectively referred to as **thyroid hormone**, are important regulators of overall basal metabolic rate. The qualitative function of these hormones is the same, but they differ in their speed and intensity of action.

Interspersed in the interstitial spaces between the follicles is another secretory cell type, the **C cells**, so-called because they secrete the peptide hormone **calcitonin**. Calcitonin plays a role in calcium metabolism and is not related in any way to the two other major thyroid hormones. We discuss calcitonin in Section 6.5, when we examine endocrine control of calcium balance. Here, we discuss $T_4$ and $T_3$.

## Synthesis and storage of the thyroid hormone

The basic ingredients for thyroid hormone synthesis are tyrosine and iodine, both of which must be taken up from the blood by the follicular cells. Tyrosine, an amino acid, is synthesized in sufficient amounts by the body, so it is not a dietary essential. By contrast, the iodine needed for thyroid hormone synthesis must be obtained from dietary intake. The dietary iodine intake needed is about 1 mg per week. In the small intestine, dietary iodine is reduced to iodide for transport in the circulation. Most Canadians exceed the daily requirement through the use of common table salt. The synthesis, storage, and secretion of thyroid hormone involve the following process.

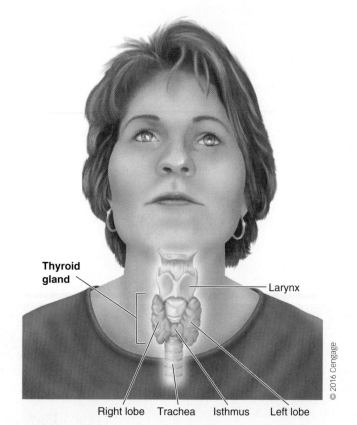

**Thyroid gland** — **Larynx**

Right lobe | Trachea | Isthmus | Left lobe

**(a)** Gross anatomy of thyroid gland

© 2016 Cengage

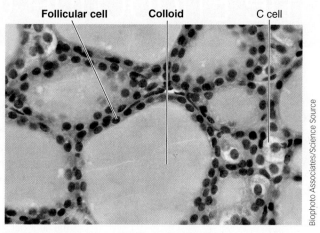

**Follicular cell** | **Colloid** | **C cell**

Biophoto Associates/Science Source

**(b)** Light-microscopic appearance of thyroid gland

> **FIGURE 6-1 Anatomy of the thyroid gland.** (a) Gross anatomy of the thyroid gland, anterior view. The thyroid gland lies over the trachea just below the larynx and consists of two lobes connected by a thin strip called the *isthmus*. (b) Light-microscope appearance of the thyroid gland. The thyroid gland is composed primarily of colloid-filled spheres enclosed by a single layer of follicular cells.

The majority of steps for thyroid hormone synthesis takes place on the thyroglobulin molecules within the colloid. Thyroglobulin itself is produced by the endoplasmic reticulum-Golgi complex of the thyroid follicular cells. Many molecules of the amino acid tyrosine become incorporated in the much larger thyroglobulin molecules as the latter are being produced. Once produced, tyrosine-containing thyroglobulin is exported from the follicular cells into the colloid by exocytosis (step **1** in › Figure 6-2).

The thyroid captures (traps) iodide from the blood and transfers it into the colloid by an *iodide pump*—the powerful, energy-requiring carrier proteins in the outer membranes of the follicular cells (step **2**). The iodide pump cotransports $Na^+$ and is driven by the concentration gradient of $Na^+$, which is established by the $Na^+-K^+$ pump located on the basolateral membrane. In moving $Na^+$ down its concentration gradient, iodide moves *against* its concentration gradient and becomes trapped in the thyroid for thyroid hormone synthesis. The iodide pump concentrates the iodide into the thyroid glandular cells and follicles to about 30 times its concentration in the blood, which is sometimes referred to as *iodide trapping*. Iodide serves no other function in the body.

Once inside the follicular cell, iodide is transported into the colloid of the follicular lumen (step **3**). Within the follicle, iodide

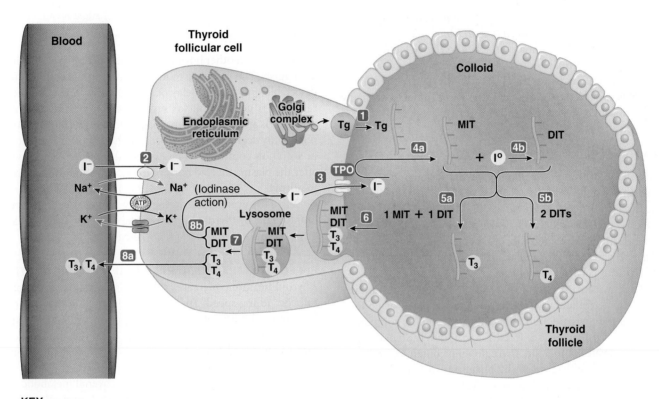

**KEY**

| | | | | | |
|---|---|---|---|---|---|
|  | = Primary active transport | $I^o$ | = Iodine | **MIT** | = Monoiodotyrosine |
| | = Secondary active transport (symporter) | **Tg** | = Thyroglobulin | **DIT** | = Di-iodotyrosine |
| | | $I^-$ | = Iodide | $T_3$ | = Tri-iodothyronine |
| | | **TPO** | = Thyroperoxidase | $T_4$ | = Tetraiodothyronine (thyroxine) |

**1** Tyrosine-containing Tg produced within the thyroid follicular cells by the endoplasmic reticulum–Golgi complex is transported by exocytosis into the colloid.

**2** Iodide is carried by secondary active transport from the blood into the colloid by symporters in the basolateral membrane of the follicular cells.

**3** Iodide exits the cell through a luminal channel to enter the colloid.

**4a** Catalyzed by TPO, iodide is simultaneously oxidized to iodine ($I^o$) and attached to tyrosine within the Tg molecule to yield MIT.

**4b** Attachment of two iodines to tyrosine yields DIT.

**5a** Coupling of one MIT and one DIT yields $T_3$.

**5b** Coupling of two DITs yields $T_4$.

**6** On appropriate stimulation, the thyroid follicular cells engulf a portion of Tg-containing colloid by phagocytosis.

**7** Lysosomes attack the engulfed vesicle and split the iodinated products from Tg.

**8a** $T_3$ and $T_4$ diffuse into the blood (secretion).

**8b** MIT and DIT are deiodinated, and the freed iodide is recycled for synthesizing more hormone.

› **FIGURE 6-2 Synthesis, storage, and secretion of thyroid hormone**

is rapidly converted into its highly reactive state, iodine, and attached to a tyrosine within the thyroglobulin molecule. This conversion (called *iodide organification*) and attachment occur almost simultaneously and is mediated by the membrane-bound enzyme **thyroperoxidase (TPO)**. Attachment of one iodine to tyrosine yields **monoiodotyrosine (MIT)** (step **4a**). Attachment of two iodines to tyrosine yields **di-iodotyrosine (DIT)** (step **4b**). A coupling process occurs between the iodinated tyrosine molecules to form the thyroid hormones. Coupling of one MIT (with one iodide) and one DIT (with two iodides) yields *triiodothyronine* ($T_3$) (with three iodides) (step **5a**). Coupling of two DITs (each bearing two iodide atoms) yields *tetraiodothyronine* ($T_4$ or thyroxine), the four-iodide form of *thyroid hormone* (step **5b**). *Coupling does not occur between two MIT molecules.*

All these products remain attached to thyroglobulin. Thyroid hormones remain stored in this form in the colloid until they are split off and secreted. The ratio of $T_3$ to $T_4$ stored in the colloid is about 1 to 10. Sufficient thyroid hormone is normally stored to supply the body's needs for several months.

## Thyroglobulin-laden colloid

The release of thyroid hormone into the systemic circulation is a rather complex process, for two reasons. First, before their release, $T_3$ and $T_4$ are still bound within the thyroglobulin molecule. Second, these hormones are stored within the follicular lumen, so they must be transported completely across the follicular cells to reach the capillaries that course through the interstitial spaces between the follicles.

The process of thyroid hormone secretion essentially involves the follicular cells "biting off" a piece of colloid, breaking the thyroglobulin molecule down into its component parts, and "spitting out" the freed $T_3$ and $T_4$ into the blood. Upon the appropriate stimulation, thyroglobulin containing colloid is taken into the follicular cells by phagocytosis (step **6**). Once inside the follicular cells, lysosomes fuse with these vesicles and release digestive enzymes that release $T_3$ and $T_4$, as well as the inactive iodotyrosines, MIT and DIT (step **7**).

The thyroid hormones, being very lipophilic, pass freely through the outer membranes of the follicular cells and into the blood (step **8a**). The MIT and DIT are of no endocrine value. The follicular cells contain an enzyme—iodinase—that swiftly removes the iodide from MIT and DIT, allowing the freed iodide to be recycled for synthesis of more hormone (step **8b**). This highly specific enzyme removes iodide only from the worthless MIT and DIT, not from the valuable $T_3$ or $T_4$.

Once released into the blood, the highly lipophilic (and, therefore, water-insoluble) thyroid hormone molecules very quickly bind with several plasma proteins. The majority of circulating $T_4$ and $T_3$ is transported by **thyroxine-binding globulin**, a plasma protein that selectively binds only thyroid hormone. The binding affinity is six times as great for $T_4$ as it is for $T_3$. Less than 0.1 percent of the $T_4$ and less than 1 percent of the $T_3$ remain in the unbound (free) form. This is remarkable, considering that only the free portion of the total thyroid hormone pool has access to the target-cell receptors and thus can exert an effect.

## $T_4$ and $T_3$

About 90 percent of the secretory product released from the thyroid gland is in the form of $T_4$, yet $T_3$ is about four times as potent in its biological activity. However, most of the secreted $T_4$ is converted into $T_3$, or *activated*, by being stripped of one of its iodines outside the thyroid gland, primarily in the liver and kidneys. About 80 percent of the circulating $T_3$ is derived from secreted $T_4$ that has been peripherally stripped. Therefore, $T_3$ is the major biologically active form of thyroid hormone at the cellular level, even though the thyroid gland secretes mostly $T_4$.

## Thyroid hormone and basal metabolic rate

Compared with other hormones, the action of thyroid hormone is sluggish. The response to an increase in thyroid hormone is detectable only after a delay of several hours, and the maximal response is not evident for several days. The duration of the response is also quite long, partially because thyroid hormone is not rapidly degraded but also because the response to an increase in secretion continues to be expressed for days or even weeks after the plasma thyroid hormone concentrations have returned to normal.

Virtually every tissue in the body is affected either directly or indirectly by thyroid hormone. The effects of $T_3$ and $T_4$ can be grouped into several overlapping categories.

### EFFECT ON METABOLIC RATE AND HEAT PRODUCTION
Thyroid hormone increases the body's overall basal metabolic rate, or "idling speed" (p. 700). It is the most important regulator of the body's rate of $O_2$ consumption and energy expenditure under resting conditions. The basal metabolic rate can increase to as much as 100 percent above normal when significant amounts of thyroid hormone are released.

Closely related to thyroid hormone's overall metabolic effect is its **calorigenic (heat-producing) effect**. Increased metabolic activity results in increased heat production.

### EFFECT ON INTERMEDIARY METABOLISM
In addition to increasing the general metabolic rate, thyroid hormone modulates the rates of many specific reactions involved in fuel metabolism. The effects of thyroid hormone on the metabolic fuels are multifaceted: Not only can it influence both the synthesis and degradation of carbohydrate, fat, and protein, but small or large amounts of the hormone may also induce opposite effects. For example, the conversion of glucose to glycogen (the storage form of glucose) is facilitated by small amounts of thyroid hormone, but the reverse—the breakdown of glycogen into glucose—occurs with large amounts of the hormone. Similarly, adequate amounts of thyroid hormone are essential for the protein synthesis needed for normal bodily growth, yet at high doses, as in thyroid hypersecretion, thyroid hormone favours protein degradation. Because thyroid hormone increases the quantity of enzymes involved in metabolism, there is also an increased need for vitamins, as vitamins are an essential element of some enzymes.

## SYMPATHOMIMETIC EFFECT

Any action similar to one produced by the sympathetic nervous system is known as a **sympathomimetic (sympathetic-mimicking) effect**. Thyroid hormone increases target-cell responsiveness to catecholamines (epinephrine and norepinephrine)—the chemical messengers used by the sympathetic nervous system and its hormonal reinforcements from the adrenal medulla. Thyroid hormone accomplishes this permissive action by causing a proliferation of specific catecholamine target-cell receptors (p. 214). Because of this action, many of the effects observed when thyroid hormone secretion is elevated are similar to those that accompany activation of the sympathetic nervous system.

## EFFECT ON THE CARDIOVASCULAR SYSTEM

Through its effect of increasing the heart's responsiveness to circulating catecholamines, thyroid hormone increases heart rate and force of contraction, thus increasing cardiac output (p. 365). In addition, in response to the heat load generated by the calorigenic effect of thyroid hormone, peripheral vasodilation occurs to carry the extra heat to the body surface for elimination to the environment (p. 715).

Aside from increasing the excitability and contractility of the heart, thyroid hormone also influences other aspects of the cardiovascular system. For example, thyroid hormone will cause an increase in blood volume and flow, but no change in arterial blood pressure.

## EFFECT ON GROWTH AND THE NERVOUS SYSTEM

Thyroid hormone is essential for normal growth because of its effects on growth hormone (GH) and IGF-I (p. 235). Thyroid hormone not only stimulates GH secretion and increases production of IGF-I by the liver but also promotes the effects of GH and IGF-I on the synthesis of new structural proteins and on skeletal growth. Thyroid-deficient children have stunted growth that can be reversed by thyroid replacement therapy. Unlike excess GH, however, excess thyroid hormone does not produce excessive growth.

Thyroid hormone plays a crucial role in the normal development of the nervous system, especially the CNS, an effect impeded in children who have thyroid deficiency from birth. Thyroid hormone is also essential for normal CNS activity in adults.

## EFFECT ON SKELETAL MUSCLE

Thyroid hormone—specifically $T_3$—has a profound influence on skeletal muscle. $T_3$ influences gene expression at transcription, posttranscription, translation, and posttranslation. Some examples of $T_3$ effects on skeletal muscle are increased muscle size, since $T_3$ influences the distribution of fibre types in the muscle (slow twitch is more sensitive to $T_3$ than fast twitch); increased calcium uptake by the sarcoplasmic reticulum; and increased maximum shortening velocity of muscle. You will learn about skeletal muscle in Chapter 7.

## The hypothalamus–pituitary–thyroid axis

Thyroid-stimulating hormone (TSH), the thyroid tropic hormone from the anterior pituitary, is the most important physiological regulator of thyroid hormone secretion (> Figure 6-3; see

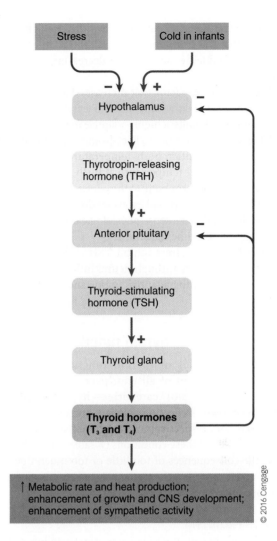

> FIGURE 6-3 **Regulation of thyroid hormone secretion**

also Chapter 5, p. 227). Almost every step of thyroid hormone synthesis and release is stimulated by TSH.

In addition to enhancing thyroid hormone secretion, TSH maintains the structural integrity of the thyroid gland. In the absence of TSH, the thyroid atrophies (decreases in size) and secretes its hormones at a very low rate. Conversely, it undergoes hypertrophy (increase in the size of each follicular cell) and hyperplasia (increase in the number of follicular cells) in response to excess TSH stimulation.

The hypothalamic thyrotropin-releasing hormone (TRH), in tropic fashion, turns on TSH secretion by the anterior pituitary (p. 229), whereas thyroid hormone, in negative-feedback fashion, turns off TSH secretion by inhibiting the anterior pituitary. Like other negative-feedback loops, the one between thyroid hormone and TSH tends to maintain a stable thyroid hormone output.

Negative feedback between the thyroid and anterior pituitary accomplishes day-to-day regulation of free thyroid hormone levels, whereas the hypothalamus mediates long-term adjustments. Unlike most other hormonal systems, the hormones in the thyroid axis in an adult normally do not undergo sudden, wide swings in secretion. The relatively steady rate of thyroid hormone secretion is in keeping with the sluggish, long-lasting responses that this hormone induces; there would be no adaptive value in suddenly increasing or decreasing plasma thyroid hormone levels.

The only known factor that increases TRH secretion (and, accordingly, TSH and thyroid hormone secretion) is exposure to cold in newborn infants, a highly adaptive mechanism. Scientists think the dramatic increase in heat-producing thyroid hormone secretion helps maintain body temperature during the abrupt drop in surrounding temperature at birth as the infant passes from the mother's warm body to the cooler environmental air. A similar TSH response to cold exposure does not occur in adults, although it would make sense physiologically and does occur in many types of experimental animals.

Various types of stress inhibit TSH and thyroid hormone secretion, presumably through neural influences on the hypothalamus, although the adaptive importance of this inhibition is unclear.

## Abnormalities of thyroid function

*Clinical Note* Abnormalities of thyroid function are among the most common of all endocrine disorders. They fall into two major categories—**hypothyroidism** and **hyperthyroidism**—reflecting deficient and excess thyroid hormone secretion, respectively. A number of specific causes can give rise to each of these conditions (▌ Table 6-1). Whatever the cause, the consequences of too little or too much thyroid hormone secretion are largely predictable, given knowledge of the functions of thyroid hormone.

### HYPOTHYROIDISM

Hypothyroidism can result (1) from primary failure of the thyroid gland itself; (2) secondary to a deficiency of TRH, TSH, or both; or (3) from an inadequate dietary supply of iodine.

The symptoms of hypothyroidism are largely caused by a reduction in overall metabolic activity. Among other things, a patient with hypothyroidism has a reduced basal metabolic rate; displays poor tolerance of cold (lack of the calorigenic effect); has a tendency to gain excessive weight (not burning fuels at a normal rate); is easily fatigued (lower energy production); has a slow, weak pulse (caused by a reduction in the rate and strength of cardiac contraction and a lowered cardiac output); and exhibits slow reflexes and slow mental responsiveness (because of the effect on the nervous system). The mental effects are characterized by diminished alertness, slow speech, and poor memory.

Another notable characteristic is an oedematous condition caused by infiltration of the skin with complex, water-retaining carbohydrate molecules, presumably as a result of altered metabolism. The resultant puffy appearance, primarily of the face, hands, and feet, is known as **myxoedema**. In fact, the term *myxoedema* is often used as a synonym for hypothyroidism in an adult, because of the prominence of this symptom.

If a person has hypothyroidism from birth, a condition known as **cretinism** develops. Because adequate levels of thyroid hormone are essential for normal growth and CNS development, cretinism is characterized by dwarfism and mental retardation, as well as other general symptoms of thyroid deficiency. The mental retardation is preventable if replacement therapy is started promptly, but it is not reversible once it has developed for a few months after birth, even with later treatment with thyroid hormone.

Treatment of hypothyroidism, with one exception, consists of replacement therapy by administering exogenous thyroid hormone. The exception is hypothyroidism caused by iodine deficiency, in which the remedy is adequate dietary iodine.

### HYPERTHYROIDISM

The most common cause of hyperthyroidism is **Graves' disease**. This is an autoimmune disease in which the body erroneously produces **long-acting thyroid stimulator (LATS)**, an antibody whose target is the TSH receptors on the thyroid cells. LATS stimulates both secretion and growth of the thyroid in a manner

### ▌ TABLE 6-1 Types of Thyroid Dysfunctions

| Thyroid Dysfunction | Cause | Plasma Concentrations of Relevant Hormones | Goitre Present? |
|---|---|---|---|
| **Hypothyroidism** | Primary failure of the thyroid gland | ↓ $T_3$ and $T_4$, ↑ TSH | Yes |
| | Secondary to hypothalamic or anterior pituitary failure | ↓ $T_3$ and $T_4$, ↓ TRH and/or ↓ TSH | No |
| | Lack of dietary iodine | ↓ $T_3$ and $T_4$, ↑ TSH | Yes |
| **Hyperthyroidism** | Abnormal presence of long-acting thyroid stimulator (LATS) (Graves' disease) | ↓ $T_3$ and $T_4$, ↓ TSH | Yes |
| | | ↓ $T_3$ and $T_4$, ↑ TRH and/or ↑ TSH | Yes |
| | Secondary to excess hypothalamic or anterior pituitary secretion | ↓ $T_3$ and $T_4$, ↓ TSH | No |
| | Hypersecreting thyroid tumour | | |

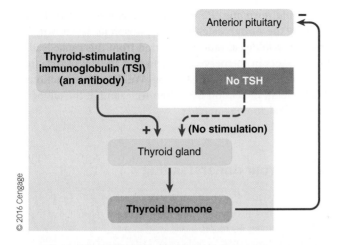

> **FIGURE 6-4 Role of long-acting thyroid stimulator in Graves' disease.** Long-acting thyroid stimulator (LATS), an antibody erroneously produced in the autoimmune condition of Graves' disease, binds with the TSH receptors on the thyroid gland and continuously stimulates thyroid hormone secretion outside the normal negative-feedback control system.

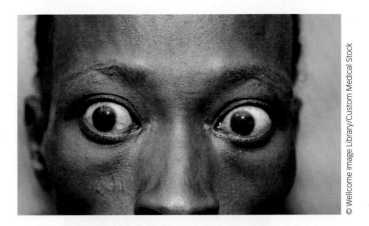

> **FIGURE 6-5 A woman with Graves' disease displaying exophthalmos.** Abnormal swelling of muscle and fat behind the eyeballs causes them to bulge forward.

**6**

similar to TSH. Unlike TSH, however, LATS is not subject to negative-feedback inhibition by thyroid hormone, so thyroid secretion and growth continue unchecked (> Figure 6-4). People with hyperthyroidism have shown secretion rates as great as 10 to 15 times the normal rate. Less frequently, hyperthyroidism occurs secondary to excess TRH or TSH or in association with a hypersecreting thyroid tumour.

As expected, the hyperthyroid patient has an elevated basal metabolic rate. The resultant increase in heat production leads to excessive perspiration and poor tolerance of heat. Despite the increased appetite and food intake that occur in response to the increased metabolic demands, body weight typically falls because the body is burning fuel at an abnormally rapid rate. Net degradation of carbohydrate, fat, and protein stores occurs. The resultant loss of skeletal muscle protein results in weakness. Various cardiovascular abnormalities are associated with hyperthyroidism, caused both by the direct effects of thyroid hormone and by its interactions with catecholamines. Heart rate and strength of contraction may increase so much that the individual has palpitations (an unpleasant awareness of the heart's activity). In severe cases, the heart may fail to meet the body's metabolic demands despite increased cardiac output. The effects on the CNS are characterized by an excessive degree of mental alertness to the point where the patient is irritable, tense, anxious, and excessively emotional.

A prominent feature of Graves' disease but not of the other types of hyperthyroidism is **exophthalmos** (bulging eyes) (> Figure 6-5). Complex, water-retaining carbohydrates are deposited behind the eyes, although why this happens is still unclear. The resulting fluid retention pushes the eyeballs forward so they bulge from their bony orbit. The eyeballs may bulge so far that the lids cannot completely close, in which case the eyes become dry, irritated, and prone to corneal ulceration. Even after correction of the hyperthyroid condition, these troublesome eye symptoms may persist.

Three general methods of treatment can suppress excess thyroid hormone secretion: surgical removal of a portion of the oversecreting thyroid gland; administration of radioactive iodine, which, after being concentrated in the thyroid gland by the iodide pump, selectively destroys thyroid glandular tissue; and use of antithyroid drugs that specifically interfere with thyroid hormone synthesis.

## Thyroid gland overstimulation

*Clinical Note* A **goitre** is an enlarged thyroid gland. Because the thyroid lies over the trachea, a goitre is readily palpable and usually highly visible (> Figure 6-6). A goitre occurs whenever either TSH or LATS excessively stimulates the thyroid gland. Note from Table 6-1 that a goitre may accompany hypothyroidism or hyperthyroidism, but it need not be present in either condition. Knowing the hypothalamus–pituitary–thyroid axis and feedback control, we can predict which types of thyroid dysfunction will produce a goitre. Let's consider hypothyroidism first.

- Hypothyroidism secondary to hypothalamic or anterior pituitary failure will not be accompanied by a goitre, because the thyroid gland is not being adequately stimulated, let alone excessively stimulated.

- With hypothyroidism caused by thyroid gland failure or lack of iodine, a goitre does develop, because the circulating level of thyroid hormone is so low that there is little negative-feedback inhibition on the anterior pituitary, and TSH secretion is therefore elevated. TSH acts on the thyroid to increase the size and number of follicular cells and to increase their rate of secretion. If the thyroid cells cannot secrete hormone because of a lack of a critical enzyme or lack of iodine, no amount of TSH will be able to induce these cells to secrete $T_3$ and $T_4$. However, TSH can still promote hypertrophy and hyperplasia of the thyroid, with a consequent paradoxical enlargement of the gland (i.e., a goitre), even though the gland is still underproducing.

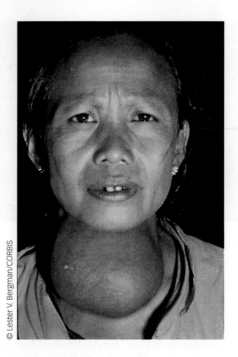

> FIGURE 6-6 Patient with a goitre

Similarly, a goitre may or may not accompany hyperthyroidism:

- Excessive TSH secretion resulting from a hypothalamic or anterior pituitary defect would obviously be accompanied by a goitre and excess $T_3$ and $T_4$ secretion because of overstimulation of thyroid growth. Because the thyroid gland in this circumstance is also capable of responding to excess TSH with increased hormone secretion, hyperthyroidism is present with this goitre.

- In Graves' disease, a hypersecreting goitre occurs because LATS promotes growth of the thyroid, as well as enhancing secretion of thyroid hormone. Because the high levels of circulating $T_3$ and $T_4$ inhibit the anterior pituitary, TSH secretion itself is low. In all other cases when a goitre is present, TSH levels are elevated and are directly responsible for excessive growth of the thyroid.

- Hyperthyroidism resulting from overactivity of the thyroid in the absence of overstimulation, such as that caused by an uncontrolled thyroid tumour, is not accompanied by a goitre. The spontaneous secretion of excessive amounts of $T_3$ and $T_4$ inhibits TSH, so there is no stimulatory input to promote growth of the thyroid.

## Parathyroid glands

There are four parathyroid glands, each the size of rice grain, located on the back surface of the thyroid gland, one in each corner. These glands share the same blood supply and venous drainage as the thyroid gland. Despite their small size, the parathyroid glands are essential for life, meaning that if the glands are removed death with follow, usually within a few days. The primary hormone secreted by the parathyroid gland is named

*parathyroid hormone (PTH)*. The primary role of PTH is to maintain the body's calcium and phosphate levels within a narrow physiological range. Deviation from this range leads to rapid changes in the nervous system and skeletal muscle function, which can be life threatening. We discuss the actions of PTH in detail in Section 6.5, when we examine the endocrine control of calcium metabolism.

---

### Check Your Understanding 6.1

1. Define *thyroid follicle, colloid, thyroglobulin, MIT, DIT, $T_3$*, and *$T_4$*.

2. Draw a flow diagram showing the effects and regulation of thyroid hormone secretion.

3. Explain how both hyperthyroidism and hypothyroidism can lead to a goitre.

---

## 6.2 | Adrenal Glands

There are two **adrenal glands** (4 g each), one embedded above each kidney in a capsule of fat (*ad* means "next to"; *renal* means "kidney") (> Figure 6-7a).

### A steroid-secreting cortex and a catecholamine-secreting medulla

Each adrenal is composed of two endocrine organs, one surrounding the other. The outer layers composing the **adrenal cortex** secrete a variety of steroid hormones; the inner portion, the adrenal medulla, secretes catecholamines. Thus, the adrenal cortex and medulla secrete hormones belonging to different chemical categories, whose functions, mechanisms of action, and regulation are entirely different. We first examine the adrenal cortex and then turn our attention to the adrenal medulla.

### Mineralocorticoids, glucocorticoids, and sex hormones

The adrenal cortex consists of three layers or zones: the **zona glomerulosa**, the outermost layer; the **zona fasciculata**, the middle and largest portion; and the **zona reticularis**, the innermost zone (> Figure 6-7b). The adrenal cortex produces a number of different **adrenocortical hormones**, all of which are steroids derived from a common precursor molecule, cholesterol (p. 216). Slight variations in structure confer different functional capabilities on the various adrenocortical hormones (corticosteroids). On the basis of their primary actions, the adrenal steroids can be divided into three categories:

1. **Mineralocorticoids**, mainly *aldosterone*, influence mineral (electrolyte) balance, specifically sodium and potassium balance.

2. **Glucocorticoids**, primarily *cortisol*, play a major role in glucose metabolism as well as in protein and lipid metabolism.

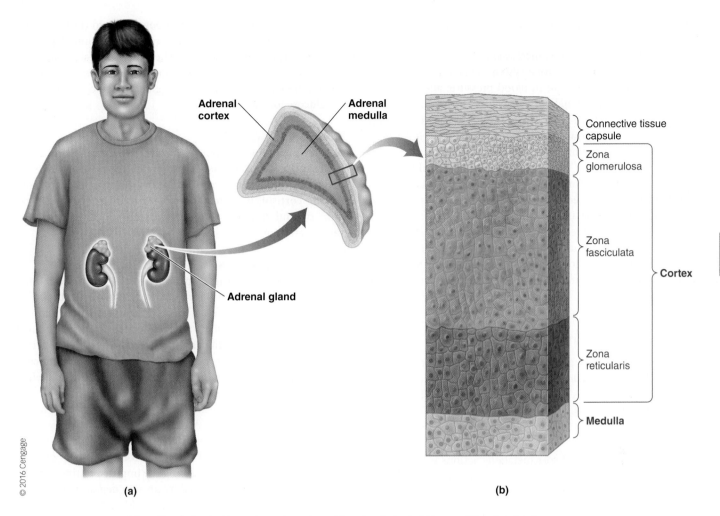

> FIGURE 6-7 **Anatomy of the adrenal glands.** (a) Location and structure of the adrenal glands. (b) Layers of the adrenal cortex

3. **Sex hormones** are identical or similar to those produced by the gonads (testes in males, ovaries in females). The most abundant and physiologically important of the adrenocortical sex hormones is *dehydroepiandrosterone*—considered a male sex hormone, although produced by both men and women.

The three categories of adrenal steroids are produced in anatomically distinct portions of the adrenal cortex. This is due to the differential distribution of the enzymes required to catalyze the different biosynthetic pathways leading to the formation of each of these steroids. Of the two major adrenocortical hormones, aldosterone is produced exclusively in the zona glomerulosa, whereas cortisol synthesis is limited to the two inner layers of the cortex—the zona fasciculata and zona reticularis—, with the former being the major source of this glucocorticoid. No other steroidogenic tissues have the capability of producing either mineralocorticoids or glucocorticoids. In contrast, the adrenal sex hormones, also produced by the two inner cortical zones, are produced in far greater abundance in the gonads.

Being lipophilic, the adrenocortical hormones are all carried in the blood extensively bound to plasma proteins. Cortisol is bound mostly to a plasma protein specific for it called **corticosteroid-binding globulin (transcortin)**, whereas aldosterone and dehydroepiandrosterone are largely bound to albumin, which nonspecifically binds with a variety of lipophilic hormones.

## Mineralocorticoids' major effects on Na⁺ and K⁺ balance

The actions and regulation of the primary adrenocortical mineralocorticoid, **aldosterone**, are described thoroughly in Chapters 13 and 14. The principal site of aldosterone action is on the distal and collecting tubules of the kidney, where it promotes $Na^+$ retention and enhances $K^+$ elimination during the formation of urine. The promotion of $Na^+$ retention by aldosterone secondarily induces osmotic retention of water, expanding the ECF volume, which is important in the long-term regulation of blood pressure.

Mineralocorticoids are *essential for life*. Without aldosterone, a person rapidly dies (in about two days to two weeks) from circulatory shock due to the marked fall in plasma volume caused by excessive losses of $H_2O$-holding $Na^+$. With most other hormonal deficiencies, death is not imminent, even though a chronic hormonal deficiency may eventually lead to a premature death.

Aldosterone secretion is increased by (1) activation of the renin–angiotensin–aldosterone system by factors related to a reduction in Na$^+$ and a fall in blood pressure; and (2) direct stimulation of the adrenal cortex by a rise in plasma K$^+$ concentration (see Figure 13-22). However, aldosterone secretion is intermingled with extracellular fluid volume, extracellular fluid electrolyte concentration, blood volume, blood pressure, and, in general, renal function. Thus, it is difficult to discuss aldosterone secretion separately from these other factors. In addition to its effect on aldosterone secretion, angiotensin promotes growth of the zona glomerulosa, in a manner similar to the effect of TSH on the thyroid. Adrenocorticotropic hormone (ACTH) from the anterior pituitary primarily promotes the secretion of cortisol, not aldosterone. Therefore, unlike cortisol regulation, the regulation of aldosterone secretion is largely independent of anterior pituitary control.

## Glucocorticoids' metabolic effects

**Cortisol**, the primary glucocorticoid, plays an important role in carbohydrate, protein, and fat metabolism; executes significant permissive actions for other hormonal activities; and helps people cope with stress.

### METABOLIC EFFECTS

The overall effect of cortisol's metabolic actions is to increase the concentration of blood glucose at the expense of protein and fat stores. Specifically, cortisol performs the following functions:

- It stimulates hepatic **gluconeogenesis**, the conversion of non-carbohydrate sources (namely, amino acids) into carbohydrate within the liver (*gluco* means "glucose"; *neo* means "new"; *genesis* means "production"). Cortisol can increase the rate of gluconeogenesis by as much as 10 times. Cortisol accomplishes this by its influence on the enzymes used by the liver to change amino acids to glucose and through its ability to increase mobilization of amino acids from muscle tissue. Between meals or during periods of fasting, when no new nutrients are being absorbed into the blood for use and storage, the glycogen (stored glucose) in the liver tends to become depleted as it is broken down to release glucose into the blood. Gluconeogenesis is an important factor in replenishing hepatic glycogen stores and thus in maintaining normal blood glucose levels between meals. This is essential because the brain can use only glucose as its metabolic fuel, yet nervous tissue cannot store glycogen to any extent. The concentration of glucose in the blood must therefore be maintained at an appropriate level to adequately supply the glucose-dependent brain with nutrients.

- It inhibits glucose uptake and use by many tissues but not by the brain, thus sparing glucose for use by the brain, which absolutely requires it as a metabolic fuel. This action contributes to the increase in blood glucose concentration brought about by gluconeogenesis.

- It facilitates lipolysis, the breakdown of lipid (fat) stores in adipose tissue, which releases free fatty acids into the blood (*lysis* means "breakdown"). The mobilized fatty acids are available as an alternative metabolic fuel for tissues that can use this energy source in lieu of glucose, thereby conserving glucose for the brain.

### PERMISSIVE ACTIONS

Cortisol is extremely important for its permissiveness (p. 214). For example, cortisol must be present in adequate amounts to permit the catecholamines to induce vasoconstriction. A person lacking cortisol, if untreated, may go into circulatory shock in a stressful situation that demands immediate widespread vasoconstriction.

### ROLE IN ADAPTATION TO STRESS

Cortisol plays a key role in adaptation to stress. Stress of any kind is one of the major stimuli for increased cortisol secretion. For example, some of the different forms of stress associated with an increase in cortisol include trauma, infection, surgery, extreme heat or cold, any debilitating disease, and fear (e.g., being attacked by a person or animal). Although cortisol's precise role in adapting to stress is not known, consider the following speculative, but plausible, explanation of what might occur if a person is faced with a life-threatening situation and must forgo eating for a period of time. A cortisol-induced shift away from protein and fat stores in favour of expanded carbohydrate stores and increased availability of blood glucose would help protect the brain from malnutrition during the imposed fasting period. Also, the amino acids liberated by protein degradation would provide a readily available supply of building blocks for tissue repair if physical injury occurred. As a result, an increased pool of glucose, amino acids, and fatty acids would be available for use as needed.

### ANTI-INFLAMMATORY AND IMMUNOSUPPRESSIVE EFFECTS

*Clinical Note* When cortisol or synthetic cortisol-like compounds are administered to yield higher-than-physiologic concentrations of glucocorticoids (i.e., *pharmacological levels*), not only are all the metabolic effects magnified but several important new actions not evidenced at normal physiologic levels are seen. The most noteworthy of glucocorticoids' pharmacological effects are *anti-inflammatory* and *immunosuppressive* (p. 471). (Although these actions are traditionally considered to occur only at pharmacologic levels, recent studies suggest cortisol may exert anti-inflammatory effects even at normal physiologic levels.) Synthetic glucocorticoids have been developed that maximize the anti-inflammatory and immunosuppressive effects of these steroids while minimizing the metabolic effects.

Administering large amounts of glucocorticoid inhibits almost every step of the inflammatory response, making these steroids effective drugs in treating conditions in which the inflammatory response itself has become destructive, such as *rheumatoid arthritis*. Glucocorticoids used in this manner do not affect the underlying disease process; they merely suppress the body's response to the disease. Because glucocorticoids also exert multiple inhibitory effects on the overall immune process, such as "knocking out of commission" the white blood cells responsible for antibody production as well as those that directly destroy foreign cells, these agents have also proved useful in managing various allergic disorders and in preventing organ transplant rejections.

When these steroids are employed therapeutically, they should be used only when warranted and, even then, only sparingly, for several important reasons. First, because they suppress the normal inflammatory and immune responses that form the backbone of the body's defence system, a glucocorticoid-treated person has limited ability to resist infections. Second, in addition to the anti-inflammatory and immunosuppressive effects readily exhibited at pharmacologic levels, other less desirable effects may also be observed with prolonged exposure to higher-than-normal concentrations of glucocorticoids. These effects include development of gastric ulcers, high blood pressure, atherosclerosis, menstrual irregularities, and bone thinning. Third, high levels of exogenous glucocorticoids act in negative-feedback fashion to suppress the hypothalamus–pituitary axis that drives normal glucocorticoid secretion and maintains the integrity of the adrenal cortex. Prolonged suppression of this axis can lead to irreversible atrophy of the cortisol-secreting cells of the adrenal gland and thus to permanent inability of the body to produce its own cortisol.

## Cortisol secretion

Cortisol secretion by the adrenal cortex is regulated by a negative-feedback system involving the hypothalamus and anterior pituitary (› Figure 6-8). ACTH from the anterior pituitary stimulates the adrenal cortex to secrete cortisol. ACTH is derived from a large precursor molecule, **pro-opiomelanocortin (POMC)**, produced within the endoplasmic reticulum of the anterior pituitary's ACTH-secreting cells. Prior to secretion, this large precursor is processed into ACTH and several other biologically active peptides, namely, *melanocyte-stimulating hormone (MSH)* and *β-endorphin*, a morphine-like substance. The possible significance of the fact that these multiple secretory products are developed from a single precursor molecule will be addressed later in this section.

Being tropic to the zona fasciculata and zona reticularis, ACTH stimulates both the growth and the secretory output of these two inner layers of the cortex. In the absence of adequate amounts of ACTH, these layers shrink considerably, and cortisol secretion is drastically reduced. Recall that angiotensin, not ACTH, maintains the size of the zona glomerulosa. Like the actions of TSH on the thyroid gland, ACTH enhances many steps in the synthesis of cortisol.

The ACTH-producing cells, in turn, secrete only at the command of corticotropin-releasing hormone (CRH) from the hypothalamus. The feedback control loop is completed by cortisol's inhibitory actions on CRH and ACTH secretion by the hypothalamus and anterior pituitary, respectively.

The negative-feedback system for cortisol maintains the level of cortisol secretion relatively constant around the set point. Superimposed on the basic negative-feedback control system

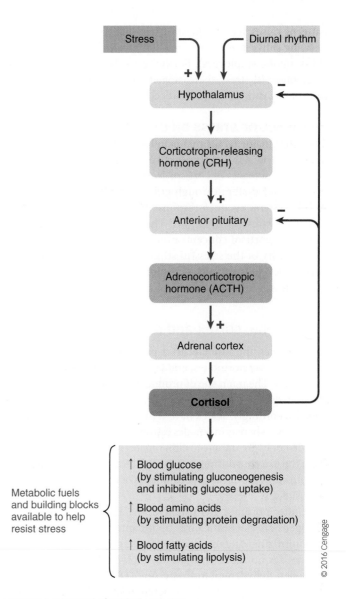

› FIGURE 6-8 **Control of cortisol secretion**

are two additional factors that influence plasma cortisol concentrations by changing the set point: *diurnal rhythm* and *stress*, both of which act on the hypothalamus to vary the secretion rate of CRH.

### INFLUENCE OF DIURNAL RHYTHM ON CORTISOL SECRETION

Recall that the plasma cortisol concentration displays a characteristic diurnal rhythm, with the highest level occurring in the morning and the lowest level at night (see Figure 5-3). This diurnal rhythm, which is intrinsic to the hypothalamus–pituitary control system, is related primarily to the sleep–wake cycle. The peak and low levels are reversed in a person who works at night and sleeps during the day. Such time-dependent variations in secretion are of more than academic interest, because it is important clinically to know at what time of day a blood sample was taken when interpreting the significance of a particular value. Also, the linking of cortisol secretion to day–night activity patterns raises serious questions about the common practice of swing shifts at work (i.e., constantly switching day and night shifts among employees). Furthermore, because cortisol helps a person resist stress, increasing attention is being given to the time of day various surgical procedures are performed.

### INFLUENCE OF STRESS ON CORTISOL SECRETION

The other major factor that is independent of, and in fact can override, the stabilizing negative-feedback control is stress. Dramatic increases in cortisol secretion, mediated by the central nervous system through enhanced activity of the CRH–ACTH system, occur in response to all kinds of mentally and physically stressful situations. The magnitude of the increase in plasma cortisol concentration is generally proportional to the intensity of the stressful stimulation; a greater increase in cortisol levels is evoked in response to severe stress than to mild stress.

## The adrenal cortex and sex hormones

In both sexes, the adrenal cortex produces both *androgens*, also called male sex hormones, and *estrogens*, also called female sex hormones. The main site of production for the sex hormones is the gonads: the testes for androgens and the ovaries for estrogens. Accordingly, males have a preponderance of circulating androgens, whereas in females estrogens predominate. However, no hormones are unique to either males or females (except those from the placenta during pregnancy), because the adrenal cortex in both sexes produces small amounts of the sex hormone of the opposite sex.

Under normal circumstances, the adrenal androgens and estrogens are not sufficiently abundant or powerful to induce masculinizing or feminizing effects, respectively. The only adrenal sex hormone that has any biological importance is the androgen **dehydroepiandrosterone (DHEA)**. The testes' primary androgen product is the potent testosterone, but the most abundant adrenal androgen is the much weaker DHEA. Adrenal DHEA is overpowered by testicular testosterone in males but is of physiological significance in females, who otherwise lack androgens. This adrenal androgen governs androgen-dependent processes in the female, such as growth of pubic and axillary (armpit) hair, enhancement of the pubertal growth spurt, and development and maintenance of the female sex drive.

Estrogens are normally produced in very small quantities in the adrenal glands because the enzymes required for their production are present in very low concentrations in the adrenocortical cells.

In addition to controlling cortisol secretion, ACTH (not the pituitary gonadotropic hormones) controls adrenal androgen secretion. In general, cortisol and DHEA output by the adrenal cortex follow the same secretion pattern. However, adrenal androgens feedback outside the hypothalamus–pituitary–adrenal cortex loop. Instead of inhibiting CRH, DHEA inhibits gonadotropin-releasing hormone, just as testicular androgens do. Furthermore, sometimes adrenal androgen and cortisol output diverge from each other. For example, at the time of puberty adrenal androgen secretion undergoes a marked surge,

but cortisol secretion does not change. This enhanced secretion initiates the development of androgen-dependent processes in females. In males, the same thing is accomplished primarily by testicular androgen secretion, which is also aroused at puberty. The nature of the pubertal inputs to the adrenals and gonads is still unresolved.

A surge in DHEA secretion begins at puberty and peaks between the ages of 25 and 30. After 30, DHEA secretion slowly tapers off until, by the age of 60, the plasma DHEA concentration is less than 15 percent of its peak level.

*Clinical Note* Some scientists suspect that the age-related decline of DHEA and other hormones, such as growth hormone (GH) (p. 232) and melatonin (p. 241), plays a role in some problems of aging. Early studies involving DHEA replacement therapy have demonstrated some physical improvement, such as an increase in lean muscle mass and a decrease in fat, but the most pronounced effect was a marked increase in psychological well-being and an improved ability to cope with stress. Advocates for DHEA replacement therapy do not suggest that maintaining youthful levels of this hormone is going to extend the lifespan, but they do propose that it may help people feel and act younger as they age. Other scientists caution that evidence supporting DHEA as an anti-aging therapy is still sparse. Also, they are concerned about DHEA supplementation until it has been thoroughly studied for possible harmful side effects. For example, some research suggests a potential increase in the risk of heart disease among women taking DHEA because of an observed reduction in HDL, the so-called good cholesterol (p. 376). Also, high doses of DHEA have been linked with increased facial hair in women. Furthermore, some experts fear that DHEA supplementation may raise the odds of acquiring ovarian or breast cancer in women and prostate cancer in men.

Although the U.S. Food and Drug Administration (FDA) banned sales of DHEA as an over-the-counter drug in 1985 because of concerns about very real risks coupled with little proof of benefits, the product is available today as an unregulated food supplement. DHEA can be marketed as a dietary supplement without approval by the U.S. FDA as long as the product label makes no specific medical claims. Health Canada, however, classifies DHEA as a controlled substance, available only by prescription.

## The adrenal cortex and hormone levels

*Clinical Note* Although uncommon, there are a number of disorders of adrenocortical function. Excessive secretion may occur with any of the three categories of adrenocortical hormones. Accordingly, three main patterns of symptoms resulting from hyperadrenalism can be distinguished, depending on which hormone type is in excess: (1) aldosterone hypersecretion, (2) cortisol hypersecretion, and (3) adrenal androgen hypersecretion.

### ALDOSTERONE HYPERSECRETION

Excess mineralocorticoid secretion may be caused by (1) a hypersecreting adrenal tumour made up of aldosterone-secreting cells (**primary hyperaldosteronism** or **Conn's syndrome**) or (2) inappropriately high activity of the renin–angiotensin system (**secondary hyperaldosteronism**). The latter may be produced by any number of conditions that cause a chronic reduction in arterial blood flow to the kidneys, thereby excessively activating the renin–angiotensin–aldosterone system. An example is atherosclerotic narrowing of the renal arteries.

The symptoms of both primary and secondary hyperaldosteronism are related to the exaggerated effects of aldosterone—namely, excessive Na$^+$ retention (*hypernatraemia*) and K$^+$ depletion (*hypokalaemia*). Also, high blood pressure (hypertension) is generally present, at least partially because of excessive Na$^+$ and fluid retention.

### CORTISOL HYPERSECRETION

Excessive cortisol secretion (**Cushing's syndrome**) can be caused by (1) overstimulation of the adrenal cortex by excessive amounts of CRH and/or ACTH; (2) adrenal tumours that uncontrollably secrete cortisol independent of ACTH; or (3) ACTH-secreting tumours located in places other than the pituitary, most commonly in the lung. Whatever the cause, the prominent characteristics of this syndrome are related to the exaggerated effects of glucocorticoid; the main symptoms are reflections of excessive gluconeogenesis. When too many amino acids are converted into glucose, the body suffers from combined glucose excess (high blood glucose) and protein shortage. Because the resultant hyperglycaemia and glucosuria (glucose in the urine) mimic diabetes mellitus, the condition is sometimes referred to as *adrenal diabetes*. One characteristic of Cushing's syndrome is the mobilization of fat from the lower extremities and the subsequent deposit of that fat into the abdominal and thoracic regions. Another feature of the elevated steroid secretion associated with Cushing's syndrome is oedema of the face. The abnormal fat distribution in the thoracic region—that is, on the back between the shoulder blades—is called a "buffalo hump," and in the face, is called a "moon face" (⟩ Figure 6-9). The appendages, in contrast, remain thin.

Besides the effects attributable to excessive glucose production, other effects arise from the widespread mobilization of amino acids from body proteins for use as glucose precursors. Loss of muscle protein leads to muscle weakness and fatigue. The protein-poor, thin skin of the abdomen becomes overstretched by the excessive underlying fat deposits, forming irregular, reddish purple linear streaks. Loss of structural protein within the walls of the small blood vessels leads to easy bruising. Wounds heal poorly, because formation of collagen, a major structural protein found in scar tissue, is depressed. Furthermore, loss of the collagen framework of bone weakens the skeleton, so fractures may result from little or no apparent injury.

### ADRENAL ANDROGEN HYPERSECRETION

Excess adrenal androgen secretion, a masculinizing condition, is more common than the extremely rare feminizing condition of excess adrenal estrogen secretion. Either condition is referred to as **adrenogenital syndrome**, emphasizing the pronounced effects that excessive adrenal sex hormones have on the genitalia and associated sexual characteristics.

The symptoms that result from excess androgen secretion depend on the sex of the individual and the age when the hyperactivity first begins.

- *In adult females*. Because androgens exert masculinizing effects, a woman with this disease tends to develop a male pattern of body hair, a condition referred to as **hirsutism**. She usually also acquires other male secondary sexual characteristics, such as deepening of the voice and more muscular arms and legs. The breasts become smaller, and menstruation may cease as a result of androgen suppression of the woman's hypothalamus–pituitary–ovarian pathway for her female sex-hormone secretion.

- *In newborn females*. Female infants born with adrenogenital syndrome manifest male-type external genitalia, because excessive androgen secretion occurs early enough during fetal life to induce development of their genitalia along male lines, similar to the development of males under the influence of testicular androgen. The clitoris, which is the female homologue of the male penis, enlarges under androgen influence and takes on a penile appearance, so in some cases it is difficult at first to determine the child's sex. Thus, this hormonal abnormality is one of the major causes of **female pseudohermaphroditism**, a condition in which female gonads (ovaries) are present but the external genitalia resemble those of a male. (A true hermaphrodite has the gonads of both sexes.)

- *In prepubertal males*. Excessive adrenal androgen secretion in prepubertal boys causes them to prematurely develop male secondary

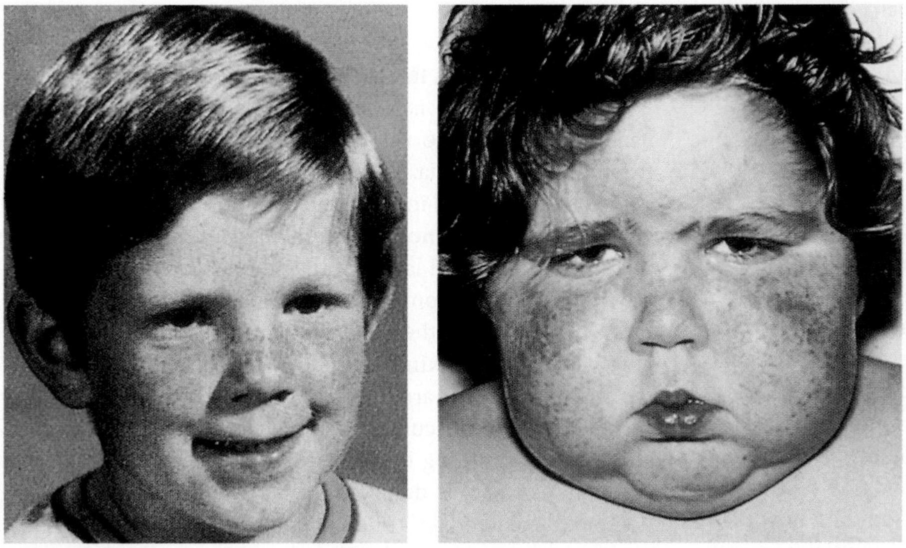

> FIGURE 6-9 Patient with Cushing's syndrome

sexual characteristics—for example, deep voice, beard, enlarged penis, and sex drive. This condition is referred to as **precocious pseudopuberty** to differentiate it from true puberty, which occurs as a result of increased testicular activity. In precocious pseudopuberty, the androgen secretion from the adrenal cortex is not accompanied by sperm production or any other gonadal activity, because the testes are still in their nonfunctional prepubertal state.

- *In adult males.* Overactivity of adrenal androgens in adult males has no apparent effect, because any masculinizing effect induced by the weak DHEA, even when in excess, is unnoticeable due to of the powerful masculinizing effects of the much more abundant and potent testosterone from the testes.

The adrenogenital syndrome is most commonly caused by an inherited enzymatic defect in the cortisol steroidogenic pathway. The pathway for synthesis of androgens branches from the normal biosynthetic pathway for cortisol. When an enzyme specifically essential for synthesis of cortisol is deficient, the result is decreased secretion of cortisol. The decline in cortisol secretion removes the negative-feedback effect on the hypothalamus and anterior pituitary so that levels of CRH and ACTH increase considerably (⟩ Figure 6-10). The defective adrenal cortex is incapable of responding to this increased ACTH secretion with cortisol output and instead shunts more of its cholesterol precursor into the androgen pathway. The result is excess DHEA production. This excess androgen does not inhibit ACTH but rather inhibits the gonadotropins. Because gamete production is not stimulated in the absence of gonadotropins, people with adrenogenital syndrome are sterile. Of course, they also exhibit symptoms of cortisol deficiency.

The symptoms of adrenal virilization, sterility, and cortisol deficiency are all reversed by glucocorticoid therapy. Administration of exogenous glucocorticoid replaces the cortisol deficit and, more dramatically, inhibits the hypothalamus and pituitary so that ACTH secretion is suppressed. Once ACTH secretion is reduced, the profound stimulation of the adrenal cortex ceases, and androgen secretion declines markedly. Removing the large quantities of adrenal androgens from circulation allows masculinizing characteristics to gradually recede and normal gonadotropin secretion to resume. Without understanding how these hormonal systems are related, it would be very difficult to comprehend how glucocorticoid administration could dramatically reverse symptoms of masculinization and sterility.

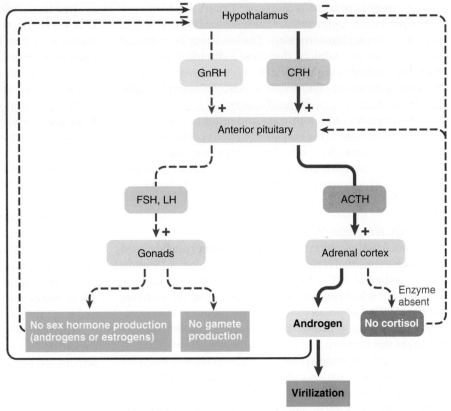

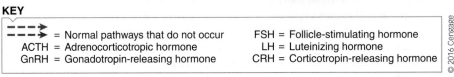

**KEY**

**- - - ▶** = Normal pathways that do not occur
**ACTH** = Adrenocorticotropic hormone
**GnRH** = Gonadotropin-releasing hormone
**FSH** = Follicle-stimulating hormone
**LH** = Luteinizing hormone
**CRH** = Corticotropin-releasing hormone

© 2016 Cengage

⟩ **FIGURE 6-10** **Hormonal interrelationships in adrenogenital syndrome.** The adrenocortical cells that are supposed to produce cortisol produce androgens instead because of a deficiency of a specific enzyme essential for cortisol synthesis. Because no cortisol is secreted to act in negative-feedback fashion, CRH and ACTH levels are elevated. The adrenal cortex responds to increased ACTH by further increasing androgen secretion. The excess androgen produces virilization and inhibits the gonadotropin pathway, with the result that the gonads stop producing sex hormones and gametes.

## ADRENOCORTICAL INSUFFICIENCY

If one adrenal gland is nonfunctional or removed, the other healthy organ can take over the function of both through hypertrophy and hyperplasia. Therefore, both glands must be affected before adrenocortical insufficiency occurs.

In **primary adrenocortical insufficiency**, also known as **Addison's disease**, all layers of the adrenal cortex are undersecreting. This condition is most commonly caused by autoimmune destruction of the cortex due to erroneous production of adrenal cortex–attacking antibodies, in which case both aldosterone and cortisol are deficient. **Secondary adrenocortical insufficiency** may occur because of a pituitary or hypothalamic abnormality, resulting in insufficient ACTH secretion. In this case, only cortisol is deficient, because aldosterone secretion does not depend on ACTH stimulation.

The symptoms associated with aldosterone deficiency in Addison's disease are the most threatening. If severe enough, the condition is fatal, because aldosterone is essential for life.

However, the loss of adrenal function may develop slowly and insidiously so that aldosterone secretion may be subnormal but not totally lacking. Patients with aldosterone deficiency display $K^+$ retention (*hyperkalaemia*), caused by reduced $K^+$ loss in the urine, and $Na^+$ depletion (*hyponatraemia*), caused by excessive urinary loss of $Na^+$. The former disturbs cardiac rhythm. The latter reduces ECF volume, including circulating blood volume, which in turn lowers blood pressure (hypotension).

Symptoms of cortisol deficiency are as would be expected: poor response to stress, hypoglycaemia (low blood glucose) caused by reduced gluconeogenic activity, and lack of permissive action for many metabolic activities. The primary form of the disease also produces hyperpigmentation (darkening of the skin), resulting from excessive secretion of ACTH. Because the pituitary is normal, the decline in cortisol secretion brings about an uninhibited elevation in ACTH output. Recall that both ACTH and melanocyte-stimulating hormone (MSH) are produced from the same large precursor molecule, pro-opiomelanocortin. As a result, levels of MSH and, accordingly, of skin-darkening melanin also rise when the blood levels of ACTH are very high.

We now shift our attention from the adrenal cortex to the adrenal medulla.

## The adrenal medulla

The adrenal medulla is actually a modified part of the sympathetic nervous system. Recall that a sympathetic pathway consists of two neurons in sequence—a preganglionic neuron originating in the CNS, whose axonal fibre terminates on a second peripherally located postganglionic neuron, which in turn terminates on the effector organ. The neurotransmitter released by sympathetic postganglionic fibres is norepinephrine, which interacts locally with the innervated organ by binding with specific target receptors known as *adrenergic receptors.*

The adrenal medulla consists of modified postganglionic sympathetic neurons. Unlike ordinary postganglionic sympathetic neurons, those in the adrenal medulla do not have axonal fibres that terminate on effector organs. Instead, on stimulation by the preganglionic fibre, the ganglionic cell bodies within the adrenal medulla release their chemical transmitter directly into the circulation (see Figure 4-43). In this case, the transmitter qualifies as a hormone instead of a neurotransmitter, and once released into circulation the hormones are carried to all body tissues. Like sympathetic fibres, the adrenal medulla does release norepinephrine (about 20 percent of its secretion), but its most abundant secretory output is a similar chemical messenger, known as epinephrine (about 80 percent). The relative contribution of these two hormones will change under different physiological conditions. Both epinephrine and norepinephrine belong to the chemical class of catecholamines, which are derived from the amino acid tyrosine. Epinephrine and norepinephrine are the same, except that epinephrine also has a methyl group.

### STORAGE OF CATECHOLAMINES
### IN CHROMAFFIN GRANULES
Catecholamine is synthesized almost entirely within the cytosol of the adrenomedullary secretory cells. Once produced, epinephrine and norepinephrine are stored in **chromaffin granules**, which are similar to the transmitter storage vesicles found in sympathetic nerve endings. Segregation of catecholamines in chromaffin granules protects them from being destroyed by cytosolic enzymes during storage.

> ■ **Clinical Connections** Pheochromocytoma is usually caused by a tumour in the adrenal medulla. However, in Megan's case, she had a paraganglioma in her celiac ganglia. In this situation, the tumour is also made up of chromaffin cells, so this tumour outside of the adrenal gland is synthesizing and storing catecholamines.

### SECRETION OF CATECHOLAMINES
### FROM THE ADRENAL MEDULLA
Catecholamines are secreted into the blood by exocytosis of chromaffin granules. Their release is analogous to the release mechanism for secretory vesicles that contain stored peptide hormones or the release of norepinephrine at sympathetic postganglionic terminals.

Epinephrine and norepinephrine are generally released by the adrenal medulla at the same time. However, epinephrine is produced exclusively by the adrenal medulla, and the bulk of norepinephrine is produced by sympathetic postganglionic fibres. Adrenomedullary norepinephrine is generally secreted in quantities too small to exert significant effects on target cells. Therefore, for practical purposes we can assume that norepinephrine effects are predominantly mediated directly by the sympathetic nervous system and that epinephrine effects are brought about exclusively by the adrenal medulla.

## Epinephrine and norepinephrine

Epinephrine and norepinephrine have varying affinities for the two major classes of receptors: alpha-adrenergic and beta-adrenergic receptors (▮ Table 6-2). There are three subclasses of beta-adrenergic receptors, $\beta_1$, $\beta_2$, and $\beta_3$, which are functionally different in different tissues. The alpha-adrenergic receptors have two subclasses, $\alpha_1$ and $\alpha_2$, which act presynaptically ($\alpha_2$) to inhibit the release of norepinephrine or postsynaptically ($\alpha_1$) to stimulate or inhibit the activity at various potassium channels. Most sympathetic target cells have $\alpha_1$ receptors—some have only $\alpha_2$ receptors, some only $\beta_2$, and some have both $\alpha_1$ and $\beta_2$, while $\beta_1$ receptors are found almost exclusively in the heart. In general, the responses elicited by activation of $\alpha_1$ and $\beta_1$ receptors are excitatory, whereas the responses to stimulation of $\alpha_2$ and $\beta_2$ receptors are typically inhibitory.

Norepinephrine binds predominantly with $\alpha_1$ and $\beta_1$ receptors located near postganglionic sympathetic-fibre terminals. Hormonal epinephrine, which can reach all $\alpha$ and $\beta_1$ receptors via its circulatory distribution, interacts with these same receptors with approximately the same potency as the neurotransmitter norepinephrine (although norepinephrine has a greater affinity than epinephrine for $\alpha$ receptors). Thus, epinephrine and norepinephrine exert similar effects in many tissues, with epinephrine generally reinforcing sympathetic nervous activity. In addition, epinephrine activates $\beta_2$ receptors, over which the sympathetic nervous system exerts little influence.

**▌TABLE 6-2** Adrenergic Receptor Types and Responses Elicited by Norepinephrine (NE) and Epinephrine (E)

| Adrenergic Receptor Type | Location | Affinity of Catecholamine for NE and E | Typical Response Elicited | Examples of Responses Elicited |
|---|---|---|---|---|
| $\alpha_1$ | Most sympathetic target cells | NE > E | Excitatory | Generalized arteriolar vasoconstriction ($\uparrow$ smooth muscle contraction) |
| $\alpha_2$ | Digestive system | NE < E | Inhibitory | Decreased motility in digestive tract ($\downarrow$ smooth muscle contraction) |
| $\beta_1$ | Heart | NE = E | Excitatory | Increased rate and strength of cardiac muscle contraction |
| $\beta_2$ | Skeletal muscle; smooth muscle of some blood vessels and organs | E only | Inhibitory | Breakdown of glycogen in skeletal muscle; bronchiolar dilation and arteriolar vasodilation in skeletal muscle and heart ($\downarrow$ smooth muscle contraction) |

Many of the essentially epinephrine-exclusive $\beta_2$ receptors are located at tissues not even supplied by the sympathetic nervous system but reached by epinephrine through the blood. An example is skeletal muscle, where epinephrine exerts metabolic effects such as promoting the breakdown of stored glycogen.

Sometimes epinephrine, through its exclusive $\beta_2$-receptor activation, brings about a different action from that elicited by norepinephrine and epinephrine action through their mutual activation of other adrenergic receptors. As an example, norepinephrine and epinephrine bring about a generalized vasoconstrictor effect that's mediated by $\alpha_1$-receptor stimulation. By contrast, epinephrine promotes vasodilation of the blood vessels that supply skeletal muscles and the heart through $\beta_2$-receptor activation.

Epinephrine functions only at the bidding of the sympathetic nervous system, however, which is solely responsible for stimulating its secretion from the adrenal medulla. Epinephrine secretion always accompanies a generalized sympathetic nervous system discharge, so sympathetic activity directly controls actions of epinephrine. By having the more versatile circulating epinephrine at its call, the sympathetic nervous system has a means of reinforcing its own neurotransmitter effects, plus a way of executing additional actions on tissues that it does not directly innervate.

## Epinephrine and the sympathetic nervous system

Adrenomedullary hormones are not essential for life, but virtually all organs in the body are affected by these catecholamines. They play important roles in mounting stress responses, regulating arterial blood pressure, and controlling fuel metabolism. The following sections discuss epinephrine's major effects, which it achieves either in collaboration with the sympathetic transmitter norepinephrine or alone to complement direct sympathetic responses.

### EFFECTS ON ORGAN SYSTEMS

Together, the sympathetic nervous system and adrenomedullary epinephrine mobilize the body's resources to support peak physical exertion in emergency or stressful situations. The sympathetic and epinephrine actions constitute a fight-or-flight response that prepares a person to combat an enemy or flee from danger (p. 191). Specifically, the sympathetic system and epinephrine increase the rate and strength of cardiac contraction, which increases cardiac output, and their generalized vasoconstrictor effects increase total peripheral resistance. Together, these effects raise arterial blood pressure, thus ensuring an appropriate driving pressure to force blood to the organs most vital for meeting the emergency. Meanwhile, vasodilation of coronary and skeletal muscle blood vessels induced by epinephrine and local metabolic factors shifts blood to the heart and skeletal muscles from other vasoconstricted regions of the body.

In addition, due to their profound influence on the heart and blood vessels, the sympathetic system and epinephrine play an important role in the ongoing maintenance of arterial blood pressure. This is explained in detail in Chapter 9.

Epinephrine (but not norepinephrine) dilates the respiratory airways to reduce the resistance encountered in moving air in and out of the lungs. Epinephrine and norepinephrine also reduce digestive activity and inhibit bladder emptying, activities that can be "put on hold" during a fight-or-flight situation.

### METABOLIC EFFECTS

Epinephrine exerts some important metabolic effects. In general, epinephrine prompts the mobilization of stored carbohydrate and fat to provide immediately available energy for use as needed to fuel muscular work. Specifically, epinephrine increases the blood glucose level by several different mechanisms. First, it stimulates both hepatic (liver) gluconeogenesis and **glycogenolysis**, the latter being the breakdown of stored glycogen into glucose, which is released into the blood. Epinephrine also stimulates glycogenolysis in skeletal muscles. Because of the difference in enzyme content between liver and muscle, however, muscle glycogen cannot be converted directly to glucose. Instead, the breakdown of muscle glycogen releases lactic acid into the blood. The liver removes lactic acid from the blood and converts it into glucose, so epinephrine's actions on skeletal muscle indirectly help raise blood glucose levels. Epinephrine and the sympathetic system may further add to this hyperglycaemic effect by inhibiting the secretion of insulin, the pancreatic hormone primarily responsible for removing glucose from the blood, and by stimulating glucagon, another pancreatic hormone that promotes hepatic glycogenolysis and gluconeogenesis. In addition to increasing blood glucose levels, epinephrine also increases the level of blood fatty acids by promoting lipolysis.

Epinephrine's metabolic effects are appropriate for fight-or-flight situations. The resulting elevated levels of glucose and fatty acids provide additional fuel to power the muscular movement required by the situation and also assure adequate nourishment for the brain during the crisis when no new nutrients are being consumed. Muscles can use fatty acids for energy production, but the brain cannot.

Because of its other widespread actions, epinephrine also increases the overall metabolic rate. Under the influence of epinephrine, many tissues metabolize faster. For example, the work of the heart and respiratory muscles increases, and the pace of liver metabolism steps up. Thus, epinephrine as well as thyroid hormone can increase the metabolic rate.

### OTHER EFFECTS

Epinephrine affects the central nervous system to promote a state of arousal and increased CNS alertness. This permits the "quick thinking" that helps a person cope with an impending emergency. Many drugs used as stimulants or sedatives exert their effects by altering catecholamine levels in the CNS.

Both epinephrine and norepinephrine cause sweating, which helps the body rid itself of extra heat generated by increased muscular activity. Also, epinephrine acts on smooth muscles within the eyes to dilate the pupil and flatten the lens. These actions adjust the eyes for more encompassing vision so that the whole threatening scene can be quickly viewed.

> **■ Clinical Connections** Because Megan's tumour was secreting catecholamines, all of her symptoms were related to what is called catecholamine excess. Under normal circumstances, the adrenal medulla releases catecholamines when the sympathetic nervous system is activated, and this acts and an amplifier of the signal. However, catecholamine-secreting tumours can act independently, causing sudden, large increases of circulating catecholamines. A key diagnostic test for pheochromocytoma involves looking for elevated urinary output of catecholamine metabolites over a 24-hour period.

## Sympathetic stimulation of the adrenal medulla

Catecholamine secretion by the adrenal medulla is controlled entirely by sympathetic input to the adrenal gland. When the sympathetic system is activated under conditions of fear or

stress, it simultaneously triggers a surge of adrenomedullary catecholamine release. The concentration of epinephrine in the blood may increase up to 300 times the normal concentration, with the amount of epinephrine released depending on the type and intensity of the stressful stimulus.

Because both components of the adrenal gland play an extensive role in responding to stress, this is an appropriate place to pull together the major factors involved in the stress response.

> ### Check Your Understanding 6.2
>
> 1. List the three categories of adrenocortical hormones, name the primary hormone in each category, and state the functions of each of these hormones.
> 2. Discuss the effect of ACTH on the adrenal cortex.
> 3. Name the two catecholamines secreted by the adrenal medulla, and describe how they are stored and released.

## 6.3 | Integrated Stress Response

**Stress** is the generalized, nonspecific response of the body to any factor that overwhelms, or threatens to overwhelm, the body's compensatory abilities to maintain homeostasis. Contrary to popular usage, the agent inducing the response is correctly called a *stressor*, whereas *stress* refers to the state induced by the stressor. The following types of noxious stimuli illustrate the range of factors that can induce a stress response: *physical* (trauma, surgery, intense heat or cold); *chemical* (reduced $O_2$ supply, acid–base imbalance); *physiological* (heavy exercise, haemorrhagic shock, pain); *infectious* (bacterial invasion); *psychological or emotional* (anxiety, fear, sorrow); and *social* (personal conflicts, change in lifestyle).

### A general reaction to stress

Different stressors may produce some specific responses characteristic of that stressor. For example, the body's specific response to cold exposure is shivering and skin vasoconstriction, whereas the specific response to bacterial invasion includes increased phagocytic activity and antibody production. In addition to their specific response, however, all stressors produce a similar nonspecific, generalized response (> Figure 6-11). This general set of responses common to exposure to all noxious stimuli is called the **general adaptation syndrome (GAS)**. The stress is first identified by the body, and then both the nervous and endocrine systems (neuroendocrine system) rally a defensive response in order to cope with the emergency (noxious stimuli). GAS is now thought to be well understood, but as with many integrative functions within the human body, there is still much to be discovered. The body reacts to stress first by releasing epinephrine and norepinephrine, and then the glucocorticoid hormone cortisol. The result is a state of intense readiness and mobilization of biochemical resources.

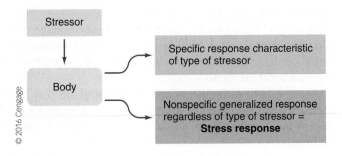

© 2016 Cengage

> FIGURE 6-11 **Action of a stressor on the body**

To appreciate the value of the multifaceted stress response, imagine a primitive cave dweller who has just seen a large wild beast lurking in the shadows. We will consider both the neural and hormonal responses that would take place in this scenario. The body responds in the same way to modern-day stressors. From the preceding discussion, you are familiar with all these responses, so now we focus on how these responses work together.

### ROLES OF THE SYMPATHETIC NERVOUS SYSTEM AND EPINEPHRINE IN STRESS

In a fight-or-flight scenario, the sympathetic nervous system (with the assistance of epinephrine from the adrenal medulla) readies the body to respond to the stressful situation. A massive discharge from the sympathetic nervous system prepares the body in various ways for the intense muscular work it needs to perform. It increases

- muscle strength;
- mental activity;
- blood pressure;
- blood flow to essential tissue, while decreasing flow to non-essential tissue; and
- cellular metabolism.

It is this sympathetic stress response that enables a person to respond to strenuous physical or emotional stress. The sympathetic stress response can occur in many emotional situations—for example, during a state of rage or fear. If you are hiking in the woods and come across a bear, your sympathetic system will be immediately heightened, providing the physiological changes listed above. The hypothalamus will be stimulated and that signal will be transmitted down to the spinal cord, providing a massive sympathetic output. If it were not for the sympathetic stress response, it would likely be impossible for you to overcome physical or emotional stress. Simultaneously, the sympathetic system calls forth hormonal reinforcements in the form of a massive outpouring of epinephrine from the adrenal medulla. Epinephrine strengthens sympathetic responses and also reaches places not innervated by the sympathetic system to perform additional functions, such as mobilizing carbohydrate and fat stores.

### ROLES OF THE CRH–ACTH–CORTISOL SYSTEM IN STRESS

Besides epinephrine, a number of other hormones are involved in the integrated stress response (▌Table 6-3). The predominant hormonal response is activation of the CRH–ACTH–cortisol system. Recall that cortisol's role in helping the body cope with stress is presumed to be related to its metabolic effects. Cortisol breaks down fat and protein stores while expanding carbohydrate stores and increasing the availability of blood glucose.

▌ **TABLE 6-3** Major Hormonal Changes during the Integrated Stress Response

| Hormone | Change | Purpose Served |
|---|---|---|
| **Epinephrine** | ↑ | Reinforces the sympathetic nervous system to prepare the body for fight-or-flight; mobilizes carbohydrate and fat energy stores; increases blood glucose and blood fatty acids |
| **CRH–ACTH–Cortisol** | ↑ | Mobilizes energy stores and metabolic building blocks for use as needed; increases blood glucose, blood amino acids, and blood fatty acids |
| | | ACTH facilitates learning and behaviour; b-endorphin co-secreted with ACTH may mediate analgesia |
| **Glucagon** | ↑ | Act in concert to increase |
| **Insulin** | ↓ | blood glucose and blood fatty acids |
| **Renin–Angiotensin–Aldosterone** | ↑ | Conserve salt and $H_2O$ to expand the plasma volume; help sustain blood pressure when acute loss of plasma volume occurs |
| **Vasopressin** | ↑ | Angiotensin II and vasopressin cause arteriolar vasoconstriction to increase blood pressure |
| | | Vasopressin facilitates learning |

From Sherwood. *Human Physiology*, 8E. © 2013 Brooks/Cole, a part of Cengage, Inc. Reproduced by permission. www.cengage.com/permissions

A logical assumption is that the increased pool of glucose, amino acids, and fatty acids is available for use as needed, such as to sustain nourishment to the brain and provide building blocks for repair of damaged tissues.

In addition to the effects of cortisol in the hypothalamus–adrenal cortex axis, ACTH may also play a role in resisting stress. ACTH is one of several peptides that facilitate learning and behaviour. Thus, an increase in ACTH during psychosocial stress may help the body cope more readily with similar stressors in the future by facilitating the learning of appropriate behavioural responses. Furthermore, ACTH is not released alone from its anterior pituitary storage vesicles. Cutting of the large pro-opiomelanocortin precursor molecule yields not only ACTH but also morphine-like $\beta$-endorphin, which is co-secreted with ACTH on stimulation by CRH during stress. As a potent endogenous opiate, $\beta$-endorphin may exert a role in mediating analgesia (reduction of pain perception) if physical injury is inflicted during stress.

### ROLE OF OTHER HORMONAL RESPONSES IN STRESS
Besides the CRH–ACTH–cortisol system, other hormonal systems play key roles in the stress response.

- *Elevation of blood glucose and fatty acids through decreased insulin and increased glucagon.* The sympathetic nervous system and the epinephrine secreted under its control both inhibit insulin and stimulate glucagon. These hormonal changes act in concert to elevate blood levels of glucose and fatty acids. Epinephrine and glucagon, whose blood levels are elevated during stress, promote hepatic glycogenolysis and (along with cortisol) hepatic gluconeogenesis. However, insulin, whose secretion is suppressed during stress, opposes the breakdown of liver glycogen stores. All these effects help increase the concentration of blood glucose. The primary stimulus for insulin secretion is a rise in blood glucose; in turn, a primary effect of insulin is to lower blood glucose. If it were not for the deliberate inhibition of insulin during the stress response, the hyperglycaemia caused by stress would stimulate secretion of glucose-lowering insulin. As a result, the elevation in blood glucose could not be sustained. Stress-related hormonal responses also promote a release of fatty acids from fat stores, because lipolysis is favoured by epinephrine, glucagon, and cortisol, but opposed by insulin.
- *Maintenance of blood volume and blood pressure through increased renin–angiotensin–aldosterone and vasopressin activity.* In addition to the hormonal changes that mobilize energy stores during stress, other hormones are simultaneously called into play to sustain blood volume and blood pressure during the emergency. The sympathetic system and epinephrine play major roles in acting directly on the heart and blood vessels to improve circulatory function. In addition, the renin-angiotensin-aldosterone system is activated as a consequence of a sympathetically induced reduction of blood supply to the kidneys (p. 576). Vasopressin secretion is also increased during stressful situations (p. 617). Collectively, these hormones expand the plasma volume by promoting retention of salt and water. Presumably, the enlarged plasma volume serves as a protective measure to

help sustain blood pressure should acute loss of plasma fluid occur through haemorrhage or heavy sweating during the impending period of danger. Vasopressin and angiotensin also have direct vasopressor effects, which would be of benefit in maintaining an adequate arterial pressure in the event of acute blood loss (p. 402). Vasopressin is further believed to facilitate learning, which has implications for future adaptation to stress.

> ### ▌ Clinical Connections
>
> Megan's diagnosed condition of pheochromocytoma mimics a chronic stress situation. Due to the elevated circulating catecholamines, heart rate and blood pressure are elevated, energy stores are mobilized to increase plasma glucose, and mental alertness is increased. The primary difference from a normal stress response, however, is that the other hormonal systems typically involved in the integrated stress response are not involved. In this situation, catecholamine release is not under regulatory control, so there is no feedback loop to shut off its inappropriate release.

## The hypothalamus

Many responses, such as the fight-or-flight response, are directly or indirectly influenced by the hypothalamus (> Figure 6-12). The hypothalamus receives and coordinates many inputs, including seasonal and circadian rhythms, complex patterns of neuroendocrine outputs, complex homeostatic mechanisms, and many important stereotyped behaviours. The hypothalamus—a very complex brain region—responds to signals generated both externally and internally. It receives input concerning physical and emotional stressors from virtually all areas of the brain and from many receptors throughout the body. In response, the hypothalamus directly activates the sympathetic nervous system, secretes CRH to stimulate ACTH and cortisol release, and triggers the release of vasopressin. Sympathetic stimulation, in turn, brings about the secretion of epinephrine, with which it has a conjoined effect on the pancreatic secretion of insulin and glucagon. Furthermore, vasoconstriction of the renal afferent arterioles by the catecholamines indirectly triggers the secretion of renin by reducing the flow of oxygenated blood through the kidneys. Renin, in turn, sets in motion the renin-angiotensin-aldosterone system. In this way, the hypothalamus integrates the responses of both the sympathetic nervous system and the endocrine system during stress.

## Chronic psychosocial stressors

Acceleration of cardiovascular and respiratory activity, retention of salt and water, and mobilization of metabolic fuels and building blocks can be of benefit in response to a physical stressor, such as an athletic competition. Most of the stressors in our everyday lives are psychosocial in nature, yet they induce these same magnified responses. Stressors, such as anxiety about an exam, conflicts with loved ones, or impatience while sitting in a traffic jam, can elicit a stress response.

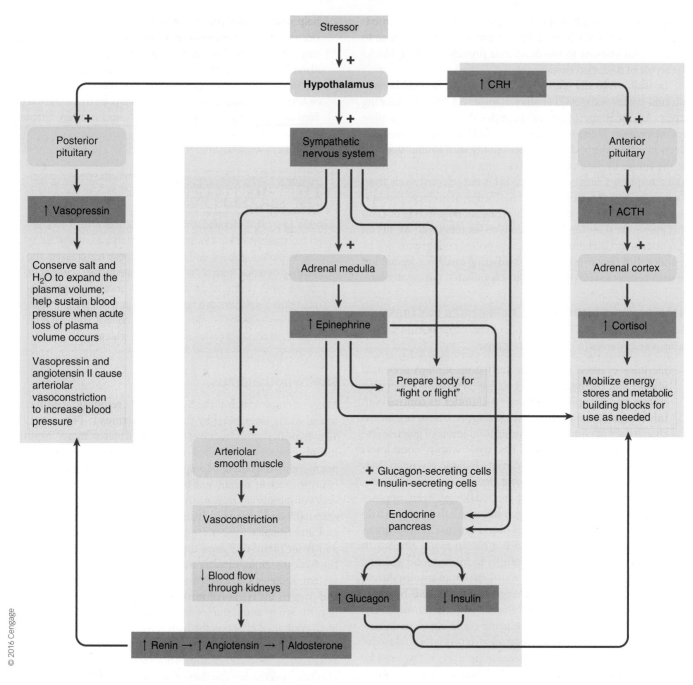

⟩ **FIGURE 6-12  Integration of the stress response by the hypothalamus**

Although the general stress response of rapid mobilization of body resources is appropriate in the face of real or threatened physical injury, it is generally inappropriate in response to nonphysical stress. If no extra energy is demanded, no tissue is damaged, and no blood lost, body stores are being broken down and fluid retained needlessly, probably to the detriment of the emotionally stressed individual. In fact, there is strong circumstantial evidence for a link between chronic exposure to psychosocial stressors and the development of pathological conditions, such as high blood pressure, although no definitive cause-and-effect relationship has been ascertained. As a result

of "unused" stress responses, could hypertension result from too much sympathetic vasoconstriction? From too much salt and water retention? From too much vasopressin and angiotensin pressor activity? A combination of these? Other factors? Recall that hypertension can develop with prolonged exposure to pharmacologic levels of glucocorticoids. Could longstanding lesser elevations of cortisol, such as might occur in the face of continual psychosocial stressors, do the same thing, only more slowly? Considerable work remains to be done to evaluate the contributions that the stressors in our everyday lives make toward disease production.

## 6.4 | Endocrine Control of Fuel Metabolism

We have just discussed the metabolic changes that are elicited during the stress response. Now we concentrate on the metabolic patterns that occur in the absence of stress, including the hormonal factors that govern normal metabolism.

### Fuel metabolism

The term **metabolism** refers to all the chemical reactions that occur within the cells of the body. Those reactions involving the degradation, synthesis, and transformation of the three classes of energy-rich organic molecules—protein, carbohydrate, and fat—are collectively known as intermediary metabolism or fuel metabolism (■ Table 6-4).

During the process of digestion, large nutrient molecules (**macromolecules**) are broken down into their smaller absorbable subunits as follows: proteins are converted into amino acids, complex carbohydrates into monosaccharides (mainly glucose), and triglycerides (dietary fats) into monoglycerides and free fatty acids. These absorbable units are transferred from the digestive tract lumen into the blood, either directly or by way of the lymph (Chapter 15).

### ANABOLISM AND CATABOLISM

These organic molecules are constantly exchanged between the blood and body cells. The chemical reactions in which the organic molecules participate within the cells are categorized into two metabolic processes: anabolism and catabolism (⟩ Figure 6-13). **Anabolism** is the buildup or synthesis of larger organic macromolecules from small organic molecular subunits and is used for cellular repair and growth. Anabolic reactions generally require energy input in the form of ATP. These reactions result in either (1) the manufacture of materials needed by the cell, such as cellular structural proteins or secretory products; or (2) storage of excess ingested nutrients not immediately needed for energy production or needed as cellular building blocks. Storage is in the form of glycogen (the storage form of glucose) or fat reservoirs.

**Catabolism** is the breakdown, or degradation, of large, energy-rich organic molecules within cells. Catabolism encompasses two levels of breakdown: (1) hydrolysis (p. 647) of large cellular organic macromolecules into their smaller subunits, similar to the process of digestion except that the reactions take place within the body cells instead of within the digestive tract lumen (e.g., release of glucose by the catabolism of stored glycogen); and (2) oxidation of the smaller subunits, such as glucose, to yield energy for ATP production (p. 24).

As an alternative to energy production, the smaller, multipotential organic subunits derived from intracellular hydrolysis may be released into the blood. These mobilized glucose, fatty acid, and amino acid molecules can then be used as needed for energy production or cellular synthesis elsewhere in the body.

In an adult, the rates of anabolism and catabolism are generally in balance, so the adult body remains in a dynamic, steady state and appears unchanged even though the organic molecules that determine its structure and function are continuously being turned over. During growth and development, anabolism exceeds catabolism.

### INTERCONVERSIONS AMONG ORGANIC MOLECULES

In addition to being able to resynthesize catabolized organic molecules back into the same type of molecules, many cells of the body, especially liver cells, can convert most types of small organic molecules into other types—as in, for example, transforming amino acids into glucose or fatty acids. Because of these interconversions, adequate nourishment can be provided by a wide range of molecules present in different types of foods. There are limits, however. **Essential nutrients**, such as the essential amino acids and vitamins, cannot be formed in the body by conversion from another type of organic molecule and therefore must be consumed in the diet.

---

**■ TABLE 6-4** Summary of Reactions in Fuel Metabolism

| Metabolic Process | Reaction | Consequence |
|---|---|---|
| **Glycogenesis** | Glucose → glycogen | ↓ Blood glucose |
| **Glycogenolysis** | Glycogen → glucose | ↑ Blood glucose |
| **Gluconeogenesis** | Amino acids → glucose | ↑ Blood glucose |
| **Protein Synthesis** | Amino acids → protein | ↓ Blood amino acids |
| **Protein Degradation** | Protein → amino acids | ↑ Blood amino acids |
| **Fat Synthesis** (Lipogenesis or Triglyceride Synthesis) | Fatty acids and glycerol → triglycerides | ↑ Blood fatty acids |
| **Fat Breakdown** (Lipolysis or Triglyceride Degradation) | Triglycerides → fatty acids and glycerol | ↑ Blood fatty acids |

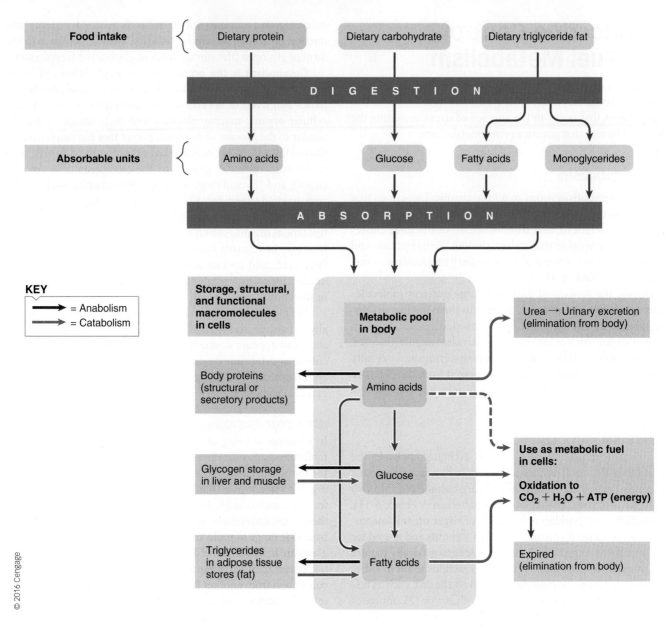

> FIGURE 6-13 **Summary of the major pathways involving organic nutrient molecules**

The major fate of both ingested carbohydrates and fats is catabolism to yield cellular energy. Amino acids are predominantly used for protein synthesis but can be used to supply energy after being converted to carbohydrate or fat by the liver. Thus, all three categories of foodstuff can be used as fuel, and excesses of any foodstuff can be deposited as stored fuel, as you will see shortly.

At a superficial level, fuel metabolism appears relatively simple: The amount of nutrients in the diet must be sufficient to meet the body's needs for energy production and cellular synthesis. This apparently simple relationship is complicated, however, by two important considerations: (1) nutrients taken in at meals must be stored and then released between meals, and (2) the brain must be continuously supplied with glucose. Let us examine the implications of each.

## Nutrients storage

Dietary fuel intake is intermittent, not continuous. As a result, excess energy must be absorbed during meals and stored for use during fasting periods between meals, when dietary sources of metabolic fuel are not available (▌ Table 6-5).

- *Excess circulating glucose* is stored in the liver and muscle as glycogen, a large molecule consisting of interconnected glucose molecules. The making of these glycogen molecules is an energy-absorbing process (endergonic) under the control of the endocrine system. Because glycogen is a relatively small energy reservoir, less than a day's energy needs can be stored in this form. Once the liver and muscle glycogen stores are filled up, additional glucose is transformed into fatty acids and glycerol, which are used to synthesize

## ■ TABLE 6-5 Stored Metabolic Fuel in the Body

| Metabolic Fuel | Circulating Form | Storage Form | Major Storage Site | Percentage of Total Body Energy Content (and Calories[1]) | Reservoir Capacity | Role |
|---|---|---|---|---|---|---|
| **Carbohydrate** | Glucose | Glycogen | Liver, muscle | 1 (1500) | Less than a day's worth of energy | First energy source; essential for the brain |
| **Fat** | Free fatty acids | Triglycerides | Adipose tissue | 77 (143 000) | About two months' worth of energy | Primary energy reservoir; energy source during a fast |
| **Protein** | Amino acids | Body proteins | Muscle | 22 (41 000) | Death results long before capacity is fully used because of structural and functional impairment | Source of glucose for the brain during a fast; last resort to meet other energy needs |

[1]Actually refers to kilocalories; see p. 700.

© 2016 Cengage

*triglycerides* (glycerol with three fatty acids attached), primarily in adipose tissue (fat).

- *Excess circulating fatty acids* derived from dietary intake also become incorporated into triglycerides.

- *Excess circulating amino acids* not needed for protein synthesis or during prolonged exercise (such as the amino acids alanine, leucine, and glutamine) are not stored as extra protein but are converted to glucose and fatty acids, which ultimately end up being stored as triglycerides.

As this list shows, the major site of energy storage for excess nutrients of all three classes is adipose tissue. Normally, enough triglyceride is stored to provide energy for about two months, more so in an overweight person. Consequently, during any prolonged period of fasting, the fatty acids released by means of triglyceride catabolism serve as the primary source of energy for most tissues. The catabolism of stored triglycerides frees glycerol as well as fatty acids, but quantitatively speaking, the fatty acids are far more important. Catabolism of stored fat yields 90 percent fatty acids and 10 percent glycerol by weight. Glycerol (but not fatty acids) can be converted to glucose by the liver and contributes in a small way to maintaining blood glucose during a fast.

As amino acids are consumed in our diet, they enter circulation, our bodies' most important amino acid pool (skeletal muscle and liver are the other two pools). Each of these pools is in equilibrium with every other pool, and therefore an increase in amino acid metabolism in one pool affects all other pools. Having these pools of amino acids to draw upon is very beneficial for sustaining life during periods when dietary intake is not consistent.

As a third energy reservoir, a substantial amount of energy is stored as structural protein, primarily in muscle, the most abundant protein mass in the body. Protein is not the first choice to tap as an energy source, however, because it serves other essential functions; in contrast, the glycogen and triglyceride reservoirs serve solely as energy depots.

## The brain and glucose

The second factor complicating fuel metabolism besides intermittent nutrient intake and the resultant necessity of storing nutrients is that the brain normally depends on the delivery of adequate blood glucose as its sole source of energy. Consequently, the blood glucose concentration must be maintained above a critical level. The units of blood glucose monitoring in Canada are reported as mmol glucose/L blood and typically is 5 mmol/L, with a normal range kept within the narrow limits of 4–6 mmol/L. (In contrast, blood glucose concentrations in the United States are given in terms of mass concentration, with a normal level hovering around 90 mg/mL.)

Liver glycogen is an important reservoir for maintaining blood glucose levels during a short fast. However, liver glycogen is depleted relatively rapidly, so during a longer fast other mechanisms must meet the energy requirements of the glucose-dependent brain. First, when no new dietary glucose is entering the blood, tissues not dependent on glucose shift their metabolism to burn fatty acids instead—sparing glucose for the brain. Fatty acids are made available by catabolism of triglyceride stores, as an alternative energy source for tissues that are not glucose dependent. Second, amino acids can be converted to glucose by gluconeogenesis, whereas fatty acids cannot. Once glycogen stores are depleted despite **glucose sparing**, new glucose supplies for the brain are provided by the catabolism of body proteins and conversion of the freed amino acids into glucose.

Amino acids therefore act as precursors for gluconeogenesis during prolonged fasting or starvation. Blood glucose homeostasis is maintained via the glucose-alanine cycle in the liver. Proteins can be broken down to provide approximately

100 grams of glucose per day. During prolonged fasting or starvation, this glucose will be used almost entirely by the brain and nervous system.

## The absorptive state and the postabsorptive state

The preceding discussion should make clear that the disposition of organic molecules depends on the body's metabolic state. The two functional metabolic states—the *absorptive state* and the *postabsorptive state*—are related to eating and fasting cycles, respectively (Table 6-6).

### ABSORPTIVE STATE

After a meal, ingested food (nutrients) is digested and absorbed into circulation. This is an absorptive (fed) state. During this time, carbohydrates absorbed as simple sugars are sent to the liver, where they are converted to glucose. The glucose then travels via circulation to be used as fuel for cellular work, or is converted to glycogen and/or fat.

The glycogen and fat is stored in the liver and adipose tissue, respectively, as reserves for the postabsorptive state. As well, some glucose is be stored as glycogen by skeletal muscle cells. Triglycerides in the form of chylomicrons, the primary result of fat digestion from the meal, are first broken down to fatty acids and glycerol through hydrolysis (lipoprotein lipase). Most of these fatty acids are re-formed as triglycerides and stored in adipose tissue. Those not stored are used for energy in the body's cells (e.g., adipose cells, skeletal muscle).

The amino acids absorbed from the protein in the meal enter the primary amino acid pools, to be used for structural purposes. Some of the amino acids (e.g., leucine, alanine, and glutamine) may be used later for fuel, if the body is in a state of fasting or starvation. Excess amino acids are stored as fat, as are excess carbohydrate and fats.

Which fuel source is used as the primary fuel depends on the intensity of the activity being performed. In the absorptive state, when there are plenty of nutrients available, the two primary fuel sources are fats and carbohydrates. In a resting state, such as when you're sitting and watching TV, the predominant fuel source is fat, followed by carbohydrate. However, as the level of activity (exertion) gradually increases, the choice of fuel gradually changes, from predominantly fat to more carbohydrate. The reason for this change is that carbohydrates are a faster form of fuel than fats, and so they can be used when the demand for energy is greater.

### POSTABSORPTIVE STATE

The average meal is absorbed roughly four hours after eating, at which point no nutrients are left in the digestive tract. So, in late morning, late afternoon, or at night when you're sleeping, the body is in a **fasted (postabsorptive) state**. The synthesis of protein and fat is curtailed as the body enters a catabolic state.

During this kind of short-term fasting, endogenous energy stores are mobilized to provide energy. The body looks for other energy sources; for example, it draws on glucose from the liver's glycogen stores and fatty acids from adipose sites. The body, brain, and nerve tissue depend on glucose for metabolism, especially when functioning at greater levels of exertion. Once the body receives its next meal, it will enter a state of anabolism.

However, if the postabsorptive state is prolonged (more than 24 hours), lipolysis becomes the primary source of fuel. This is driven by an increase in $\beta$-adrenergic sensitivity and reduced plasma insulin (hypoinsulinaemia). In the initial 24 hours of fasting there is approximately a 35 percent fall in plasma insulin, a corresponding 65 percent increase in lipolysis, and no change in glucose production. After 24 hours of fasting, the rise in lipolysis is likely a proactive mechanism to help protect blood glucose levels, which are primarily maintained by gluconeogenesis.

Note that the blood concentration of nutrients does not fluctuate markedly between the absorptive and postabsorptive states. During the absorptive state, the glut of absorbed nutrients is swiftly removed from the blood and placed into storage; during the postabsorptive state, these stores are catabolized to maintain the blood concentrations at levels necessary to fill tissue energy demands.

| ▌ TABLE 6-6 Comparison of Absorptive and Postabsorptive States | | |
| --- | --- | --- |
| **Metabolic Fuel** | **Absorptive State** | **Postabsorptive State** |
| **Carbohydrate** | Glucose and fat providing major energy source | Glycogen degradation and depletion |
| | Glycogen synthesis and storage | Glucose sparing to conserve glucose for the brain |
| | Excess converted and stored as triglyceride fat | Production of new glucose through gluconeogenesis |
| **Fat** | Triglyceride synthesis and storage | Triglyceride catabolism |
| | | Fatty acids providing the major energy source for non-glucose-dependent tissues |
| **Protein** | Protein synthesis | Protein catabolism |
| | Excess converted and stored as triglyceride fat | Amino acids used for gluconeogenesis |

Short-term fasting (seven or more days) sets off a starvation response, which encourages the body to store fat once eating is resumed. This is one of the pitfalls of so-called yo-yo dieting. The starvation response is the switching of the body from the use of carbohydrate and fat energy to amino acid and fat energy. The amino acids are synthesized from the amino acid pools (e.g., breakdown of muscle tissue). The catabolism of muscle tissue results in a reduction of the basal metabolic rate, as muscle tissue is the largest metabolically active tissue in the human body.

One of the primal effects of fasting was to reduce the body's energy needs during times of scarcity. Today, however, fasting is used as a form of weight loss. Fasting is analogous to turning the idle lower on a car. The problem with using it as a diet aid becomes apparent when food intake increases post-fasting: fewer calories are now required for basal metabolism (basal metabolic rate), and so a greater percentage of incoming calories are now stored as fat.

## ROLES OF KEY TISSUES IN METABOLIC STATES

During the alternating absorptive and postabsorptive metabolic states, various tissues play different roles, as summarized here.

- The *liver* plays the primary role in maintaining normal blood glucose levels. It stores glycogen when excess glucose is available, releases glucose into the blood when needed, and is the principal site for metabolic interconversions, such as gluconeogenesis.
- *Adipose tissue* serves as the primary energy storage site and is important in regulating fatty acid levels in the blood.
- *Muscle* is the primary site of amino acid storage and is the major energy user.
- The *brain* normally can use only glucose as an energy source, yet it does not store glycogen, making it mandatory that blood glucose levels be maintained.

## Lesser energy sources

Several other organic intermediates play a lesser role as energy sources—namely, glycerol, lactic acid, and ketone bodies.

- As mentioned earlier, *glycerol* derived from triglyceride hydrolysis (it is the backbone to which the fatty acid chains are attached) can be converted to glucose by the liver.
- Similarly, *lactic acid*, which is produced by the incomplete catabolism of glucose via glycolysis in muscle (p. 24), can also be converted to glucose in the liver.
- **Ketone bodies** are a group of compounds produced by the liver during glucose sparing. Unlike other tissues, when the liver uses fatty acids as an energy source, it oxidizes them only to acetyl coenzyme A (acetyl CoA), which it is unable to process through the citric acid cycle for further energy extraction. Thus, the liver does not degrade fatty acids all the way to carbon dioxide and water for maximum energy release. Instead, it partially extracts the available energy and converts the remaining energy-bearing acetyl CoA molecules into ketone bodies, which it releases into the blood. Ketone bodies serve as an alternative energy source for tissues capable of oxidizing them further by means of the citric acid cycle.

During long-term starvation, the brain starts using ketones instead of glucose as a major energy source. Because death resulting from starvation is usually due to protein wasting rather than hypoglycaemia (low blood glucose), prolonged survival without any caloric intake requires that gluconeogenesis be kept to a minimum as long as the energy needs of the brain are not compromised. A sizable portion of cell protein can be catabolized without serious cellular malfunction, but a point is finally reached at which a cannibalized cell can no longer function adequately. To ward off the fatal point of failure as long as possible during prolonged starvation, the brain starts using ketones as a major energy source, correspondingly decreasing its use of glucose. Use by the brain of this fatty acid "table scrap" left over from the liver's "meal" limits the necessity of mobilizing body proteins for glucose production to nourish the brain.

Both the major metabolic adaptations to prolonged starvation—a decrease in protein catabolism and use of ketones by the brain—are attributable to the high levels of ketones in the blood at the time. The brain uses ketones only when blood ketone level is high. The high blood levels of ketones also directly inhibit protein degradation in muscle. Thus, ketones spare body proteins while satisfying the brain's energy needs.

## Insulin and glucagon

How does the body know when to shift its metabolic gears from a system of net anabolism and nutrient storage to one of net catabolism and glucose sparing? The flow of organic nutrients along metabolic pathways is influenced by a variety of hormones, including insulin, glucagon, epinephrine, cortisol, and growth hormone. Under most circumstances, the pancreatic hormones, insulin and glucagon, are the dominant hormonal regulators that shift the metabolic pathways back and forth from net anabolism to net catabolism and glucose sparing, depending on whether the body is in a state of feasting or fasting, respectively.

## ISLETS OF LANGERHANS

The **pancreas** is an organ composed of both exocrine and endocrine tissues. The exocrine portion secretes a watery, alkaline solution and digestive enzymes through the pancreatic duct into the digestive tract lumen. Scattered throughout the pancreas between the exocrine cells are clusters, or islands, of endocrine cells known as the **islets of Langerhans** (see Figure 15-11). The most abundant pancreatic endocrine cells are the *β* (beta) cells, the site of *insulin* synthesis and secretion, and the *α* (alpha) cells, which produce *glucagon*. Less common are the *δ* (delta) cells, the pancreatic site of *somatostatin* synthesis. The least common islet cells, the **PP cells**, secrete *pancreatic polypeptide*, which plays a possible role in reducing appetite and food intake; these cells will not be discussed further. Here we first highlight somatostatin, and then we examine insulin and glucagon, the most important hormones in the regulation of fuel metabolism.

## SOMATOSTATIN

Pancreatic somatostatin inhibits the digestive system in a variety of ways, the overall effect of which is to inhibit digestion of nutrients and to decrease nutrient absorption. Somatostatin is released from the pancreatic δ cells in direct response to an increase in blood glucose and blood amino acids during absorption of a meal. By exerting its inhibitory effects, pancreatic somatostatin acts in negative-feedback fashion to slow down the rate at which the meal is digested and absorbed, thereby preventing excessive plasma levels of nutrients. Pancreatic somatostatin may also play a paracrine role in regulating pancreatic hormone secretion. The local presence of somatostatin decreases the secretion of insulin, glucagon, and somatostatin itself, but the importance of this function has not been determined.

Somatostatin is also produced by cells lining the digestive tract, where it acts locally as a paracrine to inhibit most digestive processes (p. 664). Furthermore, somatostatin (alias GHIH) is produced by the hypothalamus, where it inhibits the secretion of growth hormone and TSH (p. 229).

## Insulin and blood glucose, fatty acid, and amino acid levels

**Insulin** is a small protein composed of two amino acid chains that are connected to each other by linkages. These two protein chains of insulin are synthesized in the β cells of the pancreas. Insulin has been primarily associated with blood sugar and/or carbohydrates and their storage. However, insulin's effect goes far beyond carbohydrates, as it influences fats and proteins as well. Insulin is associated with a wealth of energy, especially carbohydrates. It lowers the blood levels of glucose, fatty acids, and amino acids and promotes their storage. As these nutrient molecules enter the blood during the absorptive state, insulin promotes their cellular uptake and conversion into glycogen, triglycerides, and protein, respectively. Insulin exerts its many effects either by altering transport of specific blood-borne nutrients into cells or by altering the activity of the enzymes involved in specific metabolic pathways.

The following section describes the effects of insulin on carbohydrates, fats, and proteins, and gives special attention to six forms of glucose transporters that facilitate the movement of blood glucose into specific tissues as needed. We also discuss the relationship between insulin levels and exercise.

### ACTIONS ON CARBOHYDRATES

The maintenance of blood glucose homeostasis is a particularly important function of the pancreas. Circulating glucose concentrations are determined by the balance among the following processes (> Figure 6-14): glucose absorption from the digestive tract, transport of glucose into cells, hepatic glucose production, and (abnormally) urinary excretion of glucose.

Insulin exerts four effects that lower blood glucose levels and promote carbohydrate storage:

1. Insulin facilitates glucose transport into most cells.
2. Insulin stimulates **glycogenesis**, the production of glycogen from glucose, in both skeletal muscle and the liver.
3. Insulin inhibits glycogenolysis, the breakdown of glycogen into glucose; in so doing, it favours carbohydrate storage and decreases glucose output by the liver.
4. Insulin further decreases hepatic glucose output by inhibiting gluconeogenesis, the conversion of amino acids into glucose in the liver; it does so by decreasing the amount of amino acids in the blood available to the liver for gluconeogenesis and by inhibiting the hepatic enzymes required for converting amino acids into glucose.

In these four ways, insulin decreases the concentration of blood glucose by promoting the cells' uptake of glucose from the blood for use and storage, while simultaneously blocking the two mechanisms by which the liver releases glucose into the blood (glycogenolysis and gluconeogenesis). Insulin is the only hormone capable of lowering the blood glucose level. Insulin

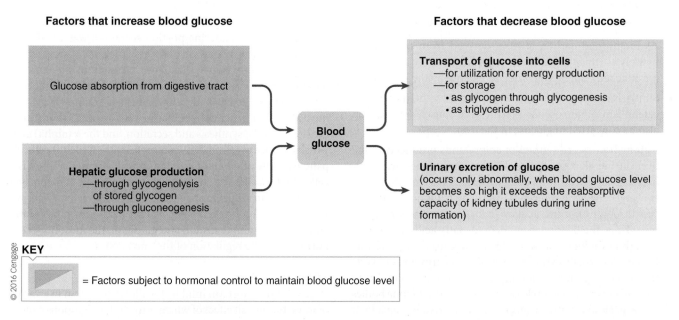

**Factors that increase blood glucose**

Glucose absorption from digestive tract

Hepatic glucose production
—through glycogenolysis of stored glycogen
—through gluconeogenesis

**Blood glucose**

**Factors that decrease blood glucose**

**Transport of glucose into cells**
—for utilization for energy production
—for storage
• as glycogen through glycogenesis
• as triglycerides

**Urinary excretion of glucose**
(occurs only abnormally, when blood glucose level becomes so high it exceeds the reabsorptive capacity of kidney tubules during urine formation)

© 2016 Cengage

**KEY**

= Factors subject to hormonal control to maintain blood glucose level

> FIGURE 6-14 **Factors that affect blood glucose concentration**

promotes the uptake of glucose by most cells through a process called glucose transporter recruitment, a topic to which we now turn our attention.

Glucose transport between the blood and cells is accomplished by means of a plasma membrane carrier known as a **glucose transporter (GLUT)**. Six forms of glucose transporters have been identified:, named in the order they were discovered—GLUT-1, GLUT-2, and so on. These glucose transporters all accomplish passive facilitated diffusion of glucose across the plasma membrane (p. 43) and are distinct from the $Na^+$-glucose cotransport carriers responsible for secondary active transport of glucose across the kidney and intestinal epithelia. Each member of the GLUT family performs slightly different functions. For example, *GLUT-1* transports glucose across the blood-brain barrier, *GLUT-2* transfers into the adjacent bloodstream the glucose that has entered the kidney and intestinal cells by means of the cotransport carriers, and *GLUT-3* is the main transporter of glucose into neurons. The glucose transporter responsible for the majority of glucose uptake by most cells of the body is *GLUT-4*, which operates only at the bidding of insulin. Glucose molecules cannot readily penetrate most cell membranes in the absence of insulin, making most tissues highly dependent on insulin for uptake of glucose from the blood and for its subsequent use. GLUT-4 is especially abundant in the tissues that account for the bulk of glucose uptake from the blood during the absorptive state—namely, skeletal muscle and adipose tissue cells

GLUT-4 is the only type of glucose transporter that responds to insulin. Unlike the other types of GLUT molecules, which are always present in the plasma membranes at the sites where they perform their functions, GLUT-4 in the absence of insulin is excluded from the plasma membrane. Insulin promotes glucose uptake by **transporter recruitment**. Insulin-dependent cells maintain a pool of intracellular vesicles containing GLUT-4. Insulin induces these vesicles to move to the plasma membrane and fuse with it, thereby inserting GLUT-4 molecules into the plasma membrane. In this way, increased insulin secretion promotes a rapid ten- to thirtyfold increase in glucose uptake by insulin-dependent cells. When insulin secretion decreases, these glucose transporters are retrieved from the membrane and returned to the intracellular pool.

Several tissues do not depend on insulin for their glucose uptake—namely, the brain, working muscles, and the liver. The brain, which requires a constant supply of glucose for its minute-to-minute energy needs, is freely permeable to glucose at all times by means of GLUT-1 and GLUT-3 molecules. Skeletal muscle cells do not depend on insulin for their glucose uptake during exercise, even though they are dependent at rest. Muscle contraction triggers the insertion of GLUT-4 into the plasma membranes of exercising muscle cells in the absence of insulin. This fact is important in managing diabetes mellitus (insulin deficiency), as described later. The liver also does not depend on insulin for glucose uptake, because it does not use GLUT-4. However, insulin does enhance the metabolism of glucose by the liver by stimulating the first step in glucose metabolism, the phosphorylation of glucose to form glucose-6-phosphate. The phosphorylation of glucose as it enters the cell keeps the intracellular concentration of "plain" glucose low so that a gradient

favouring the facilitated diffusion of glucose into the cell is maintained.

Insulin also exerts important actions on fat and protein.

### ACTIONS ON FAT

Insulin exerts multiple effects to lower blood fatty acids and promote triglyceride storage:

1. It enhances the entry of fatty acids from the blood into adipose tissue cells.

2. It increases the transport of glucose into adipose tissue cells by means of GLUT-4 recruitment. Glucose serves as a precursor for the formation of fatty acids and glycerol, which are the raw materials for triglyceride synthesis.

3. It promotes chemical reactions that ultimately use fatty acids and glucose derivatives for triglyceride synthesis.

4. It inhibits lipolysis (fat breakdown), reducing the release of fatty acids from adipose tissue into the blood.

Collectively, these actions favour removal of fatty acids and glucose from the blood and promote their storage as triglycerides.

### ACTIONS ON PROTEIN

Insulin lowers blood amino acid levels and enhances protein synthesis through several effects:

1. It promotes the active transport of amino acids from the blood into muscles and other tissues. This effect decreases the circulating amino acid level and provides the building blocks for protein synthesis within the cells.

2. It increases the rate of amino acid incorporation into protein by stimulating the cells' protein-synthesizing machinery.

3. It inhibits protein degradation.

The collective result of these actions is a protein anabolic effect. For this reason, insulin is essential for normal growth.

### SUMMARY OF INSULIN'S ACTIONS

The use of glucose by skeletal muscle depends primarily on two factors: intensity of work or exercise, and the amount of blood glucose. During periods of moderate- to high-intensity exercise, the muscles cells become highly permeable to glucose (GLUT-4), but insulin is not needed to allow the glucose to enter the muscle. As a result of muscle contraction during exercise, the muscle cell itself becomes more permeable to glucose.

In contrast, in the few hours after eating a meal, the muscle cells become much more permeable to glucose, but the permeability is the result of insulin's action on the muscle cell. The pancreas is triggered to secrete large amounts of insulin, which then act on the skeletal muscle to uptake the glucose and perform glycogenesis, and inhibit glycogenolysis (muscle and liver) and gluconeogenesis (liver). As well, the increase in blood insulin level following a meal inhibits the use of fatty acids from adipose sites for energy production. Thus, during moderate- to high-intensity exercise, and in the few hours after a meal, the muscle cells are highly permeable to glucose and they use glucose rather than fats for energy.

In short, insulin primarily exerts its effects by acting on non-working skeletal muscle, the liver, and adipose tissue. It stimulates biosynthetic pathways that lead to increased glucose use, increased carbohydrate and fat storage, and increased protein synthesis. In so doing, this hormone lowers the blood glucose, fatty acid, and amino acid levels. This metabolic pattern is characteristic of the absorptive state. Indeed, insulin secretion rises during this state and shifts metabolic pathways to net anabolism.

When insulin secretion is low, the opposite effects occur. The rate of glucose entry into cells is reduced, and net catabolism occurs rather than net synthesis of glycogen, triglycerides, and protein. This pattern is reminiscent of the postabsorptive state; indeed, insulin secretion is reduced during the postabsorptive state. However, the other major pancreatic hormone, glucagon, also plays an important role in shifting from absorptive to postabsorptive metabolic patterns (see the section Glucagon and Insulin).

### EXERCISE, INSULIN, AND EPINEPHRINE

During moderate exercise, blood glucose declines as a result of skeletal muscle uptake of glucose. As noted before, it is not necessary for the pancreas to secrete insulin in order for the skeletal muscle to uptake additional glucose during exercise, as the contraction of the skeletal muscle influences the GLUT-4 receptor facilitating glucose uptake separate from insulin. The suppression of insulin secretion, and thus the decline in blood insulin levels during exercise, is the result of increased epinephrine levels associated with exercise.

Epinephrine has a direct and indirect effect on liver glucose production. The direct effect is the stimulation of glycogenolysis in the liver; the indirect effect is seen in the inhibition of insulin secretion, which removes the inhibitory effect that insulin has on gluconeogenesis in the liver. As well, the decline in circulating insulin reduces glucose uptake by nonexercising muscle tissue and therefore spares glucose for the active muscle tissue and brain.

During more prolonged exercise, there is a decline in blood glucose and insulin, which facilitates lipolysis and the use of free fatty acids by exercising skeletal muscle. However, following intense training (over several months), a homeostatic adjustment in circulating insulin levels occurs. In trained athletes, the blood insulin levels do not decrease to the same extent as before training. Therefore, more typical blood insulin levels are found post-training, which may be associated with more normal blood glucose levels during exercise.

It is likely that increased free fatty acid metabolism and gluconeogenesis result in better control over blood glucose levels, thus not allowing them to decrease to the same extent as prior to training. Training has also been shown to increase the density of GLUT-4 receptors on skeletal muscle, thereby facilitating the uptake of blood glucose for use during exercise. Because glucose is the primary fuel during exercise, this is an important physiological adjustment.

## Regulation of blood glucose levels

The primary control of insulin secretion is a direct negative-feedback system between the pancreatic $\beta$ cells and the concentration of glucose in the blood flowing to them. An elevated blood glucose level, such as during absorption of a meal, directly stimulates the $\beta$ cells to synthesize and release insulin. The increased insulin, in turn, reduces the blood glucose to normal and promotes the use and storage of this nutrient. Conversely, a fall in blood glucose below normal, such as during fasting, directly inhibits insulin secretion. Lowering the rate of insulin secretion shifts metabolism from the absorptive to the postabsorptive pattern. This simple negative-feedback system can maintain a relatively constant supply of glucose to the tissues without requiring the participation of nerves or other hormones.

In addition to blood glucose concentration, other inputs are involved in regulating insulin secretion, as follows (⟩ Figure 6-15):

- An elevated blood amino acid level, such as after a high-protein meal, directly stimulates the $\beta$ cells to increase insulin secretion. In negative-feedback fashion, the increased insulin enhances the entry of these amino acids into the cells, lowering the blood amino acid level while promoting protein synthesis.

- Gastrointestinal hormones, secreted by the digestive tract in response to the presence of food, stimulate pancreatic insulin secretion in addition to having direct regulatory effects on the digestive system. Through this control, insulin secretion is increased in "feedforward," or anticipatory, fashion even before nutrient absorption increases the blood concentration

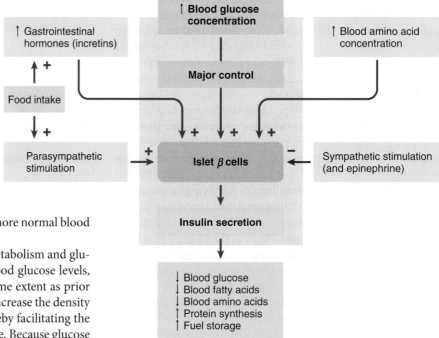

© 2016 Cengage

⟩ FIGURE 6-15 Factors controlling insulin secretion

of glucose and amino acids. Although early studies indicate a role of glucose-dependent insulinotropic peptide (GIP) (p. 693) in this effect, its exact role in humans remains controversial.

- The autonomic nervous system also directly influences insulin secretion. The islets are richly innervated by both parasympathetic (vagal) and sympathetic nerve fibres. The increase in parasympathetic activity that occurs in response to food in the digestive tract stimulates insulin release. This, too, is a feedforward response in anticipation of nutrient absorption. In contrast, sympathetic stimulation and the concurrent increase in epinephrine both inhibit insulin secretion. The fall in insulin level allows the blood glucose level to rise, an appropriate response to the circumstances under which generalized sympathetic activation occurs—namely, stress (fight-or-flight) and exercise. In both these situations, extra fuel is needed for increased muscle activity.

## Diabetes mellitus

*Clinical Note* **Diabetes mellitus** is by far the most common of all endocrine disorders. The acute symptoms of diabetes mellitus are attributable to inadequate insulin action. Because insulin is the only hormone capable of lowering blood glucose levels, one of the most prominent features of diabetes mellitus is elevated blood glucose levels, or *hyperglycaemia*. Diabetes literally means "siphon" or "running through," a reference to the large urine volume accompanying this condition. A large urine volume occurs both in diabetes mellitus (a result of insulin insufficiency) and in diabetes insipidus (a result of vasopressin deficiency). *Mellitus* means "sweet"; *insipidus* means "tasteless." The urine of patients with diabetes mellitus acquires its sweetness from excess blood glucose that spills into the urine, whereas the urine of patients with diabetes insipidus contains no sugar, so it is tasteless. (Aren't you glad you were not a health professional at the time when these two conditions were distinguished on the basis of the taste of the urine?)

Diabetes mellitus has two major variants, differing in the capacity for pancreatic insulin secretion: *type 1 diabetes*, characterized by a lack of insulin secretion, and *type 2 diabetes*, characterized by normal or even increased insulin secretion but reduced sensitivity of insulin's target cells to its presence. (For a further discussion of the distinguishing features of these two types of diabetes mellitus, see Concepts, Challenges, and Controversies.)

The acute consequences of diabetes mellitus can be grouped according to the effects of inadequate insulin action on carbohydrate, fat, and protein metabolism. These effects ultimately cause death through a variety of pathways.

### CONSEQUENCES RELATED TO EFFECTS ON CARBOHYDRATE METABOLISM

Other than hyperglycaemia, the changes induced by diabetes mellitus are an exaggerated form of low insulin activity characteristic of the postabsorptive metabolic pattern. In the usual fasting state, the blood glucose level is slightly below normal.

Hyperglycaemia—the hallmark of diabetes mellitus—arises from reduced glucose uptake by cells coupled with increased output of glucose from the liver (step **1** in Figure 6-16). As the glucose-yielding processes of glycogenolysis and gluconeogenesis proceed unchecked in the absence of insulin, hepatic output of glucose increases. Because many of the body's cells cannot use glucose without the help of insulin, paradoxically, an excess of extracellular glucose coincides with a deficiency of intracellular glucose: "starvation in the midst of plenty." Even though the non-insulin-dependent brain is adequately nourished during diabetes mellitus, further consequences of the disease lead to brain dysfunction, as you will see shortly.

When the blood glucose rises to the level where the amount of glucose filtered exceeds the tubular cells' capacity for reabsorption, glucose appears in the urine (*glucosuria*) **2**. Glucose in the urine exerts an osmotic effect that draws water with it, producing an osmotic diuresis characterized by *polyuria* (frequent urination) **3**. The excess fluid lost from the body leads to dehydration **4** which in turn can ultimately lead to peripheral circulatory failure because of the marked reduction in blood volume **5**. Circulatory failure, if uncorrected, can lead to death because of low cerebral blood flow **6** or secondary renal failure resulting from inadequate filtration pressure **7**. Furthermore, cells lose water as the body becomes dehydrated by an osmotic shift of water from the cells into the hypertonic extracellular fluid **8**. Brain cells are especially sensitive to shrinking, so nervous system malfunction ensues **9** (p. 614). Another characteristic symptom of diabetes mellitus is *polydipsia* (excessive thirst) **10** which is actually a compensatory mechanism to counteract the dehydration.

The story is not complete. In intracellular glucose deficiency, appetite is stimulated, leading to *polyphagia* (excessive food intake) **11**. Despite increased food intake, however, progressive weight loss occurs from the effects of insulin deficiency on fat and protein metabolism.

### CONSEQUENCES RELATED TO EFFECTS ON FAT METABOLISM

Triglyceride synthesis decreases while lipolysis increases, resulting in large-scale mobilization of fatty acids from triglyceride stores **12**. The increased blood fatty acids are largely used by the cells as an alternative energy source. Increased liver use of fatty acids results in the release of excessive ketone bodies into the blood, causing *ketosis* **13**. Ketone bodies include several different acids, such as acetoacetic acid, that result from incomplete breakdown of fat during hepatic energy production. The developing ketosis leads to progressive metabolic acidosis **14**. Acidosis depresses the brain and, if severe enough, can lead to diabetic coma and death **15**.

A compensatory measure for metabolic acidosis is increased ventilation to blow off extra, acid-forming carbon dioxide **16**. Exhalation of one of the ketone bodies, acetone, causes a "fruity" breath odour that smells like a combination of Juicy Fruit gum and nail polish remover. Sometimes, because of this odour, a passerby may unfortunately mistake a patient collapsed in a diabetic coma for someone passed out in a state of drunkenness. (This situation illustrates the merits of medical alert

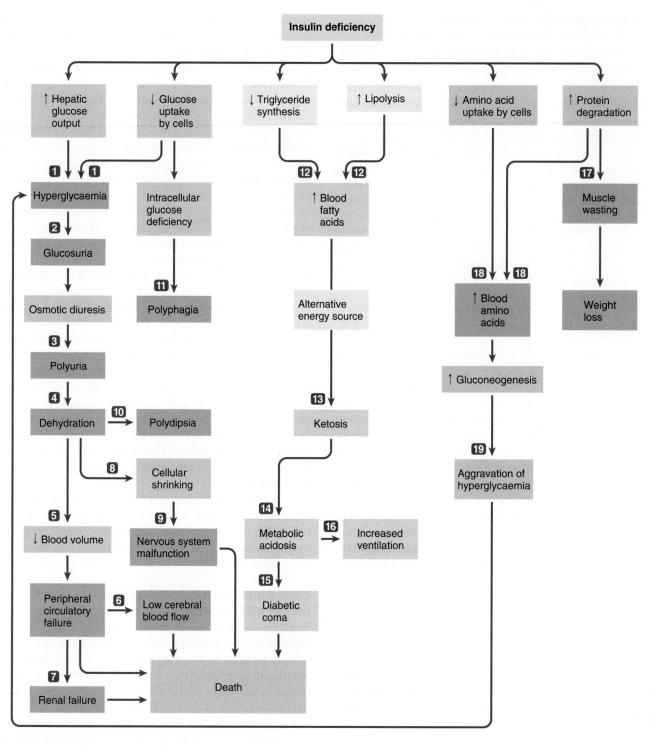

> FIGURE 6-16 **Acute effects of diabetes mellitus**

identification tags.) People with type 1 diabetes are much more prone to develop ketosis than are type 2 diabetics.

## CONSEQUENCES RELATED TO EFFECTS ON PROTEIN METABOLISM

The effects of a lack of insulin on protein metabolism result in a net shift toward protein catabolism. The net breakdown of muscle proteins leads to wasting and weakness of skeletal muscles **17** and, in child diabetics, a reduction in overall growth. Reduced amino acid uptake coupled with increased protein degradation results in excess amino acids in the blood **18**. The increased circulating amino acids can be used for additional gluconeogenesis, which further aggravates the hyperglycaemia **19**.

You can readily appreciate from this overview that diabetes mellitus is a complicated disease, one that can disturb carbohydrate, fat, and protein metabolism and also fluid and acid-base balance. It can also have repercussions on the circulatory system, kidneys, respiratory system, and nervous system.

## LONG-TERM COMPLICATIONS

In addition to these potential acute consequences of untreated diabetes, which can be explained on the basis of insulin's short-term metabolic effects, numerous long-range complications of this disease frequently occur after 15–20 years, despite treatment to prevent the short-term effects. Chronic complications, which account for the shorter life expectancy of diabetics, primarily involve degenerative disorders of the blood vessels and nervous system. Cardiovascular lesions are the most common cause of premature death in diabetics. Heart disease and strokes occur with greater incidence than in non-diabetics. Because vascular lesions often develop in the kidneys and the retinas of the eyes, diabetes is a leading cause of both kidney failure and blindness in Canada. Impaired delivery of blood to the extremities may cause these tissues to become gangrenous, and toes or even whole limbs may have to be amputated. In addition to circulatory problems, degenerative lesions in nerves lead to multiple neuropathies that result in dysfunction of the brain, spinal cord, and peripheral nerves. The latter is most often characterized by pain, numbness, and tingling, especially in the extremities.

Regular exposure of tissues to excess blood glucose over a prolonged time leads to tissue alterations responsible for the development of these long-term vascular and neural degenerative complications. Thus, to diminish the incidence of these chronic abnormalities, the best management for diabetes mellitus is to continuously keep blood glucose levels within normal limits. However, the blood glucose levels of diabetic patients on traditional therapy typically fluctuate over a broader range than normal, thereby exposing their tissues to moderately elevated blood glucose during a portion of each day. Fortunately, recent advances in understanding and learning how to manipulate underlying molecular defects in diabetes offer the hope that more effective therapies will be developed within this decade to better manage or even cure existing cases, and perhaps to prevent new cases of this devastating disease. (See Concepts, Challenges, and Controversies for current and potential future treatment strategies for diabetes.)

## Hypoglycaemia

*Clinical Note* Let's now look at the opposite of diabetes mellitus, insulin excess, which is characterized by *hypoglycaemia* (low blood glucose) and can arise in two different ways. First, insulin excess can occur in a diabetic patient when too much insulin has been injected for the person's caloric intake and exercise level, resulting in so-called **insulin shock**. Second, blood insulin level may rise abnormally high in a non-diabetic person who has a $\beta$ cell tumour or whose $\beta$ cells are overresponsive to glucose, a condition called **reactive hypoglycaemia**. Such $\beta$ cells overshoot and secrete more insulin than necessary, in response to elevated blood glucose after a high-carbohydrate meal. The excess insulin drives too much glucose into the cells, resulting in hypoglycaemia.

The consequences of insulin excess are primarily manifestations of the effects of hypoglycaemia on the brain. Recall that the brain relies on a continuous supply of blood glucose for its nourishment and that glucose uptake by the brain does not depend on insulin. With insulin excess, more glucose than necessary is driven into the other insulin-dependent cells. The result is a lowering of the blood glucose level so that not enough glucose is left in the blood to be delivered to the brain. In hypoglycaemia, the brain literally starves. The symptoms, therefore, are primarily referable to depressed brain function, which, if severe enough, may rapidly progress to unconsciousness and death. People with overresponsive $\beta$ cells usually do not become sufficiently hypoglycaemic to manifest these more serious consequences, but they do show milder symptoms of depressed CNS activity.

The true incidence of reactive hypoglycaemia is a subject of intense controversy, because laboratory measurements to confirm the presence of low blood glucose during the time of symptoms have not been performed in most people who have been diagnosed as having the condition. In mild cases, the symptoms of hypoglycaemia, such as tremor, fatigue, sleepiness, and inability to concentrate, are nonspecific. Because these symptoms could also be attributable to emotional problems or other factors, a definitive diagnosis based on symptoms alone is impossible to make.

The treatment of hypoglycaemia depends on the cause. At the first indication of a hypoglycaemic attack with insulin overdose, the diabetic person should eat or drink something sugary. Prompt treatment of severe hypoglycaemia is imperative to prevent brain damage. Note that a diabetic can lose consciousness and die from either diabetic keto-acidotic coma caused by prolonged insulin deficiency or acute hypoglycaemia caused by insulin shock. Fortunately, the other accompanying signs and symptoms differ sufficiently between the conditions to enable medical caretakers to administer appropriate therapy, either insulin or glucose. For example, ketoacidotic coma is accompanied by deep, laboured breathing (in compensation for the metabolic acidosis) and fruity breath (from exhaled ketone bodies), whereas insulin shock is not.

Paradoxically, even though reactive hypoglycaemia is characterized by a low blood glucose level, people with this disorder are treated by limiting their intake of sugar and other glucose-yielding carbohydrates to prevent their $\beta$ cells from over responding to a high glucose intake. With low carbohydrate intake, the blood glucose does not rise as much during the absorptive state. Because blood glucose elevation is the primary regulator of insulin secretion, the $\beta$ cells are not stimulated as much with a low-carbohydrate meal as with a typical meal. Accordingly, reactive hypoglycaemia is less likely to occur. Giving a symptomatic individual with reactive hypoglycaemia something sugary temporarily alleviates the symptoms. The blood glucose level is transiently restored to normal so that the brain's energy needs are once again satisfied. However, as soon as the extra glucose triggers further insulin release, the situation is aggravated.

# Diabetics and Insulin: Some Have It and Some Don't

THERE ARE TWO DISTINCT TYPES OF DIABETES MELLITUS (see the accompanying table). **Type 1 diabetes mellitus** (formerly referred to as *insulin-dependent* or *juvenile-onset*), which accounts for about 10 percent of all cases of diabetes, is characterized by a lack of insulin secretion. Because their pancreatic β cells secrete no or nearly no insulin, type 1 diabetics require exogenous insulin for survival. Hence, this form of the disease has the alternative name *insulin-dependent diabetes mellitus*. In **type 2 diabetes mellitus** (formerly referred to as *non-insulin-dependent* or *maturity-onset*), insulin secretion may be normal or even increased, but insulin's target cells are less sensitive than normal to this hormone. Ninety percent of diabetics have the type 2 form of the disease. Although either type can first manifest at any age, type 1 is more prevalent in children, whereas type 2 more generally arises in adulthood—hence, the age-related designations of the two conditions.

For information on the prevalence of diabetes in Canada as well as other facts and figures, see The Public Health Agency of Canada's website at http://www.phac-aspc.gc.ca/cd-mc/diabetes-diabete/index-eng.php, and the Canadian Diabetes Association's website at http://www.diabetes.ca/.

Statistics from 2014 suggest that the number of Canadians 12 years of age and older who have been diagnosed with diabetes is approximately 2 million, or about 6.7 percent of the Canadian population in that age range. However, the rate of diabetes within Canada's Indigenous population was triple that of the general population.

In 2013, diabetes was the sixth leading cause of death in Canada, with 7045 deaths directly associated with the disease. The actual number of deaths for which diabetes was the underlying factor is probably five times as great, since those with diabetes typically die from complications brought on by other conditions (e.g., heart disease, cardiovascular disease).

Although the direct healthcare cost of diabetes in 2015 was approximately $3 billion, the total economic burden of diabetes is much higher, at $12.2 billion, as this estimate did not include the cost of missed days of work and complications from other diseases (heart disease). Due to Canada's aging population, it is estimated that the total economic burden of diabetes will rise to over $16.9 billion annually by 2020.

In 2015, according to the International Diabetes Federation, 415 million people worldwide had diabetes, and this number is expected to increase to 642 million by 2040. Due to this prevalence of diabetes, its huge economic toll, and the fact that it forces a change in lifestyle for affected individuals and places them at increased risk for developing a variety of troublesome and even life-threatening conditions, intensive research is directed toward better understanding and controlling or preventing both types of the disease.

## Underlying Defect in Type 1 Diabetes

Type 1 diabetes is an autoimmune process involving the erroneous, selective destruction of the pancreatic β cells by inappropriately activated T lymphocytes (p. 486). The precise cause of this self-destructive immune attack remains unclear. Some individuals have a genetic susceptibility to acquiring type 1 diabetes. Environmental triggers also appear to play an important role, but investigators have not been able to definitively pin down the culprits.

In 2006, a team of doctors from the University of Alberta, led by Dr. Shapiro, Director and Head of the Clinical Islet Transplant Program, had a breakthrough in the battle against type 1 diabetes. The team developed what is known as the Edmonton Protocol, which consisted of a few adjustments to the initial protocol for islet cell transplants. Through the elimination of steroids, which are known for their toxic effect on the insulin-producing β cells, and the use of newly developed immunosuppressive drugs to prevent rejection of the graft cells, a high number of fresh islet cells were maintained. Since this adjustment, a high success rate for treating type 1 diabetes has been achieved.

## Underlying Defect in Type 2 Diabetes

Type 2 diabetics do secrete insulin, but the affected individuals exhibit *insulin* resistance. That is, the basic problem in type 2 diabetes is not lack of insulin but reduced sensitivity of insulin's target cells to its presence. Various genetic and lifestyle factors appear important in the development of type 2 diabetes. Obesity is the biggest risk factor; 90 percent of type 2 diabetics are obese.

Many type 2 diabetics have *metabolic syndrome*, or *syndrome X*, as a forerunner of diabetes. **Metabolic syndrome** encompasses a cluster of features that predispose the person to developing type 2 diabetes and atherosclerosis (p. 374). These features include obesity, large waist circumference (i.e., "apple" shapes; p. 708), high triglyceride levels, low HDL (the so-called good cholesterol; p. 376), high blood glucose, and high blood pressure. An estimated 25 percent of the Canadian population has metabolic syndrome.

The ultimate underlying cause of type 2 diabetes remains elusive despite intense investigation, but researchers have identified a number of possible links between obesity and reduced insulin sensitivity. Recent studies indicate that the responsiveness of skeletal muscle and liver to insulin can be modulated by circulating adipokines (hormones secreted by adipose cells). The implicated adipokines are distinct from leptin, the hormone secreted by adipose cells that plays a role in controlling food intake (p. 702). For example, adipose tissue secretes the hormone **resistin**, which promotes insulin resistance by interfering with insulin action. Resistin production increases in obesity. By contrast, **adiponectin**, another adipokine, increases insulin sensitivity by enhancing insulin's effects, but its production is decreased in obesity. Furthermore, free fatty acids released from adipose tissue can abnormally accumulate in muscle and interfere with insulin action in muscle. Also, evidence suggests that excessive fatty acids can indirectly trigger apoptosis (cell suicide) of $\beta$ cells.

Early in the development of the disease, the resulting decrease in sensitivity to insulin is overcome by secretion of additional insulin. However, the sustained overtaxing of the pancreas eventually exceeds the reserve secretory capacity of the genetically weak $\beta$ cells. Even though insulin secretion may be normal or somewhat elevated, symptoms of insulin insufficiency develop because the amount of insulin is still inadequate to prevent significant hyperglycaemia. The symptoms in type 2 diabetes are usually slower in onset and less severe than in type 1 diabetes.

### Treatment of Diabetes

The conventional treatment for type 1 diabetes is a controlled balance of regular insulin injections timed around meals, dietary management of the amounts and types of food consumed, and exercise. Insulin is administered by injection, because if it were swallowed, this peptide hormone would be digested by proteolytic enzymes in the stomach and small intestine. Exercise is also useful in managing both types of diabetes, because working muscles are not insulin dependent. Exercising muscles take up and use some of the excess glucose in the blood, reducing the overall need for insulin.

Whereas type 1 diabetics are permanently insulin dependent, dietary control and weight reduction may be all that is necessary to completely reverse the symptoms in type 2 diabetics. Therefore, type 2 diabetes is alternatively known as *non-insulin-dependent diabetes*. Four classes of oral medications are currently available for use if needed for treating type 2 diabetes, in conjunction with a dietary and exercise regime. These pills help the patient's body use its own insulin more effectively, each by a different mechanism, as follows:

1. By stimulating the $\beta$ cells to secrete more insulin than they do on their own (*sulphonylureas*, e.g., *Glucotrol*)
2. By suppressing the liver's output of glucose (*metformin*, e.g., *Glucophage*)
3. By blocking the enzymes that digest complex carbohydrates, thus slowing glucose absorption into the blood from the digestive tract and thereby blunting the surge of glucose immediately after a meal (*alpha-glycosidase inhibitors*, e.g., *Prandase*)
4. By making muscle and fat cells more receptive to insulin (i.e., by reducing insulin resistance) (*thiazolidinediones*, e.g., *Avandia*)

A new class of drug was recently approved for treating type 2 diabetics, an *incretin mimetic*. **Incretins** are hormones released by the digestive tract that lower blood glucose, such as GIP. Incretin mimetics are drugs that mimic these naturally occurring hormones. The first drug on the market of this type, *Byetta*, mimics the gut-released hormone *glucagon-like peptide 1 (GLP-1)*. GLP-1 is released from the small intestine L cells in response to food intake and has multiple glucose-lowering effects. GLP-1 itself is too short-lived to be suitable as a drug. Byetta, which is a version of a peptide found in the venom of a poisonous Gila monster, must be injected. Like GLP-1, this drug stimulates insulin secretion by $\beta$ cells when blood glucose is high, but not when glucose is in the normal range. It also suppresses production of glucose-raising glucagon and slows gastric emptying, thus slowing the rate at which nutrients are digested and absorbed. By promoting satiety, Byetta decreases food intake and in the long term causes weight loss. Animal studies indicate that Byetta even stimulates regeneration of pancreatic $\beta$ cells.

None of these drugs deliver new insulin to the body, so they cannot replace insulin injections for people with type 1 diabetes. Furthermore, sometimes the weakened $\beta$ cells of type 2 diabetics eventually burn out and can no longer produce insulin. In such a case, the previously non-insulin-dependent patient must be placed on insulin therapy for life.

### New Approaches to Managing Diabetes

Several new approaches are currently available for insulin-dependent diabetics that preclude the need for the one or two insulin injections daily:

- Implanted insulin pumps can deliver a prescribed amount of insulin on a regular basis, but the recipient must time meals with care to match the automatic insulin delivery.
- Pancreas transplants are also being performed more widely now, with increasing success rates. On the downside, recipients of pancreas transplants must take immunosuppressive drugs for life to prevent rejection of their donated organs. Also, donor organs are in short supply.

Current research on several fronts may dramatically change the approach to diabetic therapy in the near future. The following new treatments are on the horizon, most of which do away with the dreaded daily injections:

- Some methods under development circumvent the need for insulin injections by using alternative routes of administration that bypass the destructive digestive tract enzymes. These include the use of inhaled powdered insulin and the use of ultrasound to force insulin into the skin from an insulin-impregnated patch.
- Some researchers are seeking methods to protect swallowed insulin from destruction by the digestive tract, and others have identified a potential oral substitute for insulin—namely, a non-peptide chemical that binds with the insulin receptors and brings about the same intracellular responses as insulin does. Because this insulin mimic is not a protein, it would not be destroyed by the proteolytic digestive enzymes if taken as a pill.

*(Continued)*

- Another hope is pancreatic islet transplants. Scientists have developed several types of devices that isolate donor islet cells from the recipient's immune system. Such immunoisolation of islet cells would permit use of grafts from other animals, circumventing the shortage of human donor cells. Pig islet cells would be an especially good source, because pig insulin is nearly identical to human insulin.
- Some researchers hope they will be able to coax stem cells to develop into insulin-secreting cells that can be implanted.
- In a related approach, other scientists are turning to genetic engineering in the hope of developing surrogates for pancreatic β cells. An example is the potential reprogramming of the small-intestine endocrine cells that produce glucose-dependent insulinotropic hormone (GIP). Recall that GIP normally promotes insulin release in feedforward fashion when food is in the digestive tract. The goal is to cause these non-β cells to co-secrete both insulin and GIP on feeding.
- Another approach under development is an implanted, glucose-detecting, insulin-releasing "artificial pancreas" that would continuously monitor the patient's blood glucose level and deliver insulin in response to need.
- On another front, scientists are hopeful of one day developing immunotherapies that specifically block the attack of the immune system against the β cells, thereby curbing or preventing type 1 diabetes.
- Still other investigators are scurrying to further unravel the defective pathways underlying type 2 diabetes, with the goal of finding new therapeutic targets to prevent, halt, or reverse this condition.

### Further Reading

Chiang, J.L., Kirkman, M.S., Laffel, L.M., Peters, A.L. et al. (2014). Type 1 Diabetes through the life span: A position statement of the American Diabetes Association. *Diabetes Care*, 37(7): 2034–54.

Quirk, H., Blake, H., Tennyson, R., Randell, T.L. & Glazebrook, C. (2014). Physical activity interventions in children and young people with Type 1 diabetes mellitus: A systemic review with meta-analysis. *Diabet Med*, 10:1163–73.

Samaan, M.C. (2013). Management of pediatric and adolescent type 2 diabetes. *Int J Pediatr*, Article ID 972034, 9 pages.

## ▪ Comparison of Type 1 and Type 2 Diabetes Mellitus

| Characteristic | Type 1 Diabetes | Type 2 Diabetes |
| --- | --- | --- |
| **Level of Insulin Secretion** | None or almost none | May be normal or exceed normal |
| **Typical Age of Onset** | Childhood | Adulthood |
| **Percentage of Diabetics** | 10–20 percent | 80–90 percent |
| **Basic Defect** | Destruction of β cells | Reduced sensitivity of insulin's target cells |
| **Treatment** | Insulin injections; dietary management; exercise | Dietary control and weight reduction; exercise; sometimes oral hypoglycaemic drugs |

## Glucagon and insulin

Even though insulin plays a central role in controlling metabolic adjustments between the absorptive and postabsorptive states, the secretory product of the pancreatic islet α cells, **glucagon**, is also very important. Many physiologists view the insulin-secreting β cells and the glucagon-secreting α cells as a coupled endocrine system, whose combined secretory output is a major factor in regulating fuel metabolism.

Glucagon affects many of the same metabolic processes that insulin influences, but in most cases glucagon's actions are opposite to those of insulin. The major site of action of glucagon is the liver, where it exerts a variety of effects on carbohydrate, fat, and protein metabolism.

### ACTIONS ON CARBOHYDRATE

The overall effects of glucagon on carbohydrate metabolism result in an increase in hepatic glucose production and release and consequently an increase in blood glucose levels. Glucagon exerts its hyperglycaemic effects by decreasing glycogen synthesis, promoting glycogenolysis, and stimulating gluconeogenesis.

### ACTIONS ON FAT

Glucagon also antagonizes the actions of insulin with regard to fat metabolism by promoting fat breakdown and inhibiting triglyceride synthesis. Glucagon enhances hepatic ketone production (**ketogenesis**) by promoting the conversion of fatty acids to ketone bodies. The blood levels of fatty acids and ketones therefore increase under glucagon's influence.

### ACTIONS ON PROTEIN

Glucagon inhibits hepatic protein synthesis and promotes degradation of hepatic protein. Stimulation of gluconeogenesis further contributes to glucagon's catabolic effect on hepatic protein metabolism. Glucagon promotes protein catabolism in the liver, but it does not have any significant effect on blood amino acid levels because it does not affect muscle protein, the major protein store in the body.

## THE POSTABSORPTIVE STATE

Glucagon, like insulin, is a large string of approximately 29 amino acids. Although both glucagon and insulin are made of amino acids, they have opposite effects on the body. Considering the catabolic effects of glucagon on energy stores, you would be correct in assuming that glucagon secretion increases during the postabsorptive state and decreases during the absorptive state, just the opposite of insulin secretion. In fact, insulin is sometimes referred to as a "hormone of feasting" and glucagon as a "hormone of fasting." Insulin tends to put nutrients in storage when blood nutrient levels are high, such as after a meal, whereas glucagon promotes catabolism of nutrient stores between meals to keep up the blood nutrient levels, especially blood glucose.

As in insulin secretion, the major factor regulating glucagon secretion is a direct effect of the blood glucose concentration on the **endocrine pancreas**. In this case, the pancreatic α cells increase glucagon secretion in response to a fall in blood glucose. The hyperglycaemic actions of this hormone tend to raise the blood glucose level back to normal. Conversely, an increase in blood glucose concentration, such as after a meal, inhibits glucagon secretion, which tends to drop the blood glucose level back to normal.

### Insulin and glucagon: A team

A direct negative-feedback relationship exists between blood glucose concentration and the rates of secretion of both the β cells and the α cells, but with opposite effects. An elevated blood glucose level stimulates insulin secretion but inhibits glucagon secretion, whereas a fall in blood glucose level leads to decreased insulin secretion and increased glucagon secretion (> Figure 6-17). Because insulin lowers and glucagon raises blood glucose, the changes in secretion of these pancreatic hormones in response to deviations in blood glucose work together homeostatically to restore blood glucose levels to normal.

Similarly, a fall in blood fatty acid concentration directly inhibits insulin output and stimulates glucagon output by the pancreas, both of which are negative-feedback control mechanisms to restore the blood fatty acid level to normal.

The opposite effects exerted by blood concentrations of glucose and fatty acids on the pancreatic α and β cells are appropriate for regulating the circulating levels of these nutrient molecules because the actions of insulin and glucagon on carbohydrate and fat metabolism oppose one another. However, the effect of blood amino acid concentration on the secretion of insulin and glucagon is a different story. A rise in blood amino acid concentration stimulates *both* insulin and glucagon secretion. Why this seeming paradox, because glucagon does not exert any effect on blood amino acid concentration? The identical effect of high blood amino acid levels on both insulin and glucagon secretion makes sense when you consider the concomitant effects these two hormones have on blood glucose levels (> Figure 6-18). For example, during

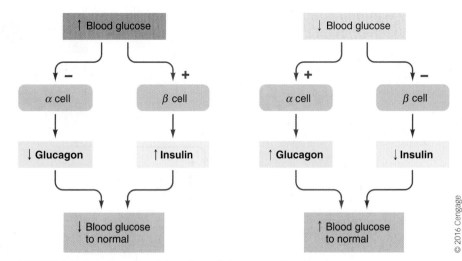

> FIGURE 6-17 **Complementary interactions of glucagon and insulin**

© 2016 Cengage

absorption of a protein-rich meal, if the rise in blood amino acids stimulated only insulin secretion, hypoglycaemia might result. Because little carbohydrate is available for absorption following consumption of a high-protein meal, the amino acid–induced increase in insulin secretion would drive too much glucose into the cells, causing a sudden, inappropriate drop in the blood glucose level. However, the simultaneous increase in glucagon secretion elicited by elevated blood amino acid levels increases hepatic glucose production. Because the hyperglycaemic effects of glucagon counteract the hypoglycaemic actions of insulin, the net result is maintenance of normal blood glucose levels (and prevention of hypoglycaemic starvation of the brain) during absorption of a meal that is high in protein but low in carbohydrates.

### Glucagon excess and diabetes mellitus

*Clinical Note* No known clinical abnormalities are caused by glucagon deficiency or excess per se. However, diabetes mellitus is frequently accompanied by excess glucagon secretion, because insulin is required for glucose to gain entry into the α cells, where it can exert control over glucagon secretion. As a result, diabetics frequently have a high rate of glucagon secretion concurrent with their insulin insufficiency because the elevated blood glucose cannot inhibit glucagon secretion as it normally would. Because glucagon is a hormone that raises blood glucose, its excess intensifies the hyperglycaemia of diabetes mellitus. For this reason, some insulin-dependent diabetics respond best to a combination of insulin and somatostatin therapy. By inhibiting glucagon secretion, somatostatin indirectly helps achieve better reduction of the elevated blood glucose concentration than can be accomplished by insulin therapy alone.

### Epinephrine, cortisol, and growth hormone

The pancreatic hormones are the most important regulators of normal fuel metabolism. However, several other hormones exert direct metabolic effects, even though control of their secretion is keyed to factors other than transitions in metabolism between feasting and fasting states (▌ Table 6-7).

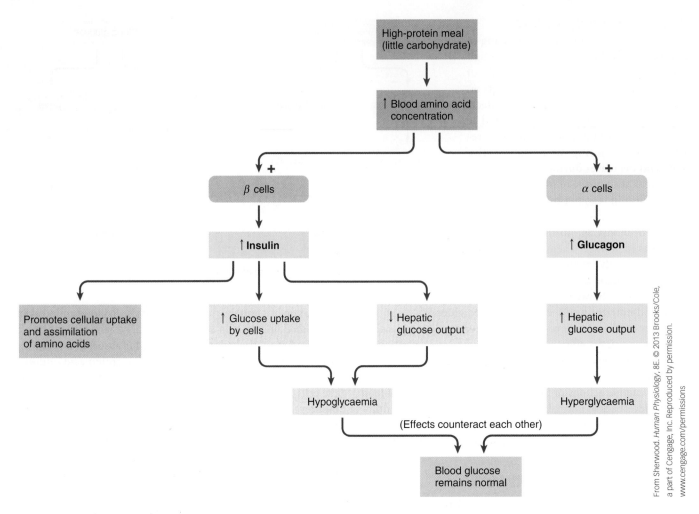

> FIGURE 6-18 **Counteracting actions of glucagon and insulin on blood glucose during absorption of a high-protein meal**

The stress hormones, epinephrine and cortisol, both increase blood levels of glucose and fatty acids through a variety of metabolic effects. In addition, cortisol mobilizes amino acids by promoting protein catabolism. Neither hormone plays an important role in regulating fuel metabolism under resting conditions, but both are important for the metabolic responses to stress. During long-term starvation, cortisol also seems to help maintain blood glucose concentration.

Growth hormone (GH) has protein anabolic effects in muscle. In fact, this is one of its growth-promoting features. Although GH can elevate the blood levels of glucose and fatty acids, it is normally of little importance to the overall regulation of fuel metabolism. Deep sleep, stress, exercise, and severe hypoglycaemia stimulate GH secretion, possibly to provide fatty acids as an energy source and spare glucose for the brain under these circumstances. GH, like cortisol, appears to help maintain blood glucose concentrations during starvation.

Although thyroid hormone increases the overall metabolic rate and has both anabolic and catabolic actions, changes in thyroid hormone secretion are usually not important for fuel homeostasis, for two reasons. First, control of thyroid hormone secretion is not directed toward maintaining nutrient levels in the blood. Second, the onset of thyroid hormone action is too slow to have any significant effect on the rapid adjustments required to maintain normal blood levels of nutrients.

Note that, with the exception of the anabolic effects of GH on protein metabolism, all the metabolic actions of these other hormones are opposite to those of insulin. Insulin alone can reduce blood glucose and blood fatty acid levels, whereas glucagon, epinephrine, cortisol, and GH all increase blood levels of these nutrients. These other hormones are therefore considered **insulin antagonists**. Thus, the main reason diabetes mellitus has such devastating metabolic consequences is that no other control mechanism is available to pick up the slack to promote anabolism when insulin activity is insufficient, so the catabolic reactions promoted by other hormones proceed unchecked. The only exception is protein anabolism stimulated by GH.

## Check Your Understanding 6.3

1. Define *glycogenesis*, *glycogenolysis*, and *gluconeogenesis*.
2. Describe the metabolic effects of insulin and glucagon.
3. Compare the effect of increased blood glucose on pancreatic β cells and α cells.

Summary of Hormonal Control of Fuel Metabolism

| | MAJOR METABOLIC EFFECTS | | | | CONTROL OF SECRETION | |
|---|---|---|---|---|---|---|
| Hormone | Effect on Blood Glucose | Effect on Blood Fatty Acids | Effect on Blood Amino Acids | Effect on Muscle Protein | Major Stimuli for Secretion | Primary Role in Metabolism |
| **Insulin** | ↓ <br><br>+Glucose uptake <br><br>+Glycogenesis <br><br>−Glycogenolysis <br><br>−Gluconeogenesis | ↓ <br><br>+ Triglyceride synthesis <br><br>−Lipolysis | ↓ <br><br>+ Amino acid uptake | ↑ <br><br>+Protein synthesis <br><br>−Protein degradation | ↑ Blood glucose <br><br>↑ Blood amino acids | Primary regulator of absorptive and postabsorptive cycles |
| **Glucagon** | ↑ <br><br>+Glycogenolysis <br><br>+Gluconeogenesis <br><br>−Glycogenesis | ↑ <br><br>+Lipolysis <br><br>− Triglyceride synthesis | No effect | No effect | ↓ Blood glucose <br><br>↑ Blood amino acids | Regulation of absorptive and postabsorptive cycles in concert with insulin; protection against hypoglycaemia |
| **Epinephrine** | ↑ <br><br>+Glycogenolysis <br><br>+Gluconeogenesis <br><br>−Insulin secretion <br><br>+Glucagon secretion | ↑ <br><br>+Lipolysis | No effect | No effect | Sympathetic stimulation during stress and exercise | Provision of energy for emergencies and exercise |
| **Cortisol** | ↑ <br><br>+Gluconeogenesis <br><br>− Glucose uptake by tissues other than brain; glucose sparing | ↑ <br><br>+Lipolysis | ↑ <br><br>+ Protein degradation | ↓ <br><br>+ Protein degradation | Stress | Mobilization of metabolic fuels and building blocks during adaptation to stress |
| **Growth Hormone** | ↑ <br><br>− Glucose uptake by muscles; glucose sparing | ↑ <br><br>+Lipolysis | ↓ <br><br>+ Amino acid uptake | ↑ <br><br>+ Protein synthesis <br><br>− Protein degradation <br><br>+ Synthesis of DNA and RNA | Deep sleep <br><br>Stress <br><br>Exercise <br><br>Hypoglycaemia | Promotion of growth; normally little role in metabolism; mobilization of fuels plus glucose sparing in extenuating circumstances |

↑ = increase ↓ = decrease

# 6.5 | Endocrine Control of Calcium Metabolism

Besides regulating the concentration of organic nutrient molecules in the blood by manipulating anabolic and catabolic pathways, the endocrine system regulates the plasma concentration of a number of inorganic electrolytes. As you already know, aldosterone controls sodium ($Na^+$) and potassium ($K^+$) concentrations in the ECF. Three other hormones—*parathyroid hormone, calcitonin,* and *vitamin D*—control calcium ($Ca^{2+}$) and phosphate ($PO_4^{3-}$) metabolism. These hormonal agents concern themselves with regulating plasma $Ca^{2+}$, and, in the process, plasma $PO_4^{3-}$ is also maintained. Plasma $Ca^{2+}$ concentration is one of the most tightly controlled variables in the body. The need for the precise regulation of plasma $Ca^{2+}$ stems from its critical influence on so many body activities (see Appendix F for reference values).

## Plasma $Ca^{2+}$

About 99 percent of the calcium in the body is in crystalline form within the skeleton and teeth. Of the remaining 1 percent, about 0.9 percent is found intracellularly within the soft tissues; less than 0.1 percent is present in the ECF. Approximately half of the ECF $Ca^{2+}$ is either bound to plasma proteins, and therefore restricted to the plasma, or complexed with $PO_4^{3-}$ and not free to participate in chemical reactions. The other half of the ECF $Ca^{2+}$ is freely diffusible and can readily pass from the plasma into the interstitial fluid and interact with the cells. The free $Ca^{2+}$ in the plasma and interstitial fluid is considered a single pool. Only this free ECF $Ca^{2+}$ is biologically active and subject to regulation; it constitutes less than one-thousandth of the total $Ca^{2+}$ in the body.

This small, free fraction of ECF $Ca^{2+}$ plays a vital role in a number of essential activities, including the following:

1. *Neuromuscular excitability.* Even minor variations in the concentration of free ECF $Ca^{2+}$ can have a profound and immediate impact on the sensitivity of excitable tissues. A fall in free $Ca^{2+}$ results in overexcitability of nerves and muscles; conversely, a rise in free $Ca^{2+}$ depresses neuromuscular excitability. These effects result from the influence of $Ca^{2+}$ on membrane permeability to $Na^+$. A decrease in free $Ca^{2+}$ increases $Na^+$ permeability, with the resultant influx of $Na^+$ moving the resting potential closer to threshold. Consequently, in the presence of *hypocalcaemia* (low blood $Ca^{2+}$), excitable tissues may be brought to threshold by normally ineffective physiologic stimuli so that skeletal muscles discharge and contract (go into spasm) "spontaneously" (in the absence of normal stimulation). If severe enough, spastic contraction of the respiratory muscles results in death by asphyxiation. *Hypercalcaemia* (elevated blood $Ca^{2+}$) is also life threatening, because it causes cardiac arrhythmias and generalized depression of neuromuscular excitability.

2. *Excitation-contraction coupling in cardiac and smooth muscle.* Entry of ECF $Ca^{2+}$ into cardiac and smooth muscle cells, resulting from increased $Ca^{2+}$ permeability in response to an action potential, triggers the contractile mechanism. Calcium is also necessary for

excitation-contraction coupling in skeletal muscle fibres, but in this case the $Ca^{2+}$ is released from intracellular $Ca^{2+}$ stores in response to an action potential. A significant part of the increase in cytosolic $Ca^{2+}$ in cardiac muscle cells also derives from internal stores.

Note that a *rise in cytosolic $Ca^{2+}$* within a muscle cell causes contraction, whereas an *increase in free ECF $Ca^{2+}$* decreases neuromuscular excitability and reduces the likelihood of contraction. Unless one keeps this point in mind, it is difficult to understand why low plasma $Ca^{2+}$ levels induce muscle hyperactivity when $Ca^{2+}$ is necessary to switch on the contractile apparatus. We are talking about two different $Ca^{2+}$ pools, which exert different effects.

3. *Stimulus-secretion coupling.* The entry of $Ca^{2+}$ into secretory cells, which results from increased permeability to $Ca^{2+}$ in response to appropriate stimulation, triggers the release of the secretory product by exocytosis. This process is important for the secretion of neurotransmitters by nerve cells and for peptide and catecholamine hormone secretion by endocrine cells.

4. *Maintenance of tight junctions between cells.* Calcium forms part of the intercellular cement that holds particular cells tightly together.

5. *Clotting of blood.* Calcium serves as a cofactor in several steps of the cascade of reactions that lead to clot formation.

In addition to these functions of free ECF $Ca^{2+}$, intracellular $Ca^{2+}$ serves as a second messenger in many cells and is involved in cell motility and cilia action. Finally, the $Ca^{2+}$ in bone and teeth is essential for the structural and functional integrity of these tissues.

Because of the profound effects of deviations in free $Ca^{2+}$, especially on neuromuscular excitability, the plasma concentration of this electrolyte is regulated with extraordinary precision. Let's see how.

## Control of $Ca^{2+}$ metabolism

Maintaining the proper plasma concentration of free $Ca^{2+}$ differs from the regulation of $Na^+$ and $K^+$ in two important ways. $Na^+$ and $K^+$ homeostasis is maintained primarily by regulating the urinary excretion of these electrolytes so that controlled output matches uncontrolled input. Although urinary excretion of $Ca^{2+}$ is hormonally controlled, in contrast to $Na^+$ and $K^+$, not all ingested $Ca^{2+}$ is absorbed from the digestive tract; instead, the extent of absorption is hormonally controlled and depends on the $Ca^{2+}$ status of the body. In addition, bone serves as a large $Ca^{2+}$ reservoir that can be drawn on to maintain the free plasma $Ca^{2+}$ concentration within the narrow limits compatible with life should dietary intake become too low. Exchange of $Ca^{2+}$ between the ECF and bone is also subject to hormonal control. Similar in-house stores are not available for $Na^+$ and $K^+$.

Regulation of $Ca^{2+}$ metabolism depends on hormonal control of exchanges between the ECF and three other compartments: bone, kidneys, and intestine. Control of $Ca^{2+}$ metabolism encompasses two aspects:

- First, regulation of **calcium homeostasis** involves the immediate adjustments required to maintain a *constant free plasma $Ca^{2+}$ concentration* on a minute-to-minute basis. This is largely accomplished by rapid exchanges between bone and ECF and to a lesser extent by modifications in urinary excretion of $Ca^{2+}$.

- Second, regulation of **calcium balance** involves the more slowly responding adjustments required to maintain a *constant total amount of $Ca^{2+}$ in the body*. Control of $Ca^{2+}$ balance ensures that $Ca^{2+}$ intake is equivalent to $Ca^{2+}$ excretion over the long term (weeks to months). Calcium balance is maintained by adjusting the extent of intestinal $Ca^{2+}$ absorption and urinary $Ca^{2+}$ excretion.

Parathyroid hormone (PTH), the principal regulator of $Ca^{2+}$ metabolism, acts directly or indirectly on all three of these effector sites. It is the primary hormone responsible for maintenance of $Ca^{2+}$ homeostasis and is essential for maintaining $Ca^{2+}$ balance, although vitamin D also contributes in important ways to $Ca^{2+}$ balance. The third $Ca^{2+}$-influencing hormone, calcitonin, is not essential for maintaining either $Ca^{2+}$ homeostasis or balance. It serves a backup function during the rare times of extreme hypercalcaemia.

We now examine the specific effects of these hormonal systems—calcium homeostasis and calcium balance—on bone, kidneys, and intestines.

## Parathyroid hormone and free plasma $Ca^{2+}$ levels

**Parathyroid hormone (PTH)** is a peptide hormone consisting of 84 amino acids secreted by the **parathyroid glands**. Like aldosterone, PTH is *essential for life*. The overall effect of PTH is to increase the $Ca^{2+}$ concentration of plasma (and, accordingly, of the entire ECF), thereby preventing hypocalcaemia. In the complete absence of PTH, death ensues within a few days, usually because of asphyxiation caused by hypocalcaemic spasm of respiratory muscles. By its actions on bone, kidneys, and intestine, PTH raises plasma $Ca^{2+}$ level when it starts to fall so that hypocalcaemia and its effects are normally avoided. This hormone also acts to lower plasma $PO_4^{3-}$ concentration. We will consider each of these mechanisms, beginning with an overview of bone remodelling and the actions of PTH on bone.

## Bone remodelling

Because 99 percent of the body's $Ca^{2+}$ is in bone, the skeleton serves as a storage depot for $Ca^{2+}$. (See Table 6–8 for other functions of the skeleton.) Bone is a living tissue, composed of an

### ▮ TABLE 6-8 Functions of the Skeleton

Support

Protection of vital internal organs

Assistance in body movement by providing site of attachment for muscles and providing leverage

Manufacture of blood cells (bone marrow)

Storage depot for $Ca^{2+}$ and $PO_4^{3-}$, which can be exchanged with the plasma to maintain plasma concentrations of these electrolytes

organic extracellular matrix impregnated with **hydroxyapatite crystals** consisting primarily of precipitated $Ca_3(PO_4)_2$ (calcium phosphate) salts. Normally, $Ca_3(PO_4)_2$ salts are in solution in the ECF, but the conditions within bone are suitable for these salts to precipitate (crystallize) around the collagen fibres in the matrix. By mobilizing some of these $Ca^{2+}$ stores in bone, PTH raises plasma $Ca^{2+}$ concentration when it starts to fall.

Despite the apparent inanimate nature of bone, its constituents are continually being turned over. **Bone deposition** (formation) and **bone resorption** (removal) normally go on concurrently so that bone is constantly being remodelled, much as people remodel buildings by tearing down walls and replacing them. Through remodelling, the adult human skeleton is completely regenerated an estimated every 10 years. Bone remodelling serves two purposes: (1) it keeps the skeleton appropriately "engineered" for maximum effectiveness in its mechanical uses, and (2) it helps maintain plasma $Ca^{2+}$ level. Let us examine in more detail the underlying mechanisms and controlling factors for each of these purposes.

Recall that three types of bone cells are present in bone. The *osteoblasts* secrete the extracellular organic matrix within which the $Ca_3(PO_4)_2$ crystals precipitate. The *osteocytes* are the retired osteoblasts imprisoned within the bony wall they have deposited around themselves. The *osteoclasts* resorb bone in their vicinity by releasing acids that dissolve the $Ca_3(PO_4)_2$ crystals and enzymes that break down the organic matrix. Thus, a constant cellular tug-of-war goes on in bone, with bone-forming osteoblasts countering the efforts of the bone-destroying osteoclasts. These construction and demolition crews, working side by side, continuously remodel bone. Throughout most of adult life, the rates of bone formation and bone resorption are about equal, so total bone mass remains fairly constant during this period.

Osteoblasts and osteoclasts both trace their origins to the bone marrow. Osteoblasts are derived from *stromal cells*, a type of connective tissue cell in the bone marrow, whereas osteoclasts differentiate from *macrophages*, which are tissue-bound derivatives of monocytes, a type of white blood cell (p. 444). In a unique communication system, osteoblasts and their immature precursors produce two chemical signals—*RANK ligand* and *osteoprotegerin*—that govern osteoclast development and activity in opposite ways, as follows (see ⟩ Figure 6-19):

- **RANK ligand (RANKL)** revs up osteoclast action. (A *ligand* is a small molecule that binds with a larger protein molecule; an example is an extracellular chemical messenger binding with a plasma membrane receptor.) As its name implies, RANK ligand binds to RANK—which stands for *receptor activator of NF-κ β*— a protein receptor on the membrane surface of nearby macrophages. This binding induces the macrophages to differentiate into osteoclasts and helps them live longer by suppressing apoptosis (cell suicide). As a result, bone resorption is stepped up and bone mass decreases.

**Osteoprotegerin (OPG)**, by contrast, suppresses osteoclast development and activity. OPG secreted into the matrix serves as a freestanding decoy receptor that binds with RANKL. By taking RANKL out of action so that it cannot bind with its intended RANK receptors, OPG prevents RANKL from revving

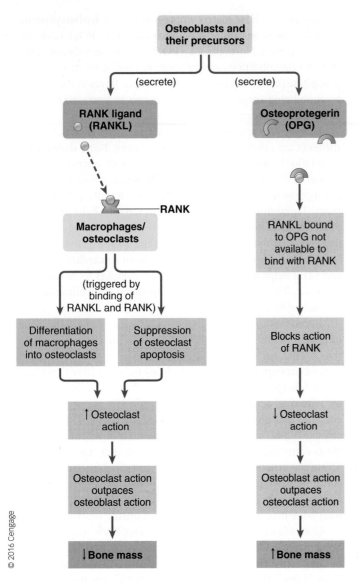

> FIGURE 6-19 **Role of osteoblasts in governing osteoclast development and activity**

Mechanical stress also tips the balance in favour of bone deposition, causing bone mass to increase and the bones to strengthen. Mechanical factors adjust the strength of bone in response to the demands placed on it. The greater the physical stress and compression to which a bone is subjected, the greater is the rate of bone deposition. For example, the bones of athletes are stronger and more massive than those of sedentary people.

By contrast, bone mass diminishes and the bones weaken when bone resorption gains a competitive edge over bone deposition in response to removal of mechanical stress. For example, bone mass decreases in people who undergo prolonged bed confinement or those in space flight. Early astronauts lost up to 20 percent of their bone mass during their time in orbit. Therapeutic exercises can limit or prevent such loss of bone.

*Clinical Note* Bone mass also decreases as a person ages. Bone density peaks when a person is in their 30s, and then starts to decline after age 40. By 50–60 years of age, bone resorption often exceeds bone formation. The result is a reduction in bone mass known as **osteoporosis** (porous bones). This bone-thinning condition is characterized by a diminished laying down of organic matrix as a result of reduced osteoblast activity and/or increased osteoclast activity rather than abnormal bone calcification. The underlying cause of osteoporosis is uncertain. Plasma $Ca^{2+}$ and $PO_4^{3-}$ levels are normal, as is PTH. Osteoporosis occurs with greatest frequency in postmenopausal women because of the associated withdrawal of bone-preserving estrogen.

## PTH and plasma $Ca^{2+}$

In addition to the factors geared toward controlling the mechanical effectiveness of bone, throughout our lifespan PTH uses bone as a "bank" from which it withdraws $Ca^{2+}$ as needed to maintain plasma $Ca^{2+}$ level. Parathyroid hormone has two major effects on bone that raise plasma $Ca^{2+}$ concentration. First, it induces a fast $Ca^{2+}$ efflux into the plasma from the small *labile pool* of $Ca^{2+}$ in the bone fluid. Second, by stimulating bone dissolution, it promotes a slow transfer into the plasma of both $Ca^{2+}$ and $PO_4^{3-}$ from the *stable pool* of bone minerals in bone itself. As a result, ongoing bone remodelling is tipped in favour of bone resorption over bone deposition. Let's examine more thoroughly PTH's actions in mobilizing $Ca^{2+}$ from its labile and stable pools in bone.

## PTH and $Ca^{2+}$ transfer

Most bone is organized into **osteon** units, each of which consists of a **central canal** surrounded by concentrically arranged **lamellae**. Lamellae are layers of osteocytes entombed within the bone they have deposited around themselves (> Figure 6-20). The osteons typically run parallel to the long axis of the bone. Blood vessels penetrate the bone from either the outer surface or the marrow cavity and run through the central canals. Osteoblasts are present along the outer surface of the bone and along the inner surfaces lining the central canals. Osteoclasts are also located on bone surfaces undergoing resorption. The surface osteoblasts and entombed osteocytes are connected by an extensive network of small, fluid-containing canals, the **canaliculi**, which allow substances to be exchanged between trapped osteocytes

up osteoclasts' bone-resorbing activity. As a result, the matrix-making osteoblasts outpace the matrix-removing osteoclasts, so bone mass increases. The balance between RANKL and OPG is therefore an important determinant of bone density. If osteoblasts produce more RANKL, the more osteoclast action, the lower the bone mass. If osteoblasts produce more OPG, the less osteoclast action, the greater the bone mass. Importantly, scientists are currently unravelling the influence of various factors on this balance. For example, the female sex hormone estrogen stimulates activity of the OPG-producing gene in osteoblasts, which is one mechanism by which this hormone preserves bone mass.

## Mechanical stress

As a child grows, the bone builders of the body keep ahead of the bone destroyers under the influence of GH and IGF-I (pp. 235).

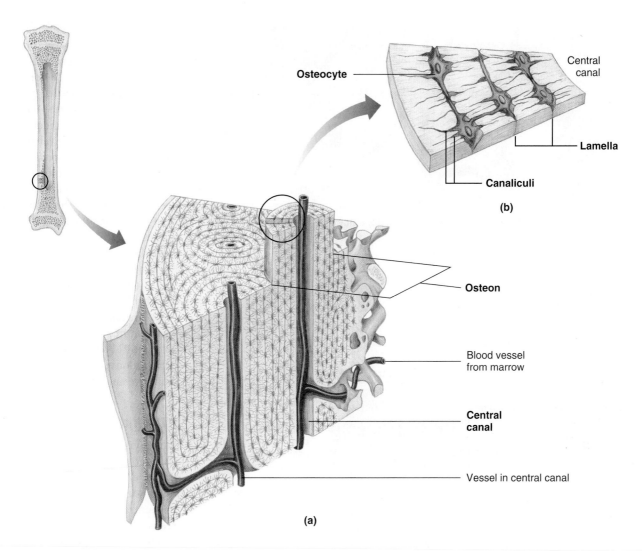

**Osteocyte**

**Central canal**

**Lamella**

**Canaliculi**

**(b)**

**Osteon**

**Blood vessel from marrow**

**Central canal**

**Vessel in central canal**

**(a)**

› **FIGURE 6-20 Organization of bone into osteons.** (a) An osteon, the structural unit of most bone, consists of concentric lamellae (layers of osteocytes entombed by the bone they have deposited around themselves) surrounding a central canal. A small blood vessel branch traverses the central canal. (b) A magnification of the lamellae within an osteon. A network of small canals, the canaliculi, interconnects the entombed osteocytes with one another and with the central canal. Long cytoplasmic processes extend from osteocyte to osteocyte within the canaliculi.

**Source:** Modified and redrawn with permission from *Human Anatomy and Physiology*, 3rd Edition, by A. Spence and E. Mason. Copyright © 1987 by The Benjamin/Cummings Publishing Company. Reprinted by permission of Pearson Education, Inc.

and the circulation. These small canals also contain long, filmy cytoplasmic extensions of osteocytes and osteoblasts that are connected to one another, much as if these cells were "holding hands." The interconnecting cell network, which is called the **osteocytic-osteoblastic bone membrane**, separates the mineralized bone itself from the plasma within the central canals (› Figure 6-21a). The small, labile pool of $Ca^{2+}$ is in the **bone fluid** that lies between this bone membrane and the adjacent bone, both within the canaliculi and along the surface of the central canal.

PTH's earliest effect is to activate membrane-bound $Ca^{2+}$ pumps located in the plasma membranes of the osteocytes and osteoblasts. These pumps promote movement of $Ca^{2+}$, without the accompaniment of $PO_4^{3-}$, from the bone fluid into these cells. From here, this $Ca^{2+}$ is transferred into the plasma within the central canal. Thus, PTH stimulates the transfer of $Ca^{2+}$ from the bone fluid across the osteocytic-osteoblastic bone membrane

into the plasma. Movement of $Ca^{2+}$ out of the labile pool across the bone membrane accounts for the fast exchange between bone and plasma (› Figure 6-21b). Because of the large surface area of the osteocytic-osteoblastic membrane, small movements of $Ca^{2+}$ across individual cells are amplified into large $Ca^{2+}$ fluxes between the bone fluid and plasma.

After $Ca^{2+}$ is pumped out, the bone fluid is replenished with $Ca^{2+}$ from the partially mineralized bone along the adjacent bone surface. Consequently, the fast exchange of $Ca^{2+}$ does not involve resorption of completely mineralized bone, and bone mass is not decreased. Through this means, PTH draws $Ca^{2+}$ out of the "quick-cash branch" of the bone bank and rapidly increases plasma $Ca^{2+}$ level without actually entering the bank (i.e., without breaking down mineralized bone itself). Under normal conditions, this exchange is much more important for maintaining plasma $Ca^{2+}$ concentration than is the slow exchange.

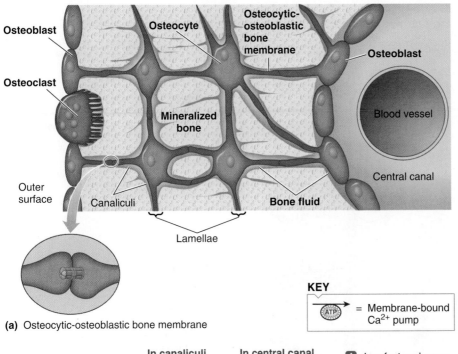

**Osteoblast**

**Osteocyte**

**Osteocytic-osteoblastic bone membrane**

**Osteoblast**

**Osteoclast**

**Mineralized bone**

**Blood vessel**

**Central canal**

**Outer surface**

**Canaliculi**

**Bone fluid**

**Lamellae**

**(a)** Osteocytic-osteoblastic bone membrane

**KEY**

= Membrane-bound Ca²⁺ pump

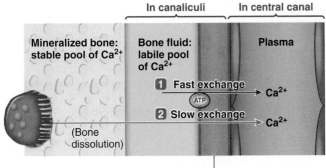

**In canaliculi**

**In central canal**

**Mineralized bone: stable pool of Ca²⁺**

**Bone fluid: labile pool of Ca²⁺**

**Plasma**

**1 Fast exchange** → Ca²⁺

**2 Slow exchange** → Ca²⁺

(Bone dissolution)

**Osteocytic-osteoblastic bone membrane** (formed by filmy cytoplasmic extensions of interconnected osteocytes and osteoblasts)

**(b)** Fast and slow exchange of Ca²⁺ between bone and plasma

**1** In a fast exchange, Ca²⁺ is moved from the labile pool in the bone fluid into the plasma by PTH-activated Ca²⁺ pumps located in the osteocytic-osteoblastic bone membrane.

**2** In a slow exchange, Ca²⁺ is moved from the stable pool in the mineralized bone into the plasma through PTH-induced dissolution of the bone by osteoclasts.

> **FIGURE 6-21 Fast and slow exchanges of Ca²⁺ across the osteocytic-osteoblastic bone membrane.** (a) Entombed osteocytes and surface osteoblasts are interconnected by long cytoplasmic processes that extend from these cells and connect to one another within the canaliculi. This interconnecting cell network, the osteocytic-osteoblastic bone membrane, separates the mineralized bone from the plasma in the central canal. Bone fluid lies between the membrane and the mineralized bone. (b) Fast exchange of Ca²⁺ between the bone and plasma is accomplished by Ca²⁺ pumps in the osteocytic-osteoblastic bone membrane that transport Ca²⁺ from the bone fluid into these bone cells, which transfer the Ca²⁺ into the plasma. Slow exchange of Ca²⁺ between the bone and plasma is accomplished by osteoclast dissolution of bone.

## PTH and localized dissolution of bone

Under conditions of chronic hypocalcaemia, such as may occur with dietary Ca²⁺ deficiency, PTH influences the slow exchange of Ca²⁺ between bone itself and the ECF by promoting actual localized dissolution of bone. It does so by stimulating osteoclasts to gobble up bone, increasing the formation of more osteoclasts, and transiently inhibiting the bone-forming activity of osteoblasts. Bone contains so much Ca²⁺ compared with the plasma (more than 1000 times as much) that even when PTH promotes increased bone resorption, there are no immediate discernible effects on the skeleton because such a tiny amount of bone is

affected. Yet the negligible amount of Ca²⁺ borrowed from the bone bank can be lifesaving in terms of restoring free plasma Ca²⁺ level to normal. The borrowed Ca²⁺ is then redeposited in the bone at another time when Ca²⁺ supplies are more abundant. Meanwhile, plasma Ca²⁺ level has been maintained without sacrificing bone integrity. However, prolonged excess PTH secretion over months or years eventually leads to the formation of cavities throughout the skeleton that are filled with very large, overstuffed osteoclasts.

When PTH promotes dissolution of the Ca₃(PO₄)₂ crystals in bone to harvest their Ca²⁺ content, both Ca²⁺ and PO₄³⁻ are released into the plasma. An elevation in plasma PO₄³⁻ is undesirable, but PTH deals with this dilemma by its actions on the kidneys, a topic to which we now turn our attention.

## PTH and the kidneys

Parathyroid hormone stimulates Ca²⁺ conservation and promotes PO₄³⁻ elimination by the kidneys during urine formation. Under the influence of PTH, the kidneys can reabsorb more of the filtered Ca²⁺, so less Ca²⁺ escapes into urine. This effect increases plasma Ca²⁺ level and decreases urinary Ca²⁺ losses. (It would be counterproductive to dissolve bone to obtain more Ca²⁺ only to lose it in urine.) By contrast, PTH decreases PO₄³⁻ reabsorption, thereby increasing urinary PO₄³⁻ excretion. As a result, PTH reduces plasma PO₄³⁻ levels at the same time as it increases plasma Ca²⁺ concentrations.

This PTH-induced removal of extra PO₄³⁻ from the body fluids is essential for preventing reprecipitation of the Ca²⁺ freed from bone. Because of the solubility characteristics of Ca₃(PO₄)₂ salt, the product of the plasma concentration of Ca²⁺ multiplied by the plasma concentration of PO₄³⁻ must remain roughly constant. Therefore, an inverse relationship exists between the plasma concentrations of Ca²⁺ and PO₄³⁻. For example, when plasma PO₄³⁻ level rises, some plasma Ca²⁺ is forced back into bone through hydroxyapatite crystal formation, thereby reducing plasma Ca²⁺ level and keeping constant the calcium phosphate product. This inverse relationship occurs because the concentrations of free Ca²⁺ and PO₄³⁻ ions in the ECF are in equilibrium with the bone crystals.

Recall that both $Ca^{2+}$ and $PO_4^{3-}$ are released from bone when PTH promotes bone dissolution. Because PTH is secreted only when plasma $Ca^{2+}$ falls below normal, the released $Ca^{2+}$ is needed to restore plasma $Ca^{2+}$ to normal, yet the released $PO_4^{3-}$ tends to raise plasma $PO_4^{3-}$ levels above normal. If plasma $PO_4^{3-}$ levels were allowed to rise above normal, some of the released $Ca^{2+}$ would have to be redeposited in the bone along with the $PO_4^{3-}$ to keep the $Ca^{2+}-PO_4^{3-}$ product constant. This self-defeating redeposition of $Ca^{2+}$ would lower plasma $Ca^{2+}$, just the opposite of the needed effect. Therefore, PTH acts on the kidneys to decrease the reabsorption of $PO_4^{3-}$ by the renal tubules. This increases urinary excretion of $PO_4^{3-}$ and lowers its plasma concentration, even though extra $PO_4^{3-}$ is released from bone into the blood.

As well as its roles in increasing $Ca^{2+}$ reabsorption and decreasing $PO_4^{3-}$ reabsorption, PTH has a role in enhancing the activation of vitamin D by the kidneys.

### PTH and intestinal absorption of $Ca^{2+}$ and $PO_4^{3-}$

Although PTH has no direct effect on the intestine, it indirectly increases both $Ca^{2+}$ and $PO_4^{3-}$ absorption from the small intestine by helping activate vitamin D. This vitamin, in turn, directly increases intestinal absorption of $Ca^{2+}$ and $PO_4^{3-}$, a topic we will discuss more thoroughly shortly.

### Plasma concentration of free $Ca^{2+}$: Regulating PTH secretion

All the effects of PTH raise plasma $Ca^{2+}$ levels. Correspondingly, PTH secretion is increased in response to a fall in plasma $Ca^{2+}$ concentration and decreased in response to a rise in plasma $Ca^{2+}$ levels. The secretory cells of the parathyroid glands are directly and exquisitely sensitive to changes in free plasma $Ca^{2+}$. Because PTH regulates plasma $Ca^{2+}$ concentration, this relationship forms a simple negative-feedback loop that controls PTH secretion without involving any nervous or other hormonal intervention (> Figure 6-22).

### Calcitonin and plasma $Ca^{2+}$ concentration

*Calcitonin* was first purified by Copp and Cheney at the University of British Columbia and thought to be a hormone of the parathyroid gland. However, it was later recognized that this hormone is produced by the C cells of the thyroid gland. Like PTH, calcitonin has two effects on bone, but in this case both effects *decrease* plasma $Ca^{2+}$ levels. First, on a short-term basis, calcitonin decreases $Ca^{2+}$ movement from the bone fluid into the plasma. Second, on a long-term basis, calcitonin decreases bone resorption by inhibiting the activity of osteoclasts. The suppression of bone resorption lowers plasma $PO_4^{3-}$ levels as well as reduces plasma $Ca^{2+}$ concentration. The hypocalcaemic and hypophosphataemic effects of calcitonin are due entirely to this hormone's actions on bone. It has no effect on the kidneys or intestine.

As with PTH, the primary regulator of calcitonin release is free plasma $Ca^{2+}$ concentration. However, in contrast to its effect on PTH release, an increase in plasma $Ca^{2+}$ stimulates calcitonin secretion, and a fall in plasma $Ca^{2+}$ inhibits calcitonin secretion

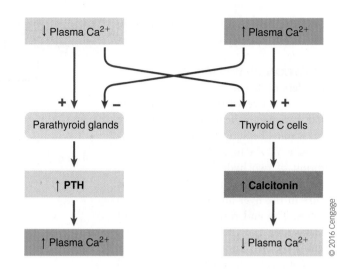

> FIGURE 6-22 **Negative-feedback loops controlling parathyroid hormone (PTH) and calcitonin secretion**

© 2016 Cengage

(Figure 6-22). Because calcitonin reduces plasma $Ca^{2+}$ levels, this system constitutes a second simple negative-feedback control over plasma $Ca^{2+}$ concentration, one opposed to the PTH system.

Most evidence suggests, however, that calcitonin plays little or no role in the normal control of $Ca^{2+}$ or $PO_4^{3-}$ metabolism. Although calcitonin protects against hypercalcaemia, this condition rarely occurs under normal circumstances. Moreover, neither thyroid removal nor calcitonin-secreting tumours alter circulating levels of $Ca^{2+}$ or $PO_4^{3-}$, which implies that this hormone is not normally essential for maintaining $Ca^{2+}$ or $PO_4^{3-}$ homeostasis. Calcitonin may, however, play a role in protecting skeletal integrity when there is a large $Ca^{2+}$ demand, such as during pregnancy or breastfeeding. Furthermore, some experts speculate that calcitonin may hasten the storage of newly absorbed $Ca^{2+}$ following a meal. Gastrointestinal hormones secreted during digestion of a meal have been shown to stimulate the release of calcitonin.

### Vitamin D and calcium absorption

The final factor involved in regulating $Ca^{2+}$ metabolism is **cholecalciferol (vitamin D)**, a steroid-like compound essential for $Ca^{2+}$ absorption in the intestine. Strictly speaking, vitamin D should be considered a hormone, because the body can produce it in the skin from a precursor related to cholesterol (7-dehydrocholesterol) on exposure to sunlight. It is subsequently released into the blood to act at a distant target site, the intestine. The skin, therefore, is actually an endocrine gland and vitamin D a hormone. Traditionally, however, this chemical messenger has been considered a vitamin, for two reasons. First, it was originally discovered and isolated from a dietary source and tagged as a vitamin. Second, even though the skin would be an adequate source of vitamin D if it were exposed to sufficient sunlight, indoor dwelling and clothing in response to cold weather and social customs preclude significant exposure of the skin to sunlight in Canada and many other parts of the world most of the time. At least part of the essential vitamin D must therefore be derived from dietary sources.

## ACTIVATION OF VITAMIN D

Regardless of its source, vitamin D is biologically inactive when it first enters the blood from either the skin or the digestive tract. It must be activated by two sequential biochemical alterations that involve the addition of two hydroxyl (−OH) groups (⟩ Figure 6-23). The first of these reactions occurs in the liver and the second in the kidneys. The end result is production of the active form of vitamin D, $1,25\text{-}(OH)_{2}\text{-}$ vitamin $D_3$, also known as *calcitriol*. The kidney enzymes involved in the second step of vitamin D activation are stimulated by PTH in response to a fall in plasma $Ca^{2+}$. To a lesser extent, a fall in plasma $PO_4^{3-}$ also enhances the activation process.

## FUNCTION OF VITAMIN D

The most dramatic effect of activated vitamin D is to increase $Ca^{2+}$ absorption in the intestine. Unlike most dietary constituents, dietary $Ca^{2+}$ is not indiscriminately absorbed by the digestive system. In fact, the majority of ingested $Ca^{2+}$ is typically not absorbed but is lost in the feces. When needed, more dietary $Ca^{2+}$ is absorbed into the plasma under the influence of vitamin D. Independently of its effects on $Ca^{2+}$ transport, the active form of vitamin D also increases intestinal $PO_4^{3-}$ absorption. Furthermore, vitamin D increases the responsiveness of bone to PTH. Thus, vitamin D and PTH are closely interdependent (⟩ Figure 6-24).

PTH is principally responsible for controlling $Ca^{2+}$ homeostasis, because the actions of vitamin D are too sluggish for it to contribute substantially to the minute-to-minute regulation of plasma $Ca^{2+}$ concentration. However, both PTH and vitamin D are essential to $Ca^{2+}$ balance—the process that ensures over the long term that $Ca^{2+}$ input into the body is equivalent to $Ca^{2+}$ output. When dietary $Ca^{2+}$ intake is reduced, the resultant transient fall in plasma $Ca^{2+}$ level stimulates PTH secretion. The increased PTH has two effects important for maintaining $Ca^{2+}$ balance: (1) it stimulates $Ca^{2+}$ reabsorption by the kidneys, thereby decreasing $Ca^{2+}$ output; and (2) it activates vitamin D, which increases the efficiency of uptake of ingested $Ca^{2+}$. Because PTH also promotes bone resorption, a substantial loss of bone minerals occurs if $Ca^{2+}$ intake is reduced for a prolonged period, even though bone is not directly involved in maintaining the balance of $Ca^{2+}$ input and output.

Recent research indicates that the functions of vitamin D are farther reaching than its effects on uptake of ingested $Ca^{2+}$ and $PO_4^{3-}$. Vitamin D, at higher blood concentrations than those sufficient to protect bone, appears to bolster muscle strength and is also an important force in energy metabolism and immune health. It helps thwart development of diabetes mellitus, fights some types of cancer, and—by presently unknown mechanisms—counters autoimmune diseases such

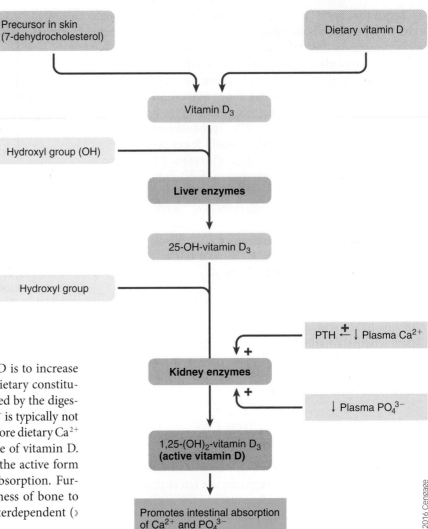

⟩ **FIGURE 6-23** **Activation of vitamin D**

as multiple sclerosis. Because of these newly found actions, scientists and dieticians are reevaluating the recommended dietary allowance (RDA) for vitamin D in the diet, especially when sufficient sun exposure is not possible. The RDA will likely be bumped up, but further study is needed to determine the optimal amount.

## Phosphate metabolism

Plasma $PO_4^{3-}$ concentration is not as tightly controlled as plasma $Ca^{2+}$ concentration. Phosphate is regulated directly by vitamin D and indirectly by the plasma $Ca^{2+}$—PTH feedback loop. To illustrate, a fall in plasma $PO_4^{3-}$ concentration exerts a twofold effect to help raise the circulating $PO_4^{3-}$ level back to normal (⟩ Figure 6-25). First, because of the inverse relationship between the $PO_4^{3-}$ and $Ca^{2+}$ concentrations in the plasma, a fall in plasma $PO_4^{3-}$ increases plasma $Ca^{2+}$, which directly suppresses PTH secretion. In the presence of reduced PTH, $PO_4^{3-}$ reabsorption by the kidneys increases, returning plasma $PO_4^{3-}$ concentration toward normal. Second, a fall in plasma $PO_4^{3-}$ also increases activation of vitamin D, which then promotes $PO_4^{3-}$ absorption

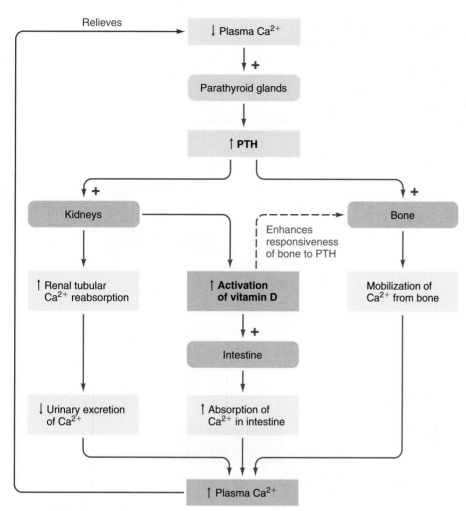

> FIGURE 6-24 **Interactions between PTH and vitamin D in controlling plasma calcium**

disorders, including decreased alertness, poor memory, and depression. Cardiac disturbances may also occur.

- Excessive mobilization of $Ca^{2+}$ and $PO_4^{3-}$ from skeletal stores leads to thinning of bone, which may result in skeletal deformities and increased incidence of fractures.

- An increased incidence of $Ca^{2+}$-containing kidney stones occurs because the excess quantity of $Ca^{2+}$ being filtered through the kidneys may precipitate and form stones. These stones may impair renal function. Passage of the stones through the ureters causes extreme pain. Because of these potential multiple consequences, hyperparathyroidism has been called a disease of "bones, stones, and abdominal groans."

- To further account for the "abdominal groans," hypercalcaemia can cause digestive disorders, such as peptic ulcers, nausea, and constipation.

### PTH HYPOSECRETION

Because of the close anatomic relation between the parathyroid glands and the thyroid, the most common cause of deficient PTH secretion—**hypoparathyroidism**—used to be inadvertent removal of the parathyroid glands (before doctors knew about their existence) during surgical removal of the thyroid gland (to treat thyroid disease). If all the parathyroid tissue was removed, these patients died, of course, because PTH is essential for life. Physicians were puzzled why some patients died soon after thyroid removal even though no surgical complications were apparent. Now that the location and importance of the parathyroid glands have been discovered, surgeons are careful to leave parathyroid tissue during thyroid removal. Rarely, PTH hyposecretion results from an autoimmune attack against the parathyroid glands.

Hypoparathyroidism leads to hypocalcaemia and hyperphosphataemia. The symptoms are mainly caused by increased neuromuscular excitability from the reduced level of free plasma $Ca^{2+}$. In the complete absence of PTH, death is imminent because respiratory muscles go into hypocalcaemic spasm. With a relative deficiency rather than a complete absence of PTH, milder symptoms of increased neuromuscular excitability become evident. Muscle cramps and twitches occur from spontaneous activity in the motor nerves, whereas tingling and pins-and-needles sensations result from spontaneous activity in the sensory nerves. Mental changes include irritability and paranoia.

### VITAMIN D DEFICIENCY

The major consequence of vitamin D deficiency is impaired intestinal absorption of $Ca^{2+}$. In the face of reduced $Ca^{2+}$ uptake, PTH maintains the plasma $Ca^{2+}$ level at the expense

in the intestine. This further helps alleviate the initial hypophosphataemia. Note that these changes do not compromise $Ca^{2+}$ balance. Although the increase in activated vitamin D stimulates $Ca^{2+}$ absorption, the concurrent fall in PTH produces a compensatory increase in urinary $Ca^{2+}$ excretion because less of the filtered $Ca^{2+}$ is reabsorbed.

## Disorders in $Ca^{2+}$ metabolism

The primary disorders that affect $Ca^{2+}$ metabolism are too much or too little PTH or a deficiency of vitamin D.

### PTH HYPERSECRETION

*Clinical Note* Excess PTH secretion, or **hyperparathyroidism**, which is usually caused by a hypersecreting tumour in one of the parathyroid glands, is characterized by hypercalcaemia and hypophosphataemia. The affected individual can be asymptomatic or symptoms can be severe, depending on the magnitude of the problem. The following are among the possible consequences:

- Hypercalcaemia reduces the excitability of muscle and nervous tissue, leading to muscle weakness and neurologic

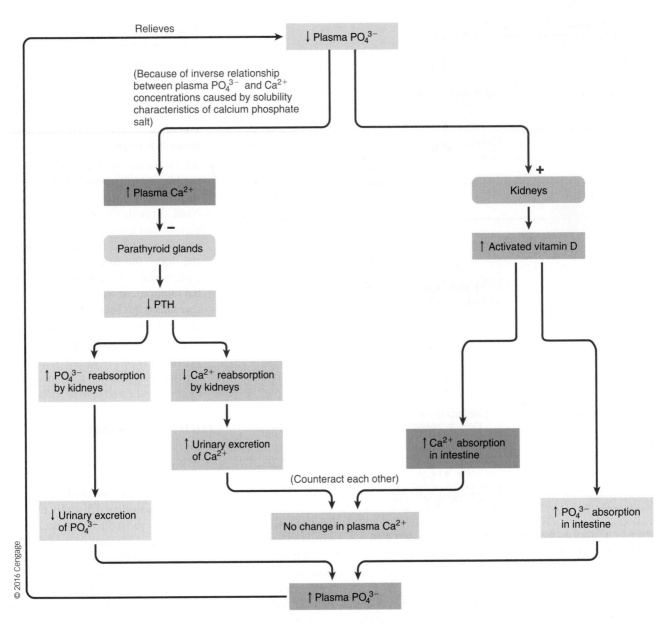

> FIGURE 6-25 **Control of plasma phosphate**

of the bones. As a result, the bone matrix is not properly mineralized, because $Ca^{2+}$ salts are not available for deposition. The demineralized bones become soft and deformed, bowing under the pressures of weight bearing, especially in children. This condition is known as **rickets** in children and **osteomalacia** in adults.

### Check Your Understanding 6.4

1. Describe the distribution of $Ca^{2+}$ in the body and the functions of free ECF $Ca^{2+}$.

2. Discuss the effects of PTH on bone, kidneys, and the intestine.

3. Explain the importance of Vitamin D in $Ca^{2+}$ homeostasis.

# Chapter in Perspective: Focus on Homeostasis

A number of peripherally located endocrine glands play key roles in maintaining homeostasis, primarily by means of their regulatory influences over the rate of various metabolic reactions and over electrolyte balance. These endocrine glands all secrete hormones in response to specific stimuli. The hormones, in turn, exert effects that act in negative-feedback fashion to resist the change that induced their secretion, thus maintaining stability in the internal environment. The specific contributions of the peripheral endocrine glands to homeostasis include the following:

- Two closely related hormones secreted by the thyroid gland, tetraiodothyronine ($T_4$) and triiodothyronine ($T_3$), increase the overall metabolic rate. Not only does this action influence the rate at which cells use nutrient molecules and oxygen within the internal environment but it also produces heat, which helps maintain body temperature.

- The adrenal cortex secretes three classes of hormones. Aldosterone, the primary mineralocorticoid, is essential for $Na^+$ and $K^+$ balance. Because of the osmotic effect of $Na^+$, $Na^+$ balance is critical to maintaining the proper ECF volume and arterial blood pressure. This action is essential for life. Without aldosterone's $Na^+$- and $H_2O$-conserving effect, so much plasma volume would be lost in the urine that death would quickly ensue. Maintaining $K^+$ balance is essential for homeostasis because changes in extracellular $K^+$ profoundly impact neuromuscular excitability, leading to detrimental effects, such as compromising normal heart function. Cortisol, the primary glucocorticoid secreted by the adrenal cortex, increases the plasma concentrations of glucose, fatty acids, and amino acids above normal. Although these actions destabilize the concentrations of these molecules in the internal environment, they indirectly contribute to homeostasis by making the molecules readily available as energy sources or building blocks for tissue repair to help the body adapt to stressful situations.

- The sex hormones secreted by the adrenal cortex do not contribute to homeostasis.

- The major hormone secreted by the adrenal medulla, epinephrine, generally reinforces activities of the sympathetic nervous system. It contributes to homeostasis directly by its role in blood pressure regulation. Epinephrine also contributes to homeostasis indirectly by helping prepare the body for peak physical responsiveness in fight-or-flight situations. This includes increasing the plasma concentrations of glucose and fatty acids above normal, thereby providing additional energy sources for increased physical activity.

- The two major hormones secreted by the endocrine pancreas—insulin and glucagon—are important in shifting metabolic pathways between the absorptive and postabsorptive states, which maintains the appropriate plasma levels of nutrient molecules.

- Parathyroid hormone from the parathyroid glands is critical to maintaining plasma concentration of $Ca^{2+}$. PTH is essential for life because of the effect of $Ca^{2+}$ on neuromuscular excitability. In the absence of PTH, death rapidly occurs from asphyxiation caused by pronounced spasms of the respiratory muscles.

## CHAPTER TERMINOLOGY

addison's disease (p. 260)
adiponectin (p. 278)
adrenal glands (p. 254)
adrenal cortex (p. 254)
adrenocortical hormones (p. 254)
adrenogenital syndrome (p. 259)
aldosterone (p. 255)
$\alpha$ (alpha) cells (p. 271)
anabolism (p. 267)
$\beta$ (beta) cells (p. 271)
Bone deposition (p. 285)
bone fluid (p. 287)
bone resorption (p. 285)
C cells (p. 248)
calcitonin (p. 248)
calcium balance (p. 285)
calcium homeostasis (p. 284)
calorigenic (heat-producing) effect (p. 250)
canaliculi (p. 286)
catabolism (p. 267)
central canal (p. 286)
cholecalciferol (vitamin D) (p. 289)
chromaffin granules (p. 261)
colloid (p. 248)
corticosteroid-binding globulin (transcortin) (p. 255)
cortisol (p. 256)
Conn's syndrome (p. 258)
cretinism (p. 252)
Cushing's syndrome (p. 259)
$\delta$ (delta) cells (p. 271)
dehydroepiandrosterone (DHEA) (p. 258)
diabetes mellitus (p. 275)

di-iodotyrosine (DIT) (p. 250)
endocrine pancreas (p. 281)
essential nutrients (p. 267)
exophthalmos (p. 253)
fasted (postabsorptive) state (p. 270)
female pseudohermaphroditism (p. 259)
follicle (p. 248)
follicular cells (p. 248)
general adaptation syndrome (GAS) (p. 263)
Graves' disease (p. 252)
goitre (p. 253)
glucagon (p. 280)
glucocorticoids (p. 254)
gluconeogenesis (p. 256)
glucose sparing (p. 269)
glucose transporter (GLUT) (p. 273)
glycogenesis (p. 272)
glycogenolysis (p. 262)
hirsutism (p. 259)
hydroxyapatite crystals (p. 285)
hyperparathyroidism (p. 291)
hyperthyroidism (p. 252)
hypoparathyroidism (p. 291)
hypothyroidism (p. 252)
incretins (p. 279)
insulin antagonists (p. 282)
insulin shock (p. 277)
islets of Langerhans (p. 271)
insulin (p. 272)
ketogenesis (p. 280)
ketone bodies (p. 271)
lamellae (p. 286)
long-acting thyroid stimulator (LATS) (p. 252)

macromolecules (p. 267)
metabolic syndrome (p. 278)
metabolism (p. 267)
mineralocorticoids (p. 254)
monoiodotyrosine (MIT) (p. 250)
myxoedema (p. 252)
osteocytic-osteoblastic bone membrane (p. 287)
osteomalacia (p. 292)
osteon (p. 286)
osteoporosis (p. 286)
osteoprotegerin (OPG) (p. 285)
pancreas (p. 271)
parathyroid glands (p. 285)
parathyroid hormone (PTH) (p. 285)
PP cells (p. 271)
precocious pseudopuberty (p. 260)
primary adrenocortical insufficiency (p. 260)
primary hyperaldosteronism (p. 258)
pro-opiomelanocortin (POMC) (p. 257)
reactive hypoglycaemia (p. 277)
resistin (p. 278)
rickets (p. 292)
secondary adrenocortical insufficiency (p. 260)
secondary hyperaldosteronism (p. 259)
sex hormones (p. 255)
stress (p. 263)
sympathomimetic (sympathetic-mimicking) effect (p. 251)
tetraiodothyronine (p. 248)
thyroid gland (p. 248)
thyroglobulin (Tg) (p. 248)
thyroperoxidase (TPO) (p. 250)

*(Continued)*

## REVIEW EXERCISES

### Objective Questions
(Answers in Appendix E, p. A-41)

1. The response to thyroid hormone is detectable within a few minutes after its secretion. *(True or false?)*

2. Male sex hormones are produced in both males and females by the adrenal cortex. *(True or false?)*

3. Adrenal androgen hypersecretion is usually due to a deficit of an enzyme crucial to cortisol synthesis. *(True or false?)*

4. Excess glucose and amino acids as well as fatty acids can be stored as triglycerides. *(True or false?)*

5. Insulin is the only hormone that can lower blood glucose levels. *(True or false?)*

6. The most life-threatening consequence of hypocalcaemia is reduced blood clotting. *(True or false?)*

7. All ingested $Ca^{2+}$ is indiscriminately absorbed in the intestine. *(True or false?)*

8. The $Ca_3(PO_4)_2$ bone crystals form a labile pool from which $Ca^{2+}$ can rapidly be extracted under the influence of PTH. *(True or false?)*

9. The lumen of the thyroid follicle is filled with _____, the chief constituent of which is a large protein molecule known as _____.

10. The common large precursor molecule that yields ACTH, MSH, and β-endorphin is known as _____.

11. _____ is the conversion of glucose into glycogen. _____ is the conversion of glycogen into glucose. _____ is the conversion of amino acids into glucose.

12. The three major tissues that are not dependent on insulin for their glucose uptake are _____, _____, and _____.

13. The three compartments with which ECF $Ca^{2+}$ is exchanged are _____, _____, and _____.

14. Which hormone does NOT exert a direct metabolic effect?
    a. epinephrine
    b. growth hormone
    c. aldosterone
    d. cortisol
    e. thyroid hormone

15. Which of the following are characteristic of the postabsorptive state? (Indicate all correct answers.)
    a. glycogenolysis
    b. gluconeogenesis
    c. lipolysis
    d. glycogenesis
    e. protein synthesis
    f. triglyceride synthesis
    g. protein degradation
    h. increased insulin secretion
    i. increased glucagon secretion
    j. glucose sparing

16. Indicate the primary circulating form and storage form of each of the three classes of organic nutrients:

|  | Primary Circulating Form | Primary Storage Form |
| --- | --- | --- |
| Carbohydrate | 1. _____ | 2. _____ |
| Fat | 3. _____ | 4. _____ |
| Protein | 5. _____ | 6. _____ |

### Written Questions

1. Describe the steps of thyroid hormone synthesis.

2. What are the effects of $T_3$ and $T_4$? Which is the more potent form? What is the source of most circulating $T_3$?

3. Describe the regulation of thyroid hormone.

4. Discuss the causes and symptoms of both hypothyroidism and hyperthyroidism. For each cause, indicate whether or not a goitre occurs, and explain why.

5. What hormones are secreted by the adrenal cortex? What are the functions and control of each of these hormones?

6. Discuss the causes and symptoms of each type of adrenocortical dysfunction.

7. What is the relationship of the adrenal medulla to the sympathetic nervous system? What are the functions of epinephrine? How is epinephrine release controlled?

8. Define *stress*. Describe the neural and hormonal responses to a stressor.

9. Define *fuel metabolism*, *anabolism*, and *catabolism*.

10. Distinguish between the absorptive and postabsorptive states with regard to the handling of nutrient molecules.

11. Name the two major cell types of the islets of Langerhans, and indicate the hormonal product of each.

12. Compare the functions and control of insulin secretion with those of glucagon secretion.

13. What are the consequences of diabetes mellitus? Distinguish between type 1 and type 2 diabetes mellitus.

14. Why must plasma $Ca^{2+}$ be closely regulated?

15. Explain how osteoblasts influence osteoclast function.

16. Discuss the contributions of parathyroid hormone, calcitonin, and vitamin D to $Ca^{2+}$ metabolism. Describe the source and control of each of these hormones.

17. Discuss the major disorders in $Ca^{2+}$ metabolism.

## POINTS TO PONDER

(Explanations in Appendix E, p. A-41)

1. Iodine is naturally present in salt water and is abundant in soil along coastal regions. Fish and shellfish living in the ocean and plants grown in coastal soil take up iodine from their environment. Fresh water does not contain iodine, and the soil becomes more iron-poor the farther inland it is. Knowing this, explain why the midwestern United States was once known as an *endemic goitre belt*. Why is this region no longer an endemic goitre belt even though the soil is still iodine poor?

2. Why do doctors recommend that people who are allergic to bee stings, and thus at risk for anaphylactic shock, carry a vial of epinephrine for immediate injection in case of a sting?

3. Why would an infection tend to raise the blood glucose level of a diabetic individual?

4. Tapping the facial nerve at the angle of the jaw in a patient with moderate hyposecretion of a particular hormone elicits a characteristic grimace on that side of the face. What endocrine abnormality could give rise to this so-called *Chvostek's sign*?

5. Soon after a technique to measure plasma $Ca^{2+}$ levels was developed in the 1920s, physicians observed that hypercalcaemia accompanied a broad range of cancers. Early researchers proposed that malignancy-associated hypercalcaemia arose from metastatic tumour cells that invaded and destroyed bone, releasing $Ca^{2+}$ into the blood. This conceptual framework was overturned when physicians noted that hypercalcaemia often appeared in the absence of bone lesions. Furthermore, cancer patients often manifested hypophosphataemia in addition to hypercalcaemia. This finding led investigators to suspect that the tumours might be producing a PTH-like substance. Explain how they reached this conclusion. In 1987, this substance was identified and named *parathyroid hormone–related peptide (PTHrP)*, which binds to and activates PTH receptors.

## CLINICAL CONSIDERATION

(Explanation in Appendix E, p. A-42)

Najma G. sought medical attention after her menstrual periods ceased and she started growing excessive facial hair. Also, she had been thirstier than usual and urinated more frequently. A clinical evaluation revealed that Najma was hyperglycaemic. Her physician told her that she had an endocrine disorder dubbed "diabetes of bearded ladies." Based on her symptoms and your knowledge of the endocrine system, what underlying defect do you think is responsible for Najma's condition?

# Muscle Physiology

Body systems maintain homeostasis

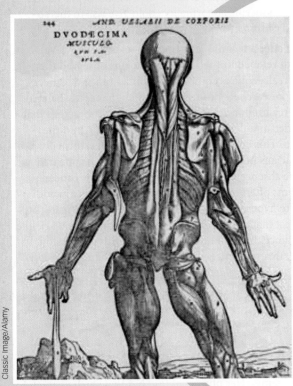

Classic Image/Alamy

Anatomical study *De Humani Corporis Fabrica*

Cells make up body systems

## Homeostasis

Skeletal muscles contribute to homeostasis by playing a major role in the procurement of food, breathing, heat generation for maintenance of body temperature, and movement away from harm.

Homeostasis is essential for survival of cells

Jubal Harshaw/shutterstock.com

Striated (voluntary) muscle

Muscles are the contraction specialists of the body. Skeletal muscles attach to the skeleton. Contraction of skeletal muscles moves the bones to which they are attached, allowing the body to perform a variety of motor activities and all the movement necessary to sustain life. Skeletal muscles that support homeostasis include those important in chewing, swallowing, and breathing. As well, heat-generating muscle contractions are important in regulating body temperature. Skeletal muscles are further used to move the body away from harm (e.g., removing your hand from a hot stove). Skeletal muscle contractions are also important for nonhomeostatic activities, such as dancing, operating a computer, or reading a book. Smooth muscle is found in the walls of hollow organs and tubes. Controlled contraction of smooth muscle, although involuntary, regulates movement of blood through blood vessels, food through the digestive tract, air through respiratory airways, and urine through the exterior. Cardiac muscle is found only in the walls of the heart, whose contractions pump life-sustaining blood throughout the body for an entire lifetime (based on 72 beats per minute to the age of 80 years, that is 3 027 456 000 beats!).

# Muscle Physiology

▌**Clinical Connections**

David is a 50-year-old male. Recently, his spouse has been trying to get him to be more active so they can exercise together. However, David finds that he cannot exercise for very long. This isn't new to him; even as a young child in school he couldn't keep up with his classmates in sports, and until recently he was content with an inactive life. Now that he is being encouraged to become active, however, this inability to exercise is becoming a concern to him. Whenever he does exercise, his legs rapidly tire and feel stiff, as if his muscles were locking on him. These symptoms go away after a period of rest, and he can continue to exercise again. Interestingly, he has greater endurance if he has recently eaten a meal.

A physical exam revealed only a generalized muscle weakness, but upon further questioning about his symptoms, David said he has on occasion noticed that his urine is darker after he has been active. Additional tests found that his creatine kinase levels were elevated, and a muscle biopsy showed excessive deposits of glycogen in his muscles. After an enzyme activity test demonstrated the absence of myophosphorylase, the doctor made a diagnosis of McArdle's disease.

## 7.1 | Introduction

By moving specialized intracellular components, muscle cells can develop tension and shorten—that is, contract. Recall that the three types of **muscle** are *skeletal muscle, cardiac muscle,* and *smooth muscle* (p. 6). Through their highly developed ability to contract, groups of muscle cells working together within a muscle can produce movement and do work. Controlled contraction of muscles allows (1) purposeful movement of the whole body or parts of the body (such as walking or waving your hand), (2) manipulation of external objects (such as driving a car or moving a piece of furniture), (3) propulsion of contents through various hollow internal organs (such as circulation of blood or movement of a meal through the digestive tract), and (4) emptying the contents of certain organs to the external environment (such as urination or giving birth).

Muscle comprises the largest group of tissues in the body, accounting for approximately half of the body's weight. Skeletal muscle alone makes up about 40 percent of body weight in men and 32 percent in women. Smooth and cardiac muscle make up another 10 percent of the total weight. Although the three muscle types are structurally and functionally distinct, they can be classified in two different ways according to their common characteristics (> Figure 7-1). First, muscles are categorized as *striated* (skeletal and cardiac muscle) or *unstriated* (smooth muscle), depending on whether alternating dark and light bands—called striations (stripes)—can be seen when the muscle is viewed under a light microscope. Second, muscles are categorized as *voluntary* (skeletal muscle) or *involuntary* (cardiac and smooth muscle), depending, respectively, on whether they are innervated by the somatic nervous system and are subject to voluntary control, or are innervated by the autonomic nervous system and are not subject to voluntary control. Although skeletal muscle is categorized as voluntary, because it can be consciously controlled, much of the skeletal muscle activity is also subject to subconscious, involuntary regulation, such as that related to posture, balance, and stereotypical movements like walking.

Most of this chapter is a detailed examination of the most abundant and best understood muscle, **skeletal muscle**. Skeletal muscles make up the muscular system. We begin with a discussion of skeletal muscle structure, and then examine how it works at the molecular level, the cell level, and in the whole muscle. The chapter also includes a discussion of the unique properties of smooth and cardiac muscle in comparison with skeletal muscle. Smooth muscle appears throughout the body systems as components of hollow organs and tubes. Cardiac muscle is found only in the heart.

## 7.2 | Structure of Skeletal Muscle

A single skeletal muscle cell, known as a muscle fibre, is relatively large, elongated, and cylinder shaped, measuring 10–100 μm (micrometre) in diameter and up to 750 000 μm, or 0.76 m, in length (1 μm = 1 millionth of a metre). In a person who is about 1.8 m tall, the longest muscle fibre in the body is about 50 cm. A skeletal muscle consists of a number of muscle fibres lying parallel to one another and bundled together by connective tissue (> Figure 7-2a). The fibres usually extend the entire length of the muscle. During embryonic development, the huge skeletal muscle fibres are formed by the fusion of many smaller cells called myoblasts (*myo* means "muscle"; *blast* means "former"); thus, a striking feature of myoblasts is the presence of multiple nuclei in a single muscle cell. Another feature is the abundance of mitochondria, the energy-generating organelles, as would be expected with the high energy demands of a tissue as active as skeletal muscle.

### Skeletal muscle fibres

The predominant structural feature of a skeletal muscle fibre is the numerous **myofibrils**. These specialized contractile elements—which constitute 80 percent of the volume of the muscle fibre—are cylindrical intracellular structures, 1 μm in diameter, that extend the entire length of the muscle fibre (> Figure 7-2b). Each myofibril consists of a regular arrangement of highly organized cytoskeletal elements: the thick and the thin filaments (> Figure 7-2c). The **thick filaments**, which are 12–18 nm (nanometre) in diameter and 1.6 μm in length, are special assemblies of the protein *myosin*. The **thin filaments**, which are 5–8 nm in diameter and 1.0 μm long, are made up primarily of the protein *actin* (> Figure 7-2d). The levels of organization in a skeletal muscle can be summarized as follows:

| Whole muscle | → | muscle fibre | → | myofibril | → | thick and thin filaments | → | myosin and actin |
|---|---|---|---|---|---|---|---|---|
| (an organ) | | (cell) | | (a specialized intracellular structure) | | (cytoskeletal elements) | | (protein molecules) |

Each muscle is covered by a dense connective tissue made primarily of collagen and to a lesser extent elastin. The connective tissue is very important in that it provides structure to the muscle and allows the transfer of force to the bone. This provides tension for stabilization and/or movement. The three anatomic connective tissues are **epimysium**, which covers the whole muscle; **perimysium**, which divides the muscle fibres into bundles or fascicles; and **endomysium**, the innermost connective tissue, which covers each muscle fibre or cell. It is the contractile components of the skeletal muscle fibres that transfer force to the connective tissue sheaths, then to the **tendon**, and finally to the bone, thereby providing the human body with the ability to stabilize body segments, manipulate objects, and locomote.

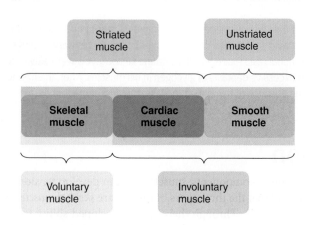

> **FIGURE 7-1 Categorization of muscle**

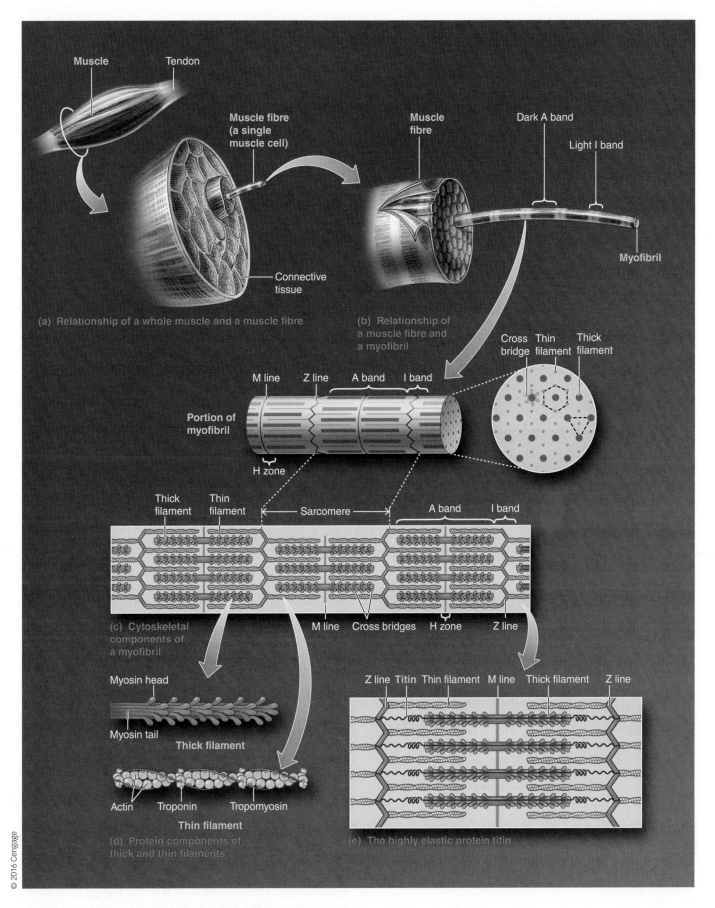

**Muscle**  **Tendon**

**Muscle fibre (a single muscle cell)**

Connective tissue

(a) Relationship of a whole muscle and a muscle fibre

**Muscle fibre**

Dark A band

Light I band

**Myofibril**

(b) Relationship of a muscle fibre and a myofibril

Cross bridge   Thin filament   Thick filament

M line   Z line   A band   I band

**Portion of myofibril**

H zone

Thick filament   Thin filament   Sarcomere   A band   I band

M line   Cross bridges   H zone   Z line

(c) Cytoskeletal components of a myofibril

Myosin head

Myosin tail   **Thick filament**

Actin   Troponin   Tropomyosin

**Thin filament**

(d) Protein components of thick and thin filaments

Z line   **Titin**   Thin filament   M line   Thick filament   Z line

(e) The highly elastic protein titin

> FIGURE 7-2 **Levels of organization in a skeletal muscle**

## A AND I BANDS

Viewed with an electron microscope, a myofibril displays alternating dark bands (the *A bands*) and light bands (the *I bands*) (› Figure 7-3a). The bands of all the myofibrils lined up parallel to one another collectively produce the striated or striped appearance of a skeletal muscle fibre visible under a light microscope (› Figure 7-3b). Alternate stacked sets of thick and thin filaments that slightly overlap one another are responsible for the A and I bands (› Figure 7-2c).

An **A band** is made up of a stacked set of thick filaments along with the portions of the thin filaments that overlap on both ends of the thick filaments. The thick filaments lie only within the A band and extend its entire width; that is, the two ends of the thick filaments within a stack define the outer limits of a given A band. The lighter area within the middle of the A band, where the thin filaments do not reach, is the **H zone**. Only the central portions of the thick filaments are found in this region. Within each stack, a system of supporting proteins holds the thick filaments together vertically. These proteins can be seen as the **M line**, which extends vertically down the middle of the A band within the centre of the H zone.

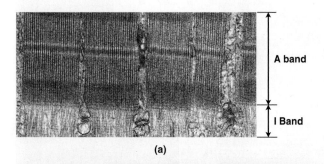

**A band**

**I Band**

**(a)**

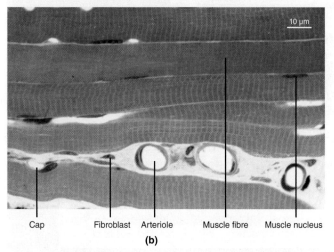

10 µm

Cap    Fibroblast  Arteriole    Muscle fibre  Muscle nucleus

**(b)**

› **FIGURE 7-3 Microscope view of skeletal muscle components.** (a) Electron microscope view of a myofibril. Note the A and I bands. (b) Low-power light-microscope view of skeletal muscle fibres. Note the striated appearance.

**Source:** Histology images were produced by Professor Stephen C. Pang, Mrs. Judy J. Janzen-Pang, and Professor Leslie W. MacKenzie of the Anatomy and Cell Biology Program in the Department of Biomedical and Molecular Sciences, Faculty of Health Sciences at Queen's University, Kingston, Ontario, Canada.

An **I band** consists of the remaining portion of the thin filaments that do not project into the A band. Visible in the middle of each I band is a dense, vertical **Z line**. The area between two Z lines is called a **sarcomere**, which is the functional unit of skeletal muscle. A **functional unit** of any organ is the smallest component that can perform all the functions of that organ. The sarcomere is the smallest component of a muscle fibre that can contract. The Z line (also known as Z discs) is a flat, cytoskeletal disc that connects the thin filaments of two adjoining sarcomeres. Each relaxed sarcomere is about 2.5 µm in width and consists of one whole A band and half of each of the two I bands located on either side. An I band contains only thin filaments from two adjacent sarcomeres, but not the entire length of these filaments. During growth, a muscle increases in length by adding new sarcomeres on the ends of the myofibrils, not by increasing the size of each sarcomere. As well, an increase in flexibility through the use of a regular stretching program is brought about by the addition of new sarcomeres.

Shown in Figure 7-2, single strands of a giant, highly elastic protein known as **titin** extend in both directions from the M line along the length of the thick filament to the Z lines at opposite ends of the sarcomere. Titin is the largest protein in the body, made up of nearly 30 000 amino acids. It serves two important roles: (1) along with the M-line proteins, titin helps stabilize the position of the thick filaments in relation to the thin filaments; and (2) by acting like a spring, it greatly augments a muscle's elasticity. That is, titin helps a muscle stretched by an external force to passively recoil or spring back to its resting length when the stretching force is removed, much like a stretched spring.

### CROSS BRIDGES

With an electron microscope, fine **cross bridges (myosin heads)** can be seen extending from each thick filament toward the surrounding thin filaments in the areas where the thick and thin filaments overlap (see › Figure 7-2c). Three-dimensionally, the thin filaments are arranged hexagonally around the thick filaments. Cross bridges project from each thick filament in all six directions toward the surrounding thin filaments. Each thin filament, in turn, is surrounded by three thick filaments. To give you an idea of the magnitude of these filaments, a single muscle fibre may contain an estimated 16 billion thick and 32 billion thin filaments, all arranged in a very precise pattern within the myofibrils.

## Myosin

Each thick filament has several hundred **myosin** molecules packed together in a specific arrangement. Myosin is considered a motor protein and is responsible for action-based motility. A myosin molecule is a protein consisting of two identical subunits, each shaped somewhat like a golf club (› Figure 7-4a). The protein's tail ends are intertwined around each other like golf-club shafts twisted together, with the two globular heads projecting out at one end. The two halves of each thick filament are mirror images made up of myosin molecules lying lengthwise

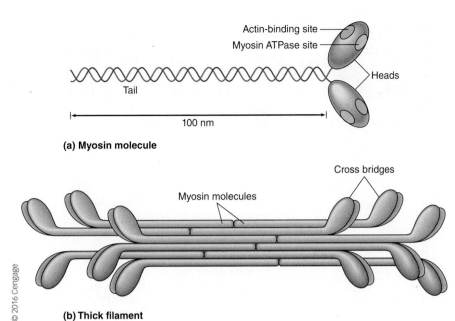

**Actin-binding site**
**Myosin ATPase site**
**Heads**
**Tail**

100 nm

**(a) Myosin molecule**

**Cross bridges**
**Myosin molecules**

© 2016 Cengage

**(b) Thick filament**

› **FIGURE 7-4** **Structure of myosin molecules and their organization within a thick filament.**
(a) Each myosin molecule consists of two identical, golf-club-shaped subunits with their tails intertwined and their globular heads, each of which contains an actin-binding site and a myosin ATPase site, projecting out at one end. (b) A thick filament is made up of myosin molecules lying lengthwise parallel to one another. Half are oriented in one direction and half in the opposite direction. The globular heads, which protrude at regular intervals along the thick filament, form the cross bridges.

in a regular, staggered array, with their tails oriented toward the centre of the filament and their globular heads protruding outward at regular intervals (› Figure 7-4b). These heads form the cross bridges between the thick and thin filaments. Each cross bridge has two important sites crucial to the contractile process: (1) an actin-binding site and (2) a myosin ATPase (ATP-splitting) site. It is the rate of ATPase activity on the head of the myosin that provides a method of muscle fibre typing (p. 319).

## Actin, tropomyosin, and troponin

Thin filaments consist of three proteins: *actin, tropomyosin,* and *troponin* (› Figure 7-5). **Actin** molecules, the primary structural proteins of the thin filament, are spherical. The backbone of a thin filament is formed by actin molecules joined into two strands and twisted together, like two intertwined strings of pearls. Each actin molecule has a special binding site for attachment with a myosin cross bridge. The binding of myosin and actin molecules at the cross bridges results in the energy-consuming contraction of the muscle fibre; this mechanism is described in Section 7.3. Accordingly, myosin and actin are often called **contractile proteins**, even though, as you will see, neither myosin nor actin actually contracts. Myosin and actin are not unique to muscle cells, but these proteins are more abundant and more highly organized in muscle cells. It is important to note that there are numerous actin-binding proteins (ABPs), an example of which is **nebulin**, but this class of proteins will not be discussed in detail.

In a relaxed muscle fibre, contraction does not take place; actin cannot bind with cross bridges, because of the way the two other types of protein—tropomyosin and troponin—are positioned within the thin filament. **Tropomyosin** molecules are threadlike proteins that lie end to end alongside the groove of the actin

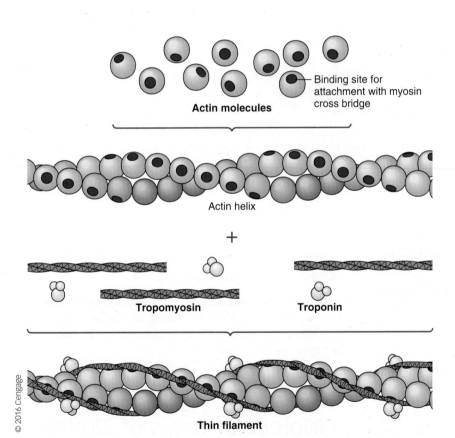

**Binding site for attachment with myosin cross bridge**

**Actin molecules**

**Actin helix**

+

**Tropomyosin**      **Troponin**

© 2016 Cengage

**Thin filament**

› **FIGURE 7-5** **Composition of a thin filament.** The main structural component of a thin filament is two chains of spherical actin molecules that are twisted together. Troponin molecules (which consist of three small, spherical subunits) and threadlike tropomyosin molecules are arranged to form a ribbon that lies alongside the groove of the actin helix and physically covers the binding sites on actin molecules for attachment with myosin cross bridges. (The thin filaments shown here are not drawn in proportion to the thick filaments in Figure 7-4. Thick filaments are two to three times larger in diameter than the thin filaments.)

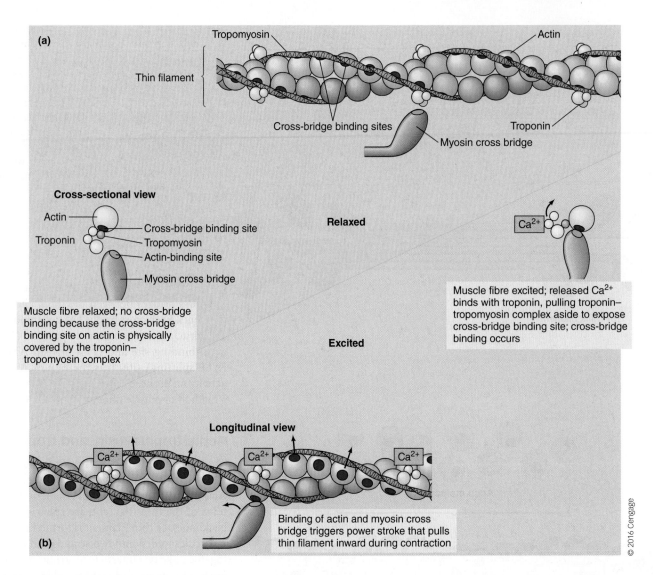

(a)

Tropomyosin

Actin

Thin filament

Cross-bridge binding sites

Troponin

Myosin cross bridge

**Cross-sectional view**

Actin

Cross-bridge binding site

Troponin

Tropomyosin

Actin-binding site

Myosin cross bridge

**Relaxed**

Ca²⁺

Muscle fibre relaxed; no cross-bridge binding because the cross-bridge binding site on actin is physically covered by the troponin–tropomyosin complex

Muscle fibre excited; released Ca²⁺ binds with troponin, pulling troponin–tropomyosin complex aside to expose cross-bridge binding site; cross-bridge binding occurs

**Excited**

**Longitudinal view**

Ca²⁺

Ca²⁺

Ca²⁺

Binding of actin and myosin cross bridge triggers power stroke that pulls thin filament inward during contraction

(b)

© 2016 Cengage

› **FIGURE 7-6** Role of calcium in turning on cross bridges

spiral. In this position, tropomyosin covers the actin sites that bind with the cross bridges, blocking the interaction that leads to muscle contraction. The other thin filament component, **troponin**, is a protein complex made of three polypeptide units: one binds to tropomyosin, one binds to actin, and a third can bind with calcium Ca²⁺.

When troponin is not bound to Ca²⁺, this protein stabilizes tropomyosin in its blocking position over actin's cross-bridge binding sites (› Figure 7-6a). When Ca²⁺ binds to troponin, the shape of this protein is changed in such a way that tropomyosin slips away from its blocking position (› Figure 7-6b). With tropomyosin out of the way, actin and myosin can bind and interact at the cross bridges, resulting in muscle contraction. Tropomyosin and troponin are often called **regulatory proteins** because of their role in covering (preventing contraction) or exposing (permitting contraction) the binding sites for cross-bridge interaction between actin and myosin.

**Check Your Understanding 7.1**

1. Compare the relationship of myofibrils and a muscle fibre with the relationship between muscle fibres and a whole muscle.

2. Describe the relationship between actin, tropomyosin, and troponin in a relaxed muscle fibre.

## 7.3 | Molecular Basis of Skeletal Muscle Contraction

Several important links in the contractile process remain to be discussed. How does cross-bridge interaction between actin and myosin bring about muscle contraction? How does a muscle action potential trigger this contractile process? What is the source of the calcium that physically repositions troponin and tropomyosin to permit cross-bridge binding? We turn our attention to these topics in this section.

## Cross bridges

Cross-bridge interaction between actin and myosin brings about muscle contraction by means of the sliding filament mechanism.

### SLIDING FILAMENT MECHANISM

During contraction, the thin filaments on each side of a sarcomere slide inward over the stationary thick filaments toward the A band's centre (⟩ Figure 7-7). As they slide inward, the thin filaments pull the Z lines to which they are attached closer together, so the sarcomere shortens. As all the sarcomeres throughout the muscle fibre's length shorten simultaneously, the entire fibre shortens. When a muscle fibre contracts, moving the Z lines closer together, the muscle fibre and thus the whole muscle shortens. This is called a *concentric contraction* (p. 315). This is the **sliding filament mechanism** of muscle contraction. The H zone, in the centre of the A band where the thin filaments do not reach, becomes smaller as the thin filaments approach each other when they slide more deeply inward. The I band, which consists of the portions of the thin filaments that do not overlap with the thick filaments, narrows as the thin filaments further overlap the thick filaments during their inward slide. The thin filaments themselves do not change length during muscle fibre shortening. The width of the A band remains unchanged during contraction, because its width is determined by the length of the thick filaments, and the thick filaments do not change length during the shortening process. Note that neither the thick nor the thin filaments decrease in length to shorten the sarcomere. Instead, contraction is accomplished by the thin filaments from the opposite sides of each sarcomere sliding closer together between the thick filaments.

### POWER STROKE

Cross-bridge activity pulls the thin filaments inward relative to the stationary thick filaments. During contraction—with the tropomyosin and troponin "chaperones" pulled out of the way by calcium—the myosin cross bridges from a thick filament can bind with the actin molecules in the surrounding thin filaments. Let's concentrate on a single cross-bridge interaction (⟩ Figure 7-8a). The two myosin heads of each myosin molecule act independently, with only one head attaching to actin at a given time. When myosin and actin make contact at a cross bridge, the bridge changes shape, bending inward as if it were on a hinge, "stroking" toward the centre of the sarcomere, like the stroking of a boat oar. This so-called **power stroke** of a cross

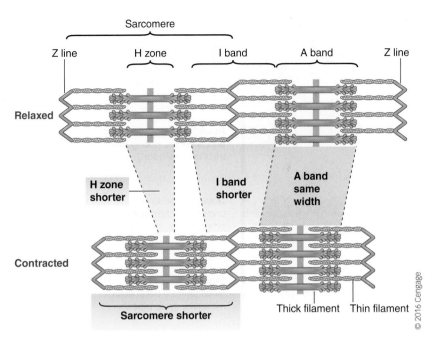

⟩ **FIGURE 7-7 Changes in banding pattern during shortening.** During muscle contraction, each sarcomere shortens as the thin filaments slide closer together between the thick filaments so that the Z lines are pulled closer together. The width of the A bands does not change as a muscle fibre shortens, but the I bands and H zones become shorter.

© 2016 Cengage

bridge pulls inward the thin filament to which it is attached. A single power stroke pulls the thin filament inward only a small percentage of the total shortening distance. Repeated cycles of cross-bridge binding and bending complete the shortening.

At the end of one cross-bridge cycle, the link between the myosin cross bridge and actin molecule breaks. The cross bridge returns to its original shape and binds to the next actin molecule behind its previous actin partner. The cross bridge bends once again to pull the thin filament in farther, then detaches and repeats the cycle. Repeated cycles of cross-bridge power strokes successively pull in the thin filaments, much like pulling in a rope hand over hand.

Because of the way myosin molecules are oriented within a thick filament (⟩ Figure 7-8b), all the cross bridges stroke toward the centre of the sarcomere so that all six of the surrounding thin filaments on each end of the sarcomere are pulled inward simultaneously (⟩ Figure 7-8c). The cross bridges aligned with given thin filaments do not stroke all in unison, however. At any time during contraction, part of the cross bridges are attached to the thin filaments and are stroking, while others are returning to their original conformation in preparation for binding with another actin molecule. Thus, some cross bridges are "holding on" to the thin filaments, whereas others "let go" to bind with new actin. Were it not for this asynchronous cycling of the cross bridges, the thin filaments would slip back toward their resting position between strokes.

How does muscle excitation switch on this cross-bridge cycling? The term **excitation–contraction coupling** refers to the series of events linking muscle excitation (the presence of an

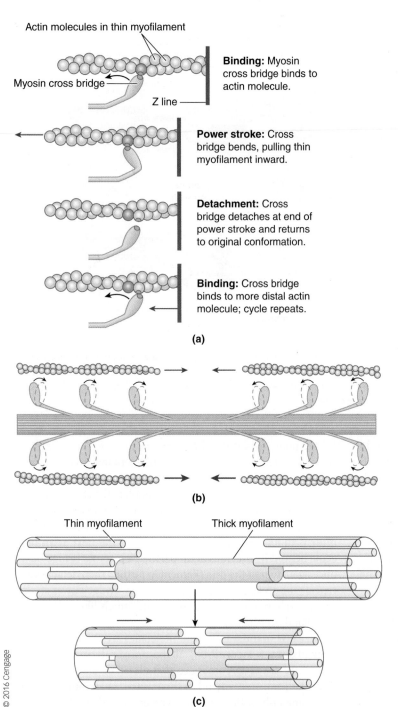

**Binding:** Myosin cross bridge binds to actin molecule.

Actin molecules in thin myofilament

Myosin cross bridge

Z line

**Power stroke:** Cross bridge bends, pulling thin myofilament inward.

**Detachment:** Cross bridge detaches at end of power stroke and returns to original conformation.

**Binding:** Cross bridge binds to more distal actin molecule; cycle repeats.

(a)

(b)

Thin myofilament    Thick myofilament

(c)

© 2016 Cengage

⟩ **FIGURE 7-8 Cross-bridge activity.** (a) During each cross-bridge cycle, the cross bridge binds with an actin molecule, bends to pull the thin filament inward during the power stroke, then detaches and returns to its resting conformation, ready to repeat the cycle. (b) The power strokes of all cross bridges extending from a thick filament are directed toward the centre of the thick filament. (c) Each thick filament is surrounded on each end by six thin filaments, all of which are pulled inward simultaneously through cross-bridge cycling during muscle contraction.

action potential in a muscle fibre) to muscle contraction (cross-bridge activity that causes the thin filaments to slide closer together to produce sarcomere shortening). We now turn our attention to excitation–contraction coupling.

## Excitation–contraction coupling

Skeletal muscles are stimulated to contract by release of acetylcholine (ACh) at neuromuscular junctions between motor-neuron terminals and muscle fibres. Recall that binding of ACh with the motor end plate of a muscle fibre brings about permeability changes in the muscle fibre, resulting in an action potential that is conducted over the entire surface of the muscle cell membrane (p. 197). Two membranous structures within the muscle fibre play an important role in linking this excitation to contraction—*transverse tubules* and the *sarcoplasmic reticulum* (SR). Let's examine the structure and function of each.

### SPREAD OF THE ACTION POTENTIAL DOWN THE T TUBULES

At each junction of an A band and I band, the surface membrane dips into the muscle fibre to form a **transverse tubule (T tubule)**, which runs perpendicularly from the surface of the muscle cell membrane into the central portions of the muscle fibre (⟩ Figure 7-9). Because the T tubule membrane is continuous with the surface membrane, an action potential on the surface membrane also spreads down into the T tubule, rapidly transmitting the surface electrical activity into the central portions of the fibre. The presence of a local action potential in the T tubules induces permeability changes in a separate membranous network within the muscle fibre—the sarcoplasmic reticulum.

### RELEASE OF CALCIUM FROM THE SARCOPLASMIC RETICULUM

The **sarcoplasmic reticulum (SR)** is a modified endoplasmic reticulum that consists of a fine network of interconnected compartments surrounding each myofibril like a mesh sleeve (⟩ Figure 7-9). This membranous network encircles the myofibril throughout its length but is not continuous. Separate segments of SR are wrapped around each A band and each I band. The ends of each segment expand to form saclike regions—the **lateral sacs** (alternatively known as **terminal cisternae**)—which are separated from the adjacent T tubules by a slight gap (⟩ Figure 7-9 and ⟩ Figure 7-10). The SR's lateral sacs store calcium. Spread of an action potential down a T tubule triggers release of calcium from the SR into the cytosol.

How is a change in T tubule potential linked with the release of calcium from the lateral sacs? An orderly arrangement of **foot proteins** extends from the SR and spans the gap between the lateral sac and T tubule. Each foot protein contains four subunits arranged in a specific pattern (⟩ Figure 7-10a). These foot proteins not only bridge the gap but also serve as *Ca²⁺-release channels*. These foot-protein Ca²⁺ channels are also known as **ryanodine receptors** because they are locked in the open position by the plant chemical ryanodine.

Half of the SR's foot proteins are "zipped together" with complementary receptors on the T tubule side of the junction. These

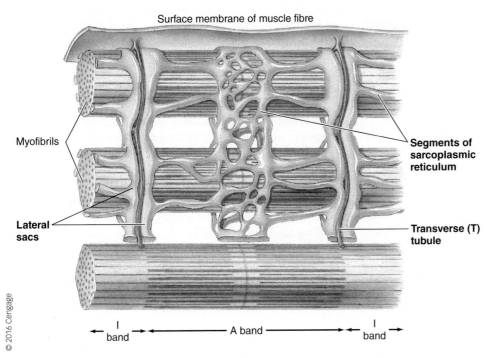

Surface membrane of muscle fibre

Myofibrils

Lateral sacs

Segments of sarcoplasmic reticulum

Transverse (T) tubule

I band — A band — I band

© 2016 Cengage

> **FIGURE 7-9 The T tubules and sarcoplasmic reticulum in relationship to the myofibrils.** The transverse (T) tubules are membranous, perpendicular extensions of the surface membrane that dip deep into the muscle fibre at the junctions between the A and I bands of the myofibrils. The SR is a fine, membranous network that runs longitudinally and surrounds each myofibril, with separate segments encircling each A band and I band. The ends of each segment are expanded to form lateral sacs that lie next to the adjacent T tubules.

T tubule receptors, which are made up of four subunits in exactly the same pattern as the foot proteins, are located like mirror images in contact with every other foot protein protruding from the SR (> Figure 7-10b and > 7-10c). These T tubule receptors are known as **dihydropyridine receptors** because they are blocked by the drug dihydropyridine. These dihydropyridine receptors are voltage-gated sensors. When an action potential is propagated down the T tubule, the local depolarization activates the voltage-gated dihydropyridine receptors. The activated T tubule receptors, in turn, trigger the opening of the directly abutting $Ca^{2+}$-release channels (alias ryanodine receptors, alias foot proteins) in the adjacent lateral sacs of the SR. Opening of the half of the $Ca^{2+}$-release channels in direct contact with the dihydropyridine receptors triggers the opening of the other half of the $Ca^{2+}$-release channels that are not directly associated with the T tubule receptors.

7

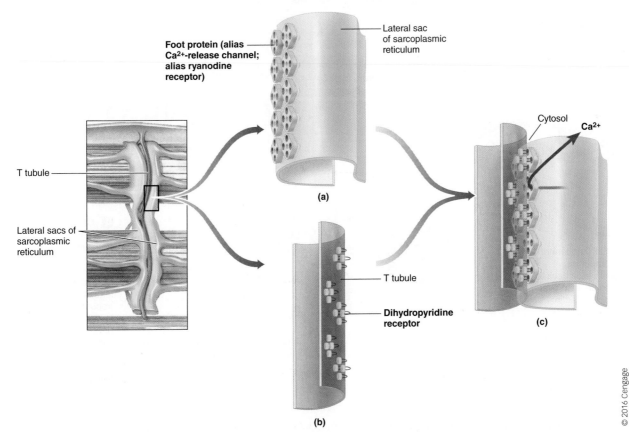

Foot protein (alias $Ca^{2+}$-release channel; alias ryanodine receptor)

Lateral sac of sarcoplasmic reticulum

T tubule

Lateral sacs of sarcoplasmic reticulum

(a)

T tubule

Dihydropyridine receptor

(b)

Cytosol

$Ca^{2+}$

(c)

© 2016 Cengage

> **FIGURE 7-10 Relationship between a T tubule and the adjacent lateral sacs of the sarcoplasmic reticulum**

Calcium is released into the cytosol from the lateral sacs through all these open $Ca^{2+}$-release channels. By slightly repositioning the troponin and tropomyosin molecules, this released calcium exposes the binding sites on the actin molecules so they can link with the myosin cross bridges at their complementary binding sites (⟩ Figure 7-11).

### ATP-POWERED CROSS-BRIDGE CYCLING

Recall that a myosin cross bridge has two special sites—an actin-binding site and an ATPase site. The latter is an enzymatic site that can bind the energy carrier *adenosine triphosphate* (ATP) and split it into *adenosine diphosphate* (ADP) and *inorganic phosphate* ($P_i$), yielding energy in the process. The breakdown of ATP occurs on the myosin cross bridge before the bridge ever links with an actin molecule (⟩ Figure 7-12, step **1**). The ADP and $P_i$ remain tightly bound to the myosin, and the generated energy is stored within the cross bridge to produce a high-energy form of myosin. To use an analogy, the cross bridge is cocked like a gun, ready to be fired when the trigger is pulled. When the muscle fibre is

excited, $Ca^{2+}$ pulls the troponin–tropomyosin complex out of its blocking position so that the energized (cocked) myosin cross bridge can bind with an actin molecule (step **2a**). This contact between myosin and actin "pulls the trigger," causing the cross-bridge bending that produces the power stroke (step **3**). Researchers have not found the mechanism by which the chemical energy released from ATP is stored within the myosin cross bridge and then translated into the mechanical energy of the power stroke. Inorganic phosphate is released from the cross bridge during the power stroke. After the power stroke is complete, ADP is released.

When the muscle is not excited and $Ca^{2+}$ is not released, troponin and tropomyosin remain in their blocking position so that actin and the myosin cross bridges do not bind and no power stroking takes place (step **2b**).

When $P_i$ and ADP are released from myosin following contact with actin and the subsequent power stroke, the myosin ATPase site is free for attachment of another ATP molecule. The actin and myosin remain linked at the cross bridge until a fresh

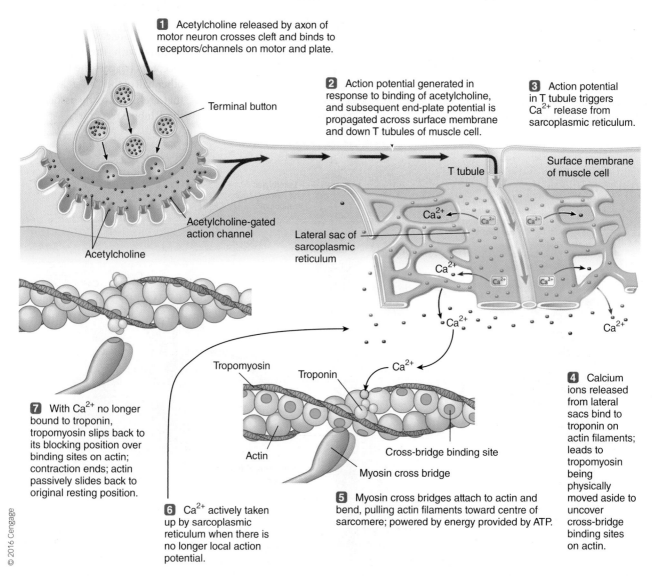

**1** Acetylcholine released by axon of motor neuron crosses cleft and binds to receptors/channels on motor and plate.

**2** Action potential generated in response to binding of acetylcholine, and subsequent end-plate potential is propagated across surface membrane and down T tubules of muscle cell.

**3** Action potential in T tubule triggers $Ca^{2+}$ release from sarcoplasmic reticulum.

Terminal button

Surface membrane of muscle cell

T tubule

$Ca^{2+}$

Acetylcholine-gated action channel

Lateral sac of sarcoplasmic reticulum

$Ca^{2+}$

Acetylcholine

$Ca^{2+}$

$Ca^{2+}$

$Ca^{2+}$

**4** Calcium ions released from lateral sacs bind to troponin on actin filaments; leads to tropomyosin being physically moved aside to uncover cross-bridge binding sites on actin.

Tropomyosin

Troponin

$Ca^{2+}$

**7** With $Ca^{2+}$ no longer bound to troponin, tropomyosin slips back to its blocking position over binding sites on actin; contraction ends; actin passively slides back to original resting position.

Actin

Cross-bridge binding site

Myosin cross bridge

**6** $Ca^{2+}$ actively taken up by sarcoplasmic reticulum when there is no longer local action potential.

**5** Myosin cross bridges attach to actin and bend, pulling actin filaments toward centre of sarcomere; powered by energy provided by ATP.

© 2016 Cengage

⟩ **FIGURE 7-11 Calcium release in excitation–contraction coupling.** Steps 1 through 5 show the events that couple neurotransmitter release and subsequent electrical excitation of the muscle cell with muscle contraction. Steps 6 and 7 show events associated with muscle relaxation.

molecule of ATP attaches to myosin at the end of the power stroke. Attachment of the new ATP molecule permits detachment of the cross bridge, which returns to its unbent form, ready to start another cycle (step **4a**). The newly attached ATP is then split by myosin ATPase, energizing the myosin cross bridge once again (step **1**). On binding with another actin molecule, the energized cross bridge again bends, and so on, successively pulling the thin filament inward to accomplish contraction.

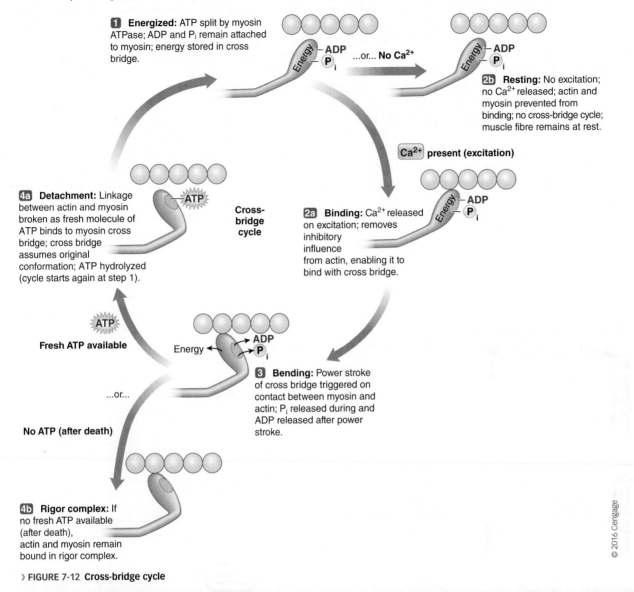

**1** **Energized:** ATP split by myosin ATPase; ADP and P$_i$ remain attached to myosin; energy stored in cross bridge.

...or... **No Ca$^{2+}$**

**2b** **Resting:** No excitation; no Ca$^{2+}$ released; actin and myosin prevented from binding; no cross-bridge cycle; muscle fibre remains at rest.

**Ca$^{2+}$** **present (excitation)**

**4a** **Detachment:** Linkage between actin and myosin broken as fresh molecule of ATP binds to myosin cross bridge; cross bridge assumes original conformation; ATP hydrolyzed (cycle starts again at step 1).

**Cross-bridge cycle**

**2a** **Binding:** Ca$^{2+}$ released on excitation; removes inhibitory influence from actin, enabling it to bind with cross bridge.

**Fresh ATP available**

...or...

**No ATP (after death)**

**3** **Bending:** Power stroke of cross bridge triggered on contact between myosin and actin; P$_i$ released during and ADP released after power stroke.

**4b** **Rigor complex:** If no fresh ATP available (after death), actin and myosin remain bound in rigor complex.

⟩ **FIGURE 7-12 Cross-bridge cycle**

© 2016 Cengage

7

**▌ Why It Matters**
**Rigor Mortis**

Note that fresh ATP must attach to myosin to permit the cross-bridge link between myosin and actin to break at the end of a cycle, even though the ATP is not split during this dissociation process. The need for ATP in separating myosin and actin is amply shown in **rigor mortis**. This "stiffness of death" is a generalized locking in place of the skeletal muscles that begins 3 to 4 hours after death and completes in about 12 hours. Following death, the cytosolic concentration of calcium begins to rise, most likely because the inactive muscle cell membrane cannot keep out extracellular Ca$^{2+}$ and perhaps also because Ca$^{2+}$ leaks out of the lateral sacs. This Ca$^{2+}$ moves the regulatory proteins aside, letting actin bind with the myosin cross bridges, which were already charged with ATP before death. Dead cells cannot produce any more ATP, so actin and myosin, once bound, cannot detach, because they lack fresh ATP. The thick and thin filaments thus stay linked by the immobilized cross bridges, leaving dead muscles stiff (⟩ Figure 7-12, step **4b**). During the next several days, rigor mortis gradually subsides as the proteins involved in the rigor complex begin to degrade.

## RELAXATION

How is **relaxation** normally accomplished in a living muscle? Just as an action potential in a muscle fibre turns on the contractile process by triggering the release of $Ca^{2+}$ from lateral sacs into the cytosol, the contractile process is turned off when $Ca^{2+}$ is returned to the lateral sacs when local electrical activity stops. The SR has an energy-consuming carrier, a *Ca²⁺–ATPase pump,* which actively transports $Ca^{2+}$ from the cytosol and concentrates it in the lateral sacs. When acetylcholinesterase (AChE) removes ACh from the neuromuscular junction, the muscle fibre action potential stops. When a local action potential is no longer in the T tubules to trigger the release of $Ca^{2+}$, the ongoing activity of the SR's $Ca^{2+}$ pump returns released $Ca^{2+}$ back into its lateral sacs. Removing cytosolic $Ca^{2+}$ lets the troponin–tropomyosin complex slip back into its blocking position, so actin and myosin can no longer bind at the cross bridges. The thin filaments, freed from cycles of cross-bridge attachment and pulling, return passively to their resting position. The muscle fibre has relaxed.

We are now going to compare the duration of contractile activity with the duration of excitation.

**Clinical Connections**

McArdle's disease is a characterized by the absence of the enzyme phosphorylase, a key enzyme required in breaking down glycogen to release glucose for ATP production. Without this enzyme, glycogen accumulates in muscles, and alternate fuel sources are required to maintain muscle activity. When David starts to exercise, he rapidly depletes his ATP pools, and he feels muscle weakness. The sensation of his legs locking is due to the lack of ATP and is the equivalent of a living rigor mortis. After a rest, his body has mobilized fatty acids and glucose that are then used by the muscle to produce ATP. (Recall from Chapter 2 how fatty acids and glucose are used to make ATP.) Then he can exercise again.

## Contractile activity

A single action potential in a skeletal muscle fibre lasts only 1–2 msec. The onset of the resulting contractile response lags behind the action potential because the entire excitation–contraction coupling must occur before cross-bridge activity begins, resulting in shortening of the sarcomere. In fact, the action potential is completed before the contractile apparatus even becomes operational. This time delay, about 0.5 msec between stimulation and the onset of contraction, is called the **latent period** (⟩ Figure 7-13).

Time is also required for generating tension within the muscle fibre, produced by the sliding interactions between the thick and thin filaments through cross-bridge activity. The time from the onset of contraction until peak tension develops—**contraction time**—averages about 40–120 msec. This time varies depending on the type of muscle fibre. Fast-twitch (FT)

muscle fibres reach peak tension faster than slow-twitch (ST) muscle fibres. This difference results from the variation in ATPase activity associated with the myosin heavy chain (for more on fibre types, see the section Muscle Fibre Types within a Single Motor Unit). Here are some examples of variation in contraction time among human skeletal muscles (approximate mean values for % ST and time in msec): frontalis oculi (15% ST), 45 msec; vastus lateralis (45% ST) and biceps brachii (40% ST), 65 msec; lateral gastrocnemius (50% ST) and soleus (80% ST), 120 msec.

The contractile response does not end until all the $Ca^{2+}$ released in response to the action potential is removed. The time from peak tension until relaxation is complete—the **relaxation time**—usually lasts another 50–200 msec. And again, the time required for the muscle fibre to relax is based on the fibre type, with FT relaxing quicker than ST. Slow-twitch fibres have a lower density of SR and slower $Ca^{2+}$ re-uptake. Remember that muscle relaxation, following a single twitch or a titanic contraction, is dictated by the efficiency of the three processes that remove $Ca^{2+}$: (1) dissociation of $Ca^{2+}$ from troponin, (2) translocation of $Ca^{2+}$ to near the site of entry into the SR, and (3) uptake of $Ca^{2+}$ into the lateral sacs of the SR by the $Ca^{2+}$ pump. Consequently, the entire contractile response to a single action potential may last up to 100 msec or more; this is much longer than the duration of the action potential that initiated it (100 msec

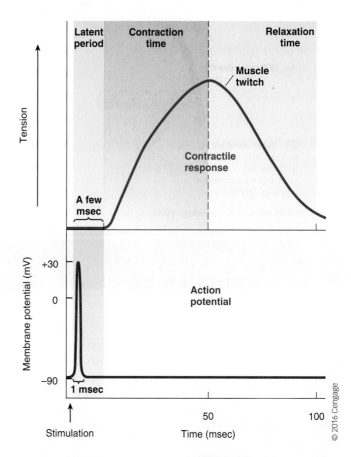

⟩ **FIGURE 7-13 Relationship of an action potential to the resultant muscle twitch**

compared with 1–2 msec). This fact is important in the body's ability to produce muscle contractions of variable strength, as you will soon discover in the next section, where we discuss skeletal muscle mechanics.

## 7.4 Skeletal Muscle Mechanics

Thus far we have described the contractile response in a single muscle fibre. In the body, groups of muscle fibres are organized into whole muscles. We now turn our attention to the contraction of whole muscles and the factors that contribute to muscle tension.

### Whole muscles

Each person has about 600 skeletal muscles, which range in size from the delicate external eye muscles that control eye movements and contain only a few hundred fibres, to the large, powerful leg muscles that contain several hundred thousand fibres.

**■ TABLE 7-1 The Number of Muscle Fibres in Various Human Skeletal Muscles**

| Muscle | Number of Muscle Fibres |
| --- | --- |
| First lumbrical | 10 250* |
| External rectus | 27 000 |
| Platysma | 27 000 |
| First dorsal interosseous | 40 500 |
| Sartorius | 128 150* |
| Brachioradialis | 129 200* |
| Tibialis anterior | 271 350 |
| Medial gastrocnemius | 1 033 000 |

**Note: Results given to nearest 50.**
*Average values. Value for sartorius from MacCallum (1898); all others from Feinstein et al. (1955).

**Source:** Reprinted with permission from B.R. MacIntosh, P.F. Gardiner, and A.J. McComas, 2006, *Skeletal Muscle: Form and Function*, 2nd ed. (Champaign, IL: Human Kinetics), p. 4.

Examples of the number of muscle fibres contained in other human muscles are provided in ▌ Table 7-1.

Each muscle is sheathed by connective tissue that penetrates from the surface into the muscle to envelop each individual fibre and divide the muscle into columns or bundles. The connective tissue extends beyond the ends of the muscle to form tough, collagenous tendons that attach the muscle to bones. A tendon may be quite long, attaching to a bone some distance from the fleshy part of the muscle. For example, some of the muscles involved in finger movement are in the forearm, with long tendons extending down to attach to the bones of the fingers. (You can see these tendons move on the posterior aspect of your hand when you wiggle your fingers.) This arrangement permits greater dexterity; the fingers would be much thicker and more awkward if all the muscles used in finger movement were actually in the fingers.

### Muscle contractions

A single action potential in a muscle fibre produces a brief, weak contraction called a **twitch**, which is too short and too weak to be useful and normally does not take place on its own. Muscle fibres are arranged into whole muscles, where they function cooperatively to produce contractions of variable grades of strength stronger than a twitch. In other words, you can vary the force you exert by the same muscle, depending on whether you are picking up a piece of paper, a book, or a 25 kg weight. Two primary factors can be adjusted to accomplish gradation of whole-muscle tension: (1) *the number of muscle fibres contracting within a muscle (i.e., motor unit recruitment)* and (2) *the firing frequency of each contracting fibre (i.e., frequency of stimulation)*.

It is worth noting that these two strategies for generating force, motor unit recruitment and frequency of stimulation, are not used equally by all muscles. Small, distal muscles (e.g., muscles of the hand) rely more on increasing the firing frequency to increase the force produced; whereas the larger, more proximal muscles rely more on motor unit recruitment. Increasing the firing frequency of the motor unit has been shown to amplify force by tenfold. Adding more motor units can be incredibly effective, as some motor units can produce 100 times more force than other motor units within the same whole muscle (e.g., quadriceps). One last comment before we examine these two force-generating strategies in more detail: it seems that when submaximal forces are required, motor unit recruitment is the more effective mechanism for increasing force, but as the muscles come closer to generating maximal force, the firing frequency becomes more important.

### Motor unit recruitment

The greater is the number of fibres recruited to contract, the greater the total muscle tension. Therefore, larger muscles consisting of more muscle fibres obviously can generate more tension than can smaller muscles with fewer fibres.

Each whole muscle is innervated by a number of different motor neurons. When a motor neuron enters a muscle, it branches, and each axon terminal supplies a single muscle fibre (❯ Figure 7-14). One motor neuron innervates a number

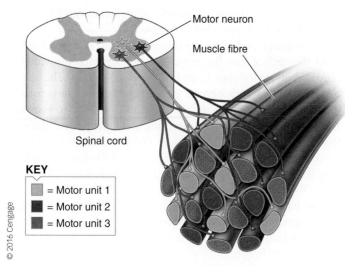

Motor neuron

Muscle fibre

Spinal cord

**KEY**

▨ = Motor unit 1

■ = Motor unit 2

■ = Motor unit 3

© 2016 Cengage

› **FIGURE 7-14  Motor units in a skeletal muscle**

of muscle fibres, but each muscle fibre is supplied by only one motor neuron. When a motor neuron is activated, all the muscle fibres it supplies are stimulated to contract simultaneously. This team of concurrently activated components—one motor neuron plus all the muscle fibres it innervates—is called a **motor unit**. The muscle fibres that compose a motor unit are dispersed throughout the whole muscle; thus, their simultaneous contraction results in an evenly distributed, although weak, contraction of the whole muscle (e.g., the biceps brachii). Each muscle consists of a number of intermingled motor units. For a weak contraction of the whole muscle, only one or a few of its motor units are activated. For stronger and stronger contractions, more and more motor units are recruited, or stimulated to contract. This phenomenon is known as **motor unit recruitment**.

The order of motor unit recruitment is based on the **size principle**: the larger the size of the motor unit, the harder it is to activate. Larger motor units are typically associated with larger motor neurons. In practical terms, it takes a larger voluntary effort, and thus a greater firing frequency, to recruit the larger motor units. As you add more weight to a barbell and lift it, with each addition of weight you will gradually recruit larger motor units to do the work. Therefore, the order of recruitment as it pertains to skeletal muscle fibre is as follows: slow-twitch → fast-twitch A → fast-twitch X. However, research has suggested that this order of recruitment may vary slightly depending on the motion at the joint. For example, the order of motor unit recruitment may change in the biceps brachii when the elbow flexes versus when it supinates. However, in most cases, the order of recruitment, smallest to largest, holds true.

How much stronger the contraction will be with the recruitment of each additional motor unit depends on the size of the motor units (i.e., the number of muscle fibres controlled by a single motor neuron). The number of muscle fibres per motor unit and the number of motor units per muscle vary widely, depending on the specific function of the muscle. For muscles that produce precise, delicate movements, such as external eye muscles and hand muscles, a single motor unit may contain as few as a dozen muscle fibres. Because so few muscle fibres are involved with each motor unit, recruitment of each additional

motor unit adds only a small increment to the whole muscle's strength of contraction. These small motor units allow very fine control over muscle tension. In contrast, in muscles designed for powerful, coarsely controlled movement, such as those of the legs, a single motor unit may contain 1500 to 2000 muscle fibres. Recruitment of each additional motor unit in these large muscles results in great incremental increases in whole-muscle tension. More powerful contractions occur at the expense of precisely controlled gradations. Thus, the number of muscle fibres participating in the whole muscle's total contractile effort depends on the number of motor units recruited and the number of muscle fibres per motor unit in that muscle.

To delay or prevent **fatigue** (the inability to maintain muscle tension at a specified level) during a sustained contraction involving only a portion of a muscle's motor units, as is necessary in muscles supporting the weight of the body against the force of gravity, **asynchronous recruitment of motor units** takes place. The body alternates motor unit activity, like shifts at a factory, to give motor units that have been active an opportunity to rest while others take over. Changing of the shifts is carefully coordinated, so the sustained contraction is smooth rather than jerky. Asynchronous motor unit recruitment is possible only for submaximal contractions, during which only some of the motor units must maintain the desired level of tension. During maximal contractions to move heavy loads, when all the muscle fibres must participate, it is impossible to alternate motor unit activity to prevent fatigue. This is one reason why you cannot support a heavy object as long as a light one.

Furthermore, the type of muscle fibre that is activated varies with the extent of gradation. Most muscles consist of a mixture of fibre types that differ metabolically, some being more resistant to fatigue than others. During weak or moderate endurance-type activities (aerobic exercise), the motor units most resistant to fatigue are recruited first (i.e., slow-twitch fibres). The last fibres to be called into play in the face of demands for further increases in tension are those that fatigue rapidly (fast-twitch A and then B). An individual can therefore engage in endurance activities for prolonged periods of time but can only briefly maintain bursts of all-out, powerful effort. Of course, even the muscle fibres most resistant to fatigue will eventually fatigue if required to maintain a certain level of sustained tension.

## Frequency of stimulation

Whole-muscle tension depends not only on the number of muscle fibres contracting but also on the tension developed by each contracting fibre. Various factors influence the extent to which tension can be developed. These factors include the following:

1. Frequency of stimulation
2. Length of the fibre at the onset of contraction
3. Extent of fatigue
4. Thickness of the fibre

We will discuss each of these factors beginning with frequency of stimulation.

## TWITCH SUMMATION AND TETANUS

Even though a single action potential (duration 1–2 msec) in a muscle fibre produces only a twitch, contractions with longer duration and greater tension can be achieved by repeated stimulation of the fibre. What happens when a second action potential occurs in a muscle fibre depends on the relaxation state of the muscle fibre. If the muscle fibre has completely relaxed (taking 80–200 msec) before the next action potential takes place, a second twitch of the same magnitude as the first occurs (❯ Figure 7-15a). The same excitation–contraction events take place each time, resulting in identical twitch responses. If, however, the muscle fibre is stimulated a second time before it has completely relaxed from the first twitch, a second action potential causes a second contractile response, which is added "piggybacked" on top of the first twitch (❯ Figure 7-15b). The two twitches from the two action potentials add together, or sum, to produce greater tension in the fibre than that produced by a single action potential. This **twitch summation** is similar to temporal summation of EPSPs at the postsynaptic neuron (p. 76).

Twitch summation is possible only because the duration of the action potential (1–2 msec) is much shorter than the duration of the resulting twitch (100 msec). Once an action potential has been initiated, a brief refractory period occurs during which another action potential cannot be initiated (p. 69). It is therefore impossible to achieve summation of action potentials. The membrane must return to resting potential and recover from its refractory period before another action potential can occur. However, because the action potential and refractory period are over long before the resulting muscle twitch is completed, the muscle fibre may be restimulated while some contractile activity still exists, to produce summation of the mechanical response.

So how many action potentials are required to induce summation—that is, produce a useful muscle contraction? Again, this depends on the muscle fibre type, but the range of firing frequencies required to initiate a twitch summation for a submaximal force output is between 8 to 12 hertz (Hz), with a moderate contractile force requiring approximately 30 Hz. To put this into perspective, 1 Hz equals one stimulation per second, so 8 Hz equals eight stimulations (action potentials) per second.

If the muscle fibre is stimulated so rapidly that it does not have a chance to relax at all between stimuli, a smooth, sustained contraction of maximal strength known as **tetanus** occurs (❯ Figure 7-15c). A tetanic contraction is usually three to four times stronger than a single twitch. (Don't confuse this normal physiological tetanus with the disease tetanus; see p. 82.) It is more difficult to determine the firing frequency necessary to produce a maximal force. The overlapping of the motor units within a whole muscle makes measurement of the firing frequency difficult. However, it is generally believed that a firing frequency of 60 Hz is sufficient to generate maximal force in a muscle (❯ Figure 7-16 shows a force-firing frequency curve for muscle), with approximately 100 Hz required for tetanus.

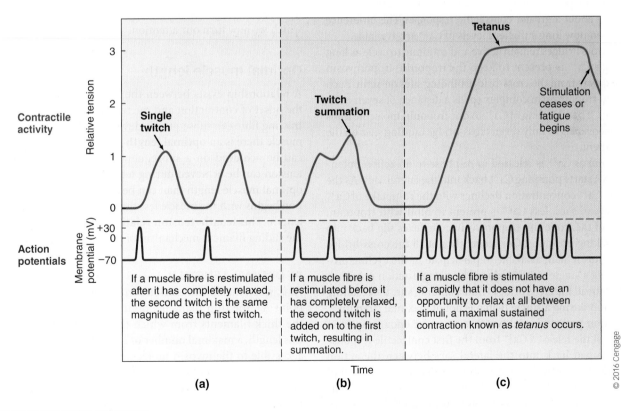

❯ FIGURE 7-15 **Summation and tetanus**

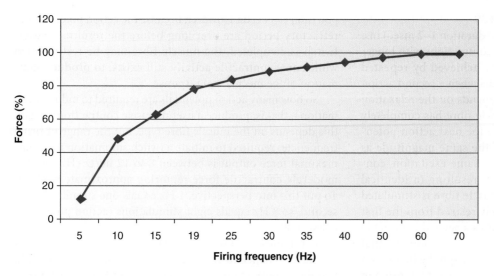

> FIGURE 7-16 **Force-firing frequency curve for human muscle.** As the firing frequency of the nerve increases, the force developed by the muscle increases, and at about 60 Hz no further increase in firing frequency will elicit a greater increase in force.

## Twitch summation at the cellular level

What is the mechanism of twitch summation and tetanus at the cell level? The tension produced by a contracting muscle fibre increases as a result of greater cross-bridge cycling. As the frequency of action potentials increases, the resulting tension development increases until a maximum tetanic contraction is achieved. Enough calcium is released in response to a single action potential to interact with all the troponin within the cell. As a result, all the cross bridges are free to participate in the contractile response. How, then, can repetitive action potentials bring about a greater contractile response? The difference depends on how long sufficient levels of $Ca^{2+}$ are available.

The cross bridges remain active and continue to cycle as long as enough $Ca^{2+}$ is present to keep the troponin–tropomyosin complex away from the cross-bridge binding sites on actin. Each troponin–tropomyosin complex spans a distance of seven actin molecules. Thus, binding of $Ca^{2+}$ to one troponin molecule leads to the uncovering of only seven cross-bridge binding sites on the thin filament.

As soon as $Ca^{2+}$ is released in response to an action potential, the SR starts pumping $Ca^{2+}$ back into the lateral sacs. As the cytosolic $Ca^{2+}$ concentration declines with the re-uptake of $Ca^{2+}$ by the lateral sacs, less $Ca^{2+}$ is present to bind with troponin, so some of the troponin–tropomyosin complexes slip back into their blocking positions. Consequently, not all the cross-bridge binding sites remain available to participate in the cycling process during a single twitch induced by a single action potential. Because not all the cross bridges find a binding site, the resulting contraction during a single twitch is not of maximal strength.

If action potentials and twitches occur far enough apart in time for all the released $Ca^{2+}$ from the first contractile response to be pumped back into the lateral sacs between the action potentials, an identical twitch response will occur as a result of the second action potential. If, however, a second action potential occurs and more $Ca^{2+}$ is released while the $Ca^{2+}$ that was released in response to the first action potential is being taken back up, the cytosolic $Ca^{2+}$ concentration remains elevated. This prolonged availability of $Ca^{2+}$ in the cytosol permits more of the cross bridges to continue participating in the cycling process for a longer time. As a result, tension development increases correspondingly. As the frequency of action potentials increases, the duration of elevated cytosolic $Ca^{2+}$ concentration increases, and contractile activity likewise increases until a maximum tetanic contraction is reached. With tetanus, the maximum number of cross-bridge binding sites remain uncovered so that cross-bridge cycling, and consequently tension development, is at its peak.

Because skeletal muscle must be stimulated by motor neurons to contract, the nervous system plays a key role in regulating contraction strength. The two main factors subject to control to accomplish gradation of contraction are the *number of motor units stimulated* and the *frequency of their stimulation*. The areas of the brain that direct motor activity combine tetanic contractions and precisely timed shifts of asynchronous motor unit recruitment to execute smooth rather than jerky contractions.

Additional factors not directly under nervous control also influence the tension developed during contraction. Among these is the length of the fibre at the onset of contraction, to which we now turn our attention.

## Optimal muscle length

A relationship exists between the length of the muscle before the onset of contraction and the tetanic tension that each contracting fibre can subsequently develop at that length. For every muscle there is an **optimal length** ($l_0$) at which maximal force can be achieved on a subsequent tetanic contraction. More tension can be achieved during tetanus when beginning at the optimal muscle length than can be achieved when the contraction begins with the muscle less than or greater than its optimal length. This **length–tension relationship** can be explained by the sliding filament mechanism of muscle contraction.

### AT OPTIMAL LENGTH ($L_0$)

At $l_0$, when maximum tension can be developed (> Figure 7-17, point A), the thin filaments optimally overlap the regions of the thick filaments from which the cross bridges project. At this length, a maximal number of cross-bridge binding sites are accessible to the myosin molecules for binding and bending. The central region of thick filaments, where the thin filaments do not overlap at $l_0$, lacks cross bridges; only myosin tails are found here.

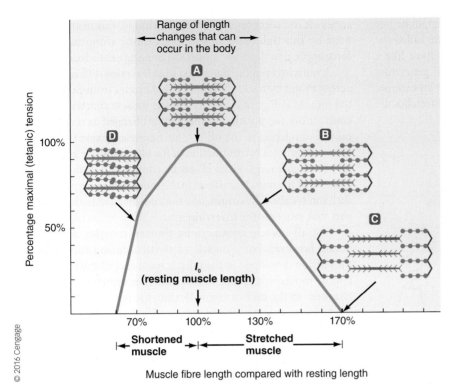

Range of length changes that can occur in the body

A

D

B

C

$l_0$ (resting muscle length)

Percentage maximal (tetanic) tension

100%

50%

70%    100%    130%    170%

Shortened muscle | Stretched muscle

Muscle fibre length compared with resting length

© 2016 Cengage

> **FIGURE 7-17 Length–tension relationship.** Maximal tetanic contraction can be achieved when a muscle fibre is at its optimal length ($l_0$) before the onset of contraction, because this is the point of optimal overlap of thick-filament cross bridges and thin-filament cross-bridge binding sites (point A). The percentage of maximal tetanic contraction that can be achieved decreases when the muscle fibre is longer or shorter than $l_0$ before contraction. When it is longer, fewer thin-filament binding sites are accessible for binding with thick-filament cross bridges, because the thin filaments are pulled out from between the thick filaments (points B and C). When the fibre is shorter, fewer thin-filament binding sites are exposed to thick-filament cross bridges because the thin filaments overlap (point D). Also, further shortening and tension development are impeded as the thick filaments become forced against the Z lines (point D). In the body, the resting muscle length is at $l_0$. Furthermore, because of restrictions imposed by skeletal attachments, muscles cannot vary beyond 30 percent of their $l_0$ in either direction (the range screened in light green). At the outer limits of this range, muscles still can achieve about 50 percent of their maximal tetanic contraction.

### AT LENGTHS GREATER THAN $L_0$

At greater lengths, as when a muscle is passively stretched (point B), the thin filaments are pulled out from between the thick filaments, decreasing the number of actin sites available for cross-bridge binding; that is, some of the actin sites and cross bridges no longer match up, so they go unused. When less cross-bridge activity can occur, less tension can develop. In fact, when the muscle is stretched to about 70 percent longer than its $l_0$ (point C), the thin filaments are completely pulled out from between the thick filaments; this prevents cross-bridge activity, and consequently, no contraction can occur.

### AT LENGTHS LESS THAN $L_0$

If a muscle is shorter than $l_0$ before contraction (point D), less tension can be developed for three reasons:

1. The thin filaments from the opposite sides of the sarcomere become overlapped, which limits the opportunity for the cross bridges to interact with actin.

2. The ends of the thick filaments become forced against the Z lines, so further shortening is impeded.

3. Besides these two mechanical factors, at muscle lengths less than 80 percent of $l_0$, not as much $Ca^{2+}$ is released during excitation–contraction coupling, for reasons unknown. Furthermore, by an unknown mechanism, the ability of $Ca^{2+}$ to bind to troponin and pull the troponin–tropomyosin complex aside is reduced at shorter muscle lengths. Consequently, fewer actin sites are uncovered for participation in cross-bridge activity.

### LIMITATIONS ON MUSCLE LENGTH

The extremes in muscle length that prevent development of tension occur only under experimental conditions, when a muscle is removed and stimulated at various lengths. In the body, the muscles are so positioned that their relaxed length is approximately their optimal length; thus, they can achieve near-maximal tetanic contraction most of the time. Because attachment to the skeleton imposes limitations, a muscle cannot be stretched or shortened more than 30 percent of its resting optimal length, and usually it deviates much less than 30 percent from normal length. Even at the outer limits (130% and 70% of $l_0$), the muscles still can generate half their maximum tension.

The factors we have discussed thus far that influence how much tension can be developed by a contracting muscle fibre—the frequency of stimulation and the muscle length at the onset of contraction—can vary from contraction to contraction. Other determinants of muscle fibre tension—the metabolic capability of the fibre relative to resistance to fatigue and the thickness of the fibre—do not vary from contraction to contraction but depend on the fibre type and can be modified over a period of time. After we complete our discussion of skeletal muscle mechanics, we will consider these metabolic factors in Section 7.5.

## Muscle tension and bone

**Tension** is produced internally within the sarcomeres, considered the **contractile component** of the muscle, as a result of cross-bridge activity and the resulting sliding of filaments. However, the sarcomeres are not attached directly to the bones. Instead, the tension generated by these contractile elements must be transmitted to the bone via the connective tissue sheaths and tendons before the bone can be moved. Connective tissue sheaths and tendon, as well as other components of the muscle, such as the intracellular titin, have a certain degree of passive elasticity. These noncontractile tissues are called the **series-elastic component** of the muscle; they behave like a stretchy spring placed between the internal tension-generating elements and the bone that is to be moved against an external load (› Figure 7-18). Shortening of the sarcomeres stretches the series-elastic component. Muscle tension is transmitted to the bone by this tightening of the series-elastic component. This force applied to the bone moves the bone against a load.

A muscle is typically attached to at least two different bones across a joint by means of tendons that extend from each end of the muscle (› Figure 7-19). When the muscle shortens during contraction, the position of the joint is changed as one bone is moved in relation to the other—for example, *flexion* (bending) of the elbow joint by contraction of the biceps muscle and *extension* (straightening) of the elbow by contraction of the triceps. The end of the muscle attached to the more stationary part of the skeleton is called the **origin**, and the end attached to the skeletal part that moves is the **insertion**.

Not all muscle contractions shorten muscles and move bones, however. For a muscle to shorten during contraction, the tension developed in the muscle must exceed the forces that oppose movement of the bone to which the muscle's insertion is attached. In the case of elbow flexion, the opposing force—the **load**—is the weight of an object being lifted. When you flex your elbow without lifting any external object, there is still a load, albeit a minimal one—the weight of your forearm being moved against the force of gravity.

Contractile component (sarcomeres)

Series-elastic component (connective tissue/tendon/intracellular titin)

Load

Load

› **FIGURE 7-18 Relationship between the contractile component and the series-elastic component in transmitting muscle tension to bone.** Muscle tension is transmitted to the bone by means of the stretching and tightening of the muscle's elastic connective tissue and tendon as a result of sarcomere shortening brought about by cross-bridge cycling.

© 2016 Cengage

## Isotonic and isometric contractions

At the level of the muscle fibre (motor unit) level, two primary types of contraction that depend on whether the muscle fibre changes length during contraction. In an **isotonic contraction** (force production is unchanged), muscle fibre tension remains constant as the muscle fibre changes length. In an *isometric contraction* (length is unchanged), the muscle fibre is prevented from shortening, so tension develops at constant muscle fibre length. The same internal events occur in both isotonic and isometric contractions: muscle fibre excitation turns on the tension-generating contractile process; the cross bridges start cycling; and filament sliding shortens the sarcomeres, which stretches the series-elastic component to exert a constant force.

## Dynamic and static contractions

In contrast to muscle contraction at the level of muscle fibres, the contraction at the level of the whole muscle (e.g., biceps) requires a different consideration because muscle force varies as the muscle contracts.

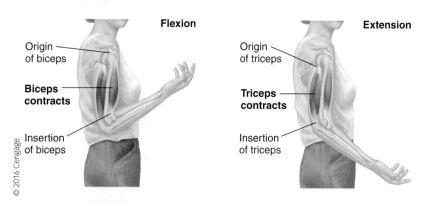

Flexion

Origin of biceps

**Biceps contracts**

Insertion of biceps

Extension

Origin of triceps

**Triceps contracts**

Insertion of triceps

© 2016 Cengage

› **FIGURE 7-19 Flexion and extension of the elbow joint**

As an example, consider your biceps. Assume you are going to lift an object. When the tension developing in your biceps becomes great enough to overcome the weight of the object in your hand, you can lift the object, and the whole muscle shortens in the process. The weight of the object does not change as it is lifted, but the force exerted by the muscle changes as the muscle shortens to accommodate the change in muscle length and the joint angle throughout the range of motion; this is a dynamic (changing-force) muscle contraction.

What happens if you try to lift an object too heavy for you (i.e., if the tension you are capable of developing in your arm muscles is less than required to lift the load)? In this case, the muscle cannot shorten and lift the object but remains at constant length despite the development of tension, so a *static* (not in motion) *contraction* occurs. A static contraction is a muscle contraction that produces an increase in muscle tension but does not result in a meaningful change in body position (limb or joint displacement). In addition to occurring when the load is too great, static contractions take place when the tension developed in the muscle is deliberately less than needed to move the load. In this case, the goal is to keep the muscle at fixed length, although it can develop more tension. These submaximal static contractions are important for maintaining posture (such as keeping the legs stiff while standing) and for supporting objects in a fixed position. During a given movement, a muscle (a whole muscle, e.g., a biceps muscle) may shift between dynamic and static contractions. For example, when you pick up a book to read, your biceps undergoes a dynamic contraction while you are lifting the book, but the contraction becomes static as you stop to hold the book in front of you.

### CONCENTRIC, ECCENTRIC, AND ISOMETRIC CONTRACTIONS

There are actually two types of dynamic contraction—*concentric* and *eccentric,* and one type of static contraction—*isometric.* In both cases of dynamic contraction, the whole muscle (e.g., biceps) changes length. The **concentric muscle contraction** is a dynamic contraction that produces tension during a shortening motion. This is the most familiar type of muscle contraction. During a concentric muscle contraction the actin filaments are pulled together by the myosin filaments, which move the Z lines closer together, shortening the sarcomere, and thus shortening the whole muscle. An example of a concentric muscle contraction is the action of the biceps muscle when you bend your arm at the elbow to raise a weight (e.g., a dumbbell or barbell). The **eccentric muscle contraction** is a dynamic contraction that produces tension while lengthening. In contrast to the concentric muscle contraction, during an eccentric muscle contraction the actin filaments are pulled apart, moving the Z lines farther from the centre and lengthening the sarcomere, which lengthens the whole muscle. An example of an eccentric muscle contraction is the action of the biceps muscle when you lower a weight by extending (i.e., straightening) your arm at the elbow.

In the case of static contraction, there is no change in muscle length. The **isometric contraction** is a contraction in which the muscle is activated, but instead of lengthening or shortening it is held at a constant length. An example of an isometric contraction is holding a box in front of you. Your arms (biceps muscles)

oppose gravity's downward pull on the box with equal upward force. Since your arms are neither raising nor lowering, your biceps are isometrically contracting, and there is no motion; hence, the term *static.*

### OTHER TYPES OF CONTRACTIONS

Skeletal muscles are not limited to pure isotonic and isometric contractions. Muscle length and tension frequently vary throughout a range of motion. Think about pulling back a bow and arrow. The tension of your biceps muscle continuously increases to overcome the progressively increasing resistance that occurs as you stretch the bow farther. At the same time, the muscle progressively shortens as you draw the bow farther back. Such a contraction occurs at neither constant tension nor constant length.

Some skeletal muscles do not attach to bones at both ends but still produce movement. For example, the tongue muscles are not attached at the free end. Isotonic contractions of the tongue muscles manoeuvre the free, unattached portion of the tongue to facilitate speech and eating. The external eye muscles attach to the skull at their origin but to the eye at their insertion. Isotonic contractions of these muscles produce the eye movements that enable us to track moving objects, read, and so on. A few skeletal muscles are completely unattached to bone and actually prevent movement. These are the voluntarily controlled rings of skeletal muscles—known as *sphincters*—that guard the exit of urine and feces from the body by isotonically contracting.

## Velocity of shortening

Force is also an important determinant of the **velocity**, or speed, of shortening (> Figure 7-20). During a concentric contraction, the greater the external force (load), the lower the velocity at which a single muscle fibre (or a constant number of contracting fibres within a muscle) shortens. The speed of shortening is maximal when there is no external force, progressively decreases with an

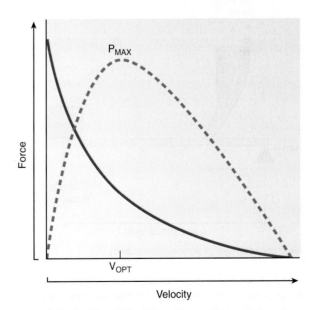

> **FIGURE 7-20 Force–velocity relationship in concentric contractions.** The velocity of shortening decreases as the force increases.

increasing force, and falls to zero (no shortening—isometric contraction) when the force cannot be overcome by maximal tetanic tension. You have frequently experienced this **force–velocity relationship**. You can lift light objects requiring little muscle tension quickly, whereas you can lift very heavy objects only slowly, if at all. This relationship between force and shortening velocity is a fundamental property of muscle, presumably because it takes the cross bridges longer to stroke against a greater force (load).

Whereas force and velocity for shortening are *inversely* related for *concentric* contractions, force and velocity for lengthening are *directly* related for *eccentric* contractions. The greater the external force stretching a muscle that is contracting to resist the stretch, the greater is the speed with which the muscle lengthens, likely because the load breaks some of the stroking cross-bridge attachments. It is the breaking of the cross bridges of the actin and myosin molecules during eccentric muscle contractions that is associated with increased muscle damage (trauma) and a phenomenon called *delayed onset muscle soreness*.

## Heat

Muscle accomplishes work in a physical sense only when an object is moved. **Work** is defined as force multiplied by distance. **Force** can be equated to the muscle tension required to overcome the load (the weight of the object). The amount of work accomplished by a contracting muscle therefore depends on how much an object weighs and how far it is moved. In an isometric contraction when no object is moved, the muscle contraction's efficiency as a producer of external work is zero. All energy consumed by the muscle during the contraction is converted to heat. In an isotonic contraction, the muscle's efficiency is about 25 percent. Of the energy consumed by the muscle during the contraction, 25 percent is realized as external work and the remaining 75 percent is converted to heat.

Much of this heat is not really wasted in a physiological sense because it is used in maintaining the body temperature. In fact, shivering—a form of involuntarily induced skeletal muscle contraction—is a well-known mechanism for increasing heat production on a cold day. Heavy exercise on a hot day, in contrast, may overheat the body because the normal heat-loss mechanisms may be unable to compensate for this increase in heat production (Chapter 16).

## Skeletal muscles, bones, and joints

Most skeletal muscles are attached to bones across joints, forming lever systems. A **lever** is a rigid structure capable of moving around a pivot point known as a **fulcrum**. In the body, the bones function as levers, the joints serve as fulcrums, and the skeletal muscles provide the force to move the bones. The portion of a lever between the fulcrum and the point where an upward force is applied is called the **power arm**; the portion between the fulcrum and the downward force exerted by a load is known as the **load arm** (> Figure 7-21a).

The most common type of lever system in the body is exemplified by flexion of the elbow joint. Skeletal muscles, such as the biceps, whose contraction flexes the elbow joint, consist of many parallel (side-by-side) tension-generating fibres that can exert a

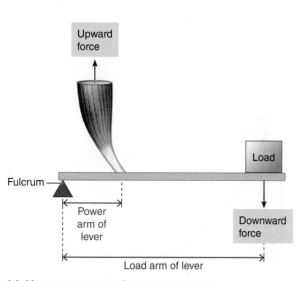

**(a)** Most common type of lever system in body

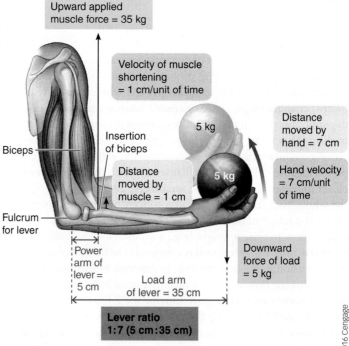

**(b)** Flexion of elbow joint as example of body lever action

© 2016 Cengage

> **FIGURE 7-21 Lever systems of muscles, bones, and joints.** (a) Schematic representation of the most common type of lever system in the body, showing the location of the fulcrum, the upward force, the downward force, the power arm, and the load arm. (b) Flexion of the elbow joint as an example of lever action in the body. Note that the lever ratio (length of the power arm to length of the load arm) is 1:7 (5 cm:35 cm), which amplifies the distance and velocity of movement seven times (distance moved by the muscle [extent of shortening] = 1 cm, distance moved by the hand = 7 cm; velocity of tening = 5 cm/unit of time), but at the expense of the muscle having to exert seven times the force of the load (muscle force = 35 kg, load = 5 kg).

The dashed line in Figure 7-20 is derived from multiplying the coordinates of each point of the force–velocity curve. Thus, despite maximal force at zero velocity, the power—the product of force and velocity—is also zero. Similarly, the power is also zero at maximal velocity because force is zero. Thus, power generation is less at slow and at very fast velocities of contraction, and it peaks at intermediate velocities. This explains why bicycles are multispeed. Cyclists, especially elite ones, maintain their pedalling frequency—equivalent to the velocity of contraction of the muscle fibres—in a range (typically 60–70 rpm) that maximizes power output. If, for example, the load increases when cycling up a hill, they shift to a lower gear ratio that allows them to maintain this pedalling frequency. Conversely, when cycling down a hill, they shift to a higher gear ratio (unless they coast to rest).

large force at their insertion but shorten only a small distance and at relatively slow velocity. The lever system of the elbow joint amplifies the slow, short movements of the biceps to produce more rapid movements of the hand that cover a greater distance.

Consider how an object weighing 5 kilograms is lifted by the hand (⟩ Figure 7-21b). When the biceps contracts, it exerts an upward force at the point where it inserts on the forearm bone about 5 cm away from the elbow joint, the fulcrum. Thus, the power arm of this lever system is 5 cm long. The length of the load arm, the distance from the elbow joint to the hand, averages 35 cm. In this case, the load arm is seven times as long as the power arm, which enables the load to be moved a distance seven times as great as the shortening distance of the muscle (while the biceps shortens a distance of 1 cm, the hand moves the load a distance of 7 cm) and at a velocity seven times as great (the hand moves 7 cm during the same length of time the biceps shortens 1 cm).

The disadvantage of this lever system is that at its insertion the muscle must exert a force seven times as great as the load. The product of the length of the power arm multiplied by the upward force applied must equal the product of the length of the load arm times the downward force exerted by the load. Because the load arm multiplied by the downward force is 35 cm × 5 kg, the power arm multiplied by the upward force must be 5 cm × 35 kg (the force that must be exerted by the muscle to be in mechanical equilibrium). Thus, skeletal muscles typically work at a mechanical disadvantage in that they must exert a considerably greater force than the actual load to be moved. Nevertheless, the amplification of velocity and distance afforded by the lever arrangement enables muscles to move loads faster over greater distances than would otherwise be possible. This amplification provides valuable manoeuvrability and speed.

Now let's shift attention from muscle mechanics to the metabolic means by which muscles power these movements.

# 7.5 | ATP and Fibre Types

## Contractile use of ATP

Three different steps in the contraction–relaxation process require ATP:

1. Splitting of ATP on the myosin head by myosin ATPase provides the energy for the power stroke of the cross bridge.

2. Binding (but not splitting) of a fresh molecule of ATP to the myosin head lets the bridge detach from the actin filament at the end of a power stroke so that the cycle can be repeated. This ATP is later split to provide energy for the next stroke of the cross bridge.

3. Active transport of $Ca^{2+}$ back into the SR during relaxation depends on energy derived from the breakdown of ATP.

Specific information on the metabolic pathways that generate ATP can be found in Chapter 2.

## Fatigue

Contractile activity in a particular skeletal muscle cannot be maintained at a given level indefinitely. Eventually, the tension in the muscle declines as fatigue sets in. There are two different types of fatigue: muscle fatigue (sometimes called peripheral fatigue) and central fatigue.

**Muscle fatigue** occurs when an exercising muscle can no longer respond to stimulation with the same degree of contractile activity. Muscle fatigue is a defence mechanism that protects a muscle from reaching a point at which it can no longer produce ATP. An inability to produce ATP would result in rigor mortis (obviously not an acceptable outcome of exercise). The underlying causes of muscle fatigue are unclear. The primary implicated factors include the following:

- The *local increase in ADP and inorganic phosphate* from ATP breakdown may directly interfere with cross-bridge cycling and/or block $Ca^{2+}$ release and uptake by the SR.

- *Accumulation of lactic acid* may inhibit key enzymes in the energy-producing pathways and/or excitation–contraction coupling process.

- The *accumulation of extracellular potassium* $(K^+)$ that occurs in the muscle when the $Na^+-K^+$ pump cannot actively transport $K^+$ back into the muscle cells as rapidly as this ion leaves during the falling phase of repeated action potentials (p. 63), causes a local reduction in membrane potential. This altered potential may decrease the release of $Ca^{2+}$ intracellularly by impairing coupling of the voltage-gated dihydropyridine receptors in the T tubules and the $Ca^{2+}$-release channels in the SR.

- *Depletion of glycogen energy reserves* may lead to muscle fatigue in exhausting exercise.

The time of onset of fatigue varies with the type of muscle fibre, as some fibres are more resistant to fatigue than others. With the intensity of the exercise, more rapid onset of fatigue is associated with high-intensity activities.

**Central fatigue** occurs when the CNS no longer adequately activates the motor neurons supplying the working muscles. The person slows down or stops exercising even though the muscles are still able to perform. Central fatigue often is psychologically based. During strenuous exercise, central fatigue may stem from discomfort associated with the activity; it takes strong motivation (a will to win) to deliberately persevere when in pain. In less strenuous activities, central fatigue may reduce physical performance in association with boredom and monotony (such as assembly-line work) or tiredness (lack of sleep). The mechanisms involved in central fatigue are poorly understood. In some cases, central fatigue may stem from biochemical insufficiencies within the brain.

*Neuromuscular fatigue* in exercise—an inability of active motor neurons to synthesize acetylcholine rapidly enough to sustain chemical transmission of action potentials from the motor neurons to the muscles—can be produced experimentally but does not occur under normal physiological conditions.

## Muscle fibre types within a single motor unit

The muscle fibres within a single motor unit are homogeneous in nature when they are compared on the basis of contractile and metabolic function. However, when muscle fibres from different motor units are compared, we find differences in contractile and metabolic function. In terms of contractility, muscle fibres can be classified as either **slow-twitch fibres** or **fast-twitch fibres,** and muscle fibre metabolic activity can be defined as either **glycolytic** or **oxidative**.

The slow-twitch muscle fibres are also known as **type I muscle fibres**. Type I fibres contract and relax more slowly than fast-twitch fibres, also called **type II muscle fibres**. When considering contractile (twitch) characteristics, we must also consider the motor input to the muscle fibre and the ATPase that is associated with the muscle fibre. Skeletal muscle fibre is innervated by alpha ($\alpha$) motor neurons, but different $\alpha$ motor neurons innervate type I and type II fibres. Type I fibres are innervated by $\alpha_2$ motor neurons, whereas type II fibres are innervated by $\alpha_1$ motor neurons, which are larger than $\alpha_2$ motor neurons. This is important because the smaller the motor neuron, the lower the activation threshold and the slower the conduction velocity of that motor unit. As a result, type I (slow-twitch) fibres are activated at lower work intensities and contract more slowly than the $\alpha_1$ innervated type II fibres. Type II (fast-twitch) muscle fibres have a maximum shortening (contraction) velocity approximately 10 times greater than type I (slow-twitch) fibres.

Additionally, the importance of motor input to the muscle has been demonstrated via cross-innervation studies. If motor input to type I ($\alpha_2$) and type II ($\alpha_1$) fibres is reversed (switched), the contractile qualities of each muscle fibre will switch as well. Thus, type I fibres will take on the qualities of type II fibres, and vice

versa. This indicates that the contractile quality of muscle fibre is strongly influenced by the type of motor input and that muscle fibre has great plasticity. However, another important piece for understanding the contractile quality of the muscle fibre is the type of myosin ATPase located on the head of the myosin molecule. The type of ATPase correlates with the maximum velocity of shortening of the fast- and slow-twitch muscle fibres. Recall that there are different isoforms (isoforms contain the same proteins but with some small differences) of ATPase located on the myosin molecule: type II fibres have the fast form, and type I have the slow form. The difference in these ATPase isoforms is helpful in the process of muscle fibre typing via the use of a **muscle biopsy**. A muscle biopsy is the process of removing cells or tissue (muscle) using a needle for the purpose of examination under a microscope.

The metabolic properties of the two primary fibre types include oxidative and glycolytic. All muscle fibres produce energy both aerobically and anaerobically, but typically one metabolic pathway predominates within each muscle fibre (and motor unit). It is best to think of the metabolic qualities of muscle fibre as a continuum between oxidative (aerobic) and glycolytic (anaerobic). The metabolic qualities of muscle fibres are determined by staining for key glycolytic and oxidative enzymes. Type I (slow-twitch) fibres are also called *slow oxidative* because they produce their energy by aerobic processes. However, type II (fast-twitch) muscle fibres are known as either *fast oxidative glycolytic* (also known as FOG, FTa, type IIa) or *fast glycolytic* (also known as FG, FTb, type IIb, or IIx), depending on the type of enzymes associated with their metabolic processes. (See Table 7-2 for an overview of the characteristics associated with the three muscle fibre types: type I, type IIa, and type IIx.)

Depending on the muscle fibre classification method used (e.g., biochemical or myosin heavy chain), research indicates that humans do not possess type IIx, but instead have a type IIx/d. Some researchers suggest that type IIx/d (or type IIx) and type IIb are the same, but that animals have type IIb and humans possess type IIx/d. (See Ennion, S., Sant'ana Pereira, J., Sargeant, A.J., et al. [1995]. Characterization of human skeletal muscle fibres according to the myosin heavy chains they express. *J Muscle Res Cell Motil,* 16[1]: 35–43; Scott, W., Stevens, J., & Binder-Macleod, S.A. [2001]. Human skeletal muscle fiber type classifications. *Phys Ther, 81*: 1810–16.)

Other related characteristics distinguish these three fibre types—slow oxidative, fast oxidative glycolytic, and fast glycolytic. Oxidative fibres, both slow and fast, contain an abundance of mitochondria, the organelles that house the enzymes involved in oxidative phosphorylation. Because adequate oxygenation is essential to support this pathway, these fibres are richly supplied with capillaries. Oxidative fibres also have a high myoglobin content. **Myoglobin** not only helps support oxygen dependency of oxidative fibres but also gives them a red colour, just as oxygenated haemoglobin produces the red colour of arterial blood. Accordingly, these muscle fibres are called **red fibres**.

In contrast, the fast fibres specialized for glycolysis contain few mitochondria but have a high content of glycolytic enzymes instead. Also, to supply the large amounts of glucose needed for glycolysis, they contain a lot of stored glycogen. Because the glycolytic fibres need relatively less oxygen to function, they have

## TABLE 7-2 The Structural and Functional Characteristics Associated with the Two Types of Skeletal Muscle Fibres

| | TYPE I | TYPE II | |
|---|---|---|---|
| | ST | FTa | FTx |
| **Contractile Metabolic** | SO | FOG | FG |
| **Structural Properties** | | | |
| *Muscle fibre diameter* | Small | Largest | Large |
| *Mitochondrial density* | High | High | Low |
| *Capillary: fibre ratio* | High | High | Low |
| *Motor units per muscle* | More, smaller | Fewer, larger | Fewer, larger |
| *SR development* | Poor | Intermediate | High |
| *Z line development* | Intermediate | Wide | Narrow |
| **Functional Properties** | | | |
| *Twitch (contraction) time* | Slow | Fast | Fast |
| *Relaxation time* | Slow | Intermediate | Fast |
| *Force production* | Low | Intermediate | High |
| *Fatigability* | Fatigue-resistant | Fatigable | Most |
| *Axonal conduction velocity* | Slow | Intermediate | Fatigable |
| *Stimulus needed for peak tension* | Low | Intermediate | Fast |
| *Sensitivity to recruitment* | High | Intermediate | High |
| **Metabolic Properties** | | | |
| *Red colour* | Dark | Dark | Pale |
| *Phosphocreatine stores* | Low | High | High |
| *Glycogen stores* | Low | Intermediate | High |
| *Triglyceride stores* | High | Intermediate | Low |
| *Glycolytic enzyme activity* | Low | Intermediate | High |
| *Krebs cycle enzyme activity* | High | Medium | Low |
| *Creatine kinase activity* | Low | Medium | High |
| *Myoglobin content* | High | Medium | Low |
| *Myosin–ATPase activity* | Low | Intermediate | High |

**Source:** S.A. Plowman and D.L. Smith, *Exercise Physiology: for health, fitness, and performance*, 2nd edition, p. 515, 2008, by Lippincott Williams & Wilkins. www.lww.com.



**Clinical Connections**

McArdle's disease patients lack the enzyme myophosphorylase that catalyzes the first step in the breakdown of glycogen. Since glycogen is the primary energy source for fast, or white, fibres, and these fibres have a lesser capillary supply to deliver energy substrates, as compared to red fibres, they are more affected by this disease.

only a meagre capillary supply compared with the oxidative fibres. The glycolytic fibres contain very little myoglobin and therefore are pale in colour, so they are sometimes called **white fibres**. (The most readily observable comparison between red and white fibres is the dark and white meat in poultry.)

### GENETIC ENDOWMENT OF MUSCLE FIBRE TYPES

In humans, most muscles contain a mixture of all three fibre types; the percentage of each type is largely determined by the type of activity for which the muscle is specialized. Accordingly, a high proportion of the slow-oxidative fibres is found in muscles specialized for maintaining low-intensity contractions for long periods of time without fatigue, such as the muscles of the back and legs that support the body's weight against the force of gravity. A preponderance of fast-glycolytic fibres are found in the arm muscles, which are adapted for performing rapid, forceful movements, such as lifting heavy objects.

The percentage of these various fibres not only differs between muscles within an individual but also varies considerably among individuals. Athletes genetically endowed with a higher percentage of the fast-glycolytic fibres are good candidates for **anaerobic (high-intensity) exercise**, such as power and sprint events, whereas those with a greater proportion of slow-oxidative fibres are more likely to succeed in endurance activities, such as marathon races.

Of course, success in any event depends on many factors other than genetic endowment, such as the extent and type of training and the level of dedication. Indeed, the mechanical and metabolic capabilities of muscle fibres can change a lot in response to the patterns of demands placed on them. Let's see how.

## Muscle fibre adaptation (plasticity)

Different types of exercise produce different patterns of neuronal discharge to the muscle involved. Depending on the pattern of neural activity, long-term adaptive changes occur in the muscle fibres, enabling them to respond most efficiently to the types of demands placed on the muscle. Therefore, skeletal muscle has a high degree of *plasticity* (p. 120). Two types of changes can be induced in muscle fibres: changes in their ATP synthesizing capacity, and changes in their diameter.

### IMPROVEMENT IN OXIDATIVE CAPACITY

Regular **aerobic (endurance) exercise**, such as long-distance jogging or cycling, induces metabolic changes within the oxidative fibres. Physiological changes associated with long, slow exercise include increases in the number of mitochondria and increases in the number of capillaries supplying blood to these fibres. Muscles so adapted can use oxygen more efficiently and therefore can better endure prolonged activity without fatiguing. However, they do not change in size and may even atrophy following prolonged (chronic) periods of endurance training. Interestingly, evidence from McMaster University in Canada suggests that the

physiological changes in muscle thought only possible from long-distance exercise, can also be gained following high-intensity interval training (HIIT) (Gibala, M.J., McGee, S.L. [2008]. Metabolic adaptations to short-term high-intensity interval training: A little pain for a lot of gain? *Exerc & Sport Sci Rev, 36*[2]: 58–63).

### MUSCLE HYPERTROPHY AND RESISTANCE TRAINING

Hypertrophy is an increase in mass or girth of a muscle and can be induced by a number of stimuli. The most familiar of these stimuli is exercise, specifically, resistance exercise (resistance training or weightlifting). For hypertrophy to occur, the message must filter down to alter the pattern of protein expression. Most research suggests that it takes approximately eight weeks of regular resistance training for actual hypertrophy to begin. However, recently, some data indicate that a small amount of hypertrophy may occur in as little as two weeks. Regardless of the precise timeline, the mechanism of hypertrophy is centred on the addition of contractile proteins to existing myofibrils. These events appear to occur within each muscle fibre (i.e., muscle cell) exercised (receives the stimuli). That is, hypertrophy results primarily from the growth of each muscle fibre, rather than an increase in the number of fibres (hyperplasia).

### MUSCLE ATROPHY AND DISUSE

 Muscle **atrophy** is the loss of muscle's mass, and therefore it is the opposite of muscle hypertrophy. More specifically it is a progressive withdrawal of anabolism and an increased catabolism. **Disuse atrophy** takes place when the skeletal muscles are not physically stressed on a regular basis (e.g., through weightlifting or some type of regular exercise) even though the nerve supply is intact, as when a cast or brace must be worn or during prolonged bed confinement. **Denervation atrophy** occurs after the nerve supply to a muscle is lost. If the muscle is stimulated electrically until innervation can be re established, such as during regeneration of a severed peripheral nerve, atrophy can be diminished, but not entirely prevented. Other factors that contribute to atrophy are hormonal, immunological, and nutritional. Atrophy is sometimes referred to as sarcopenia, but the term *sarcopenia* generally reflects reduced muscle regenerative ability (the reduced regeneration may be related to aging and a reduction in anabolic hormones [e.g., IGF] or disease [e.g., amyotrophic lateral sclerosis]). The end result of muscle atrophy is a decline in muscle force and power.

---

**▌Why It Matters**
**The Benefits of Resistance Training**

Multiple sclerosis (Ms) is a chronic disease that attacks the CNS, with symptoms ranging from mild (e.g., numbness in the limbs) to severe (e.g., paralysis). MS is unpredictable and varies from person to person in severity and symptoms. The symptoms include fatigue, occasionally severe enough that activities of daily living (e.g., shopping) cannot be performed independently, and weakness in one or more extremities. It appears that resistance training may be beneficial in the treatment of MS. Research has shown that improved lower-limb strength

via resistance training translates into improved walking gait. Strength increases in the lower limbs likely result from neurological (e.g., synchronous recruitment pattern) changes, as well as changes in skeletal muscle fibre characteristics (e.g., hypertrophic and fibre type). Following resistance training, the gait characteristics of persons with MS more closely resemble the typical gait of those without MS. Thus, chronic resistance training may be beneficial in maintaining important characteristics of skeletal muscle function.

Functionally, reduced muscular strength and power result in a failure to preserve mobility, quality of life, and the capacity to recover from illness as we age. However, current understanding of the processes that regulate muscle mass during exercise, disuse, and rehabilitation is still somewhat unclear. For example, certain poorly understood factors released from active nerve endings, perhaps packaged with the ACh vesicles, apparently contribute to the integrity and growth of muscle tissue. Research does acknowledge that regular resistance training (e.g., weightlifting) is an important corrective measure for disuse atrophy and age-related declines in skeletal muscle mass (see also Why It Matters, p. 320).

### INTERCONVERSION BETWEEN FAST MUSCLE TYPES

As previously mentioned, all the muscle fibres within a single motor unit are of the same fibre type. This pattern usually is established early in life, but the two types of fast-twitch fibres are interconvertible, depending on training efforts. Remember that fast-twitch muscle fibres are on a continuum ranging from highly glycolytic to more oxidative. That is, fast-glycolytic fibres can be converted to fast-oxidative fibres, and vice versa, depending on the types of demands repetitively placed on them. Adaptive changes in skeletal muscle gradually reverse to their original state over a period of months if the regular exercise program that induced these changes is discontinued. The default fast-twitch fibre appears to be FTb (fast-twitch glycolytic), as demonstrated in animal research (spinal transection), as well as in detraining studies (human and animal) and in the natural aging process.

Slow and fast fibres are not interconvertible, however. Although training can induce changes in muscle fibres' metabolic support systems, whether a fibre is fast or slow twitch depends on the fibre's nerve supply. Slow-twitch fibres are supplied by motor neurons that exhibit a low-frequency pattern of electrical activity, whereas fast-twitch fibres are innervated by motor neurons that display intermittent rapid bursts of electrical activity. Experimental switching of motor neurons supplying slow muscle fibres with those supplying fast fibres gradually reverses the speed at which these fibres contract.

It appears that exercise—for example, resistance training—does not increase or decrease the percentage of type I fibres a person possesses. However, in the fast-twitch (type II) muscle fibres there is some interconversion with resistance training. Resistance training has been shown to shift type IIx to type IIa fibres. Generally, this means that following chronic resistance training, the percentage of type IIa increases, whereas the percentage of type IIx decreases.

### LIMITED REPAIR OF MUSCLE

When a muscle is damaged, some limited repair is possible, even though muscle cells cannot divide mitotically to replace lost cells. A small population of inactive muscle-specific stem cells called **satellite cells** are located close to the muscle surface. When a muscle fibre is damaged, locally released factors activate the satellite cells, which divide to give rise to myoblasts. A group of myoblasts fuse to form a large, multinucleated cell, which immediately begins to synthesize and assemble the intracellular machinery characteristic of the muscle, ultimately differentiating completely into a mature muscle fibre.

Transplantation of either satellite cells or myoblasts provides one of several glimmers of hope for victims of **muscular dystrophy**—a hereditary pathological condition characterized by progressive degeneration of contractile elements, which are ultimately replaced by fibrous tissue. (See Concepts, Challenges, and Controversies, for further information on this devastating condition.)

> **▮ Clinical Connections**
>
> When tissues are injured several compounds or biomarkers are released into the blood stream. One of the biomarkers specific for muscle tissue is called myoglobin. Myoglobin is the oxygen-binding protein in muscle and is related to haemoglobin, the oxygen-binding protein in blood that we discuss in Chapter 12. Myoglobin is excreted in the urine. The darkened urine that David noticed after he exercised means he had damaged his muscles. Such repetitive injury can reduce muscle mass over time.

---

**▮ TABLE 7-3 Determinants of Whole-Muscle Tension in Skeletal Muscle**

| Number of Fibres Contracting | Tension Developed by Each Contracting Fibre |
|---|---|
| Number of motor units recruited* | Frequency of stimulation (twitch summation and tetanus)* |
| Number of muscle fibres per motor unit | Length of fibre at onset of contraction (length–tension relationship) |
| Number of muscle fibres available to contract | Extent of fatigue |
|   – Size of muscle (number of muscle fibres in muscle) |   – Duration of activity |
|   – Presence of disease (e.g., muscular dystrophy) |   – Amount of asynchronous recruitment of motor units |
|   – Extent of recovery from traumatic losses |   – Type of fibre (fatigue-resistant oxidative or fatigue-prone glycolytic) |
| | Thickness of fibre |
| |   – Pattern of neural activity (hypertrophy, atrophy) |
| |   – Amount of testosterone (larger fibres in males than females) |

*Factors controlled to accomplish gradation of contraction.

## Muscular Dystrophy: When One Small Step Is a Big Deal

**H**OPE OF TREATMENT IS ON THE HORIZON for *muscular dystrophy* (MD), a muscle-wasting disease that primarily strikes boys. The most severe form can relentlessly lead to death around age 20.

### Symptoms

Muscular dystrophy encompasses more than 30 distinct hereditary pathological conditions. It is part of a group of genetic, hereditary muscle diseases that are associated with a progressive weakening of the skeletal muscle. People with muscular dystrophy may have progressive skeletal muscle weakness, defects in muscle proteins, and the death of muscle cells. The gradual muscle wasting is characterized by progressive weakness over a period of years. Typically, a Duchenne MD patient begins to show symptoms of muscle weakness at about 2 to 3 years of age, becomes wheelchair bound when he is 10 to 12 years old, and dies within the next 10 years, either from respiratory failure when his respiratory muscles become too weak or from heart failure when his heart becomes too weak.

### Cause

MD is caused by a recessive genetic defect on the X sex chromosome, of which males have only one copy. (Males have XY sex chromosomes; females have XX sex chromosomes.) If a male inherits from his mother an X chromosome bearing the defective dystrophic gene, he is destined to develop the disease. Duchenne MD affects 1 out of every 3500 boys worldwide. To acquire the condition, females must inherit a dystrophic-carrying X gene from both parents, a much rarer occurrence.

The defective gene responsible for *Duchenne muscular dystrophy* (DMD), the most common and most devastating form of the disease, was pinpointed in 1986. The gene normally produces **dystrophin**, a large protein that provides structural stability to the muscle cell's plasma membrane. Dystrophin is part of a complex of membrane-associated proteins that form a mechanical link between actin, a major component of the muscle cell's internal cytoskeleton, and the extracellular matrix, an external support network (p. 34). This mechanical reinforcement of the plasma membrane enables the muscle cell to withstand the stresses and strains encountered during contraction and stretching.

Dystrophic muscles are characterized by a lack of dystrophin. Even though this protein represents only 0.002 percent of the total amount of skeletal muscle protein, its presence is crucial in maintaining the integrity of the muscle cell membrane. The absence of dystrophin permits a constant leakage of $Ca^{2+}$ into the muscle cells. This $Ca^{2+}$ activates proteases, protein-snipping enzymes that harm the muscle fibres. The resultant damage leads to the muscle wasting and ultimate fibrosis that characterize the disorder.

The discovery of the dystrophin gene and its deficiency in DMD brought the hope that scientists could somehow replenish this missing protein in the muscles of the disease's young victims. Although the disease is still considered untreatable and fatal, several lines of research are being pursued vigorously to intervene in the relentless muscle loss.

### Gene-Therapy Approach

One approach is a possible "gene fix." With gene therapy, healthy genes are usually delivered to the defective cells by means of viruses. Viruses operate by invading a body cell and micromanaging the cell's genetic machinery. In this way, the virus directs the host cell to synthesize the proteins needed for viral replication. With gene therapy, the desired gene is inserted into an incapacitated virus that cannot cause disease but can still enter the target cell and take over genetic commands.

One of the big challenges for gene therapy for DMD is the enormity of the dystrophin gene. This gene, more than 3 million base pairs long, is the largest gene ever found. It will not fit inside the viruses usually used to deliver genes to cells—they only have enough space for a gene one-thousandth the size of the dystrophin gene. Therefore, researchers have created a minigene that is a thousand times

---

We have now completed our discussion of all the determinants of whole-muscle tension in a skeletal muscle, which are summarized in ▌ Table 7-3. In the remaining section on skeletal muscle, we examine the central and local mechanisms involved in regulating the motor activity performed by these muscles.

### Check Your Understanding 7.4

1. Describe the differences between muscle fibres in a turkey drumstick (slow-oxidative) and muscles fibres in turkey breast meat (fast-glycolytic).

2. Discuss the relative contributions of creatine phosphate, glycolysis, and oxidative phosphorylation to the production of ATP during running of a marathon.

## 7.6 | Control of Motor Movement

Particular patterns of motor unit output govern motor activity, ranging from maintenance of posture and balance to stereotypical locomotor movements, such as walking, to individual, highly skilled motor activity, such as gymnastics. Control of any motor movement, no matter what its level of complexity, depends on converging input to the motor neurons of specific motor units. The motor neurons, in turn, trigger contraction of the muscle fibres within their respective motor units by means of the events that occur at the neuromuscular junction.

In this section we examine the neural control of motor movement. We begin with a look at the different levels of input that

smaller than the dystrophin gene but still contains the essential components for directing the synthesis of dystrophin. This stripped-down minigene can fit inside the viral carrier. Injection of these agents has stopped and even reversed the progression of MD in experimental animals. Gene-therapy clinical trials in humans have not been completed.

### Cell-Transplant Approach

Another approach involves injecting cells that can functionally rescue the dystrophic muscle tissue. Myoblasts are undifferentiated cells that fuse to form the large, multinucleated skeletal muscle cells during embryonic development. After development, a small group of stem cells known as *satellite cells* remain close to the muscle surface. Satellite cells can be activated to form myoblasts, which can fuse together to form a new skeletal muscle cell to replace a damaged cell. When the loss of muscle cells is extensive, however, as in DMD, this limited mechanism is not adequate to replace all the lost fibres.

One therapeutic approach for DMD under study involves the transplantation of dystrophin-producing myoblasts harvested from muscle biopsies of healthy donors into the patient's dwindling muscles. Other researchers are pinning their hopes on delivery of satellite cells or partially differentiated adult stem cells that can be converted into healthy muscle cells (p. 11). The aetiology and treatments for MD are under investigation by Drs. J. Tremblay (Human Genetics Institute, University of Laval) and R. Worton (Ottawa Health Research Institute). Dr. Tremblay conducts research associated with cell transplant, and Dr. Worton's research is in the area of gene therapy.

### Utrophin Approach

An alternative strategy that holds considerable promise for treating DMD is upregulation of **utrophin**, a naturally occurring protein in muscle that is closely related to dystrophin. Eighty percent of the amino acid sequence for dystrophin and utrophin is identical, but these two proteins normally have different functions. Whereas dystrophin is dispersed throughout the muscle cell's surface membrane, where it contributes to the membrane's structural stability, utrophin is concentrated at the motor end plate. Here, utrophin plays a role in anchoring the acetylcholine receptors.

When researchers genetically engineered dystrophin-deficient mice that produced extra amounts of utrophin, this utrophin upregulation compensated in large part for the absent dystrophin; that is, the additional utrophin dispersed throughout the muscle cell membrane, where it assumed dystrophin's responsibilities. The result was improved intracellular $Ca^{2+}$ homeostasis, enhanced muscle strength, and a marked reduction in the microscopic signs of muscle degeneration. Researchers are now scrambling to find a drug that will entice muscle cells to overproduce utrophin in humans, in the hopes of preventing or even repairing the muscle wasting that characterizes this devastating condition.

These steps toward an eventual treatment mean that hopefully one day the afflicted boys will be able to take steps on their own instead of being destined to wheelchairs and early death.

### Further Reading

Biggar, W.D., Klamut, H.J., Demacio, P.C., Stevens, D.J., & Ray, P.N. (2002). Duchenne muscular dystrophy: current knowledge, treatment, and future prospects. *Clin Orthop Relat Res, 401*: 88–106.

Price, F.D., Kuroda, K., & Rudnicki, M.A. (2007). Stem cell based therapies to treat muscular dystrophy. *Biochim Biophys Acta, 1772* (2): 272–83.

Roland, E.H. (2000). *Muscular dystrophy. Pediatr Rev, 21* (7): 233–37; quiz 238.

**7**

control motor output and discuss how these are integrated. Then we take a closer look at the skeletal muscle receptors that form the afferent inputs.

## Neural input

Three levels of input control motor-neuron output (❯ Figure 7-22).

1. *Input from afferent neurons*, usually through intervening interneurons, at the level of the spinal cord—that is, spinal reflexes (p. 107).

2. *Input from the primary motor cortex*. Fibres originating from neuronal cell bodies, known as **pyramidal cells,** within the primary motor cortex (p. 115) descend directly without synaptic interruption to terminate on motor neurons (or on local interneurons that terminate on motor neurons) in the spinal cord. These fibres make up the **corticospinal (pyramidal) motor system**.

3. *Input from the brain stem* as part of the multineuronal motor system. The pathways composing the **multineuronal (extrapyramidal) motor system** include a number of synapses that involve many regions of the brain (*extra* means "outside of"; *pyramidal* refers to the pyramidal system). The final link in multineuronal pathways is the brain stem, especially the reticular formation (p. 110), which in turn is influenced by motor regions of the cortex,

the cerebellum, and the basal nuclei. In addition, the motor cortex itself is interconnected with the thalamus as well as with premotor and supplementary motor areas, all part of the multineuronal system.

The only brain regions that directly influence motor neurons are the primary motor cortex and brain stem; the other involved brain regions indirectly regulate motor activity by adjusting motor output from the motor cortex and brain stem. A number of complex interactions take place among these various brain regions; the most important are represented in › Figure 7-22. (See Chapter 3 for further discussion of the specific roles and interactions of these brain regions.)

Spinal reflexes involving afferent neurons are important in maintaining posture and in executing basic protective movements, such as the withdrawal reflex. The corticospinal system

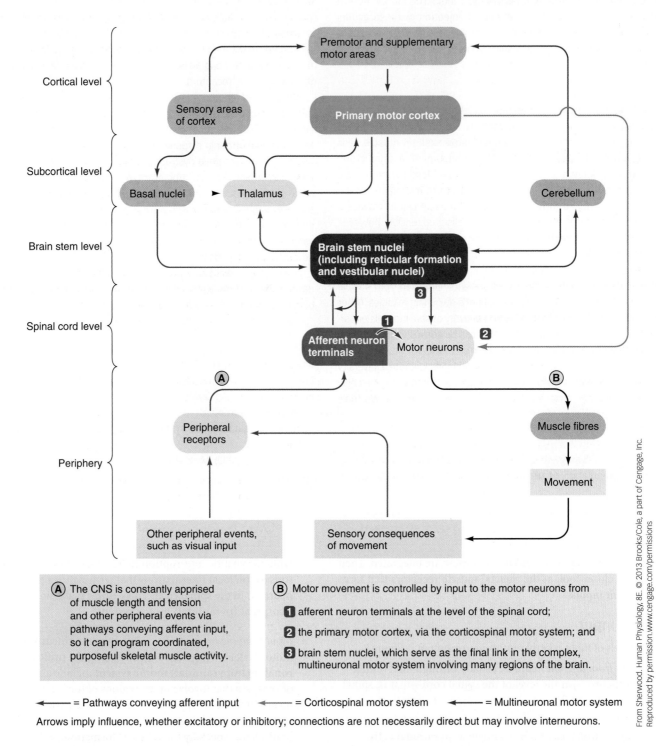

(A) The CNS is constantly apprised of muscle length and tension and other peripheral events via pathways conveying afferent input, so it can program coordinated, purposeful skeletal muscle activity.

(B) Motor movement is controlled by input to the motor neurons from

1 afferent neuron terminals at the level of the spinal cord;

2 the primary motor cortex, via the corticospinal motor system; and

3 brain stem nuclei, which serve as the final link in the complex, multineuronal motor system involving many regions of the brain.

◄——— = Pathways conveying afferent input  ◄——— = Corticospinal motor system  ◄——— = Multineuronal motor system

Arrows imply influence, whether excitatory or inhibitory; connections are not necessarily direct but may involve interneurons.

› **FIGURE 7-22 Motor control**

primarily mediates performance of fine, discrete, voluntary movements of the hands and fingers, such as those required for doing intricate needlework. Premotor and supplementary motor areas, with input from the cerebrocerebellum, plan the voluntary motor command that is issued to the appropriate motor neurons by the primary motor cortex through this descending system. The multineuronal system, in contrast, primarily regulates overall body posture involving involuntary movements of large muscle groups of the trunk and limbs. The corticospinal and multineuronal systems show considerable complex interaction and overlapping of function. To voluntarily manipulate your fingers to do needlework, for example, you subconsciously assume a particular posture of your arms that lets you hold your work.

Some of the inputs converging on motor neurons are excitatory, whereas others are inhibitory. Coordinated movement depends on an appropriate balance of activity in these inputs. The following types of motor abnormalities result from defective motor control:

- *Clinical Note* If an inhibitory system originating in the brain stem is disrupted, muscles become hyperactive because of the unopposed activity in excitatory inputs to motor neurons. This condition, characterized by increased muscle tone and augmented limb reflexes, is known as **spastic paralysis**.

- In contrast, loss of excitatory input, such as that accompanying destruction of descending excitatory pathways exiting the primary motor cortex, brings about **flaccid paralysis**. In this condition, the muscles are relaxed, and the person cannot voluntarily contract muscles, although spinal reflex activity is still present. Damage to the primary motor cortex on one side of the brain, as with a stroke, leads to flaccid paralysis on the opposite half of the body (**hemiplegia**, or paralysis of one side of the body). Disruption of all descending pathways, as in traumatic severance of the spinal cord, produces flaccid paralysis below the level of the damaged region—**tetraplegia** (*a.k.a.* quadriplegia; paralysis of all four limbs) in upper spinal cord damage and **paraplegia** (paralysis of the legs) in lower spinal cord injury.

- Destruction of motor neurons—either their cell bodies or efferent fibres—causes flaccid paralysis and lack of reflex responsiveness in the affected muscles.

- Damage to the cerebellum or basal nuclei does not result in paralysis but instead in uncoordinated, clumsy activity and inappropriate patterns of movement. These regions normally smooth out activity initiated voluntarily.

- Damage to higher cortical regions involved in planning motor activity results in the inability to establish appropriate motor commands to accomplish desired goals.

## Afferent information

Coordinated, purposeful skeletal muscle activity depends on afferent input from a variety of sources. At a simple level, afferent signals indicating that your finger is touching a hot stove trigger reflex contractile activity in appropriate arm muscles to withdraw the hand from the injurious stimulus. At a more complex level, if you are going to catch a ball, the motor systems of your brain must program sequential motor commands that will move and position your body correctly for the catch, using predictions of the ball's direction and rate of movement provided by visual input. Many muscles acting simultaneously or alternately at different joints are called into play to shift your body's location and position rapidly, while maintaining your balance in the process. It is critical to have ongoing input about your body position with respect to the surrounding environment, as well as the position of your various body parts in relationship to one another. This information is necessary for establishing a neuronal pattern of activity to perform the desired movement. To appropriately program muscle activity, your CNS must know the starting position of your body. Further, it must be constantly informed about the progression of movement it has initiated, so that it can make adjustments as needed. Your brain receives this information, which is known as proprioceptive input (p. 117), from receptors in your eyes, joints, vestibular apparatus, and skin, as well as from the muscles themselves.

You can demonstrate your joint and muscle proprioceptive receptors in action by closing your eyes and bringing the tips of your right and left index fingers together at any point in space. You can do so without seeing where your hands are, because your brain is informed of the position of your hands and other body parts at all times by afferent input from the joint and muscle receptors.

Two types of muscle receptors—*muscle spindles* and *Golgi tendon organs*—monitor changes in muscle length and tension. Muscle length is monitored by muscle spindles; changes in muscle tension are detected by Golgi tendon organs. Both these receptor types are activated by muscle stretch, but they convey different types of information. Let us see how.

### MUSCLE SPINDLE STRUCTURE

**Muscle spindles**, which are distributed throughout the fleshy part of a skeletal muscle, consist of collections of specialized muscle fibres known as **intrafusal fibres**. These fibres lie within spindle-shaped connective tissue capsules parallel to the **extrafusal fibres** (*fusus* means "spindle") (❯ Figure 7-23). Unlike an extrafusal skeletal muscle fibre, which contains contractile elements (myofibrils) throughout its entire length, an intrafusal fibre has a noncontractile central portion, with the contractile elements being limited to both ends.

Each muscle spindle has its own private efferent and afferent nerve supply. The efferent neuron that innervates a muscle spindle's intrafusal fibres is known as a **gamma motor neuron**, whereas the motor neurons that supply the extrafusal fibres are called **alpha motor neurons**. Two types of afferent sensory endings terminate on the intrafusal fibres and serve as muscle spindle receptors, both of which are activated by stretch. The **primary (annulospiral) endings** are wrapped around the central portion of the intrafusal fibres; they detect changes in the length of the fibres during stretching as well as the speed with which it occurs. The **secondary (flower-spray) endings**, which are clustered at the end segments of many of the intrafusal fibres, are sensitive only to changes in length. Muscle spindles play a key role in the stretch reflex (p. 106).

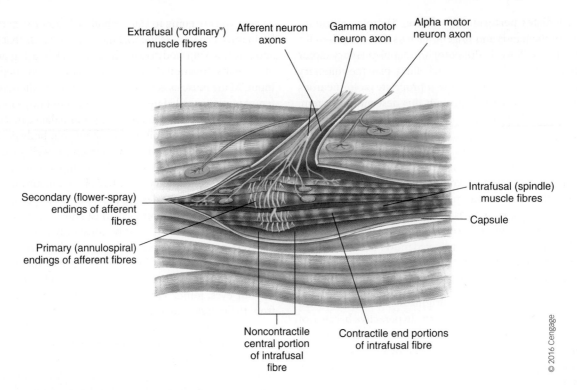

> **FIGURE 7-23 Muscle spindle.** A muscle spindle consists of a collection of specialized intrafusal fibres that lie within a connective tissue capsule parallel to the extrafusal skeletal muscle fibres. The muscle spindle is innervated by its own gamma motor neuron and is supplied by two types of afferent sensory terminals: the primary (annulospiral) endings, and the secondary (flower-spray) endings, both of which are activated by stretch.

In the figure labels:
- Extrafusal ("ordinary") muscle fibres
- Afferent neuron axons
- Gamma motor neuron axon
- Alpha motor neuron axon
- Secondary (flower-spray) endings of afferent fibres
- Primary (annulospiral) endings of afferent fibres
- Noncontractile central portion of intrafusal fibre
- Contractile end portions of intrafusal fibre
- Intrafusal (spindle) muscle fibres
- Capsule

© 2016 Cengage

### GOLGI TENDON ORGANS

In contrast to muscle spindles, which lie within the belly of the muscle, **Golgi tendon organs** are in the tendons of the muscle, where they can respond to changes in the muscle's tension rather than to changes in its length. Because a number of factors determine the tension developed in the whole muscle during contraction (e.g., frequency of stimulation or length of the muscle at the onset of contraction), it is essential that motor control systems be apprised of the tension actually achieved so that adjustments can be made as necessary.

The Golgi tendon organs consist of endings of afferent fibres entwined within bundles of connective tissue fibres that make up the tendon. When the extrafusal muscle fibres contract, the resulting pull on the tendon tightens the connective tissue bundles, which in turn increase the tension exerted on the bone to which the tendon is attached. In the process, the entwined Golgi organ afferent receptor endings are stretched, causing the afferent fibres to fire; the frequency of firing is directly related to the tension developed. This afferent information is sent to the brain for processing. Much of this information is used subconsciously for smoothly executing motor activity, but unlike afferent information from the muscle spindles, afferent information from the Golgi tendon organ reaches the level of conscious awareness. You are aware of the tension within a muscle but not of its length.

Scientists once thought the Golgi tendon organ triggered a protective spinal reflex that halted further contraction and brought about sudden reflex relaxation when the muscle tension became great enough, thus helping prevent damage to the muscle or tendon from excessive, tension-developing muscle contractions. Scientists now believe, however, that this receptor is a pure sensor and does not initiate any reflexes. Other unknown mechanisms are apparently involved in inhibiting further contraction to prevent tension-induced damage.

Having completed our discussion of skeletal muscle, we turn our attention to smooth and cardiac muscle.

## 7.7 | Smooth and Cardiac Muscle

The two other types of muscle—**smooth muscle** and **cardiac muscle**—share some basic properties with skeletal muscle, but each also displays unique characteristics ( Table 7-4). To accomplish contraction, each of the three muscle types has a specialized contractile apparatus made up of thin actin filaments that slide relative to stationary thick myosin filaments in response to a rise in cytosolic $Ca^{2+}$. As well, they all directly use ATP as the energy source for cross-bridge cycling. However, the structure and organization of fibres within these different muscle types vary, as do their mechanisms of excitation and the means by which excitation and contraction are coupled. Furthermore, there are important distinctions in the contractile response itself. In this section, we highlight the unique features of smooth and cardiac muscle, as compared with skeletal muscle. There is a detailed discussion of their function in later chapters that focus on the

| | TYPE OF MUSCLE | | | |
| --- | --- | --- | --- | --- |
| **Characteristic** | **Skeletal** | **Multiunit Smooth** | **Single-Unit Smooth** | **Cardiac** |
| **Location** | Attached to skeleton | Large blood vessels, small airways, eye, and hair follicles | Walls of hollow organs in digestive, reproductive, and urinary tracts and in small blood vessels | Heart only |
| **Function** | Movement of body in relation to external environment | Varies with structure involved | Movement of contents within hollow organs | Pumps blood out of heart |
| **Mechanism of Contraction** | Sliding filament mechanism | Sliding filament mechanism | Sliding filament mechanism | Sliding filament mechanism |
| **Innervation** | Somatic nervous system (alpha motor neurons) | Autonomic nervous system | Autonomic nervous system | Autonomic nervous system |
| **Level of Control** | Under voluntary control; also subject to subconscious regulation | Under involuntary control | Under involuntary control | Under involuntary control |
| **Initiation of Contraction** | Neurogenic | Neurogenic | Myogenic (pacemaker potentials and slow-wave potentials) | Myogenic (pacemaker potentials) |
| **Role of Nervous Stimulation** | Initiates contraction; accomplishes gradation | Initiates contraction; contributes to gradation | Modifies contraction; can excite or inhibit; contributes to gradation | Modifies contraction; can excite or inhibit; contributes to gradation |
| **Modifying Effect of Hormones** | No | Yes | Yes | Yes |
| **Presence of Thick Myosin and Thin Actin Filaments** | Yes | Yes | Yes | Yes |
| **Striated by Orderly Arrangement of Filaments** | Yes | No | No | Yes |
| **Presence of Troponin and Tropomyosin** | Yes | Tropomyosin only | Tropomyosin only | Yes |
| **Presence of T Tubules** | Yes | No | No | Yes |
| **Level of Development of Sarcoplasmic Reticulum** | Well developed | Poorly developed | Poorly developed | Moderately developed |
| **Cross Bridges Turned on by** $Ca^{2+}$ | Yes | Yes | Yes | Yes |
| **Source of Increased Cytosolic** $Ca^{2+}$ | Sarcoplasmic reticulum | Extracellular fluid and sarcoplasmic reticulum | Extracellular fluid fluid and sarcoplasmic reticulum | Extracellular fluid and sarcoplasmic reticulum |
| **Site of** $Ca^{2+}$ **Regulation** | Troponin in thin filaments | Myosin in thick filaments | Myosin in thick filaments | Troponin in thin filaments |
| **Mechanism of** $Ca^{2+}$ **Action** | Physically repositions troponin–tropomyosin complex to uncover actin cross-bridge binding sites | Chemically brings about phosphorylation of myosin cross bridges so they can bind with actin | Chemically brings about phosphorylation of myosin cross bridges so they can bind with actin | Physically repositions troponin–tropomyosin complex |

7

*(Continues)*

| | TYPE OF MUSCLE | | | |
|---|---|---|---|---|
| Characteristic | Skeletal | Multiunit Smooth | Single-Unit Smooth | Cardiac |
| **Presence of Gap Junctions** | No | Yes (very few) | Yes | Yes |
| **ATP Used Directly by Contractile Apparatus** | Yes | Yes | Yes | Yes |
| **Myosin ATPase Activity; Speed of Contraction** | Fast slow, depending | Very slow | Very slow | Slow |
| **Means by Which Gradation Accomplished** | Varying number of motor units contacting (motor unit recruitment) and frequency at which they are stimulated (twitch summation) | Varying number of muscle fibres contracting and varying cytosolic $Ca^{2+}$ concentration in each fibre by autonomic and hormonal influences | Varying cytosolic $Ca^{2+}$ concentration through myogenic activity and influences of the autonomic nervous system, hormones, mechanical stretch, and local metabolites | Varying length of fibre (depending on extent of filling of the heart chambers) and varying cytosolic $Ca^{2+}$ concentration through autonomic, hormonal, and local metabolite influences |
| **Presence of Tone in Absence of External Stimulation** | No | No | Yes | No |
| **Clear-Cut Length-Tension Relationship** | Yes | No | No | Yes |

© 2016 Cengage

organs containing these muscle types. We begin by describing the two categories of smooth muscle—multiunit and single-unit smooth muscle—that are based on differences in how the muscle fibres become excited.

## Multiunit smooth muscle

**Multiunit smooth muscle** exhibits properties partway between those of skeletal muscle and single-unit smooth muscle. As the name implies, a multiunit smooth muscle consists of multiple discrete units that function independently of one another and must be separately stimulated by nerves to contract, similar to skeletal muscle motor units. Therefore, contractile activity in both skeletal muscle and multiunit smooth muscle is **neurogenic** (nerve produced). That is, contraction in these muscle types is initiated only in response to stimulation by the nerves supplying the muscle. Whereas skeletal muscle is innervated by the voluntary somatic nervous system (motor neurons), multiunit (as well as single-unit) smooth muscle is supplied by the involuntary autonomic nervous system.

Multiunit smooth muscle is found (1) in the walls of large blood vessels; (2) in small airways to the lungs; (3) in the muscle of the eye that adjusts the lens for near or far vision; (4) in the iris of the eye, which alters the pupil size to adjust the amount of light entering the eye; and (5) at the base of hair follicles, contraction of which causes goosebumps.

## Single-unit smooth muscle cells and functional syncytia

Most smooth muscle is **single-unit smooth muscle**, alternatively called **visceral smooth muscle**, because it is found in the walls of the hollow organs or viscera (e.g., the digestive, reproductive, and urinary tracts and small blood vessels). The term *single-unit smooth muscle* derives from the fact that the muscle fibres that make up this type of muscle become excited and contract as a single unit. The muscle fibres in single-unit smooth muscle are electrically linked by gap junctions (p. 35). When an action potential occurs anywhere within a sheet of single-unit smooth muscle, it is quickly propagated via these special points of electrical contact throughout the entire group of interconnected cells, which then contract as a single, coordinated unit. Such a group of interconnected muscle cells that function electrically and mechanically as a unit is known as a **functional syncytium** (plural, *syncytia; syn* means "together"; *cyt* means "cell").

Thinking about the role of the uterus during labour can help you appreciate the significance of this arrangement. Muscle cells composing the uterine wall act as a functional syncytium. They repeatedly become excited and contract as a unit during labour, exerting a series of coordinated "pushes" that eventually deliver the baby. Independent, uncoordinated contractions of individual muscle cells in the uterine wall could not exert the uniformly applied pressure needed to expel the baby. Single-unit smooth muscle elsewhere in the body is arranged in similar functional syncytia.

## Smooth muscle cells: Small and unstriated

Most smooth muscle cells are found in the walls of hollow organs and tubes. Their contraction exerts pressure on and regulates the forward movement of the contents of these structures.

Both smooth and skeletal muscle cells are elongated, but in contrast to their large, cylindrical skeletal muscle counterparts, smooth muscle cells are spindle shaped, have a single nucleus, and are considerably smaller (2 to 10 μm in diameter and 50 to 400 μm long). Also unlike skeletal muscle cells, a single smooth muscle cell does not extend the full length of a muscle. Instead, groups of smooth muscle cells are typically arranged in sheets (> Figure 7-24a).

A smooth muscle cell has three types of filaments: (1) thick myosin filaments, which are longer than those in skeletal muscle;

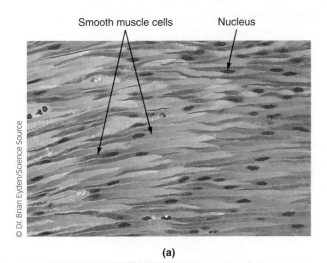

Smooth muscle cells      Nucleus

© Dr. Brian Eyden/Science Source

**(a)**

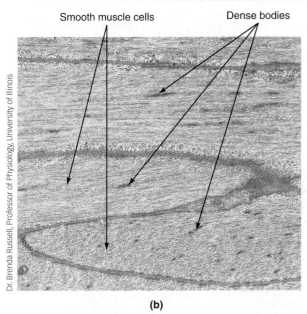

Smooth muscle cells      Dense bodies

Dr. Brenda Russell, Professor of Physiology, University of Illinois

**(b)**

> **FIGURE 7-24 Microscopic view of smooth muscle cells.** (a) Low-power light micrograph of smooth muscle cells. Note the spindle shape and single, centrally located nucleus. (b) Electron micrograph of smooth muscle cells at 14 000 × magnification. Note the presence of dense bodies and lack of banding.

(2) thin actin filaments, which contain tropomyosin but lack the regulatory protein troponin; and (3) filaments of intermediate size, which do not directly participate in contraction but are part of the cytoskeletal framework that supports the cell shape. Smooth muscle filaments do not form myofibrils and are not arranged in the sarcomere pattern found in skeletal muscle. Thus, smooth muscle cells do not show the banding or striation of skeletal muscle, hence the term *smooth* for this muscle type.

Lacking sarcomeres, smooth muscle does not have Z lines, but instead has **dense bodies** containing the same protein constituent found in Z lines (> Figure 7-24b). Dense bodies are positioned throughout the smooth muscle cell as well as attached to the internal surface of the plasma membrane. The dense bodies are held in place by a scaffold of intermediate filaments. The actin filaments are anchored to the dense bodies. Considerably more actin is present in smooth muscle cells than in skeletal muscle cells, with 10 to 15 thin filaments for each thick myosin filament in smooth muscle compared with 2 thin filaments for each thick filament in skeletal muscle.

The thick- and thin-filament contractile units are oriented slightly diagonally from side to side within the smooth muscle cell in an elongated, diamond-shaped lattice, rather than running parallel with the long axis as myofibrils do in skeletal muscle (> Figure 7-25a). Relative sliding of the thin filaments past the thick filaments during contraction causes the filament lattice to shorten and expand from side to side. As a result, the whole cell shortens and bulges out between the points where the thin filaments are attached to the inner surface of the plasma membrane (> Figure 7-25b).

## Smooth muscle cells and calcium

The thin filaments of smooth muscle cells do not contain troponin, and tropomyosin does not block actin's cross-bridge binding sites. So we have two questions: What prevents actin and myosin from binding at the cross bridges in the resting state? And how is cross-bridge activity switched on in the excited state? Light-weight chains of proteins are attached to the heads of myosin molecules. These so-called **light chains** are only of secondary importance in skeletal muscle, but they have a crucial regulatory function in smooth muscle. The smooth muscle myosin heads can interact with actin only when the myosin light chain is *phosphorylated* (i.e., has a phosphate group from ATP attached to it). During excitation, the increased cytosolic $Ca^{2+}$ acts as an intracellular messenger, initiating a chain of biochemical events that results in phosphorylation of the myosin light chain (> Figure 7-26). Smooth muscle $Ca^{2+}$ binds with **calmodulin**, an intracellular protein found in most cells that is structurally similar to troponin (p. 302). This $Ca^{2+}$–calmodulin complex binds to and activates another protein, **myosin light chain kinase (MLC kinase)**, which in turn phosphorylates the myosin light chain. Thus phosphorylated, the myosin cross bridge is able to bind with actin so that cross-bridge cycling can begin. In this way, smooth muscle is triggered to contract by a rise in cytosolic $Ca^{2+}$ similar to what happens in skeletal muscle. In smooth muscle, however, $Ca^{2+}$ ultimately turns on the cross bridges by inducing a *chemical* change in myosin in the

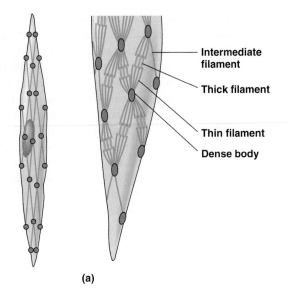

Intermediate
filament

Thick filament

Thin filament

Dense body

**(a)**

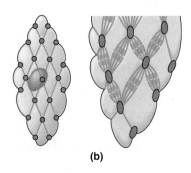

**(b)**

❯ **FIGURE 7-25 Schematic representation of the arrangement of thick and thin filaments in a smooth muscle cell in contracted and relaxed states.** (a) Relaxed smooth muscle cell. (b) Contracted smooth muscle cell

*thick* filaments, whereas in skeletal muscle it exerts its effects by invoking a *physical* change at the *thin* filaments (❯ Figure 7-27). Recall that in skeletal muscle, $Ca^{2+}$ moves troponin and tropomyosin from their blocking position, so actin and myosin are free to bind with each other.

The way excitation increases cytosolic $Ca^{2+}$ concentration in smooth muscle cells also differs from that for skeletal muscle. A smooth muscle cell has no T tubules and a poorly developed SR. The increased cytosolic $Ca^{2+}$ that triggers the contractile response comes from two sources: most $Ca^{2+}$ enters from the ECF, but some is released intracellularly from the sparse SR stores. Unlike their role in skeletal muscle cells, voltage-gated dihydropyridine receptors in the plasma membrane of smooth muscle cells function as $Ca^{2+}$ channels. When these surface-membrane channels are opened, $Ca^{2+}$ enters down its concentration gradient from the ECF. The entering $Ca^{2+}$ triggers the opening of $Ca^{2+}$ channels in the SR so that small additional amounts of $Ca^{2+}$ are released intracellularly from this meagre source. Because smooth muscle cells are so much smaller in diameter than skeletal muscle fibres, most $Ca^{2+}$ entering from the ECF can influence crossbridge activity, even in the central portions of the cell, without requiring an elaborate T tubule–SR mechanism.

Relaxation is accomplished by removal of $Ca^{2+}$ as it is actively transported out across the plasma membrane and back into the SR. When $Ca^{2+}$ is removed, myosin is dephosphorylated (the phosphate is removed) and can no longer interact with actin, so the muscle relaxes.

We are about to address the second question, How does smooth muscle become excited to contract—that is, what opens the $Ca^{2+}$ channels in the plasma membrane? First we discuss the myogenic activity of single-unit smooth muscle.

## Single-unit smooth muscle: Myogenic

Single-unit smooth muscle is **self-excitable**, so it does not require nervous stimulation for contraction. Clusters of specialized smooth muscle cells within a functional syncytium display spontaneous

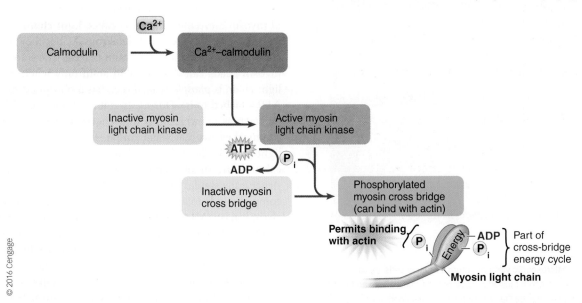

❯ **FIGURE 7-26 Calcium activation of myosin cross bridge in smooth muscle**

## Smooth muscle

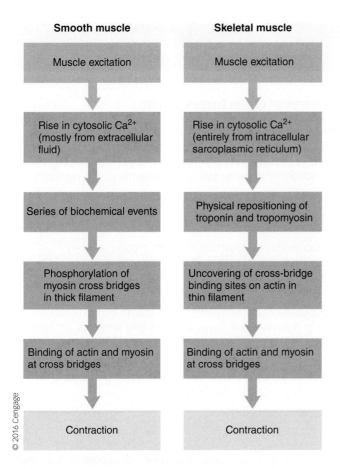

| Smooth muscle | Skeletal muscle |
|---|---|
| Muscle excitation | Muscle excitation |
| Rise in cytosolic Ca²⁺ (mostly from extracellular fluid) | Rise in cytosolic Ca²⁺ (entirely from intracellular sarcoplasmic reticulum) |
| Series of biochemical events | Physical repositioning of troponin and tropomyosin |
| Phosphorylation of myosin cross bridges in thick filament | Uncovering of cross-bridge binding sites on actin in thin filament |
| Binding of actin and myosin at cross bridges | Binding of actin and myosin at cross bridges |
| Contraction | Contraction |

© 2016 Cengage

› **FIGURE 7-27** Comparison of the role of calcium in bringing about contraction in smooth muscle and skeletal muscle

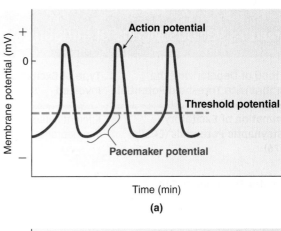

(a)

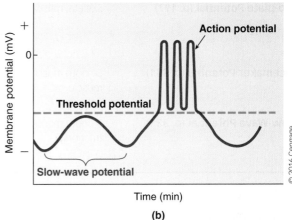

(b)

› **FIGURE 7-28** Self-generated electrical activity in smooth muscle. (a) With pacemaker potentials, the membrane gradually depolarizes to threshold on a regular periodic basis without any nervous stimulation. These regular depolarizations cyclically trigger self-induced action potentials. (b) In slow-wave potentials, the membrane gradually undergoes self-induced hyperpolarizing and depolarizing swings in potential. A burst of action potentials occurs if a depolarizing swing brings the membrane to threshold.

electrical activity; that is, they can undergo action potentials without any external stimulation. In contrast to the other excitable cells we have been discussing (such as neurons, skeletal muscle fibres, and multiunit smooth muscle), the self-excitable cells of single-unit smooth muscle do not maintain a constant resting potential. Instead, their membrane potential inherently fluctuates without any influence by factors external to the cell. Two major types of spontaneous depolarizations displayed by self-excitable cells are *pacemaker potentials* and *slow-wave potentials*.

### PACEMAKER POTENTIALS

With **pacemaker potentials**, the membrane potential gradually depolarizes on its own because of shifts in passive ionic fluxes accompanying automatic changes in channel permeability (› Figure 7-28a). When the membrane has depolarized to threshold, an action potential is initiated. After repolarizing, the membrane potential once again depolarizes to threshold, cyclically continuing in this manner to repetitively self-generate action potentials.

### SLOW-WAVE POTENTIALS

**Slow-wave potentials** are gradually alternating hyperpolarizing and depolarizing swings in potential caused by automatic cyclic changes in the rate at which sodium ions are actively transported

across the membrane (› Figure 7-28b). The potential is moved farther from threshold during each hyperpolarizing swing and closer to threshold during each depolarizing swing. If threshold is reached, a burst of action potentials occurs at the peak of a depolarizing swing. Threshold is not always reached, however, so the oscillating slow-wave potentials can continue without generating action potentials. Whether threshold is reached depends on the starting point of the membrane potential at the onset of its depolarizing swing. The starting point, in turn, is influenced by neural and local factors.

We have now discussed all the means by which excitable tissues can be brought to threshold. ▮ Table 7-5 summarizes the different triggering events that can initiate action potentials in various excitable tissues.

### MYOGENIC ACTIVITY

Self-excitable smooth muscle cells are specialized to initiate action potentials, but they are not equipped to contract. Only a very few of all the cells in a functional syncytium are

7

**▌ TABLE 7-5** Various Means of Initiating Action Potentials in Excitable Tissues

| Method of Depolarizing the Membrane to Threshold Potential | Type of Excitable Tissue Involved | Description of This Triggering Event |
| --- | --- | --- |
| **Summation of Excitatory Postsynaptic Potentials (EPSPs) (p. 76)** | Efferent neurons, interneurons | Temporal or spatial summation of slight depolarizations (EPSPs) of the dendrite/cell body end of the neuron brought about by changes in channel permeability in response to binding of excitatory neurotransmitter with surface membrane receptors |
| **Receptor Potential (p. 144)** | Afferent neurons | Typically a depolarization of the afferent neuron's receptor initiated by changes in channel permeability in response to the neuron's adequate stimulus |
| **End-plate Potential (p. 197)** | Skeletal muscle | Depolarization of the motor end plate brought about by changes in channel permeability in response to binding of the neurotransmitter acetylcholine with receptors on the end-plate membrane |
| **Pacemaker Potential (p. 331)** | Smooth muscle, cardiac muscle | Gradual depolarization of the membrane on its own because of shifts in passive ionic fluxes accompanying automatic changes in channel permeability |
| **Slow-Wave Potential (p. 331)** | Smooth muscle (in digestive tract only) | Gradual alternating hyperpolarizing and depolarizing swings in potential caused by automatic cyclical changes in active ionic transport across the membrane |

noncontractile, pacemaker cells. These cells are typically clustered in a specific location. The vast majority of smooth muscle cells in a functional syncytium are specialized to contract but cannot self-initiate action potentials. However, once an action potential is initiated by a self-excitable smooth muscle cell, it is conducted to the remaining contractile, nonpacemaker cells of the functional syncytium via gap junctions, so the entire group of connected cells contracts as a unit without any nervous input. Such nerve-independent contractile activity initiated by the muscle itself is called **myogenic** (muscle-produced) activity, in contrast to the neurogenic activity of skeletal muscle and multi-unit smooth muscle.

## Gradation of single-unit smooth muscle contraction

Single-unit smooth muscle differs from skeletal muscle in the way contraction is graded. Gradation of skeletal muscle contraction is entirely under neural control, primarily involving motor unit recruitment and twitch summation. In smooth muscle, gap junctions ensure that an entire smooth muscle mass contracts as a single unit, making it impossible to vary the number of muscle fibres contracting. Only the tension of the fibres can be modified to achieve varying strengths of contraction of the whole organ. The portion of cross bridges activated and the tension subsequently developed in single-unit smooth muscle can be graded by varying the cytosolic $Ca^{2+}$ concentration. A single excitation in smooth muscle does not cause all the cross bridges to switch on, in contrast to skeletal muscles, where a single action potential triggers release of enough $Ca^{2+}$ to permit all cross bridges to

cycle. As $Ca^{2+}$ concentration increases in smooth muscle, more cross bridges are brought into play, and greater tension develops.

### SMOOTH MUSCLE TONE
Many single-unit smooth muscle cells have sufficient levels of cytosolic $Ca^{2+}$ to maintain a low level of tension, or **tone**, even in the absence of action potentials. A sudden drastic change in $Ca^{2+}$, such as that which accompanies a myogenically induced action potential, brings about a contractile response superimposed on the ongoing tonic tension. Besides self-induced action potentials, a number of other factors, including autonomic neurotransmitters, can influence contractile activity and the development of tension in smooth muscle cells by altering their cytosolic $Ca^{2+}$ concentration.

### MODIFICATION OF SMOOTH MUSCLE ACTIVITY BY THE AUTONOMIC NERVOUS SYSTEM
Smooth muscle is typically innervated by both branches of the autonomic nervous system. In single-unit smooth muscle, this nerve supply does not *initiate* contraction, but it can *modify* the rate and strength of contraction, either enhancing or retarding the inherent contractile activity of a given organ. Recall that the isolated motor end-plate region of a skeletal muscle fibre interacts with ACh released from a single axon terminal of a motor neuron. In contrast, the receptor proteins that bind with autonomic neurotransmitters are dispersed throughout the entire surface membrane of a smooth muscle cell. Smooth muscle cells are sensitive to varying degrees and in varying ways to autonomic neurotransmitters, depending on the cells' distributions of cholinergic and adrenergic receptors (pp. 192–193).

Each terminal branch of a postganglionic autonomic fibre travels across the surface of one or more smooth muscle cells, releasing neurotransmitter from the vesicles within its multiple varicosities (bulges) as an action potential passes along the terminal (› Figure 7-29). The neurotransmitter diffuses to the many receptor sites specific for it on the cells underlying the terminal. Thus, in contrast to the discrete one-to-one relationship at motor end plates, a given smooth muscle cell can be influenced by more than one type of neurotransmitter, and each autonomic terminal can influence more than one smooth muscle cell.

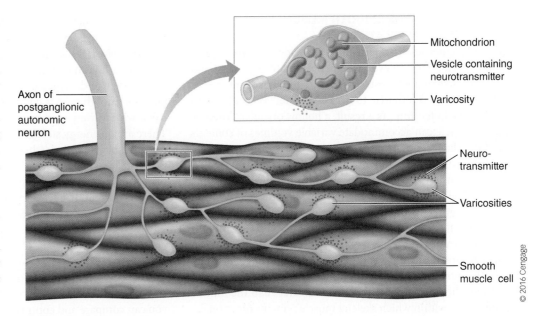

Axon of postganglionic autonomic neuron

Mitochondrion

Vesicle containing neurotransmitter

Varicosity

Neuro-transmitter

Varicosities

Smooth muscle cell

© 2016 Cengage

› **FIGURE 7-29** Schematic representation of innervation of smooth muscle by autonomic postganglionic nerve terminals

### OTHER FACTORS INFLUENCING SMOOTH MUSCLE ACTIVITY

Other factors (besides autonomic neurotransmitters) can influence the rate and strength of both multiunit and single-unit smooth muscle contraction, including mechanical stretch, certain hormones, local metabolites, and specific drugs. All these factors ultimately act by modifying the permeability of $Ca^{2+}$ channels in the plasma membrane, the SR, or both, through a variety of mechanisms. Thus, smooth muscle is subject to more external influences than skeletal muscle is, even though smooth muscle can contract on its own and skeletal muscle cannot.

Next, we look in detail at the length–tension relationship in smooth muscle, but only in regard to the effect of mechanical stretch on smooth muscle contractility. We will examine the extracellular chemical influences on smooth muscle contractility in later chapters when we discuss regulation of the various organs that contain smooth muscle.

## Smooth muscle: Tension and relaxation

The relationship between the length of the muscle fibres before contraction and the tension that can be developed on a subsequent contraction is less closely linked in smooth muscle than in skeletal muscle. Smooth muscle can have one of two different responses to being stretched: (1) being able to develop tension even when considerably stretched, and (2) inherently relaxing when stretched. The range of lengths over which a smooth muscle fibre can develop near-maximal tension is much greater than for skeletal muscle. Smooth muscle can still develop considerable tension even when stretched up to 2.5 times its resting length, for two likely reasons. First, in contrast to skeletal muscle, in which the resting length is at $l_o$, in smooth muscle the resting (nonstretched) length is much shorter than at $l_o$. Therefore, smooth muscle can be stretched considerably before reaching its optimal length. Second, the thin filaments in smooth muscle cells continue to overlap the much longer thick filaments even in the stretched-out position, so cross-bridge interaction and tension development can still take place. In contrast, when skeletal muscle is stretched only three-fourths longer than its resting length, the thick and thin filaments are completely pulled apart and can no longer interact (see › Figure 7-17).

The ability of a considerably stretched smooth muscle fibre to continue to develop tension is important, because the smooth muscle fibres within the wall of a hollow organ are progressively stretched as the volume of the organ's contents expands. Consider the urinary bladder as an example. Even though the muscle fibres in the urinary bladder are stretched as the bladder gradually fills with urine, they maintain their tone and can even develop further tension in response to inputs that regulate bladder emptying. If considerable stretching prevented tension development, as in skeletal muscle, a filled bladder would not be capable of contracting to empty.

### STRESS RELAXATION RESPONSE

When a smooth muscle is suddenly stretched, it initially increases its tension, much like the tension created in a stretched rubber band. The muscle quickly adjusts to this new length, however, and inherently relaxes to the tension level prior to the stretch, probably as a consequence of rearrangement of cross-bridge attachments. Smooth muscle cross bridges detach comparatively slowly. Some researchers speculate that, on sudden stretching of smooth muscle, any attached cross bridges would strain against the stretch, thereby contributing to a passive (not actively generated) increase in tension. As these cross bridges detach, the filaments would be able to slide into an unstrained stretched position, restoring the tension to its original level. This inherent property of smooth muscle is called the **stress relaxation response**.

These two responses of smooth muscle to being stretched—being able to develop tension even when considerably stretched and inherently relaxing when stretched—are highly advantageous. They enable smooth muscle to exist at a variety of lengths with little change in tension. As a result, a hollow organ enclosed by smooth muscle can accommodate variable volumes of contents with little change in the pressure exerted on the contents except when the contents are to be pushed out of the organ. At that time, the tension is deliberately increased by fibre shortening. It is possible for smooth muscle fibres to contract to half their normal length, enabling hollow organs to dramatically empty their contents on increased contractile activity; thus, smooth-muscled viscera can easily accommodate large volumes but can empty to practically zero volume. This length range in which smooth muscle normally functions (anywhere from 0.5 to 2.5 times the normal length) is much greater than the limited length range within which skeletal muscle remains functional.

Smooth muscle contains a lot of connective tissue, which resists being stretched. Unlike skeletal muscle, in which the skeletal attachments restrict how far the muscle can be stretched, this connective tissue prevents smooth muscle from being overstretched and thus puts an upper limit on how much a smooth-muscled hollow organ can hold.

## Smooth muscle: Slow and economical

A smooth muscle contractile response proceeds more slowly than a skeletal muscle twitch. A single smooth muscle contraction may last as long as three seconds (3000 msec), compared with the maximum of 100 msec required for a single contractile response in skeletal muscle. ATP splitting by myosin ATPase is much slower in smooth muscle, so cross-bridge activity and filament sliding occur more slowly. Smooth muscle also relaxes more slowly because of slower $Ca^{2+}$ removal. Slowness should not be equated with weakness, however. Smooth muscle can generate the same contractile tension per unit of cross-sectional area as skeletal muscle, but it does so more slowly and at considerably less energy expense. Because of slow cross-bridge cycling during smooth muscle contraction, cross bridges stay attached for more time during each cycle, compared with skeletal muscle; that is, the cross bridges latch onto the thin filaments for a longer time each cycle. This so-called **latch phenomenon** enables smooth muscle to maintain tension with comparatively less ATP consumption because each cross-bridge cycle uses up one molecule of ATP. Smooth muscle is therefore an economical contractile tissue, making it well suited for long-term sustained contractions with little energy consumption and without fatigue. In contrast to the rapidly changing demands placed on your skeletal muscles as you manoeuvre through and manipulate your external environment, your smooth muscle activities are geared for long-term duration and slower adjustments to change.

Because of its slowness and the less ordered arrangement of its filaments, smooth muscle has often been mistakenly viewed as a poorly developed version of skeletal muscle. Actually, smooth muscle is just as highly specialized for the demands placed on it—that is, being able to economically maintain tension for prolonged periods without fatigue and being able to vary considerably in the volume of contents it holds, with little change in tension. It is an extremely adaptive, efficient tissue.

Nutrient and oxygen delivery are generally adequate to support the smooth muscle contractile process. Smooth muscle can use a wide variety of nutrient molecules for ATP production. There are no energy storage pools comparable to creatine phosphate in smooth muscle; they are not necessary. Oxygen delivery is usually adequate to keep pace with the low rate of oxidative phosphorylation needed to provide ATP for the energy-efficient smooth muscle. If necessary, anaerobic glycolysis can sustain adequate ATP production if oxygen supplies are diminished.

## Cardiac muscle

Cardiac muscle functions differently from both skeletal and smooth muscle. Here we look briefly at these key differences so you can compare and contrast all three types of muscle cells. An in-depth discussion on cardiac muscle is included in the chapter on cardiac physiology (Chapter 8).

Cardiac muscle, found only in the heart, shares structural and functional characteristics with both skeletal and single-unit smooth muscle. Like skeletal muscle, cardiac muscle is striated, with its thick and thin filaments highly organized into a regular banding pattern. Cardiac thin filaments contain troponin and tropomyosin, which constitute the site of $Ca^{2+}$ action in switching on cross-bridge activity, as in skeletal muscle. Also like skeletal muscle, cardiac muscle has a clear length–tension relationship. Like the oxidative skeletal muscle fibres, cardiac muscle cells have lots of mitochondria and myoglobin. They also have T tubules and a moderately well-developed SR.

As in smooth muscle, $Ca^{2+}$ enters the cytosol from both the ECF and the SR during cardiac excitation. $Ca^{2+}$ entry from the ECF occurs through voltage-gated calcium channels. This $Ca^{2+}$ entry from the ECF triggers release of $Ca^{2+}$ intracellularly from the SR. Like single-unit smooth muscle, the heart displays pacemaker (but not slow-wave) activity, initiating its own action potentials without any external influence. Cardiac cells are interconnected by gap junctions that enhance the spread of action potentials throughout the heart, just as in single-unit smooth muscle. Also, the heart is similarly innervated by the autonomic nervous system, which, along with certain hormones and local factors, can modify the rate and strength of contraction.

Unique to cardiac muscle, the cardiac fibres are joined in a branching network, and the action potentials of cardiac muscle last much longer before repolarizing. Further details and the importance of cardiac muscle's features are addressed in Chapter 8.

# Chapter in Perspective: Focus on Homeostasis

Skeletal muscles compose the muscular system itself. Cardiac and smooth muscle are part of the organs that make up other body systems. Cardiac muscle is found only in the heart, which is part of the circulatory system. Smooth muscle is found in the walls of hollow organs and tubes, including the blood vessels in the circulatory system, airways in the respiratory system, bladder in the urinary system, stomach and intestines in the digestive system, and uterus and ductus deferens (the duct through which sperm leave the testes) in the reproductive system.

Contraction of skeletal muscles accomplishes both the movement of body parts in relation to one another and also the movement of the whole body in relation to the external environment. Thus, these muscles permit you to move through and manipulate your external environment. At a very general level, some of these movements are aimed at maintaining homeostasis, such as moving the body toward food or away from harm. Examples of more specific homeostatic functions accomplished by skeletal muscles include chewing and swallowing food for further breakdown in the digestive system into usable energy-producing nutrient molecules (the mouth and throat muscles are all skeletal muscles), and breathing to obtain oxygen and get rid of carbon dioxide (the respiratory muscles are all skeletal muscles). Heat generation by contracting skeletal muscles also is the major source of heat production in maintaining body temperature. The skeletal muscles further accomplish many nonhomeostatic activities that enable us to work and play—for example, operating a computer or riding a bicycle—so that we can contribute to society and enjoy ourselves.

All the other systems of the body, except the immune (defence) system, depend on their nonskeletal muscle components to enable them to accomplish their homeostatic functions. For example, contraction of cardiac muscle in the heart pushes life-sustaining blood forward into the blood vessels, and contraction of smooth muscle in the stomach and intestines pushes the ingested food through the digestive tract at a rate appropriate for the digestive juices secreted along the route to break down the food into usable units.

## CHAPTER TERMINOLOGY

A band (p. 300)
actin (p. 301)
aerobic (endurance) exercise (p. 320)
alpha motor neurons (p. 325)
anaerobic (high-intensity) exercise (p. 320)
asynchronous recruitment of motor units (p. 310)
atrophy (p. 320)
calmodulin (p. 329)
cardiac muscle (p. 326)
central fatigue (p. 318)
concentric muscle contraction (p. 315)
contractile component (p. 313)
contractile proteins (p. 301)
contraction time (p. 308)
corticospinal (pyramidal) motor system (p. 323)
cross bridges (myosin heads) (p. 300)
denervation atrophy (p. 320)
dense bodies (p. 329)
dihydropyridine receptors (p. 305)
disuse atrophy (p. 320)
dystrophin (p. 322)
eccentric muscle contraction (p. 315)
endomysium (p. 298)
epimysium (p. 298)
excitation–contraction coupling (p. 303)
extrafusal fibres (p. 325)
fast-twitch fibres (p. 318)
fatigue (p. 310)
flaccid paralysis (p. 325)
foot proteins (p. 304)
force (p. 316)
force–velocity relationship (p. 316)
fulcrum (p. 316)

functional syncytium (p. 328)
functional unit (p. 300)
gamma motor neuron (p. 325)
glycolytic (p. 318)
Golgi tendon organs (p. 326)
H zone (p. 300)
hemiplegia (p. 325)
I band (p. 300)
insertion (p. 314)
intrafusal fibres (p. 325)
isometric contraction (p. 315)
isotonic contraction (p. 314)
latch phenomenon (p. 334)
latent period (p. 308)
lateral sacs (p. 304)
length–tension relationship (p. 312)
lever (p. 316)
light chains (p. 329)
load (p. 314)
load arm (p. 316)
M line (p. 300)
motor unit (p. 310)
motor unit recruitment (p. 310)
multineuronal (extrapyramidal) motor system (p. 323)
multiunit smooth muscle (p. 328)
muscle (p. 297)
muscle biopsy (p. 318)
muscle fatigue (p. 317)
muscle spindles (p. 325)
muscular dystrophy (p. 321)
myofibrils (p. 298)
myogenic (p. 332)
myoglobin (p. 318)
myosin (p. 300)

myosin light chain kinase (MLC kinase) (p. 329)
nebulin (p. 301)
neurogenic (p. 328)
optimal length ($l_o$) (p. 312)
origin (p. 314)
oxidative (p. 318)
pacemaker potentials (p. 331)
paraplegia (p. 325)
perimysium (p. 298)
power arm (p. 316)
power stroke (p. 303)
primary (annulospiral) endings (p. 325)
pyramidal cells (p. 323)
red fibres (p. 318)
regulatory proteins (p. 302)
relaxation (p. 308)
relaxation time (p. 308)
rigor mortis (p. 306)
ryanodine receptors (p. 304)
sarcomere (p. 300)
sarcoplasmic reticulum (SR) (p. 304)
satellite cells (p. 321)
secondary (flower-spray) endings (p. 325)
self-excitable (p. 330)
series-elastic component (p. 314)
single-unit smooth muscle (p. 328)
size principle (p. 310)
skeletal muscle (p. 298)
sliding filament mechanism (p. 303)
slow-twitch fibres (p. 318)
slow-wave potentials (p. 331)
smooth muscle (p. 326)
spastic paralysis (p. 325)
stress relaxation response (p. 333)

tendon (p. 298)
tension (p. 313)
terminal cisternae (p. 304)
tetanus (p. 311)
tetraplegia (p. 325)
thick filaments (p. 298)
thin filaments (p. 298)
titin (p. 300)

tone (p. 332)
transverse tubule (T tubule) (p. 304)
tropomyosin (p. 301)
troponin (p. 302)
twitch (p. 309)
twitch summation (p. 311)
type I muscle fibres (p. 318)
type II muscle fibres (p. 318)

utrophin (p. 323)
velocity (p. 315)
visceral smooth muscle (p. 328)
white fibres (p. 319)
work (p. 316)
Z line (p. 300)

## REVIEW EXERCISES

### Objective Questions
### (Answers in Appendix E, p. A-42)

1. On completion of an action potential in a muscle fibre, the contractile activity initiated by the action potential ceases. *(True or false?)*

2. The velocity at which a muscle shortens depends entirely on the ATPase activity of its fibres. *(True or false?)*

3. When a skeletal muscle is maximally stretched, it can develop maximal tension on contraction, because the actin filaments can slide in a maximal distance. *(True or false?)*

4. A pacemaker potential always initiates an action potential. *(True or false?)*

5. A slow-wave potential always initiates an action potential. *(True or false?)*

6. Smooth muscle can develop tension even when considerably stretched, because the thin filaments still overlap with the long thick filaments. *(True or false?)*

7. A(n) _____ contraction is an isotonic contraction in which the muscle shortens, whereas the muscle lengthens in a(n) _____ isotonic contraction.

8. _____ motor neurons supply extrafusal muscle fibres, whereas intrafusal fibres are innervated by _____ motor neurons.

9. The two types of atrophy are _____ and _____.

10. Which of the following provide(s) direct input to alpha motor neurons? (Indicate all correct answers.)
    a. primary motor cortex
    b. brain stem
    c. cerebellum
    d. basal nuclei
    e. spinal reflex pathways

11. Which of the following is NOT involved in bringing about muscle relaxation?
    a. reuptake of $Ca^{2+}$ by the sarcoplasmic reticulum
    b. no more ATP
    c. no more action potential
    d. removal of ACh at the end plate by acetylcholinesterase
    e. filaments sliding back to their resting position

12. Match the following (with reference to skeletal muscle):

    _____ 1. $Ca^{2+}$
    _____ 2. T tubule
    _____ 3. ATP
    _____ 4. lateral sac of the sarcoplasmic reticulum
    _____ 5. myosin
    _____ 6. troponin–tropomyosin complex
    _____ 7. actin

    (a) cyclically binds with the myosin cross bridges during contraction
    (b) has ATPase activity
    (c) supplies energy for the power stroke of a cross bridge
    (d) rapidly transmits the action potential to the central portion of the muscle fibre
    (e) stores $Ca^{2+}$
    (f) pulls the troponin–tropomyosin complex out of its blocking position
    (g) prevents actin from interacting with myosin when the muscle fibre is not excited

13. Using the answer code at the right, indicate what happens in the banding pattern during contraction:

    _____ 1. thick myofilament
    _____ 2. thin myofilament
    _____ 3. A band
    _____ 4. I band
    _____ 5. H zone
    _____ 6. sarcomere

    (a) remains the same size during contraction
    (b) decreases in length (shortens) during contraction

### Written Questions

1. Describe the levels of organization in a skeletal muscle.

2. What produces the striated appearance of skeletal muscles? Describe the arrangement of thick and thin filaments that gives rise to the banding pattern.

3. What is the functional unit of skeletal muscle?

4. Describe the composition of thick and thin filaments.

5. Describe the sliding filament mechanism of muscle contraction. How do cross-bridge power strokes bring about shortening of the muscle fibre?

6. Compare the excitation–contraction coupling process in skeletal muscle with that in smooth muscle.

7. How can gradation of skeletal muscle contraction be accomplished?

8. What is a motor unit? Compare the size of motor units in finely controlled muscles with those specialized for coarse, powerful contractions. Describe motor unit recruitment.

9. Explain *twitch summation* and *tetanus*.

10. How does a skeletal muscle fibre's length at the onset of contraction affect the strength of the subsequent contraction?

11. Compare isotonic and isometric contractions.

12. Compare the three types of skeletal muscle fibres.

13. What are the roles of the corticospinal system and multineuronal system in controlling motor movement?

14. Describe the structure and function of muscle spindles and Golgi tendon organs.

15. Distinguish between multiunit and single-unit smooth muscle.

16. Differentiate between neurogenic and myogenic muscle activity.

17. How can smooth muscle contraction be graded?

18. Compare the contractile speed and relative energy expenditure of skeletal muscle with that of smooth muscle.

19. In what ways is cardiac muscle functionally similar to skeletal muscle and to single-unit smooth muscle?

## Quantitative Exercises
(Solutions in Appendix E, p. A-42)

1. Consider two individuals each throwing a baseball, one a weekend athlete and the other a professional pitcher.
   a. Given the following information, calculate the velocity of the ball as it leaves the amateur's hand:
   - The distance from his shoulder socket (humeral head) to the ball is 70 cm.
   - The distance from his humeral head to the points of insertion of the muscles moving his arm forward (we must simplify here, because the shoulder is such a complex joint) is 9 cm.
   - The velocity of muscle shortening is 2.6 m/sec.
   b. The professional pitcher throws the ball 137 km per hour. If his points of insertion are also 9 cm from the humeral head and the distance from his humeral head to the ball is 90 cm, how much faster did the professional pitcher's muscles shorten compared with the amateur's?

2. The velocity at which a muscle shortens is related to the force that it can generate in the following way:[1]

$$v = b(F_0 = F)/(F + a)$$

where $v$ is the velocity of shortening, and $F_0$ can be thought of as an "upper load limit," or the maximum force a muscle can generate against a resistance. The parameter $a$ is inversely proportional to the cross-bridge cycling rate, and $b$ is proportional to the number of sarcomeres in line in a muscle. Draw the resistance (force)–velocity curve predicted by this equation by plotting the points $F = 0$ and $F = F_0$. Values of $v$ are on the vertical axis; values of $F$ are on the horizontal axis; $a$, $b$, and $F_0$ are constants.
   a. Notice that the curve generated from this equation is the same as that in Figure 7-20, p. 315. Why does the curve have this shape? That is, what does the shape of the curve tell you about muscle performance in general?
   b. What happens to the resistance (force)–velocity curve when $F_0$ is increased? When the cross-bridge cycling rate is increased? When the size of the muscle is increased? How will each of these changes affect the performance of the muscle?

[1] F.C. Hoppensteadt & C.S. Peskin, *Mathematics in Medicine and the Life Sciences* (New York: Springer, 1992), equation 9.1.1, p. 199.

## POINTS TO PONDER

### (Explanations in Appendix E, p. A-43)

1. Why does regular aerobic exercise provide more cardiovascular benefit than weight training does? (*Hint:* The heart responds to the demands placed on it in a way similar to that of skeletal muscle.)

2. If the biceps muscle of a child inserts 4 cm from the elbow and the length of the arm from the elbow to the hand is 28 cm, how much force must the biceps generate in order for the child to lift an 8 kg stack of books with one hand?

3. Put yourself in the position of the scientists who discovered the sliding filament mechanism of muscle contraction, by considering what molecular changes must be involved to account for the observed alterations in the banding pattern during contraction. If you were comparing a relaxed and contracted muscle fibre under an electron microscope (see Figure 6-3a), how could you determine that the thin filaments do not change in length during muscle contraction? You cannot see or measure a single thin filament at this magnification. (*Hint:* What landmark in the banding pattern represents each end of the thin filament? If these

landmarks are the same distance apart in a relaxed and contracted fibre, then the thin filaments must not change in length.)

4. What type of off-the-snow training would you recommend for a competitive downhill skier versus a competitive cross-country skier? What adaptive skeletal muscle changes would you hope to accomplish in the athletes in each case?

5. When the bladder is filled and the micturition (urination) reflex is initiated, the nervous supply to the bladder promotes contraction of the bladder and relaxation of the external urethral sphincter, a ring of muscle that guards the exit from the bladder. If the time is inopportune for bladder emptying when the micturition reflex is initiated, the external urethral sphincter can be voluntarily tightened to prevent urination even though the bladder is contracting. Using your knowledge of the muscle types and their innervation, of what types of muscle are the bladder and the external urethral sphincter composed, and what branch of the efferent division of the peripheral nervous system supplies each of these muscles?

## CLINICAL CONSIDERATION

(Explanation in Appendix E, p. A-43)

Jason W. is waiting impatiently for the doctor to finish removing the cast from his leg, which Jason broke the last day of school six weeks ago. Summer vacation is half over, and he hasn't been able to swim, play baseball, or participate in any of his favourite sports. When the cast is finally off, Jason's excitement gives way to concern when he sees that the injured limb is noticeably smaller in diameter than his normal leg. What explains this reduction in size? How can the leg be restored to its normal size and functional ability?

## Normal: Human Movement

Generally, *locomotion* is simply walking, jogging, or running. Locomotion addresses the body's ability to move (i.e., locomote) from place to place. Human locomotion is bipedal (having two legs and feet). It uses the oscillating contraction–relaxation of skeletal muscles in a coordinated manner which, as a result, allows one foot to be moved in front of the other. The locomotive muscle groups are those of the hip, knee, and ankle flexors and extensors. These muscle groups are used to walk, jog, run, and stand.

To examine locomotion, we need to first discuss the human *gait* and outline a *gait cycle*. By definition, *gait* is the rhythmic alternating movement of the legs resulting in the forward movement of the body. Gait has three phases: stance (single leg support of body weight), swing (unsupported), and double support (standing). A gait cycle is the activity that takes place between heel strike of one limb and the successive heel strike of the same limb. The analysis of gait is often discussed in terms of the following components: (1) heel strike, (2) mid-stance, (3) toe off, and (4) acceleration. Many different factors influence each of these components (e.g., how much knee flexion and hip extension a person uses). Individuals have a unique characteristic gait—their gait pattern. For example, during heel strike, does the foot contact the ground more toward the heel or midfoot? For further discussion on variation and evolution in gait patterns and their relationship to running shoes and injury, Dr. Daniel Lieberman from Harvard has written some interesting articles on the topic (e.g., [2010, Jan. 28]. Foot strike patterns and collision forces in habitually barefoot versus shod runners. *Nature, 463*[7280]: 531–5; [2004, Nov 18]. Endurance running and the evolution of Homo. *Nature, 432*[7015]: 345–52). Although there is variation with each person's gait, generally all gait cycles can be broken down into the phases and components we just described. So, what controls gait and allows locomotion to occur in a smooth reflexive manner?

Several physiological components are required to accomplish locomotion. First we focus on the *control* of locomotion, which is the responsibility of the central and peripheral nervous systems (CNS and PNS; see Chapters 3 and 4). Then we discuss how skeletal muscle contractions (concentric and eccentric) enable external motion to occur.

Locomotion is executed for the most part without input from higher brain centres and requires no conscious control. Instead, the body uses learned motor programs (patterns) stored in groups of neurons in the spinal cord and brain stem; these are termed *central pattern generators (CPGs)*.

Furthermore, information from a variety of peripheral sensory receptors (e.g., muscle spindles) provides feedback to the CPGs to confirm the movement patterns and assist with corrective actions when necessary. Both afferent (sensory) and efferent (motor) neurons are needed, as well as ascending (sensory) and descending (motor) spinal pathways. Of greatest interest are the CPGs, sensory receptors, and descending motor pathways.

## Central Pattern Generators

Locomotion is complex, as there are both limb movements (e.g., leg) and segmental movements (e.g., movement of the calf independent of the thigh) that need to be coordinated during walking and running. Although sensory feedback and the mechanical properties of the musculoskeletal system contribute to this coordination, it is important to realize that the rhythms and patterns arise from the spine and brain stem (i.e., CPGs).

The body uses various CPGs to control many processes aside from locomotion—for example, breathing, chewing, and posture. CPGs are neural networks, located in both the spine and brain stem, that can produce rhythmic outputs without supraspinal input, and that bring about most rhythmic motor (movement) patterns. This suggests that rhythmic pattern generation does not rely on the entire nervous system functioning as a whole, but rather depends on smaller sections of the nervous system functioning as autonomic (automatic) neural networks. Sensory (afferent) input to the CPGs is used to make quick and complex decisions regarding repetitive movements such as when environmental changes (e.g., a change in walking surfaces—flat to uneven) are encountered.

The automatic nature of the CPGs is not fully understood, but it is believed that automaticity makes locomotion easier because the locomotive nerves (afferent and efferent) and skeletal muscles form a collective interactive system—that is, all components are interdependent. Motor pools are an example of this concept of grouping. Motor pools consist of all the motor neurons that innervate the muscle fibres within the same whole muscle. Within a motor pool there are different sizes and types of motor neurons randomly distributed within the muscle. This may seem complex, but this organization makes locomotion easier than it would be if the CNS had to coordinate each individual motor axon and muscle fibre separately. By grouping the nerves and muscles into systems, it allows the control of the movement patterns to be simplified, made formulaic, varying only in speed of movement or until sensory or supraspinal input regarding the environment (e.g., uneven surfaces) indicates an adjustment is required.

The spinal cord also interprets incoming sensory information to make the necessary adjustments in the locomotive pattern, adjusting the gait cycle. The spinal cord interprets the sensory information from all aspects of the gait during locomotion—for example, foot contact, force, changes in joint angle and acceleration—and predicts which groups of neurons (i.e., systems) to activate and which to inhibit to allow movement to continue or stop. Thus, to make adjustments in gait, sensory input from peripheral afferent receptors is paramount.

## Peripheral Sensors

The sensory receptors used in locomotion are located peripherally, and referred to as *mechanoreceptors* (see Chapter 4); they sense physiological and mechanical changes and send the information to the spinal cord for processing. Muscle spindles, Golgi tendon organs (GTOs), and free nerve endings provide much of the sensory information for *proprioception* associated with locomotion. Mechanoreceptors are located in the skeletal muscles, tendons, and joints, and provide the spinal cord with information regarding limb position and movement (kinematics). About half of the nerves in the skeletal muscle system are sensory; this provides a great deal of feedback to the CNS. To give some perspective, it is estimated that one large afferent neuron could provide in excess of 8 billion action potentials (APs) per second. Some mechanoreceptors sense force and pressure, but we will focus on the muscle spindles, which sense change in muscle fibre length.

## Muscle Spindles

Muscle spindles (see Chapter 7) contain striated muscle (i.e., intrafusal fibres) similar to those of the extrafusal fibres (i.e., skeletal muscle). The muscle spindle is surrounded by a collagen sheath that holds the contents collectively as a unit. The intrafusal fibres are innervated by $\gamma$-motor neurons (efferent) as well as sensory afferent neurons (primary and secondary). The sensory neurons generate APs that are sent to the spine when there is even a minute change in length of the intrafusal fibres; with respect to locomotion it is the primary afferent that is most important (i.e., Ia). The sensory neurons respond to passive change in length of the intrafusal fibres. This change in length is brought about by a change in length of the skeletal muscle fibres, as the extrafusal fibres and intrafusal fibres run in parallel to each other.

During locomotion, the muscle spindles send sensory input to the spine, thereby providing information about changes in fibre length as the muscles move through their normal range of motion associated with the gait cycle. So as the leg swings forward, the quadriceps muscle begins to concentrically contract (shorten) as it readies to plant (heel strike) for the stance phase and bear the weight of the body. The muscle spindles monitor both the change in length and the rate of change of the quadriceps muscle millisecond by millisecond, as well as all other muscles associated with locomotion.

## Locomotion

During walking, the muscle spindles of the lower limb musculature (e.g., quadriceps) are continuously activated, whether the leg is planted during stance phase or in the air during swing phase. It is irrelevant whether the skeletal muscles are eccentrically or concentrically contracting; the muscle spindles are activated because there is a change in length within the skeletal muscles and thus within the muscle spindles.

This muscle activity or activation pattern can be measured by electromyography (EMG), which senses the electrical activity (both the frequency and amplitude) of muscles. An interesting characteristic of muscle activation identified by EMG is that the $\gamma$-motor and $\alpha$-motor neurons activation patterns differ not only from muscle fibre to fibre, but also from step to step within the same muscle fibres. This provides a clue as to the complexity of the nervous signalling from the periphery to the spinal cord, which interprets and responds to the physiological and mechanical stimuli provided by the spindles from step to step, millisecond to millisecond.

In practical terms, the muscle spindles sense where the stance phase and swing phase of walking (or running) should begin and end. As the swing phase ends, the stance phase must begin, and thus the weight of the body must be supported. As the body weight is carried, it is sensed by the GTOs and free nerve endings to provide feedback on joint angle and the level of force needed to support the body weight. This is important for a smooth transition from swing phases to stance in the gait cycle, because it permits a smooth locomotion that is characteristic in the healthy human.

It should be noted that the GTOs, contrary to popular belief, are not simply a safety mechanism used to inhibit skeletal muscles when the muscular force generated is too great. The sensory information gathered by the GTOs is used to inhibit specific muscles during a normal walking gate, even when the forces within the muscles are relatively low. Additionally, free nerve endings are used to gather sensory information regarding mechanical stimuli such as touch, pressure, and stretch, and sends the information (e.g., sensing joint angle) to the spinal cord.

The muscle spindles, GTOs, and free nerve endings provide a wealth of proprioceptive information to the spine for incorporation and interpretation by the CPGs. It is proposed that the spine treats this incoming information as a *recognizable pattern* and compares it to known templates for locomotion. Remember, this is how we get the automaticity demonstrated by the spine (CPGs) during walking. This information is used to predictably adjust the kinematics of walking and running, which reduces the need for supraspinal input and thus conscious control.

## Supraspinal Control

As discussed, much of locomotion is directed by automaticity in the neurons associated with the CPGs of the spine and brain stem. However, evidence suggests that the *initiation and gross adjustments* of locomotion are via the supraspinal centre, the motor cortex (divisions: primary motor cortex, posterior parietal cortex, supplementary motor area, and premotor cortex). Additionally, the motor cortex is important for executing *nonrepetitive* skilled movement patterns and adjusting activation patterns of efferent motor output in unusual environments (e.g., running on uneven ground, cross-country running).

The supraspinal descending pathways used for locomotion are the reticulospinal, corticospinal, rubrospinal, and vestibulospinal tracts, which are efferent motor pathways that activate the skeletal muscles for the purposes of locomotion (among other things). Each region varies in level of activity depending on the phase of gait (e.g., swing, stance). It is fair to say that without the spinal tracts, particularly the reticulospinal tract, it is doubtful that locomotion as we know it (coordinated, smooth, modifiable, etc.) could take place.

The reticulospinal tract is formed by neuronal fibres originating in the pons and medulla (brain stem) and that descend through the spinal cord to innervate motor neurons. The cerebellum receives input from the reticulospinal tract and is vital for adjusting supraspinal input, modulating sensory feedback from the spinal cord, and adapting motor output based on need. The cerebellum also receives input from CPGs to assist in modulating motor output to the locomotor muscles. The reticulospinal tract takes the input from the brain stem, and the modulated output from the cerebellum, and continues to descend the spine to transfer the information to the locomotor muscles.

Consequently, the brain stem and cerebellum allow for unplanned modifications in locomotion such as occur for example when we encounter unforeseen objects or slippery surfaces. The combined functioning of the motor cortex, brain stem, and cerebellum enable us to plan for anticipated changes in our immediate environment that require a response. For example, when we are standing in a subway car, it is important that the motor cortex plans for the car slowing down at the next subway stop. In this case, the motor cortex initiates the transition of our body weight toward the direction opposite from where the car is moving, thereby offsetting the forward movement. Simultaneously, the brain stem and cerebellum compare the plan initiated by the motor cortex and the sensory feedback from the peripheral receptors and assess the need for adjustment: that is, did the initiated pattern match what happened? Adjustments may take place if the subway car driver stops the car at a faster rate than we anticipated, and thus our lean toward the rear of the car may need to be greater than anticipated by the motor cortex's initial plan.

When broken down into its component parts, locomotion could be seen as a complex task; in fact, it is facilitated by (1) having a lot of sensory input from a variety of sensors, (2) using motor patterns or templates for comparison, (3) grouping the neural activity into smaller systems, (4) having CPGs in the spine and brain stem make fast adjustments in neural output based on the sensory input, and (5) having automaticity so that the general process of walking and all necessary adjustments do not need to reach conscious awareness. All of these contributions reduce the need to actively think about each step of walking and running.

We take much of this integration for granted. But what happens when things go wrong? For example, what is the outcome when disease begins to disrupt the neural networking and communication between the brain, spinal cord, and skeletal muscles? We now examine abnormal function brought on by disease.

## Abnormal: Parkinson's Disease

One afternoon, 38-year-old JH sat down to read a book. After a while he noticed that both his left index finger and thumb had a slight tremor. Upon reflection, he realized that this had been going on for a month or so. Since he worked on a computer all day and was constantly texting friends and family, he thought that perhaps it was just that his fingers were stressed from constantly being in motion. A few months later, more of his fingers would now twitch on occasion and some of his toes had started

as well. He started reading about these symptoms on the Internet and decided that he should see his family physician. Two weeks later, JH was diagnosed with *Parkinson's disease*.

Parkinson's disease is a neurodegenerative disorder that has no cure. The early symptoms of Parkinson's disease are usually movement-related and manifest primarily as tremors at rest. As the disease progresses, these movement disorders worsen and can include slowness of movement, rigidity, impaired balance, and difficulty with walking. However, in addition to movement disorders, Parkinson's disease is also associated with several other symptoms such as cognitive disorders and dementia, behavioural problems, altered senses, and sleep disturbances.

Parkinson's disease is clearly a complex disease that can affect several of the body's systems. In this section, we examine the changes that occur in the central nervous system and how this affects the signals that are sent to other regions of the brain and the rest of the body. While we focus primarily on the integrative aspect of movement control, we also discuss the integrative nature of the non-movement symptoms as well.

## What Is the Cause of Parkinson's Disease?

In general, Parkinson's disease is considered to be *idiopathic*, which means there is no known cause. There is, however, a genetic link for some people and mutations in several genes can lead to the disease. Regardless of the cause, research has identified the changes that occur within the central nervous system that ultimately lead to the symptoms of the disease.

The region of the brain most affected by Parkinson's disease is within the *basal ganglia*. As the word *ganglia* suggests, the basal ganglia actually consists of several structures, including the caudate nucleus, the globus pallidus, the putamen, the subthalamic nucleus, and the *substantia nigra*. Altogether, the basal ganglia is associated with many functions, including movement, attention, and learning, so it is not surprising that alterations within these regions underlie the symptoms of Parkinson's disease.

That said, not all of these basal ganglia structures are involved. The primary symptoms of Parkinson's disease are caused by the death of dopamine-secreting neurons within the *substantia nigra pars compacta*. However, what causes these neurons to die is unknown. One of the leading theories is that, for some reason, these cells begin to produce too much of a protein called *alpha-synuclein*. This protein is not soluble and aggregates to form what are called *Lewy bodies*. It is unknown if these Lewy bodies themselves can kill neurons. Along with the Lewy bodies, iron can be concentrated, and it is thought that perhaps this excessive iron causes oxidative stress that ultimately lead to cell death.

## How Does This Affect Movement?

The control of any body system is complex and highly regulated not only to maintain homeostasis but also to respond appropriately to changes within the body or the environment. Conceptually, most systems have stimulatory and inhibitory inputs (antagonistic control), and it is the relative level of these inputs that determines the activity of a particular system. If the body needs to increase the function of a system, it can either increase the stimulatory inputs or decrease the inhibitory inputs. To decrease function, the opposite may occur. You will see this concept repeated throughout the textbook as it applies to specific systems.

This concept of control can also be applied to the control of movement. As you learned in Chapter 3, the motor division of the peripheral nervous system is composed of the autonomic nervous system, for involuntary control, and the somatic nervous system, for voluntary control. Of these, the somatic nervous system has control over skeletal muscle. In the periphery, motor neurons extend from the spinal cord to innervate muscle cells. Motor neurons themselves can only stimulate these muscle cells, so they need to be able to regulate the rate at which they fire to control contraction and, thus, movement. This regulation is accomplished by balancing the stimulatory and inhibitory inputs to the dendrites of motor neurons. Therefore, the summation of these inputs determines the relative activity of the muscle neurons.

This balance of input has three levels of control regulated by the spinal cord, the brain stem, and the primary motor cortex. Spinal cord control is responsible primarily for reflexes and rhythmic movements, whereas the brain stem and higher brain centres are responsible for voluntary movement. Afferent fibres from the primary motor cortex descend to terminate on motor neurons (or regulatory interneurons) in the spinal cord. The brain stem also projects to motor neurons but serves as a processing centre and receives input from the cerebellum, premotor cortex, and the basal ganglia. Altogether, this complex regulation of movement allows for precise movements of voluntary skeletal muscles throughout the body. However, such complexity also permits diseases that affect specific brain areas, such as Parkinson's disease, to throw this modulation of movement out of balance.

The exact role that the substantia nigra pars compacta plays in movement is not fully understood. It has been shown that direct electrical stimulation of this brain region does not result in movement. However, decreased function of these neurons, as seen in Parkinson's disease, does have an effect on movement, in particular, fine motor control.

The pars compacta receives input primarily from the caudate and putamen. Because it uses dopamine as a neurotransmitter, the pars compacta's output is considered inhibitory; this output projects back to the caudate and putamen, which in turn project to the thalamus and the cortex. Under normal circumstances, this inhibitory output is considered one of the many factors that suppress movement. This is most notable when a muscle is at rest because the inhibitory pathways then dominate the control of movement.

People with Parkinson's disease begin to lose this inhibitory output on movement, so the balance of movement control is disrupted, which unmasks the underlying stimulatory pathways. The net consequence is the occurrence of resting tremors commonly associated with this disease. In the early stages, these tremors can be masked or overridden by voluntary movement. As the disease progresses, however, tremors can appear on top of voluntary movements. The later stages of the disease are also characterized by *rigidity* of movement—both flexors and extensors are simultaneously stimulated—and *bradykinesia* (slow movement) because it becomes difficult to initiate voluntary movement.

## Non-movement Symptoms of Parkinson's Disease

Parkinson's disease is frequently considered a disease of movement, but this is certainly not the case. Given the complex organization of the brain, it should be of no surprise that Parkinson's disease also affects behaviour, mood, and cognition. Some of the other observed symptoms include anxiety, depression, behavioural changes, sleep disturbances, and dementia.

It is estimated that more than 50 percent of persons with Parkinson's disease have some form of cognitive impairment ranging from mild to *dementia*. The precise mechanisms by which this occurs is unclear, but there are similarities with a well-known dementia disease, Alzheimer's. In both Parkinson's and Alzheimer's, there is a progressive loss of memory, reasoning, and language. That said, the dementia associated with Parkinson's disease is more likely associated with the second most common form of dementia, *Lewy Body Dementia*. As already mentioned, the aggregation of alpha-synuclein leads to the presence of Lewy bodies, which is a progressive degradation of cholinergic neurons associated with cognition. However, both the role of Lewy bodies in the progressive degradation of dopaminergic neurons in the substantia nigra, and also the way that Lewy bodies contribute to cholinergic cell death remain unclear.

## Integrated Control of Movement

The purpose of this section has been to highlight the integrative nature of movement. Control of movement involves multiple steps from the initiation of movement in the primary motor cortex to the peripheral grouping of neurons (motor pools), which ultimately leads to precise skeletal muscle control and locomotion. Within this cascade of events, there are many means by which movement can be modulated. In this particular case, we used Parkinson's disease as an example of such modulation. As such, we demonstrated that the removal of one of the pathways that endogenously suppresses movement can drastically disrupt the balance of control and cause a deleterious movement disorder.

## Cardiac Physiology

Body systems maintain homeostasis

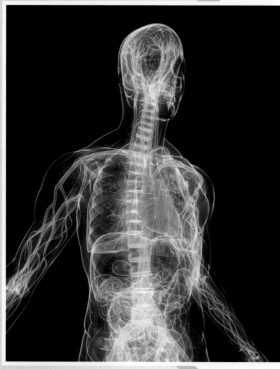

Sebastian Kaulitzki/shutterstock

Human vascular system

### Homeostasis

The circulatory system contributes to homeostasis by transporting $O_2$, $CO_2$, wastes, electrolytes, and hormones from one part of the body to another.

Homeostasis is essential for survival of cells

Cells make up body systems

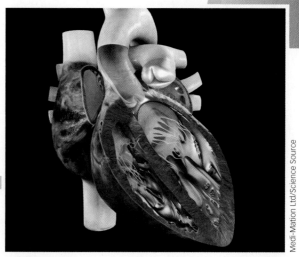

Medi-Mation Ltd/Science Source

Heart valves

The maintenance of homeostasis depends on essential materials, such as oxygen and nutrients, being continually picked up from the external environment and delivered to the cells and on waste products being continually removed. Furthermore, excess heat generated by muscles must be transported to the skin where it can be lost from the body surface to help maintain body temperature. Homeostasis also depends on the transfer of hormones, which are important regulatory chemical messengers, from their site of production to their site of action. The circulatory system, which contributes to homeostasis by serving as the body's transport system, consists of the heart, blood vessels, and blood.

All body tissues constantly depend on the life-supporting blood flow the heart provides them by contracting, or beating. The heart drives blood through the blood vessels for delivery to the tissues in sufficient amounts, whether the body is at rest or engaging in vigorous exercise.

# Cardiac Physiology

▌ **Clinical Connections**

Debbie arrived home after a long day at work and, feeling more tired than usual, sat down to relax for a few minutes. Suddenly, she felt very strange and started to feel an almost crushing pressure on her chest. She started sweating and felt as if she was unable to catch her breath. Her spouse immediately called 911, and within 40 minutes she was in the hospital emergency room.

On arrival, her electrocardiogram showed ST elevation, and she was immediately given a thrombolytic. She remained conscious, and the severity of her symptoms improved. Blood taken about two hours after her arrival showed elevated myoglobin. Subsequent blood samples also had elevated levels of troponin T. After undergoing an angiogram that showed an almost complete blockage of her left main coronary artery, she was immediately sent to surgery for a coronary artery bypass graft. Debbie had an acute myocardial infarction.

## 8.1 | Introduction

From just a matter of days following conception until death, the beat goes on. In fact, throughout an average human lifespan, the heart contracts about 3 billion times, never stopping except for a fraction of a second to fill between beats. Within the developing embryo, the heart is the first organ to become functional, which occurs about three weeks after conception. At this time, the human embryo is only a few millimetres long, about the size of a capital letter on this page.

Why does the heart develop so early, and why is it so crucial throughout life? It is this important because the circulatory system is the body's transport system. A human embryo, having very little yolk available as food, depends on promptly establishing a circulatory system that can interact with the mother's circulation to pick up and distribute to the developing tissues the supplies so critical for survival and growth. Thus begins the story of the circulatory system, which continues throughout life to be a vital pipeline for transporting materials on which the cells of the body absolutely depend.

The **circulatory system** has three basic components:

1. The **heart** serves as the pump that imparts pressure to the blood to establish the pressure gradient needed for blood to flow to the tissues. Like all liquids, blood flows down a pressure gradient from an area of higher pressure to an area of lower pressure. This chapter focuses on cardiac physiology (*cardia* means "heart").

2. The **blood vessels** serve as the passageways through which blood is directed and distributed from the heart to all parts of the body and subsequently returned to the heart (see Chapter 9).

3. **Blood** is the transport medium within which materials being moved long distances in the body (such as oxygen, carbon dioxide, nutrients, wastes, electrolytes, and hormones) are dissolved or suspended (see Chapter 10).

Blood travels continuously through the circulatory system to and from the heart through two separate vascular (blood vessel) loops, both originating and terminating at the heart (❯ Figure 8-1). The **pulmonary circulation** consists of a closed loop of vessels carrying blood between the heart and lungs (*pulmo* means "lung"). The **systemic circulation** is a circuit of vessels carrying blood between the heart and other body systems.

## 8.2 Anatomy of the Heart

The heart is a hollow, muscular organ about the size of a clenched fist. It lies in the **thoracic cavity** (chest) at about the midline, in between the **sternum** (breastbone) anteriorly and the **vertebrae** (backbone) posteriorly. Place your hand over your heart. People usually put their hand on the left side of the chest, even though the heart is actually in the middle of the chest. The heart has a broad **base** at the top and tapers to a pointed tip, the **apex**, at the bottom. It is situated at an angle under the sternum so that its base lies predominantly to the right and the apex to the left of the sternum. When the heart beats forcefully, the apex actually thumps against the inside of the chest wall on the left side. Because we become aware of the beating heart through the apex beat on the left side of the chest, we tend to think the entire heart is on the left.

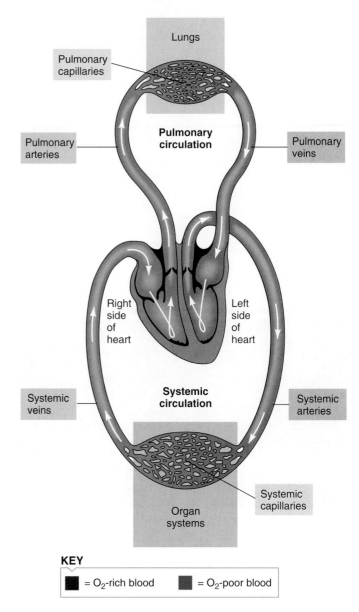

**KEY**

■ = $O_2$-rich blood ■ = $O_2$-poor blood

❯ **FIGURE 8-1 Pulmonary and systemic circulation in relation to the heart.** The circulatory system consists of two separate vascular loops: the pulmonary circulation, which carries blood between the heart and lungs; and the systemic circulation, which carries blood between the heart and organ systems.

### A dual pump

Even though anatomically the heart is a single organ, the right and left sides of the heart function as two separate pumps. The heart is divided into right and left halves and has four chambers: an upper and a lower chamber within each half (❯ Figure 8-2a). The upper chambers, the **atria** (singular, *atrium*), receive blood returning to the heart and transfer it to the lower chambers, the ventricles, which pump blood from the heart. The vessels that return blood from the tissues to the atria are **veins**, and those that carry blood away from the ventricles to the tissues are **arteries**. The two halves of the heart are separated by the **septum**, a continuous muscular partition that prevents mixture of blood from the two sides of the heart. This separation is extremely

The heart's position between two bony structures, the sternum and vertebrae, makes it possible to manually drive blood out of the heart when it is not pumping effectively; this is done by rhythmically depressing the sternum. This manoeuvre compresses the heart between the sternum and vertebrae so that blood is squeezed out

into the blood vessels, maintaining blood flow to the tissues. Often, this *external cardiac compression*, which is part of **cardiopulmonary resuscitation (CPR)**, serves as a lifesaving measure until appropriate therapy can restore the heart to normal function.

important, because the right half of the heart is receiving and pumping oxygen-poor blood, whereas the left side of the heart receives and pumps oxygen-rich blood.

### THE COMPLETE CIRCUIT OF BLOOD FLOW

Let us look at how the heart functions as a dual pump, by tracing a drop of blood through one complete circuit (⟩ Figure 8-2a and ⟩ 8-2b). Blood returning from the systemic circulation enters the right atrium via two large veins, the **venae cavae**: one returns blood from above heart level, and the other returns blood from below. The drop of blood entering the right atrium has returned from the body tissues, where oxygen has been taken from it and carbon dioxide has been added to it. This partially deoxygenated blood flows from the right atrium into the right ventricle, which pumps it out through the **pulmonary artery**. As the pulmonary artery leaves the right ventricle, it immediately forms two branches, one going to each of the two lungs. Thus, the *right side of the heart receives blood from the systemic circulation and pumps it into the pulmonary circulation.*

Within the lungs, the drop of blood loses its extra carbon dioxide and picks up a fresh supply of oxygen before being returned to the left atrium via the **pulmonary veins** coming from both lungs. This oxygen-rich blood returning to the left atrium subsequently flows into the *left ventricle, the pumping chamber that propels the blood to all body systems; that is, the left side of the heart receives blood from the pulmonary circulation and pumps it into the systemic circulation.* The single large artery carrying blood away from the left ventricle is the **aorta**. Major arteries branch from the aorta to supply the various organs of the body.

In contrast to the pulmonary circulation, in which all the blood flows through the lungs, the systemic circulation may be viewed as a series of parallel pathways. Part of the blood pumped out by the left ventricle goes to the muscles, part to the kidneys, part to the brain, and so on (Figure 8-2b). In this way, the output of the left ventricle is distributed so that each part of the body receives a fresh blood supply; the same arterial blood does not pass from organ to organ. Accordingly, the drop of blood we are tracing goes to only one of the systemic organs. Tissue cells within the organ take oxygen from the blood and use it to oxidize nutrients for energy production; in the process, the tissue cells form carbon dioxide as a waste product that is added to the blood. The drop of blood, now partially depleted of oxygen and

increased in carbon dioxide, returns to the right side of the heart, which once again pumps it to the lungs. One circuit is complete.

### COMPARISON OF THE RIGHT AND LEFT PUMPS

A critical concept is that, despite being described as two separate pumps, both sides of the heart simultaneously pump equal amounts of blood. The volume of oxygen-poor blood being pumped to the lungs by the right side of the heart soon becomes the same volume of oxygen-rich blood being delivered to the tissues by the left side of the heart. The pulmonary circulation is a low-pressure, low-resistance system, whereas the systemic circulation is a high-pressure, high-resistance system. Within the circulatory system, pressure is the force exerted on the vessel walls by the blood pumped into the vessels by the heart. Resistance is the opposition to blood flow, largely caused by friction between the flowing blood and the vessel wall. Even though the right and left sides of the heart pump the same amount of blood, the left side performs more work, because it pumps an equal volume of blood at a higher pressure into a higher-resistance and longer system. Accordingly, the heart muscle on the left side is much thicker than the muscle on the right side, making the left side a stronger pump (⟩ Figure 8-2c).

## Pressure-operated heart valves

Blood flows through the heart in one fixed direction from veins to atria to ventricles to arteries. The presence of four one-way heart valves ensures this unidirectional flow of blood. The valves are positioned so that they open and close passively because of pressure differences, similar to a one-way door (⟩ Figure 8-3). A forward pressure gradient (i.e., a greater pressure behind the valve) forces the valve open, much as you open a door by pushing on one side of it, whereas a backward pressure gradient (i.e., a greater pressure in front of the valve) forces the valve closed, just as you apply pressure to the opposite side of the door to close it. Note that a backward gradient can force the valve closed but cannot force it to swing open in the opposite direction; that is, heart valves are not like swinging, saloon-type doors.

### AV VALVES BETWEEN THE ATRIA AND VENTRICLES

Two of the heart valves, the **right and left atrioventricular (AV) valves**, are positioned between the atrium and the ventricle on the right and left sides, respectively (⟩ Figure 8-4a). These valves

8

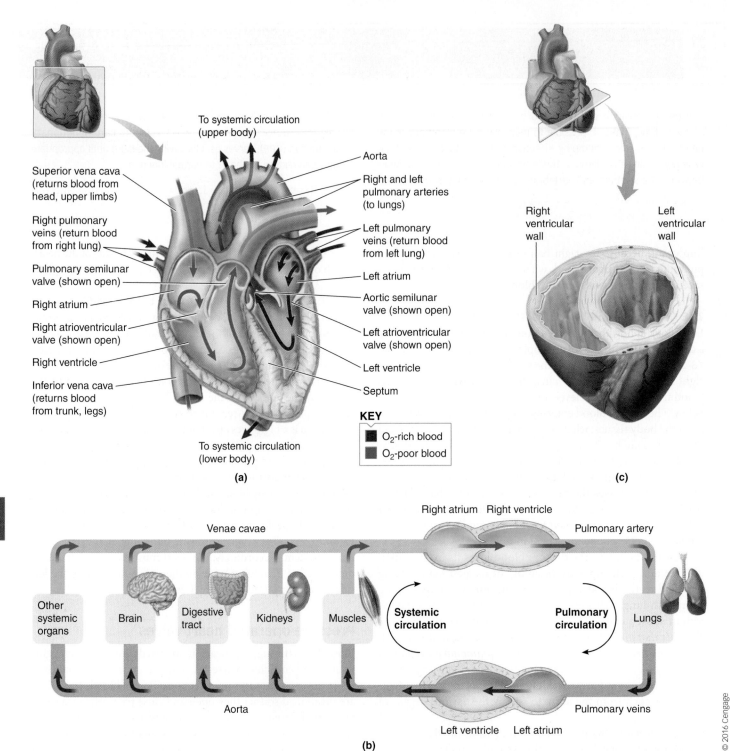

To systemic circulation
(upper body)

Aorta

Superior vena cava
(returns blood from
head, upper limbs)

Right and left
pulmonary arteries
(to lungs)

Right pulmonary
veins (return blood
from right lung)

Left pulmonary
veins (return blood
from left lung)

Pulmonary semilunar
valve (shown open)

Left atrium

Aortic semilunar
valve (shown open)

Right atrium

Right atrioventricular
valve (shown open)

Left atrioventricular
valve (shown open)

Right ventricle

Left ventricle

Inferior vena cava
(returns blood
from trunk, legs)

Septum

To systemic circulation
(lower body)

**KEY**

■ $O_2$-rich blood
■ $O_2$-poor blood

**(a)**

Right
ventricular
wall

Left
ventricular
wall

**(c)**

Venae cavae

Right atrium   Right ventricle

Pulmonary artery

Other
systemic
organs

Brain

Digestive
tract

Kidneys

Muscles

**Systemic
circulation**

**Pulmonary
circulation**

Lungs

Aorta

Pulmonary veins

Left ventricle   Left atrium

**(b)**

© 2016 Cengage

› FIGURE 8-2 **Blood flow through and pump action of the heart.** (a) Blood flow through the heart. (b) Dual pump action of the heart. The right side of the heart receives $O_2$-poor blood from the systemic circulation and pumps it into the pulmonary circulation. The left side of the heart receives $O_2$-rich blood from the pulmonary circulation and pumps it into the systemic circulation. Note the parallel pathways of blood flow through the systemic organs. (The relative volume of blood flowing through each organ is not drawn to scale.) (c) Comparison of the thickness of the right and left ventricular walls. Note that the left ventricular wall is much thicker than the right wall.

let blood flow from the atria into the ventricles during ventricular filling (when atrial pressure exceeds ventricular pressure) but prevent the backflow of blood from the ventricles into the atria during ventricular emptying (when ventricular pressure

greatly exceeds atrial pressure). If the rising ventricular pressure did not force the AV valves to close as the ventricles contracted to empty, much of the blood would inefficiently be forced back into the atria and veins instead of being pumped into the

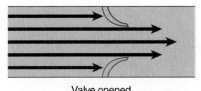

When pressure is greater behind the valve, it opens.

**Valve opened**

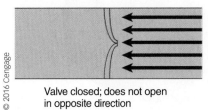

When pressure is greater in front of the valve, it closes. Note that when pressure is greater in front of the valve, it does not open in the opposite direction; that is, it is a one-way valve.

**Valve closed; does not open in opposite direction**

© 2016 Cengage

> **FIGURE 8-3 Mechanism of valve action**

arteries. The right AV valve is also called the **tricuspid valve** (*tri* means "three"), because it consists of three cusps or leaflets (⟩ Figure 8-4b). Likewise, the left AV valve, which has two cusps, is often called the **bicuspid valve** (*bi* means "two") or, alternatively, the **mitral valve** (because of its physical resemblance to a bishop's traditional hat).

The edges of the AV valve leaflets are fastened by tough, thin, fibrous cords of tendinous-type tissue, the **chordae tendineae**, which prevent the valves from being everted. That is, the chordae tendineae prevent the AV valve from being forced by the high ventricular pressure to open in the opposite direction into the atria. These cords extend from the edges of each cusp and attach to small, nipple-shaped **papillary muscles**, which protrude

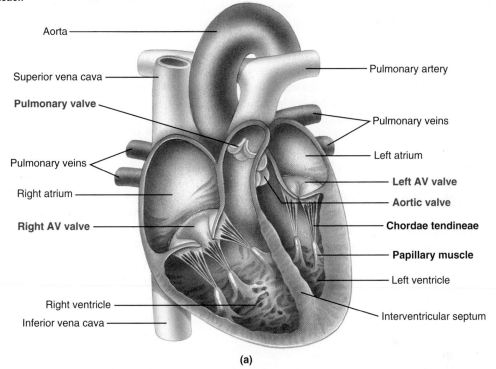

**(a)**

**8**

> **FIGURE 8-4 Heart valves.**
(a) Longitudinal section of the heart, depicting the location of the four heart valves.
(b) The heart valves in closed position, viewed from above.
(c) Prevention of eversion of the AV valves. Eversion of the AV valves is prevented by tension on the valve leaflets exerted by the chordae tendineae when the papillary muscles contract. (d) Prevention of eversion of the semilunar valves. When the semilunar valves are swept closed, their upturned edges fit together in a deep, leak-proof seam that prevents valve eversion.

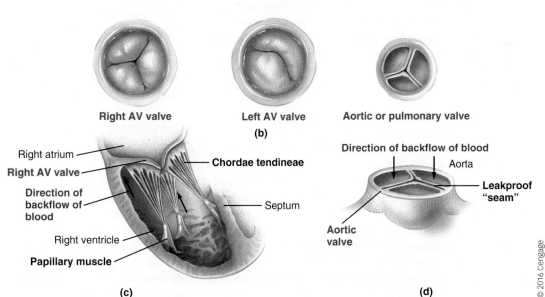

**(b)**

**(c)**

**(d)**

© 2016 Cengage

Cardiac Physiology **349**

from the inner surface of the ventricular walls (*papilla* means "nipple"). When the ventricles contract, the papillary muscles also contract, pulling downward on the chordae tendineae. This pulling exerts tension on the closed AV valve cusps and holds them in position, much like tethering ropes hold down a hot-air balloon. This action helps keep the valve tightly sealed in the face of a strong backward pressure gradient (❯ Figure 8-4c).

### SEMILUNAR VALVES BETWEEN THE VENTRICLES AND MAJOR ARTERIES

The two remaining heart valves, the **aortic valve** and **pulmonary valve**, lie at the juncture where the major arteries leave the ventricles (Figure 8-4a). They are known as **semilunar valves** because they have three cusps, each resembling a shallow half-moon-shaped pocket (*semi* means "half"; *lunar* means "moon") (Figure 8-4b). These valves are forced open when the left and right ventricular pressures exceed the pressure in the aorta and pulmonary artery, respectively, during ventricular contraction and emptying. Closure results when the ventricles relax and ventricular pressures fall below the aortic and pulmonary artery pressures. The closed valves prevent blood from flowing from the arteries back into the ventricles from which it has just been pumped.

The semilunar valves are prevented from everting by the anatomic structure and positioning of the cusps. During ventricular relaxation a backward pressure gradient is created, the back surge of blood fills the pocket-like cusps and sweeps them into a closed position, with their unattached upturned edges fitting together in a deep, leak-proof seam (❯ Figure 8-4d).

### NO VALVES BETWEEN THE ATRIA AND VEINS

Even though there are no valves between the atria and veins, backflow of blood from the atria into the veins usually is not a significant problem, for two reasons: (1) atrial pressures usually are not much higher than venous pressures, and (2) the sites where the venae cavae enter the atria are partially compressed during atrial contraction.

### FIBROUS SKELETON OF THE VALVES

Four interconnecting rings of dense connective tissue provide a firm base for attachment of the four heart valves (❯ Figure 8-5). This **fibrous skeleton**, which separates the atria from the ventricles, also provides a fairly rigid structure for attachment of the cardiac muscle. The atrial muscle mass is anchored above the rings, and the ventricular muscle mass is attached to the bottom of the rings.

It might seem rather surprising that the inlet valves to the ventricles (the AV valves) and the outlet valves from the ventricles (the semilunar valves) all lie on the same plane through the heart, as delineated by the fibrous skeleton. This relationship comes about because the heart forms from a single tube that bends on itself and twists on its axis during embryonic development. Although this turning and twisting make studying the structural relationships of the heart more difficult, the twisted structure has functional importance in that it helps the heart pump more efficiently. We will see how this happens by examining the portion of the heart that actually generates the forces responsible for blood flow: the cardiac muscle within the heart walls.

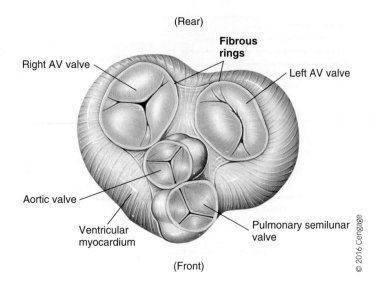

(Rear)

Fibrous rings

Right AV valve

Left AV valve

Aortic valve

Ventricular myocardium

Pulmonary semilunar valve

(Front)

© 2016 Cengage

❯ **FIGURE 8-5 Fibrous skeleton of the heart.** A view of the heart from above, with the atria and major vessels removed to show the heart valves and fibrous rings. Note that the inlet and outlet valves to the ventricles all lie on the same plane through the heart.

## The heart walls

The heart wall has three distinct layers:

- A thin inner layer, the **endothelium**, a unique type of epithelial tissue that lines the entire circulatory system
- A middle layer, the **myocardium**, which is composed of cardiac muscle and constitutes the bulk of the heart wall (*myo* means "muscle"; *cardia* means "heart")
- A thin external layer, the **epicardium**, that covers the heart (*epi* means "on")

The myocardium consists of interlacing bundles of cardiac muscle fibres arranged spirally around the circumference of the heart. The spiral arrangement is due to the heart's complex twisting during development. As a result of this arrangement, when the ventricular muscle contracts and shortens, the diameter of the ventricular chambers is reduced while the apex is simultaneously pulled upward toward the top of the heart in a rotating manner. This motion squeezes the heart, which efficiently exerts pressure on the blood within the enclosed chambers and directs it upward toward the openings of the major arteries that exit at the base of the ventricles.

## Cardiac muscle fibres

The individual cardiac muscle cells are interconnected to form branching fibres, with adjacent cells joined end to end at specialized structures known as **intercalated discs**. Within an intercalated disc, there are two types of membrane junctions: desmosomes and gap junctions (❯ Figure 8-6). A *desmosome*, a type of adhering junction that mechanically holds cells together, is particularly abundant in tissues, such as the heart, that are subject to considerable mechanical stress (p. 34). At intervals along the intercalated disc, the opposing membranes approach each other very closely to form *gap junctions*, which are areas of low electrical resistance that allow

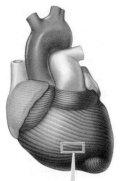

(a) Bundles of cardiac muscle are arranged spirally around the ventricle. When they contract, they "wring" blood from the apex to the base where the major arteries exit.

**Intercalated discs**

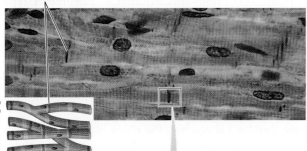

© 2016 Cengage

(b) Cardiac muscle fibres branch and are interconnected by intercalated discs.

**Desmosome**

Plasma membranes of adjacent cardiac muscle fibres

**Gap junction**          Action potential

**Intercalated disc**

© Science VU/Visuals Unlimited

(c) Intercalated discs contain two types of membrane junctions: mechanically important desmosomes and electrically important gap junctions.

> **FIGURE 8-6 Organization of cardiac muscle fibres.** Bundles of cardiac muscle fibres are arranged spirally around the ventricle. Adjacent cardiac muscle cells are joined end to end by intercalated discs, which contain two types of specialized junctions: desmosomes, which act as spot rivets mechanically holding the cells together; and gap junctions, which permit action potentials to spread from one cell to adjacent cells.

action potentials to spread from one cardiac cell to adjacent cells (p. 35). Some cardiac muscle cells can generate action potentials without any nervous stimulation. When one such cardiac cell spontaneously undergoes an action potential, the electrical impulse spreads to all the other cells that are joined by gap junctions in the surrounding muscle mass, so that they become excited and contract as a single, *functional syncytium* (p. 328). The atria and the ventricles each form a *functional syncytium* and contract as separate units. The synchronous contraction of the muscle cells that make up the walls of each of these chambers produces the pressure needed to eject the enclosed blood.

No gap junctions join the atrial and ventricular contractile cells, and, furthermore, the atria and ventricles are separated by the electrically nonconductive fibrous skeleton that surrounds and supports the valves. However, an important, specialized conduction system facilitates and coordinates transmission of electrical excitation from the atria to the ventricles to ensure synchronization between atrial and ventricular pumping.

Because of both the syncytial nature of cardiac muscle and the conduction system between the atria and ventricles, an impulse spontaneously generated in one part of the heart spreads throughout the entire heart. Therefore, unlike skeletal muscle, where graded contractions can be produced by varying the number of muscle cells contracting within the muscle (recruitment of motor units), either all the cardiac muscle fibres contract or none do. That said, cardiac contraction may be graded by varying the strength of contraction of all the cardiac muscle cells. You will learn more about this process later in Section 8.5.

## The pericardial sac

The heart is enclosed in the double-walled, membranous **pericardial sac** (*peri* means "around"). The sac consists of two layers: a tough, fibrous covering and a secretory lining. The outer fibrous covering of the sac attaches to the connective tissue partition that separates the lungs. This attachment anchors the heart so that it remains properly positioned within the chest. The sac's secretory lining secretes a thin **pericardial fluid**, which provides lubrication to prevent friction between the pericardial layers as they glide over each other with every beat of the heart.

*Clinical Note* **Pericarditis**, an inflammation of the pericardial sac that results in a painful friction rub between the two pericardial layers, occurs occasionally because of viral or bacterial infection.

Building on this foundation of heart structure, we first explain how action potentials are initiated and spread throughout the heart (Section 8.3), and then discuss how this electrical activity brings about coordinated pumping of the heart (Section 8.4).

**8**

## Check Your Understanding 8.1

1. Schematically draw the relationship of the pulmonary circulation and the systemic circulation to the chambers of the heart.
2. Name and discuss the functions of the four heart valves.
3. Describe the three layers of the heart walls.

# 8.3 | Electrical Activity of the Heart

The heart contracts, or beats, rhythmically as a result of action potentials that it generates by itself, a property called **autorhythmicity** (*auto* means "self"). These action potentials sweep across the cardiac muscle cell membranes to cause the contractions necessary to eject blood from the heart. There are two specialized types of cardiac muscle cells:

1. **Contractile cells**, which account for 99 percent of all cardiac muscle cells, do the mechanical work of pumping. These working cells normally do not initiate their own action potentials.

2. In contrast, the small but extremely important remainder of the cardiac cells, the **autorhythmic cells**, do not contract but instead are specialized for initiating and conducting the action potentials responsible for contraction of the working cells.

## Cardiac autorhythmic cells

In contrast to the cardiac contractile cells, in which the membrane remains at constant resting potential unless the cell is stimulated, the cardiac autorhythmic cells do not have a defined resting potential. Instead, they display *pacemaker activity*; that is, their membrane potential slowly depolarizes, or drifts, between action potentials until threshold is reached, at which time the membrane fires or generates an action potential. An autorhythmic cell membrane's slow drift to threshold is called the **pacemaker potential** (⟩ Figure 8-7). Through repeated cycles of drift and fire, these autorhythmic cells cyclically initiate action potentials, which then spread throughout the heart to trigger rhythmic beating without any nervous stimulation.

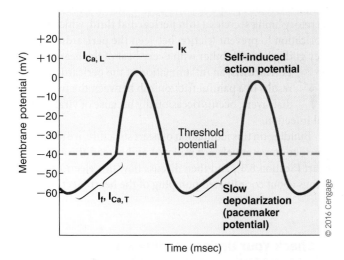

⟩ **FIGURE 8-7 Pacemaker activity of cardiac autorhythmic cells.** The first half of the pacemaker potential is due to activation of $I_f$, whereas the second half is due to opening of T-type $Ca^{2+}$ channels. Once threshold is reached, the rising phase of the action potential is due to opening of L-type $Ca^{2+}$ channels, whereas the falling phase is due to opening of $K^+$ channels.

Complex interactions of several different ionic mechanisms are responsible for the pacemaker potential. The first pacemaker current to be identified was given the name $I_f$ ("I" for current and "f" for funny). This current has very different characteristics from most cardiac ion channels in that this current is activated upon hyperpolarization. Once activated, both potassium and sodium move into the cell. This inward movement of cations causes the slow depolarization that is characteristic of pacemaker cells. $I_f$ channels have been identified as part of the hyperpolarization-activated cyclic nucleotide-gated channel (HCN) family. This is important as it not only indicates that these ion channels are hyperpolarizing but that their activity can also be modulated by cyclic nucleotides, which function as second messengers in most cell types.

Another ion channel thought to contribute to the pacemaker potential is a transient calcium ($Ca^{2+}$) channel: **T-type $Ca^{2+}$ channel ($I_{Ca, T}$)**. These ion channels differ from the longer-lasting, voltage-gated $Ca^{2+}$ channel, **L-type $Ca^{2+}$ channel ($I_{Ca, L}$)**, in that they open at lower membrane potentials. This means that the T-type $Ca^{2+}$ channels begin to open during the slow depolarization before the membrane reaches threshold. The resultant brief influx of $Ca^{2+}$ further depolarizes the membrane, helping to bring it to threshold.

Once threshold is reached, the rising phase of the action potential occurs in response to activation of L-type $Ca^{2+}$ channels and a subsequently large influx of $Ca^{2+}$. The $Ca^{2+}$-induced rising phase of a cardiac pacemaker cell differs from that in other excitable cells, where $Na^+$ influx rather than $Ca^{2+}$ influx swings the potential in the positive direction. Unlike cardiac muscle cells, true pacemaker cells, found in the core of the pacemaker region, do not have voltage-gated $Na^+$ channels. At threshold, L-type $Ca^{2+}$ channels open, causing $Ca^{2+}$ influx and depolarization of the membrane potential. The rate of this depolarization is slower than that seen in cells with $Na^+$ channels.

The falling phase is due, as usual, to the potassium ($K^+$) efflux that occurs when $K^+$ permeability increases as a result of activation of voltage-gated $K^+$ channels ($I_K$). After the action potential is over, slow closure of these $K^+$ channels coupled with the opening of $I_f$ channels initiates the next slow depolarization to threshold.

## The sinoatrial node

The specialized noncontractile cardiac cells capable of autorhythmicity lie in the following specific sites (⟩ Figure 8-8):

1. The **sinoatrial node (SA node)**, a small, specialized region in the right atrial wall near the opening of the superior vena cava.

2. The **atrioventricular node (AV node)**, a small bundle of specialized cardiac muscle cells located at the base of the right atrium near the septum, just above the junction of the atria and ventricles.

3. The **bundle of His (atrioventricular bundle)**, a tract of specialized cells that originates at the AV node and enters

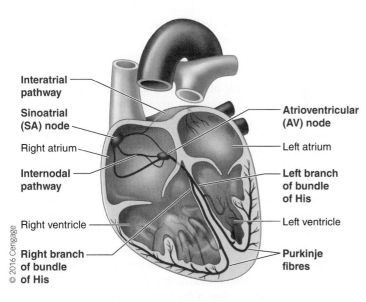

> FIGURE 8-8 **Specialized conduction system of the heart**

the interventricular septum. Here, it divides to form the right and left bundle branches that travel down the septum, curve around the tip of the ventricular chambers, and travel back toward the atria along the outer walls.

4. **Purkinje fibres**, small terminal fibres that extend from the bundle of His and spread throughout the endocardial surface of the ventricular myocardium, much like small twigs of a tree branch.

### NORMAL PACEMAKER ACTIVITY

Because these various autorhythmic cells have different rates of slow depolarization to threshold, the rates at which they are normally capable of generating action potentials also differ (▌Table 8-1). The heart cells with the fastest rate of action potential initiation are localized in the SA node. Once an action potential occurs in any cardiac muscle cell, it is propagated throughout the rest of the myocardium via gap junctions and the specialized conduction system. Therefore, the SA node, which normally has the fastest rate of autorhythmicity, at 70 to

### ▌ TABLE 8-1 Normal Rate of Action Potential Discharge in Autorhythmic Tissues of the Heart

| Tissue | Action Potentials per Minute* |
| --- | --- |
| **SA Node** (normal pacemaker) | 70–80 |
| **AV Node** | 40–60 |
| **Bundle of His and Purkinje Fibres** | 20–40 |

*In the presence of sympathetic tone; see p. 189 and p. 367.

80 action potentials per minute, drives the rest of the heart at this rate and thus is known as the **pacemaker** of the heart. That is, the entire heart becomes excited, triggering the contractile cells to contract and the heart to beat at the pace or rate set by SA node autorhythmicity, normally at 70 to 80 beats per minute. The other autorhythmic tissues cannot assume their own naturally slower rates, because they are activated by action potentials originating in the SA node before they can reach threshold at their own, slower rhythm.

The following analogy shows how the SA node drives the rest of the heart at its own pace. Suppose a train has 100 cars, 3 of which are equipped with engines and are capable of moving on their own; the other 97 cars must be pulled. One of the engine cars (the SA node) can travel at 70 km/h on its own, another engine car (the AV node) at 50 km/h, and the last engine car (the Purkinje fibres) at 30 km/h. If all these cars are joined, the engine car that can travel at 70 km/h will pull the rest of the cars at that speed. The engine cars that can travel at lower speeds on their own will be pulled at a faster speed by the car containing the fastest engine and therefore cannot assume their own slower rate as long as they are being driven by this faster car. The other 97 cars (nonautorhythmic, contractile cells), being unable to move on their own, will likewise travel at whatever speed the fastest car pulls them.

### ABNORMAL PACEMAKER ACTIVITY

*Clinical Note* If for some reason the fastest engine car breaks down (SA node damage), the next-fastest engine car (AV node) takes over, and the entire train travels at 50 km/h; that is, if the SA node becomes nonfunctional, the AV node assumes pacemaker activity. The non-SA-nodal autorhythmic tissues are **latent pacemakers** that can take over, although at a lower rate, if the normal pacemaker fails.

If impulse conduction becomes blocked between the atria and the ventricles, the atria continue to contract at the typical rate of 70 beats per minute, and the ventricular tissue, not being driven by the faster SA nodal rate, assumes its own much slower autorhythmic rate of about 30 beats per minute, initiated by the ventricular autorhythmic cells (Purkinje fibres). This situation is like a breakdown of the second engine car (AV node) so that the lead engine car (SA node) becomes disconnected from the slow third engine-containing car (Purkinje fibres) and the rest of the cars. The lead engine car continues at 70 km/h, while the rest of the train proceeds at 30 km/h. This **complete heart block** occurs when the conducting tissue between the atria and ventricles is damaged, as, for example, during a heart attack, and becomes nonfunctional. A ventricular rate of 30 beats per minute will support only a very sedentary existence; in fact, the patient usually becomes comatose.

When a person has an abnormally low heart rate, as in SA node failure or heart block, an **artificial pacemaker** can be used. Such an implanted device rhythmically generates impulses that spread throughout the heart to drive both the atria and the ventricles at the typical rate of 70 beats per minute.

Occasionally an area of the heart, such as a Purkinje fibre, becomes overly excitable and depolarizes more rapidly than the SA node (the slow engine car suddenly goes faster than the lead engine car). This abnormally excitable area, an **ectopic focus**, initiates a premature action potential that spreads throughout the rest of the heart before the SA node can initiate a normal action potential (*ectopic* means "out of place"). An occasional abnormal impulse from a ventricular ectopic focus produces a **premature ventricular contraction (PVC)**. If the ectopic focus continues to discharge at its more rapid rate, pacemaker activity shifts from the SA node to the ectopic focus. The heart rate abruptly becomes greatly accelerated and continues this rapid rate for a variable time period until the ectopic focus returns to normal. Such overly irritable areas may be associated with organic heart disease, but more frequently they occur in response to anxiety, lack of sleep, or excess caffeine, nicotine, or alcohol consumption.

We now turn our attention to how an action potential, once initiated, is conducted throughout the heart.

## Cardiac excitation

Once initiated in the SA node, an action potential spreads throughout the rest of the heart. For efficient cardiac function, the spread of excitation should satisfy three criteria:

1. *Atrial excitation and contraction should be complete before the onset of ventricular contraction.* Complete ventricular filling requires that atrial contraction precede ventricular contraction. During cardiac relaxation, the AV valves are open because pressure is lower in the ventricles compared to the atria, so venous blood entering the atria continues to flow directly into the ventricles. Almost 80 percent of ventricular filling occurs by this means before atrial contraction. When the atria do contract, more blood is squeezed into the ventricles to complete ventricular filling. Ventricular contraction then occurs to eject blood from the heart into the arteries.

   If the atria and ventricles were to contract simultaneously, the AV valves would close immediately, because ventricular pressures would greatly exceed atrial pressures. The ventricles have much thicker walls and, accordingly, can generate more pressure. Atrial contraction would be unproductive, because the atria could not squeeze blood into the ventricles through closed valves. Therefore, to ensure complete filling of the ventricles—to obtain the remaining 20 percent of ventricular filling that occurs during atrial contraction—the atria must become excited and contract before ventricular excitation and contraction. During a normal heartbeat, atrial contraction occurs about 160 msec (millisecond) before ventricular contraction.

2. Excitation of cardiac muscle fibres should be coordinated to ensure that each heart chamber contracts as a unit to pump efficiently. If the muscle fibres in a heart chamber became excited and contracted randomly rather than contracting simultaneously in a coordinated fashion, they would be unable to eject blood. A smooth, uniform ventricular contraction is essential to squeeze out the blood. As an analogy, assume you have a basting syringe full of water. If you merely poke a finger here or there into the rubber bulb of the syringe, you will not eject much water. However, if you compress the bulb in a smooth, coordinated fashion, you can squeeze out the water.

   *Clinical Note* In a similar manner, contraction of isolated cardiac muscle fibres is not successful in pumping blood. Such random, uncoordinated excitation and contraction of the cardiac cells is known as **fibrillation**. Fibrillation of the ventricles is much more serious than atrial fibrillation. Ventricular fibrillation rapidly causes death, because the heart cannot pump blood into the arteries. This condition can often be corrected by **electrical defibrillation**, in which a very strong electrical current is applied on the chest wall. When this current reaches the heart, it stimulates (depolarizes) all parts of the heart simultaneously. Usually the first part of the heart to recover is the SA node, which takes over pacemaker activity, once again initiating impulses that trigger the synchronized contraction of the rest of the heart.

3. The pair of atria and pair of ventricles should be functionally coordinated so that both members of the pair contract simultaneously. This coordination permits synchronized pumping of blood into the pulmonary and systemic circulation.

The normal spread of cardiac excitation is carefully orchestrated to ensure that these criteria are met and the heart functions efficiently (› Figure 8-9; also see Figure 8-8). We now elaborate on these three factors responsible for efficient cardiac function.

### ATRIAL EXCITATION

An action potential originating in the SA node first spreads throughout both atria, primarily from cell to cell via gap junctions. In addition, several poorly delineated, specialized

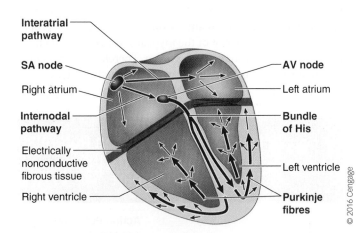

© 2016 Cengage

› **FIGURE 8-9 Spread of cardiac excitation.** An action potential initiated at the SA node first spreads throughout both atria. Its spread is facilitated by two specialized atrial conduction pathways, the interatrial and internodal pathways. The AV node is the only point where an action potential can spread from the atria to the ventricles. From the AV node, the action potential spreads rapidly throughout the ventricles, hastened by a specialized ventricular conduction system consisting of the bundle of His and Purkinje fibres.

conduction pathways speed up conduction of the impulse through the atria.

- The **interatrial pathway** extends from the SA node within the right atrium to the left atrium. Because this pathway rapidly transmits the action potential from the SA node to the pathway's termination in the left atrium, a wave of excitation can spread across the gap junctions throughout the left atrium at the same time excitation is similarly spreading throughout the right atrium. This ensures that both atria become depolarized to contract simultaneously.

- The **internodal pathway** extends from the SA node to the AV node. The AV node is the only point of electrical contact between the atria and ventricles; in other words, because the atria and ventricles are structurally connected by electrically nonconductive fibrous tissue, the only way an action potential in the atria can spread to the ventricles is by passing through the AV node. The internodal conduction pathway directs the spread of an action potential originating at the SA node to the AV node to ensure sequential contraction of the ventricles following atrial contraction. Hastened by this pathway, the action potential arrives at the AV node within 30 msec of SA node firing.

### CONDUCTION BETWEEN THE ATRIA AND THE VENTRICLES

The action potential is conducted relatively slowly through the AV node. This slowness is advantageous because it allows time for complete ventricular filling. The impulse is delayed about 100 msec (the **AV nodal delay**), which enables the atria to become completely depolarized and to contract, emptying their contents into the ventricles, before ventricular depolarization and contraction occur.

### VENTRICULAR EXCITATION

After the AV nodal delay, the impulse travels rapidly down the septum via the right and left branches of the bundle of His and throughout the ventricular myocardium via the Purkinje fibres. The network of fibres in this ventricular conduction system is specialized for rapid propagation of action potentials. Its presence hastens and coordinates the spread of ventricular excitation to ensure that the ventricles contract as a unit. The action potential is transmitted through the entire Purkinje fibre system within 30 msec.

Although this system carries the action potential rapidly to a large number of cardiac muscle cells, it does not terminate on every cell. The majority of Purkinje fibres terminate on ventricular muscle cells near the endocardial surface. The impulse quickly spreads across the ventricular wall to the epicardial surface by means of gap junctions.

The ventricular conduction system is more highly organized and more important than the interatrial and internodal conduction pathways. Because the ventricular mass is so much larger than the atrial mass, a rapid conduction system is crucial to hasten the spread of excitation in the ventricles. Purkinje fibres can transmit an action potential six times faster than the ventricular syncytium of contractile cells can. If the entire ventricular depolarization process depended on cell-to-cell spread of the impulse via gap junctions, the ventricular tissue immediately next to the AV node would become excited and contract before the impulse had even passed to the heart apex. This, of course, would not allow efficient pumping. Rapid conduction of the action potential down the bundle of His and its swift, diffuse distribution throughout the Purkinje network lead to almost simultaneous activation of the ventricular myocardial cells in both ventricular chambers, which ensures a single, smooth, coordinated contraction that can efficiently eject blood into both the systemic and pulmonary circulations at the same time.

## Cardiac contractile cells

The action potential in cardiac contractile cells, although initiated by the nodal pacemaker cells, varies considerably in its ionic mechanisms and shape from the action potentials generated at the SA node (compare Figure 8-7 and ⟩ Figure 8-10). Unlike the membrane of autorhythmic cells, the membrane of contractile cells remains essentially at rest at about −80 millivolts until excited by electrical activity propagated from the pacemaker. Once the membrane of a ventricular myocardial contractile cell is excited, an action potential is generated by a complicated interplay of permeability changes and membrane potential changes, as follows (Figure 8-10):

1. During the rising phase of the action potential, the membrane potential rapidly reverses to a positive value approaching +50 mV, due to the activation of voltage-gated $Na^+$ channels and the subsequent rapid entry of $Na^+$ into the cell, as it does in other excitable cells undergoing an action potential (p. 63).

2. The rapid depolarization brought about by activation of the "fast" $Na^+$ channels causes other voltage-dependent ion channels to be activated. The first is a transient outward

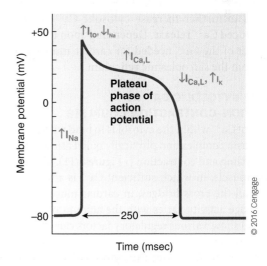

⟩ **FIGURE 8-10 Action potential in contractile cardiac muscle cells.** The action potential in cardiac contractile cells differs considerably from the action potential in cardiac autorhythmic cells (compare with Figure 8-7).

$K^+$ channel ($I_{to}$) that rapidly moves $K^+$ out of the cell to counter the influx of $Na^+$. The second is the "slow" L-type $Ca^{2+}$ channel ($I_{Ca,L}$), and the third is yet another type of voltage-gated $K^+$ channel, the delayed rectifier ($I_K$). The temporary balance of these inward and outward currents results in the characteristic plateau phase of the cardiac action potential, which can last for several hundred milliseconds.

3. Eventually, there is a time-dependent inactivation of both the transient outward $K^+$ and inward L-type $Ca^{2+}$ channels that disrupts the ionic flux balance of the plateau phase. At this time, the outward delayed rectifier $K^+$ current dominates and, as in other excitable cells, the cell returns to resting potential as $K^+$ leaves the cell.

4. The resting membrane potential is maintained by a $K^+$ channel called the inward rectifier $K^+$ channel, which is unique in that it can move $K^+$ both in and out of the cell to keep the resting membrane constant.

Let's now see how this action potential brings about contraction.

## Calcium release from the sarcoplasmic reticulum

In cardiac contractile cells, the L-type $Ca^{2+}$ channels lie primarily in the T tubules. As you learned from the preceding discussion, these voltage-gated channels open during a local action potential. Upon depolarization of the cell membrane, these L-type $Ca^{2+}$ channels open allowing the influx of $Ca^{2+}$. Once inside the cell, this $Ca^{2+}$ can directly interact with the contractile machinery or it can cause the further $Ca^{2+}$ release from internal stores. Located within the membrane of the sarcoplasmic reticulum is a protein called the *ryanodine receptor*. The ryanodine receptor is a $Ca^{2+}$ release channel that, when activated by $Ca^{2+}$ entering the cell through L-type $Ca^{2+}$ channels, causes the sarcoplasmic reticulum to release its $Ca^{2+}$ stores. This process of $Ca^{2+}$ from the ECF interacting with ryanodine receptors to further increase cytosolic $Ca^{2+}$ levels is called **$Ca^{2+}$-induced $Ca^{2+}$ release**. Depending upon the species, up to 90 percent of the $Ca^{2+}$ needed for cardiac muscle contraction comes from the sarcoplasmic reticulum.

### ROLE OF CYTOSOLIC $CA^{2+}$ IN EXCITATION–CONTRACTION COUPLING

The role of $Ca^{2+}$ within the cytosol is to bind with the troponin-tropomyosin complex and physically pull it aside to allow cross-bridge cycling and contraction (▸ Figure 8-11). However, unlike skeletal muscle, in which sufficient $Ca^{2+}$ is always released to turn on all the cross bridges, in cardiac muscle the extent of cross-bridge activity varies with the amount of cytosolic $Ca^{2+}$. As we will show, various regulatory factors can alter the amount of cytosolic $Ca^{2+}$.

Removal of $Ca^{2+}$ from the cytosol by energy-dependent mechanisms in both the plasma membrane and the sarcoplasmic reticulum restores the blocking action of troponin and tropomyosin, so contraction ceases and the heart muscle relaxes.

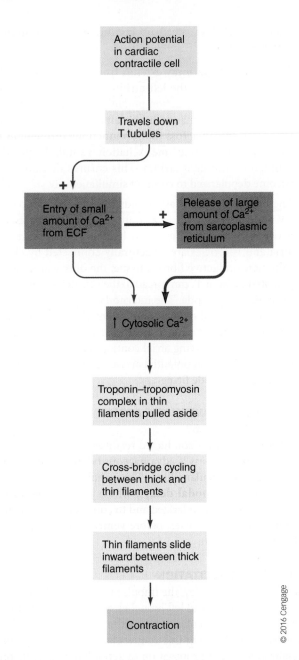

▸ **FIGURE 8-11 Excitation–contraction coupling in cardiac contractile cells**

### INFLUENCE OF ALTERED ECF $K^+$ AND $CA^{2+}$ CONCENTRATIONS

*Clinical Note* Not surprisingly, changes in the ECF concentration of potassium and calcium can have profound effects on the heart. Abnormal levels of $K^+$ are most important clinically, followed to a lesser extent by $Ca^{2+}$ imbalances. Changes in $K^+$ concentration in the ECF alter the $K^+$ concentration gradient between the ICF and ECF. Normally, there is substantially more $K^+$ inside the cells than in the ECF, but with elevated ECF $K^+$ levels, this gradient is reduced. Associated with this change is a reduction in resting potential (i.e., the membrane is less negative on the inside than normal because less $K^+$ leaves). Among the consequences is a tendency to develop ectopic foci as well as cardiac arrhythmias.

An elevated ECF $Ca^{2+}$ concentration, in contrast, augments the strength of cardiac contraction by prolonging the plateau phase of the action potential and by increasing the cytosolic concentration of $Ca^{2+}$. Contractions tend to be of longer duration, with little time to rest between contractions. Some drugs alter cardiac function by influencing $Ca^{2+}$ movement across the myocardial cell membranes. For example, $Ca^{2+}$ channel–blocking agents, such as *verapamil*, block $Ca^{2+}$ influx during an action potential, thereby reducing the force of cardiac contraction. Other drugs, such as *digitalis*, increase cardiac contractility by inducing an accumulation of cytosolic $Ca^{2+}$.

## Long refractory period

Like other excitable tissues, cardiac muscle has a refractory period. During the absolute refractory period, a second action potential cannot be triggered until an excitable membrane has recovered from the preceding action potential. In skeletal muscle, the refractory period is very short compared with the duration of the resulting contraction, so the fibre can be restimulated before the first contraction is complete to produce summation of contractions. Rapidly repetitive stimulation that does not let the muscle fibre relax between stimulations results in a sustained, maximal contraction known as *tetanus* (see Figure 7-15).

In contrast, cardiac muscle has a long refractory period that lasts about 250 msec because of the prolonged plateau phase of the action potential. This is almost as long as the period of contraction initiated by the action potential; a cardiac muscle fibre contraction averages about 300 msec ( > Figure 8-12). Consequently, cardiac muscle cannot be restimulated until contraction is almost over, which precludes summation of contractions and tetanus of cardiac muscle. This is a valuable protective mechanism, because the pumping of blood requires alternate periods of contraction (emptying) and relaxation (filling). A prolonged tetanic contraction would prove fatal. The heart chambers could not be filled and emptied again.

The chief factor responsible for the long refractory period is inactivation of the $Na^+$ channels that were activated during the initial $Na^+$ influx of the rising phase. During the plateau phase of the action potentials, $Na^+$ channels are inactivated and cannot recover from inactivation until the membrane potential has returned to near resting levels. Only then can the $Na^+$ channels be activated once again to begin another action potential. It is important to note, however, that even though $Na^+$ channels are the primary determinant for the long refractory period, factors underlying the maintenance of the plateau phase are also critical because they directly affect the time it takes for $Na^+$ channels to return to their resting state.

## The ECG

The electrical currents generated by cardiac muscle during depolarization and repolarization spread into the tissues around the heart and are conducted through the body fluids. A small part of this electrical activity reaches the body surface, where it can be detected using recording electrodes. The record produced is an **electrocardiogram (ECG)**. (Alternatively, the abbreviation EKG is often used, from the ancient Greek word *kardia* instead of the Latin *cardia* for "heart.")

Remember three important points when considering what an ECG represents:

1. An ECG is a recording of that part of the electrical activity induced in body fluids by the cardiac impulse that reaches the body surface, not a direct recording of the actual electrical activity of the heart.

2. The ECG is a complex recording representing the *overall* spread of activity throughout the heart during depolarization and repolarization. It is not a recording of a *single* action potential in a single cell at a single point in time. The record at any given time represents the sum of electrical activity in all the cardiac muscle cells, some of which may be generating action potentials while others may not yet be activated. For example, immediately after the SA node fires, the atrial cells are generating action potentials while the ventricular cells are still at resting potential. At a later point, the electrical activity will have spread to the ventricular cells while the atrial cells will be repolarizing. Therefore, the overall pattern of cardiac electrical activity varies with time as the impulse passes throughout the heart.

3. The recording represents comparisons in voltage detected by electrodes at two different points on the body surface, not the actual potential. For example, the ECG does not record a potential at all when the ventricular muscle is either completely depolarized or completely repolarized; both electrodes are viewing the same potential, so no *difference* in potential between the two electrodes is recorded.

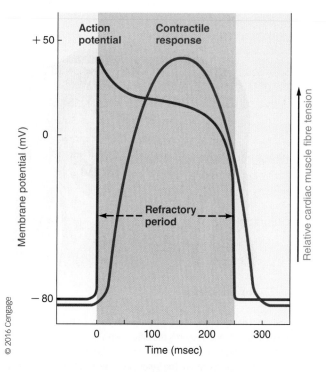

> FIGURE 8-12 **Relationship of an action potential and the refractory period to the duration of the contractile response in cardiac muscle**

The exact pattern of electrical activity recorded from the body surface depends on the orientation of the recording electrodes. Electrodes may be loosely thought of as the eyes that see the electrical activity, which then is translated into a visible recording, the ECG record. Whether an upward deflection or downward deflection is recorded is determined by the way the electrodes are oriented with respect to the current flow in the heart. For example, the spread of excitation across the heart is seen differently from the right arm, from the left leg, or from a recording directly over the heart. Even though the same electrical events are occurring in the heart, different waveforms representing the same electrical activity result when this activity is recorded by electrodes at different points on the body.

To provide standard comparisons, ECG records routinely consist of 12 conventional electrode systems, or leads. When an electrocardiograph machine is connected to pairs of recording electrodes attached to the body, the specific arrangement of each pair of connections is called a **lead**. The 12 different leads each record electrical activity in the heart from different locations—six different electrical arrangements from the limbs and six chest leads at various sites around the heart. To provide a common basis for comparison and for recognizing deviations from normal, the same 12 leads are routinely used in all ECG recordings (> Figure 8-13).

## The ECG record

Interpretation of the wave configurations recorded from each lead depends on a thorough knowledge about the sequence of cardiac excitation spread and about the position of the heart relative to electrode placement. A normal ECG has three distinct waveforms: the P wave, the QRS complex, and the T wave (> Figure 8-14).

- The **P wave** represents atrial depolarization.
- The **QRS complex** represents ventricular depolarization.
- The **T wave** represents ventricular repolarization.

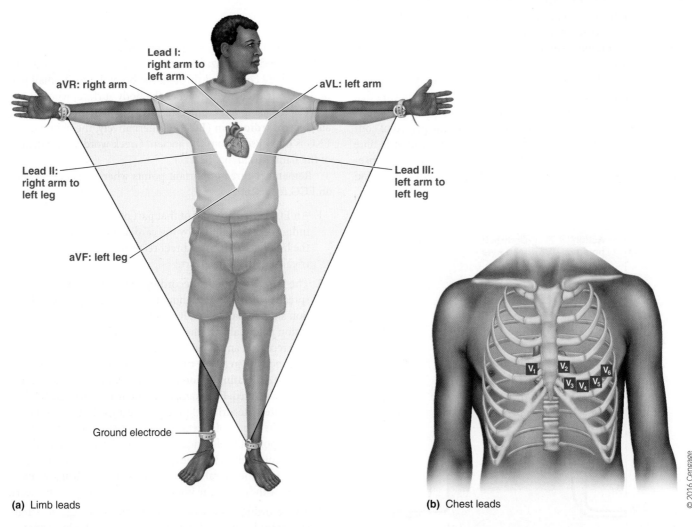

**(a)** Limb leads

**(b)** Chest leads

© 2016 Cengage

> **FIGURE 8-13 Electrocardiogram leads.** (a) The six limb leads include leads I, II, III, aVR, aVL, and aVF. Leads I, II, and III are bipolar leads because two recording electrodes are used. The tracing records the *difference* in potential between the two electrodes. For example, lead I records the difference in potential detected at the right arm and left arm. The electrode placed on the right leg serves as a ground and is not a recording electrode. The aVR, aVL, and aVF leads are unipolar leads. Even though two electrodes are used, only the actual potential under one electrode, the exploring electrode, is recorded. The other electrode is set at zero potential in comparison to the rest of the body. (b) The six chest leads, V1 through V6, are also unipolar leads. The exploring electrode mainly records the electrical potential of the cardiac musculature immediately beneath the electrode in six different locations surrounding the heart.

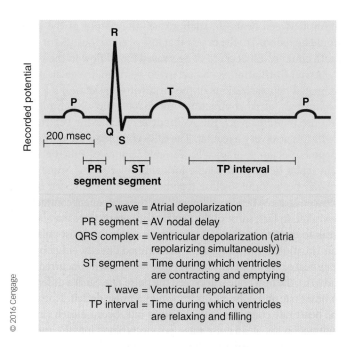

P wave = Atrial depolarization
PR segment = AV nodal delay
QRS complex = Ventricular depolarization (atria
repolarizing simultaneously)
ST segment = Time during which ventricles
are contracting and emptying
T wave = Ventricular repolarization
TP interval = Time during which ventricles
are relaxing and filling

© 2016 Cengage

> **FIGURE 8-14** Electrocardiogram waveforms in lead II

Because these shifting waves of depolarization and repolarization bring about the alternating contraction and relaxation of the heart, respectively, the cyclic mechanical events of the heart lag slightly behind the rhythmic changes in electrical activity. The following points about the ECG record should also be noted:

1. Firing of the SA node does not generate enough electrical activity to reach the body surface, so no wave is recorded for SA nodal depolarization. Therefore, the first recorded wave, the P wave, occurs when the impulse or wave of depolarization spreads across the atria.

2. In a normal ECG, no separate wave for atrial repolarization is visible. The electrical activity associated with atrial repolarization normally occurs simultaneously with ventricular depolarization and is masked by the QRS complex.

3. The P wave is much smaller than the QRS complex, because the atria have a much smaller muscle mass than the ventricles and consequently generate less electrical activity.

4. At the following three points in time, no net current flow is taking place in the heart musculature, so the ECG remains at baseline:

   a. *During the AV nodal delay.* This delay is represented by the interval of time between the end of P and the onset QRS; this segment of the ECG is known as the PR segment. (It is called the **PR segment** rather than the PQ segment because the Q deflection is small and sometimes absent, whereas the R deflection is the dominant wave of the complex.) Current is flowing through the AV node, but the magnitude is too small for the ECG electrodes to detect.

   b. *When the ventricles are completely depolarized and the cardiac contractile cells are undergoing the plateau phase of their action potential before they repolarize.* This is represented by the **ST segment**. This segment lies between QRS and T; it coincides with the time during

which ventricular activation is complete and the ventricles are contracting and emptying. Note that the ST segment is not a record of cardiac contractile activity. The ECG is a measure of the electrical activity that triggers the subsequent mechanical activity.

   c. *When the heart muscle is completely repolarized and at rest and ventricular filling is taking place, after the T wave and before the next P wave.* This period is called the **TP interval**.

## The ECG diagnosis

*Clinical Note* Because electrical activity triggers mechanical activity, abnormal electrical patterns are usually accompanied by abnormal contractile activity of the heart. So evaluation of ECG patterns can provide useful information about the status of the heart. The main deviations from normal that can be found through electrocardiography are (1) abnormalities in rate, (2) abnormalities in rhythm, and (3) cardiac myopathies (> Figure 8-15).

### ABNORMALITIES IN RATE

The heart rate can be determined from the distance between two consecutive QRS complexes on the calibrated paper used to record an ECG. A rapid heart rate of more than 100 beats per minute is called **tachycardia** (*tachy* means "fast"), whereas a slow heart rate of fewer than 60 beats per minute is called **bradycardia** (*brady* means "slow").

### ABNORMALITIES IN RHYTHM

*Rhythm* refers to the regularity or spacing of the ECG waves. Any variation from the normal rhythm and sequence of excitation of the heart is termed an **arrhythmia**. It may result from ectopic foci, alterations in SA node pacemaker activity, or interference with conduction. Heart rate is also often altered. *Extrasystoles*, or *premature ventricular contractions*, originating from an ectopic focus are common deviations from normal rhythm. Other

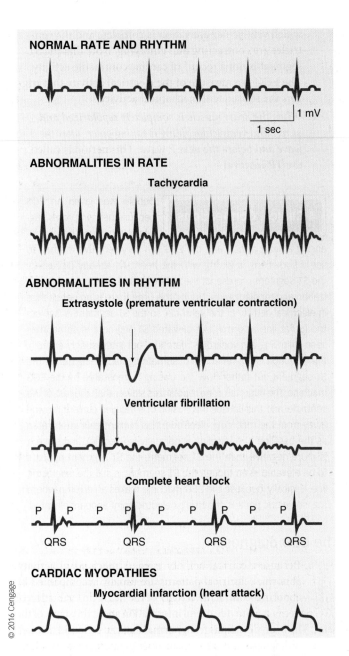

**NORMAL RATE AND RHYTHM**

1 mV

1 sec

**ABNORMALITIES IN RATE**

**Tachycardia**

**ABNORMALITIES IN RHYTHM**

**Extrasystole (premature ventricular contraction)**

**Ventricular fibrillation**

**Complete heart block**

P P P P P P P P P

QRS     QRS     QRS     QRS

**CARDIAC MYOPATHIES**

**Myocardial infarction (heart attack)**

© 2016 Cengage

› **FIGURE 8-15 Representative heart conditions detectable through electrocardiography**

abnormalities in rhythm easily detected on an ECG include atrial flutter, atrial fibrillation, ventricular fibrillation, and heart block.

**Atrial flutter** is characterized by a rapid but regular sequence of atrial depolarizations at rates between 200 and 380 beats per minute. The ventricles rarely keep pace with the racing atria. Because the conducting tissue's refractory period is longer than that of the atrial muscle, the AV node is unable to respond to every impulse that converges on it from the atria. Maybe only one out of every two or three atrial impulses successfully passes through the AV node to the ventricles. Such a situation is referred to as a *2:1 or 3:1 rhythm*. The fact that not every atrial impulse reaches the ventricle in atrial flutter is important, because it precludes a rapid ventricular rate of more than 200 beats per minute. Such a high rate would not allow adequate time for ventricular

filling between beats. In such a case, the output of the heart would be reduced to the extent that loss of consciousness or even death could result because of decreased blood flow to the brain.

**Atrial fibrillation** is characterized by rapid, irregular, uncoordinated depolarizations of the atria with no definite P waves. Accordingly, atrial contractions are chaotic and asynchronized. Because impulses reach the AV node erratically, the ventricular rhythm is also very irregular. The QRS complexes are normal in shape but occur sporadically. Variable lengths of time between ventricular beats are available for ventricular filling. Some ventricular beats come so close together that little filling can occur between beats. When less filling occurs, the subsequent contraction is weaker. In fact, some of the ventricular contractions may be too weak to eject enough blood to produce a palpable wrist pulse. In this situation, if the heart rate is determined directly, either by the apex beat or via the ECG, and the pulse rate is taken concurrently at the wrist, the heart rate will exceed the pulse rate. Such a difference in heart rate and pulse rate is known as a **pulse deficit**. Normally, the heart rate coincides with the pulse rate, because each cardiac contraction initiates a pulse wave as it ejects blood into the arteries.

**Ventricular fibrillation** is a very serious rhythmic abnormality in which the ventricular musculature exhibits uncoordinated, chaotic contractions. Multiple impulses travel erratically in all directions around the ventricles. The ECG tracing in ventricular fibrillations is very irregular, with no detectable pattern or rhythm. When contractions are so disorganized, the ventricles are ineffectual as pumps. If circulation is not restored in less than four minutes through external cardiac compression or electrical defibrillation, irreversible brain damage occurs, and death is imminent.

Another type of arrhythmia, **heart block**, arises from defects in the cardiac conducting system. The atria still beat regularly, but the ventricles occasionally fail to be stimulated and thus do not contract following atrial contraction. Impulses between the atria and ventricles can be blocked to varying degrees. In some forms of heart block, only every second or third atrial impulse is passed to the ventricles. This is known as *2:1 or 3:1 block*, which can be distinguished from the 2:1 or 3:1 rhythm associated with atrial flutter by the rates involved. In heart block, the atrial rate is normal, but the ventricular rate is considerably below normal; in atrial flutter the atrial rate is very high, and the ventricular rate is normal or above-normal. *Complete heart block* is characterized by complete dissociation between atrial and ventricular activity, which means impulses from the atria are not conducted to the ventricles at all. The SA node continues to govern atrial depolarization, but the ventricles generate their own impulses at a rate much slower than that of the atria. On the ECG, the P waves exhibit a normal rhythm. The QRS and T waves also occur regularly, but they are much slower than the P waves and completely independent of P wave rhythm. Because atrial and ventricular activity is not synchronized, waves for atrial repolarization may appear, since they are no longer masked by the QRS complex.

## CARDIAC MYOPATHIES

Abnormal ECG waves are also important for recognizing and assessing **cardiac myopathies** (damage of the heart muscle). **Myocardial ischemia** is inadequate delivery of oxygenated blood to the heart tissue. Actual death, or **necrosis**, of heart muscle cells occurs when a blood vessel supplying that area of the heart becomes blocked or ruptured. This condition is **acute myocardial infarction**, commonly called a **heart attack**. Abnormal QRS waveforms appear when part of the heart muscle becomes necrotic. In addition to ECG changes, because damaged heart muscle cells release characteristic enzymes into the blood, the level of these enzymes in the blood provides a further index of the extent of myocardial damage.

Interpretation of an ECG is a complex task requiring extensive knowledge and training. The foregoing discussion is not by any means intended to make you an ECG expert, but seeks to give you an appreciation of the ways in which the ECG can be used as a diagnostic tool, as well as to present an overview of some of the more common abnormalities of heart function.

---

### Check Your Understanding 8.2

1. Draw two graphs comparing the electrical activity in a cardiac autorhythmic cell and in a cardiac contractile cell. Label the ion movement responsible for each change in potential.
2. List the autorhythmic tissues of the heart, and indicate the normal rate of action potential discharge of each.
3. Explain why no separate wave for atrial repolarization is visible on a normal ECG.

---

## 8.4 | Mechanical Events of the Cardiac Cycle

The mechanical events of the cardiac cycle—contraction, relaxation, and the resultant changes in blood flow through the heart—are brought about by the rhythmic changes in cardiac electrical activity.

### Contracting and relaxing

The cardiac cycle consists of alternate periods of **systole** (contraction and emptying) and **diastole** (relaxation and filling). Contraction results from the spread of excitation across the heart, whereas relaxation follows the subsequent repolarization of the cardiac musculature. The atria and ventricles go through separate cycles of systole and diastole. Unless otherwise stated, the terms *systole* and *diastole* refer to what's happening with the ventricles.

The following discussion and corresponding › Figure 8-16 correlate various events that occur concurrently during the cardiac cycle, including ECG features, pressure changes, volume changes, valve activity, and heart sounds. Only the events on the left side of the heart are described, but keep in mind that identical events are occurring on the right side of the heart, except that the pressures are lower. To complete one full cardiac cycle, our discussion begins and ends with ventricular diastole.

#### MID-VENTRICULAR DIASTOLE

During most of ventricular diastole, the atrium is still also in diastole. This stage corresponds to the TP interval on the ECG—the interval after ventricular repolarization and before another atrial depolarization. Because of the continuous inflow of blood from the venous system into the atrium, atrial pressure slightly exceeds ventricular pressure even though both chambers are relaxed. As a result of this pressure differential, the AV valve is open, and blood flows directly from the atrium into the ventricle throughout ventricular diastole (**A** in Figure 8-16). As a result of this passive filling, the ventricular volume slowly continues to rise even before atrial contraction takes place.

#### LATE VENTRICULAR DIASTOLE

Late in ventricular diastole, the SA node reaches threshold and fires. The impulse spreads throughout the atria, which appears on the ECG as the P wave. Atrial depolarization brings about atrial contraction, raising the atrial pressure and squeezing more blood into the ventricle. The excitation–contraction coupling process takes place during the short delay between the P wave and the rise in atrial pressure. The corresponding rise in ventricular pressure that occurs simultaneously with the rise in atrial pressure is due to the additional volume of blood added to the ventricle by atrial contraction (**B** in Figure 8-16). Throughout atrial contraction, atrial pressure still slightly exceeds ventricular pressure, so the AV valve remains open.

#### END OF VENTRICULAR DIASTOLE

Ventricular diastole ends at the onset of ventricular contraction. By this time, atrial contraction and ventricular filling have completed. The volume of blood in the ventricle at the end of diastole is known as the **end-diastolic volume (EDV)**, which averages about 135 mL. No more blood is added to the ventricle during this cycle. Therefore, the end-diastolic volume is the maximum amount of blood that the ventricle will contain during this cycle.

#### VENTRICULAR EXCITATION AND ONSET OF VENTRICULAR SYSTOLE

After atrial excitation, the impulse travels through the AV node and specialized conduction system to excite the ventricle. Simultaneously, the atria are contracting. By the time ventricular activation is complete, atrial contraction is already over. The QRS complex represents this ventricular excitation, which induces ventricular contraction. The ventricular pressure sharply increases shortly after the QRS complex, signalling the onset of ventricular systole. The slight delay between the QRS complex and the actual onset of ventricular systole is the time required for the excitation–contraction coupling process to occur. As ventricular contraction begins, ventricular pressure immediately exceeds atrial pressure. This backward-pressure differential forces the AV valve to close.

#### ISOVOLUMETRIC VENTRICULAR CONTRACTION

After ventricular pressure exceeds atrial pressure and the AV valve has closed, to open the aortic valve, the ventricular pressure must continue to increase until it exceeds aortic pressure. Therefore, after closing of the AV valve and before opening of the aortic valve, there is a brief period of time when the ventricle remains a closed chamber (**C** in Figure 8-16). Because all valves are closed, no blood can enter or leave the ventricle during this time. This interval is termed as the period of

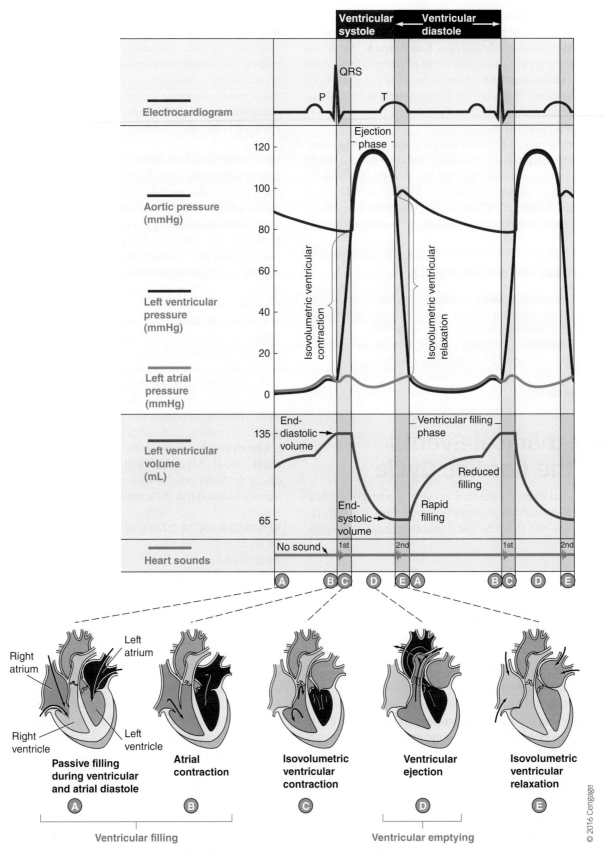

> **FIGURE 8-16 Cardiac cycle.** This graph depicts various events that occur concurrently during the cardiac cycle. Follow each horizontal strip across to see the changes that take place in the electrocardiogram; aortic, ventricular, and atrial pressures; ventricular volume; and heart sounds throughout the cycle. The last half of diastole, one full systole and diastole (one full cardiac cycle), and another systole are shown for the left side of the heart. Follow each vertical strip downward to see what happens simultaneously with each of these factors during each phase of the cardiac cycle. See pp. 361–364 for a detailed explanation of the circled letters. The sketches of the heart illustrate the flow of O₂-poor (dark blue) and O₂-rich (bright red) blood in and out of the ventricles during the cardiac cycle.

**isovolumetric ventricular contraction** (*isovolumetric* means "constant volume and length"). Because no blood enters or leaves the ventricle, the ventricular chamber stays at constant volume, and the muscle fibres stay at constant length. This isovolumetric condition is similar to an isometric contraction in skeletal muscle. During isovolumetric ventricular contraction, ventricular pressure continues to increase as the volume remains constant.

## VENTRICULAR EJECTION

When ventricular pressure exceeds aortic pressure, the aortic valve is forced open and ejection of blood begins (**D** in Figure 8-16). The amount of blood pumped out of each ventricle with each contraction is called the **stroke volume (SV)**. At rest, stroke volume is approximate 70 mL. The aortic pressure curve rises as blood becomes forced into the aorta from the ventricle faster than it can drain off into the smaller vessels at the other end. The ventricular volume decreases substantially as blood is rapidly pumped out. Ventricular systole includes both the period of isovolumetric contraction and the ventricular ejection phase.

## END OF VENTRICULAR SYSTOLE

The ventricle does not empty completely during ejection. Normally, only about half the blood within the ventricle at the end of diastole is pumped out during the subsequent systole. The amount of blood left in the ventricle at the end of systole when ejection is complete is the **end-systolic volume (ESV)**, which averages about 65 mL. This is the least amount of blood that the ventricle will contain during this cycle.

The difference between the volume of blood in the ventricle before contraction and the volume after contraction is the amount of blood ejected during the contraction; that is, EDV − ESV = SV. In our example, the end-diastolic volume is 135 mL, the end-systolic volume is 65 mL, and the stroke volume is 70 mL.

## VENTRICULAR REPOLARIZATION AND ONSET OF VENTRICULAR DIASTOLE

The T wave signifies ventricular repolarization at the end of ventricular systole. As the ventricle starts to relax, on repolarization, ventricular pressure falls below aortic pressure and the aortic valve closes. Closure of the aortic valve produces a disturbance, or notch, on the aortic pressure curve—the **dicrotic notch**. No more blood leaves the ventricle during this cycle, because the aortic valve has closed.

## ISOVOLUMETRIC VENTRICULAR RELAXATION

When the aortic valve closes, the AV valve has not yet opened, because ventricular pressure still exceeds atrial pressure, so no blood can enter the ventricle from the atrium. Therefore, all valves are once

again closed for a brief period of time known as **isovolumetric ventricular relaxation** (**E** in Figure 8-16). The muscle fibre length and chamber volume remain constant. No blood leaves or enters as the ventricle continues to relax and the pressure steadily falls.

## VENTRICULAR FILLING

When ventricular pressure falls below atrial pressure, the AV valve opens, and ventricular filling occurs again. Ventricular diastole includes both the period of isovolumetric ventricular relaxation and the ventricular filling phase.

Atrial repolarization and ventricular depolarization occur simultaneously, so the atria are in diastole throughout ventricular systole. Blood continues to flow from the pulmonary veins into the left atrium. As this incoming blood pools in the atrium, atrial pressure rises continuously. When the AV valve opens at the end of ventricular systole, blood that accumulated in the atrium during ventricular systole pours rapidly into the ventricle (**A** again). Ventricular filling thus occurs rapidly at first because of the increased atrial pressure resulting from the accumulation of blood in the atria. Then ventricular filling slows down because the accumulated blood has already been delivered to the ventricle, and atrial pressure starts to fall. During this period of reduced filling, blood continues to flow from the pulmonary veins into the left atrium and through the open AV valve into the left ventricle. During late ventricular diastole, when the ventricle is filling slowly, the SA node fires again, and the cardiac cycle starts over.

When the body is at rest, one complete cardiac cycle lasts 800 msec, with 300 msec devoted to ventricular systole and 500 msec taken up by ventricular diastole. Significantly, much of ventricular filling occurs early in diastole during the rapid-filling phase (⟩ Figure 8-17). During times of rapid heart rate, diastole length is shortened much more than systole length is. For example, if the heart rate increases from 75 to 180 beats per minute, the duration of diastole decreases about 75 percent, from 500 msec to 125 msec. This greatly reduces the time available for

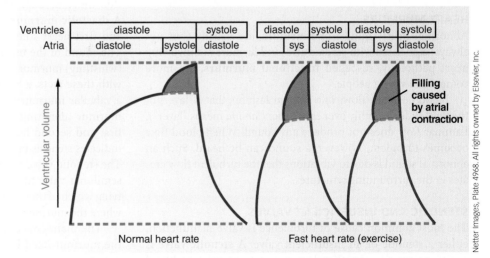

| Ventricles | diastole | | systole | diastole | systole | diastole | systole |
| Atria | diastole | systole | diastole | sys | diastole | sys | diastole |

Filling caused by atrial contraction

Ventricular volume

Normal heart rate          Fast heart rate (exercise)

⟩ **FIGURE 8-17 Ventricular filling profiles during normal and rapid heart rates.** Because much of ventricular filling occurs early in diastole during the rapid-filling phase, filling is not seriously impaired when diastolic time is reduced as a *result* of an increase in heart rate.

ventricular relaxation and filling. However, because much ventricular filling is accomplished during early diastole, filling is not necessarily impaired during periods of increased heart rate, such as during exercise. Also contributing to this is the volume of ventricular filling that is attributed to atrial contraction, the relative contribution of which increases with higher heart rates. There is a limit, however, to how rapidly the heart can beat without decreasing the period of diastole to the point that ventricular filling is severely impaired. At heart rates greater than 200 beats per minute, diastolic time is too short to allow adequate ventricular filling. With inadequate filling, the resultant cardiac output is deficient. Normally, ventricular rates do not exceed 200 beats per minute, because the relatively long refractory period of the AV node does not allow impulses to be conducted to the ventricles more frequently than this.

## Heart sounds

Two major heart sounds normally can be heard with a stethoscope during the cardiac cycle. The **first heart sound** is low-pitched, soft, and relatively long—often said to sound like "lub." The **second heart sound** has a higher pitch and is shorter and sharper—often said to sound like "dup." Thus, one normally hears "lub-dup-lub-dup-lub-dup...." The first heart sound is associated with closure of the AV valves, whereas the second sound is associated with closure of the semilunar valves (see the blue line associated with Heart sounds in Figure 8-16). Opening of valves does not produce any sound.

The sounds are caused by vibrations set up within the walls of the ventricles and major arteries during valve closure, not by the valves snapping shut. Because the AV valves close at the onset of ventricular contraction, when ventricular pressure first exceeds atrial pressure, the first heart sound signals the onset of ventricular systole. The semilunar valves close at the onset of ventricular relaxation, when the left and right ventricular pressures fall below the aortic and pulmonary artery pressures, respectively. The second heart sound, therefore, signals the onset of ventricular diastole.

### HEART MURMURS

Abnormal heart sounds, or **murmurs**, are usually (but not always) associated with cardiac disease. Murmurs not involving heart pathology, so-called **functional murmurs**, are more common in young people.

Blood normally flows in a *laminar* fashion; that is, layers of the fluid slide smoothly over each other (*lamina* means "layer"). Laminar flow does not produce any sound. When blood flow becomes turbulent, however, a sound can be heard. Such an abnormal sound is due to vibrations that the turbulent flow creates in the surrounding structures.

### STENOTIC AND INSUFFICIENT VALVES

The most common cause of turbulence is valve malfunction, either a stenotic or an insufficient valve. A **stenotic valve** is a stiff, narrowed valve that does not open completely. Blood must be forced through the constricted opening at tremendous velocity, resulting in turbulence that produces an abnormal whistling sound similar to the sound produced when you force air rapidly through narrowed lips to whistle.

An **insufficient or incompetent valve** is one that cannot close completely, usually because the valve edges are scarred and do not fit together properly. Turbulence is produced when blood flows backward through the insufficient valve and collides with blood moving in the opposite direction, creating a swishing or gurgling murmur. Such backflow of blood is known as **regurgitation**. An insufficient heart valve is often called a **leaky valve**, because it lets blood leak back through at a time when the valve should be closed.

Most often, both valvular stenosis and insufficiency are caused by **rheumatic fever**, an autoimmune (immunity against self) disease triggered by a *streptococcus* bacterial infection. Antibodies formed against toxins produced by these bacteria interact with many of the body's own tissues, resulting in immunological damage. The heart valves are among the most susceptible tissues in this regard. Large, haemorrhagic, fibrous lesions form along the inflamed edges of an affected heart valve, causing the valve to become thickened, stiff, and scarred. Sometimes the leaflet edges permanently adhere to each other. Depending on the extent and specific nature of the lesions, the valve may become either stenotic or insufficient or some degree of both.

### TIMING OF MURMURS

The valve involved and the type of defect can usually be detected by the *location* and *timing* of the murmur. Each heart valve can be heard best at a specific location on the chest. Noting where a murmur is loudest helps the diagnostician tell which valve is involved.

The timing of the murmur refers to the part of the cardiac cycle during which the murmur is heard. Recall that the first heart sound signals the onset of ventricular systole, and the second heart sound signals the onset of ventricular diastole. Therefore, a murmur between the first and second heart sounds (lub-murmur-dup, lub-murmur-dup) is a **systolic murmur**. A **diastolic murmur**, in contrast, occurs between the second and first heart sound (lub-dup-murmur, lub-dup-murmur). The sound of the murmur characterizes it as either a stenotic (whistling) murmur or an insufficient (swishy) murmur. Armed with these facts, a trained person can determine the cause of a valvular murmur (▮ Table 8-2). As an example, a whistling murmur (denoting a stenotic valve) that occurs between the first and second heart sounds (denoting a systolic murmur) indicates stenosis in a valve that should be open during systole. The stenotic valve could be either the aortic or the pulmonary semilunar valve through which blood is being ejected. Identifying which of these valves is stenotic is accomplished by finding where the murmur is best heard.

The main concern with heart murmurs, of course, is not the murmur itself but the harmful circulatory results of the defect.

## ▌TABLE 8-2 Timing and Type of Murmur Associated with Various Heart Valve Disorders

| Pattern Heard on Auscultation | Type of Valve Defect | Timing of Murmur | Valve Disorder | Comment |
|---|---|---|---|---|
| **Lub-Whistle-Dup** | Stenotic | Systolic | Stenotic semilunar valve | A whistling systolic murmur signifies that a valve that should be open during systole (a semilunar valve) does not open completely. |
| **Lub-Dup-Whistle** | Stenotic | Diastolic | Stenotic AV valve | A whistling diastolic murmur signifies that a valve that should be open during diastole (an AV valve) does not open completely. |
| **Lub-Swish-Dup** | Insufficient | Systolic | Insufficient AV valve | A swishy systolic murmur signifies that a valve that should be closed during systole (an AV valve) does not close completely. |
| **Lub-Dup-Swish** | Insufficient | Diastolic | Insufficient semilunar valve | A swishy diastolic murmur signifies that a valve that should be closed during diastole (a semi-lunar valve) does not close completely. |

### Check Your Understanding 8.3

1. Define *systole* and *diastole*.
2. State the pressure relationships among the aortic, atrial, and ventricular pressures (e.g., ventricular pressure > aortic pressure > atrial pressure) during each of these phases of the cardiac cycle: (1) ventricular filling, (2) isovolumetric ventricular contraction, (3) ventricular ejection, and (4) isovolumetric ventricular relaxation (think what pressure relationships must exist for the valves to be open or closed as appropriate in each phase).
3. How do heart sounds relate to the cardiac cycle?

## 8.5 | Cardiac Output and its Control

**Cardiac output (CO)** is the volume of blood pumped by *each ventricle* per minute (not the total amount of blood pumped by the heart). During any period of time, the volume of blood flowing through the pulmonary circulation is the same as the volume flowing through the systemic circulation. Therefore, the cardiac output from each ventricle normally is the same, although on a beat-to-beat basis, minor variations may occur (Appendix F lists reference values).

### Cardiac output

The two determinants of cardiac output are *heart rate* (beats per minute) and *stroke volume* (volume of blood pumped per beat or stroke). The average resting heart rate is 70 beats per minute, established by SA node rhythmicity; and the average resting stroke volume is 70 mL per beat, producing an average cardiac output of 4900 mL/min, or close to 5 L/min:

$$\text{Cardiac output (CO)} = \text{heart rate} \times \text{stroke volume}$$
$$= 70 \text{ beats/min} \times 70 \text{ mL/beat}$$
$$= 4900 \text{ mL/min} \approx 5 \text{ L/min}$$

Because the body's total blood volume averages 5–5.5 L (which may expand to 6 litres in some athletes), each half of the heart pumps the equivalent of the entire blood volume each minute. In other words, each minute the right ventricle normally pumps 5 litres of blood through the lungs, and the left ventricle pumps 5 litres through the systemic circulation. At this rate, each half of the heart would pump about 2.5 million litres of blood in just one year. Yet this is only the resting cardiac output! During exercise, cardiac output can increase to 20–25 L/min, and outputs as high as 40 L/min have been reported in elite endurance athletes. The difference between the cardiac output at rest and at maximum exercise is called the **cardiac reserve**.

How can cardiac output vary so tremendously, depending on the demands of the body? You can readily answer this question by thinking about how your own heart pounds rapidly (increased heart rate) and forcefully (increased stroke volume) when you engage in strenuous physical activities (need for increased cardiac output). Thus regulation of cardiac output depends on the control of both heart rate and stroke volume, topics that we discuss next.

### Heart rate

The SA node is normally the pacemaker of the heart, because it has the fastest spontaneous rate of depolarization to threshold. Recall that this automatic, gradual reduction of membrane potential between beats is due to a complex interplay of ion movements involving the inward movement of potassium and sodium through the $I_f$ channel and an increased $Ca^{2+}$ permeability. When the SA node reaches threshold, an action potential is initiated that spreads throughout the heart, inducing the heart to contract, that is, have a heartbeat. This happens about 70 times per minute, setting the average heart rate at 70 beats per minute.

The heart is innervated by both divisions of the autonomic nervous system, which can modify the rate (as well as the strength) of contraction, even though nervous stimulation is not required to initiate contraction. The parasympathetic nerve to the heart, the vagus nerve, primarily supplies the atrium, especially the SA and AV nodes. Parasympathetic innervation of the ventricles is sparse. The cardiac sympathetic nerves also supply the atria, including the SA and AV nodes, and richly innervate the ventricles as well.

Both the parasympathetic and sympathetic nervous system bring about their effects on the heart by altering the activity of the cyclic AMP second-messenger system in the innervated cardiac cells. Acetylcholine released from the vagus nerve binds to a muscarinic receptor and is coupled to an inhibitory G protein that reduces activity of the cyclic AMP pathway (p. 192) and (p. 217). By contrast, the sympathetic neurotransmitter norepinephrine binds with a $\beta_1$ adrenergic receptor and is coupled to a stimulatory G protein that accelerates the cyclic AMP pathway in the target cells (p. 193).

Let's examine the specific effects that parasympathetic and sympathetic stimulation have on the heart (▮ Table 8-3).

### EFFECT OF PARASYMPATHETIC STIMULATION ON THE HEART

- The parasympathetic nervous system's influence on the SA node is to decrease the heart rate. Acetylcholine released on increased parasympathetic activity increases the permeability

of the SA node to potassium by slowing the closure of $K^+$ channels. As a result, the rate at which spontaneous action potentials are initiated is reduced through a twofold effect:

1. Enhanced $K^+$ permeability hyperpolarizes the SA node membrane because more positive potassium ions leave than normal, making the inside even more negative. Because the resting potential starts even farther away from threshold, it takes longer to reach threshold.
2. The enhanced $K^+$ permeability induced by vagal stimulation also opposes $I_f$ current responsible for initiating the gradual depolarization of the membrane to threshold. This countering effect decreases the rate of spontaneous depolarization, prolonging the time required to drift to threshold. Therefore, the SA node reaches threshold and fires less frequently, decreasing the heart rate (❯ Figure 8-18).

- Parasympathetic influence on the AV node decreases the node's excitability, prolonging transmission of impulses to the ventricles even longer than the usual AV nodal delay. This effect is brought about by increasing $K^+$ permeability, which hyperpolarizes the membrane, thereby retarding the initiation of excitation in the AV node.
- Parasympathetic stimulation of the atrial contractile cells shortens the action potential, reducing the slow inward current carried by calcium; that is, the plateau phase is shortened. As a result, atrial contraction is weakened.
- The parasympathetic system has little effect on ventricular contraction, because of the sparseness of parasympathetic innervation to the ventricles.

Thus, the heart is more "leisurely" under parasympathetic influence. It beats less rapidly, the time between atrial and ventricular contraction is stretched out, and atrial contraction is weaker.

---

**▮ TABLE 8-3** Effects of the Autonomic Nervous System on the Heart and Structures That Influence the Heart

| Area Affected | Effect of Parasympathetic Stimulation | Effect of Sympathetic Stimulation |
| --- | --- | --- |
| **SA Node** | Decreases the rate of depolarization to threshold; decreases the heart rate | Increases the rate of depolarization to threshold; increases the heart rate |
| **AV Node** | Decreases excitability; increases the AV nodal delay | Increases excitability; decreases the AV nodal delay |
| **Ventricular Conduction Pathway** | No effect | Increases excitability; hastens conduction through bundle of His and Purkinje cells |
| **Atrial Muscle** | Decreases contractility; weakens contraction | Increases contractility; strengthens contraction |
| **Ventricular Muscle** | No effect | Increases contractility; strengthens contraction |
| **Adrenal Medulla (an endocrine gland)** | No effect | Promotes adrenomedullary secretion of epinephrine, a hormone that augments the sympathetic nervous system's actions on the heart |
| **Veins** | No effect | Increases venous return, which increases the strength of cardiac contraction through the Frank–Starling mechanism |

**KEY**

- - - = Inherent SA node pacemaker activity
——— = SA node pacemaker activity on parasympathetic stimulation
——— = SA node pacemaker activity on sympathetic stimulation

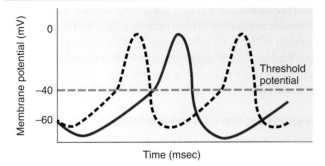

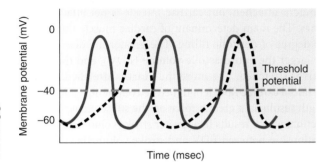

© 2016 Cengage

> **FIGURE 8-18 Autonomic control of SA node activity and heart rate.** Parasympathetic stimulation decreases the rate of SA nodal depolarization so that the membrane reaches threshold more slowly and has fewer action potentials, whereas sympathetic stimulation increases the rate of depolarization of the SA node so that the membrane reaches threshold more rapidly and has more frequent action potentials. Because each SA node action potential ultimately leads to a heartbeat, increased parasympathetic activity decreases the heart rate, whereas increased sympathetic activity increases the heart rate.

These actions are appropriate, considering that the parasympathetic system controls heart action in quiet, relaxed situations when the body is not demanding an enhanced cardiac output.

### EFFECT OF SYMPATHETIC STIMULATION ON THE HEART

- The sympathetic nervous system, which controls heart action in emergency or exercise situations, when there is a need for greater blood flow, speeds up the heart rate through its effect on the pacemaker tissue. The main effect of sympathetic stimulation on the SA node is to speed up depolarization so that threshold is reached more rapidly. Norepinephrine released from the sympathetic nerve endings enhances the pacemaker currents, reducing the time it normally takes to reach threshold. This swifter drift to threshold under sympathetic influence permits more frequent action potentials and a correspondingly faster heart rate (Figure 8-18, and Table 8-3).

- Sympathetic stimulation of the AV node reduces the AV nodal delay by increasing conduction velocity, presumably by enhancing the slow, inward $Ca^{2+}$ current.

- Similarly, sympathetic stimulation speeds up the spread of the action potential throughout the specialized conduction pathway.

- In the atrial and ventricular contractile cells, both of which have many sympathetic nerve endings, sympathetic stimulation increases contractile strength, so the heart beats more forcefully and squeezes out more blood. This effect is produced by increasing $Ca^{2+}$ permeability, which enhances the slow $Ca^{2+}$ influx and intensifies $Ca^{2+}$ participation in excitation coupling.

The overall effect of sympathetic stimulation on the heart, therefore, is to improve its effectiveness as a pump by increasing heart rate, decreasing the delay between atrial and ventricular contraction, decreasing conduction time throughout the heart, and increasing the force of contraction; that is, sympathetic stimulation revs up the heart.

### CONTROL OF HEART RATE

As you can see from the preceding discussion, the parasympathetic and sympathetic effects on heart rate are antagonistic (oppose each other), which is typical of the autonomic nervous system. At any given moment, heart rate is determined largely by the balance between inhibition of the SA node by the vagus nerve and stimulation by the cardiac sympathetic nerves. Under resting conditions, parasympathetic discharge dominates. In fact, if all autonomic nerves to the heart were blocked, the resting heart rate would increase from its average value of 70 beats per minute to about 100 beats per minute, which is the inherent rate of the SA node's spontaneous discharge when not subjected to any nervous influence. (We use 70 beats per minute as the normal rate of SA node discharge because this is the average rate under normal resting conditions when parasympathetic activity dominates.) The heart rate can be altered beyond this resting level in either direction by shifting the balance of autonomic nervous stimulation. Heart rate is speeded up by simultaneously increasing sympathetic and decreasing parasympathetic activity; heart rate is slowed by a concurrent rise in parasympathetic activity and decline in sympathetic activity. In turn, the relative level of activity in these two autonomic branches to the heart is primarily coordinated by the *cardiovascular control centre* in the brain stem.

Although autonomic innervation is the primary means by which heart rate is regulated, other factors affect it as well. The most important is epinephrine, a hormone that on sympathetic stimulation is secreted into the blood from the adrenal medulla and that acts on the heart in a manner similar to norepinephrine (the sympathetic neurotransmitter) to increase heart rate. Epinephrine therefore reinforces the direct effect that the sympathetic nervous system has on the heart.

## Stroke volume

The other component besides heart rate that determines cardiac output is stroke volume, the amount of blood pumped out by each ventricle during each beat. Two types of controls influence stroke volume: (1) *intrinsic control* related to the extent

**8**

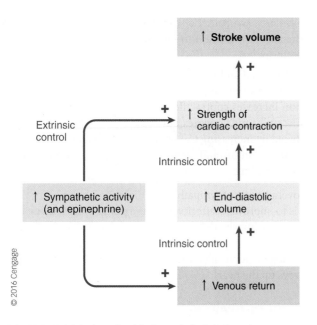

> FIGURE 8-19 **Intrinsic and extrinsic control of stroke volume**

of venous return, and (2) *extrinsic control* related to the extent of sympathetic stimulation of the heart. Both factors increase stroke volume by increasing the strength of heart contraction (› Figure 8-19). Let us examine each factor in detail to see how they influence stroke volume.

## Increased end-diastolic volume

**Intrinsic control** of stroke volume, which refers to the heart's inherent ability to vary stroke volume, depends on the direct correlation between end-diastolic volume and stroke volume. As more blood returns to the heart, the heart pumps out more blood per beat, but the relationship is not quite as simple as it might seem, because the heart does not eject all the blood it contains. This intrinsic control depends on the length–tension relationship of cardiac muscle, which is similar to that of skeletal muscle. For skeletal muscle, the resting muscle length is approximately the optimal length ($l_o$) at which maximal tension can be developed during a subsequent contraction. When the skeletal muscle is longer or shorter than $l_o$, the subsequent contraction is weaker (see › Figure 7-17). For cardiac muscle, the resting cardiac muscle fibre length is less than $l_o$. Therefore, the length of cardiac muscle fibres normally varies

along the ascending limb of the length–tension curve. An increase in cardiac muscle fibre length, by moving closer to $l_o$, increases the contractile tension of the heart on the following systole (› Figure 8-20).

Unlike in skeletal muscle, the length–tension curve of cardiac muscle normally does not operate at lengths that fall within the region of the descending limb. That is, within physiological limits, cardiac muscle does not get stretched beyond its optimal length to the point that contractile strength diminishes with further stretching.

### FRANK–STARLING LAW OF THE HEART

What causes cardiac muscle fibres to vary in length before contraction? Skeletal muscle length can vary before contraction because of the positioning of the skeletal parts to which the muscle is attached, but cardiac muscle is not attached to any bones. The main determinant of cardiac muscle fibre length is the degree of diastolic filling. The greater the diastolic filling, the larger the end-diastolic volume (EDV), and the more the heart is stretched. The more the heart is stretched, the longer the initial cardiac fibre length before contraction. The increased length results in a greater force on the subsequent cardiac contraction, which results in a greater stroke volume. This intrinsic relationship between EDV and stroke volume is known as the **Frank–Starling law of the heart**. Stated simply, the heart normally pumps out during systole the volume of blood returned to it during diastole; increased venous return results in increased stroke volume. In Figure 8-20, assume that EDV increases from point A to point B. You can see that this increase in EDV is accompanied by a corresponding increase in stroke volume from point $A^1$ to point $B^1$. The extent of filling is referred to as the

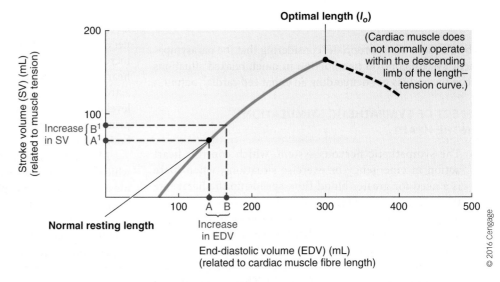

> FIGURE 8-20 **Intrinsic control of stroke volume (Frank–Starling curve).** The cardiac muscles fibre's length, which is determined by the extent of venous filling, is normally less than the optimal length for developing maximal tension. Therefore, an increase in end-diastolic volume (i.e., an increase in venous return), accomplished by moving the cardiac muscle fibre length closer to optimal length, increases the contractile tension of the fibres on the next systole. A stronger contraction squeezes out more blood. In this way, as more blood is returned to the heart and the end-diastolic volume increases, the heart automatically pumps out a correspondingly larger stroke volume.

**preload**, because it is the workload imposed on the heart before contraction begins.

### ADVANTAGES OF THE CARDIAC LENGTH–TENSION RELATIONSHIP

The built-in relationship that matches stroke volume with venous return has two important advantages. First, one of the most important functions of this intrinsic mechanism is the equalizing of output between the right and left sides of the heart so that blood pumped out by the heart is equally distributed between the pulmonary and systemic circulation. If, for example, the right side of the heart ejects a larger stroke volume, more blood enters the pulmonary circulation, so venous return to the left side of the heart increases accordingly. The increased EDV of the left side of the heart causes it to contract more forcefully, so it too pumps out a larger stroke volume. In this way, output of the two ventricular chambers is kept equal. If such equalization did not happen, too much blood would be dammed up in the venous system in front of the ventricle with the lower output.

Second, when a larger cardiac output is needed, such as during exercise, venous return is increased through action of the sympathetic nervous system and other mechanisms (see Chapter 9). The resulting increase in EDV automatically increases stroke volume correspondingly. Because exercise also increases heart rate, these two factors act together to increase the cardiac output (cardiac output = stroke volume × heart rate), and therefore more blood can be delivered to the exercising muscles.

### MECHANISM OF THE CARDIAC LENGTH–TENSION RELATIONSHIP

Although the length–tension relationship in cardiac muscle fibres depends to a degree on the extent of overlap of thick and thin filaments, the key factor relating cardiac muscle fibre length to tension development is the dependence of myofilament $Ca^{2+}$ sensitivity on the fibre's length. Specifically, as a cardiac muscle fibre's length increases along the ascending limb of the length–tension curve, the lateral spacing between adjacent thick and thin filaments is reduced. Stated differently, as a cardiac muscle fibre is stretched as a result of greater ventricular filling, its myofilaments are pulled closer together. As a result of this reduction in distance between the thick and thin filaments, more cross-bridge interactions between myosin and actin can take place when $Ca^{2+}$ pulls the troponin–tropomyosin complex away from actin's cross-bridge binding sites; that is, myofilament $Ca^{2+}$ sensitivity increases. Thus, the length–tension relationship in cardiac muscle depends not on muscle fibre length *per se* but on the resultant variations in the lateral spacing between the myosin and actin filaments.

### EFFECT OF AFTERLOAD ON STROKE VOLUME

Preload and its effect on EDV have been discussed, but another important factor affects stroke volume—**afterload**, which refers to the forces that the heart is contracting against. Afterload is a combination of the EDV, the volume of blood the ventricles are contracting on, and also the pressure that the ventricles are contracting against. In general, we consider the pressure to be the greater contributor to afterload.

For example, as discussed in Section 8.4 regarding the cardiac cycle, blood will not flow out of the left ventricle until left ventricular pressure exceeds the aortic pressure (end of C, Figure 8-16). However, in situations when aortic pressure is higher than normal, there is an increased afterload. As a result of this increased afterload, the ventricle must contract against a greater pressure and may not eject as much blood as normal. If EDV increases as a result of this decreased stroke volume then the intrinsic control of contractility will increase the strength of contraction to compensate. However, if EDV does not change then the heart will rely on extrinsic control of stroke volume.

We now shift our attention from intrinsic control to extrinsic control of stroke volume.

## Sympathetic stimulation

In addition to intrinsic control, stroke volume is also subject to extrinsic control by factors originating outside the heart, the most important of which are actions of the cardiac sympathetic nerves and epinephrine (see Table 8-3). Sympathetic stimulation and epinephrine enhance the heart's **contractility**, which is the strength of contraction at any given EDV. In other words, on sympathetic stimulation the heart contracts more forcefully and squeezes out a greater percentage of the blood it contains, leading to more complete ejection. This increased contractility is due to the increased $Ca^{2+}$ influx triggered by norepinephrine and epinephrine. The extra cytosolic $Ca^{2+}$ lets the myocardial fibres generate more force through greater cross-bridge cycling than they would without sympathetic influence. Recall that normally, the EDV is 135 mL and the end-systolic

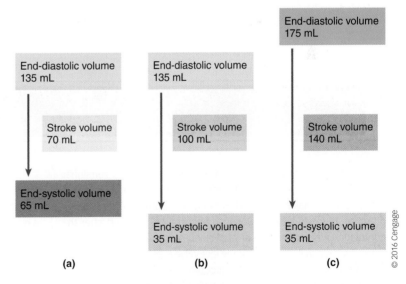

> **FIGURE 8-21 Effect of sympathetic stimulation on stroke volume.** (a) Normal stroke volume. (b) Stroke volume during sympathetic stimulation. (c) Stroke volume with combination of sympathetic stimulation and increased end-diastolic volume

© 2016 Cengage

volume (ESV) is 65 mL for a stroke volume of 70 mL (> Figure 8-21a). Another variable associated with EDV and stroke volume is **ejection fraction**, which is the fraction of blood pumped from the left ventricle with each beat of the heart. The ejection fraction is equal to stroke volume divided by EDV. It is typically expressed as a percentage and ranges between 55 and 60 percent during rest, which indicates that 40 percent is still remaining in the left ventricle. Damage to the muscle of the heart (myocardium), such as a myocardial infarction, impairs the heart's ability to eject blood and therefore reduces the ejection fraction. Consequently, the measure of ejection fraction has clinical application for such conditions as heart failure. Under sympathetic influence, for the same EDV of 135 mL, the ESV might be 35 mL and the stroke volume 100 mL (> Figure 8-21b). In effect, sympathetic stimulation shifts the Frank–Starling curve to the left (> Figure 8-22). Depending on the extent of sympathetic stimulation, the curve can be shifted to varying degrees, up to a maximal increase in contractile strength of about 100 percent greater than normal.

Sympathetic stimulation increases stroke volume not only by strengthening cardiac contractility but also by enhancing venous return (> Figure 8-21c). Sympathetic stimulation constricts the veins, and this squeezes more blood forward from the veins to the heart, increasing the EDV and subsequently increasing stroke volume even further.

### SUMMARY OF FACTORS AFFECTING STROKE VOLUME AND CARDIAC OUTPUT

The strength of cardiac muscle contraction and, accordingly, the stroke volume can be graded by (1) varying the initial length of the muscle fibres, which in turn depends on the degree of ventricular filling before contraction (intrinsic control); and (2) varying the extent of sympathetic stimulation (extrinsic control) (see Figure 8-19). This is in contrast to gradation of skeletal muscle, in which twitch summation and recruitment of motor units produce variable strength of muscle contraction. These mechanisms do not apply to cardiac muscle. Twitch summation is impossible because of the long refractory period. Recruitment of motor units is not possible because the heart muscle cells are arranged into functional syncytia where all contractile cells become excited and contract with every beat, instead of into distinct motor units that can be discretely activated.

All the factors that determine cardiac output by influencing heart rate or stroke volume are summarized in > Figure 8-23. Note that sympathetic stimulation increases cardiac output by increasing both heart rate and stroke volume. Sympathetic activity to the heart increases, for example, during exercise when the working skeletal muscles need increased delivery of oxygen-laden blood to support their high rate of ATP consumption.

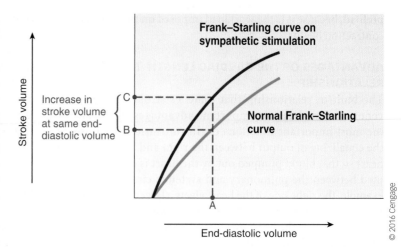

> **FIGURE 8-22 Shift of the Frank–Starling curve to the left by sympathetic stimulation.** For the same end-diastolic volume (point A), there is a larger stroke volume (from point B to point C) on sympathetic stimulation as a result of increased contractility of the heart. The Frank–Starling curve is shifted to the left by variable degrees, depending on the extent of sympathetic stimulation.

---

### ▌Why It Matters
### High Blood Pressure

When the ventricles contract, in order to force open the semilunar valves, they must generate sufficient pressure to exceed the blood pressure in the major arteries. The arterial blood pressure is called the afterload, because it is the workload imposed on the heart after the contraction has begun. If the arterial blood pressure is chronically elevated (high blood pressure) or if the exit valve is stenotic, the ventricle must generate more pressure to eject blood. For example, instead of generating the normal pressure of 120 mmHg, the ventricular pressure may need to rise as high as 400 mmHg to force blood through a narrowed aortic valve.

The heart may be able to compensate for a sustained increase in afterload by enlarging (through hypertrophy or enlargement of the cardiac muscle fibres). This enables it to contract more forcefully and maintain a normal stroke volume despite an abnormal impediment to ejection. A diseased heart or a heart weakened with age may not be able to compensate completely, however; in that case, heart failure ensues. Even if the heart is initially able to compensate for a chronic increase in afterload, the sustained extra workload placed on the heart can eventually cause pathological changes in the heart that lead to heart failure. In fact, a chronically elevated afterload is one of the two major factors that cause heart failure.

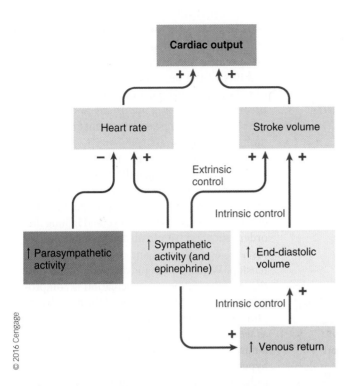

**Cardiac output**

+     +

Heart rate          Stroke volume

−   +        Extrinsic       +   +
          control

          Intrinsic control

↑ Parasympathetic activity     ↑ Sympathetic activity (and epinephrine)     ↑ End-diastolic volume

          Intrinsic control       +

          +

          → ↑ Venous return

© 2016 Cengage

❯ **FIGURE 8-23  Control of cardiac output**

---

**▮ Clinical Connections**

Myocardial infarction causes a decrease in cardiac output, and if this decrease is large enough it can cause death. In fact, many of the symptoms that can be experienced during an acute heart attack are a result of the reflex responses that try to maintain cardiac output.

During acute myocardial infarction, there is a decrease in blood delivery to the heart tissue. Since the heart requires large amounts of ATP, it is forced to shift from aerobic to anaerobic metabolism (p. 24), which is less efficient in generating ATP and leads to impaired contractile function of the heart. This decreased ability of the heart to contract is called systolic dysfunction; this means the ventricle cannot eject as much blood as normal, so stroke volume and thus cardiac output decrease. On top of this, large infarcted areas reduce ventricular compliance—its ability to expand—so during ventricular filling, ventricular pressure increases faster and EDV is decreased. This is called diastolic dysfunction, which reduces cardiac output by shifting the Frank–Starling curve to the left (see Figure 8-20 and assume B is normal and A is after the heart attack).

Next we examine how a failing heart cannot pump out enough blood.

## Heart failure

**Heart failure** is the inability of the cardiac output to keep pace with the body's demands for nutrient supplies and removal of wastes. Either one or both ventricles may progressively weaken and fail. When a failing ventricle cannot pump out all the blood returned to it, the veins behind the failing ventricle become congested with blood. Heart failure may occur for a variety of reasons, but the two most common are (1) damage to the heart muscle as a result of a heart attack or impaired circulation to the cardiac muscle; and (2) prolonged pumping against a chronically increased afterload, as with a stenotic semilunar valve or a sustained elevation in blood pressure. Recently, researchers D. Warburton, at the University of British Columbia, and M. Hay Kowsky, at the University of Alberta, have made significant contributions to understanding heart failure and to the importance of exercise in positively influencing the quality of life of those with heart failure.

### PRIME DEFECT IN HEART FAILURE

The prime defect in heart failure is a decrease in cardiac contractility; that is, weakened cardiac muscle cells contract less effectively. The intrinsic ability of the heart to develop pressure and eject a stroke volume is reduced so that the heart operates on a lower length–tension curve (❯ Figure 8-24a). The Frank–Starling curve shifts downward and to the right such that for a given EDV, a failing heart pumps out a smaller stroke volume than a normal healthy heart.

### COMPENSATORY MEASURES FOR HEART FAILURE

In the early stages of heart failure, two major compensatory measures help restore stroke volume to normal. First, sympathetic activity to the heart is increased by reflex, which increases heart contractility toward normal (❯ Figure 8-24b). Sympathetic stimulation can help compensate only for a limited period of time, however, because the heart becomes less responsive to norepinephrine after prolonged exposure, and furthermore, the stores of norepinephrine in the heart's sympathetic nerve terminals becomes depleted. Second, when cardiac output is reduced, the kidneys, in a compensatory attempt to improve their reduced blood flow, retain extra salt and water in the body during urine formation, to expand the blood volume. The increase in circulating blood volume increases the EDV. The resultant stretching of the cardiac muscle fibres enables the weakened heart to pump out a normal stroke volume (Figure 8-24b). The heart is now pumping out the blood returned to it but is operating at a greater cardiac muscle fibre length.

### DECOMPENSATED HEART FAILURE

As the disease progresses and heart contractility deteriorates further, the heart reaches a point at which it can no longer pump out a normal stroke volume (i.e., cannot pump out all the blood returned to it) despite compensatory measures. At this point, the heart slips from compensated heart failure into a state of decompensated heart failure. Now the cardiac muscle fibres are stretched to the point that they are operating in the descending limb of the length–tension curve. *Forward failure* occurs as the heart fails to pump an adequate amount of blood forward to the tissues because the stroke volume becomes progressively smaller. *Backward failure* occurs simultaneously as blood that cannot enter and be pumped out by the heart continues to dam up the venous system. The congestion in the venous system is the reason that this condition is sometimes termed **congestive heart failure**.

Left-sided failure has more serious consequences than right-sided failure. Backward failure of the left side leads to pulmonary oedema (excess tissue fluid in the lungs) because blood backs

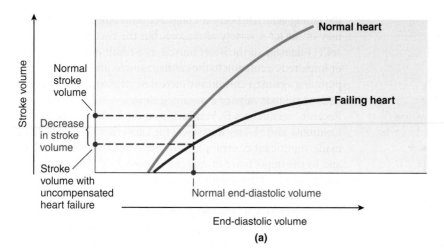

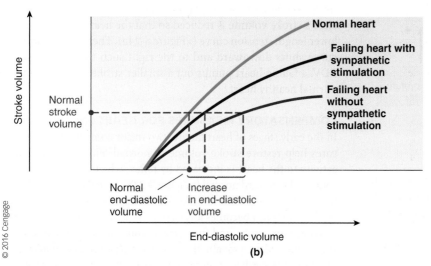

> FIGURE 8-24 **Compensated heart failure.** (a) Shift of the Frank–Starling curve downward and to the right in a failing heart. Because its contractility is decreased, the failing heart pumps out a smaller stroke volume at the same end-diastolic volume than a normal heart does. (b) Compensations for heart failure. Reflex sympathetic stimulation shifts the Frank–Starling curve of a failing heart to the left, increasing the contractility of the heart toward normal. A compensatory increase in end-diastolic volume as a result of blood volume expansion further increases the strength of contraction of the failing heart. Operating at a longer cardiac muscle fibre length, a compensated failing heart is able to eject a normal stroke volume.

up in the lungs. This fluid accumulation in the lungs reduces exchange of oxygen and carbon dioxide between the air and blood in the lungs, thereby reducing arterial oxygenation and elevating levels of acid-forming carbon dioxide in the blood. In addition, one of the more serious consequences of left-sided forward failure is an inadequate blood flow to the kidneys, which causes a twofold problem. First, vital kidney function is depressed; and second, the kidneys retain even more salt and water in the body during urine formation as they try to expand the plasma volume even further to improve their reduced blood flow. Excessive fluid retention further exacerbates the already existing problems of venous congestion.

Treatment of congestive heart failure therefore includes measures that reduce salt and water retention and increase urinary output as well as drugs that enhance the contractile ability of the weakened heart—digitalis, for example.

## OTHER TYPES OF HEART FAILURE

So far, we have discussed the chronic form of heart failure that is caused by the progressive inability of the heart to pump enough blood to satisfy the metabolic needs of the peripheral tissues. However, it is important to note there are other forms of heart failure that are just as life-threatening.

Given that the heart and lungs work in tandem to deliver oxygen throughout the body, it is clear that when the lungs are not functioning properly there can be an effect on the function of the heart. This is what is observed in lung disorders such as chronic obstructive pulmonary disease, emphysema, and pulmonary hypertension. When the lung is diseased, blood flow through the pulmonary circulation is compromised such that the right side of the heart has to pump harder to maintain blood flow. Eventually, as seen with the left ventricle in congestive heart failure, the right side of the heart will undergo hypertrophy and eventually fail. This form of right-side heart failure is called **cor pulmonale** or, more commonly, pulmonary heart disease.

Until recently, most of the research for heart failure has focused on chronic heart failure, but it is now recognized that **acute heart failure** is just as important. The main difference in these forms is that acute heart failure progresses rapidly (over days to weeks) such that the heart does not have time to try to compensate for the increased cardiac demands. Acute heart failure has many causes but always relates to something that has caused a relatively sudden inability of the heart to pump blood normally. One example of this could be in a person who has suffered an acute myocardial infarction. In such a case, if a significant part of the ventricular muscle becomes damaged, the ability of the heart to pump blood is reduced. Another example is pulmonary oedema. Similar to what is seen with cor pulmonale, pulmonary oedema causes an increase in pulmonary pressures and can rapidly affect right ventricular function.

## SYSTOLIC VERSUS DIASTOLIC HEART FAILURE

Increasingly, physicians categorize heart failure as either *systolic failure*, characterized by a decrease in cardiac contractility as just described, or *diastolic failure*, in which the heart has trouble filling. Diastolic failure is a recently recognized problem. With diastolic failure, the ventricles do not fill normally either because the heart muscle does not adequately relax between beats or because the heart muscle stiffens and cannot expand as much as usual. Because of impeded filling, a diastolic failing heart pumps out less blood than it should with each contraction. No drugs are available yet that reliably help the heart relax, so treatment is aimed at relieving symptoms or halting underlying causes of diastolic disease.

8

In Canada and around the world, heart disease is a leading cause of premature death. Recent developments in myocardial disease (cardiac muscle tissue) have led to the classification of new therapeutic targets, one of which is gene and stem cell therapies for the protection and revival of the myocardium. To date, genetic therapies have been used to treat complex diseases, such as heart failure, ischemia, and other inherited myopathies, in animals. Such progress in this field has been made that in 2013, a clinical trial led by Dr. Duncan Stewart, a cardiologist at the Ottawa Hospital Research Institute, began to study the combined effects of genetically enhanced stem cells on repairing damaged heart muscle following a heart attack.

---

### Check Your Understanding 8.4

1. Indicate the effect (increase or decrease) of parasympathetic stimulation and sympathetic stimulation on heart rate and stroke volume.
2. Draw a graph showing the relationship between end-diastolic volume and stroke volume, according to the Frank–Starling law of the heart.

---

## 8.6 | Nourishing the Heart Muscle

Cardiac muscle cells contain an abundance of mitochondria, the oxygen-dependent energy organelles. In fact, up to 40 percent of the cell volume of cardiac muscle cells is occupied by mitochondria, indicative of how much the heart depends on oxygen delivery and aerobic metabolism to generate the energy necessary for contraction. Cardiac muscle also has an abundance of myoglobin, which stores limited amounts of oxygen within the heart for immediate use.

### The heart's blood supply

Although all the blood passes through the heart, the heart muscle cannot extract oxygen or nutrients from the blood within its chambers for two reasons. First, the watertight endocardial lining does not permit blood to pass from the chamber into the myocardium. Second, the heart walls are too thick to permit diffusion of oxygen and other supplies from the blood in the chamber to the individual cardiac cells. Therefore, like other tissues of the body, heart muscle must receive blood through blood vessels, specifically via the **coronary circulation**. The coronary arteries branch from the aorta just beyond the aortic valve, and the coronary veins empty into the right atrium.

The heart muscle receives most of its blood supply during diastole. Blood flow to the heart muscle cells is substantially reduced during systole for two reasons. First, the contracting myocardium, especially in the powerful left ventricle, compresses the major branches of the coronary arteries; second, the open aortic valve partially blocks the entrance to the coronary vessels. As a result, most coronary arterial flow (about 70%) occurs during diastole, driven by the aortic blood pressure, and it declines as aortic pressure drops. Only about 30 percent of coronary arterial flow occurs during systole (>Figure 8-25).

This limited time for coronary blood flow becomes especially important during rapid heart rates, when diastolic time is much reduced. Just when increased demands are placed on the heart to pump more rapidly, it has less time to provide oxygen and nourishment to its own musculature to accomplish the increased workload.

### MATCHING OF CORONARY BLOOD FLOW TO HEART MUSCLE'S OXYGEN NEEDS

Under normal circumstances the heart muscle does receive adequate blood flow to support its activities, even during exercise, when the rate of coronary blood flow increases up to five times its resting rate. Let's see how.

Extra blood is delivered to the cardiac cells primarily by vasodilation, or enlargement, of the coronary vessels, which lets more blood flow through them, especially during diastole. The increased coronary blood flow is necessary to meet the heart's increased oxygen requirements because the heart, unlike most other tissues, is unable to remove much additional oxygen from the blood passing through its vessels to support increased metabolic activities. Most other tissues under resting conditions extract only about 25 percent of the oxygen available from the blood flowing through them, leaving a considerable oxygen reserve that can be drawn on when a tissue has increased oxygen needs; that is, the tissue can immediately increase the oxygen available to it by removing a greater percentage of oxygen from the blood passing through it. In contrast, the heart, even under resting conditions, removes up to 65 percent of the oxygen available in the coronary vessels, far more than is withdrawn by other tissues. This leaves little oxygen in reserve in the coronary blood should cardiac oxygen demands increase. Therefore, the primary means by which more oxygen can be made available to the heart muscle is by increasing coronary blood flow.

Coronary blood flow is adjusted primarily in response to changes in the heart's oxygen requirements. Among the proposed links between blood flow and oxygen needs is *adenosine*,

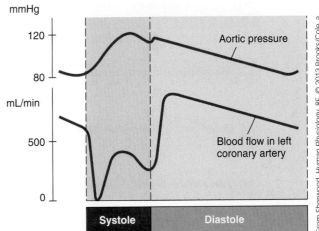

> **FIGURE 8-25 Coronary blood flow.** Most coronary blood flow occurs during diastole because the coronary vessels are compressed almost completely during systole.

which is formed from adenosine triphosphate (ATP) during cardiac metabolic activity. Cardiac cells form and release more adenosine when cardiac activity increases and the heart is using more ATP as an energy source and, accordingly, needs more oxygen. The released adenosine, acting as a paracrine (p. 51), induces dilation of the coronary blood vessels, which allows more oxygen-rich blood to flow to the more active cardiac cells to meet their increased oxygen demand (⟩ Figure 8-26). Matching oxygen delivery to oxygen needs is crucial, because heart muscle depends on oxidative processes to generate energy. The heart cannot get enough ATP through anaerobic metabolism.

### NUTRIENT SUPPLY TO THE HEART

Although the heart has little ability to support its energy needs by means of anaerobic metabolism and must rely heavily on its oxygen supply, it can tolerate wide variations in its nutrient supply. As fuel sources, the heart primarily uses free fatty acids and, to a lesser extent, glucose and lactate, depending on their availability. Because cardiac muscle is remarkably adaptable and can shift metabolic pathways to use whatever nutrient is available, the primary danger of insufficient coronary blood flow is not fuel shortage but oxygen deficiency.

## Atherosclerotic coronary artery disease

*Clinical Note* The adequacy of coronary blood flow is relative to the heart's oxygen demands at any given moment. In the normal heart, coronary blood flow increases correspondingly as oxygen demands rise. With coronary artery disease, coronary blood flow may not be able to keep pace with rising oxygen needs. The term **coronary artery disease (CAD)** refers to pathological changes within the coronary artery walls that diminish blood flow through these vessels. A given rate of

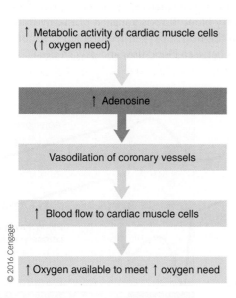

↑ Metabolic activity of cardiac muscle cells
( ↑ oxygen need)

↓

↑ Adenosine

↓

Vasodilation of coronary vessels

↓

↑ Blood flow to cardiac muscle cells

↓

↑ Oxygen available to meet ↑ oxygen need

© 2016 Cengage

⟩ **FIGURE 8-26 Matching of coronary blood flow to the oxygen need of cardiac muscle cells**

coronary blood flow may be adequate at rest but insufficient in physical exertion or other stressful situations.

Complications of CAD, including heart attacks, make it the single leading cause of death in Canada. CAD is the underlying cause of about 32 percent of all deaths in this country. CAD can cause myocardial ischemia and possibly lead to acute myocardial infarction by three mechanisms: (1) profound vascular spasm of the coronary arteries, (2) formation of atherosclerotic plaques, and (3) thromboembolism. We will discuss each in turn.

### VASCULAR SPASM

**Vascular spasm** is an abnormal spastic constriction that transiently narrows the coronary vessels. Vascular spasms are associated with the early stages of CAD and are most often triggered by exposure to cold, physical exertion, or anxiety. The condition is reversible and usually does not last long enough to damage the cardiac muscle.

When too little oxygen is available in the coronary vessels, the endothelium (blood vessel lining) releases *platelet-activating factor (PAF)*. PAF, which exerts a variety of actions, was named for its first discovered effect, activating platelets. Among its other effects, PAF, once released from the endothelium, diffuses to the underlying vascular smooth muscle and causes it to contract, bringing about vascular spasm.

### DEVELOPMENT OF ATHEROSCLEROSIS

**Atherosclerosis** is a progressive, degenerative arterial disease that gradually leads to occlusion (blockage) of affected vessels, reducing blood flow through them. Atherosclerosis is characterized by plaques forming beneath the vessel lining within arterial walls. An **atherosclerotic plaque** consists of a lipid-rich core covered by an abnormal overgrowth of smooth muscle cells, topped off by a collagen-rich connective tissue cap. As plaque forms, it bulges into the vessel lumen (⟩ Figure 8-27).

Although not all the contributing factors have been identified, in recent years investigators have sorted out the following complex sequence of events in the gradual development of atherosclerosis:

1. Atherosclerosis starts with injury to the blood vessel wall, which triggers an *inflammatory response* that sets the stage for plaque buildup. Normally, inflammation is a protective response that fights infection and promotes repair of damaged tissue (p. 466). However, when the cause of the injury persists within the vessel wall, the sustained, low-grade inflammatory response over a course of decades can insidiously lead to arterial plaque formation and heart disease. Plaque formation likely has many causes. Suspected artery-abusing agents that may set off the vascular inflammatory response include oxidized cholesterol, free radicals, high blood pressure, homocysteine, chemicals released from fat cells, or even bacteria and viruses that damage blood vessel

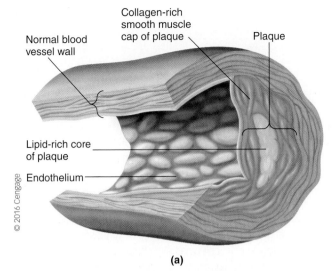

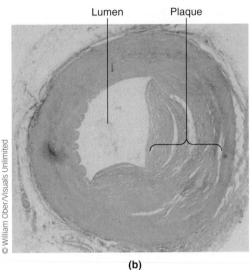

> FIGURE 8-27 **Atherosclerotic plaque.** (a) Schematic representation of the components of a plaque. (b) Photomicrograph of a severe atherosclerotic plaque in a coronary vessel

walls. The most common triggering agent appears to be oxidized cholesterol. (For a further discussion of the role of cholesterol and other factors in the development of atherosclerosis, see Concepts, Challenges, and Controversies.)

2. Typically, the initial stage of atherosclerosis is characterized by the accumulation beneath the endothelium of excessive amounts of *low-density lipoprotein (LDL)*, the so-called bad cholesterol, in combination with a protein carrier. As LDL accumulates within the vessel wall, this cholesterol product becomes oxidized, primarily by oxidative wastes produced by the blood vessel cells. These wastes are *free radicals*—very unstable electron-deficient particles that are highly reactive. Antioxidant vitamins that prevent LDL oxidation, such as *vitamin E, vitamin C,* and *beta-carotene,* have been shown to slow plaque deposition.

3. In response to the presence of oxidized LDL and/or other irritants, the endothelial cells produce chemicals that attract *monocytes,* a type of white blood cell, to the site. These immune cells trigger a local inflammatory response.

4. Once they leave the blood and enter the vessel wall, monocytes settle down permanently, enlarge, and become large phagocytic cells called *macrophages.* Macrophages voraciously phagocytose (p. 48) the oxidized LDL until these cells become so packed with fatty droplets that they appear foamy under a microscope. Now called *foam cells,* these greatly engorged macrophages accumulate beneath the vessel lining and form a visible *fatty streak,* the earliest form of an atherosclerotic plaque.

5. The earliest stage of plaque formation is characterized by accumulation beneath the endothelium of a cholesterol-rich deposit. The disease progresses as smooth muscle cells within the blood vessel wall migrate from the muscular layer of the blood vessel to a position on top of the lipid accumulation, just beneath the endothelium. This migration is triggered by chemicals released at the inflammatory site. At their new location, the smooth muscle cells continue to divide and enlarge, producing *atheromas,* which are benign (non-cancerous) tumours of smooth muscle cells within the blood vessel walls. Together the lipid-rich core and overlying smooth muscle form a maturing plaque.

6. As it continues to develop, the plaque progressively bulges into the lumen of the vessel. The protruding plaque narrows the opening through which blood can flow.

7. Further contributing to vessel narrowing, oxidized LDL inhibits release of *nitric oxide* from the endothelial cells. Nitric oxide is a local chemical messenger that relaxes the underlying layer of normal smooth muscle cells within the vessel wall. Relaxation of these smooth muscle cells dilates the vessel. Because of reduced nitric oxide release, vessels damaged by developing plaques cannot dilate as readily as normal.

8. A thickening plaque also interferes with nutrient exchange for cells located within the involved arterial wall, leading to degeneration of the wall in the vicinity of the plaque. The damaged area is invaded by *fibroblasts* (scar-forming cells), which form a connective tissue cap over the plaque. (The term *sclerosis* means "excessive growth of fibrous connective tissue"—hence, the term *atherosclerosis* for this condition that is characterized by atheromas and sclerosis, along with abnormal lipid accumulation.)

9. In the later stages of the disease, calcium often precipitates in the plaque. A vessel so afflicted becomes hard and cannot distend easily.

**Cardiac Physiology**

# Atherosclerosis: Cholesterol and Beyond

THE CAUSE OF ATHEROSCLEROSIS IS STILL NOT ENTIRELY CLEAR. Certain high-risk factors have been associated with an increased incidence of atherosclerosis and coronary heart disease. Included among them are genetic predisposition, obesity, advanced age, smoking, hypertension, lack of exercise, high blood concentrations of C-reactive protein, elevated levels of homocysteine, infectious agents, and, most notoriously, elevated cholesterol levels in the blood.

## Sources of Cholesterol

There are two sources of cholesterol for the body: (1) dietary intake of cholesterol—animal products, such as egg yolk, red meats, and butter are especially rich in this lipid (animal fats contain cholesterol, whereas plant fats typically do not); and (2) manufacture of cholesterol by cells, especially liver cells.

## Good versus Bad Cholesterol

Actually, it is not the total blood cholesterol level but the amount of cholesterol bound to various plasma protein carriers that is most important with regard to the risk of developing atherosclerotic heart disease. Because cholesterol is a lipid, it is not very soluble in blood. Most cholesterol in the blood is attached to specific plasma protein carriers in the form of lipoprotein complexes, which are soluble in blood. The three major lipoproteins are named for their density of protein as compared with lipid: (1) **high-density lipoproteins (HDL)**, which contain the most protein and least cholesterol; (2) **low-density lipoproteins (LDL)**, which have less protein and more cholesterol; and (3) **very-low-density lipoproteins (VLDL)**, which have the least protein and most lipid, but the lipid they carry is neutral fat, not cholesterol.

Cholesterol carried in LDL complexes has been termed *bad cholesterol*, because cholesterol is transported to the cells, including those lining the blood vessel walls, by LDL. The propensity toward developing atherosclerosis substantially increases with elevated levels of LDL. The presence of oxidized LDL within an arterial wall is a major trigger for the inflammatory process that leads to the development of atherosclerotic plaques (p. 375).

In contrast, cholesterol carried in HDL complexes has been dubbed *good cholesterol*, because HDL removes cholesterol *from* the cells and transports it to the liver for partial elimination from the body. Not only does HDL help remove excess cholesterol from the tissues, it also protects by inhibiting oxidation of LDL. The risk of atherosclerosis is inversely related to the concentration of HDL in the blood; that is, elevated levels of HDL are associated with a low incidence of atherosclerotic heart disease.

Some other factors known to influence atherosclerotic risk can be related to HDL levels; for example, cigarette smoking lowers HDL, whereas regular exercise raises HDL.

## Cholesterol Uptake by Cells

Unlike most lipids, cholesterol is not used as metabolic fuel by cells. Instead, it serves as an essential component of plasma membranes. In addition, a few special cell types use cholesterol as a precursor for the synthesis of secretory products, such as steroid hormones and bile salts. Although most cells can synthesize some of the cholesterol needed for their own plasma membranes, they cannot manufacture sufficient amounts and therefore must rely on supplemental cholesterol being delivered by the blood.

Cells accomplish cholesterol uptake from the blood by synthesizing receptor proteins specifically capable of binding LDL and inserting these receptors into the plasma membrane. When an LDL particle binds to one of the membrane receptors, the cell engulfs the particle by receptor-mediated endocytosis, receptor and all (p. 48). Within the cell, lysosomal enzymes break down the LDL to free the cholesterol, making it available to the cell for synthesis of new cellular membrane. The LDL receptor, which is also freed within the cell, is recycled back to the surface membrane.

If too much free cholesterol accumulates in the cell, there is a shutdown of both the synthesis of LDL receptor proteins (so that less cholesterol is taken up) and the cell's own cholesterol synthesis (so that less new cholesterol is made). Faced with a cholesterol shortage, in contrast, the cell makes more LDL receptors so that it can engulf more cholesterol from the blood.

## Maintenance of Blood Cholesterol Level and Cholesterol Metabolism

The maintenance of a blood-borne cholesterol supply to the cells involves an interaction between dietary cholesterol and the synthesis of cholesterol by the liver. When the amount of dietary cholesterol is increased, hepatic (liver) synthesis of cholesterol is turned off because cholesterol in the blood directly inhibits a hepatic enzyme essential for cholesterol synthesis. Therefore, as more cholesterol is ingested, less is produced by the liver. Conversely, when cholesterol intake from food is reduced, the liver synthesizes more of this lipid because the inhibitory effect of cholesterol on the crucial hepatic enzyme is removed. In this way, the blood concentration of cholesterol is maintained at a fairly constant level despite changes in cholesterol intake; thus, it is difficult to significantly reduce cholesterol levels in the blood by decreasing cholesterol intake.

HDL transports cholesterol to the liver. The liver secretes cholesterol as well as cholesterol-derived bile salts into the bile. Bile enters the intestinal tract, where bile salts participate in the digestive process. Most of the secreted cholesterol and bile salts are subsequently reabsorbed from the intestine into the blood to be recycled to the liver. However, the cholesterol and bile salts not reclaimed by absorption are eliminated in the feces and lost from the body.

Consequently, the liver has a primary role in determining total blood cholesterol levels, and the interplay between LDL and HDL determines the traffic flow of cholesterol between the liver and the other cells. Whenever these mechanisms are altered, blood cholesterol levels may be affected in such a way as to influence the individual's predisposition to atherosclerosis.

Varying the intake of dietary fatty acids may alter total blood cholesterol levels by influencing one or more of the mechanisms involved in cholesterol balance. The blood cholesterol level tends to be raised by ingesting saturated fatty acids found predominantly in animal fats, because these fatty acids stimulate cholesterol synthesis and inhibit its conversion to bile salts. In contrast, ingesting polyunsaturated fatty acids, the predominant fatty acids of most plants, tends to reduce blood cholesterol levels by enhancing elimination of both cholesterol and cholesterol-derived bile salts in the feces.

The treatment of elevated cholesterol has been met with moderate success. In 2008, Dr. Trisha R. Joy at the University of Western Ontario discussed the potential treatment of high cholesterol through increasing HDL (the good cholesterol), which helps remove cholesterol from the body. However, clinical trials have indicated that increasing circulating HDL as a treatment for cardiovascular disease may not be as promising as first thought. The functional quality of HDL may be more important than the circulating quantity. Thus, more research is required.

### Other Risk Factors besides Cholesterol

Despite the strong links between cholesterol and heart disease, over half of all patients with heart attacks have a normal cholesterol profile and no other well-established risk factors. Clearly, other factors are involved in the development of coronary artery disease in these people. These same factors may also contribute to development of atherosclerosis in people with unfavourable cholesterol levels. The following are among the leading other possible risk factors:

- Elevated blood levels of the amino acid **homocysteine** have recently been implicated as a strong predictor for heart disease, independent of the person's cholesterol/lipid profile. Homocysteine is formed as an intermediate product during metabolism of the essential dietary amino acid *methionine*. Investigators believe homocysteine contributes to atherosclerosis by promoting

proliferation of vascular smooth muscle cells, an early step in development of this artery-clogging condition. Furthermore, homocysteine appears to damage endothelial cells and may cause oxidation of LDL, both of which can contribute to plaque formation. Three B vitamins—*folic acid, vitamin B$_{12}$*, and *vitamin B$_6$*—all play key roles in pathways that clear homocysteine from the blood. Therefore, these B *vitamins* are all needed to keep blood homocysteine at safe levels.

- Inflammation has also been implicated in the development of atherosclerotic plaques as people with elevated levels of **C-reactive protein**, a blood-borne marker of inflammation, have a higher risk for developing coronary artery disease. In one study, people with a high level of C-reactive protein in their blood were three times more likely to have a heart attack over the next 10 years than those with a low level of this inflammatory protein. Because inflammation plays a crucial role in the development of atherosclerosis, anti-inflammatory drugs, such as Aspirin, help prevent heart attacks. Furthermore, Aspirin protects against heart attacks through its role in inhibiting clot formation.

- Accumulating data suggest that an infectious agent may be the underlying culprit in a significant number of cases of atherosclerotic disease. Among the leading suspects are respiratory infection–causing *Chlamydia pneumoniae*, cold sore–causing *herpes virus*, and gum disease–causing bacteria. Importantly, if a link between infections and coronary artery disease can be confirmed, antibiotics may be added to the regimen of heart disease prevention strategies.

As you can see, the relationship between atherosclerosis, cholesterol, and other factors is far from clear. Much research on this complex disease is currently in progress, because the incidence of atherosclerosis is so high and its consequences are potentially fatal.

### Further Reading

Austin, R.C., Lentz, S.R., & Werstuck, G.H. (2004). Role of hyperhomocysteinemia in endothelial dysfunction and atherothrombotic disease. *Cell Death Differ, 11*(Suppl 1): S56–S64.

Frohlich, J., Dobiasova, M., Lear, S., & Lee, K.W. (2001). The role of risk factors in the development of atherosclerosis. *Crit Rev Clin Lab Sci, 38*(5): 401–40.

Herzberg, G.R. (2004). Aerobic exercise, lipoproteins, and cardiovascular disease: benefits and possible risks. *Can J Appl Physiol, 29*(6): 800–7.

Mathieu, P., Pibarot, P., & Despres, J.P. (2006). Metabolic syndrome: the danger signal in atherosclerosis. *Vasc Health Risk Manag, 2*(3): 285–302.

**8**

## THROMBOEMBOLISM AND OTHER COMPLICATIONS OF ATHEROSCLEROSIS

Atherosclerosis attacks arteries throughout the body, but the most serious consequences involve damage to the vessels of the brain and heart. In the brain, atherosclerosis is the prime cause of strokes, whereas in the heart it brings about myocardial ischemia and its complications. The following are potential complications of coronary atherosclerosis:

- *Angina pectoris.* Gradual enlargement of a protruding plaque continues to narrow the vessel lumen and progressively diminishes coronary blood flow, triggering increasingly frequent bouts of transient myocardial ischemia as the ability to match blood flow with cardiac oxygen needs becomes more limited. Although the heart cannot normally be "felt," pain is associated with myocardial ischemia. Such cardiac pain, known as **angina pectoris** (pain of the chest), can be felt beneath the sternum and is often referred to (appears to come from) the left shoulder and down the left arm. The symptoms of angina pectoris recur whenever cardiac oxygen demands become too great in relation to the coronary blood flow—for example, during exertion or emotional stress. The pain is thought to result from stimulation of cardiac nerve endings by the accumulation of lactic acid when the heart shifts to its limited ability to perform anaerobic metabolism. The ischemia associated with the characteristically brief angina attacks is usually temporary and reversible and can be relieved by rest, taking vasodilator drugs, such as *nitroglycerin*, or both. Nitroglycerin brings about coronary vasodilation by being metabolically converted to nitric oxide, which in turn relaxes the vascular smooth muscle.

- *Thromboembolism.* The enlarging atherosclerotic plaque can break through the weakened endothelial lining that covers it, exposing blood to the underlying collagen in the collagen-rich connective tissue cap of the plaque. Foam cells release chemicals that can weaken the fibrous cap of a plaque by breaking down the connective tissue fibres. Plaques with thick fibrous caps are considered stable, because they are not likely to rupture. However, plaques with thinner fibrous caps are unstable, because they are likely to rupture and trigger clot formation. Blood platelets (formed elements of the blood involved in plugging vessel defects and in clot formation) normally do not adhere to smooth, healthy vessel linings. However, when platelets contact collagen at the site of vessel damage, they stick to the site and help promote the formation of a blood clot. Furthermore, foam cells produce a

potent clot promoter. Such an abnormal clot attached to a vessel wall is called a thrombus. The thrombus may enlarge gradually until it completely blocks the vessel at that site, or the continued flow of blood past the thrombus may break it loose. As it heads downstream, such a freely floating clot, or **embolus**, may completely plug a smaller vessel (> Figure 8-28). Thus, through **thromboembolism**, atherosclerosis can result in a gradual or sudden occlusion of a coronary vessel (or any other vessel).

- *Heart attack.* When a coronary vessel is completely plugged, the cardiac tissue served by the vessel soon dies from oxygen deprivation, and a heart attack occurs, unless the area can be supplied with blood from nearby vessels. Sometimes a deprived area is lucky enough to receive blood from more than one pathway. **Collateral circulation** exists when small terminal branches from adjacent blood vessels nourish the same area. These accessory vessels cannot develop suddenly after an abrupt blockage but may be lifesaving if already developed. Such alternative vascular pathways often develop over a period of time when an atherosclerotic constriction progresses slowly, or they may be induced by sustained demands on the heart through regular aerobic exercise.

In the absence of collateral circulation, the extent of the damaged area during a heart attack depends on the size of the blocked vessel: the larger the vessel occluded, the greater the area deprived of blood supply. As > Figure 8-29 illustrates, a blockage at point A in the coronary circulation would cause more extensive damage than would a blockage at point B. Because there are only two major coronary arteries, complete blockage of either one of these main branches results in extensive myocardial damage. Left coronary artery blockage is most devastating because this vessel supplies blood to 85 percent of the cardiac tissue.

A heart attack has four possible outcomes: immediate death, delayed death from complications, full functional recovery, or recovery with impaired function (▌ Table 8-4).

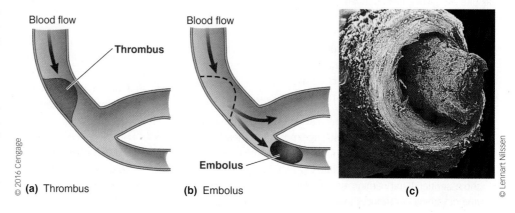

© 2016 Cengage

**(a)** Thrombus     **(b)** Embolus     **(c)**

© Lennart Nilsson

> **FIGURE 8-28 Consequences of thromboembolism.** (a) A thrombus may enlarge gradually until it completely occludes the vessel at that site. (b) A thrombus may break loose from its attachment, forming an embolus that may completely occlude a smaller vessel downstream. (c) Scanning electron microscope of a vessel completely occluded by a thromboembolic lesion

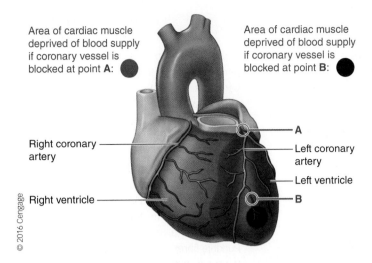

Area of cardiac muscle deprived of blood supply if coronary vessel is blocked at point **A**: ●

Area of cardiac muscle deprived of blood supply if coronary vessel is blocked at point **B**: ●

Right coronary artery

Right ventricle

A

Left coronary artery

Left ventricle

B

© 2016 Cengage

⟩ **FIGURE 8-29** **Extent of myocardial damage as a function of the size of the occluded vessel**

---

**▌ Clinical Connections**

Debbie's blockage is represented by point A in Figure 8-29, which means that she likely had an extensive area of myocardial damage. She underwent a coronary artery bypass graft that allowed blood to flow around the blockage, but by then most of the damage was irreversible. For Debbie to survive, her heart must undergo changes somewhat similar to those described for heart failure (p. 371). The surviving ventricular tissues will undergo hypertrophy to compensate for its decreased ability to contract. However, if not enough compensation occurs she will suffer heart failure and die.

---

# Chapter in Perspective: Focus on Homeostasis

Survival depends on continual delivery of needed supplies to all body cells and ongoing removal of wastes generated by the cells. Furthermore, regulatory chemical messengers, such as hormones, must be transported from their production site to their action site, where they control a variety of activities, most of which are directed toward maintaining a stable internal environment. Finally, to maintain normal body temperature, excess heat produced during muscle contraction must be carried to the skin where the heat can be lost from the body surface.

The circulatory system contributes to homeostasis by serving as the body's transport system. It provides a way to rapidly move materials from one part of the body to another. Without the circulatory system, materials would not get where they need to go to support life-sustaining activities nearly rapidly enough. For example, oxygen would take months to years to diffuse from the body surface to internal organs, yet through the heart's swift pumping action the blood can pick up and deliver oxygen and other substances to all the cells in a few seconds.

The heart serves as a dual pump to continuously circulate blood between the lungs, where oxygen is picked up, and the other body tissues, which use oxygen to support their energy-generating chemical reactions. As blood is pumped through the various tissues, other substances besides oxygen are also exchanged between the blood and tissues. For example, the blood picks up nutrients as it flows through the digestive organs, and other tissues remove nutrients from the blood as it flows through them. Even excess heat is transported by the blood from exercising muscles to the skin surface, where it is lost to the external environment.

Although all the body tissues constantly depend on the life-supporting blood flow provided to them by the heart, the heart itself is quite an independent organ. It can take care of many of its own needs without any outside influence. Contraction of this muscular organ is self-generated through a carefully orchestrated interplay of changing ionic permeabilities. Local mechanisms within the heart ensure that blood flow to the cardiac muscle normally meets the heart's need for oxygen. In addition, the heart has built-in capabilities to vary its strength of contraction, depending on the amount of blood returned to it. The heart does not act entirely autonomously, however. It is innervated by the autonomic nervous system and is influenced by the hormone epinephrine, both of which can vary heart rate and contractility, depending on the body's needs for blood delivery. Furthermore, as with all tissues, the cells that make up the heart depend on the other body systems to maintain a stable internal environment in which they can survive and function.

**8**

---

| ▌ **TABLE 8-4** Possible Outcomes of Acute Myocardial Infarction (Heart Attack) | |
|---|---|
| **Immediate Death** | **Delayed Death from Complications** |
| Acute cardiac failure occurring because the heart is too weak to pump effectively to support the body tissues | Fatal rupture of the dead, degenerating area of the heart wall |
| Fatal ventricular fibrillation brought about by damage to the specialized conducting tissue occurring because the weakened heart is deprived of oxygen | Slowly progressing congestive heart failure where the heart becomes unable to pump out all the blood returned to it |

## CHAPTER TERMINOLOGY

acute heart failure (p. 372)
acute myocardial infarction (p. 361)
afterload (p. 369)
angina pectoris (p. 378)
aorta (p. 347)
aortic valve (p. 350)
apex (p. 346)
arrhythmia (p. 359)
arteries (p. 346)
artificial pacemaker (p. 353)
atherosclerosis (p. 374)
atherosclerotic plaque (p. 374)
atria (p. 346)
atrial fibrillation (p. 360)
atrial flutter (p. 360)
atrioventricular node (AV node) (p. 352)
autorhythmic cells (p. 352)
autorhythmicity (p. 352)
AV nodal delay (p. 355)
base (p. 346)
bicuspid valve (p. 349)
blood (p. 346)
blood vessels (p. 346)
bradycardia (p. 359)
bundle of His (atrioventricular bundle)
    (p. 352)
C-reactive protein (p. 377)
$Ca^{2+}$-induced $Ca^{2+}$ release (p. 356)
cardiac myopathies (p. 360)
cardiac output (CO) (p. 365)
cardiac reserve (p. 365)
cardiopulmonary resuscitation (CPR) (p. 347)
chordae tendineae (p. 349)
circulatory system (p. 346)
collateral circulation (p. 378)
complete heart block (p. 353)
congestive heart failure (p. 371)
Contractile cells (p. 352)
contractility (p. 369)
coronary artery disease (CAD) (p. 374)
coronary circulation (p. 373)
cor pulmonale (p. 372)
diastole (p. 361)

diastolic murmur (p. 364)
dicrotic notch (p. 363)
ectopic focus (p. 354)
ejection fraction (p. 370)
electrical defibrillation (p. 354)
electrocardiogram (ECG) (p. 357)
embolus (p. 378)
end-diastolic volume (EDV) (p. 361)
end-systolic volume (ESV) (p. 363)
endothelium (p. 350)
epicardium (p. 350)
fibrillation (p. 354)
fibrous skeleton (p. 350)
first heart sound (p. 364)
Frank–Starling law of the heart (p. 368)
functional murmurs (p. 364)
heart (p. 346)
heart attack (p. 361)
heart block (p. 360)
heart failure (p. 371)
high-density lipoproteins (HDL) (p. 376)
homocysteine (p. 377)
insufficient or incompetent valve (p. 364)
interatrial pathway (p. 355)
intercalated discs (p. 350)
internodal pathway (p. 355)
intrinsic control (p. 368)
isovolumetric ventricular contraction (p. 363)
isovolumetric ventricular
    relaxation (p. 363)
latent pacemakers (p. 353)
lead (p. 358)
leaky valve (p. 364)
low-density lipoproteins (LDL) (p. 376)
L-type $Ca^{2+}$ channel ($I_{Ca, L}$) (p. 352)
mitral valve (p. 349)
murmurs (p. 364)
myocardial ischemia (p. 360)
myocardium (p. 350)
necrosis (p. 360)
pacemaker (p. 353)
pacemaker potential (p. 352)
papillary muscles (p. 349)

pericardial fluid (p. 351)
pericardial sac (p. 351)
pericarditis (p. 351)
preload (p. 369)
premature ventricular contraction (PVC)
    (p. 354)
PR segment (p. 359)
pulmonary artery (p. 347)
pulmonary circulation (p. 346)
pulmonary valve (p. 350)
pulmonary veins (p. 347)
pulse deficit (p. 360)
Purkinje fibres (p. 353)
P wave (p. 358)
QRS complex (p. 358)
regurgitation (p. 364)
rheumatic fever (p. 364)
right and left atrioventricular
    (AV) valves (p. 347)
second heart sound (p. 364)
semilunar valves (p. 350)
septum (p. 346)
sinoatrial node (SA node) (p. 352)
stenotic valve (p. 364)
sternum (p. 346)
stroke volume (SV) (p. 363)
ST segment (p. 359)
systemic circulation (p. 346)
systole (p. 361)
systolic murmur (p. 364)
T-type $Ca^{2+}$ channel ($I_{Ca, T}$) (p. 352)
tachycardia (p. 359)
thoracic cavity (p. 346)
thromboembolism (p. 378)
TP interval (p. 359)
tricuspid valve (p. 349)
T wave (p. 358)
vascular spasm (p. 374)
veins (p. 346)
venae cavae (p. 347)
ventricular fibrillation (p. 360)
vertebrae (p. 346)
very-low-density lipoproteins (VLDL) (p. 376)

## REVIEW EXERCISES

### Objective Questions (Answers in Appendix E, p. A-43)

1. Adjacent cardiac muscle cells are joined end to end at specialized structures known as _____, which contain two types of membrane junctions: _____ and _____.

2. _____ is an abnormally slow heart rate, whereas _____ is a rapid heart rate.

3. The link that coordinates coronary blood flow with myocardial oxygen needs is _____.

4. The left ventricle is a stronger pump than the right ventricle because more blood is needed to supply the body tissues than to supply the lungs. (*True or false?*)

5. The heart lies in the left half of the thoracic cavity. *(True or false?)*

6. The only point of electrical contact between the atria and ventricles is the fibrous skeletal rings. *(True or false?)*

7. The atria and ventricles each act as a functional syncytium. *(True or false?)*

8. What is the sequence of cardiac excitation?
   a. SA node → AV node → atrial myocardium → bundle of His → Purkinje fibres → ventricular myocardium.
   b. SA node → atrial myocardium → AV node → bundle of His → ventricular myocardium → Purkinje fibres.
   c. SA node → atrial myocardium → ventricular myocardium → AV node → bundle of His → Purkinje fibres.
   d. SA node → atrial myocardium → AV node → bundle of His → Purkinje fibres → ventricular myocardium.

9. What percentage of ventricular filling is normally accomplished before atrial contraction begins?
   a. 0 percent
   b. 20 percent
   c. 50 percent
   d. 80 percent
   e. 100 percent

10. What results from sympathetic stimulation of the heart?
    a. decreases in the heart rate
    b. increases in the contractility of the heart muscle
    c. shifts in the Frank–Starling curve to the right
    d. increases in the AV node delay

11. Match the following:

    | | | |
    |---|---|---|
    | _____ 1. | receives $O_2$-poor blood from venae cavae | (a) AV valves |
    | _____ 2. | prevent backflow of blood from the ventricles to the atria | (b) semilunar valves |
    | _____ 3. | pumps $O_2$-rich blood into aorta | (c) right ventricle |
    | _____ 4. | prevent backflow of blood from the arteries into the ventricles | (d) left ventricle |
    | _____ 5. | pumps $O_2$-poor blood into the pulmonary artery | (e) right atrium |

12. Circle the correct choice in each instance to complete the statement: The first heart sound is associated with closing of the *(AV/semilunar)* valves and signals the onset of *(systole/diastole)*, whereas the second heart sound is associated with closing of the *(AV/semilunar)* valves and signals the onset of *(systole/diastole)*.

13. Circle the correct choice in each instance to complete the statements: During ventricular filling, ventricular pressure must be *(greater than/less than)* atrial pressure, whereas during ventricular ejection, ventricular pressure must be *(greater than/less than)* aortic pressure. Atrial pressure is always *(greater than/less than)* aortic pressure. During isovolumetric ventricular contraction and relaxation, ventricular pressure is *(greater than/less than)* atrial pressure and *(greater than/less than)* aortic pressure.

## Written Questions

1. What are the three basic components of the circulatory system?

2. Trace a drop of blood through one complete circuit of the circulatory system.

3. Describe the location and function of the four heart valves. What keeps each of these valves from everting?

4. What are the three layers of the heart wall? Describe the distinguishing features of the structure and arrangement of cardiac muscle cells. What are the two specialized types of cardiac muscle cells?

5. Why is the SA node the pacemaker of the heart?

6. Describe the normal spread of cardiac excitation. What is the significance of the AV nodal delay? Why is the ventricular conduction system important?

7. Compare the changes in membrane potential associated with an action potential in a nodal pacemaker cell with those in a myocardial contractile cell. What is responsible for the plateau phase?

8. Why is tetanus of cardiac muscle impossible? Why is this inability advantageous?

9. Draw and label the waveforms of a normal ECG. What electrical event does each component of the ECG represent?

10. Describe the mechanical events (i.e., pressure changes, volume changes, valve activity, and heart sounds) of the cardiac cycle. Correlate the mechanical events of the cardiac cycle with the changes in electrical activity.

11. Distinguish between a stenotic and an insufficient valve.

12. Define the following: *end-diastolic volume, end-systolic volume, stroke volume, heart rate, cardiac output,* and *cardiac reserve.*

13. Discuss autonomic nervous system control of heart rate.

14. Describe the intrinsic and extrinsic control of stroke volume.

15. How is the heart muscle provided with blood? Why does the heart receive most of its own blood supply during diastole?

16. What are the pathological changes and consequences of coronary artery disease?

17. Discuss the sources, transport, and elimination of cholesterol in the body. Distinguish between "good" cholesterol and "bad" cholesterol.

## Quantitative Exercises
## (Solutions in Appendix E, p. A-43)

1. During heavy exercise, the cardiac output of a trained athlete may increase to 40 L/min. If stroke volume could not increase above the normal value of 70 mL, what heart rate would be necessary to achieve this cardiac output? Is such a heart rate physiologically possible?

2. How much blood remains in the heart after systole if the stroke volume is 85 mL and the end-diastolic volume is 125 mL?

8

(Explanations in Appendix E, p. A-44)

1. The stroke volume ejected on the next heartbeat after a premature ventricular contraction (PVC) is usually larger than normal. Can you explain why? (*Hint*: At a given heart rate, the interval between a PVC and the next normal beat is longer than the interval between two normal beats.)

2. Trained athletes usually have lower resting heart rates than normal (e.g., 50 beats/min in an athlete compared with 70 beats/min in a sedentary individual). Considering that the resting cardiac output is 5000 mL/min in both trained athletes and sedentary people, how are trained athletes able to generate the same cardiac output in the face of a reduced heart rate?

3. During fetal life, because of the tremendous resistance offered by the collapsed, nonfunctioning lungs, the pressures in the right half of the heart and pulmonary circulation are higher than in the left half of the heart and systemic circulation, a situation that reverses after birth. Also in the fetus, a vessel called the *ductus arteriosus* connects the pulmonary artery and aorta as these major vessels both leave the heart. The blood pumped out by the heart into the pulmonary circulation is shunted from the pulmonary artery into the aorta through the ductus arteriosus, bypassing the nonfunctional lungs. What force is driving blood to flow in this direction through the ductus arteriosus?

4. At birth, the ductus arteriosus normally collapses and eventually degenerates into a thin, ligamentous strand. On occasion, this fetal bypass fails to close properly at birth, leading to a *patent* (open) *ductus arteriosus*. In what direction would blood flow through a patent ductus *arteriosus*? What possible outcomes do you predict might occur as a result of this blood flow?

5. Through what regulatory mechanisms can a transplanted heart, which does not have any innervation, adjust cardiac output to meet the body's changing needs?

6. There are two branches of the bundle of His, the right and left bundle branches, each of which travels down its respective side of the ventricular septum (see Figure 8-8). Occasionally, conduction through one of these branches becomes blocked (so-called bundle-branch block). In this case, the wave of excitation spreads out from the terminals of the intact branch and eventually depolarizes the whole ventricle, but the normally stimulated ventricle completely depolarizes a considerable time before the ventricle on the side of the defective bundle branch. For example, if the left bundle branch is blocked, the right ventricle will be completely depolarized two to three times more rapidly than the left ventricle. How would this defect affect the heart sounds?

## CLINICAL CONSIDERATION

(Explanation in Appendix E, p. A-44)

1. In a physical exam, Rachel B.'s heart rate was rapid and very irregular. Furthermore, her heart rate, determined directly by listening to her heart with a stethoscope, exceeded the pulse rate taken concurrently at her wrist. No definite P waves could be detected on Rachel's ECG. The QRS complexes were normal in shape but occurred sporadically. Given these findings, what is the most likely diagnosis of Rachel's condition? Explain why the condition is characterized by a rapid, irregular heartbeat. Would cardiac output be seriously impaired by this condition? Why or why not? What accounts for the pulse deficit?

# Vascular Physiology

Body systems maintain homeostasis

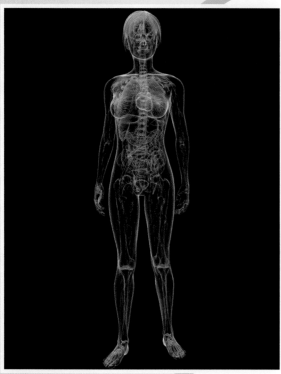

Vascular system

Cells make up body systems

### Homeostasis

The circulatory system contributes to homeostasis by transporting $O_2$, $CO_2$, wastes, electrolytes, and hormones from one part of the body to another.

Homeostasis is essential for survival of cells

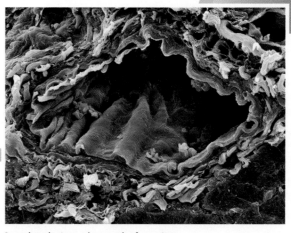

Scanning electron micrograph of an artery

The circulatory system contributes to homeostasis by serving as the body's transport system. The blood vessels transport and distribute blood pumped through them by the heart to meet the body's needs for oxygen and nutrient delivery, waste removal, and hormonal signalling. The highly elastic arteries transport blood from the heart to the organs and serve as a pressure reservoir to continue driving blood forward when the heart is relaxing and filling. The mean arterial blood pressure is closely regulated to ensure adequate blood delivery to the organs. The amount of blood that flows through a given organ depends on the calibre (internal diameter) of the highly muscular arterioles that supply the organ. Arteriolar calibre is subject to control so that cardiac output can be constantly readjusted to best serve the body's needs at each moment. The thin-walled, pore-lined capillaries are the actual site of exchange between blood and the surrounding tissue cells. The highly distensible veins return blood from the organs to the heart and also serve as a blood reservoir.

# Vascular Physiology

**▌ Clinical Connections**

John is 63 years old and recently retired. He was diagnosed with hypertension at age 40, but his blood pressure has been well maintained. He has been a pack-a-day smoker since the age of 15. His weight has always been on the heavy side, and his family physician has recommended he lose weight. He tries to walk regularly, but finds that after he walks a short distance, he starts to feel pain and sometimes cramps in his right calf. When this happens he stops and the symptoms quickly go away. He can then walk again, but the symptoms return. Some days are better than others, and he has developed a walk–rest pattern for his daily exercise. When his physician next asks how his exercise is going, John describes what has been happening.

Upon recognizing the symptoms, his physician conducted an ankle-brachial index, which involves comparing the systolic blood pressure measured in the right ankle to blood pressure measured in the right arm. John's ankle-brachial index was found to be 0.85. Subsequently, an ultrasound revealed significant blockage in an artery, and John was diagnosed with intermittent claudication, a symptom of peripheral artery disease. John underwent angioplasty, quit smoking, and is now able to exercise without any problem.

# 9.1 Introduction

Most body cells are not in direct contact with the external environment, yet these cells must make exchanges with this environment, such as picking up oxygen and nutrients and eliminating wastes. Furthermore, chemical messengers must be transported between cells to accomplish integrated activity. To achieve these long-distance exchanges, cells are linked with one another and with the external environment by vascular (blood vessel) highways. Blood is transported to all parts of the body through a system of vessels that brings fresh supplies to the vicinity of all cells while removing their wastes.

To review cardiac circulation, all blood pumped by the right side of the heart passes through the pulmonary circulation to the lungs for oxygen pickup and carbon dioxide removal. The blood pumped by the left side of the heart (ventricle) enters the systemic circulation for distribution to the systemic organs. This delivery of oxygenated blood occurs through a parallel arrangement of vessels that branch from the aorta (⟩ Figure 9-1), which ensures that all organs receive blood of the same composition $(O_2, CO_2)$. Because of this parallel arrangement, blood flow through each systemic organ can be independently adjusted as needed.

In this chapter, we first examine some general principles regarding blood flow patterns and the physics of blood flow. Then we turn our attention to the roles of the various types of blood vessels through which blood flows. We end the chapter by discussing how blood pressure is regulated to ensure adequate delivery of blood to the tissues.

## Maintenance of homeostasis

Metabolically active cells of the body are constantly exchanging oxygen and nutrients for carbon dioxide and waste products. As a result, blood constantly needs to be reconditioned so that its composition remains relatively stable, despite an ongoing drain of supplies to support metabolic activities and despite the continual addition of wastes from the tissues. Organs that recondition the blood (e.g., the kidneys) normally receive much more blood than necessary to meet their basic metabolic needs; the extra blood is used to perform their physiological role in maintaining homeostasis. For example, large percentages of the cardiac output are distributed to the digestive tract (to pick up nutrient supplies), to the kidneys (to eliminate metabolic wastes and adjust water and electrolyte composition), and to the skin (to eliminate heat). Blood flow to the other organs—heart, skeletal muscles, and so on—is solely for filling these organs' metabolic needs, and this flow can be adjusted according to the level of activity in these organs. For example, during exercise, additional blood is delivered to the active muscles to meet their increased metabolic needs.

Because the reconditioning organs—digestive organs, kidneys, and skin—receive more blood flow than they need for their own functioning, they can withstand temporary reductions in blood flow much better than can other organs that do not have this extra margin of blood supply. The brain in particular suffers irreparable damage when transiently deprived of blood supply. After only four minutes without oxygen, permanent

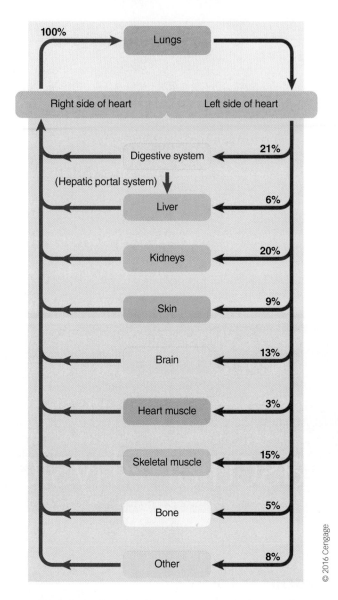

⟩ **FIGURE 9-1 Distribution of cardiac output at rest.** The lungs receive all the blood pumped out by the right side of the heart, whereas the systemic organs each receive some of the blood pumped out by the left side of the heart. The percentage of pumped blood received by the various organs under resting conditions is indicated. This distribution of cardiac output can be adjusted as needed.

brain damage may occur. Therefore, a high priority in the overall operation of the circulatory system is the constant delivery of adequate blood to the brain, which can least tolerate disrupted blood supply.

In Section 8.3, you will see how the distribution of cardiac output is adjusted according to the body's immediate needs. For now, we are going to concentrate on the factors that influence blood flow through a given blood vessel.

## Blood flow and the pressure gradient

The **flow rate** of blood through a vessel (i.e., the volume of blood passing through per unit of time) is directly proportional to the pressure gradient and inversely proportional to vascular resistance:

$$F = \frac{\Delta P}{R}$$

where

$F$ = flow rate of blood through a vessel

$\Delta P$ = pressure gradient

$R$ = resistance of blood vessels

### PRESSURE GRADIENT

The **pressure gradient** is the difference in pressure between the beginning and end of a vessel. Blood flows from an area of higher pressure to an area of lower pressure down a pressure gradient. Contraction of the heart imparts pressure to the blood, which is the main driving force for flow through a vessel. Because of frictional losses (resistance), the pressure drops as blood flows throughout the vessel's length. Accordingly, pressure is higher at the beginning than at the end of the vessel, and this difference establishes a pressure gradient for forward flow of blood through the vessel. The greater the pressure gradient forcing blood through a vessel, the greater the flow rate through that vessel (› Figure 9-2a). Think of a garden hose attached to a faucet. If you turn on the faucet slightly, a small stream of water flows out of the end of the hose, because the pressure is slightly greater at the beginning than at the end of the hose. If you open the faucet all the way, the pressure gradient increases tremendously, so that water flows through the hose much faster and spurts from the end of the hose. Note that the *difference* in pressure between the two ends of a vessel, not the absolute pressures within the vessel, determines flow rate (› Figure 9-2b).

### RESISTANCE

The other factor influencing flow rate through a vessel is **resistance**, which is a measure of the hindrance or opposition to blood flow through a vessel, caused by friction between the moving fluid and the stationary vascular walls. As resistance to flow increases, it is more difficult for blood to pass through the vessel, so flow rate decreases (as long as the pressure gradient remains unchanged). When resistance increases, the pressure gradient must increase correspondingly to maintain the same flow rate. Accordingly, when the vessels offer more resistance to flow, the heart must work harder to maintain adequate circulation.

Resistance to blood flow depends on three factors: (1) viscosity of the blood, (2) vessel length, and (3) vessel radius. The last of these three is by far the most important. The term **viscosity** (designated as $\eta$) refers to the friction developed between the molecules of a fluid as they slide over each other during flow of the fluid. The greater the viscosity is, the greater the resistance to flow. In general, the thicker a liquid, the more viscous it is. For example, molasses flows more slowly than water, because molasses has greater viscosity. Blood viscosity is determined primarily by the number of circulating red blood cells. Normally, this factor is relatively constant and thus not important in controlling resistance. Occasionally, however, blood viscosity and, accordingly, resistance to flow, are altered by an abnormal

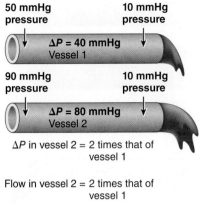

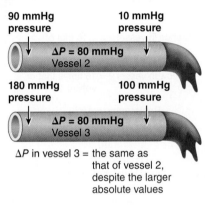

› FIGURE 9-2 **Relationship of flow to the pressure gradient in a vessel.** (a) As the difference in pressure (P) between the two ends of a vessel increases, the flow rate increases proportionally. (b) Flow rate is determined by the *difference* in pressure between the two ends of a vessel, not the magnitude of the pressures at each end.

© 2016 Cengage

number of red blood cells. When excessive red blood cells are present, blood flow is more sluggish than normal.

Because blood "rubs" against the lining of the vessels as it flows past, the greater the vessel surface area in contact with the blood, the greater the resistance to flow. Surface area is determined by both the length ($L$) and the radius ($r$) of the vessel. At a constant radius, the longer the vessel is, the greater the surface area and the greater the resistance to flow. Because vessel length remains constant in the body, it is not a variable factor in the control of vascular resistance.

Therefore, the major determinant of resistance to flow is the vessel's radius. Fluid passes more readily through a large vessel than through a smaller vessel. This occurs because a given volume of blood comes in contact with much more of the surface area in a small-radius vessel than it does in a large-radius vessel, and this increased contact results in greater resistance (› Figure 9-3a).

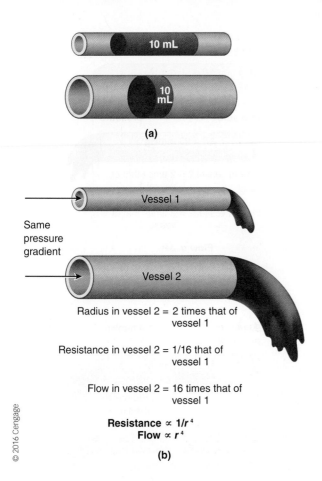

**(a)**

Same
pressure
gradient

Vessel 1

Vessel 2

Radius in vessel 2 = 2 times that of
vessel 1

Resistance in vessel 2 = 1/16 that of
vessel 1

Flow in vessel 2 = 16 times that of
vessel 1

**Resistance ∝ 1/$r^4$**
**Flow ∝ $r^4$**

© 2016 Cengage

**(b)**

> FIGURE 9-3 **Relationship of resistance and flow to the vessel radius.**
(a) The same volume of blood comes into contact with a greater surface area of a small-radius vessel compared with a large-radius vessel. Accordingly, the small-radius vessel offers more resistance to blood flow, because the blood "rubs" against a larger surface area. (b) Doubling the radius decreases the resistance to 1/16 and increases the flow 16 times, because the resistance is inversely proportional to the fourth power of the radius.

Furthermore, a slight change in the radius of a vessel brings about a notable change in flow, because the resistance is inversely proportional to the fourth power of the radius (multiplying the radius by itself four times):

$$R \propto \frac{1}{r^4}$$

Doubling the radius therefore reduces the resistance to 1/16th its original value ($r^4 = 2 \times 2 \times 2 \times 2 = 16; R \propto 1/16$) and brings about a sixteenfold increase in flow through the vessel (at the same pressure gradient) (> Figure 9-3b). The converse is also true. Only 1/16th as much blood flows through a vessel at the same driving pressure as when its radius is halved. Importantly, the radius of arterioles can be regulated; this is the most important factor in controlling resistance to blood flow throughout the vascular circuit.

**POISEUILLE'S LAW**

The factors that affect the flow rate through a vessel are integrated in **Poiseuille's law**, as follows:

$$\text{Flow rate} = \frac{\pi \Delta P r^4}{8 \eta L}$$

The significance of the relationships among flow, pressure, and resistance—as largely determined by vessel radius—will become more apparent as we embark on a voyage through the vessels in the next section, The Vascular Tree.

> ■ **Clinical Connections** John has peripheral artery disease due to atherosclerosis (see Chapter 8, p. 372). Tissue blood flow remains relatively constant unless there is an increase in pressure. At rest, pressure is relatively constant, and a partial arterial blockage caused by atherosclerosis causes an increase in resistance, which means there is decreased blood flow at rest. During exercise, the plaque restricts dilation of the artery, so blood flow is further compromised. The consequences of this decreased flow during exercise are discussed later.

## The vascular tree

The systemic and pulmonary circulations each consist of a closed system of vessels (> Figure 9-4). (For the history leading up to the conclusion that blood vessels form a closed system, see Concepts, Challenges, and Controversies.) These vascular loops each consist of a continuum of different blood vessel types that begins and ends with the heart. Here, we examine this system of vessels by looking specifically at the systemic circulation. Arteries, which carry blood from the heart to the organs, branch, or diverge, into a tree of progressively smaller vessels, with the various branches delivering blood to different regions of the body. When a small artery reaches the organ it supplies, it branches into numerous arterioles. The volume of blood flowing through an organ can be adjusted by regulating the internal diameter of the organ's arterioles. Arterioles branch further within the organs into capillaries, the smallest vessels, across which all exchanges are made with surrounding cells. Capillary exchange is the entire purpose of the **circulatory system**; all other activities of the system are directed toward ensuring an adequate distribution of replenished blood to capillaries for exchange with all cells. Capillaries rejoin to form small **venules**, which further merge to form small **veins** that leave the organs. The small veins progressively unite or converge to form larger veins that eventually empty into the heart. The arterioles, capillaries, and venules are collectively referred to as the **microcirculation**, because they are only visible through a microscope. The microcirculatory vessels are all located within the organs. The pulmonary circulation consists of the same vessel types, except that all the blood in this loop goes between the heart and lungs. If all of the vessels in the body were strung end to end, they could circle the circumference of Earth twice!

## The structure of blood vessels

Before learning about the major types of blood vessels in the body, you need to have an understanding of blood vessel structure, which is important for understanding blood vessel function.

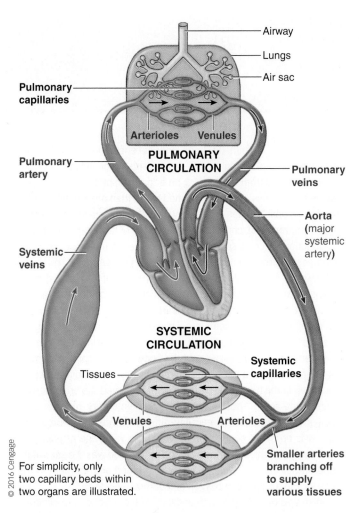

Pulmonary capillaries

Pulmonary artery

Systemic veins

Airway

Lungs

Air sac

Arterioles    Venules

**PULMONARY CIRCULATION**

Pulmonary veins

**Aorta** (major systemic artery)

**SYSTEMIC CIRCULATION**

Tissues

Venules

Systemic capillaries

Arterioles

Smaller arteries branching off to supply various tissues

© 2016 Cengage

For simplicity, only two capillary beds within two organs are illustrated.

› **FIGURE 9-4 Basic organization of the cardiovascular system.** Arteries progressively branch as they carry blood from the heart to the organs. A separate small arterial branch delivers blood to each of the various organs. As a small artery enters the organ it is supplying, it branches into arterioles, which further branch into an extensive network of capillaries. The capillaries rejoin to form venules, which further unite to form small veins that leave the organ. The small veins progressively merge as they carry blood back into the heart.

Blood vessel walls are made up of alternating layers of connective tissue (both fibrous and elastic), smooth muscle cells, and epithelial cells. The outermost layer is called the *tunica adventitia* and mainly consists of collagen-rich connective tissue. The middle layer, the *tunica media*, is the thickest and contains circular arrangements of connective tissue and smooth muscle cells. In some vessels, an elastic fibre-rich area called the **external elastic lamina** may be found between the *tunica adventitia* and the *tunica media*. The innermost layer is the *tunica intima* and consists of a single layer of endothelial cells (the endothelium) surrounded by a basement membrane of connective tissues and, in some vessels, an additional layer of elastic fibres called the **internal elastic lamina**.

Each of these layers serves distinct physiological roles depending on the function of the blood vessel. Consequently, the relative thicknesses and even the presence of these layers vary dramatically throughout the vascular tree. ▌Table 9-1 summarizes the features of the blood vessel types, which are further discussed in Sections 9.2 through 9.4.

### Check Your Understanding 9.1

1. Describe the distribution of blood flow at rest, and postulate how this might change during exercise.

2. Name the reconditioning organs.

3. Give the equation showing the relationship among flow rate of blood through a vessel, the pressure gradient, and the resistance to flow, and give the equation for the relationship between resistance and vessel radius.

## 9.2 | Arteries

### Rapid-transit passageways and pressure reservoirs

Arteries are specialized to (1) serve as rapid-transit passageways for blood from the heart to the organs (because of their large radius, arteries offer little resistance to blood flow), and (2) act as a **pressure reservoir** to provide the driving force for blood when the heart is relaxing.

We discussed the first point in the section on blood vessels and pressure gradient. Here we expand on the arteries' second specialization, as pressure reservoir. The heart alternately contracts to pump blood into the arteries and then relaxes to refill with blood from the veins. When the heart is relaxing and refilling, no blood is pumped out. However, capillary flow does not fluctuate between cardiac systole and diastole; that is, blood flow is continuous through the capillaries supplying the organs. The driving force for the continued flow of blood to the organs during cardiac relaxation is provided by the elastic properties of the arterial walls.

All vessels are lined with a thin layer of smooth, flat, endothelial cells that are continuous with the endothelial lining (the endocardium) of the heart. A thick wall made up of smooth muscle and connective tissue surrounds the arteries' endothelial lining. Arterial connective tissue contains an abundance of two types of connective tissue fibres: *collagen fibres*, which provide tensile strength against the high driving pressure of blood ejected from the heart; and *elastin fibres*, which give the arterial walls elasticity so that they behave much like a balloon (› Figure 9-5).

As the heart pumps blood into the arteries during ventricular systole, a greater volume of blood enters the arteries from the heart than leaves them to flow into smaller vessels downstream, because the smaller vessels have a greater resistance to flow. The arteries' elasticity enables them to expand to temporarily hold this excess volume of ejected blood, so they store some of the pressure energy imparted by cardiac contraction in their stretched walls— just as a balloon expands to accommodate the extra volume of air you blow into it (› Figure 9-6a). When the heart relaxes and ceases pumping blood into the arteries, the stretched arterial walls passively recoil, like an inflated balloon that is released. This recoil pushes the excess blood contained in the arteries into the vessels downstream, ensuring continued blood flow to the organs when the heart is relaxing (› Figure 9-6b).

**9**

# From Humours to Harvey: Historical Highlights in Circulation

TODAY, EVEN GRADE-SCHOOL CHILDREN KNOW that blood is pumped by the heart and continually circulates throughout the body in a system of blood vessels. Furthermore, people accept without question that blood picks up oxygen in the lungs from the air we breathe and delivers it to the various organs. This common knowledge was unknown for most of human history, however. Even though the function of blood was described as early as the 5th century BC, our modern concept of circulation did not develop until 1628, more than 2000 years later, when William Harvey published his now classical study on the circulatory system.

Ancient Greeks believed everything material in the universe consisted of just four elements: earth, air, fire, and water. Extending this view to the human body, they thought these four elements took the form of four "humours": *black bile* (representing earth), *blood* (representing air), *yellow bile* (representing fire), and *phlegm* (representing water). According to the Greeks, disease resulted when one humour was out of normal balance with the rest. The "cure" was logical: to restore normal balance, drain off whichever humour was in excess. Because the easiest humour to drain off was the blood, bloodletting became standard procedure for treating many illnesses—a practice that persisted well into the Renaissance (which began in the 1300s and extended into the 1600s).

Although the ancient Greeks' notion of the four humours was erroneous, their concept of the necessity of balance within the body was remarkably accurate. As we now know, life depends on homeostasis, maintenance of the proper balance among all elements of the internal environment.

Aristotle (384–322 BC), a biologist as well as a philosopher, was among the first to correctly describe the heart as the centre of a system of blood vessels. However, he thought the heart was both the seat of intellect (the brain was not identified as the seat of intellect until more than a century later) and a furnace that heated the blood. He considered this warmth the vital force of life, because the body cools quickly at death. Aristotle also erroneously theorized that breathing ventilated the "furnace," with air serving as a cooling agent. Aristotle could observe with his eyes the arteries and veins in cadavers but did not have a microscope with which to observe capillaries. (The microscope was not invented until the 17th century.) Thus, he did not think arteries and veins were directly connected.

In the 3rd century BC, Erasistratus, a Greek many considered the first physiologist, proposed that the liver used food to make blood, which the veins delivered to the other organs. He believed the arteries contained air, not blood. According to his view, *pneuma* (air), a living force, was taken in by the lungs, which transferred it to the heart. The heart transformed the air into a "vital spirit" that the arteries carried to the other organs.

Galen (AD 130–206), a prolific, outspoken, dogmatic Roman physician, philosopher, and scholar, expanded on the work of Erasistratus and others who had preceded him. Galen further elaborated on the pneumatic theory. He proposed three fundamental members in the body, from lowest to highest: liver, heart, and brain. Each was dominated by a special *pneuma*, or spirit. (In Greek, *pneuma* encompassed the related ideas of air, breath, and spirit.) Like Erasistratus, Galen believed that the liver made blood from food, taking on a "natural spirit" (*pneuma physicon*) in the process. The newly formed blood then proceeded through veins to organs. The natural spirit, which Galen considered a vapour rising from the blood, controlled the functions of nutrition, growth, and reproduction. Once its spirit supply was depleted, the blood moved in the opposite direction through the same venous pathways, returning to the liver to be replenished. When the natural spirit was carried in the venous blood to the heart, it mixed with air that was breathed in and transferred from the lungs to the heart. Contact with air in the heart transformed the natural spirit into a higher-level spirit, the "vital spirit" (*pneuma zotikon*). The vital spirit, which was carried by the arteries, conveyed heat and life throughout the body. The vital spirit was transformed further into a yet higher animal, or psychical, spirit (*pneuma psychikon*) in the brain. This ultimate spirit regulated the brain, nerves, feelings, and so on. Thus, according to Galenic theory, the veins and arteries were conduits for carrying different levels of pneuma, and there was no direct connection between the veins and arteries. The heart was not involved in moving blood but instead was the site where blood and air mixed. (We now know that blood and air meet in the lungs for the exchange of oxygen and carbon dioxide.)

Galen was one of the first to understand the need for experimentation, but unfortunately, his impatience and his craving for philosophical and literary fame led him to expound comprehensive theories that were not always based on the time-consuming collection of evidence. Even though his assumptions about bodily structure and functions often were incorrect, his theories were convincing because they seemed a logical way of pulling together what was known at the time. Furthermore, the sheer quantity of his writings helped establish him as an authority. In fact, his writings remained the anatomic and physiological "truth" for nearly 15 centuries, throughout the Middle Ages and well into the Renaissance. So firmly entrenched was Galenic doctrine that people who challenged its accuracy risked their lives by being declared secular heretics.

Not until the Renaissance and the revival of classical learning did independent-minded European investigators begin to challenge Galen's theories. Most notably, the English physician William Harvey (1578–1657) revolutionized the view of the roles played by the heart, blood vessels, and blood. Through careful observations, experimentation, and deductive reasoning, Harvey was the first to correctly identify the heart as a pump that repeatedly moves a small volume of blood forward in one fixed direction in a circular path through a closed system of blood vessels (the *circulatory system*). He also correctly proposed that blood travels to the lungs to mix with air (instead of air travelling to the heart to mix with blood). Even though he could not see physical connections between arteries and veins, he speculated on their existence. Not until the discovery of the microscope later in the century was the existence of these connections—capillaries—confirmed by Marcello Malpighi (1628–1694).

| | VESSEL TYPE | | | |
| --- | --- | --- | --- | --- |
| **FEATURE** | **Arteries** | **Arterioles** | **Capillaries** | **Veins** |
| **Number** | Several hundred* | Half a million | Ten billion | Several hundred* |
| **Special Features** | Thick, highly elastic walls; large radii* | Highly muscular, well-innervated walls; small radii | Very thin-walled; large total cross-sectional area | Thin-walled compared to arteries; highly distensible; large radii* |
| **Functions** | Passageway from heart to organs; serve as pressure reservoir | Primary resistance vessels; determine distribution of cardiac output | Site of exchange; determine distribution of extracellular fluid between plasma and interstitial fluid | Passageway to heart from organs; serve as blood reservoir |
| **Structure** | | | | |

Large artery — Arteriole — Capillary — Large vein

Relative Thickness of Layers in Wall
Endothelium
Elastic fibres
Smooth muscle
Collagen fibres

© 2016 Cengage

*These numbers and special features refer to the large arteries and veins, not to the smaller arterial branches or venules.

## Arterial pressure and diastole

**Blood pressure**, the force exerted by the blood against a vessel wall, depends on the volume of blood contained within the vessel and the **compliance**, or **distensibility**, of the vessel walls (how easily they can be stretched). If the volume of blood entering the arteries were equal to the volume of blood leaving the arteries during the same period, arterial blood pressure would remain constant. This is not the case, however. During ventricular systole, a stroke volume of blood enters the arteries from the ventricle, but only about one-third as much blood leaves the arteries to enter the arterioles. During diastole, no blood enters the arteries, although blood continues to leave them, driven by elastic recoil. The maximum pressure exerted in the arteries when blood is ejected into them during systole, the **systolic pressure**, averages 120 mmHg. The minimum pressure within the arteries when blood is draining off into the rest of the vessels during diastole, the **diastolic pressure**, averages 80 mmHg (⟩ Figure 9-7; see also ⟩ Figure 8-16).

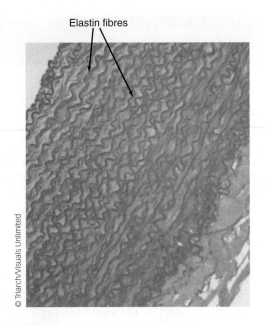

Elastin fibres

> **FIGURE 9-5 Elastin fibres in an artery.** Light micrograph of a portion of the aorta wall in cross-section, showing the numerous wavy elastin fibres common to all arteries.

## Blood pressure can be measured using a sphygmomanometer

The changes in arterial pressure throughout the cardiac cycle can be measured directly by connecting a pressure-measuring device to a needle inserted in an artery. However, it is more convenient and reasonably accurate to measure the pressure indirectly with a **sphygmomanometer**: an externally applied inflatable

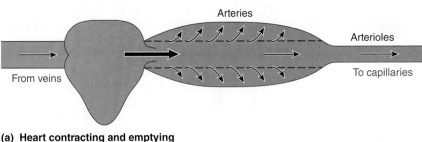

Arteries

Arterioles

From veins

To capillaries

**(a) Heart contracting and emptying**

Arteries

Arterioles

From veins

To capillaries

**(b) Heart relaxing and filling**

> **FIGURE 9-6 Arteries as a pressure reservoir.** Because of their elasticity, arteries act as a pressure reservoir. (a) The elastic arteries distend during cardiac systole as more blood is ejected into them than drains off into the narrow, high-resistance arterioles downstream. (b) The elastic recoil of arteries during cardiac diastole continues driving the blood forward when the heart is not pumping.

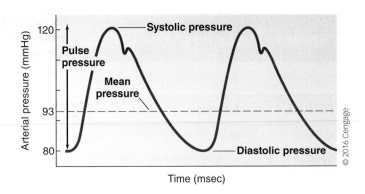

> **FIGURE 9-7 Arterial blood pressure.** The systolic pressure is the peak pressure exerted in the arteries when blood is pumped into them during ventricular systole. The diastolic pressure is the lowest pressure exerted in the arteries when blood is draining off into the vessels downstream during ventricular diastole. The pulse pressure is the difference between systolic and diastolic pressure. The mean pressure is the average pressure throughout the cardiac cycle.

cuff attached to a pressure gauge. When the cuff is wrapped around the upper arm and then inflated with air, the pressure of the cuff is transmitted through the tissues to the underlying brachial artery, the main vessel carrying blood to the forearm (> Figure 9-8). The technique involves balancing the pressure in the cuff against the pressure in the artery. When cuff pressure is greater than the pressure in the vessel, the vessel is pinched closed so that no blood flows through it. When blood pressure is greater than cuff pressure, the vessel is open and blood flows through.

### DETERMINATION OF SYSTOLIC AND DIASTOLIC PRESSURE

During a test to determine blood pressure, a stethoscope is placed over the brachial artery at the inside bend of the elbow just below the cuff. No sound can be detected either when blood is not flowing through the vessel or when blood is flowing in the normal, smooth laminar flow (p. 364). Turbulent blood flow, in contrast, creates vibrations that can be heard. The sounds heard when determining blood pressure, known as **Korotkoff sounds**, are distinct from the heart sounds associated with valve closure heard when listening to the heart with a stethoscope.

At the onset of a blood pressure determination, the cuff is inflated to a pressure greater than systolic blood pressure so that the brachial artery collapses. Because the externally applied pressure is greater than the peak internal pressure, the artery remains completely pinched closed throughout the entire cardiac cycle; no sound can be heard, because no blood is passing through (point **1** in > Figure 9-8b). As air in the cuff is slowly released, the pressure in the cuff is gradually reduced. When the cuff pressure falls to just below the peak systolic pressure, the artery transiently opens a bit when the

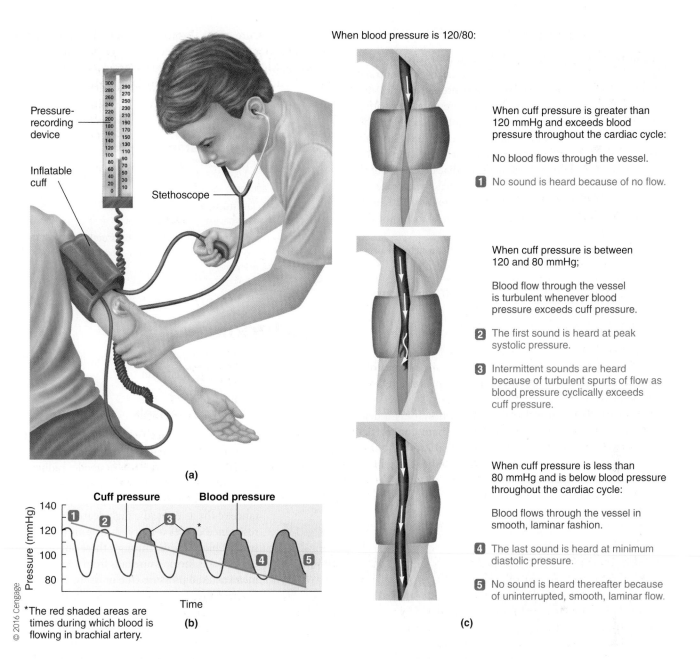

When blood pressure is 120/80:

When cuff pressure is greater than 120 mmHg and exceeds blood pressure throughout the cardiac cycle:

No blood flows through the vessel.

**1** No sound is heard because of no flow.

When cuff pressure is between 120 and 80 mmHg;

Blood flow through the vessel is turbulent whenever blood pressure exceeds cuff pressure.

**2** The first sound is heard at peak systolic pressure.

**3** Intermittent sounds are heard because of turbulent spurts of flow as blood pressure cyclically exceeds cuff pressure.

When cuff pressure is less than 80 mmHg and is below blood pressure throughout the cardiac cycle:

Blood flows through the vessel in smooth, laminar fashion.

**4** The last sound is heard at minimum diastolic pressure.

**5** No sound is heard thereafter because of uninterrupted, smooth, laminar flow.

Pressure-recording device

Inflatable cuff

Stethoscope

**(a)**

Cuff pressure    Blood pressure

*The red shaded areas are times during which blood is flowing in brachial artery.

**(b)**

**(c)**

> FIGURE 9-8 **Sphygmomanometry.** (a) Use of a sphygmomanometer in determining blood pressure. The pressure in the inflatable cuff can be varied to prevent or permit blood flow in the underlying brachial artery. Turbulent blood flow can be detected with a stethoscope, whereas smooth laminar flow and no flow are inaudible. (b) Patterns of sounds in relation to cuff pressure compared with blood pressure. (c) Blood flow through the brachial artery in relation to cuff pressure and sounds

blood pressure reaches this peak. Blood escapes through the partially occluded artery for a brief interval before the arterial pressure falls below the cuff pressure and the artery collapses once again. This spurt of blood is turbulent, so it can be heard. So the highest cuff pressure at which the *first sound* can be heard indicates the *systolic pressure* (point **2**). As the cuff pressure continues to fall, blood intermittently spurts through the artery and produces a sound with each subsequent cardiac cycle whenever the arterial pressure exceeds the cuff pressure (point **3**).

When the cuff pressure finally falls below diastolic pressure, the brachial artery is no longer pinched closed during any part of the cardiac cycle, and blood can flow uninterrupted through the vessel (point **5**). With the return of non-turbulent blood

flow, no further sounds can be heard. Therefore, the highest cuff pressure at which the *last sound* can be detected indicates the *diastolic pressure* (point **4**).

In clinical practice, arterial blood pressure is expressed as systolic pressure over diastolic pressure, with the cut-off for desirable blood pressure being less than 120/80 (120 over 80) mmHg.

**PULSE PRESSURE**

The pulse that can be felt in an artery lying close to the surface of the skin is due to the difference between systolic and diastolic pressures. This pressure difference is known as the **pulse pressure**. When blood pressure is 120/80, pulse pressure is 40 mmHg (120 – 80 mmHg).

## Mean arterial pressure

The **mean arterial pressure** is the *average pressure* driving blood forward into the tissues throughout the cardiac cycle. Contrary to what you might expect, mean arterial pressure is not the halfway value between systolic and diastolic pressure (e.g., with a blood pressure of 120/80, mean pressure is not 100 mmHg). The reason is that arterial pressure remains closer to diastolic than to systolic pressure for a longer portion of each cardiac cycle. At resting heart rate, about two-thirds of the cardiac cycle is spent in diastole, and only one-third in systole. As an analogy, if a race car travelled 80 km per hour (km/h) for 40 minutes and 120 km/h for 20 minutes, its average speed would be 93 km/h, not the halfway value of 100 km/h.

Similarly, a good approximation of the mean arterial pressure can be determined using the following formula:

Mean arterial pressure = diastolic pressure + 1/3 pulse pressure

At 120/80, mean arterial pressure = 80 mmHg + (1/3) 40 mmHg

= 93 mmHg

The mean arterial pressure, not the systolic or diastolic pressure, is monitored and regulated by blood pressure reflexes (described in Section 8.6).

Because arteries offer little resistance to flow, only a negligible amount of pressure energy is lost in them due to friction. Therefore, arterial pressure—systolic, diastolic, pulse, or mean—is essentially the same throughout the arterial tree (⟩ Figure 9-9).

Blood pressure exists throughout the entire vascular tree. However, when discussing a person's blood pressure without qualifying which blood vessel type is referred to, the term is tacitly understood to mean the pressure in the arteries.

## 9.3 | Arterioles

When an artery reaches the organ it is supplying, it branches into numerous arterioles within the organ.

### The major resistance vessels

**Arterioles** are the major resistance vessels in the vascular tree because their radius is small enough to offer considerable resistance to flow. (Even though capillaries have a smaller radius than arterioles, collectively they do not offer as much resistance to flow as the arterioles do.) In contrast to the low resistance of the arteries, the high degree of arteriolar resistance causes a marked drop in mean pressure as blood flows through these small vessels. On average, the pressure falls from 93 mmHg, the mean arterial pressure (the pressure of the blood entering the arterioles), to 37 mmHg, the pressure of the blood leaving the arterioles and entering the capillaries (⟩ Figure 9-9). This decline in pressure helps establish the pressure differential that encourages the flow of blood from the heart to the various organs downstream. If no pressure drop occurred in the arterioles, the pressure at the end of the arterioles (i.e., at the beginning of the capillaries) would be equal to the mean arterial pressure. No pressure gradient would exist to drive blood from the heart to the tissue capillary beds.

Arteriolar resistance also converts the pulsatile systolic-to-diastolic pressure swings in the arteries into the nonpulsatile pressure present in the capillaries. The non-fluctuating pressure that the arterioles create is a key component in mean arterial pressure (p. 419). Another variable in the determination of mean arterial pressure is total peripheral resistance, which again is influenced by arteriolar resistance. This illustrates the influence of arteriolar resistance in determining the overall pressure in our vascular system.

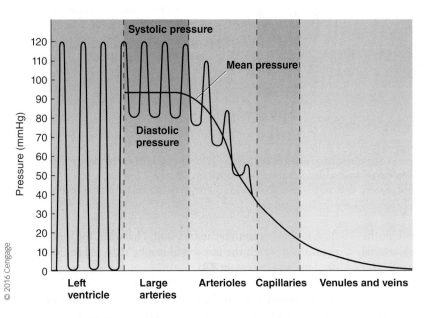

© 2016 Cengage

⟩ **FIGURE 9-9 Pressures through the systemic circulation.** Left ventricular pressure swings between a low pressure of 0 mmHg during diastole to a high pressure of 120 mmHg during systole. Arterial blood pressure, which fluctuates between a peak systolic pressure of 120 mmHg and a low diastolic pressure of 80 mmHg each cardiac cycle, is of the same magnitude throughout the large arteries. Because of the arterioles' high resistance, the pressure drops precipitously, and the systolic-to-diastolic swings in pressure are converted to a nonpulsatile pressure when blood flows through the arterioles. The pressure continues to decline, but at a slower rate, as blood flows through the capillaries and venous system.

Cross-section
of arteriole

**(b) Normal arteriolar tone**

Caused by:
- ↑ Myogenic activity
- ↑ Oxygen (O₂)
- ↓ Carbon dioxide (CO₂) and other metabolites
- ↑ Endothelin
- ↑ Sympathetic stimulation Vasopressin; angiotensin II Cold

**(c) Vasoconstriction** (increased contraction of circular smooth muscle in the arteriolar wall, which leads to increased resistance and decreased flow through the vessel)

Caused by:
- ↓ Myogenic activity
- ↓ O₂
- ↑ CO₂ and other metabolites
- ↑ Nitric oxide
- ↓ Sympathetic stimulation Histamine release Heat

© 2016 Cengage

**(d) Vasodilation** (decreased contraction of circular smooth muscle in the arteriolar wall, which leads to decreased resistance and increased flow through the vessel)

> **FIGURE 9-10 Arteriolar vasoconstriction and vasodilation.** (a) A scanning electron micrograph of an arteriole showing how the smooth muscle cells run circularly around the vessel wall. (b) Schematic representation of an arteriole in cross-section, showing normal arteriolar tone. (c) Outcome of and factors causing arteriolar vasoconstriction. (d) Outcome of and factors causing arteriolar vasodilation

The radius (and, accordingly, the resistances) of arterioles supplying individual organs can be adjusted independently to accomplish two functions: (1) to variably distribute the cardiac output among the systemic organs, depending on the body's momentary needs; and (2) to help regulate arterial blood pressure. Before considering how such adjustments are important in accomplishing these two functions, we will discuss the mechanisms involved in adjusting arteriolar resistance.

Smooth muscle cells

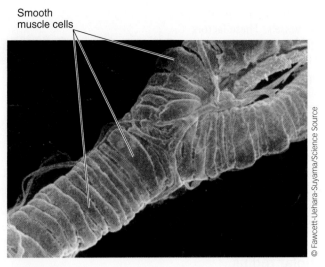

**(a)** Scanning electron micrograph of an arteriole showing how the smooth muscle cells run circularly around the vessel wall

## VASOCONSTRICTION AND VASODILATION

Unlike arteries, arteriolar walls contain very little elastic connective tissue. However, they do have a thick layer of smooth muscle that is richly innervated by sympathetic nerve fibres. The smooth muscle is also sensitive to many local chemical changes and to a few circulating hormones. The smooth muscle layer runs circularly around the arteriole (› Figure 9-10a), so when the smooth muscle layer contracts, the vessel's circumference (and its radius) becomes smaller, which increases resistance and decreases flow through that vessel. **Vasoconstriction** is the term applied to such narrowing of a vessel (› Figure 9-10c). The term **vasodilation** refers to enlargement in the circumference and radius of a vessel as a result of its smooth muscle layer relaxing (› Figure 9-10d). Vasodilation leads to decreased resistance and increased flow through that vessel.

## VASCULAR TONE

Arteriolar smooth muscle normally displays a state of partial constriction known as **vascular tone**, which establishes a baseline of arteriolar resistance (› Figure 8-10b). Two factors are responsible for vascular tone. First, arteriolar smooth muscle has considerable myogenic activity; that is, its membrane potential fluctuates independent of any neural or hormonal influences, leading to self-induced contractile activity (p. 332). Second, the sympathetic fibres supplying most arterioles continually release norepinephrine, which further enhances vascular tone.

This ongoing tonic activity makes it possible to either increase or decrease the level of contractile activity to accomplish vasoconstriction or vasodilation, respectively. Were it not for tone, it would be impossible to reduce the tension in an arteriolar wall to accomplish vasodilation; only varying degrees of vasoconstriction would be possible.

A variety of factors can influence the level of contractile activity in arteriolar smooth muscle, thereby substantially changing resistance to flow in these vessels. These factors fall into two categories: local (intrinsic) controls, which are important in determining the distribution of cardiac output; and extrinsic controls, which are important in blood pressure regulation. Now we begin our discussion of local controls; the discussion of extrinsic control begins in the section Extrinsic Sympathetic Control of Arteriolar Radius.

## Arteriolar radius

Local (intrinsic) controls are changes within an organ that alter the radius of the vessels and hence adjust blood flow through the organ by directly affecting the smooth muscle of the organ's arterioles. Local influences may be either chemical or physical. Local chemical influences on arteriolar radius include (1) local metabolic changes and (2) histamine release. Local physical influences include (1) local application of heat or cold, (2) chemical response to shear stress, and (3) myogenic response to stretch. These local influences, chemical and physical, play essential roles in regulating arteriolar radius.

The fraction of the total cardiac output delivered to each organ is not always constant: it varies, depending on the demands for blood at the time. The amount of the cardiac output received by each organ is determined by the number and calibre of the arterioles supplying that area. Recall that $F = \Delta P/R$. Because blood is delivered to all organs at the same mean arterial pressure, the driving force for flow is identical for each organ. Therefore, differences in flow to various organs are completely determined by differences in the extent of vascularization and by differences in resistance offered by the arterioles supplying each organ. On a moment-to-moment basis, the distribution of cardiac output can be varied by differentially adjusting arteriolar resistance in the various vascular beds.

As an analogy, consider a water pipe that has a number of adjustable valves located throughout its length (> Figure 9-11). Assuming that water pressure in the pipe is constant, differences in the amount of water flowing into a beaker under each valve depend entirely on which valves are open and to what extent. No water enters beakers under closed valves (high resistance), and more water flows into beakers under valves that are opened completely (low resistance) than into beakers under valves that are only partially opened (moderate resistance).

Similarly, more blood flows to areas whose arterioles offer the least resistance to its passage. During exercise, for example,

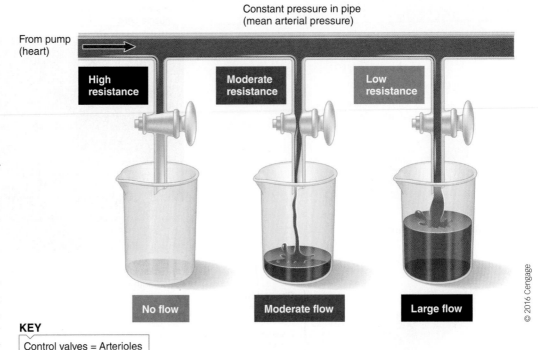

**KEY**

Control valves = Arterioles

> FIGURE 9-11 **Flow rate as a function of resistance**

not only is cardiac output increased, but, because of vasodilation in skeletal muscle and in the heart, a greater percentage of the pumped blood is diverted to these organs to support their increased metabolic activity. Simultaneously, blood flow to the digestive tract and kidneys is reduced as a result of arteriolar vasoconstriction in these organs (> Figure 9-12). Only the blood supply to the brain remains remarkably constant no matter what a person is doing, be it vigorous physical activity, intense mental concentration, or sleep. Although total blood flow to the brain remains constant, imaging (MRI) and noninvasive measurement (Doppler) techniques demonstrate that regional blood flow varies within the brain in close correlation with local neural activity patterns.

We now examine the role and mechanism of the local influences on arteriolar radius.

## Local influences on arteriolar radius

The most important local chemical influences on arteriolar smooth muscle are related to metabolic changes within a given organ. The influence of these local changes on arteriolar radius is important in matching blood flow through an organ with the organ's metabolic needs, ultimately with the intent of maintaining homeostasis within that tissue and the whole body. Local metabolic controls are especially important in skeletal muscle and the heart—organs whose metabolic activity and need for blood supply normally vary most extensively. Local metabolic controls are also important in the brain, whose overall metabolic activity and need for blood supply remain constant. Local controls help maintain the constancy of blood flow to the brain.

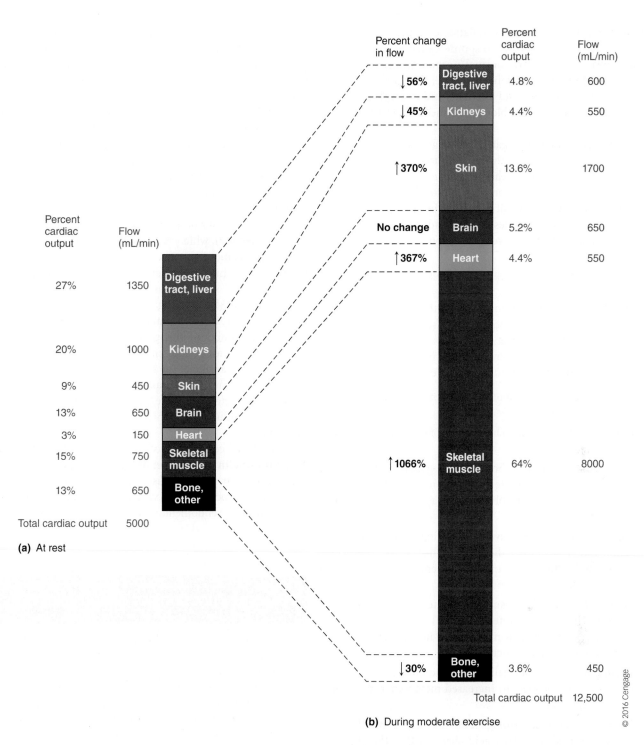

**> FIGURE 9-12 Magnitude and distribution of the cardiac output at rest and during moderate exercise.** Not only does cardiac output increase during exercise, but the distribution of cardiac output is also adjusted to support the heightened physical activity. The percentage of cardiac output going to the skeletal muscles and heart rises, thereby delivering extra oxygen and nutrients needed to support these muscles' stepped-up rate of ATP consumption. The percentage going to the skin increases as a way to eliminate from the body surface the extra heat generated by the exercising muscles. The percentage going to most other organs shrinks. Only the magnitude of blood flow to the brain remains unchanged while the distribution of cardiac output is readjusted during exercise.

## ACTIVE HYPERAEMIA

Arterioles lie within the organ they are supplying and can be acted on by local factors within that organ. During increased metabolic activity, such as when a skeletal muscle is contracting during exercise, local concentrations of a number of the organ's chemicals change. For example, the local oxygen concentration decreases as the actively metabolizing cells use up more oxygen to support oxidative phosphorylation for ATP production. This and other local chemical changes produce local arteriolar dilation by triggering relaxation of the arteriolar smooth muscle in the vicinity. Local arteriolar vasodilation then increases blood flow to that particular area, a response called **active hyperaemia** (*hyper* means "above normal"; *aemia* means "blood"). When cells are more metabolically active, they require more blood to bring in oxygen and nutrients and to remove metabolic wastes. The increased blood flow meets these increased local needs.

Conversely, when an organ, such as a relaxed muscle, is less metabolically active and thus has reduced needs for blood delivery, the resultant local chemical changes (e.g., increased local oxygen concentration) bring about local arteriolar vasoconstriction and a subsequent reduction in blood flow to the area. Local metabolic changes can thus adjust blood flow as needed without involving nerves or hormones.

### LOCAL METABOLIC CHANGES THAT INFLUENCE ARTERIOLAR RADIUS

A variety of local chemical changes act together in a cooperative, redundant manner to bring about these "selfish" local adjustments in arteriolar calibre that match a tissue's blood flow with its needs. Specifically, the following local chemical factors produce relaxation of arteriolar smooth muscle (vasodilation):

- *Decreased oxygen*, a result of increased oxidative metabolism and thus oxygen uptake by cells.

- *Increased carbon dioxide*, a by-product of increased oxidative phosphorylation.

- *Increased acid*. More carbonic acid is generated from the increased carbon dioxide production. Also, lactic acid accumulates if the glycolytic pathway is used for ATP production.

- *Increased potassium*. In actively contracting muscles, repeated action potentials outpace the ability of the $Na^+$ −$K^+$ pump to restore the resting concentration gradients and result in an increase in $K^+$ in the tissue fluid.

- *Increased osmolarity*. Osmolarity (the concentration of osmotically active solutes/particles) increases during elevated cellular metabolism because of increased formation of osmotically active particles.

- *Adenosine release*. Especially in cardiac muscle, adenosine is released in response to increased metabolic activity or oxygen deprivation.

- *Prostaglandin release*. Prostaglandins are local chemical messengers derived from fatty acid chains within the plasma membrane.

The relative contributions of the aforementioned chemicals (and possibly others) in the local metabolic control of arteriole blood flow are currently being investigated.

### LOCAL VASOACTIVE MEDIATORS

These local chemical changes do not act directly on vascular smooth muscle to change its contractile state. Instead, **endothelial cells**, the single layer of specialized epithelial cells that line the lumen of all blood vessels, release chemical mediators that play a key role in locally regulating arteriolar calibre. Until recently, scientists regarded endothelial cells as little more than a passive barrier between the blood and the rest of the vessel wall. It is now known that endothelial cells are active participants in a variety of vessel-related activities, some of which are described in ▮ Table 9-2. Among these functions, endothelial cells release locally acting chemical messengers in response to chemical changes in their environment (such as a reduction in oxygen) or physical changes (such as stretching of the vessel wall). These local vasoactive (acting on vessels) mediators act on the underlying smooth muscle to alter its state of contraction.

Among the best studied of these local vasoactive mediators is **nitric oxide (NO)**, which causes local arteriolar vasodilation by inducing relaxation of arteriolar smooth muscle in the vicinity. It does so by inhibiting the entry of contraction-inducing calcium into these smooth muscle cells. Nitric oxide is a small, highly reactive, short-lived gas molecule that once was known primarily as a toxic air pollutant. Yet studies have revealed an astonishing number of biological roles for NO, which occur in many other tissues besides endothelial cells. In fact, it appears that NO serves as one of the body's most important messenger molecules, as shown by the range of functions identified for this chemical and listed in ▮ Table 9-3. As you can see, most areas of the body are influenced by this versatile intercellular messenger molecule.

Endothelial cells release other important chemicals besides NO. **Endothelin**, another endothelial vasoactive substance, causes arteriolar smooth muscle contraction and is one of the most potent vasoconstrictors yet identified. Still other chemicals, released from the endothelium in response to chronic changes in

### ▮ TABLE 9-2 Functions of Endothelial Cells

- Line the blood vessels and heart chambers; serve as a physical barrier between the blood and the remainder of the vessel wall

- Secrete vasoactive substances in response to local chemical and physical changes; these substances cause relaxation (vasodilation) or contraction (vasoconstriction) of the underlying smooth muscle

- Secrete substances that stimulate new vessel growth and proliferation of smooth muscle cells in vessel walls

- Participate in the exchange of materials between the blood and surrounding tissue cells across capillaries through vesicular transport

- Influence formation of platelet plugs, clotting, and clot dissolution (see Chapter 10)

- Participate in the determination of capillary permeability by contracting to vary the size of the pores between adjacent endothelial cells

© 2016 Cengage

## ■ TABLE 9-3 Functions of Nitric Oxide (NO)

- Causes relaxation of arteriolar smooth muscle. By means of this action, NO plays an important role in controlling blood flow through the tissues and in maintaining mean arterial blood pressure.
- Dilates the arterioles of the penis and clitoris, thus serving as the direct mediator of erection of these reproductive organs. Erection is accomplished by rapid engorgement of these organs with blood.
- Acts as chemical warfare against bacteria and cancer cells via release from macrophages, large phagocytic cells of the immune system
- Interferes with platelet function and blood clotting at sites of vessel damage
- Serves as a novel type of neurotransmitter in the brain and elsewhere
- Plays a role in the changes underlying memory
- By promoting relaxation of digestive tract smooth muscle, helps regulate peristalsis, a type of contraction that pushes digestive tract contents forward
- Relaxes the smooth muscle cells in the airways of the lungs, helping keep these passages open to facilitate movement of air in and out of the lungs
- Modulates the filtering process involved in urine formation
- Directs blood flow to oxygen-starved tissues
- May play a role in relaxation of skeletal muscle

© 2016 Cengage

blood flow to an organ, trigger long-term vascular changes that permanently influence blood flow to a region. Some chemicals, for example, stimulate new vessel growth, a process known as **angiogenesis**.

## Local histamine release

Histamine is another local chemical mediator that influences arteriolar smooth muscle, but it is not released in response to local metabolic changes and is not derived from endothelial cells.

### ■ Why It Matters
### Local Heat or Cold Application

Heat application, by causing localized arteriolar vasodilation, is a useful therapeutic agent for promoting increased blood flow to an area. Conversely, applying ice packs to an inflamed area produces vasoconstriction, which reduces swelling by counteracting histamine-induced vasodilation.

Although histamine normally does not participate in controlling blood flow, it is important in certain pathological conditions.

Histamine is synthesized and stored within special connective tissue cells in many organs and in certain types of circulating white blood cells. When organs are injured or during allergic reactions, histamine is released and acts as a paracrine in the damaged region. By promoting relaxation of arteriolar smooth muscle, histamine is the major cause of vasodilation in an injured area. The resultant increase in blood flow into the area produces the redness and contributes to the swelling seen with inflammatory responses (see Chapter 10 for further detail).

### ■ Clinical Connections

During exercise, the muscles require more ATP, so blood flow increases to increase the delivery rate of oxygen and ATP precursors such as glucose. However, when blood flow to a muscle is compromised, such as what happens during peripheral artery disease, the muscle turns to anaerobic metabolism. As a consequence, metabolic wastes accumulate and can cause pain and cramping. Within the arteries and arterioles downstream of the plaque, local vasodilating factors accumulate to try to cause dilation and increase blood flow to the skeletal muscle. However, blood flow upstream continues to be restricted due to the atherosclerosis, so supply does not match demand. When John stopped walking, his reduced blood flow to his calves was able to again deliver sufficient oxygen to match muscle activity and to clear out the local mediators of vasodilation.

## Local physical influences

Among the physical influences on arteriolar smooth muscle, the effect of temperature changes is exploited clinically, but the chemically mediated response to shear stress and the myogenic response to stretch are most important physiologically. Let's examine each of these effects.

### SHEAR STRESS

Because of friction, blood flowing over the surface of the vessel lining creates a longitudinal force, known as **shear stress**, on the endothelial cells. An increase in shear stress causes the endothelial cells to release nitric oxide, which diffuses to the underlying smooth muscle and promotes vasodilation. The resultant increase in arteriolar diameter reduces shear stress in the vessel. In response to shear stress on a long-term basis, endothelial cells orient themselves parallel to the direction of blood flow.

### MYOGENIC RESPONSES TO STRETCH

Arteriolar smooth muscle responds to being passively stretched by myogenically increasing its tone via vasoconstriction, thereby acting to resist the initial passive stretch. Conversely, a decrease in arteriolar stretching induces a reduction in myogenic vessel tone by promoting vasodilation. Endothelial-derived vasoactive substances may also contribute to these mechanically induced responses. The extent of passive stretch varies with the volume of blood delivered to the arterioles from the arteries. An increase in mean arterial pressure drives more blood forward into the

9

arterioles and stretches them further, whereas arterial occlusion blocks blood flow into the arterioles and reduces arteriolar stretch.

Myogenic responses, coupled with metabolically induced responses, are important in reactive hyperaemia and autoregulation, topics to which we now turn our attention.

### REACTIVE HYPERAEMIA

When the blood supply to a region is completely occluded, arterioles in the region dilate because of (1) myogenic relaxation, which occurs in response to the diminished stretch accompanying no blood flow, and (2) changes in local chemical composition. Many of the chemical changes that occur in a blood-deprived tissue are the same as those that occur during metabolically induced active hyperaemia. When a tissue's blood supply is blocked, oxygen levels decrease in the deprived tissue; the tissue continues to consume oxygen, but no fresh supplies are being delivered. Meanwhile, the concentrations of carbon dioxide, acid, and other metabolites rise. Even though their production does not increase as it does when a tissue is more active metabolically, these substances accumulate in the tissue when the normal amounts produced are not washed away by blood.

After the occlusion is removed, blood flow to the previously deprived tissue is transiently much higher than normal because the arterioles are widely dilated. This postocclusion increase in blood flow, called **reactive hyperaemia**, can take place in any tissue. Such a response is beneficial for rapidly restoring the local chemical composition to normal. Of course, prolonged blockage of blood flow leads to irreversible tissue damage.

### AUTOREGULATION

When mean arterial pressure falls (e.g., because of haemorrhage or a weakened heart), the driving force is reduced, so blood flow to organs decreases. This is a milder version of what happens during vessel occlusion. The resultant changes in local metabolites and the reduced stretch in the arterioles collectively bring about arteriolar dilation to help restore tissue blood flow to normal, despite the reduced driving pressure. On the negative side, widespread arteriolar dilation reduces the mean arterial pressure still further, which aggravates the problem. Conversely, in the presence of sustained elevations in mean arterial pressure (hypertension), local chemical and myogenic influences triggered by the initial increased flow of blood to tissues bring about an increase in arteriolar tone and resistance. This greater degree of vasoconstriction subsequently reduces tissue blood flow toward normal, despite the elevated blood pressure (> Figure 9-13). **Autoregulation** (self-regulation) is the term for these local arteriolar mechanisms that keep tissue blood flow fairly constant despite rather wide deviations in mean arterial driving pressure. Not all organs autoregulate equally. As examples, the brain autoregulates best, the kidneys are good at autoregulation, and skeletal muscle has poor autoregulatory abilities.

Active hyperaemia, reactive hyperaemia, and histamine release all deliberately increase blood flow to a particular tissue for a specific purpose by inducing local arteriolar vasodilation.

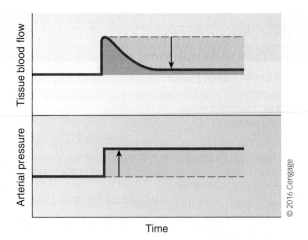

> **FIGURE 9-13 Autoregulation of tissue blood flow.** Even though blood flow through a tissue immediately increases in response to a rise in arterial pressure, the tissue blood flow is reduced gradually as a result of autoregulation within the tissue, despite a sustained increase in arterial pressure.

In contrast, autoregulation is a means by which each tissue resists alterations in its own blood flow secondary to changes in mean arterial pressure by making appropriate adjustments in arteriolar radius.

As an example of autoregulation, consider the effects of postural changes on blood flow to the brain. When you suddenly go from lying down to standing up, the body undergoes several responses in order to maintain blood flow to the brain. Within the brain itself, this sudden decrease in pressure, and thus blood flow, cause arteries and arterioles to constrict to increase pressure and maintain blood flow. This compensatory response reverses as systemic arterial pressure stabilizes to the new posture. Although this example has been oversimplified, it does highlight the ability of an organ to autoregulate its blood flow to ensure that there is a constant supply of oxygen.

This completes our discussion of the local control of arteriolar radius. Now let's examine extrinsic control of arteriolar radius.

## Extrinsic sympathetic control of arteriolar radius

Extrinsic control of arteriolar radius includes both neural and hormonal influences, the effects of the sympathetic nervous system being the most important. Sympathetic nerve fibres supply arteriolar smooth muscle everywhere in the systemic circulation, except in the brain. Recall that a certain level of ongoing sympathetic activity contributes to vascular tone. Increased sympathetic activity produces generalized arteriolar vasoconstriction, whereas decreased sympathetic activity leads to generalized arteriolar vasodilation. These widespread changes in arteriolar resistance bring about changes in mean arterial pressure because of their influence on total peripheral resistance, as we describe next.

## INFLUENCE OF TOTAL PERIPHERAL RESISTANCE ON MEAN ARTERIAL PRESSURE

To find the effect of changes in arteriolar resistance on mean arterial pressure, the formula $F = \Delta P/R$ applies to the entire circulation as well as to a single vessel:

- *F:* Looking at the circulatory system as a whole, flow ($F$) through all the vessels in either the systemic or pulmonary circulation is equal to the cardiac output.

- *P:* The pressure gradient ($\Delta P$) for the entire systemic circulation is the *mean arterial pressure*. (($\Delta P$) equals the difference in pressure between the beginning and the end of the systemic circulatory system. The beginning pressure is the mean arterial pressure as the blood leaves the left ventricle at an average of 93 mmHg. The end pressure in the right atrium is 0 mmHg. Therefore, $\Delta P = 93\,\text{mmHg} - 0\,\text{mmHg} = 93\,\text{mmHg}$, which is equivalent to the mean arterial pressure.)

- *R:* The total resistance ($R$) offered by all the systemic peripheral vessels together is the **total peripheral resistance**. By far the greatest percentage of the total peripheral resistance is due to arteriolar resistance, because arterioles are the primary resistance vessels.

Therefore, for the entire systemic circulation, rearranging

$$F = \Delta P/R$$

to

$$\Delta P = F \times R$$

gives us the equation

Mean arterial pressure = cardiac output $\times$ total peripheral resistance

Thus, the extent of total peripheral resistance offered collectively by all the systemic arterioles influences the mean arterial pressure immensely. A dam provides an analogy to this relationship. At the same time a dam restricts the flow of water downstream, it increases the pressure upstream by elevating the water level in the reservoir behind the dam. Similarly, generalized, sympathetically induced vasoconstriction reduces blood flow downstream to the organs while elevating the upstream mean arterial pressure, thereby increasing the main driving force for blood flow to all the organs.

These effects seem counterproductive. Why increase the driving force for flow to the organs by increasing arterial blood pressure while reducing flow to the organs by narrowing the vessels supplying them? In effect, the sympathetically induced arteriolar responses help maintain the appropriate driving pressure head (i.e., the mean arterial pressure) to all organs. The extent to which each organ actually receives blood flow is determined by local arteriolar adjustments that override the sympathetic constrictor effect. If all arterioles were dilated, total tissue perfusion would be greater than cardiac output, resulting in blood pressure to fall substantially, so there would not be an adequate driving force for blood flow. An analogy is the pressure head for water in the pipes in your home. If the water pressure is adequate, you can selectively obtain satisfactory water flow at any of the faucets by turning the appropriate handle to the open position. If the water pressure in the pipes is too low, however, you cannot obtain satisfactory flow at any faucet, even if you turn the handle to the maximally open position. So tonic sympathetic activity constricts most vessels (with the exception of those in the brain) and thereby helps maintain a pressure head, which organs can draw upon as needed by means of local mechanisms that control arteriolar radius.

## NOREPINEPHRINE'S INFLUENCE ON ARTERIOLAR SMOOTH MUSCLE

The norepinephrine released from sympathetic nerve endings combines with $\alpha_1$-adrenergic receptors on arteriolar smooth muscle to bring about vasoconstriction (p. 193). Cerebral (brain) arterioles are the only ones that do not have $\alpha_1$ receptors, so no vasoconstriction occurs in the brain. It is important that cerebral arterioles are not constricted by neural influences, because brain blood flow must remain constant to meet the brain's continual need for oxygen, no matter what is going on elsewhere in the body. Cerebral vessels are almost entirely controlled by local mechanisms that maintain a constant blood flow to support a constant level of brain metabolic activity. In fact, reflex vasoconstrictor activity in the remainder of the cardiovascular system is aimed at maintaining an adequate pressure head for blood flow to the vital brain.

Sympathetic activity therefore contributes in an important way to maintaining mean arterial pressure; it does so by ensuring an adequate driving force for blood flow to the brain at the expense of organs that can better withstand reduced blood flow. Other organs that really need additional blood, such as active muscles (including active heart muscle), obtain it through local controls that override the sympathetic effect.

## LOCAL CONTROLS OVERRIDING SYMPATHETIC VASOCONSTRICTION

Skeletal and cardiac muscles have the most powerful local control mechanisms with which to override generalized sympathetic vasoconstriction. For example, if you are pedalling a bicycle, the increased activity in the skeletal muscles of your legs brings about an overriding local, metabolically induced vasodilation in those particular muscles, despite the generalized sympathetic vasoconstriction that accompanies exercise. As a result, more blood flows through your leg muscles but not through your inactive arm muscles.

## NO PARASYMPATHETIC INNERVATION TO ARTERIOLES

There is no significant parasympathetic innervation to arterioles, with the exception of the abundant parasympathetic vasodilator supply to the arterioles of the penis and clitoris. The rapid, profuse vasodilation induced by parasympathetic stimulation in these organs (by means of promoting release of NO) is largely responsible for accomplishing erection. Vasodilation elsewhere is produced by decreasing sympathetic vasoconstrictor activity below its tonic level. When mean arterial pressure rises above normal, reflex reduction in sympathetic vasoconstrictor activity accomplishes generalized arteriolar vasodilation to help bring the driving pressure down toward normal.

## The medullary cardiovascular control centre

The main region of the brain that adjusts sympathetic output to the arterioles is the **cardiovascular control centre** in the medulla of the brain stem. This is the integrating centre for blood pressure regulation (described in Section 8.6). Several other brain regions also influence blood distribution, the most notable being the hypothalamus, which, as part of its temperature-regulating function, controls blood flow to the skin to adjust heat loss to the environment.

In addition to neural reflex activity, several hormones also extrinsically influence arteriolar radius. These hormones include the adrenal medullary hormones *epinephrine* and *norepinephrine*, which generally reinforce the sympathetic nervous system in most organs, as well as *vasopressin (ADH)* and *angiotensin II*, which are important in controlling fluid balance.

### INFLUENCE OF EPINEPHRINE AND NOREPINEPHRINE

Sympathetic stimulation of the adrenal medulla causes this endocrine gland to release epinephrine and norepinephrine. Adrenal medullary norepinephrine combines with the same $\alpha_1$ receptors as sympathetically released norepinephrine to produce generalized vasoconstriction. However, epinephrine, the more abundant of the adrenal medullary hormones, combines with both $\beta_2$ and $\alpha_1$ receptors, but has a much greater affinity for the $\beta_2$ receptors (Table 9-4). Activation of $\beta_2$ receptors produces vasodilation, but not all tissues have $\beta_2$ receptors; they are most abundant in the arterioles of the heart and skeletal muscles. During sympathetic discharge, the released epinephrine combines with the $\beta_2$ receptors in the heart and skeletal muscles to reinforce local vasodilatory mechanisms in these tissues. Arterioles in digestive organs and kidneys, in contrast, are equipped only with $\alpha_1$ receptors. Therefore, the arterioles of these organs undergo more profound vasoconstriction during generalized sympathetic discharge than those in the heart and skeletal muscle do. Lacking $\beta_2$ receptors, the digestive organs and kidneys do not experience an overriding vasodilatory response on top of the $\alpha_1$ receptor–induced vasoconstriction.

### INFLUENCE OF VASOPRESSIN AND ANGIOTENSIN II

The two other hormones that extrinsically influence arteriolar tone are vasopressin and angiotensin II. Vasopressin is primarily involved in maintaining water balance by regulating the amount of water the kidneys retain for the body during urine formation (p. 592). Angiotensin II is part of a hormonal pathway, the *renin-angiotensin–aldosterone system*, which is important in regulating the body's salt balance. This pathway promotes salt conservation during urine formation and also leads to water retention, because salt exerts a water-holding osmotic effect in the ECF (p. 575). Consequently, both these hormones play important roles in maintaining the body's fluid balance, which in turn is an important determinant of plasma volume and blood pressure.

In addition, both vasopressin and angiotensin II are potent vasoconstrictors. Their role in this regard is especially crucial during haemorrhage. A sudden loss of blood reduces the plasma volume, which triggers increased secretion of both these hormones to help restore plasma volume. Their vasoconstrictor effects also help maintain blood pressure despite abrupt loss of plasma volume. (The functions and control of these hormones are discussed more thoroughly in Chapters 5 and 6.)

This completes our discussion of the various factors that affect total peripheral resistance, the most important of which are controlled adjustments in arteriolar radius. These factors are summarized in ⟩ Figure 9-14.

We now turn our attention to the next vessels in the vascular tree, the capillaries.

## 9.4 | Capillaries

**Capillaries** branch extensively to bring blood within the reach of every cell and, therefore, are considered the primary site for exchange of materials between blood and tissue cells. However, it is important to note that the entire vasculature is a continuum and does not abruptly change

---

**▌ TABLE 9-4 Arteriolar Smooth Muscle Adrenergic Receptors**

| | RECEPTOR TYPE | |
| --- | --- | --- |
| CHARACTERISTIC | $\alpha_1$ | $\beta_2$ |
| **Location of the Receptor** | All arteriolar smooth muscle except in the brain | Arteriolar smooth muscle in the heart and skeletal muscles |
| **Chemical Mediator** | Norepinephrine from sympathetic fibres and the adrenal medulla | Epinephrine from the adrenal medulla (greater affinity for this receptor) |
| | Epinephrine from the adrenal medulla (less affinity for this receptor) | |
| **Arteriolar Smooth Muscle Response** | Vasoconstriction | Vasodilation |

9

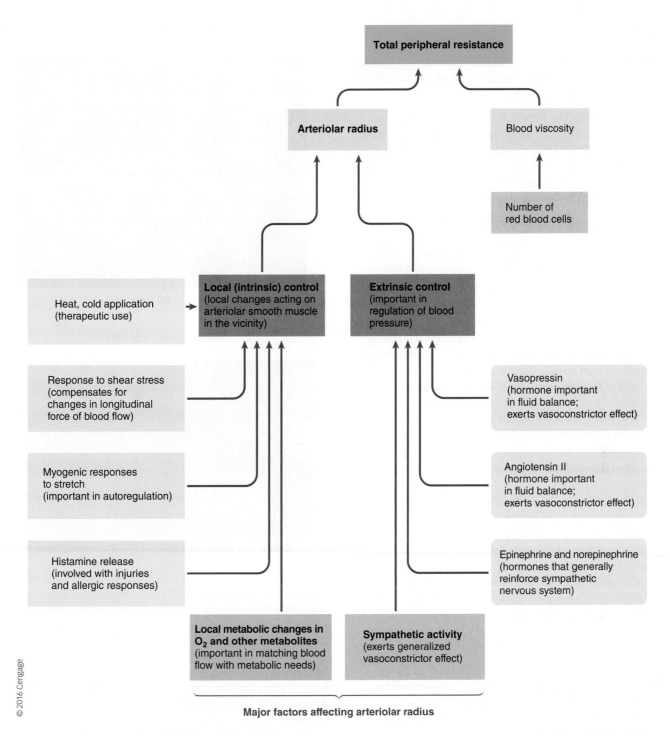

**Major factors affecting arteriolar radius**

> **FIGURE 9-14 Factors affecting total peripheral resistance.** The primary determinant of total peripheral resistance is the adjustable arteriolar radius. Two major categories of factors influence arteriolar radius: (1) local (intrinsic) control, which is primarily important in matching blood flow through a tissue with the tissue's metabolic needs and is mediated by local factors acting on the arteriolar smooth muscle; and (2) extrinsic control, which is important in regulating blood pressure and is mediated primarily by sympathetic influence on arteriolar smooth muscle.

from one vascular type to another. Consequently, some exchange takes place across the other microcirculatory vessels, especially the postcapillary venules. When the term *capillary exchange* is used, it tacitly refers to all exchange at the microcirculatory level, the majority of which occurs across the capillaries.

## Capillaries: Sites of exchange

There are no carrier-mediated transport systems across capillaries, with the exception of those in the brain that play a role in the blood–brain barrier (p. 100). Materials are exchanged across capillary walls mainly by diffusion.

## FACTORS THAT ENHANCE DIFFUSION ACROSS CAPILLARIES

Capillaries are ideally suited to enhance diffusion. They minimize diffusion distances while maximizing surface area and time available for exchange, as follows:

1. Diffusing molecules have only a short distance to travel between blood and surrounding cells because of the thin capillary wall and small capillary diameter, coupled with the close proximity of every cell to a capillary. This short distance is important because the rate of diffusion slows down as the diffusion distance increases.

   a. Capillary walls are very thin (1 μm in thickness; in contrast, the diameter of a human hair is 100 μm). Capillaries consist of only a single layer of flat endothelial cells—essentially the lining of the other vessel types. No smooth muscle or connective tissue is present (see ▌Table 9-1).

   b. Each capillary is so narrow (7 μm average diameter) that red blood cells (8 μm diameter) have to squeeze through single file (› Figure 9-15b). Consequently, plasma contents are either in direct contact with the inside of the capillary wall or are only a short diffusing distance from it.

   c. Researchers estimate that because of extensive capillary branching, no cell is farther than 0.01 cm (4/1000 in.) from a capillary.

2. Because capillaries are distributed in such incredible numbers (estimates range from 10 billion to 40 billion capillaries), a tremendous total surface area is available for exchange (an estimated 600 m²). Despite this large number of capillaries, at any point in time they contain only 5 percent of the total blood volume (250 mL out of a total of 5000 mL). As a result, a small volume of blood is exposed to an extensive surface area. If all the capillary surfaces were stretched out in a flat sheet and the volume of blood contained within the capillaries were spread over the top, this would be roughly equivalent to spreading a half pint (237 mL) of paint over the floor of a high school gymnasium. Imagine how thin the paint layer would be!

3. Blood flows more slowly in the capillaries than elsewhere in the circulatory system. The extensive capillary branching is responsible for this slow velocity of blood flow through the capillaries. Let's see why blood slows down in the capillaries.

### SLOW VELOCITY OF FLOW THROUGH CAPILLARIES

First, we need to clarify a potentially confusing point. The term *flow* can be used in two different contexts—flow rate and velocity of flow. The *flow rate* refers to the *volume* of blood per unit of time flowing through a given segment of the circulatory system (this is the flow we have been talking about in relation to the pressure gradient and resistance). The *velocity of flow* is the linear *speed*, or distance per unit of time, with which blood flows forward through a given segment of the circulatory system. Because the circulatory system is a closed system, the volume of blood flowing through any level of the system must equal the cardiac

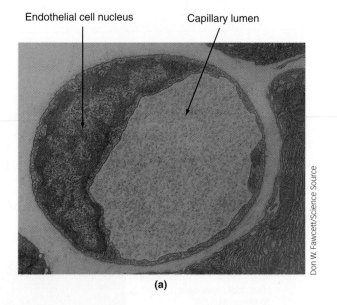

Endothelial cell nucleus     Capillary lumen

(a)

Don W. Fawcett/Science Source

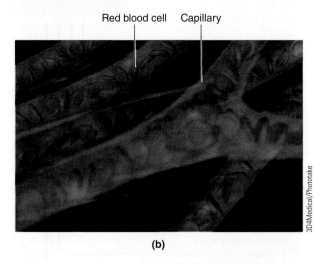

Red blood cell    Capillary

(b)

3D4Medical/Phototake

› **FIGURE 9-15 Capillary anatomy.** (a) Electron micrograph of a cross-section of a capillary. The capillary wall consists of a single layer of endothelial cells. The nucleus of one of these cells is shown. (b) Photograph of a capillary bed. The capillaries are so narrow that the red blood cells must pass through single file.

output. For example, if the heart pumps out 5 L of blood per minute, and 5 L/min return to the heart, then 5 L/min must flow through the arteries, arterioles, capillaries, and veins. Therefore, the flow rate is the same at all levels of the circulatory system.

However, the velocity with which blood flows through the different segments of the vascular tree varies, because velocity of flow is inversely proportional to the total cross-sectional area of all the vessels at any given level of the circulatory system. Even though the cross-sectional area of each capillary is extremely small compared to that of the large aorta, the total cross-sectional area of all capillaries added together is about 1300 times greater than the cross-sectional area of the aorta—because there are so many capillaries. Accordingly, blood slows considerably as it passes through the capillaries (› Figure 9-16). This slow velocity allows adequate time for exchange of nutrients and metabolic end products between blood and tissue cells. As the

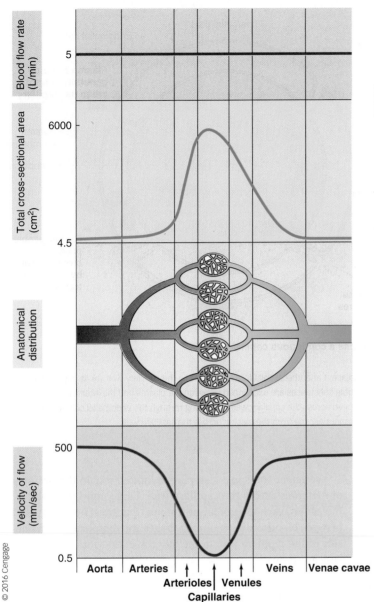

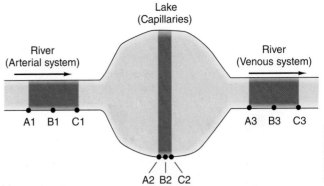

> FIGURE 9-17 **Relationship between total cross-sectional area and velocity of flow.** The three dark blue areas represent equal volumes of water. During one minute, this volume of water moves forward from points A to points C. Therefore, an identical volume of water flows past points B1, B2, and B3 during this minute; that is, the flow rate is the same at all points along the length of this body of water. However, during that minute the identical volume of water moves forward a much shorter distance in the wide lake (A2 to C2) than in the much narrower river (A1 to C1 and A3 to C3). Thus, the velocity of flow is much slower in the capillaries than in the arterial and venous systems.

volumes of water, now spread out over a larger cross-sectional area, move forward a much shorter distance in the wide lake than in the narrow river during a given period of time. You could readily observe the forward movement of water in the swift-flowing river, but the forward motion of water in the lake would be unnoticeable.

As well, because of the capillaries' tremendous total cross-sectional area, the resistance offered by all the capillaries is much lower than that offered by all the arterioles, even though each capillary has a smaller radius than each arteriole. For this reason, the arterioles contribute more to total peripheral resistance. Furthermore, arteriolar calibre (and, accordingly, resistance) is subject to control, whereas capillary calibre cannot be adjusted.

## Capillary pores

Diffusion across capillary walls also depends on the walls' permeability to the materials being exchanged. The endothelial cells forming the capillary walls fit together in a jigsaw-puzzle fashion, but the closeness of the fit varies considerably between organs. In most capillaries, narrow, water-filled gaps, or **pores**, lie at the junctions between the cells (> Figure 9-18). These pores permit passage of small water-soluble substances. Lipid-soluble substances, such as oxygen and carbon dioxide, can readily pass through the endothelial cells themselves by dissolving in the lipid bilayer barrier.

The size of the capillary pores varies from organ to organ. At one extreme, the endothelial cells in brain capillaries are joined by tight junctions so that pores are nonexistent. These junctions prevent transcapillary passage of materials between the cells and thus constitute part of the protective blood–brain barrier (p. 100). In most tissues, small, water-soluble substances, such as ions, glucose, and amino acids, can readily pass through the

> FIGURE 9-16 **Comparison of blood flow rate and velocity of flow in relation to total cross-sectional area.** The blood flow rate (red curve) is identical through all levels of the circulatory system and is equal to the cardiac output (5 L/min at rest). The velocity of flow (purple curve) varies throughout the vascular tree and is inversely proportional to the total cross-sectional area (green curve) of all the vessels at a given level. Note that the velocity of flow is slowest in the capillaries, which have the largest total cross-sectional area.

capillaries rejoin to form veins, the total cross-sectional area is once again reduced, and the velocity of blood flow increases as blood returns to the heart.

As an analogy, consider a river (the arterial system) that widens into a lake (the capillaries), then narrows into a river again (the venous system) (> Figure 9-17). The flow rate is the same throughout the length of this body of water; that is, identical volumes of water are flowing past all the points along the bank of the river and lake. However, the velocity of flow is slower in the wide lake than in the narrow river because the identical

**9**

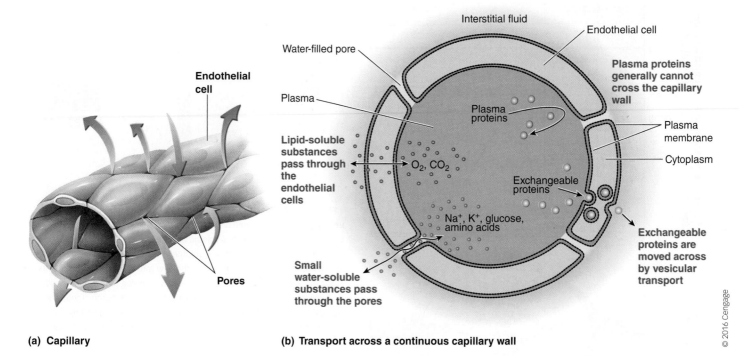

**(a) Capillary**

**(b) Transport across a continuous capillary wall**

> **FIGURE 9-18 Exchanges across the capillary wall.** (a) Slit-like gaps between adjacent endothelial cells form pores within the capillary wall. (b) As depicted in this schematic representation of a cross-section of a capillary wall, small water-soluble substances are exchanged between the plasma and the interstitial fluid by passing through the water-filled pores, whereas lipid-soluble substances are exchanged across the capillary wall by passing through the endothelial cells. Proteins to be moved across are exchanged by vesicular transport. Plasma proteins generally cannot escape from the plasma across the capillary wall.

water-filled pores, but large, non-lipid-soluble materials that cannot fit through the pores, such as plasma proteins, are kept from passing. At the other extreme, liver capillaries have such large pores that even proteins pass through readily. This is adaptive, because the liver's functions include synthesis of plasma proteins and the metabolism of protein-bound substances, such as cholesterol. These proteins must all pass through the liver's capillary walls. The leakiness of various capillary beds is therefore a function of how tightly the endothelial cells are joined, which varies according to the different organs' needs.

Scientists traditionally considered the capillary wall a passive sieve, like a brick wall with permanent gaps in the mortar acting as pores. Recent studies, however, suggest that endothelial cells can actively change to regulate capillary permeability; that is, in response to appropriate signals, the "bricks" can readjust themselves to vary the size of the holes. So the degree of leakiness does not necessarily remain constant for a given capillary bed. For example, histamine increases capillary permeability by triggering contractile responses in endothelial cells to widen the intercellular gaps. This is not a muscular contraction, because no smooth muscle cells are present in capillaries. It is due to an actin–myosin contractile apparatus in the nonmuscular capillary endothelial cells. Because of these enlarged pores, the affected capillary wall is leakier. As a result, normally retained plasma proteins escape into the surrounding tissue, where they exert an osmotic effect. Along with histamine-induced vasodilation, the resulting additional local fluid retention contributes to inflammatory swelling.

Vesicular transport also plays a limited role in the passage of materials across the capillary wall. Large non-lipid-soluble molecules, such as protein hormones, that must be exchanged between blood and surrounding tissues are transported from one side of the capillary wall to the other in endocytotic-exocytotic vesicles (p. 46).

## Capillaries under resting conditions

The branching and reconverging arrangement within capillary beds varies somewhat, depending on the tissue. Capillaries typically branch either directly from an arteriole or from a thoroughfare channel known as a **metarteriole**, which runs between an arteriole and a venule. Likewise, capillaries may rejoin at either a venule or a metarteriole (> Figure 9-19a).

Unlike the true capillaries within a capillary bed, metarterioles are sparsely surrounded by wisps of spiralling smooth muscle cells. These cells also form **precapillary sphincters**, each of which consists of a ring of smooth muscle around the entrance to a capillary as it arises from a metarteriole. However, it is important to note that although generally accepted, the existence of precapillary sphincters in humans has not been conclusively established.

### ROLE OF PRECAPILLARY SPHINCTERS

Precapillary sphincters are not innervated, but they have a high degree of myogenic tone and are sensitive to local metabolic changes. They act as stopcocks to control blood flow through

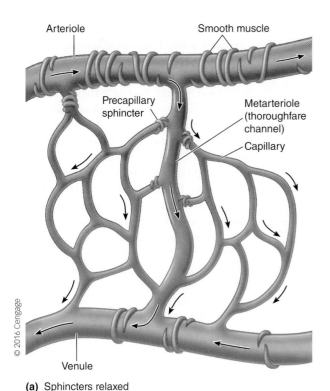

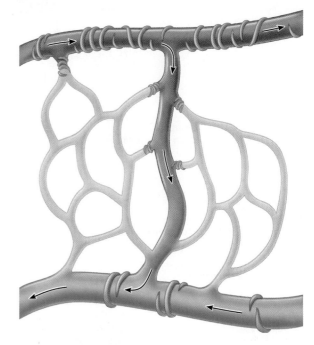

**(a)** Sphincters relaxed

**(b)** Sphincters contracted

© 2016 Cengage

> **FIGURE 9-19 Capillary bed.** Capillaries branch either directly from an arteriole or from a metarteriole, a thoroughfare channel between an arteriole and venule. Capillaries rejoin at either a venule or a metarteriole. Smooth muscle cells form precapillary sphincters that encircle capillaries as they arise from the metarteriole. (a) When the precapillary sphincters are relaxed, blood flows through the entire capillary bed. (b) When the precapillary sphincters are contracted, blood flows only through the metarteriole, bypassing the capillary bed.

the particular capillary that each one guards. Arterioles perform a similar function for a small group of capillaries. Capillaries themselves have no smooth muscle, so they cannot actively participate in regulating their own blood flow.

Generally, tissues that are more metabolically active have a greater density of capillaries. For example, skeletal muscle has more capillaries (greater density) than the tendons that join muscle to bone. Only about 10 percent of the precapillary sphincters in a resting muscle are open at any moment, so blood flow through the muscle is reduced to about 10 percent of maximum, and the rest flows through the metarterioles (> Figure 9-19b). As muscle metabolism increases, the need for oxygen and the production of carbon dioxide increase, and other chemical concentrations change, thus facilitating the relaxation of precapillary sphincters and arterioles in that region. Restoration of the chemical concentrations to normal as a result of increased blood flow to that region removes the impetus for vasodilation, so the precapillary sphincters close once again, and the arterioles return to normal tone. In this way, blood flow through any given capillary is often intermittent due to arteriolar and precapillary sphincter action working in concert.

When the muscle as a whole becomes more active, a greater percentage of the precapillary sphincters relax, simultaneously opening up more capillary beds, while concurrent arteriolar vasodilation increases total flow to the organ. As a result of more blood flowing through more open capillaries, the total

volume and surface area available for exchange increase, and the diffusion distance between the cells and an open capillary decreases (> Figure 9-20). Therefore, blood flow through a particular tissue (assuming a constant blood pressure) is regulated by (1) the degree of resistance offered by the arterioles in the organ, controlled by sympathetic activity and local factors; and (2) the number of open capillaries, controlled by action of the same local metabolic factors on precapillary sphincters.

## Interstitial fluid: A passive intermediary

Exchanges between blood and tissue cells are not made directly. Interstitial fluid, the true internal environment in immediate contact with the cells, acts as the go-between. Only 20 percent of the ECF circulates as plasma. The remaining 80 percent consists of interstitial fluid, which bathes all the cells in the body. Cells exchange materials directly with interstitial fluid, with the type and extent of exchange governed by the properties of cellular plasma membranes. Movement across the plasma membrane may be either passive (i.e., by diffusion down electrochemical gradients or by facilitated diffusion) or active (i.e., by active carrier-mediated transport or by vesicular transport) (see ▮ Table 2-5).

In contrast, exchanges across the capillary wall between plasma and interstitial fluid are largely passive. The only transport across this barrier that requires energy is the limited vesicular transport. Because capillary walls are highly permeable,

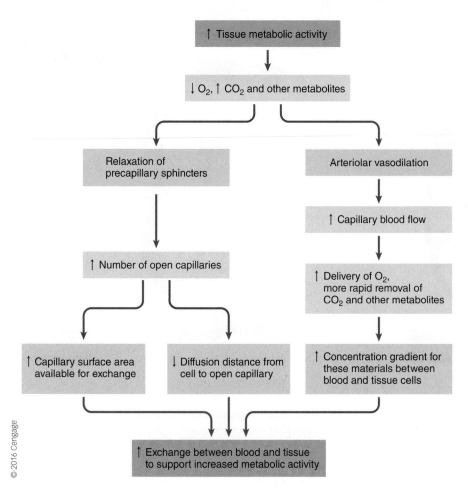

↑ Tissue metabolic activity

↓ O₂, ↑ CO₂ and other metabolites

Relaxation of precapillary sphincters

Arteriolar vasodilation

↑ Number of open capillaries

↑ Capillary blood flow

↑ Capillary surface area available for exchange

↓ Diffusion distance from cell to open capillary

↑ Delivery of O₂, more rapid removal of CO₂ and other metabolites

↑ Concentration gradient for these materials between blood and tissue cells

↑ Exchange between blood and tissue to support increased metabolic activity

© 2016 Cengage

> FIGURE 9-20 **Complementary action of precapillary sphincters and arterioles in adjusting blood flow through a tissue in response to changing metabolic needs**

in the appropriate direction across the capillary walls. The reconditioning organs continuously add nutrients and oxygen and remove carbon dioxide and other wastes as blood passes through them. Meanwhile, cells constantly use up supplies and generate metabolic wastes. As cells use up oxygen and glucose, the blood constantly brings in fresh supplies of these vital materials, maintaining concentration gradients that favour the net diffusion of these substances from blood to cells. Simultaneously, ongoing net diffusion of carbon dioxide and other metabolic wastes from cells to blood is maintained by the continual production of these wastes at the cell level and by their constant removal by the circulating blood (> Figure 9-21).

Because the capillary wall does not limit the passage of any constituent except plasma proteins, the extent of exchanges for each solute is independently determined by the magnitude of its concentration gradient between blood and surrounding cells. As cells increase their level of activity, they use up more oxygen and produce more carbon dioxide, among other things. This creates larger concentration gradients for oxygen and carbon dioxide between cells and blood, so more oxygen diffuses out of the blood into the cells, and more carbon dioxide proceeds in the opposite direction to help support the increased metabolic activity.

exchange is so thorough that the interstitial fluid takes on essentially the same composition as incoming arterial blood, with the exception of the large plasma proteins that usually do not escape from the blood. Therefore, when we speak of exchanges between blood and tissue cells, we tacitly include interstitial fluid as a passive intermediary.

Exchanges between blood and surrounding tissues across the capillary walls are accomplished in two ways: (1) passive diffusion down concentration gradients, the primary mechanism for exchanging individual solutes; and (2) bulk flow, a process that fills the totally different function of determining the distribution of the ECF volume between the vascular and interstitial fluid compartments. Now let's examine each of these mechanisms in more detail, starting with diffusion.

## Diffusion across the capillary walls

There are no carrier-mediated transport systems in most capillary walls; therefore, solutes cross primarily by diffusion down concentration gradients. The chemical composition of arterial blood is carefully regulated to maintain the concentrations of individual solutes at levels that promote each solute's movement

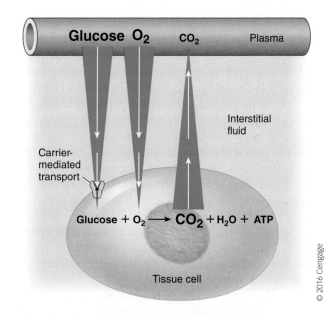

> FIGURE 9-21 **Independent exchange of individual solutes down their own concentration gradients across the capillary wall**

## Bulk flow across the capillary walls

The second means by which exchange is accomplished across capillary walls is bulk flow. A volume of protein-free plasma actually filters out of the capillary, mixes with the surrounding interstitial fluid, and then is reabsorbed. This process is called **bulk flow**, because the various constituents of the fluid are moving together in bulk, or as a unit, in contrast to the discrete diffusion of individual solutes down concentration gradients.

The capillary wall acts like a sieve, with fluid moving through its water-filled pores. When pressure inside the capillary exceeds pressure on the outside, fluid is pushed out through the pores in a process known as **ultrafiltration**. Due to the pores' filtering effect, most plasma proteins are retained on the inside during this process, although a few do escape. All other constituents in the plasma are dragged as a unit along with the volume of fluid leaving the capillary, so the filtrate is essentially a protein-free plasma. When inward-driving pressures exceed outward pressures across the capillary wall, net inward movement of fluid from the interstitial fluid into the capillaries takes place through the pores, a process known as **reabsorption**.

### FORCES INFLUENCING BULK FLOW

Bulk flow occurs because of differences in the hydrostatic and colloid osmotic pressures between plasma and interstitial fluid. These pressures are what are called the *Starling forces* and are used in the Starling equation that defines the movement of fluid across a capillary membrane. Even though pressure differences exist between plasma and surrounding fluid elsewhere in the circulatory system, only the capillaries have pores that let fluids pass through. Four forces influence fluid movement across the capillary wall (> Figure 9-22):

1. **Capillary blood pressure ($P_c$)** is the fluid or hydrostatic pressure exerted on the inside of the capillary walls by blood. This pressure tends to force fluid *out of* the capillaries into the interstitial fluid. By the level of the capillaries, blood pressure has dropped substantially because of frictional losses in pressure in the high-resistance

arterioles upstream. On average, the hydrostatic pressure is 37 mmHg at the arteriolar end of a tissue capillary (compared with a mean arterial pressure of 93 mmHg). It declines even further, to 17 mmHg, at the capillary's venular end because of further frictional loss coupled with the exit of fluid through ultrafiltration along the capillary's length (see > Figure 9-9).

2. **Plasma-colloid osmotic pressure ($\pi_p$)**, also known as *oncotic pressure*, is a force caused by colloidal dispersion of plasma proteins; it encourages fluid movement *into* the capillaries. Because plasma proteins remain in the plasma rather than entering the interstitial fluid, a protein concentration difference exists between plasma and interstitial fluid. Accordingly, there is also a water concentration difference between these two regions. Plasma has a higher protein concentration and a lower water concentration than interstitial fluid does. This difference exerts an osmotic effect that tends to move water from the area of higher water concentration in interstitial fluid to the area of lower water concentration (or higher protein concentration) in plasma (p. 38). The other plasma constituents do not exert an osmotic effect, because they readily pass through the capillary wall, so their concentrations are equal in plasma and interstitial fluid. Plasma-colloid osmotic pressure averages 25 mmHg.

3. **Interstitial fluid hydrostatic pressure ($P_{IF}$)** is the fluid pressure exerted on the outside of the capillary wall by interstitial fluid. This pressure tends to force fluid *into* the capillaries. Because of difficulties encountered in measuring interstitial fluid hydrostatic pressure, the actual value of the pressure is a controversial issue. It is either at, slightly above, or slightly below atmospheric pressure. For purposes of illustration, we will say it is 1 mmHg above atmospheric pressure.

4. **Interstitial fluid–colloid osmotic pressure ($\pi_{IF}$)** is another force that does not normally contribute significantly to bulk flow. The small fraction of plasma proteins

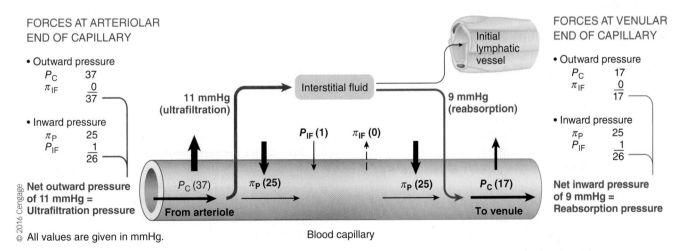

FORCES AT ARTERIOLAR
END OF CAPILLARY

- Outward pressure

| $P_C$ | 37 |
| $\pi_{IF}$ | 0 |
| | 37 |

- Inward pressure

| $\pi_P$ | 25 |
| $P_{IF}$ | 1 |
| | 26 |

**Net outward pressure of 11 mmHg = Ultrafiltration pressure**

11 mmHg (ultrafiltration)

$P_{IF}$ (1)   $\pi_{IF}$ (0)

9 mmHg (reabsorption)

Initial lymphatic vessel

Interstitial fluid

$P_C$ (37)   $\pi_P$ (25)   **From arteriole**   $\pi_P$ (25)   $P_C$ (17)   **To venule**

Blood capillary

FORCES AT VENULAR
END OF CAPILLARY

- Outward pressure

| $P_C$ | 17 |
| $\pi_{IF}$ | 0 |
| | 17 |

- Inward pressure

| $\pi_P$ | 25 |
| $P_{IF}$ | 1 |
| | 26 |

**Net inward pressure of 9 mmHg = Reabsorption pressure**

© 2016 Cengage

© All values are given in mmHg.

> **FIGURE 9-22 Bulk flow across the capillary wall.** Schematic representation of ultrafiltration and reabsorption as a result of imbalances in the physical forces acting across the capillary wall

that leak across the capillary walls into the interstitial spaces are normally returned to the blood by means of the lymphatic system. Therefore, the protein concentration in the interstitial fluid is extremely low, and the interstitial fluid–colloid osmotic pressure is very close to zero. If plasma proteins pathologically leak into the interstitial fluid, however, as they do when histamine widens the capillary pores during tissue injury, the leaked proteins exert an osmotic effect that tends to promote movement of fluid *out of* the capillaries into the interstitial fluid.

Therefore, the two pressures that tend to force fluid out of the capillary are capillary blood pressure and interstitial fluid–colloid osmotic pressure. The two opposing pressures that tend to force fluid into the capillary are plasma-colloid osmotic pressure and interstitial fluid hydrostatic pressure. Now let's analyze the fluid movement that occurs across a capillary wall because of imbalances in these opposing physical forces (> Figure 9-22).

## NET EXCHANGE OF FLUID ACROSS THE CAPILLARY WALL

Net exchange at a given point across the capillary wall can be calculated using the following:

$$\text{Net exchange pressure} = \underset{\substack{\text{(outward} \\ \text{pressure)}}}{(P_c + \pi_{IF})} - \underset{\substack{\text{(inward} \\ \text{pressure)}}}{(\pi_p + P_{IF})}$$

A positive net exchange pressure (when the outward pressure exceeds the inward pressure) represents a filtration pressure. A negative net exchange pressure (when the inward pressure exceeds the outward pressure) represents a reabsorption pressure.

At the arteriolar end of the capillary, the outward pressure totals 37 mmHg, whereas the inward pressure totals 26 mmHg, for a net outward pressure of 11 mmHg. Filtration takes place at the beginning of the capillary as this outward pressure gradient forces a protein-free filtrate through the capillary pores.

By the time the venular end of the capillary is reached, the capillary blood pressure has dropped, but the other pressures have remained essentially constant. At this point, the outward pressure has fallen to a total of 17 mmHg, whereas the total inward pressure is still 26 mmHg, for a net inward pressure of 9 mmHg. Reabsorption of fluid takes place as this inward pressure gradient forces fluid back into the capillary at its venular end.

Filtration and reabsorption, collectively known as *bulk flow*, are thus due to a shift in the balance between the passive physical forces acting across the capillary wall. No active forces or local energy expenditures are involved in the bulk exchange of fluid between the plasma and surrounding interstitial fluid. With only minor contributions from the interstitial fluid forces, filtration occurs at the beginning of the capillary because capillary blood pressure exceeds plasma-colloid osmotic pressure, whereas by the end of the capillary, reabsorption takes place because blood pressure has fallen below osmotic pressure.

It is important to realize that we have taken "snapshots" at two points—at the beginning and at the end—in a hypothetical capillary. Actually, blood pressure gradually diminishes along

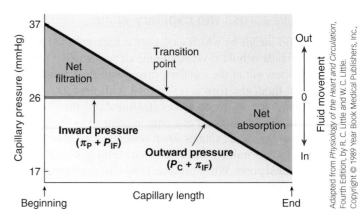

> FIGURE 9-23 **Net filtration and net reabsorption along the vessel length.** The inward pressure ($\pi_P + P_{IF}$) remains constant throughout the length of the capillary, whereas the outward pressure ($P_C + \pi_{IF}$) progressively declines throughout the capillary's length. In the first half of the vessel, where the declining outward pressure still exceeds the constant inward pressure, progressively diminishing quantities of fluid are filtered out (*upward red arrow*). In the last half of the vessel, the progressively increasing quantities of fluid are reabsorbed (*downward blue arrow*) as the declining outward pressure falls farther below the constant inward pressure.

the length of the capillary, so that progressively diminishing quantities of fluid are filtered out in the first half of the vessel and progressively increasing quantities of fluid are reabsorbed in the last half (> Figure 9-23). Even this situation is idealized. The pressures used in this figure are average values and controversial at that. Some capillaries have such high blood pressure that filtration actually occurs throughout their entire length, whereas others have such low hydrostatic pressure that reabsorption takes place throughout their length.

In fact, a recent theory that has received considerable attention is that net filtration occurs throughout the length of all *open* capillaries, whereas net reabsorption occurs throughout the length of all *closed* capillaries. According to this theory, when the precapillary sphincter is relaxed, capillary blood pressure exceeds the plasma osmotic pressure even at the venular end of the capillary, thereby promoting filtration throughout the length. When the precapillary sphincter is closed, the reduction in blood flow through the capillary diminishes capillary blood pressure below the plasma osmotic pressure even at the beginning of the capillary, so reabsorption takes place all along the capillary. Whichever mechanism is involved, the net effect is the same. A protein-free filtrate exits the capillaries and is ultimately reabsorbed.

## ROLE OF BULK FLOW

Bulk flow does not play an important role in the exchange of individual solutes between blood and tissues. This is because the quantity of solutes moved across the capillary wall by bulk flow is extremely small compared with the much larger transfer of solutes by diffusion. Thus, ultrafiltration and reabsorption are not important in the exchange of nutrients and wastes. Bulk flow is extremely important, however, in regulating the distribution of ECF between the plasma and interstitial fluid. Maintenance of proper arterial blood pressure depends in part on an appropriate volume of circulating blood. If plasma volume is reduced (e.g., by haemorrhage), blood pressure falls. The resultant lowering of capillary blood pressure alters the balance of forces across the capillary walls. Because the net outward pressure is decreased while the net inward pressure remains unchanged, extra fluid is shifted from the interstitial compartment into the plasma as a result of reduced filtration and increased reabsorption. The extra fluid soaked up from the interstitial fluid provides additional fluid for the plasma, temporarily compensating for the loss of blood. Meanwhile, reflex mechanisms acting on the heart and blood vessels also come into play to help maintain blood pressure until long-term mechanisms, such as thirst (and its satisfaction) and reduction of urinary output, can restore the fluid volume to completely compensate for the loss.

Conversely, if the plasma volume becomes overexpanded, as with excessive fluid intake, the resulting rise in capillary blood pressure forces extra fluid from the capillaries into the interstitial fluid, temporarily relieving the expanded plasma volume until the excess fluid can be eliminated from the body by long-term measures, such as increased urinary output.

These internal fluid shifts between the two ECF compartments occur automatically and immediately whenever the balance of forces acting across the capillary walls is changed; they provide a temporary mechanism to help keep plasma volume fairly constant. In the process of restoring plasma volume to an appropriate level, interstitial fluid volume fluctuates, but it is much more important that plasma volume be kept constant, to ensure that the circulatory system functions effectively.

## The lymphatic system

Even under normal circumstances, slightly more fluid is filtered out of the capillaries into the interstitial fluid than is reabsorbed from the interstitial fluid back into the plasma. The extra fluid filtered out as a result of this filtration–reabsorption imbalance is picked up by the **lymphatic system**. This extensive network of one-way vessels provides an accessory route by which fluid can be returned from the interstitial fluid to the blood. The lymphatic system functions much like a storm sewer that picks up and carries away excess rainwater so that the water does not accumulate and flood an area.

### PICKUP AND FLOW OF LYMPH

Small, blind-ended, terminal lymph vessels known as **initial lymphatics** permeate almost every tissue of the body () Figure 9-24a). The endothelial cells forming the walls of initial lymphatics slightly overlap like shingles on a roof, with their overlapping edges being free instead of attached to the surrounding cells. This arrangement creates one-way, valvelike openings in the vessel wall. Fluid pressure on the outside of the vessel pushes the innermost edge of a pair of overlapping edges inward, creating a gap between the edges (i.e., opening the valve). This opening permits interstitial fluid to enter () Figure 9-24b). Once interstitial fluid enters a lymphatic vessel, it is called **lymph**. Fluid pressure on the inside forces the overlapping edges together, closing the valves so that lymph does not escape. These lymphatic valvelike openings are much larger than the pores in blood capillaries. Consequently, large particles in the

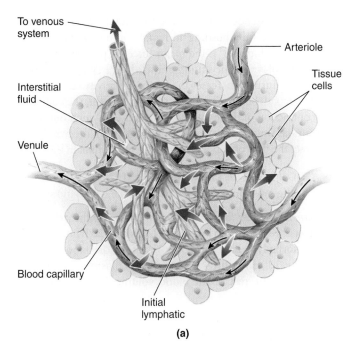

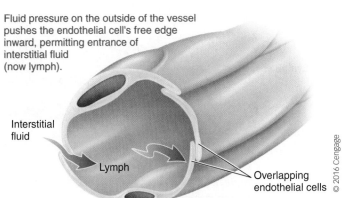

> **FIGURE 9-24 Initial lymphatics.** (a) Relationship between initial lymphatics and blood capillaries. Blind-ended initial lymphatics pick up excess fluid that has been filtered by blood capillaries and return it to the venous system in the chest. (b) Arrangement of endothelial cells in an initial lymphatic. Note that the overlapping edges of the endothelial cells create valvelike openings in the vessel wall.

interstitial fluid, such as escaped plasma proteins and bacteria, can gain access to initial lymphatics but are excluded from blood capillaries.

Initial lymphatics converge to form larger and larger **lymph vessels**, which eventually empty into the venous system near where the blood enters the right atrium () Figure 9-25a). Because there is no "lymphatic heart" to provide driving pressure, you may wonder how lymph is directed from the tissues toward the venous system in the thoracic cavity. Lymph flow is accomplished by two mechanisms. First, lymph vessels beyond the initial lymphatics are surrounded by smooth muscle, which contracts rhythmically as a result of myogenic activity. When this muscle is stretched because the vessel is distended with lymph, the muscle inherently contracts more forcefully, pushing the lymph through the vessel. This intrinsic "lymph pump" is the major force for propelling lymph. Stimulation of lymphatic smooth muscle by the sympathetic nervous system further increases the pumping activity of the lymph vessels. Second, because lymph vessels lie between skeletal muscles, contraction of these muscles squeezes the lymph out of the vessels. One-way valves spaced at intervals within the lymph vessels direct the flow of lymph toward its venous outlet in the chest.

### FUNCTIONS OF THE LYMPHATIC SYSTEM

Here are the most important functions of the lymphatic system:

- *Return of excess filtered fluid.* Normally, capillary filtration exceeds reabsorption by about 3 L per day (20 L filtered, 17 L reabsorbed) (› Figure 9-25b). Yet the entire blood volume is only 5 L, and only 2.75 L of that is plasma. (Blood cells make up the rest of the blood volume.) With an average cardiac output, 7200 L of blood pass through the capillaries daily under resting conditions (more when cardiac output increases). Even though only a small fraction of the filtered fluid is not reabsorbed by the blood capillaries, the cumulative effect of this process being repeated with every heartbeat results in the equivalent of more than the entire plasma volume left behind in the interstitial fluid each day. Obviously, this fluid must be returned to the circulating plasma, and this task is accomplished by the lymph vessels. The average rate of flow through the lymph vessels is 3 L per day, compared with 7200 L per day through the circulatory system.

- *Defence against disease.* The lymph percolates through **lymph nodes** located en route within the lymphatic system. Passage of this fluid through the lymph nodes is an important aspect of the body's defence mechanism against disease. For example, bacteria picked up from the interstitial fluid are destroyed by special phagocytes within the lymph nodes (see Chapter 11).

- *Transport of absorbed fat.* The lymphatic system is important in the absorption of fat from the digestive tract. The end products of the digestion of dietary fats are packaged by cells lining the digestive tract into fatty particles that are too large to gain access to the blood capillaries but can easily enter the initial lymphatics (see Chapter 15).

- *Return of filtered protein.* Most capillaries permit leakage of some plasma proteins during filtration. These proteins cannot readily be reabsorbed back into the blood capillaries but can easily gain access to the initial lymphatics. If the proteins were allowed to accumulate in the interstitial fluid rather than being returned to the circulation via the lymphatics, the interstitial fluid–colloid osmotic pressure (an outward pressure) would progressively increase while the plasma-colloid osmotic pressure (an inward pressure) would progressively fall. As a result, filtration forces would gradually increase and reabsorption forces would gradually decrease, resulting in progressive accumulation of fluid in the interstitial spaces at the expense of loss of plasma volume.

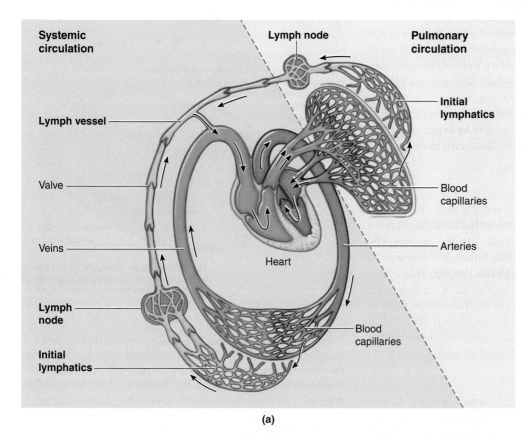

(a)

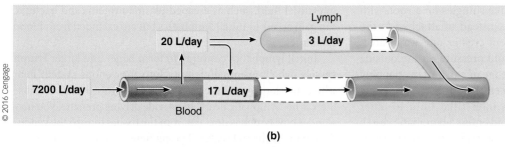

(b)

› **FIGURE 9-25 Lymphatic system.** (a) Lymph empties into the venous system near its entrance to the right atrium. (b) Lymph flow averages 3 L per day, whereas blood flow averages 7200 L per day.

## Oedema and interstitial fluid accumulation

*Clinical Note* Occasionally, excessive interstitial fluid does accumulate when one of the physical forces acting across the capillary walls becomes abnormal for some reason. Swelling of the tissues because of excess interstitial fluid is known as **oedema**. The causes of oedema can be grouped into four general categories:

1. *A reduced concentration of plasma proteins* decreases plasma-colloid osmotic pressure. Such a drop in the major inward pressure lets excess fluid filter out, whereas less-than-normal amounts of fluid are reabsorbed; hence, extra fluid remains in the interstitial spaces. Oedema can be caused by a decreased concentration of plasma proteins in several different ways: excessive loss of plasma proteins in the urine, from kidney disease; reduced synthesis of plasma proteins, from liver disease (the liver synthesizes almost all plasma proteins); a diet deficient in protein; or significant loss of plasma proteins from large burned surfaces.

2. *Increased permeability of the capillary walls* allows more plasma proteins than usual to pass from the plasma into the surrounding interstitial fluid—for example, via histamine-induced widening of the capillary pores during tissue injury or allergic reactions. The resultant fall in plasma-colloid osmotic pressure decreases the effective inward pressure, whereas the resultant rise in interstitial fluid–colloid osmotic pressure caused by excess protein in the interstitial fluid increases the effective outward force. This imbalance contributes in part to the localized oedema associated with injuries (e.g., blisters) and allergic responses (e.g., hives).

3. *Increased venous pressure*, as when blood dams up in the veins, is accompanied by an increased capillary blood pressure, because the capillaries drain into the veins. This elevation in outward pressure across the capillary walls is largely responsible for the oedema seen with congestive heart failure (p. 371). Regional oedema can also occur because of localized restriction of venous return. An example is the swelling that often occurs in the legs and feet during pregnancy. The enlarged uterus compresses the major veins that drain the lower extremities as these vessels enter the abdominal cavity. The resultant damming of blood in these veins raises blood pressure in the capillaries of the legs and feet, which promotes regional oedema of the lower extremities.

4. *Blockage of lymph vessels* produces oedema because the excess filtered fluid is retained in the interstitial fluid rather than returned to the blood through the lymphatics. Protein accumulation in the interstitial fluid compounds the problem through its osmotic effect. Local lymph blockage can occur, for example, in the arms of women whose major lymphatic drainage channels from the arm have been blocked as a result of lymph node removal during surgery for breast cancer. More widespread lymph blockage occurs with *filariasis*, a mosquito-borne parasitic

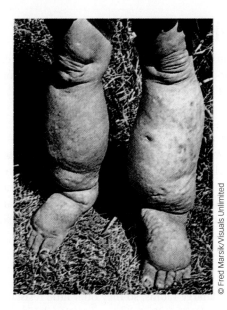

> **FIGURE 9-26 Elephantiasis.** This tropical condition is caused by a mosquito-borne parasitic worm that invades the lymph vessels. As a result of the interference with lymph drainage, the affected body parts, usually the extremities, become grossly oedematous, appearing elephant-like.

disease found predominantly in tropical coastal regions. In this condition, small, threadlike filaria worms infect the lymph vessels, where their presence prevents proper lymph drainage. The affected body parts, particularly the scrotum and extremities, become grossly oedematous. The condition is often called *elephantiasis* because of the elephant-like appearance of the swollen extremities (> Figure 9-26).

*Clinical Note* D.C. McKenzie and colleagues at the University of British Columbia studied the relationship between breast cancer, the lymphatic system, and exercise. The lymphatic system has not been found to limit exercise performance in a healthy population, but the function of the lymphatic system can be impaired in a large percentage of women (about 35%) who have survived breast cancer. The chief role of the lymphatic system during exercise is to help regulate tissue pressure and volume via the movement of fluid and plasma proteins from the interstitial space back into the cardiovascular system. Breast cancer–related lymphoedema (i.e., lymphatic obstruction) results in a chronic swelling, which typically occurs in the ipsilateral arm of those treated for breast cancer. Exercise was initially thought to facilitate the development of breast cancer–related lymphoedema through damage to the axillary lymphatic from breast cancer treatment, but the current view is that the cause is multifactorial and not simply a lymphatic obstruction.

Whatever the cause of oedema, an important consequence is reduced exchange of materials between the blood and cells. As excess interstitial fluid accumulates, the distance between the blood and cells across which nutrients, oxygen, and wastes must diffuse increases, so the rate of diffusion decreases. Therefore, cells within oedematous tissues may not be adequately supplied.

## 9.5 | Veins

The venous system completes the circulatory circuit. Blood leaving the capillary beds enters the venous system for transport back to the heart.

### Venules

At the microcirculatory level, capillaries drain into *venules*, which progressively converge to form small veins that exit the organ. In contrast to arterioles, venules have little tone and resistance. Extensive communication takes place via chemical signals between venules and nearby arterioles. This venuloarteriolar signalling is vital to matching capillary inflow and outflow within an organ.

### Veins: A blood reservoir

Veins have a large radius, so they offer little resistance to flow. Furthermore, because the total cross-sectional area of the venous system gradually decreases as smaller veins converge into progressively fewer but larger vessels, blood flow speeds up as blood approaches the heart.

In addition to serving as low-resistance passageways to return blood from the tissues to the heart, systemic veins also serve as a *blood reservoir*. Because of their storage capacity, veins are often called **capacitance vessels**. Veins have much thinner walls with less smooth muscle than arteries have. Also, in contrast to arteries, veins have very little elasticity, because venous connective tissue contains considerably more collagen fibres than elastin fibres. Unlike arteriolar smooth muscle, venous smooth muscle has little inherent myogenic tone. Because of these features, veins are highly distensible, or stretchable, and have little elastic recoil. They easily distend to accommodate additional volumes of blood, even with only a small increase in venous pressure. Arteries stretched by an excess volume of blood recoil because of the elastic fibres in their walls, driving the blood forward. Veins containing an extra volume of blood simply stretch to accommodate the additional blood without tending to recoil. In this way, veins serve as a **blood reservoir**; that is, when demands for blood are low, the veins can store extra blood in reserve because of their passive distensibility. Under resting conditions, the veins contain more than 60 percent of the total blood volume (> Figure 9-27).

Let's clarify a possible point of confusion. Contrary to a common misconception, blood stored in the veins is not being held in a stagnant holding tank. Normally all the blood is circulating all the time. When the body is at rest and many of the capillary beds are closed, the capacity of the venous reservoir is increased as extra blood bypasses the capillaries and enters the veins. When this extra volume of blood stretches the veins, the blood moves forward through the veins more slowly because the total cross-sectional area of the veins has been increased as a result of the stretching. Therefore, the blood spends more time in the veins. As a result of this slower transit time through the veins, the veins are essentially storing the extra volume of blood because it is not moving forward as quickly to the heart to be pumped out again.

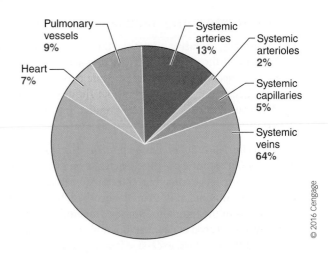

> **FIGURE 9-27 Percentage of total blood volume in different parts of the circulatory system**

© 2016 Cengage

When the stored blood is needed, such as during exercise, extrinsic factors (soon to be described) reduce the capacity of the venous reservoir and drive the extra blood from the veins to the heart so it can be pumped to the tissues. Increased venous return leads to an increased cardiac stroke volume, in accordance with the Frank–Starling law of the heart (p. 368). In contrast, if too much blood pools in the veins instead of being returned to the heart, cardiac output is abnormally diminished. In this way, a delicate balance exists between the capacity of the veins, the extent of venous return, and the cardiac output. We now turn our attention to the factors that affect venous capacity and contribute to venous return.

### Venous return

**Venous capacity**, the volume of blood that the veins can accommodate, depends on both the distensibility of the vein walls and the influence of any externally applied pressure that squeezes inwardly on the veins. At a constant blood volume, as venous capacity increases, more blood remains in the veins instead of being returned to the heart. Such venous storage decreases the effective circulating blood volume—that is, the volume of blood returned to and pumped out of the heart. Conversely, when venous capacity decreases, more blood returns to the heart and is subsequently pumped out. Thus, changes in venous capacity directly influence the magnitude of venous return, which in turn is an important (although not the only) determinant of effective circulating blood volume. The effective circulating blood volume is also influenced on a short-term basis by passive shifts in bulk flow between the vascular and interstitial fluid compartments and on a long-term basis by factors that control total ECF volume, such as salt and water balance.

In Chapter 8, we discussed the role of the heart as a pump and all the factors that influence the pump to regulate cardiac output. However, what was missing from that discussion was the influence of the venous side of the circulatory system on cardiac function. The blood that returns to the heart, or **venous return**, also plays a key role in determining cardiac output. The following discussion outlines the importance of venous return.

The term *venous return* refers to the volume of blood entering each atrium per minute from the veins. Recalling that the circulatory system is a closed loop, it is correct to state that under steady-state conditions, venous return must be equal to cardiac output and that any imbalance of this relationship causes blood to accumulate in either the pulmonary or systemic circulations. An example of an imbalance is what is observed during heart failure (p. 371) or any other disease that increases right atrial pressure.

The venous system is characterized as the low pressure, high capacitance part of the circulation. By the time the blood enters the venous system, blood pressure averages only 17 mmHg (see ⟩ Figure 9-9). However, because atrial pressure is near 0 mmHg, a small but adequate driving pressure still exists to promote the flow of blood through the large-radius, low-resistance veins. The other determining factor of venous return is venous capacitance. Unlike the elastic properties of arteries, increasing the volume of blood in the veins does not necessarily lead to a significant rise in pressure. Instead, increasing the volume of blood stored in the veins leads to a decrease in the amount of blood returned to the right atrium and, thus, the right ventricle. According to the Frank–Starling relationship, this reduction in end-diastolic volume (EDV) leads to a decreased cardiac output (see ⟩ Figure 8-20). The opposite is also true if the ability of the veins to store blood decreases. Decreased venous capacitance causes more blood to be sent back to the heart, which causes an increase in EDV and, therefore, increases cardiac output.

*Clinical Note* In addition to the driving pressure imparted by cardiac contraction, there are five main factors that enhance venous return: (1) sympathetically induced venous vasoconstriction, (2) skeletal muscle activity, (3) effect of venous valves, (4) respiratory activity, and (5) effect of cardiac suction (⟩ Figure 9-28). Most of these secondary factors affect venous return by influencing the pressure gradient between the veins and the heart. We will examine each in turn.

## EFFECT OF SYMPATHETIC ACTIVITY ON VENOUS RETURN

Veins are not very muscular and have little inherent tone, but venous smooth muscle is abundantly supplied with sympathetic nerve fibres. Sympathetic stimulation produces venous vasoconstriction, which modestly elevates venous pressure; this, in turn, increases the pressure gradient to drive more of the stored blood from the veins into the right atrium, thereby enhancing venous return. Veins normally have such a large radius that the moderate vasoconstriction from sympathetic stimulation has little effect on resistance to flow. Even when constricted, veins still have a relatively large radius and are still low-resistance vessels.

In addition to mobilizing the stored blood, venous vasoconstriction enhances venous return by decreasing venous capacity. With the filling capacity of the veins reduced, less of the blood

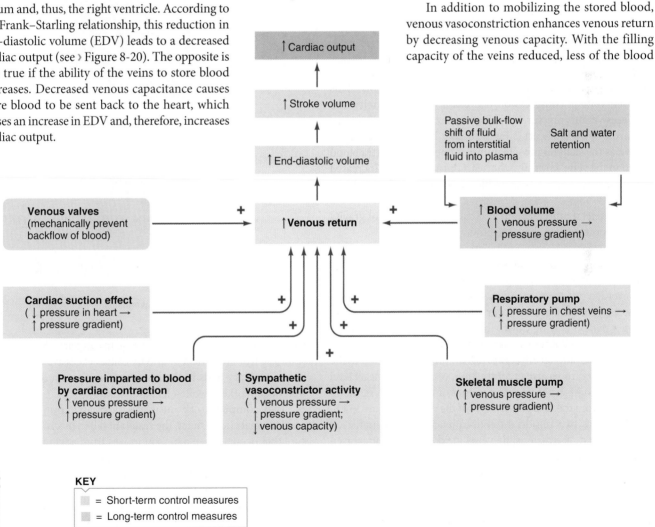

⟩ FIGURE 9-28 Factors that facilitate venous return

draining from the capillaries remains in the veins; instead, it continues to flow toward the heart. The increased venous return initiated by sympathetic stimulation leads to increased cardiac output because end-diastolic volume has increased. Sympathetic stimulation of the heart also increases cardiac output by increasing the heart rate and increasing the heart's contractility (pp. 367 and 369). As long as sympathetic activity remains elevated, as during exercise, the increased cardiac output, in turn, helps sustain the increased venous return initiated in the first place by sympathetically induced venous vasoconstriction. More blood being pumped out by the heart means greater return of blood to the heart, because the reduced-capacity veins do not stretch to store any of the extra blood being pumped into the vascular system.

It is important to recognize the different outcomes of vasoconstriction in arterioles and veins. Arteriolar vasoconstriction immediately *reduces* flow through these vessels because of their increased resistance (less blood can enter and flow through a narrowed arteriole), whereas venous vasoconstriction immediately *increases* flow through these vessels because of their decreased capacity (narrowing of veins squeezes out more of the blood already in the veins, increasing blood flow through these vessels).

### EFFECT OF SKELETAL MUSCLE ACTIVITY ON VENOUS RETURN

Many of the large veins in the extremities lie between skeletal muscles, so muscle contraction compresses the veins. This external venous compression decreases venous capacity and increases venous pressure, in effect squeezing fluid in the veins forward toward the heart (› Figure 9-29). This pumping action, known as the **skeletal muscle pump**, is one way that extra blood stored in the veins returns to the heart during exercise. Increased muscular activity pushes more blood out of the veins and into the heart.

The skeletal muscle pump also counters the effect of gravity on the venous system. Let's see how.

### COUNTERING THE EFFECTS OF GRAVITY ON THE VENOUS SYSTEM

The average pressures mentioned thus far for various regions of the vascular tree are for a person in the horizontal position. When a person is lying down, the force of gravity is uniformly applied, so it need not be considered. When a person stands up, however, gravitational effects are not uniform. In addition to the usual pressure from cardiac contraction, vessels below heart level are subject to pressure from the weight of the column of blood extending from the heart to the level of the vessel (› Figure 9-30).

There are two important consequences of this increased pressure. First, the distensible veins yield under the increased hydrostatic pressure; they further expand and thereby increase their capacity. Even though the arteries are subject to the same gravitational effects, they are not nearly as distensible and do not expand like the veins. Much of the blood entering from the capillaries tends to pool in the expanded lower leg veins instead of returning to the heart. Because venous return is reduced, cardiac output decreases, and the effective circulating volume

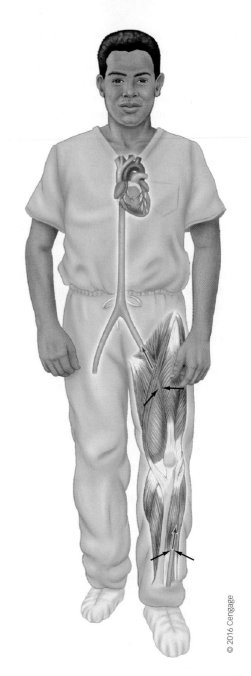

› FIGURE 9-29 **Skeletal muscle pump enhancing venous return**

shrinks. Second, the marked increase in capillary blood pressure resulting from the effect of gravity causes excessive fluid to filter out of capillary beds in the lower extremities, producing localized oedema (i.e., swollen feet and ankles).

Two compensatory measures normally counteract these gravitational effects. First, the resultant fall in mean arterial pressure that occurs when a person moves from a lying-down to an upright position triggers sympathetically induced venous vasoconstriction, which drives some of the pooled blood forward. Second, the skeletal muscle pump "interrupts" the column of blood by completely emptying certain vein segments intermittently so that a particular portion of a vein is not subjected to the weight of the entire venous column from the heart to that portion's level (› Figure 9-29, and › Figure 9-31). Reflex venous

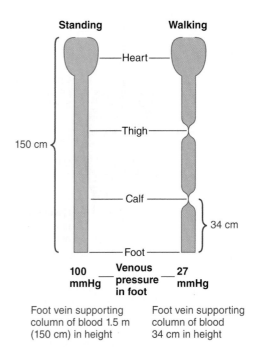

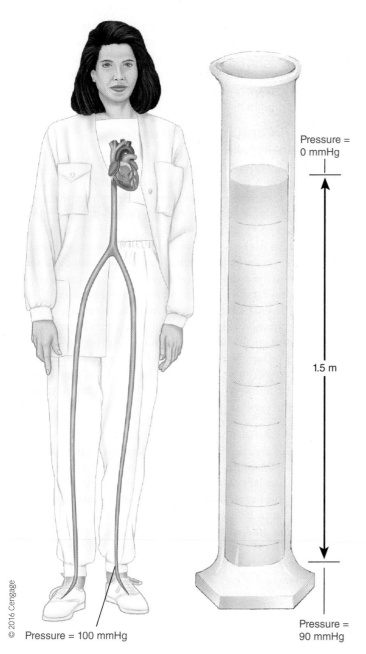

> FIGURE 9-31 **Effect of contraction of the skeletal muscles of the legs in counteracting the effects of gravity.** Contraction of skeletal muscles (as in walking) completely empties given vein segments, interrupting the column of blood that the lower veins must support.

**Source:** Adapted from *Physiology of the Heart and Circulation*, Fourth Edition, by R.C. Little and W.C. Little. Copyright © 1989 Year Book Medical Publishers, Inc., with permission from Elsevier.

reflexes aimed at maintaining mean arterial pressure. Reduced flow of blood to the brain, in turn, leads to fainting, which returns the person to a horizontal position, eliminating the gravitational effects on the vascular system and restoring effective circulation. For this reason, it is counterproductive to try to hold upright someone who has fainted. Fainting is a remedy to the problem, not the problem itself.

Because the skeletal muscle pump facilitates venous return and helps counteract the detrimental effects of gravity on the circulatory system, when you are working at a desk, it's a good idea to get up periodically, and when you are on your feet, to move around. The mild muscular activity "gets the blood moving." It is further recommended that people who must be on their feet for long periods of time use elastic stockings that apply a continuous gentle external compression, similar to the effect of skeletal muscle contraction, to further counter the effect of gravitational pooling of blood in the leg veins.

## EFFECT OF VENOUS VALVES ON VENOUS RETURN

Venous vasoconstriction and external venous compression both drive blood toward the heart. Yet if you squeeze a fluid-filled tube in the middle, fluid is pushed in both directions from the point of constriction (› Figure 9-32a). So why isn't blood driven backward as well as forward by venous vasoconstriction and the skeletal muscle pump? Blood can only be driven forward because the large veins are equipped with one-way valves spaced at 2 to 4 cm intervals; these valves let blood move forward toward the heart but keep it from moving back toward the tissues

90 mmHg caused by gravitational effect
10 mmHg caused by pressure imparted by cardiac contraction

> FIGURE 9-30 **Effect of gravity on venous pressure.** In an upright adult, the blood in the vessels extending between the heart and foot is equivalent to a 1.5 m column of blood. The pressure exerted by this column of blood as a result of the effect of gravity is 90 mmHg. The pressure imparted to the blood by the heart declines to about 10 mmHg in the lower leg veins because of frictional losses in preceding vessels. Together these pressures produce a venous pressure of 100 mmHg in the ankle and foot veins. Similarly, the capillaries in the region are subjected to these same gravitational effects.

vasoconstriction cannot completely compensate for gravitational effects without skeletal muscle activity. When a person stands still for a long time, therefore, blood flow to the brain is reduced because of the decline in effective circulating volume, despite

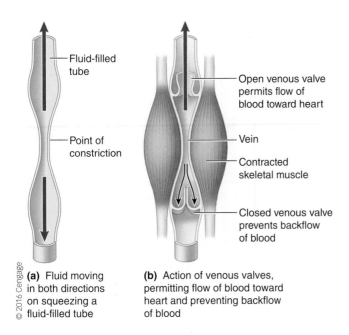

- Fluid-filled tube
- Point of constriction

- Open venous valve permits flow of blood toward heart
- Vein
- Contracted skeletal muscle
- Closed venous valve prevents backflow of blood

**(a)** Fluid moving in both directions on squeezing a fluid-filled tube

**(b)** Action of venous valves, permitting flow of blood toward heart and preventing backflow of blood

› **FIGURE 9-32 Function of venous valves.** (a) When a tube is squeezed in the middle, fluid is pushed in both directions. (b) Venous valves permit the flow of blood only toward the heart.

(› Figure 9-32b). These venous valves also play a role in counteracting the gravitational effects of upright posture by helping minimize the backflow of blood that tends to occur when a person stands up and by temporarily supporting portions of the column of blood when the skeletal muscles are relaxed.

*Clinical Note* **Varicose veins** occur when the venous valves become incompetent and can no longer support the column of blood above them. People predisposed to this condition usually have inherited an overdistensibility and weakness of their vein walls. Aggravated by frequent, prolonged standing, the veins become so distended as blood pools in them that the edges of the valves can no longer meet to form a seal. Varicosed superficial leg veins become visibly overdistended and tortuous. Contrary to what might be expected, chronic pooling of blood in the pathologically distended veins does not reduce cardiac output, because there is a compensatory increase in total circulating blood volume. Instead, the most serious consequence of varicose veins is the possibility of abnormal clot formation in the sluggish, pooled blood. Particularly dangerous is the risk that these clots may break loose and block small vessels elsewhere, especially the pulmonary capillaries.

## EFFECT OF RESPIRATORY ACTIVITY ON VENOUS RETURN

As a result of respiratory activity, the pressure within the chest cavity averages 5 mmHg less than atmospheric pressure. As the venous system returns blood to the heart from the lower regions of the body, it travels through the chest cavity, where it is exposed to this subatmospheric pressure. Because the venous system in the limbs and abdomen is subject to normal atmospheric pressure, an externally applied pressure gradient exists between the lower veins (at atmospheric pressure) and the chest veins (at 5 mmHg less than atmospheric pressure). This pressure difference squeezes blood from the lower veins to the chest veins, promoting increased venous return (› Figure 9-33). This mechanism of facilitating venous return is called the **respiratory pump**, because it results from respiratory activity. Increased respiratory activity and the effects of the skeletal muscle pump and venous vasoconstriction all enhance venous return during exercise.

## EFFECT OF CARDIAC SUCTION ON VENOUS RETURN

The extent of cardiac filling does not depend entirely on factors affecting the veins. The heart plays a role in its own filling. During ventricular contraction, the AV valves are drawn downward, enlarging the atrial cavities. As a result, the atrial pressure transiently drops below 0 mmHg, thus increasing the vein-to-atria pressure gradient so that venous return is enhanced. In addition, the rapid expansion of the ventricular chambers during ventricular relaxation creates a transient negative pressure in the ventricles so that blood is sucked in from the atria and veins; that is, the negative ventricular pressure increases the vein-to-atria-to-ventricle pressure gradient, further enhancing venous return. Thus, the heart functions like a suction pump to facilitate cardiac filling.

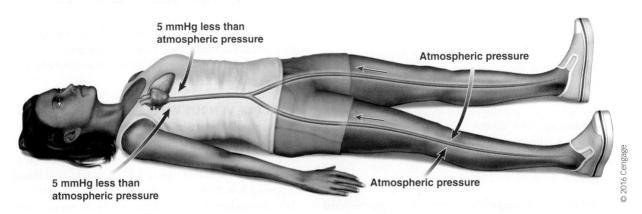

- 5 mmHg less than atmospheric pressure
- Atmospheric pressure
- 5 mmHg less than atmospheric pressure
- Atmospheric pressure

› **FIGURE 9-33 Respiratory pump enhancing venous return.** As a result of respiratory activity, the pressure surrounding the chest veins is lower than the pressure surrounding the veins in the extremities and abdomen. This establishes an externally applied pressure gradient on the veins, which drives blood toward the heart.

## 9.6 | Blood Pressure

The blood pressure that is monitored and regulated in the body is the mean arterial pressure. It is not the arterial systolic or diastolic or pulse pressure, nor the pressure in any other part of the vascular tree. Routine blood pressure measurements record the arterial systolic and diastolic pressures, which can be used as a yardstick for assessing mean arterial pressure. The cut-off for normal blood pressure has recently been designated by the American National Institutes of Health (NIH) as being less than 120/80 mmHg (Appendix G lists reference values), and it is recommended under the 2016 Canadian Hypertension Education Program that patients be managed to maintain blood pressure below 140/90 mmHg (http://www.hypertension.ca).

### Blood pressure regulation

Mean arterial pressure is the main driving force for propelling blood to the tissues. This pressure must be closely regulated for two reasons. First, it must be high enough to ensure sufficient driving pressure; without this pressure, the brain and other organs do not receive adequate flow, no matter what local adjustments are made in the resistance of the arterioles supplying them. Second, the pressure must not be so high that it creates extra work for the heart and increases the risk of vascular damage and possible rupture of small blood vessels.

#### DETERMINANTS OF MEAN ARTERIAL PRESSURE

Elaborate mechanisms involving the integrated action of the various components of the circulatory system and other body systems are vital in regulating this all-important mean arterial pressure (› Figure 9-34). Remember that the two determinants of mean arterial pressure are cardiac output and total peripheral resistance:

$$\text{Mean arterial pressure} = \text{cardiac output} \times \text{total peripheral resistance}$$

(Do not confuse this equation, which indicates the *determinants* of mean arterial pressure, namely, the magnitude of both the cardiac output and total peripheral resistance, with the equation used to *calculate* mean arterial pressure, namely, mean arterial pressure = diastolic pressure + 1/3 pulse pressure.)

Recall that a number of factors determine cardiac output (see › Figure 8-23) and total peripheral resistance (see › Figure 9-14). You can quickly appreciate the complexity of blood pressure regulation. Let's work through › Figure 9-34, reviewing all the factors that affect mean arterial pressure. Even though we covered all these factors earlier, it is useful to pull them all together. The circled numbers in the text correspond to the numbers in the figure.

- Mean arterial pressure depends on cardiac output and total peripheral resistance (**1** in › Figure 9-34).
- Cardiac output depends on heart rate and stroke volume **2**.
- Heart rate depends on the relative balance of parasympathetic activity **3**, which decreases heart rate, and sympathetic activity (tacitly including epinephrine throughout this discussion) **4**, which increases heart rate.

- Stroke volume increases in response to sympathetic activity **5** (extrinsic control of stroke volume).
- Stroke volume also increases as venous return increases **6** (intrinsic control of stroke volume according to the Frank–Starling law of the heart).
- Venous return is enhanced by sympathetically induced venous vasoconstriction **7**, the skeletal muscle pump **8**, the respiratory pump **9**, and cardiac suction **10**.
- The effective circulating blood volume also influences how much blood is returned to the heart **11**. The blood volume depends in the short term on the size of passive bulk-flow fluid shifts between plasma and interstitial fluid across the capillary walls **12**. In the long term, the blood volume depends on salt and water balance **13**, which are hormonally controlled by the renin–angiotensin–aldosterone system and vasopressin, respectively **14**.
- The other major determinant of mean arterial blood pressure—total peripheral resistance—depends on the radius of all arterioles as well as blood viscosity **15**. The main factor determining blood viscosity is the number of red blood cells **16**. However, arteriolar radius is the more important factor determining total peripheral resistance.
- Arteriolar radius is influenced by local (intrinsic) metabolic controls that match blood flow with metabolic needs **17**. For example, local changes that take place in active skeletal muscles cause local arteriolar vasodilation and increased blood flow to these muscle **18**.
- Arteriolar radius is also influenced by sympathetic activity **19**, an extrinsic control mechanism that causes arteriolar vasoconstriction **20** to increase total peripheral resistance and mean arterial blood pressure.
- Arteriolar radius is also extrinsically controlled by the hormones vasopressin and angiotensin II, which are potent vasoconstrictors **21** as well as being important in salt and water balance **22**.

Altering any of the pertinent factors that influence blood pressure will change blood pressure, unless a compensatory change in another variable keeps the blood pressure constant. Blood flow to any given organ depends on the driving force of the mean arterial pressure and on the degree of vasoconstriction of the organ's arterioles. Because mean arterial pressure depends on the cardiac output and the degree of arteriolar vasoconstriction, if the arterioles in one organ dilate, the arterioles in other organs must constrict to maintain an adequate arterial blood pressure. An adequate pressure is needed to provide a driving force to push blood not only to the vasodilated organ but also to the brain, which depends on a constant blood supply. Therefore, the cardiovascular variables must be continuously juggled to maintain a constant blood pressure, despite organs' varying needs for blood.

#### SHORT-TERM AND LONG-TERM CONTROL MEASURES

Mean arterial pressure is constantly monitored by **baroreceptors** (pressure sensors) within the circulatory system. When deviations from normal are detected, multiple reflex responses are initiated to return mean arterial pressure to its normal

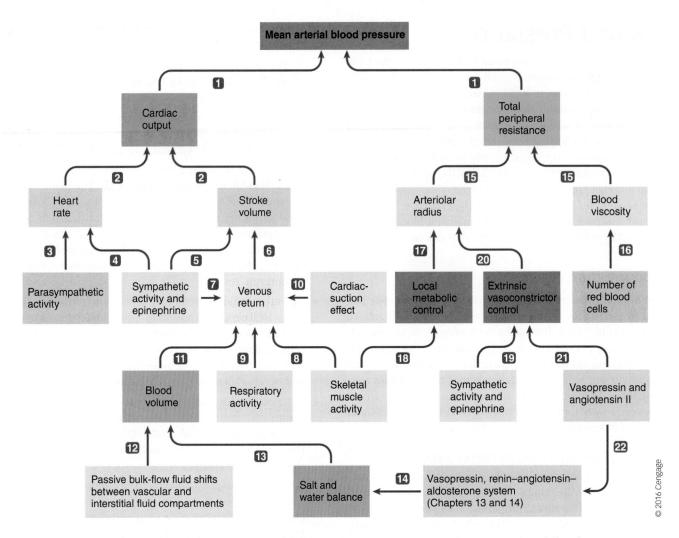

> **FIGURE 9-34 Determinants of mean arterial blood pressure.** Note that this figure is basically a composite of › Figure 8-23, Control of cardiac output; › Figure 9-14, Factors affecting total peripheral resistance; and › Figure 9-28, Factors that facilitate venous return. See the text for a discussion of the circled numbers.

value. *Short-term* (within seconds) adjustments are made by alterations in cardiac output and total peripheral resistance, mediated by means of autonomic nervous system influences on the heart, veins, and arterioles. *Long-term* (requiring minutes to days) control involves adjusting total blood volume by restoring normal salt and water balance through mechanisms that regulate urine output and thirst (Chapters 13 and 14). The size of the total blood volume, in turn, has a profound effect on cardiac output and mean arterial pressure. Let's now examine the short-term mechanisms involved in ongoing regulation of this pressure.

## The baroreceptor reflex

Any change in mean arterial pressure triggers an autonomic system–mediated **baroreceptor reflex**, which influences the heart and blood vessels to adjust cardiac output and total peripheral resistance in an attempt to restore blood pressure to normal. Like any reflex, the baroreceptor reflex includes a receptor, an afferent pathway, an integrating centre, an efferent pathway, and effector organs.

The most important receptors involved in the moment-to-moment regulation of blood pressure, the **carotid sinus** and **aortic arch baroreceptors**, are mechanoreceptors sensitive to changes in both mean arterial pressure and pulse pressure. Their responsiveness to fluctuations in pulse pressure enhances their sensitivity as pressure sensors, because small changes in systolic or diastolic pressure may alter the pulse pressure without changing the mean pressure. These baroreceptors are strategically located (› Figure 9-35) to provide critical information about arterial blood pressure in the vessels leading to the brain (the carotid sinus baroreceptor) and in the major arterial trunk before it branches to supply the rest of the body (the aortic arch baroreceptor).

The baroreceptors constantly provide information about mean arterial pressure; in other words, they continuously generate action potentials in response to the ongoing pressure within the arteries. When arterial pressure (either mean or pulse pressure) increases, the receptor potential of these baroreceptors increases, thus increasing the rate of firing in the corresponding afferent neurons. Conversely, a decrease in the mean arterial pressure slows the rate of firing generated in the afferent neurons by the baroreceptors (› Figure 9-36).

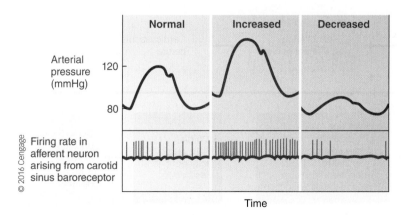

and vasodilator areas. Because these regions are highly inter-connected and functionally interrelated, we refer to them collectively as the *cardiovascular control centre*.

To review how autonomic changes alter arterial blood pressure, study › Figure 9-37, which summarizes the major effects of parasympathetic and sympathetic stimulation on the heart and blood vessels.

To fit all the pieces of the baroreceptor reflex together, let's trace the reflex activity that compensates for an elevation or fall in blood pressure. If for any reason mean arterial pressure rises above normal (› Figure 9-38a), the carotid sinus and aortic arch baroreceptors increase the rate of firing in their respective afferent neurons. On being informed by increased afferent firing that the blood pressure has become too high, the cardiovascular control centre responds by decreasing sympathetic and increasing parasympathetic activity to the cardiovascular system. These efferent signals decrease heart rate, decrease stroke volume, and produce arteriolar and venous vasodilation, which in turn lead to a decrease in cardiac output and a decrease in total peripheral resistance, with a subsequent fall in blood pressure back toward normal.

Conversely, when blood pressure falls below normal (› Figure 9-38b), baroreceptor activity decreases, inducing the cardiovascular centre to increase sympathetic cardiac and vasoconstrictor nerve activity while decreasing its parasympathetic output. This efferent pattern of activity leads to an increase in heart rate and stroke volume, coupled with arteriolar and venous vasoconstriction. These changes increase both cardiac output and total peripheral resistance, raising blood pressure back toward normal.

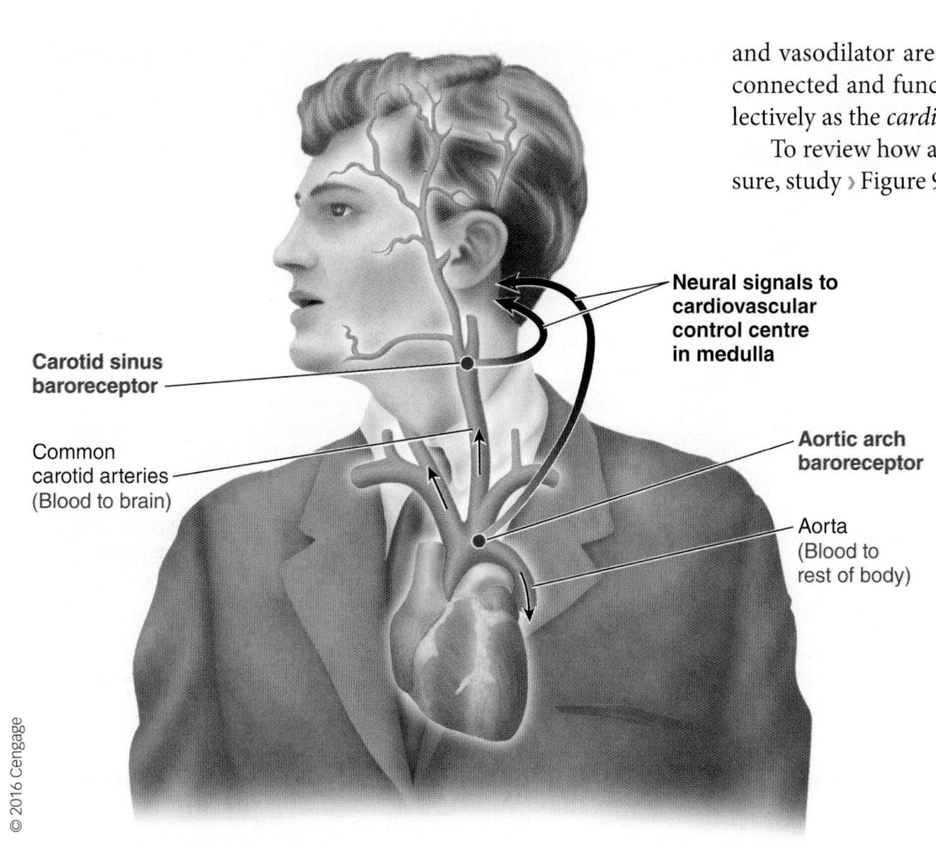

> **FIGURE 9-35** **Location of the arterial baroreceptors.** The arterial baroreceptor are strategically located to monitor the mean arterial blood pressure in the arteries that supply blood to the brain (carotid sinus baroreceptor) and to the rest of the body (aortic arch baroreceptor).

The integrating centre that receives the afferent impulses about the state of mean arterial pressure is the *cardiovascular control centre*, located in the medulla within the brain stem. The efferent pathway is the autonomic nervous system. The cardiovascular control centre alters the ratio between sympathetic and parasympathetic activity to the effector organs (the heart and blood vessels). The cardiovascular control centre is sometimes divided into cardiac and vasomotor centres, which are occasionally further classified into smaller subdivisions, such as cardio-accelleratory and cardioinhibitory centres and vasoconstrictor

### Other reflexes and responses

Besides the baroreceptor reflex, whose sole function is blood pressure regulation, several other reflexes and responses influence the cardiovascular system even though they primarily regulate other body functions. Some of these other influences deliberately move arterial pressure away from its normal value temporarily, overriding the baroreceptor reflex to accomplish a particular goal. These factors include the following:

1. Left atrial volume receptors and hypothalamic osmoreceptors are primarily important in water and salt balance in the body; they affect the long-term regulation of blood pressure by controlling the plasma volume.

2. Chemoreceptors located in the carotid and aortic arteries, in close association with but distinct from the baroreceptors, are sensitive to low oxygen or

> **FIGURE 9-36** **Firing rate in an afferent neuron from the carotid sinus baroreceptor in relation to the magnitude of mean arterial pressure**

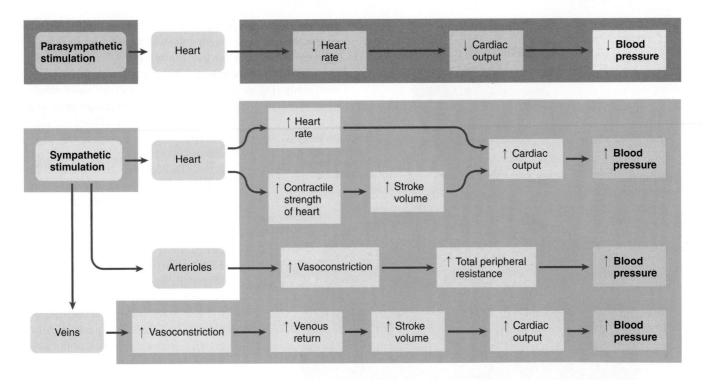

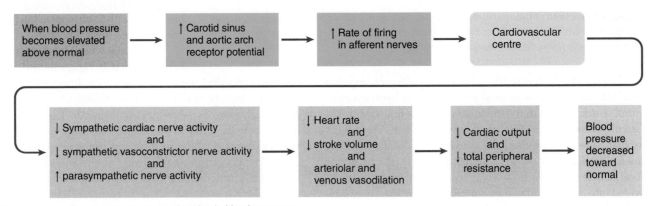

> FIGURE 9-37 Summary of the effects of the parasympathetic and sympathetic nervous systems on factors that influence mean arterial blood pressure

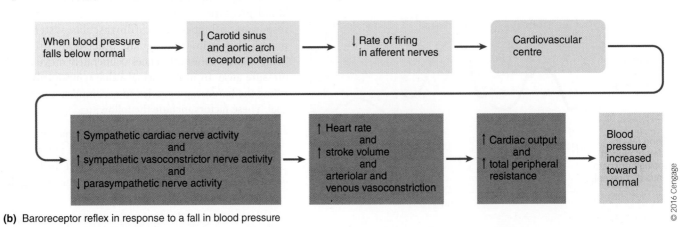

> FIGURE 9-38 Baroreceptor reflexes to restore blood pressure to normal

high acid levels in the blood. These chemoreceptors' main function is to increase respiratory activity to bring in more oxygen or to blow off more acid-forming carbon dioxide, but they also increase blood pressure by sending excitatory impulses to the cardiovascular centre.

3. Cardiovascular responses associated with certain behaviours and emotions are mediated through the cerebral cortex–hypothalamic pathway and appear pre-programmed. These responses include the widespread changes in cardiovascular activity that accompany the generalized sympathetic fight-or-flight response, the characteristic marked increase in heart rate and blood pressure associated with sexual orgasm, and the localized cutaneous vasodilation characteristic of blushing.

4. Pronounced cardiovascular changes accompany exercise, including a substantial increase in skeletal muscle blood flow (see › Figure 9-12), a significant increase in cardiac output, a fall in total peripheral resistance (because of widespread vasodilation in skeletal muscles despite generalized arteriolar vasoconstriction in most organs), and a modest increase in mean arterial pressure (❚ Table 9-5). Evidence suggests that discrete exercise centres yet to be identified in the brain induce the appropriate cardiac

and vascular changes at the onset of exercise or even in anticipation of exercise. These effects are then reinforced by afferent inputs to the medullary cardiovascular centre from chemoreceptors in exercising muscles as well as by local mechanisms important in maintaining vasodilation in active muscles. The baroreceptor reflex further modulates these cardiovascular responses.

5. Hypothalamic control over cutaneous (skin) arterioles for the purpose of temperature regulation takes precedence over control that the cardiovascular centre has over these same vessels for the purpose of blood pressure regulation. As a result, blood pressure can fall when the skin vessels are widely dilated to eliminate excess heat from the body, even though the baroreceptor responses are calling for cutaneous vasoconstriction to help maintain adequate total peripheral resistance.

6. Vasoactive substances released from the endothelial cells play a role in regulating blood pressure. For example, nitric oxide normally exerts an ongoing vasodilatory effect.

Despite these control measures, sometimes blood pressure is not maintained at the appropriate level. We next examine blood pressure abnormalities.

## ❚ TABLE 9-5 Cardiovascular Changes during Exercise

| Cardiovascular Variable | Change | Comment |
|---|---|---|
| **Heart Rate** | Increases | Occurs as a result of an increased sympathetic and a decreased parasympathetic activity to the SA node |
| **Venous Return** | Increases | Occurs as a result of sympathetically induced venous vasoconstriction and increased activity of the skeletal muscle pump and respiratory pump |
| **Stroke Volume** | Increases | Occurs both as a result of increased venous return by means of the Frank–Starling mechanism (unless diastolic filling time is significantly reduced by a high heart rate) and as a result of a sympathetically induced increase in myocardial contractility |
| **Cardiac Output** | Increases | Occurs as a result of increases in both heart rate and stroke volume |
| **Blood Flow to Active Skeletal Muscles and Heart Muscle** | Increases | Occurs as a result of locally controlled arteriolar vasodilation, which is reinforced by the vasodilatory effects of epinephrine and overpowers the weaker sympathetic vasoconstrictor effect |
| **Blood Flow to the Brain** | Unchanged | Occurs because sympathetic stimulation has no effect on brain arterioles; local control mechanisms maintain constant cerebral blood flow, whatever the circumstances |
| **Blood Flow to the Skin** | Increases | Occurs because the hypothalamic temperature control centre induces vasodilation of skin arterioles; increased skin blood flow brings heat produced by exercising muscles to the body surface where the heat can be lost to the external environment |
| **Blood Flow to the Digestive System, Kidneys, and Other Organs** | Decreases | Occurs as a result of generalized sympathetically induced arteriolar vasoconstriction |
| **Total Peripheral Resistance** | Decreases | Occurs because resistance in the skeletal muscles, heart, and skin decreases to a greater extent than resistance in the other organs increases |
| **Mean Arterial Blood Pressure** | Increases (modest) | Occurs because cardiac output increases to a greater extent than total peripheral resistance decreases |

9

# Hypertension

Clinical Note — Sometimes blood pressure control mechanisms do not function properly or are unable to completely compensate for changes that have taken place. Blood pressure may be too high (**hypertension** if above 140/90 mmHg) or too low (**hypotension** if below 100/60 mmHg). Hypotension in its extreme form is called *circulatory shock*. We will first examine hypertension, which is by far the most common of blood pressure abnormalities, and then conclude this chapter with a discussion of hypotension and shock.

There are two broad classes of hypertension: secondary hypertension and primary hypertension, depending on the cause.

## SECONDARY HYPERTENSION

A definite cause for hypertension can be established in only 10 percent of cases: specifically, those called **secondary hypertension**, which occurs secondary to another known primary problem. Here are some examples of secondary hypertension:

1. *Renal hypertension*. For example, atherosclerotic lesions protruding into the lumen of a renal artery or external compression of the vessel by a tumour may reduce blood flow through the kidney. The kidney responds by initiating the hormonal pathway involving angiotensin II. This pathway promotes salt and water retention during urine formation, thereby increasing the blood volume to compensate for the reduced renal blood flow. Recall that angiotensin II is also a powerful vasoconstrictor. Although these two effects (increased blood volume and angiotensin-induced vasoconstriction) are compensatory mechanisms to improve blood flow through the narrowed renal artery, they also are responsible for elevating the arterial pressure as a whole.

2. *Endocrine hypertension*. For example, a *pheochromocytoma* is an adrenal medullary tumour that secretes excessive epinephrine and norepinephrine. Abnormally elevated levels of these hormones bring about a high cardiac output and generalized peripheral vasoconstriction, both of which contribute to the hypertension characteristic of this disorder.

3. *Neurogenic hypertension*. An example is the hypertension caused by erroneous blood pressure control arising from a defect in the cardiovascular control centre.

## PRIMARY HYPERTENSION

In the remaining 90 percent of hypertension cases the underlying cause is unknown. Such hypertension is known as **primary (essential or idiopathic) hypertension**. Primary hypertension is a catchall category for blood pressure elevated by a variety of unknown causes rather than by a single disease entity. People show a strong genetic tendency to develop primary hypertension, which can be hastened or worsened by contributing factors, such as obesity, stress, smoking, or dietary habits. Consider the following range of potential causes for primary hypertension currently being investigated.

- *Defects in salt management by the kidneys.* Disturbances in kidney function too minor to produce outward signs of renal disease could nevertheless insidiously lead to gradual accumulation of salt and water in the body, resulting in progressive elevation of arterial pressure.

- *Excessive salt intake.* Because salt osmotically retains water, thus expanding the plasma volume and contributing to the long-term control of blood pressure, excessive ingestion of salt could theoretically contribute to hypertension. Yet controversy continues over whether or not restricting salt intake should be recommended as a means of preventing and treating high blood pressure. The research data to date have been inconclusive and subject to conflicting interpretations.

- *Diets low in fruits, vegetables, and dairy products (i.e., low in $K^+$ and $Ca^{2+}$).* Dietary factors other than salt have been shown to markedly affect blood pressure. The DASH (Dietary Approaches to Stop Hypertension) studies found that a low-fat diet rich in fruits, vegetables, and dairy products could lower blood pressure in people with mild hypertension as much as any single drug treatment. Research indicates that the high potassium intake associated with eating abundant fruits and vegetables may lower blood pressure by relaxing arteries. Furthermore, inadequate calcium intake from dairy products has been identified as the most prevalent dietary pattern among individuals with untreated hypertension, although the role of calcium in regulating blood pressure is unclear.

- *Plasma membrane abnormalities, such as defective $Na^+-K^+$ pumps.* Such defects, by altering the electrochemical gradient across plasma membranes, could change the excitability and contractility of the heart and the smooth muscle in blood vessel walls in such a way as to lead to high blood pressure. In addition, the $Na^+-K^+$ pump is crucial to salt management by the kidneys. A genetic defect in the $Na^+-K^+$ pump of hypertensive-prone laboratory rats was the first gene–hypertension link to be discovered.

- *Variation in the gene that encodes for angiotensinogen.* Angiotensinogen is part of the hormonal pathway that produces the potent vasoconstrictor angiotensin II and promotes salt and water retention. One variant of the gene in humans appears to be associated with a higher incidence of hypertension. Researchers speculate that the suspect version of the gene leads to a slight excess production of angiotensinogen, which increases activity of this blood pressure–raising pathway. This is the first gene–hypertension link discovered in humans.

- *Endogenous digitalis-like substances.* Such substances act in much the same way as the drug digitalis (p. 357) to increase cardiac contractility and also to constrict blood vessels and reduce salt elimination in the urine, all of which could cause chronic hypertension.

- *Abnormalities in nitric oxide, endothelin, or other locally acting vasoactive chemicals.* For example, a shortage of nitric oxide has been discovered in the blood vessel walls of some hypertensive patients, leading to an impaired ability to accomplish blood pressure–lowering vasodilation. Furthermore, an underlying abnormality in the gene that codes for endothelin, a locally acting vasoconstrictor, has been strongly implicated as a possible cause of hypertension, especially among those of African descent.

- *Excess vasopressin.* Recent experimental evidence suggests that hypertension may result from a malfunction of the vasopressin-secreting cells of the hypothalamus. Vasopressin is a potent vasoconstrictor and also promotes water retention.

Whatever the underlying defect, once initiated, hypertension appears to be self-perpetuating. Constant exposure to elevated blood pressure predisposes vessel walls to the development of atherosclerosis, which further raises blood pressure. (But acute, transient rises in blood pressure can be found in some exercise conditions.)

### ADAPTATION OF BARORECEPTORS DURING HYPERTENSION

The baroreceptors do not respond to bring the blood pressure back to normal during hypertension because they adapt, or become reset, to operate at a higher level. In the presence of chronically elevated blood pressure, the baroreceptors still function to regulate blood pressure, but they maintain it at a higher mean pressure.

### COMPLICATIONS OF HYPERTENSION

Hypertension imposes stresses on both the heart and the blood vessels. The heart has an increased workload because it has to pump against an increased total peripheral resistance, whereas blood vessels may be damaged by the high internal pressure, particularly when the vessel wall is weakened by the degenerative process of atherosclerosis. Complications of hypertension include congestive heart failure caused by the heart's inability to pump continuously against a sustained elevation in arterial pressure, strokes caused by rupture of brain vessels, and heart attacks caused by rupture of coronary vessels. Spontaneous haemorrhage due to the bursting of small vessels elsewhere in the body may also occur, but with less serious consequences; an example is the rupture of blood vessels in the nose, resulting in nosebleeds. Another serious complication of hypertension is renal failure caused by progressive impairment of blood flow through damaged renal blood vessels. Furthermore, retinal damage from changes in the blood vessels supplying the eyes may result in progressive loss of vision.

Until complications occur, hypertension is symptomless, because the tissues are adequately supplied with blood. Therefore, unless blood pressure measurements are made on a routine basis, the condition can go undetected until a precipitous complicating event. When you become aware of these potential complications of hypertension and consider that 22 percent of all adults in Canada are estimated to have chronically elevated blood pressure, you can appreciate the magnitude of this national health problem.

### TREATMENT OF HYPERTENSION

Once hypertension is detected, therapeutic intervention can reduce the course and severity of the problem. Dietary management, including weight loss, along with a variety of drugs that manipulate salt and water management or autonomic activity on the cardiovascular system can be used to treat hypertension. No matter what the original cause, agents that reduce plasma volume or total peripheral resistance (or both) will decrease blood pressure toward normal. Furthermore, a regular aerobic exercise program can be employed to help reduce high blood pressure.

## Prehypertension

In its recent guidelines, the U.S. National Institute of Health identified **prehypertension** as a new category for blood pressures in the range between normal and hypertension (between 120/80 and 139/89). Blood pressures in the prehypertension range can usually be reduced by appropriate dietary and exercise measures, whereas those in the hypertension range typically must be treated with blood pressure medication in addition to changing health habits. The goal in managing blood pressures in the prehypertension range is to take action before the pressure climbs into the hypertension range, where serious complications may develop.

Now we look at the other extreme, hypotension: first, transient orthostatic hypotension, and then, the (more serious) circulatory shock.

## Orthostatic hypotension

Hypotension, or low blood pressure, occurs either when there is a disproportion between vascular capacity and blood volume (in essence, too little blood to fill the vessels) or when the heart is too weak to drive the blood.

The most common situation in which hypotension occurs transiently is orthostatic hypotension. **Orthostatic (postural) hypotension** is a transient hypotensive condition resulting from insufficient compensatory responses to the gravitational shifts in blood when a person moves from a horizontal to a vertical position, especially after prolonged bed rest. When a person moves from lying down to standing up, pooling of blood in the leg veins from gravity reduces venous return, decreasing stroke volume, and thus lowering cardiac output and blood pressure. This fall in blood pressure is normally detected by the baroreceptors, which initiate immediate compensatory responses to restore blood pressure to its proper level. When a long-bedridden patient first starts to rise, however, these reflex compensatory adjustments are temporarily lost or reduced because of disuse. Sympathetic control of the leg veins is inadequate, so when the patient first stands up, blood pools in the lower extremities. The condition is further aggravated by the decrease in blood volume that typically accompanies prolonged bed rest. The resultant orthostatic hypotension and decrease in blood flow to the brain cause dizziness or actual fainting. Because postural compensatory mechanisms are depressed during prolonged bed confinement, patients sometimes are put on a tilt table so that they can be moved gradually from a horizontal to an upright position. This allows the body to adjust slowly to the gravitational shifts in blood.

9

## Circulatory shock

When blood pressure falls so low that adequate blood flow to the tissues can no longer be maintained, the condition known as **circulatory shock** occurs. Circulatory shock is categorized into four main types (› Figure 9-39):

1. *Hypovolemic (low-volume) shock* is caused by a fall in blood volume, which occurs either directly through severe haemorrhage or indirectly through loss of fluids derived from the plasma (e.g., severe diarrhoea, excessive urinary losses, or extensive sweating).

2. *Cardiogenic (heart-produced) shock* is due to a weakened heart's failure to pump blood adequately.

3. *Vasogenic (vessel-produced) shock* is caused by widespread vasodilation triggered by the presence of vasodilator substances. There are two types of vasogenic shock: septic and anaphylactic. *Septic shock*, which may accompany massive infections, is due to vasodilator substances released from the infective agents. Similarly, extensive histamine release in severe allergic reactions can cause widespread vasodilation in *anaphylactic shock*.

4. *Neurogenic (nerve-produced) shock* also involves generalized vasodilation, but not by means of the release of vasodilator substances. In this case, loss of sympathetic vascular tone leads to generalized vasodilation. This undoubtedly is responsible for the shock accompanying crushing injuries when blood loss has not been sufficient to cause hypovolemic shock. Deep, excruciating pain apparently inhibits sympathetic vasoconstrictor activity.

Using haemorrhage as an example (› Figure 9-40), we now examine the consequences of and compensations for shock. This figure may look intimidating, but we will work through it step by step. It is an important example that pulls together many of the principles discussed in this chapter. As before, the squared numbers in the text correspond to the numbers in the figure.

### CONSEQUENCES AND COMPENSATIONS OF SHOCK

- Following severe loss of blood, the resultant reduction in circulating blood volume leads to a decrease in venous return **1** and a subsequent fall in cardiac output and arterial blood pressure. (Note the blue boxes, which indicate consequences of haemorrhage.)

- Compensatory measures immediately attempt to maintain adequate blood flow to the brain. (Note the pink boxes, which indicate compensations for haemorrhage.)

- The baroreceptor reflex response to the fall in blood pressure brings about increased sympathetic and decreased

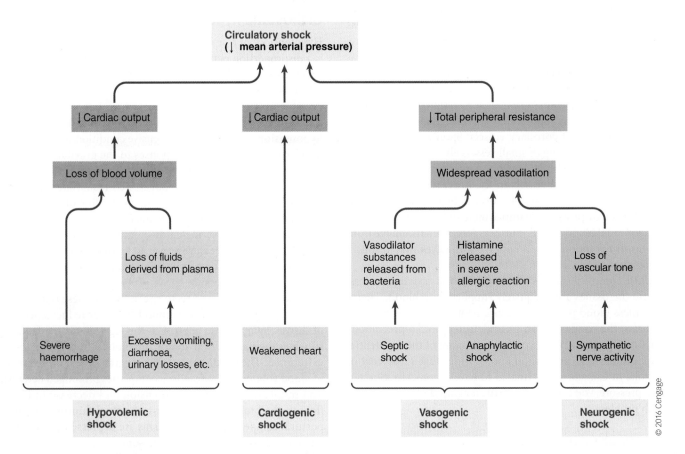

› **FIGURE 9-39 Causes of circulatory shock.** Circulatory shock, which occurs when mean arterial blood pressure falls so low that adequate blood flow to the tissues can no longer be maintained, may result from (1) extensive loss of blood volume (hypovolemic shock), (2) failure of the heart to pump blood adequately (cardiogenic shock), (3) widespread arteriolar vasodilation (vasogenic shock), or (4) neurally defective vasoconstrictor tone (neurogenic shock).

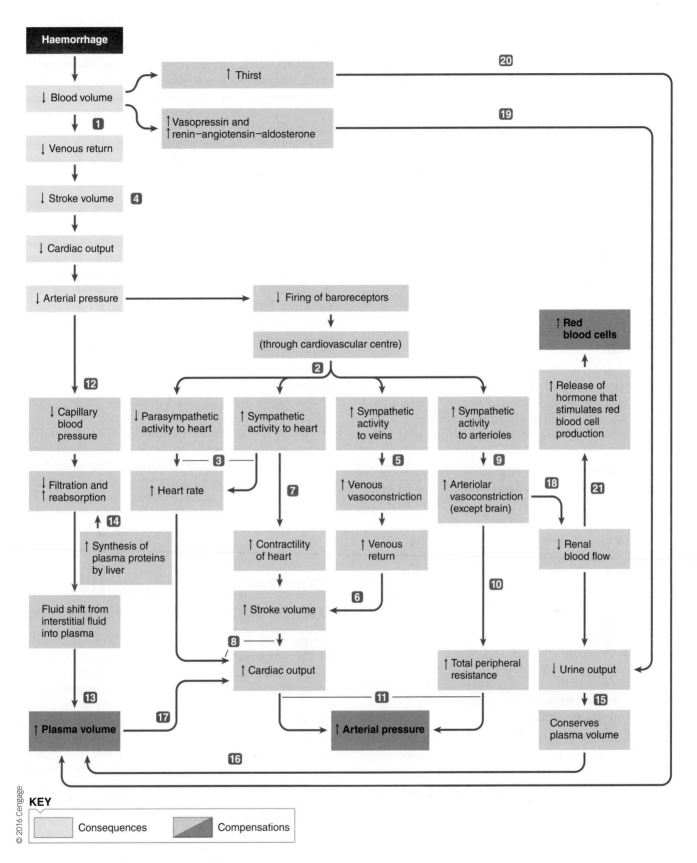

> FIGURE 9-40 **Consequences and compensations of haemorrhage.** The reduction in blood volume resulting from haemorrhage leads to a fall in arterial pressure. (Note the *blue boxes*, representing consequences of haemorrhage.) A series of compensations ensue (*light pink boxes*) that ultimately restore plasma volume, arterial pressure, and the number of red blood cells toward normal (*dark pink boxes*). Refer to the text for an explanation of the circled numbers and a detailed discussion of the compensations.

parasympathetic activity to the heart **2**. The result is an increase in heart rate **3** to offset the reduced stroke volume **4** brought about by the loss of blood volume. With severe fluid loss, the pulse is weak because of the reduced stroke volume but rapid because of the increased heart rate.

- Increased sympathetic activity to the veins produces generalized venous vasoconstriction **5**, increasing venous return by means of the Frank–Starling mechanism **6**.

- Simultaneously, sympathetic stimulation of the heart increases the heart's contractility **7** so that it beats more forcefully and ejects a greater volume of blood, likewise increasing the stroke volume.

- The increase in heart rate and in stroke volume collectively increase cardiac output **8**.

- Sympathetically induced generalized arteriolar vasoconstriction **9** leads to an increase in total peripheral resistance **10**.

- Together, the increase in cardiac output and total peripheral resistance bring about a compensatory increase in arterial pressure **11**.

- The original fall in arterial pressure is also accompanied by a fall in capillary blood pressure **12**, which results in fluid shifts from the interstitial fluid into the capillaries to expand the plasma volume **13**. This response is sometimes termed *autotransfusion*, because it restores the plasma volume as a transfusion does.

- This ECF shift is enhanced by plasma protein synthesis by the liver during the next few days following haemorrhage **14**. The plasma proteins exert a colloid osmotic pressure that helps retain extra fluid in the plasma.

- Urinary output is reduced, thereby conserving water that normally would have been lost from the body **15**. This additional fluid retention helps expand the reduced plasma volume **16**. Expansion of plasma volume further augments the increase in cardiac output brought about by the baroreceptor reflex **17**. Reduction in urinary output results from decreased renal blood flow caused by compensatory renal arteriolar vasoconstriction **18**. The reduced plasma volume also triggers increased secretion of the hormone vasopressin and activation of the salt- and water-conserving renin–angiotensin–aldosterone hormonal pathway, which further reduces urinary output **19**.

- Increased thirst is also stimulated by a fall in plasma volume **20**. The resultant increased fluid intake helps restore plasma volume.

- Over a longer course of time (a week or more), lost red blood cells are replaced through increased red blood cell production, which is triggered by a reduction in oxygen delivery to the kidneys **21**.

### IRREVERSIBLE SHOCK

These compensatory mechanisms are often not enough to counteract substantial fluid loss. Even if they can maintain an adequate blood pressure level, the short-term measures cannot continue indefinitely. Ultimately, fluid volume must be replaced from the outside through drinking, transfusion, or a combination of both. Blood supply to the kidneys, digestive tract, skin, and other organs can be compromised to maintain blood flow to the brain only so long before organ damage begins to occur. A point may be reached at which blood pressure continues to drop rapidly because of tissue damage, despite vigorous therapy. This condition is frequently termed *irreversible shock*, in contrast to *reversible shock*, which can be corrected by compensatory mechanisms and effective therapy.

*Clinical Note* Although the exact mechanism underlying irreversibility is not currently known, many logical possibilities could contribute to the unrelenting, progressive circulatory deterioration that characterizes irreversible shock. Metabolic acidosis arises when lactic acid production increases as blood-deprived tissues resort to anaerobic metabolism. Acidosis deranges the enzymatic systems responsible for energy production, and this limits the capability of the heart and other tissues to produce ATP. Prolonged depression of kidney function results in electrolyte imbalances that may lead to cardiac arrhythmias. The blood-deprived pancreas releases a chemical that is toxic to the heart (**myocardial toxic factor**), further weakening the heart. Vasodilator substances build up within ischemic organs, inducing local vasodilation that overrides the generalized reflex vasoconstriction. As cardiac output progressively declines because of the heart's diminishing effectiveness as a pump and total peripheral resistance continues to fall, hypotension becomes increasingly more severe. This causes further cardiovascular failure, which leads to a further decline in blood pressure. Consequently, when shock progresses to the point that the cardiovascular system itself starts to fail, a vicious positive-feedback cycle ensues that ultimately results in death.

# Chapter in Perspective: Focus on Homeostasis

Homeostatically, the blood vessels serve as passageways to transport blood to and from the cells for oxygen and nutrient delivery, waste removal, distribution of fluid and electrolytes, elimination of excess heat, and hormonal signalling, among other things. Cells soon die if deprived of their blood supply; brain cells succumb within four minutes. Blood is constantly recycled and reconditioned as it travels through the various organs via the vascular highways. Hence, the body needs only a very small volume of blood to maintain the appropriate chemical composition of the entire internal fluid environment on which the cells depend for their survival. For example, oxygen is continually picked up by blood in the lungs and constantly delivered to all the body cells.

The smallest blood vessels, the capillaries, are the actual site of exchange between the blood and surrounding cells. Capillaries bring homeostatically maintained blood within 0.01 cm of every cell in the body; this proximity is critical, because beyond a few centimetres materials cannot diffuse rapidly enough to support life-sustaining activities. Oxygen that would take months to years to diffuse from the lungs to all the cells of the body is continuously delivered at the "doorstep" of every cell, where diffusion can efficiently accomplish short local exchanges between the capillaries and surrounding cells. Likewise, hormones must be rapidly transported through the circulatory system from their sites of production in endocrine glands to their sites of action in other parts of the body. These chemical messengers could not diffuse nearly rapidly enough to their target organs to effectively exert their controlling effects, many of which are aimed toward maintaining homeostasis.

The rest of the circulatory system is designed to transport blood to and from the capillaries. The arteries and arterioles distribute blood pumped by the heart to the capillaries for life-sustaining exchanges to take place, and the venules and veins collect blood from the capillaries and return it to the heart, where the process is repeated.

## CHAPTER TERMINOLOGY

active hyperaemia (p. 398)
angiogenesis (p. 399)
aortic arch baroreceptors (p. 420)
arterioles (p. 394)
autoregulation (p. 400)
baroreceptor reflex (p. 420)
baroreceptors (p. 419)
blood pressure (p. 391)
blood reservoir (p. 414)
bulk flow (p. 409)
capacitance vessels (p. 414)
capillaries (p. 402)
capillary blood pressure ($P_C$) (p. 409)
cardiovascular control centre (p. 402)
carotid sinus (p. 420)
circulatory shock (p. 426)
circulatory system (p. 388)
compliance (p. 391)
diastolic pressure (p. 391)
distensibility (p. 391)
endothelial cells (p. 398)
endothelin (p. 398)
external elastic lamina (p. 389)
flow rate (p. 386)
hypertension (p. 424)

hypotension (p. 424)
initial lymphatics (p. 411)
internal elastic lamina (p. 389)
interstitial fluid hydrostatic pressure ($P_{IF}$) (p. 409)
interstitial fluid–colloid osmotic pressure ($\pi_{IF}$) (p. 409)
Korotkoff sounds (p. 392)
lymph (p. 412)
lymphatic system (p. 411)
lymph nodes (p. 412)
lymph vessels (p. 411)
mean arterial pressure (p. 394)
metarteriole (p. 406)
microcirculation (p. 388)
myocardial toxic factor (p. 428)
nitric oxide (NO) (p. 398)
oedema (p. 413)
orthostatic (postural) hypotension (p. 425)
plasma-colloid osmotic pressure ($\pi_p$) (p. 409)
poiseuille's law (p. 388)
pores (p. 405)
precapillary sphincters (p. 406)
prehypertension (p. 425)
pressure gradient (p. 387)

pressure reservoir (p. 389)
primary (essential or idiopathic) hypertension (p. 424)
pulse pressure (p. 393)
reabsorption (p. 409)
reactive hyperaemia (p. 400)
resistance (p. 387)
respiratory pump (p. 418)
secondary hypertension (p. 424)
shear stress (p. 399)
skeletal muscle pump (p. 416)
sphygmomanometer (p. 392)
systolic pressure (p. 391)
total peripheral resistance (p. 401)
ultrafiltration (p. 409)
varicose veins (p. 418)
vascular tone (p. 395)
vasoconstriction (p. 395)
vasodilation (p. 395)
veins (p. 388)
venous capacity (p. 414)
venous return (p. 414)
venules (p. 388)
viscosity (p. 387)

## REVIEW EXERCISES

### Objective Questions (Answers in Appendix E, p. A-44)

1. In general, the parallel arrangement of the vascular system enables each organ to receive its own separate arterial blood supply. (*True or false?*)

2. More blood flows through the capillaries during cardiac systole than during diastole. (*True or false?*)

3. The capillaries contain only 5 percent of the total blood volume at any point in time. (*True or false?*)

4. The same volume of blood passes through the capillaries in a minute as passes through the aorta, even though blood flow is much slower in the capillaries. (*True or false?*)

5. Because capillary walls have no carrier transport systems, all capillaries are equally permeable. (*True or false?*)

6. Because of gravitational effects, venous pressure in the lower extremities is greater when a person is standing up than when the person is lying down. (*True or false?*)

7. Which change would increase the resistance in an arteriole?
   a. a shorter length
   b. a wider calibre
   c. increased sympathetic stimulation
   d. decreased blood viscosity

8. Which of the following functions is (are) attributable to arterioles? (*Indicate all correct answers.*)
   a. produce a significant decline in mean pressure, which helps establish the driving pressure gradient between the heart and organs
   b. serve as site of exchange of materials between blood and surrounding tissue cells

9

c. act as main determinant of total peripheral resistance
d. determine the pattern of distribution of cardiac output
e. help regulate mean arterial blood pressure
f. convert the pulsatile nature of arterial blood pressure into a smooth, nonfluctuating pressure in the vessels farther downstream
g. act as a pressure reservoir

9. Using the answer code on the right, indicate whether the following factors increase or decrease venous return:

_____ 1. sympathetically induced venous vasoconstriction

_____ 2. skeletal muscle activity

_____ 3. gravitational effects on the venous system

_____ 4. respiratory activity

_____ 5. increased atrial pressure associated with a leaky AV valve

_____ 6. ventricular pressure change associated with diastolic recoil

(a) increases venous return

(b) decreases venous return

(c) has no effect on venous return

10. Using the answer code on the right, indicate what kind of compensatory changes occur in the factors in question to restore blood pressure to normal in response to hypovolemic hypotension resulting from severe haemorrhage:

_____ 1. rate of afferent firing generated by the carotid sinus and aortic arch baroreceptors

_____ 2. sympathetic output by the cardiovascular centre

_____ 3. parasympathetic output by the cardiovascular centre

_____ 4. heart rate

_____ 5. stroke volume

_____ 6. cardiac output

_____ 7. arteriolar radius

_____ 8. total peripheral resistance

_____ 9. venous radius

_____ 10. venous return

_____ 11. urinary output

_____ 12. fluid retention within the body

_____ 13. fluid movement from interstitial fluid into plasma across the capillaries

(a) increased

(b) decreased

(c) no effect

## Written Questions

1. Compare blood flow through reconditioning organs and through organs that do not recondition the blood.

2. Discuss the relationships among flow rate, pressure gradient, and vascular resistance. What is the major determinant of resistance to flow?

3. Describe the structure and major functions of each segment of the vascular tree.

4. How do the arteries serve as a pressure reservoir?

5. Describe the indirect technique of measuring arterial blood pressure by means of a sphygmomanometer.

6. Define *vasoconstriction* and *vasodilation*.

7. Discuss the local and extrinsic controls that regulate arteriolar resistance.

8. What is the primary means by which individual solutes are exchanged across capillary walls? What forces produce bulk flow across capillary walls? Of what importance is bulk flow?

9. How is lymph formed? What are the functions of the lymphatic system?

10. Define *oedema*, and discuss its possible causes.

11. How do veins serve as a blood reservoir?

12. Compare the effect of vasoconstriction on the rate of blood flow in arterioles and veins.

13. Discuss the factors that determine mean arterial pressure.

14. Review the effects on the cardiovascular system of parasympathetic and sympathetic stimulation.

15. Differentiate between secondary hypertension and primary hypertension. What are the potential consequences of hypertension?

16. Define *circulatory shock*. What are its consequences and compensations? What is irreversible shock?

## Quantitative Exercises
## (Solutions in Appendix E, p. A-44)

1. Recall that the flow rate of blood equals the pressure gradient divided by the total peripheral resistance of the vascular system. The conventional unit of resistance in physiological systems is expressed in PRU (peripheral resistance unit), which is defined as (1 L/min)/(1 mmHg). At rest, Tom's total peripheral resistance is about 20 PRU. Last week while playing racquetball, his cardiac output increased to 30 L/min and his mean arterial pressure increased to 120 mmHg. What was his total peripheral resistance during the game?

2. Systolic pressure rises as a person ages. By age 85, an average male has a systolic pressure of 180 mmHg and a diastolic pressure of 90 mmHg.

   a. What is the mean arterial pressure of this average 85-year-old male?

   b. From your knowledge of capillary dynamics, predict the result at the capillary level of this age-related change in mean arterial pressure if no homeostatic mechanisms were operating. (Recall that mean arterial pressure is about 93 mmHg at age 20.)

9

3. Compare the flow rates in the systemic and the pulmonary circulations of an individual with the following measurements:

systemic mean arterial pressure = 95 mmHg
systemic resistance = 19 PRU
pulmonary mean arterial pressure = 20 mmHg
pulmonary resistance = 4 PRU

4. Assume a person has a blood pressure recording of 125/77:
   a. What is the systolic pressure?
   b. What is the diastolic pressure?
   c. What is the pulse pressure?
   d. What is the mean arterial pressure?
   e. Would any sound be heard when the pressure in an external cuff around the arm was 130 mmHg? (*Yes or no?*)
   f. Would any sound be heard when cuff pressure was 118 mmHg?
   g. Would any sound be heard when cuff pressure was 75 mmHg?

## POINTS TO PONDER

(Explanations in Appendix E, p. A-45)

1. During coronary bypass surgery, a piece of vein is often removed from the patient's leg and surgically attached within the coronary circulatory system so that blood detours, through the vein, around an occluded coronary artery segment. Why must the patient wear, for an extended period of time after surgery, an elastic support stocking on the limb from which the vein was removed?

2. A classmate who has been standing still for several hours working on a laboratory experiment suddenly faints. What is the probable explanation? What would you do if the person next to him tried to get him up?

3. A drug applied to a piece of excised arteriole causes the vessel to relax, but an isolated piece of arteriolar muscle stripped from the other layers of the vessel fails to respond to the same drug. What is the probable explanation?

4. Explain how each of the following antihypertensive drugs would lower arterial blood pressure:
   a. drugs that block $\alpha_1$-adrenergic receptors (e.g., *phentolamine*)
   b. drugs that block $\beta_1$-adrenergic receptors (e.g., *metoprolol*) (*Hint*: Review adrenergic receptors on p. 193.)
   c. drugs that directly relax arteriolar smooth muscle (e.g., *hydralazine*)
   d. diuretic drugs that increase urinary output (e.g., *furosemide*)
   e. drugs that block release of norepinephrine from sympathetic endings (e.g., *guanethidine*)
   f. drugs that act on the brain to reduce sympathetic output (e.g., *clonidine*)
   g. drugs that block $Ca^{2+}$ channels (e.g., *verapamil*)
   h. drugs that interfere with the production of angiotensin II (e.g., *captopril*)

9

## CLINICAL CONSIDERATION

(Explanation in Appendix E, p. A-45)

Li-Ying C. has just been diagnosed as having hypertension secondary to a *pheochromocytoma*, a tumour of the adrenal medulla that secretes excessive epinephrine. Explain how this condition leads to secondary hypertension, by describing the effect that excessive epinephrine would have on various factors that determine arterial blood pressure.

# The Blood

Body systems maintain
homeostasis

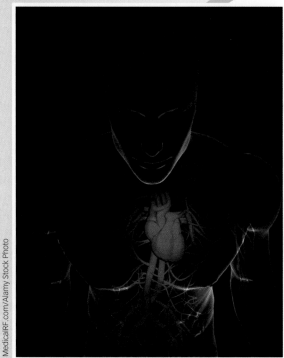

Cardiovascular system

## Homeostasis

Blood contributes to homeostasis by serving as the vehicle for transporting materials to and from the cells, buffering changes in pH, carrying excess heat to the body surface for elimination, playing a major role in the body's defence system, and minimizing blood loss when a blood vessel is damaged.

Homeostasis is essential
for survival of cells

Cells make up body systems

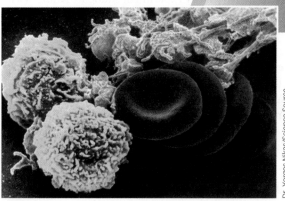

Scanning electron micrograph showing red blood cells,
a T lymphocyte, and platelets

Blood is the vehicle for long-distance, mass transport of materials between cells and the external environment or between the cells themselves. Such transport is essential for maintaining homeostasis. Blood consists of a complex liquid plasma in which the cellular elements—*erythrocytes, leukocytes*, and *platelets*—are suspended. Erythrocytes (red blood cells or RBCs) are essentially plasma membrane–enclosed bags of haemoglobin that transport oxygen in the blood. Leukocytes (white blood cells or WBCs), the immune system's mobile defence units, are transported in the blood to sites of injury or of invasion by disease-causing microorganisms. Platelets (thrombocytes) are important in haemostasis, specifically, in the stopping of bleeding from an injured vessel.

# The Blood

## CONTENTS AT A GLANCE

**❚ Clinical Connections**

Christine is 55 and has been an avid runner for 20 years. For the past month or so she has been finding her runs harder and harder. It has come to the point that she finds herself struggling to maintain her regular pace for even a 5 km run. While running, she feels her exertion level is much higher because she finds herself more breathless than usual and recovery after the run takes longer. Most days she generally feels tired all day long and goes to bed early. Both her children have grown and no longer living at home, so she finds herself working longer and not necessarily eating balanced meals as well as she used to. Concerned about fatigue, she met with her physician, who ordered standard blood work. The most overt result was a decreased hematocrit. After ruling out the potential for internal haemorrhage, Christine was put on iron supplements. Within a few weeks she was back to normal. Her diagnosis was iron deficiency anemia due to an iron-poor diet.

## 10.1 | Introduction

Blood represents about 8 percent of total body weight and has an average volume of 5 L in women and 5.5 L in men. It consists of three types of specialized cellular elements, *erythrocytes (red blood cells), leukocytes (white blood cells)*, and *platelets (thrombocytes)*, suspended in the complex liquid *plasma* (❚ Table 10-1). Erythrocytes and leukocytes are both whole cells, whereas platelets are cell fragments. For convenience, we refer collectively to all these blood cellular elements as *blood cells*.

## ∎ TABLE 10-1 Blood Constituents and Their Functions

| Constituent | Functions |
|---|---|
| **Plasma** | |
| **Water** | Transport medium; carries heat |
| **Electrolytes** | Membrane excitability; osmotic distribution of fluid between ECF and ICF; buffer pH changes |
| **Nutrients, wastes, gases, hormones** | Transported in blood; the blood gas $CO_2$ plays a role in acid–base balance |
| **Plasma proteins** | In general, exert an osmotic effect important in the distribution of ECF between the vascular and interstitial compartments; buffer pH changes |
| *Albumins* | Transport many substances; contribute most to colloid osmotic pressure |
| *Globulins* | |
| Alpha and beta | Transport many water-insoluble substances; clotting factors; inactive precursor molecules |
| Gamma | Antibodies |
| *Fibrinogen* | Inactive precursor for the fibrin meshwork of a clot |
| **Cellular Elements** | |
| **Erythrocytes** | Transport $O_2$ and $CO_2$ (mainly $O_2$) |
| **Leukocytes** | |
| **Neutrophils** | Phagocytes that engulf bacteria and debris |
| **Eosinophils** | Attack parasitic worms; important in allergic reactions |
| **Basophils** | Release histamine, which is important in allergic reactions, and heparin, which helps clear fat from the blood |
| **Monocytes** | In transit to become tissue macrophages |
| **Lymphocytes** | |
| B lymphocytes | Produce antibodies |
| T lymphocytes | Cell-mediated immune responses |
| **Platelets** | Haemostasis |

The constant movement of blood as it flows through the blood vessels keeps the blood cells rather evenly dispersed within the plasma. However, if you put a sample of whole blood in a test tube and treat it to prevent clotting, the heavier cells slowly settle to the bottom and the lighter plasma rises to the top. This process can be speeded up by centrifuging, which quickly packs the cells in the bottom of the tube (› Figure 10-1). Because more than 99 percent of the cells are erythrocytes, the **haematocrit**, or **packed cell volume**, essentially represents the percentage of erythrocytes in the total blood volume. The haematocrit averages 42 percent for women and slightly higher, 45 percent, for men. Plasma accounts for the remaining volume. Accordingly, the average volume of plasma in the blood is 58 percent for women and 55 percent for men. White blood cells and platelets, which are colourless and less dense than red cells, are packed in a thin, cream-coloured layer, the *buffy coat*, on top of the packed red cell column. They account for less than 1 percent of the total blood volume.

Let's first consider the properties of the largest portion of the blood, the plasma, before turning our attention to the cellular elements.

## 10.2 | Plasma

Plasma, being a liquid, consists of 90 percent water.

### Plasma water

Plasma water serves as a medium for materials that are carried in the blood. Also, because water has a high capacity to hold heat, plasma can absorb and distribute much of the heat generated metabolically within tissues, whereas the temperature of the blood itself undergoes only small changes. As blood travels close to the surface of the skin, heat energy not needed to maintain body temperature is eliminated to the environment.

A large number of inorganic and organic substances are dissolved in the plasma. Inorganic constituents account for about 1 percent of plasma weight. The most abundant electrolytes (ions) in the plasma are sodium and chloride, the components of common salt. There are smaller amounts of bicarbonate, potassium, calcium, and others. The most notable functions of these ions are their roles in membrane excitability, osmotic distribution of fluid between the ECF and cells, and buffering of pH changes.

The most plentiful organic constituents by weight are the plasma proteins, which compose 6–8 percent of plasma's total weight. The remaining small percentage of plasma consists of other organic substances, including nutrients (such as glucose, amino acids, lipids, and vitamins), waste products (creatinine, bilirubin, and nitrogenous substances, such as urea), dissolved gases $(O_2$ and $CO_2)$, and hormones. Most of these substances are merely being transported in the plasma. For example, endocrine glands secrete hormones into the plasma, which transports these chemical messengers to their sites of action.

We now examine the plasma proteins.

## Plasma proteins

**Plasma proteins** are the one group of plasma constituents that are not just along for the ride. These important components normally stay in the plasma, where they perform many valuable functions. Here are the most important of these functions:

1. Unlike other plasma constituents that are dissolved in the plasma water, plasma proteins are dispersed as a colloid (Appendix B, p. A-10). Furthermore, because they are the largest of the plasma constituents, plasma proteins usually do not exit through the narrow pores in the capillary walls to enter the interstitial fluid. By their presence as a colloidal dispersion in the plasma and their absence in the interstitial fluid, plasma proteins establish an osmotic gradient between blood

and interstitial fluid. This colloid osmotic pressure is the primary force that prevents excessive loss of plasma from the capillaries into the interstitial fluid, which helps maintain plasma volume (p. 409).

2. Plasma proteins are partially responsible for plasma's capacity to buffer changes in pH (p. 625).

3. The three groups of plasma proteins—*albumins, globulins,* and *fibrinogen*—are classified according to their various physical and chemical properties (Appendix F lists normal values). In addition to the general functions just listed, each type of plasma protein performs specific tasks, as follows:

   a. **Albumins**, the most abundant plasma proteins, contribute most extensively to the colloid osmotic pressure by virtue of their numbers. They also nonspecifically bind many substances that are poorly soluble in plasma (such as bilirubin, bile salts, and penicillin) for transport in the plasma.

   b. There are three subclasses of globulins: **alpha (α) globulins, beta (β) globulin**s, and **gamma (γ) globulins**.

      1 Like albumins, some alpha and beta globulins bind poorly water-soluble substances for transport in the plasma, but these globulins are highly specific as to which passenger they will bind and carry. Examples of substances carried by specific globulins include thyroid hormone (p. 248), cholesterol (p. 216), and iron (p. 685).

      2 Many of the factors involved in the blood-clotting process are alpha or beta globulins.

      3 Inactive, circulating proteins, which are activated as needed by specific regulatory inputs, belong to the alpha globulin group (e.g., the alpha globulin *angiotensinogen* is activated to *angiotensin*, which plays an important role in regulating salt balance in the body; p. 576).

      4 The gamma globulins are the *immunoglobulins* (antibodies), which are crucial to the body's defence mechanism (p. 475).

      5 **Fibrinogen** is a key factor in blood clotting.

      Plasma proteins are synthesized by the liver, with the exception of gamma globulins, which are produced by lymphocytes, one of the types of white blood cells.

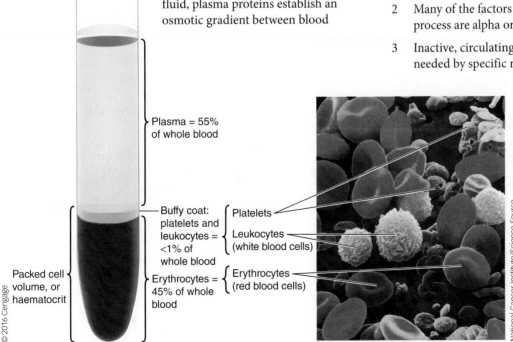

Plasma = 55% of whole blood

Buffy coat: platelets and leukocytes = <1% of whole blood

Packed cell volume, or haematocrit

Erythrocytes = 45% of whole blood

Platelets

Leukocytes (white blood cells)

Erythrocytes (red blood cells)

© 2016 Cengage

National Cancer Institute/Science Source

> **FIGURE 10-1** **Haematocrit and types of blood cells.** The values given are for men. The average haematocrit for women is 42 percent, with plasma occupying 58 percent of the blood volume. Note the biconcave shape of the erythrocytes.

**Source:** From *Textbook of Medical Physiology* 8th edition, by Guyton et al., Figure 14-11, p. 157. Copyright Elsevier, 1990.

## 10.3 | Erythrocytes

Each millilitre of blood contains about 5 billion **erythrocytes (red blood cells or RBCs)** on average, which is commonly reported clinically in a **red blood cell count** as 5 million cells per cubic millimetre.

### Structure and function

The shape and content of erythrocytes are ideally suited to carry out their primary function—namely, transporting oxygen and to a lesser extent carbon dioxide and hydrogen ion in the blood.

#### ERYTHROCYTE STRUCTURE

Erythrocytes are flat, disc-shaped cells indented in the middle on both sides, like a doughnut with a flattened centre instead of a hole (i.e., they are biconcave discs, 8 μm in diameter, 2 μm thick at the outer edges, and 1 μm thick in the centre) (see Figure 10-1). This unique shape contributes in two ways to the efficiency with which red blood cells perform their main function of oxygen transport in the blood: (1) the biconcave shape provides a larger surface area for diffusion of oxygen across the membrane than would a spherical cell of the same volume; (2) the thinness of the cell enables oxygen to diffuse rapidly between the exterior and innermost regions of the cell.

Another structural feature that facilitates RBCs' transport function is the flexibility of their membrane. Red blood cells, whose diameter is normally 8 μm, can deform amazingly as they squeeze single file through capillaries as narrow as 3 μm in diameter. Because they are extremely pliant, RBCs can travel through the narrow, tortuous capillaries to deliver their oxygen cargo at the tissue level without rupturing in the process.

The most important anatomic feature that enables RBCs to transport oxygen is the haemoglobin they contain. Let's look at this unique molecule in more detail.

#### PRESENCE OF HAEMOGLOBIN

**Haemoglobin** is found only in red blood cells. A haemoglobin molecule has two parts: (1) the **globin** portion, a protein made up of four highly folded polypeptide chains; and (2) four iron-containing, non-protein groups known as **heme groups**, each of which is bound to one of the polypeptides (> Figure 10-2). Each of the four iron atoms can combine reversibly with one $O_2$ molecule; thus, each haemoglobin molecule can pick up four $O_2$ passengers in the lungs. Because oxygen is poorly soluble in the plasma, 98.5 percent of the oxygen carried in the blood is bound to haemoglobin (p. 532).

Haemoglobin is a pigment (i.e., it is naturally coloured). Because of its iron content, it appears reddish when combined with oxygen and bluish when deoxygenated. Fully oxygenated arterial blood is therefore red in colour, and venous blood, which has lost some of its $O_2$ load at the tissue level, appears much darker, but is still red.

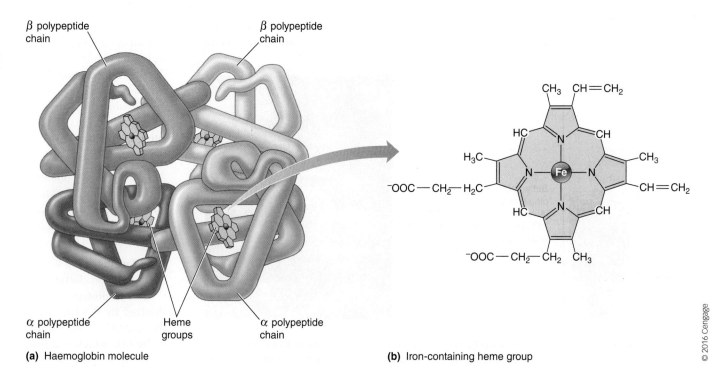

**(a)** Haemoglobin molecule

**(b)** Iron-containing heme group

© 2016 Cengage

> **FIGURE 10-2 Haemoglobin molecule.** A haemoglobin molecule consists of four highly folded polypeptide chains (the globin portion) and four iron-containing heme groups.

In addition to carrying oxygen, haemoglobin can combine with the following:

1. *Carbon dioxide.* Haemoglobin helps transport this gas from the tissue cells back to the lungs (p. 538).

2. *The acidic hydrogen ion* $(H^+)$ portion of ionized carbonic acid, which is generated at the tissue level from carbon dioxide. Haemoglobin buffers this acid so that it minimally alters the pH of the blood (p. 625).

3. *Carbon monoxide (CO).* This gas is not normally in the blood, but if inhaled, it preferentially occupies the $O_2$-binding sites on haemoglobin, causing CO poisoning.

4. *Nitric oxide (NO).* In the lungs, the vasodilator nitric oxide binds to haemoglobin. This NO is released at the tissues, where it relaxes and dilates the local arterioles (p. 399). This vasodilation helps ensure that the oxygen-rich blood can make its vital rounds and also helps stabilize blood pressure.

Therefore, haemoglobin plays the key role in $O_2$ transport and also contributes significantly to $CO_2$ transport and the pH-buffering capacity of blood. Furthermore, by toting along its own vasodilator, haemoglobin helps deliver the oxygen it is carrying. (Chapter 12 provides a detailed explanation of haemoglobin and its role in gas transport.)

---

**▌Clinical Connections**

Haemoglobin is the iron-containing molecule that transports oxygen throughout the body. When someone has iron deficient anemia, there is a decrease in production of haemoglobin and, over time, this can lead to a decrease in the total amount of haemoglobin in the blood. This decreases the oxygen-carrying capacity of the blood. At rest, this may or may not be felt, but during exercise there is insufficient oxygen delivery to the working muscles, and you will fatigue faster than normal.

---

### LACK OF NUCLEUS AND ORGANELLES

To maximize its haemoglobin content, a single erythrocyte is stuffed with more than 250 million haemoglobin molecules, excluding almost everything else. (That means each RBC can carry more than a billion $O_2$ molecules!) Red blood cells contain no nucleus, organelles, or ribosomes. During the cell's development, these structures are extruded to make room for more haemoglobin. Thus, a RBC is mainly a plasma membrane–enclosed sac full of haemoglobin.

### KEY ERYTHROCYTE ENZYMES

Only a few crucial, non-renewable enzymes remain within a mature erythrocyte: glycolytic enzymes and carbonic anhydrase. The **glycolytic enzymes** are necessary for generating the energy needed to fuel the active-transport mechanisms involved in maintaining proper ionic concentrations within the cell. Paradoxically, even though erythrocytes are the vehicles for transporting oxygen to all other tissues of the body, for energy production they themselves cannot use the oxygen they are carrying. Lacking the mitochondria that house the enzymes for oxidative phosphorylation, erythrocytes must rely entirely on glycolysis for ATP formation.

The other important enzyme within RBCs, **carbonic anhydrase**, is critical in $CO_2$ transport. This enzyme catalyzes a key reaction that ultimately leads to the conversion of metabolically produced carbon dioxide into **bicarbonate ion** $(HCO_3^-)$, which is the primary form in which carbon dioxide is transported in the blood. Thus, erythrocytes contribute to $CO_2$ transport in two ways: by means of its carriage on haemoglobin and its carbonic anhydrase–induced conversion to bicarbonate.

## Bone marrow

Each of us has a total of 25 trillion to 30 trillion RBCs streaming through our blood vessels at any given time (750 000 times more in number than the entire Canadian population)! Yet these vital gas-transport vehicles are short lived and must be replaced at the average rate of 2 million to 3 million cells per second (Appendix F lists normal reference values).

### ERYTHROCYTES' SHORT LIFESPAN

The price erythrocytes pay for their generous content of haemoglobin, which requires the exclusion of the usual specialized intracellular machinery, is a short lifespan. Without DNA and RNA, red blood cells cannot synthesize proteins for cell repair, growth, and division or for renewing enzyme supplies. Equipped only with initial supplies synthesized before they extrude their nucleus, organelles, and ribosomes, RBCs survive an average of only 120 days, in contrast to nerve and muscle cells, which last a person's entire life. During its short lifespan of four months, each erythrocyte travels about 1125 km as it circulates through the vasculature.

As a red blood cell ages, its nonreparable plasma membrane becomes fragile and prone to rupture as the cell squeezes through tight spots in the vascular system. Most old RBCs meet their final demise in the **spleen**, because this organ's narrow, winding capillary network is a tight fit for these fragile cells. The spleen lies in the upper left part of the abdomen. In addition to removing most of the old erythrocytes from circulation, the spleen has a limited ability to store healthy erythrocytes in its pulpy interior, serves as a reservoir site for platelets, and contains an abundance of lymphocytes, a type of white blood cell.

### ERYTHROPOIESIS

Erythrocytes cannot divide to replenish their own numbers, so the old ruptured cells must be replaced by the new cells produced in an erythrocyte factory—the **bone marrow**—which is the soft, highly cellular tissue that fills the internal cavities of bones. The bone marrow normally generates new red blood cells—by a process known as **erythropoiesis**—at a rate to keep pace with the demolition of old cells.

During intrauterine development, erythrocytes are produced first by the yolk sac and then by the developing liver and spleen, until the bone marrow is formed and takes over erythrocyte production exclusively. In children, most bones are filled with **red bone marrow**—capable of blood cell production. As a person matures fatty **yellow bone marrow**—incapable of erythropoiesis—gradually replaces red marrow, which remains only in a few isolated places, such as the sternum (breastbone), ribs, and upper ends of the long limb bones.

**10**

Red marrow not only produces RBCs but also is the ultimate source for leukocytes and platelets. Undifferentiated **pluripotent stem cells** reside in the red marrow, where they continuously divide and differentiate to give rise to each type of blood cells. These stem cells, the source of all blood cells, have now been isolated. The search was difficult, because stem cells are less than 0.1 percent of all cells in the bone marrow. Although a great deal of research remains to be done, this recent discovery may hold a key to a cure for a host of blood and immune disorders, as well as numerous other diseases.

The different types of immature blood cells, along with the stem cells, are intermingled in the red marrow at various stages of development. Once mature, the blood cells are released into the rich supply of capillaries that permeate the red marrow. Regulatory factors act on the *haemopoietic* (blood-producing) red marrow to govern the type and number of cells generated and discharged into the blood. Of the blood cells, the mechanism for regulating RBC production is the best understood. We consider this next.

## Erythropoietin

$O_2$ transport in the blood is the erythrocytes' primary function. So you might logically suspect that the primary stimulus for increased erythrocyte production would be reduced $O_2$ delivery to the tissues. You would be correct; however, low $O_2$ levels do not stimulate erythropoiesis by acting directly on the red bone marrow. Instead, reduced $O_2$ delivery to the kidneys stimulates them to secrete the hormone **erythropoietin (EPO)** into the blood, and this hormone, in turn, stimulates erythropoiesis by the bone marrow (› Figure 10-3).

Erythropoietin acts on derivatives of undifferentiated stem cells that are already committed to becoming RBCs, stimulating their proliferation and maturation into mature erythrocytes. This increased erythropoietic activity elevates the number of circulating RBCs, thereby increasing $O_2$-carrying capacity of the blood and restoring $O_2$ delivery to the tissues to normal. Once normal $O_2$ delivery to the kidneys is achieved, erythropoietin secretion is reduced until needed again. In this way, erythrocyte production is normally balanced against the destruction or loss of these cells so that $O_2$-carrying capacity in the blood stays fairly constant. In severe loss of RBCs, as in haemorrhage or abnormal destruction of young circulating erythrocytes, the rate of erythropoiesis can be increased to more than six times the normal level.

The preparation of an erythrocyte for its departure from the marrow involves several steps, such as synthesis of haemoglobin and extrusion of the nucleus and organelles. Cells closest to maturity need a few days to reach maturity and be released into the blood in response to erythropoietin, and the less-developed and newly proliferated cells may take up to several weeks to reach maturity. Therefore, the time required for complete replacement of lost RBCs depends on how many are needed to return the number to normal. (When you donate blood, your circulating erythrocyte supply is replenished in less than a week.)

## Blood types

There are a minimum of 30 relatively universal antigens (proteins) and many uncommon antigens associated with human blood cells. Many of these are unable to cause an antigen–antibody reaction. However, the most common and important blood typing is the A, B, and O antigen system (**ABO system**), which is associated with the antigens present on the surface of the red blood cell (erythrocytes). Scientist Karl Landsteiner is credited as having discovered the ABO blood typing

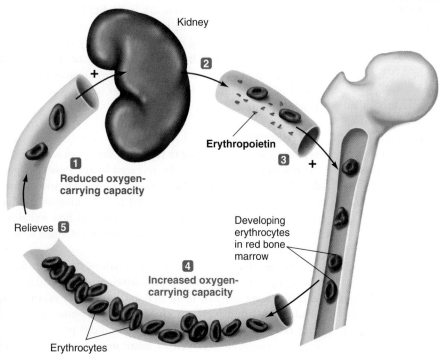

**1** The kidneys detect reduced $O_2$-carrying capacity of the blood.

**2** When less $O_2$ is delivered to the kidneys, they secrete the hormone erythropoietin into the blood.

**3** Erythropoietin stimulates erythropoiesis (erythrocyte production) by the bone marrow.

**4** The additional circulating erythrocytes increase the $O_2$-carrying capacity of the blood.

**5** The increased $O_2$-carrying capacity relieves the initial stimulus that triggered erythropoietin secretion.

Kidney

Erythropoietin

**1** Reduced oxygen-carrying capacity

Relieves **5**

Developing erythrocytes in red bone marrow

**4** Increased oxygen-carrying capacity

Erythrocytes

© 2016 Cengage

› **FIGURE 10-3 Control of erythropoiesis**

system in about 1900 and later was awarded the Nobel Prize in Physiology or Medicine for his work.

A second system of blood typing is the Rh (Rhesus) system. In this system, the transfusion reaction occurs at a slower rate. Antibodies associated with blood types are the classic example of "natural antibodies," although natural immunity is actually a special case of active immunity. Let's see how blood typing works.

## ABO BLOOD TYPES

The surface membranes of human erythrocytes contain inherited antigens that vary depending on blood type; because these antigens are inherited, they can also be used to study parentage. Within the major blood group system, the ABO system, the erythrocytes of people with type A blood contain A antigens, those with type B blood contain B antigens, those with type AB blood have both A and B antigens, and those with type O blood do not have any A or B red blood cell surface antigens.

Antibodies against erythrocyte antigens not present on the body's own erythrocytes begin to appear in human plasma after a baby is about six months of age. Accordingly, the plasma of type A blood contains anti-B antibodies, type B blood contains anti-A antibodies, no antibodies related to the ABO system are present in type AB blood, and both anti-A and anti-B antibodies are present in type O blood. These antigens and antibodies are able to cause a blood transfusion reaction, which typically occurs in the plasma of the blood. The antibodies bind with the erythrocytes' antigens, causing agglutination of the red cells (see Transfusion Reaction). Because of this process, the antigens associated with the erythrocytes are called *agglutinogens*, and the antibodies are called *agglutinins*.

Typically, one would expect antibody production against A or B antigen to be induced only if blood containing the alien antigen were injected into the body. However, high levels of these antibodies are found in the plasma of people who have never been exposed to a different type of blood. Consequently, these were considered naturally occurring antibodies—that is, produced without any known exposure to the antigen.

Scientists now realize that people are unknowingly exposed at an early age to small amounts of A- and B-like antigens associated with common intestinal bacteria. The bacteria are believed to make their way into the body via ingested food from our daily diet. Antibodies produced against the foreign antigen will then react when the wrong blood type (and associated surface antigen) is transfused into the body. This is considered a secondary immune response by the adapted division of the immune system; the first response occurred during the initial exposure to the intestinal bacteria that generated the antibodies (producing the memory cells). Your own blood type and its surface antigens are not identified as foreign (i.e., the A antigen in a type A person) because of clonal deletion.

Therefore, the test to establish blood type begins with a blood sample being drawn from a vein and mixed with antibodies against types A and B blood, then the sample is checked to see whether or not the blood cells agglutinate. If agglutination takes place, it means the blood reacted with one of the antibodies. In the second step, known as back typing, the non-cell liquid part of your blood is mixed with blood that is known to be type A and type B. These two steps can accurately determine your ABO blood type. Blood typing then proceeds to testing for Rh factor to tell whether or not you have an Rh antigen (to be discussed next) on the surface of your erythrocytes.

## RH BLOOD TYPES

The Rh blood group system (an erythrocyte antigen first observed in Rhesus [Rh] monkeys) consists of about 50 defined blood-group antigens; the with six primary antigen groups are D, C, E, d, c, and e. This classification system is termed the CDE system developed by Fisher and Race (Fisher–Race), and is the most commonly used nomenclature. A person with C antigen will not have c antigen, and the same pattern follows for the other antigens. The D antigen is frequently found in the population and is the most antigenic, and thus most important. The terms **Rh factor**—Rh positive (+) and Rh negative (−)—refer to the D antigen. The essence of the Fisher–Race system is that persons either have, or do not have, Rh factor on the surface of their erythrocytes. If they have the $Rh^+$ factor, they are considered $Rh^-$, and those without it are termed $Rh^-$.

In contrast to the ABO blood type, immunization against Rh can usually only occur through blood transfusion or placental exposure during pregnancy. This is because it requires an enormous exposure to the Rh factor to elicit a reaction. Anti-Rh antibodies are produced only by $Rh^-$ individuals, and if such people are exposed to foreign Rh antigens present in $Rh^+$ blood, they become sensitized to the $Rh^+$ antigen. A subsequent transfusion of $Rh^-$ blood could produce a transfusion reaction. $Rh^+$ individuals, in contrast, never produce antibodies against the $Rh^-$ factor that they themselves possess. Therefore, $Rh^-$ people should be given only $Rh^-$ blood, whereas $Rh^+$ people can safely receive either $Rh^-$ or $Rh^+$ blood. The Rh factor is of particular medical importance when an $Rh^-$ mother develops antibodies against the erythrocytes of an $Rh^+$ fetus she is carrying: a condition known as erythroblastosis fetalis, or haemolytic disease of the newborn. It was in 1939 that Drs. Levine and Stetson published the clinical consequences of nonrecognized Rh factor, resulting in a haemolytic transfusion reaction in a newborn.

*Clinical Note* Once a person's blood is typed as A, B, O, or AB, each blood type is further characterized as $Rh^+$ or $Rh^-$. For example, if you have blood type B and are $Rh^+$, you are said to be B-positive ($B^+$). Generally, about 85 percent of all people are $Rh^+$, and the remaining are said to be $Rh^-$.

## TRANSFUSION REACTION

*Clinical Note* If a person is given blood of an incompatible type, two different antigen–antibody interactions take place. By far the more serious consequences arise from the effect of the antibodies in the recipient's plasma on the incoming donor erythrocytes. The effect of the donor's antibodies on the recipient's erythrocyte-bound antigens is less important, unless a large amount of blood is transfused, because the donor's antibodies are so diluted by the recipient's plasma that little red blood cell damage takes place in the recipient.

Antibody interaction with erythrocyte-bound antigen may result in agglutination (clumping) or haemolysis (rupture) of the attacked red blood cells. Agglutination and haemolysis of

10

donor red blood cells by antibodies in the recipient's plasma can lead to a sometimes fatal **transfusion reaction** (⟩ Figure 10-4). Agglutinated clumps of incoming donor cells can plug small blood vessels. In addition, one of the most lethal consequences of mismatched transfusions is acute kidney failure, caused by the release of large amounts of haemoglobin from damaged donor erythrocytes. If the free haemoglobin in the plasma rises above a critical level, it will precipitate in the kidneys and block the urine-forming structures, leading to acute kidney shutdown.

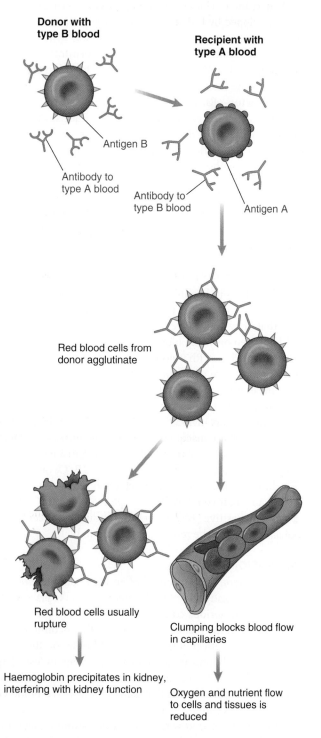

Donor with type B blood

Recipient with type A blood

Antigen B

Antibody to type A blood

Antibody to type B blood

Antigen A

Red blood cells from donor agglutinate

Red blood cells usually rupture

Clumping blocks blood flow in capillaries

Haemoglobin precipitates in kidney, interfering with kidney function

Oxygen and nutrient flow to cells and tissues is reduced

© 2016 Cengage

⟩ **FIGURE 10-4 Transfusion reaction.** A transfusion reaction resulting from type B blood being transfused into a recipient with type A blood

## UNIVERSAL BLOOD DONORS AND RECIPIENTS

Because type O individuals have no A or B antigens, their erythrocytes will not be attacked by either anti-A or anti-B antibodies, so they are considered **universal donors**. Their blood can be transfused into people of any blood type. However, type O individuals can receive only type O blood, because the anti-A and anti-B antibodies in their plasma will attack either A or B antigens in incoming blood. In contrast, type AB individuals are called **universal recipients**. Lacking both anti-A and anti-B antibodies, they can accept donor blood of any type, although they can donate blood only to other AB people. Because their erythrocytes have both A and B antigens, their cells would be attacked if transfused into individuals with antibodies against either of these antigens.

The terms *universal donor* and *universal recipient* are misleading, however. In addition to the ABO system, many other erythrocyte antigens and plasma antibodies can cause transfusion reactions, the most important of which is the Rh factor. Remember that those with Rh⁻ blood are *universal donors*, but those with Rh⁺ blood can only donate to other Rh⁺ persons.

## RETICULOCYTES

When demands for RBC production are high (e.g., following haemorrhage), the bone marrow may release large numbers of immature erythrocytes, known as **reticulocytes**, into the blood to quickly meet the need. Reticulocytes develop and mature in the red bone marrow, following which the reticulocytes circulate for about 24 hours in the bloodstream before transforming into a mature RBC. These immature cells can be recognized by staining techniques that make visible the residual ribosomes and organelle remnants that have not yet been extruded. Automated laser-marking counters are used with a fluorescent dye to accurately count the reticulocytes. Their presence above the normal level of 0.5 to 1.5 percent of the total number of circulating erythrocytes indicates a high rate of erythropoietic activity. At very rapid rates, more than 30 percent of the circulating red cells may be at the immature reticulocyte stage.

## SYNTHETIC ERYTHROPOIETIN

*Clinical Note* Researchers have identified the gene that directs erythropoietin synthesis, so this hormone can now be produced in a laboratory. Laboratory-produced erythropoietin has currently become biotechnology's single-biggest money-maker, with sales exceeding $1 billion annually. This hormone is often used to boost RBC production in patients with suppressed erythropoietic activity, such as those with kidney failure or those undergoing chemotherapy for cancer. (Chemotherapy drugs interfere with the rapid cell division characteristic of both cancer cells and the developing RBCs.) Furthermore, the ready availability of this hormone has diminished the need for blood transfusions. For example, transfusion of a surgical patient's own previously collected blood, coupled with erythropoietin to stimulate further RBC production, has reduced the use of donor blood by as much as 50 percent in some hospitals. (See Concepts Challenges, and Controversies for an update on other alternatives to whole-blood transfusions under investigation.)

## Anaemia

*Clinical Note* Despite control measures, oxygen-carrying capacity cannot always be maintained to meet tissue needs. The term **anaemia** refers to a below-normal $O_2$-carrying capacity of the blood and is characterized by a low haematocrit ( > Figure 10-5a and 10-5b). It can be brought about by a decreased rate of erythropoiesis, excessive losses of erythrocytes, or a deficiency in the haemoglobin content of erythrocytes. The various causes of anaemia can be grouped into six categories:

1. **Nutritional anaemia** is caused by a dietary deficiency of a factor needed for erythropoiesis. The production of RBCs depends on an adequate supply of essential raw ingredients, some of which are not synthesized in the body but must be provided by dietary intake. For example, *iron deficiency anaemia* (see Clinical Connection in Section 10.3) occurs when not enough iron is available for synthesizing haemoglobin.

2. **Pernicious anaemia** is caused by an inability to absorb enough ingested vitamin $B_{12}$ from the digestive tract. Vitamin $B_{12}$ is essential for normal RBC production and maturation. It is abundant in a variety of foods. The problem is a deficiency of *intrinsic factor*, a special substance secreted by the lining of the stomach (p. 664). Vitamin $B_{12}$ can be absorbed from the intestinal tract only when this nutrient is bound to intrinsic factor. When intrinsic factor is deficient, not enough ingested vitamin $B_{12}$ is absorbed. The resulting impairment of RBC production and maturation leads to anaemia.

3. **Aplastic anaemia** is caused by failure of the bone marrow to produce enough RBCs, even though all ingredients necessary for erythropoiesis are available. Reduced erythropoietic capability can be caused by destruction of red bone marrow by toxic chemicals (such as benzene), heavy exposure to radiation (e.g., fallout from a nuclear bomb explosion, or excessive exposure to X-rays), invasion of the marrow by cancer cells, or chemotherapy for cancer. The destructive process may selectively reduce the marrow's output of erythrocytes, or it may reduce the productive capability for leukocytes and platelets as well. The anaemia's severity depends on the extent to which erythropoietic tissue is destroyed; severe losses are fatal.

4. **Renal anaemia** may result from kidney disease. Because erythropoietin from the kidneys is the primary stimulus for promoting erythropoiesis, inadequate erythropoietin secretion by diseased kidneys leads to insufficient RBC production.

5. **Haemorrhagic anaemia** is caused by losing a lot of blood. The loss can be either acute, such as a bleeding wound, or chronic, such as excessive menstrual flow.

6. **Haemolytic anaemia** is caused by the rupture of too many circulating erythrocytes. **Haemolysis**, the rupture of RBCs, occurs either because otherwise normal cells are induced to rupture by external factors, as in the invasion of RBCs by *malaria* parasites, or because the cells are defective, as in sickle cell disease. **Sickle cell disease** is the best-known example among various hereditary abnormalities of erythrocytes that make these cells very fragile. It affects about 1 in 650 persons of African descent. In this condition, a defective type of haemoglobin joins together to form rigid chains that make the RBC stiff and unnaturally shaped, like a crescent or sickle ( > Figure 10-6). Unlike normal erythrocytes, these deformed RBCs tend to clump together. The resultant logjam blocks blood flow through small vessels, leading to pain and tissue damage at the affected site. Furthermore, the defective erythrocytes are fragile and prone to rupture, even as young cells, as they travel through the narrow splenic capillaries. Despite an accelerated rate of erythropoiesis triggered by the constant excessive loss of RBCs, production may not be able to keep pace with the rate of destruction, and anaemia may result.

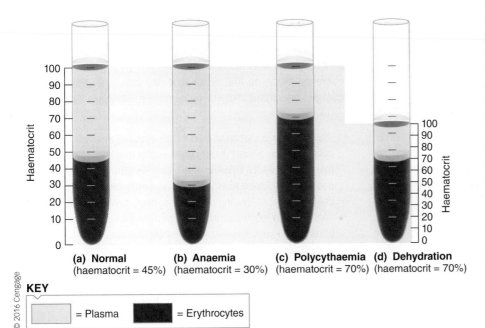

**(a) Normal** (haematocrit = 45%) **(b) Anaemia** (haematocrit = 30%) **(c) Polycythaemia** (haematocrit = 70%) **(d) Dehydration** (haematocrit = 70%)

**KEY**

☐ = Plasma   ■ = Erythrocytes

© 2016 Cengage

> **FIGURE 10-5 Haematocrit under various circumstances** (a) Normal haematocrit. (b) The haematocrit is lower than normal in anaemia because of too few circulating erythrocytes; and (c) above normal in polycythaemia because of excess circulating erythrocytes. (d) The haematocrit can also be elevated during dehydration, when the normal number of circulating erythrocytes is concentrated in a reduced plasma volume.

## Polycythaemia

*Clinical Note* Poiseuille's law states that the greater the viscosity of blood, the greater the reduction in flow within a vessel, if all other variables remain constant. We can use the

# In Search of a Blood Substitute

ONE OF THE HOTTEST MEDICAL CONTESTS of the last two decades has been the race to develop a universal substitute for human blood that is safe, inexpensive, and disease free, and that has a long shelf life.

### Need for a Blood Substitute

In Canada, over 2000 units of blood are required in hospitals every day. With only 3.5 percent of eligible Canadians actually donating blood, regional shortages of particular blood types occur. Such shortages are only expected to become more widespread as the number of blood donors continues to decline at the same time that the number of senior citizens, the group of people who most often need transfusions, continues to grow. The benefits for society of a safe blood substitute that could be administered without regard for the recipient's blood type are great, as will be the profits for the manufacturer of the first successful product. Experts estimate that the world market for a good blood substitute, more accurately called an *oxygen therapeutic*, may be as much as $10 billion per year.

Researchers began working on blood substitutes in the 1960s, but the search for an alternative to whole-blood transfusions was given new impetus in the 1980s by the rising incidence of AIDS, the tainted blood scandal, and the concomitant concern over the safety of the nation's blood supply. Infectious diseases—such as AIDS, viral hepatitis, and West Nile virus—can be transmitted from infected blood donors to recipients of blood transfusions. Furthermore, restrictions have been placed on potential donors who lived or travelled in Europe during the time mad cow disease hit the beef industry. Although careful screening of our blood supply minimizes the possibility that infectious diseases will be transmitted through transfusion, the public remains wary and would welcome a safe substitute.

Eliminating the risk of disease transmission is only one advantage of finding an alternative to whole-blood transfusion. Whole blood must be kept refrigerated, and even then it has a shelf life of only 42 days. Also, transfusion of whole blood requires blood typing and cross-matching, which cannot be done at the scene of an accident or on a battlefield.

### Major Approaches

The goal is not to find a replacement for whole blood but to duplicate its $O_2$-carrying capacity. The biggest need for blood transfusions is to replace acute blood loss in accident victims, surgical patients, and wounded soldiers. These individuals require short-term replenishment of blood's $O_2$-carrying capacity until their own bodies synthesize replacement erythrocytes. The many other important elements in blood are not as immediately critical in sustaining life as is the haemoglobin within the RBCs. Problematically, red blood cells are the whole-blood component that requires refrigeration, has a short shelf life, and bears the markers for the various blood types.

Therefore, the search for a blood substitute has focused on two major possibilities: (1) haemoglobin products that exist outside a RBC and can be stored at room temperature for up to six months to a year, and (2) chemically synthesized products that serve as artificial haemoglobin by dissolving large amounts of oxygen when $O_2$ levels are high (as in the lungs) and releasing it when $O_2$ levels are low (as in the tissues). A variety of potential blood substitutes are in various stages of development. Some have reached the stage of clinical trials, but no products have yet reached the market, although they are getting close. Let's examine each of these major approaches.

### Haemoglobin Products

By far the greatest number of research efforts has focused on manipulating the structure of haemoglobin so that it can be effectively and safely administered as a substitute for whole-blood transfusions. If appropriately stabilized and suspended in saline solution, haemoglobin could be injected to bolster the $O_2$-carrying capacity of the recipients' blood no matter what their blood type. The following strategies are among those being pursued to develop a haemoglobin product:

- One problem is that haemoglobin behaves differently when it is outside RBCs. "Naked" haemoglobin splits into halves that do not release oxygen for tissue use as readily as normal haemoglobin does. As well, these haemoglobin fragments can cause kidney damage. A cross-binding reagent has been developed that keeps haemoglobin molecules intact when they are outside the confines of red blood cells, thereby surmounting one major obstacle to administering free haemoglobin.

viscosity of water as a touchstone for that of blood, as blood is approximately three times as viscous as water. The viscosity of blood is largely dependent on the number of suspended RBCs in the blood. Consequently, if the number of RBCs increases, the frictional drag between the RBCs and the vessel's wall increases. It is this frictional drag that determines viscosity; the greater the blood's viscosity, the larger the reduction in the blood flow.

A common method used to measure the number of RBCs in blood is haematocrit. If a healthy person has a haematocrit of 42 percent, then 42 percent of the person's blood volume consists of cells (primarily RBCs), and the remaining portion

- Some products under investigation are derived from outdated, donated human blood. Instead of the blood being discarded, its haemoglobin is extracted, purified, sterilized, and chemically stabilized. However, this strategy relies on the continued practice of collecting human blood donations.
- Several products use cows' blood as a starting point. Bovine haemoglobin is readily available from slaughterhouses, is cheap, and can be treated for administration to humans. A big concern with these products is the potential of introducing into humans unknown disease-causing microbes that might be lurking in the bovine products.
- A potential candidate as a blood substitute is genetically engineered haemoglobin that bypasses the ongoing need for human donors or the risk of spreading disease from cows to humans. Genetic engineers can insert the gene for human haemoglobin into bacteria, which act as a factory to produce the desired haemoglobin product. A drawback for genetically engineered haemoglobin is the high cost involved in operating the facilities.
- One promising strategy encapsulates haemoglobin within liposomes— membrane-wrapped containers—similar to real haemoglobin-stuffed, plasma membrane–enclosed RBCs. These so-called *neo red cells* await further investigation.

### Synthetic Oxygen Carriers

Other researchers are pursuing the development of chemical-based strategies that rely on *perfluorocarbons (PFCs)*, which are synthetic $O_2$-carrying compounds. PFCs are completely inert, chemically synthesized molecules that can dissolve large quantities of oxygen in direct proportion to the amount of oxygen breathed in. Because they are derived from a nonbiological source, PFCs cannot transmit disease. This, coupled with their low cost, makes them attractive as a blood substitute. Yet use of PFCs is not without risk. Their administration can cause flu-like symptoms, and because of poor excretion, they may be retained and accumulate in the body. Ironically, PFC administration poses the danger of causing $O_2$ toxicity by delivering too much oxygen to the tissues in uncontrolled fashion (p. 541).

### Tactics to Reduce Need for Donated Blood

Other tactics besides blood substitutes aimed toward reducing the need for donated blood include the following:

- By changing surgical practices, the medical community has reduced the need for transfusions. These blood-saving methods include recycling a patient's own blood during surgery (collecting lost blood, then reinfusing it); using less invasive and therefore less bloody surgical techniques; and treating the patient with blood-enhancing erythropoietin prior to surgery.
- The necessity of matching blood types for transfusions is a major reason for waste at blood banks. Transfusion of mismatched blood causes a serious, even fatal reaction (p. 439). Therefore, a blood bank may be discarding stocks of one blood type that has gone unused while facing a serious shortage of another type. The various blood types are distinguished by differences in the short sugar-chain markers that project from the red cell's plasma membrane (p. 439). Mismatched identity markers are the target for attack in a transfusion reaction. Researchers are making considerable progress in their search for enzymes that can cleave the identity markers away from RBCs, thereby converting them all into a type that could be safely transfused into anyone. Such a product would reduce the current wastage.
- Other investigators are seeking ways to prolong the life of RBCs, either in a blood bank or in patients, thus reducing the need for fresh, transfusable blood.

As this list of strategies attests, considerable progress has been made toward developing a safe, effective alternative to whole-blood transfusions. Yet after more than three decades of intense effort, considerable challenges remain, and no ideal solution has been found.

**10**

### Further Reading

Chang, T.M. (2006). Evolution of artificial cells using nanobiotechnology of hemoglobin based RBC blood substitute as an example. *Artif Cell Blood Sub*, 34(6): 551–66.

Chang, T.M. (2004). Hemoglobin-based red blood cell substitutes. *Artif Organs*, 28(9): 789–94.

Lee, D.H., & Blajchman, M.A. (2000). Platelet substitutes and novel platelet products. *Expert Opin Inv Drug*, 9(3): 457–69.

is plasma. Normal haematocrit values for males and females are 42 percent and 38 percent, respectively. Haematocrit values may change under certain circumstances, such as disease or injection of recombinant EPO, as seen in some cases of blood doping. There is a benefit to increasing haematocrit in the form of increased haemoglobin and $O_2$-carrying capacity, but as haematocrit increases so does blood viscosity (>Figure 10-7). One negative aspect of increased haematocrit is a reduced ability of the heart to circulate the blood (because of increased friction), thereby decreasing the timely delivery of oxygen to the tissues. Thus, there is a fine balance between the benefit of increased $O_2$-carrying capacity and the reduced ability to circulate the

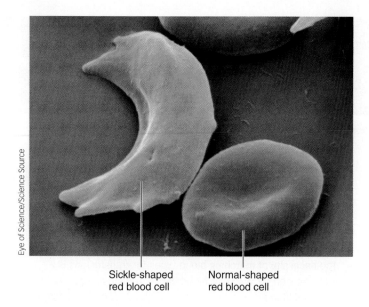

Sickle-shaped
red blood cell

Normal-shaped
red blood cell

› FIGURE 10-6 **Sickle-shaped red blood cell**

oxygen. The condition of polycythaemia is associated with an increased haematocrit.

**Polycythaemia**, in contrast to anaemia, is characterized by too many circulating RBCs and an elevated haematocrit (see › Figure 10-5c). There are two general types of polycythaemia, depending on the circumstances triggering the excess RBC production: primary polycythaemia and secondary polycythaemia.

**Primary polycythaemia** is caused by a tumour-like condition of the bone marrow in which erythropoiesis proceeds at an excessive, uncontrolled rate instead of being subject to the normal erythropoietin regulatory mechanism. The RBC count may reach 11 million cells/mm$^3$ (normal is 5 million cells/mm$^3$), and the haematocrit may be as high as 70 to 80 percent (normal is 42 to 45 percent). Inappropriate polycythaemia has harmful effects, however. The excessive number of red cells increases blood's viscosity up to five to seven times normal, causing the blood to flow very sluggishly, which may actually reduce $O_2$

delivery to the tissues (p. 387). The increased viscosity also increases the total peripheral resistance, which may elevate the blood pressure, thus increasing the workload of the heart, unless blood pressure control mechanisms can compensate (see › Figure 9-14).

**Secondary polycythaemia**, in contrast, is an appropriate erythropoietin-induced adaptive mechanism to improve the blood's $O_2$-carrying capacity in response to a prolonged reduction in $O_2$ delivery to the tissues. It occurs normally in people living at high altitudes, where less oxygen is available in the atmospheric air, or people in whom $O_2$ delivery to the tissues is impaired by chronic lung disease or cardiac failure. The red cell count in secondary polycythaemia is usually lower than that in primary polycythaemia, typically averaging 6–8 million cells/mm$^3$. The price paid for improved $O_2$ delivery is an increased viscosity of the blood.

An elevated haematocrit can occur when the body loses fluid but not erythrocytes, as in dehydration that accompanies heavy sweating or profuse diarrhoea (see › Figure 10-5d). This is not a true polycythaemia, however, because the number of circulating RBCs is not increased. A normal number of erythrocytes is simply concentrated in a smaller plasma volume. This condition is sometimes termed **relative polycythaemia**.

Relative polycythaemia frequently occurs in weight-class sports, such as boxing, wrestling, judo, and rowing. In high school or university wrestling, average weekly body weight fluctuations commonly exceed 2.5 kg, 15 times a season over a 24-to-48-hour period, mostly because of dehydration. There have been reports of athletes decreasing body weight by 5 kg in less than five hours, all via dehydration. Typically, dehydration methods include diuretics, saunas, rubber suits, and running. The physiological effects of dehydration include reductions in plasma volume, cardiac output, peripheral blood flow (muscle and skin), sweating, thermoregulatory ability, and rate of stomach emptying. Physical performance also suffers. However, once rehydration takes place via the consumption of fluids, primarily water, these negative physiological changes are reversed as the body moves toward homeostasis (e.g., normal haematocrit and plasma volume).

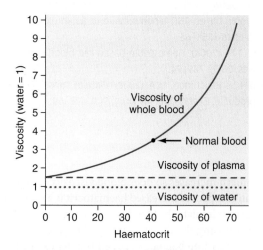

› FIGURE 10-7 **The influence of haematocrit on blood viscosity**

### Check Your Understanding 10.2

1. Describe the anatomic features of erythrocytes that contribute to the efficiency with which they transport oxygen.

2. List the chemicals with which haemoglobin can combine.

3. Discuss the source, control, and function of erythropoietin.

## 10.4 | Leukocytes

**Leukocytes (white blood cells or WBCs)** are the mobile units of the body's immune defence system (Appendix G lists normal reference values). **Immunity** is the body's ability to resist or eliminate potentially harmful foreign materials or abnormal cells. Leukocytes and their derivatives, along with a variety of plasma proteins, make up the **immune system**, an internal defence system that recognizes and either destroys or neutralizes materials within the body that are foreign to the "normal self." Specifically, the immune system (1) defends against invading **pathogens** (disease-producing microorganisms, such as bacteria and viruses); (2) identifies and destroys cancer cells that arise in the body; and (3) removes worn-out cells (such as aged red blood cells) and tissue debris (e.g., tissue damaged by trauma or disease). The latter is essential for wound healing and tissue repair.

## Defence agents

To carry out their functions, leukocytes largely use a "seek out and attack" strategy; that is, they go to sites of invasion or tissue damage. The main reason WBCs are in the blood is to be rapidly transported from their site of production or storage to wherever they are needed.

## Five types

Leukocytes lack haemoglobin, so they are colourless (i.e., white) unless specifically stained for microscopic visibility. Unlike erythrocytes, which are of uniform structure, identical function, and constant number, leukocytes vary in structure, function, and number. There are five different types of circulating leukocytes—neutrophils, eosinophils, basophils, monocytes, and lymphocytes—each with a characteristic structure and function. They are all somewhat larger than erythrocytes.

The five types of leukocytes fall into two main categories, depending on the appearance of their nuclei and the presence or absence of granules in their cytoplasm when viewed microscopically (> Figure 10-8). Neutrophils, eosinophils, and basophils are categorized as **polymorphonuclear granulocytes** (*polymorphonuclear* means "many-shaped nucleus"; *granulocyte* means "granule-containing cell"). Their nuclei are segmented into several lobes of varying shapes, and their cytoplasm contains an abundance of membrane-enclosed granules. The three types of granulocytes are distinguished on the basis of the varying affinity of their granules for dyes: *eosinophils* have an affinity for the red dye eosin; *basophils* preferentially take up a basic blue dye; and *neutrophils* are neutral, showing no dye preference. Monocytes and lymphocytes are known as **mononuclear agranulocytes** (*mononuclear* means "single nucleus"; *agranulocyte* means "cell lacking granules"). Both have a single, large, non-segmented nucleus and few granules. Monocytes are the larger of the two and have an oval or a kidney-shaped nucleus. *Lymphocytes*, the smallest of the leukocytes, characteristically have a large, spherical nucleus that occupies most of the cell.

## Production

All leukocytes ultimately originate from the same undifferentiated multipotent (having many potentials) stem cells in the red bone marrow that also give rise to erythrocytes and platelets (> Figure 10-9). The cells destined to become leukocytes eventually differentiate into various committed cell lines and proliferate under the influence of appropriate stimulating factors. Granulocytes and monocytes are produced only in the bone marrow, which releases these mature leukocytes into the blood. Lymphocytes are originally derived from precursor cells in the bone marrow, but most new lymphocytes are actually produced by lymphocytes already in the **lymphoid tissues** (lymphocyte-containing), such as the lymph nodes and tonsils.

The total number of leukocytes normally ranges from 5 million to 10 million cells per millilitre of blood, with an average of 7 million cells/mL—expressed as an average **white blood cell count** of 7000/mm³ (▌ Table 10-2). Leukocytes are the least numerous of the blood cells (about 1 white blood cell for every 700 red blood cells), not because fewer are produced but because they are merely in transit while in the blood. Normally, about two-thirds of the circulating leukocytes are granulocytes,

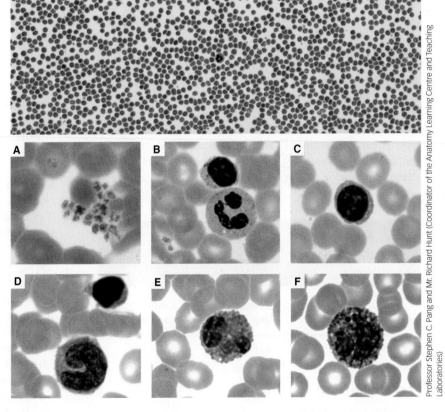

Professor Stephen C. Pang and Mr. Richard Hunt (Coordinator of the Anatomy Learning Centre and Teaching Laboratories)

> **FIGURE 10-8 Peripheral blood smear.** The relative abundance of erythrocytes (red blood cells or RBCs) and leukocytes (white blood cells or WBCs) can be visualized in this blood smear. Background view is a blood smear under low magnification showing a singular leukocyte in the centre. (a) A clump of platelets (thrombocytes); (b) a neutrophil with lobulated nucleus (bottom) and a lymphocyte; (c) a lymphocyte with condensed chromatin and very little cytoplasm; (d) a monocyte with horseshoe-shaped nucleus; (e) an eosinophil with lobulated nucleus and pink granules in the cytoplasm; and (f) a basophil with lobulated nucleus and densely stained granules. Scale bar = 7 μm.

**10**

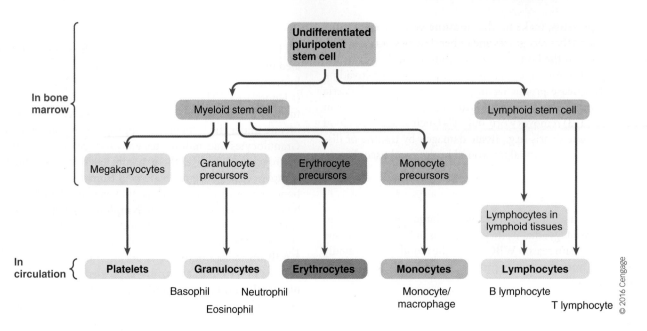

> **FIGURE 10-9 Blood cell production (haemopoiesis).** All the blood cell types ultimately originate from the same undifferentiated pluripotent stem cells in the red bone marrow.

mostly neutrophils, whereas one-third is agranulocytes, predominantly lymphocytes (› Figure 10-8). However, the total number of white cells and the percentage of each type may vary considerably to meet changing defence needs. Depending on the type and extent of assault the body is combating, different types of leukocytes are selectively produced at varying rates. Chemical messengers arising from invaded or damaged tissues or from activated leukocytes themselves govern the rates of production of the various leukocytes. Specific hormones analogous to erythropoietin direct the differentiation and proliferation of each cell type. Some of these hormones have been identified and can be produced in the laboratory; an example is **granulocyte colony–stimulating factor**, which stimulates increased replication and release of granulocytes, especially neutrophils, from the bone marrow. This achievement has opened up the ability to administer these hormones as a powerful new therapeutic tool to bolster a person's normal defence against infection or cancer.

## FUNCTIONS AND LIFESPANS OF LEUKOCYTES

The following are the functions and lifespans of the **granulocytes**:

- **Neutrophils**—major function: phagocytosis. Furthermore, scientists recently discovered that neutrophils release a web of extracellular fibres, dubbed *neutrophil extracellular traps (NETs)*. These fibres contain bacteria-killing chemicals, enabling NETs to trap and then destroy bacteria extracellularly. As a result, neutrophils can destroy bacteria both intracellularly by phagocytosis and extracellularly via the NETs they release. Neutrophils are the first defenders against bacterial invasion, are very important in inflammatory responses, and act as scavengers to clean up debris.

- *Clinical Note* As might be expected in view of these functions, an increase in circulating neutrophils (**neutrophilia**) typically accompanies acute bacterial infections. In fact, a differential WBC count (a determination of the proportion of each type of leukocyte present) can be useful in making an immediate, reasonably accurate prediction of whether an infection, such as pneumonia or meningitis, is of bacterial or viral origin. Obtaining a definitive answer as to the causative agent by culturing a sample of the infected tissue's fluid takes several days. Because an elevated neutrophil count is highly indicative of bacterial infection, it is appropriate to initiate antibiotic therapy long before the true causative agent is actually known. (Bacteria generally succumb to antibiotics, whereas viruses do not.)

- **Eosinophils**—major function: phagocytosis of parasites. An increase in circulating eosinophils (**eosinophilia**) is

### ▌TABLE 10-2 Typical Human Blood Cell Counts

| Cell Type | Concentration (per mL blood) | Count (per mm³) |
|---|---|---|
| **Erythrocytes** | 5 billion | 5 million |
| **Platelets** | 250 million | 250 000 |
| **Leukocytes (WBC)** | 7 million | 7000 |
| *Neutrophils (60–70%)* | | |
| *Lymphocytes (25–33%)* | | |
| *Monocytes (2–6%)* | | |
| *Eosinophils (1–4%)* | | |
| *Basophils (0.25–0.5%)* | | |

associated with allergic conditions (such as asthma) and with internal parasite infestations (e.g., worms). Eosinophils attach to the worm and secrete substances that kill it.

- **Basophils**—major function: chemotactic factor production. They are quite similar structurally and functionally to **mast cells**, which never circulate in the blood, but instead are dispersed in connective tissue throughout the body. Researchers have shown that basophils arise from the bone marrow, whereas mast cells are derived from precursor cells in connective tissue. Both basophils and mast cells synthesize and store *histamine* and *heparin*: powerful chemical substances that can be released on appropriate stimulation. Histamine release is important in allergic reactions, whereas heparin speeds up removal of fat particles from the blood after a fatty meal. Heparin can also prevent clotting (coagulation) of blood samples drawn for clinical analysis and is used extensively as an anticoagulant drug, but whether it plays a physiological role in clot prevention is still debated.

Once released into the blood from the bone marrow, a granulocyte usually stays in transit in the blood for less than a day before leaving the blood vessels to enter the tissues, where it survives another three to four days unless it dies sooner in the line of duty.

By comparison, the functions and lifespans of the agranulocytes are as follows:

- **Monocytes**—major function: phagocytosis, antigen presentation, cytokine production, and cytotoxicity. They emerge from the bone marrow while still immature, and circulate for only a day or two before settling down in various tissues throughout the body. At their new residences, monocytes mature and greatly enlarge, becoming the large tissue phagocytes known as **macrophages** (*macro* means "large"; *phage* means "eater"). A macrophage's lifespan may range from months to years, unless it is destroyed sooner while performing its phagocytic activity. A phagocytic cell can ingest only a limited amount of foreign material before it succumbs.

- **Lymphocytes**—major function: lymphocyte activation, cytokine production, antigen recognition, antibody production, memory, and cytotoxicity. There are two types of lymphocytes: B lymphocytes and T lymphocytes. **B lymphocytes (B cells)** produce *antibodies*, which circulate in the blood and are responsible for *antibody-mediated*, or *humoural*, *immunity*. An antibody binds with and marks for destruction (by phagocytosis or other means) the specific kinds of foreign matter, such as bacteria, that induced production of the antibody. **T lymphocytes (T cells)** do not produce antibodies; instead, they directly destroy their specific target cells by releasing chemicals that punch holes in the victim cells, a process called *cell-mediated immunity*. The target cells of T cells include body cells invaded by viruses and cancer cells. Lymphocytes live for about 100 to 300 days. During this period, most of them continually recycle among the lymphoid tissues, lymph, and blood, spending only a few hours at a time in the blood. Therefore, only a small part of the total lymphocytes are in transit in the blood at any given moment.

## ABNORMALITIES IN LEUKOCYTE PRODUCTION

*Clinical Note* Even though levels of circulating leukocytes may vary, changes in these levels are normally controlled and adjusted according to the body's needs. However, abnormalities in leukocyte production can occur that are not subject to control; that is, either too few or too many WBCs can be produced. The bone marrow can greatly slow down or even stop production of white blood cells when it is exposed to certain toxic chemical agents (such as benzene and anticancer drugs) or to excessive radiation. The most serious consequence is the reduction in professional phagocytes (neutrophils and macrophages), which greatly reduces the body's defence capabilities against invading microorganisms. The only defence still available when the bone marrow fails is the immune capabilities of the lymphocytes produced by the lymphoid organs.

In **infectious mononucleosis**, not only does the number of lymphocytes (but not other leukocytes) in the blood increase, but also many of the lymphocytes are atypical in structure. This condition, which is caused by the *Epstein-Barr virus*, is characterized by pronounced fatigue, a mild sore throat, and low-grade fever. Full recovery usually requires a month or more.

Surprisingly, one of the major consequences of **leukaemia**, a cancerous condition that involves uncontrolled proliferation of WBCs, is inadequate defence capabilities against foreign invasion. In leukaemia, the WBC count may reach as high as 500 000/mm$^3$, compared with the normal 7000/mm$^3$, but because most of these cells are abnormal or immature, they cannot perform their normal defence functions. Another devastating consequence of leukaemia is displacement of the other blood cell lines in the bone marrow. This results in anaemia due to a reduction in erythropoiesis and in internal bleeding caused by a deficit of platelets. Platelets play a critical role in preventing bleeding from the myriad tiny breaks that normally occur in small blood vessel walls. Consequently, overwhelming infections and haemorrhage are the most common causes of death in leukemic patients. In the next section we examine the role of platelets in detail to show how they normally minimize the threat of haemorrhage.

### Check Your Understanding 10.3

1. Name the polymorphonuclear granulocytes and the mononuclear agranulocytes.
2. State the function of each type of leukocyte.

## 10.5 | Platelets and Haemostasis

In addition to erythrocytes and leukocytes, **platelets (thrombocytes)** are a third type of cellular element in blood. An average of 250 million platelets are normally present in each millilitre of blood (range of 150 000 to 350 000 / m$^3$).

## Platelets

Platelets are not whole cells; instead, they are small cell fragments (about 2–4 μm in diameter) shed from the outer edges of extraordinarily large (up to 60 μm in diameter) bone marrow–bound cells known as **megakaryocytes** (〉 Figure 10-10; also see Figure 10-9). A single megakaryocyte typically produces about 1000 platelets. Megakaryocytes are derived from the same undifferentiated stem cells that give rise to the erythrocytic and leukocytic cell lines. Platelets are essentially detached vesicles containing pieces of megakaryocyte cytoplasm wrapped in plasma membrane.

Platelets remain functional for an average of 10 days, at which time they are removed from circulation by the tissue macrophages, especially those in the spleen and liver, and are replaced by newly released platelets from the bone marrow. The hormone **thrombopoietin**, produced by the liver, increases the number of megakaryocytes in the bone marrow and stimulates each megakaryocyte to produce more platelets. The factors that control thrombopoietin secretion and regulate the platelet level are currently under investigation.

Platelets do not leave the blood as WBCs do, but at any given time about one-third of them are in storage in blood-filled spaces in the spleen. These stored platelets can be released from the spleen into the circulating blood as needed (e.g., during haemorrhage) by sympathetically induced splenic contraction.

Because platelets are cell fragments, they lack nuclei. However, they have organelles and cytosolic enzyme systems that generate energy and synthesize secretory products, which they store in numerous granules dispersed throughout the cytosol. Furthermore, platelets contain high concentrations of actin and myosin, which enable them to contract. Their secretory and contractile abilities are important in haemostasis, the topic we now examine.

Megakaryocyte
Clusters of platelets about to shed off
Cluster of developing erythrocytes
Developing leukocyte

Biophoto Associates/Science Source

〉 **FIGURE 10-10 Photomicrograph of a megakaryocyte forming platelets**

## Haemostasis

**Haemostasis** is the arrest of bleeding from a broken blood vessel—that is, the stopping of haemorrhage (*haemo* means "blood"; *stasis* means "standing"). (Be sure not to confuse this with the term *homeostasis*.) For bleeding to take place from a vessel, there must be a break in the vessel wall and the pressure inside the vessel must be greater than the pressure outside it to force blood out through the defect.

The small capillaries, arterioles, and venules are often ruptured by minor traumas of everyday life; such traumas are the most common source of bleeding, although we often are not even aware that any damage has taken place. The body's inherent haemostatic mechanisms normally are adequate to seal defects and stop blood loss through these small microcirculatory vessels. The much rarer occurrence of bleeding from medium- or large-size vessels usually cannot be stopped by the body's haemostatic mechanisms alone. Bleeding from a severed artery is more profuse and therefore more dangerous than venous bleeding, because the outward driving pressure is greater in the arteries (i.e., arterial blood pressure is much higher than venous pressure). First-aid measures for a severed artery include applying to the wound external pressure that is greater than the arterial blood pressure to temporarily halt the bleeding until the torn vessel can be surgically closed. Haemorrhage from a traumatized vein can often be stopped simply by elevating the bleeding body part to reduce gravity's effects on pressure in the vein (p. 416). If the accompanying drop in venous pressure is not enough to stop the bleeding, mild external compression is usually adequate.

Haemostasis involves three major steps: (1) *vascular spasm*, (2) *formation of a platelet plug*, and (3) *blood coagulation (clotting)*. Platelets play a pivotal role in haemostasis. They obviously play a major part in forming a platelet plug, but they also contribute significantly to the other two steps.

### Vascular spasm

A cut or torn blood vessel immediately constricts. The underlying mechanism is unclear but is thought to be an intrinsic response triggered by a paracrine released locally from the inner lining (endothelium) of the injured vessel. This constriction, or **vascular spasm**, slows blood flow through the defect and thus minimizes blood loss. Also, as the opposing endothelial surfaces of the vessel are pressed together by this initial vascular spasm, they become sticky and adhere to each other, further sealing off the damaged vessel. These physical measures alone cannot completely prevent further blood loss, but they minimize blood flow through the break in the vessel until the other haemostatic measures can actually plug up the hole.

### Platelet aggregation

Platelets normally do not stick to the smooth endothelial surface of blood vessels, but when this lining is disrupted because of vessel injury, platelets become activated by the exposed *collagen*, which is a fibrous protein in the underlying connective tissue.

© 2016 Cengage

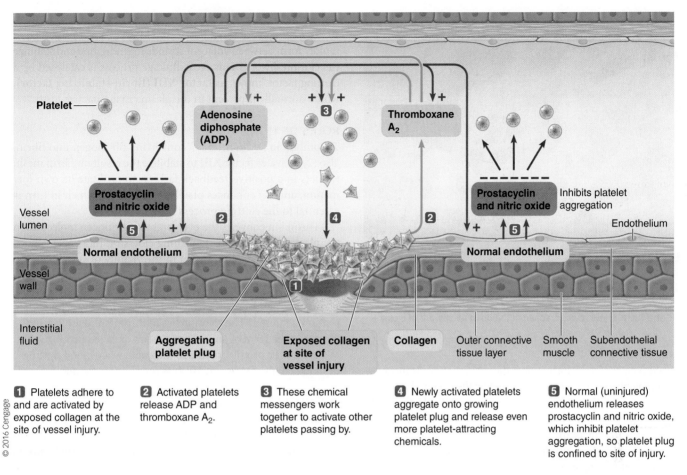

**1** Platelets adhere to and are activated by exposed collagen at the site of vessel injury.

**2** Activated platelets release ADP and thromboxane A₂.

**3** These chemical messengers work together to activate other platelets passing by.

**4** Newly activated platelets aggregate onto growing platelet plug and release even more platelet-attracting chemicals.

**5** Normal (uninjured) endothelium releases prostacyclin and nitric oxide, which inhibit platelet aggregation, so platelet plug is confined to site of injury.

› **FIGURE 10-11 Formation of a platelet plug.** Platelets aggregate at a vessel defect through a positive-feedback mechanism involving the release of adenosine diphosphate (ADP) and thromboxane **A₂** from platelets, which stick to exposed collagen at the site of the injury. Platelets are prevented from aggregating at the adjacent normal vessel lining by the release of prostacyclin and nitric oxide from the undamaged endothelial cells.

When activated, platelets quickly adhere to the collagen and form a haemostatic **platelet plug** at the site of the defect. Once platelets start aggregating, they release several important chemicals from their storage granules. Among these chemicals is *adenosine diphosphate (ADP)*, which causes the surface of nearby circulating platelets to become sticky so that they adhere to the first layer of aggregated platelets. These newly aggregated platelets release more ADP, which causes more platelets to pile on, and so on. In this way, a plug of platelets rapidly builds up at the defect site in a positive-feedback fashion (› Figure 10-11). Activation of platelets by ADP also stimulates platelets to release another prothrombotic factor, *thromboxane A₂*. This released thromboxane A₂ enhances the positive-feedback pathway by further enhancing platelet recruitment and ADP release, as well as by directly stimulating platelet aggregation.

Given the self-perpetuating nature of platelet aggregation, why doesn't the platelet plug continue to develop and expand over the surface of the adjacent normal vessel lining? A key reason is that ADP and other chemicals released by the activated platelets stimulate the release of *prostacyclin* and *nitric oxide* from the adjacent normal endothelium. Both these chemicals profoundly inhibit platelet aggregation. Thus, the platelet plug is limited to the defect and does not spread to the nearby undamaged vascular tissue (› Figure 10-11).

The aggregated platelet plug not only physically seals the break in the vessel but also performs three other important roles. (1) The actin-myosin complexes within the aggregated platelets contract to compact and strengthen what was originally a fairly loose plug. (2) The chemicals released from the platelet plug include several powerful vasoconstrictors (serotonin, epinephrine, and thromboxane A₂), which induce profound constriction of the affected vessel to reinforce the initial vascular spasm. (3) The platelet plug releases other chemicals that enhance blood coagulation, the next step of haemostasis. Although the platelet-plugging mechanism alone is often enough to seal the myriad minute tears in capillaries and other small vessels that occur many times daily, larger holes in vessels require the formation of a blood clot to completely stop the bleeding.

## Clot formation

**Blood coagulation**, or **clotting**, is the transformation of blood from a liquid into a solid gel. Formation of a clot on top of the platelet plug strengthens and supports the plug, reinforcing the seal over a break in a vessel. Furthermore, as blood in the vicinity of the vessel defect solidifies, it can no longer flow. Clotting is the body's most powerful haemostatic mechanism. It is required to stop bleeding from all but the most minute defects.

10

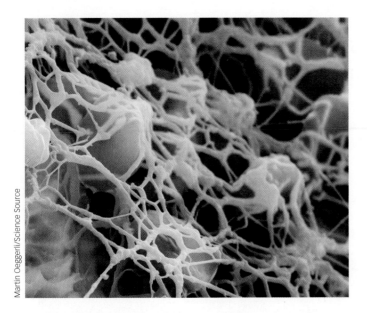

> FIGURE 10-12 **Erythrocytes trapped in the fibrin meshwork of a clot**

## CLOT FORMATION

The ultimate step in clot formation is the conversion of fibrinogen—a large, soluble plasma protein produced by the liver and normally always present in the plasma—into **fibrin**, an insoluble, threadlike molecule. This conversion into fibrin is catalyzed by the enzyme **thrombin** at the site of the injury. Fibrin molecules adhere to the damaged vessel surface, forming a loose, netlike meshwork that traps blood cells, including aggregating platelets. The resulting mass, or **clot**, typically appears red because of the abundance of trapped RBCs, but the foundation of the clot is formed of fibrin derived from the plasma (› Figure 10-12). Except for platelets, which play an important role in ultimately bringing about the conversion of fibrinogen to fibrin, clotting can take place in the absence of all other blood cells.

The original fibrin web is rather weak, because the fibrin strands are only loosely interlaced. However, chemical linkages rapidly form between adjacent strands to strengthen and stabilize the clot meshwork. This cross-linkage process is catalyzed by a clotting factor known as **factor XIII (fibrin-stabilizing factor)**, which normally is present in the plasma in an inactive form.

### ROLES OF THROMBIN

Thrombin, in addition to (1) converting fibrinogen into fibrin, also (2) activates factor XIII to stabilize the resultant fibrin mesh, (3) acts in a positive-feedback fashion to facilitate its own formation, and (4) enhances platelet aggregation, which in turn is essential to the clotting process (› Figure 10-13).

Thrombin converts the ever-present fibrinogen molecules in the plasma into a blood-staunching clot, so it must normally be absent from the plasma, except in the vicinity of vessel damage. Otherwise, blood would always be coagulated—a situation incompatible with life. How can thrombin normally be absent from the plasma, yet be readily available to trigger fibrin formation when a vessel is injured? The solution lies in thrombin's existence in the plasma in the form of an inactive precursor called **prothrombin**. What converts prothrombin into thrombin when blood clotting is desirable? This conversion involves the clotting cascade.

### THE CLOTTING CASCADE

Yet another activated plasma clotting factor, **factor X**, converts prothrombin to thrombin; factor X itself is normally present in the blood in an inactive form and must be converted into its active form by still another activated factor, and so on. Altogether, 12 plasma clotting factors participate in essential steps that lead to the final conversion of fibrinogen into a stabilized fibrin mesh (› Figure 10-14). These factors are designated by Roman numerals in the order they were discovered, not in the order they participate in the clotting process. Most of these clotting factors are plasma

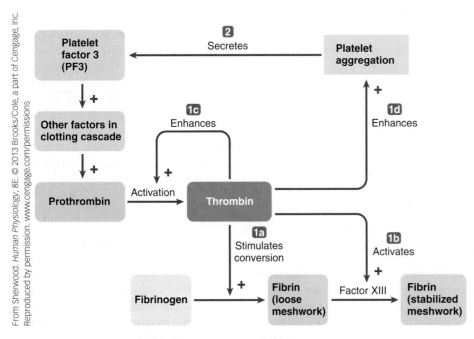

> FIGURE 10-13 **Roles of thrombin in haemostasis**

**1** Thrombin, a component of the clotting cascade, plays multiple roles in haemostasis:

**1a** It stimulates conversion of fibrinogen to fibrin.

**1b** It activates the factor that stabilizes the fibrin meshwork of the clot.

**1c** It enhances activation of more prothrombin into thrombin through positive feedback.

**1d** It enhances platelet aggregation.

**2** Through positive feedback, aggregated platelets secrete PF3, which stimulates the clotting cascade that results in thrombin activation.

**10**

proteins synthesized by the liver. One consequence of liver disease is that clotting time is prolonged due to a reduced production of clotting factors. Normally, they are always present in the plasma in an inactive form, such as fibrinogen and prothrombin. In contrast to fibrinogen, which is converted into insoluble fibrin strands, prothrombin and the other precursors act as proteolytic (protein-splitting) enzymes when converted to their active form. These enzymes activate other specific factors in the clotting sequence. Once the first factor in the sequence is activated, it activates the next factor, and so on, in a series of sequential reactions known as the **clotting cascade**; this continues until thrombin catalyzes the final conversion of fibrinogen into fibrin. Several of these steps require the presence of plasma calcium and *platelet factor 3 (PF3)*, a phospholipid secreted by the aggregated platelet plug. Thus, platelets also contribute to clot formation.

## INTRINSIC AND EXTRINSIC PATHWAYS

The clotting cascade may be triggered by the *intrinsic pathway* or the *extrinsic pathway*:

- The **intrinsic pathway** precipitates clotting within damaged vessels as well as clotting of blood samples in test tubes. All elements necessary to bring about clotting by means of the intrinsic pathway are present in the blood. This pathway, which involves seven separate steps (shown in blue in Figure 10-14), is set off when **factor XII (Hageman factor)**

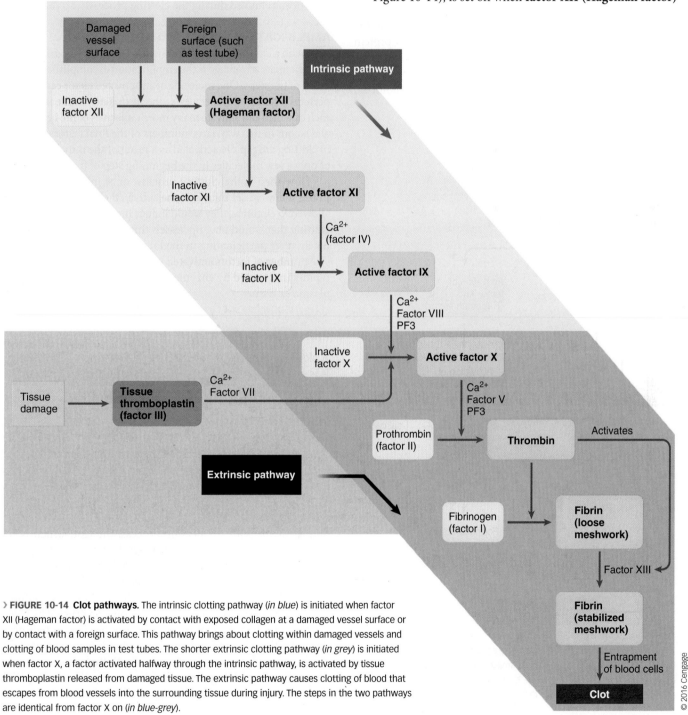

> **FIGURE 10-14 Clot pathways.** The intrinsic clotting pathway (*in blue*) is initiated when factor XII (Hageman factor) is activated by contact with exposed collagen at a damaged vessel surface or by contact with a foreign surface. This pathway brings about clotting within damaged vessels and clotting of blood samples in test tubes. The shorter extrinsic clotting pathway (*in grey*) is initiated when factor X, a factor activated halfway through the intrinsic pathway, is activated by tissue thromboplastin released from damaged tissue. The extrinsic pathway causes clotting of blood that escapes from blood vessels into the surrounding tissue during injury. The steps in the two pathways are identical from factor X on (*in blue-grey*).

is activated by coming into contact with either exposed collagen in an injured vessel or a foreign surface, such as a glass test tube. Remember that exposed collagen also initiates platelet aggregation. In this way, the formation of a platelet plug and the chain reaction leading to clot formation are simultaneously set in motion when a vessel is damaged. Furthermore, these complementary haemostatic mechanisms reinforce each other. The aggregated platelets secrete PF3, which is essential for the clotting cascade that, in turn, enhances further platelet aggregation (› Figure 10-15; also see Figure 10-13).

- The **extrinsic pathway** takes a shortcut and requires only four steps (shown in grey in Figure 10-14). This pathway, which requires contact with tissue factors external to the blood, initiates clotting of blood that has escaped into the tissues. When a tissue is traumatized, it releases a protein complex known as **tissue thromboplastin**. Tissue thromboplastin directly activates factor X, thereby bypassing all preceding steps of the intrinsic pathway. From this point on, the two pathways are identical.

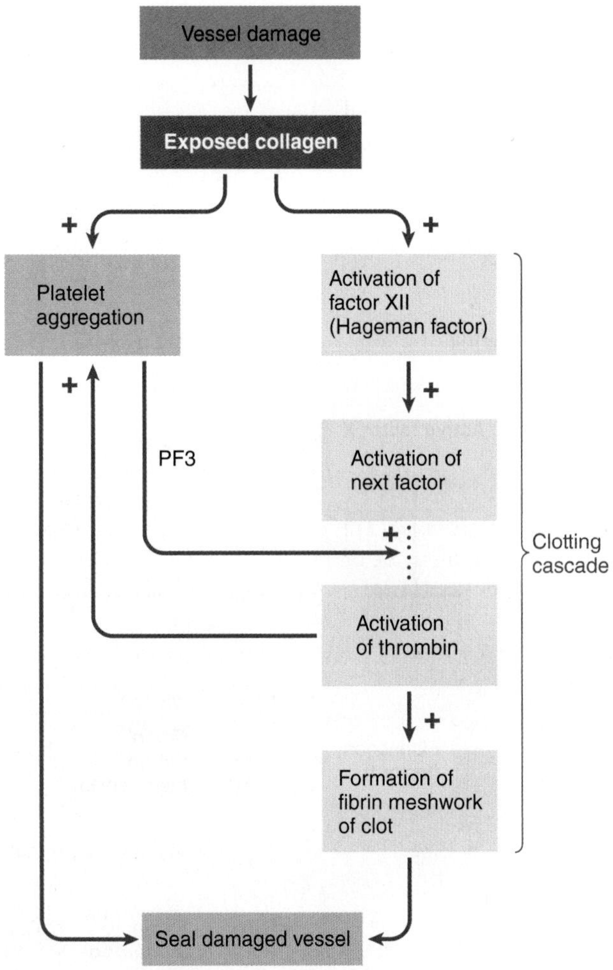

> **FIGURE 10-15 Concurrent platelet aggregation and clot formation.**
Exposed collagen at the site of vessel damage simultaneously initiates platelet aggregation and the clotting cascade. These two haemostatic mechanisms positively reinforce each other as they seal the damaged vessel.

The intrinsic and extrinsic mechanisms usually operate simultaneously. When tissue injury involves rupture of vessels, the intrinsic mechanism stops blood in the injured vessel, whereas the extrinsic mechanism clots blood that escaped into the tissue before the vessel was sealed off. Typically, clots are fully formed in three to six minutes.

### CLOT RETRACTION

Once a clot is formed, contraction of the platelets trapped within the clot shrinks the fibrin mesh, pulling the edges of the damaged vessel closer together. During **clot retraction**, fluid is squeezed from the clot. This fluid, called **serum**, is essentially plasma minus the fibrinogen and other clotting precursors that have been removed during the clotting process.

### AMPLIFICATION DURING CLOTTING PROCESS

Although a clotting process that involves so many steps may seem inefficient, its advantage is the amplification accomplished during many of the steps. One molecule of an activated factor can activate perhaps a hundred molecules of the next factor in the sequence, each of which can activate many more molecules of the next factor, and so on. In this way, large numbers of the final factors involved in clotting are rapidly activated as a result of the initial activation of only a few molecules in the beginning step of the sequence.

How then is the clotting process, once initiated, confined to the site of vessel injury? If the activated clotting factors were allowed to circulate, they would induce inappropriate widespread clotting that would plug up vessels throughout the body. Fortunately, after participating in the local clotting process, the massive number of factors activated in the vicinity of vessel injury is rapidly inactivated by enzymes and other factors present in the plasma or tissue.

## Fibrinolytic plasmin

A clot is not meant to be a permanent solution to vessel injury. It is a transient device to stop bleeding until the vessel can be repaired.

### VESSEL REPAIR

The aggregated platelets secrete a chemical that helps promote the invasion of fibroblasts (fibre formers) from the surrounding connective tissue into the wounded area of the vessel. Fibroblasts form a scar at the vessel defect.

### CLOT DISSOLUTION

Simultaneous with the healing process, the clot, which is no longer needed to prevent haemorrhage, is slowly dissolved by a fibrinolytic (fibrin-splitting) enzyme called **plasmin**. If clots were not removed after they performed their haemostatic function, the vessels, especially the small ones that endure tiny ruptures on a regular basis, would eventually become obstructed by clots.

Plasmin, like the clotting factors, is a plasma protein produced by the liver and present in the blood in an inactive precursor form, **plasminogen**. Plasmin is activated in a fast cascade of reactions involving many factors, among them factor XII (Hageman factor), which also triggers the chain reaction leading

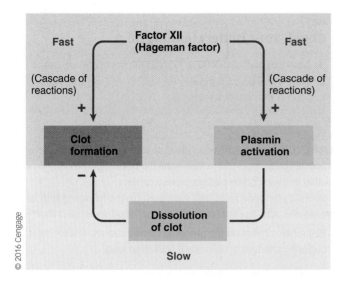

> **FIGURE 10-16** **Role of factor XII in clot formation and dissolution.** Activation of factor XII (Hageman factor) simultaneously initiates a fast cascade of reactions that results in clot formation and a fast cascade of reactions that results in plasmin activation. Plasmin, which is trapped in the clot, subsequently slowly dissolves the clot. This action removes the clot when it is no longer needed after the vessel has been repaired.

to clot formation (> Figure 10-16). When a clot is rapidly being formed, activated plasmin becomes trapped in the clot and later dissolves it by slowly breaking down the fibrin meshwork.

Phagocytic white blood cells gradually remove the products of clot dissolution. Following an injury, you have likely observed the black-and-blue marks of bruised skin. The coloration results from the slow removal of the deoxygenated clotted blood that escaped into the tissue layers of your skin. This blood is eventually cleared by plasmin action, followed by the phagocytic cleanup crew.

### PREVENTING INAPPROPRIATE CLOT FORMATION

In addition to removing clots that are no longer needed, plasmin functions continually to prevent clots from forming inappropriately. Throughout the vasculature, small amounts of fibrinogen are constantly converted into fibrin, triggered by unknown mechanisms. Clots do not develop, however, because plasmin, activated by **tissue plasminogen activator (tPA)**, quickly gets rid of the fibrin from the tissues, especially the lungs. Normally, the low level of fibrin formation is counterbalanced by a low level of fibrinolytic activity, so inappropriate clotting does not occur. Only when a vessel is damaged do additional factors precipitate the explosive chain reaction that leads to more extensive fibrin formation and results in local clotting at the site of injury.

Genetically engineered tPA and other similar chemicals that trigger clot dissolution are frequently used to limit damage to cardiac muscle during heart attacks. Administering a clot-busting drug within the first hours after a clot has blocked a coronary (heart) vessel often dissolves the clot in time to restore blood flow to the cardiac muscle that is supplied by the blocked vessel before the muscle dies of oxygen deprivation. In recent years, tPA and related drugs have also been used successfully to promptly dissolve a stroke-causing clot within a cerebral

(brain) blood vessel, thereby minimizing loss of irreplaceable brain tissue after a stroke (p. 114).

### Inappropriate clotting

Despite protective measures, clots occasionally form in intact vessels. Abnormal or excessive clot formation within blood vessels—what has been dubbed "haemostasis in the wrong place"—can compromise blood flow to vital organs. The body's clotting and anticlotting systems normally function in a check-and-balance manner. Acting in concert, they permit prompt formation of "good" blood clots, thus minimizing blood loss from damaged vessels, while preventing "bad" clots from forming and blocking blood flow in intact vessels. An abnormal intravascular clot attached to a vessel wall is known as a **thrombus**, and a freely floating clot is called **embolus**. An enlarging thrombus narrows and can eventually completely occlude the vessel in which it forms. By entering and completely plugging a smaller vessel, a circulating embolus can suddenly block blood flow (see Figure 8-29).

Several factors, acting independently or simultaneously, can cause *thromboembolism*: (1) Roughened vessel surfaces associated with atherosclerosis can lead to thrombus formation (p. 378). (2) Imbalances in the clotting–anticlotting systems can trigger clot formation. (3) Slow-moving blood is more apt to clot, probably because small quantities of fibrin accumulate in the stagnant blood, for example, in blood pooled in varicosed leg veins (p. 418). (4) Widespread clotting is occasionally triggered by the release of tissue thromboplastin into the blood from large amounts of traumatized tissue. Similar widespread clotting can occur in **septicaemic shock**, in which bacteria or their toxins initiate the clotting cascade.

### Haemophilia

In contrast to inappropriate clot formation in intact vessels, the opposite haemostatic disorder is the failure of clots to form promptly in injured vessels, resulting in life-threatening haemorrhage even from relatively mild traumas. The most common cause of excessive bleeding is **haemophilia**, which is caused by a deficiency of one of the factors in the clotting cascade. Although a deficiency of any of the clotting factors could block the clotting process, 80 percent of all haemophiliacs lack the genetic ability to synthesize factor VIII.

People with a platelet deficiency, in contrast to the more profuse bleeding that accompanies defects in the clotting mechanism, continuously develop hundreds of small, confined haemorrhagic areas throughout the body tissues as blood leaks from tiny breaks in the small blood vessels before coagulation takes place. Platelets normally are the primary sealers of these ever-occurring minute ruptures. In the skin of a platelet-deficient person, the diffuse capillary haemorrhages are visible as small, purplish blotches, giving rise to the term **thrombocytopenia purpura** (the purple of thrombocyte deficiency) for this condition. (Recall that *thrombocyte* is another name for platelet.)

Vitamin K deficiency can also cause a bleeding tendency. Vitamin K, commonly known as the blood-clotting vitamin, is essential for normal clot formation. Researchers recently figured out vitamin K's role in the clotting process. In a complex sequence of biochemical events, vitamin K combines with oxygen, releasing free energy that is ultimately used in the activation processes of the clotting cascade.

**Check Your Understanding 10.4**

1. Describe how a platelet plug forms at the site of a vessel defect.
2. Draw a flow diagram showing the intrinsic and extrinsic pathways of the clotting cascade.

# Chapter in Perspective: Focus on Homeostasis

Blood contributes to homeostasis in a variety of ways. First, the composition of interstitial fluid—the true internal environment that surrounds and directly exchanges materials with the cells—depends on the composition of blood plasma. Because of the thorough exchange that occurs between the interstitial and vascular compartments, interstitial fluid has the same composition as plasma with the exception of plasma proteins, which cannot escape through the capillary walls. Thus, blood serves as the vehicle for rapid, long-distance, mass transport of materials to and from the cells, and interstitial fluid serves as the go-between.

Homeostasis depends on the blood carrying materials such as oxygen and nutrients to the cells as rapidly as the cells consume these supplies, and carrying materials such as metabolic wastes away from the cells as rapidly as the cells produce these products. It also depends on the blood carrying hormonal messengers from their site of production to their distant site of action. Once a substance enters the blood, it can be transported throughout the body within seconds, whereas diffusion of the substance over long distances in a large multicellular organism, such as a human, would take months to years—a situation incompatible with life. Diffusion can, however, effectively accomplish short local exchanges of materials between blood and surrounding cells through the intervening interstitial fluid.

Blood has special transport capabilities that enable it to move its cargo efficiently throughout the body. For example, life-sustaining oxygen is poorly soluble in water, but blood is equipped with $O_2$-carrying specialists, the erythrocytes (red blood cells), which are stuffed full of haemoglobin, a complex molecule that transports oxygen. Likewise, homeostatically important water-insoluble hormonal messengers are shuttled in the blood by plasma protein carriers.

Specific components of the blood perform the following additional homeostatic activities that are unrelated to blood's transport function:

- Blood helps maintain the proper pH in the internal environment by buffering changes in the body's acid–base load.

- Blood helps maintain body temperature by absorbing heat produced by heat-generating tissues such as contracting skeletal muscles and distributing it throughout the body. Excess heat is carried by the blood to the body surface for elimination to the external environment.

- Electrolytes in the plasma are important in membrane excitability, which is critical for nerve and muscle function.

- Electrolytes in the plasma are important in osmotic distribution of fluid between the extracellular and intracellular fluid. Plasma proteins play a critical role in distributing extracellular fluid between the plasma and interstitial fluid.

- Through their haemostatic functions, the platelets and clotting factors minimize the loss of life-sustaining blood after vessel injury.

- The leukocytes (white blood cells), their secretory products, and certain types of plasma proteins, such as antibodies, constitute the immune defence system. This system defends the body against invading disease-causing agents, destroys cancer cells, and paves the way for wound healing and tissue repair by clearing away debris from dead or injured cells. These actions indirectly contribute to homeostasis by helping the organs that directly maintain homeostasis stay healthy. We could not survive beyond early infancy were it not for the body's defence mechanisms.

## CHAPTER TERMINOLOGY

ABO system (p. 438)
albumins (p. 435)
alpha ($\alpha$) globulins (p. 435)
anaemia (p. 441)
aplastic anaemia (p. 441)
B lymphocytes (B cells) (p. 447)
basophils (p. 447)
beta ($\beta$) globulins (p. 435)
bicarbonate ion ($HCO_3^-$) (p. 437)
blood coagulation (p. 449)
bone marrow (p. 437)
carbonic anhydrase (p. 437)
clot (p. 450)
clot retraction (p. 452)
clotting (p. 449)

clotting cascade (p. 451)
embolus (p. 453)
eosinophilia (p. 446)
eosinophils (p. 446)
erythrocytes (red blood cells or RBCs) (p. 436)
erythropoiesis (p. 437)
erythropoietin (EPO) (p. 438)
extrinsic pathway (p. 452)
factor X (p. 450)
factor XII (Hageman factor) (p. 451)
factor XIII (fibrin-stabilizing factor) (p. 450)
fibrin (p. 450)
fibrinogen (p. 435)
gamma ($\gamma$) globulins (p. 435)

globin (p. 436)
glycolytic enzymes (p. 437)
granulocyte colony–stimulating factor (p. 446)
granulocytes (p. 446)
haematocrit (p. 434)
haemoglobin (p. 436)
haemolysis (p. 441)
haemolytic anaemia (p. 441)
haemophilia (p. 453)
haemorrhagic anaemia (p. 441)
Haemostasis (p. 448)
heme groups (p. 436)
immune system (p. 445)
Immunity (p. 444)
infectious mononucleosis (p. 447)

10

## REVIEW EXERCISES

### Objective Questions (Answers in Appendix E, p. A-45)

1. Blood can absorb metabolic heat while undergoing only small changes in temperature. *(True or false?)*

2. Haemoglobin can carry only oxygen. *(True or false?)*

3. Erythrocytes, leukocytes, and platelets all originate from the same undifferentiated stem cells. *(True or false?)*

4. Erythrocytes are unable to use the oxygen they contain for their own ATP formation. *(True or false?)*

5. White blood cells spend the majority of their time in the blood. *(True or false?)*

6. Which type of leukocyte is produced primarily in lymphoid tissue? _____

7. Most clotting factors are synthesized by the _____.

8. Which of the following is NOT a function of plasma proteins?
   a. facilitating retention of fluid in the blood vessels
   b. playing an important role in blood clotting
   c. transporting water-insoluble substances in the blood
   d. transporting oxygen in the blood
   e. serving as antibodies
   f. contributing to the buffering capacity of the blood

9. Which of the following is NOT directly triggered by exposed collagen in an injured vessel?
   a. initial vascular spasm
   b. platelet aggregation
   c. activation of the clotting cascade
   d. activation of plasminogen

10. Match the following (an answer may be used more than once):
   ___ 1. causes platelets to aggregate in feedback fashion
   ___ 2. activates prothrombin
   ___ 3. fibrinolytic enzyme
   ___ 4. inhibits platelet aggregation

   ___ 5. first factor activated in intrinsic clotting pathway
   ___ 6. forms meshwork of the clot
   ___ 7. stabilizes the clot
   ___ 8. activates fibrinogen
   ___ 9. activated by tissue thromboplastin

   (a) prostacyclin
   (b) plasmin
   (c) ADP
   (d) fibrin
   (e) thrombin
   (f) factor X
   (g) factor XII
   (h) factor XIII

11. Match the following blood abnormalities with their causes:
   ___ 1. deficiency of intrinsic factor
   ___ 2. insufficient amount of iron to synthesize adequate haemoglobin
   ___ 3. destruction of bone marrow
   ___ 4. abnormal loss of blood
   ___ 5. tumour-like condition of bone marrow
   ___ 6. inadequate erythropoietin secretion
   ___ 7. excessive rupture of circulating erythrocytes
   ___ 8. associated with living at high altitudes

   (a) haemolytic anaemia
   (b) aplastic anaemia
   (c) nutritional anaemia
   (d) haemorrhagic anaemia
   (e) pernicious anaemia
   (f) renal anaemia
   (g) primary polycythaemia
   (h) secondary polycythaemia

### Written Questions

1. What is the average blood volume in women and in men?

2. What is the normal percentage of blood occupied by erythrocytes and by plasma? What is the haematocrit? What is the buffy coat?

3. What is the composition of plasma?

4. List the three major groups of plasma proteins, and state their functions.

5. Describe the structure and functions of erythrocytes.

6. Why can erythrocytes survive for only about 120 days?

7. Describe the process and control of erythropoiesis.

**10**

8. Compare the structures, functions, and lifespans of the five types of leukocytes.

9. Discuss the derivation of platelets.

10. Describe the three steps of haemostasis, including a comparison of the intrinsic and extrinsic pathways by which the clotting cascade is triggered.

11. Compare plasma and serum.

12. What normally prevents inappropriate clotting in vessels?

## Quantitative Exercises
### (Solutions in Appendix E, p. A-45)

1. The normal concentration of haemoglobin in blood (as measured clinically) is 15 g/100 mL of blood.

    a. Given that 1 mole of haemoglobin weighs 66 000 g, what is the concentration of haemoglobin in millimoles (mM)?

    b. Each haemoglobin molecule can bind four molecules of oxygen. What is the concentration of oxygen bound to haemoglobin at maximal saturation (in mM)?

    c. Given that 1 mole of an ideal gas occupies 22.4 L, what is the maximal carrying capacity of normal blood for oxygen (usually expressed in mL of $O_2$/litre of blood)?

2. Assume that the blood sample in ⟩ Figure 10-5b, is from a patient with haemorrhagic anaemia. Given a normal blood volume of 5 L, a normal red blood cell concentration of 5 billion/mL, and an RBC production rate of 3 million cells/second, how long will it take the body to return the haematocrit to normal?

3. Note that in the blood sample (in Figure 10-5c) from a patient with polycythaemia, the haematocrit has increased to 70 percent. An increased haematocrit increases blood viscosity, which in turn increases total peripheral resistance and increases the workload on the heart. The effect of haematocrit ($h$) on relative blood viscosity ($v$, viscosity relative to that of water) is given approximately by the following equation:

$$v = 1.5 \times \exp(2h)$$

Note that in this equation, $h$ is the haematocrit as a fraction, not a percentage. Given a normal haematocrit of 0.40, what percent increase in viscosity would result from the polycythaemia in Figure 10-5c? What percentage change would this cause in total peripheral resistance?

## POINTS TO PONDER

### (Explanations in Appendix E, p. A-46)

1. A person has a haematocrit of 62. Can you conclude from this finding that the person has polycythaemia? Explain.

2. There are different forms of haemoglobin. *Haemoglobin A* is normal adult haemoglobin. The abnormal form *haemoglobin S* causes RBCs to warp into fragile, sickle-shaped cells. Fetal RBCs contain *haemoglobin F*, the production of which stops soon after birth. Now researchers are trying to goad the genes that direct haemoglobin *F* synthesis back into action as a means of treating sickle cell anaemia. Explain how turning on these fetal genes could be a useful remedy. (Indeed, the first effective drug therapy recently approved for treating sickle cell anaemia, *hydroxyurea*, acts on the bone marrow to boost production of fetal haemoglobin.)

3. Low on the list of popular animals are vampire bats, leeches, and ticks, yet these animals may someday indirectly save your life. Scientists are currently examining the "saliva" of these blood-sucking creatures in search of new chemicals that might limit cardiac muscle damage in heart attack victims. What do you suspect is the nature of these sought-after chemicals?

4. *Porphyria* is a genetic disorder that shows up in about 1 in every 25 000 individuals. Affected individuals lack certain enzymes that are part of a metabolic pathway leading to formation of heme, which is the iron-containing group of haemoglobin. An accumulation of porphyrins, which are intermediates of the pathway, causes a variety of symptoms, especially after exposure to sunlight. Lesions and scars form on the skin. Hair grows thickly on the face and hands. As gums retreat from teeth, the elongated canine teeth take on a fanglike appearance. Symptoms worsen on exposure to a variety of substances, including garlic and alcohol. Affected individuals avoid sunlight and aggravating substances and get injections of heme from normal red blood cells. If you are familiar with vampire stories, which date from the Middle Ages or earlier, speculate on how they may have evolved among superstitious folk who did not have medical knowledge of porphyria.

## CLINICAL CONSIDERATION

### (Explanation in Appendix E, p. A-46)

Linda P. has just been diagnosed as having pneumonia. Her white blood cell count is 7200/mm³, and 67 percent of the white blood cells are neutrophils. It will take several days to obtain a definitive answer as to the causative agent by culturing a sample of discharges from her respiratory system. Based on the WBC count, do you think that Linda should be given antibiotics immediately, long before the causative agent is actually known? Are antibiotics likely to combat her infection?

# Body Defences

Body systems maintain homeostasis

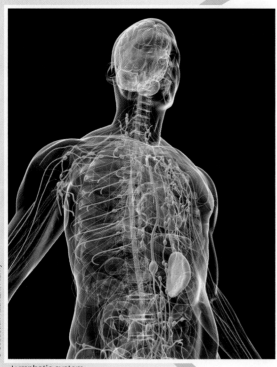

Lymphatic system

## Homeostasis

The immune system defends against foreign invaders and cancer cells and paves the way for tissue repair. The integumentary system (skin) prevents the entrance of external agents and the loss of internal fluid by serving as a protective barrier between the external environment and the remainder of the body.

Homeostasis is essential for survival of cells

Cells make up body systems

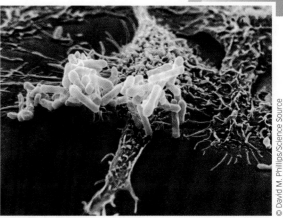

Macrophage ingesting pseudomonas

Humans constantly come into contact with external agents that could be harmful if they entered the body. The most serious are disease-causing microorganisms. If bacteria or viruses do gain entry, the body is equipped with a complex, multifaceted, internal defence system—the immune system—that provides continual protection against invasion by foreign agents. Furthermore, body surfaces exposed to the external environment, such as the integumentary system (skin), serve as a first line of defence to resist penetration by foreign microorganisms. The immune system also protects against cancer and paves the way for repair of damaged tissues.

The immune system indirectly contributes to homeostasis by helping maintain the health of organs that directly contribute to homeostasis.

# 11

# Body Defences

## CONTENTS AT A GLANCE

▌ **Clinical Connections**

Caitlin is a 14-year-old-girl who has had a history of seasonal hay fever. For the past day she has complained about the symptoms being worse than usual. During the night she woke as she was having problems drawing in breath. Gasping and wheezing she alerted her parents, who took her to the hospital where she was immediately treated for what appeared to be an asthma attack. Her blood work showed an elevated white cell count with normal haemoglobin, and there were eosinophils in her sputum. Her chest X-ray was normal. Once her breathing was under control she was discharged with asthma medication and a list of follow-up tests to be done. A pulmonary function test found that she had an obstructed airway, which was reversed with the b2-adrenergic stimulant normally given for asthma. At an allergy specialist, she had several positive skin tests. Ultimately, her diagnosis was changed to allergic asthma.

## 11.1 | Introduction

Immunity is the body's ability to resist or eliminate potentially harmful foreign materials or abnormal cells. The following activities are attributable to the **immune system**, an internal defence system that plays a key role in recognizing and either destroying or neutralizing materials within the body that are foreign to the **body's own cells**:

1. Defending against invading pathogens (microorganisms, e.g., bacteria and viruses)

2. Removing worn-out cells damaged by trauma or disease, and facilitating wound healing and tissue repair

3. Identifying and destroying abnormal or mutant cells that originated in the body, a task (*immune surveillance*) that is the chief internal-defence mechanism against cancer

4. Mounting inappropriate immune responses that lead either to *allergies*, which occur when the body turns against a normally harmless environmental chemical entity, or to *autoimmune diseases*, which happen when the defence system erroneously produces antibodies against a particular type of the body's own cells

Many actions we undertake throughout our daily lives influence the function of our immune systems and thus our immunity. However, only one action in particular has been shown to have a profound effect on the short-term (acute) and long-term (chronic) functioning of our immune system: exercise.

### Pathogenic bacteria and viruses

The primary foreign enemies that the immune system defends against are bacteria and viruses. **Bacteria** (which are large) are non-nucleated, single-celled microorganisms self-equipped with all the machinery essential for their own survival and reproduction. Pathogenic bacteria that invade the body cause tissue damage and produce disease largely by releasing enzymes or toxins that physically injure or functionally disrupt affected cells and organs. The disease-producing power of a pathogen is known as its **virulence**.

In contrast to bacteria, **viruses** (which are small) are not self-sustaining cellular entities. They consist only of nucleic acids (genetic material—DNA or RNA) enclosed by a protein coat. Because they lack cellular machinery for energy production and protein synthesis, viruses cannot carry out metabolism and reproduce unless they invade a **host cell** (a body cell of the infected individual) and take over the cell's biochemical facilities for their own uses. Not only do viruses sap the host cell's energy resources, but the viral nucleic acids also direct the host cell to synthesize proteins needed for viral replication.

When a virus becomes incorporated into a host cell, the body's own defence mechanisms may destroy the cell because they no longer recognize it as a "self" cell (they see the cell as foreign). Other ways in which viruses can lead to cell damage or death are by depleting essential cell components, dictating that the cell produce substances toxic to the cell, or transforming the cell into a cancer cell. Let's now take a brief look at the pathology of a virus.

Some common viruses are SARS (severe acute respiratory syndrome), H1N1 (influenza A virus), and avian (bird) flu. The H1N1 virus was responsible for the 2009 pandemic (epidemic of wide geography) that started in Mexico, while the avian flu was caused by a virus adapted to birds. Each is a viral infection that presents its own unique challenge to immunologists (those that study immunology) and virologists (those that study viruses).

#### SARS VIRUS

Here is a closer look at the SARS virus: it is caused by the coronavirus (common cold family) and spread by airborne respiratory secretions (coughing or sneezing infected droplets). SARS is contracted (caught) by breathing in or touching the respiratory secretions of an infected person. The virus may live on hands, tissues, and surfaces (e.g., door knob) for six hours in wet-droplet form and three hours in dried-droplet form. The virus has even been found to live for four days in feces, and might live for months when the temperature is below freezing. Many viruses are spread over large distances via world travel. If we take the first case of SARS as an example, it was spread by a traveller; a middle-age businessman travelling from China to Vietnam via Hong Kong. The businessman later died as a result of the virus. Outside of Asia, SARS had the greatest impact on Toronto, Canada, where over 25 000 persons were placed in quarantine. In total, 400 Canadians became ill with the disease and 44 died. The total death rate in 2003 was about 15 percent of cases, but in cases of persons over 65 years of age the death rate was about 50 percent.

Symptoms present in 2–10 days following contact with the virus, but there are cases outside this range. Persons with SARS symptoms are contagious, but a person may be contagious before symptoms present or after they subside. Typical symptoms include fever ($\geq 38.0°C$) and cough, difficulty breathing, or other respiratory symptoms. Treatment is commonly antibiotics, antiviral medications, high doses of steroids, and oxygen or mechanical ventilation. To reduce the risk of contracting SARS or other viruses, people should cover their mouth and nose in their arm when sneezing or coughing, and not share food, beverages, or utensils with others. Commonly touched surfaces should be cleaned with disinfectant. Although SARS is not a major health risk at this time, taking such basic precautions reduce the risk of infection and outbreaks.

#### BODY DEFENCES

When considering how the body defends against foreign pathogens, we can broadly classify the mechanisms as either **external defences** (the integumentary system) or **internal defences** (the immune system). We begin our discussion of body defences with an overview of the **integumentary system**. This is followed by a detailed description of how the immune system functions and its role in homeostasis.

## 11.2 | External Defences

The **skin (integument)**, which is the largest organ of the body, not only serves as a mechanical barrier between the external environment and the underlying tissues but also is dynamically involved in defence mechanisms and other important functions. The skin consists of two layers, an outer *epidermis* and an inner *dermis* (› Figure 11-1).

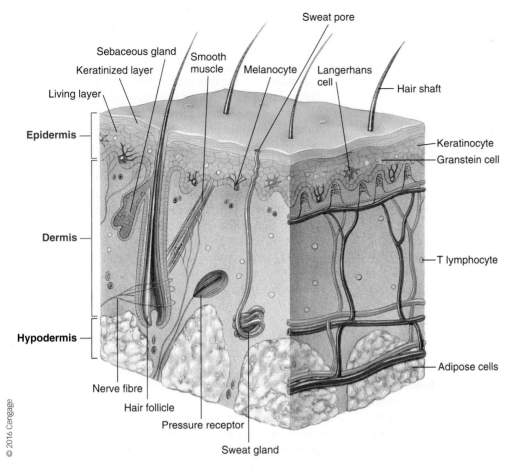

Labels in figure (clockwise/positional):
- Sweat pore
- Sebaceous gland
- Keratinized layer
- Living layer
- Smooth muscle
- Melanocyte
- Langerhans cell
- Hair shaft
- Keratinocyte
- Granstein cell
- T lymphocyte
- Adipose cells
- Nerve fibre
- Hair follicle
- Pressure receptor
- Sweat gland
- **Epidermis**
- **Dermis**
- **Hypodermis**

© 2016 Cengage

〉 **FIGURE 11-1 Anatomy of the skin.** The skin consists of two layers, a keratinized outer epidermis and a richly vascularized inner connective tissue dermis. Special infoldings of the epidermis form the sweat glands, sebaceous glands, and hair follicles. The epidermis contains four types of cells: keratinocytes, melanocytes, Langerhans cells, and Granstein cells. The skin is anchored to underlying muscle or bone by the hypodermis, a loose, fat-containing layer of connective tissue.

## Epidermis

The **epidermis** consists of numerous layers of epithelial cells. On average, the epidermis replaces itself about every two and a half months. The inner epidermal layers are composed of cube-shaped cells that are living and rapidly dividing, whereas the cells in the outer layers are dead and flattened. The epidermis has no direct blood supply. Its cells are nourished only by diffusion of nutrients from a rich vascular network in the underlying dermis. The newly forming cells in the inner layers constantly push the older cells closer to the surface, farther and farther from their nutrient supply. This, coupled with the fact that the outer layers are continuously subjected to pressure and "wear and tear," causes these older cells to die and become flattened. Epidermal cells are riveted tightly together by desmosomes (p. 34), which interconnect with intracellular keratin filaments (p. 34) to form a strong, cohesive covering. During maturation of a keratin-producing cell, keratin filaments progressively accumulate and cross-link with one another within the cytosol. As the outer cells die, this fibrous keratin core remains, forming flattened, hardened scales that provide a tough, protective, **keratinized layer**. As the scales of the outermost keratinized

layer slough or flake off through abrasion, they are continuously replaced by means of cell division in the deeper epidermal layers. The rate of cell division, and consequently the thickness of this keratinized layer, varies in different regions of the body. It is thickest in the areas where the skin is subjected to the most pressure, such as the bottom of the feet. The keratinized layer is air-tight, fairly waterproof, and impervious to most substances. It resists anything passing in either direction between the body and external environment. For example, it minimizes loss of water and other vital constituents from the body and prevents most foreign material from penetrating into the body.

*Clinical Note* This protective layer's value in holding in body fluids becomes obvious after severe burns. Bacterial infections can occur in the unprotected underlying tissue, but even more serious are the systemic consequences of loss of body water and plasma proteins, which escape from the exposed, burned surface. The resulting circulatory disturbances can be life threatening.

Likewise, the skin barrier impedes passage into the body of most materials that come into contact with the body surface, including bacteria and toxic chemicals. In many instances, the skin modifies compounds that come into contact with it. For example, epidermal enzymes can convert many potential carcinogens into harmless compounds. Some materials, however, especially lipid-soluble substances, can penetrate intact skin through the lipid bilayers of the plasma membranes of the epidermal cells. Drugs that can be absorbed by the skin, such as nicotine or estrogen, are sometimes used in the form of a cutaneous patch impregnated with the drug.

## Dermis

Under the epidermis is the **dermis**, a connective tissue layer that contains many elastin fibres (for stretch) and collagen fibres (for strength), as well as an abundance of blood vessels and specialized nerve endings. The dermal blood vessels not only supply both the dermis and epidermis but also play a major role in temperature regulation. The calibre of these vessels, and hence the volume of blood flowing through them, is subject to control to vary the amount of heat exchange between these skin surface vessels and the external environment (Chapter 16). Receptors

11

at the peripheral endings of afferent nerve fibres in the dermis detect pressure, temperature, pain, and other somatosensory input. Efferent nerve endings in the dermis control blood vessel calibre, hair erection, and secretion by the skin's exocrine glands.

## Skin's exocrine glands and hair follicles

Special infoldings of the epidermis into the underlying dermis form the skin's exocrine glands—the sweat glands and sebaceous glands—as well as the hair follicles. **Sweat glands**, which are distributed over most of the body, release a dilute salt solution through small openings, the sweat pores, onto the skin surface. Evaporation of this sweat cools the skin and is important in regulating temperature.

The amount of sweat produced is subject to regulation and depends on the environmental temperature, the amount of heat-generating skeletal muscle activity, and various emotional factors (e.g., a person often sweats when nervous). A special type of sweat gland located in the axilla (armpit) and pubic region produces a protein-rich sweat that supports the growth of surface bacteria, which give rise to a characteristic odour. In contrast, most sweat, as well as the secretions from the sebaceous glands, contains chemicals that are generally highly toxic to bacteria.

The cells of the **sebaceous glands** produce **sebum**, an oily secretion released into adjacent hair follicles. From there, sebum flows to the skin surface, oiling both the hairs and the outer keratinized layers of the skin, which helps to waterproof them and prevent them from drying and cracking. Chapped hands or lips indicate insufficient protection by sebum. The sebaceous glands are particularly active during adolescence, causing the oily skin common among teenagers.

Each **hair follicle** is lined with special keratin-producing cells, which secrete keratin and other proteins that form the hair shaft. Hairs increase the sensitivity of the skin's surface to tactile (touch) stimuli. In some other species, this function is more finely tuned. For example, the whiskers on a cat are exquisitely sensitive in this regard. An even more important role of hair in hairy species is heat conservation, a function that's not significant in humans. Like hair, the **nails** are another special keratinized product derived from living epidermal structures, the nail beds.

## Hypodermis

The skin is anchored to the underlying tissue (muscle or bone) by the **hypodermis** (*hypo* means "below"), also known as **subcutaneous tissue** (*sub* means "under"; *cutaneous* means "skin"), a loose layer of connective tissue. Most fat cells are housed within the hypodermis. These fat deposits throughout the body are collectively referred to as **adipose tissue**.

## Specialized cells

The epidermis contains four distinct resident cell types—*melanocytes, keratinocytes, Langerhans cells,* and *Granstein cells*—plus transient T lymphocytes that are scattered throughout the epidermis and dermis. Each of these resident cell types performs specialized functions.

### MELANOCYTES

**Melanocytes** produce the pigment melanin, which they disperse to surrounding skin cells. The amount and type of melanin, which can vary among black, brown, yellow, and red pigments, are responsible for the differences in human skin coloration. Fair-skinned people have about the same number of melanocytes as dark-skinned people; the difference in skin colour depends on the amount of melanin produced by each melanocyte. Melanin is produced through complex biochemical pathways in which the melanocyte enzyme *tyrosinase* plays a key role. Most people, regardless of skin colour, have enough tyrosinase that, if fully functional, could result in enough melanin to make their skin very black. In those with lighter skin, however, two genetic factors prevent this melanocyte enzyme from functioning at full capacity: (1) much of the tyrosinase produced is in an inactive form, and (2) various inhibitors that block tyrosinase action are produced. As a result, less melanin is produced.

In addition to hereditary determination of melanin content, the amount of this pigment can be increased transiently in response to exposure to ultraviolet (UV) light rays from the sun. This additional melanin provides the protective function of absorbing harmful UV rays—and gives the appearance of a suntan. For more information, see Why It Matters.

### KERATINOCYTES

The most abundant epidermal cells are the **keratinocytes**. As they die, they form the outer protective keratinized layer. They also generate hair and nails. A surprising, recently discovered function is that keratinocytes are also important immunologically. They secrete interleukin 1 (a product also secreted by macrophages), which influences the maturation of T cells that tend to localize in the skin. Interestingly, the epithelial cells of the thymus have been shown to have anatomic, molecular, and functional similarities to those of the skin. This suggests keratinocytes have an influence on some post-thymic steps in T-cell maturation that take place in the skin.

### OTHER IMMUNE CELLS OF THE SKIN

As well as keratinocytes, the Langerhans cells and the Granstein cells are the two other epidermal cell types that play a role in immunity. **Langerhans cells**, which migrate to the skin from the bone marrow, are dendritic cells that serve as antigen-presenting cells. Thus, the skin not only is a mechanical barrier but also alerts lymphocytes if the barrier is breached by invading microorganisms. Langerhans cells present antigen to helper T cells, facilitating their responsiveness to skin-associated antigens. In contrast, **Granstein cells** seem to act to hinder skin-activated immune responses. These cells are the most recently discovered and least understood of the skin's immune cells. Significantly, Langerhans cells are more susceptible to damage by UV radiation (as from the sun) than Granstein cells are. Losing Langerhans cells as a result of exposure to UV radiation can detrimentally lead to a predominant suppressor signal rather than the normally dominant helper signal, leaving the skin more vulnerable to microbial invasion and cancer cells.

The various epidermal components of the immune system are collectively termed **skin-associated lymphoid tissue (SALT)**.

The two major forms of ultraviolet (UV) light that are of concern are called UVA and UVB, both of which are of major health concern. Both UVA and UVB are capable of penetrating the skin. The longer wavelength UVA rays can penetrate deeper into the skin, but are potentially less reactive. In contrast, the shorter wavelength UVB rays cannot penetrate as deeply, but are more reactive. When melanin is present in the skin, it can act to neutralize the effects of these harmful rays. However, when the protective effects of melanin are overwhelmed, UV light can have many affects, ranging from sunburn to the development of skin cancer.

One of UV light's most deleterious actions is to damage DNA, which can be done directly or indirectly. Direct DNA damage occurs when UVB light is absorbed by DNA and disrupts the structure of the strands. Such damages are thought to underlie sunburn as well as lead to some types of cancers. The impact of indirect DNA damage is harder to assess. It is caused by UV light interacting with molecules within the skin to form reactive oxygen species, which in turn can directly damage nearby DNA as well as travel within the body to have deleterious effects elsewhere.

Melanin itself cannot always protect your skin from damage. Fortunately, the development of sunscreens and sunblocks can pick up where your natural defences drop off. Sunscreen works by absorbing the UV rays, but some of these rays still penetrate. Sunblocks are more effective and work by reflecting or scattering the UV rays so that they do not penetrate into the skin. For maximum protection, make sure you choose a lotion that contains both a sunscreen and a sunblock.

---

Recent research suggests that the skin probably plays an even more elaborate role in adaptive immune defence than described here. This is appropriate, because the skin serves as a major interface with the external environment.

### VITAMIN D SYNTHESIS BY THE SKIN

The epidermis also synthesizes vitamin D in the presence of sunlight. The cell type that produces vitamin D is undetermined. Vitamin D, which is derived from a precursor molecule closely related to cholesterol, promotes the absorption of calcium from the digestive tract into the blood (Chapter 15). Dietary supplements of vitamin D are usually required because typically the skin is not exposed to sufficient sunlight to produce adequate amounts of this essential chemical.

## Protective measures

The human body's defence system guards against entry of potential pathogens not only through the outer surface of the body but also through the internal cavities that communicate directly with the external environment—namely, the digestive system, the genitourinary system, and the respiratory system. These systems use various tactics to destroy microorganisms entering through these routes.

### DEFENCES OF THE DIGESTIVE SYSTEM

Saliva secreted into the mouth at the entrance to the digestive system contains an enzyme that lyses certain ingested bacteria. The so-called friendly bacteria that live on the back of the tongue convert food-derived nitrate into nitrite, which is swallowed. Acidification of nitrite on reaching the highly acidic stomach generates nitric oxide, which is toxic to a variety of microorganisms. Furthermore, many of the surviving bacteria that are swallowed are killed directly by the strongly acidic gastric juice in the stomach. Farther down the tract, the intestinal lining is endowed with gut-associated lymphoid tissue. These defensive mechanisms are not 100 percent effective, however. Some bacteria do manage to survive and reach the large intestine (the last portion of the digestive tract), where they continue to flourish. Surprisingly, this normal microbial population provides a natural barrier against infection within the lower intestine. These harmless resident microbes competitively suppress the growth of potential pathogens that have managed to escape the antimicrobial measures of earlier parts of the digestive tract.

*Clinical Note* On occasion, orally administered antibiotic therapy against an infection elsewhere within the body may actually induce an intestinal infection. By knocking out some of the normal intestinal bacteria, an antibiotic may permit an antibiotic-resistant pathogenic species to overgrow in the intestine.

### DEFENCES OF THE GENITOURINARY SYSTEMS

Within the genitourinary (reproductive and urinary) system, would-be invaders encounter hostile conditions in the acidic urine and acidic vaginal secretions. The genitourinary organs also produce a sticky mucus, which, like flypaper, entraps small invading particles. Subsequently, the particles are either engulfed by phagocytes or are swept out as the organ empties (e.g., flushed out with urine flow).

### DEFENCES OF THE RESPIRATORY SYSTEM

The respiratory system is likewise equipped with several important defence mechanisms against inhaled particulate matter. The respiratory system is the largest surface of the body that comes into direct contact with the increasingly polluted external environment. The surface area of the respiratory system exposed to the air is 30 times that of the skin. Larger airborne particles are filtered out of the inhaled air by hairs at the entrance of the nasal passages. Lymphoid tissues, the *tonsils* and *adenoids*, provide immunological protection against inhaled pathogens near the beginning of the respiratory system. Farther down the respiratory airways, millions of tiny hairlike projections known as *cilia* constantly beat in an outward direction. The respiratory airways are coated with a layer of thick, sticky mucus secreted by epithelial cells within the airway lining. This mucous sheet, laden with any inspired particulate debris (such as dust) that adheres to it, is constantly moved upward to the throat by ciliary action. This movement of mucus is known as the **mucociliary escalator**. The dirty mucus is either expectorated (spit out) or typically swallowed without the person being aware of it; any indigestible foreign particulate matter is later eliminated in the feces. As well as keeping the lungs clean, this mechanism is an important defence against bacterial infection, because many bacteria enter the body on dust particles. Also contributing to defence against respiratory infections are antibodies secreted in the mucus. In addition, an abundance of phagocytic specialists called the **alveolar macrophages** scavenge within the air sacs (alveoli) of the lungs. Further respiratory defences include coughs and sneezes. These commonly experienced reflex mechanisms involve forceful outward expulsion of material in an attempt to remove irritants from the trachea (*coughs*) or nose (*sneezes*).

| ■ Clinical Connections | Caitlin had suffered from hay fever for years and was used to the coughing and |
|---|---|

sneezing that occurs in an attempt to remove the irritant. However, her condition had progressed to allergic asthma, a condition in which the mucus-secreting cells become larger and secrete greater quantities of mucus, which in turn triggers more coughing. Although asthma is commonly associated with the characteristic wheezing in the airways, the most common symptom is periodic, persistent coughing.

Cigarette smoking suppresses these normal respiratory defences. The smoke from a single cigarette can paralyze the cilia for several hours, and repeated exposure eventually leads to ciliary destruction. Failure of ciliary activity to sweep out a constant stream of particulate-laden mucus enables inhaled carcinogens to remain in contact with the respiratory airways for prolonged periods. Furthermore, cigarette smoke incapacitates alveolar macrophages. Not only do particulates in cigarette smoke overwhelm the macrophages but certain components of cigarette smoke have a direct toxic effect on the macrophages, reducing their ability to engulf foreign material. In addition, noxious agents in tobacco smoke irritate the mucous linings of the respiratory tract, resulting in excess mucus production, which may partially obstruct the airways. "Smoker's cough" is an attempt to dislodge this excess stationary mucus. These and other direct toxic effects on lung tissue lead to the increased incidence of lung cancer and chronic respiratory diseases associated

### Check Your Understanding 11.2

1. Describe the epidermis, dermis, and hypodermis.

2. List and state the functions of the four resident cell types in the skin.

with cigarette smoking. Air pollutants include some of the same substances found in cigarette smoke and can similarly affect the respiratory system.

## 11.3 | Internal Defences

### Leukocytes

Leukocytes (white blood cells, or WBCs) and their derivatives, along with a variety of plasma proteins, are responsible for the different immune defence strategies.

#### LYMPHOCYTE FUNCTIONS

As a brief review, the functions of the five types of leukocytes are as follows (pp. 446–447):

1. Neutrophils are highly mobile phagocytic specialists that engulf and destroy unwanted materials.

2. Eosinophils secrete chemicals that destroy parasitic worms and are involved in allergic reactions.

3. Basophils release histamine and heparin and also are involved in allergic reactions.

4. Monocytes are transformed into **macrophages**, which are large, tissue-bound phagocytic specialists.

5. Lymphocytes are of two types.

   a. B lymphocytes (B cells) are transformed into plasma cells, which secrete antibodies that indirectly lead to the destruction of foreign material (antibody-mediated immunity).

   b. T lymphocytes (T cells) directly destroy virus-invaded cells and mutant cells by releasing chemicals that punch lethal holes in the victim cells (cell-mediated immunity).

A given leukocyte is present in the blood only transiently. Most leukocytes are out in the tissues, on defence missions. As a result, the immune system's effector cells are widely dispersed throughout the body and can defend in any location.

#### LYMPHOID TISSUES

Almost all leukocytes originate from common precursor stem cells in the bone marrow and are subsequently released into the blood. The only exception is lymphocytes, which arise in part from lymphocyte colonies in various lymphoid tissues originally populated by cells derived from bone marrow (p. 446).

Lymphoid tissues, collectively, are the tissues that produce, store, or process lymphocytes. These include the bone marrow, lymph nodes, spleen, thymus, tonsils, adenoids, appendix, and aggregates of lymphoid tissue in the lining of the digestive tract called **Peyer's patches** or **gut-associated lymphoid tissue (GALT)** (⟩ Figure 11-2). Lymphoid tissues are strategically located to intercept invading microorganisms before they have a chance to spread very far. For example, lymphocytes populating the *tonsils* and *adenoids* are situated advantageously to respond to inhaled microbes, whereas microorganisms invading through the digestive system immediately encounter lymphocytes in the *appendix* and GALT. Potential pathogens that gain access to lymph are filtered through *lymph nodes*, where they are exposed

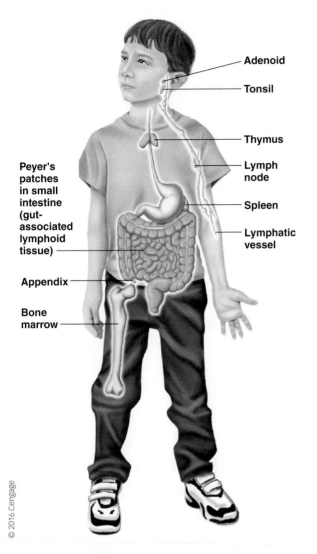

Adenoid

Tonsil

Thymus

Lymph node

Spleen

Lymphatic vessel

Peyer's patches in small intestine (gut-associated lymphoid tissue)

Appendix

Bone marrow

© 2016 Cengage

> FIGURE 11-2 **Lymphoid tissues.** The lymphoid tissues, which are dispersed throughout the body, produce, store, or process lymphocytes.

| Lymphoid Tissue | Function |
| --- | --- |
| **Bone Marrow** | Origin of all blood cells |
| | Site of maturational processing for B lymphocytes |
| **Lymph Nodes, Tonsils, Adenoids, Appendix, Gut-Associated Lymphoid** | Exchange lymphocytes with the lymph (remove, store, produce, and add them) |
| | Resident lymphocytes produce antibodies and sensitized T cells, which are released into the lymph. |
| **Tissue** | Resident macrophages remove microbes and other particulate debris from the lymph. |
| **Spleen** | Exchanges lymphocytes with the blood (removes, stores, produces, and adds them) |
| | Resident lymphocytes produce antibodies and sensitized T cells, which are released into the blood. |
| | Resident macrophages remove microbes and other particulate debris, most notably worn-out red blood cells, from the blood. |
| | Stores a small percentage of red blood cells, which can be added to the blood by splenic contraction as needed |
| **Thymus** | Site of maturational processing for T lymphocytes |
| | Secretes the hormone thymosin |

**TABLE 11-1** Functions of Lymphoid Tissues

© 2016 Cengage

to lymphocytes as well as to macrophages that line lymphatic passageways. The *spleen*, the largest lymphoid tissue, performs immune functions on blood similar to those that lymph nodes perform on lymph. Through actions of its lymphocyte and macrophage population, the spleen clears microorganisms and other foreign matter from the blood that passes through it and also removes worn-out red blood cells. The *thymus* and *bone marrow* play important roles in processing T and B lymphocytes, respectively, to prepare them to carry out their specific immune strategies. Table 11-1 summarizes the major functions of the various lymphoid tissues; some of these are described in Chapter 10 and others are discussed in Sections 11.5–11.7 of this chapter.

We now turn our attention to the two major components of the immune system's response to foreign invaders and other targets—the innate and the adaptive immune responses—and in the process further examine the roles of each type of leukocyte.

## Immune responses

Protective immunity is conferred by the complementary actions of two separate but interdependent components of the immune system: the *innate immune system* and the *adaptive,*

or *acquired, immune system.* The responses of these two systems differ in timing and in the selectivity of the defence mechanisms.

The **innate immune system** encompasses the body's *nonspecific* immune responses that come into play immediately on exposure to a threatening agent. These nonspecific responses are inherent (innate or built-in) defence mechanisms that nonselectively defend against foreign or abnormal material of any type, even on initial exposure to it. Such responses provide a first line of defence against a wide range of threats, including infectious agents, chemical irritants, and tissue injury from mechanical trauma and burns. Everyone is born with essentially the same innate immune-response mechanisms, although there are some subtle genetic differences. The **adaptive or acquired immune system**, in contrast, relies on *specific* immune responses selectively targeted against a particular foreign material to which the body has already been exposed and has had opportunity to prepare for making an attack that's discriminatingly aimed at

11

this foreign material. The adaptive immune system thus takes considerably more time to mount an immune response and takes on specific foes. The innate and adaptive immune systems work in harmony to contain, then eliminate, harmful agents.

### INNATE IMMUNE SYSTEM

The components of the innate system are always on guard, ready to unleash a limited, rather crude, repertoire of defence mechanisms at any and every invader. Of the immune effector cells, the neutrophils and macrophages—both phagocytic specialists—are especially important in innate defence. Several groups of plasma proteins also play key roles. The various nonspecific immune responses are set in motion in response to generic molecular patterns associated with threatening agents, such as cell surface carbohydrates found on bacteria but not on human cells. The responding phagocytic cells are studded with a recently discovered type of plasma membrane protein known as **toll-like receptors (TLRs)**. TLRs have been dubbed the "eyes of the innate immune system" because these immune sensors recognize and bind with the telltale bacterial markers, allowing the effector cells of the innate system to recognize pathogens as distinct from self-cells. A TLR's recognition of a pathogen triggers the phagocyte to engulf and destroy the infectious microorganism. Moreover, activation of the TLR induces the phagocytic cell to secrete chemicals, some of which contribute to inflammation—an important innate response to microbial invasion.

TLRs link the innate and adaptive branches of the immune system, because still other chemicals secreted by the phagocytes are important in activating cells of the adaptive immune system. Furthermore, foreign particles are deliberately marked for phagocytic ingestion by being coated with antibodies produced by the B cells of the adaptive immune system—another link between the innate and adaptive branches. These are but a few examples of how various components of the immune system are highly interactive and interdependent. The most significant cooperative relationships among the immune effectors are pointed out throughout this chapter.

The innate mechanisms give us all a rapid but limited and nonselective response to unfriendly challenges of all kinds, whereas innate immunity largely contains and limits the spread of infection. These nonspecific responses are important for keeping the invading pathogen under control until the adaptive immune system can be prepared to take over and mount strategies to eliminate the pathogen.

### ADAPTIVE IMMUNE SYSTEM

The responses of the adaptive or acquired immune system are mediated by the B and T lymphocytes. Each B and T cell can recognize and defend against only one particular type of foreign material, such as one kind of bacterium. Among the millions of B and T cells in the body, only the ones specifically equipped to recognize the unique molecular features of a particular infectious agent are called into action to defend against this particular agent. This specialization is similar to modern military personnel who are specially trained to accomplish a very specific task. The chosen lymphocytes multiply, expanding the pool of specialists that can launch a highly targeted attack against the invading pathogen.

The adaptive immune system is the ultimate weapon against most pathogens. The repertoire of activated and expanded B and T cells is constantly changing in response to the pathogens encountered. In this way, each person's adaptive immune system adapts to the specific pathogens in their environment. The targets of the adaptive immune system vary among people, depending on the types of immune assaults each individual meets. Furthermore, the adaptive immune system can remember each pathogen, so that when rechallenged by that pathogen, its response is much quicker. This is accomplished by forming a pool of memory cells.

We will first examine in more detail the innate immune responses and then look more closely at adaptive immunity.

---

**Check Your Understanding 11.3**

1. Define *immunity*.
2. Explain why innate immune responses are considered nonspecific and adaptive immune responses are said to be specific.

---

## 11.4 | Innate Immunity

Innate defences include the following features:

1. *Inflammation*, a nonspecific response to tissue injury in which the phagocytic specialists—neutrophils and macrophages—play a major role, along with supportive input from other immune cell types

2. *Interferon*, a family of proteins that nonspecifically defend against viral infection

3. *Natural killer cells*, a special class of lymphocyte—like cells that spontaneously and nonspecifically lyse (rupture) and thereby destroy virus-infected host cells and cancer cells

4. The *complement system*, a group of inactive plasma proteins that, when sequentially activated, bring about destruction of foreign cells by attacking their plasma membranes

In this section, we discuss each of these in turn, beginning with inflammation.

### Inflammation

The term **inflammation** refers to an innate, nonspecific series of highly interrelated events set into motion in response to foreign invasion, tissue damage, or both. The ultimate goal of inflammation is to bring to the invaded or injured area phagocytes and plasma proteins that can (1) isolate, destroy, or inactivate the invaders; (2) remove debris; and (3) prepare for subsequent healing and repair. The overall inflammatory response is remarkably similar no matter what the triggering event (bacterial invasion, chemical injury, or mechanical trauma), although some subtle differences may be evident, depending on the injurious agent or the site of damage. The following sequence of events typically occurs during the inflammatory response. We use the example of bacterial entry into a break in the skin, and ⟩ Figure 11-3 corresponds to our discussion.

### DEFENCE BY RESIDENT TISSUE MACROPHAGES

When bacteria invade through a break in the external barrier of skin, macrophages already in the area immediately begin phagocytizing the foreign microbes. Although usually not enough resident macrophages are present to meet the challenge alone, they defend against infection during the first hour or so, before other mechanisms can be mobilized. Macrophages are usually rather stationary—gobbling debris and contaminants that come their way—but when necessary, they become mobile and migrate to sites of battle against invaders (⟩ Figure 11-3, step **1**).

### LOCALIZED VASODILATION

Almost immediately on microbial invasion, arterioles within the area dilate. The vasodilation, increased permeability, and slowing of blood velocity are induced by the actions of various inflammatory mediators, such as histamine released from the mast cells

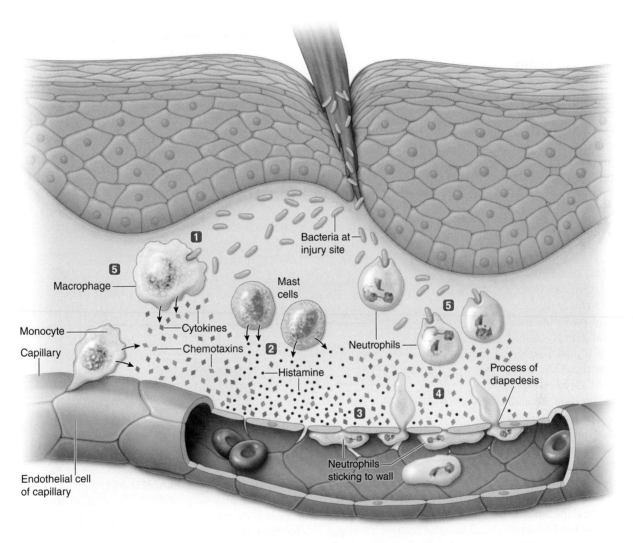

**1** A break in the skin introduces bacteria, which reproduce at the wound site. Activated resident macrophages engulf the pathogens and secrete cytokines and chemotaxins.

**2** Activated mast cells release histamine.

**3** Histamine dilates local blood vessels and widens the capillary pores. The cytokines cause neutrophils and monocytes to stick to the blood vessel wall.

**4** Chemotaxins attract neutrophils and monocytes, which squeeze out between cells of the blood vessel wall, a process called diapedesis, and migrate to the infection site.

**5** Monocytes enlarge into macrophages. Newly arriving macrophages and neutrophils engulf the pathogens and destroy them.

⟩ **FIGURE 11-3 Steps producing inflammation.** Chemotaxins released at the site of damage attract phagocytes to the scene. Note the leukocytes emigrating from the blood into the tissues by assuming an amoeba-like behaviour and squeezing through the capillary pores. Mast cells secrete vessel-dilating, pore-widening histamine. Macrophages secrete cytokines that exert multiple local and systemic effects.

© 2016 Cengage

(the connective tissue—bound "cousins" of circulating basophils) (steps **2** and **3**). In the damaged tissue area, vasodilation occurs first in the arterioles, then progresses to the capillaries, and brings about an increase in the amount of blood present, causing redness and heat. Increased local delivery of blood also brings to the site more phagocytic leukocytes and plasma proteins, both crucial to the defence response.

### INCREASED CAPILLARY PERMEABILITY

Released histamine also increases the capillaries' permeability by enlarging the capillary pores (the spaces between the endothelial cells), so the plasma proteins normally prevented from leaving the blood can escape into the inflamed tissue.

### LOCALIZED OEDEMA

Accumulation of leaked plasma proteins in the interstitial fluid raises the local interstitial fluid—colloid osmotic pressure. Furthermore, the increased local blood flow elevates capillary blood pressure. Because both these pressures tend to move fluid out of the capillaries, these changes favour enhanced ultrafiltration and reduced reabsorption of fluid across the involved capillaries. The end result of this shift in fluid balance is localized oedema (p. 413). Thus, the familiar swelling that accompanies inflammation is due to histamine-induced vascular changes. Likewise, the other well-known gross manifestations of inflammation, such as redness and heat, are largely caused by the enhanced flow of warm arterial blood to the damaged tissue. Pain is caused by local distension within the swollen tissue and also by the direct effect of locally produced substances on the receptor endings of afferent neurons that supply the area. These observable characteristics of the inflammatory process (swelling, redness, heat, and pain) are coincidental to the primary purpose of the vascular changes in the injured area—that is, to increase the number of leukocytic phagocytes and crucial plasma proteins in the area (❯ Figure 11-4).

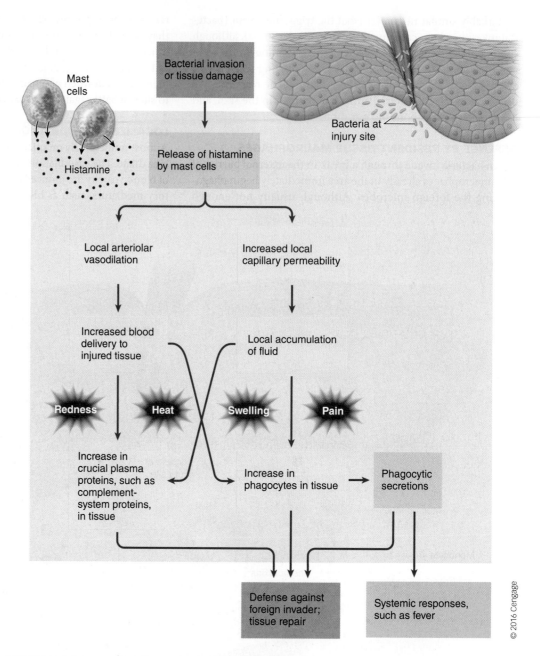

❯ FIGURE 11-4 **Gross manifestations and outcomes of inflammation**

© 2016 Cengage

### WALLING OFF THE INFLAMED AREA

The leaked plasma proteins most critical to the immune response are those in the complement system and also the clotting and anticlotting factors. On exposure to tissue thromboplastin in the injured tissue and to specific chemicals secreted by phagocytes on the scene, fibrinogen—the final factor in the clotting system—is converted into fibrin (p. 449). Fibrin forms interstitial fluid clots in the spaces around the bacterial invaders and damaged cells. This walling off of the injured region from the surrounding tissues prevents, or at least delays, the spread of bacterial invaders and their toxic products. Later, the more slowly activated anticlotting factors gradually dissolve the clots after they are no longer needed (p. 452).

11

## EMIGRATION OF LEUKOCYTES

Within an hour after injury, the area is teeming with leukocytes that have left the vessels. Neutrophils arrive first, followed during the next 8 to 12 hours by the slower-moving monocytes. The latter swell and mature into macrophages during the next 8 to 12 hours. Once neutrophils or monocytes leave the bloodstream, they never recycle back to the blood.

Leukocytes can emigrate from the blood into the tissues by means of the following steps:

- Blood-borne leukocytes, especially neutrophils and monocytes, stick to the inner endothelial lining of capillaries in the affected tissue, a process called **margination** (› Figure 11-3, step ❸). *Selectins*, a type of cell adhesion molecule (CAM; p. 32) that protrudes from the inner endothelial lining, cause leukocytes flowing by in the blood to slow down and roll along the interior of the vessel, much as the nap of a carpet slows down a child's rolling toy car. This slowing down allows the leukocytes enough time to check for local activating factors—"SOS signals" from nearby injured or infected tissues. When present, these activating factors cause the leukocytes to adhere firmly to the endothelial lining by means of interaction with another type of CAM, the *integrins*.

- Soon, the adhered leukocytes start leaving by a mechanism known as **diapedesis**. Assuming *amoeba-like behaviour*, an adhered leukocyte pushes a long, narrow projection through a capillary pore; then the remainder of the cell flows forward into the projection. In this way, the leukocyte is able to wriggle its way through the capillary pore, even though it is much larger than the pore (› Figure 11-3, step ❹). Outside the vessel, the leukocyte crawls toward the injured area. Neutrophils arrive on the inflammatory scene earliest because they are more mobile than monocytes.

- **Chemotaxis** guides phagocytic cells in the direction of migration; that is, the cells are attracted to certain chemical mediators, known as *chemotaxins* or *chemokines*, released at the site of damage. Binding of chemotaxins with protein receptors on the plasma membrane of a phagocytic cell increases calcium entry into the cell. Calcium, in turn, switches on the cellular contractile apparatus that leads to the amoeba-like crawling. Because the concentration of chemotaxins progressively increases toward the site of injury, phagocytic cells move unerringly toward this site along a chemotaxin concentration gradient.

### LEUKOCYTE PROLIFERATION

Resident tissue macrophages as well as leukocytes that exited from the blood and migrated to the inflammatory site are soon joined by new phagocytic recruits from the bone marrow. Within a few hours after the onset of the inflammatory response, the number of neutrophils in the blood may increase up to four to five times that of normal. This increase is due partly to the transfer into the blood of large numbers of preformed neutrophils stored in the bone marrow and partly to increased production of new neutrophils by the bone marrow. A slower-commencing but longer-lasting increase in monocyte production by the bone marrow also occurs, making available larger numbers of these macrophage precursor cells (› Figure 11-3, step ❺). In addition, the multiplication of resident macrophages adds to the pool of these important immune cells. Proliferation of new neutrophils, monocytes, and macrophages and mobilization of stored neutrophils are stimulated by various chemical mediators released from the inflamed region.

### MARKING OF BACTERIA FOR DESTRUCTION BY OPSONINS

Phagocytes must be able to distinguish between normal cells and foreign or abnormal cells before accomplishing their destructive mission. Otherwise, they could not selectively engulf and destroy only the unwanted materials. First, as you have learned, phagocytes, by means of their TLRs, recognize and subsequently engulf infiltrators that have standard bacterial cell wall components not found in human cells. Second, foreign particles are deliberately marked for phagocytic ingestion by being coated with chemical mediators generated by the immune system. Such body-produced chemicals that make bacteria more susceptible to phagocytosis are known as **opsonins**. Opsonins are antibodies, and they are one of the activated proteins of the complement system.

An opsonin enhances phagocytosis by linking the foreign cell to a phagocytic cell (› Figure 11-5). One portion of an opsonin molecule binds nonspecifically to the surface of an invading bacterium, whereas another portion of the opsonin molecule binds to receptor sites specific for it on the phagocytic cell's plasma membrane. This link ensures that the bacterial victim does not have a chance to get away before the phagocyte can perform its lethal attack.

### LEUKOCYTIC DESTRUCTION OF BACTERIA

Neutrophils and macrophages clear the inflamed area of infectious and toxic agents as well as tissue debris by both phagocytic and nonphagocytic means; this clearing action is the main function of the inflammatory response.

Phagocytosis involves the engulfment and intracellular degradation (breakdown) of foreign particles and tissue debris. Macrophages can engulf a bacterium in less than 0.01 second. Recall that phagocytic cells contain an abundance of lysosomes,

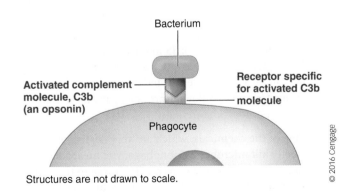

Structures are not drawn to scale.

› **FIGURE 11-5 Mechanism of opsonin action.** One of the activated complement molecules, C3b, links a foreign cell, such as a bacterium, and a phagocytic cell by nonspecifically binding with the foreign cell and specifically binding with a receptor on the phagocyte. This link ensures that the foreign material does not escape before it can be engulfed by the phagocyte.

**11**

which are organelles filled with hydrolytic enzymes. After a phagocyte has internalized targeted material, a lysosome fuses with the membrane that encloses the engulfed matter and releases its hydrolytic enzymes within the confines of the vesicle, where the enzymes begin breaking down the entrapped material (p. 48). Phagocytes eventually die from the accumulation of toxic by-products from foreign particle degradation or from inadvertent release of destructive lysosomal chemicals into the cytosol. Neutrophils usually succumb after phagocytizing from 5 to 25 bacteria, whereas macrophages survive much longer and can engulf up to 100 or more bacteria. Indeed, the longer-lived macrophages even clear the area of dead neutrophils and other tissue debris. The **pus** that forms in an infected wound is a collection of these phagocytic cells, both living and dead; necrotic (dead) tissue liquefied by lysosomal enzymes released from the phagocytes; and bacteria.

## MEDIATION OF THE INFLAMMATORY RESPONSE BY PHAGOCYTE-SECRETED CHEMICALS

Microbe-stimulated phagocytes release many chemicals that function as mediators of inflammation. All chemicals other than antibodies that leukocytes secrete are collectively called **cytokines**. Macrophages, monocytes, neutrophils, a type of T cell called a helper T cell, and some nonimmune cells, such as endothelial cells (cells lining blood vessels) and fibroblasts (fibre-formers in the connective tissue), all secrete cytokines. More than 100 cytokines have been identified, and the list continues to grow as researchers unravel the complicated chemical means by which immune effector cells communicate with one another to coordinate their activities. Unlike antibodies, cytokines do not interact directly with the antigen (foreign material) that induces their production. Instead, cytokines largely spur other immune cells into action to help ward off the invader. Cytokines typically act locally as paracrines on cells in the vicinity, but some circulate in the blood to exert endocrine effects at distant sites. The cytokines released by phagocytes induce a range of interrelated immune activities, varying from local responses to the systemic manifestations that accompany microbe invasion. Some cytokines have names related to their first identified or most important function—examples include specific *colony-stimulating factors*. Other cytokines are designated as specific numbered *interleukins*, such as interleukin 1 (IL-1), interleukin 2 (IL-2), and so on (*interleukin* means "between leukocytes"); 35 interleukins have been identified to date, numbered in the order of their discovery. Interleukins orchestrate a wide variety of independent and overlapping immune activities. The following are among the most important functions of phagocytic secretions:

1. Some of the chemicals, which are very destructive, directly kill microbes by nonphagocytic means. For example, macrophages secrete *nitric oxide (NO)*, a multipurpose chemical that is toxic to nearby microbes. As a more subtle means of destruction, neutrophils secrete **lactoferrin**, a protein that tightly binds with iron, making it unavailable for use by invading bacteria. Bacterial multiplication depends on high concentrations of available iron.

2. Several chemicals released by macrophages, namely **interleukin 1 (IL-1)**, **interleukin 6 (IL-6)**, and **tumour necrosis factor (TNF)**, collectively act to bring about a diverse array of effects locally and throughout the body, all of which are geared toward defending the body against infection or tissue injury. They promote inflammation and are largely responsible for the systemic manifestation accompanying an infection. (Tumour necrosis factor is named for its role in killing cancer cells, but it also exerts other effects.)

3. IL-1, IL-6, and TNF also function together as **endogenous pyrogen (EP)**, which induces the development of fever (*endogenous* means "from within the body"; *pyro* means "fire" or "heat"; and *gen* means "production"). This response occurs especially when the invading organisms have spread into the blood. Endogenous pyrogen causes release within the hypothalamus of *prostaglandins*; these are locally acting chemical messengers that turn up the hypothalamic "thermostat" that regulates body temperature. The function of the resulting elevation in body temperature in fighting infection remains unclear. The fact that fever is such a common systemic manifestation of inflammation suggests the raised temperature plays an important, beneficial role in the overall inflammatory response, as supported by recent evidence. For example, higher temperatures augment phagocytosis and increase the rate of the many enzyme-dependent inflammatory activities. Furthermore, an elevated body temperature may interfere with bacterial multiplication by increasing bacterial requirements for iron. Resolving the controversial issue of whether a fever can be beneficial is extremely important, given the widespread use of drugs that suppress fever.

   *Clinical Note* Although a mild fever may possibly be beneficial, there is no doubt that an extremely high fever can be detrimental, particularly by harming the central nervous system. Young children, whose temperature-regulating mechanisms are not as stable as those of more mature individuals, occasionally have convulsions in association with high fevers.

4. An additional role for IL-1, IL-6, and TNF is to decrease the plasma concentration of iron by altering iron metabolism within the liver, spleen, and other tissues. This action reduces the amount of iron available to support bacterial multiplication.

5. Furthermore, IL-1, IL-6, and TNF stimulate the release of **acute phase proteins** from the liver. This collection of proteins, which have not yet been sorted out by scientists, exerts a multitude of wide-ranging effects associated with the inflammatory process, tissue repair, and immune cell activities. One of the better-known acute phase proteins is *C-reactive protein*, considered clinically as a blood-borne marker of inflammation (p. 376).

6. TNF stimulates the release of histamine from mast cells in the vicinity. Histamine, in turn, promotes the local vasodilation and increased capillary permeability of inflammation.

7. A chemical secreted by neutrophils, **kallikrein**, converts specific plasma protein precursors produced by the liver into activated **kinins**. Activated kinins augment a variety of inflammatory events. For example, the end product of the kinin cascade, **bradykinin**, activates nearby pain receptors and thus partially produces the soreness associated with inflammation. Bradykinin also contributes to blood vessel dilation, reinforcing the effects of histamine. In a positive-feedback fashion, kinins also act as powerful chemotaxins to entice more neutrophils to join the battle.

8. Some phagocytic chemical mediators trigger both the *clotting* and *anticlotting systems* to first enhance the walling-off process and then facilitate gradual dissolution of the fibrous clot after it is no longer needed.

This list of events augmented by phagocyte-secreted chemicals is not complete, but it illustrates the diversity and complexity of responses these mediators elicit. Furthermore, there are other important macrophage—lymphocyte interactions that do not depend on the release of chemicals from phagocytic cells. Thus, the effect that phagocytes, especially macrophages, ultimately have on microbial invaders far exceeds their engulf-and-destroy tactics.

### TISSUE REPAIR

The ultimate purpose of the inflammatory process is to isolate and destroy injurious agents and to clear the area for tissue repair. In some tissues (e.g., skin, bone, and liver), the healthy organ-specific cells surrounding the injured area undergo cell division to replace the lost cells, often repairing the wound perfectly. In typically non-regenerative tissues, such as nerve and muscle, however, lost cells are replaced by **scar tissue**. Fibroblasts, a type of connective tissue cell, start to divide rapidly in the vicinity and secrete large quantities of the protein collagen, which fills in the region vacated by the lost cells and results in the formation of scar tissue. Even in a tissue as readily replaceable as skin, scars sometimes form when complex underlying structures, such as hair follicles and sweat glands, are permanently destroyed by deep wounds.

### INNATE IMMUNITY AND EXERCISE

Acute (short-term) exercise in untrained persons has been shown to stimulate phagocytic activity by increasing the activity of the monocytes, neutrophils, and macrophages. Well-trained athletes tend to demonstrate clinically normal monocyte and neutrophil counts, whereas complement levels may be suppressed. Also, there is evidence of suppressed neutrophil function at rest and following intense exercise in well-trained athletes (research has focused on endurance athletes). It has been proposed that the mild innate immune suppression (e.g., down-regulation of neutrophils) observed in some well-trained athletes may be due to a down-regulation of the inflammatory response as a result of the chronic tissue damage associated with high training volumes.

## NSAIDs and glucocorticoids

 Many drugs can suppress the inflammatory process. The most effective are the *nonsteroidal anti-inflammatory drugs*, or *NSAIDs* (aspirin, ibuprofen, and related

compounds) and *glucocorticoids* (drugs similar to the steroid hormone cortisol, which is secreted by the adrenal cortex; p. 256). For example, aspirin interferes with the inflammatory response by decreasing histamine release, thus reducing pain, swelling, and redness. Furthermore, aspirin reduces fever by inhibiting production of prostaglandins, the local mediators of endogenous pyrogen-induced fever.

Glucocorticoids, which are potent anti-inflammatory drugs, suppress almost every aspect of the inflammatory response. In addition, they destroy lymphocytes within lymphoid tissue and reduce antibody production. These therapeutic agents are useful for treating undesirable immune responses, such as allergic reactions (e.g., poison ivy rash and asthma) and the inflammation associated with arthritis. However, by suppressing inflammatory and other immune responses that localize and eliminate bacteria, such therapy also reduces the body's ability to resist infection. For this reason, glucocorticoids should be used discriminatingly.

Is naturally secreted cortisol likewise counterproductive to the immune defence system? Traditionally, cortisol has not been considered to display anti-inflammatory activity at normal blood concentrations. Instead, anti-inflammatory action has been attributed only to blood concentrations that are higher than the normal physiological range brought about by the administration of exogenous (from outside the body) cortisol-like drugs. Recent evidence, however, suggests that cortisol, whose secretion is increased in response to any stressful situation, does exert anti-inflammatory activity even at normal physiological levels. According to this proposal, the anti-inflammatory effect of cortisol modulates stress-activated immune responses, preventing them from overshooting, and thus protecting us against damage by potentially overreactive defence mechanisms.

Now let's shift attention from inflammation to interferon, another component of innate immunity.

## Interferon and viruses

Besides the inflammatory response, another innate defence mechanism is the release of **interferon** from virus-infected cells. Interferon briefly provides nonspecific resistance to viral infections by transiently interfering with replication of the same or unrelated viruses in other host cells. In fact, interferon was named for its ability to *interfere* with viral replication.

### ANTIVIRAL EFFECT OF INTERFERON

When a virus invades a cell, in response to being exposed to viral nucleic acid, the cell synthesizes and secretes interferon. Once released into the ECF from a virus-infected cell, interferon binds with receptors on the plasma membranes of healthy neighbouring cells or even distant cells that it reaches through the blood, signalling these cells to prepare for possible viral attack. Interferon thus acts as a "whistle-blower," forewarning healthy cells of potential viral attack and helping them prepare to resist. Interferon does not have a direct antiviral effect; instead, it triggers the production of virus-blocking enzymes by potential host cells. When interferon binds with these other cells, they synthesize enzymes that can break down viral messenger RNA (p. A-22) and inhibit protein synthesis. Both these processes are essential

11

for viral replication. Although viruses are still able to invade these forewarned cells, these pathogens cannot govern cellular protein synthesis for their own replication (> Figure 11-6).

The newly synthesized inhibitory enzymes remain inactive within the tipped-off potential host cell unless it is actually invaded by a virus, at which time the enzymes are activated by the presence of viral nucleic acid. This activation requirement protects the cell's own messenger RNA and protein-synthesizing machinery from unnecessary inhibition by these enzymes should viral invasion not occur. Because activation can take place only during a limited time span, this is a short-term defence mechanism.

Interferon is released nonspecifically from any cell infected by any virus and, in turn, can induce temporary self-protective activity against many different viruses in any other cells that it reaches. In this way, it provides a general, rapidly responding defence strategy against viral invasion until more specific, but slower-responding, immune mechanisms come into play.

In addition to facilitating inhibition of viral replication, interferon reinforces other immune activities. For example, it enhances macrophage phagocytic activity, stimulates production of antibodies, and boosts the power of killer cells.

## ANTICANCER EFFECTS OF INTERFERON

Interferon exerts anticancer as well as antiviral effects. It markedly enhances the actions of cell-killing cells—the natural killer cells and a special type of T lymphocyte, *cytotoxic T cells*—which attack and destroy both virus-infected cells and cancer cells. Furthermore, interferon itself slows cell division and suppresses tumour growth.

## Natural killer cells

As well as the functions of inflammation and interferon, **natural killer (NK) cells** are naturally occurring, lymphocyte-like cells that nonspecifically destroy virus-infected cells and cancer cells by directly lysing the membranes of such cells on first exposure to them. Their mode of action and major targets are similar to those of cytotoxic T cells, but the latter can fatally attack only the specific types of virus-infected cells and cancer cells to which they have been previously exposed. Furthermore, after exposure, cytotoxic T cells require a maturation period before they can launch their lethal assault. NK cells provide an immediate, nonspecific defence against virus-invaded cells and cancer cells before the more specific and more abundant cytotoxic T cells become functional.

> **FIGURE 11-6 Mechanism of action of interferon in preventing viral replication.** Interferon, which is released from virus-infected cells, binds with other uninvaded host cells and induces these cells to produce inactive enzymes capable of blocking viral replication. These inactive enzymes are activated only if a virus subsequently invades one of these prepared cells.

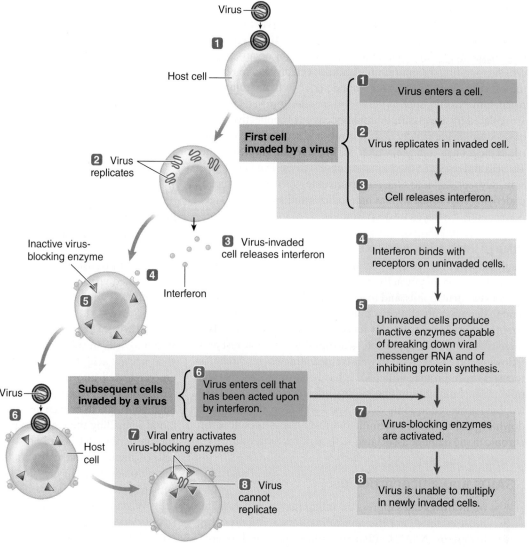

© 2016 Cengage

## The complement system

The **complement system**, the last of the four innate immunity defences, is brought into play nonspecifically in response to invading organisms. This system can be activated in two ways:

1. By exposure to particular carbohydrate chains present on the surfaces of microorganisms, but not found on human cells, generating a nonspecific innate immune response
2. By exposure to antibodies produced against a specific foreign invader, generating an adaptive immune response

In fact, the system derives its name from the fact that it *complements* the action of antibodies; it is the primary mechanism activated by antibodies to kill foreign cells. It does so by forming membrane attack complexes that punch holes in the victim cells. In addition to bringing about direct lysis of the invader, the powerful complement cascade reinforces other general inflammatory tactics.

### FORMATION OF THE MEMBRANE ATTACK COMPLEX

In the same mode as the clotting and anticlotting systems, the complement system consists of plasma proteins that are produced by the liver and circulate in the blood in inactive form. Once the first component, C1, is activated, it activates the next component, C2, and so on, in a cascade sequence of activation reactions. The five final components, C5 through C9, assemble into a large, doughnut-shaped protein complex, the **membrane attack complex (MAC)**, which embeds itself in the surface membrane of nearby microorganisms, creating a large channel through the membrane (› Figure 11-7). In other words, the parts make a hole. This hole-punching technique makes the membrane extremely leaky; the resulting osmotic flux of water into the victim cell causes it to swell and burst. This complement-induced lysis is the major means of directly killing microbes without phagocytizing them.

### AUGMENTING INFLAMMATION

Unlike the other cascade systems, in which the sole function of the various components leading up to the end step is activation of the next precursor in the sequence, several activated proteins in the complement cascade also act on their own to augment the inflammatory process—by the following methods:

- *Serving as chemotaxins,* which attract and guide professional phagocytes to the site of complement activation (i.e., the site of microbial invasion)
- *Acting as opsonins,* by binding with microbes and thereby enhancing their phagocytosis
- *Promoting vasodilation and increased vascular permeability,* thus increasing blood flow to the invaded areas
- *Stimulating release of histamine* from mast cells in the vicinity, which in turn enhances the local vascular changes characteristic of inflammation
- *Activating kinins,* which further reinforce inflammatory reactions

Several activated components in the cascade are very unstable. Because these unstable components can carry out the sequence only in the immediate area in which they are activated before they decompose, the complement attack is confined to the surface membrane of the microbe whose presence initiated activation of the system. Nearby host cells are thus spared from lytic attack.

We have now completed our discussion of innate immunity and are going to turn our attention to adaptive immunity.

---

### Check Your Understanding 11.4

1. List and briefly describe the four types of innate defence.
2. What are the gross manifestations of inflammation, and what are their underlying physiological mechanisms?
3. Define *cytokines*.

---

## 11.5 | Adaptive Immunity: General Concepts

A specific adaptive immune response is a selective attack aimed at limiting or neutralizing a particular offending target for which the body has been specially prepared after previous exposure to it.

### Antibody-mediated immunity and cell-mediated immunity

There are two classes of adaptive immune responses: **antibody-mediated (humoural) immunity**, involving production of antibodies by B lymphocyte derivatives known as *plasma cells*; and **cell-mediated immunity**, involving production of *activated T lymphocytes*, which directly attack unwanted cells.

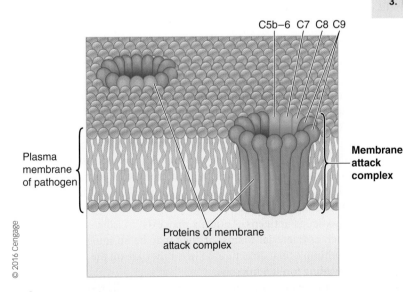

C5b–6  C7  C8  C9

Plasma membrane of pathogen

Membrane attack complex

Proteins of membrane attack complex

› **FIGURE 11-7** **Membrane attack complex (MAC) of the complement system.** Activated complement proteins C5, C6, C7, C8, and a number of C9s aggregate to form a porelike channel in the plasma membrane of the target cell. The resulting leakage leads to destruction of the cell.

11

Lymphocytes can specifically recognize and selectively respond to an almost limitless variety of foreign agents as well as cancer cells. The recognition and response processes are different in B and in T cells. In general, B cells recognize free-existing foreign invaders, such as bacteria and their toxins, and a few viruses, which they combat by secreting antibodies specific for the invaders. T cells specialize in recognizing and destroying body cells gone awry, including virus-infected cells and cancer cells. We are now going to explore the different life histories of B and T cells.

### ORIGINS OF B AND T CELLS

Both types of lymphocytes, like all blood cells, are derived from common stem cells in the bone marrow. Whether a lymphocyte and all its progeny are destined to be B or T cells depends on the site of final differentiation and maturation of the original cell in the lineage (› Figure 11-8). B cells differentiate and mature

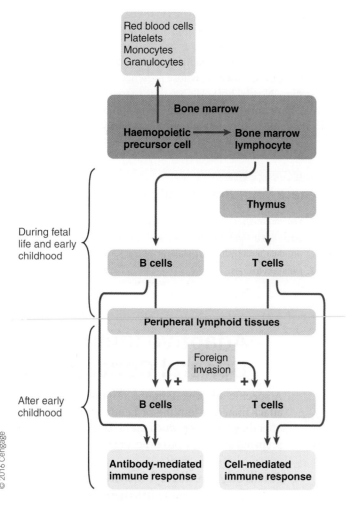

> FIGURE 11-8 **Origins of B and T cells.** B cells are derived from lymphocytes that matured and differentiated in the bone marrow, whereas T cells are derived from lymphocytes that originated in the bone marrow but matured and differentiated in the thymus. After early childhood, new B and T cells are produced primarily by colonies of B and T cells established in peripheral lymphoid tissues during fetal life and early childhood.

in the bone marrow. As for T cells, during fetal life and early childhood, some of the immature lymphocytes from the bone marrow migrate through the blood to the thymus, where they undergo further processing to become T lymphocytes (named for their site of maturation). The *thymus* is a lymphoid tissue located midline within the chest cavity above the heart in the space between the lungs (see › Figure 11-2).

On being released into the blood from either the bone marrow or the thymus, mature B and T cells take up residence and establish lymphocyte colonies in the peripheral lymphoid tissues. Here, on appropriate stimulation, they undergo cell division to produce new generations of either B or T cells, depending on their ancestry. After early childhood, most new lymphocytes are derived from these peripheral lymphocyte colonies rather than from the bone marrow.

Each of us has an estimated 2 trillion lymphocytes, which, if aggregated in a mass, would be about the size of the brain. At any one time, most of these lymphocytes are concentrated in the various strategically located lymphoid tissues, but both B and T cells continually circulate among the lymph, blood, and body tissues, where they remain on constant surveillance.

### ROLE OF THYMOSIN

Because most of the migration and differentiation of T cells occurs early in development, the thymus gradually atrophies and becomes less important as the person matures. It does, however, continue to produce **thymosin**, a hormone important in maintaining the T-cell lineage. Thymosin enhances proliferation of new T cells within the peripheral lymphoid tissues and augments the immune capabilities of existing T cells. Secretion of thymosin decreases after about 30 to 40 years of age. This decline has been suggested as a contributing factor in aging. Scientists further speculate that diminishing T-cell capacity with advancing age may be linked to increased susceptibility to viral infections and cancer, because T cells play an especially important role in defence against viruses and cancer.

Let's now see how lymphocytes detect their selected target.

## Antigens

Both B and T cells must be able to specifically recognize the unwanted cells and other material to be destroyed or neutralized as being distinct from the body's own normal cells. The presence of antigens enables lymphocytes to make this distinction. An **antigen** is a large, foreign, unique molecule that triggers a specific immune response against itself when it gains entry into the body. In general, the more complex a molecule is, the greater its antigenicity. Foreign proteins are the most common antigens because of their size and structural complexity, although other macromolecules, such as large polysaccharides and lipids, can also act as antigens. Antigens may exist as isolated molecules, such as bacterial toxins, or they may be an integral part of a multimolecular structure, as when they are on the surface of an invading foreign microbe.

In the next section we examine how B cells respond to their targeted antigen. In Section 11.7, we look at the response of T cells to their antigen.

**Check Your Understanding 11.5**

1. State the two classes of adaptive immunity, and indicate which type of lymphocyte accomplishes each.

2. Define *antigen*.

## 11.6 | B Lymphocytes: Antibody-Mediated Immunity

Each B and T cell has receptors—**B-cell receptors (BCRs)** and **T-cell receptors (TCRs)**—on its surface for binding with one particular type of the multitude of possible antigens (› Figure 11-9). These receptors are the "eyes of the adaptive immune system," although a given lymphocyte can see only one unique antigen. This is in contrast to the TLRs of the innate effector cells, which recognize generic trademarks found on the surface of all microbial invaders. Two important things to know about B cells are as follows:

- *The antigens to which B cells respond can be T-independent or T-dependent.* B cells can bind with and be directly activated by polysaccharide antigens without any assistance from T cells. These antigens, known as **T-independent antigens**, stimulate production of antibodies without any T-cell involvement. By contrast, **T-dependent antigens**, which are typically protein antigens, do not directly stimulate the production of antibodies without the help of a special type of T cell, known as a *helper T cell*. The majority of antigens to which B cells respond are T-dependent antigens.

- *Antigens stimulate B cells to convert into plasma cells that produce antibodies.* When BCRs (› Figure 11-9a) bind with an antigen, most B cells differentiate into active *plasma cells,* whereas others become dormant *memory cells.*

We first examine the role of plasma cells and their antibodies and then turn our attention to memory cells.

### Plasma cells

A **plasma cell** produces **antibodies** that can combine with the specific type of antigen that stimulated activation of the plasma cell. During differentiation into a plasma cell, a B cell swells as the rough endoplasmic reticulum (the site for synthesis of proteins to be exported) greatly expands (› Figure 11-10). Because antibodies are proteins, plasma cells essentially become prolific protein factories, producing up to 2000 antibody molecules per second. So great is the commitment of a plasma cell's protein-synthesizing machinery to antibody production that it cannot maintain protein synthesis for its own viability and growth. Consequently, it dies after a brief (five- to seven-day), highly productive lifespan.

Antibodies are directly secreted into the blood or lymph, depending on the location of the activated plasma cells, but all antibodies eventually gain access to the blood, where they are known as gamma globulins (immunoglobulins) (p. 435).

### Antibody subclasses

Antibodies are grouped into five subclasses based on differences in their biological activity:

1. **IgM** immunoglobulin serves as the B-cell surface receptor for antigen attachment and is secreted in the early stages of plasma cell response.

2. **IgG**, the most abundant immunoglobulin in the blood, is produced copiously when the body is subsequently exposed to the same antigen. Together, IgM and IgG antibodies produce most specific immune responses against bacterial invaders and a few types of viruses.

3. **IgE** helps protect against parasitic worms and is the antibody mediator for common allergic responses, such as hay fever, asthma, and hives.

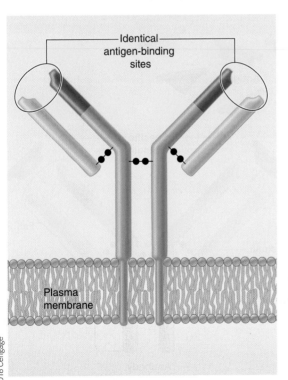

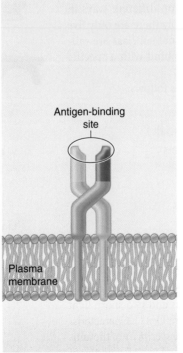

**(a)** B-cell receptor (BCR)

**(b)** T-cell receptor (TCR)

© 2016 Cengage

› **FIGURE 11-9 B-cell and T-cell receptors**

**Plasma cell**

**Unactivated B cell**

Endoplasmic
reticulum

(a)                                            (b)

⟩ **FIGURE 11-10 Comparison of an unactivated B cell and a plasma cell.** Electron micrograph of (a) an unactivated B cell, or small lymphocyte, and (b) a plasma cell. A plasma cell is an activated B cell. It is filled with an abundance of rough endoplasmic reticulum distended with antibody molecules.

4. **IgA** immunoglobulins are found in secretions of the digestive, respiratory, and genitourinary systems, as well as in milk and tears.

5. **IgD** is present on the surface of many B cells, but its function is uncertain.

Note that this classification is based on different ways in which antibodies function. It does not imply there are only five different antibodies. Within each functional subclass are millions of different antibodies, each able to bind with a specific antigen only.

The basic functions of antibodies are as follows:

- binding antigens
- activating the complement system (cascade)
- binding and neutralizing bacterial toxins
- inhibiting bacterial access to host cells
- inhibiting viral entry into host cells
- assisting phagocytosis against parasites
- aiding cytotoxicity of T and NK cells

## Antibody structure

Antibodies of all five subclasses are composed of four interlinked polypeptide chains—two long, heavy chains and two short, light chains—arranged in the shape of a Y (⟩ Figure 11-11). Characteristics of the arm regions of the Y determine the *specificity* of the antibody (i.e., with what antigen the antibody can bind). Properties of the tail portion of the antibody determine the *functional properties* of the antibody (what the antibody does once it binds with antigen).

An antibody has two identical antigen-binding sites, one at the tip of each arm. These **antigen-binding fragments (Fab)** are

unique for each different antibody, so that each antibody can interact only with an antigen that specifically matches it, much like a lock and key. The tremendous variation in the antigen-binding fragments of different antibodies leads to the extremely large number of unique antibodies that can bind specifically with millions of different antigens.

In contrast to these variable Fab regions at the arm tips, the tail portion of every antibody within each immunoglobulin subclass is identical. The tail, the antibody's so-called **constant (Fc) region**, contains binding sites for particular mediators of antibody-induced activities, which vary among the different subclasses. In fact, differences in the constant region are the basis for distinguishing between the different immunoglobulin subclasses. For example, the constant tail region of IgG antibodies, when activated by antigen binding in the Fab region, binds with phagocytic cells and serves as an opsonin to enhance phagocytosis. In comparison, the constant tail region of IgE antibodies attaches to mast cells and basophils, even in the absence of antigen. When the appropriate antigen gains entry to the body and binds with

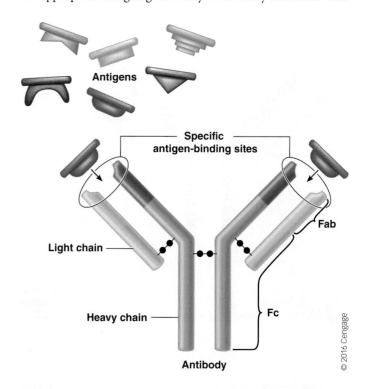

⟩ **FIGURE 11-11 Antibody structure.** An antibody is Y-shaped. It is able to bind only with the specific antigen that "fits" its antigen-binding sites (Fab) on the arm tips. The tail region (Fc) binds with particular mediators of antibody-induced activities.

© 2016 Cengage

the attached antibodies, this triggers the release of histamine from the affected mast cells and basophils. Histamine, in turn, induces the allergic manifestations that follow.

## Antibodies and immune responses

Immunoglobulins cannot directly destroy foreign organisms or other unwanted materials on binding with antigens on their surfaces. Instead, antibodies exert their protective influence by physically hindering antigens or, more commonly, by amplifying innate immune responses (⟩ Figure 11-12).

## Neutralization and agglutination

Antibodies can physically hinder some antigens from exerting their detrimental effects. For example, by combining with bacterial toxins, antibodies can prevent these harmful chemicals from interacting with susceptible cells. This process is known

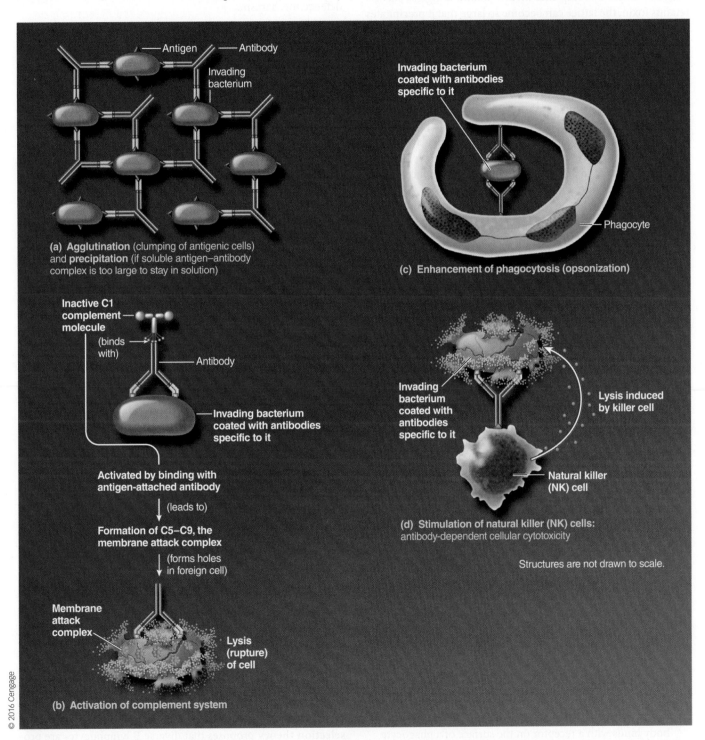

(a) **Agglutination** (clumping of antigenic cells) and **precipitation** (if soluble antigen–antibody complex is too large to stay in solution)

(c) **Enhancement of phagocytosis (opsonization)**

(b) **Activation of complement system**

(d) **Stimulation of natural killer (NK) cells:** antibody-dependent cellular cytotoxicity

Structures are not drawn to scale.

© 2016 Cengage

⟩ **FIGURE 11-12 How antibodies help eliminate invading microbes.** Antibodies physically hinder antigens through neutralization or (a) agglutination and precipitation. Antibodies amplify innate immune responses by (b) activating the complement system, (c) enhancing phagocytosis by acting as opsonins, and (d) stimulating killer cells.

as **neutralization**. Similarly, antibodies can bind with surface antigens on some types of viruses, preventing these viruses from entering cells, where they could exert their damaging effects. Sometimes multiple antibody molecules can cross-link numerous antigen molecules into chains or lattices of antigen—antibody complexes. The process in which foreign cells, such as bacteria or mismatched transfused red blood cells, bind together in such a clump is known as **agglutination**. When linked antigen—antibody complexes involve soluble antigens, such as tetanus toxin, the lattice can become so large that it precipitates out of solution. (**Precipitation** is the process in which a substance separates from a solution.) Within the body, these physical hindrance mechanisms play only a minor protective role against invading agents.

*Clinical Note* However, the tendency for certain antigens to agglutinate or precipitate on forming large complexes with antibodies specific for them is useful clinically and experimentally for detecting the presence of particular antigens or antibodies. Pregnancy diagnosis tests, for example, use this principle to detect, in urine, the presence of a hormone secreted soon after conception.

### AMPLIFICATION OF INNATE IMMUNE RESPONSES

Antibodies' most important function by far is to profoundly augment the innate immune responses already initiated by the invaders. Antibodies mark foreign material as targets for destruction by the complement system, phagocytes, or natural killer (NK) cells. At the same time they enhance the activity of these defence systems by the following methods:

1. *Activating the complement system.* When an appropriate antigen binds with an antibody, receptors on the tail portion of the antibody bind with and activate C1, the first component of the complement system. This sets off the cascade of events leading to formation of the membrane attack complex, which is specifically directed at the membrane of the invading cell that bears the antigen that initiated the activation process. In fact, antibodies are the most powerful activators of the complement system. The biochemical attack subsequently unleashed against the invader's membrane is the most important mechanism by which antibodies exert their protective influence. Furthermore, various activated complement components enhance virtually every aspect of the inflammatory process. The same complement system is activated by an antigen—antibody complex regardless of the type of antigen. Although the binding of antigen to antibody is highly specific, the outcome, which is determined by the antibody's constant tail region, is identical for all activated antibodies within a given subclass; for example, all IgG antibodies activate the same complement system.

2. *Enhancing phagocytosis.* Antibodies, especially IgG, act as opsonins. The tail portion of an antigen-bound IgG antibody binds with a receptor on the surface of a phagocyte and subsequently promotes phagocytosis of the antigen-containing victim attached to the antibody.

3. *Stimulating NK cells.* Binding of antibody to antigen induces the attack of an antigen-bearing cell by NK cells. NK cells attack and destroy both antibody-coated cells and also MHC class I target cells.

In these ways, antibodies, although unable to directly destroy invading bacteria or other undesirable material, bring about destruction of the antigens to which they are specifically attached: specifically, by amplifying other nonspecific lethal defence mechanisms.

### IMMUNE COMPLEX DISEASE

*Clinical Note* Occasionally an overzealous antigen—antibody response inadvertently causes damage to normal cells as well as to invading foreign cells. Typically, antigen—antibody complexes, formed in response to foreign invaders, are removed by phagocytic cells after having revved up nonspecific defence strategies. If large numbers of these complexes are continuously produced, however, the phagocytes cannot clear away all the immune complexes formed. Antigen—antibody complexes that are not removed continue to activate the complement system, among other things. Excessive amounts of activated complement and other inflammatory agents may "spill over," damaging surrounding normal cells as well as the unwanted cells. Furthermore, destruction is not necessarily restricted to the initial site of inflammation. antigen—antibody complexes may circulate freely and become trapped in the kidneys, joints, brain, small vessels of the skin, and elsewhere, causing widespread inflammation and tissue damage. Such damage produced by immune complexes is referred to as an **immune complex disease**, which can be a complicating outcome of bacterial, viral, or parasitic infection.

More insidiously, immune complex disease can also stem from overzealous inflammatory activity prompted by immune complexes formed by *self-antigens* (proteins synthesized by the person's own body) and antibodies erroneously produced against them. *Rheumatoid arthritis* develops in this way.

## Clonal selection

Consider the diversity of foreign molecules a person can potentially encounter during a lifetime. Yet each B cell is preprogrammed to respond to only one of these millions of different antigens. Other antigens cannot combine with the same B cell and induce it to secrete different antibodies. The astonishing implication is that each of us is equipped with millions of different preformed B lymphocytes, at least one for every possible antigen that we might ever encounter—including those specific for synthetic substances that do not exist in nature. The clonal selection theory proposes how a matching B cell responds to its antigen.

Early researchers in immunological theory believed antibodies were "made to order" whenever a foreign antigen gained entry to the body. In contrast, the currently accepted **clonal selection theory** proposes that diverse B lymphocytes are produced during fetal development, each capable of synthesizing an antibody against a particular antigen before ever being exposed to it. All offspring of a particular ancestral B lymphocyte form a family of identical cells, or a **clone**, that is committed to producing the same specific antibody. B cells remain dormant, not actually secreting their particular antibody product nor undergoing rapid division until (or unless) they come into contact with the appropriate antigen. Lymphocytes that have not yet been exposed to their specific antigen are known as **naive lymphocytes**. When an antigen gains entry to the body, the particular clone of B cells that bear receptors on their surface uniquely specific for that antigen is activated, or selected, by the antigen binding with the receptors, hence the term *clonal selection theory* (⟩ Figure 11-13).

The first antibodies produced by a newly formed B cell are IgM immunoglobulins, which are inserted into the cell's plasma membrane rather than secreted. Here, they serve as receptor sites for binding with a specific kind of antigen, almost like advertisements for the kind of antibody the cell can produce. The binding of the appropriate antigen to a B cell amounts to placing an order for the manufacture and secretion of large quantities of that particular antibody.

## Selected clones

Antigen binding causes the activated B-cell clone to multiply and differentiate into two cell types: plasma cells and memory cells. Most progeny are transformed into active plasma cells, which are prolific producers of customized antibodies that contain the same antigen-binding sites as the surface receptors. However, plasma cells switch to producing IgG antibodies, which are secreted rather than remaining membrane bound. In the blood, the secreted antibodies combine with the invading free antigen (not bound to lymphocytes), marking it for destruction by the complement system, phagocytic ingestion, or other means.

### MEMORY CELLS

Not all the new B lymphocytes produced by the specifically activated clone differentiate into antibody-secreting plasma cells. A small proportion of them become **memory cells**, which do not participate in the current immune attack against the antigen, but instead remain dormant and expand the specific clone. If the person is ever re-exposed to the same antigen, these memory cells are primed and ready for even more immediate action than were the original lymphocytes in the clone.

Even though each of us has essentially the same original pool of different B-cell clones, the pool gradually becomes appropriately biased to respond most efficiently to each person's particular antigenic environment. Those clones specific for antigens to which a person is never exposed remain dormant for life, whereas those specific for antigens in the individual's environment typically become expanded and enhanced by forming highly responsive memory cells. The different naive clones provide protection against unknown new pathogens, and the evolving populations of memory cells protect against the recurrence of infections encountered in the past.

### PRIMARY AND SECONDARY RESPONSES

During initial contact with a microbial antigen, the antibody response is delayed for several days until plasma cells are formed and does not reach its peak for a couple of weeks (⟩ Figure 11-14). This response is known as the **primary response**. Meanwhile, symptoms characteristic of the particular microbial invasion persist until either the invader succumbs to the mounting specific immune attack against it or the infected person dies. After reaching the peak, the antibody levels gradually decline over a period of time, although some circulating antibody from

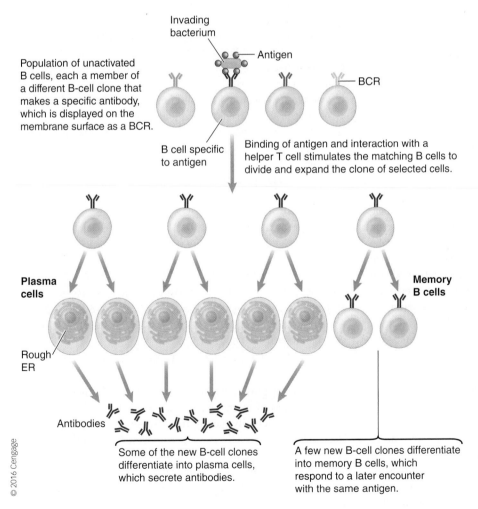

Population of unactivated B cells, each a member of a different B-cell clone that makes a specific antibody, which is displayed on the membrane surface as a BCR.

Invading bacterium
Antigen
BCR

B cell specific to antigen

Binding of antigen and interaction with a helper T cell stimulates the matching B cells to divide and expand the clone of selected cells.

Plasma cells

Memory B cells

Rough ER

Antibodies

Some of the new B-cell clones differentiate into plasma cells, which secrete antibodies.

A few new B-cell clones differentiate into memory B cells, which respond to a later encounter with the same antigen.

© 2016 Cengage

⟩ **FIGURE 11-13 Clonal selection theory.** The B-cell clone specific to the antigen proliferates and differentiates into plasma cells and memory cells. Plasma cells secrete antibodies that bind with free antigen not attached to B cells. Memory cells expand the specific clone and are primed and ready for subsequent exposure to the same antigen.

# Vaccination: A Victory over Many Dreaded Diseases

**M**ODERN SOCIETY HAS COME TO HOPE and even expect that vaccines can be developed to protect us from almost any dreaded infectious disease. This expectation has been brought into sharp focus by our current frustration over the inability to date to develop a successful vaccine against HIV, the virus that causes AIDS.

Nearly 2500 years ago, our ancestors were aware of the existence of immune protection. Writing about a plague that struck Athens in 430 BCE, Thucydides observed that the same person was never attacked twice by this disease. However, the ancients did not understand the basis of this protection, so they could not manipulate it to their advantage.

Early attempts to deliberately acquire lifelong protection against smallpox, a dreaded disease that was highly infectious and frequently fatal (up to 40% of the sick died), consisted of intentionally exposing oneself by coming into direct contact with a person suffering from a milder form of the disease. The hope was to protect against a future fatal bout of smallpox by deliberately inducing a mild case of the disease. By the beginning of the 17th century, this technique had evolved into using a needle to extract small amounts of pus from active smallpox pustules (the fluid-filled bumps on the skin, which leave a characteristic depressed scar or pock mark after healing) and introducing this infectious material into healthy individuals. This inoculation process was done by applying the pus directly to slight cuts in the skin or by inhaling dried pus.

Edward Jenner, an English physician, was the first to demonstrate that immunity against cowpox, a disease similar to but less serious than smallpox, could also protect humans against smallpox. Having observed that milkmaids who got cowpox seemed to be protected from smallpox, Jenner in 1796 inoculated a healthy boy with pus he had extracted from cowpox boils. After the boy recovered, Jenner (not being restricted by modern ethical standards of research on human subjects) deliberately inoculated him with what was considered a normally fatal dose of smallpox infectious material. The boy survived.

Jenner's results were not taken seriously, however, until a century later when, in the 1880s, Louis Pasteur, the first great experimental immunologist, extended Jenner's technique. Pasteur demonstrated that the disease-inducing capability of organisms could be greatly reduced (attenuated) so they could no longer produce disease but would still induce antibody formation when introduced into the body—the basic principle of modern vaccines. His first vaccine was against anthrax, a deadly disease of sheep and cows. Pasteur isolated and heated anthrax bacteria, then injected these attenuated organisms into a group of healthy sheep. A few weeks later at a gathering of fellow scientists, Pasteur injected these vaccinated sheep as well as a group of unvaccinated sheep with fully potent anthrax bacteria. The result was dramatic—all the vaccinated sheep survived, but all the unvaccinated sheep died. Pasteur's notorious public demonstrations, such as this, coupled with his charismatic personality, caught the attention of physicians and scientists of the time, sparking the development of modern immunology.

Today, vaccinations work using the same principles as those used by Jenner and Pasteur, but we have developed other methods of delivering the foreign antigen. Three common methods of introducing the antigen follow:

1. Inserting an inactivated vaccine consisting of virus particles that have been grown in culture and then killed (e.g., by heat)
2. Using an attenuated vaccine consisting of live virus particles with low virulence
3. Establishing a subunit vaccine that presents an antigen to the immune system without introducing viral particles

## Further Reading

Crockett, M., & Keystone, J. (2005). "I hate needles" and other factors impacting on travel vaccine uptake. *J Travel Med, 12*(Suppl 1): S41—S46.

Thomas, R.E., Jefferson, T.O., Demicheli, V. & Rivetti, D. (2006). Influenza vaccination for health-care workers who work with elderly people in institutions: A systematic review. *Lancet Infect Dis, 6*(5): 273–79.

Leung, A.K., Kellner, J.D., & Davies, H.D. (2005). Hepatitis A: A preventable threat. *Adv Ther, 22*(6): 578–86.

---

this primary response may persist for a prolonged period. Long-term protection against the same antigen, however, is primarily attributable to the memory cells. If the same antigen ever reappears, the long-lived memory cells launch a more rapid, more potent, and longer-lasting **secondary response** than occurred during the primary response. This swifter, more powerful immune attack is frequently adequate to prevent or minimize overt infection on subsequent exposures to the same microbe, forming the basis of long-term immunity against a specific disease.

*Clinical Note* The original antigenic exposure that induces the formation of memory cells can occur through the person either actually having the disease or being vaccinated. Vaccination deliberately exposes the person to a pathogen that has been stripped of its disease-inducing capability but that can still induce antibody formation against itself. (For the early history of vaccination development, see Concepts, Challenges, and Controversies.)

Memory cells are not formed for some diseases, so lasting immunity is not conferred by an initial exposure, as in the case

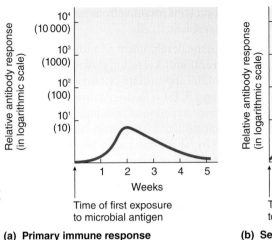

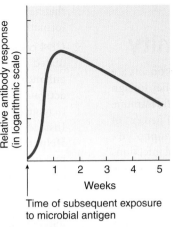

(a) **Primary immune response**

(b) **Secondary immune response**

> FIGURE 11-14 **Primary and secondary immune responses.** (a) Primary response on first exposure to a microbial antigen. (b) Secondary response on subsequent exposure to the same microbial antigen. The primary response does not peak for a couple of weeks, whereas the secondary response peaks in a week. The magnitude of the secondary response is 100 times greater than the primary response. (The relative antibody response is in the logarithmic scale.)

of strep throat. The course and severity of the disease are the same each time a person is reinfected with a microbe that the immune system does not "remember," regardless of the number of prior exposures.

## B cells and gene fragments

Considering the millions of different antigens against which each of us has the potential to actively produce antibodies, how is it possible for an individual to have such a tremendous diversity of B lymphocytes, each capable of producing a different antibody? Antibodies are proteins synthesized in accordance with a nuclear DNA blueprint. However, because all cells of the body, including the antibody-producing cells, contain the same nuclear DNA, it is hard to imagine how enough DNA could be packaged within the nuclei of every cell to code for the millions of different antibodies (a different portion of the genetic code being used by each B-cell clone), along with all the other genetic instructions used by other cells. Actually, only a relatively small number of gene fragments code for antibody synthesis, but during B-cell development these fragments are cut, reshuffled, and spliced in a vast number of different combinations. Each different combination gives rise to a unique B-cell clone. Antibody genes are later even further diversified by somatic mutation (Appendix C, p. A-30). The antibody genes of already-formed B cells are highly prone to mutations in the region that codes for the variable antigen-binding sites on the antibodies. Each different mutant cell in turn gives rise to a new clone. Thus, the great diversity of antibodies is made possible by the reshuffling of a small set of gene fragments during B-cell development, as well as by further somatic mutation

in already-formed B cells. In this way, a huge antibody repertoire is possible using only a modest share of the genetic blueprint.

## Active and passive immunity

The production of antibodies as a result of exposure to an antigen is referred to as **active immunity** against that antigen. A second way in which an individual can acquire antibodies is by the direct transfer of antibodies actively formed by another person (or animal). The immediate "borrowed" immunity conferred upon receipt of preformed antibodies is known as **passive immunity**. Such transfer of antibodies of the IgG class normally occurs from the mother to the fetus across the placenta during intrauterine development. In addition, a mother's colostrum (first milk) contains IgA antibodies that provide further protection for breastfed babies. Passively transferred antibodies are usually broken down in less than a month, but meanwhile newborns are provided important immune protection (essentially the same as their mother's) until they can begin actively mounting their own immune responses. Antibody-synthesizing ability does not develop for about a month after birth.

Passive immunity is sometimes employed clinically to provide immediate protection or to bolster resistance against an extremely virulent infectious agent or potentially lethal toxin to which a person has been exposed: for example, rabies virus, tetanus toxin in nonimmunized individuals, and poisonous snake venom. Typically, the administered preformed antibodies have been harvested from another source (often nonhuman) that has been exposed to an attenuated form of the antigen. Frequently, horses or sheep are used in the deliberate production of antibodies to be collected for passive immunizations. Although injection of serum containing these antibodies—**antiserum (antitoxin)**—is beneficial in providing immediate protection against the specific disease or toxin, the recipient may develop an immune response against the injected antibodies themselves, because they are foreign proteins. The result may be a severe allergic reaction to the treatment, a condition known as **serum sickness**.

**11**

## Check Your Understanding 11.6

1. Describe the structure and function of the Fab and Fc regions of an antibody.

2. List the means by which antibodies help to eliminate invading microbes.

3. According to the clonal selection theory, draw a flow diagram showing the formation of plasma cells and memory cells in response to the binding of an antigen to its specific B-cell receptor.

© 2016 Cengage

## 11.7 | T Lymphocytes: Cell-Mediated Immunity

As important as B lymphocytes and their antibody products are in specific defence against invading bacteria and other foreign material, they represent only half of the body's specific immune defences. T cells account for approximately 50 to 70 percent of the total lymphocyte number in circulation. The T lymphocytes are equally as important in defence against most viral infections as B lymphocytes are and also play an important regulatory role in immune mechanisms.

### T cells and targets

Whereas B cells and antibodies defend against conspicuous invaders in the ECF, T cells defend against covert invaders that hide inside cells where antibodies and the complement system cannot reach them. Unlike B cells, which secrete antibodies that can attack antigen at long distances, T cells do not secrete antibodies. Instead, they must directly contact their targets, a process known as *cell-mediated immunity*. T cells of the killer type release chemicals that destroy targeted cells that they contact, such as virus-infected cells and cancer cells.

Like B cells, T cells are clonal and exquisitely antigen specific. On its plasma membrane, each T cell bears unique receptor proteins called T-cell receptors, similar although not identical to the surface receptors on B cells. Immature lymphocytes acquire their T-cell receptors in the thymus during their differentiation into T cells. Unlike B cells, T cells are activated by a foreign antigen only when it is on the surface of a cell that also carries a marker of the individual's own identity; that is, both foreign antigens and **self-antigens** known as **major histocompatibility complex (MHC) molecules** must be on a cell's surface before a T cell can bind with it. T cells learn to recognize foreign antigens only in combination with the person's own tissue antigens. This "thymic education," lets T cells pass on what they've learned to all T cells' future progeny. The importance of this dual antigen requirement and the nature of the MHC self-antigens are described in the section Class I and Class II MHC Glycoproteins.

A delay of a few days generally follows exposure to the appropriate antigen before **sensitized (activated) T cells** are prepared to launch a cell-mediated immune attack. When exposed to a specific antigen combination, cells of the complementary T-cell clone proliferate and differentiate for several days, a process that yields large numbers of activated effector T cells that carry out various cell-mediated responses.

### Cytotoxic T cells, helper T cells, and regulatory T cells

There are three main subpopulations of T cells, depending on their roles when activated by antigens:

1. **CD8 cells (cytotoxic, or killer, T cells)** destroy host cells harbouring anything foreign and thus bearing foreign antigen, such as body cells invaded by viruses, cancer cells that have mutated proteins resulting from malignant transformations, and transplanted cells.

2. **CD4 cells** enhance the development of antigen-stimulated B cells into antibody-secreting plasma cells, enhance activity of the appropriate cytotoxic cells, and activate macrophages. CD4 cells do not directly participate in immune destruction of invading pathogens. Instead, they modulate activities of other immune cells. **Helper T cells** are by far the most numerous T cells, making up 60–80 percent of circulating T cells. Because of the important role these cells play in turning on the full power of all the other activated lymphocytes and macrophages, helper T cells constitute the immune system's master switch.

3. **Regulatory T cells** ($T_{reg}$), originally called **suppressor T cells**, are a recently identified small subset of CD4 cells. In addition, they also have CD25, a component of a receptor for IL-2, which promotes $T_{reg}$ activities. Thus, these cells are also referred to as **CD4+CD25+T cells**.

*Clinical Note* **Acquired immunodeficiency syndrome (AIDS)**, caused by the **human immunodeficiency virus (HIV)**, is extremely devastating to the immune defence system. The AIDS virus selectively invades helper T cells, destroying or incapacitating the cells that normally orchestrate much of the immune response (› Figure 11-15). The virus also invades macrophages, further crippling the immune system, and sometimes enters brain cells as well, leading to the dementia (severe impairment of intellectual capacity) noted in some AIDS victims.

### MEMORY T CELLS

Like B cells, T cells form a memory pool and display both primary and secondary responses. Primary responses tend to be initiated in the lymphoid tissues, where naive lymphocytes and antigen-presenting cells interact. For a few weeks after the infection is cleared, more than 90 percent of the huge number of effector T cells generated during the primary response die by means of *apoptosis* (cell suicide). To stay alive, activated

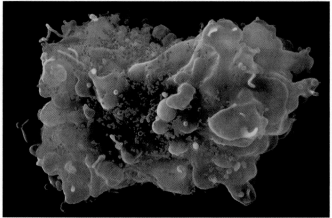

› FIGURE 11-15 **AIDS virus.** Human immunodeficiency virus (HIV) (in grey), the AIDS-causing virus, on a helper T lymphocyte, HIV's primary target

© Lennart Nilsson/Scanpix

11

T lymphocytes require the continued presence of their specific antigen and appropriate stimulatory signals. Once the foe succumbs, the vast majority of the now superfluous T lymphocytes commit suicide, because their supportive antigen and stimulatory signals are withdrawn. Elimination of most of the effector T cells following a primary response is essential to prevent congestion in the lymphoid tissues. (Such paring down is not needed for B cells because those that become plasma cells rather than memory B cells upon antigen stimulation rapidly work themselves to death while producing antibodies.) The remaining surviving effector T cells become long-lived memory T cells that migrate to all areas of the body, where they are poised for a swift secondary response to the same pathogen in the future.

Let's examine the T cells in further detail.

## Cytotoxic T cells

Cytotoxic T cells are microscopic "hit men." The targets of these destructive cells most frequently are host cells infected with viruses. When a virus invades a body cell, as it must to survive, the infected cell breaks down the envelope of proteins surrounding the virus and then loads a fragment of the viral antigen piggyback-style onto a newly synthesized self-antigen. This self-antigen and viral antigen complex is inserted into the host cell's surface membrane, where it serves as a red flag that indicates the cell is harbouring the invader (> Figure 11-16, steps **1** and **2**). To attack the intracellular virus, cytotoxic T cells must destroy the infected host cell in the process. Cytotoxic T cells of the clone that's specific for this particular virus recognize and bind to the viral antigen and self-antigen on the surface of an infected cell (Figure 11-16, step **3**). Thus sensitized by viral antigen, a cytotoxic T cell can kill the infected cell by either direct or indirect means, depending on the type of lethal chemicals the activated T cell releases. Let's elaborate.

- An activated cytotoxic T cell may directly kill the victim cell by releasing chemicals that lyse the attacked cell before viral replication can begin (Figure 11-16, step **4**). Specifically, cytotoxic T cells as well as NK cells destroy a targeted cell by releasing **perforin** molecules, which penetrate the target cell's surface membrane and join to form porelike channels (> Figure 11-17). This technique of killing a cell by punching holes in its membrane is similar to the method employed by the membrane attack complex of the complement cascade. This contact-dependent mechanism of killing has been nicknamed the "kiss of death."

- A cytotoxic T cell can also indirectly bring about death of an infected host cell by releasing **granzymes**, which are enzymes similar to digestive enzymes. Granzymes enter the target cell through the perforin channels. Once inside, these chemicals trigger the virus-infected cell to self-destruct through apoptosis (cell suicide).

The virus released on destruction of the host cell by either of these methods is directly destroyed in the ECF by phagocytic cells, neutralizing antibodies, and the complement system. Meanwhile, the cytotoxic T cell, which has not been harmed in the process, can move on to kill other infected host cells.

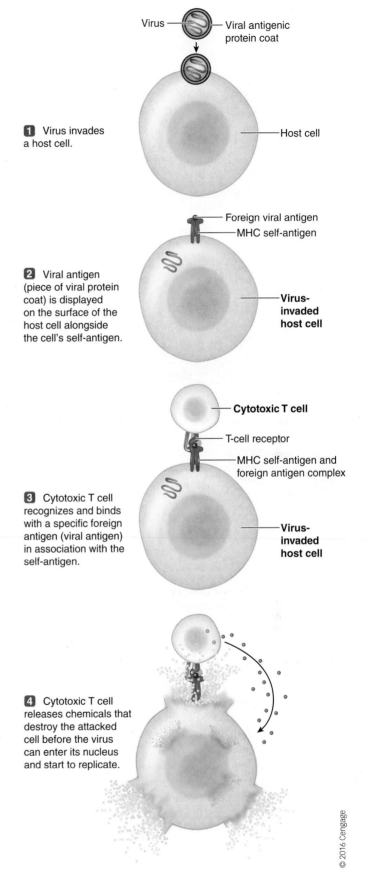

**1** Virus invades a host cell.

Virus — Viral antigenic protein coat

Host cell

**2** Viral antigen (piece of viral protein coat) is displayed on the surface of the host cell alongside the cell's self-antigen.

Foreign viral antigen
MHC self-antigen
**Virus-invaded host cell**

**3** Cytotoxic T cell recognizes and binds with a specific foreign antigen (viral antigen) in association with the self-antigen.

**Cytotoxic T cell**
T-cell receptor
MHC self-antigen and foreign antigen complex
**Virus-invaded host cell**

**4** Cytotoxic T cell releases chemicals that destroy the attacked cell before the virus can enter its nucleus and start to replicate.

© 2016 Cengage

> **FIGURE 11-16 A cytotoxic T cell lysing a virus-invaded cell**

11

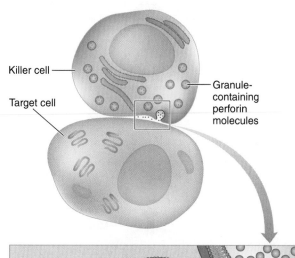

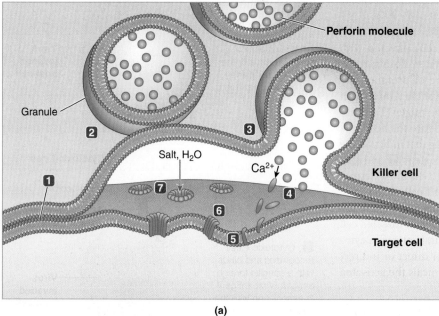

**1** The killer cell binds to its target cell.

**2** As a result of this binding, the killer cell's perforin-containing granules fuse with the plasma membrane.

**3** The granules disgorge their perforin by exocytosis into a small pocket of intercellular space between the killer cell and its target.

**4** On exposure to $Ca^{2+}$ in this ECF space, the individual perforin molecules change from a spherical to a cylindrical shape.

**5** The remodelled perforin molecules bind to the target cell membrane and insert into it.

**6** Individual perforin molecules group together like staves of a barrel to form pores.

**7** The pores admit salt and $H_2O$, causing the target cell to swell and burst.

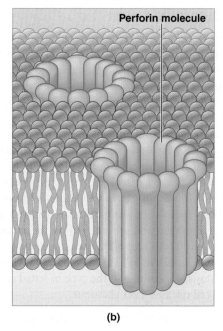

(a)     (b)

› **FIGURE 11-17 Mechanism of killing by killer cells.** (a) Details of the killing process. (b) Enlargement of perforin-formed pores in a target cell. Note the similarity to the membrane attack complex formed by complement molecules (see › Figure 11-7).

The surrounding healthy cells replace the lost cells by means of cell division. Usually, to halt a viral infection, only some of the host cells must be destroyed. If the virus has had a chance to multiply, however, with the replicated virus leaving the original cell and spreading to other host cells, the cytotoxic T-cell defence mechanism may sacrifice so many of the host cells that serious malfunction may ensue.

Recall that other nonspecific defence mechanisms also come into play to combat viral infections, most notably NK cells, interferon, macrophages, and the complement system. As usual, an intricate web of interplay exists among the immune defences that are launched against viral invaders (❙ Table 11-2).

### ANTIVIRAL DEFENCE IN THE NERVOUS SYSTEM

The usual method of destroying virus-infected host cells is not appropriate for the nervous system. If cytotoxic T cells destroyed virus-infected neurons, the lost cells could not be replaced, because neurons cannot reproduce. Fortunately, virus-infected neurons are spared from extermination by the immune system. But how are neurons protected from viruses? Immunologists long thought that the only antiviral defences for neurons were those aimed at free viruses in the extracellular fluid. Surprising new research has revealed, however, that antibodies not only target viruses for destruction in the extracellular fluid but can also eliminate viruses inside neurons. It is unclear whether antibodies actually enter the neurons and interfere directly with viral replication (neurons have been shown to take up antibodies near their synaptic endings) or bind with the surface of nerve cells and trigger intracellular changes that stop viral replication.

*Clinical Note* The fact that some viruses, such as the herpes virus, persist for years in nerve cells, occasionally flaring up to produce symptoms, demonstrates that the antibodies' intraneuronal mechanism does not provide a foolproof antiviral defence for neurons.

**When the virus is free in the ECF**

Macrophages

Destroy the free virus by phagocytosis

Process and present the viral antigen to both B and T cells

Secrete interleukin 1, which activates B- and T-cell clones specific to the viral antigen

Plasma Cells Derived from B Cells Specific to the Viral Antigen Secrete Antibodies That

Neutralize the virus to prevent its entry into a host cell

Activate the complement cascade that directly destroys the free virus and enhances phagocytosis of the virus by acting as an opsonin

**When the virus has entered a host cell (which it must do to survive and multiply, with the replicated viruses leaving the original host cell to enter the ECF in search of other host cells)**

Interferon

Is secreted by virus-infected cells

Binds with and prevents viral replication in other host cells

Enhances the killing power of macrophages, natural killer cells, and cytotoxic T cells

Natural Killer Cells

Nonspecifically lyse virus-infected host cells

Cytotoxic T Cells

Are specifically sensitized by the viral antigen and lyse the infected host cells before the virus has a chance to replicate

Helper T Cells

Secrete cytokines, which enhance cytotoxic T-cell activity and B-cell antibody production

**When a virus-infected cell is destroyed, the free virus is released into the ECF, where it is attacked directly by macrophages, antibodies, and the activated complement components.**

## Helper T cells

In contrast to cytotoxic T cells, helper T cells are not killer cells. Instead, helper T cells secrete chemicals classified as *cytokines* that help, or augment, nearly all aspects of the immune response. Cytokines are growth factors produced by leukocytes and other cells and have many functions, some of which are discussed next.

### CYTOKINES

Exposure to antigens frequently activates both the B- and T-cell mechanisms simultaneously. Just as helper T cells can modulate the secretion of antibodies by B cells, antibodies may influence cytotoxic T cells' ability to destroy a victim cell. Most effects that lymphocytes exert on other immune cells are mediated by secretion of chemical messengers. All chemicals other than the antibodies that leukocytes secrete are collectively called *cytokines*, most of which are produced by helper T cells. Unlike antibodies, cytokines do not interact directly with the antigen that induces their production. Instead, cytokines spur other immune cells into action to help ward off the invader. The following are among the best known of helper T-cell cytokines:

1. Helper T cells secrete several interleukins (*IL-4*, *IL-5*, and *IL-6*) that serve collectively as a **B-cell growth factor**, which contributes to B-cell function in concert with the IL-1 secreted by macrophages. Antibody secretion is greatly reduced or absent without the assistance of helper T cells, especially in defence against T-dependent antigen.

2. Helper T cells similarly secrete **T-cell growth factor** (*IL-2*), which augments the activity of cytotoxic T cells and even of other helper T cells responsive to the invading antigen. In typical interplay fashion, IL-1 secreted by macrophages not only enhances the activity of both the appropriate B- and T-cell clones but also stimulates secretion of IL-2 by activated helper T cells.

3. Some chemicals secreted by T cells act as *chemotaxins* to lure more neutrophils and macrophages-to-be to the invaded area.

4. Once macrophages are attracted to the area, **macrophage-migration inhibition factor**, another important cytokine released from helper T cells, keeps these large phagocytic cells in the region by inhibiting their outward migration. As a result, a great number of chemotactically attracted macrophages accumulate in the infected area. This factor also confers greater phagocytic power to the gathered macrophages. These so-called **angry macrophages** have more powerful destructive ability. They are especially important in defending against the bacteria that cause tuberculosis, because such microbes can survive simple phagocytosis by nonactivated macrophages.

5. One cytokine secreted by helper T cells (IL-5) activates eosinophils, and another (IL-4) promotes the development of IgE antibodies for defence against parasitic worms.

### T HELPER 1 AND T HELPER 2 CELLS

Not all helper cells secrete the same cytokines. Two subsets of helper T cells—**T helper 1** ($T_H1$) cells and **T helper 2** ($T_H2$) cells—augment different patterns of immune responses by secreting different types of cytokines. $T_H1$ cells rally a cell-mediated (cytotoxic T-cell) response, which is appropriate for infections with intracellular microbes, such as viruses, whereas $T_H2$ cells promote antibody-mediated immunity by B cells and rev up eosinophil activity for defence against parasitic worms.

Helper T cells produced in the thymus are in a naive state until they encounter the antigen they are primed to recognize. Whether a naive helper T cell becomes a $T_H1$ or $T_H2$ cell depends on which cytokines are secreted by macrophages and dendritic cells (macrophage-like cells) of the innate immune system that present the antigen to the uncommitted T cell. **Interleukin 12 (IL-12)**

**11**

drives a naive T cell specific for the antigen to become a $T_H1$ cell, whereas **interleukin 4 (IL-4)** favours the development of a naive cell into a $T_H2$ cell. Thus, the **antigen-presenting cells** of the nonspecific immune system can influence the whole tenor of the specific immune response by determining whether the $T_H1$ or $T_H2$ cellular subset dominates. In the usual case, the secreted cytokines promote the appropriate specific immune response against the particular threat at hand.

Scientists have also recently discovered much smaller subsets of helper T cells: $T_H17$ cells and T follicular helper ($T_{FH}$) cells. **Helper T 17 ($T_H17$) cells** produce *IL-17* (accounting for the name of this subset). They promote inflammation and are effector molecules in the development of inflammatory autoimmune diseases such as multiple sclerosis. The newest discovered **T follicular helper ($T_{FH}$) cells** specifically interact with B cells in lymph node follicles (hence the name of this subset) to help them secrete antibodies in response to T-dependent antigens.

> ▍ **Clinical Connections**
>
> In the case of allergic asthma, the allergen causes macrophages to secrete cytokines that are presented to the naive helper T cell (Th0). These Th0 cells are transformed into Th2 cells, and these cells can then activate the humoural immune system to produce antibodies against the inhaled allergen. When the allergen is again inhaled, an adaptive immune response occurs, which results in the recruitment and infiltration of eosinophils.

## The immune system and self-antigens

The term tolerance in this context refers to the phenomenon of preventing the immune system from attacking the person's own tissues. During the genetic "cut, shuffle, and paste process" that goes on during lymphocyte development, some B and T cells are by chance formed that could react against the body's own tissue antigens. If these lymphocyte clones were allowed to function, they would destroy the individual's body. Fortunately, the immune system normally does not produce antibodies or activated T cells against the body's own self-antigens, but instead directs its destructive tactics at foreign antigens only.

At least six different mechanisms are involved in tolerance:

1. *Clonal deletion.* In response to continuous exposure to body antigens early in development, lymphocyte clones specifically capable of attacking these self-antigens in most cases are permanently destroyed. This **clonal deletion** is accomplished by triggering apoptosis of immature cells that would react with the body's own proteins. This physical elimination is the major mechanism by which tolerance is developed.

2. *Clonal anergy.* The premise of **clonal anergy** is that a lymphocyte must receive two specific simultaneous signals to be activated (turned on), one from its compatible antigen and another from a stimulatory cosignal molecule known as **B7**, which is found only on the surface of an antigen-presenting cell. Both signals are present for foreign antigens, which are introduced to lymphocytes by antigen-presenting cells. Once a B or T cell is turned on by finding

its matching antigen in accompaniment with the cosignal, the cell no longer needs the cosignal to interact with other cells. For example, an activated cytotoxic T cell can destroy any virus-invaded cell that bears the viral antigen, even though the infected cell does not have the cosignal. In contrast, these dual signals—antigen plus cosignal—are never present for self-antigens because these antigens are not handled by cosignal-bearing, antigen-presenting cells. The first exposure to a single signal from a self-antigen turns *off* the compatible T cell, rendering the cell unresponsive to further exposure to the antigen instead of spurring the cell to proliferate. This reaction is referred to as *clonal anergy* (*anergy* means "lack of energy") because T cells are being inactivated ("become lazy") rather than activated by their antigens. Clonal anergy is a backup to clonal deletion. Anergized lymphocyte clones survive, but can't function.

3. *Receptor editing.* A newly identified means of ridding the body of self-reactive B cells is **receptor editing**. With this mechanism, once a B cell that bears a receptor for one of the body's own antigens encounters the self-antigen, the B cell escapes death or a lifetime of anergy by swiftly changing its antigen receptor to a nonself version. In this way, an originally self-reactive B cell survives but is "rehabilitated" so that it will never target the body's own tissues again.

4. *Inhibition by regulatory T cells.* These suppressor cells may play a role in tolerance by inhibiting throughout life some lymphocyte clones specific for the body's own tissues.

5. *Immunological ignorance,* alternatively known as *antigen sequestering.* Some self-molecules are normally hidden from the immune system, because they never come into direct contact with the ECF in which the immune cells and their products circulate. An example of such a segregated antigen is thyroglobulin, a complex protein sequestered within the hormone-secreting structures of the thyroid gland (p. 248).

6. *Immune privilege.* A few tissues, most notably the testes and the eyes, have **immune privilege**, because they escape immune attack even when they are transplanted in an unrelated individual. Scientists recently discovered that the cellular plasma membranes in these immune-privileged tissues have a specific molecule that triggers apoptosis of approaching activated lymphocytes that could attack the tissues.

## Autoimmune diseases

*Clinical Note* Occasionally, the immune system fails to distinguish self-antigens from foreign antigens, which results in an attack on its own healthy cells. The nonrecognition of self-antigens is termed an **autoimmune disease**. An example of such a disease is lupus. There are different kinds of lupus, of which the most common is systemic lupus erythematosus, which affects many parts of the body, such as the heart, joints, skin, lungs, blood vessels, liver, kidneys, and nervous system. Systemic lupus erythematosus is unpredictable, with periods of illness (known as flares) alternating with periods of relief (remission). Systemic lupus erythematosus can occur at any age and is more common in women. Other forms of lupus are discoid lupus, which causes a persistent rash; subacute cutaneous lupus, which is associated with sores after exposure to the sun; and neonatal lupus, which affects newborns. Currently, there is no cure for lupus, but lifestyle adjustments can help fight the disease and give an improved sense of well-being. Autoimmunity underlies more than 80 diseases, many of which are well known. Examples include multiple sclerosis, rheumatoid arthritis, and type 1 diabetes mellitus.

Autoimmune diseases may arise from a number of different causes:

1. Exposure of normally inaccessible self-antigens sometimes induces an immune attack against these antigens. Because the immune system is usually never exposed to hidden self-antigens, it does not learn to tolerate them. Inadvertent exposure of these normally inaccessible antigens to the immune system because of tissue disruption caused by injury or disease can lead to a rapid immune attack against the affected tissue, just as if these self-proteins were foreign invaders. *Hashimoto's disease*, which involves the production of antibodies against thyroglobulin and the destruction of the thyroid gland's hormone-secreting capacity, is one such example.

2. Normal self-antigens may be modified by factors, such as drugs, environmental chemicals, viruses, or genetic mutations, so that they are no longer tolerated by the immune system.

3. Exposure of the immune system to a foreign antigen structurally almost identical to a self-antigen may induce the production of antibodies or activated T lymphocytes that not only interact with the foreign antigen but also cross-react with the closely similar body antigen. An example of this molecular mimicry is the streptococcal bacteria responsible for strep throat. The bacteria possess antigens structurally very similar to self-antigens in the tissue covering the heart valves of some individuals, in which case the antibodies produced against the streptococcal organisms may also bind with this heart tissue. The resultant inflammatory response is responsible for the heart-valve lesions associated with *rheumatic fever*.

4. New studies hint at another possible trigger of autoimmune diseases, one that could explain why a whole host of these disorders are more common in women than in men. Traditionally, scientists have speculated that the sex bias of autoimmune diseases was somehow related to hormonal differences. Recent findings suggest, however, that the higher incidence of these self-destructive conditions in females may be a legacy of pregnancy. Researchers have learned that fetal cells, which often gain access to the mother's bloodstream during the trauma of labour and delivery, sometimes linger in the mother for decades after the pregnancy. The immune system typically clears these cells from the mother's body following childbirth, but studies involving one particular autoimmune disease demonstrated that those women with the condition were more likely than healthy women to have persistent fetal cells in their blood. The persistence of similar but not identical fetal antigens that were not wiped out early on as being foreigners may somehow trigger a gradual, more subtle immune attack that eventually turns against the mother's own closely related antigens.

What is the nature of the self-antigens that the immune system learns to recognize as markers of a person's own cells? That is the topic of the next section.

## The major histocompatibility complex

Self-antigens are plasma membrane-bound glycoproteins (proteins with sugar attached), known as *MHC molecules* because their synthesis is directed by a group of genes called the *major histocompatibility complex (MHC)*. These are the same MHC molecules that escort engulfed foreign antigen to the cell surface for presentation by antigen-presenting cells. The MHC genes are the most variable ones in humans. More than 100 different MHC molecules have been identified in human tissue, but each individual has a code for only three to six of these possible antigens. Because of the tremendous number of different combinations possible, the exact pattern of MHC molecules varies from one individual to another, much like a biochemical fingerprint or molecular identification card—except in identical twins, who have the same MHC self-antigens.

The major histocompatibility (*histo* means "tissue"; *compatibility* means "ability to get along") complex was so named because these genes and the self-antigens they encode were first discerned in relation to tissue typing (similar to blood typing), which is done to obtain the most compatible matches for tissue grafting and transplantation. However, the transfer of tissue from one individual to another does not normally occur in nature. The natural function of MHC antigens lies in their ability to direct the responses of T cells, not in their artificial role in rejecting transplanted tissue.

MHC molecules on a cell's surface, which are always combined with a foreign antigen, signal to the immune cells. T cells typically bind with MHC self-antigens only when associated with a foreign antigen, such as a viral protein, also displayed on the cell surface in a groove on the top of the MHC molecule. Thus, T-cell receptors bind only with body cells, making the statement—by bearing both self- and nonself antigens on their surface—"I, one of your own kind, have been invaded. Here's a description of the enemy I am housing within." Only T cells that specifically match up with both the self- and foreign antigen can bind with the infected cell.

### LOADING OF FOREIGN PEPTIDE ON MHC MOLECULE
Unlike B cells, T cells cannot bind with foreign antigen that is not in association with self-antigen. It would be futile for T cells to bind with free, extracellular antigens; they cannot defend against foreign material unless it is intracellular. A foreign protein first must be enzymatically broken down within a body cell into small fragments known as peptides. These antigenic peptides are inserted into the binding groove of a newly synthesized MHC

molecule before the MHC-foreign antigen complex travels to the surface membrane. Once displayed at the cell surface, the combined presence of these self- and nonself antigens alerts the immune system to the presence of an undesirable agent within the cell. Highly specific T-cell receptors fit a particular MHC-foreign antigen complex in complementary fashion. This binding arrangement can be likened to a hot dog in a bun, with the MHC molecule being the bottom of the bun, the T-cell receptor the bun's top, and the foreign antigen the hot dog. In the case of cytotoxic T cells, the outcome of this binding is the destruction of the infected body cell. Because cytotoxic T cells do not bind to MHC self-antigens in the absence of foreign antigen, normal body cells are protected from a lethal immune attack.

## CLASS I AND CLASS II MHC GLYCOPROTEINS

T cells become active only when they match a given MHC peptide combination. In addition to having to fit a specific foreign peptide, the T-cell receptor must also match the appropriate MHC protein. Each individual has two main classes of MHC-encoded molecules that are differentially recognized by cytotoxic T and helper T cells: class I and class II MHC glycoproteins, respectively (> Figure 11-18). The class I and II markers serve as signposts to guide cytotoxic and helper T cells to the precise cellular locations where their immune capabilities can be most effective.

Cytotoxic T cells can respond to foreign antigen only in association with **class I MHC glycoproteins**, which are found on the surface of virtually all nucleated body cells. To carry out their role of dealing with pathogens that have invaded host cells, it is appropriate that cytotoxic T cells bind only with cells of the organism's own body that viruses have infected—that is, with foreign antigen in association with self-antigen. Furthermore, these deadly T cells can also link up with any cancerous body cell, because class I MHC molecules also display mutated cellular proteins characteristic of these abnormal cells. Because any

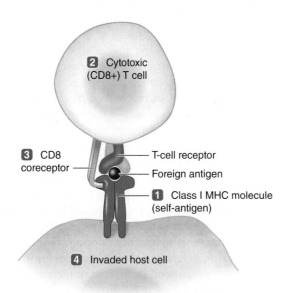

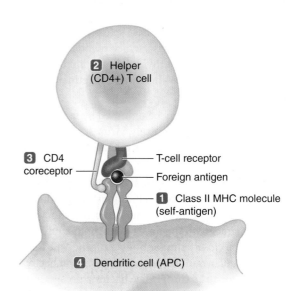

1️⃣ Class I MHC molecules are found on the surface of all cells.

2️⃣ They are recognized only by cytotoxic (CD8+) T cells.

3️⃣ CD8 coreceptor links the two cells together.

4️⃣ Linked in this way, cytotoxic T cells can destroy body cells if invaded by foreign (viral) antigen.

1️⃣ Class II MHC molecules are found on the surface of immune cells with which helper T cells interact: dendritic cells, macrophages, and B cells.

2️⃣ They are recognized only by helper (CD4+) T cells.

3️⃣ CD4 coreceptor links the two cells together.

4️⃣ To be activated, helper T cells must bind with a class II MHC-bearing APC (dendritic cell or macrophage). To activate B cells, helper T cell must bind with a class II MHC-bearing B cell with displayed foreign antigen.

(a) **Class I MHC self-antigens**

(b) **Class II MHC self-antigens**

> FIGURE 11-18 **Distinctions between class I and class II major histocompatibility complex (MHC) glycoproteins.** Specific binding requirements for the two types of T cells ensure that these cells bind only with the targets that they can influence. Cytotoxic (CD8+) T cells can recognize and bind with foreign antigen only when the antigen is in association with class I MHC glycoproteins, which are found on the surface of all body cells. This requirement is met when a virus invades a body cell, whereupon the cell is destroyed by the cytotoxic T cells. Helper (CD4+) T cells, which are activated by or enhance the activities of dendritic cells, macrophages, and B cells, can recognize and bind with foreign antigen only when it is in association with class II MHC glycoproteins, which are found only on the surface of these other immune cells. The CD8+ or CD4+ T-cell's coreceptor CD8 or CD4 links these cells to the target cell's class I or class II MHC molecules, respectively.

nucleated body cell can be invaded by viruses or become cancerous, essentially all cells display class I MHC glycoproteins, thereby enabling cytotoxic T cells to attack any virus-invaded host cell or any cancer cell.

In contrast, **class II MHC glycoproteins**, which are recognized by helper T cells, are restricted to the surface of a few special types of immune cells. That is, a helper T cell can bind with foreign antigen only when it is found on the surfaces of immune cells with which the helper T cell interacts. These include the macrophages, which present antigen to helper T cells, as well as B cells and cytotoxic T cells, whose activities are enhanced by cytokines secreted by helper T cells. The capabilities of helper T cells would be squandered if these cells were able to bind with body cells other than these special immune cells. In this way, the specific binding requirements for the two types of T cells help ensure the appropriate T-cell responses.

### TRANSPLANT REJECTION

*Clinical Note* T cells do bind with MHC antigens present on the surface of *transplanted cells* in the absence of foreign viral antigen. The ensuing destruction of the transplanted cells triggers the rejection of transplanted or grafted tissues. Presumably, some of the recipient's T cells mistake the MHC antigens of the donor cells for a closely resembling combination of a conventional viral foreign antigen complexed with the recipient's MHC self-antigens.

To minimize the rejection phenomenon, technicians match the tissues of donor and recipient according to MHC antigens as closely as possible. Therapeutic procedures to suppress the immune system then follow. In the past, the primary immunosuppressive tools included radiation therapy and drugs aimed at destroying the actively multiplying lymphocyte populations, plus anti-inflammatory drugs that suppressed growth of all lymphoid tissue. However, these measures not only suppressed the T cells that were primarily responsible for rejecting transplanted tissue but also depleted the antibody-secreting B cells. Unfortunately, the treated individual was left with little specific immune protection against bacterial and viral infections. In recent years, new therapeutic agents have become extremely useful in selectively depressing T-cell-mediated immune activity while leaving B-cell humoural immunity essentially intact. For example, *cyclosporin* blocks interleukin 2, the cytokine secreted by helper T cells that is required for expansion of the selected cytotoxic T-cell clone. Furthermore, a new technique under investigation may completely prevent rejection of transplanted tissues even from an unmatched donor. This technique involves the use of tailor-made antibodies that block specific facets of the rejection process. If proven safe and effective, the technique will have a tremendous impact on tissue transplantation.

Let's now look in more detail at the role of T cells in defending against cancer.

## Immune surveillance

Besides destroying virus-infected host cells, another important function of the T-cell system is recognizing and destroying newly arisen, potentially cancerous tumour cells before they have a chance to multiply and spread, a process known as **immune surveillance**. At least once a day, on average, your immune system destroys a mutated cell that could potentially become cancerous. Any normal cell may be transformed into a cancer cell if mutations occur within its genes that govern cell division and growth. Such mutations may occur by chance alone or, more frequently, by exposure to **carcinogenic** (cancer-causing) factors such as ionizing radiation, certain environmental chemicals, or physical irritants. Alternatively, a few cancers are caused by tumour viruses, which turn the cells they invade into cancer cells. Presumably the immune system recognizes cancer cells because they bear new and different surface antigens—due to either genetic mutation or invasion by a tumour virus—alongside the cell's normal self-antigens.

### BENIGN AND MALIGNANT TUMOURS

*Clinical Note* Cell multiplication and growth are normally under strict control, but the regulatory mechanisms are largely unknown. Cell multiplication in an adult is generally restricted to replacing lost cells. Furthermore, cells normally respect their own place and space in the body's society of cells. If a cell that has transformed into a tumour cell manages to escape immune destruction, however, it defies the normal controls on its proliferation and position. Unrestricted multiplication of a single cancer cell results in a **tumour** that consists of a clone of cells identical to the original mutated cell.

If the mass is slow growing, stays put in its original location, and does not infiltrate the surrounding tissue, it is considered a **benign tumour**. In contrast, the transformed cell may multiply rapidly and form an invasive mass that lacks the "altruistic" behaviour characteristic of normal cells. Such invasive tumours are **malignant tumours**, or **cancer**. Malignant tumour cells usually do not adhere well to the neighbouring normal cells, so often some of the cancer cells break away from the parent tumour. These "emigrant" cancer cells are transported through the blood to new territories, where they continue to proliferate, forming multiple malignant tumours. The term **metastasis** is applied to this spreading of cancer to other parts of the body.

If a malignant tumour is detected early, before it has metastasized, it can be removed surgically. Once cancer cells have dispersed and seeded multiple cancerous sites, surgical elimination of the malignancy is impossible. In this case, agents that interfere with rapidly dividing and growing cells, such as certain chemotherapeutic drugs, are used in an attempt to destroy the malignant cells. Unfortunately, these agents also harm normal body cells, especially rapidly proliferating cells, such as blood cells and the cells lining the digestive tract.

Untreated cancer is eventually fatal in most cases, for several interrelated reasons. The uncontrollably growing malignant mass crowds out normal cells by vigorously competing with them for space and nutrients, yet the cancer cells cannot take over the functions of the cells they are destroying. Cancer cells typically remain immature and do not become specialized; they often resemble embryonic cells instead (> Figure 11-19). Such poorly differentiated malignant cells lack the ability to perform the specialized functions of the normal cell type from which

11

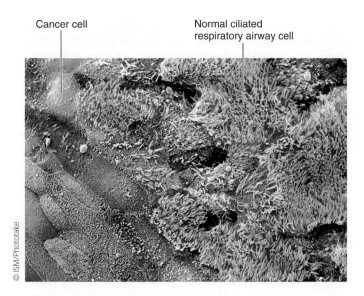

Cancer cell

Normal ciliated respiratory airway cell

© ISM/Phototake

> FIGURE 11-19 **Comparison of normal and cancerous cells in the large respiratory airways.** The normal cells display specialized cilia, which constantly contract in whiplike motion to sweep debris and microorganisms from the respiratory airways so they do not gain entrance to the deeper portions of the lungs. The cancerous cells are not ciliated, so they are unable to perform this specialized defence task.

they mutated. Affected organs gradually become disrupted to the point that they can no longer perform their life-sustaining functions, and the person dies.

## GENETIC MUTATIONS THAT DO NOT LEAD TO CANCER

Even though many body cells undergo mutations throughout a person's lifetime, most of these mutations do not result in malignancy, for three reasons:

1. Only a fraction of the mutations involve loss of control over the cell's growth and multiplication. More frequently, other facets of cellular function are altered.

2. A cell usually becomes cancerous only after an accumulation of multiple independent mutations. This requirement contributes at least in part to the much higher incidence of cancer in older individuals, in whom mutations have had more time to accumulate in a single-cell lineage.

3. Potentially cancerous cells that do arise are usually destroyed by the immune system early in their development.

## EFFECTORS OF IMMUNE SURVEILLANCE

Immune surveillance against cancer depends on an interplay among three types of immune cells—*cytotoxic T cells*, *NK cells*, and *macrophages*—as well as *interferon*. Not only can all three of these immune cell types attack and destroy cancer cells

directly, but all three also secrete interferon. Interferon, in turn, inhibits multiplication of cancer cells and increases the killing ability of the immune cells (> Figure 11-20).

Because NK cells do not require prior exposure and sensitization to a cancer cell before being able to launch a lethal attack, they are the first line of defence against cancer. In addition, cytotoxic T cells take aim at cancer cells after being sensitized by mutated surface proteins alongside normal class I MHC molecules. On contacting a cancer cell, both these killer cells release perforin and other toxic chemicals that destroy the targeted mutant cell. Macrophages, in addition to clearing away the remains of the dead victim cell, can engulf and destroy cancer cells intracellularly.

The fact that cancer does sometimes occur means that cancer cells occasionally escape these immune mechanisms. Some cancer cells are believed to survive by evading immune detection, for example, by failing to display identifying antigens on their surface or by being surrounded by counterproductive **blocking antibodies** that interfere with T-cell function. Although B cells and antibodies are not believed to play a direct role in cancer defence, B cells may produce antibodies against a mutant cancer cell when they recognize it is not a normal self-cell. These antibodies, for unknown reasons, do not activate the complement system, which could destroy the cancer cells. Instead, the antibodies bind with the antigenic sites on the cancer cell, thereby hiding these sites from recognition by cytotoxic T cells. The coating of a tumour cell by blocking antibodies thus protects the harmful cell from attack by deadly T cells. A new finding reveals that still other successful cancer cells thwart immune attack by turning on their pursuers. They induce the T cells that bind with them to commit suicide.

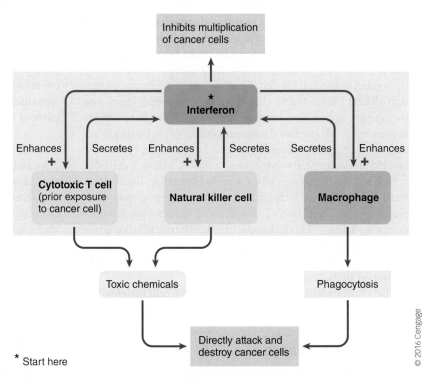

> FIGURE 11-20 **Immune surveillance against cancer.** Anticancer interactions of cytotoxic T cells, natural killer cells, macrophages, and interferon

© 2016 Cengage

## Regulatory loops

From the preceding discussion, it is obvious that complex controlling factors operate within the immune system itself. Until recently, the immune system was believed to function independently of other control systems in the body. Investigations now indicate, however, that the immune system both influences and is influenced by the two major regulatory systems—the nervous and endocrine systems. For example, interleukin 1 can turn on the stress response by activating a sequence of nervous and endocrine events that result in the secretion of cortisol, one of the major hormones released during stress. This linkage between a mediator of the immune response and a mediator of the stress response is appropriate. Cortisol mobilizes the body's nutrient stores so that metabolic fuel is readily available to keep pace with the body's energy demands at a time when the person is sick and may not be eating enough (or, in the case of an animal, may not be able to search for food). Furthermore, cortisol mobilizes amino acids, which serve as building blocks to repair any tissue damage sustained during the encounter that triggered the immune response.

In the reverse direction, lymphocytes and macrophages are responsive to blood-borne signals from the nervous system and from certain endocrine glands. These important immune cells possess receptors for a wide variety of neurotransmitters, hormones, and other chemical mediators. For example, cortisol and other chemical mediators of the stress response have a profound immunosuppressive effect, inhibiting many functions of lymphocytes and macrophages and decreasing the production of cytokines. Thus, a negative-feedback loop appears to exist between the immune system and the nervous and endocrine systems. Cytokines released by immune cells enhance the neurally and hormonally controlled stress response, whereas cortisol and related chemical mediators released during the stress response suppress the immune system. In large part because stress suppresses the immune system, stressful physical, psychological, and social life events are linked with increased susceptibility to infections and cancer. Thus, the body's resistance to disease *can* be influenced by the person's mental state—a case of "mind over matter."

There are other important links between the immune and nervous systems in addition to the cortisol connection. For example, many immune system organs, such as the thymus, spleen, and lymph nodes, are innervated by the sympathetic nervous system, the branch of the nervous system called into play during stress-related fight-or-flight situations (p. 142). In the reverse direction, immune system secretions act on the brain to produce fever and other general symptoms that accompany infections.

---

### Check Your Understanding 11.7

1. List the three major types of T cells, and state their functions.
2. Describe the role of antigen-presenting cells.
3. Define *immune surveillance*, and discuss how it is accomplished.

---

## 11.8 | Immune Diseases

Abnormal functioning of the immune system can lead to immune diseases in two general ways: immunodeficiency diseases (too little immune response) and inappropriate immune attacks (too much or mistargeted immune response).

### Immunodeficiency

*Clinical Note* Immunodeficiency diseases occur when the immune system fails to respond adequately to foreign invasion. The condition may be congenital (present at birth) or acquired (nonhereditary), and it may specifically involve impairment of antibody-mediated immunity, or cell-mediated immunity, or both.

In a rare hereditary condition known as **severe combined immunodeficiency**, both B and T cells are lacking. Such people have extremely limited defences against pathogenic organisms and die in infancy unless maintained in a germ-free environment (i.e., live in a bubble). However, that verdict has changed with recent successes using gene therapy to cure the disease in some patients.

Acquired (nonhereditary) immunodeficiency states can arise from inadvertent destruction of lymphoid tissue during prolonged therapy with anti-inflammatory agents, such as cortisol derivatives, or from cancer therapy aimed at destroying rapidly dividing cells (which unfortunately include lymphocytes as well as cancer cells). The most recent, and tragically the most common, acquired immunodeficiency disease is AIDS, which, as described earlier, is caused by HIV, a virus that invades and incapacitates the critical helper T cells.

Let's now look at inappropriate immune attacks.

### Allergies

*Clinical Note* Inappropriate adaptive immune attacks cause reactions harmful to the body. These include (1) *autoimmune responses*, in which the immune system turns against one of the body's own tissues; (2) *immune complex diseases*, which involve overexuberant antibody responses that "spill over" and damage normal tissue; and (3) *allergies*. The first two conditions have been described earlier in this chapter, so here we concentrate on allergies.

An **allergy** is the acquisition of an inappropriate specific immune reactivity, or **hypersensitivity**, to a normally harmless environmental substance, such as dust or pollen. The offending agent is known as an **allergen**. Subsequent re-exposure of a sensitized individual to the same allergen elicits an immune attack, which may vary from a mild, annoying reaction to a severe, body-damaging reaction that may even be fatal.

Allergic responses are classified into two categories: immediate hypersensitivity and delayed hypersensitivity. In **immediate hypersensitivity**, the allergic response appears within about 20 minutes after a sensitized person is exposed to an allergen. In **delayed hypersensitivity**, the reaction does not generally show up until a day or so following exposure. The difference in timing is due to the different mediators involved. A particular allergen may activate either a B-cell or a T-cell response. Immediate allergic reactions involve B cells and are elicited by antibody interactions with an allergen; delayed reactions involve T cells and the more slowly responding process of cell-mediated immunity against the allergen. Let us examine the causes and consequences of each of these reactions in more detail.

### TRIGGERS FOR IMMEDIATE HYPERSENSITIVITY

In immediate hypersensitivity, the antibodies involved and the events that ensue on exposure to an allergen differ from the typical antibody-mediated response to bacteria. The most common allergens that provoke immediate hypersensitivities are pollen grains, bee stings, penicillin, certain foods, moulds, dust, feathers, and animal fur. (Actually, people allergic to cats are not allergic to the fur itself. The true allergen is in the cat's saliva, which is deposited on the fur during licking. Likewise, people are not allergic to dust or feathers *per se*, but to tiny mites that inhabit the dust or feathers and eat the scales constantly being shed from the skin.) For unclear reasons, these allergens bind to and elicit the synthesis of IgE antibodies rather than the IgG antibodies associated with bacterial antigens. IgE antibodies are the least plentiful immunoglobulin, but their presence spells trouble. Without IgE antibodies, there would be no immediate hypersensitivity. When a person with an allergic tendency is first exposed to a particular allergen, compatible helper T cells secrete *interleukin 4*, a cytokine that prods compatible B cells to synthesize IgE antibodies specific for the allergen. During this initial **sensitization period**, no symptoms are evoked, but memory cells form that are primed for a more powerful response on subsequent re-exposure to the same allergen.

In contrast to the antibody-mediated response elicited by bacterial antigens, IgE antibodies do not freely circulate. Instead, their tail portions attach to mast cells and basophils, both of which produce and store an arsenal of potent inflammatory chemicals, such as histamine, in preformed granules. Mast cells are most plentiful in regions that come into contact with the external environment, such as the skin, the outer surface of the eyes, and the linings of the respiratory system and digestive tract. The binding of an appropriate allergen with the outreached arm regions of IgE antibodies lodged tail first in a mast cell or basophil triggers the rupture of the cell's granules. As a result, histamine and other chemical mediators spew forth into the surrounding tissue.

A single mast cell (or basophil) may be coated with a number of different IgE antibodies, each able to bind with a different allergen. Thus, the mast cell can be triggered to release its chemical products by any one of a number of different allergens (> Figure 11-21).

### CHEMICAL MEDIATORS OF IMMEDIATE HYPERSENSITIVITY

The chemicals released by the mast cells cause the reactions that characterize immediate hypersensitivity. The most important chemicals released during immediate allergic reactions, and their functions, are as follows:

1. *Histamine* brings about vasodilation and increased capillary permeability as well as increased mucus production.

2. **Slow-reactive substance of anaphylaxis (SRS-A)** induces prolonged and profound contraction of smooth muscle, especially of the small respiratory airways. SRS-A is a leukotriene, a locally acting mediator similar to prostaglandins (p. 741).

3. **Eosinophil chemotactic factor** specifically attracts eosinophils to the area. Interestingly, eosinophils release enzymes that inactivate SRS-A and may also inhibit histamine, perhaps serving as an off-switch to limit the allergic response.

### SYMPTOMS OF IMMEDIATE HYPERSENSITIVITY

Symptoms of immediate hypersensitivity vary depending on the site, allergen, and mediators involved. Most frequently, the reaction is localized to the body site in which the IgE-bearing cells first come into contact with the allergen. If the reaction is limited to the upper respiratory passages after a person inhales an allergen, such as ragweed pollen, the released chemicals bring about the symptoms characteristic of **hay fever**—for example, nasal congestion caused by histamine-induced localized oedema and sneezing and runny nose caused by increased mucus secretion. If the reaction is concentrated primarily within the bronchioles (the small respiratory airways that lead to the tiny air sacs within the lungs), **asthma** results. Contraction of the smooth muscle in the walls of the bronchioles in response to SRS-A narrows or constricts these passageways, making breathing difficult. Localized swelling in the skin because of allergy-induced histamine release causes **hives**. An allergic reaction in the digestive tract in response to an ingested allergen can lead to diarrhoea.

### TREATMENT OF IMMEDIATE HYPERSENSITIVITY

Treatment of localized immediate allergic reactions with antihistamines often offers only partial relief of the symptoms, because some of the manifestations are invoked by other chemical mediators not blocked by these drugs. For example, antihistamines are not particularly effective in treating asthma, the most serious symptoms of which are invoked by SRS-A. Adrenergic drugs (which mimic the sympathetic nervous system) are helpful through their vasoconstrictor-bronchodilator actions in counteracting the effects of both histamine and SRS-A. Anti-inflammatory drugs, such as cortisol derivatives, are often used as the primary treatment for ongoing allergen-induced inflammation, such as that associated with asthma. Newer drugs that inhibit leukotrienes, including SRS-A, have been added to the arsenal for combating immediate allergies.

### ANAPHYLACTIC SHOCK

A life-threatening systemic reaction can occur if the allergen becomes blood borne or if very large amounts of chemicals are released from the localized site into the circulation. When large amounts of these chemical mediators gain access to the blood, the extremely serious systemic (involving the entire body) reaction known as **anaphylactic shock** occurs. Severe hypotension that can lead to circulatory shock (p. 426) results from widespread vasodilation and a massive shift of plasma fluid into the interstitial spaces due to a generalized increase in capillary

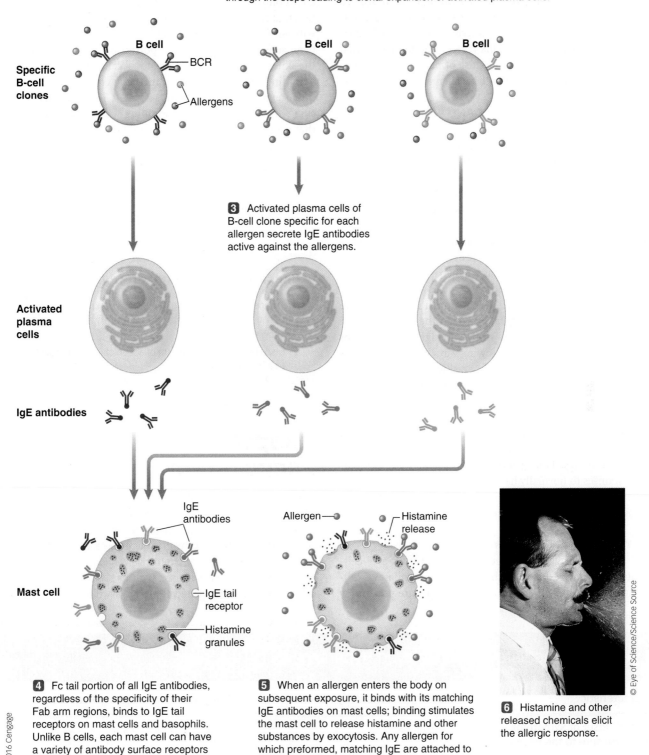

① Allergens (antigens) enter the body for the first time.

② Allergens bind to matching BCRs; B cells now process the allergens and, with stimulation by helper T cells (not shown), proceed through the steps leading to clonal expansion of activated plasma cells.

**Specific B-cell clones**

B cell — BCR — Allergens

B cell

B cell

③ Activated plasma cells of B-cell clone specific for each allergen secrete IgE antibodies active against the allergens.

**Activated plasma cells**

**IgE antibodies**

IgE antibodies

Allergen — Histamine release

**Mast cell** — IgE tail receptor — Histamine granules

④ Fc tail portion of all IgE antibodies, regardless of the specificity of their Fab arm regions, binds to IgE tail receptors on mast cells and basophils. Unlike B cells, each mast cell can have a variety of antibody surface receptors for binding different allergens.

⑤ When an allergen enters the body on subsequent exposure, it binds with its matching IgE antibodies on mast cells; binding stimulates the mast cell to release histamine and other substances by exocytosis. Any allergen for which preformed, matching IgE are attached to the mast cell can trigger histamine release.

⑥ Histamine and other released chemicals elicit the allergic response.

© Eye of Science/Science Source

© 2016 Cengage

> **FIGURE 11-21 Role of IgE antibodies and mast cells in immediate hypersensitivity.** B-cell clones are converted into plasma cells, which secrete IgE antibodies on contact with the allergen for which they are specific. The Fc tail portion of all IgE antibodies, regardless of the specificity of their Fab arm regions, binds to receptor proteins specific for IgE tails on mast cells and basophils. Unlike B cells, each mast cell bears a variety of antibody surface receptors for binding different allergens. When an allergen combines with the IgE receptor specific for it on the surface of a mast cell, the mast cell releases histamine and other chemicals by exocytosis. These chemicals elicit the allergic response.

permeability. Concurrently, pronounced bronchiolar constriction occurs and can lead to respiratory failure. The person may suffocate from an inability to move air through the narrowed airways. Unless countermeasures, such as injecting a vasoconstrictor-bronchodilator drug, are undertaken immediately, anaphylactic shock is often fatal. This reaction is the reason that even a single bee sting or a single dose of penicillin can be so dangerous in people sensitized to these allergens.

### IMMEDIATE HYPERSENSITIVITY AND ABSENCE OF PARASITIC WORMS

Although the immediate hypersensitivity response differs considerably from the typical IgG antibody response to bacterial infections, it is strikingly similar to the immune response elicited by parasitic worms. Shared characteristics of the immune reactions to allergens and parasitic worms include the production of IgE antibodies and increased basophil and eosinophil activity. This finding has led to the proposal that harmless allergens somehow trigger an immune response designed to fight worms. Mast cells are concentrated in areas where parasitic worms (and allergens) could contact the body. Parasitic worms can penetrate the skin or digestive tract or can attach to the digestive tract lining. Some worms migrate through the lungs during a part of their life cycle. Scientists suspect the IgE response helps ward off these invaders in the following way. The inflammatory response in the skin could wall off parasitic worms attempting to burrow in. Coughing and sneezing could expel worms that have migrated to the lungs. Diarrhoea could help flush out worms before they could penetrate or attach to the digestive tract lining. Interestingly, epidemiological studies suggest that the incidence of allergies in a country rises as the presence of parasites decreases. Thus, superfluous immediate hypersensitivity responses to normally harmless allergens may represent a pointless marshalling of a honed immune-response system "with nothing better to do" in the absence of parasitic worms.

### DELAYED HYPERSENSITIVITY

Some allergens invoke delayed hypersensitivity, a T-cell-mediated immune response, rather than an immediate, B cell-IgE antibody response. Among these allergens are poison ivy toxin and certain chemicals to which the skin is frequently exposed, such as cosmetics and household cleaning agents. Most commonly, the response is characterized by a delayed skin eruption that reaches its peak intensity one to three days after contact with an allergen to which the immune system has previously been sensitized. To illustrate, poison ivy toxin does not harm the skin on contact, but it activates T cells specific for the toxin, including the formation of a memory component. On subsequent exposure to the toxin, activated T cells diffuse into the skin within a day or two, combining with the poison ivy toxin that is present. The resulting interaction gives rise to the tissue damage and discomfort typical of the condition. The best relief is obtained from application of anti-inflammatory preparations, such as those containing cortisol derivatives.

▌ Table 11-3 summarizes the distinctions between immediate and delayed hypersensitivities.

### Check Your Understanding 11.8

1. Compare the type of immune response and immune effectors involved in immediate hypersensitivity and delayed hypersensitivity.

2. Discuss how the mechanism of action of IgE antibodies differs from that of IgG antibodies.

# Chapter in Perspective: Focus on Homeostasis

We could not survive beyond early infancy were it not for the body's defence mechanisms. These mechanisms resist and eliminate potentially harmful foreign agents that we continually come in contact with in our external environment and also destroy abnormal cells that often arise within the body. Homeostasis can be optimally maintained, and thus life sustained, only if the body cells are not physically injured or functionally disrupted by pathogenic

▌ **TABLE 11-3** Immediate versus Delayed Hypersensitivity Reactions

| Characteristic | Immediate Hypersensitivity Reaction | Delayed Hypersensitivity Reaction |
| --- | --- | --- |
| **Time of Onset of Symptoms after Exposure to the Allergen** | Within 20 minutes | Within one to three days |
| **Type of Immune Response** | Antibody-mediated immunity against the allergen | Cell-mediated immunity against the allergen |
| **Immune Effectors Involved** | B cells, IgE antibodies, mast cells, basophils, histamine, slow-reactive substance of anaphylaxis, eosinophil chemotactic factor | T cells |
| **Allergies Commonly Involved** | Hay fever, asthma, hives, anaphylactic shock in extreme cases | Contact allergies, such as allergies to poison ivy, cosmetics, and household cleaning agents |

microorganisms or are not replaced by abnormally functioning cells, such as traumatized cells or cancer cells. The immune defence system—a complex, multifaceted, interactive network of leukocytes, their secretory products, and plasma proteins—contributes indirectly to homeostasis by keeping other cells alive so they can perform their specialized activities to maintain a stable internal environment. The immune system protects the other healthy cells from foreign agents that have gained entrance to the body, eliminates newly arisen cancer cells, and clears away dead and injured cells to pave the way for replacement with healthy new cells.

The skin contributes indirectly to homeostasis by serving as a protective barrier between the external environment and the rest of the body cells. It helps prevent harmful foreign agents, such as pathogens and toxic chemicals from entering the body and helps prevent the loss of precious internal fluids from the body. The skin also contributes directly to homeostasis by helping maintain body temperature by means of the sweat glands and adjustments in skin blood flow. The amount of heat carried to the body surface for dissipation to the external environment is determined by the volume of warmed blood flowing through the skin.

Other systems that have internal cavities in contact with the external environment, such as the digestive, genitourinary, and respiratory systems, also have defence capabilities to prevent harmful external agents from entering the body through these avenues.

## CHAPTER TERMINOLOGY

acute phase proteins (p. 470)

acquired immunodeficiency syndrome (AIDS) (p. 482)

active immunity (p. 481)

adaptive or acquired immune system (p. 465)

adipose tissue (p. 462)

agglutination (p. 478)

allergen (p. 491)

allergy (p. 491)

alveolar macrophages (p. 464)

anaphylactic shock (p. 492)

angry macrophages (p. 485)

antibodies (p.475)

antibody-mediated (humoural) immunity (p. 473)

antigen (p. 474)

antigen-binding fragments (Fab) (p. 476)

antigen-presenting cells (p. 486)

antiserum (antitoxin) (p. 481)

asthma (p. 492)

autoimmune disease (p. 486)

bacteria (p. 460)

B-cell growth factor (p. 485)

B-cell receptors (BCRs) (p. 475)

benign tumour (p. 489)

blocking antibodies (p. 490)

body's own cells (p. 460)

bradykinin (p. 471)

B7 (p. 486)

carcinogenic (p. 489)

cancer (p. 489)

CD4 cells (p. 482)

CD4+CD25+T cells (p. 482)

CD8 cells (cytotoxic, or killer, T cells) (p. 482)

cell-mediated immunity (p. 473)

chemotaxis (p. 469)

class I MHC glycoproteins (p. 488)

class II MHC glycoproteins (p. 489)

clonal anergy (p. 486)

clonal deletion (p. 486)

clonal selection theory (p. 478)

clone (p. 479)

complement system (p. 473)

constant (Fc) region (p. 476)

cytokines (p. 470)

delayed hypersensitivity (p. 491)

dermis (p. 461)

diapedesis (p. 469)

endogenous pyrogen (EP) (p. 470)

eosinophil chemotactic factor (p. 492)

external defences (p. 460)

Granstein cells (p. 462)

granzymes (p. 483)

gut-associated lymphoid tissue (GALT) (p. 464)

hair follicle (p. 462)

hay fever (p. 492)

helper T cells (p. 482)

helper T 17 ($T_H$17) cells (p. 486)

hives (p. 492)

host cell (p. 460)

human immunodeficiency virus (HIV) (p. 482)

hypersensitivity (p. 491)

hypodermis (p. 462)

IgA (p. 476)

IgD (p. 476)

IgE (p. 475)

IgG (p. 475)

IgM (p. 475)

immediate hypersensitivity (p. 491)

immune complex disease (p. 478)

immune privilege (p. 486)

immune surveillance (p. 489)

immune system (p. 460)

inflammation (p. 466)

innate immune system (p. 465)

integumentary system (p. 460)

interferon (p. 471)

interleukin 1 (IL-1) (p. 470)

interleukin 4 (IL-4) (p. 486)

interleukin 6 (IL-6) (p. 470)

interleukin 12 (IL-12) (p. 485)

internal defences (p. 460)

kallikrein (p. 471)

keratinized layer (p. 461)

keratinocytes (p. 462)

kinins (p. 471)

lactoferrin (p. 470)

Langerhans cells (p. 462)

macrophage-migration inhibition factor (p. 485)

macrophages (p. 464)

major histocompatibility complex (MHC) molecules (p. 482)

malignant tumours (p. 489)

margination (p. 469)

melanocytes (p. 462)

membrane attack complex (MAC) (p. 473)

memory cells (p. 479)

metastasis (p. 489)

mucociliary escalator (p. 464)

nails (p. 462)

naive lymphocytes (p. 479)

natural killer (NK) cells (p. 472)

neutralization (p. 478)

opsonins (p. 469)

passive immunity (p. 481)

perforin (p. 483)

Peyer's patches (p. 464)

plasma cell (p. 475)

pus (p. 470)

precipitation (p. 478)

primary response (p. 479)

receptor editing (p. 486)

regulatory T cells ($T_{reg}$) (p. 482)

sebaceous glands (p. 462)

sebum (p. 462)

scar tissue (p. 471)

secondary response (p. 480)

sensitized (activated) T cells (p. 482)

skin (integument) (p. 460)

self-antigens (p. 482)

## REVIEW EXERCISES

### Objective Questions
(Answers in Appendix E, p. A-47)

1. The complement system can be activated only by antibodies. *(True or false?)*

2. Specific adaptive immune responses are accomplished by neutrophils. *(True or false?)*

3. Damaged tissue is always replaced by scar tissue. *(True or false?)*

4. Active immunity against a particular disease can be acquired only by actually having the disease. *(True or false?)*

5. A secondary response has a more rapid onset, is more potent, and has a longer duration than a primary response. *(True or false?)*

6. _____ are receptors on the plasma membrane of phagocytes that recognize and bind with telltale molecular patterns present on the surface of microorganisms but absent from human cells.

7. The complement system's _____ forms a doughnut-shaped complex that embeds in a microbial surface membrane, causing osmotic lysis of the victim cell.

8. _____ is a collection of phagocytic cells, necrotic tissue, and bacteria.

9. _____ is the localized response to microbial invasion or tissue injury that is accompanied by swelling, heat, redness, and pain.

10. A chemical that enhances phagocytosis by serving as a link between a microbe and the phagocytic cell is known as a(n) _____.

11. _____, collectively, are all the chemical messengers other than antibodies secreted by lymphocytes.

12. Which statement concerning leukocytes is NOT correct?
    a. Monocytes are transformed into macrophages.
    b. T lymphocytes are transformed into plasma cells that secrete antibodies.
    c. Neutrophils are highly mobile phagocytic specialists.
    d. Basophils release histamine.
    e. Lymphocytes arise in large part from lymphoid tissues.

13. Match the following with the answer code provided below:

    ___ 1. a family of proteins that nonspecifically defend against viral infection

    ___ 2. a response to tissue injury in which neutrophils and macrophages play a major role

    ___ 3. a group of plasma proteins that, when activated, bring about destruction of foreign cells by attacking their plasma membranes

    ___ 4. lymphocyte-like entities that spontaneously lyse tumour cells and virus-infected host cells

    (a) complement system
    (b) natural killer cells
    (c) interferon
    (d) inflammation

14. Using the answer code on the right, indicate whether the numbered characteristics of the adaptive immune system apply to antibody-mediated immunity, cell-mediated immunity, or both:

    ___ 1. involves secretion of antibodies

    ___ 2. mediated by B cells

    ___ 3. mediated by T cells

    ___ 4. accomplished by thymus- educated lymphocytes

    ___ 5. triggered by the binding of specific antigens to complementary lymphocyte receptors

    ___ 6. involves formation of memory cells in response to initial exposure to an antigen

    ___ 7. primarily aimed against virus-infected host cells

    ___ 8. protects primarily against bacterial invaders

    ___ 9. directly destroys targeted cells

    ___ 10. involved in rejection of transplanted tissue

    ___ 11. requires binding of a lymphocyte to a free extracellular antigen

    ___ 12. requires dual binding of a lymphocyte with both foreign antigen and self-antigens present on the surface of a host cell

    (a) antibody-mediated immunity
    (b) cell-mediated immunity
    (c) both antibody-mediated and cell-mediated immunity

15. Using the answer code on the right, indicate whether the numbered characteristics apply to the epidermis or dermis:

___ 1. is the inner layer of skin
___ 2. has layers of epithelial cells that are
___ 3. has no direct blood supply
___ 4. contains sensory nerve endings
___ 5. contains keratinocytes
___ 6. contains melanocytes
___ 7. contains rapidly dividing cells
___ 8. is mostly connective tissue

(a) epidermis
(b) dermis dead and flattened

## Written Questions

1. Distinguish between bacteria and viruses.
2. Summarize the functions of each of the lymphoid tissues.
3. Distinguish between innate and adaptive immune responses.
4. Compare the life history of B cells and of T cells.
5. What is an antigen?
6. Describe the structure of an antibody. List and describe the five subclasses of immunoglobulins.
7. In what ways do antibodies exert their effects?
8. Describe the clonal selection theory.
9. Compare the functions of B cells and T cells. What are the roles of the two main types of T cells?
10. Summarize the functions of macrophages in immune defence.
11. What mechanisms are involved in tolerance?
12. What is the importance of class I and class II MHC glycoproteins?
13. Describe the factors that contribute to immune surveillance against cancer cells.
14. Distinguish among immunodeficiency disease, autoimmune disease, immune complex disease, immediate hypersensitivity, and delayed hypersensitivity.
15. What are the immune functions of the skin?

## Quantitative Exercise (Solution in Appendix E, p. A-47)

1. As a result of the innate immune response to an infection (e.g., from a cut on the skin), capillary walls near the site of infection become very permeable to plasma proteins that normally remain in the blood. These proteins diffuse into the interstitial fluid, raising the interstitial fluid-colloid osmotic pressure. This increased colloid osmotic pressure causes fluid to leave the circulation and accumulate in the tissue, forming a welt, or wheal. This process is referred to as the *wheal response*. The wheal response is mediated in part by histamine secreted from mast cells in the area of infection. The histamine binds to receptors, called *H-1 receptors*, on capillary endothelial cells. The histamine signal is transduced via the $Ca^{2+}$ second-messenger pathway involving phospholipase C (p. 218). In response to this signal, the capillary endothelial cells contract (via internal actin-myosin interaction), which causes a widening of the intercellular gaps (pores) between the capillary endothelial cells (p. 405). In addition, substance P (p. 150) also contributes to pore widening. Plasma proteins can pass through these widened pores and leave the capillaries. Looking at Figure 11-20, compare the magnitude of the wheal response (i.e., the extent of localized oedema) if PIF were raised (a) from 0 mmHg to 5 mmHg and (b) from 0 mmHg to 10 mmHg. In both cases, compare the net exchange pressure (NEP) at the arteriolar end of the capillary, the venular end of the capillary, and the average NEP. (Assume the other forces acting across the capillary wall remain unchanged.)

## POINTS TO PONDER

### (Explanations in Appendix E, p. A-47)

1. Compare the defence mechanisms that come into play in response to bacterial and viral pneumonia.
2. Why does the frequent mutation of HIV (the AIDS virus) make it difficult to develop a vaccine against this virus?
3. What impact would failure of the thymus to develop embryonically have on the immune system after birth?
4. Medical researchers are currently working on ways to "teach" the immune system to view foreign tissue as "self." What useful clinical application will the technique have?
5. When someone looks at you, are the cells of your body that person is viewing dead or alive?

11

## CLINICAL CONSIDERATION

### (Explanations in Appendix E, p. A-48)

1. Christine W. has a blood disorder called *idiopathic thrombocytopenic purpura*. This is a condition in which the patient has an increased tendency to bleed due to a low platelet count. This low platelet count is caused by the generation of antibodies against platelets, which results in the phagocytosis of the platelets by macrophages in the spleen. In Christine's case, her platelet count dropped very low and was not responsive to treatment with steroids or immunosupressants. It was decided that her spleen needed to be removed.

2. What are some of the major functions of the spleen?
3. How will Christine's immune system be affected without a spleen?
4. What will happen to Christine's platelet count after the splenectomy?

# The Respiratory System

Body systems maintain homeostasis

High-altitude mountaineer

Harry Kikstra/Getty Images

## Homeostasis

The respiratory system contributes to homeostasis by obtaining $O_2$ from and eliminating $CO_2$ to the external environment. It helps regulate the pH of the internal environment by adjusting the rate of removal of acid-forming $CO_2$.

Homeostasis is essential for survival of cells

Cells make up body systems

Alveoli

Science Photo Library/Alamy Stock Photo

Energy is essential for sustaining such life-supporting cellular activities as protein synthesis and active transport across membranes. Cells need a continual supply of oxygen to support the metabolic reactions that produce biochemical energy—typically in the form of adenosine triphosphate (ATP; see Chapter 2)—needed for these activities. This process is known as **internal (cellular) respiration**. The carbon dioxide produced during these reactions must be eliminated from the body at the same rate as it is produced. This prevents the excessive increases in acid (necessary to maintain acid–base balance) that result from the formation of carbonic acid.

**External respiration**, in contrast, involves all processes that accomplish movement of oxygen from the atmosphere to the alveoli, where it diffuses into the pulmonary capillary blood. In exchange, carbon dioxide produced by the tissues diffuses from the pulmonary capillaries into the alveoli for excretion to the atmosphere. (Transportation of oxygen from the lung to the tissues and of carbon dioxide from the tissues to the lung is the responsibility of the cardiovascular system.) In no other system is the need for specialization to overcome the limitations of diffusion associated with large multicellular organisms better illustrated than in the respiratory system.

The movement of oxygen from room air into the alveoli and carbon dioxide in the reverse direction depends on two factors: (1) the ability of the respiratory muscles to generate pressure to move air through the airways (overcoming the resistance to flow) and inflate the lung (overcoming the elastance of the lung tissue), and (2) the ability of oxygen and carbon dioxide to diffuse across the alveolar–capillary barrier. The bulk flow and diffusion of a gas are governed by a simple equation:

$$\text{flow (or diffusion)} = \text{pressure} / \text{resistance}$$

In this chapter, we discuss how this simple equation explains the flow of gas, the diffusion of oxygen and carbon dioxide, and how various pathologies illustrate these physiological principles. Before doing so, we introduce basic respiratory anatomy, lung volumes, and measurements of ventilatory function. We then examine how the pressure gradients to inflate and deflate the lungs are generated, how diffusion across the alveolar–capillary membrane is maximized, and how oxygen and carbon dioxide are transported in blood.

# 12

# The Respiratory System

**▌ Clinical Connections**

Andrea, an active, 20-year-old female, presented at hospital with shortness of breath and agitation. Her patient history reveals that she suffers from hay fever, which she controls with over-the-counter medication. She had recently moved in with a new roommate, who has two cats, and her hay fever has been worse than usual for this time of year. Her initial vital signs were a pulse of 115 beats/minute, blood pressure of 145/90 mmHg, respiratory rate of 26 breaths/minute, and oxygen saturation of 90 percent on room air. There is bilateral wheezing during expiration and increased use of accessory muscles during exhalation. Her arterial blood gas measurements were pH of 7.0, $P_{AO_2}$ of 65, $P_{CO_2}$ of 50, and $HCO_3$ of 24. A chest X-ray indicated lung hyperinflation and a pulmonary function test showed a reduced forced expiratory flow rate and a decreased FEV1 with an increased total lung capacity. Treatment with bronchodilators brought about relief of her symptoms. She was ultimately diagnosed with a cat allergy, which had progressed to allergic asthma. She was given a suitable course of treatment and advised to find a cat-free environment to live.

## 12.1 | Introduction

The primary function of respiration—to obtain oxygen for use by the body's cells and to eliminate the carbon dioxide the cells produce—is the focus of this chapter (⟩ Figure 12-1). The respiratory system, however, has many other functions not directly related to gas exchange (▮ Table 12-1).

## External respiration

External respiration refers to the entire sequence of events in the exchange of oxygen and carbon dioxide between the external environment and the cells of the body. External respiration, the primary topic of this chapter, encompasses four steps (⟩ Figure 12-1):

1. Air is alternately moved in and out of the lungs so that air can be exchanged between the atmosphere (external environment) and air sacs (alveoli) of the lungs. This exchange is accomplished by the mechanical act of **breathing**, or **ventilation**. Ventilation is regulated according to the body's needs for oxygen uptake and carbon dioxide removal. (See ▌ Table 12-1 for exceptions.)

2. Oxygen diffuses from the air in the alveoli to the blood within the pulmonary capillaries; carbon dioxide diffuses in the opposite direction.

3. The blood transports oxygen from the lungs to the tissues and carbon dioxide from the tissues to the lungs.

4. Oxygen and carbon dioxide are exchanged between tissues and blood by the process of diffusion across the systemic (tissue) capillaries.

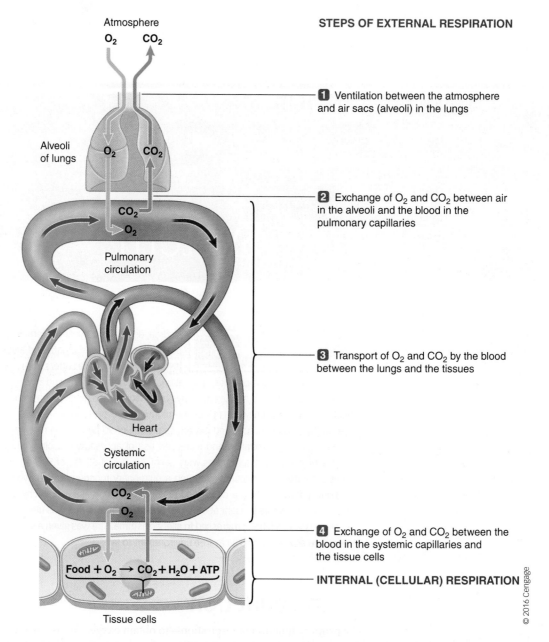

Atmosphere

$O_2$  $CO_2$

**STEPS OF EXTERNAL RESPIRATION**

Alveoli of lungs

$O_2$  $CO_2$

**1** Ventilation between the atmosphere and air sacs (alveoli) in the lungs

$CO_2$
$O_2$

Pulmonary circulation

**2** Exchange of $O_2$ and $CO_2$ between air in the alveoli and the blood in the pulmonary capillaries

Heart

Systemic circulation

**3** Transport of $O_2$ and $CO_2$ by the blood between the lungs and the tissues

$CO_2$
$O_2$

**4** Exchange of $O_2$ and $CO_2$ between the blood in the systemic capillaries and the tissue cells

Food + $O_2$ → $CO_2$ + $H_2O$ + ATP

**INTERNAL (CELLULAR) RESPIRATION**

Tissue cells

© 2016 Cengage

⟩ **FIGURE 12-1 External and internal respiration.** External respiration encompasses the steps involved in the exchange of oxygen and carbon dioxide between the external environment and the tissues (steps 1 through 4). Internal respiration encompasses the intracellular metabolic reactions involving the use of oxygen to derive energy (ATP) from food, producing carbon dioxide as a by-product. The respiratory system does not accomplish all the steps of respiration; it is involved only with ventilation and the exchange of oxygen and carbon dioxide between the lungs and blood (steps 1 and 2). The circulatory system carries out steps 3 and 4.

12

- It is a route for water loss and heat elimination. Humidification of inspired air is essential to prevent the alveolar linings from drying out. Because water has a high heat capacity (i.e., considerable energy is required to increase its temperature), heat is lost when air is expired. This is readily apparent when expired water vapour condenses on a cold day.
- It enhances venous return (see discussion of the respiratory pump, p. 418).
- It helps maintain normal acid–base balance by altering the amount of $H^+$-generating carbon dioxide exhaled (p. 626).
- It enables speech, singing, and other vocalization.
- It defends against inhaled foreign matter (p. 463). The lungs, along with the gastrointestinal tract, provide a huge surface for foreign agents to enter the body, and so both systems need to be immunologically competent. The respiratory system does this not only with alveolar macrophages but also through protective reflexes such as coughing and sneezing. It has a protective lining of mucus and underlying cilia that transport mucus with any deposited material to the esophagus, where it is swallowed. In addition, the respiratory system is responsible for generating the large pressures needed for vomiting, a reflex that protects against ingestion of noxious foods.
- It removes, modifies, activates, or inactivates many biological substances passing through the pulmonary circulation. All blood returning to the heart from the tissues must pass through the lungs before returning to the systemic circulation. The lungs, therefore, are uniquely situated to act on specific substances that have been added to the blood at the tissue level before they have a chance to reach other parts of the body via the arterial system. For example, prostaglandins, a collection of chemical messengers released in many tissues to mediate particular local responses (p. 741), may spill into the blood, but they are *inactivated* during passage through the lungs so that they cannot exert systemic effects. In contrast, the lungs convert angiotensin I to angiotensin II, a hormone that plays an important role in regulating the concentration of sodium in the extracellular fluid (p. 576).
- The nose, a part of the respiratory system, serves as the organ of smell (p. 184).
- The respiratory muscles generate the large pressures needed during childbirth (parturition).
- The same respiratory muscles are used during defecation.
- The lung also provides a reservoir of blood for the left heart, helping to equalize the outputs of the left and right ventricles.

© 2016 Cengage

## Anatomy of the respiratory system

The **respiratory system** (› Figure 12-2) consists of two components: the lungs and the chest wall. The lungs have two components: airways and alveoli. The chest wall includes other parts of the body involved in respiration: the thorax and the abdomen, which contain the muscles that generate air flow.

### THE LUNGS

The lungs occupy most of the thoracic (chest) cavity; the only other structures in the chest are the heart and associated vessels, the esophagus, the thymus, and some nerves. There are two **lungs**, each divided into several lobes and each supplied by one of the bronchi. The lung tissue consists of a series of highly branched airways, alveoli, pulmonary blood vessels, and large quantities of elastic connective tissue. The upper airway contains skeletal muscles that regulate airflow by changing the diameter of the pharynx (throat) and **larynx (voice box)**, prevent aspiration of food into the lungs, and participate in vocalization. All airways, especially the smaller ones, also contain smooth muscle within their walls. The activation or relaxation of smooth muscle—resulting from changes in activity of the autonomic nervous system—affects airway dimensions and, in turn, the resistance to airflow (p. 514). In contrast, the walls of the alveoli have no muscles to either inflate or deflate them; if there were, they would interfere with diffusion. Instead, changes in lung volume (and accompanying changes in alveolar volume) result from changes in the dimensions of the thoracic cavity due to activation of pump muscles: the diaphragm, internal and external intercostal muscles, and abdominal muscles.

### AIRWAYS

The **airways** carry air between the atmosphere and the alveoli, where oxygen and carbon dioxide are exchanged. They begin at the **nasal passages (nose)** (› Figure 12-2a) that open into the pharynx, a common passageway for both the respiratory and digestive systems. Two tubes lead from the pharynx: the **trachea**, through which air is conducted to the alveoli; and the **esophagus**, the tube through which food passes to the stomach. Air normally enters the pharynx through the nose, but it can enter by the mouth as well, especially when the nasal passages are congested (you breathe through your mouth when you have a cold) and when ventilation increases during exercise. Because the pharynx serves as a common passageway for food and air, reflex mechanisms close the trachea during swallowing so that food enters the esophagus and not the airways. The esophagus stays closed except during swallowing to keep air from entering the stomach during breathing. Control of this important junction originates in the brain stem, where the breathing rhythm is generated.

The larynx is located at the entrance of the trachea. The anterior protrusion of the larynx forms the "Adam's apple." The **vocal folds**, two bands of elastic tissue, lie across the opening of the larynx; the laryngeal muscles control their length and shape. As air is moved past the taut vocal folds, they vibrate to produce the different sounds of speech. The lips,

12

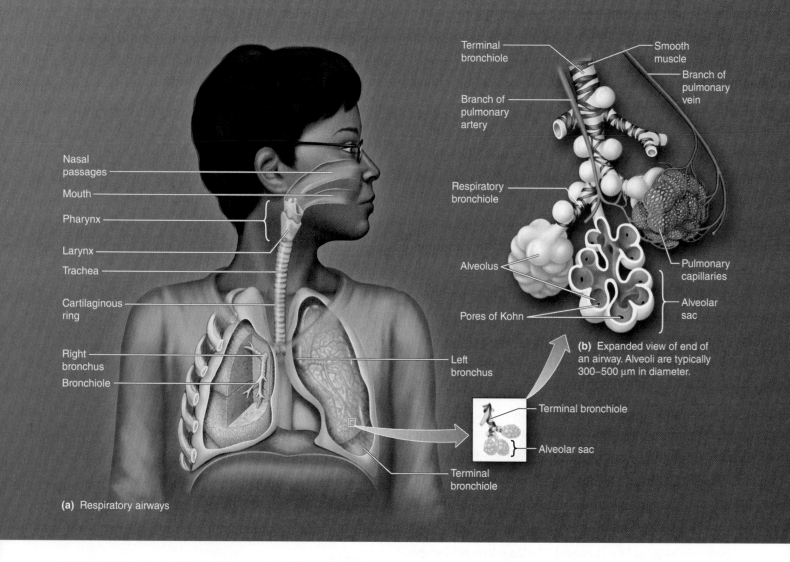

(a) Respiratory airways

Terminal bronchiole — Smooth muscle

Branch of pulmonary vein

Branch of pulmonary artery

Respiratory bronchiole

Alveolus

Pulmonary capillaries

Pores of Kohn

Alveolar sac

**(b)** Expanded view of end of an airway. Alveoli are typically 300–500 μm in diameter.

Terminal bronchiole

Alveolar sac

Left bronchus

Terminal bronchiole

Nasal passages

Mouth

Pharynx

Larynx

Trachea

Cartilaginous ring

Right bronchus

Bronchiole

> FIGURE 12-2 **Anatomy of the respiratory system.** (a) The respiratory airways. (b) Enlargement of the alveoli (air sacs) at the terminal ends of the airways. Most alveoli are clustered in grapelike arrangements at the end of the terminal bronchioles.

© 2016 Cengage

tongue, and soft palate modify the vibrations into recognizable sounds. During swallowing, the vocal folds are brought into tight apposition to close the entrance to the trachea, preventing aspiration of food.

Beyond the larynx, the trachea divides into two main branches, the right and left **bronchi**, which enter the right and left lungs, respectively. Within each lung, the bronchus continues to branch into progressively narrower, shorter, and more numerous airways, similar to the branching of a tree. The smaller branches are known as **bronchioles**; the smallest are named respiratory bronchioles because, at this stage, the walls of the airways are so thin that some gas exchange can occur. Clustered at the ends of the terminal bronchioles are the **alveoli**, the tiny air sacs where gases are exchanged between air and blood (> Figure 12-2b).

To permit flow in and out of the alveoli, the airways must remain open. The trachea and larger bronchi are fairly rigid tubes encircled by a series of cartilaginous rings that help the airways resist compression during coughing, for example. In the trachea, these rings are incomplete, resembling a horseshoe

with the two ends on the back (dorsum) joined by smooth muscle. The smaller bronchioles have no cartilage to hold them open, but their walls contain smooth muscle innervated by the autonomic nervous system; the smooth muscle is also sensitive to certain hormones and local chemicals. These factors, by varying the degree of contraction (tone) of bronchiolar smooth muscle and hence the calibre of these small terminal airways, regulate the resistance to airflow between the atmosphere and alveoli. In fact, as you will learn, regulation of airway calibre is the most important factor regulating resistance to flow (Poiseuille's law, p. 514).

The two anatomical components of the lung, airways and alveoli, have corresponding physical functions: **convection** and diffusion, respectively (> Figure 12-3). The balance between the volume dedicated to airways (large airways reduce resistance to flow, small ones increase it) and to alveoli (a larger surface area enhances diffusion) represents a compromise between these two functions. As described for the vascular system (Chapter 9), blood exiting the left heart travels through a single aorta that then branches into smaller and smaller blood vessels until reaching

12

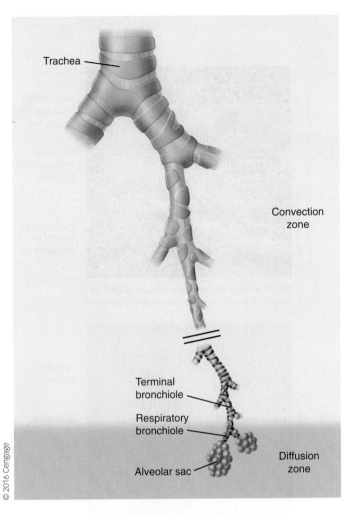

**FIGURE 12-3 Schematic of airways.** Movement of gas is purely convective until the gas arrives at the respiratory bronchioles, where, because of the increase in total cross-sectional area of the airways, motion of the gas molecules starts the transition from convection to diffusion. By the time gas arrives in the alveolar ducts and alveoli, the motion of gas molecules is purely diffusive.

systemic capillaries; likewise, the air entering the lung follows a similar path and with similar results. Although each successive airway is smaller, their number increases so dramatically that the total cross-sectional area of all airways at a given stage increases as gas flows into the lung. Because the flow at any given point is constant (i.e., the flow through the trachea equals the flow through all the bronchioles), this increase in total cross-sectional area means that the linear velocity of the gas slows. As a result, there is a transition from convection to diffusion, starting in the respiratory bronchioles (> Figure 12-3). By the time the gas reaches the alveoli, its motion is no longer convective (a process that requires energy produced by the muscles that generate flow) but diffusive (a process that does not require energy).

### ALVEOLI

The alveoli—about 300 million of them, each about 300 μg (micrometre) in diameter—are clusters of thin-walled, inflatable, grapelike sacs at the terminal branches of the conducting airways (> Figure 12-4b) and are ideally structured for gas exchange. They

are composed of two types of epithelial cells: Type I and Type II. Alveolar walls consist of a single layer of flattened **Type I alveolar cells** (> Figure 12-4a). The walls of the dense sheetlike network of pulmonary capillaries encircling each alveolus (> Figure 12-4b) are also only one-cell thick. Thus, the interstitial space between an alveolus and the surrounding capillary network is extremely thin; only 0.5 μm separates air in the alveoli from blood in the pulmonary capillaries. (A sheet of tracing paper is about 50 times as thick as this air-to-blood barrier.) Moreover, the total alveolar surface area available for diffusion of oxygen and carbon dioxide is huge, typically 40–100 m², depending on height. (If the lungs consisted of a single hollow chamber of the same dimensions rather than millions of alveoli, its surface area would be only about 0.01 m².) These two factors, along with the differences in partial pressures of oxygen and carbon dioxide across this membrane (pp. 525–526), ensure that diffusion, under all but extreme environmental or pathological conditions, is both rapid and complete as blood flows through the pulmonary capillaries.

Minute **pores of Kohn** in the walls between adjacent alveoli (see > Figure 12-2b) permit airflow between adjoining alveoli; this process is known as **collateral ventilation**. These pores are important as they allow fresh air to enter an alveolus when the terminal conducting airway is blocked because of disease.

### Check Your Understanding 12.1

1. Relate the two anatomical components of the lung to their respective contributions to movement of gases.

2. Discuss how the alveolar air–pulmonary blood interface is ideally structured for gas exchange.

3. State the functions of type I alveolar cells.

### THE CHEST WALL

The chest wall (**thorax**) is formed by 12 pairs of ribs, which join the **thoracic vertebrae** (backbone) posteriorly; ribs 1–7 join the **sternum** (breastbone) anteriorly. The **rib cage** protects the lungs and heart. The chest wall contains the muscles involved in generating the pressures that cause airflow. Like the lungs, it also contains considerable amounts of elastic connective tissue. At the neck, muscles and connective tissue enclose the thoracic cavity. In the absence of trauma, the only connection between the alveoli and the atmosphere is through the airways.

The main **inspiratory muscles** are the diaphragm and external intercostal muscles. The **diaphragm**, a large sheet of skeletal muscle, forms the floor of the thoracic cavity and separates it from the abdominal cavity (> Figure 12-5). The diaphragm is penetrated only by the esophagus and the blood vessels traversing the thoracic and abdominal cavities. It is innervated by the **phrenic nerves** that originate from cervical spinal cord segments 3, 4, and 5. The relaxed diaphragm has a dome shape that protrudes upward into the thoracic cavity. It is 1.7 mm thick at rest (up to 6 mm when contracting) and

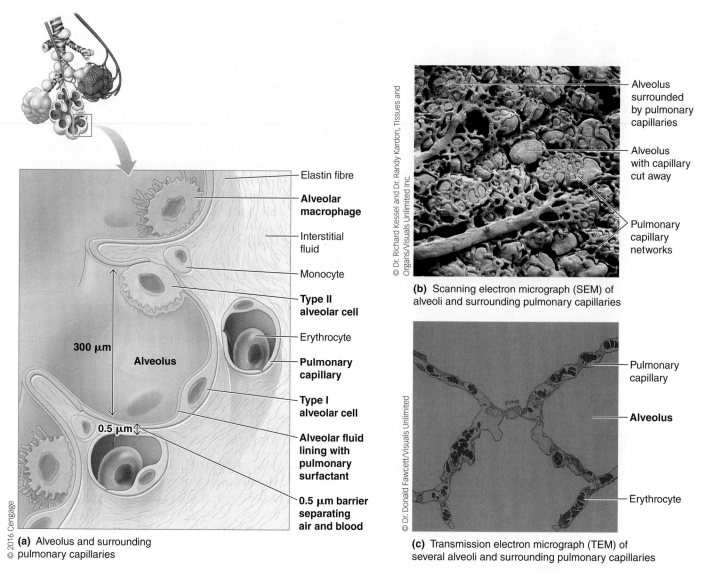

**Elastin fibre**

**Alveolar macrophage**

Interstitial fluid

Monocyte

**Type II alveolar cell**

Erythrocyte

**Pulmonary capillary**

**Type I alveolar cell**

**Alveolar fluid lining with pulmonary surfactant**

**0.5 μm barrier separating air and blood**

300 μm

Alveolus

0.5 μm

© 2016 Cengage

**(a)** Alveolus and surrounding pulmonary capillaries

© Dr. Richard Kessel and Dr. Randy Kardon, Tissues and Organs/Visuals Unlimited Inc.

Alveolus surrounded by pulmonary capillaries

Alveolus with capillary cut away

Pulmonary capillary networks

**(b)** Scanning electron micrograph (SEM) of alveoli and surrounding pulmonary capillaries

© Dr. Donald Fawcett/Visuals Unlimited

Pulmonary capillary

**Alveolus**

Erythrocyte

**(c)** Transmission electron micrograph (TEM) of several alveoli and surrounding pulmonary capillaries

› **FIGURE 12-4 Alveolus and associated pulmonary capillaries.** (a) A schematic representation of an electron microscope view of an alveolus and surrounding capillaries. This representation is *not* to scale; the diameter of an alveolus is actually about 600 times larger than the space between the alveolus and a pulmonary capillary. A single layer of flattened Type I alveolar cells forms the alveolar walls. Type II alveolar cells in the alveolar wall secrete pulmonary surfactant. Alveolar macrophages wander within the alveolar lumen. (b) A transmission electron micrograph showing several alveoli and their surrounding capillaries. (c) Note that each alveolus is surrounded by an almost continuous sheet of blood vessels.

**12**

consists of a central part, the crus, and a costal (lateral) portion on each side. At rest, the costal parts are active, but as respiratory demands increase, additional motor units of the costal diaphragm and then the crus are recruited. At the onset of **inspiration**, the diaphragm descends (moves caudally), thereby enlarging the volume of the thoracic cavity by increasing the cavity's vertical (top-to-bottom) dimension. The abdominal wall, if relaxed, moves outward during inspiration as the descending diaphragm pushes the abdominal contents downward and forward. Fifty to 75 percent of the enlargement of the thoracic cavity during quiet inspiration is accomplished by contraction of the diaphragm.

Two sets of **intercostal muscles** lie between the ribs (*inter* means "between"; *costa* means "rib"). The **external intercostal muscles** lie over the internal intercostal muscles. Contraction

of these muscles, whose fibres run downward and forward between adjacent ribs, enlarges the thoracic cavity in both the lateral (side-to-side) and anteroposterior (front-to-back) dimensions. When the external intercostals contract, they elevate the ribs, moving the sternum upward and outward. **Intercostal nerves** (T1 to T12) activate these intercostal muscles. Many respiratory physiologists think the main result of activation of the external intercostal muscles is to stabilize the chest wall, preventing it from being sucked in during inspiration. This increases the efficiency of the diaphragm because its contraction is used to inflate the lung rather than being wasted on moving the chest wall.

The expiratory muscles include the **internal intercostal muscles** and the **abdominal muscles**, consisting of the rectus abdominis, transverse abdominis, and the external and internal

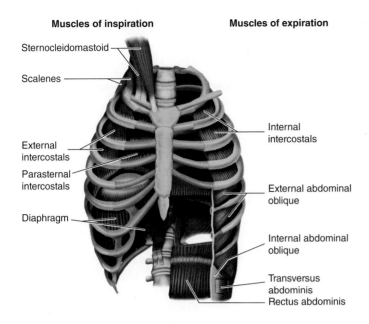

Muscles of inspiration    Muscles of expiration

Sternocleidomastoid

Scalenes

External intercostals

Parasternal intercostals

Diaphragm

Internal intercostals

External abdominal oblique

Internal abdominal oblique

Transversus abdominis

Rectus abdominis

> FIGURE 12-5 **Anatomy of the respiratory muscles**

obliques. In humans, these are inactive at rest in healthy individuals but are recruited when ventilatory demands increase, as during exercise. They are also recruited during such protective reflexes as coughing, sneezing, and vomiting. Because these reflexes are so important, the expiratory muscles can generate much higher pressures than the inspiratory muscles. The internal intercostals are innervated by branches of the same thoracic nerves that project to the external intercostals. The abdominal muscles are innervated by nerves originating from spinal cord segments T7 through L1.

### THE PLEURAL SPACE

The lung and other internal structures are covered by the visceral membrane. The inside wall of the thorax is lined by the parietal membrane. The space between these two membranes—which is very small, containing only a few millilitres of fluid—is called the **pleural space (cavity or sac)** (> Figure 12-6). The **pleural fluid**

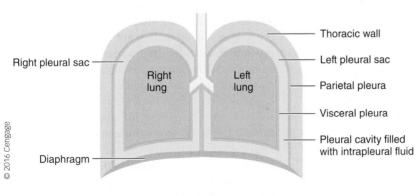

Right pleural sac

Right lung

Left lung

Thoracic wall

Left pleural sac

Parietal pleura

Visceral pleura

Pleural cavity filled with intrapleural fluid

Diaphragm

© 2016 Cengage

> FIGURE 12-6 **Pleural space.** Schematic representation of the relationship of the pleural space to the lungs and thorax. The inside membrane of the pleural space, the *visceral pleura*, closely adheres to the surface of the lung and then reflects back on itself to form the other membrane, the *parietal pleura*, which lines the interior surface of the thoracic wall. The size of the pleural space between these two layers is grossly exaggerated here for the purpose of visualization.

**▮ Why It Matters**

**P**leurisy, an inflammation of the pleural membrane, is accompanied by painful breathing, because each inflation and deflation of the lungs cause a "friction rub" between the pleural and visceral membranes.

lubricates the surfaces of the two membranes as they slide past each other during breathing.

## 12.2 | **Respiratory Mechanics**

As with blood flow through the circulatory system (p. 386), airflow through the airways depends on a pressure gradient, according to the equation flow = pressure/resistance, or $\dot{V} = \Delta P/R$. The pressure gradient is used to overcome the elastance (stiffness) of the respiratory system, the resistance to airflow, and the inertia of the system. The first two account for almost all of the pressure that must be generated. For flow to occur between the airway opening (nose, mouth) and alveoli, pressure in the alveoli must be less than pressure at the mouth. For expiration to occur, pressure in the alveoli must be greater than the pressure at the mouth. To understand how our body changes alveolar pressure requires an introduction to the mechanics of the respiratory system.

### Check Your Understanding 12.2

1. List the three factors opposing changes in volume of the respiratory system, and indicate their relative importance.

### Pressures in the respiratory system

Four different pressures are critical to ventilation (> Figure 12-7): (1) atmospheric or barometric pressure ($P_B$); (2) alveolar pressure ($P_A$); (3) pleural pressure ($P_{pl}$); and (4) transpulmonary pressure ($P_{tp}$), which is also referred to as lung recoil pressure ($P_l$) (> Figure 12-8). ▮ Table 12-2 explains the units used to measure these pressures.

### Mechanical properties of the respiratory system

Each of the two components of the respiratory system—the lung and the chest wall—has its own pressure–volume relationship (> Figure 12-9). (Appendix G explains the derivation of the pressure–volume relationships of the lung and chest wall.) It is important for clinicians to understand the mechanical properties of these two components because the diseases that affect one or both of them impair respiratory function.

**12**

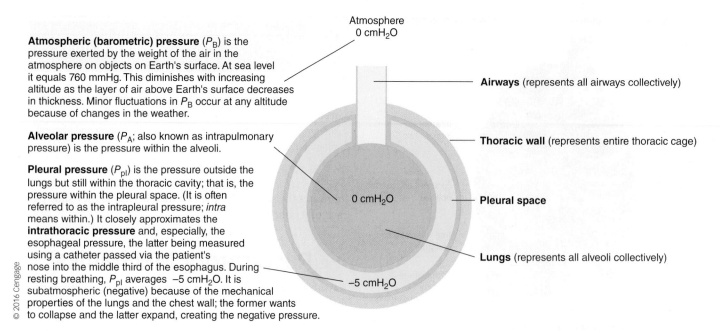

**Atmospheric (barometric) pressure** ($P_B$) is the pressure exerted by the weight of the air in the atmosphere on objects on Earth's surface. At sea level it equals 760 mmHg. This diminishes with increasing altitude as the layer of air above Earth's surface decreases in thickness. Minor fluctuations in $P_B$ occur at any altitude because of changes in the weather.

**Alveolar pressure** ($P_A$; also known as intrapulmonary pressure) is the pressure within the alveoli.

**Pleural pressure** ($P_{pl}$) is the pressure outside the lungs but still within the thoracic cavity; that is, the pressure within the pleural space. (It is often referred to as the intrapleural pressure; *intra* means within.) It closely approximates the **intrathoracic pressure** and, especially, the esophageal pressure, the latter being measured using a catheter passed via the patient's nose into the middle third of the esophagus. During resting breathing, $P_{pl}$ averages −5 cmH₂O. It is subatmospheric (negative) because of the mechanical properties of the lungs and the chest wall; the former wants to collapse and the latter expand, creating the negative pressure.

**Atmosphere**
**0 cmH₂O**

**Airways** (represents all airways collectively)

**Thoracic wall** (represents entire thoracic cage)

**0 cmH₂O**

**Pleural space**

**Lungs** (represents all alveoli collectively)

**−5 cmH₂O**

> **FIGURE 12-7 Pressures important in respiration**

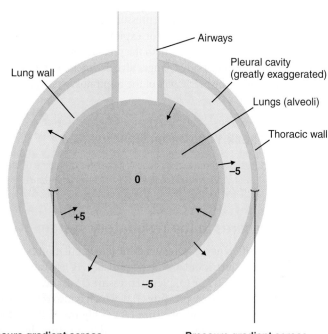

Airways

Lung wall

Pleural cavity (greatly exaggerated)

Lungs (alveoli)

Thoracic wall

0

−5

+5

−5

Pressure gradient across lung wall = alveolar pressure − pleural pressure. Here, $P_{tp} = 0 − (−5) = 5$ cmH₂O.

Pressure gradient across thoracic wall = pleural pressure − atmospheric pressure. Here, $P_{tw} = −5 − 0 = −5$ cmH₂O.

Pressures are in cmH₂O.

> **FIGURE 12-8 Transmural pressure gradient.** In this depiction, flow is zero, and all pressures are in cmH₂O, relative to atmospheric pressure set to 0 cmH₂O. The pressure gradient across a structure is called the transmural pressure (*trans* is Latin for "across"; mural is the adjective of *murus*, Latin for "wall"). A transmural pressure difference is *always* calculated as the inside pressure minus the outside pressure. When referring to the lung, the transmural pressure is called the transpulmonary pressure ($P_{tp}$). This pressure is also the recoil pressure of the lung ($P_l$). Therefore, transpulmonary pressure = alveolar pressure − pleural pressure, or as an equation: $P_{tp} = P_A − P_{pl}$.

The line labelled $P_l$ in > Figure 12-9 shows the pressure–volume relationship of the lung. As lung volume increases, its recoil pressure increases from ~0 cmH₂O at residual volume to about 30 cmH₂O at total lung capacity (p. 518). The lung therefore behaves like a balloon or a bubble, always exerting a positive (deflating) pressure.

The line labelled $P_w$ in > Figure 12-9 shows the pressure–volume relationship of the chest wall. It is like a spring that can be either compressed or expanded. Above approximately 65 percent of total volume (less when expressed as a percent of vital capacity; > Figure 12-9), when it is stretched beyond its equilibrium volume, it exerts a deflating pressure. Below 65 percent, however, it is like a compressed spring, exerting negative (inflating) pressures.

The solid line labelled $P_{rs}$ shows the pressure–volume relationship of the respiratory system. Here, the pressures represent the sums of the pressures of the lung and chest wall at any given volume. At 100 percent of the vital capacity (p. 518), both the lung and chest wall "want" to deflate, and the pressure exerted by the respiratory system is the sum of the deflationary pressures exerted by both. Conversely, at residual volume (0 percent of vital capacity), the chest wall's tendency to inflate reflects only the passive inflationary pressure of the chest wall. At approximately 40 percent of the vital capacity, however, the net pressure exerted by the respiratory system is 0 cmH₂O because, at this volume, the deflating pressure of the lung (about +5 cmH₂O) is equal and opposite to the inflating pressure (−5 cmH₂O) of the chest wall. This volume is the *functional residual capacity* (FRC) (p. 518) and is the volume the lungs return to at the end of a breath during resting breathing.

© 2016 Cengage

**12**

Pressures can be expressed in many different units. Ideally, one would use those approved by the Système Internationale (SI)—in the case of atmospheric pressure, 1 Atm or 1000 kilopascals (kPa)—but for respiratory physiology we use traditional units based on the height of a column of mercury (mmHg) or water (cmH₂O):

- mmHg for the partial pressures of gases when discussing diffusion
- cmH₂O when discussing bulk flow (convection)

For the latter, we use cmH₂O rather than mmHg because the pressures needed to generate flow are typically small, only a few cmH₂O. For example, a pressure of 5 cmH₂O, which can easily cause a high flow, equals ~0.37 cmHg or 3.7 mmHg (1 mmHg equals 13.6 mmH₂O or 1.36 cmH₂O); 3.7 mm is too small to be measured accurately when using a mercury manometer. In addition, pressures related to convective flow are expressed *relative to atmospheric pressure*. Thus, if atmospheric pressure were 1034 cmH₂O (usually given as 760 mmHg), an alveolar pressure of 1039 cmH₂O is referred to as +5 cmH₂O, and an alveolar pressure of 1029 cmH₂O as −5 cmH₂O. (Note that "negative" pressures do not exist; they are negative only because they are less than the barometric pressure to which they are referred.)

---

The shape of the pressure–volume curve of the respiratory system is important. First, like all curves, it is characterized by its slope, which represents the change in volume per unit change in pressure. This slope, similar to the curve for the lung, has the units of L/cmH₂O, and is called the compliance. The compliance is greatest at functional residual capacity, the lung volume around which we breathe, and least at low or high lung volumes. Therefore, at functional residual capacity, the need to generate pressure to stretch the respiratory system is minimized, which decreases the energy (work) needed to breathe.

It is important to note that the zero pressure condition that occurs at the functional residual capacity is above the minimum lung volume which the person can achieve upon maximal forceful expiration—that is, the residual volume. On average, the maximum volume that the lungs can hold—the total lung capacity—is about 5.7 L in healthy, young adult males and 4.2 L in females. Anatomic build (especially height), age, the distensibility of the lungs, and the presence or absence of respiratory disease all affect total lung capacity. Normally, during quiet breathing, lung volume at end-expiration is about 50 percent of the total lung capacity in a healthy, young adult. Thus, at functional residual capacity, the lungs still contain about 2.2 L of air. At rest, a typical tidal volume (p. 518) is about 0.5 L, so during quiet breathing the lung volume varies between 2.2 L at the end of expiration and

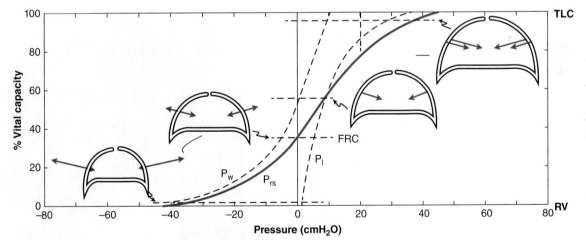

> **FIGURE 12-9 Pressure–volume relationships of the lung, chest wall, and respiratory system.** As lung volume increases, the recoil (passive) pressure of the lung (P₁), which is always positive (expiratory), also increases. In contrast, the recoil pressure of the chest wall (Pᵥᵥ) is most negative (inspiratory) at minimum volume, increases as volume increases, exerts no recoil pressure at approximately 60 percent of total volume, and is positive (expiratory) above this volume. The recoil pressure of the respiratory system (Pᵣₛ) is the sum of the pressures generated by the lung and chest wall. At functional residual capacity (FRC), the pressures of the lung and chest wall are equal and opposite; therefore, the net pressure exerted by the respiratory system is nil. Consequently, at the end of a passive expiration, the respiratory system is at FRC, approximately 50 percent of total lung capacity. This ensures that gas exchange continues during expiration. Note too that the slope of the pressure–volume relationship of the respiratory system is maximal at FRC; as a result, system compliance is maximal, which minimizes the work of breathing. For more information, see Appendix G.

E. Agostoni and J. Mead. Statics of the Respiratory System. In: *Handbook of Physiology*. Section 3, Respiration, vol. 1, pp. 387–409 (1964). Eds. W.O. Fenn and H. Rahn. Washington, DC. American Physiological Society.

2.7 L at the end of inspiration (see › Figure 12-21). This presence of at least 2 L of gas in the lung even after expiration means there is sufficient oxygen in the lung to maintain its diffusion into the blood in the pulmonary capillaries, thereby preventing swings in blood oxygen levels during the respiratory cycle. During maximal expiration, lung volume can decrease to 1.2 L in males (1 L in females), but the lungs can never be completely deflated because the small airways collapse during forced expirations at low lung volumes, blocking further outflow (see › Figure 12-18).

## Check Your Understanding 12.3

1. Describe the respective contributions of the lung and chest wall to the generation of pressure by the respiratory system at residual volume, functional residual capacity, and total lung capacity.

2. What are the physiological benefits of breathing from functional residual capacity?

### SOURCE OF THE LUNG'S ELASTIC RECOIL

The lung "wants" to collapse for two reasons: the **elastic recoil** of the tissues (pulmonary connective tissues contain large quantities of elastin fibres), and the **surface tension** of the liquid lining the inside surface of the alveoli.

Of these two factors, surface tension contributes about 70 percent of the lung's elastic recoil pressure. This was first established in the 1920s by the Swiss physiologist von Neergard who compared the pressure–volume curves of isolated lungs filled with saline and then with air (› Figure 12-10). (For a history of pulmonary mechanics, see Macklem, P.T. [2004]. A century of the mechanics of breathing, *American Journal of Respiratory and Critical Care Medicine, 170*: 10–15; › Figure 12-11.) When the lung is filled with saline, it relies on the elastic properties of the lung and generates very little recoil pressure. In contrast, when it is inflated with air, its recoil pressure is much greater.

Why does the air-filled lung generate so much more recoil pressure than a saline-filled lung? At an air–water interface, water molecules at the surface are more strongly attracted to other surrounding water molecules than to any water molecules in the air above the surface. This is because there are far more water molecules in the water than in the water vapour above the surface. This unequal attraction, due to the polarity (unequal distribution of charge) of the water molecules, produces a force known as surface tension. Surface tension has a twofold effect. First, the liquid layer resists any force that increases its surface area; that is, it opposes expansion of the alveolus, because the surface water molecules oppose being pulled apart. Accordingly, the greater the *surface tension*, the less compliant are the lungs. Second, the surface area of the liquid tends to shrink as much as possible because the surface water molecules, being preferentially attracted to one another, try to get as close together as possible. Thus, the surface tension of the liquid lining an alveolus tends to reduce alveolar size and expel alveolar gas. This property, along with the rebound of the stretched elastin fibres, produces the lung's elastic recoil that provides the pressure for passive expiration.

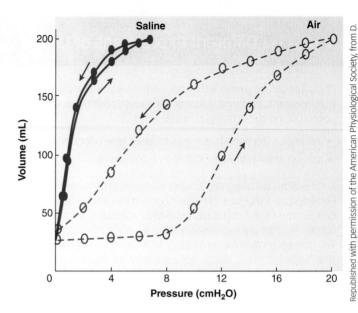

Republished with permission of the American Physiological Society, from D. F. Tierney, "Lung surfactant: some historical perspectives leading to its cellular and molecular biology," *American Journal of Physiology* 257 (2): L1–L12, 1989; permission conveyed through Copyright Clearance Center, Inc.

› FIGURE 12-10 **Pressure–volume relationship of saline- and air-filled lungs.** At a given volume, the saline-filled lung generates much less recoil pressure than the air-filled lung, because the former lacks an air–liquid interface and, therefore, surface tension. In an air-filled lung, surface tension accounts for most of the lung's recoil pressure; the rest comes from the elastic nature of the lung tissue. Note the space (or hysteresis) between the inflation and deflation limbs of the air-filled (but not saline-filled) lung. The hysteresis is exaggerated because this is the first inflation with air starting from a degassed state, as would occur when a baby takes its first breath. Over subsequent breaths, the inflation limb moves closer to the deflation limb.

› FIGURE 12-11 **Peter Macklem (1931–2011).** He was one of the world's most prominent respiratory researchers (note the Order of Canada), and was still active at the time of his death in early 2011. His achievements and those of others are recounted in his excellent review of Canadian contributions to respiratory research in the *Canadian Respiratory Journal* [2007] 14: 383–92.

In emphysema, the breakdown of the alveolar walls causes both the loss of elastin fibres and the reduction in alveolar surface tension, which decrease elastic recoil. This, along with increased airway resistance, contributes to the patient's difficulty in breathing out. Moreover, the reduced elastic recoil of the lung, in the face of unchanged outward recoil of the chest wall, means that the patient's equilibrium point—the functional residual capacity (FRC)—increases, accounting for a barrel-chested appearance in these patients.

## Alveolar stability

The division of the lung into millions of alveoli provides the advantage of a huge surface area for gas exchange, but it also presents the problem of maintaining alveolar stability. Recall that the pressure generated by alveolar surface tension is directed inward. If you visualize the alveoli as spherical bubbles, according to the **law of LaPlace**, the magnitude of this collapsing pressure is directly proportional to the surface tension and inversely proportional to the radius of the bubble:

$$P = \frac{2T}{r}$$

where

$P$ = collapsing pressure

$T$ = surface tension

$r$ = radius of bubble (alveolus)

Because the collapsing pressure is inversely proportional to the radius, the smaller the alveolus, the smaller is its radius and the greater its tendency to collapse at a given surface tension. Accordingly, if two alveoli of unequal size but with the same surface tension are connected by the same terminal airway, the smaller alveolus—because it generates a larger collapsing pressure—will collapse and empty into the larger alveolus (❯ Figure 12-12a). Two factors oppose the tendency of alveoli to collapse: pulmonary surfactant and alveolar interdependence.

### PULMONARY SURFACTANT

**Pulmonary surfactant** is a complex mixture of lipids and proteins secreted by the Type II alveolar cells (see ❯ Figure 12-4a). It disperses between the water molecules in the fluid lining the alveoli and lowers **alveolar surface tension**, because the cohesive force between a water molecule and an adjacent molecule of surfactant is very low. By lowering alveolar surface tension, surfactant provides two important benefits: (1) it increases pulmonary compliance, reducing

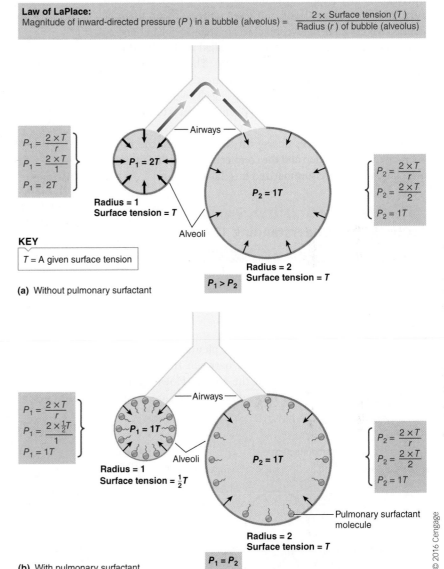

**Law of LaPlace:**
Magnitude of inward-directed pressure ($P$) in a bubble (alveolus) = $\dfrac{2 \times \text{Surface tension }(T)}{\text{Radius }(r)\text{ of bubble (alveolus)}}$

$P_1 = \dfrac{2 \times T}{r}$
$P_1 = \dfrac{2 \times T}{1}$
$P_1 = 2T$

$P_1 = 2T$

**Radius = 1**
**Surface tension = $T$**

**KEY**

$T$ = A given surface tension

**(a)** Without pulmonary surfactant

$P_2 = \dfrac{2 \times T}{r}$
$P_2 = \dfrac{2 \times T}{2}$
$P_2 = 1T$

$P_2 = 1T$

Airways
Alveoli

**Radius = 2**
**Surface tension = $T$**

$P_1 > P_2$

$P_1 = \dfrac{2 \times T}{r}$
$P_1 = \dfrac{2 \times \frac{1}{2}T}{1}$
$P_1 = 1T$

$P_1 = 1T$

**Radius = 1**
**Surface tension = $\frac{1}{2}T$**

**(b)** With pulmonary surfactant

$P_2 = \dfrac{2 \times T}{r}$
$P_2 = \dfrac{2 \times T}{2}$
$P_2 = 1T$

$P_2 = 1T$

Airways
Alveoli

Pulmonary surfactant molecule

**Radius = 2**
**Surface tension = $T$**

$P_1 = P_2$

❯ **FIGURE 12-12 Role of pulmonary surfactant in counteracting the tendency for small alveoli to collapse into larger alveoli.** (a) According to the law of LaPlace, if two alveoli of unequal size but the same surface tension are connected by the same terminal airway, the smaller alveolus generates a larger inward-directed pressure than the larger alveolus. In the absence of surfactant, it will collapse and empty its air into the larger alveolus. (b) Pulmonary surfactant reduces the surface tension of a smaller alveolus more than that of a larger alveolus. This reduction in surface tension offsets the effect of the smaller radius, and reduces the inward-directed pressure. Consequently, the collapsing pressures of the small and large alveoli are comparable and the small alveolus does not collapse and empty into the larger one.

**Source**: From Sherwood. *Human Physiology*, 8E. © 2013 Brooks/Cole, a part of Cengage, Inc. Reproduced by permission. www.cengage.com/permissions.

**12**

© 2016 Cengage

the work of inflating the lungs; and (2) it reduces the recoil pressure of smaller alveoli more than that of larger alveoli; this allows alveoli of different sizes to coexist, thereby keeping them open and able to participate in gas exchange. The net effect of surfactant, therefore, is to equalize the pressures within alveoli of different sizes, minimizing the tendency of small ones to empty into larger ones (⟩ Figure 12-12b). This helps stabilize alveoli and maintains gas exchange.

### ALVEOLAR INTERDEPENDENCE

The second factor that contributes to alveolar stability is the interdependence among neighbouring alveoli. Each alveolus is surrounded by other alveoli and interconnected with them by connective tissue. If an alveolus starts to collapse, neighbouring alveoli are stretched as their walls are pulled in the direction of the collapsing alveolus (⟩ Figure 12-13a). These neighbouring alveoli then recoil in response to being stretched, and therefore exert an expanding force on the collapsing alveolus, which helps it stay open (⟩ Figure 12-13b). This phenomenon is termed **alveolar interdependence**.

The opposing forces acting on the lung (i.e., the forces keeping the alveoli open and the countering forces that promote alveolar collapse) are summarized in ▍ Table 12-3.

---

### Check Your Understanding 12.4

1. Describe the forces that keep the alveoli open and those that promote alveolar collapse.

---

### PNEUMOTHORAX

*Clinical Note* Normally, air does not enter the pleural cavity because there is no communication between the cavity and either the atmosphere or the alveoli. However, if the chest wall is punctured, air flows down its pressure gradient

---

**▍ TABLE 12-3 Opposing Forces Acting on the Lung**

| Forces Keeping the Alveoli Open | Forces Promoting Alveolar Collapse |
|---|---|
| Positive transmural pressure | Elasticity of stretched connective tissue fibres |
| Pulmonary surfactant (decreases alveolar surface tension) | Alveolar surface tension |
| Alveolar interdependence | |

---

from the atmosphere into the pleural space (⟩ Figure 12-14a). The abnormal condition of air entering the pleural space is known as **pneumothorax** (from the Greek, "air in the chest"). Both pleural and alveolar pressure now equal or actually exceed atmospheric pressure. With no opposing negative pleural pressure to keep it inflated, the lung collapses to its unstretched size (⟩ Figure 12-14b). Likewise, with no opposing negative pleural pressure to hold it in, the thoracic wall expands to its unrestricted size (see the zero pressure point in ⟩ Figure 12-9), but this has much less serious consequences than collapse of the lung. Pneumothorax and lung collapse can also occur if air enters the pleural cavity through a hole in the lung— produced, for example, by a disease process or by overinflation of the lung (*barotrauma*) (⟩ Figure 12-14c). The latter reason explains why respiratory therapists and anesthetists carefully monitor the pressure and volume generated by mechanical ventilators.

Having completed our discussion of the physical properties of the respiratory system's two components, the lung and chest wall, we can now turn to the main question: how does alveolar pressure change so that we can breathe in and out?

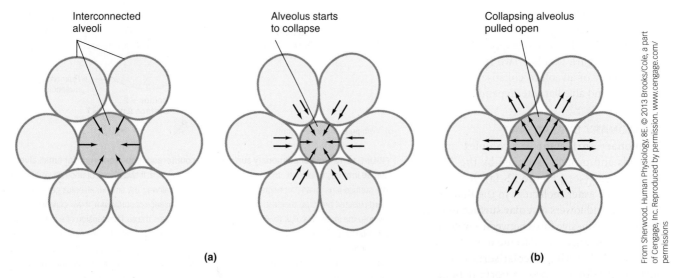

**⟩ FIGURE 12-13 Alveolar interdependence.** (a) When an alveolus (*in pink*) in a group of interconnected alveoli starts to collapse, the neighbouring alveoli become stretched by the collapsing alveolus. (b) After stretching, these alveoli recoil, pulling the collapsing alveolus outward. This helps prevent the alveolus from collapsing.

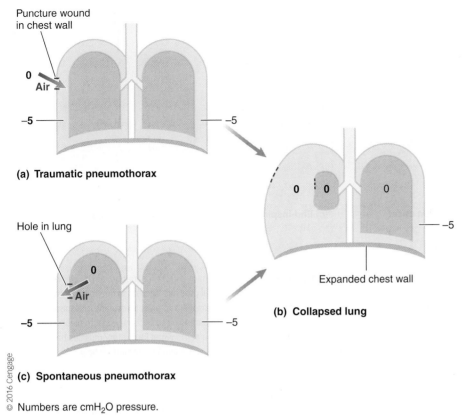

Puncture wound
in chest wall

0
Air

−5                                        −5

**(a) Traumatic pneumothorax**

0 ⦙ 0                    0

                                    −5

Expanded chest wall

**(b) Collapsed lung**

Hole in lung

0

Air

−5                                  −5

© 2016 Cengage

**(c) Spontaneous pneumothorax**

Numbers are cmH₂O pressure.

> **FIGURE 12-14 Pneumothorax.** (a) *Traumatic pneumothorax*: Puncture of the chest wall permits air from the atmosphere to flow down its pressure gradient into the pleural space, abolishing the transmural pressure gradient. (b) *Collapsed lung*: When the transmural pressure gradient is abolished, the lung collapses, and the chest wall expands. (c) *Spontaneous pneumothorax*: A hole in the lung permits air to move down its pressure gradient and enter the pleural cavity via the lungs, abolishing the transmural pressure gradient. As with traumatic pneumothorax, the lung collapses and the chest wall expands.

## How alveolar pressure changes

For air to flow into the lung, alveolar pressure must be less than barometric (atmospheric) pressure. For air to flow out of the lung, alveolar pressure must exceed barometric pressure. Because barometric pressure is reasonably constant, the alveolar pressure must change to establish the pressure gradient needed to generate airflow. The pressure across the lung, the transpulmonary pressure (identical to the lung's recoil pressure), equals the inside (alveolar) pressure minus the outside (pleural) pressure (see > Figure 12-7), as follows:

lung recoil pressure = alveolar pressure − pleural pressure

Because alveolar pressure needs to change in order for the lung to inflate or deflate, we rearrange this equation by placing *alveolar pressure* on the left side:

alveolar pressure = lung recoil pressure + pleural pressure

To change alveolar pressure, therefore, requires a change in the recoil pressure of the lung, the pleural pressure, or both. Because the recoil pressure of the lung depends on lung volume, in this equation we cannot change the lung volume by changing a pressure that reflects lung volume; this is a circular argument. Therefore, the pleural pressure has to change, and this happens when the muscles in the chest wall are activated. Activating the inspiratory muscles decreases pleural pressure, whereas activating the expiratory muscles increases it.

### ONSET OF INSPIRATION: CONTRACTION OF INSPIRATORY MUSCLES

Before inspiration, at the end of the preceding expiration, alveolar pressure equals atmospheric pressure, so no air flows into or out of the lungs (> Figures 12-15. As the pleural pressure decreases and the thoracic cavity enlarges, the pressure within the alveoli drops because of decompression (i.e., the same number of molecules are now present inside a larger container; see **Boyle's/Mariotte's law,** > Figure 12-16). Because alveolar pressure is now less than atmospheric pressure, air flows down the pressure gradient, inflating the lungs (> Figure 12-15 and > Figure 12-17), until no further gradient exists—that is, until alveolar pressure again equals atmospheric pressure. In a typical inspiratory excursion, the alveolar pressure drops only about 1 cmH₂O. (For a more detailed explanation of how this happens, see Appendix G.)

The trajectory of pleural pressure from start to end of inspiration is not linear but curved (> Figure 12-17). The "extra" pressure (hatched area) is used to overcome airflow resistance. If there were no flow resistance (as in an infinitely slow inspiration), the only pressure needed would be that needed to overcome the elastic recoil (the straight line, AC, between the start and end of inspiration).

### ONSET OF EXPIRATION: RELAXATION OF INSPIRATORY MUSCLES

At the end of inspiration, the inspiratory muscles relax, decreasing their ability to expand the thorax. As a result, pleural pressure becomes less negative (i.e., more positive) (> Figure 12-15 and > Figure 12-17). Then, because alveolar pressure is the algebraic sum of the lung recoil and pleural pressures, the alveolar pressure is now positive and expiration starts. Note that activation of the expiratory muscles is not necessary for expiration to occur; all that's required is reduction and then cessation of activity of the inspiratory muscles. Expiration stops when alveolar pressure again equals atmospheric pressure.

### ROLE OF ACCESSORY INSPIRATORY MUSCLES

Deeper inspirations can be accomplished by contracting the diaphragm and external intercostal muscles more forcefully and by recruiting the previously inactive **accessory inspiratory muscles**. Contracting these accessory muscles in the neck (see > Figure 12-5), raises the sternum and elevates the first two ribs, enlarging the upper portion of the thoracic cavity. As the volume of the thoracic cavity increases even more, making the pleural pressure and, therefore, the alveolar pressure even more negative, both inspiratory flow and tidal volume increase.

**12**

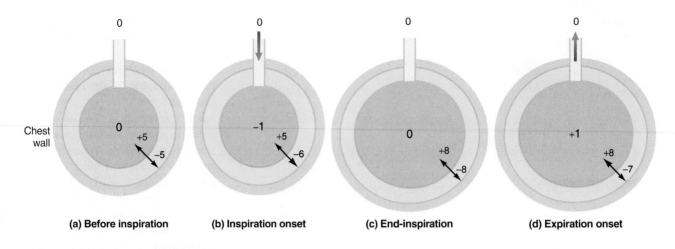

**(a) Before inspiration**  **(b) Inspiration onset**  **(c) End-inspiration**  **(d) Expiration onset**

Numbers are pressures, in cmH₂O.

> **FIGURE 12-15** **Changes in lung volume and alveolar and pleural pressures during inspiration and expiration.** (a) *Before inspiration, at the end of the preceding expiration*: Because alveolar pressure equals atmospheric pressure, there is no flow. (b) *Inspiration*: As pleural pressure decreases due to inspiratory muscle contraction, alveolar pressure decreases due to decompression. This establishes a pressure gradient from atmosphere to alveoli, resulting in inspiratory flow. (c) *End of inspiration*: Inspiration ends when the contraction of inspiratory muscles decreases, allowing lung recoil pressure to become equal to pleural pressure. As a result, alveolar pressure again equals atmospheric pressure and flow stops. (d) *Onset of expiration*: Expiration starts when inspiratory muscles have stopped contracting. Lung recoil pressure is now greater than pleural pressure, resulting in a positive alveolar pressure and, therefore, expiratory flow.

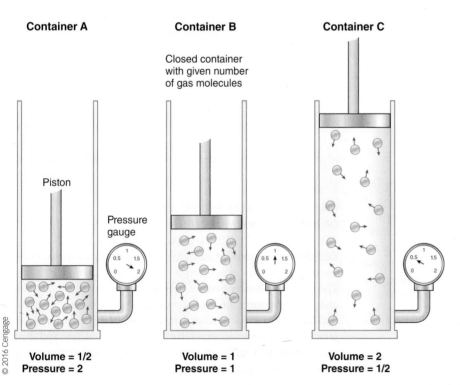

> **FIGURE 12-16** **Boyle's/Mariotte's law.** Each container has the same number of molecules. Given the random motion of gas molecules, the likelihood of a gas molecule striking the interior wall of the container and exerting pressure varies inversely with the volume of the container at a given temperature. The gas in container B exerts more pressure than the same number of the same gas molecules in container C, but less pressure than the same number of gas molecules in container A. The relation is stated as Boyle's/Mariotte's law: $P_1V_1 = P_2V_2$. As the volume of gas increases, the pressure of the gas decreases proportionately; conversely, the pressure increases proportionately as the volume decreases.

## SUMMARY OF INSPIRATION

Inspiration occurs when alveolar pressure is less than barometric pressure. This drop in alveolar pressure is the result of activation of the inspiratory muscles that lower pleural pressure (which, before inspiration starts, is about −5 cmH₂O). Because alveolar pressure is the algebraic sum of the pleural pressure and the elastic recoil pressure of the lung, and the lung's recoil pressure cannot change as quickly as the pleural pressure (it takes time for gas to flow into the lung, thereby increasing its recoil), alveolar pressure decreases. As a result, air flows into the lung (alveoli). As long as the inspiratory muscles contract and keep pleural pressure less (more negative) than the recoil pressure of the lung, inflation continues. Toward the end of inspiration, the activity of the inspiratory muscles declines. As a result, the recoil pressure of the lung "catches up" to the pleural pressure. When the two pressures are equal and opposite (e.g., pleural pressure = −8 cmH₂O and lung recoil pressure = +8 cmH₂O), inspiration stops. The "equal and opposite" situation is identical to that before inspiration started, at functional residual capacity, but the lung volume before inspiration started is greater than at the end of inspiration because the recoil pressure of the lung is greater.

For expiration to start, the cessation of inspiratory muscle activity results in pleural pressure starting to increase—that is, becomes more positive—in this case, increasing from −8 cmH₂O to −7 cmH₂O. Because the elastic recoil pressure of the lung is

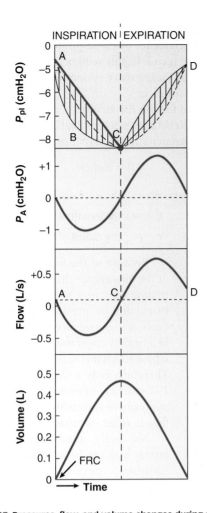

> FIGURE 12-17 **Pressures, flow, and volume changes during a respiratory cycle.** Before the onset of inspiration, pleural pressure ($P_{pl}$) is approximately $-5\,cmH_2O$, equal and opposite to the lung's recoil pressure ($+5\,cmH_2O$). Therefore, the alveolar pressure ($P_A$) is $0\,cmH_2O$ (relative to atmospheric), and there is no flow. As the inspiratory muscles contract, $P_{pl}$ decreases and so does $P_A$ (due to decompression). There is now a pressure gradient driving air from the airway opening to the alveoli. As long as $P_A$ is less than atmospheric pressure (i.e., negative), inspiratory flow continues. As the inspiratory muscles start to relax toward the end of inspiration, lung recoil pressure ($R$) "catches up" to $P_{pl}$, and $P_A$ returns to $0\,cmH_2O$, but this time at a more negative $P_{pl}$ ($-8\,cmH_2O$) and, therefore, at a higher lung volume. The straight line between points A and C indicates the trajectory of $P_{pl}$ if inspiration were infinitely slow—that is, if all the pressure required to inflate the lungs were used only to overcome elastic recoil. The hatched area between this straight line and the actual trajectory of $P_{pl}$ during inspiration indicates the additional pressure needed to overcome flow resistance.

now greater than the pleural pressure, the alveolar pressure is positive, and expiration starts. Expiration ends when the recoil pressure of the lung again equals the pleural pressure.

## ACTIVE EXPIRATION

In healthy individuals at rest, expiration is passive due to elastic recoil of the lungs. In contrast, inspiration is *always* active (i.e., requires energy expenditure), because it occurs only as a result of contraction of inspiratory muscles. **Forced or active expiration** empties the lungs more rapidly, occasionally more completely (i.e., to a lower end-expiratory lung volume). At sufficiently high levels of ventilation during exercise, the activation of expiratory muscles

can reduce the end-expiratory lung volume to below the functional residual capacity. This increases tidal volume independent of the inspiratory muscles and allows the lung volume to stay on the more compliant part of the pressure-volume curve (see › Figure 12-9).

To breathe out more forcefully, alveolar pressure must be increased even more above atmospheric pressure than can be accomplished by relaxation of the inspiratory muscles and elastic recoil of the lungs (see › Figure 12-9). The most important **expiratory muscles** are those of the abdominal wall. As they contract, the resulting increase in abdominal pressure becomes transferred to the pleural space, thereby increasing pleural pressure and generating a more forceful expiration (› Figure 12-18). The other expiratory muscles are the internal intercostal muscles. Their contraction pulls the ribs downward and inward, flattening the chest wall and further decreasing the size of the thoracic cavity— the opposite of what results from activation of the external intercostals. Aside from decreasing the volume of the thoracic cavity, the muscles of active expiration stabilize the abdomen; this aids breathing because energy is now used to move air rather than displace the abdominal contents. This is most clearly seen during heavy physical labour and aerobic exercise.

During forceful expiration, pleural pressure becomes positive, but the lungs do not collapse because alveolar pressure increases correspondingly, according to the relationship: alveolar pressure = recoil pressure + pleural pressure. As expiratory flow occurs, however, alveolar pressure gradually decreases, because energy (pressure) is lost in overcoming flow resistance. At some point along the airways, this pressure will drop until it equals and then falls below the pressure outside the airways, the pleural pressure. At this point, the **equal pressure point**, the **transmural pressure** switches from positive (inside pressure > outside pressure) to negative (inside pressure < outside pressure), a situation in which the airway is under compression (see › Figure 11-18c). How much compression occurs depends on the intrinsic strength of the airway wall (cartilage) and the degree of contraction (*tone*) of the airway smooth muscle.

■ **Clinical Connections**

Asthma is an obstructive airway disease due to bronchoconstriction of the smaller airways and an inflammatory response in the airways that leads to an increase in mucus secretion. Patients with asthma generally have normal inhalation; however, exhalation is hindered due to the obstruction. In an effort to enhance exhalation against the obstruction, there is an increased use of accessory muscles for forced exhalation. With greater recruitment of these muscles, there is an increase in pleural pressure that causes the equal pressure point to be reached earlier in the exhalation, which further constricts the bronchioles due to compression of the lungs and results in the characteristic wheezing sound associated with asthma. Exhalation is terminated early, which causes air trapping because the next inspiration cycle begins at a higher lung volume. This air trapping further increases the amount of work needed for exhalation, and this cycle continues to worsen until the obstruction can be reduced.

**12**

## Airway resistance and airflow

We have discussed airflow in and out of the lungs as a function of the magnitude of the pressure gradient between the alveoli and the atmosphere, and you have learned how this

pressure gradient is generated and used to overcome, during inspiration, the elastance of the respiratory system, especially the lung's elastic recoil. In this section, we focus on the other component of the respiratory system that needs to be overcome: airway resistance.

Resistance to flow in the airways is governed by the equation $V = \Delta P/R$. Resistance is determined by, **Poiseuille's law**, the same equation that determines resistance to blood flow (p. 388):

$$R = 8n\ell/r^4$$

where: 
$\eta$ = the viscosity of the fluid
$\ell$ = the length of the tube
$r$ = the radius of the tube

Because of the fourth power, the radii of the airways are the primary factor determining resistance to flow. In our preceding discussion of pressure gradient–induced flows, we ignored airway resistance because, in a healthy respiratory system, the radii are large enough that resistance is extremely low. Therefore, only a small pressure gradient, 1–2 $cmH_2O$, between the alveoli and the atmosphere is usually required to generate flow in either direction.

Normally, modest adjustments in airway size can be achieved by regulating autonomic nervous system activity to suit metabolic and, therefore, ventilatory demands. Parasympathetic activity,

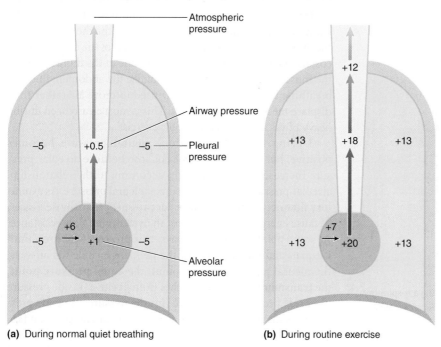

**(a)** During normal quiet breathing

**(b)** During routine exercise

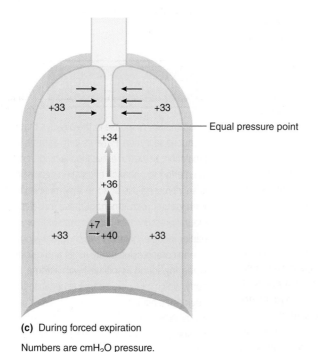

**(c)** During forced expiration

Numbers are $cmH_2O$ pressure.

> FIGURE 12-18 **Airway collapse during forced expiration.** (a) Normal, quiet expiration, during which airway resistance is low, so there is little frictional loss of pressure within the airways. Pleural pressure remains less than airway pressure throughout the airways, so they remain open. (b) In routine exercise, even though pleural pressure is elevated during the active expiration accompanying routine vigorous activity, alveolar pressure is also elevated and airway resistance is still low, so the resistance-induced drop in airway pressure does not fall below the elevated pleural pressure until gas reaches airways that are held open by cartilaginous rings. Therefore, airway collapse does not occur. (c) In forced expiration, both alveolar and pleural pressures are markedly increased. When resistive losses cause pressure in the airway to fall below the surrounding elevated pleural pressure (i.e., the transmural pressure becomes negative at the equal pressure point), the small, nonrigid airways are compressed or closed, preventing increases in expiratory flow. In healthy individuals, this dynamic compression of airways occurs only at very low lung volumes.

dominant at rest when ventilatory demands are low, promotes bronchiolar smooth muscle contraction (**bronchoconstriction**); this increases airway resistance by decreasing the radii of airways. Conversely, withdrawal of parasympathetic activity, as in exercise, causes airway smooth muscle relaxation (**bronchodilation**). Sympathetic stimulation and to a greater extent its associated hormone, epinephrine, also cause bronchodilation and decreased airway resistance via activation of $\beta_2$-adrenergic receptors (Table 12-4). Therefore, during periods of sympathetic domination, when actual or potential increased demands for oxygen uptake occur, bronchodilation ensures that the pressure gradients generated by respiratory pump muscle activity can achieve maximum flows with minimal resistance. A low resistance also helps reduce the work of breathing. Because of this bronchodilator action, epinephrine or similar drugs are useful therapeutic tools to counteract airway constriction in patients with bronchial spasm.

The situation is, however, considerably more complex than this (> Figure 12-19). Airways receive a rich innervation by postganglionic parasympathetic nerve fibres originating from the vagus nerves; the release of acetylcholine causes bronchoconstriction. Tonic activity of these fibres accounts for a resting level of contraction, known as tone. In contrast, at least in humans, innervation of the airways by postganglionic sympathetic (adrenergic) fibres from the paravertebral ganglia is much less.

Dilation can be elicited in two ways by the adrenergic system: (1) directly by the terminals releasing norepinephrine, which activates $\beta_2$-receptors on bronchial smooth muscle; or (2) indirectly by the adrenal medulla releasing epinephrine, which is carried by the circulation to its site of action—the airway smooth muscle. (This is why people with intense reactions to allergens, such as bee venom, carry syringes loaded with epinephrine.) In addition, activation of $a$-adrenergic receptors located in parasympathetic ganglia may suppress parasympathetic activity, thereby decreasing tone (indicating the importance of having resting levels of tone) and causing bronchodilation.

The peptidergic system also affects airway resistance. Release of vasoactive intestinal peptide (VIP) by the inhibitory non-adrenergic-non-cholinergic parasympathetic (i-NANC) system causes dilation (possibly mediated by nitric oxide), whereas the complementary excitatory (e-NANC) system causes bronchoconstriction via the release of such tachykinins as substance P. In addition, many other bronchoconstrictor agents (histamine, thromboxane $A_2$, prostaglandin $F_2$, and leukotrienes) are released by mast and epithelial cells or cells recruited as part of an immunological response to allergens or infections. Given the prevalence of asthma, the amount of research on this topic—both academic and commercial/pharmaceutical—is not surprising.

Airway smooth muscle controls the resistance primarily in intrathoracic airways. Skeletal muscles in the upper airway (larynx, pharynx) also control resistance to flow, but because they are skeletal, they are not under autonomic control. Instead, they are under the control of motor neurons located in the part of the brain stem that contains the neurons that generate the respiratory rhythm (p. 542).

| ▌ **TABLE 12-4** Factors Affecting Airway Resistance | | |
| --- | --- | --- |
| **Status of Airways** | **Effect on Resistance** | **Factors Producing the Effect** |
| **Bronchoconstriction** | ↓ radius, ↑ resistance to airflow | *Pathological factors:*<br>Allergy-induced spasm of the airways caused by slow-reactive substance of anaphylaxis<br>Histamine<br>Physical blockage of the airways caused by excess mucus<br>oedema of the walls<br>Airway collapse<br><br>*Physiological control factors:*<br>Neural control: parasympathetic stimulation<br>Local chemical control: ↓ $CO_2$ concentration |
| **Bronchodilation** | ↑ radius, ↓ resistance to airflow | *Pathological factors*: none<br><br>*Physiological control factors:*<br>Neural control: sympathetic stimulation (minimal effect)<br>Hormonal control: epinephrine<br>Local chemical control: ↑ $CO_2$ concentration |

12

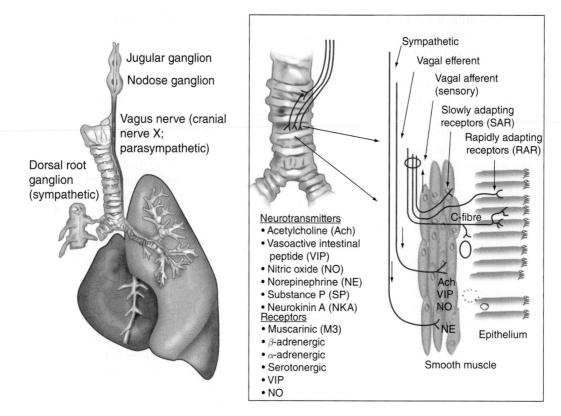

**Jugular ganglion**

**Nodose ganglion**

**Vagus nerve (cranial nerve X; parasympathetic)**

**Dorsal root ganglion (sympathetic)**

**Sympathetic**

**Vagal efferent**

**Vagal afferent (sensory)**

**Slowly adapting receptors (SAR)**

**Rapidly adapting receptors (RAR)**

C-fibre

Neurotransmitters
• Acetylcholine (Ach)
• Vasoactive intestinal peptide (VIP)
• Nitric oxide (NO)
• Norepinephrine (NE)
• Substance P (SP)
• Neurokinin A (NKA)
Receptors
• Muscarinic (M3)
• β-adrenergic
• α-adrenergic
• Serotonergic
• VIP
• NO

Ach
VIP
NO

NE

Epithelium

Smooth muscle

› **FIGURE 12-19 Airway innervation.** The lung is innervated by both branches of the autonomic nervous system. Parasympathetic neurotransmitters to airway smooth muscle include acetylcholine (Ach), vasoactive intestinal peptide (VIP), and nitric oxide (NO). The sympathetic neurotransmitter is norepinephrine (NE). Other transmitters include substance P (SP) and neurokinin A (NKA). Receptor endings located in both the epithelium and smooth muscle include slowly adapting (SAR) and rapidly adapting (RAR) receptors, along with endings of unmyelinated fibres (C-fibres). JG: jugular ganglion; NG: nodose ganglion; DRG: dorsal root ganglion.

Last, resistance can be affected by swelling of the mucosa, in the nose and elsewhere, as anyone with a "stuffy" nose can attest. Agents (sympathetic agonists) that narrow mucosal blood vessels reduce this swelling, which is why they are known as *decongestants*.

## Airway resistance and chronic pulmonary diseases

Chronic pulmonary diseases are often characterized by increased resistance resulting from narrowing of the lumen of the lower airways. When resistance increases, a larger pressure gradient must be established to maintain even a normal airflow. For example, if resistance doubles due to narrowing of the airways (which, according to Poiseuille's law, requires only a 16 percent decrease in radius), the pressure gradient must be doubled through increased activation of the respiratory pump muscles to produce the same flow. Accordingly, people with pulmonary disease work harder to breathe. Chronic (long-term) pulmonary diseases include asthma and chronic obstructive pulmonary disease (COPD).

### ASTHMA

In asthma (see www.lung.ca), airway obstruction is due to (1) thickening of airway walls, brought about by inflammation

and histamine-induced oedema (p. 467); (2) plugging of the airways by excessive secretion of very thick mucus; and (3) airway hyper-responsiveness, characterized by constriction of the smaller airways due to spasm of the smooth muscle in their walls. Triggers that lead to these inflammatory changes and the exaggerated bronchoconstrictor response include repeated exposure to allergens (such as dust mites or pollen), irritants (as in cigarette smoke or pollution), and infections. In severe attacks, pronounced clogging and narrowing of the airways can prevent all airflow, leading to death. An estimated 3 million people in Canada have asthma, and the number is steadily climbing. Asthma is the most common chronic childhood disease. Scientists are unsure why the incidence of asthma is increasing.

### CHRONIC OBSTRUCTIVE PULMONARY DISEASE

*Clinical Note* **Chronic obstructive pulmonary disease (COPD)** (the current term for emphysema and chronic bronchitis) slowly damages the airways, usually as a result of cigarette smoking. (A world authority is Dr. James C. Hogg of Vancouver, the 2013 recipient of the Gairdner Wightman Award.) The disease can also result from other airborne irritants and occupational pollutants, such as coal dust, asbestos, and silica. Combining exposure to pollutants with cigarette smoking greatly increases the likelihood of a person developing COPD.

According to the Canadian Lung Association (see www.lung.ca), approximately 600 million people worldwide have COPD, including about 1.5 million Canadians. COPD is the fourth-leading cause of death in Canada, and experts believe that, by 2020, it will be the third-largest cause of mortality worldwide, in large part because cigarette smoking is very prevalent in underdeveloped and developing countries where governments are susceptible to the influence (or lawsuits) of multinational manufacturers and dependent on taxes derived from sales of cigarettes. A good prognosis for COPD relies on early detection and diagnosis, prompt cessation of smoking when applicable, and reduction of environmental exposure to particulate pollution. Most patients demonstrate improvement in lung function at the onset of treatment, but eventually COPD progresses, and the symptoms become worse.

### CHRONIC BRONCHITIS

**Chronic bronchitis** is a long-term inflammatory condition of the lower airways, generally triggered by frequent exposure to cigarette smoke, polluted air, or allergens. In response to the chronic irritation, the airways become narrowed by prolonged oedematous thickening of the airway linings, coupled with over-production of thick mucus. Despite frequent coughing associated with the chronic irritation, the plugged mucus often cannot be removed, especially because the irritants immobilize the ciliary mucus escalator (p. 464). Pulmonary bacterial infections frequently occur because the accumulated mucus serves as an excellent medium for bacterial growth.

### EMPHYSEMA

**Emphysema** is characterized by (1) collapse of the smaller airways and (2) breakdown of alveolar walls (i.e., loss of alveolar tissue, including pulmonary capillaries). This irreversible condition can arise in two ways. Most commonly, emphysema results from the excessive release of destructive enzymes, such as *trypsin*, from alveolar macrophages as a defensive mechanism in response to chronic exposure to inhaled cigarette smoke or other irritants. A protein that inhibits trypsin—$\alpha_1$-*antitrypsin*—normally protects the lungs from damage caused by the release of the destructive enzymes. However, excessive secretion of destructive enzymes in response to chronic irritation can overwhelm the protective capability of $\alpha_1$-antitrypsin, with the result that these enzymes destroy not only foreign materials but lung tissue as well. Loss of lung tissue leads to the breakdown of alveolar walls and collapse of small airways.

Less frequently, emphysema arises from a genetic inability to produce $\alpha_1$-antitrypsin so that the lung tissue has no protection from trypsin. The unprotected lung tissue gradually disintegrates under the influence of even small amounts of macrophage-released enzymes, even in the absence of chronic exposure to inhaled irritants. The only treatment currently available is lung transplantation.

### DIFFICULTY BREATHING OUT

When pulmonary disease of any type increases airway resistance, expiration is more difficult than inspiration. The smaller airways, lacking the cartilaginous rings that hold the larger ones open, are held open by a positive transmural pressure gradient—that is, by the negative pleural pressure. Inspiration not only increases the volume of the alveoli but also dilates the airways passively because of the greater positive transmural pressure gradient. The resulting dilation of the airways decreases flow resistance (Poiseuille's law). In a healthy individual, however, airway resistance is so low that these respiratory phase–related changes in airway resistance cannot be readily detected. When airway resistance increases substantially, however, as during an asthma attack, the difference is noticeable. Thus, a person with asthma has more difficulty expiring than inspiring, giving rise to the characteristic wheezing that happens as air is forced out through the narrowed airways.

Surprisingly, patients experiencing an acute exacerbation of asthma or COPD do not complain of difficulty breathing out; instead, they complain that they cannot get enough air in! Why? Although the initial problem is obstructive ("can't breathe out"), it quickly becomes restrictive ("can't breathe in"). Because the person has trouble breathing out, they do not return to the usual end-expiratory lung volume. Therefore, the next inspiration starts at a higher lung volume. This is good because airway resistance is lower at higher lung volumes. But at this elevated lung volume, the person still cannot breathe out completely. Over time, and the process can be rapid and the person hyperinflates—a phenomenon known as **dynamic hyperinflation**. But at high lung volumes, not only is the inspiratory reserve volume reduced (often to less than a litre) but so too is system compliance (see ❭ Figure 12-9). This makes inspiration difficult and uncomfortable; the patient wants to breathe in but cannot achieve a satisfactory tidal volume. The discrepancy between what is wanted and what is achieved is thought to be the basis for the debilitating sensation of **dyspnea**, or **breathlessness**, during these acute exacerbations.

Normally, the smaller airways stay open during quiet breathing and even during active expiration when pleural pressure is elevated, as during exercise (see ❭ Figure 12-18b,). In people without pulmonary disease, the smaller airways collapse and further outflow of air stops only at very low lung volumes during a maximal forced expiration (see ❭ Figure 12-22). Because the small airways collapse, the lungs can never be emptied completely, which accounts for the residual volume. By contrast, in people who have pulmonary disease, especially emphysema, the smaller airways often collapse even during expiration at rest, preventing further outflow. Such individuals breathe at higher end-expiratory lung volumes that help hold the airways open. They may also purse their lips during expiration; by limiting flow at the mouth, they shift the equal pressure point toward the mouth, and the increased pressure inside the airways helps to keep the airways open.

## Lung volumes and capacities

A spirometer is a device that measures the volume of air breathed in and out; it consists of an air-filled drum floating in a water-filled chamber (❭ Figure 12-20). As a person breathes air in and out of the drum through a connecting tube, the rise and the fall of the drum, to which a pen is attached, are recorded as a **spirogram**,

**12**

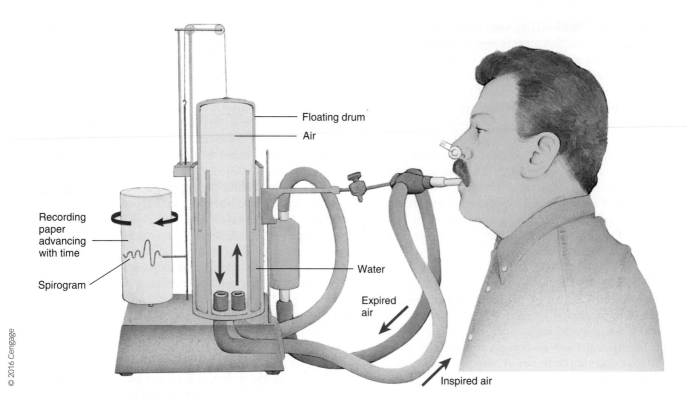

> **FIGURE 12-20 A spirometer**

which is calibrated to measure the volume (cross-sectional area multiplied by height). Water-filled spirometers have been replaced with devices that measure flow (often using a turbine); the output signal is electronically integrated to obtain volume.

> Figure 12-21 gives an example of a spirogram for a healthy, young, adult male. Generally, the values are lower for females. The following lung volumes and lung capacities can be determined. (Note a lung capacity is the sum of two or more lung volumes.)

- **Tidal volume ($V_T$).** The volume of air entering or leaving the lungs during a single breath. Average value under resting conditions = 500 mL.

- **Inspiratory reserve volume (IRV).** The extra volume that can be maximally inspired over and above the typical resting tidal volume. Average value = 3000 mL.

- **Inspiratory capacity (IC).** The maximum volume that can be inspired starting from the end of a normal quiet expiration (IC = IRV + $V_T$). Average value = 3500 mL.

- **Expiratory reserve volume (ERV).** The maximum volume that can be actively expired starting from the end of a typical resting tidal volume. Average value = 1000 mL.

- **Residual volume (RV).** The volume of air remaining in the lungs after a maximal expiration. Average value = 1200 mL. The residual volume cannot be measured directly with a spirometer, because this volume of air does not move in and out of the lungs. It can be determined indirectly, however, through gas-dilution techniques involving inspiration of a known quantity of a harmless tracer gas, such as helium.

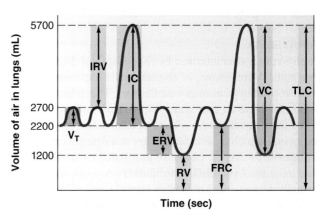

**KEY**

$V_T$ = Tidal volume (500 mL)
IRV = Inspiratory reserve volume (3000 mL)
IC = Inspiratory capacity (3500 mL)
ERV = Expiratory reserve volume (1000 mL)
RV = Residual volume (1200 mL)
FRC = Functional residual capacity (2200 mL)
VC = Vital capacity (4500 mL)
TLC = Total lung capacity (5700 mL)

> **FIGURE 12-21 Subdivisions of lung volumes.** Normal spirogram of a healthy, young, adult male. (The residual volume, RV, cannot be measured with a spirometer but can by other means, such as helium dilution.) Values for females are, on average, lower.

- **Functional residual capacity (FRC).** The volume of air in the lungs at the end of a normal passive expiration (FRC = ERV + RV). Average value = 2200 mL.

- **Vital capacity (VC).** The maximum volume of air that can be moved out during a single breath following a maximal inspiration. The person first inspires maximally, then expires maximally ($VC = IRV + V_T + ERV$). Average value = 4500 mL.

- **Total lung capacity (TLC).** The maximum volume of air that the lungs can hold ($TLC = VC + RV$) Average. value = 5700 mL.

- **Forced expiratory volume in one second ($FEV_1$).** The volume of air expired during the first second of a maximal expiratory effort starting from TLC. It is usually expressed as a ratio of the forced vital capacity—that is, the $FEV_1/FVC$—or converted to a percentage. Young, healthy people have values of 80 percent or more, but the test must be done correctly. The $FEV_1/FVC$ decreases with age. One can estimate a person's maximum voluntary ventilation (MVV, in L/min) by multiplying the ($FEV_1$) by 35.

## DYNAMIC LUNG FUNCTION AND RESPIRATORY DYSFUNCTION

*Clinical Note* As well as the measurements of their lung volumes, individuals suspected of having pulmonary problems will have their dynamic lung function assessed. Two general categories of respiratory dysfunction yield abnormal results during spirometry: *obstructive lung disease* and *restrictive lung disease* (see › Figure 12-22 and › Figure 12-23).

› Figure 12-22 illustrates a **forced vital capacity (FVC)** manoeuvre in which the person inhales maximally to total lung capacity (identical to 100 percent of the vital capacity) and then breathes out as hard, fast, and completely as possible. From this manoeuvre, considerable insight into the status of the respiratory system, especially the lungs, is obtained. The healthy person in › Figure 12-22a can exhale approximately 80 percent of her vital capacity in the first second (the forced expired volume, $FEV_1$), but individuals with pulmonary diseases have very different results. The patient with an obstructive lung disease has a lower $FEV_1$ and cannot breathe out to a low lung volume because her airways collapse, trapping gas (› Figure 12-22b). Consequently, her FRC and residual volume (RV) are greater, but vital capacity (VC) smaller. In contrast, the person with restrictive lung disease breathes at a lower lung volume but can still breathe out a normal or even greater than normal fraction of her vital capacity in the first second (› Figure 12-22c).

Another way of presenting this manoeuvre is not with a tracing of volume versus time (› Figure 12-22 and › Figure 12-23a), in which the rate of volume change indicates flow) but with a plot of flow versus volume. The flow–volume tracings from a healthy individual, a patient with obstructive, and another with restrictive lung disease are depicted in › Figure 12-23b. During the manoeuvre, expiratory flow reaches an early peak and then declines until residual volume is reached. The first part of the curve depends on the effort made and is therefore called *effort-dependent*. In contrast, the last part of the curve is *effort-independent*. In other words, even if a healthy individual tried to breathe out harder, he could not generate a higher flow. This is because a greater effort causes increased pleural pressure that compresses the smaller airways, increasing their resistance. This increase in resistance offsets the increased driving pressure.

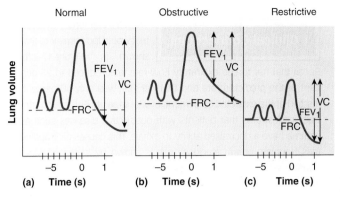

> **FIGURE 12-22 Spirometry (flow versus time) tracings.** (a) The tracing from the healthy individual shows that the volume she can breathe out in the first second of a forced expiration (the $FEV_1$) is more than 80 percent of her vital capacity, and she can exhale to a low lung volume (the residual volume). (b) In contrast, the patient with an obstructive lung disease has a lower $FEV_1$; in other words, the slope of the flow tracing during the forced expiration is reduced. Moreover, even at rest, there is air trapping in the lung during expiration due to collapsible airways, so her functional residual capacity (FRC) and residual volume (RV) are greater, but the vital capacity (VC) is smaller. Nevertheless, total lung capacity (TLC) is normal or even elevated because it includes trapped gas. Although both VC and $FEV_1$ are reduced, the $FEV_1$ is reduced more than the forced vital capacity (FVC). As a result, the $FEV_1 / FVC$ percentage is much lower than the normal 80 percent. (c) In restrictive lung disease, the lungs are less compliant than normal. The patient breathes at a lower lung volume, has a greatly reduced VC, a normal RV, but the $FEV_1 / FVC$ percentage is normal or even elevated. Therefore, changes in $FEV_1 / FVC$ are particularly useful in distinguishing between obstructive and restrictive lung disease.

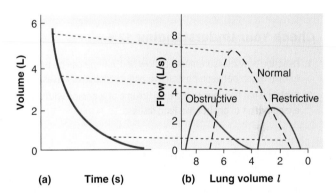

> **FIGURE 12-23 Derivation of expiratory flow–volume curves.** (a) Volume versus time tracing of a normal individual during a maximal forced expiration from total lung capacity. (b) At each volume, the slope (volume/time, or flow) is calculated (three such points are shown in the volume-time tracing) and plotted versus volume. Note the rapid rise to a peak flow of ~7 L/s, followed by a linear decrease in flow as lung volume decreases. In contrast, an individual with obstructive lung disease (left tracing) breathes at a higher lung volume but reaches a lower peak flow; throughout expiration, flows are much less than those in a normal individual, and the tracing is slightly concave to the x-axis ("scooping"). The person also cannot breathe out as much; that is, he has an elevated residual volume. His vital capacity is also slightly reduced. The person with a restrictive disease (right tracing) breathes at a lower lung volume, has a greatly reduced vital capacity, but can, at a given lung volume, generate expiratory flows greater than those of the healthy individual. These curves illustrate how such tests can help diagnose respiratory diseases.

PHYSIOLOGY OF RESPIRATION 2e by Michael P. Hlastala & Albert J. Berger (1996): Figure 3.11 (p. 58) © 1996 by Oxford University Press, Inc. By Permission of Oxford University Press, USA.

NEL                                     The Respiratory System   **519**

Andrea's pulmonary function test would look similar to that shown in ⟩ Figure 12-22b. Recall that her forced expiratory flow rate was reduced to 25 percent of the predicted rate and that her FEV₁ was only 40 percent, instead of the expected 80 percent, of vital capacity. This spirometry tracing illustrates that patients with obstructive diseases such as asthma have a decreased ability to exhale. This causes air trapping (gas trapping; breath stacking); this is shown by increased functional residual capacity and residual volume, yet the vital capacity is smaller.

In contrast, in a person with COPD, flow again reaches an early peak, but the value is much less than that predicted for a healthy person of the same age, gender, and height. Moreover, the maximal expiratory flow rapidly decreases to a low value and stays low throughout the rest of the expiration, despite the person making a maximal effort. Although not shown, the maximum flow during this effort is the same as that used by the person at rest. Last, expiratory flow terminates prematurely, at a higher volume, because of closure of the airways. In this person, the forced vital capacity is less than predicted.

The third person has a restrictive disease. The most obvious difference between this individual and the other two is the reduced lung volume (smaller lateral dimension). The person cannot inhale as much (a smaller distance along the *x*-axis between the end of a normal inspiration and the end of a maximal inspiration) and, because the recoil of the lung is reduced, peak expiratory flow is less.

### Check Your Understanding 12.6

1. Describe how measurements of flow and volume can be used to distinguish between obstructive and restrictive diseases.

2. Explain what happens to the lung volume of a person with an exacerbation of obstructive lung disease.

### WHY WE MEASURE EXPIRATORY FLOWS

Considering that inspiration is the active, energy-consuming part of breathing, why do we measure events during *expiration* to assess pulmonary function? Indeed, to assess a person's ability to breathe, why not simply ask the person to breathe as hard and fast as possible for several minutes and measure the maximum voluntary ventilation? Aside from the difficulty of knowing whether the person is making a maximal effort, the major problem is that an increase in ventilation in the absence of an increase in carbon dioxide production means the person is hyperventilating, and the partial pressure of carbon dioxide in the arterial blood ($Pa_{CO_2}$) will decrease. This, in turn, will cause systemic vasoconstriction, reducing blood flow to the periphery (accounting for numbness in the extremities), including the brain (accounting for feeling "light-headed"). It is therefore difficult for a person to sustain the effort, and the procedure would not be ethical for patients.

A simple experiment can provide part of the answer to why we measure expiratory events as an index of pulmonary function. Breathe out all the way to residual volume, and then inhale as hard and as fast as you can to total lung capacity. How long did it take? Perhaps 0.5 seconds or less? Now breathe in all the way to total lung capacity, and then breathe out as hard, fast, and completely as you can to residual volume. If you did it correctly, the expiration would take at least four to six seconds. In other words, the limitation to ventilation in a healthy individual is expiratory, not inspiratory.

There are two reasons for this. First, during expiration, lung volume decreases and the radii of the airways decrease, thereby increasing flow resistance. Second, during expiration, the expiratory muscles contract and, therefore, shorten. According to the force–length relationship (Chapter 7), this reduces the ability of skeletal muscles to generate force—in the case of expiratory muscles, the ability to generate pressure. Recall that

$$\dot{V} = \Delta P / R$$

During expiration, especially a forced expiration, the ability to generate pressure decreases because of the force–length relationship, and resistance increases because of reduced lung volume and compression of the airways due to the higher pleural pressure outside the intrathoracic airways. The combination of a decreased numerator ($\Delta P$) and an increased denominator ($R$) is responsible for the reduction in airflow as lung volume decreases.

During inspiration, the ability of the inspiratory muscles to generate pressure also decreases because they shorten. But this is offset by the decrease in resistance as lung volume increases. Therefore, inspiratory flow is limited not by the resistance but by the contractile properties of the inspiratory muscles.

Why not simply measure the maximal voluntary ventilation (MVV)? In fact, we do not need to perform this test. Instead, we can multiply the forced expired volume in the first second ($FEV_1$) of a forced vital capacity by 35 to obtain a value close to the actual MVV.

### OTHER CAUSES OF RESPIRATORY DYSFUNCTION

Obstructive and restrictive diseases are not the only categories of respiratory dysfunction, and spirometry is not the only way to assess pulmonary function. Other conditions affecting respiratory function include the following:

- Diseases impairing diffusion of oxygen and carbon dioxide across the pulmonary alveolar–capillary membranes
- Reduced ventilation because of mechanical failure, as with neuromuscular diseases affecting the respiratory muscles
- Inadequate pulmonary blood flow
- Ventilation/perfusion abnormalities involving a poor matching of air to blood (and vice versa) so that gas exchange is impaired

Some lung diseases are actually a complex mixture of different types of functional disturbances. To determine what abnormalities are present, a diagnostician relies on a variety of tests in addition to spirometry, including X-rays, blood-gas measurements, and tests to measure the diffusion capacity for oxygen across the alveolar capillary membrane.

**12**

## Pulmonary and alveolar ventilation

The volume of gas breathed during one minute is called the **minute ventilation**. It is calculated by multiplying the average tidal volume ($V_T$) over one minute by the **respiratory frequency** ($f$). Thus, minute ventilation (L/min) equals tidal volume (L/breath) multiplied by respiratory rate (breaths/min). In symbols,

$$\dot{V}_E = V_T \times f$$

where the E after $\dot{V}$ indicates the measurement was made on expired gas. At rest, an average tidal volume is 0.5 L and the respiratory frequency is 12, so the minute ventilation is 6.0 L/min. For a brief period, a healthy, young, adult male can voluntarily increase his minute ventilation twenty-five-fold, to 150 L/min. To increase ventilation, both tidal volume and respiratory frequency increase; at the start, the tidal volume increases more than the frequency, but once a limit to tidal volume is reached (large tidal volumes increase the work of breathing because system compliance is low at high volumes),

frequency increases. But not all of this ventilation can be used for gas exchange because of the anatomic dead space.

### ANATOMIC DEAD SPACE

Not all the inspired air reaches the alveoli; some remains in the conducting airways. This volume is considered anatomic dead space ($V_D$) because air within these conducting airways cannot participate in gas exchange. The volume of the airways in an adult averages about 150 mL. Anatomic dead space influences the efficiency of pulmonary ventilation. Although 500 mL of air move in and out with each breath, only 350 mL actually participate in gas exchange due to the anatomic dead space.

To understand how the dead space affects gas exchange, examine › Figure 12-24. During inspiration, the person inspires 500 mL. Most (350 mL) enters the alveoli, but 150 mL remains in the conducting airways. Therefore, at the end of inspiration, the airways are filled with 150 mL of fresh air. During the subsequent expiration, 500 mL of air are expired. The first 150 mL

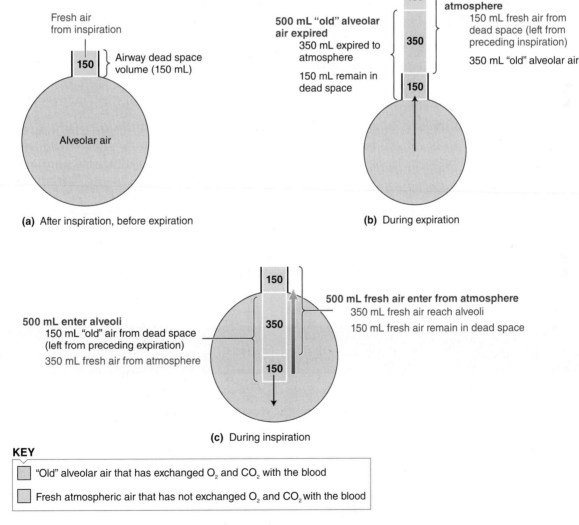

(a) After inspiration, before expiration

(b) During expiration

(c) During inspiration

**KEY**

"Old" alveolar air that has exchanged $O_2$ and $CO_2$ with the blood

Fresh atmospheric air that has not exchanged $O_2$ and $CO_2$ with the blood

© 2016 Cengage

› **FIGURE 12-24 Effect of dead space on gas exchange.** Even though 500 mL of air moves in and out between the atmosphere and the respiratory system and 500 mL moves in and out of the alveoli with each breath, only 350 mL is actually exchanged between the atmosphere and the alveoli because of the anatomic dead space (the volume of air in the airways).

is the fresh air that stayed in the airways and never participated in gas exchange. The remaining 350 mL is gas from the alveoli. But 500 mL also had to leave the alveoli. The first 350 mL are expired to the atmosphere; the other 150 mL remain in the conducting airways.

On the next inspiration, 500 mL of gas enter the alveoli. The first 150 mL to enter the alveoli is the old alveolar air that remained in the dead space during the preceding expiration. In effect, the subject is rebreathing his own gas at the start of inspiration. The other 350 mL entering the alveoli is fresh air. But because the subject's VT is 500 mL, the last 150 mL remain in the conducting airways, to be expired without benefit of being exchanged with the blood, during the next expiration.

### ALVEOLAR VENTILATION

To optimize oxygen transfer to the blood in the capillaries, we want as much as possible of the lung to be devoted to gas exchange—that is, to alveoli. But this would require the airways (which do not participate in gas exchange because their walls are too thick) to be small. But if the airways are narrow, flow resistance increases inordinately—due to the radius to the fourth power ($r^4$) term in Poiseuille's law (p. 514)—and the resistive work of breathing would be too high. The anatomy of the respiratory system is a compromise between airway (resistance) and alveolar (gas exchange) volumes. As just mentioned, during inspiration, some of the inspired air remains in the airways and never reaches the alveoli: the anatomic dead space. The dead space volume has important consequences for **alveolar ventilation** ($\dot{V}_A$). If the volume of gas a person breathes in with each breath (the tidal volume) is the same as the volume of the dead space, then the alveolar ventilation must be zero. In other words, all the inspired gas stays in the anatomic dead space. To have an effective alveolar ventilation (in terms of gas exchange), tidal volume must exceed dead space volume. It would seem, therefore, that the ideal breathing pattern to maximize alveolar ventilation is one in which an individual uses a slow, deep breathing pattern, when tidal volume is much greater than dead space volume. The effects of changes in tidal volume and respiratory frequency on minute and alveolar ventilation, at rest and during exercise, are shown in ▮ Table 12-5.

At rest, ventilation is more effective with larger tidal volumes and lower respiratory frequencies. During exercise, and assuming dead space volume is fixed (in mL, it is approximately twice the person's weight in kg), the proportion of "wasted ventilation" ($\dot{V}_D / \dot{V}_E$) decreases, and the proportion of ventilation directed to the alveoli ($\dot{V}_A / \dot{V}_E$) increases, which makes breathing more efficient. This is because the primary way to increase ventilation is to increase the tidal volume. One cannot, however, increase tidal volume too much because that places the subject on the low compliance part of the pressure–volume curve (see ﹥ Figure 12-9) and this disproportionately increases the work of breathing. As a result, after the initial increase in tidal volume, the person increases frequency as ventilatory demands continue to increase.

## Work of breathing

During normal quiet breathing, the inspiratory muscles overcome two opposing forces: the elastic recoil of the lung (reflecting the elastic nature of tissues and, mainly, surface tension) and airway resistance. Moreover, expiration is passive; the energy needed to overcome flow resistance comes from the energy stored in the expanded lung (lung recoil). Normally, the lungs are highly compliant and airway resistance is low, so only about 3 percent of the total energy expended by the body is used for quiet breathing.

Note the situation in which a tidal volume (0.15 L) equal to the dead space volume (0.15 L) results in an alveolar ventilation of 0 L/min. Although even smaller tidal volumes would seem to be incompatible with gas exchange, this is not the case. It is possible—and even desirable in several clinical situations—to ventilate individuals, particularly infants, with tiny volumes but at very high frequencies (~600/min). These high frequencies enhance gas diffusion, and carbon dioxide is literally "shaken" out of the lungs. The advantage of this technique, called *high frequency oscillatory ventilation*, is that it does not produce the high pressures typically needed to inflate damaged lungs. It therefore avoids the possibility of barotrauma. Much of this work was done in the 1970s and 1980s at Toronto's Hospital for Sick Children.

**12**

▮ **TABLE 12-5** Effect of Changes in Tidal Volume and Frequency on Alveolar Ventilation

|  | $V_T$ (L) | $V_D$ (L) | $f$ (bpm) | $\dot{V}_E$ (L/min) | $\dot{V}_A$ (L/min) | $\dot{V}_D/\dot{V}_E$ | $\dot{V}_A/\dot{V}_E$ |
|---|---|---|---|---|---|---|---|
| Rest | 0.5 | 0.15 | 12 | 6.0 | 4.2 | 0.30 | 0.70 |
| Rest | 0.3 | 0.15 | 20 | 6.0 | 3.0 | 0.50 | 0.50 |
| Rest | 0.75 | 0.15 | 8 | 6.0 | 4.8 | 0.20 | 0.80 |
| Rest | 0.15 | 0.15 | 40 | 6.0 | 0 | 1.00 | 0.00 |
| **Exercise** | | | | | | | |
| Moderate | 1.0 | 0.15 | 20 | 20 | 17 | 0.15 | 0.85 |
| Heavy | 1.5 | 0.15 | 30 | 40 | 40.5 | 0.10 | 0.90 |

(For a description of the development of this technique, see "The Oscillations of HFO" by A.C. Bryan. [2001]. *American Journal of Respiratory and Critical Care Medicine, 163:* 816–17.)

An almost infinite combination of tidal volumes and respiratory frequencies can be involved in the work of breathing for a given type of alveolar ventilation (see ▮ Table 12-5). However, we actually use only a limited range in the work of breathing. The combination of a low frequency and a high tidal volume is associated with a low resistive component but a high elastic component of the work of breathing (❯ Figure 12-25); in contrast, the combination of a high frequency and a low tidal volume is associated with a high resistive but a low elastic component of the work of breathing. In breathing, we use the particular combination of respiratory frequency and tidal volume that results in a minimal *total* of these two components. However, the range of frequencies over which we can breathe at rest and that results in the minimal work of breathing is wide: from 8 to 16 breaths per minute. At rest, we rarely use frequencies outside this range.

The work of breathing increases in four different situations:

1. *When pulmonary compliance is decreased*, such as with pulmonary fibrosis, more work is required to expand the lungs.

2. *When airway resistance is increased*, such as with COPD or asthma (especially during acute exacerbations), more work is required to achieve the greater pressures necessary to overcome increased flow resistance.

3. *When elastic recoil is decreased*, as with emphysema, passive expiration may be inadequate to expel the volume

of air normally exhaled during quiet breathing. Thus, the expiratory muscles must work to aid in emptying the lungs, even at rest.

4. *When there is a need for increased ventilation*, such as during exercise, more work is required to generate larger tidal volumes and faster breathing.

Strenuous exercise can increase the energy cost of respiration twenty-five-fold, with the respiratory muscles accounting for 10–15 percent of total oxygen consumption, and even more in elite athletes. The oxygen consumption of the respiratory muscles during intense exercise may be so great that insufficient oxygen may be available to the exercising limb muscles. Rarely, however, in healthy individuals do the demands of the respiratory muscles for oxygen become so great that they compromise blood flow and, therefore, oxygen delivery to the exercising limb muscles (a phenomenon known as *steal*). But this situation can be exacerbated in hot and/or humid environments, when the body must direct some blood flow (cardiac output) to the skin to reduce core temperature. If there is competition by different muscle groups (respiratory, locomotive) for oxygenated blood, this will accelerate the onset of fatigue and either shorten exercise time or reduce exercise intensity.

In persons with poorly compliant lungs or chronic airflow limitation, the work of breathing—even at rest—may increase to 30 percent of the total energy expenditure. During moderate-intensity exercise, the respiratory muscles of patients with chronic airflow limitations may require 35–40 percent of the body's oxygen consumption, severely limiting the ability of these individuals to exercise. Most recent work indicates that exercise training, including training of the respiratory muscles, can help such patients improve their quality of life. It is uncertain whether respiratory muscle training can, however, improve exercise performance in healthy individuals in whom exercise is limited by the ability of the cardiovascular system to deliver oxygen to working muscles.

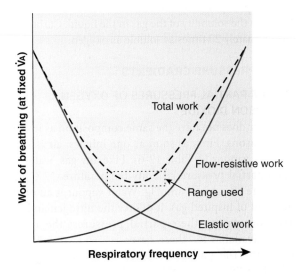

❯ **FIGURE 12-25 Work of breathing.** For a given alveolar ventilation, an increase in respiratory frequency requires a concomitant decrease in tidal volume. Therefore, as respiratory frequency increases, the flow resistive component of the work of breathing increases, but the elastic work of breathing (reflecting the tidal volume) must decrease. The total work of breathing (dashed line) is the sum of the elastic and flow-resistive components. People breathe in a frequency range—a wide trough—that minimizes the work of breathing. Behavioural aspects of breathing easily override this factor, in part because the work of breathing at rest constitutes only a few percent of the total energy expenditure.

### Check Your Understanding 12.7

1. Explain the best way to increase alveolar ventilation during exercise. What limitations apply to your explanation?

## 12.3 | Gas Exchange

The ultimate purpose of alveolar ventilation is to provide oxygen to the alveoli—where it is taken up by the blood—and to remove the carbon dioxide from that blood and excrete it to the atmosphere. Blood acts as a transport system for oxygen and carbon dioxide between the lungs and tissues, with the cells extracting oxygen from the blood and eliminating carbon dioxide into it.

Even those of us with normal lung function may have abnormal levels of oxygen and carbon dioxide in the blood. This can be explained by impaired diffusion of gases at the

**12**

alveolar–capillary membrane or poor matching of alveolar ventilation to perfusion. Again, an understanding of normal physiology provides insights into abnormal physiology, and vice versa.

Gas exchange involves diffusion of oxygen from the alveoli to the blood in the pulmonary capillaries and of carbon dioxide in the reverse direction. The factors determining diffusion are the *same* as those involved in convective flow: a pressure gradient ($\Delta P$), and a resistance ($R$) related to the physical properties of the gas and the structures through which it diffuses. For diffusion, the pressure gradient is partial pressure, not hydraulic pressure.

## Diffusion at the alveolar–capillary membrane

How much oxygen ($\dot{V}_{O_2}$) or carbon dioxide ($\dot{V}_{CO_2}$) diffuses across the alveolar–capillary membrane depends on two factors: (1) the partial pressure gradient of the gas across the membrane, and (2) the resistance to diffusion of the gas across the same membrane. For oxygen, the high end of the partial pressure gradient is set by its partial pressure in the alveoli ($P_{A_{O_2}}$), and the low end is set by its partial pressure in the pulmonary artery, which contains mixed blood ($P_{\bar{V}_{O_2}}$), because it comes from all systemic tissues. For carbon dioxide, the high end of the gradient is its partial pressure in the pulmonary arterial (mixed venous) blood ($P_{\bar{V}_{CO_2}}$), and its low end is its partial pressure in the alveoli ($P_{A_{CO_2}}$). The resistance to diffusion, in turn, depends on three factors: (1) the surface area ($A$) of the membrane across which diffusion occurs, (2) its thickness ($T$), and (3) the diffusibility ($D$) of the gas (reflecting its solubility and molecular weight). Because diffusibility is a constant, it can be ignored. The equations (see also Appendix H) are as follows:

$$\dot{V}_{O_2} = \left( P_{A_{O_2}} - P_{\bar{V}_{O_2}} \right) \cdot A/T$$

and

$$\dot{V}_{CO_2} = \left( P_{A_{CO_2}} - P_{\bar{V}_{CO_2}} \right) \cdot A/T$$

In summary, assuming the time available for diffusion is adequate, the *only* factors that can affect diffusion of either gas are *changes in their partial pressure gradients*, and the *surface area* and *diffusion distance* across which diffusion occurs. In the following sections, we explore how these factors affect diffusion.

## Partial pressure gradients

Gas exchange at both pulmonary and tissue capillary levels involves simple passive diffusion of oxygen and carbon dioxide down **partial pressure gradients**. No active transport mechanisms exist (or are needed) for these gases. What determines these gradients?

### PARTIAL PRESSURES

Atmospheric air is a mixture of gases; typical dry air contains about 79 percent nitrogen, 21 percent oxygen, and almost negligible percentages of carbon dioxide, water vapour, other gases, and pollutants. Together, these gases exert a total atmospheric pressure of 760 mmHg at sea level. This pressure is equal to the sum of the pressures contributed by each gas. Every gas molecule,

| | Dry Air | Inspired Gas | Alveolar Gas |
|---|---|---|---|
| $P_{N_2}$ (mmHg) | 600 | 563 | 573 |
| $P_{O_2}$ (mmHg) | 160 | 150 | 100 |
| $P_{H_2O}$ (mmHg) | 0 | 47 | 47 |
| $P_{CO_2}$ (mmHg) | 0.23 | 0.23 | 40 |
| Total (mmHg) | 760 | 760 | 760 |

no matter its size, exerts the same pressure; for example, a molecule of nitrogen exerts the same pressure as a molecule of oxygen. The pressure exerted by a gas is directly proportional to the percentage of that gas in the mixture (**Dalton's law**), and is therefore referred to as its **partial pressure**. Because 79 percent of the air consists of nitrogen, it is responsible for 79 percent of 760 mmHg, or 600 mmHg. Similarly, because oxygen constitutes 21 percent of the air, its partial pressure is 160 mmHg (■ Table 12-6). Thus, at sea level, the **partial pressure of oxygen** ($P_{O_2}$) in atmospheric air is normally 160 mmHg, and the **partial pressure of carbon dioxide** ($P_{CO_2}$) is a negligible 0.23 mmHg.

Gases dissolved in a liquid, such as blood or another bodily fluid, also exert a partial pressure. The greater its partial pressure, the more gas is dissolved. How much is dissolved also depends on the solubility of the gas in the liquid. Carbon dioxide is approximately 20 times as soluble as oxygen.

### PARTIAL PRESSURE GRADIENTS

### ALVEOLAR PARTIAL PRESSURES OF OXYGEN AND CARBON DIOXIDE

Alveolar air does not have the same composition as inspired air for two reasons. First, as soon as one inhales, air is saturated with water vapour (■ Table 12-6). Like any gas, water vapour exerts a partial pressure; at body temperature, 37°C, its partial pressure ($P_{H_2O}$) is 47 mmHg. Water vapour is an obligatory component of inspired gas. It dilutes the nitrogen and oxygen, thereby decreasing their partial pressures; the remaining pressure available for these gases is $760 - 47 = 713$ mmHg. This means the partial pressure of nitrogen in inspired gas ($P_{N_2}$) is $0.79 \times 713$ mmHg = 563 mmHg and that of oxygen ($P_{I_{O_2}}$) is $0.21 \times 713$ mmHg = 150 mmHg. Therefore, as a result of humidifying inspired air, $P_{O_2}$ drops by 10 mmHg, from 160 to 150 mmHg.

The alveolar partial pressure of oxygen ($P_{A_{O_2}}$) is also lower than atmospheric $P_{O_2}$. This is because inspired air mixes with the large volume of unexpelled air remaining in the lungs, including the dead space, at the end of the preceding expiration (the functional residual capacity). At the end of an inspiration (at resting ventilation), less than 15 percent of the air in the alveoli is fresh air. As a result, there is a further drop in $P_{O_2}$ such that $P_{A_{O_2}}$ is 100 mmHg. This can be calculated using the **alveolar gas equation** (simplified here):

$$P_{A_{O_2}} = P_{I_{O_2}} - P_{A_{CO_2}}/R$$

where $R$ is the **respiratory quotient**, that is, the ratio of metabolic carbon dioxide production to oxygen consumption ($\dot{V}_{CO_2}/\dot{V}_{O_2}$). At rest in a healthy individual with a typical Western diet, this ratio is 0.8. The effect of a ratio less than 1 is to magnify the effect of the carbon dioxide, which replaces some of the oxygen. Dividing a $P_{A_{CO_2}}$ of 40 mmHg by 0.8 gives 50 mmHg; subtracting this from the $P_{I_{O_2}}$ of 150 mmHg results in a $P_{A_{O_2}}$ of 100 mmHg (▮ Table 12-6).

You might think that the $P_{O_2}$ in the alveoli would increase during inspiration with the arrival of fresh air and then decrease during expiration. However, only small fluctuations occur, for two reasons. First, only a small proportion of the alveolar air is exchanged with each breath. The relatively small volume of inspired, high $P_{O_2}$ air mixes quickly with the much larger volume of retained alveolar air, which has a lower $P_{O_2}$. Thus, the oxygen in the inspired air only slightly elevates the alveolar $P_{O_2}$. Second, even this potentially small elevation of $P_{O_2}$ is diminished because of the continuous diffusion of oxygen down its partial pressure gradient from the alveoli into the blood. The oxygen arriving in the alveoli in the newly inspired air simply replaces that diffusing out of the alveoli into the pulmonary capillaries. Therefore, alveolar $P_{O_2}$ remains relatively constant at about 100 mmHg throughout the respiratory cycle. This can be explained as follows: the arterial $P_{O_2}$ —that is, the $P_{O_2}$ in the blood leaving the lung—equilibrates with the alveolar $P_{O_2}$, so it remains fairly constant at this same value. Accordingly, the arterial $P_{O_2}$ —that is, the $P_{O_2}$ responsible for the diffusion of oxygen into the systemic tissues— varies only slightly during the respiratory cycle.

A similar situation exists in reverse for carbon dioxide. Produced by the tissues as a metabolic waste product, it is constantly added to the blood in the systemic capillaries. In the pulmonary capillaries, carbon dioxide diffuses down its partial pressure gradient from the blood into the alveoli and is subsequently excreted during expiration. Like the alveolar $P_{O_2}$, the alveolar $P_{CO_2}$ remains fairly constant throughout the respiratory cycle but at a lower value of 40 mmHg.

### Check Your Understanding 12.8

1. Describe the changes in the composition and partial pressures of nitrogen, oxygen, carbon dioxide, and water vapour as air moves from the atmosphere to the alveoli.

## PARTIAL PRESSURE GRADIENTS OF OXYGEN AND CARBON DIOXIDE ACROSS THE PULMONARY CAPILLARIES

As blood passes through the lungs, it acquires oxygen and gives up carbon dioxide, due to diffusion down partial pressure gradients. Ventilation constantly replenishes alveolar oxygen and removes carbon dioxide, thereby maintaining the appropriate partial pressure gradients. The blood entering the pulmonary capillaries via the pulmonary arteries is systemic venous blood; it therefore has a relatively low partial pressure of oxygen (the mixed venous $P_{O_2}$, $P_{\bar{V}_{O_2}}$) of 40 mmHg, and a relatively high mixed venous partial pressure of carbon dioxide ($P_{\bar{V}_{CO_2}}$) of 46 mmHg. As this blood flows through the pulmonary capillaries, it is exposed to alveolar air (⟩ Figure 12-26). Because the alveolar $P_{O_2}$ (100 mmHg) is greater than the mixed venous $P_{O_2}$ (40 mmHg), oxygen diffuses down an initial 60 mmHg partial pressure gradient from the alveoli into the blood until no further gradient exists.

The partial pressure gradient for carbon dioxide is in the opposite direction. The mixed venous $P_{CO_2}$ of 46 mmHg is 6 mmHg greater than the alveolar $P_{CO_2}$. Carbon dioxide therefore diffuses from the blood into the alveoli until blood $P_{CO_2}$ equilibrates with alveolar $P_{CO_2}$ at 40 mmHg.

After leaving the lungs, the arterialized blood has a $P_{O_2}$ of 100 mmHg and a $P_{CO_2}$ of 40 mmHg, and these are the values of the blood pumped to the systemic tissues by the left heart (⟩ Figure 12-26).

Note that blood returning to the heart from the tissues still contains oxygen (mixed venous $P_{O_2}$ = 40 mmHg), and blood leaving the lungs still contains carbon dioxide (arterial $P_{CO_2}$ = 40 mmHg).

The mixed venous $P_{O_2}$ of 40 mmHg can be expressed as mixed venous oxygen content, which is 150 mL $O_2$/L of blood. Because arterial blood contains approximately 200 mL $O_2$/L, this means that at rest only 50 of the 200 mL $O_2$/L, or 25 percent, of the oxygen delivered to the tissues has been used. The remaining 150 mL $O_2$/L represents an immediately available reserve that can be used by tissues whenever their oxygen demands increase. The carbon dioxide remaining in the arterial blood plays an important role in acid–base balance because it generates carbonic acid (see Chapter 14). Furthermore, arterial $P_{CO_2}$ is important in stimulating respiration (discussed in Section 12.5).

The amount of oxygen taken up in the lungs matches the amount extracted and used by the tissues in steady-state conditions. (The two can differ, transiently, during sudden changes in metabolism, as occurs at the onset or cessation of exercise.) When tissues are more metabolically active, as in exercise, they extract more oxygen from the blood, reducing the mixed venous $P_{O_2}$ to, for example, 30 mmHg. When this blood returns to the lungs, the gradient becomes 70 (100 – 30) mmHg, rather than the 60 mmHg at rest. By lowering the mixed venous $P_{O_2}$, the gradient for diffusion is increased. Note that while this increases the amount of oxygen transferred—because the mixed venous blood contains less oxygen—it does not increase arterial oxygen content much because the arterial oxygen content was already at or close to the maximum of 200 mL $O_2$/L. (Arterial oxygen content can be increased if the concentration of haemoglobin in the blood is higher, typically by augmenting the production of red blood cells with erythropoietin; p. 438.) The larger diffusion gradient across the alveolar–capillary membrane helps ensure that diffusion is more rapid and that blood $P_{O_2}$ equilibrates with alveolar $P_{O_2}$. This occurs despite there being less time available for diffusion, because blood flows more quickly through the pulmonary capillaries during exercise.

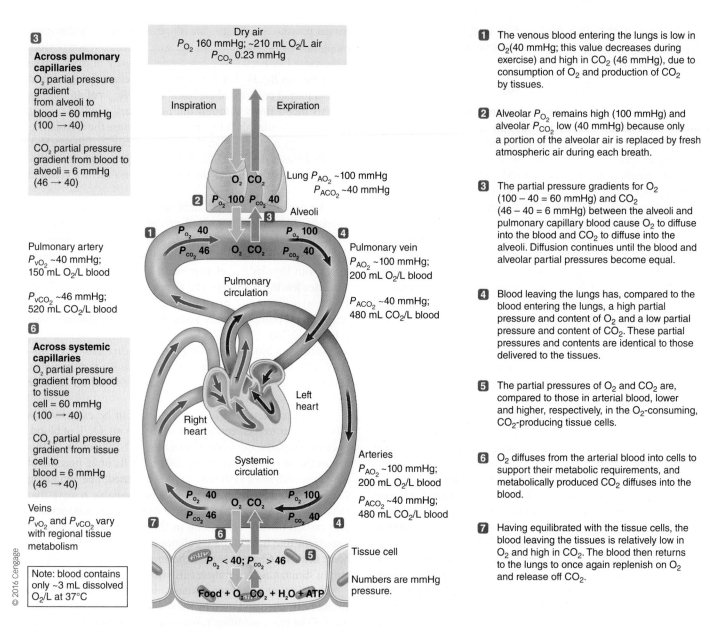

**3**

**Across pulmonary capillaries**
$O_2$ partial pressure gradient from alveoli to blood = 60 mmHg ($100 \rightarrow 40$)

$CO_2$ partial pressure gradient from blood to alveoli = 6 mmHg ($46 \rightarrow 40$)

Pulmonary artery
$P_{vO_2}$ ~40 mmHg;
150 mL $O_2$/L blood

$P_{vCO_2}$ ~46 mmHg;
520 mL $CO_2$/L blood

**6**

**Across systemic capillaries**
$O_2$ partial pressure gradient from blood to tissue cell = 60 mmHg ($100 \rightarrow 40$)

$CO_2$ partial pressure gradient from tissue cell to blood = 6 mmHg ($46 \rightarrow 40$)

Veins
$P_{vO_2}$ and $P_{vCO_2}$ vary with regional tissue metabolism

Note: blood contains only ~3 mL dissolved $O_2$/L at 37°C

© 2016 Cengage

Dry air
$P_{O_2}$ 160 mmHg; ~210 mL $O_2$/L air
$P_{CO_2}$ 0.23 mmHg

Inspiration    Expiration

$O_2$  $CO_2$    Lung $P_{AO_2}$ ~100 mmHg
$P_{ACO_2}$ ~40 mmHg

**2** $P_{O_2}$ 100 $P_{CO_2}$ 40    Alveoli
**3**

**1** $P_{O_2}$ 40    $P_{O_2}$ 100 **4**
$P_{CO_2}$ 46  $O_2$ $CO_2$  $P_{CO_2}$ 40    Pulmonary vein
$P_{AO_2}$ ~100 mmHg;
Pulmonary circulation    200 mL $O_2$/L blood

$P_{ACO_2}$ ~40 mmHg;
480 mL $CO_2$/L blood

Left heart

Right heart

Arteries
Systemic circulation    $P_{AO_2}$ ~100 mmHg;
200 mL $O_2$/L blood

**7** $P_{O_2}$ 40    $P_{O_2}$ 100 **4**
$P_{CO_2}$ 46  $O_2$ $CO_2$  $P_{CO_2}$ 40    $P_{ACO_2}$ ~40 mmHg;
480 mL $CO_2$/L blood
**6**

Tissue cell

$P_{O_2}$ < 40; $P_{CO_2}$ > 46 **5**

Food + $O_2$ $CO_2$ + $H_2O$ + ATP

Numbers are mmHg pressure.

**1** The venous blood entering the lungs is low in $O_2$(40 mmHg; this value decreases during exercise) and high in $CO_2$ (46 mmHg), due to consumption of $O_2$ and production of $CO_2$ by tissues.

**2** Alveolar $P_{O_2}$ remains high (100 mmHg) and alveolar $P_{CO_2}$ low (40 mmHg) because only a portion of the alveolar air is replaced by fresh atmospheric air during each breath.

**3** The partial pressure gradients for $O_2$ (100 – 40 = 60 mmHg) and $CO_2$ (46 – 40 = 6 mmHg) between the alveoli and pulmonary capillary blood cause $O_2$ to diffuse into the blood and $CO_2$ to diffuse into the alveoli. Diffusion continues until the blood and alveolar partial pressures become equal.

**4** Blood leaving the lungs has, compared to the blood entering the lungs, a high partial pressure and content of $O_2$ and a low partial pressure and content of $CO_2$. These partial pressures and contents are identical to those delivered to the tissues.

**5** The partial pressures of $O_2$ and $CO_2$ are, compared to those in arterial blood, lower and higher, respectively, in the $O_2$-consuming, $CO_2$-producing tissue cells.

**6** $O_2$ diffuses from the arterial blood into cells to support their metabolic requirements, and metabolically produced $CO_2$ diffuses into the blood.

**7** Having equilibrated with the tissue cells, the blood leaving the tissues is relatively low in $O_2$ and high in $CO_2$. The blood then returns to the lungs to once again replenish on $O_2$ and release off $CO_2$.

> **FIGURE 12-26** Oxygen and carbon dioxide exchange across pulmonary and systemic capillaries caused by partial pressure gradients

**12**

The other major factors influencing the high end of the gradient for diffusion of oxygen are the relative levels of ventilation (V) and perfusion (Q) to lung regions. (See the discussion of alveolar dead space, p. 529.) In regions with a high ratio of alveolar ventilation to perfusion (i.e., high $\dot{V}_A/\dot{Q}$), the alveolar $P_{O_2}$ will be higher, increasing the gradient for diffusion; in regions with a low $\dot{V}_A/\dot{Q}$ ratio, the alveolar $P_{O_2}$ will be lower, reducing the gradient.

In contrast to the 60 mmHg gradient favouring diffusion of oxygen, the gradient for the diffusion of carbon dioxide from mixed venous blood into the alveoli is only 6 mmHg (46 – 40). Carbon dioxide does not need as large a partial pressure gradient for diffusion as that for oxygen because it is approximately 20 times more soluble and, therefore, more diffusible than oxygen (see equation, p. 524). During exercise, as more carbon dioxide is released into the venous blood, its partial pressure in venous blood (the mixed venous $P_{CO_2}$) increases, and this increases the gradient for its diffusion from the blood into the alveoli for excretion to the atmosphere.

**■ Clinical Connections** Even though asthma primarily affects the bronchioles, it can also affect gas exchange at the alveoli. Normally ventilation and perfusion are matched so that adequate gas exchange can occur. However, in an obstructive disease such as asthma, inhaled air follows the path of least resistance, meaning that alveoli behind obstructed bronchioles will receive less ventilation. Consequently, there is a mismatch of ventilation and perfusion, such that the poorly ventilated areas of the lung prevent $O_2$ saturation of the blood returning to the systemic circulation. The result is hypoxia.

■ **TABLE 12-7** Factors That Influence the Rate of Gas Transfer across the Alveolar–Capillary Membrane

| Factor | Influence on Diffusion across the Alveolar–Capillary Membrane | Comments |
|---|---|---|
| **Partial pressure gradients of oxygen and carbon dioxide** | Diffusion ↑ as partial pressure gradient ↑ | Major determinant of the rate of transfer |
| **Surface area of the alveolar–capillary membrane** | Diffusion ↑ as surface area ↑ | Surface area is constant under resting conditions. Surface area ↑ during exercise as more pulmonary capillaries are recruited when cardiac output increases and the alveoli expand more as tidal volumes increase. Surface area ↓ in such pathological conditions as emphysema and lung collapse. |
| **Thickness of the barrier separating the air and blood across the alveolar–capillary membrane** | Diffusion ↓ as thickness ↑ | Thickness (distance over which diffusion occurs) normally remains constant. Thickness ↑ in such pathological conditions as pulmonary oedema, pulmonary fibrosis, and pneumonia. But it ↓ in exercise as previously closed pulmonary capillaries are recruited (opened) by the increase in pulmonary arterial pressure. |
| **Diffusion coefficient (solubility of the gas in the membrane)** | Diffusion ↑ as diffusion coefficient ↑. For a given gas, however, the diffusion coefficient is constant. | Diffusion coefficient for carbon dioxide is 20 times that for oxygen, offsetting the smaller partial pressure gradient for carbon dioxide; therefore, approximately equal amounts of carbon dioxide and oxygen diffuse across the membrane. |

## Other factors

According to **Fick's law of diffusion**, the rate of diffusion of a gas also depends on the surface area and thickness of the membrane through which the gas is diffusing and on the diffusion coefficient of the particular gas (▌ Table 12-7). In healthy individuals at rest, these other factors are relatively constant. But there are circumstances when they change, affecting the rate of gas transfer in the lungs.

### EFFECT OF SURFACE AREA AND MEMBRANE THICKNESS ON GAS EXCHANGE

At rest, some pulmonary capillaries, especially those at the apex (top) of the lung, are closed because the low blood pressure in the pulmonary circulation cannot keep them open. During exercise, however, cardiac output increases and this increases pulmonary arterial pressure. This increased pressure opens (recruits) many of the previously closed pulmonary capillaries, thereby increasing the surface area ($A$) available for gas exchange. Moreover, recruiting previously closed pulmonary capillaries now means that a given molecule of oxygen may not have to diffuse as far to reach the blood, effectively reducing diffusion distance ($T$). In addition, alveolar membranes are stretched more during exercise because of the larger tidal volumes, thereby increasing the alveolar surface area and decreasing the thickness ($T$) of the membrane. Collectively, these changes improve gas exchange during exercise.

This recruitment of pulmonary capillaries to enhance gas exchange is very important during exercise. In the pulmonary capillaries, the equilibration of $P_{O_2}$ and $P_{CO_2}$ in the blood with those in the alveoli is dependent not only on the pressure gradients and the surface area (p. 524) but also on the duration of exposure of the capillary blood to alveolar gas. This is called the **capillary transit time** (⟩ Figure 12-27). At rest, blood remains in the pulmonary capillaries for approximately 0.75 seconds; hence, the capillary transit time is 0.75 seconds. This is approximately triple the time needed for equilibration of $P_{O_2}$ and much more than that needed for equilibration of $P_{CO_2}$. During maximal aerobic exercise, however, the capillary transit time is reduced to about 0.4 seconds. This is still enough time to allow equilibration of $P_{O_2}$ and $P_{CO_2}$ because the partial pressure gradients are maintained or even increased and because alveolar surface area increases. Indeed, equilibration of the two gases may occur in as little as 0.25 seconds—and elite endurance athletes may push this limit of pulmonary diffusion even further.

The capillary transit time is a major reason why exercise at altitude, especially very high altitude, is so difficult. The lower atmospheric pressure reduces the alveolar $P_{O_2}$, which is the high end of the gradient for diffusion. Although the low end of the gradient—the mixed venous $P_{O_2}$—also decreases due to a combination of exercise and altitude, this is not enough to compensate. During extreme exercise, the arterial $P_{O_2}$ will not equilibrate with the alveolar $P_{O_2}$. At very high altitudes, such as near (8400 m) the peak of Everest (8848 m), the arterial $P_{O_2}$ of four climbers averaged 25 mmHg, and arterial $P_{CO_2}$ averaged 13 mmHg; arterial oxygen content was about 146 mL $O_2$/L (Grocott et al. [2009]. *New England Journal of Medicine, 360*: 140–9). As we shall see, this arterial oxygen content is much higher than predicted by the oxygen dissociation curve (see ⟩ Figure 12-32), and the reasons why are explored in Section 11.4.

12

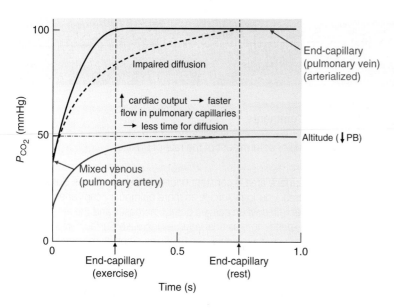

> **FIGURE 12-27 Dynamics of diffusion of oxygen across the alveolar–capillary membrane.** At sea level (black line), diffusion of oxygen is complete (i.e., the blood in the pulmonary capillary attains a partial pressure of oxygen equivalent to that in the alveolus) within approximately 0.25 seconds, leaving a "safety margin" of about 0.5 seconds. If diffusion is impaired (dashed line), equilibration still occurs before the red blood cells reach the end of the capillary, but only at rest. If the time available for diffusion decreases, for example, due to an increase in cardiac output, the person with normal diffusion is unaffected (equilibration still occurs), but the person with diffusion impairment will experience a drop in arterial $P_{O_2}$ (desaturation). At altitude (red line), the driving pressure for diffusion—the alveolar $P_{O_2}$—is reduced because of the fall in barometric pressure. This reduces the rate of oxygen diffusion, but in the healthy person equilibration is still complete (albeit at a lower alveolar $P_{O_2}$). In the patient with impaired diffusion, however, equilibration does not occur even at rest and is severely compromised during exercise.

Several pathological conditions can markedly reduce pulmonary surface area and, in turn, decrease diffusion. Most notably, in *emphysema*, surface area is reduced because many alveolar walls, with their capillaries, are lost, resulting in larger but fewer chambers (> Figure 12-28). Loss of surface area is likewise associated with collapsed regions of the lung (**atelectasis**) and also results when part of the lung is surgically removed—for example, in treating lung cancer. Compared to carbon dioxide, oxygen has lower solubility and, therefore, diffusibility (see p. 529), so reductions in pulmonary surface area decrease the diffusion of oxygen more than that of carbon dioxide.

Inadequate gas exchange can also occur when the thickness of the barrier separating the air and blood is pathologically increased. This decreases the rate of gas transfer because a gas takes longer to diffuse. Thickness increases in (1) *pulmonary oedema*, when excess interstitial fluid accumulates between the alveoli and pulmonary capillaries due to pulmonary inflammation or left-sided congestive heart failure (p. 370); (2) *pulmonary fibrosis*, which involves replacement of delicate lung tissue with thick, fibrous tissue in response to certain chronic irritants; and (3) *pneumonia*, which is characterized by inflammation-induced accumulation of fluid within or around the alveoli. Most commonly, pneumonia is due to bacterial or viral infection of the lungs, but it may also arise from accidental *aspiration* (breathing in) of food, vomitus, or chemical agents.

In a diseased lung in which diffusion is impeded because of a reduction in surface area or a thickening of the alveolar–capillary membrane, oxygen transfer is usually more seriously impaired than transfer of carbon dioxide, because of oxygen's lower diffusibility (reduced solubility). By the time the blood reaches the end of a pulmonary capillary, $P_{CO_2}$ is more likely than $P_{O_2}$ to have equilibrated with its counterpart in the alveolar compartment because carbon dioxide diffuses more rapidly. In less stressful conditions, diffusion of both gases may be adequate at rest, but during exercise, when pulmonary transit time is decreased, the $P_{O_2}$ at the end of the pulmonary capillary may be less than the $P_{O_2}$ in the alveolus, indicating failure of equilibration.

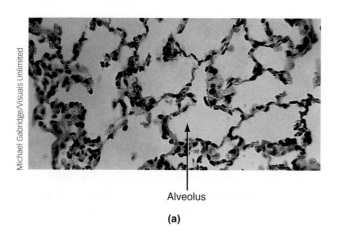

Alveolus

(a)

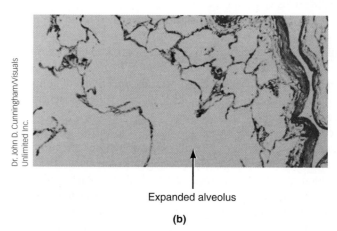

Expanded alveolus

(b)

> **FIGURE 12-28 Comparison of normal and emphysematous lung tissue.** (a) Photomicrograph of lung tissue from a healthy individual. Each of the smallest clear spaces is an alveolar lumen. (b) Photomicrograph of lung tissue from a patient with emphysema. Note the loss of alveolar walls in the emphysematous lung tissue, resulting in larger but fewer alveoli. Alveolar capillaries are also lost.

## EFFECT OF DIFFUSION COEFFICIENT ON GAS EXCHANGE

The rate of gas transfer is directly proportional to the diffusion coefficient ($D$), a constant directly related to the solubility of a particular gas in the lung tissues (mostly water) and inversely proportional to its molecular weight. The diffusion coefficient for carbon dioxide is 20 times that of oxygen because carbon dioxide is much more soluble in tissues. The rate of diffusion of carbon dioxide across the respiratory membranes is therefore approximately 20 times more rapid than that of oxygen per mmHg. The lower diffusion coefficient of oxygen is partially offset by its greater partial pressure gradient (alveolar − mixed venous = 100 − 40 = 60 mmHg), compared to that for carbon dioxide (mixed venous − alveolar = 46 − 40 = 6 mmHg).

---

### Check Your Understanding 12.9

1. List and explain the importance of the factors that determine the diffusion of oxygen and carbon dioxide across the alveolar–capillary membrane.

---

### ALVEOLAR DEAD SPACE

Not all alveoli participate equally well in gas exchange. We have assumed that all the atmospheric air entering the alveoli participates in the exchange of oxygen and carbon dioxide with pulmonary blood. However, gas exchange can only occur when gas "meets" blood. Ideally, ventilation should match perfusion and vice versa. Because, however, not all alveoli are ventilated and perfused equally, gas exchange is less than perfect. At one extreme, some alveoli (or lung regions) are perfused but not ventilated; they are called **shunts**. At the other extreme, some lung regions are ventilated but not perfused; their alveoli cannot participate in gas exchange and are referred to as **alveolar dead space**. In healthy people, alveolar dead space is small and of little importance, but it can increase to lethal levels in several types of pulmonary disease. The combination of anatomic and alveolar dead space is referred to as the **physiological dead space**.

Let's now examine why alveolar and, therefore, physiological dead space, are minimal in healthy individuals.

## Regional control of ventilation and perfusion

In our discussion of the role of airway resistance during inspiration and expiration, we referred to the overall resistance of all the airways. However, the resistance of individual airways supplying a specific region can be adjusted independently in response to regional changes in $P_{CO_2}$. Similarly, the resistance of individual blood vessels supplying a region is influenced by the regional $P_{O_2}$. This situation is analogous to the control of systemic arterioles. Recall that overall systemic arteriolar resistance (i.e., total peripheral resistance) is an important determinant of the blood flow through the systemic circulatory system (p. 401). Yet the radii of individual arterioles supplying various tissues can be adjusted locally to match the tissues' differing metabolic needs (p. 398). In the lung, this regional control can be modelled using a simple analogy: a sink.

### EFFECT OF OXYGEN

The oxygen delivered by ventilation to a lung region is analogous to the water flowing into a sink (❯ Figure 12-29). When you open the faucet (increasing ventilation), delivering more water (oxygen), the water level (the $P_{O_2}$) rises in the sink (lung region); when you close the faucet, the water level (and $P_{O_2}$) falls. How fast the oxygen (water) is removed (emptied) from the lung region depends on its perfusion; this is modelled by the degree of opening of the sink's drain. Unlike a sink's drain, however, the lung region has a built-in control to regulate the opening of the drain. If ventilation to the lung region is inadequate, the regional $P_{O_2}$ falls. This low $P_{O_2}$ (hypoxia) causes constriction of the vascular smooth muscle of the blood vessels (vasoconstriction) supplying that lung region. This is analogous to partially closing the sink's drain. As a result, the regional $P_{O_2}$ rises. And because all of the blood ejected from the right ventricle must go to the lungs, this means that the perfusion to the hypoxic lung region has been diverted to other, better ventilated (nonhypoxic) regions.

Does the reverse happen? In other words, does an increase in regional ventilation and the resulting increase in regional $P_{O_2}$ cause vasodilation of the blood vessels supplying that region? The evidence for this effect is weaker. As already discussed, the alveolar $P_{O_2}$ in a normal person at sea level is approximately 100 mmHg. The only way to exceed this value is to increase

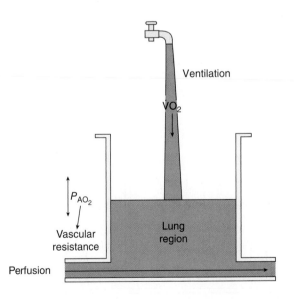

❯ **FIGURE 12-29 Regional alveolar gas pressures: Oxygen and hypoxic vasoconstriction.** The level of water in a sink is determined by how fast water is delivered to the sink through the faucet and how fast it is taken away through the drain. Similarly, the level (partial pressure) of oxygen in a given lung region—the regional alveolar $P_{O_2}$ —is determined by how much oxygen is delivered by the ventilation to that region and how much oxygen is taken up by the blood flowing to that region. Unlike a sink, the lung region can regulate its perfusion. If regional alveolar $P_{O_2}$ falls, the resulting hypoxia causes vasoconstriction to that region, thereby diverting blood to other, presumably better ventilated (nonhypoxic), regions.

ventilation disproportionate to metabolic demands (i.e., hyperventilation) and thereby decrease alveolar (and arterial) $P_{CO_2}$. According to the alveolar gas equation (p. 525), this will increase the alveolar $P_{O_2}$. This effect, however, is modest, perhaps 20 mmHg at the most. The other way to increase alveolar $P_{O_2}$ is artificial: inhalation of supplemental oxygen. This, however, is not a circumstance that would drive the evolution of a mechanism to cause vasodilation in response to high alveolar $P_{O_2}$. Consequently, the response to higher levels of regional alveolar $P_{O_2}$ should really be considered a reduction of any prevailing vasoconstriction rather than an actual vasodilation.

### EFFECT OF CARBON DIOXIDE

The carbon dioxide delivered by perfusion (output of the right ventricle) to a lung region is also analogous to the water flowing into a sink ( Figure 12-30); how fast carbon dioxide (water) is removed depends on the ventilation (degree of opening of the drain). When you open the faucet—thereby increasing perfusion and therefore delivery of carbon dioxide, which reflects metabolic production of carbon dioxide $\dot{V}_{CO_2}$—the water level $\left(P_{CO_2}\right)$ rises in the sink (lung region); when you close the faucet, the water level $P_{A_{CO_2}}$ falls. This same sink (lung region) has a built-in control to regulate the opening of the drain. If perfusion to the lung region is low, the regional $P_{CO_2}$ falls. This low $P_{CO_2}$ (hypocapnia) constricts the smooth muscle of the airways (bronchoconstriction) supplying that lung region. As a result, ventilation of that region decreases, returning $P_{CO_2}$ toward its control level. Because all the ventilation goes to the lungs, any ventilation of

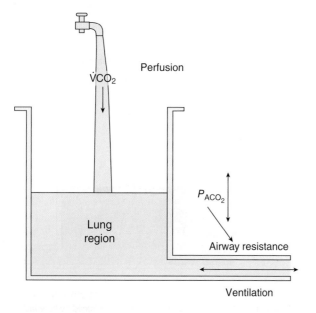

> **FIGURE 12-30 Regional alveolar gas pressures: Carbon dioxide and bronchoconstriction/bronchodilation.** The regional alveolar $P_{CO_2}$ is determined by how much carbon dioxide is delivered by the perfusion to that region and how much carbon dioxide is excreted via its ventilation. If regional alveolar $P_{CO_2}$ falls, the resulting hypocapnia causes bronchoconstriction to that region, thereby diverting ventilation to other lung regions with higher alveolar $P_{CO_2}$. Conversely, if regional alveolar $P_{CO_2}$ rises, the resulting hypercapnia causes bronchodilation in that region, thereby reducing flow resistance and increasing regional ventilation to eliminate more carbon dioxide.

an underperfused region is diverted to other better perfused regions. Conversely, an increase in perfusion to a region increases regional $P_{CO_2}$; this causes dilation of the regional airways, leading to increased regional ventilation and, as a result, a return of $P_{CO_2}$ back toward control levels.

The two mechanisms for matching airflow and blood flow function simultaneously, so normally very little air or blood is wasted in the lung ( Figure 12-31). In this way, local control mechanisms operate to match ventilation and perfusion in a particular region so as to optimize overall gas exchange.

### Check Your Understanding 12.10

1. Explain how changes in the partial pressures of oxygen and carbon dioxide in the lung operate to optimize the matching of regional alveolar ventilation to regional alveolar perfusion.

### GAS EXCHANGE ACROSS THE SYSTEMIC CAPILLARIES

Just as they do at the pulmonary capillaries, oxygen and carbon dioxide move between the systemic capillary blood and the tissue cells by simple diffusion down partial pressure gradients (see  Figure 12-26). The arterial blood that reaches the systemic capillaries is essentially the same blood that left the lungs in the pulmonary veins, because the only two places in the entire circulatory system at which gas exchange can take place are the pulmonary capillaries and the systemic capillaries. The arterial $P_{O_2}$ is 100 mmHg, and the arterial $P_{CO_2}$ is 40 mmHg, the same as the alveolar $P_{O_2}$ and $P_{CO_2}$.

### $P_{O_2}$ AND $P_{CO_2}$ GRADIENTS ACROSS THE SYSTEMIC CAPILLARIES

Cells constantly consume oxygen and produce carbon dioxide through oxidative metabolism. Cellular $P_{O_2}$ averages about 40 mmHg, and $P_{CO_2}$ about 46 mmHg; these values are highly variable depending on the tissue and the level of cellular metabolic activity. Oxygen moves by diffusion down its partial pressure gradient from the entering systemic capillary blood ($P_{O_2}$ = 100 mmHg) into the adjacent cells until equilibrium is reached. Therefore, the $P_{O_2}$ of venous blood leaving the systemic capillaries is equal to the tissue $P_{O_2}$, an average of 40 mmHg.

The reverse situation exists for carbon dioxide. Carbon dioxide rapidly diffuses out of the cells ($P_{CO_2}$ = 46 mmHg) into the entering capillary blood ($P_{CO_2}$ = 40 mmHg) and down the partial pressure gradient created by its ongoing production. Transfer of carbon dioxide continues until blood $P_{CO_2}$ equilibrates with tissue $P_{CO_2}$. Accordingly, blood leaving the systemic capillaries, pumped by the right ventricle into the pulmonary capillaries, has an average of 46 mmHg.

Actually, the partial pressures of the systemic blood gases never completely equilibrate with tissue $P_{O_2}$ and $P_{CO_2}$. Because the cells are constantly consuming oxygen and producing carbon dioxide, tissue $P_{O_2}$ is always slightly less than the $P_{O_2}$ of the blood leaving the systemic capillaries, and the tissue $P_{CO_2}$ is always slightly exceeds the systemic venous $P_{CO_2}$.

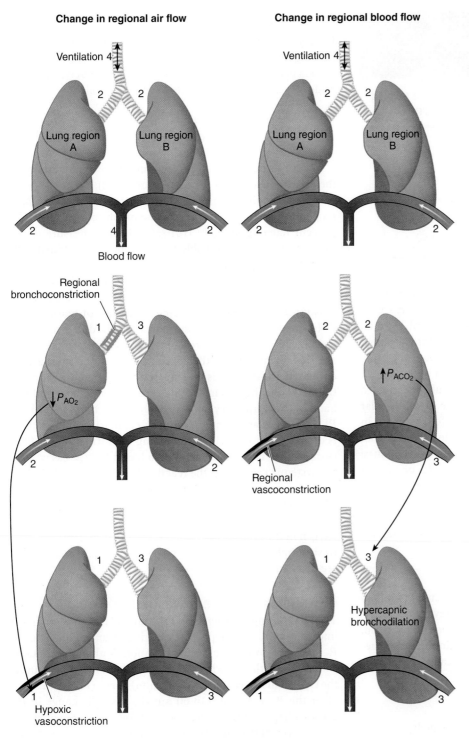

**Change in regional air flow**

Ventilation 4

2    2

Lung region A    Lung region B

2    4    2

Blood flow

Regional bronchoconstriction

1    3

↓$P_{AO_2}$

2    2

1    3

Hypoxic vasoconstriction

1    3

**Change in regional blood flow**

Ventilation 4

2    2

Lung region A    Lung region B

2    2

2    2

↑$P_{ACO_2}$

1    3

Regional vasoconstriction

1    3

Hypercapnic bronchodilation

1    3

> **FIGURE 12-31 Two-compartment model of the lung.** Top panel shows a perfect lung receiving 4 L/min of air and 4 L/min of blood, 2 L/min of air and blood to each of the two lung regions, A and B. The left middle panel shows a reduction of ventilation (V) to 1 L/min to region A; at constant total ventilation, ventilation to region B increases to 3 L/min. Thus, region A is underventilated and region B is overventilated. As a result (bottom left), there is regional vasoconstriction, reducing perfusion to region A. Because cardiac output does not change, this means that perfusion is diverted to the better ventilated region B. The right middle panel illustrates a reduction in perfusion to region A to 1 L/min; at constant cardiac output, perfusion to region B increases to 3 L/min. Thus, region A is underperfused, and region B is overperfused. As a result (bottom right), there is hypercapnia-induced bronchodilation of region B, thereby increasing ventilation to the better perfused region. Not shown here are hypocapnia-induced bronchoconstriction of region A on the left and region B on the right. In all cases, local controls help match ventilation ($\dot{V}_A$) to perfusion (Q), and vice versa, thereby preventing extremes of $\dot{V}_A$ / Q.

As a tissue's metabolism increases, its $P_{O_2}$ falls and its $P_{CO_2}$ rises. Because this increases the partial pressure gradients for diffusion, more oxygen diffuses from the blood into the tissue and more carbon dioxide diffuses out of the tissue before blood $P_{O_2}$ and $P_{CO_2}$ equilibrate with those in the tissue. Therefore, the amount of oxygen transferred to the cells and the amount of carbon dioxide carried away from the cells both depend on the rate of cellular metabolism.

### NET DIFFUSION OF OXYGEN AND CARBON DIOXIDE BETWEEN THE ALVEOLI AND TISSUES

Net diffusion of oxygen occurs first between the alveoli and blood and then between the blood and tissues, because of the partial pressure gradients of oxygen created by its continuous replenishment by alveolar ventilation and its continuous uptake by cells. Net diffusion of carbon dioxide occurs in the reverse direction, first between the tissues and blood and then between the blood and alveoli, because of the partial pressure gradients of carbon dioxide created by its continuous production in the cells and its continuous removal from the lung by alveolar ventilation (see › Figure 12-26).

We now examine how oxygen and carbon dioxide are transported in the blood between the alveoli and tissues.

## 12.4 | Gas Transport

Oxygen acquired by the blood in the lungs must be transported to the tissues. Conversely, carbon dioxide produced by cells must be transported to the lungs for elimination.

### Blood, oxygen, and haemoglobin

Oxygen is present in the blood in two forms: one physically dissolved and the other chemically bound to haemoglobin (see ▌Table 12-8).

### PHYSICALLY DISSOLVED OXYGEN

Very little oxygen dissolves in plasma water because oxygen is poorly soluble. At a normal arterial $P_{O_2}$ of 100 mmHg, only

**12**

## TABLE 12-8 Methods of Gas Transport in the Blood

| Gas | Method of Transport in Blood | Percentage Carried in This Form |
|---|---|---|
| **Oxygen** | Physically dissolved | 1.5 |
| | Bound to haemoglobin | 98.5 |
| **Carbon Dioxide** | Physically dissolved | 5–10 |
| | Bound to haemoglobin | 5–10 |
| | As bicarbonate ($HCO_3^-$) | 80–90 |

3 mL of oxygen can dissolve in 1 L of blood. At a resting cardiac output of 5 L/min, only 15 mL of $O_2$/min can dissolve. Under resting conditions, however, the body consumes 250 mL $O_2$/min. If oxygen were carried only in dissolved form, the cardiac output needed to sustain resting metabolic activity would therefore have to be 83.3 L/min (250 mL $O_2$/min divided by 3 mL $O_2$/L blood), and this assumes that all the oxygen would be extracted from the blood. It should be obvious that the cardiovascular system cannot accommodate a twenty-five-fold increase in consumption during strenuous exercise. To do so requires an additional mechanism for transporting oxygen to the tissues: haemoglobin (Hb). Haemoglobin transports 98.5 percent of the oxygen carried by the blood. Because *the oxygen bound to haemoglobin does not contribute to the $P_{O_2}$ of the blood*, the $P_{O_2}$ of the blood provides no information about its oxygen content. This is critical in understanding the difference between saturation and content.

### OXYGEN BOUND TO HAEMOGLOBIN

Haemoglobin, an iron-bearing protein molecule contained within the red blood cells, can form an easily reversible combination with oxygen (p. 436). When not combined with oxygen, haemoglobin is referred to as **reduced haemoglobin (deoxyhaemoglobin)**; when combined with oxygen, it is called **oxyhaemoglobin ($HbO_2$)**:

$$(Hb + O_2 \Leftrightarrow HbO_2)$$

Reduced haemoglobin          Oxyhaemoglobin

We need to answer several questions about the role of haemoglobin in oxygen transport. What determines whether oxygen and haemoglobin are combined or dissociated (separated)? Why does haemoglobin combine with oxygen in the lungs and release it at the tissues? How can a variable amount of oxygen be released at the tissues, depending on the level of tissue activity? How can we talk about oxygen transfer between blood and surrounding tissues in terms of the partial pressure gradients of oxygen when 98.5 percent of the oxygen is bound to haemoglobin and thus does not contribute to the $P_{O_2}$ of the blood?

## $P_{O_2}$ and haemoglobin saturation

Each of the four atoms of iron within the heme portions of a haemoglobin molecule can combine with an oxygen molecule, so each haemoglobin molecule can carry up to four oxygen molecules. As a result, the actual reactions are as follows:

$$Hb + O_2 \Leftrightarrow HbO_2 + O_2 \Leftrightarrow Hb(O_2)_2 + O_2 \Leftrightarrow Hb(O_2)_3 + O_2 \Leftrightarrow Hb(O_2)_4$$

These sequential reactions account for the sigmoid shape of the oxygen dissociation curve (> Figure 12-32).

Haemoglobin is considered fully saturated when all the haemoglobin present is carrying its maximum oxygen load. The **percent haemoglobin (% Hb) saturation**, a measure of the extent to which the available haemoglobin is combined with oxygen, can vary from 0 to 100 percent.

The most important factor determining the % Hb saturation is the $P_{O_2}$ of the blood, which is determined by how much oxygen is physically dissolved. According to the **law of mass action**, if the concentration of one substance involved in a reversible reaction is increased, the reaction is driven toward the opposite side. Conversely, if the concentration of one substance is decreased, the reaction is driven toward the side of that concentration. Applying this law to the reversible reaction involving haemoglobin and oxygen ($Hb + O_2 \Leftrightarrow HbO_2$), when blood $P_{O_2}$ increases, as in the pulmonary capillaries, the reaction is driven toward the right, increasing the formation of $HbO_2$ (increased % Hb saturation). When blood $P_{O_2}$ decreases, as in the systemic capillaries, the reaction is driven toward the left, and oxygen is released from haemoglobin as $HbO_2$ dissociates (decreased % Hb saturation). Thus, because of the difference in $P_{O_2}$ at the lungs and other tissues, haemoglobin acquires oxygen in the lungs, where ventilation is continually providing fresh supplies of oxygen, and releases it to the tissues, which are constantly consuming oxygen.

### OXYGEN–HAEMOGLOBIN DISSOCIATION CURVE

The relationship between blood $P_{O_2}$ and % Hb saturation is not linear, a feature of great physiological importance. Doubling the $P_{O_2}$ does not double the % Hb saturation. Rather, the relationship between these variables follows a sigmoid-shaped curve, the **$O_2$–Hb dissociation (or saturation) curve** (> Figure 12-32). Over the $P_{O_2}$ range of 0 to 60 mmHg, a small change in $P_{O_2}$ causes a large change in the % Hb saturation, as shown by the steep lower part of the curve. In contrast, at the upper end, between a blood $P_{O_2}$ of 60 and 100 mmHg, the curve flattens to a plateau. Over this 40 mmHg range, a rise in $P_{O_2}$ produces only a small increase in the % Hb saturation. Both portions of the curve, the lower steep part and the upper plateau, are physiologically significant.

### SIGNIFICANCE OF THE PLATEAU PORTION OF THE $O_2$–HB CURVE

The plateau is present over the blood $P_{O_2}$ range that exists in the pulmonary capillaries where haemoglobin acquires oxygen. The systemic arterial blood leaving the lungs, having equilibrated

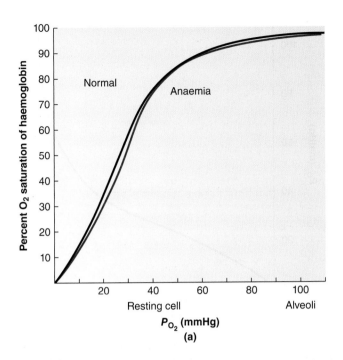

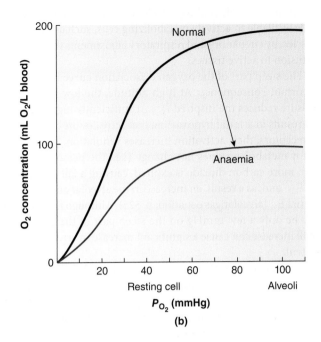

> FIGURE 12-32 **Oxygen–haemoglobin (O₂–Hb) dissociation (saturation) curve.** Relation between (a) % Hb saturation and (b) oxygen content (or concentration) versus $P_{O_2}$. The relationship between these two variables is depicted by a sigmoid-shaped curve with a steep portion between 0 and 60 mmHg and a plateau above a $P_{O_2}$ of 60 mmHg. A reduction by half of the number of red blood cells or the concentration of haemoglobin in the red blood cells (anaemia) does not affect the position of the curve (blue line) when the y-axis is (a) % Hb saturation because only the percent, not the absolute number, of binding sites occupied by oxygen is measured. When the y-axis is (b) content (or concentration), however, the anaemia curve is reduced by half.

with alveolar $P_{O_2}$, normally has a $P_{O_2}$ of 100 mmHg. According to the O₂–Hb dissociation curve, the haemoglobin of blood at a $P_{O_2}$ of 100 mmHg is 97.5 percent saturated. Therefore, haemoglobin in the systemic arterial blood is normally almost completely saturated.

If the alveolar $P_{O_2}$ and, consequently, the arterial $P_{O_2}$, fall below normal, the plateau ensures that the Hb saturation does not decrease greatly until the $P_{O_2}$ falls below 60 mmHg, at which point haemoglobin is still about 90 percent saturated. Conversely, if the blood $P_{O_2}$ is greatly increased above 100 mmHg—say, to 600 mmHg by breathing pure oxygen—the oxygen content of the blood increases very little because it was already almost completely saturated. What little increase in content does occur is due to the extra oxygen dissolved in blood. Therefore, in the $P_{O_2}$ range between 60 and 600 mmHg or even higher, there is only a 10 percent difference in the amount of oxygen carried by haemoglobin. In this way, the plateau portion of the O₂–Hb curve provides a good margin of safety in the oxygen-carrying capacity of the blood.

*Clinical Note* Arterial $P_{O_2}$ may be reduced by pulmonary diseases accompanied by inadequate ventilation or defective gas exchange, or by circulatory disorders that result in inadequate blood flow to the lungs. It may also fall in healthy people under two circumstances: (1) at high altitudes, where the $P_{O_2}$ of the inspired air is reduced because of the lower atmospheric pressure; or (2) in oxygen-deprived environments at sea level, such as if someone were accidentally locked in an air-tight vault. Unless the arterial $P_{O_2}$ becomes markedly reduced (falls below 60 mmHg) in either pathological conditions or abnormal

circumstances, near-normal amounts of oxygen can still be carried to the tissues.

## SIGNIFICANCE OF THE STEEP PORTION OF THE O₂-HB CURVE

The steep portion of the curve between 0 and 60 mmHg occurs in the blood $P_{O_2}$ range that exists at the systemic capillaries, where oxygen is unloaded from haemoglobin. In the systemic capillaries, the blood equilibrates with the surrounding tissue cells at an average (but highly variable) $P_{O_2}$ of 40 mmHg. Note in > Figure 12-32 that at a $P_{O_2}$ of 40 mmHg, the % Hb saturation is 75 percent. The blood arrives in the tissue capillaries at a $P_{O_2}$ of 100 mmHg and with 97.5 percent Hb saturation. Because haemoglobin is 75 percent saturated at a $P_{O_2}$ of 40 mmHg in the systemic capillaries, nearly 25 percent of the HbO₂ must dissociate, yielding reduced haemoglobin and oxygen. This released oxygen is free to diffuse down its partial pressure gradient from the red blood cells through the plasma and interstitial fluid into the tissue.

The normal 60 mmHg drop in $P_{O_2}$ from 100 to 40 mmHg in the systemic capillaries causes unloading of about 25 percent of the oxygen combined with haemoglobin, sufficient to sustain resting metabolism. But an additional drop in $P_{O_2}$ of only 20 mmHg from 40 mmHg unloads an additional 45 percent of the total oxygen because this change in partial pressure occurs over the steep portion of the curve. In this range, only small decreases in systemic capillary $P_{O_2}$ are required to unload considerable oxygen to meet increased metabolic demands. During strenuous exercise, for example, as much as 85 percent of the

**12**

oxygen may be released to muscle. Even more oxygen can be made available to actively metabolizing cells, such as exercising muscles, by circulatory and respiratory adjustments that increase perfusion to active tissues.

The steep part of the oxygen dissociation curve has another important consequence. At high altitude, the low barometric pressure reduces the inspired $P_{O_2}$. At sufficiently high altitudes, this results in arterial hypoxaemia that activates the carotid chemoreceptors; their activation increases ventilation. This means that if metabolism does not change (i.e., the person is still at rest), more carbon dioxide is expired, causing a fall in the arterial $P_{CO_2}$ and, as a result, an increase in the alveolar and therefore arterial $P_{O_2}$ (alveolar gas equation; p. 525). Although this increase may be only a few mmHg on the steep part of the curve, this small increase can cause a significant increase in arterial oxygen content.

The function of the steep portion of the oxygen dissociation curve may, on the basis of the previous paragraph, appear obvious—that is, augment unloading of oxygen to metabolically active tissues or augment arterial oxygen content as a result of altitude-induced hyperventilation. However, recall that oxygen diffuses to tissues down a partial pressure gradient. Over the steep part of the curve, the high end of the gradient decreases, and this would reduce the gradient that favours diffusion. Why would evolution conserve an oxygen dissociation curve that reduces the driving pressure for diffusion into cells over such a critical range? The answer is that the oxygen dissociation curve, as depicted, applies to conditions in the lung, where the $P_{O_2}$ determines the saturation or content of the arterial blood. In tissues, however, oxygen is consumed, and this means the oxygen content determines the local arterial $P_{O_2}$. When oxygen content becomes the independent variable and $P_{O_2}$ the dependent variable, a different interpretation is apparent (> Figure 12-33): as oxygen is consumed and the saturation drops, the systemic $P_{O_2}$ does not fall as much (reduced slope). In other words, the shape of the curve preserves the $P_{O_2}$, thereby maintaining the partial pressure gradient for diffusion.

## Haemoglobin and oxygen transfer

There is more to say about the role of haemoglobin in gas exchange. In our discussion of oxygen being driven from the alveoli to the blood by a $P_{O_2}$ gradient, we ignored the oxygen bound to haemoglobin because blood $P_{O_2}$ depends entirely on the *dissolved* oxygen. However, haemoglobin plays a crucial role in permitting the transfer of large quantities of oxygen to the tissues before blood $P_{O_2}$ equilibrates with the surrounding tissues (> Figure 12-34).

### ROLE OF HAEMOGLOBIN AT THE ALVEOLAR LEVEL
Haemoglobin acts as a storage depot for oxygen, removing it from solution as soon as it enters the blood from the alveoli. Because only dissolved oxygen contributes to the $P_{O_2}$, the oxygen stored in haemoglobin cannot contribute to blood $P_{O_2}$. When systemic venous blood enters the pulmonary capillaries, its $P_{O_2}$ is considerably lower than alveolar $P_{O_2}$, so oxygen immediately diffuses into the blood, raising blood $P_{O_2}$. As soon as

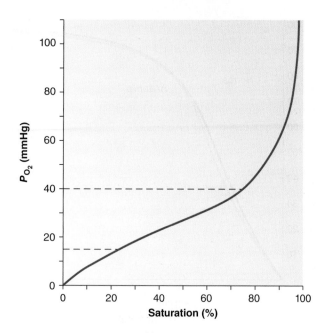

Republished with permission of Daedalus Enterprises Inc., from Downs, John, B. "Has Oxygen Administration Delayed Appropriate Respiratory Care? Fallacies Regarding Oxygen Therapy," *Respiratory Care* 48 (6), 1956; permission conveyed through Copyright Clearance Center, Inc.

> FIGURE 12-33 **Another approach to the oxygen dissociation curve.** Here, the axes have been reversed from those shown in Figure 12-32. In tissues, oxygen is consumed, and, therefore, saturation becomes the independent variable. So when the oxygen dissociation curve is plotted with percentage saturation on the x-axis and $P_{O_2}$ on the y-axis, a fall in saturation (i.e., uptake of oxygen by the tissues) does not cause as large a drop in $P_{O_2}$, thereby preserving the partial pressure responsible for diffusion of oxygen into the cells.

the blood $P_{O_2}$ increases, the percentage of haemoglobin that can bind with oxygen likewise increases, as indicated by the oxygen dissociation curve. Consequently, most of the oxygen that has diffused into the blood combines with haemoglobin and no longer contributes to blood $P_{O_2}$. As oxygen is removed from solution by combining with haemoglobin, blood $P_{O_2}$ falls to about the same level it was when the blood entered the lungs, even though the oxygen content of the blood has increased. Because blood $P_{O_2}$ is still less than alveolar $P_{O_2}$, oxygen continues to diffuse from the alveoli into the blood, only to bind to more haemoglobin.

Even though we described this process as a series of steps, the net diffusion of oxygen from alveoli to blood occurs continuously until haemoglobin becomes as completely saturated with oxygen as it can be at that particular $P_{O_2}$. At a normal $P_{O_2}$ of 100 mmHg, haemoglobin is 97.5 percent saturated. Thus, by binding oxygen, haemoglobin keeps blood $P_{O_2}$ low, preserving the partial pressure gradient so that oxygen continues to diffuse from the alveolar gas into the blood. Not until haemoglobin is maximally saturated for that $P_{O_2}$ does all the oxygen transferred into the blood remain dissolved and directly contribute to the $P_{O_2}$. Only now does blood $P_{O_2}$ rapidly equilibrate with alveolar $P_{O_2}$ and bring further oxygen transfer to a halt—but this point is not reached until haemoglobin is already loaded with oxygen to the maximum extent possible. Once blood $P_{O_2}$ equilibrates with alveolar $P_{O_2}$, no further transfer of oxygen occurs, no matter how little or how much total oxygen has already been transferred.

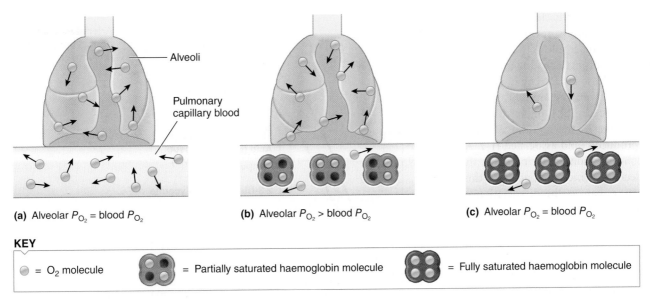

**(a)** Alveolar $P_{O_2}$ = blood $P_{O_2}$    **(b)** Alveolar $P_{O_2}$ > blood $P_{O_2}$    **(c)** Alveolar $P_{O_2}$ = blood $P_{O_2}$

**KEY**

= $O_2$ molecule    = Partially saturated haemoglobin molecule    = Fully saturated haemoglobin molecule

© 2016 Cengage

> **FIGURE 12-34 Haemoglobin facilitates transfer of oxygen by acting as a storage depot to keep $P_{O_2}$ low.** (a) In the hypothetical situation in which no haemoglobin is present in the blood, the alveolar $P_{O_2}$ and the pulmonary capillary blood $P_{O_2}$ are at equilibrium. (b) Haemoglobin has been added to the pulmonary capillary blood. As the Hb starts to bind with oxygen, it removes oxygen from solution. Because only dissolved oxygen contributes to blood $P_{O_2}$, the blood $P_{O_2}$ remains below that of the alveoli, even though the same number of oxygen molecules are present in the blood as in (a). By binding some of the dissolved oxygen, Hb favours the net diffusion of more oxygen down its partial pressure gradient from the alveoli to the blood. (c) Haemoglobin is fully saturated with oxygen, and the alveolar and blood $P_{O_2}$ are at equilibrium again. The blood $P_{O_2}$ resulting from dissolved oxygen is equal to the alveolar $P_{O_2}$, despite the fact that the total oxygen content in the blood is much greater than in (a), in which blood $P_{O_2}$ was equal to alveolar $P_{O_2}$ in the absence of Hb.

## ROLE OF HAEMOGLOBIN AT THE TISSUE LEVEL

The reverse situation occurs at the tissue level. Because the $P_{O_2}$ of blood entering the systemic capillaries is considerably higher than the $P_{O_2}$ of the surrounding tissue, oxygen immediately diffuses from the blood into the tissues, lowering blood $P_{O_2}$. When blood $P_{O_2}$ falls, haemoglobin must unload some stored oxygen. As the oxygen released from haemoglobin dissolves in the blood, blood $P_{O_2}$ increases and once again exceeds the $P_{O_2}$ of the surrounding tissues. This favours further movement of oxygen out of the blood, although the arterial oxygen content has already fallen. Only when haemoglobin can no longer release any more oxygen into solution (when Hb is unloaded to the greatest extent possible for the $P_{O_2}$ existing at the systemic capillaries) can blood $P_{O_2}$ fall as low as that in surrounding tissue. At this time, further transfer of oxygen stops. Haemoglobin, because it stores a large quantity of oxygen that can be freed by a slight reduction in $P_{O_2}$ at the systemic capillary level, permits the transfer of much more oxygen from the blood into the cells than would be possible in its absence.

Haemoglobin is crucial for oxygen transport. Because oxygen bound to haemoglobin does not contribute to the blood $P_{O_2}$, it keeps the blood $P_{CO_2}$ low and thereby maintains the partial pressure gradient needed to ensure transfer of oxygen from alveolar gas to blood until haemoglobin is saturated and the $P_{O_2}$ in the blood equilibrates with the $P_{O_2}$ in the alveoli. This ensures that enough oxygen (its concentration, or content, as distinct from its partial pressure) is transported to the tissues to meet resting and increased metabolic demands. If haemoglobin levels fall to half of normal, as in a severely anaemic patient (p. 441), the oxygen-carrying capacity of the blood falls by 50 percent, even

though the arterial $P_{O_2}$ is still the normal 100 mmHg, and the Hb is 97.5 percent saturated. An oximeter (a device that measures oxygen saturation) reading of 97.5 percent tells you only what percentage of the binding sites on haemoglobin are occupied by oxygen; it does *not* indicate how many binding sites are available (i.e., the concentration of Hb in the blood) or the arterial oxygen content. Similarly, if the of a blood sample is 100 mmHg, this reveals nothing about the concentration (or content) of oxygen in the blood; a $P_{O_2}$ of 100 mmHg is just as compatible with a concentration of 200 mL $O_2$/L blood (the maximum) as it is with 3 mL $O_2$/L blood (the minimum). In summary, an oximeter cannot detect anaemia.

## Unloading of oxygen

Even though the main factor determining % Hb saturation is the $P_{O_2}$ of the blood, other factors affect the affinity, or bond strength, between haemoglobin and oxygen. Accordingly, they shift the oxygen dissociation curve (i.e., change the % Hb saturation at a given $P_{O_2}$).

Both disease and exercise influence the oxygen dissociation curve. The by-products associated with energy turnover during exercise are heat, carbon dioxide, and lactic acid. All three enhance the release of oxygen from haemoglobin; that is, at a given $P_{O_2}$, haemoglobin holds less oxygen. Another factor that affects the dissociation of oxygen from haemoglobin is the pH. One of the forms in which carbon dioxide is transported in the plasma is as a bicarbonate ion ($HCO_3^-$), while it goes through the carbonic anhydrase–carbonic acid cycle. As bicarbonate is produced, a hydrogen ion ($H^+$) is released as a by-product.

**12**

Additionally, in muscle, lactic acid is produced because some of the energy turnover is anaerobic. This increased production of lactic acid is associated with the release of H⁺. The intensity of exercise dictates the amount of lactic acid produced. Thus, higher-intensity work increases the amount of lactic acid produced (and associated H⁺). As the concentration of H⁺ increases, the pH decreases (increased acidity); this, in turn, enhances the dissociation of oxygen from haemoglobin, a phenomenon known as the **Bohr effect**.

### BOHR EFFECT

Both carbon dioxide (Haldane effect, p. 536) and the H⁺ component of acids can combine reversibly with haemoglobin at sites other than those binding oxygen. The result is a change in the molecular shape (conformation) of haemoglobin that reduces its affinity for oxygen. Note that the % Hb saturation refers only to the extent to which haemoglobin is combined with oxygen, not the extent to which it is bound with carbon dioxide, H⁺, or other molecules. Indeed, the % Hb saturation decreases when carbon dioxide and H⁺ bind with haemoglobin, because their presence on haemoglobin facilitates release of oxygen.

In cells, oxygen is used for energy production, and carbon dioxide and lactic acid are produced as by-products. The H⁺ that accumulates because of increases in lactic acid (generated during anaerobic metabolism), and $HCO_3^-$ production lowers the blood pH and enhances the dissociation of oxygen. This is particularly important for metabolically active cells, such as muscle during exercise.

Haemoglobin binds with most of the H⁺ that accumulates within the erythrocytes when carbonic acid dissociates into H⁺ and $HCO_3^-$. As with carbon dioxide, reduced haemoglobin has a greater affinity for H⁺ than oxygenated haemoglobin. Therefore, oxygen unloading facilitates the pickup by haemoglobin of carbon dioxide–generated H⁺. Because only free, dissolved H⁺ contributes to the acidity of a solution, the venous blood would be considerably more acidic than the arterial blood if haemoglobin did not bind most of the H⁺ generated at the tissue level. Haemoglobin is therefore an important buffer.

The Bohr effect acts in reverse at the alveoli. The carbonic anhydrase–carbonic acid reaction reverses in the pulmonary capillaries. This reversal in turn decreases the amount of carbonic acid ($H_2CO_3$), bicarbonate $HCO_3^-$, and hydrogen ion (H⁺), thereby increasing pH and decreasing the dissociation of oxygen, which facilitates its loading onto haemoglobin. Thus, the Bohr effect, as represented by the sigmoidal shape of the oxygen dissociation curve (> Figure 12-35), does not influence the combination of oxygen with haemoglobin in the lung during exercise. However, at the tissue level, the Bohr effect acts in an opposite

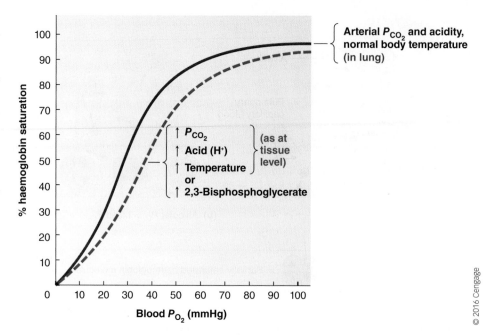

> **FIGURE 12-35 Effect of $P_{CO_2}$ increased, H⁺, temperature, and 2,3-bisphosphoglycerate on the oxygen dissociation curve.** Increased $P_{CO_2}$, acid, temperature, and 2,3-bisphosphoglycerate, as occurs in red blood cells, shift the curve to the right. Because less oxygen can bind to haemoglobin at a given $P_{O_2}$, more oxygen is unloaded from haemoglobin for use by the tissues.

fashion. In skeletal muscle during exercise, the ability of haemoglobin to bind oxygen is reduced by about 10–15 percent; as a result, more oxygen is available to the tissue, facilitating aerobic metabolism.

Other factors that influence the binding of oxygen to haemoglobin are carbon dioxide, acidity, temperature, and 2,3-bisphosphoglycerate; they will be examined separately. The oxygen dissociation curve you are already familiar with (see > Figure 12-32) is a typical sigmoid-shaped curve at normal arterial $P_{CO_2}$ and acidity levels, normal body temperature, and normal 2,3-bisphosphoglycerate concentration.

### HALDANE EFFECT

An increase in $P_{CO_2}$ shifts the oxygen dissociation curve to the right (> Figure 12-35) and vice versa. The % Hb saturation still depends on the $P_{O_2}$, but for any given $P_{O_2}$, less oxygen binds to haemoglobin as the $P_{CO_2}$ increases, and more oxygen binds to haemoglobin as the $P_{CO_2}$ decreases. This effect is important, because the $P_{CO_2}$ of the blood increases in the systemic capillaries as carbon dioxide diffuses down its gradient from the cells into the blood. The presence of this additional carbon dioxide in the blood in effect decreases the affinity of haemoglobin for oxygen, so haemoglobin unloads even more oxygen at the tissue level than it would if the reduction in $P_{O_2}$ in the systemic capillaries were the only factor affecting % Hb saturation. Note, too, that this effect of carbon dioxide is lost at high $P_{O_2}$ (as when supplemental oxygen is provided), but this beneficial effect is not needed since arterial oxygen content is high.

### EFFECT OF TEMPERATURE ON % HB SATURATION

In a similar manner, a rise in temperature shifts the oxygen dissociation curve to the right, resulting in more unloading of oxygen at a given $P_{O_2}$. An exercising muscle or other actively metabolizing tissue produces heat. The local rise in temperature enhances oxygen release from haemoglobin for use by more active tissues.

### COMPARISON OF THESE FACTORS AT THE TISSUE AND PULMONARY LEVELS

As you just learned, increases in carbon dioxide, acidity, and temperature at the tissue level—all of which are associated with increased cellular oxygen consumption—enhance the effect of a drop in $P_{O_2}$ in facilitating the release of oxygen from haemoglobin. These effects are largely reversed in the lung, where the extra acid-forming carbon dioxide is blown off and the local environment is cooler. (Although body temperature is 37°C, regional differences occur; working muscle can be as warm as 40°C.) Appropriately, therefore, haemoglobin has a higher affinity for oxygen in the pulmonary capillaries, enhancing the effect of the high $P_{O_2}$ on the uptake of oxygen by haemoglobin.

### EFFECT OF 2,3-BISPHOSPHOGLYCERATE ON % HB SATURATION

The preceding changes take place in the *environment* of the red blood cells, but a factor *inside* the red blood cells, **2,3-bisphosphoglycerate (BPG)**, can also affect the degree of binding of oxygen to haemoglobin. This erythrocyte constituent, which is produced during red blood cell metabolism, can bind reversibly with haemoglobin and reduce its affinity for oxygen, just as carbon dioxide and $H^+$ do. Thus, an increased level of BPG, like the other factors, shifts the curve to the right, enhancing oxygen unloading as blood flows through tissues.

The production of BPG by red blood cells gradually increases whenever haemoglobin in the arterial blood is chronically undersaturated—that is, when arterial oxygen saturation is below normal (hypoxemia). This condition may occur in people living at high altitudes or in those suffering from certain types of circulatory or respiratory diseases or anaemia. By helping unload oxygen from haemoglobin at the tissue level, increased BPG helps maintain oxygen availability for tissue use despite chronic reductions in arterial oxygen content.

However, unlike the other factors—which normally are present only at the tissue level and thus shift the curve to the right only in the systemic capillaries, where the shift is advantageous in unloading oxygen—BPG is present in the red blood cells and, accordingly, shifts the curve to the right to the same degree in both the tissues and the lungs. As a result, BPG decreases the ability to load oxygen in the lung, which is the negative aspect of its increased production.

### Haemoglobin and carbon monoxide

*Clinical Note* **Carbon monoxide (CO)** and oxygen compete for the same binding sites on haemoglobin, but haemoglobin's affinity for carbon monoxide is 240 times greater than that for oxygen. The combination of carbon monoxide and haemoglobin is known as **carboxyhaemoglobin (HbCO)**. Because of this much greater affinity, even small amounts of carbon monoxide can occupy a disproportionately large share of haemoglobin's binding sites, making haemoglobin unavailable for oxygen transport. Even though the Hb concentration and $P_{O_2}$ are normal (because only dissolved oxygen is responsible for the $P_{O_2}$, and the presence of carbon monoxide does not affect how much oxygen dissolves in the plasma), the oxygen content of the blood is seriously reduced. In addition, carbon monoxide also shifts the oxygen dissociation curve to the left. As a result, a much greater drop in $P_{O_2}$ is required to unload oxygen to the tissues, increasing the likelihood of cellular anoxia.

Fortunately, carbon monoxide is not a normal constituent of inspired air. It is a poisonous gas produced during the incomplete combustion (burning) of carbon products, such as gasoline, coal, wood, and tobacco. Carbon monoxide is especially dangerous because it is so insidious. If it is produced in a closed environment and its concentration continues to increase (e.g., in a parked car with the motor running and windows closed), it can reach lethal levels without a person ever being aware of the danger. People cannot detect carbon monoxide because it is odourless, colourless, tasteless, and nonirritating. Furthermore, a person with carbon monoxide poisoning has no sensation that arterial oxygen levels are low and makes no attempt to increase ventilation, even though the cells are anoxic (lack oxygen). (For a history of the treatment of carbon monoxide poisoning, and how an effective treatment was replaced by a less effective one that is still in use, see J.A. Fisher et al. [2011]. *Experimental Physiology*, 96: 1262–69.)

### Carbon dioxide

When arterial blood flows through the tissue capillaries, carbon dioxide diffuses down its partial pressure gradient from the cells into the blood. Carbon dioxide is transported in the blood in three ways (see > Figure 12-36 and also ▌Table 12-7):

1. *Physically dissolved.* As with dissolved oxygen, the amount of carbon dioxide physically dissolved in the blood depends on the $P_{CO_2}$. Because carbon dioxide is more soluble than oxygen, a greater proportion of the total carbon dioxide (compared to oxygen) is physically dissolved in the blood. Even so, only 5–10 percent of the blood's total carbon dioxide content is carried this way at the normal systemic venous $P_{CO_2}$ of 46 mmHg.

2. *Bound to haemoglobin.* Another 5–10 percent of the carbon dioxide combines with haemoglobin to form **carbamino haemoglobin (HbCO$_2$)**. Carbon dioxide binds with the globin portion of haemoglobin, unlike oxygen that combines with the heme portions. Reduced haemoglobin (i.e., haemoglobin without oxygen) has a greater affinity for carbon dioxide than does oxygenated haemoglobin. The unloading of oxygen from haemoglobin in the tissue capillaries therefore facilitates the uptake of carbon dioxide by haemoglobin.

3. *As bicarbonate.* By far the most important means of carbon dioxide transport is as **bicarbonate (HCO$_3^-$)**, with

**12**

80–90 percent of the carbon dioxide converted into bicarbonate by the following chemical reaction, which takes place inside red blood cells:

$$CO_2 + H_2O \Leftrightarrow H_2CO_3 \Leftrightarrow H^+ + HCO_3^-$$

In the first step, carbon dioxide combines with water to form **carbonic acid** $(\mathbf{H_2CO_3})$. This reaction occurs very slowly in plasma, but it proceeds swiftly inside red blood cells because of the presence of the enzyme carbonic anhydrase, which catalyzes (accelerates) the reaction. As is characteristic of acids, some of the carbonic acid molecules spontaneously dissociate into hydrogen ions $(H^+)$ and bicarbonate ions $(HCO_3^-)$. The one carbon atom and two oxygen atoms of the original carbon dioxide molecule are therefore present in the blood as an integral part of bicarbonate. This is beneficial because bicarbonate is more soluble than carbon dioxide in blood.

### CHLORIDE SHIFT

As this reaction proceeds, bicarbonate and hydrogen ions start to accumulate inside red blood cells in the systemic capillaries. The red blood cell membrane has a bicarbonate–chloride $(HCO_3^- - Cl^-)$ carrier that passively facilitates the diffusion of these ions in opposite directions across the membrane. The membrane is relatively impermeable to $H^+$. Consequently, $HCO_3^-$, but not $H^+$, diffuses down its concentration gradient out of the erythrocytes into the plasma. Because bicarbonate is a negatively charged ion, its efflux—unaccompanied by a comparable outward diffusion of positively charged ions—creates an electrical gradient (p. 38). Chloride ions $(Cl^-)$, the dominant plasma anions, diffuse into the red blood cells down this electrical gradient and restore electrical neutrality. This inward shift of chloride in exchange for the efflux of carbon dioxide–generated bicarbonate ion is known as the **chloride (Hamburger) shift**.

### REVERSE HALDANE EFFECT

Removing oxygen from haemoglobin increases the ability of haemoglobin to bind carbon dioxide and the carbon dioxide–generated $H^+$; this is known as the **Haldane effect**. The Haldane and Bohr effects work in synchrony to facilitate the liberation of oxygen and the uptake of carbon dioxide and carbon dioxide–generated $H^+$ at the tissue level. The reduced haemoglobin transports the bound carbon dioxide back to the lungs, where it is excreted (› Figure 12-36).

There is also a phenomenon known as the **reverse Haldane effect**. Increases in arterial $P_{O_2}$, as occur during inhalation of supplementary oxygen, reduce or prevent haemoglobin from binding the carbon dioxide released from tissues. As a result, carbon dioxide must be carried back to the lungs primarily as

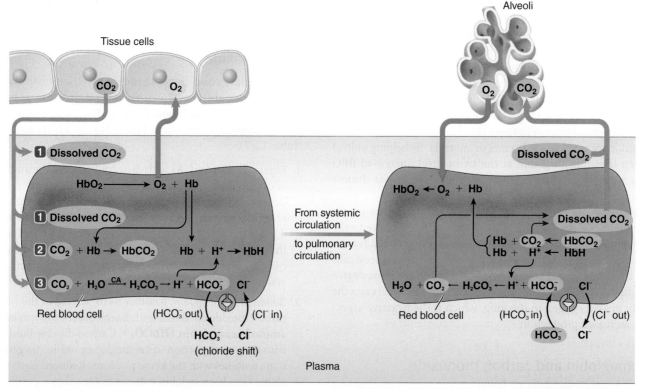

CA = Carbonic anhydrase

› **FIGURE 12-36 Carbon dioxide transport in the blood.** Carbon dioxide diffusing from tissues is transported in the blood to the lungs in three ways: (1) physically dissolved; (2) bound to haemoglobin (Hb); and (3) as bicarbonate ion $(HCO_3^-)$. Haemoglobin is present only in the red blood cells, as is carbonic anhydrase, the enzyme that catalyzes the production of $(HCO_3^-)$. The $H^+$ generated during the production of $(HCO_3^-)$ also binds to Hb. Bicarbonate moves by facilitated diffusion down its concentration gradient out of the red blood cell into the plasma, and chloride $(Cl^-)$ moves by means of the same passive carrier into the red blood cell down the electrical gradient created by the outward diffusion of $(HCO_3^-)$. The reactions that occur at the tissue level are reversed in the lung.

bicarbonate and dissolved carbon dioxide. The increase in $P_{CO_2}$ would be of no consequence except that some researchers believe the increase in $P_{CO_2}$ at the site of the central chemoreceptors (p. 547), is responsible for the increase in ventilation associated with breathing supplemental oxygen, not a decrease, as commonly thought. This causes *hypocapnia*, something that may have serious consequences for the delivery of oxygen to tissues (see Why It Matters, p. 541).

(see Why It Matters, p. 541).

### Check Your Understanding 12.11

1. Compare the transport of oxygen and carbon dioxide in the blood.

## Abnormal blood-gas levels

Table 12-9 provides a glossary of terms used to describe various states associated with respiratory abnormalities.

### ABNORMALITIES IN ARTERIAL $P_{O_2}$

The term **hypoxia** refers to the condition of insufficient oxygen at the cellular level. There are four general categories of hypoxia:

1. *Hypoxic hypoxia* is characterized by a low arterial blood $P_{O_2}$ accompanied by inadequate Hb saturation. It is caused by (a) inadequate gas exchange, typified by a normal alveolar $P_{O_2}$ but a reduced arterial $P_{O_2}$; or (b) exposure to high altitude or to a suffocating environment where atmospheric $P_{O_2}$ is reduced so that alveolar and arterial $P_{O_2}$ are likewise reduced.

2. *Anaemic hypoxia* is a reduced oxygen-carrying capacity of the blood. It can result from (a) a decrease in the number of circulating red blood cells, (b) an inadequate amount of Hb within the red blood cells, or (c) carbon monoxide poisoning. In all cases of anaemic hypoxia, the arterial $P_{O_2}$ is normal, but the oxygen content of the arterial blood is lower than normal because less haemoglobin is available to carry oxygen.

3. *Circulatory (or stagnant) hypoxia* arises when too little oxygenated blood is delivered to the tissues. Circulatory hypoxia to a limited area can result from a local vascular spasm, blockage, or injury. Alternatively, circulatory hypoxia may be general, from congestive heart failure or circulatory shock. In this case, the arterial $P_{O_2}$ and oxygen content are typically normal, but too little oxygenated blood reaches the cells.

4. In *histotoxic hypoxia*, oxygen delivery to the tissues is normal, but the cells cannot use the oxygen available to them. The classic example is *cyanide poisoning*. Cyanide blocks cellular enzymes essential for internal respiration.

**Hyperoxia**, an above-normal arterial $P_{O_2}$, cannot occur when a person is breathing air at sea level. However, breathing supplemental oxygen can increase alveolar and, consequently, arterial $P_{O_2}$. Because more of the inspired gas is oxygen, the $P_{O_2}$ is greater; as a result, more oxygen dissolves in the blood before arterial $P_{O_2}$ equilibrates with alveolar $P_{O_2}$. Even though arterial $P_{O_2}$ increases, the total blood oxygen content does not increase significantly because haemoglobin is nearly fully saturated at a normal arterial $P_{O_2}$. In certain pulmonary diseases in which the patient has a reduced arterial $P_{O_2}$, however, breathing supplemental oxygen can help establish a larger alveoli-to-blood partial pressure gradient, improving arterial $P_{O_2}$. Far from being advantageous, in contrast, a markedly elevated arterial $P_{O_2}$ can be dangerous. If the arterial $P_{O_2}$ is too high, **oxygen toxicity** can occur. Even though the total oxygen content of the blood

### ▮ TABLE 12-9 Mini-Glossary of Clinically Important Respiratory States

**Apnea**   Transient cessation of breathing

**Asphyxia**   Oxygen starvation of tissues, caused by a lack of oxygen in the air, respiratory impairment, or inability of the tissues to use oxygen; always accompanied by a rise in carbon dioxide levels

**Cyanosis**   Blueness of the skin resulting from insufficiently oxygenated blood in the arteries

**Dyspnea**   Difficult or laboured breathing

**Eupnea**   Normal breathing

**Hypercapnia**   (or hypercarbia) Excess carbon dioxide in the arterial blood

**Hyperpnea**   Increased pulmonary ventilation that matches increased metabolic demands (carbon dioxide production), as in exercise

**Hyperventilation**   Increased pulmonary ventilation in excess of metabolic requirements, resulting in decreased arterial $P_{CO_2}$ and respiratory alkalosis

**Hypocapnia**   Below-normal $P_{CO_2}$ in the arterial blood

**Hypoventilation**   Underventilation in relation to metabolic requirements, resulting in increased arterial $P_{CO_2}$ and respiratory acidosis

**Hypoxaemia**   Below normal $P_{O_2}$ in the arterial blood

**Hypoxia**   Insufficient oxygen at the cellular level

**Anaemic hypoxia**   Reduced oxygen-carrying capacity of the blood

**Circulatory hypoxia**   Inadequate oxygenated blood delivered to tissues; also called *stagnant hypoxia*

**Histotoxic hypoxia**   Inability of the cells to use available oxygen

**Hypoxic hypoxia**   Low arterial $P_{O_2}$ accompanied by inadequate Hb saturation

**Respiratory arrest**   Permanent cessation of breathing (unless clinically corrected)

**Suffocation**   Oxygen deprivation as a result of an inability to breathe

**12**

is only slightly increased, exposure to a high $P_{O_2}$ can damage some cells due to the production of reactive oxygen species and free radicals that can damage membranes. In particular, brain damage and blindness-causing damage to the retina are associated with oxygen toxicity. In addition, through mechanisms still being investigated, a high $P_{O_2}$ increases vascular resistance in many organs. If perfusion decreases more than arterial oxygen content increases, regional delivery of oxygen may actually decrease even though the patient is breathing or being ventilated with 100 percent oxygen. Therefore, oxygen must be administered cautiously; it is not needed, and may even be dangerous, in individuals who are not hypoxemic.

### ABNORMALITIES IN ARTERIAL $P_{CO_2}$

*Clinical Note* Because of carbon dioxide's high solubility and, therefore, diffusibility, it is not readily affected by changes in the resistance ($R$) term of the equation for diffusion. The alveolar $P_{CO_2}$ is, however, directly related to the metabolic production of carbon dioxide and inversely related to the alveolar ventilation, as described by the alveolar ventilation equation: $P_{A_{CO_2}} = k \times V_{CO_2}/\dot{V}_A$ (where $k$, a constant, is 863 mmHg; see Quantitative Exercises, p. 553). Thus, at a given metabolism, the greater the alveolar ventilation, the lower is the alveolar $P_{CO_2}$, and vice versa.

The term **hypercapnia** (sometimes called *hypercarbia*) refers to excess carbon dioxide in arterial blood; it is caused by **hypoventilation** (ventilation inadequate to meet metabolic needs for oxygen delivery and carbon dioxide removal). With most pulmonary diseases, carbon dioxide accumulates in arterial blood concurrently with an oxygen deficit, because the exchanges of both gases between lungs and atmosphere are equally affected ( > Figure 12-37). However, when a decrease in arterial $P_{O_2}$ is due to reduced pulmonary diffusing capacity—as in pulmonary oedema or emphysema—oxygen transfer suffers more than that of carbon dioxide, because oxygen is 20 times less diffusible than carbon dioxide. As a result, in these circumstances hypoxic hypoxia occurs much more readily than hypercapnia. (This can also occur in cases of ventilation/perfusion mismatch in the lung.)

**Hypocapnia** (or *hypocarbia*), below-normal arterial $P_{CO_2}$ levels, results from **hyperventilation**, a condition in which alveolar ventilation is greater than that needed to excrete metabolically produced carbon dioxide. Hyperventilation can be triggered by anxiety, fever, and aspirin poisoning. It also occurs during high levels of exercise when there is a shift to anaerobic metabolism. Alveolar $P_{O_2}$ increases during hyperventilation as more fresh oxygen is delivered to the alveoli from the atmosphere than the blood extracts from the alveoli for tissue consumption, and arterial $P_{O_2}$ increases correspondingly ( > Figure 12-37). However, because haemoglobin is almost fully saturated at the normal arterial $P_{O_2}$, very little additional oxygen is added to the blood. Nor can the $P_{O_2}$ increase very much; it can never exceed the partial pressure of inspired oxygen (about 150 mmHg, at sea level). Except for the small, extra amount of dissolved oxygen, blood oxygen content in a normal person remains essentially unchanged during hyperventilation.

Increased ventilation is not synonymous with hyperventilation. Increased ventilation that matches an increased metabolic demand, such as that which occurs during exercise, is termed **hyperpnea**. (Recall that increased ventilation

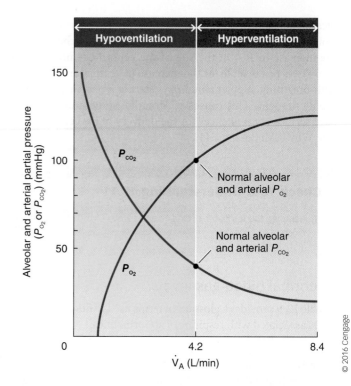

> FIGURE 12-37 **Effects of hyperventilation and hypoventilation on arterial $P_{O_2}$ and $P_{CO_2}$.** Alveolar and arterial $P_{CO_2}$ at a given carbon dioxide production ($\dot{V}_{CO_2}$) are inversely related to alveolar ventilation, as indicated by the alveolar ventilation equation. From the change in alveolar $P_{CO_2}$, one can compute the alveolar $P_{O_2}$ using the alveolar gas equation (p. 525). Doubling alveolar ventilation cannot double alveolar $P_{O_2}$ to 200 mmHg because the upper limit of alveolar $P_{O_2}$ is the inspired $P_{O_2}$, approximately 150 mmHg at sea level.

that exceeds metabolic demand is called hyperventilation.) During exercise below the threshold at which the exercising muscles switch to anaerobic metabolism, alveolar and arterial $P_{O_2}$ and $P_{CO_2}$ remain constant, with the increased alveolar ventilation matching the increased oxygen consumption and carbon dioxide production. Above the anaerobic threshold, the increased production of acid (lactic acid) causes alveolar ventilation to exceed that required for excretion of carbon dioxide, and the person becomes hypocapnic.

### CONSEQUENCES OF ABNORMALITIES IN ARTERIAL BLOOD GASES

*Clinical Note* The consequences of reduced oxygen availability to tissues during hypoxia are apparent. The cells need adequate oxygen to sustain energy-generating metabolism. The consequences of abnormal levels of carbon dioxide in the blood are less obvious. Changes in blood concentration of carbon dioxide primarily affect acid–base balance. Hypercapnia elevates production of carbonic acid. The subsequent generation of excess $H^+$ produces a condition termed *respiratory acidosis*. Conversely, less-than-normal amounts of $H^+$ are generated through carbonic acid formation in conjunction with hypocapnia. The resultant condition is called *respiratory alkalosis* (Chapter 14). (To learn about the effects of altitude and exercise training at high altitude, see Chapter 14, Integration of Human Physiology.)

Oxygen is the most commonly administered drug—and it is a drug—in the world. It is given to patients who have, or are suspected of having, hypoxic or anoxic tissues. Some of the most common clinical situations are acute exacerbations of chronic obstructive pulmonary disease, myocardial infarctions (heart attacks), and strokes. Many of these patients are not, however, hypoxic, based on measurements of arterial oxygen saturation using an oximeter: a small device typically placed over a finger or earlobe. Guidelines for emergency medical personnel typically stipulate that, regardless of the oximeter reading, patients be given oxygen by facemask or nasal tubes.

Is there any evidence that oxygen is clinically beneficial in these nonhypoxemic patients? Surprisingly, the answer is no. Instead, control trials, in which patients are randomly assigned to treatment groups and the outcomes later evaluated "blind" (i.e., without knowing the group to which the patient was assigned) demonstrate either that the extra oxygen provides no benefit or that those given oxygen have greater mortality than those given less oxygen or even air.

The reasons for the worse outcomes are uncertain but there are two main possibilities. The first is that the high levels of oxygen cause the formation of reactive oxygen species that damage membranes and interfere with cellular function, and may even activate processes that lead to cell death (apoptosis). The other major possibility, and one that predates the first, is that oxygen increases vascular resistance in many tissues, including the brain and heart. Although the relative contributions of the possible mechanisms (e.g., interference by reactive oxygen species with endothelial nitric oxide, hyperoxia-induced hypocapnia) may be controversial, the physiological reasoning is incontrovertible.

A nonhypoxemic patient is, by definition, almost fully saturated and has, therefore, a near-maximum concentration of oxygen in the blood (200 mL $O_2$/L blood). Breathing 100 percent oxygen, therefore, cannot increase haemoglobin saturation; the only way to increase arterial oxygen content is by increasing the amount of oxygen dissolved in the blood. But because oxygen is relatively insoluble, breathing 100 percent oxygen will provide no more than an 8 percent increase in content. In contrast, breathing 100 percent oxygen commonly provokes decreases of cerebral or coronary blood flow of 20 percent or more. If the perfusion decreases more than the oxygen concentration increases, a paradox results: breathing 100 percent oxygen actually decreases oxygen delivery to tissues. Although the arterial $P_{O_2}$ may increase to as much as 600 mmHg during inhalation of oxygen, this increased driving pressure for diffusion is of limited value if perfusion (which transports the amount of oxygen that cells need to consume) is inadequate.

Even before these recent (as late as 2013) findings of the effects of administration of oxygen to nonhypoxemic adults appeared, a major change in the guidelines about the use of oxygen for resuscitating newborns had occurred. Administering 100 percent oxygen to newborns in difficulty (a low Apgar [Appearance, Pulse, Grimace, Activity, Respiration] score—introduced by Virginia Apgar, an American physician who founded the discipline of neonatology) delays the onset of spontaneous breathing and generally results in worse outcomes. Some have linked the administration of oxygen to neurological disorders, but this is contentious. Readers interested in this fascinating story, and the difficulties overcoming the inherent bias in favour of oxygen, can start by checking the scientific literature for publications by the Norwegian neonatologist, O.D. Saugstad (see also [2010]. *Lancet, 376*[9757]: 1970–71).

When there is clear evidence of hypoxia, either at the systemic or tissue level, oxygen is a lifesaver. One example is the use of high-pressure (hyperbaric) oxygen for treating gangrene or foot ulcers in people with diabetes. However, individuals promoting the use of hyperbaric chambers often claim unproven benefits on the grounds that hyperbaric oxygen delivers more oxygen to tissues. It is true that the arterial $P_{O_2}$ can be as high as 2000 mmHg, but this does not ensure that more oxygen is actually delivered—and it is uncertain that this would be beneficial—because they fail to take into account that a high $P_{O_2}$ reduces perfusion. As is true for any drug, oxygen has its uses, but it must not be used indiscriminately.

**12**

### Check Your Understanding 12.12

1. Describe the effects of changes in alveolar ventilation on the contents and partial pressures of oxygen and carbon dioxide in arterial blood.

## 12.5 | Control of Breathing

Like the heartbeat, breathing must occur in a continuous, cyclic pattern to sustain life. Just as the heart rhythmically contracts and relaxes to alternately pump blood from the heart and fill it again, the inspiratory muscles rhythmically contract and relax to alternately fill the lungs with air and empty them. Both these activities are accomplished automatically, without conscious effort. However, the underlying mechanisms and control of these two systems differ markedly.

### Respiratory centres in the brain

Whereas the heart can generate its own rhythm by means of its intrinsic pacemaker activity, the respiratory muscles, being skeletal muscles, contract only when stimulated. The rhythmic pattern of breathing is established by their cyclic neural activation. The neuronal discharges that establish breathing rhythm, and reflexively adjust the level of ventilation to match uptake of oxygen to removal of carbon dioxide, originate in the respiratory control centres in the brain (specifically the medulla), not in the lungs or respiratory muscles. This contrasts with the heart, where external input is not needed to initiate a heartbeat, but rather external input can modify the rate and strength of cardiac contraction. Furthermore, unlike cardiac activity, which is not subject to voluntary control, respiratory activity can be voluntarily modified to accomplish speaking, singing, whistling, playing a wind instrument, or holding one's breath while swimming.

## COMPONENTS OF NEURAL CONTROL OF RESPIRATION

Neural control of respiration involves three distinct components: (1) generation of the alternating inspiratory/expiratory rhythm, (2) regulation of the level of ventilation (i.e., the rate and depth of breathing) to match metabolism, and (3) modification of respiratory activity to serve other purposes. The latter modifications may be either voluntary, as in speech, or involuntary, as in cough or sneeze.

Respiratory control centres in the brain stem generate the rhythmic pattern of breathing. The primary respiratory control centre, the *medullary respiratory centre*, consists of several aggregations of cells within the medulla that generate output to the respiratory muscles. (These observations about the location of the neurons are based on recordings from animals.) In addition, other respiratory centres are present in the pons, a more rostral region of the brain stem. These centres influence output from the medullary respiratory centre. Following is a description of how these various regions interact to establish respiratory rhythmicity.

## INSPIRATORY AND EXPIRATORY NEURONS IN THE MEDULLARY CENTRE

We rhythmically breathe in and out during quiet breathing because of alternate contraction and relaxation of the inspiratory muscles—the diaphragm and external intercostal muscles—which are supplied by the phrenic nerve and intercostal nerves, respectively. The cell bodies of the axons projecting to these muscles are located in the spinal cord. When activated, they stimulate motor units of the inspiratory muscles, causing inspiration; when inactive, the inspiratory muscles relax, and expiration takes place. These spinal motor neurons receive inputs originating from pre-inspiratory neurons (i.e., neurons that do not project directly to the muscles) of the medullary respiratory centre.

The **medullary respiratory centre** consists of two clusters known as the *dorsal respiratory group* and the *ventral respiratory group*. The **dorsal respiratory group (DRG)** contains primarily *inspiratory neurons* whose descending fibres terminate on the spinal motor neurons that supply the inspiratory muscles. When neurons in the DRG fire, inspiration occurs; when they cease firing, expiration occurs. Expiration ends when synaptic input to the inspiratory neurons is again sufficient to bring them to threshold and make them discharge. Many neurons of the DRG project to neurons of the ventral respiratory group (VRG); if these projections are their only ones, they are referred to as *respiratory interneurons* because their connections remain within the respiratory control centres. They may also project to neurons in the VRG via collaterals from their axons that project to inspiratory motor neurons in the spinal cord.

The **ventral respiratory group (VRG)** is composed of *inspiratory neurons* and *expiratory neurons*, some of which are inactive during resting breathing. Many neurons that discharge in phase with inspiration, especially those in the nucleus ambiguus, are actually inspiratory motor neurons because they project directly to the many skeletal muscles, including the tongue, that widen (abduct) the upper airway (larynx and oropharynx) and reduce resistance to flow during inspiration.

Neurons that discharge in phase with expiration (defined as the absence of activity of the inspiratory muscles) are premotor expiratory neurons and expiratory motor neurons. The former project to motor neurons in the spinal cord that innervate the expiratory pump muscles (abdominal and internal intercostal muscles). The latter innervate muscles that narrow (adduct) the upper airway. Many neurons in this region are recruited by increased input from neurons of the DRG during periods of increased ventilatory demand, and this accounts for the recruitment of expiratory pump muscles when expiration becomes active.

## GENERATION OF RESPIRATORY RHYTHM

Until recently, the DRG was generally thought to generate the basic rhythm of ventilation. Then, the generation of respiratory rhythm was attributed to pacemaker-like neurons in the **pre-Bötzinger complex**, a region located near the rostral end of the VRG. The issue of whether or not neurons in this complex display pacemaker activity is contentious. To be a true pacemaker, a cell has to display rhythmic changes in membrane potential in the absence of synaptic input, and this has not always been observed. (Moreover, the first preparations in which activity was recorded from these neurons were probably hypoxic, if not anoxic; the respiratory activity more closely resembled gasping, not normal breathing.) The most recent data indicate that respiratory rhythm arises from synaptic interactions between multiple structures, specifically the VRG, the Bötzinger and pre-Bötzinger complexes, and the pons. There may well be a pacemaker component in the pre-Bötzinger complex, but it is one in which its state is determined by inputs from other structures.

## Influences from higher centres

It is a truism that every neuron in the brain is connected, inevitably, with every other neuron. It is therefore not surprising that structures elsewhere in the brain stem affect respiratory pattern. In the 1920s, an English investigator, Thomas Lumsden, observed that destruction of certain areas in the pons dramatically altered breathing pattern in anesthetized cats. One area, called the **apneustic centre** (from the Greek, "to hold one's breath") was associated with prolonged and deep inspirations; another, called the **pneumotaxic centre** (from the Greek, "to regulate respiratory rate") stopped inspiration and caused prolonged expirations. These terms—apneustic and pneumotaxic—are still used by some investigators, but they have little physiological meaning because they describe breathing patterns only rarely observed, as during pathological conditions such as coma. More importantly, although lesions of these areas cause abnormal breathing patterns while the experimental animals are anesthetized, their breathing appears normal after recovery from the anesthetic.

## DISCHARGE PATTERNS OF RESPIRATORY NEURONS AND MOTOR NEURONS

Respiratory neurons in the brain stem can be characterized by their different discharge patterns, as defined by timing and discharge frequency. In terms of timing, many neurons fire only during one phase (i.e., during inspiration or expiration); others fire only during a particular phase (e.g., during late inspiration) or during the transition between phases. In terms of discharge patterns, common ones include augmenting (i.e., discharge frequency starts at a low value and increases during the time the neuron is active), decrementing (i.e., high initial onset frequency that decreases thereafter), or plateau (rapid increase to a peak followed by a steady discharge until firing ceases). Many investigators have modelled these discharge patterns to reveal how the interconnections between these various neuronal types can generate the oscillating pattern of discharge that defines respiration.

These discharge patterns can be most readily understood by linking them to the functions of the respiratory muscles that generate pressure and those that regulate resistance of the upper airway. During inspiration, the elastic recoil of the respiratory system, especially the lung, increases (see ⟩ Figure 12-10). To overcome this increasing recoil, the discharge of inspiratory pump muscles increases (⟩ Figure 12-38). By both increasing the frequency of discharge and recruiting previously inactive motor neurons during an inspiration, inspiratory muscles generate more pressure to overcome the elastic recoil. These inspiratory motor neurons all stop firing at the end of inspiration regardless of when they started to fire during inspiration.

The discharge pattern of expiratory motor neurons can be similarly understood in terms of their function. During resting breathing, expiratory flow is generated through passive recoil;

as a result, expiratory motor neurons are often inactive. (This does not mean that expiratory premotor neurons in the brain stem are inactive; they may be discharging at levels too low to activate the motor neurons to which they project. And expiratory interneurons—those that project to other neurons participating in the generation of respiratory rhythm—are still active.) But when ventilation increases in response to metabolic or other demands, expiratory motor neurons are recruited; in addition to the pressure generated by elastic recoil, this increases expiratory flow. Unlike inspiratory motor neurons, however, expiratory motor neurons often stop discharging during the expiratory phase before the onset of the next inspiration because the last part of expiration can include, especially at rest, an expiratory pause.

The discharge patterns of motor neurons that adjust the calibre of the upper airway differ markedly from those providing input to the muscles that generate airflow. During inspiration, motor neurons innervating the upper airway abductors often start to fire before those innervating the inspiratory pump muscles, rapidly reach a peak discharge frequency, and maintain this level of activity until the end of inspiration. This pattern results in a rapid opening of the upper airway that keeps it open during inspiration, thereby decreasing flow resistance and, consequently, the energy expended to generate inspiratory flow.

At high levels of ventilatory demand, as in exercise, some of these motor neurons become tonically (continuously) active, keeping the airways open and helping to reduce flow resistance during both phases of respiration. In contrast, motor neurons innervating upper airway muscles that are active during expiration often display activity at low ventilatory levels—especially at the start of expiration—then a decrease in activity during the expiratory phase. This results in narrowing of the upper airway, increasing flow resistance and impeding expiratory flow, especially at the start of expiration. This helps maintain lung volume and gas exchange. As ventilatory demands increase, however, their activity decreases, ensuring that the upper airway is dilated and therefore offers minimal resistance to flow.

In summary, the discharge patterns of the motor neurons innervating the respiratory muscles—pump muscles (flow generation) and airway muscles (control of flow resistance)—are appropriate for their respective functions.

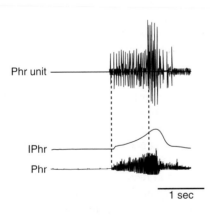

Republished with permission of the American Physiological Society, from S. Milano, Gretlot, A. L., and S. Iscoe, "Discharge patterns of phrenic motoneurons during fictive coughing and vomiting in decerebrate cats," *Journal of Applied Physiology*, 73 (4), 1992; permission conveyed through Copyright Clearance Center, Inc.

⟩ **FIGURE 12-38 Phrenic motor neuron discharge patterns.** Traces, from top down: a small filament from the phrenic nerve with activity of two distinct motor axons (Phr unit); the smoothed (integrated) activity of the contralateral whole phrenic nerve (IPhr); and the raw activity of the contralateral phrenic nerve (Phr). During inspiration in a decerebrate cat, the amplitude of the discharge of all phrenic motor neurons (which innervate motor units of the diaphragm) increases in a ramplike pattern. This is due to both an increase in the discharge frequency of individual motor axons and the recruitment of additional motor axons throughout inspiration. Later-recruited motor axons innervate larger motor units that generate more force. The increased discharge overcomes the increased elastic recoil of the lung as lung volume increases.

### Check Your Understanding 12.13

1. Describe the differences in discharge patterns between motor neurons innervating pump muscles (e.g., diaphragm) and motor neurons innervating muscles controlling the diameter of the upper airway

2. How are these discharge patterns affected by increases in the chemical drive to breathing?

## Control of respiratory pattern

Two main types of receptors, mechanical and chemical, control respiratory pattern and, therefore, the level of ventilation.

## MECHANORECEPTORS

Any receptor that detects a change in pressure, flow, or displacement of a structure in the respiratory system is a *mechanoreceptor*. These receptors are found in both components of the respiratory system, the lung and chest wall.

## PULMONARY RECEPTORS

The lungs contain some of the best-studied mechanoreceptors in physiology, in part because their reflex effects can be easily measured and because the effects of their activation can be readily eliminated by section of the vagus nerves (cranial nerve X), the nerves carrying their afferents to the brain stem. They include slowly and rapidly adapting receptors (with myelinated afferents) and C-fibres (with unmyelinated afferents). (See › Figure 12-19.)

- *Slowly adapting receptors*: **Slowly adapting receptors**, the endings of which are located in the smooth muscle of the airways respond to changes in lung volume (actually, changes in transmural pressure). Inflation of the lung activates them, if previously inactive, or increases their discharge if they were already active. As their name implies, they maintain their discharge frequency if lung inflation is sustained. They are the receptors responsible for one of the best known reflexes, the Breuer–Hering reflex.

- *Rapidly adapting receptors*: **Rapidly adapting receptors**, in contrast, have endings in the airway epithelia, primarily in larger airways. In response to chemical or mechanical stimuli, they exhibit a rapid burst of high frequency action potentials that quickly adapts (ceases). They used to be called irritant receptors because they could be activated by such stimulants as smoke or histamine. Their activation can lead to airway narrowing and cough, part of a protective reflex that limits inhalation of irritants into the lung. Their activation can also elicit mucus production that helps eliminate inhaled particles. Also, because they respond to deflation of lung regions, they initiate sighs, which help open regions of the lung that are in the process of closing (atelectasis) and thereby help maintain gas exchange.

- *C-fibres (J receptors)*: The endings of these receptors are close (or juxtaposed; J) to the pulmonary capillaries. As such, they can detect increases in pulmonary arterial pressure and exudation of fluid into the interstitial space (i.e., pulmonary oedema). They also respond to chemical stimuli, such as capsaicin (the pungent extract of chili peppers), that signal inflammation. When activated, they cause bronchoconstriction and rapid shallow breathing (tachypnea) typical of dyspnea. Their activation also causes inhibition of spinal motor neurons; this would reduce motor activity associated with exercise and, hence, the need to increase cardiac output to the pulmonary circulation.

## BREUER–HERING REFLEX

The first description of a negative biological feedback loop was made in the 1860s by Joseph Breuer (the student) and Ewald Hering (the professor) in Austria. In rabbits, they observed that inflation of the lungs inhibited inspiration and prolonged expiration, whereas deflation had the opposite effects. (They could easily monitor the rabbit's breathing by watching the opening and closing of its nostrils.) They referred to this phenomenon as *selbststeurung* (German for "self-steering"). In hundreds of subsequent studies, researchers have confirmed these findings, in ever greater detail, in anaesthetized animals. In brief, the discharge of brain stem inspiratory neurons activates inspiratory motor neurons, causing lung inflation. Inflation activates the pulmonary slowly adapting receptors. The discharge of their afferents, in the vagus nerve projecting to the medulla, activates neurons in the brain stem that inhibit inspiratory premotor neurons, thereby stopping inspiration. This is *the* classic negative feedback loop.

This feedback loop controls the depth and duration of inspiration in spontaneously breathing *anaesthetized* animals. These studies show that more feedback from slowly adapting receptors is required to terminate inspiration soon after the onset of inspiration, but less feedback is required as inspiration continues. The central (brain stem respiratory centre) threshold, in terms of feedback from slowly adapting receptors, for terminating inspiration therefore falls as inspiration proceeds. Thus, large tidal volumes are associated with short inspiratory durations, and small tidal volumes are associated with longer inspiratory durations. Studies in human infants also suggest that this reflex is operative.

There is, however, little evidence that the Breuer–Hering reflex operates in *conscious* animals and, especially, in conscious humans. Indeed, in conscious adults, the longer the inspiration (within limits), the larger is the tidal volume. If one adds carbon dioxide to the inspirate, this relation still holds, but the relation is steeper. That is, for a given duration of inspiration, a larger tidal volume is achieved due to an increased flow. This is consistent with carbon dioxide increasing the drive to breathe (i.e., production of a greater inspiratory flow). The Breuer–Hering reflex does appear to operate at higher tidal volumes, when inspiration becomes more difficult because the system is less compliant.

The Breuer–Hering reflex may, however, operate *during* inspiration as opposed to terminating inspiration. In lightly anaesthetized animals, activity in the phrenic nerves (that innervate the diaphragm) is greater during unimpeded inspirations than during inspiratory phases when lung inflation is prevented. In other words, the slowly adapting receptors provide *positive feedback* during inspiration. The probable explanation is that increased activity (higher discharge frequencies of already active motor units and recruitment of previously inactive ones) is needed to overcome progressively greater elastic recoil as lung volume increases. If the level of anesthesia is slightly increased, this effect disappears, accounting for the failure to observe this phenomenon in earlier studies in more deeply anesthetized animals.

## CHEST WALL RECEPTORS

- *Diaphragm:* The diaphragm contains relatively few receptors—spindles and Golgi tendon organs—with myelinated afferents, and the effects of their activation are poorly understood. The lack of these receptors, so prevalent in limb muscles, may reflect the fact that the diaphragm's role

12

is largely limited to respiration and such protective reflexes as coughing, sneezing, and vomiting. During these reflexes, these receptors may act to limit its contraction and, therefore, the high intrathoracic and abdominal pressures that are generated. In contrast, the diaphragm has many receptors with small myelinated and unmyelinated afferents, the receptors of which respond to local metabolic conditions (*metaboreceptors*). Their activation may elicit a reflex inhibition of the diaphragm, via either spinal or supraspinal interneurons, thereby reducing its activity and the production of catabolites. Their activation may also contribute to the sense of dyspnea.

- *Rib cage:* Unlike the diaphragm, the muscles of the chest wall, especially the intercostal muscles, are richly supplied with muscle spindles and, to a lesser degree, Golgi tendon organs. What is unclear, however, is their role in respiration. The spindles are identical to those in limb muscles and are therefore able to detect discrepancies between expected outcome (shortening of the intrafusal fibres due to activation of the $\gamma$ efferents) and achieved outcome (shortening of the extrafusal fibres via $\alpha$ efferents). A discrepancy, such as a smaller increase in chest wall distension than expected, would cause the spindle receptor to discharge, eliciting a reflex increase in $\alpha$ efferent activity in an attempt to "unload" the spindle. Some researchers believe that this failure to meet contractile expectations—in this case, a person's awareness of not getting the anticipated tidal volume—is the underlying cause of breathlessness (dyspnea). This is plausible, but recall that the intercostals and abdominal muscles are not only respiratory muscles; they also have an important role in posture or stabilization of the thorax during such movements as lifting a heavy object. Therefore, the spindles could also have a role in postural adjustments.

## CHEMICAL CONTROL

### VENTILATION ADJUSTMENT

No matter how much oxygen is extracted from the blood or how much carbon dioxide is added to it at the tissue level, the $P_{O_2}$ and $P_{CO_2}$ of the arterial blood are held remarkably constant, indicating that arterial blood-gas tensions are precisely regulated.

Arterial blood gases are maintained within the normal range by varying the rate and depth of breathing (and so too is alveolar ventilation) to match the body's metabolism (oxygen uptake and carbon dioxide production). If more oxygen and carbon dioxide are extracted from and unloaded to the alveoli, respectively, because of increased tissue metabolism, alveolar ventilation increases to maintain arterial blood gases within a normal range.

The medullary respiratory centres receive many inputs. One of the most important is information about the chemical composition of the blood. The two most obvious signals to increase ventilation are a decreased arterial $P_{O_2}$ and an increased arterial $P_{CO_2}$. Intuitively, you would suspect that if oxygen levels in the arterial blood declined or if carbon dioxide accumulated, ventilation would be stimulated to obtain more oxygen or to eliminate excess carbon dioxide. These two factors do indeed influence the magnitude of ventilation, but not to the same degree nor through the same pathway(s). Also, a third chemical factor, $H^+$, influences the level of respiratory activity. Next we examine the role of each of these important chemical factors in the control of ventilation (see ▌ Table 12-10).

### DECREASED ARTERIAL $P_{O_2}$

Arterial levels of oxygen are monitored by **peripheral chemoreceptors**, the **carotid bodies** and **aortic bodies**, which lie at the forks of the left and right common carotid arteries and in the arch of the aorta, respectively (❯ Figure 12-39). These chemoreceptors respond to specific changes in the chemical content of the arterial blood that bathes them. (They should not be confused with the carotid sinus and aortic arch baroreceptors, located in the same vicinity, which help regulate systemic arterial blood pressure (p. 419). Those interested in a more detailed description of the versatility of both sets of chemoreceptors, and the role of ATP, should consult work by Colin Nurse at McMaster University (Piskuric & Nurse [2013]. *Journal of Physiology, 591*: 415–22).

### EFFECT OF A LARGE DECREASE IN $P_{O_2}$ ON THE PERIPHERAL CHEMORECEPTORS

In terms of respiratory control, the carotid chemoreceptors are more important than the aortic chemoreceptors. This is

▌ **TABLE 12-10** Influence of Chemical Factors on Respiration

| Chemical Factor | Effect on the Carotid Chemoreceptors | Effect on the Central Chemoreceptors |
|---|---|---|
| ↓ $P_{A_{O_2}}$ | Activated only when arterial $P_{O_2} < 60$ mmHg; an emergency mechanism | Directly depresses central chemoreceptors and respiratory centre when < 60 mmHg, but may activate other neurons in brain stem |
| ↑ $P_{A_{CO_2}}$ ( = ↑ $H^+$ in brain ECF) | Weakly stimulates; sensitizes receptors to hypoxia | Strongly stimulates; is the dominant control of ventilation (levels > 70–80 mmHg depress the respiratory centre and central chemoreceptors) |
| ↑ arterial [$H^+$] | Stimulates; important in acid–base balance; sensitizes receptors to hypoxia | Does not affect; cannot penetrate blood–brain barrier |

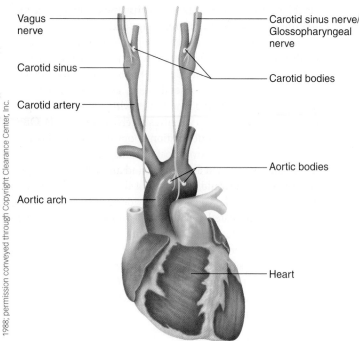

Vagus nerve

Carotid sinus

Carotid artery

Aortic arch

Carotid sinus nerve/ Glossopharyngeal nerve

Carotid bodies

Aortic bodies

Heart

› **FIGURE 12-39** **Discharge patterns of pump and airway muscles**

because the former respond to changes in arterial $P_{O_2}$, whereas the latter respond to changes in arterial oxygen content. The reasoning is as follows: increases and decreases in alveolar ventilation increase and decrease, respectively, the alveolar $P_{O_2}$. Assuming equilibration between alveolar and arterial pressures, this means that alveolar ventilation controls arterial $P_{O_2}$. But the carotid chemoreceptors are relatively insensitive to modest decreases in arterial $P_{O_2}$ until the $P_{O_2}$ falls below 60 mmHg, the level at which desaturation starts to fall to levels that could impair oxygenation of peripheral tissues. When arterial $P_{O_2}$ does fall below 60 mmHg, the receptors are activated and reflexively increase ventilation and, therefore, alveolar $P_{O_2}$. Activating the carotid chemoreceptors at $P_{O_2}$ s > 60 mmHg is unnecessary because an increase in alveolar $P_{O_2}$ would not increase arterial oxygen content much more since the blood is already well saturated.

In contrast, the aortic chemoreceptors respond not to arterial $P_{O_2}$ but to arterial oxygen content. When content falls, their activation causes an increase not in ventilation but in cardiac output, thereby increasing systemic oxygen delivery (the product of arterial oxygen content and cardiac output: $mL\,O_2/L\,blood \times L\,blood/min = mL\,O_2/min$).

You can appreciate the distinction between the two stimuli by considering an individual with anaemia. This person has a normal arterial $P_{O_2}$ but a reduced arterial oxygen content. Consequently, there is no need for a reflex increase in ventilation. An increase would marginally increase arterial $P_{O_2}$ but would not increase content; only correction of the anaemia would do that. An anaemic person would, however, benefit from an increased cardiac output because this compensates for the reduced arterial oxygen content, thereby maintaining oxygen delivery to the periphery.

In summary, the two sets of peripheral arterial chemoreceptors work together, when needed, to ensure adequate oxygen delivery to the tissues. The carotid chemoreceptors respond to and control arterial $P_{O_2}$ and therefore arterial oxygen content, by adjusting ventilation. The aortic chemoreceptors respond to decreases in arterial oxygen content and adjust cardiac output to maintain oxygen delivery to the tissues.

### DIRECT EFFECT OF A LARGE DECREASE IN $P_{O_2}$ ON THE RESPIRATORY CENTRES

*Clinical Note* In general, hypoxia depresses cellular activity, particularly the maintenance of ionic gradients. Were it not for the intervention of the peripheral chemoreceptors when the arterial $P_{O_2}$ falls below 60 mmHg, a vicious cycle ending in cessation of breathing would ensue. Direct depression of the respiratory centre by a markedly low arterial $P_{O_2}$ would further reduce ventilation, leading to an even greater fall in arterial $P_{O_2}$ that would depress the respiratory centre even more until ventilation ceased and death occurred. (This is an example of positive feedback; most positive feedbacks destabilize a system, resulting in a bad outcome.)

Until recently, the only cells thought capable of responding to hypoxia were the carotid chemoreceptors. Although most cells in the brain decrease their metabolic activity during exposure to hypoxia, some neurons in the caudal hypothalamus and rostral ventrolateral medulla of rats increase their activity during hypoxia. Nevertheless, in animals lacking feedback from the periphery (cut vagus and carotid sinus nerves), progressive hypoxia depresses inspiratory activity, so the hypoxia-induced activation of these neurons has little effect.

There is another related issue. Newborn animals are extremely resistant to hypoxia; they can recover completely from prolonged anoxia or asphyxiation that would kill an adult. This feature has obvious evolutionary value during and immediately after birth until breathing starts.

### CARBON DIOXIDE–GENERATED H⁺

In contrast to arterial $P_{O_2}$, which does not contribute to the minute-to-minute regulation of ventilation, arterial $P_{CO_2}$ is the most important factor regulating ventilation under resting conditions. This role is appropriate because changes in alveolar ventilation have an immediate and pronounced effect on arterial (see alveolar ventilation equation, p. 540). Even slight alterations from normal in arterial $P_{CO_2}$ induce significant reflex changes in ventilation. An increase in arterial $P_{CO_2}$ stimulates the respiratory centre, and the resulting increase in ventilation promotes elimination of the excess carbon dioxide to the atmosphere. Conversely, a fall in arterial $P_{CO_2}$ reduces respiratory drive and, therefore, ventilation; as a result, metabolically produced carbon dioxide accumulates, returning to $P_{CO_2}$ normal.

### EFFECT OF INCREASED $P_{CO_2}$ ON THE CENTRAL CHEMORECEPTORS

Surprisingly, given the key role of arterial $P_{CO_2}$ in regulating respiration, no important receptors monitor arterial $P_{CO_2}$ *per se*. The carotid bodies respond weakly to changes in arterial $P_{CO_2}$, so they play only a minor role in reflexively stimulating ventilation in response to an elevation in arterial $P_{CO_2}$. However, they do sensitize the carotid chemoreceptors to hypoxia, such that for a given arterial $P_{O_2}$, the response is greater during hypercarbia. More important in linking changes in arterial $P_{CO_2}$ to compensatory adjustments in ventilation are the **central chemoreceptors**, located in the medulla near the respiratory centres. Originally, these receptors were thought to lie on or just beneath the ventral surface of the medulla, but later work indicates that they are present in many areas of the brain stem, often close to capillaries in regions containing neurons that generate the respiratory rhythm. These central chemoreceptors do not monitor carbon dioxide itself; however, they are sensitive to carbon dioxide–induced changes in $H^+$ concentration in the brain extracellular fluid that bathes them.

Movement of materials across the brain capillaries is restricted by the blood–brain barrier (p. 100). Because this barrier is readily permeable to carbon dioxide, any increase in arterial $P_{CO_2}$ causes a similar rise in brain extracellular fluid $P_{CO_2}$ as carbon dioxide diffuses down its partial pressure gradient from the cerebral capillaries into the brain extracellular fluid. The increased $P_{CO_2}$ within the brain extracellular fluid correspondingly raises the concentration of $H^+$ according to the law of mass action as it applies to this reaction: $CO_2 + H_2O \Leftrightarrow H_2CO_3 \Leftrightarrow H^+ + HCO_3^-$. An elevation of $H^+$ concentration in the brain extracellular fluid directly stimulates the central chemoreceptors, which in turn increase ventilation by stimulating the respiratory centre through synaptic connections (Figure 12-40). As the excess carbon dioxide is subsequently blown off, the arterial $P_{CO_2}$ and the $P_{CO_2}$ and $H^+$ concentration of the brain extracellular fluid return to normal. Conversely, a decline in arterial $P_{CO_2}$ below normal is paralleled by a fall in $P_{CO_2}$ and $H^+$ in the extracellular fluid; the result of this is a central chemoreceptor–mediated decrease in ventilation. As carbon dioxide produced by cell metabolism is consequently allowed to accumulate, arterial $P_{CO_2}$ and $P_{CO_2}$ and $H^+$ of the brain extracellular fluid return toward normal.

Unlike carbon dioxide, $H^+$ cannot readily permeate the blood–brain barrier, so $H^+$ in the plasma cannot gain access to the central chemoreceptors. Accordingly, the central chemoreceptors respond only to $H^+$ generated within the brain extracellular fluid itself as a result of the entry of carbon dioxide. Thus, the major mechanism controlling ventilation under resting conditions is specifically aimed at regulating $H^+$ concentration in the brain extracellular fluid, which in turn directly reflects the arterial $P_{CO_2}$. Unless there are extenuating circumstances, such as reduced availability of oxygen in the inspired air, arterial $P_{O_2}$ is coincidentally maintained at its normal value by receptors responding to brain extracellular fluid $H^+$.

The powerful influence of the central chemoreceptors on the respiratory centre is responsible for your inability to hold your breath for more than about a minute. While you hold your breath, metabolically produced carbon dioxide continues to accumulate in your blood and then to build up the $P_{CO_2}$ / $H^+$ concentration in your brain extracellular fluid. Finally, the increased $H^+$ stimulus to respiration becomes so powerful that central chemoreceptor excitatory input overrides the voluntary inhibition of breathing. This happens long before arterial $P_{O_2}$ falls to the threateningly low levels that trigger the carotid chemoreceptors. Therefore, you cannot deliberately hold your breath long enough to create a dangerously high level of carbon dioxide or low level of oxygen in the arterial blood.

But activation of the central chemoreceptors is not the sole determinant of breath-hold duration. If, for example, a person at the breakpoint of a breath-hold is allowed to inhale a gas mixture that does not improve the blood gases, the person can continue to hold his or her breath. Moreover, blowing air in through the mouth and out through the nostrils while breath-holding also allows the breath-hold to continue; the sensation of airflow through the nose apparently fools people into believing they are breathing. Consider the experimental situation in which a person agrees be paralyzed except for an arm used to signal the experimenter (the arm is spared because a tourniquet prevents its paralysis). This paralyzed person can hold his or her breath

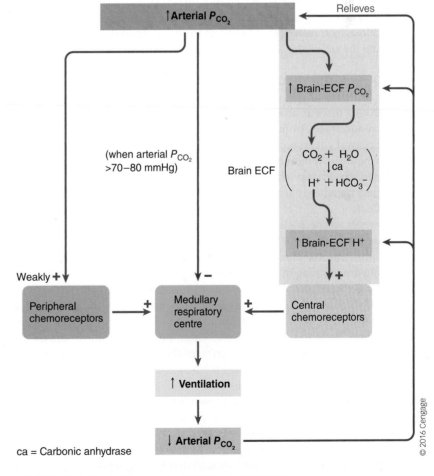

ca = Carbonic anhydrase

> **FIGURE 12-40** Effect of increased arterial $P_{CO_2}$ on ventilation

much longer than others who can voluntarily breathe. These results indicate that unrewarded breathing efforts, particularly contractions of the diaphragm, are a source of sensory input that causes the person to stop the breath-hold. The relevance of these experiments is that they provide information related to the sensation of breathlessness (dyspnea), a debilitating symptom in many people with respiratory diseases.

### DIRECT EFFECT OF A LARGE INCREASE IN $P_{CO_2}$ ON THE RESPIRATORY CENTRES

*Clinical Note* In contrast to the normal reflex stimulatory effect of the increased $P_{CO_2}$-$H^+$ mechanism on respiratory activity, very high levels of carbon dioxide directly depress the entire brain, including the respiratory centre, just as very low levels of oxygen do. Up to a $P_{CO_2}$ of 70–80 mmHg, progressively higher $P_{CO_2}$ levels induce correspondingly more vigorous respiratory efforts in an attempt to blow off the excess carbon dioxide. A further increase in $P_{CO_2}$ beyond 70–80 mmHg, however, does not increase ventilation more. but actually depresses (hyperpolarizes) respiratory neurons. For this reason, carbon dioxide must be removed and oxygen supplied in closed environments, such as closed-system anesthesia machines, submarines, or space capsules.

Surprisingly, however, high levels of carbon dioxide are not lethal if the $P_{CO_2}$ is adequate. There are reports of individuals who have, accidentally, experienced extremely high levels of arterial $P_{O_2}$ (supercarbia) but have recovered with no neurological deficits. The likely explanation is that the body can limit the intracellular acidosis by transferring bicarbonate from the extracellular space into the cell.

### LOSS OF SENSITIVITY TO $P_{CO_2}$ WITH LUNG DISEASE

During prolonged hypoventilation caused by certain types of chronic lung disease, an elevated $P_{CO_2}$ occurs simultaneously with a reduced $P_{O_2}$. In most cases, the elevated $P_{CO_2}$ (acting via the central chemoreceptors) and the reduced (acting via the peripheral chemoreceptors) are *synergistic*; that is, the combined stimulatory effect on respiration exerted by these two inputs together is greater than the sum of their independent effects.

However, some patients with severe chronic lung disease lose their sensitivity to an elevated arterial $P_{CO_2}$. Chronic retention of carbon dioxide results in an increased generation of hydrogen ions in the brain extracellular fluid, which is eventually neutralized by the influx of bicarbonate across the blood–brain barrier. The additional bicarbonate combines with the excess $H^+$, removing it from solution so that it no longer contributes to the free $H^+$ concentration. As a result, the $H^+$ concentration in the brain extracellular fluid returns to normal, although $P_{CO_2}$ in the arterial blood and brain extracellular fluid remains high. Because the central chemoreceptors respond to $H^+$, not $P_{CO_2}$, there is no reflex stimulation of the respiratory centres. Consequently, the drive to eliminate carbon dioxide is blunted in such patients; that is, their level of ventilation is low considering their high arterial $P_{CO_2}$. In these patients, the hypoxic drive to ventilation is the primary respiratory stimulus, whereas in normal individuals the arterial

$P_{CO_2}$ is the dominant factor governing the level of ventilation. Paradoxically, administering oxygen to such patients to relieve the hypoxia can markedly depress ventilation. This leads to acute worsening of carbon dioxide retention because it removes the stimulus to breathe. In addition, administration of oxygen also worsens the mismatching of ventilation to perfusion ($\dot{V}_A / \dot{Q}$ see ⟩ Figure 12-31) due to the release of hypoxic vasoconstriction. This increases perfusion to alveoli that are relatively underventilated, thereby leading to an increase in arterial $P_{CO_2}$. Last, the higher $P_{O_2}$ displaces carbon dioxide (reverse Haldane effect) from haemoglobin, increasing arterial $P_{CO_2}$. Thus, supplemental oxygen must be administered cautiously to patients with chronic pulmonary diseases.

## Acid–base balance

Changes in arterial $H^+$ concentration cannot influence the central chemoreceptors because $H^+$ does not readily cross the blood–brain barrier. However, the carotid chemoreceptors do respond to fluctuations in arterial $H^+$ concentration, in contrast to their weak sensitivity to changes in arterial $P_{CO_2}$ and their unresponsiveness to arterial $P_{O_2}$ until it falls below approximately 60 mmHg.

Any change in arterial $P_{CO_2}$ brings about a corresponding change in the $H^+$ concentration of the blood as well as of the brain extracellular fluid. These carbon dioxide–induced changes of $H^+$ in the arterial blood are detected by the peripheral chemoreceptors; the result is a reflex increase in ventilation in response to increased arterial $H^+$ concentration (decreased pH) and a reflex decrease in response to a decreased arterial $H^+$ concentration (increased pH). However, these changes in ventilation mediated by the carotid chemoreceptors are far less important than the powerful central chemoreceptor mechanism(s) that adjust ventilation in response to changes in carbon dioxide–generated $H^+$ concentration.

The peripheral chemoreceptors do play a major role in adjusting ventilation in response to alterations in arterial $H^+$ concentration unrelated to fluctuations in $P_{CO_2}$. In many situations, even though $P_{CO_2}$ is normal, arterial $H^+$ concentration is changed by the addition or loss of noncarbonic acids from the body. For example, arterial $H^+$ concentration increases during diabetes mellitus because of the addition of excess $H^+$-generating keto acids to the blood. A rise in arterial $H^+$ concentration reflexly stimulates ventilation by means of the carotid chemoreceptors. Conversely, the carotid chemoreceptors reflexly suppress respiratory activity in response to a fall in arterial $H^+$ concentration resulting from nonrespiratory causes. (Nonrespiratory causes of a change in $H^+$ are generally referred to as *metabolic*—that is, metabolic acidosis and, rarely, metabolic alkalosis.) Changes in ventilation by this mechanism are extremely important in regulating the body's acid–base balance because they determine how much of the carbon dioxide generated by $H^+$ is eliminated. The resulting adjustment in the amount of $H^+$ added to the blood from carbon dioxide can compensate for the metabolic-induced abnormality in arterial $H^+$ concentration that first elicited the respiratory response. (See Chapter 14 for further details.)

**12**

## Effect of exercise on ventilation

Alveolar ventilation may increase up to twentyfold during heavy exercise (90 to 195 L/min) to keep pace with the increased demand for oxygen uptake and carbon dioxide elimination. ( Table 12-11 highlights changes in oxygen- and carbon dioxide–related variables during exercise.) The cause of increased ventilation during exercise is still largely speculative. Although it would seem logical that decreased arterial $P_{O_2}$ and increased arterial $P_{CO_2}$ and $H^+$ would account for the increase in ventilation, this does not appear to be the case.

- Despite the marked increase in oxygen use during exercise, arterial $P_{O_2}$ does not decrease but remains normal or may actually increase slightly, because the increase in alveolar ventilation keeps pace with or even slightly exceeds the elevated rate of oxygen consumption.

- Likewise, despite the marked increase in carbon dioxide production during exercise, arterial $P_{CO_2}$ does not increase but remains normal or even decreases slightly, because the extra carbon dioxide produced is removed as rapidly or even more rapidly by the increase in ventilation.

- During mild or moderate exercise, $H^+$ concentration does not increase because $H^+$-generating carbon dioxide is held constant. During heavy exercise, $H^+$ concentration does increase somewhat due to the release of $H^+$-generating lactic acid into the blood by anaerobic metabolism in the exercising muscles. Even so, the elevation in $H^+$ concentration resulting from lactic acid formation is not enough to account for the large increase in ventilation that accompanies exercise. This anaerobic production of $H^+$ is, however, sufficient to cause an increase in alveolar ventilation disproportionate to the production of carbon dioxide; as a result, the person actually hyperventilates (defined as a decrease in arterial $P_{CO_2}$), and this is responsible for the increase in arterial $P_{O_2}$ during exercise above the anaerobic threshold.

Some investigators argue that the constancy of the three chemical regulatory factors during exercise shows that ventilatory responses to exercise are actually controlled by these

## ▌ TABLE 12-11 Oxygen- and Carbon Dioxide–Related Variables during Exercise

| Oxygen- or Carbon Dioxide–Related Variables | Change | Comment |
|---|---|---|
| Oxygen Use | Marked ↑ | Active muscles oxidize nutrient molecules more rapidly to meet ↑ energy needs. |
| Carbon Dioxide Production | Marked ↑ | More actively metabolizing muscles produce more carbon dioxide. |
| Alveolar Ventilation | Marked ↑ | By mechanisms not completely understood, alveolar ventilation keeps pace with or even slightly exceeds the increased metabolic demands during exercise. |
| Arterial $P_{O_2}$ | Normal or slight ↑ | Despite a marked ↑ in oxygen use and carbon dioxide production during exercise, alveolar ventilation keeps pace with or even slightly exceeds the elevated rates of oxygen consumption and carbon dioxide production. |
| Arterial $P_{CO_2}$ | Normal or slight ↓ | |
| Oxygen Delivery to Muscles | Marked ↑ | Although arterial $P_{CO_2}$ remains normal, oxygen delivery to muscles is greatly ↑ by the ↑ blood flow to exercising muscles, accomplished by ↑ cardiac output coupled with local vasodilation of arterioles in active muscles. |
| Oxygen Extraction by Muscles | Marked ↑ | ↑ use of oxygen ↓ the $P_{O_2}$ at the tissue level, which results in more oxygen unloading from haemoglobin; this is enhanced by ↑ $P_{CO_2}$, and ↑ $H^+$ and ↑ temperature. |
| Carbon Dioxide Removal from Muscles | Marked ↑ | ↑ blood flow to exercising muscles removes the excess carbon dioxide produced by these more actively metabolizing tissues. |
| Arterial H¹ Concentration | | |
| Mild-to-Moderate Exercise | Normal | Because carbonic acid–generating carbon dioxide is held constant in arterial blood, arterial $H^+$ concentration does not change. |
| Heavy Exercise | Modest ↑ | In heavy exercise, when muscles resort to anaerobic metabolism, lactic acid is added to the blood. |

© 2016 Cengage

12

factors—particularly by $P_{CO_2}$, because it is normally the dominant variable controlled at rest. According to this reasoning, how else could alveolar ventilation be increased in exact proportion to carbon dioxide production, thereby keeping the $P_{CO_2}$ constant? This proposal, however, cannot account for the observation that during heavy exercise, alveolar ventilation often increases more than carbon dioxide production increases, thereby actually causing a slight decline in $P_{CO_2}$. Also, ventilation increases abruptly at the onset of exercise (within seconds), long before changes in arterial blood gases could influence the respiratory centre (which requires a matter of minutes).

## Other factors that may increase ventilation during exercise

A number of other factors, including the following, play a role in the ventilatory response to exercise:

1. *Reflexes originating from body movements.* Joint and muscle receptors (mechanoreceptors) excited during muscle contraction reflexly stimulate the respiratory centre, abruptly increasing ventilation. Golgi tendon organs, muscle spindles, and skeletal joint receptors send afferent signals to the sensory cortex, which then relays the information to the respiratory centre. Even passive movement of the limbs (e.g., someone else alternately flexing and extending a person's knee) may increase ventilation several fold through activation of these receptors, although no actual exercise is occurring. Thus, the mechanical events of exercise are believed to play an important role in coordinating respiratory activity with the increased metabolic requirements of the active muscles.

2. *Increase in body temperature.* Much of the energy generated during muscle contraction is converted to heat rather than to actual mechanical work. Heat-loss mechanisms, such as sweating, frequently cannot keep pace with the increased heat production that accompanies increased physical activity, so body temperature often rises slightly during exercise (p. 711). Because raised body temperature stimulates ventilation, this exercise-related heat production undoubtedly contributes to the respiratory response to exercise. For the same reason, increased ventilation often accompanies a fever.

3. *Epinephrine release.* The adrenal medullary hormone epinephrine also stimulates ventilation. The level of circulating epinephrine rises during exercise in response to the sympathetic nervous system discharge that accompanies increased physical activity.

4. *Impulses from the cerebral cortex.* Especially at the onset of exercise, the motor areas of the cerebral cortex are believed to simultaneously stimulate the medullary respiratory neurons and activate the motor neurons of the exercising muscles. This is similar to the cardiovascular adjustments initiated by the motor cortex at the onset of exercise. In this way, the motor control region of the brain elicits increased ventilatory and circulatory responses to support the increased physical activity it is about to orchestrate. These anticipatory adjustments are feedforward regulatory mechanisms; that is, they occur *before* any homeostatic factors actually change. This contrasts with the more common case in which regulatory adjustments to restore homeostasis take place *after* a factor has become altered.

Experiments in animals show that stimulation of a region of the brain that elicits locomotion also increases ventilation. But if an experimenter paralyzes the animal (and places it on mechanical ventilation), the same stimulation now elicits activity in the nerves projecting to the leg muscles and to the diaphragm. This activation occurs in the absence of any movement and, therefore, of any feedback from the periphery.

None of these factors or combinations of factors fully explains the abrupt and profound effect exercise has on ventilation, nor can they completely account for the high degree of correlation between respiratory activity and the body's needs for gas exchange during exercise.

## Apnea and dyspnea

**Apnea** is the transient interruption of ventilation, with breathing resuming spontaneously. If breathing does not resume, the condition is called **respiratory arrest**. Because ventilation normally decreases and the central chemoreceptors are less sensitive to the $P_{CO_2}$ drive during sleep, especially paradoxical sleep (p. 134), apnea (known as **sleep apnea**) is most likely to occur during this time.

### DYSPNEA

*Clinical Note* People who have dyspnea have the subjective sensation that they are not getting enough air; that is, they feel short of breath. *Dyspnea* (also called *breathlessness*) is the anguish associated with the unsatiated desire for more adequate ventilation. It often accompanies the laboured breathing characteristic of obstructive lung disease or the pulmonary oedema associated with congestive heart failure. In contrast, during exercise, a person can breathe very hard without experiencing dyspnea, because such exertion is not accompanied by a sense of anxiety over the adequacy of ventilation. Surprisingly, dyspnea is not directly related to chronic elevation of arterial $P_{CO_2}$ or reduction of $P_{O_2}$. The subjective feeling of air hunger may occur even when alveolar ventilation and the blood gases are normal. Some people experience dyspnea when they *perceive* that they are short of air, even though this is not actually the case, such as in a crowded elevator. A generally accepted theory, first advanced by E.J.M. Campbell (1925–2004) at McMaster University, is that the discrepancy between the respiratory drive—the "want"—and the tidal volume achieved—the "get"—accounts for the disabling sensation of dyspnea.

Many people with chronic obstructive pulmonary disease (COPD) experience dyspnea, especially when they increase their ventilation during exercise. Because expiration is difficult (the obstruction), they do not have time to breathe out completely to functional residual capacity, even to the elevated functional residual capacity characteristic of people with COPD. Consequently, they inhale more air than they exhale and, over time, they increase their end-expiratory lung volume, a process referred to as *dynamic hyperinflation* (p. 517). This limits how much they can inspire because the inspiratory reserve volume (the difference between total lung capacity and end-inspired lung volume) decreases. Thus, despite the disease being one that limits expiration, these people complain of an inability to breathe in adequately, especially during exercise or acute exacerbations of COPD. A fall in the inspiratory reserve volume below a certain value, typically about 0.8 L, is associated with the onset of severe dyspnea. One of the world's leading researchers in this area is Denis O'Donnell at Queen's University, Kingston, Ontario.

# Chapter in Perspective: Focus on Homeostasis

The respiratory system contributes to homeostasis by obtaining oxygen from, and eliminating carbon dioxide to, the external environment. All cells ultimately need an adequate supply of oxygen to use in oxidizing nutrient molecules to generate ATP. Adult brain cells that depend on a continual supply of oxygen die if deprived of oxygen for more than four minutes. Even cells that can resort to anaerobic metabolism for energy production, such as strenuously exercising muscles, can do so only transiently by incurring an oxygen deficit that ultimately must be made up during the period of excess post-exercise oxygen consumption.

As a result of these energy-yielding metabolic reactions, the body produces large quantities of carbon dioxide that must be eliminated. Because carbon dioxide and water form carbonic acid, adjustments in the rate of carbon dioxide elimination by the respiratory system are important in regulating acid–base balance in the internal environment.

## CHAPTER TERMINOLOGY

abdominal muscles (p. 504)
accessory inspiratory muscles (p. 511)
airways (p. 501)
alveolar dead space (p. 529)
alveolar gas equation (p. 525)
alveolar interdependence (p. 510)
alveolar surface tension (p. 509)
alveolar ventilation (p. 522)
alveoli (p. 502)
aortic bodies (p. 545)
apnea (p. 550)
apneustic centre (p. 542)
atelectasis (p. 528)
2,3-bisphosphoglycerate (BPG) (p. 537)
bicarbonate (p. 537)
Bohr effect (p. 536)
Boyle's/Mariotte's law (p. 511)
breathing (p. 500)
breathlessness (p. 517)
bronchi (p. 502)
bronchioles (p. 502)
bronchoconstriction (p. 515)
bronchodilation (p. 515)
capillary transit time (p. 527)
carbamino haemoglobin ($HbCO_2$) (p. 537)
carbon monoxide (CO) (p. 537)
carbonic acid ($H_2CO_3$) (p. 538)
carboxyhaemoglobin (HbCO) (p. 537)
carotid bodies (p. 545)
central chemoreceptors (p. 547)
chloride (Hamburger) shift (p. 538)
chronic bronchitis (p. 517)

chronic obstructive pulmonary disease (COPD) (p. 516)
collateral ventilation (p. 503)
convection (p. 502)
Dalton's law (p. 524)
diaphragm (p. 503)
dorsal respiratory group (DRG) (p. 542)
dynamic hyperinflation (p. 517)
dyspnea (p. 517)
elastic recoil (p. 505)
emphysema (p. 517)
equal pressure point (p. 513)
esophagus (p. 501)
expiratory muscles (p. 513)
expiratory reserve volume (ERV) (p. 518)
external intercostal muscles (p. 504)
external respiration (p. 498)
Fick's law of diffusion (p. 526)
forced expiratory volume in one second ($FEV_1$) (p. 519)
forced or active expiration (p. 513)
forced vital capacity (FVC) (p. 519)
functional residual capacity (FRC) (p. 518)
Haldane effect (p. 538)
hyperoxia (p. 539)
hyperpnea (p. 540)
hyperventilation (p. 540)
hypocapnia (p. 540)
hypoventilation (p. 540)
hypoxia (p. 539)
inspiration (p. 504)
inspiratory muscles (p. 503)

inspiratory reserve volume (IRV) (p. 518)
inspiratory capacity (IC) (p. 518)
intercostal muscles (p. 504)
intercostal nerves (p. 504)
internal (cellular) respiration (p. 498)
internal intercostal muscles (p. 504)
intrathoracic pressure (p. 506)
larynx (voice box) (p. 501)
law of LaPlace (p. 509)
law of mass action (p. 532)
lungs (p. 501)
medullary respiratory centre (p. 542)
minute ventilation (p. 521)
nasal passages (nose) (p. 501)
$O_2$–Hb dissociation (or saturation) curve (p. 532)
oxyhaemoglobin ($HbO_2$) (p. 532)
partial pressure (p. 524)
oxygen toxicity (p. 539)
partial pressure gradients (p. 524)
partial pressure of carbon dioxide ($Pco_2$) (p. 524)
partial pressure of oxygen ($Po_2$) (p. 524)
percent haemoglobin (% Hb) saturation (p. 532)
peripheral chemoreceptors (p. 545)
phrenic nerves (p. 503)
physiological dead space (p. 529)
pleural fluid (p. 505)
pleural space (cavity or sac) (p. 505)
pneumotaxic centre (p. 542)
pneumothorax (p. 510)

12

## REVIEW EXERCISES

### Objective Questions
### (Answers in Appendix E, p. A-48)

1. Breathing is accomplished by alternate contraction and relaxation of muscles within the lung tissue. *(True or false?)*

2. Normally the alveoli empty completely during maximal expiratory efforts. *(True or false?)*

3. Alveolar ventilation does not always increase when pulmonary ventilation increases. *(True or false?)*

4. Oxygen and carbon dioxide have equal diffusion coefficients. *(True or false?)*

5. Haemoglobin has a higher affinity for oxygen than for any other substance. *(True or false?)*

6. Rhythmicity of breathing is brought about by pacemaker activity displayed by the respiratory muscles. *(True or false?)*

7. The expiratory neurons send impulses to the motor neurons controlling the expiratory muscles during normal quiet breathing. *(True or false?)*

8. The three forces that tend to keep the alveoli open are _____, _____, and _____.

9. The two forces that promote alveolar collapse are _____ and _____.

10. _____ is a measure of the magnitude of change in lung volume accomplished by a given change in the transmural pressure gradient.

11. _____ is the phenomenon of the lungs snapping back to their resting size after having been stretched.

12. _____ is the erythrocytic enzyme that catalyzes the conversion of $CO_2$ into $HCO_3^-$.

13. Which reaction takes place in the pulmonary capillaries?

$$Hb + O_2 \rightarrow HbO_2$$
$$CO_2 + H_2O \rightarrow H_2CO_3 : H^+ + HCO_3^-$$
$$Hb + CO_2 \rightarrow HbCO_2$$
$$HbH \rightarrow Hb + H^+$$

14. Indicate the oxygen and carbon dioxide partial pressure relationships important in gas exchange by circling > *(greater than)*, < *(less than)*, or = *(equal to)* as appropriate in each of the following statements:

a. $P_{O_2}$ in blood entering the pulmonary capillaries is (>, <, or =) $P_{O_2}$ in the alveoli.

b. $P_{CO_2}$ in blood entering the pulmonary capillaries is (>, <, or =) $P_{CO_2}$ in the alveoli.

c. $P_{O_2}$ in the alveoli is (>, <, or =) $P_{O_2}$ in blood leaving the pulmonary capillaries.

d. $P_{CO_2}$ in the alveoli is (>, <, or =) $P_{CO_2}$ in blood leaving the pulmonary capillaries.

e. $P_{O_2}$ in blood leaving the pulmonary capillaries is (>, <, or =) $P_{O_2}$ in blood entering the systemic capillaries.

f. $P_{CO_2}$ in blood leaving the pulmonary capillaries is (>, <, or =) in $P_{CO_2}$ blood entering the systemic capillaries.

g. $P_{O_2}$ in blood entering the systemic capillaries is (>, <, or =) $P_{O_2}$ in the tissues.

h. $P_{CO_2}$ in blood entering the systemic capillaries is (>, <, or =) $P_{CO_2}$ in the tissues.

i. $P_{O_2}$ in the tissues is (>, <, or approximately =) $P_{O_2}$ in blood leaving the systemic capillaries.

j. $P_{CO_2}$ in the tissues is (>, <, or approximately =) $P_{CO_2}$ in blood leaving the systemic capillaries.

k. $P_{O_2}$ in blood leaving the systemic capillaries is (>, <, or =) $P_{O_2}$ in blood entering the pulmonary capillaries.

l. $P_{CO_2}$ in blood leaving the systemic capillaries is (>, <, or =) $P_{CO_2}$ in blood entering the pulmonary capillaries.

15. Using the answer code on the right, indicate which chemo-receptors are being described:

___ 1. stimulated by an arterial $P_{O_2}$ of 80 mmHg

___ 2. stimulated by an arterial $P_{O_2}$ of 55 mmHg

___ 3. directly depressed by an arterial of 55 $P_{O_2}$ mmHg

___ 4. weakly stimulated by an elevated arterial $P_{CO_2}$

___ 5. strongly stimulated by an elevated brain ECF $H^+$ concentration induced by an elevated arterial $P_{CO_2}$

___ 6. stimulated by an elevated arterial $H^+$ concentration

(a) peripheral chemoreceptors

(b) central chemoreceptors

(c) both peripheral and central chemoreceptors

(d) neither peripheral nor central chemoreceptors

## Written Questions

1. Distinguish between internal and external respiration. List the steps in external respiration.

2. Describe the two anatomical components of the respiratory system. What is the site of gas exchange?

3. Describe the two physical processes involved in ventilation/gas exchange.

4. Compare atmospheric, alveolar, and pleural pressures.

5. Why are the lungs normally stretched even during expiration?

6. Explain why air enters the lungs during inspiration and leaves during expiration.

7. Why does a forced maximal expiration take much longer than a forced maximal inspiration?

8. Why is inspiration normally active and expiration normally passive?

9. Why does airway resistance become an important determinant of flow in chronic obstructive pulmonary disease?

10. Explain pulmonary elasticity in terms of compliance and elastic recoil.

11. State the source and function of pulmonary surfactant.

12. Define the various lung volumes and capacities.

13. Compare pulmonary ventilation and alveolar ventilation. What is the consequence of anatomic and alveolar dead space?

14. What determines the partial pressures of a gas in air and in blood?

15. List how oxygen and carbon dioxide are transported in blood.

16. What is the primary factor that determines the percent haemoglobin saturation? What is the significance of the plateau and the steep portions of the $O_2$–Hb dissociation curve?

17. How does haemoglobin promote the net transfer of oxygen from the alveoli to the blood?

18. Explain the Bohr and Haldane effects.

19. Define the following: *hypoxic hypoxia, anemic hypoxia, circulatory hypoxia, histotoxic hypoxia, hypercapnia, hypocapnia, hyperventilation, hypoventilation, hyperpnea, apnea,* and *dyspnea.*

20. Distinguish between the DRG and the VRG.

21. What brain region establishes the rhythmicity of breathing?

22. Describe the functional significance of the discharge properties of motor neurons of inspiratory pump muscles and of the skeletal muscles of the upper airway.

## Quantitative Exercises
## (Solutions in Appendix E, p. A-48)

1. The two curves in ⟩ Figure 12-37 show partial pressures for oxygen and carbon dioxide at various alveolar ventilations. These curves can be calculated from the following two equations:

$$P_{A_{O_2}} = P_{I_{O_2}} - \left(\dot{V}_{O_2}/\dot{V}A\right) \times 863 \text{ mmHg}$$
$$P_{A_{O_2}} = \left(\dot{V}_{CO_2}/\dot{V}_A\right) \times 863 \text{ mmHg}$$

2. In these equations, $P_{A_{O_2}}$ equals the partial pressure of oxygen in the alveoli; $P_{A_{CO_2}}$ equals the partial pressure of carbon dioxide in the alveoli; $P_{I_{O_2}}$ equals the partial pressure of oxygen in the inspired air; $\dot{V}_{O_2}$ equals the rate of oxygen consumption by the body; $\dot{V}_{CO_2}$ equals the rate of carbon dioxide production by the body; VA equals the alveolar ventilation; and 863 mmHg is a constant that accounts for atmospheric pressure and temperature.

3. John is training for a marathon tomorrow and just ate a meal of pasta (assume this is pure carbohydrate, which is metabolized with an R of 1). His alveolar ventilation $\left(\dot{V}_A\right)$ is 3.0 L/min, and he is consuming oxygen at a rate of 300 mL/min. What is his $P_{A_{CO_2}}$?

4. Assume you are flying in an airplane that is cruising at 5500 m (18 000 ft.), where the pressure outside the plane is 380 mmHg.

   a. Calculate the $P_{O_2}$ in the air outside the plane, ignoring water vapour pressure.

   b. If the plane depressurized, what would be your $P_{A_{O_2}}$? Assume that the ratio of your oxygen consumption $\left(\dot{V}_{O_2}\right)$ to ventilation does not change (i.e., equals 0.06), and note that under these conditions the constant in the equation that accounts for atmospheric pressure and temperature decreases from 863 to 431.5 mmHg.

   c. Calculate your $P_{A_{CO_2}}$, assuming that your carbon dioxide production $\left(\dot{V}_{CO_2}\right)$ and alveolar ventilation $\left(\dot{V}_A\right)$ remained unchanged at 200 mL/min and 4.2 L/min, respectively.

5. A student has a tidal volume of 350 mL. While breathing at 12 breaths/min, her alveolar ventilation $\left(\dot{V}_A\right)$ is 80 percent of her minute ventilation $\left(\dot{V}_E\right)$. What is the volume of her anatomic dead space?

12

(Explanations on p. A-48)

1. Why is it important that airplane interiors are maintained at, for example, a pressure equivalent to that at an altitude of 2100 m when the plane is at an altitude of 12 000 m? Explain the physiological value of using oxygen masks when the pressure in the airplane interior cannot be maintained.

2. Would hypercapnia accompany the hypoxia produced in each of the following situations? Explain why or why not.
   a. cyanide poisoning
   b. pulmonary oedema
   c. restrictive lung disease
   d. high altitude
   e. severe anaemia
   f. congestive heart failure
   g. obstructive lung disease

3. If a person lives 1.6 km above sea level in Denver, Colorado, where the atmospheric pressure is 630 mmHg, what would the $P_{O_2}$ of the inspired air be once it is humidified in the upper airway?

4. Based on what you know about the control of respiration, explain why it is dangerous to voluntarily hyperventilate to lower the arterial $P_{O_2}$ before going underwater. The purpose of the hyperventilation is to stay under longer before $P_{O_2}$ rises above normal and drives the swimmer to surface for a breath of air.

5. If a person whose alveolar membranes are thickened by disease has an alveolar $P_{O_2}$ of 100 mmHg and an alveolar $P_{CO_2}$ of 40 mmHg, which of the following values of systemic arterial blood gases are most likely?

$$P_{O_2} = 105 \, \text{mmHg}, P_{CO_2} = 35 \, \text{mmHg}$$
$$P_{O_2} = 100 \, \text{mmHg}, P_{CO_2} = 40 \, \text{mmHg}$$
$$P_{O_2} = 90 \, \text{mmHg}, P_{CO_2} = 45 \, \text{mmHg}$$

6. If the person is administered 100 percent oxygen, will the arterial $P_{O_2}$ increase, decrease, or remain the same? Will the arterial $P_{CO_2}$ increase, decrease, or remain the same?

(Explanation in Appendix E, p. A-49)

1. Keith M., a former heavy cigarette smoker, has severe emphysema. How does this condition affect his airway resistance? How does this change in airway resistance influence Keith's inspiratory and expiratory efforts? Describe how his respiratory muscle activity and intra-alveolar pressure changes compare with normal to accomplish a normal tidal volume. How would his spirogram compare with that of a normal individual? What influence would Keith's condition have on gas exchange in his lungs? What blood-gas abnormalities are likely present? Would it be appropriate to administer oxygen to Keith to relieve his hypoxia?

## Exercise and Obstructive Sleep Apnea

## Normal: Exercise

When exercise begins, are the cardiovascular (e.g., heart rate, blood pressure, and cardiac output) and respiratory (e.g., tidal volume and breathing frequency) systems coordinated to meet the increased metabolic needs of the skeletal muscles? Do these two systems respond independently or together? If they respond together, is there a common control system? The answers to these questions have taken many years to unravel (Johansson J.E. [1895]. *Scandinavian Archives of Physiology*, 5: 20–66) and continue to be studied. However, it is generally agreed that these two systems respond as one (the cardiorespiratory system), and are under a common control—central command. We now explore the reasoning behind the concept of central command.

## Evidence of Cardiorespiratory Integration during Exercise

There is a parallel between an increase in ventilation and cardiac output and an increase in oxygen uptake, which suggests a common control system and integrated response. The integrators consist of motor and sensory areas of the brain, with the brain stem linking respiratory and cardiovascular controls. Neural outflow from the motor cortex initiates exercise as well as changes in respiration and cardiovascular function; this ties together all three systems: skeletal muscle, respiratory, and cardiovascular. This activation is a feedforward process.

Central command is demonstrated by the fact that the ventilatory and cardiac responses are similar over a range of submaximal exercise intensities. Both ventilation and cardiac output increase abruptly at the onset of exercise, followed by a more gradual increase over the next one to three minutes, until a steady state is achieved between minutes three and five. How quickly a steady state is attained in a healthy individual depends on exercise intensity, movement economy, and fitness level. *Economy* is considered the amount of energy required to exercise at a specific intensity. If one person uses less energy while exercising at the same intensity as another person, the person using less energy is said to be more economical (their economy is greater), or efficient. At the termination of exercise, ventilation and cardiac output both rapidly decline in the first minute, followed by a more gradual decline over the next several minutes.

The cardiorespiratory adjustments to exercise result from the integration of both neural and humoural factors (i.e., blood and body fluid–borne factors), but the neural component is more important to overall control and integration. Signals from the brain to the active muscles—the motor outflow called central command—pass through the reticular activating system of the medulla, and this command signal from the motor cortex activates (irradiates) the respiratory and cardiovascular control centres in the medulla. The theory of central command gained

momentum through research conducted by August Krogh and Johannes Lindhard ([1913–14]. *Journal of Physiology*, 47: 112–36), who believed that the exercise-induced rise in heart rate and ventilation was produced by irradiation (neural impulses) from the motor cortex. More specifically, the supraspinal centres believed to be involved in central command are the primary motor cortex, posterior hypothalamus, subthalamus, and mesencephalon. We next examine these neural elements more closely.

## The Central and Peripheral Neural Elements

During exercise, there is a major neural outflow to the skeletal, cardiovascular, and respiratory systems. Central command produces descending neural activation of motor nerves, causing muscular contraction and movement and also stimulation of the supraspinal centres responsible for autonomic nervous system (ANS) outflow to the periphery. As the muscle contractions associated with exercise begin, descending signals from supraspinal centres—in conjunction with afferent neural feedback from activated proprioceptors (e.g., muscle spindles), chemoreceptors, baroreceptors, and metaboreceptors—act to control and modulate the cardiorespiratory system in relation to muscle activation and exercise intensity.

Supraspinal activation of the ANS increases heart rate and myocardial contractility (facilitating the Frank–Starling mechanism); vasodilates peripheral muscle arterioles; and vasoconstricts nonworking vasculature (e.g., in the stomach). However, central command has little direct influence on the sympathetic nervous system (SNS) and primarily acts through withdrawal of parasympathetic tone. Thus, the initial increase in heart rate occurs via the withdrawal of parasympathetic (vagal) tone. Sympathetic activity is influenced more by neural reflexes (e.g., baroreceptors) during exercise. The activation (irradiation) hypothesis of central command is considered a feedforward mechanism, which accelerates the cardiorespiratory variables until they are turned off or modified by feedback.

Afferent feedback from proprioceptors (e.g., muscle spindles) in the active skeletal muscles is relayed via spinal afferents to central command. This verifies that exercise has begun and skeletal muscle metabolism has increased, which changes the chemical milieu of muscle, interstitial fluid, and blood. The central (throughout the medulla) and peripheral (aortic and carotid bodies) chemoreceptors, along with metaboreceptors in skeletal muscles respond to changes in the levels of carbon dioxide and $H^+$, by-products of metabolism, thereby providing feedback for modulation of central command; this feedback is likely one of the most important modifiers of cardiorespiratory function.

The arterial $P_{O_2}$ and $P_{CO_2}$ are typically within normal range below the anaerobic threshold, and during this time the central chemoreceptors dominate. The central chemoreceptors are protected by the blood–brain barrier, but recall that carbon dioxide can readily cross the blood–brain barrier and act on the chemoreceptors by increasing the local concentration of $H^+$. During submaximal steady-state exercise, respiration is adjusted in parallel to skeletal muscle production of carbon dioxide.

In contrast, this is not the case during transitions from rest to exercise and during intense (but still submaximal) exercise when the peripheral chemoreceptors and muscle metaboreceptors are stimulated. The basis for this stimulation of ventilation is unclear because, below the anaerobic threshold, arterial $P_{O_2}$ and $P_{CO_2}$ do not change. (Note, however, that taking a blood sample to measure $P_{O_2}$ and $P_{CO_2}$ eliminates any respiration-related oscillations in $P_{O_2}$ and $P_{CO_2}$; if the receptors are sensitive to the *rate* of change of $P_{O_2}$ and $P_{CO_2}/H^+$ during the respiratory cycle, and these changes are faster and larger during exercise, then we could not, on the basis of a blood sample drawn over several seconds, detect their contributions to the increase in ventilation.) Exercising muscles release potassium $(K^+)$ that activates the peripheral chemoreceptors; $K^+$ could therefore be an important factor contributing to the increase in ventilation. Above the anaerobic threshold, the production of metabolic acids (e.g., lactic acid) stimulates the peripheral chemoreceptors. As a result, the subject hyperventilates; arterial $P_{O_2}$ increases and $P_{CO_2}$ decreases, blunting the response to metabolic acids. The peripheral chemoreceptors provide feedback to the respiratory centres to increase respiration. Activated central and peripheral chemoreceptors and/or muscle metaboreceptors communicate with central command to modulate respiration. In essence, chemoreceptors provide feedback for control of breathing (rate and depth of respiration) by central command, as they detect mismatches between respiration and muscular work.

Similarly, activation of the baroreceptors provides afferent feedback to central command to modify heart rate, on a beat-to-beat basis, and vascular resistance through the SNS. Recall that baroreceptors are mechanoreceptors that sense change in pressure within the walls of the vessels. Changes in pressure in the aortic arch and carotid sinus are detected by mechanoreceptors and signal central command to adjust cardiac vagal tone and SNS outflow accordingly, which change heart rate and the diameters of the arterioles that determine peripheral resistance. Likewise, the cardiopulmonary baroreflex senses changes in blood volume and filling pressure in the pulmonary arteries and veins. Again, afferent feedback to central command allows for the appropriate modification of vagal tone and SNS outflow to adjust heart rate and the diameters of arterioles, which increases or decreases blood pressure.

Note that systolic and arterial pulse pressures increase during rhythmical dynamic exercise (e.g., running), and there is an adjustment of the baroreflex to accommodate the higher pressure. This is accomplished by resetting the baroreflex to regulate arterial blood pressure at a higher operating point (new set point) during exercise. If the set point of the baroreflex were not increased, the ANS could not make the necessary cardiovascular adjustments; instead, the baroreceptors would attempt to reduce cardiac output, heart rate, and blood pressure, all of which would oppose the needed responses to exercise. The resetting of the arterial baroreflex is believed to be accomplished by central command.

It can be difficult to definitively distinguish between the central and peripheral neural contributions to the cardiorespiratory response to exercise. However, many animal and human studies of these responses to exercise support the theory of central command initiating and coordinating the cardiorespiratory responses to exercise, with the peripheral elements providing information to central command to modify its control.

## Peripheral Humoural Elements

The humoural factors that influence skeletal muscle blood flow, cardiac output, and ventilation are metabolic vasodilators (e.g., ADP) and hormones (e.g., epinephrine). The focus here will be on the metabolic vasodilators that influence muscle metaboreceptors. Carbon dioxide has already been mentioned and will not be discussed further. Accordingly, the focus of this section is on the response of skeletal muscle metaboreceptors to chemical factors other than carbon dioxide.

The muscle metaboreceptors are local sensors that monitor chemical changes. For example, if oxygen delivery to the quadriceps muscle is inadequate at the onset of exercise, the concentrations of chemical factors (e.g., AMP and ADP) increase in the quadriceps muscle, interstitial fluid, and blood, causing local vasodilation. Local interstitial factors thought to influence metaboreceptor activity include adenosine, AMP, ADP, Pi, $H^+$, lactic acid, bradykinin, prostaglandins, $K^+$, and nitric oxide (NO); however, adenosine is considered a weak vasodilator. These interstitial factors stimulate the nerve endings surrounding the skeletal muscle fibres; activation of their afferents influences the ANS.

The local changes are relayed to central command and modify ANS activity. The resulting increase in SNS outflow vasoconstricts blood vessels in nonworking tissue but vasodilates those in working tissue. Some experimental evidence even suggests that the resulting increase in SNS outflow from the stimulation of muscle metaboreceptors increases heart rate via β-adrenergic modulation. Nevertheless, increased vasodilation associated with stimulation of metaboreceptors is specific to the site of working muscle, making this response local. The end result ensures sufficient delivery of oxygen to active muscle tissue, while reducing delivery of oxygen to tissues not essential to exercise.

The preceding discussion supports the concept that at the onset of exercise, neural outflow from central command increases, and this activates the cardiorespiratory system. Feedback from central and peripheral chemoreceptors, metaboreceptors, baroreceptors, and peripheral proprioceptors help match cardiorespiratory output and muscle perfusion with exercise intensity and the mass of muscle activated. This matching supports the delivery of blood, oxygen, and other nutrients (e.g., glucose) to the working limb muscles and the removal of metabolic by-products (e.g., $H^+$).

## Abnormal: Obstructive Sleep Apnea (OSA)

Obstructive sleep apnea (OSA) is a disease that illustrates the interactions between the respiratory and cardiovascular systems. Although the problem is respiratory—cessation of breathing for a few seconds or up to a minute as many as 500 times a night—the pathological consequences are primarily cardiovascular.

OSA occurs in approximately 4 percent of males and 2 percent of females. Snoring, a frequent precursor to OSA, is even more common. Contributing factors include age and obesity, mostly because the extra weight, especially in the neck, predisposes the upper airway to collapse. This effect is exacerbated when the neural drive to the muscles in the upper airway that dilate it during inspiration decreases during sleep. As a result, the increased inspiratory efforts can suck the upper airway closed, an effect exacerbated by surface tension between the opposing walls of the collapsed airway.

## The Effects of OSA

The effects of OSA have profound effects on health. Each episode of OSA-induced asphyxia (particularly the hypoxia) stimulates the sympathetic nervous system, activating the heart and vascular smooth muscle at a time when oxygen delivery to tissues decreases because of the asphyxia. This is especially stressful to the heart because even at rest it consumes approximately 70 percent of the oxygen delivered to it. Therefore, unless blood flow to the heart increases, there is a risk that it will receive insufficient oxygen during the apnea to meet its increased metabolic demands—caused by the surge in sympathetic activity—and the increase in afterload—caused by peripheral vasoconstriction. Moreover, if the asphyxia is severe enough, cardiac arrhythmias can occur, possibly leading to a heart attack. These factors may account for the increased rates of death at night, typically attributed to heart attacks, reported in patients with OSA.

Aside from these acute but repeated effects of OSA, there are insidious chronic effects. Over time, the acute and repeated episodes of increases in sympathetic activity become permanent (chronic), persisting even when the individuals are awake. The resulting effects on the heart and vasculature (the walls of the blood vessels become stiffer) include hypertension, a major factor causing heart attacks and strokes. (That obstructive sleep apnea *causes*, and is not just associated with, hypertension was first demonstrated using ingenious experiments in dogs by a Toronto group headed by E.A. Phillipson: D. Brooks et al. [1997]. *Journal of Clinical Investigation*, 99: 106–109.) Moreover, recent evidence shows that OSA increases the coagulability of the blood. Both hypertension and hypercoagulability, along with a host of others factors (including diabetes—often caused by obesity) contribute to, if not cause, the high rates of myocardial infarctions and strokes in patients with OSA. Not surprisingly, the cost to the Canadian healthcare system is high—about double that of age-matched controls (approximately $4000 versus $2000) over the 10 years before a diagnosis of OSA—because of the associated comorbidities (Ronald, J., et al. [1998]. *Sleep Research Online, 1:* 71–74). This is true even if one compares obese individuals with OSA to weight-matched controls without OSA (Banno, K., et al. [2009]. Sleep, 32: 247–52).

## Treatment of OSA

Treatment of OSA is relatively straightforward. The patient wears a mask that blows air into the airway at a pressure slightly above atmospheric pressure; this is called *continuous positive airway pressure* (CPAP). The pressure splints the airway open, preventing the recurrent episodes of airway collapse and, therefore, apnea. Unfortunately, many people have difficulty tolerating the device, and so adherence can be low. Treating OSA will become increasingly important as the incidence of obesity continues to increase as the population ages.

# The Urinary System

Body systems maintain homeostasis

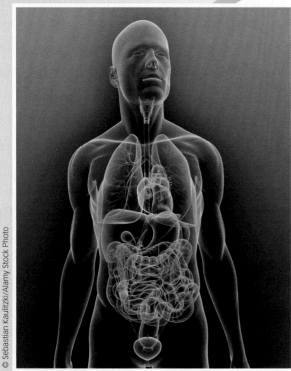

Urinary system

## Homeostasis

The urinary system contributes to homeostasis by helping regulate the volume, electrolyte composition, and pH of the internal environment and by eliminating metabolic waste products.

Homeostasis is essential for survival of cells

Cells make up body systems

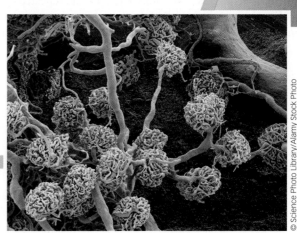

Kidney glomeruli

The survival and proper functioning of cells depend on the maintenance of stable concentrations of salt, acids, and other electrolytes in the internal fluid environment. Cell survival also depends on continual removal of toxic metabolic wastes that cells produce as they perform life-sustaining chemical reactions. The kidneys play a major role in maintaining homeostasis by regulating the concentration of many plasma constituents, especially electrolytes and water, and by eliminating all metabolic wastes (except carbon dioxide, which is removed by the lungs). As plasma repeatedly filters through the kidneys, they retain constituents of value for the body and eliminate undesirable or excess materials in the urine. Of special importance is the kidneys' ability to regulate the volume and osmolarity (solute concentration) of the internal fluid environment by controlling salt and water balance. Also crucial is their ability to help regulate pH by controlling elimination of acid and base in the urine.

# 13

# The Urinary System

**▌ Clinical Connections**

Jacob is 45 years old and an avid cyclist. On a daily basis, he cycles to and from work, a trip of 25 km each way. For the past two weeks, the average daytime high has exceeded 27°C. While mowing his lawn on yet another hot and sunny Saturday, he suddenly had an intense pain in his lower left side that radiated around his lower back. Upon arriving at the emergency room, the intensity of the pain was making him feel nauseous, and he was having problems sitting comfortably. Blood and urine samples were taken. His urine sample had a pinkish hue to it and upon analysis was found to have an osmolality of 1200 mOsm/l and contained small calcium oxalate crystals. Suspecting kidney stones as the source of the pain, an abdominal CT was ordered. The diagnosis of kidney stones was confirmed as the CT scan showed a few stones of varying size in his left kidney. He was given medication to help with the pain and instructed to drink lots of water to help pass the stones.

## 13.1 | Introduction

Exchanges between the cells and the ECF could notably alter the composition of the ECF if there were no mechanisms to keep it stable.

## The kidneys

The kidneys are involved in homeostatic function, assisting in electrolyte regulation (concentrations), acid–base balance, blood volume control, and blood pressure control. This is achieved both independently and through coordination with other organs; especially those of the endocrine system (Chapter 6).

The kidneys, controlled by hormonal and neural input, are the organs primarily responsible for maintaining the stability of ECF volume, electrolyte composition, and osmolarity (solute concentration). The kidneys maintain water and electrolyte balance within a narrow range, despite a wide range of intake and losses of electrolytes through other avenues. Kidneys not only adjust for wide variations in the ingestion of water ($H_2O$), salt, and other electrolytes, but also adjust urinary output of these ECF constituents to compensate for abnormal losses through heavy sweating, vomiting, diarrhoea, or haemorrhage.

When the ECF has a surplus of water or a particular electrolyte, such as salt (NaCl), it can eliminate the excess in the urine. In a deficit state, the kidneys cannot make up for a depleted constituent, but it can limit further urinary loss of the constituent and conserve it until the person can take in more. Consequently, the kidneys can compensate more efficiently for excesses than for deficits. The kidneys cannot completely halt the loss of a valuable substance in the urine, even though the substance may be in short supply. A prime example is the case of an $H_2O$ deficit. Even if a person is not consuming any $H_2O$, the kidneys must put out a minimum about 500 mL of $H_2O$ in the urine each day to fill another major role as the body's cleaners. Because $H_2O$ eliminated in the urine is derived from the blood plasma, a person stranded without $H_2O$ eventually urinates to death.

Additionally, the kidneys are the main route for eliminating potentially toxic metabolic wastes and foreign compounds from the body. These wastes cannot be eliminated as solids, and thus must be excreted in solution as waste-filled urine.

### OVERVIEW OF KIDNEY FUNCTIONS

The kidneys perform the following functions, most of which help preserve the constancy of the internal fluid environment:

1. *Maintaining $H_2O$ balance in the body* (Chapter 14)

2. *Maintaining the proper osmolarity of body fluids*, primarily through regulating $H_2O$ balance. This is important to prevent osmotic fluxes into or out of the cells, which could lead to detrimental swelling or shrinking of the cells, respectively (Chapter 14).

3. *Regulating the quantity and concentration of most ECF ions*, including sodium ($Na^+$), chloride ($Cl^-$), potassium ($K^+$), calcium ($Ca^{2+}$), hydrogen ion ($H^+$), bicarbonate ($HCO_3^-$), phosphate ($PO_4^{3-}$), sulphate ($SO_4^{2-}$), and magnesium ($Mg^+$). Even minor fluctuations in the ECF concentrations of some of these electrolytes can have profound influences. For example, changes in the ECF concentration of $K^+$ can potentially lead to fatal cardiac dysfunction (p. 356).

4. *Maintaining proper plasma volume*, which is important in long-term regulation of arterial blood pressure. This is accomplished through the kidneys' regulatory role in salt ($Na^+$ and $Cl^-$; NaCl) and $H_2O$ balance (Chapter 14).

5. *Helping maintain the proper acid–base balance of the body* by adjusting urinary output of $H^+$ and $HCO_3^-$ (Chapter 14)

6. *Excreting (eliminating) the end products (wastes) of bodily metabolism*, such as urea, uric acid, and creatinine. These wastes are toxic, especially to the brain.

7. *Excreting many foreign compounds*, such as drugs, food additives, pesticides, and other exogenous non-nutritive materials that have entered the body

8. *Producing erythropoietin*, a hormone that stimulates red blood cell production (Chapter 10)

9. *Producing renin*, an enzymatic hormone that triggers a chain reaction important in salt conservation by the kidneys

10. *Converting vitamin D into its active form* (Chapter 6)

## The kidneys and urine

The kidneys are bean-shaped organs, with a concave side that faces medially (see ⟩ Figure 13-1a). They are located in the posterior aspect of the abdomen on each side of the spine and surrounded by two layers of fat. The top is approximately at the level of the twelfth thoracic vertebrae. An adult kidney is about 10 cm long, 5 cm wide, 3 cm thick, and weighs about 150 g. The right kidney sits below the liver and the left below the diaphragm.

Above each kidney is an adrenal gland. The size and location of the liver in the abdominal cavity causes the right kidney to be placed slightly lower than the left kidney, and the left kidney to be placed more medial compared with the right. The kidneys lie behind the peritoneum lining of the abdominal cavity. The renal artery, vein, nerves, and ureter are located on the medial side of each kidney. *The outer surface of the kidney is called the renal cortex, and deep to the cortex is the renal medulla.* Urine empties into the renal pelvis, which is the medial inner core of each kidney (⟩ Figure 13-1b). The urine is then channelled into the ureter, a smooth muscle–walled duct that exits at the medial boarder. The blood supply for each kidney comes from the renal arteries, which branch out from the abdominal aorta.

The **urinary bladder**, which temporarily stores urine, is a hollow, distensible, smooth muscle–walled sac. Periodically, as a result of bladder contraction, urine is emptied from the bladder via the **urethra** to the outside. The urethra in females is straight and short, passing directly from the neck of the bladder to the outside (⟩ Figure 13-2a; see also Figure 17-2). In males, the urethra is much longer and follows a curving course from the bladder to the outside, passing through both the prostate gland and the penis (Figure 13-1a and 13-2b; see also Figure 17-1). The male urethra serves the dual function of providing both a route for eliminating urine from the bladder and a passageway for semen from the reproductive organs. The prostate gland lies below the neck of the bladder and completely encircles the urethra.

 *Clinical Note* Prostatic enlargement, which often occurs during middle to older age, can partially or completely occlude the urethra, impeding the flow of urine.

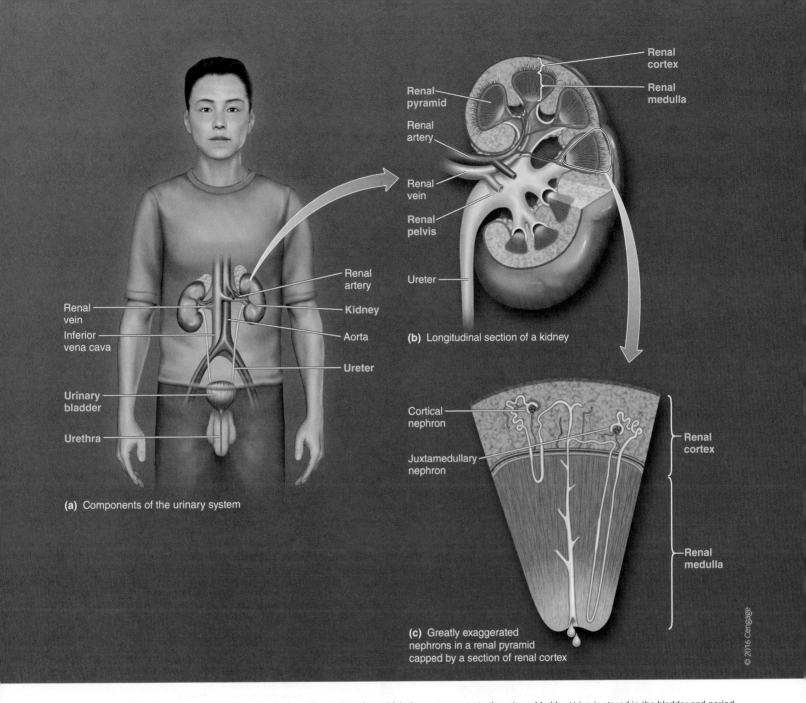

**(a)** Components of the urinary system

**(b)** Longitudinal section of a kidney

**(c)** Greatly exaggerated nephrons in a renal pyramid capped by a section of renal cortex

© 2016 Cengage

> **FIGURE 13-1 The urinary system.** (a) The pair of kidneys forms the urine, which the ureters carry to the urinary bladder. Urine is stored in the bladder and periodically emptied to the exterior through the urethra. (b) The kidney consists of an outer, granular-appearing renal cortex and an inner, striated-appearing renal medulla. The renal pelvis at the medial inner core of the kidney collects urine after it is formed. (c) Each kidney has a million nephrons. The two types of these microscopic functional units are shown here, greatly exaggerated, in a medullary renal pyramid capped by a section of renal cortex.

**13**

The parts of the urinary system beyond the kidneys merely serve as ductwork to transport urine to the outside. Once formed by the kidneys, urine is not altered in composition or volume as it moves downstream through the rest of the tract.

## The nephron

The basic functional unit of the kidney is the nephron. A functional unit is the smallest unit within an organ that is capable of performing all of that organ's functions; for the kidney, that means the formation of urine. There are more than 1 million nephrons within the cortex and medulla of each healthy human adult kidney. Nephrons regulate water and solute (especially electrolytes) by filtering the blood under pressure and then reabsorbing necessary fluid and molecules back into the blood, and by secreting other unneeded molecules. The main function of the kidneys is to produce urine and by doing so maintain consistency in the ECF composition.

The arrangement of nephrons within the kidneys gives rise to two distinct regions: an outer region called the **renal cortex**,

Ureter
Smooth muscle of bladder wall
Ureteral openings
Internal sphincter
Pelvic diaphragm
External sphincter
Urethra
External urethral orifice
(a) Female
Prostate gland (an accessory sex gland)
Bulbourethral glands (accessory sex glands)
Urethra
External urethral orifice
(b) Male



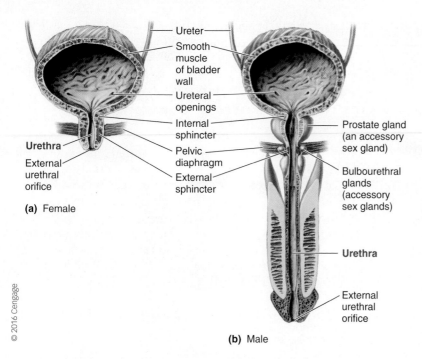

› **FIGURE 13-2 Comparison of the urethra in females and males.** (a) In females, the urethra is straight and short. (b) In males, the urethra, which is much longer, passes through the prostate gland and penis.

which looks granular, and an inner region, the **renal medulla**, which is made up of striated triangles, the **renal pyramids** (Figure 13-1b and 13-1c).

Knowledge of the structural arrangement of an individual nephron is essential for understanding the distinction between the cortical and medullary regions of the kidney and, more important, for understanding renal function. Each nephron consists of a *vascular component* and a *tubular component*, both of which are intimately related structurally and functionally (› Figure 13-3).

### VASCULAR COMPONENT OF THE NEPHRON

The dominant parts of the nephron's vascular component are the **glomeruli** (singular, *glomerulus*). These ball-like tufts of capillaries filter some of the water and solutes from the blood passing through them. This filtered fluid, which is almost identical in composition to plasma, then passes through the nephron's tubular component, where various transport processes convert it into urine.

On entering the kidney, the renal artery subdivides to ultimately form many small vessels known as **afferent arterioles**, one of which supplies each

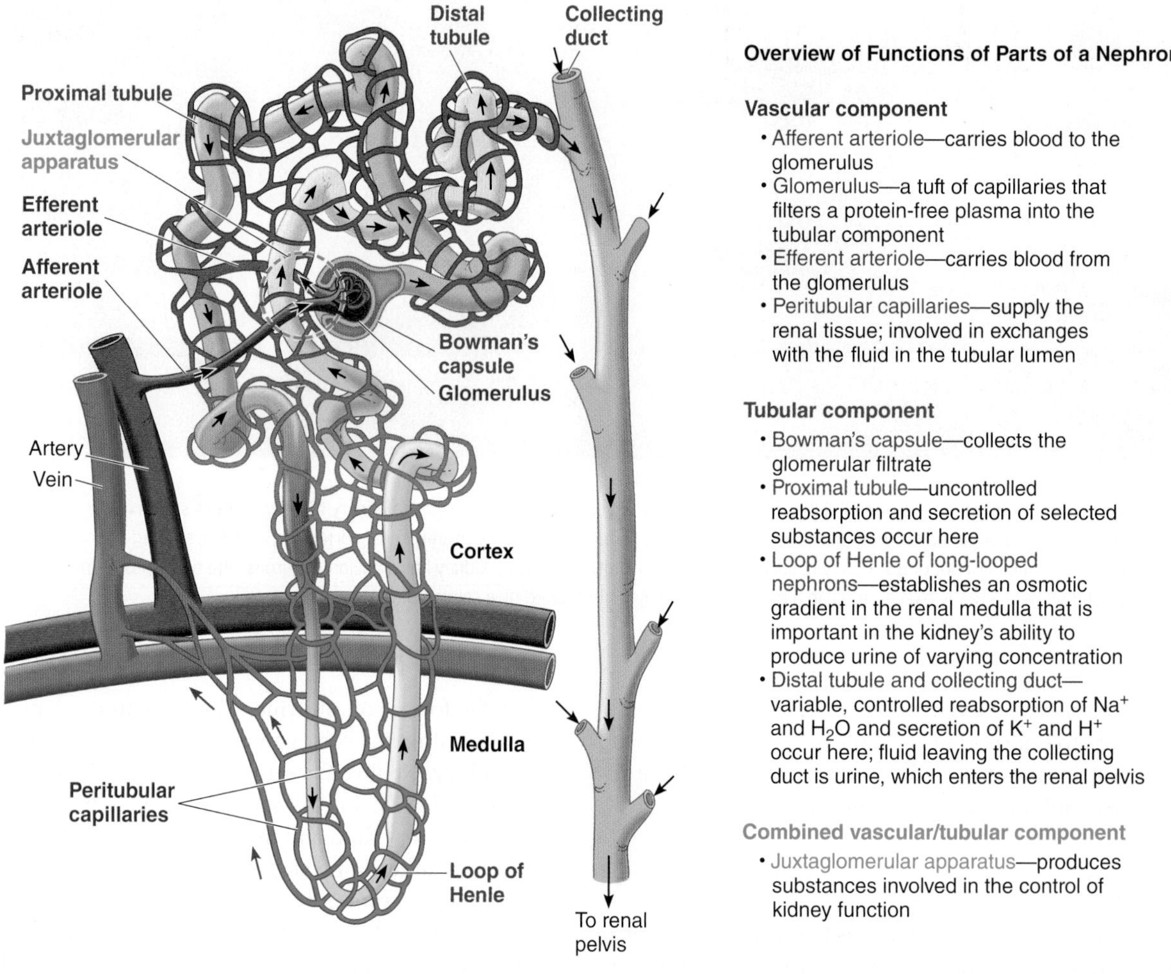

Distal tubule
Collecting duct
Proximal tubule
Juxtaglomerular apparatus
Efferent arteriole
Afferent arteriole
Bowman's capsule
Glomerulus
Artery
Vein
Cortex
Medulla
Peritubular capillaries
Loop of Henle
To renal pelvis
Overview of Functions of Parts of a Nephron
Vascular component
• Afferent arteriole—carries blood to the glomerulus
• Glomerulus—a tuft of capillaries that filters a protein-free plasma into the tubular component
• Efferent arteriole—carries blood from the glomerulus
• Peritubular capillaries—supply the renal tissue; involved in exchanges with the fluid in the tubular lumen
Tubular component
• Bowman's capsule—collects the glomerular filtrate
• Proximal tubule—uncontrolled reabsorption and secretion of selected substances occur here
• Loop of Henle of long-looped nephrons—establishes an osmotic gradient in the renal medulla that is important in the kidney's ability to produce urine of varying concentration
• Distal tubule and collecting duct—variable, controlled reabsorption of $Na^+$ and $H_2O$ and secretion of $K^+$ and $H^+$ occur here; fluid leaving the collecting duct is urine, which enters the renal pelvis
Combined vascular/tubular component
• Juxtaglomerular apparatus—produces substances involved in the control of kidney function



› **FIGURE 13-3 A nephron.** A schematic representation of a cortical nephron, the most abundant type of nephron in humans

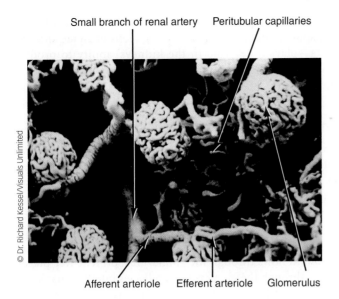

Small branch of renal artery · Peritubular capillaries

© Dr. Richard Kessel/Visuals Unlimited

Afferent arteriole · Efferent arteriole · Glomerulus

› **FIGURE 13-4 Glomerulus vasculature**

nephron. The afferent arteriole delivers blood to the glomerulus. The glomerular capillaries rejoin to form another arteriole, the **efferent arteriole**, through which blood that was not filtered into the tubular component leaves the glomerulus (Figure 13-3 and › Figure 13-4). The efferent arterioles are the only arterioles in the body that drain from capillaries. Typically, arterioles break up into capillaries that rejoin to form venules. At the glomerular capillaries, no oxygen or nutrients are extracted from the blood for use by the kidney tissues nor are waste products picked up from the surrounding tissue. Thus, arterial blood enters the glomerular capillaries through the afferent arteriole, and arterial blood leaves the glomerulus through the efferent arteriole.

The efferent arteriole quickly subdivides into a second set of capillaries, the **peritubular capillaries**, which supply the renal tissue with blood and are important in exchanges between the tubular system and blood during conversion of the filtered fluid into urine. These peritubular capillaries, as their name implies, are intertwined around the tubular system (*peri* means "around"). The peritubular capillaries rejoin to form venules that ultimately drain into the renal vein, by which blood leaves the kidney.

## TUBULAR COMPONENT OF THE NEPHRON

The nephron's tubular component is a hollow, fluid-filled tube formed by a

single layer of epithelial cells. Even though the tubule is continuous from its beginning near the glomerulus to its ending at the renal pelvis, it is arbitrarily divided into various segments based on differences in structure and function along its length (Figure 13-3 and › Figure 13-5). The tubular component begins with **Bowman's capsule**, an expanded, double-walled invagination that cups around the glomerulus to collect the fluid filtered from the glomerular capillaries.

From the Bowman's capsule, the filtered fluid passes into the **proximal tubule**, which lies entirely within the cortex and is highly coiled or convoluted throughout much of its course. The next segment, the **loop of Henle**, forms a sharp U-shaped or hairpin loop that dips into the renal medulla. The *descending limb* of the loop of Henle plunges from the cortex into the medulla; the *ascending limb* traverses back up into the cortex. The ascending limb returns to the glomerular region of its own nephron, where it passes through the fork formed by the afferent and efferent arterioles. Both the tubular and vascular cells at this point are specialized to form the **juxtaglomerular apparatus**, a structure that lies next to the glomerulus (*juxta* means "next to"). This specialized region plays an important role in regulating kidney function. Beyond the juxtaglomerular

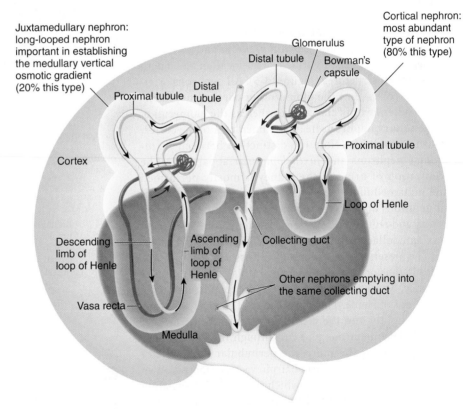

For better visualization, the nephrons are grossly exaggerated in size, and the peritubular capillaries have been omitted, except for the vasa recta.

© 2016 Cengage

› **FIGURE 13-5 Comparison of juxtamedullary and cortical nephrons.** The glomeruli of cortical nephrons lie in the outer cortex, whereas the glomeruli of juxtamedullary nephrons lie in the inner part of the cortex next to the medulla. The loops of Henle of cortical nephrons dip only slightly into the medulla, but the juxtamedullary nephrons have long loops of Henle that plunge deep into the medulla. The juxtamedullary nephrons' peritubular capillaries form hairpin loops known as *vasa recta*.

**13**

The nephron is very important to the body's fluid regulation, and thus it is a target for drugs used to treat hypertension (high blood pressure). These drugs, called *diuretics,* inhibit the ability of the nephron to reabsorb water, which then increases the volume of urine produced and excreted each day. Recall that maintaining proper plasma volume is important in long-term regulation of blood pressure, accomplished through the kidneys' regulation of salt and water balance.

apparatus, the tubule once again coils tightly to form the **distal tubule**, which also lies entirely within the cortex. The distal tubule empties into a **collecting duct (tubule)**, with each collecting duct draining fluid from up to eight separate nephrons. Each collecting duct plunges down through the medulla to empty its fluid contents (now converted into urine) into the renal pelvis.

### CORTICAL AND JUXTAMEDULLARY NEPHRONS

The two types of nephrons—*cortical nephrons* and *juxtamedullary nephrons*—are distinguished by the location and length of some of their structures (Figure 13-5). All nephrons originate in the cortex, but the glomeruli of **cortical nephrons** lie in the outer layer of the cortex, whereas the glomeruli of **juxtamedullary nephrons** lie in the inner layer of the cortex, next to the medulla. (Note the distinction between *juxtamedullary* nephrons and *juxtaglomerular* apparatus.) The concentration of urine in the kidney is mostly performed by the juxtamedullary nephron. Approximately 80 percent of all nephrons are cortical nephrons; these mostly perform excretory and regulatory functions. The remaining 20 percent are juxtamedullary nephrons, which concentrate and dilute urine. The presence of all glomeruli and associated Bowman's capsules in the cortex is responsible for this region's granular appearance. These two nephron types differ most markedly in their loops of Henle. The hairpin loop of cortical nephrons dips only slightly into the medulla. In contrast, the loop of juxtamedullary nephrons plunges through the entire depth of the medulla. Furthermore, the peritubular capillaries of juxtamedullary nephrons form hairpin vascular loops known as **vasa recta** (straight vessels), which run in close association with the long loops of Henle. In cortical nephrons, the peritubular capillaries do not form vasa recta, but instead entwine around these nephrons' short loops of Henle. As they course through the medulla, the collecting ducts of both cortical and juxtamedullary nephrons run parallel to the ascending and descending limbs of the juxtamedullary nephrons' long loops of Henle and vasa recta. The parallel arrangement of tubules and vessels in the medulla creates this region's striated appearance. More important, as you will see in Section

13.6, this arrangement—coupled with the permeability and transport characteristics of the long loops of Henle and vasa recta—plays a key role in the kidneys' ability to produce urine of varying concentrations, depending on the needs of the body.

## The three basic renal processes

Three basic processes are involved in forming urine: *glomerular filtration, tubular reabsorption,* and *tubular secretion.* To aid in visualizing the relationships among these renal processes, it is useful to unwind the nephron schematically, as shown in ⟩ Figure 13-6.

### GLOMERULAR FILTRATION

As blood flows through the glomerulus, protein-free plasma filters through the glomerular capillaries into Bowman's capsule. Normally, about 20 percent of the plasma that enters the glomerulus is filtered. This process, known as **glomerular**

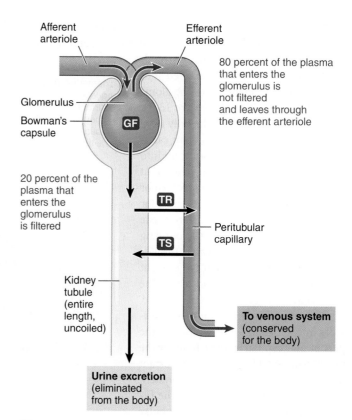

⟩ **FIGURE 13-6 Basic renal processes.** Anything filtered or secreted but not reabsorbed is excreted in the urine and lost from the body. Anything filtered and subsequently reabsorbed, or not filtered at all, enters the venous blood and is saved for the body.

**filtration**, is the first step in urine formation. On average, 125 mL of glomerular filtrate (filtered fluid) are formed collectively through all the glomeruli each minute. This amounts to 180 L (about 47.5 gallons) each day. Considering that the average plasma volume in an adult is 2.75 L, this means that the kidneys filter the entire plasma volume about 65 times per day. If everything filtered passed out in the urine, the total plasma volume would be urinated in less than half an hour! This does not happen, however, because the kidney tubules and peritubular capillaries are intimately related throughout their lengths, so that materials can be transferred between the fluid inside the tubules and the blood within the peritubular capillaries.

### TUBULAR REABSORPTION

As the filtrate flows through the tubules, substances of value to the body are returned to the peritubular capillary plasma. This selective movement of substances from inside the tubule (the tubular lumen) into the blood is called **tubular reabsorption**. Reabsorbed substances are not lost from the body in the urine, but instead are carried by the peritubular capillaries to the venous system and then to the heart to be recirculated. Of the 180 L of plasma filtered per day, 178.5 L, on average, are reabsorbed. The remaining 1.5 L left in the tubules pass into the renal pelvis to be eliminated as urine. In general, substances the body needs to conserve are selectively reabsorbed, whereas unwanted substances that must be eliminated stay in the urine.

### TUBULAR SECRETION

The third renal process, **tubular secretion**, is the selective transfer of substances from the peritubular capillary blood into the tubular lumen. It provides a second route for substances to enter the renal tubules from the blood, the first being by glomerular filtration. Only about 20 percent of the plasma flowing through the glomerular capillaries is filtered into Bowman's capsule; the remaining 80 percent flows on through the efferent arteriole into the peritubular capillaries. In contrast, tubular secretion provides a mechanism for more rapid elimination of selected substances from the plasma. This is accomplished by extracting an additional quantity of a particular substance from the 80 percent of unfiltered plasma in the peritubular capillaries and adding it to the quantity of the substance already present in the tubule as a result of filtration.

### URINE EXCRETION

**Urine excretion** is the elimination of substances from the body in the urine. It is not really a separate process but the result of the first three processes. All plasma constituents filtered or secreted but not reabsorbed remain in the tubules and pass into the renal pelvis to be excreted as urine and eliminated from the body (Figure 13-6). Note that anything filtered and subsequently reabsorbed, or not filtered at all, enters the venous blood from the peritubular capillaries and is therefore conserved for the body instead of being excreted in urine, despite passing through the kidneys.

### THE BIG PICTURE OF THE BASIC RENAL PROCESSES

Glomerular filtration is largely an indiscriminate process. With the exception of blood cells and plasma proteins, all constituents within the blood—water, nutrients, electrolytes, wastes, and so on—nonselectively enter the tubular lumen as a bulk unit during filtration. That is, of the 20 percent of the plasma filtered at the glomerulus, everything in that part of the plasma enters Bowman's capsule, except for the plasma proteins. The highly discriminating tubular processes then work on the filtrate to return to the blood a fluid that has the composition and volume necessary to maintain the constancy of the internal fluid environment. The unwanted filtered material is left behind in the tubular fluid to be excreted as urine. Glomerular filtration can be thought of as pushing a part of the plasma—with all its essential components as well as those that need to be eliminated from the body—onto a tubular "conveyor belt" that terminates at the renal pelvis, which is the collecting point for urine within the kidney. All plasma constituents that enter this conveyor belt and are not subsequently returned to the plasma by the end of the line are spilled out of the kidney as urine. It is up to the tubular system to salvage by reabsorption the filtered materials that need to be preserved for the body, while leaving behind substances that must be excreted. In addition, some substances are not only filtered but also secreted onto the tubular conveyor belt, so the amounts of these substances excreted in the urine are greater than the amounts that were filtered. For many substances, these renal processes are subject to physiological control. Thus, the kidneys handle each constituent in the plasma in a characteristic manner by a particular combination of filtration, reabsorption, and secretion.

The kidneys act only on the plasma, yet the ECF consists of both plasma and interstitial fluid. The interstitial fluid is actually the true internal fluid environment of the body, because it is the only component of the ECF that comes into direct contact with the cells. However, because of the free exchange between plasma and interstitial fluid across the capillary walls (with the exception of plasma proteins), interstitial fluid composition reflects the composition of plasma. By performing their regulatory and excretory roles on the plasma, the kidneys maintain the proper interstitial fluid environment for optimal cell function. Most of the rest of this chapter is devoted to considering how the basic renal processes are accomplished and the mechanisms by which they are carefully regulated to help maintain homeostasis.

## 13.2 | Renal Blood Flow

The path of renal blood flow (RBF) is renal arteries → afferent arterioles → glomerular capillaries → efferent arterioles → proximal peritubular capillaries → vasa recta → distal peritubular capillaries → capillaries associated with the collecting duct → renal venules. The highest RBF occurs at rest, and is about 1200 mL/min, which is about 22 percent of cardiac output. The purpose of this relatively high blood flow is to supply the kidney with sufficient plasma for filtration, secretion, and reabsorption. During exercise, RBF is reduced. Under extreme conditions—for example, shock—RBF can approach zero. Another purpose of the blood supply is to deliver oxygen for energy to perform active transport (e.g., of $Na^+$). However, resting RBF is more than adequate to satisfy the kidneys' $O_2$ requirements. Generally, blood flow could be reduced, and $O_2$ requirements would still be met.

Control of RBF is largely impacted by the ANS—specifically, the sympathetic division and the associated vasoconstriction. As well, the adrenal medulla reinforces the sympathetic vasoconstriction. Interestingly, although the nerves innervating the kidney aid in regulation of RBF, they are not essential for normal renal function. The kidney also has an intrinsic ability to control blood flow, which is termed autoregulation. It is thought that RBF autoregulation is based primarily on two mechanisms: the myogenic response and the tubuloglomerular feedback. The myogenic response is a function of smooth muscle to contract in response to external stretching force. In vascular smooth muscle this causes vasoconstriction on a rise in arterial pressure, thereby allowing for autoregulation. Tubuloglomerular feedback is a mechanism specific to the kidney that leads to constriction of the afferent arteriole in response to an increase in NaCl concentration in the distal tubule. These mechanisms allow the kidney to adjust resistance in order to maintain blood flow even when perfusion pressure changes.

## 13.3 | Glomerular Filtration

Fluid filtered from the glomerulus into Bowman's capsule must pass through the following three layers that make up the **glomerular membrane** (› Figure 13-7): (1) the glomerular capillary wall, (2) the basement membrane, and (3) the inner layer of Bowman's capsule. Collectively, these layers function as a fine molecular sieve that retains the blood cells and plasma proteins but permits water and solutes of small molecular dimension to filter through. Let's consider each layer in more detail.

### The glomerular membrane

The *glomerular capillary wall* consists of a single layer of flattened endothelial cells. It is perforated by many large pores that

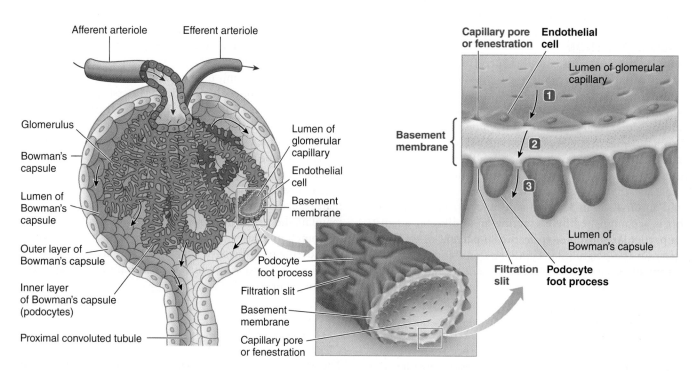

To be filtered, a substance must pass through

**1** the pores between and the fenestrations within the endothelial cells of the glomerular capillary,

**2** an acellular basement membrane, and

**3** the filtration slits between the foot processes of the podocytes in the inner layer of Bowman's capsule.

› **FIGURE 13-7 Layers of the glomerular membrane**

© 2016 Cengage

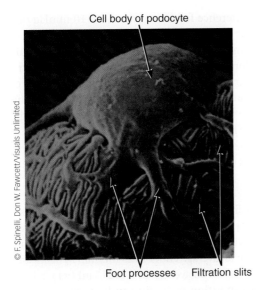

Cell body of podocyte

Foot processes    Filtration slits

> FIGURE 13-8 **Bowman's capsule podocytes with foot processes and filtration slits.** Note the filtration slits between adjacent foot processes on this scanning electron micrograph. The podocytes and their foot processes encircle the glomerular capillaries.

make it more than 100 times as permeable to water and solutes as capillaries elsewhere in the body.

The *basement membrane* is an acellular (lacking cells) gelatinous layer composed of collagen and glycoproteins that is sandwiched between the glomerulus and Bowman's capsule. The collagen provides structural strength, and the glycoproteins discourage the filtration of small plasma proteins. The larger plasma proteins cannot be filtered, because they cannot fit through the capillary pores, but the pores are just barely large enough to permit passage of albumin, the smallest of plasma proteins. However, because the glycoproteins are negatively charged, they repel albumin and other plasma proteins, which are also negatively charged. Therefore, plasma proteins are almost completely excluded from the filtrate, with less than 1 percent of the albumin molecules escaping into Bowman's capsule.

*Clinical Note* Some renal diseases characterized by excessive albumin in the urine (*albuminuria*) are due to disruption of the negative charges within the basement membrane, which makes the glomerular membrane more permeable to albumin, even though the size of the capillary pores remains constant.

The final layer of the glomerular membrane is the inner layer of Bowman's capsule. It consists of **podocytes**, octopus-like cells that encircle the glomerular tuft. Each podocyte bears many elongated foot processes (*podo* means "foot"; a *process* is a projection or appendage) that interdigitate with foot processes of adjacent podocytes, much as you interlace your fingers between each other when you cup your hands around a ball (Figure 13-8). The narrow slits between adjacent foot processes, known as **filtration slits**, provide a pathway through which fluid leaving the glomerular capillaries can enter the lumen of Bowman's capsule.

Thus, the route that filtered substances take across the glomerular membrane is completely extracellular—first through capillary pores, then through the acellular basement membrane, and finally through capsular filtration slits (Figure 13-7).

## Glomerular capillary blood pressure

To accomplish glomerular filtration, a force must drive a part of the plasma in the glomerulus through the openings in the glomerular membrane. No active transport mechanisms or local energy expenditures are involved in moving fluid from the plasma across the glomerular membrane into Bowman's capsule. Passive physical forces similar to those acting across capillaries elsewhere accomplish glomerular filtration.

Because the glomerulus is a tuft of capillaries, the same principles of fluid dynamics that cause ultrafiltration across other capillaries (p. 409) apply here, except for two important differences: (1) the glomerular capillaries are much more permeable than capillaries elsewhere, so more fluid is filtered for a given filtration pressure; and (2) the balance of forces across the glomerular membrane is such that filtration occurs throughout the entire length of the capillaries. In contrast, the balance of forces in other capillaries shifts so that filtration occurs in the beginning part of the vessel but reabsorption occurs toward the vessel's end (see Figure 9-23).

### FORCES INVOLVED IN GLOMERULAR FILTRATION

Three physical forces are involved in glomerular filtration (▮ Table 13-1): (1) glomerular capillary blood pressure, (2) plasma-colloid osmotic pressure, and (3) Bowman's capsule hydrostatic pressure. Let's examine the role of each.

1. *Glomerular capillary blood pressure* is the fluid pressure exerted by the blood within the glomerular capillaries. It ultimately depends on the contraction of the heart (the source of energy that produces glomerular filtration) and the resistance to blood flow offered by the afferent and efferent arterioles. Glomerular capillary blood pressure, at an estimated average value of 55 mmHg, is higher than the capillary blood pressure (18 mmHg on average) found in other capillaries. The reason for the higher pressure in the glomerular capillaries is the larger diameter of the afferent arteriole compared with that of the efferent arteriole. Because blood can more readily enter the glomerulus through the wide afferent arteriole than it can leave through the narrower efferent arteriole, glomerular capillary blood pressure is maintained high as a result of blood damming up in the glomerular capillaries. Furthermore, because of the high resistance offered by the efferent arterioles, blood pressure does not have the same tendency to fall along the length of the glomerular capillaries as it does along other capillaries. This elevated, nondecremental glomerular blood pressure tends to push fluid out of the glomerulus into Bowman's capsule along the glomerular capillaries' entire length, and it is the major force producing glomerular filtration.

   Whereas glomerular capillary blood pressure favours filtration, the two other forces acting across the glomerular membrane (plasma-colloid osmotic pressure and Bowman's capsule hydrostatic pressure) oppose filtration.

**The Urinary System    567**

## TABLE 13-1 Forces Involved in Glomerular Filtration

| Force | Effect | Magnitude (MMHG) |
|---|---|---|
| **Glomerular Capillary Blood Pressure** | Favours filtration | 55 |
| **Plasma-colloid Pressure** | Opposes filtration | 30 |
| **Bowman's Capsule Hydrostatic Pressure** | Opposes filtration | 15 |
| **Net Filtration Pressure** (Difference between force favouring filtration and forces opposing filtration) | Favours filtration | 10 |

55−(30+15)=10

© 2016 Cengage

2. *Plasma-colloid osmotic pressure* is caused by the unequal distribution of plasma proteins across the glomerular membrane. Because plasma proteins cannot be filtered, they are in the glomerular capillaries, but not in Bowman's capsule. Accordingly, the concentration of $H_2O$ is higher in Bowman's capsule than in the glomerular capillaries. The resulting tendency for $H_2O$ to move by osmosis down its own concentration gradient from Bowman's capsule into the glomerulus opposes glomerular filtration. This opposing osmotic force averages 30 mmHg, which is slightly higher than across other capillaries. It is higher because much more $H_2O$ is filtered out of the glomerular blood, so the concentration of plasma proteins is higher than elsewhere.

3. *Bowman's capsule hydrostatic pressure*, the pressure exerted by the fluid in this initial part of the tubule, is estimated to be about 15 mmHg. This pressure, which tends to push fluid out of Bowman's capsule, opposes the filtration of fluid from the glomerulus into Bowman's capsule.

### GLOMERULAR FILTRATION RATE

As you can see in Table 13-1, the forces acting across the glomerular membrane are not in balance. The total force favouring filtration is the glomerular capillary blood pressure at 55 mmHg. The total of the two forces opposing filtration is 45 mmHg.

The net difference favouring filtration (10 mmHg of pressure) is called the **net filtration pressure**. This modest pressure forces large volumes of fluid from the blood through the highly permeable glomerular membrane. The actual rate of filtration, the **glomerular filtration rate (GFR)**, depends not only on the net filtration pressure but also on how much glomerular surface area is available for penetration and how permeable the glomerular membrane is (i.e., how "holey" it is). These properties of the glomerular membrane are collectively referred to as the **filtration coefficient** ($K_f$). Accordingly,

$$GFR = K_f \times \text{net filtration pressure}$$

Normally, about 20 percent of the plasma that enters the glomerulus is filtered at the net filtration pressure of 10 mmHg. This produces collectively through all glomeruli 180 L of glomerular filtrate each day for an average GFR of 125 mL/min in males and 160 L of filtrate per day for an average GFR of 115 mL/min in females.

*Clinical Note* An injection of inulin can be used to determine glomerular filtration rate. Inulin is a naturally occurring oligosaccharide produced by some plants and stored for energy. It is completely filtered by the nephron at the glomerulus but is then neither secreted nor reabsorbed by the tubules. This means it can be used as a clinical and highly accurate measure of the glomerular filtration rate.

### Changes in GFR

Because the net filtration pressure that accomplishes glomerular filtration is simply due to an imbalance of opposing physical forces between the glomerular capillary plasma and Bowman's capsule fluid, alterations in any of these physical forces can affect the GFR. We now examine the effect that changes in each of these physical forces have on the GFR.

#### UNREGULATED INFLUENCES ON THE GFR

Plasma-colloid osmotic pressure and Bowman's capsule hydrostatic pressure are not subject to regulation and, under normal conditions, do not vary much.

*Clinical Note* However, they can change pathologically and thus inadvertently affect the GFR. Because plasma-colloid osmotic pressure opposes filtration, a decrease in plasma protein concentration, by reducing this pressure, leads to an increase in the GFR. An uncontrollable reduction in plasma protein concentration might occur, for example, in severely burned patients who lose a large quantity of protein-rich, plasma-derived fluid through the exposed burned lesions of their skin. Conversely, in situations in which the plasma-colloid osmotic pressure is elevated, such as in cases of dehydrating diarrhoea, the GFR is reduced.

Bowman's capsule hydrostatic pressure can become uncontrollably elevated, and filtration subsequently can decrease, given a urinary tract obstruction, such as a kidney stone or enlarged prostate. The damming up of fluid behind the obstruction elevates capsular hydrostatic pressure.

#### CONTROLLED ADJUSTMENTS IN THE GFR

Unlike plasma-colloid osmotic pressure and Bowman's capsule hydrostatic pressure—which may be uncontrollably altered

**13**

in various disease states and thereby inappropriately alter the GFR—glomerular capillary blood pressure can be controlled to adjust the GFR to suit the body's needs. Assuming that all other factors stay constant, as the glomerular capillary blood pressure goes up, the net filtration pressure increases and the GFR increases correspondingly. The magnitude of the glomerular capillary blood pressure depends on the rate of blood flow within each of the glomeruli. The amount of blood flowing into a glomerulus per minute is determined largely by the magnitude of the mean systemic arterial blood pressure and the resistance offered by the afferent arterioles. If resistance increases in the afferent arteriole, less blood flows into the glomerulus, decreasing the GFR. Conversely, if afferent arteriolar resistance is reduced, more blood flows into the glomerulus and the GFR increases. Two major control mechanisms regulate the GFR, both directed at adjusting glomerular blood flow by regulating the radius and thus the resistance of the afferent arteriole. These mechanisms are (1) autoregulation, which is aimed at preventing spontaneous changes in GFR; and (2) extrinsic sympathetic control, which is aimed at long-term regulation of arterial blood pressure.

## MECHANISMS RESPONSIBLE FOR AUTOREGULATION OF THE GFR

Because arterial blood pressure is the force that drives blood into the glomerulus, the glomerular capillary blood pressure and, accordingly, the GFR would increase in direct proportion to an increase in arterial pressure if everything else remained constant (> Figure 13-9). Similarly, a fall in arterial blood pressure would be accompanied by a decline in GFR. Such spontaneous, inadvertent changes in GFR are largely prevented by intrinsic regulatory mechanisms initiated by the kidneys themselves, a process known as **autoregulation** (*auto* means "self"). The kidneys can, within limits, maintain a constant blood flow into the glomerular capillaries (and thus a constant glomerular capillary blood pressure and a stable GFR), despite changes in the driving arterial pressure. They do so primarily by altering afferent arteriolar calibre, thereby adjusting resistance to flow through these vessels. For example, if the GFR increases as a direct result of a rise in arterial pressure, the net filtration pressure and GFR can be reduced to normal by constriction of the afferent arteriole, which decreases the flow of blood into the glomerulus

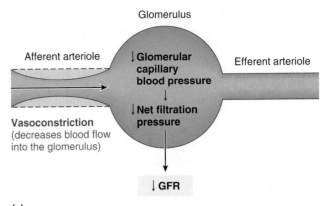

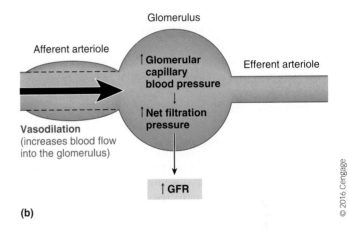

> FIGURE 13-10 **Adjustments of afferent arteriole calibre to alter the GFR.** (a) Arteriolar vasoconstriction reduces the GFR. (b) Arteriolar vasodilation increases the GFR

(> Figure 13-10a). This local adjustment lowers the glomerular blood pressure and the GFR to normal.

Conversely, when GFR falls in the presence of a decline in arterial pressure, glomerular pressure can be increased to normal by vasodilation of the afferent arteriole, which allows more blood to enter despite the reduction in driving pressure (> Figure 13-10b). The resultant buildup of glomerular blood volume increases glomerular blood pressure, which in turn brings the GFR back up to normal.

Two intrarenal mechanisms contribute to autoregulation: (1) a *myogenic* mechanism, which responds to changes in pressure within the nephron's vascular component; and (2) a *tubuloglomerular feedback* mechanism, which senses changes in the salt level in the fluid flowing through the nephron's tubular component.

1. The **myogenic** mechanism is a common property of vascular smooth muscle (*myogenic* means "muscle-produced"). Arteriolar vascular smooth muscle contracts inherently in response to the stretch accompanying

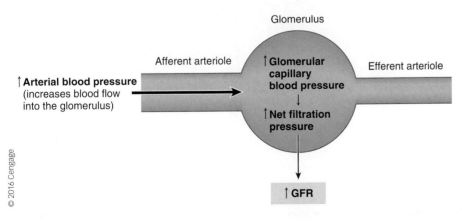

> FIGURE 13-9 **Direct effect of arterial blood pressure on the glomerular filtration rate (GFR)**

increased pressure within the vessel (p. 399). Accordingly, the afferent arteriole automatically constricts on its own when it is stretched because of an increased arterial driving pressure. This response helps limit blood flow into the glomerulus to normal, despite the elevated arterial pressure. Conversely, inherent relaxation of an unstretched afferent arteriole when pressure within the vessel is reduced increases blood flow into the glomerulus, despite the fall in arterial pressure.

2. The **tubuloglomerular feedback (TGF)** mechanism involves the *juxtaglomerular apparatus*, which is the specialized combination of tubular and vascular cells where the tubule, after having bent back on itself, passes through the angle formed by the afferent and efferent arterioles as they join the glomerulus ( > Figure 13-11; see also Figure 13-3). Some of the pericytes within the wall of the afferent arteriole in this region are specialized to form **granular cells**, so called because they contain many secretory granules. Specialized tubular cells in this region are collectively known as the **macula densa**. The macula densa cells detect changes in the salt level of the fluid flowing past them through the tubule. Other mediators of TGF are ATP and adenosine.

If the GFR is increased due to an elevation in arterial pressure, more fluid than normal is filtered and flows through the distal tubule. In response to the resultant rise in salt delivery to the distal tubule, the macula densa cells release ATP. Extracellular

degradation of the ATP forms adenosine, which acts locally as a paracrine on the adjacent afferent arteriole; this causes vasoconstriction, which reduces glomerular blood flow and ultimately returns GFR to normal. In the opposite situation, when less salt is delivered to the distal tubule due to a spontaneous decline in GFR accompanying a fall in arterial pressure, less ATP is released by macula densa cells, and this results in less extracellular adenosine. The resultant afferent arteriolar vasodilation increases the glomerular flow rate, restoring the GFR to normal. By means of the TGF mechanism, the nephron is able to monitor the salt level in the fluid flowing through it and also use ATP and adenosine to regulate the rate of filtration through its glomerulus to keep the early distal tubular fluid and salt delivery constant.

## IMPORTANCE OF AUTOREGULATION OF THE GFR

The myogenic and TGF mechanisms work in unison to autoregulate the GFR within the mean arterial blood pressure range of 80 to 180 mmHg. Within this wide range, intrinsic autoregulatory adjustments of afferent arteriolar resistance can compensate for changes in arterial pressure, thereby preventing inappropriate fluctuations in GFR, even though glomerular pressure tends to change in the same direction as arterial pressure. Normal mean arterial pressure is 93 mmHg, so the range of 80–180 mmHg encompasses the transient changes in blood pressure that accompany daily activities unrelated to the need for the kidneys to regulate water and salt excretion, such as the normal elevation in blood pressure during exercise. Autoregulation is important because unintentional shifts in GFR could

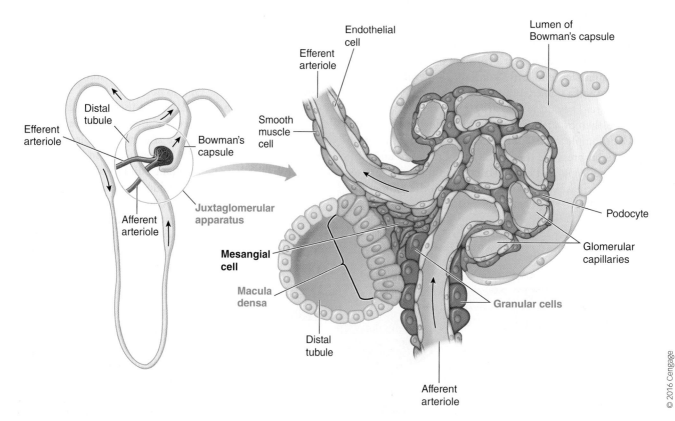

> **FIGURE 13-11 The juxtaglomerular apparatus.** The juxtaglomerular apparatus consists of specialized vascular cells (the granular cells) and specialized tubular cells (the macula densa) at a point where the distal tubule passes through the fork formed by the afferent and efferent arterioles of the same nephron.

lead to dangerous imbalances of fluid, electrolytes, and wastes. Because at least a certain portion of the filtered fluid is always excreted, the amount of fluid excreted in the urine is automatically increased as the GFR increases. If autoregulation did not occur, the GFR would increase and water and solutes would be lost needlessly as a result of the rise in arterial pressure during heavy exercise. If, by contrast, the GFR were too low, the kidneys could not eliminate enough wastes, excess electrolytes, and other materials that need to be excreted. Autoregulation thus greatly blunts the direct effect that changes in arterial pressure would otherwise have on GFR and subsequently on water, solute, and waste excretion.

When changes in mean arterial pressure fall outside the autoregulatory range, these mechanisms cannot compensate. Therefore, dramatic changes in mean arterial pressure (<80 mmHg or >180 mmHg) directly cause the glomerular capillary pressure and, accordingly, the GFR to decrease or increase in proportion to the change in arterial pressure.

## IMPORTANCE OF EXTRINSIC SYMPATHETIC CONTROL OF THE GFR

In addition to the intrinsic autoregulatory mechanisms designed to keep the GFR constant in the face of fluctuations in arterial blood pressure, the GFR can be *changed on purpose*—even when the mean arterial blood pressure is within the autoregulatory range—by extrinsic control mechanisms that override the autoregulatory responses. Extrinsic control of GFR, which is mediated by sympathetic nervous system input to the afferent arterioles, is aimed at regulating arterial blood pressure. The parasympathetic nervous system does not exert any influence on the kidneys.

If plasma volume is decreased—for example, by haemorrhage—the resulting fall in arterial blood pressure is detected by the arterial carotid sinus and aortic arch baroreceptors, which initiate neural reflexes to raise blood pressure toward normal (p. 420). These reflex responses are coordinated by the cardiovascular control centre in the brain stem and are mediated primarily through increased sympathetic activity to the heart and blood vessels. Although the resulting increase in both cardiac output and total peripheral resistance helps raise blood pressure toward normal, plasma volume is still reduced. In the long term, plasma volume must be restored to normal. One compensation for a depleted plasma volume is reduced urine output so that more fluid than normal is conserved for the body. Urine output is reduced in part

by reducing the GFR; if less fluid is filtered less is available to excrete.

## ROLE OF THE BARORECEPTOR REFLEX IN EXTRINSIC CONTROL OF THE GFR

No new mechanism is needed to decrease the GFR. It is reduced by the baroreceptor reflex response to a fall in blood pressure (> Figure 13-12). During this reflex, sympathetically induced vasoconstriction occurs in most arterioles throughout the body (including the afferent arterioles) as a compensatory mechanism to increase total peripheral resistance. The afferent arterioles are innervated with sympathetic vasoconstrictor fibres to a far greater extent than are the efferent arterioles. When the afferent arterioles carrying blood to the glomeruli constrict from increased sympathetic activity, less blood flows into the

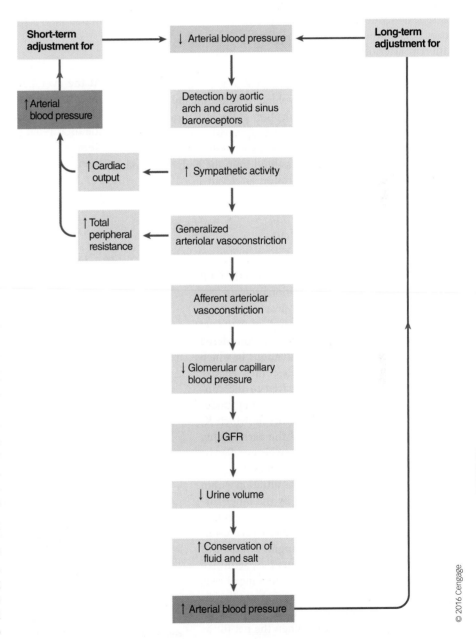

> **FIGURE 13-12** Baroreceptor reflex influence on the GFR in long-term regulation of arterial blood pressure

glomeruli than normal, lowering glomerular capillary blood pressure (see ⟩ Figure 13-10a). The resulting decrease in GFR, in turn, reduces urine volume. In this way, some of the water and salt that would otherwise have been lost in the urine are saved for the body, helping in the long term to restore plasma volume to normal so that short-term cardiovascular adjustments that have been made are no longer necessary. Other mechanisms, such as increased tubular reabsorption of water and salt as well as increased thirst, also contribute to long-term maintenance of blood pressure, despite a loss of plasma volume, by helping to restore plasma volume.

Conversely, if blood pressure is elevated (e.g., because of an expansion of plasma volume following ingestion of excessive fluid), the opposite responses occur. When the baroreceptors detect a rise in blood pressure, sympathetic vasoconstrictor activity to the arterioles, including the renal afferent arterioles, is reflexly reduced, allowing afferent arteriolar vasodilation to occur. As more blood enters the glomeruli through the dilated afferent arterioles, glomerular capillary blood pressure rises, increasing the GFR (see Figure 13-10b). As more fluid is filtered, more fluid is available to be eliminated in the urine. Contributing to the increase in urine volume is a hormonally adjusted reduction in the tubular reabsorption of water and salt. These two renal mechanisms—increased glomerular filtration and decreased tubular reabsorption of water and salt—increase urine volume and eliminate the excess fluid from the body. Reduced thirst and fluid intake also help restore an elevated blood pressure to normal.

### The GFR and the filtration coefficient

Up to now we have discussed changes in the GFR as a result of changes in net filtration pressure. The rate of glomerular filtration, however, depends on the filtration coefficient $(K_f)$ as well as on the net filtration pressure. For years, $K_f$ was considered a constant, except in disease situations in which the glomerular membrane becomes leakier than usual. Exciting new research to the contrary indicates that $K_f$ is subject to change under physiological control. Both factors on which $K_f$ depends—the surface area and the permeability of the glomerular membrane—can be modified by contractile activity within the membrane.

The surface area available for filtration within the glomerulus is represented by the inner surface of the glomerular capillaries that comes into contact with blood. Each tuft of glomerular capillaries is held together by **mesangial cells**. These cells contain contractile elements (i.e., actin-like filaments). Contraction of these mesangial cells closes off a portion of the filtering capillaries, reducing the surface area available for filtration within the glomerular tuft. When the net filtration pressure remains unchanged, this reduction in $K_f$ decreases GFR. Sympathetic stimulation

causes the mesangial cells to contract—a second mechanism (besides promoting afferent arteriolar vasoconstriction) by which sympathetic activity can decrease the GFR.

Podocytes also possess actin-like contractile filaments. Their contraction or relaxation can, respectively, decrease or increase the number of open filtration slits in the inner membrane of Bowman's capsule by changing the shapes and proximities of the foot processes (⟩ Figure 13-13). The number of slits is a determinant of permeability: the more open slits, the greater the permeability. Contractile activity of the podocytes, which in turn affects permeability and $K_f$, is under physiological control by mechanisms that are incompletely understood.

Before turning our attention to the process of tubular reabsorption, we are first going to examine the percentage of cardiac output that goes to the kidneys. This will reinforce the concept of how much blood flows through the kidneys and how much of that fluid is filtered and subsequently acted on by the tubules.

### The kidneys and cardiac output

At the average net filtration pressure and $K_f$, 20 percent of the plasma that enters the kidneys is converted into glomerular filtrate. That means at an average GFR of 125 mL/min, the total renal plasma flow must average about 625 mL/min. Because 55 percent of whole blood consists of plasma (i.e., haematocrit = 45; p. 434), the total flow of blood through the

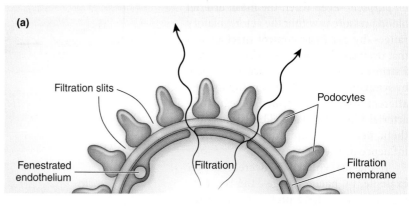

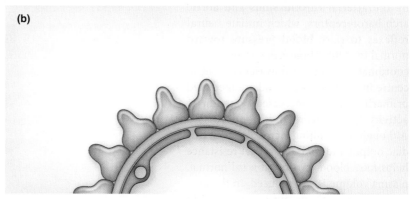

⟩ **FIGURE 13-13 Change in the number of open filtration slits caused by podocyte relaxation and contraction.** (a) Podocyte relaxation narrows the bases of the foot processes, increasing the number of fully open, intervening filtration slits that span a given area. (b) Podocyte contraction flattens the foot processes and thus decreases the number of intervening filtration slits.

kidneys averages 1140 mL/min. This quantity is about 22 percent of the total cardiac output of 5 L (5000 mL)/min, although the kidneys compose less than 1 percent of total body weight.

The kidneys need to receive such a seemingly disproportionate share of the cardiac output because they must continuously perform their regulatory and excretory functions on the huge volumes of plasma delivered to them to maintain stability in the internal fluid environment. Most of the blood goes to the kidneys not to supply the renal tissue but to be adjusted and purified by the kidneys. On average, 20–25 percent of the blood pumped out by the heart each minute "goes to the cleaners" instead of serving its normal purpose of exchanging materials with the tissues. Only by continuously processing such a large proportion of the blood can the kidneys precisely regulate the volume and electrolyte composition of the internal environment and adequately eliminate the large quantities of metabolic waste products that are constantly produced.

### KIDNEY FUNCTION DURING EXERCISE

Exercise produces considerable changes in RBF and in electrolyte and protein excretion. RBF is reduced in proportion to exercise intensity; as exercise intensity increases, RBF decreases. Renal blood flow during maximal exercise may decrease by about 300 mL, receiving less than 5 percent of cardiac output.

This results from the redistribution of blood flow away from the renal, splanchnic, liver, and stomach vascular beds, which are nonessential during exercise, to the skeletal muscle, heart, and skin. During vigorous exercise, the body increases the reabsorption of water and salt in the proximal and distal tubules, and gradually returns them to the circulatory system. The elevated levels of vasopressin in the blood facilitate the reabsorption of water.

The conservation of body water and electrolytes during exercise is not substantial, as the maximum volume of water that can be reabsorbed during exercise is only 30–45 mL/h. When you compare the volume of water reabsorption by the kidneys with typical skin sweat rates (1000–2000 mL/h), it is clear the kidney cannot keep pace. The kidney water reabsorption is important in the first 48 hours post-exercise, when fluid restoration can gradually occur.

### Check Your Understanding 13.2

1. Prepare a table showing the effect and magnitude of the physical forces involved in glomerular filtration.
2. Discuss how extrinsic control of the GFR is accomplished and the importance of this mechanism.

## 13.4 | Tubular Reabsorption

All plasma constituents except the proteins are indiscriminately filtered together through the glomerular capillaries. In addition to waste products and excess materials that the body must eliminate, the filtered fluid contains nutrients, electrolytes, and other substances that the body cannot afford to lose in the urine. The essential materials that are filtered are returned to the blood by *tubular reabsorption*. Tubular reabsorption is the method by which solutes and water are removed from the tubular fluid and transported into the blood. Reabsorption is a two-step process, beginning with the active or passive extraction of substances from the tubule fluid into the renal interstitium (connective tissue surrounding the nephrons) and then the transport (active and passive transport) of these substances from the interstitium into the bloodstream.

### Highly selective and variable

Tubular reabsorption is a highly selective process. All constituents except plasma proteins are at the same concentration in the glomerular filtrate as in plasma. In most cases, the quantity reabsorbed of each substance is the amount required to maintain the proper composition and volume of the internal fluid environment. In general, the tubules have a high reabsorptive capacity for substances needed by the body and little or no reabsorptive capacity for substances of no value ( Table 13-2). Accordingly, only a small percentage, if any, of filtered plasma constituents that are useful to the body are present in the urine, most having been reabsorbed and returned to the blood. Only excess amounts of essential materials, such as electrolytes, are excreted in the urine. For the essential plasma constituents regulated by the kidneys, absorptive capacity may vary depending on the body's needs. In contrast, a large percentage of filtered waste products is present in the urine. These wastes, which are useless or even potentially harmful to the body if allowed to accumulate, are not reabsorbed to any extent. Instead, they stay in the tubules, to be eliminated in the urine. As water and other valuable constituents are reabsorbed, the waste products remaining in the tubular fluid become highly concentrated.

Of the 125 mL/min filtered, typically 124 mL/min are reabsorbed. Considering the magnitude of glomerular filtration, the extent of tubular reabsorption is tremendous: the tubules

### ▌ TABLE 13-2 Fate of Various Substances Filtered by the Kidneys

| Substance | Average Percentage of Filtered Substance Reabsorbed | Average Percentage of Filtered Substance Excreted |
|---|---|---|
| **Water** | 99 | 1 |
| **Sodium** | 99.5 | 0.5 |
| **Glucose** | 100 | 0 |
| **Urea** (a waste product) | 50 | 50 |
| **Phenol** (a waste product) | 0 | 100 |

typically reabsorb 99 percent of the filtered water (210 L/day), 100 percent of the filtered sugar (1 kg/day), and 99.5 percent of the filtered salt (165 g/day).

## Transepithelial transport

**Transepithelial transport** (*transepithelial* means "across the epithelium") is also referred to as *transcellular transport*, and can be defined as solute movement across an epithelial cell layer through the cell. Throughout its entire length, the tubule wall is one cell thick and in close proximity to a surrounding peritubular capillary (> Figure 13-14). Adjacent tubular cells do not come into contact with each other, except where they are joined by tight junctions at their lateral edges near their *luminal membranes*, which face the tubular lumen. Interstitial fluid lies in the gaps between adjacent cells—the **lateral spaces**—as well as between the tubules and capillaries. The *basolateral membrane* faces the interstitial fluid at the base and lateral edges of the cell. The tight junctions largely prevent substances from moving *between* the cells, so generally materials must pass *through* the cells to leave the tubular lumen and gain entry to the blood.

### STEPS OF TRANSEPITHELIAL TRANSPORT

To be reabsorbed, a substance must traverse five distinct barriers (Figure 13-14):

- Step 1: It must leave the tubular fluid by crossing the luminal membrane of the tubular cell.
- Step 2: It must pass through the cytosol from one side of the tubular cell to the other.
- Step 3: It must cross the basolateral membrane of the tubular cell to enter the interstitial fluid.

- Step 4: It must diffuse through the interstitial fluid.
- Step 5: It must penetrate the capillary wall to enter the blood plasma.

In contrast to transcellular (transepithelial) transport is the concept of paracellular transport. Paracellular transport refers to the movement of solvent across an epithelial cell layer through the tight junctions (between the epithelial cells). The primary role of the proximal tubule is fluid and electrolyte transport. As the proximal tubule epithelial cells have leaky tight junctions, fluid and solutes can cross this nephron segment not only transcellularly, but also paracellularly. However, the amount of transcellular transport (~95%) is much greater than paracellular transport, with paracellular transport suggested to be about 2–5 percent of total transport. Regulation of paracellular transport in the proximal tubule is yet to be fully identified, but some possible regulators of the tight junctions include influences of pressure, hormones, and cyclic nucleotides.

### PASSIVE VERSUS ACTIVE REABSORPTION

The two types of tubular reabsorption—*passive reabsorption* and *active reabsorption*—depend on whether local energy expenditure is needed for reabsorbing a particular substance. In **passive reabsorption**, all steps in the transepithelial transport of a substance from the tubular lumen to the plasma are passive; that is, no energy is spent for the substance's net movement, which occurs down electrochemical or osmotic gradients (p. 38). In contrast, **active reabsorption** takes place if any one of the steps in the transepithelial transport of a substance requires energy, even if the four other steps are passive. With active reabsorption, net movement of the substance from the tubular lumen to the plasma occurs *against* an electrochemical gradient. Substances

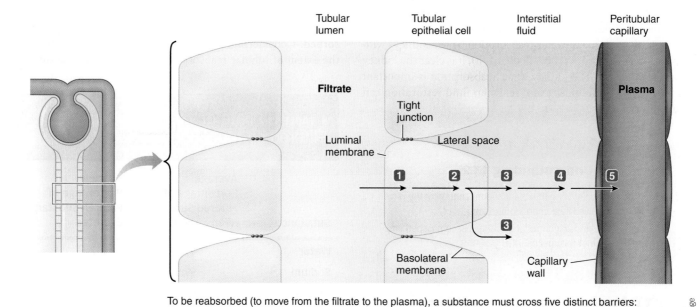

To be reabsorbed (to move from the filtrate to the plasma), a substance must cross five distinct barriers:

**1** the luminal cell membrane  **3** the basolateral cell membrane  **5** the capillary wall

**2** the cytosol  **4** the interstitial fluid

© 2016 Cengage

> **FIGURE 13-14 Steps of transepithelial transport**

that are actively reabsorbed are of particular importance to the body, such as glucose, amino acids, and other organic nutrients, as well as $Na^+$ and other electrolytes, such as $PO_4^{3-}$. Rather than specifically describe the reabsorptive process for each of the many filtered substances returned to the plasma, we will provide illustrative examples of the general mechanisms involved. But first we highlight the unique and important case of $Na^+$ reabsorption.

### The $Na^+$–$K^+$ ATPase pump

Sodium reabsorption is unique and complex. Of the total energy spent by the kidneys, 80 percent is used for $Na^+$ transport, indicating the importance of this process. Unlike most filtered solutes, $Na^+$ is reabsorbed throughout most of the tubule, but to varying extents in different regions. Of the $Na^+$ filtered, 99.5 percent is normally reabsorbed. Of the $Na^+$ reabsorbed, on average 67 percent is reabsorbed in the proximal tubule, 25 percent in the loop of Henle, and 8 percent in the distal and collecting tubules. Sodium reabsorption plays different important roles in each of these segments, as will become apparent as our discussion continues. Here is a preview of these roles.

- Sodium reabsorption in the *proximal tubule* plays a pivotal role in reabsorbing glucose, amino acids, $H_2O$, $Cl^-$, and urea.

- Sodium reabsorption in the ascending limb of the *loop of Henle*, along with $Cl^-$ reabsorption, plays a critical role in the kidneys' ability to produce urine of varying concentrations and volumes, depending on the body's need to conserve or eliminate $H_2O$.

- Sodium reabsorption in the *distal and collecting tubules* is variable and subject to hormonal control. It plays a key role in regulating ECF volume, which is important in long-term control of arterial blood pressure and is also linked in part to $K^+$ secretion and $H^+$ secretion.

Sodium is reabsorbed throughout the tubule with the exception of the descending limb of the loop of Henle. You will learn about the significance of this exception later. Throughout all $Na^+$ reabsorbing segments of the tubule, the active step in $Na^+$ reabsorption involves the energy-dependent $Na^+$–$K^+$ ATPase carrier located in the tubular cell's basolateral membrane ( ) Figure 13-15). This carrier is the same $Na^+$–$K^+$ pump present in all cells that actively extrudes $Na^+$ from the cell (p. 45). As this basolateral pump transports $Na^+$ out of the tubular cell into the lateral space, it keeps the intracellular $Na^+$ concentration low while simultaneously building up the concentration of $Na^+$ in the lateral space; that is, it moves $Na^+$ against a concentration gradient. Because the intracellular $Na^+$ concentration is kept low by basolateral pump activity, a concentration gradient is established that favours the passive movement

of $Na^+$ from its higher concentration in the tubular lumen across the luminal border into the tubular cell. The nature of the luminal $Na^+$ channels and/or transport carriers that permit movement of $Na^+$ from the lumen into the cell varies for different parts of the tubule, but in each case, movement of $Na^+$ across the luminal membrane is always a passive step. For example, in the proximal tubule, $Na^+$ crosses the luminal border by a cotransport carrier that simultaneously moves $Na^+$ and an organic nutrient, such as glucose, from the lumen into the cell. You will learn more about this cotransport process shortly. By contrast, in the collecting duct, $Na^+$ crosses the luminal border through a $Na^+$ channel. Once $Na^+$ enters the cell across the luminal border by whatever means, it is actively extruded to the lateral space by the basolateral $Na^+$–$K^+$ pump. This step is the same throughout the tubule. Sodium continues to diffuse down a concentration gradient from its high concentration in the lateral space into the surrounding interstitial fluid and finally into the peritubular capillary blood. Thus, net transport of $Na^+$ from the tubular lumen into the blood occurs at the expense of energy.

First let's consider the importance of regulating $Na^+$ reabsorption in the distal portion of the nephron and examine how this control is accomplished. Later, in Section 13.6, we will explore in further detail the roles of $Na^+$ reabsorption in the proximal tubule and in the loop of Henle.

### Aldosterone and $Na^+$ reabsorption

In the proximal tubule and loop of Henle, a constant percentage of the filtered $Na^+$ is reabsorbed regardless of the **$Na^+$ load**

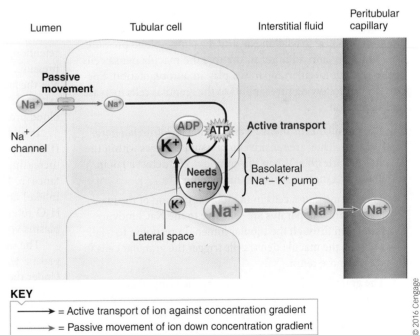

**KEY**

⟶ = Active transport of ion against concentration gradient

⟶ = Passive movement of ion down concentration gradient

> **FIGURE 13-15 Sodium reabsorption.** The basolateral $Na^+$–$K^+$ ATPase carrier actively transports $Na^+$ from the tubular cell into the interstitial fluid within the lateral space. This process establishes a concentration gradient for passive movement of $Na^+$ from the lumen into the tubular cell and from the lateral space into the peritubular capillary, accomplishing net transport of $Na^+$ from the tubular lumen into the blood at the expense of energy.

**13**

(*total amount* of Na$^+$ in the body fluids, *not the concentration* of Na$^+$ in the body fluids). In the distal part of the tubule, the reabsorption of a small percentage of the filtered Na$^+$ is subject to hormonal control. The extent of this controlled, discretionary reabsorption is inversely related to the magnitude of the Na$^+$ load in the body. If there is too much Na$^+$, little of this controlled Na$^+$ is reabsorbed; instead, it is lost in the urine, thereby removing excess Na$^+$ from the body. If Na$^+$ is depleted, however, most or all of this controlled Na$^+$ is reabsorbed, conserving for the body Na$^+$ that otherwise would be lost in the urine.

The Na$^+$ load in the body is reflected by the ECF volume. Sodium and its accompanying anion Cl$^-$ account for more than 90 percent of the ECF's osmotic activity. (NaCl is common table salt.) Recall that osmotic pressure can be thought of loosely as a force that attracts and holds H$_2$O (p. 38). When the Na$^+$ load is above normal and the ECF's osmotic activity is therefore increased, the extra Na$^+$ holds extra H$_2$O, expanding the ECF volume. Conversely, when the Na$^+$ load is below normal, thereby decreasing ECF osmotic activity, less H$_2$O than normal can be held in the ECF, so the ECF volume is reduced. Because plasma is part of the ECF, the most important result of a change in ECF volume is the matching change in blood pressure with expansion (↑ blood pressure) or reduction (↓ blood pressure) of the plasma volume. Thus, long-term control of arterial blood pressure ultimately depends on Na$^+$-regulating mechanisms. We now turn our attention to these mechanisms.

### ACTIVATION OF THE RENIN–ANGIOTENSIN SYSTEM

The most important and best-known hormonal system involved in regulating Na$^+$ is the **renin–angiotensin–aldosterone system (RAAS)**. The granular cells of the juxtaglomerular apparatus (see Figure 13-11) secrete an enzymatic hormone, **renin**, into the blood in response to a fall in NaCl/ECF volume/blood pressure. This function is in addition to the role the macula densa cells of the juxtaglomerular apparatus play in autoregulation. Specifically, the following three inputs to the granular cells increase renin secretion:

1. The granular cells themselves function as *intrarenal baroreceptors*. They are sensitive to pressure changes within the afferent arteriole. When the granular cells detect a fall in blood pressure, they secrete more renin.

2. The macula densa cells in the tubular portion of the juxtaglomerular apparatus are sensitive to the NaCl moving past them through the tubular lumen. In response to a fall in NaCl, the macula densa cells trigger the granular cells to secrete more renin.

3. The granular cells are innervated by the sympathetic nervous system. When blood pressure falls below normal, the baroreceptor reflex increases sympathetic activity. As part of this reflex response, increased sympathetic activity stimulates the granular cells to secrete more renin.

These interrelated signals for increased renin secretion all indicate the need to expand the plasma volume to increase the arterial pressure to normal on a long-term basis. Through a complex series of events involving the RAAS, increased renin secretion brings about increased Na$^+$ reabsorption by the distal and collecting tubules. Chloride always passively follows Na$^+$ down the electrical gradient established by sodium's active movement. The ultimate benefit of this salt retention is that it osmotically promotes H$_2$O retention, which helps restore the plasma volume and, therefore, is important in the long-term control of blood pressure.

Let's examine in further detail the RAAS mechanism by which renin secretion ultimately leads to increased Na$^+$ reabsorption (⟩ Figure 13-16). Once secreted into the blood, renin acts as an enzyme to activate **angiotensinogen** into **angiotensin I**. Angiotensinogen, a plasma protein synthesized by the liver, is always present in the plasma in high concentration. On passing through the lungs via the pulmonary circulation, angiotensin I is converted into **angiotensin II** by **angiotensin-converting enzyme (ACE)**, which is abundant in the pulmonary capillaries. Angiotensin II is the main stimulus for secretion of the hormone *aldosterone* from the adrenal cortex. The *adrenal cortex* is an endocrine gland that produces several different hormones, each secreted in response to different stimuli.

### FUNCTIONS OF THE RENIN–ANGIOTENSIN–ALDOSTERONE SYSTEM

Among its actions, aldosterone increases Na$^+$ reabsorption by the distal and collecting tubules. It does so by promoting the insertion of additional Na$^+$ channels into the luminal membranes and additional Na$^+$–K$^+$ ATPase carriers into the basolateral membranes of the distal and collecting tubular cells. The net result is a greater passive inward flux of Na$^+$ into the tubular cells from the lumen and increased active pumping of Na$^+$ out of the cells into the plasma—that is, an increase in Na$^+$ reabsorption, with Cl$^-$ following passively. RAAS thus promotes salt retention and a resulting H$_2$O retention and rise in arterial blood pressure. Acting in negative-feedback fashion, this system alleviates the factors that triggered the initial release of renin: namely, salt depletion, plasma volume reduction, and decreased arterial blood pressure (Figure 13-16).

In addition to stimulating aldosterone secretion, angiotensin II is a potent constrictor of the systemic arterioles, directly increasing blood pressure by increasing total peripheral resistance (p. 401). Furthermore, it stimulates thirst (increasing fluid intake) and stimulates vasopressin (a hormone that increases H$_2$O retention by the kidneys), both of which contribute to plasma volume expansion and elevation of arterial pressure.

The opposite situation exists when the Na$^+$ load, ECF and plasma volume, and arterial blood pressure are above normal. Under these circumstances, renin secretion is inhibited. Therefore, because angiotensinogen is not activated to angiotensin I and II, aldosterone secretion is not stimulated. Without aldosterone, the small aldosterone-dependent part of Na$^+$ reabsorption in the distal segments of the tubule does not occur. Instead, this nonreabsorbed Na$^+$ is lost in the urine. In the absence of aldosterone, the ongoing loss of this small percentage of filtered Na$^+$ can rapidly remove excess Na$^+$ from the body. Even though only about 8 percent of the filtered Na$^+$ depends on aldosterone for reabsorption, this small loss, multiplied many times as the entire

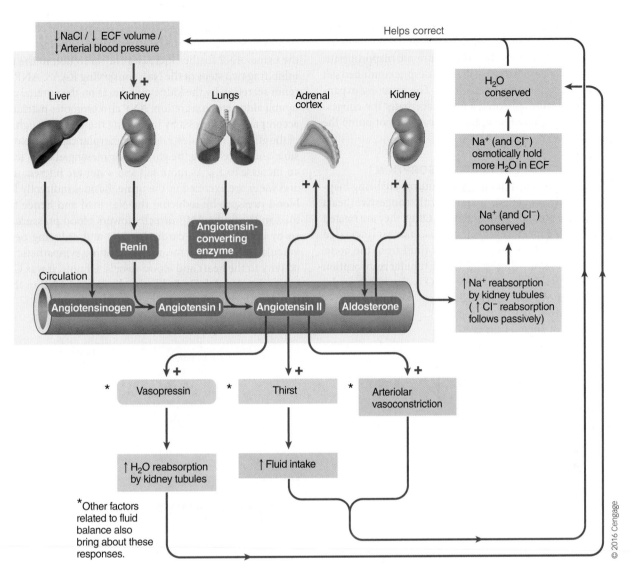

> FIGURE 13-16 **Renin–angiotensin–aldosterone system (RAAS).** The kidneys secrete the hormone renin in response to reduced NaCl, ECF volume, and arterial blood pressure. Renin activates angiotensinogen, a plasma protein produced by the liver, into angiotensin I. Angiotensin I is converted into angiotensin II by angiotensin-converting enzyme (ACE) produced in the lungs. Angiotensin II stimulates the adrenal cortex to secrete the hormone aldosterone, which stimulates $Na^+$ reabsorption by the kidneys. The resulting retention of $Na^+$ exerts an osmotic effect that holds more $H_2O$ in the ECF. Together, the conserved $Na^+$ and $H_2O$ help correct the original stimuli that activated this hormonal system. Angiotensin II also exerts other effects that help rectify the original stimuli, such as by promoting arteriolar vasoconstriction.

plasma volume is filtered through the kidneys many times per day, can lead to a sizable loss of $Na^+$.

In the complete absence of aldosterone, 20 g of salt may be excreted per day. With maximum aldosterone secretion, all the filtered $Na^+$ (and, accordingly, all the filtered $Cl^-$) is reabsorbed, so salt excretion in the urine is zero. The amount of aldosterone secreted, and consequently the relative amount of salt conserved versus salt excreted, usually varies between these extremes, depending on the body's needs. For example, an average salt consumer typically excretes about 10 g of salt per day in the urine, a heavy salt consumer excretes more, and someone who has lost considerable salt during heavy sweating excretes less urinary salt. By varying the amount of renin and aldosterone secreted in accordance with the salt-determined fluid load in

the body, the kidneys can finely adjust the amount of salt conserved or eliminated. In doing so, they maintain the salt load and ECF volume/arterial blood pressure at a relatively constant level, despite wide variations in salt consumption and abnormal losses of salt-laden fluid.

## ROLE OF THE RENIN–ANGIOTENSIN–ALDOSTERONE SYSTEM IN VARIOUS DISEASES

*Clinical Note* Some cases of hypertension (high blood pressure) are due to abnormal increases in RAAS activity. This system is also responsible in part for the fluid retention and oedema accompanying congestive heart failure. Because of the failing heart, cardiac output is reduced and arterial blood pressure is low, despite a normal or even expanded plasma volume.

When a fall in blood pressure is due to a failing heart rather than a reduced salt/fluid load in the body, the salt- and fluid-retaining reflexes triggered by the low blood pressure are inappropriate. Sodium excretion may fall to virtually zero despite continued salt ingestion and accumulation in the body. The resulting expansion of the ECF produces oedema and exacerbates the congestive heart failure because the weakened heart cannot pump the additional plasma volume.

### DRUGS THAT AFFECT SODIUM REABSORPTION

*Clinical Note* Because their salt-retaining mechanisms are being inappropriately triggered, patients with congestive heart failure are placed on low-salt diets. Often they are treated with **diuretics**, therapeutic agents that cause *diuresis* (increased urinary output) and thus promote loss of fluid from the body. Many of these drugs function by inhibiting tubular reabsorption of $Na^+$. As more $Na^+$ is excreted, more $H_2O$ is also lost from the body, helping remove the excess ECF.

**ACE inhibitor drugs**, which block the action of angiotensin-converting enzyme (ACE), and **aldosterone receptor blockers** are both also beneficial in treating hypertension and congestive heart failure. By blocking the generation of angiotensin II or by blocking the binding of aldosterone with its renal receptors, respectively, these two classes of drugs halt the ultimate salt- and fluid-conserving actions and arteriolar-constrictor effects of RAAS.

## Atrial natriuretic peptide and reabsorption

Atrial natriuretic peptide (ANP) is a polypeptide hormone involved in the homeostatic control of body $H_2O$ and $Na^+$. ANP reduces $H_2O$ and $Na^+$ loads on the circulatory system and thus lowers blood pressure, which is the opposite effect of aldosterone and the RAAS. ANP is released by atrial myocytes (muscle cells) from granules in the atria of the heart in response to high blood pressure, hypervolaemia, and exercise. When blood volume is increased or there is increased venous return (exercise), stretch receptors in the left atria, aortic arch, and carotid sinus stimulate the release of ANP (> Figure 13-17). Other potential stimulants for the release of ANP are sympathetic stimulation, elevated $Na^+$ concentration, angiotensin II, and endothelin (vasoconstrictor).

The main action of ANP is to directly inhibit $Na^+$ reabsorption in the distal nephron, thereby increasing $Na^+$ excretion in the urine. ANP further increases $Na^+$ excretion in the urine by inhibiting two steps of the $Na^+$-conserving RAAS. ANP inhibits renin secretion by the kidneys and acts on the adrenal cortex to inhibit aldosterone secretion. ANP also promotes natriuresis and accompanying diuresis by increasing the GFR through dilation of the afferent arterioles, raising glomerular capillary blood pressure, and by relaxing the glomerular mesangial cells, leading to an increase in $K_f$. As more salt and water are filtered, more salt and water are excreted in the urine. Besides indirectly lowering blood pressure by reducing the $Na^+$ load and hence the fluid load in the body, ANP directly lowers blood pressure. It does so by decreasing the cardiac output and reducing peripheral vascular resistance by means of inhibiting sympathetic nervous activity to the heart and blood vessels. Of note, it was Canadian researcher Dr. A.J. De Bold who discovered ANP in the 1980s while at Queen's University in Kingston, Ontario.

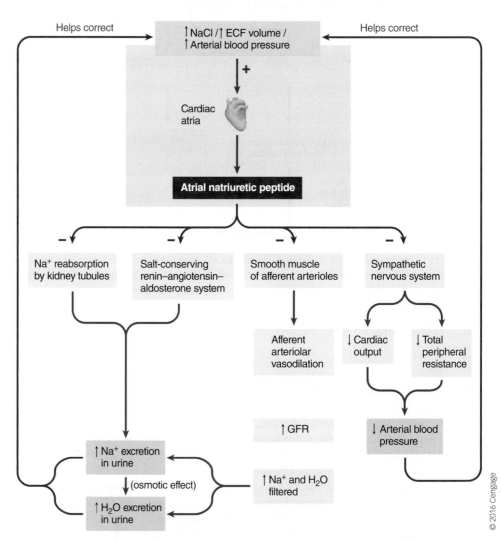

> **FIGURE 13-17 Atrial natriuretic peptide.** The atria secrete the hormone atrial natriuretic peptide (ANP) in response to being stretched by $Na^+$ retention, expansion of the ECF volume, and increase in arterial blood pressure. ANP, in turn, promotes natriuretic, diuretic, and hypotensive effects to help correct the original stimuli that resulted in its release.

The relative contributions of ANP in maintaining salt and water balance and blood pressure regulation are presently being intensively investigated. Importantly, derangements of this system could logically contribute to hypertension. In fact, recent studies suggest that a deficiency of the counterbalancing natriuretic system may underlie some cases of long-term hypertension by leaving the powerful $Na^+$-conserving system unopposed. The resulting $Na^+$ retention, especially in association with high salt intake, could expand ECF volume and elevate blood pressure.

We now shift our attention to the reabsorption of other filtered solutes. However, we will continue to discuss $Na^+$ reabsorption, because the reabsorption of many other solutes is linked in some way to $Na^+$ reabsorption.

## Glucose and amino acids

Large quantities of nutritionally important organic molecules, such as glucose and amino acids, are filtered each day. Normally these substances are completely reabsorbed back into the blood by energy- and $Na^+$-dependent mechanisms located in the proximal tubule, so none of these materials are usually excreted in the urine. This rapid and thorough reabsorption early in the tubules protects against the loss of these important organic nutrients.

Even though glucose and amino acids are moved uphill against their concentration gradients from the tubular lumen into the blood until their concentration in the tubular fluid is virtually zero, no energy is directly used to operate the glucose or amino acid carriers. Glucose and amino acids are transferred by secondary active transport. With this process, specialized *cotransport carriers* located only in the proximal tubule simultaneously transfer both $Na^+$ and the specific organic molecule from the lumen into the cell (p. 44). This luminal cotransport carrier is the means by which $Na^+$ passively crosses the luminal membrane in the proximal tubule. The lumen-to-cell $Na^+$ concentration gradient maintained by the energy-consuming basolateral $Na^+-K^+$ pump drives this cotransport system and pulls the organic molecule against its concentration gradient without the direct expenditure of energy. Because the overall process of glucose and amino acid reabsorption depends on the use of energy, these organic molecules are considered to be actively reabsorbed, even though energy is not used directly to transport them across the membrane. In essence, glucose and amino acids get a "free ride" at the expense of energy already used in the reabsorption of $Na^+$. Secondary active transport requires the presence of $Na^+$ in the lumen; without $Na^+$, the cotransport carrier is inoperable. Once transported into the tubular cells, glucose and amino acids passively diffuse down their concentration gradients across the basolateral membrane into the plasma, facilitated by a carrier that is not dependent on energy.

## Actively reabsorbed substances

All actively reabsorbed substances bind with plasma membrane carriers that transfer them across the membrane against a concentration gradient. Each carrier is specific for the types of substances it can transport; for example, the glucose cotransport carrier cannot transport amino acids, or vice versa. Because a limited number of each carrier type is present in the cells lining the tubules, there is an upper limit on how much of a particular substance can be actively transported from the tubular fluid in a given period of time. The maximum reabsorption rate is reached when all the specific carriers for a particular substance are fully occupied or saturated, so they cannot handle any additional passengers at that time. This transport maximum is designated as the **tubular maximum ($T_m$)**. Any quantity of a substance filtered beyond its $T_m$ is not reabsorbed and escapes instead into the urine. With the exception of $Na^+$, all actively reabsorbed substances have a tubular maximum. (Even though individual $Na^+$ transport carriers can become saturated, the tubules as a whole do not display a tubular maximum for $Na^+$, because aldosterone promotes the synthesis of more active $Na^+-K^+$ carriers in the distal and collecting tubular cells as needed.)

The plasma concentrations of some, but not all, substances that display carrier-limited reabsorption are regulated by the kidneys. How can the kidneys regulate some actively reabsorbed substances but not others, when the renal tubules limit the quantity of each of these substances that can be reabsorbed and returned to the plasma? We will compare glucose, a substance that has a $T_m$ but is *not regulated* by the kidneys, with phosphate, a $T_m$-limited substance that is *regulated* by the kidneys.

## Glucose and the kidneys

The normal plasma concentration of glucose is 100 mg of glucose/100 mL of plasma. Because glucose is freely filterable at the glomerulus, it passes into Bowman's capsule at the same concentration it has in the plasma. Accordingly, 100 mg of glucose are present in every 100 mL of plasma filtered. With 125 mL of plasma normally being filtered each minute (average GFR = 125 mL/min), 125 mg of glucose pass into Bowman's capsule with this filtrate every minute. The quantity of any substance filtered per minute, known as its **filtered load**, can be calculated as follows:

Filtered load of a substance = plasma concentration of the
substance $\times$ GFR

$$\text{Filtered load of glucose} = 100\,\text{mg}/100\,\text{mL} \times 125\,\text{mL}/\text{min}$$
$$= 125\,\text{mg}/\text{min}$$

At a constant GFR, the filtered load of glucose is directly proportional to the plasma glucose concentration. Doubling the plasma glucose concentration to 200 mg/100 mL doubles the filtered load of glucose to 250 mg/min, and so on (> Figure 13-18).

### TUBULAR MAXIMUM FOR GLUCOSE

The $T_m$ for glucose averages 375 mg/min; that is, the glucose carrier mechanism is capable of actively reabsorbing up to 375 mg of glucose per minute before it reaches its maximum transport capacity. At a normal plasma glucose concentration of 100 mg/100 mL, the 125 mg of glucose filtered per minute can readily be reabsorbed by the glucose carrier mechanism, because the filtered load is well below the $T_m$ for glucose. Ordinarily, therefore, no glucose appears in the urine. Not until the filtered load of glucose exceeds 375 mg/min is the $T_m$ reached. When more

13

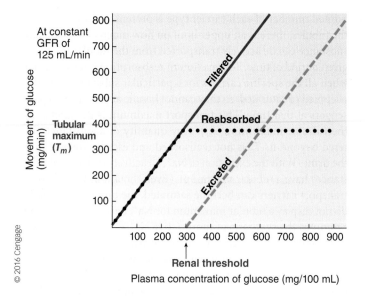

**At constant GFR of 125 mL/min**

Movement of glucose (mg/min)

Tubular maximum ($T_m$)

Renal threshold

Plasma concentration of glucose (mg/100 mL)

› **FIGURE 13-18 Renal handling of glucose as a function of plasma glucose concentration.** At a constant GFR, the quantity of glucose filtered per minute is directly proportional to the plasma concentration of glucose. All the filtered glucose can be reabsorbed up to the tubular maximum ($T_m$). If the amount of glucose filtered per minute exceeds the $T_m$, the maximum amount of glucose is reabsorbed (a $T_m$ worth), and the rest stays in the filtrate to be excreted in urine. The renal threshold is the plasma concentration at which the $T_m$ is reached and glucose first starts appearing in the urine.

© 2016 Cengage

glucose is filtered per minute than can be reabsorbed because the $T_m$ is exceeded, the maximum amount is reabsorbed, while the rest stays in the filtrate to be excreted. Accordingly, the plasma glucose concentration must be greater than 300 mg/100 mL—more than three times the normal value—before the amount filtered exceeds 375 mg/min and glucose starts spilling into the urine.

### RENAL THRESHOLD FOR GLUCOSE

The plasma concentration at which the $T_m$ of a particular substance is reached and the substance first starts appearing in urine is called the **renal threshold**. At the average $T_m$ of 375 mg/min and GFR of 125 mL/min, the renal threshold for glucose is 300 mg/100 mL.[1] Beyond the $T_m$, reabsorption stays constant at its maximum rate, and any further increase in the filtered load leads to a directly proportional increase in the amount of the substance excreted. For example, at a plasma glucose concentration of 400 mg/100 mL, the filtered load of glucose is 500 mg/min, 375 mg/min of which can be reabsorbed (a $T_m$

---

[1]This is an idealized situation. In reality, glucose often starts spilling into the urine at glucose concentrations of 180 mg/100 mL and above. Glucose is often excreted before the average renal threshold of 300 mg/100 mL is reached for two reasons. First, not all nephrons have the same $T_m$, so some nephrons may have exceeded their $T_m$ and be excreting glucose, while others have not yet reached their $T_m$. Second, the efficiency of the glucose cotransport carrier may not be working at its maximum capacity at elevated values less than the true $T_m$, so some of the filtered glucose may fail to be reabsorbed and spill into the urine even though the average renal threshold has not been reached.

worth) and 125 mg/min of which are excreted in the urine. At a plasma glucose concentration of 500 mg/100 mL, the filtered load is 625 mg/min, but only 375 mg/min can be reabsorbed, and 250 mg/min spill into the urine (Figure 13-18).

*Clinical Note* The plasma glucose concentration can become extremely high in *diabetes mellitus*, an endocrine disorder involving inadequate insulin action. Insulin is a pancreatic hormone that facilitates transport of glucose into many body cells. When cellular glucose uptake is impaired, glucose that cannot gain entry into cells stays in the plasma, elevating the plasma glucose concentration. Consequently, although glucose does not normally appear in urine, it is found in the urine of people with diabetes when the plasma glucose concentration exceeds the renal threshold, even though renal function has not changed.

What happens when plasma glucose concentration falls below normal? The renal tubules, of course, reabsorb all the filtered glucose, because the glucose reabsorptive capacity is far from being exceeded. The kidneys cannot do anything to raise a low plasma glucose level to normal. They simply return all the filtered glucose to the plasma.

### REASON THAT THE KIDNEYS DO NOT REGULATE GLUCOSE

The kidneys do not influence plasma glucose concentration over a wide range of values from abnormally low levels up to three times the normal level. Because the $T_m$ for glucose is well above the normal filtered load, the kidneys usually conserve all the glucose, thereby protecting against loss of this important nutrient in the urine. The kidneys do not regulate glucose, because they do not maintain glucose at some specific plasma concentration. Instead, this concentration is normally regulated by endocrine and liver mechanisms, with the kidneys merely maintaining whatever plasma glucose concentration is set by these other mechanisms (except when excessively high levels overwhelm the kidneys' reabsorptive capacity). The same general principle holds true for other organic plasma nutrients, such as amino acids and water-soluble vitamins.

## Phosphate and the kidneys

The kidneys do directly contribute to the regulation of many electrolytes, such as phosphate ($PO_4^{3-}$) and calcium ($Ca^{2+}$), because the renal thresholds of these inorganic ions equal their normal plasma concentrations. The transport carriers for these electrolytes are located in the proximal tubule. Here we use $PO_4^{3-}$ as an example. Our diets are generally rich in $PO_4^{3-}$, but because the tubules can reabsorb up to the normal plasma concentration's worth of $PO_4^{3-}$ and no more, the excess ingested $PO_4^{3-}$ is quickly spilled into the urine, restoring the plasma concentration to normal. The greater the amount of $PO_4^{3-}$ ingested beyond the body's needs, the greater is the amount excreted. In this way, the kidneys maintain the desired plasma $PO_4^{3-}$ concentration while eliminating any excess $PO_4^{3-}$ ingested.

Unlike the reabsorption of organic nutrients, the reabsorption of $PO_4^{3-}$ and $Ca^{2+}$ is also subject to hormonal control. Parathyroid hormone can alter the renal thresholds for $PO_4^{3-}$ and

Ca$^{2+}$, thus adjusting the quantity of these electrolytes conserved, depending on the body's momentary needs (Chapter 6).

## Responsibility of Na$^+$ reabsorption for Cl$^-$, H$_2$O, and urea reabsorption

Not only is secondary active reabsorption of glucose and amino acids linked to the basolateral Na$^+$−K$^+$ pump, but passive reabsorption of Cl$^-$, H$_2$O, and urea also depends on this active Na$^+$ reabsorption mechanism.

### CHLORIDE REABSORPTION

The negatively charged chloride ions are passively reabsorbed down the electrical gradient created by the active reabsorption of the positively charged sodium ions. For the most part, chloride ions pass between, not through, the tubular cells. The amount of Cl$^-$ reabsorbed is determined by the rate of active Na$^+$ reabsorption, instead of being directly controlled by the kidneys.

### WATER REABSORPTION

Water is passively reabsorbed throughout the length of the tubule because it osmotically follows the Na$^+$ that is actively reabsorbed. Of the H$_2$O filtered, 65 percent—117 L per day—is passively reabsorbed by the end of the proximal tubule. Another 15 percent of the filtered H$_2$O is obligatorily reabsorbed from the loop of Henle. This 80 percent of the filtered H$_2$O is reabsorbed in the proximal tubule and Henle's loop, regardless of the H$_2$O load in the body, and is not subject to regulation. Variable amounts of the remaining 20 percent are reabsorbed in the distal portions of the tubule; the extent of reabsorption in the distal and collecting tubules is under direct hormonal control, depending on the body's state of hydration. No part of the tubule directly requires energy for this tremendous reabsorption of H$_2$O.

During reabsorption, H$_2$O passes through aquaporins (water channels), formed by specific plasma membrane proteins in the tubular cells. Different types of water channels are present in various parts of the nephron. The water channels in the proximal tubule are always open, which accounts for the high H$_2$O permeability of this region. The channels in the distal parts of the nephron, in contrast, are regulated by the hormone *vasopressin*, which accounts for the variable H$_2$O reabsorption in this region.

The main driving force for H$_2$O reabsorption in the proximal tubule is a compartment of hypertonicity in the lateral spaces between the tubular cells established by the basolateral pump's active extrusion of Na$^+$ (⟩ Figure 13-19). As a result of this pump activity, the concentration of Na$^+$ rapidly diminishes in the tubular fluid and tubular cells, while it simultaneously increases in the localized region within the lateral spaces. This osmotic gradient induces the passive net flow of H$_2$O from the lumen into the lateral spaces, either through the cells or intercellularly through leaky tight junctions. The accumulation of fluid in the lateral spaces results in a build-up of hydrostatic (fluid) pressure, which flushes H$_2$O out of the lateral spaces into the interstitial fluid and finally into the peritubular capillaries. Water also osmotically follows other preferentially reabsorbed solutes, such as glucose (which is also Na$^+$ dependent). However, the direct influence of Na$^+$ reabsorption on passive H$_2$O reabsorption is quantitatively more important.

This return of filtered H$_2$O to the plasma is enhanced by the fact that the plasma-colloid osmotic pressure is greater in the peritubular capillaries than elsewhere. The concentration of plasma proteins, which is responsible for the plasma osmotic pressure, is elevated in the blood entering the peritubular capillaries because of the extensive filtration of H$_2$O through the glomerular capillaries upstream. The plasma proteins left behind in the glomerulus are concentrated into a smaller volume of plasma H$_2$O, increasing the plasma-colloid osmotic pressure of the unfiltered blood that leaves the glomerulus and enters the peritubular capillaries. This force tends to "pull" H$_2$O into the peritubular capillaries simultaneously with the push of the hydrostatic pressure in the lateral spaces that drives H$_2$O toward the capillaries. By these means, 65 percent of the filtered H$_2$O—117 L per day—is passively reabsorbed by the end of the proximal tubule.

The mechanisms responsible for H$_2$O reabsorption beyond the proximal tubule are described later.

### UREA REABSORPTION

Passive reabsorption of urea, in addition to Cl$^-$ and H$_2$O, is also indirectly linked to active Na$^+$ reabsorption. **Urea** is a waste product from the breakdown of protein. The osmotically induced

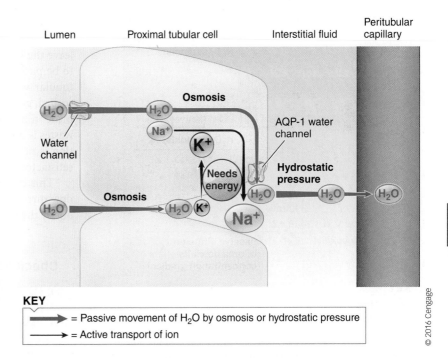

**KEY**

⟹ = Passive movement of H$_2$O by osmosis or hydrostatic pressure

⟶ = Active transport of ion

© 2016 Cengage

⟩ **FIGURE 13-19 Water reabsorption in the proximal tubule.** The force for H$_2$O reabsorption is the compartment of hypertonicity in the lateral spaces established by active extrusion of Na$^+$ by the basolateral pump. The resultant accumulation of H$_2$O in the lateral spaces creates a hydrostatic pressure that drives the H$_2$O into the peritubular capillaries.

13

reabsorption of $H_2O$ in the proximal tubule secondary to active $Na^+$ reabsorption produces a concentration gradient for urea that favours passive reabsorption of this waste (> Figure 13-20). Extensive reabsorption of $H_2O$ in the proximal tubule gradually reduces the original 125 mL/min of filtrate until only 44 mL/min of fluid remain in the lumen by the end of the proximal tubule (65% of the $H_2O$ in the original filtrate, or 81 mL/min, has been reabsorbed). Substances that have been filtered but not reabsorbed become progressively more concentrated in the tubular fluid because $H_2O$ is reabsorbed, and they are left behind. Urea is one such substance. Urea's concentration as it is filtered at the glomerulus is identical to its concentration in the plasma entering the peritubular capillaries. The quantity of urea present in the 125 mL of filtered fluid at the beginning of the proximal tubule, however, is concentrated almost threefold in the 44 mL left at the end of the proximal tubule. As a result, the urea concentration within the tubular fluid becomes much greater than the urea concentration in the adjacent capillaries. Therefore, a concentration gradient is created for urea to passively diffuse from the tubular lumen into the peritubular capillary plasma. Because the walls of the proximal tubules are only somewhat permeable to urea, only about 50 percent of the filtered urea is passively reabsorbed by this means.

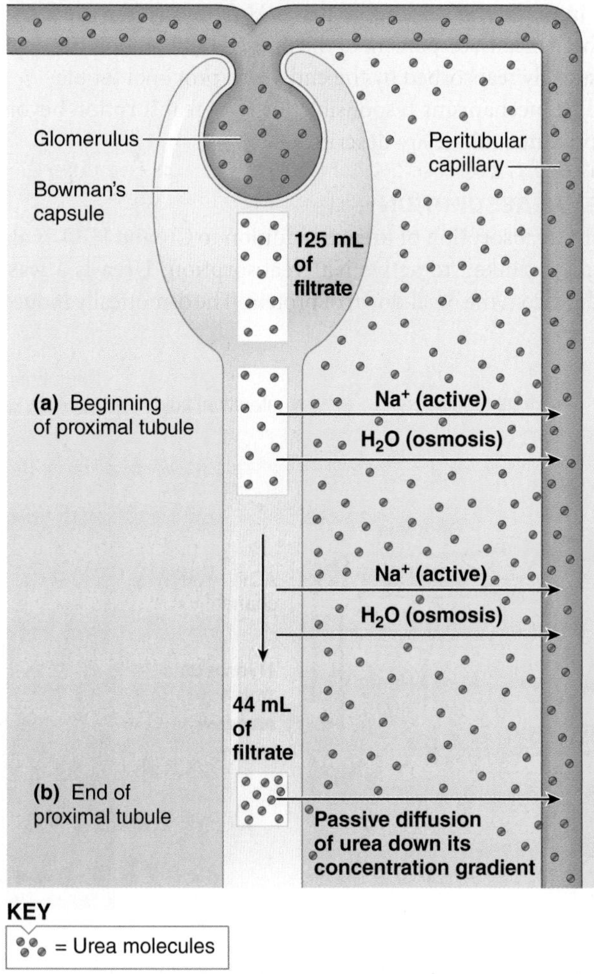

Clinical Note Even though only half of the filtered urea is eliminated from the plasma with each pass through the nephrons, this removal rate is adequate. The urea concentration in the plasma becomes elevated only in impaired kidney function, when much less than half of the urea is removed. An elevated urea level was one of the first chemical characteristics to be identified in the plasma of patients with severe renal failure. Accordingly, clinical measurement of **blood urea nitrogen (BUN)** came into use as a crude assessment of kidney function. It is now known that the most serious consequences of renal failure are not attributable to the retention of urea, which itself is not especially toxic, but rather to the accumulation of other substances that are not adequately excreted because of their failure to be properly secreted—most notably $H^+$ and $K^+$. Health professionals still often refer to renal failure as **uraemia** (urea in the blood), indicating excess urea in the blood, even though urea retention is not this condition's major threat.

## Unwanted waste products

The filtered waste products other than urea—such as *phenol* (mainly from ingested sources, but also produced within the body) and *creatinine*—are likewise concentrated in the tubular fluid as $H_2O$ leaves the filtrate to enter the plasma. However, they are not passively reabsorbed, as urea is. Urea molecules, the smallest of the waste products, are the only wastes passively reabsorbed by this concentrating effect. Even though the other wastes are also concentrated in the tubular fluid, they cannot leave the lumen and move down their concentration gradients to be passively reabsorbed, because they cannot permeate the tubular wall. Therefore, the waste products, not being reabsorbed, generally remain in the tubules and are excreted in the urine in highly concentrated form. This excretion of metabolic wastes is not subject to physiological control. When renal function is normal, however, the excretory processes proceed at a satisfactory rate even though they are not controlled.

This completes our discussion of tubular reabsorption. We now turn our attention to the other basic renal process carried out by the tubules: tubular secretion.

© 2016 Cengage

**KEY**

= Urea molecules

> **FIGURE 13-20 Passive reabsorption of urea at the end of the proximal tubule.** (a) In Bowman's capsule and at the beginning of the proximal tubule, urea is at the same concentration as in the plasma and surrounding interstitial fluid. (b) By the end of the proximal tubule, 65 percent of the original filtrate has been reabsorbed, concentrating the filtered urea in the remaining filtrate. This establishes a concentration gradient favouring passive reabsorption of urea.

Labels within figure:
- Glomerulus
- Bowman's capsule
- Peritubular capillary
- 125 mL of filtrate
- **(a)** Beginning of proximal tubule
- $Na^+$ (active)
- $H_2O$ (osmosis)
- $Na^+$ (active)
- $H_2O$ (osmosis)
- 44 mL of filtrate
- **(b)** End of proximal tubule
- **Passive diffusion of urea down its concentration gradient**

---

### Check Your Understanding 13.3

1. Show the steps of transepithelial transport on a sketch you make of a kidney tubule and associated peritubular capillary.

2. Describe the sequence of events that take place in the renin–angiotensin–aldosterone system in response to a fall in NaCl, ECF volume, and arterial blood pressure.

3. Explain how the kidneys regulate the plasma concentration of phosphate, but not of glucose, when the kidney tubules display a transport maximum ($T_m$) for both of these substances.

# 13.5 | Tubular Secretion

Like tubular reabsorption, tubular secretion involves transepithelial transport, but now the steps are reversed. *Tubular secretion* refers to the discrete transfer of substances from the peritubular capillaries into the tubular lumen. As a second route of entry into the tubules for selected substances, it hastens elimination of these compounds from the body. Anything that gains entry to the tubular fluid—whether by glomerular filtration or tubular secretion—and fails to be reabsorbed is eliminated in the urine.

The most important substances secreted by the tubules are hydrogen ion $(H^+)$, potassium ion $(K^+)$, and organic anions and cations, many of which are compounds foreign to the body.

## Hydrogen ion secretion

Renal $H^+$ secretion is extremely important in regulating acid–base balance in the body. Hydrogen ions secreted into the tubular fluid are eliminated from the body in the urine. Hydrogen ions can be secreted by the proximal, distal, and collecting tubules, and the amount of this secretion depends on the acidity of the body fluids. When the body fluids are too acidic, $H^+$ secretion increases. Conversely, $H^+$ secretion is reduced when the $H^+$ concentration in the body fluids is too low. (See Chapter 14 for further detail.)

## Potassium ion secretion

Potassium ions are selectively moved in opposite directions in different parts of the tubule; they are actively reabsorbed in the proximal tubule and actively secreted in the distal and collecting tubules. Early in the tubule, $K^+$ is reabsorbed in a constant, unregulated fashion, whereas later in the tubule, $K^+$ secretion is variable and subject to regulation. The filtered $K^+$ is almost completely reabsorbed in the proximal tubule, so most $K^+$ in the urine is derived from a controlled $K^+$ secretion in the distal parts of the nephron, rather than from filtration.

During $K^+$ depletion, $K^+$ secretion in the distal parts of the nephron is reduced to a minimum, so only the small percentage of filtered $K^+$ that escapes reabsorption in the proximal tubule is excreted in the urine. In this way, $K^+$ that normally would have been lost in urine is conserved for the body. Conversely, when plasma $K^+$ levels are elevated, $K^+$ secretion is adjusted so that just enough $K^+$ is added to the filtrate for elimination to reduce the plasma $K^+$ concentration to normal. Thus, $K^+$ secretion, not the filtration or reabsorption of $K^+$, is varied in a controlled fashion to regulate the rate of $K^+$ excretion and maintain the desired plasma $K^+$ concentration.

### MECHANISM OF K⁺ SECRETION

Potassium ion secretion in the distal and collecting tubules is coupled to $Na^+$ reabsorption by the energy-dependent basolateral $Na^+{-}K^+$ pump

($\rangle$ Figure 13-21). This pump not only moves $Na^+$ out of the cell into the lateral space but also transports $K^+$ from the lateral space into the tubular cells. The resulting high intracellular $K^+$ concentration favours net movement of $K^+$ from the cells into the tubular lumen. Movement across the luminal membrane occurs passively through the large number of $K^+$ channels in this barrier in the distal and collecting tubules. By keeping the interstitial fluid concentration of $K^+$ low as it transports $K^+$ into the tubular cells from the surrounding interstitial fluid, the basolateral pump encourages passive movement of $K^+$ out of the peritubular capillary plasma into the interstitial fluid. Potassium ion leaving the plasma in this manner is later pumped into the cells, from which it passively moves into the lumen. In this way, the basolateral pump actively induces the net secretion of $K^+$ from the peritubular capillary plasma into the tubular lumen in the distal parts of the nephron.

$K^+$ secretion is linked with $Na^+$ reabsorption by the $Na^+{-}K^+$ pump, so why isn't $K^+$ secreted throughout the $Na^+$-reabsorbing segments of the tubule, instead of taking place only in the distal parts of the nephron? The answer lies in the location of the passive $K^+$ channels. In the distal and collecting tubules, the $K^+$ channels are concentrated in the luminal membrane, providing a route for $K^+$ pumped into the cell to exit into the lumen, thus being secreted. In the other tubular segments, the $K^+$ channels are located primarily in the basolateral membrane. As a result, $K^+$ pumped into the cell from the lateral space by the $Na^+{-}K^+$ pump simply moves back out into the lateral space through these channels. This $K^+$ recycling permits the ongoing operation of the $Na^+{-}K^+$ pump to accomplish $Na^+$ reabsorption with no local net effect on $K^+$.

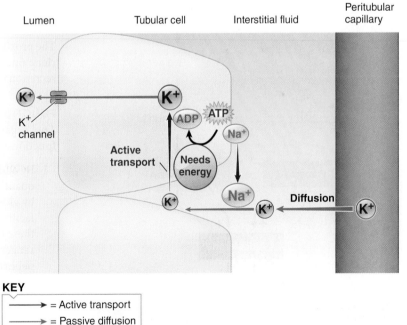

**KEY**

→ = Active transport

→ = Passive diffusion

$\rangle$ **FIGURE 13-21 Potassium ion secretion.** The basolateral pump simultaneously transports $Na^+$ into the lateral space and $K^+$ into the tubular cell. In the parts of the tubule that secrete $K^+$, this ion leaves the cell through channels located in the luminal border, thus being secreted. (In the parts of the tubule that do not secrete $K^+$, the $K^+$ pumped into the cell during $Na^+$ reabsorption leaves the cell through channels located in the basolateral border, thus being retained in the body.)

## CONTROL OF K⁺ SECRETION

Several factors can alter the rate of K⁺ secretion, the most important being aldosterone. This hormone stimulates K⁺ secretion by the tubular cells late in the nephron, while simultaneously enhancing these cells' reabsorption of Na⁺. A rise in plasma K⁺ concentration directly stimulates the adrenal cortex to increase its output of aldosterone, which in turn promotes the secretion and ultimate urinary excretion and elimination of excess K⁺. Conversely, a decline in plasma K⁺ concentration causes a reduction in aldosterone secretion and a corresponding decrease in aldosterone-stimulated renal K⁺ secretion.

Note that a rise in plasma K⁺ concentration directly stimulates aldosterone secretion by the adrenal cortex, whereas a fall in plasma Na⁺ concentration stimulates aldosterone secretion by means of the complex RAAS pathway. Thus, aldosterone secretion can be stimulated by two separate pathways (› Figure 13-22). No matter what the stimulus, however, increased aldosterone secretion always promotes simultaneous Na⁺ reabsorption and K⁺ secretion. For this reason, K⁺ secretion can be inadvertently stimulated as a result of increased aldosterone activity brought about by Na⁺ depletion, ECF volume reduction, or a fall in arterial blood pressure totally unrelated to K⁺ balance. The resulting inappropriate loss of K⁺ can lead to K⁺ deficiency.

## EFFECT OF H⁺ SECRETION ON K⁺ SECRETION

Another factor that can inadvertently alter the magnitude of K⁺ secretion is the acid–base status of the body. The basolateral pump in the distal portions of the nephron can secrete either K⁺ or H⁺ in exchange for reabsorbed Na⁺. An increased rate of secretion of either K⁺ or H⁺ is accompanied by a decreased rate of secretion of the other ion. Normally, the kidneys secrete a preponderance of K⁺, but when the body fluids are too acidic and H⁺ secretion is increased as a compensatory measure, K⁺ secretion is correspondingly reduced. This reduced secretion leads to inappropriate K⁺ retention in the body fluids.

## IMPORTANCE OF REGULATING PLASMA K⁺ CONCENTRATION

Except in the overriding circumstances of K⁺ imbalances inadvertently induced during renal compensations for Na⁺ or ECF volume deficits or acid–base imbalances, the kidneys usually exert a fine degree of control over plasma K⁺ concentration. This is extremely important, because even minor fluctuations in plasma K⁺ concentration can have detrimental consequences. Potassium plays a key role in the membrane electrical activity of excitable tissues. Both increases and decreases in the plasma (ECF) K⁺ concentration can alter the intracellular-to-extracellular K⁺ concentration gradient, which in turn can change the resting membrane potential. A rise in ECF K⁺ concentration leads to a reduction in resting potential and a subsequent increase in excitability, especially of heart muscle. This cardiac overexcitability can lead to a rapid heart rate and even fatal cardiac arrhythmias. Conversely, a fall in ECF K⁺ concentration results in hyperpolarization of nerve and muscle cell membranes, which reduces their excitability. The manifestations of ECF K⁺ depletion are skeletal muscle weakness, diarrhoea, and abdominal distension caused by smooth muscle dysfunction, and abnormalities in cardiac rhythm and impulse conduction.

## Organic anion and cation secretion

The proximal tubule contains two distinct types of secretory carriers: one for the secretion of organic anions and another for secretion of organic cations.

### FUNCTIONS OF ORGANIC ION SECRETORY SYSTEMS

The organic ion secretory systems serve three important functions:

1. By adding more of a particular type of organic ion to the quantity that has already gained entry to the tubular fluid by glomerular filtration, these organic secretory pathways facilitate excretion of these substances. Included among these organic ions are certain blood-borne chemical messengers, such as prostaglandins, histamine, and norepinephrine, which having served their purpose, must be rapidly removed from the blood so that their biological activity is not unduly prolonged.

2. In some important instances, organic ions are poorly soluble in water. To be transported in blood, they are extensively but not irreversibly bound to plasma proteins. Because they are attached to plasma proteins, these substances cannot be filtered through the glomeruli. Tubular secretion facilitates elimination of these nonfilterable organic ions in urine. Even though a given organic ion is

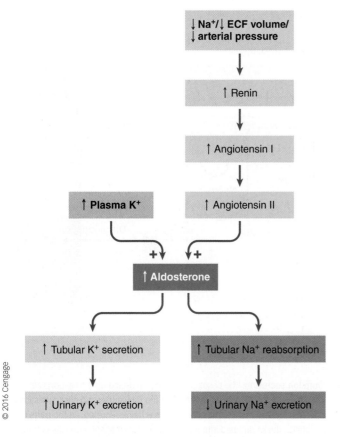

› **FIGURE 13-22** Dual control of aldosterone secretion by K⁺ and Na⁺

largely bound to plasma proteins, a small percentage of the ion always exists in free or unbound form in the plasma. Removal of this free organic ion by secretion permits the unloading of some of the bound ion, which is then free to be secreted. This, in turn, encourages the unloading of even more organic ion, and so on.

3. Most importantly, the proximal tubule organic ion secretory systems play a key role in eliminating many foreign compounds from the body. These systems can secrete a large number of different organic ions, both those produced endogenously (within the body) and those foreign organic ions that have gained access to the body fluids. This nonselectivity permits these organic ion secretory systems to hasten removal of many foreign organic chemicals, including food additives, environmental pollutants (e.g., pesticides), drugs, and other non-nutritive organic substances that have entered the body. Even though this mechanism helps rid the body of potentially harmful foreign compounds, it is not subject to physiological adjustments. The carriers cannot increase their secretory pace when confronted with an elevated load of these organic ions.

The liver plays an important role in helping rid the body of foreign compounds. Many foreign organic chemicals are not ionic in their original form, so they cannot be secreted by the organic ion systems. The liver converts these foreign substances into an anionic form that facilitates their secretion by the organic anion system and, thus, accelerates their elimination.

*Clinical Note* Many drugs, such as penicillin and nonsteroidal anti-inflammatory drugs (NSAIDs), are eliminated from the body by the organic ion secretory systems. To keep the plasma concentration of these drugs at effective levels, the dosage must be repeated on a regular, frequent basis to keep pace with the rapid removal of these compounds in the urine.

## Summary of reabsorptive and secretory processes

This completes our discussion of the reabsorptive and secretory processes that occur across the proximal and distal portions of the nephron. These processes are summarized in ▌Table 13-3. To generalize, the proximal tubule does most of the reabsorbing. This mass reabsorber transfers much of the filtered water and needed solutes back into the blood in unregulated fashion. Similarly, the proximal tubule is the major site of secretion, with the exception of $K^+$ secretion. The distal and collecting tubules then determine the final amounts of $H_2O$, $Na^+$, $K^+$, and $H^+$ excreted in the urine and thus eliminated from the body. They do so by fine-tuning the amount of $Na^+$ and $H_2O$ reabsorbed and the amount of $K^+$ and $H^+$ secreted. These processes in the distal part of the nephron are all subject to control, depending on the body's momentary needs. The unwanted filtered waste products are left behind to be eliminated in the urine, along with excess amounts of filtered or secreted nonwaste products that fail to be reabsorbed.

▌**TABLE 13-3** Summary of Transport across Proximal and Distal Portions of the Nephron

### PROXIMAL TUBULE

| Reabsorption | Secretion |
|---|---|
| 67 percent of filtered $Na^+$ actively reabsorbed; not subject to control; $Cl^-$ follows passively | Variable $H^+$ secretion, depending on acid–base status of body |
| All filtered glucose and amino acids reabsorbed by secondary active transport; not subject to control | Organic ion secretion; not subject to control |
| Variable amounts of filtered $PO_3^{4-}$ and other electrolytes reabsorbed; subject to control; 65 percent of filtered $H_2O$ osmotically reabsorbed; not subject to control | |
| 50 percent of filtered urea passively reabsorbed; not subject to control | |
| Almost all filtered $K^+$ reabsorbed; not subject to control | |

### DISTAL TUBULE AND COLLECTING DUCT

| Reabsorption | Secretion |
|---|---|
| Variable $Na^+$ reabsorption, controlled by aldosterone; $Cl^-$ follows passively | Variable $H^+$ secretion, depending on acid–base status of body |
| Variable $H_2O$ reabsorption, controlled by vasopressin | Variable $K^+$ secretion, controlled by aldosterone |

Next we focus on the end result of the basic renal processes—what's left in the tubules to be excreted in urine, and, as a consequence, what has been cleared from plasma.

### Check Your Understanding 13.4

1. List the three secretory processes accomplished by the kidney tubules.
2. Explain how most foreign organic compounds are eliminated from the body.

## 13.6 | Urine Excretion and Plasma Clearance

Of the 125 mL of plasma filtered per minute, typically 124 mL/min are reabsorbed, so the final quantity of urine formed averages 1 mL/min. Thus, of the 180 L filtered per day, 1.5 L of urine are excreted.

13

Urine contains high concentrations of various waste products plus variable amounts of the substances regulated by the kidneys—any excess having spilled into the urine. Useful substances are conserved by reabsorption, so they do not appear in the urine.

A relatively small change in the quantity of filtrate reabsorbed can bring about a large change in the volume of urine formed. For example, a reduction of less than 1 percent in the total reabsorption rate, from 124 to 123 mL/min, increases the urinary excretion rate by 100 percent, from 1 to 2 mL/min.

## Plasma clearance: Volume

Compared with plasma entering the kidneys through the renal arteries, plasma leaving the kidneys through the renal veins lacks the materials that were left behind to be eliminated in the urine. By excreting substances in the urine, the kidneys clean, or clear, the plasma flowing through them of these substances. The **plasma clearance** of any substance is defined as the volume of plasma completely cleared of that substance by the kidneys per minute.[2] It refers not to the *amount of the substance* removed but to the *volume of plasma* from which that amount was removed. Plasma clearance is actually a more useful measure than urine excretion; it is more important to know what effect urine excretion has on removing materials from body fluids than to know the volume and composition of discarded urine. Plasma clearance expresses the kidneys' effectiveness in removing various substances from the internal fluid environment.

Plasma clearance can be calculated for any plasma constituent as follows:

$$\text{Clearance rate of a substance (mL/min)} = \frac{\substack{\text{urine concentration} \\ \text{of the substance} \\ \text{(quantity / mL urine)}} \times \substack{\text{urine flow rate} \\ \text{(mL / min)}}}{\substack{\text{plasma concentration of the substance} \\ \text{(quantity / mL plasma)}}}$$

The plasma clearance rate varies for different substances, depending on how the kidneys handle each substance.

## Plasma clearance rate if a substance is filtered but not reabsorbed

Assume that a plasma constituent, substance X, is freely filterable at the glomerulus, but is not reabsorbed or secreted. As 125 mL/min of plasma are filtered and subsequently reabsorbed, the quantity of substance X originally contained within the 125 mL is left behind in the tubules to be excreted. Thus, 125 mL of plasma are cleared of substance X each minute (❯ Figure 13-23). (Of the 125 mL/min of plasma filtered, 124 mL/min of the filtered fluid are returned, through reabsorption, to the plasma minus substance X, thereby clearing this 124 mL/min of substance X. In addition, the 1 mL/min of fluid lost in urine is eventually replaced by an equivalent volume of ingested $H_2O$ that is already clear of substance X. Therefore, 125 mL of plasma cleared of substance X are, in effect, returned to the plasma for every 125 mL of plasma filtered per minute.)

*Clinical Note* There is no endogenous chemical with the characteristics of substance X. All substances naturally present in the plasma, even wastes, are reabsorbed or secreted to some extent. However, **inulin** (do not confuse with insulin)—a harmless foreign carbohydrate produced by Jerusalem artichokes, onions, and garlic—is freely filtered and not reabsorbed or secreted: an ideal substance X. Inulin can be injected and its plasma clearance determined as a clinical means of ascertaining the GFR. All glomerular filtrate formed is cleared of inulin at the following rate:

$$\begin{aligned} \text{Clearance rate for inulin} &= \frac{30\,\text{mg / mL urine} \times 1.25\,\text{mL urine / min}}{0.30\,\text{mg / mL plasma}} \\ &= 125\,\text{mL plasma / min} \end{aligned}$$

So the volume of plasma cleared of inulin per minute equals the volume of plasma filtered per minute—that is, the GFR.

Although determination of inulin plasma clearance is accurate and straightforward, it is not very convenient, because inulin must be infused continuously throughout the determination to maintain a constant plasma concentration. Therefore, the plasma clearance of an endogenous substance, **creatinine**, is often used instead to find a rough estimate of the GFR. Creatinine, an end product of muscle metabolism, is produced at a relatively constant rate. It is freely filtered and not reabsorbed but is slightly secreted. Accordingly, creatinine clearance is not a completely accurate reflection of the GFR, but it does provide a close approximation and can be more readily determined than inulin clearance.

## Plasma clearance rate if a substance is filtered and reabsorbed

Some or all of a reabsorbable substance that has been filtered is returned to the plasma. Because less than the filtered volume of plasma will have been cleared of the substance, the plasma clearance rate of a reabsorbable substance is always less than the GFR. For example, the plasma clearance for glucose is normally zero. All the filtered glucose is reabsorbed along with the rest of the returning filtrate, so none of the plasma is cleared of glucose (❯ Figure 13-23b).

For a substance that is partially reabsorbed, such as urea, only part of the filtered plasma is cleared of that substance. With about 50 percent of the filtered urea passively reabsorbed, only half of the filtered plasma, or 62.5 mL, is cleared of urea each minute (❯ Figure 13-23c).

---

[2]Actually, plasma clearance is an artificial concept, because when a particular substance is excreted in the urine, that substance's concentration in the plasma as a whole is uniformly decreased as a result of thorough mixing in the circulatory system. However, it is useful for comparative purposes to consider clearance in effect as the volume of plasma that would have contained the total quantity of the substance (at the substance's concentration prior to excretion) that the kidneys excreted in one minute; that is, the hypothetical volume of plasma completely cleared of that substance per minute.

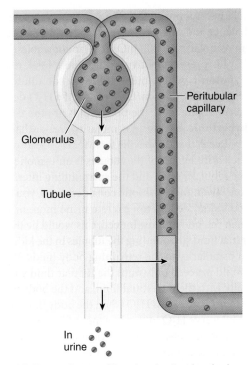

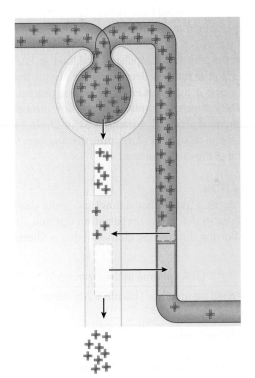

Glomerulus

Peritubular capillary

Tubule

In urine

**(a)** For a substance filtered and not reabsorbed or secreted, such as inulin, all of the filtered plasma is cleared of the substance.

**(b)** For a substance filtered, not secreted, and completely reabsorbed, such as glucose, none of the filtered plasma is cleared of the substance.

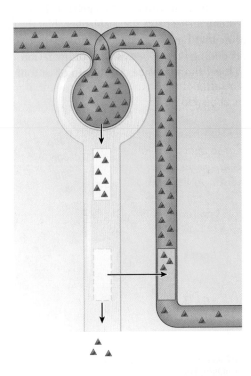

**(c)** For a substance filtered, not secreted, and partially reabsorbed, such as urea, only a portion of the filtered plasma is cleared of the substance.

**(d)** For a substance filtered and secreted but not reabsorbed, such as hydrogen ion, all of the filtered plasma is cleared of the substance, and the peritubular plasma from which the substance is secreted is also cleared.

› **FIGURE 13-23 Plasma clearance for substances handled in different ways by the kidneys**

**13**

## Plasma clearance rate if a substance is filtered and secreted but not reabsorbed

Tubular secretion allows the kidneys to clear certain materials from the plasma more efficiently. Only 20 percent of the plasma entering the kidneys is filtered. The remaining 80 percent passes unfiltered into the peritubular capillaries. The only means by which this unfiltered plasma can be cleared of any substance during this trip through the kidneys before being returned to the general circulation is by secretion. An example is hydrogen ion. Not only is filtered plasma cleared of nonreabsorbable $H^+$, but the plasma from which $H^+$ is secreted is also cleared of $H^+$. For example, if the quantity of $H^+$ secreted is equivalent to the quantity of $H^+$ present in 25 mL of plasma, the clearance rate for $H^+$ will be 150 mL/min at the normal GFR of 125 mL/min. Every minute 125 mL of plasma will lose its $H^+$ through filtration and failure of reabsorption, and 25 mL more of plasma will lose its $H^+$ through secretion. The plasma clearance for a secreted substance is always greater than the GFR (⟩ Figure 13-23d).

*Clinical Note* Just as inulin can be used clinically to determine the GFR, plasma clearance of another foreign compound, the organic anion **para-aminohippuric acid (PAH)**, can be used to measure renal plasma flow. Like inulin, PAH is freely filterable and nonreabsorbable. It differs, however, in that all the PAH in the plasma that escapes filtration is secreted from the peritubular capillaries by the organic anion secretory pathway in the proximal tubule. In this way, PAH is removed from all the plasma that flows through the kidneys—both from plasma that is filtered and subsequently reabsorbed without its PAH, and from unfiltered plasma that continues on in the peritubular capillaries and loses its PAH by active secretion into the tubules. Because all the plasma that flows through the kidneys is cleared of PAH, the plasma clearance for PAH is a reasonable estimate of the rate of plasma flow through the kidneys. Typically, renal plasma flow averages 625 mL/min, for a RBF (plasma plus blood cells) of 1140 mL/min; this amounts to more than 20 percent of the cardiac output.

### FILTRATION FRACTION

If you know PAH clearance (renal plasma flow) and inulin clearance (GFR), you can easily determine the **filtration fraction**, the fraction of plasma flowing through the glomeruli that is filtered into the tubules.

$$\text{Filtration fraction} = \frac{\text{GFR (plasma inulin clearance)}}{\text{renal plasma flow (plasma PAH clearance)}}$$

$$= \frac{125\ \text{mL/min}}{625\ \text{mL/min}} = 20\ \text{percent}$$

Thus, 20 percent of the plasma that enters the glomeruli is typically filtered.

## The kidneys and urine of varying concentrations

Having considered how the kidneys deal with a variety of solutes in the plasma, we now examine renal handling of plasma $H_2O$. The ECF osmolarity (solute concentration) depends on the relative amount of $H_2O$ compared with solute. At normal fluid balance and solute concentration, the body fluids are **isotonic** at an osmolarity of 300 mOsm/L (milliosmoles/litre). If too much $H_2O$ is present relative to the solute load, the body fluids are **hypotonic**, which means they are too dilute at an osmolarity less than 300 mOsm/L. However, if a $H_2O$ deficit exists relative to the solute load, the body fluids are too concentrated or are **hypertonic**, having an osmolarity greater than 300 mOsm/L.

Recall that the driving force for $H_2O$ reabsorption throughout the entire length of the tubules is an osmotic gradient between the tubular lumen and the surrounding interstitial fluid. Therefore, given these osmotic considerations, you would expect that the kidneys could not excrete urine more or less concentrated than the body fluids. Indeed, this would be the case if the interstitial fluid surrounding the tubules in the kidneys were identical in osmolarity to the remaining body fluids. Water reabsorption would proceed only until the tubular fluid equilibrated osmotically with the interstitial fluid, and the body would have no way to eliminate excess $H_2O$ when the body fluids were hypotonic or to conserve $H_2O$ in the presence of hypertonicity.

Fortunately, a large **vertical osmotic gradient** is uniquely maintained in the interstitial fluid of the medulla of each kidney. The concentration of the interstitial fluid progressively increases from the cortical boundary down through the depth of the renal medulla until it reaches a maximum of 1200 mOsm/L in humans at the junction with the renal pelvis (⟩ Figure 13-24).

The vertical osmotic gradient enables the kidneys to produce urine that ranges in concentration from 100 to 1200 mOsm/L, depending on the body's state of hydration. When the body is in ideal fluid balance, 1 mL/min of isotonic urine is formed.

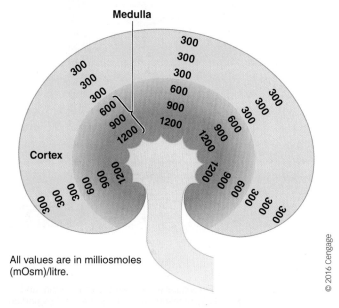

All values are in milliosmoles (mOsm)/litre.

⟩ **FIGURE 13-24 Vertical osmotic gradient in the renal medulla.** Schematic representation of the kidney rotated 90 degrees from its normal position in an upright person for better visualization of the vertical osmotic gradient in the renal medulla. The osmolarity of the interstitial fluid throughout the renal cortex is isotonic at 300 mOsm/L, but the osmolarity of the interstitial fluid in the renal medulla increases progressively from 300 mOsm/L at the boundary with the cortex to a maximum of 1200 mOsm/L at the junction with the renal pelvis.

© 2016 Cengage

When the body is overhydrated (too much $H_2O$), the kidneys can produce a large volume of dilute urine (up to 25 mL/min and hypotonic at 100 mOsm/L), eliminating the excess $H_2O$ in the urine. Conversely, the kidneys can put out a small volume of concentrated urine (down to 0.3 mL/min and hypertonic at 1200 mOsm/L) when the body is dehydrated (too little $H_2O$), conserving $H_2O$ for the body.

Unique anatomic arrangements and complex functional interactions between the various nephron components in the renal medulla establish and use the vertical osmotic gradient. Recall that the hairpin loop of Henle dips only slightly into the medulla in cortical nephrons, but in juxtamedullary nephrons the loop plunges through the entire depth of the medulla so that the tip of the loop lies near the renal pelvis (see Figure 13-5). Also, the vasa recta of juxtamedullary nephrons follow the same deep hairpin loop as the long loop of Henle. Flow in both the long loops of Henle and the vasa recta is considered countercurrent, because the flow in the two closely adjacent limbs of the loop moves in opposite directions. Also running through the medulla in the descending direction only, on their way to the renal pelvis, are the collecting ducts that serve both types of nephrons. This arrangement, coupled with the permeability and transport characteristics of these tubular segments, plays a key role in the kidneys' ability to produce urine of varying concentrations, depending on the body's needs for water conservation or elimination. Briefly, the juxtamedullary nephrons' long loops of Henle *establish* the vertical osmotic gradient; their vasa recta *preserve* this gradient while providing blood to the renal medulla; and the collecting ducts of all nephrons *use* the gradient, in conjunction with the hormone vasopressin, to produce urine of varying concentrations. Collectively, this entire functional organization is known as the **medullary countercurrent system**. We will examine each of its facets in detail.

*Clinical Note* The colour of human urine can vary with hydration level and other factors. A typical urine chart is set up on an eight-step scale, with the colour of urine ranging from a very light yellow (well hydrated) to a very dark yellow (severe dehydration). Hydration levels may vary based on fluid intake and sweat rate. The colours 1 through 3 show a balance in hydration. Colours 4 and 5 suggest mild dehydration. The colours 6 through 8 indicate severe dehydration. In general, colours 4 through 8 suggest that fluid intake needs to be increased.

Other factors, such as the removal of excess B vitamins from the bloodstream, may cause a yellowing of the urine. If blood is present in the urine (haematuria), there may be damage to the kidney, and immediate medical attention is required. A dark-orange to brown urine can be a sign of jaundice. Dark-coloured (melanuria) urine may be caused by a melanoma (a malignant tumour of melanocytes). Reddish urine is associated with porphyria, which is a disorder of certain enzymes associated with heme and its biochemical pathway. Milky white urine is a condition called *chyluria*, caused by the presence of chyle that consists of lymph fluid and fats.

## The medullary vertical osmotic gradient

Here we follow the filtrate through a long-looped nephron to show how this structure establishes a vertical osmotic gradient in the medulla. Immediately after the filtrate is formed, uncontrolled osmotic reabsorption of filtered $H_2O$ occurs in the proximal tubule secondary to active $Na^+$ reabsorption. As a result, by the end of the proximal tubule about 65 percent of the filtrate has been reabsorbed, but the 35 percent remaining in the tubular lumen still has the same osmolarity as the body fluids. Therefore, the fluid entering the loop of Henle is still isotonic. The additional 15 percent of the filtered $H_2O$ is obligatorily reabsorbed from the loop of Henle during the establishment and maintenance of the vertical osmotic gradient, thereby altering the osmolarity of the tubular fluid.

### PROPERTIES OF THE DESCENDING AND ASCENDING LIMBS OF A LONG HENLE'S LOOP

The following functional distinctions between the descending limb of a long Henle's loop (which carries fluid from the proximal tubule down into the depths of the medulla) and the ascending limb (which carries fluid up and out of the medulla into the distal tubule) are crucial for establishing the incremental osmotic gradient in the medullary interstitial fluid.

The *descending limb* (1) is highly permeable to $H_2O$, and (2) does not actively extrude $Na^+$—that is, does not reabsorb $Na^+$. (It is the only segment of the entire tubule that does not do so.)

The *ascending limb* (1) actively transports NaCl out of the tubular lumen into the surrounding interstitial fluid, and (2) is always impermeable to $H_2O$. Therefore, salt leaves the tubular fluid without $H_2O$ osmotically following along.

### MECHANISM OF COUNTERCURRENT MULTIPLICATION

The close proximity and countercurrent flow of the two limbs allow important interactions between them. Even though the flow of fluids is continuous through the loop of Henle, we will see what happens by going step by step, much like making an animated film run so slowly that each frame can be viewed. (See ⟩ Figure 13-25 to follow each of these steps.)

- Initial scene: Before the vertical osmotic gradient is established, the medullary interstitial fluid concentration is uniformly 300 mOsm/L, as are the rest of the body fluids.

- Step 1: The active NaCl pump in the ascending limb can transport NaCl out of the lumen until the surrounding interstitial fluid is 200 mOsm/L more concentrated than the tubular fluid in this limb. When the ascending limb pump starts actively extruding salt, the medullary interstitial fluid becomes hypertonic. Water cannot follow osmotically from the ascending limb, because this limb is impermeable to $H_2O$. However, net diffusion of $H_2O$ does occur from the descending limb into the interstitial fluid. The tubular fluid entering the descending limb from the proximal tubule is isotonic. Because the descending limb is highly permeable to $H_2O$, net diffusion of $H_2O$ occurs by osmosis out of the descending limb into the more concentrated interstitial fluid. The passive movement of $H_2O$ out of the descending limb continues until the osmolarities of the fluid in the descending limb and interstitial fluid become equilibrated. Thus, the tubular fluid entering the loop of Henle immediately starts to become more concentrated as it loses $H_2O$.

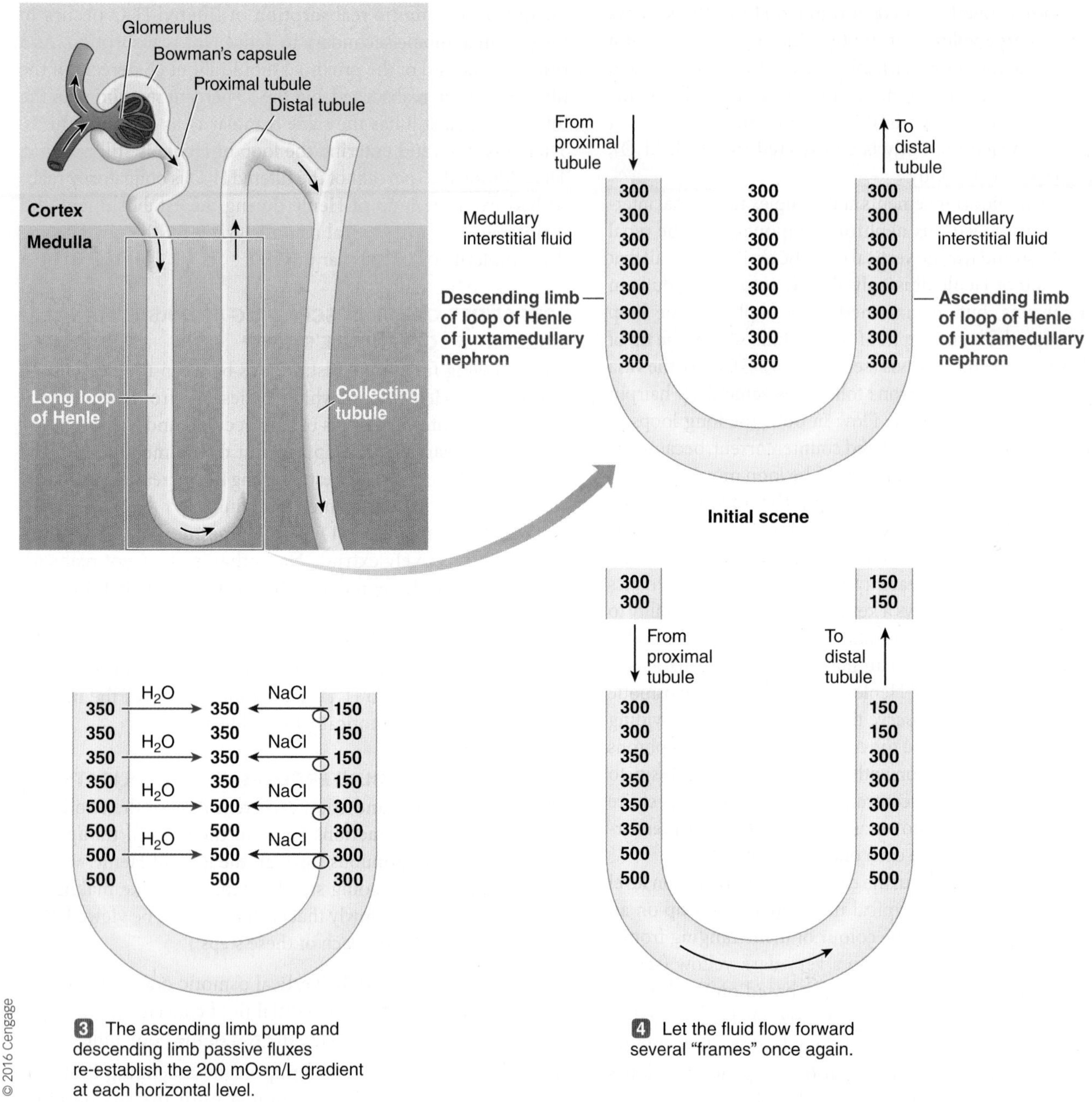

Medullary interstitial fluid — **Descending limb of loop of Henle of juxtamedullary nephron**

Medullary interstitial fluid — **Ascending limb of loop of Henle of juxtamedullary nephron**

**Initial scene**

**3** The ascending limb pump and descending limb passive fluxes re-establish the 200 mOsm/L gradient at each horizontal level.

**4** Let the fluid flow forward several "frames" once again.

> FIGURE 13-25 **Countercurrent multiplication in the renal medulla**

At equilibrium, the osmolarity of the ascending limb fluid is 200 mOsm/L, and the osmolarities of the interstitial fluid and descending limb fluid are equal at 400 mOsm/L.

- Step 2: Advancing the entire column of fluid in the loop of Henle several frames, we see that a mass of 200 mOsm/L fluid exits from the top of the ascending limb into the distal tubule, and a new mass of isotonic fluid at 300 mOsm/L enters the top of the descending limb from the proximal tubule. At the bottom of the loop, a comparable mass of 400 mOsm/L fluid from the descending limb moves forward around the tip into the ascending limb, placing it opposite a

400 mOsm/L region in the descending limb. Note that the 200 mOsm/L concentration difference has been lost at both the top and the bottom of the loop.

- Step 3: The ascending limb pump again transports NaCl out, while $H_2O$ passively leaves the descending limb until a 200 mOsm/L difference is reestablished between the ascending limb and both the interstitial fluid and descending limb at each horizontal level. Note, however, that the concentration of tubular fluid is progressively increasing in the descending limb and progressively decreasing in the ascending limb.

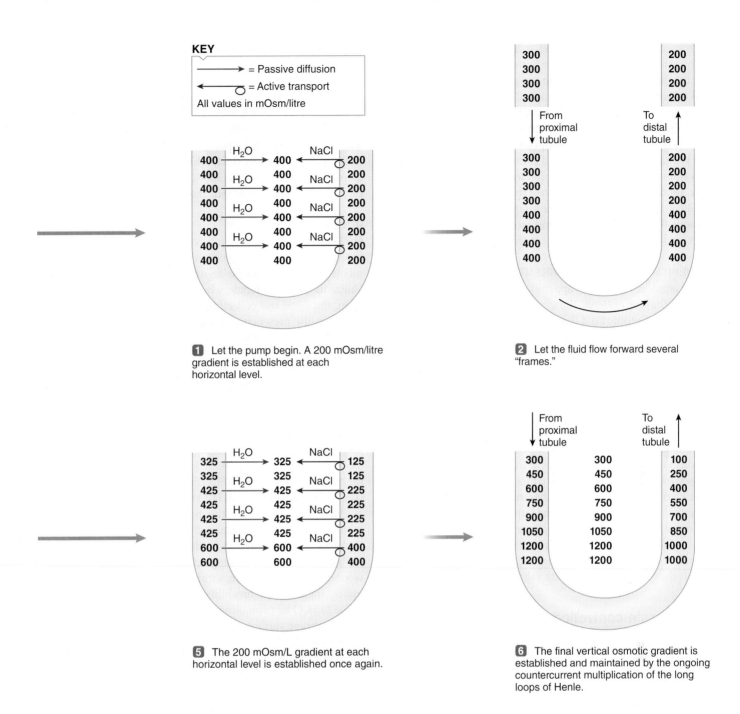

**KEY**

→ = Passive diffusion

← ⟲ = Active transport

All values in mOsm/litre

**1** Let the pump begin. A 200 mOsm/litre gradient is established at each horizontal level.

**2** Let the fluid flow forward several "frames."

From proximal tubule    To distal tubule

**5** The 200 mOsm/L gradient at each horizontal level is established once again.

**6** The final vertical osmotic gradient is established and maintained by the ongoing countercurrent multiplication of the long loops of Henle.

From proximal tubule    To distal tubule

- Step 4: As the tubular fluid is advanced still further, the 200 mOsm/L concentration gradient is disrupted once again at all horizontal levels.

- Step 5: Again, active extrusion of NaCl from the ascending limb, coupled with the net diffusion of $H_2O$ out of the descending limb, reestablishes the 200 mOsm/L gradient at each horizontal level.

- Step 6 and on: As the fluid flows slightly forward again and this stepwise process continues, the fluid in the descending limb becomes progressively more hypertonic until it reaches a maximum concentration of 1200 mOsm/L at the bottom of the loop; this is four times the normal concentration of body

fluids. Because the interstitial fluid always achieves equilibrium with the descending limb, an incremental vertical concentration gradient, ranging from 300 to 1200 mOsm/L, is likewise established in the medullary interstitial fluid. In contrast, the concentration of the tubular fluid progressively decreases in the ascending limb because salt is pumped out, but $H_2O$ is unable to follow. In fact, the tubular fluid even becomes hypotonic before it leaves the ascending limb to enter the distal tubule at a concentration of 100 mOsm/L—one-third the normal concentration of body fluids.

Note that although a gradient of only 200 mOsm/L exists between the ascending limb and the surrounding fluids at

13

each medullary horizontal level, a much larger vertical gradient exists from the top to the bottom of the medulla. Even though the ascending limb pump can generate a gradient of only 200 mOsm/L, this effect is multiplied into a large vertical gradient because of the countercurrent flow within the loop. This concentrating mechanism accomplished by the loop of Henle is known as **countercurrent multiplication**.

We have artificially described countercurrent multiplication in a stop-and-flow, stepwise fashion to facilitate understanding. It is important to realize that once the incremental medullary gradient is established, it stays constant. This is due to the continuous flow of fluid, the ongoing ascending limb active transport, and the accompanying descending limb passive fluxes.

### BENEFITS OF COUNTERCURRENT MULTIPLICATION

If you consider only what happens to the tubular fluid as it flows through the loop of Henle, the whole process seems an exercise in futility. The isotonic fluid that enters the loop becomes progressively *more concentrated* as it flows down the descending limb, achieving a maximum concentration of 1200 mOsm/L, only to become progressively *more dilute* as it flows up the ascending limb, finally leaving the loop at a minimum concentration of 100 mOsm/L. What is the point of concentrating the fluid fourfold and then turning around and diluting it until it leaves at one-third the concentration at which it entered? Such a mechanism offers three benefits. First, it establishes a vertical osmotic gradient in the medullary interstitial fluid. This gradient, in turn, is used by the collecting ducts to concentrate the tubular fluid so that a urine *more concentrated* than normal body fluids can be excreted. Second, the fact that the fluid is hypotonic as it enters the distal parts of the tubule enables the kidneys to excrete a urine *more dilute* than normal body fluids. And third, it allows for the overall volume of urine to be significantly reduced, which also allows the body to conserve both salt and water. Let's see how.

## Vasopressin-controlled H$_2$O reabsorption

Vasopressin is secreted from the posterior pituitary gland in response to a reduction in blood plasma volume or an increase in plasma osmolarity. Secretion in response to reduced plasma volume is activated by pressure receptors in the veins, atria, and carotids. Secretion in response to increases in plasma osmotic pressure is mediated by osmoreceptors in the hypothalamus.

Many factors can reduce the secretion of vasopressin from the posterior pituitary gland, including the consumption of caffeine. The resulting decrease in water reabsorption by the kidneys may lead to a higher urinary production. Thus, caffeine causes the body to lose more water and may lead to dehydration if it is consumed excessively. As well, angiotensin II, which causes vasoconstriction and the release of aldosterone from the adrenal cortex, also stimulates the release of vasopressin.

After the obligatory H$_2$O reabsorption from the proximal tubule (65% of the filtered H$_2$O) and loop of Henle (15% of the filtered H$_2$O), 20 percent of the filtered H$_2$O remains in the lumen to enter the distal and collecting tubules for variable reabsorption that is under hormonal control. This is still a large volume of filtered H$_2$O subject to regulated reabsorption; 20 percent $\times$ GFR $(180\,\text{L}/\text{day}) = 36\,\text{L}/\text{day}$ to be reabsorbed to

varying extents, depending on the body's state of hydration. This is more than 13 times the amount of plasma H$_2$O in the entire circulatory system.

The fluid leaving the loop of Henle enters the distal tubule at 100 mOsm/L, so it is hypotonic to the surrounding isotonic (300 mOsm/L) interstitial fluid of the renal cortex, through which the distal tubule passes. The distal tubule then empties into the collecting duct, which is bathed by progressively increasing concentrations (300–1200 mOsm/L) of surrounding interstitial fluid as it descends through the medulla.

### ROLE OF VASOPRESSIN

For H$_2$O absorption to occur across a segment of the tubule, two criteria must be met: (1) an osmotic gradient must exist across the tubule, and (2) the tubular segment must be permeable to H$_2$O. Unlike the proximal tubule, where the aquaporins are present and open, the distal and collecting tubules are *impermeable* to H$_2$O except in the presence of **vasopressin**—also known as **antidiuretic hormone** (*anti* means "against"; *diuretic* means "increased urine output")[3]—which increases their permeability to H$_2$O. Vasopressin is produced by several specific neuronal cell bodies in the hypothalamus part of the brain, then stored in the posterior pituitary gland, which is attached to the hypothalamus by a thin stalk. The hypothalamus controls release of vasopressin from the posterior pituitary into the blood. In negative-feedback fashion, vasopressin secretion is stimulated by a H$_2$O deficit when the ECF is too concentrated (i.e., hypertonic) and H$_2$O must be conserved for the body; it is inhibited by a H$_2$O excess when the ECF is too dilute (i.e., hypotonic) and surplus H$_2$O must be eliminated in urine.

Vasopressin reaches the basolateral membrane of the tubular cells lining the distal and collecting tubules through the circulatory system. Here it binds with receptors specific for it (› Figure 13-26). This binding activates the cyclic AMP (cAMP) second-messenger system within the tubular cells, which ultimately increases permeability of the opposite luminal membrane to H$_2$O by promoting the translocation of aquaporins to the membrane surface; vasopressin's action is similar to that of insulin's on the GLUT4 receptor involved in glucose uptake by the skeletal muscle cell. Without these aquaporins, the luminal membrane is impermeable to H$_2$O. Once H$_2$O enters the tubular cells from the filtrate through these vasopressin-regulated luminal water channels, it passively leaves the cells down the osmotic gradient across the cells' basolateral membrane to enter the interstitial fluid. The H$_2$O channels in the basolateral membrane are always present, so this membrane is always permeable to H$_2$O. By permitting more H$_2$O to permeate from the lumen into the tubular cells, the additional vasopressin-regulated luminal channels increase H$_2$O reabsorption from the filtrate into the interstitial fluid. The tubular response to vasopressin is graded: the more vasopressin present, the more luminal water channels are inserted, and the greater the permeability of the distal and collecting tubules to H$_2$O. The increase in luminal membrane water channels is not permanent, however. The channels are retrieved when vasopressin secretion decreases and

---

[3]Even though textbooks traditionally have tended to use the name *antidiuretic hormone* for this hormone, especially when discussing its actions on the kidney, investigators in the field now prefer *vasopressin*.

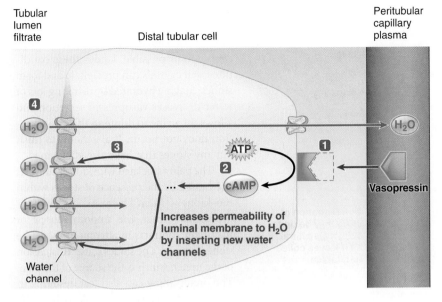

Tubular lumen filtrate

Distal tubular cell

Peritubular capillary plasma

**1** Blood-borne vasopressin binds with its receptor sites on the basolateral membrane of a distal or collecting tubule cell.

**2** This binding activates the cyclic AMP second-messenger system within the cell.

**3** Cyclic AMP increases the opposite luminal membrane's permeability to $H_2O$ by promoting the insertion of water channels in this membrane. This membrane is impermeable to water in the absence of vasopressin.

**4** Water enters the tubular cell from the tubular lumen through the inserted water channels and subsequently enters the blood, in this way being reabsorbed. Water exits the cell through a different water channel permanently positioned at the basolateral border.

© 2016 Cengage

⟩ **FIGURE 13-26** Mechanism of action of vasopressin

cAMP activity is similarly decreased. Accordingly, $H_2O$ permeability is reduced when vasopressin secretion decreases.

Vasopressin influences $H_2O$ permeability only in the distal part of the nephron, especially the collecting ducts. It has no influence over the 80 percent of the filtered $H_2O$ that is obligatorily reabsorbed without control in the proximal tubule and loop of Henle. The ascending limb of Henle's loop is always impermeable to $H_2O$, even in the presence of vasopressin.

### REGULATION OF $H_2O$ REABSORPTION IN RESPONSE TO A $H_2O$ DEFICIT

When vasopressin secretion increases in response to a $H_2O$ deficit and the permeability of the distal and collecting tubules to $H_2O$ accordingly increases, the hypotonic tubular fluid entering the distal part of the nephron can lose progressively more $H_2O$ by osmosis into the interstitial fluid. This occurs as the tubular fluid first flows through the isotonic cortex and then is exposed to the ever-increasing osmolarity of the medullary interstitial fluid when it plunges toward the renal pelvis (⟩ Figure 13-27a). As the 100 mOsm/L tubular fluid enters the distal tubule and is exposed to a surrounding interstitial fluid of 300 mOsm/L, $H_2O$ leaves the tubular fluid by osmosis across the now permeable tubular cells until the tubular fluid reaches a maximum concentration of 300 mOsm/L by the end of the distal tubule. As this 300 mOsm/L tubular fluid progresses farther into the collecting duct, it is exposed to even higher osmolarity in the surrounding

medullary interstitial fluid. Consequently, the tubular fluid loses more $H_2O$ by osmosis and becomes further concentrated, only to move farther forward and be exposed to an even higher interstitial fluid osmolarity and lose even more $H_2O$, and so on.

Under the influence of maximum levels of vasopressin, it is possible to concentrate the tubular fluid up to 1200 mOsm/L by the end of the collecting ducts. No further modification of the tubular fluid occurs beyond the collecting duct, so what remains in the tubules at this point is urine. As a result of this extensive vasopressin-promoted reabsorption of $H_2O$ in the late segments of the tubule, a small volume of urine concentrated up to 1200 mOsm/L can be excreted. As little as 0.3 mL of urine may be formed each minute, less than one-third the normal urine flow rate of 1 mL/min. The reabsorbed $H_2O$ entering the medullary interstitial fluid is picked up by the peritubular capillaries and returned to the general circulation, thus being conserved for the body.

Although vasopressin promotes $H_2O$ conservation by the body, it cannot completely halt urine production, even when a person is not taking in any $H_2O$, because a minimum volume of $H_2O$ must be excreted with the solute wastes. Collectively, the waste products and other constituents eliminated in the urine average 600 mOsm each day. Because the maximum urine concentration is 1200 mOsm/L, the minimum volume of urine required to excrete these wastes is 500 mL/day (600 mOsm of waste / day ÷ 1200 mOsm / L of urine = 0.5L, or 500mL / day, or 0.3mL / minof urine = 0.5 L, or 500 mL/day, or 0.3 mL/min). So under maximal vasopressin influence, 99.7 percent of the 180 L of plasma $H_2O$ filtered per day is returned to the blood, with an obligatory $H_2O$ loss of 0.5 L.

The kidneys' ability to tremendously concentrate urine to minimize $H_2O$ loss when necessary is possible only because of the presence of the vertical osmotic gradient in the medulla. If this gradient did not exist, the kidneys could not produce a urine more concentrated than the body fluids, no matter how much vasopressin was secreted, because the only driving force for $H_2O$ reabsorption is a concentration differential between the tubular fluid and the interstitial fluid.

**▌ Clinical Connections** Jacob had a prolonged period of continuous exercise in very hot conditions and did not stay adequately hydrated. Excessive sweating increases the osmolality in all of the body's fluid compartments. Ultimately, this leads to the secretion of vasopressin (antidiuretic hormone) to enhance water reabsorption in the distal tubules. Given his urine osmolality

*Continued*

**13**

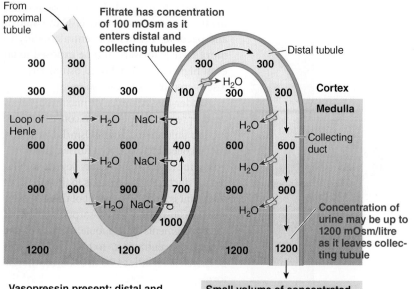

**Vasopressin present: distal and collecting tubules permeable to H₂O**

Small volume of concentrated urine excreted; reabsorbed H₂O picked up by peritubular capillaries and conserved for body

**(a)** In the face of a water deficit

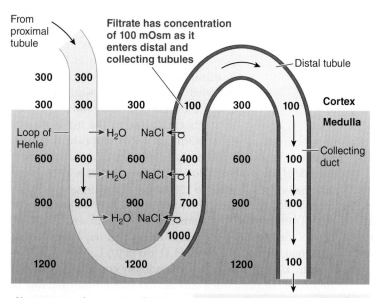

**No vasopressin present: distal and collecting tubules impermeable to H₂O**

Large volume of dilute urine; no collecting tubule H₂O reabsorbed in distal portion of nephron; excess H₂O eliminated

**(b)** In the face of a water excess

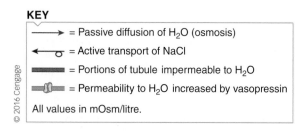

**KEY**

| | |
|---|---|
| → | = Passive diffusion of $H_2O$ (osmosis) |
| ←○ | = Active transport of NaCl |
| ▬ | = Portions of tubule impermeable to $H_2O$ |
| ▭ | = Permeability to $H_2O$ increased by vasopressin |

All values in mOsm/litre.

© 2016 Cengage

> **FIGURE 13-27 Excretion of urine of varying concentration depending on the body's needs**

of 1200 mOsm/L, his urine was as concentrated as possible. Under these conditions, salt crystals can precipitate and form stones. Staying hydrated by drinking lots of water decreases vasopressin secretion and allows the urine to dilute so that precipitation does not occur. It also helps to flush out any stones that have already formed.

The pain associated with kidney stones is not caused by the presence of stones within the kidneys themselves. As stones enter the urinary tract, the smooth muscle of the ureter contracts in attempt to move the stones forward. The stones cause a distension of the ureter, which is the source of the pain. The ureter lining can also tear, resulting in bleeding, which gives a pink hue to the urine.

## REGULATION OF H₂O REABSORPTION IN RESPONSE TO A H₂O EXCESS

Conversely, when a person consumes large quantities of $H_2O$, the excess $H_2O$ must be removed from the body without simultaneously losing solutes that are critical for maintaining homeostasis. Under these circumstances, no vasopressin is secreted, so the distal and collecting tubules remain impermeable to $H_2O$. The tubular fluid entering the distal tubule is hypotonic (100 mOsm/L), having lost salt without an accompanying loss of $H_2O$ in the ascending limb of Henle's loop. As this hypotonic fluid passes through the distal and collecting tubules (> Figure 13-27b), the medullary osmotic gradient cannot exert any influence because of the late tubular segments' impermeability to $H_2O$. In other words, none of the $H_2O$ remaining in the tubules can leave the lumen to be reabsorbed, even though the tubular fluid is less concentrated than the surrounding interstitial fluid. Therefore, in the absence of vasopressin, the 20 percent of the filtered fluid that reaches the distal tubule is not reabsorbed. Meanwhile, excretion of wastes and other urinary solutes remains constant. The net result is a large volume of dilute urine, which helps rid the body of excess $H_2O$. Urine osmolarity may be as low as 100 mOsm/L, the same as in the fluid entering the distal tubule. Urine flow may be increased up to 25 mL/min in the absence of vasopressin, compared with the normal urine production of 1 mL/min.

The ability to produce urine less concentrated than the body fluids depends on the fact that the tubular fluid is hypotonic as it enters the distal part of the nephron. This dilution is accomplished in the ascending limb when NaCl is actively extruded, but $H_2O$ cannot follow. Therefore, the loop of Henle, by simultaneously establishing the medullary osmotic gradient and diluting the tubular fluid before it enters the distal segments, plays a key role in allowing the kidneys to excrete urine that ranges in concentration from 100 to 1200 mOsm/L.

## Countercurrent exchange within the vasa recta

The vasa recta supply the renal medulla with blood to nourish its tissues and also to transport the water reabsorbed by the loops of Henle and collecting ducts back to the general circulation.

Another key contribution of the vasa recta is to support the countercurrent multiplier mechanism, which produces a high concentration of solutes in the interstitial fluid. It accomplishes this due to the following important characteristics: (1) the hairpin (U-shape) construction of the vasa recta loops back through the concentration gradient; (2) the blood flow in the vasa recta is opposite that of the fluid movement through the loop of Henle; (3) the vasa recta lie in close proximity to the loop of Henle; (4) the two arms of the vasa recta lie in close proximity to each other; and (5) the vasa recta are highly permeable (NaCl and $H_2O$). These characteristics allow the rapid exchange of fluid and solutes ($H_2O$ and NaCl) in the two parallel streams and thereby maintain a large concentration difference between the two ends of the vasa recta. This enables the blood in the vasa recta to leave the medulla and enter the renal vein essentially isotonic (~300–320 mOsm/L) to incoming arterial blood (> Figure 13-28).

As blood passes down the descending limb of the vasa recta, it equilibrates with the progressively increasing concentration of the surrounding interstitial fluid. It picks up $Na^+$ and $Cl^-$ (NaCl) and some urea, and loses $H_2O$ until it is very hypertonic (1200 mOsm/L) by the bottom of the loop. Then, as blood flows up the ascending limb, NaCl diffuses back out into the interstitium, and $H_2O$ reenters the vasa recta because the surrounding interstitial fluid has progressively decreasing concentrations; the passive exchange allows the blood leaving the vasa recta to be isotonic (~300–320 mOsm/L). This passive exchange of solutes and $H_2O$ between the two limbs of the vasa recta and the interstitial fluid is known as **countercurrent exchange**. Unlike countercurrent multiplication, it does not establish the concentration gradient. Rather, it preserves (*prevents the dissolution of*) the gradient. Because blood enters and leaves the medulla at the same osmolarity as a result of countercurrent exchange, the medullary tissue is nourished with blood, yet the incremental gradient of hypertonicity in the medulla is preserved.

## Water reabsorption

It is important to distinguish between the reabsorption of $H_2O$ that must follow solute reabsorption and the reabsorption of "free" $H_2O$ that is not linked to solute reabsorption (i.e., *free water clearance*). Here are the significant differences.

- *In the tubular segments permeable to $H_2O$, solute reabsorption is always accompanied by comparable $H_2O$ reabsorption* because of osmotic considerations. Therefore, the total volume of $H_2O$ reabsorbed is determined in large part by the total mass of solute reabsorbed; this is especially true of NaCl, because it is the most abundant solute in the ECF.

- *Solute excretion is always accompanied by comparable $H_2O$ excretion* because of osmotic considerations. This fact is responsible for the obligatory excretion of at least a minimal volume of $H_2O$, even when a person is severely dehydrated.

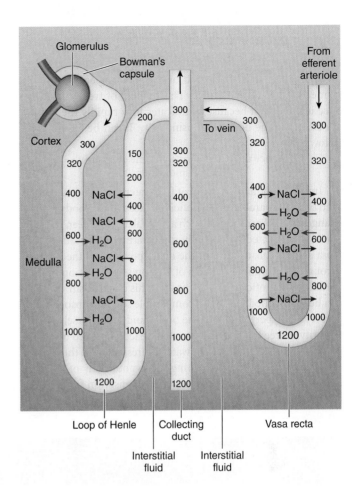

> FIGURE 13-28 **Countercurrent exchange in the renal medulla.** Hypothetical pattern of blood flow. If the blood supply to the renal medulla flowed straight through from the cortex to the inner medulla, the blood would be isotonic on entering but very hypertonic on exiting, having picked up salt and lost water as it equilibrated with the surrounding interstitial fluid at each incremental horizontal level. It would be impossible to maintain the vertical osmotic gradient, because the salt pumped out by the ascending limb of Henle's loop would be continuously flushed away by blood flowing through the medulla. However, the actual shape of the vasa recta mimics the loop of Henle, but the blood flow is reversed. Blood equilibrates with the interstitial fluid at each incremental horizontal level in both the descending limb and the ascending limb of the vasa recta, so blood is isotonic as it enters and leaves the medulla. This countercurrent exchange prevents dissolution of the medullary osmotic gradient, while providing blood to the renal medulla.

For the same reason, when excess unreabsorbed solute is present in the tubular fluid, its presence exerts an osmotic effect to hold excessive $H_2O$ in the lumen. This phenomenon is known as *osmotic diuresis*. Diuresis is increased urinary excretion, of which there are two types: osmotic diuresis and water diuresis.

*Clinical Note* **Osmotic diuresis** involves the increased excretion of both $H_2O$ and solute due to an excess of unreabsorbed solute in the tubular fluid, such as occurs in diabetes mellitus. The large quantity of unreabsorbed glucose that remains in the tubular fluid in people with diabetes osmotically drags $H_2O$ with it into the urine. Some diuretic drugs act by blocking specific solute reabsorption so that extra $H_2O$ spills into the urine along with the unreabsorbed solute.

13

**Water diuresis**, in contrast, involves the increased urinary output of $H_2O$ with little or no increase in excretion of solutes. Excess water that is excreted is termed **free water**. The total plasma volume that is cleared of excess water each minute is termed **free water clearance**. The concept of free water clearance is important because it indicates how rapidly the kidneys are changing the body fluid osmolarity. It can be calculated as follows:

$$\text{Clearance of } H_2O = \text{Urine volume per minute} - \text{Osmolar substances}$$

Importantly, a loss or gain of free water that is not accompanied by a comparable solute deficit or excess in the body leads to changes in ECF osmolarity. Such an imbalance between $H_2O$ and solute is corrected by partially dissociating $H_2O$ reabsorption from solute reabsorption in the distal portions of the nephron through the combined effects of vasopressin secretion and the medullary osmotic gradient. Through this mechanism, free $H_2O$ can be reabsorbed without a comparable solute reabsorption to correct for hypertonicity of the body fluids. Conversely, a large quantity of free $H_2O$ can be excreted unaccompanied by a comparable solute excretion (i.e., water diuresis) to rid the body of excess pure $H_2O$, thus correcting for hypotonicity of the body fluids. Water diuresis is normally a compensation for ingesting too much $H_2O$.

Excessive water diuresis follows alcohol ingestion. Because alcohol inhibits vasopressin secretion, the kidneys inappropriately lose too much $H_2O$. Typically, more fluid is lost in the urine than is consumed in the alcoholic beverage, so the body becomes dehydrated despite substantial fluid ingestion.

▌Table 13-4 summarizes how various tubular segments of the nephron handle $Na^+$ and $H_2O$ and the significance of these processes.

## Renal failure

1. Urine excretion and the resulting clearance of wastes and excess electrolytes from the plasma are crucial for maintaining homeostasis. When the functions of both kidneys are so disrupted that they cannot perform their regulatory and excretory functions sufficiently to maintain homeostasis, renal failure has set in. **Renal failure** can be described physiologically as a decrease in the glomerular filtration rate, with a biochemical symptom of elevated serum creatinine. One of the most widely used blood chemistry tests of renal function is the serum creatinine test. Serum creatinine levels can be influenced by the amount of skeletal muscle mass, however, and thus age, sex, and race need to be considered when using this test

---

**▌ TABLE 13-4 Handling of Sodium and Water by Various Tubular Segments of the Nephron**

| Tubular Segment | Na⁺ REABSORPTION | | H₂O REABSORPTION | |
| --- | --- | --- | --- | --- |
| | Percentage of Reabsorption in This Segment | Distinguishing Features | Percentage of Reabsorption in This Segment | Distinguishing Features |
| **Proximal Tubule** | 67 | Active; uncontrolled; plays a pivotal role in the reabsorption of glucose, amino acids, Cl⁻, H₂O, and urea | 65 | Passive; obligatory osmotic reabsorption following active Na⁺ reabsorption |
| **Loop of Henle** | 25 | Active, uncontrolled; Na⁺ along with Cl⁻ reabsorption from the ascending limb helps establish the medullary interstitial vertical osmotic gradient, which is important in the kidneys' ability to produce urine of varying concentrations and volumes, depending on the body's needs | 15 | Passive; obligatory osmotic reabsorption from the descending limb as the ascending limb extrudes NaCl into the interstitial fluid (i.e., reabsorbs NaCl) |
| **Distal and Collecting Tubules** | 8 | Active; variable and subject to aldosterone control; important in the regulation of ECF volume and long-term control of blood pressure; linked to K⁺ secretion and H⁺ secretion | 20 | Passive; not linked to solute reabsorption; variable quantities of "free" H₂O reabsorption subject to vasopressin control; driving force is the vertical osmotic gradient in the medullary interstitial fluid established by the long loops of Henle; important in regulating ECF osmolarity |

as a marker of renal function. More accurate methods of assessing renal function tend to be more expensive. Renal failure has a variety of causes, some of which begin elsewhere in the body and affect renal function secondarily. Among the causes are the following: *Infectious organisms*, either blood-borne or gaining entrance to the urinary tract through the urethra

2. *Toxic agents*, such as lead, arsenic, pesticides, or even long-term exposure to high doses of aspirin

3. *Inappropriate immune responses*, such as *glomerulonephritis*, which occasionally follow streptococcal throat infections, as antigen–antibody complexes leading to localized inflammatory damage are deposited in the glomeruli (p. 478)

4. *Obstruction of urine flow* by kidney stones, tumours, or an enlarged prostate gland, with back pressure reducing glomerular filtration as well as damaging renal tissue

5. *An insufficient renal blood supply* that leads to inadequate filtration pressure, which can occur secondary to circulatory disorders, such as heart failure, haemorrhage, shock, or narrowing and hardening of the renal arteries by atherosclerosis

Although these conditions may have different origins, almost all can cause some degree of nephron damage. The glomeruli and tubules may be independently affected, or both may be dysfunctional. Regardless of cause, renal failure can manifest itself either as (1) *acute renal failure*, characterized by a sudden onset with rapidly reduced urine formation until less than the essential minimum of around 500 mL of urine is produced per day; or (2) *chronic renal failure*, characterized by slow, progressive, insidious loss of renal function. A person may die from acute renal failure, or the condition may be reversible and lead to full recovery. Chronic renal failure, in contrast, is not reversible. Gradual, permanent destruction of renal tissue eventually proves fatal. Chronic renal failure is insidious, because up to 75 percent of the kidney tissue can be destroyed before the loss of kidney function is even noticeable. Because of the abundant reserve of kidney function, only 25 percent of the kidney tissue is needed to adequately maintain all the essential renal excretory and regulatory functions. With less than 25 percent of functional kidney tissue remaining, however, renal insufficiency becomes apparent. *End-stage renal failure* ensues when 90 percent of kidney function has been lost.

We will not be describing the different stages and symptoms associated with various renal disorders, but ▌ Table 13-5 summarizes the potential consequences of renal failure to give you an idea of the broad effects of kidney impairment. The extent of these effects should not be surprising, considering the central role the kidneys play in maintaining homeostasis. When the kidneys cannot maintain a normal internal environment, widespread disruption of cell activities can bring about abnormal function in other organ systems as well. By the time end-stage renal failure occurs, literally every body system has become impaired to some extent.

Because chronic renal failure is irreversible and eventually fatal, treatment is aimed at maintaining renal function by

alternative methods, such as dialysis and kidney transplantation. (For further explanation of these procedures, see Concepts, Challenges, and Controversies.)

This completes our discussion of kidney function. In the remainder of the chapter, we focus on the plumbing that stores and carries the urine formed by the kidneys to the outside.

## Check Your Understanding 13.5

1. State how the plasma clearance rate for each of the following substances compares with the GFR: (a) a substance that is filtered but not reabsorbed or secreted, (b) a substance that is filtered and reabsorbed but not secreted, and (c) a substance that is filtered and secreted but not reabsorbed.

2. Identify and explain which nephron component establishes, which component preserves, and which component uses the vertical osmotic gradient in the renal medulla.

3. Explain how vasopressin increases the permeability of the distal and collecting tubules to $H_2O$.

## Urine storage

Once urine has been formed by the kidneys, it is transmitted through the ureters to the urinary bladder. Urine does not flow through the ureters by gravitational pull alone. Peristaltic (forward-pushing) contractions of the smooth muscle within the ureteral wall propel the urine forward from the kidneys to the bladder. The ureters penetrate the wall of the bladder obliquely, coursing through the wall several centimetres before they open into the bladder cavity. This anatomic arrangement prevents backflow of urine from the bladder to the kidneys when pressure builds up in the bladder. As the bladder fills, the ureteral ends within its wall are compressed closed. Urine can still enter, however, because ureteral contractions generate enough pressure to overcome the resistance and push urine through the occluded ends.

### ROLE OF THE BLADDER

The bladder can accommodate large fluctuations in urine volume. The bladder wall consists of smooth muscle lined by a special type of epithelium. It was once assumed that the bladder was an inert sac. However, both the epithelium and the smooth muscle actively participate in the bladder's ability to accommodate large changes in urine volume. The epithelial lining can increase and decrease in surface area by the orderly process of membrane recycling as the bladder alternately fills and empties. Membrane-enclosed cytoplasmic vesicles are inserted by exocytosis into the surface area during bladder filling; then the vesicles are withdrawn by endocytosis to shrink the surface area following emptying (p. 46). As is characteristic of smooth muscle, bladder muscle can stretch tremendously without building up bladder wall tension (p. 333). In addition, the highly folded bladder wall flattens out during filling to increase bladder storage capacity. Because the kidneys continuously form urine, the bladder must have enough storage capacity to preclude the need to continually get rid of the urine.

13

**Uremic toxicity** caused by retention of waste products

Nausea, vomiting, diarrhoea, and ulcers caused by a toxic effect on the digestive system

Bleeding tendency arising from a toxic effect on platelet function

Mental changes—such as reduced alertness, insomnia, and shortened attention span, progressing to convulsions and coma—caused by toxic effects on the central nervous system

Abnormal sensory and motor activity caused by a toxic effect on the peripheral nerves

**Metabolic acidosis\*** caused by the inability of the kidneys to adequately secrete $H^+$ that is continually being added to the body fluids as a result of metabolic activity

Altered enzyme activity caused by the action of too much acid on enzymes

Depression of the central nervous system caused by the action of too much acid interfering with neuronal excitability

**Potassium retention\*** resulting from inadequate tubular secretion of $K^+$

Altered cardiac and neural excitability as a result of changing the resting membrane potentials of excitable cells

**Sodium imbalances** caused by the inability of the kidneys to adjust $Na^+$ excretion to balance changes in $Na^+$ consumption

Elevated blood pressure, generalized oedema, and congestive heart failure if too much $Na^+$ is consumed

Hypotension and, if severe enough, circulatory shock if too little $Na^+$ is consumed

**Phosphate and calcium imbalances** arising from impaired reabsorption of these electrolytes

Disturbances in skeletal structures caused by abnormalities in deposition of calcium phosphate crystals, which harden bone

**Loss of plasma proteins** as a result of increased "leakiness" of the glomerular membrane

Oedema caused by a reduction in plasma-colloid osmotic pressure

**Inability to vary urine concentration** as a result of impairment of the countercurrent system

Hypotonicity of body fluids if too much $H_2O$ is ingested

Hypertonicity of body fluids if too little $H_2O$ is ingested

**Hypertension** arising from the combined effects of salt and fluid retention and vasoconstrictor action of excess angiotensin II

**Anaemia** caused by inadequate erythropoietin production

**Depression of the immune system**, most likely caused by toxic levels of wastes and acids

Increased susceptibility to infections

\* Among the most life-threatening consequences of renal failure.

© 2016 Cengage

---

The bladder smooth muscle is richly supplied by parasympathetic fibres, stimulation of which causes bladder contraction. If the passageway through the urethra to the outside is open, bladder contraction empties urine from the bladder. The exit from the bladder, however, is guarded by two sphincters: the *internal urethral sphincter* and the *external urethral sphincter*.

### ROLE OF THE URETHRAL SPHINCTERS

A **sphincter** is a ring of muscle that, when contracted, closes off passage through an opening. The **internal urethral sphincter**—which is smooth muscle and, accordingly, is under involuntary control—is not really a separate muscle, but instead consists of the last part of the bladder. Although it is not a true sphincter, it performs the same function as a sphincter. When the bladder is relaxed, the anatomic arrangement of the internal urethral sphincter region closes the outlet of the bladder.

Farther down the passageway, the urethra is encircled by a layer of skeletal muscle, the **external urethral sphincter**. This sphincter is reinforced by the entire **pelvic diaphragm**, a skeletal muscle sheet that forms the floor of the pelvis and helps support the pelvic organs (see › Figure 13-2). The motor neurons that supply the external sphincter and pelvic diaphragm fire continuously at a moderate rate, unless they are inhibited, keeping these muscles tonically contracted so they prevent urine from escaping through the urethra. Normally, when the bladder is relaxed and filling, both the internal and external urethral sphincters are closed to keep urine from dribbling out. Furthermore, because the external sphincter and pelvic diaphragm are skeletal muscle and thus under voluntary control, the person can deliberately tighten them to prevent urination from occurring even when the bladder is contracting and the internal sphincter is open.

### MICTURITION REFLEX

**Micturition (urination)**, the process of bladder emptying, is governed by two mechanisms: the micturition reflex and voluntary control. The **micturition reflex** is initiated when stretch

13

# Dialysis: Cellophane Tubing or Abdominal Lining as an Artificial Kidney

**BECAUSE CHRONIC RENAL FAILURE IS IRREVERSIBLE** and eventually fatal, treatment is aimed at maintaining renal function by alternative methods, such as dialysis and kidney transplantation. The process of dialysis bypasses the kidneys to maintain normal fluid and electrolyte balance and remove wastes artificially. In the original method of dialysis, **haemodialysis**, a patient's blood is pumped through cellophane tubing that is surrounded by a large volume of fluid similar in composition to normal plasma. After dialysis, the blood is returned to the patient's circulatory system. Like capillaries, cellophane is highly permeable to most plasma constituents but is impermeable to plasma proteins. As blood flows through the tubing, solutes move across the cellophane down their individual concentration gradients; plasma proteins, however, stay in the blood. Urea and other wastes, which are absent in the dialysis fluid, diffuse out of the plasma into the surrounding fluid, cleaning the blood of these wastes. Plasma constituents that are not regulated by the kidneys and are at normal concentration, such as glucose, do not move across the cellophane into the dialysis fluid, because there is no driving force to produce their movement. (The dialysis fluid's glucose concentration is the same as normal plasma glucose concentration.) Electrolytes, such as $K^+$ and $PO_4^{3-}$, are higher than their normal plasma concentrations because the diseased kidneys cannot eliminate excess quantities of these substances. So they move out of the plasma until equilibrium is achieved between the plasma and the dialysis fluid. Because the dialysis fluid's solute concentrations are maintained at normal plasma values, the solute concentration of the blood returned to the patient after dialysis is essentially normal. Haemodialysis is repeated as often as necessary to maintain the plasma composition within an acceptable level. Typically, it is done three times per week for several hours at each session.

In a more recent method of dialysis, **continuous ambulatory peritoneal dialysis (CAPD)**, the peritoneal membrane (the lining of the abdominal cavity) is used as the dialysis membrane. With this method, two litres of dialysis fluid are inserted into the patient's abdominal cavity through a permanently implanted catheter. Urea, $K^+$, and other wastes and excess electrolytes diffuse from the plasma across the peritoneal membrane into the dialysis fluid, which is drained off and replaced several times a day. The CAPD method offers several advantages: The patient can self-administer, the patient's blood is continuously purified and adjusted, and the patient can engage in normal activities while dialysis is being accomplished. One drawback is the increased risk of peritoneal infections.

Although dialysis can remove metabolic wastes and foreign compounds and help maintain fluid and electrolyte balance within acceptable limits, this plasma-cleansing technique cannot make up for the failing kidneys' reduced ability to produce hormones (erythropoietin and renin) and to activate vitamin D. One new experimental technique incorporates living kidney cells derived from pigs within a dialysis-like machine. Standard ultrafiltration technology like that used in haemodialysis purifies and adjusts the plasma as usual. Importantly, the living cells not only help maintain even better control of plasma constituents, especially $K^+$, but also add the deficient renal hormones to the plasma passing through the machine and activate vitamin D. This promising new technology has not yet been tested in large-scale clinical trials.

For now, transplanting a healthy kidney from a donor is another option for treating chronic renal failure. A kidney is one of the few transplants that can be provided by a living donor. Because 25 percent of the total kidney tissue can maintain the body, both the donor and the recipient have ample renal function with only one kidney each. The biggest problem with transplants is the possibility that the patient's immune system will reject the organ. Risk of rejection can be minimized by matching the tissue types of the donor and the recipient as closely as possible (the best donor choice is usually a close relative), coupled with immunosuppressive drugs.

Another new technique on the horizon for treating end-stage renal failure is a continuously functioning artificial kidney that mimics natural renal function. Using nanotechnology (very small-scale devices), researchers are working on a device that contains two membranes, the first for filtering blood as the glomerulus does and the second for mimicking the renal tubules by selectively altering the filtrate. The device, which will directly process the blood on an ongoing basis without using dialysis fluid, will return important substances to the body while discharging unneeded substances to a discardable bag that will serve as an external bladder. Scientists have developed computer models for such a device and thus far have created the filtering membrane.

**13**

receptors within the bladder wall are stimulated (⟩ Figure 13-29). The bladder in an adult can accommodate approximately 250–400 millilitres of urine before the tension within its walls begins to rise sufficiently to activate the stretch receptors (⟩ Figure 13-30). The greater the distension beyond this, the greater is the extent of receptor activation. Afferent fibres from the stretch receptors carry impulses into the spinal cord and eventually, via interneurons, stimulate the parasympathetic supply to the bladder and inhibit the motor-neuron supply to the external sphincter. Parasympathetic stimulation of the bladder causes it to contract. No special mechanism is required to open the internal sphincter; changes in the shape of the bladder during contraction mechanically pull the internal sphincter open. Simultaneously, the external sphincter relaxes as its motor

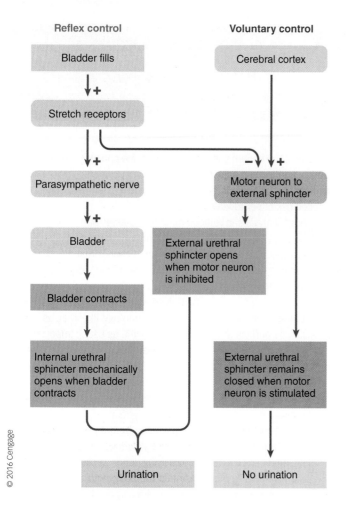

Reflex control

Bladder fills → **+** Stretch receptors → **+** Parasympathetic nerve → **+** Bladder → Bladder contracts → Internal urethral sphincter mechanically opens when bladder contracts → Urination

Voluntary control

Cerebral cortex → **−** / **+** Motor neuron to external sphincter

External urethral sphincter opens when motor neuron is inhibited

External urethral sphincter remains closed when motor neuron is stimulated → No urination

© 2016 Cengage

> FIGURE 13-29 **Reflex and voluntary control of micturition**

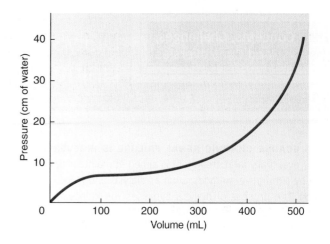

© 2016 Cengage

> FIGURE 13-30 **Pressure changes within the urinary bladder as the bladder fills with urine**

neuron supply is inhibited. Now both sphincters are open, and urine is expelled through the urethra by the force of bladder contraction. This micturition reflex, which is entirely a spinal reflex, governs bladder emptying in infants. As soon as the bladder fills enough to trigger the reflex, the baby automatically wets.

**VOLUNTARY CONTROL OF MICTURITION**
In addition to triggering the micturition reflex, bladder filling also gives rise to the conscious urge to urinate. The perception of bladder fullness appears before the external sphincter reflexly

relaxes, warning that micturition is imminent. As a result, voluntary control of micturition, learned during toilet training in early childhood, can override the micturition reflex so that bladder emptying can take place at the person's convenience rather than when bladder filling first activates the stretch receptors. If the time when the micturition reflex is initiated is inopportune for urination, the person can voluntarily prevent bladder emptying by deliberately tightening the external sphincter and pelvic diaphragm. Voluntary excitatory impulses from the cerebral cortex override the reflex inhibitory input from the stretch receptors to the involved motor neurons (the relative balance of EPSPs and IPSPs), keeping these muscles contracted so that no urine is expelled.

Urination cannot be delayed indefinitely. As the bladder continues to fill, reflex input from the stretch receptors increases with time. Finally, reflex inhibitory input to the external sphincter motor neuron becomes so powerful that it can no longer be overridden by voluntary excitatory input, so the sphincter relaxes and the bladder uncontrollably empties.

Micturition can also be deliberately initiated, even though the bladder is not distended, by voluntarily relaxing the external sphincter and pelvic diaphragm. Lowering of the pelvic floor allows the bladder to drop downward, which simultaneously pulls open the internal urethral sphincter and stretches the

**13**

**? \***

**▐ Why It Matters**
**Urinary Incontinence**

Urinary incontinence, or inability to prevent discharge of urine, occurs when descending pathways in the spinal cord that mediate voluntary control of the external sphincter and pelvic diaphragm are disrupted, as in spinal cord injury. Because the components of the micturition reflex arc are still intact in the lower spinal cord, bladder emptying is governed by an uncontrollable spinal reflex, as

in infants. A lesser degree of incontinence characterized by urine escaping when bladder pressure suddenly increases transiently, such as during coughing or sneezing, can result from impaired sphincter function. This is common in women who have borne children or in men whose sphincters have been injured during prostate surgery.

bladder wall. The subsequent activation of the stretch receptors brings about bladder contraction by the micturition reflex. Voluntary bladder emptying may be further assisted by contracting the abdominal wall and respiratory diaphragm. The resulting increase in intra-abdominal pressure squeezes down on the bladder to facilitate its emptying.

# Chapter in Perspective: Focus on Homeostasis

The kidneys contribute to homeostasis more extensively than any other single organ. They regulate the electrolyte composition, volume, osmolarity, and pH of the internal environment and eliminate all the waste products of bodily metabolism, except for respiration-removed carbon dioxide. They accomplish these regulatory functions by eliminating in the urine substances the body doesn't need, such as metabolic wastes and excess quantities of ingested salt or water, while conserving useful substances. The kidneys can maintain the plasma constituents they regulate within the narrow range compatible with life, despite wide variations in intake and losses of these substances through other avenues. Illustrating the magnitude of the kidneys' task, about 25 percent of the blood pumped into the systemic circulation goes to the kidneys to be adjusted and purified, and 75 percent of the blood is used to supply all the other tissues.

The kidneys contribute to homeostasis in the following specific ways:

## REGULATORY FUNCTIONS

- The kidneys regulate the quantity and concentration of most ECF electrolytes, including those important in maintaining proper neuromuscular excitability.
- They help maintain proper pH by eliminating excess $H^+$ (acid) or $HCO_3^-$ (base) in the urine.
- They help maintain proper plasma volume, which is important in long-term regulation of arterial blood pressure, by controlling salt balance in the body. The ECF volume, including plasma volume, reflects total salt load in the ECF, because $Na^+$ and its attendant anion, $Cl^-$, are responsible for more than 90 percent of the ECF's osmotic (water-holding) activity.
- The kidneys maintain water balance in the body, which is important in maintaining proper ECF osmolarity (concentration of solutes). This role is important in maintaining stability of cell volume by keeping water from osmotically moving into or out of the cells, thereby preventing them from swelling or shrinking, respectively.

## EXCRETORY FUNCTIONS

- The kidneys excrete the end products of metabolism in urine. If allowed to accumulate, these wastes are toxic to cells.
- The kidneys also excrete many foreign compounds that enter the body.

## HORMONAL FUNCTIONS

- The kidneys produce erythropoietin, the hormone that stimulates the bone marrow to produce red blood cells. This action contributes to homeostasis by helping maintain the optimal $O_2$ content of blood. More than 98 percent of $O_2$ in the blood is bound to haemoglobin within red blood cells.
- They also produce renin, the hormone that initiates the renin–angiotensin–aldosterone pathway for controlling renal tubular $Na^+$ reabsorption, which is important in long-term maintenance of plasma volume and arterial blood pressure.

## METABOLIC FUNCTIONS

- The kidneys help convert vitamin D into its active form. Vitamin D is essential for $Ca^{2+}$ absorption from the digestive tract. Calcium, in turn, exerts a wide variety of homeostatic functions.

# CHAPTER TERMINOLOGY

ACE inhibitor drugs (p. 578)
active reabsorption (p. 574)
afferent arterioles (p. 562)
aldosterone receptor blockers (p. 578)
anaemia (p. 598)
angiotensin-converting enzyme (ACE) (p. 576)
angiotensin I (p. 576)
angiotensin II (p. 576)
angiotensinogen (p. 576)
antidiuretic hormone (p. 592)
autoregulation (p. 569)
blood urea nitrogen (BUN) (p. 582)
Bowman's capsule (p. 563)
collecting duct (tubule) (p. 564)

continuous ambulatory peritoneal dialysis (CAPD) (p. 599)
cortical nephrons (p. 564)
countercurrent exchange (p. 595)
countercurrent multiplication (p. 592)
creatinine (p. 586)
depression of the immune system (p. 598)
distal tubule (p. 564)
diuretics (p. 578)
efferent arteriole (p. 563)
external urethral sphincter (p. 598)
filtered load (p. 579)
filtration coefficient $(K_f)$ (p. 568)
filtration fraction (p. 588)

filtration slits (p. 567)
free water (p. 596)
free water clearance (p. 596)
glomerular filtration (p. 564)
glomerular filtration rate (GFR) (p. 568)
glomerular membrane (p. 566)
glomeruli (p. 562)
granular cells (p. 570)
haemodialysis (p. 599)
hypertension (p. 598)
hypertonic (p. 588)
hypotonic (p. 588)
inability to vary urine concentration (p. 598)
internal urethral sphincter (p. 598)

**13**

## REVIEW EXERCISES

### Objective Questions
(Answers in Appendix E, p. A-49)

1. Part of the kidneys' energy supply is used to accomplish glomerular filtration. *(True or false?)*

2. Sodium reabsorption is under hormonal control throughout the length of the tubule. *(True or false?)*

3. Glucose and amino acids are reabsorbed by secondary active transport. *(True or false?)*

4. Solute excretion is always accompanied by comparable $H_2O$ excretion. *(True or false?)*

5. Water excretion can occur without comparable solute excretion. *(True or false?)*

6. The functional unit of the kidneys is the _____

7. _____ is the only ion actively reabsorbed in the proximal tubule and actively secreted in the distal and collecting tubules.

8. The daily minimum volume of obligatory $H_2O$ loss that must accompany excretion of wastes is _____ mL.

9. Indicate whether each of the factors would (a) increase or (b) decrease the GFR, if everything else remained constant.

_____ 1. a rise in Bowman's capsule pressure resulting from ureteral obstruction by a kidney stone

_____ 2. a fall in plasma protein concentration resulting from loss of these proteins from a large burned surface of skin

_____ 3. a dramatic fall in arterial blood pressure following severe haemorrhage (80 mmHg)

_____ 4. afferent arteriolar vasoconstriction

_____ 5. tubuloglomerular feedback response to decreased salt delivery to the distal tubule

_____ 6. myogenic response of an afferent arteriole stretched as a result of an increased driving blood pressure

_____ 7. increased sympathetic activity to the afferent arterioles

_____ 8. contraction of mesangial cells

_____ 9. contraction of podocytes

10. Which filtered substance is normally NOT present in the urine at all?
   a. $Na^+$
   b. $PO_4^{3-}$
   c. urea
   d. $H^+$
   e. glucose

11. Reabsorption of which substance is NOT linked in some way to active $Na^+$ reabsorption?
   a. glucose
   b. $PO_4^{3-}$
   c. $H_2O$
   d. urea
   e. $Cl^-$

In questions 12–14, indicate, by writing the identifying letters in the proper order in the blanks, the proper sequence through which fluid flows as it traverses the structures in question.

12. a. ureter          ____ ____ ____ ____ ____
    b. kidney
    c. urethra
    d. bladder
    e. renal pelvis

13. a. efferent arteriole
    b. peritubular capillaries
    c. renal artery
    d. glomerulus
    e. afferent arteriole
    f. renal vein

    — — — — — —

14. a. loop of Henle
    b. collecting duct
    c. Bowman's capsule
    d. proximal tubule
    e. renal pelvis
    f. distal tubule
    g. glomerulus

    — — — — — —

15. Using the answer code on the right, indicate what the osmolarity of the tubular fluid is at each of the designated points in a nephron:

    ___ 1. Bowman's capsule
    ___ 2. end of proximal tubule
    ___ 3. tip of Henle's loop of juxtamedullary nephron (at the bottom of the U-turn)
    ___ 4. end of Henle's loop of juxtamedullary nephron (before entry into distal tubule)
    ___ 5. end of collecting duct

    (a) isotonic (300 mOsm/L)
    (b) hypotonic (100 mOsm/L)
    (c) hypertonic (1200 mOsm/L)
    (d) ranging from hypotonic to hypertonic (100 mOsm/L to 1200 mOsm/L)

## Written Questions

1. List the functions of the kidneys.
2. Describe the anatomy of the urinary system. Describe the components of a nephron.
3. Describe the three basic renal processes, and indicate how they relate to urine excretion.
4. Distinguish between *secretion* and *excretion*.
5. Discuss the forces involved in glomerular filtration. What is the average GFR?
6. How is GFR regulated as part of the baroreceptor reflex?
7. Why do the kidneys receive a seemingly disproportionate share of the cardiac output? What percentage of RBF is normally filtered?
8. List the steps in transepithelial transport.
9. Distinguish between active and passive reabsorption.
10. Describe all the tubular transport processes that are linked to the basolateral $Na^+-K^+$ ATPase carrier.
11. Describe the renin–angiotensin–aldosterone system. Discuss the source and function of atrial natriuretic peptide.
12. To what do the terms *tubular maximum* $(T_m)$ and *renal threshold* refer? Compare two substances that display a $T_m$: one substance that *is* and one that *is not* regulated by the kidneys.

13. What is the importance of tubular secretion? What are the most important secretory processes?
14. What is the average rate of urine formation?
15. Define *plasma clearance*.
16. What establishes a vertical osmotic gradient in the medullary interstitial fluid? Of what importance is this gradient?
17. Discuss the function and mechanism of action of vasopressin.
18. Describe the transfer of urine to, the storage of urine in, and the emptying of urine from the bladder.

## Quantitative Exercises
### (Solutions in Appendix E, p. A-50)

1. Two patients are voiding protein in their urine. To determine whether or not this proteinuria indicates a serious problem, a physician injects small amounts of inulin and PAH into each patient. Recall that inulin is freely filtered and neither secreted nor reabsorbed in the nephron and that PAH at this concentration is completely removed from the blood by tubular secretion. The data collected are given in the following table, where $[X]_u$ is the concentration of substance X (either inulin or PAH) in the urine (in mM); $[X]_p$ is the concentration of X in the plasma, and $v_u$ is the flow rate of urine (in mL/min).

| Patient | $[I]_u$ | $[I]_p$ | $[PAH]_u$ | $[PAH]_p$ | $[V]_u$ |
|---------|---------|---------|-----------|-----------|---------|
| 1 | 25 | 2 | 186 | 3 | 10 |
| 2 | 31 | 1.5 | 300 | 4.5 | 6 |

   a. Calculate each patient's GFR and renal plasma flow.
   b. Calculate the RBF for each patient, assuming both have a haematocrit of 0.45.
   c. Calculate the filtration fraction for each patient.
   d. Which of the values calculated for each patient are within the normal range? Which values are abnormal? What could be causing these deviations from normal?

2. What is the filtered load of sodium if inulin clearance is 125 mL/min and the sodium concentration in plasma is 145 mM?

3. Calculate a patient's rate of urine production, given that his inulin clearance is 125 mL/min and his urine and plasma concentrations of inulin are 300 mg/L and 3 mg/L, respectively.

4. If the urine concentration of a substance is 7.5 mg/mL of urine, its plasma concentration is 0.2 mg/mL of plasma, and the urine flow rate is 2 mL/min, what is the clearance rate of the substance? Is the substance being reabsorbed or secreted by the kidneys?

13

(Explanations in Appendix E, p. A-50)

1. The juxtamedullary nephrons of animals adapted to survive with minimal water consumption, such as desert rats, have relatively much longer loops of Henle than humans have. Of what benefit would these longer loops be?

2. a. If the plasma concentration of substance X is 200 mg/100 mL and the GFR is 125 mL/min, how much is the filtered load of this substance?

   b. If the $T_m$ for substance X is 200 mg/min, how much of the substance will be reabsorbed at a plasma concentration of 200 mg/100 mL and a GFR of 125 mL/min?

   c. How much of substance X will be excreted?

3. *Conn's syndrome* is an endocrine disorder brought about by a tumour of the adrenal cortex that secretes excessive aldosterone in uncontrolled fashion. Given what you know about the functions of aldosterone, describe the most prominent features of this condition.

4. Because of a mutation, a child was born with an ascending limb of Henle that was water permeable. What would be the minimum/maximum urine osmolarities (in units of mOsm/L) the child could produce?

   a. 100/300

   b. 300/1200

   c. 100/100

   d. 1200/1200

   e. 300/300

5. An accident victim suffers permanent damage of the lower spinal cord and is paralyzed from the waist down. Describe what governs bladder emptying in this individual.

## CLINICAL CONSIDERATION

(Explanation in Appendix E, p. A-50)

1. Marcus T. has noted a gradual decrease in his urine flow rate and is now experiencing difficulty in initiating micturition. He needs to urinate frequently, and often he feels as if his bladder is not empty even though he has just urinated. Analysis of Marcus's urine reveals no abnormalities. Are his urinary tract symptoms most likely caused by kidney disease, a bladder infection, or prostate enlargement?

## Fluid and Acid–Base Balance

Body systems maintain homeostasis

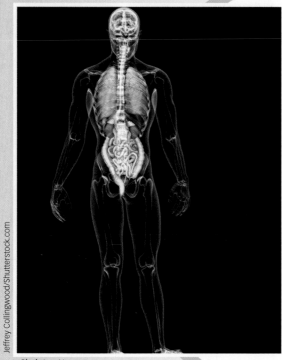

Jeffrey Collingwood/Shutterstock.com

Skeleton X-ray

### Homeostasis

The kidneys, in conjunction with hormones involved in salt and water balance, are responsible for maintaining the volume and osmolarity of the extracellular fluid (internal environment). Along with the respiratory system and chemical buffer systems in the body fluids, the kidneys also contribute to homeostasis by maintaining the proper pH in the internal environment.

Homeostasis is essential for survival of cells

Cells make up body systems

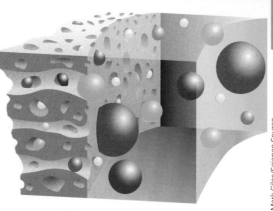

Mark Giles/Science Source

Osmosis

Homeostasis depends on maintaining a balance between the input and the output of all constituents in the internal fluid environment. Regulation of fluid balance involves two separate components:

1. *control of ECF volume*, which includes circulating plasma volume; and
2. *control of ECF osmolarity*, which includes the solute concentration.

The kidneys control ECF volume by regulating salt balance and ECF osmolarity by regulating water balance. They maintain this balance by adjusting the output of salt and water in the urine, as needed to compensate for variations in the input and output of these constituents.

Similarly, the kidneys help maintain acid–base balance by adjusting the urinary output of hydrogen ions (acid) and bicarbonate ions (base) as needed. Also contributing to acid–base balance in the body fluids are the chemical buffer systems (e.g., bicarbonate buffer system) and the respiratory system (i.e., the lungs). For example, the lungs can adjust the rate at which hydrogen ion–generating carbon dioxide is eliminated by increasing or decreasing the rate and depth of breathing.

# Fluid and Acid–Base Balance

## CONTENTS AT A GLANCE

**▮ Clinical Connections**

Diane, a 22-year-old female university student, has been healthy until recently. Despite having an increased appetite and no change in her level of activity, she had a weight loss of 5 kg (11 lb) over the past 15 days. She has also felt intensely thirsty and has been producing frequent, large volumes of urine. When Diane experienced generalized abdominal pain, vomiting, and increasing lethargy, her friend brought her immediately to the emergency room for an assessment.

Diane was found to have an elevated heart rate (tachycardia) and decreased blood pressure (hypotensive). Her plasma glucose level was extremely elevated at 25 mmol/L (Normal: ≤ 11.1 mmol/L), and ketoacids were present in her blood. Further bloodwork revealed the following:

| | |
|---|---|
| Serum pH: 7.25 (LOW) | (Normal: 7.35–7.45) |
| Arterial pCO2: 30 mmHg (LOW) | (Normal: 35–40 mmHg) |
| Plasma $HCO_3^-$: 13 mEq/L (LOW) | (Normal: 23–29 mEq/L) |
| Anion Gap: 22 mEq/L (HIGH) | (Normal: 8–16 mEq/L) |

Diane was diagnosed as having diabetic ketoacidosis (DKA), and treatment was initiated.

## 14.1 | Balance Concept

The cells of complex multicellular organisms are able to survive and function only within a very narrow range of composition of the extracellular fluid (ECF)—that is, the internal fluid environment that bathes them.

### The internal pool of a substance

The quantity of any particular substance in the ECF is considered a readily available internal **pool**. The amount of the substance in the pool may be increased either by transferring more in from the external environment (most commonly by ingestion) or by metabolically producing it within the body (⟩ Figure 14-1). Substances may be removed from the body by being excreted to the outside or by being used up in a metabolic reaction. If the quantity of a substance is to remain stable within the body, its **input** through ingestion or metabolic production must be balanced by an equal **output** through excretion or metabolic consumption. This relationship, known as the **balance concept**, is extremely important in maintaining homeostasis. Not all input and output pathways are applicable for each body-fluid constituent. For example, salt is not synthesized or consumed by the body, so the stability of salt concentration in the body fluids depends entirely on a balance between salt ingestion and salt excretion.

#### EXCHANGES BETWEEN THE POOL AND OTHER INTERNAL SITES

The ECF pool can be altered by transferring a particular ECF constituent into storage within the body. If the body as a whole has a surplus or deficit of a particular stored substance, the storage site can be expanded or partially depleted to maintain the ECF concentration of the substance within homeostatically prescribed limits. For example, after absorption of a meal, when more glucose is entering the plasma than is being consumed by the cells, the extra glucose can be temporarily stored, in the form of glycogen, in muscle and liver cells. This storage depot can then be tapped between meals as needed to maintain the plasma glucose level when no new nutrients are being added to the blood by eating. This internal storage capacity is limited, however. Although an internal exchange between the ECF and a storage depot can temporarily restore the plasma concentration of a particular substance to normal, in the long run any excess or deficit of that constituent must be compensated for by appropriate adjustments in total body input or output.

Another possible internal exchange between the pool and the rest of the body is the reversible incorporation of certain plasma constituents into more complex molecular structures. For example, iron is incorporated into haemoglobin within the red blood cells during their synthesis, but it is released intact back into the body fluids when the red cells degenerate. This process differs from the metabolic consumption of a substance, in which the substance is irretrievably converted into another form—for example, glucose converted into carbon dioxide plus water plus energy. It also differs from storage in that the latter serves no purpose other than storage. By contrast, reversible incorporation into a more complex structure serves a specific purpose.

### Maintenance of a balanced ECF constituent

When total body input of a particular substance equals its total body output, a **stable balance** exists. When the gains via input for a substance exceed its losses via output, a **positive balance** exists. The result is an increase in the total amount of the substance in the body. In contrast, when losses for a substance exceed its gains, a **negative balance** exists, and the total amount of the substance in the body decreases.

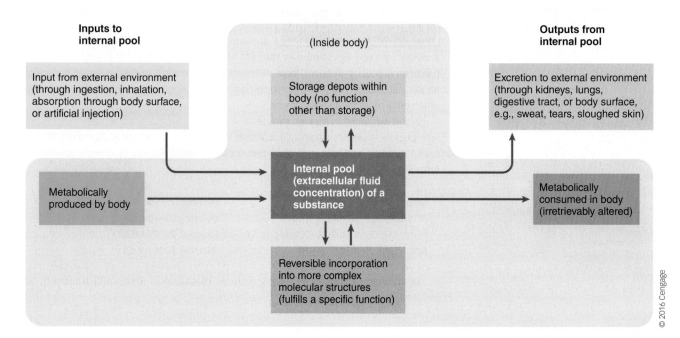

⟩ **FIGURE 14-1** Inputs to and outputs from the internal pool of a body constituent

Changing the magnitude of any of the input or output pathways for a given substance can alter its plasma concentration. To maintain homeostasis, any change in input must be balanced by a corresponding change in output (e.g., increased salt intake must be matched by a corresponding increase in salt output in the urine), and, conversely, increased losses must be compensated for by increased intake. Thus, maintaining a stable balance requires control. However, not all input and output pathways are regulated to maintain balance. Generally, input of various plasma constituents is poorly controlled or not controlled at all. We frequently ingest salt and water, for example, not because we need them but because we want them, so the intake of salt and water is highly variable. Likewise, hydrogen ion ($H^+$) is uncontrollably generated internally and added to the body fluids. Salt, water, and $H^+$ can also be lost to the external environment to varying degrees through the digestive tract (vomiting), skin (sweating), and elsewhere without regard for the salt, water, or $H^+$ balance in the body. Compensatory adjustments in the urinary excretion of these substances maintain the body fluids' volume and salt and acid composition within the extremely narrow homeostatic range compatible with life, despite the wide variations in input and unregulated losses of these plasma constituents. It is important to note that although excretion of solutes by the kidney is extremely efficient, it does have a maximum concentrating ability. Ingesting solutes in concentrations above this maximum results in an inability to fully eliminate them. In these situations, it is possible for solute concentrations to rise to a level that, without treatment, can be fatal.

The rest of this chapter is devoted to discussing the regulation of fluid balance (maintaining salt and water balance) and acid–base balance (maintaining $H^+$ balance).

### Check Your Understanding 14.1

1. List the possible inputs to and outputs from the internal pool of a given body constituent. Define *stable balance*, *positive balance*, and *negative balance*.

## 14.2 | Fluid Balance

Water is by far the most abundant component of the human body, constituting 60 percent of body weight, on average, but ranging from 40 to 80 percent. The $H_2O$ content of an individual remains fairly constant, largely because the kidneys efficiently regulate **water balance**, but the percentage of body $H_2O$ varies from person to person. The reason for the wide range in body $H_2O$ among individuals is the variability in the amount of their adipose tissue (fat). Adipose tissue has a low $H_2O$ percentage compared with other tissues. Plasma, as you might suspect, is more than 90 percent $H_2O$. Even the soft tissues, such as skin, muscles, and internal organs, consist of 70 to 80 percent $H_2O$. The relatively drier skeleton is only 22 percent $H_2O$. Fat, however, is the driest tissue of all, having only 10 percent $H_2O$ content. Accordingly, a high body $H_2O$ percentage is associated with

leanness, and a low body $H_2O$ percentage with obesity, because a larger proportion of the overweight body consists of relatively dry fat.

The percentage of body $H_2O$ is also influenced by the sex and age of the individual. Women have a lower body $H_2O$ percentage than men, primarily because the female sex hormone, estrogen, promotes fat deposition in the breasts, buttocks, and elsewhere. This not only gives rise to the typical female figure but also endows women with a higher proportion of adipose tissue and, therefore, a lower body $H_2O$ proportion. The percentage of body $H_2O$ also decreases progressively with age.

### Body water distribution

Body water is distributed between two major fluid compartments: fluid within the cells, **intracellular fluid (ICF)**, and fluid surrounding the cells, **extracellular fluid (ECF)** (Table 14-1). (The terms $H_2O$ and *fluid* are commonly used interchangeably. Although this usage is not entirely accurate, because it ignores the solutes in body fluids, it is acceptable when discussing total volume of fluids, because the major proportion of these fluids consists of $H_2O$.)

#### PROPORTION OF $H_2O$ IN THE MAJOR FLUID COMPARTMENTS

The ICF compartment composes about two-thirds of the total body $H_2O$. Even though each cell contains its own unique mixture of constituents, these trillions of minute fluid compartments are similar enough to be considered collectively as one large fluid compartment.

The remaining third of the body $H_2O$ found in the ECF compartment is further subdivided into plasma and interstitial fluid. The **plasma**, which makes up about 25 percent of the ECF volume, is the fluid portion of blood. The **interstitial fluid**,

### ■ TABLE 14-1 Classification of Body Fluid

| Compartment | Volume of Fluid (in Litres) | Percentage of Body Fluid | Percentage of Body Weight |
| --- | --- | --- | --- |
| **Total Body Fluid** | 42 | 100 | 60 |
| **Intracellular Fluid (ICF)** | 28 | 67 | 40 |
| **Extracellular Fluid (ECF)** | 14 | 33 | 20 |
| *Plasma* | 3.5 | 8.3 (25% of ECF) | 5 |
| *Interstitial fluid* | 10.5 | 25 (75% of ECF) | 15 |
| *Lymph* | Negligible | Negligible | Negligible |
| *Transcellular fluid* | Negligible | Negligible | Negligible |

© 2016 Cengage

which represents the other 75 percent of the ECF compartment, is the fluid in the spaces between cells. It bathes and makes exchanges with tissue cells.

### MINOR ECF COMPARTMENTS

Two other minor categories are included in the ECF compartment: lymph and transcellular fluid. **Lymph** is the fluid being returned from the interstitial fluid to the plasma by means of the lymphatic system, where it is filtered through lymph nodes for immune defence purposes (pp. 411 and 464). **Transcellular fluid** consists of a number of small, specialized fluid volumes, all of which are secreted by specific cells into a particular body cavity to perform some specialized function. Transcellular fluid includes *cerebrospinal fluid* (the brain and spine); *intraocular fluid* (the eye); *synovial fluid* (the joints); *pericardial, intrapleural, and peritoneal fluids* (the heart, lungs, and intestines, respectively); and the *digestive juices* (the stomach).

Although these fluids are extremely important functionally, they represent an insignificant fraction of total body $H_2O$. Furthermore, the transcellular compartment as a whole usually does not reflect changes in the body's fluid balance. For example, cerebrospinal fluid does not decrease in volume when the body as a whole is experiencing a negative $H_2O$ balance. This is not to say that these fluid volumes never change. Localized changes in a particular transcellular fluid compartment can occur pathologically (such as too much intraocular fluid accumulating in the eyes of people with glaucoma; p. 153), but such a localized fluid disturbance does not affect the fluid balance of the body. Therefore, the transcellular compartment can usually be ignored when dealing with problems of fluid balance. The main exception to this generalization occurs when digestive juices are abnormally lost from the body during heavy vomiting or diarrhoea, which can bring about a fluid imbalance.

*Clinical Note* Bioelectrical impedance can be a useful technique for body composition analysis in healthy individuals and in those with a number of chronic conditions, such as mild to moderate obesity, diabetes mellitus, and other medical conditions in which major disturbances of water distribution are not prominent.

## Plasma and interstitial fluid

Several barriers separate the body-fluid compartments, and they limit the movement of $H_2O$ and solutes between the various compartments to differing degrees.

### THE BARRIER BETWEEN PLASMA AND INTERSTITIAL FLUID: BLOOD VESSEL WALLS

The two components of the ECF—plasma and interstitial fluid—are separated by the walls of the blood vessels. However, $H_2O$ and all plasma constituents except for plasma proteins are continuously and freely exchanged between plasma and interstitial fluid by passive means across the thin, pore-lined capillary walls. Accordingly, plasma and interstitial fluid are nearly identical in composition, except that interstitial fluid lacks plasma proteins. Any change in one of these ECF compartments is quickly reflected in the other compartment, because they are constantly mixing.

### THE BARRIER BETWEEN THE ECF AND ICF: CELLULAR PLASMA MEMBRANES

In contrast to the very similar solute concentrations of the plasma and interstitial fluid compartments, the ECF differs significantly from that of the ICF ( > Figure 14-2). The physical barrier that separates the ECF and ICF is the cell membrane. Unlike the constant mixing between the plasma and interstitial fluid compartments, the cell membrane is highly selective and only permits the passage of certain materials. Due to this high specificity for movement there is a pronounced difference in the composition of the ECF and ICF compartments.

There are several differences among the ECF and ICF. First, the ICF has proteins that cannot permeate the enveloping membranes to leave the cells. Like most proteins, they are negatively charged and therefore can attract positively charged ions. Second there is an unequal distribution of $Na^+$ and $K^+$ and their attendant anions. This is the result of the membrane-bound $Na^+-K^+$ ATPase pump that is present in all cells. It actively transports three $Na^+$ out of and two $K^+$ into cells. As a result, $Na^+$ is the primary ECF cation, and $K^+$ is the primary ICF cation.

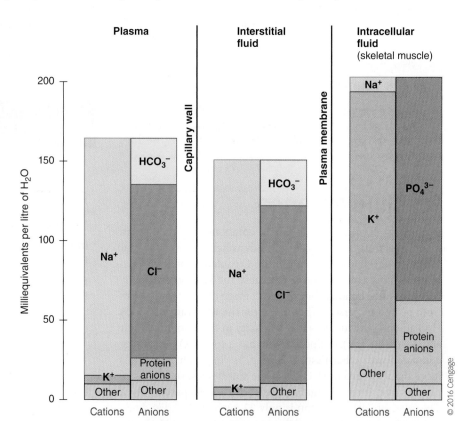

> FIGURE 14-2 **Ionic composition of the major body-fluid compartments**

All of these substances can result in the movement of $H_2O$ into the cell, which will cause the cell to swell and lyse if not checked. The $Na^+-K^+$ ATPase pump is also a mechanism to prevent this. Additionally, this unequal distribution of $Na^+$ and $K^+$, coupled with differences in membrane permeability to these ions, is crucial for the electrical properties of cells. These include both the initiation and the propagation of action potentials in excitable tissues (see Chapters 2 and 3). The concept of having an unequal distribution of solutes in different fluid compartments is known as **chemical disequilibrium**.

Except for the extremely small, electrically unbalanced portion of the total intracellular and extracellular ions involved in membrane potential, most ECF and ICF ions are electrically balanced. In the ECF, for example, $Na^+$ is accompanied primarily by the chloride anion $(Cl^-)$ and to a lesser extent by bicarbonate $(HCO_3^-)$. In the ICF, $K^+$ is accompanied by phosphate $(PO_4^{3-})$ and the negatively charged proteins trapped within the cell.

Most cells are freely permeable to water. Recall (p. 409) that the movement of water between the ECF compartments (plasma and the interstitial fluid) across capillary walls is governed by Starling forces. These forces, resulting from the imbalances between capillary hydrostatic pressure and colloid osmotic pressure, dictate whether water is absorbed into a capillary from the interstitial space or filtered out of a capillary into the interstitial space. In contrast, the movement of water across the cell membrane (between the interstitial fluid and the ICF) occurs as a result of osmotic effects alone. The hydrostatic pressures of the interstitial fluid and ICF are both extremely low and fairly constant, and therefore do not significantly contribute to the movement of water.

## ECF volume and osmolarity

Extracellular fluid serves as an intermediary between the cells and the external environment. All exchanges of water and other constituents between the ICF and external world must occur through the ECF. Water added to the body fluids always enters the ECF compartment first, and fluid always leaves the body via the ECF.

Plasma is the only fluid that can be acted on directly to control its volume and composition. Plasma circulates through all of the reconditioning organs that perform homeostatic adjustments (p. 386). Any adjustments made to the plasma equally effect the interstitial fluid, as a result of the free exchange between these compartments across the capillary walls. As a general rule of thumb, any control mechanism that changes plasma, will similarly change the interstitial fluid compartment. In contrast to this, the degree to which the ICF is influenced by the ECF is dictated by the permeability of cell membrane.

Two factors are regulated to maintain **fluid balance** in the body: ECF volume and ECF osmolarity. Although regulation of these two factors is interrelated, both being dependent on the relative NaCl and $H_2O$ load in the body, the reasons that they are closely controlled are different:

1. *ECF volume* must be closely regulated to help *maintain blood pressure*. Maintaining *salt balance* is of primary importance in the long-term regulation of ECF volume.

2. *ECF osmolarity* must be closely regulated to *prevent swelling or shrinking of cells*. Maintaining *water balance* is of primary importance in regulating ECF osmolarity.

## Control of ECF volume

ECF volume acts directly on blood pressure by changing plasma volume. For example, expanding ECF volume raises arterial blood pressure by increasing plasma volume. Two compensatory measures come into play to transiently adjust blood pressure until the ECF volume can be restored to normal. Let's review them.

### REVIEW OF SHORT-TERM CONTROL MEASURES TO MAINTAIN BLOOD PRESSURE

1. *The baroreceptor reflex alters both cardiac output and total peripheral resistance* to adjust blood pressure in the proper direction through autonomic nervous system effects on the heart and blood vessels (p. 420). Cardiac output and total peripheral resistance both increase to raise blood pressure when it falls too low, and conversely, both decrease to reduce blood pressure when it rises too high.

2. *Fluid shifts occur temporarily and automatically between plasma and interstitial fluid* as a result of changes in the balance of hydrostatic and osmotic forces (Starling forces) acting across the capillary walls when plasma volume deviates from normal (p. 409). For example, a reduction in plasma volume is partially compensated for by a shift of fluid out of the interstitial compartment and into the blood vessels. This helps to expand the circulating plasma volume at the expense of the interstitial compartment. Conversely, when plasma volume is too large, much of the excess fluid shifts into the interstitial compartment.

These two measures are short-term adjustment to help keep blood pressure fairly constant, but they are not long-term solutions. Furthermore, these short-term compensatory measures have a limited ability to minimize a change in blood pressure. For example, if plasma volume is too inadequate, blood pressure remains too low no matter how vigorous the pump action of the heart, how constricted the resistance vessels, or what proportion of interstitial fluid shifts into the blood vessels.

### LONG-TERM CONTROL MEASURES TO MAINTAIN BLOOD PRESSURE

It is important that other compensatory measures come into play in the long run to restore the ECF volume to normal. Long-term regulation of blood pressure rests with the kidneys and the thirst mechanism, controlling urinary output and fluid intake, respectively. Although these two measures play an important long-term influence on arterial blood pressure, urinary output by the kidneys is most crucial for maintaining blood pressure. You will see why as we discuss these long-term mechanisms in more detail.

14

## Control of salt

By way of review, sodium and its attendant anions account for more than 90 percent of the ECF's osmotic activity. As the kidneys conserve salt, they automatically conserve water because water follows $Na^+$ osmotically. As a result, this retained salt solution is isotonic (p. 40). In other words, changing the amount of salt in the ECF does not change its concentration; it merely changes the volume of the ECF. An increase to salt in the ECF leads to an increase in water retention, so the ECF remains isotonic, but increased in volume. Similarly, a reduction in salt leads to decreased water retention, so the ECF remains isotonic, but reduced in volume. The total mass of $Na^+$ salts in the ECF is referred to as the **$Na^+$ load**, and determines the ECF's volume. Appropriately, regulation of ECF volume therefore depends primarily on controlling salt balance.

To maintain **salt balance** at a set level, salt input must equal salt output in order to prevent salt accumulation or deficit in the body. Athletes often challenge the body's ability to manage salt balance during ultra-endurance sporting events.

### POOR CONTROL OF SALT INTAKE

The only avenue for salt input is ingestion, which typically is well in excess of the body's need for replacing obligatory salt losses. In our example of a typical daily salt balance ( Table 14-2), salt intake is 10.5 g per day. (The average Canadian salt intake is around 3.5 g per day.) Only 0.5 g of salt per day is adequate to replace the small amounts of salt usually lost in the feces and sweat, not considering salt lost during regular physical activity of moderate to high intensity.

The salt intake of humans is not well controlled and as a result, we typically consume salt in excess of our needs. Carnivores (meat eaters) and omnivores (eaters of meat and plants, like humans), naturally get enough salt in fresh meat (meat contains an abundance of salt-rich ECF). From a physiologic standpoint, humans normally do not display a regulatory appetite to seek additional salt. In fact, humans generally only have a hedonistic (pleasure-seeking) reason for consuming salt—we consume it because we like it. In contrast, herbivores (plant eaters) lack salt in their diet and develop salt hunger from a regulatory standpoint. Herbivores will travel many kilometres to a salt lick to meet their physiologic need for salt.

### PRECISE CONTROL OF SALT OUTPUT IN THE URINE

To maintain salt balance, excess ingested salt must be excreted in the urine. The three avenues for salt output are the obligatory loss of salt in *sweat and feces* and the controlled excretion of salt in *urine* (Table 14-2). The total amount of sweat produced is unrelated to salt balance and determined rather by factors that control body temperature. The small amount of salt normally lost in feces is also not subject to control. This uncontrollable loss of salt is referred to as **obligatory loss**. Except during heavy sweating or diarrhoea, this obligatory loss amounts to only about 0.5 g of salt per day. This represents the only salt that normally needs to be replaced by salt intake.

Since human salt consumption typically exceeds the amount that needs to be replaced from obligatory losses, the kidneys precisely excrete excess salt in the urine to maintain salt balance. In our example, 10 g of salt are eliminated in the urine per day so that total salt output exactly equals salt input. By regulating the rate of urinary salt excretion (i.e., by regulating the rate of $Na^+$ excretion), the kidneys normally keep the total $Na^+$ mass in the ECF constant, despite any notable changes in dietary intake or unusual losses through sweating or diarrhoea. In keeping the total $Na^+$ mass in the ECF constant, the ECF volume, in turn, is maintained within the narrowly prescribed limits essential for normal circulatory function.

Deviations in the ECF volume accompanying changes in the salt load trigger renal compensatory responses that quickly bring the $Na^+$ load and ECF volume back to normal. Sodium is freely filtered at the glomerulus and actively reabsorbed, but it is not secreted by the tubules. Therefore, the amount of $Na^+$ excreted in the urine represents the amount of $Na^+$ filtered minus the amount reabsorbed.

$$Na^+ \text{ excreted} = Na^+ \text{ filtered} - Na^+ \text{ reabsorbed}$$

The kidneys accordingly adjust the amount of salt excreted by controlling two processes: (1) the glomerular filtration rate (GFR), and (2) tubular reabsorption of $Na^+$. The regulation of these two processes are discussed in Chapter 13, so here we examine how these processes relate to the long-term control of ECF volume and blood pressure.

- *The amount of $Na^+$ filtered is controlled by regulating the GFR.* The amount of $Na^+$ filtered is equal to the plasma $Na^+$ concentration multiplied by the GFR. At any given plasma $Na^+$ concentration, a change in the GFR will correspondingly change the amount of $Na^+$ and the accompanying fluid filtered. Thus, control of the GFR can adjust the amount of $Na^+$ filtered each minute. Recall that the GFR is deliberately changed to alter the amount of salt and fluid filtered as part of the general baroreceptor reflex response to change in blood pressure (see Figure 13-12). The amount of salt filtered is therefore adjusted as part of the general blood pressure–regulating reflexes. It is important to note that the $Na^+$ load in the body is not sensed as such. Rather, it is monitored indirectly through the effect that $Na^+$ ultimately has on blood

---

| **█ TABLE 14-2** Daily Salt Balance | | | |
|---|---|---|---|
| **SALT INPUT** | | **SALT OUTPUT** | |
| Avenue | Amount (g/day) | Avenue | Amount (g/day) |
| **Ingestion** | 10.5 | Obligatory loss in sweat and feces | 0.5 |
| | | Controlled excretion in urine | 10.0 |
| **Total Input** | 10.5 | Total output | 10.5 |

pressure via its role in determining the ECF volume. Fittingly, baroreceptors that monitor fluctuations in blood pressure bring about adjustments in the amounts of Na⁺ filtered and eventually excreted.

- *The amount of Na⁺ reabsorbed is controlled through the renin–angiotensin–aldosterone system.* The amount of Na⁺ reabsorbed also depends on regulatory systems that play an important role in controlling blood pressure. Although Na⁺ is reabsorbed throughout most of the length of the nephron, only its reabsorption in the distal aspects are subject to control. The main factor controlling the extent of Na⁺ reabsorption in the distal tubule and collecting ducts is the important renin–angiotensin–aldosterone system (RAAS). Activation of the RAAS, specifically aldosterone, promotes Na⁺ reabsorption and thereby Na⁺ retention. Na⁺ retention, in turn, promotes osmotic retention of $H_2O$ and the subsequent expansion of plasma volume and elevation of arterial blood pressure. Appropriately, this Na⁺− conserving system can be activated by a number of means. Namely, by a reduction in salt, ECF volume, and/or arterial blood pressure (see Figure 13-16).

Thus, control of GFR and Na⁺ reabsorption are interrelated and intimately tied to long-term regulation of ECF volume as reflected by blood pressure. For example, a fall in arterial blood pressure carries with it two very important effects. First, there is a reflex reduction in the GFR to decrease the amount of Na⁺ filtered. Second, there is a hormonally adjusted increase in the amount of Na⁺ reabsorbed by the action of aldosterone (⟩ Figure 14-3). Together, these effects conserve the amount of Na⁺ by reducing the amount of Na⁺ excreted. This is accompanied by a reduction in water excretion, which helps compensate for the fall in arterial pressure.

## Control of ECF osmolarity

Maintaining fluid balance depends on regulating both ECF volume and ECF osmolarity. Whereas regulating ECF volume is important in long-term control of blood pressure, regulating ECF osmolarity is important in preventing changes in cell volume. The **osmolarity** of a fluid is a measure of the concentration of the individual solute particles dissolved in it. The higher the osmolarity is, the higher the concentration of solutes. Recall that water moves by osmosis down its concentration gradient from an area of lower solute (higher water) concentration to an area of higher solute (lower water) concentration.

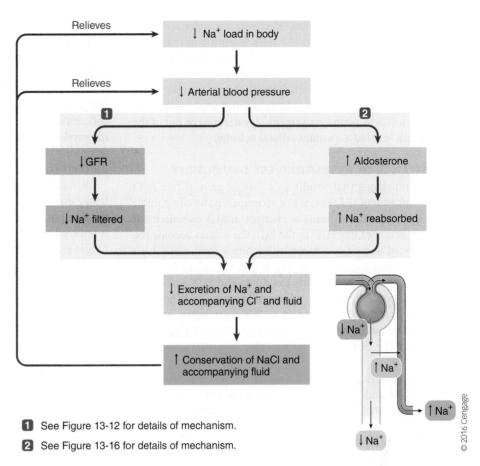

**1** See Figure 13-12 for details of mechanism.

**2** See Figure 13-16 for details of mechanism.

⟩ **FIGURE 14-3** Dual effect of a fall in arterial blood pressure on renal handling of Na⁺

### IONS RESPONSIBLE FOR ECF AND ICF OSMOLARITY

Osmosis occurs across the cellular plasma membranes only when a difference in concentration of nonpenetrating solutes exists between the ECF and ICF. Solutes that can penetrate a barrier separating two fluid compartments quickly become equally distributed between the two compartments and thus do not contribute to osmotic differences.

Sodium and its attendant anions are by far the most abundant nonpenetrating solutes in the ECF. Since they are nonpenetrating, they account for the vast majority of the osmotic activity of the ECF. In contrast, K⁺ and its accompanying anions are the most abundant nonpenetrating solutes of the ICF and are therefore responsible for its osmotic activity. Even though small amounts of Na⁺ and K⁺ passively diffuse across the plasma membrane all the time, these ions behave as if they were nonpenetrating because of Na⁺−K⁺ pump activity. Any Na⁺ that passively diffuses down its electrochemical gradient into the cell is promptly pumped back outside, so the result is the same as if Na⁺ were barred from the cells. In reverse, K⁺, in effect, remains trapped within the cells. The resulting unequal distribution of Na⁺ and K⁺ and their accompanying anions between the ECF and ICF is responsible for the osmotic activity of these two fluid compartments.

Normally, the osmolarities of the ECF and ICF are the same, because the total concentration of K⁺ and other effectively nonpenetrating solutes inside the cells is equal to the total concentration of $Na^+$ and other effectively nonpenetrating solutes in the fluid surrounding the cells. Even though nonpenetrating solutes in the ECF and ICF differ, their concentrations are normally identical. Note that it is the number, not the nature of the unequally distributed particles per volume that determines the osmolarity of the fluid. The identical osmolarities of the ECF and ICF results in no net movement of water into or out of the cells, which leads to a constant cellular volume.

### IMPORTANCE OF REGULATING ECF OSMOLARITY

Any circumstance that results in a loss or gain of *free $H_2O$* (i.e., loss or gain of $H_2O$ that is not accompanied by comparable solute deficit or excess) leads to changes in ECF osmolarity. If there is a deficit of free $H_2O$ in the ECF, the solutes become too concentrated and ECF osmolarity becomes abnormally high (i.e., it becomes *hypertonic*; p. 41). If there is excess free $H_2O$ in the ECF, the solutes become too dilute and ECF osmolarity becomes abnormally low (i.e., it becomes *hypotonic*). When ECF osmolarity changes with respect to ICF osmolarity, osmosis takes place. Water either leaves or enters the cells, depending on whether the ECF is more or less concentrated, respectively, than the ICF.

The osmolarity of the ECF must therefore be regulated to prevent these undesirable shifts of $H_2O$ out of or into the cells. It is crucial that the ECF osmolarity be maintained within a very narrow range to prevent cell shrinking (osmotic fluid shift from the ICF to the ECF) or swelling (osmotic fluid shift from the ECF to the ICF).

The following sections examine the fluid shifts that occur between the ECF and ICF when ECF osmolarity becomes either hypertonic or hypotonic relative to the ICF. Then we consider how water balance and, consequently, ECF osmolarity are normally maintained to minimize harmful changes in cell volume.

## ECF hypertonicity and shrinking cells

**Hypertonicity** of the ECF, the excessive concentration of ECF solutes, is often associated with **dehydration**—that is, a negative free water balance.

### CAUSES OF HYPERTONICITY

Hypertonicity can be brought about in four major ways:

1. *Insufficient* water intake, such as might occur during desert travel or might accompany difficulty in swallowing

2. *Excessive* water loss, such as might occur during heavy sweating, vomiting, or diarrhoea. Even though both water and solutes are lost during these conditions, relatively more water is lost, resulting in more concentrated solutes in the plasma.

3. **Diabetes insipidus**, a disease characterized by a deficiency of vasopressin (see Why It Matters, p. 615).

4. **Iatrogenic** hypertonicity, caused, for example, by intravenous fluid replacement with hypertonic saline. Iatrogenic means the condition results from medical intervention.

### DIRECTION AND RESULTING SYMPTOMS OF $H_2O$ MOVEMENT DURING HYPERTONICITY

*Clinical Note* Whenever the ECF compartment becomes hypertonic, water moves out of the cells by osmosis into the more concentrated ECF until the ICF osmolarity equilibrates with the ECF. As water leaves the cells, they shrink. Of particular concern is that considerable shrinking of brain neurons disturbs the myelin sheath and effects overall brain function. The manifestations of this include mental confusion and irrationality, delirium, convulsions, coma, or even death.

Rivalling the neural symptoms in seriousness are the circulatory disturbances. Since most cases of ECF hypertonicity are the result of dehydration, the extent to which the plasma volume is reduced determines the severity in physiologic changes. This ranges from a slight reduction in blood pressure to circulatory shock and death.

Other more common signs and symptoms are evident even in very mild cases of dehydration. These include dry skin, sunken eyeballs, a parched tongue, dry lips, more concentrated urine, and reduced urine volume.

## ECF hypotonicity and swelling cells

**Hypotonicity** of the ECF is often associated with excess free $H_2O$, sometimes referred to as **overhydration**. When a positive free $H_2O$ balance exists, the ECF is less concentrated (more dilute) than normal.

### CAUSES OF HYPOTONICITY (OVERHYDRATION)

Under normal circumstances, hypotonicity of the ECF does not occur because any surplus free water is promptly excreted in the urine. Despite this, hypotonicity can still occur and the major causes are as follows:

1. Patients with *renal failure* who cannot excrete a dilute urine become hypotonic when they consume relatively more water than solutes.

2. Hypotonicity can occur transiently in healthy people *if* water *is rapidly ingested* to such an excess that the kidneys can't respond quickly enough to eliminate the extra water.

3. Hypotonicity can occur when excess water without solute is retained in the body as a result of *inappropriate secretion of vasopressin*.

4. Similar to hypertonicity, iatrogenic hypotonicity can occur by intravenous fluid replacement with hypotonic fluids (e.g., hypotonic saline).

Vasopressin is normally secreted in response to a water deficit, which is relieved by increasing water reabsorption in the distal part of the nephrons. However, vasopressin release occurs for other reasons, even when there is no water deficit. These include pain, acute infections, trauma, and other stressful situations. The increased vasopressin secretion and resulting water retention elicited by stress are appropriate in anticipation of potential blood loss in the stressful situation. The extra, retained water could minimize the effect a loss of blood volume would have on blood pressure. However, because modern-day stressful

14

Vasopressin (*a.k.a.* antidiuretic hormone) increases the permeability of the distal and collecting tubules to $H_2O$ and thus enhances $H_2O$ conservation by reducing urinary output of $H_2O$ (p. 592). This is accomplished via the translocation of aquaporins (water channels) to the surface of the luminal side of various parts of the nephron—specifically, to the distal and collecting tubules. The aquaporins allow for $H_2O$ to be reabsorbed in these parts of the nephron. In a condition called *diabetes insipidus*, this normal response of water conservation by vasopressin does not occur. This condition is not to be confused with diabetes mellitus, where the major problem is with insulin and glucose regulation. Patients with *diabetes insipidus* can produce up to 20 L of very dilute urine daily, compared with the normal average of 1.5 L/day.

In general, individuals with diabetes insipidus have their symptoms for one of two reasons. Some patients release a normal amount of vasopressin from the posterior pituitary gland, but the kidneys are unable to respond to it. This variant of diabetes insipidus is referred to as *nephrogenic diabetes insipidus* because the problem originates from the level of the kidney. Nephrogenic diabetes insipidus can be caused by medications (e.g., lithium carbonate, a mood stabilizer).

Other patients do not release vasopressin from the posterior pituitary despite a normally functioning kidney. This variant is referred to as *central diabetes insipidus* because the problem originates from the level of the pituitary gland. Some infections, inflammatory conditions or even a tumour pressing on the pituitary gland can lead to central diabetes insipidus. Central diabetes insipidus is treated by replacing vasopressin, often in the form of a nasal spray. Those with diabetes insipidus must keep pace with the tremendous loss of $H_2O$ in the urine to avoid quickly becoming dehydrated. Frequently, those with *diabetes insipidus* complain that they spend an extraordinary amount of time day and night going to the bathroom and getting drinks.

---

situations generally do not involve blood loss, the increased vasopressin secretion is inappropriate as far as the body's fluid balance is concerned. The reabsorption and retention of too much water dilute the body's solutes.

### DIRECTION AND RESULTING SYMPTOMS OF WATER MOVEMENT DURING HYPOTONICITY

*Clinical Note* Excess free water retention first dilutes the ECF compartment, making it hypotonic. The resulting difference in osmotic potential between the ECF and ICF induces water to move from the more dilute ECF and into the ICF, resulting in cellular swelling. In contrast to the shrinking of cerebral neurons in hypertonic states, pronounced swelling of brain cells also leads to brain dysfunction in hypotonic states. Symptoms include confusion, irritability, lethargy, headache, dizziness, vomiting, drowsiness, convulsions, coma, and even death.

Non-neural symptoms of overhydration include muscle weakness (from muscle cell swelling), hypertension (from expansion of plasma volume), and oedema (from increased hydrostatic pressure).

The condition of overhydration, hypotonicity, and cellular swelling due to excess free water retention is sometimes referred to as **water intoxication**. This should not be confused with the fluid retention that occurs with excess salt retention. In the case of fluid retention, the ECF is still isotonic because the increase in salt is matched by a corresponding increase in water. Since the interstitial fluid is isotonic, no osmotic gradient exists to drive extra $H_2O$ into the cells. The excess salt and $H_2O$ burden is therefore confined to the ECF compartment, with circulatory consequences being the most important concern. When water intoxication occurs, there is excess free water in the ECF. Therefore, an osmotic gradient exists between the ECF and the ICF, and in addition to any circulatory disturbances, symptoms caused by cellular swelling also become a problem.

Let us now contrast the situations of hypertonicity and hypotonicity with what happens as a result of isotonic fluid gain or loss.

### MAINTENANCE OF CELL WATER WHEN ECF IS ISOTONIC

*Clinical Note* An example of an isotonic fluid gain is therapeutic intravenous administration of an isotonic solution, such as isotonic saline (also known as *normal saline*). When an isotonic fluid is injected into the ECF compartment, ECF volume increases, but the concentration of ECF solutes remains unchanged; in other words, the ECF is still isotonic. Since the osmolarity of the ECF has not changed, there is no osmotic gradient between the ECF and the ICF. Therefore, no net fluid shift occurs between the two compartments. The ECF compartment has increased in volume without shifting water into the cells. Thus, for intravenous fluid therapy, unless a qualified health professional is trying to correct an osmotic imbalance, the fluid of choice is isotonic with the ECF. This prevents fluctuations in intracellular volume and possible neural symptoms.

Similarly, in an isotonic fluid loss, such as haemorrhage, the loss is confined to the ECF compartment, with no corresponding loss of fluid from the ICF. Fluid does not shift out of the cells because the ECF remaining within the body is still isotonic. Of course, many other mechanisms counteract loss of blood, but the ICF compartment is not directly affected by the loss.

Thus, when the ECF and ICF are in osmotic equilibrium, no net movement of $H_2O$ into or out of the cells occurs. This is true regardless of whether the ECF volume increases or decreases. Shifts in water between the ECF and ICF take place only when the concentration of solutes in the ECF changes due to a loss or gain of free water.

Now let's look at how free water balance is normally maintained.

**14**

## CONTROL OF WATER BALANCE

Control of free water balance is crucial for regulating ECF osmolarity. Increases in free water cause the ECF to become too dilute, and deficits of free water cause the ECF to become too concentrated. These types of changes must immediately be corrected in order to avoid harmful osmotic fluid shifts into or out of the cells.

To maintain a stable water balance, water input must equal water output.

### SOURCES OF WATER INPUT

- In a person's typical daily water balance (▌Table 14-3), a little more than a litre of water is added to the body by *drinking liquids.*

- Surprisingly, an amount almost equal to that is obtained from *eating solid food.* Recall that muscles consist of about 75 percent water; meat (animal muscle) is therefore 75 percent water. Fruits and vegetables consist of 60–90 percent water. Therefore, people normally get almost as much water from solid foods as from the liquids they drink.

- The third source of water input is *metabolically produced* water. Chemical reactions within the cells convert food and $O_2$ into energy, producing $CO_2$ and $H_2O$ in the process (e.g., the electron transport train). This **metabolic water** produced during cell metabolism averages about 350 mL/day.

The average water intake from these three sources totals 2600 mL/day. Medically, another source of water is intravenous infusion of fluid.

### SOURCES OF WATER OUTPUT

- The body loses close to a litre of water daily without being aware of it. This is referred to as **insensible loss** (i.e., the person has no sensory awareness of it) and occurs from the

*lungs* and the *nonsweating skin.* During respiration, inspired air becomes saturated with water within the airways. This water is lost when the moistened air is subsequently expired. Normally we are not aware of this water loss, but on cold days when water vapour condenses, we can "see our breath." The other insensible loss is continual loss of water from the skin even in the absence of sweating. Water molecules can diffuse through skin cells and evaporate without being noticed. Fortunately, the skin is fairly waterproofed by its keratinized exterior layer, which protects against a much greater loss of water by this avenue (p. 461). If this protective surface layer is lost, for example in the case of skin burns, there is a significantly increased fluid loss from the burned surface. Those with serious burns are at a much higher risk of severe dehydration. As a result, one critically important component of burn management is adequate fluid replacement therapy to counteract this increased water loss.

- Sensible loss (i.e., the person has sensory awareness) of water from the skin occurs through *sweating.* At an air temperature of 20°C, an average of 100 mL of water is lost daily through sweating. Loss of water through sweating can vary substantially depending on the environmental temperature, the humidity, and the degree of physical activity. It may range from zero millilitres to as much as several litres per hour in very hot weather.

- Another avenue for water loss is through the *feces.* Normally, only about 100 mL of water are lost through this route each day. During fecal formation in the large intestine, most water is absorbed out of the digestive tract lumen into the blood, thereby conserving fluid and solidifying the digestive tract's contents for elimination. Significantly increased water loss through the digestive tract can occur in the case of vomiting and/or diarrhoea.

- By far the most important output mechanism for water is *urine excretion,* with 1500 mL (1.5 L) of urine being produced daily on average.

The example in Table 14-3 shows that the total water input is the same as the total water output (2600 mL/day). This balance is not by chance. Water input must match water output in order for the body to maintain both water balance and a normal solute concentration of the ECF.

### FACTORS REGULATED TO MAINTAIN WATER BALANCE

Of the many sources of water input and output, only two can be regulated to maintain water balance. Water input is regulated by thirst, and this influences the amount of fluid ingested. Also, water output regulated by the kidneys influences how much urine is formed. The most important mechanism controlling water balance is urine output.

Other factors that involve water input and output are also regulated, but not for the purpose of maintaining water balance. For example, the food we eat contains water, but the amount of food we ingest is regulated to maintain energy balance. Furthermore, we regulate the amount we sweat, but this is to help us maintain a constant body temperature. Metabolic water production and insensible losses are completely unregulated.

---

▌ **TABLE 14-3** Daily Water Balance

| WATER INPUT | | WATER OUTPUT | |
|---|---|---|---|
| Avenue | Quantity (mL/day) | Avenue | Quantity (mL/day) |
| **Fluid Intake** | 1250 | Insensible loss (from lungs and nonsweating skin) | 900 |
| **Water in Food Intake** | 1000 | | |
| **Metabolically Produced Water** | 350 | Sweat | 100 |
| | | Feces | 100 |
| | | Urine | 1500 |
| **Total Input** | 2600 | Total output | 2600 |

© 2016 Cengage

## CONTROL OF WATER OUTPUT IN THE URINE BY VASOPRESSIN

Fluctuations in ECF osmolarity caused by imbalances between water input and output are quickly compensated for by changes in the amount of urinary excretion of water. This can happen without a change in the usual excretion of salt. This means that water balance can be regulated independently of solute regulation, so that the amount of free water retained or eliminated can be varied to quickly restore ECF osmolarity to normal. Free water reabsorption and excretion are adjusted through changes in vasopressin secretion (p. 592). Throughout most of the nephron, $H_2O$ reabsorption is important in regulating ECF volume; this is because salt reabsorption (regulated by aldosterone) is accompanied by comparable $H_2O$ reabsorption. In contrast, the distal and collecting tubules have variable free water reabsorption without comparable salt reabsorption. This is the result of the increasing vertical osmotic gradient in the renal medulla. Vasopressin increases the permeability of the distal and collecting tubules to water. Depending on the amount of vasopressin present, the amount of free water reabsorbed can be adjusted as necessary to restore ECF osmolarity to normal.

Vasopressin is produced by the hypothalamus and stored in the posterior pituitary gland. It is released from the posterior pituitary on command from the hypothalamus.

## CONTROL OF $H_2O$ INPUT BY THIRST

*Clinical Note* **Thirst** is the subjective sensation that drives you to ingest $H_2O$. A **thirst centre** is located in the hypothalamus in close proximity to the vasopressin-secreting cells. Excessive thirst is known as *polydipsia*; excessive urination is known as *polyuria*. Both of these signs are often present in individuals with untreated diabetes mellitus and are the result of high glucose concentration in the plasma and urine.

## Vasopressin secretion and thirst: Largely triggered simultaneously

The hypothalamic control centres that regulate vasopressin secretion and thirst act simultaneously. Both centres are stimulated by a free water deficit and suppressed by a free water excess. Appropriately, the same circumstances that call for reducing urinary output to conserve body water also give rise to the sensation of thirst to replenish body water.

### ROLE OF HYPOTHALAMIC OSMORECEPTORS

The predominant excitatory input for both vasopressin secretion and thirst comes from **hypothalamic osmoreceptors** located near

the vasopressin-secreting cells and thirst centre. These osmoreceptors monitor the osmolarity of fluid surrounding them. Because this fluid reflects the concentration of the entire internal fluid environment, as the osmolarity increases and the need for water conservation increases, vasopressin secretion and thirst are both stimulated (> Figure 14-4). As a result, reabsorption of water in the distal and collecting tubules is increased. This leads to a decreased urine output and increased water conservation. At the same time, the thirst centre is stimulated, and water intake is encouraged. These actions restore depleted water stores and relieve hypertonicity by diluting the solutes to their normal concentration. In contrast, water excess manifests as a reduced ECF osmolarity, and this prompts increased urine output (through decreased vasopressin release) and suppression of thirst. Together, these factors reduce free water in the body and restore normal ECF osmolarity.

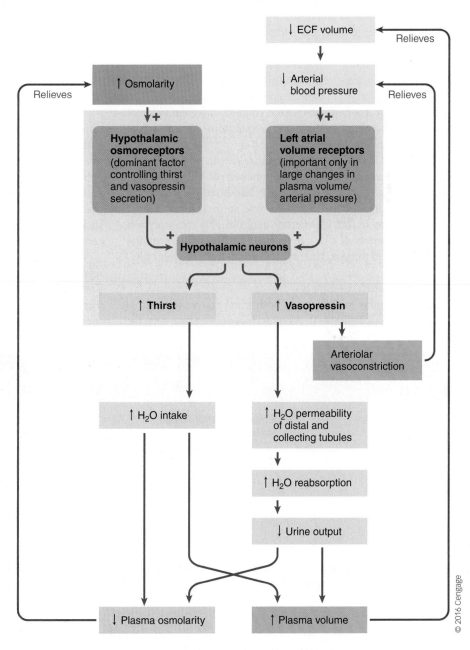

> FIGURE 14-4 **Control of increased vasopressin secretion and thirst during a water deficit**

## ROLE OF LEFT ATRIAL VOLUME RECEPTORS

Although the major stimulus for vasopressin secretion and thirst is an increase in ECF osmolarity, the vasopressin-secreting cells and thirst centre are both influenced to a moderate extent by an overall change to ECF volume. This is mediated by input from the **left atrial volume receptors**. Located in the left atrium, these volume receptors monitor the pressure of blood flowing through, which reflects the ECF volume. In response to a major reduction in ECF volume (>7% loss of volume) and therefore a concordant change to arterial pressure, the left atrial volume receptors reflexively stimulate both vasopressin secretion and thirst. This degree of volume loss is often seen in the case of severe haemorrhage. The outpouring of vasopressin and the increased thirst lead to decreased urine output and increased fluid intake, respectively. Furthermore, the concentration of vasopressin at the circulating levels, elicited by a large decline in ECF volume and arterial pressure, exerts a potent vasoconstrictor effect on arterioles (giving rise to its name; p. 402). By helping increase both the plasma volume and the total peripheral resistance, vasopressin helps relieve the low blood pressure that elicited vasopressin secretion. Conversely, vasopressin and thirst are both inhibited when ECF/plasma volume and arterial blood pressure are elevated. The resultant suppression of water intake, coupled with elimination of excess plasma volume in the urine, helps restore a normal blood pressure and plasma volume.

Recall that low ECF/plasma volume and low arterial blood pressure also reflexively increase aldosterone secretion. The resulting increase in $Na^+$ reabsorption ultimately leads to osmotic retention of water, expansion of ECF volume, and an increase in arterial blood pressure. In fact, aldosterone-controlled $Na^+$ reabsorption is the most important factor in regulating ECF volume, with the vasopressin and thirst mechanism playing a supportive role only.

## ROLE OF ANGIOTENSIN II

Another stimulus for increasing both thirst and vasopressin is angiotensin II (▌ Table 14-4). As all hormones need receptors, the receptor for angiotensin is a G protein, which has two structures often termed A1 and A2. Angiotensin II is the primary ligand, or binding molecule, of both A1 and A2. When the renin–angiotensin–aldosterone mechanism is activated to conserve $Na^+$, angiotensin II acts directly on the brain to give rise to the urge to drink. A molecule that increases the urge to drink is known as a **dipsogen**. Concurrently, angiotensin II stimulates vasopressin secretion, and this reduces urine output by increasing water reabsorption in the kidney (p. 576). The resultant increased water intake and decreased urine output help correct the reduction in ECF volume that triggered the renin–angiotensin–aldosterone system.

## REGULATORY FACTORS THAT DO NOT LINK VASOPRESSIN AND THIRST

Several factors increase release of vasopressin secretion, but not thirst. Vasopressin can be stimulated by stress-related inputs—such as pain, fear, and trauma. These do not directly relate to water balance. In fact, water retention from the inappropriate secretion of vasopressin can bring about hypotonicity in the ECF fluid, and an overall water surplus. In contrast, alcohol acts to inhibit vasopressin secretion and can lead to ECF hypertonicity by promoting excessive free water excretion.

One stimulus that promotes thirst but not vasopressin secretion is the direct effect of dryness of the mouth. Nerve endings in the mouth are directly stimulated by dryness, which causes an intense sensation of thirst. This can often be relieved by moistening the mouth, even if no water is actually ingested. A dry mouth, known as **xerostomia**, can exist when salivation is suppressed by factors unrelated to the water content of the body, such as nervousness, excessive smoking, medications, or even certain medical conditions (e.g., Sjögren's syndrome).

Factors that affect vasopressin secretion or thirst and are not related to the water balance in the body are usually short-lived. The dominant, long-standing control of vasopressin and thirst is directly correlated with the state of water in the body—namely, the status of ECF osmolarity and ECF volume.

---

**▌ TABLE 14-4** Factors Controlling Vasopressin Secretion and Thirst

| Factor | Effect on Vasopressin Secretion | Effect on Thirst | Comment |
|---|---|---|---|
| ↑ **ECF Osmolarity** | ↑ | ↑ | Major stimulus for vasopressin secretion and thirst |
| ↓ **ECF Volume** | ↑ | ↑ | Important only in large changes in ECF volume/arterial blood pressure |
| **Angiotensin II** | ↑ | ↑ | Part of dominant pathway for promoting compensatory salt and $H_2O$ retention when ECF volume/arterial blood pressure are reduced |
| **Pain, Fear, Trauma, and Other Stress-Related Inputs** | Inappropriate ↑ unrelated to body's $H_2O$ balance | No effect | Promotes excess $H_2O$ retention and ECF hypotonicity (resultant $H_2O$ retention of potential value in maintaining arterial blood pressure in case of blood loss in the stressful situation) |
| **Alcohol** | Inappropriate ↓ unrelated to body's $H_2O$ balance | No effect | Promotes excess $H_2O$ loss and ECF hypertonicity |

Diane developed diabetic ketoacidosis (DKA) as a result of undiagnosed Type I diabetes mellitus (DM). This disease usually affects individuals under the age of 30 and is the result of autoimmune destruction of pancreatic insulin-producing beta islet cells. Without insulin, glucose cannot be transported into the cells of the body from the blood. Therefore, it accumulates in the bloodstream and creates severe hyperglycaemia.

Hyperglycaemia causes an increased osmolarity of the blood, and this, in turn, activates central osmoreceptors and the thirst centre. Patients, therefore, often experience extreme thirst and excessive oral fluid intake, known as *polydipsia*.

Under normal circumstances, glucose is reabsorbed into the bloodstream in the kidney via protein-mediated transport. In untreated Type I DM, the level of hyperglycaemia often surpasses the ability of the kidney to reabsorb all the glucose. The glucose that remains in the renal tubules increases the osmolarity of the filtrate. Consequently, less water is reabsorbed from the nephron, and more frequent, larger volumes of urine are produced. This is known as *polyuria*.

The polyuria eventually reduces the volume of intravascular fluid and causes dehydration. To maintain a constant cardiac output and blood pressure, the heart rate increases (tachycardia). Without treatment, the level of dehydration can eventually become so profound that a normal cardiac output cannot be maintained. At this point, a decrease in blood pressure occurs (hypotension). One very important initial step in the management of new onset Type I DM and DKA is rehydration in the form of intravenous fluid replacement.

### ORAL WATER METERING

"Oral water metering" appears to exist in most animals. For example, a thirsty animal will rapidly drink only enough water to satisfy its water deficit. It will generally stop drinking before the ingested $H_2O$ has even had time to be absorbed from the digestive tract and return the ECF compartment to normal. Exactly what factors are involved in signalling that enough water has been consumed is still uncertain. This mechanism seems to be less effective in humans because we frequently drink more than is necessary to meet the needs of our body or may not drink enough to make up a deficit.

### NONPHYSIOLOGICAL INFLUENCES ON FLUID INTAKE

Even though the thirst mechanism exists to control water intake, fluid consumption by humans is often complex and also influenced by habit and sociological factors. Thus, even though water intake is critical in maintaining fluid balance, it is not precisely controlled in humans. We usually drink when we are thirsty, but we often drink even when we are not thirsty because, for example, we are on a coffee break.

With water intake being inadequately controlled and sometimes even contributing to water imbalances in the body, the primary factor involved in maintaining water balance is urinary output regulated by the kidneys. Accordingly, *vasopressin-controlled $H_2O$ reabsorption is of primary importance in regulating ECF osmolarity.*

Before we shift attention to acid–base balance, examine ▌Table 14-5, which summarizes the regulation of ECF volume and osmolarity, the two factors important in maintaining fluid balance.

### Check Your Understanding 14.2

1. Make a chart showing the percentage of body $H_2O$ distributed among the major body fluid compartments.

2. Compare how ECF volume and ECF osmolarity are regulated, and discuss why it is important that each is regulated.

3. Compare the effect of ECF hypertonicity and hypotonicity on cell volume.

---

### ▌ TABLE 14-5 Summary of the Regulation of ECF Volume and Osmolarity

| Regulated Variable | Need to Regulate the Variable | Outcomes If the Variable Is Not Normal | Mechanism for Regulating the Variable |
|---|---|---|---|
| **ECF Volume** | Important in the long-term control of arterial blood pressure | ↓ ECF volume → <br> ↓ arterial blood pressure <br><br> ↑ ECF volume → <br> ↑ arterial blood pressure | Maintenance of salt balance; salt osmotically "holds" $H_2O$, so the $Na^+$ load determines the ECF volume <br><br> Accomplished primarily by aldosterone-controlled adjustments in urinary $Na^+$ excretion |
| **ECF Osmolarity** | Important to prevent detrimental osmotic movement of $H_2O$ between the ECF and ICF | ↑ ECF osmolarity (hypotonicity) → <br> $H_2O$ enters the cells → cells swell <br><br> ↓ ECF osmolarity (hypertonicity) → <br> $H_2O$ leaves the cells → cells shrink | Maintenance of free $H_2O$ balance <br><br> Accomplished primarily by vasopressin-controlled adjustments in excretion of $H_2O$ in the urine |

14

## 14.3 | Acid–Base Balance

The term **acid–base balance** refers to the precise regulation of **free hydrogen ion ($H^+$) concentration** in the body fluids. To indicate the concentration of a chemical, its symbol is enclosed in square brackets: [ ]. Thus, [$H^+$] designates $H^+$ concentration.

### Acids liberate $H^+$; bases accept them

**Acids** are a special group of hydrogen-containing substances that *dissociate,* or separate, when in solution, liberating free $H^+$ and anions. Many other substances (e.g., carbohydrates) also contain hydrogen, but they are not classified as acids, because the hydrogen is tightly bound within their molecular structure and is never liberated as free $H^+$.

A strong acid has a greater tendency to dissociate in solution than a weak acid. This means that a greater percentage of strong acid molecules separate into free $H^+$ and anions. Hydrochloric acid (HCl) is an example of a strong acid. Every HCl molecule dissociates into free $H^+$ and $Cl^-$ (chloride) when dissolved in $H_2O$. Carbonic acid ($H_2CO_3$) is an example of a weak acid, and only a portion of these molecules dissociate in solution into $H^+$ and $HCO_3^-$ (bicarbonate anions). The rest of the $H_2CO_3$ molecules remain intact. *Only the free $H^+$ contribute to the acidity of a solution.* This makes $H_2CO_3$ a weaker acid than HCl, since only a portion of $H_2CO_3$ molecules dissociate in solution to make free $H^+$ (> Figure 14-5).

The extent of dissociation for a given acid is always constant. When in solution, the same proportion of the molecules of the acid always separate to liberate free $H^+$, and the other portion always remains intact. The constant degree of dissociation for a particular acid (in this example, $H_2CO_3$) is expressed by the **dissociation constant, K**. Consider the following chemical equation:

$$H_2CO_3 \rightarrow H^+ + HCO_3^-$$

The dissociation constant is calculated as

$$[H^+][HCO_3^-]/[H_2CO_3] = K$$

where $[H^+][HCO_3^-]$ represents the concentration of ions resulting from [$H_2CO_3$] dissociation, and [$H_2CO_3$] represents the concentration of intact (undissociated) molecules. This dissociation constant varies for different acids.

A **base** is a substance that can combine with a free $H^+$ and thus remove it from solution. The stronger the base, the more readily it can bind to free $H^+$ in solution.

### The pH designation

The [$H^+$] in the ECF is normally $4 \times 10^{-8}$ or 0.00000004 equivalents per litre. The concept of pH was developed to express [$H^+$] more conveniently. Specifically, **pH** equals the logarithm (log) to the base 10 of the reciprocal of the hydrogen ion concentration:

$$pH = \log 1/[H^+] \quad (or\ pH = -\log[H^+])$$

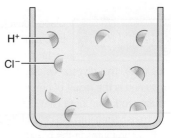

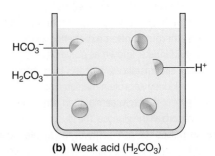

**(a)** Strong acid (HCl)

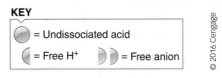

**(b)** Weak acid ($H_2CO_3$)

**KEY**

- = Undissociated acid
- = Free $H^+$
- = Free anion

© 2016 Cengage

> **FIGURE 14-5 Comparison of a strong and a weak acid.** (a) Five molecules of a strong acid. A strong acid, such as HCl (hydrochloric acid), completely dissociates into free $H^+$ and anions in solution. (b) Five molecules of a weak acid. A weak acid, such as $H_2CO_3$ (carbonic acid), only partially dissociates into free $H^+$ and anions in solution.

Two important points should be noted about this formula:

1. Since [$H^+$] is in the denominator, a high [$H^+$] corresponds to a low pH, and a low [$H^+$] corresponds to a high pH.

2. Since pH represents a base-10 logarithmic relationship, every unit change in pH actually represents a tenfold change in free [$H^+$]. A log to the base 10 indicates how many times 10 must be multiplied by itself to produce a given number. For example, the log of 10 = 1, whereas the log of 100 = 2. The number 10 must be multiplied by itself twice to yield 100 ($10 \times 10 = 100$). Numbers less than 10 have a log of less than 1. Numbers between 10 and 100 have logs between 1 and 2, and so on. Accordingly, each unit of change in pH indicates a tenfold change in free [$H^+$]. For example, a solution with a pH of 7 has a [$H^+$] that is $10 (10^1)$ times less than that of a solution with a pH of 6 (a 1 pH-unit difference) and $100 (10^2)$ times less than that of a solution with a pH of 5 (a 2 pH-unit difference). For further discussion on pH, see Concepts, Challenges, and Controversies.

### ACIDIC AND BASIC SOLUTIONS IN CHEMISTRY

The pH of pure water is 7.0, which is considered chemically neutral. An extremely small proportion of $H_2O$ molecules dissociate into $H^+$ and $OH^=$ (hydroxide ions). Although $OH^-$ is

# Acid–Base Balance: The Physico-chemical Approach

**A**CID–BASE BALANCE IS ONE OF THE MORE—if not most—difficult topics in physiology. Not only is it seemingly arbitrary and empirical, it is also highly quantitative. Moreover, it relies on pH, a system in which the $H^+$ concentration $[H^+]$ and its changes are not intuitively comprehensible. For example, if the control pH is 7.4 and the $[H^+]$ doubles, what is the new pH? If you are still trying to calculate this, try this version: if the control $[H^+]$ is 40 nEq/L and the $[H^+]$ doubles, what is the new $[H^+]$? The relative ease of answering the latter question illustrates this point.

In the early 1980s, Peter Stewart (1921–1993), a Winnipeg-born physiologist working at Brown University in Rhode Island, changed our approach to acid–base physiology, using the principles of physico-chemistry. One of Stewart's most important contributions is that he distinguishes between independent and dependent variables. In traditional acid–base physiology, if one adds hydrochloric acid to water, the pH decreases (becomes more acidic) because the acid immediately dissociates into $H^+$ and $Cl^-$, and the increased $[H^+]$ accounts for the increased acidity (drop in pH). But because $[H^+]$ is the dependent variable, how can the addition of the dependent variable (the $H^+$ in HCl) change itself? In physico-chemistry, it is the addition of $Cl^-$ that changes the acidity of the solution.

Stewart uses physico-chemistry to show that three independent variables determine the acidity of a physiological solution: the $P_{CO_2}$, the net charge of strong ions (i.e., ones that easily dissociate: $Na^+$, $K^+$, $Cl^-$, and lactate), and total weak acid (typically proteins, such as albumin). Note that $H^+$ is not listed because it is, as mentioned, the dependent variable.

Stewart's approach is unassailably correct, even though some traditional physiologists protested. Nevertheless, the pH approach is so thoroughly entrenched that it remains the one described by all current texts, including this edition.

Those interested in Stewart's approach can consult the Internet; start with his Wikipedia entry, and then www.acidbase.org, where you click on "read the book." It is a challenging read, but Stewart's work is of enormous perception and importance.

---

considered basic since it has the ability to bind with $H^+$, an equal number of $H^+$ are produced from dissociation of water. This makes water neutral: neither acidic nor basic. Solutions having a pH less than 7.0 contain a higher $[H^+]$ than pure water and are considered **acidic**. Conversely, solutions with a pH value of greater than 7.0 have a lower $[H^+]$ and are considered **basic**, or **alkaline** (› Figure 14-6a). › Figure 14-7 compares the pH values of common solutions.

### ACIDOSIS AND ALKALOSIS IN THE BODY

The pH of arterial blood is normally 7.45, and the pH of venous blood is 7.35, for an average blood pH of 7.4. The pH of skeletal muscle is about 7.15, making it slightly more acidic than either arterial or venous blood. The pH value of skeletal muscle decreases with greater intensity of exercise and is attributed to an increase in lactate generated during increased muscular effort. Venous blood is normally slightly more acidic than arterial blood because it contains more $CO_2$. This is because $CO_2$ reacts with $H_2O$ to generate more $H_2CO_3$, carbonic acid—a weak acid. As a result, there is slightly more free $H^+$ in solution and, therefore, an increase in acidity (decreased pH).

Many disease processes cause deviations from the normal plasma pH range of 7.35–7.45. The term **acidaemia** is used to describe a plasma pH that is below 7.35. The suffix *-aemia* means blood, so this term literally means "acidic blood." The general term **acidosis** refers to an overall acidic state in the body. Likewise, the term **alkalaemia** is used to describe a plasma pH that is above 7.45. Similar to acidaemia, alkalaemia literally means "alkaline blood." The general term **alkalosis** refers to an overall alkalotic or basic state in the body. In most instances, a state of acidosis results in an acadaemia, and a state of alkalosis results in an alkalaemia. For this reason, in the rest of the chapter we use the terms acidosis and alkalosis when referring to abnormalities in plasma pH below 7.35 or above 7.45, respectively. (See › Figure 14-6b.) Note that the reference point for determining the body's acid–base status is not the chemically neutral pH of 7.0; rather, it is the normal plasma pH of 7.4. Thus, a plasma pH of 7.2, for example, is considered a state of acidosis in the body, even though in chemistry a pH of 7.2 is considered basic or alkaline.

Even very minor deviations (few hundredths of a point) from the normal pH range can result in serious illness. An arterial pH of less than 6.8 or greater than 8.0 signifies an emergency and must be corrected immediately, as it is not compatible with life for any extended period of time. Since death will occur with an arterial pH outside the range of 6.8–8.0, the $[H^+]$ in the body fluids must be carefully regulated.

### FLUCTUATIONS IN $[H^+]$

Even small changes in $[H^+]$ have dramatic effects on normal cell function. The main consequences of fluctuations in $[H^+]$ outside of the narrow normal range (7.35–7.45) include the following:

1. Changes in excitability of nerve and muscle cells

   - States of acidosis in the body lead to an overall depression of the central nervous system. Individuals in a state of acidosis often become disoriented and, in severe cases, eventually die in a comatose state.

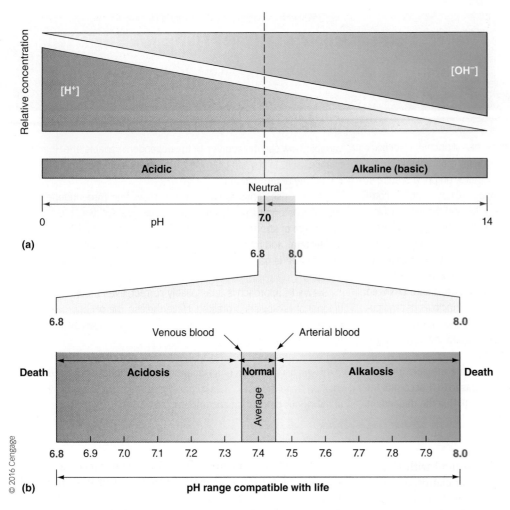

> **FIGURE 14-6 pH considerations in chemistry and physiology.** (a) Relationship of pH to the relative concentrations of H⁺ and base (OH⁻) under chemically neutral, acidic, and alkaline conditions. (b) Plasma pH range under normal, acidosis, and alkalosis conditions

3. *Influence on K⁺ levels* in the body. When reabsorbing Na⁺ from the filtrate, the renal tubular cells secrete either K⁺ or H⁺ in exchange (p. 583). Normally, they secrete a preponderance of K⁺ compared with H⁺. As a result of this intimate relationship between secretion of H⁺ and K⁺ by the kidneys, an increased rate of secretion of one of these ions is accompanied by a decreased rate of secretion of the other. For example, if more H⁺ than normal is eliminated by the kidneys (as occurs in states of acidosis), less K⁺ than usual can be excreted. The resulting K⁺ retention can affect cardiac function, among other detrimental consequences.

## Hydrogen ions

As with any other constituent, the input of hydrogen ions must be balanced by an equal output to maintain a constant [H⁺] in the body fluids. Only a small percentage of the acid in our body comes from the foods we eat or fluids we drink. Most ingested acids are weak and therefore only release a small amount of H⁺ as a result of dissociation. An example of this is the weak citric acid found in oranges. Most H⁺ in the body fluids is generated internally from metabolic activities.

### SOURCES OF H⁺ IN THE BODY

Normally, H⁺ is continually added to the body fluids from the three following sources:

1. *Carbonic acid formation.* The major source of H⁺ is through $H_2CO_3$ formation from metabolically produced $CO_2$. Cellular oxidation of nutrients yields energy, but also $CO_2$ and $H_2O$ as end products. $CO_2$ and $H_2O$, in the presence of *carbonic anhydrase (ca)*, forms $H_2CO_3$. This partially dissociates to liberate free H⁺ and $HCO_3^-$.

$$CO_2 + H_2O \xrightarrow{ca} H_2CO_3 \rightarrow H^+ + HCO_3^-$$

This reaction is reversible because it can proceed in either direction. The direction depends on the concentrations of the substrates involved, as dictated by the *law of mass action* (p. 532). Within the systemic capillaries, the $CO_2$ level in the blood increases as metabolically produced $CO_2$ enters

- States of alkalosis lead to an overall hyperexcitability of the nervous system. The sensory component of the peripheral nervous system is often affected first. Patients often complain of **paraesthesias** (a sensation of pins-and-needles) in the extremities. This is thought to be the result of the nerves becoming so excitable that they fire even in the absence of normal stimuli. Motor nerves can also be affected and begin to twitch and, in severe cases, will spasm. Death can occur in extreme states of alkalosis because muscle spasms in the muscles of respiration lead to a serious impairment to breathing. Also, patients in severe states of alkalosis may die from uncontrollable convulsions as a result of hyperexcitability of the central nervous system (CNS).

2. A *marked influence on enzyme activity.* Most enzymes have very narrow pH range where enzyme activity is optimal. Changes in pH on either side of this narrow range produce low or high reaction rates. This occurs because some of the forces holding the enzyme (protein) in its normal conformation depend on charged groups, such as H⁺.

**14**

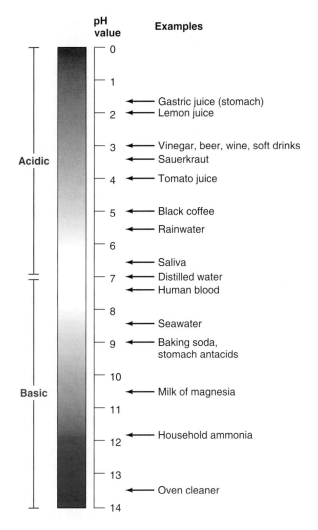

| pH value | Examples |
|---|---|
| 0 | |
| 1 | |
| 2 | Gastric juice (stomach) / Lemon juice |
| 3 | Vinegar, beer, wine, soft drinks / Sauerkraut |
| 4 | Tomato juice |
| 5 | Black coffee |
| | Rainwater |
| 6 | |
| 7 | Saliva / Distilled water / Human blood |
| 8 | |
| | Seawater |
| 9 | Baking soda, stomach antacids |
| 10 | |
| | Milk of magnesia |
| 11 | |
| 12 | Household ammonia |
| 13 | |
| | Oven cleaner |
| 14 | |

Acidic (pH 0–7), Basic (pH 7–14)

> FIGURE 14-7 Comparison of pH values of common solutions

from the tissues. Initially, this drives the reaction to the acid side, generating $H^+$ and $HCO_3^-$. However, in the lungs, the $CO_2$ diffuses from the blood in pulmonary capillaries into the alveoli (air sacs). The resultant reduction in blood $CO_2$ reverses the initial change and drives the reaction back toward the $CO_2$ side. Overall, there is no change in overall blood $[H^+]$ and therefore no change in pH. When the rate of $CO_2$ removal by the lungs does not match the rate of $CO_2$ production at the tissue level, however, the resulting change in $CO_2$ leads to an overall change in the free $H^+$ in the body fluids.

2. *Inorganic acids produced during breakdown of nutrients.* Dietary proteins found abundantly in meat protein contain a large quantity of sulphur and phosphorus. When these nutrient molecules are broken down, sulphuric acid and phosphoric acid are produced as by-products. Being moderately strong acids, these two inorganic acids largely dissociate, liberating free $H^+$ into the body fluids. In contrast, breakdown of fruits and vegetables produce various bases that neutralize the acids derived from protein metabolism. Generally, more acids than bases are produced during breakdown of ingested food, leading to an excess of these acids.

3. *Organic acids resulting from intermediary metabolism.* Numerous organic acids are produced during normal intermediary metabolism. For example, fatty acids are produced during fat metabolism, and lactic acid is produced by muscles during heavy exercise. These acids partially dissociate to yield free $H^+$.

Hydrogen ion generation therefore normally goes on continuously as a result of metabolism. In certain disease states, additional acids may be produced that further contribute to the total body pool of $H^+$. In untreated insulin-dependent diabetes mellitus, large quantities of ketoacids are produced and released into the plasma from abnormal fat metabolism. Also, certain acid-producing medications can add to the total $[H^+]$ in the body.

The input of $H^+$ is unceasing, highly variable, and essentially unregulated. For this reason, the body has several mechanisms to help maintain a normal plasma pH.

### THREE LINES OF DEFENCE AGAINST CHANGES IN [H⁺]

The generated free $H^+$ must be largely removed from solution in the body and ultimately must be eliminated so that the pH of body fluids can remain within the narrow normal range of 7.35–7.45. Mechanisms must also exist to compensate rapidly for the occasional situation in which the ECF becomes too alkaline.

Three lines of defence against changes in $[H^+]$ operate to maintain $[H^+]$ of body fluids at a nearly constant level despite unregulated input. These include (1) the *chemical buffer systems,* (2) the *respiratory mechanism of pH control,* and (3) the *renal mechanism of pH control.*

### Chemical buffer systems

A **chemical buffer system** is a mixture in a solution of two chemical compounds that minimize pH changes when either an acid or a base is added to or removed from the solution. A buffer system consists of a pair of substances involved in a reversible reaction. One of these substances must yield free $H^+$ if the $[H^+]$ falls, and the other must bind with free $H^+$ if the $[H^+]$ rises.

An important example of such a buffer system is the carbonic acid:bicarbonate ($H_2CO_3 : HCO_3^-$) buffer pair involved in the following reversible reaction:

$$H_2CO_3 \rightarrow H^+ + HCO_3^-$$

When a strong acid, such as HCl, is added to an unbuffered solution, all the dissociated $H^+$ remains free in the solution (> Figure 14-8a). In contrast, when HCl is added to a solution containing the $H_2CO_3 : HCO_3^-$ buffer pair, the $HCO_3^-$ immediately binds with the free $H^+$ to form $H_2CO_3$ (> Figure 14-8b). This weak $H_2CO_3$ dissociates only slightly compared with the complete dissociation of HCl that occurred when the buffer system was not present. In the opposite case, when the pH of the solution starts to rise due to the addition of base or loss of acid, $H_2CO_3$ releases $H^+$ to minimize the rise in pH.

**14**

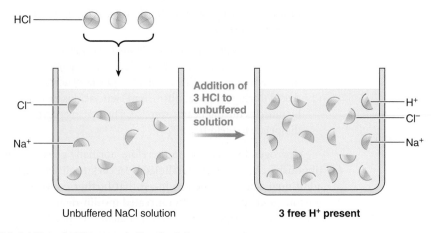

**(a)** Addition of HCl to an unbuffered solution

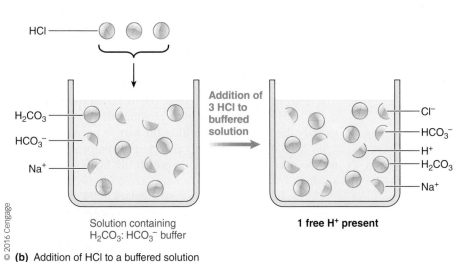

© 2016 Cengage

**(b)** Addition of HCl to a buffered solution

> **FIGURE 14-8 Action of chemical buffers.** (a) All the added hydrogen ions ($H^+$) remain free and contribute to the acidity of the solution. (b) Bicarbonate ions ($HCO_3^-$), the basic member of the buffer pair, bind with some of the added $H^+$ and remove them from solution so that they do not contribute to its acidity.

The body has four buffer systems and each serves a different important role in the body: (1) the ($H_2CO_3$:$HCO_3^-$) buffer system, (2) the protein buffer system, (3) the haemoglobin buffer system, and (4) the phosphate buffer system. (Table 14-6).

### The $H_2CO_3$:$HCO_3^-$ buffer pair

The ($H_2CO_3$:$HCO_3^-$) buffer pair is the most important buffer system in the ECF for buffering pH changes brought about by causes other than fluctuations in $CO_2$-generated $H_2CO_3$. This buffer pair is a very effective ECF buffer system for two reasons. First, $H_2CO_3$ and $HCO_3^-$ are abundant in the ECF and are therefore readily available to resist changes in pH. Second, each component of this buffer system is closely regulated. The respiratory system closely regulates $CO_2$ (ultimately generating $H_2CO_3$), and the kidneys closely regulate [$HCO_3^-$]. Thus, the ($H_2CO_3$:$HCO_3^-$) buffer system involves $CO_2$, via the following reaction:

$$CO_2 + H_2O \rightarrow H_2CO_3 \rightarrow H^+ + HCO_3^-$$

When new $H^+$ is added to the plasma from any source other than $CO_2$ (e.g., lactic acid release into the ECF from exercising muscles), the preceding reaction is driven toward the left side of the equation. The extra $H^+$ binds with $HCO_3^-$ and therefore no longer contributes to the acidity of body fluids. In contrast, a decrease in [$H^+$] for a reason other than a reduction in $CO_2$ (e.g., loss of plasma-derived HCl in the gastric juices during vomiting), the reaction is driven toward the right side of the equation. Dissolved $CO_2$ and $H_2O$ in the plasma form $H_2CO_3$, which generates additional $H^+$ to make up for the $H^+$ deficit. In so doing, the $H_2CO_2$:$HCO_3^-$ buffer system resists the fall in [$H^+$].

It is very important to note that this system cannot buffer changes in pH induced by fluctuations in $HCO_3^-$, $CO_2$, or $H_2CO_3$. A buffer system cannot buffer itself. Consider, for example, the situation in which the plasma [$H^+$] is elevated by $CO_2$ retention due to a breathing problem. According to the law of mass action, the rise in $CO_2$ drives the reaction to the right, thereby elevating [$H^+$]. The increase in [$H^+$] occurs because the reaction is driven to the *right* by an increase in $CO_2$. The elevated [$H^+$] cannot drive the reaction to the *left* to buffer the increase in [$H^+$] because the elevated [$H^+$] is due to a rightward shift of the reaction. The $H_2CO_3$:$HCO_3^-$ buffer is only effective against increases [$H^+$] brought about by mechanisms other than an increase $CO_2$ content. In the opposite situation, the $H_2CO_3$:$HCO_3^-$ buffer system cannot compensate for a reduction in [$H^+$] from

■ **TABLE 14-6** Chemical Buffers and Their Primary Roles

| Buffer System | Major Functions |
|---|---|
| **Carbonic acid: Bicarbonate buffer system** | Primary ECF buffer against noncarbonic acid changes |
| **Protein buffer system** | Primary ICF buffer; also buffers ECF |
| **Haemoglobin buffer system** | Primary buffer against carbonic acid changes |
| **Phosphate buffer system** | Important urinary buffer; also buffers ICF |

© 2016 Cengage

a deficit of $CO_2$, for the similar reason outlined above. Changes in pH caused by fluctuations in $CO_2$ levels can be regulated, but a different mechanism is required.

## THE HENDERSON–HASSELBALCH EQUATION AND NORMAL PLASMA PH OF 7.4

The relationship between $[H^+]$ and the members of a buffer pair can be expressed according to the **Henderson–Hasselbalch equation**. For the $H_2CO_3:HCO_3^-$ buffer system, the equation is as follows:

$$pH = pK + \log[HCO_3^-]/[H_2CO_3]$$

Although you do not need to know the mathematical manipulations involved, it is helpful to understand how this formula is derived. Recall that the dissociation constant K for $H_2CO_3$ is

$$[H^+][HCO_3^-]/[H_2CO_3] = K$$

and the relationship between pH and $[H^+]$ is

$$pH = \log 1/[H^+] \text{ or } pH = -\log[H^+]$$

Like pH, pK is the negative logarithm of K. Notice that in both cases, the negative log function is replaced by the letter *p*. Solving the dissociation constant equation for $[H^+]$ gives the following:

$$[H^+] = K \times [H_2CO_3]/[HCO_3^-]$$

By taking the negative log of each side, the formula becomes

$$-\log[H^+] = -\log\{(K) \times [H_2CO_3]/[HCO_3^-]\}$$

Using the mathematical principles of logarithms, this equation can be simplified to

$$-\log[H^+] = -\log(K) + \log([HCO_3^-]/[H_2CO_3])$$

The final manipulation of the above expression required in order to obtain the Henderson-Hasselbalch equation is to replace the negative log functions with the letter *p*:

$$pH = pK + \log[HCO_3^-]/[H_2CO_3]$$

Practically speaking, $[H_2CO_3]$ directly reflects the concentration of dissolved $CO_2$ because $CO_2$ in the plasma is converted into $H_2CO_3$. Replacing $[H_2CO_3]$ with $[CO_2]$, the equation becomes

$$pH = pK + \log[HCO_3^-]/[CO_2]$$

The pK value always remains a constant for any given acid. For example, the pK of $H_2CO_3$ is 6.1. Since pK is always a constant, changes in pH are associated with changes in the ratio between $[HCO_3^-]$ and $[CO_2]$.

- Under normal circumstances, the ratio between $[HCO_3^-]$ and $[CO_2]$ in the ECF is 20:1. If we substitute the ratio of 20:1 and solve for pH, we obtain the following:

$$\begin{aligned} pH &= pK + \log[HCO_3^-]/[CO_2] \\ &= 6.1 + \log 20/1 \end{aligned}$$

- $\log(20) = 1.3$. Therefore, $pH = 6.1 + 1.3 = 7.4$, which is the normal pH of plasma.

- When the ratio of $[HCO_3^-]:[CO_2]$ increases to values above 20:1, pH rises. This occurs with either a rise in $[HCO_3^-]$ or a fall in $[CO_2]$.

- In contrast, when the ratio of $[HCO_3^-]:[CO_2]$ decreases to values below 20:1, pH falls. This occurs with either a decrease in $[HCO_3^-]$ or increase in $[CO_2]$.

The $[HCO_3^-]$ is regulated by the kidneys and $[CO_2]$ is regulated by the lungs. This means that the pH of the plasma can be shifted up and down by altering $[HCO_3^-]$ at the level of the kidneys and/or the $[CO_2]$ at the level of the lungs. The kidneys and lungs regulate pH (and thus free $[H^+]$) largely by controlling plasma $[HCO_3^-]$ and $[CO_2]$, respectively, to restore their ratio to normal. Accordingly, the pH of the plasma is related to the plasma $[HCO_3^-]$ and $[CO_2]$ based on the following relationship:

$$pH \propto \frac{[HCO_3^-] \text{controlled by kidney function}}{[CO_2] \text{controlled by respiratory function}}$$

Although the kidneys and the lungs are crucial to maintaining normal plasma pH, there are disease states where the kidney or respiratory tract are unable to properly regulate $[HCO_3^-]$ or $[CO_2]$. This leads to conditions that are collectively referred to as **acid-base disorders**. They occur as a result of alterations in the ratio of $[HCO_3^-]:[CO_2]$. We explore these disorders in relation to respiratory and renal control of pH and acid–base disturbances in the section on Acid-Base Imbalances.

## The protein buffer system

The most plentiful buffers of the body fluids are the proteins. These include both intracellular proteins and the plasma proteins. Proteins are excellent buffers because they contain both acidic and basic groups that can give up or take up $H^+$. Quantitatively, due to the sheer abundance of intracellular proteins, the protein system is most important in buffering changes in $[H^+]$ in the ICF.

## The haemoglobin buffer system

Haemoglobin buffers the $H^+$ generated from the metabolically produced $CO_2$ that travels between the tissues and lungs. At the level of the systemic capillaries, $CO_2$ continuously diffuses into the blood from the tissue cells where it is produced. The greatest percentage of this $CO_2$ forms $H_2CO_3$, which partially dissociates into $H^+$ and $HCO_3^-$. As discussed, free $H^+$ produced from $H_2CO_3$ cannot be buffered by $HCO_3^-$. Instead, the free $[H^+]$ becomes bound to Hb and no longer contributes to the acidity of body fluids. Were it not for Hb, blood would become much too acidic after picking up $CO_2$ at the tissues. With the tremendous buffering capacity of the Hb system, venous blood is only slightly more acidic than arterial blood, despite the large volume of free $H^+$ produced from $CO_2$ carried in venous blood. At the lungs, the reactions are reversed, and the resulting $CO_2$ is exhaled.

## The phosphate buffer system

Since phosphates are most abundant intracellularly, this system contributes significantly to intracellular buffering—rivalled only by the more plentiful intracellular proteins.

14

The phosphate buffer system consists of an acidic phosphate salt ($NaH_2PO_4$) that can donate a free $H^+$ when the $[H^+]$ falls and a basic phosphate salt ($Na_2HPO_4$) that can accept a free $H^+$ when the $[H^+]$ rises. The buffer pair $NaH_2PO_4$:$Na_2HPO_4$ can alternately switch a $H^+$ for a $Na^+$ as demanded by the $[H^+]$:

$$Na_2HPO_4 + H^+ \rightarrow NaH_2PO_4 + Na^+$$

Despite the phosphate salt pair being a good buffer, its concentration in the ECF is rather low, so it is not very important as an ECF buffer.

Even more importantly, the phosphate system serves as an excellent urinary buffer. Humans normally consume more phosphate than needed. Any excess phosphate filtered through the kidneys is not reabsorbed because the renal threshold for phosphate has been exceeded (p. 580). Therefore, the excess phosphate remains in the tubular fluid and can buffer any $H^+$ secreted into the tubular fluid until it is eventually excreted in the urine. This ensures that the free $[H^+]$ in the tubular fluid remains low, and the pH of the urine does not change significantly. The excess $H^+$ ingested and created in the body can therefore continue to be secreted and excreted in the amounts required to maintain a normal plasma pH. Unlike the other buffers discussed, this is the only one present in the tubular fluid that buffers urine during its formation.

## Chemical buffer systems: The first line of defence

All chemical buffer systems act immediately (within fractions of a second) to minimize changes in pH. Accordingly, the buffer systems are considered the *first line of defence* against changes in $[H^+]$ in the body because they are the first mechanisms to respond.

Through the mechanism of buffering, most excess $H^+$ seem to disappear from the body fluids. It must be emphasized, however, that none of the chemical buffer systems actually eliminate $H^+$ from the body. Buffered $H^+$ is merely removed from solution by being incorporated within one member of the buffer pair, preventing $H^+$ from contributing to the acidity of the body's fluid. Although effective, each buffer system has a limited capacity to soak up $H^+$. Since $H^+$ is unceasingly produced in the body, it must ultimately be removed. If $H^+$ were not eventually eliminated, all of the buffers would become bound with $H^+$, effectively stopping all further buffering ability. The body therefore requires other lines of defence against changes in $[H^+]$, specifically, lines of defence with an ability to eliminate excess $H^+$.

The respiratory and renal systems can very effectively eliminate acid from the body, instead of merely suppressing it, but the trade-off is their slower response compared to chemical buffer systems.

## The respiratory system and $[H^+]$

The respiratory system plays an important role in acid–base balance through its ability to alter pulmonary ventilation and consequently to alter excretion of $CO_2$. The level of respiratory activity is governed in part by arterial $[H^+]$ through the following mechanism ( Table 14-7):

- When arterial $[H^+]$ increases as the result of a *nonrespiratory* cause, the respiratory centre in the brain stem is reflexively stimulated to increase pulmonary ventilation (the rate at which gas is exchanged between the lungs and the atmosphere) (p. 548). As the rate and depth of breathing increase, more $CO_2$ than usual is exhaled. This means that less $CO_2$ is available to react with $H_2O$ in the blood and less $H_2CO_3$ is produced. Therefore, removal of $CO_2$ helps to remove excess acid from nonrespiratory sources in the body.

- Conversely, when arterial $[H^+]$ falls, pulmonary ventilation is reduced. As a result of slower and shallower breathing, less $CO_2$ is removed from the body by the lungs. Metabolically produced $CO_2$ therefore accumulates in the blood. This $CO_2$ reacts with $H_2O$ to produce $H_2CO_3$, restoring $[H^+]$ toward normal.

The lungs are extremely important in maintaining $[H^+]$. Every day they remove from body fluids what amounts to 100 times more $H^+$ derived from carbonic acid as the kidneys remove from sources other than carbonic acid. Furthermore, the

**TABLE 14-7** Respiratory Adjustments to Acidosis and Alkalosis Induced by Nonrespiratory Causes

| Respiratory Compensations | Normal (pH 7.4) | ACID–BASE STATUS | |
| --- | --- | --- | --- |
| | | Nonrespiratory (metabolic) Acidosis (pH 7.1) | Nonrespiratory (metabolic) Alkalosis (pH 7.7) |
| Ventilation | Normal | ↑ | ↓ |
| Rate of $CO_2$ Removal | Normal | ↑ | ↓ |
| Rate of $H_2CO_3$ Formation | Normal | ↓ | ↑ |
| Rate of $H^+$ Generation from $CO_2$ | Normal | ↓ | ↑ |

respiratory system, through its ability to regulate arterial $[CO_2]$, can adjust the amount of $H^+$ added to body fluids as needed to restore a normal pH.

## The respiratory system: The second line of defence

If deviations in $[H^+]$ are not swiftly and completely corrected by the buffer systems, the respiratory system comes into action. Unlike the immediate response of the buffer systems of the body, adjustments in ventilation require a few minutes to be initiated. This makes it the *second line of defence* against changes in $[H^+]$ in the body.

The respiratory system can compensate only for non-respiratory changes to plasma pH. Furthermore, it can only partially correct these abnormalities in pH (50 to 75 percent). There are two main reasons for this. First, the peripheral chemoreceptors and central chemoreceptors change pulmonary ventilation using different mechanisms. The peripheral chemoreceptors increase ventilation in response to an elevated arterial $[H^+]$, and the central chemoreceptors increase ventilation in response to a rise in $[CO_2]$ (p. 545). Consider the example of an acidosis arising from a nonrespiratory cause. When the peripheral chemoreceptors detect the increase in arterial $[H^+]$, they reflexively *stimulate* the respiratory centre to increase ventilation, allowing for more $CO_2$ to be blown off. The $[CO_2]$ therefore decreases, and the central chemoreceptors *inhibit* the respiratory centre. By opposing the action of the peripheral chemoreceptors, the central chemoreceptors stop the compensatory increase in ventilation short of restoring pH back to the normal range.

Second, as the pH moves toward normal, the driving force for the compensatory increase in ventilation is diminished. Ventilation is increased by the peripheral chemoreceptors in response to a rise in arterial $[H^+]$, but as the $[H^+]$ is gradually reduced, so is the enhanced response in ventilation.

When changes in $[H^+]$ stem from $[CO_2]$ fluctuations arising from respiratory abnormalities, the respiratory mechanism does not contribute to pH control. Take for instance an acidosis that exists because of $CO_2$ accumulation due to a lung disease. It was the initial impairment in the lungs that led to an increased $CO_2$, and so the lungs cannot possibly compensate by increasing $CO_2$ removal. The buffer systems (other than the $H_2CO_3 : HCO_3^-$ pair) plus renal regulation are the only mechanisms available for defending against respiratory-induced acid–base disturbances.

## The kidneys and plasma pH control

The kidneys control the pH of body fluids by adjusting three interrelated factors: (1) $H^+$ excretion, (2) $HCO_3^-$ excretion, and (3) ammonia $NH_3$ secretion.

Acids are continuously being added to body fluids as a result of metabolic activities, yet the generated $H^+$ must not be allowed to accumulate. Although the body's buffer systems can resist changes in pH by removing $H^+$ from solution, the persistent production of acidic metabolic products would eventually overwhelm the capacity of these buffers. Therefore, the constantly generated $H^+$ must ultimately be eliminated from the body. The lungs can remove only carbonic acid by eliminating $CO_2$. The task of eliminating $H^+$ derived from sulphuric, phosphoric, lactic, and other acids rests with the kidneys. Furthermore, the kidneys can also eliminate extra $H^+$ derived from carbonic acid.

### MECHANISM OF RENAL $H^+$ EXCRETION AND SECRETION

Recall that the filtration rate of $H^+$ equals plasma $[H^+]$ multiplied by GFR. Since normal plasma $[H^+]$ is extremely low (less than in pure $H_2O$), the filtration rate of $H^+$ is also extremely low. Although the filtered $H^+$ is normally eliminated in the urine, it only accounts for a very minute fraction of the excreted $H^+$. The vast majority of excreted $H^+$ is secreted into the filtrate at various places across the nephron. These locations include the proximal tubule, distal tubule, and collecting duct. Since the normal tendency of the kidney is to remove excess $H^+$ by excretion, urine is usually acidic (pH ~ 6.0).

The $H^+$ secretory process begins in the tubular cells with $CO_2$ from three sources: $CO_2$ diffused into the tubular cells from (1) plasma, (2) tubular fluid, or (3) $CO_2$ metabolically produced within the tubular cells. Catalyzed by carbonic anhydrase, $CO_2$ and $H_2O$ form $H_2CO_3$, which dissociates into $H^+$ and $HCO_3^-$. The produced $H^+$ is secreted by two main mechanisms. First, an energy-dependent carrier on the luminal surface can transport $H^+$ out of the cell into the tubule lumen. Second, the proximal tubule utilizes a *luminal $Na^+$–$H^+$ antiporter*. The net effects of this antiporter include $H^+$ secretion into the lumen and $Na^+$ reabsorption into the peritubular capillaries (> Figure 14-9). Furthermore, the $HCO_3^-$ produced is reabsorbed into the peritubular capillaries. This means that $Na^+$ reabsorption and plasma pH regulation (by way of renal $H^+$ secretion) are partially linked. In fact, this proximal tubule antiporter is responsible for the majority of the $Na^+$ reabsorption in this region of the kidneys.

### FACTORS INFLUENCING THE RATE OF $H^+$ SECRETION

The magnitude of $H^+$ secretion depends primarily the acid–base status of the plasma and the effect it has on the tubular cells of the kidney (> Figure 14-10). There are no neural or hormonal controls involved.

- When the $[H^+]$ of the plasma passing through the peritubular capillaries is elevated above normal, the tubular cells respond by secreting a greater-than-usual amount of $H^+$ from the plasma into the tubular fluid.

- Conversely, when plasma $[H^+]$ is lower than normal, the kidneys conserve $H^+$ by reducing secretion and the subsequent excretion in the urine.

- Since $H^+$ secretion in the kidneys begins with $CO_2$, its rate of secretion is also influenced by the $[CO_2]$.

14

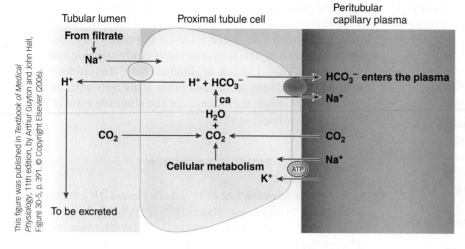

Tubular lumen   Proximal tubule cell   Peritubular capillary plasma

**From filtrate**

> FIGURE 14-9 The Na⁺–H⁺ exchanger (NHE) of the proximal tubule

- When plasma $[CO_2]$ increases, these reactions proceed more rapidly, and the rate of H⁺ secretion also increases (Figure 14-10).

- Conversely, when plasma $[CO_2]$ decreases, these reactions proceed less rapidly and the rate of H⁺ secretion also decreases.

These responses are extremely important in terms of how the kidney compensates for acid–base abnormalities involving a change in $H_2CO_3$ caused by respiratory dysfunction. Therefore, the kidneys can adjust H⁺ excretion to compensate for changes in both carbonic and noncarbonic acids.

## The kidneys and HCO₃⁻ excretion

Before being eliminated by the kidneys, the H⁺ generated from noncarbonic acids is buffered to a large extent by plasma $HCO_3^-$ (see Why It Matters, p. 629). Therefore, renal handling

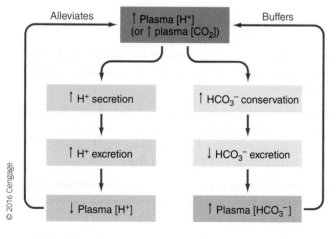

> FIGURE 14-10 Control of the rate of tubular H⁺ secretion

of acid–base balance also involves adjustment of $HCO_3^-$ excretion, depending on the H⁺ load in the plasma (Figure 14–10).

The kidneys regulate plasma $[HCO_3^-]$ by two interrelated mechanisms: (1) variable reabsorption of filtered $HCO_3^-$ back into the plasma, and (2) variable addition of new $HCO_3^-$ to the plasma. Both of these mechanisms are inextricably linked with H⁺ secretion by the kidney tubules. Each H⁺ secreted into the tubular fluid results in the simultaneous transfer of a $HCO_3^-$ into the peritubular capillary plasma. Whether a filtered $HCO_3^-$ is reabsorbed or a new $HCO_3^-$ is added to the plasma in accompaniment with H⁺ secretion depends on whether filtered $HCO_3^-$ is present in the tubular fluid to react with the secreted H⁺.

### COUPLING OF HCO₃⁻ REABSORPTION WITH H⁺ SECRETION

Bicarbonate is freely filtered, but because the luminal membranes of tubular cells are impermeable to filtered $HCO_3^-$, it cannot diffuse into these cells. In fact, $HCO_3^-$ is not freely permeable to membranes in general. Therefore, reabsorption of $HCO_3^-$ must occur via an indirect mechanism (> Figure 14-11). H⁺ secreted into the tubular fluid combines with filtered $HCO_3^-$ to form $H_2CO_3$. Under the influence of carbonic anhydrase, which is present on the surface of the luminal membrane, $H_2CO_3$ decomposes into $CO_2$ and $H_2O$ within the filtrate. Unlike $HCO_3^-$, the $CO_2$ can easily penetrate tubular cell membranes. Once it is intracellular, cytoplasmic carbonic anhydrase helps the $CO_2$ react with $H_2O$ to form $H_2CO_3$ which dissociates into H⁺ and $HCO_3^-$. Since $HCO_3^-$ can permeate the basolateral membrane of the tubule cells, it passively diffuses out of the cells and into the peritubular capillary plasma. Meanwhile, the generated H⁺ is actively secreted. Notice that the disappearance of a $HCO_3^-$ from the tubular fluid is coupled with the appearance of another $HCO_3^-$ in the plasma. In effect, a $HCO_3^-$ has been "reabsorbed." Even though the $HCO_3^-$ entering the plasma is not the same one that was filtered, the net result is the same as if $HCO_3^-$ were directly reabsorbed.

Under normal circumstances, the amount of H⁺ secreted is greater than the amount of $HCO_3^-$ filtered. Accordingly, all the filtered $HCO_3^-$ is usually reabsorbed. By far, most of the secreted H⁺ combines with $HCO_3^-$ and is "used up" in $HCO_3^-$ reabsorption rather than excreted. However, the slight excess of secreted H⁺ that is not matched by filtered $HCO_3^-$ is excreted in urine.

Secretion H⁺ that is *excreted* is generally coupled with the *addition of new* $HCO_3^-$ to the plasma. This H⁺ is generated by $CO_2$ from cellular metabolism, or by the $CO_2$ that diffuses from the plasma and into the tubular cell (> Figure 14-12). The $CO_2$ reacts with $H_2O$ in the cell to produce H⁺ and $HCO_3^-$. The $HCO_3^-$ is absorbed into the peritubular capillaries. This $HCO_3^-$ is considered "new" because it is not associated with the reabsorption of filtered $HCO_3^-$. The H⁺ is secreted, combines with urinary buffers, especially basic phosphate ($HPO_4^{2-}$), and is ultimately excreted in the urine.

**14**

This figure was published in *Textbook of Medical Physiology*, 11th edition, by Arthur Guyton and John Hall, Figure 30-5, p. 391. © Copyright Elsevier (2006).

© 2016 Cengage

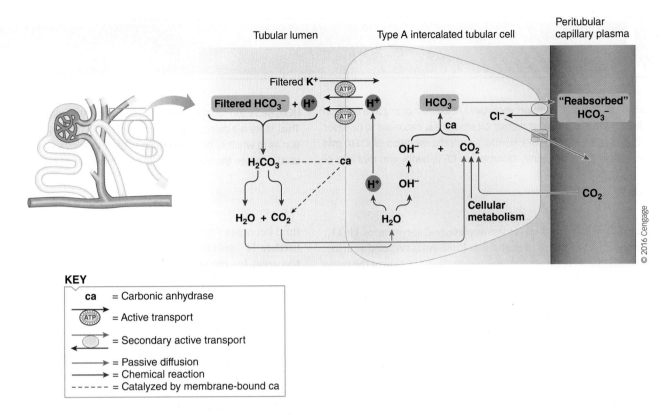

**KEY**

**ca** = Carbonic anhydrase

(ATP) = Active transport

= Secondary active transport

⟶ = Passive diffusion

⟶ = Chemical reaction

- - - - - = Catalyzed by membrane-bound ca

> **FIGURE 14-11 Hydrogen ion secretion coupled with bicarbonate reabsorption.** Because the disappearance of a filtered $HCO_3^-$ from the tubular fluid is coupled with the appearance of another $HCO_3^-$ in the plasma, $HCO_3^-$ is considered to have been reabsorbed.

## RENAL HANDLING OF $H^+$ AND $HCO_3^-$ DURING ACIDOSIS AND ALKALOSIS

When plasma $[H^+]$ is elevated during acidosis, more $H^+$ is secreted than normal. At the same time, less plasma $[HCO_3^-]$ is filtered than normal because more of it is used up in buffering the excess $H^+$ in the ECF. This imbalance between filtered $HCO_3^-$ and secreted $H^+$ has two consequences. First, more of the secreted $H^+$ is excreted in the urine because more $H^+$ are entering the tubular fluid at a time when fewer are needed to reabsorb the reduced quantities of filtered $HCO_3^-$. This extra $H^+$ makes the urine more acidic than normal. Second, the increased excretion of $H^+$ is linked with the addition of new $HCO_3^-$ to the plasma. This additional $HCO_3^-$ is available to buffer excess $H^+$ present in the body.

In the situation of alkalosis, the rate of $H^+$ secretion diminishes, and the rate of $HCO_3^-$ filtration increases. When plasma $[H^+]$ is below normal, a smaller proportion of the $HCO_3^-$ pool is tied up buffering $H^+$. This causes the plasma $[HCO_3^-]$ to increase. As a result, the rate of $HCO_3^-$

filtration correspondingly increases. Not all of the filtered $HCO_3^-$ is reabsorbed because $HCO_3^-$ is in excess of secreted $H^+$ in the tubular fluid. Since $HCO_3^-$ cannot be reabsorbed without first

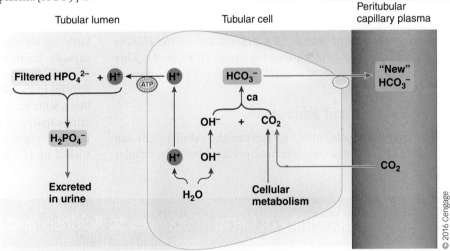

> **FIGURE 14-12 Hydrogen ion secretion and excretion coupled with the addition of new $HCO_3^-$ to the plasma.** Secreted $H^+$ does not combine with filtered $HPO_4^{2-}$ and is not subsequently excreted until all the filtered $HCO_3^-$ has been reabsorbed, as depicted in Figure 14-11. Once all the filtered $HCO_3^-$ has combined with secreted $H^+$, further secreted $H^+$ is excreted in the urine, primarily in association with urinary buffers such as basic phosphate. Excretion of $H^+$ is coupled with the appearance of new $HCO_3^-$ in the plasma. The new $HCO_3^-$ represents a net gain rather than merely a replacement for filtered $HCO_3^-$.

**14**

Why is bicarbonate transport important? Aside from acting as a buffer in skeletal muscle and other tissues, bicarbonate is important to diseases such as cystic fibrosis (CF). Normal pancreatic secretion of a 140 mM $HCO_3^-$ solution is frequently disrupted in CF patients, and that loss of

$HCO_3^-$ secretion capacity correlates strongly with disease severity in CF. Thus, there is a defect in the transport of $HCO_3^-$. This has raised the question as to whether delivering $HCO_3^-$ to diseased tissues can someday be used to lessen the effects of CF in patients and even extend their lives.

reacting with $H^+$, there is incomplete reabsorption of $HCO_3^-$. Excess $HCO_3^-$ remains in the tubular fluid and is excreted in urine. The net effects of this excretion include alkalinization of the urine and an overall reduction in plasma $HCO_3^-$.

In summary, when plasma $[H^+]$ increases above normal during *acidosis,* renal compensation includes the following (❚ Table 14-8):

1. Increased secretion and excretion of $H^+$ in the urine, which eliminates the excess $H^+$ and decreases plasma $[H^+]$

2. Reabsorption of all the filtered $HCO_3^-$ and the addition of new $HCO_3^-$ to the plasma, resulting in increased plasma $[HCO_3^-]$

When plasma $[H^+]$ falls below normal during *alkalosis,* renal compensation includes the following:

1. Decreased secretion and excretion of $H^+$ in the urine, which conserves $H^+$ and increases plasma $[H^+]$

2. Incomplete reabsorption of filtered $HCO_3^-$ and subsequent increased excretion of $HCO_3^-$, resulting in decreased plasma $[HCO_3^-]$.

Note that in states of acidosis, the kidneys acidify urine (by secreting extra $H^+$) and alkalinize plasma (by conserving $HCO_3^-$) to bring pH back to normal. In states of alkalosis, the kidneys alkalinize the urine (by eliminating excess $HCO_3^-$) and acidify plasma (by reducing secretion of $H^+$) to bring pH back to normal.

## The kidneys and ammonia

The energy-dependent $H^+$ carriers in the tubular cells can secrete $H^+$ against a concentration gradient until the tubular

fluid becomes 800 times as acidic as the plasma. At this point, further $H^+$ secretion stops because the gradient becomes too great for the secretory process to continue. The kidneys cease to secrete $H^+$ when urinary pH becomes 4.5. Only about 1 percent of the excess free $H^+$ secreted daily can acidify the urine to this pH. This means the other 99 percent of the normally secreted $H^+$ load could not be eliminated—an intolerable situation. For $H^+$ secretion to proceed, therefore, most secreted $H^+$ must be buffered in the tubular fluid so that it does not exist as free $H^+$ and, accordingly, not contribute to tubular acidity.

Bicarbonate is unable to buffer urinary $H^+$ as it does $H^+$ in the ECF because it is not excreted in the urine simultaneously with $H^+$. (Whichever of these substances is in excess in the plasma is excreted in the urine.) There are, however, two important urinary buffers that help remove free $H^+$: (1) filtered phosphate buffers, and (2) secreted ammonia.

### FILTERED PHOSPHATE AS A URINARY BUFFER
Normally, secreted $H^+$ is first buffered by the phosphate buffer system. Phosphate is present in the tubular fluid only because of dietary excess, not because of any specific mechanism for buffering secreted $H^+$. Phosphate just happens to have buffering capacity. Furthermore, tubular fluid contains phosphate levels above the transport maximum. Above this level excess phosphate remains in the tubular fluid. The basic member of this buffer pair binds with secreted $H^+$. When $H^+$ secretion is high, the buffering capacity of urinary phosphates is exceeded. As soon as all the basic phosphate ions that are coincidentally excreted have soaked up $H^+$, the acidity of the tubular fluid quickly rises as

**14**

## ❚ TABLE 14-8 Summary of Renal Responses to Acidosis and Alkalosis

| Acid–Base Abnormality | $H^+$ Secretion | $H^+$ Excretion | $HCO_3^-$ Reabsorption and Addition of New $HCO_3^-$ to Plasma | $HCO_3^-$ Excretion | pH of Urine | Compensatory Change in Plasma pH |
|---|---|---|---|---|---|---|
| **Acidosis** | ↑ | ↑ | ↑ | Normal (zero; all filtered is reabsorbed) | Acidic | Alkalinization toward normal |
| **Alkalosis** | ↓ | ↓ | ↓ | ↑ | Alkaline | Acidification toward normal |

© 2016 Cengage

more H⁺ is secreted. Without additional buffering capacity from another source, H⁺ secretion would soon halt abruptly because the free $[H^+]$ in the tubular fluid can quickly rise to the critical limiting level (pH 4.5).

## SECRETED NH₃ AS A URINARY BUFFER

When acidosis exists, the tubular cells secrete **ammonia (NH₃)** into the tubular fluid once the normal urinary phosphate buffers are saturated. This NH₃ enables the kidneys to continue secreting additional H⁺ ions because NH₃ combines with free H⁺ in the tubular fluid to form **ammonium ion** ($NH_4^+$), as follows:

$$NH_3 + H^+ \rightarrow NH_4^+$$

The tubular membranes are not very permeable to $NH_4^+$, so $NH_4^+$ remains in the tubular fluid and is subsequently lost in urine. Each molecule of $NH_4^+$ eliminated takes an H⁺ with it. NH₃ secreted during acidosis allows for the necessary increased buffering capacity in the tubular fluid, so large amounts of H⁺ can be secreted. Were it not for NH₃ secretion, the extent of H⁺ secretion would be limited to whatever phosphate-buffering capacity coincidentally happened to be present as a result of dietary excess.

In contrast to the urinary phosphate buffer system, which is in the tubular fluid because the phosphate have been filtered but not reabsorbed, $NH_4^+$ is deliberately synthesized from the amino acid *glutamine* within the proximal tubular cells (⟩ Figure 14-13a). Once synthesized, $NH_4^+$ is absorbed into the tubular lumen. In the thick ascending limb of the loop of Henle, $NH_4^+$ can take the place of K⁺ and be reabsorbed into the renal interstitium via the Na⁺–K⁺–2 Cl⁻ transporter. The relative pH of the renal interstitium is high. $NH_4^+$ therefore liberates

a H⁺ to the interstitium and becomes NH₃. Unlike $NH_4^+$, NH₃ readily diffuses passively into cells down its concentration gradient. This allows NH₃ to diffuse into the collecting duct of the nephron. Here it combines with a free H⁺ and is excreted as $NH_4^+$ (⟩ Figure 14-13b). The rate of NH₃ secretion by the tubular cells varies directly based on the excess H⁺ needing to be secreted. When someone has been in a state of acidosis for more than two or three days, the rate of NH₃ production increases substantially. This extra NH₃ provides additional buffering capacity and contributes to renal compensation for acidosis.

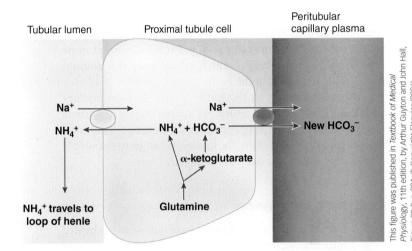

⟩ **FIGURE 14-13a Ammonium production from glutamine metabolism in the proximal tubule cell.** $NH_4^+$ from glutamine metabolism is secreted into the tubule lumen and travels to the loop of Henle. The bicarbonate molecules produced in the process are reabsorbed into the peritubular capillaries and constitute new bicarbonate ions.

**\*\*The high pH of the interstitium causes NH₄⁺ to dissociate.**

⟩ **FIGURE 14-13b Ammonium and ammonia movement through the loop of Henle and distal nephron.** Ammonium takes the place of K⁺ on the Na⁺–K⁺–2Cl⁻ transporter and enters the renal interstitium. The relatively high pH of the renal interstitium causes $NH_4^+$ to liberate a H⁺ The resultant NH₃ diffuses through the collecting duct cell and into the tubular lumen, where it combines with a metabolically produced free H⁺ The $NH_4^+$ becomes trapped in the lumen and is excreted in the urine.

Fluid and Acid–Base Balance **631**

## The kidneys: A powerful third line of defence

The kidneys require hours to days to compensate for changes in body-fluid pH, compared with the immediate responses of the buffer systems and the few minutes of delay before the respiratory system responds. Therefore, they are the *third line of defence* against $[H^+]$ changes in body fluids. However, the kidneys are considered the most potent acid–base regulatory organs. Not only can they vary the degree to which removal of $H^+$ occurs from any source, they can also variably conserve or eliminate $HCO_3^-$, depending on the acid–base status of the body. For example, during renal compensation for acidosis, each $H^+$ excreted in urine results in a new $HCO_3^-$ added to the plasma to buffer (by means of the $H_2CO_3 : HCO_3^-$ system) another $H^+$ that remains in body fluids. By simultaneously removing acid ($H^+$) from and adding base ($HCO_3^-$) to body fluids, the kidneys can restore the pH toward normal more effectively than the lungs, which adjust only the amount of $H^+$-generating $CO_2$ in the body.

Also contributing to the kidneys' acid–base regulatory potency is their ability to return pH almost exactly to normal. Compared to the respiratory system that can only partially compensate for a pH abnormality, the kidneys continue to change pH until compensation is essentially complete.

## Acid–base imbalances

*Clinical Note* Deviations from normal acid–base status can be divided into four general categories based on the source and direction of the abnormal change in $[H^+]$. These categories are *respiratory acidosis, respiratory alkalosis, metabolic acidosis,* and *metabolic alkalosis.*

As you have learned, changes in plasma $[H^+]$ are reflected by changes in the ratio of $[HCO_3^-]$ to $[CO_2]$. Recall that the normal ratio is 20/1. Using the Henderson-Hasselbalch equation with this ratio and the pK value of 6.1, we arrive at a normal plasma pH of 7.4. Compared to taking direct measurements of $[H^+]$ alone, we can find more important clues about the underlying factors responsible for acid–base disturbances by focusing on the changes in specific values for $[HCO_3^-]$ and $[CO_2]$. The following rules of thumb apply when examining acid–base imbalances *before any compensations take place:*

1. A change in pH that has an underlying respiratory cause is associated with a change in $[CO_2]$. For example, a respiratory acidosis is caused by an increase in $[CO_2]$, and a respiratory alkalosis is caused by a decrease in $[CO_2]$. Both of these scenarios result in a change in free $H^+$ that is derived from $H_2CO_3$. In contrast, a pH deviation of metabolic origin is associated with an abnormal $[HCO_3^-]$. A metabolic acidosis results in a low plasma $[HCO_3^-]$, and a metabolic alkalosis results in a high $[HCO_3^-]$. These changes are caused by an inequality between the amount of $HCO_3^-$ available and the amount of $H^+$ generated from noncarbonic acid sources that the $HCO_3^-$ must buffer.

2. Observing how the overall ratio changes $[HCO_3^-]/[CO_2]$ helps us determine whether an acidotic or alkalotic state exists. When the ratio falls below 20/1, an acidosis exists. Mathematically, this makes sense. The log of any number lower than 20 is less than 1.3 and, when added to the pK of 6.1, yields an acidotic pH < 7.4. Likewise, when the ratio exceeds 20/1, an alkalosis exists. The log of any number greater than 20 is more than 1.3 and, when added to the pK of 6.1, yields an alkalotic pH > 7.4.

3. We now have enough information to determine which of the four categories an acid–base disturbance falls into:

   - *Respiratory acidosis* has a ratio of less than 20/1 due to an elevation in $[CO_2]$.

   - *Respiratory alkalosis* has a ratio greater than 20/1 due to a decrease in $[CO_2]$.

   - *Metabolic acidosis* has a ratio of less than 20/1 due to a decrease in $[HCO_3^-]$.

   - *Metabolic alkalosis* has a ratio greater than 20/1 due to an elevation in $[HCO_3^-]$.

Next we examine each of these categories in more detail, paying particular attention to possible causes and compensations, and using the examples in ⟩ Figure 14-14 (a–i). In this figure the "balance beam" concept and the Henderson–Hasselbalch equation will help you better visualize the contributions of the lungs and kidneys to the causes of and compensations for various acid–base disorders. The normal situation is represented in ⟩ Figure 14-14a.

## Respiratory acidosis and an increase in $[CO_2]$

Respiratory acidosis occurs when $CO_2$ builds up in the blood as a result of hypoventilation (p. 540). This leads to an increased production of $H_2CO_3$ and therefore increased free $H^+$ in the plasma, causing the pH to become more acidic (pH < 7.4).

### CAUSES OF RESPIRATORY ACIDOSIS

Respiratory acidosis is commonly caused by pulmonary conditions (p. 516–517), including emphysema, chronic bronchitis, asthma, and severe pneumonia. It may also be triggered by metabolic alkalosis.

In uncompensated respiratory acidosis, $[CO_2]$ is elevated, whereas $[HCO_3^-]$ is normal. In our example in ⟩ Figure 14-14b, the $[CO_2]$ is doubled, so the ratio becomes 20/2 (or 10/1). It may appear that $[H^+]$ and $[HCO_3^-]$ should increase proportionally because an elevated $[CO_2]$ drives the reaction $CO_2 + H_2O \rightarrow H_2CO_3 \rightarrow H^+ + HCO_3^-$ to the right. This is not the case, however, because the $[HCO_3^-]$ is normally 600 000 times that of $[H^+]$. Take for example, a situation where there is 1 $H^+$ and 600 000 $HCO_3^-$ present in the ECF. The generation of 1 additional $H^+$ will double its concentration, but one extra $HCO_3^-$ only increases the $HCO_3^-$ by 0.00017 percent (from 600 000 to 600 001 ions). Therefore, an elevation in

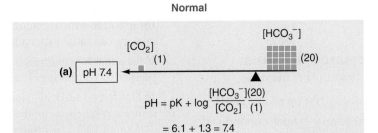

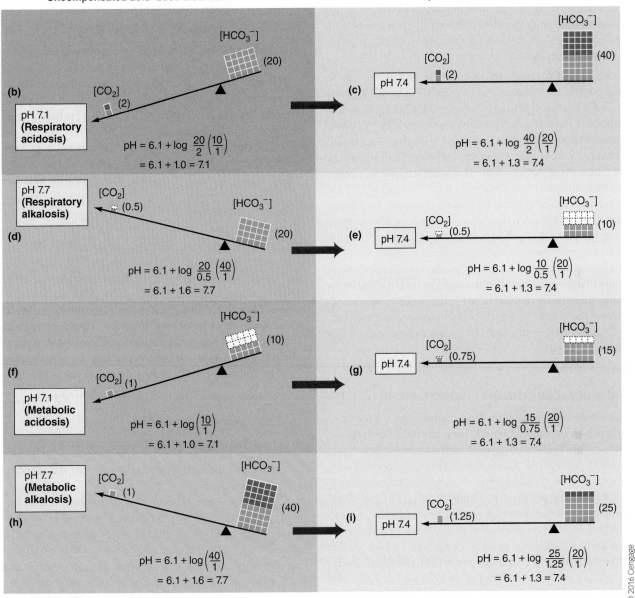

The lengths of the arms of the balance beams are not to scale.

> **Figure 14-14 Schematic representation of the relationship of [HCO₃⁻] and [CO2] to pH in various acid–base statuses.** (a) Normal acid–base balance. The $[HCO_3^-] / [CO_2]$ ratio is 20/1. (b) Uncompensated respiratory acidosis. The $[HCO_3^-] / [CO_2]$ ratio is reduced (20/2), because $CO_2$ has accumulated. (c) Compensated respiratory acidosis. Compensatory retention of $[HCO_3^-]$ to balance the $CO_2$ accumulation restores the $[HCO_3^-] / [CO_2]$ ratio to a normal equivalent (40/2). (d) Uncompensated respiratory alkalosis. The $[HCO_3^-] / [CO_2]$ ratio is increased (20/0.5) by a reduction in $CO_2$. (e) Compensated respiratory alkalosis. Compensatory elimination of $[HCO_3^-]$ to balance the $CO_2$ deficit restores the $[HCO_3^-] / [CO_2]$ ratio to a normal equivalent (10/0.5). (f) Uncompensated metabolic acidosis. The $[HCO_3^-] / [CO_2]$ ratio is reduced (10/1) by a $[HCO_3^-]$ deficit. (g) Compensated metabolic acidosis. Conservation of $[HCO_3^-]$, which partially makes up for the $[HCO_3^-]$ deficit, and a compensatory reduction in $CO_2$ restore the $[HCO_3^-] / [CO_2]$ to a normal equivalent (15/0.75). (h) Uncompensated metabolic alkalosis. The $[HCO_3^-] / [CO_2]$ ratio is increased (40/1) by excess $[HCO_3^-]$. (i) Compensated metabolic alkalosis. Elimination of some of the extra $[HCO_3^-]$ and a compensatory increase in $CO_2$ restore the ratio $[HCO_3^-] / [CO_2]$ to a normal equivalent (25/1.25).

**14**

$[CO_2]$ brings about a pronounced increase in $[H^+]$, but $[HCO_3^-]$ remains essentially normal.

### COMPENSATIONS FOR RESPIRATORY ACIDOSIS

Compensatory measures act to restore pH to normal.

- The chemical buffers immediately take up additional $H^+$.

- The respiratory mechanism usually cannot respond with compensatory increased ventilation, because impaired respiration is the reason for the abnormality in the first place.

- The kidneys are therefore most important in compensating for respiratory acidosis. They reabsorb all the filtered $HCO_3^-$ and add new $HCO_3^-$ to the plasma, and simultaneously they secrete and excrete more $H^+$.

As a result, $HCO_3^-$ stores in the body rise. In ⟩ Figure 14-14c, the plasma $[HCO_3^-]$ is doubled, so the $[HCO_3^-]/[CO_2]$ ratio becomes 40/2 rather than 20/2. A ratio of 40/2 is equivalent to a normal 20/1 ratio, so pH returns to the normal 7.4. Enhanced renal conservation of $HCO_3^-$ and new $HCO_3^-$ formation has fully compensated for $CO_2$ accumulation, although both $[CO_2]$ and $[HCO_3^-]$ are now distorted. Note that maintenance of normal pH depends on preserving a normal ratio between $[HCO_3^-]$ and $[CO_2]$, no matter what the absolute values of each of these buffer components are. Although the occurrence of complete compensation to the point that pH returns exactly to the normal state of 7.4 is extremely rare, in each example in ⟩ Figure 14-14 we assume full compensation for ease of mathematical calculations. Also note that the values used in these examples represent hypothetical situations. Deviations in pH actually occur over a wide range, and the degree to which compensation can be accomplished varies considerably (e.g., $HCO_3^-$ may rise or fall only by one or two points).

## Respiratory alkalosis and a decrease in $[CO_2]$

Respiratory alkalosis results from hyperventilation, and is a condition in which the amount of $CO_2$ normally found in the blood decreases. This leads to an overall decrease in $H^+$ and hence an elevated plasma pH. Common causes of hyperventilation include fever, anxiety, and serious infections. Other stresses to the body, including pregnancy or high altitude, can cause respiratory alkalosis. It may also be triggered by metabolic acidosis.

Looking at the biochemical abnormalities in uncompensated respiratory alkalosis, as shown in Figure 14-14d, we find the increase in pH reflects a reduction in $[CO_2]$ (half the normal value in this example), whereas $[HCO_3^-]$ remains normal. This yields a ratio of 20/0.5 (or 40/1), which will cause an overall increase in plasma pH.

### COMPENSATIONS FOR RESPIRATORY ALKALOSIS

Compensatory measures act to shift pH back toward normal.

- The chemical buffer systems begin to liberate $H^+$ to diminish the severity of the alkalosis.

- As plasma $[CO_2]$ and $[H^+]$ fall below normal because of excessive ventilation, two of the normally potent stimuli for

driving ventilation are removed. This effect tends to limit the extent to which a nonrespiratory factor, such as fever or anxiety, can cause hyperventilation.

- If the situation continues for a few days, the kidneys compensate by conserving $H^+$ and excreting more $HCO_3^-$.

In Figure 14-14e, $HCO_3^-$ stores are reduced to half (by loss in the urine) to compensate for the loss in $[CO_2]$. The $[HCO_3^-]/[CO_2]$ ratio becomes 10/0.5 (equivalent to the normal 20/1), and pH is restored to normal.

## Metabolic acidosis and a fall in $[HCO_3^-]$

**Metabolic acidosis** (also known as **nonrespiratory acidosis**) encompasses all types of acidosis besides that caused by excess $CO_2$ in body fluids. In the uncompensated state, metabolic acidosis is characterized by a reduction in plasma $[HCO_3^-]$, whereas $[CO_2]$ remains normal. In ⟩ Figure 14-14f, $[HCO_3^-]$ is halved and $[CO_2]$ remains normal, producing a $[HCO_3^-]/[CO_2]$ ratio of 10/1. A metabolic acidosis may arise from an excessive loss of body fluids rich in $HCO_3^-$ or from an accumulation of noncarbonic acids. In the latter case, plasma $HCO_3^-$ is used up in buffering the additional $H^+$.

### SERUM ANION GAP

A relatively simple measurement can be used to determine the cause of metabolic acidosis: the **anion gap**. The serum is uncharged or electro-neutral, meaning that the number of anions and number of cations are equal. The anion gap is a mathematical way of determining the difference in laboratory-measured anions (negatively charged ions) and laboratory-measured cations (positively charged ions) in serum. Therefore, to determine the anion gap, look at the ions that are *measured* in the serum, and this gives you an idea of the amount of *unmeasured* anions.

The equation for calculating the anion gap is as follows:

$$\text{anion gap} = ([Na^+] + [K^+]) - ([Cl^-] + [HCO_3^-])$$

This is generally shortened to

$$\text{anion gap} = [Na^+] - ([Cl^-] + [HCO_3^-])$$

As serum potassium is generally very low, it does not greatly affect the calculation.

The normal value of serum anion gap is considered 12 mEq/L, but typically its range is 8–16 mEq/L. This number represents the main unmeasured anions that are normally present in the plasma: phosphate, citrate, sulphate, and protein (⟩ Figure 14-15). When the anion gap increases, this generally indicates there is an excess of one or more of these unmeasured anions. It can also mean that other anions, such as lactate or keto acids, have been added to the plasma. Some diseases, extreme exercise, alcohol abuse, or even septic shock can cause unmeasured anions to increase. These anions are generally acidic and, as such, cause a decrease in serum bicarbonate by the action of chemical buffering.

In cases of a normal anion-gap metabolic acidosis, the disturbance is often due to a loss of bicarbonate (e.g., diarrhoea and some renal diseases). To maintain an electrically neutral serum, the loss in bicarbonate is generally countered by an increase in serum chloride.

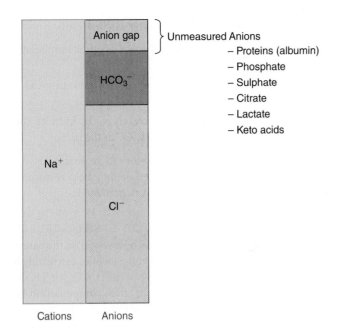

> **FIGURE 14-15** Serum anion gap

Unmeasured Anions
- Proteins (albumin)
- Phosphate
- Sulphate
- Citrate
- Lactate
- Keto acids

Cations    Anions

## CAUSES OF METABOLIC ACIDOSIS

*Severe diarrhoea.* During digestion, a digestive juice rich in $HCO_3^-$ is normally secreted into the digestive tract and is later reabsorbed back into the plasma when digestion is completed. During diarrhoea, this $HCO_3^-$ is lost from the body before it can be reabsorbed. Consequently, less $HCO_3^-$ is available to buffer $H^+$, which leads to more free $H^+$ in the body fluids. Also, the loss of $HCO_3^-$ shifts the $CO_2 + H_2O \rightarrow H^+ + HCO_3^-$ reaction to the right to compensate for the $HCO_3^-$ deficit. This increases $[H^+]$ above normal.

1. *Diabetes mellitus.* The absence of insulin results in an inability for cells to use glucose for metabolic processes. As a result, abnormal fat metabolism leads to the formation of excess keto acids whose dissociation increases plasma $[H^+]$. These excess keto acids cause a high anion gap.

2. *Strenuous exercise.* When muscles resort to anaerobic glycolysis during strenuous exercise, excess lactic acid is produced, raising plasma $[H^+]$ (p. 317). The excess lactic acid causes a high anion gap.

3. *Uraemic acidosis.* In severe kidney failure (uraemia), the kidneys cannot rid the body of normal amounts of metabolically generated $H^+$ from noncarbonic acid sources. $H^+$ therefore begins to accumulate in the body fluids. Furthermore, the dysfunction results in an inability to conserve adequate amounts of $HCO_3^-$.

## COMPENSATIONS FOR METABOLIC ACIDOSIS

Except in uraemic acidosis, metabolic acidosis is compensated for by both respiratory and renal mechanisms as well as by chemical buffers.

- The buffers take up extra $H^+$.
- The lungs blow off additional $H^+$-generating $CO_2$.
- The kidneys excrete more $H^+$ and conserve more $HCO_3^-$.

In > Figure 14-14f, g), these compensatory measures restore the ratio to normal by reducing $[CO_2]$ and by raising $[HCO_3^-]$. In this example, the ratio is brought to 15/0.75, which is equivalent to 20/1.

Note that in compensating for metabolic acidosis, the lungs deliberately change $[CO_2]$ to levels that deviate from normal in an attempt to restore a normal $[H^+]$. In respiratory-induced acid–base disorders, on the other hand, the abnormal $[CO_2]$ is the *cause* of the $[H^+]$ imbalance.

When kidney disease causes metabolic acidosis, complete compensation is not possible because the renal mechanism is not available for pH regulation. Recall that the respiratory system can compensate only up to 75 percent of the way toward normal. Uraemic acidosis is very serious because the kidneys cannot help restore pH all the way to normal.

> ### ▌Clinical Connections
>
> Without insulin production, Diane's energy metabolism shifts to a catabolic state. This leads to the breakdown of lipids and proteins and induces an overall energy deficient state, increasing hunger and food intake (*polyphagia*). Although caloric intake increases, it is usually insufficient to counterbalance the overall catabolic state, and weight loss ensues.
>
> The fatty acids from lipid catabolism are converted in the liver to the ketones ß-hydroxybutyric acid and acetoacetic acid. These two metabolically produced compounds are acidic, which leads to a decrease in the plasma pH and a state of metabolic acidosis. Furthermore, these compounds are anionic and increase the number of unmeasured anions in the plasma, thereby elevating the plasma anion gap. These two features help define the specific acid–base disturbance seen in DKA: an elevated plasma anion-gap metabolic acidosis.
>
> In DKA, the increase in buffering activity related to the high $[H^+]$ in the plasma results in a decrease in plasma bicarbonate. Furthermore, the respiratory system attempts to compensate for the high $[H^+]$ by increasing the rate and depth of breathing to remove $CO_2$. This is reflected in a decreased pCO2 level. Although these compensatory mechanisms attempt to correct the acid–base disturbance, the condition will worsen without emergent medical intervention.
>
> Definitive treatment of DKA requires fluid replacement, correction of electrolyte disturbances, administration of insulin, and correction of the acidotic state in the body. Prior to exogenous insulin injections, new-onset Type I DM was a universally fatal condition.
>
> Now, with injectable exogenous insulin available, the prompt identification of new-onset Type I DM and effective treatment of DKA can help make Type I DM a chronic illness that can be effectively managed with insulin and careful blood glucose monitoring.

## Metabolic alkalosis and an elevation in $[HCO_3^-]$

**Metabolic alkalosis** is a reduction in plasma $[H^+]$ caused by a relative deficiency of noncarbonic acids. This acid–base disturbance is associated with an increase in $[HCO_3^-]$, which, in the uncompensated state, is not accompanied by a change in $[CO_2]$. In Figure 14-14h, $[HCO_3^-]$ is doubled, producing an alkalotic ratio of 40/1.

## CAUSES OF METABOLIC ALKALOSIS

1. *Vomiting* causes abnormal loss of $[H^+]$ from the body as a result of lost acidic gastric juices. Hydrochloric acid is secreted into the stomach lumen in preparation for and during digestion. Bicarbonate is added to the plasma during gastric HCl secretion. This $HCO_3^-$ is neutralized by $H^+$ as the gastric secretions are eventually reabsorbed back into the plasma, so normally there is no net addition of $HCO_3^-$ to the plasma from this source. However, when this acid is lost from the body during vomiting, not only is plasma $[H^+]$ decreased, but reabsorbed $H^+$ is no longer available to neutralize the $HCO_3^-$ entering the plasma during gastric HCl secretion. Thus, loss of HCl in effect increases plasma $[HCO_3^-]$. Less frequently, vomiting can have deeper effects; that is, it can cause the loss of $HCO_3^-$ from the digestive juices in the upper intestine. In these instances, vomiting can result in a metabolic acidosis instead of metabolic alkalosis.

2. *Ingestion of alkaline drugs* can produce alkalosis, such as when baking soda ($NaHCO_3$, which dissociates in solution into $Na^+$ and $HCO_3^-$) is used as a self-administered remedy for treating gastric hyperacidity. Although $HCO_3^-$ neutralizes excess acid in the stomach and helps relieve the symptoms heartburn, there is a risk that more $HCO_3^-$ than needed may be ingested. The extra $HCO_3^-$ absorbed from the digestive tract increases the plasma $[HCO_3^-]$. Ultimately, the increased $HCO_3^-$ binds with more of the free $H^+$ than usual, and so the amount of free $[H^+]$ decreases. Commercially available alkaline products for treating gastric hyperacidity are generally not absorbed from the digestive tract to any extent and therefore do not alter the body's acid–base status.

## COMPENSATIONS FOR METABOLIC ALKALOSIS

- In metabolic alkalosis, chemical buffer systems immediately liberate $H^+$.
- Ventilation is reduced so that extra $H^+$-generating $CO_2$ is retained in the body fluids.
- If the condition persists for several days, the kidneys conserve $H^+$ and excrete the excess $HCO_3^-$ in the urine.

In ⟩ Figure 14-14i, there is a compensatory increase in $[CO_2]$ and a partial reduction in $[HCO_3^-]$. Together they help restore the $[HCO_3^-]/[CO_2]$ ratio back to the equivalent of 20/1.

## OVERVIEW OF COMPENSATED ACID–BASE DISORDERS

An individual's acid–base status cannot be assessed on the basis of pH alone. Uncompensated acid–base abnormalities can readily be distinguished on the basis of deviations of either $[CO_2]$ or $[HCO_3^-]$ from normal (▯ Table 14-9). However, when compensation has been accomplished and pH is essentially normal, determinations of $[HCO_3^-]$ and $[HCO_3^-]$ can reveal an acid–base disorder, but the type of disorder is more difficult to distinguish. The procedures for making these distinctions are beyond the scope of this textbook. For example, in both compensated respiratory acidosis and compensated metabolic alkalosis, $[CO_2]$ and $[HCO_3^-]$ are both above normal. Similarly, compensated respiratory alkalosis and compensated metabolic acidosis share similar patterns of $[CO_2]$ and $[HCO_3^-]$. A full medical history and examination are often extremely helpful in determining the particular type of acid–base disorder that exists. In addition, many clinicians use a specific device called a nomogram to determine which type of acid–base disorder their patients have (⟩ Figure 14-16). With this device, the arterial blood levels for pH, $pCO_2$, and $HCO_3^-$ are mapped on a graph. The type of acid–base disorder is determined by the point at which the test results show up on the graph—that is, within which of the highlighted sections as shown in Figure 14-16.

▮ **TABLE 14-9** Summary of $[CO_2]$, $[HCO_3^-]$ and pH in Uncompensated and Compensated Acid–Base Abnormalities

| Acid–Base Status | pH | $[CO_2]$ (Compared with Normal) | $[HCO_3^-]$ (Compared with Normal) | $HCO_3^-/[CO_2]$ |
|---|---|---|---|---|
| **Normal** | Normal | Normal | Normal | 20/1 |
| **Uncompensated Respiratory Acidosis** | Decreased | Increased | Normal | 20/2 (10/1) |
| **Compensated Respiratory Acidosis** | Normal | Increased | Increased | 40/2 (20/1) |
| **Uncompensated Respiratory Alkalosis** | Increased | Decreased | Normal | 20/0.5 (40/1) |
| **Compensated Respiratory Alkalosis** | Normal | Decreased | Decreased | 10/0.5 (20/1) |
| **Uncompensated Metabolic Acidosis** | Decreased | Normal | Decreased | 10/1 |
| **Compensated Metabolic Acidosis** | Normal | Decreased | Decreased | 15/0.75 (20/1) |
| **Uncompensated Metabolic Alkalosis** | Increased | Normal | Increased | 40/1 |
| **Compensated Metabolic Alkalosis** | Normal | Increased | Increased | 25/1.25 (20/1) |

14

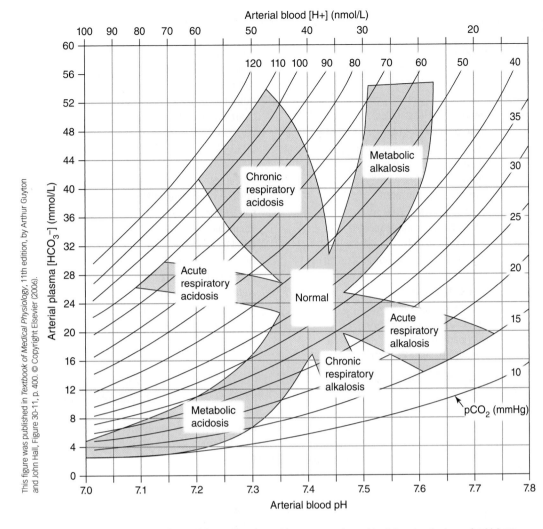

This figure was published in *Textbook of Medical Physiology*, 11th edition, by Arthur Guyton and John Hall, Figure 30-11, p. 400. © Copyright Elsevier (2006).

**FIGURE 14-16 A nomogram for acid–base disorders.** This nomogram is used to determine the type of acid–base disorder based on the patient's arterial pH, $pCO_2$, and $HCO_3^-$.

body's salt load affects the osmotic determination of the ECF volume, of which plasma volume is a part. An increased salt load in the ECF leads to an expansion in ECF volume, including plasma volume, which in turn causes a rise in blood pressure. Conversely, a reduction in the ECF salt load brings about a fall in blood pressure. Salt balance is maintained by constantly adjusting salt output in the urine to match unregulated, variable salt intake.

Control of water balance is important in preventing changes in ECF osmolarity, which induces detrimental osmotic shifts of $H_2O$ between the cells and ECF. Such shifts of $H_2O$ into or out of the cells cause the cells to swell or shrink, respectively. Cells, especially brain neurons, do not function normally when swollen or shrunken. Water balance is largely maintained by controlling the volume of free $H_2O$ ($H_2O$ not accompanied by solute) lost in the urine to compensate for uncontrolled losses of variable volumes of $H_2O$ from other avenues, such as through sweating or diarrhoea, and for poorly regulated $H_2O$ intake. Even though a thirst mechanism exists to control $H_2O$ intake based on need, the amount a person drinks is often influenced by social custom and habit instead of thirst alone.

## Check Your Understanding 14.3

1. Explain why only a narrow pH range is compatible with life.
2. If a person has severe diarrhoea, what type of acid–base abnormality will likely result? Describe the compensatory responses of the three lines of defence against this change in [H+].

# Chapter in Perspective: Focus on Homeostasis

Homeostasis depends on maintaining a balance between the input and output of all constituents present in the internal fluid environment. Regulation of fluid balance involves two separate components: control of salt balance and control of water balance. Control of salt balance is primarily important in the long-term regulation of arterial blood pressure, because the

A balance between input and output of H$^+$ is critical to maintaining the body's acid–base balance within the narrow limits compatible with life. Deviations in pH of the internal fluid environment lead to altered neuromuscular excitability, changes in enzymatically controlled metabolic activity, and [K$^+$] imbalances. Imbalances in [K$^+$] can cause cardiac arrhythmias. These effects are fatal if the pH falls outside the range of 6.8–8.0.

Hydrogen ions are uncontrollably and continually added to the body fluids as a result of ongoing metabolic activities, yet the ECF's pH must be kept constant at a slightly alkaline level of 7.4 for optimal body function. Like salt and water balance, control of H$^+$ output by the kidneys is the main regulatory factor in achieving H$^+$ balance. The lungs, which can adjust their rate of excretion of H$^+$-generating $CO_2$, also help eliminate H$^+$ from the body. Furthermore, chemical buffer systems can take up or liberate H$^+$, transiently keeping its concentration constant within the body until its output can be brought into line with its input. Such a mechanism is not available for salt or water balance.

14

## CHAPTER TERMINOLOGY

acidaemia (p. 31)
acid–base balance (p. 620)
acid-base disorders (p. 620)
acidic (p. 621)
acidosis (p. 621)
acids (p. 620)
alkalaemia (p. 621)
alkaline (p. 621)
alkalosis (p. 621)
ammonia ($NH_3$) (p. 631)
ammonium ion (p. 631)
anion gap (p. 634)
balance concept (p. 608)
base (p. 620)
basic (p. 621)
chemical buffer system (p. 623)
chemical disequilibrium (p. 611)
dehydration (p. 614)
diabetes insipidus (p. 614)
dipsogen (p. 619)

dissociation constant, K (p. 620)
extracellular fluid (ECF) (p. 609)
fluid balance (p. 611)
free hydrogen ion ($H^+$) concentration (p. 620)
Henderson–Hasselbalch equation (p. 625)
hypertonicity (p. 614)
hypothalamic osmoreceptors (p. 617)
hypotonicity (p. 614)
input (p. 608)
interstitial fluid (p. 609)
insensible loss (p. 616)
intracellular fluid (ICF) (p. 609)
Iatrogenic (p. 614)
left atrial volume receptors (p. 618)
lymph (p. 610)
metabolic acidosis 634
metabolic alkalosis (p. 635)
metabolic water (p. 616)
$NA^+$ load (p. 611)
negative balance (p. 608)

nonrespiratory acidosis (p. 634)
obligatory loss (p. 612)
osmolarity (p. 613)
output (p. 608)
overhydration (p. 614)
paraesthesias (p. 622)
pH (p. 620)
plasma (p. 609)
Pool (p. 608)
positive balance (p. 608)
salt balance (p. 612)
stable balance (p. 608)
thirst (p. 617)
thirst centre (p. 617)
transcellular fluid (p. 610)
water balance (p. 609)
water intoxication (p. 615)
xerostomia (p. 618)

## REVIEW EXERCISES

### Objective Questions
### (Answers in Appendix E, p. A-50)

1. The only avenue by which materials can be exchanged between the cells and the external environment is the ECF. *(True or false?)*

2. Water is driven into the cells when the ECF volume is expanded by an isotonic fluid gain. *(True or false?)*

3. Salt balance in humans is poorly regulated because of our hedonistic salt appetite. *(True or false?)*

4. An unintentional increase in $CO_2$ is a cause of respiratory acidosis, but a deliberate increase in $CO_2$ compensates for metabolic alkalosis. *(True or false?)*

5. The largest body-fluid compartment is the _____

6. Of the two members of the $H_2CO_3 : HCO_3^-$ buffer system, _____ is regulated by the lungs and _____ is regulated by the kidneys.

7. Which individuals would have the lowest percentage of body $H_2O$?
   a. a chubby baby
   b. a well-proportioned female university student
   c. a well-muscled male university student
   d. an obese elderly woman
   e. a lean elderly man

8. Which of the following factors does NOT increase vasopressin secretion?
   a. ECF hypertonicity
   b. alcohol
   c. stressful situations
   d. ECF volume deficit
   e. angiotensin II

9. Which of the following statements about pH are correct? *(Indicate all correct answers.)*
   a. It equals $\log 1/[H^+]$.
   b. It equals $pK + \log[CO_2]/[HCO_3^-]$.
   c. It is high in acidosis.
   d. It falls as $[H^+]$ increases.
   e. It is normal when the $[HCO_3^-]/[CO_2]$ ratio is 20/1.

10. Which of the following statements about acidosis are correct? *(Indicate all correct answers.)*
    a. It causes overexcitability of the nervous system.
    b. It exists when the plasma pH falls below 7.35.
    c. It occurs when the $[HCO_3^-]/[CO_2]$ ratio exceeds 20/1.
    d. It occurs when $CO_2$ is blown off more rapidly than it is being produced by metabolic activities.
    e. It occurs when excessive bicarbonate is lost from the body, as in diarrhoea.

11. Why do the kidney tubular cells secrete $NH_3$? *(Indicate all correct answers.)*
    a. because the urinary pH becomes too high
    b. because the body is in a state of alkalosis
    c. to enable further renal secretion of $H^+$
    d. to buffer excess filtered $HCO_3^-$
    e. because there is excess $NH_3$ in the body fluids

12. Complete the following chart:

| $\dfrac{[HCO_3^-]}{[CO_2]}$ | Uncompensated Abnormality | Possible Cause | pH |
|---|---|---|---|
| 10/1 | 1. _____ | 2. _____ | 3. _____ |
| 20/0.5 | 4. _____ | 5. _____ | 6. _____ |
| 20/2 | 7. _____ | 8. _____ | 9. _____ |
| 40/1 | 10. _____ | 11. _____ | 12. _____ |

14

## Written Questions

1. Explain the balance concept.
2. Outline the distribution of body $H_2O$.
3. Define *transcellular fluid*, and identify its components. Does the transcellular compartment as a whole reflect changes in the body's fluid balance?
4. Compare the ionic composition of plasma, interstitial fluid, and intracellular fluid.
5. What factors are regulated to maintain the body's fluid balance?
6. Why is regulation of ECF volume important? How is it regulated?
7. Why is regulation of ECF osmolarity important? How is it regulated? What are the causes and consequences of ECF hypertonicity and ECF hypotonicity?
8. Outline the sources of input and output in a daily salt balance and a daily $H_2O$ balance. Which are subject to control to maintain the body's fluid balance?
9. Distinguish between an acid and a base.
10. What is the relationship between $[H^+]$ and pH?
11. What is the normal pH of body fluids? How does this compare with the pH of $H_2O$?
12. Define *acidosis* and *alkalosis*.
13. What are the consequences of fluctuations in $[H^+]$?
14. What are the body's sources of $H^+$?
15. Describe the three lines of defence against changes in $[H^+]$ in terms of their mechanisms and speed of action.
16. List and indicate the functions of each of the body's chemical buffer systems.
17. What are the causes of the four categories of acid–base imbalances?
18. Why is uraemic acidosis so serious?

## Quantitative Exercises
### (Solutions in Appendix E, p. A-50)

1. Given that plasma pH 5 7.4, arterial $P_{CO_2} = 40$ mmHg, and each mmHg partial pressure of $CO_2$ is equivalent to a plasma $[CO_2]$ of 0.03 mM, what is the value of plasma $[HCO_3^-]$?
2. Death occurs if the plasma pH falls outside the range of 6.8–8.0 for an extended time. What is the concentration range of $H^+$ represented by this pH range?
3. A person drinks 1 L of distilled water. Use the data in Table 14-1 to calculate the resulting percentage increase in total body water (TBW), ICF, ECF, plasma, and interstitial fluid. Repeat the calculations for ingestion of 1 L of isotonic NACl. Which solution would be better at expanding plasma volume in a patient who has just haemorrhaged?

## POINTS TO PONDER

### (Explanations in Appendix E, p. A-51)

1. Alcoholic beverages inhibit vasopressin secretion. Given this fact, predict the effect of alcohol on the rate of urine formation. Predict the actions of alcohol on ECF osmolarity. Explain why a person still feels thirsty after excessive consumption of alcoholic beverages.
2. If a person loses 1500 mL of salt-rich sweat and drinks 1000 mL of water during the same time period, what will happen to vasopressin secretion? Why is it important to replace both the water and the salt?
3. If a solute that can penetrate the plasma membrane, such as dextrose (a type of sugar), is dissolved in sterile water at a concentration equal to that of normal body fluids and then is injected intravenously, what is the impact on the body's fluid balance?
4. Explain why it is safer to treat gastric hyperacidity with antacids that are poorly absorbed from the digestive tract than with baking soda, which is a good buffer for acid but is readily absorbed.
5. Which would buffer the acidosis accompanying severe pneumonia?
   a. $H^+ + HCO_3^- \rightarrow H_2CO_3 \rightarrow CO_2 + H_2O$
   b. $CO_2 H_2O \rightarrow H_2CO_3 \rightarrow H^+ + HCO_3^-$
   c. $H^+ + Hb \rightarrow HHb$
   d. $HHb \rightarrow H^+ + Hb$
   e. $NaH_2PO_4 + Na^+ \rightarrow Na_2HPO_4 + H^+$

## CLINICAL CONSIDERATION

### (Explanation in Appendix E, p. A-51)

Marilyn Y. has had pronounced diarrhoea for more than a week as a result of having acquired *salmonellosis*, a bacterial intestinal infection, from improperly handled food. What impact has this prolonged diarrhoea had on her fluid and acid–base balance? In what ways has Marilyn's body been trying to compensate for these imbalances?

14

## Altitude Adaptation

When a person travels to or lives at altitude, the cardiorespiratory system adapts. These changes can occur quickly (seconds) or take many months; some changes eventually become part of the genetic makeup of individuals whose ancestors have lived at altitude for generations.

Altitude generally refers to an elevation of ≥1500 m (4921 ft). As altitude increases, barometric pressure falls. As a result, so does the prevailing partial pressure of oxygen ($P_{O_2}$). At altitudes greater than 2500 m, this results in alveolar hypoxia and, assuming gas exchange across the alveolar–capillary membrane is normal, arterial hypoxaemia. Because a major function of the cardiorespiratory system is to deliver oxygen to tissues to maintain cellular metabolism, the fall of arterial $P_{O_2}$ reduces $O_2$ delivery to tissues, thereby compromising cellular function. To prevent this, the body compensates for (adapts to) altitude-induced hypoxia.

How much oxygen is delivered to the tissues depends on two factors: the concentration (or content) of oxygen in the blood and perfusion (for the whole body, this is the cardiac output). Cardiorespiratory adjustments, therefore, represent the body's efforts to increase these two factors. The study of responses to altitude is clinically relevant because these responses are similar in patients whose disease state is hypoxaemic.

In terms of the arterial $O_2$ content, adaptation starts immediately upon ascent to altitude and is especially noticeable if the ascent is rapid, as when one travels by air. The carotid chemoreceptors are immediately activated upon exposure to a hypoxic environment: for example, at 3000 m, where barometric $P_{O_2}$ is 525 mmHg, and ambient, inspired, and alveolar $P_{O_2}$ are approximately 110, 100, and 50 mmHg, respectively, in a healthy person. Due to the arterial hypoxaemia, the person hyperventilates. (By definition, a drop in the alveolar $P_{O_2}$ and arterial $P_{O_2}$ at the same metabolic $CO_2$ production indicates hyperventilation.) The fall in alveolar $P_{O_2}$, according to the alveolar gas equation (p. 525), increases alveolar $P_{O_2}$. Even if the increase is only a few mmHg, this can substantially increase the $O_2$ saturation and, therefore, arterial $O_2$ content when the body is functioning on the steep part of the $O_2$ dissociation curve (p. 532). The hypoxia-induced increase in ventilation, however, is blunted by the accompanying hypocapnia and the resulting respiratory alkalosis (pp. 540).

Over the next few days of acclimatization, ventilation gradually increases, for reasons that are still incompletely understood. The classical explanation is that the kidneys compensate for the alkalosis by not reabsorbing bicarbonate ($HCO_3^-$). The loss of $HCO_3^-$ occurs in all body compartments, including the cerebral spinal fluid (CSF) to which the central chemoreceptors are exposed. Thus, as the pH of the CSF gradually returns toward normal (i.e., re-acidifies), ventilation increases and, accompanying it, the arterial $P_{O_2}$ and arterial $O_2$ content. An alternate explanation for this acclimatization is an increase in peripheral chemoreceptor sensitivity to hypoxia. This is based on findings (primarily in research on goats) that the acclimatization occurs even when only the carotid bodies, not the rest of the body (including the brain), are exposed to hypoxic blood. In humans, acclimatization still occurs even if hypocapnia and respiratory alkalosis are prevented through the addition of carbon dioxide to the inspired gas.

The arterial $O_2$ content also increases rapidly at altitude because of a decrease in plasma volume, which is increased urine production and/or a shift of fluid to the interstitial and intracellular compartments. If a greater proportion of the blood is occupied by red blood cells (i.e., an increase in haematocrit), then for a given blood flow, more oxygen is delivered to the tissues.

The arterial hypoxaemia associated with altitude also stimulates, via activation of hypoxia-inducible factor, the release of erythropoietin, the protein that controls the production of red blood cells. Although earlier work indicated that the primary site of erythropoietin production was the kidneys and liver, more recent work indicates that many tissues, including the heart and diaphragm, contain the mRNA for erythropoietin. Although the red blood cells that are released may be immature, the haematocrit increases, thereby increasing the $O_2$-carrying capacity of the blood. This response takes about three to four weeks to become fully effective.

Increases in arterial $O_2$ content are, however, ineffective in maintaining $O_2$ delivery if perfusion to tissues is inadequate. At altitude, aortic chemoreceptors are stimulated by the decreased arterial $O_2$ content (not by the decreased arterial $P_{O_2}$) and reflexively stimulate the heart (increased heart rate), thereby increasing its output.

Many vascular beds of the systemic vasculature, especially those in the brain and heart, are sensitive to the prevailing blood gases. Hypoxaemia causes arterioles to dilate, increasing perfusion. This effect, however, is blunted by the hypocapnia resulting from altitude-induced hyperventilation; as a result, vascular resistance depends on the net effect of the opposing vasodilator and vasoconstrictor stimuli. In addition to arterial $P_{O_2}$ and $P_{O_2}$, other circulating factors (pH, adrenaline, angiotensin II, and adenosine) determine the vascular resistance. Many factors derived from the endothelium affect vascular smooth muscle, including nitric oxide and prostaglandins (vasodilators) and endothelin-1 and superoxides (vasoconstrictors). To these must be added the mechanical effects of changes in blood pressure and viscosity. Research to determine how all these factors interact—and change over time as the cardiorespiratory system responds—to control vascular resistance at altitude is in the early stages.

The vasculature in the lung responds differently to changes in arterial $P_{O_2}$. As described elsewhere (p. 539), hypoxia causes constriction, not dilation, of the pulmonary vasculature. Ordinarily, this is useful because *regional* constriction would force pulmonary blood flow to other, presumably better-ventilated regions with higher regional alveolar $P_{O_2}$ s. At altitude, however, the entire lung will be hypoxic (even though regional variations in ventilation/perfusion ratio will persist and, along with it, regional variations in alveolar $P_{O_2}$). As a result, pulmonary vascular resistance increases, causing an increase in the pressure generated by the right ventricle. At moderate altitudes, this is innocuous, but at higher altitudes the increased pressure can cause fluid to leak from the pulmonary blood vessels into the lung: a condition known as *high altitude pulmonary edema* (HAPE). There is considerable variation among individuals in terms of susceptibility to HAPE and, as yet, no way to predict who will suffer from it. The best way to prevent HAPE is by gradual ascent to altitude. Treatment involves administration of supplemental oxygen and descent.

Long-term adaptations once thought to include increased capillary density and mitochondria—both of which would improve $O_2$ delivery and utilization—have not been confirmed by later research. Indeed, mountain climbers, experience decreases in both body and muscle mass. Because the number of muscle capillaries does not change, the increase in capillary density is an artefact related to the loss of muscle mass. In fact, the ratio of capillaries to muscle fibres does not change.

Understandably, researchers and clinicians concentrate on oxygen and its delivery to tissues. As a result, much less attention has been given to the effects of changes in carbon dioxide on adaptation to altitude. Among the Quechua people of South America, who reside above 4000 m, new mothers swaddle their newborns, including their heads, even though this decreases the $P_{O_2}$ of the air inhaled by the infants (E.Z. Tronick et al. [1994]. The Quechua manta pouch: A caretaking practice for buffering the Peruvian infant against the multiple stressors of high altitude, *Child Dev*, 65: 1005–13). Although newborns have fetal haemoglobin, which has an $O_2$ dissociation curve displaced to the left (i.e., it binds oxygen more avidly than adult haemoglobin, making it better at taking up oxygen in the lung), the investigators did not measure the $P_{O_2}$ of the inhaled gas. It is possible that a higher $P_{O_2}$ would help dilate (or prevent as much vasoconstriction of) systemic blood vessels, thereby increasing blood flow and, therefore, $O_2$ delivery, despite a lower inspired $P_{O_2}$. A higher arterial $P_{O_2}$ would also help unload oxygen to tissues. It is unclear if the practice is actually beneficial; it has not been described in the few other societies that live at high altitude. Moreover, if this practice is beneficial, why do adults not use it? Perhaps they no longer need it because they are fully adapted.

## Altitude Training for Performance Enhancement

Formal altitude training has been used since the 1968 Olympic Games in Mexico City, Mexico. The many benefits associated with training at altitude (i.e., altitude training) result from the reduction in air density and partial pressure of oxygen ($P_{O_2}$), which stimulates increased red blood cell (RBC) production and leads to improved sea-level endurance performance.

Those who live well above sea level ($\geq$2500 m) have an advantage over athletes who reside at sea level, and their acclimatization is due to their chronic exposure to moderate-to-high-altitude

hypoxia. The acclimatized person has high $O_2$-carrying capacity (high haematocrit, haemoglobin, and erythrocyte count), similar to that observed in chronic polycythaemia. Increases in the levels of endogenous erythropoietin (EPO)—released principally from the kidneys, but with small amounts from the liver—increase the RBC volume and gradually reduce the increased ventilation and cardiac frequency that typically accompany acute moderate-to-high-altitude exposure. In response to hypoxic pulmonary vasoconstriction, other long-term physiological changes associated with the cardiovascular system include increases in blood volume, pulmonary arterial wall thickness, and right ventricular muscle mass. If the person does not respond well to altitude, the hypoxic vasoconstriction may lead to pulmonary hypertension, and right ventricular failure can develop, as seen in chronic mountain illness.

However, for most athletes, exposure to an altitude of 1500–2500 m for training purposes is physiologically manageable and has been shown by some research to be beneficial to athletic performance. Many sports teams train at an altitude of 2500 m. Two of the most common altitude training methodologies are "live high, train high" and "live high, train low." The latter is arguably the most frequently used and effective method even though its effectiveness has recently been questioned. Nonetheless, our discussion will focus on this strategy.

The basic guidelines for altitude-training camp (for eating and sleeping) state that an elevation of approximately 2500 m above sea level is required to stimulate an increase in RBC production and thus improve aerobic performance. At this height, the amount of oxygen available is about 25 percent less than at sea level. A minimum altitude exposure of about 8 to 10 hours per day, for 10 to 14 days, is needed in order to stimulate the adaptations necessary to increase the number of RBCs and thus changes in haematocrit and haemoglobin, which are essential to increasing the $O_2$ transport ability of the blood. These changes, along with changes in skeletal muscle mitochondria and oxidative enzymes improve endurance performance. Four weeks of training is typically recommended.

Recent research has advanced the development of successful altitude training protocols. Benjamin D. Levine and James Stray-Gundersen (1997) developed the basic protocol for the "live high, train low" approach (Living high-training low: Effect of moderate-altitude acclimatization with low-altitude training on performance. *J Appl Physiol*, 83: 102–12). They used college-level competitive runners as the participants. The benefit associated with their method was not only to expose the athletes to the hypoxic environment long enough to stimulate an increase in haematocrit but also to allow the athletes to maintain training intensity by training at low altitude. Recall that upon exposure to moderate to high altitude, many physiological and performance measures decrease, an effect known as *detraining*; this is due to the reduced inspired $P_{O_2}$. Upon acute exposure to altitude (2300 m), athletes demonstrated a 14 percent drop in maximum oxygen consumption ($\dot{V}O_2$ max), an 8 percent slower 1.5 km run time, and a 10 percent slower 5 km run time.

After four weeks of "live high, train low," Levine and Stray-Gundersen recorded an improvement in sea-level 5000 m time trials (13.4 s), and increases in RBC volume (9%), $O_2$ max

(5%), and maximal running speed at $O_2$ max. The improvements in $O_2$ max, RBC volume, and time trial performance were roughly proportional, indicating the importance of haematocrit to endurance performance.

These two researchers later repeated the study using elite male and female runners ([2001] *J App Physiol, 91*: 1113–20). The purpose was to determine if elite runners who were likely near their maximal structural and functional adaptive capacity (near their genetic ceiling) would be able to improve performance similarly to that of the college-level distance runners in the earlier 1997 study. The results indicated that elite distance runners improved their 3000 m run time by an average of 1.1 percent, which is a meaningful improvement at the world-class level. $O_2$ max also improved by 3 percent. The physiological changes following acclimatization to moderate altitude were an almost doubling of circulating EPO levels and a 10 g/L increase in haemoglobin concentration. These results indicate that the natural stimulation of the kidneys, liver, heart, and diaphragm to produce endogenous EPO could improve sea-level endurance performance in both well-trained and elite distance runners. Other researchers have observed similar improvements. However, as previously mentioned, a recent study suggests that part or all of the improvement in performance may be related to a placebo effect (C. Siebenmann et al., "Live high-train low" using normobaric hypoxia: A double-blinded, placebo-controlled study. *J Appl Physiol* [2011] doi:10.1152/japplphysiol.00388.2011).

On return to sea level, the body returns to homeostatic conditions associated with sea level. In the first two weeks, there is, typically, a decrease in haematocrit and an increase in plasma volume, and erythrocyte (RBC) formation in bone marrow is depressed. However, for competitive purposes, coaches suggest that when altitude-trained endurance athletes travel to sea-level competitions, they will have a higher RBC concentration for about 10 to 14 days, and this gives the athlete a competitive advantage. Approximately two months following the return from moderate-to-high-altitude training, total blood volume decreases and haematocrit is normal or slightly below normal. These adjustments gradually return the body to sea-level homeostasis.

# The Digestive System

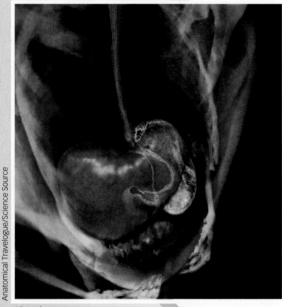

Digestive system

Body systems maintain homeostasis

## Homeostasis

The digestive system contributes to homeostasis by transferring nutrients, water, and electrolytes from the external environment to the internal environment.

Homeostasis is essential for survival of cells

Cells make up body systems

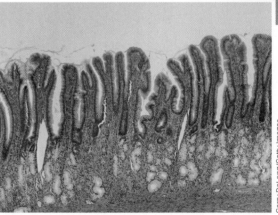

Photomicrograph of the pylorus

To maintain homeostasis, nutrient molecules used for energy production must continually be replaced by new, energy-rich nutrients. As well, nutrient molecules, especially proteins, are needed for ongoing synthesis of new cells and cell parts in the course of tissue turnover and growth. Similarly, water and electrolytes constantly lost in urine and sweat and through other avenues must be replenished regularly. The digestive system contributes to homeostasis by transferring nutrients, water, and electrolytes from the external environment to the internal environment. The digestive system does not directly regulate the concentration of any of these constituents in the internal environment. It does not vary nutrient, water, or electrolyte uptake based on body needs (with few exceptions); rather, it optimizes conditions for digesting and absorbing what is ingested.

# 15

# The Digestive System

▌ **Clinical Connections**

Cathy is 40 years old and has recently complained of a burning sensation in her abdomen. The pain, which occurs between meals and generally subsides while she is eating, has not been relieved by NSAIDs but is partially relieved by antacids. She has frequent belching, feels bloated, and has a general fatigue. She has also recently experienced a noticeable weight loss. When her vitals were taken, she was hypotensive and had tachycardia. A fecal occult blood test was positive. She underwent endoscopy that visualized a peptic ulcer, and a biopsy from the site tested positive for *H. pylori*.

## 15.1 | Introduction

The primary function of the **digestive system** is to transfer nutrients, water, and electrolytes from the food we eat into the body's internal environment. The digestive system is a series of hollow organs that are joined together. It begins with the mouth and terminates at the anus. Ingested food is essential as an energy source, or fuel, from which the cells can generate ATP to carry out their particular energy-dependent activities, such as active transport, contraction, synthesis, and secretion. Food is also a source of building supplies for the renewal and addition of body tissues.

The act of eating does not automatically make the preformed organic molecules in food available to the body cells as a source of fuel or as building blocks. First, the food must be digested, that is, biochemically broken down, into small, simple molecules that can be absorbed from the digestive tract into the circulatory system for distribution to the cells. Normally, about 95 percent of the ingested food is made available for the body's use. The sequence in nutrient acquisition is ingestion, digestion, absorption, distribution, and usage.

We first provide an overview of the digestive system, examining the common features of the various components of the system, and then we begin a detailed tour of the tract from beginning to end.

## Four digestive processes

There are four basic digestive processes: *motility, secretion, digestion*, and *absorption*.

### MOTILITY

The term **motility** refers to the muscular contractions that mix and move forward the contents of the digestive tract. Like vascular smooth muscle, the smooth muscle in the walls of the digestive tract maintains a constant low level of contraction known as tone. Tone is important in maintaining a steady pressure on the contents of the digestive tract as well as in preventing its walls from remaining permanently stretched following distension.

Two basic types of digestive motility are superimposed on this ongoing tonic activity: propulsive movements and mixing movements. *Propulsive movements* propel, or push, the contents forward through the digestive tract; the rate of propulsion varies depending on the functions of the different regions. That is, the contents move forward in a given segment at a velocity that's appropriate for that segment to fulfill its role in digestion. *Mixing movements* serve a twofold function. First, by mixing food with the digestive juices, these movements promote digestion of the food. Second, they facilitate absorption by exposing all parts of the intestinal contents to the absorbing surfaces of the digestive tract.

Contraction of the smooth muscle within the walls of the digestive organs accomplishes movement of material through most of the digestive tract. The exceptions occur at the ends of the tract—the mouth through the early part of the oesophagus at the beginning and the external anal sphincter at the end. In these regions, the acts of chewing, swallowing, and defecation have voluntary components, and thus use skeletal muscle. By contrast, motility accomplished by smooth muscle throughout the rest of the tract is controlled by complex involuntary mechanisms.

### SECRETION

A number of digestive juices are secreted into the digestive tract lumen by exocrine glands (p. 7) along the route, each with its own specific secretory product. Each **digestive secretion** consists of water, electrolytes, and specific organic constituents important in the digestive process, such as enzymes, bile salts, or mucus. The secretory cells extract from the plasma large volumes of water and the raw materials necessary to produce their particular secretion.

Secretion of all digestive juices requires energy, both for active transport of some of the raw materials into the cell (others diffuse in passively) and for synthesis of secretory products by the endoplasmic reticulum. On appropriate neural or hormonal stimulation, the secretions are released into the digestive tract lumen. Normally, there is no net loss of water as the digestive secretions are reabsorbed in one form or another back into the blood after their participation in digestion. However, in the case of vomiting or diarrhoea, the digestive secretions are not fully reabsorbed, resulting in a net loss of the water that had initially been borrowed from the plasma. This can lead to dehydration.

Furthermore, endocrine cells located in the digestive tract wall secrete into the blood the gastrointestinal hormones (e.g., gastrin, secretin, motilin) that help control digestive motility and exocrine gland secretion.

### DIGESTION

Humans consume three different biochemical categories of energy-rich foodstuffs: *carbohydrates, proteins*, and *fats*. These large molecules cannot cross plasma membranes intact to be absorbed from the lumen of the digestive tract into the blood or lymph. The term **digestion** refers to the biochemical breakdown of the structurally complex foodstuffs of the diet into smaller, absorbable units by the enzymes produced within the digestive system, as follows:

1. The simplest form of **carbohydrates** is the simple sugars or **monosaccharides** (one-sugar molecules), such as **glucose, fructose**, and **galactose**, very few of which are normally found in the diet (Appendix B, p. A-11). Most ingested carbohydrate is in the form of **polysaccharides** (many-sugar molecules), which consist of chains of interconnected glucose molecules. The most common polysaccharide consumed is **starch** derived from plant sources. In addition, meat contains **glycogen**, the polysaccharide storage form of glucose in muscle. **Cellulose**, another dietary polysaccharide, found in plant walls, cannot be digested into its constituent monosaccharides by the digestive juices humans secrete; it represents the indigestible *fibre*, or the bulk, in our diets. Besides polysaccharides, a lesser source of dietary carbohydrate is in the form of **disaccharides** (two-sugar molecules), including **sucrose** (table sugar, which consists of one glucose and one fructose molecule) and **lactose** (milk sugar made up of one glucose and one galactose molecule). Through the process of digestion, starch, glycogen, and disaccharides are converted into monosaccharides, principally glucose with small amounts of fructose and galactose. These monosaccharides are the absorbable units for carbohydrates.

2. The sources of dietary **protein** are meats, legumes, eggs, grains, and dairy products. Dietary proteins consist of various combinations of **amino acids** held together by peptide bonds (Appendix B, p. A-14). Through the process of digestion, proteins are degraded primarily into their constituent amino acids as well as a few **small polypeptides** (several amino acids linked by peptide bonds), both of which are the absorbable units for protein.

3. Most dietary **fats** are in the form of **triglycerides,** which are neutral fats, each consisting of a glycerol with three **fatty acid** molecules attached (*tri* means "three") (Appendix B, p. A-12). During digestion, two of the fatty acid molecules are split off, leaving a **monoglyceride**—a glycerol molecule with one fatty acid molecule attached (*mono* means "one"). Thus, the end products of fat digestion are monoglycerides and free fatty acids, which are the absorbable units of fat.

Digestion is accomplished by enzymatic **hydrolysis** (breakdown by water; Appendix B, p. A-15). By adding water at the bond site, enzymes in the digestive secretions break down the bonds that hold the small molecular subunits within the nutrient molecules together, thus setting the small molecules free (⟩ Figure 15-1). The removal of water at the bond sites (dehydration synthesis) originally joined these small subunits to form nutrient molecules. Hydrolysis replaces the water and frees the small absorbable units. Digestive enzymes are specific in the type of bonds they can hydrolyze. As food moves through the digestive tract, it is subjected to various enzymes, each of which breaks down the food molecules even further. In this way, large food molecules are converted to simple absorbable units in a progressive, stepwise fashion, like an assembly line in reverse, as the digestive tract contents are propelled forward.

### ABSORPTION

In the small intestine, digestion is completed and most absorption occurs. Through the process of **absorption**, the small absorbable units that result from digestion, along with water, vitamins, and electrolytes, are transferred from the digestive tract lumen into the blood or lymph. As we examine the digestive tract from beginning to end, we will discuss the four processes of motility, secretion, digestion, and absorption as they take place within each digestive organ (see ▌Table 15-1).

## The digestive tract and accessory organs

The digestive system consists of the *digestive* (or *gastrointestinal*) *tract* plus the accessory digestive organs (*gastro* means "stomach"). The **accessory digestive organs** include the *salivary glands,* the *exocrine pancreas,* and the *biliary system,* which is composed of the *liver* and *gallbladder*. These exocrine organs lie outside the digestive tract and empty their secretions through ducts into the digestive tract lumen.

The **digestive tract** is a tube about 4.5 m in length in its normal contractile state (in the uncontracted state of a cadaver the tract is 9 m long). Running through the middle of the body, the digestive tract includes the following organs (▌Table 15-1): *mouth; pharynx* (throat); *oesophagus; stomach; small intestine* (consisting of the *duodenum, jejunum,* and *ileum*); *large intestine* (the *cecum, appendix, colon,* and *rectum*); and *anus*. Although these organs are continuous with one another, they are considered separate entities because of regional modifications that allow for specialized activities. Because the digestive tract is continuous from the mouth to the anus, like the lumen of a straw, it is continuous with the external environment. As a result, the contents within the lumen of the digestive tract are technically outside the body, just as the liquid you suck through a straw is not a part of the straw. Only after a substance has been absorbed from the lumen across the digestive tract wall is it considered part of the body. This is important, because conditions essential to digestion can be tolerated in the digestive tract lumen, but not in the body proper. Consider the following examples:

- The pH of the stomach contents is ≤ 2 as a result of the gastric secretion of hydrochloric acid (HCl), yet the body fluid's pH in other regions ranges between 6.8 and 8.0.

- The digestive enzymes that hydrolyze the protein in food could also destroy the body's own tissues (proteins) that produce them. Therefore, once these enzymes are synthesized in inactive form, they are not activated until they reach the lumen, where they actually attack the food outside the body (i.e., within the lumen), thereby protecting the body tissues against self-digestion.

- In the lower intestine exist quadrillions of living microorganisms that are normally harmless and even beneficial, yet if these same microorganisms enter the body proper (as may happen with a ruptured appendix), they may be extremely harmful or even lethal.

## The digestive tract wall

The digestive tract wall has the same general structure throughout most of its length, from oesophagus to anus, with some local variations characteristic of each region. A cross-section of the digestive tube reveals four major tissue layers (⟩ Figure 15-2). From the innermost layer outward they are the *mucosa*, the *submucosa*, the *muscularis externa*, and the *serosa*.

### MUCOSA

The **mucosa** lines the luminal surface of the digestive tract, as well as other body cavities that are exposed to the external environment and internal organs. It is divided into three layers:

- The primary component is the **mucous membrane,** an inner epithelial layer that serves as a protective surface. It is modified in particular areas for secretion and absorption. The mucous membrane contains

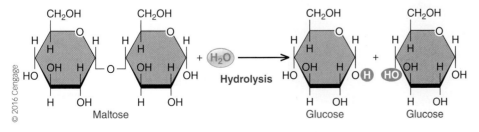

⟩ **FIGURE 15-1 An example of hydrolysis.** In this example, the disaccharide maltose (the intermediate breakdown product of polysaccharides) is broken down into two glucose molecules by the addition of $H_2O$ at the bond site.

© 2016 Cengage

15

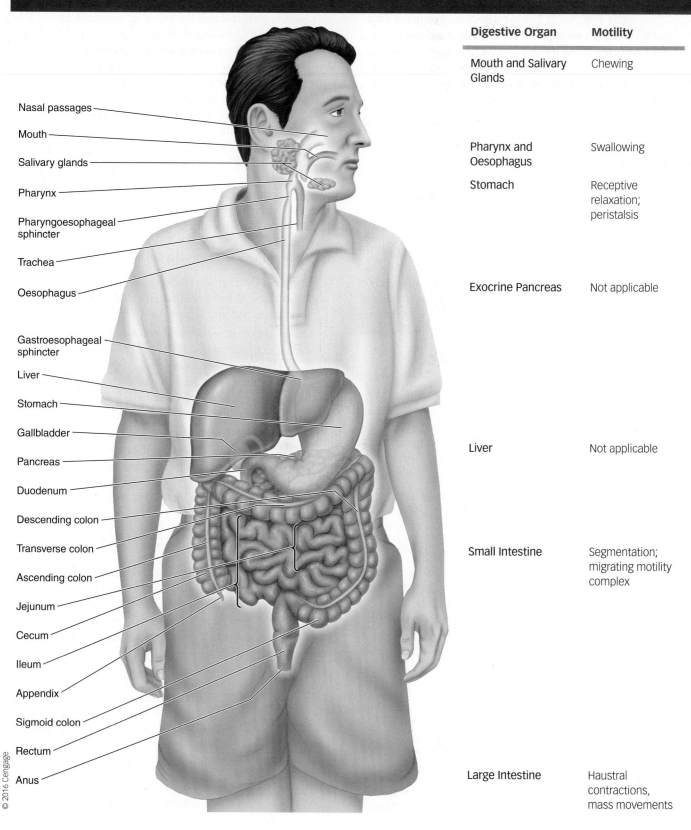

| Digestive Organ | Motility |
|---|---|
| Mouth and Salivary Glands | Chewing |
| Pharynx and Oesophagus | Swallowing |
| Stomach | Receptive relaxation; peristalsis |
| Exocrine Pancreas | Not applicable |
| Liver | Not applicable |
| Small Intestine | Segmentation; migrating motility complex |
| Large Intestine | Haustral contractions, mass movements |

Nasal passages

Mouth

Salivary glands

Pharynx

Pharyngoesophageal sphincter

Trachea

Oesophagus

Gastroesophageal sphincter

Liver

Stomach

Gallbladder

Pancreas

Duodenum

Descending colon

Transverse colon

Ascending colon

Jejunum

Cecum

Ileum

Appendix

Sigmoid colon

Rectum

Anus

© 2016 Cengage

**15**

| Secretion | Digestion | Absorption |
|---|---|---|
| **Saliva**<br>• Amylase<br>• Mucus<br>• Lysozyme | Carbohydrate digestion begins | No foodstuffs; a few medications— for example, nitroglycerin |
| **Mucus** | None | None |
| **Gastric Juice**<br>• HCl<br>• Pepsin<br>• Mucus<br>• Intrinsic factor | Carbohydrate digestion continues in body of stomach; protein digestion begins in antrum of stomach | No foodstuffs; a few lipid-soluble substances, such as alcohol and aspirin |
| **Pancreatic Digestive Enzymes**<br>• Trypsin, chymotrypsin, carboxypeptidase<br>• Amylase<br>• Lipase<br>**Pancreatic aqueous NaHCO₃ secretion** | These pancreatic enzymes accomplish digestion in duodenal lumen | Not applicable |
| **Bile**<br>• Bile salts<br>• Alkaline secretion<br>• Bilirubin | Bile does not digest anything, but bile salts facilitate fat digestion and absorption in duodenal lumen | Not applicable |
| **Succus Entericus**<br>**Mucus**<br>**Salt**<br>(Small intestine enzymes— disaccharidases and aminopeptidases— are not secreted but function within the brush-border membrane.) | In lumen, under influence of pancreatic enzymes and bile, carbohydrate and protein digestion continues, and fat digestion is completely accomplished; in brush border, carbohydrate and protein digestion is completed. | All nutrients, most electrolytes, and water |
| **Mucus** | None | Salt and water, converting contents to feces |

*exocrine gland cells* for secretion of digestive juices, *endocrine gland cells* for secretion of blood-borne gastrointestinal hormones, and *epithelial cells* specialized for absorbing digested nutrients.

- The **lamina propria** is a thin middle layer of connective tissue on which the epithelium rests. It houses the **gut-associated lymphoid tissue (GALT)**, which is important in the defence against disease-causing intestinal bacteria (p. 464).
- The **muscularis mucosa**, a sparse layer of smooth muscle, is the outermost mucosal layer that lies adjacent to the submucosa.

The mucosal surface is generally highly folded (ridged), which greatly increases the surface area available for absorption. The degree of folding varies depending on the region of the tract; it is most extensive in the small intestine, where maximum absorption occurs, and least in the oesophagus, which merely serves as a transit tube. The pattern of surface folding can be modified by contraction of the muscularis mucosa.

### SUBMUCOSA

The **submucosa** (under the mucosa) is a thick layer of connective tissue that supports the mucosa and provides the digestive tract with its distensibility and elasticity. It contains larger blood and lymph vessels, both of which send branches inward to the mucosal layer and outward to the surrounding thick muscle layer. Also, a nerve network known as the *submucosal plexus* lies within the submucosa (*plexus* means "network").

### MUSCULARIS EXTERNA

The **muscularis externa**, the major smooth muscle coat of the digestive tube, surrounds the submucosa. In most parts of the tract, the muscularis externa consists of two layers: an *inner circular layer* (encircle) and an *outer longitudinal layer.* Contraction of these circular fibres decreases the lumen diameter, constricting the tube at the point of contraction. Contraction of the fibres in the outer layer (longitudinally), shortens the tube. Together, contractile activity of these smooth muscle layers produces the propulsive and mixing movements. Another nerve network, the *myenteric plexus*, lies between the two muscle layers (*myo* means "muscle"; *enteric* means "intestine"). Together the submucosal and myenteric plexuses, along with hormones and local chemical mediators, help regulate local gut activity.

### SEROSA

The outer connective tissue covering of the digestive tract is the **serosa (serous membrane)**, which secretes a watery fluid (**serous fluid**) that lubricates to reduce friction between the digestive organs and surrounding viscera. Throughout much of the tract, the serosa is continuous with the **mesentery**, which suspends the digestive organs from the inner wall of the abdominal cavity like a sling (› Figure 15-2). This attachment provides relative fixation, supporting the digestive organs in proper position, while still allowing them freedom for mixing and propulsive movements.

**15**

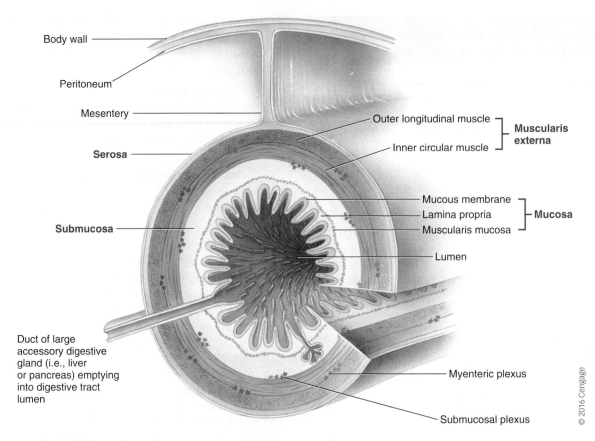

⟩ **FIGURE 15-2 Layers of the digestive tract wall.** The digestive tract wall consists of four major layers: from the innermost out, they are the mucosa, submucosa, muscularis externa, and serosa.

## Regulation of digestive function

Digestive motility and secretion are carefully regulated to maximize digestion and absorption of ingested food. Four factors are involved in regulating digestive system function: (1) autonomous smooth muscle function, (2) intrinsic nerve plexuses, (3) extrinsic nerves, and (4) gastrointestinal hormones.

### AUTONOMOUS SMOOTH MUSCLE FUNCTION

Similar to some self-excitable cardiac muscle cells, some smooth muscle cells are pacemaker cells that display rhythmic, spontaneous variations in membrane potential. The prominent type of self-induced electrical activity in digestive smooth muscle is slow-wave potentials (p. 331), alternatively referred to as the digestive tract's **basic electrical rhythm (BER)**. Muscle-like but noncontractile cells known as the **interstitial cells of Cajal** are the pacemaker cells that instigate cyclic slow-wave activity. These pacemaker cells lie at the boundary between the longitudinal and circular smooth muscle layers. Slow waves are not action potentials and do not directly induce muscle contraction; they are rhythmic, wavelike fluctuations in membrane potential that cyclically bring the membrane closer to or farther from threshold potential. These slow-wave oscillations are believed to be due to cyclic variations in calcium release from the endoplasmic reticulum and calcium uptake by the mitochondria of the pacemaker cell. If these waves reach threshold at the peaks of

depolarization, a volley of action potentials is triggered at each peak, resulting in rhythmic cycles of muscle contraction.

Like cardiac muscle, smooth muscle cells are connected by gap junctions, through which charge-carrying ions can flow (p. 36). In this way, electrical activity initiated in a digestive tract pacemaker cell spreads to the adjacent contractile smooth muscle cells. New evidence suggests that this electrical activity may also spread via the enteric nervous system, described shortly. Thus, the whole muscle sheet behaves like a functional syncytium, becoming excited and contracting as a unit at threshold (p. 328). If threshold is not achieved, the oscillating slow-wave sweeps across the muscle sheet without being accompanied by contractile activity.

Whether threshold is reached depends on various mechanical, neural, and hormonal factors that influence the starting point around which the slow-wave rhythm oscillates. If the starting point is nearer the threshold level, as it is when food is present in the digestive tract, the depolarizing slow-wave peak reaches threshold, so action potential frequency and its accompanying contractile activity increase. Conversely, if the starting point is farther from threshold, as when no food is present, there is a lower likelihood of reaching threshold.

The *rate* (frequency) of self-induced rhythmic digestive contractile activities, such as peristalsis in the stomach, segmentation in the small intestine, and haustral contractions in the large intestine, depends on the inherent rate established by the involved pacemaker cells. The *intensity* (strength) of these contractions depends on the number of action potentials that occur when the slow-wave potential reaches threshold, which in turn depends on how long threshold is sustained. At threshold, voltage-gated $Ca^{2+}$ channels are activated (p. 58), resulting in $Ca^{2+}$ influx into the smooth muscle cell. The resultant $Ca^{2+}$ entry has two effects: (1) it is responsible for the rising phase of an action potential, with the falling phase being brought about as usual by $K^+$ efflux; and (2) it triggers a contractile response (p. 329). The greater the number of action potentials that take place, the higher the cytosolic $Ca^{2+}$ concentration, the greater the cross-bridge activity, and the stronger the contraction. Other factors that influence contractile activity also do so by altering the cytosolic $Ca^{2+}$ concentration. Consequently, the level of contractility can range from low-level tone to vigorous mixing and propulsive movements due to the variations in cytosolic $Ca^{2+}$ concentration.

### INTRINSIC NERVE PLEXUSES

The **intrinsic nerve plexuses** are the two major networks of nerve fibres—the submucosal plexus and the **myenteric plexus**—that lie entirely within the digestive tract wall and run its entire length. This means the digestive tract has its own intramural nervous system, which contains as many neurons as the spinal cord and endows the tract with a considerable degree of self-regulation. These two plexuses are often termed the **enteric nervous system**. However, the enteric nervous system does receive substantial input from the autonomic nervous system.

The intrinsic plexuses influence all facets of digestive tract activity. Various types of neurons are present in the intrinsic plexuses. Some are sensory, and these have receptors that respond to specific local stimuli in the digestive tract. Other local neurons innervate the smooth muscle cells and the exocrine and endocrine cells of the digestive tract to directly affect digestive tract motility, secretion of digestive juices, and secretion of gastrointestinal hormones. As with the CNS, the enteric nervous system is linked by interneurons; some neurons are excitatory, and some are inhibitory. For example, neurons that release *acetylcholine* promote smooth muscle contraction, whereas *nitric oxide* and *vasoactive intestinal peptide* cause relaxation. These intrinsic nerve networks primarily coordinate local activity within the digestive tract. To illustrate, if a large piece of food gets stuck in the oesophagus, the intrinsic plexuses coordinate local responses to push the food forward. Intrinsic nerve activity can, in turn, be influenced by the extrinsic nerves.

### EXTRINSIC NERVES

The **extrinsic nerves** are the nerve fibres from both branches of the ANS that originate outside the digestive tract and innervate the various digestive organs. The autonomic nerves influence digestive tract motility and secretion either by modifying ongoing activity in the intrinsic plexuses, altering the level of gastrointestinal hormone secretion, or, in some instances, acting directly on the smooth muscle and glands.

Recall that, in general, the sympathetic and parasympathetic nerves supplying any given tissue exert opposing actions on that tissue. The sympathetic nervous system tends to inhibit or slow down digestive tract contraction and secretion. This action is appropriate, considering that digestive processes are not of highest priority when the body faces an emergency. The parasympathetic nervous system dominates when general maintenance types of activities, such as digestion, can proceed optimally. Accordingly, the parasympathetic nerve fibres supplying the digestive tract tend to increase smooth muscle motility and promote secretion of digestive enzymes and hormones. A unique parasympathetic nerve supply of the digestive tract—the postganglionic parasympathetic nerve fibres—is actually a part of the intrinsic nerve plexuses, which secrete acetylcholine within the plexuses. Accordingly, acetylcholine is released in response to local reflexes coordinated entirely by the intrinsic plexuses as well as to vagal stimulation, which acts through the intrinsic plexuses.

In addition to being called into play during generalized sympathetic or parasympathetic discharge, the autonomic nerves, especially the vagus nerve, can be discretely activated to modify digestive activity only. One of the major purposes of specific activation of extrinsic innervation is to coordinate activity between different regions of the digestive system. For example, the act of chewing food reflexly increases not only salivary secretion but also stomach, pancreatic, and liver secretion via vagal reflexes in anticipation of the arrival of food.

*Clinical Note* Swallowing can become a great concern for the elderly, because such diseases as Alzheimer's can interfere with the autonomic nervous system. It is important to correct this condition, since it affects the same neuromuscular structures that influence speech. Hospitals often use speech therapists to work with Alzheimer's patients.

### GASTROINTESTINAL HORMONES

Tucked within the mucosa of certain regions of the digestive tract are endocrine gland cells that release hormones into the blood. These **gastrointestinal hormones** are carried through the blood to other areas of the digestive tract, where they exert either excitatory or inhibitory influences on smooth muscle and exocrine gland cells; for example, gastrin stimulates the release of gastric juices (acid) by the stomach. Interestingly, many of these same hormones are released from neurons in the brain, where they act as neurotransmitters and neuromodulators. During embryonic development, certain cells of the developing neural tissue migrate to the digestive system, where they become endocrine cells.

## Receptor activation

The digestive tract wall contains three types of sensory receptors that respond to local changes in the digestive tract: (1) *chemoreceptors* sensitive to chemical components within the lumen, (2) *mechanoreceptors* sensitive to stretch or tension within the wall, and (3) *osmoreceptors* sensitive to the osmolarity of the luminal contents.

Stimulation of these receptors elicits neural reflexes or secretion of hormones, both of which alter the level of activity in the digestive system's effector cells. These effector cells include smooth muscle cells (for modifying motility),

15

exocrine gland cells (for controlling secretion of digestive juices), and endocrine gland cells (for varying secretion of gastrointestinal hormones (> Figure 15-3). Receptor activation may bring about two types of neural reflexes: short reflexes and long reflexes. When the intrinsic nerve networks influence local motility or secretion in response to specific local stimulation, all elements of the reflex are located within the wall of the digestive tract itself; that is, a **short reflex** takes place. Extrinsic autonomic nervous activity can be superimposed on the local controls to modify smooth muscle and glandular responses, either to correlate activity between different regions of the digestive system or to modify digestive system activity in response to external influences. Because the autonomic reflexes involve long pathways between the central nervous system and digestive system, they are known as **long reflexes**. In addition to these neural reflexes, digestive system activity is coordinated by the gastrointestinal hormones, which are released in response to local changes in the digestive tract or by short or long reflexes.

Sensory receptors within the digestive tract wall monitor luminal content and wall tension. Receptor proteins within the plasma membranes of the digestive system's effector cells bind with and respond to gastrointestinal hormones, neurotransmitters, and local chemical mediators.

From this overview, you can see that regulation of gastrointestinal function is very complex, influenced by many synergistic, interrelated pathways that are designed to ensure that the appropriate responses occur to digest and absorb the ingested food. In no other system of the body is there so much overlapping control.

Next we take a tour of the digestive tract, beginning with the mouth and ending with the anus. Along the way, we will examine each digestive organ in terms of the four basic digestive processes of motility, secretion, digestion, and absorption. Table 15-1 summarizes these activities and serves as a useful reference throughout the rest of the chapter.

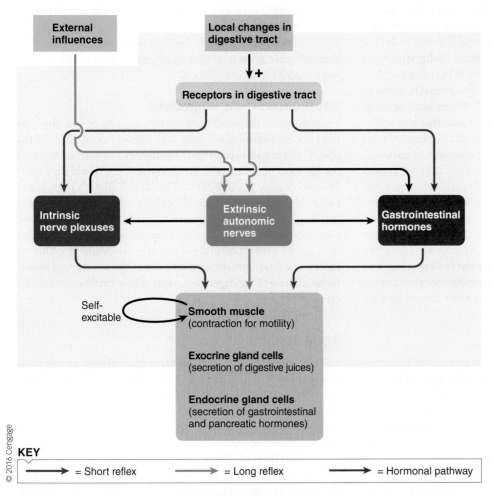

> FIGURE 15-3 **Summary of pathways controlling digestive system activities**

KEY

⟶ = Short reflex    ⟶ = Long reflex    ⟶ = Hormonal pathway

© 2016 Cengage

## 15.2 | Mouth

### The oral cavity

Entry to the digestive tract is through the **mouth (oral cavity)**. The opening is formed by the muscular **lips**, which help procure, guide, and hold the food in the mouth. The lips also serve nondigestive functions: for example, in speech, which depends on particular lip formations, and in interpersonal relationships, such as in kissing when the lips serve as a sensory receptor.

The **palate**, which forms the arched roof of the oral cavity, separates the mouth from the nasal passages. It allows breathing and chewing or sucking to take place simultaneously. Hanging down from the palate in the rear of the throat is the **uvula**, which plays an important role in sealing off the nasal passages during swallowing.

The **tongue** forms the floor of the oral cavity, is composed of voluntarily controlled skeletal muscle, and contains the majority of the **taste buds** (p. 181). Movements of the tongue guide food within the mouth for chewing and swallowing, and are also important in speech.

The **pharynx** is the cavity at the rear of the throat. It acts as a common passageway for both the digestive system (links the mouth and oesophagus) and the respiratory system. This arrangement necessitates mechanisms to guide food and air into the proper passageways beyond the pharynx. Within the side walls of the pharynx are the **tonsils**, lymphoid tissues that are part of the immune system.

## The teeth

**Mastication (chewing)** is the motility of the mouth that involves the slicing, tearing, grinding, and mixing of ingested food by the **teeth**. The teeth are embedded in and protrude from the jawbones. The exposed part of a tooth is covered by **enamel**, the hardest structure of the body, which forms before the tooth's eruption.

The upper and lower teeth normally fit together when the jaws are closed. This **occlusion** allows food to be ground and crushed between the tooth surfaces. The teeth can exert great forces during eating; an adult male can exert up to 90 kg. This is sufficient to crack a hard nut. However, the degree of occlusion is most important in determining the efficiency of chewing.

*Clinical Note* When the teeth do not make proper contact (**malocclusion**) they cannot accomplish their normal cutting and grinding action. This occurs due to abnormal positioning of the teeth, often caused by either overcrowding of teeth or displacement of the jaw in relation one another. Malocclusion can cause abnormal wearing of affected tooth surfaces and dysfunction and pain of the **temporomandibular joint (TMJ)**, where the jawbones articulate with each other. Malocclusion is often corrected by braces, which exert prolonged gentle pressure against the teeth to move them gradually into position.

The functions of chewing are (1) to grind and break food up for swallowing and increase the food surface area on which salivary enzymes act, (2) to mix food with saliva, and (3) to stimulate the taste buds. The third function not only gives rise to the pleasurable subjective sensation of taste, but also reflexively increases salivary, gastric, pancreatic, and bile secretion to prepare for the arrival of food.

The act of chewing can be voluntary, but most chewing during a meal is a rhythmic reflex brought about by activation of the skeletal muscles of the jaws, lips, cheeks, and tongue in response to the pressure of food against the oral tissues.

## Saliva

**Saliva** secretion is produced largely by three major pairs of salivary glands that lie outside the oral cavity and discharge saliva through short ducts into the mouth. It is stimulated by both the parasympathetic and sympathetic nervous systems.

Saliva is about 99.5 percent water and 0.5 percent electrolytes and protein, and approximately 1 to 1.5 L/day is secreted. Saliva secretions range from a basal rate of 0.5 mL/min to a maximum flow rate of about 5 mL/min in response to a potent stimulus (e.g., sucking a lemon). The salivary salt (NaCl) concentration is only one-seventh of that in the plasma, which is important in perceiving salty tastes. Similarly, discrimination of sweet tastes

is enhanced by the absence of glucose in the saliva. The most important salivary proteins are *amylase, mucus,* and *lysozyme.* They contribute to the functions of saliva as follows:

1. Saliva begins digestion of carbohydrate in the mouth via **salivary amylase**, an enzyme that breaks polysaccharides down into **maltose**, a disaccharide (see › Figure 15-1).

2. Saliva facilitates swallowing by moistening food particles, holding them together, and by providing lubrication by the **mucus**.

3. Saliva exerts some antibacterial action—first, by **lysozyme**, an enzyme that destroys certain bacteria by breaking down their cell walls; and second, by rinsing away material that may serve as a food source for bacteria.

4. Saliva serves as a solvent for molecules; only molecules in solution can react with taste bud receptors. Self-demonstration: Dry your tongue and then drop some sugar on it—you cannot taste the sugar until it is moistened.

5. Saliva aids speech by facilitating movements of the lips and tongue.

6. Saliva plays an important role in oral hygiene by helping keep the mouth and teeth clean. The constant flow of saliva helps flush away food residues, foreign particles, and old epithelial cells that have shed from the oral mucosa.

7. Saliva is rich in bicarbonate buffers, which neutralize acids in food as well as acids produced by bacteria in the mouth, thereby helping prevent dental caries (this is the correct term for what is commonly referred to as tooth decay or dental cavities).

Despite these many functions, saliva is not essential for digesting and absorbing foods, because enzymes produced by the pancreas and small intestine can complete food digestion even in the absence of salivary and gastric secretion.

*Clinical Note* The main problems associated with diminished salivary secretion, a condition known as **xerostomia**, are difficulty in chewing and swallowing, inarticulate speech unless frequent sips of water are taken when talking, and a rampant increase in dental caries unless special precautions are taken.

## Salivary secretion

Basal secretions brought on by low-level stimulation by the parasympathetic nervous system are important in keeping the mouth and throat moist at all times. In addition to this continuous, low-level secretion, salivary secretion may be increased by two types of salivary reflexes: the simple and conditioned salivary reflexes (› Figure 15-4).

### SIMPLE AND CONDITIONED SALIVARY REFLEXES

The **simple salivary reflex** occurs when chemoreceptors and pressure receptors within the oral cavity respond to the presence of food. On activation, these receptors initiate impulses in afferent nerve fibres that carry the information to the **salivary centre**, which is located in the medulla of the brain stem, as are

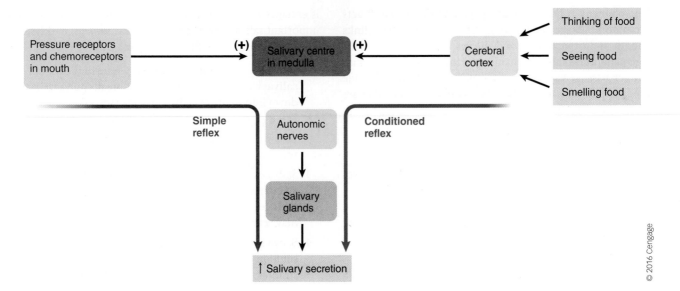

> FIGURE 15-4 **Control of salivary secretion**

all the brain centres that control digestive activities. The salivary centre, in turn, sends impulses via the extrinsic autonomic nerves to the salivary glands to promote increased salivation.

With the **conditioned (acquired) salivary reflex**, salivation occurs without oral stimulation. Just thinking about, seeing, smelling, or hearing the preparation of pleasant food initiates salivation through this reflex. All of us have experienced such "mouth-watering" in anticipation of something delicious to eat. This reflex is a learned response based on previous experience. Through functions of the cerebral cortex, our mental associations with the pleasure of eating stimulate the medullary salivary centre.

### AUTONOMIC INFLUENCE ON SALIVARY SECRETION

The salivary centre controls the degree of salivary output by means of the autonomic nerves that supply the salivary glands. Unlike the autonomic nervous system elsewhere in the body, sympathetic and parasympathetic responses in the salivary glands are not antagonistic. Both sympathetic and parasympathetic stimulation increase salivary secretion, but the quantity, characteristics, and mechanisms are different. Parasympathetic stimulation, which exerts the dominant role in salivary secretion, produces a prompt and abundant flow of watery saliva that is rich in enzymes. Sympathetic stimulation, by contrast, produces a much smaller volume of thick saliva that is rich in mucus. Because sympathetic stimulation elicits a smaller volume of saliva, the mouth feels drier than usual during circumstances when the sympathetic system is dominant, such as stress situations (giving a speech).

Salivary secretion is the only digestive secretion entirely under neural control. All other digestive secretions are regulated by both nervous system reflexes and hormones.

### Digestion

Digestion in the mouth involves the hydrolysis of polysaccharides into disaccharides by amylase. However, most digestion by this enzyme is accomplished in the body of the stomach after the food mass and saliva have been swallowed. Acid inactivates amylase, but in the centre of the food mass, where stomach acid has not yet reached, this salivary enzyme continues to function for several more hours.

No absorption of foodstuff occurs from the mouth. Importantly, some drugs can be absorbed by the oral mucosa; a prime example is *nitroglycerin*, a vasodilator drug sometimes used by cardiac patients to relieve anginal attacks (p. 378) associated with myocardial ischemia (p. 360).

---

#### Check Your Understanding 15.2

1. State the functions of salivary mucus, amylase, and lysozyme.
2. Compare the effects of parasympathetic versus sympathetic stimulation of the salivary glands.

---

## 15.3 | Pharynx and Oesophagus

The motility associated with the pharynx and oesophagus is **swallowing**. Most of us think of swallowing as the limited act of moving food out of the mouth into the oesophagus. However, swallowing actually is the entire process of moving food from the mouth through the oesophagus into the stomach.

### Swallowing: A programmed all-or-none reflex

Swallowing is initiated when a **bolus**, or ball of chewed or liquid food, is voluntarily forced by the tongue to the rear of the mouth into the pharynx (> Figure 15-5a). Swallowing is a complex task requiring a coordinated effort of 25 pairs of muscles. The pressure of the bolus stimulates pharyngeal pressure receptors, which send afferent impulses to the **swallowing centre** located in the medulla of the brain stem. The swallowing centre then reflexly activates in the appropriate sequence the muscles involved in swallowing. Swallowing is the most complex reflex in the body. Multiple highly coordinated responses are triggered in a specific all-or-none pattern over a period of time to accomplish the act of swallowing. Swallowing is initiated voluntarily, but once begun it cannot be stopped. Perhaps you have experienced this when a large piece of hard candy inadvertently slipped to the rear of your throat, triggering an unintentional swallow.

## The stages of swallowing

Swallowing is divided into the oropharyngeal stage and the esophageal stage. The **oropharyngeal stage** lasts about one second and consists of moving the bolus from the mouth through the pharynx and into the oesophagus. When the bolus enters the pharynx, it must be directed into the oesophagus and prevented from entering the other openings that communicate with the pharynx (i.e., the trachea). All of this is managed by the following coordinated activities (⟩ Figure 15-5b):

- The position of the tongue against the hard palate keeps food from reentering the mouth during swallowing.
- The uvula is elevated and lodges against the back of the throat, sealing off the nasal passages.
- Food is prevented from entering the trachea primarily by elevation of the larynx and tight closure of the vocal folds across the laryngeal opening, or **glottis** (⟩ Figure 15-5c). The first part of the trachea is the *larynx*, or *voice box*, across which are stretched the *vocal folds*. Contraction of laryngeal muscles aligns the vocal folds in tight apposition to each other, thus sealing the glottis entrance. Also, the bolus tilts a small flap of cartilage, the **epiglottis** (*epi* means "upon"), backward over the closed glottis, thereby protecting the respiratory airways. It is the glottis that is closed by the epiglottis during the valsalva manoeuvre, which is used to stabilize the core area during heavy lifting.
- The respiratory passages are temporarily sealed off during swallowing, because the swallowing centre briefly inhibits the nearby respiratory centre.
- With the larynx and trachea sealed off, pharyngeal muscles contract to force the bolus into the oesophagus.

## The pharyngoesophageal sphincter

The oesophagus is a fairly straight muscular tube that extends between the pharynx and stomach (see ▮ Table 15-1). Lying for the most part in

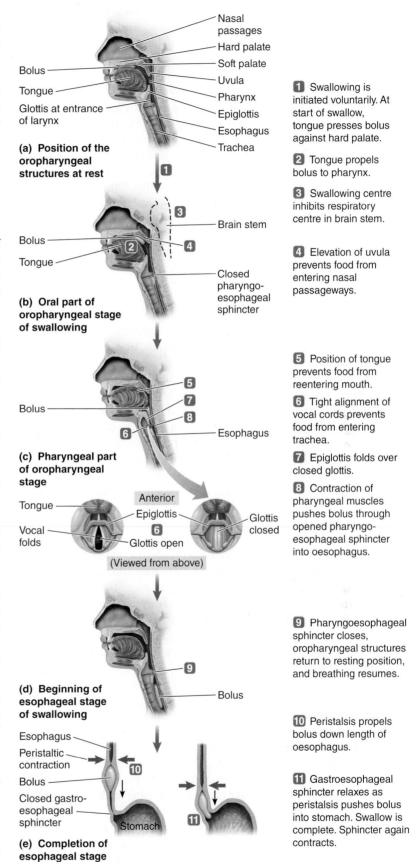

(a) Position of the oropharyngeal structures at rest

(b) Oral part of oropharyngeal stage of swallowing

(c) Pharyngeal part of oropharyngeal stage

(d) Beginning of esophageal stage of swallowing

(e) Completion of esophageal stage

1 Swallowing is initiated voluntarily. At start of swallow, tongue presses bolus against hard palate.

2 Tongue propels bolus to pharynx.

3 Swallowing centre inhibits respiratory centre in brain stem.

4 Elevation of uvula prevents food from entering nasal passageways.

5 Position of tongue prevents food from reentering mouth.

6 Tight alignment of vocal cords prevents food from entering trachea.

7 Epiglottis folds over closed glottis.

8 Contraction of pharyngeal muscles pushes bolus through opened pharyngo-esophageal sphincter into oesophagus.

9 Pharyngoesophageal sphincter closes, oropharyngeal structures return to resting position, and breathing resumes.

10 Peristalsis propels bolus down length of oesophagus.

11 Gastroesophageal sphincter relaxes as peristalsis pushes bolus into stomach. Swallow is complete. Sphincter again contracts.

⟩ FIGURE 15-5 Oropharyngeal and esophageal stages of swallowing

© 2016 Cengage

15

the thoracic cavity, it penetrates the diaphragm and joins the stomach in the abdominal cavity. The oesophagus is guarded at both ends by sphincters (ringlike muscle). The sphincter when closed prevents passage through the tube. The upper esophageal sphincter is the *pharyngoesophageal sphincter*, and the lower esophageal sphincter is the *gastroesophageal sphincter*.

Because the oesophagus is exposed to subatmospheric intrapleural pressure as a result of respiratory activity (p. 513), a pressure gradient exists between the atmosphere and the oesophagus. Except during a swallow, the **pharyngoesophageal sphincter** keeps the entrance to the oesophagus closed to prevent air from entering the oesophagus and stomach during breathing. Otherwise, the digestive tract would be subjected to large volumes of gas, which would lead to excessive **eructation** (burping). During swallowing, this sphincter opens and allows the bolus to pass into the oesophagus. Once the bolus has entered the oesophagus, the pharyngoesophageal sphincter closes and breathing resumes.

### Peristaltic waves

The **esophageal stage** of the swallow now begins (⟩ Figure 15-5d). The swallowing centre triggers a **primary peristaltic wave** that sweeps from the beginning to the end of the oesophagus, forcing the bolus ahead of it through the oesophagus to the stomach. The term **peristalsis** refers to ringlike contractions of the circular smooth muscle that move progressively forward (⟩ Figure 15-5e). The peristaltic wave takes about five to nine seconds to reach the lower end of the oesophagus. Progression of the wave is controlled by the swallowing centre, with innervation by means of the vagus.

If a large or sticky swallowed bolus, such as a bite of peanut butter sandwich, fails to be carried along to the stomach, the lodged bolus distends the oesophagus, stimulating pressure receptors within its walls. A second more forceful peristaltic wave is initiated, mediated by the intrinsic nerve plexuses at the level of the distension. These **secondary peristaltic waves** do not involve the swallowing centre, nor is the person aware of their occurrence. Distension of the oesophagus also reflexly increases salivary secretion. The trapped bolus is eventually dislodged and moved forward. Esophageal peristalsis is so effective you could eat an entire meal while you were upside down and it would all promptly be pushed to the stomach.

#### THE GASTROESOPHAGEAL SPHINCTER

The **gastroesophageal sphincter** is a highly muscular sphincter in the lower oesophagus that remains constricted at all times, except during either swallowing or vomiting. As the peristaltic wave sweeps down the oesophagus, the gastroesophageal sphincter relaxes reflexly so that the bolus can pass into the stomach (⟩ Figure 15-5e). After the bolus has entered the stomach, the gastroesophageal sphincter again contracts.

It is essential that the gastroesophageal sphincter stays contracted to maintain a barrier between the stomach and oesophagus, thereby reducing the risk of reflux of acidic gastric contents into the oesophagus. If gastric contents do flow backward past the sphincter—called gastroesophageal reflux (**gastric reflux**)—the acidity of these contents irritates the oesophagus,

causing discomfort known as **heartburn**. (The heart itself is not involved.) The acidity of the contents can damage the esophageal lining if the reflux occurs on a regular basis (i.e., >1/week). Regular gastric reflux can lead to serious disorders, such as esophageal ulcers or esophageal cancer.

#### ESOPHAGEAL SECRETION

Mucus is secreted throughout the length of the digestive tract by mucus-secreting glands in the mucosa. By lubricating the passage of food, esophageal mucus reduces the likelihood of damage by sharp edges in the entering food. It also protects the esophageal wall from acid and enzymes in gastric juice if gastric reflux occurs.

The entire transit time in the pharynx and oesophagus averages a mere 6 to 10 seconds, too short a time for any digestion or absorption to occur. Let's now move on to examine the stomach.

---

### Check Your Understanding 15.3

1. Describe how the pharynx prevents food entry into the trachea during a swallow.
2. State the function of the pharyngoesophageal sphincter.

---

## 15.4 | Stomach

The **stomach** is a J-shaped saclike chamber lying between the oesophagus and the small intestine. It has three sections based on anatomic, histological, and functional distinctions (⟩ Figure 15-6). The **fundus** is the part of the stomach that lies above the esophageal opening. The middle or main part of the stomach is the **body**. The smooth muscle layers in the fundus and body are relatively thin, but the lower part of the stomach, the **antrum**, has much heavier musculature. This difference in muscle thickness plays an important role in gastric motility in these two regions. There are also glandular differences in the mucosa of these regions. The terminal portion of the stomach is the **pyloric sphincter**, which acts as a barrier between the stomach and the upper part of the small intestine, the duodenum.

### Food storage and protein digestion

The stomach performs three main functions:

1. The stomach's most important function is to store ingested food until it can be emptied into the small intestine at a rate appropriate for optimal digestion and absorption. Because the small intestine is the primary site for this digestion and absorption, it is important that the stomach store the food and meter it into the duodenum at a rate that does not exceed the small intestine's capacities.

2. The stomach secretes hydrochloric acid (HCl) and enzymes that begin protein digestion. The stomach is capable of secreting about 3 L of gastric juice each day.

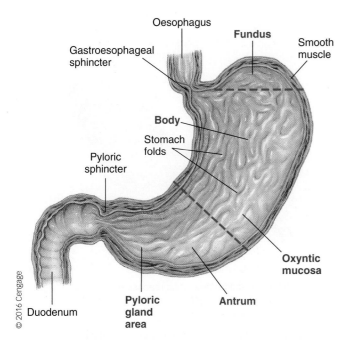

Oesophagus

Gastroesophageal sphincter

**Fundus**

Smooth muscle

**Body**

Stomach folds

Pyloric sphincter

**Oxyntic mucosa**

© 2016 Cengage

Duodenum

**Pyloric gland area**

**Antrum**

⟩ **FIGURE 15-6 Anatomy of the stomach.** The stomach is divided into three sections based on structural and functional distinctions: the fundus, body, and antrum. Based on differences in glandular secretion, the mucosal lining of the stomach is divided into the oxyntic mucosa and the pyloric gland area.

3. Through the stomach's mixing movements, the ingested food is pulverized and mixed with gastric secretions to produce a thick liquid known as **chyme.** The stomach contents must be converted to chyme before they can be emptied into the duodenum.

We next discuss these stomach functions in terms of the four basic digestive processes—motility, secretion, digestion, and absorption. Starting with motility, gastric motility is complex and subject to multiple regulatory inputs. The four aspects of gastric motility are (1) filling, (2) storage, (3) mixing, and (4) emptying. We begin with gastric filling.

## Gastric filling

When empty, the stomach's volume is about 50 mL, but it can expand to a capacity of about 1000 mL (twentyfold increase) during a meal, and can be physically distended to as much as 4000 mL. The stomach accommodates this change in volume with little change in tension and intragastric pressure by means of the following mechanism. The interior of the stomach has deep folds. During a meal, the folds get smaller and nearly flatten out as the stomach relaxes with each mouthful. This reflex relaxation of the stomach is called **receptive relaxation**; it enhances the stomach's ability to accommodate the extra volume of food with little rise in stomach pressure. Receptive relaxation is triggered by eating and is mediated by the vagus nerve. If too much food is consumed, the stomach becomes overdistended and intragastric pressure rises along with discomfort.

## Gastric storage

Pacemaker cells located in the fundus region generate slow-wave potentials that sweep down the length of the stomach toward the pyloric sphincter at a rate of three per minute. This rhythmic spontaneous depolarization—the basic electrical rhythm (BER) of the stomach—occurs continuously and may or may not be accompanied by contraction of the stomach's circular smooth muscle layer. Depending on the level of smooth muscle excitability, smooth muscle cells may be brought to threshold by this flow of current and undergo action potentials, which in turn initiate peristaltic waves that sweep over the stomach in pace with the BER.

Once initiated, the peristaltic wave spreads over the fundus and body to the antrum and pyloric sphincter. When the waves reach the antrum they become much stronger and more vigorous, because the muscle there is much thicker than that of the fundus and body. Because weak mixing movements occur in the fundus and body, food delivered to the stomach is stored in the relatively quiet body without being mixed. The fundic area contains only a pocket of gas. Food is gradually fed from the body into the antrum, where mixing does take place.

## Gastric mixing

The strong antral peristaltic contractions mix the food with gastric secretions to produce chyme. Each antral peristaltic wave propels chyme forward toward the pyloric sphincter. Tonic contraction of the pyloric sphincter keeps it almost, but not completely, closed. The small opening is large enough for water and other fluids to pass through but too small for the thicker chyme, except when a strong antral peristaltic contraction pushes it through. Before more chyme can be squeezed out, the peristaltic wave reaches the pyloric sphincter and causes it to contract more forcefully, closing the exit and blocking further passage into the duodenum. The bulk of the antral chyme that was being propelled forward but failed to be pushed into the duodenum is stopped at the closed sphincter and folds back into the antrum, only to be propelled forward and tumbled back again as the new peristaltic wave advances (⟩ Figure 15-7). This process, called **retropulsion**, thoroughly mixes the chyme in the antrum and thoroughly shears and grinds the chyme until the particles are small enough for emptying.

> **■ Clinical Connections** — Cathy has a peptic ulcer, which essentially is damage to the stomach lining. While her stomach is empty she feels pain because the stomach acids irritate the ulcer area. With food in her stomach, the acids are essentially neutralized, so the irritation is not felt. The ulcer, if severe enough, can cause bleeding. This bleeding can be detected with a fecal occult blood test. The loss of blood could also account for her hypertension and reflex tachycardia.

## Gastric emptying

In addition to mixing gastric contents, the antral peristaltic contractions are the driving force for gastric emptying. Of the 30 mL of chyme that the antrum can hold, usually only a few millilitres

15

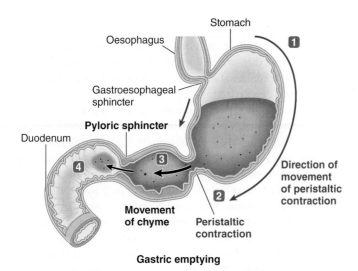

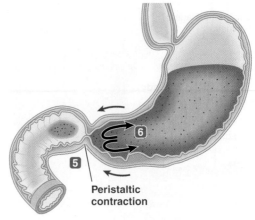

**Gastric emptying**

**1** A peristaltic contraction originates in the upper fundus and sweeps down toward the pyloric sphincter.

**2** The contraction becomes more vigorous as it reaches the thick-muscled antrum.

**3** The strong antral peristaltic contraction propels the chyme forward.

**4** A small portion of chyme is pushed through the partially open sphincter into the duodenum. The stronger the antral contraction, the more chyme is emptied with each contractile wave.

**Gastric mixing**

**5** When the peristaltic contraction reaches the pyloric sphincter, the sphincter is tightly closed and no further emptying takes place.

**6** When chyme that was being propelled forward hits the closed sphincter, it is tossed back into the antrum. Mixing of chyme is accomplished as chyme is propelled forward and tossed back into the antrum with each peristaltic contraction, a process called retropulsion.

> FIGURE 15-7 **Gastric emptying and mixing as a result of antral peristaltic contractions**

of antral contents are pushed into the duodenum with each peristaltic wave. However, the amount of chyme that escapes into the duodenum before the pyloric sphincter tightly closes depends largely on the strength of peristalsis. The intensity of antral peristalsis can vary markedly under the influence of different signals from both the stomach and the duodenum; thus, gastric emptying is regulated by both gastric and duodenal factors (▮ Table 15-2). These factors influence the stomach's excitability by slightly depolarizing or hyperpolarizing the gastric smooth muscle. This excitability is a determinant of the degree of antral peristaltic activity. The greater the excitability, the more frequently the BER will generate action potentials, the greater the degree of peristaltic activity in the antrum, and the faster the rate of gastric emptying.

### FACTORS IN THE STOMACH THAT INFLUENCE THE RATE OF GASTRIC EMPTYING

The main gastric factor that influences the strength of contraction is the amount of chyme in the stomach. Other things being equal, the stomach empties at a rate proportional to the volume of chyme in it at any given time. Stomach distension triggers increased gastric motility through a direct effect of stretch on the smooth muscle as well as through involvement of the intrinsic plexuses, the vagus nerve, and the stomach hormone *gastrin*. (The source, control, and other functions of this hormone are described in Section 15.8.)

Furthermore, the degree of fluidity of the chyme in the stomach influences gastric emptying. The stomach contents must be converted into a finely divided, thick liquid form before emptying. The sooner the appropriate degree of fluidity can be achieved, the more rapidly the contents are ready to be evacuated.

### FACTORS IN THE DUODENUM THAT INFLUENCE THE RATE OF GASTRIC EMPTYING

Despite these gastric influences, factors in the duodenum are of primary importance in controlling the rate of gastric emptying. The duodenum must be ready to receive the chyme and can delay gastric emptying by reducing peristaltic activity in the stomach until the duodenum is ready to accommodate more chyme. Even if the stomach is distended and its contents are in a liquid form, it cannot empty until the duodenum is ready to deal with the chyme.

The four most important duodenal factors that influence gastric emptying are *fat, acid, hypertonicity,* and *distension.* The presence of one or more of these stimuli in the duodenum activates the appropriate duodenal receptors, which trigger either a neural or a hormonal response that reduces gastric motility by reducing the excitability of the gastric smooth muscle. The subsequent reduction in antral peristaltic activity slows down the rate of gastric emptying.

- The *neural* response is mediated through both the intrinsic nerve plexuses (short reflex) and the autonomic nerves (long reflex). Collectively, these reflexes are called the **enterogastric reflex.**

## ▌ TABLE 15-2 Factors Regulating Gastric Motility and Emptying

| Factors | Mode of Regulation | Effects on Gastric Motility and Emptying |
|---|---|---|
| **Within the Stomach** | | |
| Volume of chyme | Distension has a direct effect on gastric smooth muscle excitability, as well as acting through the intrinsic plexuses, the vagus nerve, and gastrin | Increased volume stimulates motility and emptying |
| Degree of fluidity | Direct effect; contents must be in a fluid form to be evacuated | Increased fluidity allows more rapid emptying |
| **Within the Duodenum** | | |
| Presence of fat, acid, hypertonicity, or distension | Initiates the enterogastric reflex or triggers the release of enterogastrones (cholecystokinin, secretin) | These factors in the duodenum inhibit further gastric motility and emptying until the duodenum has coped with factors already present |
| **Within the Intestine** | | |
| Glucagon-like peptide-1 (GLP-1) | Inhibits gastric secretions, delays carbohydrate absorption | Slows motility and gastric emptying in the stomach |
| Gastric inhibitory peptide (GIP) | Possibly inhibits acid secretions | No effect on motility or emptying |
| **Outside the Digestive System** | | |
| Emotion | Alters autonomic balance | Stimulates or inhibits motility and emptying |
| Intense pain | Increases sympathetic activity | Inhibits motility and emptying |

© 2016 Cengage

- The *hormonal response* involves the release from the duodenal mucosa of several hormones collectively known as **enterogastrones**. The blood carries these hormones to the stomach, where they inhibit antral contractions to reduce gastric emptying. The two most important enterogastrones are **secretin** and **cholecystokinin (CCK)**. Secretin was the first hormone discovered (in 1902). Because it was a secretory product that entered the blood, it was termed *secretin*. The name *cholecystokinin* derives from the fact that this same hormone also causes contraction of the bile-containing gallbladder (*chole* means "bile"; *cysto* means "bladder"; and *kinin* means "contraction"). Secretin and CCK are major gastrointestinal hormones that perform other important functions in addition to serving as enterogastrones.

Let's examine why it is important that each of these stimuli in the duodenum (fat, acid, hypertonicity, and distension) delays gastric emptying (acting through the enterogastric reflex or one of the enterogastrones).

- *Fat*. Fat is digested and absorbed more slowly than the other nutrients. Furthermore, fat digestion and absorption take place only within the lumen of the small intestine. Therefore, when fat is already in the duodenum, further gastric emptying of more fatty stomach contents into the duodenum is prevented until the small intestine has processed the fat that's already there. In fact, fat is the most potent stimulus for inhibition of gastric motility. This is evident when you compare the rate of emptying of a high-fat meal (after six hours some of a bacon-and-eggs meal may still be in the stomach) with

that of a protein and carbohydrate meal (a meal of lean meat and potatoes may empty in three hours).

- *Acid*. Because the stomach secretes hydrochloric acid (HCl), highly acidic chyme is emptied into the duodenum, where it is neutralized by the sodium bicarbonate ($NaHCO_3$) secreted into the duodenal lumen, primarily from the pancreas. Unneutralized acid irritates the duodenal mucosa and inactivates the pancreatic digestive enzymes that are secreted into the duodenal lumen. Appropriately, therefore, unneutralized acid in the duodenum inhibits further emptying of acidic gastric contents until complete neutralization can be accomplished.

- *Hypertonicity*. As molecules of protein and starch are digested in the duodenal lumen, large numbers of amino acid and glucose molecules are released. If absorption of these amino acid and glucose molecules does not keep pace with the rate at which protein and carbohydrate digestion proceeds, these large numbers of molecules remain in the chyme and increase the osmolarity of the duodenal contents. Osmolarity depends on the number of molecules present, not on their size, and one protein molecule may be split into several hundred amino acid molecules, each of which has the same osmotic activity as the original protein molecule. The same holds true for one large starch molecule, which yields many smaller but equally osmotically active glucose molecules. Because water is freely diffusable across the duodenal wall, it enters the duodenal lumen from the plasma as the duodenal osmolarity rises. Large volumes of water entering the intestine from the plasma lead to intestinal distension, and, more important, circulatory disturbances ensue because of the reduction in plasma volume. To prevent these effects, gastric emptying is reflexly inhibited when the osmolarity of the duodenal contents starts to rise. In this way, the amount of food entering the duodenum for further digestion into a multitude of osmotically active particles is reduced until absorption processes have had an opportunity to catch up.

- *Distension.* Too much chyme in the duodenum inhibits the emptying of even more gastric contents, giving the distended duodenum time to cope with the excess volume of chyme it already contains before it gets any more.

*Clinical Note* Gastric dumping syndrome (rapid gastric emptying) occurs when there is a rapid movement of stomach contents, mainly undigested, into the small intestine after eating. There are two types: early and late gastric dumping. Early gastric dumping begins during or right after a meal, with symptoms of nausea, vomiting, bloating, diarrhoea, dizziness, and fatigue. Late gastric dumping occurs one to three hours postmeal. Its symptoms include weakness, sweating, and dizziness.

It is not uncommon to have both types of gastric dumping problems, and the syndrome is typically diagnosed on the basis of symptoms. Gastric dumping syndrome is largely preventable, by avoiding certain foods and eating a balanced diet. Treatment includes adjustment to eating habits and medication. Severe conditions may be treated with medications that slow digestion, or with surgery as a last resort. This syndrome is most commonly associated with patients who have undergone gastric bypass surgery.

## Emotions and gastric motility

Other factors unrelated to digestion, such as emotions, can also alter gastric motility by acting through the autonomic nerves to influence the degree of gastric smooth muscle excitability. Even though the effect of emotions on gastric motility varies from one person to another and is not always predictable, sadness and fear generally tend to decrease motility, whereas anger and aggression tend to increase it. In addition to emotional influences, intense pain from any part of the body tends to inhibit motility, not just in the stomach but throughout the digestive tract. This response is brought about by increased sympathetic activity.

## Vomiting

The complex act of vomiting, also known as *emesis*, is coordinated by a **vomiting centre** in the medulla of the brain stem. Vomiting begins with a deep inspiration and closure of the glottis. The contracting diaphragm descends downward on the stomach while simultaneous contraction of the abdominal muscles compresses the abdominal cavity, increasing the intra-abdominal pressure and forcing the abdominal viscera upward. As the flaccid stomach is squeezed between the diaphragm from above and the compressed abdominal cavity from below, the gastric contents are forced upward through the relaxed sphincters

and oesophagus and out through the mouth. The glottis is closed, so vomited material does not enter the respiratory airways. Also, the uvula is raised to close off the nasal cavity. The vomiting cycle may be repeated several times until the stomach is emptied. Vomiting is usually preceded by profuse salivation, sweating, rapid heart rate, and the sensation of nausea, all of which are characteristic of a generalized discharge of the autonomic nervous system.

### CAUSES OF VOMITING

Vomiting can be initiated by afferent input to the vomiting centre from a number of receptors throughout the body. The causes of vomiting include the following:

- Tactile (touch) stimulation of the back of the throat, which is one of the most potent stimuli. For example, sticking a finger in the back of the throat or even the presence of a tongue depressor or dental instrument in the back of the mouth is enough stimulation to cause gagging and even vomiting in some people.

- Irritation or distension of the stomach and duodenum

- Elevated intracranial pressure, such as that caused by cerebral haemorrhage. Thus, vomiting after a head injury is considered a bad sign; it suggests swelling or bleeding within the cranial cavity.

- Rotation or acceleration of the head producing dizziness, such as in motion sickness

- Chemical agents, including drugs or noxious substances that initiate vomiting (i.e., **emetics**) either by acting in the upper parts of the gastrointestinal tract or by stimulating chemoreceptors in a specialized **chemoreceptor trigger zone** next to the vomiting centre in the brain. Activation of this zone triggers the vomiting reflex. For example, chemotherapeutic agents used in treating cancer often cause vomiting by acting on the chemoreceptor trigger zone. As well, this region of the brain is associated with morning sickness during early pregnancy and vomiting during a hangover.

- Psychogenic vomiting induced by emotional factors, including those accompanying nauseating sights and odours and anxiety before taking an examination or in other stressful situations

### EFFECTS OF VOMITING

With excessive vomiting, the body experiences large losses of secreted fluids and acids that normally would be reabsorbed. The resulting reduction in plasma volume can lead to dehydration and circulatory problems, and the loss of acid from the stomach can lead to metabolic alkalosis (p. 636).

Vomiting is not always harmful, however. Limited vomiting triggered by irritation of the digestive tract can be useful in removing noxious material from the stomach, rather than letting it stay and be absorbed. In fact, emetics are sometimes taken after accidental ingestion of a poison to quickly remove the offending substance from the body.

We have now completed our discussion of gastric motility and will shift to gastric secretion.

**15**

## Gastric juice

The cells that secrete gastric juice are in the lining of the stomach, the gastric mucosa, which is divided into two distinct areas: (1) the **oxyntic mucosa**, which lines the body and fundus; and (2) the **pyloric gland area (PGA)**, which lines the antrum. The luminal surface of the stomach is pitted with deep pockets formed by infoldings of the gastric mucosa. The first part of these invaginations are called **gastric pits**, at the base of which lie the **gastric glands**. A variety of secretory cells line these invaginations, some exocrine and some endocrine or paracrine (▮ Table 15-3). Let's look at the gastric exocrine secretory cells first.

Three types of gastric exocrine secretory cells are found in the walls of the pits and glands in the oxyntic mucosa.

- **Mucous cells** line the gastric pits and the entrance of the glands. They secrete a thin, watery *mucus*. (*Mucous* is the adjective; *mucus* is the noun.)
- The deeper parts of the gastric glands are lined by chief and parietal cells. The more numerous **chief cells** secrete the enzyme precursor *pepsinogen*.
- The **parietal (oxyntic) cells** secrete *HCl* and *intrinsic factor* (*oxyntic* means "sharp," a reference to these cells' potent HCl secretory product).
- These exocrine secretions are all released into the gastric lumen. Collectively, they make up the gastric digestive juice.

A few **stem cells** are also found in the gastric pits. These cells rapidly divide and serve as the parent cells of all new cells of the gastric mucosa. The daughter cells that result from cell division either migrate out of the pit to become surface epithelial cells or migrate down deeper to the gastric glands, where they differentiate into chief or parietal cells. Through this activity, the entire stomach mucosa is replaced about every three days. This frequent turnover is important, because the harsh acidic stomach contents expose the mucosal cells to lots of wear and tear.

Between the gastric pits, the gastric mucosa is covered by **surface epithelial cells**, which secrete a thick, viscous, alkaline mucus that forms a visible layer several millimetres thick over the surface of the mucosa.

The gastric glands of the PGA primarily secrete mucus and a small amount of pepsinogen; no acid is secreted in this area, in contrast to the oxyntic mucosa.

Let's consider these exocrine products and their roles in digestion in further detail.

## Hydrochloric acid

The parietal cells actively secrete HCl into the lumen of the gastric pits, which in turn empty into the lumen of the stomach. As a result of this HCl secretion, the pH of the luminal contents falls as low as 2. Hydrogen ion ($H^-$) and chloride ion ($Cl^-$) are actively transported by separate pumps in the parietal cells' plasma membrane. Hydrogen ion is actively transported against a tremendous concentration gradient, with the $H^+$ concentration being as much as 3 million times greater in the lumen than in the blood. Chloride is secreted by a secondary active transport

mechanism against a much smaller concentration gradient that's only 1.5 times greater in the lumen than in the blood.

The secreted $H^+$ is not transported from the plasma but is derived instead from metabolic processes within the parietal cell (⊳ Figure 15-8). Specifically, the $H^+$ to be secreted is derived from the breakdown of $H_2O$ molecules into $H^+$ and $OH^-$ (hydroxyl ions) within the parietal cells. This $H^+$ is secreted into the lumen by $H^+ - K^+$ **ATPase** in the parietal cell's luminal membrane. This primary active-transport carrier also pumps $K^+$ into the cell from the lumen—a mechanism similar to the $Na^+ - K^+$ ATPase pump you are already familiar with. The transported $K^+$ then passively leaks back into the lumen through luminal $K^+$ channels, leaving $K^+$ levels unchanged by the process of $H^+$ secretion.

Meanwhile, the $OH^-$ generated by the breakdown of $H_2O$ becomes neutralized through combining with a new $H^+$ generated from carbonic acid ($H_2CO_3$). The parietal cells contain an abundance of the enzyme *carbonic anhydrase*. In the presence of carbonic anhydrase, $H_2O$ readily combines with $CO_2$, which either has been produced within the parietal cell by metabolic processes or has diffused in from the blood. The combination of $H_2O$ and $CO_2$ results in the formation of $H_2CO_3$, which partially dissociates to yield $H^+$ and $HCO_3^2$. The generated $H^+$ in essence replaces the one secreted.

The generated $HCO_3^-$ is moved into the plasma by a $Cl^- - HCO_3^-$ **exchanger** in the parietal cell's basolateral membrane. This exchanger transports $Cl^-$ into the parietal cell by means of secondary active transport (p. 44). Driven by the gradient, this carrier moves $HCO_3^-$ out of the cell into the plasma down its electrochemical gradient and simultaneously transports $Cl^-$ from the plasma into the parietal cell against its electrochemical gradient. This exchanger builds up the concentration of $Cl^-$ inside the parietal cell, establishing a $Cl^-$ concentration gradient between the parietal cell and gastric lumen. Because of this concentration gradient and because the cell interior is negative compared with the luminal contents, the negatively charged $Cl^-$ pumped into the cell by the basolateral exchanger diffuses out of the cell and moves down its electrochemical gradient through channels in the luminal membrane and into the gastric lumen—completing the $Cl^-$ secretory process.

Hydrochloric acid performs the following specific functions that aid digestion:

1. Activates the enzyme precursor pepsinogen to become an active enzyme, pepsin, and provides an acid medium that is optimal for pepsin activity
2. Aids in the breakdown of connective tissue and muscle fibres, reducing large food particles into smaller particles
3. Denatures protein, that is, uncoils proteins from their highly folded final form, thus exposing more of the peptide bonds for enzymatic attack
4. Along with salivary lysozyme, kills most of the microorganisms ingested with food, although some escape and continue to grow and multiply in the large intestine

15

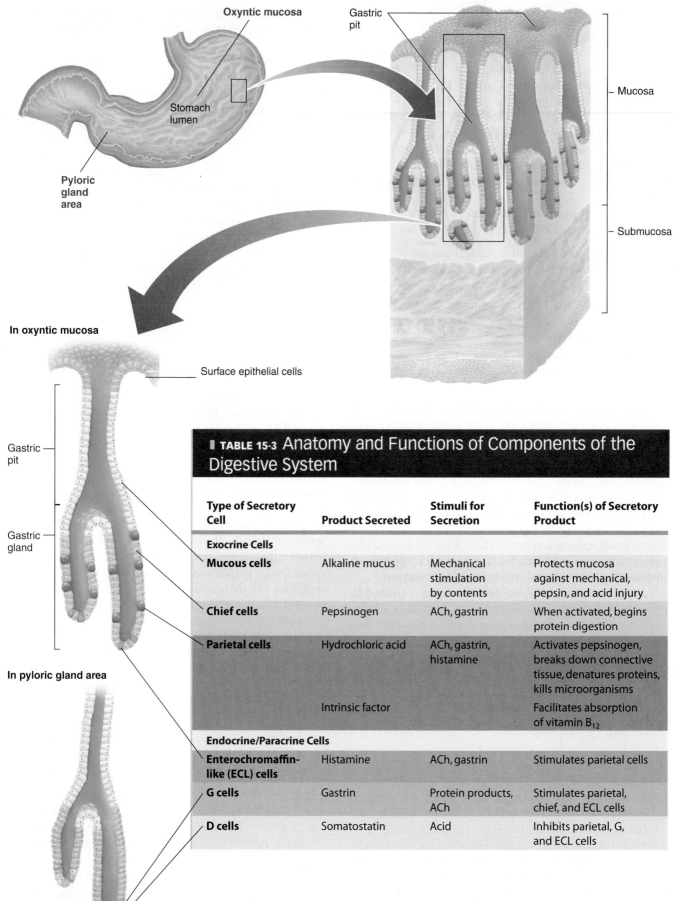

Oxyntic mucosa

Gastric pit

Stomach lumen

Pyloric gland area

Mucosa

Submucosa

In oxyntic mucosa

Surface epithelial cells

Gastric pit

Gastric gland

In pyloric gland area

| ▌TABLE 15-3 Anatomy and Functions of Components of the Digestive System | | | |
|---|---|---|---|
| **Type of Secretory Cell** | **Product Secreted** | **Stimuli for Secretion** | **Function(s) of Secretory Product** |
| **Exocrine Cells** | | | |
| **Mucous cells** | Alkaline mucus | Mechanical stimulation by contents | Protects mucosa against mechanical, pepsin, and acid injury |
| **Chief cells** | Pepsinogen | ACh, gastrin | When activated, begins protein digestion |
| **Parietal cells** | Hydrochloric acid | ACh, gastrin, histamine | Activates pepsinogen, breaks down connective tissue, denatures proteins, kills microorganisms |
| | Intrinsic factor | | Facilitates absorption of vitamin $B_{12}$ |
| **Endocrine/Paracrine Cells** | | | |
| **Enterochromaffin-like (ECL) cells** | Histamine | ACh, gastrin | Stimulates parietal cells |
| **G cells** | Gastrin | Protein products, ACh | Stimulates parietal, chief, and ECL cells |
| **D cells** | Somatostatin | Acid | Inhibits parietal, G, and ECL cells |

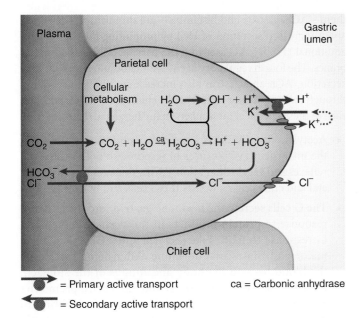

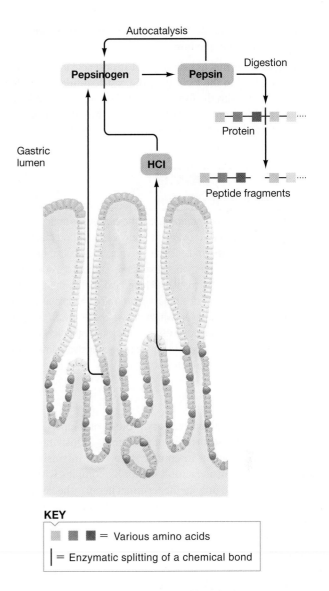

= Primary active transport     ca = Carbonic anhydrase

= Secondary active transport

› **FIGURE 15-8 Mechanism of HCl secretion.** The stomach's parietal cells actively secrete H⁺ and Cl⁻ by the actions of two separate pumps. Hydrogen ion is secreted into the lumen by a primary H⁺–K⁺ ATPase active-transport pump at the parietal cell's luminal border. The K⁺ transported into the cell by the pump promptly exits through a luminal K⁺ channel, thus being recycled between the cell and lumen. The secreted H⁺ is derived from the breakdown of $H_2O$ into H⁺ and OH⁻. The OH⁻ is neutralized by another H⁺ that's derived from the $H_2CO_3$ generated within the cell from the $CO_2$ that is either metabolically produced in the cell or diffuses in from the plasma. Chloride is secreted by secondary active transport. Driven by the concentration gradient, a Cl⁻–$HCO_3^-$ exchanger in the basolateral membrane transports $HCO_3^-$ generated from $H_2CO_3$ dissociation into the plasma and down its concentration gradient; simultaneously it transports Cl⁻ into the parietal cell against its concentration gradient. Chloride secretion is completed as the Cl⁻ that entered from the plasma diffuses out of the cell down its electrochemical gradient and through a luminal Cl⁻ channel into the lumen.

**KEY**

= Various amino acids

= Enzymatic splitting of a chemical bond

› **FIGURE 15-9 Pepsinogen activation in the stomach lumen.** In the lumen, hydrochloric acid (HCl) activates pepsinogen to its active form, pepsin, by cleaving off a small fragment. Once activated, pepsin autocatalytically activates more pepsinogen and begins protein digestion. Secretion of pepsinogen in the inactive form prevents it from digesting the protein structures of the cells in which it is produced.

## Pepsinogen

The major digestive constituent of gastric secretion is **pepsinogen**, an inactive enzymatic molecule produced by the chief cells. Chief cells release precursor enzymes. Pepsinogen is stored in the chief cell's cytoplasm within secretory vesicles known as **zymogen granules**, from which it is released by exocytosis on appropriate stimulation. When pepsinogen is secreted into the gastric lumen, HCl cleaves off a small fragment of the molecule, converting it to the active form of the enzyme, **pepsin** (› Figure 15-9). Once formed, pepsin acts on other pepsinogen molecules to produce more pepsin. A mechanism such as this, whereby an active form of an enzyme activates other molecules of the same enzyme, is called an **autocatalytic process** (*autocatalytic* means "self-activating").

Pepsin initiates protein digestion by splitting certain amino acid linkages in proteins to yield peptide fragments (small amino acid chains); it works most effectively in the acid environment provided by HCl. Because pepsin can digest protein, it must be stored and secreted in an inactive form so it does not digest the proteins of the cells in which it is formed. Therefore, pepsin is maintained in the inactive form of pepsinogen until it reaches the gastric lumen, where it is activated by HCl that's secreted into the lumen by a different cell type.

## Mucus

The surface of the gastric mucosa is covered by a layer of mucus derived from the surface epithelial cells and mucous cells. This mucus serves as a protective barrier against several forms of potential injury to the gastric mucosa:

- It acts as a lubricant and protects the gastric mucosa against mechanical injury.

- It helps protect the stomach wall from self-digestion, because pepsin is inhibited when it comes in contact with the layer

15

of mucus coating the stomach lining, but it does not affect pepsin activity in the lumen.

- Being alkaline, mucus helps protect against acid injury by neutralizing HCl in the vicinity of the gastric lining, but it does not interfere with the function of HCl in the lumen. Whereas the pH in the lumen may be as low as 2, the pH in the layer of mucus adjacent to the mucosal cell surface is about 7.

> **▌Clinical Connections**  Mucus helps to protect the stomach wall from self-digestion. However, an *H. pylori* infection can compromise the protective mucus and cause a peptic ulcer. *H. pylori* can survive in the stomach by producing urease enzymes. These enzymes metabolize urea to carbon dioxide and ammonia. Excessive carbon dioxide production can lead to frequent belching, and the ammonia produced can help neutralize stomach acid. The ammonia also can damage the epithelial layer, which allows the *H. pylori* access to the underlying connective tissues. Once the mucous layer is compromised, pepsin and HCl can cause extensive damage to the stomach wall, creating the ulcer.

## Intrinsic factor

**Intrinsic factor**, another secretory product of the parietal cells, is important in the absorption of vitamin $B_{12}$. This vitamin can be absorbed only when in combination with intrinsic factor. Binding of the intrinsic factor–vitamin $B_{12}$ complex with a special receptor located only in the terminal ileum—the last part of the small intestine—triggers the receptor-mediated endocytosis of the complex at this location.

Vitamin $B_{12}$ is essential for the normal formation of red blood cells.

## Parietal and chief cells

In addition to the gastric exocrine secretory cells, certain secretory cells in the gastric glands release endocrine and paracrine regulatory factors, instead of products involved in the digestion of nutrients in the gastric lumen. The locations and secretions of these types of cells—G cells, enterochromaffin-like (ECL) cells, and D cells—are as follows (see also ▌Table 15-3):

- Endocrine cells known as **G cells** are found in the gastric pits only in the PGA; they secrete the hormone *gastrin* into the blood.
- **Enterochromaffin-like (ECL) cells** are dispersed among the parietal and chief cells in the gastric glands of the oxyntic mucosa; they secrete the paracrine *histamine*.
- **D cells** are scattered in glands near the pylorus but are more numerous in the duodenum; they secrete the paracrine *somatostatin*.

These three regulatory factors from the gastric pits, along with the neurotransmitter *acetylcholine* (*ACh*), primarily control the secretion of gastric digestive juices. Parietal cells have separate receptors for each of these chemical messengers. Three of them—ACh, gastrin, and histamine—are stimulatory. They bring about increased secretion of HCl by promoting the insertion of additional $H^+-K^+$ ATPases into the parietal cells' plasma membranes. The fourth regulatory agent—somatostatin—inhibits HCl secretion. ACh and gastrin also increase pepsinogen secretion through their stimulatory effect on the chief cells. We now consider each of these chemical messengers in further detail (▌Table 15-3).

- Acetylcholine (ACh) is a neurotransmitter released from the intrinsic nerve plexuses in response to both short local reflexes and vagal stimulation. ACh stimulates both the parietal and chief cells as well as the G cells and ECL cells.
- The G cells secrete the hormone **gastrin** into the blood in response to protein products in the stomach lumen and in response to ACh. Like secretin and CCK, gastrin is a major gastrointestinal hormone. After being carried by the blood back to the body and fundus of the stomach, gastrin stimulates the parietal and chief cells, promoting secretion of a highly acidic gastric juice. In addition to directly stimulating the parietal cells, gastrin indirectly promotes HCl secretion by stimulating the ECL cells to release histamine. Gastrin is the main factor that brings about increased HCl secretion during meal digestion. Gastrin is also *trophic* (growth-promoting) to the mucosa of the stomach and small intestine, thereby maintaining their secretory capabilities.
- **Histamine**, a paracrine, is released from the ECL cells in response to ACh and gastrin. Histamine acts locally on nearby parietal cells to speed up HCl secretion.
- Somatostatin is released from the D cells in response to acid. It acts locally as a paracrine in negative-feedback fashion to inhibit secretion by the parietal cells, G cells, and ECL cells, thereby turning off the HCl-secreting cells and their most potent stimulatory pathway.

From this list, it is obvious not only that multiple chemical messengers influence the parietal and chief cells but also that these chemicals influence one another. Next, as we examine the phases of gastric secretion, you will see under what circumstances each of these regulatory agents is released.

## Control of gastric secretion

The rate of gastric secretion can be influenced by (1) factors arising before food ever reaches the stomach, (2) factors resulting from the presence of food in the stomach, and (3) factors in the duodenum after food has left the stomach. Accordingly, gastric secretion is divided into three phases: the cephalic, gastric, and intestinal phases.

### CEPHALIC PHASE

The *cephalic phase of gastric secretion* refers to the increased secretion of HCl and pepsinogen that occurs in feedforward fashion in response to various stimuli even before food reaches the stomach (*cephalic* means "head"). Thinking about, tasting, smelling, chewing, and swallowing food increase gastric secretion by vagal nerve activity in two ways. First, vagal stimulation

## ▌TABLE 15-4 Stimulation of Gastric Secretion

| Phase | Stimuli | Excitatory Mechanism for Enhancing Gastric Secretion |
|---|---|---|
| **Cephalic phase of gastric secretion** | Stimuli in the head—seeing, smelling, tasting, chewing, swallowing food | 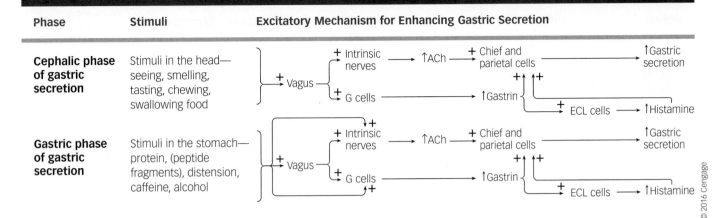 |
| **Gastric phase of gastric secretion** | Stimuli in the stomach—protein, (peptide fragments), distension, caffeine, alcohol | |

© 2016 Cengage

of the intrinsic plexuses promotes increased secretion of ACh, which in turn leads to increased secretion of HCl and pepsinogen by the secretory cells. Second, vagal stimulation of the G cells within the PGA causes the release of gastrin, which in turn further enhances secretion of HCl and pepsinogen; the effect on HCl is strengthened due to gastrin promoting the release of histamine (▌Table 15-4).

### GASTRIC PHASE

The *gastric phase of gastric secretion* begins when food actually reaches the stomach. Protein, especially peptide fragments, stomach distension, caffeine, and alcohol all act as stimuli in the stomach; they increase gastric secretion by way of overlapping efferent pathways. For example, protein in the stomach, the most potent stimulus, stimulates chemoreceptors that activate the intrinsic nerve plexuses, which in turn stimulate the secretory cells. Furthermore, protein brings about activation of the extrinsic vagal fibres to the stomach. Vagal activity further enhances intrinsic nerve stimulation of the secretory cells and triggers the release of gastrin. Protein also directly stimulates the release of gastrin. Gastrin, in turn, is a powerful stimulus for further HCl and pepsinogen secretion and also calls forth the release of histamine, which further increases HCl secretion. Through these synergistic and overlapping pathways, protein induces the secretion of a highly acidic, pepsin-rich gastric juice, which continues the digestion of the protein that first initiated the process (▌Table 15-4).

When the stomach is distended with protein-rich food that needs to be digested, these secretory responses are appropriate. Caffeine and, to a lesser extent, alcohol also stimulate the secretion of a highly acidic gastric juice, even when no food is present. This unnecessary acid can irritate the linings of the stomach and duodenum. For this reason, people with ulcers or gastric hyperacidity should avoid caffeinated and alcoholic beverages.

### INTESTINAL PHASE

The *intestinal phase of gastric secretion* encompasses the factors originating in the small intestine that influence gastric secretion. Whereas the other phases are excitatory, this phase is inhibitory.

The intestinal phase is important in helping shut off the flow of gastric juices when chyme begins to be emptied into the small intestine, a topic to which we now turn.

## Gastric secretion

You have learned about the factors that turn on gastric secretion before and during a meal, but how is the flow of gastric juices shut off when they are no longer needed? Gastric secretion is gradually reduced in three different ways as the stomach empties (▌Table 15-5):

- As the meal is gradually emptied into the duodenum, the major stimulus for enhanced gastric secretion—the presence of protein in the stomach—is withdrawn.

- After foods leave the stomach, gastric juices accumulate to such an extent that gastric pH falls very low. This fall in pH within the stomach lumen comes about largely because food proteins that had been buffering HCl are no longer present in the lumen as the stomach empties. (Recall that proteins serve as excellent buffers; p. 625.) Somatostatin is released in response to this high gastric acidity. In negative-feedback fashion, somatostatin's inhibitory effects bring about a decline in gastric secretion.

- The same stimuli that inhibit gastric motility (fat, acid, hypertonicity, or distension in the duodenum brought about by the emptying of stomach contents into the duodenum) inhibit gastric secretion as well. The enterogastric reflex and the enterogastrones suppress the gastric secretory cells, while they simultaneously reduce the excitability of the gastric smooth muscle cells. This inhibitory response is the intestinal phase of gastric secretion.

## The gastric mucosal barrier

How can the stomach contain strong acid contents and proteolytic enzymes without destroying itself? Recall that mucus provides a protective coating. In addition, the mucosal lining provides other barriers to mucosal acid damage. First, the

**15**

| Region | Stimuli | Inhibitory Mechanism for Gastric Secretion |
|---|---|---|
| **Body and antrum** | Removal of protein and distension as the stomach empties | — Intrinsic nerves / — Vagus / — G cells → ↓Gastrin → ↓Histamine ⟶ ↓Gastric secretion |
| **Antrum and duodenum** | Accumulation of acid | + D cells → ↑Somatostatin → — Parietal cells / — G cells / — ECL cells ⟶ ↓Gastric secretion |
| **Duodenum** (intestinal phase of gastric secretion) | Fat Acid Hypertonicity Distension | + Enterogastric reflex / ↑Enterogastrones (cholecystokinin and secretin) → — Parietal cells / — Chief cells / — Smooth muscle cells ⟶ ↓Gastric secretion and motility |

© 2016 Cengage

luminal membranes of the gastric mucosal cells are almost impermeable to H⁺, so acid cannot penetrate into the cells and damage them. Furthermore, the lateral edges of these cells are joined near their luminal borders by tight junctions, so acid cannot diffuse between the cells from the lumen into the underlying submucosa (see › Figure 15-2). The properties of the gastric mucosa that enable the stomach to contain acid without injuring itself constitute the **gastric mucosal barrier** (› Figure 15-10). These protective mechanisms are enhanced by the fact that the entire stomach lining is replaced every three days. Because of rapid mucosal turnover, cells are usually replaced before they are exposed to the wear and tear of harsh gastric conditions for long enough to suffer damage.

*Clinical Note* Despite the protection provided by mucus, the gastric mucosal barrier, and the frequent turnover of cells,

the protective barrier is occasionally broken, and the gastric wall is injured by the stomach's acidic and enzymatic contents. When this occurs, an erosion, or **peptic ulcer**, of the stomach wall results. Excessive gastric reflux into the oesophagus and dumping of excessive acidic gastric contents into the duodenum can lead to peptic ulcers in these locations as well. (For a further discussion of ulcers, see ◗ Concepts, Challenges, and Controversies, p. 670.)

We now turn to the remaining two digestive processes in the stomach, gastric digestion and absorption.

## Carbohydrate digestion and protein digestion

Two separate digestive processes take place within the stomach. In the body of the stomach, food remains in a

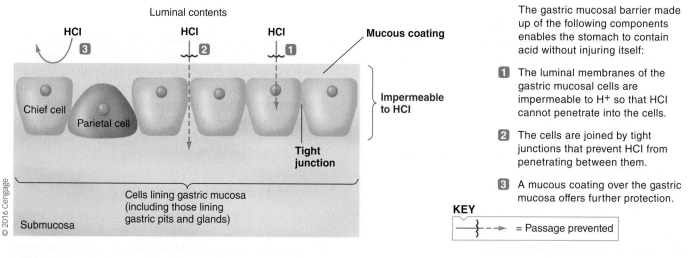

› **FIGURE 15-10** Gastric mucosal barrier

semisolid mass, because peristaltic contractions in this region are too weak for mixing to occur. Because food is not mixed with gastric secretions in the body of the stomach, very little protein digestion occurs here. In the interior of the mass, however, carbohydrate digestion continues under the influence of salivary amylase. Even though acid inactivates salivary amylase, the unmixed interior of the food mass is free of acid.

Digestion by the gastric juice itself is accomplished in the antrum of the stomach, where the food is thoroughly mixed with HCl and pepsin, which begins protein digestion.

### Alcohol and Aspirin absorption, but no food

No food or water is absorbed into the blood through the stomach mucosa. However, two noteworthy non-nutrient substances are absorbed directly by the stomach—*ethyl alcohol* and *aspirin*. Alcohol is somewhat lipid soluble, so it can diffuse through the lipid membranes of the epithelial cells that line the stomach and can enter the blood through the submucosal capillaries. Although alcohol can be absorbed by the gastric mucosa, it can be absorbed even more rapidly by the small-intestine mucosa, because the surface area for absorption in the small intestine is much greater than in the stomach. Therefore, alcohol absorption occurs more slowly if gastric emptying is delayed so that the alcohol remains longer in the more slowly absorbing stomach. Because fat is the most potent duodenal stimulus for inhibiting gastric motility, consuming fat-rich foods (e.g., whole milk, pizza, or nuts) before or during alcohol ingestion delays gastric emptying and prevents the alcohol from producing its effects as rapidly.

*Clinical Note* Another category of substances absorbed by the gastric mucosa includes weak acids, most notably *acetylsalicylic acid* (aspirin). In the highly acidic environment of the stomach lumen, weak acids are almost totally un-ionized; that is, the $H^+$ and the associated anion of the acid are bound together. In an un-ionized form, these weak acids are lipid soluble, so they can be absorbed quickly by crossing the plasma membranes of the epithelial cells that line the stomach. Most other drugs are not absorbed until they reach the small intestine, so they do not begin to take effect as quickly.

This completes our discussion of the stomach, and we now turn to the next part of the digestive tract—the small intestine and the accessory digestive organs that release their secretions into the small-intestine lumen.

---

### Check Your Understanding 15.4

1. Describe the process of retropulsion, and explain what it accomplishes.
2. Discuss how food-related stimuli induce gastric secretions during the cephalic phase of gastric secretion.
3. Describe the mechanisms that protect the gastric mucosa from acid damage.

---

## 15.5 | Pancreatic and Biliary Secretions

When gastric contents are emptied into the small intestine, they are mixed not only with juice secreted by the small-intestine mucosa but also with the secretions of the exocrine pancreas and liver that are released into the duodenal lumen. Next we discuss the roles of each of these accessory digestive organs.

### The pancreas

The pancreas is an elongated gland that lies behind and below the stomach, above the first loop of the duodenum (▸ Figure 15-11). This mixed gland contains both exocrine and endocrine tissue. The predominant exocrine part consists of grapelike clusters of secretory cells that form sacs known as **acini**, which connect to ducts that eventually empty into the duodenum. The smaller endocrine part consists of isolated islands of endocrine tissue, the islets of Langerhans, which are dispersed throughout the pancreas. The most important hormones secreted by the islet cells are insulin and glucagon (Chapter 6). The exocrine and endocrine pancreas are derived from different tissues during embryonic development and have only their location in common. Although both are involved with the metabolism of nutrient molecules, they have different functions under the control of different regulatory mechanisms.

#### THE EXOCRINE PANCREAS

The **exocrine pancreas** secretes a pancreatic juice consisting of two components: (1) *pancreatic enzymes* actively secreted by the *acinar cells* that form the acini, and (2) an *aqueous alkaline solution* actively secreted by the *duct cells* that line the pancreatic ducts. The aqueous (watery) alkaline component is rich in sodium bicarbonate ($NaHCO_3$).

Pancreatic enzymes, like pepsinogen, are stored within zymogen granules after being produced, then released by exocytosis as needed. These pancreatic enzymes are important because they can almost completely digest food in the absence of all other digestive secretions. The acinar cells secrete three different types of pancreatic enzymes capable of digesting all three categories of foodstuffs: (1) **proteolytic enzymes** for protein digestion, (2) **pancreatic amylase** for carbohydrate digestion, and (3) **pancreatic lipase** for fat digestion.

#### PANCREATIC PROTEOLYTIC ENZYMES

The three major pancreatic proteolytic enzymes are *trypsinogen*, *chymotrypsinogen*, and *procarboxypeptidase*, each of which is secreted in an inactive form. Of less importance are several nucleases and elastases. The most plentiful of these enzymes are trypsinogen (inactive) and trypsin (active). When **trypsinogen** is secreted into the duodenal lumen, it is activated to its active enzyme form, **trypsin**, by **enterokinase** (also known as **enteropeptidase**), an enzyme embedded in the luminal border of the cells that line the duodenal mucosa. Trypsin then autocatalytically activates more trypsinogen. Like pepsinogen, trypsinogen must remain inactive within the pancreas to prevent

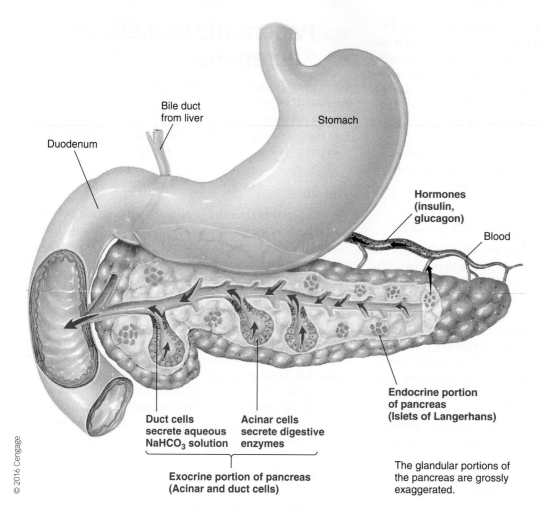

**Chymotrypsinogen** and **procarboxypeptidase**, the other pancreatic proteolytic enzymes, are converted by trypsin to their active forms, **chymotrypsin** and **carboxypeptidase**, respectively, within the duodenal lumen. Once enterokinase has activated some of the trypsin, trypsin then carries out the rest of the activation process.

Each of these proteolytic enzymes attacks different peptide linkages. The end products that result from this action are a mixture of small peptide chains and amino acids. Mucus secreted by the intestinal cells protects against digestion of the small-intestine wall by the activated proteolytic enzymes.

## PANCREATIC AMYLASE

Like salivary amylase, pancreatic amylase contributes to carbohydrate digestion by converting polysaccharides into the disaccharide maltose. Pancreatic amylase hydrolyzes starches, glycogen, and most other carbohydrates, with the exception of cellulose. Amylase is secreted in

Bile duct from liver

Stomach

Duodenum

Hormones (insulin, glucagon)

Blood

Duct cells secrete aqueous NaHCO₃ solution

Acinar cells secrete digestive enzymes

Endocrine portion of pancreas (Islets of Langerhans)

Exocrine portion of pancreas (Acinar and duct cells)

The glandular portions of the pancreas are grossly exaggerated.

© 2016 Cengage

> **FIGURE 15-11 Schematic representation of the exocrine and endocrine portions of the pancreas.** The exocrine pancreas secretes into the duodenal lumen a digestive juice composed of digestive enzymes secreted by the acinar cells and an aqueous NaHCO₃ solution secreted by the duct cells. The endocrine pancreas secretes the hormones insulin and glucagon into the blood.

this proteolytic enzyme from digesting the proteins of the cells in which it is formed. Trypsinogen remains inactive, therefore, until it reaches the duodenal lumen, where enterokinase triggers the activation process, which then proceeds autocatalytically. As further protection, the pancreas also produces a chemical known as **trypsin inhibitor**, which blocks trypsin's actions if spontaneous activation of trypsinogen inadvertently occurs within the pancreas.

the pancreatic juice in an active form, because active amylase does not endanger the secretory cells. These cells do not contain any polysaccharides.

## PANCREATIC LIPASE

Pancreatic lipase is extremely important because, throughout the entire digestive system, it is the only enzyme secreted that can digest fat. (Insignificant amounts of lipase are secreted in the

### ▌Why It Matters
## Pancreatic Insufficiency

When pancreatic enzymes are deficient, digestion of food is incomplete. Because the pancreas is the only significant source of lipase, pancreatic enzyme deficiency results in serious maldigestion of fats. The main clinical manifestation of pancreatic exocrine insufficiency is **steatorrhoea**, or excessive undigested fat in the feces. Up to 60 to 70 percent of the ingested fat may be excreted in the feces. Digestion of protein and carbohydrates is impaired to a lesser degree because salivary, gastric, and small-intestinal enzymes contribute to the digestion of these two foodstuffs.

**15**

saliva and gastric juice in humans.) Pancreatic lipase hydrolyzes dietary triglycerides into monoglycerides and free fatty acids, which are the absorbable units of fat. Like amylase, lipase is secreted in its active form because there is no risk of pancreatic self-digestion by lipase. Triglycerides are not a structural component of pancreatic cells.

### PANCREATIC AQUEOUS ALKALINE SECRETION

Pancreatic enzymes function best in a neutral or slightly alkaline environment, yet the highly acidic gastric contents are emptied into the duodenal lumen in the vicinity of pancreatic enzyme entry into the duodenum. This acidic chyme must be neutralized quickly in the duodenal lumen, not only to allow optimal functioning of the pancreatic enzymes but also to prevent acid damage to the duodenal mucosa. The alkaline ($NaHCO_3$-rich) fluid secreted by the pancreatic duct cells into the duodenal lumen serves the important function of neutralizing the acidic chyme as it empties into the duodenum from the stomach. This aqueous $NaHCO_3$ secretion is by far the largest component of pancreatic secretion. The volume of pancreatic secretion ranges between 1 and 2 L per day, depending on the type and degree of stimulation.

## Pancreatic exocrine secretion

Pancreatic exocrine secretion is regulated primarily by hormonal mechanisms. A small amount of parasympathetically induced pancreatic secretion occurs during the cephalic phase of digestion, with a further token increase occurring during the gastric phase in response to gastrin. However, the predominant stimulation of pancreatic secretion occurs during the intestinal phase of digestion when chyme is in the small intestine. The release of the two major enterogastrones, secretin and cholecystokinin (CCK), in response to chyme in the duodenum plays the central role in controlling pancreatic secretion (⟩ Figure 15-12).

### ROLE OF SECRETIN IN PANCREATIC SECRETION

Of the factors that stimulate enterogastrone release (fat, acid, hypertonicity, and distension), the primary stimulus specifically for secretin release is acid in the duodenum. Secretin is secreted by the duodenal and jejunal mucosa in response to the release of acid. Secretin, in turn, is carried by the blood to the pancreas, where it stimulates the duct cells to markedly increase their secretion of a $NaHCO_3$-rich aqueous fluid into the duodenum. Even though other stimuli may cause the release of secretin, it is appropriate that the most potent stimulus is acid, because secretin promotes the alkaline pancreatic secretion that neutralizes the acid. This mechanism provides a control system

for maintaining neutrality of the chyme in the intestine. The amount of secretin released is proportional to the amount of acid that enters the duodenum, so the amount of $NaHCO_3$ secreted parallels the duodenal acidity.

### ROLE OF CCK IN PANCREATIC SECRETION

Cholecystokinin (CCK) is important in regulating pancreatic digestive enzyme secretion. The main stimulus for release of CCK from the duodenal mucosa is the presence of fat and, to a lesser extent, protein products. The circulatory system transports CCK to the pancreas, where it stimulates the pancreatic acinar cells to increase digestive enzyme secretion. Among these enzymes are lipase and the proteolytic enzymes, which further digest the fat and protein that initiated the response and also help digest carbohydrate. In contrast to fat and protein, carbohydrate does not have any direct influence on pancreatic digestive enzyme secretion.

All three types of pancreatic digestive enzymes are packaged together in the zymogen granules, so all the pancreatic enzymes are released together on exocytosis of the granules. Therefore, even though the *total amount* of enzymes released varies depending on the type of meal consumed (the most being secreted in response to fat), the *proportion* of enzymes released does not vary on a meal-to-meal basis. That is, a high-protein meal does not cause the release of a greater proportion of proteolytic enzymes. Evidence suggests, however, that long-term adjustments in the proportion of the types of enzymes produced may occur as an adaptive response to a prolonged change in diet.

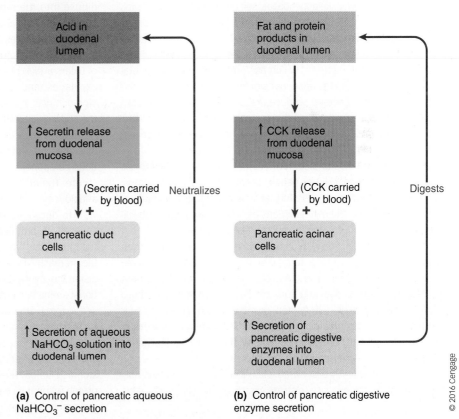

(a) Control of pancreatic aqueous $NaHCO_3^-$ secretion

(b) Control of pancreatic digestive enzyme secretion

⟩ **FIGURE 15-12** Hormonal control of pancreatic exocrine secretion

© 2016 Cengage

15

# Ulcers: When Bugs Break the Barrier

**P**EPTIC ULCERS ARE EROSIONS that typically begin in the mucosal lining of the stomach and may penetrate into the deeper layers of the stomach wall. They occur when the gastric mucosal barrier is disrupted, and thus pepsin and hydrochloric acid act on the stomach wall instead of on food in the lumen. Frequent backflow of acidic gastric juices into the oesophagus or excess unneutralized acid from the stomach in the duodenum can lead to peptic ulcers in these sites as well.

Until recently, the exact cause of ulcers was unknown, but in a surprising discovery in the early 1990s, the bacterium *Helicobacter pylori* was pinpointed as the cause of more than 80 percent of all peptic ulcers. Thirty percent of the population harbours *H. pylori*. Those who have this slow bacterium have 3 to 12 times greater risk of developing an ulcer within 10 to 20 years of acquiring the infection than those without the bacterium. They are also at increased risk of developing stomach cancer.

For years, scientists had overlooked the possibility that ulcers could be triggered by an infectious agent, because bacteria typically cannot survive in a strongly acidic environment such as the stomach lumen. An exception, *H. pylori* exploits several strategies to survive in this hostile environment. First, these organisms are motile, being equipped with four to six flagella (whiplike appendages; see the accompanying photo), which enable them to tunnel through and take up residence under the stomach's thick layer of alkaline mucus. Here they are protected from the highly acidic gastric contents. Furthermore, *H. pylori* preferentially settles in the antrum, which has no acid-producing parietal cells, although HCl from the upper parts of the stomach does reach the antrum. Also, these bacteria produce *urease*, an enzyme that breaks down urea, an end product of protein metabolism, into ammonia ($NH_3$) and $CO_2$. Ammonia serves as a buffer (p. 631) that neutralizes stomach acid locally in the vicinity of the *H. pylori*.

*H. pylori* contributes to ulcer formation in part by secreting toxins that cause a persistent inflammation, or *chronic superficial gastritis*, at the site it colonizes. *H. pylori* further weakens the gastric mucosal barrier by disrupting the tight junctions between the gastric epithelial cells, thereby making the gastric mucosa leakier than normal.

Alone or in conjunction with this infectious culprit, other factors are known to contribute to ulcer formation. Frequent exposure to some chemicals can break the gastric mucosal barrier; the most important of these are ethyl alcohol and nonsteroidal anti-inflammatory drugs (NSAIDs), such as aspirin, ibuprofen, or more potent medications for the treatment of arthritis or other chronic inflammatory processes. The barrier frequently breaks in patients with preexisting debilitating conditions, such as severe injuries or infections. Persistent stressful situations are frequently associated with ulcer formation, presumably because emotional response to stress can stimulate excessive gastric secretion.

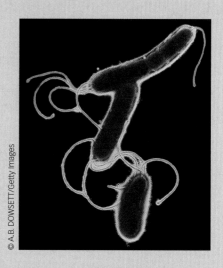

© A.B. DOWSETT/Getty Images

***Helicobacter Pylori.*** *Helicobacter pylori*, the bacterium responsible for most cases of peptic ulcers, has flagella that enable it to tunnel beneath the protective layer of mucus that coats the stomach lining.

When the gastric mucosal barrier is broken, acid and pepsin diffuse into the mucosa and underlying submucosa, with serious pathophysiological consequences. The surface erosion, or ulcer, progressively enlarges as increasing levels of acid and pepsin continue to damage the stomach wall. Two of the most serious consequences of ulcers are (1) haemorrhage, resulting from damage to submucosal capillaries, and (2) perforation, or complete erosion through the stomach wall, resulting in the escape of potent gastric contents into the abdominal cavity.

Treatment of ulcers includes antibiotics, H2 histamine receptor blockers, and proton pump inhibitors. With the discovery of the infectious component of most ulcers, antibiotics are now a treatment of choice. The other drugs are also used alone or in combination with antibiotics.

Two decades before the discovery of *H. pylori*, researchers discovered an antihistamine (*cimetidine*) that specifically blocks H2 receptors, the type of receptors that bind histamine released from the stomach. These receptors differ from H1 receptors that bind the histamine involved in allergic respiratory disorders. Accordingly, traditional antihistamines used for respiratory allergies (such as hay fever and asthma) are not effective against ulcers, nor is cimetidine useful for respiratory problems.

Another recent class of drugs used in treating ulcers inhibits acid secretion by directly blocking the pump that transports H$^+$ into the stomach lumen. These so-called proton-pump inhibitors (H$^+$ is a naked proton without its electron) help reduce the corrosive effect of HCl on the exposed tissue.

For example, with a long-term switch to a high-protein diet, a greater proportion of proteolytic enzymes are produced. CCK may play a role in the adaptation of pancreatic digestive enzymes to changes in diet.

Just as gastrin is trophic to the stomach and small intestine, CCK and secretin exert trophic effects on the exocrine pancreas to maintain its integrity.

We now look at the contributions of the remaining accessory digestive unit: the liver and gallbladder.

## The liver

Besides pancreatic juice, the other secretory product emptied into the duodenal lumen is **bile**. The **biliary system** includes the *liver*, the *gallbladder*, and associated ducts.

### LIVER FUNCTIONS

The **liver** is the largest and most important metabolic organ in the body; it can be viewed as the body's major biochemical factory. Its importance to the digestive system is its secretion of *bile salts,* which aid fat digestion and absorption. The liver also performs a wide variety of functions not related to digestion, including the following:

1. Metabolic processing of the major categories of nutrients (carbohydrates, proteins, and lipids) after their absorption from the digestive tract
2. Detoxifying or degrading body wastes and hormones, as well as drugs and other foreign compounds
3. Synthesizing plasma proteins, including those needed for blood clotting and those that transport steroid and thyroid hormones and cholesterol in the blood
4. Storing glycogen, fats, iron, copper, and many vitamins
5. Activating vitamin D, which the liver does in conjunction with the kidneys
6. Removing bacteria and "worn-out" red blood cells, thanks to its resident macrophages (p. 447)
7. Excreting cholesterol and bilirubin, the latter being a breakdown product derived from the destruction of worn-out red blood cells

Given this wide range of complex functions, there is amazingly little specialization among cells within the liver. Each liver cell, or **hepatocyte**, performs the same wide variety of metabolic and secretory tasks (*hepato* means "liver"; *cyte* means "cell"). The specialization comes from the highly developed organelles within each hepatocyte. The only liver function not accomplished by the hepatocytes is the phagocytic activity carried out by the resident macrophages, which are known as **Kupffer cells**.

### LIVER BLOOD FLOW

To carry out these wide-ranging tasks, the anatomic organization of the liver permits each hepatocyte to be in direct contact with blood from two sources: arterial blood coming from the aorta and venous blood coming directly from the digestive tract. Approximately 1 L of blood flows from the portal vein

into the liver each minute. Another 350 mL flow into the liver each minute from the hepatic artery. Like other cells, the hepatocytes receive fresh arterial blood via the hepatic artery, which supplies their oxygen and delivers blood-borne metabolites for hepatic processing. Venous blood also enters the liver by the **hepatic portal system**, a unique and complex vascular connection between the digestive tract and liver ( > Figure 15-13). The

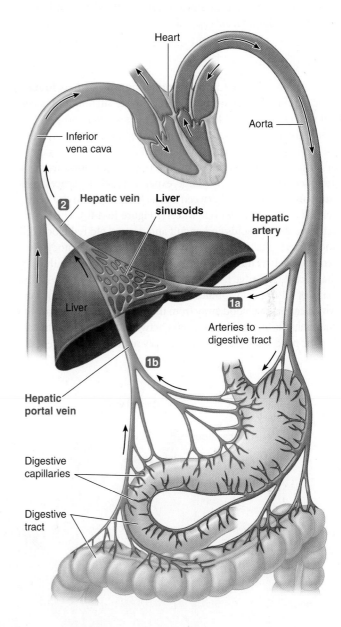

The liver receives blood from two sources:

**1a** Arterial blood, which provides the liver's $O_2$ supply and carries blood-borne metabolites for hepatic processing, is delivered by the **hepatic artery**.

**1b** Venous blood draining the digestive tract is carried by the **hepatic portal vein** to the liver for processing and storage of newly absorbed nutrients.

**2** Blood leaves the liver via the **hepatic vein**.

> FIGURE 15-13 **Schematic representation of liver blood flow**

© 2016 Cengage

15

veins draining the digestive tract do not directly join the inferior vena cava, the large vein that returns blood to the heart. Instead, the veins from the stomach and intestine enter the hepatic portal vein, which carries the products absorbed from the digestive tract directly to the liver for processing, storage, or detoxification before they gain access to the general circulation. Within the liver, the portal vein once again breaks up into a capillary network (the liver *sinusoids*) to permit exchange between the blood and hepatocytes before draining into the hepatic vein, which joins the inferior vena cava.

## Liver lobules

The liver is organized into functional units known as **lobules**, which are hexagonal arrangements of tissue surrounding a central vein (> Figure 15-14a). At each of the six outer corners of the lobule are three vessels: a branch of the hepatic artery, a branch of the hepatic portal vein, and a bile duct. Blood from the branches of both the hepatic artery and the portal vein flows from the periphery of the lobule into large, expanded capillary spaces called **sinusoids**, which run between rows of liver cells to the central vein like spokes on a bicycle wheel (> Figure 15-14b). The Kupffer cells line the sinusoids and engulf and destroy old red blood cells and bacteria that pass through in the blood. The hepatocytes are arranged between the sinusoids in plates two cell layers thick, so that each lateral edge faces a sinusoidal pool of blood. The central veins of all the liver lobules converge to form the hepatic vein, which carries the blood away from the liver. The thin bile-carrying channel, a **bile canaliculus**, runs between the cells within each hepatic plate. Hepatocytes continuously secrete bile into these thin channels, which carry the bile to a bile duct at the periphery of the lobule. The bile ducts from the various lobules converge to eventually form the *common bile duct*, which transports the bile from the liver to the duodenum. Each hepatocyte is in contact with a sinusoid on one side and a bile canaliculus on the other side.

## Bile

The opening of the bile duct into the duodenum is guarded by the **sphincter of Oddi**, also called the hepatopancreatic sphincter, which prevents bile from entering the duodenum, except during digestion of meals (> Figure 15-15). When this sphincter is closed, most of the bile secreted by the liver is diverted back up into the **gallbladder**, a small, saclike structure tucked beneath but not directly connected to the liver. The bile is stored there until needed in the duodenum. The gallbladder can hold about 50 mL. Thus, bile is not transported directly from the liver to the gallbladder. The bile is subsequently stored and concentrated in the gallbladder between meals. After a meal, bile enters the duodenum as a result of the combined effects of gallbladder emptying and increased bile secretion by the liver. The amount of bile secreted per day ranges from 250 mL to 1 L, depending on the degree of stimulation.

## Bile salts

Bile contains several organic constituents, namely *bile salts, cholesterol, lecithin,* and *bilirubin* (all derived from hepatocyte activity) in an *aqueous alkaline fluid* (added by the duct cells) similar to the pancreatic $NaHCO_3$ secretion. Even though bile does not

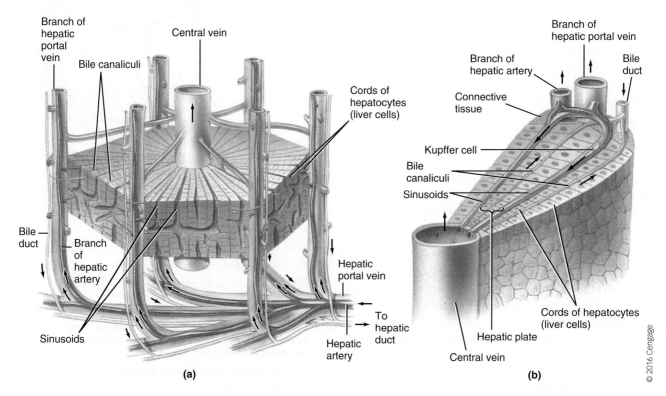

> **FIGURE 15-14 Anatomy of the liver.** (a) Hepatic lobule. (b) Wedge of a hepatic lobule

© 2016 Cengage

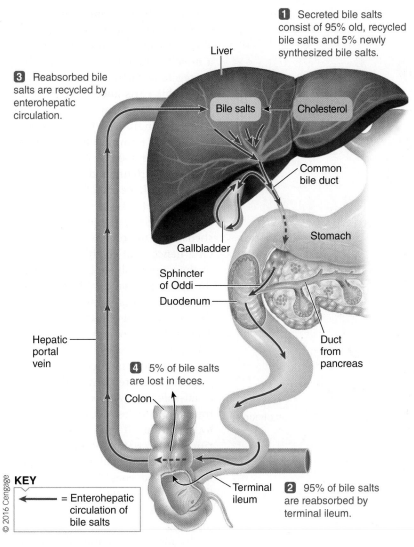

**1** Secreted bile salts consist of 95% old, recycled bile salts and 5% newly synthesized bile salts.

**3** Reabsorbed bile salts are recycled by enterohepatic circulation.

Liver

Bile salts ← Cholesterol

Common bile duct

Gallbladder

Stomach

Sphincter of Oddi

Duodenum

Duct from pancreas

Hepatic portal vein

**4** 5% of bile salts are lost in feces.

Colon

Terminal ileum

**2** 95% of bile salts are reabsorbed by terminal ileum.

© 2016 Cengage

**KEY**

⟵ = Enterohepatic circulation of bile salts

> **FIGURE 15-15 Enterohepatic circulation of bile salts.** The majority of bile salts are recycled between the liver and small intestine through the enterohepatic circulation (blue arrows). After participating in fat digestion and absorption, most bile salts are reabsorbed by active transport in the terminal ileum and returned through the hepatic portal vein to the liver, which resecretes them in the bile.

contain any digestive enzymes, it is important for the digestion and absorption of fats, primarily through the activity of bile salts.

**Bile salts** are derivatives of cholesterol. They are actively secreted into the bile and eventually enter the duodenum along with the other biliary constituents. Following their participation in fat digestion and absorption, most bile salts are reabsorbed into the blood by special active-transport mechanisms located in the terminal ileum. From here, bile salts are returned by the hepatic portal system to the liver, which resecretes them into the bile. This recycling of bile salts (and some of the other biliary constituents) between the small intestine and liver is called the **enterohepatic circulation** (*entero* means "intestine"; *hepatic* means "liver") (> Figure 15-15).

The total amount of bile salts in the body averages about 3–4 g, yet 3–15 g of bile salts may be emptied into the duodenum in a single meal. Obviously, bile salts must be recycled many

times per day. Usually, only about 5 percent of the secreted bile escapes into the feces daily. These lost bile salts are replaced by new bile salts synthesized by the liver; in this way, the size of the pool of bile salts is kept constant.

## Bile salts and fat digestion and absorption

Bile salts aid fat digestion through their detergent action (emulsification) and facilitate fat absorption by participating in the formation of micelles. Both functions are related to the structure of bile salts. Let's see how.

### DETERGENT ACTION OF BILE SALTS

The term **detergent action** refers to bile salts' ability to convert large fat globules into a **lipid emulsion** consisting of many small fat droplets, each about 1 mm in diameter. Suspended in the aqueous chyme, the lipid emulsion increases the surface area of fat droplets that are available for attack by pancreatic lipase. Fat globules, no matter their size, are made up primarily of undigested triglyceride molecules. To digest fat, lipase must come into direct contact with the triglyceride molecule. Because triglycerides are not soluble in water, they tend to aggregate into large droplets in the watery environment of the small-intestine lumen. If bile salts did not emulsify these large droplets, lipase could act on the triglyceride molecules only at the surface of the large droplets, and fat digestion would be greatly prolonged.

Bile salts exert a detergent action similar to that of the detergent you use to break up grease when you wash dishes. A bile salt molecule contains a lipid-soluble part (a steroid derived from cholesterol) plus a negatively charged, water-soluble part. Bile salts *adsorb* on the surface of a fat droplet; that is, the lipid-soluble part of the bile salt dissolves in the fat droplet, leaving the charged water-soluble part projecting from the surface of the droplet (> Figure 15-16a). This detergent-like action on the fat decreases the surface tension of the particle, increasing the ease of breakdown. Intestinal mixing movements break up large fat droplets into smaller ones. These small droplets would quickly recoalesce were it not for bile salts adsorbing on their surface and creating a shell of water-soluble negative charges on the surface of each little droplet. Because like charges repel, these negatively charged groups on the droplet surfaces cause the fat droplets to repel one another (> Figure 15-16b). This electrical repulsion prevents the small droplets from recoalescing into large fat droplets, and thus produces a lipid emulsion that increases the surface area available for lipase action.

Although bile salts increase the surface area available for attack by pancreatic lipase, lipase alone cannot penetrate the layer of bile salts adsorbed on the surface of the small emulsified fat droplets. Therefore, the pancreas secretes the polypeptide **colipase** along with lipase. Colipase binds both to lipase and to

**15**

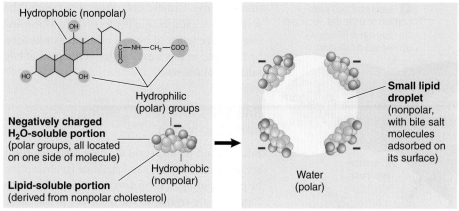

Hydrophobic (nonpolar)

OH

Hydrophilic
(polar) groups

**Negatively charged
H₂O-soluble portion**
(polar groups, all located
on one side of molecule)

Hydrophobic
(nonpolar)

**Lipid-soluble portion**
(derived from nonpolar cholesterol)

**Small lipid
droplet**
(nonpolar,
with bile salt
molecules
adsorbed on
its surface)

Water
(polar)

**(a)** Structure of bile salts and their adsorption on the surface of a small lipid droplet

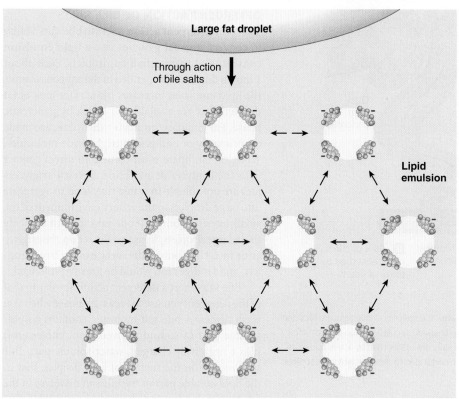

Large fat droplet

Through action
of bile salts

Lipid
emulsion

© 2016 Cengage

**(b)** Formation of a lipid emulsion through the action of bile salts

> **FIGURE 15-16 Schematic structure and function of bile salts.** (a) A bile salt consists of a lipid-soluble part that dissolves in the fat droplet and a negatively charged, water-soluble part that projects from the surface of the droplet. (b) When a large fat droplet is broken up into smaller fat droplets by intestinal contractions, bile salts adsorb on the surface of the small droplets, creating shells of negatively charged, water-soluble bile salt components that cause the fat droplets to repel one another. This emulsifying action holds the fat droplets apart and prevents them from recoalescing, thereby increasing the surface area of exposed fat available for digestion by pancreatic lipase.

the bile salts at the surface of the fat droplets, thereby anchoring lipase to its site of action.

**MICELLAR FORMATION**

Bile salts—along with cholesterol and lecithin, which are also constituents of bile—play an important role in facilitating fat absorption through micellar formation. Like bile salts, lecithin

has both a lipid-soluble and a water-soluble part, whereas cholesterol is almost totally insoluble in water. In a **micelle**, the bile salts and lecithin aggregate in small clusters with their fat-soluble parts huddled together in the middle to form a hydrophobic (water-fearing) core, while their water-soluble parts form an outer hydrophilic (water-loving) shell (› Figure 15-17). A micelle is 3–10 nm in diameter, about one-millionth the size of an emulsified lipid droplet. Micelles, being water soluble by virtue of their hydrophilic shells, can dissolve water-insoluble (and hence lipid-soluble) substances in their lipid-soluble cores. Micelles thus provide a handy vehicle for carrying water-insoluble substances through the watery luminal contents. The most important lipid-soluble substances carried within micelles are the products of fat digestion (monoglycerides and free fatty acids) as well as fat-soluble vitamins, which are all transported to their sites of absorption by this means. If they did not hitch a ride in the water-soluble micelles, these nutrients would float on the surface of the aqueous chyme (just as oil floats on top of water), never reaching the absorptive surfaces of the small intestine.

In addition, cholesterol, a highly water-insoluble substance, dissolves in the micelle's hydrophobic core. This mechanism is important in cholesterol homeostasis. The amount of cholesterol that can be carried in micellar formation depends on the relative amount of bile salts and lecithin in comparison to cholesterol.

## Bilirubin

**Bilirubin**, the other major constituent of bile, does not play a role in digestion at all; instead it is a waste product excreted in the bile. Bilirubin is the primary bile pigment derived from the breakdown of worn-out red blood cells. The typical lifespan of a red blood cell in the circulatory system is 120 days. Worn-out red blood cells are removed from the blood by the macrophages that line the liver sinusoids and reside in other areas in the body. Bilirubin is the end product from degradation of the heme (iron-containing) part of the haemoglobin contained within these old red blood cells (p. 437). This bilirubin

15

When cholesterol secretion by the liver is out of proportion to bile salt and lecithin secretion (either too much cholesterol or too little bile salts and lecithin), the excess cholesterol in the bile precipitates into microcrystals that can aggregate into **gallstones**. One treatment of cholesterol-containing gallstones involves ingestion of bile salts to increase

the bile salt pool in an attempt to dissolve the cholesterol stones. Only about 75 percent of gallstones are derived from cholesterol, however. The other 25 percent are made up of abnormal precipitates of another bile constituent, bilirubin.

is extracted from the blood by the hepatocytes and is actively excreted into the bile.

Bilirubin is a yellow pigment that gives bile its yellow colour. Within the intestinal tract, this pigment is modified by bacterial enzymes, giving rise to the characteristic brown colour of feces. When bile secretion does not occur, as when the bile duct is completely obstructed by a gallstone, the feces are greyish white. A small amount of bilirubin is normally reabsorbed by the intestine back into the blood, and when it is eventually excreted in the urine, it is largely responsible for the urine's yellow colour. The kidneys cannot excrete bilirubin until after it has been modified during its passage through the liver and intestine.

*Clinical Note* If bilirubin is formed more rapidly than it can be excreted, it accumulates in the body and causes **jaundice**. Patients with this condition appear yellowish, and this colour is

most readily seen in the whites of their eyes. Jaundice can be brought about in three different ways:

1. *Prehepatic* (the problem occurs "before the liver"), or *haemolytic, jaundice* is due to excessive breakdown (haemolysis) of red blood cells, which results in the liver being presented with more bilirubin than it is capable of excreting.

2. *Hepatic* (the problem is the liver) *jaundice* occurs when the liver is diseased and cannot deal with even the normal load of bilirubin.

3. *Posthepatic* (the problem occurs "after the liver"), or *obstructive, jaundice* occurs when the bile duct is obstructed, such as by a gallstone, so that bilirubin cannot be eliminated in the feces.

## Secretion

Bile secretion may be increased by chemical, hormonal, and neural mechanisms:

- *Chemical mechanism (bile salts).* Any substance that increases bile secretion by the liver is called a **choleretic**. The most potent choleretic is bile salts themselves. Between meals, bile is stored in the gallbladder, but during a meal, bile is emptied into the duodenum as the gallbladder contracts. After bile salts participate in fat digestion and absorption, they are reabsorbed and returned by the enterohepatic circulation to the liver, where they act as potent choleretics to stimulate further bile secretion. Therefore, during a meal, when bile salts are needed and being used, bile secretion by the liver is enhanced.

- *Hormonal mechanism (secretin).* Besides increasing the aqueous $NaHCO_3$ secretion by the pancreas, secretin stimulates an aqueous alkaline bile secretion by the liver ducts without any corresponding increase in bile salts.

- *Neural mechanism (vagus nerve).* Vagal stimulation of the liver plays a minor role in bile secretion during the cephalic phase of digestion, promoting an increase in liver bile flow before food ever reaches the stomach or intestine.

## The gallbladder

Even though the factors just described increase bile secretion by the liver during and after a meal, bile secretion by the liver occurs continuously. Between meals, the secreted bile is shunted into

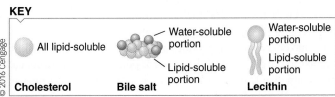

**Hydrophobic core**

**Hydrophilic shell**

**KEY**

| | | |
|---|---|---|
| All lipid-soluble | Water-soluble portion / Lipid-soluble portion | Water-soluble portion / Lipid-soluble portion |
| **Cholesterol** | **Bile salt** | **Lecithin** |

© 2016 Cengage

❯ **FIGURE 15-17 A micelle.** Bile constituents (bile salts, lecithin, and cholesterol) aggregate to form micelles that consist of a hydrophilic (water-loving) shell and a hydrophobic (water-fearing) core. Because the outer shell of a micelle is water soluble, the products of fat digestion, which are not water soluble, can be carried through the watery luminal contents to the absorptive surface of the small intestine by dissolving in the micelle's lipid-soluble core. An emulsified fat droplet ranges in diameter from 200 to 5000 nm (average 1000 nm) compared to a micelle, which is 3 to 10 nm in diameter.

**15**

the gallbladder, where it is stored and concentrated. Active transport of salt out of the gallbladder, with water following osmotically, results in a 5 to 10 times concentration of the organic constituents.

Clinical Note Because the gallbladder stores this concentrated bile, it is the primary site for precipitation of concentrated bile constituents into gallstones. Fortunately, the gallbladder does not play an essential digestive role, so its removal as a treatment for gallstones or other gallbladder disease presents no particular problem. The bile secreted between meals is stored instead in the common bile duct, which becomes dilated.

During digestion of a meal, when chyme reaches the small intestine, the presence of food, especially fat products, in the duodenal lumen triggers the release of CCK. This hormone stimulates contraction of the gallbladder and relaxation of the sphincter of Oddi, so bile is discharged into the duodenum, where it appropriately aids in the digestion and absorption of the fat that initiated the release of CCK.

## Hepatitis and cirrhosis

Clinical Note **Hepatitis** is an inflammatory disease of the liver that results from a variety of causes, including viral infection or exposure to toxic agents such as alcohol, carbon tetrachloride, and certain tranquilizers. Hepatitis ranges in severity from mild, reversible symptoms to acute, massive liver damage, with possible imminent death resulting from acute hepatic failure.

Repeated or prolonged hepatic inflammation, usually in association with chronic alcoholism, can lead to **cirrhosis**, a condition in which damaged hepatocytes are permanently replaced by connective tissue. Liver tissue has the ability to regenerate, normally undergoing a gradual turnover of cells. If part of the hepatic tissue is destroyed, the lost tissue can be replaced by an increase in the rate of cell division. There is a limit, however, to how rapidly hepatocytes can be replaced. In addition to hepatocytes, a small number of fibroblasts (connective tissue cells) are dispersed between the hepatic plates and form a supporting framework for the liver. If the liver is exposed to toxic substances, such as alcohol, so often that new hepatocytes cannot be generated rapidly enough to replace the damaged cells, the sturdier fibroblasts take advantage of the situation and overproduce. This extra connective tissue leaves little space for the hepatocytes' regrowth. Thus, as cirrhosis develops slowly over time, active liver tissue is gradually reduced, leading eventually to chronic liver failure.

This completes our discussion of the accessory digestive organs that empty their exocrine products into the small-intestine lumen. We now focus on the functions of the small intestine itself.

---

### Check Your Understanding 15.5

1. Explain the significance of some pancreatic enzymes being stored as precursors in zymogen granules.
2. Describe how bile salts contribute to dietary fat digestion.

---

## 15.6 | Small Intestine

The **small intestine** is the site where most digestion and absorption take place. No further digestion is accomplished after the luminal contents pass beyond the small intestine, nor does further absorption of ingested nutrients occur, although the large intestine does absorb small amounts of salt and water. The small intestine lies coiled within the abdominal cavity, extending between the stomach and large intestine. It is roughly divided into three segments: the **duodenum**, the **jejunum**, and the **ileum**.

In this section we examine the small intestine in terms of motility, secretion, digestion, and absorption, as we have for the other digestive organs. Small-intestine motility includes *segmentation* and the *migrating motility complex*. Let's consider segmentation first.

### Segmentation

**Segmentation**, the small intestine's primary method of motility during digestion of a meal, both mixes and slowly propels the chyme. When a portion of the small intestine becomes distended (stretched) with the chyme, the stretch of the intestine begins localized concentric contractions, which occur at regular intervals. This consists of oscillating, ringlike contractions of the circular smooth muscle along the small intestine's length; between the contracted segments are relaxed areas containing a small bolus of chyme. The contractile rings occur every few centimetres, dividing the small intestine into segments like a chain of sausages. These contractile rings do not sweep along the length of the intestine as peristaltic waves do. Rather, after a brief period of time, the contracted segments relax, and ringlike contractions appear in the previously relaxed areas (> Figure 15-18). The new contraction forces the chyme in a previously relaxed segment to move in both directions into the now relaxed adjacent segments. A newly relaxed segment therefore receives chyme from both the contracting segment immediately ahead of it and the one immediately behind it. Shortly thereafter, the areas of contraction and relaxation alternate again. In this way, the chyme is chopped, churned, and thoroughly mixed. These contractions can be compared to squeezing a pastry tube with your hands to mix the contents.

#### INITIATION AND CONTROL OF SEGMENTATION

Segmentation contractions are initiated by the small intestine's pacemaker cells, which produce a basic electrical rhythm (BER) similar to the gastric BER that governs peristalsis in the stomach. If the small-intestine BER brings the circular smooth muscle layer to threshold, segmentation contractions are induced, with the frequency of segmentation following the frequency of the BER. The frequency of contraction in the duodenum and jejunum is about 12 per minute, while in the terminal ileum the frequency drops to about 9 per minute.

The circular smooth muscle's degree of responsiveness and thus the intensity of segmentation contractions can be influenced by distension of the intestine, by the hormone gastrin, and by extrinsic nerve activity. All these factors influence the

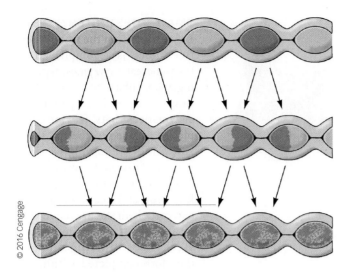

> FIGURE 15-18 **Segmentation.** Segmentation consists of ringlike contractions along the length of the small intestine. Within a matter of seconds, the contracted segments relax, and the previously relaxed areas contract. These oscillating contractions thoroughly mix the chyme within the small-intestine lumen.

excitability of the small-intestine smooth muscle cells by moving the starting potential around which the BER oscillates closer to or farther from the threshold. Segmentation is slight or absent between meals but becomes very vigorous immediately after a meal. Both the duodenum and the ileum start to segment simultaneously when the meal first enters the small intestine. The duodenum starts to segment primarily in response to local distension caused by the presence of chyme. Segmentation of the empty ileum, in contrast, is brought about by gastrin secreted in response to the presence of chyme in the stomach, a mechanism known as the **gastroileal reflex**. Extrinsic nerves can modify the strength of these contractions. Parasympathetic stimulation enhances segmentation, whereas sympathetic stimulation depresses segmental activity.

### FUNCTIONS OF SEGMENTATION

The mixing accomplished by segmentation serves the dual functions of mixing the chyme with the digestive juices secreted into the small-intestine lumen and exposing all the chyme to the absorptive surfaces of the small-intestine mucosa.

Segmentation not only accomplishes mixing but also slowly moves chyme through the small intestine. How can this be, when each segmental contraction propels chyme both forward and backward? The chyme slowly progresses forward because the frequency of segmentation declines along the length of the small intestine. The pacemaker cells in the duodenum spontaneously depolarize faster than those farther down the tract; segmentation contractions in the duodenum occur at a rate of 12 per minute, and in the terminal ileum they occur at a rate of only 9 per minute. Because segmentation occurs with greater frequency in the upper part of the small intestine than in the lower part, more chyme, on average, is pushed forward than is pushed backward. As a result, chyme is moved very slowly from the upper to the lower part of the

small intestine, being shuffled back and forth to accomplish thorough mixing and absorption in the process. This slow propulsive mechanism is advantageous because it allows ample time for the digestive and absorptive processes to take place. The contents usually take three to five hours to move through the small intestine.

## The migrating motility complex

When most of the meal has been absorbed, segmentation contractions cease and are replaced between meals by the **migrating motility complex**—the "intestinal housekeeper." This between-meal motility consists of weak, repetitive peristaltic waves that move a short distance down the intestine before dying out. The waves start at the stomach and migrate down the intestine; that is, each new peristaltic wave is initiated at a site a little farther down the small intestine. These short peristaltic waves take about 100 to 150 minutes to gradually migrate from the stomach to the end of the small intestine, with each contraction sweeping any remnants of the preceding meal plus mucosal debris and bacteria forward toward the colon, just like a good intestinal housekeeper. After the end of the small intestine is reached, the cycle begins again and continues to repeat itself until the next meal. The migrating motility complex is thought to be regulated between meals by the hormone **motilin**, which is secreted during the unfed state by endocrine cells of the small-intestine mucosa. When the next meal arrives, segmental activity is triggered again, and the migrating motility complex ceases. Motilin release is inhibited by feeding.

## The ileocecal juncture

At the juncture between the small and large intestines, the last part of the ileum empties into the cecum (> Figure 15-19). Two factors contribute to this region's ability to act as a barrier between the small and large intestines. First, the anatomic arrangement is such that valvelike folds of tissue protrude from the ileum into the lumen of the cecum. When the ileal contents are pushed forward, this **ileocecal valve** is easily pushed open, but the folds of tissue are forcibly closed when the cecal contents attempt to move backward. Second, the smooth muscle within the last several centimetres of the ileal wall is thickened, forming a sphincter that is under neural and hormonal control. Most of the time, this **ileocecal sphincter** remains at least mildly constricted. Pressure on the cecal side of the sphincter causes it to contract more forcibly; distension of the ileal side causes the sphincter to relax, a reaction mediated by the intrinsic plexuses in the area. In this way, the ileocecal juncture prevents the bacteria-laden contents of the large intestine from contaminating the small intestine and at the same time lets the ileal contents pass into the colon. If the colonic bacteria gained access to the nutrient-rich small intestine, they would multiply rapidly. Relaxation of the sphincter is enhanced by the release of gastrin at the onset of a meal, when increased gastric activity is taking place. This relaxation allows the undigested fibres and unabsorbed solutes from the preceding meal to be moved forward as the new meal enters the tract.

**15**

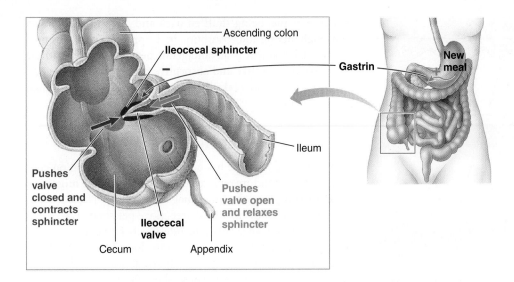

> FIGURE 15-19 **Control of the ileocecal valve/sphincter.** The juncture between the ileum and large intestine is the ileocecal valve, which is surrounded by thickened smooth muscle, the ileocecal sphincter. Pressure on the cecal side pushes the valve closed and contracts the sphincter, preventing the bacteria-laden colonic contents from contaminating the nutrient-rich small intestine. The valve/sphincter opens and allows ileal contents to enter the large intestine in response to pressure on the ileal side of the valve and to the hormone gastrin that's secreted as a new meal enters the stomach.

## Small-intestine secretions

Each day, the exocrine gland cells in the small-intestine mucosa secrete into the lumen about 1.5 L of an aqueous salt and mucous solution called **succus entericus** (juice of intestine). Secretion increases after a meal in response to local stimulation of the small-intestine mucosa by the presence of chyme.

The mucus in the secretion provides protection and lubrication. Furthermore, this aqueous secretion provides plenty of water to participate in the enzymatic digestion of food. Recall that digestion involves hydrolysis—bond breakage by reaction with water—which proceeds most efficiently when all the reactants are in solution.

No digestive enzymes are secreted into this intestinal juice. The small intestine does synthesize digestive enzymes, but they act within the brush-border membrane of the epithelial cells that line the lumen, instead of being secreted directly into the lumen.

## SMALL-INTESTINE ENZYMES

Digestion within the small-intestine lumen is accomplished by the pancreatic enzymes, with fat digestion enhanced by bile secretion. As a result of pancreatic enzymatic activity, fats are completely reduced to their absorbable units of monoglycerides and free fatty acids, proteins are broken down into small peptide fragments and some amino acids, and carbohydrates are reduced to disaccharides and some monosaccharides. Fat digestion is therefore completed within the small-intestine lumen, but carbohydrate and protein digestion have not yet been brought to completion.

Special hairlike projections on the luminal surface of the small-intestine epithelial cells, the **microvilli**, form the **brush border**. The brush-border plasma membrane contains three categories of membrane-bound enzymes:

1. *Enterokinase,* which activates the pancreatic enzyme trypsinogen

2. The **disaccharidases** (maltase, sucrase, and lactase), which complete carbohydrate digestion by hydrolyzing the remaining disaccharides (maltose, sucrose, and lactose, respectively) into their constituent monosaccharides

3. The **aminopeptidases**, which hydrolyze the small peptide fragments into their amino acid components, thereby completing protein digestion

Carbohydrate and protein digestion are therefore completed within the confines of the brush border. ( Table 15-6 provides a summary of the digestive processes for the three major categories of nutrients.)

We are now ready to discuss absorption of nutrients. Up to this point, no food, water, or electrolytes have been absorbed.

**∎ Why It Matters**
Lactose Intolerance

A fairly common disorder, **lactose intolerance**, involves a deficiency of lactase, the disaccharidase specific for the digestion of lactose, or milk sugar. Most children under 4 years of age have adequate lactase, but this may be gradually lost so that in many adults, lactase activity is diminished or absent. When lactose-rich milk or dairy products are consumed by a person with lactase deficiency, the undigested lactose remains in the lumen and has several related consequences.

First, accumulation of undigested lactose creates an osmotic gradient that draws water into the intestinal lumen. Second, bacteria living in the large intestine have lactose-splitting ability, so they eagerly attack the lactose as an energy source, producing large quantities of carbon dioxide and methane gas in the process. Distension of the intestine by both fluid and gas produces pain (cramping) and diarrhoea. Infants with lactose intolerance may also suffer from malnutrition.

| Nutrients | Enzymes for Digesting Nutrient | Source of Enzymes | Site of Action of Enzymes | Action of Enzymes | Absorbable Units of Nutrients |
|---|---|---|---|---|---|
| **Carbohydrate** | Amylase | Salivary glands | Mouth and body of stomach | Hydrolyzes polysaccharides to disaccharides | |
| | | Exocrine pancreas | Small-intestine lumen | | |
| | Disaccharidases (maltase, sucrase, lactase) | Small-intestine epithelial cells | Small-intestine brush border | Hydrolyze disaccharides to monosaccharides | Monosaccharides, especially glucose |
| **Protein** | Pepsin | Stomach chief cells | Stomach antrum | Hydrolyzes protein to peptide fragments | |
| | Trypsin, chymotrypsin carboxypeptidase | Exocrine pancreas | Small-intestine lumen | Attack different peptide fragments | |
| | Aminopeptidases | Small-intestine epithelial cells | Small-intestine brush border | Hydrolyze peptide fragments to amino acids | Amino acids and a few small peptides |
| **Fat** | Lipase | Exocrine pancreas | Small-intestine lumen | Hydrolyzes triglycerides to fatty acids and monoglycerides | Fatty acids and monoglycerides |
| | Bile salts (not an enzyme) | Liver | Small-intestine lumen | Emulsify large fat globules for attack by pancreatic lipase | |

© 2016 Cengage

## The small intestine and absorption

All products of carbohydrate, protein, and fat digestion, as well as most of the ingested electrolytes, vitamins, and water, are normally absorbed by the small intestine indiscriminately. The absorption from the small intestine each day consists of (depending on the diet) ≥ 200 g of carbohydrates, ≥ 100 g of fat, ≥ 50 g of amino acids, and 7–8 L of water. However, the capacity of the small intestine is much greater. Usually only the absorption of calcium and iron is adjusted to the body's needs. So the more food consumed, the more will be digested and absorbed— as people who are trying to control their weight are all too painfully aware.

Most absorption occurs in the duodenum and jejunum; very little occurs in the ileum, not because the ileum does not have absorptive capacity but because most absorption has already been accomplished before the intestinal contents reach the ileum. The small intestine has an abundant reserve of absorptive capacity. About 50 percent of the small intestine can be removed with little interference to absorption—with one exception. If the terminal ileum is removed, vitamin $B_{12}$ and bile salts are not properly absorbed, because the specialized transport mechanisms for these two substances are located only in this region. All other substances can be absorbed throughout the small intestine's length.

The mucous lining of the small intestine is remarkably well adapted for its special absorptive function for two reasons:

(1) it has a very large surface area, and (2) the epithelial cells in this lining have a variety of specialized transport mechanisms.

### ADAPTATIONS THAT INCREASE THE SMALL INTESTINE'S SURFACE AREA

The following special modifications of the small-intestine mucosa greatly increase the surface area available for absorption (› Figure 15-20):

- The inner surface of the small intestine, which is thrown into permanent circular folds that are visible to the naked eye, increase the surface area threefold.

- Projecting from this folded surface are microscopic finger-like projections known as **villi**, which give the lining a velvety appearance and further increase the surface area (› Figure 15-21). The surface of each villus is covered by epithelial cells interspersed occasionally with mucous cells.

- Even smaller hairlike projections, the brush border or microvilli, arise from the luminal surface of these epithelial cells, again increasing the surface area. Each epithelial cell has as many as 3000 to 6000 of these microvilli, which are visible only with an electron microscope. The small-intestine enzymes perform their functions within the membrane of this brush border.

Altogether, the folds, villi, and microvilli provide the small intestine with a luminal surface area 600 times greater than if it

15

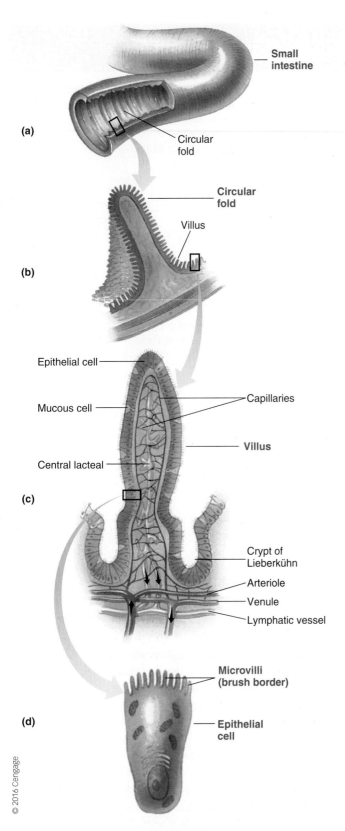

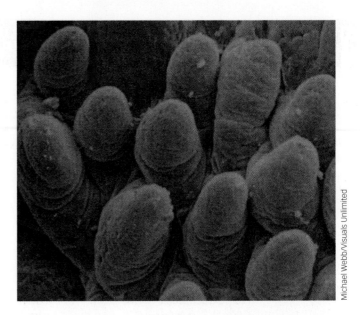

> Figure 15-21 **Scanning electron micrograph of villi projecting from the small-intestine mucosa**

> **FIGURE 15-20 Small-intestine absorptive surface.** (a) Gross structure of the small intestine. (b) One of the circular folds of the small-intestine mucosa, which collectively increase the absorptive surface area threefold. (c) Microscopic finger-like projection known as a *villus*. (d) Electron microscope view of a villus epithelial cell, depicting the presence of microvilli on its luminal border. Together, these surface modifications make the small intestine's surface area 600 times more absorptive than it would be without them.

were a tube of the same length and diameter lined by a flat surface. In fact, if the surface area of the small intestine were spread out flat, it would cover an entire tennis court.

### STRUCTURE OF A VILLUS

Absorption across the digestive tract wall involves transepithelial transport similar to movement of material across the kidney tubules (p. 574). Each villus has the following major components (see > Figure 15-20c):

- *Epithelial cells that cover the surface of the villus.* The epithelial cells are joined at their lateral borders by tight junctions, which limit passage of luminal contents between the cells, although the tight junctions in the small intestine are leakier than those in the stomach. Within their luminal brush borders, these epithelial cells have carriers for absorption of specific nutrients and electrolytes from the lumen as well as membrane-bound digestive enzymes that complete carbohydrate and protein digestion.

- *A connective tissue core.* This core is formed by the lamina propria.

- *A capillary network.* Each villus is supplied by an arteriole that breaks up into a capillary network within the villus core. The capillaries rejoin to form a venule that drains away from the villus.

- *A terminal lymphatic vessel.* Each villus is supplied by a single blind-ended lymphatic vessel known as the **central lacteal**, which occupies the centre of the villus core.

During the process of absorption, digested substances enter the capillary network or the central lacteal. To be absorbed, a substance must pass completely through the epithelial cell, diffuse through the interstitial fluid within the connective tissue core of the villus, and then cross the wall of a capillary or lymph

Malabsorption (impairment of absorption) may be caused by damage to or reduction of the surface area of the small intestine. One of the most common causes is **gluten enteropathy**, also known as **celiac disease**. In this condition, the person's small intestine is abnormally sensitive to *gluten*, a protein constituent of wheat, barley, and rye. These grain products are widely prevalent in processed foods. This condition is a complex immunological disorder in which exposure to gluten erroneously activates a T-cell response that damages the intestinal villi: the normally luxuriant array of villi is reduced, the mucosa becomes flattened, and the brush border becomes short and stubby (❭ Figure 15-22). Because this loss of villi decreases the surface area available for absorption, absorption of all nutrients is impaired. The condition is treated by eliminating gluten from the diet.

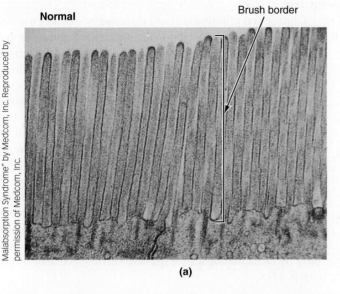

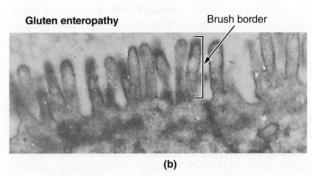

❭ **FIGURE 15-22 Reduction in the brush border with gluten enteropathy.** (a) Electron micrograph of the brush border of a small-intestine epithelial cell in a normal individual. (b) Electron micrograph of the short, stubby brush border of a small-intestine epithelial cell in a patient with gluten enteropathy

vessel. Like renal transport, intestinal absorption may be active or passive; active absorption involves energy expenditure during at least one of the transepithelial transport steps.

## The mucosal lining

Dipping down into the mucosal surface between the villi are shallow invaginations known as the **crypts of Lieberkühn** (see ❭ Figure 15-20c). Unlike the gastric pits, these intestinal crypts do not secrete digestive enzymes, but they do secrete water and electrolytes, which, along with the mucus secreted by the cells on the villus surface, constitute the succus entericus.

Furthermore, the crypts function as nurseries. The epithelial cells lining the small intestine slough off and are replaced at a rapid rate as a result of the high mitotic activity of *stem cells* in the crypts. New cells that are continually produced in the crypts migrate up the villi and, in the process, push off the older cells at the tips of the villi into the lumen. In this manner, more than 100 million intestinal cells are shed per minute. The entire trip from crypt to tip averages about three days, so the epithelial lining of the small intestine is replaced approximately every three days. Because of this high rate of cell division, the crypt stem cells are very sensitive to damage by radiation and anticancer drugs, both of which may inhibit cell division.

The new cells undergo several changes as they migrate up the villus. The concentration of brush-border enzymes increases, and the capacity for absorption improves, so the cells at the tip of the villus have the greatest digestive and absorptive capability. Just at their peak, these cells are pushed off by the newly migrating cells. Thus, the luminal contents are constantly exposed to cells that are optimally equipped to complete the digestive and absorptive functions efficiently. Furthermore, just as in the stomach, the rapid turnover of cells in the small intestine is essential because of the harsh luminal conditions. Cells exposed to the abrasive and corrosive luminal contents are easily damaged and cannot live for long, so they must be continually replaced by a fresh supply of newborn cells.

The old cells sloughed off into the lumen are not entirely lost to the body. These cells are digested, with the cell constituents

15

absorbed into the blood and reclaimed for synthesis of new cells, among other things.

In addition to the stem cells, Paneth cells are also found in the crypts. **Paneth cells** serve a defensive function, safeguarding the stem cells. They produce two chemicals that thwart bacteria: (1) *lysozyme*, the bacterial-lysing enzyme also found in saliva; and (2) *defensins*, small proteins with antimicrobial powers.

We now examine how the epithelial lining of the small intestine is specialized to accomplish absorption of luminal contents and also the mechanisms by which the specific dietary constituents are normally absorbed.

## Sodium absorption and water absorption

Sodium may be absorbed both passively and actively. When the electrochemical gradient favours movement of $Na^+$ from the lumen to the blood, passive diffusion of $Na^+$ can occur *between* the intestinal epithelial cells—through what's referred to as leaky tight junctions—and into the interstitial fluid within the villus. Movement of $Na^+$ *through* the cells is energy dependent and involves two different carriers; the process is similar to $Na^+$ reabsorption across the kidney tubules (pp. 575) (Sodium passively enters the epithelial cells across the luminal border either by itself through $Na^+$ channels or in the company of glucose or amino acid by means of a cotransport carrier. Sodium is actively pumped out of the cell at the basolateral border into the interstitial fluid in the lateral spaces between the cells where they are not joined by tight junctions. From the interstitial fluid, $Na^+$ diffuses into the capillaries.

As with the renal tubules in the early part of the nephron, the absorption of $Cl^-$, $H_2O$, glucose, and amino acids from the small intestine is linked to this energy-dependent $Na^+$ absorption. Chloride passively follows down the electrical gradient created by $Na^-$ absorption and can be actively absorbed as well if needed. Most $H_2O$ absorption in the digestive tract depends on the active carrier that pumps $Na^+$ into the lateral spaces, resulting in a concentrated area of high osmotic pressure in that localized region between the cells, similar to the situation in the kidneys (p. 581). This localized high osmotic pressure induces $H_2O$ to move from the lumen through the cell (and possibly from the lumen through the leaky tight junction) into the lateral space. Water entering the space reduces the osmotic pressure but raises the hydrostatic (fluid) pressure. As a result, $H_2O$ is flushed out of the lateral space into the interior of the villus, where it is picked up by the capillary network. Meanwhile, more $Na^+$ is pumped into the lateral space to encourage more $H_2O$ absorption.

## Carbohydrate and protein

Absorption of the digestion end products of both carbohydrates and proteins involves special carrier-mediated transport systems that require energy expenditure and $Na^+$ cotransport, and both categories of end products are absorbed into the blood.

### CARBOHYDRATE ABSORPTION

Dietary carbohydrates are generally complex and must be broken down prior to absorption. The actions of salivary amylase break down these polysaccharides into disaccharides, which are further broken down into monosaccharides by disaccharidases located in the brush borders of the small-intestine epithelial cells. All carbohydrates are absorbed as monosaccharides ( > Figure 15-23).

Very little carbohydrate is absorbed by way of basic diffusion, and instead is absorbed by active transport. The transporters that mediate the absorption of monosaccharides are (1) $Na^+$ monosaccharide cotransport and (2) $Na^+$–independent facilitated diffusion. Recall that the pores of the mucosa used for diffusion are impermeable to water-soluble solutes with larger molecular weights, including carbohydrates. As well, the active transport process is selective, with certain monosaccharides being preferentially transported before others. For example, galactose is transported before glucose, and glucose is transported before fructose.

Glucose and galactose are both absorbed by secondary active transport, in which cotransport carriers on the luminal border transport both the monosaccharide and $Na^+$ from the lumen into the interior of the intestinal cell. The operation of these cotransport carriers, which do not directly use energy themselves, depends on the $Na^+$ concentration gradient established by the energy-consuming basolateral $Na^+-K^+$ pump (p. 44). Glucose (or galactose), having been concentrated in the cell by the cotransport carriers, leaves the cell down its concentration gradient by means of a passive carrier in the basolateral border and enters the blood within the villus. In addition to glucose absorption through the cells by means of the cotransport carrier, recent evidence suggests that a significant amount of glucose crosses the epithelial barrier through the leaky tight junctions between the epithelial cells. Fructose is absorbed into the blood solely by facilitated diffusion (passive carrier-mediated transport; p. 43).

### PROTEIN ABSORPTION

Both ingested proteins and endogenous (within the body) proteins that have entered the digestive tract lumen from the three following sources are digested and absorbed:

1. Digestive enzymes—all of which are proteins—that have been secreted into the lumen

2. Proteins within the cells that are pushed off from the villi into the lumen during the process of mucosal turnover

3. Small amounts of plasma proteins that normally leak from the capillaries into the digestive tract lumen

About 20–40 g of endogenous protein enter the lumen each day from these three sources. This quantity can amount to more than the quantity of protein in ingested food. All endogenous proteins must be digested and absorbed along with the dietary proteins to prevent depletion of the body's protein stores. The amino acids absorbed from both food and endogenous protein are used primarily to synthesize new protein in the body. As the building of muscle tissue is important to athletes, amino acids are important to their diet. Just how much protein to consume and when to consume it is something that many athletes and coaches consider for optimal performance.

The protein presented to the small intestine for absorption is primarily in the form of amino acids and a few small peptide

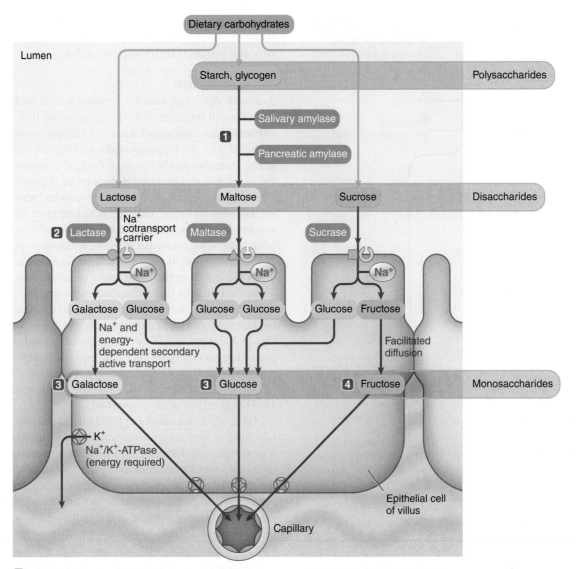

**1** The dietary polysaccharides starch and glycogen are converted into the disaccharide maltose through the action of salivary and pancreatic amylase.

**2** Maltose and the dietary disaccharides lactose and sucrose are converted to their respective monosaccharides by the disaccharidases (maltase, lactase, and sucrase) located in the brush borders of the small-intestine epithelial cells.

**3** The monosaccharides glucose and galactose are absorbed into the interior of the cell and eventually enter the blood by means of $Na^+$ and energy-dependent secondary active transport.

**4** The monosaccharide fructose is absorbed into the blood by passive facilitated diffusion.

› **FIGURE 15-23 Carbohydrate digestion and absorption**

fragments (› Figure 15-24). Amino acids are absorbed across the intestinal cells by secondary active transport, similar to glucose and galactose absorption. Thus, glucose, galactose, and amino acids all get a "free ride" in on the energy expended for $Na^+$ transport. Small peptides gain entry by means of a different carrier and are broken down into their constituent amino acids by the aminopeptidases in the brush-border membrane or by intracellular peptidases. Like monosaccharides, amino acids enter the capillary network within the villus.

## Digested fat

Fat absorption is quite different from carbohydrate and protein absorption, because the insolubility of fat in water presents a special problem. Fat must be transferred from the watery chyme through the watery body fluids, even though fat is not water soluble. Therefore, fat must undergo a series of physical and chemical transformations to circumvent this problem during its digestion and absorption (› Figure 15-25).

**15**

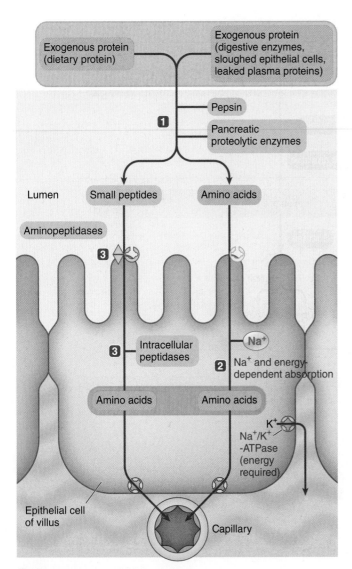

**1** Dietary and endogenous proteins are hydrolyzed to their constituent amino acids and a few small peptide fragments by gastric pepsin and pancreatic proteolytic enzymes.

**2** Amino acids are absorbed into the small-intestine epithelial cells and eventually enter the blood by means of Na⁺ and energy-dependent secondary active transport. Various amino acids are transported by carriers specific for them.

**3** The small peptides, which are absorbed by a different type of carrier, are broken down into their amino acids by aminopeptidases in the brush borders of epithelial cells or by intracellular peptidases.

⟩ **FIGURE 15-24 Protein digestion and absorption**

## A REVIEW OF FAT EMULSIFICATION AND DIGESTION

When the stomach contents are emptied into the duodenum, the ingested fat is aggregated into large, oily triglyceride droplets that float in the chyme. Recall that through the bile salts' detergent action in the small-intestine lumen, the large droplets are dispersed into a lipid emulsification of small droplets, exposing a much greater surface area of fat for digestion by pancreatic lipase. The products of lipase digestion (monoglycerides and free fatty acids) are also not very water soluble, so very little of these end products of fat digestion can diffuse

through the aqueous chyme to reach the absorptive lining. However, biliary components facilitate absorption of these fatty end products by forming micelles.

### FAT ABSORPTION

As already discussed, micelles are water-soluble particles that can carry the end products of fat digestion within their lipid-soluble interiors. Once these micelles reach the luminal membranes of the epithelial cells, the monoglycerides and free fatty acids passively diffuse from the micelles through the lipid component of the epithelial cell membranes to enter the interior of these cells. As these fat products leave the micelles and are absorbed across the epithelial cell membranes, the micelles can pick up more monoglycerides and free fatty acids, which have been produced from digestion of other triglyceride molecules in the fat emulsion.

Bile salts continuously repeat their fat-solubilizing function down the length of the small intestine until all fat is absorbed. Then the bile salts themselves are reabsorbed in the terminal ileum by special active transport. This is an efficient process, because relatively small amounts of bile salts can facilitate digestion and absorption of large amounts of fat, with each bile salt performing its ferrying function repeatedly before it is reabsorbed. When bile salts are present, about 96–98 percent of the fat is absorbed, but in the absence of bile salts only about 49–51 percent of the fat is absorbed.

Once within the interior of the epithelial cells, the monoglycerides and free fatty acids are resynthesized into triglycerides. These triglycerides conglomerate into droplets and are coated with a layer of lipoprotein (synthesized by the endoplasmic reticulum of the epithelial cell), which makes the fat droplets water soluble. The large, coated fat droplets, known as **chylomicrons**, are extruded by exocytosis from the epithelial cells into the interstitial fluid within the villus. The chylomicrons subsequently enter the central lacteals rather than the capillaries because of the structural differences between these two vessels. Capillaries have a basement membrane (an outer layer of polysaccharides) that prevents chylomicrons from entering, but the lymph vessels do not have this barrier. Therefore, fat can be absorbed into the lymphatics, but not directly into the blood.

The actual absorption or transfer of monoglycerides and free fatty acids from the chyme across the luminal membranes of the intestinal epithelial cells is a passive process, because the lipid-soluble fatty end products merely dissolve in and pass through the lipid part of the membrane. However, the overall sequence of events needed for fat absorption does require energy. For example, bile salts are actively secreted by the liver, and the resynthesis of triglycerides and formation of chylomicrons within the epithelial cells are active processes.

## Vitamin absorption

Water-soluble vitamins are primarily absorbed passively with water, whereas fat-soluble vitamins are carried in the micelles and absorbed passively with the end products of fat digestion. Some of the vitamins can also be absorbed by carriers, if necessary. Vitamin B₁₂ is unique in that it must be in combination with gastric intrinsic factor for absorption by receptor-mediated endocytosis in the terminal ileum.

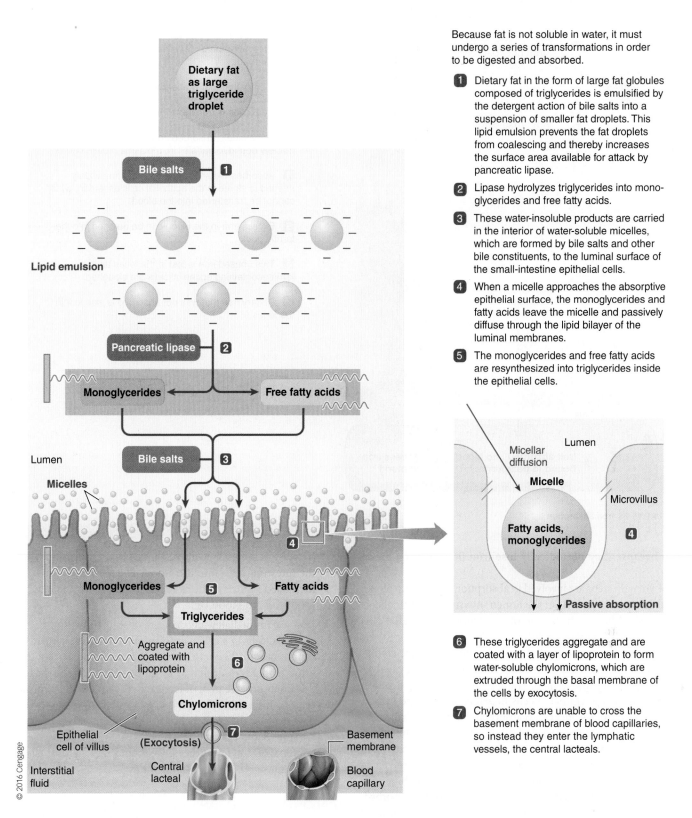

Because fat is not soluble in water, it must undergo a series of transformations in order to be digested and absorbed.

**1** Dietary fat in the form of large fat globules composed of triglycerides is emulsified by the detergent action of bile salts into a suspension of smaller fat droplets. This lipid emulsion prevents the fat droplets from coalescing and thereby increases the surface area available for attack by pancreatic lipase.

**2** Lipase hydrolyzes triglycerides into monoglycerides and free fatty acids.

**3** These water-insoluble products are carried in the interior of water-soluble micelles, which are formed by bile salts and other bile constituents, to the luminal surface of the small-intestine epithelial cells.

**4** When a micelle approaches the absorptive epithelial surface, the monoglycerides and fatty acids leave the micelle and passively diffuse through the lipid bilayer of the luminal membranes.

**5** The monoglycerides and free fatty acids are resynthesized into triglycerides inside the epithelial cells.

**6** These triglycerides aggregate and are coated with a layer of lipoprotein to form water-soluble chylomicrons, which are extruded through the basal membrane of the cells by exocytosis.

**7** Chylomicrons are unable to cross the basement membrane of blood capillaries, so instead they enter the lymphatic vessels, the central lacteals.

> FIGURE 15-25 **Fat digestion and absorption**

## Iron and calcium absorption

In contrast to the almost complete, unregulated absorption of other ingested electrolytes, dietary iron and calcium may not be absorbed completely because their absorption is subject to regulation, depending on the body's needs for these electrolytes.

### IRON ABSORPTION

Iron is essential for haemoglobin production. The normal iron intake is typically 15–20 mg/day, yet a man usually absorbs about 0.5–1 mg/day into the blood, and a woman takes up slightly more, at 1.0–1.5 mg/day (women need more

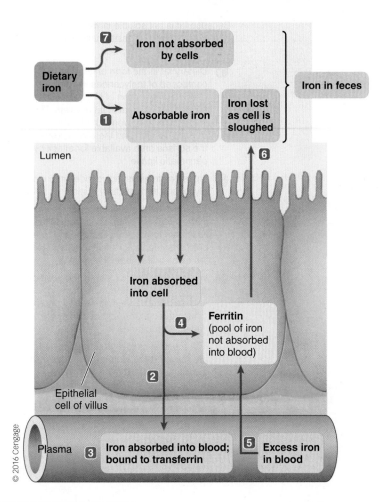

Iron not absorbed by cells

**7** Iron not absorbed by cells

Dietary iron

**1** Absorbable iron

Iron lost as cell is sloughed

Iron in feces

Lumen

**6**

Iron absorbed into cell

**4** Ferritin (pool of iron not absorbed into blood)

Epithelial cell of villus

**2**

© 2016 Cengage

Plasma **3** Iron absorbed into blood; bound to transferrin

**5** Excess iron in blood

**1** Only a portion of ingested iron is in a form that can be absorbed.

**2** Dietary iron that is absorbed into the small intestine epithelial cells and is immediately needed for red blood cell production is transferred into the blood.

**3** In the blood, absorbed iron is carried to the bone marrow bound to transferrin, a plasma protein carrier.

**4** Absorbed dietary iron that is not immediately needed is stored in the epithelial cells as ferritin, which cannot be transferred into the blood.

**5** Excess iron in the blood can be dumped into the ferritin pool.

**6** This unused iron is lost in the feces as the ferritin-containing epithelial cells are sloughed.

**7** Dietary iron that was not absorbed is also lost in the feces.

› **FIGURE 15-26** Iron absorption

iron because they periodically lose iron in menstrual blood flow).

Two main steps are involved in absorption of iron into blood: (1) absorption of iron from the lumen into small-intestinal epithelial cells, and (2) absorption of iron from the epithelial cells into the blood (› Figure 15-26).

Iron is actively transported from the lumen into the epithelial cells, with women having about four times as many active-transport sites for iron as men. The extent to which ingested iron is taken up by the epithelial cells depends on the type of iron consumed (ferrous iron, $Fe^{2+}$, is absorbed more easily than ferric iron, $Fe^{3+}$). Also, the presence of other substances in the lumen can either promote or reduce iron absorption. For example, vitamin C increases iron absorption, primarily by reducing ferric to ferrous iron. Phosphate and oxalate, in contrast, combine with ingested iron to form insoluble iron salts that cannot be absorbed.

After active absorption into the small-intestine epithelial cells, iron has two possible fates:

1. Iron needed immediately for production of red blood cells is absorbed into the blood for delivery to the bone marrow, the site of red blood cell production. Iron is transported in the blood by a plasma protein carrier known as **transferrin**. The hormone responsible for stimulating red blood cell production, erythropoietin

(p. 438), is believed to also enhance iron absorption from the intestinal cells into the blood. The absorbed iron is then used in the synthesis of haemoglobin for the newly produced red blood cells.

2. Iron not immediately needed remains stored within the epithelial cells in a granular form called **ferritin**, which cannot be absorbed into the blood. If the blood level of iron is too high, excess iron may be dumped from the blood into this unabsorbable pool of ferritin in the intestinal epithelial cells. Iron stored as ferritin is lost in the feces within three days because the epithelial cells containing these granules are sloughed off during mucosal regeneration. Large amounts of iron in the feces give them a dark, almost black colour.

**CALCIUM ABSORPTION**

The amount of calcium ($Ca^{2+}$) ions absorbed is controlled in relation to the need of the body for calcium. Calcium has to be actively absorbed, but a small amount of $Ca^{2+}$ is absorbed passively. Vitamin D greatly stimulates this active transport. Vitamin D can exert this effect only after it has been activated in the liver and kidneys—a process enhanced by parathyroid hormone. Appropriately, secretion of parathyroid hormone increases in response to a fall in $Ca^{2+}$ concentration in the blood.

Normally, of the average 1000 mg of $Ca^{2+}$ taken in daily, only about two-thirds is absorbed in the small intestine, with the rest passing out in the feces.

## Absorbed nutrients and the liver

The venules that leave the small-intestine villi, along with those from the rest of the digestive tract, empty into the hepatic portal vein, which carries the blood to the liver. Consequently, anything absorbed into the digestive capillaries first must pass through the hepatic biochemical factory before entering the general circulation. Thus, the products of carbohydrate and protein digestion are channelled into the liver, where many of these energy-rich products are subjected to immediate metabolic processing. Furthermore, harmful substances that may have been absorbed are detoxified by the liver before gaining access to the general circulation. After passing through the portal circulation, the venous blood from the digestive system empties into the vena cava and returns to the heart to be distributed throughout the body, carrying glucose and amino acids for use by the tissues.

Fat, which cannot penetrate the intestinal capillaries, is picked up by the central lacteal and enters the lymphatic system instead, bypassing the hepatic portal system. Contractions of the villi, accomplished by the muscularis mucosa, periodically compress the central lacteal and squeeze the lymph out of this vessel. The lymph vessels eventually converge to form the *thoracic duct*, a large lymph vessel that empties into the venous system within the chest. In this way, fat ultimately gains access to the blood. The absorbed fat is carried by the systemic circulation to the liver and to other tissues of the body. Therefore, the liver does have a chance to act on the digested fat, but not until the fat has been diluted by the blood in the general circulatory system. This dilution of fat protects the liver from being inundated with more fat than it can handle at one time.

## Extensive absorption and secretion

The small intestine normally absorbs about 9 L of fluid per day in the form of water and solutes, including the absorbable units of nutrients, vitamins, and electrolytes. How can that be, when humans normally ingest only about 1250 mL of fluid and consume 1250 g of solid food (80% of which is $H_2O$) per day? ▌ Table 15-7 illustrates the tremendous daily absorption performed by the small intestine. Each day about 9500 mL of water and solutes enter the small intestine. Note that of this 9500 mL, only 2500 mL are ingested from the external environment. The remaining 7000 mL (7 L) of fluid are digestive juices derived from the plasma. Recall that plasma is the ultimate source of digestive secretions, because the secretory cells extract from the plasma the necessary raw materials for their secretory product. Considering that the entire plasma volume is only about 2.75 L, absorption must closely parallel secretion to keep the plasma volume from falling sharply.

Of the 9500 mL of fluid entering the small-intestine lumen per day, about 95 percent, or 9000 mL of fluid, is normally absorbed by the small intestine back into the plasma, with only 500 mL of the small-intestine contents passing on into the colon. Thus, the body does not lose the digestive juices. After

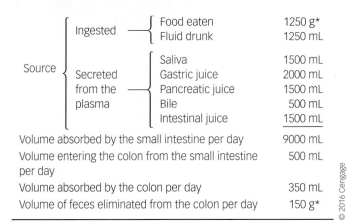

**▌ TABLE 15-7 Volumes Absorbed by the Small and Large Intestine per Day**

**Volume entering the small intestine per day**

| | | | |
|---|---|---|---|
| Source | Ingested | Food eaten | 1250 g* |
| | | Fluid drunk | 1250 mL |
| | Secreted from the plasma | Saliva | 1500 mL |
| | | Gastric juice | 2000 mL |
| | | Pancreatic juice | 1500 mL |
| | | Bile | 500 mL |
| | | Intestinal juice | 1500 mL |

| | |
|---|---|
| Volume absorbed by the small intestine per day | 9000 mL |
| Volume entering the colon from the small intestine per day | 500 mL |
| Volume absorbed by the colon per day | 350 mL |
| Volume of feces eliminated from the colon per day | 150 g* |

© 2016 Cengage

\* 1 mL of $H_2O$ weighs 1 g. Therefore, because of a high percentage of food and feces is $H_2O$, we can roughly equate grams of food or feces with millilitres of fluid.

the constituents of the juices are secreted into the digestive tract lumen and perform their function, they are returned to the plasma. The only secretory product that escapes from the body is bilirubin, a waste product that must be eliminated.

## Biochemical balance

Because the secreted juices are normally absorbed back into the plasma, the acid–base balance of the body is not altered by digestive processes. When secretion and absorption do not parallel each other, however, acid–base abnormalities can result. ⟩ Figure 15-27 provides a summary of the biochemical balance that normally exists among the stomach, pancreas, and small intestine. The arterial blood entering the stomach contains $Cl^-$, $CO_2$, $H_2O$, and $Na^+$, among other things. During HCl secretion, the gastric parietal cells extract $Cl^-$, $CO_2$, and $H_2O$ from the plasma (the $CO_2$ and $H_2O$ are essential for $H^+$ secretion) and add to it (the $HCO_3^-$ is formed in the process of generating $H^+$). The $HCO_3^-$ is transported into the plasma in exchange for the secreted $Cl^-$. Plasma $Na^+$ levels are not altered by gastric secretory processes. Because $HCO_3^-$ is an alkaline ion, venous blood leaving the stomach is more alkaline than arterial blood delivered to it.

The overall acid–base balance of the body is not altered, however, because the pancreatic duct cells extract a comparable amount of $HCO_3^-$ (along with $Na^+$) from the plasma to neutralize the acidic gastric chyme as it empties into the small intestine. Within the intestinal lumen, the alkaline pancreatic $NaHCO_3$ secretion neutralizes the gastric HCl secretion, yielding NaCl and $H_2CO_3$. The latter molecules form $Na^+$ and $Cl^-$ plus $CO_2$ and $H_2O$, respectively. All four of these constituents ($Na^+$, $Cl^-$, $CO_2$, and $H_2O$) are absorbed by the intestinal epithelium into the plasma. Note that these are exactly the same constituents present in the arterial blood entering the stomach. Through

15

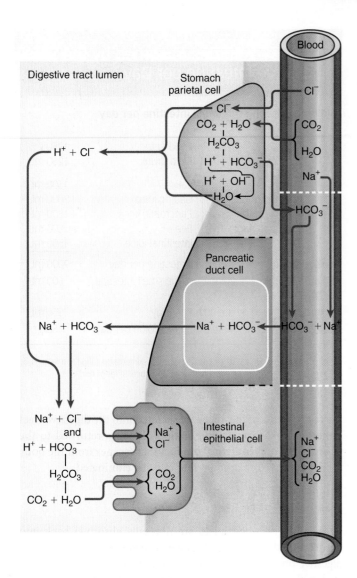

> **FIGURE 15-27 Biochemical balance among the stomach, pancreas, and small intestine.** When digestion and absorption proceed normally, no net loss or gain of acid or base or other chemicals from the body fluids occurs as a consequence of digestive secretions. The parietal cells of the stomach extract $Cl^-$, $CO_2$, and $H_2O$ from and add to the blood during HCl secretion. The pancreatic duct cells extract the $HCO_3^-$ as well as $Na^+$ from the blood during $NaHCO_3$ secretion. Within the small-intestine lumen, pancreatic $NaHCO_3$ neutralizes gastric HCl to form NaCl and $H_2CO_3$, which decomposes into $CO_2$ and $H_2O$. Subsequently, the intestinal cells absorb $Na^+$, $Cl^-$, $CO_2$, and $H_2O$ into the blood, thereby replacing the constituents that were extracted from the blood during gastric and pancreatic secretion.

these interactions, the body normally does not experience a net gain or loss of acid or base during digestion.

## Diarrhoea

**Clinical Note** When vomiting or diarrhoea occur, these normal neutralization processes cannot take place. We have already described vomiting in the section on gastric motility. The other common digestive tract disturbance that can lead to a loss of fluid and an acid–base imbalance is **diarrhoea**. This condition

is characterized by passage of a highly fluid fecal matter, often with increased frequency of defecation. Just as with vomiting, the effects of diarrhoea can be either beneficial or harmful. Diarrhoea is beneficial when rapid emptying of the intestine hastens elimination of harmful material from the body. However, not only are some of the ingested materials lost but some of the secreted materials that normally would have been reabsorbed are lost as well. Excessive loss of intestinal contents causes dehydration, loss of nutrient material, and metabolic acidosis resulting from loss of $HCO_3^-$ (p. 635). The abnormal fluidity of the feces usually occurs because the small intestine is unable to absorb fluid as extensively as normal. This extra unabsorbed fluid passes out in the feces.

The causes of diarrhoea are as follows:

1. The most common cause of diarrhoea is excessive small-intestinal motility, which arises either from local irritation of the gut wall by bacterial or viral infection of the small intestine or from emotional stress. Rapid transit of the small-intestine contents does not allow enough time for adequate absorption of fluid to occur.

2. Diarrhoea also occurs when excess osmotically active particles, such as those found in lactase deficiency, are present in the digestive tract lumen. These particles cause excessive fluid to enter and be retained in the lumen, thus increasing the fluidity of the feces.

3. Toxins of the bacterium *Vibrio cholerae* (the causative agent of cholera) and certain other microorganisms promote the secretion of excessive amounts of fluid by the small-intestine mucosa, resulting in profuse diarrhoea. Diarrhoea produced in response to toxins from infectious agents is the leading cause of death of small children in developing nations. Fortunately, a low-cost, effective *oral rehydration therapy* that takes advantage of the intestine's glucose cotransport carrier is saving the lives of millions of children. (For details about oral rehydration therapy, see ▶ Concepts, Challenges, and Controversies, p. 690.)

### Check Your Understanding 15.6

1. Describe the structural features that increase the surface area of the small intestine, and explain the significance of increasing the surface area.

2. Discuss how the $Na^+$–$K^+$ pump of mucosal epithelial cells facilitates nutrient absorption.

## 15.7 Large Intestine

The **large intestine** consists of the colon, cecum, appendix, and rectum (▶ Figure 15-28). The **cecum** forms a blind-ended pouch below the junction of the small and large intestines at the ileocecal valve. The small, finger-like projection at the bottom of the cecum is the **appendix**, a lymphoid tissue that houses lymphocytes (p. 464). The **colon**, which makes up most of the large intestine, is not coiled like the small intestine but consists of

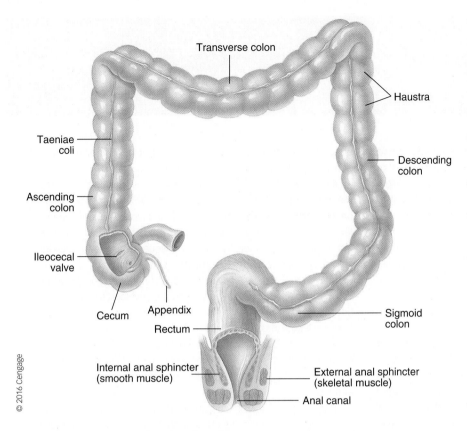

**Internal anal sphincter** (smooth muscle)

**External anal sphincter** (skeletal muscle)

**Anal canal**

© 2016 Cengage

> **FIGURE 15-28 Anatomy of the large intestine**

three relatively straight parts: the *ascending colon*, the *transverse colon*, and the *descending colon*. The end part of the descending colon becomes S-shaped, forming the *sigmoid colon* (*sigmoid* means "S-shaped"), then straightens out to form the **rectum** (*rectum* means "straight").

## A drying and storage organ

The colon normally receives about 500–1500 mL of chyme from the small intestine each day. Because most digestion and absorption have been accomplished in the small intestine, the contents delivered to the colon consist of indigestible food residues (such as cellulose), unabsorbed biliary components, and the remaining fluid. The colon extracts more water and salt from the contents. Most of the absorption occurs in the upper half of the colon, while the distal half is used as more of a storage area. What remains to be eliminated is known as **feces**. The primary function of the large intestine is to store feces before defecation. Cellulose and other indigestible substances in the diet provide bulk and help maintain regular bowel movements by contributing to the volume of the colonic contents.

## Haustral contractions

The outer longitudinal smooth muscle layer does not completely surround the large intestine. Instead, it consists only of three separate, conspicuous, longitudinal bands of muscle, the **taeniae coli**, which run the length of the large intestine. These taeniae coli are shorter than the underlying circular smooth muscle and mucosal layers would be if these layers were stretched out flat. Because of this, the underlying layers are gathered into pouches or sacs called **haustra**, much as the material of a full skirt is gathered at the narrower waistband. The haustra are not merely passive, permanent gathers, however; they actively change location as a result of contraction of the circular smooth muscle layer.

Most of the time, movements of the large intestine are slow and nonpropulsive, which is appropriate for its absorptive and storage functions. The colon's main motility is **haustral contractions** initiated by the autonomous rhythmicity of colonic smooth muscle cells. These contractions, which throw the large intestine into haustra, are similar to small-intestine segmentations, but they occur much less frequently. Thirty minutes may elapse between haustral contractions, whereas segmentation contractions in the small intestine occur at rates of between 9 and 12 per minute. The location of the haustral sacs gradually changes as a relaxed segment that has formed a sac slowly contracts, while a previously contracted area simultaneously relaxes to form a new sac. These movements are nonpropulsive; they slowly shuffle the contents in a back-and-forth mixing movement that exposes the colonic contents to the absorptive mucosa. Haustral contractions are largely controlled by locally mediated reflexes involving the intrinsic plexuses.

## Mass movements

Three to four times a day, generally after meals, a marked increase in motility takes place during which large segments of the ascending and transverse colon contract simultaneously, driving the feces one-third to three-fourths of the length of the colon in a few seconds. These massive contractions, appropriately called **mass movements**, drive the colonic contents into the distal part of the large intestine, where material is stored until defecation occurs.

When food enters the stomach, mass movements are triggered in the colon primarily by the **gastrocolic reflex**, which is mediated from the stomach to the colon by gastrin and by the extrinsic autonomic nerves. In many people, this reflex is most evident after the first meal of the day and is often followed by the urge to defecate. Thus, when a new meal enters the digestive tract, reflexes are initiated to move the existing contents farther along down the tract to make way for the incoming food. The gastroileal reflex moves the remaining small-intestine contents into the large intestine, and the gastrocolic reflex pushes the colonic contents into the rectum, triggering the defecation reflex.

**15**

# Oral Rehydration Therapy: Sipping a Simple Solution Saves Lives

**D**IARRHOEA-INDUCING MICROORGANISMS SUCH AS *Vibrio CHOLERAE*, which causes cholera, are the leading cause of death in children younger than age 5 worldwide. The problem is especially pronounced in developing countries, refugee camps, and elsewhere where poor sanitary conditions encourage the spread of the microorganisms, and medical supplies and healthcare personnel are scarce. Fortunately, a low-cost, easily obtainable, uncomplicated remedy—**oral rehydration therapy (ORT)**—has been developed to combat potentially fatal diarrhoea. This treatment exploits the membrane transporters located at the luminal border of the villus epithelial cells.

Let's examine the pathophysiology of life-threatening diarrhoea and then see how simple ORT can save lives. During digestion of a meal, the crypt cells of the small intestine normally secrete succus entericus—a solution of salt and mucus—into the lumen. These cells actively transport $Cl^-$ into the lumen, promoting the parallel passive transport of $Na^+$ and $H_2O$ from the blood into the lumen. The fluid provides the watery environment needed for enzymatic breakdown of ingested nutrients into absorbable units. Glucose and amino acids, the absorbable units of dietary carbohydrates and proteins, respectively, are absorbed by secondary active transport. This absorption mechanism uses the $Na^+$–glucose (or amino acid) cotransport carriers (SGLT) located at the luminal membrane of the villus epithelial cells. In addition, separate active $Na^+$ carriers not linked with nutrient absorption transfer $Na^+$, passively accompanied by $Cl^-$ and $H_2O$, from the lumen into the blood.

The net result of these various carrier activities is absorption of the secreted salt and $H_2O$ along with the digested nutrients. Normally, absorption of salt and $H_2O$ exceeds their secretion, so not only are the secreted fluids salvaged but also additional ingested salt and $H_2O$ are absorbed.

Cholera and most diarrhoea-inducing microbes cause diarrhoea by stimulating the secretion of $Cl^-$ or impairing the absorption of $Na^+$. As a result, more fluid is secreted from the blood into the lumen than is subsequently transferred back into the blood. The excess fluid is lost in the feces, producing the watery stool characteristic of diarrhoea. More important, the loss of fluids and electrolytes that came from the blood leads to dehydration. The subsequent reduction in effective circulating plasma volume can cause death in a matter of days or even hours, depending on the severity of the fluid loss.

In the middle of the last century, physicians learned that replacing the lost fluids and electrolytes intravenously saves the lives of most patients with diarrhoea. In many parts of the world, however, adequate facilities, equipment, and personnel are not available to administer intravenous rehydration therapy. Consequently, millions of children still succumbed to diarrhoea annually.

In 1966 researchers learned that SGLT is not affected by diarrhoea-causing microbes. This discovery led to the development of ORT. When both $Na^+$ and glucose are present in the lumen, they are cotransported from the lumen into the villus epithelial cells, from which they enter the blood. Because $H_2O$ osmotically follows the absorbed $Na^+$, ingestion of a glucose and salt solution promotes the uptake of fluid into the blood from the intestinal tract without the need for intravenous replacement of fluids.

The first proof of ORT's life-saving ability in the field came in 1971 when several million refugees poured into India from war-ravaged Bangladesh. Of the thousands of refugees who fell victim to cholera and other diarrhoeal diseases, more than 30 percent died because of the scarcity of sterile fluids and needles for intravenous therapy. In one refugee camp, however, under the supervision of a group of scientists who had been experimenting with ORT, families were taught to administer ORT to people with diarrhoea, most of whom were small children. The scarce intravenous solutions were reserved for those unable to drink. Death from diarrhoea was reduced to 3 percent in this camp, compared with a tenfold higher mortality among refugees elsewhere.

Based on this evidence, the World Health Organization (WHO) started aggressively promoting ORT. Packets of dry ingredients for ORT are now manufactured locally in more than 60 countries. The WHO estimates that about 30 percent of the world's children who contract diarrhoea are treated with the prepackaged mixture or home-prepared versions. In the United States, commercially prepared oral solutions are widely available at pharmacies and supermarkets. An estimated 1 million children worldwide are saved annually as a result of ORT.

## The defecation reflex

When mass movements of the colon move feces into the rectum, the resultant distension of the rectum stimulates stretch receptors in the rectal wall, initiating the **defecation reflex**. When fecal matter fills the rectum, sensory impulses initiated by stretch receptors due to the the distension are sent to the sacral portion of the spinal cord, and a reflex signal is sent via the parasympathetics to the distal colon. This reflex causes the **internal anal sphincter** (which is smooth muscle) to relax and the rectum and sigmoid colon to contract more vigorously. If the **external anal sphincter** (which is skeletal muscle) is also relaxed, defecation occurs. Being skeletal muscle, the external anal sphincter is under voluntary control. The initial distension of the rectal wall is accompanied by the conscious urge to defecate. If circumstances are unfavourable for defecation, voluntary tightening of the external anal sphincter can prevent defecation despite the defecation reflex. If defecation is delayed, the distended rectal wall gradually relaxes, and the urge to defecate subsides until the next mass movement propels more feces into the rectum, once again distending the rectum and triggering the defecation reflex. During periods of inactivity, both anal sphincters remain contracted to ensure fecal continence.

When defecation does occur, it is usually assisted by voluntary straining movements that involve simultaneous contraction of the abdominal muscles and a forcible expiration against a closed glottis. This is another instance where the valsalva manoeuvre is used, with the purpose of greatly increasing intra-abdominal pressure, which helps expel the feces.

## Constipation

If defecation is delayed too long, **constipation** may result. When colonic contents are retained for longer periods of time than normal, more than the usual amount of water is absorbed from the feces, so they become hard and dry. Normal variations in frequency of defecation among individuals range from after every meal to up to once a week. When the frequency is delayed beyond what is normal for a particular person, constipation and its attendant symptoms may occur. These symptoms include abdominal discomfort, dull headache, loss of appetite sometimes accompanied by nausea, and mental depression.

Contrary to popular belief, these symptoms are not caused by toxins absorbed from the retained fecal material. Although bacterial metabolism produces some potentially toxic substances in the colon, these substances normally pass through the portal system and are removed by the liver before they can reach the systemic circulation. Instead, the symptoms associated with constipation are caused by prolonged distension of the large intestine, particularly the rectum; the symptoms promptly disappear after relief from distension.

Possible causes for delayed defecation that might lead to constipation include (1) ignoring the urge to defecate; (2) decreased colon motility accompanying aging, emotion, or a low-bulk diet; (3) obstruction of fecal movement in the large bowel caused by a local tumour or colonic spasm; and (4) impairment of the defecation reflex, such as through injury of the nerve pathways involved. _Clinical Note_ If hardened fecal material becomes lodged in the appendix, it may obstruct normal circulation and mucous secretion in this narrow, blind-ended appendage. This blockage leads to inflammation of the appendix, that is, **appendicitis**. The inflamed appendix often becomes swollen and filled with pus, and the tissue may die as a result of local circulatory interference. If not surgically removed, the diseased appendix may rupture, spewing its infectious contents into the abdominal cavity.

## Large-intestine secretion

The large intestine does not secrete any digestive enzymes. None are needed, because digestion is completed before chyme ever reaches the colon. Colonic secretion consists of an alkaline ($NaHCO_3$) mucous solution, whose function is to protect the large-intestine mucosa from mechanical and chemical injury. The mucus provides lubrication to facilitate passage of the feces, whereas the $NaHCO_3$ neutralizes irritating acids produced by local bacterial fermentation. Secretion increases in response to mechanical and chemical stimulation of the colonic mucosa mediated by short reflexes and parasympathetic innervation.

No digestion takes place within the large intestine because there are no digestive enzymes. However, the colonic bacteria do digest some of the cellulose for their use.

### THE COLON

Because of slow colonic movement, bacteria have time to grow and accumulate in the large intestine. In contrast, in the small intestine the contents are normally moved through too rapidly for bacterial growth to occur. Furthermore, the mouth, stomach, and small intestine secrete antibacterial agents, but the colon does not. Not all ingested bacteria are destroyed by lysozyme and HCl, however. The surviving bacteria continue to thrive in the large intestine. The number of bacteria living in the human colon is about 10 times the number of cells in the human body. Collectively, this mass of bacteria amounts to about 1000 g. An estimated 500 to 1000 different species of bacteria typically live in the colon. These colonic microorganisms are typically harmless and, in fact, provide beneficial functions. For example, indigenous bacteria (1) enhance intestinal immunity by competing with potentially pathogenic microbes for nutrients and space (p. 463), (2) promote colonic motility, (3) help maintain colonic mucosal integrity, and (4) make nutritional contributions. For example, bacteria synthesize absorbable vitamin K and raise colonic acidity, thereby promoting the absorption of calcium, magnesium, and zinc. Furthermore, contrary to earlier assumptions, some of the glucose released during bacterial processing of dietary fibre is absorbed by the colonic mucosa.

### SALT AND WATER

Some absorption takes place within the colon, but not to the same extent as in the small intestine. Because the luminal surface of the colon is fairly smooth, it has considerably less absorptive surface area than the small intestine. Furthermore, the colon does not have the extensive, specialized transport mechanisms that the small intestine has. When excessive small-intestine motility delivers the contents to the colon before the absorption of nutrients has been completed, the colon cannot absorb most of these materials, and they are lost in diarrhoea.

The colon normally absorbs salt and water. Sodium is actively absorbed, chloride follows passively down the electrical gradient, and water follows osmotically. The colon absorbs token amounts of other electrolytes, as well as vitamin K synthesized by colonic bacteria.

Through absorption of salt and water, a firm fecal mass is formed. Of the 500 mL of material entering the colon per day from the small intestine, the colon normally absorbs about 350 mL, leaving 150 g of feces to be eliminated from the body each day (see ▌ Table 15-7). This fecal material normally consists of 100 g of water and 50 g of solid, including undigested cellulose, bilirubin, bacteria, and small amounts of salt. Thus, contrary to popular thinking, the digestive tract is not a major excretory passageway for eliminating wastes from the body. The main waste product excreted in the feces is bilirubin. The other fecal constituents are unabsorbed food residues and bacteria, which were never actually a part of the body.

15

**INTESTINAL GASES**

Occasionally, instead of feces passing from the anus, intestinal gas, or **flatus**, passes out. This gas is derived primarily from two sources: (1) swallowed air (as much as 500 mL of air may be swallowed during a meal), and (2) gas produced by bacterial fermentation in the colon. The flatus is typically a mixture of nitrogen and carbon dioxide, and small amounts of hydrogen, methane, and hydrogen sulphide. The presence of gas percolating through the luminal contents gives rise to gurgling sounds known as **borborygmi**. Eructation (burping) removes most of the swallowed air from the stomach, but some passes on into the intestine. Usually, very little gas is present in the small intestine, because the gas is either quickly absorbed or passes into the colon. Most gas in the colon is due to bacterial activity, with the quantity and nature of the gas dependent on the type of food eaten and the characteristics of the colonic bacteria. Some foods, such as beans, contain types of carbohydrates that humans cannot digest but that can be attacked by gas-producing bacteria. Much of the gas is absorbed through the intestinal mucosa. The rest is expelled through the anus.

To selectively expel gas when feces are also present in the rectum, the person voluntarily contracts the abdominal muscles and the external anal sphincter at the same time. When abdominal contraction raises the pressure against the contracted anal sphincter sufficiently, the pressure gradient forces air out at a high velocity through a slitlike anal opening that is too narrow for solid feces to escape through. This passage of air at high velocity causes the edges of the anal opening to vibrate, giving rise to the characteristic low-pitched sound accompanying passage of gas.

---

### Check Your Understanding 15.7

1. Compare haustral contractions of the large intestine to segmentation contractions of the small intestine.
2. Describe the role of $NaHCO_3$ secretions by the large intestine mucosa. Compare the function of this secretion with that of pancreatic $NaHCO_3$ secretion.

---

## 15.8 | Overview of the Gastrointestinal Hormones

Throughout our discussion of digestion, we have mentioned different functions of the three major gastrointestinal hormones: gastrin, secretin, and CCK. In this section we fit all these functions together so you can appreciate the overall adaptive importance of these interactions. Also, we introduce a more recently identified gastrointestinal hormone, GIP, as well as a hormone that connects nutrient consumption with glucose metabolism, GLP-1.

### Gastrin

Protein in the stomach stimulates the release of gastrin, which performs the following functions:

1. It acts in multiple ways to increase secretion of HCl and pepsinogen. These two substances, in turn, are of primary importance in initiating digestion of the protein that promoted their secretion.

2. It enhances gastric motility, stimulates ileal motility, relaxes the ileocecal sphincter, and induces mass movements in the colon—functions that are all aimed at keeping the contents moving through the digestive tract on the arrival of a new meal.

3. It also is trophic not only to the stomach mucosa but also to the small-intestine mucosa, which helps maintain a well-developed, functionally viable digestive tract lining.

Predictably, gastrin secretion is inhibited by an accumulation of acid in the stomach and by the presence in the duodenal lumen of acid and other constituents that necessitate a delay in gastric secretion.

### Secretin

As the stomach empties into the duodenum, the presence of acid in the duodenum stimulates the release of secretin, which performs the following interrelated functions:

1. It inhibits gastric emptying to prevent further acid from entering the duodenum until the acid already present is neutralized.

2. It inhibits gastric secretion to reduce the amount of acid being produced.

3. It stimulates the pancreatic duct cells to produce a large volume of aqueous $NaHCO_3$ secretion, which is emptied into the duodenum to neutralize the acid.

4. It stimulates secretion by the liver of a $NaHCO_3$-rich bile, which likewise is emptied into the duodenum to assist in the neutralization process. Neutralization of the acidic chyme in the duodenum helps prevent damage to the duodenal walls and provides a suitable environment for the optimal functioning of the pancreatic digestive enzymes, which are inhibited by acid.

5. Along with CCK, secretin is trophic to the exocrine pancreas.

### CCK

As chyme empties from the stomach, fat and other nutrients enter the duodenum. These nutrients, especially fat and, to a lesser extent, protein products, cause the release of CCK, which performs the following interrelated functions:

1. It inhibits gastric motility and secretion, thereby allowing adequate time for the nutrients already in the duodenum to be digested and absorbed.

2. It stimulates the pancreatic acinar cells to increase secretion of pancreatic enzymes, which continue the digestion

of these nutrients in the duodenum (this action is especially important for fat digestion, because pancreatic lipase is the only enzyme that digests fat).

3. It causes contraction of the gallbladder and relaxation of the sphincter of Oddi so that bile empties into the duodenum to aid fat digestion and absorption. Bile salts' detergent action is particularly important in enabling pancreatic lipase to perform its digestive task. Once again, the multiple effects of CCK are remarkably well adapted to dealing with the fat and other nutrients whose presence in the duodenum triggered the release of this hormone.

4. Furthermore, it is appropriate that both secretin and CCK, which have profound stimulatory effects on the exocrine pancreas, are trophic to this tissue.

5. CCK has also been implicated in long-term adaptive changes in the proportion of pancreatic enzymes produced in response to prolonged changes in diet.

6. Besides facilitating the digestion of ingested nutrients, CCK is an important regulator of food intake. It plays a key role in satiety, the sensation of having had enough to eat (p. 705).

## GIP

A more recently recognized hormone released by the K cells in the duodenum, **glucose-dependent insulinotrophic peptide (GIP)**, helps promote metabolic processing of the nutrients once they are absorbed. This hormone was originally named *gastric inhibitory peptide (GIP)* for its presumed role as an enterogastrone. It was believed to inhibit gastric motility and secretion, similar to secretin and CCK. GIP's contribution in this regard is now considered minimal. Instead, this hormone stimulates insulin release by the pancreas. To a lesser extent, GIP inhibits water and electrolyte absorption in the small intestine and the secretion of acids and pepsin. Again, this is remarkably adaptive. As soon as the meal is absorbed, the body has to shift its metabolic gears to use and store the newly arriving nutrients. The metabolic activities of this absorptive phase are largely under the control of insulin (pp. 272). Stimulated by the presence of a meal in the digestive tract, GIP initiates the release of insulin in anticipation of absorption of the meal, in a feedforward fashion. Insulin is especially important in promoting the uptake and storage of glucose. Appropriately, glucose in the duodenum increases GIP secretion.

## GLP-1

The major source of the hormone **glucagon-like peptide-1 (GLP-1)** in the body is the intestinal L cell. Secretion of GLP-1 by L cells is dependent on the presence of nutrients in the lumen of the small intestine. The nutrients that initiate secretion (secretagogues) include major nutrients such as carbohydrate, protein, and lipid. The major physiological role of GLP-1 is to connect the consumption of nutrients with glucose metabolism. In circulation, GLP-1 has a short half-life (<2 minutes) and is rapidly catabolized. It is a potent glucose-dependent stimulator (antihyperglycaemic) of insulin secretion, while suppressing glucagon secretion. The effectiveness of GLP-1 in post-meal (post-prandial) insulin secretion is easy to demonstrate, because an oral dose of glucose triggers a much greater plasma insulin response than does an intravenous glucose dose; the intravenous dose bypasses the digestive tract. When plasma glucose concentration is in the normal fasting range (~4 mmol/L) GLP-1 no longer stimulates insulin. GLP-1 appears to restore the glucose sensitivity of pancreatic $\beta$-cells, which may involve increased expression of GLUT2 receptor and glucokinase. GLP-1 inhibits gastric secretion and motility, which delays carbohydrate absorption and contributes to satiety.

Both GLP-1 and GIP use feedforward mechanisms to increase the amount of insulin released from the $\beta$-cells of the pancreas; hence, these two hormones are termed *incretin hormones*. A substantial volume of research on both of these incretins has been conducted by Canadian researchers, many associated with UBC—S.J. Kim, C.H. McIntosh, S. Widenmaier, and T.J Kieffer to name a few. The specific effects of GLP-1 are as follows:

- Promotes insulin sensitivity (potential for diabetes treatment)
- Increases insulin secretion from the pancreas
- Decreases glucagon secretion from the pancreas
- Increases insulin sensitivity in both $\alpha$-cells and $\beta$-cells
- Increases the mass $\beta$-cells and their insulin gene expression
- Inhibits gastric secretion and gastric emptying in the stomach
- Increases satiety, thereby reducing food intake

This overview of the multiple, integrated, adaptive functions of the gastrointestinal hormones provides an excellent example of the remarkable efficiency of the human body.

---

### Check Your Understanding 15.8

1. Explain the significance of some GI hormones being trophic.
2. Name the targets of secretin and of CCK.

---

# Chapter in Perspective: Focus on Homeostasis

To maintain constancy in the internal environment, materials that are used up in the body (such as energy-rich nutrients and $O_2$) or uncontrollably lost from the body (such as evaporative $H_2O$ loss from the airways or salt loss in sweat) must constantly be replaced by new supplies of these materials from the external environment. All these replacement supplies, except oxygen, are acquired through the digestive system. Fresh supplies of oxygen are transferred to the internal environment by the respiratory system, but all the nutrients, water, and various electrolytes needed to maintain homeostasis are acquired through the digestive system. The large, complex, ingested food is broken down by the digestive system into

15

small absorbable units. These small energy-rich nutrient molecules are transferred across the small-intestine epithelium into the blood for delivery to the cells to replace the nutrients constantly used for ATP production and for repair and growth of body tissues. Likewise, ingested water, salt, and other electrolytes are absorbed by the intestine into the blood.

Unlike regulation in most body systems, regulation of digestive system activities is not aimed at maintaining homeostasis. The quantity of nutrients and water ingested is subject to control, but the quantity of ingested materials absorbed by the digestive tract is not subject to control, with few exceptions. The hunger mechanism governs food intake to help maintain energy balance (Chapter 16), and the thirst mechanism controls water intake to help maintain water balance (Chapter 14). However, we do not always need these control mechanisms, and we often eat and drink when we are not hungry or thirsty.

Once these materials are in the digestive tract, the digestive system does not vary its rate of nutrient, water, or electrolyte uptake according to body needs (with the exception of iron and calcium); rather, it optimizes conditions for digesting and absorbing what is ingested. Truly, what you eat is what you get. The digestive system is subject to many regulatory processes, but these are not influenced by the nutritional or hydration state of the body. Instead, these control mechanisms are governed by the composition and volume of digestive tract contents so that the rate of motility and secretion of digestive juices are optimal for digestion and absorption of the ingested food.

If excess energy-rich nutrients are ingested and absorbed, the extra is placed in storage, such as in adipose tissue (fat), so that the blood level of nutrient molecules remains at a constant level. Excess ingested water and electrolytes are eliminated in the urine to homeostatically maintain the blood levels of these constituents.

## CHAPTER TERMINOLOGY

absorption (p. 647)
accessory digestive organs (p. 647)
acini (p. 667)
amino acids (p. 646)
aminopeptidases (p. 678)
antrum (p. 656)
appendicitis (p. 691)
appendix (p. 688)
ATPase (p. 661)
autocatalytic process (p. 663)
basic electrical rhythm (BER) (p. 650)
bile (p. 671)
bile canaliculus (p. 672)
bile salts (p. 673)
biliary system (p. 671)
bilirubin (p. 674)
body (p. 656)
bolus (p. 654)
borborygmi (p. 692)
brush border (p. 678)
carbohydrates (p. 646)
carboxypeptidase (p. 668)
cecum (p. 688)
celiac disease (p. 681)
cellulose (p. 646)
central lacteal (p. 680)
chemoreceptor trigger zone (p. 660)
chief cells (p. 661)
cholecystokinin (CCK) (p. 659)
choleretic (p. 675)
chylomicrons (p. 684)
chyme (p. 657)
chymotrypsin (p. 668)
chymotrypsinogen (p. 668)
cirrhosis (p. 676)
colipase (p. 673)
colon (p. 688)

conditioned (acquired) salivary reflex (p. 654)
constipation (p. 691)
crypts of Lieberkühn (p. 681)
D cells (p. 664)
defecation reflex (p. 690)
detergent action (p. 673)
diarrhoea (p. 688)
digestion (p. 646)
digestive secretion (p. 646)
digestive system (p. 645)
digestive tract (p. 647)
disaccharidases (p. 678)
disaccharides (p. 646)
duodenum (p. 676)
emetics (p. 660)
enamel (p. 653)
enteric nervous system (p. 651)
enterochromaffin-like (ECL) cells (p. 664)
enterogastric reflex (p. 658)
enterogastrones (p. 659)
enterohepatic circulation (p. 673)
enterokinase (p. 667)
enteropeptidase (p. 667)
epiglottis (p. 655)
eructation (p. 656)
esophageal stage (p. 656)
exchanger (p. 661)
exocrine pancreas (p. 667)
external anal sphincter (p. 690)
extrinsic nerves (p. 651)
fats (p. 647)
fatty acid (p. 647)
feces (p. 689)
ferritin (p. 686)
flatus (p. 692)
fructose (p. 646)
fundus (p. 656)

G cells (p. 664)
galactose (p. 646)
gallbladder (p. 672)
gallstones (p. 675)
gastric glands (p. 661)
gastric mucosal barrier (p. 666)
gastric pits (p. 661)
gastric reflux (p. 656)
gastrin (p. 664)
gastrocolic reflex (p. 689)
gastroesophageal sphincter (p. 656)
gastroileal reflex (p. 677)
gastrointestinal hormones (p. 651)
glottis (p. 655)
glucogon-like Peptide-1 (GLP-1) (p. 693)
glucose (p. 646)
glucose-dependent insulinotrophic peptide (GIP) (p. 693)
gluten enteropathy (p. 681)
glycogen (p. 646)
gut-associated lymphoid tissue (GALT) (p. 649)
haustra (p. 689)
haustral contractions (p. 689)
heartburn (p. 656)
hepatic artery (p. 671)
hepatic portal system (p. 671)
hepatic portal vein (p. 671)
hepatitis (p. 676)
hepatic vein (p. 671)
hepatocyte (p. 671)
histamine (p. 664)
hydrolysis (p. 647)
ileocecal sphincter (p. 677)
ileocecal valve (p. 677)
ileum (p. 676)
internal anal sphincter (p. 690)

## REVIEW EXERCISES

### Objective Questions (Answers on p. A-52)

1. The extent of nutrient uptake from the digestive tract depends on the body's needs. *(True or false?)*

2. The stomach is relaxed during vomiting. *(True or false?)*

3. Acid cannot normally penetrate into or between the cells that line the stomach, and this enables the stomach to contain acid without injuring itself. *(True or false?)*

4. Protein is continually lost from the body through digestive secretions and sloughed epithelial cells, which pass out in the feces. *(True or false?)*

5. Foodstuffs not absorbed by the small intestine are absorbed by the large intestine. *(True or false?)*

6. The endocrine pancreas secretes secretin and CCK. *(True or false?)*

7. A digestive reflex involving the autonomic nerves is known as a _____ reflex, whereas a reflex in which all elements of the reflex arc are located within the gut wall is known as a _____ reflex.

8. When food is mechanically broken down and mixed with gastric secretions, the resultant thick, liquid mixture is known as _____.

9. The entire lining of the small intestine is replaced approximately every _____ days.

10. The two substances absorbed by specialized transport mechanisms located only in the terminal ileum are _____ and _____.

11. The most potent choleretic is _____.

12. Which of the following is NOT a function of saliva?
    a. begins digestion of carbohydrate
    b. facilitates absorption of glucose across the oral mucosa
    c. facilitates speech
    d. exerts an antibacterial effect
    e. plays an important role in oral hygiene

**15**

13. Match the following:

_____ 1. prevents reentry of food into the mouth during swallowing

_____ 2. triggers the swallowing reflex

_____ 3. seals off the nasal passages during swallowing

_____ 4. prevents air from entering the oesophagus during breathing

_____ 5. closes off the respiratory airways during swallowing

_____ 6. prevents gastric contents from backing up into the oesophagus

(a) closure of the pharyngoesophageal sphincter

(b) elevation of the uvula

(c) position of the tongue against the hard palate

(d) closure of the gastroesophageal sphincter

(e) bolus pushed to the rear of the mouth by the tongue

(f) tight apposition of the vocal folds

14. Use the answer code on the right to identify the characteristics of the listed substances:

_____ 1. activates pepsinogen

_____ 2. inhibits amylase

_____ 3. is essential for vitamin $B_{12}$ absorption

_____ 4. can act autocatalytically

_____ 5. is a potent stimulant for acid secretion

_____ 6. breaks down connective tissue and muscle fibres

_____ 7. begins protein digestion

_____ 8. serves as a lubricant

_____ 9. kills ingested bacteria

_____ 10. is alkaline

_____ 11. coats the gastric mucosa

(a) pepsin

(b) mucus

(c) HCl

(d) intrinsic factor

(e) histamine

## Written Questions

1. Describe the four basic digestive processes.

2. List the three categories of energy-rich foodstuffs and the absorbable units of each.

3. List the components of the digestive system. Describe the cross-sectional anatomy of the digestive tract.

4. What four general factors are involved in regulating digestive system function? What is the role of each?

5. Describe the types of motility in each component of the digestive tract. What factors control each type of motility?

6. State the composition of the digestive juice secreted by each component of the digestive system. Describe the factors that control each digestive secretion.

7. List the enzymes involved in digesting each category of foodstuff. Indicate the source and control of secretion of each of the enzymes.

8. Why are some digestive enzymes secreted in inactive form? How are they activated?

9. What absorption processes take place within each component of the digestive tract? What special adaptations of the small intestine enhance its absorptive capacity?

10. Describe the absorptive mechanisms for salt, water, carbohydrate, protein, and fat.

11. What are the contributions of the accessory digestive organs? What are the nondigestive functions of the liver?

12. Summarize the functions of each of the three major gastrointestinal hormones.

13. What waste product is excreted in the feces?

14. How is vomiting accomplished? What are the causes and consequences of vomiting, diarrhoea, and constipation?

15. Describe the process of mucosal turnover in the stomach and small intestine.

## Quantitative Exercise (Solution on p. A-52)

1. Suppose a lipid droplet in the gut is essentially a sphere with a diameter of 1 cm.
   a. What is the surface area–to–volume ratio of the droplet? (*Hint:* The area of a sphere is $4\pi r^2$, and the volume is $4/3\pi r^3$.)
   b. Now, suppose that this sphere were emulsified into 100 essentially equal-sized droplets. What is the average surface area–to–volume ratio of each droplet?
   c. How much greater is the total surface area of these 100 droplets compared with the original single droplet?
   d. How much did the total volume change as a result of emulsification?

## 15 POINTS TO PONDER

### (Explanations on p. A-53)

1. Why do patients who have had a large part of their stomach removed for treatment of stomach cancer or severe peptic ulcer disease have to eat small quantities of food frequently instead of consuming three meals a day?

2. The number of immune cells in the *gut-associated lymphoid tissue (GALT)* housed in the mucosa is estimated to be equal to the total number of these defence cells in the rest of the body. Speculate on the adaptive significance of this extensive defence capability of the digestive system.

3. How would defecation be accomplished in a patient paralyzed from the waist down by a lower spinal cord injury?

4. After bilirubin is extracted from the blood by the liver, it is conjugated (combined) with glycuronic acid by the enzyme glucuronyl transferase within the liver. Only when conjugated can bilirubin be actively excreted into the bile. For the first few days of life, the liver does not make adequate quantities of glucuronyl transferase. Explain how this transient enzyme deficiency leads to the common condition of jaundice in newborns.

5. Explain why removal of either the stomach or the terminal ileum leads to pernicious anaemia.

## CLINICAL CONSIDERATION

(Explanation on p. A-53)

Thomas W. experiences a sharp pain in his upper-right abdomen after eating a high-fat meal. Also, he has noted that his feces are greyish white instead of brown. What is the most likely cause of his symptoms? Explain why each of these symptoms occurs with this condition.

**15**

# Energy Balance and Temperature Regulation

Body systems maintain homeostasis

## Homeostasis

Among the factors that are homeostatically maintained are the availability of energy-rich nutrients to the cells and the temperature of the internal environment. The hypothalamus helps regulate food intake, which is of primary importance in energy balance. The hypothalamus also helps maintain body temperature. It can vary heat production by the skeletal muscles and can adjust heat loss from the skin surface by varying the amount of warmed blood flowing through the skin vessels and by controlling sweat production.

Cells make up body systems

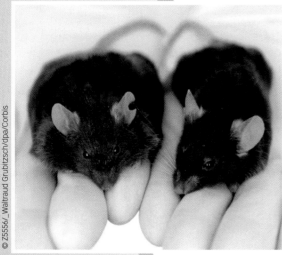

Fat mouse and thin mouse

Cells make up body systems

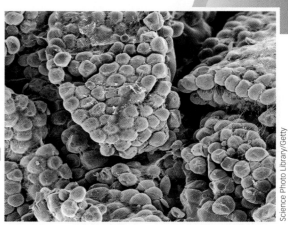

Scanning electron micrograph of fat (adipose) tissue

Food intake is essential to power cell activities. For body weight to remain constant, the caloric value of food must equal total energy needs. Energy balance and thus body weight are maintained primarily by controlling food intake. Energy expenditure generates heat, which is important in temperature regulation. Humans, usually in environments cooler than their bodies, must constantly generate heat to maintain their body temperature. Also, they must have mechanisms to cool the body if it gains too much heat from heat-generating skeletal muscle activity or from a hot external environment. Body temperature must be regulated because the rate of cellular chemical reactions depends on temperature, and overheating damages cell proteins. The hypothalamus is the major integrating centre for maintaining both energy balance and body temperature.

# 16

# Energy Balance and Temperature Regulation

**▌ Clinical Connections**

Joanne, a 42-year-old female, has been morbidly obese her entire life. Two years ago she weighed 161 kg and was 165 cm tall. This calculates to a BMI of 59.1, which is class 3 obesity. Five years earlier she was diagnosed with type 2 diabetes, likely associated with her obesity. She was also hypertensive and suffered from sleep apnea. Despite a lifetime of attempting to lose weight through diet and exercise, she was unsuccessful. At age 40, due to morbid obesity and underlying medical conditions, her physician recommended bariatric surgery, and she underwent a Roux-en-Y gastric bypass procedure.

During her recovery over the past two years she lost 56 kg and now weighs 106 kg. Although she still has type 2 diabetes mellitus, she now controls it by diet and exercise. As well, there has been a decrease in the degree of her hypertension, and she no longer requires treatment for sleep apnea. She now feels much better about herself and is able to maintain her weight through regular exercise and by eating many healthy, small meals throughout the day.

## 16.1 | Energy Balance

Each cell in the body, whether it is operating aerobically or anaerobically, needs energy to perform the functions essential for the cell's own survival (such as active transport and cellular repair) and also to carry out its specialized contributions toward maintaining homeostasis (such as gland secretion or muscle contraction). All energy used by cells is ultimately provided by food intake of carbohydrates, fats, and proteins.

In biological terms, **energy balance** can be represented in the following equation:

$$\text{Energy Intake} = \text{internal heat produced} + \text{external work} + \text{internal work} + \text{energy storage}$$

In this section we discuss the different elements that make up this energy balance equation.

## Thermodynamics

According to the **first law of thermodynamics**, energy can be neither created nor destroyed. Therefore, energy is subject to the same kind of input–output balance as are the chemical components of the body, such as water and salt (p. 608).

### ENERGY INPUT AND OUTPUT

The energy within ingested food, primarily carbohydrates, fats, and proteins, constitutes *energy input* to the body. Chemical energy locked in the bonds that hold the atoms together in nutrient molecules is released when these molecules are broken down in the body. Cells capture a portion of this nutrient energy by means of the high-energy phosphate bonds of ATP (pp. 24 and Appendix B, p. A-17). Energy harvested from biochemical processing of ingested nutrients is either used immediately to perform biological work or stored in the body (e.g., glycogen, lipids) for later use as needed during periods when food is not being digested and absorbed.

*Energy output*, or *expenditure*, by the body falls into two categories (> Figure 16-1): external work and internal work. **External work** is the energy expended when skeletal muscles contract to move external objects or to move the body in relation to the environment. **Internal work** constitutes all other forms of biological energy expenditure that do not accomplish mechanical work outside the body. Internal work encompasses two types of energy-dependent activities: (1) skeletal muscle activity used for purposes other than external work (e.g., muscle contractions for postural maintenance and shivering), and (2) all the energy-expending activities that must go on all the time just to sustain life. The latter include the work of pumping blood and breathing; the energy required for active transport of critical materials across plasma membranes; and the energy used during synthetic reactions for the maintenance, repair, and growth of cellular structures—in short, the "metabolic cost of living."

### CONVERSION OF NUTRIENT ENERGY TO HEAT

Not all energy in nutrient molecules can be harnessed to perform biological work. Energy cannot be created or destroyed, but it can be converted from one form to another. The energy in nutrient molecules not used to energize work is transformed into **thermal energy**, otherwise known as **heat**. During biochemical processing, only about 50–60 percent of the energy in nutrient molecules is transferred to ATP; the other 40–50 percent of nutrient energy is lost as heat. During ATP expenditure by the cells, another 25 percent of the energy derived from ingested food becomes heat. Therefore, only about 25–35 percent of nutrient energy is available for internal and external work. The remaining about 75 percent is lost as heat during the transfer of energy from nutrient → ATP → cellular use.

Moreover, of the energy actually captured for use by the body, almost all eventually becomes heat. For example, energy used by the heart to pump blood is gradually changed into heat by friction as blood flows through the vessels. Even in performing external work, skeletal muscles convert chemical energy (ATP) into mechanical energy inefficiently; about 75 percent of the energy is lost as heat. In this way, all energy liberated from ingested food that is not directly used for moving external objects or stored in fat (adipose tissue) deposits (or, in the case of growth, as protein) eventually becomes body heat. However, much of the heat is used to maintain body temperature.

## Metabolic rate

The rate at which energy is expended by the body during both external and internal work is known as the **metabolic rate**:

$$\text{Metabolic rate} = \text{energy expenditure/unit of time}$$

Because most of the body's energy expenditure eventually appears as heat, the metabolic rate is normally expressed as the rate of heat production in kilocalories per hour. The basic unit of heat energy is the **calorie**, which is the amount of heat needed to raise the temperature of 1 g of $H_2O$ by $1°C$. This unit is too small when discussing the human body because of the amount of heat produced, so the term **kilocalorie (kcal) or Calorie (C)** is used, which is equivalent to 1000 calories. Thus, $1\,\text{kcal} = 1C = 1000$ calories. When nutritionists speak of calories in quantifying the energy content of various foods, they are referring to a kcal or Calorie. There are 4 kcal and 9 kcal of heat energy released when 1 g of glucose and 1 g of fat are burned (oxidized), respectively. It does not matter whether the oxidation takes place inside or outside the body.

### CONDITIONS FOR MEASURING THE BASAL METABOLIC RATE

Metabolic rate and, consequently, the amount of heat produced change in relation to a variety of factors, such as exercise, daily activities, age, anxiety, body composition and size, thyroid hormone, sympathetic stimulation (release of epinephrine and norepinephrine), growth

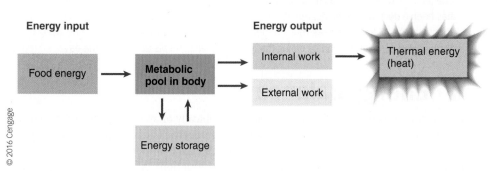

**Energy input**      **Energy output**

Food energy → Metabolic pool in body → Internal work / External work → Thermal energy (heat)

Energy storage

© 2016 Cengage

> **Figure 16-1 Energy input and output**

hormone, shivering, climate, sleep, and food intake. Increased skeletal muscle activity through exercise or other daily activities is the factor that can increase metabolic rate to the greatest extent (▮ Table 16-1). Even slight increases in muscle tone notably elevate the metabolic rate, and therefore heat production. For this reason, a person's metabolic rate is determined under standardized conditions that have been established to control factors that alter metabolic rate. In this way, the metabolic activity necessary to maintain basic body function can be determined. **Basal metabolic rate (BMR)** refers to the rate of energy use in the body during absolute rest, but while awake; it is the minimum energy (kcal) needed to maintain basic physiological function. The BMR is measured under the following conditions:

1. Physical rest is required—no exercise for at least 30 minutes prior to test to eliminate excess muscular heat production.

2. Mental rest is also needed to minimize skeletal muscle tone (sympathetic stimulation, epinephrine) associated with stress.

3. The room environment should be comfortable (20−24°C, low lights, and person laying down).

4. The person should not have eaten any food for at least 12 hours to avoid **diet-induced thermogenesis** (*thermo*

**▮ TABLE 16-1** Rate of Energy Expenditure for a 70 kg Person during Different Types of Activity*

| Form of Activity | Energy Expenditure (kcal/h) |
| --- | --- |
| Sleeping | 65 |
| Awake, lying still | 77 |
| Sitting at rest | 100 |
| Standing relaxed | 105 |
| Getting dressed | 118 |
| Keyboarding | 140 |
| Walking slowly on level (4 km/h) | 200 |
| Carpentry, painting a house | 240 |
| Sexual intercourse | 280 |
| Bicycling on level (8.8 km/h) | 304 |
| Shovelling snow, sawing wood | 480 |
| Swimming | 500 |
| Jogging (8.5 km/h) | 570 |
| Rowing (20 strokes/min) | 828 |
| Walking up stairs | 1100 |

© 2016 Cengage

*To provide some perspective when viewing this table, keep in mind that two regular Oreo cookies contain approximately 107 Calories, which translates into 107 kcals. Thus, one hour of TV watching does not expend the kilocalorie equivalent of two Oreo cookies.

means "heat"; *genesis* means "production") or the obligatory increase in metabolic rate that occurs from food intake. This short-lived (≤12 hour) rise in metabolic rate is due not to digestive activities but to the increased metabolic activity associated with the processing and storage of ingested nutrients (e.g., in the liver).

## METHODS OF MEASURING THE BASAL METABOLIC RATE

*Clinical Note* The rate of heat produced can be measured directly or indirectly. With **direct calorimetry**, the person sits in an insulated chamber with water circulating through the walls. The difference in the temperature of the water entering and leaving the chamber reflects the amount of heat liberated from the person (metabolic heat) and picked up by the water as it passes through the chamber. This provides an accurate measurement of heat production, but due to the size and cost of the chamber is not practical. A more convenient and common method of BMR measurement is **indirect calorimetry**. This method simply requires the measurement of a person's oxygen uptake per unit, which is easy and more cost effective. Recall that

$$Food + O_2 \rightarrow CO_2 + H_2O + energy \text{ (mostly heat)}$$

Accordingly, a direct relationship exists between the volume of oxygen used and the quantity of heat produced. This relationship also depends on the type of food that is oxidized. Carbohydrates (4 kcals/g), proteins (4 kcals/g), and fats (9 kcals/g) need different amounts of oxygen for their oxidation, yielding different amounts of kilocalories when oxidized. However, when averaged, we can estimate the quantity of heat produced per litre of oxygen consumed on a typical, mixed, Canadian diet. This approximate value, known as the **energy equivalent of O₂**, is 5 kcals/L of oxygen consumed. Thus, the metabolic rate of a person consuming 15 L/h of oxygen can be estimated as follows:

| 15 L/h | = | O₂ consumption |
| --- | --- | --- |
| ×5 kcals/L | = | energy equivalent of O₂ |
| 75 kcals/h | = | estimated metabolic rate |

In this way, a simple measurement of oxygen consumption can be used to reasonably estimate body heat production to determine metabolic rate.

Once the rate of heat production is determined under the prescribed basal conditions, it must be compared to norms for people of the same sex, age, height, weight, and body composition, because these factors all affect the basal rate of energy expenditure. For example, a large man actually has a higher rate of heat production than a smaller man, but expressed in terms of total surface area (which reflects height and weight), the output in kilocalories per hour per square metre of body surface area is normally similar.

### FACTORS INFLUENCING THE BASAL METABOLIC RATE

Many factors influence metabolic rate, including the following key factors:

- *Thyroid hormone.* Thyroxin is a very important determinant of BMR due to its whole-body metabolic effect. When

**16**

maximal quantities of thyroxin are secreted, metabolic rate increases 50–100 times. As thyroid hormone increases, the BMR increases correspondingly.

- *Sympathetic stimulation.* Often in concert with the release of epinephrine and norepinephrine, stimulation of the SNS increases metabolic rate by increasing cellular metabolism (e.g., skeletal smooth muscle tone). This neural-hormonal milieu also influences the rate of glycogenolysis. In addition, sympathetic stimulation is associated with the release of heat from brown fat stores. Contrary to the belief that only neonates have brown fat stores, adults have brown fat stores as well, and those who work outdoors frequently have greater brown fat stores than those who work inside. The heat released from **brown fat** results from the high number of mitochondria in the fat stores. When stimulated by the sympathetic nervous system, the oxidation of fat droplets results in large amounts of heat production. Why is all the heat produced? Because oxidative phosphorylation within the brown fat stores is uncoupled, so no ATP is generated, just heat.

- *Exercise.* By far, this is the factor that causes the greatest increase in BMR rate, due to the size and metabolic potential of the muscle tissue (45–70% of total body weight). A few seconds of maximal muscular exercise can increase overall heat production by about 50 times for a few minutes.

- *Daily activities.* Similar to exercise, the more vigorous the daily activity (e.g., moving furniture) and the larger the percentage of muscle mass involved in completing the activity, the larger the increase in metabolism.

- *Sex.* Men often have more muscle mass than women; thus, they have a greater amount of metabolically active tissue, which gives them a greater overall metabolism.

- *Age.* As we age our metabolic rate declines. From birth to about 18 years, there is a rapid decline, and from 18 to 70, there is a more gradual decline. The high metabolic rate in the early years of life results from the body's need to rapidly synthesize cellular material for growth. Obviously, this declines with age. As we age, the maintenance of muscle mass through physical activity (e.g., resistance training) is important.

Surprisingly, the BMR is not the body's lowest metabolic rate. The rate of energy expenditure during sleep is 10–15 percent lower than the BMR, presumably because of the more complete muscle relaxation that occurs during the paradoxical stage of sleep (p. 134).

## Energy input and energy output

Because energy cannot be created or destroyed, energy input must equal energy output, as follows:

$$\text{Energy input} = \text{energy output}$$

$$\text{Energy in food} = \text{external} + \text{internal heat} \pm \text{stored}$$

$$\text{consumed} \qquad \text{work} \qquad \text{production} \qquad \text{energy}$$

There are three possible states of energy balance:

1. *Neutral energy balance.* If the amount of energy in food intake equals the amount of energy expended during external and internal work, plus the basal internal energy expenditure (appearing as body heat), then energy input and output are balanced, and body weight should remain constant.

2. *Positive energy balance.* If the amount of energy in food intake is greater than the amount of energy expended during external and internal work, the extra energy consumed but not used is stored in the body (e.g., fat, glycogen), and body weight increases.

3. *Negative energy balance.* Conversely, if the energy from food intake is less than the body's immediate energy needs, the body must use stored energy (fats and glycogen) to meet energy needs, and body weight decreases accordingly.

For a person to maintain a constant body weight (with the exception of minor fluctuations caused by changes in $H_2O$ content), energy acquired through food intake must equal energy expenditure by the body. Because the average adult maintains a fairly constant weight over long periods of time, this implies that precise homeostatic mechanisms exist to maintain a long-term balance between energy intake and energy expenditure. Theoretically, total body energy content could be maintained at a constant level by regulating the magnitude of food intake, physical activity, internal work, and heat production. Control of food intake to match changing metabolic expenditures is the major means of maintaining a neutral energy balance. The level of physical activity is principally under voluntary control, and mechanisms that alter the degree of internal work and heat production are aimed primarily at regulating body temperature rather than total energy balance.

However, after several weeks of eating less or more than desired, small counteracting changes in metabolism may occur. For example, a compensatory increase in the body's efficiency of energy use in response to underfeeding partially explains why some dieters become stuck at a plateau after having lost the first five or so kilograms of weight fairly easily. Similarly, a compensatory reduction in the efficiency of energy use in response to overfeeding accounts in part for the difficulty experienced by very thin people who are deliberately trying to gain weight. Despite these modest compensatory changes in metabolism, regulation of food intake is the most important factor in the long-term maintenance of energy balance and body weight.

## Food intake

Even though food intake is adjusted to balance changing energy needs over a period of time, there are no calorie receptors per se to monitor energy input and output, or total body energy content. Instead, blood-borne chemicals that signal the body's nutritional state, such as levels of stored fat or feeding status, are important factors regulating food intake. Control of food intake relies on more than a single signal; it is established by the integration of multiple signals that provide overall information about the body's energy status. Various molecular signals (discussed in this section) ensure that feeding behaviour is synchronized with the body's short- and long-term energy needs. Some of this information is used for short-term regulation of food intake, that is, controlling meal size and frequency. Interestingly, over a 24-hour period, the energy ingested rarely matches the energy expenditure for that time. However, over the long-term (i.e., months), the correlation between total caloric intake and total energy output is good. If the relationship is undisturbed (e.g., by illness), total energy content of the body—and thus body weight—remains relatively stable. *In saying this, maintaining a healthy body weight is more complex than simply calculating total calories consumed versus total calories expended; the type of food (e.g., sugars) consumed and genetics (epigenetics) also play an important role.*

Before we focus on other factors that influence energy balance, we first turn our attention to the body's long- and short-term signals (hormones) for monitoring caloric need and energy content.

### ■ Clinical Connections

The purpose of bariatric surgery to treat morbid obesity is to reduce food intake. Although there are many different types of procedures, Joanne underwent a Roux-en-Y gastric bypass procedure. The surgery involves creating a small pouch with a volume of around 30 mL by dividing the upper end of the stomach. The small intestine is then severed and attached to the pouch. The section of small intestine still attached to the stomach is then reattached further down the small intestine to allow drainage. With the entire stomach being replaced by the small volume of the pouch, food intake is severely restricted.

### ROLE OF THE ARCUATE NUCLEUS: NEUROPEPTIDE Y AND MELANOCORTINS

Control of energy balance and food intake is primarily a function of the hypothalamus. The **arcuate nucleus** of the hypothalamus plays a central role in both the long-term control of energy balance and body weight and the short-term control of food intake on a meal-to-meal basis. The arcuate nucleus is an arc-shaped collection of neurons located adjacent to the floor of the third ventricle. Multiple, highly integrated, redundant pathways criss-cross into and out of the arcuate nucleus—indicative of this structure's complex involvement in feeding and satiety. **Feeding (appetite) signals** give rise to the sensation of **hunger**, driving us to eat. By contrast, **satiety** is the feeling of being full, telling us when we have had enough, and suppressing the desire to eat.

The arcuate nucleus has two subsets of neurons that function in an opposing manner. One subset releases *neuropeptide Y,* and the other releases *melanocortins.*[1] **Neuropeptide Y (NPY)**, a potent

---

[1]The two subsets of neurons in the arcuate nucleus are the NPY/AgRP population and the POMC/CART population. *AgRP* stands for *agouti-related protein.* Both NPY and AgRP stimulate appetite. POMC stands for *pro-opiomelanocortin,* the precursor molecule that gives rise to melanocortins. CART stands for *cocaine-and amphetamine-related transcript.* Melanocortins and CART peptide both suppress appetite. For simplicity, we discuss the role of NPY and melanocortins only but recognize that other chemical signals released from the arcuate nucleus exert similar functions.

appetite stimulator, leads to increased food intake, thus promoting weight gain. **Melanocortins**, a group of hormones typically associated with skin colour, have been shown to exert an unexpected role in energy homeostasis. Melanocortins, most notably a *melanocyte-stimulating hormone* (p. 226), suppress appetite, leading to reduced food intake and weight loss (loss of fat stores). Melanocortins do not play a direct role in determining human skin colouration; rather, this is determined by the activation of the melanocortin 1 receptor and by the particular form of melanin that's produced in melanocytes following stimulation by melanocortins.

But NPY and melanocortins are not the final effectors in appetite control. These arcuate-nucleus chemical messengers, in turn, influence the release of neuropeptides in other parts of the brain that exert more direct control over food intake. Scientists are currently trying to unravel the other factors that act upstream and downstream from NPY and melanocortins to regulate appetite. The following regulatory inputs to the arcuate nucleus and beyond are important in the long-term maintenance of energy balance and the short-term control of food intake at meals (⟩ Figure 16-2).

### LONG-TERM MAINTENANCE OF ENERGY BALANCE: LEPTIN AND INSULIN

The notion that fat cells (**adipocytes**) in adipose tissue (subcutaneous or visceral) are merely a storage site for triglycerides has undergone a shift with the discovery of their active role in energy homeostasis. Adipose tissue is an intricate, vital, and active metabolic and endocrine organ that has both local (autocrine/paracrine) and systemic (endocrine) affects. Adipose tissue contains connective and nervous tissues and immune cells, and together functions as an integrated unit. Not only does adipose tissue respond to afferent signals from hormones and the central nervous system, but also adipocytes secrete their own hormones (bioactive peptides), collectively termed **adipokines** (▌ Table 16-2). These factors include cytokines, adiponectin, plasminogen, FFA, adipsin, angiotensinogens, and resistin. We now take a closer look at the important hormone *leptin.*

**Leptin** was discovered in the mid-1990s by Dr. Friedman, and was the first blood-borne molecular satiety signal identified. Leptin is one of the most important adipokines, a hormone important in the regulation of food intake and body weight (*leptin* means "thin"). The amount of leptin in the blood is an excellent indicator of the total triglycerides stored in adipose tissue: the larger the volume of fat stores, the more leptin released into the blood. Many of the effects of leptin related to energy intake and expenditure are mediated via the arcuate nucleus of the hypothalamus, whereas other effects are mediated directly on peripheral tissues (e.g., muscle and pancreatic cells). However, leptin's primary role is to serve as a metabolic signal of energy sufficiency. It acts via negative feedback, because increased leptin from growing fat stores serves to turn off the hunger signal, thus decreasing food consumption and promoting weight loss. Its inhibitory action works by suppressing hypothalamic output of appetite-stimulating NPY and by stimulating output of appetite-suppressing melanocortins.

In contrast, a decline in leptin is linked with the adaptive physiological response to starvation, which increases appetite and decreases energy expenditure (conservation of fuel).

**16**

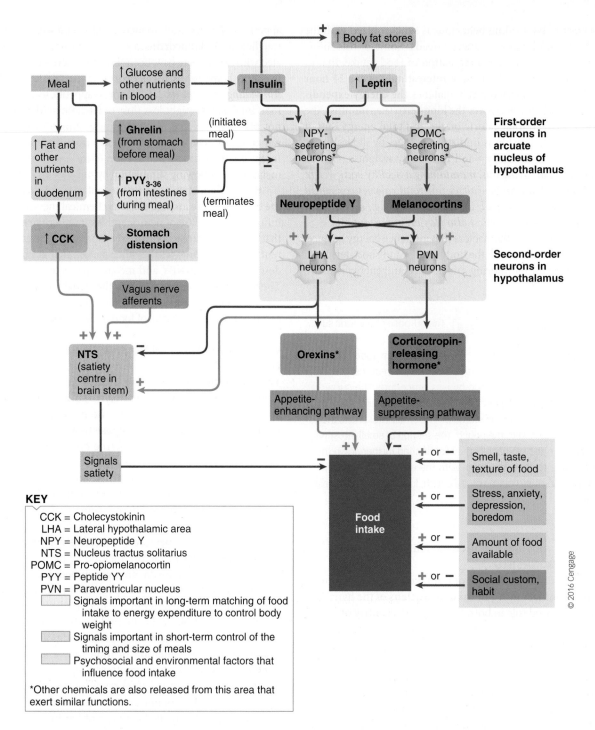

KEY

CCK = Cholecystokinin
LHA = Lateral hypothalamic area
NPY = Neuropeptide Y
NTS = Nucleus tractus solitarius
POMC = Pro-opiomelanocortin
PYY = Peptide YY
PVN = Paraventricular nucleus

Signals important in long-term matching of food intake to energy expenditure to control body weight

Signals important in short-term control of the timing and size of meals

Psychosocial and environmental factors that influence food intake

*Other chemicals are also released from this area that exert similar functions.

⟩ Figure 16-2 **Factors that influence food intake**

The starvation response can be staved off with the administration of low-dose leptin (increasing circulating leptin) replacement. This starvation response is even found in states of obesity, and interestingly, naturally high levels of leptin (endogenous) or treatment with low-dose leptin (exogenous) is effective in satiating obese persons. This is consistent with the theory of leptin resistance (discussed in the section Obesity: Kcals Required and Other Factors).

Another blood-borne signal that plays an important role in long-term control of body weight is **insulin**. This hormone is secreted by the pancreas in response to an increase in glucose concentration and other nutrients in the blood post-meal. Insulin stimulates the cellular uptake and storage of the nutrient glucose. In this way, the increase in insulin secretion that accompanies nutrient abundance, use, and storage appropriately inhibits the NPY-secreting cells of the arcuate nucleus, thereby suppressing further food intake.

In addition to the importance of leptin, insulin, and perhaps other so-called **adiposity signals** (fat-related) in the long-term control of body weight, other factors play a role in controlling the timing and size of meals. Several blood-borne messengers from the digestive tract and pancreas are important in regulating how often and how much we eat in a given day. We now look at how these messengers function.

| Adipokine | Function |
|-----------|----------|
| **Leptin** | Released from stored fat; suppresses appetite; dominant long-term regulator of energy balance and body weight |
| **Adiponectin** | Secretion from adipocytes suppressed in obesity; promotes fatty acid oxidation by muscle; increases sensitivity to insulin; decreases body weight by increasing energy expenditure; has anti-inflammatory actions |
| **Resistin** | Released primarily in obesity; leads to insulin resistance |
| **Visfatin** | Released primarily from visceral fat; stimulates glucose uptake; binds with insulin receptor at a site distinct from the insulin-binding site |
| **Tumour Necrosis Factor α (TNF-α) and Interleukin 6 (IL-6)** | Promote low-level inflammation in fat and throughout body |

**Source:** From Sherwood. *Human Physiology*, 8E. © 2013 Brooks/Cole, a part of Cengage, Inc. Reproduced by permission. www.cengage.com/permissions

## SHORT-TERM EATING BEHAVIOUR: GHRELIN AND PYY$_{3-36}$ SECRETION

Two peptides important in the short-term control of food intake have recently been identified: *ghrelin* and *peptide YY$_{3-36}$* (PYY$_{3-36}$), which signify hunger and fullness, respectively. Both are secreted by the digestive tract. **Ghrelin**, the so-called hunger hormone, is a fast-acting appetite stimulator produced by the stomach and regulated by the feeding status (*ghrelin* is the Hindu word for "growth"). Secretion of this mealtime stimulator peaks before meals and makes people feel like eating, then falls once food is eaten. Ghrelin stimulates appetite by activating the hypothalamic NPY-secreting neurons.

**Peptide YY$_{3-36}$ (PYY$_{3-36}$)** is a counterpart of ghrelin. The secretion of (PYY$_{3-36}$), which is produced by the small and large intestines, is at its lowest level before a meal but rises during meals and signals satiety. This peptide acts by inhibiting the appetite-stimulating NPY-secreting neurons in the arcuate nucleus. By thwarting appetite, PYY$_{3-36}$ is believed to be an important mealtime terminator.

We now examine several additional factors involved in signalling the body's condition on the hunger–satiety continuum.

## BEYOND THE ARCUATE NUCLEUS: OREXINS AND OTHERS

Two hypothalamic areas are richly supplied by axons from the NPY- and melanocortin-secreting neurons of the arcuate nucleus. These second-order neuronal areas involved in energy balance and food intake are the **lateral hypothalamic area (LHA)** and **paraventricular hypothalamic nucleus (PVN)**. In a recently proposed model, the LHA and PVN release chemical messengers in response to input from the arcuate nucleus neurons. These messengers act downstream from the NPY and melanocortin signals to regulate appetite. The LHA produces two closely related neuropeptides known as **orexins**, which are potent stimulators of food intake (*orexis* means "appetite"). Moreover, orexin may also stimulate wakefulness and energy expenditure. Orexin dysregulation was linked with narcolepsy, and recent animal studies indicate that a potential chief role of orexin might be to integrate metabolism, circadian rhythm, and sleep to determine periods of sleep and alertness. NPY stimulates the release of orexins, leading to an increase in appetite and greater food intake, whereas melanocortins inhibit the secretion of orexins. By contrast, the PVN releases chemical messengers—for example, **corticotropin-releasing hormone**—that decrease appetite and food intake. (As its name implies, corticotropin-releasing hormone is better known for its role as a hormone; see Chapter 6.) Melanocortins stimulate and NPY inhibits the release of these appetite-suppressing neuropeptides.

In contrast to the key role of the hypothalamus in maintaining energy balance and long-term control of body weight, there is a region in the brain stem known as the **nucleus tractus solitarius (NTS)** that processes signals important in the feeling of being full. This region is considered the *satiety centre*. Not only does the NTS receive input from the higher hypothalamic neurons involved in energy homeostasis, but it also receives afferent inputs from the digestive tract and elsewhere that signal satiety (e.g., afferent neural input indicating the extent of stomach distension). We now turn our attention to cholecystokinin, the most important of these satiety signals.

## CHOLECYSTOKININ AS A SATIETY SIGNAL

Cholecystokinin (CCK), one of the gastrointestinal hormones released from the duodenal mucosa during digestion of a meal, is an important satiety signal for regulating the size of meals. CCK is secreted in response to the presence of nutrients in the small intestine. Through multiple effects on the digestive system, CCK facilitates digestion and absorption of these nutrients (p. 692). It is appropriate that this blood-borne signal, whose rate of secretion is correlated with the amount of nutrients ingested, also contributes to the sense of being full after a meal has been consumed but before it has actually been digested and absorbed. We feel satisfied when enough food to replenish our body's energy store is in the digestive tract—even though the energy store is still low. This explains why we stop eating before the ingested food is digested and therefore available to meet the body's energy needs.

■ **Clinical Connections**

Gastric bypass surgery influences food intake not only because it physically restricts food intake. Due to the very small pouch size after the surgery, gastric distension is rapid, which activates mechanoreceptors that tell the brain that the stomach is full and to stop eating. The small stomach volume also results in food entering the small intestine faster, which activates chemical satiety signal pathways such as CCK and PPY earlier. It has been suggested that these later effects have a greater effect on post-bariatric surgery weight loss than the reduction of food volume itself.

**16**

We have described the automatic, involuntary signals that control food intake. In addition, as with water intake, there are psychological, social, and environmental factors that shape people's eating habits. Often our decision to eat or stop eating is not determined merely by whether we are hungry or full, respectively. Frequently, we eat out of habit (eating three meals a day on schedule no matter what our status on the hunger–satiety continuum) or because of social custom (food often plays a prime role in entertainment, leisure, and business activities). Even well-intentioned family pressure—"Clean your plate before you leave the table"—can have an impact on the amount consumed.

Furthermore, the amount of pleasure derived from eating can reinforce feeding behaviour. Eating foods with an enjoyable taste, smell, and texture can increase appetite and food intake. This has been demonstrated in an experiment in which rats were offered their choice of highly palatable human foods. They overate by as much as 70–80 percent and became obese. When the rats returned to eating their regular monotonous but nutritionally balanced rat chow, their obesity was rapidly reversed.

Stress, anxiety, depression, and boredom have also been shown to alter feeding behaviour in ways that are unrelated to energy needs in both experimental animals and humans. University students often eat when studying to reduce stress, or because they are tired of studying; in this instance, food satisfies a psychological need, not hunger. Thus, any comprehensive explanation of how food intake is controlled must take into account these voluntary eating acts that can reinforce or override the internal signals governing feeding behaviour.

## The gut microbiome and energy homeostasis

What foods we choose to consume and the amount of exercise we do are important factors in energy homeostasis. However, an important element often overlooked is the intestinal microbiome. Gut bacteria represent 1–3 percent of our total body mass. It is a highly active population that, because of constantly growing and dividing, likely contributes significantly to our body's total energy usage. The gut bacteria also play a significant role in regulating digestion efficiency, in that the microbiome processes certain ingested foods and molecules to make them available to us, their host. Gut bacteria are also considered a virtual endocrine organ because they secrete compounds that affect other cells within the body and may even play a role in the hypothalamic-pituitary-adrenal axis as part of the body's integrated response to stress. Such secretions potentially can affect total body energy balance and energy stores to modulate weight gain and loss. While there has been extensive research on how the microbiome can influence weight in rodent models, the direct connection in humans is still unclear.

## Obesity: kcals required and other factors

**Obesity** is defined as excessive fat content in the adipose tissue stores; being 20 percent overweight compared with normal standards is considered the arbitrary boundary for obesity. In 2014, Statistics Canada reported that 20.2 percent of Canadian adults were categorized as obese, using body mass index (BMI). This is up from 2003, when about 15 percent of the adult population was obese. These trends are the same for both males and females. If we take overweight and obese classifications together, approximately 53.6 percent of adult Canadians were in this category. (For more information on the obesity epidemic in Canada, see Concepts, Challenges, and Controversies, pp. 708–709.) Much of the world is following the same trend, recently leading the World Health Organization to coin the new word *globesity* to describe the worldwide situation.

Obesity occurs when, over a period of time, more kilocalories are ingested in food than are used to support the body's energy needs, and the excess energy is stored as triglycerides in adipose tissue. Early in the development of obesity, existing fat cells get larger (hypertrophy). An average adult has between 40 billion and 50 billion adipocytes. Each fat cell can store the maximum of about 1.2 µg of triglycerides. Once the existing fat cells are full, if people continue to consume more calories than they expend, they make more adipocytes (hyperplasia).

Although it is clear that a sedentary lifestyle is one primary reason for obesity, and that increasing physical activity levels may help maintain a normal healthy body weight, control of food intake is very important. The causes of obesity are many, and some remain obscure. However, some factors that may be involved include the following:

- *Disturbances in leptin (and potentially gherlin) signalling.* The discovery of the hormone leptin was an important milestone linking the regulation of metabolism (i.e., hypothalamus) with the levels of energy storage (adipocytes). Research in this area has been aided by the use of genetically altered mice, including db/db (i.e., diabetic) and ob/ob (i.e., obese) mice. The primary genetic defect in these mice relates to either diminished leptin production (ob/ob) or impaired leptin receptors (db/db). Ob/ob mice, for example, can display many abnormalities seen in starving animals (even though they are obese), such as hyperphagia and decreased immune function, which can be reversed with a regular injection of leptin (leptin replacement). Thus, even though the ob/ob mouse's body weight and energy balance have not changed, increased leptin levels correct these abnormalities. Similar examples of obesity in humans have been linked with an alteration of leptin or the leptin receptor. Some cases of obesity have been linked to leptin resistance, meaning the brain does not detect leptin as a signal to turn down appetite until a higher set point (fat storage point) is achieved.

Nevertheless, the mechanism for leptin resistance is unknown, but may be related to defects in leptin signalling or transport across the blood–brain barrier. In both rodent (db/db mice) and human obesity studies, plasma leptin concentrations are elevated in proportion to the degree of adiposity (volume of fat cells), supporting the concept of leptin resistance in obesity. One goal of obesity research is to find the mechanism for leptin resistance and develop drugs that bypass the resistance by targeting neural processing steps, which are downstream from leptin (e.g., release of melanocortins).

Interestingly, just as circulating leptin levels are increased in the obese person, the level of the appetite-stimulating hormone ghrelin is decreased. It is certain that ghrelin has a hunger-stimulating effect, as well as an adipogenic effect (fat-storing effect). However, the mechanism through which ghrelin contributes to the development or maintenance of obesity is not yet clear. However, it is suggested that from an overall functional perspective, ghrelin and leptin might be complementary peptides in the regulatory system that informs the central nervous system about short-term and long-term energy balance.

- *Lack of exercise.* Numerous studies have shown that, on average, overweight people do not eat any more than those of normal body weight. One possible explanation is that overweight persons do not overeat but "underexercise"—the "couch potato" syndrome. Very low levels of physical activity are typically not accompanied by comparable reductions in food intake.

Moreover, moderate- to high-intensity physical activity (e.g., exercise) has three important effects that influence caloric expenditure and body weight: (1) during exercise more kilocalories are burned compared to sitting on the couch, so extra kilocalories are not stored but expended; (2) exercise improves insulin sensitivity, which reduces insulin levels and assists the satiety signal; (3) exercise makes the TCA cycle (p. 25) run faster, decreasing the amount of citrate leaving the mitochondria, which reduces the availability of the acetyl group for the production of fat; and (4) exercise reduces stress, which reduces cortisol and appetite associated with stress.

- *Differences in the "fidget factor."* **Nonexercise activity thermogenesis (NEAT)**, or the fidget factor, might explain some variation in fat storage among people. NEAT refers to energy expended by physical activities other than planned exercise. Those who engage in toe tapping or other types of repetitive, spontaneous physical activity expend a substantial number of kilocalories throughout the day without a conscious effort.

- *Differences in extracting energy from food.* Another reason why lean people and obese people may have dramatically different body weights despite consuming the same number of kilocalories may lie in the efficiency with which each extracts energy from food. Studies suggest that lean individuals tend to derive less energy from the food they consume, because they convert more of the food's energy into heat than into energy for immediate use or for storage. For example, slim individuals have more **uncoupling proteins**, which allow their cells to convert more of the nutrient calories into heat instead of fat; these people can eat a lot without gaining weight. By contrast, obese people may have more efficient metabolic systems for extracting energy from food; this is a useful trait in times of food shortage, but a hardship when trying to maintain a desirable weight when food is plentiful.

- *Development of an excessive number of fat cells as a result of overfeeding.* One of the problems in fighting obesity is that once fat cells are created, they do not disappear with dieting and weight loss. Even if a dieter loses a large portion of the triglyceride fat stored in these cells, the depleted cells remain, ready to refill. Therefore, rebound weight gain after losing weight is difficult to avoid and discouraging for the person.

- *The existence of certain endocrine disorders, such as hypothyroidism* (p. 252). Hypothyroidism involves a deficiency of thyroid hormone, the main factor that bumps up the BMR so that the body burns more kilocalories in its resting state.

- An abundance of convenient, highly palatable, energy-dense, relatively inexpensive foods.

- The presence of brown fat stores may protect against obesity as we age. A relationship between BMI and brown fat stores has been detected. The trend indicates that a high level of brown fat stores typically correlate with a low BMI. Rodent studies have indicated that brown fat has profound effects on body weight and energy balance, and brown fat stores are found in adult humans (p. 702). Recall that one grouping of neurons in the arcuate nucleus is the NPY (potent appetite stimulator) neurons. Injection of NPY into this brain area reduces sympathetic stimulation of brown adipose tissue and, in doing so, simultaneously decreases energy expenditure and increases the expression of enzymes involved in lipogenesis in white fat (Billington, C.J., et al. [1991]. Effects of intracerebroventricular injection of neuropeptide Y on energy metabolism. *Am J Physiol Integr Comp Physiol*, 260: R321–27; G. Bray. [1992]. Peptides affect the intake of specific nutrients and the sympathetic nervous system. *Am J Clin Nutr*, 55: 265S–71S).

- The consumption of the sugar fructose may be associated with the obesity epidemic. Fructose, the sugar typically found in fruit, is also found in table sugar (sucrose) and many other foods (e.g., fast foods). Fructose is processed by the body in a different way than glucose. For example, glucose is an immediate energy source, whereas fructose must be converted in the liver to energy, fat, or glucose before it can be used by tissues. Additionally, the pathway for glucose breakdown is turned off when there is an accumulation of energy and fat, thereby preventing excess production. However, fructose breakdown does not have a feedback mechanism. Therefore, a massive ingestion of fructose would result in an overwhelming production of energy and fat. Some suggest that eating a high sugar diet—that is, a diet high in sucrose (because it is half fructose) or high-fructose corn syrup—is similar to eating a high fat diet, which results in weight gain and high cholesterol. So not all sugars are created equal!

- Genetics and inheritable traits. As discussed, the root of obesity is multifaceted, involving interactions among neuroendocrine status, environmental factors, sedentarism or poor dietary habits, and also genetics. Genetics play a large role in our lives, and thus it is no surprise that they can influence energy balance. For example, certain genetic diseases (e.g., Prader–Willi syndrome, PWS) have metabolic consequences. Prader–Willi syndrome is a genetic disorder present at birth in which seven genes on chromosome 15 are deleted or unexpressed. Those with PWS may have physical, mental, and behavioural problems—for example, the insistent feeling of hunger, which causes high levels of food consumption and thus energy balance.

16

## What Is Obesity?

Obesity means having an excessive amount of body fat. Fat is contained under the skin (subcutaneous), within the muscle (intramuscularly), and around the organs (visceral). Fat is an energy reserve, and when we consume too many kilocalories, the extra energy is stored as body fat.

The distribution (location) of body fat is also a concern. Obese people often have a significant amount of abdominal subcutaneous fat, intramuscular fat, and visceral fat deposits, all of which are associated with the development of numerous cardiovascular diseases and insulin resistance.

Two commonly used determinants of obesity are percentage of body fat and body mass index (BMI). Most healthcare professionals agree that men who have more than 25 percent body fat and women who have more than 30 percent body fat are obese. Women generally store more fat than men. Alternatively, the body weight classification of BMI (body weight in kilograms/height in metres squared) indicates that a value greater than 30 is classified as obese. Much time, energy, and money have been directed into the study of obesity, with the hope of reducing its prevalence in today's society and the prevalence of associated diseases.

## Measurement Techniques

Techniques used to measure body composition include hydrostatic weighing (underwater weighing), the Bod Pod (air displacement), and Dual Energy X-ray Absorptiometry (DEXA, an X-ray test). These methods are used primarily in research.

The most common method of indirectly determining the percentage body fat is skin-fold thickness—the measurement of the thickness of a layer of skin and subcutaneous fat in several parts of the body. Another method is bioelectrical impedance (BIA), which involves sending a low-voltage electrical current through the body. However, the results from these two methods are less accurate than hydrostatic weighing, the Bod Pod method, or DEXA.

One of the simplest and most accurate measures of whole-body composition is BMI. Canadian healthcare professionals often rely on BMI to diagnose obesity. BMI measurement is supported by the World Health Organization, the American College of Sports Medicine, Health Canada, and the Canadian Society for Exercise Physiology as a valid body-composition classification system and representation of total body fat. It is simple to perform because it requires only height and weight to calculate, and it is inexpensive to use.

If the waist circumference (WC, in centimetres) is included with BMI, a more comprehensive assessment of body composition or weight distribution can be achieved, and it provides a more complete assessment of the health risks. (Body mass index is considered an indicator of whole body composition, whereas WC is an indicator of central adiposity [abdominal subcutaneous and visceral fat]). Taken together, a better picture of fat weight distribution can be gathered.

The importance of WC has also been demonstrated in scientific research. Those who have a greater amount of abdominal fat (subcutaneous and visceral) are predisposed to numerous diseases (e.g., heart disease). Waist circumference measurement allows the fine-tuning of the assessment of body composition. For example, a man can have a normal BMI (e.g., 24) but have a large amount of abdominal adiposity (e.g., ≥ 102); he would thus be at an increased risk for various diseases, whereas someone of normal BMI (e.g., 24) and lower WC (e.g., 95) is at lower risk for certain diseases.

BMI can vary greatly from person to person, and thus the classification of obesity varies. It is generally divided into the following subcategories: Class I (BMI 30.0–34.9), with a high risk of developing health problems; Class II (BMI 35.0–39.9), with a very high risk of health problems; and Class III (BMI 40 or more), with an extremely high risk of health issues. (See table Percentage Distribution of BMI.)

The diseases associated with being overweight or obese include type 2 diabetes (insulin resistance), dyslipidaemia (high amount of blood lipids), hypertension (high blood pressure), coronary heart disease (blockage of coronary artery), gallbladder disease (gallstones), obstructive sleep apnea (disruption of breathing), and certain cancers (such as colon cancer). These are generally classified as cardiovascular diseases, and the mortality rates associated with these diseases increase with weight gain. Obviously, a person who is in the obese category has an increased risk for these diseases, compared to a person in the overweight category.

Obesity occurs when individuals consume more calories from food than they expend as energy. The human body needs calories to sustain life—to supply energy for physical activity. It is important for a person of normal BMI and WC to consume adequate calories to maintain body weight and energy balance. However, when individuals eat more calories than they burn, the energy balance is tipped toward weight gain. This imbalance between calories in and calories out may differ from one person to another, since such factors as genetics and environment also play a role.

## Genetic Factors

Obesity tends to run in families, which suggests a genetic association. However, families also tend to have the same or a similar diet and lifestyle habits that might contribute to obesity. Fettering out a genetic influence from other potential influences (environmental) on obesity is usually very difficult. Nonetheless, research does show a link between obesity and heredity.

The relationship between ZFP36 gene expression levels, obesity-related phenotypes, and adipokines is an example of genetics influencing obesity. (Dr. David Dyck at the University of Guelph has studied the role of adipokines in skeletal muscle insulin sensitivity.) The findings indicate that ZFP36 gene expression in omental adipose tissue (visceral adipose tissue around the liver) may offer partial protection against the development of insulin resistance and diabetes normally associated with obesity. There are also genetic disorders that predispose a person to obesity. An example of this is Prader–Willi syndrome, a rare disorder in which genes are missing on chromosome 15, resulting in hyperphagia (excessive hunger), leading to overeating and thus obesity.

### Environmental and Social Factors

Environment strongly influences obesity. The number of obese adult persons in Canada has risen significantly in the last 20 years. Thus, the adult population is gaining weight and changing their body composition in a negative fashion. Since the 1980s, the genetic makeup has not changed, but the environment has.

Environmental factors typically include physical activity (recreational pastimes, access to facilities, transportation), nutrition (portion size, restaurant selection, food selection), and social factors (poverty, education). We often eat out too frequently, consume too much fat, and are too physically inactive. Also, our environment does not always support healthy choices and habits. For example, work weeks are longer, and often both parents are working, which makes day-to-day activity more hectic and planning for nutritious meals more difficult.

In Canada, a 2015 Canadian Health Measures Survey (CMHS) indicates that 28.1 percent of adults 18 years or older had a BMI ≥ 30, and 36 percent were overweight.

The 2015 CMHS shows that the number of persons in Canada considered obese has increased dramatically since the late 1970s. Using height and weight data collected in the late 1970s during the Canada Health Survey, the age-adjusted obesity rate was 13.8 percent, which is below the 2015 figure of 28.1 percent. The increase is evident in each of the three obesity categories, particularly in Classes II and III.

### ■ Percentage Distribution of BMI, by Sex and Household Population Aged 18 and Older, Canada (Excluding Territories), 2015

| | BOTH SEXES | MEN | WOMEN |
| --- | --- | --- | --- |
| | Percent | Percent | Percent |
| Underweight | 1.7 | 0.8 | 2.6 |
| Normal Weight | 34.2 | 30.8 | 37.5 |
| Overweight (not obese) | 36.0 | 39.9 | 32.1 |
| Obese Class I | 16.7 | 18.1 | 15.4 |
| Obese Class II | 7.4 | 7.8 | 6.9 |
| Obese Class III | 4.0 | 2.6 | 5.5 |
| Overweight and Obese (BMI ≥25) | 64.1 | 68.4 | 59.9 |
| Obese (BMI ≥30) | 28.1 | 28.5 | 27.8 |

**Source:** Statistics Canada, CANSIM Table 117-0005, Distribution of the household population by adult body mass index (BMI) - Health Canada (HC) classification, by sex and age group. http://www5.statcan.gc.ca/cansim/a26?lang=eng&id=1170005&p2=33

Obesity in Canadian children has also increased. Between 1981 and 1996, the rate of obesity in children (7 to 13 years old) tripled. The rate of overweight and obesity combined among children also increased by more than 250 percent between 1981 and 2001. In 2015, for children ages 3 to 17, 14.2 percent were overweight, and 9.0 percent were obese.

Obesity rates are also different for various populations (cultures) and geographical locations within Canada. A population that has recently been investigated and potentially found to have a high prevalence of obesity is Indigenous children. A study completed by Dr. Noreen Willows from the University of Alberta examined the prevalence of overweight and obesity in Quebec Cree children (aged 9 to 12 years). It revealed that 33 percent of the children were overweight, and 38 percent were obese.

To fight this trend, health needs to be promoted in a variety of school subject areas, and social inequities need to be addressed, including the unequal access to physical activity facilities. The financial cost of obesity and physical inactivity varies depending on the types of costs included, but the cost is estimated to have been about $5.3 billion in 2001. This is a large economic burden on Canada's healthcare system and a particular burden on individual Canadians.

The **Canadian Obesity Network (CON)**—which consists of researchers, physicians, teachers, hospitals, industries, and others—facilitates access to information, people, funding, and the collective experience of a wide range of partners and individuals. CON is now one of Canada's leading obesity research, prevention, and management communities. It assists with the dissemination of information via news, resources, forums, and workshops, which makes possible a better understanding and management of obesity. Further information can be found on the CON website: http://www.obesitynetwork.ca.

### Other Factors

Other factors that are or may be associated with obesity are specific diseases, such as hypothyroidism (reduced thyroid hormone), which can lower metabolic rate and energy levels. As well, some medical drugs have been associated with obesity: for example, antidepressants, steroids, and psychiatric medications.

### Further Reading

Ceddia, R.B. (2005). Direct metabolic regulation in skeletal muscle and fat tissue by leptin: Implications for glucose and fatty acids homeostasis. *Int J Obesity*, 29(10): 1175–83.

Poirier, P., Giles, T.D., Bray, G.A., Hong, Y., Stern, J.S., Pi-Sunyer, F.X., et al. (2006). Obesity and cardiovascular disease: Pathophysiology, evaluation, and effect of weight loss. *Arterioscl Throm Vas*, 26(5): 968–76.

Tremblay, A., & Therrien, F. (2006). Physical activity and body functionality: Implications for obesity prevention and treatment. *Can J Physiol Pharm*, 84(2): 149–56.

16

Additionally, differences in the regulatory pathways for energy balance—either those governing food intake or those of energy expenditure—arise from genetic variations. Specifically, *epigenetics* are those inheritable changes in genetic expression that occur without changes in the sequencing of DNA. Individuals who demonstrate changes in DNA methylation (i.e., the addition of a methyl group) or other epigenetically related processes (e.g., covalent modifications to histones) could have greater susceptibility to changes in energy storage. Understanding epigenetics may eventually assist nutritionists and exercise specialists in devising the most appropriate diet and exercise plan based on the prediction of body-weight homeostasis (energy balance).

Despite this rather lengthy list, our knowledge about the causes and control of obesity is still limited, as evidenced by the number of people who are constantly trying to stabilize their weight at a more desirable level. This is important from more than an aesthetic viewpoint. It is known that obesity can predispose an individual to disease and premature death.

### Anorexia nervosa: An issue of control

*Clinical Note* The converse of obesity is generalized nutritional deficiency. The obvious causes for reduction of food intake below energy needs are lack of availability of food, interference with swallowing or digestion, and impairment of appetite.

One poorly understood disorder in which lack of appetite is a prominent feature is **anorexia nervosa**. Anorexia (anorexia nervosa) is an eating disorder focused on weight loss. However, the weight loss eventually becomes a form of control over one's body, and the desire to become thinner and more aesthetically pleasing becomes secondary. This means the pattern of dieting or eating restriction does not change, but the psychology of the condition changes to one of control.

People with anorexia persist with the endless cycle of eating restriction, frequently to the point of starvation, in an attempt to feel a sense of control over their body. This cycle is akin to any type of drug or substance addiction. Anorexia afflicts primarily females: Canadian statistics suggest that more than 95 percent of anorexics are female, but males are also at risk.

Anorexia typically becomes apparent near preadolescence or adolescence. In Canada, since 1987, hospitalizations for eating disorders have increased by approximately 35 percent among young women under the age of 15 years. Men represent only about 5 percent of those with anorexia. An Ontario study found that about 0.3 percent of men aged 15 to 64 years and about 2.1 percent of women had anorexia nervosa or bulimia (http://www.phac-aspc.gc.ca/publicat/human-humain06/10-eng.php). Closely related to anorexia nervosa is bulimia nervosa, an eating disorder whose root cause is also psychological. There are two forms or types of bulimia: purging and nonpurging. The purging type is the most common and involves self-induced vomiting, laxatives, and diuretics as a means of rapidly reducing the contents of the stomach. Nonpurging bulimia is less common but typically involves binge eating followed by excessive exercise or fasting to counter the calories.

Research suggests that people for whom thinness is highly important or a professional requirement (e.g., athletes, models) tend to be at risk for eating disorders in general. There is no research that outlines why some people develop anorexia and others do not. Societal demands and expectations may play a role, but are not thought to be the root of the disease. The disease likely begins with the basic and common pressure to be thin and attractive, as portrayed in magazines and on television, but it is accelerated by a poor self-image.

Research also suggests that genetics may play a role in determining a person's susceptibility to anorexia, and so attempts have been made to identify the gene(s) that might affect a person's tendency to develop this disorder. Additionally, some researchers suggest that anorexia may be related to dysfunction in the hypothalamus, which is responsible for regulating certain metabolic processes. Still others have suggested that imbalances in neurotransmitter levels in the brain may occur in people suffering from anorexia.

Treatment typically takes the form of a multidisciplinary or team-oriented approach. The team may consist of a physician (initial diagnosis), nutritionist (treatment protocol), and psychologist (cognitive behavioural therapy). The goal is often to improve body functioning (e.g., strength and immune system functioning) and treat the psychology of the condition.

---

### Check Your Understanding 16.1

1. Define external work, internal work, metabolic rate, appetite signals, satiety signals, adiposity signals, adipokines, visceral fat, and subcutaneous fat.

2. Explain how the basal metabolic rate can be determined indirectly.

3. Make a chart listing the involuntary regulatory signals on appetite, and indicate the source and effect of each (i.e., whether it increases or decreases appetite).

---

## 16.2 | Temperature Regulation

*Clinical Note* Humans are usually in environments cooler than their bodies, but we constantly generate heat internally, which helps maintain body temperature. Heat production ultimately depends on the oxidation of metabolic fuel derived from food.

Changes in body temperature in either direction alter cell activity; an increase in temperature speeds up cellular chemical reactions, whereas a fall in temperature slows down these reactions. Because cell function is sensitive to fluctuations in internal temperature, **temperature regulation** is important, and we homeostatically maintain body temperature at a level optimal for cellular metabolism to proceed in a stable fashion. Overheating is more serious than cooling. Even moderate elevations of body temperature begin to cause nerve malfunction and irreversible protein denaturation. Most people suffer convulsions when the internal body temperature reaches about 41°C; 43.3°C is considered the upper limit compatible with life.

By contrast, most of the body's tissues can transiently withstand substantial cooling. This characteristic is useful during cardiac surgery when the heart must be stopped. The patient's body temperature is deliberately lowered. The cooled tissues need less nourishment than they do at normal body temperature because of their reduced metabolic activity. However, a pronounced, prolonged fall in body temperature slows metabolism to a fatal level.

## Internal core temperature

The normal body temperature, taken orally (by mouth), has traditionally been considered 37°C. However, a recent study indicates that normal body temperature varies among individuals and varies throughout the day, ranging from 35.5°C in the morning to 37.7°C in the evening, with an overall average of 36.7°C.

Furthermore, there is no single body temperature, because the temperature varies from organ to organ. From a thermoregulatory viewpoint, the body may conveniently be viewed as a *central core* surrounded by an *outer shell*. The temperature within the inner core, which consists of the abdominal and thoracic organs, the central nervous system, and the skeletal muscles, generally remains fairly constant. This internal **core temperature** is subject to precise regulation to maintain its homeostatic constancy. The core tissues function best at a relatively constant temperature of around 37.8°C.

The skin and subcutaneous fat constitute the outer shell. In contrast to the constant high temperature in the core, the temperature within the shell is generally cooler and may vary substantially. For example, skin temperature may fluctuate between 20°C and 40°C without damage. In fact, the temperature of the skin is deliberately varied as a control measure to help maintain the core's thermal constancy.

### SITES FOR MONITORING BODY TEMPERATURE

Several easily accessible sites are used for monitoring body temperature. The oral and axillary (under the armpit) temperatures are comparable, whereas rectal temperature averages about 0.56°C higher. Also recently available is a temperature-monitoring instrument that scans the heat generated by the eardrum and converts this temperature into an oral equivalent. However, none of these measurements is an absolute indication of the internal core temperature, which is a bit higher, at 37.8°C, than the monitored sites.

### NORMAL VARIATIONS IN CORE TEMPERATURE

Even though the core temperature is held relatively constant, several factors cause it to vary slightly:

1. Most people's core temperature normally varies about 1°C during the day, with the lowest level occurring early in the morning before rising (6–7 a.m.) and the highest point occurring in late afternoon (5–7 p.m.). This variation is due to an innate biological rhythm, or biological clock (p. 240).

2. Women also experience a monthly rhythm in core temperature in connection with their menstrual cycle. The core temperature averages 0.5°C higher during the last half of the cycle, from the time of ovulation to menstruation. This mild sustained elevation in temperature during this period was once thought to be caused by the increased secretion of progesterone, one of the ovarian hormones, but this is no longer believed to be the case. The actual cause is still undetermined.

3. The core temperature increases during exercise because of the tremendous increase in heat production by the contracting muscles. During hard exercise, the core temperature may increase to as much as 40°C. In a resting person, this temperature would be considered a fever, but it is normal during strenuous exercise.

4. Because the temperature-regulating mechanisms are not 100 percent effective, the core temperature may vary slightly with exposure to extremes of temperature. For example, the core temperature may fall several degrees in cold weather or rise a degree or so in hot weather.

In these ways, the core temperature can vary at the extremes between about 35.6°C and 40°C, but it usually deviates less than a few degrees. This relative constancy is made possible by multiple thermoregulatory mechanisms coordinated by the hypothalamus.

## Heat input and heat output

The core temperature is a reflection of the body's total heat content. Heat input to the body must balance heat output to maintain a constant total heat content and thus a stable core temperature (⟩ Figure 16-3). *Heat input* occurs by way of heat gain from the external environment and from internal heat production, the latter being the most important source of heat for the body. Recall that most of the body's energy expenditure ultimately appears as heat. This heat is important in maintaining core temperature. In fact, usually more heat is generated than required to maintain body temperature at a normal level, so the excess heat must be eliminated from the body. *Heat output* occurs by way of heat loss from exposed body surfaces to the external environment.

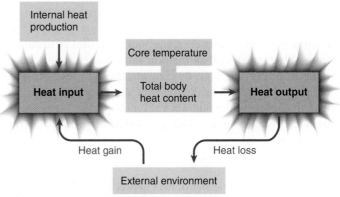

⟩ **Figure 16-3 Heat input and output**

Energy Balance and Temperature Regulation **711**

Balance between heat input and output is frequently disturbed by (1) changes in internal heat production for purposes unrelated to regulation of body temperature, most notably by exercise, which markedly increases heat production; and (2) changes in the external environmental temperature that influence the degree of heat gain or heat loss that occurs between the body and its surroundings. Compensatory adjustments must take place in heat-loss and heat-gain mechanisms to maintain body temperature within narrow limits despite changes in metabolic heat production and changes in environmental temperature. If the core temperature starts to fall, heat production is increased and heat loss is minimized so that normal temperature can be restored. Conversely, if the temperature starts to rise above normal, core temperature can be corrected by increasing heat loss while simultaneously reducing heat production.

We now elaborate on the means by which heat gains and losses can be adjusted to maintain body temperature.

## Heat exchange: Radiation, conduction, convection, and evaporation

All heat loss or heat gain between the body and external environment must take place between the body surface and its surroundings. The same physical laws of nature that govern heat transfer between inanimate objects also control the transfer of heat between the body surface and environment. The temperature of an object may be thought of as a measure of the concentration of heat within the object. Accordingly, heat always moves down its concentration gradient, that is, down a **thermal gradient** from a warmer to a cooler region (*thermo* means "heat").

The body uses four mechanisms of heat transfer: *radiation, conduction, convection*, and *evaporation*.

### RADIATION

**Radiation** is the emission of heat energy from the surface of a warm body in the form of **electromagnetic waves (heat waves)**, which travel through space (\> Figure 16-4a). When radiant energy strikes an object and is absorbed, the energy of the wave motion is transformed into heat within the object. The human body both emits (source of heat loss) and absorbs (source of heat gain) radiant energy. Whether the body loses or gains heat by radiation depends on the difference in temperature between the skin surface and the surfaces of other objects in the body's environment. Because net transfer of heat by radiation is always from warmer objects to cooler ones, the body gains heat by radiation from objects warmer than the skin surface, such as the sun, a radiator, or burning logs. By contrast, the body loses heat by radiation to objects in its environment whose surfaces are cooler than the surface of the skin, such as building walls, furniture, or trees. On average, humans lose close to half of their heat energy through radiation.

### CONDUCTION

**Conduction** is the transfer of heat between objects of differing temperatures that are in direct contact with each other, with heat moving down its thermal gradient from the warmer to the cooler object by being transferred from molecule to molecule. All molecules are constantly in vibratory motion, with warmer molecules moving faster than cooler ones. When molecules of differing heat content touch each other, the faster-moving, warmer molecule agitates the cooler molecule into more rapid motion, thereby warming up the cooler molecule. During this process, the original warmer molecule loses some of its thermal energy as it slows down and cools off a bit. Given enough time, therefore, the temperature of the two touching objects eventually equalizes.

The rate of heat transfer by conduction depends on the *temperature difference* between the touching objects and the *thermal conductivity* of the substances involved (i.e., how easily heat is conducted by the molecules of the substances). Heat can be lost or gained by conduction when the skin is in contact with a good conductor (\> Figure 16-4b). When you hold a snowball, for example, your hand becomes cold because heat moves by conduction from your hand to the snowball. Conversely, when you apply a heating pad to a body part, the part is warmed up as heat is transferred directly from the pad to the body.

Similarly, you either lose or gain heat by conduction to the layer of air in direct contact with your body. The direction of heat transfer depends on whether the air is cooler or warmer, respectively, than your skin. Only a small percentage of total heat exchange between the skin and environment takes place by conduction alone, however, because air is not a very good conductor of heat. (For this reason, swimming pool water at 26.7°C feels cooler than air at the same temperature; heat is conducted more rapidly from the body surface into the water, which is a good conductor, than into the air, which is a poor conductor.)

### CONVECTION

The term *convection* refers to the transfer of heat energy by *air* (or $H_2O$) *currents*. As the body loses heat by conduction to the surrounding cooler air, the air in immediate contact with the skin is warmed. Because warm air is lighter (less dense) than cool air, the warmed air rises while cooler air moves in next to the skin to replace the vacating warm air. The process is then repeated (\> Figure 16-4c). These air movements, known as *convection currents*, help carry heat away from the body. If it were not for convection currents, no further heat could be dissipated from the skin by conduction once the temperature of the layer of air immediately around the body equilibrated with skin temperature.

The combined conduction–convection process of dissipating heat from the body is enhanced by forced movement of air across the body surface, either by external air movements, such as those caused by the wind or a fan, or by movement of the body through the air, as during bicycle riding. Because forced air movement sweeps away the air warmed by conduction and replaces it with cooler air more rapidly, a greater total amount of heat can be carried away from the body over a given time period. Thus, wind makes us feel cooler on hot days, and windy days in the winter are more chilling than calm days at the same cold temperature. For this reason, weather forecasters have developed the concept of *wind chill factor*.

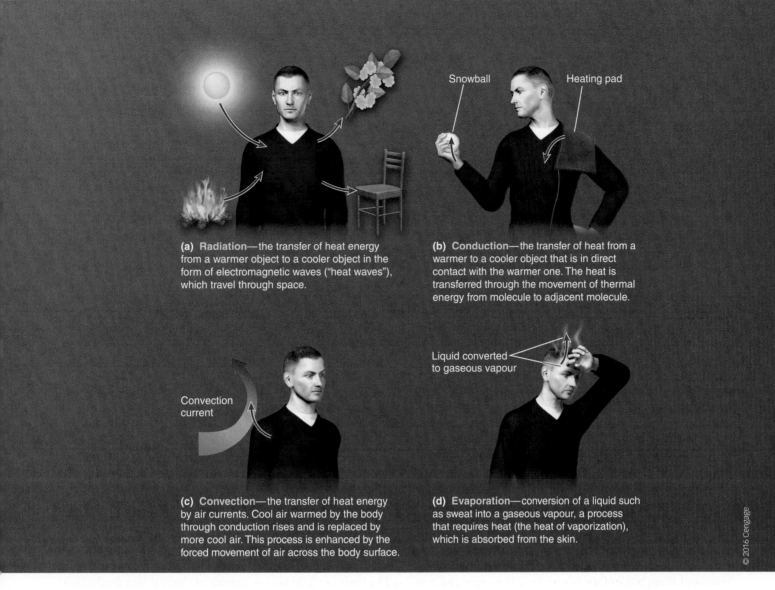

(a) **Radiation**—the transfer of heat energy from a warmer object to a cooler object in the form of electromagnetic waves ("heat waves"), which travel through space.

(b) **Conduction**—the transfer of heat from a warmer to a cooler object that is in direct contact with the warmer one. The heat is transferred through the movement of thermal energy from molecule to adjacent molecule.

(c) **Convection**—the transfer of heat energy by air currents. Cool air warmed by the body through conduction rises and is replaced by more cool air. This process is enhanced by the forced movement of air across the body surface.

(d) **Evaporation**—conversion of a liquid such as sweat into a gaseous vapour, a process that requires heat (the heat of vaporization), which is absorbed from the skin.

© 2016 Cengage

> Figure 16-4 **Mechanisms of heat transfer**

## EVAPORATION

**Evaporation** is the final method of heat transfer used by the body. When water evaporates from the skin surface, the heat required to transform water from a liquid to a gaseous state is absorbed from the skin, thereby cooling the body (> Figure 16-4d). Evaporative heat loss is what makes you feel cooler when your bathing suit is wet than when it is dry. Evaporative heat loss occurs continually from the linings of the respiratory airways and from the surface of the skin. Heat is continuously lost through the $H_2O$ vapour in the expired air as a result of the air's humidification during its passage through the respiratory system. Similarly, because the skin is not completely waterproof, $H_2O$ molecules constantly diffuse through the skin and evaporate. This ongoing evaporation from the skin is completely unrelated to the sweat glands. These passive evaporative heat-loss processes are not subject to physiological control and go on even in very cold weather, when the problem is one of conserving body heat.

**Sweating** is an active evaporative heat-loss process under sympathetic nervous control. The rate of evaporative heat loss can be deliberately adjusted by varying the extent of sweating, which is an important homeostatic mechanism to eliminate excess heat as needed. For example, see Why It Matters, which illustrates the importance of sweating. In fact, when the environmental temperature exceeds the skin temperature, sweating is the only avenue for heat loss, because the body is gaining heat by radiation and conduction under these circumstances.

**Sweat** is a dilute salt solution actively extruded to the surface of the skin by sweat glands dispersed all over the body. The sweat glands can produce up to 4 L of sweat per hour. Sweat must be evaporated from the skin for heat loss to occur. If sweat merely drips from the surface of the skin or is wiped away, no heat loss is accomplished. The most important factor determining the extent of evaporation of sweat is the *relative humidity* of the surrounding air (the percentage of $H_2O$ vapour actually present in the air compared with the greatest amount that the air can possibly hold at that temperature; e.g., a relative humidity of 70 percent means the air contains 70 percent of the $H_2O$ vapour it is capable of holding). When the relative humidity is high, the

**16**

One of the most effective heat-loss mechanisms for the purpose of thermoregulation in a warm to hot environment is evaporation from sweating. For the sweating mechanism to work effectively, you must acclimate to the new warmer environment. This takes a little time. For example, the average sweat rate for a nonacclimated person in a warmer environment is about 1.5 L/h. Following 10 days of acclimation, the person's sweat rate increases to approximately 3 L/h, and in the following six weeks may achieve a sweat rate of 3.75 L/h. The change in sweat rate associated with acclimation is one reason why it is a good idea to arrive early when you travel to a warmer environment—your physiological systems need the time to adjust. This would be good advice for Canadians travelling to warm countries during our winter months,

or for Canadian athletes travelling to warmer countries to compete in outdoor events. Note that if you decide to travel early to a warm environment for the purpose of acclimation, the next step you must take is to properly hydrate so that the sweating mechanism can work effectively. Without sufficient water intake, the sweating mechanism does not function efficiently, and your body will begin to gain and store unneeded heat; this impairs normal physiological functions. The general rule of thumb during moderate intensity work (e.g., labourer) or exercise (e.g., golfing) in a warm to hot environment is 150–350 mL of water every 15–20 minutes, and following the work or exercise about 450–675 mL for every 0.5 kg decrease in body weight. So those heading south for a winter getaway golfing trip need to acclimate and hydrate.

air is already almost fully saturated with $H_2O$, so it has limited ability to take up additional moisture from the skin. Thus, little evaporative heat loss can occur on hot, humid days. The sweat glands continue to secrete, but the sweat simply remains on the skin or drips off instead of evaporating and producing a cooling effect. As a measure of the discomfort associated with combined heat and high humidity, meteorologists have devised the *temperature–humidity index*.

### The hypothalamus and thermosensory inputs

The hypothalamus serves as the body's thermostat. The home thermostat keeps track of the temperature in a room and triggers a heating mechanism (the furnace) or a cooling mechanism (the air conditioner) as necessary to maintain the room temperature at the indicated setting. Similarly, the hypothalamus, as the body's thermoregulatory integrating centre, receives afferent information about the temperature in various regions of the body and initiates extremely complex, coordinated adjustments in heat-gain and heat-loss mechanisms as necessary to correct any deviations in core temperature from the normal setting. The hypothalamus is far more sensitive than the home thermostat, however, because it can respond to changes in blood temperature as small as 0.01°C. The degree of response to deviations in body temperature is finely matched so that precisely enough heat is lost or generated to restore the temperature to normal.

To appropriately adjust the delicate balance between the heat-loss mechanisms and the opposing heat-producing and heat-conserving mechanisms, the hypothalamus must be continuously informed of both the core and the skin temperature. This is accomplished by specialized temperature-sensitive receptors called thermoreceptors. The core temperature is monitored by *central thermoreceptors*, which are located within the hypothalamus and elsewhere in the central nervous system and the abdominal organs. *Peripheral thermoreceptors* monitor skin

temperature throughout the body and transmit information about changes in surface temperature to the hypothalamus.

Two centres for temperature regulation have been identified in the hypothalamus. The *posterior region* is activated by cold and subsequently triggers reflexes that mediate heat production and heat conservation. The *anterior region*, which is activated by warmth, initiates reflexes that mediate heat loss. Let's examine the means by which the hypothalamus fulfills its thermoregulatory functions.

### Shivering

The body can gain heat as a result of internal heat production generated by metabolic activity or from the external environment if the latter is warmer than body temperature. Because body temperature usually is higher than environmental temperature, metabolic heat production is the primary source of body heat. In a resting person, most body heat is produced by the thoracic and abdominal organs as a result of ongoing, homeostatic metabolic activities. Beyond this basal level, the rate of metabolic heat production can be variably increased primarily by changes in skeletal muscle activity or, to a lesser extent, by certain hormonal actions. Therefore, changes in skeletal muscle activity constitute the primary method of heat gain to assist temperature regulation.

#### ADJUSTMENTS IN HEAT PRODUCTION BY SKELETAL MUSCLES

In response to a fall in core temperature caused by exposure to cold, the hypothalamus takes advantage of the fact that an increase in skeletal muscle activity generates additional heat. Acting through descending pathways that terminate on the motor neurons controlling the skeletal muscles, the hypothalamus first gradually increases skeletal muscle tone (tension within the muscle). Soon shivering begins. **Shivering** consists of rhythmic, oscillating skeletal muscle contractions and relaxations at a rate of 10–20 per second. This mechanism is very effective in increasing heat production; all the energy liberated during these muscle tremors is converted to heat because no external work is accomplished. Within a matter of seconds to minutes, internal heat production may increase two- to fivefold as a result of shivering.

Frequently, these reflex changes in skeletal muscle activity are augmented by increased voluntary, heat-producing actions, such as bouncing up and down or hand clapping. Such behavioural responses appear to share neural systems in common with the involuntary physiological responses. The hypothalamus and limbic system are extensively involved with controlling motivated behaviour (p. 113).

In the opposite situation—a rise in core temperature caused by heat exposure—two mechanisms reduce heat-producing skeletal muscle activity: muscle tone is reflexly reduced, and voluntary movement is curtailed. When the air becomes very warm, people often complain it is "too hot to even move." These responses are not as effective at reducing heat production during heat exposure as are the muscular responses that increase heat production during cold exposure, for two reasons. First, because muscle tone is normally quite low, the capacity to reduce it further is limited. Second, the elevated body temperature tends to increase the rate of metabolic heat production because the temperature has a direct effect on the rate of chemical reactions.

### NONSHIVERING THERMOGENESIS

Although reflex and voluntary changes in muscle activity are the major means of increasing the rate of heat production, **nonshivering (chemical) thermogenesis** also plays a role in thermoregulation. In most experimental animals, chronic cold exposure brings about an increase in metabolic heat production that is independent of muscle contraction—that is, the increase is brought about by changes in heat-generating chemical activity. In humans, nonshivering thermogenesis is most important in newborns, because they lack the ability to shiver. Nonshivering thermogenesis is mediated by the hormones epinephrine and thyroid hormone, both of which increase heat production by stimulating fat metabolism. Newborns have deposits of a special type of adipose tissue known as *brown fat*, which is especially capable of converting chemical energy into heat. The role of nonshivering thermogenesis in adults remains controversial.

Having examined the mechanisms for adjusting heat production, we now turn to the other side of the equation—adjustments in heat loss.

## Heat loss

Heat-loss mechanisms are also subject to control, again largely by the hypothalamus. When we are hot, we want to increase heat loss to the environment; when we are cold, we want to decrease heat loss. The amount of heat lost to the environment by radiation and conduction–convection is largely determined by the temperature gradient between the skin and external environment. The body's central core is a heat-generating chamber in which the temperature must be maintained at approximately 37.8°C. Surrounding the core is an insulating shell through which heat exchanges between the body and external environment take place. To maintain a constant core temperature, the insulative capacity and temperature of the shell can be adjusted to vary the temperature gradient between the skin and external environment, thereby influencing the extent of heat loss.

The insulative capacity of the shell can be varied by controlling the amount of blood flowing through the skin. Skin blood flow serves two functions. First, it provides a nutritive blood supply to the skin. Second, as blood is pumped to the skin from the heart, it has been heated in the central core and carries this heat to the skin. Most skin blood flow is for the function of temperature regulation; at normal room temperature, 20–30 times more blood flows through the skin than is needed to meet the skin's nutritional needs.

In the process of thermoregulation, skin blood flow can vary tremendously, from 400 mL/min up to 2500 mL/min. The more blood that reaches the skin from the warm core, the closer the skin's temperature is to the core temperature. The skin's blood vessels diminish the effectiveness of the skin as an insulator by carrying heat to the surface, where it can be lost from the body by radiation and conduction–convection. Accordingly, vasodilation of the skin vessels (specifically, the arterioles), which permits increased flow of heated blood through the skin, increases heat loss. Conversely, vasoconstriction of the skin vessels, which reduces skin blood flow, decreases heat loss by keeping the warm blood in the central core, where it is insulated from the external environment. Cold, relatively bloodless skin provides excellent insulation between the core and the environment. However, the skin is not a perfect insulator, even with maximum vasoconstriction. Despite minimal blood flow to the skin, some heat can still be transferred by conduction from the deeper organs to the skin surface and then can be lost from the skin to the environment.

These skin vasomotor responses are coordinated by the hypothalamus by means of sympathetic nervous system output. Increased sympathetic activity to the skin vessels produces heat-conserving vasoconstriction in response to cold exposure, whereas decreased sympathetic activity produces heat-losing vasodilation of the skin vessels in response to heat exposure.

Recall that the cardiovascular control centre in the medulla oblongata also exerts control over the skin arterioles (as well as arterioles throughout the body) by means of adjusting sympathetic activity to these vessels for the purpose of blood pressure regulation (p. 400). Hypothalamic control over the skin arterioles for the purpose of temperature regulation takes precedence over the cardiovascular control centre's control of these same vessels (p. 402). Thus, changes in blood pressure can result from pronounced thermoregulatory skin vasomotor responses. For example, blood pressure can fall on exposure to a very hot environment, because the skin vasodilator response set in motion by the hypothalamic thermoregulatory centre overrides the skin vasoconstrictor response called forth by the medullary cardiovascular control centre.

16

## The hypothalamus, heat production, and heat loss

Let's now pull together the coordinated adjustments in heat production and heat loss in response to exposure to either a cold or a hot environment (Table 16-3).

### COORDINATED RESPONSES TO COLD EXPOSURE

In response to cold exposure, the posterior region of the hypothalamus directs the increase in heat production, such as by shivering and, simultaneously, the decrease in heat loss (i.e., conserving heat) by skin vasoconstriction and other measures.

Because there is a limit to the body's ability to reduce skin temperature through vasoconstriction, even maximum vasoconstriction is insufficient to prevent excessive heat loss when the external temperature falls extremely low. Accordingly, other measures are needed to further reduce heat loss. In animals with dense fur or feathers, the hypothalamus, acting through the sympathetic nervous system, brings about contraction of the tiny muscles at the base of the hair or feather shafts to lift the hair or feathers off the skin surface. This puffing up traps a layer of poorly conductive air between the skin surface and the environment, which increases the insulating barrier between the core and the cold air, thereby reducing heat loss. Even though the hair-shaft muscles contract in humans in response to cold exposure, this heat-retention mechanism is ineffective because of the low density and fine texture of most human body hair. The result, instead, is *goosebumps*.

After maximum skin vasoconstriction has been achieved as a result of exposure to cold, further heat dissipation in humans can be prevented only by behavioural adaptations, such as postural changes that reduce as much as possible the exposed surface area from which heat can escape. These postural changes include manoeuvres such as hunching over, clasping the arms in front of the chest, or curling up in a ball.

Putting on warmer clothing further insulates the body from too much heat loss. Clothing entraps layers of poorly conductive air between the skin surface and the environment, thereby diminishing loss of heat by conduction from the skin to the cold external air and curtailing the flow of convection currents.

### COORDINATED RESPONSES TO HEAT EXPOSURE

Under the opposite circumstance—heat exposure—the anterior part of the hypothalamus reduces heat production by decreasing skeletal muscle activity and promotes increased heat loss by inducing skin vasodilation. When even maximal skin vasodilation is inadequate to rid the body of excess heat, sweating is brought into play to accomplish further heat loss through evaporation. In fact, if the air temperature rises above the temperature of maximally vasodilated skin, the temperature gradient reverses itself so that heat is gained from the environment. Sweating is the only means of heat loss under these conditions.

We also employ voluntary measures, such as using fans, wetting the body, drinking cold beverages, and wearing cool clothing, to further enhance heat loss. Contrary to popular belief, wearing light-coloured, loose clothing is cooler than being nude. Naked skin absorbs almost all the radiant energy that strikes it, whereas light-coloured clothing reflects almost all the radiant energy that falls on it. So it is actually cooler to wear light-coloured clothing that is loose and thin enough to permit convection currents and evaporative heat loss to occur, rather than wearing no clothes.

### THERMONEUTRAL ZONE

Skin vasomotor activity is highly effective in controlling heat loss in environmental temperatures between 20°C and 29°C. Within this range—known as the **thermoneutral zone**—the core temperature can be kept constant by vasomotor responses, without the need for supplementary heat-production or heat-loss mechanisms to come into play. When external air temperature falls below the lower limits of the ability of skin vasoconstriction to reduce heat loss further, the major burden of maintaining core temperature is borne by increased heat production, especially shivering. At the other extreme, when external air temperature exceeds the upper limits of the ability of skin vasodilation to increase heat loss further, sweating becomes the dominant factor in maintaining core temperature.

### ▌ TABLE 16-3 Coordinated Adjustments in Response to Cold or Heat Exposure

| IN RESPONSE TO COLD EXPOSURE (COORDINATED BY THE POSTERIOR HYPOTHALAMUS) | | IN RESPONSE TO HEAT EXPOSURE (COORDINATED BY THE ANTERIOR HYPOTHALAMUS) | |
|---|---|---|---|
| **Increased Heat Production** | **Decreased Heat Loss (heat conservation)** | **Decreased Heat Production** | **Increased Heat Loss** |
| Increased muscle tone | Skin vasoconstriction | Decreased muscle tone | Skin vasodilation |
| Shivering Increased voluntary exercise* | Postural changes to reduce exposed surface area (hunching shoulders, etc.)* | Decreased voluntary exercise* | Sweating Cool clothing* |
| Nonshivering thermogenesis | Warm clothing* | | |

*Behavioural adaptations

© 2016 Cengage

## FEVER

*Clinical Note* The term **fever** refers to an elevation in body temperature as a result of infection or inflammation. In response to microbial invasion, certain phagocytic cells (macrophages) release a chemical known as endogenous pyrogen, which, among its many infection-fighting effects (p. 470), acts on the hypothalamic thermoregulatory centre to raise the setting of the body's thermostat (> Figure 16-5). The hypothalamus now maintains the temperature at the new set level, instead of maintaining normal body temperature. If, for example, endogenous pyrogen raises the set point to 38.9°C, the hypothalamus senses that the normal prefever temperature is too cold, so it initiates the cold-response mechanisms to raise the temperature to 38.9°C. Specifically, the hypothalamus initiates shivering to rapidly increase heat production, and promotes skin vasoconstriction to rapidly reduce heat loss, both of which drive the temperature upward. These events account for the sudden cold chills often experienced at the onset of a fever. Feeling cold, the person may put on more blankets as a voluntary mechanism that helps raise body temperature by conserving body heat. Once the new temperature is achieved, body temperature is regulated as normal in response to cold and heat but at a higher setting. Fever production in response to an infection is therefore a deliberate outcome and is not due to a breakdown of thermoregulatory mechanisms. Although the physiological significance of a fever is still unclear, many medical experts believe a rise in body temperature has a beneficial role in fighting infection. A fever augments the inflammatory response and may interfere with bacterial multiplication.

During fever production, endogenous pyrogen raises the set point of the hypothalamic thermostat by triggering the local release of *prostaglandins*, which are local chemical mediators that act directly on the hypothalamus. Aspirin reduces a fever by inhibiting the synthesis of prostaglandins. Aspirin does not lower the temperature in a person without a fever, because in the absence of endogenous pyrogen, prostaglandins are not present in the hypothalamus in appreciable quantities.

The exact molecular cause of a fever "breaking" naturally is unknown, although it presumably results from reduced pyrogen release or decreased prostaglandin synthesis. When the hypothalamic set point is restored to normal, the temperature at 38.9°C (in this example) is too high. The heat-response mechanisms cool down the body. Skin vasodilation occurs, and sweating commences. The person feels hot and throws off extra covers. The gearing up of these heat-loss mechanisms by the hypothalamus reduces the temperature to normal.

### Hyperthermia

**Hyperthermia** denotes any elevation in body temperature above the normally accepted range. The term *fever* is usually reserved for an elevation in temperature caused by the release of endogenous pyrogen, which resets the hypothalamic set point during infection or inflammation; *hyperthermia* refers to all other imbalances between heat gain and heat loss that increase body temperature. Hyperthermia has a variety of causes, some of which are normal and harmless, others pathological and fatal.

#### EXERCISE-INDUCED HYPERTHERMIA

The most common cause of hyperthermia is sustained exercise. As a physical consequence of the tremendous heat load generated by exercising muscles, body temperature rises during the initial stage of exercise because heat gain exceeds heat loss (> Figure 16-6). The elevation in core temperature reflexly triggers heat-loss mechanisms (skin vasodilation and sweating), which eliminate the discrepancy between heat production and

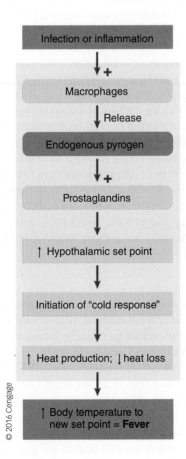

> Figure 16-5 **Fever production**

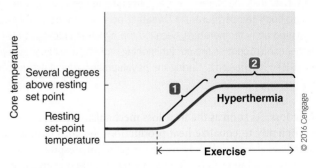

**1** At the onset of exercise, the rate of heat production initially exceeds the rate of heat loss, so the core temperature rises.

**2** When heat loss mechanisms are reflexly increased sufficiently to equalize the elevated heat production, the core temperature stabilizes slightly above the resting point for the duration of the exercise.

> Figure 16-6 **Hyperthermia in sustained exercise**

16

# The Extremes of Heat and Cold Can Be Fatal

PROLONGED EXPOSURE TO TEMPERATURE EXTREMES in either direction can overtax the body's thermoregulatory mechanisms, leading to disorders and even death.

## Heat-Related Disorders

**Heat exhaustion**, a state of collapse usually manifested by fainting, is caused by reduced blood pressure due to overtaxing the heat-loss mechanisms. Extensive sweating reduces cardiac output by depleting the plasma volume, and pronounced skin vasodilation causes a drop in total peripheral resistance. Because blood pressure is determined by cardiac output multiplied by total peripheral resistance, the following occurs: blood pressure falls, an insufficient amount of blood is delivered to the brain, and fainting takes place. Heat exhaustion is a consequence of overactivity of the heat-loss mechanisms rather than a breakdown of these mechanisms. Because the heat-loss mechanisms have been very active, body temperature is only mildly elevated in heat exhaustion. By forcing the cessation of activity when the heat-loss mechanisms are no longer able to cope with heat gain through exercise or a hot environment, heat exhaustion serves as a safety valve to help prevent the more serious consequences of heatstroke.

**Heatstroke** is an extremely dangerous situation that arises from the complete breakdown of the hypothalamic thermoregulatory systems. Heat exhaustion may progress into heatstroke if the heat-loss mechanisms continue to be overtaxed. Heatstroke is more likely to occur on overexertion during a prolonged exposure to a hot, humid environment. The elderly—in whom thermoregulatory responses are generally slower and less efficient—are particularly vulnerable to heatstroke during prolonged, stifling heat waves. So too are individuals who are taking certain common tranquilizers, such as Valium, because these drugs interfere with the hypothalamic thermoregulatory centres' neurotransmitter activity.

The most striking feature of heatstroke is a lack of compensatory heat-loss measures, such as sweating, in the face of a rapidly rising body temperature. No sweating occurs, despite a markedly elevated body temperature, because the hypothalamic thermoregulatory control centres are not functioning properly and cannot initiate heat-loss mechanisms. During the development of heatstroke, body temperature starts to climb because the heat-loss mechanisms are eventually overwhelmed by prolonged, excessive heat gain. Once the core temperature reaches the point at which the hypothalamic temperature-control centres are damaged by the heat, the body temperature rapidly rises even higher due to the complete shutdown of heat-loss mechanisms. Furthermore, as body temperature increases, the rate of metabolism increases correspondingly. This is because higher temperatures speed up the rate of all chemical reactions, resulting in even greater heat production. This positive-feedback state sends the temperature spiralling upward. Heatstroke is a very dangerous situation that is rapidly fatal if untreated. Even with treatment to halt and reverse the rampant rise in body temperature, the mortality rate is still high. The rate of permanent disability in survivors is also high because of irreversible protein denaturation caused by the high internal heat.

## Cold-Related Disorders

At the other extreme, the body can be harmed by cold exposure in two ways: frostbite and generalized hypothermia. **Frostbite** involves excessive cooling of a particular part of the body to the point where tissue in that area is damaged. When exposed tissues actually freeze, the tissue damage results from cell disruption caused by the formation of ice crystals or by a lack of liquid water.

**Hypothermia**, a fall in body temperature, occurs when generalized cooling of the body exceeds the ability of the normal heat-producing and heat-conserving regulatory mechanisms to match the excessive heat loss. As hypothermia sets in, the rate of all metabolic processes slows down due to the declining temperature. Higher cerebral functions are the first affected by body cooling, leading to loss of judgment, apathy, disorientation, and tiredness, all of which diminish the cold victim's ability to initiate voluntary mechanisms to reverse the falling body temperature. As body temperature continues to plummet, depression of the respiratory centre occurs, reducing the ventilatory drive so that breathing becomes slow and weak. Activity of the cardiovascular system is also gradually reduced. The heart is slowed and cardiac output decreased. Cardiac rhythm is disturbed, eventually leading to ventricular fibrillation and death.

heat loss. As soon as the heat-loss mechanisms have stepped up sufficiently to equalize heat production, the core temperature stabilizes at a level slightly above the set point, despite continued heat-producing exercise. Thus, during sustained exercise, body temperature initially rises and then is maintained at the higher level as long as the exercise continues.

## PATHOLOGICAL HYPERTHERMIA

 *Clinical Note* Hyperthermia can also be brought about in a completely different way: excessive heat production in connection with abnormally high circulating levels of thyroid hormone or epinephrine that result from dysfunctions of the thyroid gland or adrenal medulla, respectively. Both these hormones elevate the core temperature by increasing the overall rate of metabolic activity and heat production.

Hyperthermia can also result from malfunction of the hypothalamic control centres. Certain brain lesions, for example, destroy the normal regulatory capacity of the hypothalamic thermostat. When the thermoregulatory mechanisms are not functional, lethal hyperthermia may occur very rapidly. Normal metabolism produces enough heat to kill a person in less than five hours if the heat-loss mechanisms are completely shut down.

In addition to causing brain lesions, exposure to severe, prolonged heat stress can also break down the function of hypothalamic thermoregulation. Similarly, the body can be harmed by extreme cold exposure. (For a discussion of the effects of extreme heat or cold exposure, see Concepts, Challenges, and Controversies, p. 718.)

## Check Your Understanding 16.2

1. Describe the mechanisms of heat transfer.
2. Make a chart comparing the responses initiated by the posterior hypothalamus and also the anterior hypothalamus to maintain core body temperature when the environmental temperature becomes hot or cold.

# Chapter in Perspective: Focus on Homeostasis

Energy can be neither created nor destroyed. For body weight and body temperature to remain constant, input must equal output in the case of the body's total energy balance and its heat energy balance, respectively. If total energy input exceeds total energy output, the extra energy is stored in the body, and body weight increases. Similarly, if the input of heat energy exceeds its output, body temperature increases. Conversely, if output exceeds input, body weight decreases or body temperature falls. The hypothalamus is the major integrating centre for maintaining both a constant total energy balance (therefore a constant body weight) and a constant heat energy balance (therefore a constant body temperature).

Body temperature, which is one of the homeostatically regulated factors of the internal environment, must be maintained within narrow limits, because the structure and reactivity of the chemicals that compose the body are temperature sensitive. Deviations in body temperature outside a limited range result in protein denaturation and death of the individual if the temperature rises too high, or metabolic slowing and death if the temperature falls too low.

Body weight, in contrast, varies widely among individuals. Only the extremes of imbalances between total energy input and output become incompatible with life. For example, in the face of insufficient energy input in the form of ingested food during prolonged starvation, the body resorts to breaking down muscle protein to meet its needs for energy expenditure after the adipose stores are depleted. Body weight dwindles because of this self-cannibalistic mechanism, until death finally occurs as a result of loss of heart muscle, among other things. At the other extreme, when the food energy input greatly exceeds the energy expended, the extra energy is stored as adipose tissue, and body weight increases. The resultant gross obesity can also lead to heart failure. Not only must the heart work harder to pump blood to the excess adipose tissue but obesity also predisposes the person to atherosclerosis and heart attacks (p. 374).

## CHAPTER TERMINOLOGY

arcuate nucleus (p. 703)
adipocytes (p. 703)
adipokines (p. 703)
adiposity signals (p. 704)
anorexia nervosa (p. 701)
basal metabolic rate (BMR) (p. 701)
brown fat (p. 702)
Canadian Obesity Network (CON) (p. 709)
calorie (p. 700)
conduction (p. 712)
core temperature (p. 711)
corticotropin-releasing hormone (p. 705)
diet-induced thermogenesis (p. 701)
direct calorimetry (p. 701)
electromagnetic waves (heat waves) (p. 712)
energy balance (p. 700)
energy equivalent of $O_2$ (p. 701)
evaporation (p. 713)
external work (p. 700)
feeding (appetite) signals (p. 703)

fever (p. 717)
first law of thermodynamics (p. 700)
frostbite (p. 718)
ghrelin (p. 705)
heat (p. 700)
heat exhaustion (p. 718)
heatstroke (p. 718)
hunger (p. 703)
hyperthermia (p. 717)
hypothermia (p. 718)
indirect calorimetry (p. 701)
insulin (p. 704)
internal work (p. 700)
kilocalorie (kcal) or Calorie (C) (p. 700)
lateral hypothalamic area (LHA) (p. 705)
leptin (p. 703)
Melanocortins (p. 703)
metabolic rate (p. 700)
nucleus tractus solitarius (NTS) (p. 705)

neuropeptide Y (NPY) (p. 703)
nonexercise activity thermogenesis (NEAT) (p. 707)
nonshivering (chemical) thermogenesis (p. 715)
obesity (p. 706)
orexins (p. 705)
paraventricular hypothalamic nucleus (PVN) (p. 705)
peptide $YY_{3-36}$ ($PYY_{3-36}$) (p. 705)
radiation (p. 712)
satiety (p. 703)
shivering (p. 714)
sweat (p. 713)
sweating (p. 713)
temperature regulation (p. 710)
thermal energy (p. 700)
thermal gradient (p. 712)
thermoneutral zone (p. 716)
uncoupling proteins (p. 707)

## Objective Questions
## (Answers in Appendix E, p. A-53)

1. If more food energy is consumed than is expended, the excess energy is lost as heat. *(True or false?)*

2. All the energy within nutrient molecules can be harnessed to perform biological work. *(True or false?)*

3. Each litre of $O_2$ contains 4.8 kcal of heat energy. *(True or false?)*

4. A body temperature greater than 36.8°C is always indicative of a fever. *(True or false?)*

5. Core temperature is relatively constant, but skin temperature can vary markedly. *(True or false?)*

6. Sweat that drips off the body has no cooling effect. *(True or false?)*

7. Production of goosebumps in response to cold exposure has no value in regulating body temperature. *(True or false?)*

8. The posterior region of the hypothalamus triggers shivering and skin vasoconstriction. *(True or false?)*

9. The _____ of the hypothalamus contains two populations of neurons, one that secretes appetite-enhancing NPY and another that secretes appetite-suppressing melanocortins.

10. The primary means of involuntarily increasing heat production is _____.

11. Increased heat production independent of muscle contraction is known as _____.

12. The only means of heat loss when the environmental temperature exceeds the core temperature is _____.

13. Which statement concerning heat exchange between the body and the external environment is NOT correct?
   a. Heat gain occurs primarily by means of internal heat production.
   b. Radiation serves as a means of heat gain, but not of heat loss.
   c. Heat energy always moves down its concentration gradient from warmer to cooler objects.
   d. The temperature gradient between the skin and the external air is subject to control.
   e. Very little heat is lost from the body by conduction alone.

14. Which statement concerning fever production is NOT correct?
   a. Endogenous pyrogen is released by macrophages in response to microbial invasion.
   b. The hypothalamic set point is elevated.
   c. The hypothalamus initiates cold-response mechanisms to increase the body temperature.
   d. Prostaglandins mediate the effect.
   e. The hypothalamus is not effective in regulating body temperature during a fever.

15. Using the answer code on the right, indicate which mechanism of heat transfer is being described:

| | | |
|---|---|---|
| _____ 1. sitting on a cold metal chair | (a) | radiation |
| _____ 2. sunbathing on the beach | (b) | conduction |
| _____ 3. being in a gentle breeze | (c) | convection |
| _____ 4. sitting in front of a fireplace | (d) | evaporation |
| _____ 5. sweating | | |
| _____ 6. riding in a car with the windows open | | |
| _____ 7. lying under an electric blanket | | |
| _____ 8. sitting in a wet bathing suit | | |
| _____ 9. fanning yourself | | |
| _____ 10. immersing yourself in cold water | | |

## Written Questions

1. Differentiate between external and internal work.

2. Define *metabolic rate* and *basal metabolic rate*. Explain the process of indirect calorimetry.

3. Describe the three states of energy balance.

4. By what means is energy balance primarily maintained?

5. List the sources of heat input and output for the body.

6. Describe the source and role of the following in the long-term regulation of energy balance and the short-term control of the timing and size of meals: neuropeptide Y, melanocortins, leptin, insulin, ghrelin, $PYY_{3-36}$, orexins, corticotropin-releasing hormone, and cholecystokinin.

7. Discuss the compensatory measures that occur in response to a fall in core temperature as a result of cold exposure and in response to a rise in core temperature as a result of heat exposure.

## Quantitative Exercise
## (Solutions in Appendix E, p. A-54)

1. The basal metabolic rate (BMR) is a measure of how much energy the body consumes to maintain its rate at rest. The normal BMR for men is about 70 cal/h (p. 701). The vast majority of this energy is converted to heat. Our thermoregulatory systems function to eliminate this heat to keep body temperature constant. If our body were not able to lose this heat, our temperature would rise until we boiled (of course, a person would die before reaching that temperature). It is relatively easy to calculate how long it would take to reach the hypothetical boiling point. If an amount of energy $\Delta U$ is put into a liquid of mass $m$, the temperature change $\Delta T$ (in °C) is given by the following formula:

$$\Delta T = \Delta U / m \times C$$

2. In this equation, *C* is the specific heat of the liquid. For water, *C* is 1 kcal/kg °C. Use this information to calculate how long it would take for the heat from the BMR to boil your body fluids (assume 42 L of water in your body and a starting point of normal body temperature at 37°C). When exercising maximally, a person consumes about 1000 cal/h. How long would it take to boil in this case?

## POINTS TO PONDER

### (Explanations in Appendix E, p. A-54)

1. Explain how drugs that selectively inhibit CCK increase feeding behaviour in experimental animals.

2. What advice would you give an overweight friend who asks for your help in designing a safe, sensible, inexpensive program for losing weight?

3. Why is it dangerous to engage in heavy exercise on a hot, humid day?

4. Describe the avenues for heat loss in a person soaking in a hot bath.

5. Consider the difference between you and a fish in a local pond with regard to control of body temperature. Humans are *thermoregulators*; we can maintain a remarkably constant, rather high, internal body temperature despite the body's exposure to a wide range of environmental temperatures. To maintain thermal homeostasis, the human body physiologically manipulates mechanisms to adjust heat production, heat conservation, and heat loss. In contrast, fish are *thermoconformers*; their body temperatures conform to the temperature of their surroundings. Thus, their body temperature varies capriciously with changes in the environmental temperature. Even though fish produce heat, they cannot physiologically regulate internal heat production, nor can they control heat exchange with their environment to maintain a constant body temperature when the temperature in their surroundings rises or falls. Knowing this, do you think fish run a fever when they have a systemic infection? Why or why not?

## CLINICAL CONSIDERATION

### (Explanation in Appendix E, p. A-54)

Michael, F., a near-drowning victim, was pulled from icy water by rescuers 15 minutes after he fell through the thin ice on which he was skating. Michael is now alert and recuperating in the hospital. How do you explain his "miraculous" survival, even though he was submerged for 15 minutes, and irreversible brain damage, soon followed by death, normally occurs if the brain is deprived of its critical oxygen supply for more than four or five minutes?

16

# The Reproductive System

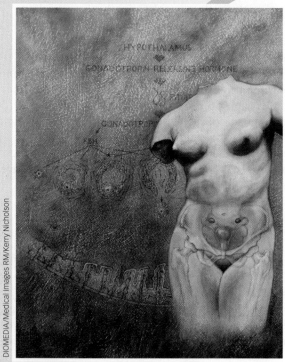

DIOMEDIA/Medical Images RM/Kerry Nicholson

Venus torso

Body systems maintain homeostasis

## Homeostasis

The reproductive system does not contribute to homeostasis but is essential for perpetuation of the species.

Homeostasis is essential for survival of cells

Cells make up body systems

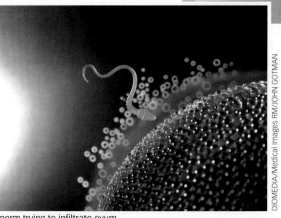

DIOMEDIA/Medical Images RM/JOHN GOTMAN

Sperm trying to infiltrate ovum

Normal functioning of the reproductive system is not aimed toward homeostasis and is not necessary for survival of an individual, but it is essential for survival of the species. Only through reproduction can the complex genetic blueprint of each species survive beyond the lives of individual members of the species.

# 17

# The Reproductive System

**❚ Clinical Connections**

Jourdain is a 19-year-old university student. For the past few days he has been experiencing some pain during urination. He has also noticed a cloudy discharge from the tip of his penis. Suspecting he has contracted a sexually transmitted disease, he went to the student health clinic on campus. His patient history revealed that he met his first girlfriend, Anita, about two months ago, and they had recently become sexually active. After providing a urine sample that tested positive for chlamydia trachomatis, he was prescribed antibiotics to clear up the infection. Concerned for Anita, the physician insisted that Jourdain tell her about the infection so that she could also be tested.

## 17.1 | Introduction

The central theme of this book has been the physiologic processes aimed at maintaining homeostasis to ensure survival of the individual. In this chapter, we depart from this theme to discuss the **reproductive system**. Sexual reproduction is the process of producing offspring for the purpose of the species' survival.

Even though the reproductive system does not contribute to homeostasis and is not essential for survival of an individual, it still plays an important role in a person's life. For example, the manner in which people relate as sexual beings contributes in significant ways to psychosocial behaviour and has important influences on how people view themselves and how they interact with others. Reproductive function also has a profound effect on society.

The universal organization of societies into family units provides a stable environment that is conducive for perpetuating our species. On the other hand, the population explosion and its resultant drain on dwindling resources have recently led to worldwide concern with the means by which reproduction can be limited.

Reproductive capability depends on intricate relationships among the hypothalamus, anterior pituitary, reproductive organs, and target cells of the sex hormones. These relationships employ many of the regulatory mechanisms used by other body systems for maintaining homeostasis, such as negative-feedback control. In addition to these basic biological processes, sexual behaviour and attitudes are deeply influenced by emotional factors and the sociocultural mores of the society in which the individual lives. Here we focus on the basic sexual and reproductive functions that are under nervous and hormonal control.

## The reproductive system

**Reproduction** depends on the union of male and female **gametes (reproductive or germ cells)**—each with a half set of chromosomes—to form a new individual with a full, unique set of chromosomes. Unlike the other body systems, which are essentially identical in the two sexes, the reproductive systems of males and females are remarkably different, befitting their different roles in the reproductive process. The **male and female reproductive systems** are designed to enable union of genetic material from the two sexual partners, and the female system is equipped to house and nourish the offspring to the developmental point at which it can survive independently in the external environment.

The **primary reproductive organs (gonads)** consist of a pair of **testes** in the male and a pair of **ovaries** in the female. In both sexes, the mature gonads perform the dual function of (1) producing gametes (**gametogenesis**), specifically, **spermatozoa (sperm)** in the male and **ova (eggs)** in the female; and (2) secreting sex hormones, specifically, **testosterone** in males and **estrogen** and **progesterone** in females.

In addition to the gonads, the reproductive system in each sex includes a **reproductive tract**, which encompasses a system of ducts that are specialized to transport or house the gametes after they are produced, plus **accessory sex glands**, which empty their supportive secretions into these passageways. In females, the *breasts* are also considered accessory reproductive organs. The externally visible portions of the reproductive system are known as **external genitalia**.

### SECONDARY SEXUAL CHARACTERISTICS

Secondary sexual characteristics distinguish males and females and are the external characteristics not directly involved in the reproductive process. Examples of these are hair-distribution patterns. In humans, for example, males have broader shoulders, whereas females have curvier hips; and males have beards, whereas females do not. Testosterone in the male and estrogen in the female govern the development and maintenance of these characteristics. Even though the growth of axillary and pubic hair at puberty is promoted in both sexes by androgens—testosterone in males and adrenocortical dehydroepiandrosterone in females (p. 258)—this

hair growth is not a secondary sexual characteristic, because both sexes display this feature. Thus, testosterone and estrogen alone govern the nonreproductive distinguishing features.

In some species, the secondary sexual characteristics are of great importance in courting and mating behaviour; for example, the rooster's headdress, or comb, attracts the female's attention, and the stag's antlers are useful to ward off other males. In humans, the differentiating marks between males and females do serve to attract a sexual partner, but attraction is also strongly influenced by the complexities of human society and cultural behaviour.

### OVERVIEW OF MALE REPRODUCTIVE FUNCTIONS AND ORGANS

The male reproductive system comprises the testes and a series of ducts and glands. The essential reproductive functions of the male are as follows:

1. Production of sperm (*spermatogenesis*)
2. Delivery of sperm to the female

The sperm-producing organs, the testes, are suspended outside the abdominal cavity in a skin-covered sac, the **scrotum**, which lies within the angle between the legs. The scrotum, which is divided internally into two sacs, is a continuation of the abdomen. Initially, the testes are in the abdomen, but at approximately the seventh month of gestation the testes drop into the scrotum, one in each sac. The male reproductive system is designed to deliver sperm to the female reproductive tract in a liquid vehicle, *semen,* which is conducive to sperm viability. The major **male accessory sex glands**, whose secretions provide the bulk of the semen, are the *seminal vesicles, prostate gland,* and *bulbourethral glands* (〉 Figure 17-1). The **penis** is the organ used to deposit semen in the female. Sperm exit each testis through the **male reproductive tract**, consisting on each side of an *epididymis, ductus (vas) deferens,* and *ejaculatory duct.* These pairs of reproductive tubes empty into a single *urethra,* the canal that runs the length of the penis and empties to the exterior. These parts of the male reproductive system are described thoroughly when their functions are discussed in Section 17.2.

### OVERVIEW OF FEMALE REPRODUCTIVE FUNCTIONS AND ORGANS

The female's role in reproduction is more complicated than the male's. The essential female reproductive functions include the following:

1. Cyclical production of ova (*oogenesis*)
2. Reception of sperm
3. Transport of the sperm and ovum to a common site for union (*fertilization* or *conception*)
4. Maintenance of the developing fetus until it can survive in the outside world (*gestation* or *pregnancy*), including formation of the *placenta,* the organ of exchange between mother and fetus
5. Giving birth to the baby (*parturition*)
6. Nourishing the infant after birth by milk production (*lactation*)

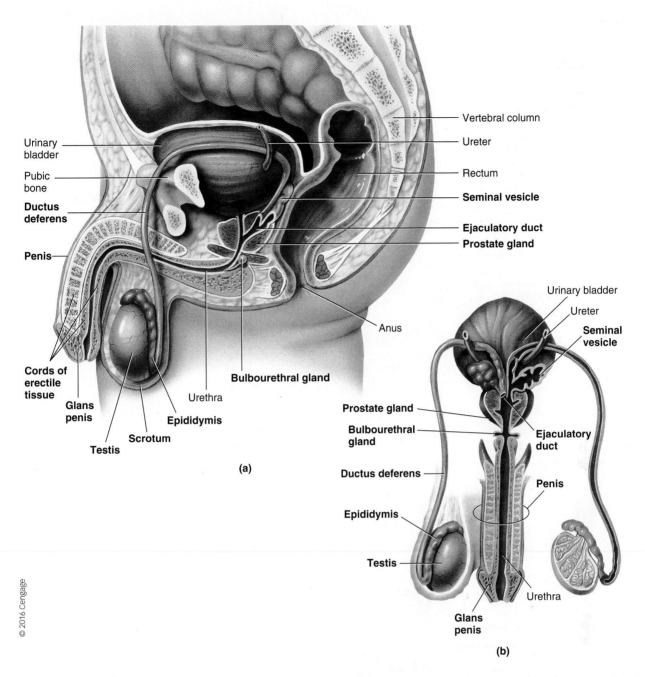

**© 2016 Cengage**

**> FIGURE 17-1 The male reproductive system.** (a) The pelvis in sagittal section. (b) Posterior view of the reproductive organs. Portions of some organs have been removed.

The product of fertilization is known as an **embryo** during the first two months of intrauterine development when tissue differentiation is taking place. Beyond this time, the developing living being is recognizable as human and is known as a **fetus** during the remainder of gestation. Although no further tissue differentiation takes place during fetal life, it is a time of tremendous tissue growth and maturation.

The ovaries and female reproductive tract lie within the pelvic cavity (> Figures 17-2a and 17-2b). The female reproductive tract consists of two *Fallopian tubes*, a *uterus*, a *cervix*, and a *vagina*. Two **oviducts (uterine or Fallopian tubes)**, which are in close association with the two ovaries, pick up ova on ovulation

(ovum release from an ovary) and serve as the site for fertilization. The thick-walled, hollow **uterus** is primarily responsible for maintaining the fetus during its development and expelling it at the end of pregnancy. The **vagina** is a muscular, expandable tube that connects the uterus to the external environment. The lowest portion of the uterus, the cervix, projects into the vagina and contains a single, small opening, the **cervical canal**. Sperm are deposited in the vagina by the penis during sexual intercourse. The cervical canal serves as a pathway for sperm to enter the uterus where it can then access the site of fertilization in the oviduct and, when greatly dilated during parturition, serves as the passageway for delivery of the baby from the uterus.

**17**

**The Reproductive System**

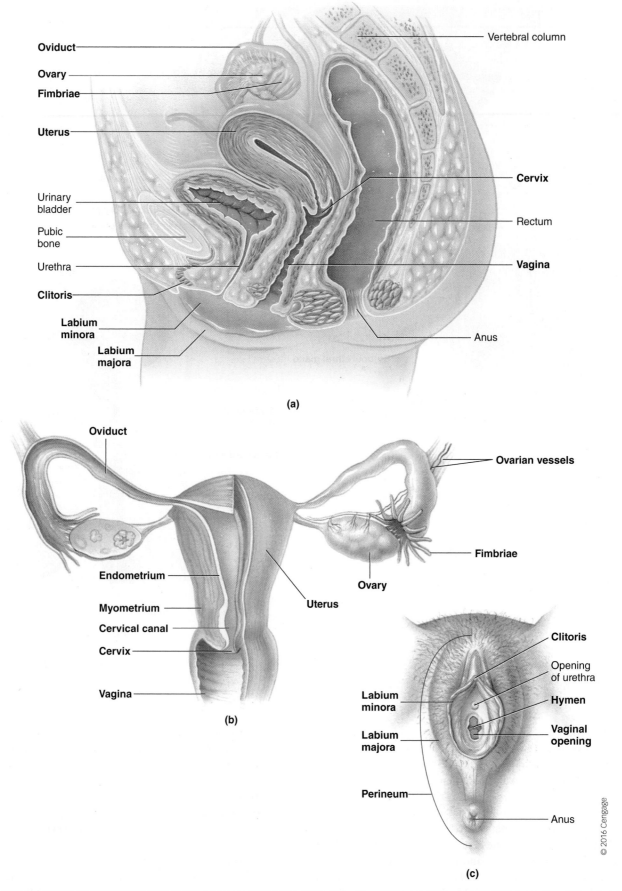

**> FIGURE 17-2 The female reproductive system.** (a) The pelvis in sagittal section. (b) Posterior view of the reproductive organs. (c) Perineal view of the external genitalia

The **vaginal opening** is located in the **perineal region** between the urethral opening anteriorly and the anal opening posteriorly (> Figure 17-2c). It is partially covered by a thin mucous membrane, the **hymen**, which typically is physically disrupted by the first sexual intercourse. The vaginal and urethral openings are surrounded laterally by two pairs of skin folds, the **labia minora** and **labia majora**. The smaller labia minora are located medially to the more prominent labia majora. The **clitoris**, a small erectile structure composed of tissue similar to that of the penis, lies at the anterior end of the folds of the labia minora. The female external genitalia are collectively referred to as the **vulva**.

## Reproductive cells

The DNA molecules that carry the cell's genetic code are not randomly crammed into the nucleus but are precisely organized into **chromosomes** (Appendix C, p. A-19). Each chromosome consists of a different DNA molecule that contains a unique set of genes. Somatic (body) cells contain 46 chromosomes (the **diploid number**), which can be sorted into 23 pairs on the basis of various distinguishing features. Chromosomes composing a matched pair are termed **homologous chromosomes**—one member of each pair having been derived from the individual's maternal parent and the other member from the paternal parent. Gametes (i.e., sperm and eggs) contain only one member of each homologous pair, for a total of 23 chromosomes (the **haploid number**).

## Gametogenesis

Most cells in the human body have the ability to reproduce themselves, a process important in growth, replacement, and repair of tissues. Cell division involves two components: division of the nucleus and division of the cytoplasm. Nuclear division in somatic cells is accomplished by **mitosis**. In mitosis, the chromosomes replicate (make duplicate copies of themselves); then the identical chromosomes are separated so that a complete set of genetic information (i.e., a diploid number of chromosomes) is distributed to each of the two new daughter cells. Nuclear division in the specialized case of gametes is accomplished by **meiosis**, in which only a half set of genetic information (i.e., a haploid number of chromosomes) is distributed to each of four new daughter cells (Appendix C, p. A-27).

During meiosis, a specialized diploid germ cell undergoes one chromosome replication followed by two nuclear divisions. In the first meiotic division, the replicated chromosomes do not separate into two individual, identical chromosomes but remain joined. The doubled chromosomes sort themselves into homologous pairs, and the pairs separate so that each of two daughter cells receives a half set of doubled chromosomes. During the second meiotic division, the doubled chromosomes within each of the two daughter cells separate and are distributed into two cells, yielding four daughter cells, each containing a half set of chromosomes, that is, a single member of each pair. During this process, the maternally and paternally derived chromosomes of each homologous pair are distributed to the daughter cells in random assortments containing one member of each chromosome pair without regard for its original derivation. That is, not all of the mother-derived chromosomes go to one daughter cell and the father-derived chromosomes to the other cell. More than 8 million ($2^{23}$) different mixtures of the 23 paternal and maternal chromosomes are possible. This genetic mixing provides novel combinations of chromosomes. Crossing-over contributes even further to genetic diversity. *Crossing-over* refers to the physical exchange of chromosome material between the homologous pairs prior to their separation during the first meiotic division (Appendix C, p. A-27).

Thus, sperm and ova each have a unique haploid number of chromosomes. When fertilization takes place, a sperm and ovum fuse to form the start of a new individual with 46 chromosomes—one member of each chromosomal pair having been inherited from the mother and the other member from the father (> Figure 17-3).

## Sex chromosomes

Whether individuals are destined to be males or females is a genetic phenomenon determined by the sex chromosomes they possess. As the 23 chromosome pairs are separated during meiosis, each sperm or ovum receives only one member of each chromosome pair. Of the chromosome pairs, 22 are **autosomal chromosomes** that code for general human characteristics as well as for specific traits, such as eye colour. The remaining pair of chromosomes consists of the **sex chromosomes**, of which

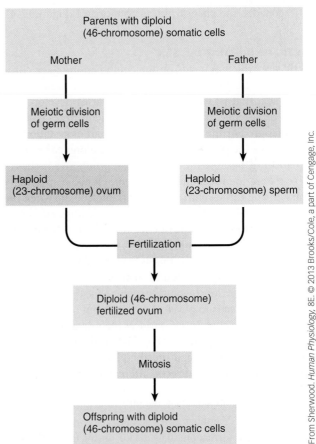

> **FIGURE 17-3 Chromosomal distribution in sexual reproduction**

**17**

there are two genetically different types: a larger **X chromosome**, and a smaller **Y chromosome**.

**Sex determination** depends on the combination of sex chromosomes. **Genetic males** have both an X and a Y sex chromosome; **genetic females** have two X sex chromosomes. Therefore, the genetic difference responsible for all the anatomic and functional distinctions between males and females is the single Y chromosome. Males have it; females do not.

As a result of meiosis during gametogenesis, all chromosome pairs are separated so that each daughter cell contains only one member of each pair, including the sex chromosome pair. When the XY sex chromosome pair separates during sperm formation, half the sperm receive an X chromosome and the other half a Y chromosome. In contrast, during oogenesis, every ovum receives an X chromosome, because separation of the XX sex chromosome pair yields only X chromosomes. During fertilization, the combination of an X-bearing sperm with an X-bearing ovum produces a genetic female, XX, whereas union of a Y-bearing sperm with an X-bearing ovum results in a genetic male, XY. Consequently, genetic sex is determined at the time of conception and depends on which type of sex chromosome is contained within the fertilizing sperm (an X or a Y).

## Sexual differentiation

Differences between males and females exist at three levels: genetic, gonadal, and phenotypic (anatomic) sex (› Figure 17-4).

### GENETIC AND GONADAL SEX

**Genetic sex**, which depends on the combination of sex chromosomes at the time of conception, in turn, determines **gonadal sex**, that is, whether testes or ovaries develop. The presence or absence of a Y chromosome determines gonadal differentiation. For the first month and a half of gestation, all embryos have the potential to differentiate along either male or female lines, because the developing reproductive tissues of both sexes are identical and indifferent. Gonadal specificity appears during the seventh week of intrauterine life when the indifferent gonadal tissue of a genetic male begins to differentiate into testes under the influence of the **sex-determining region of the Y chromosome (SRY)**—the single gene, found in the urogenital ridge cells, that is responsible for sex determination. The SRY gene is found in all mammals and shows little variation over time (evolution). This gene triggers a chain of reactions that leads to physical development of a male. SRY "masculinizes" the gonads (i.e., induces their development into testes) by stimulating production of **H-Y antigen** by primitive gonadal cells. H-Y antigen, a specific plasma membrane protein found only in males, directs differentiation of the gonads into testes.

Because genetic females lack the SRY gene and consequently do not produce H-Y antigen, their gonadal cells never receive a signal for testicular formation, so by default during the ninth week, the undifferentiated gonadal tissue starts developing into ovaries instead.

### PHENOTYPIC SEX

**Phenotypic sex**, the apparent anatomic sex of an individual, depends on the genetically determined gonadal sex. The term **sexual differentiation** refers to the embryonic development of the external genitalia and reproductive tract along either male or female lines. As with the undifferentiated gonads, embryos of both sexes have the potential to develop either male or female external genitalia and reproductive tracts. Differentiation into a male-type reproductive system is induced by **androgens**, which are masculinizing hormones secreted by the developing testes. Testosterone is the most potent androgen. The absence of these testicular hormones in female fetuses results in the development of a female-type reproductive system. By 10 to 12 weeks of gestation, the sexes can easily be distinguished by the anatomic appearance of the external genitalia.

### SEXUAL DIFFERENTIATION OF THE EXTERNAL GENITALIA

Male and female external genitalia develop from the same embryonic tissue. In both sexes, the undifferentiated external genitals consist of a *genital tubercle,* paired *urethral folds* surrounding a urethral groove, and, more laterally, *genital (labioscrotal) swellings* (› Figure 17-5). The **genital tubercle** gives rise to exquisitely sensitive erotic tissue—in males, the **glans penis** (the cap at the distal end of the penis) and in females, the clitoris. The major distinctions between the glans penis and clitoris are the smaller size of the clitoris and the urethral opening through the glans penis. The urethra is the tube through which urine is transported from the bladder to the outside and also serves in males as a passageway for exit of semen through the penis to the outside. In males, the **urethral folds** fuse around the urethral groove to form the penis, which encircles the urethra. The **genital swellings** similarly fuse to form the scrotum and **prepuce**, a fold of skin that extends over the end of the penis and more or less completely covers the glans penis. In females, the urethral folds and genital swellings do not fuse at midline but develop instead into the labia minora and labia majora, respectively. The urethral groove remains open, providing access to the interior through both the urethral opening and the vaginal opening.

### SEXUAL DIFFERENTIATION OF THE REPRODUCTIVE TRACT

Although the male and female external genitalia develop from the same undifferentiated embryonic tissue, this is not the case with the reproductive tracts. Two primitive duct systems—Wolffian ducts and the Müllerian ducts—develop in all embryos. In males, the reproductive tract develops from the **Wolffian ducts**, and the Müllerian ducts degenerate. In females the **Müllerian ducts** differentiate into the reproductive tract, and the Wolffian ducts regress (› Figure 17-6). Because both duct systems are present before sexual differentiation occurs, the early embryo has the potential to develop either a male or a female reproductive tract.

Development of the reproductive tract along male or female lines is determined by the presence or absence of two hormones secreted by the fetal testes—*testosterone* and *Müllerian-inhibiting*

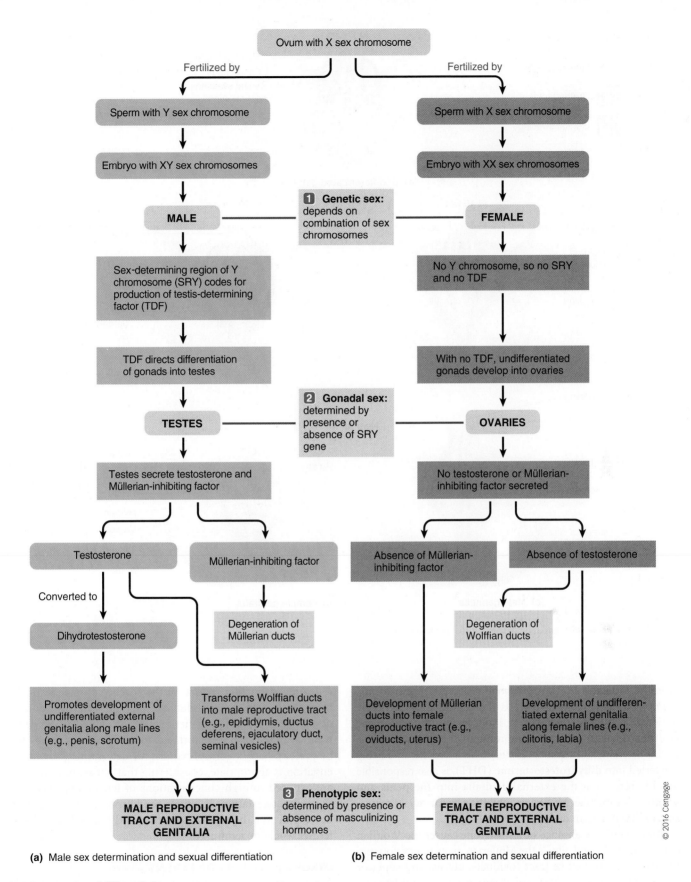

**(a)** Male sex determination and sexual differentiation

**(b)** Female sex determination and sexual differentiation

> **FIGURE 17-4 Sexual differentiation**

17

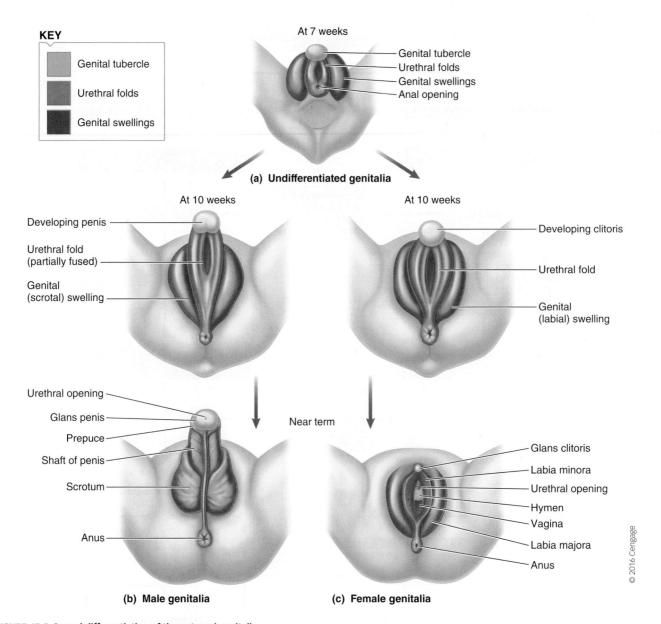

**KEY**

- Genital tubercle
- Urethral folds
- Genital swellings

At 7 weeks
- Genital tubercle
- Urethral folds
- Genital swellings
- Anal opening

**(a) Undifferentiated genitalia**

At 10 weeks
- Developing penis
- Urethral fold (partially fused)
- Genital (scrotal) swelling

At 10 weeks
- Developing clitoris
- Urethral fold
- Genital (labial) swelling

Near term

- Urethral opening
- Glans penis
- Prepuce
- Shaft of penis
- Scrotum
- Anus

- Glans clitoris
- Labia minora
- Urethral opening
- Hymen
- Vagina
- Labia majora
- Anus

© 2016 Cengage

**(b) Male genitalia**

**(c) Female genitalia**

> FIGURE 17-5 Sexual differentiation of the external genitalia

factor (see › Figure 17-4). A hormone released by the placenta, *human chorionic gonadotropin,* is the stimulus for this early testicular secretion. Testosterone induces development of the Wolffian ducts into the male reproductive tract (epididymis, ductus deferens, and seminal vesicles). This hormone, after being converted into **dihydrotestosterone (DHT)**, is also responsible for differentiating the external genitalia into the penis and scrotum. Meanwhile, Müllerian-inhibiting factor causes regression of the Müllerian ducts.

In females, the absence of testosterone causes the Wolffian ducts to degrade and allows the Müllerian ducts to develop into the female reproductive tract (oviducts, uterus, and superior portion of vagina) and the external genitalia to differentiate into the clitoris and labia.

Note that the undifferentiated embryonic reproductive tissue passively develops into a female structure unless actively acted on by masculinizing factors. In the absence of male testicular hormones, a female reproductive tract and external genitalia develop regardless of the genetic sex of the individual. For feminization of the fetal genital tissue, ovaries do not even need to be present. Such a control pattern for determining sex differentiation is appropriate, considering that fetuses of both sexes are exposed to high concentrations of female sex hormones throughout gestation. If female sex hormones influenced the development of the reproductive tract and external genitalia, all fetuses would be feminized.

**ERRORS IN SEXUAL DIFFERENTIATION**

 *Clinical Note* Genetic sex and phenotypic sex are usually compatible; that is, a genetic male anatomically appears to be a male and functions as a male, and the same compatibility holds true for females. Occasionally, however, discrepancies

17

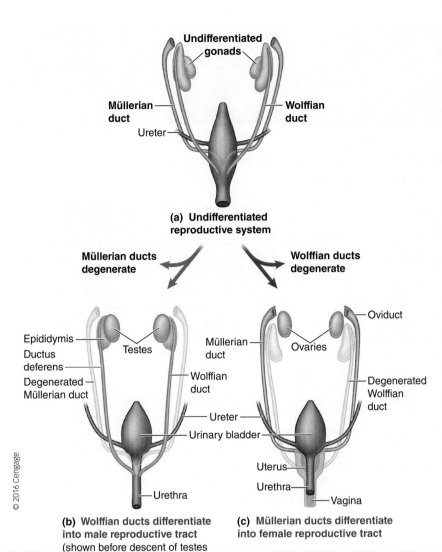

**Undifferentiated gonads**

Müllerian duct

Wolffian duct

Ureter

**(a) Undifferentiated reproductive system**

**Müllerian ducts degenerate**

**Wolffian ducts degenerate**

Epididymis

Ductus deferens

Degenerated Müllerian duct

Testes

Wolffian duct

Müllerian duct

Oviduct

Ovaries

Degenerated Wolffian duct

Ureter

Urinary bladder

Uterus

Urethra

Urethra

Vagina

**(b) Wolffian ducts differentiate into male reproductive tract** (shown before descent of testes into scrotum)

**(c) Müllerian ducts differentiate into female reproductive tract**

© 2016 Cengage

› FIGURE 17-6 **Sexual differentiation of the reproductive tract**

females. However, pathologically excessive secretion of this hormone in a genetically female fetus during critical developmental stages imposes differentiation of the reproductive tract and genitalia along male lines (see *adrenogenital syndrome*, p. 259).

Sometimes these discrepancies between genetic sex and apparent sex are not recognized until puberty, when the discovery produces a psychologically traumatic gender identity crisis. For example, a masculinized genetic female with ovaries but with male-type external genitalia may be reared as a boy until puberty, when breast enlargement (caused by estrogen secretion by the awakening ovaries) and lack of beard growth (caused by lack of testosterone secretion in the absence of testes) signal an apparent problem. Therefore, it is important to diagnose any problems in sexual differentiation in infancy. Once a sex has been assigned, it can be reinforced, if necessary, with surgical and hormonal treatment so that psychosexual development can proceed as normally as possible. Less dramatic cases of inappropriate sex differentiation often appear as sterility problems.

In some circumstances, such as athletics, it is considered performance enhancing to be a phenotypic female but genetic male due to the greater production of testosterone. Consequently, genetic testing of female athletes is sometimes required to establish genotypic sex (see ▶Concepts, Challenges, and Controversies, p. 732).

occur between genetic and anatomic sexes because of errors in sexual differentiation, as the following examples illustrate:

- If testes in a genetic male fail to properly differentiate and secrete hormones, the result is the development of an apparent anatomic female in a genetic male, who, of course, will be sterile. Similarly, genetic males whose target cells lack receptors for testosterone are feminized, even though their testes secrete adequate testosterone (see *testicular feminization syndrome*, p. 214).

- Because testosterone acts on the Wolffian ducts to convert them into a male reproductive tract but the testosterone derivative DHT is responsible for masculinization of the external genitalia, a genetic deficiency of the enzyme that converts testosterone into DHT results in a genetic male with testes and a male reproductive tract, but with female external genitalia.

- The adrenal gland normally secretes a weak androgen, *dehydroepiandrosterone*, in insufficient quantities to masculinize

**Check Your Understanding 17.1**

1. Name the primary reproductive organs in males and in females, and state the dual function of these organs in each sex.

2. Prepare a table comparing what reproductive structures are derived from the gonadal ridge, genital tubercle, urethral folds, genital swellings, Wolffian ducts, and Müllerian ducts during embryonic development in males and in females.

## 17.2 | Male Reproductive Physiology

In the embryo, the testes develop from the gonadal ridge located at the rear of the abdominal cavity. In the last months of fetal life, they begin a slow descent, passing out of the abdominal cavity through the **inguinal canal** into the scrotum; one testis drops into each pocket of the scrotal sac. Testosterone from the fetal testes induces the descent of the testes into the scrotum.

After the testes descend into the scrotum, the opening in the abdominal wall through which the inguinal canal passes closes

# Sex Testing in Sports

**S**EX VERIFICATION (SEX DETERMINATION) IN SPORTS means verifying the eligibility of an athlete to compete in a sporting event that is limited to a single sex—in essence, making sure that female athletes are actually female. Sex verification began at the 1966 European Track and Field Championships in response to suspicions that some of the top female athletes—primarily from the Soviet Union—were actually men; more suspicion was raised when some of the participants unexpectedly retired when the testing began.

The most recognized case of a male impersonating a female in an international athletic competition involved a German athlete (Hermann Ratjen), who took the name Dora and placed fourth in the high jump event in the 1936 Olympics. The deception was not discovered until 1955. Testing was introduced at the Olympics in 1968, in Mexico City. In those days, the athlete had to stand unclothed in front of a panel of gynaecologists, who would together make the determination. Later, this crude method gave way to much more sophisticated and objective laboratory tests (i.e., testing for two X chromosomes).

By the 1980s, testing methods had changed yet again. The International Olympic Committee (IOC) replaced the X chromosome test with the polymerase chain reaction test (for the Y-linked gene SRY), using the cells from a buccal smear. This test was introduced during the 1992 Winter Olympics in Albertville, France. In contrast, the International Association of Athletics Federations (IAAF) stopped genetic testing for sex verification in 1992, because mandatory urine samples made the deception virtually impossible.

Although the polymerase chain reaction test is less intrusive, many experts feel the test is meaningless and that it can cause serious psychological damage to females who may unknowingly have certain disorders of sexual differentiation. Data indicate that sex tests for approximately 1 in 500–600 athletes are abnormal and might result in their disqualification. For example, androgen insensitivity syndrome (testicular feminization) may provide a false positive test. In this case, an individual is genetically male (i.e., has both an X and a Y chromosome), but his or her tissues are unable to respond to the androgens, and thus he or she develops into a woman.

### Further Reading

Elsas, L.J., Ljungqvist, A., Ferguson-Smith, M.A., Simpson, J.L., Genel, M., Carlson, A.S., et al. (2000). Gender verification of female athletes. *Genet Med*, 2(4): 249–54.

Ferris, E.A. (1992). Gender verification testing in sport. *Brit Med Bull*, 48(3): 683–97.

Stephenson, J. (1996). Female Olympians' sex tests outmoded. *JAMA*, 276(3): 177–78.

---

snugly around the sperm-carrying duct and blood vessels that traverse between each testis and the abdominal cavity. Incomplete closure or rupture of this opening permits abdominal viscera to slip through, resulting in an **inguinal hernia**.

Although the time varies somewhat, descent is usually complete by the seventh month of gestation. As a result, descent is complete in 98 percent of full-term baby boys.

*Clinical Note* However, in a substantial percentage of premature male infants, the testes are still within the inguinal canal at birth. In most instances of retained testes, the descent occurs naturally before puberty or can be encouraged with administration of testosterone. Rarely, a testis remains undescended into adulthood, a condition known as **cryptorchidism** (hidden testis).

## Scrotal location of the testes

The temperature within the scrotum averages several degrees Celsius less than normal body (core) temperature. Descent of the testes into this cooler environment is essential, because spermatogenesis is temperature sensitive and cannot occur at normal body temperature. Therefore, a person with cryptorchidism is unable to produce viable sperm.

The position of the scrotum in relation to the abdominal cavity can be varied by two muscles and a spinal reflex mechanism that play an important role in regulating testicular temperature. Reflex contraction of the two scrotal muscles on exposure to a cold environment raises the scrotal sac to bring the testes closer to the warmer abdomen. The scrotum will take on a smaller, more wrinkled appearance. Conversely, relaxation of the muscles on exposure to heat permits the scrotal sac to become more pendulous, moving the testes farther from the warm core of the body.

## Leydig cells

The testes perform the dual function of producing sperm and secreting testosterone. About 80 percent of the testicular mass consists of highly coiled **seminiferous tubules**, within which spermatogenesis and meiosis take place. The endocrine cells that produce testosterone (androgen)—the **Leydig (interstitial) cells**—lie in the connective tissue (interstitial tissue) between the seminiferous tubules (› Figure 17-7). The production of testosterone by the Leydig cells is under the control of the pituitary gland's secretion of luteinizing hormone. Thus, the portions of the testes that produce sperm and secrete testosterone are structurally and functionally distinct.

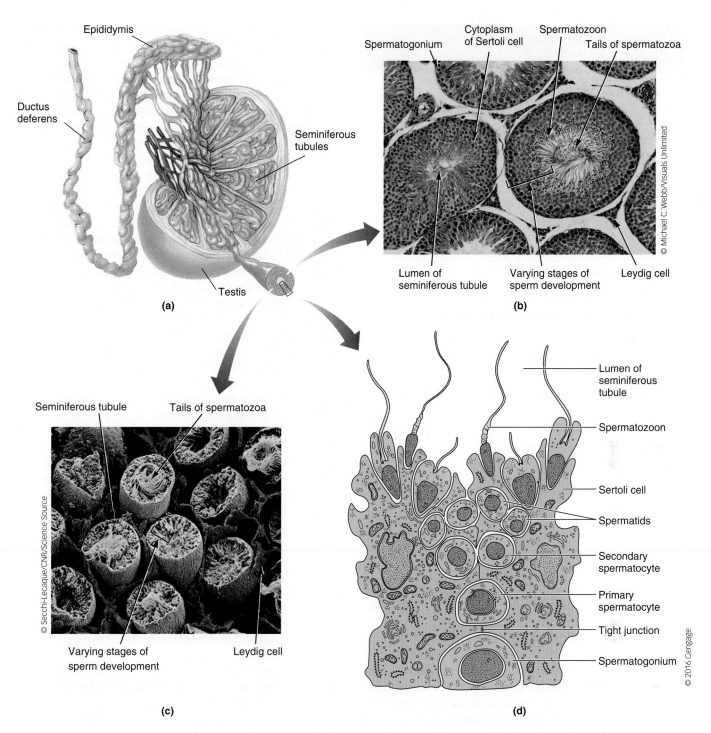

Epididymis

Ductus deferens

Seminiferous tubules

Testis

**(a)**

Spermatogonium

Cytoplasm of Sertoli cell

Spermatozoon

Tails of spermatozoa

© Michael C. Webb/Visuals Unlimited

Lumen of seminiferous tubule

Varying stages of sperm development

Leydig cell

**(b)**

Seminiferous tubule

Tails of spermatozoa

© Secchi-Lecaque/CNR/Science Source

Varying stages of sperm development

Leydig cell

**(c)**

Lumen of seminiferous tubule

Spermatozoon

Sertoli cell

Spermatids

Secondary spermatocyte

Primary spermatocyte

Tight junction

Spermatogonium

© 2016 Cengage

**(d)**

> **FIGURE 17-7 Testicular anatomy depicting the site of spermatogenesis.** (a) Longitudinal section of a testis showing the location and arrangement of the seminiferous tubules, the sperm-producing portion of the testis. (b) Light micrograph of a cross-section of a seminiferous tubule. The undifferentiated germ cells (the spermatogonia) lie in the periphery of the tubule, and the differentiated spermatozoa are in the lumen, with the various stages of sperm development in between. (c) Scanning electron micrograph of a cross-section of a seminiferous tubule. (d) Relationship of the Sertoli cells to the developing sperm cells

Testosterone is a steroid hormone derived from a cholesterol precursor molecule, as are the female sex hormones, estrogen and progesterone. Once produced, some of the testosterone is secreted into the blood, where it is transported—primarily bound to plasma proteins—to its target sites of action. A substantial portion of the newly synthesized testosterone goes into the lumen of the seminiferous tubules, where it plays an important role in sperm production.

Most but not all of testosterone's actions ultimately function to ensure delivery of sperm to the female. The effects of testosterone can be grouped into five categories: (1) effects on the reproductive system before birth, (2) effects on sex-specific

**17**

## TABLE 17-1 Effects of Testosterone

**Effects before Birth**

Masculinizes the reproductive tract and external genitalia

Promotes descent of the testes into the scrotum

**Effects on Sex-Specific Tissues after Birth**

Promotes growth and maturation of the reproductive system at puberty

Is essential for spermatogenesis

Maintains the reproductive tract throughout adulthood

**Other Reproduction-Related Effects**

Develops the sex drive at puberty

Controls gonadotropin hormone secretion

**Effects on Secondary Sexual Characteristics**

Induces the male pattern of hair growth (e.g., beard)

Causes the voice to deepen because of thickening of the vocal folds

Promotes muscle growth responsible for the male body configuration

**Nonreproductive Actions**

Exerts a protein anabolic effect

Promotes bone growth at puberty

Closes the epiphyseal plates after its conversion to estrogen by aromatase

May induce aggressive behaviour

© 2016 Cengage

tissues after birth, (3) other reproduction-related effects, (4) effects on secondary sexual characteristics, and (5) nonreproductive actions (Table 17-1).

### EFFECTS ON THE REPRODUCTIVE SYSTEM BEFORE BIRTH

Before birth, testosterone secretion by the fetal testes masculinizes the reproductive tract and external genitalia and promotes descent of the testes into the scrotum, as already described. After birth, testosterone secretion ceases, and the testes and the rest of the reproductive system remain small and nonfunctional until puberty.

### EFFECTS ON SEX-SPECIFIC TISSUES AFTER BIRTH

**Puberty** is the period of arousal and maturation of the previously nonfunctional reproductive system, culminating in sexual maturity and the ability to reproduce. It usually begins sometime between the ages of 10 and 14; on average, it begins about two years earlier in females than in males. Usually lasting three to five years, puberty encompasses a complex sequence of endocrine, physical, and behavioural events. **Adolescence** is a broad concept that refers to the entire transition period between childhood and adulthood, not just to the time of sexual maturation.

At puberty, the Leydig cells start secreting testosterone once again. Testosterone is responsible for growth and maturation of the entire male reproductive system. Under the influence of the pubertal surge in testosterone secretion, the testes enlarge and start producing sperm for the first time, the accessory sex glands enlarge and become secretory, and the penis and scrotum enlarge.

Ongoing testosterone secretion is essential for spermatogenesis and for maintaining a mature male reproductive tract throughout adulthood. Once initiated at puberty, testosterone secretion and spermatogenesis occur continuously throughout the male's life. Testicular efficiency gradually declines after 45 to 50 years of age, even though men in their 70s and beyond may continue to enjoy an active sex life, and some even father a child at this late age. The gradual reduction in circulating testosterone levels and in sperm production is not caused by a decrease in stimulation of the testes, but probably arises instead from degenerative changes associated with aging that occur in the small testicular blood vessels. This gradual decline is often termed "male menopause" or "andropause," although it is not specifically programmed, as is female menopause. Recently the androgen decline in males has been more aptly termed **androgen deficiency in aging males (ADAM)**.

 *Clinical Note* Following **castration** (surgical removal of the testes) or testicular failure caused by disease, the other sex organs regress in size and function.

### OTHER REPRODUCTION-RELATED EFFECTS

Testosterone governs the development of sexual libido at puberty and helps maintain the sex drive in the adult male. Stimulation of this behaviour by testosterone is important for facilitating delivery of sperm to females. In humans, libido is also influenced by many interacting social and emotional factors. Once libido has developed, testosterone is no longer absolutely required for its maintenance. Castrated males often remain sexually active but at a reduced level.

In another reproduction-related function, testosterone participates in the normal negative-feedback control of gonadotropin hormone secretion by the anterior pituitary, a topic covered in the section LH and FSH.

### EFFECTS ON SECONDARY SEXUAL CHARACTERISTICS

All male secondary sexual characteristics depend on testosterone for their development and maintenance. These nonreproductive male characteristics induced by testosterone include (1) the male pattern of hair growth (e.g., beard and chest hair and, in genetically predisposed men, baldness); (2) a deep voice caused by enlargement of the larynx and thickening of the vocal folds; (3) thick skin; and (4) the male body configuration (e.g., broad shoulders and heavy arm and leg musculature) as a result of protein deposition. A male castrated before puberty (a **eunuch**) does not mature sexually, nor does he develop secondary sexual characteristics.

### NONREPRODUCTIVE ACTIONS

Testosterone exerts several important effects not related to reproduction. It has a general protein anabolic (synthesis) effect and promotes bone growth, thereby contributing to the

17

more muscular physique of males and to the pubertal growth spurt. Ironically, testosterone not only stimulates bone growth but eventually prevents further growth by sealing the growing ends of the long bones (i.e., ossifying, or closing, the epiphyseal plates; p. 233). Testosterone also stimulates oil secretion by the sebaceous glands. This effect is most striking during the adolescent surge of testosterone secretion, predisposing the young man to develop acne. Testosterone also stimulates erythropoietin release, which is a contributing factor to the higher haematocrit in males.

In animals, testosterone induces aggressive behaviour, but whether it influences human behaviour other than in the area of sexual behaviour is an unresolved issue. Even though some athletes and body builders who take testosterone-like anabolic androgenic steroids to increase muscle mass have been observed to display more aggressive behaviour, it is unclear to what extent general behavioural differences between the sexes are hormonally induced or a result of social conditioning.

### CONVERSION OF TESTOSTERONE TO ESTROGEN IN MALES

Although testosterone is classically considered the male sex hormone and estrogen a female sex hormone, the distinctions are not as clear-cut as once thought. In addition to the small amount of estrogen produced by the adrenal cortex (p. 258), a portion of the testosterone secreted by the testes is converted to estrogen outside the testes by the enzyme **aromatase**, which is widely distributed. Because of this conversion, it is sometimes difficult to distinguish between the cellular effects of testosterone and those of testosterone-turned-estrogen. For example, scientists recently learned that closure of the epiphyseal plates in males is induced not by testosterone per se but by testosterone turned into estrogen by aromatization. Estrogen is also produced in adipose tissue in both sexes. Estrogen receptors have been identified in the testes, prostate, bone, and elsewhere in males. Recent findings suggest that estrogen plays an essential role in male reproductive health; for example, it is important in spermatogenesis and surprisingly contributes to male heterosexuality. Also, it likely contributes to bone homeostasis (p. 286). The depth, breadth, and mechanisms of action of estrogen in males are only beginning to be explored. (Likewise, in addition to the weak androgenic hormone DHEA produced by the adrenal cortex in both sexes, the ovaries in females secrete a small amount of testosterone, the functions of which remain unclear.)

We now shift attention from testosterone secretion to the other function of the testes—sperm production.

## Spermatogenesis

About 250 m of sperm-producing seminiferous tubules are packed within the testes (see ⟩ Figure 17-7a). Two functionally important cell types are present in these tubules: *germ cells,* most of which are in various stages of sperm development; and *Sertoli cells,* which provide crucial support for spermatogenesis (see ⟩ Figures 17-7b, 17-7c, and 17-7d). **Spermatogenesis** is a complex process by which relatively undifferentiated primordial germ cells—the **spermatogonia** (each of which contains

a diploid complement of 46 chromosomes)—proliferate and are converted into extremely specialized, motile spermatozoa (sperm), each bearing a randomly distributed haploid set of 23 chromosomes.

Microscopic examination of a seminiferous tubule reveals layers of germ cells in an anatomic progression of sperm development, starting with the least differentiated in the outer layer and moving inward through various stages of division to the lumen, where the highly differentiated sperm are ready for exit from the testis. Spermatogenesis takes 64 days for development from a spermatogonium to a mature sperm. Up to several hundred million sperm may reach maturity daily. Spermatogenesis encompasses three major stages: *mitotic proliferation, meiosis, and packaging* (⟩ Figure 17-8).

### MITOTIC PROLIFERATION

Spermatogonia located in the outermost layer of the tubule continuously divide mitotically, with all new cells bearing the full complement of 46 chromosomes identical to those of the parent cell. Such proliferation provides a continual supply of new germ cells. Following mitotic division of a spermatogonium, one of the daughter cells remains at the outer edge of the tubule as an undifferentiated spermatogonium, thus maintaining the germ-cell line. The other daughter cell starts moving toward the lumen while undergoing the various steps required to form sperm, which will be released into the lumen. In humans, the sperm-forming daughter cell divides mitotically twice more to form four identical **primary spermatocytes**. After the last mitotic division, the primary spermatocytes enter a resting phase, during which the chromosomes are duplicated and the doubled strands remain together in preparation for the first meiotic division.

### MEIOSIS

During meiosis, each primary spermatocyte (with a diploid number of 46 doubled chromosomes) forms two **secondary spermatocytes** (each with a haploid number of 23 doubled chromosomes) during the first meiotic division, finally yielding four **spermatids** (each with 23 single chromosomes) as a result of the second meiotic division.

No further division takes place beyond this stage of spermatogenesis. Each spermatid is remodelled into a single spermatozoon. Because each sperm-producing spermatogonium mitotically produces four primary spermatocytes and each primary spermatocyte meiotically yields four spermatids (spermatozoa-to-be), the spermatogenic sequence in humans can theoretically produce 16 spermatozoa each time a spermatogonium initiates this process. Usually, however, some cells are lost at various stages, so the efficiency of productivity is rarely this high.

### PACKAGING

Even after meiosis, spermatids still resemble undifferentiated spermatogonia structurally, except for their half complement of chromosomes. Production of extremely specialized, mobile spermatozoa from spermatids requires extensive remodelling, or **packaging**, of cell elements, a process known as **spermiogenesis**. Sperm are essentially stripped-down cells in which most of the cytosol and any organelles not needed for delivering the

**17**

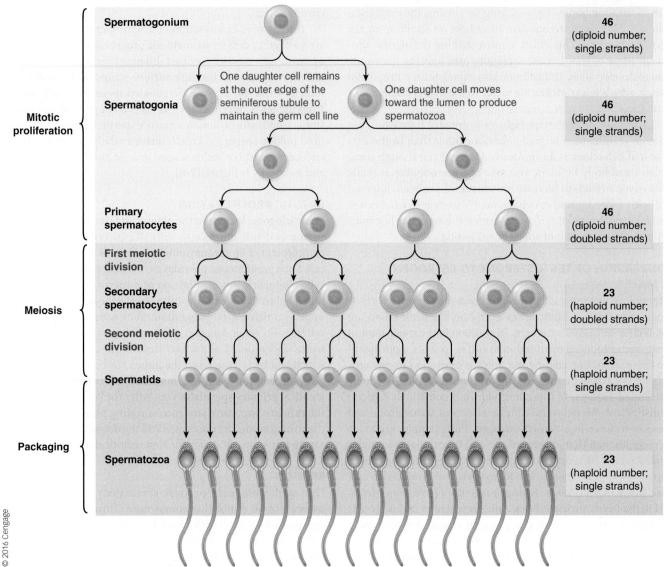

Chromosomes in each cell

Spermatogonium — 46 (diploid number; single strands)

Mitotic proliferation

Spermatogonia — One daughter cell remains at the outer edge of the seminiferous tubule to maintain the germ cell line — One daughter cell moves toward the lumen to produce spermatozoa — 46 (diploid number; single strands)

Primary spermatocytes — 46 (diploid number; doubled strands)

First meiotic division

Meiosis

Secondary spermatocytes — 23 (haploid number; doubled strands)

Second meiotic division

Spermatids — 23 (haploid number; single strands)

Packaging

Spermatozoa — 23 (haploid number; single strands)

> FIGURE 17-8 Spermatogenesis

sperm's genetic information to an ovum have been extruded. So the sperm travel with only what's essential to accomplish fertilization.

A spermatozoon has four parts (> Figure 17-9): a head, an acrosome, a midpiece, and a tail. The **head** consists primarily of the nucleus, which contains the sperm's complement of genetic information. The **acrosome**, an enzyme-filled vesicle that caps the tip of the head, is used as an "enzymatic drill" for penetrating the ovum. The acrosome is formed by aggregation of vesicles produced by the endoplasmic reticulum/Golgi complex before these organelles are discarded. Mobility for the spermatozoon is provided by a long, whiplike **tail**. Movement of the tail is powered by energy generated by the mitochondria concentrated within the **midpiece** of the sperm.

Until sperm maturation is complete, the developing germ cells arising from a single primary spermatocyte remain joined by cytoplasmic bridges. These connections, which result from incomplete cytoplasmic division, permit the four developing sperm to exchange cytoplasm. This linkage is important, because the X chromosome, but not the Y chromosome, contains genes that code for cell products essential for sperm development. (Whereas the large X chromosome contains several thousand genes, the small Y chromosome has only a few dozen, the most important of which are the SRY gene and others that play critical roles in male fertility.) During meiosis, half the sperm receive an X and the other half a Y chromosome. Were it not for the sharing of cytoplasm so that all the haploid cells are provided with the products coded for by X chromosomes until sperm development is complete, the Y-bearing, male-producing sperm could not develop and survive.

17

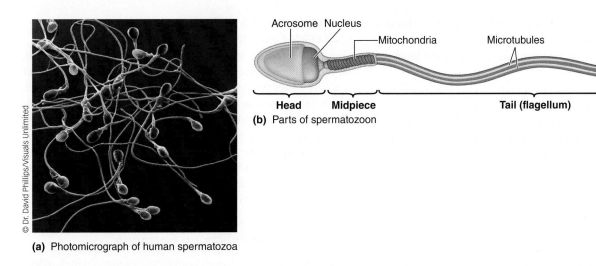

Acrosome  Nucleus
Mitochondria          Microtubules

**Head**    **Midpiece**          **Tail (flagellum)**

**(b)** Parts of spermatozoon

**(a)** Photomicrograph of human spermatozoa

> **FIGURE 17-9 Anatomy of a spermatozoon.** (a) A phase-contrast photomicrograph of human spermatozoa. (b) A spermatozoon has three functional parts: a head with its acrosome cap, a midpiece, and a tail.

**❚ Clinical Connections**

Chlamydia infections in males can lead to infertility. If the infection is asymptomatic or ignored, it can spread from the urethra further into the male reproductive tract. If it reaches the testicles, chlamydia can affect sperm in a few ways. DNA is normally packaged very tightly within the head of sperm. However, chlamydia infections can cause DNA fragmentation by reducing how tightly the DNA is packed, which can result in the breakage of DNA strands. Chlamydia can also lead to physical abnormalities in sperm that reduces their mobility. Either of these, individually or combined, can drastically reduce male fertility.

## Sertoli cells

The seminiferous tubules house the **Sertoli cells** in addition to the spermatogonia and developing sperm cells. The Sertoli cells lie side by side and form a ring that extends from the outer surface of the tubule to the lumen. Each Sertoli cell spans the entire distance from the outer surface membrane of the seminiferous tubule to the fluid-filled lumen (see > Figures 17-7b and 17-7d). Adjacent Sertoli cells are joined by tight junctions at a point slightly beneath the outer membrane. Developing sperm cells are tucked between adjacent Sertoli cells, with spermatogonia lying at the outer perimeter of the tubule, outside the tight junction (see > Figures 17-7b and 17-7d). The Sertoli cells form a barrier that prevents the immune system from becoming sensitized to antigens associated with sperm development. During spermatogenesis, the developing sperm cells arising from spermatogonial mitotic activity pass through the tight junctions, which transiently separate to make a path for them. They then migrate toward the lumen in close association with the adjacent Sertoli cells and, during this migration, undergo further divisions. The cytoplasm of the Sertoli cells envelops the migrating sperm cells, which remain buried within these cytoplasmic recesses throughout their development. At all stages of spermatogenic maturation, the developing sperm and Sertoli cells communicate with one another by means of direct cell-to-cell binding

and through paracrine secretions (p. 51). A recently identified carbohydrate on the surface membrane of the developing sperm enables them to bind to the supportive Sertoli cells.

Sertoli cells perform the following functions essential for spermatogenesis:

1. The tight junctions between adjacent Sertoli cells form a **blood–testes barrier** that prevents blood-borne substances from passing between the cells to gain entry to the lumen of the seminiferous tubule. Because of this barrier, only selected molecules that can pass through the Sertoli cells reach the intratubular fluid. As a result, the composition of the intratubular fluid varies considerably from that of the blood. The unique composition of this fluid that bathes the germ cells is critical for later stages of sperm development. The blood–testes barrier also prevents the antibody-producing cells in the ECF from reaching the tubular sperm factory, thus preventing the formation of antibodies against the highly differentiated spermatozoa.

2. Because the secluded developing sperm cells do not have direct access to blood-borne nutrients, the Sertoli cells provide nourishment for them.

3. Sertoli cells have an important phagocytic function. They engulf the cytoplasm extruded from the spermatids during their remodelling, and they destroy defective germ cells that fail to successfully complete all stages of spermatogenesis.

4. Sertoli cells secrete into the lumen **seminiferous tubule fluid**, which flushes the released sperm from the tubule into the epididymis for storage and further processing.

5. An important component of this Sertoli secretion is **androgen-binding protein (ABP)**. As the name implies, this protein binds androgens—specifically, testosterone—and thereby maintains a very high level of this hormone within the seminiferous tubule lumen. Testosterone is 100 times more concentrated in the seminiferous tubule fluid than in the blood. This high local concentration of testosterone is essential for sustaining sperm production.

**17**

ABP is necessary to retain testosterone within the lumen, because this steroid hormone is lipid soluble and could easily diffuse across the plasma membranes and leave the lumen.

6. Sertoli cells are the site of action for control of spermatogenesis by both testosterone and follicle-stimulating hormone (FSH). This production of FSH is associated with the production of ABP. The Sertoli cells themselves release another hormone, *inhibin,* which acts in negative-feedback fashion to regulate FSH secretion.

## LH and FSH

The testes are controlled by the two gonadotropic hormones secreted by the anterior pituitary, luteinizing hormone (LH) and follicle-stimulating hormone (FSH), which are named for their functions in females (p. 227). These two hormones are important to male reproductive function too. For example, testosterone is essential for maintaining spermatogenesis in the adult male, and it is under the direct control of LH.

### FEEDBACK CONTROL OF TESTICULAR FUNCTION

LH and FSH act on separate components of the testes (> Figure 17-10). LH acts on the Leydig (interstitial) cells to regulate testosterone secretion; this accounts for its alternative name in males—*interstitial cell–stimulating hormone (ICSH)*. FSH acts on the seminiferous tubules, specifically the Sertoli cells, to enhance spermatogenesis. (There is no alternative name for FSH in males.) Secretion of both LH and FSH from the anterior pituitary is stimulated in turn by a single hypothalamic hormone, **gonadotropin-releasing hormone (GnRH)** (p. 229).

Even though GnRH stimulates both LH and FSH secretion, the blood concentrations of these two gonadotropic hormones do not always parallel each other because two other regulatory factors besides GnRH—*testosterone* and *inhibin*—differentially influence the secretory rate of LH and FSH. Testosterone, the product of LH stimulation of the Leydig cells, acts in negative-feedback fashion to inhibit LH secretion in two ways. The predominant negative-feedback effect of testosterone is to decrease GnRH release by acting on the hypothalamus, which indirectly decreases both LH and FSH release by the anterior pituitary. In addition, testosterone acts directly on the anterior pituitary to reduce the responsiveness of the LH secretory cells to GnRH. The latter action explains why testosterone exerts a greater inhibitory effect on LH secretion than on FSH secretion.

The testicular inhibitory signal specifically directed at controlling FSH secretion is the peptide hormone **inhibin**, which is secreted by the Sertoli cells. Inhibin acts directly on the anterior pituitary to inhibit FSH secretion. This feedback inhibition of FSH by a Sertoli cell product is appropriate, because FSH stimulates spermatogenesis by acting on the Sertoli cells.

### ROLES OF TESTOSTERONE AND FSH IN SPERMATOGENESIS

Both testosterone and FSH play critical roles in controlling spermatogenesis, each exerting its effect by acting on the Sertoli cells. Testosterone is essential for both mitosis and meiosis of the germ cells, whereas FSH is needed for spermatid remodelling. Testosterone concentration is much higher in the testes than in the blood, because a substantial portion of this hormone produced locally by the Leydig cells is retained in the intratubular fluid complexed with androgen-binding protein secreted by the Sertoli cells. Only this high concentration of testicular testosterone is adequate to sustain sperm production.

## Gonadotropin-releasing hormone

Even though the fetal testes secrete testosterone, which directs masculine development of the reproductive system, after birth the testes become dormant until puberty. During the prepubertal period, LH and FSH are not secreted at adequate levels to stimulate any significant testicular activity. The prepubertal

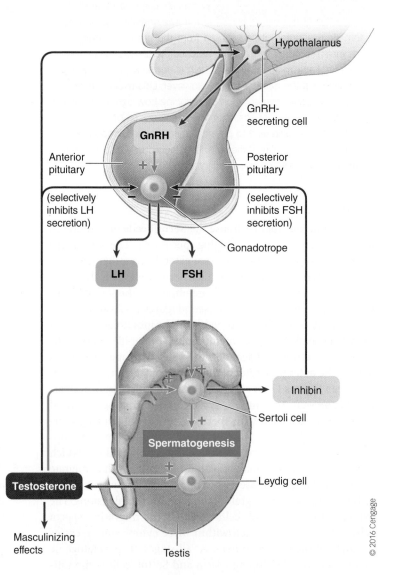

> FIGURE 17-10 **Control of testicular function**

© 2016 Cengage

17

delay in the onset of reproductive capability allows time for the individual to mature physically enough to handle child rearing. (This physical maturation is especially important in the female, whose body must support the developing fetus.)

During the prepubertal period, GnRH activity is inhibited. The pubertal process is initiated by an increase in GnRH activity sometime between 8 and 12 years of age. Early in puberty, GnRH secretion occurs only at night, causing a brief nocturnal increase in LH secretion and, accordingly, testosterone secretion. The extent of GnRH secretion gradually increases as puberty progresses until the adult pattern of GnRH, FSH, LH, and testosterone secretion is established. Under the influence of the rising levels of testosterone during puberty, the physical changes that encompass the secondary sexual characteristics and reproductive maturation become evident.

The factors responsible for initiating puberty in humans remain unclear. The leading proposal focuses on a potential role for the hormone *melatonin,* which is secreted by the *pineal gland* within the brain (p. 241). Melatonin, whose secretion decreases during exposure to the light and increases during exposure to the dark, has an antigonadotropic effect in many species. Light striking the eyes inhibits the nerve pathways that stimulate melatonin secretion. In many seasonally breeding species, the overall decrease in melatonin secretion in connection with longer days and shorter nights initiates the mating season. Some researchers suggest that an observed reduction in the overall rate of melatonin secretion at puberty in humans—particularly during the night, when the peaks in GnRH secretion first occur—is the trigger for the onset of puberty.

This completes our discussion of testicular function, and we now shift our attention to the roles of other components of the male reproductive system.

## The reproductive tract

The remainder of the male reproductive system (i.e., besides the testes) is designed to deliver sperm to the female reproductive tract. Essentially, it consists of (1) a tortuous pathway of tubes (the reproductive tract), which transports sperm from the testes to outside the body; (2) several accessory sex glands, which contribute secretions that are important to the viability and motility of the sperm; and (3) the penis, which is designed to penetrate and deposit the sperm within the vagina of the female. We will examine each of these parts in detail, beginning with the reproductive tract.

## COMPONENTS OF THE MALE REPRODUCTIVE TRACT

A comma-shaped **epididymis** is loosely attached to the rear surface of each testis (see ⟩ Figures 17-1 and 17-7a). The epididymis is about 5 m in length and tightly coiled. After sperm have been produced in the seminiferous tubules, they are swept into the epididymis due to the pressure created by the continual secretion of tubular fluid by the Sertoli cells. The spermatozoa entering the epididymis are nonmotile, which is in part a result of the low pH associated with the epididymis and vas deferens. The epididymal ducts from each testis converge to form a large, thick-walled, muscular duct called the **ductus (vas) deferens**. The ductus deferens from each testis passes up out of the scrotal sac and runs back through the inguinal canal into the abdominal cavity, where it eventually empties into the urethra at the neck of the bladder (see ⟩ Figure 17-1). The urethra carries sperm out of the penis during ejaculation, the forceful expulsion of semen from the body.

## FUNCTIONS OF THE EPIDIDYMIS AND DUCTUS DEFERENS

These ducts perform several important functions. The epididymis and ductus deferens serve as the sperm's exit route from the testis. As they leave the testis, the sperm are capable of neither movement nor fertilization. They gain both capabilities during their passage through the epididymis. This maturational process is stimulated by the testosterone retained within the tubular fluid bound to androgen-binding protein. Sperm's capacity to fertilize is enhanced even further by exposure to secretions of the female reproductive tract. This enhancement of sperm's capacity in the male and female reproductive tracts is known as **capacitation**. Scientists believe that *defensin,* a protein secreted by the epididymis that defends sperm from microorganisms, may serve a second role by boosting sperm's motility. The epididymis also concentrates the sperm a hundredfold by absorbing most of the fluid that enters from the seminiferous tubules. The maturing sperm are slowly moved through the epididymis into the ductus deferens by rhythmic contractions of the smooth muscle in the walls of these tubes.

The ductus deferens serves as an important site for sperm storage. Because the tightly packed sperm are relatively inactive and their metabolic needs are accordingly low, they can be stored in the ductus deferens for many days, even though they have no nutrient blood supply and are nourished only by simple sugars present in the tubular secretions.

### ⦚ Why It Matters
### Vasectomy

In a vasectomy, a common sterilization procedure in males, a small segment of each ductus deferens (alias vas deferens, hence the term *vasectomy*) is surgically removed. The segment removed is the part of the duct between the testis and the inguinal canal, so the surgery effectively blocks the exit of sperm from the testes. The sperm that build up behind the tied-off testicular end of the severed ductus are removed by phagocytosis. Although this procedure blocks sperm exit, it does not interfere with testosterone activity, because the Leydig cells secrete testosterone into the blood, not through the ductus deferens. Thus, testosterone-dependent masculinity or libido should not diminish after a vasectomy.

**17**

## The accessory sex glands

Several accessory sex glands—the seminal vesicles and prostate—empty their secretions into the duct system before it joins the urethra (see › Figure 17-1). A pair of saclike *seminal vesicles* empty into the last portion of the two ductus deferens, one on each side. The short segment of duct that passes beyond the entry point of the seminal vesicle to join the urethra is called the *ejaculatory duct*. The *prostate* is a large, single gland that completely surrounds the ejaculatory ducts and urethra. In a significant number of men, the prostate enlarges in middle to older age. Difficulty in urination is often encountered as the enlarging prostate impinges on the portion of the urethra that passes through the prostate. Another pair of accessory sex glands, the *bulbourethral glands,* drain into the urethra after it has passed through the prostate and just before it enters the penis. Numerous mucus-secreting glands also lie along the length of the urethra.

### SEMEN

During ejaculation, the accessory sex glands contribute secretions that provide support for the continuing viability of the sperm inside the female reproductive tract. These secretions constitute the bulk of the **semen**, which is a mixture of accessory sex gland secretions, sperm, and mucus. Sperm make up only a small percentage of the total ejaculated fluid.

### FUNCTIONS OF THE MALE ACCESSORY SEX GLANDS

Although the accessory sex gland secretions are not absolutely essential for fertilization, they do greatly facilitate the process:

- The **seminal vesicles** (1) supply fructose, which serves as the primary energy source for ejaculated sperm; (2) secrete *prostaglandins*, which stimulate contractions of the smooth muscle in both the male and female reproductive tracts, thereby helping to transport sperm from their storage site in the male to the site of fertilization in the female oviduct; (3) provide more than half the semen, which helps wash the sperm into the urethra and also dilutes the thick mass of sperm, enabling them to become mobile; and (4) secrete fibrinogen, a precursor of fibrin, which forms the meshwork of a clot (p. 449).

- The **prostate gland** (1) secretes an alkaline fluid that neutralizes the acidic vaginal secretions, an important function because sperm are more viable in a slightly alkaline environment; and (2) provides clotting enzymes and fibrinolysin. The prostatic clotting enzymes act on fibrinogen from the seminal vesicles to produce fibrin, which "clots" the semen, thus helping keep the ejaculated sperm in the female reproductive tract during withdrawal of the penis. Shortly thereafter, the seminal clot is broken down by fibrinolysin, a fibrin-degrading enzyme from the prostate, thus releasing mobile sperm within the female tract.

- During sexual arousal, the **bulbourethral glands** secrete a clear, mucus-like substance, the primary role of which is to lubricate the urethra for spermatozoa to pass through.

▌ Table 17-2 summarizes the locations and functions of the components of the male reproductive system.

Before turning to the act of delivering sperm to the female (sexual intercourse), we are going to briefly discuss the diverse roles of prostaglandins, which were first discovered in semen but are abundant throughout the body.

---

## ▌ TABLE 17-2 Location and Functions of the Components of the Male Reproductive System

| Component | Number and Location | Functions |
|---|---|---|
| **Testis** | Pair; located in the scrotum, a skin-covered sac suspended within the angle between the legs | Produce sperm<br>Secrete testosterone |
| **Epididymis and Ductus Deferens** | Pair; one epididymis attached to the rear of each testis; one ductus deferens travels from each epididymis up out of the scrotal sac through the inguinal canal and empties into the urethra at the neck of the bladder | Serve as the sperm's exit route from the testis<br>Serve as the site for maturation of the sperm for motility and fertility<br>Concentrate and store the sperm |
| **Seminal Vesicle** | Pair; both empty into the last portion of the ductus deferens, one on each side | Supply fructose to nourish the ejaculated sperm<br>Secrete prostaglandins that stimulate motility to help transport the sperm within the male and the female<br>Provide the bulk of the semen<br>Provide precursors for the clotting of semen |
| **Prostate Gland** | Single; completely surrounds the urethra at the neck of the bladder | Secretes an alkaline fluid that neutralizes the acidic vaginal secretions<br>Triggers clotting of the semen to keep the sperm in the vagina during penis withdrawal |
| **Bulbourethral Gland** | Pair; both empty into the urethra, one on each side, just before the urethra enters the penis | Secrete mucus for lubrication |

17

## Prostaglandins

Although **prostaglandins** were first identified in the semen and were believed to be of prostate gland origin (hence their name, even though they are actually secreted into the semen by the seminal vesicles), their production and actions are by no means limited to the reproductive system. These 20-carbon fatty acid derivatives are among the most ubiquitous chemical messengers in the body. They are produced in virtually all tissues from arachidonic acid, a fatty acid constituent of the phospholipids within the plasma membrane. On appropriate stimulation, arachidonic acid is split from the plasma membrane by a membrane-bound enzyme and then is converted into the appropriate prostaglandin, which acts as a paracrine locally within or near its site of production. After prostaglandins act, they are rapidly inactivated by local enzymes before they gain access to the blood; or if they do reach the circulatory system, they are swiftly degraded on their first pass through the lungs so that they are not dispersed through the systemic arterial system.

Prostaglandins are designated as belonging to one of three groups—PGA, PGE, or PGF—according to structural variations in the five-carbon ring that they contain at one end (> Figure 17-11). Within each group, prostaglandins are further identified by the number of double bonds present in the two side chains that project from the ring structure (e.g., $PGE_1$ has one double bond and $PGE_2$ has two double bonds).

Prostaglandins and other closely related arachidonic-acid derivatives—namely, *prostacyclins, thromboxanes,* and *leukotrienes*—are collectively known as **eicosanoids**, and they are among the most biologically active compounds known. Prostaglandins exert a bewildering variety of effects. Not only are slight variations in prostaglandin structure accompanied by profound differences in biological action, but the same prostaglandin molecule may even exert opposite effects in different tissues. Besides enhancing sperm transport in semen, these abundant chemical messengers are known or suspected to exert other actions in the female reproductive system and in the respiratory, urinary, digestive, nervous, and endocrine systems, in addition to affecting platelet aggregation, fat metabolism, and inflammation (▌Table 17-3). As scientists better understand the various actions of prostaglandins, new ways of manipulating them therapeutically are becoming available. A classic example is the use of Aspirin, which blocks the conversion of arachidonic acid into prostaglandins, for fever reduction and pain relief. Prostaglandin action is also therapeutically inhibited in the treatment of premenstrual symptoms and menstrual cramping. Furthermore, specific prostaglandins have been medically administered in such diverse situations as inducing labour, treating asthma, and treating gastric ulcers.

### ▌ TABLE 17-3 Actions of Prostglandins

| Body-System Activity | Actions of Prostaglandins |
|---|---|
| Reproductive System | Promote sperm transport by action on smooth muscle in the male and female reproductive tracts |
| | Play a role in ovulation |
| | Play an important role in menstruation |
| | Contribute to preparation of the maternal portion of the placenta |
| | Contribute to parturition |
| Respiratory System | Some promote bronchodilation, others bronchoconstriction |
| Urinary System | Increase the renal blood flow |
| | Increase excretion of water and salt |
| Digestive System | Inhibit HCl secretion by the stomach |
| | Stimulate intestinal motility |
| Nervous System | Influence neurotransmitter release and action |
| | Act at the hypothalamic "thermostat" to increase body temperature |
| | Exacerbate sensation of pain |
| Endocrine System | Enhance cortisol secretion |
| | Influence tissue responsiveness to hormones in many instances |
| Circulatory System | Influence platelet aggregation |
| Fat Metabolism | Inhibit fat breakdown |
| Body Defence System | Promote many aspects of inflammation, including development of fever |

© 2016 Cengage

### Check Your Understanding 17.2

1. List the functions of testosterone.
2. Define *seminiferous tubules, Leydig cells, Sertoli cells, spermatogenesis, spermiogenesis, spermatogonia, primary spermatocytes, secondary spermatocytes, spermatids,* and *spermatozoa.*

## 17.3 Sexual Intercourse between Males and Females

Ultimately, union of male and female gametes to accomplish reproduction in humans requires delivery of sperm-laden semen into the female vagina through the **sex act**, also known as **sexual intercourse**, **coitus**, or **copulation**.

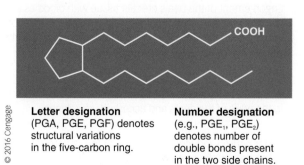

**Letter designation**
(PGA, PGE, PGF) denotes structural variations in the five-carbon ring.

**Number designation**
(e.g., $PGE_1$, $PGE_2$) denotes number of double bonds present in the two side chains.

© 2016 Cengage

> FIGURE 17-11 **Structure and nomenclature of prostaglandins**

17

## The male sex act

1. The *male sex act* involves two components: (1) **erection**, or hardening of the normally flaccid penis to permit its entry into the vagina; and (2) **ejaculation**, or forceful expulsion of semen into the urethra and out of the penis (▌ Table 17-4). In addition to these strictly reproduction-related components, the **sexual response cycle** encompasses broader physiologic responses that can be divided into four phases:

2. The *excitement phase* includes erection and heightened sexual awareness.

3. The *plateau phase* is characterized by intensification of these responses, plus more generalized body responses, such as steadily increasing heart rate, blood pressure, respiratory rate, and muscle tension.

4. The *orgasmic phase* includes ejaculation as well as other responses that culminate the mounting sexual excitement and are collectively experienced as an intense physical pleasure.

5. The *resolution phase* returns the genitalia and body systems to their prearousal state.

The human sexual response is a multicomponent experience that, in addition to these physiologic phenomena, encompasses emotional, psychological, and sociological factors. We will examine only the physiological aspects of sex.

## Erection

Erection is accomplished by engorgement of the penis with blood. The penis consists almost entirely of **erectile tissue** made up of three columns of spongelike vascular spaces extending the length of the organ (see ▸ Figure 17-1). In the absence of sexual excitation, the erectile tissues contain little blood, because the arterioles that supply these vascular chambers are constricted. As a result, the penis remains small and flaccid. During sexual arousal, these arterioles reflexly dilate and the erectile tissue fills with blood, causing the penis to enlarge both in length and width and to become more rigid. The reflexive dilation of the arterioles is initiated by the nervous system. The process typically takes about 10 seconds. During this erectile process, sympathetic input to the arterioles is inhibited, and nonadrenergic neurons release nitric oxide, thereby relaxing the arterioles. The veins that drain the erectile tissue are mechanically compressed by this engorgement and expansion of the vascular spaces; this reduces venous outflow and thereby contributes even further to the buildup of blood, or *vasocongestion*. These local vascular responses transform the penis into a hardened, elongated organ capable of penetrating the vagina.

### ERECTION REFLEX

The erection reflex is a spinal reflex triggered by stimulation of highly sensitive mechanoreceptors located in the *glans penis,* which caps the tip of the penis. A recently identified **erection-generating centre** lies in the lower spinal cord. Tactile stimulation of the glans reflexly triggers, by means of this centre, increased parasympathetic vasodilator activity and decreased sympathetic vasoconstrictor activity to the penile arterioles. The result is rapid, pronounced vasodilation of these arterioles and an ensuing erection (▸ Figure 17-12). As long as this spinal reflex arc remains intact, erection is possible even in men paralyzed by a higher spinal cord injury.

This parasympathetically induced vasodilation is the major instance of direct parasympathetic control over blood vessel diameter in the body. Parasympathetic stimulation brings about relaxation of penile arteriolar smooth muscle by nitric oxide, which causes arteriolar vasodilation in response to local tissue changes elsewhere in the body (p. 398). Arterioles are typically supplied only by sympathetic nerves; increased sympathetic activity produces vasoconstriction, and decreased sympathetic activity results in vasodilation (p. 401). Concurrent parasympathetic stimulation and sympathetic inhibition of penile arterioles accomplish vasodilation more rapidly and in greater magnitude than is possible in other arterioles supplied only by sympathetic nerves. Through this efficient means of rapidly increasing blood flow into the penis, the penis can become completely erect in as little as 5–10 seconds. At the same time, parasympathetic impulses promote secretion of lubricating mucus from the bulbo-urethral glands and the urethral glands in preparation for coitus.

---

## ▌ TABLE 17-4 Components of the Male Sex Act

| Components of Male Sex Act | Definition | How Accomplished |
| --- | --- | --- |
| **Erection** | Hardening of the normally flaccid penis to permit its entry into the vagina | Engorgement of the penis erectile tissue with blood as a result of marked parasympathetically induced vasodilation of the penile arterioles and mechanical compression of the veins |
| **Ejaculation** | | |
| *Emission phase* | Emptying of sperm and accessory sex gland secretions (semen) into the urethra | Sympathetically induced contraction of the smooth muscle in the walls of the ducts and accessory sex glands |
| *Expulsion phase* | Forceful expulsion of semen from the penis | Motor neuron–induced contraction of the skeletal muscles at the base of the penis |

© 2016 Cengage

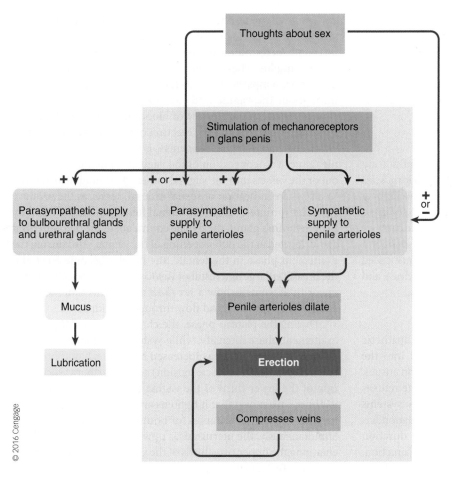

> FIGURE 17-12 **Erection reflex**

Erectile dysfunction is widespread. More than 50 percent of men between ages 40 and 70 experience some impotence, climbing to nearly 70 percent by age 70. No wonder, then, that more prescriptions were written for the much-publicized drug *sildenafil* (Viagra) during its first year on the market after its approval in 1998 for treating erectile dysfunction than for any other new drug in history. Sildenafil does not produce an erection, but it amplifies and prolongs an erectile response triggered by usual means of stimulation. Here's how the drug works: nitric oxide released in response to parasympathetic stimulation activates a membrane-bound enzyme, *guanylate cyclase,* within nearby arteriolar smooth muscle cells. This enzyme activates *cyclic guanosine monophosphate (cGMP),* an intracellular second messenger similar to cAMP (p. 217). Cyclic GMP, in turn, leads to relaxation of the penile arteriolar smooth muscle, bringing about pronounced local vasodilation. Under normal circumstances, once cGMP is activated and brings about an erection, this second messenger is broken down by the intracellular enzyme *phosphodiesterase 5 (PDE5).* Sildenafil inhibits PDE5. As a result, cGMP remains active longer, so that penile arteriolar vasodilation continues and the erection is sustained long enough for a formerly impotent man to accomplish the sex act. Just as pushing a pedal on a piano does not cause a note to be played but does prolong a played note, sildenafil cannot cause the release of nitric oxide and subsequent activation of erection-producing cGMP, but it can prolong the triggered response. The drug has no benefit for those who do not have erectile dysfunction, but its success rate has been high among sufferers of the condition.

A flurry of recent research has led to the discovery of numerous regions throughout the brain that can influence the male sexual response. The erection-influencing brain sites appear extensively interconnected and function as a unified network to either facilitate or inhibit the basic spinal erection reflex, depending on the momentary circumstances. As an example of facilitation, psychic stimuli, such as viewing something sexually exciting, can induce an erection in the complete absence of tactile stimulation of the penis. In contrast, failure to achieve an erection despite appropriate stimulation may result from inhibition of the erection reflex by higher brain centres. Let's examine erectile dysfunction in detail.

### ERECTILE DYSFUNCTION

*Clinical Note* A pattern of failing to achieve or maintain an erection suitable for sexual intercourse—**erectile dysfunction (impotence)**—may be attributable to psychological or physical factors. An occasional episode of a failed erection does not constitute impotence, but a man who becomes overly anxious about his ability to perform the sex act may well be on his way to chronic failure. Anxiety can lead to erectile dysfunction, which fuels the man's anxiety level and thus perpetuates the problem. Impotence may also arise from physical limitations, including nerve damage, certain medications that interfere with autonomic function, and problems with blood flow through the penis.

## Ejaculation

The second component of the male sex act is *ejaculation.* Like erection, ejaculation is a spinal reflex. The same types of tactile and psychic stimuli that induce erection cause ejaculation when the level of excitation intensifies to a critical peak. The overall ejaculatory response occurs in two phases: *emission* and *expulsion* (see ▮ Table 17-4).

### EMISSION

First, sympathetic impulses cause sequential contraction of smooth muscles in the prostate, reproductive ducts, and seminal vesicles. This contractile activity delivers prostatic fluid, then sperm, and finally seminal vesicle fluid (collectively, semen) into the urethra. This phase of the ejaculatory reflex is called **emission**. During this time, the sphincter at the neck of the bladder is tightly closed to prevent semen from entering the bladder and urine from being expelled along with the ejaculate through the urethra.

## EXPULSION

Second, filling of the urethra with semen triggers nerve impulses that activate a series of skeletal muscles at the base of the penis. Rhythmic contractions of these muscles occur at intervals of 0.8 seconds and increase the pressure within the penis, forcibly expelling the semen through the urethra to the exterior. This is the **expulsion** phase of ejaculation.

## ORGASM

The rhythmic contractions that occur during semen expulsion are accompanied by involuntary rhythmic throbbing of pelvic muscles and peak intensity of the overall body responses that were climbing during the earlier phases. Heavy breathing, a heart rate of up to 180 beats per minute, marked generalized skeletal muscle contraction, and heightened emotions are characteristic. These pelvic and overall systemic responses that culminate the sex act are associated with an intense pleasure characterized by a feeling of release and complete gratification, an experience known as **orgasm**.

## RESOLUTION

During the resolution phase following orgasm, sympathetic vasoconstrictor impulses slow the inflow of blood into the penis, causing the erection to subside. A deep relaxation ensues, often accompanied by a feeling of fatigue. Muscle tone returns to normal, while the cardiovascular and respiratory systems return to their prearousal level of activity. Once ejaculation has occurred, a temporary refractory period of variable duration ensues before sexual stimulation can trigger another erection. Males therefore cannot experience multiple orgasms within a matter of minutes, as females sometimes do.

## VOLUME AND SPERM CONTENT OF THE EJACULATE

The volume and sperm content of the ejaculate depend on the length of time between ejaculations. The average volume of semen is 2.75 mL, ranging from 2 mL to 6 mL, the higher volumes following periods of abstinence. An average human ejaculate contains about 180 million sperm (66 million/mL), but some ejaculates contain as many as 400 million sperm.

*Clinical Note* Both quantity and quality of sperm are important determinants of fertility. A man is considered clinically infertile if his sperm concentration falls below 20 million/mL of semen. Even though only one spermatozoon actually fertilizes the ovum, large numbers of accompanying sperm are needed to provide sufficient acrosomal enzymes to break down the barriers surrounding the ovum until the victorious sperm penetrates into the ovum's cytoplasm. The quality of sperm also must be taken into account when assessing the fertility potential of a semen sample. The presence of substantial numbers of sperm with abnormal motility or structure, such as sperm with distorted tails, reduces the chances of fertilization.

## The female sexual cycle

Both sexes experience the same four phases of the sexual cycle—excitement, plateau, orgasm, and resolution. Furthermore, the physiologic mechanisms responsible for orgasm are fundamentally the same in males and females.

The excitement phase in females can be initiated by either physical or psychological stimuli. Tactile stimulation of the clitoris and surrounding perineal area is an especially powerful sexual stimulus. These stimuli trigger spinal reflexes that bring about parasympathetically induced vasodilation of arterioles throughout the vagina and external genitalia, especially the clitoris. The resultant inflow of blood becomes evident as the swelling of the labia and the erection of the clitoris. The latter—like its male homologue, the penis—is composed largely of erectile tissue. Vasocongestion of the vaginal capillaries forces fluid out of the vessels into the vaginal lumen. This fluid, which is the first positive indication of sexual arousal, serves as the primary lubricant for intercourse. Additional lubrication is provided by the **Bartholin's gland**, the female equivalent of the bulbourethral gland, located at the outer opening of the vagina. Also during the excitement phase in the female, the nipples become erect and the breasts enlarge as a result of vasocongestion. In addition, the majority of women show a *sex flush* during this time, which is caused by increased blood flow through the skin.

During the plateau phase, the changes initiated during the excitement phase intensify, while systemic responses similar to those in the male (such as increased heart rate, blood pressure, respiratory rate, and muscle tension) occur. Further vasocongestion of the lower third of the vagina during this time reduces its inner capacity so that it tightens around the thrusting penis, heightening tactile sensation for both the female and the male. Simultaneously, the uterus raises upward, lifting the cervix and enlarging the upper two-thirds of the vagina. This ballooning, or **tenting effect**, creates a space for ejaculate deposition.

If erotic stimulation continues, the sexual response culminates in orgasm as sympathetic impulses trigger rhythmic contractions of the pelvic musculature at 0.8 second intervals, the same rate as in males. The contractions occur most intensely in the engorged lower third of the vaginal canal. Systemic responses identical to those of the male orgasm also occur. In fact, the orgasmic experience in females parallels that of males with two exceptions. First, there is no female counterpart to ejaculation. Second, females do not become refractory following an orgasm, so they can respond immediately to continued erotic stimulation and achieve multiple orgasms. If stimulation continues, the sexual intensity only diminishes to the plateau level following orgasm and can quickly be brought to a peak again. Women have been known to achieve as many as 12 successive orgasms in this manner.

During resolution, pelvic vasocongestion and the systemic manifestations gradually subside. As with males, this is a time of great physical relaxation for females.

Next we examine the female part of the reproductive process.

17

# Female Reproductive Physiology

Female reproductive physiology is more complex than male reproductive physiology.

## Complex cycling

Unlike the continuous sperm production and essentially constant testosterone secretion characteristic of the male, release of ova is intermittent, and secretion of female sex hormones displays wide cyclic swings. The tissues influenced by these sex hormones also undergo cyclic changes, the most obvious of which is the monthly menstrual cycle. During each cycle, the female reproductive tract is prepared for the fertilization and implantation of an ovum that's released from the ovary at ovulation. If fertilization does not occur, the cycle repeats. If fertilization does occur, the cycles are interrupted while the female system adapts to nurture and protect the newly conceived human being until it has developed into an individual capable of living outside the maternal environment. Furthermore, the female continues her reproductive functions after birth by producing milk (lactation) for the baby's nourishment. Thus, the female reproductive system is characterized by complex cycles that are interrupted by even more complex changes when pregnancy occurs.

The ovaries are the primary female reproductive organs, performing the dual function of producing ova (oogenesis) and secreting estrogen and progesterone. These hormones act together to promote fertilization of the ovum and to prepare the female reproductive system for pregnancy. Estrogen in the female governs many functions similar to those carried out by testosterone in the male, such as maturation and maintenance of the entire female reproductive system and establishment of female secondary sexual characteristics. In general, the actions of estrogen are important to preconception events. Estrogen is essential for ova maturation and release, development of physical characteristics that are sexually attractive to males, and transport of sperm from the vagina to the site of fertilization in the oviduct. Furthermore, estrogen contributes to breast development in anticipation of lactation. The other ovarian steroid, progesterone, is important in preparing a suitable environment for nourishing a developing embryo/fetus and for contributing to the breasts' ability to produce milk.

As in males, reproductive capability begins at puberty in females, but unlike males, who have reproductive potential throughout life, female reproductive potential ceases during middle age at menopause.

## The steps of gametogenesis

**Oogenesis** contrasts sharply with spermatogenesis in several important aspects, even though the identical steps of chromosome replication and division take place during gamete production in both sexes. The undifferentiated primordial germ cells in the fetal ovaries, the **oogonia** (comparable to the spermatogonia), divide mitotically to give rise to 6 million to 7 million oogonia by the fifth month of gestation, when mitotic proliferation ceases.

## FORMATION OF PRIMARY OOCYTES AND PRIMARY FOLLICLES

During the last part of fetal life, the oogonia begin the early steps of the first meiotic division, but do not complete it. Known now as **primary oocytes**, they contain the diploid number of 46 replicated chromosomes, which are gathered into homologous pairs, but do not separate. The primary oocytes remain in this state of **meiotic arrest** for years until they are prepared for ovulation.

Before birth, each primary oocyte is surrounded by a single layer of flat **granulosa cells**. Together, an oocyte and surrounding granulosa cells make up a **primordial follicle**. Oocytes that are not incorporated into follicles self-destruct by apoptosis. At birth, only about 2 million primary follicles remain, each containing a single primary oocyte capable of producing a single ovum. The traditional view is that no new oocytes or follicles appear after birth, with the follicles already present in the ovaries at birth serving as a reservoir from which all ova throughout the reproductive life of a female must arise. However, researchers recently discovered, in mice at least, that new oocytes and follicles are produced after birth from previously unknown ovarian stem cells capable of generating primordial germ cells (oogonia). Despite the potential for similar egg-generating stem cells in humans, the follicular pool gradually dwindles away as a result of processes that deplete the oocyte-containing follicles.

The pool of primordial follicles gives rise to an ongoing trickle of developing follicles. During this development phase, the granulosa cells become cuboidal, and the follicles are now referred to as **primary follicles**. Once it starts to develop, a follicle is destined for one of two fates: it will reach maturity and ovulate, or it will degenerate to form scar tissue, a process known as **atresia**. Until puberty, all the follicles that start to develop undergo atresia in the early stages without ever ovulating. Even for the first few years after puberty, many of the cycles are **anovulatory** (i.e., no ovum is released). Of the total pool of follicles, only about 400 will mature and release ova; 99.98 percent never ovulate, but instead undergo atresia at some stage in development. By menopause, which occurs on average in a woman's early 50s, few primary follicles remain, having either already ovulated or become atretic. From this point on, the woman's reproductive capacity ceases.

This limited gamete potential in females is in sharp contrast to the continual process of spermatogenesis in males, who have the potential to produce several hundred million sperm in a single day. Furthermore, considerable chromosome wastage occurs in oogenesis compared with spermatogenesis. Let's see how.

## FORMATION OF SECONDARY OOCYTES AND SECONDARY FOLLICLES

The primary oocyte within a primary follicle is still a diploid cell that contains 46 doubled chromosomes. From puberty until menopause, a portion of the resting pool of follicles starts developing into *secondary (antral) follicles* on a cyclic basis. The mechanisms determining which follicles in the pool will develop during a given cycle are unknown. Development of a secondary follicle is characterized by growth of the primary oocyte and by expansion and differentiation of the surrounding cell layers. The oocyte enlarges about a

**17**

thousandfold. This oocyte enlargement is caused by a buildup of cytoplasmic materials that are needed by the early embryo.

Just before ovulation, the primary oocyte, whose nucleus has been in meiotic arrest for years, completes its first meiotic division. This division yields two daughter cells, each receiving a haploid set of 23 doubled chromosomes, analogous to the formation of secondary spermatocytes (> Figure 17-13). However, almost all the cytoplasm remains with one of the daughter cells, now called the **secondary oocyte**, which is destined to become the ovum. The chromosomes of the other daughter cell together with a small share of cytoplasm form the **first polar body**. In this way, the ovum-to-be loses half of its chromosomes to form a haploid gamete but retains all of its nutrient-rich cytoplasm. The nutrient-poor polar body soon degenerates.

### FORMATION OF A MATURE OVUM

Actually, the secondary oocyte, and not the mature ovum, is ovulated and fertilized, but common usage refers to the developing female gamete as an *ovum* even in its primary and secondary oocyte stages. Sperm entry into the secondary oocyte is needed to trigger the second meiotic division. Unfertilized secondary oocytes never complete this final division. During this division, a half set of chromosomes and a thin layer of cytoplasm is extruded as the **second polar body**. The other half set of 23 unpaired chromosomes remains behind in what is now the **mature ovum**. These 23 maternal chromosomes unite with the 23 paternal chromosomes of the penetrating sperm to complete fertilization. If the first polar body has not already degenerated, it too undergoes the second meiotic division at the same time the fertilized secondary oocyte is dividing its chromosomes.

### COMPARISON OF STEPS IN OOGENESIS AND SPERMATOGENESIS

The steps involved in chromosome distribution during oogenesis parallel those of spermatogenesis, except that the cytoplasmic distribution and time span for completion sharply differ. Just as four haploid spermatids are produced by each primary spermatocyte, four haploid daughter cells are produced by each

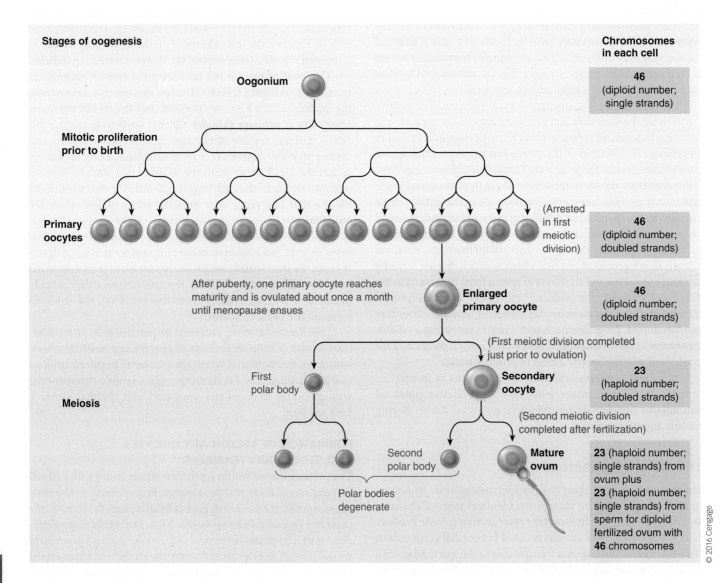

> **FIGURE 17-13 Oogenesis.** Compare with > Figure 17-8, spermatogenesis

primary oocyte (if the first polar body does not degenerate before it completes the second meiotic division). In spermatogenesis, each daughter cell develops into a highly specialized, motile spermatozoon unencumbered by unessential cytoplasm and organelles; its only destiny is to supply half the genes for a new individual. In oogenesis, however, of the four daughter cells only the one destined to become the ovum receives cytoplasm. This uneven distribution of cytoplasm is important, because the ovum, in addition to providing half the genes, provides all the cytoplasmic components needed to support early development of the fertilized ovum. The large, relatively undifferentiated ovum contains numerous nutrients, organelles, and structural and enzymatic proteins. The three other cytoplasm-scarce daughter cells, the polar bodies, rapidly degenerate—their chromosomes deliberately wasted.

Note also the considerable difference in time to complete spermatogenesis and oogenesis. It takes about two months for a spermatogonium to develop into fully remodelled spermatozoa. In contrast, development of an oogonium (present before birth) to a mature ovum requires anywhere from 11 years (beginning of ovulation at onset of puberty) to 50 years (end of ovulation at onset of menopause). The actual length of the active steps in meiosis is the same in both males and females, but in females the developing eggs remain in meiotic arrest for a variable number of years.

 The older age of ova released by women in their late 30s and 40s is believed to account for the higher incidence of genetic abnormalities, such as Down syndrome, in children born to women in this age range.

## The ovarian cycle

After the onset of puberty, the ovary constantly alternates between two phases: the **follicular phase**, which is dominated by the presence of *maturing follicles;* and the **luteal phase**, which is characterized by the presence of the *corpus luteum.* Normally, this cycle is interrupted only if pregnancy occurs and is finally terminated by menopause. The average ovarian cycle lasts 28 days, but this varies among women and among cycles in any particular woman. The follicle operates in the first half of the cycle to produce a mature egg ready for ovulation at midcycle. The corpus luteum takes over during the last half of the cycle to prepare the female reproductive tract for pregnancy if fertilization of the released egg occurs.

## The follicular phase

At any given time throughout the cycle, a portion of the primary follicles start to develop. However, only those that do so during the follicular phase, when the hormonal environment is right to promote their maturation, continue beyond the early stages of development. The others, lacking hormonal support, undergo atresia. During follicular development, as the primary oocyte is synthesizing and storing materials for future use if fertilized, important changes take place in the cells surrounding the reactivated oocyte in preparation for the egg's release from the ovary () Figure 17-14).

### PROLIFERATION OF GRANULOSA CELLS AND FORMATION OF THE ZONA PELLUCIDA

First, the single layer of granulosa cells in a primary follicle proliferates to form several layers that surround the oocyte. These granulosa cells and the oocyte secrete several glycoproteins that form a thick extracellular matrix that covers the oocyte and separates it from the surrounding granulosa cells. This intervening membrane is known as the **zona pellucida**.

Scientists have recently discovered gap junctions that penetrate the zona pellucida and extend between the oocyte and the surrounding granulosa cells in a developing follicle. Ions and small molecules can travel through these connecting tunnels. Recall that gap junctions between excitable cells permit the spread of action potentials from one cell to the next as charge-carrying ions pass through these connecting tunnels (p. 35). The cells in a developing follicle are not excitable, so gap junctions here serve a role other than transfer of electrical activity. Glucose, amino acids, and other important molecules are delivered to the oocyte from the granulosa cells through these tunnels, enabling the egg to stockpile these critical nutrients. Also, signalling molecules pass through the gap junctions in both directions, helping coordinate the changes that take place in the oocyte and the surrounding cells as both mature and prepare for ovulation.

### PROLIFERATION OF THECAL CELLS: ESTROGEN SECRETION

At the same time as the oocyte enlarges and the granulosa cells proliferate, specialized ovarian connective tissue cells in contact with the expanding granulosa cells both proliferate and differentiate to form an outer layer of **thecal cells**. The thecal and granulosa cells, collectively known as **follicular cells**, function as a unit to secrete estrogen. Of the three physiologically important estrogens—estradiol, estrone, and estriol—**estradiol** is the principal ovarian estrogen.

### FORMATION OF THE ANTRUM

The hormonal environment of the follicular phase promotes enlargement and development of the follicular cells' secretory capacity, converting the primary follicle into a **secondary (antral) follicle** capable of estrogen secretion. During this stage of follicular development, a fluid-filled cavity, or **antrum**, forms in the middle of the granulosa cells () Figures 17-14 **6** and 17-15). The follicular fluid originates partially from plasma that passes through capillary pores and partially from follicular cell secretions. As the follicular cells start producing estrogen, some of this hormone is secreted into the blood for distribution throughout the body. However, a portion of the estrogen collects in the hormone-rich antral fluid.

The oocyte has reached full size by the time the antrum begins to form. The shift to an antral follicle initiates a period of rapid follicular growth. During this time, the follicle increases in size from a diameter of less than 1 mm to 12–16 mm shortly before ovulation. Part of the follicular growth is due to continued proliferation of the granulosa and thecal cells, but most is due to a dramatic expansion of the antrum. As the follicle grows, estrogen is produced in increasing quantities.

17

NEL

The Reproductive System **747**

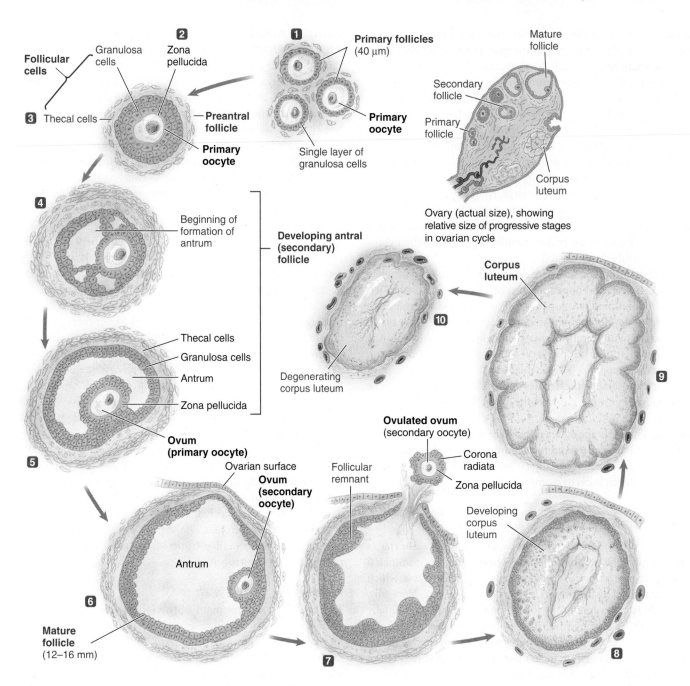

**Follicular cells**
- Granulosa cells
- Zona pellucida
- Thecal cells

**2**

**3**

**1** Primary follicles (40 μm)

Primary oocyte

Single layer of granulosa cells

Preantral follicle

Primary oocyte

Mature follicle

Secondary follicle

Primary follicle

Corpus luteum

Ovary (actual size), showing relative size of progressive stages in ovarian cycle

**4** Beginning of formation of antrum

Developing antral (secondary) follicle

Corpus luteum

**10**

Degenerating corpus luteum

Thecal cells
Granulosa cells
Antrum
Zona pellucida

**5** Ovum (primary oocyte)

**9**

Ovarian surface

Ovum (secondary oocyte)

Follicular remnant

Ovulated ovum (secondary oocyte)

Corona radiata
Zona pellucida

Developing corpus luteum

Antrum

**6**

Mature follicle (12–16 mm)

**7**

**8**

**1** In a primary follicle, a primary oocyte is surrounded by a single layer of granulosa cells.

**2** Under the influence of local paracrines, granulosa cells proliferate and form the zona pellucida around the oocyte.

**3** Surrounding ovarian connective tissue differentiates into thecal cells, converting a primary follicle into a preantral follicle.

**4** Follicles reaching the preantral stage are recruited for further development under the influence of FSH at the beginning of the follicular phase of the ovarian cycle. A recruited follicle develops into an antral, or secondary,

follicle as an estrogen-rich antrum starts to form.

**5** The antrum continues to expand as the secondary follicle rapidly grows.

**6** After about two weeks of rapid growth under the influence of FSH, the follicle has developed into a mature follicle, which has a greatly expanded antrum; the oocyte, which by now has developed into a secondary oocyte, is displaced to one side.

**7** At midcycle, in response to a burst in LH secretion, the mature follicle, bulging on the ovarian surface, ruptures and releases the

oocyte, resulting in ovulation and ending the follicular phase.

**8** Ushering in the luteal phase, the ruptured follicle develops into a corpus luteum under the influence of LH.

**9** The corpus luteum continues to grow and secrete progesterone and estrogen that prepare the uterus for implantation of a fertilized ovum.

**10** After 14 days, if a fertilized ovum does not implant in the uterus, the corpus luteum degenerates, the luteal phase ends, and a new follicular phase begins under the influence of a changing hormonal milieu.

> FIGURE 17-14 Ovarian cycle

**17**

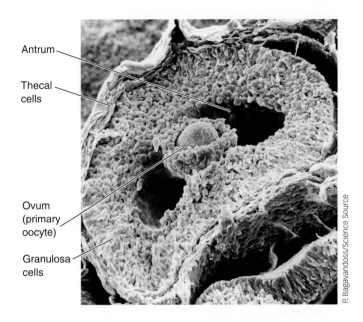

Antrum

Thecal cells

Ovum (primary oocyte)

Granulosa cells

P. Bagavandoss/Science Source

› **FIGURE 17-15** Scanning electron micrograph of a developing secondary follicle

### FORMATION OF A MATURE FOLLICLE

One of the follicles usually grows more rapidly than the others, developing into a **mature (preovulatory, tertiary, or Graafian) follicle** within about 14 days after the onset of follicular development. The antrum occupies most of the space in a mature follicle. The oocyte, surrounded by the zona pellucida and a single layer of granulosa cells, is displaced asymmetrically at one side of the growing follicle, in a little mound that protrudes into the antrum.

### OVULATION

The greatly expanded mature ovarian follicle bulges on the ovarian surface, creating a thin area that ruptures to release the oocyte at **ovulation**. The process of ovulation is controlled by the release of hormones (LH and FSH) from the anterior pituitary gland. Rupture of the follicle is facilitated by the release from the follicular cells of enzymes that digest the connective tissue in the follicular wall. The bulging wall is thus weakened so that it balloons out even farther, to the point that it can no longer contain the rapidly expanding follicular contents.

Just before ovulation, the oocyte completes its first meiotic division. The ovum (secondary oocyte), still surrounded by its tightly adhering zona pellucida and granulosa cells (now called the **corona radiata**, meaning "radiating crown"), is swept out of the ruptured follicle into the abdominal cavity by the leaking antral fluid (› Figure 17-14 **7**). The released ovum is quickly drawn into the oviduct, where fertilization may or may not take place.

The other developing follicles that failed to reach maturation and ovulate undergo degeneration, never to be reactivated. Occasionally, two (perhaps more) follicles reach maturation and ovulate at about the same time. If both are fertilized, **fraternal twins** result. Because fraternal twins arise from separate ova fertilized by separate sperm, they share no more in common than any other two siblings except for the same birth date. **Identical twins**, in contrast, develop from a single fertilized ovum

that completely divides into two separate, genetically identical embryos at a very early stage in development.

Rupture of the follicle at ovulation signals the end of the follicular phase and ushers in the luteal phase.

## The luteal phase

The ruptured follicle left behind in the ovary after release of the ovum changes rapidly. The granulosa and thecal cells remaining in the remnant follicle first collapse into the emptied antral space that has been partially filled by clotted blood.

### FORMATION OF THE CORPUS LUTEUM; ESTROGEN AND PROGESTERONE SECRETION

These old follicular cells soon undergo a dramatic structural transformation to form the **corpus luteum**, in a process called **luteinization** (› Figure 17-14 **8** and **9**). The follicular-turned-luteal cells enlarge and are converted into very active steroid hormone–producing tissue. Abundant storage of cholesterol, the steroid precursor molecule, in lipid droplets within the corpus luteum gives this tissue a yellowish appearance, hence its name (*corpus* means "body"; *luteum* means "yellow").

The corpus luteum becomes highly vascularized as blood vessels from the thecal region invade the luteinizing granulosa. These changes are appropriate for the corpus luteum's function: to secrete into the blood abundant quantities of progesterone along with smaller amounts of estrogen. Estrogen secretion in the follicular phase followed by progesterone secretion in the luteal phase is essential for preparing the uterus for implantation of a fertilized ovum. The corpus luteum becomes fully functional within four days after ovulation, but it continues to increase in size for another four or five days.

### DEGENERATION OF THE CORPUS LUTEUM

If the released ovum is not fertilized and does not implant, the corpus luteum degenerates within about 14 days after its formation (› Figure 17-14 **10**). The luteal cells degenerate and are phagocytized, the vascular supply is withdrawn, and connective tissue rapidly fills in to form a fibrous tissue mass known as the **corpus albicans** (white body). The luteal phase is now over, and one ovarian cycle is complete. A new wave of follicular development—which begins when the degeneration of the old corpus luteum has completed—signals the onset of a new follicular phase.

### CORPUS LUTEUM OF PREGNANCY

If fertilization and implantation do occur, the corpus luteum continues to grow and produce increasing quantities of progesterone and estrogen, instead of degenerating. Now called the *corpus luteum of pregnancy,* this ovarian structure persists until pregnancy ends. It provides the hormones essential for maintaining pregnancy until the developing placenta can take over this crucial function.

## The ovarian cycle

The ovary has two related endocrine units: the estrogen-secreting follicle during the first half of the cycle, and the corpus luteum, which secretes both progesterone and estrogen, during the last half of the cycle. These units are sequentially triggered by complex cyclic hormonal relationships among the hypothalamus, the anterior pituitary, and these two ovarian endocrine units.

As in the male, gonadal function in the female is directly controlled by the anterior pituitary gonadotropic hormones, namely, follicle-stimulating hormone (FSH) and luteinizing hormone (LH). These hormones, in turn, are regulated by hypothalamic gonadotropin-releasing hormone (GnRH) and feedback actions of gonadal hormones. Unlike in the male, however, control of the female gonads is complicated by the cyclic nature of ovarian function. For example, the effects of FSH and LH on the ovaries depend on the stage of the ovarian cycle. Furthermore, estrogen exerts negative-feedback effects during part of the cycle and positive-feedback effects during another part of the cycle, depending on the concentration of estrogen. Also in contrast to the male, FSH is not strictly responsible for gametogenesis, nor is LH solely responsible for gonadal hormone secretion. We will consider the control of follicular function, ovulation, and the corpus luteum separately, using ⟩ Figure 17-16 as a means of integrating the various concurrent and sequential activities that take place throughout the cycle. To facilitate correlation between this rather intimidating figure and the accompanying text description of this complex cycle, the grey numbers in the figure and its legend correspond to the grey numbers in the text description.

### CONTROL OF FOLLICULAR FUNCTION

We begin with the follicular phase of the ovarian cycle [1]. The factors that initiate follicular development are poorly understood. The early stages of preantral follicular growth and oocyte maturation do not require gonadotropic stimulation. Hormonal support is required, however, for antrum formation, further follicular development [2], and estrogen secretion [3]. Estrogen, FSH [4], and LH [5] are all needed. Antrum formation is induced by FSH. Both FSH and estrogen stimulate proliferation of the granulosa cells. Both LH and FSH are required for synthesis and secretion of estrogen by the follicle, but these hormones act on different cells and at different steps in the estrogen production pathway (⟩ Figure 17-17). Both granulosa and thecal cells participate in estrogen production. The conversion of cholesterol into estrogen requires a number of sequential steps, the last of which is conversion of androgens into estrogens. Thecal cells readily produce androgens but have limited capacity to convert them into estrogens. Granulosa cells, in contrast, contain the enzyme *aromatase,* so they can readily convert androgens into estrogens, but they cannot produce androgens in the first place. LH acts on the thecal cells to stimulate androgen production, whereas FSH acts on the granulosa cells to promote conversion of thecal androgens (which diffuse into the granulosa cells from the thecal cells) into estrogens. Because low basal levels of FSH [6] are sufficient to promote this final conversion to estrogen, the rate of estrogen secretion by the follicle primarily depends on the circulating level of LH, which continues to rise during the follicular phase [7]. Furthermore, as the follicle continues to grow, more estrogen is produced simply due to the presence of more estrogen-producing follicular cells.

Part of the estrogen produced by the growing follicle is secreted into the blood and is responsible for the steadily increasing plasma estrogen levels during the follicular phase 8.

The rest of the estrogen remains within the follicle, contributing to the antral fluid and stimulating further granulosa cell proliferation (⟩ Figure 17-17).

The secreted estrogen, in addition to acting on sex-specific tissues such as the uterus, inhibits the hypothalamus and anterior pituitary in typical negative-feedback fashion (⟩ Figure 17-18). The rising, moderate levels of estrogen that characterize the follicular phase act directly on the hypothalamus to inhibit GnRH secretion, thereby suppressing GnRH-prompted release of FSH and LH from the anterior pituitary. However, estrogen's primary effect is directly on the pituitary itself. Estrogen reduces the sensitivity to GnRH of the cells that produce gonadotropic hormones, especially the FSH-secreting cells.

This differential sensitivity of FSH- and LH-secreting cells induced by estrogen is at least in part responsible for the fact that the plasma FSH level, unlike the plasma LH concentration, declines during the follicular phase as the estrogen level rises [6]. Another contributing factor to the fall in FSH during the follicular phase is secretion of *inhibin* by the follicular cells. Inhibin preferentially inhibits FSH secretion by acting at the anterior pituitary, just as it does in the male. The decline in FSH secretion brings about atresia of all except the single most mature of the developing follicles.

In contrast to FSH, LH secretion continues to rise slowly during the follicular phase [7], despite inhibition of GnRH (and thus, indirectly, LH) secretion. This seeming paradox is due to the fact that estrogen alone cannot completely suppress **tonic LH secretion** (*tonic* here meaning "low-level, ongoing"); to completely inhibit tonic LH secretion, both estrogen and progesterone are required. Because progesterone does not appear until the luteal phase of the cycle, the basal level of circulating LH slowly increases during the follicular phase under incomplete inhibition by estrogen alone.

### CONTROL OF OVULATION

Ovulation and subsequent luteinization of the ruptured follicle are triggered by an abrupt, massive increase in LH secretion [9]. This **LH surge** brings about four major changes in the follicle:

1. It halts estrogen synthesis by the follicular cells [11].
2. It reinitiates meiosis in the oocyte of the developing follicle, apparently by blocking release of an *oocyte maturation inhibiting factor* produced by the granulosa cells. This substance is believed to be responsible for arresting meiosis in the primary oocytes once they are wrapped within granulosa cells in the fetal ovary.
3. It triggers production of locally acting prostaglandins, which induce ovulation by promoting vascular changes that cause rapid swelling of the follicle while inducing enzymatic digestion of the follicular wall. Together, these actions lead to rupture of the weakened wall that covers the bulging follicle [10].
4. It causes differentiation of follicular cells into luteal cells. Because the LH surge triggers both ovulation and luteinization, formation of the corpus luteum automatically follows ovulation [12]. Thus, the midcycle burst in LH secretion is a dramatic point in the cycle; it terminates the follicular phase and initiates the luteal phase [15].

17

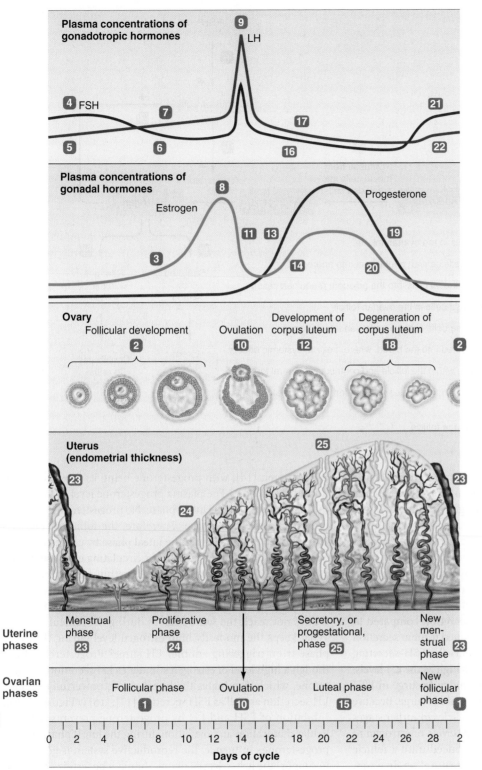

**Plasma concentrations of gonadotropic hormones**

**9** LH

**4** FSH **7** **21**

**5** **6** **17** **16** **22**

**Plasma concentrations of gonadal hormones**

**8** Progesterone

Estrogen

**11** **13** **19**

**3** **14** **20**

**Ovary**

| Follicular development | Ovulation | Development of corpus luteum | Degeneration of corpus luteum |

**2** **10** **12** **18** **2**

**Uterus (endometrial thickness)**

**25**

**23** **23**

**24**

| **Uterine phases** | Menstrual phase **23** | Proliferative phase **24** | Secretory, or progestational, phase **25** | New menstrual phase **23** |
| **Ovarian phases** | Follicular phase **1** | Ovulation **10** | Luteal phase **15** | New follicular phase **1** |

0  2  4  6  8  10  12  14  16  18  20  22  24  26  28  2

**Days of cycle**

© 2016 Cengage

**› FIGURE 17-16 Correlation between hormonal levels and cyclic ovarian and uterine changes.** During the follicular phase (the first half of the ovarian cycle **1**), the ovarian follicle **2** secretes estrogen **3** under the influence of FSH **4**, LH **5**, and estrogen **3** itself. The rising, moderate levels of estrogen inhibit FSH secretion, which declines during the last part of the follicular phase **6**, and incompletely suppress tonic LH secretion, which continues to rise throughout the follicular phase **7**. When the follicular output of estrogen reaches its peak **8**, the high levels of estrogen trigger a surge in LH secretion at midcycle **9**. This LH surge brings about ovulation of the mature follicle **10**. Estrogen secretion plummets **11** when the follicle meets its demise at ovulation.

The old follicular cells are transformed into the corpus luteum **12**, which secretes progesterone **13** as well as estrogen **14** during the luteal phase of the last half of the ovarian cycle **15**. Progesterone strongly inhibits both FSH **16** and LH **17**, which continue to decrease throughout the luteal phase. The corpus luteum degenerates **18** in about two weeks if the released ovum has not been fertilized and implanted in the uterus. Progesterone **19** and estrogen **20** levels sharply decrease when the corpus luteum degenerates, removing the inhibitory influences on FSH and LH. As these anterior pituitary hormone levels start to rise again **21**, **22** on the withdrawal of inhibition, they begin to stimulate the development of a new batch of follicles as a new follicular phase is ushered in **1**, **2**.

Concurrent uterine phases reflect the influences of the ovarian hormones on the uterus. Early in the follicular phase, the highly vascularized, nutrient-rich endometrial lining is sloughed off (the uterine menstrual phase) **23**. This sloughing results from the withdrawal of estrogen and progesterone **19**, **20** when the old corpus luteum degenerated at the end of the preceding luteal phase **18**. Late in the follicular phase, the rising levels of estrogen **3** cause the endometrium to thicken (the uterine proliferative phase) **24**. After ovulation **10**, progesterone from the corpus luteum **13** brings about vascular and secretory changes in the estrogen-primed endometrium to produce a suitable environment for implantation (the uterine secretory, or progestational, phase) **25**. When the corpus luteum degenerates **18**, a new ovarian follicular phase **1**, **2** and uterine menstrual phase **23** begin.

The two different modes of LH secretion—the tonic secretion of LH [7] responsible for promoting ovarian hormone secretion and the LH surge [9] that causes ovulation—not only occur at different times and produce different effects on the ovaries but also are controlled by different mechanisms. Tonic LH secretion is partially suppressed [7] by the inhibitory action of the rising, moderate levels of estrogen [3] during the follicular phase, and

is completely suppressed [17] by the increasing levels of progesterone during the luteal phase [13]. Because tonic LH secretion stimulates both estrogen and progesterone secretion, this is a typical negative-feedback control system.

In contrast, the LH surge is triggered by a *positive-feedback effect*. Whereas the rising, moderate levels of estrogen early in the follicular phase *inhibit* LH secretion, the high level of estrogen

**17**

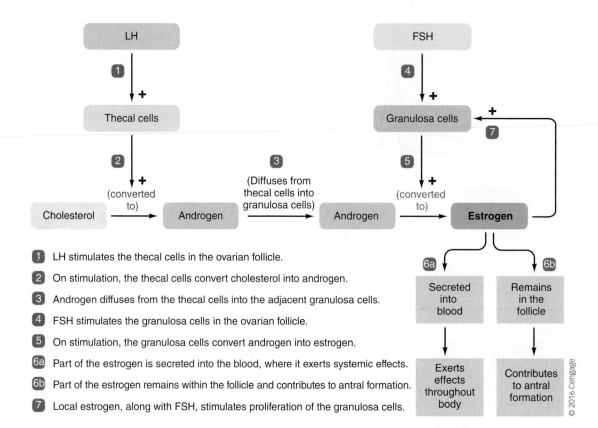

The following is a description of the figure:

| | LH | | FSH |
|---|---|---|---|

**1** LH stimulates the thecal cells in the ovarian follicle.

**2** On stimulation, the thecal cells convert cholesterol into androgen.

**3** Androgen diffuses from the thecal cells into the adjacent granulosa cells.

**4** FSH stimulates the granulosa cells in the ovarian follicle.

**5** On stimulation, the granulosa cells convert androgen into estrogen.

**6a** Part of the estrogen is secreted into the blood, where it exerts systemic effects.

**6b** Part of the estrogen remains within the follicle and contributes to antral formation.

**7** Local estrogen, along with FSH, stimulates proliferation of the granulosa cells.

Flow: Thecal cells → (converted to +) Cholesterol → Androgen → (Diffuses from thecal cells into granulosa cells) → Androgen → (converted to +) Estrogen. Granulosa cells (stimulated by FSH) +. Estrogen: 6a Secreted into blood → Exerts effects throughout body; 6b Remains in the follicle → Contributes to antral formation. 7 Local estrogen + FSH → Granulosa cells.

© 2016 Cengage

› **FIGURE 17-17** Production of estrogen by an ovarian follicle

that occurs during peak estrogen secretion late in the follicular phase [8] *stimulates* LH secretion and initiates the LH surge (› Figure 17-19). Thus, LH enhances estrogen production by the follicle, and the resultant peak estrogen concentration stimulates LH secretion. The high plasma concentration of estrogen acts directly on the hypothalamus to increase GnRH, thereby increasing both LH and FSH secretion. It also acts directly on the anterior pituitary to specifically increase the sensitivity of LH-secreting cells to GnRH. The latter effect accounts in large part for the much greater surge in LH secretion compared to FSH secretion at midcycle [9]. Also, continued inhibin secretion by the follicular cells preferentially inhibits the FSH-secreting cells, keeping the FSH levels from rising as high as the LH levels. There is no known role for the modest midcycle surge in FSH that accompanies the pronounced and pivotal LH surge. Because only a mature, preovulatory follicle, not follicles in earlier stages of development, can secrete high-enough levels of estrogen to trigger the LH surge, ovulation is not induced until a follicle has reached the proper size and degree of maturation. In a way, then, the follicle lets the hypothalamus know when it is ready to be stimulated to ovulate. The LH surge lasts for about a day at midcycle, just before ovulation.

### CONTROL OF THE CORPUS LUTEUM

LH maintains the corpus luteum; that is, after triggering development of the corpus luteum, LH stimulates ongoing steroid hormone secretion by this ovarian structure. Under the influence of LH, the corpus luteum secretes both progesterone [13]

and estrogen [14], with progesterone being its most abundant hormonal product. The plasma progesterone level increases for the first time during the luteal phase. No progesterone is secreted during the follicular phase. Therefore, the follicular phase is dominated by estrogen, and the luteal phase by progesterone.

A transitory drop in the level of circulating estrogen occurs at midcycle [11] as the estrogen-secreting follicle meets its demise at ovulation. The estrogen level climbs again during the luteal phase because of the corpus luteum's activity, although it does not reach the same peak as during the follicular phase. What keeps the modestly high estrogen level during the luteal phase from triggering another LH surge? Progesterone. Even though a high level of estrogen stimulates LH secretion, progesterone, which dominates the luteal phase, powerfully inhibits LH secretion as well as FSH secretion [17], [16] (› Figure 17-20). Inhibition of FSH and LH by progesterone prevents new follicular maturation and ovulation during the luteal phase. Under progesterone's influence, the reproductive system is gearing up to support the just-released ovum, should it be fertilized, instead of preparing other ova for release. No inhibin is secreted by the luteal cells.

The corpus luteum functions for an average of two weeks, then degenerates if fertilization does not occur [18]. The mechanisms that govern degeneration of the corpus luteum are not fully understood. The declining level of circulating LH [17], driven down by inhibitory actions of progesterone, undoubtedly contributes to the degeneration of the corpus luteum. Prostaglandins and estrogen released by the luteal cells themselves may

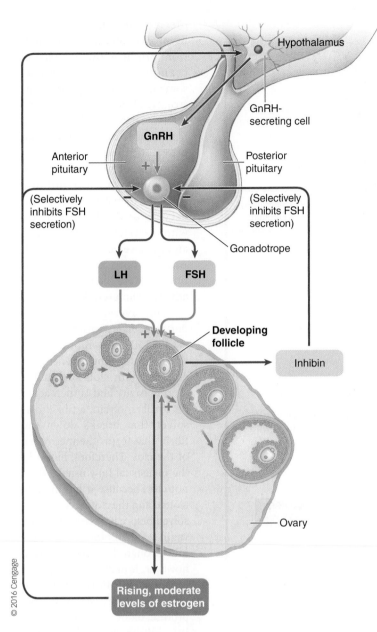

Hypothalamus

GnRH-
secreting cell

**GnRH**

Anterior
pituitary

Posterior
pituitary

(Selectively
inhibits FSH
secretion)

(Selectively
inhibits FSH
secretion)

Gonadotrope

**LH**    **FSH**

**Developing
follicle**

Inhibin

Ovary

Rising, moderate
levels of estrogen

© 2016 Cengage

› **FIGURE 17-18** Feedback control of FSH and tonic LH secretion during the follicular phase

reflects hormonal changes during the ovarian cycle, the menstrual cycle averages 28 days, as does the ovarian cycle, although even healthy women vary considerably from this mean. The outward manifestation of the cyclic changes in the uterus is the menstrual bleeding once during each menstrual cycle (i.e., once a month). Less obvious changes take place throughout the cycle, however, as the uterus is prepared for implantation should a released ovum be fertilized, then is stripped clean of its prepared lining (menstruation) if implantation does not occur, only to repair itself and start preparing for the ovum that will be released during the next cycle.

We now briefly examine the influences of estrogen and progesterone on the uterus, and then we consider the effects of cyclic fluctuations of these hormones on uterine structure and function.

### INFLUENCES OF ESTROGEN AND PROGESTERONE ON THE UTERUS

The uterus consists of two main layers: the **myometrium**, the outer smooth muscle layer; and the **endometrium**, the inner lining that contains numerous blood vessels and glands. Estrogen stimulates growth of both the myometrium and the endometrium. It also induces the synthesis of progesterone receptors in the endometrium. Therefore, the endometrium needs to be primed by estrogen before progesterone can exert an effect on it. Progesterone acts on the estrogen-primed endometrium to convert it into a hospitable and nutritious lining suitable for implantation of a fertilized ovum. Under the influence of progesterone, the endometrial connective tissue becomes loose and oedematous as a result of an accumulation of electrolytes and water, which facilitates implantation of the fertilized ovum. Progesterone further prepares the endometrium to sustain an early-developing embryo by inducing the endometrial glands to secrete and store large quantities of glycogen and by causing tremendous growth of the endometrial blood vessels. Progesterone also reduces the contractility of the uterus to provide a quiet environment for implantation and embryonic growth.

The menstrual cycle consists of three phases: the *menstrual phase*, the *proliferative phase*, and the *secretory*, or *progestational*, *phase*.

### MENSTRUAL PHASE

The menstrual phase is the most overt phase, characterized by discharge of blood and endometrial debris from the vagina. By convention, the first day of menstruation is considered the start of a new cycle. It coincides with termination of the ovarian luteal phase and onset of the follicular phase [23], › Figure 17-16. As the corpus luteum degenerates because fertilization and implantation of the ovum released during the preceding cycle did not take place [18], circulating levels of progesterone and estrogen drop precipitously [19], [20]. Because the net effect of progesterone and estrogen is

play a role. Demise of the corpus luteum terminates the luteal phase and sets the stage for a new follicular phase. As the corpus luteum degenerates, plasma progesterone [19] and estrogen [20] levels fall rapidly, because these hormones are no longer being produced. Withdrawal of the inhibitory effects of these hormones on the hypothalamus allows FSH [21] and tonic LH [22] secretion to modestly increase once again. Under the influence of these gonadotropic hormones, another batch of primary follicles [2] is induced to mature as a new follicular phase begins [1].

## Cyclic uterine changes

The fluctuations in circulating levels of estrogen and progesterone during the ovarian cycle induce profound changes in the uterus, giving rise to the **menstrual (uterine) cycle**. Because it

17

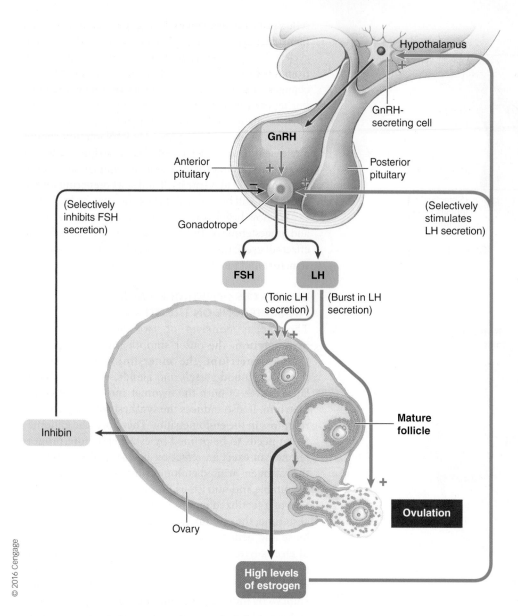

uterine contractions caused by prostaglandin overproduction produce the menstrual cramps (**dysmenorrhoea**) some women experience. Menstrual dysfunction, or abnormal menstruation, is common in female athletes. There is a high prevalence of amenorrhoea (stopping of menstrual flow), oligomenorrhoea (sporadic or slight menstrual flow), and delayed menarche in athletes who restrict caloric intake and/or have very low body fat. Amenorrhoea and other dysfunctions are likely to occur in athletes who exercise too much.

The average blood loss during a single menstrual period is 50–150 mL. Blood seeps slowly through the degenerating endometrium clots within the uterine cavity and is then acted on by fibrinolysin, a fibrin dissolver that breaks down the fibrin that forms the meshwork of the clot. Therefore, blood in the menstrual flow usually does not clot, because it has already clotted and the clot has been dissolved before it passes out of the vagina. When blood flows rapidly through the leaking vessels, however, it may not be exposed to sufficient fibrinolysin, so when the menstrual flow is most profuse, blood clots may appear. In addition to the blood and endometrial debris, large numbers of leukocytes are found in the menstrual flow. These white blood cells play an important defence role in helping the raw endometrium resist infection.

> FIGURE 17-19 **Control of the LH surge at ovulation**

to prepare the endometrium for implantation of a fertilized ovum, withdrawal of these steroids deprives the highly vascular, nutrient-rich uterine lining of its hormonal support.

The fall in ovarian hormone levels also stimulates release of a uterine prostaglandin that causes vasoconstriction of the endometrial vessels, disrupting the blood supply to the endometrium. The subsequent reduction in oxygen delivery causes death of the endometrium, including its blood vessels. The resulting bleeding through the disintegrating vessels flushes the dying endometrial tissue into the uterine lumen. Most of the uterine lining sloughs off during each menstrual period except for a deep, thin layer of epithelial cells and glands, from which the endometrium will regenerate. The same local uterine prostaglandin also stimulates mild rhythmic contractions of the uterine myometrium. These contractions help expel the blood and endometrial debris from the uterine cavity out through the vagina as **menstrual flow**. Excessive

Menstruation typically lasts for about five to seven days after degeneration of the corpus luteum, coinciding in time with the early portion of the ovarian follicular phase [23], [1]. Withdrawal of progesterone and estrogen [19], [20] on degeneration of the corpus luteum leads simultaneously to sloughing of the endometrium (menstruation) [23] and development of new follicles in the ovary [1], [2] under the influence of rising gonadotropic hormone levels [21], [22]. The drop in gonadal hormone secretion removes inhibitory influences from the hypothalamus and anterior pituitary, so FSH and LH secretion increases, and a new follicular phase begins. After five to seven days under the influence of FSH and LH, the newly growing follicles are secreting enough estrogen [3] to induce repair and growth of the endometrium.

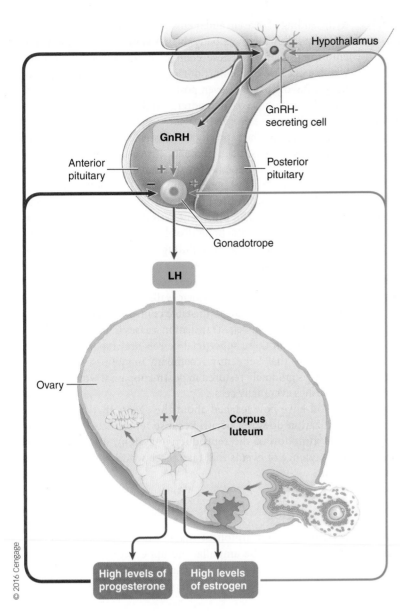

Ovary

> FIGURE 17-20 Feedback control during the luteal process

### PROLIFERATIVE PHASE

Menstrual flow ceases, and the **proliferative phase** of the uterine cycle begins concurrent with the last portion of the ovarian follicular phase as the endometrium starts to repair itself and proliferate [24] under the influence of estrogen from the newly growing follicles. When the menstrual flow ceases, a thin endometrial layer less than 1 mm thick remains. Estrogen stimulates proliferation of epithelial cells, glands, and blood vessels in the endometrium, increasing this lining to a thickness of 3–5 mm. The estrogen-dominant proliferative phase lasts from the end of menstruation to ovulation. Peak estrogen levels [8] trigger the LH surge [9] responsible for ovulation [10].

### SECRETORY, OR PROGESTATIONAL, PHASE

After ovulation, when a new corpus luteum is formed [12], the uterus enters the **secretory (progestational) phase**, which coincides in time with the ovarian luteal phase [25], [15]. The corpus

luteum secretes large amounts of progesterone [13] and estrogen [14]. Progesterone converts the thickened, estrogen-primed endometrium to a richly vascularized, glycogen-filled tissue. This period is called either the *secretory phase,* because the endometrial glands are actively secreting glycogen, or the *progestational (before pregnancy) phase,* referring to the development of a lush endometrial lining capable of supporting an early embryo. If fertilization and implantation do not occur, the corpus luteum degenerates and a new follicular phase and menstrual phase begin once again.

## Hormonal fluctuations

Hormonally induced changes also take place in the cervix during the ovarian cycle. Under the influence of estrogen during the follicular phase, the mucus secreted by the cervix becomes abundant, clear, and thin. This change—which is most pronounced when estrogen is at its peak and ovulation is approaching—facilitates passage of sperm through the cervical canal. After ovulation, under the influence of progesterone from the corpus luteum, the mucus becomes thick and sticky, essentially plugging up the cervical opening. This plug is an important defence mechanism, preventing bacteria (which might threaten a possible pregnancy) from entering the uterus from the vagina. Sperm also cannot penetrate this thick mucus barrier.

## Pubertal changes in females

Regular menstrual cycles are absent in both young and aging females, but for different reasons. The female reproductive system does not become active until puberty. Unlike the fetal testes, the fetal ovaries need not be functional, because in the absence of fetal testosterone secretion in a female, the reproductive system is automatically feminized, without requiring the presence of female sex hormones. The female reproductive system remains quiescent from birth until puberty, which occurs at about 12 years of age when hypothalamic GnRH activity increases for the first time. As in the male, the mechanisms that govern the onset of puberty are not clearly understood but are believed to involve the pineal gland and melatonin secretion.

GnRH begins stimulating release of anterior pituitary gonadotropic hormones, which in turn stimulate ovarian activity. The resulting secretion of estrogen by the activated ovaries induces growth and maturation of the female reproductive tract as well as development of the female secondary sexual characteristics. Estrogen's prominent action in the latter regard is to promote fat deposition in specific locations, such as the breasts, buttocks, and thighs, giving rise to the typical curvaceous female figure. Enlargement of the breasts at puberty is due primarily to fat deposition in the breast tissue, not to functional development of the mammary glands. The pubertal rise in estrogen also closes the epiphyseal plates, halting further growth in height; this is similar to the effect of testosterone-turned-estrogen in males. Three other pubertal

**17**

changes in females—growth of axillary and pubic hair, the pubertal growth spurt, and development of libido—are attributable to a spurt in adrenal androgen secretion at puberty, not to estrogen.

## Menopause

**Menopause** is a physiological event in which the menstrual cycle stops (pauses) permanently. Menopause is also known as the "change of life" and typically occurs between the ages of 45 and 55. It has traditionally been attributed to the limited supply of ovarian follicles present at birth. According to this proposal, once this reservoir is depleted, ovarian cycles—and hence, menstrual cycles—cease. Thus, the termination of reproductive potential in a middle-aged woman is already determined at her birth. Recent evidence suggests, however, that a midlife hypothalamic change, instead of aging ovaries, may trigger the onset of menopause. Evolutionarily, menopause may have developed as a mechanism that prevented pregnancy in women beyond the time that they could likely rear a child before their own death.

Males do not experience complete gonadal failure as females do, for two reasons. First, a male's germ cell supply is unlimited because mitotic activity of the spermatogonia continues. Second, gonadal hormone secretion in males is not inextricably dependent on gametogenesis, as in females. If female sex hormones were produced by separate tissues unrelated to those governing gametogenesis, as are male sex hormones, estrogen and progesterone secretion would not automatically stop when oogenesis stopped.

Menopause is preceded by a period of progressive ovarian failure characterized by increasingly irregular cycles and dwindling estrogen levels. This entire period of transition from sexual maturity to cessation of reproductive capability is known as the **climacteric (perimenopause)**. Ovarian estrogen production declines from as much as 300 mg per day to essentially nothing. Postmenopausal women are not completely devoid of estrogen, however, because adipose tissue, the liver, and the adrenal cortex continue to produce up to 20 mg of estrogen per day. In addition to the ending of ovarian and menstrual cycles, the loss of ovarian estrogen following menopause brings about many physical and emotional changes. These changes include vaginal dryness, which can cause discomfort during sex, and gradual atrophy of the genital organs. However, postmenopausal women still have a sex drive, because of their adrenal androgens.

*Clinical Note* Because estrogen has widespread physiological actions beyond the reproductive system, the dramatic loss of ovarian estrogen in menopause affects other body systems, most notably the skeleton and the cardiovascular system. Estrogen helps build strong bones, shielding premenopausal women from the bone-thinning condition of **osteoporosis** (p. 286). The postmenopausal reduction in estrogen increases activity of the bone-dissolving osteoclasts and diminishes activity of the bone-building osteoblasts. The result is decreased bone density and a greater incidence of bone fractures.

Estrogen also helps modulate the actions of epinephrine and norepinephrine on the arteriolar walls. The menopausal diminution of estrogen leads to unstable control of blood flow, especially in the skin vessels. Transient increases in the flow of warm blood through these superficial vessels are responsible for the **hot flashes** that frequently accompany menopause. Vasomotor stability is gradually restored in postmenopausal women so that hot flashes eventually cease.

You have now learned about the events that take place if fertilization does not occur. Because reproduction is the primary function of the reproductive system, we next turn to the sequence of events that take place when fertilization does occur.

## The oviduct

**Fertilization**, the union of male and female gametes, normally occurs in the ampulla, the upper third of the oviduct (> Figure 17-21). Both the ovum and the sperm must be transported from their gonadal site of production to the ampulla.

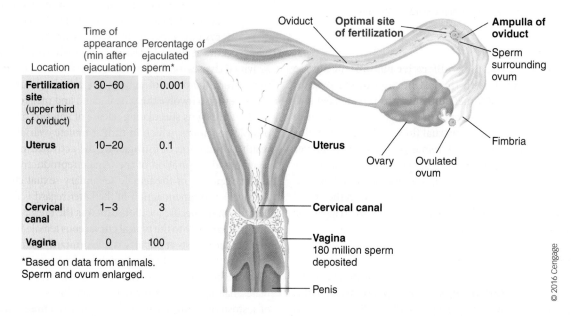

| Location | Time of appearance (min after ejaculation) | Percentage of ejaculated sperm* |
|---|---|---|
| **Fertilization site** (upper third of oviduct) | 30–60 | 0.001 |
| **Uterus** | 10–20 | 0.1 |
| **Cervical canal** | 1–3 | 3 |
| **Vagina** | 0 | 100 |

*Based on data from animals. Sperm and ovum enlarged.

© 2016 Cengage

> FIGURE 17-21 **Ovum and sperm transport to the site of fertilization**

## OVUM TRANSPORT TO THE OVIDUCT

When the ovum is released at ovulation, it is quickly picked up by the oviduct. The dilated end of the oviduct cups around the ovary and contains **fimbriae**, finger-like projections that contract in a sweeping motion to guide the released ovum into the oviduct (〉 Figure 17-21; see also 〉 Figure 17-2b). Furthermore, the fimbriae are lined by cilia—fine, hairlike projections that beat in waves toward the interior of the oviduct—further assuring the ovum's passage into the oviduct. Within the oviduct, the ovum is rapidly propelled by peristaltic contractions and ciliary action to the ampulla.

Conception can take place during a very limited time span in each cycle (the **fertile period**). If not fertilized, the ovum begins to disintegrate within 12 to 24 hours and is subsequently phagocytized by cells that line the reproductive tract. Fertilization must therefore occur within 24 hours after ovulation, when the ovum is still viable. Sperm typically survive about 48 hours but can survive up to five days in the female reproductive tract, so sperm deposited from five days before ovulation to 24 hours after ovulation may be able to fertilize the released ovum, although these times vary considerably.

*Clinical Note* Occasionally an ovum fails to be transported into the oviduct and remains instead in the peritoneal cavity. Rarely, such an ovum gets fertilized, resulting in an **ectopic abdominal pregnancy**, in which the fertilized egg implants in the rich vascular supply to the digestive organs rather than in its usual site in the uterus (*ectopic* means "out of place"). An abdominal pregnancy often leads to life-threatening haemorrhage because the digestive organ blood supply is not primed to respond appropriately to implantation as the endometrium is. If this unusual pregnancy proceeds to term, the baby must be delivered surgically, because the normal vaginal exit is not available. The probability of maternal complications at birth is greatly increased because the digestive vasculature is not designed to seal itself off after birth as the endometrium does.

## SPERM TRANSPORT TO THE OVIDUCT

After sperm are deposited in the vagina at ejaculation, they must travel through the cervical canal, through the uterus, and then to the egg in the upper third of the oviduct (〉 Figure 17-21). The first sperm arrive in the oviduct within half an hour after ejaculation. Even though sperm are mobile by means of whiplike contractions of their tails, 30 minutes is much too soon for a sperm's mobility to transport itself to the site of fertilization. To make this formidable journey, sperm need the help of the female reproductive tract.

The first hurdle is passage through the cervical canal. Throughout most of the cycle, because of high progesterone or low estrogen levels, the cervical mucus is too thick to permit sperm penetration. The cervical mucus becomes thin and watery enough to permit sperm to penetrate only when estrogen levels are high, as in the presence of a mature follicle about to ovulate. Sperm migrate up the cervical canal under their own power. The canal remains penetrable for only two or three days during each cycle, around the time of ovulation.

Once sperm have entered the uterus, contractions of the myometrium churn them around in "washing-machine" fashion. This action quickly disperses sperm throughout the uterine cavity. When sperm reach the oviduct, they are propelled to the fertilization site in the upper end of the oviduct by upward contractions of the oviduct

smooth muscle. These myometrial and oviduct contractions that facilitate sperm transport are induced by the high estrogen level just before ovulation, aided by seminal prostaglandins.

New research indicates that when sperm reach the ampulla, ova are not passive partners in conception. Mature eggs release **allurin**, a recently identified chemical that attracts sperm and causes them to propel themselves toward the waiting female gamete. Scientists also recently found the sperm receptor that detects and responds to the ovum-released chemoattractant. Interestingly, this receptor, called **hOR17-4**, is an olfactory receptor (OR) similar to those found in the nose for smell perception (p. 184). Therefore, sperm "smell" the egg. According to current thinking, the activation of the hOR17-4 receptor on binding with allurin (or another signal) from the egg triggers a second-messenger pathway in sperm that brings about intracellular $Ca^{2+}$ release. This $Ca^{2+}$ turns on the microtubule sliding that brings about tail movement and sperm swimming in the direction of the chemical signal.

Even around ovulation time, when sperm can penetrate the cervical canal, of the several hundred million sperm deposited in a single ejaculate, only a few thousand make it to the oviduct (〉 Figure 17-21). That only a very small percentage of the deposited sperm ever reach their destination is one reason why sperm concentration must be so high (20 million/mL of semen) for a man to be fertile. The other reason is that the acrosomal enzymes of many sperm are needed to break down the barriers surrounding the ovum (〉 Figure 17-22).

## FERTILIZATION

The tail of the sperm is used to manoeuvre for final penetration of the ovum. To fertilize an ovum, a sperm must first pass through the corona radiata and zona pellucida surrounding it. The sperm penetrates the corona radiata by means of membrane-bound enzymes in the surface membrane that surrounds the head (〉 Figure 17-23a, step 1; also see chapter-opening photo). Sperm can penetrate the zona pellucida only after binding with specific binding sites on the surface of this layer. The binding partners between the sperm and ovum were recently identified. **Fertilin**, a protein found on the plasma membrane of the sperm, binds with glycoproteins known as ZP3 in the outer layer of the zona pellucida. Only sperm of the same species can bind to these zona pellucida sites and pass through. Binding of sperm triggers the $Ca^{2+}$-dependent acrosome reaction, in which the acrosomal membrane disrupts and the acrosomal

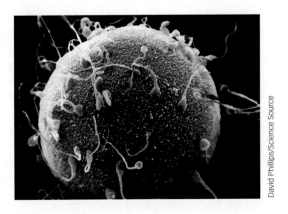

〉 **FIGURE 17-22 Scanning electron micrograph of sperm amassed at the surface of an ovum**

**17**

enzymes are released (step 2). Calcium that enters the sperm tail through the opened cation channels rapidly moves within a few seconds to the head, where it participates in the acrosome reaction. The acrosomal enzymes digest the zona pellucida, enabling the sperm, with its tail still beating, to tunnel a path through this protective barrier (step 3). The first sperm to reach the ovum itself fuses with the plasma membrane of the ovum (actually a secondary oocyte), and its head (bearing its DNA) enters the ovum's cytoplasm (step 4). The sperm's tail is frequently lost in this process, but the head carries the crucial genetic information. Sperm–egg fusion triggers a chemical change in the ovum's surrounding membrane that makes this outer layer impenetrable to the entry of any more sperm. This phenomenon is known as **block to polyspermy** (many sperm). Fertilization-induced release of intracellular $Ca^{2+}$ into the ovum cytosol triggers the exocytosis of enzyme-filled **cortical granules** from the outermost, or cortical, region of the ovum and into the space between the egg membrane and the zona pellucida (step 5). These enzymes diffuse into the zona pellucida, where they inactivate the ZP3 receptors so that other sperm reaching the zona pellucida cannot bind with it. The enzymes also harden the zona pellucida and seal off tunnels in the process to keep other penetrating sperm from advancing.

The released $Ca^{2+}$ in the ovum cytosol also triggers the second meiotic division of the egg, which is now ready to unite with the sperm to complete the fertilization process.

Within an hour, the sperm and egg nuclei fuse, thanks to a centrosome (microtubule organizing centre) provided by the sperm that forms microtubules; this brings the male and female chromosome sets together for uniting. In addition to contributing its half of the chromosomes to the fertilized ovum, now called a **zygote**, the victorious sperm also activates the ovum enzymes that are essential for early embryonic development. In this way, fertilization accomplishes the dual events of combining genes from the two parents to form a genetically unique organism and setting in motion the development of that organism.

## The blastocyst

During the first three to four days following fertilization, the zygote remains within the ampulla, because a constriction between the ampulla and the remainder of the oviduct canal prevents further movement of the zygote toward the uterus.

### THE BEGINNING STEPS IN THE AMPULLA

The zygote is not idle during this time, however. It rapidly undergoes a number of mitotic cell divisions to form a solid ball of cells called the *morula* (› Figure 17-24). Meanwhile, the rising levels of progesterone from the newly developing corpus luteum that formed after ovulation stimulate release of glycogen from the endometrium into the reproductive tract lumen for use as energy by the early embryo. The nutrients stored in the cytoplasm of the ovum can sustain the product of conception for less than a day. The concentration of secreted nutrients increases more rapidly in the small confines of the ampulla than in the uterine lumen.

### DESCENT OF THE MORULA TO THE UTERUS

About three to four days after ovulation, progesterone is produced in sufficient quantities to relax the oviduct constriction,

thereby permitting the morula to be rapidly propelled into the uterus by oviductal peristaltic contractions and ciliary activity. The temporary delay before the developing embryo passes into the uterus lets enough nutrients accumulate in the uterine lumen to support the embryo until implantation can take place. If the morula arrives prematurely, it dies.

When the morula descends to the uterus, it floats freely within the uterine cavity for another three to four days, living on endometrial secretions and continuing to divide. During the first six to seven days after ovulation, while the developing embryo is in transit in the oviduct and floating in the uterine lumen, the uterine lining is simultaneously being prepared for implantation under the influence of luteal-phase progesterone. During this time, the uterus is in its secretory, or progestational, phase, storing up glycogen and becoming richly vascularized.

Clinical Note Occasionally the morula fails to descend into the uterus and continues to develop and implant in the lining of the oviduct. This leads to an **ectopic tubal pregnancy,** which must be terminated. Ninety-five percent of ectopic pregnancies are tubal pregnancies. Such a pregnancy can never succeed, because the oviduct cannot expand as the uterus does to accommodate the growing embryo. The first warning of a tubal pregnancy is pain caused by the growing embryo stretching the oviduct. If not removed, the enlarging embryo will rupture the oviduct, causing possibly lethal haemorrhage.

### IMPLANTATION OF THE BLASTOCYST IN THE PREPARED ENDOMETRIUM

By the time the endometrium is suitable for implantation (about a week after ovulation), the morula has descended to the uterus and continued to proliferate and differentiate into a *blastocyst* capable of implantation. The week's delay after fertilization and before implantation allows time for both the endometrium and the developing embryo to prepare for implantation.

A **blastocyst** is a single-layer hollow ball of about 50 cells encircling a fluid-filled cavity, with a dense mass of cells grouped together at one side (› Figure 17-24). This dense mass, known as the **inner cell mass**, becomes the embryo/fetus itself. The rest of the blastocyst is never incorporated into the fetus; instead it serves a supportive role during intrauterine life. The thin outermost layer, the **trophoblast**, accomplishes implantation, after which it develops into the fetal portion of the placenta.

Before a blastocyst can implant, it must shed the zona pellucida and undergo a process called zona hatching. By this time the endometrium is ready to accept the early embryo. The blastocyst adheres to the uterine lining on the side of its inner cell mass (› Figure 17-25, step 1). **Implantation** begins when, on contact with the endometrium, the trophoblastic cells overlying the inner cell mass release protein-digesting enzymes. These enzymes digest pathways between the endometrial cells, permitting finger-like cords of trophoblastic cells to penetrate into the depths of the endometrium, where they continue to digest uterine cells (› Figure 17-25, step 2). Through its cannibalistic actions, the trophoblast performs the dual functions of (1) accomplishing implantation as it carves out a hole in the endometrium for the blastocyst and (2) making metabolic fuel and raw materials available for the developing embryo while the advancing trophoblastic projections break

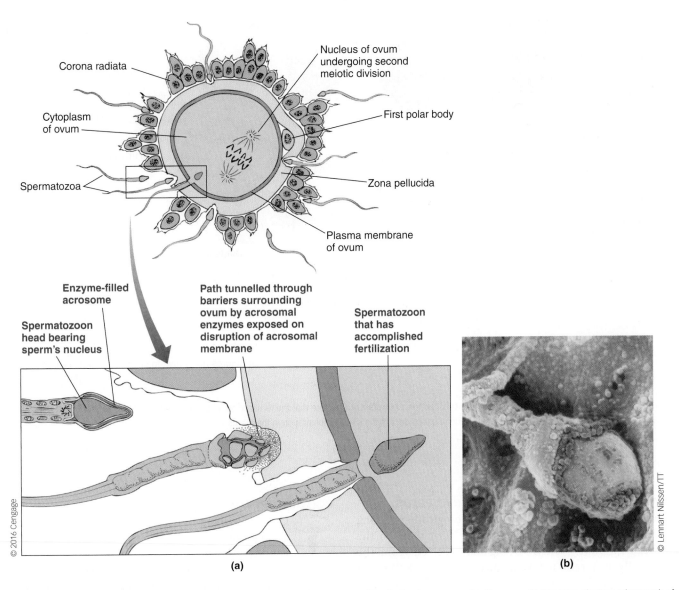

Corona radiata

Nucleus of ovum
undergoing second
meiotic division

Cytoplasm
of ovum

First polar body

Spermatozoa

Zona pellucida

Plasma membrane
of ovum

Enzyme-filled
acrosome

Path tunnelled through
barriers surrounding
ovum by acrosomal
enzymes exposed on
disruption of acrosomal
membrane

Spermatozoon
that has
accomplished
fertilization

Spermatozoon
head bearing
sperm's nucleus

© 2016 Cengage

© Lennart Nilsson/TT

(a)

(b)

> **FIGURE 17-23 Process of fertilization.** (a) Schematic representation of sperm tunnelling the barriers surrounding the ovum. (b) Scanning electron micrograph of a spermatozoon in which the acrosomal membrane has been disrupted and the acrosomal enzymes (in red) are exposed

down the nutrient-rich endometrial tissue. The cell walls of the advancing trophoblastic cells break down, forming a multinucleated syncytium that eventually becomes the fetal portion of the placenta.

Stimulated by the invading trophoblast, the endometrial tissue at the contact site undergoes dramatic changes that enhance its ability to support the implanting embryo. In response to a chemical messenger released by the blastocyst, the underlying endometrial cells secrete prostaglandins, which locally increase vascularization, produce oedema, and enhance nutrient storage. The endometrial tissue so modified at the implantation site is called the **decidua**. It is into this super-rich decidual tissue that the blastocyst becomes embedded. After the blastocyst burrows into the decidua by means of trophoblastic activity, a layer of endometrial cells covers over the surface of the hole, completely burying the blastocyst within the uterine lining (> Figure 17-25, step 3). The trophoblastic layer continues

to digest the surrounding decidual cells, providing energy for the embryo until the placenta develops.

■ **Clinical Connections**

The physician was insistant that Jourdain tell Anita right away because chlamydia infections in women are frequently (70–80 percent) asymptomatic for months or even years, but they can have serious consequences. Infections can lead to pelvic inflammatory disease, which can include infection of the uterus, the ovaries, or the fallopian tubes. Such infections can result in chronic pelvic pain and scarring of the female reproductive system. Scarring in the uterus can cause infertility by preventing the implantation of the blastocyst in the uterine endometrium. Chlamydia infections can also increase the chances of an ectopic pregnancy.

17

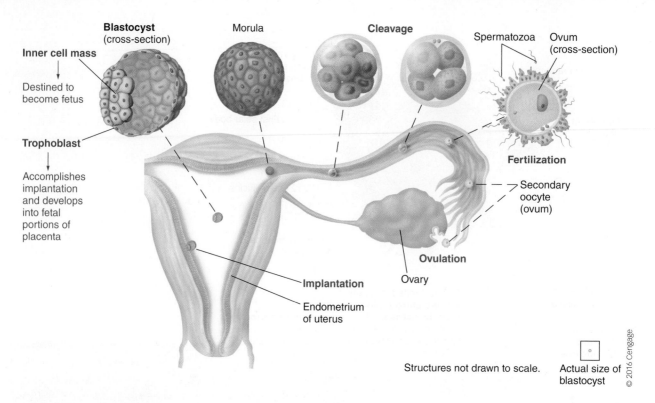

Inner cell mass
Destined to become fetus

Trophoblast
Accomplishes implantation and develops into fetal portions of placenta

Blastocyst (cross-section)

Morula

Cleavage

Spermatozoa    Ovum (cross-section)

Fertilization

Secondary oocyte (ovum)

Ovulation

Ovary

Implantation

Endometrium of uterus

Structures not drawn to scale.    Actual size of blastocyst

© 2016 Cengage

> FIGURE 17-24 **Early stages of development from fertilization to implantation.** Note that the fertilized ovum progressively divides and differentiates into a blastocyst as it moves from the site of fertilization in the upper oviduct to the site of implantation in the uterus.

## PREVENTING REJECTION OF THE EMBRYO/FETUS

What prevents the mother from immunologically rejecting the embryo/fetus, which is actually a "foreigner" to the mother's immune system—being half derived from the genetically different paternal chromosomes? Following are several proposals under investigation. New evidence indicates that the trophoblasts produce **Fas ligand**, which binds with **Fas**, a specialized receptor on the surface of approaching, activated, maternal cytotoxic T cells. Cytotoxic T cells are the immune cells that destroy foreign cells (p. 482). This binding of the trophoblasts with Fas triggers the apoptosis of the cytotoxic T immune cells that are targeted to destroy the developing foreigner, thus sparing the embryo/fetus from immune rejection. Other researchers have found that the fetal portion of the placenta, which is derived from trophoblasts, produces an enzyme, **indoleamine 2,3-dioxygenase (IDO)**, which destroys tryptophan. Tryptophan, an amino acid, is a critical factor in the activation of maternal cytotoxic T cells. So the embryo/fetus, through its trophoblast connection, is believed to defend itself against rejection by shutting down the activity of the mother's cytotoxic T cells in the placenta that would otherwise attack the developing foreign tissues. Furthermore, recent studies demonstrate that production of regulatory T cells is doubled or tripled in pregnant experimental animals. Regulatory T cells suppress maternal cytotoxic T cells that might target the fetus (p. 481).

## CONTRACEPTION

*Clinical Note* Couples wishing to engage in sexual intercourse but avoid pregnancy have a number of available methods of **contraception** (against conception). These methods act by blocking one of three major steps in the reproductive process: sperm transport to the ovum, ovulation, or implantation. (See Concepts, Challenges, and Controversies, pp. 764–765, for further details on the ways and means of contraception.)

Now let's examine the placenta in further detail.

## The placenta

The glycogen stores in the endometrium are sufficient to nourish the embryo only during its first few weeks. To sustain the growing embryo/fetus for the duration of its intrauterine life, the **placenta**, a specialized organ of exchange between the maternal and fetal blood, rapidly develops (> Figure 17-26). The placenta is derived from both trophoblastic and decidual tissue.

## FORMATION OF THE PLACENTA AND AMNIOTIC SAC

By day 12, the embryo is completely embedded in the decidua. By this time the trophoblastic layer is two cell layers thick and is called the **chorion**. As the chorion continues to release enzymes and expand, it forms an extensive network of cavities within the decidua. As the expanding chorion erodes decidual capillary walls, maternal blood leaks from the capillaries and fills these cavities. The blood is kept from clotting by an anticoagulant produced by

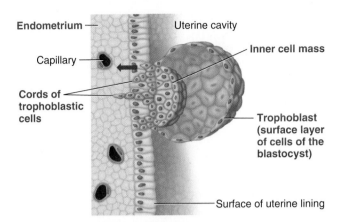

Endometrium — Uterine cavity

Capillary —

Cords of trophoblastic cells —

Inner cell mass

Trophoblast (surface layer of cells of the blastocyst)

Surface of uterine lining

**1** When the free-floating blastocyst adheres to the endometrial lining, cords of trophoblastic cells begin to penetrate the endometrium.

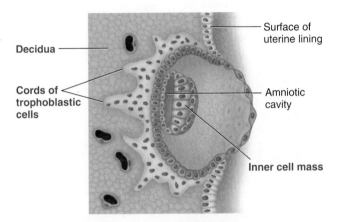

Surface of uterine lining

Decidua —

Cords of trophoblastic cells —

Amniotic cavity

Inner cell mass

**2** Advancing cords of trophoblastic cells tunnel deeper into the endometrium, carving out a hole for the blastocyst. The boundaries between the cells in the advancing trophoblastic tissue disintegrate.

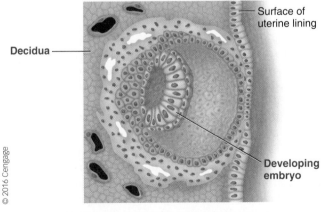

Surface of uterine lining

Decidua —

Developing embryo

**3** When implantation is finished, the blastocyst is completely buried in the endometrium.

© 2016 Cengage

> FIGURE 17-25 **Implantation of the blastocyst**

the chorion. Finger-like projections of chorionic tissue extend into the pools of maternal blood. Soon the developing embryo sends out capillaries into these chorionic projections to form **placental villi**. Some villi extend completely across the blood-filled spaces to anchor the fetal portion of the placenta to the endometrial tissue, but most simply project into the pool of maternal blood.

Each placental villus contains embryonic (later fetal) capillaries surrounded by a thin layer of chorionic tissue, which separates the embryonic/fetal blood from the maternal blood in the intervillus spaces. Maternal and fetal blood do not actually mingle, but the barrier between them is extremely thin. To visualize this relationship, think of your hands (the fetal capillary blood vessels) in rubber gloves (the chorionic tissue) immersed in water (the pool of maternal blood). Only the rubber gloves separate your hands from the water. In the same way, only the thin chorionic tissue (plus the capillary wall of the fetal vessels) separates the fetal and maternal blood. All exchanges between these two bloodstreams take place across this extremely thin barrier. This entire system of interlocking maternal (decidual) and fetal (chorionic) structures makes up the placenta.

Even though not fully developed, the placenta is well established and operational by five weeks after implantation. By this time, the heart of the developing embryo is pumping blood into the placental villi as well as to the embryonic tissues. Throughout gestation, fetal blood continuously traverses between the placental villi and the circulatory system of the fetus by means of the **umbilical arteries** (there are usually two) and **umbilical vein**, which are wrapped within the **umbilical cord**—a lifeline between the fetus and the placenta (> Figure 17-26). The maternal blood within the placenta is continuously replaced as fresh blood enters through the uterine arterioles; percolates through the intervillus spaces, where it exchanges substances with fetal blood in the surrounding villi; and then exits through the uterine vein.

Meanwhile, during the time of implantation and early placental development, the inner cell mass forms a fluid-filled **amniotic cavity** between the chorion and the portion of the inner cell mass destined to become the fetus. The epithelial layer that encloses the amniotic cavity is called the **amniotic sac (amnion)**. As it continues to develop, the amniotic sac eventually fuses with the chorion, forming a single combined membrane that surrounds the embryo/fetus. The fluid in the amniotic cavity, the **amniotic fluid**, which is similar in composition to normal ECF, surrounds and cushions the fetus throughout gestation (> Figure 17-27; also see > Figures 17-25 and 17-26).

### FUNCTIONS OF THE PLACENTA

During intrauterine life, the placenta performs the functions of the digestive system, the respiratory system, and the kidneys for the fetus. The fetus has these organ systems, but within the uterine environment they cannot (and do not need to) function. Nutrients and oxygen diffuse from the maternal blood across the thin placental barrier into the fetal blood, whereas carbon dioxide and other metabolic wastes simultaneously diffuse from the fetal blood into the maternal blood. The nutrients and oxygen brought to the fetus in the maternal blood are acquired by the mother's digestive and respiratory systems, and the carbon dioxide and wastes transferred into the maternal

**17**

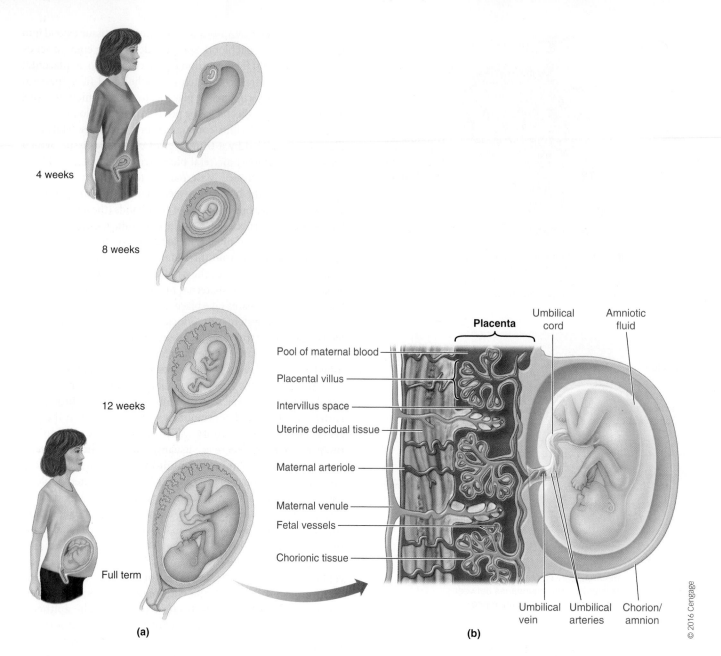

Umbilical cord

Amniotic fluid

**Placenta**

Pool of maternal blood

Placental villus

Intervillus space

Uterine decidual tissue

Maternal arteriole

Maternal venule

Fetal vessels

Chorionic tissue

Umbilical vein

Umbilical arteries

Chorion/ amnion

© 2016 Cengage

4 weeks

8 weeks

12 weeks

Full term

(a)

(b)

> **FIGURE 17-26 Placentation.** (a) Relationship between the developing fetus and uterus as pregnancy progresses. (b) Schematic representation of interlocking maternal and fetal structures that form the placenta. Finger-like projections of chorionic (fetal) tissue form the placental villi, which protrude into a pool of maternal blood. Decidual (maternal) capillary walls are broken down by the expanding chorion so that maternal blood oozes through the spaces between the placental villi. Fetal placental capillaries branch off the umbilical arteries and project into the placental villi. Fetal blood flowing through these vessels is separated from the maternal blood by the thin chorionic layer that forms the placental villi. Maternal blood enters through the maternal arterioles, then percolates through the pool of blood in the intervillus spaces. Here, exchanges are made between the fetal and maternal blood before the fetal blood leaves through the umbilical vein and maternal blood exits through the maternal venules.

**Source:** From STARR. Biology: Concepts and Applications w/CD- ROM + InfoTrac, 4E, Fig. 38.25b, p. 655. © 2000 Brooks/Cole, a part of Cengage Learning, Inc. Reproduced by permission. www.cengage.com/permissions

blood are eliminated by the mother's lungs and kidneys, respectively. Thus, the mother's digestive tract, respiratory system, and kidneys serve the fetus's needs as well as her own.

Some substances traverse the placental barrier by special mediated transport systems in the placental membranes, whereas others move across by simple diffusion. Unfortunately, many drugs, environmental pollutants, other chemical agents, and microorganisms in the mother's bloodstream can also cross

the placental barrier, and some of them may harm the developing fetus. Individuals born limbless as a result of exposure to *thalidomide*, a tranquilizer prescribed for pregnant women before this drug's devastating effects on the growing fetus were known, serve as a grim reminder of this fact. Similarly, newborns who have been exposed during gestation to their mother's abuse of a drug, such as heroin, suffer addiction withdrawal symptoms after birth. Even more common chemical agents, such as

**762 CHAPTER 17**

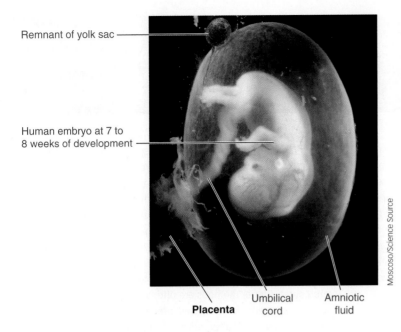

Remnant of yolk sac

Human embryo at 7 to 8 weeks of development

Moscoso/Science Source

**Placenta**  Umbilical cord  Amniotic fluid

> **FIGURE 17-27 A human fetus surrounded by the amniotic sac.** The fetus is near the end of the first trimester of development.

Aspirin, alcohol, and agents in cigarette smoke, can reach the fetus and have adverse effects. Likewise, fetuses can acquire AIDS before birth if their mothers are infected with the virus. A pregnant woman should therefore be very cautious about potentially harmful exposure from any source.

The placenta assumes yet another important responsibility; it becomes a temporary endocrine organ during pregnancy, a topic to which we now turn.

## Hormones and the placenta

The fetally derived portion of the placenta has the remarkable capacity to secrete a number of peptide and steroid hormones essential for maintaining pregnancy. The most important are *human chorionic gonadotropin, estrogen,* and *progesterone* (▌ Table 17-5). Serving as the major endocrine organ of pregnancy, the placenta is unique among endocrine tissues in two ways. First, it is a transient tissue. Second, secretion of its hormones is not subject to extrinsic control, this in contrast to the stringent, often complex mechanisms that regulate the secretion of other hormones. Instead, the type and rate of placental hormone secretion depend primarily on the stage of pregnancy.

### SECRETION OF HUMAN CHORIONIC GONADOTROPIN

One of the first endocrine events is secretion by the developing chorion of **human chorionic gonadotropin (hCG)**, a peptide hormone that prolongs the lifespan of the corpus luteum. Recall that during the ovarian cycle, the corpus luteum degenerates and the highly prepared, luteal-dependent uterine lining sloughs off if fertilization and implantation do not occur. When fertilization does occur, the implanted blastocyst saves itself from being flushed out in menstrual flow by producing hCG. This hormone is functionally similar to LH and structurally similar enough that it can also bind to LH receptors. It is because of

these similarities that hCG stimulates and maintains the corpus luteum so it does not degenerate. Now called the **corpus luteum of pregnancy**, this ovarian endocrine unit grows even larger and produces increasingly greater amounts of estrogen and progesterone for an additional 10 weeks until the placenta takes over secretion of these steroid hormones. Because of the persistence of estrogen and progesterone, the thick, pulpy endometrial tissue is maintained rather than shed. Accordingly, menstruation ceases during pregnancy.

Stimulation by hCG is necessary to maintain the corpus luteum of pregnancy because LH, which maintains the corpus luteum during the normal luteal phase of the uterine cycle, is suppressed through feedback inhibition by the high levels of progesterone.

Maintenance of a normal pregnancy depends on high concentrations of progesterone and estrogen. Therefore, hCG production is critical during the first trimester to maintain ovarian output of these hormones. In a male fetus, hCG also stimulates the precursor Leydig cells in the fetal testes to secrete testosterone, which masculinizes the developing reproductive tract.

The secretion rate of hCG increases rapidly during early pregnancy to save the corpus luteum from demise. Peak secretion of hCG occurs about 60 days after the end of the last menstrual period (> Figure 17-28). By the tenth week of pregnancy, hCG output declines to a low rate of secretion that is maintained for the duration of gestation. The fall in hCG occurs at a time when the corpus luteum is no longer needed for its steroid hormone output, because the placenta has begun to secrete substantial quantities of estrogen and progesterone. The corpus luteum of pregnancy partially regresses as hCG secretion dwindles, but it is not converted into scar tissue until after delivery of the baby.

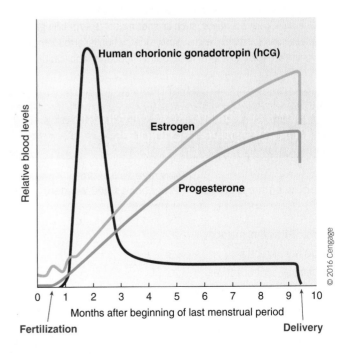

> **FIGURE 17-28 Secretion rates of placental hormones**

17

# The Ways and Means of Contraception

THE TERM *CONTRACEPTION* REFERS TO THE PROCESS of avoiding pregnancy while engaging in sexual intercourse. The available methods of contraception range in ease of use and effectiveness (see the accompanying table). These methods can be grouped into three categories based on the means by which they prevent pregnancy: (1) blockage of sperm transport to the ovum, (2) prevention of ovulation, or (3) blockage of implantation. After examining the most common ways that contraception can be accomplished by these means, we will briefly look at future contraceptive possibilities and then discuss termination of unwanted pregnancies.

### Blockage of Sperm Transport to the Ovum

■ *Natural contraception* or the *rhythm method* of birth control relies on abstinence from intercourse during the woman's fertile period. The woman can predict when ovulation will occur by keeping careful records of her menstrual cycles. Because of variability in cycles, this technique is only partially effective. The time of ovulation can be determined more precisely by recording body temperature each morning before getting up. Body temperature rises slightly about a day after ovulation has taken place. The temperature rhythm method is not useful in determining when it is safe to engage in intercourse before ovulation, but it can be helpful in determining when it is safe to resume sex after ovulation.

■ *Coitus interruptus* involves withdrawal of the penis from the vagina before ejaculation occurs. This method is only moderately effective, however, because timing is difficult, and some sperm may pass out of the urethra prior to ejaculation.

■ *Chemical contraceptives*, such as spermicidal (sperm-killing) jellies, foams, creams, and suppositories, when inserted into the vagina are toxic to sperm for about an hour after application.

## ■ Average Failure Rate of Various Contraceptive Techniques

| Contraceptive Method | Average Failure Rate (Annual Pregnancies/100 Women) |
| --- | --- |
| None | 90 |
| Natural (rhythm) methods | 20–30 |
| Coitus interruptus | 23 |
| Chemical contraceptives | 20 |
| Barrier methods | 10–15 |
| Oral contraceptives | 2–2.5 |
| Implanted contraceptives | 1 |
| Intrauterine device | 4 |

■ *Barrier methods* mechanically prevent sperm transport to the oviduct. For males, the *condom* is a thin, strong, rubber or latex sheath placed over the erect penis before ejaculation to prevent sperm from entering the vagina. For females, the *diaphragm*, which must be fitted by a trained professional, is a flexible rubber dome that is inserted through the vagina and positioned over the cervix to block sperm entry into the cervical canal. It is held in position by lodging snugly against the vaginal wall and must be left in place for at least 6 hours, but no longer than 24 hours, after intercourse. Barrier methods are often used in conjunction with spermicidal agents for increased effectiveness. The *cervical cap* is a recently developed alternative to the diaphragm. Smaller than a diaphragm, the cervical cap—coated with a film of spermicide—cups over the cervix and is held in place by suction.

■ The *female condom* (or *vaginal pouch*) is the latest barrier method developed. It is a polyurethane, cylindrical pouch, 18 cm long, closed on one end and open on the other end, with a flexible ring at both ends. The ring at the closed end of the device is inserted into the vagina and fits over the cervix, similar to a diaphragm. The ring at the open end of the pouch is positioned outside the vagina over the external genitalia.

■ *Sterilization,* which involves surgical disruption of either the ductus deferens (*vasectomy*) in men or the oviduct (*tubal ligation*) in women, is considered a permanent method of preventing sperm and ovum from uniting.

### Prevention of Ovulation

■ *Oral contraceptives,* or *birth control pills,* available only by prescription, prevent ovulation primarily by suppressing gonadotropin secretion. These pills, which contain synthetic estrogen-like and progesterone-like steroids, are taken for three weeks, either in combination or in sequence, and then withdrawn for one week. These steroids, like the natural steroids produced during the ovarian cycle, inhibit GnRH and thus FSH and LH secretion. As a result, follicle maturation and ovulation do not take place, so conception is impossible. The endometrium responds to the exogenous steroids by thickening and developing secretory capacity, just as it would to the natural hormones. When these synthetic steroids are withdrawn after three weeks, the endometrial lining sloughs and menstruation occurs, as it normally would on degeneration of the corpus luteum. In addition to blocking ovulation, oral contraceptives prevent pregnancy by increasing the viscosity of cervical mucus, which makes sperm penetration more difficult, and by decreasing muscular contractions in the female reproductive tract, which reduces sperm transport to the oviduct. Oral contraceptives have been shown to increase the risk of intravascular clotting, especially in women who also smoke tobacco.

■ Several other contraceptive methods contain synthetic female sex hormones and act similarly to birth control pills to prevent ovulation. These include (1) *long-acting subcutaneous* (under the skin) *implantation* of hormone-containing capsules that gradually release hormones at a nearly steady rate for five years and (2) *birth control patches* impregnated with hormones that are absorbed through the skin.

### Blockage of Implantation

Medically, pregnancy is not considered to begin until implantation. According to this view, any mechanism that interferes with implantation is said to prevent pregnancy. Not all hold this view, however. Some consider pregnancy to begin at time of fertilization. To these people, any interference with implantation is a form of abortion. Therefore, methods of contraception that rely on blockage of implantation are more controversial than methods that prevent fertilization from taking place.

- Blockage of implantation is most commonly accomplished by a physician inserting a small *intrauterine device (IUD)* into the uterus. The IUD's mechanism of action is not completely understood, although most evidence suggests that the presence of this foreign object in the uterus induces a local inflammatory response that prevents implantation of a fertilized ovum.

- Implantation can also be blocked by so-called *morning-after pills,* also known as *emergency contraception.* The first term is actually a misnomer, because these pills can prevent pregnancy if taken within 72 hours after, not just the morning after, unprotected sexual intercourse. The most common form of emergency contraception in Canada is called Plan B and is available, in most provinces, at pharmacies without a prescription. Plan B contains levonorgestrel, a second-generation synthetic progestogen, which acts as a hormone that binds and activates the progesterone receptor. In addition to blocking implantation, Plan B can also prevent both ovulation and fertilization, making it an effective treatment no matter where the woman is in her cycle when she takes the pills.

### Future Possibilities

On the horizon are improved varieties of currently available contraceptive techniques, such as a new birth control pill that suppresses ovulation and menstrual periods for months at a time.

- A future birth control technique is *immunocontraception*—the use of vaccines that prod the immune system to produce antibodies targeted against a particular protein critical to the reproductive process. The contraceptive effects of the vaccines are expected to last about a year. For example, in the testing stage is a vaccine that induces the formation of antibodies against human chorionic gonadotropin so that this essential corpus luteum–supporting hormone is not effective if pregnancy occurs. Another promising immunocontraception approach is aimed at blocking the acrosomal enzymes so that sperm cannot enter the ovum. Still other researchers have developed an experimental vaccine that targets a protein added by the epididymis to the surface of sperm during their maturation.

- Some researchers are exploring ways to block the union of sperm and egg by interfering with a specific interaction that normally occurs between the male and the female gametes. For example, under study are chemicals introduced into the vagina that trigger premature release of the acrosomal enzymes, depriving the sperm of a means to fertilize an ovulated egg.

- Some scientists are seeking ways to manipulate hormones to block sperm production in males without depriving the man of testosterone. One example of a male contraception under development is a combination of testosterone and progestin that inhibits GnRH and the gonadotropic hormones, thereby turning off the signals that stimulate spermatogenesis.

- Another outlook for male contraception is chemical sterilization designed to be reversed, unlike surgical sterilization, which is considered irreversible. In this experimental technique, a nontoxic polymer is injected into the ductus deferens, where the chemical interferes with sperm's fertilizing capabilities. Flushing of the polymer from the ductus deferens by a solvent reverses the contraceptive effect.

- One interesting avenue being explored holds hope for a unisex contraceptive that would stop sperm in their tracks and could be used by either males or females. Based on preliminary findings, the idea is to use $Ca^{2+}$-blocking drugs to prevent the entry of $Ca^{2+}$ into sperm tails. As in muscle cells, $Ca^{2+}$ switches on the contractile apparatus responsible for the sperm's motility. With no $Ca^{2+}$, sperm would not be able to manoeuvre to accomplish fertilization.

### Termination of Unwanted Pregnancies

- When contraceptive practices fail or are not used and an unwanted pregnancy results, women often turn to *abortion* to terminate the pregnancy. In Canada, approximately 100 000 abortions take place each year, with about half of these occurring in hospitals and the other half in clinics. Currently, the surgical removal of the embryo/fetus by a physician is legal in Canada, but the use of the drug RU-486 is not.

- There has been much debate over the use of RU-486 (the "abortion pill," or mifepristone). Clinical trials were conducted in 2000 women using both methotrexate and mifepristone. While both drugs had overall similar results, mifepristone was found to act faster. RU-486 and other similar drugs have been available in other countries (e.g., France) since 1989. The drug terminates an early pregnancy by chemical interference rather than by surgery. RU-486, a progesterone antagonist, binds tightly with the progesterone receptors on the target cells, but does not evoke progesterone's usual effects, and prevents progesterone from binding and acting. Deprived of progesterone activity, the highly developed endometrial tissue sloughs off, carrying the implanted embryo with it. RU-486 administration is followed in 48 hours by a prostaglandin that induces uterine contractions to help expel the endometrium and embryo.

**17**

| Hormone | Function |
|---|---|
| **Human Chorionic Gonadotropin (hCG)** | Maintains the corpus luteum of pregnancy; Stimulates secretion of testosterone by the developing testes in XY embryos |
| **Estrogen** (also secreted by the corpus luteum of pregnancy) | Stimulates growth of the myometrium, increasing uterine strength for parturition; Helps prepare the mammary glands for lactation |
| **Progesterone** (also secreted by the corpus luteum of pregnancy) | Suppresses uterine contractions to provide a quiet environment for the fetus |
| | Promotes formation of a cervical mucous plug to prevent uterine contamination |
| | Helps prepare the mammary glands for lactation |
| **Human Chorionic Somatomammotropin** (has a structure similar to that of both growth hormone and prolactin) | Believed to reduce maternal use of glucose and to promote the breakdown of stored fat (similar to growth hormone) so that greater quantities of glucose and free fatty acids may be shunted to the fetus |
| | Helps prepare the mammary glands for lactation (similar to prolactin) |
| **Relaxin** (also secreted by the corpus luteum of pregnancy) | Softens the cervix in preparation for cervical dilation at parturition |
| | Loosens the connective tissue between the pelvic bones in preparation for parturition |
| **Placental PTHrp** (parathyroid hormone–related peptide) | Increases maternal plasma $Ca^{2+}$ level for use in calcifying fetal bones; if necessary, promotes localized dissolution of maternal bones, mobilizing their $Ca^{2+}$ stores for use by the developing fetus |

© 2016 Cengage

Human chorionic gonadotropin is eliminated from the body in the urine. Pregnancy diagnosis tests can detect hCG in urine as early as the first month of pregnancy, about two weeks after the first missed menstrual period. Because this is before the growing embryo can be detected by physical examination, the test permits early confirmation of pregnancy.

A frequent early clinical sign of pregnancy is morning sickness, a daily bout of nausea and vomiting that often occurs in the morning but can take place at any time of day. Because this condition usually appears shortly after implantation and coincides with the time of peak hCG production, scientists speculate that this early placental hormone may trigger the symptoms, perhaps by acting on the chemoreceptor trigger zone in the vomiting centre (p. 660).

### SECRETION OF ESTROGEN AND PROGESTERONE

Why doesn't the developing placenta start producing estrogen and progesterone in the first place instead of secreting hCG, which in turn stimulates the corpus luteum to secrete these two critical hormones? The answer is that, for different reasons, the placenta cannot produce enough estrogen or progesterone in the first trimester of pregnancy. In the case of estrogen, the placenta does not have all the enzymes needed for complete synthesis of this hormone. Estrogen synthesis requires a complex interaction between the placenta and the fetus (> Figure 17-29). The placenta converts the androgen hormone produced by the fetal adrenal cortex, dehydroepiandrosterone (DHEA), into estrogen. The placenta cannot produce estrogen until the fetus has developed to the point that its adrenal cortex is secreting DHEA into the blood. The placenta extracts DHEA from the fetal blood and converts it into estrogen, which it then secretes into the maternal blood. The primary estrogen synthesized by this means is

**estriol**, in contrast to the main estrogen product of the ovaries, estradiol. Consequently, measurement of estriol levels in the maternal urine can be used clinically to assess the viability of the fetus.

In the case of progesterone, the placenta can synthesize this hormone soon after implantation. Even though the early placenta has the enzymes necessary to convert the cholesterol extracted from the maternal blood into progesterone, it does not produce much of this hormone, because the amount of progesterone produced is proportional to placental weight. The placenta is simply too small in the first 10 weeks of pregnancy to produce enough progesterone to maintain the endometrial tissue. The notable increase in circulating progesterone in the last seven months of gestation reflects placental growth during this period.

### ROLES OF ESTROGEN AND PROGESTERONE DURING PREGNANCY

As noted earlier, high concentrations of estrogen and progesterone are essential to maintain a normal pregnancy. Estrogen stimulates growth of the myometrium, which increases in size throughout pregnancy. The stronger uterine musculature is needed to expel the fetus during labour. Estriol also promotes development of the ducts within the mammary glands, through which milk is ejected during lactation.

Progesterone performs various roles throughout pregnancy. Its main function is to prevent miscarriage by suppressing contractions of the uterine myometrium. Progesterone also promotes formation of a mucous plug in the cervical canal to prevent vaginal contaminants from reaching the uterus. Finally, placental progesterone stimulates development of milk glands in the breasts, in preparation for lactation.

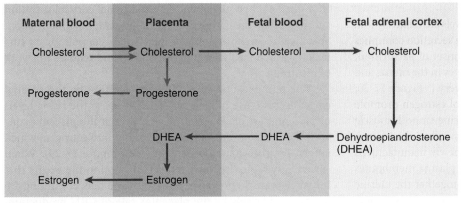

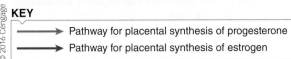

**KEY**

→ Pathway for placental synthesis of progesterone

➡ Pathway for placental synthesis of estrogen

› **FIGURE 17-29 Secretion of estrogen and progesterone by the placenta.** The placenta secretes increasing quantities of progesterone and estrogen into the maternal blood after the first trimester. The placenta itself can convert cholesterol into progesterone (*orange pathway*) but lacks some of the enzymes necessary to convert cholesterol into estrogen. However, the placenta can convert DHEA derived from cholesterol in the fetal adrenal cortex into estrogen when DHEA reaches the placenta by means of the fetal blood (*blue pathway*).

## Maternal body systems and gestation

The period of **gestation (pregnancy)** is about 38 weeks from conception (40 weeks from the end of the last menstrual period). During gestation, the embryo/fetus develops and grows to the point of being able to leave its maternal life-support system. Meanwhile, a number of physical changes within the mother accommodate the demands of pregnancy. The most obvious change is uterine enlargement. The uterus expands and increases in weight more than 20 times, exclusive of its contents. The breasts enlarge and develop the ability to produce milk. Body systems other than the reproductive system also make needed adjustments. The volume of blood increases by 30 percent, and the cardiovascular system responds to the increasing demands of the growing placental mass. Weight gain during pregnancy is due only in part to the weight of the fetus. The remainder is mostly from increased weight of the uterus, including the placenta, and increased blood volume. Respiratory activity increases by about 20 percent to handle the additional fetal requirements for oxygen utilization and carbon dioxide removal. Urinary output increases, and the kidneys excrete the additional wastes from the fetus.

The increased metabolic demands of the growing fetus increase nutritional requirements for the mother. In general, the fetus takes what it needs from the mother, even if this leaves the mother with a nutritional deficit. For example, the placental hormone **human chorionic somatomammotropin (hCS)** is thought responsible for the decreased use of glucose by the mother and the mobilization of free fatty acids from maternal adipose stores—similar to the actions of growth hormone (p. 232). (In fact, hCS has a structure similar to that of both growth hormone and prolactin and exerts similar actions.) The hCS-induced metabolic changes in the mother make available greater quantities of glucose and fatty acids for shunting to the fetus. Also, if the mother does not consume enough calcium, yet another placental hormone similar to parathyroid hormone, **parathyroid**

hormone–related peptide (PTHrp), mobilizes calcium from the maternal bones to ensure adequate calcification of the fetal bones (see ▌ Table 17-5).

## Parturition

**Parturition (labour, delivery, birth)** requires (1) dilation of the cervical canal to accommodate passage of the fetus from the uterus through the vagina and to the outside and (2) contractions of the uterine myometrium that are sufficiently strong to expel the fetus.

Several changes take place during late gestation in preparation for the onset of parturition. During the first two trimesters of gestation, the uterus remains relatively quiet, because of the inhibitory effect of the high levels of progesterone on the uterine muscle. During the last trimester, however, the uterus becomes progressively more excitable, so that mild contractions (**Braxton–Hicks contractions**) are experienced with increasing strength and frequency. Sometimes these contractions become regular enough to be mistaken for the onset of labour, a phenomenon called false labour.

Throughout gestation, the exit of the uterus remains sealed by the rigid, tightly closed cervix. As parturition approaches, the cervix begins to soften (or ripen) as a result of the dissociation of its tough connective tissue (collagen) fibres. Because of this softening, the cervix becomes malleable so that it can gradually yield, dilating the exit, as the fetus is forcefully pushed against it during labour. This cervical softening is caused largely by **relaxin**, a peptide hormone produced by the corpus luteum of pregnancy and by the placenta. Other factors to be described shortly contribute to cervical softening. Relaxin also "relaxes" the birth canal by loosening the connective tissue between pelvic bones.

Meanwhile, the fetus shifts downward (the baby "drops") and is normally oriented so that the head is in contact with the cervix in preparation for exiting through the birth canal. In a **breech birth**, any part of the body other than the head approaches the birth canal first.

## The factors that trigger parturition

Rhythmic, coordinated contractions, usually painless at first, begin at the onset of true labour. As labour progresses, the contractions increase in frequency, intensity, and discomfort. These strong, rhythmic contractions force the fetus against the cervix, dilating the cervix. Then, after having dilated the cervix enough for the fetus to pass through, these contractions force the fetus out through the birth canal.

The exact factors triggering the increase in uterine contractility and thus initiating parturition are not fully established, although much progress has been made in recent years in unravelling the sequence of events. Let's take a look at what is known about this process.

**17**

## ROLE OF HIGH ESTROGEN LEVELS

During early gestation, maternal estrogen levels are relatively low, but as gestation proceeds, placental estrogen secretion continues to rise. In the immediate days before the onset of parturition, soaring levels of estrogen bring about changes in the uterus and cervix to prepare them for labour and delivery (> Figure 17-30; also see > Figure 17-28). First, high levels of estrogen promote the synthesis of connexons within the uterine smooth muscle cells. These myometrial cells are not functionally linked to any extent throughout most of gestation. The newly manufactured connexons are inserted in the myometrial plasma membranes to form gap junctions that electrically link together the uterine smooth muscle cells so they become able to contract as a coordinated unit (p. 328).

Simultaneously, high levels of estrogen dramatically and progressively increase the concentration of myometrial receptors for oxytocin. Together, these myometrial changes collectively bring about the increased uterine responsiveness to oxytocin that ultimately initiates labour.

In addition to preparing the uterus for labour, the increasing levels of estrogen promote production of local prostaglandins that contribute to cervical ripening by stimulating cervical enzymes that locally degrade collagen fibres. Furthermore, these prostaglandins themselves increase uterine responsiveness to oxytocin.

## ROLE OF OXYTOCIN

Oxytocin is a peptide hormone produced by the hypothalamus, stored in the posterior pituitary, and released into the blood from the posterior pituitary on nervous stimulation by the hypothalamus (p. 227). A powerful uterine muscle stimulant, oxytocin plays the key role in the progression of labour. However, this hormone was once discounted as the trigger for parturition, because the circulating levels of oxytocin remain constant prior to the onset of labour. The discovery that uterine responsiveness to oxytocin is 100 times greater at term than in nonpregnant women (because of the increased concentration of myometrial oxytocin receptors) led to the now widely accepted conclusion that labour is initiated when the oxytocin receptor concentration reaches a critical threshold that permits the onset of strong, coordinated contractions in response to ordinary levels of circulating oxytocin.

## ROLE OF CORTICOTROPIN-RELEASING HORMONE

Until recently, scientists were baffled by the factors that raise levels of placental estrogen secretion. Recent research has shed new light on the probable mechanism. Evidence suggests that *corticotropin-releasing hormone (CRH)* secreted by the fetal portion of the placenta into both the maternal and fetal circulations not only drives the manufacture of placental estrogen, ultimately dictating the timing of the onset of labour, but also promotes changes in the fetal lungs needed for breathing air (> Figure 17-30). Recall that CRH is normally secreted by the hypothalamus and regulates the output of ACTH by the anterior pituitary (pp. 231). In turn, ACTH stimulates production of both cortisol and DHEA by the adrenal cortex. In the fetus, much of the CRH comes from the placenta rather than solely from the fetal hypothalamus. The additional cortisol secretion summoned by the extra CRH promotes fetal lung maturation. Specifically, cortisol stimulates the synthesis of pulmonary surfactant, which facilitates lung expansion and reduces the work of breathing.

The bumped-up rate of DHEA secretion by the adrenal cortex in response to placental CRH leads to the rising levels of placental estrogen secretion. Recall that the placenta converts DHEA from the fetal adrenal gland into estrogen, which enters the maternal bloodstream (see > Figure 17-29). When sufficiently high, this estrogen sets in motion the events that initiate labour. So pregnancy duration and delivery timing are determined largely by the placenta's rate of CRH production. That is, a "**placental clock**" ticks out the length of time until parturition. The timing of parturition is established early in pregnancy, with delivery at the end point of a maturational process that extends throughout the entire gestation. The ticking of the placental clock is measured by the rate of placental secretion. As the pregnancy progresses, CRH levels in maternal plasma rise. Researchers can accurately predict the timing of parturition by measuring the maternal plasma levels of CRH as early as the end of the first trimester. Higher-than-normal levels are associated with premature deliveries, whereas lower-than-normal levels indicate late deliveries. These and other data suggest that when a critical level of placental CRH is reached, parturition is triggered. This critical CRH level ensures that when labour begins, the infant is ready for life outside the womb. It does so by concurrently increasing the fetal cortisol needed for lung maturation and the estrogen needed for the uterine changes that bring on labour. The remaining unanswered puzzle regarding the placental clock is, what controls placental secretion of CRH?

## ROLE OF INFLAMMATION

Surprisingly, recent research suggests that inflammation plays a central role in the labour process, in the onset of both full-term labour and premature labour. Key to this inflammatory response is activation of **nuclear factor $\kappa\beta$ (NF-$\kappa\beta$)** in the uterus. NF-$\kappa\beta$ boosts production of inflammatory cytokines, such as *interleukin 8 (IL-8)*, and prostaglandins that increase the sensitivity of the uterus to contraction-inducing chemical messengers and help soften the cervix. What activates NF-$\kappa\beta$, setting off an inflammatory cascade that helps prompt labour? Various factors associated with the onset of full-term labour and premature labour can cause an upsurge in NF-$\kappa\beta$. These include stretching of the uterine muscle and the presence of the pulmonary surfactant protein SP-A in the amniotic fluid. SP-A promotes the migration of fetal macrophages to the uterus. These macrophages, in turn, produce the inflammatory cytokine *interleukin 1$\beta$ (IL-1$\beta$)* that activates NF-$\kappa\beta$. In this way, fetal lung maturation contributes to the onset of labour.

*Clinical Note* Premature labour can be incited by bacterial infections and allergic reactions that activate NF-$\kappa\beta$. Also, multiple-fetus pregnancies are at risk for premature labour, likely because the increased uterine stretching triggers earlier activation of NF-$\kappa\beta$.

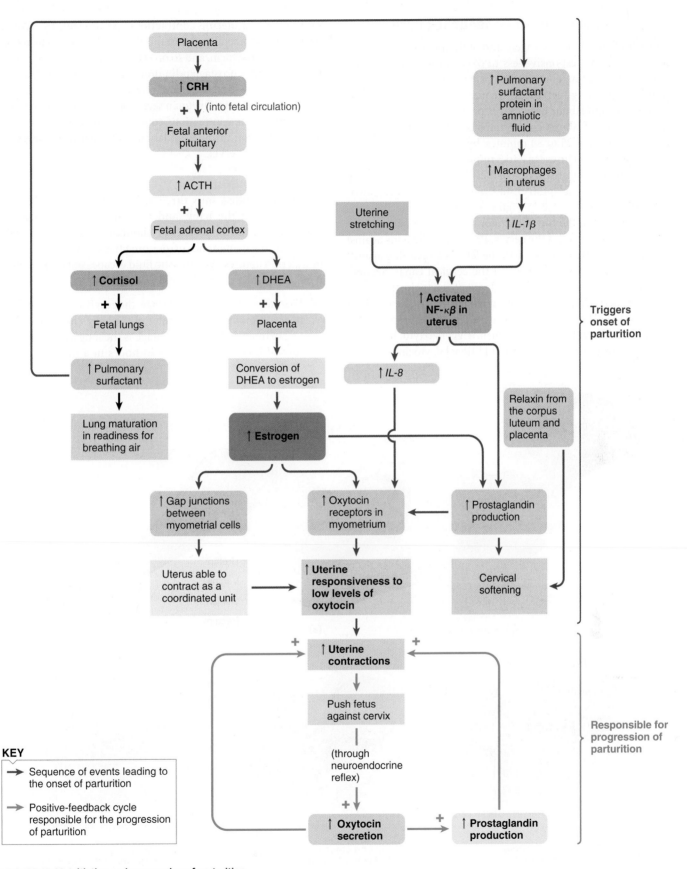

> FIGURE 17-30 **Initiation and progression of parturition**

**17**

## Parturition and a positive-feedback cycle

Once high levels of estrogen and inflammatory cytokines increase uterine responsiveness to oxytocin to a critical level and regular uterine contractions begin, myometrial contractions progressively increase in frequency, strength, and duration throughout labour until they expel the uterine contents. At the beginning of labour, contractions lasting 30 seconds or less occur about every 25 to 30 minutes; by the end, they last 60 to 90 seconds and occur every 2 to 3 minutes.

As labour progresses, a positive-feedback cycle involving oxytocin and prostaglandin ensues, incessantly increasing myometrial contractions (see ▶ Figure 17-30). Each uterine contraction begins at the top of the uterus and sweeps downward, forcing the fetus toward the cervix. Pressure of the fetus against the cervix does two things. First, the fetal head pushing against the softened cervix wedges open the cervical canal. Second, cervical stretch stimulates the release of oxytocin through a neuroendocrine reflex. Stimulation of receptors in the cervix in response to fetal pressure sends a neural signal up the spinal cord to the hypothalamus, which in turn triggers oxytocin release from the posterior pituitary. This additional oxytocin promotes more powerful uterine contractions. As a result, the fetus is pushed more forcefully against the cervix, stimulating the release of even more oxytocin, and so on. This cycle is reinforced as oxytocin stimulates prostaglandin production by the decidua. As a powerful myometrial stimulant, prostaglandin further enhances uterine contractions. Oxytocin secretion, prostaglandin production, and uterine contractions continue to increase in positive-feedback fashion throughout labour until delivery relieves the pressure on the cervix.

### STAGES OF LABOUR

Labour is divided into three stages: (1) cervical dilation, (2) delivery of the baby, and (3) delivery of the placenta (▶ Figure 17-31). At the onset of labour or sometime during the first stage, the membrane surrounding the amniotic sac, or "bag of waters," ruptures. As amniotic fluid escapes out of the vagina, it helps lubricate the birth canal.

- *First stage.* During the first stage, the cervix is forced to dilate to accommodate the diameter of the baby's head, usually to a maximum of 10 cm. This stage is the longest, lasting from several hours to as long as 24 hours in a first pregnancy.

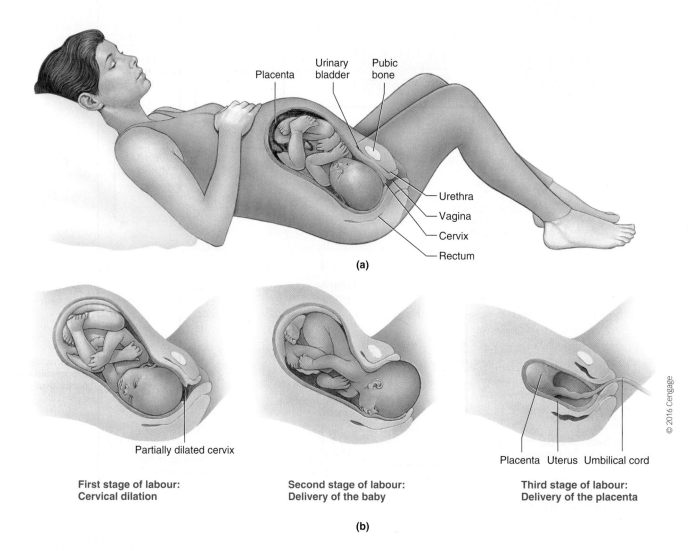

**(a)**

Placenta  Urinary bladder  Pubic bone  Urethra  Vagina  Cervix  Rectum

Partially dilated cervix

Placenta  Uterus  Umbilical cord

| First stage of labour: Cervical dilation | Second stage of labour: Delivery of the baby | Third stage of labour: Delivery of the placenta |

**(b)**

© 2016 Cengage

▶ **FIGURE 17-31 Stages of labour.** (a) Position of the fetus near the end of pregnancy. (b) Stages of labour

If another part of the fetus's body other than the head is oriented against the cervix, it is generally less effective than the head as a wedge. The head has the largest diameter of the baby's body. If the baby approaches the birth canal feet first, the feet may not dilate the cervix enough to let the head pass. In such a case, without medical intervention the baby's head would remain stuck behind the too-narrow cervical opening.

- *Second stage.* The second stage of labour, the actual birth of the baby, begins once cervical dilation is complete. When the infant begins to move through the cervix and vagina, stretch receptors in the vagina activate a neural reflex that triggers contractions of the abdominal wall in synchrony with the uterine contractions. These abdominal contractions greatly increase the force pushing the baby through the birth canal. The mother can help deliver the infant by voluntarily contracting the abdominal muscles at this time in unison with each uterine contraction, that is, by pushing with each labour pain. Stage 2 is usually much shorter than the first stage and lasts 30 to 90 minutes. The infant is still attached to the placenta by the umbilical cord at birth. The cord is tied and severed, with the stump shrivelling up in a few days to form the **umbilicus (navel)**.

- *Third stage.* Shortly after delivery of the baby, a second series of uterine contractions separates the placenta from the myometrium and expels it through the vagina. Delivery of the placenta, or **afterbirth**, constitutes the third stage of labour, typically the shortest stage, being completed within 15 to 30 minutes after the baby is born. After the placenta is expelled, continued contractions of the myometrium constrict the uterine blood vessels supplying the site of placental attachment, to prevent haemorrhage.

### UTERINE INVOLUTION

After delivery, the uterus shrinks to its pregestational size, a process known as **involution**, which takes four to six weeks to complete. During involution, the remaining endometrial tissue not expelled with the placenta gradually disintegrates and sloughs off, producing a vaginal discharge called **lochia** that continues for three to six weeks following parturition. After this period, the endometrium is restored to its nonpregnant state.

Involution occurs largely because of the precipitous fall in circulating estrogen and progesterone when the placental source of these steroids is lost at delivery. The process is facilitated in mothers who breastfeed their infants, because oxytocin is released in response to suckling. In addition to playing an important role in lactation, this periodic nursing-induced release of oxytocin promotes myometrial contractions that help maintain uterine muscle tone, enhancing involution. Involution is usually complete in about four weeks in nursing mothers but takes about six weeks in those who do not breastfeed.

## Lactation

The female reproductive system supports the new being from the moment of conception through gestation and continues to nourish it during its early life outside the supportive uterine environment. Milk (or its equivalent) is essential for survival of the newborn. Accordingly, during gestation the **mammary glands (breasts)** are prepared for **lactation** (milk production).

The breasts in nonpregnant females consist mostly of adipose tissue and a rudimentary duct system. Breast size is determined by the amount of adipose tissue, which has nothing to do with the ability to produce milk.

### PREPARATION OF THE BREASTS FOR LACTATION

Under the hormonal environment present during pregnancy, the mammary glands develop the internal glandular structure and function necessary for milk production. A breast capable of lactating has a network of progressively smaller ducts that branch out from the nipple and terminate in lobules (> Figure 17-32a). Each lobule is made up of a cluster of saclike epithelial-lined, milk-producing glands known as **alveoli**. Milk is synthesized by the epithelial cells, then secreted into the alveolar lumen, which is drained by a milk-collecting duct that transports the milk to the surface of the nipple (> Figure 17-32b).

During pregnancy, the high concentration of *estrogen* promotes extensive duct development, whereas the high level of *progesterone* stimulates abundant alveolar-lobular formation. Elevated concentrations of *prolactin* (an anterior pituitary hormone stimulated by the rising levels of estrogen) and *human chorionic somatomammotropin* (a placental hormone that has a structure similar to that of both growth hormone and prolactin) also contribute to mammary gland development by inducing the synthesis of enzymes needed for milk production.

### PREVENTION OF LACTATION DURING GESTATION

Most of these changes in the breasts occur during the first half of gestation, so the mammary glands are fully capable of producing milk by the middle of pregnancy. However, milk secretion does not occur until parturition. The high estrogen and progesterone concentrations during the last half of pregnancy prevent lactation by blocking prolactin's stimulatory action on milk secretion. Prolactin is the primary stimulant of milk secretion. Thus, even though the high levels of placental steroids induce the development of the milk-producing machinery in the breasts, they prevent these glands from becoming operational until the baby is born and milk is needed.

The abrupt decline in estrogen and progesterone that occurs with loss of the placenta at parturition initiates lactation.

This completes our discussion of the functions of estrogen and progesterone during gestation and lactation and also throughout the reproductive life of females. These functions are summarized in ▌ Table 17-6.

### STIMULATION OF LACTATION VIA SUCKLING

- Once milk production begins after delivery, two hormones are critical for maintaining lactation: (1) *prolactin,* which promotes milk secretion, and (2) *oxytocin,* which causes milk ejection. **Milk ejection (milk letdown)** refers to the forced expulsion of milk from the lumen of the alveoli out through the ducts. Release of both of these hormones is stimulated by a neuroendocrine reflex triggered by suckling (> Figure 17-33). Let's examine each of these hormones and their roles in further detail.

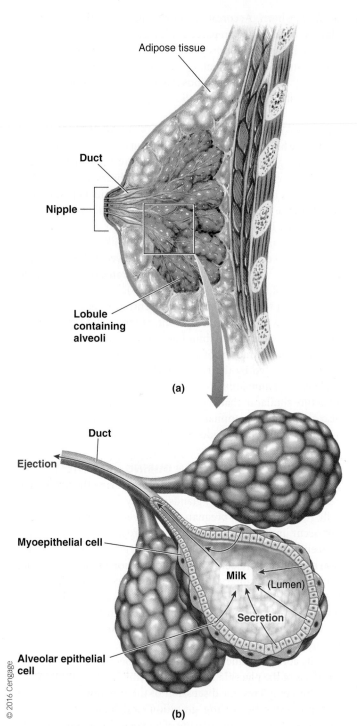

**Adipose tissue**

**Duct**

**Nipple**

**Lobule containing alveoli**

(a)

**Duct**

**Ejection**

**Myoepithelial cell**

**Milk** (Lumen)

**Secretion**

**Alveolar epithelial cell**

© 2016 Cengage

(b)

❯ **FIGURE 17-32 Mammary gland anatomy.** (a) Internal structure of the mammary gland, lateral view. (b) Schematic representation of the microscopic structure of an alveolus within the mammary gland. The alveolar epithelial cells secrete milk into the lumen. Contraction of the surrounding myoepithelial cells ejects the secreted milk out through the duct.

- *Oxytocin release and milk ejection.* The infant cannot directly suck milk out of the alveolar lumen. Instead, milk must be actively squeezed out of the alveoli into the ducts, and hence toward the nipple, by contraction of specialized myoepithelial cells (muscle-like epithelial cells) that surround each alveolus (see ❯ Figure 17-32b). The infant's suckling of the breast

stimulates sensory nerve endings in the nipple, thereby initiating action potentials that travel up the spinal cord to the hypothalamus. Once activated, the hypothalamus triggers a burst of oxytocin release from the posterior pituitary. Oxytocin, in turn, stimulates contraction of the myoepithelial cells in the breasts to induce milk ejection. Milk letdown continues only as

17

long as the infant continues to nurse. In this way, the milk-ejection reflex ensures that the breasts release milk only when and in the amount needed by the baby. Even though the alveoli may be full of milk, the milk cannot be released without oxytocin. The reflex can become conditioned to stimuli other than suckling, however. For example, the infant's cry can trigger milk letdown, causing a spurt of milk to leak from the nipples. In contrast, psychological stress, acting through the hypothalamus, can easily inhibit milk ejection. For this reason, a positive attitude toward breastfeeding and a relaxed environment are essential for successful breastfeeding.

- *Prolactin release and milk secretion.* Suckling not only triggers oxytocin release but also stimulates prolactin secretion. Prolactin output by the anterior pituitary is controlled by two hypothalamic secretions: **prolactin-inhibiting hormone (PIH)** and **prolactin-releasing hormone (PRH)**. PIH is now known to be *dopamine,* which also serves as a neurotransmitter in the brain. The chemical nature of PRH has not been identified with certainty, but scientists suspect PRH is *oxytocin* secreted by the hypothalamus into the hypothalamic–hypophyseal portal system to stimulate prolactin secretion by the anterior pituitary (p. 227). This role of oxytocin is distinct from the roles of oxytocin produced by the hypothalamus and stored in the posterior pituitary.

Throughout most of the female's life, PIH is the dominant influence, so prolactin concentrations normally remain low. During lactation, a burst in prolactin secretion occurs each time the infant suckles. Afferent impulses initiated in the nipple on suckling are carried by the spinal cord to the hypothalamus. This reflex ultimately leads to prolactin release by the anterior pituitary, although it is unclear whether this is from inhibition of PIH or stimulation of PRH secretion or both. Prolactin then acts on the alveolar epithelium to promote secretion of milk to replace the ejected milk (❭ Figure 17-33).

Concurrent stimulation by suckling of both milk ejection and milk production ensures that the rate of milk synthesis keeps pace with the baby's needs for milk. The more the infant nurses, the more milk is removed by letdown and the more milk is produced for the next feeding.

In addition to prolactin, which is the most important factor controlling synthesis of milk, at least four other hormones are essential for their permissive role in ongoing milk production: cortisol, insulin, parathyroid hormone, and growth hormone.

## Breastfeeding

Nutritionally, **milk** is composed of water, triglyceride fat, the carbohydrate lactose (milk sugar), a number of proteins, vitamins, and the minerals calcium and phosphate.

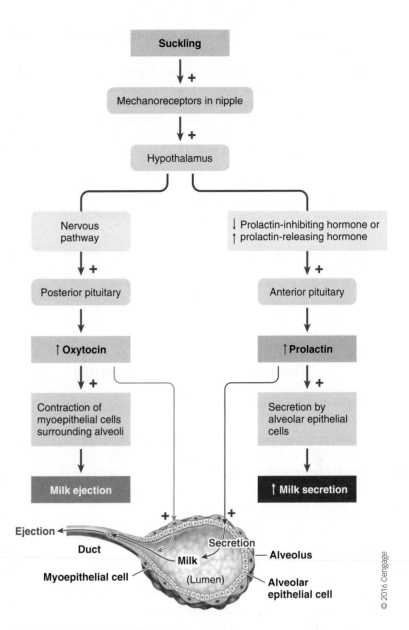

❭ FIGURE 17-33 Suckling reflexes

© 2016 Cengage

### ADVANTAGES OF BREASTFEEDING FOR THE INFANT

In addition to nutrients, milk contains a host of immune cells, antibodies, and other chemicals that help protect the infant against infection until it can mount an effective immune response on its own a few months after birth. **Colostrum**, the milk produced for the first five days after delivery, contains lower concentrations of fat and lactose but higher concentrations of immunoprotective components. All human babies acquire some passive immunity during gestation by antibodies passing across the placenta from the mother to the fetus (p. 481). These antibodies are short-lived, however, and often do not persist until the infant can fend for itself immunologically. Breastfed babies gain additional protection during this vulnerable period through a variety of mechanisms:

- Breast milk contains an abundance of *immune cells*—both B and T lymphocytes, macrophages, and neutrophils (pp. 446–447)—that produce antibodies and destroy pathogenic

17

microorganisms outright. These cells are especially plentiful in colostrum.

- *Secretory IgA,* a special type of antibody, is present in great amounts in breast milk. Secretory IgA consists of two IgA antibody molecules joined with a so-called secretory component that helps protect the antibodies from destruction by the infant's acidic gastric juice and digestive enzymes. The collection of IgA antibodies that a breastfed baby receives is specifically aimed against the particular pathogens in the environment of the mother—and, accordingly, of the infant as well. Appropriately, therefore, these antibodies protect against the infectious microbes that the infant is most likely to encounter.

- Some components in mother's milk, such as *mucus,* adhere to potentially harmful microorganisms, preventing them from attaching to and crossing the intestinal mucosa.

- *Lactoferrin* is a breast-milk constituent that thwarts growth of harmful bacteria by decreasing the availability of iron, a mineral needed for multiplication of these pathogens.

- *Bifidus factor* in breast milk, in contrast to lactoferrin, promotes multiplication of the nonpathogenic microorganism *Lactobacillus bifidus* in the infant's digestive tract. Growth of this harmless bacterium helps crowd out potentially harmful bacteria.

- Other components in breast milk promote maturation of the baby's digestive system so that it is less vulnerable to diarrhoea-causing bacteria and viruses.

- Still other factors in breast milk hasten the development of the infant's own immune capabilities.

Thus, breast milk helps protect infants from disease in a variety of ways.

Some studies hint that in addition to the benefits of breast milk during infancy, breastfeeding may reduce the risk of developing certain serious diseases later in life. Examples include allergies (such as asthma), autoimmune diseases (such as type 1 diabetes mellitus), and cancers (such as lymphoma).

Infants who are bottle-fed on a formula made from cow's milk or another substitute do not have the protective advantage provided by human milk and, accordingly, have a higher incidence of infections of the digestive tract, respiratory tract, and ears than breastfed babies do. Also, the digestive system of a newborn is better equipped to handle human milk than a cow milk–derived formula, so bottle-fed babies tend to have more digestive upsets.

### ADVANTAGES OF BREASTFEEDING FOR THE MOTHER

Breastfeeding is also advantageous for the mother. Oxytocin release triggered by nursing hastens uterine involution. In addition, suckling suppresses the menstrual cycle by inhibiting LH and FSH secretion, probably by inhibiting GnRH. Lactation, therefore, tends to prevent ovulation, decreasing the likelihood of another pregnancy (although it is not a reliable means of contraception). This mechanism permits all the mother's resources to be directed toward the newborn instead of being shared with a new embryo.

### CESSATION OF MILK PRODUCTION AT WEANING

When the infant is weaned, two mechanisms contribute to the cessation of milk production. First, without suckling, prolactin secretion is not stimulated, which removes the main stimulus for continued milk synthesis and secretion. Also, because there is no suckling and thus no oxytocin release, milk letdown does not occur. Because milk production does not immediately shut down, milk accumulates in the alveoli, engorging the breasts. The resulting pressure buildup acts directly on the alveolar epithelial cells to suppress further milk production. Therefore, the cessation of lactation at weaning results from a lack of suckling-induced stimulation of both prolactin and oxytocin secretion.

## The end and a new beginning

Reproduction is an appropriate way to end our discussion of physiology from cells to systems. The single cell resulting from the union of male and female gametes divides mitotically and differentiates into a multicellular individual made up of a number of different body systems that interact cooperatively to maintain homeostasis (i.e., stability in the internal environment). All the life-supporting homeostatic processes introduced throughout this book begin all over again at the start of a new life.

### Check Your Understanding 17.4

1. Indicate what ovarian hormones are secreted by the follicle and by the corpus luteum, state the effects of these hormones on the uterus, and indicate during which phase of the ovarian cycle each of the phases of the uterine cycle takes place.

2. Define *zygote, blastocyst, inner cell mass, trophoblast, decidua, chorion, placenta, embryo,* and *fetus.*

3. Discuss the role of oxytocin during parturition and during breastfeeding.

# Chapter in Perspective: Focus on Homeostasis

The reproductive system is unique in that it is not essential for homeostasis or for survival of the individual, but it is essential for sustaining the thread of life from generation to generation. Reproduction depends on the union of male and female gametes (reproductive cells), each with a half set of chromosomes, to form a new individual with a full, unique set of chromosomes. Unlike the other body systems, which are essentially identical in the two sexes, the reproductive systems of males and females are remarkably different, befitting their different roles in the reproductive process.

The male system is designed to continuously produce huge numbers of mobile spermatozoa that are delivered to the female during the sex act. Male gametes must be produced in abundance for two reasons: (1) Only a small percentage of them survive the hazardous journey through the female reproductive tract to the site of fertilization; and (2) the cooperative effort of many spermatozoa is required to break down the barriers surrounding the female gamete (ovum or egg) to enable one spermatozoon to penetrate and unite with the ovum.

The female reproductive system undergoes complex changes on a cyclic monthly basis. During the first half of the cycle, a single non-motile ovum is prepared for release. During the second half, the reproductive system is geared toward preparing a suitable environment for supporting the ovum if fertilization (union with a spermatozoon) occurs. If fertilization does not occur, the prepared supportive environment within the uterus sloughs off, and the cycle starts over again as a new ovum is prepared for release. If fertilization occurs, the female reproductive system adjusts to support growth and development of the new individual until it can survive on its own on the outside.

There are three important parallels in the male and female reproductive systems, even though they differ considerably in structure and function. First, the same set of undifferentiated reproductive tissues in the embryo can develop into either a male or a female system, depending on the presence or absence, respectively, of male-determining factors. Second, the same hormones—namely, hypothalamic GnRH and anterior pituitary FSH and LH—control reproductive function in both sexes. In both cases, gonadal steroids and inhibin act in negative-feedback fashion to control hypothalamic and anterior pituitary output. Third, the same events take place in the developing gamete's nucleus during sperm formation and egg formation, although males produce millions of sperm in one day, whereas females produce only about 400 ova in a lifetime.

## CHAPTER TERMINOLOGY

meiosis (p. 727)
meiotic arrest (p. 745)
menopause (p. 756)
menstrual (uterine) cycle (p. 753)
menstrual flow (p. 754)
midpiece (p. 736)
milk (p. 773)
milk ejection (milk letdown) (p. 771)
mitosis (p. 727)
Müllerian ducts (p. 728)
myometrium (p. 753)
nuclear factor $\kappa\beta$ (NF-$\kappa\beta$) (p. 768)
oogenesis (p. 745)
oogonia (p. 745)
orgasm (p. 744)
osteoporosis (p. 756)
ova (eggs) (p. 724)
ovaries (p. 724)
oviducts (uterine or Fallopian tubes) (p. 725)
ovulation (p. 749)
packaging (p. 735)
parathyroid hormone–related peptide
    (PTHrp) (p. 767)
parturition (labour, delivery, birth) (p. 767)
penis (p. 724)
perineal region (p. 727)
phenotypic sex (p. 728)
placenta (p. 760)
placental clock (p. 768)
placental villi (p. 761)
prepuce (p. 728)

primary follicles (p. 745)
primary oocytes (p. 745)
primary reproductive organs (gonads) (p. 724)
primary spermatocytes (p. 735)
primordial follicle (p. 745)
progesterone (p. 724)
prolactin-inhibiting hormone (PIH) (p. 773)
prolactin-releasing hormone (PRH) (p. 773)
proliferative phase (p. 755)
prostaglandins (p. 740)
prostate gland (p. 740)
puberty (p. 734)
relaxin (p. 767)
reproduction (p. 724)
reproductive system (p. 723)
reproductive tract (p. 724)
scrotum (p. 724)
second polar body (p. 746)
secondary (antral) follicle (p. 747)
secondary oocyte (p. 746)
secondary spermatocytes (p. 735)
secretory (progestational)
    phase (p. 755)
semen (p. 740)
seminal vesicles (p. 740)
seminiferous tubule fluid (p. 737)
seminiferous tubules (p. 732)
Sertoli cells (p. 737)
sex act (p. 741)
sex chromosomes (p. 727)
sex determination (p. 728)

sex-determining region of the Y
    chromosome (SRY) (p. 728)
sexual differentiation (p. 728)
sexual intercourse (p. 741)
sexual response cycle (p. 742)
spermatids (p. 735)
spermatogenesis (p. 735)
spermatogonia (p. 735)
spermatozoa (sperm) (p. 724)
spermiogenesis (p. 735)
tail (p. 736)
tenting effect (p. 744)
testes (p. 724)
testosterone (p. 724)
thecal cells (p. 747)
tonic LH secretion (p. 750)
trophoblast (p. 758)
umbilical arteries (p. 761)
umbilical cord (p. 761)
umbilical vein (p. 761)
umbilicus (navel) (p. 771)
urethral folds (p. 728)
uterus (p. 725)
vagina (p. 725)
vaginal opening (p. 727)
vulva (p. 727)
Wolffian ducts (p. 728)
X chromosome (p. 728)
Y chromosome (p. 728)
zona pellucida (p. 747)
zygote (p. 758)

## REVIEW EXERCISES

### Objective Questions (Answers on p. A-55)

1. It is possible for a genetic male to have the anatomic appearance of a female. *(True or false?)*

2. Testosterone secretion essentially ceases from birth until puberty. *(True or false?)*

3. Prostaglandins are derived from arachidonic acid found in the plasma membrane. *(True or false?)*

4. Females do not experience erection. *(True or false?)*

5. Most of the lubrication for sexual intercourse is provided by the female. *(True or false?)*

6. If a follicle does not reach maturity during one ovarian cycle, it can finish maturing during the next cycle. *(True or false?)*

7. Rising, moderate levels of estrogen inhibit tonic LH secretion, whereas high levels of estrogen stimulate the LH surge. *(True or false?)*

8. Spermatogenesis takes place within the _____ of the testes, stimulated by the hormones _____ and _____.

9. During estrogen production by the follicle, the _____ cells under the influence of the hormone _____ produce androgens, and the _____ cells under the influence of the hormone _____ convert these androgens into estrogens.

10. The source of estrogen and progesterone during the first 10 weeks of gestation is the _____.

11. Detection of _____ in the urine is the basis of pregnancy diagnosis tests.

12. Which statement concerning chromosomal distribution is NOT correct?
    a. All human somatic cells contain 23 chromosomal pairs for a total diploid number of 46 chromosomes.
    b. Each gamete contains 23 chromosomes, one member of each chromosomal pair.
    c. During meiotic division, the members of the chromosome pairs regroup themselves into the original combinations derived from the individual's mother and father for separation into haploid gametes.

17

d. Sex determination depends on the combination of sex chromosomes, an XY combination being a genetic male, and XX a genetic female.

e. The sex chromosome content of the fertilizing sperm determines the sex of the offspring.

13. What happens when the corpus luteum degenerates?
   a. Circulating levels of estrogen and progesterone rapidly increase.
   b. FSH and LH secretion start to rise.
   c. The endometrium thickens.
   d. Prostaglandins are released by endometrium.

Match the following:

_____1. secrete(s) prostaglandins    (a) epididymis and ductus deferens

_____2. increase(s) motility and fertility of sperm    (b) prostate gland

_____3. secrete(s) an alkaline fluid    (c) seminal vesicles

_____4. provide(s) fructose    (d) bulbourethral glands

_____5. storage site for sperm    (e) penis

_____6. concentrate(s) the sperm a hundredfold

_____7. secrete(s) fibrinogen

_____8. provide(s) clotting enzymes

_____9. contain(s) erectile tissue

Using the answer code on the right, indicate when each event takes place during the ovarian cycle:

_____1. development of antral follicles    (a) occurs during the follicular phase

_____2. secretion of estrogen

_____3. secretion of progesterone    (b) occurs during the luteal phase

_____4. menstruation

_____5. repair and proliferation of the endometrium    (c) occurs during both the follicular and luteal phases

_____6. increased vascularization and glycogen storage in the endometrium

## Written Questions

1. What are the primary reproductive organs, gametes, sex hormones, reproductive tract, accessory sex glands, external genitalia, and secondary sexual characteristics in males and in females?

2. List the essential reproductive functions of the male and of the female.

3. Discuss the differences between males and females with regard to genetic, gonadal, and phenotypic sex.

4. What parts of the male and female reproductive systems develop from each of the following: genital tubercle, urethral folds, genital swellings, Wolffian ducts, and Müllerian ducts?

5. Of what functional significance is the scrotal location of the testes?

6. Discuss the source and functions of testosterone.

7. Describe the three major stages of spermatogenesis. Discuss the functions of each part of a spermatozoon. What are the roles of Sertoli cells?

8. Discuss the control of testicular function.

9. Compare the sex act in males and females.

10. Compare oogenesis with spermatogenesis.

11. Describe the events of the follicular and luteal phases of the ovarian cycle. Correlate the phases of the uterine cycle with those of the ovarian cycle.

12. How are the ovum and spermatozoa transported to the site of fertilization? Describe the process of fertilization.

13. Describe the process of implantation and placenta formation.

14. What are the functions of the placenta? What hormones does the placenta secrete?

15. What is the role of human chorionic gonadotropin?

16. What factors contribute to the initiation of parturition? What are the stages of labour? What is the role of oxytocin?

17. Describe the hormonal factors that play a role in lactation.

18. Summarize the actions of estrogen and progesterone.

17

(Explanations on p. A-55)

1. The hypothalamus releases GnRH in pulsatile bursts once every two to three hours, with no secretion occurring in between. The blood concentration of GnRH depends on the frequency of these bursts of secretion. A promising line of research for a new method of contraception involves administration of GnRH-like drugs. In what way could such drugs act as contraceptives when GnRH is the hypothalamic hormone that triggers the chain of events leading to ovulation? (*Hint:* The anterior pituitary is programmed to respond only to the normal pulsatile pattern of GnRH.)

2. Occasionally, testicular tumours composed of interstitial cells of Leydig may secrete up to 100 times the normal amount of testosterone. When such a tumour develops in young children, they grow up much shorter than their genetic potential. Explain why. What other symptoms would be present?

3. What type of sexual dysfunction might arise in men taking drugs that inhibit sympathetic nervous system activity as part of the treatment for high blood pressure?

4. Explain the physiological basis for administering a posterior pituitary extract to induce or facilitate labour.

5. The symptoms of menopause are sometimes treated with supplemental estrogen and progesterone. Why wouldn't treatment with GnRH or FSH and LH also be effective?

## CLINICAL CONSIDERATION

(Explanation on p. A-55)

Maria A., who is in her second month of gestation, has been experiencing severe abdominal cramping. Her physician has diagnosed her condition as a tubal pregnancy: the developing embryo is implanted in the oviduct instead of in the uterine endometrium. Why must this pregnancy be surgically terminated?

# Système Internationale/Physiological Measurements

## ▌ TABLE A-1 Metric Measures and Imperial Equivalents

| Unit | Measure | Symbol | Imperial Equivalent |
|---|---|---|---|
| **Linear Measure** | | | |
| 1 kilometre | = 1000 metres | 103 m km | 0.62137 mile |
| 1 metre | | 100 m m | 39.37 inches |
| 1 decimetre | = 1/10 metre | $10^{-1}$ m dm | 3.937 inches |
| 1 centimetre | = 1/100 metre | $10^{-2}$ m cm | 0.3937 inch |
| 1 millimetre | = 1/1000 metre | $10^{-3}$ m mm | Not used |
| 1 micrometre (or micron) | = 1/1 000 000 metre | $10^{-6}$ m μm (or μ) | Not used |
| 1 nanometre | = 1/1 000 000 000 metre | $10^{-9}$ m nm | Not used |
| **Measures of Capacity (for fluids and gases)** | | | |
| 1 litre | | L | 1.0567 U.S. liquid quarts |
| 1 millilitre | = 1/1000 litre = volume of 1 g of water at STP* | mL | |
| **Measures of Volume** | | | |
| 1 cubic metre | | $m^3$ | |
| 1 cubic decimetre | = 1/1000 cubic metre = 1 litre | $dm^3$ = L L | |
| 1 cubic centimetre | = 1/1 000 000 cubic metre = 1 millilitre (mL) | $cm^3$ = mL | |
| 1 cubic millimetre | = 1/100 000 000 cubic metre | $mm^3$ | |
| **Measures of Mass** | | | |
| 1 kilogram | = 1000 grams | kg | 2.2046 pounds |
| 1 gram | | g | 15.432 grains |
| 1 milligram | = 1/1000 gram | mg | 0.01 grain (about) |
| 1 microgram | = 1/1 000 000 gram | μg (or mcg) | |

*STP = standard temperature and pressure

Comparison of human height in metres with the sizes of some biological and molecular structures

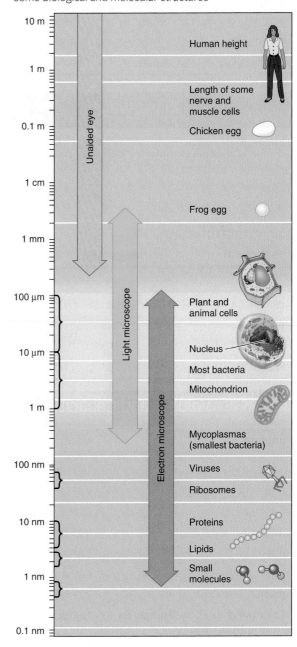

# Length

| Imperial | | Metric |
| --- | --- | --- |
| inch | = | 2.54 centimetres |
| foot | = | 0.30 metre |
| yard | = | 0.91 metre |
| mile (5280 feet) | = | 1.61 kilometre |

| To Convert | Multiply By | To Obtain |
| --- | --- | --- |
| inches | 2.54 | centimetres |
| feet | 30.00 | centimetres |
| centimetres | 0.39 | inches |
| millimetres | 0.039 | inches |

# Mass/Weight

| Imperial | | Metric |
| --- | --- | --- |
| grain | = | 64.80 milligrams |
| ounce | = | 28.35 grams |
| pound | = | 453.60 grams |
| ton (short) (2000 pounds) | = | 0.91 metric ton |

| To Convert | Multiply By | To Obtain |
| --- | --- | --- |
| ounces | 28.3 | grams |
| pounds | 453.6 | grams |
| pounds | 0.45 | kilograms |
| grams | 0.035 | ounces |
| kilograms | 2.2 | grams |

# Volume and Capacity

| Imperial | | Metric |
| --- | --- | --- |
| cubic inch | = | 16.39 cubic centimetres |
| cubic foot | = | 0.03 cubic metre |
| cubic yard | = | 0.765 cubic metres |
| ounce | = | 0.03 litre |
| pint | = | 0.47 litre |
| quart | = | 0.95 litre |
| gallon | = | 3.79 litres |

| To Convert | Multiply By | To Obtain |
| --- | --- | --- |
| fluid ounces | 30.00 | millilitres |
| quart | 0.95 | litres |
| millilitres | 0.03 | fluid ounces |
| litres | 1.06 | quarts |

**Linear Measurement Comparison**

**Fahrenheit–Celsius Temperature Comparison**

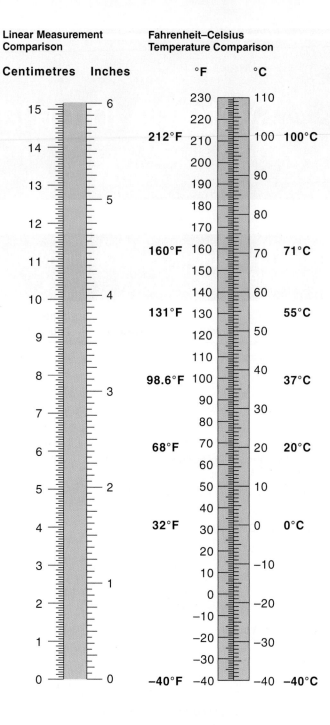

# A Review of Chemical Principles

By Spencer Seager, Weber State University, and Lauralee Sherwood

## B.1 | Chemical Level of Organization in the Body

*Matter* is anything that occupies space and has mass, including all living and nonliving things in the universe. *Mass* is the amount of matter in an object. *Weight*, in contrast, is the effect of gravity on that mass. The more gravity exerted on a mass, the greater the weight of the mass. An astronaut has the same mass whether on Earth or in space, but is weightless in the zero gravity of space.

### Atoms

All matter is made up of tiny particles called *atoms*. These particles are too small to be seen individually, even with the most powerful electron microscopes available today.

Although atoms are extremely small, they consist of three types of even smaller subatomic particles. Different types of atoms vary in the number of these various subatomic particles they contain. *Protons* and *neutrons* are particles of nearly identical mass: protons carry a positive charge and neutrons have no charge. *Electrons* have a much smaller mass than protons and neutrons and are negatively charged. An atom consists of two regions—a dense, central *nucleus* made of protons and neutrons surrounded by a three-dimensional *electron cloud*, where electrons move rapidly around the nucleus in orbitals (›Figure B-1). The magnitude of the charge of a proton exactly matches that of an electron, but it is opposite in sign, being positive. In all atoms, the number of protons in the nucleus is equal to the number of electrons moving around the nucleus, so their charges balance, and the atoms are neutral.

### Elements and atomic symbols

A pure substance composed of only one type of atom is called an **element**. A pure sample of the element carbon contains only carbon atoms, even though the atoms might be arranged in the form of diamond or in the form of graphite (pencil lead). Each element is designated by an **atomic symbol**, a one- or two-letter chemical shorthand for the element's name. Usually these symbols are easy to follow, because they are derived from the English name for the element. Thus, H stands for *hydrogen*, C for *carbon*, and O for *oxygen*. In a few cases, the atomic symbol is based on

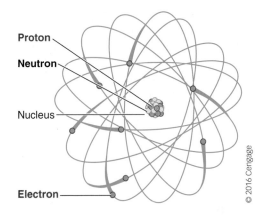

› **FIGURE B-1 The atom.** The atom consists of two regions. The central nucleus contains protons and neutrons and makes up 99.9 percent of the mass. Surrounding the nucleus is the electron cloud, where the electrons move rapidly around the nucleus. (Figure not drawn to scale.)

the element's Latin name—for example, Na for *sodium* (*natrium* in Latin) and K for *potassium* (*kalium*). Of the 109 known elements, 26 are normally found in the body. Four elements—oxygen, carbon, hydrogen, and nitrogen—compose 96 percent of the body's mass.

### Compounds and molecules

Pure substances composed of more than one type of atom are known as **compounds**. Pure water, for example, is a compound that contains atoms of hydrogen and atoms of oxygen in a 2-to-1 ratio, regardless of whether the water is in the form of liquid, solid (ice), or vapour (steam). A **molecule** is the smallest unit of a pure substance that has the properties of that substance and is capable of a stable, independent existence. For example, a molecule of water consists of two atoms of hydrogen and one atom of oxygen, held together by chemical bonds.

### Atomic number

Exactly what are we talking about when we refer to a "type" of atom? That is, what makes carbon, hydrogen, and oxygen atoms different? The answer is the number of protons in the nucleus. Regardless of where they are found, all hydrogen atoms have one proton in the nucleus, all carbon atoms have six, and all oxygen atoms have eight. Of course, these numbers also represent the number of electrons moving around each nucleus, because the number of electrons and number of protons in an atom are equal. The number of protons in the nucleus of an atom of an element is called the **atomic number** of the element.

## Atomic weight

As expected, tiny atoms have tiny masses. For example, the actual mass of a hydrogen atom is $1.67 \times 10^{-24}$ g, that of a carbon atom is $1.99 \times 10^{-23}$ g, and that of an oxygen atom is $2.66 \times 10^{-23}$ g. These very small numbers are inconvenient to work with in calculations, so a system of relative masses has been developed. These relative masses simply compare the actual masses of the atoms with each other. Suppose the actual masses of two people were determined to be 45.50 kg and 113.75 kg. Their relative masses are then determined by dividing each mass by the smaller mass of the two: $45.50/45.50 = 1.00$, and $113.75/45.50 = 2.50$. Thus, the relative masses of the two people are 1.00 and 2.50; these numbers simply express the fact that the mass of the heavier person is 2.50 times that of the other person. The relative masses of atoms are called *atomic masses*, or *atomic weights*, and are given in *atomic mass units (amu)*. In this system, hydrogen atoms, the least massive of all atoms, have an atomic weight of 1.01 amu. The atomic weight of carbon atoms is 12.01 amu, and that of oxygen atoms is 16.00 amu. Thus, oxygen atoms have a mass about 16 times that of hydrogen atoms. ▌ Table B-1 gives the atomic weights and some other characteristics of the elements that are most important physiologically.

### ▌ TABLE B-1 Characteristics of Selected Elements

| Name and Symbol | Number of Protons | Atomic Number | Atomic Weight (amu) |
|---|---|---|---|
| Hydrogen (H) | 1 | 1 | 1.01 |
| Carbon (C) | 6 | 6 | 12.01 |
| Nitrogen (N) | 7 | 7 | 14.01 |
| Oxygen (O) | 8 | 8 | 16.00 |
| Sodium (Na) | 11 | 11 | 22.99 |
| Magnesium (Mg) | 12 | 12 | 24.31 |
| Phosphorus (P) | 15 | 15 | 30.97 |
| Sulphur (S) | 16 | 16 | 32.06 |
| Chlorine (Cl) | 17 | 17 | 35.45 |
| Potassium (K) | 19 | 19 | 39.10 |
| Calcium (Ca) | 20 | 20 | 40.08 |

© 2016 Cengage

## B.2 | Chemical Bonds

Because all matter is made up of atoms, atoms must somehow be held together to form matter. The forces holding atoms together are called *chemical bonds*. Not all chemical bonds are formed in the same way, but all involve the electrons of atoms. Whether one atom will bond with another depends on the number and arrangement of its electrons. An atom's electrons are arranged in electron shells, to which we now turn our attention.

### Electron shells

Electrons tend to move around the nucleus in a specific pattern. The orbitals, or pathways, travelled by electrons around the nucleus, are arranged in an orderly series of concentric layers known as *electron shells*, which consecutively surround the nucleus. Each electron shell can hold a specific number of electrons. The first (innermost) shell closest to the nucleus can contain a maximum of only two electrons, no matter what the element is. The second shell can hold a total of eight more electrons. The third, fourth, and fifth shells can hold 18, 32, and 50 electrons, respectively. However, it is important to note that atoms are most stable when the outer shell contains only eight electrons, an exception of course are atoms with only one outer shell. As the number of electrons increases with increasing atomic number, still more electrons occupy successive shells, each at a greater distance from the nucleus. Each successive shell from the nucleus has a higher *energy level*. Because the negatively charged electrons are attracted to the positively charged nucleus, it takes more energy for an electron to overcome the nuclear attraction and orbit farther from the nucleus. Therefore, the first electron shell has the lowest energy level and the outermost shell of an atom has the highest energy level.

In general, electrons belong to the lowest energy shell possible, up to the maximum capacity of each shell. For example, hydrogen atoms have only one electron, so it is in the first shell. Helium atoms have two electrons, which are both in the first shell and fill it. Carbon atoms have six electrons, two in the first shell and four in the second shell, whereas the eight electrons of oxygen are arranged with two in the first shell and six in the second shell.

### Bonding characteristics of an atom and valence

Atoms tend to undergo processes that result in a filled outermost electron shell. So the electrons of the outer or higher-energy shell determine the bonding characteristics of an atom and its ability to interact with other atoms. Atoms that have a vacancy in their outermost shell tend to either give up, accept, or share electrons with other atoms (whichever is most favourable energetically) so that all participating atoms have filled outer shells. For example, an atom that has only one electron in its outermost shell may empty this shell so its remaining shells are completely full. By contrast, another atom that lacks only one electron in its outer shell may acquire the deficient electron from the first atom to fill all its shells to the maximum. The number of electrons an atom loses, gains, or shares to achieve a filled outer shell is known as

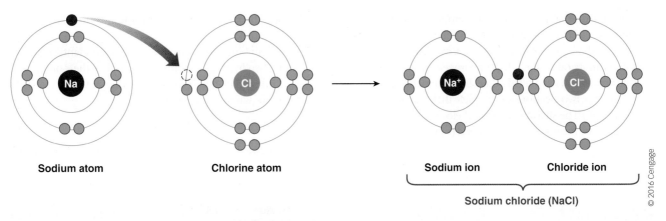

> FIGURE B-2 **Ions and ionic bonds.** Sodium (Na) and chlorine (Cl) atoms both have partially filled outermost shells. Therefore, sodium tends to give up its lone electron in the outer shell to chlorine. As a result, sodium becomes a positively charged ion, and chlorine becomes a negatively charged ion known as chloride. The oppositely charged ions attract each other, forming an ionic bond.

© 2016 Cengage

the atom's *valence*. A *chemical bond* is the force of attraction that holds participating atoms together as a result of an interaction between their outermost electrons.

Consider sodium atoms (Na) and chlorine atoms (Cl) (>Figure B-2). Sodium atoms have 11 electrons: 2 in the first shell, 8 in the second shell, and 1 in the third shell. Chlorine atoms have 17 electrons: 2 in the first shell, 8 in the second shell, and 7 in the third shell. Because 8 electrons are required to fill the second shell, sodium atoms have 1 electron more than is needed to provide a filled second shell, whereas chlorine atoms have 1 less electron than is needed to have 8 electrons in the third shell. Each sodium atom can lose an electron to a chlorine atom, leaving each sodium with 10 electrons, 8 of which are in the second shell, which is full and is now the outer shell occupied by electrons. By accepting 1 electron, each chlorine atom now has a total of 18 electrons, with 8 of them in the third, which increases its stability.

## Ions and ionic bonds

Recall that atoms are electrically neutral because they have an identical number of positively charged protons and negatively charged electrons. By giving up and accepting electrons, the sodium atoms and chlorine atoms have achieved filled outer shells, but now each atom is unbalanced electrically. Although each sodium now has 10 electrons, it still has 11 protons in the nucleus and a net electrical charge, or valence, of +1. Similarly, each chlorine now has 18 electrons, but only 17 protons. Thus each chlorine has a –1 charge. Such charged atoms are called *ions*. Positively charged ions are called *cations*; negatively charged ions are called *anions*. As a helpful hint to keep these terms straight, imagine the "t" in *cation* as standing for a "+" sign and the first "n" in *anion* as standing for "negative."

Note that both a cation and anion are formed whenever an electron is transferred from one atom to another. Because opposite charges attract, sodium ions $(Na^+)$ and charged chlorine atoms, now called *chloride* ions $(Cl^-)$, are attracted toward each other. This electrical attraction that holds cations and anions together is known as an *ionic bond*. Ionic bonds hold $Na^+$ and $Cl^-$ together in the compound *sodium chloride (NaCl)*, which is common table salt. A sample of sodium chloride actually contains sodium and chloride ions in a three-dimensional geometric arrangement called a *crystal lattice*. The ions of opposite charge occupy alternate sites within the lattice (>Figure B-3).

## Covalent bonds

It is not favourable, energywise, for an atom to give up or accept more than three electrons. Nevertheless, carbon atoms, which have four electrons in their outer shell, form compounds. They do so by another bonding mechanism, *covalent bonding*. Atoms that would have to lose or gain four or more electrons to achieve outer-shell stability usually bond by *sharing* electrons. Shared electrons actually orbit around *both* atoms. Thus a carbon atom can share its four outer electrons with the four electrons of four hydrogen atoms, as shown in Equation B-1, where the outer-shell electrons are shown as dots around the symbol of each atom. (The resulting compound is methane, $CH_4$, a gas made up of individual $CH_4$ molecules.)

**Eq. B-1**

Shared electron pairs

$$\cdot \overset{\cdot}{\underset{\cdot}{C}} \cdot + 4 \cdot H \rightarrow H : \overset{\overset{\displaystyle H}{..}}{\underset{\underset{\displaystyle H}{..}}{C}} : H$$

Shared electron pairs

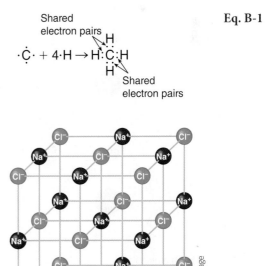

© 2016 Cengage

> FIGURE B-3 **Crystal lattice for sodium chloride (table salt)**

A Review of Chemical Principles

Each electron that is shared by two atoms is counted toward the number of electrons needed to fill the outer shell of each atom. Thus, each carbon atom shares four pairs, or eight electrons, and so has eight in its outer shell. Each hydrogen shares one pair, or two electrons, and so has a filled outer shell. (Remember, hydrogen atoms need only two electrons to complete their outer shell, which is the first shell.) The sharing of a pair of electrons by atoms binds them together by means of a *covalent bond* (›Figure B-4). Covalent bonds are the strongest of chemical bonds; that is, they are the hardest to break.

Covalent bonds also form between some identical atoms. For example, two hydrogen atoms can complete their outer shells by sharing one electron pair made from the single electrons of each atom, as shown in Equation B-2:

$$H \cdot + \cdot H \rightarrow H : H \qquad \textbf{Eq. B-2}$$

Therefore, hydrogen gas consists of individual $H_2$ molecules (›Figure B-4a). (A subscript following a chemical symbol indicates the number of that type of atom present in the molecule.) Several other nonmetallic elements also exist as molecules, because covalent bonds form between identical atoms; oxygen ($O_2$) is an example (›Figure B-4b).

Often, an atom can form covalent bonds with more than one atom. One of the most familiar examples is water ($H_2O$), consisting of two hydrogen atoms each forming a single covalent bond with one oxygen atom (›Figure B-4c). Equation B-3 represents the formation of water's covalent bonds:

$$\begin{array}{l} H \cdot \\ \quad + \cdot \ddot{O} : \rightarrow H : \ddot{O} : \qquad \textbf{Eq. B-3} \\ H \cdot \qquad\qquad\;\; H \end{array}$$

The water molecule is sometimes represented as

$$\begin{array}{l} H — O \\ \qquad | \\ \qquad H \end{array}$$

where the nonshared electron pairs are not shown and the covalent bonds, or shared pairs, are represented by dashes.

## Nonpolar and polar molecules

The electrons between two atoms in a covalent bond are not always shared equally. When the atoms sharing an electron pair are identical, such as two oxygen atoms, the electrons are attracted equally by both atoms and are therefore shared equally. The result is a *nonpolar molecule*. The term *nonpolar* implies no difference at the two ends (the two poles) of the bond. Because both atoms within the molecule exert the same pull on the shared

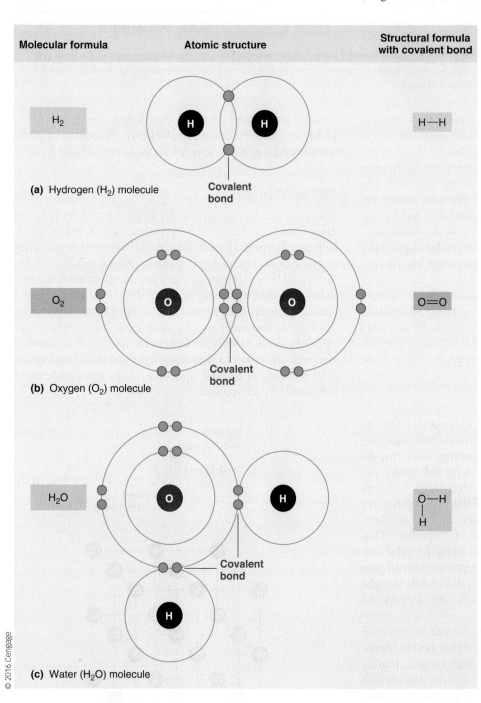

| Molecular formula | Atomic structure | Structural formula with covalent bond |
|---|---|---|

**(a)** Hydrogen ($H_2$) molecule — $H_2$ — Covalent bond — H—H

**(b)** Oxygen ($O_2$) molecule — $O_2$ — Covalent bond — O=O

**(c)** Water ($H_2O$) molecule — $H_2O$ — Covalent bond — O—H / H

› **FIGURE B-4** **A covalent bond.** A covalent bond is formed when atoms that share a pair of electrons are both attracted toward the shared pair.

electrons, each shared electron spends the same amount of time orbiting each atom. So both atoms remain electrically neutral in a nonpolar molecule such as $O_2$.

When the sharing atoms are not identical, unequal sharing of electrons occurs, because atoms of different elements do not exert the same pull on shared electrons. For example, an oxygen atom strongly attracts electrons when it is bonded to other atoms. A *polar molecule* results from the unequal sharing of electrons between different types of atoms covalently bonded together. The water molecule is a good example of a polar molecule. The oxygen atom pulls the shared electrons more strongly than do the hydrogen atoms within each of the two covalent bonds. Consequently, the electron of each hydrogen atom tends to spend more time away orbiting around the oxygen atom than at home around the hydrogen atom. Because of this nonuniform distribution of electrons, the oxygen side of the water molecule where the shared electrons spend more time is slightly negative, and the two hydrogens that are visited less frequently by the electrons are slightly more positive () Figure B-5). Note that the entire water molecule has the same number of electrons as it has protons, and so as a whole has no net charge. This is unlike ions, which have an electron excess or deficit. Polar molecules have a balanced number of protons and electrons but an unequal distribution of the shared electrons among the atoms making up the molecule.

## Hydrogen bonds

Polar molecules are attracted to other polar molecules. In water, for example, an attraction exists between the positive hydrogen ends of some molecules and the negative oxygen ends of others. Hydrogen is not a part of all polar molecules, but when it is covalently bonded to an atom that strongly attracts electrons to form a covalent molecule, the attraction of the positive (hydrogen) end of the polar molecule to the negative end of another polar molecule is called a *hydrogen bond* () Figure B-6). Thus, the polar attractions of water molecules to each other are an example of hydrogen bonding.

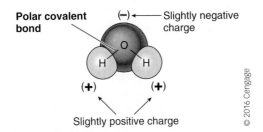

> **FIGURE B-5 A polar molecule.** A water molecule is an example of a polar molecule, in which the distribution of shared electrons is not uniform. Because the oxygen atom pulls the shared electrons more strongly than the hydrogen atoms do, the oxygen side of the molecule is slightly negatively charged, and the hydrogen sides are slightly positively charged.

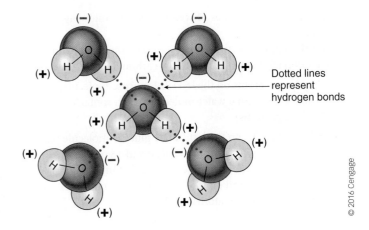

> **FIGURE B-6 A hydrogen bond.** A hydrogen bond is formed by the attraction of a positively charged hydrogen end of a polar molecule to the negatively charged end of another polar molecule.

## B.3 | Chemical Reactions

Processes in which chemical bonds are broken and/or formed are called *chemical reactions*. Reactions are represented by equations in which the reacting substances *(reactants)* are typically written on the left, the newly produced substances *(products)* are written on the right, and an arrow meaning "yields" points from the reactants to the products. These conventions are illustrated in Equation B-4:

$$\underset{\text{Reactants}}{A + B} \rightarrow \underset{\text{Products}}{C + D} \qquad \text{Eq. B-4}$$

### Balanced equations

A chemical equation is a "chemical bookkeeping" ledger that describes what happens in a reaction. By the *law of conservation of mass*, the total mass of all materials entering a reaction equals the total mass of all the products. Thus, the total number of atoms of each element must always be the same on the left and right sides of the equation, because no atoms are lost. Such equations in which the same number of atoms of each type appear on both sides are called *balanced equations*. When writing a balanced equation, the number *preceding* a chemical symbol designates the number of independent (unjoined) atoms, ions, or molecules of that type, whereas a number written as a subscript *following* a chemical symbol denotes the number of a particular atom within a molecule. The absence of a number indicates *one* of that particular chemical. Let's look at a specific example, the oxidation of glucose (the sugar that cells use as fuel), as shown in Equation B-5:

$$\underset{\text{glucose}}{C_6H_{12}O_6} + \underset{\text{oxygen}}{6O_2} \rightarrow \underset{\substack{\text{carbon} \\ \text{dioxide}}}{6CO_2} + \underset{\text{water}}{6H_2O} \qquad \text{Eq. B-5}$$

According to this equation, one molecule of glucose reacts with six molecules of oxygen to produce six molecules of carbon dioxide and six molecules of water. Note the following balance in this reaction:

- 6 carbon atoms on the left (in 1 glucose molecule) and 6 carbon atoms on the right (in 6 carbon dioxide molecules)
- 12 hydrogen atoms on the left (in 1 glucose molecule) and 12 on the right (in 6 water molecules, each containing 2 hydrogen atoms)
- 18 oxygen atoms on the left (6 in 1 glucose molecule plus 12 more in the 6 oxygen molecules) and 18 on the right (12 in 6 carbon dioxide molecules, each containing 2 oxygen atoms, and 6 more in the 6 water molecules, each containing 1 oxygen atom)

## Reversible and irreversible reactions

Under appropriate conditions, the products of a reaction can be changed back to the reactants. For example, carbon dioxide gas dissolves in and reacts with water to form carbonic acid, $H_2CO_3$:

$$CO_2 + H_2O \rightarrow H_2CO_3 \qquad \textbf{Eq. B-6}$$

Carbonic acid is not very stable, however, and as soon as some is formed, part of it decomposes to give carbon dioxide and water:

$$H_2CO_3 \rightarrow CO_2 + H_2O \qquad \textbf{Eq. B-7}$$

Reactions that go in both directions are called *reversible reactions*. They are usually represented by double arrows pointing in both directions:

$$CO_2 + H_2O \rightarrow H_2CO_3 \qquad \textbf{Eq. B-8}$$

Theoretically, every reaction is reversible. Often, however, conditions are such that a reaction, for all practical purposes, goes in only one direction; such a reaction is called *irreversible*. For example, an irreversible reaction takes place when an explosion occurs, because the products do not remain in the vicinity of the reaction site to get together to react.

## Catalysts and enzymes

The rates (speeds) of chemical reactions are influenced by a number of factors, of which catalysts are one of the most important. A *catalyst* is a "helper" molecule that speeds up a reaction without being used up in the reaction. Living organisms use catalysts known as *enzymes*. These enzymes exert amazing influence on the rates of chemical reactions that take place in the organisms. Reactions that take weeks or even months to occur under normal laboratory conditions take place in seconds under the influence of enzymes in the body. One of the fastest-acting enzymes is *carbonic anhydrase*, which catalyzes the reaction between carbon dioxide and water to form carbonic acid. This reaction is important in the transport of carbon dioxide from tissue cells, where it is produced metabolically, to the lungs, where it is excreted. The equation for the reaction is shown in Equation B-6. Each molecule of carbonic anhydrase catalyzes the conversion of 36 million $CO_2$ molecules per minute! Enzymes are important in essentially every chemical reaction that takes place in living organisms.

## B.4 | Molecular and Formula Weight and the Mole

Because molecules are made up of atoms, the relative mass of a molecule is simply the sum of the relative masses (atomic weights) of the atoms found in the molecule. The relative masses of molecules are called *molecular masses* or *molecular weights*. The molecular weight of water ($H_2O$) is the sum of the atomic weights of two hydrogen atoms and one oxygen atom, or 1.01 amu + 1.01 amu + 16.00 amu = 18.02 amu.

Not all compounds exist in the form of molecules. Ionically bonded substances such as sodium chloride consist of three-dimensional arrangements of sodium ions ($Na^+$) and chloride ions ($Cl^-$) in a 1-to-1 ratio. The formulas for ionic compounds reflect only the ratio of the ions in the compound and should not be interpreted in terms of molecules. Thus, the formula for sodium chloride (NaCl) indicates that the ions combine in a 1-to-1 ratio. It is convenient to apply the concept of relative masses to ionic compounds even though they do not exist as molecules. The *formula weight* for such compounds is defined as the sum of the atomic weights of the atoms found in the formula. Thus, the formula weight of NaCl is equal to the sum of the atomic weights of one sodium atom and one chlorine atom—22.99 amu + 35.45 amu = 58.44 amu.

As you have seen, chemical reactions can be represented by equations and discussed in terms of numbers of molecules, atoms, and ions reacting with each other. To carry out reactions in the laboratory, however, a scientist cannot count out numbers of reactant particles but instead must be able to weigh out the correct amount of each reactant. Using the mole concept makes this task possible. A *mole* (abbreviated *mol*) of a pure element or compound is the amount of material contained in a sample of the pure substance that has a mass in grams equal to the substance's atomic weight (for elements) or the molecular weight or formula weight (for compounds). Thus, 1 mol of potassium (K) would be a sample of the element with a mass of 39.10 g. Similarly, 1 mol of $H_2O$ would have a mass of 18.02 g, and 1 mol of NaCl would be a sample with a mass of 58.44 g.

The fact that atomic weights, molecular weights, and formula weights are relative masses leads to a fundamental characteristic of moles. One mole of oxygen atoms has a mass of 16.00 g, and 1 mol of hydrogen atoms has a mass of 1.01 g. Thus, the ratio of the masses of 1 mol of each element is 16.00/1.01, the same as the ratio of the atomic weights for the two elements. Recall that these atomic weights compare the relative masses of oxygen and hydrogen. Accordingly, the number of oxygen atoms present in 16 g of oxygen (1 mol of oxygen) is the same as the number of hydrogen atoms present in 1.01 g of hydrogen. Therefore, 1 mol of oxygen contains exactly the same number of oxygen atoms as the number of hydrogen atoms in 1 mol of hydrogen. Thus, it is possible and sometimes useful to think of a mole as a specific number of particles. This number, called *Avogadro's number*, is equal to $6.02 \times 10^{23}$.

## B.5 | Solutions, Colloids, and Suspensions

In contrast to a compound, a *mixture* consists of two or more types of elements or molecules physically blended together (intermixed) instead of being linked by chemical bonds. A compound has very different properties from the individual elements of which it is composed. For example, the solid, white NaCl (table salt) crystals you use to flavour your food are very different from either sodium (a silvery white metal) or chlorine (a poisonous yellow-green gas found in bleach). By comparison, each component of a mixture retains its own chemical properties. If you mix salt and sugar together, each retains its own distinct taste and other individual properties. The constituents of a compound can only be separated by chemical means—bond breakage. By contrast, the components of a mixture can be separated by physical means, such as filtration or evaporation. The most common mixtures in the body are mixtures of water and various other substances. These mixtures are categorized as *solutions*, *colloids*, or *suspensions*, depending on the size and nature of the substance mixed with water.

### Solutions

Most chemical reactions in the body take place between reactants that have dissolved to form solutions. *Solutions* are homogenous mixtures containing a relatively large amount of one substance called the *solvent* (the dissolving medium) and smaller amounts of one or more substances called *solutes* (the dissolved particles). Salt water, for example, contains mostly water, which is thus the solvent, and a smaller amount of salt, which is the solute. Water is the solvent in most solutions found in the human body.

### Electrolytes and nonelectrolytes

When ionic solutes are dissolved in water to form solutions, the resulting solution will conduct electricity. This is not true for most covalently bonded solutes. For example, a salt–water solution conducts electricity, but a sugar–water solution does not. When salt dissolves in water, the solid lattice of $Na^+$ and $Cl^-$ is broken down, and the individual ions are separated and distributed uniformly throughout the solution. These mobile, charged ions conduct electricity through the solution. Solutes that form ions in solution and conduct electricity are called *electrolytes*. Some very polar covalent molecules also behave this way. When sugar dissolves, however, individual covalently bonded sugar molecules become uniformly distributed throughout the solution. These uncharged molecules cannot conduct a current. Solutes that do not form conductive solutions are called *nonelectrolytes*.

### Measures of concentration

The amount of solute dissolved in a specific amount of solution can vary. For example, a salt–water solution might contain 1 g of salt in 100 mL of solution, or it could contain 10 g of salt in 100 mL of solution. Both solutions are salt–water solutions, but they have different concentrations of solute. The *concentration* of a solution indicates the relationship between the amount of solute and the amount of solution. Concentrations can be given in a number of different units.

### MOLARITY/MOLALITY

Concentrations given in terms of *molarity (M)* give the number of moles of solute in exactly 1 L of solution. Thus, a half molar (0.5 M) solution of NaCl would contain one-half mole, or 29.22 g, of NaCl in each litre of solution. In contrast, *molality (m)* gives the number of moles of solute in 1 kg of solvent. Converting between molality and molarity requires that you know the density of the solute. When using water as a solvent this conversion is quite simple: at room temperature, 1 kg of water occupies a volume of 1 L, meaning the molarity and molality are essential equal. Another common laboratory solvent is dimethyl-sulphoxide, which has a density of 1.100 kg/L. Considering this solvent, a 1 M solution would be 0.909 m.

### NORMALITY

When the solute is an electrolyte, it is sometimes useful to express the concentration of the solution in a unit that gives information about the amount of ionic charge in the solution. This is done by expressing concentration in terms of *normality (N)*. The normality of a solution gives the number of equivalents of solute in exactly 1 L of solution. An *equivalent* of an electrolyte is the amount that produces 1 mol of positive (or negative) charges when it dissolves. The number of equivalents of an electrolyte can be calculated by multiplying the number of moles of electrolyte by the total number of positive charges produced when one formula unit of the electrolyte dissolves. Consider NaCl and calcium chloride $(CaCl_2)$ as examples. The ionization reactions for one formula unit of each solute are as follows:

$$NaCl \rightarrow Na^+ + Cl^- \qquad \textbf{Eq. B-9}$$

$$CaCl_2 \rightarrow Ca^{2+} + 2Cl^- \qquad \textbf{Eq. B-10}$$

Thus, 1 mol of NaCl produces 1 mol of positive charges $(Na^+)$ and so contains 1 equivalent:

$$(1\,mol\,NaCl)(1) = 1\,equivalent$$

where the number 1 used to multiply the 1 mol of NaCl came from the +1 charge on $Na^+$.

One mole of $CaCl_2$ produces 1 mol of $Ca^{2+}$, which is 2 mol of positive charge. Thus 1 mol of $CaCl_2$ contains 2 equivalents:

$$(1\,mol\,CaCl_2)(2) = 2\,equivalents$$

where the number 2 used in the multiplication came from the +2 charge on $Ca^{2+}$.

If two solutions were made such that one contained 1 mol of NaCl per litre and the other contained 1 mol of $CaCl_2$ per litre, the NaCl solution would contain 1 equivalent of solute per litre and would be 1 normal (1 N). The $CaCl_2$ solution would contain 2 equivalents of solute per litre and would be 2 normal (2 N).

**OSMOLARITY/OSMOLALITY**

Another expression of concentration frequently used in physiology is *osmolarity (osm)*, which indicates the total *number* of solute particles in a litre of solution, instead of the relative weights of the specific solutes. The osmolarity of a solution is the product of M and *n*, where *n* is the number of moles of solute particles obtained when 1 mol of solute dissolves. Because nonelectrolytes such as glucose do not dissociate in solution, $n = 1$ and the osmolarity ($n \times$ M) is equal to the molarity of the solution. For electrolyte solutions, the osmolarity exceeds the molarity by a factor equal to the number of ions produced on dissociation of each molecule in solution. For example, because a NaCl molecule dissociates into two ions, $Na^+$ and $Cl^-$, the osmolarity of a 1 M solution of NaCl is $2 \times 1\,M = 2$ osm. As you saw with molarity and molality, this method of expressing concentration can also be reported as the number of osmoles of solute relative to the weight of the solvent, in which case it is called *osmolality*, with the units osm/kg.

## Colloids and suspensions

In solutions, solute particles are ions or small molecules. By contrast, the particles in colloids and suspensions are much larger than ions or small molecules. In colloids and suspensions, these particles are known as *dispersed-phase particles* instead of solutes. When the dispersed-phase particles are no more than about 100 times the size of the largest solute particles found in a solution, the mixture is called a *colloid*. The dispersed-phase particles of colloids generally do not settle out. All dispersed-phase particles of colloids carry electrical charges of the same sign. So they repel each other. The constant buffeting from these collisions keeps the particles from settling. The most abundant colloids in the body are small functional proteins dispersed in the body fluids. An example is the colloidal dispersion of the plasma proteins in the blood (p. 435).

When dispersed-phase particles are larger than those in colloids, if the mixture is left undisturbed the particles will settle out because of the force of gravity. Such mixtures are usually called *suspensions*. The major example of a suspension in the body is the mixture of blood cells suspended in the plasma. The constant movement of blood as it circulates through the blood vessels keeps the blood cells rather evenly dispersed within the plasma. However, if a blood sample is placed in a test tube and treated to prevent clotting, the heavier blood cells gradually settle to the bottom of the tube.

## B.6 | Inorganic and Organic Chemicals

Chemicals are commonly classified into two categories: inorganic and organic.

## Distinction between inorganic and organic chemicals

The original criterion used for this classification was the origin of the chemicals. Those that came from living or once-living sources were called *organic*, and those that came from other sources were *inorganic*. Today, the basis for classification is the element carbon. *Organic* chemicals are generally those that contain carbon. All others are classified as *inorganic*. A few carbon-containing chemicals are also classified as inorganic; the most common are pure carbon in the form of diamond and graphite, carbon dioxide ($CO_2$), carbon monoxide (CO), carbonates such as limestone ($CaCO_3$), and bicarbonates such as baking soda ($NaHCO_3$).

The unique ability of carbon atoms to bond to each other and form networks of carbon atoms results in an interesting fact. Even though organic chemicals all contain carbon, scientists have identified millions of these compounds. Some have been isolated from natural plant or animal sources, and many have been synthesized in laboratories. Inorganic chemicals include all the other 108 elements and their compounds. The number of known inorganic chemicals composed of all these other elements is estimated to be about 250 000, compared to millions of organic compounds composed predominantly of carbon.

## Monomers and polymers

Another result of carbon's ability to bond to itself is the large size of some organic molecules. Organic molecules range in size from methane ($CH_4$), a small, simple molecule with one carbon atom, to molecules such as DNA that contain as many as a million carbon atoms. Organic molecules that are essential for life are called *biological molecules*, or *biomolecules* for short. Some biomolecules are rather small organic compounds, including *simple sugars, fatty acids, amino acids,* and *nucleotides.* These small, single units, known as *monomers* (meaning "single unit"), are building blocks for the synthesis of larger biomolecules, including *complex carbohydrates, lipids, proteins,* and *nucleic acids,* respectively. These larger organic molecules are called *polymers* (meaning "many units"), reflecting the fact that they are made by the bonding together of a number of smaller monomers. For example, starch is formed by linking many glucose molecules together. Very large organic polymers are often referred to as *macromolecules,* reflecting their large size (*macro* means "large"). Macromolecules include many naturally occurring molecules, such as DNA and structural proteins, as well as many molecules that are synthetically produced, such as synthetic textiles (e.g., nylon) and plastics.

## B.7 | Acids, Bases, and Salts

Acids, bases, and salts may be inorganic or organic compounds.

## Acids and bases

Acids and bases are chemical opposites, and salts are produced when acids and bases react with each other. In 1887, Swedish chemist Svante Arrhenius proposed a theory defining acids and bases. He said that an *acid* is any substance that will dissociate, or break apart, when dissolved in water and, in the process, release a hydrogen ion ($H^+$). Similarly, *bases* are substances that dissociate when dissolved in water and, in the process, release a hydroxyl ion ($OH^-$). Hydrogen chloride (HCl) and sodium hydroxide (NaOH) are examples of Arrhenius acids and bases; their dissociations in water are represented in Equations B-11 and B-12, respectively:

$$HCl \rightarrow H^+ + Cl^- \qquad \text{Eq. B-11}$$

$$NaOH \rightarrow Na^+ + OH^- \qquad \text{Eq. B-12}$$

Note that the hydrogen ion is a bare proton, the nucleus of a hydrogen atom. Also note that both HCl and NaOH would behave as electrolytes.

Arrhenius did not know that free hydrogen ions cannot exist in water. They covalently bond to water molecules to form hydronium ions, as shown in Equation B-13:

$$H^+ + :\overset{\displaystyle \cdot\cdot}{\underset{\displaystyle |}{O}}-H \rightarrow \left[ H-\overset{\displaystyle \cdot\cdot}{\underset{\displaystyle |}{O}}-H \right]^+ \qquad \text{Eq. B-13}$$

In 1923, Johannes Brønsted in Denmark and Thomas Lowry in England proposed an acid–base theory that took this behaviour into account. They defined an *acid* as any hydrogen-containing substance that donates a proton (hydrogen ion) to another substance (an acid is a *proton donor*) and a *base* as any substance that accepts a proton (a base is a *proton acceptor*). According to these definitions, the acidic behaviour of HCl given in Equation B-11 is rewritten as shown in Equation B-14:

$$HCl + H_2O \rightarrow H_3O^+ + Cl^- \qquad \text{Eq. B-14}$$

Note that this reaction is reversible, and the hydronium ion is represented as $H_3O^+$. In Equation B-14, the HCl acts as an acid in the forward reaction (left to right), whereas water acts as a base. In the reverse reaction (right to left), the hydronium ion gives up a proton and thus is an acid, whereas the chloride ion, $Cl^-$, accepts the proton and so is a base. It is still a common practice to use equations such as B-11 to simplify the representation of the dissociation of an acid, even though scientists recognize that equations like B-14 are more correct.

### Salts and neutralization reactions

At room temperature, *inorganic salts* are crystalline solids that contain the positive ion (cation) of an Arrhenius base such as NaOH and the negative ion (anion) of an acid such as HCl. Salts can be produced by mixing solutions of appropriate acids and bases, allowing a neutralization reaction to occur. In *neutralization reactions*, the acid and base react to form a salt and water. Most salts that form are water soluble and can be recovered by evaporating the water. Equation B-15 is a neutralization reaction:

$$HCl + NaOH \rightarrow NaCl + H_2O \qquad \text{Eq. B-15}$$

When acids or bases are used as solutes in solutions, the concentrations can be expressed as normalities just as they were earlier for salts. An equivalent of acid is the amount that gives up 1 mol of $H^+$ in solution. Thus, 1 mol of HCl is also 1 equivalent, but 1 mol of $H_2SO_4$ is 2 equivalents. Bases are described in a similar way, but an equivalent is the amount of base that gives 1 mol of $OH^-$.

See Chapter 14 for a discussion of acid–base balance in the body.

## B.8 | Functional Groups of Organic Molecules

Organic molecules consist of carbon and one or more additional elements covalently bonded to one another in "Tinkertoy" fashion. The simplest organic molecules, *hydrocarbons*, such as methane and petroleum products, have only hydrogen atoms attached to a carbon backbone of varying lengths. All biomolecules always have additional elements besides hydrogen added to the carbon backbone. The carbon backbone forms the stable portion of most biomolecules. Other atoms covalently bonded to the carbon backbone, either alone or in clusters, form what is termed *functional groups*. All organic compounds can be classified according to the functional group or groups they contain. *Functional groups* are specific combinations of atoms that generally react in the same way, regardless of the number of carbon atoms in the molecule to which they are attached. For example, all *aldehydes* contain a functional group that contains one carbon atom, one oxygen atom, and one hydrogen atom covalently bonded in a specific way:

$$(-\overset{\displaystyle O}{\overset{\displaystyle \|}{C}}-H)$$

The carbon atom in an aldehyde group forms a single covalent bond with the hydrogen atom and a *double bond* (a bond in which two covalent bonds are formed between the same atoms, designated by a double line between the atoms) with the oxygen atom. The aldehyde group is attached to the rest of the molecule by a single covalent bond extending to the left of the carbon atom. Most aldehyde reactions are the same regardless of the size and nature of the rest of the molecule to which the aldehyde group is attached. Reactions of physiological importance often occur between two functional groups or between one functional group and a small molecule such as water.

## B.9 | Carbohydrates

Carbohydrates are organic compounds of tremendous biological and commercial importance. They are widely distributed in nature and include such familiar substances as starch, table sugar, and cellulose. Carbohydrates have five important functions in living organisms: they provide energy, serve as a stored form of chemical energy, provide dietary fibre, supply carbon atoms for the synthesis of cell components, and form part of the structural elements of cells.

### Chemical composition of carbohydrates

*Carbohydrates* contain carbon, hydrogen, and oxygen. They acquired their name because most of them contain these three elements in an atomic ratio of one carbon to two hydrogens to one oxygen. This ratio suggests that the general formula is $CH_2O$ and that the compounds are simply carbon hydrates ("watered" carbons), or carbohydrates. It is now known that they are not hydrates of carbon, but the name persists. All carbohydrates have a large number of functional groups per molecule. The most common functional groups in carbohydrates are *alcohol, ketone,* and *aldehyde*:

$$(—OH), \quad (—\overset{\displaystyle O}{\overset{\|}{C}}—), \quad (—\overset{\displaystyle O}{\overset{\|}{C}}—H)$$

Alcohol     Ketone     Aldehyde

## Types of carbohydrates

The simplest carbohydrates are simple sugars, also called *monosaccharides*. As their name indicates, they consist of single, simple-sugar units called saccharides (*mono* means "one"). The molecular structure of *glucose*, an important monosaccharide, is shown in ›Figure B-7a. In solution, most glucose molecules assume the ring form shown in ›Figure B-7b. Other common monosaccharides are *fructose, galactose,* and *ribose*—or functional groups formed by reactions between pairs of these three.

*Disaccharides* are sugars formed by linking two monosaccharide molecules together through a covalent bond (*di* means "two"). Some common examples of disaccharides are *sucrose* (common table sugar) and *lactose* (milk sugar). Sucrose molecules are formed from one glucose and one fructose molecule. Lactose molecules each contain one glucose and one galactose unit.

Because of the many functional groups on carbohydrate molecules, large numbers of simple carbohydrate molecules are able to bond together and form long chains and branched networks. The resultant substances, *polysaccharides*, contain many saccharide units (*poly* means "many"). Three common polysaccharides made up entirely of glucose units are glycogen, starch, and cellulose.

- *Glycogen* is a storage carbohydrate found in animals. It is a highly branched polysaccharide that averages a branch every 8 to 12 glucose units. The structure of glycogen is

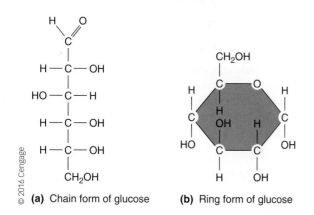

(a) Chain form of glucose     (b) Ring form of glucose

› FIGURE B-7 **Forms of glucose**

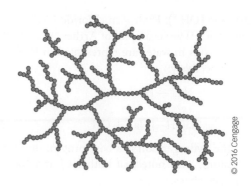

› FIGURE B-8 **A simplified representation of glycogen.** Each circle represents a glucose molecule.

represented in ›Figure B-8, where each circle represents one glucose unit.

- *Starch*, a storage carbohydrate of plants, consists of two fractions, amylose and amylopectin. Amylose consists of long, essentially unbranched chains of glucose units. Amylopectin is a highly branched network of glucose units averaging 24 to 30 glucose units per branch. Thus, it is less highly branched than glycogen.

- *Cellulose*, a structural carbohydrate of plants, exists in the form of long, unbranched chains of glucose units. The bonding between the glucose units of cellulose is slightly different from the bonding between the glucose units of glycogen and starch. Humans have digestive enzymes that catalyze the breaking (hydrolysis) of the glucose-to-glucose bonds in starch but lack the necessary enzymes to hydrolyze cellulose glucose-to-glucose bonds. So starch is a form of food for humans, but cellulose is not. Cellulose is the indigestible fibre in our diets.

## B.10 | Lipids

Lipids are a diverse group of organic molecules made up of substances with widely different compositions and molecular structures. Unlike carbohydrates, which are classified on the basis of their *molecular structure*, lipids are classified on the basis of their *solubility*. Lipids are insoluble in water but soluble in nonpolar solvents such as alcohol. Thus, lipids are the waxy, greasy, or oily compounds found in plants and animals. Lipids repel water, a useful characteristic of the protective wax coatings found on some plants. Fats and oils are energy rich and have relatively low densities. These properties account for the use of fats and oils as stored energy in plants and animals. Still other lipids occur as structural components, especially in cellular membranes. The oily plasma membrane that surrounds each cell serves as a barrier that separates the intracellular contents from the surrounding extracellular fluid (pp. 5, 29, and 30).

## Simple lipids

Simple lipids contain just two types of components: fatty acids and alcohols. *Fatty acid molecules* consist of a hydrocarbon chain with a *carboxyl* functional group (–COOH) on the end. The hydrocarbon chain can be of variable length, but natural fatty acids always contain an even number of carbon atoms. The hydrocarbon chain can also contain one or more double bonds between carbon atoms. Fatty acids with no double bonds are called *saturated fatty acids*, whereas those with double bonds are called *unsaturated fatty acids*. The more double bonds present, the higher the degree of unsaturation. Saturated fatty acids predominate in dietary animal products (e.g., meat, eggs, and dairy products), whereas unsaturated fatty acids are more prevalent in plant products (e.g., grains, vegetables, and fruits). Consumption of a greater proportion of saturated rather than unsaturated fatty acids is linked with a higher incidence of cardiovascular disease (p. 374).

The most common alcohol found in simple lipids is *glycerol* (glycerin), a three-carbon alcohol that has three alcohol functional groups (–OH).

Simple lipids called fats and oils are formed by a reaction between the carboxyl group of three fatty acids and the three alcohol groups of glycerol. The resulting lipid is an E-shaped molecule called a *triglyceride*. Such lipids are classified as fats or oils on the basis of their melting points. *Fats* are solids at room temperature, whereas *oils* are liquids. Their melting points depend on the degree of unsaturation of the fatty acids of the triglyceride. The melting point goes down with increasing degree of unsaturation. Therefore, oils contain more unsaturated fatty acids than fats do. Examples of the components of fats and oils and a typical triglyceride molecule are shown in Figure B-9.

When triglycerides form, a molecule of water is released as each fatty acid reacts with glycerol. Adipose tissue in the body contains triglycerides. When the body uses adipose tissue as an energy source, the triglycerides react with water to release free fatty acids into the blood. The fatty acids can be used as an immediate energy source by many organs. In the liver, free fatty acids are converted into compounds called *ketone bodies*. Two of the ketone bodies are acids, and one is the ketone called acetone. Excess ketone bodies are produced during *diabetes mellitus*, a condition in which most cells resort to using fatty acids as an energy source because the cells are unable to take up adequate amounts of glucose in the face of inadequate insulin action (p. 278).

## Complex lipids

Complex lipids have more than two types of components. The different complex lipids usually contain three or more of the following components: glycerol, fatty acids, a phosphate group, an alcohol other than glycerol, and a carbohydrate. Those that contain phosphate are called *phospholipids*. ›Figure B-10 contains representations of a few complex lipids; it emphasizes the components but does not give details of the molecular structures.

*Steroids* are lipids that have a unique structural feature consisting of a fused carbon ring system containing three six-membered rings and a single five-membered ring (›Figure B-11). Different steroids possess this characteristic ring structure but have different functional groups and carbon chains attached.

*Cholesterol*, a steroidal alcohol, is the most abundant steroid in the human body. It is a component of cell membranes and is used by the body to produce other important steroids that include bile salts, male and female sex hormones, and adrenocortical hormones. The structures of cholesterol and cortisol, an important adrenocortical hormone, are shown in ›Figure B-12.

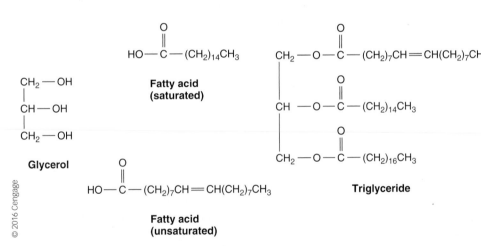

© 2016 Cengage

> **FIGURE B-9** **Triglyceride components and structure**

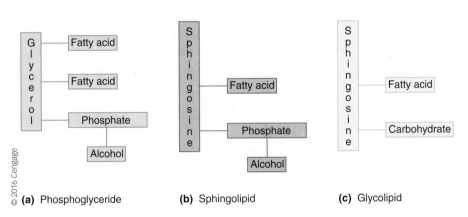

© 2016 Cengage

**(a)** Phosphoglyceride          **(b)** Sphingolipid          **(c)** Glycolipid

> **FIGURE B-10** **Examples of complex lipids.** (a) A phosphoglyceride. (b) A sphingolipid (sphingosine is an alcohol). (c) A glycolipid.

## B.11 | Proteins

The name *protein* is derived from the Greek word *proteios*, which means "of first importance." It is certainly an appropriate term for these very important biological compounds. Proteins are indispensable components of all living things, where they play

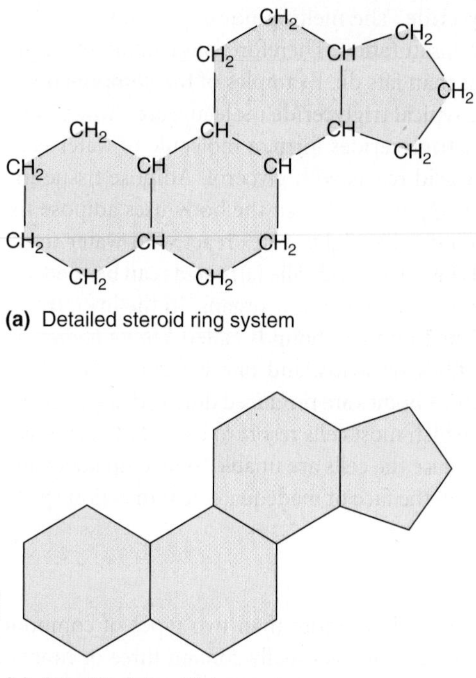

**(a)** Detailed steroid ring system

**(b)** Simplified steroid ring system

© 2016 Cengage

› **FIGURE B-11** **The steroid ring system**

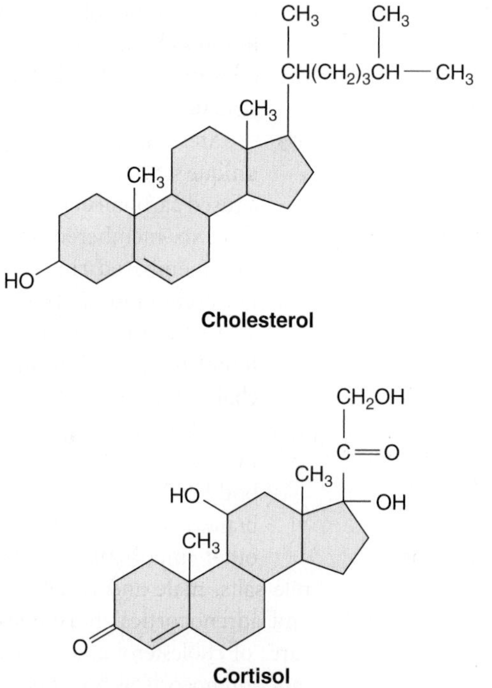

**Cholesterol**

**Cortisol**

© 2016 Cengage

› **FIGURE B-12** **Examples of steroidal compounds**

crucial roles in all biological processes. Proteins are the main structural component of cells, and all chemical reactions in the body are catalyzed by enzymes, all of which are proteins.

## Chemical composition of proteins

Proteins are macromolecules made up of monomers called *amino acids*. Hundreds of different amino acids, both natural and synthetic, are known, but only 20 are commonly found in natural proteins. From this limited pool of 20 amino acids, cells build thousands of different types of proteins, each with a distinctly different function, much the same way that composers create a diversity of unique music from a relatively small number of notes. Different proteins are constructed according to variations in the types and numbers of amino acids used and also in the order they are linked together. However, proteins are not formed haphazardly by a random linking of amino acids. Every protein in the body is deliberately and precisely synthesized under the direction of the blueprint laid down in the person's genes. Thus amino acids are assembled in a specific pattern to produce a given protein to accomplish a particular structural or functional task in the body. (More information about protein synthesis can be found in Appendix C.)

## Peptide bonds

Each amino acid molecule has three important parts: an amino functional group ($-NH_2$), a carboxyl functional group ($-COOH$), and a characteristic side chain or R group. These components are shown in expanded form in › Figure B-13. Amino acids form long chains as a result of reactions between the amino group of one amino acid and the carboxyl group of another amino acid. This reaction is illustrated in Equation B-16:

**Eq. B-16**

Notice that after the two molecules react, the ends of the product still have an amino group and a carboxyl group that can react to extend the chain length. The covalent bond formed in the reaction is called a *peptide bond* (Figure B-14).

On a molecular scale, proteins are immense molecules. Their size can be illustrated by comparing a glucose molecule to a molecule of *haemoglobin*, a protein. Glucose has a molecular weight of 180 amu and a molecular formula of $C_6H_{12}O_6$. *Haemoglobin*, a relatively small protein, has a molecular weight of 65 000 amu and a molecular formula of $C_{2952}H_{4664}O_{832}N_{812}S_8Fe_4$.

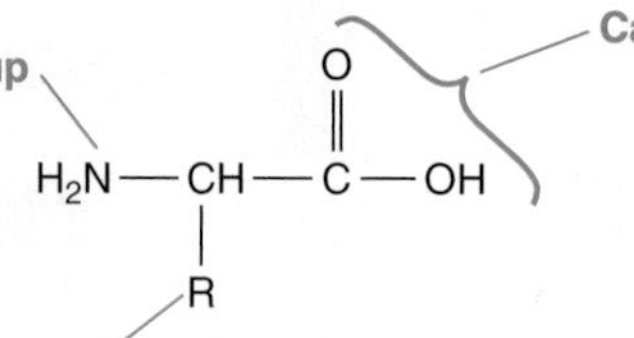

Amino group

Carboxyl group

$H_2N—CH—C—OH$

R

**Side chain** (different for each amino acid)

© 2016 Cengage

› **FIGURE B-13** **The general structure of amino acids**

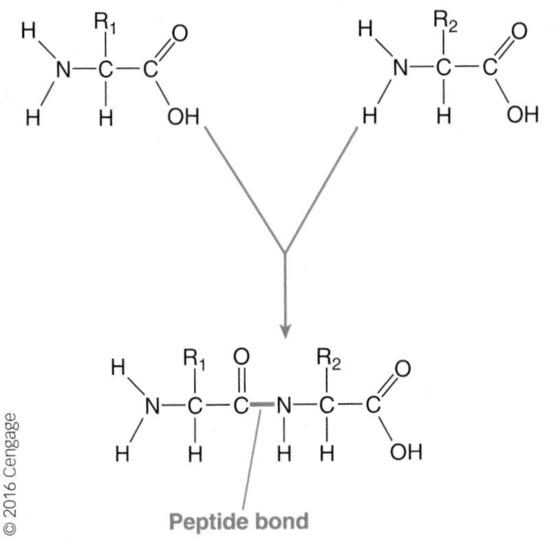

© 2016 Cengage

> **FIGURE B-14 A peptide bond.** In forming a peptide bond, the carboxyl group of one amino acid reacts with the amino group of another amino acid.

## Levels of protein structure

The many atoms in a protein are not arranged in a random way. In fact, proteins have a high degree of structural organization that plays an important role in their behaviour in the body.

### PRIMARY STRUCTURE

The first level of protein structure is called the *primary structure*. It is simply the order in which amino acids are bonded together to form the protein chain. Amino acids are frequently represented by three-letter abbreviations, such as Gly for glycine and Arg for arginine. When this practice is followed, the primary structure of a protein can be represented as in > Figure B-15, which shows part of the primary structure of human insulin, or as in > Figure B-16a, which depicts a portion of the primary structure of haemoglobin.

### SECONDARY STRUCTURE

The second level of protein structure, called the *secondary structure*, results when hydrogen bonding occurs between the amino hydrogen of one amino acid and the carboxyl oxygen

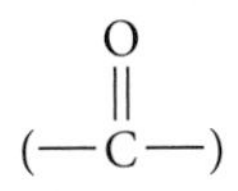

$$
\begin{array}{c}
O \\
\parallel \\
(\!-\!C\!-\!)
\end{array}
$$

of another amino acid in the same chain. As a result of this hydrogen bonding, the involved portion of the chain typically assumes a coiled, helical shape called the alpha ($\alpha$)

© 2016 Cengage

| Thr | Lys | Pro | Thr | Tyr | Phe | Phe | Gly | Arg | . . . . .

Thr—Lys—Pro—Thr—Tyr—Phe—Phe—Gly—Arg— . . . . .

> **FIGURE B-15 A portion of the primary protein structure of human insulin**

helix, which is by far the most common secondary structure found in the body (>Figure B-16b). Other secondary structures such as the beta ($\beta$) pleated sheet and random coils can also form, depending on the pattern of hydrogen bonding between amino acids located in different parts of the same chain.

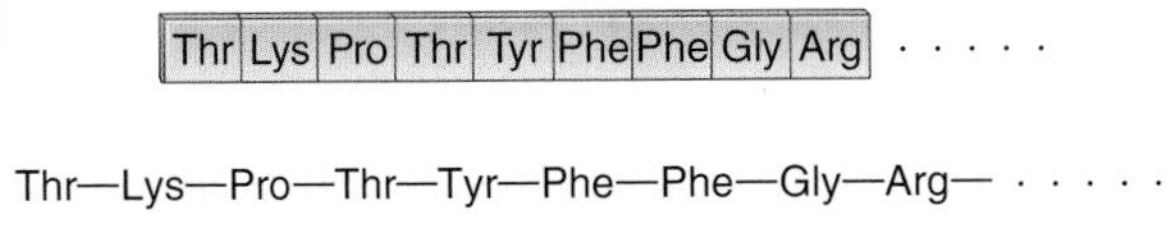

### TERTIARY AND QUATERNARY STRUCTURE

The third level of structure in proteins is the *tertiary structure*. It results when functional groups of the side chains of amino acids in the protein chain react with each other. Several different types of interactions are possible, as shown in >Figure B-17. Tertiary structures can be visualized by letting a length of wire represent the chain of amino acids in the primary structure of a protein. Next imagine that the wire is wound around a pencil to form a helix, which represents the secondary structure. The pencil is removed, and the helical structure is now folded back on itself or carefully wadded into a ball. Such folded or spherical structures represent the tertiary structure of a protein (>Figure B-16c).

All functional proteins exist in at least a tertiary structure. Sometimes, several polypeptides interact with each other to form a fourth level of protein structure, the *quaternary structure*. For example, *haemoglobin* contains four highly folded polypeptide chains (the *globin* portion) (>Figure B-16d). Four iron-containing *heme* groups, one tucked within the interior of each of the folded polypeptide subunits, complete the quaternary structure of *haemoglobin* (see Figure 9-2).

## Hydrolysis and denaturation

One of the important functions of proteins is to serve as enzymes that catalyze the many essential chemical reactions of the body. In addition to catalyzing reactions, proteins can undergo reactions themselves. Two of the most important are hydrolysis and denaturation.

### HYDROLYSIS

Notice that according to Equation B-16, the formation of peptide bonds releases water molecules. Under appropriate conditions, it is possible to reverse such reactions by adding water to the peptide bonds and breaking them. *Hydrolysis* (breakdown by $H_2O$) reactions of this type convert large proteins into smaller fragments or even into individual amino acids. Hydrolysis is the means by which digestive enzymes break down ingested food into small units that can be absorbed from the digestive tract lumen into the blood.

### DENATURATION

*Denaturation* of proteins occurs when the bonds holding a protein chain in its characteristic tertiary or secondary conformation are broken. When this happens, the protein chain takes on a random, disorganized conformation. Denaturation can result when proteins are subjected to heating (including when body temperature rises too high; p. 711), to extremes of pH (p. 621), or to treatment with specific chemicals such as alcohol or heavy metal ions. In some instances, denaturation is accompanied by

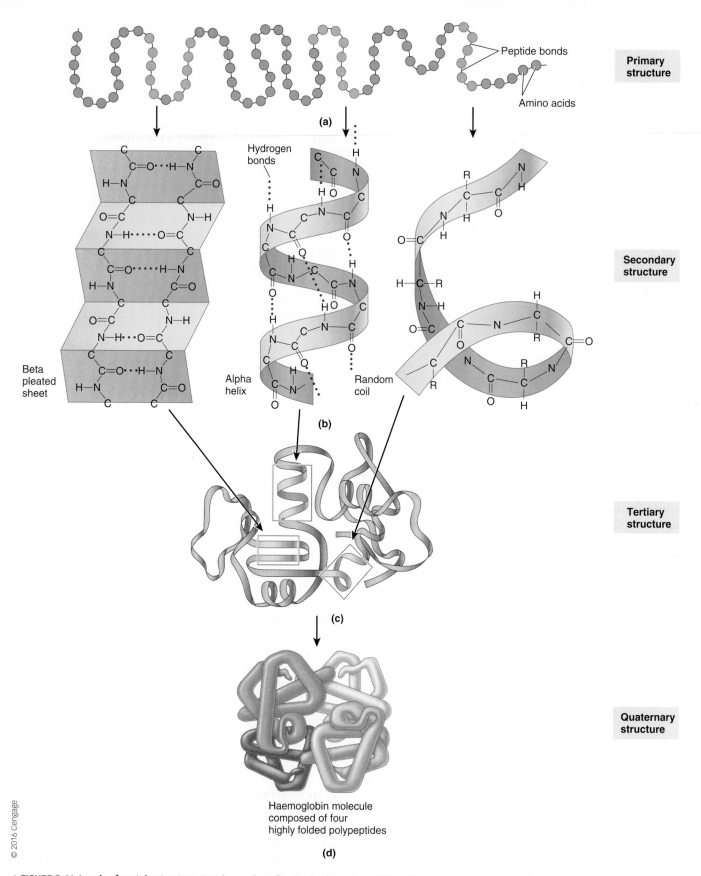

**Primary structure**

Peptide bonds

Amino acids

**(a)**

**Secondary structure**

Hydrogen bonds

Beta pleated sheet

Alpha helix

Random coil

**(b)**

**Tertiary structure**

**(c)**

**Quaternary structure**

Haemoglobin molecule composed of four highly folded polypeptides

**(d)**

© 2016 Cengage

> **FIGURE B-16 Levels of protein structure.** Proteins can have four levels of structure. (a) The primary structure is a particular sequence of amino acids bonded in a chain. (b) At the secondary level, hydrogen bonding occurs between various amino acids within the chain, causing the chain to assume a particular shape. The most common secondary protein structure in the body is the alpha helix. (c) The tertiary structure is formed by the folding of the secondary structure into a functional three-dimensional configuration. (d) Many proteins form a fourth level of structure composed of several polypeptides, as exemplified by haemoglobin.

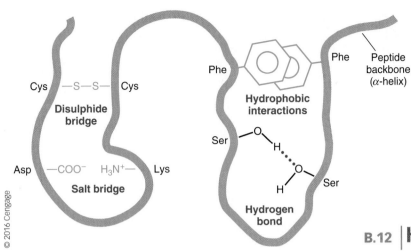

> FIGURE B-17 Side chain interactions leading to the tertiary protein structure

coagulation or precipitation, as illustrated by the changes that occur in the white of an egg as it is fried.

### Nucleic acids

*Nucleic acids* are high-molecular-weight macromolecules responsible for storing and using genetic information in living cells and passing it on to future generations. These important biomolecules are classified into two categories: *deoxyribonucleic acid (DNA)* and *ribonucleic acid (RNA)*. DNA is found primarily in the cell's nucleus, and RNA is found primarily in the cytoplasm that surrounds the nucleus.

Both types of nucleic acid are made up of units called *nucleotides*, which in turn consist of three simpler components. Each nucleotide contains an organic nitrogenous base, a sugar, and a phosphate group. The three components are chemically bonded together with the sugar molecule lying between the base and the phosphate. In RNA, the sugar is *ribose*, whereas in DNA it is *deoxyribose*. When nucleotides bond together to form nucleic acid chains, the bonding is between the phosphate of one

nucleotide and the sugar of another. The resulting nucleic acids consist of chains of alternating phosphates and sugar molecules, with a base molecule extending out of the chain from each sugar molecule (see > Figure C-1, Appendix C, p. A-20).

The chains of nucleic acid have structural features somewhat like those found in proteins. DNA takes the form of two chains that mutually coil around one another to form the well-known double helix. Some RNA occurs in essentially straight chains, whereas in other types the chain forms specific loops or helices. See Appendix C for further details.

## B.12 | High-energy Biomolecules

Not all nucleotides are used to construct nucleic acids. One very important nucleotide—*adenosine triphosphate (ATP)*—is used as the body's primary energy carrier. Certain bonds in ATP temporarily store energy that is harnessed during the metabolism of foods and made available to the parts of the cells where it is needed to do specific cellular work (pp. 24-28). Let's see how ATP functions in this role. Structurally, ATP is a modified RNA (ribose-containing) nucleotide that has adenine as its base and two additional phosphates bonded in sequence to the original nucleotide phosphate. This means adenosine triphosphate, as the name implies, has a total of three phosphates attached in a string to *adenosine*, the composite of ribose and adenine (Figure B-18). Attaching these additional phosphates requires considerable energy input. The high-energy input used to create these *high-energy phosphate bonds* is "stored" in the bonds for later use. Most energy transfers in the body involve ATP's terminal phosphate bond. When energy is needed, the third phosphate is cleaved off by hydrolysis, yielding *adenosine diphosphate (ADP)* and an inorganic phosphate ($P_i$) and releasing energy in the process (Equation B-17):

$$ATP \rightarrow ADP + P_i + \text{energy for use by cell} \qquad \textbf{Eq. B-17}$$

Why do cells use ATP as an energy currency that they can cash in by the splitting of high-energy phosphate bonds as needed? Why don't they simply and directly use the energy released during the oxidation of nutrient molecules such as glucose? The answer is that if all the energy stored in glucose were to be released at once, this amount of energy would be more than the cell could immediately use, and so much of it would be squandered. Instead, the energy trapped within the glucose bonds is gradually released and harnessed in the form of the high-energy phosphate bonds of ATP—that is, in cellular "bite-size pieces."

Under the influence of an enzyme, ATP can be converted to a cyclic form of adenosine monophosphate, which contains only one phosphate group, the other two having been cleaved off. The resultant molecule, called *cyclic* AMP (cAMP), serves as an intracellular messenger, affecting the activities of a number of enzymes involved in important reactions in the body (p. 217).

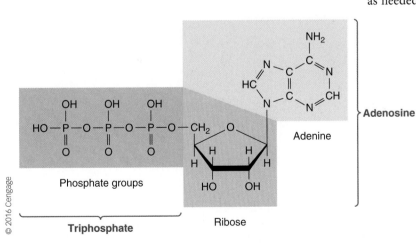

> FIGURE B-18 The structure of ATP

# Storage, Replication, and Expression of Genetic Information

## C.1 | Deoxyribonucleic Acid (DNA) and Chromosomes

The nucleus of the cell houses *deoxyribonucleic acid* (DNA), the genetic blueprint that is unique for each individual.

### Functions of DNA

As genetic material, DNA serves two essential functions. First, it contains instructions for assembling the structural and enzymatic proteins of the cell. Cellular enzymes in turn control the formation of other cellular structures and also determine the functional activity of the cell by regulating the rate at which metabolic reactions proceed. The nucleus serves as the cell's control centre by directly or indirectly controlling almost all cell activities; this is accomplished through the role of the cell's DNA in governing protein synthesis. Because cells make up the body, the DNA code determines the structure and the function of the body as a whole. The DNA of an organism not only dictates whether the organism is a human, a toad, or a pea but also determines the unique physical and functional characteristics of that individual, all of which ultimately depend on the proteins produced under DNA control.

Second, by replicating (making copies of itself), DNA perpetuates the genetic blueprint within all new cells formed within the body, and is thereby responsible for passing on genetic information from parents to children.

We will examine the structure of DNA and the coding mechanism it uses, and then turn our attention to the means by which DNA replicates itself and controls protein synthesis.

### Structure of DNA

Deoxyribonucleic acid is a huge molecule, composed in humans of millions of nucleotides arranged into two long, paired, strands that spiral around each other to form a double helix. Each *nucleotide* has three components: (1) a *nitrogenous base*, a ring-shaped organic molecule containing nitrogen; (2) a five-carbon ring-shaped sugar molecule, which in DNA is *deoxyribose*; and (3) a phosphate group. Nucleotides are joined end to end by linkages between the sugar of one nucleotide and the phosphate group of the adjacent nucleotide to form a long polynucleotide (many

nucleotide) strand with a sugar–phosphate backbone and bases projecting out one side (›Figure C-1). The four different bases in DNA are the double-ringed bases *adenine (A)* and *guanine (G)* and the single-ringed bases *cytosine (C)* and *thymine (T)*. The two polynucleotide strands within a DNA molecule are wrapped around each other so that their bases all project to the interior of the helix. The strands are held together by weak hydrogen bonds formed between the bases of adjoining strands (Appendix B, › Figure B-6). Base pairing is highly specific: Adenine pairs only with thymine and guanine pairs only with cytosine (›Figure C-2).

### Genes

The composition of the repetitive sugar–phosphate backbones that form the "sides" of the DNA "ladder" is identical for every molecule of DNA, but the sequence of the linked bases that form the "rungs" varies among different DNA molecules. The particular sequence of bases in a DNA molecule serves as instructions, or a molecular code, that dictates the assembly of amino acids into a given order for the synthesis of specific *polypeptides* (chains of amino acids linked by peptide bonds; see Appendix B, p. A-14). A *gene* is a stretch of DNA that codes for the synthesis of a particular polypeptide. Polypeptides, in turn, are folded into a three-dimensional configuration to form a functional protein. Not all portions of a DNA molecule code for structural or enzymatic proteins. Some stretches of DNA code for proteins that regulate genes. Other segments appear important in organizing and packaging DNA within the nucleus. Still other regions are known as nonsense base sequences because they have no apparent significance.

### Packaging of DNA into chromosomes

The DNA molecules within each human cell, if lined up end to end, would extend more than 2 m (2 000 000 μm), yet these molecules are packed into a nucleus that is only 5 μm in diameter. These molecules are not randomly crammed into the nucleus but are precisely organized into **chromosomes**. Each chromosome consists of a different DNA molecule and contains a unique set of genes.

*Somatic* (body) *cells* contain 46 chromosomes (the *diploid number*), which can be sorted into 23 pairs on the basis of various distinguishing features. Chromosomes composing a matched

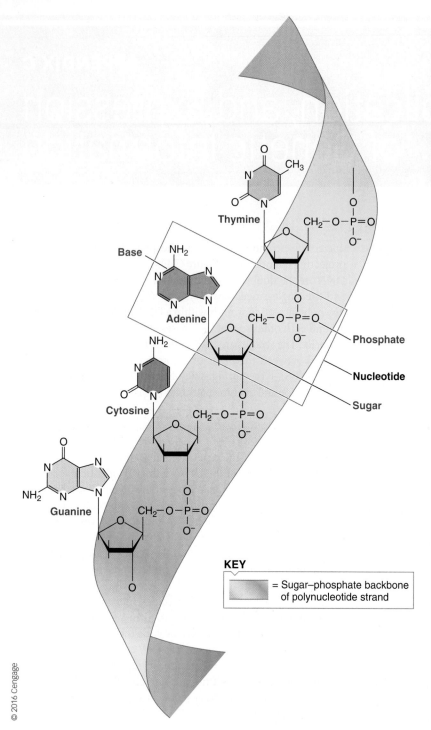

KEY

= Sugar–phosphate backbone of polynucleotide strand

› **FIGURE C-1 Polynucleotide strand.** Sugar–phosphate bonds link adjacent nucleotides together to form a polynucleotide strand with bases projecting to one side. The sugar–phosphate backbone is identical in all polynucleotides, but the sequence of the bases varies.

© 2016 Cengage

DNA molecules are packaged and compressed into discrete chromosomal units in part by nuclear proteins associated with DNA. Two classes of proteins—histone and nonhistone proteins—bind with DNA. *Histones* form bead-shaped bodies that play a key role in packaging DNA into its chromosomal structure. The *nonhistones* are important in gene regulation. The complex formed between the DNA and its associated proteins is known as *chromatin*. The long threads of DNA within a chromosome are wound around histones at regular intervals, thereby compressing a given DNA molecule to about one-sixth its fully extended length. This "beads-on-a-string" structure is further folded and supercoiled into higher and higher levels of organization to further condense DNA into rodlike chromosomes that are readily visible through a light microscope during cell division (›Figure C-3). When the cell is not dividing, the chromosomes partially decondense (i.e., unravel) to a less compact form of chromatin that is indistinct under a light microscope but appears as thin strands and clumps with an electron microscope. The decondensed form of DNA is its working form; this is the form used as a template for protein assembly. Let's turn our attention to this working form of DNA in operation.

## C.2 | Complementary Base Pairing, Replication, and Transcription

Complementary base pairing serves as the foundation for both DNA replication and the initial step of protein synthesis. We now examine the mechanism and significance of complementary base pairing in each of these circumstances, starting with DNA replication.

### DNA replication

During DNA replication, the two decondensed DNA strands "unzip" as the weak bonds between the paired bases are enzymatically broken. Then *complementary base pairing* takes place: New nucleotides present within the nucleus pair with the exposed bases from each unzipped strand (› Figure C-4). New adenine-bearing nucleotides pair with exposed thymine-bearing nucleotides in an old strand, and new guanine-bearing nucleotides pair with exposed cytosine-bearing nucleotides in an old strand. This complementary base pairing is initiated at one end of the two old strands and proceeds in an orderly fashion to the other end. The new nucleotides attracted to and thus aligned in a prescribed order by the old nucleotides are sequentially joined by sugar–phosphate linkages to form two new strands that are complementary to each of the old strands.

pair are termed *homologous chromosomes*—one member of each pair having been derived from the individual's maternal parent and the other member from the paternal parent. *Germ* (reproductive) *cells* (i.e., sperm and eggs) contain only one member of each homologous pair for a total of 23 chromosomes (the *haploid number*). Union of a sperm and an egg results in a new diploid cell with 46 chromosomes, consisting of a set of 23 chromosomes from the mother and another set of 23 from the father (p. 727).

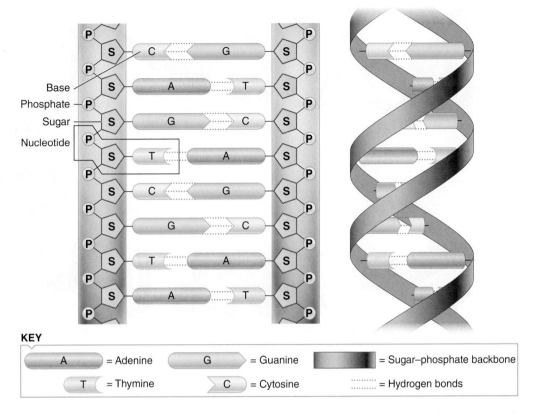

**KEY**

| | | | | | |
|---|---|---|---|---|---|
| A | = Adenine | G | = Guanine | ▬▬▬ | = Sugar–phosphate backbone |
| T | = Thymine | C | = Cytosine | ⋯⋯⋯ | = Hydrogen bonds |

❯ **FIGURE C-2 Complementary base pairing in DNA.** (a) Two polynucleotide strands held together by weak hydrogen bonds formed between the bases of adjoining strands—adenine always paired with thymine and guanine always paired with cytosine. (b) Arrangement of the two bonded polynucleotide strands of a DNA molecule into a double helix.

This replication process results in two complete double-stranded DNA molecules: one strand within each molecule having come from the original DNA molecule, and one strand having been newly formed by complementary base pairing. These two DNA molecules are both identical to the original DNA molecule; due to the imposed pattern of base pairing, the "missing" strand in each of the original separated strands has been produced anew. This replication process, which occurs only during cell division, is essential for perpetuating the genetic code in both the new daughter cells. The duplicate copies of DNA are separated and evenly distributed to the two halves of the cell before it divides. The topic of cell division is covered in detail in Section C.4.

## DNA transcription and messenger RNA

At other times, when it's not replicating in preparation for cell division, DNA serves as a blueprint for dictating cellular protein synthesis. How is this accomplished when DNA is sequestered within the nucleus and protein synthesis is carried out by ribosomes within the cytoplasm? Several types of another nucleic acid, *ribonucleic acid (RNA)*, serve as the go-between.

### STRUCTURE OF RIBONUCLEIC ACID

Ribonucleic acid differs structurally from DNA in three regards: (1) The five-carbon sugar in RNA is *ribose* instead of deoxyribose, the only difference between them being the presence in

ribose of a single oxygen atom that is absent in deoxyribose; (2) RNA contains the closely related base *uracil* instead of thymine, with the three other bases being the same as in DNA; and (3) RNA is single-stranded and not self-replicating.

All RNA molecules are produced in the nucleus using DNA as a template, and then exit the nucleus through openings in the nuclear membrane, called *nuclear pores*. These pores are large enough for passage of RNA molecules, but they block the much larger DNA molecules.

The DNA instructions for assembling a particular protein coded in the base sequence of a given gene are transcribed into a molecule of *messenger RNA (mRNA)*. The segment of the DNA molecule to be copied uncoils, and the base pairs separate to expose the particular sequence of bases in the gene. In any given gene, only one of the DNA strands is used as a template for transcribing RNA. Which strand is copied differs for each gene along the same DNA molecule. The beginning and end of a gene within a DNA strand are designated by particular base sequences that serve as start and stop signals.

### TRANSCRIPTION

*Transcription* is accomplished by complementary base pairing of free RNA nucleotides with their DNA counterparts in the exposed gene (❯ Figure C-5). The same pairing rules apply except that uracil, the RNA nucleotide substitute for thymine, pairs with adenine in the exposed DNA nucleotides. As soon as the RNA nucleotides pair

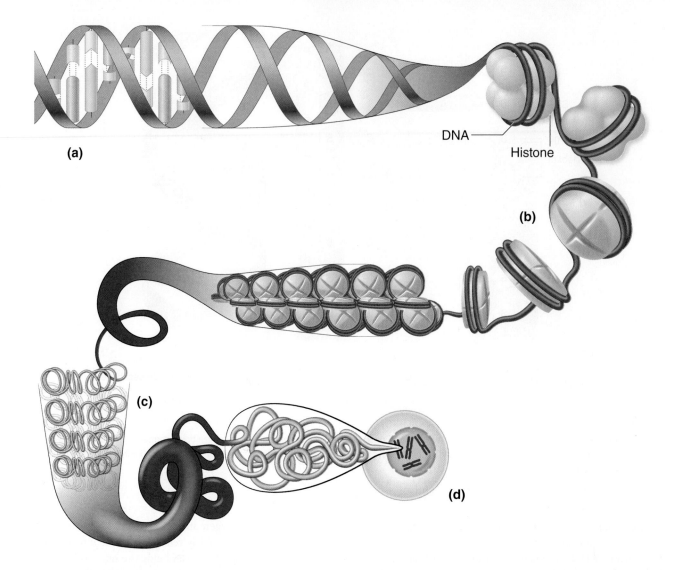

DNA

Histone

(a)

(b)

(c)

(d)

> FIGURE C-3 **Levels of organization of DNA.** (a) Double helix of a DNA molecule. (b) DNA molecule wound around histone proteins, forming a "beads-on-a-string" structure. (c) Further folding and supercoiling of the DNA–histone complex. (d) Rodlike chromosomes, the most condensed form of DNA, which are visible in the cell's nucleus during cell division.

with their DNA counterparts, sugar–phosphate bonds are formed to join the nucleotides together into a single-stranded RNA molecule that is released from DNA once transcription is complete. The original conformation of DNA is then restored. The RNA strand is much shorter than a DNA strand, because only a one-gene segment of DNA is transcribed into a single RNA molecule. The length of the finished RNA transcript varies, depending on the size of the gene. Within its nucleotide base sequence, this RNA transcript contains instructions for assembling a particular protein. Note that the message is coded in a base sequence that is *complementary to, not identical to,* the original DNA code.

Messenger RNA delivers the final coded message to the ribosomes for *translation* into a particular amino acid sequence to form a given protein. Thus, genetic information flows from DNA (which can replicate itself) through RNA to protein. This is accomplished first by *transcription* of the DNA code into a complementary RNA code, followed by *translation* of the RNA code into a specific protein. In the next section, you will learn more

about the steps in translation. The structural and functional characteristics of the cell as determined by its protein composition can be varied, subject to control, depending on which genes are switched on to produce mRNA.

Free nucleotides present in the nucleus cannot be randomly joined together to form either DNA or RNA strands, because the enzymes required to link the sugar and phosphate components of nucleotides are active only when bound to DNA. This ensures that DNA, mRNA, and protein assembly occur only according to genetic plan.

## C.3 | Translation and Protein Synthesis

Three forms of RNA participate in protein synthesis. Besides messenger RNA, two other forms of RNA are required for translation of the genetic message into cellular protein: ribosomal RNA and transfer RNA.

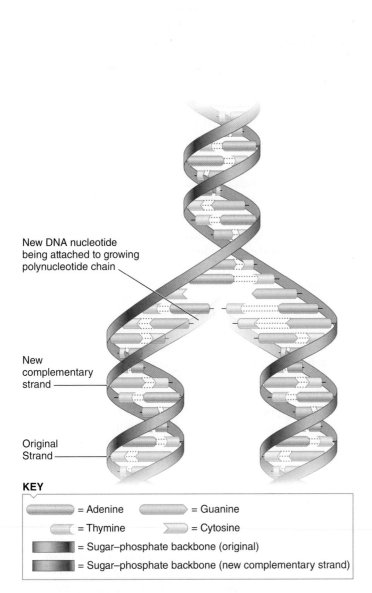

New DNA nucleotide being attached to growing polynucleotide chain

New complementary strand

Original Strand

**KEY**

| | | | |
|---|---|---|---|
| = Adenine | | = Guanine | |
| = Thymine | | = Cytosine | |
| = Sugar–phosphate backbone (original) | | | |
| = Sugar–phosphate backbone (new complementary strand) | | | |

❭ **FIGURE C-4 Complementary base pairing during DNA replication**. During DNA replication, the DNA molecule is unzipped, and each old strand directs the formation of a new strand; the result is two identical double-helix DNA molecules.

- *Messenger* RNA carries the coded message from nuclear DNA to a cytoplasmic ribosome, where it directs the synthesis of a particular protein.

- *Ribosomal* RNA (rRNA) is an essential component of *ribosomes*, the "workbenches" for protein synthesis (p. 23). Ribosomes "read" the base sequence code of mRNA and translate it into the appropriate amino acid sequence during protein synthesis.

- *Transfer* RNA (tRNA) transfers the appropriate amino acids in the cytosol to their designated site in the amino acid sequence of the protein under construction.

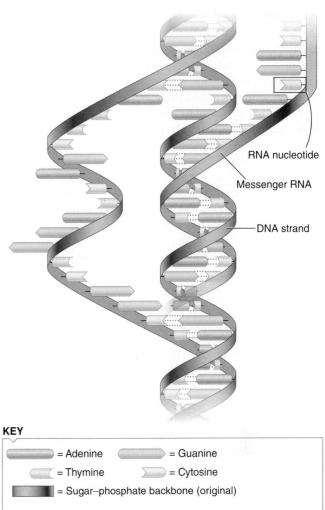

RNA nucleotide

Messenger RNA

DNA strand

**KEY**

| | | | |
|---|---|---|---|
| = Adenine | | = Guanine | |
| = Thymine | | = Cytosine | |
| = Sugar–phosphate backbone (original) | | | |

❭ **FIGURE C-5 Complementary base pairing during DNA transcription**. During DNA transcription, a messenger RNA molecule is formed as RNA nucleotides are assembled by complementary base pairing at a given segment of one strand of an unzipped DNA molecule (i.e., a gene).

## Triplet code and codon

Twenty different amino acids are used to construct proteins, yet only four different nucleotide bases are used to code for these 20 amino acids. In the "genetic dictionary," each different amino acid is specified by a *triplet code* that consists of a specific sequence of three bases in the DNA nucleotide chain. For example, the DNA sequence ACA (adenine, cytosine, adenine) specifies the amino acid cysteine, whereas the sequence ATA specifies the amino acid tyrosine. Each DNA triplet code is transcribed into mRNA as a complementary code word, or *codon*, consisting of a sequenced order of the three bases that pair with the DNA triplet. For example, the DNA triplet code ATA is transcribed as UAU (uracil, adenine, uracil) in mRNA.

Sixty-four different DNA triplet combinations (and, accordingly, 64 different mRNA codon combinations) are possible using the four different nucleotide bases ($4^3$). Of these possible combinations, 61 code for specific amino acids and the remaining 3

**Storage, Replication, and Expression of Genetic Information**

serve as "stop signals." A stop signal is like a period at the end of a sentence. The sentence consists of a series of triplet codes that specify the amino acid sequence in a particular protein. When the stop codon is reached, ribosomal RNA releases the finished polypeptide product. Because 61 triplet codes each specify a particular amino acid and there are 20 different amino acids, a given amino acid may be specified by more than one base-triplet combination. For example, tyrosine is specified by the DNA sequence ATG as well as by ATA. In addition, one DNA triplet code, TAC (mRNA codon sequence AUG) functions as a "start signal" in addition to specifying the amino acid methionine. This code marks the place on mRNA where translation is to begin; this ensures the message is started at the correct end and reads in the right direction. Interestingly, the same genetic dictionary is used universally; a given three-base code stands for the same amino acid in all living things, including microorganisms, plants, and animals.

## Ribosomes

A *ribosome* brings together all components that participate in protein synthesis—mRNA, tRNA, and amino acids—and provides the enzymes and energy required for linking the amino acids together. The nature of the protein synthesized by a given ribosome is determined by the mRNA message being translated. Each mRNA serves as a code for only one particular polypeptide.

A ribosome is an rRNA-protein structure organized into two subunits of unequal size. These subunits are brought together only when a protein is being synthesized (› Figure C-6, step **1**). During assembly of a ribosome, an mRNA molecule attaches to the smaller of the ribosomal subunits by means of a *leader sequence*, a section of mRNA that precedes the start codon. The small subunit with mRNA attached then binds to a large subunit to form a complete, functional ribosome. When the two subunits unite, a groove is formed that accommodates the mRNA molecule as it is being translated.

## Transfer RNA and anticodons

Free amino acids in the cytosol cannot recognize and bind directly with their specific codons in mRNA. Transfer RNA must bring the appropriate amino acid to its proper codon. Even though tRNA is single-stranded, as are all RNA molecules, it is folded back onto itself into a T shape with looped ends (› Figure C-7). The open-ended stem portion recognizes and binds to a specific amino acid. There are at least 20 varieties of tRNA, each able to bind with only one of the 20 kinds of amino acids. A tRNA is said to be "charged" when it is carrying its passenger amino acid. The loop end of a tRNA opposite the amino acid binding site contains a sequence of three exposed bases, known as the *anticodon*, which is complementary to the mRNA codon that specifies the amino acid being carried. Through complementary base pairing, a tRNA can bind with mRNA and insert its amino acid into the protein under construction only at the site designated by the codon for the amino acid. For example, the tRNA molecule that binds with tyrosine bears the anticodon AUA, which can pair only with the mRNA codon UAU, which specifies tyrosine.

This dual-binding function of tRNA molecules ensures that the correct amino acids are delivered to mRNA for assembly in the order specified by the genetic code. Transfer RNA can bind with mRNA only at a ribosome, so protein assembly does not occur except in the confines of a ribosome.

## Steps of protein synthesis

The three steps of protein synthesis are initiation, elongation, and termination.

1. *Initiation.* Protein synthesis is initiated when a charged tRNA molecule bearing the anticodon specific for the start codon binds at this site on mRNA (› Figure C-6, step **2**).

2. *Elongation.* A second charged tRNA bearing the anticodon specific for the next codon in the mRNA sequence then occupies the site next to the first tRNA (step **3**). At any given time, a ribosome can accommodate only two tRNA molecules bound to adjacent codons. Through enzymatic action, a peptide bond is formed between the two amino acids that are linked to the stems of the adjacent tRNA molecules (step **4**). The linkage is subsequently broken between the first tRNA and its amino acid passenger, leaving the second tRNA with a chain of two amino acids. The uncharged tRNA molecule (the one minus its amino acid passenger) is released from mRNA (step **5**). The ribosome then moves along the mRNA molecule by precisely three bases, a distance of one codon (step **6**), so that the tRNA bearing the chain of two amino acids is moved into the number one ribosomal site for tRNA. Then, an incoming charged tRNA with a complementary anticodon for the third codon in the mRNA sequence occupies the number two ribosomal site that was vacated by the second tRNA (step **7**). The chain of two amino acids subsequently binds with and is transferred to the third tRNA to form a chain of three amino acids (step **8**). Through repetition of this process, amino acids are subsequently added one at a time to a growing polypeptide chain in the order designated by the codon sequence. In this way, the ribosomal translation machinery moves stepwise along the mRNA molecule one codon at a time (step **9**). This process is rapid. Up to 10 to 15 amino acids can be added per second.

3. *Termination.* Elongation of the polypeptide chain continues until the ribosome reaches a stop codon in the mRNA molecule, at which time the polypeptide is released. The polypeptide is then folded and modified into a full-fledged protein. The ribosomal subunits dissociate and are free to reassemble into another ribosome for translation of other mRNA molecules.

## Energy cost of protein synthesis

Protein synthesis is expensive, in terms of energy. Attachment of each new amino acid to the growing polypeptide chain requires a total investment of splitting four high-energy phosphate bonds:

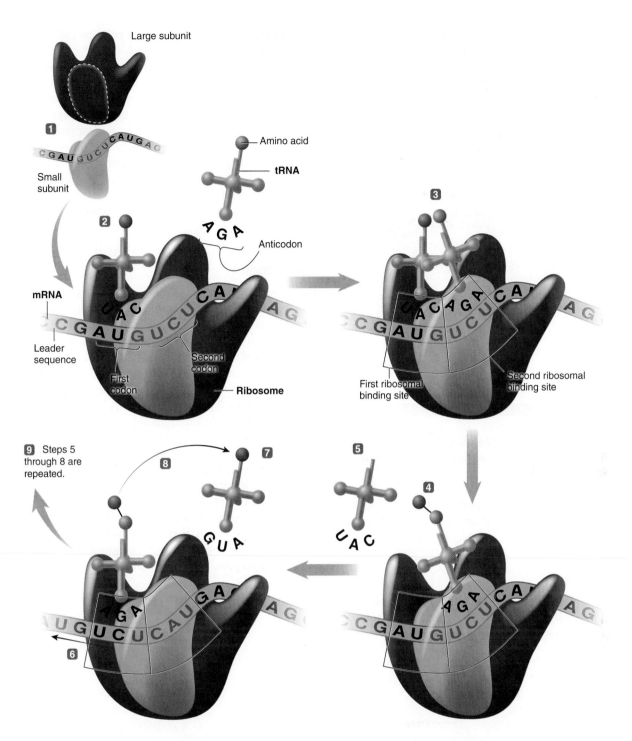

Large subunit

Small subunit

Amino acid

tRNA

Anticodon

mRNA

Leader sequence

First codon

Second codon

Ribosome

First ribosomal binding site

Second ribosomal binding site

**9** Steps 5 through 8 are repeated.

**1** On binding with a messenger RNA (mRNA) molecule, the small ribosomal subunit joins with the large subunit to form a functional ribosome.

**2** A transfer RNA (tRNA), charged with its specific amino acid passenger, binds to mRNA by means of a complementary base pairing between the tRNA anticodon and the first mRNA codon positioned in the first ribosomal binding site.

**3** Another tRNA molecule attaches to the next codon on mRNA positioned in the second ribosomal binding site.

**4** The amino acid from the first tRNA is linked to the amino acid on the second tRNA.

**5** The first tRNA detaches.

**6** The mRNA molecule shifts forward one codon (a distance of a three-base sequence).

**7** Another charged tRNA moves in to attach with the next codon on mRNA, which has now moved into the second ribosomal binding site.

**8** The amino acids from the tRNA in the first ribosomal site are linked with the amino acid in the second site.

**9** This process continues (i.e., steps 5 through 8 are repeated), with the polypeptide chain continuing to grow, until a stop codon is reached and the polypeptide chain is released.

> FIGURE C-6 **Ribosomal assembly and protein translation**

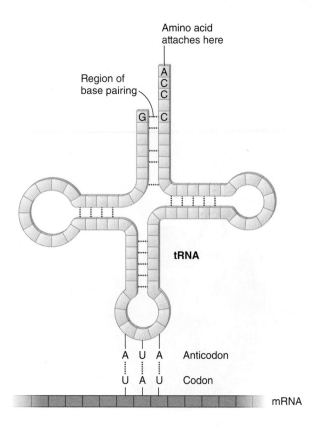

**Amino acid attaches here**

**Region of base pairing**

A
C
C

G · · · · C

tRNA

A U A  Anticodon
U A U  Codon

mRNA

> **FIGURE C-7  Structure of a tRNA molecule.** The open end of a tRNA molecule attaches to free amino acids. The anticodon loop of the tRNA molecule attaches to a complementary mRNA codon.

two to charge tRNA with its amino acid, one to bind tRNA to the ribosomal-mRNA complex, and one to move the ribosome forward one codon.

## Polyribosomes

A number of copies of a given protein can be produced from a single mRNA molecule before the latter is chemically degraded. As one ribosome moves forward along the mRNA molecule, a new ribosome attaches at the starting point on mRNA and also starts translating the message. Attachment of many ribosomes to a single mRNA molecule results in a *polyribosome*. Multiple copies of the identical protein are produced as each ribosome moves along and translates the same message (⟩ Figure C-8). The released proteins are used within the cytosol, except for the few that move into the nucleus through the nuclear pores.

In contrast to the cytosolic polyribosomes, ribosomes directed to bind with the rough endoplasmic reticulum (ER) feed their growing polypeptide chains into the ER lumen. The resultant proteins are subsequently packaged for export out of the cell or for replacement of membrane components within the cell.

## Control of gene activity and protein transcription

Because each somatic cell in the body has an identical DNA blueprint, you might assume they would all produce the same proteins. This is not the case, however, because different cell types are able to transcribe different sets of genes and thus synthesize different sets of structural and enzymatic proteins. For example, only red blood cells can synthesize haemoglobin, even though all body cells carry the DNA instructions for haemoglobin synthesis. Only about 7 percent of the DNA sequences in a typical cell are ever transcribed into mRNA for ultimate expression as specific proteins.

Control of gene expression involves gene regulatory proteins that activate (switch on) or repress (switch off) the genes that code for specific proteins within a given cell. Various DNA segments that do not code for structural and enzymatic proteins code for synthesis of these regulatory proteins. The molecular mechanisms by which these regulatory genes in turn are controlled in human cells are only beginning to be understood. In some instances, regulatory proteins are controlled by

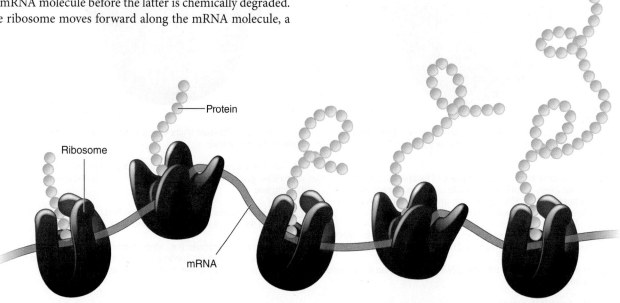

Protein

Ribosome

mRNA

> **FIGURE C-8  A polyribosome.** A polyribosome is formed by numerous ribosomes simultaneously translating mRNA.

*gene-signalling factors* that bring about differential gene activity among various cells to accomplish specialized tasks. The largest group of known gene-signalling factors in humans is the hormones. Some hormones exert their homeostatic effect by selectively altering the transcription rate of the genes that code for enzymes, which are responsible for catalyzing the reaction(s) regulated by the hormone. For example, the hormone cortisol promotes the breakdown of fat stores by stimulating synthesis of the enzyme that catalyzes the conversion of stored fat into its component fatty acids. In other cases, gene action appears to be time specific; that is, certain genes are expressed only at a certain developmental stage in the individual. This is especially important during embryonic development.

# c.4 | Cell Division

Most cells in the human body can reproduce themselves— a process important in growth, replacement, and repair of tissues. The rate at which cells divide is highly variable. Cells within the deeper layers of the intestinal lining divide every few days to replace cells that are continually sloughed off the surface of the lining into the digestive tract lumen. In this way, the entire intestinal lining is replaced about every three days (p. 681). At the other extreme are nerve cells, which permanently lose the ability to divide beyond a certain period of fetal growth and development. Consequently, when nerve cells are lost through trauma or disease, they cannot be replaced (p. 5). In between these two extremes are cells that divide infrequently, but do so when needed to replace damaged or destroyed tissue. The factors that control the rate of cell division remain obscure.

## Mitosis

Recall that cell division involves two components: nuclear division and cytoplasmic division (*cytokinesis*). Nuclear division in somatic cells is accomplished by *mitosis*, in which a complete set of genetic information (i.e., a diploid number of chromosomes) is distributed to each of two new daughter cells.

A cell capable of dividing alternates between periods of mitosis and periods of nondivision. The interval of time between cell division is known as *interphase*. Because mitosis takes less than an hour to complete, the vast majority of cells in the body at any given time are in interphase.

Replication of DNA and growth of the cell take place during interphase in preparation for mitosis. Although mitosis is a continuous process, it displays four distinct phases: *prophase, metaphase, anaphase,* and *telophase* (› Figure C-9).

### PROPHASE

1. Chromatin condenses and becomes microscopically visible as chromosomes. The condensed duplicate strands of DNA, known as *sister chromatids*, remain joined within the chromosome at a point called the *centromere* (› Figure C-10).
2. Cells contain a pair of centrioles, the short cylindrical structures that form the mitotic spindle during cell division. The

centriole pair divides, and the daughter centrioles move to opposite ends of the cell, where they assemble between them a mitotic spindle made up of microtubules.
3. The membrane surrounding the nucleus starts to break down.

### METAPHASE

1. The nuclear membrane completely disappears.
2. The 46 chromosomes, each consisting of a pair of sister chromatids, align themselves at the midline, or equator, of the cell. Each chromosome becomes attached to the spindle by means of several spindle fibres that extend from the centriole to the centromere of the chromosome.

### ANAPHASE

1. The centromeres split, converting each pair of sister chromatids into two identical chromosomes, which separate and move toward opposite poles of the spindle. Molecular motors pull the chromosomes along the spindle fibres toward the poles.
2. At the end of anaphase, an identical set of 46 chromosomes is present at each of the poles, for a transient total of 92 chromosomes in the soon-to-be-divided cell.

### TELOPHASE

1. The cytoplasm divides through formation and gradual tightening of an actin contractile ring at the midline of the cell, thus forming two separate daughter cells, each with a full diploid set of chromosomes.
2. The spindle fibres disassemble.
3. The chromosomes uncoil to their decondensed chromatin form.
4. A nuclear membrane reforms in each new cell.

Cell division is complete with the end of telophase. Each of the new cells now enters interphase.

## Meiosis

Nuclear division in the specialized case of germ cells is accomplished by *meiosis*, in which only half a set of genetic information (i.e., a haploid number of chromosomes) is distributed to each daughter cell. Meiosis differs from mitosis in several important regards (Figure C-9). Specialized diploid germ cells undergo one chromosome replication followed by two nuclear divisions to produce four haploid germ cells.

### MEIOSIS I

1. During prophase of the first meiotic division, the members of each homologous pair of chromosomes line up side by side to form a *tetrad*, which is a group of four sister chromatids with two identical chromatids within each member of the pair.
2. The process of crossing over occurs during this period, when the maternal copy and the paternal copy of each

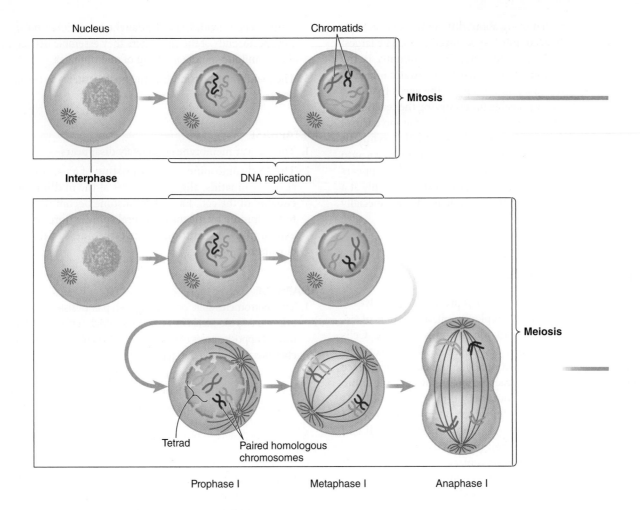

> FIGURE C-9  A comparison of events in mitosis and meiosis

chromosome are paired. *Crossing over* involves a physical exchange of chromosome material between nonsister chromatids within a tetrad (> Figure C-11). This process yields new chromosome combinations, thus contributing to genetic diversity.

3. During metaphase, the 23 tetrads line up at the equator.

4. At anaphase, homologous chromosomes, each consisting of a pair of sister chromatids joined at the centromere, separate and move toward opposite poles. Maternally and paternally derived chromosomes migrate to opposite poles in random assortments of one member of each chromosome pair without regard for its original derivation. This genetic mixing provides novel new combinations of chromosomes.

5. During the first telophase, the cell divides into two cells. Each cell contains 23 chromosomes consisting of two sister chromatids.

**MEIOSIS II**

1. Following a brief interphase in which no further replication occurs, the 23 unpaired chromosomes line up at the equator, the centromeres split, and the sister chromatids

separate for the first time into independent chromosomes that move to opposite poles.

2. During cytokinesis, each of the daughter cells derived from the first meiotic division forms two new daughter cells. The end result is four daughter cells, each containing a haploid set of chromosomes.

Union of a haploid sperm and haploid egg results in a zygote (fertilized egg) that contains the diploid number of chromosomes. Development of a new multicellular individual from the zygote is accomplished by mitosis and cell differentiation. Because DNA is normally faithfully replicated in its entirety during each mitotic division, all cells in the body possess an identical aggregate of DNA molecules. Structural and functional variations between different cell types result from differential gene expression.

## Mutations

An estimated $10^{16}$ cell divisions take place in the body during the course of a person's lifetime to accomplish growth, repair, and normal cell turnover. Because more than 3 billion nucleotides must be replicated during each cell division, it's no wonder "copying errors" occasionally occur. Any change in the DNA

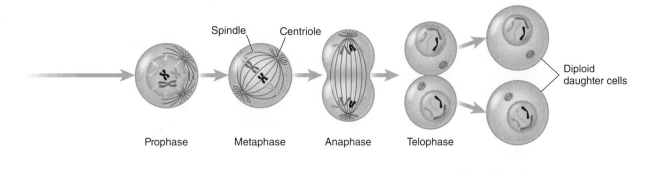

Prophase  Metaphase  Anaphase  Telophase  Diploid daughter cells

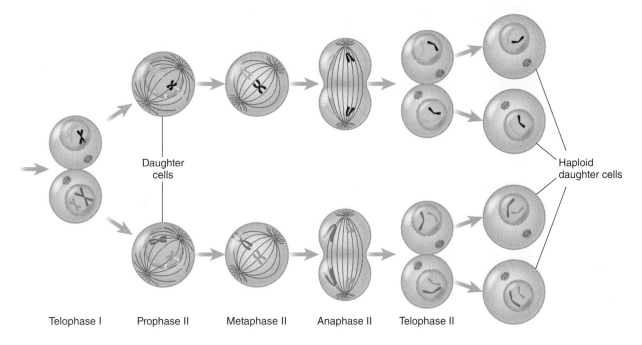

Telophase I  Prophase II  Metaphase II  Anaphase II  Telophase II

Daughter cells

Haploid daughter cells

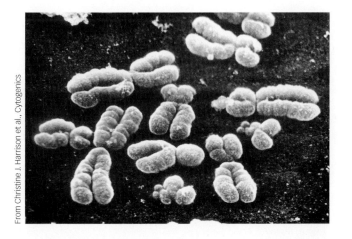

> FIGURE C-10 **A scanning electron micrograph of human chromosomes from a dividing cell**. The replicated chromosomes appear as double structures, with identical sister chromatids joined in the middle at a common centromere.

> FIGURE C-11 **Crossing over.** (a) During prophase I of meiosis, each homologous pair of chromosomes lines up side by side to form a tetrad. (b) Physical exchange of chromosome material occurs between nonsister chromatids. (c) As a result of this crossing over, new combinations of genetic material are formed within the chromosomes.

sequence is known as a *point (gene) mutation*. A point mutation arises when a base is inadvertently substituted, added, or deleted during the replication process.

When a base is inserted in the wrong position during DNA replication, the mistake can often be corrected by a built-in "proofreading" system. Repair enzymes remove the newly replicated strand back to the defective segment, at which time normal base pairing resumes to resynthesize a corrected strand. Not all mistakes can be corrected, however.

Mutations can arise spontaneously by chance alone or they can be induced by *mutagens*, which are factors that increase the rate at which mutations take place. Mutagens include various

Storage, Replication, and Expression of Genetic Information  **A-29** chemical agents as well as ionizing radiation, such as X-rays and atomic radiation. Mutagens promote mutations either by chemically altering the DNA base code through a variety of mechanisms or by interfering with the repair enzymes so that abnormal base segments cannot be cut out.

Depending on the location and nature of a change in the genetic code, a given mutation may (1) have no noticeable effect if it does not alter a critical region of a cellular protein; (2) adversely alter cell function if it impairs the function of a crucial protein; (3) be incompatible with the life of the cell, in which case the cell dies and the mutation is lost with it; or (4) in rare cases, prove beneficial if a more efficient structural or enzymatic protein results. If a mutation occurs in a body cell *(somatic mutation)*, the outcome will be reflected as an alteration in all future copies of the cell in the affected individual, but it will not be perpetuated beyond the life of the individual. If, by contrast, a mutation occurs in a sperm- or egg-producing cell *(germ cell mutation)*, the genetic alteration may be passed on to succeeding generations.

In most instances, *cancer* results from multiple somatic mutations that occur over a course of time within DNA segments known as *proto-oncogenes*. Proto-oncogenes are normal genes whose coded products are important in the regulation of cell growth and division. These genes have the potential of becoming overzealous *oncogenes* (cancer genes), which induce the uncontrolled cell proliferation characteristic of cancer. Proto-oncogenes can become cancer producing as a result of several sequential mutations in the gene itself or by changes in adjacent regions that regulate the proto-oncogenes. Less frequently, tumour viruses become incorporated in the DNA blueprint and act as oncogenes. Alternatively, cancer may arise from mutations that disable *tumour suppressor genes*, which normally restrain cell proliferation in check-and-balance fashion.

# Principles of Quantitative Reasoning

By Kim E. Cooper, Midwestern University, and John D. Nagy, Scottsdale Community College

## D.1 Introduction

Historically, as a branch of science matures, it typically becomes more precise and usually more quantitative. This trend is becoming increasingly true of biology and especially of physiology. Most students, however, are uncomfortable with quantitative reasoning. Students are usually quite capable of doing the mechanical manipulations of mathematics but have trouble translating back and forth between words, concepts, and equations. This appendix is meant to help you become more comfortable working with equations.

## D.2 Why Are Equations Useful?

A great deal of what we do in science involves establishing functional relations between variables of interest (e.g., blood pressure and heart rate, transport rate and concentration gradient). Equations are simply a compact and exact way of expressing such relationships. The tools of mathematics then allow us to draw conclusions systematically from these relationships. Mathematics is a very powerful set of tools or, more generally, a very powerful way of thinking. Mathematics allows you to think extremely precisely, and therefore clearly, about complex relationships. Equations and quantitative notions are the keys to that precision. For example, a quantitative comparison of the predictions of a theory against the results of measurement forms the basis of statistics and of much of the hypothesis testing on which science is based. A scientific conclusion without adequate quantification and statistical backing may be little more than an impression or opinion.

It may seem odd to say that mathematics allows you to think more clearly about complex ideas. People unfamiliar with mathematical thinking often complain that even simple relationships produce complicated equations and that complex relationships are mathematically intractable. Certainly, many basic concepts require considerable mathematical expertise to be handled properly, but such concepts are in fact not simple. More commonly, however, many simple equations are seen as complex because many students are poorly trained in how to think about equations.

## D.3 How to Think about an Equation

In this section we take the first, and often overlooked, step in thinking quantitatively. That is, how do we begin to think about some new equation presented to us? We start by becoming comfortable with the "meaning" of an equation. This step is absolutely necessary for you to use an equation properly. As a specific example, consider the Nernst equation (p. 55) for potassium. Various books show this equation in several different forms, but they all say essentially the same thing:

$$E_{K^+} = (RT/zF)\ln\left\{[K^+]_{out}/[K^+]_{in}\right\}$$

$$E_{K^+} = (RT/zF)2.303\log\left\{[K^+]_{out}/[K^+]_{in}\right\}$$

$$E_{K^+} = (61\text{mV}/z)\log\left\{[K^+]_{out}/[K^+]in\right\}$$

These equations may seem like meaningless strings of symbols. But what are these equations trying to tell us? What do they represent? The following four steps are intended to help you become comfortable with any new equation. Try them with the Nernst equation.

1. Be sure you can define the symbols and give dimensions and units. Check the equation for dimensional consistency.

One of the first steps is to figure out which symbols represent the variables of interest and which are simple constants. In this case, all the symbols are constants except two.

$E_{K^+}$ is the Nernst (equilibrium) potential for potassium. It represents the concentration gradient (force of diffusion) on a mole of potassium ions. $E_{K^+}$ has the dimensions of a voltage and is usually given in units of mV (millivolts). This dimension is used so that the concentration gradient is expressed in the same dimensions as the other force acting on the ions—that is, the electrical gradient. Using the same dimensions makes it possible to compare the two forces.

$[K^+]$ represents the concentration of potassium. With the subscript *out*, this symbol refers to the concentration of potassium outside the cell. With the subscript in, this symbol refers to the concentration of potassium inside the cell. $[K^+]$ has dimensions of concentration and is usually expressed in units of mM (millimolars; millimoles/litre).

2. Identify the dependent and independent variables. Try to find normal values and ranges for the variables. Before continuing, we need to define *dependent* and *independent variables*.

Recall that equations represent relationships between variables. Whenever you hear the word *relationship*, think of a graph, as in › Figure D-1, for example.

Graphs are often a good way to represent relationships and therefore equations. This graph says the value of variable 2 depends on the value of variable 1. Thus, for any value of variable 1 the corresponding value for variable 2 can be determined from the graph. In other words, variable 1 determines the value of variable 2. Because variable 2 depends on variable 1, we call variable 2 the *dependent variable*. Variable 1, in contrast is independent of variable 2, so we call variable 1 the *independent variable*. There can be any number of dependent and *independent variables*.

How do you determine which variables are dependent and which are independent? The answer usually depends on cause and effect: *Effects* are dependent variables, and *causes* are independent. For example, we know (see Chapter 9, p. 401) that mean arterial pressure (*MAP*) is the product of cardiac output (*CO*) and total peripheral resistance (*TPR*); that is,

$$MAP = CO \times TPR$$

*MAP* is on the left-hand side of this equation because we think of mean arterial pressure as a result of cardiac output and total peripheral resistance. Or, to put this another way, mean arterial pressure is a function of cardiac output and total peripheral resistance. As a cause–effect relationship, it seems backward to think of mean arterial pressure somehow "causing" cardiac output to be a certain value. Therefore, *MAP* is the effect, the dependent variable, and we place it on the left-hand side of the equality symbol. Conversely, *CO* and *TPR* are the causes, the independent variables, and we put them on the right.

In our Nernst equation example, the independent variables are the concentrations. The dependent variable is the Nernst potential because we think of the potential as being a result of the ion concentrations. We also know that $E_{K^+}$ is about −90 mV, and $[K^+]_{out}$ and $[K^+]_{in}$ are about 5 mM and 150 mM, respectively.

3. Identify the constants and know their numeric values:

- R is the gas constant. It has dimensions of energy per mole per degree of temperature and the value of 8.31 joules/kelvin · mole. It is also convenient to note that a joules = volt × coulomb.

- T is temperature, with the dimension of temperature in units of kelvins. Normal body temperature is around $37 C [= 308 \text{ kelvins} (K)]$.

- z is the valence of the ion. Valence is the charge on an ion, including the sign. For potassium, $z = +1$.

- F is Faraday's constant, which has dimensions of charge per mole, units of coulombs per mole, and a value of 96 500 C/mol.

Refer back to the Nernst equations given on the preceding page. Note that the constants just defined appear in the first two equations, but not the third. In the third equation, the quantity $RT/F$ has already been evaluated for you, as follows:

$$RT/F = [(8.31 \text{ V} \cdot \text{C/K} \cdot \text{mol})(308 \text{ K})]/(96\,500 \text{ C/mol})$$
$$= 26.5 \text{ mV}$$

We multiply this value by 2.303 to convert the natural logarithm (ln) to the base 10 logarithm. Note that $26.5 \text{ mV} \times 2.303 = 61 \text{ mV}$.

4. State the equation in words. Summarize it in a few sentences so that someone can understand what it is about. Don't just say the names of the symbols.

Just saying the names of the symbols would be equivalent to saying the following: "The Nernst potential is given by a constant times the logarithm of the ratio of the ion concentrations." This is certainly true, but does nothing to aid our intuition. A preferable statement would be "The Nernst equation allows us to calculate the force pushing ions into or out of a cell via diffusion." This is valuable, because we can compare this force to the force moving ions in and out via the membrane voltage and see which is larger and hence in what direction the ions will actually move. The force is expressed in electrical units so we can compare it directly with the membrane voltage. The constants convert from concentration to electrical units.

When you understand what an equation means you will be able to use it to answer questions. The next section gives you some guidance in taking this next step.

## D.4 | How to Think with an Equation

Before you can use an equation to help you think, you need to develop a few basic skills, and these skills are not difficult to learn.

1. Be sure you know the algebraic rules for manipulating variables within any function involved (such as $\sqrt{\phantom{x}}$, exp, or log).

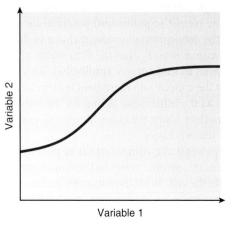

Variable 2

Variable 1

› **FIGURE D-1**

In the case of the Nernst equation, the tricky function is the logarithm. Consult a university-level algebra book if you need to brush up on the rules of working with logs or any other function. For instance, it is useful to know that

$$\log\{A\} = -\log\{1/A\}$$

and that the log operation is undone by taking it to the power of 10; that is,

$$10^{\log\{A\}} = A$$

2. Be able to solve for any variable in terms of the others.

Given just three variables ($E_{K^+}$, $[K^+]_{in}$, and $[K^+]_{out}$) only a few types of questions can be asked. Two of the three variables must be given, and you must solve for the third. If the two concentrations are given, then the formula is already set to give you the Nernst potential. If the Nernst potential and one concentration are given, however, you must be able to solve for the other concentration. Try to do this and obtain the two following equations:

$$[K^+]_{out} = [K^+]_{in} 10^{(E_K+/61mV)}$$
$$[K^+]_{in} = [K^+]_{out} 10^{(-E_K+/61mV)}$$

3. Be able to sketch, at least approximately, the dependence of any variable on any other variable.

The ability to do this is exceedingly valuable. Sketching helps you generate insight about equations; it helps you understand what an equation means. Therefore, sketching helps you understand the solution, as well as solve the problem. If you apply this technique consistently, you may find equations far simpler to handle than you previously suspected. In addition, be sure you can relate your sketch to experimental measurements and physiological situations.

For example, we can draw the relationships between the Nernst potential for potassium and the external potassium ion concentration predicted by the equations as in › Figure D-2. This sketch makes clear that the Nernst potential, which can be measured physiologically, should decrease linearly as the log $[K^+]_{out}$ increases, which can be controlled experimentally. Therefore,

this sketch suggests an experiment: vary $[K^+]_{out}$. If $E_{K^+}$ does not decrease linearly with increasing log $[K^+]_{out}$, then we would have a flaw in our understanding. The Nernst equation would not describe the real situation, as we think it should. Scientific advances are almost always heralded by such contradictions.

4. Be able to combine several equations to find new relationships.

Combining separate pieces of information is always useful. In fact, some scientists have argued that this activity is all scientists ever do. To integrate knowledge for yourself, you must be able to combine the various relations you learn about into new combinations. This allows you to solve increasingly complex problems. As an example, consider the following relation:

$$l_{K^+} = G_{K^+} \times (Vm - E_{K^+})$$

This equation describes the number of potassium ions flowing across a membrane if both a concentration gradient ($E_{K^+}$) and an electrical gradient ($V_m$) are present. This equation can be combined with the Nernst equation for potassium to answer the following question. Suppose $V_m$, $G_{K^+}$, and $[K^+]_{in}$ are fixed. What would the external potassium ion concentration have to be such that no net flux of potassium ions occurs across the membrane? No net flux implies that $l_{K^+} = 0$. But if $G_{K^+} > 0$, $l_{K^+} = 0$ only when the membrane voltage equals the Nernst potential ($V_m = E_{K^+}$). To answer the question, then, we set ($E_{K^+} = V_m$) in the Nernst equation and solve it for $[K^+]_{out}$.

5. Know the equation's underlying assumptions and limits of validity.

Every equation comes from some underlying theory or set of observations and, therefore, has some limited range of validity and rests on certain assumptions. Failure to understand this simple point often leads students to apply equations outside their realm of applicability. In that case, even though the math is done correctly, the results will be incorrect.

In the case of the Nernst equation, things are fairly simple. This equation is derived from a very powerful theory known as *equilibrium thermodynamics*, and hence it has very wide applicability. As another example, consider enzyme kinetics. The rate at which an enzyme catalyzes a reaction ($v$) is related to

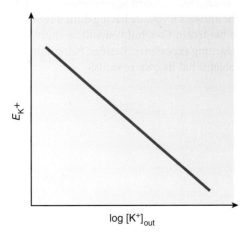

› **FIGURE D-2**

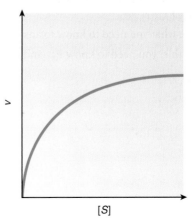

› **FIGURE D-3**

**Principles of Quantitative Reasoning**

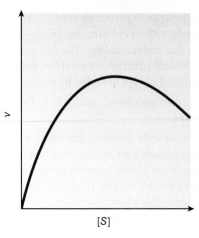

> FIGURE D-4

the concentration of substrate on which the enzyme works ([S]) by an equation called the Michaelis–Menton relationship. The graph of this relationship can be seen in > Figure D-3.

However, this relationship between reaction velocity and substrate concentration does not hold true for some real enzymes, such as lactate dehydrogenase. For this enzyme, the relationship between reaction velocity and substrate concentration is depicted in > Figure D-4.

At high substrate concentrations, the enzyme actually is inhibited by too much substrate. For such enzymes, the Michaelis–Menton theory, which works well at low [S], is invalid at higher [S].

## D.5 An Approach to Problem Solving

The final step is to apply these skills to solve a problem. As an example, calculate the concentration of potassium that must exist inside a cell if $E_{K^+} = -95$ mV and the interstitial fluid has a potassium concentration of 4 mM. Try using the following procedure to solve this problem:

1. Get a clear picture of what is being asked. State it out loud or write it down.

The question asks for the concentration of potassium in the cell, that is, $[K^+]_{in}$.

2. Determine what you need to know to answer the question.

To answer this, you need to know $E_{K^+}$ and $[K^+]_{out}$.

3. Determine what information is given. Is it sufficient? Are other relevant facts or relationships not stated in the problem? Specifically, do you need any other equations?

$E_{K^+}$ is given explicitly in the problem, but you have to translate the words to realize that $[K^+]_{out} = 4$ mM.

4. Manipulate the equation algebraically so that the unknown is on the left-hand side and everything else is on the right-hand side.

We now solve the Nernst equation for $[K^+]_{in}$. This was done in Section D.4, p. A-33:

$$[K^+]_{in} = [K^+]_{out}10^{(-E_{K^+}/61\,mV)}$$

Substituting for the values given, we obtain the following:

$$[K^+]_{in} = 4\,mM\,10^{(95\,mV/61\,mV)}$$
$$= 4\,mM\,10^{(1.56)}$$
$$= 4\,mM\,(36.3)$$
$$= 145\,mM$$

5. Is the answer dimensionally correct? Do not skip this step. It will tell you immediately if something went wrong.

Yes, the answer is in mM—the proper dimension and unit.

6. Does the answer make sense?

Yes, the value is not alarmingly low or high. Also, because the potassium ion is positively charged, if the potassium concentration outside the cell is lower than that inside the cell, then the interior of the cell would have to be negative at passive equilibrium, which it is.

Apply the approach to problem solving we have just outlined to the quantitative questions in the chapters. When you start out, apply the approach formally and carefully. For example, go through each step, and write everything out as we have done for the Nernst equation. After a time, you may not need to be so formal. Also, as you progress, you will develop your own style and approach to problem solving. Be prepared to spend some time and patience on some of the problems. Not every answer will be immediately apparent. This situation is normal. If you run into difficulties, relax, return to this appendix for guidance, and work through the problem again carefully. Do not go immediately to the answers if you are having difficulty with a problem. It may ease the frustration, but you will be cheating yourself of a valuable learning experience. Besides, being able to solve challenging problems has its own rewards.

# Answers to End-of-Chapter Objective Questions, Quantitative Exercises, Points to Ponder, and Clinical Considerations

## Chapter 1 The Foundation of Physiology

### Check Your Understanding

**1.1** (Questions on p. 8.)

1. Physiology is the study of body functions.
2. Chemical level: Various atoms and molecules make up all body structures. Cellular level: Specific chemicals are organized into living cells, which are the basic units of both structure and function. Tissue level: Groups of cells of similar specialization are organized into tissues. Organ level: An organ is made up of several tissue types that act together as a unit to perform a particular function or functions. Body system level: A body system is a collection of related organs that interact to accomplish a common activity essential for survival of the whole body. Organism level: The body systems are structurally and functionally packaged together into the whole body, which is a single, multicellular organism capable of living independently in the surrounding external environment.
3. Every cell performs basic cell functions essential for its survival, including (1) obtaining food and $O_2$ from the environment surrounding the cell; (2) performing chemical reactions using food and $O_2$ to provide energy for the cell; (3) eliminating to the surrounding environment wastes produced by these reactions; (4) synthesizing proteins and other components needed by the cell; (5) largely controlling exchanges between the cell and surrounding environment; (6) moving materials within the cell or, in the case of some cells, moving the cell; (7) being sensitive and responsive to changes in the surrounding environment; and (8) reproducing (except for nerve cells and muscle cells). Many cells in the body have additional functions that are specialized to their purpose. For example, nerve cells are specialized to transmit electrical signals, and muscle cells are specialized to contract.
4. The external environment is the surrounding environment in which an organism lives. The internal environment is the fluid inside the body and outside the cells in which the cells live. The intracellular fluid is the fluid collectively within all body cells. The extracellular fluid is the fluid inside the body and outside the cells and constitutes the internal environment. The extracellular fluid consists of the plasma, the fluid portion of the blood, and interstitial fluid, the fluid that surrounds and bathes the cells.

**1.2** (Questions on p. 16.)

1. *Intrinsic controls* are inherent compensatory responses that act locally in an organ; *extrinsic controls* are systemic controls initiated outside an organ by the regulatory systems (nervous or endocrine systems) to alter the organ's activity.
2. In *negative feedback,* the output of a control system drives a controlled variable in the opposite direction of an initial change, thereby counteracting the change. In *positive feedback,* the output of a control system drives a controlled variable in the same direction as the initial change, thus enhancing the change.
3. See ⟩ Figure 1-6a, Negative feedback, p. 15

### Objective Questions

(Questions on p. 17.)

**1.** e **2.** b **3.** c **4.** T **5.** F **6.** T **7.** muscle tissue, nervous tissue, epithelial tissue, connective tissue **8.** secretion **9.** exocrine, endocrine, hormones **10.** intrinsic, extrinsic **11.** 1. d, 2. g, 3. a, 4. e, 5. b, 6. j, 7. h, 8. i, 9. c, 10. f

### Points to Ponder

(Questions on p. 18.)

1. The respiratory system eliminates internally produced $CO_2$ to the external environment. A decrease in $CO_2$ in the internal environment brings about a reduction in respiratory activity (i.e., slower, shallower breathing) so that $CO_2$ produced within the body is allowed to accumulate, instead of being blown off as rapidly as normal to the external environment. The extra $CO_2$ retained in the body increases the $CO_2$ levels in the internal environment to normal.
2. b, c, b
3. b
4. immune defence system
5. When a person is engaged in strenuous exercise, the temperature-regulating centre in the brain brings about the widening of the blood vessels of the skin. The resultant increased blood flow through the skin carries the extra heat generated by the contracting muscles to the body surface, where it can be lost to the surrounding environment.

### Clinical Consideration

(Question on p. 18.)

Loss of fluids threatens the maintenance of proper plasma volume and blood pressure. Loss of acidic digestive juices threatens the maintenance of the proper pH in the internal fluid environment. The urinary system helps restore the proper plasma volume and pH by reducing the amount of water and acid eliminated in the urine. The respiratory system helps restore the pH by adjusting the rate of removal of acid-forming $CO_2$. Adjustments ade in the circulatory system help maintain blood pressure despite fluid loss. Increased thirst encourages increased fluid intake to help restore plasma volume. These compensatory changes in the urinary, respiratory, and circulatory systems, as well as the sensation of thirst, are all be regulated by the two major regulatory systems, the nervous and endocrine systems. Furthermore, the endocrine system makes internal adjustments to help maintain the concentration of nutrients in the internal environment even though no new nutrients are being absorbed from the digestive system.

# Chapter 2 Cell Physiology

## Check Your Understanding

**2.1** (Questions on p. 22.)
1. See ▌Table 2-1, Principles of the Cell Theory
2. These cells are all about the same size.

**2.2** (Questions on p. 29.)
1. In anaerobic conditions 2 molecules of ATP are produced (by glycolysis), and in aerobic conditions 32 molecules of ATP are produced (2 by glycolysis, 2 by the citric acid cycle, and 28 by oxidative phosphorylation) from one glucose molecule.

**2.3** (Questions on p. 36.)
1. Under an electron microscope, the plasma membrane has a sandwich appearance, with two dark layers separated by a light middle layer. The two dark layers are the hydrophilic polar regions of the lipid and protein molecules that take up a stain, whereas the light middle layer is the poorly stained hydrophobic core composed of the nonpolar regions of these molecules.
2. desmosome (adhering junction that spot-rivets two adjacent but nontouching cells, anchoring them together in tissues subject to considerable stretching); (2) tight junction (impermeable junction that joins the lateral edges of epithelial cells near their luminal borders, thus preventing movement of materials between the cells); and (3) gap junction (communicating junction made up of small connecting tunnels that permit movement of charge-carrying ions and small molecules between two adjacent cells)
3. See ❯ Figure 2-9, Desmosome, p. 35.

**2.4** (Questions on p. 46.)
1. See ❯ Figure 2-20, Comparison of carrier-mediated transport and simple diffusion down a concentration gradient, p. 43.
2. In facilitated diffusion, the carrier undergoes spontaneous changes in shape as a result of thermal energy. In primary active transport, phosphorylation (binding of the phosphate group derived from the carrier splitting ATP) increases affinity of the carrier for its passenger ion; this binding causes the carrier to change its shape. In secondary active transport, the change in shape of a cotransport carrier that binds both $Na^+$ and the transported solute is driven by a $Na^+$ concentration gradient established by a primary active transport mechanism.

**2.5** (Questions on p. 59.)
1. When a cell's electrical potential is at the resting membrane potential, the membrane is more permeable to $K^+$ than $Na^+$. Because of this, the resting membrane potential is closer to the $K^+$ equilibrium potential.
2. See ❯ Figure 2-30, Types of changes in membrane potential, p. 58.
3. Voltage-gated channels open or close in response to changes in membrane potential. Chemically gated channels change conformation in response to binding of a specific extracellular chemical messenger to a surface membrane receptor. Mechanically gated channels respond to mechanical deformation such as stretching. Thermally gated channels respond to heat or cold.

**2.6** (Questions on p. 64.)
1. Because the resting membrane is 25 to 30 times more permeable to $K^+$ than to $Na^+$, $K^+$ passes through more readily than $Na^+$. The substantially larger movement of $K^+$ out of the cell influences the resting membrane potential to a much greater extent than the smaller movement of $Na^+$ into the cell does. As a result, the resting potential (−70 mV) is closer to the equilibrium potential for $K^+$ (−70 mV) than to the equilibrium potential for $Na^+$ (+60 mV). The resting potential is less than the $K^+$ equilibrium potential because the limited entry of $Na^+$ neutralizes some of the potential that would be created by $K^+$ alone.
2. See ❯ Figure 2-37 Permeability changes and ion fluxes during a neuronal action potential, p. 63.

**2.7** (Questions on p. 69.)
1. During the absolute refractory period, all the $Na^+$ channels are in the inactivated state. However, during the relative refractory period, some $Na^+$ channels are starting to move from inactive to closed. A stimulus at this time can cause these closed $Na^+$ channels to move to the open state and, depending on how many open, fire an action potential. By the end of the relative refractory period, all the $Na^+$ channels are in the closed state.
2. During the absolute refractory period, no matter how strong the stimulus, an action potential cannot be fired. Therefore, the length of the absolute refractory period defines the maximum rate that a neuron can fire.

**2.8** (Questions on p. 79.)
1. At excitatory synapses, neurotransmitters cause the opening of nonspecific cation channels that allow cations (+ charges) to move into the cell. This charge movement depolarizes the membrane potential, making the membrane more excitable because the membrane potential is closer to threshold. In contrast, at inhibitory synapses, different neurotransmitters open certain $Cl^-$ or $K^+$ channels that cause the membrane to hyperpolarize and make the membrane less excitable.
2. Summation refers to the cumulative effects of EPSPs and IPSPs on a postsynaptic cell. In temporal summation, repetitive EPSPs from the same presynaptic neuron can sum together and potentially reach threshold. Spatial summation refers to the additive effects of several excitatory presynaptic inputs that sum to potentially reach threshold.

## Objective Questions

(Questions on pp. 85–86.)

1. plasma membrane 2. deoxyribonucleic acid (DNA), nucleus 3. organelles, cytosol, cytoskeleton 4. adenosine triphosphate (ATP) 5. F 6. T 7. F 8. 1. b, 2. c, 3. c, 4. a, 5. b, 6. c, 7. a, 8. c 9. 1. b, 2. a, 3. b, 4. a, 5. c, 6. b, 7. a, 8. b 10. 1. a, 2. a, 3. b, 4. a, 5. b, 6. a, 7. b 11. 1. c, 2. b, 3. a, 4. a, 5. c, 6. b, 7. c, 8. a, 9. b 12. F 13. T 14. F 15. F 16. refractory period 17. axon hillock 18. synapse 19. temporal summation 20. 1. b, 2. a, 3. a, 4. b, 5. b, 6. a

## Quantitative Exercises

(Questions on pp. 86–87.)

1. b
2. $24\ mol\,O_2/day \times 6\ mol\,ATP/mol\,O_2 = 144\ mol\,ATP/day$
   $144\ mol\,ATP/day \times 507\ g\,ATP/mol \times 73\,000\ g\,ATP/day$
   $1000\ g/2.2\ lb = 73\,000\ g/x\ lb$
   $1000x = 73\,000\,32.2$
   $x = $ approximately 160 lb, or 73 kg
3. $144\ mol/day\,(7300\ cal/mol) = 1\,051\,200\ cal/day\,(1051\ kilocal/day)$
4. About two-thirds of the water in the body is intracellular. Because a person's mass is about 60 percent water, for a 68 kg person

   $$68\ kg(0.6)(2/3) = 27.2\ kg$$

   is the mass of water. Assume that 1 mL of body water weighs 1 g. The total volume in the person's cells then is about 27.2L. The volume of an average cell is

   $$\frac{4}{3}\pi(1 \times 10^{-3}cm)^3 \approx 4.2 \times 10^{-9}cm^3$$
   $$= 4.2 \times 10^{-9}mL$$

   So, the number of cells in a 68 kg person is about

   $$27.2\ litres\left(\frac{1000\ mL}{1\ litre}\right)\left(\frac{1\ cell}{4.2 \times 10^{-9}mL}\right)$$
   $$= 6.476 \times 10^{12}cells$$

5. $150\ mg\left(\dfrac{1mL}{0.015\ mg}\right) = 10\,000\ mL\,(10\ L)$

6. $E = \dfrac{61\ mV}{z}\log\dfrac{C_o}{C_i}$

   **a.** $\dfrac{61\ mV}{2}\log\dfrac{1 \times 10^{-3}}{100 \times 10^{-9}} = +122\ mV$

   **b.** $\dfrac{61\ mV}{-1}\log\dfrac{110 \times 10^{-3}}{10 \times 10^{-3}} = -63.5\ mV$

7. $I_x = G_x(V_m - E_x)$

$$E_{Na^+} = 61\,mV \log\frac{145\,mM}{15\,mM} = 60.1\,mV$$

   a. $= 1ns(-70mV - 60.1mV)$

      $= 1ns(-130mV)$

      $= -130\,pA\,(A = amperes)$

   b. entering

   c. with concentration gradient

   d. with electrical gradient

8. $V_m = \dfrac{G_{Na^+}}{G_r}E_{Na^+} + \dfrac{G_{K^+}}{G_r}E_{K^+}$

   a. $G_r = 1\,nS + 5.3\,nS = 6.3\,nS$;

$$V_m = \frac{1}{6.3}59.1\,mV + \frac{5.3}{6.3}(-94.4\,mV)$$

$$= 9.4\,mV - 79.4\,mV = -70\,mV$$

   b. $E_{K^+} = 0\,mV$; $V_m = 9.4\,mV$; that is, large depolarization

## Points to Ponder

(Questions on pp. 87–88.)

1. With cyanide poisoning, the cellular activities that depend on ATP expenditure could not continue, such as synthesis of new chemical compounds, membrane transport, and mechanical work. The resultant inability of the heart to pump blood and failure of the respiratory muscles to accomplish breathing would lead to imminent death.

2. ATP is required for muscle contraction. Muscles are able to store limited supplies of nutrient fuel for use in the generation of ATP. During anaerobic exercise, muscles generate ATP from these nutrient stores by means of glycolysis, which yields two molecules of ATP per glucose molecule processed. During aerobic exercise, muscles can generate ATP by means of oxidative phosphorylation, which yields 36 molecules of ATP per glucose molecule processed. Because glycolysis inefficiently generates ATP from nutrient fuels, it rapidly depletes the muscle's limited stores of fuel, and ATP can no longer be produced to sustain the muscle's contractile activity. Aerobic exercise, in contrast, can be sustained for prolonged periods. Not only does oxidative phosphorylation use far less nutrient fuel to generate ATP, but it can be supported by nutrients delivered to the muscle by means of the blood instead of relying on stored fuel in the muscle. Intense anaerobic exercise outpaces the ability to deliver supplies to the muscle by the blood, so the muscle must rely on stored fuel and inefficient glycolysis, thus limiting anaerobic exercise to brief periods of time before energy sources are depleted.

3. d. active transport. Levelling off of the curve designates saturation of a carrier molecule, so carrier-mediated transport is involved. The graph indicates that active transport is being used instead of facilitated diffusion, because the concentration of the substance in the intracellular fluid is greater than the concentration in the extracellular fluid at all points until after the transport maximum is reached. Thus, the substance is being moved *against* a concentration gradient, so active transport must be the method of transport being used.

4. vesicular transport. The maternal antibodies in the infant's digestive tract lumen are taken up by the intestinal cells by pinocytosis and are extruded on the opposite side of the cell into the interstitial fluid by exocytosis. The antibodies are picked up from the intestinal interstitial fluid by the blood supply to the region.

5. c. The action potentials would stop as they met in the middle. As the two action potentials moving toward each other both reached the middle of the axon, the two adjacent patches of membrane in the middle would be in a refractory period, so further propagation of either action potential would be impossible.

6. A subthreshold stimulus would transiently depolarize the membrane but not sufficiently to bring the membrane to threshold, so no action potential would occur. Because a threshold stimulus would bring the membrane to threshold, an action potential would occur. An action potential of the same magnitude and duration would occur in response to a suprathreshold stimulus as to a threshold stimulus. Because of the all-or-none law, a stimulus larger than that necessary to bring the membrane to threshold would not produce a larger action potential. (The magnitude of the stimulus is coded in the *frequency* of action potentials generated in the neuron, not the *size* of the action potentials.)

7. The hand could be pulled away from the hot stove by flexion of the elbow accomplished by summation of EPSPs at the cell bodies of the neurons controlling the biceps muscle, thus bringing these neurons to threshold. The subsequent action potentials generated in these neurons would stimulate contraction of the biceps. Simultaneous contraction of the triceps muscle, which would oppose the desired flexion of the elbow, could be prevented by generation of IPSPs at the cell bodies of the neurons controlling this muscle. These IPSPs would keep the triceps neurons from reaching threshold and firing so that the triceps would not be stimulated to contract. The arm could deliberately be extended despite a painful finger prick by voluntarily generating EPSPs to override the reflex IPSPs at the neuronal cell bodies controlling the triceps while simultaneously generating IPSPs to override the reflex EPSPs at the neuronal cell bodies controlling the biceps.

8. An EPSP, being a graded potential, spreads decrementally from its site of initiation in the postsynaptic neuron. If presynaptic neuron A (near the axon hillock of the postsynaptic cell) and presynaptic neuron B (on the opposite side of the postsynaptic cell body) both initiate EPSPs of the same magnitude and frequency, the EPSPs from A will be of greater strength when they reach the axon hillock than will the EPSPs from B. An EPSP from B will decrease more in magnitude as it travels farther before reaching the axon hillock, the region of lowest threshold and thus the site of action potential initiation. Temporal summation of the larger EPSPs from A may bring the axon hillock to threshold and initiate an action potential in the postsynaptic neuron, whereas temporal summation of the weaker EPSPs from B at the axon hillock may not be sufficient to bring this region to threshold. Therefore, the proximity of a presynaptic neuron to the axon hillock can bias its influence on the postsynaptic cell.

## Clinical Consideration

(Question on p. 88.)

As $Cl^-$ is secreted by the intestinal cells into the intestinal tract lumen, $Na^+$ follows passively along the established electrical gradient. Water passively accompanies this salt ($Na^+$ and $Cl^-$) secretion by osmosis. The toxin produced by the cholera pathogen prevents the normal inactivation of cAMP in intestinal cells, so cAMP levels rise. An increase in cAMP opens the $Cl^-$ channels in the luminal membranes of these cells. Increased secretion of $Cl^-$ and the subsequent passively induced secretion of $Na^+$ and water are responsible for the severe diarrhoea that characterizes cholera.

# Chapter 3 The Central Nervous System

## Check Your Understanding

3.1 (Questions on p. 98.)

1. See ❯ Figure 3-1, Organization of the nervous system, p. 92.

2. An *afferent neuron* has a sensory receptor at its peripheral ending, a long peripheral axon (afferent fibre), a cell body devoid of presynaptic inputs located adjacent to the spinal cord, and a short central axon that terminates in the spinal cord. Afferent neurons relay signals from the periphery to the CNS. The cell body of an *efferent neuron* lies in the CNS and has many presynaptic inputs converging on it. Its long peripheral axon (efferent fibre) branches into axon terminals at the effector organ. Efferent neurons carry instructions from the CNS to effector organs. In contrast to afferent and

efferent neurons, which lie primarily in the PNS, *interneurons* lie entirely in the CNS. The cell body of an interneuron receives converging input from afferent neurons and other interneurons, and its diverging output terminates on efferent neurons or other interneurons. Interneurons are important in integrating afferent information, formulating an efferent response, and accomplishing all higher mental functions associated with the human mind.

Atrocytes are the most abundant glial cell and have a starlike shape. They function to hold neurons together, support the blood–brain barrier, help in repair of brain injuries and neural scar formation, as well as helping to regulate the extracellular environment by removing excess $K^+$ and neurotransmitters. Oligodendrocytes have elongated projections that wrap around axons to form myelin. Microglia are the immune cells of the CNS. Structurally, they have many branches. At rest, they are found in the ECF, where they secrete growth factors until they are activated, at which time they move to the affected area to perform their immune function.

**3.2** (Questions on p. 101.)
1. dura mater, arachnoid mater, and pia mater
2. Tight junctions anatomically prevent transport between the cells that form the walls of brain capillaries, and highly selective membrane-bound carriers physiologically restrict transport through these cells. Together, these mechanisms constitute the blood–brain barrier.

**3.3** (Questions on p. 109.)
1. See ⟩ Figure 3-9, Spinal cord in cross-section, p. 104.
2. A tract is a bundle of nerve fibres (axons of long interneurons) with similar function that travel up (ascending tract) or down (descending tract) in the white matter of the spinal cord. A ganglion is a collection of neuronal cell bodies located outside the CNS. A centre, or nucleus, is a functional collection of neuronal cell bodies located within the CNS. A nerve is a bundle of peripheral axons (both afferent and efferent fibres), enclosed by a connective-tissue covering and following the same pathway.

**3.4** (Questions on p. 113.)
1. All sensory input on its way to the higher cortex synapses in the thalamus, which screens out insignificant signals and routes the important signals to appropriate areas of the cortex. In this way the thalamus serves as a relay station for preliminary processing of sensory input.
2. hypothalamus

**3.5** (Questions on p. 121.)
1. See ⟩ Figure 3-24, Functional areas of the cerebral cortex, p. 117.
2. the ability of the brain to change or be functionally remodelled in response to the demands placed on it

**3.6** (Questions on p. 126.)
1. emotions, basic survival and sociosexual behavioural patterns, motivation, and learning
2. amygdala

**3.7** (Questions on p. 132.)
1. the process of transferring and fixing short-term memory traces into long-term memory stores
2. Short-term memory involves transient modifications in the function of pre-existing synapses, such as increased neurotransmitter release from a presynaptic neuron or increased responsiveness of a postsynaptic neuron to neurotransmitter. Long-term memory involves gene activation and protein synthesis that leads to relatively permanent structural or functional changes, such as formation of new synapses.
3. The hippocampus is important for declarative memories, the "what" memories of specific people, places, objects, facts, and events that often result after only one experience. The cerebellum plays an essential role in the "how to" procedural memories involving motor skills gained through repetitive training. The prefrontal association cortex is the major orchestrator of working memory, which temporarily holds currently relevant data—both new information and knowledge retrieved from memory stores—and manipulates and relates them to accomplish complex reasoning functions.

**3.8** (Questions on p. 136.)
1. refers to subjective awareness of the external world and self
2. (1) an arousal system, which is regulated by a group of neurons in the hypothalamus and involves the reticular activating system in the brain stem; (2) a slow-wave sleep centre in the hypothalamus that contains neurons that induce slow-wave sleep; and (3) a paradoxical sleep centre in the brain stem, which houses REM neurons that switch from slow-wave to REM sleep

## Objective Questions
(Questions on p. 138.)
**1.** F **2.** F **3.** T **4.** F **5.** F **6.** habituation **7.** consolidation **8.** dorsal, ventral
**9.** 1. a, 2. c, 3. a and b, 4. b, 5. a, 6. c, 7. c **10.** 1. d, 2. c, 3. f, 4. e, 5. a, 6. b

## Points to Ponder
(Questions on p. 139.)
1. Only the left hemisphere has language ability. When sharing of information between the two hemispheres is prevented as a result of severance of the corpus callosum, visual information presented only to the right hemisphere cannot be verbally identified by the left hemisphere, because the left hemisphere is unaware of the information. However, the information can be recognized by nonverbal means, of which the right hemisphere is capable.
2. c. visual disturbances. A severe blow to the back of the head is most likely to traumatize the visual cortex in the occipital lobe.
3. Salivation when seeing or smelling food, striking the appropriate letter on the keyboard when typing, and many of the actions involved in driving a car are conditioned reflexes. You undoubtedly will have many other examples.
4. Insulin excess is going to cause a lowering of blood-borne glucose concentrations. Consequently, this means there is less glucose available for maintaining brain cell function. When glucose concentrations drop enough, neurons will start to die, resulting in brain damage.

## Clinical Consideration
(Question on p. 139.)
The deficits following the stroke—numbness and partial paralysis on the upper right side of the body and inability to speak—are indicative of damage to the left somatosensory cortex and left primary motor cortex in the regions devoted to the upper part of the body, plus Broca's area.

# Chapter 4 The Peripheral Nervous System: Sensory, Autonomic, Somatic

## Check Your Understanding
**4.1** (Questions on p. 149.)
1. A stimulus is a change detectable by the body. A receptor potential is a graded potential change in a receptor in response to a stimulus. A labelled line is a committed, incoming neural pathway that carries information regarding a particular sensory modality detected by a specialized receptor type at a specific site in the periphery, and delivers it to a defined area in the somatosensory cortex. Perception is the conscious interpretation of the external world as created by the brain from the sensory input it receives.
2. See ⟩ Figure 4-4, Tonic and phasic receptors, p. 146.
3. The receptive field size for a sensory neuron on your tongue is smaller than it is for a sensory neuron on your back; this is because your tongue has greater discriminative ability than your back does.

**4.2** (Questions on p. 152.)
1. See ▌ Table 4-2, Characteristics of Pain Nerve Fibres
2. A-delta fibres constitute a fast pain pathway that carries signals arising from mechanical and thermal nociceptors. C fibres constitute a slow pain pathway that carries impulses from polymodal nociceptors.
3. Endogenous opioids serve as endogenous analgesics by binding with opiate receptors at the synaptic knob of afferent pain fibres where they inhibit release of the pain neurotransmitter, substance P, thereby blocking further transmission of the pain signal.

**4.3** (Questions on p. 169.)
1. See ⟩ Figure 4-15a, Emmetropia, myopia, and hyperopia, p. 159.
2. When a photopigment absorbs light, retinal changes to the all-trans form, activating the photopigment. The activated photopigment activates the G protein transducin, which then activates the intracellular enzyme phosphodiesterase. In the dark, the second messenger cGMP had been keeping chemically gated $Na^+$ channels open, resulting in a passive,

inward, depolarizing $Na^+$ leak (dark current). Activated phosphodiesterase degrades cGMP, permitting these chemically gated $Na^+$ channels to close, thereby stopping the depolarizing $Na^+$ leak and causing hyperpolarization of the photoreceptor (the receptor potential).

3. See Table 4-3, Properties of Rod Vision and Cone Vision.

**4.4** (Questions on p. 181.)

1. The middle ear amplifies the tympanic membrane vibrations and converts them into wavelike movements in the inner ear fluid at the same frequency as the original sound waves.

2. Pitch discrimination depends on which region of the basilar membrane naturally vibrates maximally with a given sound frequency. Loudness discrimination depends on the amplitude of the vibrations. Timbre discrimination depends on overtones of varying frequencies, which cause many points along the basilar membrane to vibrate simultaneously, but less intensely than the fundamental tone.

3. See ❯ Figure 4-33, Vestibular apparatus, p. 179.

**4.5** (Questions on p. 187.)

1. salty (stimulated by chemical salts); sour (caused by free $H^+$ in acids); sweet (evoked by the particular configuration of glucose or by artificial sweeteners, which are organic molecules similar in structure to glucose but have no calories); bitter (elicited by alkaloids and poisonous substances, thus discouraging ingestion of potentially dangerous compounds); umami (triggered by amino acids, especially glutamate, as in meat)

2. An odourant is broken down into various components, and each olfactory receptor responds to a particular odour component that multiple scents may have in common. Each glomerulus in the olfactory bulb receives signals only from receptors that detect a particular odour component. Odour discrimination is based on different patterns of glomeruli (the "smell files") activated by various scents.

**4.6** (Questions on p. 194.)

1. See ❯ Figure 4-41, Autonomic nervous system, p. 189.

2. The sympathetic nervous system dominates in emergency or stressful (fight-or-flight) situations and promotes responses that prepare the body for strenuous physical activity. The parasympathetic nervous system dominates in quiet, relaxed (rest-and-digest) situations and promotes "general housekeeping" activities such as digestion.

3. The adrenal medulla is a modified sympathetic ganglion that does not give rise to postganglionic fibres but instead, on stimulation by the preganglionic fibre, secretes the hormones epinephrine and norepinephrine into the blood.

**4.7** (Questions on p. 197.)

1. The autonomic nervous system innervates cardiac muscle, smooth muscle, most exocrine glands, some endocrine glands, and adipose tissue. The somatic nervous system innervates skeletal muscles.

2. Motor neurons are considered the final common pathway because the only way any other parts of the nervous system can influence skeletal muscle activity is by acting, in common, on these motor neurons.

**4.8** (Questions on p. 201.)

1. Acetylcholine (ACh) is the neuromuscular junction neurotransmitter. When it is released from the motor-neuron terminal button in response to an action potential, it binds with and opens nonspecific cation-receptor channels in the motor end plate of the muscle fibre. The resultant ion movement leads to an end-plate potential, which initiates a contraction-inducing action potential that is propagated throughout the muscle fibre. Acetylcholinesterase (AChE), an enzyme in the membrane of the motor end plate, inactivates ACh. By removing ACh, AChE permits the choice between allowing relaxation to take place (no more ACh released) or keeping the contraction going (more ACh released), depending on the body's momentary needs.

2. An EPP is larger than an EPSP. Because of its magnitude, an EPP is normally large enough to promote sufficient local current flow to bring the muscle membrane adjacent to the motor end plate to threshold, thereby initiating an action potential. Thus, one-to-one transmission of action potentials occurs between a motor neuron and a muscle fibre at a neuromuscular junction. By contrast, one EPSP does not have sufficient magnitude to bring the postsynaptic neuron to threshold. Summation of EPSPs arising from multiple presynaptic action potentials is needed to initiate an action potential in the postsynaptic neuron.

## Objective Questions

(Questions on pp. 204–205)

1. transduction 2. adequate stimulus 3. F 4. T 5. T 6. T 7. F 8. T 9. F 10. F 11. 1. f, 2. h, 3. l, 4. d, 5. i, 6. e, 7. b, 8. j, 9. a, 10. g, 11. c, 12. k 12. 1. a, 2. b, 3. c, 4. c, 5. c, 6. a, 7. b, 8. b 13. T 14. F 15. c 16. c 17. sympathetic, parasympathetic 18. adrenal medulla 19. 1. a, 2. b, 3. a, 4. b, 5. a, 6. a, 7. b 20. 1. b, 2. b, 3. a, 4. a, 5. b, 6. b, 7. a

## Quantitative Exercises

(Questions on pp. 205–206)

1. The slow pain pathway takes about $(1.3\,m)(1\,sec/12\,m) = 0.1083\,sec$. The fast pathway takes $(1.3\,m)(1\,sec/30\,m) = 0.0433\,sec$. The difference is $0.1083\,sec - 0.0433\,sec = 0.065\,sec = 65\,msec$.

2. **a.** The amount of light entering the eye is proportional, approximately, to the area of the open pupil. Recall that the area of a circle is $\pi r^2$. Let $r$ be the pupil radius and $A_1$ be the original pupil area. Halving the diameter also halves the radius, so the new pupil area is

$$\pi\left(\frac{1}{2}r^2\right) = \frac{1}{4}\pi r^2 = \frac{1}{4}A_1$$

   **b.** Therefore, the amount of light allowed into the eye is a quarter of what it was originally.

   The area of a rectangle is $hw$, where $h$ is the height and $w$ the width. Halving either dimension halves the area and hence the amount of light allowed into the eye.

   **c.** The cat's pupil can be considered more precise. Think about the coarse and fine adjustments on a microscope. Fine adjustment translates rotations of the knob into much smaller movement of the microscope's stage than does coarse adjustment.

3. **a.** Solve the following for I:

   $\beta = (10\,dB)\log_{10}(I/I_0)$

   $I = I_0 10^{B/10}\,W/m^2$

   Therefore,

   $I_1 = 10^{-12}(10^{20/10}) = 10^{-12}(10^2) = 10^{-10}\,W/m^2$

   $I_2 = 10^{-12}(10^{70/10}) = 10^{-12}(10^7) = 10^{-5}\,W/m^2$

   $I_3 = 10^{-12}(10^{120/10}) = 10^{-12}(10^{12}) = 1\,W/m^2$

   $I_4 = 10^{-12}(10^{170/10}) = 10^{-12}(10^{17}) = 10^{-5}\,W/m^2$

   **b.** Because of the logarithm in the definition of decibel, the sound intensity increases exponentially with respect to sound level. This fact should be clear from the definition of dB solved for I. This result implies that the human ear performs well throughout an enormous range of sound intensities.

4. $t = \dfrac{x^2}{2D} = \dfrac{(200\,nm)^2}{2 \times 10^{-5}\,cm^2/sec}$

   $= \dfrac{4 \times 10^{-14}\,m^2 \cdot sec}{2 \times 10^{-5}\,cm^2} = \left(\dfrac{10^4\,cm^2}{m^2}\right) = 20\,\mu sec$

## Points to Ponder

(Questions on p. 206.)

1. Pain is a conscious warning that tissue damage is occurring or about to occur. A patient unable to feel pain because of a nerve disorder does not consciously take measures to withdraw from painful stimuli and thus prevent more serious tissue damage.

2. Pupillary dilation (mydriasis) can be deliberately induced by ophthalmic instillation of either an adrenergic drug (such as epinephrine or related compound) or a cholinergic-blocking drug (such as atropine or related compounds). Adrenergic drugs produce mydriasis by causing contraction of the sympathetically supplied radial (dilator) muscle of the iris. Cholinergic blocking drugs cause pupillary dilation by blocking parasympathetic activity to the circular (constrictor) muscle of the iris so that action of the adrenergically controlled radial muscle of the iris is unopposed.

3. The defect would be in the left optic tract or optic radiation.

4. Fluid accumulation in the middle ear in accompaniment with middle ear infections impedes the normal movement of the tympanic membrane, ossicles, and oval window in response to sound. All these structures vibrate less vigorously in the presence of fluid, causing temporary hearing

impairment. Chronic fluid accumulation in the middle ear is sometimes relieved by surgical implantation of drainage tubes in the eardrum. Hearing is restored to normal as the fluid drains to the exterior. Usually, the tube "falls out" as the eardrum heals and pushes out the foreign object.

5. The sense of smell is reduced when you have a cold because odourants do not reach the receptor cells as readily when the mucous membranes lining the nasal passageways are swollen, and excess mucus is present.

6. By promoting arteriolar constriction, epinephrine administered in conjunction with local anaesthetics reduces blood flow to the region; this helps the anaesthetic stay in the region instead of being carried away by the blood.

7. No. Atropine blocks the effect of acetylcholine at muscarinic receptors but does not affect nicotinic receptors. Nicotinic receptors are present on the motor end plates of skeletal muscle fibres.

8. The voluntarily controlled external urethral sphincter is composed of skeletal muscle and supplied by the somatic nervous system.

9. By interfering with normal acetylcholine activity at the neuromuscular junction, α-bungarotoxin leads to skeletal muscle paralysis, with death ultimately occurring as a result of an inability to contract the diaphragm and breathe.

10. If the motor neurons that control the respiratory muscles, especially the diaphragm, are destroyed by poliovirus or amyotrophic lateral sclerosis, the person is unable to breathe and dies (unless breathing is assisted by artificial means).

## Clinical Consideration

(Question on p. 206.)

Syncope most frequently occurs as a result of inadequate delivery of blood carrying sufficient oxygen and glucose supplies to the brain. Possible causes include circulatory disorders, such as impaired pumping of the heart or low blood pressure; respiratory disorders, resulting in poorly oxygenated blood; anaemia, in which the oxygen-carrying capacity of the blood is reduced; or low blood glucose, due to improper endocrine management of blood glucose levels. Vertigo, in contrast, typically results from a dysfunction of the vestibular apparatus, arising, for example, from viral infection or trauma, or abnormal neural processing of vestibular information, as, for example, with a brain tumour.

# Chapter 5 Principles of Endocrinology: The Central Endocrine Glands

## Check Your Understanding

**5.1** (Questions on p. 218.)

1. In general, peptide hormones are synthesized and stored within vesicles. Upon the appropriate stimulation, the vesicles undergo exocytosis to release the peptide hormones into the blood. Peptide hormones are usually hydrophilic so they do not require a transport protein in the blood to reach their target. In contrast, because steroid hormones are lipophilic, once they are synthesized they are not stored, but will leave the cell. In the plasma, they are generally bound to transport proteins to help them reach their targets.

2. Different cells have different proteins available for phosphorylation by protein kinase A. The particular cellular action of cAMP depends on what these proteins do once they are phosphorylated.

**5.2** (Questions on p. 222.)

1. (1) Negative feedback maintains the plasma concentration of a hormone at a given set level because the output of the hormonal control system counteracts a change in the input. For example, when the concentration of a hormone falls, the control system triggers increased secretion of the hormone, which feeds back to shut off the stimulatory response when the set level is achieved. (2) Neuroendocrine reflexes produce a sudden increase in hormone secretion (that is, "turn up the thermostat setting") in

response to a specific stimulus, which is often external to the body. (3) Diurnal (circadian) rhythms are rhythmic fluctuations up and down in the secretion rate of many hormones as a function of the time of day.

2. Down regulation is a reduction in the number of target-cell receptors in the face of a prolonged increase in a hormone. Permissiveness refers to the need for one hormone to be present in adequate amounts to permit another hormone to fully exert its effects. Synergism results when the combined effect of two hormones is greater than the sum of their separate effects. Antagonism takes place when one hormone decreases the effectiveness of another hormone.

3. See ▶ Figure 5-8, Activation of genes by lipophilic hormones, p. 221.

**5.3** (Questions on p. 231.)

1. See ▶ Figure 5-12, Hierarchic chain of command and negative feedback in endocrine control, p. 230.

2. (a) Posterior pituitary hormones—vasopressin (antidiuretic hormone, ADH) and oxytocin. (b) Anterior pituitary hormones—growth hormone (GH, somatotropin), thyroid-stimulating hormone (TSH, thyrotropin), adrenocorticotropic hormone (ACTH, corticotropin), follicle-stimulating hormone (FSH), luteinizing hormone (LH), and prolactin (PRL). (c) Hypophysiotropic hormones—thyrotropin-releasing hormone (TRH), corticotropin-releasing hormone (CRH), gonadotropin-releasing hormone (GnRH), growth hormone–releasing hormone (GHRH), somatostatin (growth hormone–inhibiting hormone, GHIH), prolactin-releasing peptide (PrRP), and dopamine (prolactin-inhibiting hormone, PIH)

3. The hypothalamus controls hormonal output from the posterior pituitary by a neural connection and controls hormonal output from the anterior pituitary by a vascular connection. The posterior pituitary is a neural extension of the hypothalamus. Neuronal cell bodies in the hypothalamus produce vasopressin and oxytocin, which are transported down the axons that pass through the connecting stalk to the posterior pituitary, where these hormones are stored in the neuronal terminals. When stimulated by the hypothalamus, these hormones are independently released into the systemic blood. The hypothalamus secretes hypophysiotropic hormones into the hypothalamic–hypophyseal portal system, a capillary-to-capillary vascular link that transports them through the connecting stalk to the anterior pituitary, where they control the secretion of hormones produced by the anterior pituitary into the systemic blood.

**5.4** (Questions on p. 240.)

1. GH directly increases fatty acid levels in the blood by enhancing the breakdown of triglyceride stores in adipose tissue and increases blood-glucose levels by decreasing glucose uptake by muscles and increasing glucose output by the liver, thus mobilizing fat stores as a major energy source for muscle, while conserving glucose for the glucose-dependent brain. GH also directly and indirectly (via insulin-like growth factor-I [IGF-I]) brings about protein synthesis, decreasing blood amino acids in the process.

2. GH does not act directly to bring about most of its growth-producing actions (increased cell division, enhanced protein synthesis, and bone growth). Instead, GH stimulates the liver to release IGF-I, which directly mediates these growth-promoting actions. GH's only direct growth-promoting action is to stimulate protein synthesis, which it does in conjunction with IGF-I.

3. Growth hormone has a well-characterized diurnal rhythm for secretion. It is primarily under the control of growth hormone–releasing hormone and growth hormone–inhibiting hormone. However, its secretion can be increased by exercise, stress, and low blood glucose.

## Objective Questions

(Questions on p. 244.)

**1.** T **2.** T **3.** F **4.** T **5.** T **6.** T **7.** tropic **8.** down regulation **9.** epiphyseal plate **10.** suprachiasmatic nucleus **11.** 1. c, 2. b, 3. b, 4. a, 5. a, 6. c, 7. c

## Points to Ponder

(Questions on p. 244.)

1. The concentration of hypothalamic releasing and inhibiting hormones would be considerably lower (in fact, almost nonexistent) in a systemic venous blood sample compared to the concentration of these hormones

in a sample of hypothalamic–hypophyseal portal blood. These hormones are secreted into the portal blood for local delivery between the hypothalamus and anterior pituitary. Any portion of these hormones picked up by the systemic blood at the anterior pituitary capillary level is greatly diluted by the much larger total volume of systemic blood, compared to the extremely small volume of blood within the portal vessel.

2. Above normal. Without sufficient iodine, the thyroid gland is unable to synthesize enough thyroid hormone. The resultant reduction in negative-feedback activity by the reduced level of thyroid hormone would lead to increased TSH secretion. Despite the elevated TSH, however, the thyroid gland still could not secrete adequate thyroid hormone because of the iodine deficiency.

3. If CRH and/or ACTH is elevated in accompaniment with the excess cortisol secretion, the condition is secondary to a defect at the hypothalamic/anterior pituitary level. If CRH and ACTH levels are below normal in accompaniment with the excess cortisol secretion, the condition is due to a primary defect at the adrenal cortex level, with the excess cortisol inhibiting the hypothalamus and anterior pituitary in negative-feedback fashion.

4. Males with testicular feminization syndrome would be unusually tall because of the inability of testosterone to promote closure of the epiphyseal plates of the long bones in the absence of testosterone receptors.

5. Full-grown athletes sometimes illegally take supplemental doses of growth hormone because it promotes increased skeletal muscle mass through its protein anabolic effect. However, excessive growth hormone can have detrimental side effects, such as possibly causing diabetes or high blood pressure.

## Clinical Consideration

(Question on p. 245.)
Hormonal replacement therapy following pituitary gland removal should include thyroid hormone (the thyroid gland will not produce sufficient thyroid hormone in the absence of TSH) and glucocorticoid (because of the absence of ACTH), especially in stress situations. If indicated, male or female sex hormones can be replaced, even though these hormones are not essential for survival. For example, testosterone in males plays an important role in libido. Growth hormone and prolactin need not be replaced because their absence will produce no serious consequences in this individual. Vasopressin may have to be replaced if insufficient quantities of this hormone are picked up by the blood at the hypothalamus in the absence of the posterior pituitary.

# Chapter 6 The Endocrine Glands

## Check Your Understanding

**6.1** (Questions on p. 254.)
1. A thyroid follicle, the functional unit of the thyroid gland, is a hollow sphere consisting of a single layer of follicular cells enclosing an inner lumen filled with colloid, which serves as an extracellular storage site for thyroid hormone. Colloid is filled with thyroglobulin, a large glycoprotein within which are incorporated the thyroid hormones in their various stages of synthesis. During synthesis, attachment of one iodide to tyrosine yields MIT (monoiodotyrosine), and attachment of two iodides to tyrosine yields DIT (di-iodotyrosine). Coupling of one MIT and one DIT yields T3 (tri-iodothyronine), and coupling of two DITs yields T4 (tetraiodothyronine or thyroxine). T3 and T4 are collectively referred to as thyroid hormone.
2. See ❯ Figure 6-3, Regulation of thyroid hormone secretion, p. 251.
3. See ❚ Table 6.1, Types of Thyroid Dysfunctions
**6.2** (Questions on p. 263.)
1. (a) Mineralocorticoids—aldosterone (promotes $Na^+$ retention and $K^+$ elimination during urine formation). (b) Glucocorticoids—cortisol (increases blood glucose at the expense of protein and fat stores and helps the body adapt to stress). (c) Sex hormones—dehydroepiandrosterone

(governs androgen-dependent processes in the female, such as growth of pubic and axillary hair, enhancing pubertal growth spurt, and maintaining the sex drive)
2. ACTH stimulates the adrenal cortex to secrete cortisol but has no effect on aldosterone.
3. epinephrine and norepinephrine, which are stored in chromaffin granules, from which they are secreted into the blood by exocytosis on stimulation by preganglionic sympathetic fibres.
**6.3** (Questions on p. 282.)
1. Glycogenesis is the conversion of glucose to glycogen, glycogenolysis is the conversion of glycogen to glucose, and gluconeogenesis is the conversion of amino acids to glucose.
2. Insulin ↓ blood glucose and promotes carbohydrate storage by facilitating glucose transport into most cells via transporter recruitment and by stimulating glycogenesis and inhibiting glycogenolysis and gluconeogenesis. It ↓ blood fatty acids and promotes triglyceride storage by increasing the entry of fatty acids and glucose into adipose cells, thereby promoting triglyceride synthesis and inhibiting lipolysis. It ↓ blood amino acids and enhances protein synthesis by promoting active transport of amino acids into muscles, stimulating protein synthesis, and inhibiting protein degradation. Glucagon ↑ blood glucose and reduces carbohydrate stores by increasing hepatic glucose production via decreasing glycogenesis and by promoting glycogenolysis and gluconeogenesis. It ↑ blood fatty acids and reduces fat stores by promoting lipolysis and inhibiting triglyceride synthesis. It ↑ blood ketone levels by enhancing conversion of fatty acids to ketone bodies by the liver. Glucagon inhibits protein synthesis and promotes protein degradation in the liver but not in muscle, the body's major protein store, so it does not have any significant effect on blood amino acid levels.
3. Increased blood glucose stimulates the pancreatic β cells, thus increasing insulin secretion, and inhibits the pancreatic α cells, thus decreasing glucagon secretion.
**6.4** (Questions on p. 292.)
1. Following is the distribution of body $Ca^{2+}$: 99 percent in the skeleton and teeth, 0.9 percent in the cells, and 0.1 percent in the ECF. Of the $Ca^{2+}$ in the ECF, half is bound in complexes and not available to participate in chemical reactions, and the other half is free $Ca^{2+}$, which is biologically active. This free ECF $Ca^{2+}$ plays a vital role in neuromuscular excitability, excitation–contraction coupling in cardiac and smooth muscle, stimulus–secretion coupling, excitation–secretion coupling, maintenance of tight junctions, and blood clotting.
2. PTH acts on bone, kidneys, and the intestine to increase plasma $Ca^{2+}$ as follows: PTH stimulates a fast exchange of $Ca^{2+}$ from the small labile pool of $Ca^{2+}$ in the bone fluid into the plasma by activating $Ca^{2+}$ pumps in the osteocytic–osteoblastic bone membrane. It induces a slow exchange of $Ca^{2+}$ from the stable pool of $Ca^{2+}$ in the bone minerals of the bone itself into the plasma by stimulating osteoclasts to dissolve bone. PTH acts on the kidneys to conserve $Ca^{2+}$ and eliminate $PO_4^{3-}$ during urine formation and to activate vitamin D. PTH indirectly increases $Ca^{2+}$ and $PO_4^{3-}$ absorption from the small intestine via its role in activating vitamin D, which directly promotes the intestinal absorption of these ingested electrolytes.
3. Vitamin D is essential for the absorption of $Ca^{2+}$ from the intestine. Without adequate vitamin D, the body will activate the process to mobilize $Ca^{2+}$ from internal stores, such as the bones.

## Objective Questions

(Questions on p. 294.)
**1.** F **2.** T **3.** T **4.** T **5.** T **6.** F **7.** F **8.** F **9.** colloid, thyroglobulin
**10.** pro-opiomelanocortin **11.** glycogenesis, glycogenolysis, gluconeogenesis
**12.** brain, working muscles, liver **13.** bone, kidneys, digestive tract **14.** c
**15.** a, b, c, g, i, j **16.** 1. glucose, 2. glycogen, 3. free fatty acids, 4. triglycerides, 5. amino acids, 6. body proteins

## Points to Ponder

(Questions on p. 295.)
1. The midwestern United States is no longer an endemic goitre belt even though the soil is still iodine poor, because individuals living in this

region obtain iodine from iodine-supplemented nutrients, such as iodinated salt, and from seafood and other naturally iodine-rich foods shipped from coastal regions.

2. Anaphylactic shock is an extremely serious allergic reaction brought about by massive release of chemical mediators in response to exposure to a specific allergen—such as one associated with a bee sting—to which the individual has been highly sensitized. These chemical mediators bring about circulatory shock (severe hypotension) through a twofold effect: (1) by relaxing arteriolar smooth muscle, thus causing widespread arteriolar vasodilation and a resultant fall in total peripheral resistance and arterial blood pressure; and (2) by causing a generalized increase in capillary permeability, resulting in a shift of fluid from the plasma into the interstitial fluid. This shift decreases the effective circulating volume, further reducing arterial blood pressure. Additionally, these chemical mediators bring about pronounced bronchoconstriction, making it impossible for the victim to move sufficient air through the narrowed airways. Because these responses take place rapidly and can be fatal, people allergic to bee stings are advised to keep injectable epinephrine in their possession. By promoting arteriolar vasoconstriction through its action on $\alpha_1$ receptors in arteriolar smooth muscle and by promoting bronchodilation through its action on $\beta_2$ receptors in bronchiolar smooth muscle (see ▌Table 6-2), epinephrine counteracts the life-threatening effects of the anaphylactic reaction to the bee sting.

3. An infection elicits the stress response, which brings about increased secretion of cortisol and epinephrine, both of which increase the blood glucose level. This becomes a problem for diabetic patients who have to bring down the elevated blood glucose by injecting additional insulin or, preferably, by reducing carbohydrate intake and/or exercising to use up some of the extra blood glucose. In a normal individual, the check-and-balance system between insulin and the other hormones that oppose insulin's actions helps maintain the blood glucose within reasonable limits during the stress response.

4. The presence of Chvostek's sign is due to increased neuromuscular excitability caused by moderate hyposecretion of parathyroid hormone.

5. If malignancy-associated hypercalcaemia arose from metastatic tumour cells that invaded and destroyed bone, both hypercalcaemia and hyperphosphataemia would result as calcium phosphate salts were released from the destructed bone. The fact that hypophosphataemia, not hyperphosphataemia, often accompanies malignancy-associated hypercalcaemia led investigators to rule out bone destruction as the cause of the hypercalcaemia. Instead, they suspected that the tumours produced a substance that mimics the actions of PTH in promoting concurrent hypercalcaemia and hypophosphataemia.

## Clinical Consideration

(Question on p. 295.)

"Diabetes of bearded ladies" is descriptive of both excess cortisol and excess adrenal androgen secretion. Excess cortisol secretion causes hyperglycaemia and glucosuria. Glucosuria promotes osmotic diuresis, which leads to dehydration and a compensatory increased sensation of thirst. All these symptoms—hyperglycaemia, glucosuria, polyuria, and polydipsia—mimic diabetes mellitus. Excess adrenal androgen secretion in females promotes masculinizing characteristics, such as beard growth. Simultaneous hypersecretion of both cortisol and adrenal androgen most likely occurs secondary to excess CRH/ACTH secretion, because ACTH stimulates both cortisol and androgen production by the adrenal cortex.

# Chapter 7 Muscle Physiology

## Check Your Understanding

**7.1** (Questions on p. 302.)

1. A *muscle fibre* is composed of *myofibrils* that extend the entire length of the muscle fibre; in general, the larger-diameter *muscle fibres* have a greater number of myofibrils. A whole muscle is composed of muscle fibres that extend the entire length of the muscle; in general, the larger-diameter muscles have more muscle fibres.

2. The regulatory protein, troponin, binds to both actin and tropomyosin. In the relaxed state, troponin assumes a conformation that causes tropomyosin to cover the myosin cross-bridge binding sites on the actin molecules.

**7.2** (Questions on p. 309.)

1. See ❯ Figure 7-7, Changes in banding pattern during shortening, p. 303.

2. The dihydropyridine receptors serve as voltage-gated sensors that are activated by an action potential as it propagates along the T tubule. The activated dihydropyridine receptors trigger the opening of $Ca^{2+}$-release channels (ryanodine receptors) in the adjacent lateral sacs of the sarcoplasmic reticulum, thereby permitting $Ca^{2+}$ release from the lateral sacs. This released $Ca^{2+}$ repositions the troponin–tropomyosin complex so that actin and the myosin cross bridges can interact to accomplish contraction.

3. See the cross-bridge cycle in ❯ Figure 7-12. ATP binds to the myosin head and causes the head to detach from the actin molecule. During the cocking of the myosin head, ATP is hydrolyzed to ADP and $P_i$. When the myosin head binds to actin, $P_i$ is released from the head during the power stroke. ADP is released from the myosin head after the power stroke.

**7.3** (Questions on p. 314.)

1. Greater strength of contraction can be achieved through motor unit recruitment, twitch summation, positioning the muscle at its optimal length, the absence of fatigue, and hypertrophy of a muscle (strength training).

2. In twitch summation, the level of cytosolic $Ca^{2+}$ is increased by repeated release of $Ca^{2+}$ from the lateral sacs. In addition, with repeated excitation of a skeletal muscle cell, insufficient time is available between action potentials for the sarcoplasmic reticulum to pump all the released $Ca^{2+}$ back into the lateral sacs. The sustained, elevated cytosolic $Ca^{2+}$ leads to prolonged exposure of myosin cross-bridge binding sites for interaction with actin and thus greater power stroke opportunities.

3. When skeletal muscle is stretched beyond its optimal length, fewer actin sites are available for cross-bridge cycling, which reduces tension generation.

**7.4** (Questions on p. 322.)

1. The leg (drumstick) muscles of a turkey consist primarily of red muscle fibres, which have a large number of mitochondria, high levels of myoglobin, low glycogen content, and relatively few glycolytic enzymes. The turkey's leg muscles are built for endurance, not for speed or power. In contrast, the turkey's breast muscles, composed primarily of white muscle fibres, have relatively few mitochondria, low levels of myoglobin, high glycogen content, and an abundance of glycolytic enzymes. The breast muscles are built for speed and power, but lack endurance (for example, turkeys can fly only a very short distance).

2. ATP comes from all of these sources during the course of the race, but oxidative phosphorylation is responsible for generating the largest amount of ATP expended in this event.

## Objective Questions

(Questions on pp. 336–337.)

**1.** F **2.** F **3.** F **4.** T **5.** F **6.** T **7.** concentric, eccentric **8.** alpha, gamma
**9.** denervation atrophy, disuse atrophy **10.** a, b, e **11.** b **12.** 1. f, 2. d, 3. c, 4. e, 5. b, 6. g, 7. a **13.** 1. a, 2. a, 3. a, 4. b, 5. b, 6. b

## Quantitative Exercises

(Questions on p. 337.)

**1. a.** For the weekend athlete, the lever ratio is 70 cm/9 cm. So the velocity at the end of the arm is 2.6 m/sec (70/9) = 20.2 m/sec (about 72 km/h).

**b.** For the professional ballplayer, the lever ratio is 90 cm/9 cm. So

$$10x = 137 \text{ km/h}$$
$$x = 13 \cdot 7 \text{ km/h} \left(1 \text{ hr}/36\,000 \text{ sec}\right)$$
$$= 3 \cdot 8 \text{ m/sec}$$

2. The force–velocity curve is as follows:

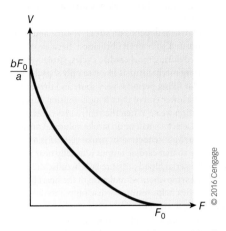

© 2016 Cengage

**a.** The shape of the curve indicates that it takes time to develop force and that the greater the force developed, the more time is needed.

**b.** The maximum velocity does not change when $F_0$ is increased, but the muscle can lift heavier loads or generate more force. The maximum load does not increase when the cross-bridge cycling rate increases, but the muscle can lift lighter loads faster. If the muscle increases in size, $b$ increases, and the entire curve shifts up with respect to the $v$ axis.

## Points to Ponder
(Questions on pp. 337–338.)

1. By placing increased demands on the heart to sustain increased delivery of $O_2$ and nutrients to working skeletal muscles, regular aerobic exercise induces changes in cardiac muscle that enable it to use $O_2$ more efficiently, such as increasing the number of capillaries supplying blood to the heart muscle. Intense exercise of short duration, such as weight training, in contrast, does not induce cardiac efficiency. Because this type of exercise relies on anaerobic glycolysis for ATP formation, no demands are placed on the heart for increased delivery of blood to the working muscles.

2. The power arm of the lever is 4 cm, and the load arm is 28 cm for a lever ratio of 1:7 (4 cm:28 cm). Thus, to lift an 8 kg stack of books with one hand, the child must generate an upward applied force in the biceps muscle of 56 kg. (With a lever ratio of 1:7, the muscle must exert seven times the force of the load; $7 \times 8\,kg = 56\,kg$.)

3. The length of the thin filaments is represented by the distance between a Z line and the edge of the adjacent H zone. This distance remains the same in a relaxed and contracted myofibril, which leads to the conclusion that the thin filaments do not change in length during muscle contraction.

4. Regular bouts of anaerobic, short-duration, high-intensity resistance training would be recommended for competitive downhill skiing. By promoting hypertrophy of the fast glycolytic fibres, such exercise better adapts the muscles to activities that require intense strength for brief periods, such as a swift, powerful descent downhill. In contrast, regular aerobic exercise would be more beneficial for competitive cross-country skiers. Aerobic exercise induces metabolic changes within the oxidative fibres that enable the muscles to use $O_2$ more efficiently. These changes, which include an increase in mitochondria and capillaries within the oxidative fibres, adapt the muscles to better endure the prolonged activity of cross-country skiing without fatiguing.

5. Because the site of voluntary control to overcome the micturition reflex is at the external urethral sphincter and not the bladder, the external urethral sphincter must be skeletal muscle, which is innervated by the voluntarily controlled somatic nervous system, and the bladder must be smooth muscle, which is innervated by the involuntarily controlled autonomic nervous system. The only other type of involuntarily controlled muscle besides smooth muscle is cardiac muscle, which is found only in the heart. Therefore, the bladder must be smooth, not cardiac, muscle.

## Clinical Consideration
(Question on p. 338.)

The muscles in the immobilized leg have undergone disuse atrophy. The physician or physical therapist can prescribe regular resistance-type exercises that specifically use the atrophied muscles to help restore them to their normal size.

# Chapter 8 Cardiac Physiology

## Check Your Understanding

**8.1** (Questions on p. 351.)
1. See ❯ Figure 8-1, Pulmonary and systemic circulations in relation to the heart, p. 346.
2. The *right* and *left atrioventricular (AV valves)* let blood flow from the atria to the ventricles during ventricular filling, but prevent backflow of blood from the ventricles into the atria during ventricular emptying. The *aortic* and *pulmonary semilunar valves* let blood flow from the ventricles into the aorta and pulmonary arteries during ventricular emptying, but prevent backflow of blood from these major arteries into the ventricles during ventricular filling.
3. The endocardium is the thin inner layer. The middle myocardium mainly consists of the cardiac muscle cells. The epicardium is the thin outer layer.

**8.2** (Questions on p. 361.)
1. See ❯ Figure 8-7, Pacemaker activity of cardiac autorhythmic cells, p. 352 and ❯ Figure 8-10, Action potential of cardiac contractile muscle cells, p. 355.
2. *SA node* (70–80 action potentials/min), *AV node* (40–60 action potentials/min), and *Bundle of His and Purkinje fibres* (20–40 action potentials/min)
3. The electrical activity associated with atrial repolarization occurs simultaneously with ventricular depolarization and is masked by the QRS complex on a normal ECG.

**8.3** (Questions on p. 365.)
1. *Systole* is the period of contraction and emptying, and *diastole* is the period of relaxation and filling during the cardiac cycle.
2. (1) aortic pressure > atrial pressure > ventricular pressure, (2) aortic pressure > ventricular pressure > atrial pressure, (3) ventricular pressure > aortic pressure > atrial pressure, (4) aortic pressure > ventricular pressure > atrial pressure
3. The first heart sound is associated with the closure of the AV valves; this occurs at the beginning of isovolumetric ventricular contraction. The second heart sound is the semilunar valves; this occurs at the beginning of isovolumetric ventricular relaxation.

**8.4** (Questions on p. 373.)
1. *Parasympathetic stimulation* decreases heart rate and has no effect on stroke volume. *Sympathetic stimulation* increases heart rate and increases stroke volume by increasing the contractile strength of the heart.
2. See ❯ Figure 8-20, Intrinsic control of stroke volume (Frank-Starling curve), p. 368.

## Objective Questions
(Questions on pp. 380–381.)

1. intercalated discs, desmosomes, gap junctions 2. bradycardia, tachycardia 3. adenosine 4. F 5. F 6. F 7. T 8. d 9. d 10. b 11. 1. e, 2. a, 3. d, 4. b, 5. c, 12. AV, systole, semilunar, diastole 13. less than, greater than, less than, greater than, less than

## Quantitative Exercises
(Questions on pp. 381–382.)

1. CO = HR × SV
   40L/min = HR × 0.07L

   HR = (40L/min)/(0.07 L) = 571 beats/min
   This rate is not physiologically possible.

2. ESV = EDV − SV
    = 125mL − 85mL
    = 40mL

## Points to Ponder

(Questions on p. 382.)

1. Because, at a given heart rate, the interval between a premature beat and the next normal beat is longer than the interval between two normal beats, the heart fills for a longer period of time following a premature beat before the next period of contraction and emptying begins. Because of the longer filling time, the end-diastolic volume is larger, and, according to the Frank–Starling law of the heart, the subsequent stroke volume will be correspondingly larger.

2. The heart of a trained athlete is stronger and can pump blood more efficiently, so the resting stroke volume is larger than in an untrained person. For example, if the resting stroke volume of a strong-hearted athlete is 100 mL, a resting heart rate of only 50 beats/minute produces a normal resting cardiac output of 5000 mL/minute. An untrained person with a resting stroke volume of 70 mL, in contrast, must have a heart rate of about 70 beats/minute to produce a comparable resting cardiac output.

3. The direction of flow through a patent ductus arteriosus is the reverse of the flow that occurs through this vascular connection during fetal life. With a patent ductus arteriosus, some of the blood present in the aorta is shunted into the pulmonary artery because, after birth, the aortic pressure is greater than the pulmonary artery pressure. This abnormal blood flow produces a "machinery murmur," which lasts throughout the cardiac cycle but is more intense during systole and less intense during diastole. Thus, the murmur waxes and wanes with each beat of the heart, sounding somewhat like a washing machine as the agitator rotates back and forth. The murmur is present throughout the cardiac cycle because a pressure differential between the aorta and pulmonary artery is present during both systole and diastole. The murmur is more intense during systole because more blood is diverted through the patent ductus arteriosus as a result of the greater pressure differential between the aorta and pulmonary artery during ventricular systole than during ventricular diastole. Typically, the systolic aortic pressure is 120 mmHg, and the systolic pulmonary arterial pressure is 24 mmHg, for a pressure differential of 96 mmHg. By contrast, the diastolic aortic pressure is normally 80 mmHg, and the diastolic pulmonary arterial pressure is 8 mmHg, for a pressure differential of 72 mmHg.

4. A transplanted heart that does not have any innervation adjusts the cardiac output to meet the body's changing needs by means of both intrinsic control (the Frank–Starling mechanism) and extrinsic hormonal influences, such as the effect of epinephrine on the rate and strength of cardiac contraction.

5. In left bundle-branch block, the right ventricle becomes completely depolarized more rapidly than the left ventricle. As a result, the right ventricle contracts before the left ventricle, and the right AV valve is forced closed prior to closure of the left AV valve. Because the two AV valves do not close in unison, the first heart sound is "split"; that is, two distinct sounds in close succession can be detected as closure of the left valve lags behind closure of the right valve.

## Clinical Consideration

(Question on p. 382.)

The most likely diagnosis is atrial fibrillation. This condition is characterized by rapid, irregular, uncoordinated depolarizations of the atria. Many of these depolarizations reach the AV node at a time when it is not in its refractory period, thereby bringing about frequent ventricular depolarizations and a rapid heartbeat. However, because impulses reach the AV node erratically, the ventricular rhythm and thus the heartbeat are also very irregular as well as being rapid.

Ventricular filling is only slightly reduced despite the fact that the fibrillating atria are unable to pump blood because most ventricular filling occurs during diastole prior to atrial contraction. Because of the erratic heartbeat, variable lengths of time are available between ventricular beats for ventricular filling. However, the majority of ventricular filling occurs early in ventricular diastole after the AV valves first open, so even though the filling period may be shortened, the extent of filling may be near normal. Only when the ventricular filling period is very short is ventricular filling substantially reduced.

Cardiac output, which depends on stroke volume and heart rate, usually is not seriously impaired with atrial fibrillation. Because ventricular filling is only slightly reduced during most cardiac cycles, stroke volume, as determined by the Frank–Starling mechanism, is likewise only slightly reduced. Only when the ventricular filling period is very short and the cardiac muscle fibres are operating on the lower end of their length–tension curve is the resultant ventricular contraction weak. When the ventricular contraction becomes too weak, the ventricles eject a small or no stroke volume. During most cardiac cycles, however, the slight reduction in stroke volume is often offset by the increased heart rate so that cardiac output is usually near normal. Furthermore, if the mean arterial blood pressure falls because the cardiac output does decrease, increased sympathetic stimulation of the heart brought about by the baroreceptor reflex helps restore cardiac output to normal by shifting the Frank–Starling curve to the left.

On those cycles when ventricular contractions are too weak to eject enough blood to produce a palpable wrist pulse, if the heart rate is determined directly, either by the apex beat or via the ECG, and the pulse rate is taken concurrently at the wrist, the heart rate will exceed the pulse rate, producing a pulse deficit.

# Chapter 9 Vascular Physiology

## Check Your Understanding

**9.1** (Questions on p. 389.)

1. At rest, blood flow is widely distributed and includes the digestive system and the kidneys. During exercise, blood flow is diverted to organs such as skeletal muscles for increase oxygen delivery and the skin to help remove excess heat. Blood flow to the digestive system and kidneys will be reduced as they are not important at this time.

2. digestive tract, kidneys, and skin

3. $F = \Delta P/R$
   $R \alpha 1/r^4$

**9.2** (Questions on p. 394.)

1. See ▌Table 9-1, Features of Blood Vessels

2. Arteries have an abundance of elastin fibres that allow them to stretch to accommodate the extra volume of blood pumped into them during systole, and then to recoil and drive the extra blood forward into the remaining vasculature during diastole

3. See ❭ Figure 9-7a, Arterial blood pressure, p. 392.

## Objective Questions

(Questions on pp. 429–430.)

**1.** T **2.** F **3.** T **4.** T **5.** F **6.** T **7.** c **8.** a, c, d, e, f **9.** 1. a, 2. a, 3. b, 4. a, 5. b, 6. a
**10.** 1. b, 2. a, 3. b, 4. a, 5. a, 6. a, 7. b, 8. a, 9. b, 10. a, 11. b, 12. a, 13. a

## Quantitative Exercises

(Questions on pp. 430–431.)

1. (120mmHg)/(30L/min) = 4PRU
2. a. 90mmHg + (180mmHg − 90mmHg)/3 = 120mmHg
   b. Because the other forces acting across the capillary wall, such as plasma colloid osmotic pressure, typically do not change with age, one would suspect fluid loss from the capillaries into the tissues would occur as a result of the increase in capillary blood pressure.
3. systemic: (95 mmHg)/(19 PRU)
   = 95mmHg/(19mmHg/L/min) = 5L/min
   pulmonary: (20mmHg)/(4PRU) = 5L/min
4. a. 125 mmHg
   b. 77 mmHg
   c. 48mmHg(125mmHg − 77 mmHg = 48mmHg)
   d. $93\,\text{mmHg}\left[77 + \dfrac{1}{3}(48) = 77 + 16 = 93\,\text{mmHg}\right]$

f. Yes; blood would flow through the brachial artery when the arterial pressure was between 118 and 125 mmHg and would not flow through when the arterial pressure fell below 118 mmHg. The turbulence created by this intermittent blood flow would produce sounds.

g. No; blood would flow continuously through the brachial artery in smooth, laminar fashion, so no sound would be heard.

## Points to Ponder

(Questions on p. 431.)

1. An elastic support stocking increases external pressure on the remaining veins in the limb to produce a favourable pressure gradient that promotes venous return to the heart and minimizes swelling that would result from fluid retention in the extremity.

2. The classmate has apparently fainted because of insufficient blood flow to the brain as a result of pooling of blood in the lower extremities brought about by standing still for a prolonged time. When the person faints and assumes a horizontal position, the pooled blood will quickly return to the heart, improving cardiac output and blood flow to his brain. Trying to get the person up would be counterproductive, so the classmate trying to get him up should be advised to let him remain lying down until he recovers on his own.

3. The drug is apparently causing the arteriolar smooth muscle to relax by causing the release of a local vasoactive chemical mediator from the endothelial cells that induces relaxation of the underlying smooth muscle.

4. a. Because activation of $\alpha_1$-adrenergic receptors in vascular smooth muscle brings about vasoconstriction, blockage of $\alpha_1$-adrenergic receptors reduces vasoconstrictor activity, thereby lowering the total peripheral resistance and arterial blood pressure.

   b. Because activation of $\beta_1$-adrenergic receptors, which are found primarily in the heart, increases the rate and strength of cardiac contraction, drugs that block $\beta_1$-adrenergic receptors reduce cardiac output and thus arterial blood pressure by decreasing the rate and strength of the heartbeat.

   c. Drugs that directly relax arteriolar smooth muscle lower arterial blood pressure by promoting arteriolar vasodilation and reducing total peripheral resistance.

   d. Diuretic drugs reduce the plasma volume, thereby lowering arterial blood pressure, by increasing urinary output. Salt and water that normally would have been retained in the plasma are excreted in the urine.

   e. Because sympathetic activity promotes generalized arteriolar vasoconstriction, thereby increasing total peripheral resistance and arterial blood pressure, drugs that block the release of norepinephrine from sympathetic endings lower blood pressure by preventing this vasoconstrictor effect.

   f. Similarly, drugs that act on the brain to reduce sympathetic output lower blood pressure by preventing the effect of sympathetic activity on promoting arteriolar vasoconstriction and the resultant increase in total peripheral resistance and arterial blood pressure.

   g. Drugs that block $Ca^{2+}$ channels reduce the entry of $Ca^{2+}$ into the vascular smooth muscle cells from the ECF in response to excitatory input. Because the level of contractile activity in vascular smooth muscle cells depends on their cytosolic $Ca^{2+}$ concentration, drugs that block $Ca^{2+}$ channels reduce the contractile activity of these cells by reducing $Ca^{2+}$ entry and lowering their cytosolic $Ca^{2+}$ concentration. Total peripheral resistance and, accordingly, arterial blood pressure are decreased as a result of reduced arteriolar contractile activity.

   h. Drugs that interfere with the production of angiotensin II block activation of the hormonal pathway that promotes salt and water conservation (the renin–angiotensin–aldosterone system). As a result, more salt and water are lost in the urine, and less fluid is retained in the plasma. The resultant reduction in plasma volume lowers the arterial blood pressure.

## Clinical Consideration

(Question on p. 431.)

The abnormally elevated levels of epinephrine found with a pheochromocytoma bring about secondary hypertension by (1) increasing the heart rate;

(2) increasing cardiac contractility, which increases stroke volume; (3) causing venous vasoconstriction, which increases venous return and subsequently stroke volume by means of the Frank–Starling mechanism; and (4) causing arteriolar vasoconstriction, which increases total peripheral resistance. Increased heart rate and stroke volume both lead to increased cardiac output. Increased cardiac output and increased total peripheral resistance both lead to increased arterial blood pressure.

# Chapter 10 The Blood

## Check Your Understanding

**10.1** (Questions on p. 436.)

1. See ⟩ Figure 10-1, Haematocrit and types of blood cells, p. 435.

2. Plasma proteins (1) exert an osmotic effect important in the distribution of ECF between the vascular and interstitial compartments, (2) buffer pH changes, (3) transport many substances that are poorly soluble in plasma, (4) include clotting factors, (5) include inactive precursor molecules, and (6) include antibodies.

**10.2** (Questions on p. 444.)

1. Four anatomic features of erythrocytes contribute to the efficiency with which they transport $O_2$. (1) Most importantly, they are completely full of $O_2$-carrying haemoglobin, and 98.5 percent of the $O_2$ in the blood is bound to haemoglobin. (2) Their unique biconcave shape provides a larger surface area for diffusion of $O_2$ across the membrane than would a spherical cell of the same volume. (3) Their thinness enables $O_2$ to diffuse rapidly from the exterior to the innermost regions of the cell. (4) Their membrane flexibility enables them to deform so that they can squeeze through capillaries less than half their diameter on their $O_2$-delivery route.

2. $O_2$, $CO_2$, $H^+$, CO, and NO

3. *Erythropoietin*, which the kidneys secrete into the blood in response to reduced $O_2$ delivery, stimulates increased erythrocyte production by the red bone marrow.

**10.3** (Questions on p. 447.)

1. See ⟩ Figure 10-8, Peripheral blood smear, p. 445.

2. Neutrophils are phagocytic specialists, act like suicide bombers by releasing neutrophil extracellular traps (NETs) during a unique type of programmed cell death, are the first defenders on the scene of bacterial invasion, and are important in inflammatory responses. Eosinophils are important in allergic reactions and attack parasitic worms. Basophils release histamine that is important in allergic reactions and heparin that speeds up removal of fat particles from the blood. Monocytes leave the blood and set up residence in tissues throughout the body, where they mature into large tissue phagocytes known as macrophages. B lymphocytes produce antibodies. T lymphocytes accomplish cell-mediated immunity by releasing chemicals that punch holes in virally invaded body cells and in cancer cells.

**10.4** (Questions on p. 454.)

1. The plasma protein von Willebrand factor (vWF) adheres to exposed collagen at a vessel defect. Circulating platelets attach to binding sites on vWF. Collagen activates the bound platelets, which release ADP that causes passing-by platelets to become sticky and pile on top of the growing *platelet plug* at the defect site in a positive-feedback fashion.

2. See ⟩ Figure 10-14, Clot pathways, p. 451.

## Objective Questions

(Questions on pp. 455–456.)

**1.** T **2.** F **3.** T **4.** T **5.** F **6.** lymphocytes **7.** liver **8.** d **9.** a **10.** 1. c, 2. f, 3. b, 4. a, 5. g, 6. d, 7. h, 8. e, 9. f 11. 1. e, 2. c, 3. b, 4. d, 5. g, 6. f, 7. a, 8. h

## Quantitative Exercises

(Questions on p. 456.)

1. a. $(15\,g)/(100\,mL) = (150\,g/L)$
   $(150\,g/L) \times (1\,mol/66 \times 10^3\,g) = 2.27\,mM$
   b. $(2.27\,mM) \times (4\,O_2/Hb) = 9.09\,mM$

**c.** $(9.09 \times 10^{-3} \text{ mol } O_2/\text{L blood}) \times (22.4 \text{L } O_2/\text{l mol } O_2) = 204 \text{ mL } O_2/\text{L}$ blood

2. Normal blood contains $5 \times 10^9$ RBCs/mL.

   Normal blood volume is 5 L.

   Thus, a normal person has $(5 \times 10^9 \text{ RBCs/mL}) \times (5000 \text{ mL}) = 25 \times 10^{12}$ RBCs.

   The normal haematocrit (Ht) is 45 percent, whereas the anaemic has a Ht of 30 percent. This represents a loss of 1/3 of the RBCs, that is, $8.3 \times 10^{12}$ RBCs. If RBCs are produced at a rate of $3 \times 10^6$ RBCs/ sec, then the time to re-establish the Ht is $8.3 \times 10^{12}$ RBCs/($3310^6$ RBCs/sec) = $2.77310^6$ sec = 32 days.

   Thus, it takes about a month to replace a haemorrhagic loss of RBCs of this magnitude.

3. $v = 1.5 \times \exp(2h)$; calculate $v$ for $h = 0.4$ and $h = 0.7$.

   When $h = 0.4$, $v = 1.5 \times \exp(0.8) = 3.3$.

   When $h = 0.7$, $v = 1.5 \times \exp(1.4) = 6.1$.

   $6.1/3.3 = 1.85$, that is, an 85 percent increase in viscosity. Because resistance is directly proportional to viscosity, the resistance will also increase by 85 percent.

## Points to Ponder

(Questions on p. 456.)

1. No, you cannot conclude that a person with a haematocrit of 62 definitely has polycythaemia. With 62 percent of the whole-blood sample consisting of erythrocytes (normal being 45%), the number of erythrocytes compared to the plasma volume is definitely elevated. However, the person *may* have polycythaemia, in which the number of erythrocytes is abnormally high, or may be dehydrated, in which case a normal number of erythrocytes is concentrated in a smaller-than-normal plasma volume.

2. If the genes that direct fetal haemoglobin-F synthesis could be reactivated in a patient with sickle cell anaemia, a portion of the abnormal haemoglobin S that causes the erythrocytes to warp into defective sickle-shaped cells would be replaced by "healthy" haemoglobin F, thus sparing a portion of the RBCs from premature rupture. Haemoglobin F would not completely replace haemoglobin S because the gene for synthesis of haemoglobin S would still be active.

3. Most heart attack deaths are attributable to the formation of abnormal clots that prevent normal blood flow. The sought-after chemicals in the "saliva" of bloodsucking creatures are particular agents that break up or prevent the formation of these abnormal clots.

   Although genetically engineered tissue–plasminogen activator (tPA) is already being used as a clot-busting drug, this agent brings about degradation of fibrinogen as well as fibrin. Thus, even though the life-threatening clot in the coronary circulation is dissolved, the fibrinogen supplies in the blood are depleted for up to 24 hours until new fibrinogen is synthesized by the liver. If the patient sustains a ruptured vessel in the interim, insufficient fibrinogen might be available to form a blood-staunching clot. For example, many patients treated with tPA suffer haemorrhagic strokes within 24 hours of treatment due to incomplete sealing of a ruptured cerebral vessel. Therefore, scientists are searching for better alternatives to combat abnormal clot formation by examining the naturally occurring chemicals produced by bloodsucking creatures that permit them to suck a victim's blood without the blood clotting.

4. When considering the symptoms of porphyria, one could imagine how tales of vampires—blood-craving, hairy, fanged, monstrous-looking creatures who roamed in the dark and were warded off by garlic— might easily have evolved from people's encounters with victims of this condition. This possibility is especially likely when considering how stories are embellished and distorted as they get passed along by word of mouth.

## Clinical Consideration

(Question on p. 456.)

Because the white blood cell count is within the normal range, the patient's pneumonia is most likely not caused by a bacterial infection. Bacterial infections are typically accompanied by an elevated total white blood cell count and an increase in percentage of neutrophils. Therefore, the pneumonia is probably caused by a virus. Because antibiotics are more useful in combating bacterial than viral infections, antibiotics are not likely to be useful in combating this patient's pneumonia.

# Chapter 11 Body Defences

## Check Your Understanding

**11.1** (Questions on p. 464.)

1. The skin consists of an outer *epidermis* and inner *dermis*. The epidermis has an inner layer of living cube-shaped cells and an outer keratinized layer of dead, flattened cells. The epidermis has no direct blood supply, instead it gets its nourishment from the underlying dermis, a connective tissue layer that has an abundance of blood vessels. The skin is anchored to muscle or bone by the *hypodermis*: a layer of loose connective tissue that often contains an abundance of fat cells, in which case it is called *adipose tissue*.

2. (1) *Melanocytes* produce the pigment melanin, which is responsible for different skin colours. (2) *Keratinocytes* produce keratin, which gives rise to the outer, protective, keratinized layer of the epidermis. (3) *Langerhans cells* present antigen to helper T cells. (4) *Granstein cells* are immune-suppressive cells.

**11.2** (Questions on p. 466.)

1. *Immunity* is the body's ability to resist or eliminate potentially harmful foreign materials or abnormal cells.

2. *Innate immune responses* are inherent defense mechanisms that nonselectively defend against foreign or abnormal material of any type, even on initial exposure to it. *Adaptive immune responses* are selectively targeted against a particular foreign material to which the body has already been exposed and had an opportunity to prepare for an attack aimed discriminatingly at the enemy.

**11.3** (Questions on p. 473.)

1. (1) *Inflammation* is a nonspecific response to foreign invasion or tissue damage mediated largely by phagocytes and macrophages that destroy or incapacitate the invaders, remove debris, and prepare for subsequent healing and repair. (2) *Interferon* is released by virally invaded cells and, nonspecifically, transiently interferes with the replication of the same or unrelated viruses in other host cells. (3) *Natural killer cells* nonspecifically destroy virally invaded cells and cancer cells by releasing chemicals that directly lyse these cells on first exposure to them. (4) The *complement system* is a group of inactive plasma proteins that, when sequentially activated, destroy foreign cells by forming holes in their plasma membranes.

2. The gross manifestations of inflammation are redness, heat, swelling, and pain. Redness and heat are due to the increased blood flow. Fluids accumulate, and this causes the swelling. Pain is caused by chemicals that activate pain receptors.

3. *Cytokines* refer to all the chemicals, other than antibodies, secreted by leukocytes. Cytokines help enhance the function of other immune cells as well as many other immune functions.

**11.4** (Questions on p. 475.)

1. *Antibody mediated immunity* is accomplished by B lymphocytes, and *cell-mediated immunity* is accomplished by T lymphocytes.

2. An *antigen* is a large, foreign, unique molecule that triggers a specific immune response against itself.

**11.5** (Questions on p. 481.)

1. An *antibody* is Y-shaped. The *Fab* regions on the tip of each arm are variable among antibodies and determine with which specific antigen the antibody can bind in a lock-and-key fashion. The *Fc* tail region, which is constant for a given antibody class, binds with a particular antibody mediator, thereby determining what the antibody does once it binds with antigen.

2. neutralization of bacterial toxins, agglutination, activation of the complement system, enhancement of phagocytosis by acting as opsonins, and stimulation of natural killer cells

3. See ⟩ Figure 11-13, Clonal selection theory, p. 479.

**11.6** (Questions on p. 491.)

1. *Cytotoxic*, or *killer, T cells* destroy virus-invaded host cells and cancer cells by secreting (1) perforin, which forms hole-punching complexes in the victim cell; or (2) granzymes, which trigger the victim cell to self-destruct through apoptosis. *Helper T cells* secrete cytokines that amplify the activities of other immune cells. *Regulatory T cells* suppress immune responses; they inhibit both innate and adaptive immune responses in check-and-balance fashion to minimize harmful immune pathology.

2. *Antigen-presenting cells* process and present antigen, complexed with MHC molecules (self-antigens), on their surface to T cells. T cells cannot interact with antigen without this "formal introduction."

3. In the process of *immune surveillance*, natural killer cells, cytotoxic T cells, macrophages, and the interferon they collectively secrete normally eradicate newly arisen cancer cells before they have a chance to multiply and spread.

**11.7** (Questions on p. 494.)

1. See ▌Table 11-3, Immediate versus Delayed Hypersensitivity Reactions.

2. In contrast to *IgG antibodies*, which freely circulate and immediately amplify innate defense mechanisms on binding with their specific antigen, *IgE antibodies* specific for different antigens attach by their tail portions to mast cells and basophils in the absence of antigen. The binding of an appropriate antigen triggers the rupture of the cell's granules, which contain histamine and other inflammatory chemical mediators that spew forth into the surrounding tissue, where they cause the allergic response.

## Objective Questions

(Questions on pp. 496–497.)

**1.** F **2.** F **3.** F **4.** F **5.** T **6.** toll-like receptors **7.** membrane-attack complex **8.** pus **9.** inflammation **10.** opsonin **11.** cytokines **12.** b **13.** 1. c, 2. d, 3. a, 4. b **14.** 1. a, 2. a, 3. b, 4. b, 5.c, 6. c, 7. b, 8. a, 9. b, 10. b, 11. a, 12. b **15.** 1. b, 2. a, 3. a, 4. b, 5. a, 6. a, 7. a, 8. b

## Quantitative Exercise

(Question on p. 497.)

1. $NEP$ = net outward pressure − net inward pressure

$NEP = (P_C + \pi_{IF}) - (P_{IF} + \pi_P)$

Note for this problem, $(P_{IF} + \pi_P) = (25\,mmHg + 1\,mmHg) = 26\,mmHg$, is constant for all cases.

## Points to Ponder

(Questions on p. 497.)

1. Against bacteria, defences begin with the innate immune mechanism of inflammation. Resident macrophages engulf invading bacteria, while histamine-induced vascular responses increase blood flow. This increased blood flow brings additional immune-effector cells such as neutrophils and monocytes to help engulf and destroy foreign invaders and debris. These innate immune mechanisms are coupled with adaptive immune mechanisms. Plasma cells secrete customized antibodies, which specifically bind to invading bacteria, further leading to the bacteria's destruction.

2. A vaccine against a particular microbe can be effective only if it induces formation of antibodies and/or activated T cells against a stable antigen that is present on all microbes of this type. It has not been possible to produce a vaccine against HIV because it frequently mutates. Specific immune responses induced by vaccination against one form of HIV may prove to be ineffective against a slightly modified version of the virus.

3. Failure of the thymus to develop would lead to an absence of T lymphocytes and no cell-mediated immunity after birth. This outcome would seriously compromise the individual's ability to defend against viral invasion and cancer.

4. Researchers are working on ways to "teach" the immune system to view foreign tissue as "self" as a means to prevent the immune system of an organ-transplant patient from rejecting the foreign tissue, while leaving the patient's capacity for immune defence fully intact. The immunosuppressive drugs now used to prevent transplant rejection cripple the recipients' immune defence systems and leave the patients more vulnerable to microbial invasion.

5. The skin cells visible on the body's surface are all dead.

## Clinical Consideration

(Questions on p. 497.)

1. The spleen primarily acts as a blood filter. It is responsible for the removal of old blood cells and platelets, the storage of red blood cells, and the production, storage, and release of lymphocytes and macrophages.

2. Patients who have had their spleen—an important element of the immune system—removed can develop infections much quicker than a person with a spleen. The severity of infections also tends to be worse, such that infections that would normally be considered mild can become life threatening.

3. Christine's platelet count will increase. Because the spleen normally removes old platelets from the blood, her platelet count can increase above normal and increase the risk of blood clots.

# Chapter 12 The Respiratory System

## Check Your Understanding

**12.1** (Questions on p. 503.)

1. See ❭ Figure 12-3, Schematic of airways, p. 503.

2. Only 0.5 μm separates the air in the alveoli from the blood in the pulmonary capillaries, and the alveolar air-blood interface presents a tremendous surface area (75 m²) for exchange. The thinness and extensive surface area of the alveolar membrane facilitate gas exchange because the rate of diffusion is inversely proportional to the thickness and directly proportional to the surface area of this interface.

3. *Type I alveolar cells* form the walls of the alveoli.

**12.2** (Questions on p. 505.)

1. The three factors are the pressure gradient, resistance to flow, and the inertia of the system. Of these, the first two are most important.

**12.3** (Questions on p. 508.)

1. At residual volume, the deflating actions of the lungs do not oppose the inflating actions of the chest wall, so the pressure reflects the passive inflation of the chest wall. At functional residual capacity, the pressure of the lungs to deflate is equal to the pressure of the chest wall to inflate, so the net pressure is 0. At total lung capacity, both the lungs and the chest wall exert pressure to deflate.

2. At functional residual capacity, the need to generate pressure is minimized, which decreases the amount of work (energy) needed to breathe.

**12.4** (Questions on p. 510.)

1. The forces that keep the alveoli open are the transmural pressure gradient and pulmonary surfactant (which opposes alveolar surface tension), and the forces that promote alveolar collapse are elastic recoil and alveolar surface tension.

**12.5** (Questions on p. 514.)

1. Alveolar pressure must decrease to below atmospheric pressure.

2. During inspiration pleural pressure decreases, transpulmonary pressure decreases, whereas alveolar pressure initially decreases and then increases to 0 again. During expiration, pleural pressure increases, transpulmonary pressure increases, and alveolar pressure first increases then decreases to 0.

**12.6** (Questions on p. 520.)

1. In the case of obstructive disease, lung volume is greater and peak flow is achieved earlier, but it is less than for a normal person, and forced vital capacity is less than expected. In the case of restrictive lung disease, lung volume is smaller and peak flow is reduced.

2. With an exacerbation of obstructive lung disease, a person cannot breathe out due to airway collapse. This results in trapping gas and increasing lung volume.

**12.7** (Questions on p. 523.)

1. To increase alveolar ventilation during exercise you need to breathe faster and generate a higher tidal volume. However, this capability is limited by the delivery of oxygen to the respiratory muscles; without enough oxygen delivery, fatigue will occur.

**12.8** (Questions on p. 525.)

1. See ▌ Table 12.6, Partial Pressures of Gases in Different Parts of the Respiratory System.

**12.9** (Questions on p. 529.)

1. See ▌ Table 12-7, Factors that Influence the Rate of Gas Transfer across the Alveolar-Capillary Membrane.

**12.10** (Questions on p. 530.)

1. If ventilation to a region is not sufficient, there will be a regional decrease in $P_{O_2}$. This will cause a constriction of the vascular smooth muscle, thereby increasing vascular resistance and decreasing blood flow, which allows a rise in $P_{O_2}$. If $P_{CO_2}$ is high, it will cause a dilation of the regional airways to increase regional ventilation to decrease $P_{CO_2}$.

**12.11** (Questions on p. 539.)

1. Oxygen can be transported both as dissolved gas (1–2%) and bound to haemoglobin (98–99%). Carbon dioxide can be transported as dissolved gas (5–10%), bound to haemoglobin (5–10%), or as bicarbonate (80–90%).

**12.12** (Questions on p. 541.)

1. See ❯ Figure 12-37, Effects of hyperventilation and hypoventilation on arterial $P_{O_2}$ and $P_{CO_2}$, p. 540.

**12.13** (Questions on p. 543.)

1. The discharge of motor neurons that innervate pump muscles change with inspiration and expiration. In contrast, the discharge patterns for motor neurons that innervate muscles controlling the diameter of the upper airway are relatively constant.

2. When the chemical drive to breathing is increased, motor neurons innervating muscles of the upper airway are recruited, and their discharge pattern resembles those for the pump neurons.

**12.14** (Questions on p. 549.)

1. Potential factors are $P_{O_2}$, $P_{CO_2}$, arterial $H^+$. $P_{CO_2}$ is likely the most important due to its actions on the central chemoreceptors of the respiratory centre.

## Objective Questions

(Questions on p. 552.)

**1.** F **2.** F **3.** T **4.** F **5.** F **6.** F **7.** F **8.** transmural pressure gradient, pulmonary surfactant action, alveolar interdependence **9.** pulmonary elasticity, alveolar surface tension **10.** compliance **11.** elastic recoil **12.** carbonic anhydrase **13.** a **14.** a. <, b. >, c. =, d. =, e. =, f. =, g. >, h. <, i. approximately =, j. approximately =, k. =, l. = **15.** 1. d, 2. a, 3. b, 4. a, 5. b, 6. a

## Quantitative Exercises

(Questions on p. 553.)

For general reference for questions 1 and 2:

$$P_{A_{O_2}} = P_{I_{O_2}} - (\dot{V}_{O_2}/\dot{V}_A) \times 863 \, mmHg$$

$$P_{A_{O_2}} = (\dot{V}_{CO_2}/\dot{V}_A) \times 863 \, mmHg$$

**1.** $\dot{V}_A = 3 \, L/min$

$\dot{V}_{O_2} = 0.3 \, L/min, R = 1,$ therefore $\dot{V}_{CO_2} = 0.3 \, L/min$

$P_{A_{CO_2}} = (0.3 \, L/min / 3 \, L/min) \times 863 \, mmHg = 86.3 \, mmHg$

**2. a.** $380 \, mmHg \times 0.21 = 79.8 \, mmHg$

**b.** $P_{A_{O_2}} = 79.8 \, mmHg - (0.06) \times 431.5 \, mmHg$

$= 79.8 \, mmHg - 25.8 \, mmHg$

$= 57 \, mmHg$

**c.** $P_{A_{CO_2}} = (0.2 \, L/min / 4.2 \, L/min) \times 431.5 \, mmHg$

$= 20.5 \, mmHg$

**3.** $V_T = 350 \, mL, f = 12/min, \dot{V}_A = 0.8 \times \dot{V}_E, V_D = ?$

$\dot{V}_A = f \times (V_T - V_D)$

$\dot{V}_E = f \times V_T$

$0.8 = \dot{V}_A/\dot{V}_E = f \times [(V_T - V_D)]/(f \times V_T)$

$\quad = 1 - (V_D/V_T)$

$0.8 = 1 - V_D/350 \, mL)$

$V_D/350 _{mL} = 0.2$

$V_D = 0.2(350 \, mL) = 70 \, mL$

## Points to Ponder

(Questions on p. 554.)

1. Total atmospheric pressure decreases with increasing altitude, yet the percentage of $O_2$ in the air remains the same. At an altitude of 9144 m (30 000 feet), the atmospheric pressure is only 226 mmHg. Because 21 percent of atmospheric air consists of $O_2$, the $P_{O_2}$ of inspired air at 9144 m is only 47.5 mmHg, and alveolar $P_{O_2}$ is even lower, at about 20 mmHg. At this low $P_{O_2}$, haemoglobin is only about 30 percent saturated with $O_2$; this is much too low to sustain tissue needs for $O_2$.

   The $P_{O_2}$ of inspired air can be increased by two means when flying at high altitude. First, by pressurizing the plane's interior to a pressure comparable to that of atmospheric pressure at sea level, the $P_{O_2}$ of inspired air within the plane is 21 percent of 760 mmHg, or the normal 160 mmHg. Accordingly, alveolar and arterial $P_{O_2}$ and percentage of Hb saturation are likewise normal. In the emergency situation of failure to maintain internal cabin pressure, breathing pure $O_2$ can raise the $P_{O_2}$ considerably above that accomplished by breathing normal air. When a person is breathing pure $O_2$, the entire pressure of inspired air is attributable to $O_2$. For example, with a total atmospheric pressure of 226 mmHg at an altitude of 9144 m, the $P_{O_2}$ of inspired pure $O_2$ is 226 mmHg, which is more than adequate to maintain normal arterial Hb saturation.

2. **a.** Hypercapnia would not accompany the hypoxia associated with cyanide poisoning. In fact, $CO_2$ levels decline, because oxidative metabolism is blocked by the tissue poisons so that $CO_2$ is not being produced.

   **b.** Hypercapnia could but may not accompany the hypoxia associated with pulmonary oedema. Pulmonary diffusing capacity is reduced in pulmonary oedema, but $O_2$ transfer suffers more than $CO_2$ transfer because the diffusion coefficient for $CO_2$ is 20 times that for $O_2$. As a result, hypoxia occurs much more readily than hypercapnia in these circumstances. Hypercapnia does occur, however, when pulmonary diffusing capacity is severely impaired.

   **c.** Hypercapnia would accompany the hypoxia associated with restrictive lung disease because ventilation is inadequate to meet the metabolic needs for both $O_2$ delivery and $CO_2$ removal. Both $O_2$ and $CO_2$ exchange between the lungs and atmosphere are equally affected.

   **d.** Hypercapnia would not accompany the hypoxia associated with high altitude. In fact, arterial $P_{CO_2}$ levels actually decrease. One of the compensatory responses in acclimatization to high altitudes is reflex stimulation of ventilation as a result of the reduction in arterial $P_{O_2}$. This compensatory hyperventilation to obtain more $O_2$ blows off too much $CO_2$ in the process, so arterial $P_{CO_2}$ levels decline below normal.

   **e.** Hypercapnia would not accompany the hypoxia associated with severe anaemia. Reduced $O_2$-carrying capacity of the blood has no influence on blood $CO_2$ content, so arterial $P_{CO_2}$ levels are normal.

   **f.** Hypercapnia would accompany the circulatory hypoxia associated with congestive heart failure. Just as the diminished blood flow fails to deliver adequate $O_2$ to the tissues, it also fails to remove sufficient $CO_2$.

   **g.** Hypercapnia would accompany the hypoxia associated with obstructive lung disease, because ventilation would be inadequate to meet the metabolic needs for both $O_2$ delivery and $CO_2$ removal. Both $O_2$ and $CO_2$ exchange between the lungs and atmosphere would be equally affected.

3. $P_{CO_2} = 122 \, mmHg$

   0.21(atmospheric pressure − partial pressure of $H_2O$)

   $= 0.21(630 \, mmHg - 47 \, mmHg)$

   $= 0.21(583 \, mmHg) = 122 \, mmHg$

4. Voluntarily hyperventilating before going underwater lowers the arterial $P_{CO_2}$ but does not increase the $O_2$ content in the blood. Because the $P_{CO_2}$ is below normal, the swimmers can hold their breath longer than usual before the arterial $P_{O_2}$ increases to the point that they are driven to surface for a breath. Therefore, a swimmer can stay underwater longer. The risk, however, is that the $O_2$ content of the blood, which was normal, not increased, before going underwater, continues to fall. Therefore, the $O_2$ level in the

blood can fall dangerously low before the $CO_2$ level builds to the point of driving the swimmer to take a breath. Low arterial $P_{O_2}$ does not stimulate respiratory activity until it has fallen to 60 mmHg. Meanwhile, the person may lose consciousness (known as shallow water blackout) and drown due to inadequate $O_2$ delivery to the brain. If the person does not hyperventilate so that both the arterial $P_{CO_2}$ and $O_2$ content are normal before going underwater, the buildup of $CO_2$ will drive the person to the surface for a breath before the $O_2$ levels fall to a dangerous point.

5. c. The arterial $P_{O_2}$ will be less than the alveolar $P_{CO_2}$, and the arterial will be greater than the alveolar $P_{CO_2}$. Because pulmonary diffusing capacity is reduced, arterial $P_{CO_2}$ and $P_{O_2}$ do not equilibrate with alveolar $P_{CO_2}$ and $P_{O_2}$.

If the person is administered 100 percent $O_2$, the alveolar $P_{O_2}$ will increase, and the arterial $P_{O_2}$ will increase accordingly. Even though $P_{O_2}$ arterial will not equilibrate with alveolar $P_{O_2}$, it will be higher than when the person is breathing atmospheric air.

The arterial $P_{CO_2}$ will remain the same whether the person is administered 100 percent $O_2$ or is breathing atmospheric air. The alveolar and thus the blood-to-alveolar $P_{CO_2}$ gradient are not changed by breathing 100 percent $O_2$ because the $P_{CO_2}$ in atmospheric air and 100 percent $O_2$ are both essentially zero ( $P_{CO_2}$ in atmospheric air 0.3 mmHg).

## Clinical Consideration

(Question on p. 554.)

Emphysema is characterized by a collapse of the smaller respiratory airways and a breakdown of alveolar walls. Because of the collapse of smaller airways, airway resistance is increased with emphysema. As with other chronic obstructive pulmonary diseases, expiration is impaired to a greater extent than inspiration because airways are naturally dilated slightly more during inspiration than expiration as a result of the greater transmural pressure gradient during inspiration. Because airway resistance is increased, a patient with emphysema must produce larger-than-normal intra-alveolar pressure changes to accomplish a normal tidal volume. Unlike quiet breathing in a normal person, the accessory inspiratory muscles (neck muscles) and the muscles of active expiration (abdominal muscles and internal intercostal muscles) must be brought into play to inspire and expire a normal tidal volume of air.

The spirogram would be characteristic of chronic obstructive pulmonary disease. Because the patient experiences more difficulty emptying the lungs than filling them, the total lung capacity would be essentially normal, but the functional residual capacity and the residual volume would be elevated as a result of the additional air trapped in the lungs following expiration. Because the residual volume is increased, the inspiratory capacity and vital capacity will be reduced. Also, the $FEV_1$ will be markedly reduced because the expiratory flow is decreased by the airway obstruction. The $FEV_1$-to-vital capacity ratio will be much lower than the normal 80 percent.

Because of the reduced surface area for exchange as a result of a breakdown of alveolar walls, gas exchange would be impaired. Therefore, arterial $P_{CO_2}$ would be elevated and arterial $P_{O_2}$ reduced compared to normal.

Paradoxically, administering $O_2$ to this patient to relieve his hypoxic condition would markedly depress his drive to breathe by elevating the arterial $P_{O_2}$ and removing the primary driving stimulus for respiration. In addition, changes that occur in the distribution of ventilation and perfusion impair gas exchange. Because of this danger, $O_2$ therapy should either not be administered or administered extremely cautiously.

# Chapter 13 The Urinary System

## Check Your Understanding

**13.1** (Questions on p. 566.)
1. See ❯ Figure 13-6, Basic renal processes, p. 564.
2. *Cortical nephrons* are the most abundant type of nephron. The *juxtamedullary nephrons* are important in establishing the medullary vertical osmotic gradient. The glomeruli of cortical nephrons lie in the outer cortex, whereas the glomeruli of juxtamedullary nephrons lie in the

inner part of the cortex next to the medulla. The loops of Henle of cortical nephrons dip only slightly into the medulla, but the juxtamedullary nephrons have long loops of Henle that plunge deep into the medulla. The juxtamedullary nephrons' peritubular capillaries form hairpin loops known as vasa recta. See ❯ Figure 13-5, Comparison of juxtamedullary and cortical nephrons, p. 563.
3. At rest there is more than adequate renal blood flow to supply $O_2$ to the kidneys. However, during exercise, other organs, such as skeletal muscle, have a higher requirement for blood flow, and the temporary shunting of blood flow to those regions does not greatly impact the kidneys. As well, decreased renal blood flow decreases urine production.

**13.2** (Questions on p. 573.)
1. See ❙ Table 13-1, Forces Involved in Glomerular Filtration.
2. Extrinsic control of the GFR is accomplished by sympathetic nervous system input to the afferent arterioles aimed at regulating arterial blood pressure. When arterial blood pressure falls, sympathetically induced afferent arteriolar vasoconstriction decreases glomerular capillary blood pressure, causing a decrease in GFR. The resultant reduction in urine volume conserves fluid and salt that would have been lost from the body in urine. This saved fluid helps restore plasma volume and blood pressure to normal in the long term. The opposite compensatory effect takes place when arterial blood pressure is too high.

**13.3** (Questions on p. 582.)
1. See ❯ Figure 13-14, Steps of transepithelial transport, p. 574.
2. The kidneys secrete the enzymatic hormone renin in response to reduced NaCl, ECF volume, and arterial blood pressure. Renin activates angiotensinogen, a plasma protein produced by the liver, into angiotensin I, which is converted into angiotensin II by angiotensin-converting enzyme produced in the lungs. Angiotensin II stimulates the adrenal cortex to secrete the hormone aldosterone, which stimulates $Na^+$ reabsorption by the kidneys. The resulting retention of $Na^+$ exerts an osmotic effect that holds more $H_2O$ in the ECF. Together, the conserved $Na^+$ and $H_2O$ help correct the original stimuli that activated this hormonal system.
3. For phosphate, the $T_m$ is fixed at its normal plasma concentration, whereas for glucose the $T_m$ is greater than the normal plasma concentration. Any excess phosphate in the plasma is immediately secreted. However, with slight increases in plasma glucose concentrations, there is no increase in glucose excretion until the $T_m$ is reached.

**13.4** (Questions on p. 585.)
1. Secretion of $H^+$, $K^+$, and organic anions and cations
2. The removal of foreign organic compounds from the body is hastened via secretion by the organic ion secretory systems in the proximal tubule. The liver helps rid the body of foreign compounds that are not ionic in their original form by converting them into an anionic form.

**13.5** (Questions on p. 597.)
1. (1) clearance rate = GFR, (2) clearance rate, GFR, (3) clearance rate. GFR
2. The loop of Henle establishes the gradient by reabsorbing Na and Cl, the vasa recta preserve the gradient by countercurrent exchange, and the collecting ducts use the gradient to concentrate urine.
3. *Vasopressin*, which is secreted into the blood by the posterior pituitary in response to a $H_2O$ deficit, increases the permeability of the distal and collecting tubules to $H_2O$. Vasopressin activates the cAMP pathway within the principal cells lining these tubules when it binds with V2 receptors specific for it at the basolateral membrane of these cells. cAMP increases the opposite luminal membrane's permeability to $H_2O$ by promoting the insertion of aquaporin water channels into the membrane. Water enters the cell from the tubular lumen through the inserted water channels and exits the cell through different, permanently positioned water channels at the basolateral border to enter the blood, in this way being reabsorbed to correct the $H_2O$ deficit.

## Objective Questions

(Questions on pp. 602–603.)
**1.** F **2.** F **3.** T **4.** T **5.** T **6.** nephron **7.** potassium **8.** 500 **9.** 1. b, 2. a, 3. b, 4. b, 5. a, 6. b, 7. b, 8. b, 9. b 10. e 11. b **12.** b, e, a, d, c **13.** c, e, d, a, b, f **14.** g, c, d, a, f, b, e **15.** 1. a, 2. a, 3. c, 4. b, 5.d

## Quantitative Exercises

(Questions on p. 603)

1.

|  | Patient 1 | Patient 2 |
|---|---|---|
| a. GFR | 125 mL/min | 124 mL/min |
| RPF | 620 mL/min | 400 mL/min |
| b. RBF | 1127 mL/min | 727 mL/min |
| c. FF | 0.20 | 0.31 |

d. All of patient 1's values are within the normal range. Patient 2's GFR is normal, but he has a low renal plasma flow and a high filtration fraction. Therefore, his GFR is too high for that RPF. This might imply enlarged filtration slits or a leaky glomerulus, in general. The low RPF may imply low renal blood pressure, perhaps from a partially blocked renal artery.

2. filtered load = GFR × plasma concentration

$$= (0.125 \text{ litre/min}) \times (145 \text{ mmol/litre})$$

$$= 8.125 \text{ mmol/min}$$

3. GFR = $(U \times [I]_U)/[I]_B$

$U = (\text{GFR} \times [I]_B)/[I]_U = (125 \text{ mL/min})(3 \text{ mg/L})/(300 \text{ mg/L})$

$= 1.25 \text{ mL/min}$

$$\text{Clearance rate of a substance} = \frac{\text{urine concentration of a substance} \times \text{urine flow rate}}{\text{plasma concentration of the substance}}$$

4. $= \dfrac{7.5 \text{ mg/mg} \times 2 \text{ mL/min}}{0.2 \text{ mg/mL}}$

$= 75 \text{ mL/min}$

Because a clearance rate of 75 mL/min is less than the average GFR of 125 mL/min, the substance is being reabsorbed.

## Points to Ponder

(Questions on p. 604.)

1. The longer loops of Henle in desert rats (known as *kangaroo rats*) permit a greater magnitude of countercurrent multiplication and thus a larger medullary vertical osmotic gradient. As a result, these rodents can produce urine that is concentrated up to an osmolarity of almost 6000 mOsm/L, which is five times more concentrated than maximally concentrated human urine at 1200 mOsm/L. Because of this tremendous concentrating ability, kangaroo rats never have to drink; the $H_2O$ produced metabolically within their cells during oxidation of foodstuff (food + $O_2 \rightarrow CO_2 + H_2O$ + energy) is sufficient for their needs.

2. a. 250 mg/min filtered

filtered load of substance = plasma concentration of substance × GFR

filtered load of substance = 200 mg/100 mL × 125 mL/min

$= 250 \text{ mg/min}$

b. 200 mg/min reabsorbed; a $T_m$'s worth of the substance will be reabsorbed

c. 50 mg/min excreted amount of substance excreted = amount of substance filtered − amount of substance reabsorbed = 250 mg/min − 200 mg/min = 50 mg/min

3. Aldosterone stimulates $Na^+$ reabsorption and $K^+$ secretion by the renal tubules. Therefore, the most prominent features of Conn's syndrome (hypersecretion of aldosterone) are hypernatraemia (elevated $Na^+$ levels in the blood) caused by excessive $Na^+$ reabsorption, hypokalemia (below-normal $K^+$ levels in the blood) caused by excessive $K^+$ secretion, and hypertension (elevated blood pressure) caused by excessive salt and water retention.

4. e. 300/300. If the ascending limb were permeable to water, it would not be possible to establish a vertical osmotic gradient in the interstitial fluid of the renal medulla, nor would the ascending-limb fluid become hypotonic before entering the distal tubule. As the ascending limb pumped NaCl into the interstitial fluid, water would osmotically follow, so both the interstitial fluid and the ascending limb would remain isotonic at 300 mOsm/L. With the tubular fluid entering the distal tubule being 300 mOsm/L, instead of the normal 100 mOsm/L, it would not be possible to produce urine with an osmolarity less than 300 mOsm/L. Likewise, in the absence of the medullary vertical osmotic gradient, it would not be possible to produce urine more concentrated than 300 mOsm/L, no matter how much vasopressin was present.

5. Because the descending pathways between the brain and the motor neurons that supply the external urethral sphincter and pelvic diaphragm are no longer intact, the accident victim can no longer voluntarily control micturition. Therefore, bladder emptying in this individual is governed entirely by the micturition reflex.

## Clinical Consideration

(Question on p. 604.)

prostate enlargement

# Chapter 14 Fluid and Acid-Base Balance

## Check Your Understanding

14.1 (Questions on p. 609.)

1. *inputs* = ingestion or metabolic production; *outputs* = excretion or metabolic consumption

2. *Stable balance*: when total body input of a substance equals its total body output. *Positive balance*: when the gains via input for a substance exceed its losses via output. *Negative balance*: when losses for a substance exceed its gains

14.2 (Questions on p. 619.)

1. See Table 14-1, Classification of Body Fluid

2. *ECF volume* is regulated by maintaining salt balance and is important in maintaining blood pressure. *ECF osmolarity* is regulated by maintaining free water balance and is important for preventing the swelling or shrinking of cells.

3. When the ECF is hypertonic, water osmotically leaves the cells, causing them to shrink. When the ECF is hypotonic, water osmotically enters the cells, causing them to swell.

14.3 (Questions on p. 637.)

1. Only a narrow pH range is compatible with life because even small changes in $[H^+]$ have dramatic effects on normal cell function. Changes in $[H^+]$ (1) lead to changes in excitability of nerve and muscle cells, which can cause death due to either severe acidosis by depressing the CNS or severe alkalosis by causing spasm of the respiratory muscles; (2) alter enzyme activity, thus disturbing life-supporting metabolic activity catalyzed by these enzymes; and (3) influence $[K^+]$, which can lead to fatal cardiac abnormalities.

2. metabolic acidosis. To compensate, the ECF buffers take up extra $H^+$, the lungs blow off additional $H^+$-generating $CO_2$, and the kidneys excrete more $H^+$ and conserve more $HCO_3^-$

## Objective Questions

(Questions on p. 638.)

1. T 2. F 3. F 4. T 5. T 6. intracellular fluid 7. $H_2CO_3$, $HCO_3^-$ 8. d 9. b 10. a, d, e 11. b, e 12. c 13. 1. metabolic acidosis, 2. diabetes mellitus, 3. pH = 7.1, 4. respiratory alkalosis, 5. anxiety, 6. pH = 7.7, 7. respiratory acidosis, 8. pneumonia, 9. pH = 7.1, 10. metabolic alkalosis, 11. vomiting, 12. pH = 7.7

## Quantitative Exercises

(Questions on p. 639.)

1. pH = 6.1 + log$[HCO_3^-]/(0.03 \text{ mM/mmHg} \times 40 \text{ mmHg})$

$7.4 = 6.1 + \log[HCO_3^-]/1.2 \text{ mM}$

$\log[HCO_3^-]/1.2 \text{ mM} = 7.4 - 6.1 = 1.3$

$[HCO_3^-] = 1.2 \text{ mM} \times (10^{1.3}) = 24 \text{ mM}$

| Ingested Fluid | Compartment | Size of Compartment before Ingestion (litres) | Size of Compartment after Ingestion (litres) | Percent Increase in Size of Compartment after Ingestion |
|---|---|---|---|---|
| **Distilled** | TBW | 42 | 43 | 2 |
| **Water** | ICF (2/3 TBW) | 28 | 28.667 | 2 |
| | ECF (1/3 TBW) | 14 | 14.333 | 2 |
| | plasma (20% ECF) | 2.8 | 2.866 | 2 |
| | ISF (80% ECF) | 11.2 | 11.466 | 2 |
| **Saline** | TBW | 42 | 43 | 2 |
| | ICF | 28 | 28 | 0 |
| | ECF | 14 | 15 | 7 |
| | plasma | 2.8 | 3 | 7 |
| | ISF | 11.2 | 12 | 7 |

2. $pH = -\log[H^+], [H^+] = 10^{-pH}$

$[H^+] = 10^{-6.8} = 158\,nM$ for $pH = 6.8$

$[H^+] = 10^{-8.0} = 10\,nM$ for $pH = 8.0$

3. Note that distilled water is permeable across all barriers, so it distributes equally among all compartments. However, the saline does not enter cells, so it stays in the ECF. The resultant distributions are summarized in the chart below. Clearly, saline is better at expanding the plasma volume.

## Points to Ponder

(Questions on p. 639.)

1. The rate of urine formation increases when alcohol inhibits vasopressin secretion and the kidneys are unable to reabsorb water from the distal and collecting tubules. Because extra free water that normally would have been reabsorbed from the distal parts of the tubule is lost from the body in the urine, the body becomes dehydrated, and the ECF osmolarity increases following alcohol consumption. That is, more fluid is lost in the urine than is consumed in the alcoholic beverage as a result of alcohol's action on vasopressin. Thus, the imbibing person experiences a water deficit and still feels thirsty, despite the recent fluid consumption.

2. If a person loses 1500 mL of salt-rich sweat and drinks 1000 mL of water without replacing the salt during the same time period, there will still be a volume deficit of 500 mL, and the body fluids will have become hypotonic (the remaining salt in the body will be diluted by the ingestion of 1000 mL of free water). As a result, the hypothalamic osmoreceptors (the dominant input) will signal the vasopressin-secreting cells to *decrease* vasopressin secretion and thus increase urinary excretion of the extra free water that is making the body fluids too dilute. Simultaneously, the left atrial volume receptors will signal the vasopressin-secreting cells to *increase* vasopressin secretion to conserve water during urine formation and thus help relieve the volume deficit. These two conflicting inputs to the vasopressin-secreting cells are counterproductive. For this reason, it is important to replace both water and salt following heavy sweating or abnormal loss of other salt-rich fluids. If salt is replaced along with water intake, the ECF osmolarity remains close to normal and the vasopressin-secreting cells receive signals only to increase vasopressin secretion to help restore the ECF volume to normal.

3. When a dextrose solution equal in concentration to that of normal body fluids is injected intravenously, the ECF volume is expanded, but the ECF and ICF are still osmotically equal. Therefore, no net movement of water occurs between the ECF and ICF. When the dextrose enters the cell and is metabolized, however, the ECF becomes hypotonic as this solute leaves the plasma. If the excess free water is not excreted in the urine rapidly enough, water will move into the cells by osmosis.

4. Because baking soda ($NaHCO_3$) is readily absorbed from the digestive tract, treatment of gastric hyperacidity with baking soda can lead to metabolic alkalosis as too much ($HCO_3^-$) is absorbed. Treatment with antacids that are poorly absorbed is safer because these products remain in the digestive tract and do not produce an acid-base imbalance.

5. c. The haemoglobin buffer system buffers carbonic acid-generated hydrogen ion. In the case of respiratory acidosis accompanying severe pneumonia, the $H^+ + Hb \rightarrow HHb$ reaction shifts toward the HHb side, thus removing some of the extra free $H^+$ from the blood.

## Clinical Consideration

(Question on p. 639.)

The resultant prolonged diarrhoea leads to dehydration and metabolic acidosis due, respectively, to excessive loss in the feces of fluid and $NaHCO_3$ that normally would have been absorbed into the blood.

Compensatory measures for dehydration have included increased vasopressin secretion, resulting in increased water reabsorption by the distal and collecting tubules and a subsequent reduction in urine output. Simultaneously, fluid intake has been encouraged by increased thirst. The metabolic acidosis has been combated by removal of excess $H^+$ from the ECF by the $[HCO_3^-]$ member of the $H_2CO_3$: $[HCO_3^-]$ buffer system, by increased ventilation to reduce the amount of acid-forming $CO_2$ in the body fluids, and by the kidneys excreting extra $H^+$ and conserving $[HCO_3^-]$.

# Chapter 15 The Digestive System

## Check Your Understanding

15.1 (Questions on p. 652.)

1. See ⟩ Figure 15-2, Layers of the digestive tract wall, p. 650.

2. Pacemaker cells called interstitial cells of Cajal (ICC) generate slow-wave potentials that cyclically bring the membrane closer to or farther from threshold potential. Slow-wave potentials spread via gap junctions from ICCs to the surrounding smooth muscle cells, which themselves are interconnected by gap junctions into functional syncytia. If the slow waves reach threshold at the peaks of depolarization, a volley of action potentials is triggered at each peak, resulting in rhythmic cycles of contraction in the sheet of smooth muscle cells driven by the pacemaker. The slow-wave frequency varies regionally in the digestive tract. Contractions within the tract occur at the rate of slow-wave frequency within a given section of tract.

3. The sympathetic nervous system tends to slow the digestive tract and secretion. The parasympathetic nervous system increases smooth muscle motility and secretion of digestive enzymes.

**15.2** (Questions on p. 654.)

1. *mucus*: facilitates swallowing by moistening food particles to hold them together and by providing lubrication; *amylase*: begins digestion of dietary starches; *lysosyme*: exerts antibacterial actions by lysing bacteria, binding IgA antibodies, secreting lactoferrin that binds iron needed for bacterial multiplication, and rinsing away material that may serve as a food source for bacteria

2. *Parasympathetic stimulation* induces production of a large volume of watery saliva that is rich in enzymes. *Sympathetic stimulation* produces a small volume of thick saliva that is rich in mucus.

**15.3** (Questions on p. 656.)

1. During swallowing, respiration is temporarily inhibited while the entrance to the trachea is closed off to prevent food from entering the respiratory airways. Contraction of the laryngeal muscles tightly aligns the vocal folds across the glottis. For further protection, the epiglottis folds downward over the closed glottis.

2. The *pharyngoesophageal sphincter* at the upper end of the esophagus separates the pharynx from the esophagus and prevents air from entering the esophagus and stomach during breathing. This sphincter remains tonically contracted except during a swallow, when it opens. The *gastroesophageal sphincter* at the lower end of the esophagus separates the esophagus from the stomach and prevents reflux of gastric contents into the esophagus. It relaxes during a swallow.

**15.4** (Questions on p. 667.)

1. A strong peristaltic contraction in the stomach antrum pushes the luminal contents downward toward a slightly open pyloric sphincter. A small portion of the chyme is forced through the sphincter before the peristaltic wave reaches the sphincter and closes it tightly. When food that is moving forward hits the closed sphincter, it is tossed backward, only to be propelled forward again by the next peristaltic wave. This *retropulsion* (cycles of food being moved forward and backward in the antrum) shears and grinds the food into smaller pieces and thoroughly mixes it with gastric secretions, converting it into chime before emptying.

2. The cephalic phase of gastric secretion occurs in response to food-related stimuli such as thinking about, tasting, smelling, seeing, chewing, and swallowing food. The response is mediated by vagal activation of the chief and parietal secretory cells (which secrete HCl and pepsinogen, respectively) and the endocrine G cells (which secrete the hormone gastrin that further stimulates the chief and parietal cells).

3. The following components of the gastric mucosal barrier enable the stomach to contain potent HCl without injuring itself: (1) The luminal membranes of the gastric mucosal cells are impermeable to $H^+$, so HCl cannot penetrate into the cells; (2) tight junctions join the cells and prevent HCl from penetrating between them; (3) the surface mucous cells secrete mucus, which serves as a physical barrier to acid penetration; and (4) $HCO_3^-$ secreted with the mucus serves as a chemical barrier that neutralizes acid in the vicinity of the mucosa. Finally, the entire stomach lining undergoes significant turnover, with the cells being replaced every three days.

**15.5** (Questions on p. 676.)

1. The pancreas produces proteolytic enzymes for digesting food proteins, but these enzymes could also digest the proteins of the cells that produce them if they were not stored in inactive form until they are secreted. These inactive enzymes are stored in secretory zymogen granules in the pancreatic acinar cells until appropriate stimuli induce their secretion. They are activated to protein-digesting enzymes (trypsin, chymotrypsin, and carboxypeptidase) only when they reach the duodenal lumen, where they act on food, not the pancreatic cells.

2. Bile salts contribute to dietary fat digestion by their detergent action (fat emulsification) in the small intestine. Intestinal mixing movements break up large fat globules into smaller droplets. The droplets do not recoalesce into the large globule because bile salts adsorb on the surface of the smaller droplets, creating a *lipid emulsion* consisting of many small fat droplets suspended in the watery chyme. This action increases the surface area of fat available for attack by pancreatic lipase.

**15.6** (Questions on p. 688.)

1. Three major structural features increase the surface area of the small intestine available for absorption: (1) Large *circular folds* on the small-intestine luminal surface increase the surface area threefold; (2) microscopic finger-like projections, known as *villi*, that sit atop the circular folds increase the surface area another tenfold; and (3) a multitude of hairlike protrusions, referred to as *microvilli* or the *brush border* on each villus surface, collectively increase the surface area another twentyfold. Together these three adaptations increase the absorptive surface area of the small intestine approximately 600 times greater than compared to a completely smooth tube of the same length and diameter.

2. Absorption of most dietary carbohydrate and protein is accomplished by secondary active transport that involves the cotransport of $N^+$ and the nutrient molecule into the absorptive cell. Although the cotransporter does not use ATP directly, the $Na^+-K^+$ pump (which directly uses energy) establishes a $N^+$ concentration gradient that drives the cotransporter to move $Na^+$ downhill and move the nutrient molecule uphill into the cell across the luminal membrane. The nutrient molecules move out of the cell into the blood across the basal membrane by facilitated diffusion, completing the absorptive process.

**15.7** (Questions on p. 692.)

1. Large-intestine *haustral contractions* are very similar to small intestine *segmentation contractions* in that both are rhythmic, autonomous, oscillating ringlike contractions initiated by pacemaker cells, but haustral contractions occur much less frequently (once every 30 minutes) than segmentation contractions (9 to 12 per minute). Haustral contractions, which are nonpropulsive, shuffle the colonic content back and forth, facilitating salt and water absorption, which compact the content and form a firm fecal mass. In contrast, segmentation in the small intestine is both a slowly propulsive and mixing movement.

2. The $NaHCO_3$ produced by the *large intestine* mainly serves to protect the large intestine from acid produced during colonic fermentation by the intestinal bacteria. The primary role of pancreatic $NaHCO_3$ secreted into the duodenal lumen is to neutralize the acidic chyme that empties from the stomach, thereby protecting the small intestine from acid damage in addition to creating an alkaline environment that optimizes pancreatic digestive enzyme activity.

**15.8** (Questions on p. 693.)

1. *Trophic hormones* induce growth of their target tissues. For example, gastrin is trophic to the gastric mucosa, in addition to stimulating secretion by the parietal and chief cells. The significance is that the hormone helps maintain the gastric mucosa, despite harsh conditions, thereby maintaining the secretory capabilities of this stomach lining.

2. *secretin*: stomach smooth muscle and parietal cells (acts as an enterogastrone to inhibit gastric motility and acid secretion), pancreatic duct cells (stimulates $NaHCO_3$ secretion); *CCK*: stomach smooth muscle and parietal cells (acts as an enterogastrone), pancreatic acinar cells (stimulates secretion of pancreatic digestive enzymes), gallbladder (causes contraction), sphincter of Oddi (causes relaxation), and brain (signals satiety)

## Objective Questions

(Questions on pp. 695–696.)

1. F 2. T 3. T 4. F 5. F 6. F 7. long, short 8. chyme 9. three 10. vitamin $B_{12}$, bile salts 11. bile salts 12. b 13. 1. c, 2. e, 3. b, 4. a, 5. f, 6. d 14. 1. c, 2. c, 3. d, 4. a, 5. e, 6. c, 7. a, 8. b, 9. c, 10. b, 11. b

## Quantitative Exercise

(Question on p. 696.)

1. **a.** $r = 0.5$ cm, area $= 4\pi(0.25) = \pi\, cm^2$

$$\text{volume} = \left(\frac{4}{3}\pi\right)(0.5)^3 = 0.5236\ cm^3$$

area/volume $= 6$

**b.** Each new droplet's volume is $5.236 \times 10^{-3}$ cm³, so the average radius is therefore 0.1077 cm. The area of a sphere with that radius is $4\pi(0.1077 \text{ cm})^2 = 0.1458$ cm²
area/volume = 27.8

**c.** Area emulsified/area droplet = $(100)(0.1458 \text{ cm}^2)/\pi \text{ cm}^2 = 4.64$. Thus, the total surface area of all 100 emulsified droplets is 4.64 times the area of the original larger lipid droplet.

**d.** Volume emulsified/volume droplet = $(100)(5.236 \times 10^{-3} \text{ cm}^3)/0.5236 \text{ cm}^3 = 1.0$.
Therefore, the total volume did not change as a result of emulsification, as would be expected, because the total volume of the lipid is conserved during emulsification. The volume originally present in the large droplet is divided up among the 100 emulsified droplets.

## Points to Ponder
(Questions on pp. 696–697.)

1. Patients who have had their stomachs removed must eat small quantities of food frequently instead of consuming the typical three meals a day because they have lost the ability to store food in the stomach and meter it into the small intestine at an optimal rate. If a person without a stomach consumed a large meal that entered the small intestine all at once, the luminal contents would quickly become too hypertonic as digestion of the large nutrient molecules into a multitude of small, osmotically active, absorbable units outpaced the more slowly acting process of absorption of these units. As a consequence of this increased luminal osmolarity, water would enter the small intestine lumen from the plasma by osmosis, resulting in circulatory disturbances as well as intestinal distension. To prevent this "dumping syndrome" from occurring, the patient must "feed" the small intestine only small amounts of food at a time so that absorption of the digested end products can keep pace with their rate of production. The person has to consciously take over metering the delivery of food into the small intestine because the stomach is no longer present to assume this responsibility.

2. The gut-associated lymphoid tissue launches an immune attack against any pathogenic (disease-causing) microorganisms that enter the readily accessible digestive tract and escape destruction by salivary lysozyme or gastric HCl. This action defends against entry of these potential pathogens into the body proper. The large number of immune cells in the gut-associated lymphoid tissue is adaptive as a first line of defence against foreign invasion, considering that the surface area of the digestive tract lining represents the largest interface between the body proper and the external environment.

3. Defecation would be accomplished entirely by the defecation reflex in a patient paralyzed from the waist down because of lower spinal cord injury. Voluntary control of the external anal sphincter would be impossible because of interruption in the descending pathway between the primary motor cortex and the motor neuron supplying this sphincter.

4. When insufficient glucuronyl transferase is available in the neonate to conjugate all the bilirubin produced during erythrocyte degradation with glycuronic acid, the extra unconjugated bilirubin cannot be excreted into the bile. Therefore, this extra bilirubin remains in the body, giving rise to mild jaundice in the newborn.

5. Removal of the stomach leads to pernicious anaemia because of the resultant lack of intrinsic factor, which is necessary for absorption of vitamin $B_{12}$. Removal of the terminal ileum leads to pernicious anaemia because this is the only site where vitamin $B_{12}$ can be absorbed.

## Clinical Consideration
(Question on p. 697.)
A person whose bile duct is blocked by a gallstone experiences a painful gallbladder attack after eating a high-fat meal because the ingested fat triggers the release of cholecystokinin, which stimulates gallbladder contraction. As the gallbladder contracts and bile is squeezed into the blocked bile duct, the duct becomes distended prior to the blockage. This distension is painful.

The feces are greyish white because no bilirubin-containing bile enters the digestive tract when the bile duct is blocked. Bilirubin, when acted on by bacterial enzymes, is responsible for the brown colour of feces, which are greyish white in its absence.

# Chapter 16 Energy Balance and Temperature Regulation

## Check Your Understanding
**16.1** (Questions on p. 710.)

1. *External work* is the energy expended by skeletal muscles to move external objects or to move the body in relation to the environment. *Internal work* constitutes all biological energy expenditure that does not accomplish mechanical work outside the body; it includes skeletal muscle activity associated with postural maintenance and shivering and all energy-expending activities essential for sustaining life, such as pumping blood or breathing. *Metabolic rate* is the rate at which energy is expended during both external and internal work: metabolic rate = energy expenditure/unit of time. *Appetite signals* give the sensation of hunger, driving us to eat. *Satiety signals* give the sensation of being full, suppressing the desire to eat. *Adiposity signals* are indicative of the size of fat stores in adipose tissue and are important in the long-term control of body weight. *Adipokines* refer to hormones secreted by adipocytes. *Visceral fat* is the deep, "bad" fat that surrounds the abdominal organs and is likely to be chronically inflamed and secrete harmful adipokines. *Subcutaneous fat* is deposited under the skin (it is the fat you can pinch) and is less harmful than visceral fat.

2. The basal metabolic rate can be determined indirectly by measuring a person's $O_2$ uptake per unit of time and multiplying this value by the energy equivalent of $O_2$, which is 4.8 kcal of energy liberated per litre of $O_2$ consumed. This technique is based on the fact that a direct relationship exists between the volume of $O_2$ used and the quantity of heat produced as follows: Food + $O_2 \rightarrow CO_2 + H_2O$ + energy (mostly transformed into heat).

3. Neuropeptide Y, from the arcuate nucleus of hypothalamus, increases appetite.
Melanocortins, from the arcuate nucleus of hypothalamus, decreases appetite.
Leptin, from adipose tissue, decreases appetite.
Insulin, from the pancreas, decreases appetite.
Orexins, from the lateral hypothalamus, increase appetite.
Corticotrophin-releasing hormone, from the paraventricular nucleus of hypothalamus, decreases appetite.
Ghrelin, from the stomach, increases appetite.
PPY, from the small and large intestines, decreases appetite.
CCK, from the small intestine, decreases appetite.
Stomach distension, from the stomach, decreases appetite.

**16.2** (Questions on p. 719.)

1. *radiation*: transfer of heat energy from a warmer object to a cooler object in the form of heat waves that travel through space; *conduction*: transfer of heat between objects of differing temperatures that are in direct contact with each other; *convection*: transfer of heat energy by air currents; *evaporation*: conversion of a liquid such as sweat to a gaseous vapour, a process that requires heat, which is absorbed from the skin

2. See ▮ Table 16-4, Coordinated Adjustments in Response to Heat or Cold Exposure.

## Objective Questions
(Questions on p. 720.)

1. F **2.** F **3.** F **4.** F **5.** T **6.** T **7.** T **8.** T **9.** arcuate nucleus **10.** shivering **11.** nonshivering thermogenesis **12.** sweating **13.** b **14.** e **15.** 1. b, 2. a, 3. c, 4. a, 5. d, 6. c, 7. b, 8. d, 9. c, 10. b

## Quantitative Exercise

(Question on pp. 720–721.)

1. From physics we know that $\Delta T(°C) = \Delta U/(m \times C)$.

   Also note that $\Delta U/t = BMR$; that is, the rate of using energy is the basal metabolic rate. $m$ represents the mass of body fluid; for a typical person, this is 42 L.

   $(42 \text{ L}) \times (1 \text{ kg/L}) = 42 \text{ kg}$

   $C = 1.0 \text{ kcal}/(\text{kg}°C)$

   Given that water boils at 100°C and normal body temperature is 37°C, we need to change the temperature by 63°C. Thus,

   $t = (\Delta T \times C \times m)/BMR = (63°C)[1.0 \text{ kcal/kg}°C](42 \text{ kg})/(75 \text{ kcal/h}) = 35 \text{ h}$

   At the higher metabolic rate during exercise,
   $t = (63°C)[1.0 \text{ kcal}/(\text{kg}°C)](42 \text{ kg})/(1000 \text{ kcal/h}) = 2.6 \text{ h}$

## Points to Ponder

(Questions on p. 721.)

1. Evidence suggests that CCK serves as a satiety signal. It is believed to serve as a signal to stop eating when enough food has been consumed to meet the body's energy needs, even though the food is still in the digestive tract. Therefore, when drugs that inhibit CCK release are administered to experimental animals, the animals overeat because this satiety signal is not released.

2. Don't go on a "crash diet." Be sure to eat a nutritionally balanced diet that provides all essential nutrients, but reduce total caloric intake, especially by cutting down on high-fat foods. Spread out consumption of the food throughout the day instead of just eating several large meals. Avoid bedtime snacks. Burn more calories through a regular exercise program.

3. Engaging in heavy exercise on a hot day is dangerous because of problems arising from trying to eliminate the extra heat generated by the exercising muscles. First, there are conflicting demands for distribution of the cardiac output—temperature-regulating mechanisms trigger skin vasodilation to promote heat loss from the skin surface, whereas metabolic changes within the exercising muscles induce local vasodilation in the muscles to match the increased metabolic needs with increased blood flow. Further exacerbating the problem of conflicting demands for blood flow is the loss of effective circulating plasma volume resulting from the loss of a large volume of fluid through another important cooling mechanism: sweating. Therefore, it is difficult to maintain an effective plasma volume and blood pressure and simultaneously keep the body from overheating when engaging in heavy exercise in the heat, so heat exhaustion is likely to ensue.

4. When a person is soaking in a hot bath, loss of heat by radiation, conduction, convection, and evaporation is limited to the small surface area of the body exposed to the cooler air. Heat is being gained by conduction at the larger skin surface area exposed to the hotter water.

5. The thermoconforming fish would not run a fever when it has a systemic infection because it has no mechanisms for regulating internal heat production or for controlling heat exchange with its environment. The body temperature of fish varies capriciously with the external environment, no matter whether the fish has a systemic infection or not. It is not able to maintain body temperature either at some normal set point or at an elevated set point (i.e., a fever).

## Clinical Consideration

(Question on p. 721.)

Cooled tissues need less nourishment than tissues at normal body temperature because of their pronounced reduction in metabolic activity. The lower $O_2$ need of cooled tissues accounts for the occasional survival of drowning victims who have been submerged in icy water considerably longer than one could normally survive without $O_2$.

# Chapter 17 The Reproductive System

## Check Your Understanding

17.1 (Questions on p. 731.)

1. *testes* in males and *ovaries* in females. In both sexes, these primary reproductive organs perform the dual function of producing gametes (sperm in males and ova in females) and secreting sex hormones (testosterone in males and estrogen and progesterone in females).

2. See table below.

17.2 (Questions on p. 741.)

1. See Table 17-1, Effects of Testosterone.

2. The *seminiferous tubules* are the highly coiled tubular component of the testes where spermatogenesis takes place. The *Leydig cells*, which lie in the interstitial spaces between the seminiferous tubules, secrete testosterone. *Sertoli cells* are epithelial cells that lie in close association with and protect, nurse, and enhance the developing sperm cells throughout their development in the seminiferous tubules. Sertoli cells also secrete inhibin and are the site of action for control of *spermatogenesis* by both testosterone and FSH. *Spermatogenesis* is the sequence of steps by which relatively undifferentiated primordial germ cells proliferate and are converted into extremely specialized, motile spermatozoa. *Spermiogenesis* refers to the remodelling or packaging of the haploid spermatids into spermatozoa. *Spermiation* is the final release of a mature spermatozoon from the Sertoli cell to which it has been attached throughout development. *Spermatogonia* are the undifferentiated primordial germ cells that have a diploid number of single-strand chromosomes. Two mitotic divisions of a spermatogonium yield four *primary spermatocytes*, each with a diploid number of double-strand chromosomes. The first meiotic division of these primary spermatocytes yields eight *secondary spermatocytes*, each with a haploid number of

| Embryonic Structure | Reproductive Structure Derived from this Embryonic Structure in Males | Reproductive Structure Derived from this Embryonic Structure in Females |
|---|---|---|
| Gonadal Ridge | Testes | Ovaries |
| Genital Tubercle | Penis | Clitoris |
| Urethral Folds | Cord of erectile tissue that surrounds the urethra in the penis | Labia minora |
| Genital Swellings | Scrotum and prepuce | Labia majora |
| Wolffian Ducts | Epididymis, ductus deferens, ejaculatory duct, and seminal vesicles | Nothing; ducts degenerate |
| Müllerian Ducts | Nothing; ducts degenerate | Oviducts, uterus, cervix, and upper vagina |
| Urogenital Sinus | Prostate gland, bulbourethral gland | Lower vagina |

double-strand chromosomes. The second meiotic division of these secondary spermatocytes yields 16 spermatids, each with a haploid number of single-strand chromosomes. The *spermatids* are packaged into highly specialized *spermatozoa*, each with a haploid number of single-strand chromosomes.

**17.3** (Questions on p. 744.)

1. (1) *excitement phase* (erection and heightened sexual awareness); (2) *plateau phase* (intensification of sexual responses and systemic responses, such as increased heart rate, blood pressure, respiratory rate, and muscle tension); (3) *orgasmic phase* (ejaculation in males, orgasm in both sexes); and (4) *resolution phase* (return to prearousal state)

2. The male sex act includes erection and ejaculation. *Erection*, hardening of the normally flaccid penis to permit its entry into the vagina, is accomplished by engorgement of the penis erectile tissue with blood as a result of marked parasympathetically induced vasodilation of the penile arterioles and mechanical compression of the veins. *Ejaculation* occurs in two phases: The *emission phase*, emptying of sperm and accessory sex gland secretions (semen) into the urethra, is accomplished by sympathetically induced contraction of smooth muscle in the walls of the reproductive ducts and accessory sex glands. The *expulsion phase*, the forceful expulsion of semen from the penis, is accomplished by motor neuron –induced contraction of the skeletal muscles at the base of the penis.

**17.4** (Questions on p. 774.)

1. The ovarian *follicle* secretes estrogen; the corpus luteum secretes progesterone (most) and estrogen. *Estrogen* stimulates growth of the endometrium and myometrium and induces synthesis of progesterone receptors in the endometrium. *Progesterone* acts on the estrogen-primed endometrium to convert it into a hospitable and nutritious lining (with loose, edematous connective tissue, glycogen stores, and increased blood supply) suitable for implantation. The *menstrual phase* of the uterine cycle takes place during the first half of the ovarian follicular phase, after the end of the last cycle when the estrogen and progesterone supply was withdrawn upon degeneration of the corpus luteum. The highly developed uterine lining sloughs (menstruation) simultaneously with the beginning of a new follicular phase, which is under the influence of rising FSH and LH upon withdrawal of the inhibitory actions of estrogen and progesterone. The *proliferative phase* of the uterine cycle occurs during the last half of the ovarian follicular phase, as the endometrium stops sloughing and starts to repair itself and proliferate under the influence of rising levels of estrogen from the newly recruited, rapidly growing antral follicles. Peak estrogen secretion causes the LH surge, which triggers ovulation, ending both the ovarian follicular phase and the uterine proliferative phase. The *secretory*, or *progestational, phase* of the uterine cycle begins concurrent with the ovarian luteal phase, as progesterone from the corpus luteum converts the thickened, estrogen-primed endometrium to a lush environment capable of supporting an early embryo, should the released egg be fertilized and implant.

2. The *zygote* is the fertilized ovum, following union of the male and female chromosomes. The *blastocyst* is a single-layer hollow ball of about 50 cells resulting from mitotic cell divisions of the zygote; the blastocyst is the developmental stage that implants in the endometrium. The *inner cell mass* is a dense mass on one side of the blastocyst that is destined to become the fetus. The *trophoblast* is the thin outermost layer of the blastocyst that accomplishes implantation, after which it develops into the fetal part of the placenta. The *decidua* is the endometrial tissue that's modified at the implantation site to enhance its ability to support the implanting embryo. The *chorion* is the thickened trophoblastic layer after implantation is complete. The *placenta* is the specialized organ of exchange between the maternal and fetal blood that is derived from both trophoblastic embryonic tissue and decidual maternal tissue and that secretes peptide and steroid hormones essential for maintaining

pregnancy. The *embryo* is the product of fertilization during the first two months of intrauterine development, when tissue differentiation is taking place. The *fetus* is the product of fertilization during the last seven months of gestation, after differentiation is complete and tremendous tissue growth and maturation occur.

3. During *parturition*, a positive-feedback loop involving oxytocin, a powerful uterine muscle stimulant, is responsible for the progression of labour. Oxytocin promotes uterine muscle contractions that force the fetus against the cervix, dilating it and triggering a neuroendocrine reflex that results in secretion of even more oxytocin, which stimulates even stronger contractions, and so on as labour progresses until the cervix is dilated sufficiently for the baby to be pushed out. During *breastfeeding*, oxytocin causes milk ejection (milk letdown) by stimulating contraction of the myoepithelial cells surrounding the milk-secreting alveoli.

## Objective Questions

(Questions on pp. 779–777.)

**1.** T **2.** T **3.** T **4.** F **5.** T **6.** F **7.** T **8.** seminiferous tubules, FSH, testosterone **9.** thecal, LH, granulosa, FSH **10.** corpus luteum of pregnancy **11.** human chorionic gonadotropin **12.** c **13.** b **14.** 1. c, 2. a, 3. b, 4. c, 5. a, 6.a, 7. c, 8. b, 9. e **15.** 1. a, 2. c, 3. b, 4. a, 5. a, 6. b

## Points to Ponder

(Questions on p. 778.)

1. The anterior pituitary responds only to the normal pulsatile pattern of GnRH and does not secrete gonadotropins in response to continuous exposure to GnRH. In the absence of FSH and LH secretion, ovulation and other events of the ovarian cycle do not ensue, so continuous GnRH administration may find use as a contraceptive technique.

2. Testosterone hypersecretion in a young boy causes premature closure of the epiphyseal plates so that he stops growing before he reaches his genetic potential for height. The child would also display signs of precocious pseudopuberty, characterized by premature development of secondary sexual characteristics, such as deep voice, beard, enlarged penis, and sex drive.

3. A potentially troublesome side effect of drugs that inhibit sympathetic nervous system activity as part of the treatment for high blood pressure is males' inability to carry out sexual activity. Both divisions of the autonomic nervous system are required for the male sexual activity. Parasympathetic activity is essential for accomplishing erection, and sympathetic activity is important for ejaculation.

4. Posterior pituitary extract contains an abundance of stored oxytocin, which can be administered to induce or facilitate labour by increasing uterine contractility. Exogenous oxytocin is most successful in inducing labour if the woman is near term, presumably because of the increasing concentration of myometrial oxytocin receptors at that time.

5. GnRH or FSH and LH are not effective in treating the symptoms of menopause because the ovaries are no longer responsive to the gonadotropins. Thus, treatment with these hormones would not cause estrogen and progesterone secretion. In fact, GnRH, FSH, and LH levels are already elevated in postmenopausal women because of lack of negative feedback by the ovarian hormones.

## Clinical Consideration

(Question on p. 778.)

The first warning of a tubal pregnancy is pain caused by stretching of the oviduct by the growing embryo. A tubal pregnancy must be surgically terminated because the oviduct cannot expand as the uterus does to accommodate the growing embryo. If not removed, the enlarging embryo will rupture the oviduct, causing possibly lethal haemorrhage.

# Reference Values for Commonly Measured Variables in Blood and Commonly Measured Cardiorespiratory Variables

| Blood Gases | Normal values |
|---|---|
| $P_{O_2}$ (arterial) | 11–13 kPa (kilopascals) |
| $P_{CO_2}$ (arterial) | 4.7–5.9 kPa |
| **Electrolytes** | |
| $Ca^{2+}$ (total) | 2.2–2.6 mmol/L |
| $Cl^-$ | 97–110 mmol/L |
| $K^+$ | 3.5–5.0 mmol/L |
| $Na^+$ | 135–146 mmol/L |
| **Hormones** | |
| Aldosterone | 83–277 pmol/L |
| Cortisol | |
|   8:00 a.m. | 140–690 nmol/L |
|   4:00 p.m. | 40–330 nmol/L |
| Estradiol | |
|   Women (early follicular phase) | 73–367 pmol/L |
|   Women (midcycle peak) | 551–2753 pmol/L |
|   Men | 37–184 pmol/L |
| Insulin (fasted) | 43–186 pmol/L |
| Insulin-like growth factor I | |
|   16–24 years old | 182–780 µg/L |
|   25–50 years old | 114–492 µg/L |
| Parathyroid hormone | 10–75 ng/L |
| Progesterone | |
|   Women (luteal phase) | 6–81 nmol/L |
|   Women (pregnancy) | 15–770 nmol/L |
|   Men | 0.6–4.3 nmol/L |
| Testosterone | |
|   Women | <3.5 nmol/L |
|   Men | 9–35 nmol/L |
| Thyroid-stimulating hormone | 0.3–4.0 mIU/L |
| Thyroxine (adults) | 64–140 nmol/L |

| Blood Gases | Normal values |
|---|---|
| **Nutrients (fasting)** | |
| Glucose | 4–6 mmol/L |
| Free fatty acids | 0.3–1.0 mmol/L |
| Triglycerides | <1.8 mmol/L |
| **Proteins** | |
| Albumin | 35–55 g/L |
| Globulins | 20–35 g/L |
| Fibrinogens | 2–4 g/L |
| **Red and White Blood Cells** | |
| Red blood cell count | $4.1–5.4 \times 10^{12}$/L |
| Haematocrit | |
|   Male | 0.42–0.52 |
|   Female | 0.37–0.48 |
| Haemoglobin | |
|   Male | 140–180 g/L |
|   Female | 120–160 g/L |
| Iron | 9–27 µmol/L |
| Leukocytes (total) | $4.3–10.8 \times 10^{9}$/L |
| Osmolarity | 285–295 mosmol/L |
| pH | 7.38–7.45 |

| Cardiovascular Response | | Units |
|---|---|---|
| **Systolic Blood Pressure (rest)** | | |
| Male | 20–30 years | 120 mmHg |
| | 50–60 years | 134 mmHg |
| Female | 20–30 years | 120 mmHg |
| | 50–60 years | 130 mmHg |
| **Diastolic Blood Pressure (rest)** | | |
| Male | 20–30 years | 80 mmHg |
| | 50–60 years | 84 mmHg |

| Cardiovascular Response | | Units |
| --- | --- | --- |
| Female | 20–30 years | 74 mmHg |
| | 50–60 years | 84 mmHg |
| **Systolic Blood Pressure (maximal exercise)** | | |
| Male | 20–30 years | 190 mmHg |
| | 50–60 years | 200 mmHg |
| Female | 20–30 years | 190 mmHg |
| | 50–60 years | 200 mmHg |
| **Diastolic Blood Pressure (maximal exercise)** | | |
| Male | 20–30 years | 70 mmHg |
| | 50–60 years | 84 mmHg |
| Female | 20–30 years | 64 mmHg |
| | 50–60 years | |
| **Stroke Volume (rest)** | | |
| Male | 20–30 years | 90 mL/beat |
| | 50–60 years | 70 mL/beat |
| Female | 20–30 years | 75 mL/beat |
| | 50–60 years | 62 mL/beat |
| **Stroke Volume (maximal exercise)** | | |
| Male | 20–30 years | 128 mL/beat |
| | 50–60 years | |
| Female | 20–30 years | 92 mL/beat |
| | 50–60 years | 75 mL/beat |
| **Heart Rate (rest)** | | |
| Male | 20–30 years | 75 b/min |
| | 50–60 years | 80 b/min |
| Female | 20–30 years | 75 b/min |
| | 50–60 years | 80 b/min |
| **Heart Rate (maximal exercise)** | | |
| Male | 20–30 years | 195 b/min |
| | 50–60 years | 155 b/min |
| Female | 20–30 years | 195 b/min |
| | 50–60 years | 155 b/min |
| **Cardiac Output (rest)** | | |
| Male | 20–30 years | 6.5 L/min |
| | 50–60 years | 5.5 L/min |
| Female | 20–30 years | 5.5 L/min |
| | 50–60 years | 5.0 L/min |

| Cardiovascular Response | | Units |
| --- | --- | --- |
| **Cardiac Output (maximal exercise)** | | |
| Male | 20–30 years | 25 L/min |
| | 50–60 years | 16 L/min |
| Female | 20–30 years | 18 L/min |
| | 50–60 years | 12 L/min |
| **Oxygen Consumption (rest)** | | |
| Male | 20–30 years | 0.35 L/min (3.5 mL/kg/min) |
| | 50–60 years | 0.35 L/min (3.5 mL/kg/min) |
| Female | 20–30 years | 0.35 L/min (3.5 mL/kg/min) |
| | 50–60 years | 0.35 L/min (3.5 mL/kg/min) |
| **Oxygen Consumption (maximal exercise)** | | |
| Male | 20–30 years | 3.5 L/min |
| | 50–60 years | 2.5 L/min |
| Female | 20–30 years | 2.5 L/min |
| | 50–60 years | 1.5 L/min |
| **Breathing Pattern (rest)** | | |
| $V_E$ (minute ventilation) | 6.0 L/min | Values for 30-year-old untrained males |
| $F_b$ (breathing frequency) | 15 breaths/min | |
| $V_A$ (alveolar ventilation) | 3.7 L/min | |
| $V_T$ (title volume) | 0.4 L/min | |
| **Breathing Pattern (maximal exercise)** | | |
| $V_E$ | 130 L/min | Values for 30-year-old untrained males |
| $F_b$ | 45 breaths/min | |
| $V_A$ | 96 L/min | |
| $V_T$ | 2.90 L/min | |

# A Deeper Look into Chapter 12, The Respiratory System

## G.1 Terminology for Respiratory Physiology

### Primary symbols

C    Content (or concentration) in blood

D    Delivery (as in $\dot{D}O_2$)

F    Fractional concentration in dry gas

P    Pressure or partial pressure

Q    Volume of blood

$\dot{Q}$    Volume of blood per unit time (also cardiac output)

R    Respiratory exchange ratio

S    Saturation of haemoglobin with $O_2$

V    Volume of gas

$\dot{V}$    Volume of gas per unit time (flow or ventilation)

### Secondary symbols for gas phase

A    alveolar

B    barometric

D    dead space

E    expired

ET    end-tidal

I    inspired

L    lung

T    tidal

### Secondary symbols for blood phase

a    arterial

c    capillary

c'    end capillary

i    ideal

v    venous

$\bar{v}$    mixed venous

A time derivative is indicated by a dot over the symbol, a mean (or average) by a bar. The location of a parameter in the gas phase (e.g., alveoli, inspiratory gas) is denoted by smaller upper case letters, whereas those in the blood phase are indicated by lower case letters.

### Other abbreviations

ATPS    ambient temperature, pressure, saturated with water vapour; used for measurements of gas volumes under ambient conditions.

BTPS    body temperature $(37°C)$, pressure (ambient $P_B$), saturated; used for volumes and flows in the body.

STPD    standard temperature $(0°C)$, pressure $(760\ mmHg)$, dry; used for measurements of $O_2$ consumption $(\dot{V}_{O_2})$ and $CO_2$ production $(\dot{V}_{CO_2})$.

Note: Measurements of gas volumes under ambient conditions (e.g., in a bag) are given as ATPS. Volumes or flows in the body are always given as BTPS because they always exist at this condition. Volumes of gases diffusing or consumed $(\dot{V}_{O_2})$ or produced $(\dot{V}_{CO_2})$ are always given as STPD to enable comparisons of measurements made under different ambient conditions.

### Examples

$Ca_{O_2}$    $O_2$ content (or concentration) of arterial blood

$F_{E_{N_2}}$    Fractional concentration of $N_2$ in expired gas

$P_{\bar{V}_{CO_2}}$    Partial pressure of $CO_2$ in mixed venous blood

$P_{ET_{CO_2}}$    Partial pressure of $CO_2$ in end-tidal gas

$\dot{V}$    Volume of gas per minute (or second), that is, flow

$\dot{V}_{O_2}$    Consumption or diffusion (context determines which) of $O_2$

## G.2 Gas Laws and Other Variables

1. Gay-Lussac's/Charles's law: Volume and temperature

$$V_1/V_2 = T_1/T_2 \text{ and } P_1/P_2 = T_1/T_2$$

where $V$ is volume, $T$ is temperature, $P$ is pressure, and 1 and 2 are conditions 1 and 2, respectively.

2. Boyle's/Mariotte's law: Volume and pressure

$$P \times V = \text{constant, at constant temperature}$$

or

$$P_1V_1 = P_2V_2$$

where $P$ is the pressure, and $V$ is the volume of a perfect gas.

3. Dalton's law of partial pressures: The partial pressure of a given gas $x$ is equal to the product of the total pressure, $P_B$ multiplied by $F$, the fractional concentration of the gas in the mixture.

$$P_x = P_B \times F$$

For inspired/alveolar gases, when $T = 37°C$ and $P_{H_2O} = 47$ mmHg,

$$P_x = (P_B - 47) \times F$$

where $F$ is the fractional concentration of $O_2$, $CO_2$, or $N_2$ in the dry gas mixture.

4. Fick's law of diffusion of a gas $x$:

$$\dot{V}_x = A \times D \times \Delta P/T$$

where $\dot{V}_x$ = the amount (mL) of substance $x$ diffused per unit time

$A$ = surface area for diffusion

$D$ = the diffusion constant for substance $x$

$\Delta P$ = the pressure gradient between the two points across which diffusion occurs

$T$ = the distance over which diffusion occurs

5. Graham's law:

$$\dot{V}_x \ \alpha \ 1/\sqrt{MW}.$$

where $\dot{V}_x$ is the amount (mL) of substance $x$ diffused per unit time, and MW is its molecular weight.

# G.3 Pressure–Volume Relationship of the Respiratory System

Neither the elastic recoil pressure of the lung $(P_l)$ nor the pleural pressure $(P_{pl})$ has a unique value; both depend on the volume of the respiratory system. We can determine the relation between system volume and pressure by measuring mouth pressure $(P_m)$ at different lung volumes under conditions when the mouth and nose are closed (or a valve through which the subject is breathing is closed) and there is no activity of the respiratory muscles. At the end of a normal expiration, at the functional residual capacity (FRC), a valve is closed temporarily (so the subject cannot breathe), and she relaxes her respiratory muscles. We measure $P_m$ "inside" the valve. Because there is no tendency to either inhale or exhale, $P_m$ is 0 cm $H_2O$—that is, atmospheric pressure $(P_B)$. The valve is then opened, the subject breathes in until she cannot inhale any more (total lung capacity, TLC), and again relaxes her respiratory muscles against the reclosed valve.

She will now feel herself wanting to breathe out; in other words, mouth pressure is greater than atmospheric pressure, or positive. If she now repeats this, but this time after breathing all the way to residual volume (RV), she will feel herself wanting to breathe in; in other words, mouth pressure is less than atmospheric pressure, or negative. Recall that these pressures are generated only by the *passive* properties of the respiratory system, not by the respiratory pump muscles. Moreover, because she relaxes her respiratory pump muscles and her upper airway (pharynx) is open, mouth pressure must equal alveolar pressure $(P_A)$; if they were different, flow would occur between the upper airway and alveoli, and that is not happening. If we repeat this experiment at different lung volumes, not just FRC, TLC, and RV, we obtain the relation depicted in Figure 12-10 (p. 508). At lung volumes above the end of a normal expiration (FRC), the respiratory system passively generates positive pressures. At volumes below FRC, the pressures are negative.

You may be puzzled by the arrangement of the axes, with the independent variable, volume $(V)$, on the $y$-axis and the dependent variable, mouth pressure $(P_m)$, on the $x$-axis. Under normal circumstances, we change lung (system) volume by changing the pressure, either through changes in the degree of activation of the respiratory pump muscles or, in ventilated individuals, by changing the pressure generated by the ventilator. These pressure changes increase or decrease lung volume.

Mouth pressure $(P_m)$ is the pressure generated by the entire respiratory system $(P_{rs})$ (i.e., $P_m = P_{rs}$) and the respiratory system is composed of the lung and chest wall (Figure 12-9, p. 507). In other words, the pressure generated by the respiratory system $(P_{rs})$ equals the sum of the pressure generated by the lung $(P_l)$ and the pressure generated by the chest wall $(P_w)$:

$$P_{rs} = P_l + P_w$$

We know $P_{rs}$ because under the experimental conditions, it is $P_m$. But how can we measure the recoil pressures of the lung $(P_l)$ and of the chest wall $(P_w)$? We can measure $P_l$ indirectly by inserting a catheter via the nose into the middle third of the esophagus and measuring its pressure; this, as verified experimentally, provides an excellent approximation of the pleural pressure $(P_{pl})$. We now repeat the procedures just described and measure pleural pressure at each lung volume. How does this help us? Recall from the rearranged equation describing the transpulmonary pressure that

$$P_l = P_A - P_{pl}$$

We know $P_A$ because, under the conditions of this experiment (valve closed, airways open, and respiratory pump muscles relaxed), $P_m = P_A$. Thus, we know both terms on the right side of the equation and can now calculate $P_l$ at different lung volumes to get the second curve labelled $P_l$ in Figure 12-9.

In addition, because

$$P_{rs} = P_l + P_w$$

and we know both $P_{rs}$ and $P_l$, we can calculate $P_w$. The passive recoil pressure of the chest wall at different lung volumes is given by the curve designated $P_w$ in Figure 12-9.

# G.4 How We Inspire

At the end of a normal expiration, we return to a lung volume (FRC) at which the tendency of the chest wall to expand is equal and opposite to the tendency of the lung to collapse. This relation can be described in terms of the two pressures acting at the surface of the lung:

- $P_A$ is the pressure inside the alveolus ($= P_B$ when flow is zero and the airway is open)
- $P_{pl}$ is the pleural pressure

The difference between the two pressures is called the transpulmonary pressure $(P_{tp})$. *This pressure is also referred to as the recoil pressure of the lung* $(P_l)$. Note too that the difference across a structure, whether the lung or an airway (in which case, we use the term *transmural pressure*) is *always* determined as the pressure inside the structure minus the pressure outside the structure.

Or, using abbreviations,

$$P_{tp} = P_A = P_{pl}$$

At end-expiration when no flow occurs $(\dot{V} = 0)$, $P_A$ *must* equal zero (0) cmH$_2$O: if it were any other value, flow would occur. Since $P_{pl}$ is –5 cmH$_2$O,

$$P_{tp} = 0 - (-5) = +5 \text{ cmH}_2\text{O}$$

The positive transpulmonary pressure only means that the lung is inflated—the normal state of affairs. Note too that the positive $P_{tp}$ results from only one component, the negative (subatmospheric) pressure generated by the tendency of the chest wall to expand. The alveolar pressure, as we shall see, is purely passive, determined by the algebraic sum of the pressure generated by the elastic recoil pressure of the lung $(P_l)$ as it "tries" to collapse and the outward recoil of the chest wall as it "tries" to expand.

Note that the recoil pressure of the lung $(P_l)$ *must* be *equal and opposite* to the pleural pressure $(P_{pl})$ to give a $P_A = 0$ cmH$_2$O, which is required for the zero flow condition to be true.

For flow to occur, $P_A$ has to change. Therefore, we rearrange the equation to place $P_A$ on the left side. (We do this because, as indicated earlier, this is the variable that has to change in order to produce flow, since pressure at the airway opening $[P_B]$ is always set to zero.) Therefore, rearranging the equation:

$$P_B = P_{tp} + P_{pl}$$

Stated in words, alveolar pressure $(P_A)$ is the algebraic sum of transpulmonary pressure $(P_{tp})$ and pleural pressure $(P_{pl})$.

Note that we could also express it slightly differently using the following equation:

$$P_A = P_l + P_{pl}$$

where alveolar pressure is the algebraic sum of the lung recoil and the pleural pressures.

To produce inspiratory flow, $P_A$ has to be negative (subatmospheric). To do this, the negative pressure pulling open the lung $(P_{pl})$ must be more negative than the transpulmonary pressure is positive (e.g., $-6$ cmH$_2$O is more negative than $+5$ cmH$_2$O is positive). This is done by contracting the inspiratory muscles, mainly the diaphragm. The diaphragm descends, lowering $P_{pl}$, decompressing the gas in the lungs and, according to Boyle's/Mariotte's law, decreasing $P_A$. This creates the negative $P_A$ required for inspiratory flow.

The lung behaves mechanically much like a balloon placed inside a bottle. To inflate the balloon a pressure must be generated to overcome the tendency of the balloon to deflate (i.e., elastic recoil). In this model, the balloon (lung) is connected to the atmosphere via a tube that passes through a stopper. A catheter is placed inside the jar to measure the equivalent of pleural pressure (in cmH$_2$O), registered on the meter above the jar. As we pull down on the rubber sheet (diaphragm) across the bottom of the jar, the pressure inside the jar (equivalent to pleural pressure) becomes more subatmospheric and the balloon inflates (right-hand figure). Because it takes time for air to flow, due to the resistance to airflow offered by the tubing (airways) and by the gas itself, the balloon (lung) expansion and, therefore, its elastic recoil $(P_l)$ cannot catch up to the more negative $P_{pl}$ as long as we continues to contract the diaphragm more. When, eventually, we stop pulling on the rubber sheet (equivalent to the diaphragm no longer contracting), $P_l$ catches up to and equals $P_{pl}$, but at greater absolute values (e.g., $P_{pl} = -8$ cmH$_2$O and $P_l = +8$ cmH$_2$O). At this point, inspiration stops.

For expiration to occur at rest, all we have to do is relax the diaphragm. The energy stored in the lung, its elastic recoil (due to stretching of the elastic fibres and, more importantly, increased surface tension), is sufficient to generate expiratory flow. Thus, at the onset of expiration, $P_{pl}$ becomes slightly less negative, say $-7.9$ cmH$_2$O. But $P_l$ is still 8 cmH$_2$O. Therefore,

$$P_A = P_{tp} + P_{pl} = +8 + (-7.9) = +0.1 \text{ cmH}_2\text{O}$$

In other words, $P_A$ is now greater than $P_B$ (atmospheric = mouth pressure) and expiratory flow starts.

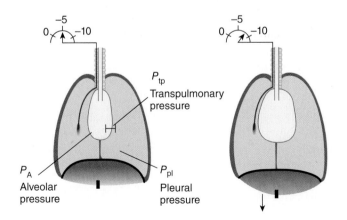

$P_{tp}$
Transpulmonary pressure

$P_A$
Alveolar pressure

$P_{pl}$
Pleural pressure

# Glossary

Definitions are provided for selected key terms.

## A

**A band** One of the dark bands that alternate with light (I) bands to create a striated appearance in a skeletal or cardiac muscle fibre when these fibres are viewed with a light microscope (p. 300)

**absolute refractory period** Along with the *relative refractory period*, one of the two components of the refractory period. These two periods occur as a result of the changing status of the voltage-gated $Na^+$ and $K^+$ channels during and after an action potential. (p. 67)

**accessory digestive organs** Exocrine organs outside the wall of the digestive tract that empty their secretions through ducts into the digestive tract lumen (p. 647)

**accessory sex glands** Glands that empty their secretions into the reproductive tract (p. 724)

**accommodation** The ability to adjust the strength of the lens in the eye so that both near and far sources can be focused on the retina (p. 157)

**acetylcholine (ACh)** (as′-uh-teal-KŌ-lēn) The neurotransmitter released from all autonomic preganglionic fibres, parasympathetic postganglionic fibres, and motor neurons (p. 188)

**acetylcholinesterase (AChE)** (as′-uh-teal-kō-luh-NES-tuh-rās) An enzyme present in the motor-end-plate membrane of a skeletal muscle fibre that inactivates acetylcholine (p. 199)

**ACh** See acetylcholine. (p. 188)

**AChE** See *acetylcholinesterase*. (p. 199)

**acidaemia** A plasma pH that is below 7.35 (p. 621)

**acid-base balance** The precise regulation of free hydrogen ion concentration in the body fluids (p. 620)

**acidic** A solution with a pH of less than 7.0 or with a $[H^+]$ greater than that of pure water (p. 621)

**acidosis** (as-i-DŌ-sus) Blood pH of less than 7.35 (p. 621)

**acids** Hydrogen-containing substance that yields a free hydrogen ion and anion on dissociation (p. 620)

**acini** (ĀS-i-nī) The secretory component of the saclike exocrine glands, such as digestive enzyme–producing pancreatic glands or milk-producing mammary glands (p. 667)

**acquired (conditioned) reflexes** A new or modified response elicited by a stimulus after conditioning (p. 105)

**acquired immunodeficiency syndrome (AIDS)** A set of symptoms and infections resulting from damage to the immune system caused by the human immunodeficiency virus (HIV) (p. 482)

**acromegaly** A disproportionate growth pattern producing a disfiguring condition ; bone thickening in the extremities and the face. (p. 238)

**ACTH** See *adrenocorticotropic hormone*. (p. 227)

**active forces** Forces that require expenditure of cellular energy (ATP) in the transport of a substance across the plasma membrane (p. 37)

**active hyperaemia** The increased blood flow resulting when a tissue is active (p. 398)

**active reabsorption** A type of tubular reabsorption in which any one of the five steps in the transepithelial transport of a substance reabsorbed across the kidney tubules requires energy expenditure (p. 574)

**active transport** Active carrier-mediated transport involving transport of a substance against its concentration gradient across the plasma membrane (p. 43)

**acuity** Discriminative ability; the ability to discern between two different points of stimulation (p. 147)

**adaptation** A reduction in receptor potential despite sustained stimulation of the same magnitude (p. 145)

**addicted** Compulsive seeking out and taking a drug at all costs, first to experience the pleasurable sensations and later to avoid the negative withdrawal symptoms, even when the drug no longer provides pleasure (p. 82)

**A-delta (δ) fibre** Type of sensory fibre associated with cold and pressure. Nociceptor stimulation is interpreted as fast pain information. (p. 149)

**adenohypophysis** The anterior pituitary (p. 226)

**adenosine diphosphate (ADP)** (uh-DEN-uh-sēn) The two-phosphate product formed from the splitting of ATP to yield energy for the cell's use (p. 24)

**adenosine triphosphate (ATP)** The body's common energy currency, which consists of an adenosine with three phosphate groups attached; splitting of the high-energy, terminal phosphate bond provides energy to power cellular activities (p. 24)

**adenylyl cyclase** (ah-DEN-il-il sī-klās) The membrane-bound enzyme that is activated by a G protein intermediary in response to the binding of an extracellular messenger with a surface membrane receptor and that, in turn, activates cyclic AMP, an intracellular second messenger (p. 217)

**adequate stimulus** A property of a sensory receptor that determines the type of energy to which a sensory receptor responds with the initiation of sensory transduction (p. 143)

**ADH** Antidiuretic hormone. See *vasopressin*. (p. 82)

**adipocytes** Fat cells in adipose tissue; store triglyceride fat and secrete hormones called *adipokines* (p. 703)

**adipokines** Hormones secreted by adipose tissue that play important roles in energy balance and metabolism (p. 703)

**adipose tissue** The tissue specialized for storage of triglyceride fat; found under the skin in the hypodermis (p. 462)

**adolescence** The stage of physical and mental human development that occurs between childhood and adulthood; involves biological, social, and psychological changes (p. 734)

**ADP** See *adenosine diphosphate*. (p. 24)

**adrenal cortex** (uh-DRĒ-nul) The outer portion of the adrenal gland; secretes three classes of steroid hormones: glucocorticoids, mineralocorticoids, and sex hormones (p. 254)

**adrenal medulla** (muh-DUL-uh) The inner portion of the adrenal gland; an endocrine gland that is a modified sympathetic ganglion that secretes the hormones epinephrine and norepinephrine into the blood in response to sympathetic stimulation (p. 192)

**adrenergic fibres** (ad′-ruh-NUR-jik) Nerve fibres that release norepinephrine as their neurotransmitter (p. 188)

**aerobic** Condition in which oxygen is available (p. 28)

**aerobic (endurance) exercise** Exercise that can be supported by ATP formation, accomplished by oxidative phosphorylation because adequate oxygen is available to support the muscle's modest energy demands; also called endurance-type exercise (p. 320)

**afferent arterioles** (AF-er-ent ar-TIR-ē-ōl) Vessels that carry blood into the glomerulus of the kidney's nephron (p. 562)

**afferent neurons** Neurons that possess a sensory receptor at their peripheral ending and carry information to the central nervous system (p. 92)

**afferent pathway** [neuron] Nerve pathway that carries nerve impulses from receptors toward the central nervous system (p. 106)

**after hyperpolarization** (hī′-pur-pō-luh-ruh-ZA-shun) A slight, transient hyperpolarization that sometimes occurs at the end of an action potential (p. 62)

**afterload** The pressure that the left ventricle of the heart must generate in order to eject blood from the chamber (p. 369)

**agonists** Substances that initiate a physiological response when combined with a receptor (p. 193)

**AIDS** See *acquired immunodeficiency syndrome*. (p. 482)

**albumins** (al-BEW-min) Smallest and most abundant of the plasma proteins; binds and transports many water-insoluble substances in the blood; contributes extensively to plasma-colloid osmotic pressure (p. 435)

**aldosterone** (al-dō-steer-OWN) or (al-DOS-tuh-rōn) The adrenocortical hormone that stimulates $Na^+$ reabsorption by the distal and collecting tubules of the kidney's nephron during urine formation (p. 255)

**alkalaemia** A plasma pH that is above 7.45 (p. 621)

**alkalosis** (al′-kuh-LŌ-sus) Blood pH that is greater than 7.45 (p. 621)

**allergy** Acquisition of an inappropriate, specific immune reactivity to a normally harmless environmental substance (p. 491)

**all-or-none law** An excitable membrane either responds to a stimulus with an action potential that spreads nondecrementally throughout the membrane or does not respond with an action potential at all. (p. 69)

**α (alpha) cells** The endocrine pancreatic cells that secrete the hormone glucagon (p. 271)

**alpha motor neurons** Motor neurons that innervate ordinary skeletal muscle fibres (p. 325)

**alveolar dead space** (al-VE-o-lur) The difference between physiological dead space and anatomical dead space, which represents that part of the physiological dead space resulting from ventilation of relatively non-perfused alveoli (p. 529)

**alveolar surface tension** (al-VE-o-lur) The surface tension of the fluid lining the alveoli in the lungs (p. 509)

**alveolar ventilation** ($V_A$) The volume of air exchanged between the atmosphere and alveoli per minute; equals (tidal volume minus dead space volume) multiplied by respiratory rate (p. 522)

**alveoli** [of lungs] (al-VĒ-ō-lī) The air sacs across which oxygen and carbon dioxide are exchanged between the blood and air in the lungs (p. 502)

**amniotic cavity** The fluid-filled cavity surrounding the developing embryo (p. 761)

**amniotic fluid** The fluid within the amnion that surrounds, nourishes, and protects the fetus from injury (p. 761)

**AMPA receptors** Primarily responsible for generating EPSPs in response to glutamate activation (p. 130)

**ampulla** A dilated portion of a canal or duct (p. 178)

**amyotrophic lateral sclerosis (ALS)** A neurological disorder characterized by progressive degeneration of motor cells in the spinal cord and brain, often referred to as Lou Gehrig's disease (p. 195)

**anabolic** See *anabolism*. (p. 22)

**anabolism** (ah-NAB-ō-li-zum) The buildup, or synthesis, of larger organic molecules from the small organic molecular subunits (p. 267)

**anaemia** A reduction below normal in $O_2$-carrying capacity of the blood (p. 441)

**anaerobic** (an´-uh-RŌ-bik) Condition in which oxygen is not present (p. 28)

**anaerobic (high-intensity) exercise** High-intensity exercise that can be supported by ATP formation accomplished by anaerobic glycolysis for brief periods of time when $O_2$ delivery to a muscle is inadequate to support oxidative phosphorylation (p. 28)

**analgesic system** Neural pathway that includes the secretion of endorphins by the brain in response to the perception of pain (p. 150)

**anatomic dead space** ($V_d$) The volume of the conducting airways, including the nose, mouth, trachea, and alveoli; represents that portion of inspired gas that is unavailable for exchange with pulmonary capillary blood (p. 521)

**anatomy** A branch of biology that considers the structure of living things; closely interrelated with physiology because body functions are highly dependent on the structure of the body parts that carry them out (p. 4)

**androgens** Masculinizing, male sex hormone; includes testosterone from the testes and dehydroepiandrosterone from the adrenal cortex (p. 232)

**angiogenesis** A physiological process involving the formation and growth of new blood vessels from pre-existing vessels (p. 399)

**anion gap** The unmeasured negatively charged ions in the plasma, calculated by taking the difference between the laboratory measured cations and the laboratory anions (p. 634)

**antagonism** Actions opposing each other; in the case of hormones, when one hormone causes the loss of another hormone's receptors, reducing the effectiveness of the second hormone (p. 214)

**antagonistic** Of or relating to an antagonist or its action (p. 192)

**antagonists** Substances that interfere with or inhibit the physiological action of another (p. 193)

**anterior pituitary gland** The glandular portion of the pituitary that synthesizes, stores, and secretes six different hormones: growth hormone, TSH, ACTH, FSH, LH, and prolactin (p. 225)

**antibodies** Immunoglobulins produced by a specific activated B lymphocyte (plasma cell) against a particular antigen; bind with the specific antigen against which they are produced and promote the antigenic invader's destruction by augmenting nonspecific immune responses already initiated against the antigen (p. 475)

**antibody-mediated (humoural) immunity** A specific immune response accomplished by antibody production by B cells (p. 473)

**antidiuretic hormone** j(an´-ti-dī´-yū-RET-ik) See *vasopressin*. (p. 592)

**antigen** A large, complex molecule that triggers a specific immune response against itself when it gains entry into the body (p. 474)

**antioxidant** A substance that helps inactivate biologically damaging free radicals (p. 242)

**antrum** [of ovary] The fluid-filled cavity formed within a developing ovarian follicle (p. 656)

**antrum** [of stomach] The lower portion of the stomach (p. 656)

**aorta** (a-OR-tah) The large vessel that carries blood from the left ventricle (p. 347)

**aortic valve** A one-way valve that permits the flow of blood from the left ventricle into the aorta during ventricular emptying but prevents the backflow of blood from the aorta into the left ventricle during ventricular relaxation (p. 350)

**aphasias** Language disorders caused by damage to specific cortical areas, mostly resulting from strokes; not to be confused with speech impediments, which are caused by a defect in the mechanical aspect of speech, such as weakness or incoordination of the muscles controlling the vocal apparatus (p. 132)

**apnea** The cessation of breathing (p. 550)

**aquaporins (water channels)** Membrane proteins that form channels used for the passage of water. About a billion water molecules can pass single file through an aquaporin channel in one second. (p. 38)

**aqueous humour** (Ā-kwē-us) The clear, watery fluid in the anterior chamber of the eye; provides nourishment for the cornea and lens (p. 153)

**arachnoid mater** A delicate, richly vascularized layer with a cobweb appearance. The space between the arachnoid layer and the underlying pia mater, the subarachnoid space, is filled with CSF. (p. 100)

**arachnoid villi** Protrusions of arachnoid tissue that penetrate through gaps in the overlying dura and project into the dural sinuses. CSF is reabsorbed across the surfaces of these villi into the blood circulating within the sinuses. (p. 100)

**arcuate nucleus** (ARE-kyou-it´) Subcortical brain region that houses neurons that secrete appetite-enhancing neuropeptide Y and those that secrete appetite-suppressing melanocortins (p. 703)

**arrhythmia** Variation from the normal rhythm of the heartbeat, encompassing abnormalities of rate, regularity, origin of impulse, and sequence of activation (p. 359)

**arterioles** (ar-TIR-ē-ōlz) The highly muscular, high-resistance vessels, the calibre of which can be changed subject to control to determine how much of the cardiac output is distributed to each of the various tissues (p. 394)

**arteries** Vessels that carry blood away from the heart (p. 346)

**ascending tracts** Bundles of nerve fibres of similar function that travel up the spinal cord to transmit signals derived from afferent input to the brain (p. 103)

**association areas** Areas of the brain that make up much of the cerebral cortex and are free to process all kinds of information; enable the brain to produce behaviours requiring the coordination of many brain areas (p. 133)

**asthma** An obstructive pulmonary disease characterized by profound constriction of the smaller airways caused by allergy-induced spasm of the smooth muscle in the walls of these airways (p. 492)

**astigmatism** A defect in the eye or lens caused by a deviation from spherical curvature, resulting in distorted images, as light rays are prevented from meeting at a common focus (p. 156)

**astrocytes** A type of glial cell in the brain, whose major functions include holding the neurons together in proper spatial relationship, inducing the brain capillaries to form tight junctions important in the blood–brain barrier, and enhancing synaptic activity (p. 95)

**asynchronous recruitment of motor units** Condition in which the body alternates motor unit activity to give motor units that have been active an opportunity to rest while others take over; involves a carefully coordinated shift of motor unit activity so that sustained contraction is smooth rather than jerky (p. 310)

**atherosclerosis** (ath-uh-rō-skluh-RŌ-sus) A progressive, degenerative arterial disease that leads to gradual blockage of affected vessels, reducing blood flow through them (p. 374)

**ATP** See *adenosine triphosphate*. (p. 24)

**ATP synthase** Converts ADP 1 Pi to ATP, providing a rich yield of 32 more ATP molecules for each glucose molecule processed (p. 27)

**atria** (*atrium*, singular) (Ā-tree-a) Upper chambers of the heart that receive blood from the veins and transfers it to the ventricles (p. 346)

**atrioventricular node (AV node)** (ā´-trē-ō-ven-TRIK-yuh-lur) A small bundle of specialized cardiac cells at the junction of the atria and ventricles that is the only site of electrical contact between the atria and ventricles (p. 352)

**atrophy** (AH-truh-fē) Decrease in mass of an organ (p. 320)

**auditory (cochlear) nerve** A bundle of nerve fibres that carries hearing information between the cochlea and the brain (p. 174)

**autoimmune disease** Disease characterized by erroneous production of antibodies against one of the body's own tissues (p. 486)

**autonomic nervous system (ANS)** The portion of the efferent division of the peripheral nervous system that innervates smooth and cardiac muscle and exocrine glands; composed of two subdivisions—the sympathetic nervous system and the parasympathetic nervous system (p. 92)

**autorhythmicity** The ability of an excitable cell to rhythmically initiate its own action potentials (p. 352)

**AV nodal delay** The delay in impulse transmission between the atria and ventricles at the AV node to allow enough time for the atria to become completely depolarized and contract, emptying their contents into the ventricles, before ventricular depolarization and contraction occur (p. 355)

**axon** A single, elongated tubular extension of a neuron that conducts action potentials away from the cell body; also known as a *nerve fibre* (p. 65)

**axon hillock** The first portion of a neuronal axon plus the region of the cell body from which the axon leaves; the site of action-potential initiation in most neurons (p. 65)

**axon terminals** The branched endings of a neuronal axon, which release a neurotransmitter that influences target cells in close association with the axon terminals (p. 66)

## B

**baroreceptor reflex** An autonomically mediated reflex response that influences the heart and blood vessels to oppose a change in mean arterial blood pressure (p. 420)

**baroreceptors** Receptors located within the circulatory system that monitor blood pressure (p. 419)

**basal metabolic rate (BMR)** (BĂ-sul) The minimal waking rate of internal energy expenditure; occurs when the body is at rest (p. 701)

**basal ganglia** Several masses of grey matter located deep within the white matter of the cerebrum of the brain; play an important inhibitory role in motor control (p. 123)

**base** A substance that can combine with a free hydrogen ion and remove it from solution (p. 346)

**basic behavioural patterns** Behaviours controlled at least in part by the limbic system; include those aimed at individual survival (attack, searching for food) and those directed toward perpetuating the species (sociosexual behaviours conducive to mating) (p. 124)

**basic electrical rhythm (BER)** Self-induced electrical activity of the digestive-tract smooth muscle (p. 650)

**basic or alkaline** A solution with a pH of greater than 7.0 or with a $[H^+]$ less than that of pure water (p. 621)

**basilar membrane** (BAS-ih-lar) The membrane that forms the floor of the middle compartment of the cochlea and bears the organ of Corti, the sense organ for hearing (p. 172)

**basophils** (BAY-so-fills) White blood cells that synthesize, store, and release histamine, which is important in allergic responses, and heparin, which hastens the removal of fat particles from the blood as well as prevents blood clotting (p. 447)

**BER** See *basic electrical rhythm.* (p. 650)

**β (beta) cells** The endocrine pancreatic cells that secrete the hormone insulin (p. 271)

**bicarbonate ion** ($HCO_3^-$) The anion resulting from dissociation of carbonic acid, $H_2CO_3$ (p. 437)

**bile salts** Cholesterol derivatives secreted in the bile that facilitate fat digestion through their detergent action and facilitate fat absorption through their micellar formation (p. 673)

**biliary system** (BIL-ē-air´-ē) The bile-producing system, consisting of the liver, gallbladder, and associated ducts (p. 671)

**bilirubin** (bill-eh-RU-bin) A bile pigment, which is a waste product derived from the degradation of haemoglobin during the breakdown of old red blood cells (p. 674)

**binocular field of vision** The field of vision simultaneously received by both eyes; varies depending on the position of the eyes in the skull (p. 167)

**bipolar cells** Specialized sensory neurons that display graded potentials and have action potentials that do not originate until the ganglion cells propagate the sensory message over long distances to the brain; includes smell, sight, taste, hearing, and vestibular functions (p. 157)

**bitter taste** The most sensitive of the tastes, and perceived by most to be unpleasant, sharp, or disagreeable; includes tastes associated with unsweetened cocoa, coffee, beer, and olives (p. 184)

**blastocyst** The developmental stage of the fertilized ovum by the time it is ready to implant; consists of a single-layered sphere of cells encircling a fluid-filled cavity (p. 758)

**blind spot** An area that cannot be directly observed under existing circumstances; physiologically, the small, circular, optically insensitive region in the retina, where fibres of the optic nerve emerge from the eyeball (p. 158)

**blood–brain barrier (BBB)** Special structural and functional features of the brain capillaries that limit access of materials from the blood into the brain tissue (p. 100)

**B lymphocytes (B cells)** White blood cells that produce antibodies against specific targets to which they have been exposed (p. 447)

**body** [of the stomach] The main, or middle, part of the stomach (p. 7)

**body systems** Collection of organs that perform related functions and interact to accomplish a common activity that is essential for survival of the whole body; for example, the digestive system (p. 7)

**bone marrow** The soft, highly cellular tissue that fills the internal cavities of bones; the source of most blood cells (p. 437)

**botulism** Food poisoning caused by a bacterium (*botulinum*) growing on improperly sterilized meats and preserved foods (p. 199)

**Bowman's capsule** The beginning of the tubular component of the kidney's nephron that cups around the glomerulus and collects the glomerular filtrate as it is formed (p. 563)

**Boyle's/Mariotte's law** (boils) At any constant temperature, the pressure exerted by a gas varies inversely with the volume of the gas (p. 511)

**bradycardia** A resting heart rate less than 60 beats per minute; seldom symptomatic until the rate drops below 50 beats per minute (p. 359)

**bradykinin** A compound sometimes released in the blood that causes contraction of smooth muscle and vasodilation; increases capillary permeability, constricts smooth muscle, and stimulates pain receptors (p. 150)

**brain stem** The portion of the brain that is continuous with the spinal cord; serves as an integrating link between the spinal cord and higher brain levels, and controls many life-sustaining processes, such as breathing, circulation, and digestion (p. 109)

**bronchioles** (BRONG-kē-ōlz) The small, branching airways within the lungs (p. 502)

**bronchoconstriction** Narrowing of the respiratory airways (p. 515)

**bronchodilation** Widening of the respiratory airways (p. 515)

**brush border** The collection of microvilli projecting from the luminal border of epithelial cells lining the digestive tract and kidney tubules (p. 678)

**bulbourethral glands** (bul-bo-you-RĔTH-ral) Male accessory sex glands that secrete mucus for lubrication (p. 740)

**bulk flow** Movement in bulk of a protein-free plasma across the capillary walls between the blood and surrounding interstitial fluid; encompasses ultrafiltration and reabsorption (p. 409)

**bundle of His (atrioventricular bundle)** (hiss) A tract of specialized cardiac cells that rapidly transmits an action potential down the interventricular septum of the heart (p. 352)

## C

**cadherins** Connective tissue fibres that interlace on the surface of adjacent cells and interlock in zipper fashion to help hold the cells within tissues and organs together (p. 32)

**calcitonin** (kal´-suh-TŌ-nun) A hormone secreted by the thyroid C cells that lowers plasma $Ca^{2+}$ levels (p. 248)

**calcium balance** Maintenance of a constant total amount of $Ca^{2+}$ in the body; accomplished by slowly responding adjustments in intestinal $Ca^{2+}$ absorption and in urinary $Ca^{2+}$ excretion (p. 285)

**calcium homeostasis** Maintenance of a constant free plasma $Ca^{2+}$ concentration, accomplished by rapid exchanges of $Ca^{2+}$ between the bone and ECF and, to a lesser extent, by modifications in urinary $Ca^{2+}$ excretion (p. 284)

**calmodulin** (kal´-MA-jew-lin) An intracellular $Ca^{2+}$–binding protein that, on activation by $Ca^{2+}$, induces a change in structure and function of another intracellular protein; especially important in smooth-muscle excitation–contraction coupling (p. 219)

**CAMs** See *cell adhesion molecules.* (p. 32)

**capacitance vessels** The distensibility of blood vessels, mainly veins, which are capable of holding and storing approximately 60–70 percent of the body's blood volume (p. 414)

**capillaries** The thin-walled, pore-lined smallest of blood vessels, across which exchange between the blood and surrounding tissues takes place (p. 441)

**capsaicin** A colourless, pungent compound derived from capsicum; the source of the hotness of hot peppers (p. 149)

**carbonic anhydrase** (an-HĪ-drās) The enzyme that catalyzes the conversion of $CO_2$ and $H_2O$ into carbonic acid, $H_2CO_3$ (p. 437)

**cardiac muscle** The specialized muscle found only in the heart (p. 326)

**cardiac output (CO)** The volume of blood pumped by each ventricle each minute; equals stroke volume multiplied by heart rate (p. 365)

**cardiovascular control centre** The integrating centre located in the medulla of the brain stem that controls mean arterial blood pressure (p. 402)

**carrier-mediated transport** Transport of a substance across the plasma membrane, facilitated by a carrier molecule (p. 41)

**carrier molecules** Membrane proteins, which, by undergoing reversible changes in shape so that specific binding sites are alternately exposed at either side of the membrane, can bind with and transfer particular substances otherwise unable to cross the plasma membrane on their own (p. 31)

**cartilage** Tough and flexible connective tissue found in many areas of the body for structure and support (p. 233)

**cascade** A series of sequential reactions that culminates in a final product, such as a clot (p. 220)

**catabolic** See *catabolism*. (p. 267)

**catabolism** (kuh-TAB-ō-li-zum) The breakdown, or degradation, of large, energy-rich molecules within cells (p. 267)

**cataract** A cloudiness or opacity in the normally transparent lens of the eye that can cause a decrease in vision and lead to blindness (p. 157)

**cauda equina** The thick bundle of elongated nerve roots within the lower vertebral canal; called the cauda equina ("horse's tail") because of its appearance (p. 102)

**C cells** The thyroid cells that secrete calcitonin (p. 248)

**CD4 cells** See *helper T cells*. (p. 482)

**CD8 cells (cytotoxic, or killer, T cells)** The population of T cells that destroys host cells bearing foreign antigen, such as body cells invaded by viruses or cancer cells (p. 482)

**cell** The smallest unit capable of carrying out the processes associated with life; the basic unit of both structure and function of living organisms (p. 5)

**cell adhesion molecules (CAMs)** Proteins that protrude from the surface of the plasma membrane and form loops or other appendages that the cells use to grip one another and the surrounding connective tissue fibres (p. 32)

**cell body** The portion of a neuron that houses the nucleus and organelles (p. 64)

**cell differentiation** The process by which less specialized cells become more specialized (p. 5)

**cell-mediated immunity** A specific immune response accomplished by activated T lymphocytes, which directly attack unwanted cells (p. 473)

**central chemoreceptors** (kē-mō-rē-SEP-turz) Receptors located in the medulla near the respiratory centre; respond to changes in ECF H⁺ concentration resulting from changes in arterial $P_{CO_2}$ and adjust respiration accordingly (p. 547)

**central fatigue** Occurs when the CNS no longer adequately activates the motor neurons supplying the working muscles; causes a person to slow down or stop exercising even though the muscles can still perform; often psychologically based (p. 318)

**central lacteal** (LAK-tē-ul) The initial lymphatic vessel that supplies each of the small-intestinal villi (p. 680)

**central nervous system (CNS)** The brain and spinal cord (p. 92)

**central sulcus** (SUL-kus) A deep infolding of the brain surface that runs roughly down the middle of the lateral surface of each cerebral hemisphere and separates the parietal and frontal lobes (p. 115)

**cerebellum** (ser´-uh-BEL-um) The part of the brain attached at the rear of the brain stem; concerned with maintaining proper position of the body in space and subconscious coordination of motor activity (p. 121)

**cerebral cortex** The outer shell of grey matter in the cerebrum; site of initiation of all voluntary motor output and final perceptual processing of all

sensory input, as well as integration of most higher neural activity (p. 114)

**cerebral hemispheres** The cerebrum's two halves, which are connected by a thick band of neuronal axons (p. 113)

**cerebrocerebellum** Pertaining to the cerebrum and the cerebellum (p. 122)

**cerebrospinal fluid (CSF)** (ser´-uh-brō-SPĪ-nul) or (sah-REE-brō-SPĪ-nul) A special cushioning fluid that is produced by, surrounds, and flows through the central nervous system (p. 100)

**cerebrovascular accident (CVA)** Brain damage that occurs when a brain (cerebral) blood vessel is blocked by a clot or ruptures and the brain tissue supplied by that vessel loses its vital oxygen and glucose supply; results in damage and usually death of the deprived tissue (p. 114)

**cerebrum** (SER-uh-brum) or (sah-REE-brum) The division of the brain that consists of the basal nuclei and cerebral cortex (p. 94)

**cerumen** A brownish yellow, waxy secretion of the ceruminous glands of the external auditory meatus; commonly known as earwax (p. 172)

**C fibres** Afferent fibres in the peripheral nerves of the somatic sensory system; convey input signals from the periphery to the central nervous system (p. 149)

**channels** Small, water-filled passageways through the plasma membrane; formed by membrane proteins that span the membrane and provide highly selective passage for small water-soluble substances such as ions (p. 58)

**chemical buffer system** A mixture in a solution of two or more chemical compounds that minimize pH changes when either an acid or a base is added to or removed from the solution (p. 623)

**chemical disequilibrium** The concept of having an unequal distribution of solutes in different fluid compartments (p. 611)

**chemical gradient** A passive process in which ions move down an electrochemical gradient through open channels (from high to low concentration and by attraction of ion to area of opposite charge) (p. 37)

**chemically gated channels** Channels in the plasma membrane that open or close in response to the binding of a specific chemical messenger with a membrane receptor site that is in close association with the channel (p. 58)

**chemical nociceptors** Pain receptors (e.g., TRP) that respond to noxious chemical stimuli (p. 149)

**chemiosmotic mechanism** A four-step process that collectively harnesses the energy stored in the H⁺ gradient across the inner mitochondrial membrane to synthesize ATP (p. 27)

**chemoreceptors** (kē-mo-rē-sep'-tur) Sensory receptors sensitive to specific chemicals (p. 144)

**chief cells** The cells in the gastric pits that secrete pepsinogen (p. 661)

**cholecystokinin (CCK)** (kō´-luh-sis-tuh-kī-nun) A hormone released from the duodenal mucosa primarily in response to the presence of fat; inhibits gastric motility and secretion, stimulates pancreatic enzyme secretion, stimulates gallbladder contraction, and acts as a satiety signal (p. 659)

**cholesterol** A type of fat molecule that serves as a precursor for steroid hormones and bile salts and is a stabilizing component of the plasma membrane (p. 29)

**cholinergic fibres** (kō´-lin-ER-jik) Nerve fibres that release acetylcholine as their neurotransmitter (p. 188)

**choroid** The pigmented vascular layer of the eyeball between the retina and the sclera (p. 153)

**choroid plexuses** Tissue found in particular regions of the ventricle cavities of the brain; consist of richly vascularized, cauliflower-like masses of pia mater tissue that dip into pockets formed by ependymal cells; have a key role in the production of cerebrospinal fluid as a result of selective transport mechanisms across their membranes (p. 100)

**chronic obstructive pulmonary disease (COPD)** A group of lung diseases characterized by increased airway resistance resulting from narrowing of the lumen of the lower airways; includes asthma, chronic bronchitis, and emphysema (p. 516)

**chyme** (kīm) A thick liquid mixture of food and digestive juices (p. 657)

**ciliary body** The portion of the eye that produces aqueous humour and contains the ciliary muscle (p. 153)

**ciliary muscle** A circular ring of smooth muscle within the eye whose contraction increases the strength of the lens to accommodate for near vision (p. 157)

**circular (constrictor) muscle** Circularly oriented muscle layer contained within the iris of the eye that contracts in response to parasympathetic stimulation to decrease the size of the pupil and reduce the amount of light entering the eye (p. 155)

**circulatory shock** Condition that occurs when mean arterial blood pressure falls so low that adequate blood flow to the tissues can no longer be maintained (p. 426)

**circumventricular organs** Highly vascularized structures in the brain lacking a blood–brain barrier (p. 101)

**citric acid cycle** A cyclic series of biochemical reactions that involves the further processing of intermediate breakdown products of nutrient molecules, resulting in the generation of carbon dioxide and the preparation of hydrogen-carrier molecules for entry into the high-energy-yielding electron transport chain (p. 25)

**classical neurotransmitters** Synthesized in nerve terminals (acetylcholine, GABA, etc.), and generally present within the neuron; stimulus-dependent for release, act on postsynaptic cells, and have a mechanism of removal (p. 80)

**CNS** See *central nervous system*. (p. 92)

**cochlea** (KOK-lē-uh) The snail-shaped portion of the inner ear that houses the receptors for sound (p. 172)

**cochlear duct** An endolymph-filled cavity inside the cochlea, between the scala tympani and the scala vestibule; separated from these compartments by the basilar membrane and Reissner's membrane, respectively (p. 172)

**cochlear implants** Surgically implanted electronic device that provides a sense of sound to a person who is well below normal hearing (p. 177)

**cognition** The act or process of knowing, including both awareness and judgment (p. 93)

**collagen** Protein that forms cable-like fibres or sheets that provide tensile strength (resistance to longitudinal stress) (p. 34)

**collateral circulation** Occurs when an area of tissue or an organ has been cut off from its normal blood supply; forms new pathways (i.e., anastamoses) to deliver blood to the tissue or organ (p. 378)

**collateral ganglia** Sympathetic ganglia that lie about halfway between the CNS and the target organ (p. 188)

**collecting duct (tubule)** The last portion of tubule in the kidney's nephron that empties into the renal pelvis (p. 564)

**colloid** (KOL-oid) The thyroglobulin-containing substance enclosed within the thyroid follicles (p. 248)

**colour blindness** Condition that occurs in individuals who lack a particular type of cone, so their colour vision is a product of the differential sensitivity of only two types of cones; causes a different perception of colours and an inability to distinguish as many varieties of colours (p. 165)

**colour vision** The ability to distinguish objects that are observed when incoming light reacts with the several types of cone photoreceptors in the eye; a process in which the brain and the nervous system respond to the stimuli (p. 165)

**competition** Occurs when several closely related compounds compete for a ride across the membrane on the same carrier. If a given binding site can be occupied by more than one type of molecule, the rate of transport of each substance is less when both molecules are present than when either is present by itself. (p. 42)

**complement system** A collection of plasma proteins activated in cascade fashion on exposure to invading microorganisms, ultimately producing a membrane attack complex that destroys the invaders (p. 473)

**compliance** The distensibility of a hollow, elastic structure, such as a blood vessel or the lungs; a measure of how easily the structure can be stretched (p. 391)

**concave** Curved in, as a surface of a lens that diverges light rays (p. 155)

**concentration gradient** A difference in concentration of a particular substance between two adjacent areas (p. 37)

**concentric muscle contraction** A dynamic contraction that produces tension during a shortening motion; most familiar type of muscle contraction; occurs when actin filaments are pulled together by the myosin filaments, which move the Z lines closer together, shortening the sarcomere, and thus shortening the whole muscle (p. 315)

**conductive deafness** Condition that occurs when sound waves are not adequately conducted through the external and middle portions of the ear to set the fluids in the inner ear in motion (p. 177)

**cones** The eye's photoreceptors used for colour vision in the light (p. 153)

**congestive heart failure** The inability of the cardiac output to keep pace with the body's needs for blood delivery, with blood damming up in the veins behind the failing heart (p. 371)

**connective tissue** Tissue that serves to connect, support, and anchor various body parts; distinguished by relatively few cells dispersed within an abundance of extracellular material (p. 7)

**connexon** Six protein subunits arranged in a hollow tubelike structure. Two connexons extend outward, one from each of the plasma membranes of two adjacent cells, and join end to end to form the connecting tunnel between the two cells (p. 35)

**consciousness** The subjective awareness of both the external world and self, including awareness of thoughts, perceptions, and dreams; involves conscious experience that depends on the integrated functioning of many parts of the nervous system, even though the final level of awareness resides in the cerebral cortex and a crude sense of awareness is detected by the thalamus (p. 134)

**consolidation** Stabilization of memory by a process of transferring and fixing short-term memory traces into long-term memory stores (p. 126)

**contiguous conduction** The means by which an action potential is propagated throughout an unmyelinated nerve fibre; occurs when local current flow between an active and an adjacent inactive area brings the inactive area to threshold, triggering an action potential in a previously inactive area (p. 66)

**contractile component** The cross bridges formed by the actin and myosin molecules contained within the smallest functional unit of the muscle cell, the sarcomere (p. 313)

**contractile proteins** Myosin and actin, whose interaction brings about the shortening (contraction) of a muscle fibre (p. 301)

**contraction time** The period of time it takes for a muscle to generate force; varies among human skeletal muscle fibre types, for example, fast twitch and slow twitch (p. 308)

**control centre** The operation centre that integrates incoming information (p. 14)

**controlled variable** Some factor that can vary, but is controlled to reduce the amount of variability and keep the factor at a relatively steady state (p. 14)

**convection** Transfer of heat energy by air or water currents (p. 502)

**convergence** The converging of many presynaptic terminals from thousands of other neurons on a single neuronal cell body and its dendrites so that activity in the single neuron is influenced by the activity in many other neurons (p. 83)

**convex** Curved out, as a surface in a lens that converges light rays (p. 155)

**core temperature** The temperature within the inner core of the body (abdominal and thoracic organs, central nervous system, and skeletal muscles) that is homeostatically maintained at about 37.8°C (p. 711)

**cornea** (KOR-nee-ah) The clear, most anterior, outer layer of the eye through which light rays pass to the interior of the eye (p. 153)

**coronary artery disease (CAD)** Atherosclerotic plaque formation and narrowing of the coronary arteries that supply the heart muscle (p. 374)

**coronary circulation** The blood vessels that supply the heart muscle (p. 373)

**corpus callosum** The neural connection that enables the brain's two hemispheres to communicate and cooperate with each other by means of constant exchange of information; the body's "information superhighway" (p. 113)

**corpus luteum** (LOO-tē-um) The ovarian structure that develops from a ruptured follicle after ovulation (p. 749)

**corticospinal (or pyramidal) motor system** Efferent motor neurons that terminate within the spinal cord (p. 323)

**cortisol** (KORT-uh-sol) The adrenocortical hormone that plays an important role in carbohydrate, protein, and fat metabolism and helps the body resist stress (p. 256)

**cotransport carriers** Carriers, such as the luminal carriers in intestinal and kidney cells that have two binding sites, each of which interacts with a different molecule to be transported; for example, the $Na^+$ and nutrient cotransporters, or the $Na^+$–$K^+$ ATPase (p. 44)

**cranial nerves** The 12 pairs of peripheral nerves, the majority of which arise from the brain stem (p. 110)

**creatine kinase (CK)** An enzyme expressed by various tissues and cell types; catalyzes the conversion of creatine and ATP to phosphocreatine (p. 24)

**creatine phosphate** A phosphorylated creatine molecule that acts as a rapidly mobilizable reserve of high-energy phosphates in skeletal muscle and brain tissues (p. 24)

**CREB** Refers to cAMP response element-binding, which is a cellular transcription factor; binds to particular DNA sequences, called cAMP response elements, increasing or decreasing their transcription of the downstream genes; commonly associated with the formation of long-term memory (p. 131)

**CREB2** This second messenger plays a regulatory role in LTP as well as in simpler forms of short-term memory, such as sensitization (p. 131)

**cross bridges (myosin heads)** The myosin molecules' globular heads that protrude from a thick filament within a muscle fibre and interact with the actin molecules in the thin filaments to bring about shortening of the muscle fibre during contraction (p. 300)

**crossed extensor reflex** Ensures that the opposite limb will be in a position to bear the weight of the body as an injured limb is withdrawn from a stimulus (p. 109)

**cupula** An overlying, caplike, gelatinous layer, in which the imbedded hairs protrude into the ampulla; sways in the direction of the surrounding endolymph movement, much like seaweed leaning in the direction of the prevailing tide (p. 178)

**curare** An alkaloid that will interfere with neuromuscular junction activity by blocking the effect of released ACh; reversibly binds to the ACh receptor sites on the motor end plate (p. 199)

**current** A flow of electrical charges; by convention, the direction of current flow is always designated by the direction in which the positive charges are moving (p. 59)

**cyclic adenosine monophosphate (cyclic AMP or cAMP)** An intracellular second messenger derived from adenosine triphosphate (ATP) (p. 217)

**cyclic AMP** See *cyclic adenosine monophosphate*. (p. 217)

**cyclic guanosine monophosphate (cyclic GMP or cGMP)** A second messenger in a system analogous to the cAMP system (p. 220)

**cytokines** All chemicals other than antibodies that are secreted by lymphocytes (p. 470)

**cytoplasm** (SĪ-tō-plaz´-um) Portion of the cell interior not occupied by the nucleus (p. 22)

## D

**deafness** Classified into two types—*conductive deafness* and *sensorineural deafness*—depending on the part of the hearing mechanism that fails to function adequately (p. 177)

**declarative memories** A capacity that requires conscious recall and involves the significant role of the hippocampus and associated temporal/limbic structures in maintaining a durable record of the everyday episodic events (p. 128)

**dehydration** A water deficit in the body (p. 614)

**dehydroepiandrosterone (DHEA)** (dē-HĪ-drō-ep-i-and-row-steer-own) The androgen (masculinizing hormone) secreted by the adrenal cortex in both sexes (p. 258)

**denervation atrophy** Condition that occurs after the nerve supply to a muscle is lost (p. 320)

**dense-core vesicles** Vesicles that undergo Ca²⁺-induced exocytosis and release neuropeptides at the same time that the neurotransmitter is released from the synaptic vesicles (p. 80)

**depolarization** (de´-pō-luh-ruh_ZĀ-shun) A reduction in membrane potential from resting potential; movement of the potential from resting toward 0 mV (p. 58)

**depolarization block** A prolonged depolarization of the diaphragm, resulting in respiratory paralysis; occurs when the voltage-gated Na⁺ channels are trapped in an inactivated state, thereby prohibiting the initiation of new action potentials and resultant contraction of the diaphragm; results in the inability to breathe (p. 199)

**depression** A mental disorder associated with defects in limbic system neurotransmitters, specifically, a functional deficiency of serotonin or norepinephrine or both; characterized by a pervasive negative mood accompanied by a generalized loss of interests, an inability to experience pleasure, and suicidal tendencies (p. 126)

**depth perception** The slight disparity in the information received from the two eyes allows the brain to estimate distance, which enables perception of three-dimensional objects in spatial depth; some depth perception possible using only one eye, based on experience and comparison with other cues (p. 167)

**dermatome** A specific region of the body surface that is supplied by a particular spinal nerve; concerns sensory input (p. 103)

**dermis** The connective tissue layer that lies under the epidermis in the skin; contains the skin's blood vessels and nerves (p. 461)

**descending tracts** Bundles of nerve fibres of similar function that travel down the spinal cord to relay messages from the brain to efferent neurons (p. 103)

**desmosomes** (dez´-muh-sōm) Adhering junctions between two adjacent, but nontouching, cells formed by the extension of filaments between the cells' plasma membranes; most abundant in tissues that are subject to considerable stretching (p. 34)

**DHEA** See *dehydroepiandrosterone.* (p. 258)

**diabetes insipidus** (in-SIP´-ud-us) An endocrine disorder characterized by a deficiency of vasopressin (p. 614)

**diabetes mellitus** (muh-LĪ-tus) An endocrine disorder characterized by inadequate insulin action (p. 275)

**diacylglycerol (DAG)** The product of PIP₂ breakdown, a component of the tails of the phospholipid molecules within the cell membrane itself (p. 219)

**diaphragm** (DIE-uh-fram) A dome-shaped sheet of skeletal muscle that forms the floor of the thoracic cavity; the major inspiratory muscle (p. 503)

**diaphysis** The fairly uniform cylindrical shaft of a long bone (p. 233)

**diastole** (dī-AS-tō-lē) The period of cardiac relaxation and filling (p. 361)

**diastolic pressure** The lowest pressure during the resting phase of the cardiac cycle; the period of time when the heart fills with blood following systole (contraction), and thus the ventricles are relaxed (p. 391)

**diencephalon** (dī´-un-SEF-uh-lan) The division of the brain that consists of the thalamus and hypothalamus (p. 112)

**diffusion** Random collisions and intermingling of molecules as a result of their continuous thermally induced random motion (p. 37)

**digestion** The breaking-down process whereby the structurally complex foodstuffs of the diet are converted into smaller absorbable units by the enzymes produced within the digestive system (p. 646)

**diploid number** (DIP-loid) A complete set of 46 chromosomes (23 pairs), as found in all human somatic cells (p. 727)

**dipsogen** A molecule that increases the urge to drink (p. 618)

**dissociation constant, K** The number that represents the proportion of molecules of a particular acid that separate to liberate free H⁺ (p. 620)

**distal tubule** A highly convoluted tubule that extends between the loop of Henle and the collecting duct in the kidney's nephron (p. 564)

**diurnal (circadian) rhythm** (dī-URN´-ul) (sir-KĀ-dē-un) Oscillations in the set point of various body activities, such as hormone levels and body temperature, that are very regular and have a frequency of one cycle every 24 hours, usually linked to the light–dark cycle; also known as biological rhythm (p. 212)

**divergence** The diverging, or branching, of a neuron's axon terminals so that activity in this single neuron influences the many other cells with which its terminals synapse (p. 83)

**docking marker** A protein located at a specific destination within the cell where an appropriate vesicle can "dock" and "unload" its cargo (p. 31)

**docking-marker acceptors** A protein that binds lock-and-key fashion with the docking markers of secretory vesicles (p. 31)

**dorsal (posterior) horn** The posterior column of grey matter in the spinal cord when viewed in cross-section; receives several types of sensory information, including light touch, proprioception, and vibration (p. 103)

**dorsal root** Area where afferent fibres carrying incoming signals from peripheral receptors enter the spinal cord (p. 103)

**dorsal root ganglion** A cluster of afferent neuronal cell bodies located adjacent to the spinal cord (p. 103)

**down regulation** A reduction in the number of receptors for (and thereby the target cell's sensitivity to) a particular hormone as a direct result of the effect that an elevated level of the hormone has on its own receptors (p. 214)

**dura mater** A tough, inelastic covering consisting of two layers (*dura* means "tough") (p. 98)

**dural sinuses** The smaller of the two types of cavities that receive venous blood from the brain and from which the blood returns to the heart. The larger cavities are known as venous sinuses. (p. 98)

**duration** A continuance of length of time (p. 144)

**dwarfism** Hyposecretion of GH in a child is one cause of dwarfism, where the predominant feature is short stature caused by retarded skeletal growth. GH deficiency may be caused by a pituitary defect (lack of GH) or may occur secondary to hypothalamic dysfunctions (lack of GHRH). (p. 238)

**dyslexia** Stems from a deficit in phonological processing, meaning an impaired ability to break down written words into their underlying phonetic components. Dyslexics have difficulty decoding and thus identifying and assigning meaning to words. The condition is in no way related to intellectual ability. (p. 132)

**E**

**ear** A prominent skin-covered flap of cartilage; collects sound waves and channels them down the external ear canal. The ear consists of three parts: the *external*, the *middle*, and the *inner ear.* (p. 169)

**ear canal** A tube running from the outer ear to the middle ear; also called the external auditory meatus; (p. 172)

**ECG** See *electrocardiogram.* (p. 357)

**EDV** See *end-diastolic volume.* (p. 361)

**EEG** See *electroencephalogram.* (p. 357)

**effector** The component of the physiological system used to bring about the desired effect as determined by the control centre (p. 14)

**effector organs** The muscles or glands that are innervated by the nervous system and that carry out the nervous system's orders to bring about a desired effect, such as a particular movement or secretion (p. 92)

**efferent division** (EF-er-ent) The portion of the peripheral nervous system that carries instructions from the central nervous system to effector organs (p. 92)

**efferent neurons** Neurons that carry information from the central nervous system to an effector organ (p. 93)

**efferent pathway** Nerve pathways that carry nerve impulses away from the central nervous system to effectors such as muscles or glands (p. 106)

**ejection fraction** The fraction of blood pumped out of the ventricle with each heart beat (p. 370)

**elastic recoil** Rebound of the lungs after having been stretched (p. 508)

**elastin** A rubber-like protein fibre most abundant in tissues that must be capable of easily stretching and then recoiling after the stretching force is removed; found, for example, in the lungs, which stretch and recoil as air moves in and out (p. 34)

**electrical gradient** A difference in charge between two adjacent areas (p. 38)

**electrocardiogram (ECG)** The graphic record of the electrical activity that reaches the surface of the body as a result of cardiac depolarization and repolarization (p. 357)

**electrochemical gradient** The simultaneous existence of an electrical gradient and a concentration (chemical) gradient for a particular ion (p. 38)

**electroencephalogram (EEG)** (i-lek´-trō-in-SEF-uh-luh-gram´) A graphic record of the collective postsynaptic potential activity in the cell bodies and dendrites located in the cortical layers under a recording electrode (p. 357)

**electron transport chain** Consists of electron carrier molecules located in the inner mitochondrial membrane lining the cristae. The high-energy electrons extracted from the hydrogens held in NADH and FADH2 are transferred through a series of steps from one electron-carrier molecule to another, within the cristae membrane, in a kind of assembly line. (p. 27)

**embolus** (emboli, plural) (EM-bō-lus) A freely floating clot (p. 378)

**emotion** A complex experience involving biochemical and environmental processes within the limbic association cortex, which is located mostly on the bottom and adjoining inner portion of each temporal lobe (p. 124)

**emphysema** (em´-fuh-ZĒ-muh) A pulmonary disease characterized by collapse of the smaller airways and a breakdown of alveolar walls (p. 517)

**end-diastolic volume (EDV)** The volume of blood in the ventricle at the end of diastole, when filling is complete (p. 361)

**endocrine glands** Ductless glands that secrete hormones into the blood (p. 7)

**endocrine system** One of the body's two major regulatory systems; the other is the nervous system. Through its relatively slowly acting hormone messengers, the endocrine system generally regulates activities that require duration rather than speed. (p. 209)

**endocrinology** The study of the homeostatic chemical adjustments and other activities that hormones accomplish (p. 210)

**endocytosis** (en´-dō-sī-TŌ-sis) Internalization of extracellular material within a cell as a result of the plasma membrane forming a pouch that contains the extracellular material, then sealing at the surface of the pouch to form a small, intracellular, membrane-enclosed vesicle with the contents of the pouch trapped inside (p. 46)

**endocytotic vesicle** A vesicle within the cell that is formed during the process of endocytosis and contains substances being imported from outside the cell (p. 48)

**endogenous pyrogen (EP)** (pī´-ruh-jun) A chemical released from macrophages during inflammation that acts by means of local prostaglandins to raise the set point of the hypothalamic thermostat to produce a fever (p. 470)

**endolymph** The fluid within the cochlear duct (p. 172)

**endometrium** (en´-dō-MĒ-trĒ-um) The lining of the uterus (p. 753)

**endothelium** (en´-dō-THĒ-le-um) The thin, single-celled layer of epithelial cells that lines the entire circulatory system (p. 350)

**end-plate potential (EPP)** The graded receptor potential that occurs at the motor end plate of a skeletal muscle fibre in response to binding with acetylcholine (p. 197)

**end-systolic volume (ESV)** The volume of blood in the ventricle at the end of systole, when emptying is complete (p. 363)

**enterogastrones** (ent´-uh-rō-GAS-trōnz) Hormones secreted by the duodenal mucosa that inhibit gastric motility and secretion; include secretin and cholecystokinin (p. 659)

**enterohepatic circulation** (en´-tur-ō-hi-PAT-ik) The recycling of bile salts and other bile constituents between the small intestine and liver by means of the hepatic portal vein (p. 673)

**entrained** To pull along after itself (p. 212)

**eosinophils** (ē´-uh-SIN-uh-fils) White blood cells that are important in allergic responses and in combating internal parasite infestations (p. 446)

**ependymal cells** Cells that line the internal, fluid-filled cavities of the CNS (p. 98)

**epidermis** (ep´-uh-DER-mus) The outer layer of the skin, consisting of numerous layers of epithelial cells, with the outermost layers being dead and flattened (p. 461)

**epilepsy** A neurological disorder characterized by seizures (p. 120)

**epinephrine (adrenaline)** (ep´-uh-NEF-rin) The primary hormone secreted by the adrenal medulla; important in preparing the body for fight-or-flight responses and in regulating arterial blood pressure; also known as adrenaline (p. 192)

**epiphyseal plate** (eh-pif-i-SEE-al) A layer of cartilage that separates the diaphysis (shaft) of a long bone from the epiphysis (flared end); the site of growth of bones in length before the cartilage ossifies (turns into bone) (p. 233)

**epiphysis** The flared articulating knob at either end of a long bone (p. 233)

**epithelial tissue** (ep´-uh-THĒ-lē-ul) A functional grouping of cells specialized in the exchange of materials between the cell and its environment; lines and covers various body surfaces and cavities and forms secretory glands (p. 6)

**EPSP** See *excitatory postsynaptic potential*. (p. 76)

**equilibrium potential** The membrane voltage at which there is no net movement of a particular ion across the membrane (p. 55)

**erythrocytes (red blood cells or RBCs)** (i-RITH-ruh-sīts) Red blood cells, which are plasma membrane-enclosed bags of haemoglobin that transport oxygen and, to a lesser extent, carbon dioxide and hydrogen ions in the blood (p. 436)

**erythropoiesis** (i-rith´-rō-poi-Ē-sus) Erythrocyte production by the bone marrow (p. 437)

**erythropoietin (EPO)** The hormone released from the kidneys in response to a reduction in O₂ delivery to the kidneys; stimulates the bone marrow to increase erythrocyte production (p. 438)

**esophagus** (i-SOF-uh-gus) A straight muscular tube that extends between the pharynx and stomach (p. 501)

**estrogen** Feminizing, female sex hormone (p. 724)

**ESV** See *end-systolic volume*. (p. 363)

**eustachian (auditory) tube** Connects the middle ear to the pharynx; exposes the inside of the ear drum facing the middle ear cavity to atmospheric pressure (p. 172)

**excitable tissues** Tissues capable of producing electrical signals when excited; include nervous and muscle tissue (p. 57)

**excitation–contraction coupling** The series of events linking muscle excitation (the presence of an action potential) to muscle contraction (filament sliding and sarcomere shortening) (p. 303)

**excitatory postsynaptic potential (EPSP)** (pōst´-si-NAP-tik) A small depolarization of the postsynaptic membrane in response to neurotransmitter binding, bringing the membrane closer to threshold (p. 76)

**excitatory synapse** (SIN-aps´) Synapse in which the postsynaptic neuron's response to neurotransmitter release is a small depolarization of the postsynaptic membrane, bringing the membrane closer to threshold (p. 76)

**exocrine glands** Glands that secrete through ducts to the outside of the body or into a cavity that communicates with the outside (p. 7)

**exocytosis** (eks´-ō-sī-TŌ-sis) Fusion of a membrane-enclosed intracellular vesicle with the plasma membrane, followed by the opening of the vesicle and the emptying of its contents to the outside (p. 46)

**expiratory muscles** The skeletal muscles whose contraction reduces the size of the thoracic cavity and lets the lungs recoil to a smaller size, bringing about movement of air from the lungs to the atmosphere (p. 513)

**expiratory reserve volume (ERV)** The amount of additional air that can be breathed out after the end-expiratory level of normal breathing; if a person exhales as much as possible, only the residual volume remains (p. 518)

**external eye muscles** A set of six external eye muscles, for each eye, that position and move the eye so that it can better locate, see, and track objects. Eye movements are among the fastest, most discretely controlled movements of the body. (p. 169)

**external intercostal muscles** Inspiratory muscles whose contraction elevates the ribs, thereby enlarging the thoracic cavity (p. 504)

**external work** Energy expended by contracting skeletal muscles to move external objects or to move the body in relation to the environment (p. 700)

**extracellular fluid (ECF)** All the body's fluid outside the cells; consists of interstitial fluid and plasma (p. 9)

**extracellular matrix (ECM)** An intricate meshwork of fibrous proteins embedded in a watery, gel-like substance; secreted by local cells (p. 34)

**extrafusal fibres** Class of muscle fibre innervated by alpha motor neurons (p. 325)

**extrinsic controls** Regulatory mechanisms initiated outside an organ that alter the activity of the organ; accomplished by the nervous and endocrine systems (p. 14)

**extrinsic nerves** The nerves that originate outside the digestive tract and innervate the various digestive organs (p. 651)

**eye** A sense organ that enables vision by reacting to light; rod and cone cells within the retina allow vision (p. 191)

**eyelashes** Protection for the eyes that trap fine, airborne debris, such as dust, before it can fall into the eye (p. 153)

**eyelids** Act like shutters to protect the anterior portion of the eye from environmental insults; close reflexly to cover the eye under threatening circumstances, such as rapidly approaching objects, dazzling light, and instances when the exposed surface of the eye or eyelashes are touched (p. 152)

## F

**facilitated diffusion** Passive carrier-mediated transport involving transport of a substance down its concentration gradient across the plasma membrane (p. 43)

**fasted (postabsorptive) state** The metabolic state after a meal is absorbed during which endogenous energy stores must be mobilized and glucose must be spared for the glucose-dependent brain; fasting state (p. 270)

**fast pain** Pain that's easily localized and typically perceived initially as a brief, sharp, prickling sensation. The fast pain pathway originates from specific mechanical or heat nociceptors. (p. 149)

**fatigue** Inability to maintain muscle tension at a given level despite sustained stimulation (p. 310)

**feedback** Responses made after a change has been detected (p. 14)

**feedforward** A response designed to prevent an anticipated change in a controlled variable (p. 14)

**feeding or appetite signals** Appetite signals that give rise to the sensation of hunger and promote the desire to eat (p. 703)

**fibrinogen** (fi-BRIN-uh-jun) A large, soluble plasma protein converted into an insoluble, thread-like molecule that forms the meshwork of a clot during blood coagulation (p. 435)

**fibronectin** Protein fibres that promote cell adhesion and hold cells in position (p. 34)

**Fick's law of diffusion** The rate of net diffusion of a substance across a membrane is directly proportional to the substance's concentration gradient, the membrane's permeability to the substance, and the surface area of the membrane and inversely proportional to the substance's molecular weight and the diffusion distance. (p. 527)

**fight-or-flight response** The changes in activity of the various organs innervated by the autonomic nervous system in response to sympathetic stimulation, which collectively prepare the body for strenuous physical activity in the face of an emergency or stressful situation, such as a physical threat from the environment (p. 191)

**fire** State of an excitable cell when it undergoes an action potential (p. 62)

**first messenger** An extracellular messenger, such as a hormone, that binds with a surface membrane receptor and activates an intracellular second messenger to carry out the desired cellular response (p. 53)

**flavine adenine dinucleotide (FAD)** Redox cofactor involved in several important metabolic reactions (p. 26)

**fluid mosaic model** A model of membrane structure, in which the lipid bilayer is embedded with membrane proteins (p. 30)

**follicular cells** [of ovary] (fah-LIK-you-lar) Collectively, the granulosa and thecal cells (p. 248)

**follicular cells** [of thyroid gland] The cells that form the walls of the colloid-filled follicles in the thyroid gland and secrete thyroid hormone (p. 248)

**forced or active expiration** Emptying of the lungs more completely than when at rest by contracting the expiratory muscles; also called *forced expiration* (p. 513)

**fovea** The small central pit in the retina with a high density of cones (p. 158)

**Frank–Starling law of the heart** Intrinsic control of the heart such that increased venous return resulting in increased end-diastolic volume leads to an increased strength of contraction and increased stroke volume—that is, the heart normally pumps out all the blood returned to it (p. 368)

**free radicals** Very unstable electron-deficient particles that are highly reactive and destructive (p. 114)

**frontal lobes** The lobes of the cerebral cortex that lie at the top of the brain in front of the central sulcus and that are responsible for voluntary motor output, speaking ability, and elaboration of thought (p. 115)

**FSH** See *follicle-stimulating hormone*. (p. 227)

**functional residual capacity (FRC)** Volume of air present in the lungs at the end of passive expiration (p. 518)

**functional syncytium** (sin-sish´-ē-um) A group of smooth or cardiac muscle cells that are interconnected by gap junctions and function electrically and mechanically as a single unit (p. 328)

**functional unit** The smallest component of an organ that can perform all the functions of the organ (p. 300)

## G

**gametes (reproductive or germ cells)** (GAM-ētz) Reproductive, or germ, cells, each containing a haploid set of chromosomes; sperm and ova (p. 724)

**gamma (γ) globulins** Antibody proteins produced by the immune system to identify and destroy foreign objects; also called immunoglobulins (im´-ū-nō-GLOB-yū-lunz) (p. 435)

**gamma motor neuron** A motor neuron that innervates the fibres of a muscle spindle receptor (p. 325)

**ganglion cells** The nerve cells in the outermost layer of the retina and whose axons form the optic nerve (p. 157)

**gap junction** A communicating junction formed between adjacent cells by small connecting tunnels that permit passage of charge-carrying ions between the cells, so that electrical activity in one cell is spread to the adjacent cell (p. 35)

**gastrin** A hormone secreted by the pyloric gland area of the stomach that stimulates the parietal and chief cells to secrete a highly acidic gastric juice (p. 664)

**gated channels** Channels that have gates that can alternately be open, permitting ion passage through them, or closed, preventing ion passage through them. The opening and closing of gates results from a change in the three-dimensional conformation (shape) of the protein that forms the gated channel, (p. 58)

**gestation (pregnancy)** The period during which an embryo develops in the uterus (p. 767)

**gigantism** A rapid growth in height without distortion of body proportions; occurs when overproduction of GH begins in childhood before the epiphyseal plates close (p. 238)

**glands** Epithelial tissue derivatives that are specialized for secretion (p. 6)

**glial cells** (glē-ul) Serve as the connective tissue of the CNS and help support the neurons both physically and metabolically; include astrocytes, oligodendrocytes, ependymal cells, and microglia (p. 95)

**glomerular filtration** (glow-MER-yū-lur) Filtration of a protein-free plasma from the glomerular capillaries into the tubular component of the kidney's nephron as the first step in urine formation (p. 568)

**glomerular filtration rate (GFR)** The rate at which glomerular filtrate is formed (p. 568)

**glomeruli** (*glomerulus*, singular) (glow-MER-yū-li) Ball-like tuft of capillaries in the kidney's nephron that filters water and solute from the blood as the first step in urine formation (p. 185)

**glucagon** (GLOO-kuh-gon) The pancreatic hormone that raises blood glucose and blood fatty-acid levels (p. 280)

**glucagon-like peptide I (GLP-I)** Gastrointestinal hormone involved in digestion; primarily secreted by intestinal L cells; subject of research as a potential treatment for diabetes mellitus. (p. 693)

**glucocorticoids** (gloo´-kō-KOR-ti-koidz) The adrenocortical hormones that are important in intermediary metabolism and in helping the body resist stress; primarily cortisol (p. 254)

**gluconeogenesis** (gloo´-kō-nē-ō-JEN-uh-sus) The conversion of amino acids into glucose (p. 256)

**glycogen** (GLĪ-kō-jen) The storage form of glucose in the liver and muscle (p. 646)

**glycogenesis** (glī´-kō-JEN-i-sus) The conversion of glucose into glycogen (p. 272)

**glycogenolysis** (glī´-kō-juh-NOL-i-sus) The conversion of glycogen to glucose (p. 262)

**glycolysis** (glī-KOL-uh-sus) A biochemical process that takes place in the cell's cytosol and involves the breakdown of glucose into two pyruvic acid molecules (p. 25)

**GnRH** See *gonadotropin-releasing hormone*. (p. 738)

**Golgi tendon organs** Proprioceptive sensor receptor organ located at the insertion of skeletal muscle fibres into thes of skeletal muscle (p. 326)

**gonadotropin-releasing hormone (GnRH)** (gō-nad´-uh-TRŌ-pin) The hypothalamic hormone that stimulates the release of FSH and LH from the anterior pituitary (p. 738)

**gonadotropins** FSH and LH; hormones that are tropic to the gonads (p. 227)

**G protein** A membrane-bound intermediary, which, when activated on binding of an extracellular first messenger to a surface receptor, activates the enzyme adenylyl cyclase on the intracellular side of the membrane in the cAMP second-messenger system (p. 217)

**graded potentials** Local change in membrane potential that occurs in varying grades of magnitude; serves as a short-distance signal in excitable tissues (p. 59)

**grand postsynaptic potential (GPSP)** Spatial or temporal summation of many small potentials; dictated by the rates of firing of many presynaptic neurons that jointly control the (grand) membrane potential in the body of a single postsynaptic cell. (p. 78)

**granular cells** Pericytes modified to become secretory cells; also called juxtaglomerular cells; contain large quantities of secretory vesicles that contain renin (p. 570)

**granulocytes** (gran´-yuh-lō-sīts) Leukocytes that contain granules, including neutrophils, eosinophils, and basophils (p. 446)

**granulosa cells** (gran´-yuh-LO-suh) The layer of cells immediately surrounding a developing oocyte within an ovarian follicle (p. 745)

**grey matter** The portion of the central nervous system composed primarily of densely packaged neuronal cell bodies and dendrites (p. 102)

**growth hormone (GH, somatotropin)** An anterior pituitary hormone primarily responsible for regulating overall body growth and also important in intermediary metabolism (p. 227)

**growth hormone–releasing hormone (GHRH)** Regulatory hormone that stimulates growth hormone secretion; one of two antagonistic hormones from the hypothalamus involved in controlling growth hormone secretion. (p. 227)

**guanosine diphosphate (GDP)** A nucleotide sugar that is a substrate for glycosyltransferase reaction in metabolism (p. 27)

**guanosine triphosphate (GTP)** A high-energy molecule (similar to ATP) whose energy is transferred to ATP (p. 27)

**gustation** The sensation of taste (p. 181)

## H

**habituation** A decreased responsiveness to repetitive presentations of an indifferent stimulus—that is, one that's neither rewarding nor punishing (p. 128)

**haematocrit** (hi-MAT´-uh-krit) The percentage of blood volume occupied by erythrocytes when they are packed down in a centrifuged blood sample (p. 434)

**haemoglobin (Hb)** (HĒ-muh-glō´-bun) A large, iron-bearing protein molecule found within erythrocytes that binds with and transports most oxygen in the blood; also carries some of the carbon dioxide and hydrogen ions in the blood (p. 27)

**haemolysis** (hē-MOL-uh-sus) Rupture of red blood cells (p. 441)

**haemostasis** (hē´-mō-STĀ-sus) The stopping of bleeding from an injured vessel (p. 448)

**haploid number** (HAP-loid) The number of chromosomes found in gametes; a half set of chromosomes, one member of each pair, for a total of 23 chromosomes in humans (p. 727)

**Hb** See *haemoglobin.* (p. 27)

**hCG** See *human chorionic gonadotropin.* (p. 763)

**hearing** The sense by which sound is perceived (p. 169)

**heart failure** An inability of the cardiac output to keep pace with the body's demands for supplies and for removal of wastes (p. 371)

**helper T cells** The population of T cells that enhances the activity of other immune-response effector cells; also called *CD4 cells* (p. 482)

**Henderson-Hasselbalch equation** The mathematical relationship between [H⁺] and the members of a buffer pair (p. 625)

**hepatic portal system** (hi-PAT-ik) A complex vascular connection between the digestive tract and liver, such that venous blood from the digestive system drains into the liver for processing of absorbed nutrients before being returned to the heart (p. 671)

**hippocampus** (hip-oh-CAM-pus) The elongated, medial portion of the temporal lobe that is a part of the limbic system and crucial for forming long-term memories (p. 128)

**histamine** A chemical released from mast cells or basophils that brings about vasodilation and increased capillary permeability; important in allergic responses and inflammation (p. 664)

**homeostasis** (hō´-mē-ō-STĀ-sus) The maintenance by the highly coordinated, regulated actions of the body systems of relatively stable chemical and physical conditions in the internal fluid environment that bathes the body's cells (p. 9)

**homeostatic drives** The subjective urges associated with specific bodily needs that motivate appropriate behaviour to satisfy those needs (p. 125)

**hormone response element (HRE)** The specific attachment site on DNA for a given steroid hormone and its nuclear receptor (p. 221)

**hormones** Long-distance chemical mediators secreted by an endocrine gland into the blood, which transports them to their target cells (p. 52)

**host cell** A body cell infected by a virus (p. 460)

**HRE** See *hormone response element.* (p. 221)

**human chorionic gonadotropin (hCG)** (kō-rē-ON-ik gō-nad´-uh-TRŌ-pin) A hormone secreted by the developing placenta that stimulates and maintains the corpus luteum of pregnancy (p. 763)

**human physiology** (fiz-ē-OL-ō-gē) The study of body functions (p. 3)

**hydrocephalus** A condition in which there is an abnormal accumulation of cerebrospinal fluid (p. 100)

**hydrolysis** (hī-DROL-uh-sis) The digestion of a nutrient molecule by the addition of water at a bond site (p. 647)

**hydrophilic** "Water loving"; substances that can interact with water molecules. Phospholipids have hydrophilic, polar (electrically charged) heads, containing a negatively charged phosphate group; the polar heads line up on the outsides of a lipid bilayer. (p. 29)

**hydrophilic hormones** Hormones that bind with surface membrane receptors and thereby trigger a chain of intracellular events by means of a second-messenger system that ultimately alters pre-existing cell proteins, usually enzymes, which exert the effect leading to the target cell's response to the hormone (p. 214)

**hydrophobic** "Water fearing"; substances that will not mix with water. The hydrophobic tails of the molecules in a lipid bilayer bury themselves in the centre away from the water, whereas the hydrophilic heads line up on both sides of the lipid bilayer, in contact with the water (p. 29)

**hydrostatic (fluid) pressure** (hī-dro-STAT-ik) The pressure exerted by fluid on the walls that contain it (p. 39)

**hypercapnia** A condition in which there is too much carbon dioxide in the blood (p. 540)

**hyperopia** Condition in which the eyeball is too short or the lens is too weak. Far objects are focused by accommodation, whereas near objects are focused behind the retina. (p. 157)

**hyperplasia** (hī-pur-PLĀ-zē-uh) An increase in the number of cells (p. 233)

**hyperpnea** Abnormally deep or rapid breathing (p. 540)

**hyperpolarization** An increase in membrane potential from resting potential; potential becomes even more negative than at resting potential (p. 62)

**hypersecretion** Too much of a particular hormone secreted (p. 213)

**hypertension** (hī´-pur-TEN-chun) Sustained, above-normal mean arterial blood pressure (p. 424)

**hypertonicity** The excessive concentration of solutes in the extracellular fluid compartment (p. 614)

**hypertonic solution** (hī-pur-TON-ik) A solution having an osmolarity greater than that of normal body fluids; more concentrated than normal (p. 41)

**hypertrophy** (hī-PUR-truh-fē) Increase in the size of an organ as a result of an increase in the size of its cells (p. 233)

**hyperventilation** Overbreathing; rate of ventilation in excess of the body's metabolic needs for $CO_2$ removal (p. 540)

**hypophysiotropic hormones** (hi-PO-fiz-e-oh-) Hormones secreted by the hypothalamus that regulate the secretion of anterior pituitary hormones; see also *releasing hormone* and *inhibiting hormone* (p. 229)

**hyposecretion** Too little of a particular hormone secreted (p. 213)

**hypotension** (hī-pō-TEN-chun) Sustained, below-normal mean arterial blood pressure (p. 424)

**hypothalamic-hypophyseal portal system** (hī-pō-thuh-LAM-ik hī-pō-FIZ-ē-ul) The vascular connection between the hypothalamus and anterior pituitary gland used for the pickup and delivery of hypophysiotropic hormones (p. 229)

**hypothalamic osmoreceptors** The specialized sensory receptor located near the vasopressin-secreting cells and thirst centre; the predominant excitatory input for both vasopressin secretion and thirst (p. 617)

**hypothalamus** (hī´-pō-THAL-uh-mus) The brain region located beneath the thalamus that is concerned with regulating many aspects of the internal fluid environment, such as water and salt balance and food intake; serves as an important link between the autonomic nervous system and endocrine system (p. 113)

**hypotonicity** A lower than normal concentration of solutes in the extracellular fluid compartment, often associated with an excess of free (p. 614)

**hypotonic solution** (hī´-pō-TON-ik) A solution having an osmolarity less than that of normal body fluids; more dilute than normal (p. 41)

**hypoventilation** Underbreathing; ventilation inadequate to meet the metabolic needs for $O_2$ delivery and $CO_2$ removal (p. 540)

**hypoxia** (hī-POK-sē-uh) Insufficient oxygen at the cellular level (p. 539)

**I**

**iatrogenic** A condition that arises as a result of medical intervention (p. 614)

**I band** One of the light bands that alternate with dark (A) bands to create a striated appearance in a skeletal or cardiac muscle fibre when these fibres are viewed with a light microscope (p. 300)

**insensible loss** The elimination of water from the body that a person has no sensory awareness of, for example, from the lungs and nonsweating skin (p. 616)

**IGF-I** A protein hormone similar to insulin (hence the name) (p. 235)

**IGF-II** A growth-promoting hormone that promotes the growth of bone and soft tissues (p. 235)

**immediate early genes (IEGs)** Genes that play a critical role in memory consolidation, governing the synthesis of the proteins that encode long-term memory (p. 132)

**immune surveillance** Recognition and destruction of newly arisen cancer cells by the immune system (p. 489)

**immunity** The body's ability to resist or eliminate potentially harmful foreign materials or abnormal cells (p. 444)

**impermeable** Prohibiting passage of a particular substance through the plasma membrane (p. 36)

**implantation** The burrowing of a blastocyst into the endometrial lining (p. 758)

**inflammation** An innate, nonspecific series of highly interrelated events, especially involving neutrophils, macrophages, and local vascular changes, that are set into motion in response to foreign invasion or tissue damage (p. 466)

**inhibin** (in-HIB-un) A hormone secreted by the Sertoli cells of the testes or by the ovarian follicles that inhibits FSH secretion (p. 738)

**inhibiting hormones** Hypothalamic hormone that inhibits the secretion of a particular anterior pituitary hormone (p. 229)

**inhibitory postsynaptic potential (IPSP)** (pōst´-si-NAP-tik) A small hyperpolarization of the postsynaptic membrane in response to neurotransmitter binding, thereby moving the membrane farther from threshold (p. 76)

**inhibitory synapse** (SIN-aps´) Synapse in which the postsynaptic neuron's response to neurotransmitter release is a small hyperpolarization of the postsynaptic membrane, moving the membrane farther from threshold (p. 76)

**innate immune system** Inherent defence system that nonselectively defends against foreign or abnormal material, even on initial exposure to it; see also *inflammation, interferon, natural killer cells*, and *complement system* (p. 465)

**innervate** When a neuron terminates on a muscle or a gland, the neuron is said to innervate the structure. (p. 74)

**inositol trisphosphate (IP$_3$)** A secondary messenger molecule involved in signal transduction (p. 219)

**inspiration** A breath in (p. 504)

**inspiratory capacity (IC)** The volume that can be inhaled after a tidal breath out (p. 518)

**inspiratory muscles** The skeletal muscles whose contraction enlarges the thoracic cavity, bringing about lung expansion and movement of air into the lungs from the atmosphere (p. 503)

**insulin** (IN-suh-lin) The pancreatic hormone that lowers blood levels of glucose, fatty acids, and amino acids and promotes their storage (p. 272)

**insulin-like growth factor (IGF)** See *somatomedins*. (p. 235)

**integrating centre** A region that determines efferent output based on the processing of afferent input (p. 106)

**integrins** Membrane proteins that serve as a structural link between the outer membrane surface and its extracellular surroundings and also connect the inner membrane surface to the intracellular cytoskeletal scaffolding (p. 32)

**intercostal muscles** (int-ur-KOS-tul) The muscles that lie between the ribs; see also *external intercostal muscles* and *internal intercostal muscles* (p. 504)

**interferon** (in´-tur-FĒR-on) A chemical released from virus-invaded cells that provides nonspecific resistance to viral infections by transiently interfering with replication of the same or unrelated viruses in other host cells (p. 471)

**intermediary metabolism** The collective set of intracellular chemical reactions that involve the degradation, synthesis, and transformation of small nutrient molecules; also known as fuel metabolism (p. 22)

**internal environment** The body's aqueous extracellular environment, which consists of the plasma and interstitial fluid and which must be homeostatically maintained for the cells to make life-sustaining exchanges with it (p. 9)

**internal intercostal muscles** Expiratory muscles whose contraction pulls the ribs downward and inward, thereby reducing the size of the thoracic cavity (p. 504)

**internal (cellular) respiration** The intracellular metabolic reactions and processes by which living organisms (cells) convert biochemical energy (e.g., glucose) from nutrients into adenosine triphosphate (ATP) for energy (p. 498)

**internal work** All forms of biological energy expenditure that do not accomplish mechanical work outside the body (p. 700)

**interneurons** Neurons that lie entirely within the central nervous system and are important for integrating peripheral responses to peripheral information as well as for the abstract phenomena associated with the human mind (p. 93)

**interstitial cell–stimulating hormone (ICSH)** In males, the hormone that stimulates the interstitial Leydig cells in the testes to secrete the male sex hormone, testosterone, a chemical produced by the body that regulates the activity of certain cells or certain organs (p. 227)

**interstitial fluid** (in´-tur-STISH-ul) The portion of the extracellular fluid that surrounds and bathes all the body's cells (p. 10)

**intracellular fluid (ICF)** The fluid collectively contained within all the body's cells (p. 9)

**intrinsic (local) controls** Local control mechanisms inherent to an organ (p. 14)

**intrinsic factor** A special substance secreted by the parietal cells of the stomach that must be combined with vitamin B$_{12}$ for this vitamin to be absorbed by the intestine. Deficiency produces pernicious anaemia. (p. 664)

**intrinsic nerve plexuses** Interconnecting networks of nerve fibres within the digestive tract wall (p. 651)

**ion concentration gradient** A gradient that results from an unequal distribution of ions across the cell membrane; for example, an Na$^+$ gradient (p. 44)

**IPSP** See *inhibitory postsynaptic potential.* (p. 76)

**islets of Langerhans** (LAHNG-er-honz) The endocrine portion of the pancreas that secretes the hormones insulin and glucagon into the blood (p. 271)

**isometric contraction** (ī´-sō-MET-rik) A muscle contraction in which the development of tension occurs at constant muscle length (p. 315)

**isotonic contraction** A muscle contraction in which muscle tension remains constant as the muscle fibre changes length (p. 314)

**isotonic solution** (ī´-sō-TON-ik) A solution having an osmolarity equal to that of normal body fluids (p. 40)

## J

**jet lag** When one's inherent rhythm (dictated by the SCN and pineal gland) is out of step with external cues; typically a temporary disorder that causes fatigue, insomnia, and other symptoms as a result of flight across time zones; disrupts circadian rhythm, which disrupts sleep (p. 240)

**juxtaglomerular apparatus** (juks´-tuh-glō-MER-yū-lur) A cluster of specialized vascular and tubular cells at a point where the ascending limb of the loop of Henle passes through the fork formed by the afferent and efferent arterioles of the same nephron in the kidney (p. 563)

## K

**Krebs cycle** Consists of a series of eight separate biochemical reactions directed by the enzymes of the mitochondrial matrix, a series of biochemical reactions (p. 25)

## L

**lacrimal gland** The gland located in the upper lateral corner under the eyelid; produces tears to moisten and cleanse the eye (p. 153)

**lactation** Milk production by the mammary glands (p. 771)

**language** The method of human communication, either spoken or written, consisting of the use of words in a structured and conventional way; an excellent example of early cortical plasticity coupled with later permanence (p. 132)

**Laron dwarfism** Condition that involves abnormally short stature, despite normal levels of growth hormone, due to inability of tissues to respond normally to growth hormone. This condition is in contrast with the more typical dwarfism that involves abnormally short stature due to growth hormone deficiency. (p. 238)

**larynx (voice box)** (LARE-inks) The structure at the entrance of the trachea that contains the vocal cords (p. 501)

**lateral horn** The lateral column of grey matter of the spinal cord when viewed in cross-section; involved with sympathetic activity of the autonomic nervous system (p. 103)

**lateral inhibition** The phenomenon in which the most strongly activated signal pathway originating from the centre of a stimulus area inhibits the less excited pathways from the fringe areas by means of lateral inhibitory connections within sensory pathways (p. 147)

**lateral sacs** The expanded saclike regions of a muscle fibre's sarcoplasmic reticulum; store and release calcium, which plays a key role in triggering muscle contraction (p. 304)

**law of mass action** If the concentration of one of the substances involved in a reversible reaction is increased, the reaction is driven toward the opposite side, and if the concentration of one of the substances is decreased, the reaction is driven toward that side. (p. 532)

**leak channels** Ion channels that are more frequently open than closed (p. 58)

**learning** The acquisition of knowledge or skills as a consequence of experience, instruction, or both. It is widely believed that rewards and punishments are integral parts of many types of learning. (p. 126)

**left atrial volume receptors** The receptors in the left atrium of the heart that monitor the pressure of blood flowing through the left atrium (p. 618)

**left cerebral hemisphere** Region of the brain that excels in logical, analytic, sequential, and verbal tasks, such as math, language forms, and philosophy (p. 134)

**length–tension relationship** The relationship between the length of a muscle fibre at the onset of contraction and the tension the fibre can achieve on a subsequent tetanic contraction (p. 312)

**lens** A transparent, biconvex structure of the eye that refracts (bends) light rays, and whose strength can be adjusted to accommodate for vision at different distances (p. 153)

**leptin** A hormone released from adipose tissue that plays a key role in long-term regulation of body weight, by acting on the hypothalamus to suppress appetite (p. 703)

**leukocytes (white blood cells or WBCs)** (LOO-kuh-sīts) White blood cells, which are the immune system's mobile defence units (p. 444)

**Leydig (interstitial) cells** (LI-dig) The interstitial cells of the testes that secrete testosterone (p. 732)

**LH** See *luteinizing hormone.* (p. 227)

**LH surge** The burst in LH secretion that occurs at midcycle of the ovarian cycle and triggers ovulation (p. 750)

**limbic association cortex** Area of the brain located mostly on the bottom and adjoining inner portion of each temporal lobe; concerned primarily with motivation and emotion and extensively involved in memory (p. 134)

**limbic system** (LIM-bik) A functionally interconnected ring of forebrain structures that surrounds the brain stem and is concerned with emotions, basic survival and sociosexual behavioural patterns, motivation, and learning (p. 124)

**lipid bilayer** A non-rigid plasma membrane structure made of two layers of lipid molecules. Lipid bilayers are flat sheets that form a continuous barrier around cells, and that surround the nucleus, organelles, and other subcellular structures. (p. 29)

**lipid emulsion** A suspension of small fat droplets held apart as a result of adsorption of bile salts on their surface (p. 673)

**lipophilic hormones** Hormones (steroids and thyroid hormone) that bind with intracellular receptors and primarily produce their effects in their target cells by activating specific genes to cause the synthesis of new enzymatic or structural proteins (p. 215)

**long-term memory** Capacity to process, codify, and store for the long-term informational aspects of memory along with other memories of the same type; a form of organization that facilitates future searching of memory stores to retrieve desired information (p. 126)

**long-term potentiation (LTP)** Molecular mechanism of short-term memory that involves a prolonged increase in the strength of existing synaptic connections in activated pathways following brief periods of repetitive stimulation (p. 130)

**loop of Henle** (HEN-lē) A hairpin loop that extends between the proximal and distal tubule of the kidney's nephron (p. 563)

**lumen** (LOO-men) The interior space of a hollow organ or tube (p. 6)

**luteal phase** (LOO-tē-ul) The phase of the ovarian cycle dominated by the presence of a corpus luteum (p. 747)

**luteinization** (loot´-ē-un-uh-ZĀ-shun) Formation of a postovulatory corpus luteum in the ovary (p. 749)

**luteinizing hormone (LH)** An anterior pituitary hormone that stimulates ovulation, luteinization, and secretion of estrogen and progesterone in females, and stimulates testosterone secretion in males (p. 227)

**lymph** Interstitial fluid that is picked up by the lymphatic vessels and returned to the venous system, meanwhile passing through the lymph nodes for defence purposes (p. 411)

**lymphocytes** White blood cells that provide immune defence against targets for which they are specifically programmed (p 447)

**lymphoid tissues** Tissues that produce and store lymphocytes, such as lymph nodes and tonsils (p. 445)

## M

**macrophages** (MAK-ruh-fĀjs) Large, tissue-bound phagocytes (p. 464)

**major histocompatibility complex (MHC) molecules** Major histocompatibility complex molecules are found on cell surfaces and mediate interactions with leukocytes; also called human leukocyte antigen (HLA) (p. 482)

**mast cells** Cells located within connective tissue that synthesize, store, and release histamine, as during allergic responses (p. 447)

**mean arterial pressure** The average pressure responsible for driving blood forward through the arteries into the tissues throughout the cardiac cycle; equals cardiac output multiplied by total peripheral resistance (p. 394)

**mechanically gated channels** Channels that open or close in response to stretching or other mechanical deformation (p. 58)

**mechanical nociceptors** Pain receptors that respond to mechanical damage (p. 149)

**mechanistic approach** Explanation of body functions in terms of mechanisms of action—that is, the "how" of events that occur in the body (p. 4)

**mechanoreceptors** (meh-CAN-oh-rē-SEP-tur) or (mek´-uh-nō-rē-SEP-tur) A sensory receptor sensitive to mechanical energy, such as stretching or bending (p. 143)

**medicine** The art and science of healing (p. 3)

**medullary respiratory centre** (MED-you-LAIR-ē) Several aggregations of neuronal cell bodies within the medulla that provide output to the respiratory muscles and receive input important for regulating the magnitude of ventilation (p. 542)

**meiosis** (mī-ō-sis) Cell division in which the chromosomes replicate followed by two nuclear divisions so that only a half set of chromosomes is distributed to each of four new daughter cells (p. 727)

**Meissner's corpuscles** A type of mechanoreceptor located in the skin; responds to light touch and very sensitive to vibration. (p. 148)

**melanin** A pigment that is the primary determinant of skin colour; also found in hair, the pigmented tissue underlying the iris of the eye, and the stria vascularis of the inner ear. In the skin melanin is produced by melanocytes located in the basal layer of the epidermis. (p. 226)

**melanocyte-stimulating hormone (MSH)** (mel-AH-nō-sīt) A hormone produced by the anterior pituitary in humans and by the intermediate lobe of the pituitary in lower vertebrates; regulates skin colouration by controlling the dispersion of melanin granules in lower vertebrates; involved with control of food intake and possibly memory and learning in humans (p. 226)

**melanopsin** A protein found in a special retinal ganglion cell; the receptor for light that keeps the body in tune with external time (p. 242)

**melatonin** (mel-uh-TŌ-nin) A hormone secreted by the pineal gland during darkness that helps entrain the body's biological rhythms with the external light/dark cues (p. 240)

**membrane attack complex (MAC)** A collection of the five, final, activated components of the complement system that aggregate to form a porelike channel in the plasma membrane of an invading microorganism, with the resultant leakage leading to destruction of the invader (p. 473)

**membrane-bound enzymes** Surface-located proteins that control specific chemical reactions at either the inner or the outer cell surface. Cells are specialized in the types of enzymes embedded within their plasma membranes. (p. 31)

**membrane potential** A separation of charges across the membrane; most often a slight excess of negative charges lined up along the inside of the plasma membrane and separated from a slight excess of positive charges on the outside (p. 54)

**membrane proteins** Proteins with unique combinations of sugar chains projecting from them that serve as the trademark of a particular cell type, enabling a cell to recognize others of its own kind (p. 30)

**memory cells** B or T cells that are newly produced in response to a microbial invader, but do not participate in the current immune response against the invader; remain dormant, ready to launch a swift, powerful attack should the same microorganism invade again in the future (p. 479)

**meninges** The three membranes (dura mater, arachnoid, and pia mater) that line the skull and vertebral canal and enclose the brain and spinal cord (p. 98)

**menstrual (uterine) cycle** (men´-stroo-ul) The cyclic changes in the uterus that accompany the hormonal changes in the ovarian cycle (p. 753)

**menstrual phase** The phase of the menstrual cycle characterized by sloughing of endometrial debris and blood out through the vagina (p. 753)

**Merkel's discs** The most sensitive of the mechanoreceptors to low frequency vibrations (5–15 Hz), with a receptive field of about 2–3 mm; classified as slowly adapting type I myelinated mechanoreceptors, which provide touch information (p. 148)

**metabolic acidosis** (met-uh-bol´-ik) Acidosis resulting from any cause other than excess accumulation of carbonic acid in the body (p. 598)

**metabolic alkalosis** (al´-kuh-LŌ-sus) Alkalosis caused by a relative deficiency of noncarbonic acid (p. 635)

**metabolic rate** Energy expenditure per unit of time (p. 700)

**metabolic water** The $H_2O$ produced as a result of biochemical reactions within the cells of the body (p. 616)

**MHC molecules** See *Major histocompatibility complex (MHC) molecules.* (p. 482)

**micelle** (mī-SEL) A water-soluble aggregation of bile salts, lecithin, and cholesterol that has a hydrophilic shell and a hydrophobic core; carries the water-insoluble products of fat digestion to their site of absorption (p. 674)

**microglia** The immune cells of the CNS; scavengers cells that are "cousins" of monocytes, a type of white blood cell that leaves the blood and sets up residence as a front-line (i.e., with innate immunity) defence agent in various tissues throughout the body (p. 97)

**microvilli** (mī´-krō-VIL-ī) Actin-stiffened, nonmotile, hairlike projections from the luminal surface of epithelial cells lining the digestive tract and kidney tubules; tremendously increase the surface area of the cell exposed to the lumen (p. 678)

**micturition (urination)** (mik-too-RISH-un) or (mik-chuh-RISH-un) The process of bladder emptying (p. 598)

**middle ear** The portion of the ear internal to the eardrum and external to the oval window of the cochlea (p. 172)

**milk ejection (milk letdown)** The squeezing out of milk produced and stored in the alveoli of the breasts by means of contraction of the myoepithelial cells that surround each alveolus (p. 771)

**millivolt (mV)** 1 millivolt = 1/1000 volt (p. 54)

**mineralocorticoids** (min-uh-rul-ō-KOR-ti-koidz) The adrenocortical hormones that are important in $Na^+$ and $K^+$ balance; primarily aldosterone (p. 254)

**mitosis** (mī-TŌ-sis) Cell division in which the chromosomes replicate before nuclear division so that each of the two daughter cells receives a full set of chromosomes (p. 727)

**mitral cells** Cells on which the olfactory receptors terminate in the glomeruli that refine the smell signals and relay them to the brain for further processing (p. 185)

**modality** Determined by the type of sensory neuron that's activated and its point of termination in the brain. (p. 143)

**monocytes** (MAH-nō-sīts) White blood cells that emigrate from the blood, enlarge, and become macrophages; large-tissue phagocytes (p. 447)

**monosaccharides** (mah´-nō-SAK-uh-rīdz) Simple sugars, such as glucose; the absorbable unit of digested carbohydrates (p. 646)

**monosynaptic reflex** A reflex arc that contains only one synapse (p. 106)

**motion sickness** Sensitivity to particular motions that activate the vestibular apparatus and cause symptoms of dizziness and nausea (p. 180)

**motivation** The ability to direct behaviour toward specific goals, some of which are aimed at satisfying specific identifiable physical needs related to homeostasis (p. 125)

**motor end plate** The specialized portion of a skeletal muscle fibre that lies immediately underneath the terminal button of the motor neuron and possesses receptor sites for binding acetylcholine released from the terminal button (p. 197)

**motor homunculus** Visual representation of the connection between different body parts and the areas in brain hemispheres responsible for movement within those body parts (p. 116)

**motor neurons** The neurons that innervate skeletal muscle and whose axons constitute the somatic nervous system (p. 194)

**motor program** The different, related functions carried out by the three higher motor areas of the cortex and the cerebellum; important in programming and coordinating complex movements involving simultaneous contraction of many muscles; includes signals transmitted through efferent and afferent pathways that allow the central nervous system to anticipate and plan movement (p. 119)

**motor unit** One motor neuron plus all the muscle fibres it innervates (p. 310)

**motor unit recruitment** The progressive activation of a muscle fibre's motor units to accomplish increasing gradations of contractile strength (p. 310)

**mucosa** (mew-KŌ-sah) The innermost layer of the digestive tract that lines the lumen (p. 184)

**multiunit smooth muscle** A smooth muscle mass that consists of multiple discrete units that function independently of one another and that must be separately stimulated by autonomic nerves to contract (p. 328)

**muscarinic receptors** (MUS-ka-rin´-ik) Type of cholinergic receptors found at the effector organs of all parasympathetic postganglionic fibres (p. 192)

**muscle biopsy** Procedure in which a piece of muscle tissue is removed with the use of a needle for the purpose of examination (p. 318)

**muscle fibre** A single muscle cell, which is relatively long and cylindrical in shape (p. 197)

**muscle spindles** Sensory receptors within the belly of a muscle; primarily detect changes in the length of the muscle, which is sent by sensory neurons to the central nervous system (p. 325)

**muscle tissue** A functional grouping of cells specialized for contraction and force generation (p. 6)

**muscular dystrophy** A degenerative disease in which muscles get progressively weaker (p. 321)

**myasthenia gravis** A disease involving the neuromuscular junction, characterized by extreme muscular weakness; an autoimmune condition in which the body erroneously produces antibodies against its own motor-end-plate ACh receptors (p. 201)

**myelin** (MĪ-uh-lun) An insulative lipid covering that surrounds myelinated nerve fibres at regular intervals along the axon's length (p. 70)

**myelinated fibres** Neuronal axons covered at regular intervals with insulative myelin (p. 70)

**myoblast** Undifferentiated cell in the mesoderm of the embryo; precursor of a muscle cell (p. 22)

**myocardial ischemia** (mī-ō-KAR-dē-ul is-KĒ-mē-uh) Inadequate blood supply to the heart tissue (p. 360)

**myocardium** (mī´-ō-KAR-dē-um) The cardiac muscle within the heart wall (p. 350)

**myofibrils** (mī´-ō-FĪ–B-rul) Specialized intracellular structures of muscle cells that contain the contractile apparatus (p. 298)

**myoglobin** Iron-containing protein found in muscle fibres; structurally similar to haemoglobin but having a higher affinity for oxygen than blood haemoglobin (p. 318)

**myometrium** (mī´-ō-mē-TRĒ-um) The smooth muscle layer of the uterus (p. 753)

**myosin** (MĪ-uh-sun) The contractile protein that forms the thick filaments in muscle fibres (p. 300)

**myopia** Condition in which the eyeball is too long or the lens is too strong. A near light source is focused on the retina without accommodation (even though accommodation is normally used for near vision), whereas a far light source is focused in front of the retina and is blurry. (p. 157)

## N

**pump** See $Na^+–K^+$ ATPase pump. (p. 44)

**ATPase pump** A carrier that actively transports $Na^+$ out of the cell and $K^+$ into the cell; also known as $Na^+–K^+$ ATPase pump (p. 44)

**$Na^+$ load** The total mass of $Na^+$ salts in the ECF

**$Na^+$ permeability** ($P_{Na^+}$) The relative ability for $Na^+$ diffusing down its chemical gradient from a higher concentration outside the cell to a lower concentration inside the cell. During an action potential, once threshold is met, enough $Na^+$ gates have opened to allow a significant influx of $Na^+$ into the cell. (p. 63)

**natural killer (NK) cells** Naturally occurring, lymphocyte-like cells that nonspecifically destroy virus-infected cells and cancer cells by directly lysing their membranes on first exposure to them (p. 472)

**negative balance** Situation in which the losses for a substance exceed its gains, so that the total amount of the substance in the body decreases (p. 608)

**negative feedback** A regulatory mechanism in which a change in a controlled variable triggers a response that opposes the change, thus maintaining a relatively steady set point for the regulated factor (p. 14)

**neostigmine** A medication in the category of anti-cholinesterases; prolongs the action of acetylcholine, found naturally in the body (p. 201)

**nerve** A bundle of peripheral neuronal axons, some afferent and some efferent, enclosed by a connective tissue covering and following the same pathway (p. 103)

**nervous system** One of the two major regulatory systems of the body; in general, coordinates rapid activities of the body, especially those involving interactions with the external environment (p. 91)

**nervous tissue** A functional grouping of cells specialized for initiation and transmission of electrical signals (p. 6)

**net diffusion** The difference between two opposing movements (p. 37)

**net filtration pressure** The net difference in the hydrostatic and osmotic forces acting across the glomerular membrane that favours the filtration of a protein-free plasma into Bowman's capsule (p. 568)

**neuroendocrine reflexes** Cells that receive neuronal input (neurotransmitters released by nerve cells) and, as a result, release message hormones to the blood for communication (p. 212)

**neuroendocrinology** The study of the interactions between the nervous system and the endocrine system (p. 222)

**neuroglia** See *glial cells.* (p. 95)

**neuroglobin** An oxygen-binding protein in the brain (p. 101)

**neurohormone** A substance released by a neurosecretory cell directly into the blood that functions as a hormone; generally structurally identical to neurotransmitters (p. 53)

**neurohypophysis** An alternative name for the posterior pituitary, which is composed of nervous tissue (p. 225)

**neuromodulators** (ner´ō-MA-jew-lā´-torz) Chemical messengers that bind to neuronal receptors at nonsynaptic sites (i.e., not at the subsynaptic membrane), and bring about long-term changes that subtly depress or enhance synaptic effectiveness (p. 80)

**neuromuscular junction** The juncture between a motor neuron and a skeletal muscle fibre (p. 197)

**neuron** (NER-on) A nerve cell, typically consisting of a cell body, dendrites, and an axon and specialized to initiate, propagate, and transmit electrical signals (p. 64)

**neuropeptides** Large, slowly acting peptide molecules released from axon terminals along with classical neurotransmitters; most neuropeptides function as neuromodulators (p. 80)

**neurotransmitters** Chemical messengers released from the axon terminal of a neuron in response to an action potential; influence another neuron or an effector with which the neuron is anatomically linked (p. 80)

**neutrophils** (new´-truh-filz) White blood cells that are phagocytic specialists and important in inflammatory responses and defence against bacterial invasion (p. 446)

**nicotinamide adenine dinucleotide (NAD⁺)** Coenzyme found in all living cells; involved in metabolism, carrying electrons from one reaction to another (p. 26)

**nicotinic receptors** (nick´-o-TIN-ik) Type of cholinergic receptors found at all autonomic ganglia and the motor end plates of skeletal muscle fibres (p. 192)

**night blindness** Condition of sight related to dietary deficiency of vitamin A; reduced photopigment concentrations in both rods and cones, although sufficient cone photopigment exists to allow response to the intense stimulation of bright light, except in the most severe cases (p. 165)

**nitric oxide** A recently identified local chemical mediator released from endothelial cells and other tissues; exerts a wide array of effects, which include causing local arteriolar vasodilation, acting as a toxic agent against foreign invaders, and serving as a unique type of neurotransmitter (p. 130)

**NMDA receptors** A glutamate receptor, the predominant molecular device for controlling synaptic plasticity and memory function (p. 114)

**nociceptor** (nō-sē-SEP-tur) A pain receptor, sensitive to tissue damage (p. 149)

**nodes of Ranvier** (RAN-vē-ā) The portions of a myelinated neuronal axon between the segments of insulative myelin; the axonal regions where the axonal membrane is exposed to the ECF and membrane potential exists (p. 70)

**noradrenaline** Chemically very similar to *adrenaline (epinephrine)*, the primary hormone product secreted by the adrenal medulla gland (p. 188)

**norepinephrine** (nor´-ep-uh-NEF-run) The neurotransmitter released from sympathetic postganglionic fibres; noradrenaline (p. 188

**nucleus** [of cells] A distinct spherical or oval structure that is usually located near the centre of a cell and that contains the cell's genetic material, deoxyribonucleic acid (DNA) (p. 22)

**nursing** A healthcare profession that works collaboratively with other healthcare professionals for the care of an individual (p. 3)

## O

**obligatory loss** The loss of a substance from the body that cannot be avoided by any control mechanisms (p. 612)

**O₂–Hb dissociation curve** A graphic depiction of the relationship between arterial $P_{O_2}$ and percent haemoglobin saturation (p. 532)

**occipital lobes** (ok-SIP´-ut-ul) The lobes of the cerebral cortex that are located posteriorly and that initially process visual input (p. 115)

**odourants** Molecules that yield odours or fragrance and that act through second-messenger systems to trigger receptor potentials. The afferent signals arising from the olfactory receptors are sorted according to scent component by the glomeruli within the olfactory bulb. (p. 184)

**oedema** (i-DĒ-muh) Swelling of tissues as a result of excess interstitial fluid (p. 413)

**off response** Situation in which the receptor rapidly adapts by no longer responding to a maintained stimulus, but typically responds with a slight depolarization when the stimulus is removed (p. 145)

**olfaction** The sense of smell (p. 181)

**olfactory bulb** A complex neural structure containing several different layers of cells that are functionally similar to the retinal layers of the eye; two olfactory bulbs, one on each side of the head, are about the size of small grapes (p. 185)

**olfactory mucosa** A 3 cm² patch of mucosa in the ceiling of the nasal cavity; contains three cell types: *olfactory receptor cells*, *supporting cells*, and *basal cells* (p. 184)

**olfactory nerve** The axons of the olfactory receptor cells collectively form the olfactory nerve (p. 184)

**olfactory receptor cell** An afferent neuron whose receptor portion lies in the olfactory mucosa in the nose and whose afferent axon traverses into the brain (p. 184)

**oligodendrocytes** (ol-i-gō´-DEN-drō-sitz) The myelin-forming cells of the central nervous system (p. 70)

**on-centre and off-centre ganglion cells** Ganglion cells that respond in opposite ways, depending on the relative illumination of the centre and the periphery of their receptive fields (p. 166)

**opiate receptors** A type of protein found in the central nervous system and the gastrointestinal tract. Opiates activate pain relief and stimulate pleasure centres of the brain. (p. 150)

**opsin** A group of light-sensitive membrane-bound proteins found in photoreceptor cells of the retina (p. 161)

**opsonins** (OP´-suh-nuns) Body-produced chemical that links bacteria to macrophages, thereby making the bacteria more susceptible to phagocytosis (p. 469)

**optic chiasm** Area beneath the hypothalamus where the optic fibres from the medial half of each retina cross to the opposite side, and those from the lateral half remain on the original side (p. 166)

**optic disc** The point on the retina at which the optic nerve leaves and through which blood vessels pass; the region often called the blind spot since it has no rods and cones and no image can be detected in this area(p. 158)

**optic nerve** The bundle of nerve fibres that leave the retina, relaying information about visual input (p. 158)

**optic radiations** Fibre bundles that relay information received from the lateral geniculate nucleus in the thalamus to different zones in the cortex, each of which processes different aspects of the visual stimulus (e.g., colour, form, depth, movement) (p. 167)

**optimal length (lo)** The length before the onset of contraction of a muscle fibre at which maximal force can be developed on a subsequent tetanic contraction (p. 312)

**organ of Corti** (KOR-tē) The sense organ of hearing within the inner ear; contains hair cells whose hairs are bent in response to sound waves, setting up action potentials in the auditory nerve (p. 173)

**organophosphates** A group of chemicals that modify neuromuscular junction activity by irreversibly inhibiting AChE, which prevents the inactivation of released ACh (p. 200)

**organism** A living entity, either single celled (unicellular) or made up of many cells (multicellular) (p. 5)

**organs** Distinct structural units composed of two or more types of primary tissue organized to perform one or more particular functions; for example, the stomach (p. 7)

**osmolarity** (oz´-mō-LAR-ut-ē) A measure of the concentration of solute molecules in a solution (p. 613)

**osmosis** (os-MO-sis) Movement of water across a membrane down its own concentration gradient toward the area of higher solute concentration (p. 38)

**osmotic pressure** A measure of the tendency for water to move into a solution because of its relative concentration of nonpenetrating solutes and water. Net movement of water continues until the opposing hydrostatic pressure exactly counterbalances the osmotic pressure. (p. 40)

**ossification** The natural process of bone formation; the hardening or calcification of soft tissue into a bonelike material (p. 233)

**osteoblasts** (OS-tē-ō-blasts´) Bone cells that produce the organic matrix of bone (p. 233)

**osteoclasts** Bone cells that dissolve bone in their vicinity (p. 233)

**osteocytes** Osteoblasts that retire from active bone-forming duty, because their imprisonment prevents them from laying down new bone; also involved in the hormonally regulated exchange of calcium between bone and the blood (p. 233)

**osteoporosis** A disease of bone that leads to an increased risk of fracture (p. 286)

**outer hair cells** Sound receptor cells found in the organ of Corti. The 16 000 hair cells within each cochlea are arranged in four parallel rows along the length of the basilar membrane: one row of inner hair cells, and three rows of outer hair cells. (p. 173)

**oval window** The membrane-covered opening that separates the air-filled middle ear from the upper compartment of the fluid-filled cochlea in the inner ear (p. 172)

**overhydration** Water excess in the body (p. 614)

**overshoot** When a signal exceeds its steady-state value (p. 62)

**ovulation** (ov´-yuh-LĀ-shun) Release of an ovum from a mature ovarian follicle (p. 749)

**oxidative phosphorylation** (fos´-fōr-i-LĀ-shun) The entire sequence of mitochondrial biochemical reactions that uses oxygen to extract energy from the nutrients in food and transforms it into ATP, producing carbon dioxide and water in the process (p. 27)

**oxyhaemoglobin (HbO2)** (ok-si-HĒ-muh-glō-bun) Haemoglobin combined with oxygen (p. 532)

**oxyntic mucosa** (ok-SIN-tic) The mucosa that lines the body and fundus of the stomach; contains gastric pits that lead to the gastric glands lined by mucous neck cells, parietal cells, and chief cells (p. 661)

## P

**Pacinian corpuscles** Mechanoreceptors located in the skin that respond to touch and deep pressure; useful in detecting moderately rough to rougher surfaces and are very sensitive to vibration (p. 147)

**pancreas** (PAN-krē-us) A mixed gland composed of an exocrine portion that secretes digestive enzymes and an aqueous alkaline secretion into the duodenal lumen and an endocrine portion that secretes the hormones insulin and glucagon into the blood (p. 271)

**paracrines** (PEAR-uh-krin) Local chemical messengers whose effect is exerted only on neighbouring cells in the immediate vicinity of its site of secretion (p. 51)

**paradoxical sleep** A 10- to 15-minute episode that punctuates the end of each slow-wave sleep cycle (p. 134)

**parasympathetic nervous system** (pear´-uh-sim-puh-THET-ik) The subdivision of the autonomic nervous system that dominates in quiet, relaxed situations and promotes body maintenance activities such as digestion and emptying of the urinary bladder (p. 92)

**parathyroid glands** (pear´-uh-THĪ-roid) Four small glands located on the posterior surface of the thyroid gland that secrete parathyroid hormone (p. 285)

**parathyroid hormone (PTH)** A hormone that raises plasma Ca²⁺ levels (p. 285)

**paresthesias** The sensation of numbness and/or tingling in the skin (p. 622)

**parietal (oxyntic) cells** (puh-RĪ-ut-ul) The stomach cells that secrete hydrochloric acid and intrinsic factor (p. 661)

**parietal lobes** The lobes of the cerebral cortex that lie at the top of the brain behind the central sulcus and contain the somatosensory cortex (p. 115)

**parietal–temporal–occipital association cortex** A region concerned with integrating such sensory inputs as sight, sound, and touch (p. 133)

**Parkinson's disease (PD)** Condition associated with a deficiency of dopamine, an important neurotransmitter in the basal ganglia (p. 82)

**partial pressure** The individual pressure exerted independently by a particular gas within a mixture of gases (p. 524)

**partial pressure gradients** Difference in the partial pressure of a gas between two regions that promotes the movement of the gas from the region of higher partial pressure to the region of lower partial pressure (p. 524)

**parturition (labour, delivery, birth)** (par´-too-RISH-un) Delivery of a baby (p. 767)

**passive forces** Forces that do not require expenditure of cellular energy to accomplish transport of a substance across the plasma membrane (p. 37)

**passive reabsorption** When none of the steps in the transepithelial transport of a substance reabsorbed across the kidney tubules requires energy expenditure (p. 574)

**pathogens** (PATH-uh-junz) Disease-causing microorganisms, such as bacteria or viruses (p. 445)

**pathophysiology** (path´-ō-fiz-ē-OL-ō-gē) Abnormal functioning of the body associated with disease (p. 16)

**pepsin** An enzyme involved in protein digestion (p. 663)

**pepsinogen** (pep-SIN-uh-jun) A peptide secreted by the stomach that is the inactive precursor of pepsin (p. 663)

**percent haemoglobin (% Hb) saturation** A measure of the extent to which the haemoglobin present in the blood is combined with oxygen (p. 532)

**perception** The conscious interpretation of the external world as created by the brain from a pattern of nerve impulses delivered to it from sensory receptors (p. 143)

**perilymph** Fluid contained in the scala vestibuli and scala tympani (p. 172)

**peripheral chemoreceptors** (kē´-mō-rē-SEP-turz) The carotid and aortic bodies, which respond to changes in arterial $P_{O_2}$, $P_{CO_2}$, and $H^+$ and adjust respiration accordingly (p. 545)

**peripheral nervous system (PNS)** Nerve fibres that carry information between the central nervous system and other parts of the body (p. 92)

**peristalsis** (per´-uh-STOL-sus) Ringlike contractions of the circular smooth muscle of a tubular organ that move progressively forward with a stripping motion, pushing the contents of the organ ahead of the contraction (p. 656)

**peritubular capillaries** (per´-i-TŪ-bū-lur) Capillaries that intertwine around the tubules of the kidney's nephron; supply the renal tissue and participate in exchanges between the tubular fluid and blood during the formation of urine (p. 563)

**permeable** Permitting passage of a particular substance through a plasma membrane (p. 36)

**permissiveness** Situation in which one hormone must be present in adequate amounts for the effect of another hormone to become fully exerted (p. 214)

**pernicious anaemia** (per-NEE-shus) The anaemia produced as a result of intrinsic factor deficiency (p. 441)

**pH** The logarithm to the base 10 of the reciprocal of the hydrogen ion concentration; $pH = \log 1/[H^+]$ or $pH = -\log[H^+]$ (p. 620)

**phagocytosis** (fag´-oh-sī-TŌ-sus) A type of endocytosis in which large, multimolecular, solid particles are engulfed by a cell (p. 48)

**phantom pain** An experience of pain that may arise due to extensive remodelling of the brain region that originally handled sensations from a severed limb (p. 147)

**pharynx** (FARE-inks) The back of the throat, which serves as a common passageway for the digestive and respiratory systems (p. 172)

**phasic receptors** Receptors that are useful in situations where it is important to signal a change in stimulus intensity rather than to relay status quo information (p. 145)

**pheromones** Chemical factor that is secreted to induce sexual response (p. 186)

**phosphatidylinositol bisphosphate (PIP$_2$)** A component of the tails of the phospholipid molecules within the cell membrane (p. 219)

**phospholipase C** An enzyme that cleaves phospholipids just in front of the phosphate group (p. 218)

**phospholipids** A class of lipids that are a major component of all cell membranes because they form lipid bilayers (p. 29)

**photons** Particle-like individual packets of electromagnetic radiation that travel in wave-like fashion (p. 155)

**photopigments** Molecules that undergo chemical alterations when activated by light. Through a series of steps, this light-induced change and subsequent activation of photopigments brings about a receptor potential that ultimately leads to the generation of action potentials, which transmit this information to the brain for visual processing. (p. 161)

**photoreceptors** Sensory receptors responsive to light (p. 143)

**phototransduction** The mechanism of converting light stimuli into electrical activity by the rods and cones of the eye (p. 161)

**physical therapy** The treatment of disease, injury, or deformity by physical methods such as active mobilization (p. 3)

**physiological range** The normal range for a physiological variable (e.g., temperature, pH) found in the population of interest (e.g., humans) that will be maintained most of the time (>85%) (p. 9)

**physiologist** Person who specializes in the study of physiology (p. 4)

**pia mater** The innermost layer of the meninges (*mater* means "mother," indicative of this membrane's protective and supportive role) (p. 100)

**pineal gland** (PIN-ē-ul) A small endocrine gland located in the centre of the brain that secretes the hormone melatonin (p. 240)

**pinna** A prominent skin-covered flap of cartilage that collects sound waves and channels them down the external ear canal. Many species (dogs, for

example) can cock their ears in the direction of sound to collect more sound waves, but human ears are relatively immobile. (p. 171)

**pinocytosis** (pin-oh-cī-TŌ-sus) Type of endocytosis in which the cell internalizes fluid (p. 48)

**pitch** The tone of a sound, determined by the frequency of vibrations (e.g., whether a sound is a C or G note) (p. 171)

**pitch discrimination** The ability to distinguish between various frequencies of incoming sound waves; depends on the shape and properties of the basilar membrane, which is narrow and stiff at its oval window end and wide and flexible at its helicotrema end (p. 174)

**pituitary gland (hypophysis)** (pih-TWO-ih-tair-ee) A small endocrine gland connected by a stalk to the hypothalamus; consists of the anterior pituitary and posterior pituitary (p. 225)

**placenta** (plah-SEN-tah) The organ of exchange between the maternal and fetal blood; also secretes hormones that support the pregnancy (p. 761)

**plasma** The liquid portion of the blood (p. 10)

**plasma cell** An antibody-producing derivative of an activated B lymphocyte (p. 475)

**plasma clearance** The volume of plasma that is completely cleared of a given substance by the kidneys per minute (p. 586)

**plasma-colloid osmotic pressure (Πp)** (KOL-oid os-MOT-ik) The force caused by the unequal distribution of plasma proteins between the blood and surrounding fluid; encourages fluid movement into the capillaries (p. 409)

**plasma membrane** A protein-studded lipid bilayer that encloses each cell, separating it from the extracellular fluid (p. 5)

**plasma proteins** The proteins that remain within the plasma, where they perform important functions; include albumins, globulins, and fibrinogen (p. 435)

**plasticity** (plas-TIS-uh-tē) The ability of portions of the brain to assume new responsibilities in response to the demands placed on it (p. 120)

**platelets (thrombocytes)** (PLA–T-lets) Specialized cell fragments in the blood that participate in haemostasis by forming a plug at a vessel defect (p. 447)

**pluripotent stem cells** Precursor cells; for example, those that reside in the bone marrow and continuously divide and differentiate to give rise to each of the types of blood cells (p. 438)

**polarization** Any time at which the membrane potential is not 0 mV (p. 62)

**poliovirus** A contagious causative agent; a human enterovirus that destroys motor neuron cells (p. 195)

**polycythaemia** (pol-i-sī-THĒ-mē-uh) Excess circulating erythrocytes, accompanied by an elevated haematocrit (p. 444)

**polymodal nociceptors** Nociceptors that respond to more than one type of noxious stimuli (p. 150)

**polysaccharides** (pol´-ī-SAK-uh-rīdz) Complex carbohydrates, consisting of chains of interconnected glucose molecules (p. 646)

**polysynaptic** Term than means "many synapses" (p. 106)

**positive balance** Situation in which the gains via input for a substance exceed its losses via output, so that the total amount of the substance in the body increases (p. 608)

**positive feedback** A regulatory mechanism in which the input and the output in a control system continue to enhance each other, so that the controlled variable is progressively moved farther from a steady state (p. 15)

**posterior parietal cortex** Part of the brain that plays a key role in producing planned movements. Before an effective movement can be started, the central nervous system must know the original positions of the body parts that are to be moved, and the positions of any external objects that the body may contact. (p. 119)

**posterior pituitary gland** The neural portion of the pituitary that stores and, upon hypothalamic stimulation, releases into the blood two hormones produced by the hypothalamus—vasopressin and oxytocin (p. 225)

**postganglionic fibre** (pōst´-gan-glē-ON-ik) The second neuron in the two-neuron autonomic nerve pathway; originates in an autonomic ganglion and terminates on an effector organ (p. 188)

**postsynaptic neuron** (pōst´-si-NAP-tik) The neuron that conducts its action potentials away from a synapse (p. 74)

**prefrontal association cortex** The front portion of the frontal lobe just anterior to the premotor cortex; the part of the brain involved in brainstorming and, specifically, in planning for voluntary activity, decision making, creativity, and personality traits (p. 133)

**preganglionic fibre** The first neuron in the two-neuron autonomic nerve pathway; originates in the central nervous system and terminates on an autonomic ganglion (p. 188)

**preload** The pressure stretching the ventricle of the heart, after atrial contraction and subsequent passive filling of the ventricle; describes the initial stretching of a single cardiac myocyte prior to contraction (p. 369)

**premotor cortex** Part of the brain that works in coordination with the motor cortex (a few millimetres anterior of the primary motor cortex) to plan and execute movements (p. 119)

**pressure gradient** A difference in pressure between two regions that drives the movement of blood or air from the region of higher pressure to the region of lower pressure (p. 387)

**presynaptic facilitation** Enhanced release of neurotransmitter from a presynaptic axon terminal as a result of excitation of another neuron that terminates on the axon terminal (p. 81)

**presynaptic inhibition** A reduction in the release of a neurotransmitter from a presynaptic axon terminal as a result of excitation of another neuron that terminates on the axon terminal (p. 81)

**presynaptic neuron** (prē-si-NAP-tik) The neuron that conducts its action potentials toward a synapse (p. 74)

**primary active transport** A carrier-mediated transport system in which energy is directly required to operate the carrier and move the transported substance against its concentration gradient (p. 44)

**primary auditory cortex** The part of the brain located in the temporal lobe that is responsible for the processing of auditory information (p. 176)

**primary follicles** Primary oocyte surrounded by a single layer of granulosa cells in the ovary (p. 745)

**primary motor cortex** The portion of the cerebral cortex that lies anterior to the central sulcus and is responsible for voluntary motor output (p. 115)

**primary olfactory cortex** A portion of the cortex involved in olfaction (p. 185)

**primary reproductive organs (gonads)** The primary reproductive organs, which produce the gametes and secrete the sex hormones; testes and ovaries (p. 724)

**primary tastes** Salty, sour, sweet, bitter, umami, and the recently added fifth taste—a meaty, "amino-acid" taste (p. 182)

**procedural memories** Type of memory associated with motor skills gained through repetitive training, such as memorizing a dance routine. The cortical areas important for a given procedural memory are the specific motor or sensory systems engaged in performing the routine. (p. 128)

**prolactin (PRL)** (prō-LAK-tun) An anterior pituitary hormone that stimulates breast development and milk production in females (p. 227)

**prolactin-inhibiting hormone (PIH)** Hormone that inhibits the release of prolactin; acts as a neurotransmitter and as a neuromodulator. For example, PIH is identical to *dopamine*, a major neurotransmitter in the basal nuclei and elsewhere. (p. 229)

**proliferative phase** The phase of the menstrual cycle during which the endometrium repairs itself and thickens following menstruation; lasts from the end of the menstrual phase until ovulation (p. 755)

**pro-opiomelanocortin (POMC)** (prō-op´Ē-ō-ma-LAN-oh-kor´-tin) A large precursor molecule produced by the anterior pituitary that is cleaved into adrenocorticotropic hormone, melanocyte-stimulating hormone, and endorphin (p. 257)

**proprioception** (prō´-prē-ō-SEP-shun) Awareness of position of body parts in relation to one another and to surroundings (p. 117)

**prostaglandins** (pros´-tuh-GLAN-dins) Local chemical mediators derived from a component of the plasma membrane, arachidonic acid (p. 741)

**prostate gland** A male accessory sex gland that secretes an alkaline fluid, which neutralizes acidic vaginal secretions (p. 740)

**protein kinase** A (KŪ-nase) An enzyme that phosphorylates and thereby induces a change in the shape and function of a particular intracellular protein (p. 217)

**proteolytic enzymes** (prōt´-ē-uh-LIT-ik) Enzymes that digest protein (p. 667)

**proximal tubule** (PROKS-uh-mul) A highly convoluted tubule that extends between Bowman's capsule and the loop of Henle in the kidney's nephron (p. 563)

**pseudopods** Surface projections from a white blood cell that completely surround or engulf large multimolecular particles, such as a bacterium or tissue debris, and trap it within an internalized vesicle (p. 48)

**psychoactive drugs** Drugs that affect human moods; some have been shown to influence self-stimulation in experimental animals (p. 125)

**PTH** See *parathyroid hormone.* (p. 285)

**pulmonary artery** (PULL-mah-nair-ē) The large vessel that carries blood from the right ventricle to the lungs (p. 347)

**pulmonary circulation** The closed loop of blood vessels carrying blood between the heart and lungs (p. 346)

**pulmonary surfactant** (sur-FAK-tunt) A phospholipoprotein complex secreted by the Type II alveolar cells that intersperses between the water molecules that line the alveoli, thereby lowering the surface tension within the lungs (p. 509)

**pulmonary valve** A one-way valve that permits the flow of blood from the right ventricle into the pulmonary artery during ventricular emptying, but prevents the backflow of blood from the pulmonary artery into the right ventricle during ventricular relaxation (p. 350)

**pulmonary veins** The large vessels that carry blood from the lungs to the heart (p. 347)

**pump** An active transport mechanism in living cells by which specific ions are moved through the cell membrane against a concentration or electrochemical gradients (p. 44)

**pupil** An adjustable round opening in the centre of the iris through which light passes to the interior portions of the eye (p. 154)

**Purkinje fibres** (pur-KIN-jē) Small terminal fibres that extend from the bundle of His and rapidly transmit an action potential throughout the ventricular myocardium (p. 353)

**pyloric gland area (PGA)** (pī-LŌR-ik) The specialized region of the mucosa in the antrum of the stomach that secretes gastrin (p. 661)

**pyloric sphincter** (pī-lōr´-ik SFINGK-tur) The juncture between the stomach and duodenum (p. 656)

**pyramidal cells** Neurons with cone-shaped cell bodies (p. 114)

## R

**radial (dilator) muscle** Muscle fibres that project out from the pupil and, when stimulated by the autonomic nervous system, contract to increase the size of the pupil (p. 155)

**radiation** Emission of heat energy from the surface of a warm body in the form of electromagnetic waves (p. 712)

**reabsorption** The net movement of interstitial fluid into the capillary (p. 409)

**reactive hyperaemia** Hyperaemia (an increase in blood flow) in a tissue or organ, resulting from the restoration of temporarily occluded (blocked) blood flow (p. 400)

**receptive field** Refers to the area surrounding a receptor within which the receptor can detect stimuli (p. 147)

**receptor** See *sensory receptor or receptor site.* (p. 31)

**receptor-mediated endocytosis** A highly selective process that enables cells to import specific large molecules that it needs from its environment; triggered by the binding of a specific molecule, such as a protein, to a surface membrane receptor site specific for that protein (p. 48)

**receptor potential** The graded potential change that occurs in a sensory receptor in response to a stimulus; generates action potentials in the afferent neuron fibre (p. 144)

**receptor sites** Membrane proteins that bind with a specific extracellular chemical messenger, thereby bringing about a series of membrane and intracellular events that alter the activity of the particular cell (p. 31)

**reconsolidated** To combine into one system, back into a restabilized, inactive state (p. 126)

**red, green, and blue cones** Photoreceptor cells that respond selectively to various wavelengths of light, making colour vision possible (p. 161)

**reduced haemoglobin (deoxyhaemoglobin)** Haemoglobin that is not combined with oxygen (p. 532)

**referred pain** Pain that appears in one part of the body, but originates from another. For example, pain originating in the heart may appear to come from the left shoulder and arm, suggesting that inputs arising from the heart may share a pathway to the brain in common with inputs from the left upper extremity, but the mechanism responsible for referred pain is not completely understood. (p. 103)

**reflex** Any response that occurs automatically without conscious effort; controlled by a reflex arc that includes a receptor, afferent pathway, integrating centre, efferent pathway, and effector (p. 105)

**reflex arc** A neural pathway that controls an action reflex (e.g., patellar reflex). In humans, most sensory neurons do not pass directly into the brain, but synapse in the spinal cord and include afferent neurons, interneurons, and efferent neurons to complete the arc. (p. 105)

**refraction** Bending of a light ray (p. 155)

**refractory period** (rē-FRAK-tuh-rē) The time period when a recently activated patch of membrane is refractory (unresponsive) to further stimulation, preventing the action potential from spreading backward into the area through which it has just passed, thereby ensuring the unidirectional propagation of the action potential away from the initial site of activation (p. 67)

**regeneration tube** Formed by Schwann cells to guide a regenerating nerve fibre to its proper destination. The remaining portion of the axon connected to the cell body starts to grow and move forward within the Schwann cell column by amoeboid movement. (p. 72)

**regulatory proteins** Troponin and tropomyosin, which play a role in regulating muscle contraction by either covering or exposing the sites of interaction between the contractile proteins (p. 302)

**regulatory T cells (T$_{reg}$)** A class of T lymphocytes that suppresses the activity of other lymphocytes (p. 482)

**relative refractory period** The interval immediately following the absolute refractory period, during which initiation of a second action potential is inhibited but not impossible (p. 68)

**releasing hormones** Hypothalamic hormone that stimulates the secretion of a particular anterior pituitary hormone (p. 229)

**renal cortex** An outer, granular-appearing region of the kidney (p. 254)

**renal medulla** (RĒ-nul muh-DUL-uh) An inner, striated-appearing region of the kidney (p. 192)

**renal threshold** The plasma concentration at which the $T_m$ of a particular substance is reached and the substance first starts appearing in the urine (p. 580)

**renin** (RĒ-nin) An enzymatic hormone released from the kidneys in response to a decrease in NaCl/ECF volume/arterial blood pressure; activates angiotensinogen (p. 576)

**renin–angiotensin–aldosterone system (RAAS)** (an´jē-ō-TEN-sun al-dō-steer-OWN) The salt-conserving system triggered by the release of renin from the kidneys, which activates angiotensin, which stimulates aldosterone secretion, which stimulates Na$^+$ reabsorption by the kidney tubules during the formation of urine (p. 576)

**repolarization** (rē´-pō-luh-ruh-ZĀ-shun) Return of membrane potential to resting potential following a depolarization (p. 58)

**reproductive tract** The system of ducts specialized to transport or house the gametes after they are produced (p. 724)

**residual volume (RV)** The minimum volume of air remaining in the lungs even after a maximal expiration (p. 518)

**resistance** Hindrance of flow of blood or air through a passageway (blood vessel or respiratory airway, respectively) (p. 60)

**rest-and-digest** Bodily functions promoted by the parasympathetic system; coincides with a slowing down of activities enhanced by the sympathetic system (p. 192)

**resting membrane potential** The membrane potential that exists when an excitable cell is not displaying an electrical signal (p. 54)

**reticular activating system (RAS)** (ri-TIK-ū-lur) Ascending fibres that originate in the reticular formation and carry signals upward to arouse and activate the cerebral cortex (p. 112)

**reticular formation** A network of interconnected neurons that runs throughout the brain stem and initially receives and integrates all synaptic input to the brain (p. 110)

**retina** The innermost layer in the posterior region of the eye that contains the eye's photoreceptors, the rods and cones (p. 153)

**retinene** The light-absorbing part of the photopigment. There are four different photopigments, one in the rods and one in each of three types of cones. (p. 161)

**retino-hypothalamic tract** A photic input pathway involved in the circadian rhythms. The message regarding light status relayed from the SCN to the pineal gland is the major way the internal clock is coordinated to a 24-hour day. (p. 242)

**rhodopsin** A pigment of the retina that is responsible for both the formation of the photoreceptor cells and the first events in the perception of light (p. 161)

**right and left atrioventricular (AV) valves** One-way valves that permit the flow of blood from the atrium to the ventricle during filling of the heart, but prevent the backflow of blood from the ventricle to the atrium during emptying of the heart (p. 347)

**right cerebral hemisphere** Part of the brain that excels in nonlanguage skills, especially spatial perception and artistic and musical talents; generates a big-picture, holistic view, in contrast to the left hemisphere that tends to process information in a fine-detail, fragmentary way (p. 134)

**rods** The eye's photoreceptors used for night vision (p. 153)

**round window** The membrane-covered opening that separates the lower chamber of the cochlea in the inner ear from the middle ear (p. 172)

**Ruffini corpuscles** Myelinated, slow-adapting nerve endings found in the deep layers of the skin (subcutaneous tissue) and joints; encapsulated, tied into the local collagen matrix, and respond to stretch (deformation) and torque; register mechanical deformation within joints (sensitivity of 2 degrees), as well as continuous pressure (p. 148)

## S

**saccule** Contains hairs that respond selectively to tilting of the head away from a horizontal position (such as getting up from bed) and to vertically directed linear acceleration and deceleration (p. 178)

**salivary amylase** (AM-uh-lās´) An enzyme produced by the salivary glands that begins carbohydrate digestion in the mouth and continues in the body of the stomach after the food and saliva have been swallowed (p. 653)

**saltatory conduction** (SAL-tuh-tōr´-ē) The means by which an action potential is propagated throughout a myelinated fibre, with the impulse jumping over the myelinated regions from one node of Ranvier to the next (p. 70)

**salt balance** The equalization of the body's salt input and output to prevent salt accumulation or deficit (p. 612)

**salty taste** One of the five basic tastes; stimulated by chemical salts, especially NaCl (table salt). Direct entry of positively charged Na$^+$ through specialized Na$^+$ channels in the receptor cell membrane, a movement that reduces the cell's internal negativity, is responsible for receptor depolarization in response to salt. (p. 182)

**SA node** See *sinoatrial node*. (p. 352)

**sarcomere** (SAR-kō-mir) The functional unit of skeletal muscle; the area between two Z lines within a myofibril (p. 300)

**sarcoplasmic reticulum (SR)** (ri-TIK-yuh-lum) A fine meshwork of interconnected tubules that surrounds a muscle fibre's myofibrils; contains expanded lateral sacs, which store calcium that is released into the cytosol in response to a local action potential (p. 304)

**satellite cells** Small mononuclear progenitor cells with virtually no cytoplasm; found in mature muscle (p. 321)

**saturation** State in which all binding sites on a carrier molecule are occupied (p. 41)

**scala media** A blind-ended cochlear duct that constitutes the middle compartment of the cochlea (p. 172)

**scala tympani** The lower compartment of the cochlea; follows the outer contours of the spiral (p. 172)

**scala vestibuli** The upper compartment of the cochlea; follows the inner contours of the spiral (p. 172)

**Schwann cells** (shwahn) The myelin-forming cells of the peripheral nervous system (p. 70)

**sclera** The visible white part of the eye (p. 153)

**secondary active transport** A transport mechanism in which a carrier molecule for glucose or an amino acid is driven by a Na$^+$ concentration gradient established by the energy-dependent Na$^+$ pump to transfer the glucose or amino acid uphill without directly expending energy to operate the carrier (p. 44)

**secondary (antral) follicle** A developing ovarian follicle that is secreting estrogen and forming an antrum (p. 747)

**second messenger** An intracellular chemical that is activated by binding of an extracellular first messenger to a surface receptor site, and that triggers a preprogrammed series of biochemical events, which result in altered activity of intracellular proteins to control a particular cellular activity (p. 53)

**second-order sensory neuron** Neurons located in the medulla oblongata that receive input from primary sensory neurons, and thereby form the nucleus of the solitary tract that integrates all visceral information (p. 146)

**secretin** (si-KRĒT-′n) A hormone released from the duodenal mucosa primarily in response to the presence of acid; inhibits gastric motility and secretion and stimulates secretion of a NaHCO₃ solution from the pancreas and from the liver (p. 659)

**secretion** The release to a cell's exterior, on appropriate stimulation, of substances that have been produced by the cell (p. 6)

**secretory (progestational) phase** The phase of the menstrual cycle characterized by the development of a lush endometrial lining capable of supporting a fertilized ovum; also known as the progestational phase (p. 755)

**secretory vesicles** (VES-i-kuls) Membrane-enclosed sacs containing proteins that have been synthesized and processed by the endoplasmic reticulum/Golgi complex of the cell, and which are released to the cell's exterior by exocytosis on appropriate stimulation (p. 48)

**segmentation** The small intestine's primary method of motility; consists of oscillating, ringlike contractions of the circular smooth muscle along the small intestine's length (p. 676)

**selectively permeable** Permitting some particles to pass through the plasma membrane while excluding others (p. 36)

**self-antigens** Antigens that are characteristic of a person's own cells (p. 482)

**semen** (SĒ-men) A mixture of accessory sex gland secretions and sperm (p. 740)

**semicircular canals** Sense organ in the inner ear that detects rotational or angular acceleration or deceleration of the head (p. 178)

**semilunar valves** (sem′-ī-LEW-nur) The aortic and pulmonary valves (p. 740)

**seminal vesicles** (VES-i-kuls) Male accessory sex glands that supply fructose to ejaculated sperm and secrete prostaglandins (p. 740)

**seminiferous tubules** (sem′-uh-NIF-uh-rus) The highly coiled tubules within the testes that produce spermatozoa (p. 732)

**sensitization** Increased responsiveness to mild stimuli following a strong or noxious stimulus (p. 129)

**sensitization period** The period of first exposure to a particular allergen, but no symptoms evoked; includes formation of memory cells that can elicit an immune response upon subsequent exposure to the allergen (p. 492)

**sensor** A cell (receptor) that detects or measures a physical property and records, indicates, or otherwise responds to it (p. 14)

**sensorineural deafness** Condition wherein sound waves are transmitted to the inner ear but are not translated into nerve signals that are interpreted by the brain as sound sensations (p. 177)

**sensory afferent** Pathway coming into the central nervous system that carries information that reaches the level of consciousness (p. 142)

**sensory homunculus** Orderly somatotopic representation of the sensitivity of body parts; gives an upside down and unequal representation on the somatosensory cortex; the larger the cortical representation, the higher the sensitivity and use (p. 118)

**sensory receptor** An afferent neuron's peripheral ending, which is specialized to respond to a particular stimulus in its environment (p. 92)

**Sertoli cells** (sur-TŌ-lē) Cells located in the seminiferous tubules that support spermatozoa during their development (p. 737)

**set point** The desired level at which homeostatic control mechanisms maintain a controlled variable, (nonsteady) physiological state (p. 9)

**short-term memory** The capacity to hold a small amount (approximately eight elements) of information in mind in a readily available state for a short period of time (seconds) (p. 126)

**signal transduction** The sequence of events in which incoming signals (instructions from extracellular chemical messengers such as hormones) are conveyed to the cell's interior for execution (p. 53)

**simple (basic) reflexes** Built-in reflexes requiring no conscious thought (p. 105)

**single-unit smooth muscle** The most abundant type of smooth muscle; made up of muscle fibres that are interconnected by gap junctions so that they become excited and contract as a unit; also known as *visceral smooth muscle* (p. 328)

**sinoatrial node (SA node)** (sī-nō-Ā-trē-ul) A small, specialized, autorhythmic region in the right atrial wall of the heart that has the fastest rate of spontaneous depolarizations and serves as the normal pacemaker of the heart (p. 352)

**skeletal muscle** Striated muscle, which is attached to the skeleton and is responsible for movement of the bones in purposeful relation to one another; innervated by the somatic nervous system and under voluntary control (p. 298)

**skin (integument)** The covering for the outside of the body, along with the underlying connective tissue; also called integument (in-TEG-yuh-munt) (p. 460)

**slow pain** Type of persistent pain characterized as dull or aching; originates from chemical nociceptors (p. 149)

**slow pain pathway** Impulses carried on unmyelinated slow C fibres that carry slow pain, heat, cold, and mechanical stimuli at a much slower rate of <1–2 m/s (p. 150)

**slow-wave potentials** Self-excitable activity of an excitable cell in which its membrane potential undergoes gradually alternating depolarizing and hyperpolarizing swings (p. 331)

**smell** The ability to perceive odours; olfaction (p. 181)

**smooth muscle** Involuntary muscle innervated by the autonomic nervous system and found in the walls of hollow organs and tubes (p. 326)

**somaesthetic sensations** (SŌ-mes-THEH-tik) Awareness of sensory input such as touch, pressure, temperature, and pain from the body's surface (p. 116)

**somatic nervous system** The portion of the efferent division of the peripheral nervous system that innervates skeletal muscles; consists of the axonal fibres of the alpha motor neurons (p. 92)

**somatic sensation** Sensory information arising from the body surface, including somaesthetic sensation and proprioception (p. 142)

**somatomedins** (sō′-mat-uh-MĒ-dinz) Hormones secreted by the liver or other tissues, in response to growth hormone, that act directly on the target cells to promote growth; insulin-like growth factors (IGF) (p. 235)

**somatosensory cortex** The region of the parietal lobe immediately behind the central sulcus; the site of initial processing of somaesthetic and proprioceptive input (p. 117)

**somatosensory pathways** Consists of components of the central and peripheral nervous systems that receive and interpret sensory information from receptors in the joints, ligaments, muscles, and skin (p. 146)

**somatosensory receptors** Free-nerve endings consisting of a neuron with an exposed receptor; see also *special sense receptor* (p. 143)

**somatostatin** Hormone secreted by the hypothalamus and the pancreas; acts as a paracrine in the stomach (p. 229)

**sound waves** Travelling vibrations of air that consist of regions of high pressure caused by compression of air molecules alternating with regions of low pressure caused by rarefaction of the molecules (p. 169)

**sour taste** Taste caused by acids containing a free hydrogen ion. The citric acid content of lemons, for example, accounts for their distinctly sour taste. (p. 182)

**spatial summation** The summing of several post-synaptic potentials arising from the simultaneous activation of several excitatory (or several inhibitory) synapses (p. 79)

**special senses** Vision, hearing, taste, and smell (p. 143)

**special senses receptor** Used in the ear; turns a mechanical stimulation (non-neural) into a neural signal by synapsing onto a sensory neuron (p. 143)

**specificity** Ability of carrier molecules to transport only specific substances across the plasma membrane (p. 41)

**spermatogenesis** (spur′-mat-uh-JEN-uh-sus) Sperm production (p. 735)

**sphincter** (sfink-tur) A voluntarily controlled ring of skeletal muscle that controls passage of contents through an opening into or out of a hollow organ or tube (p. 598)

**spike** Recorded appearance of an action potential; a sharp rapid rise followed by a rapid decrease (p. 62)

**spinal cord** Tubular bundle of nervous tissue and cells connected to the medulla oblongata and extending down the vertebral column (p. 101)

**spinal nerves** Paired spinal nerves emerge from the spinal cord through spaces formed between the bony, winglike arches (pedicle and lamina) of adjacent vertebrae; named according to the region of the vertebral column from which they originate (p. 101)

**spinal reflex** A reflex that is integrated by the spinal cord (p. 107)

**spinocerebellum** Part of the brain that enhances muscle tone and helps coordinate voluntary movement, especially fast, phasic motor activities (p. 121)

**spleen** A lymphoid tissue in the upper left part of the abdomen that stores lymphocytes and platelets and destroys old red blood cells (p. 437)

**spoken languages** Capacity for speaking developed in the early years of human life. A hearing impairment can prevent not only the ability to talk but also the ability to understand the spoken word (p. 177)

**stable balance** The equalization of the total body input of a particular substance and its total body output (p. 608)

**stapes** The small ossicle in the middle ear, shaped like a stirrup (p. 172)

**steady state** A dynamic equilibrium (p. 37)

**stellate cells** Neurons with a starlike shape (p. 114)

**Stem Cell Network** A Canadian venture that fosters ethical stem cell research by bringing together over 50 leading scientists to investigate the potential of stem cells (p. 11)

**stem cells** Relatively undifferentiated precursor cells that give rise to highly differentiated, specialized cells (p. 661)

**stereocilia** Actin-stiffened microvilli that protrude from the surface of each hair cell (p. 173)

**stimulus** A detectable physical or chemical change in the environment of a sensory receptor (p. 106)

**stress** The generalized, nonspecific response of the body to any factor that overwhelms, or threatens to overwhelm, the body's compensatory abilities to maintain homeostasis (p. 263)

**stretch reflex** A monosynaptic reflex in which an afferent neuron originating at a stretch-detecting receptor in a skeletal muscle terminates directly on the efferent neuron supplying the same muscle and causes it to contract and counteract the stretch (p. 106)

**stroke volume (SV)** The volume of blood pumped out of each ventricle with each contraction, or beat, of the heart (p. 363)

**subarachnoid space** The space between the arachnoid mater and the pia mater (p. 100)

**submucosa** The connective tissue layer of the digestive tract that lies under the mucosa and contains the larger blood and lymph vessels and a nerve network (p. 649)

**substance P** The neurotransmitter released from pain fibres (p. 150)

**subsynaptic membrane** (sub-sih-NAP-tik) The portion of the postsynaptic cell membrane that lies immediately underneath a synapse and contains receptor sites for the synapse's neurotransmitter (p. 75)

**supplementary motor area** Part of the sensorimotor cerebral cortex, located on the medial face of the hemisphere, just in front of primary motor cortex (p. 119)

**suprachiasmatic nucleus (SCN)** (soup´-ra-kī-as-MAT-ik) A cluster of nerve cell bodies in the hypothalamus that serves as the master biological clock, acting as the pacemaker that establishes many of the body's circadian rhythms (p. 240)

**suspensory ligaments** Structure that attaches the ciliary muscle to the lens (p. 157)

**sweet taste** Taste evoked by the particular configuration of glucose. From an evolutionary perspective, a craving for sweet foods is associated with the supply of necessary calories in a readily usable form. (p. 184)

**sympathetic ganglion chain** A chain of ganglia situated ventral and lateral to the spinal cord and extending from the upper neck down to the coccyx (p. 188)

**sympathetic nervous system** The subdivision of the autonomic nervous system that dominates in emergency (fight-or-flight) or stressful situations and prepares the body for strenuous physical activity (p. 92)

**synapse** (SIN-aps´) The specialized junction between two neurons where an action potential in the presynaptic neuron influences the membrane potential of the postsynaptic neuron by means of the release of a chemical messenger that diffuses across the small cleft that separates the two neurons (p. 76)

**synaptic cleft** The space between two nerve cells in which neurotransmitters cross over (p. 74)

**synaptic knob** The bulbous structures on the end of an axon, each of which contains many synaptic vesicles (p. 74)

**synaptic vesicles** Vesicles that store various neurotransmitters that are released at the synapse. The release is regulated by voltage-dependent calcium channels (p. 74)

**synergism** (SIN-er-jiz´-um) Several complementary actions that produce a combined effect that is greater than the sum of their separate effects (p. 214)

**systemic circulation** (sis-TEM-ik) The closed loop of blood vessels carrying blood between the heart and body systems (p. 346)

**systole** (SIS-tō-lē) The period of cardiac contraction and emptying (p. 361)

**systolic pressure** Peak pressure in the arteries; occurs near the beginning of the cardiac cycle when the ventricles are contracting (p. 391)

## T

**T₃** See *triiodothyronine*. (p. 248)

**T₄** See *tetraiodothyronine*. (p. 248)

**tachycardia** The cardiac rhythm, which produces a ventricular rate greater than 100 beats per minute (p. 359)

**target cells** The cells that a particular extracellular chemical messenger, such as a hormone or a neurotransmitter, influences (p. 51)

**tastant** A taste-provoking chemical (p. 182)

**taste** One of the five senses; detected through sensory organs, the taste buds (p. 181)

**taste pore** A small opening that houses the taste bud, through which fluids in the mouth come into contact with the surface of its receptor cells (p. 182)

**taste receptor cells** Modified epithelial cells with many surface folds, or microvilli, that protrude slightly through the taste pore, greatly increasing the surface area exposed to the oral contents (p. 182)

**tears** Eye-washing fluid that is lubricating, cleansing, and bactericidal; produced continuously by the lacrimal gland in the upper lateral corner under the eyelid (p. 153)

**tectorial membrane** A gelatinous structure that is a highly hydrated, non-cellular matrix found on the cells of the cochlea (p. 173)

**teleological approach** (tē´-lē-ō-LA-ji-kul) Explanation of body functions in terms of their particular purpose in fulfilling a bodily need—that is, the "why" of body processes (p. 4)

**temporal lobes** The cerebral cortex lobes that are located laterally and that initially process auditory input (p. 115)

**temporal summation** The summing of several postsynaptic potentials occurring very close together in time because of successive firing of a single presynaptic neuron (p. 78)

**tendon** The tough, fibrous tissue that connects muscle to bone (p. 298)

**tension** The force produced during muscle contraction due to the shortening of the sarcomeres, resulting in stretching and tightening of the muscle's elastic connective tissue and tendon, which transmit the tension to the bone to which the muscle is attached (p. 313)

**terminal button** A motor neuron's enlarged knob-like ending that terminates near a skeletal muscle fibre and releases acetylcholine in response to an action potential in the neuron (p. 197)

**terminal ganglia** Very short postganglionic fibres that lie in or near an effector organ and end on the cells of that organ (p. 188)

**testosterone** (tes-TOS-tuh-rōn) The male sex hormone, secreted by the Leydig cells of the testes (p. 724)

**tetanus** (TET´-n-us) A smooth, maximal, muscle contraction that occurs when the fibre is stimulated so rapidly that it does not have a chance to relax at all between stimuli (p. 311)

**tetraiodothyronine (T₄ or thyroxine)** (thī-ROCKS-in) The most abundant hormone secreted by the thyroid gland; important in the regulation of overall metabolic rate; also known as thyroxine or T₄ (p. 248)

**thalamus** (THAL-uh-mus) The brain region that serves as a synaptic integrating centre for preliminary processing of all sensory input on its way to the cerebral cortex (p. 112)

**thecal cells** (THAY-kel) The outer layer of specialized ovarian connective tissue cells in a maturing follicle (p. 747)

**thermally gated channels** Gated channels that respond to local changes in temperature (heat or cold) (p. 58)

**thermal nociceptors** Nociceptors formed late during neurogenesis, developing from neural crest stem cells, that respond to temperature extremes, especially heat (p. 149)

**thermoreceptors** (thur´-mō-rē-SEP-tur) Sensory receptors sensitive to heat and cold (p. 144)

**thick filaments** Specialized cytoskeletal structures within skeletal muscle that are made up of myosin molecules and interact with the thin filaments to shorten the fibre during muscle contraction (p. 298)

**thin filaments** Specialized cytoskeletal structures within skeletal muscle that are made up of actin, tropomyosin, and troponin molecules and interact with the thick filaments to shorten the fibre during muscle contraction (p. 298)

**third-order sensory neuron** The long sensory tract that ascends within the spinal cord; found in the thalamic nuclei and transmits information from the thalamus to the primary sensory cortex (p. 146)

**thirst** The subjective sensation that drives you to drink H₂O (p. 617)

**thirst centre** An area of the hypothalamus in close proximity to vasopressin secreting cells (p. 617)

**thoracic cavity** (chest) (thō-RAS-ik) The cavity that lies above the diaphragm and contains the heart and lungs (p. 346)

**threshold potential** The critical membrane potential that must be reached before an action potential is initiated in an excitable cell (p. 61)

**thromboembolism** A general term describing both thrombosis and its main complication, embolization; the formation of a blood clot (a thrombus) inside a blood vessel (p. 378)

**thrombus** An abnormal clot attached to the inner lining of a blood vessel (p. 453)

**thymus** (THĪ-mus) A lymphoid gland located midline in the chest cavity that processes T lymphocytes and produces the hormone thymosin, which maintains the T-cell lineage (p. 224)

**thyroglobulin (Tg)** (thī´-rō-GLOB-yuh-lun) A large, complex molecule on which all steps of thyroid hormone synthesis and storage take place (p. 248)

**thyroid gland** A bi-lobed endocrine gland that lies over the trachea and secretes three hormones: tetraiodothyronine (thyroxine) and triiodothyronine, which regulate overall basal metabolic rate, and

**calcitonin,** which contributes to control of calcium balance (p. 248)

**thyroid hormone** Collectively, the hormones secreted by the thyroid follicular cells, namely, tetraiodothyronine (thyroxine) and triiodothyronine (p. 248)

**thyroid-stimulating hormone (TSH, thyrotropin)** An anterior pituitary hormone that stimulates secretion of thyroid hormone and promotes growth of the thyroid gland (p. 227)

**thyrotropin-releasing hormone (TRH)** Hormone that stimulates the release of TSH (alias thyrotropin) from the anterior pituitary (p. 229)

**tidal volume ($V_T$)** The volume of air entering or leaving the lungs during a single breath (p. 518)

**tight junctions** Impermeable junctions between two adjacent epithelial cells formed by the sealing together of the cells' lateral edges near their luminal borders; prevent passage of substances between the cells (p. 34)

**timbre** Quality of a sound that is determined by its characteristic overtones (p. 171)

**tip links** Structure that pulls on mechanically gated ion channels in the hair cell when the stereocilia are deflected by endolymph movement (p. 178)

**tissues** (1) Functional aggregation of cells of a single specialized type, such as nerve cells forming nervous tissue; (2) the aggregate of various cellular and extracellular components that make up a particular organ, such as lung tissue (p. 6)

**T lymphocytes (T cells)** White blood cells that accomplish cell-mediated immune responses against targets to which they have been previously exposed; see also *cytotoxic T cells, helper T cells,* and *regulatory T cells* (p. 447)

**tolerance** Refers to the desensitization to an addictive drug so that the user needs greater quantities of the drug to achieve the same effect (p. 82)

**tone** [muscle] The ongoing baseline of activity in a given system or structure, as in muscle tone, sympathetic tone, or vascular tone (p. 171)

**tone** [sound] The quality or character of sound. The characteristic quality or timbre of a particular sound (p. 41)

**tonic activity** The typically constant activity of some neurons, in contrast to the phasic activity (pulsitile firing) of other neurons. Slowly adapting, or tonic, receptors respond to steady stimulus and produce a steady rate of firing; they most often respond to increased intensity of stimulus by increasing their firing frequency (impulses per second). (p. 192)

**tonicity** The effect a solution has on cell volume—whether the cell remains the same size, swells, or shrinks—when the solution surrounds the cell (p. 40)

**tonic receptors** Receptors that adapt slowly or do not adapt at all; important in situations where it is valuable to maintain information about a stimulus; Examples include muscle stretch receptors, which monitor muscle length, and joint proprioceptors, which measure the degree of joint flexion (p. 145)

**total lung capacity (TLC)** The volume of gas contained in the lung at the end of maximal inspiration (p. 519)

**total peripheral resistance** The resistance offered by all the peripheral blood vessels, with arteriolar resistance contributing most extensively (p. 401)

**trachea** (TRĂ-kē-uh) The conducting airway that extends from the pharynx and branches into two

bronchi, each entering a lung; known as the windpipe (p. 501)

**transcellular fluid** A number of small, specialized fluid volumes, all of which are secreted by specific cells into a particular body cavity to perform some specialized function (p. 610)

**transducin** A G protein found in rod and cone cells (p. 161)

**transduction** Conversion of stimuli into action potentials by sensory receptors (p. 143)

**transepithelial transport** (tranz-ep-i-THĒ-lē-al) The entire sequence of steps involved in the transfer of a substance across the epithelium between either the renal tubular lumen or digestive tract lumen and the blood (p. 574)

**transporter recruitment** The phenomenon of inserting additional transporters (carriers) for a particular substance into the plasma membrane, thereby increasing membrane permeability to the substance in response to an appropriate stimulus (p. 273)

**transport maximum ($T_m$)** The maximum rate of a substance's carrier-mediated transport across the membrane when the carrier is saturated; known as *tubular maximum* in the kidney tubules (p. 41)

**transverse tubule (T tubule)** A perpendicular infolding of the surface membrane of a muscle fibre; rapidly spreads surface electric activity into the central portions of the muscle fibres (p. 304)

**tricarboxylic acid cycle** The second stage of cellular respiration, which is the three-stage process by which living cells break down organic fuel molecules in the presence of oxygen (p. 25)

**triglycerides** (trī-GLIS-uh-rīdz) Neutral fats composed of one glycerol molecule with three fatty acid molecules attached (p. 647)

**triiodothyronine ($T_3$)** (trī-ī-ō-dō-THĪ-rō-nēn) The most potent hormone secreted by the thyroid follicular cells; important in the regulation of overall metabolic rate (p. 248)

**trilaminar structure** The plasma membrane's three-layered appearance due to the specific arrangement of its molecules; consists of two dark layers separated by a light middle layer (p. 29)

**trophoblast** (TRŌF-uh-blast′) The outer layer of cells in a blastocyst that is responsible for accomplishing implantation and developing the fetal portion of the placenta (p. 758)

**tropic hormone** (TRŌ-pik) A hormone that regulates the secretion of another hormone (p. 210)

**tropomyosin** (trōp′-uh-MĪ-uh-sun) One of the regulatory proteins in the thin filaments of muscle fibres (p. 301)

**troponin** (tro-PŌ-nun) One of the regulatory proteins found in the thin filaments of muscle fibres (p. 302)

**TSH** See *thyroid-stimulating hormone.* (p. 227)

**T tubule** See *transverse tubule.* (p. 304)

**tubular maximum ($T_m$)** The maximum amount of a substance that the renal tubular cells can actively transport within a given time period; the kidney cells' equivalent of transport maximum (p. 579)

**tubular reabsorption** The selective transfer of substances from tubular fluid into peritubular capillaries during urine formation (p. 565)

**tubular secretion** The selective transfer of substances from peritubular capillaries into the tubular lumen during urine formation (p. 565)

**twitch** A brief, weak contraction that occurs in response to a single action potential in a muscle fibre (p. 309)

**twitch summation** The addition of two or more muscle twitches as a result of rapidly repetitive stimulation, resulting in greater tension in the fibre than that produced by a single action potential (p. 311)

**tympanic membrane** (tim-PAN-ik) The eardrum, which is stretched across the entrance to the middle ear and which vibrates when struck by sound waves funnelled down the external ear canal (p. 172)

**Type I alveolar cells** (al-VĒ-ō-lur) The single layer of flattened epithelial cells that forms the wall of the alveoli within the lungs (p. 503)

**Type 1 diabetes mellitus** A form of diabetes mellitus; an autoimmune disease that results in the permanent destruction of insulin-producing beta cells of the pancreas; also known as insulin-dependent, or juvenile onset, diabetes (p. 278)

**Type 2 diabetes mellitus** A metabolic disorder characterized by insulin resistance, relative insulin deficiency, and hyperglycaemia; also known as non-insulin-dependent, or maturity-onset, diabetes (p. 278)

## U

**ultrafiltration** The net movement of a protein-free plasma out of the capillary into the surrounding interstitial fluid (p. 409)

**umami taste** A taste, first identified and named by a Japanese researcher, that is triggered by amino acids, especially glutamate. The presence of amino acids, as found in meat, for example, serves as a marker for a desirable, nutritionally protein-rich food. (p. 184)

**urethra** (yū-RĒ-thruh) A tube that carries urine from the bladder to outside the body (p. 560)

**urine excretion** The elimination of substances from the body in the urine; anything filtered or secreted and not reabsorbed is excreted (p. 565)

**use-dependent competition** Competition for a binding site that is based on use (p. 119)

**utricle** One of two otolith organs housed within a bony chamber situated between the semicircular canals and the cochlea (p. 178)

## V

**vagus nerve** (VĂ-gus) The tenth cranial nerve, which serves as the major parasympathetic nerve (p. 110)

**varicosities** Swellings on the terminal branches of autonomic fibres that simultaneously release neurotransmitter over a large area of the innervated organ, rather than on single cells (p. 188)

**vasoconstriction** (vă-zō-kun-STRIK-shun) The narrowing of a blood vessel lumen due to contraction of the vascular circular smooth muscle (p. 395)

**vasodilation** The enlargement of a blood vessel lumen due to relaxation of the vascular circular smooth muscle (p. 395)

**vasopressin** (vă-zō-PRES-sin) A hormone secreted by the hypothalamus, then stored and released from the posterior pituitary; increases the permeability of the distal and collecting tubules of the kidneys to water and promotes arteriolar vasoconstriction; also known as *antidiuretic hormone (ADH)* (p. 592)

**veins** Vessels that carry blood toward the heart (p. 346)

**venae cavae** (*vena cava*, singular) (VĒ-nah CĀV-ah; VĒ-nē cāv-ē) Two large veins (superior and inferior) that empty blood into the right atrium (p. 347)

**venous return** (VĒ-nus) The volume of blood returned to each atrium per minute from the veins (p. 414)

**venous sinuses** Cavities between the different layers of the brain through which venous blood drains (p. 98)

**ventilation** The mechanical act of moving air in and out of the lungs; breathing (p. 500)

**ventral corticospinal tract** Descending pathway that originates in the motor region of the cerebral cortex, then travels down the ventral portion of the spinal cord, and terminates in the spinal cord on the cell bodies of efferent motor neurons supplying skeletal muscles (p. 103)

**ventral (anterior) horn** The ventral column of grey matter in the spinal cord when viewed in cross-section; contains cell bodies of the efferent motor neurons supplying skeletal muscles (p. 103)

**ventral root** Area where efferent fibres carrying outgoing signals to skeletal muscles exit the spinal cord (p. 103)

**ventral spinocerebellar tract** Ascending pathway that originates in the spinal cord, runs up the ventral margin of the cord, has several synapses along the way, and eventually terminates in the cerebellum (p. 103)

**ventricles** (VEN-tri-kul) Lower chambers of the heart that pump blood into the arteries (p. 98)

**vesicular transport** Movement of large molecules or multimolecular materials into or out of the cell by being enclosed in a vesicle, as in endocytosis or exocytosis (p. 46)

**vestibular apparatus** (veh-STIB-yuh-lur) The component of the inner ear that provides information essential for the sense of equilibrium and for coordinating head movements with eye and postural movements; consists of the semicircular canals, utricle, and saccule (p. 178)

**vestibular membrane** Forms the ceiling of the cochlear duct and separates it from the scala vestibuli (p. 172)

**vestibular nerve** One of the two branches of the vestibulocochlear nerve, the cochlear nerve being the other; transmits information from the semicircular canals to the vestibular ganglion; receives positional information (p. 178)

**vestibular nuclei** A cluster of neuronal cell bodies in the brain stem to which signals arising from the various components of the vestibular apparatus are carried through the vestibulocochlear nerve (p. 180)

**vestibulocerebellum** Part of the brain that helps maintain balance and controls eye movements (p. 121)

**vestibulocochlear nerve** Nerve that contains both the auditory nerve and the vestibular nerve (p. 178)

**villi** (*villus*, singular) (VIL-us) Microscopic finger-like projections from the inner surface of the small intestine (p. 679)

**virulence** (VIR-you-lentz) The disease-producing power of a pathogen (p. 460)

**visceral afferent** A pathway coming into the central nervous system that carries subconscious information derived from the internal viscera (p. 142)

**visceral smooth muscle** (VIS-uh-rul) See *single-unit smooth muscle.* (p. 328)

**viscosity** (vis-KOS-i-tē) The friction developed between molecules of a fluid as they slide over each other during flow of the fluid; the greater the viscosity, the greater the resistance to flow (p. 387)

**visual field** Area that the eyes view, with each eye having a slightly different vantage point and also tremendous overlap at the same time; overlap giving rise to binocular field of vision, important for depth perception (p. 165)

**vital capacity (VC)** The maximum volume of air that can be moved out during a single breath following a maximal inspiration (p. 519)

**vitreous humour** A semifluid, jelly-like substance located in the larger posterior cavity between the lens and retina; important in maintaining the spherical shape of the eyeball (p. 153)

**vomeronasal organ (VNO)** Organ located about 15 mm inside the human nose next to the vomer bone; common in other mammals, and until recently thought to be nonexistent in humans (p. 186)

**voltage-gated channels** Channels in the plasma membrane that open or close in response to changes in membrane potential (p. 58)

## W

**wavelength** The distance between two wave peaks (p. 155)

**Wernicke's area** Receives input from the visual cortex in the occipital lobe, a pathway important in reading comprehension and in describing objects seen, as well as from the auditory cortex in the temporal lobe, a pathway essential for understanding spoken words. Wernicke's area also receives input from the somatosensory cortex, a pathway important in the ability to read Braille (p. 132)

**white matter** The portion of the central nervous system composed of myelinated nerve fibres (p. 103)

**withdrawal reflex** A polysynaptic spinal reflex intended to protect the body from damaging stimuli; causes stimulation of sensory, association, and motor neurons (p. 107)

**working memory** Capacity to temporarily hold and interrelate various pieces of information relevant to a current mental task; involves briefly holding and processing data for immediate use—both newly acquired information and related, previously stored, transiently brought forth, knowledge—and enables the evaluation of incoming data in context (p. 126)

## Z

**Z line** A flattened disclike cytoskeletal protein that connects the thin filaments of two adjoining sarcomeres (p. 300)

**zona fasciculata** (zō-nah fa-SIK-ū-lah-ta) The middle and largest layer of the adrenal cortex; major source of cortisol (p. 254)

**zona glomerulosa** (glō-MER-yū-lō-sah) The outermost layer of the adrenal cortex; sole source of aldosterone (p. 254)

**zona reticularis** (ri-TIK-yuh-lair-us) The innermost layer of the adrenal cortex; produces cortisol, along with the zona fasciculata (p. 254)

# Index

adenylyl cyclase, 217
adipocytes, 703
adipokines, 225, 703, 705
adipose tissue, 462
adiposity signals, 704
adolescence, 734
ADP. *See* adenosine diphosphate (ADP)
adrenal cortex, 192, 254, 258, 576
  cortisol secretion, 227
  enzyme for producing cortisol, 216
  growth, 227
adrenal diabetes, 259
adrenal glands
  adrenal androgen hypersecretion, 259–260
  adrenal cortex and sex hormones, 258
  adrenal medulla, 261, 263
  adrenocortical insufficiency, 260–261
  aldosterone hypersecretion, 258–259
  catecholamine-secreting medulla, 254
  cortisol hypersecretion, 259
  cortisol secretion, 257–258
  epinephrine, 261–263
  glucocorticoids, 254, 256–258
  mineralocorticoids, 254–256
  norepinephrine, 261–262
  sex hormones, 255
  steroid-secreting cortex, 254
adrenaline, 192
adrenal medulla, 192, 261
adrenergic drugs, 492
adrenergic fibres, 188
adrenergic system, 515
adrenocortical hormones, 254
adrenocortical insufficiency, 260–261
adrenocorticotropic hormone (ACTH), 227, 231, 256, 257
adrenogenital syndrome, 259, 260, 731
adult stem cells, 10, 72
aerobic (endurance) exercise, 320, 425
aerobic metabolism, 24, 28
afferent artirioles, 562
afferent division, 92
afferent fibres, 103
afferent nerve fibres and acupuncture analgesia (AA), 153
afferent neurons, 92, 93, 107, 117, 128, 176, 178, 187
  information, 325–326
  input, 323
  spinal reflexes, 324–325
afferent pain fibres, 149–150
afferent pathway, 106
afterbirth, 771
after hyperpolarization, 62
afterload, 369
agglutination, 477–478
agglutinins, 439
agglutinogens, 439
agonists, 193–194
agranulocyte, 445
airflow and airway resistance, 514–516
airway resistance, 529–530

airflow, 514–516
  bronchoconstriction, 515
  bronchodialation, 515
  chronic pulmonary diseases, 516
airways, 501–503
  atomic dead space ($V_D$), 521–522
  smooth muscle and airway resistance, 515
albumins, 435
albuminuria, 567
alcohol
  absorption, 667
  gastric secretion, 665
aldosterone, 255–256
  tubular reabsorption, 575–578
aldosterone hypersecretion, 258–259
aldosterone receptor blockers, 578
alkalaemia, 621
alkaline, 621
alkalosis, 621
  handling hydrogen and $HCO_3^-$ during, 629–630
allergen, 491
allergies, 460, 491–492
allied health professions, 3
all-or-none law, 69
allurin, 757
alpha (α) globulins, 435
alpha motor neurons, 325
alpha (α) receptors, 193
alphasynuclein, 341
altitude, physiological response to, 640–642
alveolar-capillary membrane diffusion, 524
alveolar dead space gas exchange, 529
alveolar gas equation, 525
alveolar interdependence, 510
alveolar macrophages, 464
alveolar partial pressures of oxygen and carbon dioxide, 524–525
alveolar pressure ($P_A$), 505, 511, 513
alveolar ventilation ($V_A$), 522
alveoli, 502–503, 771
  alveolar interdependence, 510
  Bohr effect, 536
  collateral ventilation, 503
  lowering surface tension, 509
  net diffusion of oxygen and carbon dioxide, 531
  pores of Kohn, 503
  pressure changes, 511
  pulmonary surfactant, 504, 509–510
  shunts, 529
  stability, 509–510
  surface tension, 508, 509
  Type I alveolar cells, 503
  Type II alveolar cells, 503
Alzheimer's
  progressive loss of memory, 342
  swallowing, 651
*American Journal of Respiratory and Critical Care Medicine*, 508
amine hormones, 215
amino acids, 28, 269, 646

growth hormone (GH), 236
  tubular reabsorption, 579
aminopeptidases, 678
ammonia ($NH_3$) and kidneys, 630–631
ammonium ion ($NH_4^+$), 632
amnesia, 127–128
amniotic cavity, 761
amniotic fluid, 761
amniotic sac (amnion), 761
amoeba-like behaviour, 469
amoebas, 5
AMPA receptors, 130
  glutamate and, 150
amphetamines, 125
ampulla, 178
amygdala, 124
amyotrophic lateral sclerosis (ALS), 195
anabolic processes, 22
anabolism, 267
anaemia, 441, 598
anaemic hypoxia, 539
anaerobic (high-intensity) exercise, 320
anaerobic glycolysis, 24
anaerobic metabolism, 27
analgesic system, 150, 182
anaphylactic shock, 426, 492–494
anatomy, 4
androgen-binding protein (ABP), 737
androgen deficiency in aging males (ADAM), 734
androgens, 232, 239–240, 728
angina pectoris, 378
angiogenesis, 399
angiotensin, 435
angiotensin-converting enzyme (ACE), 576
angiotensin I, 576
angiotensin II, 402, 424, 576, 618
angiotensinogen, 576
  variation in gene encoding for, 424
angiotensinogen alpha goblin, 435
angry macrophages, 485
anion gap, 634–635
anions, 38
anorexia nervosa, 710
anovulatory, 745
antagonism, 214
antagonistic control, 192
antagonists, 193–194
anterior pituitary, 211
anterior pituitary gland, 225–226
  hormones, 226–229
  hypophysiotropic hormones, 229
  hypothalamic hormones, 229
  hypothalamic-hypophyseal portal system, 229
  target-gland hormones feedback, 227
anterior pituitary tropic hormone, 231
anterograde amnesia, 127
anthrax, 480
antibiotics and intestinal infections, 463
antibodies, 447

eosinophil chemotactic factor, 492
eosinophilia, 446–447
eosinophils, 445, 464
   helper T cells, 485
ependymal cells, 98
epicardium, 350
epidermal cells, 461
epidermal growth factor, 225
epidermis, 460, 461
epididymis, 724, 739
epigenetics, 710
epiglottis, 655
epilepsy, 120
epimysium, 298
epinephrine (adrenaline), 192, 214,
     424, 449, 515, 516
   adrenergic receptors, 193
   arteriolar radius, 402
   release during exercise, 550
epinephrine (E)
   alpha-adrenergic and beta-adrenergic
     receptors, 261, 262
   CNS, 263
   fuel metabolism, 281–283
   metabolic effects, 262–263
   organ systems, 262
epiphyseal plate, 233, 234
epiphysis, 233
epithelial cells, 6, 649
   epidermis, 461
   villi, 680
epithelial glands, 7
epithelial sheets, 35
epithelial tissue, 6–7
equal pressure point, 513
equilibrium, 169–181
Erasistratus, 390
erectile dysfunction (impotence), 743
erectile tissue, 742
erection, 742
   erectile dysfunction, 743
   erection reflex, 742–743
erection-generating centre, 742
eructation (burping), 656
erythroblastosis fetalis, 439
erythrocytes (red blood cells), 433, 436.
     *See also* red blood cells
   ABO blood types, 439
   agglutinins, 439
   agglutinogens, 439
   antibodies against antigens, 439
   biconcave shape, 435, 436
   bone marrow producing, 437
   carbonic anhydrase, 437
   colour, 436
   erythropoietin (EPO), 437
   fibrinogen, 435
   flexibility of membrane, 436
   glycolytic enzymes, 437
   haemoglobin, 436
   intrauterine development, 437
   key enzymes, 437

   lack of nucleus and organelles, 437
   nitric oxide (NO), 437
   nonreparable, 437
   oxygen transport, 436
   preparation for departure from red bone
     marrow, 438
   Rh blood types, 439
   rupturing, 441
   short lifespan, 437
   spleen, 437
   structure and function, 436–437
   thinness, 436
   without DNA and RNA, 437
erythropoietin (EPO), 225, 437
esophageal stage, 656
esophagus, 501, 647, 654–656
essential nutrients, 267
estradiol, 215, 747
estriol, 766
estrogen, 221, 227, 232, 240, 461, 724
estrogen-like (phytoestrogen)
    compounds, 221
ETC. *See* electron transport chain (ETC)
eunuch, 734
eustachian (auditory) tube, 172
evaporation, 713–714
excessive salt intake, 424
exchanger, 661
excitable tissues, 54, 57
excitation-contraction coupling, 284
   action potential spread down transverse
     tubules (T tubules), 304, 305
   adenosine triphosphate (ATP) powering
     cross-bridge cycling, 306–307
   dihydropyridine receptors, 305
   foot proteins, 305
   lateral sacs, 305
   relaxation, 308
   release of calcium from sarcoplasmic
     reticulum (SR), 304
   rigor mortis, 307
   ryanodine receptors, 305
   terminal cisternae, 304
excitatory postsynaptic potential (EPSP),
    76–79
excitatory synapses, 76
excitement phase, 742, 744
exercise
   aerobic (endurance) exercise, 320
   anaerobic (high-intensity) exercise, 320
   blood pressure, 423
   body temperature increase, 550
   cardiac output (CO), 365
   cardiovascular changes during, 423
   cerebral cortex impulses, 550
   effect on ventilation, 549–550
   ephinephrine release, 550
   high-intensity interval training (HIIT), 320
   innate immunity, 471
   kidney function during, 572–573
   lymphatic system, 413
   oxygen dissociation curve, 536

   reflexes originating from body
     movements, 550
   resistance training, 320
   respiration energy cost, 523
exercise-induced hyperthermia, 717–718
exercise physiology, 3
exocrine gland cells, 649
exocrine glands, 7, 462
exocrine pancreas, 647, 667
exocytosis, 46, 48, 75, 216
   balancing with endocytosis, 50
exophthalmos, 253
expiration
   active, 513
   alveolar pressure ($P_A$), 513
   decreasing size of thoracic cavity, 513
   equal pressure point, 513
   expiratory muscles, 513
   force–length relationship, 520
   measuring expiratory flows, 520
   onset, 511
   transmural pressure, 513
expiratory muscles, 513
expiratory neurons, 542
expiratory reserve volume (ERV), 518
expression, 132
expulsion, 744
extension, 314
external anal sphincter, 690
external auditory meatus, 171
external cardiac compression, 347
external defences, 460–464
external ear, 169, 171–172
external elastic lamina, 389
external genitalia, 724
external intercostal muscles, 504
external respiration, 498, 500–501
external urethral sphincter, 598
external work, 700
extracellular chemical messengers, 52–53
extracellular fluid (ECF), 9–11, 29, 608, 609
   balanced constituent, 608–609
   cell water maintenance, 615
   controlling volume, 611
   exchanges between other sites
    and pool, 608
   hypertonicity, 614
   hypotonicity, 614–617
   interstitial fluid, 10
   intracellular fluid (ICF) and, 610–611
   lymph, 561
   minor compartments, 610
   normal alkalinity, 623
   osmolarity, 611, 613–614
   plasma, 10
   plasma and interstitial fluid barrier, 610
   pool, 608
   shrinking cells, 614
   swelling cells, 614–617
   transcellular fluid, 610
extracellular matrix (ECM), 34
extrafusal fibres, 325, 340

heart (*continued*)

cardiac muscle, 334
cardiac muscle fibres, 350
cardiac myopathies, 360–361
as centre of blood vessels, 390
centre of intellect, 390
comparing left and right pumps, 347
complete heart block, 353, 360
contracting or beating, 352
coronary circulation, 373
cytosolic $Ca^{2+}$ in excitation–contraction coupling, 356
dual pump, 346–347
ectopic focus, 354
electrical activity of, 352–361
endocardial lining, 373
enhancing contractility, 369
external cardiac compression, 347
extrasystoles, 359
fibrillation, 354
free fatty acids, 374
furnace to heat blood, 390
glucose, 374
heart attacks, 361, 378
heart block, 360
heart walls, 373
hypertension, 425
impulses spreading throughout, 351
inability to extract oxygen or nutrients from blood, 373
lactate, 374
left atrium, 347
left ventricle, 347
murmurs, 364
myocardial infarction, 370
myocardial ischemia, 360
nutrient supply to, 374
pacemaker, 352–353
parasympathetic stimulation effects on, 366–367
pericardial sac, 351
position of, 347
premature ventricular contraction (PVC), 354, 359
pressure-operated heart valves, 347–350
pulmonary artery, 347
pulmonary veins, 347
Purkinje fibres, 353
right atrium, 347
right ventricle, 347
sinoatrial node (SA node), 352–354
sympathetic stimulation effects on, 367
tachycardia, 359
thoracic cavity, 346
3:1 block, 360
3:1 rhythm, 360
thromboembolism, 378
thyroid hormone, 251
twisted structure, 350
2:1 block, 360
2:1 rhythm, 360
vagus nerve, 366

vascular inflammatory response, 374
veins, 346
venae cavae, 347
ventricles, 346
ventricular fibrillation, 360
heart attacks, 361, 378
clot-busting drugs, 453
coronary artery disease (CAD), 374–378
c-reactive protein, 377
extent of damaged area, 378
hypertension, 425
possible outcomes, 378
heart block, 360
heartburn, 656
heart failure, 371–373
heart murmurs, 364
heart muscle, 373–379
heart muscle cells
necrosis, 360
vasodilation, 373
heart rate, 365–366, 419
heart sounds, 364–365
heart valves, 347–350
insufficient or incompetent valves, 364
leaky valves, 364
stenotic valve, 364
heart wall, 350
heat, 700
and skeletal muscle, 316
heat exchange, 712–714
heat exhaustion, 718
heat input, 711
heat loss mechanisms
cold exposure, 716
fever, 717
heat exposure, 716
hypothalamus and heat production, 716
thermoneutral zone, 716
heat output, 711
heat-related disorders, 718
heatstroke, 718
heat transfer mechanisms, 713
heel strike, 339
*Helicobacter pylori*, 670
helicotrema, 172
helper T cells, 475, 482–483, 485–486, 489
helper T 17 ($T_H17$) cells, 486
heme groups, 436
hemiplegia, 325
Henderson–Hasselbalch equation, 625
Henle's loop, 589
heparin, 447
hepatic artery, 671
hepatic jaundice, 675
hepatic portal system, 229, 671
hepatic portal vein, 671
hepatic vein, 671
hepatitis, 676
heptaocytes, 671
Hering, Ewald, 544
herpes virus, 377
high blood pressure, 370

high-density lipoproteins (HDL), 376–377
higher centres, 542–543
higher cortex role in basic behavioural patterns, 125
high frequency oscillatory ventilation, 522
high-intensity interval training (HIIT), 320
highly lipid-soluble particles, 36
hippocampus, 128
hirsutism, 259
histamine, 52, 470, 664
arterioles, 399
basophils, 447
complement system stimulating release, 473
immmediate hypersensitivity, 492
increasing blood flow to tissues, 400
mast cells, 447
histotoxic hypoxia, 539
hives, 492
$H^+–K^+$ ATPase, 661
H1N1 (influenza A virus), 460
Hodgkin–Huxley model, 62
Hodgkin, Lloyd, 62
Hogg, James C., 516
homeostasis, 9–13, 83, 136, 390
body systems, 11–13
disruptions in, 16
glial cells, 129
maintenance of, 386
homeostatic control systems, 14–16
homeostatic drives, 125
homocysteine, 377
homologous chromosomes, 727
hOR17-4, 757
hormonal control systems, 211–212
hormonal responses *versus* neural responses, 222
hormone response element (HRE), 221
hormones, 210–211
activating, 212
adipokines, 225
adrenocortcotropic hormone (ACTH, adrenocorticotropin), 227
aldosterone, 576
amine hormones, 215
androgens, 232, 239–240
angiotensin II, 402
antagonism, 214
anterior and posterior pituitary gland, 225–226
anterior pituitary gland, 227
anterior pituitary tropic hormone, 230
arteriolar radius, 402
atrial natriuretic peptide, 225
as chemical messengers, 211
chemical properties, 216–217
communication principles, 214–222
cyclical secretion of, 211
deliberately altered target-cell receptors, 214
diurnal (circadian) rhythm, 212
endocrine glands, 7, 211
endocrine rhythm, 212

large intestine (*continued*)

drying and storage organ, 689

external anal sphincter, 690

feces, 689

haustra, 689

haustral contractions, 689

internal anal sphincter, 690

intestinal gases, 692

mass movements, 689

salt absorption, 691

secretion, 691–692

taeniae coli, 689

water ($H_2O$) absorption, 691

Laron dwarfism, 221, 238

larynx (voice box), 501

laser eye surgery, 157

latch phenomenon, 334

latent pacemakers, 353

latent period, 308

lateral corticospinal tract, 116

lateral horn, 103

lateral hypothalamic area (LHA), 705

lateral inhibition, 147

lateral sacs, 305

lateral spaces, 574

Iatrogenic, 614

law of LaPlace, 509

law of mass action, 24, 532

leads, 358

leak channels, 58

leaky valves, 364

learned motor programs (patterns), 339

learning

habituation, 129

limbic system, 126

lecithin, 672

left atrial volume receptors, 421, 668

left cerebral hemisphere, 134

legal determination of brain death, 120

length–tension relationship, 312–313

lens, 153, 155–159

leptin, 703

leukaemia, 447

leukocytes (white blood cells), 433,
444–447, 464

abnormalities in production, 447

amoeba-like behaviour, 469

basophils, 445, 464

chemokines, 469

chemotaxins, 469

chemotaxis, 469

cytokines, 470

defence agents, 445

destruction of bacteria, 469–470

emigration of, 469

eosinophils, 445–447, 464

functions and life spans, 446–447

lymphocytes, 445–447, 464

lymphoid tissues, 464–465

monocytes, 445, 447, 464

mononuclear agranulocytes, 445

neutrophils, 445–446, 464

polymorphonuclear granulocytes, 445

production, 445–446

proliferation, 469

red bone marrow, 437

types, 445

leukotrienes, 741

levels of organization in body

body system level, 7–8

cellular level, 5

chemical level, 4–5

organism level, 8

organ level, 7

tissue level, 6–7

levers, 316–317

Levine, Benjamin D., 641

levodopa (L-dopa), 82

Lewy bodies, 341, 342

Lewy Body Dementia, 342

Leydig, 227

Leydig cells, 732–735

after birth, sex-specific tissues, 734

before birth, testosterone secretion, 734

nonreproductive actions, 734–735

reproduction-related effects, 734

secondary sexual characteristics, 734

testosterone production, 733–734

testosterone-turned-estrogen, 735

Leydig (interstitial) cells, 732

LH surge, 750

Lieberman, Daniel, 339

light, 154–155

light adaptation, 164

light chains, 329

light ray, 1155

light sleep, 134

light waves, 155–156

limbic association cortex, 134

limbic system

basal ganglia, 124

basic behavioural patterns, 124–125

emotion, 124

hypothalamus, 125

learning, 126

lobes of cerebral cortex, 124

memory, 126–128

modifying responses, 125

motivated behaviours, 125

neurotransmitters, 125–126

punishment centres, 125

reward centres, 125

thalamus, 112

lipase, 673

lipid bilayer, 29–31, 36

lipid emulsion, 673

lipids

plasma membrane, 29–30

lipid-soluble extracellular chemical
messengers, 53

lipid-soluble molecules, 41

lipophilic hormones, 215–216

activating specific genes in target cell, 217

binding to plasma proteins, 211, 213

cholesterol, 216

crossing biological membranes, 221

estrogen, 221

estrogen-like (phytoestrogen)
compounds, 221

forming intracellular proteins, 217

general means of action, 217

processing, 216

protein synthesis, 221–222

receptors location, 217

steroid hormones, 215

thyroid hormones, 215

transporting, 215–216

Lippershey, Hans, 22

lips, 652

liver, 390, 647, 671

absorbed nutrients and, 687

aqueous alkaline fluid, 672

bile, 671, 672

bile canaliculus, 672

bile salts, 673

bilirubin, 672, 674–675

blood flow, 671–672

cholesterol, 672

functions, 671

glycogen, 269

hepatic portal system, 671

heptaocytes, 671

Kupffer cells, 671

lecithin, 672

lobules, 672

plasma proteins, 212

sinusoids, 672

sphincter of Oddi, 672

total blood cholesterol, 377

load, 314

load arm, 316

lobules, 672

lochia, 771

lockjaw, 82

locomotion, 339–341

long-acting thyroid stimulator (LATS),
252–253

long reflexes, 652

long-term memory, 126–127

loop of Henle, 563, 575

loss of plasma proteins, 598

loudness, 171

Lou Gehrig's disease, 195

low-density lipoproteins (LDL), 376

low lipid-soluble particles, 36

L-type $Ca^{2+}$ channel ($I_{Ca, L}$), 352

lumbar nerves, 101

lumen, 6

luminal membranes, 574

lymph vessels blockage, 413

Lumsden, Thomas, 542

lung disease, 548

lung disorders

chronic obstructive pulmonary disease, 372

dynamic lung function and respiratory
dysfunction, 519–521

# Study Card

## 1.1 | Introduction to Physiology (pp. 3–4)

- Physiology is the study of body functions.
- Physiologists explain body function in terms of the mechanisms of action involving cause-and-effect sequences of physical and chemical processes.
- Physiology and anatomy are closely interrelated because body functions are highly dependent on the structures of the body parts that carry them out.

## 1.2 | Levels of Organization in the Body (pp. 4–8)

- The human body is a complex combination of specific atoms and molecules.
- These nonliving chemical components are organized in a precise way to form cells, the smallest entities capable of carrying out the processes associated with life. Cells are the basic structural and functional living building blocks of the body. *(Review Figure 1-1.)*
- The basic functions performed by each cell for its own survival include (1) obtaining oxygen and nutrients, (2) performing energy-generating chemical reactions, (3) eliminating wastes, (4) synthesizing proteins and other cell components, (5) controlling movement of materials between the cell and its environment, (6) moving materials throughout the cell, (7) responding to the environment, and (8) reproducing.
- In addition to its basic functions, each cell in a multicellular organism performs a specialized function.
- Combinations of cells of similar structure and specialized function form the four primary tissues of the body: muscle, nervous, epithelial, and connective.
- Glands are derived from epithelial tissue and specialized for secretion. Exocrine glands secrete through ducts to the body surface or a cavity that communicates with the outside; endocrine glands secrete hormones into the blood. *(Review Figure 1-2.)*
- Organs are combinations of two or more types of tissues that act together to perform one or more functions. An example is the stomach.
- Body systems are collections of organs that perform related functions and interact to accomplish a common activity essential for survival of the whole body. An example is the digestive system. *(Review Figure 1-3.)*
- Organ systems combine to form the organism, or whole body.

## 1.3 | Concepts of Homeostasis (pp. 9–13)

- The fluid inside the cells of the body is intracellular fluid; the fluid outside the cells is extracellular fluid.
- Because most body cells are not in direct contact with the external environment, cell survival depends on maintaining a relatively stable internal fluid environment with which the cells directly make life-sustaining exchanges.
- The extracellular fluid serves as the internal environment. It consists of the plasma and interstitial fluid. *(Review Figure 1-4.)*
- Homeostasis is the maintenance of a dynamic, steady state in the internal environment.
- The factors of the internal environment that must be homeostatically maintained are its (1) concentration of nutrient molecules; (2) concentration of oxygen and carbon dioxide; (3) concentration of waste products; (4) pH; (5) concentration of water, salt, and other electrolytes; (6) volume and pressure; and (7) temperature.
- The functions performed by the body systems are directed toward maintaining homeostasis. The body systems' functions ultimately depend on the specialized activities of the cells that make up each system. Thus, homeostasis is essential for each cell's survival, and each cell contributes to homeostasis. *(Review Figure 1-5.)*

## 1.4 | Homeostatic Control Systems (pp. 14–16)

- A homeostatic control system is a network of body components working together to maintain a given factor in the internal environment relatively constant near an optimal set level.
- Homeostatic control systems can be classified as (1) intrinsic (local) controls, which are inherent compensatory responses of an organ to a change; and (2) extrinsic controls, which are responses of an organ triggered by factors external to the organ, namely by the nervous and endocrine systems.
- Both intrinsic and extrinsic control systems generally operate on the principle of negative feedback: A change in a controlled variable triggers a response that drives the variable in the opposite direction of the initial change, thus opposing the change. *(Review Figure 1-6.)*
- In positive feedback, a change in a controlled variable triggers a response that drives the variable in the same direction as the initial change, thus amplifying the change.
- Feedforward mechanisms are compensatory responses that occur in anticipation of a change.
- Pathophysiological states ensue when one or more of the body systems fail to function properly and, therefore, an optimal internal environment can no longer be maintained. Serious homeostatic disruption leads to death.

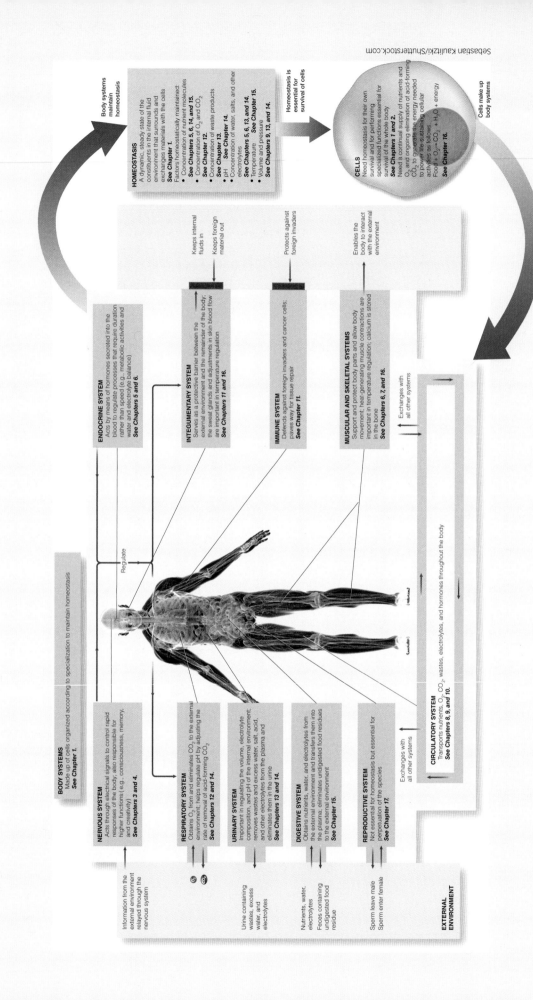

- The complex organization and interaction of the chemicals within a cell confer the unique characteristics of life.
- Cells are the living building blocks of the body.

## 2.1 | Observation of Cells (p. 22)

- Cells are too small for the unaided eye to see.
- Using early microscopes, investigators learned that all plant and animal tissues consist of individual cells.
- Through more sophisticated techniques, scientists now know that a cell is a complex, highly organized, compartmentalized structure.

## 2.2 | An Overview of Cell Structure (p. 22)

- Cells have three major subdivisions: the plasma membrane, the nucleus, and the cytoplasm. *(Review Table 2-2.)*
- The plasma membrane encloses the cell and separates the intracellular and extracellular fluid.
- The cytoplasm consists of cytosol, a complex gel-like mass laced with a cytoskeleton, and organelles, which are highly organized, membrane-enclosed structures dispersed within the cytosol.

## 2.3 | Cellular Metabolism (pp. 22–29)

- The rod-shaped mitochondria are the energy organelles of the cell. They house the enzymes of the citric acid cycle (in the mitochondrial matrix) and electron transport chain (on the cristae of the mitochondrial inner membrane). Together, these two biochemical processes efficiently convert the energy in food molecules to the usable energy stored in ATP molecules. *(Review Figures 2-1, 2-2, and 2-3.)*
- During this process, which is known as *oxidative phosphorylation*, the mitochondria use molecular $O_2$ and produce $CO_2$ and $H_2O$ as by-products.
- A cell is more efficient at converting food energy into ATP when oxygen is available. In an anaerobic (without $O_2$) condition, a cell can produce only two molecules of ATP for every glucose molecule processed by means of glycolysis, which takes place in the cytosol. In an aerobic (with $O_2$) condition, the mitochondria further degrade the products of glycolysis to yield another 34 molecules of ATP for every glucose molecule processed.
- Cells use ATP as an energy source for synthesis of new chemical compounds, for membrane transport, and for mechanical work.

## 2.4 | The Plasma Membrane (pp. 29–33)

- All cells are bounded by a plasma membrane, a thin lipid bilayer in which proteins are interspersed and to which carbohydrates are attached on the outer surface.
- The electron microscopic appearance of the plasma membrane as a trilaminar structure (two dark lines separated by a light interspace) is caused by the arrangement of the molecules composing it. The phospholipids orient themselves to form a bilayer with a hydrophobic interior (light interspace) sandwiched between the hydrophilic outer and inner surfaces (dark lines). *(Review Figures 2-6, 2-7, and 2-8.)*

- This lipid bilayer forms the structural boundary of the cell, serving as a barrier for water-soluble substances and being responsible for the fluid nature of the membrane.
- Cholesterol molecules tucked between the phospholipids contribute to the fluidity and stability of the membrane.
- According to the fluid mosaic model of membrane structure, the lipid bilayer is embedded with proteins. *(Review Figure 2-8.)* Membrane proteins, which vary in type and distribution among cells, serve as (1) channels for passage of small ions across the membrane, (2) carriers for transport of specific substances in or out of the cell, (3) docking-marker acceptors for fusion with and subsequent exocytosis of secretory vesicles, (4) membrane-bound enzymes that govern specific chemical reactions, (5) receptors for detecting and responding to chemical messengers that alter cell function, and (6) cell adhesion molecules that help hold cells together and serve as a structural link between the extracellular surroundings and intracellular cytoskeleton.
- The membrane carbohydrates, short sugar chains that project from the outer surface only, serve as self-identity markers. *(Review Figure 2-8.)* They are important in the recognition of "self" in cell-to-cell interactions, such as tissue formation and tissue growth.

## 2.5 | Cell-to-Cell Adhesions (pp. 34–36)

- Special cells locally secrete a complex extracellular matrix (ECM), which serves as a biological glue between the cells of a tissue.
- The ECM consists of a watery, gel-like substance interspersed with three major types of protein fibres: collagen, elastin, and fibronectin.
- Many cells are further joined by specialized cell junctions, of which there are three types: desmosomes, tight junctions, and gap junctions.
- Desmosomes serve as adhering junctions to hold cells together mechanically and are especially important in tissues subject to a great deal of stretching. *(Review Figure 2-9.)*
- Tight junctions actually fuse cells together to seal off passage between cells, thereby permitting only regulated passage of materials through the cells. These impermeable junctions are found in the epithelial sheets that separate compartments with very different chemical compositions. *(Review Figure 2-10.)*
- Gap junctions are communicating junctions between two adjacent, but not touching, cells. Cells joined by gap junctions are connected by small tunnels that permit exchange of ions and small molecules between the cells. Such movement of ions plays a key role in the spread of electrical activity to synchronize contraction in heart and smooth muscle. *(Review Figure 2-11.)*

## 2.6 | Overview of Membrane Transport (pp. 36–37)

- Materials can pass between the ECF and ICF by unassisted and assisted means.
- Transport mechanisms may also be passive (the particle moves across the membrane without the cell expending energy) or active (the cell expends energy to move the particle across the membrane).

## 2.7 | Unassisted Membrane Transport (pp. 37–41)

- Lipid-soluble particles and ions can cross the membrane unassisted. Nonpolar (lipid-soluble) molecules of any size can dissolve in and passively pass through the lipid bilayer down concentration gradients by means of the process of diffusion. (*Review Figures 2-12 and 2-13.*) Small ions traverse the membrane passively down electrochemical gradients through open protein channels specific for the ion.
- Osmosis is a special case of water passively moving down its own concentration gradient to an area of higher solute concentration. (*Review Figures 2-14 through 2-18.*)

## 2.8 | Assisted Membrane Transport (pp. 41–51)

- Carrier mechanisms are important for the assisted transfer of small polar molecules and for selected movement of ions across the membrane.
- In carrier-mediated transport, the particle is transported across by specific membrane carrier proteins. Carrier-mediated transport may be passive and move the particle down its concentration gradient (facilitated diffusion), or active and move the particle against its concentration gradient (active transport). (*Review Figures 2-19 through 2-22.*)
- A given carrier can move a single specific substance in one direction, two substances in opposite directions, or two substances in the same direction. Primary active transport requires the direct use of ATP to drive the pump, whereas secondary active transport is driven by an ion concentration gradient established by a primary active transport system. (*Review Figures 2-21 and 2-22.*)
- Large polar molecules and multimolecular particles can leave or enter the cell by being wrapped in a piece of membrane to form vesicles that can be internalized (endocytosis) or externalized (exocytosis). (*Review Figures 2-23 and 2-24.*)
- Cells are differentially selective in what enters or leaves because they possess varying numbers and kinds of channels, carriers, and mechanisms for vesicular transport.
- Large polar molecules (too large for channels and not lipid soluble) for which there are no special transport mechanisms are unable to permeate.

## 2.9 | Intercellular Communication and Signal Transduction (pp. 51–54)

- The coordination of the diverse activities of cells throughout the body to accomplish life-sustaining and other desired activities depends on the ability of cells to communicate with one another.

## 2.10 | Membrane Potential (pp. 54–59)

- This potential enables cellular communication in nervous tissue and muscle, and is made possible via the separation of opposite charges. The separation of charges is based on the difference in the relative number of cations (+) and anions (–) in the ICF and ECF.

## 2.11 | Graded Potentials (pp. 59–61)

- Graded potentials are local changes in membrane potential that occur in varying grades of magnitude or strength, do not travel long distances, and diminish in strength as they travel. Graded potentials may or may not result in an action potential.

## 2.12 | Action Potentials (pp. 61–71)

- Action potentials are brief, rapid, large changes in membrane potential during which the potential actually reverses, so that the inside of the excitable cell transiently becomes more positive than the outside.

## 2.13 | Regeneration of Nerve Fibres (pp. 72–73)

- Peripheral nervous system axons that are cut can regenerate, whereas those in the central nervous system cannot.
- Regeneration of peripheral axons is made possible by Schwann cells. In a peripheral nerve axon, the portion of the axon farthest from the cell body degenerates and is removed by the Schwann cells, which remain and form a regeneration tube to guide the restoration of the new nerve fibre to its proper destination.

## 2.14 | Synapses and Neuronal Integration (pp. 74–83)

- The axon, or nerve fibre (the conducting zone), conducts action potentials in undiminished fashion from the axon hillock to the axon terminals. The axon terminal (the output zone) serves as the presynaptic component, releasing a neurotransmitter that influences other postsynaptic cells in response to action potential propagation down the axon.
- The released neurotransmitter combines with receptor channels on the postsynaptic neuron. (*Review Figure 2-44.*) (1) If nonspecific cation channels that permit passage of both $Na^+$ and $K^+$ are opened, the resultant ionic fluxes cause an EPSP, a small depolarization that brings the postsynaptic cell closer to threshold. (2) If either $K^+$ or $Cl^-$ channels are opened, the likelihood that the postsynaptic neuron will reach threshold is diminished when an IPSP, a small hyperpolarization, is produced. (*Review Figure 2-45.*)
- If the dominant activity is in its excitatory inputs, the postsynaptic cell will likely be brought to threshold and have an action potential. This can be accomplished by either (1) temporal summation (EPSPs from a single, repetitively firing, presynaptic input occurring so close together in time that they add together) or (2) spatial summation (adding of EPSPs occurring simultaneously from several different presynaptic inputs). (*Review Figure 2-46.*) If inhibitory inputs dominate, the postsynaptic potential is brought farther than usual from threshold. If excitatory and inhibitory activity to the postsynaptic neuron is balanced, the membrane remains close to resting.

## 3.1 | Organization of the Nervous System (pp. 92–93)

■ The nervous system consists of the central nervous system (CNS), which includes the brain and spinal cord, and the peripheral nervous system, which includes the nerve fibres carrying information to (afferent division) and from (efferent division) the CNS. *(Review Figure 3-1.)*

■ Three functional classes of neurons—afferent neurons, efferent neurons, and interneurons—compose the excitable cells of the nervous system. *(Review Figure 3-2.)* (1) Afferent neurons inform the CNS about conditions in both the external and internal environment; (2) efferent neurons carry instructions from the CNS to effector organs—namely, muscles and glands; (3) interneurons are responsible for integrating afferent information and formulating an efferent response, as well as for all higher mental functions associated with the human mind.

## 3.2 | Structure and Function of the Central Nervous System (pp. 93–94)

■ Even though no part of the brain acts in isolation of other brain regions, the brain is organized into networks of neurons within discrete locations that are ultimately responsible for carrying out particular tasks.

■ The parts of the brain from the lowest, most primitive level to the highest, most sophisticated level are the brain stem, cerebellum, hypothalamus, thalamus, basal ganglia, and cerebral cortex. *(Review Table 3-1 and Figure 3-7.)*

## 3.3 | Protection of the Central Nervous System (pp. 95–101)

■ Glial cells form the connective tissue within the CNS and physically, metabolically, and functionally support the neurons.

■ The four types of glial cells are the astrocytes, oligodendrocytes, microglia, and ependymal cells. *(Review Figure 3-3 and Table 3-2.)*

■ The brain has several protective devices, which is important because neurons cannot divide to replace damaged cells. (1) The brain is wrapped in three layers of protective membranes—the meninges—and is further surrounded by a hard, bony covering. (2) Cerebrospinal fluid flows within and around the brain to cushion it against physical jarring *(review Figure 3-6)*. (3) Protection against chemical injury is conferred by a blood–brain barrier that limits access of blood-borne substances to the brain.

■ The brain depends on a constant blood supply for delivery of oxygen and glucose because it cannot generate ATP in the absence of either of these substances.

## 3.4 | Spinal Cord (pp. 101–109)

■ The spinal cord has two vital functions. (1) It serves as the neuronal link between the brain and the peripheral nervous system. All communication up and down the spinal cord occurs through ascending and descending tracts in the cord's outer white matter. *(Review Figures 3-10 and 3-11.)* (2) It is the integrating centre for spinal reflexes, including some of the basic protective and postural reflexes and those involved with the emptying of the pelvic organs. *(Review Figures 3-15 and 3-17.)*

■ The basic reflex arc includes a receptor, an afferent pathway, an integrating centre, an efferent pathway, and an effector. *(Review Figure 3-15.)* The centrally located grey matter of the spinal cord contains the interneurons interposed between the afferent input and efferent output as well as the cell bodies of efferent neurons. *(Review Figures 3-9 and 3-12.)*

## 3.5 | Brain Stem (pp. 109–112)

■ The brain stem is an important link between the spinal cord and higher brain levels.

■ The brain stem is the origin of the cranial nerves. *(Review Figure 3-18.)* It also contains centres that control cardiovascular, respiratory, and digestive function; regulates postural muscle reflexes; controls the overall degree of cortical alertness; and plays a key role in the sleep–wake cycle.

■ The prevailing state of consciousness depends on the cyclical interplay among (1) an arousal system (the reticular activating system) originating in the brain stem, (2) a slow-wave sleep centre in the hypothalamus, and (3) a paradoxical sleep centre in the brain stem. *(Review Figure 3-19.)*

## 3.6 | Thalamus and Hypothalamus (pp. 112–113)

■ The thalamus and hypothalamus are deep subcortical brain structures that interact extensively with the cortex in performing their functions. *(Review Figure 3-20)*

■ The thalamus serves as a relay station for preliminary processing of sensory input on its way to the cortex. It also accomplishes a crude awareness of sensation and some degree of consciousness.

■ The hypothalamus regulates many homeostatic functions, in part through its extensive control of the autonomic nervous system and endocrine system.

## 3.7 | Cerebral Cortex (pp. 113–121)

■ The cerebral cortex is the outer shell of grey matter that caps an underlying core of white matter. The white matter consists of bundles of nerve fibres that interconnect various cortical regions with other areas. *(Review Figure 3-28.)* The cortex itself consists primarily of neuronal cell bodies, dendrites, and glial cells.

■ Ultimate responsibility for many discrete functions is localized in particular regions of the cortex as follows: (1) the occipital lobes house the visual cortex; (2) the auditory cortex is in the temporal lobes; (3) the parietal lobes are responsible for

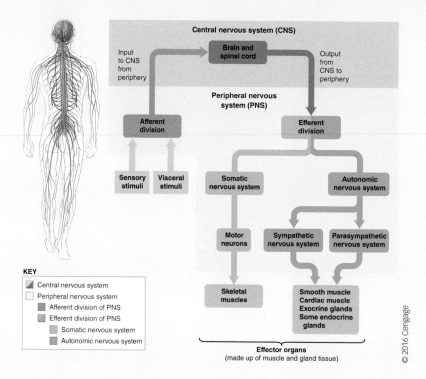

Central nervous system (CNS)

Brain and spinal cord

Input to CNS from periphery

Output from CNS to periphery

Peripheral nervous system (PNS)

Afferent division

Efferent division

Sensory stimuli | Visceral stimuli

Somatic nervous system

Autonomic nervous system

Motor neurons

Sympathetic nervous system

Parasympathetic nervous system

Skeletal muscles

Smooth muscle
Cardiac muscle
Exocrine glands
Some endocrine glands

**KEY**
▨ Central nervous system
☐ Peripheral nervous system
▨ Afferent division of PNS
▨ Efferent division of PNS
☐ Somatic nervous system
▨ Autonomic nervous system

Effector organs
(made up of muscle and gland tissue)

© 2016 Cengage

reception and perceptual processing of somatosensory (somaesthetic and proprioceptive) input; and (4) voluntary motor movement is set into motion by the frontal lobe, where the primary motor cortex and higher motor areas are located. *(Review Figures 3-22 through 3-24.)*

## 3.8 | Cerebellum (pp. 121–122)

■ The vestibulocerebellum helps maintain balance and controls eye movements.

■ The spinocerebellum enhances muscle tone and helps coordinate voluntary movement, especially fast, phasic motor activities.

■ The cerebrocerebellum plays a role in initiating voluntary movement and in storing procedural memories. *(Review Figure 3-27.)*

## 3.9 | Basal Ganglia (pp. 123–124)

■ The basal ganglia consist of several masses of grey matter and are associated with motor control, cognition, emotions, and learning. *(Review Figure 3-28)*

■ The basal ganglia inhibit muscle tone; coordinate slow, sustained postural contractions; and suppress useless patterns of movement.

## 3.10 | The Limbic System and Its Functional Relations with the Higher Cortex (pp. 124–126)

■ The limbic system, which includes portions of the hypothalamus and other forebrain structures that encircle the brain stem, is responsible for emotion as well as for basic, inborn behavioural patterns related to survival and perpetuation of the species. It also plays an important role in motivation and learning. *(Review Figure 3-29.)*

## 3.11 | Learning, Memory, and Language (pp. 126–134)

■ There are two types of memory: (1) a short-term memory with limited capacity and brief retention, coded by modification of activity at pre-existing synapses; and (2) a long-term memory with large storage capacity and enduring memory traces, involving relatively permanent structural or functional changes, such as the formation of new synapses, between existing neurons. *(Review Table 3-3 and Figures 3-30 and 3-31.)*

■ The hippocampus and associated structures are especially important in declarative, or "what," memories of specific objects, facts, and events.

■ The cerebellum and associated structures are especially important in procedural, or "how to," memories of motor skills gained through repetitive training.

■ The prefrontal association cortex is the site of working memory, which temporarily holds currently relevant data—both new information and knowledge retrieved from memory stores. It manipulates and relates this information to accomplish the higher-reasoning processes of the brain.

■ Language ability depends on the integrated activity of two primary language areas—Broca's area and Wernicke's area—typically located only in the left cerebral hemisphere. *(Review Figure 3-32.)*

## 3.12 | Sleep (pp. 134–136)

■ Sleep is an active process, not just the absence of wakefulness. While sleeping, a person cyclically alternates between slow-wave sleep and paradoxical sleep. *(Review Table 3-4.)*

■ Slow-wave sleep is characterized by slow waves on the EEG and little change in behaviour pattern from the waking state, except for not being aware of the external world. *(Review Figure 3-34.)*

■ Paradoxical, or REM, sleep is characterized by an EEG pattern similar to that of an alert, awake individual. Rapid eye movements, dreaming, and abrupt changes in behaviour pattern occur. *(Review Figure 3-34.)*

## 4.1 | Introduction (p. 141)

- The peripheral nervous system links the periphery and the central nervous system. It is divided into the afferent division and efferent division.

## 4.2 | The Peripheral Nervous System: Afferent Division (pp. 142–143)

- The afferent division of the peripheral nervous system carries information about the internal and external environment to the CNS.

- Sensory information, afferent information that reaches the level of conscious awareness, includes (1) somatic sensation (somaesthetic sensation and proprioception) and (2) special senses.

- Perception is the conscious interpretation of the external world that the brain creates from sensory input.

## 4.3 | Receptor Physiology (pp. 143–149)

- A stimulus brings about a graded, depolarizing receptor potential by altering the receptor's membrane permeability. Receptor potentials, if of sufficient magnitude, generate action potentials in the afferent neuronal membrane next to the receptor by opening $Na^+$ channels in this region. These action potentials self-propagate along the afferent neuron to the CNS. *(Review Figure 4-2.)*

- Receptor potential size is also influenced by the extent of receptor adaptation, which is a reduction in receptor potential despite sustained stimulation. (1) Tonic receptors adapt slowly or not at all and thus provide continuous information about the stimuli they monitor. (2) Phasic receptors adapt rapidly and frequently exhibit off responses, thereby providing information about changes in the energy form they monitor. *(Review Figure 4-4.)*

- The term *receptive field* refers to the area surrounding a receptor within which the receptor can detect stimuli. The acuity, or discriminative ability, of a body region varies inversely with the size of its receptive fields and also depends on the extent of lateral inhibition in the afferent pathways arising from receptors in the region. *(Review Figure 4-5.)*

## 4.4 | Pain (pp. 149–152)

- Painful experiences are elicited by noxious mechanical, thermal, or chemical stimuli and consist of two components: (1) the perception of pain and (2) the emotional and behavioural responses to it.

- Three categories of nociceptors, or pain receptors, respond to these stimuli: mechanical nociceptors, thermal nociceptors, and chemical nociceptors.

- Pain signals are transmitted over two afferent pathways: (1) a fast pain pathway that carries sharp, prickling pain signals; and (2) a slow pain pathway that carries dull, aching, persistent pain signals. *(Review Table 4-2.)*

- Afferent pain fibres terminate in the spinal cord on ascending pathways that transmit the signal to the brain for processing. *(Review Figure 4-6.)*

- Descending pathways from the brain use endogenous opiates to suppress the release of substance P, a pain-signalling neurotransmitter from the afferent pain-fibre terminal. Thus, these descending pathways block further transmission of the pain signal and serve as a built-in analgesic system. *(Review Figure 4-6.)*

## 4.5 | Eye: Vision (pp. 152–169)

- Light is a form of electromagnetic radiation that travels in wavelike fashion, with visible light being only a small band in the total electromagnetic spectrum. *(Review Figures 4-10 and 4-11.)*

- The eye is a specialized structure housing the light-sensitive receptors essential for vision perception—namely, the rods and cones found in its retinal layer. *(Review Table 4-4 and Figures 4-7 and 4-19.)*

- The cornea and lens are the primary refractive structures that bend the incoming light rays to focus the image on the retina. The cornea contributes most to the total refractive ability of the eye. The strength of the lens can be adjusted through action of the ciliary muscle to accommodate for differences in near and far vision. *(Review Figures 4-12 through 4-14.)*

- The rod and the cone photoreceptors are activated when the photopigments they contain differentially absorb various wavelengths of light. Light absorption causes a change in the rate of action potential propagation in the visual pathway leaving the retina. The conversion of light stimuli into electrical signals is known as *phototransduction*. *(Review Figures 4-16, 4-19, 4-20, and 4-21.)*

- The visual message is transmitted via a complex crossed and uncrossed pathway to the visual cortex in the occipital lobe of the brain for perceptual processing. *(Review Figure 4-24.)*

- The eyes' sensitivity is increased during dark adaptation, by the regeneration of rod photopigments that had been broken down during preceding light exposure. Sensitivity is decreased during light adaptation by the rapid breakdown of cone photopigments.

## 4.6 | Ear: Hearing and Equilibrium (pp. 169–181)

- Sound waves consist of high-pressure regions of compression alternating with low-pressure regions of rarefaction of air molecules. The pitch (tone) of a sound is determined by the frequency of its waves, the loudness (intensity) by the amplitude of the waves, and the timbre (quality) by its characteristic overtones. *(Review Figures 4-26 and 4-27 and Table 4-5.)*

- Sound waves are funnelled through the external ear canal to the tympanic membrane, which vibrates in synchrony with the waves.

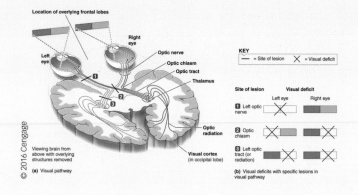

Location of overlying frontal lobes

Right eye

Left eye

Optic nerve
Optic chiasm
Optic tract
Thalamus

**KEY**
— = Site of lesion    X = Visual deficit

Optic radiation

Visual cortex (in occipital lobe)

Viewing brain from above with overlying structures removed

**(a)** Visual pathway

© 2016 Cengage

| Site of lesion | Visual deficit | |
|---|---|---|
| | Left eye | Right eye |
| **1** Left optic nerve | | |
| **2** Optic chiasm | | |
| **3** Optic tract (or radiation) | | |

**(b)** Visual deficits with specific lesions in visual pathway

■ Middle ear bones bridging the gap between the tympanic membrane and the inner ear amplify the tympanic movements and transmit them to the oval window, whose movement sets up travelling waves in the cochlear fluid. *(Review Figures 4-28 and 4-29.)* This sets the basilar membrane in motion. *(Review Figure 4-29.)*

■ On top of the basilar membrane are the receptive inner hair cells of the organ of Corti, whose hairs bend when the basilar membrane deflects up and down in relation to the overhanging stationary tectorial membrane that the hairs contact. *(Review Figures 4-28 and 4-30.)*

■ Neural signals are generated in response to the mechanical deformation of hair cells by specific movement of fluid and related structures within these vestibular sense organs. This information is important for the sense of equilibrium and for maintaining posture. *(Review Figures 4-34 and 4-35.)*

■ Vestibular input goes to the vestibular nuclei in the brain stem and to the cerebellum for use in maintaining balance and posture, controlling eye movement, and perceiving motion and orientation. *(Review Figure 4-36.)*

## 4.7 | Chemical Senses: Taste and Smell (pp. 181–187)

■ Taste receptors are housed in taste buds on the tongue; olfactory receptors are located in the olfactory mucosa in the upper part of the nasal cavity. *(Review Figures 4-37 and 4-38.)*

■ The five primary tastes are salty, sour, sweet, bitter, and umami. The recently added fifth taste is a meaty, "amino-acid" taste. Taste discrimination beyond the primary tastes depends on the patterns of stimulation of the taste buds, each of which responds in varying degrees to the different primary tastes.

■ There are 1000 different types of olfactory receptors, each of which responds to only one discrete component of an odour, an odourant. The afferent signals arising from the olfactory receptors are sorted according to scent component by the glomeruli within the olfactory bulb. Odour discrimination depends on the patterns of activation of the glomeruli. *(Review Figure 4-39.)*

## 4.8 | The Peripheral Nervous System: Efferent Division (p. 187)

■ There are two types of efferent output: (1) the autonomic nervous system, which is under involuntary control and supplies cardiac and smooth muscle as well as most exocrine and some endocrine glands; and (2) the somatic nervous system, which

is subject to voluntary control and supplies skeletal muscle. *(Review Tables 4-10 and 4-11.)*

## 4.9 | Autonomic Nervous System (pp. 187–194)

■ The ANS consists of two subdivisions: the sympathetic and parasympathetic nervous systems. *(Review Figures 4-41 and 4-42 and Tables 4-8 and 4-9.)*

■ An autonomic nerve pathway consists of a two-neuron chain. The preganglionic fibre originates in the CNS and synapses with the cell body of the postganglionic fibre in a ganglion outside the CNS. The postganglionic fibre ends on the effector organ. *(Review Figures 4-40 and 4-41 and Table 4-9.)*

■ Tissues innervated by the ANS possess one or more of several different receptor types for the postganglionic chemical messengers. Cholinergic receptors include nicotinic and muscarinic receptors; adrenergic receptors include $\alpha_1$, $\alpha_2$, $\beta_1$, and $\beta_2$ receptors. *(Review Figure 4-45 and Table 4-9.)*

■ The sympathetic system dominates in emergency or stressful (fight-or-flight) situations and promotes responses that prepare the body for strenuous physical activity. The parasympathetic system dominates in quiet, relaxed (rest-and-digest) situations and promotes body-maintenance activities, such as digestion. *(Review Tables 4-8 and 4-9.)*

■ Autonomic activities are controlled by multiple areas of the CNS, including the spinal cord, medulla, hypothalamus, and prefrontal association cortex.

## 4.10 | Somatic Nervous System (pp. 194–196)

■ The somatic nervous system consists of the axons of motor neurons, which originate in the spinal cord or brain stem and end on skeletal muscle. *(Review Table 4-10.)*

■ Acetylcholine, the neurotransmitter released from a motor neuron, stimulates muscle contraction.

## 4.11 | Neuromuscular Junction (pp. 197–201)

■ Each axon terminal of a motor neuron forms a neuromuscular junction with a single muscle cell (fibre). *(Review Figure 4-44 and Table 4-12.)*

■ An action potential in the axon terminal causes the release of ACh from its storage vesicles. The released ACh diffuses across the space separating the nerve and muscle cell and binds to special receptor sites on the underlying motor end plate of the muscle cell membrane. This combination of ACh with the receptors triggers the opening of specific channels in the motor end plate. The subsequent ion movements depolarize the motor end plate, producing the end-plate potential (EPP). *(Review Figure 4-45.)*

■ Local current flow between the depolarized end plate and adjacent muscle cell membrane brings these adjacent areas to threshold, initiating an action potential that is propagated throughout the muscle fibre.

■ This muscle action potential triggers muscle contraction.

■ Acetylcholinesterase inactivates ACh, ending the EPP and, subsequently, the action potential and resultant contraction. *(Review Figure 4-45.)*

## 5.1 | General Principles of Endocrinology (pp. 209–214)

- Hormones are long-distance chemical messengers secreted by the ductless endocrine glands into the blood, which transports the hormones to specific target sites where they control a particular function by altering protein activity within target cells.
- Hormones are grouped into two categories based on differences in their solubility and are further grouped according to their chemical structure: hydrophilic hormones (peptide hormones and catecholamines), and lipophilic hormones (steroid hormones and thyroid hormone).
- The endocrine system is especially important in regulating organic metabolism, water and electrolyte balance, growth, and reproduction and in helping the body cope with stress. *(Review Figure 5-1 and Table 5-2.)*
- Some hormones are tropic, meaning their function is to stimulate and maintain other endocrine glands.
- The effective plasma concentration of each hormone is normally controlled by regulated changes in the rate of hormone secretion. Secretory output of endocrine cells is primarily influenced by two types of direct regulatory inputs: (1) neural input, which increases hormone secretion in response to a specific need and governs diurnal variations in secretion; and (2) input from another hormone, which involves either stimulatory input from a tropic hormone or inhibitory input from a target-cell hormone in negative-feedback fashion. *(Review Figures 5-2, 5-3, and 5-12.)*
- The effective plasma concentration of a hormone can also be influenced by its rate of removal from the blood by metabolic inactivation and excretion and, for some hormones, by its rate of activation or its extent of binding to plasma proteins.
- Endocrine dysfunction arises when too much or too little of any particular hormone is secreted or when there is decreased target-cell responsiveness to a hormone. *(Review Table 5-1.)*

## 5.2 | Principles of Hormonal Communication (pp. 214–222)

- Hormones are long-distance chemical messengers secreted by the endocrine glands into the blood, which transports the hormones to specific target sites where they control a particular function by altering protein activity within the target cells.
- Hormones are grouped into two categories based on their solubility differences: (1) hydrophilic (water-soluble) hormones, which include peptides (the majority of hormones) and catecholamines (secreted by the adrenal medulla); and (2) lipophilic (lipid-soluble) hormones, which include steroid hormones (the sex hormones and those secreted by the adrenal cortex) and thyroid hormone. *(Review Table 5-2.)*

- Hydrophilic peptide hormones are synthesized and packaged for export by the endoplasmic reticulum/Golgi complex, stored in secretory vesicles, and released by exocytosis on appropriate stimulation.
- Hydrophilic peptide hormones dissolve freely in the plasma for transport to their target cells.
- At their target cells, hydrophilic hormones bind with surface membrane receptors. On binding, a hydrophilic hormone triggers a chain of intracellular events by means of a second-messenger system that ultimately alters pre-existing cell proteins, usually enzymes, which exert the effect leading to the target cell's response to the hormone. *(Review Figures 5-5 and 5-6.)*
- Steroids are synthesized by modifications of stored cholesterol by means of enzymes specific for each steroidogenic tissue. *(Review Figure 5-4.)*
- Steroids are not stored in the endocrine cells. Being lipophilic, they diffuse out through the lipid membrane barrier as soon as they are synthesized. Control of steroids is directed at their synthesis.
- Lipophilic steroids and thyroid hormone are both transported in the blood largely bound to carrier plasma proteins, with only free, unbound hormone being biologically active.
- Lipophilic hormones readily enter through the lipid membrane barriers of their target cells and bind with nuclear receptors. Hormonal binding activates the synthesis of new enzymatic or structural intracellular proteins that carry out the hormone's effect on the target cell. *(Review Figure 5-8.)*

## 5.3 | Comparison of the Nervous and Endocrine Systems (p. 222)

- The nervous and endocrine systems are the two main regulatory systems of the body. In general, the nervous system coordinates rapid responses, whereas the endocrine system regulates activities that require duration rather than speed.

## 5.4 | The Endocrine Tissues (pp. 222–225)

- The endocrine system consists of many tissues that are critical for maintaining homeostasis. Review Table 5-3 for a summary of the major endocrine tissues and their hormones.

## 5.5 | Hypothalamus and Pituitary (pp. 225–231)

- The pituitary gland consists of two distinct lobes, the posterior pituitary and the anterior pituitary. *(Review Figure 5-9.)*
- The hypothalamus, a portion of the brain, secretes nine peptide hormones. Two are stored in the posterior pituitary, and seven are carried through a special vascular link—the hypothalamic-hypophyseal portal system—to the anterior pituitary, where they regulate the release of particular anterior pituitary hormones. *(Review Figures 5-10 and 5-13.)*

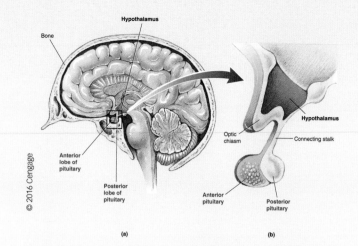

Bone

Hypothalamus

Optic chiasm

Hypothalamus

Connecting stalk

Anterior lobe of pituitary

Posterior lobe of pituitary

Anterior pituitary

Posterior pituitary

(a)

(b)

■ The posterior pituitary is essentially a neural extension of the hypothalamus. Two small peptide hormones, vasopressin and oxytocin, are synthesized within the cell bodies of neurosecretory neurons located in the hypothalamus, from which they pass down the axon to be stored in nerve terminals within the posterior pituitary. These hormones are independently released from the posterior pituitary into the blood in response to action potentials originating in the hypothalamus. (*Review Figure 5-10.*)
(1) Vasopressin conserves water during urine formation.
(2) Oxytocin stimulates uterine contraction during childbirth and milk ejection during breastfeeding.

■ The anterior pituitary secretes six different peptide hormones that it produces itself. Five anterior pituitary hormones are tropic.
(1) Thyroid-stimulating hormone (TSH) stimulates secretion of thyroid hormone. (2) Adrenocorticotropic hormone (ACTH) stimulates secretion of cortisol by the adrenal cortex. (3 and 4) The gonadotropic hormones—follicle-stimulating hormone (FSH) and luteinizing hormone (LH)—stimulate production of gametes (eggs and sperm) as well as secretion of sex hormones. (5) Growth hormone (GH) stimulates growth indirectly by stimulating secretion of somatomedins, which in turn promote growth of bone and soft tissues. GH exerts metabolic effects as well. (6) Prolactin stimulates milk secretion and is not tropic to another endocrine gland. (*Review Figure 5-11.*)

■ The anterior pituitary releases its hormones into the blood at the bidding of releasing and inhibiting hormones from the hypothalamus. The hypothalamus, in turn, is influenced by a variety of neural and hormonal controlling inputs. (*Review Table 5-4 and Figures 5-12 and 5-13.*)

■ Both the hypothalamus and anterior pituitary are inhibited in negative-feedback fashion by the product of the target endocrine gland in the hypothalamus–anterior pituitary–target gland axis. (*Review Figure 5-12.*)

## 5.6 | Endocrine Control of Growth (pp. 231–240)

■ Growth depends not only on growth hormone and other growth-influencing hormones, such as thyroid hormone, insulin, and the sex hormones, but also on genetic determination, an

adequate diet, and freedom from chronic disease or stress. Major growth spurts occur the first few years after birth and during puberty. (*Review Figure 5-14.*)

■ Growth hormone (GH) primarily promotes growth indirectly by stimulating the liver's production of somatomedins. The major somatomedin, or insulin-like growth factor, is IGF-I, which acts directly on bone and soft tissues to bring about most growth-promoting actions. The GH/IGF-I pathway causes growth by stimulating protein synthesis, cell division, and the lengthening and thickening of bones. (*Review Figures 5-15 and 5-16.*)

■ Growth hormone also directly exerts metabolic effects unrelated to growth, such as conservation of carbohydrates and mobilization of fat stores. (*Review Figure 5-16.*)

■ Growth hormone secretion by the anterior pituitary is regulated in negative-feedback fashion by two hypothalamic hormones: growth hormone–releasing hormone, and growth hormone–inhibiting hormone (somatostatin). (*Review Figure 5-16.*)

■ Growth hormone levels are not highly correlated with periods of rapid growth. The primary signals for increased growth hormone secretion are related to metabolic needs rather than growth, namely, deep sleep, exercise, stress, low blood glucose, increased blood amino acids, or decreased blood fatty acids. (*Review Figure 5-16.*)

## 5.7 | Pineal Gland and Circadian Rhythms (pp. 240–242)

■ The suprachiasmatic nucleus (SCN) is the body's master biological clock. Self-induced cyclic variations in the concentration of clock proteins within the SCN bring about cyclic changes in neural discharge from this area. Each cycle takes about a day and drives the body's circadian (daily) rhythms.

■ The inherent rhythm of this endogenous oscillator is a bit longer than 24 hours. Therefore, each day the body's circadian rhythms must be entrained or adjusted to keep pace with environmental cues so that the internal rhythms are synchronized with the external light–dark cycle.

■ In the eyes, special photoreceptors that respond to light but are not involved in vision send input to the SCN. Acting through the SCN, the pineal gland's secretion of the hormone melatonin rhythmically fluctuates with the light–dark cycle, decreasing in the light and increasing in the dark. Melatonin, in turn, is believed to synchronize the body's natural circadian rhythms, such as diurnal (day–night) variations in hormone secretion and body temperature, with external cues such as the light–dark cycle.

■ Other proposed roles for melatonin include (1) promoting sleep; (2) influencing reproductive activity, including the onset of puberty; (3) acting as an antioxidant to remove damaging free radicals; and (4) enhancing immunity.

## 6.1 | Thyroid Gland (pp. 248–254)

■ The thyroid gland contains two types of endocrine secretory cells: (1) follicular cells, which produce the iodine-containing hormones, $T_4$ (thyroxine or tetraiodothyronine) and $T_3$ (triiodothyronine), collectively known as thyroid hormones; and (2) C cells, which synthesize a $Ca^{2+}$-regulating hormone, calcitonin. (Review Figure 6-1.)

■ All steps of thyroid hormone synthesis take place on the large thyroglobulin molecules within the colloid—an extracellular site located within the interior of the thyroid follicles. Thyroid hormone is secreted by means of the follicular cells phagocytizing a piece of colloid and freeing $T_4$ and $T_3$, which diffuse across the plasma membrane and enter the blood. (Review Figures 6-1 and 6-2.)

■ Thyroid hormone is the primary determinant of the overall metabolic rate of the body. By accelerating the metabolic rate of most tissues, it increases heat production. Thyroid hormone also enhances the actions of the chemical mediators of the sympathetic nervous system. Through this and other means, thyroid hormone indirectly increases cardiac output. Finally, thyroid hormone is essential for normal growth as well as the development and function of the nervous system.

■ Thyroid hormone secretion is regulated by a negative-feedback system between hypothalamic TRH, anterior pituitary TSH, and thyroid gland $T_3$ and $T_4$. The feedback loop maintains thyroid hormone levels relatively constant. Cold exposure in newborn infants is the only input to the hypothalamus known to be effective in increasing TRH and thereby thyroid hormone secretion. (Review Figure 6-3.)

## 6.2 | Adrenal Glands (pp. 254–263)

■ Each adrenal gland (of the pair) consists of two separate endocrine organs: (1) an outer, steroid-secreting adrenal cortex; and (2) an inner, catecholamine-secreting adrenal medulla. (Review Figure 6-7.)

■ The adrenal cortex secretes three different categories of steroid hormones: (1) mineralocorticoids (primarily aldosterone); (2) glucocorticoids (primarily cortisol); and (3) adrenal sex hormones (primarily the weak androgen, dehydroepiandrosterone).

■ Aldosterone regulates $Na^+$ and $K^+$ balance and is important for blood pressure homeostasis, which is achieved secondarily by the osmotic effect of $Na^+$ in maintaining the plasma volume—a lifesaving effect.

■ Control of aldosterone secretion is related to $Na^+$ and $K^+$ balance and to blood pressure regulation and is not influenced by ACTH.

■ Cortisol helps regulate fuel metabolism and is important in stress adaptation. It increases blood levels of glucose, amino acids, and fatty acids and spares glucose for use by the glucose-dependent brain. The mobilized organic molecules are available for use as needed for energy or for repair of injured tissues. (Review Figure 6-8 and Table 6-3.)

■ Cortisol secretion is regulated by a negative-feedback loop involving hypothalamic CRH and pituitary ACTH. Stress is the most potent stimulus for increasing activity of the CRH–ACTH–cortisol axis. Cortisol also displays a characteristic diurnal rhythm. (Review Figures 6-8, and 6-12.)

■ Dehydroepiandrosterone (DHEA) governs the sex drive and growth of pubertal hair in females.

■ The adrenal medulla consists of modified sympathetic postganglionic neurons, which secrete the catecholamine epinephrine into the blood in response to sympathetic stimulation. For the most part, epinephrine reinforces the sympathetic system in mounting general systemic fight-or-flight responses and in maintaining arterial blood pressure. Epinephrine also exerts important metabolic effects, namely, increasing blood glucose and blood fatty acids. (Review Figure 4-43, and Table 6-2.)

■ The primary stimulus for increased adrenomedullary secretion is activation of the sympathetic system by stress. (Review Table 6-3, and Figure 6-12.)

## 6.3 | Integrated Stress Response (pp. 263–266)

■ The term *stress* refers to the generalized nonspecific response of the body to any factor that overwhelms, or threatens to overwhelm, the body's compensatory ability to maintain homeostasis. The term *stressor* refers to any noxious stimulus that elicits the stress response.

■ In addition to specific responses to various stressors, all stressors produce the following similar generalized stress response: (1) activation of the sympathetic nervous system accompanied by epinephrine secretion, which together prepare the body for a fight-or-flight response; (2) activation of the CRH–ACTH–cortisol system, which helps the body cope with stress primarily by mobilizing metabolic resources; (3) elevation of blood glucose and fatty acids through decreased insulin and increased glucagon secretion; and (4) maintenance of blood volume and blood pressure through increased activity of the renin–angiotensin–aldosterone system along with increased vasopressin secretion. All these actions are coordinated by the hypothalamus. (Review Figures 6-11 and 6-12, and Table 6-3.)

## 6.4 | Endocrine Control of Fuel Metabolism (pp. 267–283)

■ Intermediary or fuel metabolism is, collectively, the synthesis (anabolism), breakdown (catabolism), and transformation of the three classes of energy-rich organic nutrients—carbohydrate, fat, and protein—within the body. (Review Table 6-4 and Figure 6-13.)

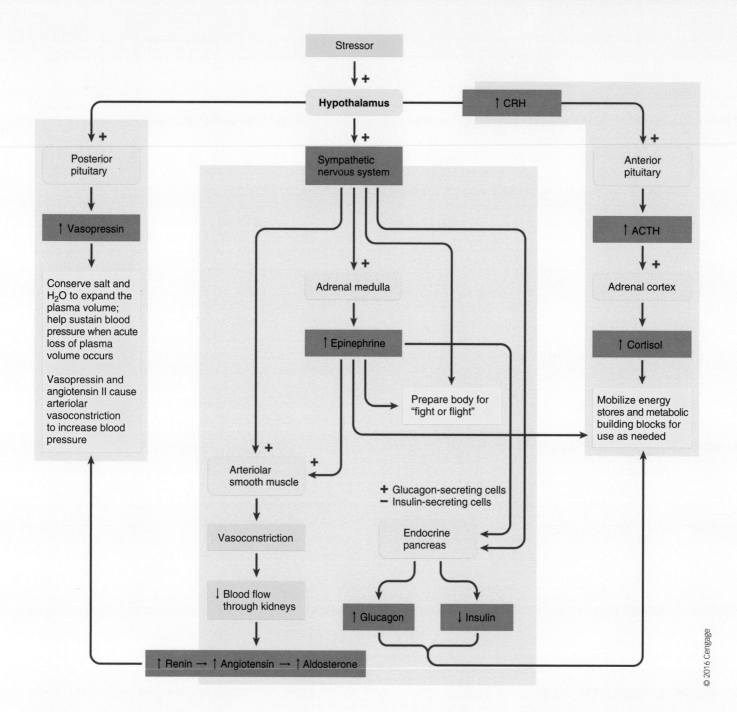

- Glucose and fatty acids derived from carbohydrates and fats, respectively, are primarily used as metabolic fuels, whereas amino acids derived from proteins are primarily used for synthesis of structural and enzymatic proteins. *(Review Table 6-5.)*

- During the absorptive state following a meal, excess absorbed nutrients not immediately needed for energy production or protein synthesis are stored to a limited extent as glycogen in the liver and muscle but mostly as triglycerides in adipose tissue. *(Review Table 6-6.)*

- During the postabsorptive state between meals when no new nutrients are entering the blood, the glycogen and triglyceride stores are catabolized to release nutrient molecules into

the blood. If necessary, body proteins are degraded to release amino acids for conversion into glucose. The blood glucose concentration must be maintained above a critical level even during the postabsorptive state, because the brain depends on blood-delivered glucose as its energy source. Tissues not dependent on glucose switch to fatty acids as their metabolic fuel, sparing glucose for the brain. *(Review Table 6-6.)*

- These shifts in metabolic pathways between the absorptive and postabsorptive state are hormonally controlled. The most important hormone in this regard is insulin. Insulin is secreted by the $\alpha$ cells of the islets of Langerhans, the endocrine portion of the pancreas. The other major pancreatic hormone, glucagon, is secreted by the $\beta$ cells of the islets. *(Review Table 6-7.)*

- Insulin is an anabolic hormone; it promotes the cellular uptake of glucose, fatty acids, and amino acids and enhances their conversion into glycogen, triglycerides, and proteins, respectively. In so doing, it lowers the blood concentrations of these small organic molecules.
- Insulin secretion is increased during the absorptive state, primarily by a direct effect of an elevated blood glucose on the cells, and is largely responsible for directing the organic traffic into cells during this state. *(Review Figures 6-14 through 6-18.)*
- Glucagon mobilizes the energy-rich molecules from their stores during the postabsorptive state. Glucagon, which is secreted in response to a direct effect of a fall in blood glucose on the pancreatic $\alpha$ cells, in general opposes the actions of insulin. *(Review Figures 6-17 and 6-18.)*

## 6.5 | Endocrine Control of Calcium Metabolism (pp. 284–292)

- Changes in the concentration of free, diffusible plasma $Ca^{2+}$, the biologically active form of this ion, produce profound and life-threatening effects, most notably on neuromuscular excitability. Hypercalcaemia reduces excitability, whereas hypocalcaemia brings about overexcitability of nerves and muscles. If the overexcitability is severe enough, fatal spastic contractions of respiratory muscles can occur.
- Three hormones regulate the plasma concentration of $Ca^{2+}$ (and concurrently regulate $PO_4^{3-}$)—parathyroid hormone (PTH), calcitonin, and vitamin D.
- PTH, whose secretion is directly increased by a fall in plasma $Ca^{2+}$ concentration, acts on bone, kidneys, and the intestine to raise the plasma $Ca^{2+}$ concentration. In so doing, it is essential for life by preventing the fatal consequences of hypocalcaemia. The specific effects of PTH on bone promote $Ca^{2+}$ movement from the bone fluid into the plasma in the short term and promote localized dissolution of bone by enhancing activity of the osteoclasts (bone-dissolving cells) in the long term. *(Review Figures 6-19 through 6-22.)*
- Dissolution of the calcium phosphate bone crystals releases $PO_4^{3-}$ as well as $Ca^{2+}$ into the plasma. PTH acts on the kidneys to enhance the reabsorption of filtered $Ca^{2+}$, thereby reducing the urinary excretion of $Ca^{2+}$ and increasing its plasma concentration. Simultaneously, PTH reduces renal $PO_4^{3-}$ reabsorption, in this way increasing $PO_4^{3-}$ excretion and lowering plasma $PO_4^{3-}$ levels. This is important because a rise in plasma $PO_4^{3-}$ would force the redeposition of some of the plasma $Ca^{2+}$ back into the bone.
- Furthermore, PTH facilitates the activation of vitamin D, which in turn stimulates $Ca^{2+}$ and $PO_4^{3-}$ absorption from the intestine. *(Review Figures 6-23 and 6-24.)*
- Vitamin D can be synthesized from a cholesterol derivative in the skin when exposed to sunlight, but frequently this endogenous source is inadequate, so vitamin D must be supplemented by dietary intake. From either source, vitamin D must be activated first by the liver and then by the kidneys (the site of PTH regulation of vitamin D activation) before it can exert its effect on the intestine. *(Review Figure 6-23.)*
- Calcitonin, a hormone produced by the C cells of the thyroid gland, is the third factor that regulates $Ca^{2+}$. In negative-feedback fashion, calcitonin is secreted in response to an increase in plasma $Ca^{2+}$ concentration and acts to lower plasma $Ca^{2+}$ levels by inhibiting activity of bone osteoclasts. Calcitonin is unimportant except during the rare condition of hypercalcaemia. *(Review Figure 6-22.)*

# NOTES

## 7.1 | Introduction (pp. 297–298)

■ Muscles are specialized for contraction. Through their highly developed ability to move specialized cytoskeletal components, they are able to develop tension, shorten, produce movement, and accomplish work. The three types of muscle—skeletal, cardiac, and smooth—are categorized in two different ways according to common characteristics. (1) Skeletal muscle and cardiac muscle are striated, whereas smooth muscle is unstriated. (2) Skeletal muscle is voluntary, whereas cardiac muscle and smooth muscle are involuntary. (*Review Figure 7-1 and Table 7-4.*)

## 7.2 | Structure of Skeletal Muscle (pp. 298–302)

■ Skeletal muscles are made up of bundles of long, cylindrical muscle cells known as muscle fibres, wrapped in connective tissue.

■ Muscle fibres are packed with myofibrils, each myofibril consisting of alternating, slightly overlapping stacked sets of thick and thin filaments. This arrangement leads to a skeletal muscle fibre's striated microscopic appearance, which consists of alternating dark A bands and light I bands. (*Review Figures 7-2, 7-3, and 7-4.*)

■ Thick filaments consist of the protein myosin. Cross bridges made up of the myosin molecules' globular heads project from each thick filament toward the surrounding thin filaments. (*Review Figures 7-2 and 7-5.*)

■ Thin filaments consist primarily of the protein actin, which can bind and interact with the myosin cross bridges to bring about contraction. Accordingly, myosin and actin are known as contractile proteins. However, in the resting state two other regulatory proteins, tropomyosin and troponin, lie across the surface of the thin filament to prevent this cross-bridge interaction. (*Review Figure 7-2.*)

## 7.3 | Molecular Basis of Skeletal Muscle Contraction (pp. 302–309)

■ Excitation of a skeletal muscle fibre by its motor neuron brings about contraction through a series of events that results in the thin filaments sliding closer together between the thick filaments. (*Review Figure 7-7.*)

■ This sliding filament mechanism of muscle contraction is switched on by the release of calcium ions from the lateral sacs of the sarcoplasmic reticulum. (*Review Figures 7-9, 7-10, and 7-11.*)

■ Calcium release occurs in response to the spread of a muscle fibre action potential into the central portions of the fibre via the T tubules. (*Review Figures 7-9 and 7-11.*)

■ Released $Ca^{2+}$ binds to the troponin–tropomyosin complex of the thin filament, slightly repositioning the complex to uncover actin's cross-bridge binding sites. (*Review Figures 7-6 and 7-11.*)

■ After the exposed actin attaches to a myosin cross bridge, molecular interaction between actin and myosin releases energy within the myosin head that was stored from the prior splitting of ATP by the myosin ATPase site. This released energy powers cross-bridge stroking. (*Review Figure 7-12.*)

■ During a power stroke, an activated cross bridge bends toward the centre of the thick filament, pulling in the thin filament to which it is attached. (*Review Figure 7-8.*)

■ With the addition of a fresh ATP molecule to the myosin cross bridge, myosin and actin detach, the cross bridge returns to its original shape, and the cycle is repeated. (*Review Figure 7-12.*)

■ Repeated cycles of cross-bridge activity slide the thin filaments inward step by step. (*Review Figure 7-8.*)

■ When there is no longer a local action potential, the lateral sacs actively take up the $Ca^{2+}$, troponin and tropomyosin slip back into their blocking position, and relaxation occurs. (*Review Figure 7-11.*)

■ The entire contractile response lasts about 100 times longer than the action potential. (*Review Figure 7-13.*)

## 7.4 | Skeletal Muscle Mechanics (pp. 309–317)

■ Gradation of whole-muscle contraction can be accomplished by (1) varying the number of muscle fibres contracting within the muscle and (2) varying the tension developed by each contracting fibre. (*Review Table 7-3.*)

■ The greater the tension developed by each contracting fibre, the stronger the contraction of the whole muscle. Two readily variable factors that affect fibre tension are (1) frequency of stimulation (firing frequency), which determines the extent of twitch summation; and (2) length of the fibre before the onset of contraction. (*Review Figure 7-16 and Table 7-3.*)

■ The term *twitch summation* refers to the increase in tension accompanying repetitive stimulation of the muscle fibre. After undergoing an action potential, the muscle cell membrane recovers from its refractory period and can be restimulated while some contractile activity triggered by the first action potential still remains. As a result, the contractile responses (twitches) induced by the two rapidly successive action potentials can sum, thereby increasing the tension developed by the fibre. If the muscle fibre is stimulated so rapidly that it does not have a chance to start relaxing between stimuli, a smooth, sustained maximal (maximal for the fibre at that length) contraction known as tetanus occurs. (*Review Figure 7-15.*)

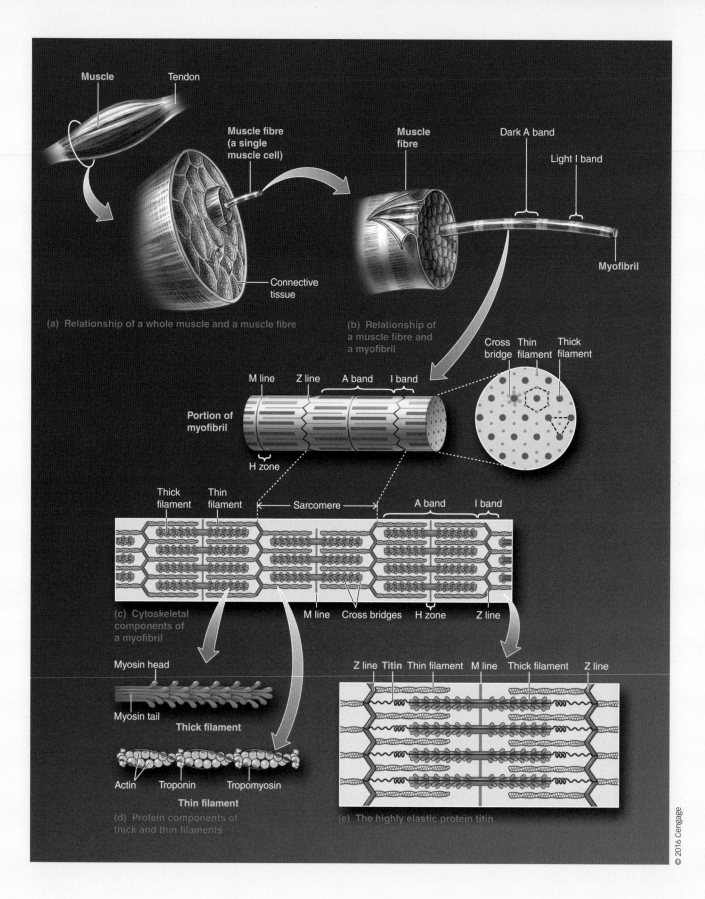

(a) Relationship of a whole muscle and a muscle fibre

(b) Relationship of a muscle fibre and a myofibril

(c) Cytoskeletal components of a myofibril

(d) Protein components of thick and thin filaments

(e) The highly elastic protein titin

■ The two primary types of muscle contraction—isometric (constant length) and isotonic (constant tension)—depend on the relationship between muscle tension and the load. The load is the force opposing contraction: that is, the weight of an object being lifted. (1) If tension is less than the load, the muscle cannot shorten and lift the object but remains at constant length, producing an isometric contraction. (2) In an isotonic contraction, the tension exceeds the load so the muscle can shorten and lift the object, maintaining constant tension throughout the period of shortening. *(Review Figure 7-18.)*

■ The velocity, or speed, of shortening is inversely proportional to the load. *(Review Figure 7-20.)*

■ The bones, muscles, and joints form lever systems. The most common type amplifies the velocity and distance of muscle shortening to increase the speed and range of motion of the body part moved by the muscle. This increased manoeuvrability is accomplished at the expense of the muscle having to exert considerably more force than the load. *(Review Figure 7-21.)*

## 7.5 | ATP and Fibre Types (pp. 317–322)

■ The transfer of high-energy phosphates from stored creatine phosphate to ADP provides the first source of ATP at the onset of exercise.

■ Fatigue is of two types: (1) muscle fatigue, which occurs when an exercising muscle can no longer respond to neural stimulation with the same degree of contractile activity; and (2) central fatigue, which occurs when the CNS no longer adequately activates the motor neurons.

■ The three types of muscle fibres are classified by the pathways they use for ATP synthesis (oxidative or glycolytic) and the rapidity with which they split ATP and subsequently contract (slow-twitch or fast-twitch): (1) slow-oxidative fibres, (2) fast-oxidative fibres, and (3) fast-glycolytic fibres. *(Review Table 7-2.)*

■ Muscle fibres adapt in response to different demands placed on them. Regular endurance exercise promotes improved oxidative capacity in oxidative fibres, whereas high-intensity resistance training promotes hypertrophy of fast glycolytic fibres. The two types of fast-twitch fibres are interconvertible, depending on the type and extent of training. Fast- and slow-twitch fibres are not interconvertible. Muscles atrophy when not used.

## 7.6 | Control of Motor Movement (pp. 322–326)

■ Control of any motor movement depends on the activity levels of the presynaptic inputs that converge on the motor neurons supplying various muscles. These inputs come from three sources: (1) spinal reflex pathways, which originate with afferent neurons; (2) the corticospinal (pyramidal) motor system, which originates at the pyramidal cells in the primary motor cortex and is concerned primarily with discrete, intricate movements of the hands; and (3) the multineuronal (extrapyramidal) motor system, which originates in the brain stem and is mostly involved with postural adjustments and involuntary movements of the trunk and limbs. The final motor output from the brain stem is influenced by the cerebellum, basal nuclei, and cerebral cortex. *(Review Figure 7-22.)*

■ Establishment and adjustment of motor commands depend on continuous afferent input, especially feedback about changes in muscle length (monitored by muscle spindles) and muscle tension (monitored by Golgi tendon organs). *(Review Figure 7-23.)*

## 7.7 | Smooth and Cardiac Muscle (pp. 326–334) (Review Table 7–4.)

■ The thick and thin filaments of smooth muscle are not arranged in an orderly pattern, so the fibres are not striated. *(Review Figures 7-24 and 7-25.)*

■ In smooth muscle, cytosolic $Ca^{2+}$, which enters from the extracellular fluid and is also released from sparse intracellular stores, activates cross-bridge cycling by initiating a series of biochemical reactions that result in phosphorylation of the myosin cross bridges to enable them to bind with actin. *(Review Figures 7-26 and 7-27.)*

■ Multiunit smooth muscle is neurogenic, requiring stimulation of individual muscle fibres by its autonomic nerve supply to trigger contraction.

■ Single-unit smooth muscle is myogenic; it can initiate its own contraction without any external influence, as a result of spontaneous depolarizations to threshold potential brought about by automatic shifts in ionic fluxes. Only a few of the smooth muscle cells in a functional syncytium are self-excitable. The two major types of spontaneous depolarizations displayed by self-excitable smooth muscle cells are pacemaker potentials and slow-wave potentials. *(Review Figure 7-28 and Table 7-5.)*

■ The autonomic nervous system, as well as hormones and local metabolites, can modify the rate and strength of the self-induced smooth muscle contractions. All these factors influence smooth muscle activity by altering cytosolic $Ca^{2+}$ concentration.

■ Smooth muscle does not have a clear-cut length–tension relationship. It can develop tension when considerably stretched and inherently relaxes when stretched.

■ Smooth muscle contractions are energy efficient, enabling this type of muscle to economically sustain long-term contractions without fatigue. This economy, coupled with the fact that single-unit smooth muscle can exist at a variety of lengths with little change in tension, makes single-unit smooth muscle ideally suited for its task of forming the walls of hollow organs that can distend.

■ Cardiac muscle is found only in the heart. It has highly organized striated fibres, like skeletal muscle. Like single-unit smooth muscle, some cardiac muscle fibres can generate action potentials, which are spread throughout the heart with the aid of gap junctions.

# NOTES

## 8.1 | Introduction (pp. 345–346)

- The circulatory system is the transport system of the body.
- The three basic components of the circulatory system are the heart (the pump), the blood vessels (the passageways), and the blood (the transport medium).

## 8.2 | Anatomy of the Heart (pp. 346–351)

- The heart is basically a dual pump that provides the driving pressure for blood flow through the pulmonary and systemic circulations. *(Review Figures 8-1 and 8-2.)*
- The heart has four chambers: each half of the heart consists of an atrium, or venous input chamber, and a ventricle, or arterial output chamber. *(Review Figures 8-2 and 8-4.)*
- Four heart valves direct the blood in the right direction and keep it from flowing in the other direction. *(Review Figures 8-3, 8-4, and 8-5.)*
- Contraction of the spirally arranged cardiac muscle fibres produces a wringing effect important for efficient pumping. It is also important for efficient pumping that the muscle fibres in each chamber act as a functional syncytium, contracting as a coordinated unit. *(Review Figure 8-6.)*

## 8.3 | Electrical Activity of the Heart (pp. 352–361)

- The heart is self-excitable, initiating its own rhythmic contractions.
- Autorhythmic cells are 1 percent of the cardiac muscle cells; they do not contract but are specialized to initiate and conduct action potentials. Autorhythmic cells display a pacemaker potential, a slow drift to threshold potential, as a result of a complex interplay of inherent changes in ion movement across the membrane. *(Review Figure 8-7.)* The other 99 percent of cardiac cells are contractile cells that contract in response to the spread of an action potential initiated by autorhythmic cells.
- The cardiac impulse originates at the SA node, the pacemaker of the heart, which has the fastest rate of spontaneous depolarization to threshold. *(Review Table 8-1 and Figure 8-8.)*
- Once initiated, the action potential spreads throughout the right and left atria, partially facilitated by specialized conduction pathways, but mostly by cell-to-cell spread of the impulse through gap junctions. *(Review Figures 8-8 and 8-9.)*
- The impulse passes from the atria into the ventricles through the AV node, the only point of electrical contact between these chambers. The action potential is delayed briefly at the AV node, ensuring that atrial contraction precedes ventricular contraction to allow complete ventricular filling. *(Review Figures 8-8 and 8-9.)*

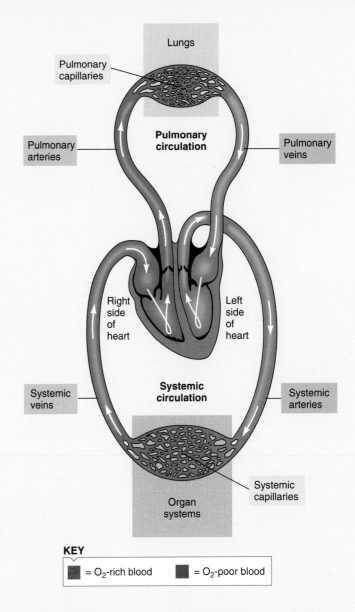

**KEY**

■ = $O_2$-rich blood    ■ = $O_2$-poor blood

- The impulse then travels rapidly down the interventricular septum via the bundle of His and rapidly disperses throughout the myocardium by means of the Purkinje fibres. The rest of the ventricular cells are activated by cell-to-cell spread of the impulse through gap junctions. *(Review Figures 8-8 and 8-9.)*
- Thus, the atria contract as a single unit, followed after a brief delay by a synchronized ventricular contraction. The action potentials of cardiac contractile cells exhibit a prolonged positive phase, or plateau, accompanied by a prolonged period of contraction, which ensures adequate ejection time. This plateau is primarily due to activation of slow L-type $Ca^{2+}$ channels. *(Review Figure 8-10.)*

- Calcium entry though the L-type channels in the T tubules triggers a much larger release of $Ca^{2+}$ from the sarcoplasmic reticulum. This $Ca^{2+}$-induced $Ca^{2+}$ release leads to cross-bridge cycling and contraction. *(Review Figure 8-11.)*
- Because a long refractory period occurs in conjunction with this prolonged plateau phase, summation and tetanus of cardiac muscle are impossible, ensuring the alternate periods of contraction and relaxation essential for pumping of blood. *(Review Figure 8-12.)*
- The spread of electrical activity throughout the heart can be recorded from the body surface. This record, the ECG, can provide useful information about the status of the heart. *(Review Figures 8-13, 8-14, and 8-15.)*

## 8.4 | Mechanical Events of the Cardiac Cycle (pp. 361–365)

- The cardiac cycle consists of three important events:
  1. The generation of electrical activity as the heart auto-rhythmically depolarizes and repolarizes
  2. Mechanical activity consisting of alternate periods of systole (contraction and emptying) and diastole (relaxation and filling), which are initiated by the rhythmic electrical cycle
  3. Directional flow of blood through the heart chambers, guided by valve opening and closing induced by pressure changes that are generated by mechanical activity
- Valve closing gives rise to two normal heart sounds. The first heart sound is caused by closing of the atrioventricular (AV) valves and signals the onset of ventricular systole. The second heart sound is due to closing of the aortic and pulmonary valves at the onset of diastole.
- The atrial pressure curve remains low throughout the entire cardiac cycle, with only minor fluctuations (normally varying between 0 and 8 mmHg). The aortic pressure curve remains high the entire time, with moderate fluctuations (normally varying between a systolic pressure of 120 mmHg and a diastolic pressure of 80 mmHg). The ventricular pressure curve fluctuates dramatically, because ventricular pressure must be below the low atrial pressure during diastole to allow the AV valve to open for filling, and it must be above the high aortic pressure during systole to force the aortic valve open to allow emptying. Therefore, ventricular pressure normally varies from 0 mmHg during diastole to slightly more than 120 mmHg during systole. *(Review Figure 8-16.)*
- The end-diastolic volume is the volume of blood in the ventricle when filling is complete at the end of diastole. The end-systolic volume is the volume of blood remaining in the ventricle when ejection is complete at the end of systole. The stroke volume is the volume of blood pumped out by each ventricle each beat. *(Review Figure 8-16.)*
- Defective valve function produces turbulent blood flow, which is audible as a heart murmur. Abnormal valves may be either stenotic and not open completely or insufficient and not close completely. *(Review Table 8-2.)*

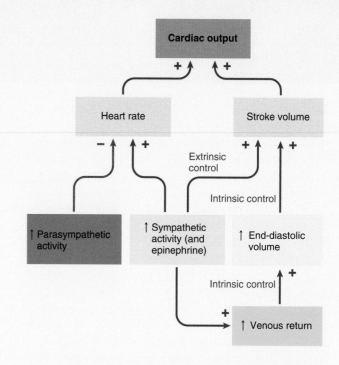

## 8.5 | Cardiac Output and Its Control (pp. 365–373)

- Cardiac output, the volume of blood ejected by each ventricle each minute, is determined by the heart rate multiplied by the stroke volume. *(Review Figure 8-23.)*
- Heart rate is varied by altering the balance of parasympathetic and sympathetic influence on the SA node. Parasympathetic stimulation slows the heart rate, and sympathetic stimulation speeds it up. *(Review Figure 8-18 and Table 8-3.)*
- Stroke volume depends on (1) the extent of ventricular filling, with an increased end-diastolic volume resulting in a larger stroke volume by means of the length–tension relationship (Frank–Starling law of the heart, a form of intrinsic control); and (2) the extent of sympathetic stimulation, with increased sympathetic stimulation resulting in increased contractility of the heart—that is, increased strength of contraction and increased stroke volume at a given end-diastolic volume (extrinsic control). *(Review Figures 8-19 through 8-22.)*

## 8.6 | Nourishing the Heart Muscle (pp. 373–379)

- Cardiac muscle is supplied with oxygen and nutrients by blood delivered to it by the coronary circulation, not by blood within the heart chambers.
- Most coronary blood flow occurs during diastole, because during systole the contracting heart muscle compresses the coronary vessels. *(Review Figure 8-25.)*
- Coronary blood flow is normally varied to keep pace with cardiac oxygen needs. *(Review Figure 8-26.)*
- Coronary blood flow may be compromised by development of atherosclerotic plaques, which can lead to ischemic heart disease, ranging in severity from mild chest pain on exertion to fatal heart attacks. *(Review Figures 8-27 through 8-29 and Table 8-4.)*

## 9.1 | Introduction (pp. 386–389)

- Materials can be exchanged between various parts of the body and with the external environment by means of the blood vessel network that transports blood to and from all organs. *(Review Figure 9-1.)*

- Organs that replenish nutrient supplies and remove metabolic wastes from the blood receive a greater percentage of the cardiac output than is warranted by their metabolic needs. These "reconditioning" organs can better tolerate reductions in blood supply than can organs that receive blood solely for meeting their own metabolic needs.

- The brain is especially vulnerable to reductions in its blood supply. Therefore, maintaining adequate flow to this vulnerable organ is a high priority in circulatory function.

- The flow rate of blood through a vessel is directly proportional to the pressure gradient and inversely proportional to the resistance. The higher pressure at the beginning of a vessel is established by the pressure imparted to the blood by cardiac contraction. The lower pressure at the end is due to frictional losses as flowing blood rubs against the vessel wall. *(Review Figure 9-2.)*

- Resistance, the hindrance to blood flow through a vessel, is influenced most by the vessel's radius. Resistance is inversely proportional to the fourth power of the radius, so small changes in radius profoundly influence flow. As the radius increases, resistance decreases and flow increases. *(Review Figure 9-3.)*

- Blood flows in a closed loop between the heart and the organs. The arteries transport blood from the heart throughout the body. The arterioles regulate the amount of blood that flows through each organ. The capillaries are the actual site where materials are exchanged between blood and surrounding tissue cells. The veins return blood from the tissue level back to the heart. *(Review Figure 9-4 and Table 9-1.)*

## 9.2 | Arteries (pp. 389–394)

- Arteries are large-radius, low-resistance passageways from the heart to the organs.

- They also serve as a pressure reservoir. Because of their elasticity, arteries expand to accommodate the extra volume of blood pumped into them by cardiac contraction and then recoil to continue driving the blood forward when the heart is relaxing. *(Review Figures 9-5 and 9-6.)*

- Systolic pressure is the peak pressure exerted by the ejected blood against the vessel walls during cardiac systole. Diastolic pressure is the minimum pressure in the arteries when blood is draining off into the vessels downstream during cardiac diastole. *(Review Figures 9-7 and 9-8.)*

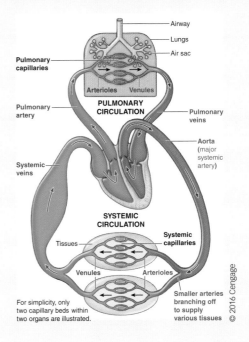

- The average driving pressure throughout the cardiac cycle is the mean arterial pressure, which can be estimated using the following formula: mean arterial pressure = diastolic pressure + 1/3 pulse pressure. *(Review Figure 9-9.)*

## 9.3 | Arterioles (pp. 394–402)

- Arterioles are the major resistance vessels. Their high resistance produces a large drop in mean pressure between the arteries and capillaries. This decline enhances blood flow by contributing to the pressure differential between the heart and organs. *(Review Figure 9-9.)*

- Tone, a baseline of contractile activity, is maintained in arterioles at all times.

- Arteriolar vasodilation, an expansion of arteriolar calibre above tonic level, decreases resistance and increases blood flow through the vessel, whereas vasoconstriction, a narrowing of the vessel, increases resistance and decreases flow. *(Review Figure 9-10.)*

- Arteriolar calibre is subject to two types of control mechanisms: local (intrinsic) controls and extrinsic controls.

- Local controls primarily involve local chemical changes associated with changes in the level of metabolic activity in an organ. These changes in local metabolic factors cause the release of vasoactive mediators from the endothelial cells in the vicinity. Examples include vasodilating nitric oxide and vasoconstricting endothelin. These vasoactive mediators act on the underlying arteriolar smooth muscle to bring about an appropriate change in the calibre of the arterioles supplying the organ. By adjusting the resistance to blood flow in this manner, the local control mechanism adjusts blood flow to the organ to match the momentary metabolic needs of the organ. (*Review Figures 9-10, 9-11, and 9-14, and Tables 9-2 and 9-3.*)

■ Arteriolar calibre can be adjusted independently in different organs by local control factors. Such adjustments are important in variably distributing cardiac output. (*Review Figure 9-12.*)

■ Extrinsic control is accomplished primarily by sympathetic nerve influence and to a lesser extent by hormonal influence over arteriolar smooth muscle. Extrinsic controls are important in maintaining mean arterial pressure. Arterioles are richly supplied with sympathetic nerve fibres, whose increased activity produces generalized vasoconstriction and a subsequent increase in total peripheral resistance, thus increasing mean arterial pressure. Decreased sympathetic activity produces generalized arteriolar vasodilation, which lowers mean arterial pressure. These extrinsically controlled adjustments of arteriolar calibre help maintain the appropriate pressure head for driving blood forward to the tissues. (*Review Figure 9-14.*)

## 9.4 | Capillaries (pp. 402–413)

■ The thin-walled, small-radius, extensively branched capillaries are ideally suited to serve as sites of exchange between the blood and surrounding tissue cells. Anatomically, the surface area for exchange is maximized and diffusion distance is minimized in the capillaries. Furthermore, because of their large total cross-sectional area, the velocity of blood flow through capillaries is relatively slow, providing adequate time for exchanges to take place. (*Review Figures 9-15 through 9-17.*)

■ Two types of passive exchanges—diffusion and bulk flow—take place across capillary walls.

■ Individual solutes are exchanged primarily by diffusion down concentration gradients. Lipid-soluble substances pass directly through the single layer of endothelial cells lining a capillary, whereas water-soluble substances pass through water-filled pores between the endothelial cells. Plasma proteins generally do not escape. (*Review Figures 9-18 and 9-21.*)

■ Imbalances in physical pressures acting across capillary walls are responsible for bulk flow of fluid through the pores back and forth between plasma and interstitial fluid. (1) Fluid is forced out of the first portion of the capillary (ultrafiltration), where outward pressures (mainly capillary blood pressure) exceed inward pressures (mainly plasma-colloid osmotic pressure). (2) Fluid is returned to the capillary along its last half, when outward pressures fall below inward pressures. The reason for the shift in balance down the length of the capillary is the continuous decline in capillary blood pressure while the plasma-colloid osmotic pressure remains constant. (*Review Figures 9-9, 9-22, and 9-23.*)

■ Bulk flow is responsible for the distribution of extracellular fluid between plasma and interstitial fluid.

■ Normally, slightly more fluid is filtered than is reabsorbed. The extra fluid, any leaked proteins, and bacteria in the tissue are picked up by the lymphatic system. Bacteria are destroyed as lymph passes through the lymph nodes en route to being returned to the venous system. (*Review Figures 9-24 and 9-25.*)

## 9.5 | Veins (pp. 414–418)

■ Veins are large-radius, low-resistance passageways through which blood returns from the organs to the heart.

■ In addition, veins can accommodate variable volumes of blood and therefore act as a blood reservoir. The capacity of veins to hold blood can change markedly with little change in venous pressure. Veins are thin-walled, highly distensible vessels that can passively stretch to store a larger volume of blood. (*Review Figure 9-27.*)

■ The primary force that produces venous flow is the pressure gradient between the veins and atrium (i.e., what remains of the driving pressure imparted to the blood by cardiac contraction). (*Review Figures 9-9 and 9-28.*)

■ Venous return is enhanced by sympathetically induced venous vasoconstriction and by external compression of the veins from contraction of surrounding skeletal muscles, both of which drive blood out of the veins. These actions help counter the effects of gravity on the venous system. (*Review Figures 9-28 through 9-31.*)

■ One-way venous valves ensure that blood is driven toward the heart and kept from flowing back toward the tissues. (*Review Figure 9-32.*)

■ Venous return is also enhanced by the respiratory pump and the cardiac suction effect. Respiratory activity produces a less-than-atmospheric pressure in the chest cavity, thus establishing an external pressure gradient that encourages flow from the lower veins that are exposed to atmospheric pressure to the chest veins that empty into the heart. (*Review Figure 9-33.*)

■ In addition, slightly negative pressures created within the atria during ventricular systole and within the ventricles during ventricular diastole exert a suctioning effect that further enhances venous return and facilitates cardiac filling.

## 9.6 | Blood Pressure (pp. 419–428)

■ Regulation of mean arterial pressure depends on control of its two main determinants: cardiac output and total peripheral resistance. (*Review Figure 9-34.*)

■ Control of cardiac output, in turn, depends on regulation of heart rate and stroke volume, whereas total peripheral resistance is determined primarily by the degree of arteriolar vasoconstriction. (*Review Figures 8-23 and 9-14.*)

■ Short-term regulation of blood pressure is accomplished mainly by the baroreceptor reflex. Carotid sinus and aortic arch baroreceptors continuously monitor mean arterial pressure. When they detect a deviation from normal, they signal the medullary cardiovascular centre, which responds by adjusting autonomic output to the heart and blood vessels to restore the blood pressure to normal. (*Review Figures 9-35 through 9-38.*)

■ Long-term control of blood pressure involves maintaining proper plasma volume through the kidneys' control of salt and water balance. (*Review Figure 9-34.*)

■ Blood pressure can be abnormally high (hypertension) or abnormally low (hypotension). Severe, sustained hypotension resulting in generalized inadequate blood delivery to the tissues is known as circulatory shock. (*Review Figures 9-39 and 9-40.*)

## 10.1 | Introduction (pp. 433–434)

■ Blood consists of three types of cellular elements—erythrocytes (red blood cells), leukocytes (white blood cells), and platelets (thrombocytes)—suspended in the liquid plasma. (Review Table 10-1.)

■ The 5–5.5 L volume of blood in an adult consists of 42–45 percent erythrocytes, less than 1 percent leukocytes and platelets, and 55–58 percent plasma. The percentage of whole-blood volume occupied by erythrocytes is the haematocrit. (Review Figure 10-1.)

## 10.2 | Plasma (pp. 434–435)

■ Plasma is a complex liquid consisting of 90 percent water that serves as a transport medium for substances being carried in the blood.

■ The most abundant inorganic constituents in plasma are sodium and chloride. The most plentiful organic constituents in plasma are plasma proteins.

■ All plasma constituents are freely diffusible across the capillary walls, except the plasma proteins, which remain in the plasma, where they perform a variety of important functions. Plasma proteins include the albumins, globulins, and fibrinogen. (Review Table 10-1.)

## 10.3 | Erythrocytes (pp. 436–444)

■ Erythrocytes are specialized for their primary function of $O_2$ transport in the blood. Their biconcave shape maximizes the surface area available for diffusion of oxygen into cells of this volume. (Review Figure 10-1.) Erythrocytes do not contain a nucleus, organelles, or ribosomes, but instead are packed full of haemoglobin—an iron-containing molecule that can loosely and reversibly bind with oxygen. Because oxygen is poorly soluble in blood, haemoglobin is indispensable for oxygen transport. (Review Figure 10-2.)

■ Haemoglobin also contributes to carbon dioxide transport and buffering of blood by reversibly binding with carbon dioxide and hydrogen ions.

■ Unable to replace cell components, erythrocytes are destined to a short lifespan of about 120 days.

■ Undifferentiated multipotent stem cells in the red bone marrow give rise to all cellular elements of the blood. (Review Figure 10-8.) Erythrocyte production (erythropoiesis) by the marrow normally keeps pace with the rate of erythrocyte loss, keeping the red cell count constant. Erythropoiesis is stimulated by erythropoietin, a hormone secreted by the kidneys in response to reduced $O_2$ delivery. (Review Figure 10-3.)

## 10.4 | Leukocytes (pp. 444–447)

■ Leukocytes are the defence corps of the body. They attack foreign invaders (the most common of which are bacteria and viruses), destroy cancer cells that arise in the body, and clean up cellular debris. Leukocytes as well as certain plasma proteins make up the immune system.

■ Each of the five types of leukocytes has a different task: (1) Neutrophils, the phagocytic specialists, are important in engulfing bacteria and debris. (2) Eosinophils specialize in attacking parasitic worms and play a role in allergic responses. (3) Basophils release two chemicals: histamine, which is also important in allergic responses; and heparin, which helps clear fat particles from the blood. (4) Monocytes, on leaving the blood, set up residence in the tissues and greatly enlarge to become the large-tissue phagocytes known as macrophages. (5) Lymphocytes provide immune defence against bacteria, viruses, and other targets for which they are specifically programmed. Their means of defence include the production of antibodies that mark the victim for destruction by phagocytosis or other processes (for B lymphocytes) and the release of chemicals that punch holes in the victim (for T lymphocytes). (Review Figure 10-8 and Table 10-1.)

■ Leukocytes are present in the blood only while in transit from their site of production and storage in the bone marrow (and also in the lymphoid tissues in the case of the lymphocytes) to their site of action in the tissues. (Review Figure 10-9.) At any given time, most leukocytes are out in the tissues on surveillance missions or performing actual combat missions.

■ All leukocytes have a limited lifespan and must be replenished by ongoing differentiation and proliferation of precursor cells. The total number and percentage of each of the different types of leukocytes produced vary depending on the momentary defence needs of the body.

## 10.5 | Platelets and Haemostasis (pp. 447–454)

■ Platelets are cell fragments derived from large megakaryocytes in the bone marrow. (Review Figures 10-8, 10-9, and 10-10.)

■ Platelets play a role in haemostasis, the arrest of bleeding from an injured vessel. The three main steps in haemostasis are (1) vascular spasm, (2) platelet plugging, and (3) clot formation.

■ Vascular spasm reduces blood flow through an injured vessel.

■ Aggregation of platelets at the site of vessel injury quickly plugs the defect. Platelets start to aggregate on contact with exposed collagen in the damaged vessel wall. (Review Figures 10-11 and 10-15.)

■ Clot formation reinforces the platelet plug and converts blood in the vicinity of a vessel injury into a nonflowing gel.

- Most factors necessary for clotting are always present in the plasma in inactive precursor form. When a vessel is damaged, exposed collagen initiates a cascade of reactions involving successive activation of these clotting factors, ultimately converting fibrinogen into fibrin via the intrinsic clotting pathway. *(Review Figures 10-13, 10-14, and 10-15.)*
- Fibrin, an insoluble threadlike molecule, is laid down as the meshwork of the clot; the meshwork in turn entangles blood cellular elements to complete clot formation. *(Review Figure 10-12.)*
- Blood that has escaped into the tissues clots on exposure to tissue thromboplastin, which sets the extrinsic clotting pathway into motion. *(Review Figure 10-14.)*
- When no longer needed, clots are dissolved by plasmin, a fibrinolytic factor also activated by exposed collagen. *(Review Figure 10-16.)*

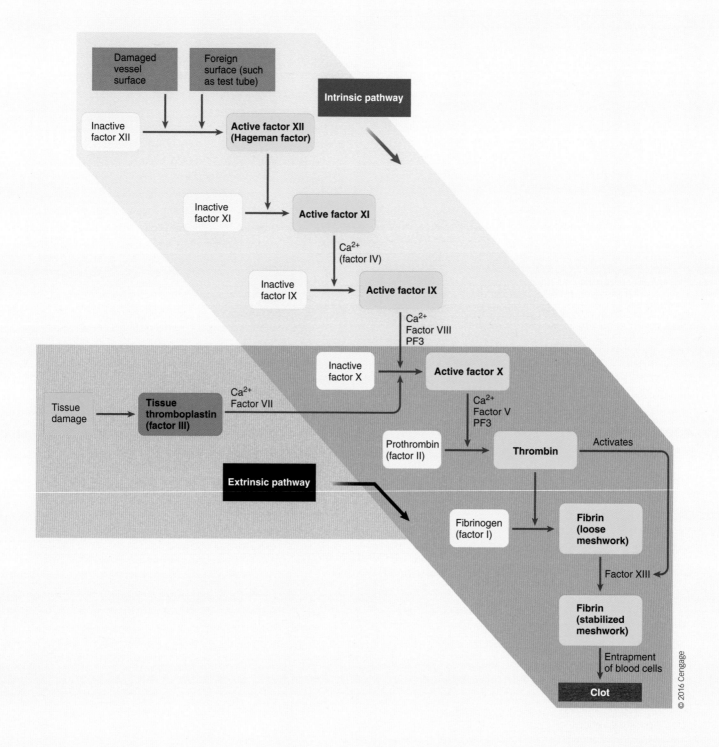

© 2016 Cengage

## 11.1 | Introduction (p. 460)

■ The most common invaders are bacteria and viruses. Bacteria are self-sustaining, single-celled organisms, which produce disease by virtue of the destructive chemicals they release. Viruses are protein-coated nucleic acid particles, which invade host cells and take over the cellular metabolic machinery for their own survival to the detriment of the host cell.

## 11.2 | External Defences (pp. 460–464)

■ The body surfaces exposed to the outside environment— both the outer covering of skin and the linings of internal cavities that communicate with the external environment—not only serve as mechanical barriers to deter would-be pathogenic invaders but also play an active role in thwarting the entry of bacteria and other unwanted materials.

■ The epidermis contains four cell types:

1. Melanocytes produce a pigment, melanin, the colour and amount of which is responsible for the varying shades of skin colour. Melanin protects the skin by absorbing harmful UV radiation.

2. The most abundant cells are the keratinocytes, producers of the tough keratin that forms the outer, protective layer of the skin. This physical barrier discourages bacteria and other harmful environmental agents from entering the body and prevents water and other valuable body substances from escaping. Keratinocytes further serve immunologically by secreting interleukin 1, which enhances post-thymic T-cell maturation within the skin.

3. Langerhans cells also function in specific immunity by presenting antigen to helper T cells.

4. Granstein cells suppress skin-activated immune responses.

■ The dermis contains (1) blood vessels, (2) sensory nerve endings, and (3) several exocrine glands and hair follicles. (Review Figure 11-1.)

■ The skin's exocrine glands include sebaceous glands, which produce sebum, an oily substance that softens and waterproofs the skin, and sweat glands, which produce cooling sweat. Hair follicles produce hairs, the distribution and function of which are minimal in humans.

■ In addition, the skin synthesizes vitamin D in the presence of sunlight.

■ Besides the skin, the other main routes by which potential pathogens enter the body are (1) the digestive system, which is defended by an antimicrobial salivary enzyme, destructive acidic gastric secretions, gut-associated lymphoid tissue, and harmless colonic resident flora; (2) the genitourinary system, which is protected by destructive acidic and particle-entrapping mucus secretions; and (3) the respiratory system, whose defence depends on alveolar macrophage activity and on secretion of a sticky mucus that traps debris, which is subsequently swept out by ciliary action.

## 11.3 | Internal Defences (pp. 464–466)

■ Leukocytes and their derivatives are the major effector cells of the immune system and are reinforced by a number of different plasma proteins. Leukocytes include neutrophils, eosinophils, basophils, monocytes, and lymphocytes.

■ Leukocytes are primarily produced in the bone marrow, but some lymphocytes are also produced and differentiated and perform their defence activities within lymphoid tissues strategically located at likely points of foreign infiltration. (Review Figure 11-2 and Table 11-1.)

## 11.4 | Innate Immunity (pp. 466–473)

■ Innate immune responses, which form a first line of defence against atypical cells (foreign, mutant, or injured cells) within the body, include inflammation, interferon, natural killer cells, and the complement system.

■ Inflammation is a nonspecific response to foreign invasion or tissue damage mediated largely by the professional phagocytes (neutrophils and monocytes-turned-macrophages) and their secretions. (Review Figure 11-3.)

■ The phagocytic cells destroy foreign and damaged cells both by phagocytosis and by the release of lethal chemicals.

■ Histamine-induced vasodilation and increased permeability of local capillaries at the site of invasion or injury permit enhanced delivery of more phagocytic leukocytes and inactive plasma protein precursors crucial to the inflammatory process, such as clotting factors and components of the complement system. These vascular changes also largely produce the observable local manifestations of inflammation—swelling, redness, heat, and pain. (Review Figure 11-4.)

■ Interferon is nonspecifically released by virus-infected cells and transiently inhibits viral multiplication in other cells to which it binds. (Review Figure 11-6.) Interferon further exerts anticancer effects by slowing division and growth of tumour cells as well as by enhancing the power of killer cells. (Review Figure 11-20.)

■ Natural killer (NK) cells nonspecifically lyse and destroy virus-infected cells and cancer cells on first exposure to them.

## 11.5 | Adaptive Immunity: General Concepts (pp. 473–475)

■ There are two broad classes of adaptive immune responses: antibody-mediated immunity and cell-mediated immunity. In both instances, the ultimate outcome of a particular lymphocyte binding with a specific antigen is destruction of the antigen, but the effector cells, stimuli, and tactics involved are different. Plasma cells derived from B lymphocytes (B cells) are responsible for antibody-mediated immunity, whereas T lymphocytes (T cells) accomplish cell-mediated immunity. *(Review Figure 11-8.)*

■ B cells develop from a lineage of lymphocytes that originally matured within the bone marrow. The T-cell lineage arises from lymphocytes that migrated from the bone marrow to the thymus to complete their maturation. *(Review Figure 11-8.)*

## 11.6 | B Lymphocytes: Antibody-Mediated Immunity (pp. 475–481)

■ Each B cell recognizes specific free extracellular antigen that is not associated with cell-bound self-antigens, such as that found on the surface of bacteria.

■ After being activated by binding with its specific antigen, a B cell rapidly proliferates, producing a clone of its own kind that can specifically wage battle against the invader. Most lymphocytes in the expanded B-cell clone become plasma cells that participate in the primary response against the invader. *(Review Figures 11-10 and 11-13.)*

■ Antibodies are Y-shaped molecules. The antigen-binding sites on the tips of each arm of the antibody determine with what specific antigen the antibody can bind. Properties of the antibody's tail portion determine what the antibody does once it binds with antigen. *(Review Figure 11-11.)*

■ There are five subclasses of antibodies, depending on differences in the biological activity of their tail portion: IgM, IgG, IgE, IgA, and IgD immunoglobulins.

■ Antibodies do not directly destroy material. Instead, they exert their protective effect by physically hindering antigens through neutralization or agglutination or by intensifying lethal innate immune responses already called into play by the foreign invasion. Antibodies activate the complement system, enhance phagocytosis, and stimulate killer cells. *(Review Figure 11-12.)*

■ Some of the newly developed lymphocytes in the activated B-cell clone do not participate in the attack but become memory cells that lie in wait, ready to launch a swifter and more forceful secondary response should the same foreigner ever invade the body again. *(Review Figures 11-13 and 11-14.)*

## 11.7 | T Lymphocytes: Cell-Mediated Immunity (pp. 482–491)

■ T cells accomplish cell-mediated immunity by being in direct contact with their targets.

■ There are three main subpopulations of T cells: cytotoxic T cells, helper T cells, and regulatory T cells.

■ The targets of cytotoxic T cells are virally invaded cells and cancer cells, which they destroy by releasing perforin molecules that form a lethal hole-punching complex that inserts into the membrane of the victim cell or by releasing granzymes that trigger the victim cell to undergo apoptosis. *(Review Figures 11-16 and 11-17, and Table 11-2.)*

■ Like B cells, T cells bear receptors that are antigen specific, undergo clonal selection, exert primary and secondary responses, and form memory pools for long-lasting immunity against targets to which they have already been exposed.

■ Those lymphocytes produced by chance that can attack the body's own antigen-bearing cells are eliminated or suppressed so that they are prevented from functioning. In this way, the body is able to "tolerate" (not attack) its own antigens. Tolerance is accomplished by clonal deletion, clonal anergy, receptor editing, regulatory (suppressor) T cells, immunological ignorance, and immune privilege.

■ The major self-antigens on the surface of body cells are known as MHC molecules, which are coded for by the major histocompatibility complex (MHC), a group of genes with DNA sequences unique for each individual.

■ The presence of class I or class II MHC self-antigens on the surface of these foreign antigen-bearing host cells causes the two different types of T cells to differentially interact with them. *(Review Figure 11-18.)*

1. Cytotoxic T cells are able to bind only with virus-infected host cells or cancer cells, which always bear the class I MHC self-antigen in association with foreign or abnormal antigen. On binding with the abnormal host cell, these T cells release toxic substances that kill the dangerous body cell.

2. Helper T cells can bind only with other T cells, B cells, and macrophages that have encountered foreign antigen. These immune cells bear the class II MHC self-marker in association with foreign antigen. Subsequently, helper T cells enhance the immune powers of these other effector cells by secreting specific chemical mediators.

■ In the process of immune surveillance, natural killer cells, cytotoxic T cells, macrophages, and the interferon they collectively secrete normally eradicate newly arisen cancer cells before they have a chance to spread. *(Review Figure 11-20.)*

## 11.8 | Immune Diseases (pp. 491–494)

■ Immune diseases are of two types: immunodeficiency diseases (insufficient immune responses) or inappropriate immune attacks (excessive or mistargeted immune responses).

■ With immunodeficiency diseases, the immune system fails to defend normally against bacterial or viral infections through a deficit of B or T cells, respectively.

■ With inappropriate immune attacks, the immune system becomes overzealous. There are three categories of inappropriate attacks: (1) autoimmune diseases; (2) immune complex diseases; and (3) allergies or hypersensitivities.

## 12.1 | Introduction (pp. 499–505)

- Internal respiration encompasses the intracellular metabolic reactions that use oxygen and produce carbon dioxide during energy-yielding oxidation of nutrient molecules.

- Respiratory airways conduct air from the atmosphere to the alveoli, the gas-exchanging portion of the lungs. (*Review Figure 12-1.*)

- External respiration involves the transfer of oxygen and carbon dioxide between the external environment and tissue cells. The respiratory and circulatory systems function together to accomplish external respiration. (*Review Figure 12-2.*)

- The respiratory system moves air between the atmosphere and lungs through the process of ventilation.

- In the lungs, oxygen and carbon dioxide diffuse across the extremely thin walls of the alveoli for exchange between the gas in the alveoli and the blood in the pulmonary capillaries. The alveolar walls are formed by Type I alveolar cells. Type II alveolar cells secrete pulmonary surfactant. (*Review Figure 12-4.*)

- The lungs are housed within the thorax, the volume of which is changed by contraction of respiratory muscles.

- Each lung is surrounded by a double-walled, closed sac, the pleural space. (*Review Figure 12-6.*)

## 12.2 | Respiratory Mechanics (pp. 505–523)

- Ventilation, or breathing, is the process of cyclically moving air in and out of the lungs so that alveolar air that has already participated in the exchange of oxygen and carbon dioxide with the pulmonary capillary blood is exchanged for fresh atmospheric air.

- The respiratory system consists of two components: the lungs and the chest wall. The lungs always exert an expiratory (deflationary) pressure, whereas the chest wall exerts an inspiratory (inflationary) pressure at lung volumes below about 60 percent of its capacity and expiratory (deflationary) pressures above this volume. These two components account for the pressure–volume relationship of the respiratory system. At approximately 40 percent of the vital capacity, the inspiratory and expiratory pressures are in equilibrium due to the equal and opposite pressures of the lung and chest wall. (*Review Figure 12-8*).

- Ventilation is accomplished by alternating the pressure gradient for airflow between the atmosphere and alveoli. When alveolar pressure decreases below atmospheric pressure as a result of decompression of the pleural space, air flows into the lungs. When alveolar pressure increases above atmospheric as a result of increased lung recoil at higher lung volumes, and this recoil pressure is not opposed by an equal and opposite pleural pressure (due to contraction of inspiratory muscles), air flows out of the lungs. (*Review Figures 12-7, 12-15, 12-16, and 12-18.*)

- Alternate contraction and relaxation of the inspiratory muscles (primarily the diaphragm) indirectly produce periodic inflation and deflation of the lungs by cyclically decompressing and compressing the pleural space, and therefore reducing and increasing intrathoracic pressure; the lungs passively follow the movements of the thorax. (*Review Figures 12-7, 12-16, and 12-18.*)

- The lungs can be stretched to varying degrees during inspiration; they then recoil to their pre-inspiratory size during expiration because of their elastic recoil.

- The term *compliance* refers to the distensibility of the lungs—the change in volume per unit change in the pressure gradient across the lung wall.

- Elastic recoil depends on the connective tissue meshwork within the lungs and, mostly, on alveolar surface tension. Surface tension, due to the attractive forces between the surface water molecules in the liquid film lining each alveolus, opposes alveolar expansion (decreases compliance). (*Review Table 12-3.*)

- Because energy is required for contracting the inspiratory muscles, inspiration is an active process, but expiration is passive during quiet breathing because it is accomplished by elastic recoil of the lungs on relaxing inspiratory muscles.

- For more forceful active expiration, contraction of the expiratory muscles (primarily the abdominal muscles) increases the alveolar-to-atmospheric pressure gradient. (*Review Figures 12-15.*)

- The larger the gradient between the alveoli and atmosphere in either direction, the greater the airflow. Air continues to flow until the alveolar pressure equilibrates with atmospheric pressure. (*Review Figures 12-16 and 12-18.*)

- Besides being directly proportional to the pressure gradient, airflow is also inversely proportional to airway resistance. (*Review Table 12-4.*) Because airway resistance, which depends on the calibre of the conducting airways, is normally very low, airflow depends primarily on the pressure gradient between the alveoli and atmosphere.

- If airway resistance is increased by disease, either the respiratory muscles must generate more pressure to maintain airflow or the person must breathe more slowly.

- If the alveoli were lined by water alone, the surface tension would be so great that the lungs would not be compliant and would tend to collapse. Pulmonary surfactant interspersed between the water molecules lowers the alveolar surface tension, thereby increasing the compliance of the lungs and counteracting the tendency for alveoli to collapse. Alveolar independence also counteracts the tendency for alveoli to collapse, because a collapsing alveolus is pulled open by the recoil of surrounding alveoli that have been stretched by the collapsing alveolus. (*Review Figures 12-10 and 12-12 and Table 12-3.*)

- Normally, the lungs operate at "half full." Lung volume typically varies from about 2 to 2.5 L as an average tidal volume of 500 mL of air is moved in and out with each breath. (*Review Figures 12-20 and 12-21.*)

- The amount of air moved in and out of the lungs in one minute, the pulmonary ventilation, is equal to tidal volume multiplied by respiratory frequency.

- Not all the air moved in and out is available for $O_2$ and $CO_2$ exchange with the blood, because part occupies the conducting airways, known as the *anatomic dead space*. Alveolar ventilation, the volume of air exchanged between the atmosphere and alveoli in one minute, is a measure of the air actually available for gas exchange with the blood. Alveolar ventilation = (tidal volume − dead space volume) respiratory rate. (*Review Figure 12-23 and Table 12-4.*)

## 12.3 | Gas Exchange (pp. 523–531)

- Oxygen and carbon dioxide move across membranes by passive diffusion down partial pressure gradients.

- The partial pressure of a gas is that fraction of the total atmospheric pressure contributed by this individual gas, which in turn is directly proportional to the percentage of this gas in the air. The partial pressure of a gas in blood depends on the amount of this particular gas dissolved in the blood. (*Review Table 12-6.*)

- Net diffusion of $O_2$ occurs first between the alveoli and blood and then between the blood and tissues as a result of the $O_2$ partial pressure gradients created by continuous use of $O_2$ in the cells and continuous replenishment of fresh alveolar $O_2$ provided by ventilation. (*Review Figure 12-26.*)

- Net diffusion of $CO_2$ occurs in the reverse direction, first between the tissues and blood and then between the blood and alveoli, as a result of the $CO_2$ partial pressure gradients created by continuous production of $CO_2$ in the cells and continuous removal of alveolar $CO_2$ through ventilation. (*Review Figure 12-26.*)

- Factors other than the partial pressure gradient that influence the rate of gas exchange are the surface area and thickness of the membrane across which the gas is diffusing and the diffusion coefficient of the gas in the membrane, according to Fick's law of diffusion. (*Review Table 12-7.*)

## 12.4 | Gas Transport (pp. 531–541)

- Because carbon dioxide and especially oxygen are not very soluble in blood, they must be transported primarily by mechanisms other than simply being physically dissolved. (*Review Table 12-8.*)

- Only 1.5 percent of the oxygen is physically dissolved in the blood; the 98.5 percent is chemically bound to haemoglobin (Hb). (*Review Figure 12-35.*)

- The primary factor that determines the extent to which haemoglobin and oxygen are combined (the % Hb saturation) is the $P_{O_2}$ of the blood. This relation is depicted by an S-shaped curve known as the $O_2$–Hb dissociation curve. (*Review Figure 12-33.*)

  1. In the $P_{O_2}$ range of the pulmonary capillaries (the plateau portion of the curve), haemoglobin is still almost fully saturated even if the blood $P_{O_2}$ falls to almost 60 mmHg. This provides a margin of safety by ensuring near-normal oxygen delivery to the tissues, despite a substantial reduction in arterial $P_{O_2}$.

  2. In the $P_{O_2}$ range in the systemic capillaries (the steep portion of the curve), Hb unloading increases greatly in response to a small local decline in blood $P_{O_2}$ associated with increased cellular metabolism. In this way, more oxygen is provided to match the increased tissue needs. (*Review Figure 12-33.*) The shape of the oxygen dissociation curve, when viewed from the perspective of the tissues, helps maintain the gradient for diffusion despite oxygen uptake by the tissues. (*Review Figure 12-31.*)

- Carbon dioxide picked up at the systemic capillaries is transported in the blood in three forms: (1) 10 percent is physically dissolved, (2) 30 percent is bound to haemoglobin, and (3) 60 percent takes the form of bicarbonate $HCO_3^-$. (*Review Table 12-8.*)

- The erythrocyte enzyme carbonic anhydrase catalyzes conversion of $CO_2$ to $HCO_3^-$ according to the reaction $CO_2 + H_2O \Leftrightarrow H_2CO_3 \Leftrightarrow H^+ + HCO_3^-$. The carbon and oxygen originally present in $CO_2$ are now part of the $HCO_3^-$. The generated $H^+$ binds to Hb. These reactions are all reversed in the lungs as $CO_2$ is eliminated to the alveoli and then to the atmosphere. (*Review Figure 12-37.*)

## 12.5 | Control of Breathing (pp. 541–551)

- Ventilation involves three aspects, all subject to neural control: (1) rhythmic cycling between inspiration and expiration; (2) how frequently this cycling occurs (respiratory frequency); and (3) the amplitude of the signal sent to the respiratory muscles (tidal volume).

- Respiratory rhythm is established by a network of neurons in the medulla. The outputs are directed to both respiratory pump muscles and those that control the size of the upper airway. (*Review Figures 12-39 and 12-41.*)

- When the inspiratory neurons stop firing, the inspiratory muscles relax and expiration takes place. For active expiration to occur, the expiratory muscles are activated by output from the medullary expiratory neurons in the ventral respiratory group (VRG) of the medullary respiratory control centre.

- Three chemical factors play a role in determining the magnitude of ventilation: $P_{CO_2}$, $P_{O_2}$, and $H^+$ concentration of the arterial blood. (*Review Table 12-10.*)

- The dominant factor in the minute-to-minute regulation of ventilation is the arterial $P_{CO_2}$. An increase in arterial $P_{CO_2}$ is the most potent chemical stimulus for increasing ventilation. Changes in arterial $P_{CO_2}$ alter ventilation primarily by bringing about corresponding changes in the brain ECF $H^+$ concentration, to which the central chemoreceptors are exquisitely sensitive. (*Review Figure 12-39.*)

- The carotid chemoreceptors reflexly stimulate the respiratory centre in response to a marked reduction in arterial $P_{CO_2}$ (< 60 mmHg).

- The carotid chemoreceptors respond to (or are sensitized by) an increase in arterial $H^+$ concentration, which reflexly increases ventilation. The resulting adjustment in arterial $H^+$–generating $CO_2$ is important in maintaining the acid–base balance of the body. (*Review Figure 12-40.*)

## 13.1 | Introduction (pp. 559–565)

- Each of the pair of kidneys consists of an outer renal cortex and inner renal medulla. (Review Figure 13-1.)
- The kidneys form urine. They eliminate unwanted plasma constituents in the urine while conserving materials of value to the body.
- The urine-forming functional unit of the kidneys is the nephron, which is composed of interrelated vascular and tubular components. (Review Figure 13-3.)
- The vascular component consists of two capillary networks in series: (1) the glomerulus, and (2) the peritubular capillaries. (Review Figures 13-3 and 13-4.)
- The tubular component begins with Bowman's capsule, which cups around the glomerulus to catch the filtrate, then continues a specific tortuous course to ultimately empty into the renal pelvis. (Review Figure 13-3.)
- The kidneys perform three basic processes in carrying out their regulatory and excretory functions: (1) glomerular filtration, the nondiscriminating movement of protein-free plasma from the blood into the tubules; (2) tubular reabsorption, the selective transfer of specific constituents in the filtrate back into the blood of the peritubular capillaries; and (3) tubular secretion, the highly specific movement of selected substances from peritubular capillary blood into the tubular fluid. Everything filtered or secreted but not reabsorbed is excreted as urine. (Review Figure 13-6.)

## 13.2 | Renal Blood Flow (pp. 565–566)

- RBF is greatest at rest (1200 mL/min) and becomes reduced during exercise. The purpose of this relatively high blood flow is to supply the kidney with sufficient plasma for filtration, secretion, and reabsorption.
- Control of RBF is primarily via the ANS, with the sympathetic division exerting the greatest control through vasoconstriction. The adrenal medulla reinforces the sympathetic vasoconstriction.

## 13.3 | Glomerular Filtration (pp. 566–573)

- Glomerular filtrate is produced when part of the plasma flowing through each glomerulus is passively forced under pressure through the glomerular membrane into the lumen of the underlying Bowman's capsule. (Review Figure 13-7.)
- The net filtration pressure that induces filtration is caused by an imbalance in the physical forces acting across the glomerular membrane. A high glomerular capillary blood pressure favouring filtration outweighs the combined opposing forces of plasma-colloid osmotic pressure and Bowman's capsule hydrostatic pressure. (Review Table 13-1.)
- Of the plasma flowing through the kidneys, normally 20 percent is filtered through the glomeruli, producing an average glomerular filtration rate (GFR) of 125 mL/min. This filtrate is identical in composition to plasma, except for the plasma proteins, which are held back by the glomerular membrane.
- The GFR can be deliberately altered through a change in the glomerular capillary blood pressure by means of sympathetic influence on the afferent arterioles as part of the baroreceptor reflex response that compensates for changed arterial blood pressure.
- As the GFR is altered, the amount of fluid lost in urine changes correspondingly, and plasma volume adjusts as needed to help restore blood pressure to normal on a long-term basis. (Review Figure 13-12.)

## 13.4 | Tubular Reabsorption (pp. 573–582)

- After a protein-free plasma is filtered through the glomerulus, the tubules handle each substance discretely, so that even though the concentrations of all constituents in the initial glomerular filtrate are identical to their concentrations in the plasma (with the exception of plasma proteins), the concentrations of different constituents are variously altered as the filtered fluid flows through the tubular system. (Review Tables 13-2 and 13-3.)
- The reabsorptive capacity of the tubular system is tremendous. More than 99 percent of the filtered plasma is returned to the blood through reabsorption. On average, 124 mL out of the 125 mL filtered per minute are reabsorbed. (Review Table 13-2.)
- Tubular reabsorption involves transepithelial transport from the tubular lumen into the peritubular capillary plasma. This process may be active (requiring energy) or passive (using no energy). (Review Figure 13-14.)
- The pivotal event to which most reabsorptive processes are somehow linked is the active reabsorption of $Na^+$, driven by an energy-dependent $Na^+ - K^+$ ATPase carrier located in the basolateral membrane of almost all tubular cells. The transport of $Na^+$ out of the cells into the lateral spaces between adjacent cells by this carrier induces the net reabsorption of $Na^+$ from the tubular lumen to the peritubular capillary plasma. (Review Figure 13-15.)
- Most $Na^+$ reabsorption takes place early in the nephron in constant unregulated fashion, but in the distal and collecting tubules, the reabsorption of a small percentage of the filtered $Na^+$ is variable and subject to control, depending primarily on the complex renin–angiotensin–aldosterone system. (Review Table 13-4.)
- Because $Na^+$ and its attendant anion, $Cl^-$, are the major osmotically active ions in the ECF, the ECF volume is determined by the $Na^+$ load in the body. In turn, the plasma volume, which reflects the total ECF volume, is important in the long-term determination of arterial blood pressure. Whenever the $Na^+$ load, ECF volume, plasma volume, and arterial blood pressure are below normal, the juxtaglomerular apparatus of the kidneys secretes renin, an enzymatic hormone that triggers a series of events

ultimately leading to increased secretion of aldosterone from the adrenal cortex. Aldosterone increases Na$^+$ reabsorption from the distal portions of the tubule, thereby correcting for the original reduction in Na$^+$, ECF volume, and blood pressure. *(Review Figures 13-11 and 13-16.)*

■ By contrast, Na$^+$ reabsorption is inhibited by atrial natriuretic peptide, a hormone released from the cardiac atria in response to expansion of the ECF volume and a subsequent increase in blood pressure. *(Review Figure 13-17.)*

■ In addition to driving the reabsorption of Na$^+$, the energy used to supply the Na$^+$−K$^+$ ATPase carrier is ultimately responsible for the reabsorption of organic nutrient molecules from the proximal tubule by secondary active transport. Specific cotransport carriers located at the luminal border of the proximal tubular cell are driven by the Na$^+$ concentration gradient to selectively transport glucose or an amino acid from the luminal fluid into the tubular cell, from which the nutrient eventually enters the plasma.

■ The other electrolytes besides Na$^+$ that are actively reabsorbed by the tubules, such as PO$_4^{3-}$ and Ca$^{2+}$, have their own independently functioning carrier systems within the proximal tubule.

■ Because these carriers, like the organic-nutrient cotransport carriers, can become saturated, each exhibits a maximal carrier-limited transport capacity ($T_m$). Once the filtered load of an actively reabsorbed substance exceeds the $T_m$, reabsorption proceeds at a constant maximal rate, and the additional filtered quantity of the substance is excreted in urine. *(Review Figure 13-18.)*

■ Active Na$^+$ reabsorption also drives the passive reabsorption of Cl$^-$ (via an electrical gradient), H$_2$O (by osmosis), and urea (down a urea concentration gradient created as a result of extensive osmotic-driven H$_2$O reabsorption). Sixty-five percent of the filtered H$_2$O is reabsorbed from the proximal tubule in unregulated fashion, driven by active Na$^+$ reabsorption. Reabsorption of H$_2$O increases the concentration of other substances remaining in the tubular fluid, most of which are filtered waste products. The small urea molecules are the only waste products that can passively permeate the tubular membranes. Accordingly, urea is the only waste product partially reabsorbed as a result of being concentrated. About 50 percent of the filtered urea is reabsorbed. *(Review Figures 13-19 and 13-20 and Table 13-4.)*

■ The other waste products, which are not reabsorbed, remain in the urine in highly concentrated form.

## 13.5 | Tubular Secretion (pp. 583–585)

■ By tubular secretion, the kidney tubules can selectively add some substances to the quantity already filtered. Secretion of substances hastens their excretion in the urine.

■ The most important secretory systems are for (1) H$^+$, which is important in regulating acid–base balance; (2) K$^+$, which keeps the plasma K$^+$ concentration at an appropriate level to maintain normal membrane excitability in muscles and nerves; and (3) organic ions, which accomplish more efficient elimination of foreign organic compounds from the body. *(Review Figures 13-21 and 13-22.)*

■ H$^+$ is secreted in the proximal, distal, and collecting tubules. K$^+$ is secreted only in the distal and collecting tubules under control of aldosterone. Organic ions are secreted only in the proximal tubule. *(Review Table 13-3.)*

## 13.6 | Urine Excretion and Plasma Clearance (pp. 585–601)

■ Of the 125 mL/min of filtrate formed in the glomeruli, normally only 1 mL/min remains in the tubules to be excreted as urine.

■ Only wastes and excess electrolytes not wanted by the body are left behind, dissolved in a given volume of H$_2$O to be eliminated in the urine.

■ Because the excreted material is removed, or cleared, from the plasma, the term *plasma clearance* refers to the volume of plasma cleared of a particular substance each minute by renal activity. *(Review Figure 13-23.)*

■ The kidneys can excrete urine of varying volumes and concentrations to either conserve or eliminate H$_2$O, depending on whether the body has a H$_2$O deficit or excess, respectively. The kidneys can produce urine ranging from 0.3 mL/min at 1200 mOsm/L to 25 mL/min at 100 mOsm/L by reabsorbing variable amounts of H$_2$O from the distal portions of the nephron.

■ This variable reabsorption is made possible by a vertical osmotic gradient in the medullary interstitial fluid, which is established by the long loops of Henle of the juxtamedullary nephrons via countercurrent multiplication and preserved by the vasa recta of these nephrons via countercurrent exchange. *(Review Figures 13-5, 13-24, 13-25, and 13-28.)* This vertical osmotic gradient, to which the hypotonic (100 mOsm/L) tubular fluid is exposed as it passes through the distal portions of the nephron, establishes a passive driving force for progressive reabsorption of H$_2$O from the tubular fluid, but the actual extent of H$_2$O reabsorption depends on the amount of vasopressin (antidiuretic hormone) secreted. *(Review Figure 13-27.)*

■ Vasopressin increases the permeability of the distal and collecting tubules to H$_2$O; they are impermeable to H$_2$O in its absence. *(Review Figure 13-26.)* Vasopressin secretion increases in response to a H$_2$O deficit, and H$_2$O reabsorption increases accordingly. Vasopressin secretion is inhibited in response to a H$_2$O excess, reducing H$_2$O reabsorption. In this way, adjustments in vasopressin-controlled H$_2$O reabsorption help correct any fluid imbalances.

■ The bladder can accommodate approximately 250–400 mL of urine before stretch receptors within its wall initiate the micturition reflex. *(Review Figure 13-30.)* This reflex causes involuntary emptying of the bladder by simultaneous bladder contraction and the opening of both the internal and external urethral sphincters. Micturition can transiently be voluntarily prevented until a more opportune time by deliberate tightening of the external sphincter and surrounding pelvic diaphragm. *(Review Figure 13-29.)*

## 14.1 | Balance Concept (pp. 608–609)

- The internal pool of a substance is the quantity of that substance in the ECF.

- Inputs to the pool occur by way of ingestion or metabolic production of the substance. Outputs from the pool occur by way of excretion or metabolic consumption of the substance. (*Review Figure 14-1.*)

- Input must equal output to maintain a stable balance of the substance.

## 14.2 | Fluid Balance (pp. 609–619)

- On average, the body fluids compose 60 percent of total body weight. This figure varies among individuals, depending on how much fat (a tissue with a low $H_2O$ content) a person has.

- Two-thirds of the body $H_2O$ is found in the intracellular fluid (ICF). The remaining third present in the extracellular fluid (ECF) is distributed between plasma (20% of ECF) and interstitial fluid (80% of ECF). (*Review Table 14-1.*)

- Because all plasma constituents are freely exchanged across the capillary walls, the plasma and interstitial fluid are nearly identical in composition, except for the lack of plasma proteins in the interstitial fluid. In contrast, the ECF and ICF have markedly different compositions, because the plasma membrane barriers are highly selective as to what materials are transported into or out of the cells. (*Review Figure 14-2.*)

- The essential components of fluid balance are control of ECF volume by maintaining salt balance and control of ECF osmolarity by maintaining water balance. (*Review Tables 14-2, 14-3, and 14-5.*)

- Because of the osmotic holding power of $Na^+$, the major ECF cation, a change in the body's total $Na^+$ content brings about a corresponding change in ECF volume, including plasma volume, which, in turn, alters arterial blood pressure in the same direction. Appropriately, in the long run, $Na^+$-regulating mechanisms compensate for changes in ECF volume and arterial blood pressure. (*Review Table 14-5.*)

- Salt intake is not controlled in humans, but control of salt output in the urine is closely regulated. Blood pressure–regulating mechanisms can vary the GFR, and, accordingly, the amount of $Na^+$ filtered, by adjusting the radius of the afferent arterioles that supply the glomeruli. Simultaneously, blood pressure–regulating

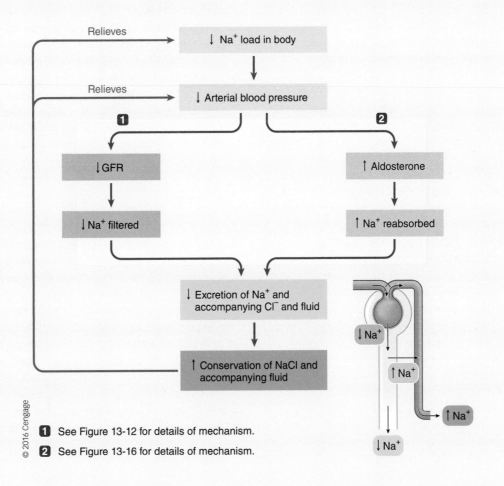

© 2016 Cengage

**1** See Figure 13-12 for details of mechanism.

**2** See Figure 13-16 for details of mechanism.

mechanisms can vary the secretion of aldosterone to adjust $Na^+$ reabsorption by the renal tubules. Varying $Na^+$ filtration and $Na^+$ reabsorption can adjust how much $Na^+$ is excreted in the urine to regulate plasma volume and, consequently, arterial blood pressure in the long term. *(Review Figure 14-3.)*

■ Changes in ECF osmolarity are primarily detected and corrected by the systems that maintain $H_2O$ balance.

■ ECF osmolarity must be closely regulated to prevent osmotic shifts of $H_2O$ between the ECF and ICF, because cell swelling or shrinking is harmful, especially to brain neurons. Excess free $H_2O$ in the ECF dilutes ECF solutes; the resulting ECF hypotonicity drives $H_2O$ into the cells. An ECF free $H_2O$ deficit, by contrast, concentrates ECF solutes, so $H_2O$ leaves the cells to enter the hypertonic ECF. *(Review Table 14-5.)*

■ To prevent these harmful fluxes, free $H_2O$ balance is regulated largely by vasopressin and, to a lesser degree, by thirst.

■ Changes in vasopressin secretion and thirst are both governed primarily by hypothalamic osmoreceptors, which monitor ECF osmolarity. The amount of vasopressin secreted determines the extent of free $H_2O$ reabsorption by distal portions of the nephrons, thereby determining the volume of urinary output. *(Review Figure 14-4 and Table 14-4.)*

■ Simultaneously, intensity of thirst controls the volume of fluid intake. However, because the volume of fluid drunk is often not directly correlated with the intensity of thirst, control of urinary output by vasopressin is the most important regulatory mechanism for maintaining $H_2O$ balance.

## 14.3 | Acid–Base Balance (pp. 620–637)

■ Acids liberate free hydrogen ions ($H^+$) into solution; bases bind with free $H^+$ ions and remove them from solution. *(Review Figure 14-5.)*

■ *Acid–base balance* refers to regulation of $H^+$ concentration ($[H^+]$) in the body fluids. To precisely maintain $[H^+]$, input of $H^+$ by metabolic production of acids within the body must continually be matched with $H^+$ output by urinary excretion of $H^+$ and respiratory removal of $H^+$-generating $CO_2$. Furthermore, between the time of its generation and its elimination, $H^+$ must be buffered within the body to prevent marked fluctuations in $[H^+]$.

■ Hydrogen ion concentration is often expressed in terms of pH, which is the logarithm of $1/[H^+]$.

■ The normal pH of the plasma is 7.4, slightly alkaline compared to neutral $H_2O$, which has a pH of 7.0. A pH lower than normal (higher $[H^+]$ than normal) indicates a state of acidosis. A pH higher than normal (lower $[H^+]$ than normal) characterizes a state of alkalosis. *(Review Figure 14-6.)*

■ Fluctuations in $[H^+]$ have profound effects on body chemistry, most notably (1) changes in neuromuscular excitability, with acidosis depressing excitability, especially in the central nervous system, and alkalosis producing overexcitability of both the peripheral and the central nervous systems; (2) disruption of normal metabolic reactions by altering the structure and function of all enzymes; and (3) alterations in plasma $[K^+]$ brought about by $H^+$-induced changes in the rate of $K^+$ elimination by the kidneys.

■ The primary challenge in controlling acid–base balance is maintaining normal plasma alkalinity despite continual addition of $H^+$ to the plasma from ongoing metabolic activity. The major source of $H^+$ is from the dissociation of $CO_2$-generated $H_2CO_3$.

■ The three lines of defence for resisting changes in $[H^+]$ are (1) the chemical buffer systems, (2) respiratory control of pH, and (3) renal control of pH.

■ Chemical buffer systems, the first line of defence, each consist of a pair of chemicals involved in a reversible reaction, one that can liberate $H^+$ and the other that can bind $H^+$. By acting according to the law of mass action, a buffer pair acts immediately to minimize any changes in pH. *(Review Figure 14-8 and Table 14-6.)*

■ The respiratory system, the second line of defence, normally eliminates the metabolically produced $CO_2$ so that $H_2CO_3$ does not accumulate in the body fluids.

■ When chemical buffers alone have been unable to immediately minimize a pH change, the respiratory system responds within a few minutes by altering its rate of $CO_2$ removal. An increase in $[H^+]$ from sources other than carbonic acid stimulates respiration so that more $H_2CO_3$-forming $CO_2$ is blown off, thereby compensating for acidosis by reducing the generation of $H^+$ from $H_2CO_3$. Conversely, a fall in $[H^+]$ depresses respiratory activity so that $CO_2$ and thus $H^+$-generating $H_2CO_3$ can accumulate in the body fluids to compensate for alkalosis. *(Review Table 14-7.)*

■ The kidneys are the third and most powerful line of defence. They require hours to days to compensate for a deviation in body-fluid pH. However, they not only eliminate the normal amount of $H^+$ produced from non-$H_2CO_3$ sources but they can also alter their rate of $H^+$ removal in response to changes in both non-$H_2CO_3$ and $H_2CO_3$ acids. In contrast, the lungs can adjust only $H^+$ generated from $H_2CO_3$. Furthermore, the kidneys can regulate $[HCO_3^-]$ in body fluids as well.

■ The kidneys compensate for acidosis by secreting the excess $H^+$ in the urine while adding new $HCO_3^-$ to the plasma to expand the $HCO_3^-$-buffer pool. During alkalosis, the kidneys conserve $H^+$ by reducing its secretion in urine. They also eliminate $HCO_3^-$ which is in excess because less $HCO_3^-$ than usual is tied up buffering $H^+$ when $H^+$ is in short supply. *(Review Figures 14-9, 14-10, and 14-11 and Table 14-8.)*

■ Secreted $H^+$ must be buffered in the tubular fluid to prevent the $H^+$ concentration gradient from becoming so great that it blocks further $H^+$ secretion. Normally, $H^+$ is buffered by the urinary phosphate buffer pair, which is abundant in the tubular fluid because excess dietary phosphate spills into the urine to be excreted from the body.

■ In acidosis, when all the phosphate buffer is already used up in buffering the extra secreted $H^+$, the kidneys secrete $NH_3$ into the tubular fluid to serve as a buffer so that $H^+$ secretion can continue.

■ The four types of acid–base imbalances are respiratory acidosis, respiratory alkalosis, metabolic acidosis, and metabolic alkalosis. Respiratory acid–base disorders stem from deviations from normal $[CO_2]$, whereas metabolic acid imbalances include all deviations in pH other than those caused by abnormal $[CO_2]$. *(Review Figure 14-12 and Table 14-9.)*

## 15.1 | Introduction (pp. 645–652)

- The four basic digestive processes are motility, secretion, digestion, and absorption.
- The three classes of energy-rich nutrients are digested into absorbable units as follows: (1) Dietary carbohydrates in the form of the polysaccharides starch and glycogen are digested into their absorbable units of monosaccharides, especially glucose. (Review Figure 15-1.) (2) Dietary proteins are digested into their absorbable units of amino acids and a few small polypeptides. (3) Dietary fats in the form of triglycerides are digested into their absorbable units of monoglycerides and free fatty acids.
- The digestive system consists of the digestive tract and accessory digestive organs (salivary glands, exocrine pancreas, and biliary system). (Review Table 15-1.)
- The lumen of the digestive tract is continuous with the external environment, so its contents are technically outside the body; this arrangement permits digestion of food without self-digestion occurring in the process.
- The digestive tract wall has four layers throughout most of its length. From innermost outward, they are the mucosa, submucosa, muscularis externa, and serosa. (Review Figure 15-2.)

## 15.2 | Mouth (pp. 652–654)

- **Motility:** Food enters the digestive system through the mouth, where it is chewed and mixed with saliva to facilitate swallowing.
- **Secretion:** The salivary enzyme, amylase, begins the digestion of carbohydrates. More important than its minor digestive function, saliva is essential for articulating speech and plays an important role in dental health. Salivary secretion is controlled by a salivary centre in the medulla, mediated by autonomic innervation of the salivary glands. (Review Figure 15-4.)
- **Digestion:** Salivary amylase begins to digest polysaccharides into the disaccharide maltose, a process that continues in the stomach after the food has been swallowed and until amylase is eventually inactivated by the acidic gastric juice. (Review Figure 15-1, and Table 15-6.)
- **Absorption:** No absorption of nutrients occurs from the mouth.

## 15.3 | Pharynx and Esophagus (pp. 654–656)

- **Motility:** Following chewing, the tongue propels the bolus of food to the rear of the throat, which initiates the swallowing reflex. The swallowing centre in the medulla coordinates a complex group of activities that result in closure of the respiratory passages and propulsion of food through the pharynx and esophagus into the stomach. (Review Figure 15-5.)
- **Secretion:** The esophageal secretion, mucus, is protective in nature.

- **Digestion and absorption:** No nutrient digestion or absorption occurs in the pharynx or esophagus.

## 15.4 | Stomach (pp. 656–667)

- **Motility:** The four aspects of gastric motility are gastric filling, storage, mixing, and emptying.
- Gastric filling is facilitated by vagally mediated receptive relaxation of the stomach muscles.
- Gastric storage takes place in the body of the stomach, where peristaltic contractions of the thin muscle walls are too weak to mix the contents.
- Gastric mixing in the thick-muscled antrum results from vigorous peristaltic contractions. (Review Figure 15-7.)
- Gastric emptying is influenced by the following factors in the stomach and duodenum: (1) the volume and fluidity of chyme in the stomach; and (2) the duodenal factors, which are the dominant factors controlling gastric emptying. The specific factors in the duodenum that delay gastric emptying are fat, acid, hypertonicity, and distension. (Review Figure 15-7 and Table 15-2.)
- **Secretion:** Gastric secretions into the stomach lumen include (1) HCl (from the parietal cells), which activates pepsinogen, denatures protein, and kills bacteria; (2) pepsinogen (from the chief cells), which, once activated, initiates protein digestion; (3) mucus (from the mucous cells), which provides a protective coating to supplement the gastric mucosal barrier, enabling the stomach to contain harsh luminal contents without self-digestion; and (4) intrinsic factor (from the parietal cells), which plays a vital role in vitamin B12 absorption, a constituent essential for normal red blood cell production. (Review Table 15-3 and Figures 15-8, 15-9, and 15-10.)
- The stomach also secretes the following endocrine and paracrine regulatory factors: (1) the hormone gastrin (from the G cells), which plays a dominant role in stimulating gastric secretion; (2) the paracrine histamine (from the ECL cells), a potent stimulant of acid secretion by the parietal cells; and (3) the paracrine somatostatin (from the D cells), which inhibits gastric secretion. (Review Table 15-3.)
- Gastric secretion is under complex control mechanisms. Gastric secretion is increased during the cephalic and gastric phases of gastric secretion before and during a meal by mechanisms involving excitatory vagal and intrinsic nerve responses along with the stimulatory actions of gastrin and histamine. (Review Tables 15-4 and 15-5.)
- **Digestion:** Carbohydrate digestion continues in the body of the stomach under the influence of the swallowed salivary amylase. Protein digestion is initiated by pepsin in the antrum of the stomach, where vigorous peristaltic contractions mix the food

with gastric secretions, converting it to a thick liquid mixture known as *chyme. (Review Table 15-6.)*

- **Absorption:** No nutrients are absorbed from the stomach.

## 15.5 | Pancreatic and Biliary Secretions (pp. 667–676)

- Pancreatic exocrine secretions and bile from the liver both enter the duodenal lumen.

- Pancreatic secretions include (1) potent digestive enzymes from the acinar cells, which digest all three categories of foodstuff; and (2) an aqueous $NaHCO_3$ solution from the duct cells, which neutralizes the acidic contents emptied into the duodenum from the stomach. This neutralization is important to protect the duodenum from acid injury and to allow pancreatic enzymes, which are inactivated by acid, to perform their important digestive functions. *(Review Figure 15-11.)*

- The pancreatic digestive enzymes include (1) the proteolytic enzymes trypsinogen, chymotrypsinogen, and procarboxypeptidase, which are secreted in inactive form and are activated in the duodenal lumen on exposure to enterokinase and activated trypsin; (2) pancreatic amylase, which continues carbohydrate digestion; and (3) lipase, which accomplishes fat digestion. *(Review Table 15-6.)*

- Pancreatic secretion is primarily under hormonal control, which matches composition of the pancreatic juice with the needs in the duodenal lumen. Secretin stimulates the pancreatic duct cells, and cholecystokinin (CCK) stimulates the acinar cells. *(Review Figure 15-12.)*

- The liver, the body's largest and most important metabolic organ, performs many varied functions. *(Review Figure 15-14.)* Its contribution to digestion is the secretion of bile, which contains bile salts.

- Bile salts aid fat digestion through their detergent action and facilitate fat absorption by forming water-soluble micelles that can carry the water-insoluble products of fat digestion to their absorption site. *(Review Figures 15-13 and 15-15 to 15-17.)*

- Bile also contains bilirubin, a derivative of degraded haemoglobin, which is the major excretory product in the feces.

## 15.6 | Small Intestine (pp. 676–688)

- The small intestine is the main site for digestion and absorption.

- **Motility:** Segmentation, the small intestine's primary motility during digestion of a meal, thoroughly mixes the food with pancreatic, biliary, and small-intestine juices to facilitate digestion; it also exposes the products of digestion to the absorptive surfaces. *(Review Figure 15-18.)*

- **Secretion:** The juice secreted by the small intestine does not contain any digestive enzymes.

- **Digestion:** The pancreatic enzymes continue carbohydrate and protein digestion in the small-intestine lumen. The small-intestine brush-border enzymes complete the digestion of carbohydrates and protein. Fat is digested entirely in the small-intestine lumen, by pancreatic lipase. *(Review Figure 15-22a and Table 15-6.)*

- **Absorption:** The small-intestine lining is remarkably adapted for its digestive and absorptive functions. Its folds bear a rich array of finger-like projections, the villi, which have a multitude of even smaller hairlike protrusions, the microvilli. Together, these surface modifications tremendously increase the area available to house the membrane-bound enzymes and to accomplish both active and passive absorption. *(Review Figures 15-20, 15-21, and 15-22.)*

- The energy-dependent process of $Na^+$ absorption provides the driving force for $Cl^-$, water, glucose, and amino acid absorption. All these absorbed products enter the blood. *(Review Figures 15-23 and 15-24.)*

- Because they are not soluble in water, the products of fat digestion must undergo a series of transformations that enable them to be passively absorbed, eventually entering the lymph.

## 15.7 | Large Intestine (pp. 688–692)

- The colon serves primarily to concentrate and store undigested food residues (fibre, the indigestible cellulose in plant walls) and bilirubin until they can be eliminated from the body as feces. *(Review Figure 15-28.)*

- **Motility:** Haustral contractions slowly shuffle the colonic contents back and forth to mix and facilitate absorption of most of the remaining fluid and electrolytes. Mass movements several times a day, usually after meals, propel the feces long distances. Movement of feces into the rectum triggers the defecation reflex, which the person can voluntarily prevent by contracting the external anal sphincter if the time is inopportune for elimination.

- **Secretion:** The alkaline mucous secretion of the large intestine is primarily protective in function.

- **Digestion and absorption:** No secretion of digestive enzymes or absorption of nutrients takes place in the colon, as all nutrient digestion and absorption was previously completed in the small intestine. Absorption of some of the remaining salt and water converts the colonic contents into feces.

## 15.8 | Overview of the Gastrointestinal Hormones (pp. 692–693)

- Each of these hormones performs multiple interrelated functions.

- Gastrin is released primarily in response to the presence of protein products in the stomach, and its effects promote digestion of protein, movement of materials through the digestive tract, and maintenance of the integrity of the stomach and small-intestine mucosa.

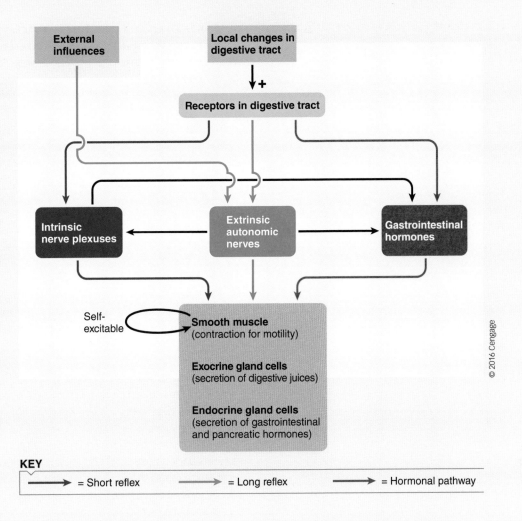

**KEY**

————▶ = Short reflex     ————▶ = Long reflex     ————▶ = Hormonal pathway

© 2016 Cengage

■ Secretin is released primarily in response to the presence of acid in the duodenum, and its effects neutralize the acid in the duodenal lumen and maintain the integrity of the exocrine pancreas.

■ Cholecystokinin is released primarily in response to the presence of fat products in the duodenum, and its effects optimize conditions for digesting fat and other nutrients and for maintaining the integrity of the exocrine pancreas.

# NOTES

## 16.1 | Energy Balance (pp. 699–710)

- Energy input to the body in the form of food energy must equal energy output, because energy cannot be created or destroyed.

- Energy output or expenditure includes (1) external work, performed by skeletal muscles to move an external object or move the body through the external environment; and (2) internal work, which consists of all other energy-dependent activities that do not accomplish external work, including active transport, smooth and cardiac muscle contraction, glandular secretion, and protein synthesis. (Review Figure 16-1.)

- Only about 25 percent of the chemical energy in food is harnessed to do biological work. The rest is immediately converted to heat. Furthermore, all the energy expended to accomplish internal work is eventually converted into heat, and 75 percent of the energy expended by working skeletal muscles is lost as heat. Therefore, most of the energy in food ultimately appears as body heat.

- The metabolic rate, which is energy expenditure per unit of time, is measured in kilocalories of heat produced per hour.

- The basal metabolic rate (BMR) is a measure of the body's minimal waking rate of internal energy expenditure.

- For a neutral energy balance, the energy in ingested food must equal energy expended in performing external work and transformed into heat. If more energy is consumed than is expended, the extra energy is stored in the body, primarily as adipose tissue, so body weight increases. By contrast, if more energy is expended than is available in the food, body energy stores are used to support energy expenditure, so body weight decreases.

- Usually, body weight remains fairly constant over a prolonged period of time (except during growth), because food intake is adjusted to match energy expenditure on a long-term basis.

- Food intake is controlled primarily by the hypothalamus by means of complex regulatory mechanisms in which hunger and satiety are important components. Feeding or appetite signals give rise to the sensation of hunger and promote eating, whereas satiety signals lead to the sensation of fullness and suppress eating.

- The arcuate nucleus of the hypothalamus plays a key role in energy homeostasis by virtue of the two clusters of appetite-regulating neurons it contains: neurons that secrete neuropeptide Y (NPY), which increases appetite and food intake; and neurons that secrete melanocortins, which suppress appetite and food intake. (Review Figure 16-2.)

- Adipocytes in fat stores secrete the hormone leptin, which reduces appetite and decreases food consumption by inhibiting the NPY-secreting neurons and stimulating the melanocortins-secreting neurons of the arcuate nucleus. This mechanism is primarily important in the long-term matching of energy intake with energy output, thereby maintaining body weight over the long term. (Review Table 16-2 and Figure 16-2.)

- Insulin released by the endocrine pancreas in response to increased glucose and other nutrients in the blood also inhibits the NPY-secreting neurons and contributes to long-term control of energy balance and body weight.

- NPY and melanocortins bring about their effects by acting on the lateral hypothalamus area (LHA) and paraventricular nucleus (PVN) to alter the release of chemical messengers from these areas. The LHA secretes orexins that are potent stimulators of food intake, whereas the PVN releases neuropeptides such as corticotropin-releasing hormone that decrease food intake. (Review Figure 16-2.)

- Short-term control of the timing and size of meals is mediated primarily by the actions of signals arising from the digestive tract and pancreas. Of major importance are two peptides secreted by the digestive tract. (1) Ghrelin, a mealtime initiator, is secreted by the stomach before a meal and signals hunger. Its secretion drops when food is consumed. Ghrelin stimulates appetite and promotes feeding behaviour by stimulating the NPY-secreting neurons. (2) $PYY_{3-36}$, a mealtime terminator, is secreted by the small and large intestines during a meal and signals satiety. Its secretion is lowest before a meal. $PYY_{3-36}$ inhibits the NPY-secreting neurons. (Review Figure 16-2.)

- The nucleus tractus solitarius (NTS) in the brain stem serves as the satiety centre and in this capacity also plays a key role in the short-term control of meals. The NTS receives input from the higher hypothalamic areas concerned with control of energy balance and food intake as well as input from the digestive tract and pancreas. Satiety signals acting through the NTS to inhibit further food intake include stomach distension and increased cholecystokinin, a hormone released from the duodenum in response to the presence of nutrients, especially fat, in the digestive tract lumen. (Review Figure 16-2.)

- Psychosocial and environmental factors can also influence food intake above and beyond the internal signals that govern feeding behaviour.

## 16.2 | Temperature Regulation (pp. 710–719)

- The body can be thought of as a heat-generating core (internal organs, CNS, and skeletal muscles) surrounded by a shell of variable insulating capacity (the skin).

- The skin exchanges heat energy with the external environment, with the direction and amount of heat transfer depending on the environmental temperature and the momentary insulating capacity of the shell.

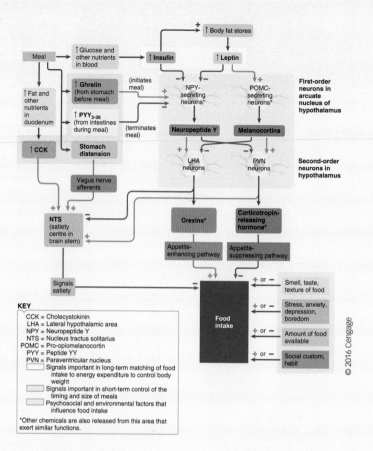

**KEY**

CCK = Cholecystokinin
LHA = Lateral hypothalamic area
NPY = Neuropeptide Y
NTS = Nucleus tractus solitarius
POMC = Pro-opiomelanocortin
PYY = Peptide YY
PVN = Paraventricular nucleus

Signals important in long-term matching of food intake to energy expenditure to control body weight

Signals important in short-term control of the timing and size of meals

Psychosocial and environmental factors that influence food intake

*Other chemicals are also released from this area that exert similar functions.

© 2016 Cengage

- The four physical means by which heat is exchanged between the body and external environment are (1) radiation (net movement of heat energy via electromagnetic waves); (2) conduction (exchange of heat energy by direct contact); (3) convection (transfer of heat energy by means of air currents); and (4) evaporation (extraction of heat energy from the body by the heat-requiring conversion of liquid $H_2O$ to $H_2O$ vapour). Because heat energy moves from warmer to cooler objects, radiation, conduction, and convection can be channels for either heat loss or heat gain, depending on whether surrounding objects are cooler or warmer, respectively, than the body surface.

- Normally, they are avenues for heat loss, along with evaporation resulting from sweating. (Review Figure 16-4.)

- To prevent serious cell malfunction, the core temperature must be held constant at about 36.8°C (equivalent to an average oral temperature of 36.8°C) by continuously balancing heat gain and heat loss despite changes in environmental temperature and variation in internal heat production. (Review Figure 16-3.)

- This thermoregulatory balance is controlled by the hypothalamus. Peripheral thermoreceptors inform the hypothalamus of the skin temperature, and central thermoreceptors—the most important of which are located in the hypothalamus itself—inform the hypothalamus of the core temperature.

- The primary means of heat gain is heat production by metabolic activity; the biggest contributor is skeletal muscle contraction.

- Heat loss is adjusted by sweating and by controlling to the greatest extent possible the temperature gradient between the skin and surrounding environment. The latter is accomplished by regulating the diameter of the skin's blood vessels. (1) Vasoconstriction of the skin vessels reduces the flow of warmed blood through the skin so that skin temperature falls. The layer of cool skin between the core and environment increases the insulating barrier between the warm core and the external air. (2) Conversely, skin vasodilation brings more warmed blood through the skin so that skin temperature approaches the core temperature, thus reducing the insulative capacity of the skin.

- On exposure to cool surroundings, the core temperature starts to fall as heat loss increases, because of the larger-than-normal skin-to-air temperature gradient. The hypothalamus responds to reduce the heat loss by inducing skin vasoconstriction, while simultaneously increasing heat production through heat-generating shivering. (Review Table 16-4.)

- Conversely, in response to a rise in core temperature (resulting either from excessive internal heat production accompanying exercise or from excessive heat gain on exposure to a hot environment), the hypothalamus triggers heat-loss mechanisms, such as skin vasodilation and sweating, while simultaneously decreasing heat production, such as by reducing muscle tone. (Review Table 16-4.)

- In both cold and heat responses, voluntary behavioural actions also contribute importantly to maintenance of thermal homeostasis.

- A fever occurs when endogenous pyrogen released from macrophages in response to infection raises the hypothalamic set point. An elevated core temperature develops as the hypothalamus initiates cold-response mechanisms to raise the core temperature to the new set point. (Review Figure 16-5.)

## 17.1 | Introduction (pp. 723–731)

- Both sexes produce gametes (reproductive cells), sperm in males and ova (eggs) in females, each of which bears one member of each of the 23 pairs of chromosomes present in human cells. Union of a sperm and an ovum at fertilization results in the beginning of a new individual that has 23 complete pairs of chromosomes, half from the father and half from the mother. (*Review Figure 17-3.*)
- In both sexes, the reproductive system consists of (1) a pair of gonads, testes in males and ovaries in females, which are the primary reproductive organs that produce the gametes and secrete sex hormones; (2) a reproductive tract composed of a system of ducts that transport and/or house the gametes after they are produced; and (3) accessory sex glands that provide supportive secretions for the gametes. The externally visible portions of the reproductive system constitute the external genitalia. (*Review Figures 17-1 and 17-2.*)
- Secondary sexual characteristics are the distinguishing features between males and females not directly related to reproduction.
- Sex determination is a genetic phenomenon dependent on the combination of sex chromosomes at the time of fertilization: an XY combination is a genetic male, and an XX combination, a genetic female. (*Review Figure 17-4.*)
- The term *sex differentiation* refers to the embryonic development of the gonads, reproductive tract, and external genitalia along male or female lines, which gives rise to the apparent anatomic sex of the individual. In the presence of masculinizing factors, a male reproductive system develops; in their absence, a female system develops. (*Review Figures 17-4, 17-5, and 17-6.*)

## 17.2 | Male Reproductive Physiology (pp. 731–741)

- The testes are located in the scrotum. The cooler temperature in the scrotum than in the abdominal cavity is essential for spermatogenesis.
- Spermatogenesis (sperm production) occurs in the testes' highly coiled seminiferous tubules. (*Review Figures 17-7 and 17-8.*)
- Leydig cells in the interstitial spaces between these tubules secrete the male sex hormone, testosterone, into the blood. (*Review Figure 17-7.*)
- Testosterone is secreted before birth to masculinize the developing reproductive system; then its secretion ceases until puberty, at which time it begins once again and continues throughout life. Testosterone is responsible for maturation and maintenance of the entire male reproductive tract, for development of secondary sexual characteristics, and for stimulating libido. (*Review Table 17-1.*)
- The testes are regulated by the anterior pituitary hormones: luteinizing hormone (LH) and follicle-stimulating hormone (FSH). These gonadotropic hormones, in turn, are under control of hypothalamic gonadotropin-releasing hormone (GnRH). (*Review Figure 17-10.*)
- Testosterone secretion is regulated by LH stimulation of the Leydig cells, and, in negative-feedback fashion, testosterone inhibits gonadotropin secretion. (*Review Figure 17-10.*)
- Spermatogenesis requires both testosterone and FSH. Testosterone stimulates the mitotic and meiotic divisions required to transform the undifferentiated diploid germ cells, the spermatogonia, into undifferentiated haploid spermatids. FSH stimulates the remodelling of spermatids into highly specialized motile spermatozoa. (*Review Figure 17-8.*)
- Also present in the seminiferous tubules are Sertoli cells, which protect, nurse, and enhance the germ cells throughout their development. Sertoli cells also secrete inhibin, a hormone that inhibits FSH secretion, completing the negative-feedback loop. (*Review Figures 17-7b and 17-7d and 17-10.*)
- The still immature sperm are flushed out of the seminiferous tubules into the epididymis by fluid secreted by the Sertoli cells.
- The epididymis and ductus deferens store and concentrate the sperm and increase their motility and fertility prior to ejaculation. (*Review Table 17-2 and Figure 17-7.*)
- During ejaculation, the sperm are mixed with secretions released by the accessory glands. (*Review Table 17-2.*)
- The seminal vesicles supply fructose for energy and prostaglandins, which promote smooth muscle motility in both the male and female reproductive tracts, for enhancement of sperm transport. The seminal vesicles also contribute the bulk of the semen.
- Prostaglandins are produced throughout the body, not just in the reproductive tract. These ubiquitous chemical messengers are derived from arachidonic acid, a component of the plasma membrane. By acting as paracrines, specific prostaglandins exert a variety of local effects. (*Review Figure 17-11 and Table 17-3.*)
- The prostate gland contributes an alkaline fluid for neutralizing the acidic vaginal secretions.
- The bulbourethral glands release lubricating mucus.

## 17.3 | Sexual Intercourse between Males and Females (pp. 741–744)

- The male sex act consists of erection and ejaculation, which are part of a much broader systemic, emotional response that typifies the male sexual response cycle. (*Review Table 17-4.*)
- Erection is a hardening of the normally flaccid penis that enables it to penetrate the female vagina. Erection is accomplished by marked vasocongestion of the penis brought about

by reflexly induced vasodilation of the arterioles supplying the penile erectile tissue. (Review Figure 17-12.)

■ When sexual excitation reaches a critical peak, ejaculation occurs. It consists of two stages: (1) emission, the emptying of semen (sperm and accessory sex gland secretions) into the urethra; and (2) expulsion of semen from the penis. The latter is accompanied by a set of characteristic systemic responses and intense pleasure referred to as *orgasm*. (Review Table 17-4.)

■ Females experience a sexual cycle similar to that of males, with both having excitation, plateau, orgasmic, and resolution phases. The major difference is that women do not ejaculate.

■ During the female sexual response, the outer portion of the vagina constricts to grip the penis, whereas the inner part expands to create space for sperm deposition.

## 17.4 | Female Reproductive Physiology (pp. 745–774)

■ The ovaries perform the dual and interrelated functions of oogenesis (producing ova) and secretion of estrogen and progesterone. (Review Table 17-6.) Two related ovarian endocrine units sequentially accomplish these functions: the follicle and the corpus luteum.

■ The same steps in chromosome replication and division take place in oogenesis as in spermatogenesis, but the timing and end result are markedly different. Spermatogenesis is accomplished within two months, whereas the similar steps in oogenesis take anywhere from 12 to 50 years to complete on a cyclic basis from the onset of puberty until menopause. A female is born with a limited, largely nonrenewable supply of germ cells, whereas postpubertal males can produce several hundred million sperm each day. Each primary oocyte yields only one cytoplasm-rich ovum along with three doomed cytoplasm-poor polar bodies that disintegrate, whereas each primary spermatocyte yields four equally viable spermatozoa. (Review Figure 17-13 and Figure 17-8.)

■ Oogenesis and estrogen secretion take place within an ovarian follicle during the first half of each reproductive cycle (the follicular phase) under the influence of FSH, LH, and estrogen. (Review Figures 17-14 through 17-18.)

■ At approximately midcycle, the maturing follicle releases a single ovum (ovulation). Ovulation is triggered by an LH surge brought about by the high level of estrogen produced by the mature follicle. (Review Figures 17-14, 17-16, and 17-19.)

■ Under the influence of LH, the empty follicle is then converted into a corpus luteum, which produces progesterone as well as estrogen during the last half of the cycle (the luteal phase). This endocrine unit prepares the uterus for implantation if the released ovum is fertilized. (Review Figures 17-14, 17-16, and 17-20.)

■ If fertilization and implantation do not occur, the corpus luteum degenerates. The consequent withdrawal of hormonal support for the highly developed uterine lining causes it to disintegrate and slough, producing menstrual flow. Simultaneously, a new follicular phase is initiated. (Review Figures 17-14 and 17-16.)

■ Menstruation ceases and the uterine lining (endometrium) repairs itself under the influence of rising estrogen levels from the newly maturing follicle. (Review Figure 17-16.)

■ If fertilization does take place, it occurs in the oviduct as the released egg and sperm deposited in the vagina are both transported to this site. (Review Figures 17-21 through 17-23.)

■ The fertilized ovum begins to divide mitotically. Within a week it grows and differentiates into a blastocyst capable of implantation. (Review Figure 17-24.)

■ Meanwhile, the endometrium has become richly vascularized and stocked with stored glycogen under the influence of luteal-phase progesterone. (Review Figure 17-16.) Into this especially prepared lining the blastocyst implants by means of enzymes released by the trophoblasts, which form the blastocyst's outer layer. These enzymes digest the nutrient-rich endometrial tissue, accomplishing the dual function of carving a hole in the endometrium for implantation of the blastocyst while simultaneously releasing nutrients from the endometrial cells for use by the developing embryo. (Review Figure 17-25.)

■ After implantation, an interlocking combination of fetal and maternal tissues— the placenta—develops. The placenta is the organ of exchange between the maternal and fetal blood and also acts as a transient, complex endocrine organ that secretes a number of hormones essential for pregnancy. Human chorionic gonadotropin, estrogen, and progesterone are the most important of these hormones. (Review Figures 17-26, 17-28, and 17-29, and Table 17-5.)

■ Human chorionic gonadotropin maintains the corpus luteum of pregnancy, which secretes estrogen and progesterone during the first trimester of gestation until the placenta takes over this function in the last two trimesters. High levels of estrogen and progesterone are essential for maintaining a normal pregnancy. (Review Figure 17-28.)

■ Parturition is initiated by a complex interplay of multiple maternal and fetal factors. Once the contractions are initiated at the onset of labour, a positive-feedback cycle is established that progressively increases their force. As contractions push the fetus against the cervix, secretion of oxytocin, a powerful uterine muscle stimulant, is reflexly increased. The extra oxytocin causes stronger contractions, giving rise to even more oxytocin release, and so on. This positive-feedback cycle progressively intensifies until cervical dilation and delivery are complete. (Review Figure 17-30.)

■ Prolactin stimulates the synthesis of enzymes essential for milk production by the alveolar epithelial cells. However, the high gestational level of estrogen and progesterone prevents prolactin from promoting milk production. Withdrawal of the placental steroids at parturition initiates lactation.

■ Lactation is sustained by suckling, which triggers the release of oxytocin and prolactin. Oxytocin causes milk ejection by stimulating the myoepithelial cells surrounding the alveoli to squeeze the secreted milk out through the ducts. Prolactin stimulates the secretion of more milk to replace the milk ejected as the baby nurses. (Review Figures 17-32 and 17-33.)